ANIMAL

SMITHSONIAN

ANIMAL

Editors-in-Chief
David Burnie and Don E. Wilson

LONDON, NEW YORK, MUNICH, MELBOURNE, AND DELHI

DORLING KINDERSLEY, LONDON

SENIOR EDITORS Angeles Gavira, Peter Frances
PROJECT EDITORS David Summers, Sean O'Connor
EDITORS Alison Copland, Lesley Riley, Polly Boyd
ZOOLOGICAL EDITORS Kim Dennis-Bryan, Stephen Parker
INDEXER Jane Parker
PRODUCTION CONTROLLERS Elizabeth Cherry, Michelle Thomas
CATEGORY PUBLISHER Jonathan Metcalf

SENIOR ART EDITORS Ina Stradins, Vanessa Hamilton
ART EDITORS Kirsten Cashman, Philip Ormerod, Sara Freeman
DESIGNER Amir Reuveni
DTP DESIGNERS Louise Waller, Rajen Shah, Martin Nilsson, Simon Longstaff
PICTURE RESEARCHERS Cheryl Dubyk-Yates, Kate Duncan, Sean Hunter
ILLUSTRATORS Richard Tibbets, Evi Antoniou, Paul Banville
ART DIRECTOR Bryn Walls

DORLING KINDERSLEY, DELHI

MANAGING EDITORS Ira Pande, Prita Maitra
PROJECT EDITORS Ranjana Saklani, Atanu Raychaudhuri
EDITORS Rimli Borooah, Kajori Aikat
DTP COORDINATORS Jacob Joshua, Rajesh Bisht

MANAGING ART EDITOR Shuka Jain
PROJECT DESIGNER Shefali Upadhyay
DESIGNER Pallavi Narain
DESIGN ASSISTANCE Elizabeth Thomas, Suresh Kumar
DTP DESIGNER Pankaj Sharma

STUDIO CACTUS
SENIOR DESIGNER Sharon Moore
PROJECT EDITOR Lizzie Mallard-Shaw

SCHERMULY DESIGN COMPANY
SENIOR DESIGNER Hugh Schermuly
DESIGNERS Phil Gamble, Nick Buzzard, Sally Geeve, Paul Stork
SENIOR EDITOR Cathy Meeus
EDITORS Joanna Chisholm, Jackie Jackson, Mary Pickles

FOR THIS EDITION

DORLING KINDERSLEY, LONDON

ZOOLOGICAL EDITOR Kim Dennis-Bryan
SENIOR EDITOR Jemima Dunne
US EDITORS Jill Hamilton, Jane Perlmutter, Rebecca Warren
MANAGING EDITOR Camilla Hallinan
SENIOR PRODUCTION EDITOR Jennifer Murray
PRODUCTION CONTROLLER Sophie Argyris
PUBLISHING DIRECTOR Jonathan Metcalf

ART EDITORS Anna Hall, Amy Orsborne
SENIOR ART EDITOR Gadi Farfour, Helen Spencer
MANAGING ART EDITOR Karen Self
PICTURE RESEARCHER Jo Walton
DK PICTURE LIBRARY Martin Copeland
JACKET DESIGNER Mark Cavanagh
ART DIRECTOR Philip Ormerod

DORLING KINDERSLEY, DELHI

MANAGING EDITOR Rohan Sinha
SENIOR EDITOR Alka Ranjan
EDITORS Dharini, Suefa Lee, Rupa Rao, Priyaneet Singh, Catherine Thomas
PRODUCTION MANAGER Pankaj Sharma
DTP DESIGNERS Neeraj Bhatia, Vishal Bhatia, Jaypal Singh Chauhan, Shanker Prasad, Arjinder Singh, Tanveer Zaidi

MANAGING ART EDITOR Arunesh Talapatra
CONSULTANT ART DIRECTOR Shefali Upadhyay
SENIOR ART EDITOR Mitun Banerjee
DESIGNERS Shriya Parameswaran, Arijit Ganguly, Nidhi Mehra, Shreya Anand Virmani, Zaurin Thoidingjam
DTP MANAGER/CTS Balwant Singh
SENIOR DTP DESIGNER Harish Aggarwal

First American edition published in 2001
This American edition published in 2011 in the United States by
DK Publishing, 375 Hudson Street, New York, New York 10014

2 4 6 8 10 9 7 5 3 1
001—177859—Sept/2011

A catalog record for this book is available from the Library of Congress.
ISBN 978-0-7566-8677-2

Color reproduction by Colourscan, Singapore
Printed and bound by Hung Hing, China

EDITORS-IN-CHIEF **DAVID BURNIE & DON E. WILSON**

MAIN CONSULTANTS

MAMMALS

DR. JULIET CLUTTON-BROCK
&
DR. DON E. WILSON

BIRDS

DR. FRANÇOIS VUILLEUMIER
&
CARLA DOVE

REPTILES

CHRIS MATTISON
&
RONALD CROMBIE

AMPHIBIANS

PROFESSOR TIM HALLIDAY
&
RONALD CROMBIE

FISH

PROFESSOR RICHARD ROSENBLATT
&
CAROLE BALDWIN
DR. STANLEY WEITZMAN

INVERTEBRATES

DR. GEORGE C. MCGAVIN, DR. RICHARD BARNES, DR. FRANCES DIPPER
&
DR. STEPHEN CAIRNS, TIMOTHY COFFER, DR. KRISTIAN FAUCHALD, DR. M.G. HARASEWYCH, GARY F. HEVEL, DR. W. DUANE HOPE, DR. BRIAN F. KENSLEY, DR. DAVID PAWSON, DR. KLAUS RUETZLER

CONTRIBUTORS AND CONSULTANTS

Dr. Richard Barnes
Dr. Paul Bates
Dr. Simon K. Bearder
Deborah Behler
John Behler
Keith Betton
Dr. Michael de L. Brooke
Dr. Charles R. Brown
Dr. Donald Bruning
George H. Burgess
Dr. Kent E. Carpenter
Norma G. Chapman
Ben Clemens
Dr. Malcolm C. Coulter
Dominic M. Couzens
Dr. Timothy M. Crowe
Dr. Kim Dennis-Bryan
Dr. Christopher Dickman
Joseph A. DiCostanzo
Prof. Philip Donoghue
Dr. Nigel Dunstone
Dr. S. Keith Eltringham
Prof. Brock Fenton
Joseph Forshaw
Susan D. Gardieff
Dr. Anthony Gill
Dr. Joshua Ginsberg
Prof. Colin Groves
Dr. Jurgen H. Haffer
Prof. Tim Halliday
Dr. Gavin Hanke
Dr. Cindy Hull
Dr. Barry J. Hutchins
Dr. Paul A. Johnsgard
Dr. Angela Kepler
Dr. Jiro Kikkawa
Prof.Nigel Leader-Williams
Dr. Douglas Long
Dr. Manuel Marin
Chris Mattison
Dr. George C. McGavin
Dr. Jeremy McNeil
Dr. Rodrigo A. Medellin
Dr. Fridtjof Mehlum
Chris Morgan
Rick Morris
Dr. Bryan G. Nelson
Dr. Gary L. Nuechterlein
Jemima Parry-Jones
Malcolm Pearch
Prof. Christopher Perrins
Prof. Ted Pietsch
Dr. Tony Prater
Dr. Galen B. Rathbun
Dr. Ian Redmond
Dr. James D. Rising
Robert H. Robins
Jeff Sailer
Dr. Scott A. Schaeffer
Dr. Karl Schuchmann
Prof. John D. Skinner
Dr. Andrew Smith
Dr. Ronald L. Smith
Dr. David D. Stone
Dr. Mark Taylor
Dr. David H. Thomas
Dr. Dominic Tollit
Dr. Jane Wheeler
Dr. Ben Wilson
Dr. David B. Wingate
Dr. Hans Winkler
Dr. Kevin Zippel

ADDITIONAL CONSULTANTS FOR THIS EDITION
Derek Harvey
Darren Mann

FOR THE SMITHSONIAN INSTITUTION

REPTILES AND AMPHIBIANS
Jeremy F. Jacobs

FISH
Dr. Jeffrey T. Williams

INVERTEBRATES
Dr. Rafael Lemaitre
Christopher Mayer
Jonathan Norenburg
Michael Vecchione

CONTENTS

BIRDS 262

REPTILES 370

AMPHIBIANS 436

FISHES 466

INVERTEBRATES 530

PREFACE

First published in 2001, *Animal* set a new standard in reference guides to the natural world. That standard continues today in this new edition, which has been revised and updated to reflect the scientific developments in the past 10 years. Compiled by a team of over 70 zoologists, biologists, and naturalists, it forms a complete survey of the animal world, from familiar species that are seen almost every day to ones that only experts can recognize and name. Its nearly 2,000 animal profiles—and much more besides—have been verified by the Smithsonian Institution, one of the leading scientific research bodies in the world. It is a partnership of which the publishers continue to be proud.

Ten years is a tiny amount of time in the history of our planet, but it has seen some key developments in biosciences. Thanks mainly to genetic research, understanding of animal classification is undergoing rapid change, as scientists find out with greater precision how different animals are related. This new edition of *Animal* reflects these changes in the way animals are grouped, and sometimes in the way they are named. Genetic research has also identified new species among ones that were known before, while scientific expeditions—traveling through forests, oceans, and even sifting through the soil—have discovered an extraordinary number of new species every year. From the golden palace monkey to the bizarre yeti lobster, some of the most exciting of these discoveries can be found in the pages of this new edition.

Paradoxically, this ongoing process of discovery occurs at a time when the world's animals are threatened as never before. Some species—such as the red fox—are remarkably adaptable, and can make themselves at home in a wide range of different habitats, from city centers to semidesert, often far outside their native range. But many are not like this. Instead, home for them is a particular habitat, with a precise range of climatic conditions, living space, and food. If just one factor alters—even very slightly—a species is slowly, but surely, put at risk.

This is the prospect facing many of the world's animals as the human race grows, and our influence continues to be felt planet-wide. Since *Animal* was first published, the human population has

swollen by an extra billion, and our demand for resources—and our effect on the world's habitats—makes this pressure greater still. Pollution and habitat destruction have an almost immediate effect on animals, but global warming, already under way a decade ago, is likely to affect most species on land and in freshwater, as well as many in the seas. For some animals, the threat is so severe that they face imminent risk of extinction.

However, as *Animal* enters its second edition, there are grounds for hope. Thanks to modern technology, we have never been in a better position to know what lives on our planet, how it is changing, and which species need the most urgent protection. History shows that—given the opportunity—animals have the ability to bounce back. When they do, the results can be extraordinary. The Hawaiian goose, once down to just 30 birds, has now recovered to nearly 1,000 in the wild, with the same again in captivity. The southern white rhino, once on the critical list, now stands at nearly 18,000 animals, and the population is still growing strong. But even this is eclipsed by the Antarctic fur seal, hunted almost to extinction by the early 20th century. Today, 6 million of them crowd the shores of South Georgia—the largest collection of marine mammals anywhere on earth.

Today, international conservation has never been better organized, or more serious about addressing the task ahead. At the UN Convention on Biodiversity, held in Japan in 2010, participating countries recognized the vital role played by natural habitats, and pledged themselves to protect 17 percent of the world's land surface by 2020, and 10 percent of the seas. Already, the International Union for Conservation of Nature (IUCN) collates the work of thousands of scientists in documenting species under threat, while the Convention on International Trade in Endangered Species (CITES) monitors cross-border traffic in hundreds of threatened animals, from insects to elephants. Other organizations monitor countless more aspects of the natural world, from bird migration routes to life in coral reefs and the open seas. Together, it adds up to a staggering information-gathering system that is truly planet-wide.

Biodiversity is important for all of us, and conservation keeps it intact. Animals are a key element of biodiversity, so they play a key part in our world. With its unrivaled range of species, from the biggest mammals to the smallest invertebrates, *Animal* celebrates the richness and beauty of animal life, at a time when it needs our help.

David Burnie & Don E. Wilson
Editors-in-Chief

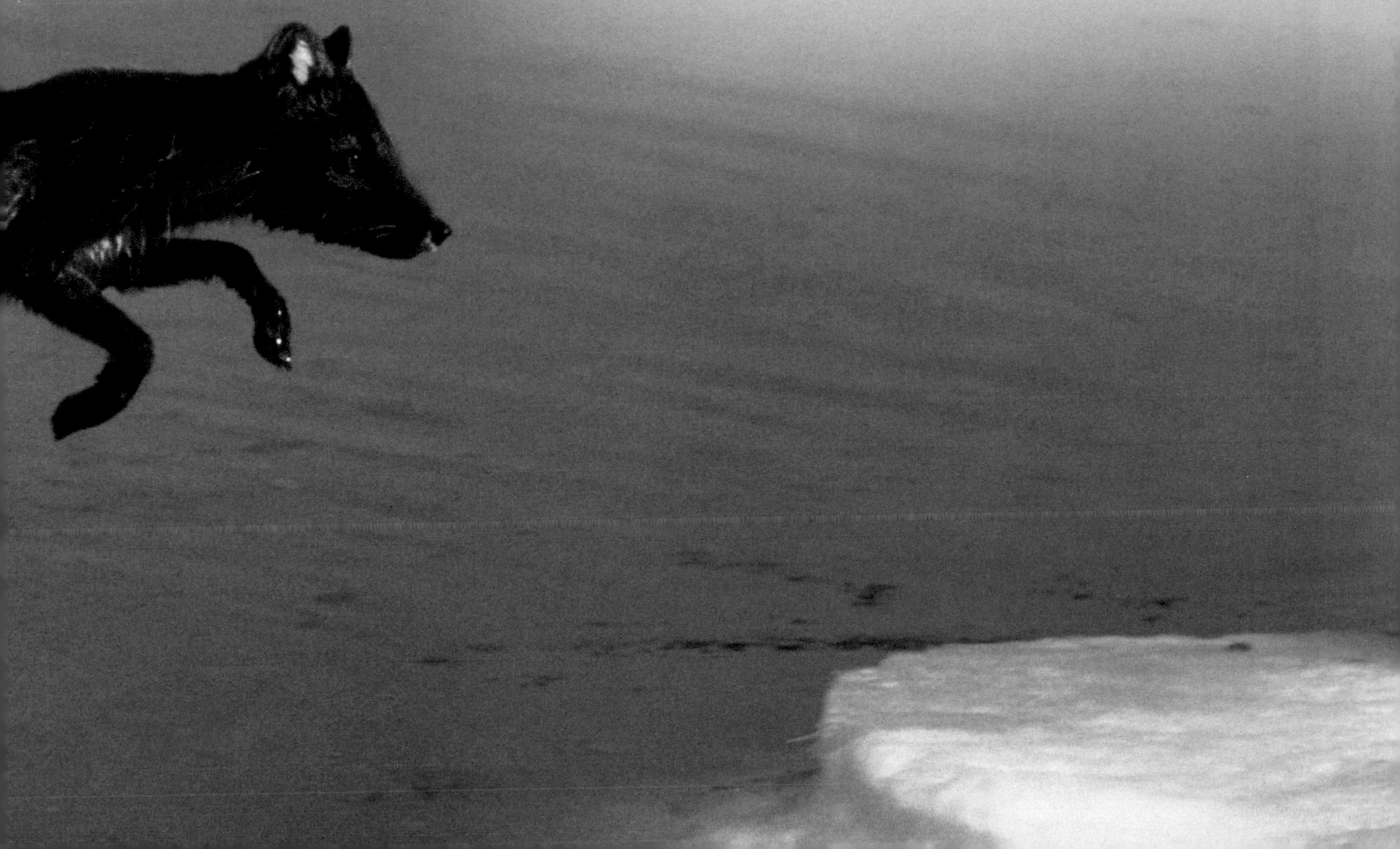

ABOUT THIS BOOK

Animal is organized into 3 main sections: a general **Introduction** to animals and their lives; a section on the world's main **Habitats**; and the main part of the book, **The Animal Kingdom**, which is dedicated to the description of animal groups and species. At the end of the book, a detailed glossary defines all zoological and technical terms used, while a full index lists all the groups and species featured, by both Latin name and common name, including alternative common names.

INTRODUCTION

Animal begins with an overview of all aspects of animal life. This includes an account of what an animal is—and how it differs from other living things. It also examines animal anatomy, life cycles, evolution, behavior, and conservation. In addition, there is a comprehensive presentation of the classification scheme that underpins the species profiles in **The Animal Kingdom**.

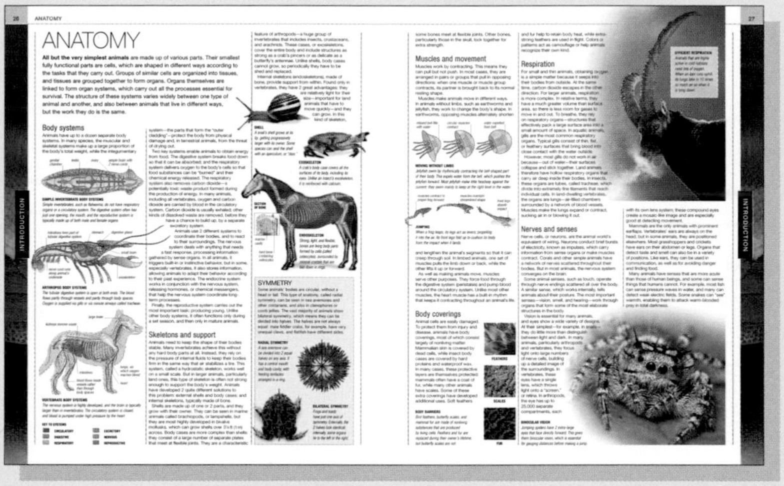

HABITATS

This section looks at the world's main animal habitats. Coverage of each habitat is divided into 2 parts. The first (illustrated below) describes the habitat itself, including its climate and plant life, and the types of animals found there. The pages that immediately follow describe how the anatomy and behavior of animals are adapted to suit the conditions in which they live.

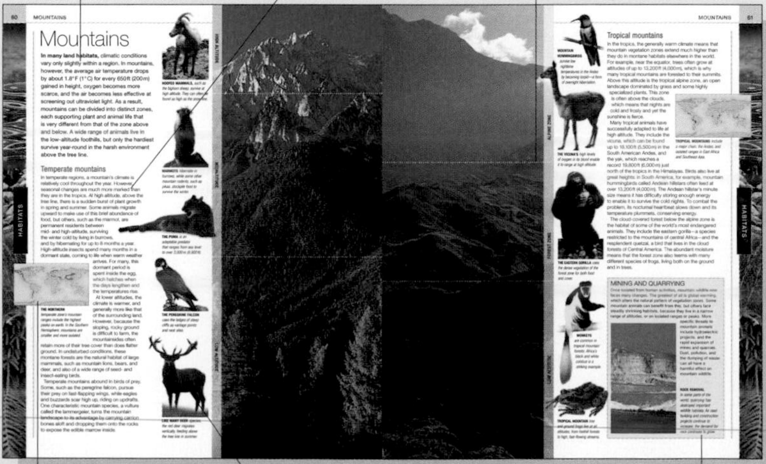

THE ANIMAL KINGDOM

This section of *Animal* is divided into 6 chapters: mammals, birds, reptiles, amphibians, fish, and invertebrates. Each chapter begins with an introduction to the animal group. Lower-ranked groups such as orders and families are then introduced, followed by profiles of the species classified within them. The invertebrates are organized slightly differently (see opposite), while passerine birds are profiled at family level, with only representative species shown.

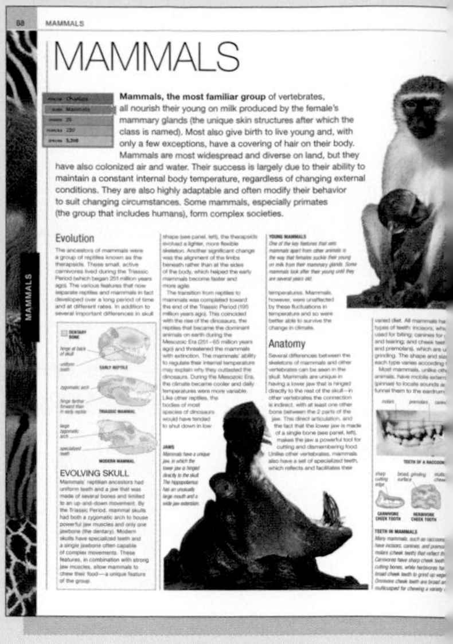

HABITAT SYMBOLS

Symbols in profiles are listed as shown below, not in order of main occurrence.

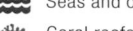

- Temperate forest, including woodland
- Coniferous forest, including woodland
- Tropical forest and rain forest
- Mountains, highlands, scree slopes
- Desert and semidesert
- Open habitats, including grassland, moor, heath, savanna, fields, scrub
- Wetlands and all still bodies of water, including lakes, ponds, pools, marshes, bogs, and swamps
- Rivers, streams, and all flowing water
- Mangrove swamps, above or below the waterline
- Coastal areas, including beaches and cliffs, areas just above high tide, in the intertidal zone, and in shallow, offshore waters
- Seas and oceans
- Coral reefs and waters immediately around them
- Polar regions, including tundra and icebergs
- Urban areas, including buildings, parks, and gardens
- Parasitic; living on or inside another animal

DATA FIELD

Summary information is given at the start of each profile. Measurements are for adult males of the species and may be a typical range, single-figure average, or maximum, depending on available records.

LENGTH (all groups, except Invertebrates)
MAMMALS: head and body. **BIRDS:** tip of bill to tip of tail. **REPTILES:** length of carapace for tortoises and turtles; head and body, including tail, for all other species. **FISHES & AMPHIBIANS:** head and body, including tail.

TAIL (Mammals only) Length.

WEIGHT (Mammals, Birds, and Fishes only) Body weight.

SOCIAL UNIT (Mammals only)
Whether a species lives mainly alone (*Individual*), in a *Group*, in a *Pair*, or varies between these units (*Variable*).

PLUMAGE (Birds only)
Whether sexes are *alike* or *different*.

MIGRATION (Birds only)
Whether a bird is a *Migrant*, *Partial migrant*, *Nonmigrant*, or *Nomadic*.

BREEDING (Reptiles and Fishes only)
Whether the species is *Viviparous*, *Oviparous*, or *Ovoviviparous*.

HABIT (Reptiles and Amphibians only)
Whether the species is partly or wholly *Terrestrial*, *Aquatic*, *Burrowing*, or *Arboreal*.

BREEDING SEASON (Amphibians only)
The time of year in which breeding occurs.

SEX (Fishes only)
Whether the species has separate males and females (*Male/Female*), is a *Hermaphrodite*, or a *Sequential hermaphrodite*.

OCCURRENCE (Invertebrates only)
Number of species in the family, class, or phylum; their distribution and the microhabitats they can be found in.

STATUS (all groups)
Animal uses The IUCN Red List (see p.33) and other threat categories, as follows:

EXTINCT IN THE WILD (IUCN) Known only to survive in captivity, as a naturalized population outside its natural range, or as a small, managed population that has been reintroduced.

CRITICALLY ENDANGERED (IUCN) Facing an extremely high risk of extinction in the wild in the immediate future.

ENDANGERED (IUCN) Facing a very high risk of extinction in the wild in the near future.

VULNERABLE (IUCN) Facing a high risk of extinction in the wild in the medium-term future.

NEAR THREATENED (IUCN) Strong possibility of becoming endangered in the near future.

LEAST CONCERN (IUCN) Low-risk category that includes widespread and common species.

DATA DEFICIENT (IUCN) Not a threat category. Population and distribution data is insufficient for assessment.

NOT EVALUATED Data not yet assessed against the 5 IUCN criteria.

For fuller details see The IUCN Red List web site (www.iucnredlist.org)

◁ **MAJOR GROUPS**
Each of the 6 major animal groups is introduced with an overview of its defining characteristics. These pages also cover key aspects of evolution, life cycle, and behavior.

◁ **PHYLA, CLASSES, AND ORDERS**
Within each major group, lower-ranked groups such as orders are described in separate introductions. These identify the animals found in the group, describe their defining properties, and explain many of the terms used in the species profiles.

INVERTEBRATE ANIMALS

There are so many invertebrate species that in order to provide representative coverage of the group as a whole, this book focuses on profiling the larger taxonomic groups—phylum, class, order, and family—rather than on individual species. Groups are presented in 3 ways: a brief introduction to an order or class, supported by a description of one or more constituent families, with each family illustrated by representative species (e.g., insects); a profile of a class, illustrated by species that represent distinct groups or types within it (e.g., crustaceans); a description of a phylum illustrated by a representative species of that phylum (used for minor phyla only).

◁ **LOWER-RANKED GROUPS**
In many chapters, animal groups are divided into smaller groups, such as families. An introduction to each of these smaller sections describes common anatomical features. Relevant aspects of reproduction and behavior are also introduced.

▽ **SPECIES PROFILES**
Over 2,000 wild animals are profiled in The Animal Kingdom. Every profile contains a text summary and, in most cases, a color illustration and a distribution map.

CLASSIFICATION KEY

In each animal group introduction, a color-coded panel shows the position in the animal kingdom of the group being described. The group is identified by a white outline and an arrow. The taxonomic ranks above the group are named in the upper part of the table, while the number of lower ranks that it contains are shown below it. In some introductions, an extra Classification Note is included, to list subgroups or identify areas of debate.

LATIN NAME
The species' Latin name appears in a colored band at the beginning of each entry

DATA FIELD
Core information is provided in summary form at the beginning of each species entry. The categories vary between groups (see opposite)

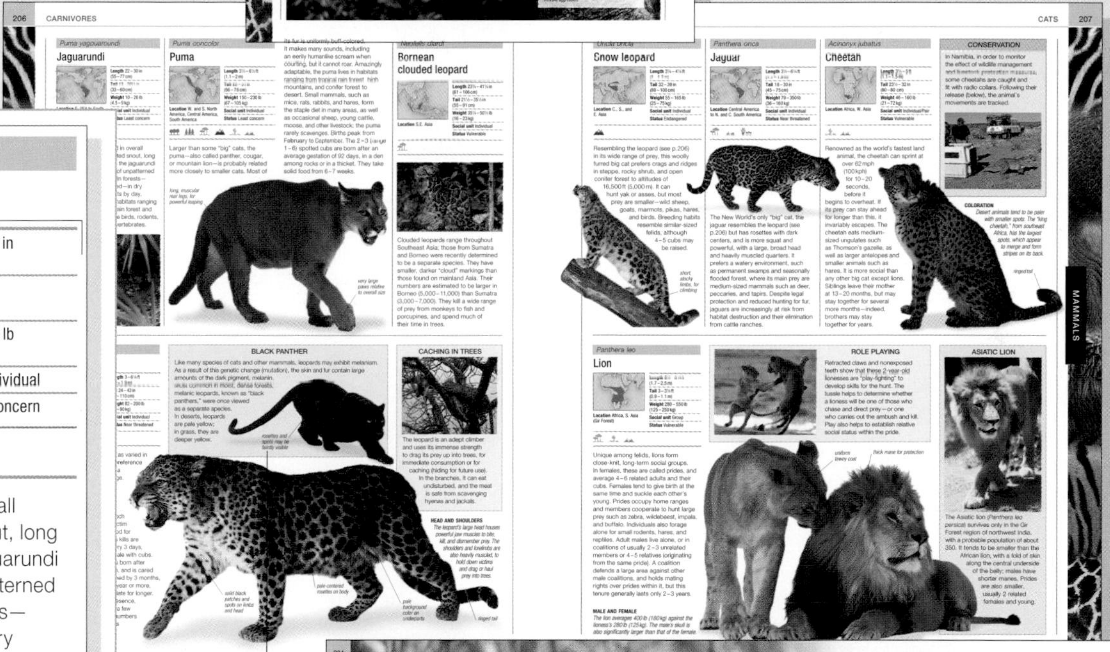

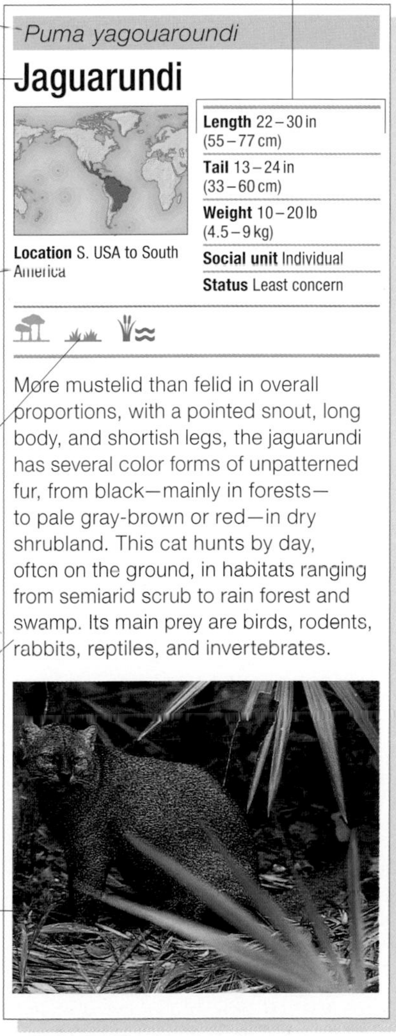

Puma yagouaroundi
Jaguarundi
Length 22–30 in (55–77 cm)
Tail 13–24 in (33–60 cm)
Weight 10–20 lb (4.5–9 kg)
Location S. USA to South America
Social unit Individual
Status Least concern
More mustelid than felid in overall proportions, with a pointed snout, long body, and shortish legs, the jaguarundi has several color forms of unpatterned fur, from black—mainly in forests—to pale gray-brown or red—in dry shrubland. This cat hunts by day, often on the ground, in habitats ranging from semiarid scrub to rain forest and swamp. Its main prey are birds, rodents, rabbits, reptiles, and invertebrates.

COMMON NAME
The species' common name appears below the Latin name, with alternative common names given in the text profile below

LOCATION
Color maps indicate the world distribution for each species with further details given in the caption below

HABITAT SYMBOLS
These define the environment in which the animal is found (see opposite for key). Additional information may be given in the text profile below

TEXT PROFILE
Each entry has a description of the most characteristic and noteworthy features of the species

ILLUSTRATIONS
Most entries include a color photograph or artwork of the species. The animals pictured are normally adult males

FEATURE BOXES
Profiles may include features on anatomy or species variation (blue box), behavior (yellow box), or conservation (green box)

FEATURE PROFILES ▷
Species of particular interest are given a double-page entry. These consist of a profile with feature boxes, a spectacular, close-up photograph, and, in some cases, an action sequence.

INTRODUCTION

Animals form the largest of the natural world's 6 kingdoms. Although they evolved after other living things, they are now the dominant form of life on earth. Among the attributes that have made them so successful are the abilities to adapt their behavior and to move. This section looks at the features that set them apart from other life forms, how they have evolved, their often complex responses to the world around them, and the way biologists classify them. It also examines the many threats they face.

WHAT ARE ANIMALS?

With almost 2 million species identified to date, and even more than that awaiting discovery, animals are the most varied living things on the planet. For over a billion years, they have adapted to the changing world around them, developing a vast array of different lifestyles in the struggle to survive. At one extreme, animals include fast-moving predators, such as sharks, big cats, and birds of prey; at the other, there are the inconspicuous sorters and sifters of the animal world's leftovers, living unseen in the soil or on the deep seabed. Together, they make up the animal kingdom—a vast collection of living things that are linked by a shared biology and that occupy a dominant place in life on Earth.

Characteristics of animals

Animals are usually easy to distinguish from other forms of life because most of them can move. However, while this rule works for most animals that live on land, it does not apply to many that live in water. Here, many animals spend their adult lives in one place, and some have trailing arms or tentacles that make them look very much like plants. A more reliable way of identifying animals is by their basic biological features: their bodies are composed of many cells, and they have nerves and muscles that enable them to respond to the world around them. Most important of all, they get the energy they need by taking in food.

Animals are highly complex and remarkably responsive, compared with other forms of life. Even the simplest animals react quickly to changes around them, shrinking away from potential danger or reaching out for food. Animals with well-developed nervous systems can go much further: they can learn from experience—an ability that is unique to the animal world.

STATIC LIFE
Sea squirts are typical "sessile" animals: they spend their adult lives fixed to a solid surface. The young, which resemble tadpoles, can move freely, enabling sea squirts to spread.

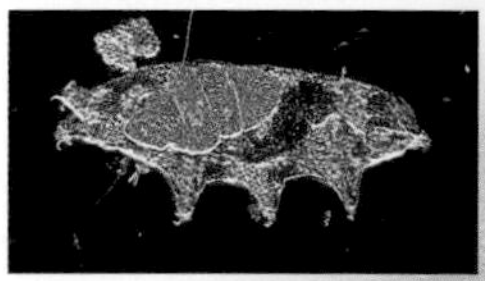

MINIATURE ANIMALS
Microscopic tardigrades (also called water bears) move about on tiny legs. Their behavior is as complex as that of animals thousands of times their size.

The scale of animal life

The world's largest living animals, baleen whales, can be up to 82 ft (25 m) long and weigh 132 tons (120 tonnes). At the other end of the spectrum are microscopic organisms—rotifers and tardigrades are only 1/500 in (0.05 mm) long—and submicroscopic flies and beetles about 1/125 in (0.2 mm) long. These animals are so tiny that their weight is negligible. Even so, they possess all the body systems needed for survival.

Different body sizes allow animals to live in different ways. Whales have few natural predators; and the same is true of elephants, the largest land animals. Their massive bodies are highly energy-efficient because they process food on such a large scale. However, they take a long time to reach maturity, which means they are slow to reproduce. Insects, on the other hand, are easy prey for many animals, and their small size means that their bodies are not as energy-efficient as those of large animals. But, since they can breed rapidly when conditions are favorable, their numbers can climb at a prodigious rate.

Chordates and invertebrates

Almost all the world's largest and most familiar animals are vertebrates—animals that have bony internal skeletons. They include the fastest animals on land, in water, and in the air, and also many species with elaborate behavior and highly developed brains. Vertebrates themselves belong to a group of animals called chordates, which share a common ancestry stretching back millions of years. However, despite leading the animal kingdom in many fields, chordates make up a tiny minority of the animal species alive today. The vast majority of species are invertebrates, or animals without backbones.

Unlike chordates, invertebrates often have very little in common with each other, apart from their lack of a backbone. The giant squid, the largest invertebrate,

OCEAN GIANTS
A humpback whale bursts out of the sea. Animals this large depend on aquatic habitats because water can support most of their weight.

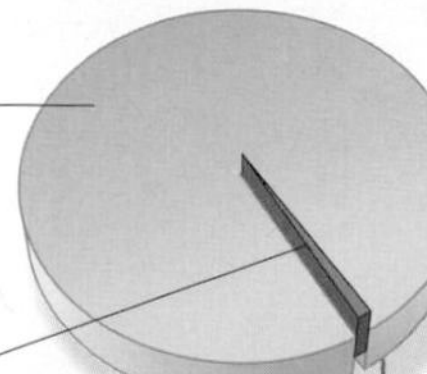

INVERTEBRATE MAJORITY
Chordates make up less than 3 percent of the world's animals. The remaining 97 percent are invertebrates.

KINGDOMS OF LIFE

Biologists classify all living things into overall groups, called kingdoms. The members of each kingdom are alike in fundamental ways, such as in the nature of their cells or in the way they obtain energy. In the most widely used system of classification there are 6 kingdoms, of which the animal kingdom is the largest. In recent years, a new classification system has been proposed. In this, there are 3 "superkingdoms": Archaebacteria, Eubacteria, and Eukaryota. The first and second reflect chemical and physical differences within bacteria. The third contains the living things that, unlike bacteria, have complex cells: protists, fungi, plants, and animals.

ANIMALS

Animals are multicellular organisms that obtain energy by ingesting food. All animals are capable of moving at least some parts of themselves, and many can move from place to place.

PLANTS

Plants are multicellular organisms that grow by harnessing the energy in light. Through a process called photosynthesis, they use this energy to build up organic matter from simple materials, creating most of the food on which animals rely.

FUNGI

Most fungi are multicellular. They collect energy from organic matter, which they do not ingest but break down externally using microscopic threads that spread throughout their food. Many fungi are too small to be seen, but some form large fruiting bodies.

PROTISTS

Protists are single-celled organisms that typically live in water, or in permanently moist habitats. Their cells are larger and more complex than those of bacteria. Some protists behave like plants, collecting energy from sunlight; others, known as protozoa, are more like animals, acquiring energy by ingesting food.

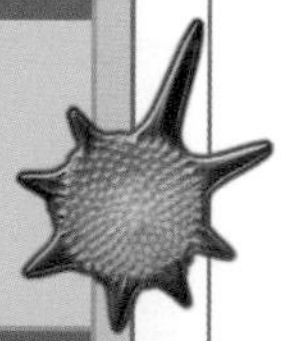

BACTERIA

Together with Archaea, bacteria are the simplest fully independent living things. Their cells are prokaryotic, meaning they lack organelles—the specialized structures that more complex cells use for carrying out different tasks. Bacteria gather energy from various sources, including organic and inorganic matter, and sunlight.

ARCHAEA

Single-celled archaeans used to be classified among bacteria, and like them they lack cellular organelles. But details of their chemical makeup link them more closely to higher organisms. Many archaeans are extremophiles: they are adapted to live in extreme environments, such as hot volcanic springs and super-saline lakes.

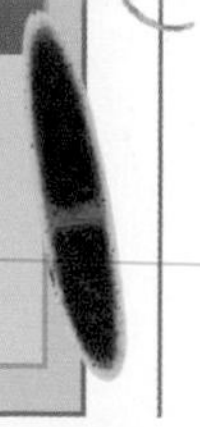

INTRODUCTION

SUPPORT SYSTEMS
Many invertebrates—such as leeches—have no hard body parts; to keep their shape they rely on the pressure of their body fluids. Simple chordates, which include lancelets, have a strengthening rod, or notochord, that runs along the length of their bodies. Vertebrates, which are more advanced chordates, are the only animals that have internal skeletons made of bone.

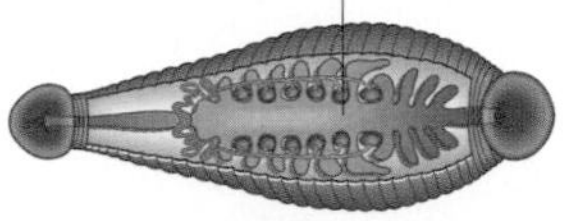
fluid prevents leech's body from collapsing
LEECH

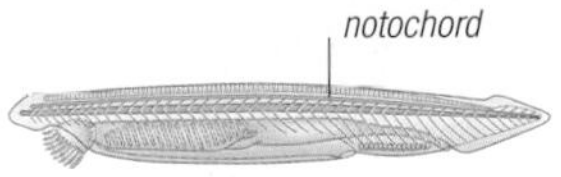
notochord
LANCELET

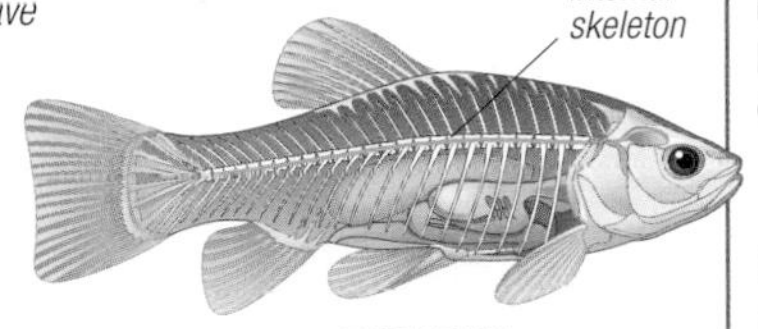
internal skeleton
BONY FISH

can measure over 52 ft (16 m) long, but it is very much an exception. Most invertebrates are tiny, and many live in inaccessible habitats, which explains why they are still poorly known compared with chordates.

Warm- and cold-blooded animals

Most animals are cold-blooded (or ectothermic), which means that their body temperature is determined by that of their surroundings. Birds and mammals are warm-blooded (or endothermic), which means that they generate their own heat and maintain a constant internal temperature regardless of the conditions outside.

This difference in body temperature has some far-reaching effects on the way animals live, because animal bodies work best when they are warm. Cold-blooded animals, such as reptiles, amphibians, and insects, operate very effectively in warm conditions, but they slow down if the temperature drops. They can absorb some heat by basking in sunshine, but if the temperature falls below about 50°F (10°C) their muscles work so slowly that they find it difficult to move. Birds and mammals, on the other hand, are barely affected by this kind of temperature change. Their internal heat and good insulation enable them to remain active even when temperatures fall below freezing.

HEAT CONTROL
A basking butterfly soaks up the sunshine. By basking, or by hiding in the shade, butterflies and other cold-blooded animals can adjust their body temperature. Even so, they have difficulty coping with extreme temperatures, especially severe cold.

Individuals and colonies

Physically, most chordates function as separate units, even though they may live together in families or larger groups. In the invertebrate world, it is not unusual for animals to be permanently linked together, forming clusters called colonies. Colonies often look and behave like single animals. Most are static, but some—particularly ones that live in the sea—can move about. Colonial species include some of the world's most remarkable invertebrates. Pyrosomes, for example, form colonies shaped like test tubes that are large enough for a diver to enter. But, in ecological terms, the most important colonial animals are reef-building corals, which create complex structures that provide havens for a range of other animals. In reef-building corals, the members of each colony are usually identical. But in some colonial species, the members have different shapes designed for different tasks. For example, the Portuguese man-of-war—an oceanic drifter that has a highly potent sting—looks like a jellyfish but consists of separate animals, called polyps, that capture food, digest it, or reproduce. They dangle beneath a giant, gas-filled polyp that acts as the colony's float.

LIVING TOGETHER
This branching coral is covered with a living "skin" that connects its individual animals, or polyps. The polyps are in constant contact with each other but otherwise lead separate lives. Each has a set of stinging tentacles and catches its own food.

Fuel for life

Animals obtain their energy from organic matter, or food. They break food up by digesting it, and then they absorb the substances that are released. These substances are carried into the animal's cells, where they are combined with oxygen to release energy. This process—called cellular respiration—is like a highly controlled form of burning, with food acting as the fuel.

The majority of animals are either herbivores, which eat plants, or carnivores, which eat other animals. Carnivores include predators, which hunt and kill prey, and parasites, which feed in or on the living bodies of other animals. There are also omnivores, which eat both animal and plant food, and scavengers, which feed on dead matter—from decaying leaves and corpses to fur and even bones.

All animals, regardless of lifestyle, ultimately provide food for other animals. All are connected by food chains, which pass food—and its energy—from one species to another. However, individual food chains are rarely more than 5 or 6 links long. This is because up to 90 percent of an animal's energy cannot be passed on: it is used up in making the animal's own body work.

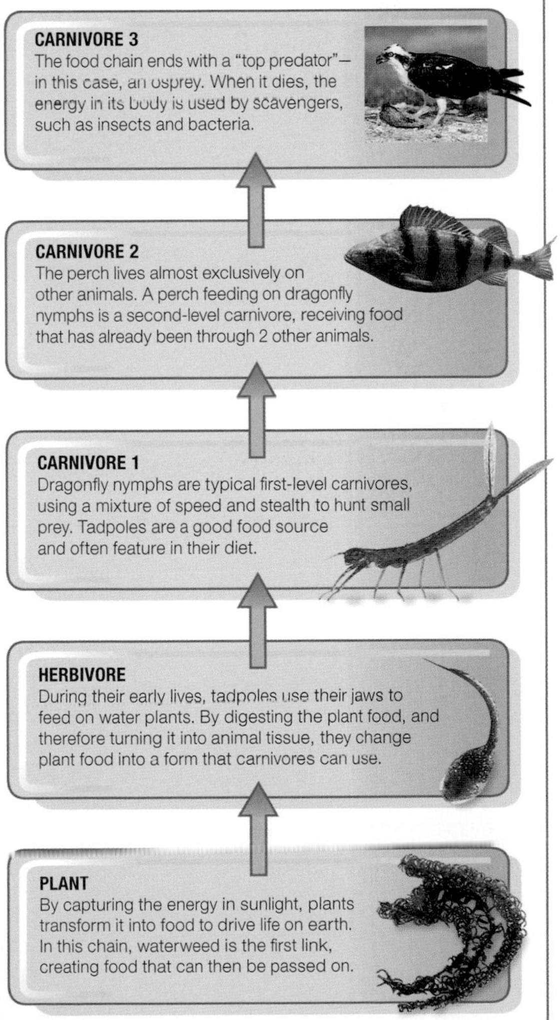
CARNIVORE 3
The food chain ends with a "top predator"—in this case, an osprey. When it dies, the energy in its body is used by scavengers, such as insects and bacteria.
CARNIVORE 2
The perch lives almost exclusively on other animals. A perch feeding on dragonfly nymphs is a second-level carnivore, receiving food that has already been through 2 other animals.
CARNIVORE 1
Dragonfly nymphs are typical first-level carnivores, using a mixture of speed and stealth to hunt small prey. Tadpoles are a good food source and often feature in their diet.
HERBIVORE
During their early lives, tadpoles use their jaws to feed on water plants. By digesting the plant food, and therefore turning it into animal tissue, they change plant food into a form that carnivores can use.
PLANT
By capturing the energy in sunlight, plants transform it into food to drive life on earth. In this chain, waterweed is the first link, creating food that can then be passed on.

FOOD CHAIN
This is a typical 5-stage food chain from a freshwater habitat. Food and energy move upward through the chain, being passed on every time one organism eats another. The chain ends when it reaches an animal that has no natural predators. The energy in the final animal's body is ultimately passed on to scavengers and to recyclers, known as decomposers, many of which live in soil.

INSULATION
In sub-zero conditions, superb insulation keeps the body temperature of these young emperor penguins at an almost constant 104°F (40°C).

EVOLUTION

Like all living things, animals undergo changes as each new generation succeeds the one before. These changes are usually so slight that they are very difficult to see, but over thousands or millions of years, they can completely alter the way animals look and also the way they behave. This process of change is called evolution. It allows animals to exploit new opportunities and to adapt to changes that take place in the world around them. Evolution works by modifying existing characteristics, usually through a series of extremely small mutations. The result of this is that every animal is a living store of evolutionary history—one that helps to show how different species are related.

Animal adaptation

Evolution is made possible by the variations that exist within animals, and it occurs mainly because animals compete with each other for limited resources, such as space and food. In this competition, some characteristics prove to be more useful than others, which means that their owners are more likely to thrive, and to produce the most young. Animals with less useful features face more difficulties and find it harder to breed. The least successful animals are therefore gradually weeded out, while those with "winning" characteristics become widespread. This weeding-out process is called natural selection.

TAWNY OWL

SNOWY OWL

BLENDING IN
Most owls have brown plumage, which helps them hide in trees. However, the snowy owl, from the treeless Arctic tundra, is mainly white—an adaptation that increases its chances of catching food and raising young.

A HISTORY OF LIFE
Geologists divide earth's history into periods characterized by major physical changes, such as bursts of volcanic activity, collisions between continents, or alterations in climate. Many of these periods have ended in worldwide mass extinctions, both on land and in the sea. For each period described below, the date shown indicates the period's end.

Natural selection operates all the time, invisibly screening all the subtle variations that come about when animals reproduce. For example, for many animals, camouflage is a valuable aid to survival. Natural selection ensures that any improvements in an animal's camouflage—a slight change in color, pattern, or behavior—are passed on to the next generation, increasing its chances of survival and therefore its chances of producing young.

Adaptations such as camouflage last only as long as they are useful: if an animal's lifestyle changes, the path that evolution follows changes, too. This has happened many times with birds: some lineages have evolved the power of flight only to lose it when they take up life on land.

Species and speciation

A species is any group of animals that has the potential to interbreed and that, under normal circumstances, does not breed with any other group. Speciation is one of the evolutionary processes that brings about new species. It usually occurs when an existing species becomes split up into 2 or more isolated groups, which are kept apart either by physical barriers, such as seas and mountain ranges, or by changes in behavior. If these groups remain separate for long enough, they evolve their own characteristic adaptations and become so different that they can no longer interbreed.

Speciation is difficult to observe because it occurs so slowly, but evidence of it is not hard to find. Many animals—from butterflies to freshwater fish—show distinct regional differences. In time, these local forms, or subspecies, can become species in their own right.

GERMAN FORM

JAPANESE FORM

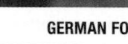

SPECIATION
The Apollo butterfly is a highly variable species. Many forms are restricted to specific parts of the Northern Hemisphere. Since the Apollo often lives on mountains, its different forms tend to remain apart. In time, these forms could evolve into new species.

Extinction

Extinction is a natural feature of evolution because for some species to succeed, others must fail. Since life began, about 99 percent of the Earth's species have disappeared and, on at least 5 occasions, huge numbers have died out in a relatively short time. The most recent of these mass extinctions, about 65 million years ago, swept away the dinosaurs and many other forms of life. However, despite such catastrophes, the total number of living species has, until recently, followed a generally upward trend. Today, the extinction rate is increasing rapidly as a result of human interference in natural ecosystems. Primates, tropical birds, and many amphibians are particularly threatened. For the foreseeable future, this decline is set to continue because evolution generates new species far more slowly than the current rate of extinction.

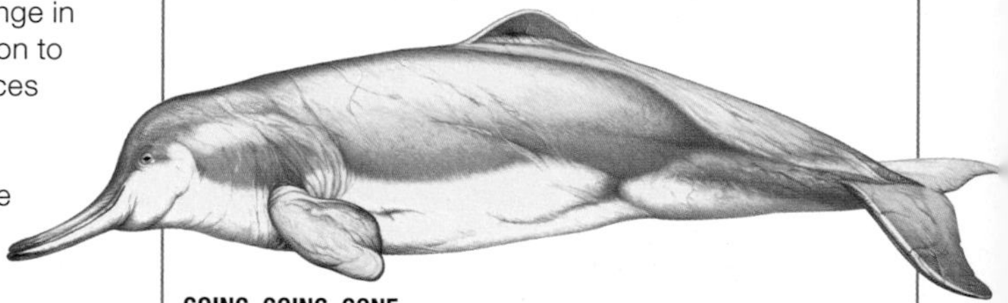

GOING, GOING, GONE
Declared extinct in August 2007, the Chinese river dolphin lived only in the Yangtze river basin. Increasing human activity caused the population to decline drastically from 400 in the 1980s to 100 in the mid-1990s. None has been found in recent surveys.

Convergence

Unrelated animals often develop very noticeable similarities. For example, sharks and dolphins are fundamentally very different, but both have streamlined bodies with an upright dorsal fin—a shape that gives them speed and stability underwater—while moles and marsupial moles share a range of adaptations for life underground. Amphisbaenians and caecilians also look very similar, although the former are reptiles and the latter are highly unusual amphibians. Such similarities are the result of convergent evolution,

NUMBER OF EXTINCT FAMILIES	PRECAMBRIAN	CAMBRIAN	ORDOVICIAN	SILURIAN	DEVONIAN	CARBONIFEROUS
800 700 600 500 400 300 200 100 0	This vast expanse of geological time stretches from when the continents first formed to when animals with hard body parts first appeared in the fossil record. Life itself emerged near the beginning of Precambrian time, about 3.8 billion years ago. The date for the first animals is less certain: being soft-bodied, they left few traces, although their existence is evident from fossils of burrows and tracks (about one billion years old).	The start of the Cambrian period was marked by an extraordinary explosion of animal life in the seas. Cambrian animals included mollusks, echinoderms, and arthropods. These were among the first creatures to have hard, easily fossilized body parts. By the end of this period, all the major divisions of animal life, or phyla, that exist today had been established.	During this period, life was still confined to the sea. Animals included the earliest crustaceans and some of the earliest jawless fish. Trilobites (arthropods with a 3-lobed body) were numerous, as were long-shelled nautiloids (predatory mollusks with sucker-bearing arms). Like the Cambrian period, the Ordovician ended in a mass extinction, probably caused by climatic changes.	Silurian times saw the evolution of the first fish with jaws and of giant sea scorpions—relatives of today's arachnids. The first land plants appeared.	In Devonian times, jawed and jawless fish diversified rapidly, which is why the period is known as the "Age of Fishes." Many fish lived in freshwater, where warm conditions and falling water levels encouraged the evolution of primitive lungs. As a result, amphibians evolved, becoming the first vertebrates to live on land. On land, insects became widespread, and the first true forests began to form. The Devonian ended with the third mass extinction, which killed up to 70 percent of animal species.	During this period, which is also called the "Coal Age," a warm global climate encouraged the growth of forests on swampy ground—home to amphibians and flying insects, including dragonflies with wingspans of up to 23½in (60cm). In the sea, ammonoids (mollusks related to today's nautiluses), were common.
YEARS AGO (MILLIONS):	542	488	444	416	359	29

PALAEOZOIC ERA

a process in which natural selection comes up with the same set of adaptations to a particular way of life, reshaping body parts or whole animals until they outwardly look the same. Convergence is responsible for a whole series of striking similarities in the animal kingdom. It can make the task of tracing evolution extremely difficult, which is why the theories regarding animal classification often change.

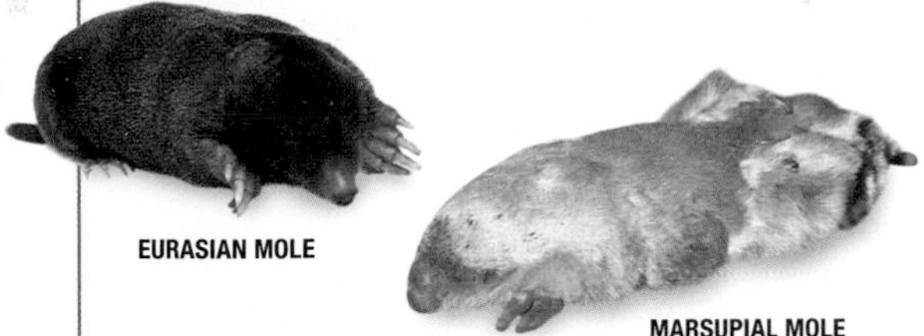

EURASIAN MOLE

MARSUPIAL MOLE

SIMILAR SHAPES

Apart from their difference in color, Eurasian moles and marsupial moles are alike in many ways. However, they are not at all closely related; their similarities are a result of convergence.

Animal partnerships

Over millions of years, animals have evolved complex partnerships with each other and with other forms of life. In one common form of partnership, called mutualism, both species benefit from the arrangement. Examples are oxpeckers and large mammals, and corals and microscopic algae. Many partnerships are loose ones but, as with pollinating insects and plants, some are so highly evolved that the 2 partners cannot survive without each other. By contrast, a commensal partnership—such as that between remoras and their host fish (see p.522)—is one in which one species gains but the other neither gains nor loses.

Partnerships may appear to be mutually obliging, but each partner is driven purely by self-interest. If one partner can tip the balance in its favor, natural selection will lead it to do so. The ultimate outcome is parasitism, in which one animal, the parasite, lives on or inside another, entirely at the host's expense.

MIXED BLESSING

Oxpeckers eat ticks and other parasites that live on the skin of large animals, but they also feed on blood from wounds—a habit that is less helpful to their hosts.

Biogeography

The present-day distribution of animals is the combined result of many factors. Among them are continental drift and volcanic activity, which constantly reshape the surface of the earth. By splitting up groups of animals, and creating completely new habitats, these geological processes have had a profound impact on animal life. One of the most important effects can be seen on remote islands, such as Australia and Madagascar, which have been isolated from the rest of the world for millions of years. Until humans arrived, their land-based animals lived in total seclusion, unaffected by competitors from outside. The result is a whole range of indigenous species, such as kangaroos and lemurs, which are found nowhere outside their native homes.

Animals are separated when continents drift apart, and they are brought together when they collide. The distribution of animals is evidence of such events long after they occur. For example, Australasia and Southeast Asia became close neighbors long ago, but their wildlife remains entirely different: it is divided by "Wallace's line," an invisible boundary that indicates where the continents came together.

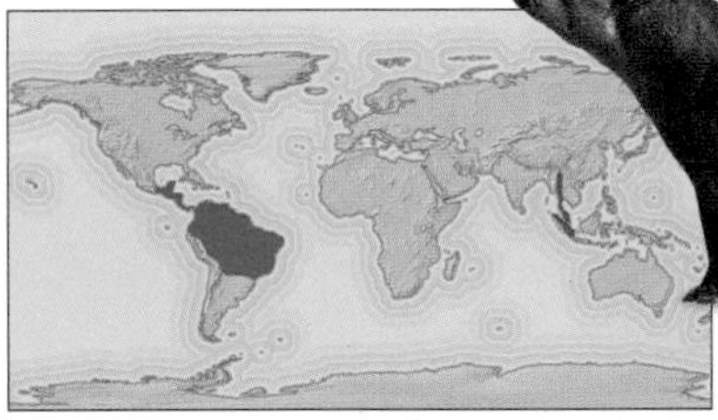

BRAZILIAN TAPIR

BRAZILIAN TAPIR DISTRIBUTION

MALAYSIAN TAPIR DISTRIBUTION

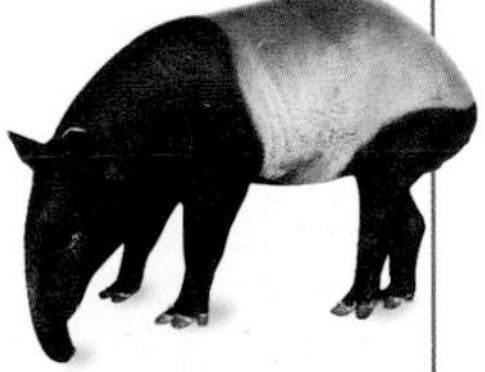

SPLIT BY CONTINENTAL DRIFT

One species of tapir lives in Southeast Asia; another 3 live in Central and South America. This indicates that these 2 landmasses were once joined.

MALAYSIAN TAPIR

ARTIFICIAL SELECTION

The variations that natural selection works on are often difficult to see. One herring, for example, looks very much like another, while starlings in a flock are almost impossible to tell apart. This is because natural selection operates on a huge range of features among many individuals in a species. However, when animals are bred in controlled conditions, their hidden variations are very easy to bring out. Animal breeders rigorously concentrate on the reproduction of particular features, such as a specific size or color, and by selecting only those animals with the desired features they can exaggerate those features with remarkable speed. This process, called artificial selection, is responsible for all the world's domesticated animals and all cultivated plants.

FAMILY LIKENESS

All domesticated dogs are descended from the gray wolf. Through artificial selection, individual breeds can be established in a very short time.

CHIHUAHUA

GRAY WOLF

PERMIAN	TRIASSIC	JURASSIC	CRETACEOUS	TERTIARY	QUARTERNARY
eptiles became the ominant land animals. ne continents formed single landmass. ne period ended with e mass extinction—robably from climate nange and volcanic ctivity—of over 75 ercent of land species nd over 90 percent marine species.	During the Triassic period, reptiles dominated life on Earth. They included flying pterosaurs, swimming forms such as nothosaurs and ichthyosaurs, and the first true dinosaurs. Early mammals also existed, but with reptiles in the ascendant they made up only a minor part of Earth's land-based fauna.	During this period, known as the "Age of Dinosaurs," reptiles strengthened their position as the dominant form of animal life. They included a wide range of plant- and meat-eating species. There was a rich variety of vegetation and, toward the end of the Jurassic, flowering plants appeared, creating new opportunities for animals—particularly pollinating insects. Birds gradually evolved from dinosaurs. The first known bird, *Archaeopteryx*, probably took to the air over 150 million years ago.	The Cretaceous period saw rapid evolution in flowering plants and in animals that could use them as food. But the most striking feature of this period was the rise of the most gigantic land animals that have ever lived. They were all dinosaurs, and they included immense plant-eating sauropods, perhaps weighing up to 88 tons (80 tonnes), as well as colossal 2-legged carnivores, such as *Tyrannosaurus*, and its even bigger relative, *Carcharodontosaurus*. Toward the end of the Cretaceous, dinosaurs were already in decline, but they were wiped out by a mass extinction that also erased many other forms of life.	Mammals existed long before this period began, but the disappearance of the dinosaurs enabled mammals to evolve rapidly, becoming larger and much more diverse. By this time in the Earth's history, the continents had split apart, allowing different groups of mammals to evolve in isolation from each other.	This period has seen abrupt changes in climate. Mammals have become supreme, and humans have become the dominant form of life.
251	200	145	65	1.8	PRESENT
	MESOZOIC ERA			CENOZOIC ERA	

CLASSIFICATION

Nearly 2 million kinds of animals have been identified by scientists, and thousands more are discovered every year. The real total may be 10 times this number, which means that the task of locating and identifying all the world's animals will never be complete. To make sense of this bewildering diversity, biologists use classification—a system that works like a gigantic filing system of all the past and present life on our planet. In classification, the entire living world is divided into groups, known as taxa. Each type of animal is given a unique scientific name, and is placed with its closest relatives, reflecting the path that evolution is thought to have followed.

Principles of classification

Scientific classification dates back over 300 years, to the work of John Ray (1627–1705). He classified plants and animals, using names that described features such as leaf shape. He also realized the importance of species, an idea that is still key to classification today. In the 18th century, another botanist—Carl Linnaeus (1707–1778)—laid the foundations of modern classification by devising classification hierarchies and 2-part scientific names. Like Ray, Linnaeus worked long before the discovery of evolution. He saw species as fixed, with unique shapes and lifestyles that were part of a divine plan.

At first sight, Linnaeus's binomial names may seem cumbersome, even though they are far more concise than the ones that were used before. They have 2 immense advantages. Unlike common names, they are unique to a particular species and they act like signposts, showing exactly where a species fits into life as a whole.

The study of classification, known as taxonomy, has rules for the way scientific names are devised and used. Each species has a genus name and a specific name by which it can be identified. The tiger, for example, is *Panthera tigris*, showing that its closest relatives are other members of the genus *Panthera*—the lion, the leopard, and the jaguar. The wild cat, on the other hand, belongs to the genus *Felis*, which includes most of the world's smaller cats. At a glance, this shows that there are differences between tigers and wild cats, in the way they have evolved as well as in their size.

Classification levels

In Linnaean classification, the species level is the basic starting point. Species are organized into groups of increasing size, starting with genera and working upward through families, orders, classes, and phyla, and then into kingdoms and domains, which are the largest groups of all.

These groups work like flexible folders, and even equivalent levels can vary enormously in size. The cat family, for example, contains 41 species, while the aardvark family contains just one. At the other extreme, in the insect world, the weevil family currently contains 50,000 species, and there are probably many more. Not all animals fit neatly into groups that Linnaeus originally devised and so, over 2 centuries of study, taxonomists have created a wide array of intermediate levels. These include superfamilies, infraclasses, suborders, and subphyla, each fitting into the hierarchy at different points. At a much finer level, many species are divided into variants called subspecies—the milksnake, for example, has about 24, each with its own distinctive markings and geographic range.

These intermediate levels do not mean that Linnaeus's system does not work. Rather, it simply reflects the fact that classification levels are really convenient labels, rather than things that have a physical existence of their own. The only category that really exists is the species, and even this level can be difficult to define. Traditionally, a species is defined as a group of living things that breed exclusively with their own kind to produce fertile young. This works well enough with many animals, but does not always hold true.

Gathering evidence

Identifying which group an animal belongs to is vital to the classification process, and anatomical features play a key part. They can be very useful in tracing the path of evolution, often showing how body parts, such as limbs or teeth, have been developed and modified for different ways of life. The limbs of 4-legged vertebrates are one of the best-known examples of this kind of evolutionary improvization. The basic limb pattern, dating back

PHYLUM	Chordata
CLASS	Mammalia
ORDER	Carnivora
FAMILY	Felidae
SPECIES	*Panthera tigris*

TIGER CLASSIFICATION
In this book, panels such as the one above are used to identify the position of animal groups in the taxonomic hierarchy. The larger panel on the right defines the various taxonomic ranks, starting at the top with the kingdom (one of the highest ranks) and ending with the species. Taking the tiger as an example, it also shows how a particular animal's physical characteristics are used to determine its place in the classification system.

KINGDOM A kingdom is an overall division containing organisms that work in fundamentally similar ways.	**Animalia** *The kingdom Animalia contains multicellular organisms that obtain energy by eating food. Most have nerves and muscles, and are mobile.*	
PHYLUM A phylum is a major subdivision of a kingdom, and it contains one or more classes and their subgroups.	**Chordata** *Animals of the phylum Chordata have a strengthening rod or notochord running the length of their bodies, for all or part of their lives.*	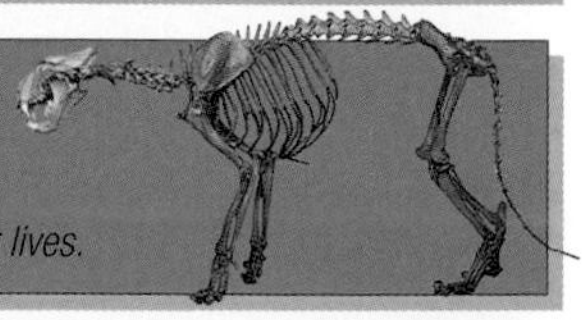
CLASS A class is a major subdivision of a phylum, and it contains one or more orders and their subgroups.	**Mammalia** *The class Mammalia contains chordates that are warm-blooded, have hair, and suckle their young. The majority of them give birth to live young.*	
ORDER An order is a major subdivision of a class, and it contains one or more families and their subgroups.	**Carnivora** *The order Carnivora contains mammals that have teeth specialized for biting and shearing. Many of them, including the tiger, live primarily on meat.*	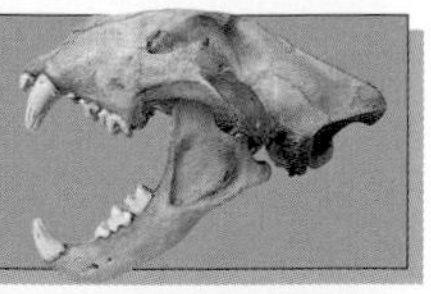
FAMILY A family is a subdivision of an order, and it contains one or more genera and their subgroups.	**Felidae** *The family Felidae contains carnivores with short skulls and well-developed claws. In most cases, the claws are retractable.*	
GENUS A genus is a subdivision of a family, and it contains one or more species and their subgroups.	***PANTHERA*** *The genus* Panthera *contains large cats that have a specialized larynx with elastic ligaments. Unlike other cats, they can roar as well as purr.*	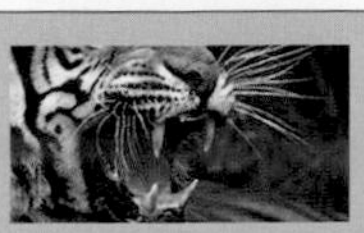
SPECIES A species is a group of similar individuals that are able to interbreed in the wild.	***PANTHERA TIGRIS*** *The tiger is the only member of the genus* Panthera *that has a striped coat when adult. There are several varieties, or subspecies.*	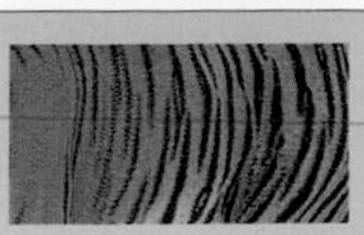

MICRO-ANIMALS

A large part of the animal kingdom consists of animals that are too small to be seen with the naked eye. Some kinds, such as the tardigrades and rotifers, are almost always microscopic, while others have close relatives that are much easier to see. Whatever their size, most have a clearly recognizable body plan, which can be used in their classification. However, occasionally, a micro-animal appears that does not fit any recognized group. One example is the phylum Loricifera—the most recent phylum of animals, which was added in the late 20th century.

SAMPLING A STREAM
Aquatic habitats often teem with tiny animals, which can be trapped with ultrafine nets. Many are straightforward to classify, but larval forms can cause problems—in the past, they have sometimes been classified as species in their own right.

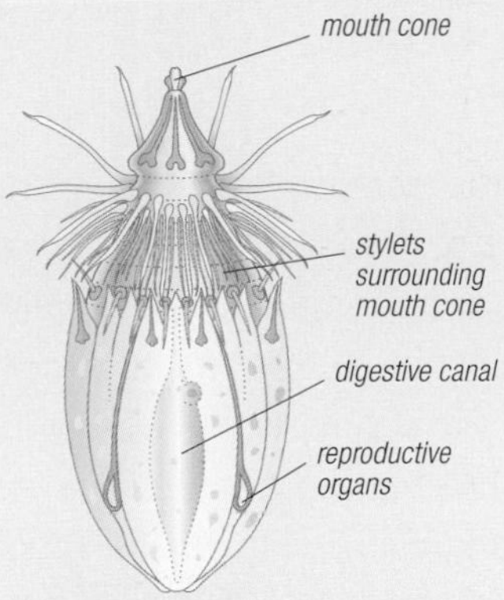

A NEW FORM OF LIFE
Loriciferans were first described in 1983. Measuring about (1mm) long, they live in marine sediments, and have a mouth surrounded by sharp stylets, or spines. They probably live worldwide.

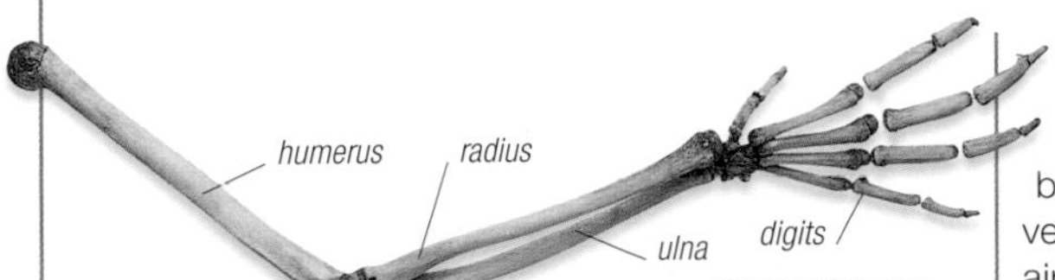

CHIMPANZEE ARM

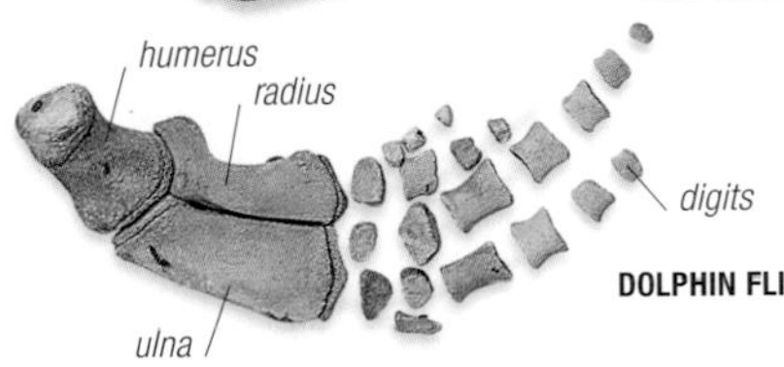

DOLPHIN FLIPPER

ARMS AND FLIPPERS
A human arm and a dolphin's flipper look very different in life, but they contain the same arrangement of bones. This similarity is strong evidence that chimpanzees (and humans) and dolphins evolved from a common ancestor.

over 300 million years, is built around 3 main sets of bones: a single bone near the body, 2 bones farther out, and 5 sets of smaller bones at the limb's farthest point. As 4-legged vertebrates spread from the water to the land and air, their limbs became very different in shape and size, and in the way they worked, however, the underlying pattern of bones was retained, making it an identifying feature that remains to this day. Because this pattern is unlikely to have evolved more than once, it provides important evidence that these animals evolved from a shared ancestor.

Conflicting clues

Taxonomy is rarely straightforward, because evolution provides a mix of useful and confusing clues. Useful ones include the underlying patterns of bones or body segments, and microscopic

SUPERFICIALLY SIMILAR
Although they look similar, sharks and bony fish differ in many ways. Sharks have skeleton made of cartilage, and are covered with denticles, or rough scales. Bony fish have a calcified skeleton, a swim bladder, and a flap that covers their gills.

features, such as the way a single cell divides when a new animal starts to take shape. They also include vestigial organs that have lost their original function over time, and become smaller as a result. Among them are the hind limb bones of some snakes that play no part in movement, but indicate a 4-legged ancestry.

Of the confusing factors, the most difficult is convergence (p.16)—the process that results in unrelated species developing similar adaptations for similar ways of life. Convergence can make them look superficially very similar, a case in point being cartilaginous and bony fishes (see above), both of which are adapted to living in water. However, when examined in more detail, it becomes clear they have converged, and there are fundamental differences that show they have evolved from separate branches of the evolutionary tree.

Convergence can make the relationships between organisms extremely difficult to unravel. A record of the past is built into every living thing in the form of DNA. Since the 1990s, technological advances in both computing and DNA extraction have helped explain convergence and many other questions. By studying the similarities and differences in genetic material, relationships between different organisms can be established regardless of their appearance.

NEW SPECIES

Every year, the world's taxonomists examine thousands of new species that have been collected or photographed in many different parts of the world. Most of them are invertebrates, particularly from the sea, but there are also reptiles, birds, and mammals, often from remote areas of the tropics. Each new species is formally identified, named, and described as part of the classification process.

ONE-OFF
Described in 2004, the Laotian rock rat Laonastes aenigmamus *is so different from other rodents that it is classified in a family of its own.*

New discoveries

Since classification first began, taxonomists have identified and catalogued species from all over the world. Today, that process continues. Most large, land-dwelling animals are known, but huge numbers of invertebrates, particularly in the deep ocean, have yet to be discovered and described.

Victorian zoologists—including famous ones such as Charles Darwin (1809–1882)—thought little about trapping new species, or shooting them out of the trees. In today's more enlightened times, technology plays a role instead. Automatic cameras photograph animals on rain

HIDDEN DEPTHS
Discovered in 2003, the red jellyfish Tiburonia granrojo *grows up to 10ft (3m) across. It was found in deep water in the northern Pacific.*

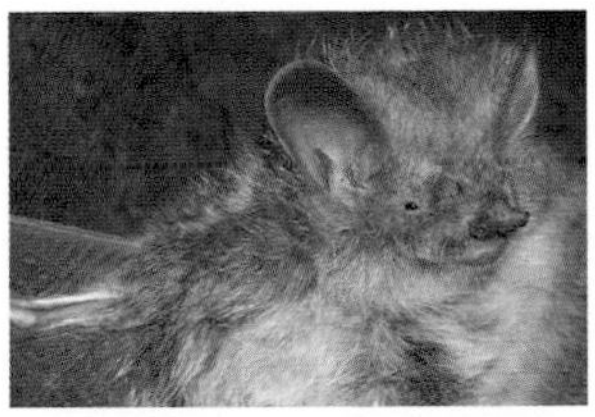

UNUSUAL BAT
*Elery's tube-nosed bat (*Murina eleryi*) was described in 2009. This tiny North Vietnamese bat lives in deciduous and dry evergreen forest that grows on limestone.*

forest floors, and deep in the oceans where there is no natural light. Thanks to these, some highly secretive mammals have been observed, such as the Bornean clouded leopard (p.206)—a nocturnal rain forest predator. Giant squid (p.599), long known from dead remains, have been seen in their natural habitat for the very first time.

Splitting species

New species can also be found in another way—by reexamining ones that are already known. Sometimes, this shows that 2 species are actually one, but it can also have the opposite effect. Two species may look similar, but their anatomy, behavior, and genes may show that this similarity is just skin-deep. When this happens, the species is split into 2, and one of them renamed.

There are many examples of new species being "discovered" in this way. For example, the greater flamingo (p.285) is now recognized as being different from the American flamingo, whereas formerly they were thought to be one. The tuatara (p.383)—long treated as a unique species of reptile—is currently classified as 2. "Splitting" is particularly common among forest primates, accounting for a considerable growth in the number of species, even though their numbers are often in serious decline.

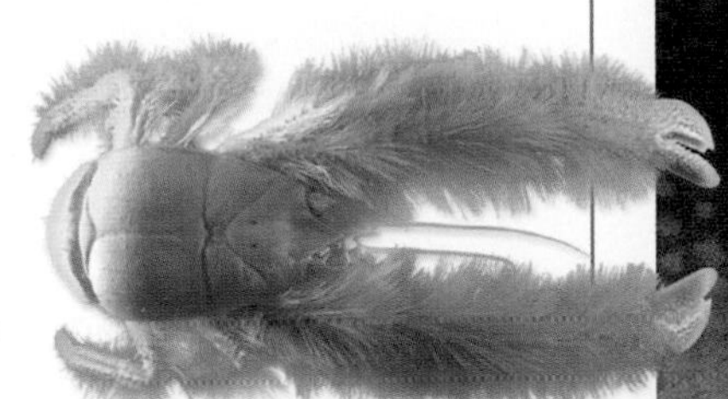

HAIRY LOBSTER
The remarkable yeti lobster Kiwi hirsuta *was found in 2005, during a census of marine life. It lives in the South Pacific, around deep-sea vents.*

CLADISTICS

During the 1950s, a new classification system was proposed by the German entomologist Willi Hennig (1913–1976). Unlike traditional Linnaean classification, Hennig's system—called cladistics, or phylogenetics—works solely with hierarchical groups called clades. Clades can contain a handful of species, or hundreds of thousands, but each is a single branch from the tree of life, with an ancestor and all of its descendants. Cladistics was controversial when it was first introduced, but with the advance of DNA analysis in the 1990s, it has moved to the forefront of modern classification.

Clades

Cladistics investigates the sequence in which features have evolved. It makes 2 basic assumptions: that living things are all related—however distantly—and that features evolve over time. By taking a group of species, and then comparing the features that they do or do not share, taxonomists generate cladograms, or "family trees," showing the most likely path that evolution has followed.

Cladistics differs from Linnaean classification in several ways. Firstly, it is a statistical science, which aims to be as objective as possible about the species being studied. Secondly, the only groups it recognizes are clades. These can be of any size, but by definition they are monophyletic, meaning that they contain a complete evolutionary line, with all the offshoots that it has produced. Linnaean classification, on the other hand, long predates the discovery of evolution. Many of its groups are indeed clades, but some are polyphyletic, meaning that they contain several evolutionary lines treated as if they were one. Others are paraphyletic, which means that they contain some branches of a clade, but not all. For example, Linnaean classification recognizes reptiles as a separate class, and it gives the same treatment to birds. Cladists recognize them both as members of the same clade, which also includes the dinosaurs.

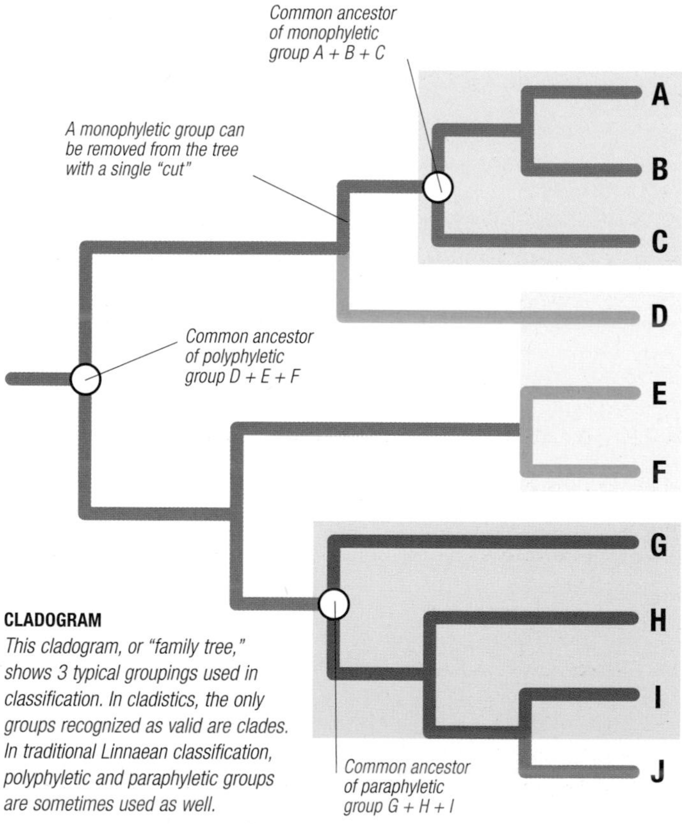

CLADE
Also known as a monophyletic group, this contains a single shared ancestor, and all its descendants, both alive and extinct. A clade can be of any size, and may contain further clades nested within it.

POLYPHYLETIC GROUP
Unlike a clade, this contains a number of species, but not their shared ancestor. Warm-blooded animals, for example, form a group like this: they consist of mammals and birds, but do not form a clade.

PARAPHYLETIC GROUP
A paraphyletic group contains an ancestral species, but not all of its descendants. As a result, it does not form a clade. Reptiles are an example, because, traditionally, they do not include birds.

CLADOGRAM
This cladogram, or "family tree," shows 3 typical groupings used in classification. In cladistics, the only groups recognized as valid are clades. In traditional Linnaean classification, polyphyletic and paraphyletic groups are sometimes used as well.

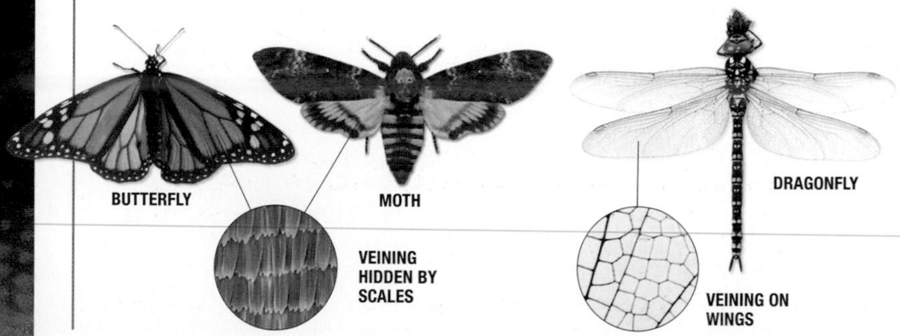

REVEALING WINGS
All flying insects have evolved from a single common ancestor that had 6–8 strengthening veins along its wings. In dragonflies, the primitive pattern of veins has been subdivided many times. But in butterflies and moths, it is usually hidden by scales; these scales point to their shared ancestry.

GENETIC INSIGHTS

When used with cladistics, genetic evidence often confirms traditional classification, but in some cases it refutes existing ideas. For example, snakes and lizards are traditionally classified in the order Squamata, which also includes the amphisbaenians; some snakes have the vestiges of pelvic bones, which strongly suggests that snakes had lizardlike ancestors. However, DNA evidence shows that the picture may be more complex; some snakes are more closely related to lizards than they are to each other, suggesting that a snakelike body form has evolved several times. Based on this evidence, snakes may not form an individual clade.

LIZARD

SNAKE

LOSING LEGS
Lizards typically have well developed legs, many rows of scales on their undersides, and eyelids that can blink. By contrast, snakes are legless, with a single row of belly scales, and eyes covered by a single transparent scale, or brille. These adaptations may have evolved more than once, originally for a burrowing lifestyle.

Cladistic analysis

When cladistics was first introduced, it relied on structural features, such as bones, feathers, or vein patterns on insects' wings. It still uses these today, but in addition, modern cladistics works by comparing genes. Whatever the features, the procedure is the same: the "primitive" (ancestral) or "derived" (altered) state of various features is scored and compared for a particular group of species. Those sharing the most derived features are likely to be most closely related—a principle that is used to build up the branch-points of a family tree. For example, moths and butterflies share numerous features with other insects, but they differ in being covered in microscopic scales. This derived feature—together with many others—strongly suggests that they are more closely related to each other than to other insects. Initially, cladists processed their data by hand, but today's computers can deal with thousands of features, analyzing huge numbers of outcomes to produce the most probable result.

Two systems

Linnaean classification has a specific hierarchy of named levels. Cladistics does not work like this, because branch-points can occur at any stage in a family tree. As a result, cladists prefer not to give clades different levels of their own. Instead, they are often referred to as "unranked clades," followed by a definition or shortened name.

Taken to its logical conclusion, cladistics rewrites much of animal classification, creating a plethora of unranked clades in the place of traditional Linnaean groups. However, the Linnaean system is very practical, which is why the 2 systems are often used side by side. Cladistics provides a powerful way of investigating the past, and determining how closely different species are related, while the Linnaean system provides a concise way of identifying and dealing with familiar groups and their names.

ANIMAL GROUPS AND NAMES

The following 5 pages summarize the classification system used in this book. Like all classification systems, it represents current thinking, which is always liable to change. Groups are "nested" to show how they are related; informal groups with no distinct biological identity are bounded by dotted lines. Some of the species totals, particularly for the invertebrates, are based on estimates; as with new discoveries, these can be subject to rapid change.

Animal groups

The animal kingdom is divided into more than 30 phyla. Some of the smallest phyla contain less than 100 species, while at the other extreme, the arthropod phylum contains as least 1.2 million, making these arguably the most successful animals of all time. Between these 2 extremes are the chordates—animals that usually have a backbone, and that make up a large part of this book.

Although not front runners in terms of species, chordates are exceptional in many ways. They show an extraordinary range in dimensions, from fish the size of a fingernail to majestic 100-ton whales. They can be found on land, at sea, and in the air, where they outrun, outswim, or outfly all other forms of life. Since antiquity, they have been much observed, which is why many names for animal groups have their origin in those times.

Group names

Like species names, group names often look obscure but they have a language and logic of their own. Many of them are named after particular body parts that make a group distinct. Chordates, for example, get their name from the notochord—a strengthening rod that runs down their backs for part or all of their lives. Arthropods get their name from their joint-bearing legs, one of the features that accounts for their success. Among chordates, frogs and toads make up the order Anura, which simply means "without tails," while elephants make up the order Proboscidea, after their long proboscis, or nose.

Sometimes, ancient folklore also plays a part in names. For example, nightjars and frogmouths, in the avian order Caprimulgiformes, got their name from the belief that nightjars sucked milk from goats. Dugongs and manatees are more fanciful still. They belong to the order Sirenia, from their supposed resemblance to the three sirens of Greek mythology—mythical figures who seduced sailors into shipwreck on the rocks.

An ongoing process

Classification changes all the time as more is discovered about animals and their evolutionary past. Some changes affect life at its highest levels: today, for example, it is generally accepted that there are 6 kingdoms of life instead of the previous 5 (see p.14). Changes like this are highly unusual, but changes to lower-level classification happen every day. Through cladistics and DNA analysis, much more information about the relationships between species is becoming available, and as a result, groups at different taxonomic levels are often moved, split, or joined together.

Seals, for example, were formerly considered to be a separate group of mammals, the order Pinnipedia. Today, most zoologists classify them as a family within the order Carnivora, together with terrestrial carnivores such as cats, dogs, and bears. Among birds, the hammerkop and shoebill were formerly classified with herons and their relatives in the order Ciconiiformes; today, they are classified with pelicans in the order Pelecaniformes instead. With moves like these, the animals themselves do not change. What does change is our understanding of them: how they have evolved, and which animals are their closest relatives.

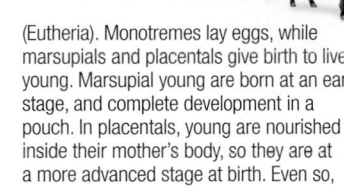

VERTEBRATES

GREVY'S ZEBRA

MAMMALS

CLASS Mammalia

Mammals are chordates characterized by having fur or hair, and raising their young on milk. There are over 5,000 living species, in 29 orders, ranging from a single species to over 2,000. Mammals are divided into 3 groups, depending on their reproductive strategy: monotremes, marsupials (Metatheria), and placental mammals (Eutheria). Monotremes lay eggs, while marsupials and placentals give birth to live young. Marsupial young are born at an early stage, and complete development in a pouch. In placentals, young are nourished inside their mother's body, so they are at a more advanced stage at birth. Even so, parental care may last at least a year.

EGG-LAYING MAMMALS
ORDER Monotremata — FAMILIES 2 — SPECIES 5

MARSUPIALS
INFRACLASS Marsupialia — FAMILIES 20 — SPECIES 319

AMERICAN OPOSSUMS
ORDER Didelphimorphia — FAMILIES 1 — SPECIES 86

AUSTRALASIAN CARNIVOROUS MARSUPIALS
ORDER Dasyuromorphia — FAMILIES 2 — SPECIES 70

BANDICOOTS
ORDER Paramelemorphia — FAMILIES 3 — SPECIES 19

MARSUPIAL MOLES
ORDER Notoryctemorphia — FAMILIES 1 — SPECIES 2

KANGAROOS AND RELATIVES
ORDER Diprotodontia — FAMILIES 11 — SPECIES 136

SHREW OPOSSUMS
ORDER Paucituberculata — FAMILIES 1 — SPECIES 6

MONITO DEL MONTE
ORDER Microbiotheria — FAMILIES 1 — SPECIES 1

SENGIS
ORDER Macroscelidea — FAMILIES 1 — SPECIES 15

TENRECS AND GOLDEN MOLES
ORDER Afrosoricidae — FAMILIES 2 — SPECIES 53

AARDVARK
ORDER ORYCTEROPODIDAE — FAMILIES 1 — SPECIES 1

DUGONG AND MANATEES
ORDER Sirenia — FAMILIES 2 — SPECIES 4

ELEPHANTS
ORDER Proboscidea — FAMILIES 1 — SPECIES 3

HYRAXES
ORDER Hyracoidea — FAMILIES 1 — SPECIES 4

ARMADILLOS
ORDER Cingulata — FAMILIES 1 — SPECIES 21

INTRODUCTION

SLOTHS AND ANTEATERS
ORDER Pilosa — FAMILIES 4 — SPECIES 11

RABBITS, HARES, AND PIKAS
ORDER Lagomorpha — FAMILIES 3 — SPECIES 92

RODENTS
ORDER Rodentia — FAMILIES 33 — SPECIES 2,272

- **SQUIRREL-LIKE RODENTS** — SUBORDER Sciuromorpha — FAMILIES 3 — SPECIES 307
- **BEAVERLIKE RODENTS** — SUBORDER Castorimorpha — FAMILIES 3 — SPECIES 102
- **MOUSELIKE RODENTS** — SUBORDER Myomorpha — FAMILIES 7 — SPECIES 1,570
- **CAVYLIKE RODENTS** — SUBORDER Hystricomorpha — FAMILIES 18 — SPECIES 284
- **SPRINGHARES AND RELATIVES** — SUBORDER Anomaluromorpha — FAMILIES 2 — SPECIES 9

COLUGOS
ORDER Dermoptera — FAMILIES 1 — SPECIES 2

TREE SHREWS
ORDER Scandentia — FAMILIES 2 — SPECIES 20

PRIMATES
ORDER Primates — FAMILIES 15 — SPECIES 382

- **PROSIMIANS** — SUBORDER Strepsirrhini — FAMILIES 7 — SPECIES 90
- **MONKEYS AND APES** — SUBORDER Haplorrhini
  - **MONKEYS** — FAMILIES 6 — SPECIES 271
  - **APES** — FAMILIES 2 — SPECIES 21

BATS
ORDER Chiroptera — FAMILIES 18 — SPECIES 1,117

HEDGEHOGS AND RELATIVES
ORDER Erinaceomorpha — FAMILIES 1 — SPECIES 24

SHREWS AND RELATIVES
ORDER Soricomorpha — FAMILIES 3 — SPECIES 418

PANGOLINS
ORDER Pholidota — FAMILIES 1 — SPECIES 8

CARNIVORES
ORDER Carnivora — FAMILIES 15 — SPECIES 285

- **DOGS AND RELATIVES** — FAMILY Canidae — SPECIES 34
- **BEARS** — FAMILY Ursidae — SPECIES 8
- **SEA LIONS AND FUR SEALS** — FAMILY Otariidae — SPECIES 16
- **WALRUS** — FAMILY Odobenidae — SPECIES 1
- **EARLESS SEALS** — FAMILY Phocidae — SPECIES 18
- **SKUNKS** — FAMILY Mephitidae — SPECIES 12
- **RACCOONS AND RELATIVES** — FAMILY Procyonidae — SPECIES 14
- **RED PANDA** — FAMILYAiluridae — SPECIES 1
- **MUSTELIDS** — FAMILY Mustelidae — SPECIES 58
- **MALAGASY CARNIVORES** — FAMILY Eupleridae — SPECIES 9
- **AFRICAN PALM CIVET** — FAMILY Nandiniidae — SPECIES 1
- **MONGOOSES** — FAMILY Herpestidae — SPECIES 33
- **CIVETS AND RELATIVES** — FAMILY Viverridae — SPECIES 35
- **CATS** — FAMILY Felidae — SPECIES 41
- **HYENAS AND AARDWOLF** — FAMILY Hyaenidae — SPECIES 4

HOOFED MAMMALS

ODD-TOED HOOFED MAMMALS
ORDER Perissodactyla — FAMILIES 3 — SPECIES 17

- **HORSES AND RELATIVES** — FAMILY Equidae — SPECIES 8
- **RHINOCEROSES** — FAMILY Rhinocerotidae — SPECIES 5
- **TAPIRS** — FAMILY Tapiridae — SPECIES 4

EVEN-TOED HOOFED MAMMALS
ORDER Artiodactyla — FAMILIES 10 — SPECIES 376

- **PIGS** — FAMILY Suidae — SPECIES 19
- **PECCARIES** — FAMILY Tayassuidae — SPECIES 3
- **HIPPOPOTAMUSES** — FAMILY Hippopotamidae — SPECIES 2
- **CAMELS AND RELATIVES** — FAMILY Camelidae — SPECIES 4
- **DEER** — FAMILY Cervidae — SPECIES 51
- **MUSK DEER** — FAMILY Moschidae — SPECIES 7
- **CHEVROTAINS** — FAMILY Tragulidae — SPECIES 8
- **PRONGHORN** — FAMILY Antilocapridae — SPECIES 1
- **GIRAFFE AND OKAPI** — FAMILY Giraffidae — SPECIES 2
- **CATTLE AND RELATIVES** — FAMILY Bovidae — SPECIES 279

CETACEANS
ORDER Cetacea — FAMILIES 11 — SPECIES 85

- **BALEEN WHALES** — SUBORDER Mysticeti — FAMILIES 4 — SPECIES 14
- **TOOTHED WHALES** — SUBORDER Odontoceti — FAMILIES 7 — SPECIES 71

BIRDS

BULLOCK'S ORIOLE

CLASS **Aves**

Birds are the only members of the phylum Chordata that possess feathers. They use them to keep warm and, in most cases, to fly. Birds are closely related to reptiles, and form a single clade (p.20) with them and the now-extinct dinosaurs. In recent years, comparison of DNA from different species of birds has led to many changes in avian classification, but most systems still separate the class into 29 orders—the system used in this book. Some bird orders are very small: the ostrich order, for example, contains just one species, while that containing the mousebirds has only 6. By comparison, the order containing the passeriformes, or perching birds, contains as many species (over 6,000) as all the other orders of birds combined, and it includes all the world's songbirds. Within this huge order, there is considerable disagreement about how many families of passerine birds there are. Some ornithologists consider that there are just 60 families, but many now put the total at about 100 or even more.

	ORDER	FAMILIES	SPECIES
TINAMOUS	Tinamiformes	1	45
KIWIS	Apterygiformes	1	3
CASSOWARIES AND EMUS	Casuariiformes	2	4
OSTRICH	Struthioniformes	1	1
RHEAS	Rheiformes	1	2
GAMEBIRDS	Galliformes	5	290
WATERFOWL	Anseriformes	3	174
PENGUINS	Sphenisciformes	1	17
ALBATROSSES AND PETRELS	Procellariiformes	4	133
LOONS	Gaviiformes	1	5
GREBES	Podicipediformes	1	22
FLAMINGOS	Phoenicopteriformes	1	6
HERONS AND RELATIVES	Ciconiiformes	3	121
PELICANS AND RELATIVES	Pelecaniformes	8	67
BIRDS OF PREY	Falconiformes	3	319
CRANES AND RELATIVES	Gruiformes	11	228
WADERS, GULLS, AND AUKS	Charadriiformes	19	379
PIGEONS	Columbiformes	2	321
SANDGROUSE	Pterocliditormes	1	16
PARROTS	Psittaciformes	1	375
CUCKOOS AND TURACOS	Cuculiformes	3	170
OWLS	Strigiformes	2	202
NIGHTJARS AND FROGMOUTHS	Caprimulgiformes	5	125
HUMMINGBIRDS AND SWIFTS	Apodiformes	3	447
MOUSEBIRDS	Coliiformes	1	6
TROGONS	Trogoniformes	1	40
KINGFISHERS AND RELATIVES	Coraciiformes	10	218
WOODPECKERS AND TOUCANS	Piciformes	5	411
PASSERINES	Passeriformes	c.96	c.6,000

INTRODUCTION

INTRODUCTION

REPTILES

AFRICAN STRIPED SKINK

CLASS Reptilia

Chordates with scaly skin, or reptiles, were the first 4-legged animals to be fully at home on dry land. This is because their skin is waterproof. Snakes and lizards make up over 90 percent of living reptile species, although the largest species are tortoises and crocodiles. Most reptiles lay eggs, but a small minority give birth to live young.

TORTOISES AND TURTLES
ORDER Chelonia — FAMILIES 13 — SPECIES 317

TUATARAS
ORDER Rhyncocephalia — FAMILIES 1 — SPECIES 2

SQUAMATES
ORDER Squamata — FAMILIES 48 — SPECIES c.9,000

SNAKES
SUBORDER Serpentes — FAMILIES 18 — SPECIES c.3,400

BOAS, PYTHONS, AND RELATIVES
SUPERFAMILY Henophidia — FAMILIES 12 — SPECIES 186

COLUBRIDS AND RELATIVES
SUPERFAMILY Caenophidia — FAMILIES 3 — SPECIES c.2,800

- COLUBRIDS — FAMILY Colubridae — SPECIES c.2,100
- VIPERS — FAMILY Viperidae — SPECIES 292
- ELAPIDS — FAMILY Elapidae — SPECIES 347

BLIND AND THREAD SNAKES
SUPERFAMILY Scolecophidia — FAMILIES 3 — SPECIES 407

LIZARDS
SUBORDER Lacertilia — FAMILIES 24 — SPECIES c.5,500

IGUANAS AND RELATIVES
SUPERFAMILY Iguania — FAMILIES 3 — SPECIES 1,607

GECKOS AND SNAKE LIZARDS
SUPERFAMILY Gekkota — FAMILIES 7 — SPECIES 1,369

SKINKS AND RELATIVES
SUPERFAMILY Scincomorpha — FAMILIES 7 — SPECIES 2,252

ANGUIMORPH LIZARDS
SUPERFAMILY Anguimorpha — FAMILIES 7 — SPECIES 224

AMPHISBAENIANS
SUBORDER Amphisbaenia — FAMILIES 6 — SPECIES 181

CROCODILIANS
ORDER Crocodilia — FAMILIES 3 — SPECIES 24

AMPHIBIANS

PACIFIC GIANT SALAMANDER

CLASS Amphibia

Frogs and toads make up the largest order of amphibians, and show the widest range of adaptations for terrestrial life. Newts and salamanders most closely resemble ancestral amphibians; caecilians are an aberrant and relatively little-known group.

NEWTS AND SALAMANDERS
ORDER Caudata — FAMILIES 9 — SPECIES 597

CAECILIANS
ORDER Gymnophiona — FAMILIES 3 — SPECIES 183

FROGS AND TOADS
ORDER Anura — FAMILIES 49 — SPECIES 5,858

FISHES

AMERICAN PADDLEFISH

Despite superficial similarities, fish are a varied collection of animals with very different evolutionary histories. Lampreys have no jaws—a feature they share with the earliest vertebrates, which evolved over 500 million years ago. Cartilaginous fishes and bony fishes have jaws, skulls, and skeletons, but their anatomy is very different, and their lifestyles are usually quite distinct as well. Today, bony fishes make up by far the largest class (about 96 percent of species) with representatives in all kinds of aquatic habitats, ranging from the salty open ocean and brackish estuaries to isolated desert pools and seasonally flowing rivers. The major subclass of this group contains so many orders and species that it is dealt with at superorder level in this book.

JAWLESS FISHES

LAMPREYS
CLASS Cephalaspidomorphi — ORDERS 1 — FAMILIES 1 — SPECIES 43

CARTILAGINOUS FISHES
CLASS Chondrichthyes — ORDERS 12 — FAMILIES 51 — SPECIES c.1,200

SHARKS AND RAYS
SUBCLASS Elasmobranchii

SHARKS
ORDERS 8 — FAMILIES 31 — SPECIES c.500

SKATES AND RAYS
ORDERS 3 — FAMILIES 17 — SPECIES c.650

CHIMAERAS
SUBCLASS Holocephali — ORDERS 1 — FAMILIES 3 — SPECIES 49

BONY FISHES
CLASS Osteichthyes — ORDERS 48 — FAMILIES 482 — SPECIES c.31,000

FLESHY-FINNED FISHES
SUBCLASS Sarcopterygii — ORDERS 3 — FAMILIES 4 — SPECIES 8

RAY-FINNED FISHES
SUBCLASS Actinopterygii

PRIMITIVE RAY-FINNED FISHES
ORDERS 4 — FAMILIES 5 — SPECIES 48

BONY-TONGUED FISHES
ORDER Osteoglossiformes — FAMILIES 7 — SPECIES 219

TARPONS AND EELS
SUPERORDER Elopomorpha — ORDERS 5 — FAMILIES 24 — SPECIES 1,000

HERRINGS AND RELATIVES
SUPERORDER Clupeomorpha — ORDERS 1 — FAMILIES 6 — SPECIES 392

CATFISH AND RELATIVES
SUPERORDER Ostariophysi — ORDERS 5 — FAMILIES 73 — SPECIES c.9,600

SALMON AND RELATIVES
SUPERORDER Protacanthopterygii — ORDERS 3 — FAMILIES 16 — SPECIES 538

DRAGONFISHES AND RELATIVES
SUPERORDER Stenopterygii — ORDERS 2 — FAMILIES 5 — SPECIES c.430

LANTERNFISHES AND RELATIVES
SUPERORDER Scolepomorpha — ORDERS 2 — FAMILIES 19 — SPECIES c.520

COD AND ANGLERFISHES
SUPERORDER Paracanthopterygii — ORDERS 5 — FAMILIES 37 — SPECIES c.1,600

SPINY-RAYED FISHES
SUPERORDER Acanthopterygii — ORDERS 17 — FAMILIES 286 — SPECIES c.16,500

INVERTEBRATES

SPONGES

PHYLUM Porifera CLASSES 3 ORDERS 24 FAMILIES 127 SPECIES c.10,000

CNIDARIANS

PHYLUM Cnidaria CLASSES 6 ORDERS 24 FAMILIES 300 SPECIES c.11,000

FLATWORMS

PHYLUM Platyhelminthes CLASSES 6 ORDERS 41 FAMILIES 424 SPECIES c.20,000

SEGMENTED WORMS

PHYLUM Annelida CLASSES 4 ORDERS 17 FAMILIES 130 SPECIES c.21,000

ROUNDWORMS

PHYLUM Nematoda CLASSES 2 ORDERS 17 FAMILIES 160 SPECIES c.20,000

MINOR PHYLA

ROTIFER

Invertebrates are classified in about 30 phyla, which vary considerably in size. In this book, major phyla are treated separately, but a selection of minor phyla also appears on pages 544–5. Almost all these minor phyla contain marine animals or ones that live in damp habitats.

COMB JELLIES
PHYLUM Ctenophora SPECIES c.200

PEANUT WORMS
PHYLUM Sipuncula SPECIES c.150

BRYOZOANS
PHYLUM Bryozoa SPECIES c.6,000

ROTIFERS
PHYLUM Rotifera SPECIES c.2,000

RIBBON WORMS
PHYLUM Nemertea SPECIES c.1,400

BRACHIOPODS
PHYLUM Brachiopoda SPECIES c.400

HORSESHOE WORMS
PHYLUM Phorona SPECIES c.20

ARROW WORMS
PHYLUM Chaetognatha SPECIES c.150

WATER BEARS
PHYLUM Tardigrada SPECIES c.1,000

VELVET WORMS
PHYLUM Onychophora SPECIES c.180

SPOONWORMS
PHYLUM Echiura SPECIES c.200

HEMICHORDATES
PHYLUM Hemichordata SPECIES c.130

10 OTHER MINOR INVERTEBRATE GROUPS

MOLLUSKS

PHYLUM Mollusca CLASSES 7 ORDERS 53 FAMILIES 609 SPECIES c.110,000

ECHINODERMS

PHYLUM Echinodermata CLASSES 5 ORDERS 38 FAMILIES 173 SPECIES c.7,000

INVERTEBRATE CHORDATES

TUNICATE

Invertebrate chordates are animals that share some characteristics with vertebrates but lack a bony skeleton. There are 3 subphyla—tunicates (the majority of species), lancelets, and hagfish. Tunicates have a swimming tadpole stage but are baglike as adults; lancelets and hagfish are mobile and in their internal anatomy bear strong resemblances to vertebrates.

TUNICATES
SUBPHYLUM Urochordata
CLASSES 3 ORDERS 7 FAMILIES 36 SPECIES c.2,900

LANCELETS
SUBPHYLUM Cephalochordata
CLASSES 1 ORDERS 1 FAMILIES 1 SPECIES 30

HAGFISH
CLASSES 1 ORDERS 1 FAMILIES 1 SPECIES 77

ARTHROPODS

SPIDER-HUNTING WASP

PHYLUM **Arthropoda**

Arthropods form the largest phylum in the animal kingdom. Insects make up the biggest subgroup, but the phylum also contains 2 other giant classes—crustaceans and arachnids—which dwarf many phyla in the invertebrate world.

MANDIBULATES
SUBPHYLUM Mandibulata CLASSES 16 ORDERS 109 FAMILIES c.2,230 SPECIES c.1.2 million

HEXAPODS
SUPERCLASS Hexapoda CLASSES 4 ORDERS 32 FAMILIES c.1,047 SPECIES c.1.1 million

SPRINGTAILS
CLASS Collembola ORDERS 1 FAMILIES 32 SPECIES c.8,100

PROTURANS
CLASS Protura ORDERS 1 FAMILIES 7 SPECIES c.760

DIPLURANS
CLASS Diplura ORDERS 1 FAMILIES 8 SPECIES c.975

INSECTS
CLASS Insecta ORDERS 29 FAMILIES 1005 SPECIES c.1.1 million

Group	Order	Species
BRISTLETAILS	Archaeognatha	c.470
SILVERFISH	Thysanura	c.570
MAYFLIES	Ephemeroptera	c.3,000
DAMSELFLIES AND DRAGONFLIES	Odonata	c.5,600
CRICKETS AND GRASSHOPPERS	Orthoptera	c.10,500
STONEFLIES	Plecoptera	c.3,000
ROCK CRAWLERS	Grylloblattodea	30
STICK AND LEAF INSECTS	Phasmatodea	c.2,500
EARWIGS	Dermaptera	c.1,900
MANTIDS	Mantodea	c.2,300
COCKROACHES	Blattodea	c.4,600
TERMITES	Isoptera	c.2,900
WEB-SPINNERS	Embioptera	c.400
ANGEL INSECTS	Zoraptera	43
BARKLICE AND BOOKLICE	Psocoptera	c.5,600
PARASITIC LICE	Phthiraptera	c.5,200
BUGS	Hemiptera	c.88,000
THRIPS	Thysanoptera	c.7,400
DOBSONFLIES AND ALDERFLIES	Megaloptera	c.300
SNAKEFLIES	Rapdhidioptera	c.200
ANTLIONS, LACEWINGS, AND RELATIVES	Neuroptera	c.11,000
BEETLES	Coleoptera	c.370,000
STREPSIPTERANS	Strepsiptera	c.580
SCORPIONFLIES	Mecoptera	c.550
FLEAS	Siphonaptera	c.2,400
FLIES	Diptera	c.150,000
CADDISFLIES	Trichoptera	c.10,000
MOTHS AND BUTTERFLIES	Lepidoptera	c.165,000
BEES, WASPS, ANTS, AND SAWFLIES	Hymenoptera	c.198,000

MYRIAPODS
SUPERCLASS Myriapoda CLASSES 2 ORDERS 21 FAMILIES 171 SPECIES c.13,150

CRUSTACEANS
SUPERCLASS Crustacea CLASSES 7 ORDERS 56 FAMILIES c.1,000 SPECIES c.70,000

CHELICERATES
SUBPHYLUM Chelicerata CLASSES 3 ORDERS 14 FAMILIES 675 SPECIES c.104,350

SEA SPIDERS
CLASS Pycnogonida ORDERS 1 FAMILIES 13 SPECIES c.1,330

HORSESHOE CRABS
CLASS Merostomata ORDERS 1 FAMILIES 1 SPECIES 4

ARACHNIDS
CLASS Arachnida ORDERS 12 FAMILIES 661 SPECIES c.103,000

ANATOMY

All but the very simplest animals are made up of various parts. Their smallest fully functional parts are cells, which are shaped in different ways according to the tasks that they carry out. Groups of similar cells are organized into tissues, and tissues are grouped together to form organs. Organs themselves are linked to form organ systems, which carry out all the processes essential for survival. The structure of these systems varies widely between one type of animal and another, and also between animals that live in different ways, but the work they do is the same.

Body systems

Animals have up to a dozen separate body systems. In many species, the muscular and skeletal systems make up a large proportion of the body's total weight, while the integumentary

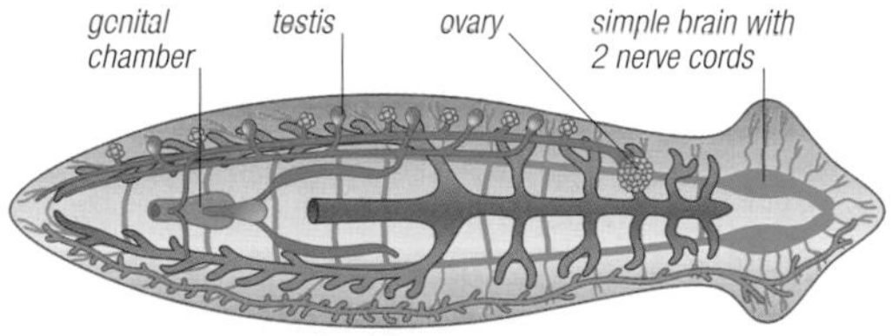

SIMPLE INVERTEBRATE BODY SYSTEMS
Simple invertebrates, such as flatworms, do not have respiratory organs or a circulatory system. The digestive system often has just one opening, the mouth, and the reproductive system is typically made up of both male and female organs.

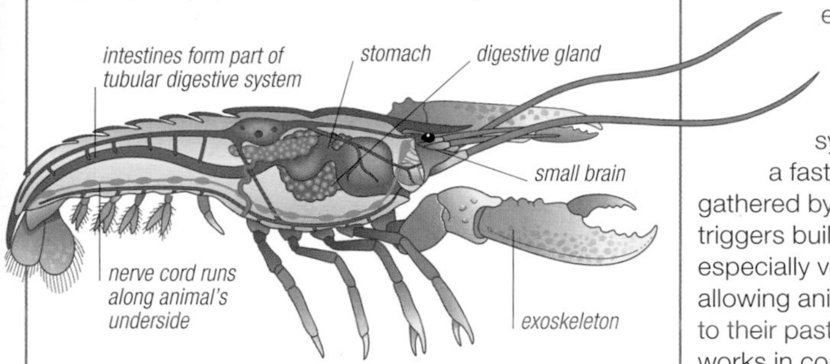

ARTHROPOD BODY SYSTEMS
The tubular digestive system is open at both ends. The blood flows partly through vessels and partly through body spaces. Oxygen is supplied via gills or via minute airways called tracheae.

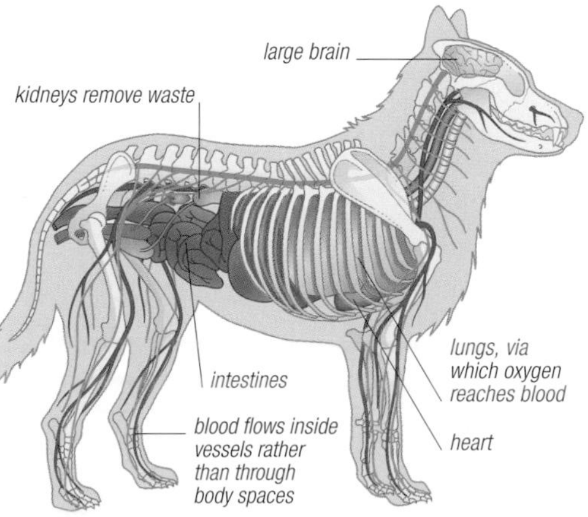

VERTEBRATE BODY SYSTEMS
The nervous system is highly developed, and the brain is typically larger than in invertebrates. The circulatory system is closed, and blood is pumped under high pressure by the heart.

KEY TO SYSTEMS

CIRCULATORY	EXCRETORY
DIGESTIVE	NERVOUS
RESPIRATORY	REPRODUCTIVE

system—the parts that form the "outer cladding"—protect the body from physical damage and, in terrestrial animals, from the threat of drying out.

Two key systems enable animals to obtain energy from food. The digestive system breaks food down so that it can be absorbed; and the respiratory system delivers oxygen to the body's cells so that food substances can be "burned" and their chemical energy released. The respiratory system also removes carbon dioxide—a potentially toxic waste product formed during the production of energy. In many animals, including all vertebrates, oxygen and carbon dioxide are carried by blood in the circulatory system. Carbon dioxide is usually exhaled; other kinds of dissolved waste are removed, before they have a chance to build up, by a separate excretory system.

Animals use 2 different systems to coordinate their bodies, and to react to their surroundings. The nervous system deals with anything that needs a fast response, processing information gathered by sense organs. In all animals, it triggers built-in or instinctive behavior, but in some, especially vertebrates, it also stores information, allowing animals to adapt their behavior according to their past experience. The endocrine system works in conjunction with the nervous system, releasing hormones, or chemical messengers, that help the nervous system coordinate long-term processes.

Finally, the reproductive system carries out the most important task: producing young. Unlike other body systems, it often functions only during a set season, and then only in mature animals.

Skeletons and support

Animals need to keep the shape of their bodies stable. Many invertebrates achieve this without any hard body parts at all. Instead, they rely on the pressure of internal fluids to keep their bodies firm in the same way that air stabilizes a tire. This system, called a hydrostatic skeleton, works well on a small scale. But in larger animals, particularly land ones, this type of skeleton is often not strong enough to support the body's weight. Animals have developed 2 quite different solutions to this problem: external shells and body cases; and internal skeletons, typically made of bone.

Shells are made up of one or 2 parts, and they grow with their owner. They can be seen in marine animals called brachiopods, or lampshells, but they are most highly developed in bivalve mollusks, which can grow shells over 3¼ ft (1 m) across. Body cases are more complex than shells: they consist of a large number of separate plates that meet at flexible joints. They are a characteristic feature of arthropods—a huge group of invertebrates that includes insects, crustaceans, and arachnids. These cases, or exoskeletons, cover the entire body and include structures as strong as a crab's pincers or as delicate as a butterfly's antennae. Unlike shells, body cases cannot grow, so periodically they have to be shed and replaced.

Internal skeletons (endoskeletons), made of bone, provide support from within. Found only in vertebrates, they have 2 great advantages: they are relatively light for their size—important for land animals that have to move quickly—and they can grow. In this kind of skeleton,

SHELL
A snail's shell grows at its lip, getting progressively larger with its owner. Some species can seal the shell with an operculum, or "door."

EXOSKELETON
A crab's body case covers all the surfaces of its body, including its eyes. Unlike an insect's exoskeleton, it is reinforced with calcium.

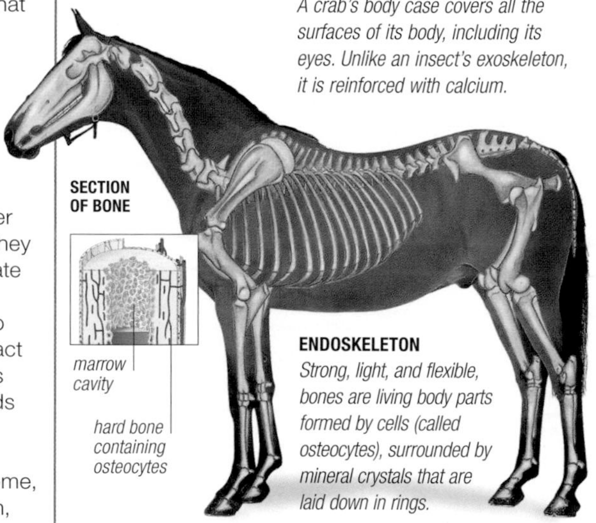

ENDOSKELETON
Strong, light, and flexible, bones are living body parts formed by cells (called osteocytes), surrounded by mineral crystals that are laid down in rings.

SYMMETRY

Some animals' bodies are circular, without a head or tail. This type of anatomy, called radial symmetry, can be seen in sea anemones and other cnidarians, and also in ctenophores or comb jellies. The vast majority of animals show bilateral symmetry, which means they can be divided into halves. The halves are not always equal: male fiddler crabs, for example, have very unequal claws, and flatfish have different sides.

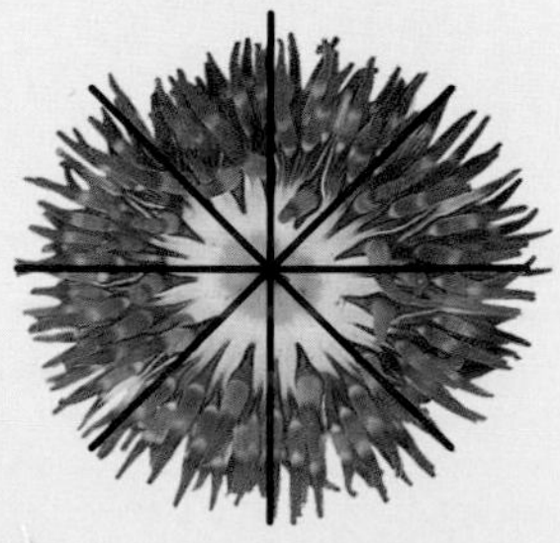

RADIAL SYMMETRY
A sea anemone can be divided into 2 equal halves on any axis. It has a central mouth and body cavity, with feeding tentacles arranged in a ring.

BILATERAL SYMMETRY
Frogs and toads have just one axis of symmetry. Externally, the 2 halves look identical; internally, some organs lie to the left or the right.

some bones meet at flexible joints. Other bones, particularly those in the skull, lock together for extra strength.

Muscles and movement

Muscles work by contracting. This means they can pull but not push. In most cases, they are arranged in pairs or groups that pull in opposing directions: when one muscle or muscle group contracts, its partner is brought back to its normal resting shape.

Muscles make animals move in different ways. In animals without limbs, such as earthworms and jellyfish, they work to change the body's shape. In earthworms, opposing muscles alternately shorten

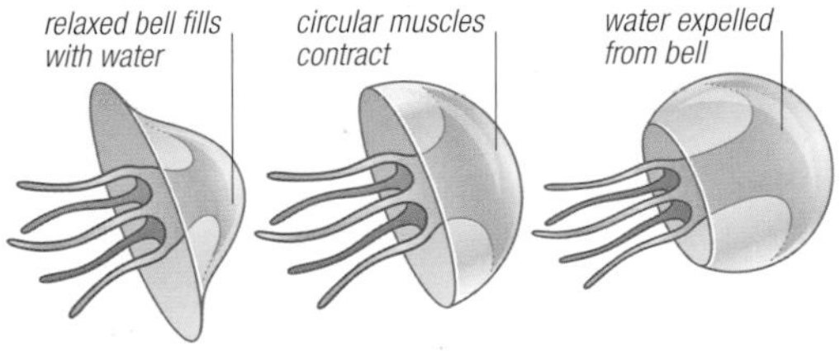

MOVING WITHOUT LIMBS
Jellyfish swim by rhythmically contracting the bell-shaped part of their body. This expels water from the bell, which pushes the jellyfish forward. Most jellyfish make little headway against the current: they swim mainly to keep at the right level in the water.

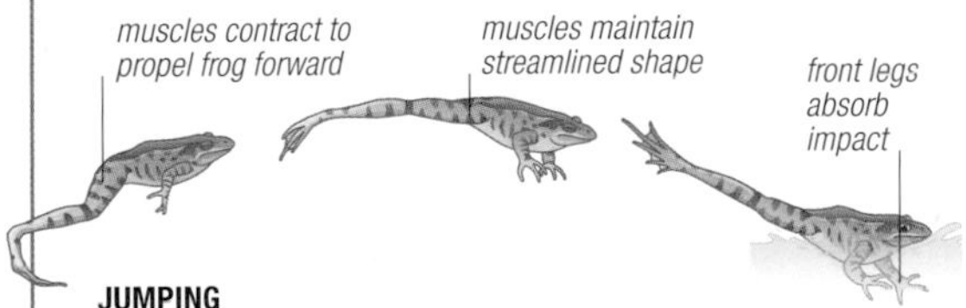

JUMPING
When a frog leaps, its legs act as levers, propelling it into the air. Its front legs fold up to cushion its body from the impact when it lands.

and lengthen the animal's segments so that it can creep through soil. In limbed animals, one set of muscles pulls the limb down or back, while the other lifts it up or forward.

As well as making animals move, muscles serve other purposes. They force food through the digestive system (peristalsis) and pump blood around the circulatory system. Unlike most other muscles, the heart muscle has a built-in rhythm that keeps it contracting throughout an animal's life.

Body coverings

Animal cells are easily damaged. To protect them from injury and disease, animals have body coverings, most of which consist largely of nonliving matter. Mammalian skin is covered by dead cells, while insect body cases are covered by hard proteins and waterproof wax. In many cases, these protective layers are themselves protected: mammals often have a coat of fur, while many other animals have scales. Some of these extra coverings have developed additional uses. Soft feathers and fur help to retain body heat, while extra-strong feathers are used in flight. Colors or patterns act as camouflage or help animals recognize their own kind.

FEATHERS

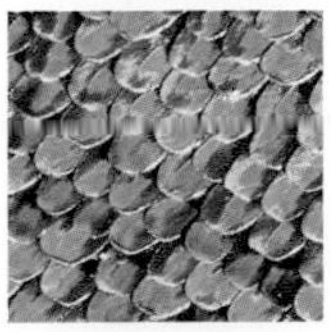

SCALES

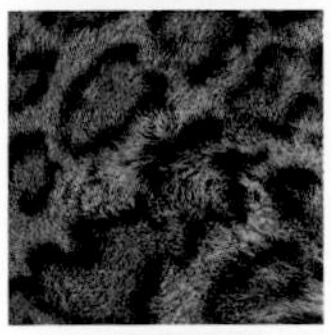

FUR

BODY BARRIERS
Bird feathers, butterfly scales, and mammal fur are made of nonliving substances that are produced by living cells. Feathers and fur are replaced during their owner's lifetime, but butterfly scales are not.

Respiration

For small and thin animals, obtaining oxygen is a simple matter because it seeps into their bodies from outside. At the same time, carbon dioxide escapes in the other direction. For larger animals, respiration is more complex. In relative terms, they have a much greater volume than surface area, so there is less room for gases to move in and out. To breathe, they rely on respiratory organs—structures that effectively pack a large surface area into a small amount of space. In aquatic animals, gills are the most common respiratory organs. Typical gills consist of thin, flat, or feathery surfaces that bring blood into close contact with the water outside.

However, most gills do not work in air because—out of water—their surfaces collapse and stick together. Land animals therefore have hollow respiratory organs that carry air deep inside their bodies. In insects, these organs are tubes, called tracheae, which divide into extremely fine filaments that reach individual cells. In land-dwelling vertebrates, the organs are lungs—air-filled chambers surrounded by a network of blood vessels. Muscles make the lungs expand or contract, sucking air in or blowing it out.

EFFICIENT RESPIRATION
Animals that are highly active in cold habitats need lots of oxygen. When an ibex runs uphill, its lungs take in 10 times as much air as when it is lying down.

Nerves and senses

Nerve cells, or neurons, are the animal world's equivalent of wiring. Neurons conduct brief bursts of electricity, known as impulses, which carry information from sense organs or make muscles contract. Corals and other simple animals have a network of nerves scattered throughout their bodies. But in most animals, the nervous system converges on the brain.

Some animal senses, such as touch, operate through nerve endings scattered all over the body. A similar sense, which works internally, tells animals about their posture. The most important senses—vision, smell, and hearing—work through organs that form some of the most elaborate structures in the body.

Vision is essential for many animals, and eyes show a wide variety of designs. At their simplest—for example, in snails—they do little more than distinguish between light and dark. In many animals, particularly arthropods and vertebrates, they focus light onto large numbers of nerve cells, building up a detailed image of the surroundings. In vertebrates, these eyes have a single lens, which throws light onto a "screen," or retina. In arthropods, the eye has up to 25,000 separate compartments, each with its own lens system; these compound eyes create a mosaic-like image and are especially good at detecting movement.

Mammals are the only animals with prominent earflaps. Vertebrates' ears are always on the head, but in some animals, they are positioned elsewhere. Most grasshoppers and crickets have ears on their abdomen or legs. Organs that detect taste and smell can also be in a variety of positions. Like ears, they can be used in communication, as well as for avoiding danger and finding food.

Many animals have senses that are more acute than those of human beings, and some can sense things that humans cannot. For example, most fish can sense pressure waves in water, and many can detect weak electric fields. Some snakes can "see" warmth, enabling them to attack warm-blooded prey in total darkness.

BINOCULAR VISION
Jumping spiders have 2 extra-large eyes that face directly forward. This gives them binocular vision, which is essential for gauging distances before making a jump.

BEHAVIOR

An animal's behavior encompasses all the things that it does as well as the way that it does them. Behavior ranges from simple actions, such as eating or keeping clean, to highly elaborate activities, such as hunting in a pack, courting a mate, or building a nest. In some animals, behavior is almost entirely predictable; in others, it develops with experience, so the more an animal does something, the more skilled it becomes. As with all aspects of animal biology, behavior is the product of evolution, which means that it gradually changes as time goes by. These changes enable species to react in the most effective way to the opportunities and dangers that they encounter in daily life.

Instinct and learning

In simple animals, behavior is "hard wired"—governed by inherited instincts—which means that it consists of fixed sequences of actions prompted by triggers. For example, day-old birds instinctively beg for food when their parents appear at the nest. At this stage, they are usually blind: their behavior is triggered by noise and movement rather than by the sight of food itself.

Instinctive behavior may seem basic, but it can produce quite remarkable results. The structures that animals build—from nests to dams—are the results of inborn behavioral impulses. When beavers set out to make a dam, they do so without any knowledge of engineering principles. Yet the structure they make is shaped to withstand water pressure as if it had been scientifically designed.

Beavers do not have to think how to build, just as spiders do not need to work out how to weave webs. Even so, the results of instinctive behavior can change. As animals repeat certain tasks, such as making a nest, their performance often improves. This is particularly important for some animals—such as male weaverbirds—which use their nest-building skills to attract a mate.

Apart from octopuses and their relatives, most invertebrates have narrow limits when it comes to learning. For vertebrates, on the other hand, learned behavior is often extremely important. Frogs and toads quickly learn to avoid animals that taste unpleasant, while mammals acquire a wide range of skills from their parents, including how to hunt. Among primates, individuals very occasionally "invent" new behavior, which is then copied by their neighbors. This copying process produces culture—patterns of behavior that are handed on down the generations. Culture is something that humans, as a species, have developed to a unique degree.

LEARNING TO FEED
Eurasian oystercatchers learn how to feed by watching their parents. Some birds hammer at shells to break them, while others stab at the shells' hinges to force them to open. Once a bird has learned one technique, it uses it for life.

INSTINCTIVE WEAVING
Spiders often build highly complex webs, but they always produce them to one particular design. As a result, it is often possible to identify a spider from its web alone. This garden spider will use its web for just one day; then it will eat it before starting on a replacement.

ANIMAL INTELLIGENCE

At one time, intelligence was thought to be rare in animals, with some exceptions such as dolphins, monkeys, and chimpanzees. Mammals include gifted communicators, but they are not alone in having problem-solving skills. Many other animals use simple tools, and some even shape them as well. Among the most impressive tool-makers are crows and their relatives. In captivity, one New Caledonian crow was seen bending a straight piece of wire, which it successfully turned into a hook for reaching food. It did this despite never having seen wire before—a remarkable example of insight that even a primate would find hard to match.

THE SIGNS OF INTELLIGENCE
Chimpanzees can solve problems, and they are able to learn sign language to communicate with humans. They can learn symbols for objects and actions, and they occasionally combine the symbols in ways that resemble spoken phrases.

Communication

For most animals, keeping in touch with their own kind is essential to their survival. Animals communicate with each other for a range of reasons, including finding food, attracting a mate, and bringing up their young. Different methods of communication have their own advantages and drawbacks. Body language—which includes facial expressions and physical displays—works well at close quarters but is ineffective at a distance and in habitats where dense vegetation gets in the way. In such cases, communication by sound is much more practical. Whales call to each other over immense distances, while some small animals produce remarkably loud sounds for their size. Treefrogs, cicadas, and mole crickets, for example, can often be heard at a distance of over 1 mile (2 km). Each species uses its own distinctive "call sign," and many behave like ventriloquists, pitching their calls in a way that throws predators off their track.

Animals that are capable of producing light also use identifying call signs. These can consist of specific sequences of flashes or—in many deep-sea fishes—illuminated body patterns. But, like body language and sound, this form of communication works only when the signaller is actively signaling. Scent communication is quite

NIGHT LIGHT
Female glow worms use light to signal their presence to the males, which fly overhead. Signaling with light can be dangerous because it can attract predators as well as potential mates. If a glow worm senses danger, she quickly "switches off" her light.

JOINING THE CHORUS
By howling, wolves advertise their ownership of a hunting territory to any other wolves that may be in the area. Wolves often howl at night, after they have made a successful kill.

different because the signal lingers long after the animal that made it has moved on. Animal scents are specific, allowing animals to lay trails and to advertise their presence to potential mates. Some male insects are able to respond to individual molecules of airborne scent, allowing them to track down females far upwind.

Living in groups

Some animals spend all their lives alone and never encounter another member of their species. But, for many, getting together is an important part of life. Animal groups vary in size as well as in how long they last: mayflies, for example, form mating swarms that last just a few hours, while migrating birds often assemble for several weeks. Many other animals, including fish and grazing mammals, form groups that are maintained for life.

GROUP FORMATION
By forming a V (or skein), geese can reduce the amount of energy needed to migrate by flying in the leading birds' slipstream. They take turns to lead.

SAFETY IN NUMBERS
In open habitats, such as grassland, where predators tend to be fast and hiding places scarce, prey mammals typically live in large groups. This group of zebras has further increased its security by grazing near a herd of wildebeest.

Groups of animals may seem to be easy targets for predators, but the opposite is usually true. Predators find it difficult to single out individuals from a group, so living together gives animals a better chance of survival. Groups are also more difficult to catch by surprise because there is always more than one animal on the alert for signs of danger.

In most animal groups, the members belong to a single species but do not necessarily share the same parents. However, in the most tight-knit groups, the members are all closely related. Examples of such extended families include wolf packs and kookaburra "clans", where the young remain with their parents instead of setting up independently. This kind of group-living reaches its extreme in social insects, such as termites and ants, which cannot survive alone.

Defence and attack

Both predatory and prey animals use specialized behavior to help them survive. For example, while many prey animals simply try to escape, others keep perfectly still, relying on camouflage to protect them. A wide range of species, from moths to lizards, try to make themselves appear dangerous by exaggerating their size or by revealing colored spots that look like eyes. Sometimes such threats are real: for example, the brilliant colors of poison-arrow frogs indicate that they contain some of the animal kingdom's most potent poisons.

RESPONDING TO THREAT
When threatened, puffer fish enlarge themselves by gulping water. Once distended, they can barely move, but their spines make them practically impossible to attack.

Predatory animals use one of 2 techniques to catch prey: they either wait for it to come their way or they track it down. "Sit-and-wait" predators are often camouflaged, and some actively entice their victims within range. In anglerfish, for example, the snout has a long, luminous protuberance, called a lure, which the fish dangle appetizingly in front of their mouths: anything swimming close to inspect this lure is snapped up whole. For active hunting, nature puts a premium on speed and keen senses, which is why animals such as cheetahs, peregrine falcons, and blue marlins are among the fastest in the world. Some active predators operate in groups. By working together, gray wolves, African wild dogs, and lions can tackle prey much larger than themselves.

PACK HUNTING
A pack of African wild dogs pulls down a wildebeest that they have run to exhaustion. Once their victim is dead, the dogs will devour it. On returning to their den, they regurgitate some of the meat for any pups that have been left behind.

Behavioral cycles

Some kinds of behavior, including self-defense, can be provoked at any time. Others are cyclical, triggered by cues that keep animals in step with changes around them. One of the most important cycles is the alternation between night and day. Others include the rise and fall of the tide and the annual sequence of changing seasons.

Cyclical behaviors are all instinctive. They may be stimulated by external changes, by built-in "biological clocks," or by a combination of the 2. Birds, for example, often gather to roost late in the day, a form of cyclical behavior that is triggered by falling light levels as the sun nears the horizon. On a much longer time scale, ground squirrels show an annual cycle in body weight, getting heavier before they enter hibernation. However, ground squirrels maintain their cycle even if kept in conditions of constant temperature and day length, which shows that the rhythm is controlled biologically. Biological clocks often involve hormones, but the way they work is not yet fully understood.

FEEDING TIME
Fiddler crabs emerge from their burrows to pick over the nutrient-rich sediment for small particles of food. They cannot feed underwater, so their feeding behavior is governed by the daily rise and fall of the tide.

LIFE CYCLES

An animal's life cycle consists of all the stages between the beginning of one generation and the beginning of the next. In some species—especially insects and other small invertebrates—the entire cycle is completed within a few weeks; in much larger animals, it often takes many years. Regardless of how long it takes to complete, an animal's life cycle always involves 2 main steps: a period of growth and development, followed by reproduction. Some animals reproduce once and then die: for them, reproduction marks the end of life as well as the completion of the life cycle. For many, reproduction continues throughout adulthood, giving animals more than one chance to produce young.

Reproduction

The ability to reproduce is the cornerstone of life because it allows living things to multiply, exploit new opportunities around them, and evolve. Animals reproduce in one of 2 ways: asexually (without sex) or sexually.

In asexual reproduction, a single parent partitions off part of itself to form a new animal. The partitioning process can happen in a variety of ways. Hydras, for example, produce small buds that grow into complete new animals, while sea anemones literally tear themselves in 2. Some animals produce eggs that develop without fertilization – a process called parthenogenesis. This is common in aphids and other sap-sucking insects, but it is rare in vertebrates (whiptail lizards are one of the few examples). Asexual reproduction is relatively quick and simple, but it has one important disadvantage: since only one parent is involved, the offspring are either genetically identical to that parent or very similar to it. As a result, parent and offspring are equally vulnerable to threats such as disease: if one animal dies, the rest will often follow suit.

Sexual reproduction gets around this problem because the involvement of 2 parents produces offspring that are genetically varied. Each one has a unique combination of characteristics, which allows the fittest to survive and the species slowly evolves. However, the disadvantage of sexual reproduction is that it is much more complicated: the parents must be of the right species and correct sex, and in most cases they must cooperate to breed. In addition, only one parent—the female—actually produces young, so some reproductive potential is lost.

SEXUAL REPRODUCTION
As a female common frog lays her eggs, her mate sprays them with sperm to fertilize them. Each tadpole will be genetically unique.

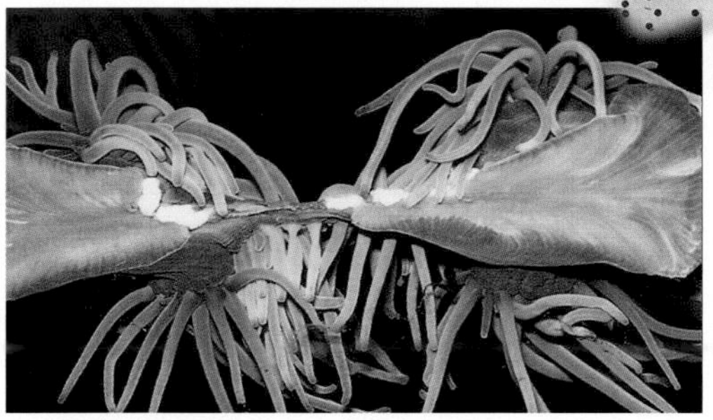

ASEXUAL REPRODUCTION
This sea anemone is in the final stages of reproducing, which it does by tearing itself in half. The result will be 2 individuals that have exactly the same genetic makeup.

LIFESPANS

In general, animal lifespans are directly related to adult body size: the larger the animal, the longer it is likely to live, although there are exceptions to this rule. One important factor affecting lifespan is metabolic rate—the rate at which an animal uses energy to make its body work. "Cold-blooded" animals, such as amphibians and reptiles, have a relatively low metabolic rate and tend to be long-lived, while "warm-blooded" animals, such as birds and mammals, have a high metabolic rate and tend to be relatively short-lived. This is especially true of small species because a low body mass means that heat escapes quickly from the body and has to be constantly replaced by food. Environmental factors, such as temperature and humidity, also have an effect. Houseflies, for example, often die within 6 weeks in warm conditions but may survive for many months if it is cool, and some microscopic animals survive for decades if they remain in a dormant state. In general, animals rarely live for long after their reproductive life has come to an end.

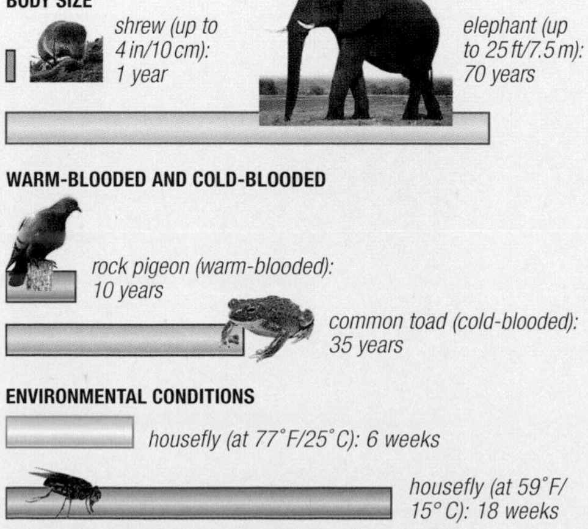

FACTORS DETERMINING LIFESPAN
Three of the major factors influencing an animal's lifespan are illustrated above. The life expectancy of each animal is represented by a colored bar, indicating the degree to which a factor can affect its lifespan.

Despite these difficulties, sexual reproduction is widespread throughout the animal world, which demonstrates its long-term value. Even species that normally reproduce without sex periodically include a sexual phase in their life cycle, thereby getting the best of both worlds.

Animal sexes

All animals that reproduce sexually with a partner of the opposite sex show dimorphism—that is, the males and females are anatomically different and behave in different ways. In some species, the differences are not obvious, but in others, they are quite distinct. Dimorphism exists because the sexes have developed different roles in reproduction and need different body forms to carry them out.

However, not all sexually reproducing animals are of opposite sexes. Some—earthworms and terrestrial snails, for example—are hermaphrodite (they have both male and female sex organs). This simplifies sexual reproduction because any adult of a species is potentially a suitable mate for another. A further variation is that some species have separate sexes but individuals can change sex during adult life. Parrotfishes, for example, often live in small schools dominated by a single male. If the male dies, a female changes sex and takes his place.

DIMORPHIC PARTNERS
In some animals, sex differences are extreme. Here, a wingless female vapourer moth (seen on the right) has attracted a winged male. After mating, the female will crawl away to lay her eggs.

Courtship

Before an animal can mate, it has to find a partner. This is easy enough for species that live in groups, but for animals that live alone, it poses problems. Solitary animals locate potential mates by sending out signals, such as sounds or airborne scents. Each species has its own characteristic "call sign," ensuring that it finds others of its own kind.

Once the sexes are in contact, one partner—usually the male—has to overcome the other's wariness and demonstrate his suitability as a mate. This process is known as courtship. It often takes the form of ritualized behavior that displays the male's physical fitness or his ownership of a good provision of food. If the female is sufficiently impressed, she will accept him as her mate.

While some species form lifelong partnerships, many go their separate ways after mating. In the latter case, the males typically mate with several females but take no part in raising the young. More rarely, things work in the opposite way, with one female mating with several males. Where this happens—for example, in phalaropes (see p.312)—the female is more brightly colored than the male and often takes the lead in courtship. In general, these females take little or no part in rearing young.

COURTSHIP RITUAL
A male frigatebird inflates his throat pouch to attract the attention of a female. Birds have good color vision, which explains why males often have striking plumage.

Fertilization

Marking the start of a new life, fertilization occurs when a male sperm and female egg cell fuse. In animals, it takes place in one of 2 ways: either outside or inside the female's body.

External fertilization is used by many animals that either live permanently in water or return to it to breed. In the simplest version of this process—seen in static invertebrates such as corals—vast numbers of the male and female sex cells are shed into the water, where they mingle so that fertilization can occur. A more advanced version of this, shown by animals that can move around, such as frogs, involves 2 partners pairing up. Although they appear to mate, fertilization nevertheless takes place in the water rather than inside the female's body.

External fertilization does not work on land because sex cells soon dry out and die when exposed to air. Most terrestrial animals therefore use internal fertilization. In general, this involves the male injecting sperm into the female. However, some terrestrial animals, such as salamanders and newts, do not copulate. Instead, the male deposits a package of sperm (a spermatophore) near the female; she then collects it with her reproductive organ so that internal fertilization can take place.

INTERNAL FERTILIZATION
Like all insects, flat-footed bugs have to pair up so that the female's eggs can be fertilized. Mating takes several hours.

Starting life

Most animals—apart from the ones that use asexual reproduction—start life as a single fertilized egg cell. If the egg has been fertilized externally, it will already be outside the mother's body, perhaps drifting in the sea or glued to seabed plants or sand. If the egg has been fertilized internally, it will either be laid, to hatch afterward, or it will be retained inside the mother while it begins to develop into a young animal.

The degree of development that takes place at any one stage varies from one type of animal to another. Oviparous species, such as birds, lay their fertilized eggs before fetal development begins. In birds, development is often deferred for several more days until the clutch is complete; it begins as soon as the parent starts to incubate the eggs. Ovoviviparous species, which include many reptiles and sharks, incubate their eggs internally, "giving birth" at the moment when the eggs are about to hatch. Viviparous species—which include virtually all mammals, as well as some reptiles, amphibians, and fish—give birth to live young.

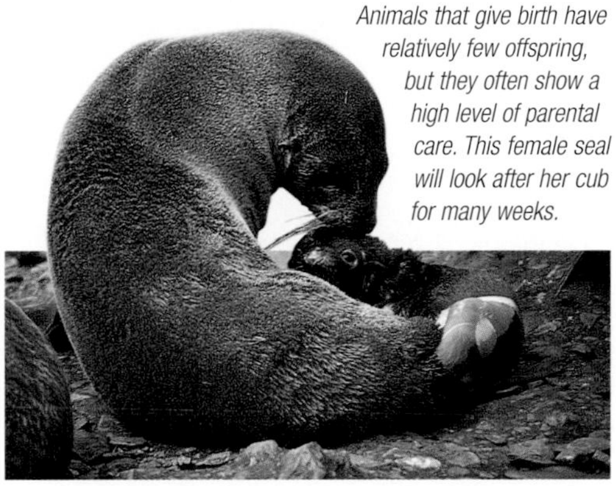

LIVE BIRTH
Animals that give birth have relatively few offspring, but they often show a high level of parental care. This female seal will look after her cub for many weeks.

Metamorphosis

All animals change shape as they grow and develop. In some, the changes are gradual and relatively minor, but in others, they are so far-reaching that the animal is completely transformed. This transformation is called metamorphosis. It allows animals to live in different ways—and often in different habitats—during their young and adult lives.

EXTERNAL FERTILIZATION
Corals release their sex cells into the water. They are triggered to do this by the changing phases of the moon—a natural clock that many animals use to synchronize their breeding behavior.

LIVING TOGETHER

Many animals associate with their own kind but continue to lead independent lives. However, eusocial species, such as termites, ants, and bees, form permanent groups or colonies in which just one member—the queen—produces all the colony's young. As a result, the colony's members are closely related and in many ways behave like a single organism. The great success of the system is apparent in the fact that these insects are among the most numerous on earth.

QUEEN TERMITE
Hidden deep inside a termite nest, where she is attended by her workers, a queen termite lays up to 30,000 eggs a day. In her complete dependence on the workers for food, she represents the ultimate form of reproductive specialization.

Although metamorphosis is most common in invertebrates, it does occur in amphibians and some fish. Animals that undergo the metamorphic process spend the early part of their lives as larvae. In the sea, larvae often drift near the surface as part of the plankton, and because they are carried far and wide they play an important role in helping their species spread.

In the insect world, metamorphosis occurs in 2 ways. Incomplete metamorphosis, shown by grasshoppers and bugs, involves a series of gradual changes that are made as the young insect, or nymph, matures. Complete metamorphosis—shown by butterflies, beetles, and flies—involves more drastic changes, which occur during a resting stage, called pupation, when the body is broken down and rearranged.

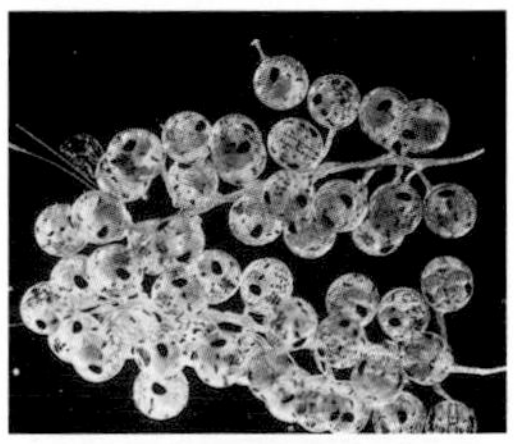

CRAB EGGS

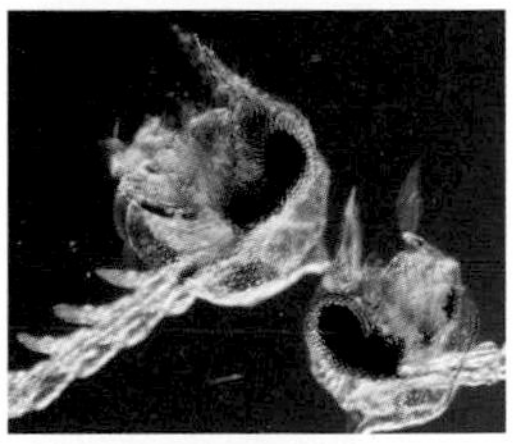

ZOEA LARVA

FROM LARVA TO ADULT
Like most crustaceans, shore crabs start life as eggs (top). The eggs hatch to produce the first larval stage, called a zoea (center), which floats in plankton. This changes into a megalopa larva, then sinks to the seabed and finally changes into an adult crab.

ANIMALS IN DANGER

Until a century ago, the earth still contained large areas of wilderness, where animals had little or no contact with human beings. Since then, the human population has soared to 7 billion, and our increasing use of energy and raw materials affects the entire planet. Natural habitats are disappearing, and the earth's biodiversity—the sum total of all living species—is in sharp decline. This is a dangerous situation for humanity, because it reduces the earth's biological resources, and makes the world a less stable place. For animals, the results can be disastrous. Some changes are local ones, which threaten individual species, but others, particularly climate change, are global in their reach.

Habitat change

Humans first started to alter habitats when they discovered fire, but with the start of farming, about 10,000 years ago, habitat destruction rapidly increased. Agriculture has been the main driving force behind deforestation, which swept across the Northern Hemisphere in historical times, and continues in the tropics today. It has also been responsible for the destruction of some of the world's major natural grasslands, and of marshes and other freshwater wetlands that supply many wild animals with their food. In recent times, urbanization has become almost as important as a threat—towns and cities take up space, and the roads between them use up even more.

The pattern of habitat change is often as important as its scale. For example, if large areas are preserved, the habitat can often function as before, although on a reduced scale. But if the same amount of habitat is divided up into smaller isolated fragments, the effect on animals is much more severe. This is because many species—particularly predators at the top of food chains—need extensive territories to survive. These fragmented habitats are also exposed to more intrusion and disturbance from humans and domestic animals, making it much harder for wild animals to feed and breed.

NATURAL PRAIRIE

CEREAL FIELD

FROM PRAIRIE TO FARMLAND
North America's forests and prairies underwent an explosive burst of habitat change with the arrival of European settlers. Both were originally home to a wide variety of animals, such as bears and bison, which were hunted and then displaced as agriculture expanded. Much of the forest has grown back, but the prairies are now almost entirely devoted to cereal farming: the original grassland habitat has largely disappeared.

Pollution

Pollution occurs when chemicals or other agents disrupt natural ecosystems. Sometimes it has a natural origin, but in most cases it is the result of human activity. It can affect animals physically—for example, entangling them in waste, or clogging them with oil—but its chemical effects are often more serious, and harder to identify or predict.

The most problematic chemical pollutants are synthetic organic (carbon-containing) substances, such as solvents, pesticides, and herbicides. Hundreds of thousands of these chemicals now exist, and new kinds are produced every year. Their chemical structure means that they are often absorbed by living tissue, where they are ideally placed to cause the most damage. Some of these substances are toxic to all forms of life, but others are more selective. They are passed on when predators eat their prey, and as a result, they accumulate in species at the top of food chains, such as whales, polar bears, and birds of prey. As well as dissolved pollution, marine animals have to cope with solid plastic waste. Taking decades or even centuries to degrade, this forms huge eddies or gyres in the world's oceans, which can often be hundreds of miles across. Small particles of these plastics are often ingested by animals, weakening them or killing them outright.

Animals are also affected by air pollution, which is created mainly when fuels are burned. Air pollution causes localized problems such as acid rain, which can have a highly damaging effect on freshwater fish. On a much broader scale, it is also responsible for global warming—the biggest environmental change of all.

DEADLY SLICK
Oil spills devastate wildlife. These pelicans were victims of the Deepwater Horizon spill of 2010 in the Gulf of Mexico. Oil is toxic and destroys the waterproofing of plumage; without treatment, affected birds die.

Hunting, fishing, and collecting

Unlike many of the world's other resources, animals can reproduce. This means that—in theory—useful species can be harvested without them ever running out. Unfortunately, many species have been overexploited, with the result that some have died out, while others are now in serious danger.

The list of past casualties from hunting includes the African blue buck, which died out in about 1800, and the North American passenger pigeon, which became extinct in 1914, despite formerly being the most numerous bird in the world. These animals were killed primarily for food, a practice that continues today in a more diverse way, in the bushmeat trade. Once a subsistence activity, the bushmeat trade has recently become a global business, focusing on all kinds of forest animals that can be caught and sold as food. Primates are particularly threatened, but the trade also endangers many other animals, from snakes to pangolins. Animals are also hunted to meet less pressing needs. Elephants are in demand for their ivory, and rhinos for their horns. Tigers are hunted for their fur and body parts, which fetch increasingly inflated prices as the number of surviving animals falls.

AWAITING EXPORT
Packed in tiny wire cages, these parakeets are destined to supply the caged-bird trade. The trade is driven by money from dealers overseas. However, in the birds' native countries, collecting birds may sometimes be the only way of earning a living.

At sea, fish have become victim to the kind of overexploitation once reserved for animals on land. Plummeting stocks of once-common species, such as tuna and cod, are typical of a resource that is often only weakly regulated, or not at all. Some fish breed at an early age, and can recover from overfishing if the pressure is reduced. But with species like tuna, maturity takes time, so adult fish can become too rare to guarantee a future supply of young. Comparatively little is known about the effect of this relentless harvesting on marine and coastal life. However, fish play a key part in many food chains, and when their numbers fall, the effects are felt by countless other animals, from seabed invertebrates to fish-eating birds.

Introduced species

Even before Columbus discovered America, explorers and colonists had spread animals to new parts of the world. The process increased rapidly with the Age of Exploration, and the result—hundreds of years later—is that the wildlife of isolated regions has been overwhelmed by a host of intruders, from rats and cats to mosquitoes. Some of these introduced species cause problems by actively preying on local wildlife. Others harm native animals indirectly, by competing with them for food, or by transmitting diseases such as avian malaria.

In Australia, introduced species have disrupted the ecology of an entire continent. Kangaroos still thrive, but many small marsupials now live in a tiny fraction of their original range, in marginal habitats that introduced species find difficult to reach. Similar problems affect New Zealand and Madagascar, and on much smaller oceanic islands, the situation is often more severe. Their native birds are often wiped out by cats and rats—tenacious newcomers that are extremely difficult to eradicate. In this age of rapid travel and expanding tourism, the threat from introduced species is never far away.

UNWELCOME ARRIVALS
Rabbits were introduced to Australia in the mid-19th century for food and their fur. They quickly spread inland, displacing native animals and destroying vegetation. In dry areas, soil erosion set in, permanently changing the landscape.

Global warming

The world's climate has always changed, but the current period of rapid warming is without precedent in modern times. Most scientists believe that the cause is increasing levels of atmospheric greenhouse gases, caused by human activities. Greenhouse gases include water vapor, chlorofluorocarbons (CFCs), and carbon dioxide. They make the atmosphere trap outgoing heat, warming up the earth.

Wildlife has coped with changes in the past, but the speed and severity of this episode could result in extinction on a worldwide scale. Atmospheric warming, in itself, is only part of the story, because global warming has dozens of knock-on effects. It makes sea ice melt and changes the pattern of oceanic currents, altering climatic conditions on land. Over the longer term, it makes the oceans more acid, threatening shelled animals and coral reefs. It also makes sea levels rise, both by melting ice caps, and by making seawater expand. This expansion happens very slowly, but once started, will take centuries to reverse.

SHRINKING ICE
Polar bears use winter sea ice as a platform for catching seals. With global warming, the Arctic Ocean's ice cover is diminishing, making it more difficult for polar bears to stock up on food during this crucial time of year.

Attempts are under way to limit greenhouse gas emissions, following research collated by the International Panel on Climate Change (IPCC). So far, these efforts have had some results, but the upward trend in the production of greenhouse gases will take decades to reverse.

Animals on the brink

The International Union for Conservation of Nature (IUCN) maintains The IUCN Red List of Threatened Species™ (see panel, below), which is constantly updated by the work of scientists worldwide. In 2010, nearly 56,000 species were assessed, representing about 3 percent of those that have been formally classified. Although this is only a small portion of the world's species, this sample indicates how life on earth is faring, how little is known, and how urgent the need is to assess more species. Comprehensive assessments have been carried out for birds, mammals, amphibians, sharks, reef-building coral, cycads, and conifers, and the statistics are disturbing, with 1 in 8 birds, 1 in 4 mammals, 1 in 3 corals, and more than 1 in 3 amphibians at risk of extinction.

Being "on the brink" means different things for different species. Some animals—particularly invertebrates—can reproduce rapidly when conditions are good, which means that they have the potential to make a fast comeback. But many species on the IUCN Red List are slow breeders, and take a long time to recover if their numbers fall. Albatrosses are typical examples: they take up to 7 years to become mature, lay just one egg, and often breed only in alternate years.

To make matters more complex, animals cannot necessarily breed if they find a suitable habitat, and a partner of the opposite sex. This is because many species breed in groups, and rely on the stimulus of others around them to trigger essential behavior, such as courtship and nest-building. The passenger pigeon was a classic example of a communal breeder, nesting in colonies many square miles in extent. Even when many thousands were left, it had already stepped over the threshold into oblivion.

THREAT CATEGORIES
The IUCN Red List of Threatened Species places animals in one of 8 categories according to the degree of risk they face: the most threatened species (such as the tiger and the black rhinoceros) are "critically endangered"; the next category (including the orangutan and Queen Alexandra's birdwing) are "endangered," and so on. Details of the categories—which are also used in this book—can be found on page 10.

THE IUCN RED LIST

The IUCN Red List of Threatened Species is published by the International Union for Conservation of Nature (IUCN). The IUCN, founded in 1948 by the United Nations, carries out a range of activities aimed at safeguarding the natural world. Part of its work is the regular compilation of the IUCN Red List, which draws together information provided by over 10,000 scientists from all over the world; this list has become a global directory to the state of living things on our planet.

The current IUCN Red List shows that threatened species are often grouped in particular parts of the world. Today's "hot spots" include East Africa, Southeast Asia, and the American tropics. One of the reasons for this is that these regions have a much greater diversity of species than regions farther north or south: the American tropics, for example, are particularly rich in bird species. In recent years, these areas have seen rapid habitat change—particularly deforestation—which has come about partly because an expanding human population needs more land on which to grow food.

CONSERVATION

The negative impact that human beings have on wildlife grows day by day, but so, too, does the impact of conservation. Across the world, organizations big and small are engaged in a concerted effort to protect nature in its original state, or to ensure that we use it in a sustainable way. It is a huge task, and one that raises some difficult practical and philosophical questions. Which is the best way of safeguarding species? How do you go about saving an animal that is on the verge of extinction? And, if resources are limited, are some animals more "important" than others? Experts do not always agree on the answers, but there is no doubt that conservation is an urgent priority if today's threatened species are to survive.

Habitat protection

By far the most effective way of safeguarding animals is to protect their natural habitats. An animal's habitat provides everything necessary for its survival, and in its natural state it can continue to do this indefinitely as food and energy is passed from one species to another.

This is the thinking behind national parks and wildlife reserves. Even small parks can be effective—particularly when they protect breeding grounds—but, in general, the larger the area that is protected, the more species benefit and the greater are the chances that the habitat is truly self-sustaining. For example, Manú National Park—the largest in Peru—includes an extraordinary range of habitats from high-altitude Andean grassland to lowland Amazonian rainforest. It is home to at least 150 species of mammals, over 1,000 species of birds, and even more species of butterflies, making it one of the richest tropical reserves in the world. Its success is partly due to its remote location, which has restricted human settlement, unlike many parts of the Amazon farther east.

WATERSIDE VANTAGE POINT
Specially constructed blinds allow visitors to watch birds in a wetland reserve. After centuries of drainage for agriculture, reserves like these are vitally important to many wetland species.

In other parts of the world, national parks and reserves can suffer from their own popularity, and also from the pressure for resources. In the Galapagos Islands, for example, conservationists are engaged in an often difficult struggle to balance the needs of wildlife against the needs of an expanding human population, and increasing numbers of visitors.

Techniques and technology

When a species is in immediate danger of extinction, captive breeding can be a highly effective way of bringing it back from the brink. In 1982, this was the situation with the California condor, when only about 24 birds were left in the wild. During the 1980s, a breeding program was initiated, and all the remaining birds were caught—a drastic measure that caused considerable controversy at the time. Three decades later, the intervention has been vindicated: the total population has reached about 400, with around half this number flying free.

In recent years, new technology has played an increasing part in this kind of conservation work. Satellite tracking of released animals helps show where they feed and where they breed.

READY FOR RELEASE
Raised in captivity, this California condor may one day help swell the population in the wild. However, in comparison with life in captivity, life in the wild can be difficult and even hazardous.

Using DNA technology, there is even a possibility that recently extinct species could be "brought back to life". However, most conservationists believe that these techniques—on their own—are not long-term routes to survival. This is partly because they require a large commitment of time, money, and space. But a more significant problem lies in their outcome: if a species' natural habitat is disappearing, captive animals will have no home to go to if they are released.

Controlling incomers

In isolated parts of the world, introduced, or "alien," species make life extremely difficult for native animals. Cats, foxes, and rats head the list of these problematic incomers, although plant-eating mammals can also cause immense damage. In some of the worst-affected regions, such as Australia and New Zealand, conservation programs are now under way to reduce this threat.

In an island as vast as Australia, eradicating feral cats or foxes is not a feasible goal. But in some parts of the country, large areas have been fenced off to protect bandicoots, bilbies, and other vulnerable marsupials. In these giant enclosures, alien species are either trapped or controlled by poison bait. The poisons are substances

EXCLUDING INTRUDERS
In Western Australia, this electric fence protects the Peron Peninsula from introduced mammals, such as cats. The entire peninsula—covering 390 square miles (1,000 square km)—is to become an "alien-free" haven for endangered marsupials.

ANIMAL APPEAL

One problem with animal conservation is that our reactions to species differ. For example, everyone loves giant pandas, but far fewer like "creepy crawlies"—the invertebrates that underpin every ecosystem on land, and also in the sea. Invertebrates are essential for making life work, particularly as many of them recycle nutrients on which so many other living things depend. To be effective, conservation has to protect all animals in a habitat. These range from so-called "charismatic megafauna"—big animals with star-appeal—right down to animals that few people see, and that even fewer can name.

POLES APART
The giant panda and the European longhorn beetle are at opposite ends of the spectrum of public interest and concern. While the panda attracts funds and media attention, the beetle and its like rarely arouse comment.

from native plants, which affect alien species, but leave native ones unharmed. Killing for conservation is a difficult and divisive issue, particularly when the victims are cats that have run wild. However, there is no doubt—as far as Australian marsupials are concerned—that it is a highly effective measure.

Introduced species are even more of a problem on offshore islands, where they can devastate land animals and colonies of nesting birds. Many of the world's remotest islands, such as Kerguelen in the southern Indian Ocean, have been overrun by rats, which arrived aboard ships several centuries ago. Rats can be extremely difficult to control and on Kerguelen at least, eradication programs have not succeeded. However, on several islands off the coast of New Zealand, rats have been eradicated to create safe havens for tuataras—among the most endangered reptiles in the world. The small size of these islands makes them ideal "arks," because they are relatively easy to keep alien-free.

Legal protection

After centuries of indiscriminate exploitation, endangered animals are now protected by a host of international agreements and national laws. One of the most important of these is the Convention on International Trade in Endangered Species, or CITES, which came into force in 1975. Other international agreements protect particular habitats—such as wetlands—or entire continents, as in the case of the Antarctic Treaty. Some international bodies promote conservation as a way of managing wildlife resources. For example, the International Whaling Commission (IWC), which was set up in 1946, originally supervised the "sustainable harvesting" of whales. When it became clear that numbers of most great whales were plummeting, whaling limits were gradually tightened, until a moratorium on commercial whaling was introduced in 1986.

Legal protection is an essential part of wildlife conservation, and in the case of whales, has almost certainly saved some species from extinction. However, its effectiveness is sometimes undermined by loopholes, or by illegal activity. Two notorious examples of the latter are the poaching of black rhinoceroses, whose numbers have collapsed from about 100,000 in the early 1960s to about 4,000 today, and the tiger, which has been reduced to a total population of about 3,500. Both these animals are killed for their body parts, which fetch extremely high prices in the Far East.

TOURIST ATTRACTION
Watched by a group of tourists, a cheetah relaxes in the evening sunshine. Its tameness is unusual and is a sign that tourism's intrusion into its habitat is affecting its natural behavior.

Commercial exploitation

Few people would condone the sale of rhino horn or tiger bones, but some conservationists do believe that—where possible—wild animals should be made to "pay their way". According to this viewpoint, animals are best conserved if they generate income, because this provides an incentive for protecting them. There are 2 main ways by which this can happen; wildlife tourism can be encouraged, with some of the revenue being used for conservation work; alternatively, animals themselves can be managed as a resource.

Wildlife tourism is a booming business, although it has undeniable drawbacks, such as increasing habitat disturbance. But wildlife experts are often sharply divided about the use of animals as a resource. In recent years, the African elephant has been a case in point, with different conservation bodies at odds about the exploitation of ivory. In this debate, one side believes that the legal sale of ivory is bound to have a damaging effect on elephant numbers. The other side believes that if it is carefully controlled, the sale of ivory could actually safeguard the species by generating money to protect it.

At present, no one knows whether commerce has a real place in wildlife conservation. If it does, one factor is certain: the income generated by wild animals will have to benefit local people, as they are the ones who can make conservation work.

WINNERS AND LOSERS
Almost extinct in the 1930s, the Antarctic fur seal now numbers 1.5 million. Its recovery is owed not only to protection from hunting but also to reduced numbers of whales, which compete for krill.

CITES

The Convention on International Trade in Endangered Species (CITES) has over 120 signatories, and is the most important piece of international legislation governing the movement of live animals and animal products across international borders. CITES completely prohibits trade in over 400 species, and requires special permits for trade in others. Some illegally traded objects are easy to identify, but others can be distinguished by DNA analysis—a relatively new technique increasingly used by customs officials. CITES has been successful in some areas, but despite increasing vigilance at ports and airports, smuggling is still a problem.

BANNED GOODS
All the items shown in this photograph are made of turtle shell—an animal product that cannot be exported under CITES regulations. In some countries, national laws make it illegal to own animals or objects listed by CITES.

INTRODUCTION

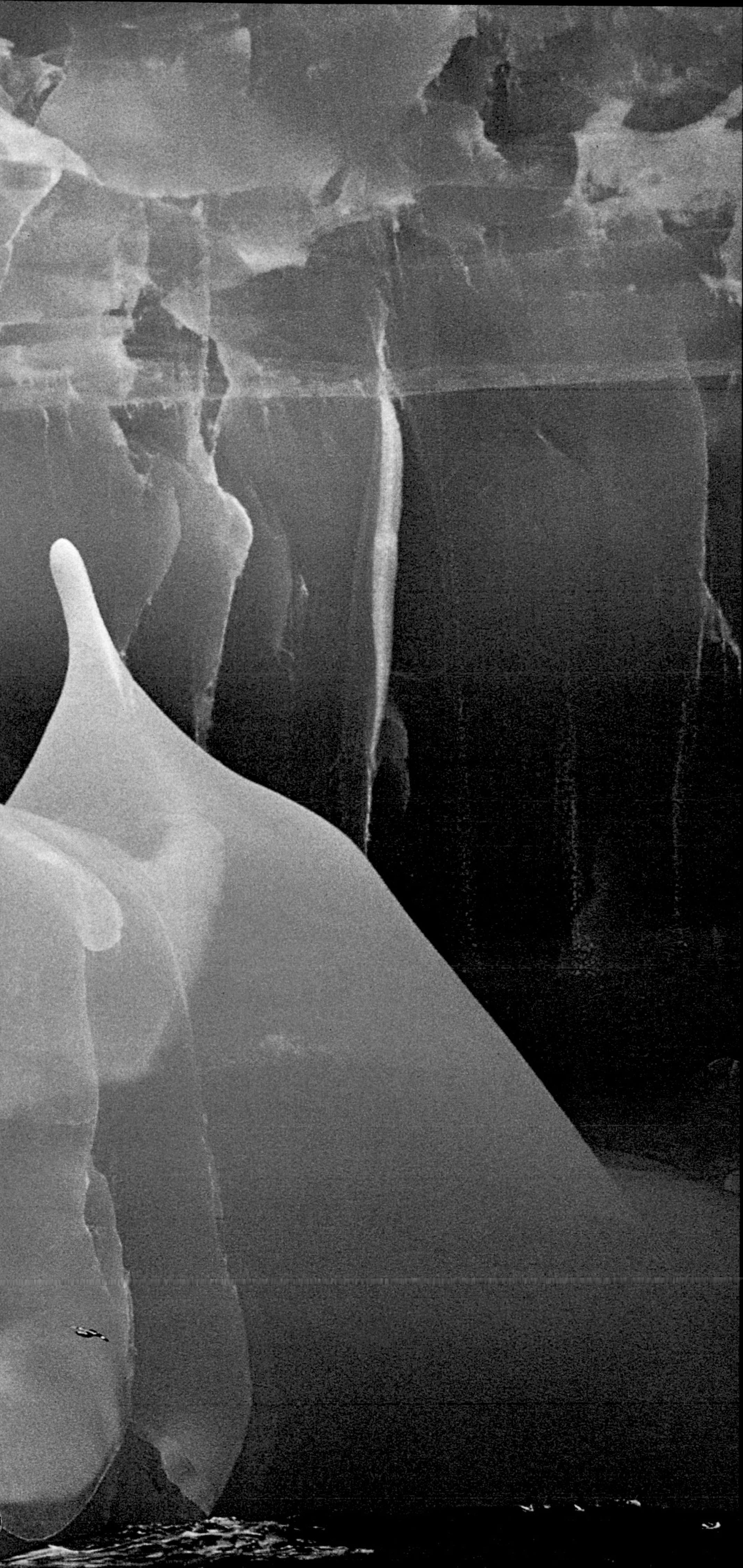

HABITATS

The earth is the only place in the universe known to support life. Like all other living things, animals occupy a zone between the lower part of the earth's atmosphere and the floors of its oceans. Although animal life is richest in the warm, wet conditions of the tropics, some animals have adapted to life in the hostile conditions found in arid deserts, high mountains, and the polar ice caps. Wherever they live, animals interact with each other, with other living things, and with their nonliving surroundings to produce complex, constantly changing environments known as habitats. This section looks at the habitats of the world and the animals that live in them.

WORLD HABITATS

Seen from space, the most striking feature of the earth is its sheer physical variety. Not only is there dry land and open sea, but there are also mountains, plains, rivers, coastal shelves, and deep oceanic trenches. The earth also varies in its climate: in some parts of the world, weeks or months pass by under almost cloudless skies, while in others, the ground is scoured by icy winds or soaked by intense tropical storms. Differences like these create a complex jigsaw puzzle of varied habitats, enabling the Earth to support a rich diversity of animal life. Some species are highly adaptable, and can survive in a wide range of conditions, but the vast majority are found in one kind of habitat and nowhere else.

What are habitats?

In its narrowest sense, a habitat is the environment in which something lives. For some animals, a habitat might be as restricted as a temporary pool in a desert or as small as a piece of decaying wood. In a broader sense—the one used in this book—a habitat can mean a characteristic grouping of living things, together with the setting in which they are found. In ecology, a habitat defined in this way is known as a biome.

Habitats contain both living and nonliving matter. In some—for example, true desert—living things are thinly scattered, so the nonliving part of the environment is dominant. In others, such as forest and coral reefs, living things are so abundant that they fill all the available space and create habitats for each other. In these habitats, huge numbers of species exist side by side, forming extraordinarily complex webs of life.

EFFECTS OF CLIMATE
These two habitats—in Argentina (left) and Chile (right)—are at the same latitude, but they have very different climates, and therefore very different plant and animal life. The contrast is created by the Andes, a mountain barrier that blocks rain-bearing winds. Chile is on the windward side of the Andes.

Factors that shape habitats

Geology plays a part in shaping habitats, but by far the most important factor is climate. As a result, differences in climate—which sometimes occur over remarkably small distances—can have a huge effect on plant and animal life. A classic example of this occurs where mountain ranges intercept rain-bearing winds. On the windward side of the mountains, heavy rainfall often creates lush forests teeming with all kinds of animal life. But in the "rainshadow," to the lee of the mountains, low rainfall can produce desert or scrub, where only drought-tolerant animals can survive.

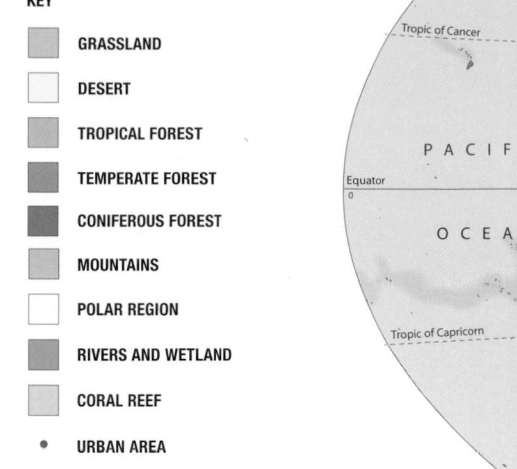

Temperature is another climatic factor that has an important effect both on land and in the sea. For example, in the far north, coniferous forest eventually peters out in the face of biting winter frosts. This northern tree line, which runs like a ragged ring around the Arctic, marks the outer range of crossbills, wood wasps, and many other animals that depend on conifers for survival. On coasts and at sea, temperature changes are usually more gradual than they are inland. However, warmth—or lack of it—still determines where some habitats are found. For example, reef-building corals do not thrive at temperatures of less than about 68°F (20°C), so most reefs are found in the tropics. However, on the west coast of Africa and the Americas, reefs are rare because, although the climate is warm, cold currents pass close to the shore. Mangrove swamps present a similar pattern: in the Southern Hemisphere, they reach as far as South Australia; in the Northern Hemisphere, they extend only just out of the tropics.

CHEMICAL CYCLES

In all habitats, living things take part in cycles that shuttle chemical elements between living and nonliving matter. About 25 elements are essential to life and, of these, just 4 make up the bulk of living things. These are hydrogen, oxygen, nitrogen, and—the key element—carbon. In the nonliving world, carbon can be found in the atmosphere (as a gas), in water (in dissolved form), and in the ground (in rocks and fossil fuels). Plants absorb carbon dioxide from the atmosphere, and most other forms of life give off carbon dioxide when they break down carbon-containing substances to release energy. Carbon is also released by burning fossil fuels.

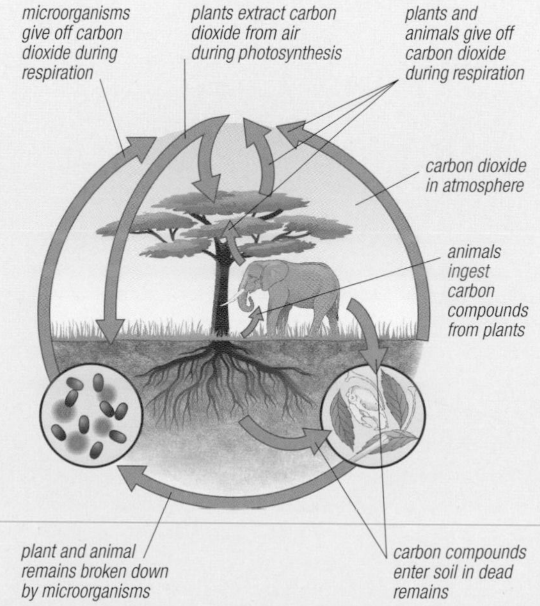

THE CARBON CYCLE
This diagram shows some of the main pathways in the carbon cycle. The time taken for each part of the cycle to be completed varies greatly. Carbon may stay in living things for only a few days, but it can remain locked up underground for thousands of years.

HABITATS OF THE WORLD
This map shows the distribution of major habitats across the world, and also cities with populations of 5 million or more. The habitat distribution shown here is the pattern that would exist if man-made changes, resulting from urbanization and the spread of agriculture, had not occurred.

KEY
- GRASSLAND
- DESERT
- TROPICAL FOREST
- TEMPERATE FOREST
- CONIFEROUS FOREST
- MOUNTAINS
- POLAR REGION
- RIVERS AND WETLAND
- CORAL REEF
- • URBAN AREA

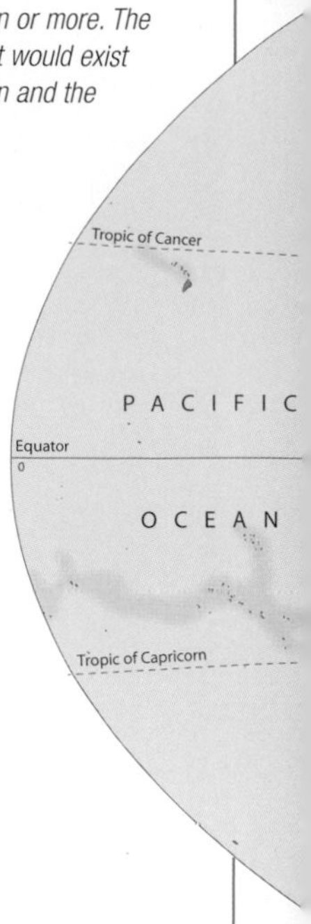

Biodiversity

From the earliest days of scientific exploration, naturalists noticed great variations in biodiversity, or species richness. In the far north and south, species totals are low compared with the numbers found near the equator. Arctic tundra, for example, is inhabited by just a few hundred species of insects, while in tropical forests, the total is probably at least a million. A similar picture—albeit on a smaller scale—is true for mammals and birds. However, high-latitude habitats make up for their lack of biodiversity by having some phenomenally large species populations. The seas around Antarctica, for example, harbour perhaps 40 million crabeater seals—the most numerous large wild mammals on earth.

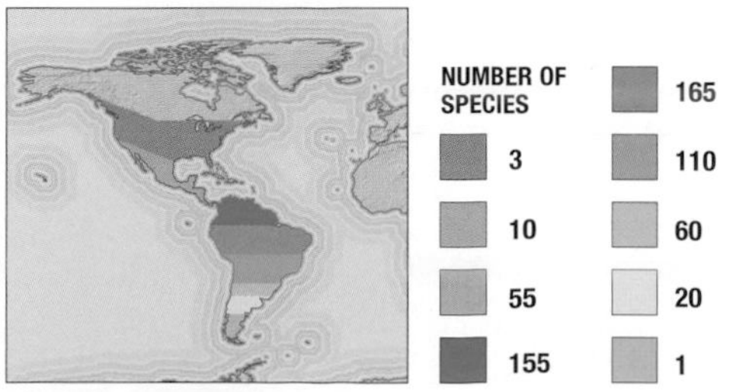

TROPICAL TREASURY
The hummingbird family vividly demonstrates the high level of species diversity that is to be found in the tropics. Only a handful of hummingbird species live at high latitudes—and most of these are migrants—but on the equator, the number of species rises to over 150.

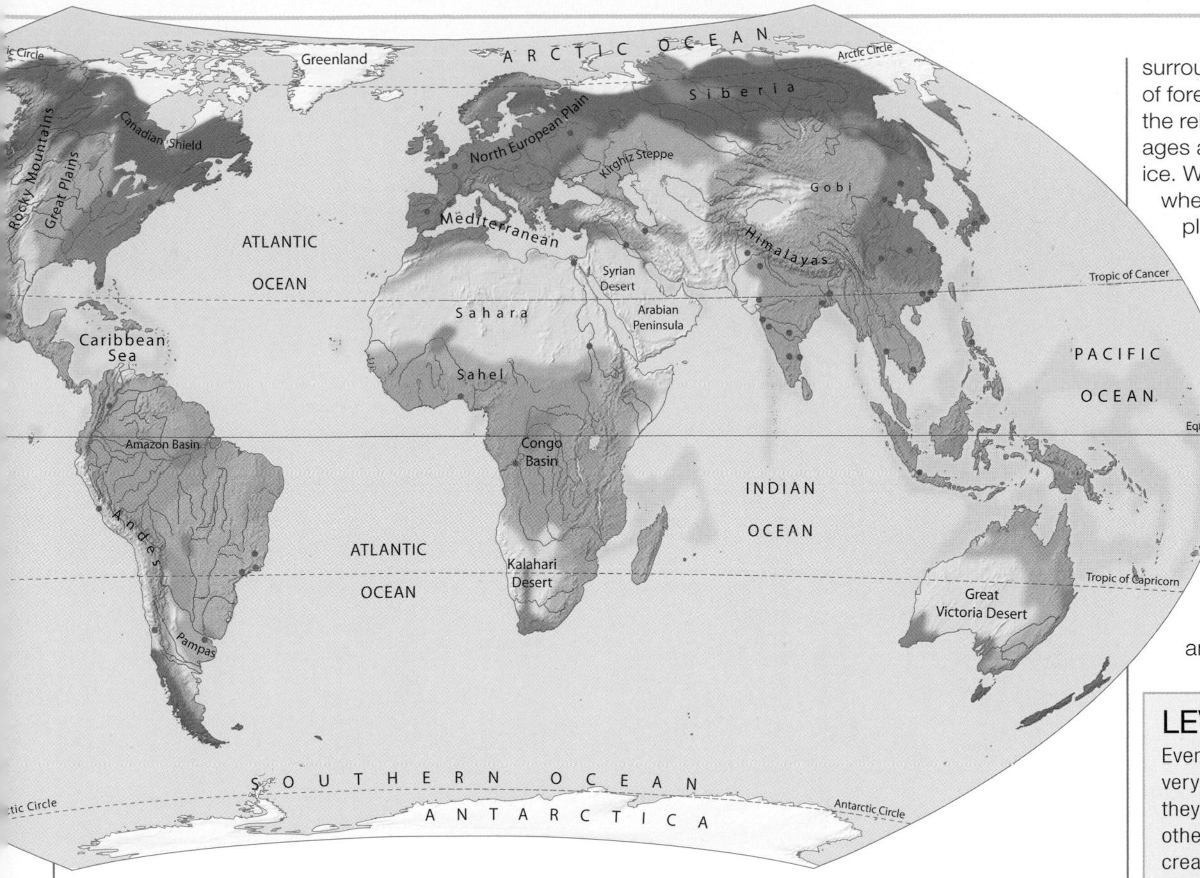

The reasons for such variation in biodiversity are still not fully understood, although climate almost certainly plays a part. However, in an age in which many animal species are endangered, biodiversity—and ways of maintaining it—has become an important topic. Tropical forests and coral reefs are especially rich in species, which is why so much attention is currently focused on preserving them and their animal life.

Animal distributions

A glance at the map on this page shows that various types of habitat are spread across large expanses of the world. However, with a few exceptions, most of their animals are not. Instead, each species has a characteristic distribution, which comes about partly through its evolutionary history (see p.17), and partly as a result of its way of life.

In many cases, an animal's lifestyle shapes its distribution in unexpectedly subtle ways. For example, in the Americas, the brown pelican is found all along the western coast, apart from the far north and south; in the east, it does not reach south of the Caribbean. The reason for this is that, unlike its relatives, the brown pelican feeds by diving for fish, and needs clear water to spot its prey. The Caribbean is clear, but farther south lies the Amazon River, which pours vast amounts of mud-laden water into the sea. For the pelican, this muddy water is a barrier that cannot be crossed. Many animals' distributions are linked to those of particular plants. Extreme examples include the yucca moth, which depends on yucca plants; the fig wasp, which develops inside figs; and countless bees that depend on particular flowers. Not all plant-dependent animals are insects, however. The robber crab—the largest and heaviest terrestrial crustacean—feeds predominantly on rotting coconuts, which it scavenges along the shore. As a result, it is found only where coconuts grow. Mammals can be just as particular. The giant panda—one of the most famous examples—depends on about 2 dozen species of bamboo, which are found only at mid-altitudes in the mountains of central China.

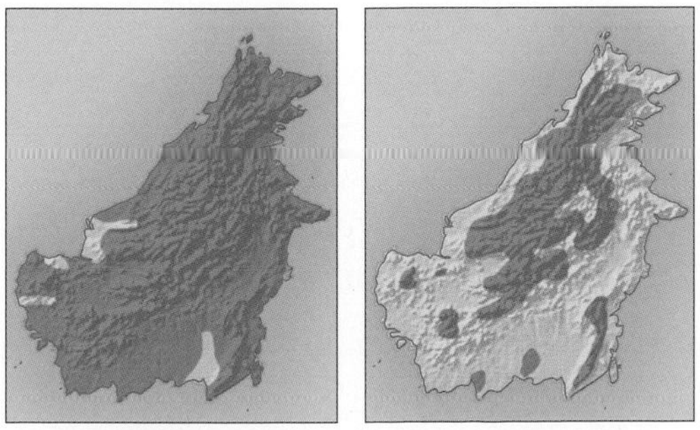

IMPOVERISHED RAINFOREST
Borneo—an island of extraordinary biological richness—had extensive rainforest 60 years ago (left). Deforestation (projected 2020 cover, right) not only endangers many species and renders others extinct, but can cause erosion and even change local climate.

Changing habitats

In nature, habitats change all the time. Forest and grassland catch fire, rivers burst their banks, and storms batter coral reefs and coasts. Such unpredictable occurrences are facts of life, and animals—along with other living things—have evolved ways of surviving them. Habitats can also change in much more profound ways, over much longer periods of time. Here, the driving force is usually climate change, a natural process that is triggered by a host of factors, including continental drift. On several occasions in the distant past—most recently about 12,000 years ago—the polar ice caps expanded, destroying existing habitats and evicting their animals. On each occasion, when the ice eventually melted, plants moved back into the empty landscape, and animals followed suit.

The world's climates are interrelated, which means that changes in one area can have long-term effects all over the globe. For example, during the last ice age, the climate in the tropics became drier, and the Amazon rainforest shrank to form scattered "refugia"—islands of forest surrounded by grassland. Even today, these areas of forest still contain a wider variety of birds than the relatively new forest that has grown back. Ice ages also affect sea levels, by locking up water as ice. When sea levels fall, land habitats expand; when they rise, the land is drowned again and plants and animals are forced to retreat.

Since the last ice age ended, natural changes have not been the only ones that have affected the world's habitats. On a local and global scale, human activity has had an increasing impact and, as a result, the pattern we now see is partly man-made. This is especially true of forests, which have been cut back to make space for agriculture, but it is also true of some grasslands, wetlands, and even deserts. In some remote regions—particularly in the far north—the original pattern still remains, but in populated regions, it has been transformed, creating a world where wild animals can have difficulty finding a home.

LEVELS OF LIFE

Even in the remotest places on earth, animals very rarely live entirely on their own. Instead, they interact with other individuals, and with other species. Taken together, these interactions create a range of different ecological levels, from local populations, communities, and ecosystems to the whole biosphere—the sum total of all the places where living things can be found. Because microorganisms are so widespread, the biosphere extends high into the atmosphere, and probably several miles underground.

INDIVIDUAL
An individual animal is normally an independent unit that finds its own food. It often lives within a set home range.

POPULATION
A population is a group of individuals that belong to the same species, live in one area, and interbreed.

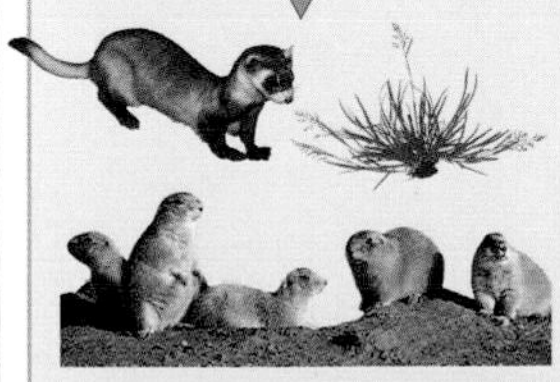

COMMUNITY
A community is a collection of populations. Although they belong to different species, they depend on each other for survival.

ECOSYSTEM
An ecosystem is made up of a community and its physical surroundings. Characteristic ecosystem types are known as biomes (or habitats).

BIOSPHERE
The biosphere consists of all the ecosystems on Earth, and therefore all the places—from the Earth's crust to the atmosphere—that living things inhabit.

HABITATS

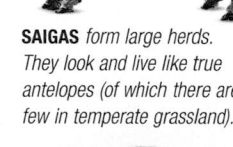

Grassland

In parts of the world where it is too dry for trees to grow, yet moist enough to prevent the land from becoming desert, grasses are the dominant plants. Grasses are unusual in that their stems grow from a point near the ground. This means that unlike most other plants, which grow from their tips, grasses are unharmed by grazing. In fact, grazing animals help grasses to maintain their dominance by stunting the growth of competing plants. This creates a vast, open habitat in which there is plenty of plant food—for those that can digest it—but little shelter from the elements.

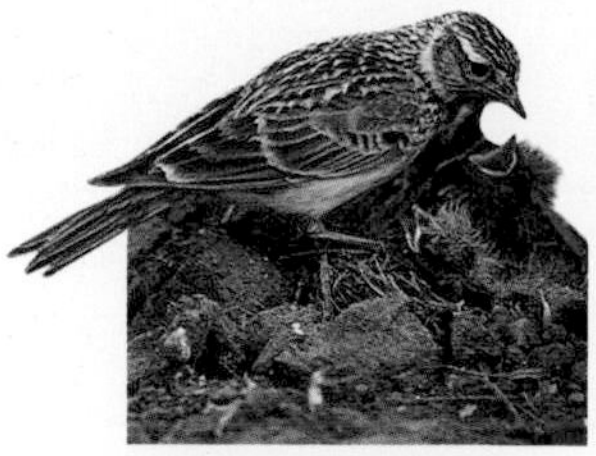

GRASSLAND SONGBIRDS, *such as the skylark, have no trees to perch on, so they broadcast their songs and calls from the air.*

SAIGAS *form large herds. They look and live like true antelopes (of which there are few in temperate grassland).*

Temperate grassland

Before the advent of farming, grassland covered large parts of the temperate world, notably in the northern hemisphere. These vast grasslands—which include the prairies of North America, and the steppes of Europe and central Asia—are nearly all in the center of large landmasses, far away from coasts and their moisture-laden winds. Summers are often warm, but winters can be long and cold, with biting winds.

An unusual feature of this kind of habitat is that the majority of the plant matter is hidden away below ground—the exact opposite of the situation elsewhere on land. This is because grass plants direct much more energy into growing roots than into producing leaves, and their roots form a continuous mat that protects the surface of the ground by holding the soil in place. If grassland is burned, or hit by drought, it can soon recover because the grass can draw on its buried reserves in order to start growing again. The root mat makes a useful source of food for insects and other small animals. It is also a perfect medium for burrowers, because it is easy to dig through and, unlike loose soil, rarely caves in.

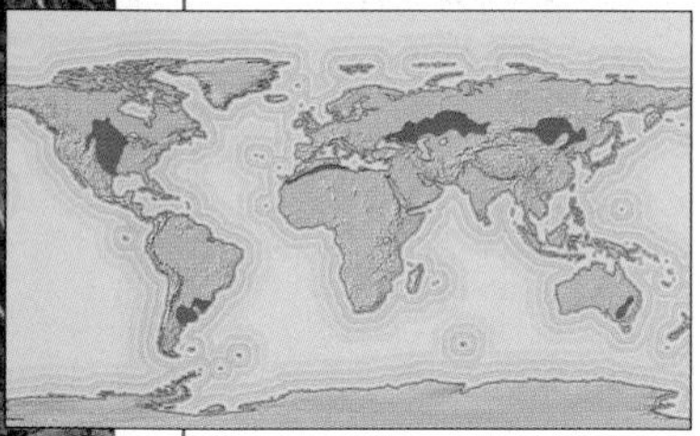

THE LARGEST AREAS *of temperate grassland are found in North America, South America, eastern Europe, and central and eastern Asia.*

Above ground, the food supply is closely tied to the seasons. In temperate grassland, most of the year's water usually comes in the form of spring rain or melting snow. This creates a flush of growth during spring and early summer, which is the time that most grazing animals breed. By late summer, the grass is brown and dry, although for a while grass seeds make a valuable fall harvest. Winter is a difficult time for all grassland animals, but particularly for grazers because they often have to survive on low-grade food that is hidden under snow.

THE AMERICAN BISON, *now relatively rare, is a large, heavy grazer that once had a profound effect on the ecology of the prairies.*

GRASSLAND REPTILES, *such as this sand lizard, spend the winter underground and are active in summer.*

SOME BIRDS, *such as this burrowing owl, lay their eggs in underground burrows because there are no trees to provide nest holes.*

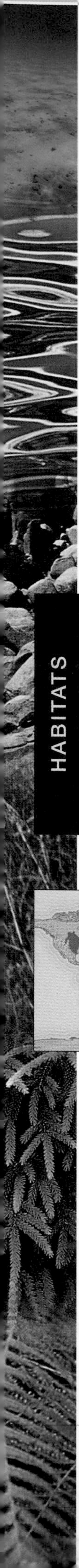

TREES AND SHRUBS

AERIAL SCAVENGERS, *such as this African white-backed vulture, exploit the savanna's strong thermals, uninterrupted views, and abundance of wildlife.*

THE GIRAFFE'S *extraordinary reach is responsible for the umbrella-like shape of many African grassland trees.*

FLIGHTLESS *birds, such as the ostrich, thrive in savanna, where running is an effective means of escape.*

GROUND

THE CHEETAH *attacks in the open, so it relies on speed, rather than stalking, to enable it to make a kill.*

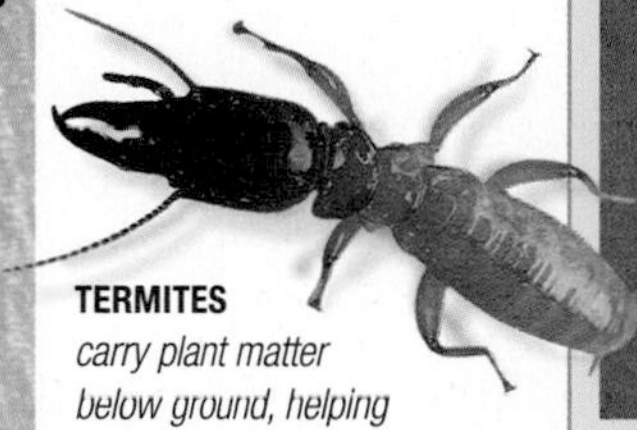

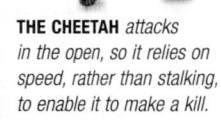

TERMITES *carry plant matter below ground, helping to recycle vital nutrients.*

Savanna

Savanna is tropical or subtropical grassland that contains scattered shrubs and trees. The grasslands of east Africa are a familiar example, with their diverse wildlife and distinctive vegetation (particularly the flat-topped acacia trees). Compared with temperate grassland, savanna is very variable: in some savanna habitats, trees are few and far between; in others, they form scattered thickets, merging into open woodland.

Trees have a major impact on the savanna's animal life. This is because they produce a wide variety of food, including wood, leaves, flowers, and seeds; and they also create shelter and breeding sites for animals that live off the ground. The balance between trees and grass is a delicate one that is sometimes changed by the animals themselves. For example, elephants destroy trees by pushing them over so that they can reach their leaves. However, elephants also help trees reproduce because they ingest the trees' seeds, which are then passed in their dung—an ideal medium for promoting seed growth. Browsing mammals often keep trees in check by nibbling saplings before they have had a chance to become established. Fire also helps hold back trees, and its effect is most apparent in places where trees grow close together.

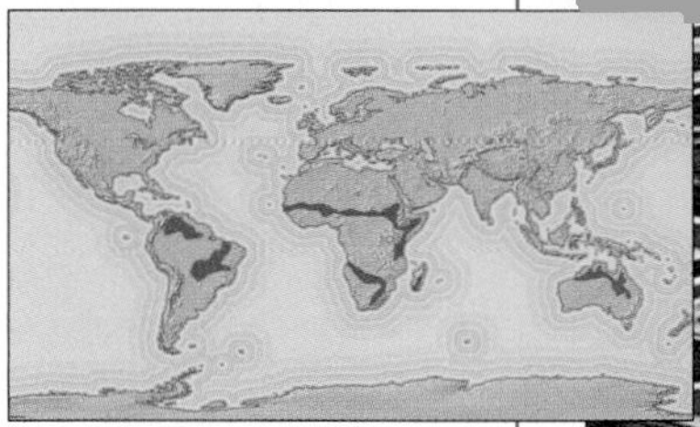

SAVANNA IS FOUND *mainly in Central and South America, tropical Africa, southern Asia, and northern Australia.*

Unlike temperate grassland habitats, savanna is usually warm all year round. There is often a long dry season, when most trees lose their leaves, followed by a wet, or "rainy," season, which produces a rapid burst of growth that turns the landscape green. During this wet season, plant-eating animals rarely have to contend with a shortage of food; in the dry season, the threat of starvation is never far away, and many animals travel long distances to find water and food.

SOIL EROSION

At one time, natural grassland covered about two-fifths of the earth's land surface. With the spread of agriculture, much of this has been taken over for growing crops or raising livestock, leaving only small remnants with their original vegetation and wildlife. One of the side effects of this process has been a great increase in soil erosion. Grasses gradually build up soil, and hold it together with their fibrous, matlike roots. If they are removed, the soil is exposed to the wind and rain, which can strip it away. Erosion is a major problem for agriculture, because soil takes so long to form. In the short term, it cannot be replaced.

DEEP CUTS *Seen here in Africa, soil erosion can form characterstic V-shaped gullies several yards deep. Trees fall in when their roots are undermined.*

HABITATS

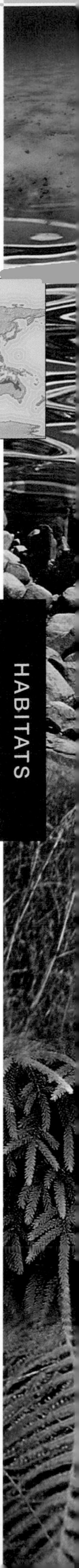

Life in grassland

Despite centuries of human disturbance, grassland supports some of the largest concentrations of animal life on earth. Survival in grassland habitats is far from easy, however: aside from the lack of shelter and plant diversity, there are hazards such as drought and fire to contend with. Added to this is the ever-present risk of attack by some of the world's fastest and most powerful predators.

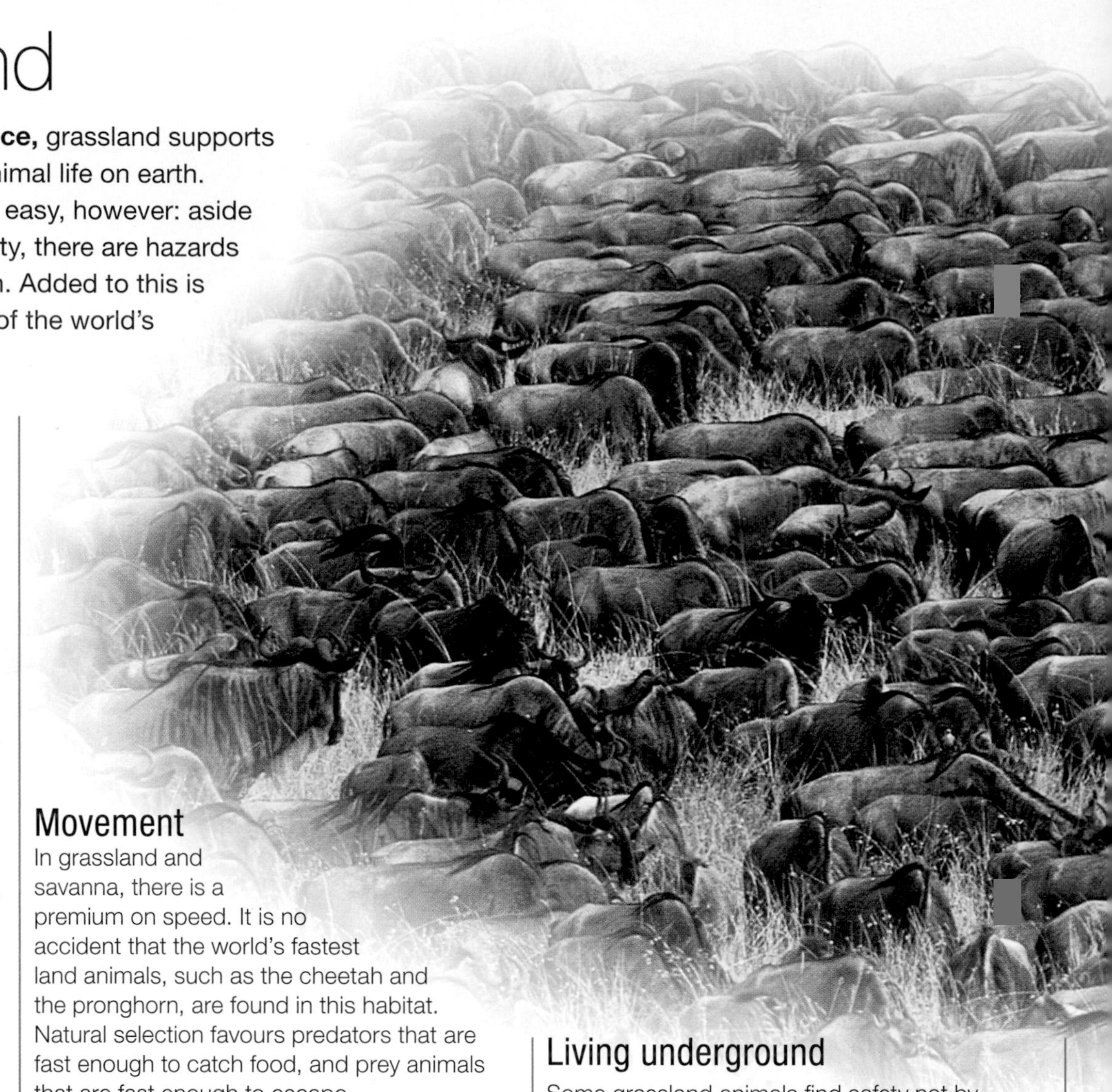

Herding

Life in open grassland is often dangerous because there are few places to hide. To increase their chances of survival, many large plant eaters live in herds. This makes it more difficult for predators to attack, because while most members of the herd are eating, some are always on the lookout for danger.

Today's largest herds are found on Africa's plains. Here, migrating wildebeest can form herds over a quarter of a million strong, and 25 miles (40 km) long, although even these herds are small compared with some that existed in the past. During the 19th century, springbok herds in southern Africa sometimes contained more than 10 million animals. In North America, bison herds probably reached similar sizes before hunting brought the species to the edge of extinction.

Life in herds does have its problems, one of which is the risk that an animal might wander off and become lost. Most herding species have scent glands on their hooves so that if an animal becomes isolated it can follow the scent tracks to rejoin the herd. Another problem is giving birth. To prevent their young from being trampled or attacked, many grazers give birth in cover, and rejoin the herd a few weeks later. Some, however, are born in the open and have to be able to keep up with the herd when they are just a few hours old.

Movement

In grassland and savanna, there is a premium on speed. It is no accident that the world's fastest land animals, such as the cheetah and the pronghorn, are found in this habitat. Natural selection favours predators that are fast enough to catch food, and prey animals that are fast enough to escape.

Most of the fast runners are mammals; however, grassland also has nature's fastest-running birds, including ostriches, rheas, and emus—giant species that have lost the ability to fly. These birds can reach speeds of up to 44 mph (70 kph). More importantly, they are able to maintain such speeds for up to 30 minutes—long enough to outrun most of their enemies unless a predator launches an attack from a very close range.

Despite the many fast runners, grassland life often appears tranquil. This is because running is extremely energy-intensive, and animals run only when they absolutely have to. Prey animals have invisible "security thresholds" that vary according to the threat they face. For example, gazelles often let lions approach to within about 650 ft (200 m) because they are instinctively aware that lions that are visible at this distance are unlikely to be stalking prey.

A solitary cheetah, on the other hand, will send a gazelle herd sprinting, even if it is seen to be 4 times farther away.

FLIGHTLESS BIRDS
Standing guard over its chicks, a greater rhea watches for danger. Like other flightless birds that live in grassland, it uses its height and large eyes to spot potential predators at a distance so that it has time to run for safety.

Living underground

Some grassland animals find safety not by running away but by retreating into burrows below ground. There, they can stay out of reach of most predators and find some protection from the worst of the elements. Subterranean animals include a wide variety of species, ranging from mammals to insects. Some animals, particularly snakes, do not excavate their own burrows; instead, they adopt existing ones. The largest burrows, made by African aardvarks, are big enough to accomodate a person, and are a serious hazard to vehicles; the most extensive are made by prairie dogs and other rodents. Before farming became widespread in North

MAKING A HOME BELOW GROUND
A black-tailed prairie dog collects grass to line a nesting chamber. The colony, or town, consists of tunnels up to 16 ft (5 m) deep.

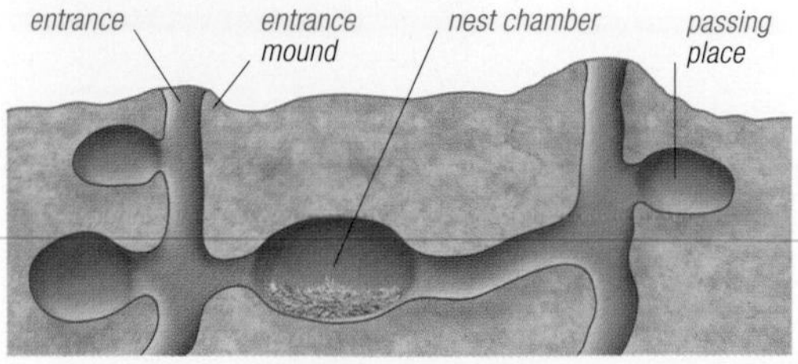

PRAIRIE-DOG BURROWS

MIGRATING WILDEBEEST
In Tanzania's Serengeti National Park, wildebeest migrate in a cyclical path to take advantage of fresh plant growth at different times of the year. They spend the wet season in open grassland, and the dry season in wooded savanna.

FIRE

Fires, ignited by lightning, are a natural feature of grassland life, clearing away dead growth, and allowing fresh grass to sprout. In the long run, fires help grassland wildlife, but while they are burning they can be lethal. As the flame front advances, most animals react by running or flying for safety, often abandoning their usual caution in an urgent bid to escape. Some animals, such as bustards and storks, have learned to make the most of this frenzied exodus from the flames. They gather close to fires and snap up insects and other small animals as they scurry away; and once the fire has moved on, they pick over the charred ground for casualties.

FEEDING BY THE FLAMES
A European white stork searches for small animals fleeing before an advancing fire. Soaring high over the plains, the stork will have been attracted by the fire's smoke.

America's prairies, some prairie-dog burrow systems covered several thousand square miles and housed millions of animals.

Termites are also accomplished builders, constructing giant, elaborate, subterranean nests that extend high above ground level. These nests house large, cooperative communities that can contain over 30 million inhabitants. Along with ants, they make up a very large part of the habitat's animal life, and provide food for the large insect eaters.

Feeding

Although grass is rich in nutrients and easy to find, it is difficult to digest. Many mammals, including humans, cannot break it down at all because it contains large amounts of cellulose—a carbohydrate that most animals cannot digest. Grazing mammals, however, have special microorganisms in the gut that break down cellulose so that the body is able to use it. Some nonmammal species also use microbes to digest plant material. In tropical savanna, for example, termites rely on them to break down dead leaves and wood.

The animals that are most efficient at using cellulose are ruminant mammals—antelopes, buffaloes, and giraffes, for example—which helps to explain why these animals dominate grasslands. The ruminant's complex stomach acts like a fermentation tank, working to extract the maximum amount of nutrients from food. The animal assists in the process by regurgitating its food and chewing it a second time, making it even easier to break down. Nonruminant plant eaters, such as zebras, have less efficient digestive systems and must therefore eat more to survive.

In pure grassland, plant eaters compete for the same food, although each may have a preference for a different type of grass. In savanna, the presence of trees and shrubs makes for a wider range of food, and browsing mammals have minimized competition further by evolving specific ways of feeding. This means that a remarkable number of species can live side by side. For example, the small Kirk's dik dik antelope feeds on shoots and fruit, and rarely touches grass, while the much larger eland will eat almost anything from fruit and seeds to roots scraped up from the ground. Scavengers, which also play an important part in the habitat's ecology, include birds as well as coyotes, jackals, and hyenas. Most airborne scavengers are vultures, but there are several species of storks, one of which, the marabou stork, rivals the Andean condor for the title of the world's largest flying animal.

REACHING FOR FOOD
Unlike most antelopes, the gerenuk can stand on its back legs. This means that although it measures only about 39 in (1 m) at the shoulder, it can browse leaves over 6½ ft (2 m) up—much higher than the reach of other antelopes of similar size.

INSECT EATER
The giant anteater uses its powerful claws to break open termite mounds and reach the insects inside.

COLLECTING DUNG
Dung beetles make use of the large amount of dung produced by grazing mammals. They gather the dung into balls, which they then roll away and bury below ground where it is used as food for the beetles' grubs.

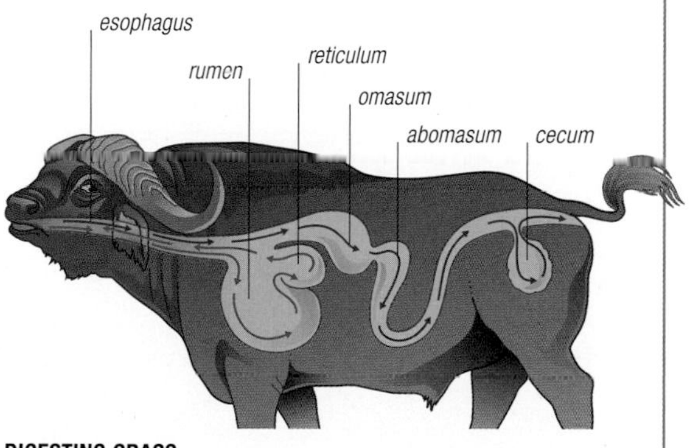

DIGESTING GRASS
The largest of the 4 chambers in a buffalo's stomach, the rumen, contains millions of bacteria and protozoans. These simple organisms produce cellulose-splitting enzymes that turn cellulose into simple nutrients that can then be absorbed.

Desert

Deserts are places of extremes. Besides being dry, they experience intense sunshine, and a greater daily temperature range than any other land habitat. Rain—when it comes—typically falls in brief but torrential downpours, while strong winds pick up sand and grit, carrying it almost horizontally through the air. Although no 2 deserts are identical, true desert is usually defined as having less than 6 in (15 cm) of rainfall a year. Semidesert has more rainfall—up to 16 in (40 cm) a year—which typically falls during a relatively short spring or wet season followed by months of drought.

SANDGROUSE *overcome the problem of supplying water to nestlings by transporting it in their breast feathers.*

THE SCIMITAR-HORNED ORYX, *like other nondrinkers, can get all the water it needs from its food.*

True desert

Most of the world's true desert is found in 2 belts, one straddling each of the tropics. Here, zones of high atmospheric pressure persist for months at a time, preventing low-pressure air from bringing in rain. Desert also forms where mountains block rain-bearing winds, and where cold, coastal currents chill the air so that it carries very little moisture inland.

In true desert, the amount of rain is so meager, and so unpredictable, that very few plants can survive. The ones that do—such as cacti and other succulents—are highly effective at collecting and conserving what little water nature provides: they have large networks of shallow roots, which drain the surrounding ground so thoroughly that, often, nothing else can grow near by.

For animals, this arid environment creates some interesting effects. With so few plants, there is very little soil, which severely limits invertebrate life. Most small animals, such as insects, are found either on the plants themselves or in the debris that accumulates immediately beneath them. Larger animals, such as reptiles and rodents, venture away from these pockets of greenery, but even they have to be careful to avoid the worst of the daytime heat.

Lack of vegetation means that most of the ground is exposed. Bare ground absorbs warmth very quickly when the sun rises, and reradiates it once the sun has set. The dry air accentuates this effect, allowing daytime surface temperatures to soar to over 158°F (70°C). As a result, most animals living in true desert are active after dark. During the day, they hide away, leaving little sign of themselves other than their tracks.

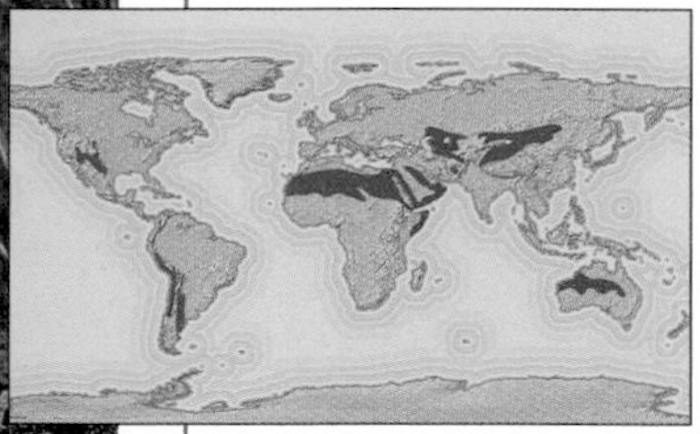

TRUE DESERT *occurs at midlatitudes in the northern and southern hemispheres, and it can have less than 2 in (5 cm) of rain a year.*

THE GOLDEN JACKAL *is able to live nocturnally, which means that it can avoid the worst of the desert's daytime heat.*

THE THORNY DEVIL *has protective spines in addition to the good camouflage that other slow-moving ground dwellers often depend on for survival.*

DESERT SNAKES, *such as the desert horned viper, will pursue their prey into their burrows.*

COUCH'S SPADEFOOT TOAD *is one of the desert frogs and toads that spend most of their lives underground.*

GROUND

PLANT LAYER

GROUND

DESERT BATS *play a vital ecological role because they feed on insects and pollinate flowers.*

SEMIDESERT BIRDS, *such as the roadrunner, often nest among spiny plants to protect their young from predators.*

THE MEERKAT *lives in large colonies. Its varied diet and cooperative foraging technique help it survive when food is scarce.*

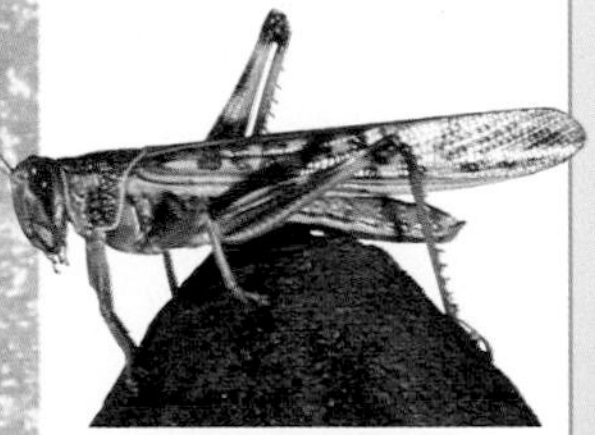

THE LOCUST'S *migratory lifestyle is an adaptation to a habitat where food supplies are erratic and unpredictable.*

TARANTULAS *hunt mainly by touch, enabling them to find their prey after dark.*

Semidesert

Compared with true desert, semidesert is more widespread, and it is also much more biologically productive. It is found in every continent, including some regions far outside the tropics.

The modest but nevertheless reliable rainfall that semidesert receives has a dramatic effect on the landscape and the types of animals that it can support. Plants often grow in profusion, creating tangled thickets of vegetation that provide plenty of cover. There are woody species, which store most of their water in underground roots, and fleshy succulents, which store it above ground in their stems and leaves. Most of these desert species are well protected from plant eaters—cacti, for example, have extremely sharp spines, while spurges exude a poisonous milky sap when they are damaged—but, for animals that can overcome these defenses, they are an important source of food. Semidesert also has plants known as desert ephemerals, which spring up rapidly after rain, flower, set seed, and then die. This short life cycle produces extra fresh food for animals, and adds to the stock of seeds scattered over the desert floor.

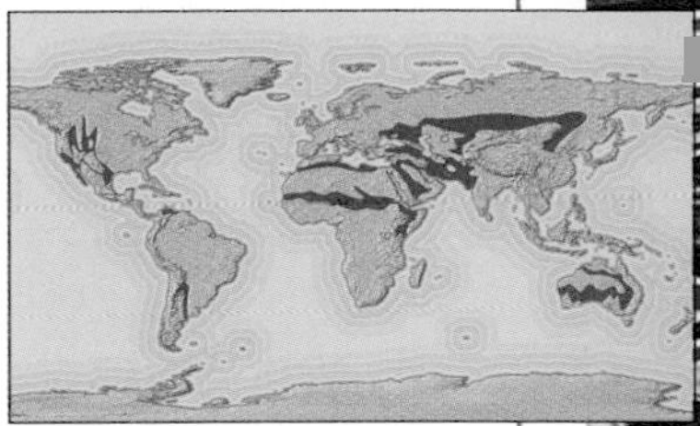

SEMIDESERT *is most widespread in the world's major landmasses, and it extends well into the temperate zone in both hemispheres.*

While some semideserts are warm or hot throughout the year, others are surprisingly cold in winter. In the deserts of central Asia, and in the northern parts of America's Great Basin—the desert region between the Rocky Mountains and the coastal ranges farther west—temperatures can fall to -22°F (-30°C). In these areas, animals need protection against winter cold as well as against summer heat: small animals, such as insects, usually become dormant in winter, and many burrowing mammals hibernate until the spring.

DESERTIFICATION

The world's desert regions constantly shift, because rainfall patterns change as time goes by. These changes are generally very slow, which gives wildlife time to adjust. However, deserts can also be created by human activities—particularly poor farming practices, such as overstocking with cattle or goats. This result, called desertification, currently affects many parts of the world, from China and California to the Sahel—the arid region to the south of the Sahara Desert. Desertification reduces plant cover and speeds up erosion, driving out animals that normally survive in dry habitats. Once it has occurred, it is difficult to reverse.

SAND STORM
Looming high into the sky, a vast cloud of sand approaches a desert town. A powerful storm can displace millions of tons of sand.

HABITATS

Life in desert

In a habitat where moisture is scarce, obtaining and conserving water are every animal's top priorities. Desert animals practice a tight "water economy," which means collecting water wherever they can, and minimizing water loss wherever possible. However, being economical with water is not in itself enough to guarantee survival: desert species have had to evolve various other adaptations to enable them to cope with a wide range of temperatures and the ever-present threat of food shortage. As a result, these animals are able to live in some of the driest places on earth.

Conserving water

Most deserts have a scattering of oases, where animals gather to drink. Some species need to drink daily, which restricts how far they can roam from an oasis. Others can survive on their on-board reserves for days or even weeks, depending on the temperature. A remarkable feature of desert life is that some animals can manage without drinking at all. Instead, they get all their water from their food. Some extract it from the moisture contained in food; but most use the food to manufacture metabolic water, which is created by chemical reactions when the energy in food is released. Seed-eating rodents are expert at this: although their food looks dry, they are able to metabolize all the water they need.

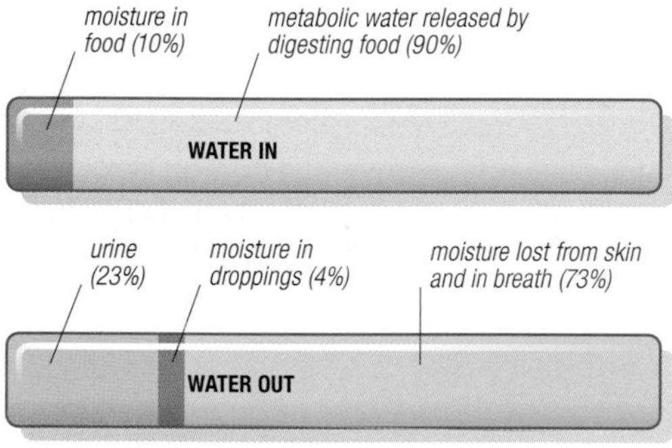

WATER BALANCE
This diagram illustrates how a kangaroo rat survives entirely on the water in its food. The water taken in has to balance that which is lost to prevent the animal from becoming dehydrated.

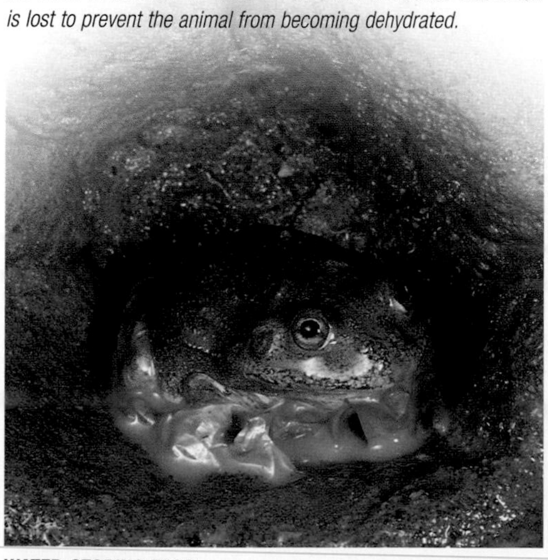

WATER-STORING FROG
The Australian water-holding frog stores water in its bladder and beneath its skin. To prevent this water from drying out, the frog then seals itself in a semipermeable cocoon underground.

For drinkers and nondrinkers alike, water has to be eked out to make sure that it lasts. Compared with animals from other habitats, desert species lose very little moisture in their urine and droppings, and only a small amount is released from their skin and in their breath. Desert species are also good at withstanding dehydration. The dromedary, or one-humped camel, can lose nearly one half of its body water and survive. For humans, losing just a fifth can be fatal.

Storing food

To enable them to cope with erratic food supplies, many animals keep their own food reserves. Some do this by hiding food away. The North American kangaroo rat, for example, constructs underground granaries that contain up to 11 lb (5 kg) of seeds. But for predators, and for animals that browse on shrubs, creating such larders is not possible. Their food is difficult to collect and to transport, and even if it could be hoarded it would be unlikely to remain usable for more than just a few days. The answer is to store food inside the body. The classic example of this is the camel, which stores surplus food, in the form of fat, in its hump. Several other species, such as the Gila monster and fat-tailed dunnart, store food in their tails.

RAPID REFILL
After going without water for several days, a camel can drink over 11 gallons (50 liters) in just a few minutes. It also metabolizes water from surplus food, laid down as fat in its hump. Its salt tolerance is high, which is useful in a habitat where water is often brackish.

Coping with heat and cold

In desert, the temperature rarely stays steady for more than a few hours, and it can reach extremes of both heat and cold very quickly. Humans lose excess heat by sweating, but at very high temperatures, this cooling system can use as much as 35 fl. oz (1 liter) of water an hour—far more than any desert animal could afford.

Desert animals tackle the heat problem in 2 ways: by reducing the heat they absorb, and by increasing the heat they give out. Light-colored skin or fur reflects some of the sun's rays, minimizing heat absorption; but a much more effective method—used by many desert animals—is to avoid the most intense heat by being nocturnal, spending the day sheltering underground. Burrows do not have to be very deep to make a difference: while the desert surface may be too hot to touch, the ground just a few inches below it will be relatively cool.

Getting rid of excess heat is more difficult, particularly when an animal's body temperature is dangerously high. Lizards and snakes are

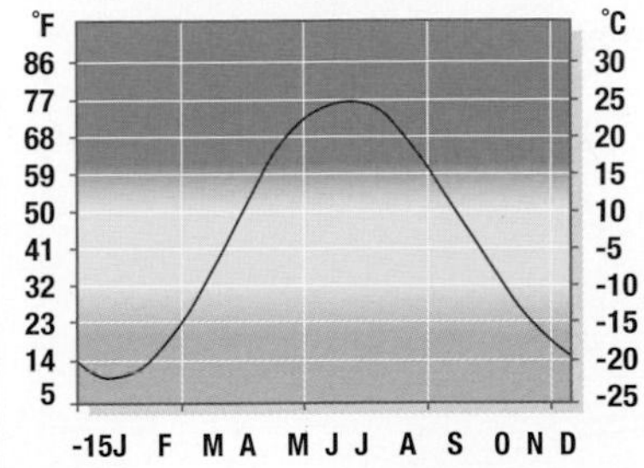

COLD DESERT
This graph shows the average annual temperature on the western edge of the Gobi Desert. From November through to March, temperatures fall below freezing.

CHANGING COAT
In the deserts of central Asia, freezing winters are followed by soaring temperatures in spring. The Bactrian camel manages by growing a thick coat that falls off as soon as the spring warm up begins. This camel is about halfway through this process.

often described as "cold-blooded," but this actually means that their body temperature rises and falls with that of their surroundings. Although they thrive in warm habitats, and can survive with a body temperature of up to 111° F (44° C), they often have to sit out the hottest part of the day in shade. Some desert birds cool down by panting, which involves fluttering the flap of skin over their throats. Desert kangaroos and wallabies lick their front legs, covering them with saliva. As the saliva evaporates, the animal's blood cools down.

In high-latitude desert regions, such as the Gobi Desert of central Asia and the Great Basin Desert of North America, winter can be extremely cold. Animals have various ways of coping with this. Most reptiles hibernate, while birds often fly to warmer climates. Mammals keep warm by growing thick fur, or by sheltering underground.

Explosive breeding

Desert animals often have highly variable breeding seasons. Instead of reproducing at a fixed time of year, many produce young when there is the best chance of finding food. Female kangaroos, for example, give birth extremely regularly when food is plentiful, but when food is scarce they stop breeding entirely. This flexible system is an efficient way of using resources because it prevents parents having to tend hungry youngsters when they are hungry themselves.

Some desert species carry irregular breeding to extremes. Desert wildlife includes a number of animals that, paradoxically, live or breed in water and, for those species, reproducing is a highly unpredictable and time-sensitive business. Such animals include burrowing frogs and toads, and also freshwater shrimps that live in temporary pools. For months or even years at a time, they are an invisible part of desert wildlife, with the amphibian species lying hidden underground, and the shrimps present only as eggs in dried-up ground. But immediately after a heavy storm, the frogs and toads dig their way to the surface, and the shrimp eggs hatch. Once active, these animals immediately set about finding mates because they have to complete their life cycles before the pools dry up again.

LIFE IN BRIEF
Trapped in a rapidly shrinking desert pool, these adult tadpole shrimps have only a few days to live, but the eggs they leave behind can survive in a dried-out state for several years—long enough to last until the next heavy storm, when they will hatch.

MOVING ON SAND
Flaps of skin between the web-footed gecko's toes make for snowshoelike feet that allow it to run across dunes.

ATTACK FROM BELOW
Guided by vibrations overhead, this desert golden mole has emerged from its burrow to ambush a gecko.

Movement

Desert sand makes life difficult for animals on the move. Large animals sink into it, while small ones struggle to climb up and down slopes of shifting grains. To combat the problem, some animals, such as golden and marsupial moles, move through the sand rather than above it. Others, such as camels and geckos, have extra large feet, which help to spread their body weight over the surface of the sand and so increase stability. Sidewinding snakes have a different solution: they throw themselves forwards in a succession of sideways jumps, leaving a characteristic pattern of J-shaped tracks. In addition to saving energy, this method helps to minimize contact with hot ground.

Some insects and lizards have learned to tolerate hot ground by alternating the feet that are in contact with the ground at any one time. Having long legs also helps as they hold the animal's body away from the sand's surface, where the heat is fiercest.

NOCTURNAL ACTIVITY
Desert jerboas are typical of the small rodents that live in dry habitats. They are entirely nocturnal, and feed mainly on seeds. Jerboas can travel over 6 miles (10 km) in their search for food, hopping on their long back legs and balancing with their tails.

NOMADIC ANIMALS

Where food and water supply is patchy, some animals adopt a wandering lifestyle. This is common in desert habitats, especially in animals that can fly. Desert locusts are famous for their huge nomadic swarms, and some desert birds, particularly seed eaters, form large nomadic flocks. Unlike migrants, nomadic animals do not follow fixed routes—the weather often dictates their course—and they breed erratically, wherever they find a good food supply.

WILD BUDGERIGARS
Australia's nomadic birds include the budgerigar, the cockatiel, and several pigeon species. Budgerigars can breed when just one month old; and, since parents can raise several families in quick succession, flocks can build to prodigious numbers.

HABITATS

Tropical forest

Forests have flourished in the tropics for longer than they have existed anywhere else on earth, which helps to explain why the animal species that live there outnumber those of all other land habitats combined. Most large tropical forest animals have been identified and classified, but the invertebrate life is so diverse that the task of cataloguing it will never be complete. There are 2 main types of tropical forests: rain forest, which is closest to the equator; and seasonal, or monsoon, forest, which grows toward the edges of the tropical belt.

Tropical rain forest

Near the equator, the climate is warm and moist all year round, creating ideal conditions for plant growth. As a result, trees and other forest plants grow almost incessantly in an endless competition for light. Some plants put all their resources into growing towering trunks, while others are adapted for survival in partial shade. As a result of these different growth patterns, the forest is divided into clearly defined layers, each with its own characteristic animal species.

The highest layer, at about 245 ft (75 m), consists of giant, isolated trees called emergents. These provide nest sites for predatory birds and feeding platforms for monkeys. Beneath this level is the canopy, where copious light, combined with some protection provided by the emergents, results in a continuous layer of branches and lush foliage up to 65 ft (20 m) deep. This layer feeds or harbors most of the forest's animal life. Below the canopy is the understory—a more open layer made up of shade-tolerant trees. On the forest floor, leaf litter is food for some very small animals as well as support for plants and saplings that grow where sufficient light filters through from above.

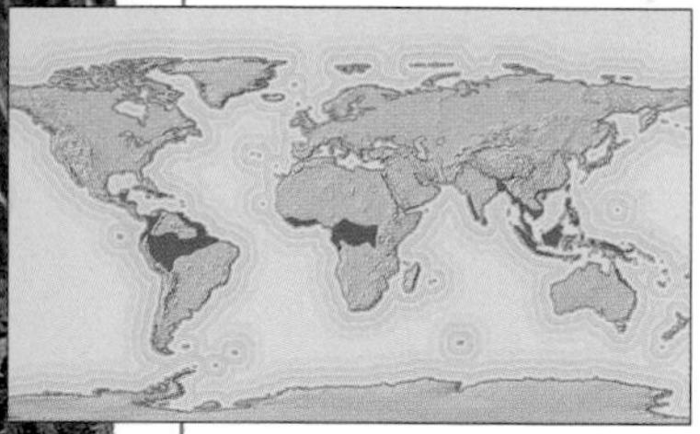

TROPICAL RAIN FOREST *is found near the equator, where annual rainfall exceeds 8¼ ft (2.5 m) and is spread throughout the year.*

This zonal pattern is characteristic of lowland rain forest (the most common rain forest type). At higher altitudes, the trees are lower and the layers are more compressed—an effect that is exacerbated as altitude increases until eventually the trees form elfin forests little more than head high. Soil is also an important factor in shaping the forest. In some parts of the tropics, such as the Rio Negro region of South America, infertile sand results in the growth of stunted trees with leathery leaves.

FOREST EAGLES *have broad wings that enable them to glide through the canopy in their search for prey.*

SLOTHS *spend most of their lives suspended from branches in the forest canopy, relying on camouflage to avoid attack.*

TROPICAL RAIN FOREST *harbors a greater variety of tree frogs than any other habitat.*

HOWLER MONKEYS *of South America's rain forest are among the few primates that survive on leaves.*

LEAFCUTTER ANTS *are found at all levels in rain forest, from the ground to the highest tree tops.*

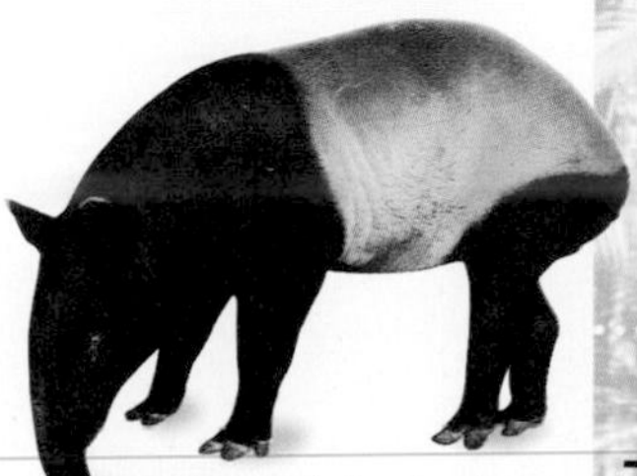

TAPIRS, *and other hoofed mammals, gather fallen fruit from the tropical forest floor.*

EMERGENTS

CANOPY

UNDERSTORY

FOREST FLOOR

HABITATS

HORNBILLS *use their long bill to collect fruit growing on the high branches in seasonal forest. In tropical rain forest, toucans feed in the same way.*

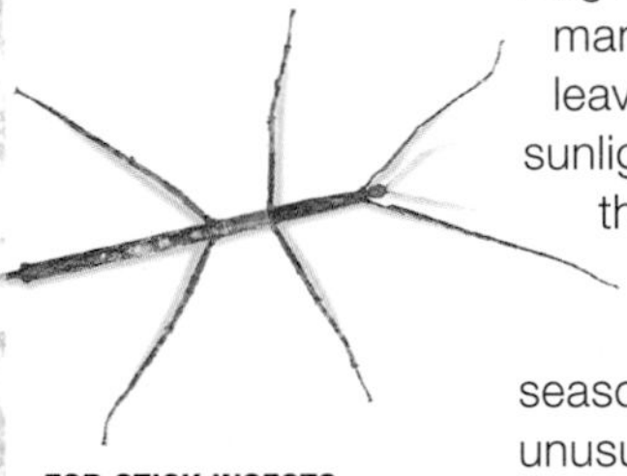
FOR STICK INSECTS *and leaf insects, highly developed camouflage is the key to survival in seasonal forest.*

FRUIT BATS, *the world's largest bats, are almost all found in tropical forest habitats.*

THE ATLAS MOTH *is found in seasonal forest, as are many of the world's other large, winged insects.*

FOREST CATS *make up more than half of the world's cat species. The tiger is the largest and one of the most endangered.*

SNAKES *in seasonal forest include species that hunt on the ground, and those that climb trees to search for prey.*

Seasonal (monsoon) forest

Unlike rain forest, where the climate is very stable, seasonal forest grows where rainfall is concentrated into a wet, or rainy season, which is known as a monsoon. Up to 8¼ ft (2.5 m) of rain can fall in just 3 months—as much as some tropical rain forests receive throughout the whole year. As a consequence, seasonal forest is not as tall as tropical rain forest and, typically, the canopy is more open and extends farther toward the forest floor.

Immediately after the monsoon, seasonal forest is lush and green; but in the long dry season that follows, many of the trees shed their leaves, and the piercing sunlight is able to reach through the bare branches to the ground. Some seasonal forest trees are unusual in that they flower and fruit after losing their leaves. Where this happens, birds, insects, and mammals congregate in large numbers to feed. In the rainy season, the forest's animals are well hidden by the foliage; once the leaves have fallen, they become much easier to find.

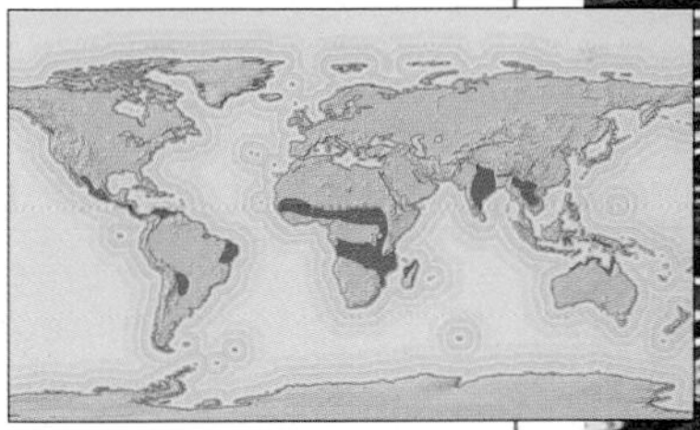
SEASONAL FOREST *grows on either side of the equator. The dry season lasts longer the farther the region is from the equator.*

Despite the yearly cycle of deluge and drought, the animal life of seasonal forest is some of the most numerous and varied in the world. In southern Asia, which has the largest area of this type of forest, the habitat supports elephants, monkeys, leopards, and also tigers. In Asia's seasonal forests, there are some spectacular birds, including giant hornbills, and some of the world's largest snakes. In Africa, seasonal forests abound with browsing antelopes while, in Central America, they are inhabited by pumas, coatis, and white-tailed deer. Most of these animals breed during the wet season, when they can take advantage of the abundant supply of fresh leaves.

DEFORESTATION

The world's tropical forests are being cleared at a rate unparalleled in human history. About half the original cover has been destroyed in the last 50 years, reducing the amount of carbon locked up in trees, which in turn boosts climate change. In many tropical countries, large areas of forest now have legal protection, but with growing pressure for farmland and lumber, controls are often difficult to enforce.

Tropical deforestation is also fueled by the international demand for commodities such as cocoa, rubber, and palm oil—an ingredient of biofuels, soap, and many processed foods. These are often grown in plantations on formerly forested land.

FORMER FOREST *Tropical forest clearance follows the pattern set in forests in other parts of the world, but its impact on plants and animals is greater because there are so many species at stake.*

Life in tropical forest

Some tropical forest animals spend all their time on the ground. For most, however, daily life involves getting around among trees. The canopy holds most of the forest's food, so an animal that is good at moving around in the tree tops has the greatest chance of thriving. Some animals are so well adapted to life in the trees—breeding as well as feeding there—that they very rarely have to visit the forest floor.

Moving in trees

Small animals need relatively few special adaptations for moving about in trees. Ants, for example, are so light that it makes little difference to them whether they are traveling up and down trees or across the forest floor. But for larger animals, such as apes, monkeys, and other primates, climbing is a dangerous occupation: if they lose their grip—as occasionally happens—they risk a fatal fall.

Most primates climb by running or leaping along the tops of branches, often using familiar routes that act like highways through the trees. Monkeys follow these routes mainly by sight, but many of the more primitive primates, such as bushbabies, move around after dark, identifying their pathways partly by smell. Gibbons are different again: they travel underneath the branches by swinging hand-over-hand in a breathtakingly acrobatic manner. This unusual but highly effective form of movement is called brachiation.

PREHENSILE TAILS
Climbing snakes—such as this emerald tree boa—use their tails to clamp themselves to branches. The front of the snake's body folds up, ready to straighten out as it strikes anything that comes within reach.

Tropical forest harbors a huge variety of flying animals—birds, bats, and flying insects—that swoop or hover among the tree tops. However, during the course of evolution, many unrelated animals, including mammals, frogs, and even snakes, have developed winglike flaps of skin that enable them to glide. Some of these gliders can travel over 330 ft (100 m) from tree to tree and, remarkably, many of them are most active after dark.

Communication

In any kind of forest, animals face problems keeping in touch. In the canopy, leaves and branches make it difficult to see for more than a few yards, while tree trunks get in the way on the ground. As a result, many forest animals rely on sound and scent, rather than visual signals, to claim territories and attract partners. Some of the loudest animals in the world live in tropical forest. They include howler monkeys, bellbirds, parrots, cicadas, and an enormous variety of tree frogs. Like mammals and birds, each species of tree frog has its own characteristic call: some produce a short metallic "tink"; others generate a sustained trilling that sounds like machinery.

ARM OVER ARM
Swinging, or "brachiation," is a highly efficient way of moving around in trees. This lar gibbon can leap over 16 ft (5 m) between hand holds and can easily overtake someone running on the ground.

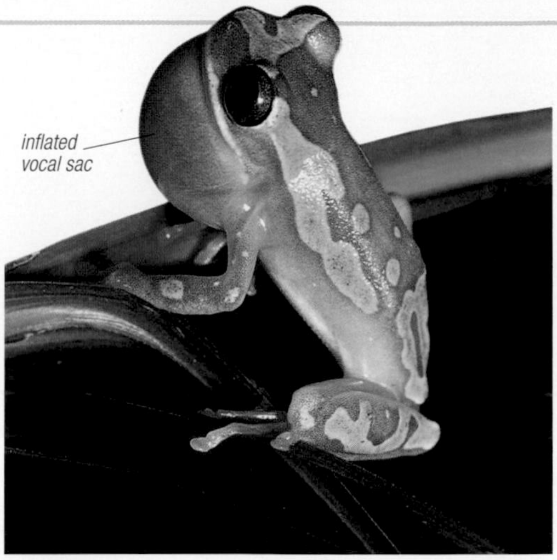

PRIVATE CALL
With its vocal sac inflated, a male tree frog calls to deter rival males and to attract potential mates. Female frogs react only to the call sign of their own species; and the louder and longer a male calls, the more likely females are to respond.

Signaling with sound can be dangerous because it can attract predators as well as potential mates. Tree frogs and cicadas minimize the problem by pitching their calls so that the source is very difficult to locate. Other animals, including many mammals and flying insects, avoid the problem by using scent to stay in touch. One great advantage of scent is that it lingers: for example, in marking its territory, a jaguar or okapi leaves a signal that will last for several days.

Keeping out of danger

Tropical forest abounds with camouflaged animals as well as species that mimic others. Animals that use camouflage—chiefly insects and spiders, but also snakes, lizards, frogs, and toads—resemble a huge variety of inanimate objects, from bark, thorns, and bird droppings to branches and fallen leaves. Many animals use camouflage to avoid being spotted and eaten, but some predators also use it to enable them to ambush their prey.

Mimicry, in which one species "pretends" to be another, is a subtler means of avoiding attack. It involves a relatively harmless species evolving to look like one that is dangerous, and it is most common in invertebrates.

Some tropical forest spiders, for example, closely resemble stinging ants and even move like them. Matters are complicated where several species come to look alike. Some groups of unrelated butterflies, which contain poisons that are distasteful to birds, imitate each other; thus they have evolved the insect equivalent of a shared warning trademark.

Warning signals are most developed in extremely toxic animals. For example, unlike other frogs, tiny poison-arrow frogs hop nonchalantly about the forest floor, relying on their extraordinarily vivid colors to warn other animals that they are not merely unpalatable but highly dangerous to eat.

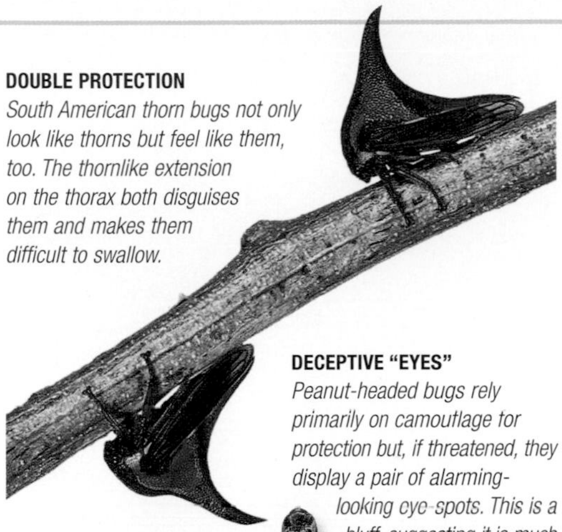

DOUBLE PROTECTION
South American thorn bugs not only look like thorns but feel like them, too. The thornlike extension on the thorax both disguises them and makes them difficult to swallow.

DECEPTIVE "EYES"
Peanut-headed bugs rely primarily on camouflage for protection but, if threatened, they display a pair of alarming-looking eye-spots. This is a bluff, suggesting it is much larger than it actually is.

Feeding

Near the equator, trees grow, flower, and set seed all year round, generating a nonstop supply of food. Many forest animals—including bats, birds, and insects—live almost exclusively on the abundant nectar and fruit. Some of these animals help trees spread their pollen and seeds. Quetzals, for example, swallow fruit whole, and then regurgitate the pits onto the forest floor where they can germinate.

Compared with flowers and fruit, tropical forest leaves are difficult to digest. Animals that feed on these leaves generally pick them while they are still young—before protective toxins have had a chance to build up inside them. Insects are the most prolific leaf eaters, but some of the forest's larger animals also rely on this difficult diet. They include several kinds of monkeys and sloths, and the hoatzin—a highly unusual bird from South America. The hoatzin processes its food much as a grazing mammal does; after eating, it is often so heavy that it can barely fly.

Tropical forest predators range from some of the world's smallest insects to the largest cats. In an environment that provides lots of cover, most of them stalk their prey rather than running it down. Army and driver ants are the most remarkable exceptions: they hunt in "packs" over 50,000 strong, overpowering and eating anything that cannot escape.

Reproduction

For animals that live in trees, breeding can sometimes involve unusual adaptations. Some tree frogs come down to the forest floor to lay their eggs, but many lay them high up in the canopy, either in water-filled tree holes or in the pools of water that gather in plants. Some frog species are more creative, laying their eggs in nests of foam that keep their eggs moist until the tadpoles are ready to hatch.

Many tropical birds start life in the safety of tree holes, but climbing mammals rarely build nests, and many of their young start life in the open. Young monkeys often cling to their mothers' chests, keeping a tight grip as the parent runs along branches or leaps through the air. For young murine (mouselike) opossums, which live in the American tropics, early life is even more precarious because their mothers do not have well-developed pouches. This means that until their feet have developed the ability to cling, these tiny marsupials hang from their mothers' teats by their mouths, their legs dangling in the air.

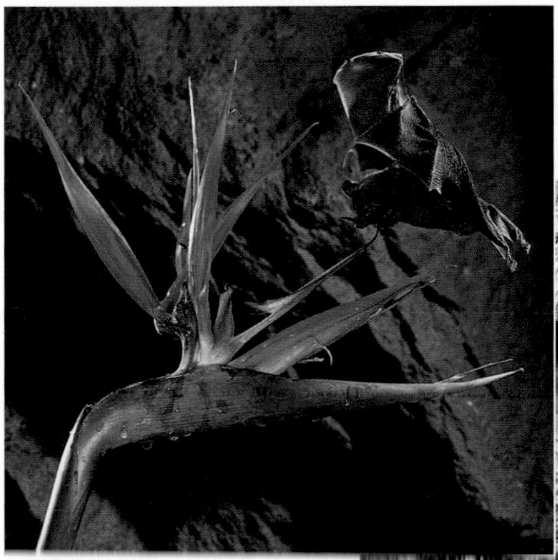

NECTAR FEEDERS
A bat laps up flower nectar. Flowers pollinated by bats and birds have to be robust if they are not to be damaged by their visitors. Birds are attracted to certain flowers mainly by their bright colors, bats by their pungent scent.

RODENT OPPORTUNISTS
Agoutis and other small, ground-dwelling mammals often follow the progress of parrots and monkeys through the trees, gathering the pieces of food that accidentally fall to the ground.

PROTECTIVE NEST
These African treefrogs have grouped together to make a giant foam nest hanging from a branch. The nest's exterior hardens to protect the eggs and to keep the interior foam moist. When the tadpoles hatch, they break out and drop into water below.

ANT FOLLOWERS

In Central and South American forests, a single column of marauding army ants attracts up to 30 different kinds of birds, all swooping down to catch the tiny animals that burst out of the leaf litter to escape the ants. Some birds flutter ahead of the ant column; others dart among the ants themselves. Some are full-time ant followers, while others follow the throng only for as long as the ants are marching through their territories. A few small animals, including wasps and rove beetles, travel with the army on the ground. Although this sounds dangerous, these animals mimic the ants very closely—an adaptation that prevents them from being attacked while gaining them protection and a supply of free food. Army ants also attract lizards and frogs, as well as parasitic flies that lay their eggs on the fleeing animals.

ARMY ANTS

AERIAL ATTACK
The blue-crowned motmot, a part-time ant follower, is distantly related to kingfishers. Like them, it swoops down on its prey, carries it back to a perch, and hits it against a branch before eating it.

HABITATS

Temperate forest

Temperate forest grows in regions that have a wide range of climates. In some, winters are cold and summers are cool; in others, the winter is relatively mild, and the summer heat rivals that in the tropics. Where winters are cold, temperate forest trees are usually deciduous, shedding their leaves in winter and growing a new set in spring; in warmer regions, many trees keep their leaves all year. Although temperate forest does not have as many animal species as tropical forest, it is still among the richest wildlife habitats on land.

Deciduous forest

In the depths of winter, deciduous forest can seem gaunt and empty, and largely devoid of animal life. But as the days lengthen in spring, and buds begin to burst, the habitat becomes alive with birdsong and animals on the move. This transformation is triggered by a sudden abundance of plant food—one that nourishes large numbers of both the plant-eating insects as well as the animals that feed on them. Many of these forest animals are permanent residents, but they also include migratory birds that fly in from distant parts of the world.

Compared with tropical forest, temperate deciduous forest has relatively few tree species: the maximum number—found in some of the forests of eastern North America—is several hundred, while tropical forest might contain several thousand. Nevertheless, temperate forest trees are powerhouses of life. Large oak trees, for example, can produce over a quarter of a million leaves a year—enough to sustain the army of weevils, gall wasps, and moth caterpillars that feed rapidly in spring and early summer while the leaf crop is at its freshest and most nutritious.

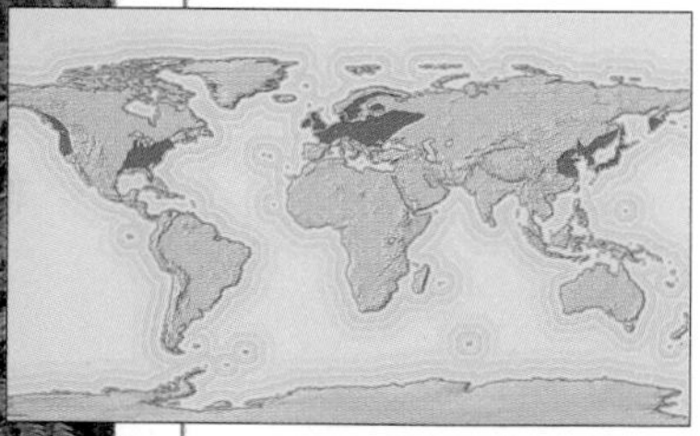

LARGE AREAS *of temperate deciduous forest are found in the Northern Hemisphere; most temperate forest in the Southern Hemisphere is evergreen.*

Like tropical forest, deciduous forest has a clear, vertically layered structure, but there are some important differences. The trees are rarely more than 100 ft (30 m) tall, and the canopy layer is usually deep but open, allowing light to reach the understory and encourage plant growth. Fallen leaves rot slowly in cool conditions, so deciduous forest has an unusually deep layer of leaf litter that insects, woodlice, and millipedes use as food and cover. This means that while many small animals live in the cracks and crevices in bark, the place that is richest in invertebrate life is not the trees but the ground.

MOTHS *lay their eggs on leaves, buds, or bark. Their caterpillars are often the most numerous leaf eaters in the forest.*

TREECREEPERS *patrol tree trunks in search of small insects hidden in bark crevices.*

SQUIRRELS *spend the autumn collecting food, which they then store away for use in the winter.*

DEER *feed on leaves in summer; in winter, they often strip bark from shrubs and young understory trees.*

WILD BOAR *dig up the leaf litter with their snouts, feeding on animals, nuts, and roots.*

IN THE MOIST *conditions of the forest floor, lungless salamanders can absorb oxygen through their skin.*

CANOPY

UNDERSTORY

GROUND

HABITATS

CANOPY

THE RINGTAILED POSSUM *has a prehensile tail, which helps it climb along high branches to reach flowers and fruit.*

HOOPOES *swoop down from the canopy to catch animals on the ground.*

KOALAS *live in the forest canopy, but also walk across the ground to reach isolated clumps of trees.*

UNDERSTORY

THE AUSTRALIAN GREEN TREEFROG *has exceptionally thick skin to minimize moisture loss and enable it to deal with the dry conditions in evergreen forest.*

BUSHCRICKETS *and other insects, such as beetles, provide food for reptiles and mammals that live on the ground.*

GROUND

SUN-SEEKING *lizards and snakes often bask in the warm sunlight that bathes the forest floor.*

Evergreen forest

In warm parts of the temperate world, many broad-leaved trees are evergreen. Unlike trees of deciduous forest, which grow in spring and summer, evergreens grow in winter and spring, when temperatures are low but not cold, and when water is readily available. Described by botanists as sclerophyllous (meaning hard-leaved) forest, this habitat is found in several widely scattered regions of the world, including parts of California and western South America, the Mediterranean region in Europe, and large areas of eastern and southwestern Australia. In some of these places, the forest is low-growing, but in Australia, where eucalyptus is the dominant species, it includes the tallest broad-leaved trees in the world.

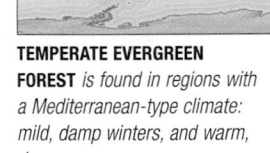

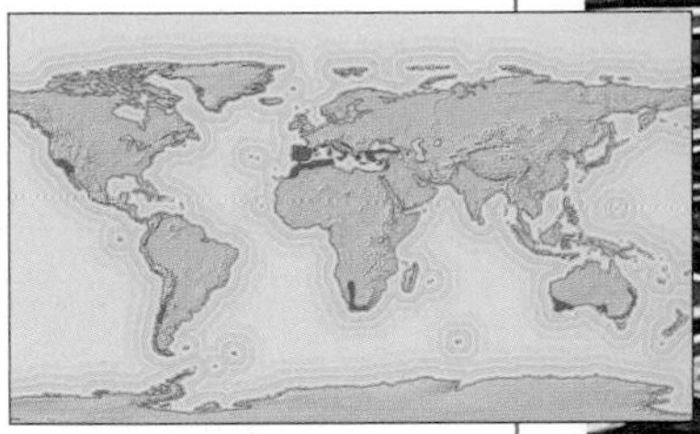

TEMPERATE EVERGREEN FOREST *is found in regions with a Mediterranean-type climate: mild, damp winters, and warm, dry summers.*

Temperate evergreen trees usually have open crowns, which means that the vertical layers are often less pronounced than they are in forests in cooler regions, and plenty of light is able to reach the forest floor. As a result, these forests are rich in ground-based wildlife, and warmth-loving animals—such as lizards and butterflies, which are usually associated with higher levels—can often be seen sunbathing on the floor. The open structure also makes it easy for birds, such as kookaburras and other forest kingfishers, rollers, and hoopoes, to swoop down on animals moving around on the ground.

The air in evergreen forest often smells pleasantly aromatic because most of the leaves are filled with pungent oils. These oils help to stop the leaves from drying out, and they also protect them from animals. They are a highly effective deterrent, as relatively few animals—apart from specialists such as the koala—include these leaves in their diet.

SALINIZATION

In Australia, the arrival of Europeans has brought huge changes to the continent's forests, with their unique plants and wildlife. Millions of acres have been cut for lumber, claiming some of the world's tallest broad-leaved trees, and displacing woodland animals. Even more land has been cleared for grazing and for crops, sometimes with unintended consequences. One of them is salinization—a process that mobilizes subterranean salts when native vegetation is removed. Salinization often kills surviving trees, turning them into gaunt skeletons that take many years to decay.

SALTY SOIL *These Australian eucalyptuses have been killed by salinization—rising salt levels in the ground. Salinization is caused by forest clearance, which changes the water balance of the ground. Salinized land is useless for agriculture.*

Life in temperate forest

The factor that most affects life in temperate forest is the variable food supply. At all levels—from the tree tops to the forest floor—the life cycles of forest animals living in temperate forest move in step with the seasons so that they produce their young when food is easiest to find. Life is relatively easy in spring and summer; but in winter, when the food supply is at its lowest, special adaptations are needed for survival.

Feeding

As a habitat, deciduous temperate forest—the kind of forest found across much of the northern hemisphere—has one very useful feature. The trees that grow in it, such as oak and beech, produce leaves that are designed to last for just one growing season. As a result, these leaves are usually thin and easy to eat, which is why vast numbers of insects feast on them from the moment they begin to appear in spring.

This sudden explosion of insect life attracts an army of highly specialized avian predators. In Europe, northern Asia, and North America, dozens of warbler species migrate north as the buds open. These birds have extremely acute eyesight, enabling them to scour leaves for the tiniest grubs and caterpillars, which they then pick up with their tweezerlike beaks. Other birds, including treecreepers, woodpeckers, and nuthatches, concentrate on the bark, seeking out and pecking at the tiny animals hidden among the crevices.

By midsummer, leaves stop growing and animal feeding behavior changes. Most temperate trees are pollinated by wind, which means that they do not produce enticing nectar-rich flowers. However, these trees do produce large crops of nuts and other seeds, which are extremely important foods for animals because, unlike leaves, they can be stored away and used when other food is scarce.

WINTER GRANARY
In western North America, acorn woodpeckers store acorns in trees, cramming them into holes that they have previously drilled in the bark. These storage trees are called granaries. A single granary may hold up to 50,000 acorns and may be stocked by up to a dozen woodpeckers working and nesting together. The woodpeckers will also store acorns in fence posts and telephone poles.

Food storage, or "caching," is practiced by many forest birds and mammals. Jays bury acorns in the ground, while acorn woodpeckers store them in trees. Squirrels bury seeds of all kinds, and red foxes bury anything that is even faintly edible, from half-eaten remains to food wrappers and discarded shoes. Some animals locate their stores by scent, but most are very good at pinpointing them by memory alone, finding and digging up their food even when it is covered by snow.

Seed-caching has an important impact on forest ecology. Although animals that bury seeds have good memories, some of what they hide is always forgotten about. This means that provided the seeds are not discovered by rodents or other animals, they remain effectively planted and ready to germinate, helping the forest's trees to reproduce.

TREE GALLS

Common in broad-leaved forest, galls are growths whose development is triggered by insects and other organisms. Galls often look like berries or buttons on leaves and twigs. They act as both home and larder, or nursery, allowing their occupants to feed unseen. In broad-leaved forest, most galls are produced by tiny wasps and flies. Each species of insect affects a particular tree, and each produces galls of a characteristic shape, making them easy to identify.

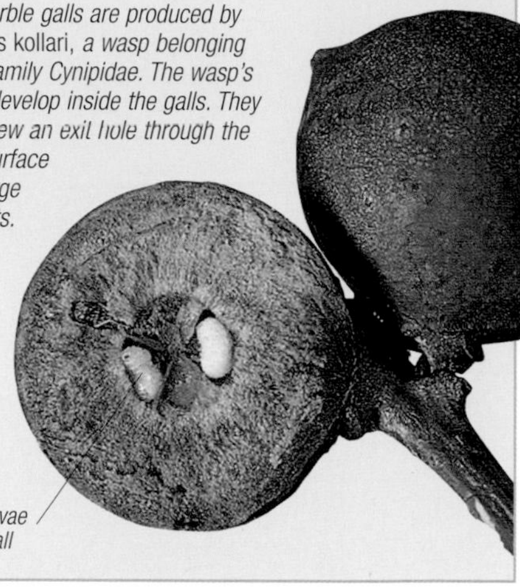

OAK GALLS
Oak marble galls are produced by Andricus kollari, *a wasp belonging to the family Cynipidae. The wasp's larvae develop inside the galls. They then chew an exit hole through the outer surface to emerge as adults.*

FOREST-FLOOR FORAGERS
Watching over their piglets, adult wild boars root in the leaf litter for food. Wild boars feed on acorns and other nuts, and they also use their spadelike snouts to dig up roots, fungi, and small animals hidden in soil or among fallen leaves.

Hibernation

In the fall, many insect-eating birds migrate to warmer climes, leaving the forest's remaining animals to face the winter cold. Animals that store food can remain active throughout this difficult time of year, but others use a very different survival strategy: they hibernate, living on the fat reserves they have built up during the summer months.

How long and how deeply an animal hibernates depends on where it lives. In the forests of northwest Europe, hedgehogs may hibernate for up to 6 months, whereas farther south, their winter sleep is much shorter. In eastern North America, woodchucks—or groundhogs—typically hibernate from October to February; their wanderings early in the year are a traditional sign that spring is not far off.

Some hibernating animals, such as the common dormouse, hardly ever interrupt their winter break, even if they are picked

SLEEPING THROUGH THE COLD
Common dormice use leaves and moss to make well-insulated winter nests among brambles and other plants. Here, part of the nest has been temporarily removed to reveal the hibernating animal, fast asleep with its tail wrapped tightly around its body.

up. However, many hibernators behave in a different way. If the weather turns warm, they briefly rouse themselves: bats, for example, take to the air for feeding flights, while hedgehogs often move out of one hibernation nest and into another. But forest hibernators have to be careful not to do this too often: activity uses up their bodily food reserves, and it therefore puts them at risk of running out before the winter is truly over. Many insects also hibernate, often hidden under bark; but in some species, the adults die out, leaving behind tough, overwintering eggs that will hatch in spring.

NIGHT FLIGHT
Unlike most squirrels, North American flying squirrels feed at night. The elastic skin flaps between their legs enable them to glide for over 165 ft (50 m). They steer with their front legs, and use their large eyes to navigate safely in the dark.

Movement

While monkeys and gibbons are the most impressive climbers in tropical forest, squirrels are the experts in temperate forest. Unlike many climbing mammals, they can run head-first down tree trunks, as well as up them, by hooking their long, curved hind claws into bark. Squirrels have excellent eyesight, and they instinctively scuttle to the back of a tree if they spot a potential predator—a simple behavior that makes them difficult to catch.

Temperate forest is inhabited by gliding rodents, and also—in Australasia—by gliding marsupials. But for precise maneuvering among trees, owls and birds of prey are unrivaled. Unlike their relatives in open habitats, most of these aerial hunters have relatively short, broad wings that enable them to twist and turn effectively. A prime example of this adaptation to woodland life is the Eurasian sparrowhawk: rather than soaring and then swooping, it speeds among trees and along hedgerows—sometimes only a yard or so above ground—ambushing small birds in midair and carrying them away in its talons.

Temperate forest provides ground-dwelling animals with lots of cover. As a result, small mammals, such as voles and shrews, abound on the forest floor. To avoid being seen as they move around, these animals often use runways partly covered by grass or fallen leaves. Voles use a combination of vision, smell, and touch to find their way along these runs, but shrews, which have very poor eyesight, navigate partly by emitting high-pitched pulses of sound. These signals bounce back from nearby objects in the same way as those sent out by a bat's sonar system.

Living in leaf litter

The leaf litter in temperate forest is one of the world's richest animal micro-habitats. This deep layer of decomposing matter harbors vertebrates, including mammals and salamanders, but its principal inhabitants are invertebrates that feed on leaf fragments, on fungi and bacteria, or on each other. Some of these animals—such as centipedes and woodlice—are large enough to be easily seen, but many others are microscopic. Animals that live deep in leaf litter exist in total darkness, so most of them rely on their sense of touch to find food. This is especially true of predators: centipedes locate their prey with long antennae, while tiny pseudoscorpions use the sensory hairs that cover their pincers. Like true scorpions, pseudoscorpions are venomous; but they are so small that they pose no threat to anything much bigger than themselves. This is fortunate because in just a few square yards of leaf litter their numbers can run into millions. Dead leaves are a useful screen, hiding leaf-litter dwellers from other animals foraging on the forest floor. However, it is not totally secure. Some temperate forest birds, particularly thrushes, pick up leaves and toss them aside, snapping up leaf-litter animals as they try to rush away from the light.

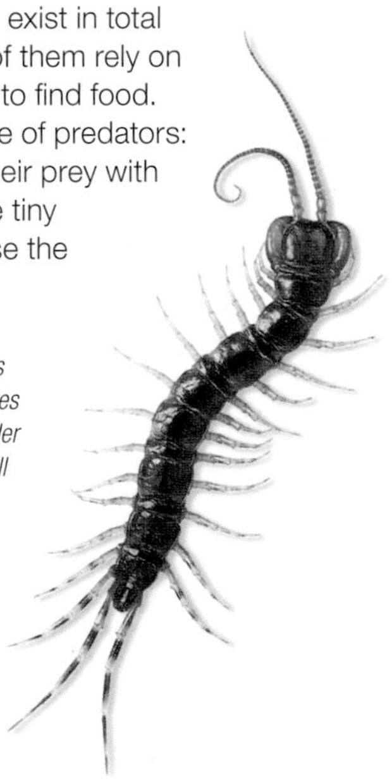

HUNTER IN THE LEAVES
Lithobiid centipedes, such as this one, have flattened bodies that allow them to crawl under leaves and fallen logs as well as on the ground surface. Geophilid centipedes live permanently underground, and therefore have narrow, almost wormlike bodies.

INSECT ENGRAVERS

Female bark beetles tunnel through the sapwood beneath living bark, laying eggs at intervals along the way. The hatched larvae eat their way out at right angles to the original tunnel, creating distinctive "galleries" that can be seen when dead bark falls away. The side tunnels end in exit holes, from which the developed beetle emerges and flies away. Common in deciduous forest, bark beetles can be highly destructive because they often infect trees with fungi. One species—the elm bark beetle—spreads Dutch elm disease, a fungus that has wiped out elms in parts of Europe and North America.

BARK GALLERIES
The adult bark beetle has a cylindrical body and a round-fronted thorax hiding most of the head. The gallery patterns vary from one species to another.

Coniferous forest

Conifers are the world's toughest trees. Their small, needle-shaped leaves can withstand extreme cold and are impervious to strong sunshine and wind, and their relatively narrow, upright habit enables them to grow closely together to form dense, sheltered forest. As a result, conifers thrive where few broad-leaved trees can survive, such as the far north and in mountain ranges. They also flourish in places that have very heavy rainfall. In such areas they form temperate rain forest, home to some of the largest trees in the world.

THE ANIMAL LIFE *in coniferous forest provides birds of prey, such as the northern goshawk, with a year-round food supply.*

TREE TRUNKS *provide the growing larvae of horntails and other wood wasps with protection from the worst of the winter cold.*

Boreal forest

Named after Boreas, the Greek god of the north wind, boreal forest, or taiga, is the largest continuous expanse of forest on earth. It covers about 6 million square miles (15 million square km) and stretches in an almost unbroken belt across the far north, often reaching deep into the Arctic. In some places, the belt is over 1,000 miles (1,600 km) wide. Across the boreal forest belt as a whole, winter temperatures routinely drop below -13°F (-25°C), but in some of the coldest regions, such as northeast Siberia, temperatures can fall below -49°F (-45°C). Summers in boreal forest are brief but can be warm.

Compared with the types of forest that occur at lower latitudes, boreal forest has only a handful of tree species and therefore provides only a limited variety of food for herbivores. Plant diversity is also restricted both by the amount of light that can reach the forest floor and by the high acidity of pine needles. Even in summer, the interior of the forest is often dark, with a thick layer of dead needles carpeting the floor. Fungi thrive in these conditions, but the only forest-floor plants that live here are the ones that can tolerate low light levels and acidic soil conditions.

Other than insects, few animals can digest conifer leaves or wood, so most plant eaters concentrate on seeds, buds, and bark, or on berries from low-growing shrubs. However, what this habitat lacks in variety, it more than makes up for in quantity, especially as there is relatively little competition for food. This is one of the reasons why many boreal forest animals, from birds to bears, have far more extensive ranges than species that live in warmer parts of the world.

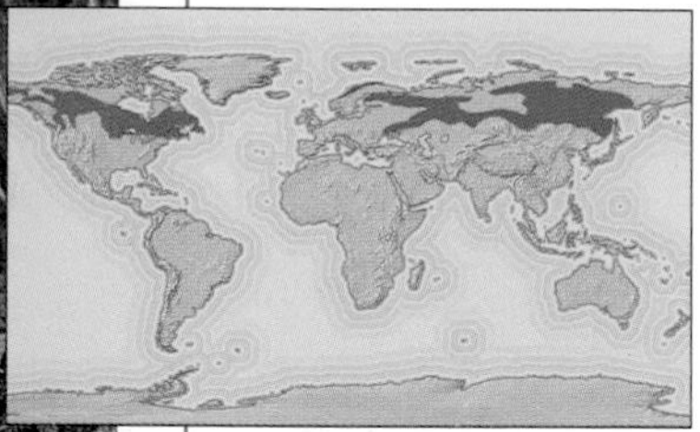

BOREAL FOREST *stretches across much of the far north. There is no equivalent habitat in the Southern Hemisphere.*

THE BROWN BEAR *can climb trees, but it finds most of its food on the ground.*

GRAY WOLVES *hunt in packs, a strategy that enables them to kill animals larger than themselves. This is particularly important during winter, when food is scarce.*

FOR MOOSE *and other hoofed mammals, boreal forest offers protection from the cold winds and blizzard conditions of winter.*

HABITATS

Temperate rain forest

The world's largest areas of rain forest are found in the tropics, but rain forest also exists in parts of the temperate world. It grows where west-facing mountains intercept moist air blowing in from the sea and, unlike boreal forest, it experiences relatively mild temperatures all year round.

Compared with boreal forest, temperate rain forest is a rare habitat, occurring in a few widely separated areas. In the Southern Hemisphere, it is found in the South Island of New Zealand, and in parts of southern Chile. In both of these places, most rain forest trees are broad-leaved species, but in America's Pacific northwest—where the largest temperate rain forest in the world can be found—the trees are almost entirely conifers. Some are over nearly 250 ft (75 m) high, and more than 500 years old. This kind of coniferous forest looks unlike any other. On the ground, and in the understory, every surface is draped with ferns or waterlogged moss. Densely packed trunks, some over 9¾ ft (3 m) across, rise up to the canopy high overhead, where the sky is always laden with rain.

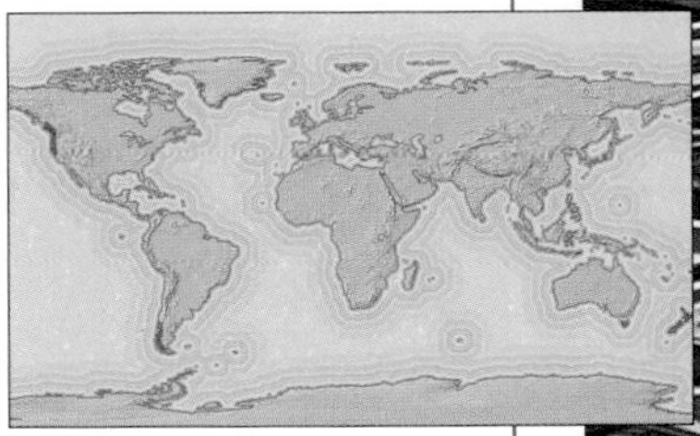

TEMPERATE RAIN FOREST *is found on west-facing coasts, where heavy rain falls throughout much of the year.*

Temperate rain forest supports many animals that are found in coniferous forest all over the world, but it has some additional features that set it apart: the mild, damp conditions, which make it a haven for slugs and salamanders, and the immense amount of fallen lumber, which creates opportunities for insects that feed on dead wood. In its natural state, the forest teems with mammals, as well as with owls and other birds that need large, old trees as nest sites. Unfortunately, these trees are in great demand by the lumber industry and, as a result, untouched temperate rain forest is increasingly rare.

THE GOLDEN-CROWNED KINGLET *is one of many small insect eaters that feed and nest high above the ground.*

OWLS, *such as this great horned owl, fly at night, hunting the many small mammals and birds that live in temperate rain forest.*

THE NORTH AMERICAN PORCUPINE *is a slow and awkward tree climber, but it manages to reach the tree tops to eat buds and bark.*

THE FISHER *belies its name in that it rarely catches fish. It is one of the few animals that includes porcupines among its prey.*

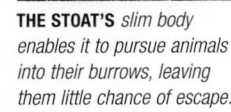

THE STOAT'S *slim body enables it to pursue animals into their burrows, leaving them little chance of escape.*

SLUGS *in temperate rain forest include plant-eating species and some that prey on other slugs.*

CLEAR CUTTING

Conifers are essential in lumber production and paper making, but harvesting them has far-reaching effects on wildlife. Many animals—particularly hole-nesting birds—rely on old trees, and are rare in planted forests that are routinely clear cut. On the other hand, where smaller blocks are cut at different times, clear cutting can be used to create a patchwork of forest habitats. This helps forest animals, although plantations are still no match for original or "old-growth" forest, which has never been felled.

COMPLETE CLEARANCE *Total clear cutting removes all the trees in large areas of forest, forcing the animal inhabitants to move out.*

Life in coniferous forest

For animals and trees alike, life in boreal forest is dominated by the need to survive long and extremely cold winters. Animals that remain active in winter, such as wolves, need a constant supply of food simply to avoid freezing. Conifers are difficult to exploit for food, which means that animals that rely on them have developed some highly specialized physical and behavioral characteristics.

Feeding

Compared with many broad-leaved trees, conifers are well protected against attack. In addition to tough leaves, they often have oily resins that make both their leaves and their wood difficult to digest. Furthermore, if the sapwood is injured, this resin oozes out and traps insects and spiders as effectively as glue.

Despite these defenses, some animals manage to live entirely on coniferous trees. Among the most successful are sawfly larvae. These caterpillarlike grubs bore deep inside the trunks, leaving cylindrical tunnels in the wood. They do not eat the wood itself; instead, they feed on a fungus that grows on the walls of the tunnels they have built. Female sawflies carry small amounts of this fungus with them when they emerge as adults, and they infect new trees when they lay their eggs. This kind of symbiotic partnership is vital to sawflies, but it is not entirely unique: in other habitats, particularly in the tropics, ants and termites also "cultivate" fungi as food.

Although wood-boring grubs are safe from most predators, they are not entirely immune from attack. Coniferous forest is the habitat of some of the world's largest woodpeckers, which hammer their way into tree trunks to reach the grubs inside. Sawflies also face a threat from the ichneumon wasp, which drills through the wood with its long ovipositor to lay its eggs on the sawfly larvae. It is thought that the wasp locates the larvae by smell. When the eggs hatch, the ichneumon grubs eat their host larvae alive.

Some animals, such as the capercaillie and North American porcupine, eat large quantities of conifer needles, but moth caterpillars are the leading leaf eaters: as always in coniferous forest, the number of species involved is small, but the damage they inflict can be vast. This is especially true of species such as the gypsy moth, which has been accidentally introduced to many parts of the world.

MOTH ATTACK
Although the gypsy moth itself does no damage, its caterpillars feed voraciously on conifers as well as on a range of deciduous trees.

DRILLING THROUGH WOOD
Bracing itself with its feet, a female ichneumon wasp bores into a branch to lay eggs on a sawfly grub.

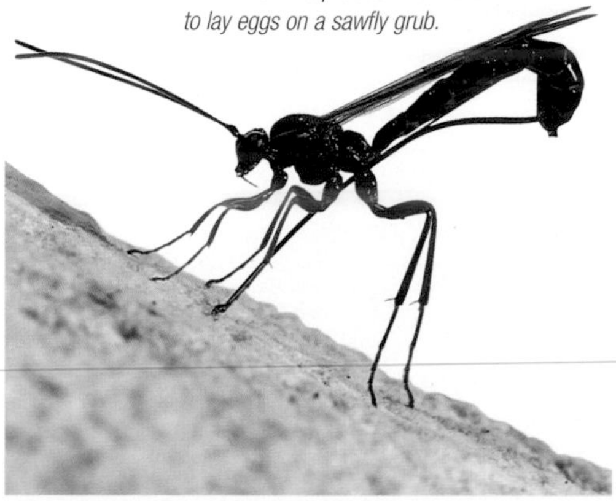

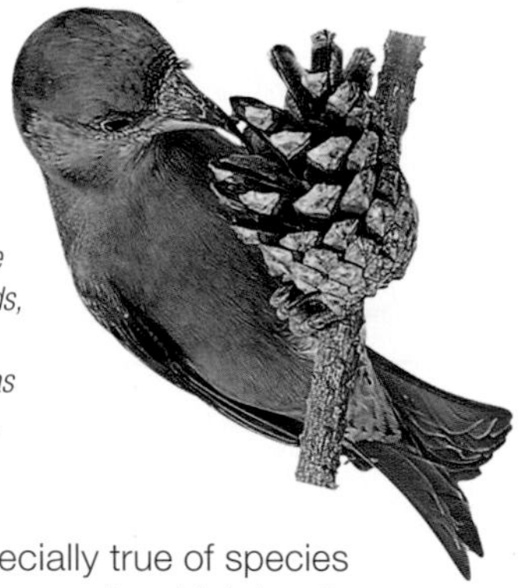

EXTRACTING SEEDS
A crossbill's beak overlaps at its tips, making it an ideal tool for extracting cone seeds. For short periods, crossbill nestlings can survive temperatures as low as -31°F (-35°C).

Finding winter food

Conifers do not have true flowers, but they nevertheless produce seeds. For birds and small mammals, this seed crop is a valuable winter fuel. However, accessing conifer seeds is not easy: they develop inside woody cones, and the cones stay tightly closed until the seeds inside are mature.

Coniferous forest animals have developed a variety of ways of removing these seeds before the trees scatter them on the forest floor. Squirrels gnaw through the soft, unripe cone while it is still attached to the branch, eating the seeds and dropping the remains of the cone on the ground below. Woodpeckers often take fallen cones and wedge them into tree holes or broken stumps, using these to hold the cone firm while they peck out all the seeds. Crossbills are even more proficient: their beaks are uniquely adapted for dealing with cones, enabling them to extract the seeds with surgical precision.

Compared with seeds, bark is a low-quality food, but in winter it is vital to some species' survival. Deer strip it away from the base of young saplings, while bank voles and porcupines often climb trees to attack the bark higher up. Bark stripping often stunts a tree's growth, and a severe attack can kill it.

COUP DE GRACE
At the end of a high-speed chase, a Canadian lynx catches a snowshoe hare. Population cycles of both species are closely linked.

BROWSING DEER
Deer have a major impact on coniferous forest by eating young tree shoots and bark. Where deer are numerous, such browsing can kill small saplings (inset). However, this creates open glades, which are useful to other animals because they let in light, allowing fruit-forming plants—a good food source—to grow.

Coping with cold

Coniferous forest has its share of seasonal visitors: principally insect-eating birds that arrive in spring and then leave for the south once they have raised their young. But for the rest of its animals, long winters are an inescapable fact of forest life. Some species hibernate, but many remain active even during the coldest months, relying on their insulation for survival.

One group of forest animals, mustelids, have coats that are exceptionally well insulating. This group includes pine martens, wolverines, minks, and sables, all of them agile hunters renowned for their thick and luxurious fur. As with most mammals, their coats contain two different kinds of hair: long, outer guard hairs

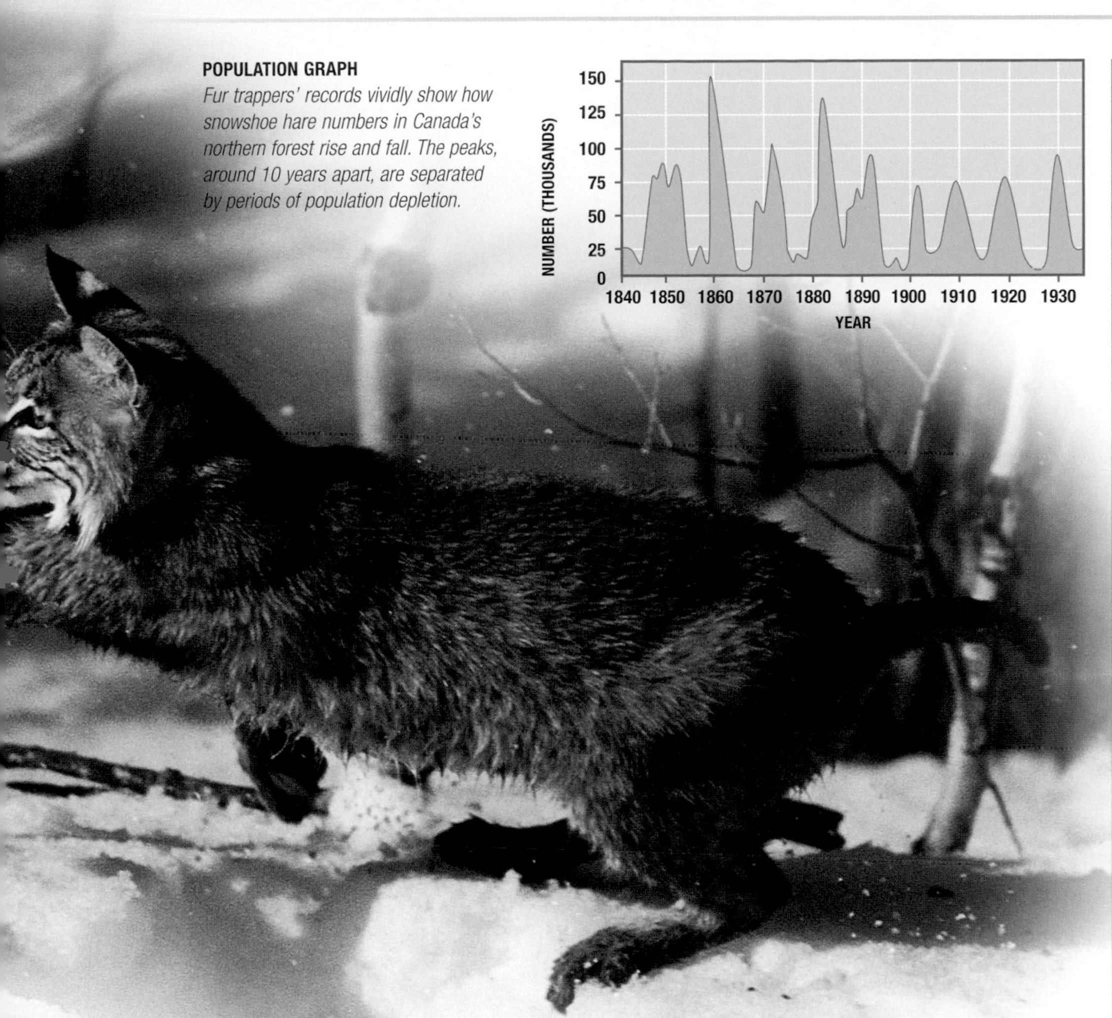

POPULATION GRAPH
Fur trappers' records vividly show how snowshoe hare numbers in Canada's northern forest rise and fall. The peaks, around 10 years apart, are separated by periods of population depletion.

form the coat's water-repellent surface, while shorter, much denser, hairs—the underfur—trap a layer of air close to the body, keeping the animal warm. Northern species all grow an extra-thick coat after their late-summer molt; and some species, such as the stoat, use this molt to change color, developing a white coat that provides better camouflage for the winter.

Keeping warm is relatively easy for large mammals because their bodies contain a large store of heat. But for the smallest warm-blooded inhabitants of coniferous forest, winter conditions test their cold tolerance to its limits. Voles and other rodents can hide in burrows, but birds spend most of their lives in the open. For wrens and tits, which often weigh less than ⅜ oz (10 g), winter nights are a particularly dangerous time. With such minute bodies, their fuel reserves are tiny, and so they must make special provision if they are to stay alive until dawn when the search for food can resume. Some of them make the most of what body heat they have by huddling together in tree holes, but a few, such as the Siberian tit, bed down in the snow, using it as an insulating material.

RUNNING ON SNOW
Cold winters actually make it easier for the wolverine to catch food: its ability to run at high speed over frozen snow enables it to catch large animals, such as reindeer, which can outpace it during the summer months.

Population cycles

Since there are relatively few animal species in the northern coniferous forests, the lives of predators and prey are very closely linked. During mild years, strong tree growth can trigger a population explosion among small animals; as a result, the predators that feed on them begin to increase in number. These conditions never last for long, though: as the plant eaters begin to outstrip the food supply, their numbers start to fall again. And as the rate of the fall accelerates, the predators soon follow suit.

Despite the unpredictability of the northern climate, these ups and downs occur with surprising regularity. In North America, fur trappers' records dating back over a century provide some long-term evidence of population swings. For example, they show that the snowshoe hare population roughly follows a 10-year cycle, with 2 or 3 good years, followed by a lengthy slump. The Canadian lynx—one of the snowshoe hare's main predators—follows the same pattern, but with a one- or 2-year time-lag. Similar cycles involving lemmings and other small mammals take place in tundra (see p.67).

While there is little that they can do to prevent this boom-and-bust pattern from occurring, animals like the snowshoe hare are able to make a fairly fast recovery from a population slump by breeding quickly when conditions are favorable.

IRRUPTIONS

In the northern coniferous forests, food supplies are affected by the weather and the degree of competition. In winters that follow cool summers, the supply of seeds and berries can be thin, leaving seed- and fruit-eating birds with little to survive on. Rather than starve, these birds fly south in waves, called irruptions, which may involve traveling beyond their normal winter range by as much as 930 miles (1,500 km). Species that frequently irrupt include crossbills, waxwings, and tits, as well as nutcrackers and other seed-eating members of the crow family.

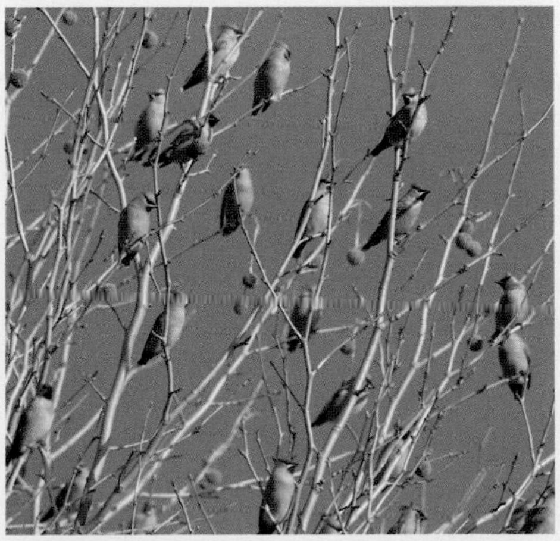

FLYING SOUTH
A flock of Bohemian waxwings pause during their southward flight. Like other irrupting species, waxwings usually spread south when their population reaches a high level. Shortage of winter food can trigger a mass exodus.

Mountains

In many land habitats, climatic conditions vary only slightly within a region. In mountains, however, the average air temperature drops by about 1.8° F (1° C) for every 650 ft (200 m) gained in height, oxygen becomes more scarce, and the air becomes less effective at screening out ultraviolet light. As a result, mountains can be divided into distinct zones, each supporting plant and animal life that is very different from that of the zone above and below. A wide range of animals live in the low-altitude foothills, but only the hardiest survive year-round in the harsh environment above the tree line.

Temperate mountains

In temperate regions, a mountain's climate is relatively cool throughout the year. However, seasonal changes are much more marked than they are in the tropics. At high altitude, above the tree line, there is a sudden burst of plant growth in spring and summer. Some animals migrate upward to make use of this brief abundance of food, but others, such as the marmot, are permanent residents between mid- and high-altitude, surviving the winter cold by living in burrows, and by hibernating for up to 8 months a year. High-altitude insects spend many months in a dormant state, coming to life when warm weather arrives. For many, this dormant period is spent inside the egg, which hatches when the days lengthen and the temperatures rise.

At lower altitudes, the climate is warmer, and generally more like that of the surrounding land. However, because the sloping, rocky ground is difficult to farm, the mountainsides often retain more of their tree cover than does flatter ground. In undisturbed conditions, these montane forests are the natural habitat of large mammals, such as mountain lions, bears, and deer, and also of a wide range of seed- and insect-eating birds.

Temperate mountains abound in birds of prey. Some, such as the peregrine falcon, pursue their prey on fast-flapping wings, while eagles and buzzards soar high up, riding on updrafts. One characteristic mountain species, a vulture called the lammergeier, turns the mountain landscape to its advantage by carrying carrion bones aloft and dropping them onto the rocks to expose the edible marrow inside.

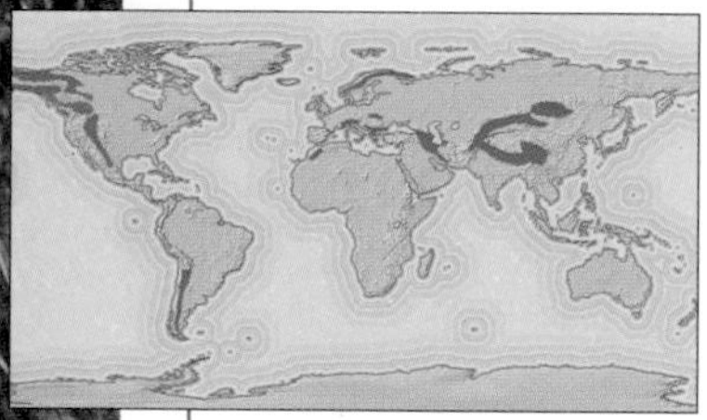

THE NORTHERN *temperate zone's mountain ranges include the highest peaks on earth. In the Southern Hemisphere, mountains are smaller and more isolated.*

HOOFED MAMMALS, *such as the bighorn sheep, survive at high altitude. They can often be found as high as the snow line.*

MARMOTS *hibernate in burrows, while some other mountain rodents, such as pikas, stockpile food to survive the winter.*

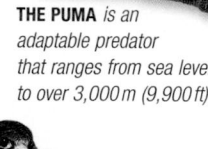

THE PUMA *is an adaptable predator that ranges from sea level to over 3,000 m (9,900 ft).*

THE PEREGRINE FALCON *uses the ledges of steep cliffs as vantage points and nest sites.*

LIKE MANY DEER *species, the red deer migrates vertically, feeding above the tree line in summer.*

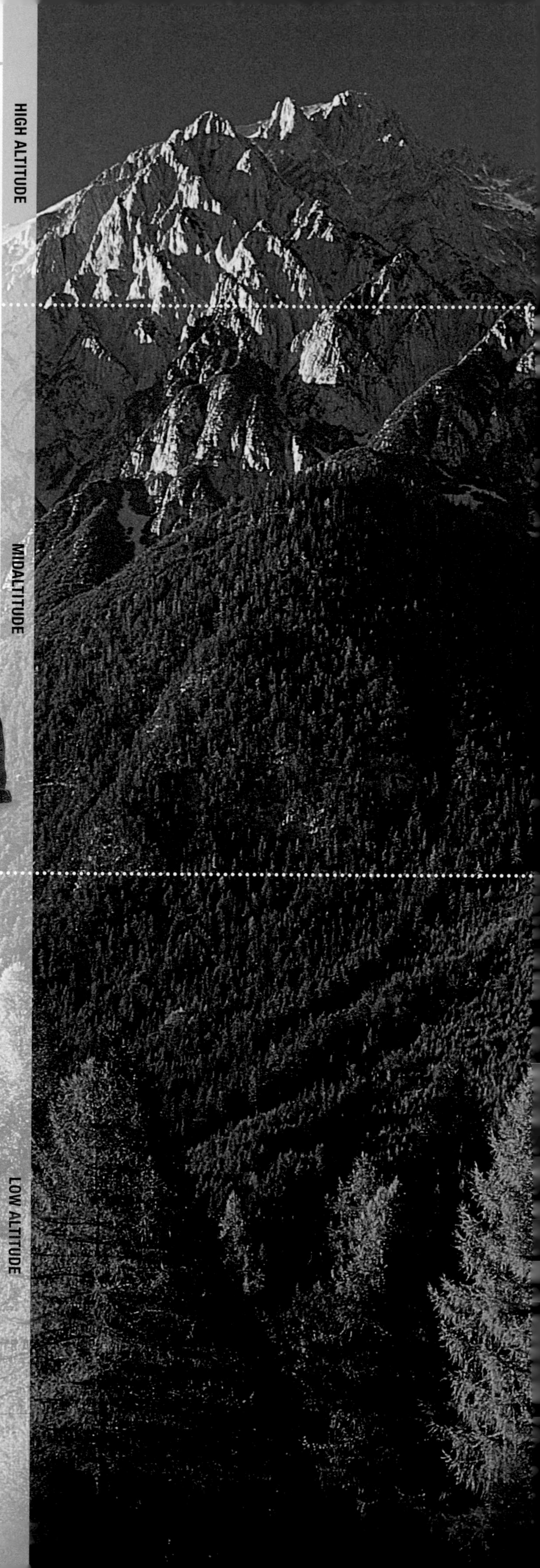

MOUNTAIN HUMMINGBIRDS *survive low nighttime temperatures in the Andes by becoming torpid—a form of overnight hibernation.*

THE VICUNA'S *high levels of oxygen in its blood enable it to range at high altitude.*

THE EASTERN GORILLA *uses the dense vegetation of the forest zone for both food and cover.*

MONKEYS *are common in tropical mountain forests. Africa's black and white colobus is a striking example.*

TROPICAL MOUNTAIN *tree and ground frogs live at all altitudes, from foothill forests to high, fast-flowing streams.*

Tropical mountains

In the tropics, the generally warm climate means that mountain vegetation zones extend much higher than they do in montane habitats elsewhere in the world. For example, near the equator, trees often grow at altitudes of up to 13,200 ft (4,000 m), which is why many tropical mountains are forested to their summits. Above this altitude is the tropical alpine zone, an open landscape dominated by grass and some highly specialized plants. This zone is often above the clouds, which means that nights are cold and frosty and yet the sunshine is fierce.

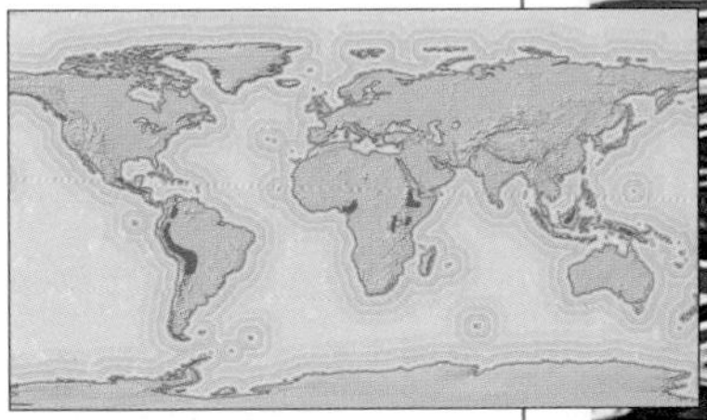

TROPICAL MOUNTAINS *include a major chain, the Andes, and isolated ranges in East Africa and Southeast Asia.*

Many tropical animals have successfully adapted to life at high altitude. They include the vicuna, which can be found up to 18,100 ft (5,500 m) in the South American Andes, and the yak, which reaches a record 19,800 ft (6,000 m) just north of the tropics in the Himalayas. Birds also live at great heights: in South America, for example, mountain hummingbirds called Andean hillstars often feed at over 13,200 ft (4,000 m). The Andean hillstar's minute size means it has difficulty storing enough energy to enable it to survive the cold nights. To combat the problem, its nocturnal heartbeat slows down and its temperature plummets, conserving energy.

The cloud-covered forest below the alpine zone is the habitat of some of the world's most endangered animals. They include the eastern gorilla—a species restricted to the mountains of central Africa—and the resplendent quetzal, a bird that lives in the cloud forests of Central America. The abundant moisture means that the forest zone also teems with many different species of frogs, living both on the ground and in trees.

MINING AND QUARRYING

Once isolated from human activities, mountain wildlife now faces many changes. The greatest of all is global warming, which alters the natural pattern of vegetation zones. Some mountain animals can benefit from this, but others face steadily shrinking habitats, because they live in a narrow range of altitudes, or on isolated ranges or peaks. More specific threats to mountain animals include hydroelectric projects, and the rapid expansion of mines and quarries. Dust, pollution, and the dumping of waste can all have a harmful effect on mountain wildlife.

ROCK REMOVAL *In some parts of the world, quarrying has destroyed important wildlife habitats. As road building and construction projects continue to increase, the demand for rock continues to grow.*

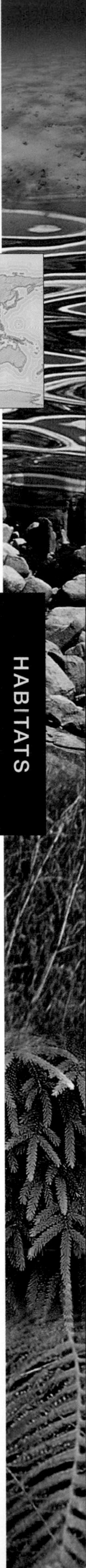

Life in mountains

Life at high altitude can be harsh. Food is often scarce, the weather can be treacherous, and the thin air can make it difficult to breathe. However, there is also more space, relatively little interference from humans, and fewer predators than there are lower down. Many animals are "incomers," using high ground as an extension of their normal range, but there are also some that live only on high ground. In large mountain chains, many animals have a wide distribution; but isolated peaks are often inhabited by animals that are found nowhere else.

Breathing thin air

At 19,800 ft (6,000 m), air is half as dense as it is at sea level. As a result, it contains only half the normal amount of oxygen—so little that anyone trying to breathe at this height would have difficulty remaining conscious. Yet some mountain animals live even higher than this because they have evolved specialized body systems that enable them to get the maximum amount of oxygen into their blood.

In the vertebrate world, birds are the unrivaled experts at high-altitude living. This is because air passes through their lungs in only one direction, not in and out, which ensures that a high proportion of the air's oxygen enters the blood—far more than enters a mammal's bloodstream in the same conditions. This fact is apparent from the height at which birds are capable of flying. In the Himalayas, choughs have been seen fluttering around campsites at over 26,400 ft (8,000 m), and there are records of vultures colliding with planes at over 36,300 ft (11,000 m)—far higher than Mount Everest.

Birds are unusual in being able to cope with rapid changes in altitude without experiencing any ill effects. For mammals, moving from one altitude to another necessitates special adjustment by the body, which is achieved by acclimatization—a process that can take several weeks to complete. During acclimatization, the number of red cells in the blood slowly increases, boosting its oxygen-carrying capacity. This physical adjustment, which is shown by a broad range of mammals—including humans—is temporary. If an animal moves back to lower ground, the process is reversed.

However, for mountain mammals, such as the vicuna and ibex, adaptation to life high up is a permanent state, not something that can be switched on and off. When measured as a proportion of volume, vicunas have 3 times more red cells in their blood than most other mammals, and the hemoglobin in their red cells is unusually good at collecting oxygen. As a result, vicunas can run almost effortlessly on the altiplano—the high-altitude plateau that runs the length of the Andes.

SOARING
Updrafts provide an almost effortless means of travel for birds like the Andean condor, which can cover hundreds of miles in a day by riding waves of rising air.

Compared with mammals, cold-blooded animals, such as reptiles, have fewer problems with thin air because they use oxygen more slowly. For them, the main problem with mountain life is cold: if the temperature is too low, their body processes slow down, and their muscles have difficulty working.

Movement

Mountains seem almost purpose made for soaring birds because strong air currents make it easy for them to gain height. For animals on or close to the ground, moving around is not so easy. Many insects are wingless, and species that do fly usually keep close to the rock to reduce the risk of being blown away by the wind.

For larger animals, the situation is even more hazardous, for a single misjudged move can lead to a fatal fall. Many rock-dwelling mammals therefore have feet designed to prevent slipping. In Australia, rock wallabies use their large back feet to take leaps of up to 13 ft (4 m), with their long tails helping them to balance. In Africa and the Middle East, hyraxes run over rocks and boulders with the help of specialized soles, which work like suction cups. But in most of the world's mountains, the most adept climbers are hoofed mammals.

NONSLIP FEET
The hyrax's small, adhesive feet are perfect for negotiating rock, although some species also use them for climbing trees. In addition to living high up, rock hyraxes inhabit kopjes—miniature mountains of eroded boulders scattered on the African savanna.

Hooves may seem far from ideal for climbing, and it is true that some hoofed mammals, such as horses, have great difficulty moving around on rocky slopes. But in climbing mammals, such as mountain goats and klipspringers, hooves have evolved into perfect aids for moving about in mountains: they are small and compact, allowing them to fit onto narrow ledges, and they have hard edges surrounding rough, nonslip pads. These combined characteristics make for good grip in all conditions, including rain and snow.

Getting a firm grip is essential for moving on rock, but equally important is a strong sense of balance and head for heights. Most terrestrial mammals are instinctively afraid of steep drops but, from an early age, mountain dwellers show what appears to be a reckless disregard for their own safety. Adult chamois take 20 ft (6 m)

VERTICAL MIGRANTS
In mountainous regions, red deer spend the summer high up, where food is plentiful and there are relatively few biting flies. Their downhill migration in fall often coincides with the start of the rutting season, when males grow a mane of hair on the neck and compete with each other for the right to mate.

leaps, and can run down nearly vertical slopes as easily as they can run up them; and their young are able to keep up with them when just a few weeks old.

Coping with winter

In tropical mountains, conditions are often much the same all year-round, which means that animals can stay at one altitude all their lives. But in temperate mountains, seasonal changes affect the food supply. Winter is the critical time: anything that cannot survive the cold weather conditions and the shortage of food has to move to lower ground or hibernate until the return of spring.

Animals that are resident at high altitude have a variety of ways of dealing with the changes. Insects often enter a dormant state, called diapause, which puts their development on hold. Many small mammals, such as marmots, survive mountain winters by hibernating, while many of those that remain active live on food stores accumulated earlier in the year. Pikas, for example, gather up leaves and grass and build them into "haystacks" among the broken rocks around their homes. Before adding fresh supplies to a stack, they sometimes spread them out to dry in the sunshine, which reduces the chances of the food from rotting.

For other animals, the first fall snows are the signal to move downhill. These vertical migrations are a common feature of mountain life in temperate regions, and they are demonstrated by a wide range of mammals and birds, from mountain sheep and deer to choughs and grouse. In many cases, the migration involves moving from the exposed mountaintop to the forests lower down, but some mountain forest species also migrate. Among these latter migrants are birds such as nutcrackers, which feed on conifer seeds. If the seed crop fails, they fly downhill in a form of sporadic migration called an irruption (see p.59). Clark's nutcracker, from the Rocky Mountains, is a typical example: normally found at up to 8,200 ft (2,500 m), it descends as low as sea level when food becomes hard to find.

FLEXIBLE MIGRANT
Like many temperate mountain animals, the nutcracker lives at various altitudes, depending on the conditions: severe cold drives it downhill, but because there is more competition for food on low ground it returns uphill as soon as the weather improves.

Finding food

As in most land habitats, a mountain's animal life depends ultimately on plants, for plants provide food for herbivorous animals, which, in turn, are eaten by a wide range of predators. However, some mountain animals make use of a very different food source—the cargo of small animals, mainly insects, that are carried uphill by the wind to be stranded among the rocks, snow, and ice.

Most of these wind-blown animals are so tiny that they are practically invisible; yet they provide useful nourishment for scavengers that live above the snow line. They consist almost entirely of invertebrates, such as springtails and snow fleas, which can survive the very low temperatures of high-altitude winters. During the depths of winter, they hide among rocks and moss, but when the weather warms up, they can often be seen hopping across banks of snow, feasting on the debris that the wind has brought up from lower ground.

LIFE IN CAVES

Many animals use caves temporarily, but some have adapted to spend their entire lives in them. These permanent cave dwellers, called troglodytes, feed either on each other or on the droppings deposited by roosting bats and birds.

As a habitat, deep caves have the advantage that temperatures remain fairly constant throughout the year. However, they are also completely dark, which means that eyes are useless. Bats, oilbirds, and swiftlets use echolocation to navigate while underground, but permanent cave dwellers sense their surroundings largely by touch, often using smell to track down food. Cave crickets detect food using antennae, while spiders and harvestmen use their feet. In subterranean streams and pools, cave salamanders, such as the olm, sense vibrations in the water. The cave fish has a row of pressure sensors along each side of its body that enables it to detect other animals several yards away.

Although food is scarce, animals exist many miles underground and have even been found in pothole systems that have no direct contact with the surface except via water trickling its way underground.

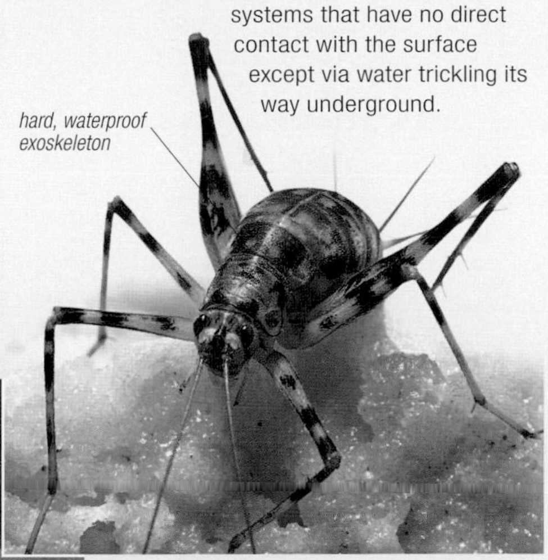

GIANT ANTENNAE
Cave crickets' antennae, which can be 2 or 3 times the body length, locate the dead remains and bat droppings that make up the animal's diet.

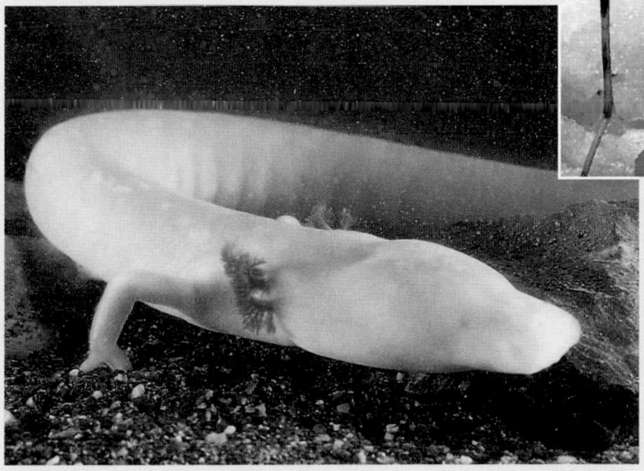

UNDERWATER LIFE
Like most aquatic vertebrates in caves, the olm has very little pigment in its skin. Its eyes are vestigial and covered with skin. Olms breed in water and remain in it throughout adult life.

Polar regions

SNOWY OWLS *are forced to hunt by daylight during the weeks of high summer.*

The Arctic and Antarctic are the coldest places on earth. The Arctic is a partly frozen ocean, hemmed in by large expanses of windswept tundra; the Antarctic is an ice-covered continent, surrounded by the world's stormiest seas. They are similar to each other—and unlike any other habitat—in that they have 24-hour daylight in summer and perpetual darkness in winter, but they are physically different in ways that have important effects on animal life. In the Arctic, many animals live on land; in the Antarctic, animal life is based almost entirely in the ocean.

WILLOW GROUSE, *like many other animals, turn white in winter to blend in with the snow.*

Arctic and tundra

Covering about 4.6 million square miles (12 million square km), the Arctic Ocean is both the smallest and the shallowest ocean in the world. For several months in summer, permanent daylight produces a constant supply of energy, which is harnessed by vast quantities of planktonic algae. These form the first link in the Arctic Ocean's food chain, which ultimately nourishes animals as large as whales and polar bears.

Sea ice—or the lack of it—is a major factor in determining where large mammals live, especially during winter when the surface area of the ice is at its greatest. Polar bears and Arctic foxes can traverse the ice to find food, but seals and some other marine mammals must maintain breathing holes to survive.

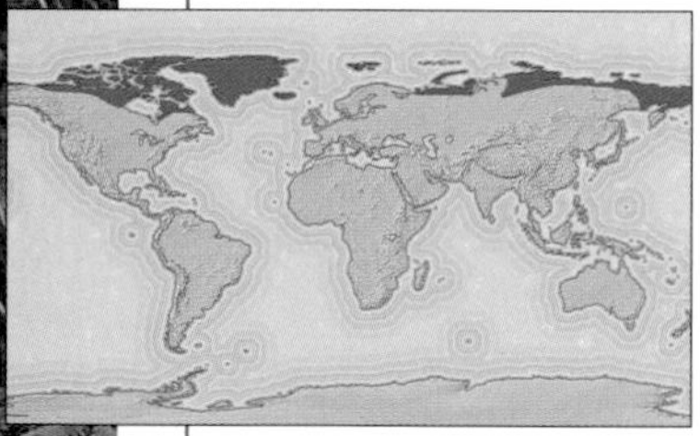

TRUE TUNDRA *is found north of the Arctic Circle, but tundralike conditions exist on some mountains elsewhere.*

Despite the icy conditions, sea life is plentiful in the Arctic because cold water is rich in oxygen and the seabed sediment is rich in nutrients. On land, though, intense winter cold means that trees cannot survive. The result is tundra—an open, often featureless, landscape, scraped smooth by glaciers during the last ice age. Today, Arctic glaciers are restricted mainly to mountains and to the ice cap that covers Greenland, but large areas of tundra remain permanently frozen underground. This frozen zone—the permafrost layer—prevents spring melt water from draining away, creating waterlogged landscapes in a region where rainfall, or snow, is paradoxically very low.

In late spring and early summer, tundra plants grow, and flower, very rapidly. Geese and other migratory birds arrive to breed, and vast numbers of mosquitoes emerge from tundra pools. The migrants' departure, when the short summer draws to a close, marks the end of another biological year.

MUSK OXEN *and other tundra grazers reach food by scraping away snow with their hooves.*

LEMMINGS *provide food for many of the Arctic's predators, including foxes.*

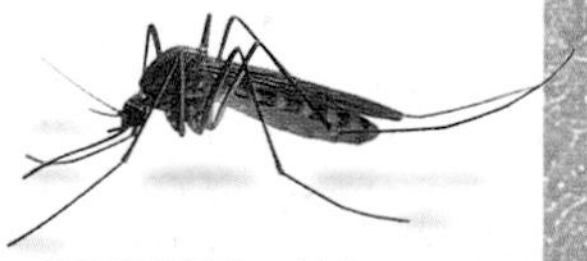

MOSQUITOES, *which emerge in early summer, make life uncomfortable for mammals and birds.*

THE POLAR BEAR *is a superb swimmer, and it is as much at home on shifting pack ice as it is on tundra.*

Antarctic

Unlike the Arctic, mainland Antarctica is isolated from the rest of the world. It is covered with ice, up to 13,200 ft (4,000 m) thick, which continues out to sea forming large ice shelves. On the Antarctic Peninsula—a finger of land pointing toward South America—summer temperatures rise to a few degrees above freezing point, but in the rest of the continent, average temperatures are below freezing all year- round.

Algae and lichens grow on bare rocks in many parts of the Antarctic coastline, but the Antarctic Peninsula is the only part of the continent where terrestrial plants can survive. This is also the only place that has a significant range of terrestrial animals, although these are chiefly springtails, mites, and nematode worms—few of which are over ⅕ in (5 mm) long. The rest of Antarctica's land-based animal life consists of species that feed in the sea and come ashore to breed, such as penguins, or those that scavenge food at these animals' breeding grounds, such as skuas. With the exception of emperor penguins, vertebrates desert the ice at the end of summer to spend the winter at sea.

The Southern Ocean, which surrounds Antarctica, is one of the most biologically productive seas in the world. Although species numbers are relatively low, population sizes are often enormous because the nonstop summer daylight generates a vast food supply. Krill—small crustaceans that form the diet of seals and whales—are especially prolific: some of their swarms are estimated to weigh in excess of 10 million tons and are large enough to be seen by satellites in space. Although the Southern Ocean is always cold, it maintains a minimum temperature of about 28.8°F (-1.8°C); below this, sea water freezes. As a result, the ocean is quite warm compared with Antarctica itself.

SKUAS, *found in both the Arctic and the Antarctic, feed on carrion and on other birds' eggs and chicks.*

NESTING COLONIES *of some birds, such as Adelie penguins, can contain over a million members.*

SPRINGTAILS *eat decaying plant remains—unlike most Antarctic animals, which rely on food from the sea.*

KRILL *harvest plankton, forming the basis of many Antarctic food chains.*

LEOPARD SEALS *feed on krill, fish, and seabirds. This predatory lifestyle has no direct equivalent in the Arctic.*

BALEEN WHALES *play a major part in the Southern Ocean's ecology by eating vast amounts of plankton.*

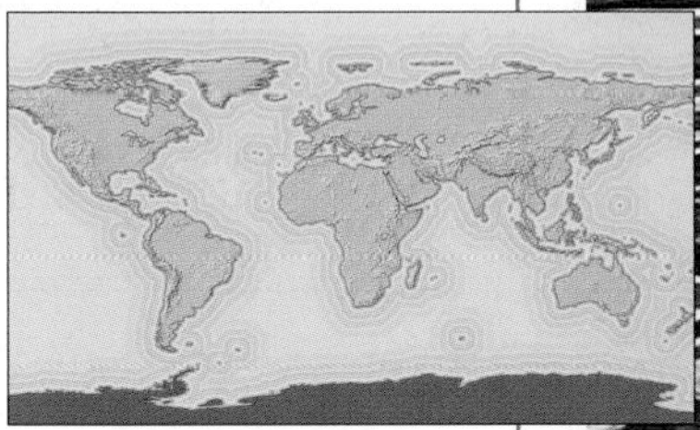

MOST OF ANTARCTICA, *excluding the relatively mild Antarctic Peninsula, lies south of the Antarctic Circle.*

CLIMATE CHANGE

The polar regions are uniquely sensitive to climate change, because global warming—or cooling—has a direct effect on their ice cover, both on land and at sea. Sea ice is rarely more than 16½ ft (5 m) thick, and spreads and retreats with the seasons. It is a vital platform for "ice seals," the approximately 10 species that haul out onto ice to breed, and it is also used by polar bears and some penguins to get near their food.

GAPS IN THE ICE *The Arctics's summer sea ice is currently shrinking fast, and may disappear altogether by 2060. Paradoxically, Antarctica's sea ice is slowly increasing—a trend that may be linked to changes in wind patterns.*

Life in polar regions

Although they live at opposite ends of the earth, the animals that inhabit the Arctic and Antarctic share many adaptations. Resilience to extreme cold is first and foremost among these, but almost as important is the ability to cope with a highly seasonal food supply. For some animals, winter is a good time for catching food but, for most, hunger and cold make the long winter months a critical time of year. Such testing conditions mean that in comparison with other parts of the world the poles are inhabited by very few animal species. However, those that do thrive can be extraordinarily numerous.

Coping with cold

Warm-blooded animals have to maintain a constant body temperature, which means that combating heat loss is a major priority in the polar environment. Cold-blooded animals can function with a fluctuating body temperature, but even they have limits—in subzero conditions, they can freeze solid. Fish are particularly at risk of freezing, for while their body fluids normally freeze at about 30.6°F (-0.8°C) polar sea water is often slightly colder still. To help prevent freezing, the blood of many cold-blooded species contains proteins that lower its normal freezing point. Some insects can survive at -49°F (-45°C) without any ice forming in their bodies.

Since mammals and birds cannot afford to let their internal temperature fall even slightly, they need insulation to keep warm. Fur and feathers are among the finest insulating materials that nature has devised, but many polar animals, such as whales, seals, and penguins, have additional insulation in the form of blubber—a layer of yellowish fat that is laid down under the skin. Blubber, which can be up to 12 in (30 cm) thick, is such an effective insulator that all these animals feel cold to the touch even when their internal body temperature is above 100.5°F (38°C). Blubber is particularly useful in the sea because water conducts heat away from the body 25 times faster than air. It also has another valuable function: because fat contains lots of energy, blubber can be used as a food reserve when supplies are low.

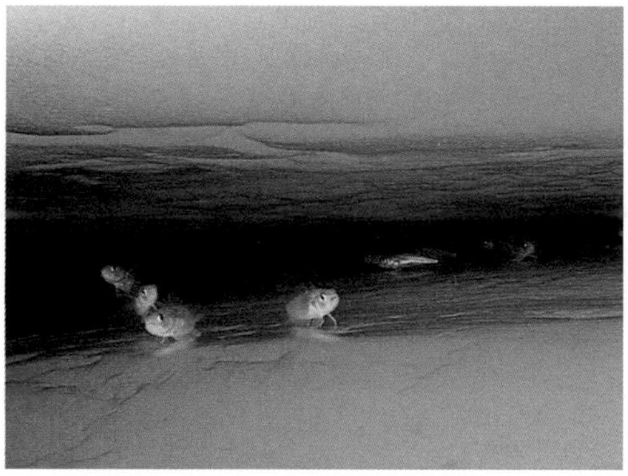

THE ANTARCTIC'S ICE FISH
In the Antarctic, fish live in a world that is often roofed by ice. The high level of oxygen in cold water allows some Antarctic fish to survive without any hemoglobin in their blood. As a result, their blood is almost clear rather than red.

SLEEPING ON ICE
A polar bear cub rests at its mother's side. Polar bears have long body fur, and their feet have furry soles. This keeps them warm on the ice, while a thick layer of blubber enables them to retain body heat while swimming in the ocean.

Surviving under ice

Whales and seals face problems during the long polar winter because sea ice restricts their access to air. They dive under the ice to feed, but they must then surface to breathe. Some species avoid the problem by moving to lower latitudes. Those that remain behind survive either by maintaining breathing holes or by congregating in polynia—areas where the wind and currents keep the water ice free.

Seals start making breathing holes when the ice is thin, rasping away at it with their teeth. As the depth of the ice increases with the progress of winter, they continue to visit and work on their holes to ensure they remain clear. The Weddell seal, which lives farther south than any other species, spends so much time keeping its breathing holes open that its teeth develop distinctive patterns of wear, and by late winter, its breathing holes can be 6½ ft (2 m) deep. The seals have to find their holes in almost complete darkness because during the Antarctic winter the sun stays below the horizon for weeks.

PAUSE FOR BREATH
A Weddell seal heads up toward a breathing hole in the ice. Weddell seals can stay underwater for over an hour, making return trips of up to 6 miles (10 km) before they have to come up for air.

Whales rarely make breathing holes. Instead, they head for polynia, where they can come up for air in open water. This less laborious strategy means they are not tied to one place, but it does have its dangers: groups of whales can become ensnared in shrinking polynia, unable to reach the next stretch of open water. There are records of narwhals—the world's most northerly whales—being trapped in their hundreds, making them easy targets for hunters.

Changing color

In the treeless Arctic tundra, camouflage is one of the most effective ways both of avoiding attack and of making an attack unseen. The summer and winter landscapes look so different that many tundra animals change their camouflage twice a year. The Arctic fox is a classic example: its summer coat is usually brownish gray, but in early fall it turns white; in spring, the process is reversed so that the fox blends with the rapidly thawing tundra. In some parts of the far north—particularly western Alaska and northern Greenland—Arctic foxes develop a bluish winter coat instead. Some researchers have suggested that this blue coat is an adaptation to coastal landscapes, where there is less winter snow, but as these foxes have been widely introduced by fur farmers, the theory is difficult to prove.

The least weasel, the world's smallest carnivore, changes its coat in a similar way, as do ptarmigans and many other tundra birds. Some, such as the snowy owl, keep their white plumage all year round, which suggests that good camouflage is most important in winter and less so in summer when food is easier to find.

LONG-DISTANCE MIGRANTS
In Alaska and northern Canada, vast herds of reindeer migrate between their summer grounds on open tundra and their winter grounds in coniferous forest. Some travel over 620 miles (1,000 km) each way, swimming across rivers and sea inlets en route. Pregnant females lead the herd during the spring leg of the journey.

SEASONAL COATS
When the Arctic fox changes color, the consistency of its coat changes as well. Its white winter coat has long guard hairs and thick underfur, making for superb insulation. Its brown summer coat is shorter, with thinner underfur, to help prevent overheating.

Summer migrants

Near the poles, 24-hour daylight in summer creates the ideal conditions for rapid plant growth. This short-lived but profuse supply of food has a dramatic effect on tundra life, attracting vast numbers of migrants. Geese come to crop the plants with their beaks, waders arrive to feed on worms and insects that live in swampy ground, while terns find food both on the tundra and in water close to the shore. This annual influx of visitors is mirrored in the ocean. Many of the world's baleen whales head toward polar waters during summer to make use of the annual upsurge in planktonic life. However, unlike migratory birds, these huge mammals do not breed at high latitudes. Instead, they put on weight and then return to warmer waters to give birth. During the breeding season, they often do not feed at all.

Winter food

Although the food supply in polar seas slowly falls in the fall, there is still a reasonable amount for animals to eat. On land, life is not so easy. The growth of tundra plants comes to a complete halt and, to make matters more difficult, the plants themselves are often covered by deep snow. For herbivores, this lack of accessibility is a major problem at a critical time of year. In the Arctic tundra, plant-eating animals reach plants in one of 2 different ways. Reindeer (caribou) and musk oxen use their hooves to clear away the snow to reveal the lichens and dwarf willows underneath. Lemmings turn the snow to their advantage by burrowing in it. The snow protects them from predators and from the weather outside: no matter how cold or windy it is on the surface, the lemmings enjoy a benign micro-climate that allows them to feed all year. In Antarctica, there are very few terrestrial plants, which means that almost no animals stay active in winter on food gathered from land. With so much ice, even food from the ocean can be difficult to reach. Male emperor penguins, guarding their eggs, do not even attempt to find it: huddling on the ice through the long night of winter, they go without food until spring.

RUNWAYS UNDER THE SNOW
Lemmings need cover to survive. In winter, they develop enlarged front claws that enable them to tunnel in snow. The tunnels keep them secure, although they must still contend with sparse winter food supplies. When the snow melts, they move underground.

TUNDRA INSECTS

Unlike Antarctica, the Arctic tundra teems with insect life, including aphids, bumble bees, damselflies, and unimaginable numbers of mosquitoes. The mosquitoes spend their larval and pupal stages in tundra pools, emerging as adults in early summer. For warm-blooded animals, tundra mosquitoes are a serious problem. The females need blood before they can breed, and they are relentless in their attempts to get it. When the mosquito season is at its height, some animals head for high ground, but most have no alternative but to sit it out. Mosquitoes do bring some benefits: the larvae and pupae are a useful source of protein for waterfowl and waders.

CLOUD OF MOSQUITOES
Adult tundra mosquitoes live for only a few weeks once they have emerged from their aquatic nurseries—just long enough to feed, mate, and lay eggs, before fall frosts glaze their breeding pools with ice.

Freshwater

Every year, about 24,000 cubic miles (100,000 cubic km) of water evaporates from the world's oceans, condenses, and then falls as rain or snow. Most of this water disappears back into the atmosphere to continue this cycle, but about a third returns to the oceans by flowing either over ground or beneath the land surface. This steady supply of freshwater sustains all the world's land-based life, as well as creating highly diverse habitats—from streams, rivers, and lakes to reedbeds, marshes, and swamps—in which a wide range of different animal and plant life can thrive.

HERONS *hunt by stealth, waiting along riverbanks and lake shores for fish to come within range.*

DRAGONFLIES, *like many other insects, spend the early part of their lives in freshwater, leaving it only when they are ready to breed.*

Lakes and rivers

For permanent water dwellers, life in lakes and rivers is shaped by many different factors. One of these is the water's chemical makeup, which is often dictated by the type of rock that forms the bed of a river or lake. Hard water, for example, is good for animals that grow shells because it contains calcium that can be used as shell-building material, while water that is especially rich in oxygen is important for highly active predatory fish, such as salmon and trout. Water that is very deficient in oxygen, on the other hand, provides a poor environment for animal life as a whole because relatively few aquatic species, other than specialized worms, can survive in it.

In lakes and rivers, aquatic animals usually occupy clearly defined zones. The brightly lit water close to the surface often teems with water-fleas, copepods, and other microscopic forms of animal life. They live here in order to feed on phytoplankton—the microscopic algae that become extremely abundant during the summer months. Feeding in the middle zone and near the water's surface are larger animals, such as fish, that are able to hold their own against the strength of the current. Weak swimmers live near the banks, where the current is slower, or among stones and sediment on the bottom. In still and slow-flowing water, surface tension supports insects that hunt by walking or running over the water.

Some animals that are associated with freshwater habitats are not necessarily permanent water dwellers; instead, they divide their time between water and the adjacent land, entering the lake or river to hunt or breed, or using it as a nursery for their young.

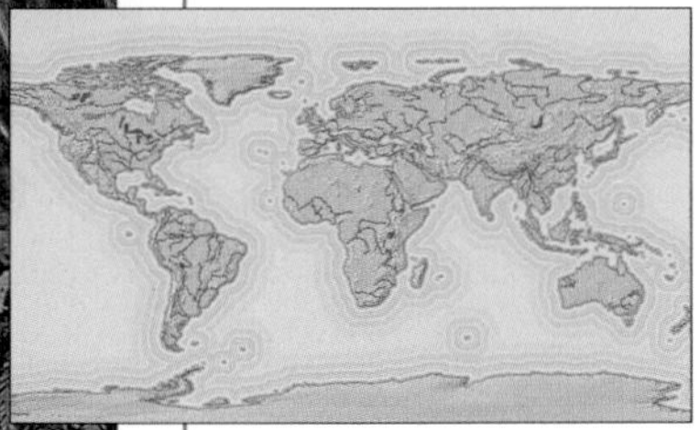

IN GEOLOGICAL TERMS, *lakes and rivers are highly changeable. Lakes gradually fill with sediment, while rivers often change course.*

FROGS *vary in their affinity for water. The American bullfrog rarely leaves it, even as an adult.*

MANY AMPHIBIANS, *such as the great crested newt, use still or slow-flowing water containing plenty of plants as a repository for their eggs.*

SMALL CRUSTACEANS *are part of the animal plankton living at all levels in ponds, and at the surface in lakes.*

PIKE *are sit-and-wait predators, using dense waterside vegetation as cover from which to ambush their prey.*

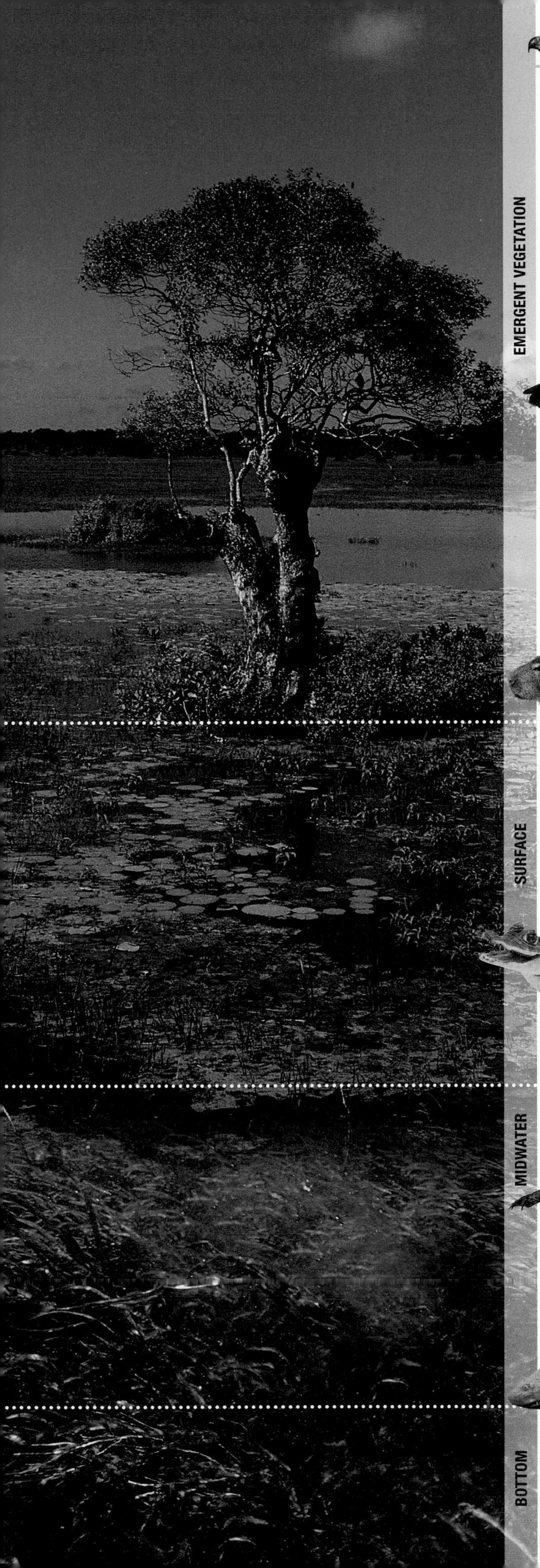

FISH EAGLES *use their long talons to snatch fish from the surface of the water.*

THE JACANA, *or lilytrotter, has long toes that spread the bird's weight so that it can stand on floating leaves.*

THE CAPYBARA, *the world's largest rodent, feeds on land, but uses water as a refuge from attack by predators.*

CAIMANS, *and many other crocodilians, can survive for many weeks when wetlands dry out by wallowing in moist mud.*

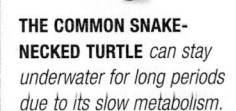

THE COMMON SNAKE-NECKED TURTLE *can stay underwater for long periods due to its slow metabolism.*

LUNGFISH *survive times of drought by lying dormant in mud, protected by their waterproof cocoons.*

Wetland

A wetland is any waterlogged or flooded area with a covering of water plants. In some wetlands—reedbeds and bogs, for example—the plants hide the water completely. However, in most wetland, areas of open water and dense vegetation are mixed, creating a rich and complex habitat that can be exploited by animals of almost every kind.

Biologically, wetland is among the most productive inland habitats, sometimes surpassing even rain forest in the amount of food that it generates for animals. In temperate parts of the world, this productivity reaches a peak during spring and summer, but in the tropics and subtropics, it is more affected by the water supply. Some tropical wetlands—South America's Pantanal, for example—largely dry out during the dry season, but then look like vast lakes once it has rained.

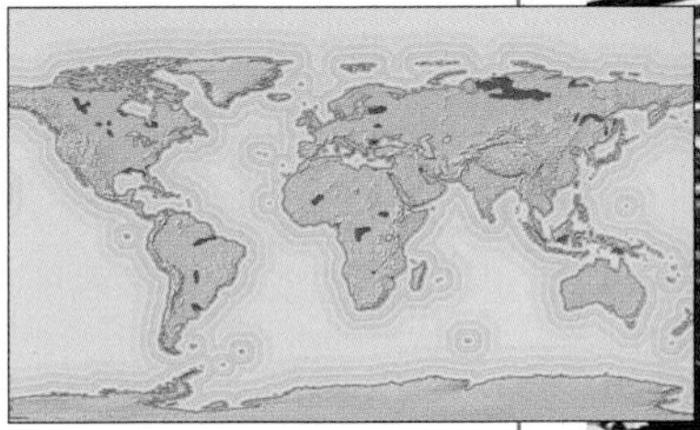

FRESHWATER WETLANDS *include the Pantanal, South America, and the Okavango, an inland delta in southern Africa.*

In many wetlands, the water is no more than a yard (3¼ft) or so deep, which means that bottom-living animals and surface dwellers are rarely far apart. This kind of environment is ideal for air-breathing swimmers, such as snakes and turtles, and also for land-based animals that use water as a temporary refuge from danger. Unlike large lakes, wetlands have an extra dimension in the form of emergent plants, which grow up through the water's surface and into the air. These plants range in size from small grasses and rushes just tall enough to keep insects clear of the water, to water-loving trees that grow to over 115ft (35m) high. Trees act as important roosting and breeding sites for waterbirds, providing them with shelter and relative safety from predators as well as keeping them close to the source of their food.

POLLUTION AND ABSTRACTION

Freshwater habitats are affected by human activities in many ways. Water pollution is a general problem that has profound consequences for a wide range of animal life. However, in addition, large areas of wetland have been drained for agricultural use, often because wetland conditions produce highly fertile soil. Water abstraction, both for agriculture and for domestic use, is also an increasing threat to freshwater habitats. Efforts to protect wetlands have produced several international agreements. One of these, the Ramsar Convention, adopted in 1971, focuses on wetlands used by migrating waterfowl.

HIGH AND DRY *In central Asia, the Aral Sea—has a giant inland lake—has shrunk dramatically because incoming river water has been diverted to irrigate fields.*

Life in freshwater

Freshwater is an essential resource for all land-dwelling animals, but it is also an important habitat in its own right. Freshwater habitats vary immensely, from temporary pools to giant lakes, and from tiny streams to rivers thousands of miles long. As a result, the problems that animals must overcome in order to survive in their environment can be very different from one freshwater habitat to another. Strong currents, periodic drought, and intense competition for food are some of the difficulties for which animals must find solutions.

Staying in place

Animals living in fast-flowing rivers face a constant battle with the current, and they deal with it in one of 2 main ways. The first is to avoid the problem by staying close to the riverbed, where the current is relatively slow. Many invertebrates, such as stonefly and mayfly larvae, never venture into open water; and to improve their staying power they often have a flat profile, so that if they position themselves with their heads facing upstream, the current presses down on their backs, helping them to stay in place. Remarkably, the same technique is used by dippers—the only songbirds that feed underwater: they are naturally buoyant, but as they walk upstream over the riverbed, the force of the current keeps them submerged.

The second solution is to compete with the current by swimming against it. Species that cannot avoid the current match their swimming speed to the water flow, enabling them to stay in place—and they keep swimming even when they are asleep.

COPING WITH THE CURRENT
Brown trout live in cool, fast-flowing streams, where it takes a lot of energy simply to stay in place. Trout can hold their own because fast-moving water is usually rich in oxygen—exactly what the fish need to keep their muscles working.

Migration

In addition to supporting freshwater residents, rivers and lakes receive visitors from the sea. These are migratory fish, which divide their time between fresh- and saltwater habitats. Anadromous species, such as salmon, breed in rivers but spend most of their adult lives offshore. Catadromous species, such as some eels, do exactly the opposite: they live in freshwater, but swim out to sea to breed.

For most fish, freshwater makes a much safer nursery than the ocean, and this justifies their long journey upriver to lay their eggs. However, because the food supply in freshwater is limited, the young eventually make for the much more dangerous, but also more fruitful, environment of the open sea. The advantages of this are apparent where some, but not all, members of a species migrate: those that swim out to sea usually grow much bigger than those that stay behind.

Like migrating birds, migrating fish are often remarkably accurate in pinpointing the place of their birth, returning there to spawn even though it might mean an inland journey of over 1,500 miles (2,500 km). Each river has its own characteristic chemical fingerprint and, with their acute sense of smell, migratory fish are able to identify the estuary that they left as young fish. By monitoring the scent of the water as they progress upriver, they home in on their own spawning ground.

Migrating fish often meet barriers to their progress. Salmon are famous for jumping waterfalls and rapids, while eels tackle obstacles by slithering around them over land. Eels usually do this after dark, in damp conditions, when they can survive out of water by breathing through their skin.

Living in and out of water

Most amphibians, such as frogs and toads, develop in freshwater habitats and then, as they approach adulthood, take up life on land, returning to water to breed. However, insects evolved amphibious lifestyles long before the first true amphibians appeared. Today, almost every patch of freshwater, from the smallest puddle to the largest lake, is inhabited by insects. Mosquito larvae feed on microscopic freshwater life, whereas the larvae of dragonflies and damselflies stalk larger prey, catching it with a set of extensible jaws called a mask. These insects leave the water when they become adult, but water beetles and bugs remain in it for life, although their ability to fly makes it easy for them to spread from one pool to another.

Some larger animals have developed lifestyles that straddle water and land. Snakes are good swimmers, and a number of species,

A DAYTIME REFUGE
Hippos lounge in rivers and lakes by day and emerge at night, sometimes traveling over 9 miles (15 km) to find good grazing. Their skin is thin and almost hairless, but it exudes a special secretion that helps to protect them from the daytime sun.

JOURNEY'S END
Most migratory fish travel upriver to breed several times during their lives. Sockeye salmon, however, almost always die once the adults have spawned. Sockeyes that are ready to breed typically develop hooked jaws and humped backs.

DRY TIMES
Two staring yellow eyes identify a common caiman in the mud of a dried-out lake. During drought, caimans can survive being entombed for many months as long as the mud stays moist.

such as the grass snake, specialize in catching aquatic animals. The anaconda, the heaviest snake in the world, uses water for cover and for support: despite its size, water buoys it up, reducing its effective weight to almost nothing and enabling it to swim at considerable speed.

In addition to providing food, freshwater is a valuable resource for some land-based animals seeking refuge from predators or

ROOSTING AMONG THE REEDS
Swallows, starlings, and other flock-forming songbirds often use reeds as overnight roosts. The sleeping birds are relatively safe from predators because reeds are usually surrounded by water.

daytime heat. At dusk, birds often roost on lakes and reservoirs while, during the day, hippos, capybaras, and beavers spend a lot of time in the water, emerging under the cover of darkness to feed on land.

Surviving drought

In warm parts of the world—particularly in the tropics—rivers and lakes often dry out completely for several months each year. In these conditions, freshwater animals need some unusual adaptations to survive.

When a lake's water level drops, its temperature increases. Warm, stagnant water often contains so little oxygen that many fish suffocate, but lungfish—characteristic inhabitants of tropical lakes and wetlands—are experts at dealing with drought. They gulp air at the surface and, when their home starts to dry out, they burrow into the mud, sealing themselves inside mucous cocoons. Later in the year, when heavy rain falls and soaks the mud, the cocoon breaks down, and the fish wiggle away.

Caimans and turtles also hide away in this manner, although their scaly, waterproof skins mean that cocoons are not needed. Being "cold-blooded," or ectothermic, they need relatively little energy to stay alive, so they can survive drought-induced food shortages for months.

Animals that cannot survive drought often leave behind drought-resistant eggs, which hatch when water returns—an effective way of bridging the gap between one wet period and the next. This survival strategy is used by water fleas, rotifers, tardigrades, and many other "micro-animals" that live in temporary pools or in the film of freshwater that covers mosses and other plants.

BREATHING AT THE SURFACE
Mosquito larvae hang upside down at the water's surface, breathing through snorkel-like tubes. The tubes are tipped with water-repellent hairs that break through the surface film.

Coming up for air

Freshwater animals need to breathe oxygen. Fish use gills to extract it from the water, but many invertebrates collect it from the air. These air breathers include water snails, many insects, and the water spider—the only spider to have evolved a fully aquatic lifestyle.

For animals living at the water's surface, air is easy to reach. For fully submerged ones, breathing requires periodic trips to the surface to replenish air reserves, which are stored in or on the insect's body, often forming a film-like bubble that gives the animal a silvery sheen. The water spider has an exceptionally elaborate storage system. It constructs a "diving bell" from strands of silk, trapping a large bubble of air inside it. The bell acts as both lair and nursery—a unique example of an animal creating a submerged habitat that resembles dry land.

LIVING IN SALT LAKES

Salt and soda lakes are the most saline habitats on earth, with up to 10 times as much dissolved salt as the sea. They form where rainfall is low and evaporation rates are high because temperatures often climb to over 104° F (40° C). Relatively few animals can deal with these extreme conditions, but the ones that do can be extremely numerous because they face very little competition. Salt-lake food chains are based on cyanobacteria—plantlike microorganisms that harness the energy in sunlight, turning it into food.

LIVING IN BRINE
Brine shrimps are small crustaceans that thrive in salt-rich desert lakes. They breed rapidly, producing eggs that can withstand several years of drought.

FILTER FEEDERS
Lesser flamingos use their beaks to filter microorganisms out of the water. A large colony can consume more than one ton of cyanobacteria a day.

Oceans

The oceans form by far the largest continuous habitat on earth, and they were almost certainly the environment in which life first evolved. The underwater landscape is made up of mountains and volcanoes, cliffs, deep valleys, and vast, flat plains, many of them far larger than any found on land. The oceans are so immense—they cover more than three-quarters of the earth's surface—and difficult to explore that scientific knowledge of ocean wildlife lags behind that of life on land. However, research has shown that life is found at all levels, from the sunlit surface to the deepest trenches over 7 miles (11 km) down.

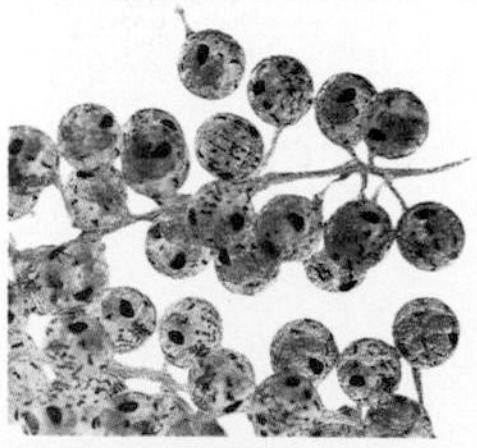

PLANKTONIC LIFE *near the surface of inshore waters consists partly of sea animals starting their lives as larvae.*

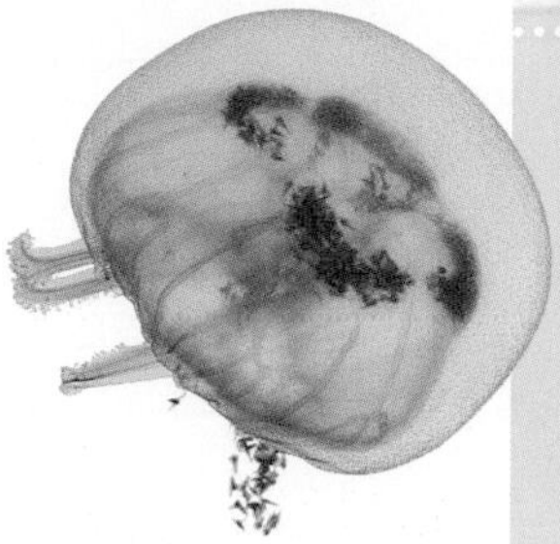

JELLYFISH, *the largest individual planktonic animals, can swim as well as drift near the surface. Some are over 6½ ft (2 m) across.*

Inshore waters

Some inshore waters are so shallow that if the world's oceans were lowered by just 245 ft (75 m), huge areas of seabed would be exposed. Off Western Europe, for example, the coast would be extended by about 125 miles (200 km), and in parts of Siberia, by more than 435 miles (700 km). These shallow waters owe their existence to continental shelves—the gently sloping plateaux that flank many of the deep ocean basins.

MOST SHARKS *live in the relatively shallow water of the continental shelves rather than in the deep sea.*

Continental shelves are a key habitat for sea life, supporting large schools of fish and a diverse collection of other animals, from lobsters and crabs to mollusks and burrowing worms. This wealth of life is possible because, in shallow water, sunlight can reach the seabed, promoting the growth of algae, seagrasses, and countless other organisms that need energy from light to survive. Just like plants on land, these provide animals with a year-round supply of food, as well as with plenty of cover and places to breed.

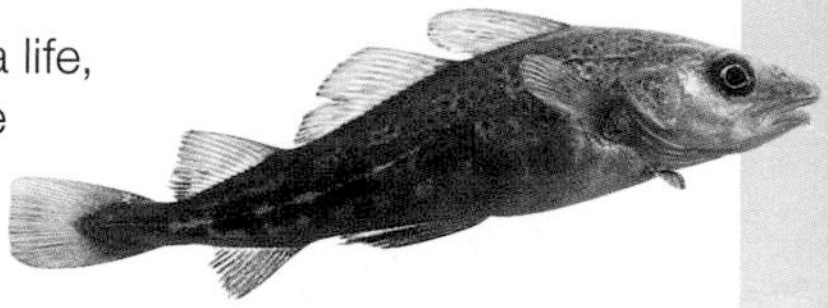

FISH *such as cod shed vast numbers of eggs near the seabed. The eggs float gradually upward, and the young fish live near the surface.*

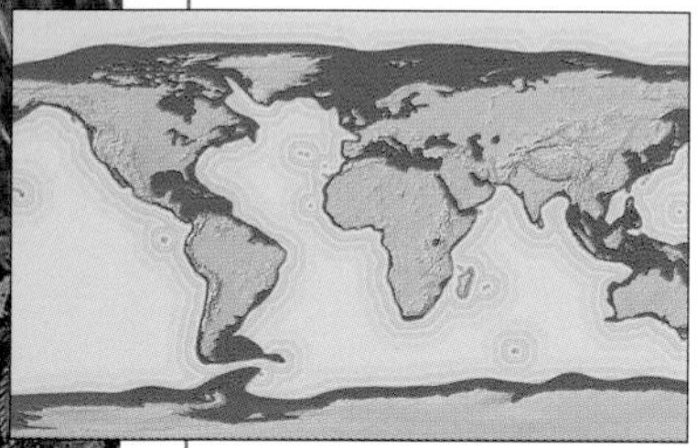

CONTINENTAL SHELVES *vary in width from a few miles—in coasts close to deep-sea trenches—to over 620 miles (1,000 km).*

Some inshore animals, such as lobsters and flatfish, spend most of their lives on the seabed, while a few live entirely in midwater or at the surface. Others live in quite different habitats at different stages of their lives. For example, some larvae develop as part of the surface plankton, then move to midwater or the seabed as adults.

LOBSTERS *are common seabed scavengers, crawling over the bottom in search of live prey and dead remains.*

Inshore waters are also visited by animals from the open sea—most often by passive drifters, such as jellyfish, but also by powerful swimmers, such as whales and sharks. Some come to breed, while others arrive by accident and then return to deeper water. Occasionally, the latter get into difficulties: jellyfish often end up on the shore, and whales can become stranded when their navigation systems guide them into shallow water instead of safely out to sea.

LIKE MANY FLATFISH, *the European plaice lives on the seabed. It is camouflaged to match the sediment.*

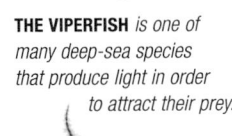

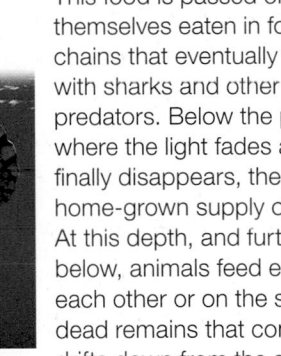

OCEAN WANDERERS, *such as tropicbirds, spend most of their lives far out at sea, returning to land only to breed.*

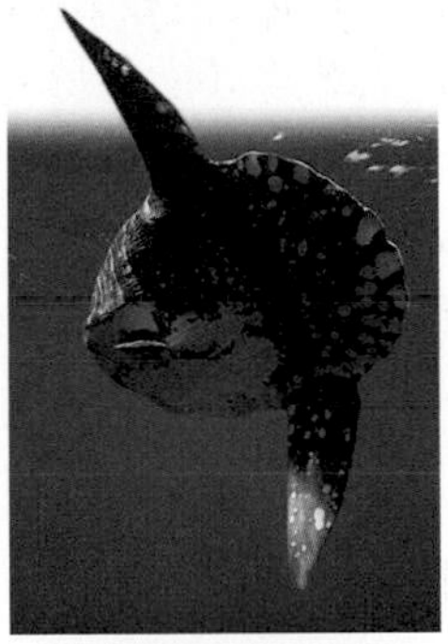

THE HUGE SUNFISH *is a typical pelagic species: it spends its life in open water, often hundreds of miles from land.*

MARINE TURTLES *often migrate long distances between their feeding grounds and the beaches where they breed.*

DOLPHINS *have streamlined bodies that give them exceptional swimming ability. They often roam far out of sight of land.*

THE VIPERFISH *is one of many deep-sea species that produce light in order to attract their prey.*

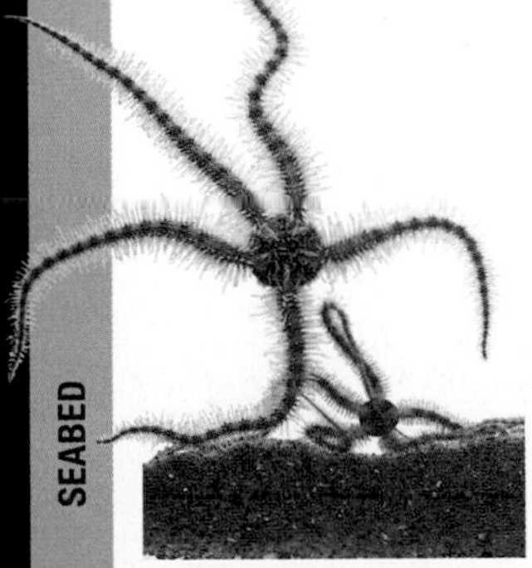

DEEP-SEA SCAVENGERS, *such as brittlestars, feed on organic matter that drifts down to the seabed.*

Open sea

Even in the clearest sea water, light penetrates no farther than about 825 ft (250 m) below the surface. More than any other factor, this has a crucial effect on sea life because it determines what there is to eat.

In the brightly lit surface zone, microscopic algae grow by harnessing the energy in sunlight, creating an invisible harvest for planktonic animals. This food is passed on when the plankton are themselves eaten in food chains that eventually end with sharks and other large predators. Below the point where the light fades and finally disappears, there is no home-grown supply of food. At this depth, and further below, animals feed either on each other or on the supply of dead remains that constantly drifts down from the surface.

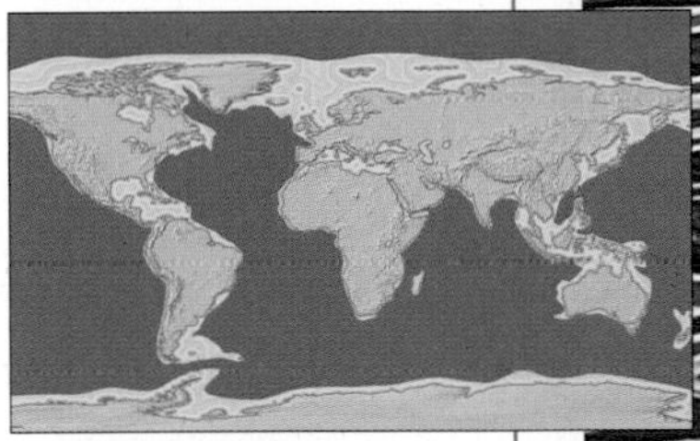

IF THE CONTINENTAL *shelves are excluded, the average depth of the world's ocean basins is about 13,200 ft (4,000 m).*

Despite the unimaginable volume of the oceans, few of the world's animal species—perhaps 5 percent—live in open water. Since most of these stay near the surface, where they can take advantage of the relatively plentiful food supply, animal life in the huge midwater zone and the deeper abyssal zone is comparatively sparse.

In contrast with these central zones, much of the ocean floor abounds with animals. Deep-sea creatures, which are known collectively as benthic animals, include species that swim or crawl over the seabed as well as those that burrow through it, mining the soft sediment for food. Many of these animals appear to have changed little over millions of years, for although the water is extremely cold, and the pressure intense, the deep seabed is not subjected to the changeable conditions that can affect the surface. It is therefore one of the most stable habitats on earth.

OVERFISHING

The world's total fish catch currently stands at nearly 98 million tons (100 million tonnes) a year, with about nine-tenths coming from the sea. This colossal harvest has a major impact on marine ecosystems, with fishing spreading from traditional grounds to polar waters and the deep seabed. Managed appropriately, marine fish can be sustainably harvested, whereas overfishing can disrupt breeding, pushing stocks to the point of collapse. One of the latest victims is the bluefin tuna in the western Atlantic: its numbers have fallen by over 70 percent in 40 years. As a result, it is now listed as critically endangered.

SURE CATCH
Using sonar and global positioning systems, modern trawlers can track down schools with pinpoint accuracy. Without strict controls, far fewer adult fish are left to breed.

Life in oceans

In the billion or more years since the first animal species evolved, competition for survival in the sea has become ever more intense. Today, the oceans are home to the largest predators on the planet, as well as to vast numbers of microscopic animals that drift by unseen. As on land, ocean wildlife is affected by local conditions, the most important being the supply of food. For many, survival also depends on being able to defend themselves against attack. In some regions, life is thinly spread, but in others, animals are found in greater numbers than they are anywhere else on earth.

Feeding at sea

The oceans are so huge that, although they contain plenty of nutritious food, marine animals face a challenge in finding enough to eat without expending too much energy in the process.

Some concentrate on large prey. The sperm whale, for example, hunts giant squid at depths of over 3,300 ft (1,000 m), although most pursuit hunters search for food near the surface rather than in the depths. Other large sea animals eat smaller fare, scooping it up in huge amounts, often straining it with their gills. This technique, known as filter feeding, is used by baleen whales and some of the largest sharks and rays. Most filter feeders live on plankton, which is so abundant that it allows them to reach a gigantic size. Drifting animals also feed on plankton, although on a much smaller scale. Comb jellies or sea gooseberries, for example, haul it in with stinging tentacles that work like fishing nets.

On the seabed, animal life depends almost entirely on the dead organic matter that drifts steadily down from above. Brittlestars are typical of these scavengers, collecting food particles with their arms. However, there are also predators—bizarre fish, for example—that hunt in the total darkness, on or near the seabed. Animal life at the bottom of the sea can be sparse, so these predators cannot afford to miss any opportunity to feed. Many of them therefore have gigantic mouths and elastic stomachs that enable them to swallow prey that is almost as large as themselves.

Controlling buoyancy

Very few marine animals—with the exception of deep-diving mammals—are found at all levels in the sea. Instead, most are adapted for life at a particular depth, and have buoyancy devices that help to keep them there. Surface drifters, such as the violet sea snail, have simple floats: filled with material that is lighter than water, they ensure that the animal stays at the surface, even in a heavy swell.

For animals that spend their lives fully immersed, remaining at one level requires more complex apparatus. They have to be neutrally buoyant at their optimal depth and yet able to rise or sink as the need arises. To do this, some use adjustable buoyancy aids hidden inside their bodies. Bony fish, for example, have a gas-filled chamber, called the swim bladder, just below the backbone. If the fish needs to sink, it removes some of the gas from the swim bladder by pumping it into the bloodstream; if it needs to rise, it pumps it back into the bladder. Cartilaginous fish, such as sharks, do not have swim bladders; instead, they rely on their large, oily liver to keep them afloat. Many sharks are actually slightly heavier than sea water: swimming provides the lift that allows them to control their depth.

ANIMALS ADRIFT
The violet sea snail hangs from a raft made of hardened bubbles of mucus. To improve buoyancy, its shell is unusually thin. It drifts with the currents, feeding on other surface animals.

FILTER FEEDING
Using large flaps on either side of its head, a manta ray channels plankton into its mouth. Filter feeders like the manta are indiscriminate eaters, swallowing anything that becomes trapped by their strainerlike gills.

Avoiding predators

In the open sea, there is nowhere to hide, which leaves animals highly vulnerable to predators. To survive, some rely on camouflage or disguise; others behave in ways that make them difficult to attack.

For slow-moving invertebrates, such as those that make up plankton, one of the most effective disguises is transparency. Planktonic animals are often as clear as glass, which makes them difficult to see, even at close quarters. Most of these animals are only a few millimeters long, although some tunicates form translucent, tube-shaped colonies that can be over 9¾ ft (3 m) in length.

Some fish are transparent when they are very small, but then use camouflage of a different kind as they get older. Almost all species that live in brightly lit, open water have dark backs but much paler undersides. This pattern, known as countershading, protects fish in 2 ways: it hides them from predators deeper down by disguising their silhouette against the

BIOLUMINESCENCE

A variety of marine animals produce light—some to maintain contact with their own kind, others to lure prey. In some species, notably planktonic invertebrates, the value of light is less easy to explain. Bioluminescence is most common in bathypelagic species (those that live in very deep, open water). Typically, the light is produced by skin organs called photophores. Light production is not always confined to the animal itself: some fish can eject luminous clouds to distract predators while they make an escape.

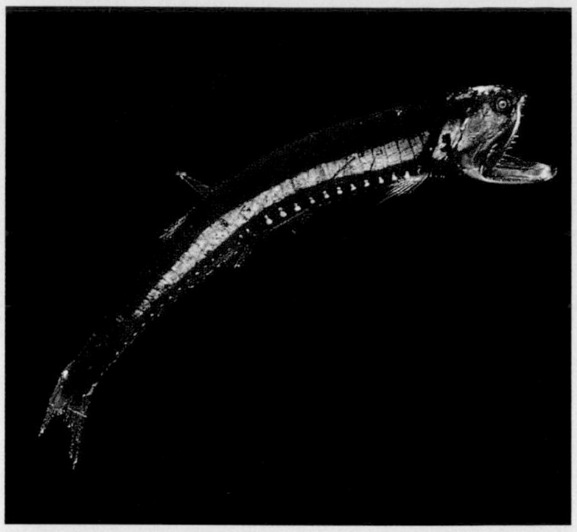

LIGHTS IN THE DARK
Bristlemouths have photophores that emit a yellowish green light. As in most luminous fish, the photophores are arranged in characteristic patterns that enable members of a species to recognize each other in the dark. Photophores may shine for extended periods or they may be flashed on and off.

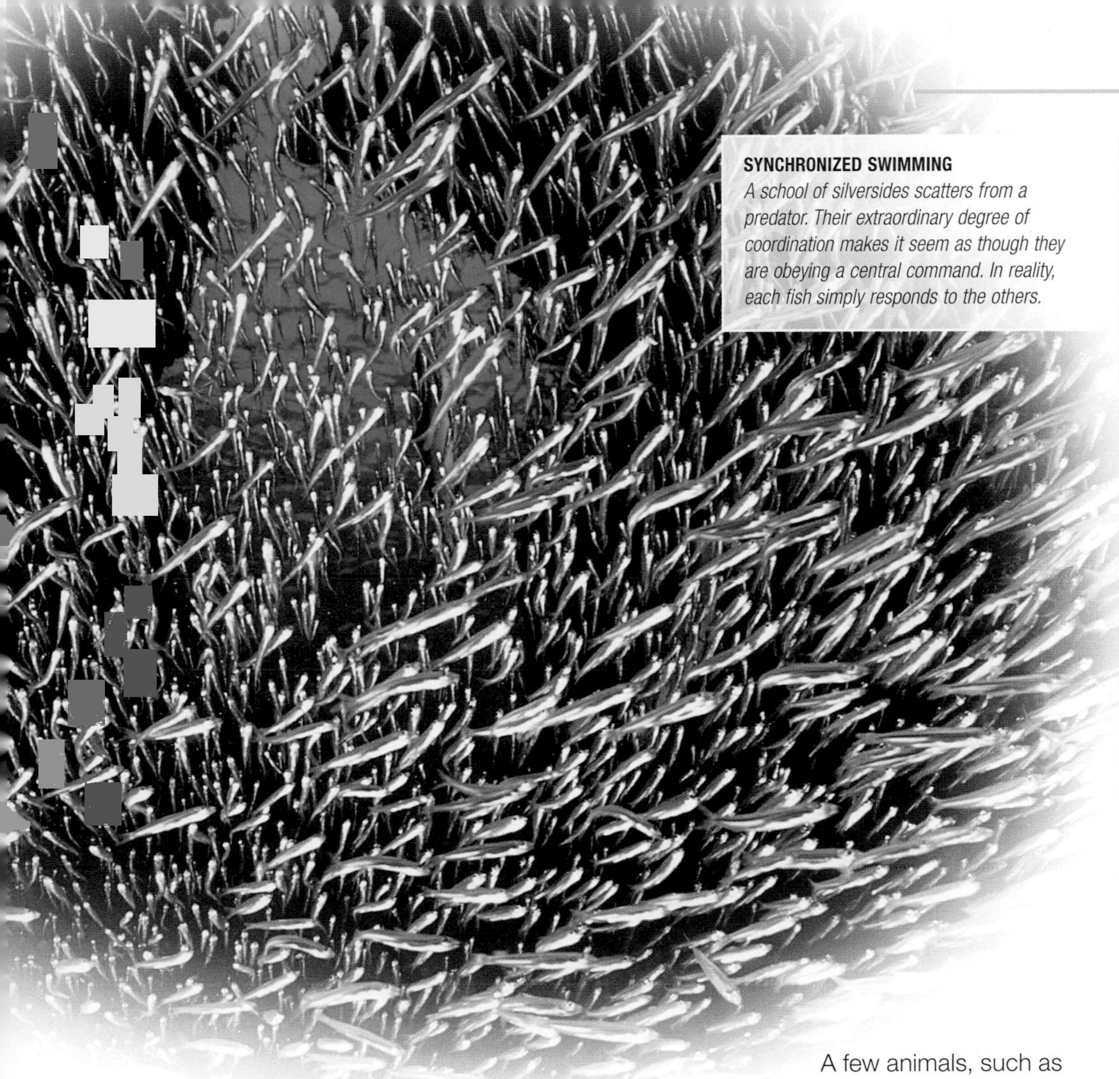

SYNCHRONIZED SWIMMING
A school of silversides scatters from a predator. Their extraordinary degree of coordination makes it seem as though they are obeying a central command. In reality, each fish simply responds to the others.

bright sky, and it conceals them from surface hunters, including seabirds, by making them blend in with the dark water beneath.

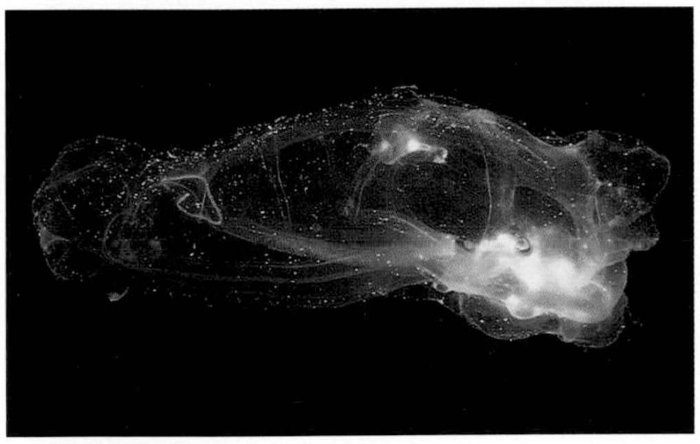

TRANSPARENT LIFE
Barrel-shaped salps have clear, jellylike bodies that are difficult for predators to see. Up to 4 in (10 cm) long, they live in swarms that can extend hundreds of miles across the ocean's surface.

Living in groups

On land, animal groups can be very large, but none rivals the size of those that can occur at sea. Fish often live in schools thousands or even millions strong, while some planktonic animals form swarms that can be over 60 miles (100 km) long.

These giant aggregations of underwater life often seem easy targets for predators. Whales gorge on krill with almost nonchalant ease, while other predators make huge inroads into schools of fish. But, in general, animals living in groups are safer than they would be alone: they are more difficult to single out and much more difficult to take by surprise.

A few animals, such as dolphins, live in sophisticated social groups. Dolphins use cholocation to locate prey, to warn each other of danger, and to organize themselves during hunting.

Migration

Many marine animals—including most of the largest whales—migrate between breeding grounds in the tropics and feeding grounds at higher latitudes. The gray whale probably travels the greatest distances: its lifetime annual migrations total up to 500,000 miles (800,000 km)—twice the distance to the moon. Tuna are also known to undertake immense journeys: some schools travel the length of the Mediterranean; in the Pacific, one fish tagged in Mexico was recovered off Japan. Turtles show very precise migration patterns, returning to the same stretch of beach year after year to lay their eggs. This is all the more remarkable because it takes 2 or 3 decades for the animal to mature: during this time, a turtle remembers precisely where it hatched so that when it is ready to breed it can make the long journey back.

On a smaller scale, many fish swim inshore to spawn, while other slow-moving animals migrate across the seabed. Caribbean spiny lobsters travel between shallow reefs, where they breed, and deeper water, where they overwinter; they set off in single file, each one following the tail of the animal in front.

Journeys like these are usually annual events. But some animals migrate daily to feed. Planktonic animals often rise to the surface at night, sinking back into the depths by day. Some of the plankton's predators copy this pattern, creating a 24-hour cycle involving many animals. These vertical migrations are clearly revealed by shipboard sonar, which shows a reflective layer rising at sunset and sinking at dawn.

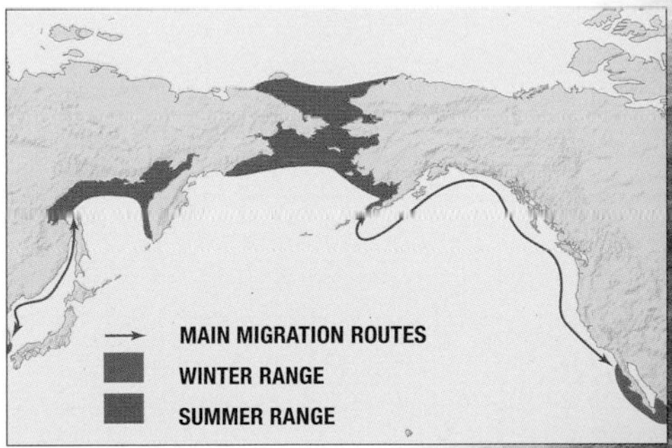

WHALES ON THE MOVE
North American gray whales make an annual return trip between their winter breeding grounds off northwest Mexico and their summer feeding grounds around the Bering Sea. A smaller number migrate between Korea and eastern Siberia.

HYDROTHERMAL VENTS

First seen in 1977, off the Galapagos Islands, hydrothermal vents are remarkable ecological oases on the deep seabed. Vents are created by volcanic activity releasing streams of intensely hot, mineral-laden water into the ocean. They are among the few habitats where life does not rely on energy from the sun—specialized bacteria use the minerals to produce energy, and vent animals either consume food created by the bacteria or eat each other. Over 300 animal species have been discovered around vents; most of these are not found elsewhere.

GHOSTLY ANIMALS
Like most vent animals, these crabs and clams—caught in the powerful lights of a submersible—have very little pigmentation, which explains their white appearance.

HABITATS

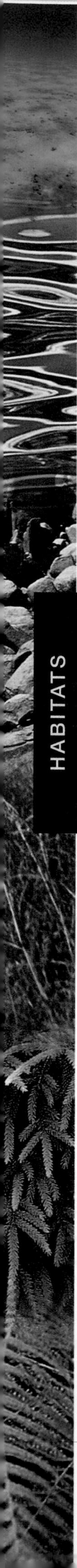

HABITATS

Coasts and coral reefs

In the natural world, the boundaries between different environments are often rich habitats for wildlife. The world's coasts are the ultimate example of this meeting of habitats because they bring together animals that live on the land and those that live in the sea. Coastal wildlife varies according to local conditions but, on rocky coasts as well as sandy ones, shore animals are associated with clear-cut zones that are usually determined by the tides. Coral reefs are a special kind of coastal habitat. Famed for their spectacular shape and color, they can grow to vast proportions and are unrivaled in the immense variety of aquatic life that they support. Reef-building corals require specific conditions, the most important being warmth and bright sunlight all year round. As a result, coral reefs are largely restricted to the tropics.

MARINE POLLUTION

With their abundant wildlife, coasts and coral reefs are easily damaged by pollution. Among the most visible pollutants are crude oil, which can be fatal to surface-dwelling animals, and nonbiodegradable plastics, which litter shallows and shores. Pollution can also affect the water invisibly, when contaminants such as industrial effluents are washed into the sea. On a global scale, marine life faces larger problems: the gradual warming and acidification of seawater, caused by increasing levels of carbon dioxide in the earth's atmosphere. In coral reefs, periodic warming sometimes triggers bleaching, which occurs when corals expel the symbiotic algae that provide them with much of their food. Corals often recover from bleaching, but acidification is a more long-term and pervasive change. It makes it harder for marine animals to extract calcium carbonate—the key mineral that they use for building skeletons and shells.

BLEACHED CORALS
Reaching upward like skeletal fingers, bleached Acropora *coral contrasts with the living coral around it. Bleaching is not necessarily fatal to corals, but it is a sign of a reef under stress—something that is becoming more common as global sea temperatures rise.*

Coasts

Some habitats look much the same from one year to the next, but the seashore is always changing. Waves pound away at rocks, undermining them and breaking them up, while coastal currents reshape the shore in a less dramatic way by moving pebbles and sand. Superimposed on this is the rhythmic movement of the tides—a twice-daily cycle that has a profound impact on seashore wildlife.

Tides vary enormously in different parts of the world. Around islands in mid-ocean, the total rise and fall is often less than 12 in (30 cm), while in deep bays and inlets on continental coasts, it can be over 33 ft (10 m). Whatever its height range, the tide divides the shore into 3 different zones, each with its own distinctive animal life.

The highest of these zones is the supralittoral, which is the part of the shore just beyond the reach of the highest tides. Although this zone is never actually submerged, it is affected by salt spray, which means that animals that are sensitive to salt—and also salt-intolerant plants—are rarely found in this zone. Below this is the littoral, a zone that is regularly covered and then exposed as the tide floods and recedes. The animals found here, such as mussels and limpets, lead a double life in that they have to be able to survive both in water and in air. The next zone, or infralittoral, is always submerged, even during the lowest tides. Most of the animals that live in this zone are fully marine, although a few leave the water to breed.

Coastal wildlife is also affected by the geology of the shore. Many animals live on rocky coasts, while others specialize in living in sand or coastal mud. Compared with these, shingle is a difficult habitat for animals, although some shoreline waders use it as a place to nest.

CLIFFS *are important nesting sites for many seabirds, although some, such as terns, nest on the ground above the high-tide line.*

SUPRALITTORA

Coral reefs

There are 2 main types of corals: hard and soft. Coral reefs are formed by hard corals. The individual coral animals, called polyps, secrete external skeletons that persist after they die. Soft corals are found all over the world, but hard, reef-building corals grow only in clear, nutrient-poor water, where their symbiotic algae can harness the energy in sunlight. Such is the richness of the habitat created by coral reefs that a huge number of animals are able to live side by side without competing for the same food.

There are 3 main types of reefs: fringing reefs, which grow close to the shore; barrier reefs, which are separated from the coast by deep channels, sometimes over 62 miles (100 km) wide; and atolls, which are ring- or horseshoe-shaped reefs that grow around oceanic islands, often where volcanoes have subsided into the sea.

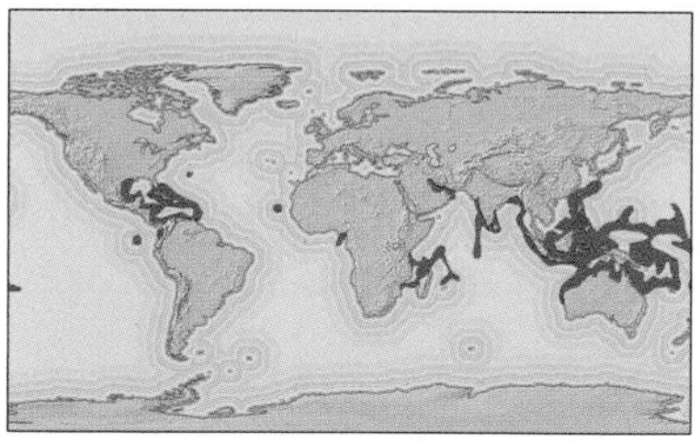

CORAL REEFS *are found in parts of the world where the sea temperature never drops below about 64°F (18°C). The world's largest reefs are in the Indo-Pacific region.*

Although every reef is unique, reefs share a common structure. Depth and exposure are 2 important factors that influence this because corals vary in their need for light and in their ability to withstand the force of the waves. The fastest-growing corals, which need lots of light and relatively calm water, typically form the central zone of the reef, projecting just above the surface at the lowest tides. On its inshore edge, the central zone is often backed by lagoons—large pools of open water lying over coral sand. By contrast, the seaward edge of the reef often forms a submarine cliff that drops steeply into the depths. The corals in this zone are solid and resilient because they have to withstand heavy breakers rolling in from the open sea.

LAGOON

RAYS, *and many other botto dwelling fish, feed on anima buried in the coral sand of reef lagoons. Most are well camouflaged.*

THE LUMPFISH *stays in place by sticking itself to submerged rocks with its suckerlike pelvic fins.*

STARFISH *are common predators in the littoral zone, particularly on rocky shores with plenty of bivalves to eat.*

BIVALVES *are found on all types of shore; some, such as mussels, fasten themselves to rocks, while others live buried in mud.*

SEALS *spend most of their time in the water but have to come ashore to breed. Some breeding colonies contain thousands of animals.*

SEASHORE WORMS, *including ragworms, hunt and scavenge on the seabed. Many species burrow, filtering food from the water.*

TURTLES *use their toothless beaks to feed on algae and the various small animals that live in all actively growing parts of reefs.*

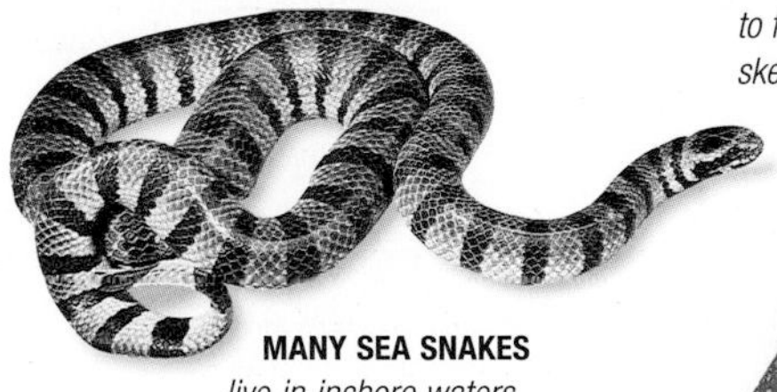

MANY SEA SNAKES *live in inshore waters. Some hunt among coral reefs, preying upon small fish and eels.*

PARROTFISH *use beaklike jaws to feed on corals. The chewed-up skeletons are excreted as sand, which builds up on the reef.*

BRANCHING CORALS, *mainly of the genus* Acropora, *are characteristic of the central zone of most reefs. Some can gain 6 in (15 cm) in height each year.*

SPONGES *filter food from water around reefs. Some are minute, but deep-water forms can reach 39 in (1 m) in height.*

Life on coasts and coral reefs

Compared with the open sea, coasts and coral reefs are disproportionately rich in marine wildlife. Rocky coasts and mudflats abound with seabirds and invertebrates, while coral reefs probably contain at least a third of the world's fish species. Conditions in coasts and reefs are variable, so the animals that use them are often extremely specialized. Some roam over large areas of the shore, but for most, a distance of just a few yards makes the difference between an ideal habitat and one in which it is impossible to survive.

Adapting to tides

For coastal animals, life is governed by the rhythm of the tides. As well as adapting to the twice-daily ebb and flow, they have to adapt to the drawn-out rhythm of extra-large spring tides, which are normally 14 days apart. Knowing where the tide stands is important because animals caught unprepared run the risk of either drying out or drowning.

Some shore animals, such as barnacles, adjust their behavior according to whether or not they are submerged. Most aquatic animals, however, are much more sophisticated, reacting to their internal biological clocks, which march in step with the tides. Even if they are taken away from the shore and placed in a tank, their tidal clocks continue to tick.

CLOSED ANEMONE OPEN ANEMONE

PREVENTING WATER LOSS
To protect themselves from drying out at low tide, many sea anemones withdraw their tentacles, turning into jellylike blobs that can be seen on coastal rocks. They often live under overhangs, which keeps them out of direct sunlight.

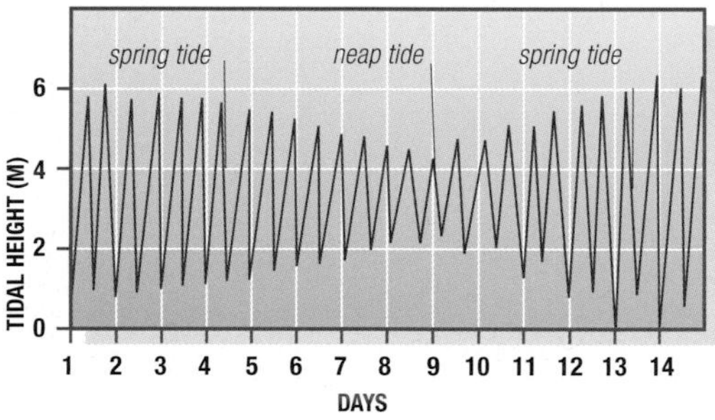

TIDAL RHYTHMS
When the sun and moon line up with the earth, extra high and low "spring" tides occur. When they are at right angles, the result is a neap tide, with minimal variation.

This built-in ability to keep time enables animals to anticipate events. For example, submerged limpets crawl over rocks to graze on microscopic algae. But since they are vulnerable away from their niche on the rock, they have to return home before the sea ebbs away. Their biological clocks tell them when to head back, ensuring that they are securely in place, and airtight, before the tide goes out. Conversely, the fiddler crab's biological clock prompts it to come out and feed at low tide and then return to its underground burrow before the incoming tide engulfs it.

MANGROVE SWAMPS

Mangrove swamps are formed by trees that are adapted for life in salty intertidal mud. Found only in the tropics and subtropics, they play an important ecological role by stabilizing the coast and providing inshore nurseries for fish and other marine animals. Most mangroves develop arching prop roots, which are exposed when the tide is out and submerged when it is in. These provide anchorage for mollusks, and act as convenient perches for mudskippers—the air-breathing fish that thrive in this habitat. Mangrove foliage is tough and leathery, and eaten mainly by insects; but the dense canopy attracts large numbers of birds, which use mangroves as breeding sites and overnight roosts. The mud in mangrove swamps is deep and often foul smelling, but it is rich in organic matter that is replenished each day by the tide.

Until the late 20th century, mangroves escaped much of the large-scale deforestation in the tropics. But, with the increase in shrimp farming, large areas of mangroves have been cut down, raising concerns about the long-term impact on coastal wildlife.

SWAMP VEGETATION
Mangroves are the only trees that can grow in mud periodically flooded by the tide. Few animals eat the trees' leaves, partly because they are often rich in salt, but these coastal forests are important breeding grounds for birds and marine animals.

Feeding on the shore

Coasts provide a much greater variety of food than the open sea. High up the shore, an abundance of animal and plant remains left in strandlines by the retreating tide are consumed by beach-hoppers, springtails, and other scavengers, which are then preyed upon by gulls and waders.

In the intertidal zone, filter feeders are common. Unlike the giant filter feeders of the open sea (see p.74), coastal ones are generally small and often spend their entire adult lives fixed in one place. Mussels and other bivalves, which filter particles of food using modified gills, are examples of these. Barnacles have a different filtering technique: despite their resemblance to mollusks, barnacles are crustaceans, with a set of feathery legs. At high tide, the legs protrude from the barnacle's case to collect food particles.

Rocky shores are home to some fast-moving swimmers as well as to many other animals that take a more leisurely approach to finding food. Starfish and sea urchins are among the slowest, crawling over rocks on hundreds of fluid-filled feet.

FEEDING AT THE SHORE
Sandpipers wade into the sea in their search for food. Like many coastal waders, they follow each wave as it retreats, and then they scuttle back up the shore as the next one breaks.

Living in mud and sand

Compared with rocky shores, coastal mud and sand seem to harbor only a limited amount of animal life. But appearances are deceptive. Mud in particular often teems with hidden life feeding on organic matter brought in by the tide.

One of the advantages of living below ground is that despite the attacks of curlews and other long-billed birds, it provides good protection from predators. The chief disadvantage is that the surface is constantly shifting, cutting off buried animals from the water above—and from oxygen and food.

Some of these buried animals have specialized body parts that enable them to connect with the surface. Clams, for example, have leathery tubes or siphons. In many species, the siphons can be retracted inside the shell, but in some, they are too long to be stowed away. Species that do not have such accessories often live in burrows. The lugworm

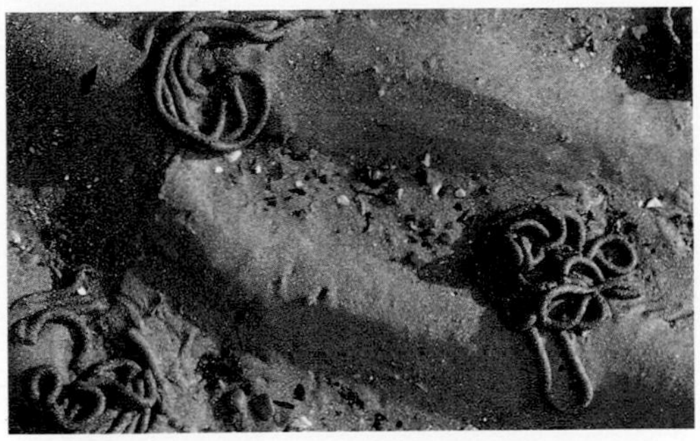

LUGWORM CASTS
Lugworm burrows have 2 holes—one for incoming water and another for ejected waste. The waste forms a pile, or "cast," that looks like muddy toothpaste squirted out of a tube.

is an example: it makes a U-shaped burrow, lining it with mucus to keep it intact so that seawater does not make it collapse.

Symbiotic partnerships

Symbiosis is a feature of life in all habitats, but it is particularly apparent on coasts. In some cases, the symbiotic partners are animals; but, in many others, one is an animal and the other is microscopic algae that live inside its body. These algae, known as zooxanthellae, can be found in thousands of coastal animals, including corals, jellyfish, and giant clams.

Zooxanthellae live by photosynthesis—the same process by which plants grow. Through a complex series of chemical reactions, they harness the energy in sunlight and use it to build up organic matter. The host animal provides the zooxanthellae with protection from the outside and, in return, the algae surrender some of their manufactured food.

These partnerships are very important to reef-building corals because they allow them to live in places where the supply of food is otherwise low. For the partnership between coral and algae to work, the corals have to encourage algae growth, which means they must live in bright sunshine near the water surface. However, this limits their upward growth because few corals can survive more than an hour of exposure at low tide.

NEW GENERATION
A giant clam expels eggs into the sea. The larvae will drift away to settle on distant parts of the reef. Giant clams live on the food generated by the symbiotic algae harboring in their fleshy lips.

MUTUAL BENEFIT
Rather than harming clownfish, the sea anemone's stinging tentacles protect them from predators, while the fish probably help keep the anemone clean. However, the partnership is not entirely equal, for while anemones can live alone, clownfish cannot.

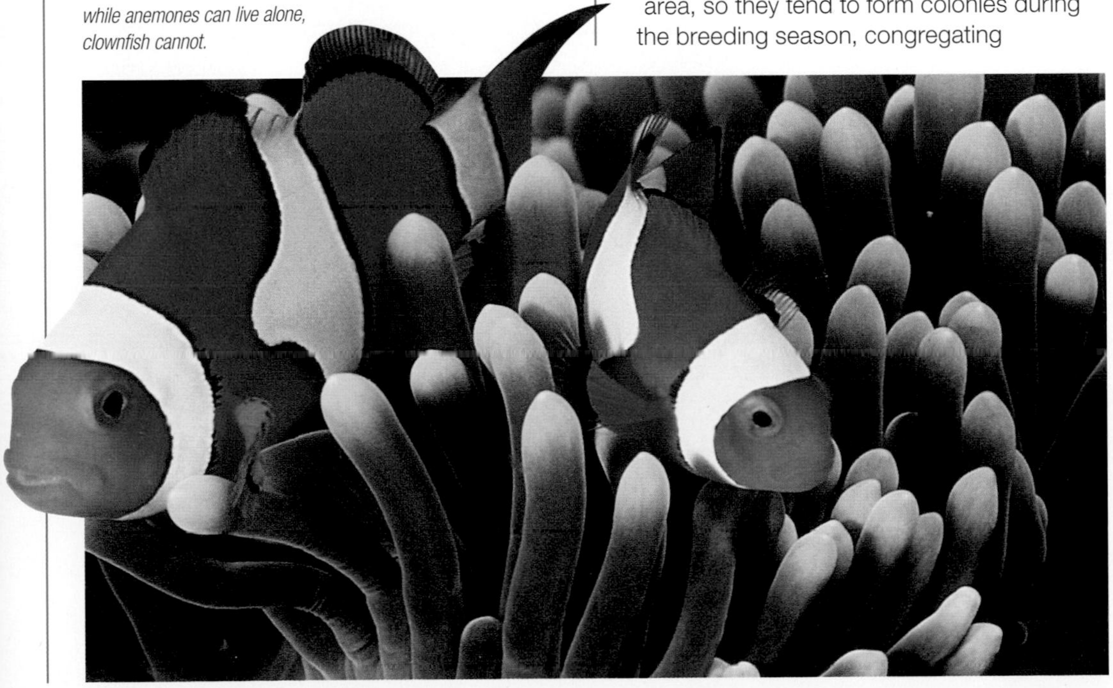

Reproduction

Many land animals have adapted to life at sea. Some are now fully marine, but others—such as turtles, some sea snakes, and seals and their relatives—must come ashore to breed. These animals are often scattered over a wide area, so they tend to form colonies during the breeding season, congregating in the same place each year to maximize their chances of finding a mate. Many coastal invertebrates, on the other hand, spend their entire adult lives in one place. For them, reproduction is an opportunity not only to multiply but also to disperse. Their eggs hatch into planktonic larvae, which may then drift long distances in coastal currents before eventually settling down. For some species, such as barnacles, choosing a home is an irrevocable decision because a larva cannot detach itself once it has settled. Chemical cues help it to "make up its mind" before it takes this momentous step.

Some fish come inshore to breed because the shoreline offers plenty of hiding places for their eggs. An extreme example of such a fish is the California grunion, which lays its eggs not in water but in damp sand on beaches. Grunions stage mass spawnings during high spring tides at night. At the next spring tide, the eggs hatch and the young are washed into the sea.

A LIFE IN THE SEA
The sea otter is the only otter to live entirely in water: it gives birth in the sea, and floats upside down to suckle its young. This female is swimming with her cub on her chest.

HABITATS

Urban areas

Two hundred years ago, only about 3 percent of the world's population lived in cities. Today, the figure is rapidly approaching 50 percent even in previously rural countries, and the human population has increased nearly sevenfold. This phenomenal growth in urban living has transformed large areas of the planet. It has created a wide range of artificial habitats—both in and out of doors—that animals can use as their homes, as well as vast amounts of waste that form the basis of animal food chains. As a result, there is a wealth of wildlife living with and among us.

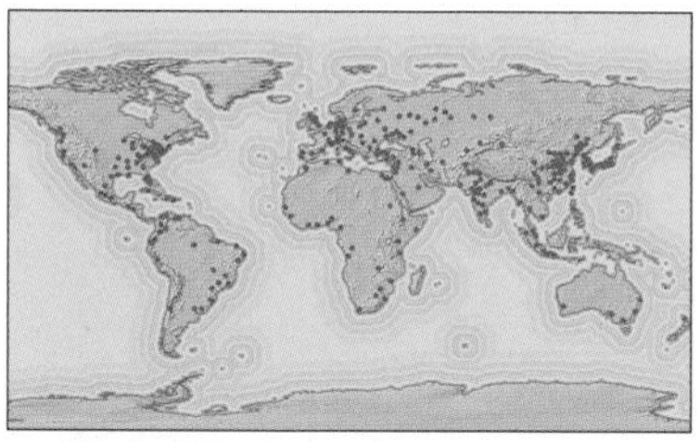

CITIES *that have a population of more than one million people are spread throughout the world. There are more than 250 of these cities, and the total is rapidly increasing.*

Outdoors

For animals that can deal with disturbance, cities and towns can be good places to live. They have plenty of suitable places for sheltering or raising young, from trees and window ledges to underground passageways, and for omnivorous species, they provide a constant supply of leftover food. In winter, the artificial heat that escapes from buildings offers additional benefit. Even better, cities are relatively safe: cats and dogs aside, they are free of many of the predators that animals would encounter in their natural homes.

Animals have adapted to urban expansion with different degrees of success. Some species are never found in cities, always retreating as the concrete advances. Others, such as migrating birds and insects, are occasional visitors, touching down briefly before moving on. More adaptable animals—raccoons and red foxes, for example—are equally at home in town or country, and treat built-up areas as extensions of their natural habitat. True urban specialists, such as the ubiquitous feral pigeon and the house sparrow, are now so fully adapted to city living that they are rarely seen anywhere else.

While feral pigeons can survive in the busiest city centers, many urban animals are found chiefly in parks and gardens—the small-scale versions of their habitats in the wild. These animals vary from one part of the world to another, but they include tree-dwelling mammals, such as squirrels and opossums, and a wide range of birds. The spread of suburbia is normally a threat to wildlife, but for these species it can actually be a help because it creates a patchwork of suitable habitats, sometimes with the bonus of food hand outs.

FERAL PIGEONS, *a familiar sight in the city, are descended from domestic birds that escaped into the wild.*

THE HOUSE SPARROW'S *exploitation of urban areas has enabled it to spread from Europe, Africa, and Asia, to other parts of the world.*

THE BRUSH-TAILED POSSUM *thrives in gardens and parks, and sometimes breeds in buildings.*

MANY BUTTERFLIES *breed in urban areas, but the Camberwell beauty is only a visitor in search of nectar and fallen fruit.*

THE RED FOX *is active at night, reducing the risk of its being disturbed while foraging.*

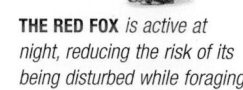

THE BROWN RAT'S *burrowing lifestyle enables it to live close to people without being easily seen.*

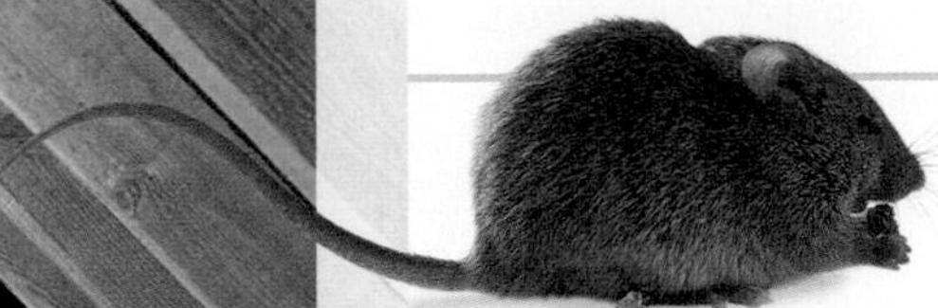

THE HOUSE MOUSE, *originally from central Asia, has spread all over the world. Other household rodents include rats and some species of dormice.*

FOR GECKOS, *house walls make good nocturnal hunting grounds—especially if there are lights nearby to attract flying insects.*

HOUSEFLIES *enter buildings to find food, but some other species—such as the cluster fly—use attics and undisturbed empty rooms for hibernation.*

COCKROACHES *can become a serious pest in warm parts of the world and in houses with central heating.*

SILVERFISH *are widespread but largely harmless. They feed nocturnally on all kinds of starchy substances, from flour to wallpaper paste.*

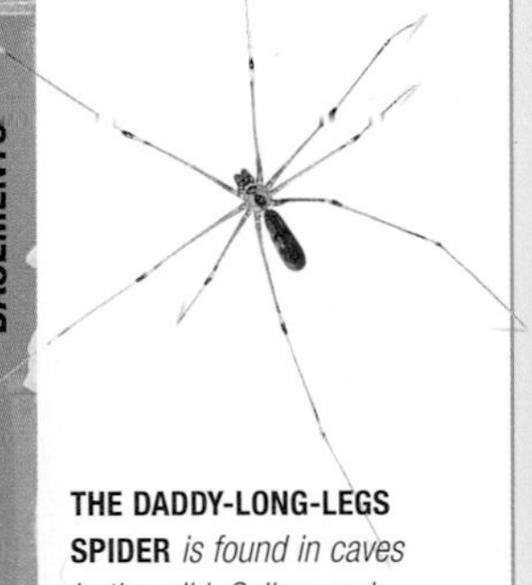

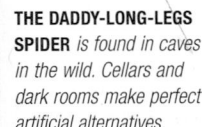

THE DADDY-LONG-LEGS SPIDER *is found in caves in the wild. Cellars and dark rooms make perfect artificial alternatives.*

Indoors

In the natural world, many animals inadvertently create habitats for other species when they build their nests. Humans do exactly the same. However, because our "nests" are so extensive and complex, they can host an exceptionally wide range of animal life. Much of it is harmless, but some can cause problems or at least inconvenience.

Most indoor animals are small and nocturnal, which helps them to avoid being noticed by their human hosts. This is especially true of species that share daily living areas and that scavenge leftover food. Silverfish, for example, emerge after dark to search for flour and other starchy produce, scuttling away if cupboards or drawers are suddenly opened, exposing them to the light. Cockroaches behave in a similar way, but they are more of a nuisance because they spread disease. At dawn, nocturnal animals hide away, leaving the day shift to take over. Houseflies, for example, are most active during the day because they navigate by sight. In basements and attics, wildlife is less affected by the cycle of light and dark, and it is less frequently disturbed by human comings and goings. For wild animals, attics resemble extra-large tree holes, while basements resemble caves. Wasps, birds, and house mice will all nest in attics—if they can get in—and they sometimes share this habitat with roosting bats. Basements and cellars provide a haven for spiders, which can survive for long periods without food and, in many cases, catch their prey in total darkness.

The advent of central heating has been an important factor in the increase in the number of animals that choose to share our homes. For example, cockroaches, which were originally found mainly in warm parts of the world, are now widespread in cooler regions. Soft furnishings and carpeting also play a part: as well as helping keep the home warm, they provide hiding places and nesting material for various animals.

URBAN HAZARDS

Urban animals face a form of "unnatural selection" in that any species that are not suited to city life are ruthlessly weeded out, regardless of how successful they would be in the wild. Shortage of space, combined with noise and pollution prevent many animals from establishing themselves in cities, while other environmental factors, such as bright street lighting, deter many more. Such are the hazards of city life that even experienced, and otherwise successful, urban animals sometimes succumb to the dangers. Traffic, of course, is a major hazard that claims the lives of thousands of animals (including humans) every day.

PARTING OF THE WAYS *Urban traffic is a major hazard to animals, but roads are just as much of a threat in rural areas. They divide natural habitats, separating animal populations that normally interbreed.*

Life in urban areas

Animals have had millions of years to adapt to earth's natural habitats, but only a fraction of that time to adjust to life in cities. Despite this, animals are never far away in built-up places. Their success is due mainly to "preadaptation"—characteristics that evolved to suit one way of life, or habitat, but that accidentally turn out to be useful for another. Thus, some animals thrive in man-made habitats that resemble the ones they would use in nature. Others succeed because they are highly adaptable and can exploit the opportunities that we inadvertently provide.

Feeding

Some outdoor urban animals live on the same foods that they eat in the wild, but for scavenging species—such as raccoons, foxes, and pigeons—the daily fare is often very different than that of their natural homes. These versatile creatures will try any kind of leftover food, however unfamiliar it looks and smells, and this highly opportunistic streak is the secret of their great success. Modern food packaging can sometimes present problems, but they quickly learn how to tear or peck away at plastic and paper to get at the edible contents within.

Indoor animals get their food from one of 3 sources: the things we eat, the animals that eat those things, and the fabric of our homes. The first category contains a wide range of household pests, such as rats, mice, houseflies, and cockroaches; the second category consists chiefly of spiders, but also centipedes and geckos in warm parts of the world. Spiders are almost perfectly adapted to indoor life, and although widely disliked, they make a positive contribution by keeping indoor insect numbers in check. Animals in the third category are the least welcome of these uninvited guests. They include wood eaters such as termites and beetles, as well as insects that attack other organic materials, such as wool. In many parts of the world, these animals are serious pests.

Light and warmth

In cities, streetlights light up the night sky, while heat from buildings and traffic makes them far warmer than nearby countryside. Artificial lights confuse insect navigation systems and interfere with birds' biological clocks. As a result, songbirds sometimes sing late at night, and some species start building nests in winter, convinced by the bright light that it is spring. Extra warmth is appreciated by a range of animals, from butterflies to birds. In some regions, starlings commute into cities on winter afternoons to roost on buildings where they are relatively warm.

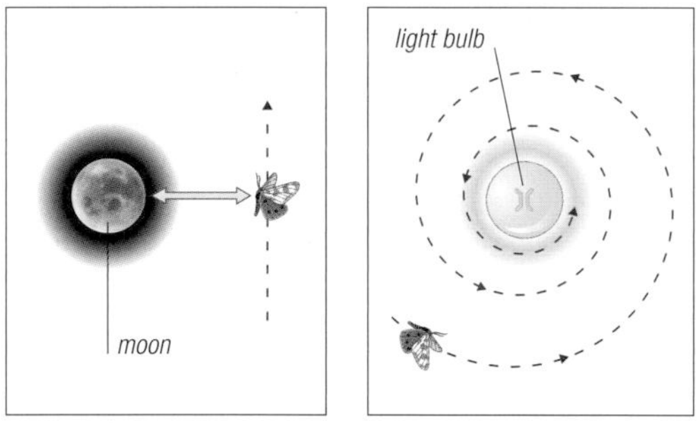

DRAWN TO THE LIGHT
Moths navigate by maintaining a set angle to the moon, which, because it is so far away, works like a compass, enabling the moth to follow a straight line. With closer lights, such as street lamps, the "compass" appears to the moth to drift. The moth adjusts its path accordingly, and ends up spiraling inward.

DAWN RAID
Alert for signs of danger, and using its keen sense of smell and highly dexterous front paws, a raccoon sorts through trash from a container that it has overturned.

INTRODUCED SPECIES

Species that have set up home in unfamiliar parts of the world are plentiful in urban areas. Some have been introduced deliberately, while others arrive with imported food. Pets, also, are sometimes released, or manage to escape, into the wild. Some of these animals remain urban, but a few—such as the starling in North America—have gone on to colonize entire continents.

FAR FROM HOME
The rose-ringed parakeet from tropical Africa and southern Asia is good at dealing with cold, so it can thrive in parts of North America and Europe.

Rooftop animals

For birds and bats, the tops of buildings can make ideal homes. High above the ground, and relatively undisturbed by people, animals feed and breed undisturbed. Some species nest in attics or under eaves, while others favor the tops of chimneys. Swifts, swallows, and martins are foremost among rooftop dwellers and are prime examples of preadaptation at work: they naturally nest on cliffs or in crevices, but the rapid spread of towns and cities has provided alternatives that have enabled them to extend their range to places where they would otherwise be rare.

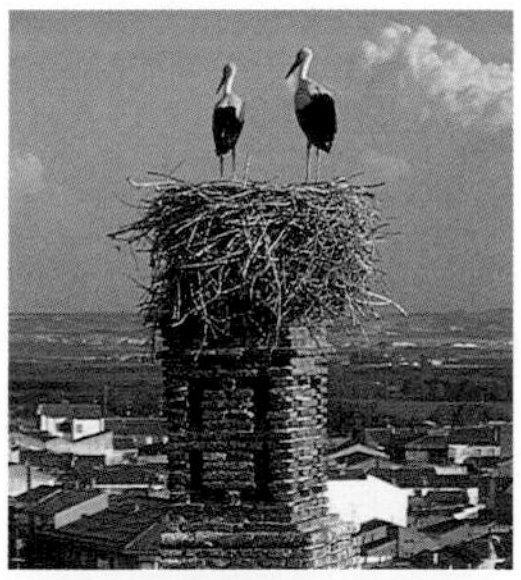

LIFE ON THE ROOFTOPS
A pair of European white storks look down on a town from their chimney-top nest. Although they often nest on buildings, they feed in fields. They were once common urban visitors in northern Europe, but changing agricultural practices have reduced their numbers.

STRUCTURAL PESTS

Since humans first started to build wood houses, wood-eating insects have been a problem. Wood-eating beetles are common in temperate regions, while termites attack timber in the tropics. At one time, little could be done to control severe attacks; today, insecticides are effective at keeping them in check. Even so, many of these animals, especially wood-boring beetles, are far more widespread than they were because they have been exported in the lumber shipped all over the world.

EATEN FROM INSIDE
Deathwatch beetle larvae chew tunnels through old lumber. During the 5 years they take to turn into adults, they can weaken lumber so much that it eventually collapses. A related species attacks furniture wood.

URBAN HUNTER
The peregrine falcon is one of the most adaptable of all birds of prey. It feeds mainly on other birds, which it usually catches in midair. City rooftops and ledges provide an excellent vantage point for the falcon to pick out its quarry before launching into a high-speed downward dive, known as a stoop.

THE ANIMAL KINGDOM

Almost two million species of animals have so far been identified—and it is thought that the true number may be several times greater than this. Members of the animal kingdom range from invertebrates that are too small to be seen

MAMMALS

MAMMALS

PHYLUM	Chordata
CLASS	Mammalia
ORDERS	29
FAMILIES	220
SPECIES	5,398

Mammals, the most familiar group of vertebrates, all nourish their young on milk produced by the female's mammary glands (the unique skin structures after which the class is named). Most also give birth to live young and, with only a few exceptions, have a covering of hair on their body. Mammals are most widespread and diverse on land, but they have also colonized air and water. Their success is largely due to their ability to maintain a constant internal body temperature, regardless of changing external conditions. They are also highly adaptable and often modify their behavior to suit changing circumstances. Some mammals, especially primates (the group that includes humans), form complex societies.

YOUNG MAMMALS
One of the key features that sets mammals apart from other animals is the way that females suckle their young on milk from their mammary glands. Some mammals look after their young until they are several years old.

Evolution

The ancestors of mammals were a group of reptiles known as the therapsids. These small, active carnivores lived during the Triassic Period (which began 251 million years ago). The various features that now separate reptiles and mammals in fact developed over a long period of time and at different rates. In addition to several important differences in skull shape (see panel, left), the therapsids evolved a lighter, more flexible skeleton. Another significant change was the alignment of the limbs beneath rather than at the sides of the body, which helped the early mammals become faster and more agile.

The transition from reptiles to mammals was completed toward the end of the Triassic Period (195 million years ago). This coincided with the rise of the dinosaurs, the reptiles that became the dominant animals on earth during the Mesozoic Era (251–65 million years ago) and threatened the mammals with extinction. The mammals' ability to regulate their internal temperature may explain why they outlasted the dinosaurs. During the Mesozoic Era, the climate became cooler and daily temperatures were more variable. Like other reptiles, the bodies of most species of dinosaurs would have tended to shut down in low temperatures. Mammals, however, were unaffected by these fluctuations in temperature and so were better able to survive the change in climate.

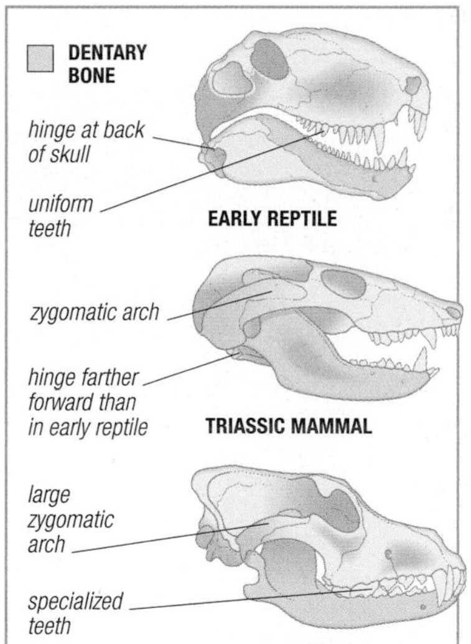

EVOLVING SKULL

Mammals' reptilian ancestors had uniform teeth and a jaw that was made of several bones and limited to an up-and-down movement. By the Triassic Period, mammal skulls had both a zygomatic arch to house powerful jaw muscles and only one jawbone (the dentary). Modern skulls have specialized teeth and a single jawbone often capable of complex movements. These features, in combination with strong jaw muscles, allow mammals to chew their food—a unique feature of the group.

JAWS
Mammals have a unique jaw, in which the lower jaw is hinged directly to the skull. The hippopotamus has an unusually large mouth and a wide jaw extension.

Anatomy

Several differences between the skeletons of mammals and other vertebrates can be seen in the skull. Mammals are unique in having a lower jaw that is hinged directly to the rest of the skull—in other vertebrates the connection is indirect, with at least one other bone between the 2 parts of the jaw. This direct articulation, and the fact that the lower jaw is made of a single bone (see panel, left), makes the jaw a powerful tool for cutting and dismembering food. Unlike other vertebrates, mammals also have a set of specialized teeth, which reflects and facilitates their varied diet. All mammals have 3 types of teeth: incisors, which are used for biting; canines for gripping and tearing; and cheek teeth (molars and premolars), which are used for grinding. The shape and size of each type varies according to diet.

Most mammals, unlike other animals, have mobile external ears (pinnae) to locate sounds and then funnel them to the eardrum, where

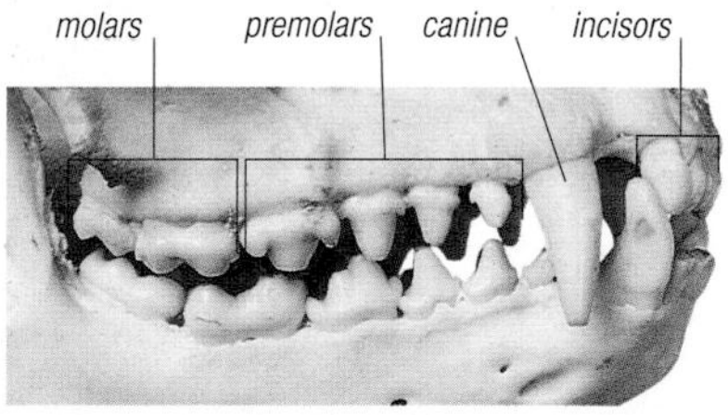

TEETH OF A RACCOON

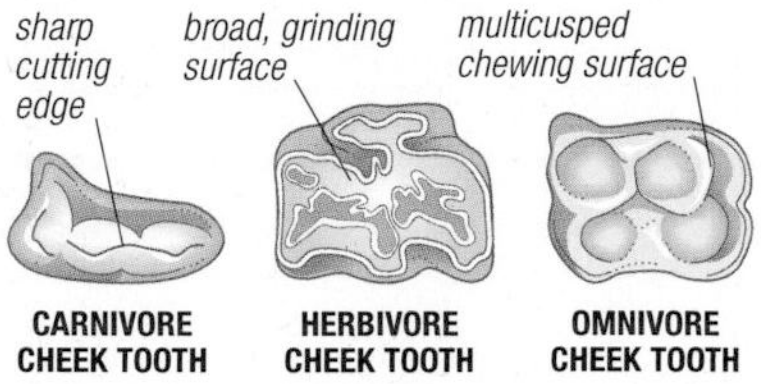

TEETH IN MAMMALS
Many mammals, such as raccoons (top), have incisors, canines, and premolars and molars (cheek teeth) that reflect their diet. Carnivores have sharp cheek teeth for cutting bones, while herbivores have broad cheek teeth to grind up vegetation. Omnivore cheek teeth are broad and multicusped for chewing a variety of foods.

HAIR

Mammals are the only animals with a covering of hair on the body. A hair consists of a rod of cells strengthened by the protein keratin. Hair can take several forms, including whiskers, spines, prickles, and even horns (as in rhinoceroses). Most common is fur, which usually consists of an insulating layer of underfur and a projecting guard coat that protects the skin and gives hair its color (which may aid camouflage). Hairs such as whiskers may also have a sensory function.

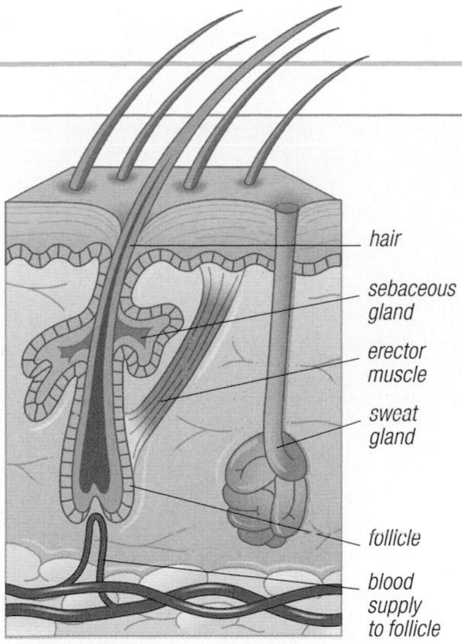

SKIN SECTION
Each hair arises from a pocket in the skin called the follicle. Next to the follicle is an erector muscle that raises or lowers the hair, changing the insulating properties of the coat.

MODIFIED HAIR
Porcupines have whiskers, long body hairs, and defensive spines (which are modified hairs).

they are transmitted to the inner ear by 3 tiny bones, and so to the brain. The fennec fox has enormous, sensitive pinnae; by contrast, true seals have lost their pinnae.

One of the most distinctive parts of a mammal's body is its skin. This consists of 2 layers: a protective outer layer of dead cells (the epidermis) and an inner layer (the dermis) that contains blood vessels, nerve-endings, and glands. It is the glands in the dermis that are particularly unusual: the sebaceous (or scent) glands secrete chemicals that mammals use to communicate with one another, the mammary glands produce the milk used to nourish newborn young, and the sweat glands—together with the hair, which also arises from the dermis (see panel, above)—play an important part in regulating temperature.

Reproduction

Depending on the way they reproduce, mammals are divided into 3 groups—the Prototheria, Marsupialia or Metatheria, and Eutheria. In all 3, fertilization is internal (see p.31). The monotremes that form the infraclass Protheria, lay eggs. Members of the other 2 groups have live young. Mammals of the infraclass Marsupialia have no true placenta (see below) and produce young at a very early stage of their development. In some species, the young are kept in a pouch on the outside of the mother's body until they are more fully grown. The largest reproductive group is the infraclass Eutheria, which contains the placental mammals. The unborn young develop in the mother's uterus. During pregnancy, food and oxygen pass from mother to fetus through an organ known as the placenta, while waste substances move in the opposite direction. When born, infant placental mammals are more highly developed than those of marsupials.

The juveniles of all mammals are fed on milk secreted by the mother's mammary glands, which become active after the young are born. Except in monotremes, the milk is delivered through teats. As well as providing nourishment (it is rich in proteins and fats), the milk contains antibodies that help establish resistance to infection. Being nourished on milk during the early weeks of life also means that young mammals do not have to forage for their own food, which greatly increases their chances of survival.

Litter sizes vary from 20 (Virginia opossum) to only 1 (orangutan); and gestation periods range from 12 days (short-nosed bandicoot) to 22 months (African elephant).

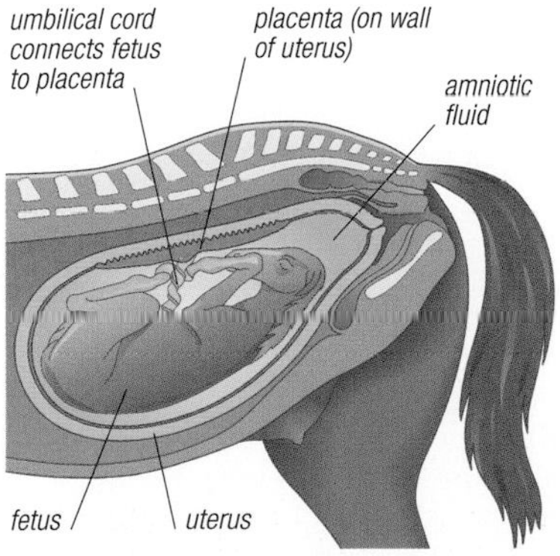

PLACENTAL MAMMALS
The fertilized egg of a placental mammal, such as a horse, divides many times and eventually becomes a fetus. As the fetus grows, the uterus expands in size and weight. In the horse, pregnancy lasts about 11 months.

Temperature control

Mammals, like birds, are endothermic, meaning that they maintain a constant body temperature and can therefore remain active at extremely high or low external temperatures. This is why mammals are able to occupy every major habitat and are more widespread than any other vertebrates (except birds). Many species, such as the seals and whales of the Antarctic, live in regions where the temperature is well below freezing for much or all of the year. An area of the brain known as the hypothalamus monitors body temperature and adjusts it if necessary. Body temperature can be altered by increasing or decreasing the metabolic rate, widening or constricting blood vessels that carry heat to the skin's surface, raising or lowering body hair to trap or release an insulating layer of air, and causing shivering (to gain heat) or evaporation via sweating or panting (to lose heat). Mammals can also control their body temperature by adopting special body postures: a monkey, for example, will hunch up in the cold (and many mammals huddle together in small groups to keep warm); lemurs warm up in the early morning sun by sitting up and spreading their arms to expose their thinly haired undersurface.

Behavioral patterns also help regulate body temperature. For

STAYING WARM
Large mammals, such as red deer, that do not hibernate in winter, survive and keep warm by using fat reserves built up during the summer. Over time, this leads to weight loss.

KEEPING COOL
In hot climates, mammals avoid overheating by resting in the hottest part of the day. They may also keep cool by panting, as this puma is doing. Panting helps lower the body temperature through the evaporation of water from internal surfaces, such as the tongue.

HIBERNATION

Some mammals, especially small species, conserve energy during the cold months by hibernating, just as some reptiles (for example, the garter snake) do. Their temperature falls, their breathing slows, their metabolism drops to almost imperceptible levels, and they fast, drawing on stored fat reserves. When hibernating, the animal is torpid and difficult to rouse. A West European hedgehog, for example, begins hibernation when the outside temperature falls below 59° F (15° C), and in midwinter its body temperature lowers to about 43° F (6° C). In some bats, rectal temperatures of 32° F (0° C) have been recorded during hibernation. Larger mammals, such as the American black bear, do not truly hibernate—they sleep and become cold but are roused more easily. Related to hibernation is estivation, which is torpidity during the summer. Like hibernation, estivation saves energy when food is short.

COLD SLEEP
Many bats that live in temperate regions hibernate over the winter. While sleeping, the temperature of their body falls to that of the roost site, as shown by the dew on this Daubenton's bat, right.

WINTER HIDEAWAY
For hedgehogs, effective camouflage is an important part of winter hibernation.

AQUATIC MAMMALS

Three groups of mammals have adapted to aquatic life, developing a streamlined body and the ability to stay underwater for long periods (although all of them return to the surface to breathe). The largest group, the whales and dolphins, are the most specialized. They have lost their body hair and spend their entire life in water. Like whales, seals and sea lions rely on subcutaneous fat to keep warm, but have retained their fur, which is kept waterproof by an oily secretion from the sebaceous glands. Sirenians (the manatees and dugong) live in warm coastal waters and estuaries and are the only herbivorous aquatic mammals. The sea otter is the only other mammal that spends most of its life in water. It lacks subcutaneous fat and instead keeps itself warm by trapping air in its dense fur.

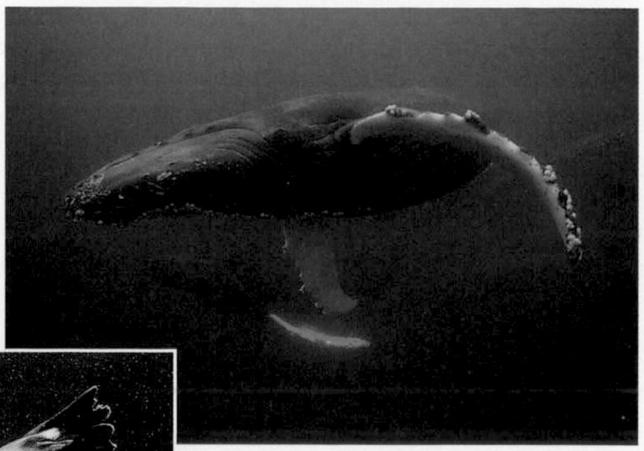

LIVING IN THE SEA
Aquatic mammals, such as this humpback whale, evolved from land-living mammals. Although the back limbs have been lost, some whales retain a vestigial pelvic girdle.

GRACEFUL MOVERS
Seals, such as this harbor seal, are highly acrobatic swimmers, with front and back limbs that have been modified into webbed flippers. Ungainly on land, they rarely venture far from the water, even to give birth to their young.

example, in the desert and in tropical grasslands, in the heat of the day rodents retreat into cool, moist burrows, while larger mammals rest in the shade in cool depressions in the ground. Body coloring is another important factor: dark colors absorb heat; light colors reflect it. Desert mammals are therefore often light fawn in color, while those living in cool climates are dark. This arrangement may conflict with the need for camouflage—where there is winter snow, mammals may turn white (stoats) or have a permanently white coloration (polar bears), and possess extra-thick fur to compensate.

Feeding

To maintain a high, constant body temperature is "energy expensive"—mammals therefore need a nutritious and plentiful diet. While the earliest mammals were probably predators, different species have since adapted to meet their dietary requirements in a variety of ways. Some eat animal prey—this is a carnivorous diet (and includes insect-eating). Other mammals, called herbivores, eat plants. A herbivorous diet includes sub-types such as fruit-eating and grass-eating. An omnivore eats both animal prey and plants.

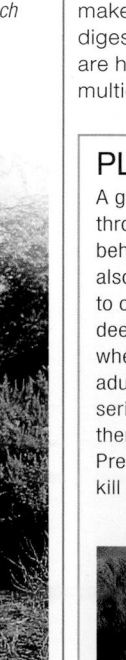

Carnivorous mammals have a simple digestive tract, because the proteins, lipids, and minerals found in meat require little in the way of specialized digestion. Plants, on the other hand, contain complex carbohydrates, such as cellulose. The digestive tract of a herbivore is therefore host to bacteria that ferment these substances and make them available for digestion. The bacteria are housed either in a multichambered stomach or in a large cecum (a blind-ending sac in the large intestine). The size of an animal is a factor in determining diet type. Small mammals have a high ratio of heat-losing surface area to heat-generating volume, which means that they tend to have high energy requirements and a high metabolic rate. Because they cannot tolerate the slow, complex digestive process of a herbivore, mammals that weigh less than about 18 oz (500 g) are mostly insectivorous. Larger mammals, on the other hand, generate more heat and less of this heat is lost. They can therefore tolerate either a slower collection process (those that prey on large vertebrates) or a slower digestive process (herbivores). Furthermore, mammals that weigh more than about 18 oz (500 g) usually cannot collect enough insects during their waking hours to sustain themselves. The only large insectivorous mammals are those that feed on huge quantities of colonial insects (ants or termites).

FEEDING HIGH UP
Mammals avoid competing with one another by eating different foods or by getting the same food in different ways. For example, the long neck of a giraffe allows it to feed at a height beyond the reach of other herbivores.

Social structures

Mammals communicate socially by scent, either from glands (which may be located in the face, in the feet, or in the groin) or in their urine or feces (which contain sexual hormones).

FEAR SUBMISSION EXCITEMENT

FACIAL EXPRESSION
Some mammals communicate using facial expressions. This ability is well developed in primates, such as the chimpanzee. Fear is shown by baring the teeth, submission by a pouting smile, and excitement by exposed teeth and an open mouth.

PLAY

A great deal of a young mammal's learning occurs through play, when infants experiment with adult behaviors, such as fighting and hunting. Play can also take the form of exploring and displaying to one another. Hoofed mammals, such as deer, establish dominance rankings when young to avoid conflict as adults. This reduces the risk of serious injury that would make them vulnerable to predators. Predators must learn to stalk and kill prey to survive as adults.

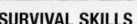

SURVIVAL SKILLS
Potential predators, such as these lion cubs, use play to practice pouncing, biting, and raking with the back feet, stopping before real damage is done.

FINDING THEIR PLACE
Like many herbivores, these elephant calves use play to establish their rank in the herd. Young elephants also use their trunks to examine novel objects.

INTELLIGENCE

In mammals thought to be intelligent—such as primates—the "thinking" part of the brain, the cerebrum, is larger relative to the rest of the brain. Intelligence itself is not easy to define, but indications of intelligence include the ability to learn, matched with behavioral flexibility. Rats, for example, are considered to be highly intelligent because they can learn to perform new tasks, an ability that may be important when they first colonize a fresh habitat. In some mammals, food-gathering appears to be related to intelligence: a deer feeding on plants has a brain relatively smaller than a cat that must "think" to outwit its prey.

RAT IN A MAZE
Animal intelligence is often assessed by a creature's ability to complete simple tasks, usually using a food incentive. This albino brown rat is finding its way through a maze.

FOOD FOR THOUGHT
Primates are rare among mammals in that some can use a variety of objects as tools, while others clean their food prior to eating it. This Japanese macaque has learned to separate wheat grains from sand by immersing the food in water. The sand sinks; the wheat grains float on the surface.

They also communicate by body posture or facial gesture, by touch, and by sound, which may develop into complex messages.

Socialization begins at birth, when signals between a mother and her infant facilitate suckling. The process continues throughout the juvenile period, when the young interact with each other through play (see panel, opposite below) and gradually learn to interpret the behavior of adults.

Some mammals only interact to mate or to nurture young, but many form temporary or permanent social groupings beyond these minimal needs. The hoofed mammals, for example, form herds for safety, because predators are more likely to be spotted.

BENEFITS OF COMMUNAL LIVING
Living in a group has several advantages for small mammals, such as these dwarf mongooses. For example, there is more chance of spotting predators, the burden of rearing young can be shared, and the territory can be defended collectively.

Interwoven with social behavior is the use of space: most mammals have a home range (the area within which an animal, or group, performs most of its activities), and in some these areas are developed into territories (an area that the animal, or group, defends against others of the same species).

Solitary mammals, such as aardvarks and most cats, usually occupy a territory that they defend against members of the same sex. A tigress, for example, will not permit other females to enter her territory, whereas the territory of a male tiger overlaps those of several females.

Some mammals, such as gibbons, live in monogamous pairs (with their immature offspring), and each pair occupies and defends a territory. However, recent studies indicate that the assumption of "lifelong fidelity" is wrong: surreptitious copulations outside the pair are frequent, and many pairs break up and find new mates after some years.

Other mammals form larger social groups, of varied composition. In some species, such as seals and elephants, the sexes live separately for most of the year. Related females form social units, and the males either live alone or form small bachelor groups. In mammals living under this system there is ferocious competition among the males for the right to mate. The most successful males are usually the largest and strongest individuals that also have the best weapons (antlers, horns, or tusks). In elephant seals and sea lions, as the females become sexually receptive, the males bellow and display to establish dominance.

Other species, such as Burchell's zebras, form small, permanent harem groups, in which one male leads and mates with a group of females. The subordinate males are relegated to bachelor bands unless one can successfully challenge and expel a harem male.

The final type of larger social group exhibited by mammals is one in which the collection of individuals consists of several males and females. Such societies are restricted to primates and the social carnivores (although some other species, such as migrating wildebeests, form temporary mixed-sex groups). Among baboons, for example, several adult males join a troop of females related to each other (but not to the males) and compete for dominance and the right to mate with the females. These power struggles result in a continual change in the dominance hierarchy. In lion prides, the males are related to each other and cooperate in the defense of the females rather than competing for them. In wolf packs, only the dominant pair breeds; the other adults are the offspring from previous years who, instead of leaving to form their own packs, stay with their parents and help rear their younger siblings.

Locomotion

Mammals dwell in a huge variety of habitats and have accordingly developed a number of different ways of moving around. While a quadrupedal (4-footed) gait is the most common form of mammal locomotion, some species, such as kangaroos, have a bipedal (2-footed) gait. As mammals move, the way in which the foot comes into contact with the ground varies in one of 3 different ways (see below). Locomotion is also related to the lifestyle of the animal. Predators, which require short, explosive bursts of speed to catch their prey, have a flexible spine. For example, in the cheetah—the fastest terrestrial animal—the backbone coils and uncoils with every stride, propeling the animal forward. It has been estimated that these movements increase the cheetah's speed by about 19 mph (30 kph). In contrast, prey species, such as gazelles, rely on endurance to outrun predators. They have a rigid back, and the "energy cost" of their movement is reduced as a result of a lengthening of the lower segments of the limbs and a concentration of the muscles close to the body.

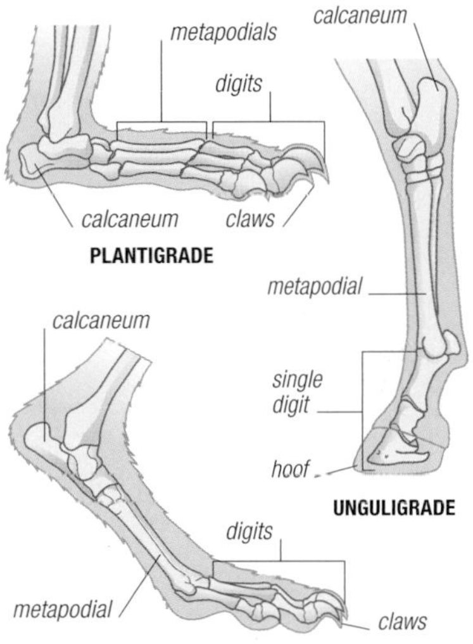

WALKING GAITS
When moving, plantigrade mammals, such as bears, keep the calcaneum (heel bone), metapodials, and digits (toes) of each foot on the ground. Digitigrade mammals, such as dogs, move with only the digits touching the ground. Unguligrade mammals, such as horses, walk and run on the tip of each digit.

AERIAL MAMMAL: LONG-EARED BAT

ARBOREAL MAMMAL: BLACK LEMUR

TERRESTRIAL MAMMAL: SNOW LEOPARD

AQUATIC MAMMAL: DOLPHIN

ADAPTABLE MAMMALS
Over time, the limbs of some mammals have evolved into wings, flippers, or grasping hands and feet, enabling bats to take to the air, whales and dolphins to adopt an aquatic existence, and primates to move freely through the trees. Predatory land mammals, such as cats, have developed a flexible spine for speed and maneuverability, and a long tail for balance.

Egg-laying mammals

PHYLUM	Chordata
CLASS	Mammalia
ORDER	Monotremata
FAMILIES	2
SPECIES	5

Also known as monotremes, this group comprises the duck-billed platypus and the echidnas. Monotremes are the only mammals that lay eggs. The duck-billed platypus has a ducklike bill, a beaverlike tail, and webbed feet. Echidnas have a tubular beak, huge claws for digging, and spines. The duck-billed platypus lives in freshwater in eastern Australia; echidnas exist in varied habitats in Australia and New Guinea.

AQUATIC LIFE
The duck-billed platypus is a strong swimmer. Its coat is waterproof and has a dense underfur for insulation. When diving, the eyes and ears are closed.

Anatomy

Monotremes are short-legged animals with a small head and tiny eyes. In all species, the digestive, urinary, and reproductive tracts empty into a common chamber (the cloaca), which terminates in a single exit ("monotreme" means "one hole").

DIFFERENT SNOUTS
The platypus (top) has a flattened bill, covered with sensitive skin. Echidnas, however, have a cylindrical beak. In both, the snout is used to probe for and locate food.

WEBBED FEET
In water, the duck-billed platypus is propelled by powerful thrusts of its fully webbed front feet.

Feeding

The platypus uses its sensitive bill to probe the beds of rivers or lakes for crustaceans and insect larvae. The short-nosed echidna digs with its powerful claws for ants, termites, or earthworms; long-nosed echidnas mainly eat earthworms. Adult monotremes lack teeth: food is ground between plates or spines in the mouth.

STICKY CATCH
The short-nosed echidna uses its long, slender tongue to catch termites and ants. A sticky coating of saliva ensures that the insects do not escape.

Tachyglossus aculeatus

Short-beaked echidna

Location Australia (including Tasmania), New Guinea

Length 12–18 in (30–45 cm)
Tail ⅜ in (1 cm)
Weight 5½–15 lb (2.5–7 kg)
Social unit Individual
Status Least concern

The spines of this species, also called the spiny anteater, are longer than the fur between them. Active both day and night, this echidna is solitary and can become torpid in very cold or hot weather, when its temperature falls from the normal 88–92° F (31–33° C) to as low as 39° F (4° C). It eats a variety of ants, termites, grubs, and worms. These are detected by smell and perhaps by sensors on the long snout that detect electric signals. The small head joins the shoulders with no external neck.

Zaglossus bartoni

Eastern long-beaked echidna

Location New Guinea

Length 23½–39 in (60–100 cm)
Tail None
Weight 11–22 lb (5–10 kg)
Social unit Individual
Status Critically endangered

The downcurved snout of this species may exceed 8 in (20 cm) in length, with a tiny mouth at the tip and small, close-set eyes at its base. As in other echidna species, the female eastern long-beaked echidna digs a burrow for her egg, but in this species she carries and suckles the hatched infant in her pouch. The largest monotreme, it is slow moving, rolling into a spiny ball for defense. The spines are often only just visible through its long, black fur.

Ornithorhynchus anatinus

Duck-billed platypus

Location E. Australia, Tasmania

Length 16–23½ in (40–60 cm)
Tail 3¼–6 in (8.5–15 cm)
Weight 1¾–5½ lb (0.8–2.5 kg)
Social unit Individual
Status Least concern

The ducklike, beak-shaped mouth, sprawling, reptile-type gait, and flattened, almost scaly, beaverlike tail make the duck-billed platypus an unmistakable animal. Its waterproof plum-colored body fur has a plush texture that is reminiscent of that of a mole. The probing bill is very sensitive to both touch and water-borne electrical signals from the muscles of its small aquatic prey. It has an unusually high density of red blood cells to enable it to make deep dives in pursuit of its quarry. Platypuses are solitary, yet occupy overlapping home ranges. However, males in the breeding season defend these territories. The male has a poisonous spur on each hind foot, with which he attempts to wound his rivals. Home is a bankside burrow, usually about 16 ft (5 m) long but it may reach up to 98 ft (30 m). After a gestation period of one month, the female incubates her clutch of 1–2 soft, leathery-shelled eggs here for 10 days. After the young hatch, she suckles them for 3–4 months in the burrow, leaving them walled in for periods of up to 38 hours while she forages.

MAMMALS

Marsupials

PHYLUM	Chordata
CLASS	Mammalia
INFRACLASS	Marsupialia
ORDERS	7
FAMILIES	20
SPECIES	319

CLASSIFICATION NOTE

The taxonomic rank of Marsupialia has changed. It is now an infraclass containing 7 orders: Didelphinomorphia (the American opossums), Dasyuromorphia (the Australian carnivorous marsupials), Peramelemorphia (bandicoots and bilbies), Notoryctemorphia (marsupial "moles"), Diprotodontia (kangaroos, koala, and relatives), Paucituberculata ("shrew" opossums), and Microbiotheria (the monito del monte). Because of their shared reproductive biology, the orders are not being treated separately but as infraclass Marsupialia.

Like other mammals (apart from the monotremes), marsupials bear live offspring. They are, however, distinct from all other live-bearers (together described as placental mammals), in that they give birth at a very early stage of the embryo's development and nourish the newborn on milk rather than by a placenta. Marsupials are, therefore, classified as infraclass Marsupialia (or Metatheria), while the placental mammals are placed in the infraclass Eutheria. Marsupials are amazingly diverse, including animals such as kangaroos, possums, and bandicoots. The Australasian marsupials, through a lack of competing species, have diversified and become specialized insectivores, carnivores, and herbivores. In South America, the marsupials are small and mostly arboreal; only one species, the Virginia opossum, has spread to North America.

Early life

Marsupial offspring are born in an almost embryonic state after a very short gestation (only 12 days in some bandicoot species). The newborn makes its way to one of the mother's nipples, where it remains attached for several weeks. Larger species have single births, but the small quolls and dunnarts have litters of up to 8. In many kangaroo species, the female mates again while pregnant, but the new embryo remains dormant until the previous young leaves the pouch.

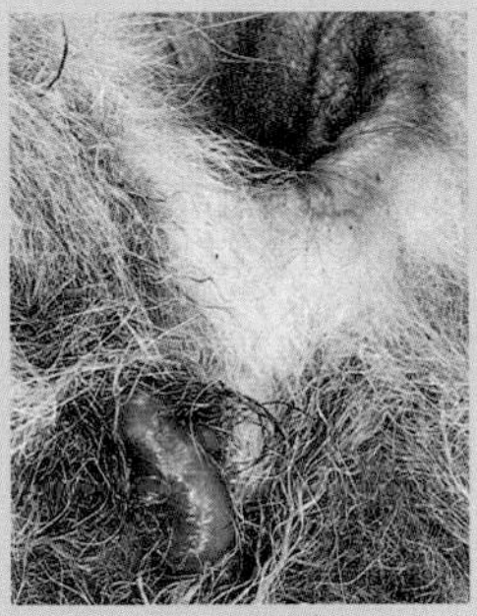

AMAZING JOURNEY
This tiny tammar wallaby baby is hauling itself over its mother's fur to reach a nipple, to which it will firmly attach itself. As in most marsupials, the nipples are protected by a pouch. The joey will not relinquish the nipple until, months later, it begins to explore the outside world.

OVERCROWDING
A mother's pouch may be too small to carry several offspring. This mouse opossum has no option but to carry her babies externally.

SPECIAL POUCHES
Most marsupials have a pouch to carry their young. This western grey kangaroo has a forward-facing pouch, from which her joey's head is protruding. Some pouches face backward.

Anatomy

Externally, marsupials are highly varied, although many have long back legs and feet (for example, kangaroos) and elongated snouts; almost all have large eyes and ears. Female marsupials have a unique "doubled" reproductive tract (see below), and in some males the penis is forked. The testes are held in a pendulous scrotum with a long, thin stalk, which swings in front of the penis. Apart from their specialized reproductive system, marsupials also differ from placental mammals in that their brain is relatively smaller and lacks a corpus callosum (the nerve tract connecting the two cerebral hemispheres). The group that contains the kangaroos, possums, wombats, and the koala, and the group containing the bandicoots, have an arrangement called syndactyly: the second and third toes of the back foot are combined to form a single digit with 2 claws.

FEMALE REPRODUCTIVE SYSTEM
Unlike placental mammals, which have a single uterus and vagina (far right), female marsupials have a double system with 2 uteri, each with its own lateral vagina (right). The young is born through a separate, central birth canal. In some marsupials, this canal forms before each birth. In others, it remains after the first birth (but sometimes fills with connective tissue).

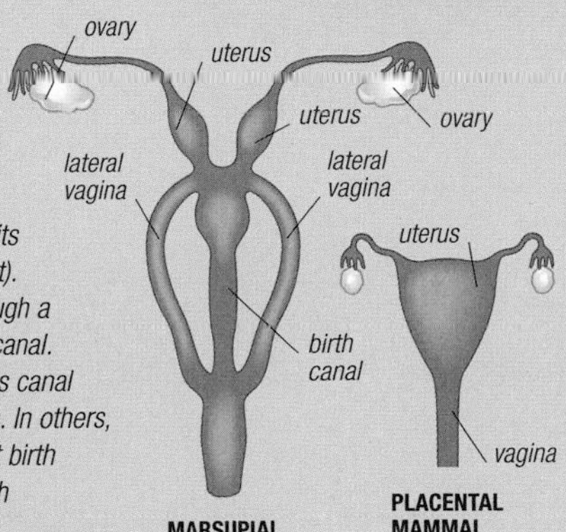

Movement

Although most marsupials run or scurry, there are several variations. For example, wombats waddle, while koalas and possums climb. Kangaroos and wallabies hop on their long back legs, using the extended middle toe as an extra limb segment. Although hopping at low speeds uses more energy than running on all fours, above approximately 6 ft (1.8 m) per second, the larger species begin to conserve energy. This is because energy is stored in the tendons of the foot, and the heavy tail swings up and down like a pendulum, providing momentum.

HOPPING
The faster a large kangaroo (such as this red kangaroo) moves, the more energy efficient it becomes.

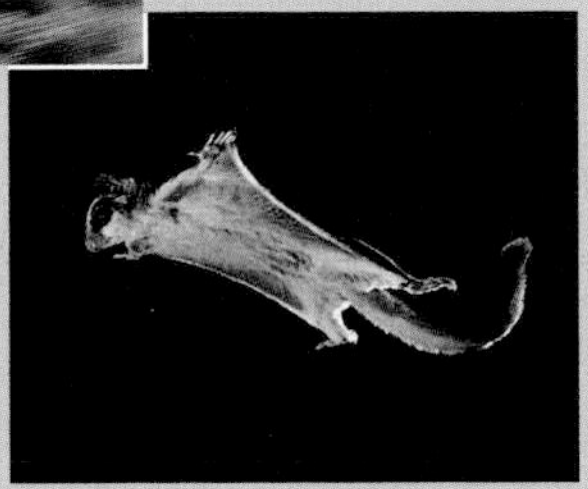

GLIDING
Some possums, known as gliders, have a membrane between their front and back legs, which they use as a parachute.

Marmosa murina

Common mouse opossum

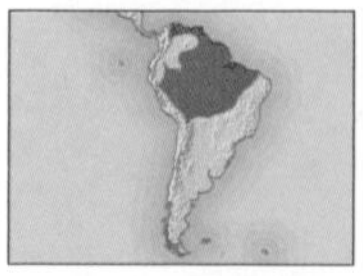

Location N. and C. South America

Length 4¼–5¾ in (11–14.5 cm)
Tail 5¼–8½ in (13.5–21 cm)
Weight $^{5}/_{16}$–1⅝ oz (15–45 g)
Social unit Individual
Status Least concern

Found near forest streams and human habitation, this opossum is pale buff to gray on the upperparts, creamy white below, with a black face mask. Its diet includes small invertebrates, such as insects and spiders, small vertebrates such as lizards, birds' eggs, and chicks, and some fruit. It is a fast, agile climber and rests by day in a tree-hole, old bird's nest, or tangle of twigs among the branches. The 5–8 young are weaned at 60–80 days.

Chironectes minimus

Water opossum

Location S. Mexico to C. South America

Length 10–16 in (26–40 cm)
Tail 12–17 in (31–43 cm)
Weight 20–29 oz (550–800 g)
Social unit Individual
Status Least concern

Also called the yapok, this is the only aquatic marsupial. It has fine, dense, water-repellent fur, long, webbed toes on its rear feet, and—in both male and female—a pouch with a muscular opening that can close tight underwater. The diet includes fish, frogs, and similar freshwater prey, detected and grabbed by the dextrous, clawless front toes. It rests by day in a leaf-lined river-bank den.

Caluromysiops irrupta

Black-shouldered opossum

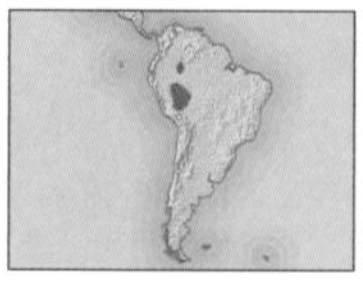

Location W. South America

Length 8½–10 in (21–26 cm)
Tail 9–12 in (22–31 cm)
Weight 7–21 oz (200–600 g)
Social unit Variable
Status Least concern

A broad, black stripe from shoulders to front legs distinguishes this woolly furred opossum. It forages by night in trees for a wide diet, including fruits and grubs. Its densely furred tail is gray, turning white at the end, with a hairless underside. Gestation is 13–14 days and the litter size is 2.

Neophascogale lorentzii

Long-clawed marsupial mouse

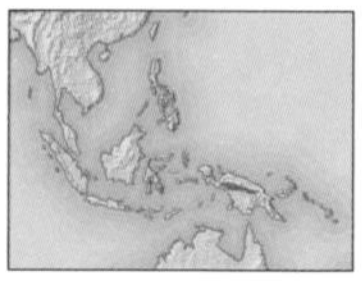

Location New Guinea

Length 6½–9 in (16–23 cm)
Tail 6½–9 in (17–22 cm)
Weight 7–9 oz (200–250 g)
Social unit Individual
Status Least concern

Also called the speckled dasyure because of the sprinkling of long, white hairs in the dark gray upper fur, this species has short, powerful limbs with very long claws on all toes. It digs by day for grubs, worms, and similar prey. Information on nesting and breeding is very sparse; 4 young have been recorded in the female's pouch.

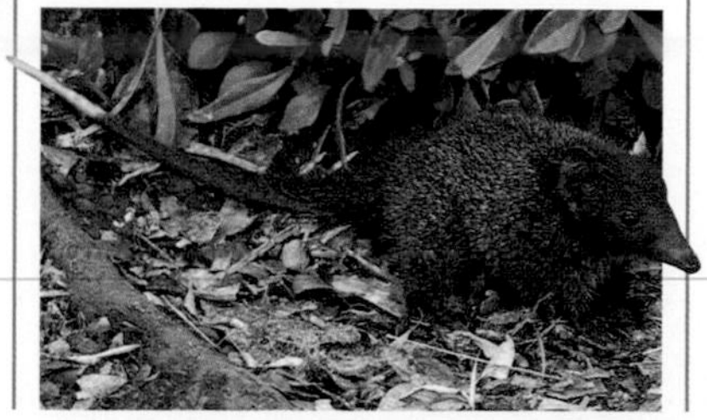

Didelphis virginiana

Virginia opossum

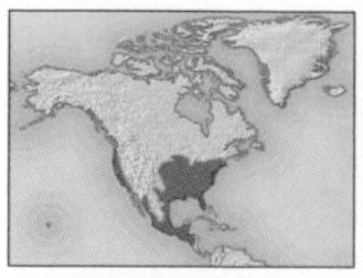

Location W., C., and E. USA, Mexico, Central America

Length 13–20 in (33–50 cm)
Tail 10–21½ in (25–54 cm)
Weight 4½–12 lb (2–5.5 kg)
Social unit Individual
Status Least concern

The highly adaptable Virginia opossum is actively expanding its range in North America. It benefits from human habitation, both for shelter because it nests in piles of debris or outbuildings, and for food because it scavenges for scraps. A true omnivore, its diet ranges from grubs and eggs to flowers, fruits, and carrion; it may raid farm poultry and damage garden plants. The largest American marsupial, the Virginia opossum is nocturnal and usually terrestrial, but also climbs well and swims strongly. Although not territorial, individuals avoid each other except when breeding. The female bears up to 18 per litter, yet has only enough teats for 13. The survivors attach to the teats for 50 days, leaving the pouch at 70 days.

"PLAYING POSSUM"

When threatened, the Virginia opossum feigns death in an attempt to make predators lose interest. It may lie immobile, partly curled on its side, for up to 6 hours. The mouth and eyes are open, and it does not respond to stimuli, including touch. It may void a foul-smelling fluid from its anus for added effect.

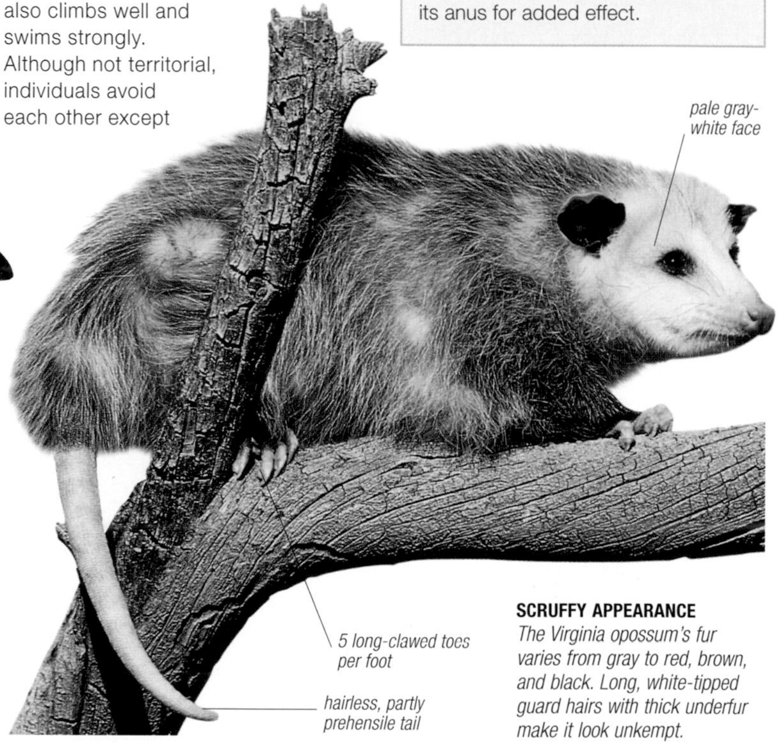

SCRUFFY APPEARANCE
The Virginia opossum's fur varies from gray to red, brown, and black. Long, white-tipped guard hairs with thick underfur make it look unkempt.

Ningauri ridei

Inland ningaui

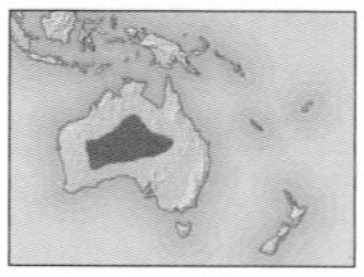

Location C. Australia

Length 2–3 in (5–7.5 cm)
Tail 2–2¾ in (5–7 cm)
Weight $^{7}/_{32}$–$^{7}/_{16}$ oz (6–12 g)
Social unit Individual
Status Least concern

Formally described for science only in 1975, the inland or wongai ningaui is a small, fierce, solitary, nocturnal, shrewlike predator of invertebrates, such as beetles, crickets, and spiders, mostly less than ⅜ in (1 cm) long. It hunts mainly by smell and hearing in clumps of spinifex (hummock grass), resting by day in thick undergrowth or an old hole dug by a lizard, rodent, or large spider. After an incubation of 13–21 days, 6–7 newborn ningaui attach to the female's teats in her relatively open pouch area for 6 weeks. They are weaned by 13 weeks.

Sminthopsis crassicaudata

Fat-tailed dunnart

Location Australia

- **Length** 2¼–3½ in (6–9 cm)
- **Tail** 1½–2¾ in (4–7 cm)
- **Weight** 3/8–11/16 oz (10–20 g)
- **Social unit** Individual
- **Status** Least concern

Widespread in wood, scrub, and dry grassland, the fat-tailed dunnart has benefited from the spread of grazing farmland. It has fawn or brown upper fur and white underparts. It actively searches by night across open soil and leaf litter for grubs, worms, and similar small invertebrates. It is solitary during the breeding season of August to March, but at other times small groups, of less than 10, may huddle together in sheltered crevices, logs, or burrows during cold periods.

wide tail base to store fat

Antechinomys laniger

Kultarr

Location C. and S. Australia

- **Length** 2¾–4 in (7–10 cm)
- **Tail** 4–6 in (10–15 cm)
- **Weight** 7/8–1¼ oz (25–35 g)
- **Social unit** Individual
- **Status** Least concern

Well adapted to a range of habitats, from scattered woodland to scrubby semidesert, the kultarr preys at night on grubs and similar small invertebrates. It bounds on its relatively large, 4-toed rear feet, but lands on its front legs (like a rabbit). It has large ears and dark-ringed eyes, a thin, pointed face, and a compact body. Its long tail has a dark-tufted tip. The fur is fawn or brown above and white underneath.

Pseudantechinus macdonnellensis

Fat-tailed false antechinus

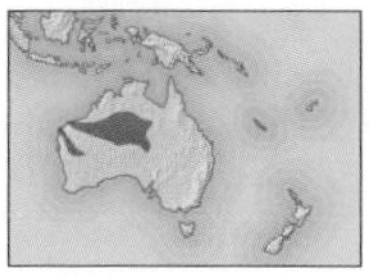

Location W. and C. Australia

- **Length** 3¾–4¼ in (9.5–10.5 cm)
- **Tail** 3–3¼ in (7.5–8.5 cm)
- **Weight** 11/16–1⅝ oz (20–45 g)
- **Social unit** Individual
- **Status** Least concern

When well fed, this insectivorous, nocturnal species, also called the red-eared antechinus, has a carrot-shaped tail swollen with fat (as a food store). Its fur is gray-brown, with red patches behind the ears and gray-white underparts. It prefers dry, rocky areas with clumped bushes and grass. After a gestation of 45–55 days, the female carries 6 young, one on each teat in her pouch, for 40 days. She suckles them in a nest until 14 weeks.

Parantechinus apicalis

Dibbler

Location S.W. Australia

- **Length** 4–6½ in (10–16 cm)
- **Tail** 3–4¾ in (7.5–12 cm)
- **Weight** 1 7/16–4 oz (40–125 g)
- **Social unit** Individual
- **Status** Endangered

After a gap of some 80 years, this species was rediscovered in 1967 in extreme southwest Australia and in 1985 on 2 offshore islands. Distinctive features include a white eye-ring, white-flecked "grizzled" gray-brown fur, a well-haired, short, tapering tail, and large canine teeth for killing prey. This includes small vertebrates, such as mice, birds, and lizards, as well as insects and other invertebrates.

Dasycercus byrnei

Kowari

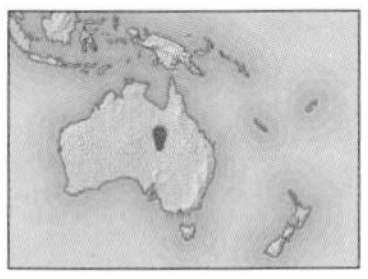

Location C. Australia

- **Length** 5¼–7 in (13.5–18 cm)
- **Tail** 4¼–5½ in (11–14 cm)
- **Weight** 2½–5 oz (70–150 g)
- **Social unit** Individual
- **Status** Vulnerable

Despite a squirrel-like appearance and bushy, black-tufted tail, the kowari is a burrow-dwelling carnivore. It kills large prey, such as lizards, birds, and rodents, by holding them in its front paws and biting repeatedly and rapidly. It also eats insects and similar creatures, mostly more than 3/8 in (1 cm) long. The kowari is solitary and extends its burrow after rain, which softens the stony soil in its harsh, sparsely vegetated desert habitat. It marks its burrow and home range with urine, feces, and chin- and chest-gland scents, and communicates with neighbors through chatters, hisses, and flicks of its tail.

Dasyurus viverrinus

Eastern quoll

Location Tasmania

- **Length** 11–18 in (28–45 cm)
- **Tail** 6½–11 in (17–28 cm)
- **Weight** 1½–4½ lb (0.7–2 kg)
- **Social unit** Individual
- **Status** Near threatened

Last seen on the Australian mainland in the 1960s, in Sydney's suburbs, this catlike marsupial (sometimes called the spotted native cat) now survives only in Tasmania. It hunts on the ground by night for small mammals, birds, and other vertebrates, and large insects; it also eats fruit, grass, and carrion. Quolls occupy varied woody, shrubby, and grassy habitats, as provided by mixed agriculture. The male is longer and 50 percent heavier than the female.

MATERNAL CARE

Like many female marsupials, the quoll has a large litter, up to 24, but only 6 nipples in her pouch, so only 6 young can survive. They attach for 8 weeks, then remain in the den as the mother goes off to feed and returns to suckle. She regularly takes them to new dens, often by piggyback.

SPOTTED COAT
Slim and agile, the eastern quoll has white spots on its brown or black body, but not on the tail. The ears are large and erect.

Dasycercus cristicauda

Mulgara

Location C. Australia

- **Length** 4¾–8 in (12–20 cm)
- **Tail** 3–4¼ in (7.5–11 cm)
- **Weight** 1¾–5 oz (50–150 g)
- **Social unit** Individual
- **Status** Least concern

This inhabitant of sandy desert with scattered spinifex (hummock grass) resembles the kowari (above) in body form and, like several similar species, has a fat-swollen tail base. Its fur is sandy brown on the upperparts, off-white below, with a crest of black hairs on the tail's upper surface toward the tip. Its burrow varies from single, simple, and short to a system of tunnels with 5–6 entrances. At night, the mulgara hunts small rodents, lizards, birds, and large insects, and may also tackle large centipedes.

Sarcophilus harrisii

Tasmanian devil

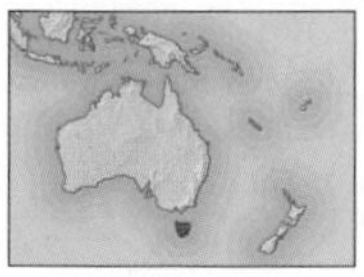

Location Tasmania

Length 20½–32 in (52–80 cm)
Tail 9–12 in (23–30 cm)
Weight 8¾–26 lb (4–12 kg)
Social unit Individual
Status Endangered

The largest marsupial carnivore, the "devil" hunts prey of varied sizes from insects to wallabies, as well as scavenging. Nocturnal, it screeches and rears up if alarmed. It makes its den in a burrow or among rocks or roots. Born after a gestation of 30–31 days, the young attach to the female's 4 teats in her shallow, rear-opening pouch. Disease has caused a sharp decline in numbers in recent years.

WHITE ON BLACK
The devil has long, white patches on its chest and rump.

COMPETING AT A KILL

The Tasmanian devil is normally solitary, but its excellent sense of smell leads to many converging on a large carcass. Its powerful jaws and sharp, sturdy teeth rip up the hide and crush gristle and bone. The devils growl and snarl at each other for the prime parts, but stop short of physical clashes.

Myrmecobius fasciatus

Numbat

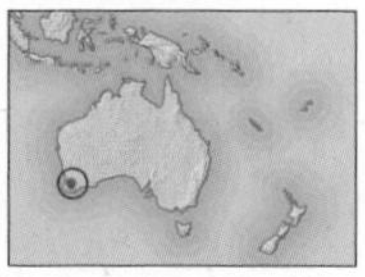

Location S.W. Australia

Length 8–11 in (20–28 cm)
Tail 6½–8½ in (16–21 cm)
Weight 11–26 oz (300–725 g)
Social unit Individual
Status Endangered

The numbat, or banded anteater, has a long (4 in/10 cm) tongue to lick up termites, or very rarely ants, after it has ripped open the nest using the large-clawed, powerful forefeet. It also has more teeth—52—than any other land mammal, although these are very small. Day-active and solitary, the numbat chases same-sex intruders from its territory. Offspring attach to the female's 4 teats for 4 months, then are suckled in the nest for a further 2–3 months.

Perameles gunnii

Eastern barred bandicoot

Location S.E. Australia, Tasmania

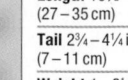

Length 10½–14 in (27–35 cm)
Tail 2¾–4¼ in (7–11 cm)
Weight 1–3¼ lb (0.5–1.5 kg)
Social unit Individual
Status Near threatened

A few hundred of these bandicoots survive in mainland Australia, in Victoria, the majority being found in Tasmania. This omnivorous opportunist forages alone at night and shelters by day in a simple nest of grasses, leaves, and twigs. The eastern barred bandicoot has rabbitlike ears, 3–4 whitish back stripes, and a white tail. Its life cycle is rapid: 12-day gestation, weaning by 60 days, and sexual maturity at 3 months.

Echymipera kalubu

New Guinean spiny bandicoot

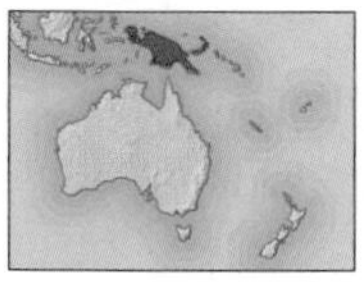

Location New Guinea and surrounding islands

Length 8–20 in (20–50 cm)
Tail 2–5 in (5–12.5 cm)
Weight 1–3¼ lb (0.5–1.5 kg)
Social unit Individual
Status Least concern

Bandicoots are mostly insectivorous, rat- to rabbit-sized marsupials. The New Guinean spiny bandicoot has a long, mobile snout, an outer coat of stiff, stout hairs in shades of brown, copper, yellow, and black, buff underparts, and a hairless tail. It forages alone at night for prey, takes fruits, berries, and other plant matter, and aggressively repels others of its kind. It shelters by day in a hollow log, leaf pile, or self-dug burrow.

Macrotis lagotis

Greater bilby

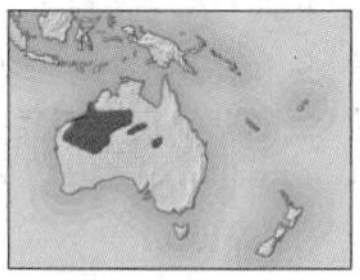

Location W. and C. Australia

Length 12–22 in (30–55 cm)
Tail 8–11½ in (20–29 cm)
Weight 2¼–5½ lb (1–2.5 kg)
Social unit Pair
Status Vulnerable

Also known as the rabbit-eared bandicoot, this omnivorous species is recognized by its huge ears, long hind feet, and tricolored tail: blue-gray at the base (as on the body), then black, with the last half white and feathery. It digs powerfully to shelter by day in a burrow 10 ft (3 m) long and some 7 ft (2 m) deep. Pairs associate for the breeding season and the 2 offspring, born after 13–16 days' gestation, leave the pouch after 80 days.

Notoryctes typhlops

Southern marsupial mole

Location S. Australia

Length 4¾–7 in (12–18 cm)
Tail ¾–1 in (2–2.5 cm)
Weight 1⁷⁄₁₆–2½ oz (40–70 g)
Social unit Individual
Status Data deficient

The southern marsupial mole tunnels down to 8 ft (2.5 m) in sandy deserts, loose-soiled grassland, and scrubby bush. It "swims" through light sand, which collapses behind leaving no permanent tunnel. Food includes fungi and tubers as well as animal prey, which is eaten whenever encountered. The silky, off-white to cinnamon fur is rubbed shiny by burrowing, and may be stained deep red by iron minerals in the soil. The female's pouch, in which she carries 1–2 young, opens rearward so it does not fill with soil.

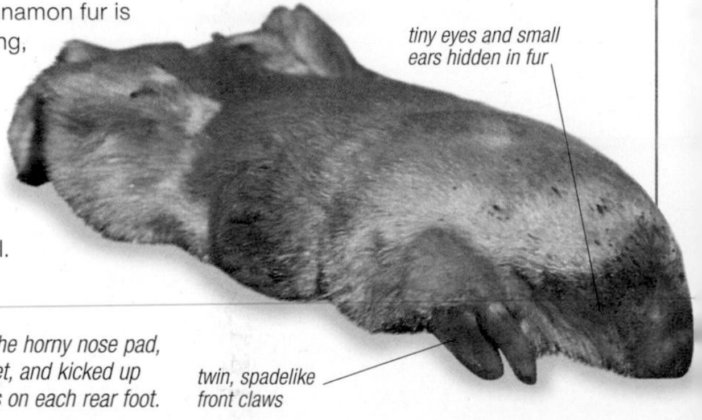

BURROWING MACHINE
Sand and soil are probed by the horny nose pad, scooped aside by the front feet, and kicked up and back by the 3 large claws on each rear foot.

MOLE FOOD

The southern marsupial mole eats soil-dwelling worms, grubs, centipedes, and even small lizards, such as the gecko held here by the huge front claws. Using smell and touch, it follows small tunnels made by potential prey in order to catch its victims. It may forage on the surface after rain.

MAMMALS

Phascolarctos cinereus

Koala

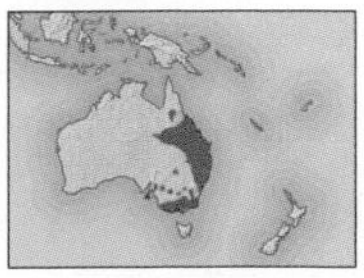

Location E. Australia

Length 26–32 in (65–82 cm)
Tail ⅜–¾ in (1–2 cm)
Weight 8¾–33 lb (4–15 kg)
Social unit Individual
Status Least concern

The koala lives nearly all its life in eucalyptus trees. It feeds for about 4 hours at night, eating some 1 lb (500 g) of leaves, and dozes for the remaining time, wedged securely in a branch fork. Occasionally, it descends to change trees or promote digestion by eating soil, bark, and gravel. Individuals bellow to each other during the breeding season and dominant males mate with more females than do junior males. Copulation is brief and often accompanied by bites and scratches. After a gestation of 35 days, the single offspring crawls into the pouch, where it is suckled for 6 months. It then clings to the mother's back. Among its early solid food is the mother's droppings, which carry helpful food-digesting and disease-fighting microbes. Despite a placid appearance, the koala readily bites and scratches.

large, rounded, white-tufted ears

BEARLIKE MARSUPIAL
The koala's face is large and wide, like that of a bear, with a smooth black muzzle; the body is compact and thick set. The fur is soft, long, and mainly gray to gray-brown, with slightly paler underparts, and mottling on the rump.

CONSERVATION

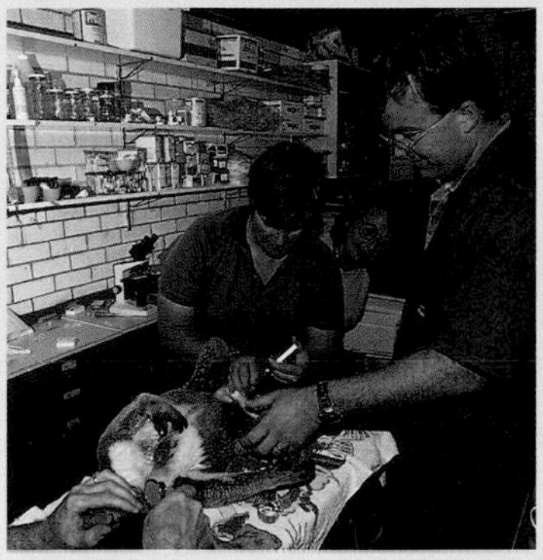

Once common in Australia's eastern forests, the koala is in decline in some regions, mainly as a result of forest fragmentation, parasitic disease, and traffic accidents. Veterinary aid (see left) can be vital in saving lives. However, conservation measures have to be carried out with care, as koala populations can rise steeply given sufficient food. On Kangaroo Island, where koalas have been introduced, the species is now a major forest pest.

HOLDING ON

The koala's short, powerful limbs have sharp claws (except on the first rear toe) and granular paw pads, to grip bark and branches. Toes 1 and 2 on the front foot can oppose the other 3, allowing a pincerlike grasp of thin boughs.

Vombatus ursinus

Common wombat

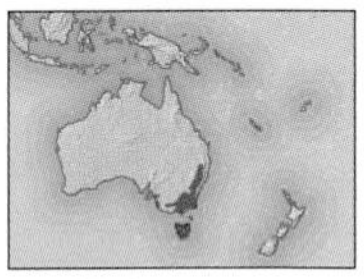

Location E. Australia, Tasmania

Length 28–47 in (70–120 cm)
Tail ¾–1¼ in (2–3 cm)
Weight 55–88 lb (25–40 kg)
Social unit Individual
Status Least concern

The common wombat has a hairless nose and remarkably bearlike form. A prolific burrower, its tunnel system generally has only one entrance but many underground branches, with a total length of up to 650 ft (200 m). Preferred sites are slopes above creeks and valleys, where the wombat grazes, mainly at night, on grasses, sedges, roots, and tubers. In winter, it may graze or sunbathe by day. Although wombats are generally solitary, and adult males tend to chase intruders from their home areas, they also seem to visit each other's burrows on occasion. After the offspring has left its mother's pouch, it stays at first in a nest chamber lined with dry grass and leaves, inside the burrow. The common wombat is not dangerous to humans, but its holes may trip horses, harbor dingoes, undermine banks, and damage rabbit-proof fences.

OUT OF THE POUCH

The common wombat usually produces a single offspring, which remains in the mother's pouch for 6–7 months. It returns there occasionally over the following 3 months to suckle or seek shelter. Weaning occurs when it is 15 months old.

coarse, dense, gray-tinged, brown fur

BROAD BURROWER
The common wombat's broad, angular head, compact body, and strong limbs with wide, large-clawed feet make it an accomplished digger.

Lasiorhinus latifrons

Southern hairy-nosed wombat

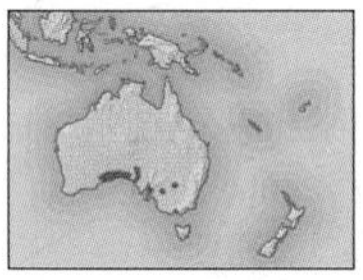

Location S. Australia

Length 30–37 in (77–95 cm)
Tail 1¼–2¼ in (3–6 cm)
Weight 42–71 lb (19–32 kg)
Social unit Group
Status Least concern

Even stockier and shorter-limbed than the common wombat (see left), this species has longer ears, long, silky fur mottled brown and gray, and gray-white hairs on the snout. It is colonial, groups of 5–10 occupying a burrow system (warren) covering several hundred square yards. This wombat grazes at night on grass and other low vegetation. Senior males repel strangers of their kind unless these are receptive females at breeding time.

MAMMALS

Trichosurus vulpecula

Common brush-tailed possum

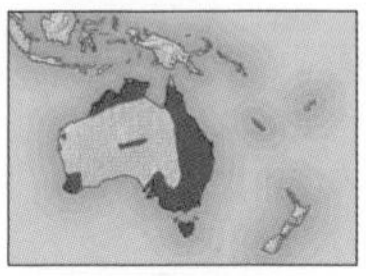

Location Australia (including Tasmania)

Length 14–15¾ in (35–58 cm)
Tail 10–16 in (25–40 cm)
Weight 3¼–10 lb (1.5–4.5 kg)
Social unit Individual
Status Least concern

Familiar in many habitats, including parks and gardens, the common brush-tailed possum has mainly silver-gray body fur, which is short and tinged red in the north of its range, and longer and darker gray in the south. It bounds and climbs with ease as it forages for eucalypt, acacia, and other leaves, as well as flowers, fruit, and, occasionally, birds' eggs and chicks. This vocal possum, with its range of hisses, chitters, grunts, and growls, is solitary apart from the brief mating season. It lives in a crevice or hollow in a tree, log, rock, or roof. The usually single young remains in the pouch for 5 months.

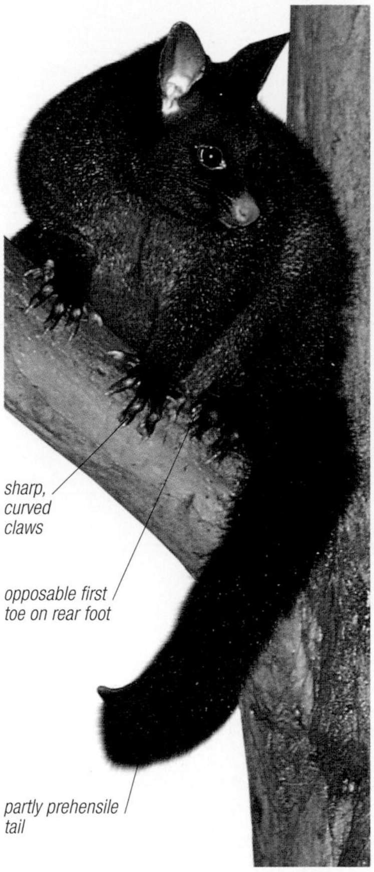

Phalanger orientalis

Common cuscus

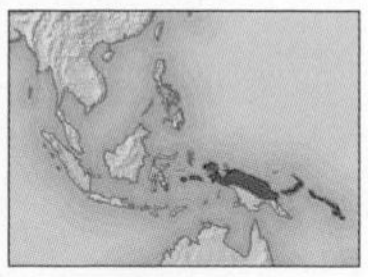

Location New Guinea, Solomon Islands

Length 15–19 in (38–48 cm)
Tail 11–17 in (28–43 cm)
Weight 3¼–7¾ lb (1.5–3.5 kg)
Social unit Individual
Status Least concern

Also known as the northern common cuscus, the coloring of this species varies across its range of many islands, from almost white to black. There is generally, however, a stripe along the back, and part of the tail is always hairless and white. Resembling a combination of sloth and monkey, it has deliberate but agile climbing habits as it forages for leaves, fruit, and other plant material. Docile by nature, it is a common household pet.

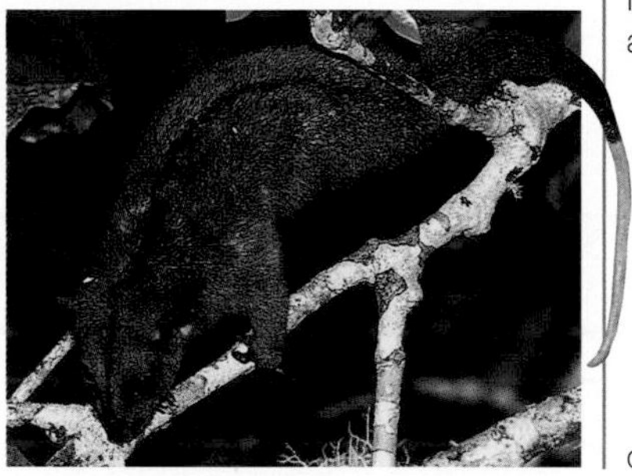

Gymnobelideus leadbeateri

Leadbeater's possum

Location S.E. Australia (Central Highlands, Victoria)

Length 6–7 in (15–17 cm)
Tail 6–8 in (14.5–20 cm)
Weight 3⅝–6 oz (100–175 g)
Social unit Group
Status Endangered

Notable for its rediscovery in 1961, after 52 years of presumed extinction, Leadbeater's possum has a gray back with a dark central stripe from head to tail, and off-white underside. Speedy and elusive in trees, it feeds at night on small invertebrates and tree juices, such as gum, sap, and nectar. Colonies of up to 8 comprise a senior breeding pair and their offspring, mainly males. Females defend the group's territory.

Petaurus norfolcensis

Squirrel glider

Location E. Australia

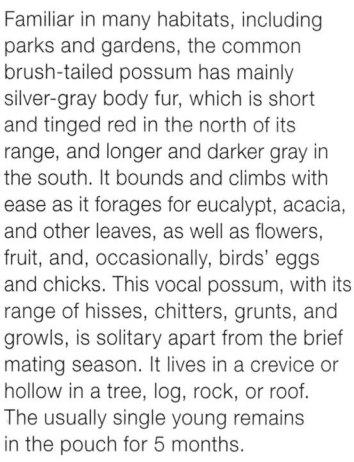

Length 7–9 in (18–23 cm)
Tail 8¾–12 in (22–30 cm)
Weight 7–11 oz (200–300 g)
Social unit Group
Status Least concern

The squirrel glider's furry gliding membrane extends from each fifth front toe to the back foot. The long, bushy, soft-furred, squirrel-like tail acts as a rudder when parachuting as far as 165 ft (50 m) to other trees or to escape enemies. This possum lives in small groups of one adult male, 1–3 females, and their offspring of the season. It feeds on insects and similar small tree creatures, as well as sap, gum, pollen, and seeds. The 1–2 young remain in the pouch for up to 3 months and are weaned by 4 months.

Dactylopsila trivirgata

Striped possum

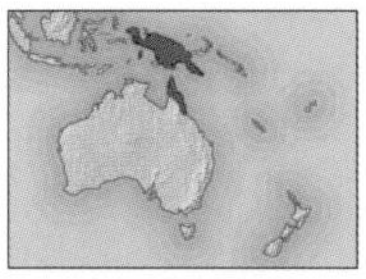

Location N.E. Australia, New Guinea

Length 9½–11 in (24–28 cm)
Tail 12–15⅓ in (30–39 cm)
Weight 9–19 oz (250–525 g)
Social unit Variable
Status Least concern

Skunklike black and white stripes and a bushy, black, white-tipped tail are the distinguishing features of the striped possum. Like the skunk (see panel, p.186), it can emit a foul, penetrating odor from glands in the genital region. At night, it forages alone through tree branches, probing for wood-boring grubs, ants, and termites with the extra-long claw on its front fourth toe. It also eats fruit, birds, and small mammals.

Hemibelideus lemuroides

Brushy-tailed ringtail

Location N.E. Australia

Length 12–16 in (30–40 cm)
Tail 12½–14½ in (32–37 cm)
Weight 29–36 oz (800–1,000 g)
Social unit Variable
Status Near threatened

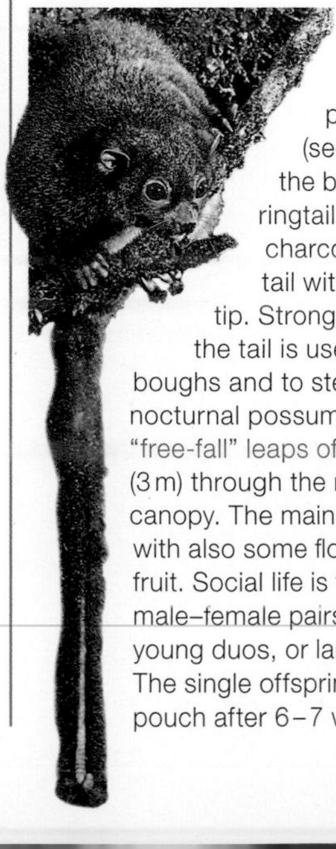

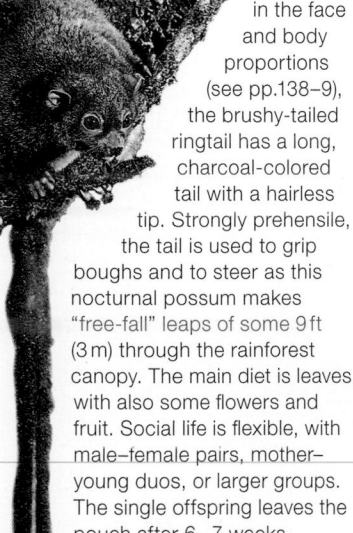

Lemurlike in the face and body proportions (see pp.138–9), the brushy-tailed ringtail has a long, charcoal-colored tail with a hairless tip. Strongly prehensile, the tail is used to grip boughs and to steer as this nocturnal possum makes "free-fall" leaps of some 9 ft (3 m) through the rainforest canopy. The main diet is leaves, with also some flowers and fruit. Social life is flexible, with male–female pairs, mother–young duos, or larger groups. The single offspring leaves the pouch after 6–7 weeks.

Pseudocheirus peregrinus

Common ringtail

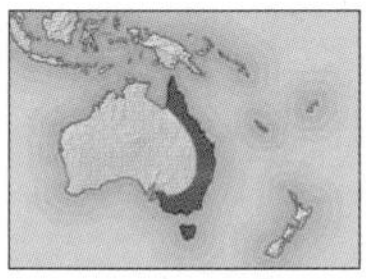

Location E. Australia, Tasmania

Length 12–14 in (30–35 cm)
Tail 12–14 in (30–35 cm)
Weight 25–39 oz (700–1,100 g)
Social unit Group
Status Least concern

The common ringtail's reddish or gray-brown fur extends to the base of the tapering, strongly prehensile tail, becomes darker along the first half of the tail, then changes to white. Found in many habitats, including parks and gardens, the main food of this group-dwelling mammal is leaves, especially eucalypts and acacia. In the north of its range, it nests in a tree-hole; in the south, in a squirrel-like drey.

Petauroides volans

Greater glider

Location E. Australia

Length 14–19 in (35–48 cm)
Tail 18–23½ in (45–60 cm)
Weight 2–3½ lb (0.9–1.5 kg)
Social unit Pair
Status Least concern

This nocturnal forager feeds in eucalypt woodland (not rain forest), preferring the tender young leaves of just a few eucalypt species. It forms female–male pairs, which share the same tree-hollow den for most of the year. The single offspring stays in the pouch for 5 months, then remains in the den or rides on the mother's back for another 1–2 months. By 10 months, young males are driven away by the father.

Groups of scratch marks on tree trunks betray regular landing sites of the greater glider, the largest gliding marsupial. Like other tree-dwelling marsupials, it has sharp claws to grip bark, and 2 toes on the front foot oppose the other 3, giving a pincerlike grip.

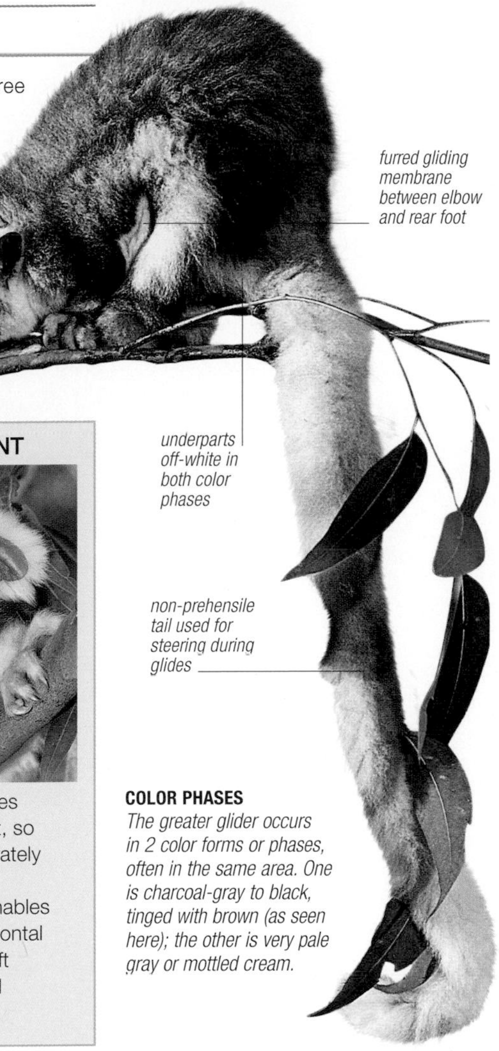

COLOR PHASES
The greater glider occurs in 2 color forms or phases, often in the same area. One is charcoal-gray to black, tinged with brown (as seen here); the other is very pale gray or mottled cream.

FACING THE FRONT

The greater glider's large eyes and huge ears face the front, so it can judge distances accurately by stereoscopic vision and stereophonic sound. This enables the glider to parachute horizontal distances of more than 330 ft (100 m) at night and yet land precisely on a tree trunk.

Distoechurus pennatus

Feather-tailed possum

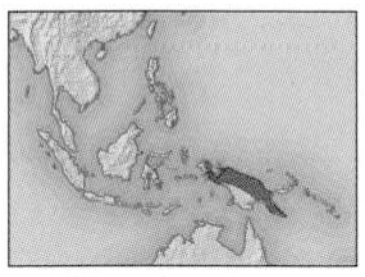

Location New Guinea

Length 4⅛–5½ in (10.5–13.5 cm)
Tail 5–6½ in (12.5–15.5 cm)
Weight 1 7/16–2⅛ oz (40–60 g)
Social unit Group
Status Least concern

The feather-tailed possum is identified by its white face, with 4 black stripes, and its feather- or quill-like tail with a prehensile tip. It moves through the tree branches by darting leaps, gripping with the sharp claws on all toes, and consumes insects (especially cicadas), as well as flowers, nectar, and fruit. These possums appear to live in groups of 2–3, and the female carries 1–2 young in her pouch, then riding on her back near the nest. Details of the social and breeding habits are, however, lacking.

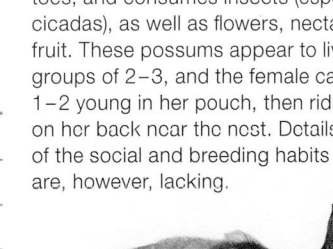

Cercartetus lepidus

Little pygmy-possum

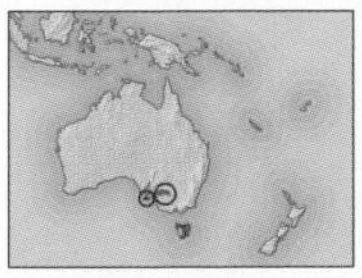

Location S.E. Australia, Tasmania

Length 2–2¾ in (5–6.5 cm)
Tail 2½–3 in (6–7.5 cm)
Weight 3/16–11/32 oz (5–9 g)
Social unit Individual
Status Least concern

The smallest possum species, with a thumb-sized body, this is the only pygmy-possum with gray fur on the underside; the upperparts are fawn or brown. It has a short, blunt face and large, erect ears. The prehensile tail can support the animal's entire weight, and expands at its base to store excess food as fat. Nocturnal and usually solitary, this marsupial feeds in low bushes and shrubs, or on the ground, on a variety of small animals from insects to lizards.

Acrobates pygmaeus

Pygmy glider

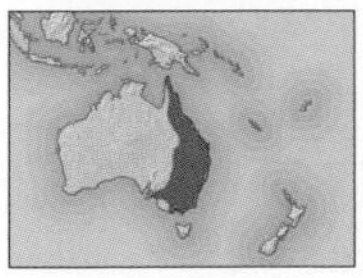

Location E. Australia

Length 2¾–3 in (6.5–8 cm)
Tail 2¾–3 in (7–8 cm)
Weight 11/32–9/16 oz (9–15 g)
Social unit Group
Status Least concern

Also called the feather-tailed glider, this tiny, agile marsupial has a long tail with a row of stiff hairs on either side. The gliding membrane extends between the front and rear limbs, and the "dual-purpose" toes have sharp claws to dig into bark. Expanded, padlike tips on the toes grip smooth, shiny surfaces such as leaves—even glass windows. The tongue is long and brush-tipped, to gather nectar, pollen, and small insects from flowers.

Hypsiprymnodon moschatus

Musky rat-kangaroo

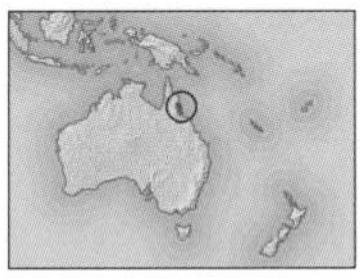

Location N.E. Australia

Length 6½–11 in (16–28 cm)
Tail 4¾–7 in (12–17 cm)
Weight 13–24 oz (375–675 g)
Social unit Individual
Status Least concern

Neither rat nor kangaroo, this is a potoroid marsupial (see right). Its preferred habitat is thick rain forest. It forages alone by day, mainly on the ground for fallen fruit such as figs, as well as palm nuts, seeds, and fungi. It also hoards food at scattered sites—unusual behavior among marsupials—and bounds on all fours, gripping with the opposable big toe on each rear foot. Both sexes produce a musky odor, particularly during breeding.

Potorous longipes

Long-footed potoroo

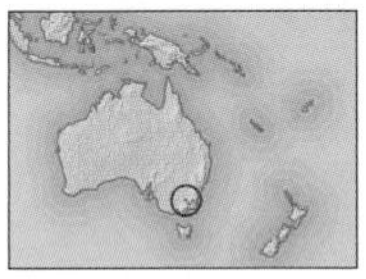

Location S.E. Australia

Length 15–16½ in (38–42 cm)
Tail 12–13 in (30–33 cm)
Weight 3¼–4½ lb (1.5–2 kg)
Social unit Individual
Status Endangered

This potoroo ("rat-kangaroo") is an active, solitary, nocturnal fungus-eater. It bounds at high speed, kangaroolike, on its large back feet, and scrabbles for food with the shorter but strong front limbs. Some 30 fungus species make up four-fifths of its diet. It also eats insects and green plant matter. After an incubation period of 38 days, the single young suckles in the pouch for up to 5 months, and remains with its mother for a further 2–3 months.

ALERT TO DANGER
Red kangaroos usually browse and graze head down, but remain alert, looking and listening for predators. Their sharp eyes can spot a dingo 1,150 ft (350 m) away, and their large ears are very sensitive.

Macropus rufus

Red kangaroo

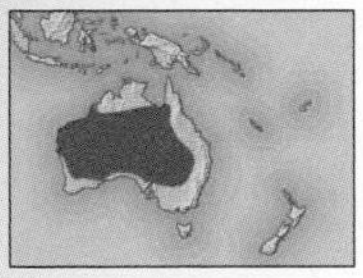

Location Australia

Length 3¼–5¼ ft (1–1.6 m)
Tail 2½–4 ft (75–120 cm)
Weight 55–200 lb (25–90 kg)
Social unit Group
Status Least concern

The largest living marsupial, the red kangaroo is found over much of Australia, with the highest numbers living in open savanna woodland. Its population varies greatly from year to year: when rainfall is high, numbers may reach as many as 12 million, falling to 5 million in times of drought. Females will conceive only if there has been enough rain to produce plenty of green vegetation. In a prolonged drought, the males no longer produce sperm. Red kangaroos use a highly developed sense of smell to detect water, and if it is scarce will migrate up to 125 miles (200 km) from their usual grazing grounds to find it. They forage mainly at night, eating succulent grass shoots, herbs, and leaves. They live in groups of 2 to 10 animals, although 1,500 or so may gather at a water hole during a drought. The dominant male of the group mates with several females. The red kangaroo is regarded as a pest in its native Australia, and is hunted for its meat and skin. Apart from humans and occasionally the wedge-tailed eagle, its only predator is the dingo.

DEFENSIVE MEASURES

The dominant male of a group of red kangaroos will fight off any challenge for supremacy from other males. If one of a group spies a potential predator, it will warn the rest by stamping its foot or thumping its tail on the ground. The group will then flee, taking refuge in water if possible.

PLAY FIGHTING
When playing or fighting among themselves, kangaroos may stand up and spar, but their normal defense is to deliver a powerful kick.

IN FULL FLIGHT
Fleeing from danger, the red kangaroo bounds on its hind legs. It can reach 30 mph (50 kph) for short periods.

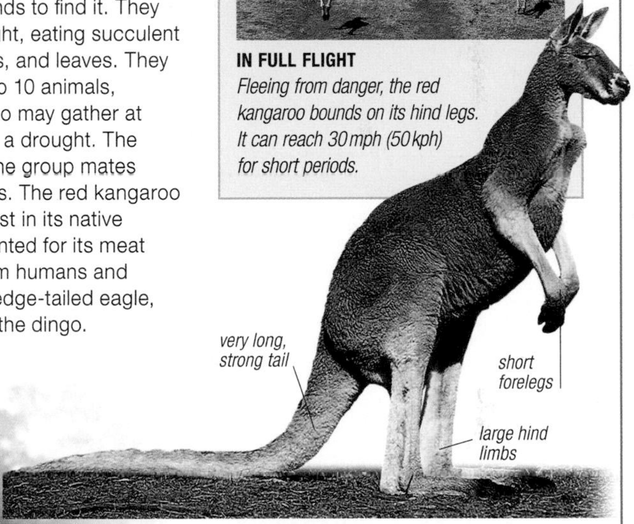

RUFOUS MALE
Males, which may be more than double the weight of females, are usually orange-red, while females are blue-gray – although coloration varies.

IN MOTHER'S POUCH
The infant kangaroo spends 190 days entirely in its mother's pouch, before making its first foray outside. It becomes independent when a year old.

MAMMALS

Bettongia penicillata

Brush-tailed bettong

Length 12–15 in (30–38 cm)
Tail 11½–14 in (29–36 cm)
Weight 2¼–3¼ lb (1–1.5 kg)
Social unit Individual
Status Critically endangered

Location S.W. Australia

Like the potoroo family (see p.99), the brush-tailed bettong—also called the woylie—is a fungivore. It forages in woodland soil by night, scraping earth to find the fungi, which form 90 percent of its diet. The remainder of its diet comprises roots, bulbs, tubers, and worms. This bettong has a tail almost as long as its head and body, with a crest of black fur along the upper side. By day, it shelters in a large, domed nest of bark, leaves, and grass.

Aepyprymnus rufescens

Rufous bettong

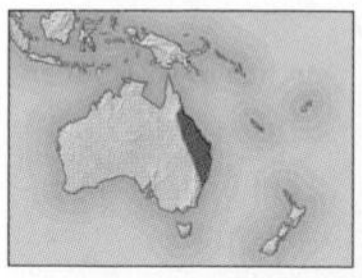

Length 14½–20½ in (37–52 cm)
Tail 13½–16 in (34–40 cm)
Weight 5½–7¾ lb (2.5–3.5 kg)
Social unit Variable
Status Least concern

Location E. Australia

Grizzled red-brown fur over the body gives this rat-kangaroo its name of rufous bettong, although some fur is white. It uses its grasping tail to gather grasses and stems for its tall, cone-shaped nest, which it builds on the woodland floor, often against a log or tree; one individual may have 5 such refuges. The diet is fungi, grass, roots, leaves, flowers, seeds, and small invertebrates. The single young leaves the pouch after 16 weeks.

Thylogale stigmatica

Red-legged pademelon

Length 15–23 in (38–58 cm)
Tail 12–18½ in (30–47 cm)
Weight 5½–15 lb (2.5–7 kg)
Social unit Individual
Status Least concern

Location N. and E. Australia, New Guinea

This slender-headed, stout-bodied, thick-tailed pademelon (a type of small wallaby) tends to be brown-gray in rain forests but paler fawn in open woodland. Active both day and night, it is usually solitary but may gather in groups at a fruiting tree to feed. Other foods include leaves and seeds.

Petrogale penicillata

Brush-tailed rock wallaby

Length 20–23½ in (50–60 cm)
Tail 20–28 in (50–70 cm)
Weight 11–24 lb (5–11 kg)
Social unit Group
Status Near threatened

Location S.E. Australia

The rock wallaby is specialized for leaping and scrambling over boulders, cliffs, ledges, and outcrops. It can make single bounds of 13 ft (4 m), and the soles of the rear feet are enlarged, padded, and roughened, providing excellent grip. By day, this wallaby rests in a cool rock crevice or cave, occasionally sunbathing. At night, it feeds on grasses, ferns, bush leaves, and fruit. Colonies may exceed 50 in number.

Lagorchestes conspicillatus

Spectacled hare wallaby

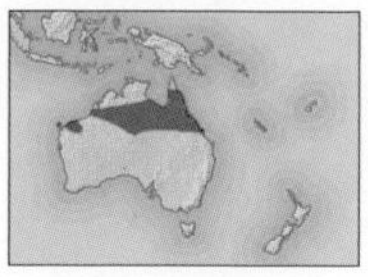

Length 40–48 cm (16–19 in)
Tail 37–50 cm (14½–20 in)
Weight 1.5–4.5 kg (3¼–10 lb)
Social unit Individual
Status Least concern

Location N. Australia

A conspicuous orange eye patch in the generally shaggy, white-grizzled, gray-brown coat gives this small member of the kangaroo family the name of spectacled hare wallaby. It grazes by night, usually alone, on grasses and herbs, and hides in a burrow or thicket by day. After 29–31 days' gestation, the single young stays in the pouch for 5 months, and is weaned by 7 months.

Setonix brachyurus

Quokka

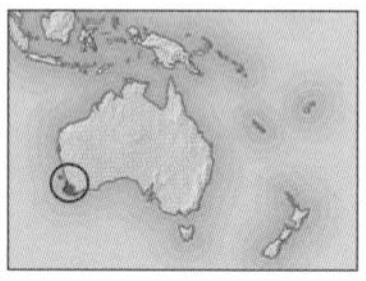

Length 16–21½ in (40–54 cm)
Tail 10–14 in (25–35 cm)
Weight 5½–11 lb (2.5–5 kg)
Social unit Group
Status Vulnerable

Location S.W. Australia (Rottnest and Bald islands)

Very rare on mainland Australia, the quokka survives on 2 islands off the southwest coast, mainly because introduced predators such as foxes are absent. This small wallaby frequents thick forest, open woodland, low scrub, and swamp edges or river banks where available. After resting by day in dense vegetation, it feeds at night on leaves, grasses, and fruit. Small family groups associate as larger gatherings, which maintain a group territory. Gestation is 27 days; the single young leaves the pouch after 6 months.

CONSERVATION

Between 5,000 and 15,000 quokkas live on Rottnest Island. As well as being a popular tourist attraction, they have enabled scientists to carry out detailed research into marsupial biology—essential knowledge for their future conservation.

COMPACT SHAPE
The quokka has a rounded body, short ears and snout, and a stout tail. Its dense, coarse fur is brown, tinged red around the face and neck.

Petrogale concinna

Little rock wallaby

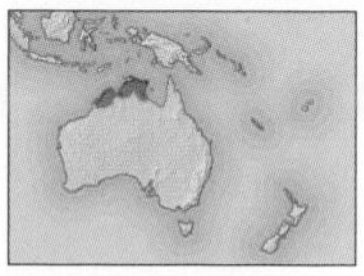

Length 11½–14 in (29–35 cm)
Tail 9–12 in (22–31 cm)
Weight 2¼–3¼ lb (1–1.5 kg)
Social unit Individual
Status Data deficient

Location N. Australia

Similar to the brush-tailed rock wallaby (see left) in form and habits, this species—also called the nabarlek—has short, silky, reddish brown fur with a dark shoulder stripe and brush-tipped tail. It grazes alone by night on grass and sedge, but its main food in the dry season is nardoo (a tough fern). To cope with this abrasive food, the wallaby's molar teeth slowly move forward in the jaw as they wear, and are then replaced by another set—a process of repeated tooth replacement unique among marsupials.

Dendrolagus dorianus

Doria's tree kangaroo

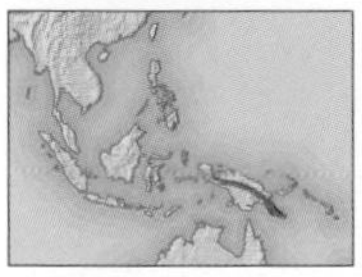

Location New Guinea

- **Length** 20–31 in (51–78 cm)
- **Tail** 17½–26 in (44–66 cm)
- **Weight** 14–32 lb (6.5–14.5 kg)
- **Social unit** Individual
- **Status** Vulnerable

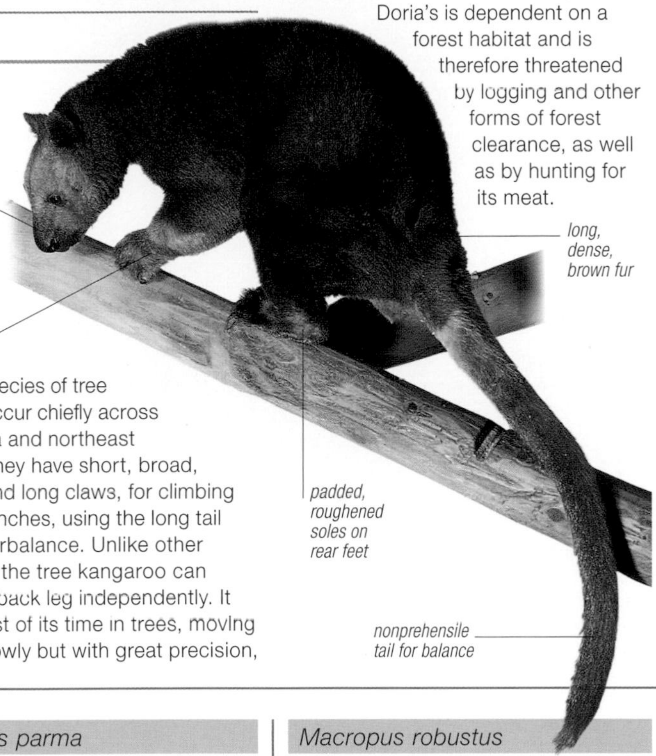

Some 10 species of tree kangaroo occur chiefly across New Guinea and northeast Australia. They have short, broad, stout feet and long claws, for climbing through branches, using the long tail as a counterbalance. Unlike other kangaroos, the tree kangaroo can move each back leg independently. It spends most of its time in trees, moving relatively slowly but with great precision, although it can walk and bound quickly on the ground. Doria's tree kangaroo is one of the largest and, like most other species in its genus, is mostly solitary and nocturnal, with a diet of various leaves, buds, flowers, and fruit. It has black ears, a whorled fur pattern in the middle of the back, and a pale brown or cream tail. After a gestation period of approximately 30 days, the single young attaches to the mother's teat in her pouch and suckles for up to 10 months. Like other tree kangaroos, Doria's is dependent on a forest habitat and is therefore threatened by logging and other forms of forest clearance, as well as by hunting for its meat.

Macropus fulginosus

Western grey kangaroo

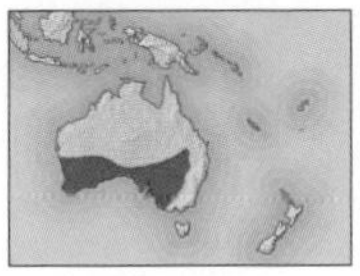

Location S. Australia

- **Length** 3–4½ ft (0.9–1.4 m)
- **Tail** 30–39 in (75–100 cm)
- **Weight** 33–120 lb (15–54 kg)
- **Social unit** Group
- **Status** Least concern

One of the largest, most abundant kangaroos, the western grey kangaroo has thick, coarse fur that varies from pale gray-brown to chocolate-brown, with a paler chest and belly. It lopes like a rabbit when moving slowly, using all 4 limbs, with the tail as a brace, but bounds on its back legs at high speed. Males are up to twice the size of females and can cover 33 ft (10 m) in one leap. This species grazes primarily at night, mainly on grasses, but it also browses on leafy shrubs and low trees. It lives in stable groups of up to 15. The dominant male is usually the only male to breed. The incubation period is 30–31 days.

"BOXING" KANGAROOS

Male kangaroos battle for females during the mating season, and also for food or resting sites if these are limited. The antagonists lock arms and attempt to push each other over. They may also lean back on their tails and kick with the rear feet. As in most such contests, serious injury is rare.

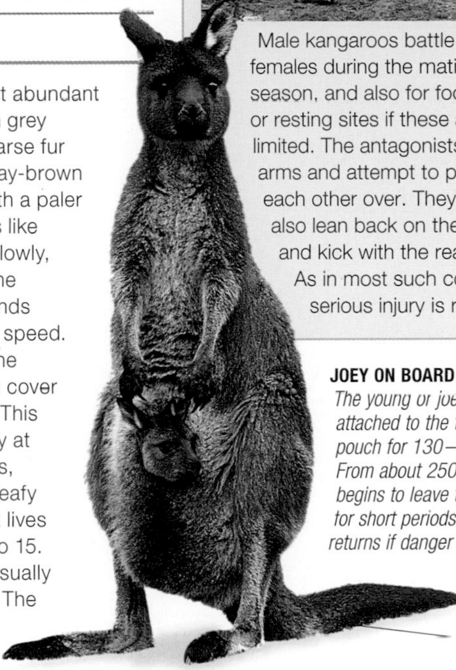

JOEY ON BOARD
The young or joey remains attached to the teat in the pouch for 130–150 days. From about 250 days, it begins to leave the pouch for short periods, but quickly returns if danger threatens.

Macropus parma

Parma wallaby

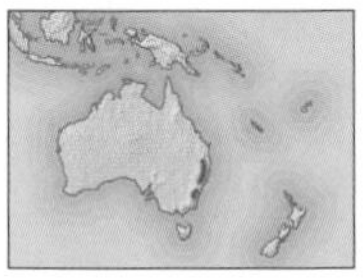

Location E. Australia

- **Length** 18–21 in (45–53 cm)
- **Tail** 16–21½ in (41–54 cm)
- **Weight** 7¾–13 lb (3.5–6 kg)
- **Social unit** Individual
- **Status** Near threatened

The parma wallaby's distinguishing marks are a black stripe in the center of the red- or gray-brown fur on the back, from the neck to the middle of the back, and a white stripe on the side of the muzzle and cheek. Solitary, shy, and well camouflaged in dense vegetation, for a century it was believed to be extinct on the Australian mainland, but was rediscovered in 1967. This night-time grazer and browser feeds on a very wide range of plants.

Macropus robustus

Wallaroo

Location Australia

- **Length** 2½–4½ ft (0.8–1.4 m)
- **Tail** 23½–35 in (60–90 cm)
- **Weight** 33–105 lb (15–47 kg)
- **Social unit** Individual
- **Status** Least concern

Also called the euro or hill kangaroo, the wallaroo is found in a range of habitats, but usually in and around rocky outcrops, cliffs, and boulder piles. It shelters there by day, thereby surviving very hot, dry conditions, and by late afternoon moves out to forage on grasses, sedges, and other leafy foods. The wallaroo resembles other brown wallabies but may adopt a distinctive pose with shoulders back, elbows together, and wrists raised.

Wallabia bicolor

Swamp wallaby

Location E. Australia

- **Length** 26–34 in (66–85 cm)
- **Tail** 26–34 in (66–85 cm)
- **Weight** 23–45 lb (10.5–20.5 kg)
- **Social unit** Individual
- **Status** Least concern

Unlike other wallabies, this species moves with its head low and its tail held out straight behind. It has coarse, brown-black fur and a much darker face, snout, front and back feet, and tail. Also called the stinker or black wallaby, it feeds at night on a wide diet of plant material, including toxic types such as hemlock.

Tarsipes rostratus

Honey possum

Location S.W. Australia

- **Length** 2½–3½ in (6.5–9 cm)
- **Tail** 2¾–4¼ in (7–10.5 cm)
- **Weight** ¼–9/16 oz (7–16 g)
- **Social unit** Group
- **Status** Least concern

One of the smallest possums, this tiny, agile, nocturnal species has a long, pointed, well-whiskered snout and a long, prehensile tail. It lives in small groups, and its toes have soft, padlike tips and small, sharp claws, to grip both bark and glossy leaves. The 1-in (2.5-cm) long bristle-tipped tongue gathers pollen and nectar ("honey") from flowers, which are then scraped onto the much-reduced teeth (upper canines and lower incisors only).

MAMMALS

Sengis

PHYLUM	Chordata
CLASS	Mammalia
ORDER	Macroscelidea
FAMILIES	1 (Macroscelididae)
SPECIES	15

The long, pointed snout, which is flexible and highly sensitive, gives sengis their alternative name of elephant-shrew. They have keen senses of hearing and vision, and long, powerful back legs for running swiftly around their territory. Sengis are found only in Africa, where their habitat ranges from stony ground and grassland to forest undergrowth. They forage mostly by day, feeding on invertebrates.

BURROWS AND TRAILS
Sengis are strictly terrestrial and live in a variety of habitats. The Karoo rock sengi (right) favors rocky areas where it will either dig its own burrow or occupy one that has been abandoned. It maintains a system of trails that lead from the burrow to feeding areas. This behavioral pattern is seen in several sengi species.

Elephantulus rufescens

Rufous sengi

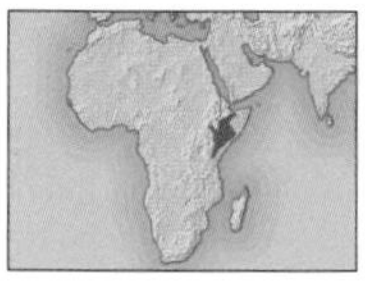

Location E. Africa

Length 4¾–5 in (12–12.5 cm)
Tail 5–5½ in (13–13.5 cm)
Weight 1¾–2⅛ oz (50–60 g)
Social unit Individual/Pair
Status Least concern

Gray to brown with white underparts, this species has a white eye-ring with a dark patch on the outer edge. In addition to small creatures, it eats soft fruit, seeds, and buds. The 1–2 young are born after a gestation period of 60 days. A male–female pair defend their territory by drumming their back feet and chasing away the intruder. The male chases out other males, and the female other females.

Rhynchocyon chrysopygus

Golden-rumped sengi

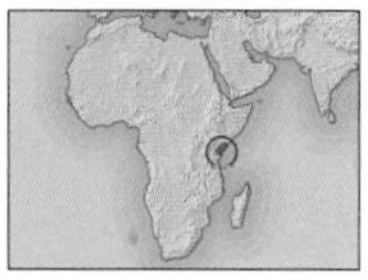

Location E. Africa

Length 10½–11½ in (27–29 cm)
Tail 9–10 in (23–26 cm)
Weight 1¼ lb (525–550 g)
Social unit Individual/Group
Status Endangered

This colorful species has hairless black feet, legs, and ears, a mainly black, white-tipped, sparsely furred tail, russet-colored head and body, and a golden patch on the rump. Its diet consists of small invertebrates, such as worms, insects, and centipedes. When threatened, this sengi slaps its tail on the leaves to warn others, then bounds away at great speed, often with a stiff-legged leap to demonstrate its vigor.

Elephantulus pilicaudus

Karoo rock sengi

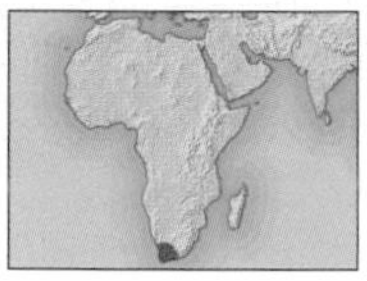

Location South Africa

Length 10.5–14.5 cm (4–5½ in)
Tail 11–13.5 cm (4¼–5¼ in)
Weight 50 g (1¾ oz)
Social unit Individual/Pair
Status Data deficient

This species was distinguished from the closely allied Cape sengi in 2008. It is restricted to the Nama-Karoo region of South Africa, and is distinguished by a tufted tail and subtle features of its body pattern, such as mottled underparts and buff-colored cheeks. It lives among boulders in scattered rocky habitats on mountain slopes above 4,265 ft (1,300 m)—a landscape unsuitable for agriculture or development. However, it is known from only 17 specimens, so probably has low population density.

Rhynchocyon udzungwensis

Grey-faced sengi

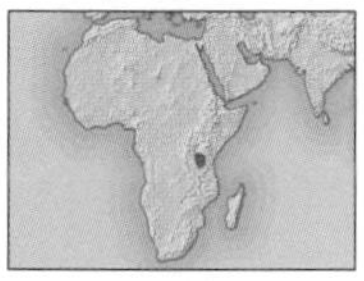

Location E. Africa

Length 11¾–12½ in (30–32 cm)
Tail 9½–10 in (24–26 cm)
Weight 1½ lb (710 g)
Social unit Individual/Pair
Status Vulnerable

Initially found in 2005 by camera-trapping, this animal is one of many new species discovered in the exceptionally diverse communities of Udzungwa Mountain in Tanzania. It is the largest member of a genus of "giant sengis" and is distinguished by a gray face, yellow chest, orange sides, maroon back, and black thighs. It is known from only 2 populations, covering a combined forest area of scarcely 115 square miles (300 square km)—with around 50–80 individuals per square km. This makes this species vulnerable to natural disasters—such as fires (whether natural or human-caused). Global warming could reduce its cool montane habitat too. Like related species, it builds a nest consisting of a cup in soil lined with dry leaves, topped with a cryptic dome of leaf litter.

Rhynchocyon petersi

Black and rufous sengi

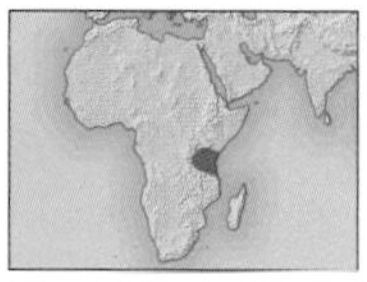

Location E. Africa

Length 9–12½ in (23–32 cm)
Tail 7½–10½ in (19–27 cm)
Weight c.⅞ lb (400 g)
Social unit Individual/Pair
Status Vulnerable

One of four brightly colored "giant" or "checkered" sengis, this species has an orange head and upper body, graduating through red on the back to black on the rump—and with an orange-brown tail. It occurs in southeastern Kenya and northeastern Tanzania, including Zanzibar Island. "Giant" sengis of the genus *Rhynchocyon* are nervous, twitchy forest-dwelling animals and do not bound as much as other members of the family.

MAMMALS

Tenrecs and golden moles

PHYLUM	Chordata
CLASS	Mammalia
ORDER	Afrosoricida
FAMILIES	2
SPECIES	53

Genetic studies show that tenrecs and golden moles, which were once classified with shrews, moles, and hedgehogs, are actually unrelated to them. The Madagascan tenrecs (and semiaquatic African otter shrews) include species that are unusual among mammals in having a reptilian cloaca: a common opening to the anus and the urinogenital tract. They feed mainly on invertebrates. The African golden moles have strong limbs for burrowing, tiny skin-covered eyes, and lack external ears. They prey on underground animals, such as earthworms and burrowing lizards.

NOCTURNAL HUNTER
At night, Grant's golden moles forage for food on the surface, feeding mainly on termites. They will, however, eat a wide variety of invertebrates, and small lizards too, if they can catch them. Here, Grant's golden mole feasts on a locust.

Tenrec ecaudatus

Common tenrec

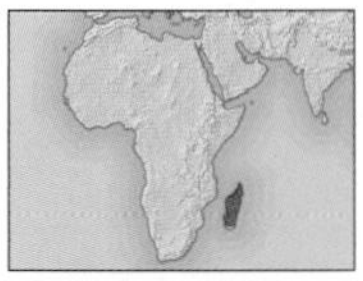

Location Madagascar

Length 10–15½ in (26–39 cm)
Tail ⅜–½ in (1–1.5 cm)
Weight 3¼–5½ lb (1.5–2.5 kg)
Social unit Individual
Status Least concern

The 30 species of tenrecs are mainly from central Africa and Madagascar. Most resemble a combination of shrew and hedgehog. The nocturnal, common tenrec is the largest land-dwelling species. It has coarse, gray to reddish gray fur and sharp spines. Using its long, mobile snout, it grubs among leaves for worms and other small creatures. It also scavenges and hunts frogs and mice. In defense, a tenrec squeals, erects the spiny hairs on its neck into a crest, jumps and bucks, and readily bites. It shelters by day in a nest of grass and leaves under a log, rock, or bush. After a gestation of 50–60 days, a litter of 10–12 is born. When young, they are striped black and white.

Limnogale mergulus

Web-footed tenrec

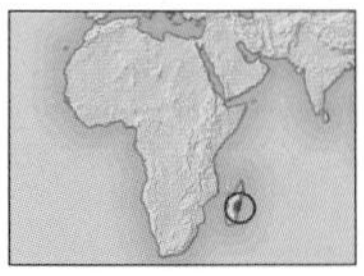

Location E. Madagascar

Length 4¾–6½ in (12–17 cm)
Tail 4¾–6½ in (12–16 cm)
Weight 1⁷⁄₁₆–2⅛ oz (40–60 g)
Social unit Individual
Status Vulnerable

Once believed extinct, the only aquatic species of tenrec has been relocated in the wild but information is very scarce. It has an otterlike form, with a broad, round-muzzled head, long whiskers, small high-set eyes and ears, and webbed rear feet with sharp claws, used to grip slippery rocks and prey. The short, dense, water-repellent fur is reddish brown. The web-footed tenrec noses among stones and weeds for water insects, crabs, and crayfish, propelled mainly by means of its long, scantily haired tail.

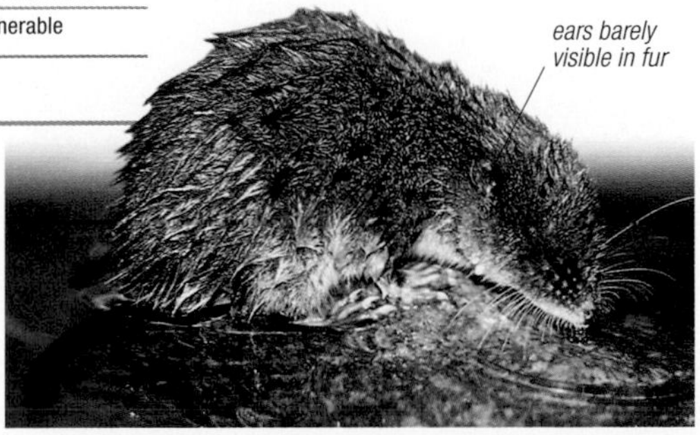

Micropotamogale lamottei

Nimba otter-shrew

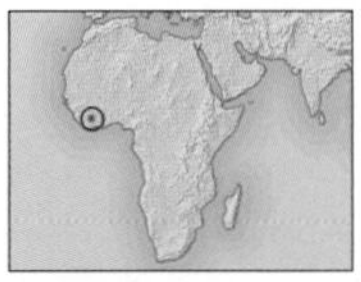

Location W. Africa (Mount Nimba area)

Length 4¾–8 in (12–20 cm)
Tail 4–6 in (10–15 cm)
Weight 4 oz (125 g)
Social unit Individual
Status Endangered

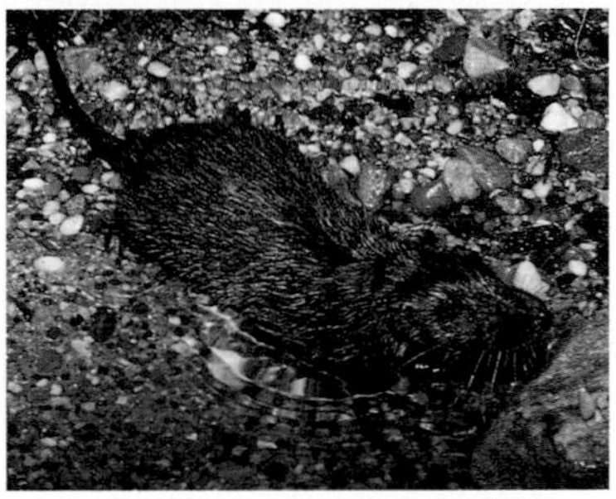

This otter-shrew is largely restricted to upland forest streams in an area of about 580 square miles (1,500 square km) around Mount Nimba in West Africa. It has a fleshy nose, rounded head, compact body, and long tail. Long, gray or dark brown fur usually hides the eyes and most of the ears. It is mostly nocturnal, catching small fish, crabs, water insects, and similar prey on short dives or along river banks, and eating them on land. It digs a short nesting burrow in soft soil.

Hemicentetes semispinosus

Lowland streaked-tenrec

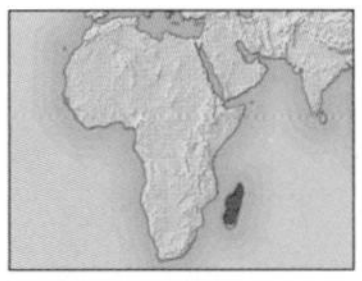

Location Madagascar

Length 6½–7½ in (16–19 m)
Tail None
Weight 2⅞–10 oz (80–275 g)
Social unit Group
Status Least concern

This species is distinguished from other tenrecs by its two-tone color of background black with variable stripes of white, yellow, or brown. The fur is coarse with scattered spines and a patch of spiky hairs on the crown, which is erect as a crest. The main foods are worms and grubs. The streaked tenrec lives in groups of 15 or more, and all help to protect each female's litter of 2–4 young.

Potamogale velox

Giant otter-shrew

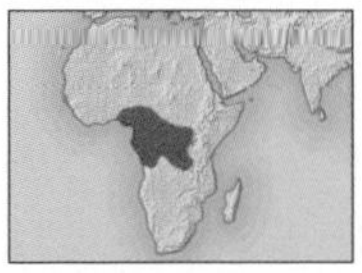

Location W. and C. Africa

Length 11½–14 in (29–35 cm)
Tail 9½–35 in (24–90 cm)
Weight 13 oz (350 g)
Social unit Individual
Status Least concern

Easily mistaken at a glance for a small otter (which is a mustelid rather than an insectivore), the giant otter-shrew has a rounded muzzle, a long, flexible body, and a long, muscular tail flattened from side to side, which it uses to propel itself through water. This species inhabits a variety of freshwater habitats, from still pools to mountain torrents at altitudes of 6,000 ft (1,800 m). The eyes and ears are small and high-set, for swimming low in the water; yet, unlike an otter, the toes lack webs. Mainly nocturnal and solitary outside the breeding season, the giant otter-shrew hunts primarily at night for fish, frogs, shellfish, and other freshwater animals. Its bankside burrow, where it shelters by day, has an underwater entrance.

MAMMALS

Setifer setosus

Greater hedgehog-tenrec

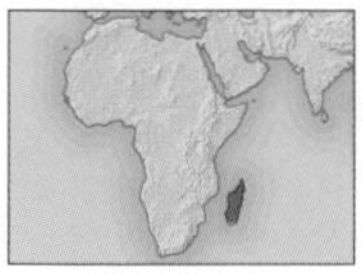

Location Madagascar

Length 6–9 in (15–22 cm)
Tail ½ in (1.5 cm)
Weight 6–10 oz (175–275 g)
Social unit Individual
Status Least concern

This hedgehoglike tenrec has short, pointed, white-tipped spines on its long body and coarse hair, varying from gray to black, on the head and legs. Unlike true hedgehogs, it is active by day but, like them, it rolls into a prickly ball when threatened. It climbs well and eats a wide variety of worms, amphibians, reptiles, insects, carrion, fruit, and berries. It becomes torpid for several weeks in adverse conditions.

Chrysochloris asiatica

Cape golden mole

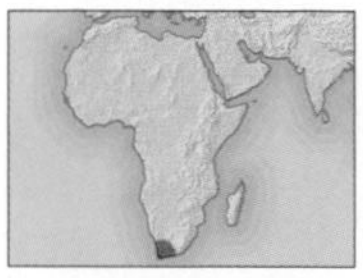

Location Southern Africa

Length 3½–5½ in (9–14 cm)
Tail None
Weight Not recorded
Social unit Individual
Status Least concern

African golden moles (chrysochlorids) belong to a different order from the true moles (Soricomorpha). The Cape golden mole has the soft, dense fur typical of moles, which may appear olive, brown, or gray depending on the direction of the light. Its snout has a hairless nose pad, its eyes and ears are tiny, and each front leg has two large digging claws—all adaptations for a tunneling lifestyle. It is solitary and eats worms, grubs and other soil creatures found when burrowing or that have fallen into its more permanent tunnels.

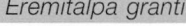

Eremitalpa granti

Grant's golden mole

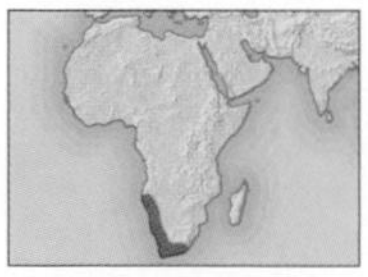

Location Southern Africa

Length 2¾–3¼ in (7–8 cm)
Tail None
Weight 9/16–1 1/16 lb (15–30 g)
Social unit Individual
Status Least concern

Long, soft, silky fur covers almost the entire body of Grant's golden mole, and varies from steel-gray to buff or white. This mole has tiny, almost invisible eyes and ears, a hard, hairless nose pad, and 3 long, broad claws on each foot. It pushes through loose sand as though "swimming," making more permanent tunnels deeper down in sand or near the surface in harder, more compacted soil. The main components of its diet are various small desert animals, from ants, termites, and beetles to lizards and snakes. Grant's golden mole is solitary, and probably active for short periods throughout the day and night. It hardly, if ever, comes to the surface except to locate a mate. The specialized desert habitat of this species is under increasing threat from mining and other human activities.

Aardvark

PHYLUM	Chordata
CLASS	Mammalia
ORDER	Tubulidentata
FAMILIES	1 (Orycteropodidae)
SPECIES	1

The only surviving member of the order Tubulidentata, the aardvark is a solitary animal found in Africa. It is characterized by nonfunctional, columnar cheek teeth, a long snout, large ears, a piglike body, and powerful limbs and shovel-shaped claws for digging. Ears that can be folded back (and a profusion of nostril hairs) help keep out dirt when burrowing. The aardvark has a primitive brain and poor eyesight. It does, however, have an excellent sense of smell. This is used to locate termites and ants, which are then captured, using the animal's long, sticky tongue.

MASTER BURROWER
The aardvark is a fast and prolific burrower, using its strong, clawed front limbs to dig and its back feet to push away excavated soil. Some aardvark burrows consist of an extensive tunnel network; others are shorter and provide temporary refuge. The aardvark always exits its burrow head first, as shown here.

Orycteropus afer

Aardvark

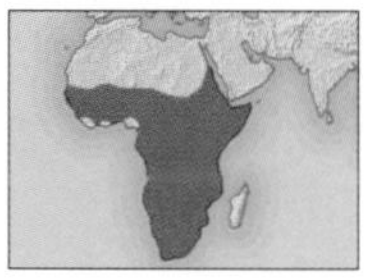

Location Africa (south of the Sahara)

Length 5¼ ft (1.6 m)
Tail 22 in (55 cm)
Weight 84–140 lb (38–64 kg)
Social unit Individual
Status Least concern

Also known as ant-bear, the aardvark is one of the most powerful mammal diggers, excavating burrows up to 33 ft (10 m) long around its home range of ¾–2 square miles (2–5 square km). The single young is born after a gestation period of 243 days and weighs 3¾ lb (1.7 kg). Aardvarks chew one species of ant with their molar teeth, but other species of ant and termites are swallowed whole and probably ground up in their muscular stomach.

HUNCHED BACK
The aardvark has a distinctive curved back, and its snout, ears, and tail are long and tapering. The bristly, scant, brown fur is tinged with yellows and grays.

SEASONAL DIET

The aardvark favors ants as its food, which are more abundant in summer, but it also eats termites at times when ants are not available. It breaks into a nest or mound using its front feet, which are armed with long, sharp claws. The dense mat of hairs that surrounds the aardvark's nostrils effectively filters dust as it digs.

Dugong and manatees

PHYLUM	Chordata
CLASS	Mammalia
ORDER	Sirenia
FAMILIES	2
SPECIES	4

Sirenians—the dugong and manatees—are large, slow-moving creatures with a streamlined body. They are the only marine mammals that feed primarily on plants. Sirenians must rise to the surface to breathe, but they can remain submerged for up to 20 minutes. Even though they have no enemies apart from humans, sirenians may number fewer than 150,000—making them among the least abundant of any mammal order.

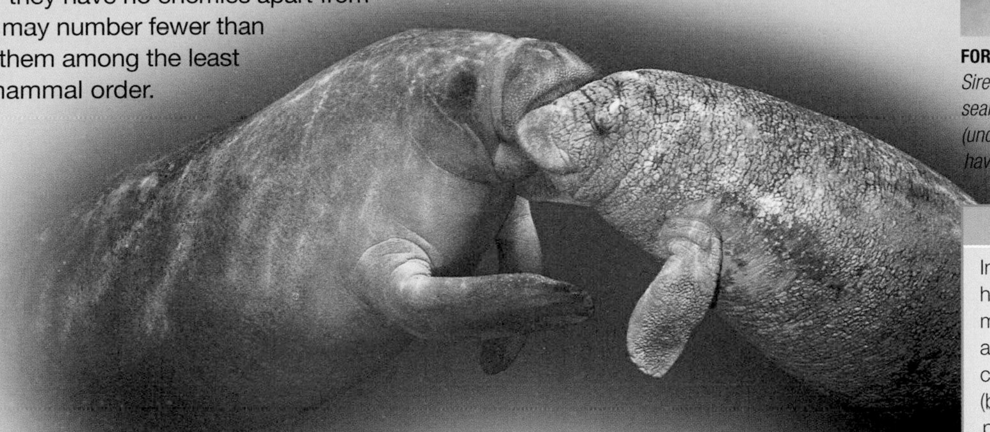

Anatomy

Sirenians have paddlelike front limbs and a flat tail to aid propulsion. Their skin is thick and tough, and they have a relatively small brain. Due to the large volume of gas given off during the digestion of plant matter, sirenians are highly buoyant. To compensate, their bones are heavy and dense. Underwater, sirenians can close their nostrils and contract their eyelids.

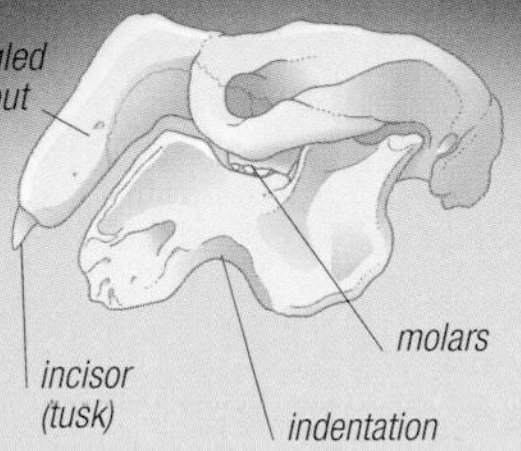

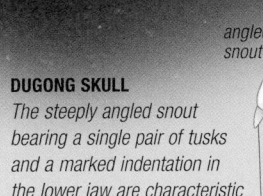

DUGONG SKULL
The steeply angled snout bearing a single pair of tusks and a marked indentation in the lower jaw are characteristic features of a dugong skull.

MOTHER AND CALF
Sirenians are slow breeders: usually only one calf is produced every 2 years. The "mouthing" contact between this manatee mother and calf helps preserve the family bond.

Feeding

Sirenians strip vegetation using their large, mobile upper lip. Food is then crushed between horny plates on the front part of the palate and on the lower jaw, and finally ground between the teeth.

FORAGING ON THE SEA BED
Sirenians, such as this dugong, frequently search the sea bed for the rhizomes (underground stems) of sea grasses, which have high concentrations of carbohydrates.

CONSERVATION

In the past, sirenians were hunted extensively for their meat, hide, and oil. Today, they are vulnerable to injury or death caused by boat propellers (below), fishing nets, and the pollution of coastal waters.

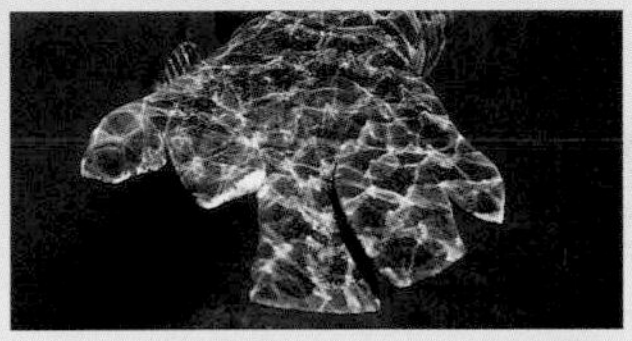

Dugong dugon

Dugong

Length 8¼–13 ft (2.5–4 m)
Weight 550–1,985 lb (250–900 kg)
Social unit Group
Status Vulnerable

Location E. Africa, W. Asia, S. Asia, S.E. Asia, Australia, Pacific islands

The dugong has a crescent-shaped tail, and short foreflippers. Primarily diurnal, it moves regularly each day between on- and offshore areas depending on the tides and food supply; in some areas, it undertakes longer seasonal migrations—perhaps hundreds of kilometres—to follow marine plant growth and avoid cold-water currents. Some dugongs are solitary but most form loose groups averaging 10–20, exceptionally 100 or more, with little social structure. Members may gather to intimidate and butt predators such as sharks. Males compete for females by sounds and pushes. Courtship and mating (which is monogamous) are similarly auditory and tactile. One offspring, up to 4 ft (1.2 m) long and 77 lb (35 kg) in weight, is born after a gestation of 13–14 months. It is cared for by the mother, with help from older siblings and female relatives, and is weaned by 18 months. A dugong may live for about 60 years.

Trichechus manatus

West Indian manatee

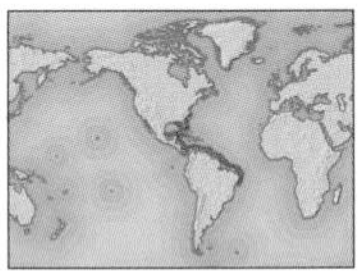

Length 8¼–15 ft (2.5–4.5 m)
Weight 440–1,320 lb (200–600 kg)
Social unit Variable
Status Vulnerable

Location S.E. USA to N.E. South America, Caribbean

Best known of the 3 manatee species (the other 2 being the Amazonian and West African), the West Indian manatee lives along shallow shores and estuaries, and in nearby rivers and freshwater lagoons. Groups of up to 20 swell to 100 or more where food is plentiful; however, there is little cohesion, and individuals come and go, ranging widely. Reproduction is similar to the dugong's (see left), although the manatee is polygynous.

FOOD INTAKE

Feeding occurs from the surface down about 13 ft (4 m). The manatee holds food with its flippers, and directs it into the mouth using its flexible lips. Daily food intake is up to a quarter of its body weight, and may include a few fish (for protein).

BLUNT FRONT
Like other sirenians, this species has tiny eyes and no external ear flaps. Its gray or gray-brown skin, paler below, may harbor growths of algae.

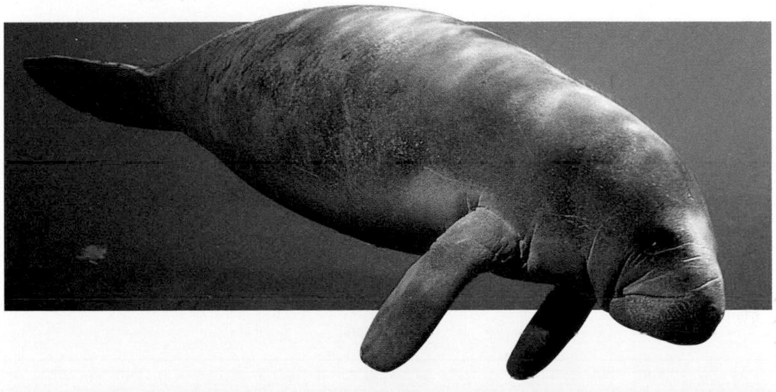

Elephants

PHYLUM	Chordata
CLASS	Mammalia
ORDER	Proboscidea
FAMILIES	1 (Elephantidae)
SPECIES	3

The largest living animals on land—the male African elephant may be as tall as 13ft (4m) and weigh nearly 11 tons (10tonnes)—elephants are characterized by pillarlike legs, a thick-set body with a convexly curved spine, large ears (relatively smaller in Asian elephants), and a heavy head with a long, mobile trunk. African and Asian elephants live in savanna and light forest; African forest elephants (recently given species status) mainly live deep in the African rain forest (they occasionally venture on to the savanna). Elephants live for about 65–70 years—longer than any other mammals except humans. Males (and females to a lesser degree) grow throughout life: 50-year-old males are noticeably larger than 20 year olds.

Anatomy

Perhaps the most distinctive feature is the trunk, a flexible elongation of the upper lip and nose that consists of thousands of muscle pairs. It is used like a "5th limb" to pluck grass, pull down branches, lift logs, or squirt water or dust. Also immediately noticeable are the tusks (upper incisors), which are large, thick, and curved in most bull elephants; cows have smaller tusks (in female Asian elephants they do not protrude beyond the lip). The skeleton consists of thick, heavy bones, which are able to support the animal's great weight. The large, fan-shaped ears, which contain a network of blood vessels, are constantly in motion to aid heat loss. In aggressive displays, the ears are spread sideways. The skin is thick, finely wrinkled, and sparsely haired.

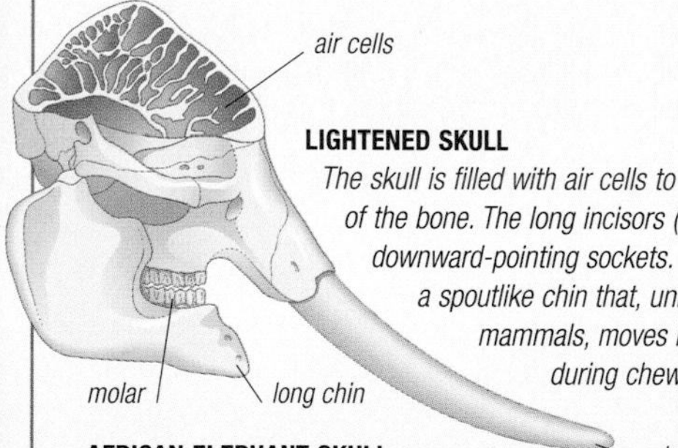

LIGHTENED SKULL
The skull is filled with air cells to lighten the weight of the bone. The long incisors (tusks) have deep, downward-pointing sockets. The lower jaw has a spoutlike chin that, unlike in most mammals, moves horizontally during chewing.

AFRICAN ELEPHANT SKULL

Feeding

Elephants have large, ridged cheek teeth (molars and premolars) to deal with their coarse diet of bark, leaves, branches, and grass (African forest elephants also eat fruit). In eating these foods, elephants cause enormous damage: grass is pulled up in tufts, branches are broken off, bark is stripped, and small trees are sometimes uprooted. Some areas have alternated between closed woodland and open savanna, depending on the number of elephants living there.

STRIPPING FOLIAGE
Elephants use their mobile trunk to pull down branches. An adult needs to eat about 350lb (160kg) of food daily.

DIGGING FOR SALT
Elephants often need to supplement their diet with extra salt. This juvenile African elephant is loosening fragments of salt-rich soil with its tusks. Juveniles learn from older members of the herd where to find salt.

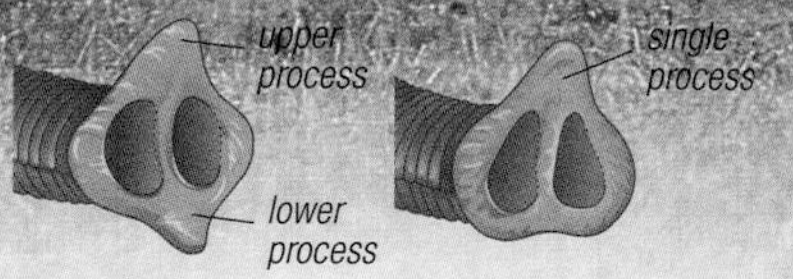

TRUNK SHAPE
African elephants have 2 opposing, fingerlike outgrowths called processes at the tip of the trunk; Asian elephants have one. In both, the processes are used to pick up small objects.

TAKING A DUST-BATH

DAILY ROUTINE
To keep the skin healthy, African elephants take a daily dust-bath.

SUCKING UP THE DUST
Dust is sucked up into the trunk, which is a tubular extension of the upper lip.

PROTECTION
The dust acts as a sunscreen, protecting the elephant's skin from the direct rays of the sun.

REPELLING INSECTS
Dust is also a good insect repellent, deterring insects from biting the sensitive skin.

MAMMALS

Social group

Elephants live in family groups that consist of the oldest, most experienced female (the matriarch) and other females of various ages (and their young). For protection, or when feeding in lush areas, small herds of African elephants may join together to form groups made up of several hundred individuals. African forest elephants and Asian elephants live in small family groups only. Males, however, only join the herd when a female is sexually receptive and are otherwise either solitary (older bulls) or live in bachelor groups (young bulls). Adult bull Asian elephants have annual periods of sexual excitement, called "musth" (bull African elephants have an equivalent condition, about which less is known).

PROTECTING YOUNG
Elephant calves are protected from predators and other dangers by all members of the herd, which are usually blood relatives. This Asian elephant calf is only a few weeks old and remains close to its mother. Two smaller females are close by, ready to assist the mother if necessary.

LIVING TOGETHER
This is a typical African elephant family group. Communication within a herd takes many forms, including vocalizations (some of which are below the range of human hearing), touch, foot stamping, and body postures. Cooperative behavior—such as employing a system of lookouts while bathing—is common.

CONSERVATION

The Asian elephant is endangered due to competition with a growing human population, and its expanding need for land. The same is true of African elephants, where their original habitat of forest and savanna is becoming increasingly fragmented and farmed. African elephants also face the hazard of hunting—a traditional practice that is a source of meat, and of highly valuable ivory. During the 20th century, ivory prices soared, and commercial hunting became widespread. Elephant numbers crashed—an effect spurred on by periodic droughts. In 1989, the Kenyan authorities acted, burning a stockpile of seized ivory that sent a worldwide message. In 1990, the international sale of ivory was banned, although in the 21st century, strong demand for it remains.

5

EPOSITING THE DUST
e elephant blows out the dust through the unk, depositing it on the back and head.

6

KEEPING THE SKIN HEALTHY
To maintain good skin condition, regular dust-baths are as important as water-baths.

Loxodonta africana

African savanna elephant

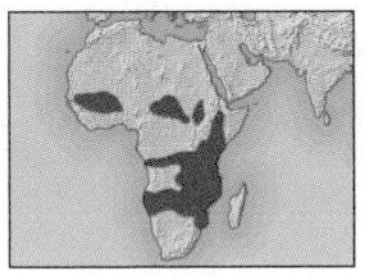

Location Africa

Length 13–16 ft (4–5 m)
Tail 3¼–5 ft (1–1.5 m)
Weight 4⅜–7¾ tons (4–7 tonnes)
Social unit Group
Status Vulnerable

Although it is also called the African bush or savanna elephant, this species—the largest of the 3 elephant species—lives in varied habitats from desert to high rain forest. It has larger ears than the Asian elephant, a concave curve to its back, and 2 processes on the tip of the trunk rather than one (see opposite). Both male and female African savanna elephants have forward-curving tusks (incisor teeth), which are sometimes used as tools to loosen mineral-rich soil that is then eaten. Requiring substantial amounts of food and a large area in which to forage for it, a herd of African savanna elephants may cause dramatic changes to the environment, especially during prolonged periods of drought.

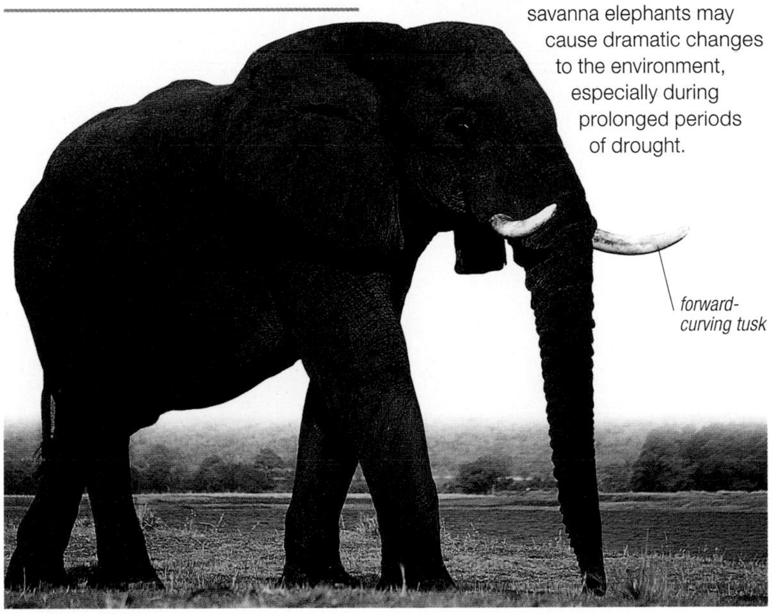

Loxodonta cyclotis

African forest elephant

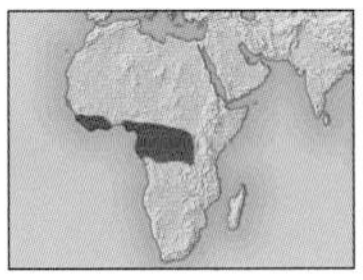

Location W. and C. Africa

Length 9¾–13 ft (3–4 m)
Tail 2½–4 ft (50–120 cm)
Weight 1–3¼ tons (0.9–3 tonnes)
Social unit Group
Status Endangered

Formerly regarded as a subspecies of the African savanna elephant, this species is smaller, has darker skin, more rounded ears, and a hairier trunk. The yellow or brownish tusks are parallel and point downward, adaptations that allow the African forest elephant to move freely through dense vegetation.

Elephas maximus

Asiatic elephant

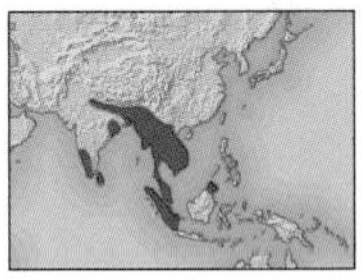

Location S. and S.E. Asia

Length Up to 11 ft (3.5 m)
Tail 3¼–5 ft (1–1.5 m)
Weight 2¼–5½ tons (2–5 tonnes)
Social unit Group
Status Endangered

Asiatic elephants have smaller ears than the African species, and a unique tip to their trunk (see opposite). The tusks are small and may be absent in females (above). The molar teeth are very like those of the extinct mammoth, which suggests a close relationship between them. The Asiatic elephant has a long association with humans, and animals from all 4 subspecies—Malaysian, Sumatran, mainland Indian, and Sri Lankan—have been domesticated.

Hyraxes

PHYLUM	Chordata
CLASS	Mammalia
ORDER	Hyracoidea
FAMILIES	1 (Procaviidae)
SPECIES	4

Although hyraxes resemble rabbits in size and shape, genetic evidence suggests that they are more closely related to primitive hoofed mammals. The pads of their feet are moistened by glandular secretions, which make the soles more adhesive. This, along with the opposable toe on each back foot, enables hyraxes to climb steep rock faces. They are found in Africa and parts of the Middle East. Some species inhabit rocky outcrops; others are arboreal. They can survive food shortages because they eat almost every type of vegetation and require very little water (their kidneys are highly efficient).

KEEPING WARM
Hyraxes have poor temperature control. These rock hyraxes are huddling together for warmth.

Procavia capensis

Rock hyrax

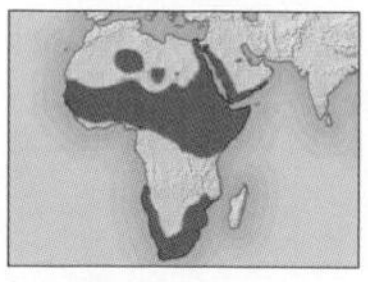

Location W., S., and E. Africa, W. Asia

Length	12–23 in (30–58 cm)
Tail	8–12 in (20–31 cm)
Weight	$6\frac{1}{2}$–11 lb (3–5 kg)
Social unit	Group
Status	Least concern

The rock hyrax has a plump body and short, dense, gray or gray-brown fur, which is paler below. Also called the rock dassie, it lives in colonies of 4–40, usually of one dominant male, other males, and females and young. Although found in a wide variety of habitats, it usually lives in rocky outcrops and crags, among boulders, where it makes a grass-lined nest.

Dendrohyrax arboreus

Southern tree hyrax

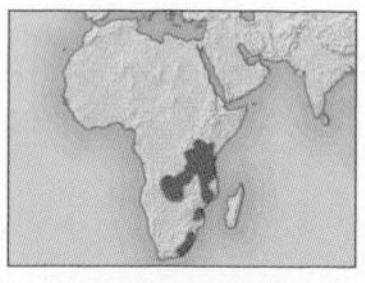

Location E. and southern Africa

Length	16–28 in (40–70 cm)
Tail	$\frac{3}{8}$–$1\frac{1}{4}$ in (1–3 cm)
Weight	$3\frac{1}{4}$–10 lb (1.5–4.5 kg)
Social unit	Variable
Status	Least concern

The southern tree hyrax or tree dassie has a yellowish patch on the back near the rump, but is otherwise gray-brown with buff underparts. The head, legs, and tail seem almost too small for the stocky body. True to its name, the southern tree hyrax lives among trees, shrubs, and creepers, and nests in a tree hole. It feeds on the ground only occasionally. Like all hyraxes, this species has poor internal control of body temperature, so groups often sunbathe to warm up, or rest in shade to cool down. Litters of 1–3 are born after a gestation period of 7–8 months.

Armadillos

PHYLUM	Chordata
CLASS	Mammalia
ORDER	Cingulata
FAMILIES	1
SPECIES	21

Instantly recognized by their protective armorlike covering of hardened skin on their head, back, sides, and limbs, and reduced fur, armadillos are ground-living animals allied to anteaters and sloths. They have simple peglike teeth and short, strong limbs for digging; extra support for excavating burrows is provided by special articulations of the lower vertebral column (a feature shared with nonburrowing anteaters and sloths). Armadillos are mainly insect-eaters, but also eat a wide range of other items, including plant materials, eggs, and carrion. Most species occur in open habitats of South America, but one can also be found in the United States.

ARMADILLOS
Contrary to popular belief, not all armadillos can roll in a ball to defend themselves. Only species of the genus Tolypeutes *can do this.*

Priodontes maximus

Giant armadillo

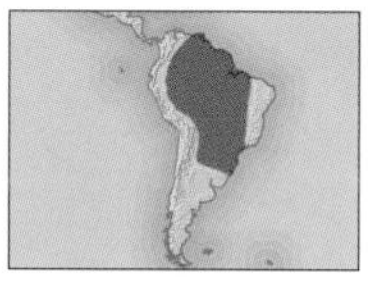

Location N. and C. South America

Length	30–39 in (75–100 cm)
Tail	20 in (50 cm)
Weight	66 lb (30 kg)
Social unit	Individual
Status	Vulnerable

By far the largest armadillo, this species has 11–13 bands of slightly flexible, hinged plates over its body, and 3–4 bands over its neck. The long, tapering tail is likewise armored. The main body color is brown, with a pale yellow-white head, tail, and band along the lower edges of the plates. The especially large third front claw is used to rip up soil for small food items—mainly termites, ants, worms, spiders, small snakes, and lizards. The front claws also dig a burrow in which the giant armadillo shelters by day. It feeds in an area for 2–3 weeks, then moves on. Like most other armadillos, it shows little social or territorial behavior. The gestation period is 4 months, and the 1–2 offspring are weaned by 6 weeks and are sexually mature by 12 months.

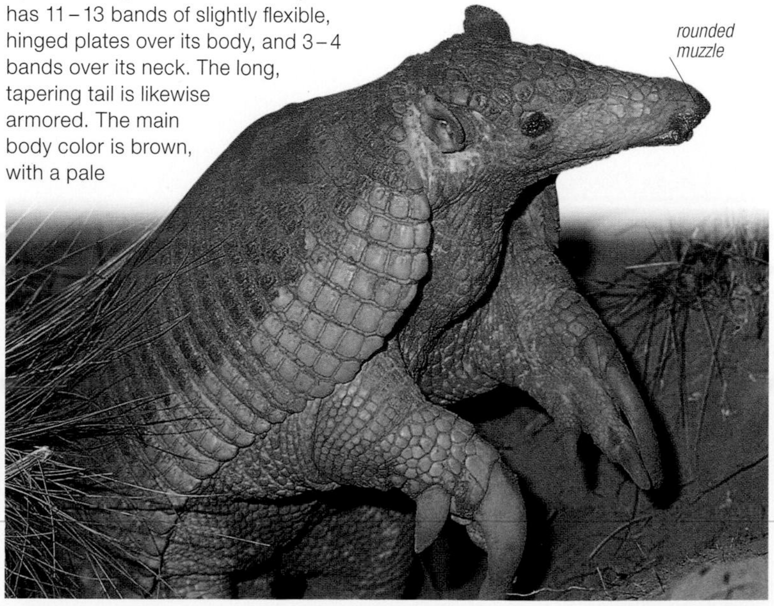

Chaetophractus villosus

Big hairy armadillo

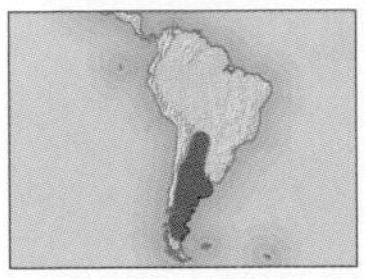

Location S. South America

Length 9–16 in (22–40 cm)
Tail 3½–6½ in (9–17 cm)
Weight 2¼–6½ lb (1–3 kg)
Social unit Individual
Status Least concern

This armadillo of arid habitats has long, coarse hairs projecting between the 18 or so distinct bands of bony, skin-covered armor on its body. Some of the 7 or 8 bands are hinged. When threatened and unable to burrow, it tucks in its feet and presses its armored body into the ground. This protects its softer, white- or brown-furred underside and provides an effective defense against both canids and birds of prey. In summer, it is mainly nocturnal and eats varied small prey, from grubs to rodents. In winter, activity is mainly by day and the diet includes more plant matter.

Dasypus novemcinctus

Nine-banded armadillo

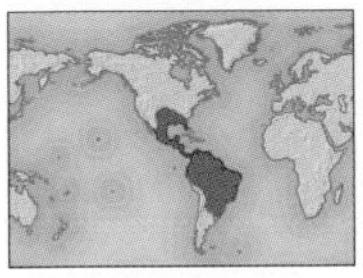

Location S. USA, Mexico, Caribbean, Central America, South America

Length 14–22½ in (35–57 cm)
Tail 9½–18 in (24–45 cm)
Weight 5½–14 lb (2.5–6.5 kg)
Social unit Individual
Status Least concern

The most commonly seen armadillo, this species has 8–10 bony bands around its middle, which allows some flexibility. The bony armor and leathery skin account for one-sixth of the total weight. Like most armadillos, it digs an extensive burrow system. It takes most foods, from ants and birds to fruit and roots, and is solitary yet may share a burrow with others of its kind. Offspring are nearly always quadruplets—4 of the same sex.

Zaedyus pichiy

Pichi

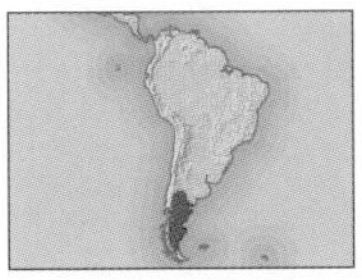

Location S. South America

Length 10–13½ in (26–34 cm)
Tail 4–4¾ in (10–12 cm)
Weight 2¼–4½ lb (1–2 kg)
Social unit Individual
Status Near threatened

When threatened, the pichi squats down and draws its sharp-clawed feet under its body, relying on its wide, low, domed body armor for protection, or it wedges itself into a burrow with its armor facing outward. This small armadillo digs a short tunnel for shelter and eats various small insects, worms, other invertebrates, and sometimes also carrion.

Cabassous centralis

Northern naked-tailed armadillo

Location Central America and N. South America

Length 12–16 in (30–40 cm)
Tail 5–7 in (13–18 cm)
Weight 2–3.5 kg (4½–7¾ lb)
Social unit Individual
Status Data deficient

Occupying a wide range of habitats, this big-eared armadillo has large claws, especially the middle forefoot claw, for digging up prey and making a burrow to shelter in by day. It also has a long, sticky tongue to lick up termites and ants in the manner of an anteater. In common with other armadillos, it is mainly silent, perhaps growling when threatened, as it tries to dig itself into the ground so only the armored upper body is exposed.

Sloth and anteaters

PHYLUM	Chordata
CLASS	Mammalia
ORDER	Pilosa
FAMILIES	4
SPECIES	11

Sloths and anteaters are a tropical American group of mammals united by a coarse coat and a highly specialized diet: tubular-snouted anteaters are insectivorous and slow-moving. Sloths are vegetarian, mainly eating leaves of a limited number of tree species. They have reduced dentition: sloths have only cheek teeth, and anteaters lack teeth altogether. Anteaters rip apart nests of ants and termites using enlarged front claws, and lap up prey with a long tongue and sticky saliva. Giant anteaters are ground-dwellers, but other species climb. Sloths are entirely arboreal: they have long, slender forelimbs for climbing through branches, long claws for grasping, and have complex digestive systems for digesting foliage.

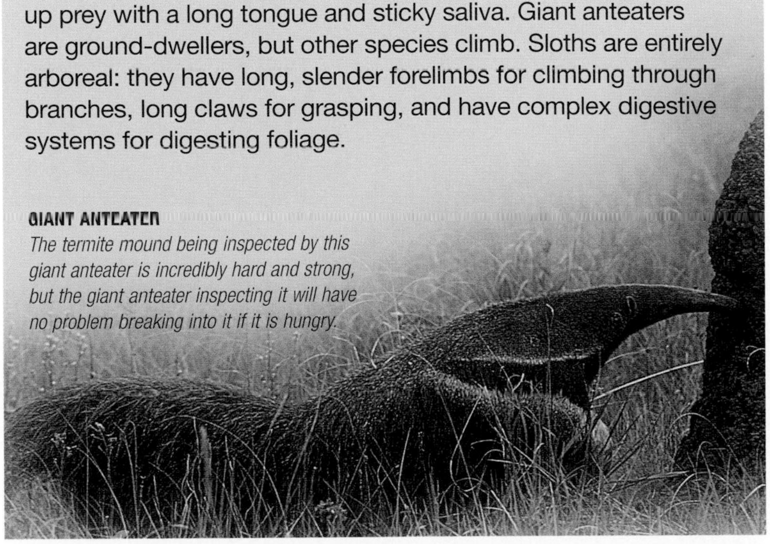

GIANT ANTEATER
The termite mound being inspected by this giant anteater is incredibly hard and strong, but the giant anteater inspecting it will have no problem breaking into it if it is hungry.

Bradypus torquatus

Maned sloth

Location E. South America

Length 18–20 in (45–50 cm)
Tail 1½–2 in (4–5 cm)
Weight 7¾–8¾ lb (3.5–4 kg)
Social unit Individual
Status Endangered

This species has the typical sloth's small head, tiny eyes and ears, and small tail hidden in the fur, which contrasts with the large body and powerful limbs. Algae, mites, ticks, beetles, and even moths live in the coarse outer coat, which is longer, darker, and manelike around the head, neck, and shoulders. The underfur is fine, dense, and pale. The maned sloth eats the leaves, buds, and soft twigs of a few forest trees, notably *Cecropia*. It comes to the ground only to defecate, or to move to another tree if it cannot travel through the branches. On the ground, the sloth drags itself along by its longer, stronger front legs and claws. Surprisingly, it can swim well. In addition to its physical slowness, the sloth's muscles are small and weak for its overall body size, and even its metabolism is less rapid than most other mammals, giving a low body temperature of just above 86° F (30° C). Its main defense is to stay still and unnoticed or to lash out with its formidable claws. After 5–6 months' gestation, one offspring is born, weighing about 9 oz (250 g). It clings to its mother's abdomen with its well-formed, hook-shaped claws. The young suckles for up to 4 weeks and after weaning stays with her, being carried and learning feeding patterns, for a further 6 months.

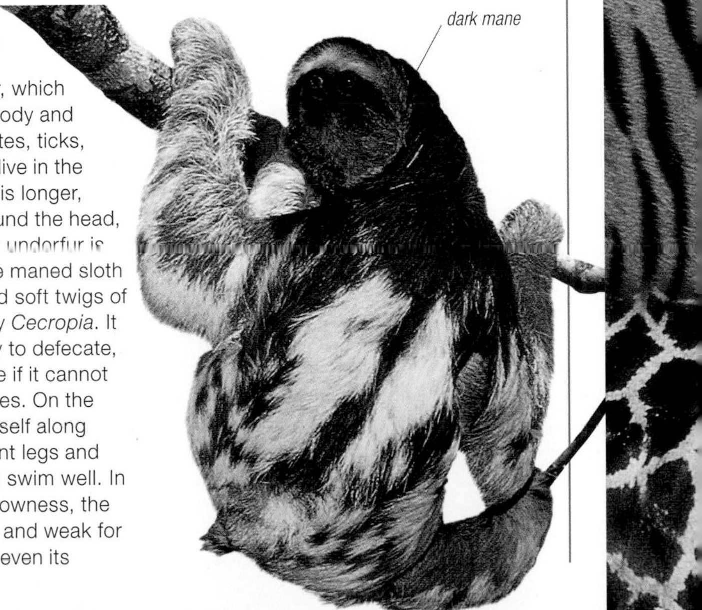

Bradypus pygmaeus

Pygmy three-toed sloth

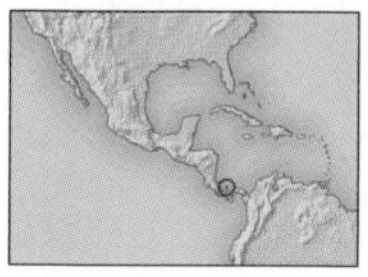

Length 18¾–20½ in (48–53 cm)
Tail 1¾–2¼ in (4.5–6 cm)
Weight 5½–7¾ lb (2.5–3.5 kg)
Social unit Individual
Status Critically endangered

Location Central America

Confined to Isla Escudo de Veraguas off the Caribbean coast of Panama, this diminutive sloth evolved in isolation from its larger mainland ancestors. It is the only sloth that subsists entirely on red mangrove leaves—an adaptation that restricts it to mangrove forest beside the sea. This critically endangered species is threatened by poaching from visiting fishermen and potential tourism development.

Choloepus didactylus

Linnaeus' two-toed sloth

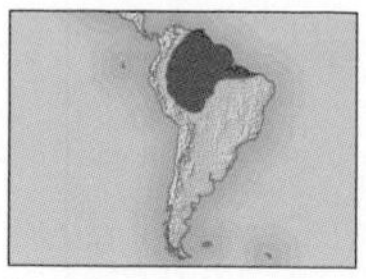

Length 18–34 in (46–86 cm)
Tail ½–1½ in (1.5–3.5 cm)
Weight 8¾–19 lb (4–8.5 kg)
Social unit Individual
Status Least concern

Location N. South America

Linnaeus' two-toed sloth has 2 hook-clawed toes on each front foot but 3 on each rear one. Its coarse fur is gray-brown, paler on the face, but, like other sloths, it may be tinged green by algae growing on the hairs. It eats a typical sloth's diet of leaves and fruit and, in common with all sloths, is solitary, moves extremely slowly, and descends to the ground only to defecate (about once a week).

Tamandua tetradactyla

Southern tamandua

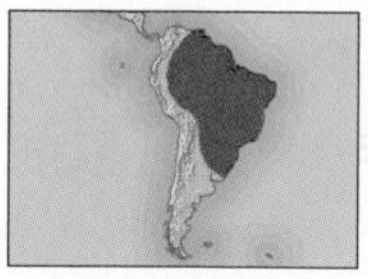

Length 21–35 in (53–88 cm)
Tail 16 in (40 cm)
Weight 7¾–19 lb (3.5–8.5 kg)
Social unit Individual
Status Least concern

Location N. and E. South America

Also called the collared anteater, the southern tamandua, has a long, narrow head and sparsely haired, prehensile tail. It climbs well and feeds both in the branches and on the ground, breaking into the nests of ants, termites, and bees. It is active for 8-hour periods, day or night. A single offspring is born after 4–5 months' gestation and rides on the mother.

SELF-DEFENSE

When threatened, a southern tamandua backs against a trunk or rock, rears up onto its hind legs, props itself up by its tail, and holds its powerful front legs outstretched. In this position it can slash out at its attacker with the long, sharp claws on its front feet.

BLACK VEST
The southern tamandua is pale yellow with a black "vest" over the shoulders, chest, sides, and lower back.

Cyclopes didactylus

Silky anteater

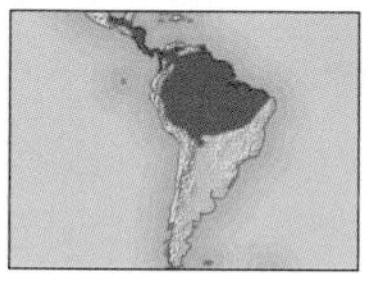

Length 6½–8½ in (16–21 cm)
Tail 6½–9 in (16–23 cm)
Weight 5–10 oz (150–275 g)
Social unit Individual
Status Least concern

Location Central America to N. South America

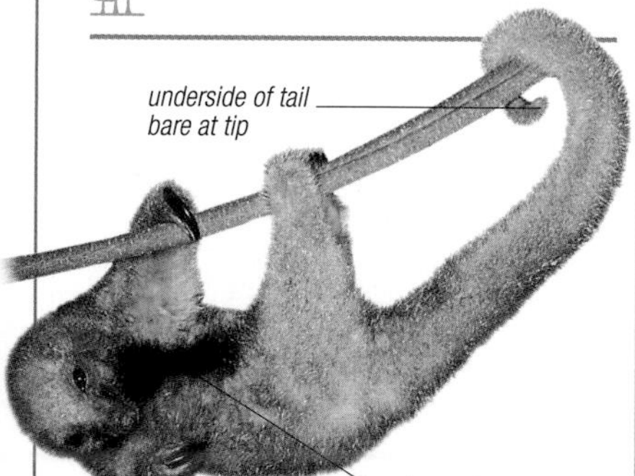

Solitary and nocturnal, this anteater has long, dense, fine fur, usually smoky gray with a silver sheen and a variable brown stripe on the side of the body. Specialized for arboreal life, it grips strongly with its feet and hooklike claws, supported by its prehensile tail, which is bare on the underside near the tip. It breaks open tree-ant nests and licks out the ants with its long, saliva-coated tongue.

Myrmecophaga tridactyla

Giant anteater

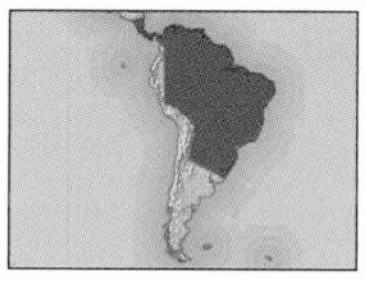

Length 3¼–6½ ft (1–2 m)
Tail 25–35 in (64–90 cm)
Weight 49–86 lb (22–39 kg)
Social unit Individual
Status Vulnerable

Location Central to South America

ANTS'-EYE VIEW

This anteater rips open ant nests and termite mounds with its big-clawed front feet, and then uses its tongue, which can protrude more than 2 ft (60 cm), to take its prey. The tongue is covered with minute, backward-pointing spines and sticky saliva, to which tiny prey adhere. The remaining termites then repair the nest.

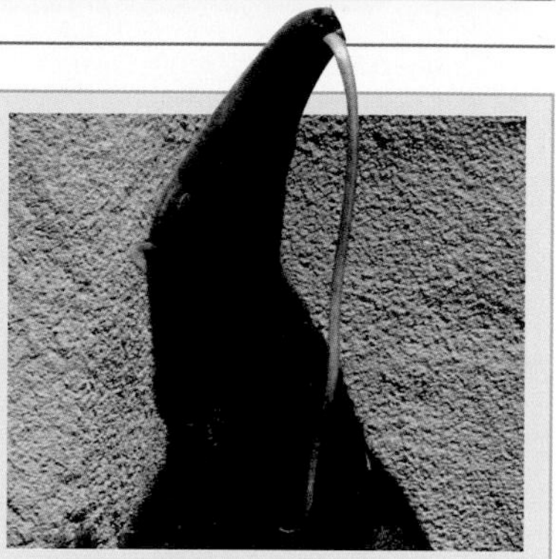

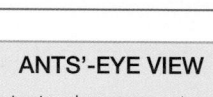

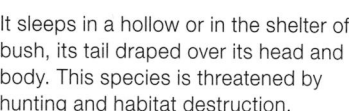

The giant anteater has a long and tubular snout that widens to a small face with tiny ears and eyes. With its massive front legs and smaller rear legs, it walks with an ambling gait, protecting its large front claws from wear by walking on its knuckles. Active day and night, this species wanders its home range, which may cover up to 10 square miles (25 square km), depending on the availability of food. It sleeps in a hollow or in the shelter of bush, its tail draped over its head and body. This species is threatened by hunting and habitat destruction.

NARROW BODY AND BUSHY TAIL
The giant anteater has long, coarse fur on its narrow body. Its coloration is mainly gray with black and white markings. Its brown tail is very bushy.

Rabbits, hares, and pikas

PHYLUM	Chordata
CLASS	Mammalia
ORDER	Lagomorpha
FAMILIES	3
SPECIES	92

Although these small to medium-sized gnawing animals are similar to rodents in many ways (for example, both groups have large incisors), they differ in several respects, including the presence of a second set of upper incisors and a lighter skull structure. Rabbits, hares, and pikas (also termed "calling hares" on account of their varied vocalizations)—the lagomorphs—are among the most hunted of all animals. Their natural predators are carnivores and birds, and they are also hunted by humans for sport, for food, and for their fur. All species are terrestrial and are found worldwide (except the West Indies, the southern parts of South America, Madagascar, and several islands of Southeast Asia) in habitats ranging from arctic tundra to semidesert.

Feeding

Lagomorphs are herbivorous and generally eat grass and other succulent plants. Matter that cannot be digested initially is expelled in the form of a moist pellet and eaten, usually straight from the anus. It is then held in the stomach to be mixed with other food for second digestion before being excreted as a dry pellet. In this way, most food travels through the digestive system twice, enabling the animal to derive maximum nutrition from its diet. This process is called refection.

PLANT-EATERS
All lagomorphs, such as this North American pika, spend much of their time feeding. During the summer, the pika also gathers and stores food for winter, creating a "haystack" of dried foliage outside its rocky shelter.

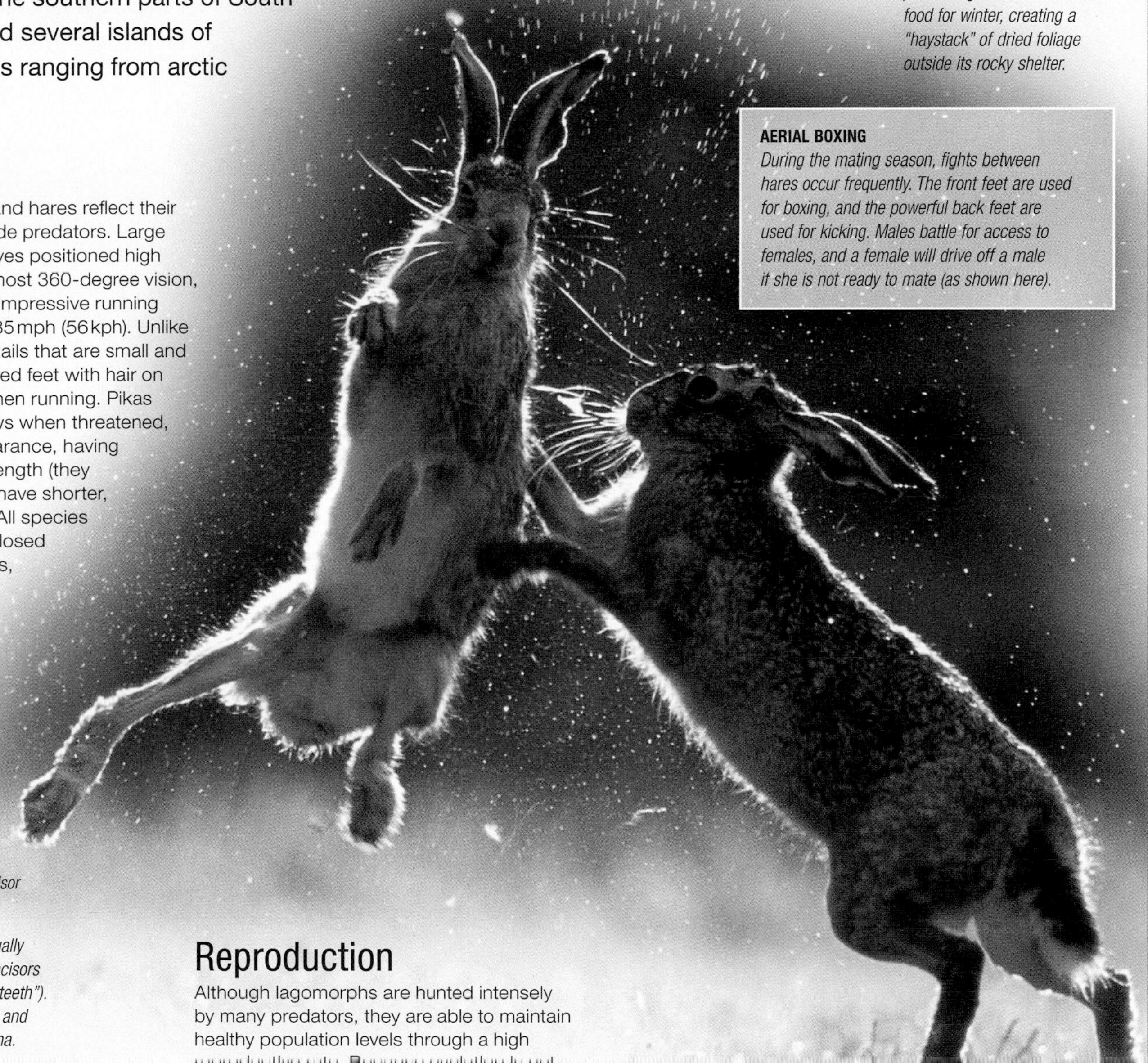

AERIAL BOXING
During the mating season, fights between hares occur frequently. The front feet are used for boxing, and the powerful back feet are used for kicking. Males battle for access to females, and a female will drive off a male if she is not ready to mate (as shown here).

Anatomy

The physical features of rabbits and hares reflect their need to perceive danger and elude predators. Large ears provide excellent hearing, eyes positioned high on each side of the head give almost 360-degree vision, and elongated back legs enable impressive running speeds—hares can reach up to 35 mph (56 kph). Unlike rodents, rabbits and hares have tails that are small and spherical, and they have well-furred feet with hair on the soles, which provides grip when running. Pikas tend to hide in crevices or burrows when threatened, and are more mouselike in appearance, having legs of approximately the same length (they cannot run as swiftly). They also have shorter, rounded ears and no visible tail. All species have slitlike nostrils that can be closed completely. Unusual for mammals, the females of some species are larger than the males.

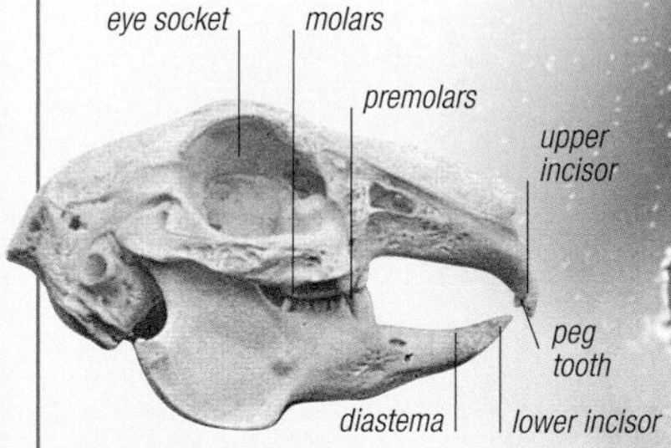

RABBIT SKULL
Lagomorphs have well-developed, continually growing incisor teeth. Behind the upper incisors is a second pair of smaller incisors ("peg teeth"). There is a large gap between the incisors and premolars in both jaws, called the diastema.

FLEXIBLE NECK
In the lagomorphs, self-grooming is important: mutual grooming between individuals is rare. Great flexibility in the neck—this European rabbit is able to rotate its head through 180 degrees—allows them to reach the fur on the back.

Reproduction

Although lagomorphs are hunted intensely by many predators, they are able to maintain healthy population levels through a high reproductive rate. Because ovulation is not cyclical but is instead triggered in response to copulation, females can become pregnant directly after giving birth. Some species may even conceive a second litter before giving birth to the first. Rabbits, the most prolific breeders of all the lagomorphs, can produce litters of up to 12 young 6 times annually. Furthermore, rabbits are sexually mature at a young age (the European rabbit is able to conceive when only 3 months old), and the gestation period may be very short (the Florida cottontail rabbit, for example, gestates for as little as 26 days).

FEEDING TIME
Although some rabbits give birth in their burrows, all hares are born above ground. These young brown hares remain hidden during the day but are collected together at sunset when the mother visits to nurse them.

Ochotona princeps

American pika

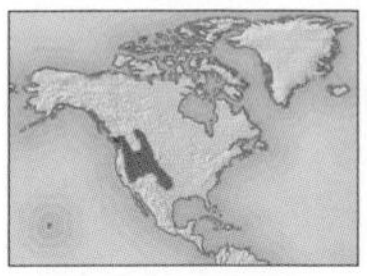

Location S.W. Canada, W. USA

Length 6½–9 in (16–22 cm)
Tail None
Weight 4–6 oz (125–175 g)
Social unit Individual
Status Least concern

Found at high altitudes, this pika lives on a talus—an area of piled, broken rocks fringed by alpine meadows and similar low, grassy vegetation. At each talus, solitary pikas use whistling calls to defend territories that alternate across the area by gender, giving a female–male patchwork. A typical territory is 6,500 square ft (600 square m), and it has a foraging area and a den in a burrow or rock crevice.

READY FOR WINTER

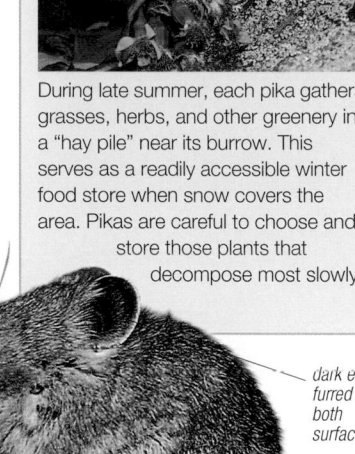

During late summer, each pika gathers grasses, herbs, and other greenery in a "hay pile" near its burrow. This serves as a readily accessible winter food store when snow covers the area. Pikas are careful to choose and store those plants that decompose most slowly.

EGG-SHAPED MAMMAL
A crouched pika has a rounded outline resembling an egg. The fur is varying shades of brown.

Ochotona curzoniae

Black-lipped pika

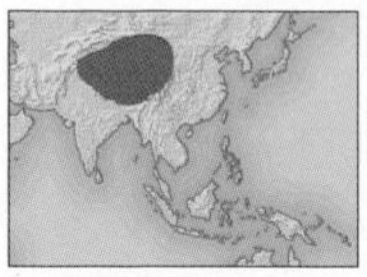

Location E. Asia

Length 5½–7½ in (14–18.5 cm)
Tail None
Weight 4–6 oz (125–175 g)
Social unit Group
Status Least concern

This pika is sandy brown above and dull yellow-white on the underside, with a rust-hued patch behind the ear and a dark nose and lips. An extended family occupies each burrow system and members are sociable. In some areas this pika is so numerous as to be considered a pest. Females can have up to 5 litters of 8 young per year, which are cared for by both parents.

Romerolagus diazi

Volcano rabbit

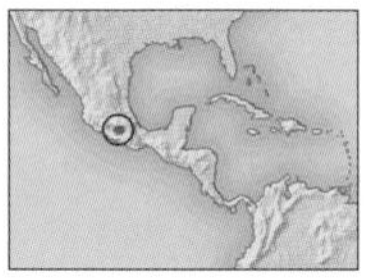

Location C. Mexico

Length 23–32 cm (9–12½ in)
Tail 1–3 cm (⅜–1¼ in)
Weight 375–600 g (13–21 oz)
Social unit Group
Status Endangered

Restricted to open pine forests on volcanic peaks near Mexico City, the volcano rabbit lives in groups of 2–5. It has very short, rounded ears for a rabbit, relatively small back legs and feet, and communicates by means of penetrating whistles. Its diet is mainly the tall, dense grass in which it also makes its burrow.

Pentalagus furnessi

Amami rabbit

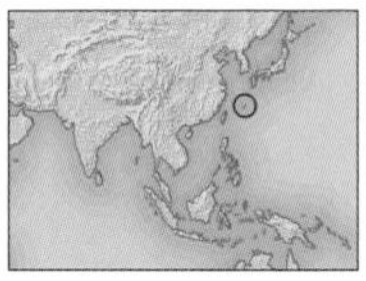

Location Amami and Tokuno islands (Japan)

Length 16½–20 in (42–50cm)
Tail ⅜–1½ in (1–3.5 cm)
Weight 4½–6½ lb (2–3 kg)
Social unit Individual/Group
Status Endangered

Found only on 2 small Japanese islands, this rabbit's many distinctive features include an all-black coat, pointed snout, small eyes and ears, and short, long-nailed limbs for digging nest-holes. Nocturnal in habit, it eats forest plants, such as pampas grass leaves, sweet potato runners, bamboo sprouts, nuts, and bark. Social and breeding habits are little known, although it communicates by means of clicking sounds. The female has 2 litters of 2–3 young each year.

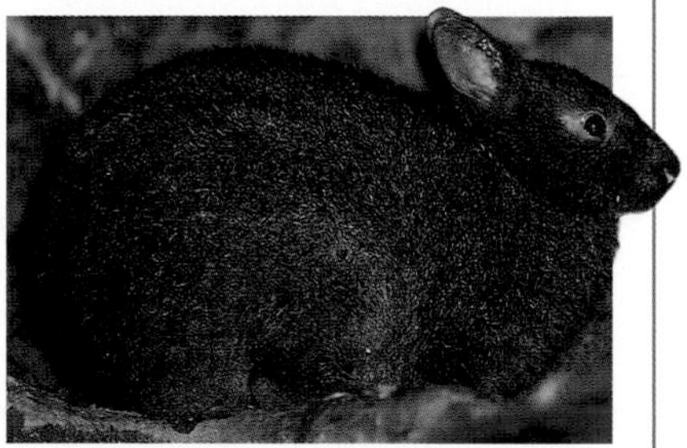

Brachylagus idahoensis

Pygmy rabbit

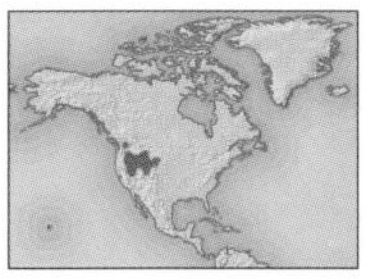

Location W. USA

Length 9–11½ in (22–29 cm)
Tail ½–1 in (1.5–2.5 cm)
Weight 13–16 oz (350–450 g)
Social unit Individual
Status Least concern

The world's smallest rabbit, this arid-adapted species digs a large burrow system and feeds on big sagebrush and closely related species. The long, silky fur on its back is gray in winter and brown in summer; the abdomen is whitish. Unlike most rabbit, its climbs well into bushes to feed, and, although solitary, it also makes pikalike whistling calls to warn neighbors of approaching predators. Breeding details are poorly known: the gestation period may be 26–28 days, and litter size 4–8, with up to 3 litters per year.

Sylvilagus aquaticus

Swamp rabbit

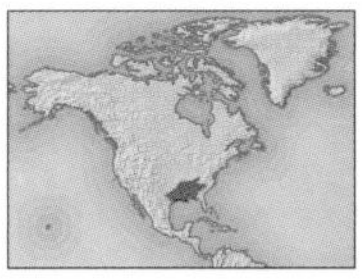

Location S.E. USA

Length 18–22 in (45–55 cm)
Tail 1½ in (4 cm)
Weight 3¼–5½ lb (1.5–2.5 kg)
Social unit Group
Status Least concern

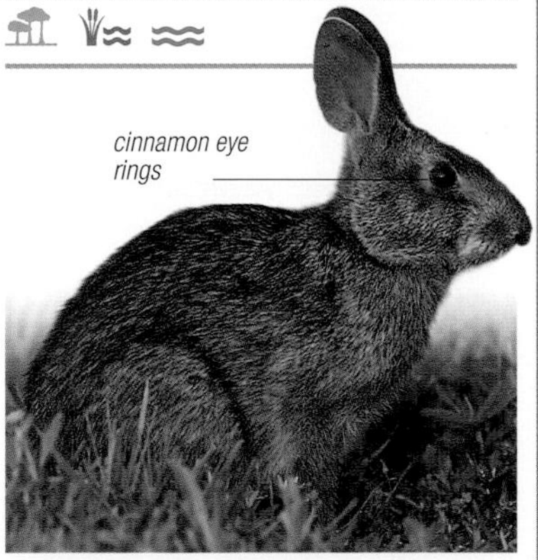

Always associated with water, in marshes, creeks, and pools, the swamp rabbit has black to rusty brown fur. It swims well and readily, especially if threatened, and feeds by day or night on sedges, rushes, and other aquatic plants, including swamp bamboos (*Arundinaria*). The swamp rabbit lives in small groups, usually controlled by a dominant territorial male, and builds an above-ground nest of plant stalks and stems. The female lines the breeding nest with her fur, in the typical rabbit manner. The average litter size is 3.

Sylvilagus floridanus

Eastern cottontail

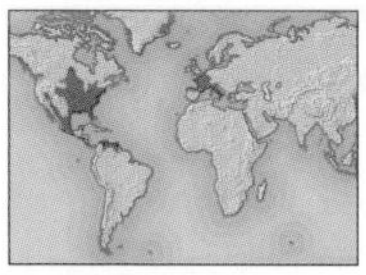

Location S.E. Canada to Mexico, Central America, N. South America, Europe

Length 15–19½ in (38–49 cm)
Tail 1–2¾ in (2.5–7 cm)
Weight 2¼–3¼ lb (1–1.5 kg)
Social unit Group
Status Least concern

Widely distributed in many habitats over its natural range, and introduced to new areas of North America and Europe, the cottontail has the typical rabbit body form with a reddish-topped white tail. In summer, it feeds on lush green vegetation; bark and twigs predominate in winter. Groups have established dominance hierarchies. Average litter size ranges from 5 in North America to 2 in South America.

Oryctolagus cuniculus

European rabbit

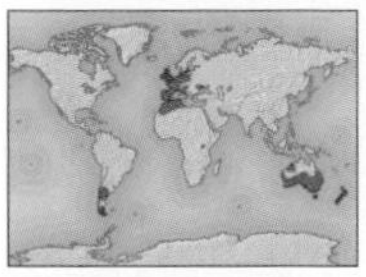

Location Europe, N.W. Africa, Australia, New Zealand, S. South America

Length 13½–20 in (34–50 cm)
Tail 1½–3¼ in (4–8 cm)
Weight 2¼–5½ lb (1–2.5 kg)
Social unit Group
Status Near threatened

Originally from southwest Europe and perhaps northwest Africa, where it is now rare, this species has been introduced to many other regions and has reached severe pest status in some, devastating farmland and wildlife. It is the ancestor of all breeds of domestic rabbits. Nocturnal and very sociable, this European rabbit lives in colonies and digs complex tunnel systems (warrens) with many entrances and exits. It eats grass, herbs, twigs, and some bark. Senior females nest in the main warren but lower-ranking mothers may dig separate short burrows (stops) for their young.

BUFF PATCHES
The European rabbit is buff-coloured between the shoulders, and has a pale buff eye-ring, inner limb surfaces, and underside.

FAST BREEDER

The rabbit's legendary powers of reproduction include a gestation period of 28–33 days and, in good conditions, a litter size of up to 8 (average 5), with as many as 6 litters a year. The newborn are helpless with eyes closed, and for warmth the mother lines the nursery chamber with dry grass, moss, and fur plucked from her own belly. She visits to suckle them for only a few minutes daily.

Caprolagus hispidus

Hispid hare

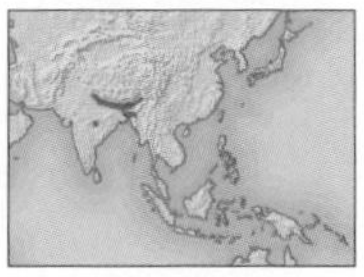

Location S. Asia

Length 15–20 in (38–50 cm)
Tail 1–2¼ in (2.5–5.5 cm)
Weight 4½–5½ lbs (2–2.5 kg)
Social unit Individual/Pair
Status Endangered

Also called the bristly rabbit after its coarse, dark brown fur, the hispid hare lives in tall "elephant grass" country, feeding by night on the soft shoots and roots of the grass. It has short ears, and its back legs are not much larger than its front legs. It does not burrow but shelters in surface vegetation, living alone or as a female–male pair. Reproductive information is scant but the suspected litter size is small for a lagomorph, perhaps 2–5, with 2 or possibly 3 litters produced each year.

Lepus europaeus

Brown hare

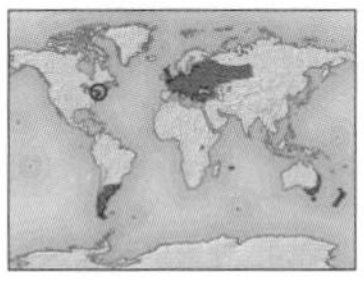

Location Europe, Australia, New Zealand, North and South America

Length 19–28 in (48–70 cm)
Tail 2¾–5 in (7–13 cm)
Weight 5½–15 lb (2.5–7 kg)
Social unit Individual
Status Least concern

This hare has a conspicuous tail, black on top and white below. The gray ears have a black patch near the outer tip. This species has been introduced from Europe and West Asia to many other regions, and adapts to open woods, bush, mixed farmland, and even scrubby semi-desert. Its diet is grass, herbs, bark, and, rarely, carrion. Nocturnal and solitary, hares gather in courting pairs or groups in late winter and spring. At this time, "boxing" may occur between rival males or unreceptive females and rejected males. The young (litter size 1–10) are reared in a shallow depression (form) in grass or bushes.

Lepus californicus

Black-tailed jackrabbit

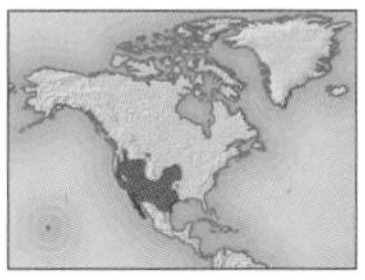

Location W., C., and S. USA to N. Mexico

Length 18½–25 in (47–63 cm)
Tail 2–4¼ in (5–11 cm)
Weight 3¼–7¾ lb (1.5–3.5 kg)
Social unit Individual
Status Least concern

This lean, long-legged hare's huge ears (up to 15 cm/6 in) detect the faint sounds of predators, and also rid the body of excess warmth in the hot summers of its generally arid habitat. It prefers succulent grass and herbs, but can survive by nibbling woody twigs during winter or drought. One of the speediest lagomorphs, it can run at 35 mph (56 kph). The complex courtship involves the pair jumping, chasing, and fighting.

Lepus arcticus

Arctic hare

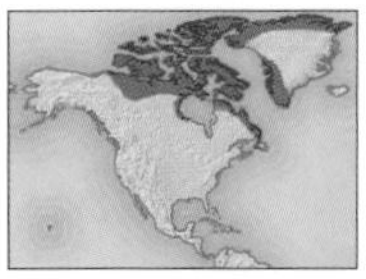

Location N. Canada, Greenland

Length 17–26 in (43–66 cm)
Tail 1¾–4 in (4.5–10 cm)
Weight 6½–15 lb (3–7 kg)
Social unit Variable
Status Least concern

The Arctic hare (sometimes confused with a similar white-in-winter species, the snowshoe hare, *Lepus americanus*) is a true tundra species. It can survive in an open, treeless habitat through the long and bitter cold season. Preferred sites are rocky outcrops or hillsides with crevices and crannies for shelter. This hare may be solitary but, especially in winter, and uniquely among lagomorphs, it shows "flocking" behaviour in which large groups of up to 300 gather, move, run, and change direction almost as one. The diet is a variety of low-growing grasses, herbs, and shrubs, including lichens, mosses, and most parts of the arctic dwarf willow. However, these opportunistic hares may also eat small animals or larger carrion. During the aggressive spring courtship, the male follows the female and may bite her neck so hard that bleeding occurs. Litter size is 1–8 with 1–3 litters per year. The young hares (leverets) stay in their nest (form), a hollow near rocks, lined with grass, moss, and fur. The mother visits to suckle them for only 2 minutes every 18 hours.

WINTER COAT

The Arctic hare's thick winter coat is almost pure white with black ear tips. It provides both warmth and camouflage in snow and ice. In most regions the spring molt produces the gray-brown summer coat, although in some areas this is also white. The timing of the molt depends on the number of daylight hours, detected by the eyes and then controlled via the body's hormonal (endocrine) system.

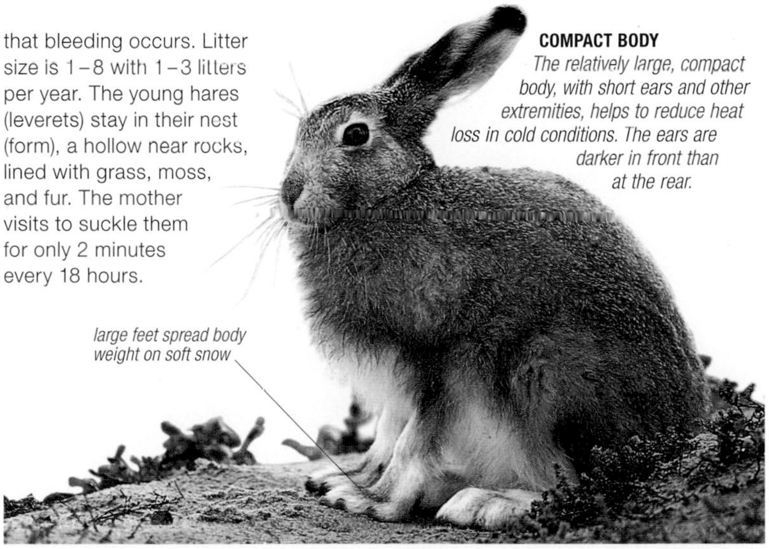

COMPACT BODY
The relatively large, compact body, with short ears and other extremities, helps to reduce heat loss in cold conditions. The ears are darker in front than at the rear.

Rodents

PHYLUM	Chordata
CLASS	Mammalia
ORDER	Rodentia
FAMILIES	33
SPECIES	2,272

CLASSIFICATION NOTE

Some mammalogists prefer to divide the order Rodentia into 2 suborders: (Sciurognathi and Hystricognathi). Others advocate a division into 5 suborders, which are defined by jaw musculature: Sciuromorpha (squirrel-like rodents), Castorimorpha (beaverlike rodents), Myomorpha (mouselike rodents), Histricomorpha (cavylike rodents), and Anomaluromorpha (springhare). For greater ease of reference, the latter division is used here.

Squirrel-like rodents see pp.118–21
Beaverlike rodents see pp.121–2
Mouselike rodents see pp.123–8
Cavylike rodents see pp.129–31
Springhares see p.132

Representing over 40 percent of all mammal species, rodents form a successful and highly adaptable order. They are found worldwide (except Antarctica) in almost every habitat: lemmings, for example, favor the cold climate of the arctic tundra, while gundis prefer the heat of African desert regions. Despite the variety of lifestyles and habitats exhibited by members of this order, there are many common characteristics: most rodents are small quadrupeds with a long tail, clawed feet, long whiskers, and teeth (especially the long incisors) and jaws specialized for gnawing. Although generally terrestrial, some species are arboreal (such as tree squirrels), burrowing (mole-rats, for instance, live almost wholly underground), or semiaquatic (such as beavers). Some species, such as the woodchuck, are solitary, but most are highly social and form large communities.

Senses

Most rodents enjoy acute senses of smell and hearing, which, in combination with their long and numerous touch-sensitive whiskers, provide them with a heightened awareness of their surroundings. Nocturnal species have larger eyes than diurnal species, to maximize the amount of light received by the retina (the greater the amount of light, the brighter and clearer the image). Rodents communicate by smell (odors are secreted from scent glands on the body) and by an extensive range of vocalizations.

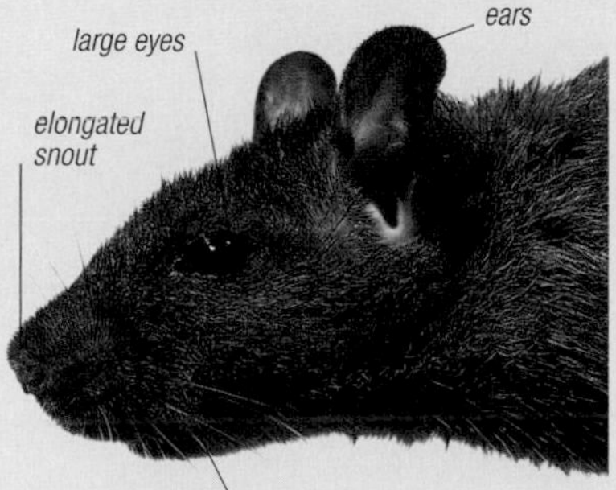

HIGHLY TUNED
Well-developed sense organs are present in most rodents and may contribute to the adaptability of species, such as this brown rat. The large eyes and ears, elongated snout, and long whiskers are typical of many rodents.

Anatomy

While the anatomy of rodents is more uniform than that of most other orders of mammals, some characteristics, such as a compact body and a long tail and whiskers, are shared by many species. The front foot usually has 5 digits (although the thumb may be vestigial or absent), the back foot has 3–5 digits, and the method of locomotion is generally plantigrade. Different species use their tail to perform distinct functions: beavers have a flat, wide tail for propulsion when swimming, while the Eurasian harvest mouse uses its prehensile tail when climbing in long grass. In some species, part of the tail skin, or the tail itself, will break off if caught, enabling the animal to escape. Because rodent anatomy is more generalized than that of other mammals, they can adapt easily and are able to thrive in many different habitats.

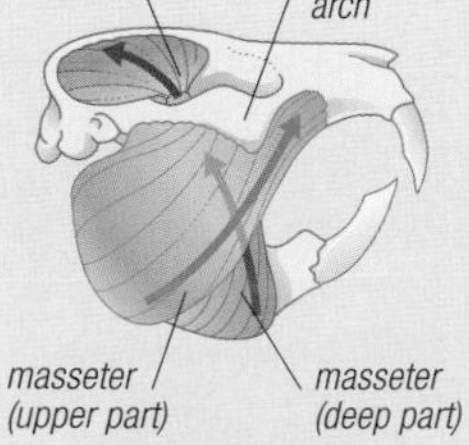

SQUIRREL-LIKE

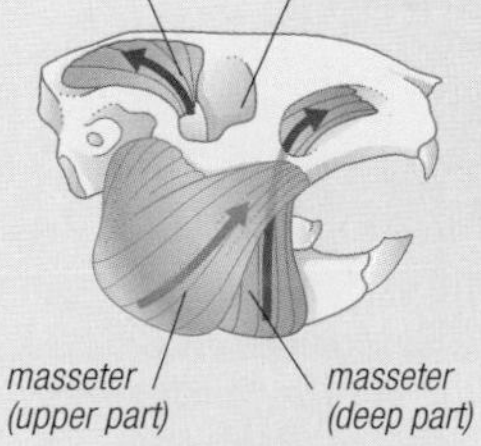

CAVYLIKE

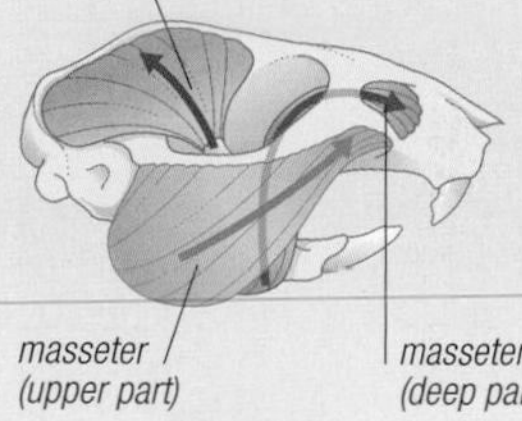

MOUSELIKE

JAW MUSCLES
Rodents have an enlarged chewing muscle (the masseter), which permits both a vertical and a back and forth motion of the lower jaw. In squirrel- and beaverlike rodents, the upper part of the masseter reaches the back of the skull, the deep part extends to the zygomatic arch, and the temporal muscle is small. This system allows a strong forward motion when biting. In mouselike rodents, the deep part of the masseter extends onto the upper jaw, the upper part is located forward, and the temporal is large. This permits a versatile chewing action. In cavylike rodents and the springhare, the deep part of the masseter extends in front of the eye, and the temporal is small. This gives a strong forward bite.

LARGE INCISORS
The 4 huge incisors (seen here in a marmot) distinguish rodents from most other mammal orders. These teeth are long, curved, and grow continually. Only the front surface of these teeth has enamel, however—the back surface consists of softer dentine, which is eroded by constant gnawing to ensure that the teeth remain sharp.

Feeding

Most rodents have a plant-based diet that may include leaves, fruit, seeds, and roots. However, many species have alternative diets: water rats and the wood mouse eat snails; rice rats take young turtles; muskrats eat clams and crayfish; the southern grasshopper mouse eats ants and scorpions; and the black rat scavenges in human food supplies. To assist digestion, rodents have a large cecum, a blind-ending sac in the large intestine. This contains bacteria that split cellulose, the main component of plant cell walls, into digestible carbohydrates. In some rodents, after food is processed in the cecum, it is ejected from the anus and is eaten again. Once in the stomach, the carbohydrates (amounting to 80 percent of the energy contained in the food) are absorbed. This highly efficient process, known as refection, leaves only a dry fecal pellet to be excreted.

OMNIVOROUS DIET
Some rodents are omnivorous: they feed on both plants and animals, depending on availability. This spiny mouse eats mainly vegetation but will also feed on insects.

HERBIVOROUS DIET
Most rodents are herbivorous, eating only plants. The European water vole shown here feeds on aquatic and land plants. Food may be stored for consumption during winter shortages.

Reproduction

The high birth rate among rodents enables them to maintain stable population levels in adverse conditions. This means that predation and human controls (such as poisoning) have little effect on the survival of a species, and in favorable conditions numbers may increase rapidly. A brown rat, for example, is able to breed at only 2 months of age and may yield litters of more than 10 young every month or so. Voles are also prolific breeders: some species may produce more than 13 litters annually. Smaller rodents tend to produce more young than larger species (such as the capybara)—as a result, small rodents form the staple diet of a wider range of animals. In rodents, the complete cycle of reproduction, from sexual attraction right through to raising young, is influenced by the emission of pungent glandular secretions. Female rats, for example, produce a pheromone about 8 days after giving birth. This scented chemical is secreted into the mother's feces and helps prevent the offspring from becoming separated from her.

CAPYBARA FAMILY
Not all rodent species breed as prolifically as mice and rats do. The capybara, for example, usually produces only one litter a year unless conditions are particularly favorable. Litter size varies between one and 8 but is usually 5. Capybara offspring are well developed at birth and are soon able to follow their mother and eat solid food.

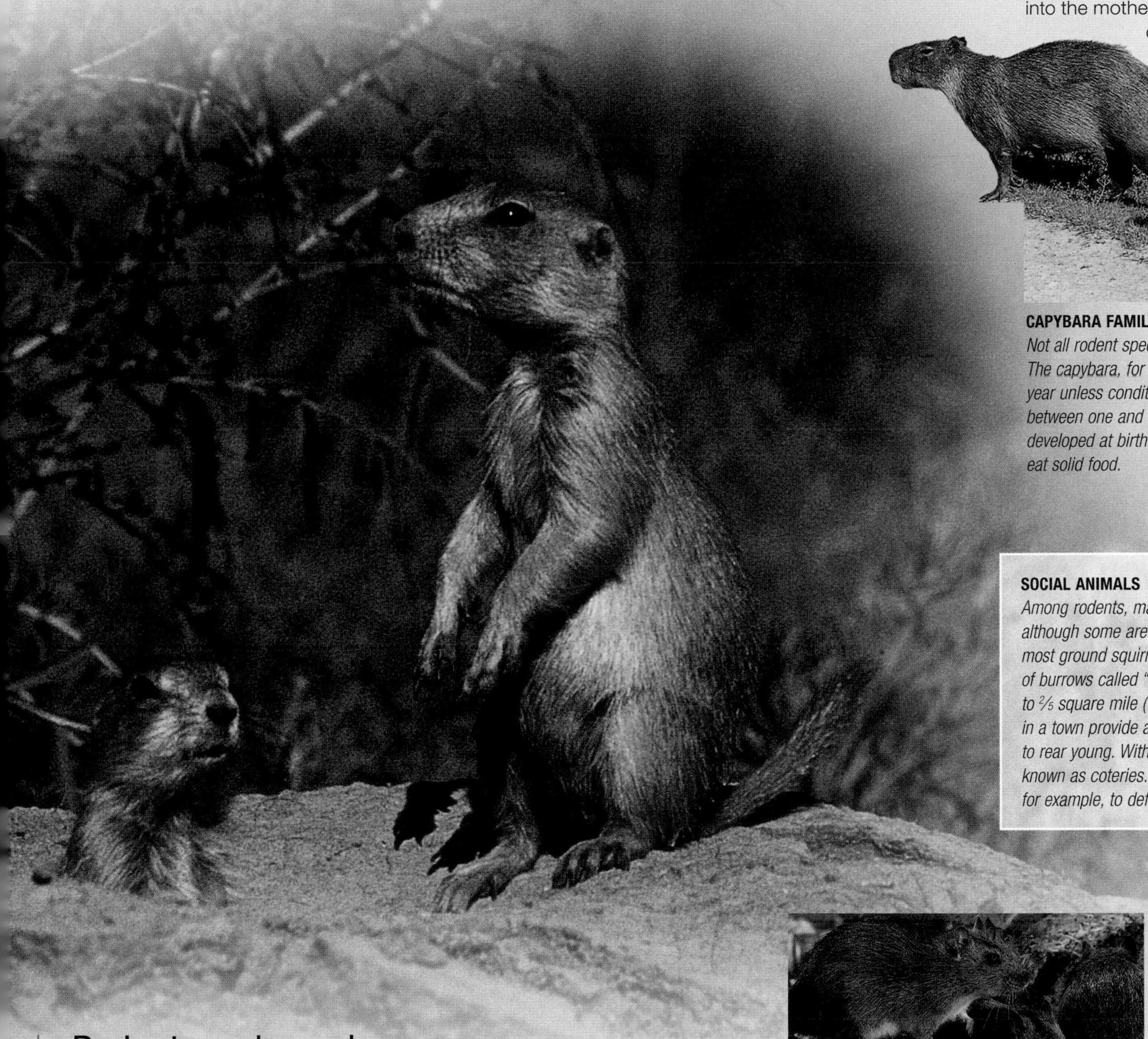

SOCIAL ANIMALS
Among rodents, many species live in organized communities, although some are solitary. These black-tailed prairie dogs, like most ground squirrels, are highly sociable. They live in a system of burrows called "towns," each of which may cover an area up to ⅖ square mile (1 square km). The interconnected burrows in a town provide a refuge from predators and a safe place to rear young. Within the town, prairie dogs form subgroups known as coteries. Members of a coterie act cooperatively; for example, to defend their territory.

GREAT OPPORTUNISTS
Rodents are highly gregarious animals and have successfully colonized many habitats, especially those created by human settlement, such as garbage dumps and sewers. These brown rats are scavenging for food among the garbage.

Rodents and people

Some rodents, mainly rats and mice, are considered pests by people because they are often in direct competition with humankind (they occupy the same habitats and eat the same food) and are highly adaptable. Rodents consume over 39 million tons (40 million tonnes) of human food every year, contaminate stored food with their urine and fecal pellets, and are known to transmit more than 20 disease-causing organisms. Although some control of rodent populations is brought about by the use of traps and poisons, many species are sufficiently intelligent to learn to avoid such measures. Only a few of the 2,272 species of rodents, however, are genuine pests: many benefit people, for example, by destroying insects and weeds or by maintaining the health of forests by spreading fungi. Beavers and chinchillas are farmed for their fur, while rats, mice, and guinea pigs are kept as pets and are used extensively in medical research.

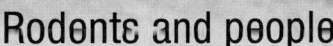

GNAWING DAMAGE
All rodents use their well-developed incisor teeth to gnaw. A beaver can fell trees, such as this birch, by gnawing through the trunk. Branches and smaller trunks from the tree are used to build a lodge or to dam a river.

MAMMALS

Squirrel-like rodents

PHYLUM	Chordata
CLASS	Mammalia
ORDER	Rodentia
SUBORDER	Sciuromorpha
FAMILIES	3
SPECIES	307

This group, which is defined by the arrangement of the jaw muscles (see p.116), embraces a variety of rodents, such as marmots and squirrels. Members of the squirrel family all have long whiskers, a cylindrical body, and a well-haired tail. Species from other families vary anatomically from the ground-dwelling mountain beaver to the mouselike dormouse. Squirrel-like rodents are distributed worldwide, in a variety of habitats.

BALANCING ACT
The Eurasian red squirrel uses its tail for balance when running along branches.

Aplodontia rufa

Mountain beaver

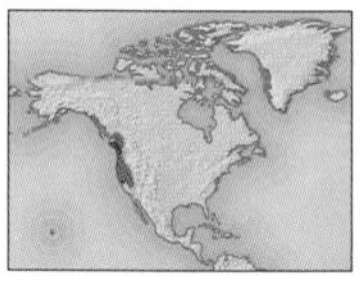

Location S.W. Canada to S.W. USA

Length 12–18 in (30–46 cm)
Tail ¾–1½ in (2–4 cm)
Weight 1¾–3¼ lb (0.8–1.5 kg)
Social unit Individual
Status Least concern

This rodent digs its tunnels and home (lodge) under felled trees, and so is increasing where commercial logging occurs. It lives alone and its tunnel openings lead directly to food such as bark, twigs, shoots, and soft plants, which it brings back to eat or store. Also known as the sewellel, it climbs well and destroys many small trees. The long fur is black to red-brown above, yellow-brown beneath, with a white spot below each ear.

Marmota monax

Woodchuck

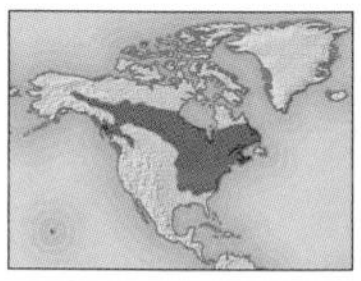

Location Alaska and W. Canada to E. Canada and E. USA

Length 12½–20½ in (32–52 cm)
Tail 3–4½ in (7.5–11.5 cm)
Weight ½–11 lb (63–5 kg)
Social unit Variable
Status Least concern

Also called the groundhog, this is one of the largest, strongest ground squirrels. It feeds mainly in the afternoon, often in a loose group with others of its kind, on a variety of seeds, grasses, clovers, fruit, and small animals such as grasshoppers and snails. Despite its size, it is an able climber and swims well. In autumn, it excavates a deeper burrow for its long winter hibernation. During this period it survives on stored body fat, which makes up about 20 percent of its body weight. In North America, February 2 is Groundhog Day, when the woodchuck supposedly peers from its winter burrow to assess the weather. The woodchuck regularly shows aggressive behavior to its own kind, especially as males fight for dominance in the spring mating season. It also vigorously defends its burrow by arching its back, jumping, flicking its stiffened tail, and chattering its bared teeth. When frightened, it makes a sharp whistling alarm call.

POWERFUL BUILD
The woodchuck has a stout body, small ears, short legs, and a bushy tail.

Marmota flaviventris

Yellow-bellied marmot

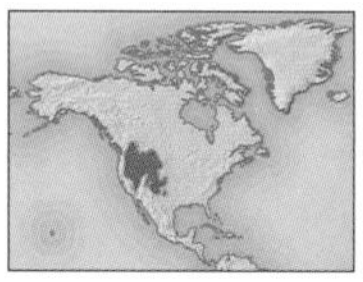

Location S.W. Canada to W. USA

Length 13½–20 in (34–50 cm)
Tail 5–9 in (13–22 cm)
Weight 3¼–11 lb (1.5–5 kg)
Social unit Group
Status Least concern

Adaptable in habitat, the yellow-bellied marmot takes a wide diet of grasses, flowers, herbs, and seeds. It feeds mainly in the morning and late afternoon, then grooms with others in its colony—usually one male and several females. Its long hibernation lasts up to 8 months. The 3–8 young are weaned after 20–30 days.

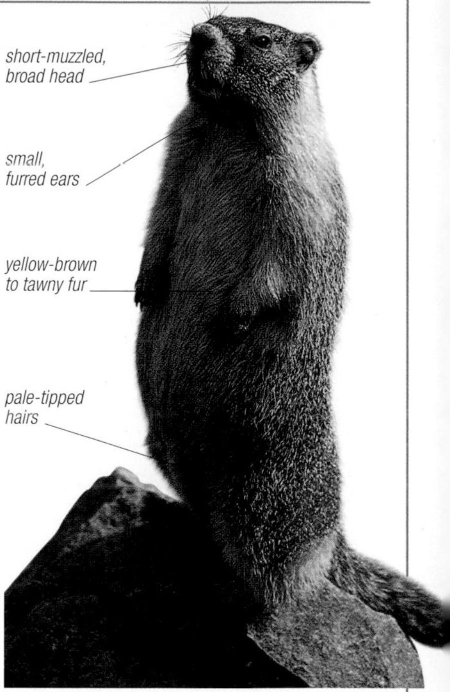

Tamias striatus

Eastern chipmunk

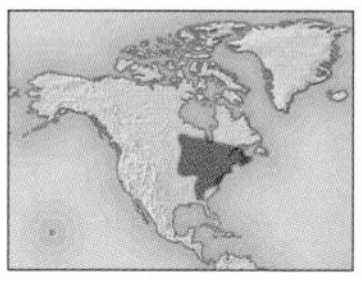

Location S.E Canada to C. and E. USA

Length 6–6½ in (15.5–16.5 cm)
Tail 2¾–4 in (7–10 cm)
Weight 2⅞–4 oz (80–125 g)
Social unit Individual
Status Least concern

The eastern chipmunk, occasionally kept as a pet, is also familiar in the wild, as a bold visitor to picnic sites. It frequents mainly deciduous woods, especially birch, as well as woody areas containing abundant rock crevices. It can climb but forages mainly on the ground for seeds and nuts, especially during mid-morning and midafternoon, carrying food items in its cheek pouches. The basic coloring of the eastern chipmunk is grayish or reddish brown, becoming paler red on the rump. It lives alone in its burrow system and hibernates during winter, although it may emerge on mild days to feed. The noisy "chip" and "cuk" calls act as alarm signals for fellow chipmunks and other small animals living nearby.

Spermophilus columbianus

Columbian ground squirrel

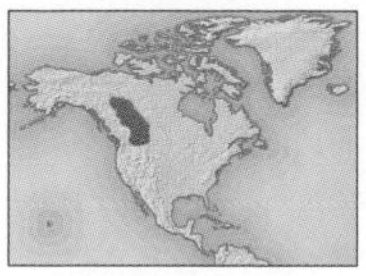

Location W. Canada to N.W. USA

Length 10–11½ in (25–29 cm)
Tail 3¼–4½ in (8–11.5 cm)
Weight 30–36 oz (850–1,000 g)
Social unit Group
Status Least concern

The Columbian ground squirrel ("Columbia" being British Columbia, Canada) comes from mountain meadows and hilly grassland. Its diet consists of fruit, seeds, flowers, bulbs, and on occasions insects caught in midair. Living in a small colony, it "kisses" when it meets other Columbian ground squirrels, tilting its head to sniff the oral scent glands near the mouth.

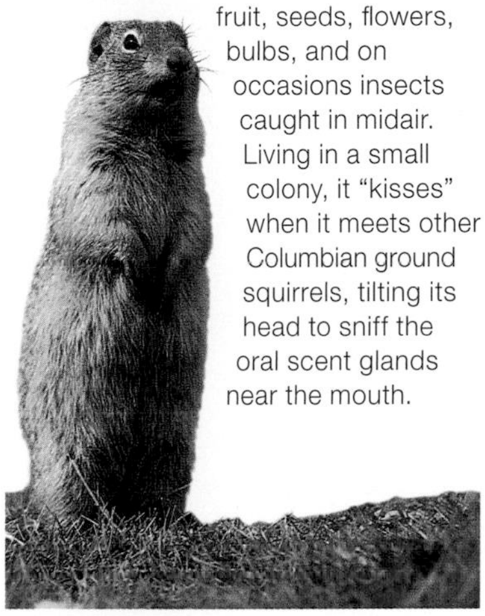

Cynomys ludovicianus

Black-tailed prairie dog

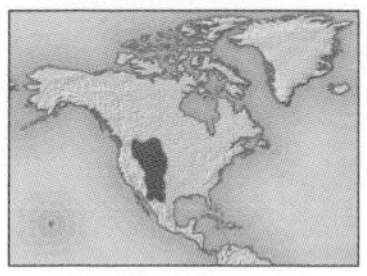

Location S.W. Canada to N. Mexico

Length 11–12 in (28–30 cm)
Tail 2¾–4½ in (7–11.5 cm)
Weight 1½–3¼ lb (0.7–1.5 kg)
Social unit Group
Status Least concern

Preference for a grassy habitat and a barklike, doggy "yip" give 5 species of ground squirrels their common name of prairie dogs. The black-tailed or plains prairie dog dwells at altitudes of 4,300–6,600 ft (1,300–2,000 m) across the Great Plains of North America and south into the more arid, extreme north of Mexico. The body hairs of the black-tailed prairie dog are tipped black in winter but white in summer, and the whiskers and end third of the tail are black. This rodent eats a range of seasonal plants, such as wheat grass, buffalo grass, globe mallow, and rabbit brush in summer, and thistles, cacti such as prickly pear, and underground roots and bulbs in winter. Prairie dogs breed rapidly, with up to 8 young born after a gestation period of 33–38 days. In the past, their feeding habits led to massive destruction of wheat and other cereal crops, and their burrows tripped horses and farm stock. As a result, prairie dogs were subject to extermination campaigns, which were very successful. In recent times, the rodents have been restricted mainly to parks and reserves. Their drastic reduction in numbers has seriously threatened the black-footed ferret (see p.190), for which they were virtually sole prey.

COLORATION
The black-tailed prairie dog is generally brown or reddish brown on the upperparts, shading to white on the underparts. The whiskers and tail tip are black.

COTERIES AND WARD

The basic prairie dog social unit is the coterie of one male, several females, and their young. Several coteries form a ward, members guarding their territories and burrows with an energetic "jump-yip" display, bared chattering teeth, and fluffed-up tails. Many ward form a township of up to 160 acres (65 hectares).

Xerus inauris

Cape ground squirrel

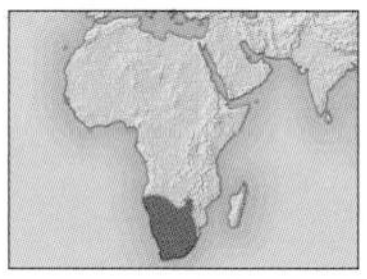

Location Southern Africa

Length 8–12 in (20–30 cm)
Tail 7–10 in (18–26 cm)
Weight 21 oz (575 g)
Social unit Group
Status Least concern

The large claws of the Cape ground squirrel can burrow in hard, dry, stony soil. The upper fur is brownish pink with a white flank stripe and belly. The prominent eyes are circled with white and the muzzle and feet are also white. There are black bands near the base and tip of the tail. Its diet is opportunistic, ranging from seeds, bulbs, and roots to insects and birds' eggs. This ground squirrel lives in colonies of 6–10, and in some cases up to 30.

Heliosciurus gambianus

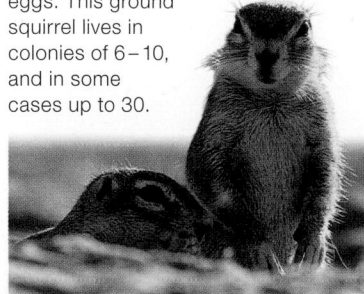

Gambian sun squirrel

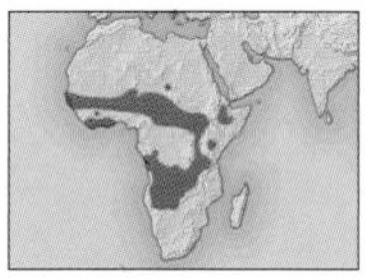

Location Africa

Length 6–8½ in (15.5–21 cm)
Tail 6–12 in (15.5–30 cm)
Weight 9–13 oz (250–350 g)
Social unit Individual/Pair
Status Least concern

Banded hairs in yellows, browns, and grays give the sun squirrel a speckled olive-brown appearance. The tail has 14 rings along its length and the eyes are white ringed. In habit, this squirrel is a typical ground-and-tree species, with a diet ranging from seeds to birds' eggs. Its distinctive behaviors include "basking" on sunny branches, as its name suggests, and relining its nest each night with freshly plucked leaves.

Ratufa indica

Indian giant squirrel

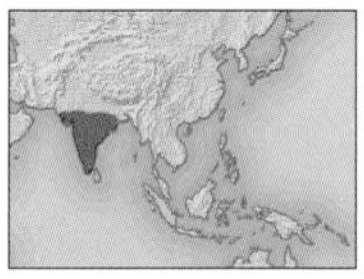

Location S. Asia

Length 14–16 in (35–40 cm)
Tail 14–23½ in (35–60 cm)
Weight 3¼–4½ lb (1.5–2 kg)
Social unit Individual/Pair
Status Least concern

The huge, bushy tail of the Indian giant squirrel is usually longer than the head and body combined. The upperparts are dark, the head and limbs red-brown, and the underside whitish. This squirrel, alert and wary, makes massive leaps of 20 ft (6 m) among branches as it forages for fruit, nuts, bark, insects, and eggs. Its characteristic feeding posture is not upright but leaning forward or down, perched on its back legs with the tail as a counterbalance. Its short, broad thumbs help it to manipulate food. It builds the typical squirrel-type nest (drey) for resting and rearing young. Reproduction in this species is not well known but they are thought to breed throughout the year. Females usually produce 1–2 offspring in each litter after a gestation period of 28–35 days.

FEEDING POSITION
This squirrel shows the typical feeding posture of this species. Its long tail provides stability while it eats its food, which is skillfully manipulated by dexterous front paws.

Callosciurus prevosti

Prevost's squirrel

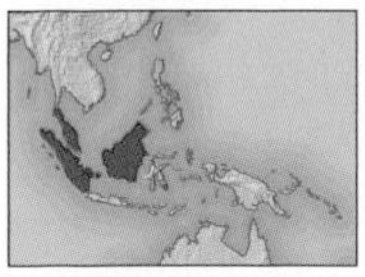

Location S.E. Asia

Length	5–11 in (13–28 cm)
Tail	3¼–10 in (8–26 cm)
Weight	5–18 oz (150–500 g)
Social unit	Variable
Status	Least concern

One of the most brightly colored mammals, this tree-dwelling, twilight-active squirrel is black on its upperparts and vivid chestnut-red on the underparts, with a broad, gleaming white band along each side from nose to thigh. It lives alone or in small family groups, communicating by birdlike calls and visual displays of the bushy tail. The diet includes fruit, soft seeds, oily nuts, and buds, as well as termites, ants, grubs, and birds' eggs. After 46–48 days' gestation, the 2–3 young are born in a large nest of twigs and leaves, in a tree hole or among branches.

Sciurus vulgaris

Eurasian red squirrel

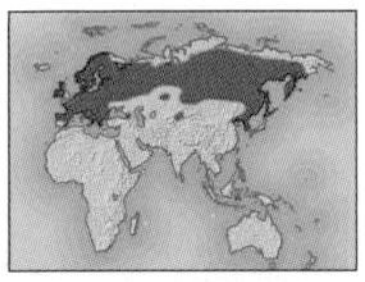

Location W. Europe to E. Asia

Length	8–10 in (20–25 cm)
Tail	6–8 in (15–20 cm)
Weight	7–17 oz (200–475 g)
Social unit	Individual
Status	Least concern

The "red" squirrel varies from red to brown, gray, or black on its back, and may turn gray-brown in winter. The underparts are always pale or white. An excellent climber and leaper, it feeds on the ground and in branches on seeds (especially those of conifers), nuts, mushrooms and other fungi, shoots, fruit, soft bark, and sap. It lives alone except when a female is nursing young, which she usually does twice a year. There are 5–7 young in each litter.

BREEDING NEST

The female red squirrel gives birth to 2–5 blind, naked babies. She nurses them for 12 weeks in the breeding nest, which may be a larger version of the ball-shaped, twig drey in a branch fork, or in a tree hole. The nest is lined with soft, fine material, to keep the young warm in her absence.

BUSHY TAIL, TUFTED EARS
Although the red squirrel moults twice a year, the tail hair is only replaced once. Its ears are always tufted, particularly in winter.

Sciurus carolinensis

Eastern gray squirrel

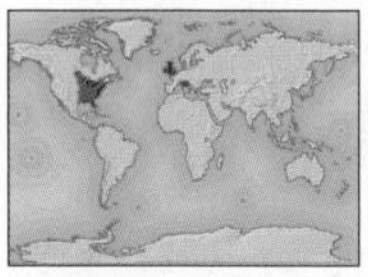

Location S. and S.E. Canada to S. USA, Europe

Length	9–11 in (23–28 cm)
Tail	6–10 in (15–26 cm)
Weight	11–25 oz (300–700 g)
Social unit	Variable
Status	Least concern

Introduced from North America into parts of Europe, the eastern gray squirrel has a gray back and white or pale underparts. The face, back, and forelegs are brown-tinged. This opportunistic feeder takes nuts, seeds, flowers, fruit, buds, and fungi. It may emerge from its grass- and bark-lined twig nest (drey) in winter to forage. It also eats food that it stored during the autumn, locating it from memory and smell. Breeding occurs twice a year. The 2–4 young are helpless at birth, but grow quickly and are fully weaned at 10 weeks.

COLOR VARIATION

Despite their name, gray squirrels exhibit variation in fur color: some are black; others are reddish. These color forms belong to the same species and all can freely interbreed; coat color is passed on to offspring following a simple pattern of heredity. Fur color is determined by the distribution and types of chemical pigment in hair—including eumelanin (black pigment) and pheomelanin (reddish pigment). Reddish and black squirrels can arise anywhere in a "wild-type" gray population by random genetic mutation, and the color can spread as they pass their genes on when they breed. In North America, black squirrels are more common farther north because they retain heat better in cold climates.

SITTING PRETTY
When feeding on the ground, gray squirrels typically sit up in this way so that their forepaws are free to handle food items.

Petaurista elegans

Giant flying squirrel

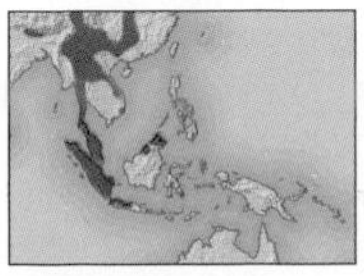

Location S.E. Asia

Length	12–18 in (30–45 cm)
Tail	12½–24 in (32–61 cm)
Weight	1–4½ lb (0.5–2 kg)
Social unit	Pair
Status	Least concern

This squirrel stretches out its limbs to extend large patches of thin, loose, furred skin between front and rear legs, in a parachute-like shape. A glider rather than a true flier, it can travel more than 1,300 ft (400 m) at a 3-in-1 glide ratio (covering 3 times the distance that it loses in height). It changes the glide angle using its front legs. Glides are generally from high in one tree to lower in another, rather than to the ground, and are often to escape danger. Unlike most squirrels, this species is nocturnal, leaving its tree hole at night to find conifer seeds, nuts, fruit, leaves, shoots, and buds. Fur color varies but is usually tawny to reddish brown above, with black rings around the large, "night-vision" eyes and hairless ears.

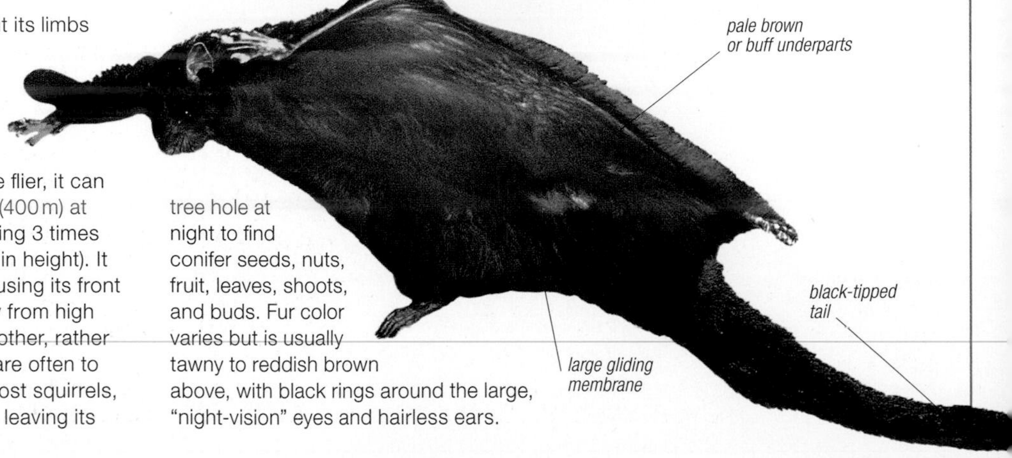

Muscardinus avellanarius

Hazel dormouse

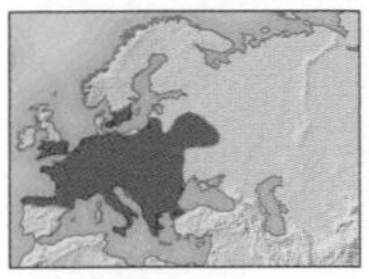

Location Europe

Length 2½–3¼ in (6.5–8.5 cm)
Tail 2¼–3 in (5.5–8 cm)
Weight $\frac{3}{16}$–1¼ oz (15–35 g)
Social unit Individual
Status Least concern

About the size of a house mouse (see p.128), the common dormouse is an excellent climber and jumper. Feeding mainly in trees, it changes its diet with the season from flowers, grubs, and birds' eggs in spring and summer to seeds, berries, fruit, and nuts in autumn. Dormice are the only rodents that do not have a cecum (a part of the large intestine), which may indicate that their diet is low in cellulose. The nest is made of grass in a thick bush or tree hole. Several individuals may live nearby and share the feeding area. They communicate using a wide range of whistles and growling noises. The female is pregnant for 22–24 days and has a litter of 2–7 young, with up to 2 litters per year. The tail is bushy and its skin can detach if seized by a predator.

DEEP WINTER SLEEP

Most dormice hibernate deeply in winter ("dor" meaning "sleeping"). The hazel dormouse rests for about 7 months in a nest about 4¾ in (12 cm) in diameter—larger than its summer quarters. This nest may be in a burrow or under moss or leaves. It stores food both in its nest and in its body as fat, to enable it to survive throughout the cold season.

VARIABLE COLORATION
The hazel or common dormouse has yellow, red, orange, or brown upperparts and a white underside.

Glis glis

Edible dormouse

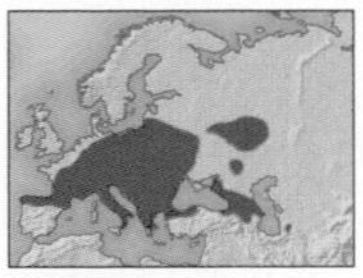

Location C. and S. Europe to W. Asia

Length 5–8 in (13–20 cm)
Tail 4–7 in (10–18 cm)
Weight 2½–5 oz (70–150 g)
Social unit Group
Status Least concern

This species inhabits woods and out-buildings, nesting in tree holes or crevices in roofs and under floors. Native to mainland Europe, it was introduced to Britain in 1902. It is usually nocturnal but may also be active at dawn and dusk (crepuscular). As the days become shorter in autumn, it lays down fat to provide reserves. The edible dormouse eats leaves, seeds, fruit, nuts, bark, mushrooms, creatures such as insects, and birds' eggs and chicks. Like other dormice, it forms loose social groups and communicates by squeaks and twitters. While pregnant, the female is solitary, giving birth to 1–11 young after a pregnancy of 30–32 days. They grow quickly and hibernate for the first time aged 8–9 weeks.

SQUIRREL-LIKE LEAPER

The edible dormouse resembles a squirrel with its long, bushy tail, large back legs for leaping in branches, and semiupright posture. Rough pads on its hands and feet also aid climbing. It is highly arboreal and can leap over 23 ft (7 m) between branches. Its fine dense fur is brown or silver-gray with dark eye patches and white underparts.

ONCE "EDIBLE"
This dormouse was kept in ancient Rome and fattened with extra food in autumn, to be served at the dinner table—hence its name.

Beaverlike rodents

PHYLUM	Chordata
CLASS	Mammalia
ORDER	Rodentia
SUBORDER	Castorimorpha
FAMILIES	3
SPECIES	102

Despite their superficial differences, beavers, pocket gophers, and kangaroo rats are classified together on the basis of anatomical details, such as skull structure and evidence from DNA. Beavers are large, semiaquatic rodents with webbed back feet and a flattened scaly tail and occur across the Northern Hemisphere. Pocket gophers, kangaroo rats, and allies have fur-lined cheek pouches ("pockets") and are confined to the Americas. Pocket gophers are short-legged for burrowing, but kangaroo rats are bipedal with a bounding gait.

DANGER SIGNAL
By slapping their broad, flat tails on the surface of the water, beavers warn others of potential danger.

Orthogeomys grandis

Large pocket gopher

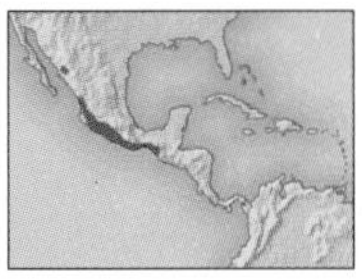

Location Mexico to Central America

Length 4–14 in (10–35 cm)
Tail 1½–5½ in (4–14 cm)
Weight 11–32 oz (300–900 g)
Social unit Individual
Status Least concern

Like other pocket gophers, the large pocket gopher digs a burrow system (lodge) using its strong, large-clawed forefeet. It feeds on roots, bulbs, and other underground plant parts and also comes above ground at night to forage for stems and shoots, which it carries to its lodge in its fur-lined cheek pouches. Normally solitary, during the breeding season the large pocket gopher forms groups of one male and 4 females. Two or more young are born to each female in a grass-lined nesting chamber at the lowest level of the lodge.

Cratogeomys merriami

Merriam's pocket gopher

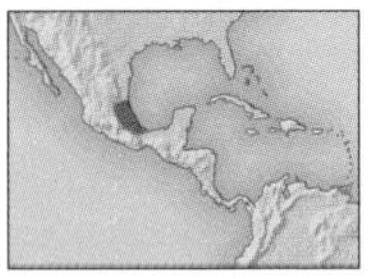

Location E. Mexico

Length 5½–10 in (14–26 cm)
Tail 2½–5 in (6–13 cm)
Weight 8–32 oz (225–900 g)
Social unit Individual
Status Least concern

The pocket gophers have a pocketlike pouch of furred skin on each cheek, for carrying food back to the nest. This species occupies a variety of habitats from sea level up to almost 13,200 ft (4,000 m). Its front incisor teeth are very long and can cope with many kinds of plant foods, from spiky cacti and farm crops (gophers sometimes become a pest) to fir tree needles and seeds. The upperparts are yellow, brown, or almost black; the underside is paler.

MAMMALS

Thomomys bottae

Botta's pocket gopher

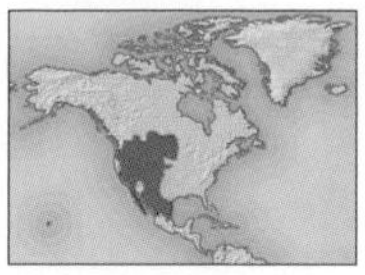

Location W. USA to N. Mexico

- **Length** 4½–12 in (11.5–30 cm)
- **Tail** 1½–3¾ in (4–9.5 cm)
- **Weight** 1⅝–2 oz (45–55 g)
- **Social unit** Individual
- **Status** Least concern

Botta's pocket gopher lives mainly alone. It digs an extensive burrow system (lodge) in loose soil, using its strong forelegs equipped with large claws. This gopher stays underground most of its life, consuming roots, tubers, bulbs, and other subsurface plant parts. It is grayish brown above and browny orange below, and has adaptations typical of burrowers, such as a flat head, long whiskers, small eyes, and small ears closed by flaps.

Heteromys catopterius

Overlook spiny pocket mouse

Location N. South America

- **Length** 5–5¼ in (13–13.5 cm)
- **Tail** 6½ in (16.5 cm)
- **Weight** 2¾–3 oz (78–86 g)
- **Social unit** Group
- **Status** Not evaluated

Discovered in 2010 in the high, wet mountain forests of coastal Venezuela, the overlook spiny pocket mouse occurs from 1,148 to 8,063 ft (350–2,450 m). It belongs to a group of spiny mice so-called because of their coarse, spiny fur. This species has darker fur than more widespread relatives of the lowlands, and is adapted to cool, high altitudes. Scientists speculate that it may have had a more extensive distribution during the last Ice Age. Although it lives in a remote location, it is possible that its survival will be threatened by global warming. Its habits are probably similar to other *Heteromys* spiny mice, emerging from burrows at night to take seeds and other plant material—gathered in cheek pouches for storage in underground caches.

Microdipodops megacephalus

Dark kangaroo mouse

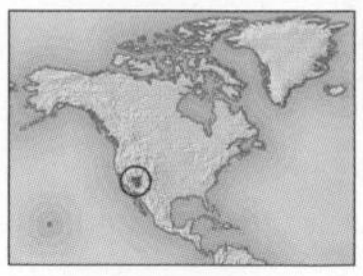

Location S.W. USA

- **Length** 2½–3 in (6.5–7.5 cm)
- **Tail** 2½–4¼ in (6.5–10.5 cm)
- **Weight** ⅜–⅝ oz (10–17 g)
- **Social unit** Individual
- **Status** Least concern

The dark kangaroo mouse is named after its dark-furred, large back legs, which it uses for hopping and leaping over sand dunes and soft, dry soil. The upperparts are gray-brown, the underparts pure white. The prominent eyes, large ears, long snout, and bushy whiskers indicate adaptation for a nocturnal lifestyle. Like many small desert rodents, this species is a seasonal opportunist. In summer, it consumes mainly insects; in winter, it switches to being a seed-hoarder in its burrow nest, carrying food items in its external cheek pouches. It also stores food as body fat in its dumpy tail. Each male dark kangaroo mouse aggressively defends a territory of up to 7,900 square yd (6,600 square m) against others of its species. Female territories tend to be much smaller, only about 480 square yd (400 square m). Each female produces a litter of 2–7 offspring.

Dipodomys merriami

Merriam's kangaroo rat

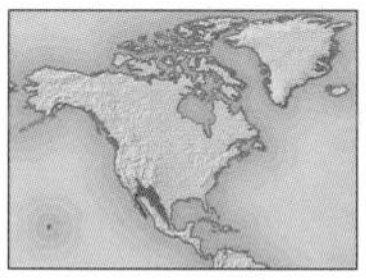

Location S.W. USA to N. Mexico

- **Length** 3¼–5½ in (8–14 cm)
- **Tail** 5½–6½ in (14–16 cm)
- **Weight** 7/16–1⅝ oz (40–45 g)
- **Social unit** Group
- **Status** Least concern

Relatively large back legs and feet, for kangaroo-like hopping, and a very long, slender, tufted tail, for balance, allow Merriam's kangaroo rat to move at high speed over sandy soil in its desert habitat. The silky fur is gray above and white on the underparts, with narrow dark gray and white stripes along the flanks. This kangaroo rat regularly has energetic dust-baths to keep its fur and skin clean. It digs, burrows, or searches for food, especially cockle- and sand-burr seeds in winter, and cactus seeds in summer.

Castor fiber

Eurasian beaver

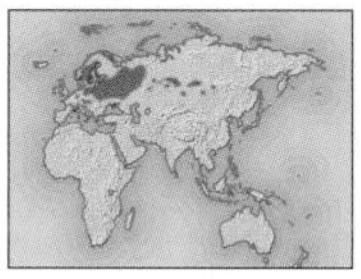

Location Europe to C. Asia

- **Length** 33–39 in (83–100 cm)
- **Tail** 12–15 in (30–38 cm)
- **Weight** 51–77 lb (23–35 kg)
- **Social unit** Group
- **Status** Least concern

Similar to the American beaver (see right) in appearance, habits, and lifestyle, the Eurasian beaver is usually heavier. As in its close cousin, glands at the base of the tail produce an oily, waterproofing secretion that is spread through the fur when grooming. In areas with many natural waterways, the Eurasian beaver does not build a sticks-and-mud lodge but digs tunnels in the bank, with underwater entrances. It eats bark, leaves, and plants, and can stay submerged for 20 minutes.

Castor canadensis

American beaver

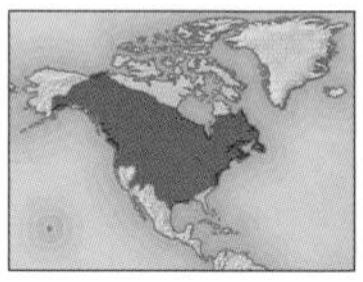

Location North America

- **Length** 29–35 in (74–88 cm)
- **Tail** 10–13 in (26–33 cm)
- **Weight** 24–57 lb (11–26 kg)
- **Social unit** Group
- **Status** Least concern

The American beaver is well adapted to aquatic life. Its feet are webbed for swimming and the flat, scaly tail slaps the surface as a loud alarm signal. Underwater, the ears and nose shut with valvelike flaps and the lips close behind the incisor teeth, which can then be used for nibbling and gnawing. The eyes have a third, transparent eyelid (known as the nictitating membrane) to see below the surface. Long whiskers feel the way in the dark. The American beaver feeds on the leaves, twigs, and bark of bankside trees and water plants. It also gnaws and fells small trees to eat the tender shoots and leaves. It uses the fallen branches and small trunks for building its lodge (see panel) and for dam construction, dragging them to the dam site in their strong jaws. Following a gestation period of 107 days, the 3–4 kits (young beavers) are born fully furred and are weaned within 2 months.

LAKESIDE LODGE

Beavers rest by day in their lodge—a pile of mud and sticks built in a pool or lake. The lodge's underwater entrances keep out land-based predators. The beaver digs channels and builds dams of mud, stones, and branches to maintain a system of waterways. These activities are believed by some to harm crops and trees, and affect local wildlife. An alternate view is that beavers reduce local floods, and help return the habitat to its natural state.

TWO COATS
The American beaver's long guard coat (outer fur) varies from yellowish brown to black, although it is usually reddish brown. The dense underfur is dark gray and retains body heat even in freezing water.

Mouselike rodents

PHYLUM	Chordata
CLASS	Mammalia
ORDER	Rodentia
SUBORDER	Myomorpha
FAMILIES	7
SPECIES	1,570

This suborder—which is distinguished from the other 4 rodent suborders by the way the jaw muscles are arranged (see p.116)—constitutes over a quarter of all mammal species. Within the group are rats and mice (including voles, lemmings, hamsters, and gerbils), and jerboas. Mouselike rodents often have a pointed face and long whiskers, and are usually small, nocturnal seed-eaters. They are found worldwide (except Antarctica), in almost all terrestrial habitats. Some species, such as the naked mole-rat, live underground; others, such as the water vole, occupy aquatic habitats. Species that live in open areas may have longer legs and feet (for quick escapes), and larger ears to detect danger from a distance.

PROLIFIC BREEDERS
Mouselike rodents, such as this harvest mouse, usually have large litters.

Oxymycterus nasutus

Long-nosed hocicudo

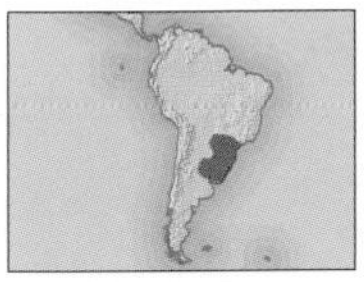

Location E. South America

Length 3¾–6½ in (9.5–17 cm)
Tail 2¾–5¾ in (7–14.5 cm)
Weight 3¼ oz (90 g)
Social unit Individual
Status Least concern

This large mouse rarely comes out into the open. It forages by day under leaves, logs, and stones for creatures such as grubs and worms, sniffing with its long, flexible, shrewlike snout and scrabbling up prey with its large-clawed forefeet. Its back is black, tinged with red or yellow, and is darker along the midline; its flanks are yellow-brown; the underside is yellow-orange mixed with gray; and the short, scaly tail is sparsely haired. It rarely excavates its own tunnels or pathways, using those of other rodents.

Baiomys taylori

Northern pygmy mouse

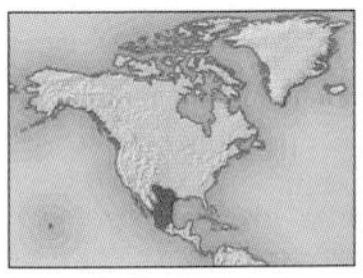

Location S. USA to C. Mexico

Length 2–2½ in (5–6.5 cm)
Tail 1½–1¾ in (3.5–4.5 cm)
Weight ¼–11/32 oz (7–9 g)
Social unit Individual
Status Least concern

The female of this species, North America's smallest rodent, can become pregnant at the youngest age of any New World mouse—just 4 weeks. The fur is mid-brown on the back and gray underneath. This mouse occupies a territory up to 100 ft (30 m) across and feeds at twilight on plants and seeds. Its nest is in a burrow under logs or plants.

Rhizomys sinensis

Chinese bamboo rat

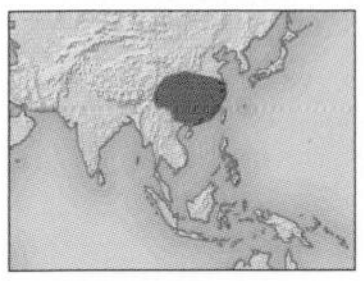

Location E. Asia

Length 9–16 in (22–40 cm)
Tail 2–3¾ in (5–9.5 cm)
Weight 2¼–6½ lb (1–3 kg)
Social unit Individual
Status Least concern

This beaverlike rat has thick gray-brown fur, a squat snout, and short, scaly tail. It lives in bamboo thickets, digging tunnels and nest chambers between the roots. An excellent climber, it feeds on bamboo as well as seeds and fruit.

Reithrodontomys raviventris

Saltmarsh harvest mouse

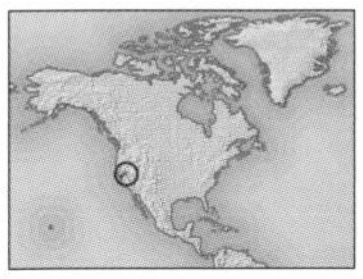

Location W. USA (San Francisco Bay area)

Length 2¾–3 in (7–7.5 cm)
Tail 1¾–4½ in (4.5–11.5 cm)
Weight ¼–⅔ oz (6–20 g)
Social unit Individual
Status Endangered

Similar to the house mouse (see p.128), the saltmarsh harvest mouse has large ears and a long tail. It builds a summer nest of grass above ground in a bush or undergrowth, and feeds on seeds, shoots, and insects. In winter, it moves into a burrow that has been dug, but then deserted, by another rodent.

Tachyoryctes macrocephalus

Ethiopian African mole rat

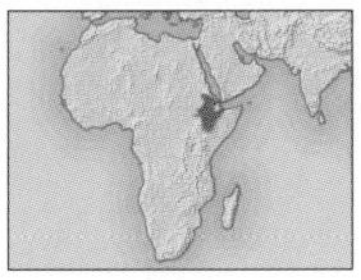

Location E. Africa

Length 12 in (31 cm)
Tail 3½–4 in (9–10 cm)
Weight 13–36 oz (350–1,000 g)
Social unit Variable
Status Endangered

This giant African mole rat lives like a mole, spending a great deal of time in its burrow. This may exceed 165 ft (50 m) in length and houses one rat. Burrowing adaptations include a blunt and rounded head, robust body, short limbs, small eyes and ears, and thick fur. This mole rat gnaws roots and other plant parts with its large, projecting, orange-yellow incisors, and digs its extensive tunnel system in the same way.

Peromyscus leucopus

White-footed mouse

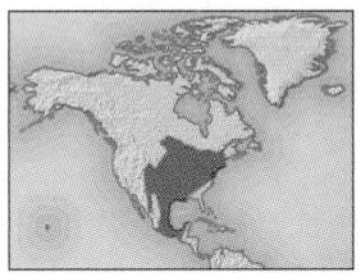

Location S.E. Canada to Mexico

Length 3½–4¼ in (9–10.5 cm)
Tail 2½–4 in (6–10 cm)
Weight ½–1 1/16 oz (14–30 g)
Social unit Paired
Status Least concern

A widespread and common species, this mouse resembles others in the deer mouse group, with its white feet and underparts, and brown fur on the back. White-footed mice usually live in pairs, which occupy a small den in a sheltered place under tree roots, below a log or stone, or in a thicket; or they may dig a den in soil, or take over an abandoned burrow. The nest is made from soft, dry plant matter, including shredded stems. The pair forage principally at night for fruit, berries, seeds, and insects, staying mainly on the ground even though they climb well. Food is stored in the den near the nest, covered with soil. In cold weather, the white-footed mouse may hibernate for a few hours each day. The female is pregnant for 22–23 days, and the average litter size is 4–5.

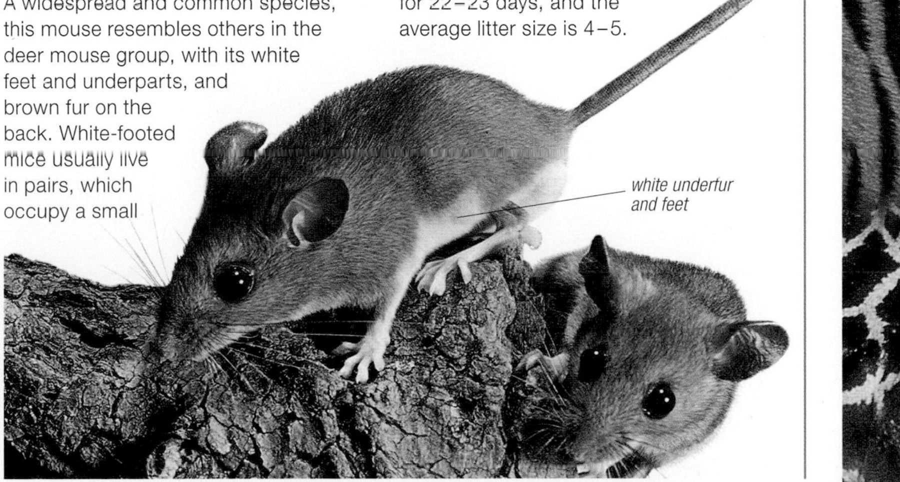

MAMMALS

Calomys laucha

Little lauca

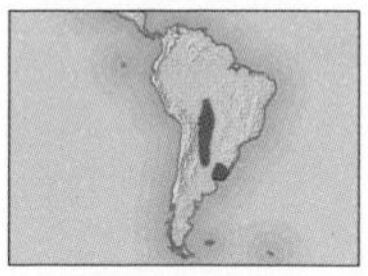

Location C. and E. South America

Length 2¾ in (7 cm)
Tail 2¼ in (5.5 cm)
Weight 7/16 oz (13 g)
Social unit Variable
Status Least concern

The little lauca resembles the house mouse (see p.128) not only in appearance but also in living near human dwellings and in its occasional population surges, when it may become a pest. It is gray-white above and pale to dark brown on the underside, with a white patch behind each medium-sized ear, and a long, sparsely furred tail. It builds a grassy nest in any crevice, such as under a log, rock, floorboards, or even—being a skillful climber—in a tree fork. The main diet is plants of all kinds, supplemented by a few insects such as beetles and caterpillars.

Sigmodon hispidus

Hispid cotton rat

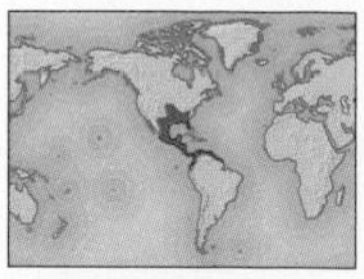

Location S. USA to N. South America

Length 5–8 in (13–20 cm)
Tail 3¼–6½ in (8–16.5 cm)
Weight 3⅝–8 oz (100–225 g)
Social unit Individual
Status Least concern

This is one of 13 cotton rat species in the Americas. Some are extremely rare, others locally numerous, even increasing to pest status when food is plentiful. The hispid cotton rat feeds on many foods, including plants (sometimes ruining crops such as sweet potato and sugar cane), insects, and grubs. Being a good swimmer, it also takes freshwater crabs, crayfish, and frogs, and climbs reeds to eat birds' eggs and chicks. It is active day and night, usually living alone in a grassy nest in a sheltered depression in the ground or in a burrow up to 30 in (75 cm) deep. The hispid cotton rat digs a shallow burrow as it feeds, and establishes well-worn foraging runways. Its stiff fur is brown to brownish gray on the back, gray-white underneath. The female becomes sexually mature at 6–8 weeks, and, after 27 days' gestation, she gives birth to up to 12 young.

Kunsia tomentosus

Giant South American water rat

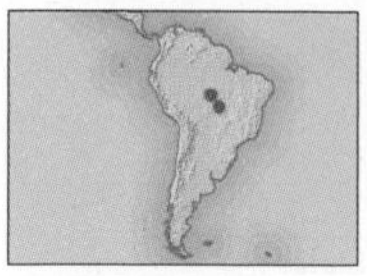

Location C. South America

Length 11½ in (29 cm)
Tail 6 in (15 cm)
Weight Not recorded
Social unit Individual
Status Least concern

A little-known species, this large, small-eared, short-tailed rat spends much time burrowing, using its strong feet equipped with long, curved claws. Its upper fur is dark brown with stiff hairs, and the underneath hairs are gray with white tips. As the giant South American water rat tunnels, it consumes roots, tubers, and other underground plant parts. During the flood season, however, water fills its tunnels, so this rat stays mainly above ground, where its diet changes to mostly grasses and shoots.

Phodopus roborovskii

Roborovski's desert hamster

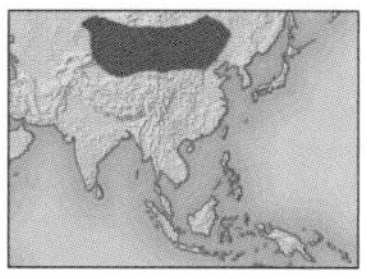

Location E. Asia

Length 2¼–4 in (5.5–10 cm)
Tail 2¾–4¼ in (7–11 cm)
Weight ⅞–1¾ oz (25–50 g)
Social unit Individual
Status Least concern

Also called the dwarf hamster, this small, short-tailed, prominent-eared rodent is pale brown on its upperparts with pure white underparts. Its rear feet are short and broad, with dense fur on the underside for jumping across hot, loose desert sand. Like other hamsters, the Roborovski's desert hamster crams seeds into its internal cheek pouches and takes them back to its burrow for storage. It also eats insects such as beetles, locusts, and earwigs. The nesting burrow is dug in firm sand and is lined with hair shed by camels and sheep.

Mesocricetus auratus

Golden hamster

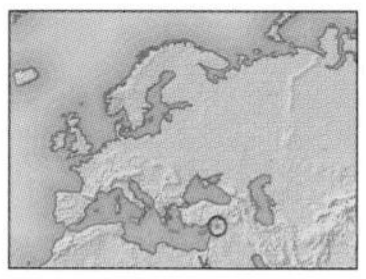

Location W. Asia

Length 5–5¼ in (13–13.5 cm)
Tail ½ in (1.5 cm)
Weight 3⅝–4 oz (100–125 g)
Social unit Individual
Status Vulnerable

The golden hamster, familiar around the world as a pet, is restricted in the wild to a small area of western Asia. Its golden coat may show a darker patch on the forehead and a black stripe from each cheek to the upper neck. The fur pales to gray-white on the underside. Grooming is important to keep the coat in good condition, and both front teeth and claws are used for this purpose. This hamster excavates a burrow down to 6½ ft (2 m), which it rarely leaves except to feed on a diet of seeds, nuts, and small creatures such as ants, flies, cockroaches, bugs, and even wasps. It is aggressive toward other hamsters.

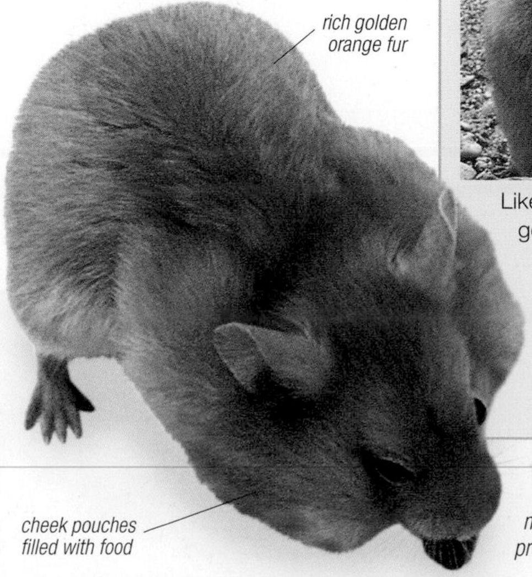

rich golden orange fur

cheek pouches filled with food

ROBUST RODENT
The golden hamster has a blunt muzzle, broad face, small eyes, prominent ears, and tiny tail.

SOLITARY GROOMING

Like many other hamsters, the golden hamster lives alone, so it has to rid its own coat of dirt, old fur, tangles, and pests such as fleas, rather than relying on mutual grooming like many social rodents.

Cricetus cricetus

Common hamster

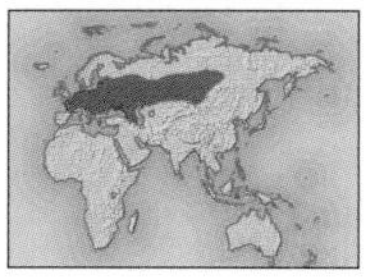

Location Europe to C. Asia

Length 8–13½ in (20–34 cm)
Tail 1½–2¼ in (4–6 cm)
Weight 3⅝–32 oz (100–900 g)
Social unit Individual
Status Least concern

The largest hamster, this species has distinctively thick fur, which is red-brown on the back and mainly black on the underside, with white patches on the nose, cheeks, throat, flanks, and paws—darker underside coloration compared to the upperparts is very unusual among mammals. In autumn, the common hamster hoards seeds, roots, and other plant matter, carrying items in its large, external cheek pouches back to its burrow. It then hibernates until spring, waking every 5–7 days to feed. In summer, it also eats grubs, worms, and other creatures. After 18–20 days' gestation, the female produces a litter of up to 12, which are weaned by 3 weeks and fully grown by 8 weeks.

MAMMALS

Microtus arvalis

Common vole

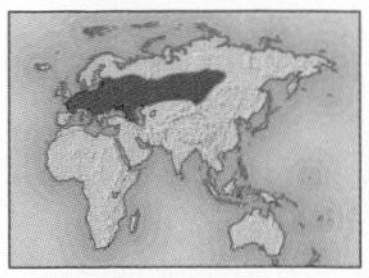

Length $3\frac{1}{2}$–$4\frac{3}{4}$ in (9–12 cm)
Tail $1\frac{1}{4}$–$1\frac{3}{4}$ in (3–4.5 cm)
Weight $1\frac{1}{16}$–$1\frac{5}{8}$ oz (20–45 g)
Social unit Group
Status Least concern

Location W. Europe to W. and C. Asia

This medium-sized vole is one of the most numerous rodents in grassy and farmland habitats. It has short fur, gray-brown or sandy on the back changing to gray underneath, a blunt snout, small eyes and ears, stocky body, and short, furred tail. It digs burrows to make nest chambers and food stores, and eats chiefly green plant parts such as grass blades. In winter, it may take refuge in a barn or haystack, and gnaw soft bark.

Ondatra zibethicus

Muskrat

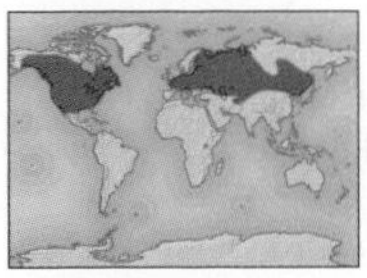

Length 10–14 in (25–35 cm)
Tail 8–10 in (20–25 cm)
Weight $1\frac{1}{4}$–$4\frac{1}{2}$ lb (0.6–2 kg)
Social unit Group
Status Least concern

Location North America, W. Europe to N. and E. Asia

The muskrat usually lives in a group of up to 10, digging tunnels in the bank or building a beaverlike home (lodge) from mud, plant stems, and twigs. It eats reeds and other water plants, and occasionally hunts for crayfish, frogs, fish, and mollusks. The female builds a nest in a dry tunnel chamber or lodge platform, where the litter of 1–3 young are born. Musk secretions are used, with droppings and urine, to mark out territories.

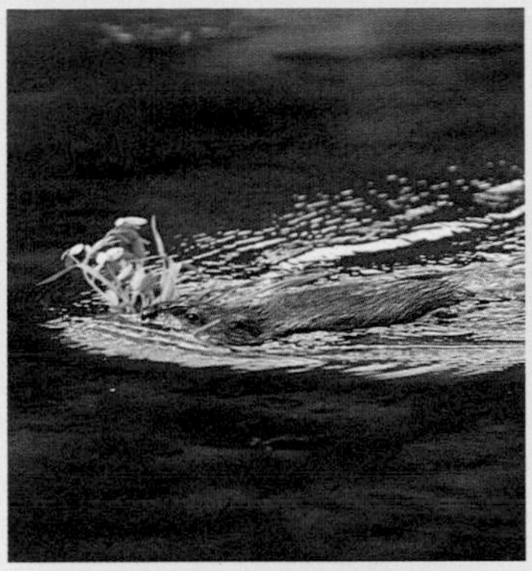

EXPERT SWIMMER

The largest species of burrowing vole, the muskrat is well adapted to swimming with large back feet that have small webs between the toes and a row of stiff hairs along one edge, forming a "swimming fringe." Its long, furless tail is flattened from side to side, which enables it to be used as a rudder. The nostrils and small ears are closed by flaps during dives, which may last 20 minutes. It can swim up to 330 ft (100 m) underwater without surfacing.

MUSKY SMELL
The muskrat is named after the musky-smelling secretions from glands around its genital and anal region. The glands are especially prominent in males and enlarge at breeding time.

Myodes glareolus

Bank vole

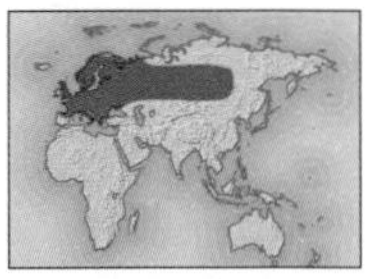

Length $2\frac{3}{4}$–$5\frac{1}{4}$ in (7–13.5 cm)
Tail $1\frac{1}{2}$–$2\frac{1}{2}$ in (3.5–6.5 cm)
Weight $\frac{7}{16}$–$1\frac{1}{4}$ oz (12–35 g)
Social unit Group
Status Least concern

Location W. Europe to N. Asia

Typically blunt-headed, the bank vole's upperparts vary from yellowish to reddish or brown, with gray flanks, gray-white rump, white feet, and a slightly bushy-tipped tail. The species also varies greatly in size, being twice as long and 3 times as heavy in some regions compared to others. The bank vole is very adaptable, nesting in burrows, thickets, and tree stumps, and eating a huge range of foods, from fungi and mosses to seeds, buds, insects, and birds' eggs.

Arvicola amphibius

Eurasian water vole

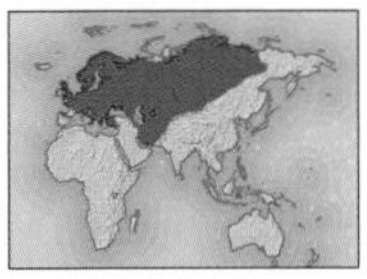

Length $4\frac{3}{4}$–9 in (12–23 cm)
Tail $2\frac{3}{4}$–$4\frac{1}{4}$ in (7–11 cm)
Weight $2\frac{1}{8}$–11 oz (60–300 g)
Social unit Individual/Pair
Status Least concern

Location W. Europe to W. and N. Asia

Water voles that mainly burrow in meadows and woods are almost half the size of those that live near rivers, lakes, and marshes. Both types eat plant foods and have thick fur, which is gray, brown, or black on the upperparts and dark gray to white below. The rounded tail is half the body length. This vole is threatened by pollution, loss of habitat, and an introduced predator, mink.

Lemmus sibiricus

Brown lemming

Length $4\frac{3}{4}$–6 in (12–15 cm)
Tail $\frac{3}{8}$–$\frac{1}{2}$ in (1–1.5 cm)
Weight $1\frac{5}{8}$–$5\frac{1}{2}$ oz (45–150 g)
Social unit Group
Status Least concern

Location N. Asia

The brown lemming lives in large colonies, breeds prolifically, and makes small-scale seasonal migrations between high, shrubby grassland and moors, and sheltered lowlands for winter. Its migrations are much less spectacular than those of the Norway lemming (*Lemmus lemmus*), which is sometimes driven by instinct to try to swim rivers or scramble down cliffs. The brown lemming eats mosses, sedges, herbs, and soft twigs, and sometimes birds' eggs. The female builds a nest from grass and her own fur and, after a gestation period of 18 days, produces up to 12 young.

Lagurus lagurus

Steppe vole

Length $3\frac{1}{4}$–$4\frac{3}{4}$ in (8–12 cm)
Tail $\frac{1}{3}$–$\frac{3}{4}$ in (0.7–2 cm)
Weight $\frac{7}{8}$–$1\frac{1}{4}$ oz (25–35 g)
Social unit Variable
Status Least concern

Location E. Europe to E. Asia

Long, waterproof fur, even covering feet and ears, keeps the steppe vole warm in the cold north Asian steppes. It has a black stripe along the middle of its pale gray or cinnamon back, and pale underparts. Burrows up to 12 in (30 cm) deep give temporary shelter, while more permanent ones—3 times as deep—house grass-lined nests. There are 5 litters per year of up to 12 young.

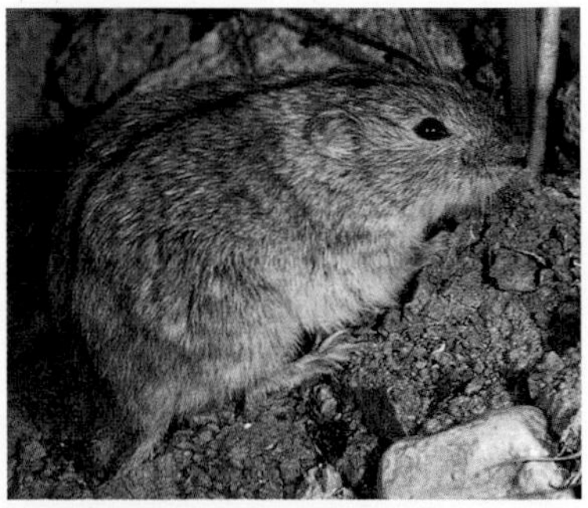

Nyctomys sumichrasti

Sumichrast's vesper rat

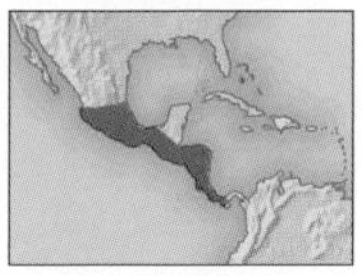

Length 4¼–5 in (11–13 cm)
Tail 3¼–6 in (8.5–15.5 cm)
Weight Not recorded
Social unit Group
Status Least concern

Location S. Mexico to S. Central America

One of the more arboreal and brightly colored rats, this species has tawny or pinkish brown upperparts with darker hairs along the centre of the back. It has pale flanks, white underparts, a dark ring around the eye, and a brown, scaly, hairy tail. The ears are short and finely furred. On each foot the first toe is almost thumblike, developed for gripping twigs. Sumichrast's vesper rat lives in a colony, builds squirrel-like nests of twigs, leaves, and creepers, and rarely descends to the ground. It is active mainly at night, eating a variety of plants, including figs and avocados.

Parotomys brantsii

Brants's whistling rat

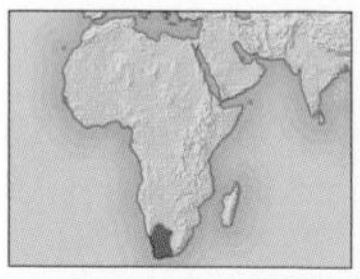

Length 5–6½ in (2.5–16.5 cm)
Tail 3–4¼ in (7.5–10.5 cm)
Weight 3–4 oz (85–125 g)
Social unit Group
Status Least concern

Location Southern Africa

Also known as the karoo rat from its arid, rocky, saltbrush habitat, this wary, diurnal species feeds on grass and other low-growing vegetation. It never wanders far from its extensive tunnel system, which can be accessed via numerous entrances. If danger threatens, it whistles loudly to warn other members of the colony. In favorable conditions, it can breed up to 4 times a year, females producing up to 4 young in each litter. Offspring mature quickly and are capable of breeding themselves when about 3 months old. Brants's whistling rat has a reddish orange nose and small yellow ears. The fur is patchy yellow and brown-black, with gray-white underparts.

Pachyuromys duprasi

Fat-tailed jird

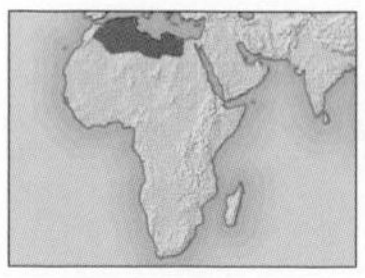

Length 3¾–5 in (9.5–13 cm)
Tail 2¼–6½ in (5.5–16.5 cm)
Weight 11/16–1¾ oz (20–50 g)
Social unit Variable
Status Least concern

Location N. Africa

A Sahara dweller, the fat-tailed jird has long, soft fur, a pointed snout, and long rear feet. Coloration is chestnut-cinnamon on the back and sides, with black tips to the hairs, shading to white underneath. The club-shaped tail contains a store of body fat for nourishment and water. This jird emerges from its burrow at dusk to search for insects such as crickets. It also eats leaves, seeds, and other plant matter.

Meriones unguiculatus

Mongolian jird

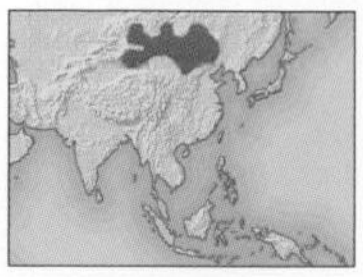

Length 4–5 in (10–13 cm)
Tail 3¾–4¼ in (9.5–11 cm)
Weight 2–2⅛ oz (50–60 g)
Social unit Group
Status Least concern

Location C. Asia

Familiar as a pet, this is one of some 17 jird species and is native to Central Asia. It burrows in dry steppes and is active by day and night, summer and winter. It eats mainly seeds, storing excess in its elaborate burrow, which it may share with its mate and up to 12 young. Family members groom each other but are quick to attack a stranger. Mainly brown haired with black tips, the underparts are gray or white.

Hypogeomys antimena

Votsovotsa

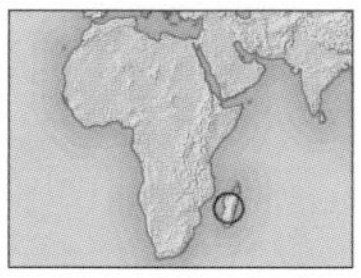

Length 12–14 in (30–35 cm)
Tail 8½–10 in (21–25 cm)
Weight 2¼–3¼ lb (1–1.5 kg)
Social unit Group
Status Endangered

Location W. Madagascar

Known locally as the votsotsa, this rat has tall, rabbitlike ears and large rear feet. Similar to a rabbit in behavior, too, it hops rather than runs and lives in a family group of male, female, and offspring of the past 2–3 years. The votsovotsa digs a burrow system with up to 6 entrances in the sandy soil of coastal forests. It eats fruit, shoots, and soft bark, holding them in the forefeet. This species is threatened by habitat loss and competition from introduced roof rats.

Cricetomys gambianus

Northern giant pouched rat

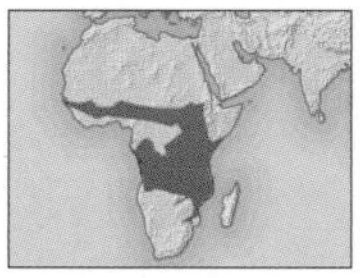

Length 14–16 in (35–40 cm)
Tail 14½–18 in (37–45 cm)
Weight 2¼–3¼ lb (1–1.5 kg)
Social unit Variable
Status Least concern

Location W., C., E., and southern Africa

This rat is solitary, active mainly at night, and climbs and swims well. It eats a great variety of moist or fleshy foods, from termites to avocados, and also peanuts and corn. Its huge cheek pouches carry items back to its burrow, which has extensive chambers for food, resting, breeding, and defecation. Its bristly hair is buff-brown on the back, fading to white on the throat and underside, with a dark brown eye-ring. This large, big-eared, docile rat is kept both as a pet and for meat.

Jaculus jaculus

Lesser Egyptian jerboa

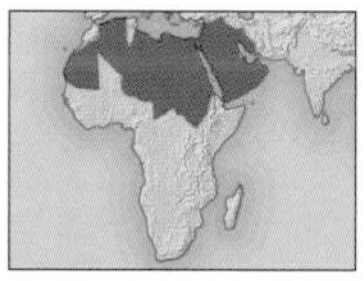

Length 4–4¾ in (10–12 cm)
Tail 6½–8 in (16–20 cm)
Weight 1⅝–2⅝ oz (45–75 g)
Social unit Individual
Status Least concern

Location N. Africa to W. Asia

Well adapted for desert sand, this jerboa has very long hind legs, each with 3 toes on a pad of hairs. It hops at high speed, balancing with its very long tail, which has a black band near the fluffy white tip. The fur is brown-orange on the back, gray-orange along the sides, and white below with a whitish hip band. The lesser Egyptian jerboa feeds at night on seeds, roots, and leaves. By day, it plugs the entrances to its burrow to keep out the heat, predators, and other animals.

Allactaga tetradactyla

Four-toed jerboa

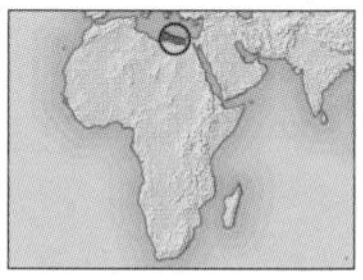

Length 4–4¾ in (10–12 cm)
Tail 6–7 in (15.5–18 cm)
Weight 1¾–2 oz (50–55 g)
Social unit Individual
Status Vulnerable

Location N. Africa

Each back foot of this jerboa has an extra, fourth toe, which is small compared to the 3 functional toes. In other respects it is a typical jerboa, with its huge, hopping back feet and tall, rabbitlike ears. The upperparts are speckled black and orange, the rump orange, the sides gray, and the underparts white. The long, balancing tail has a black band near the white, feathery tip. Emerging at night, the jerboa eats grass, leaves, and soft seeds.

MAMMALS

Apodemus flavicollis

Field mouse

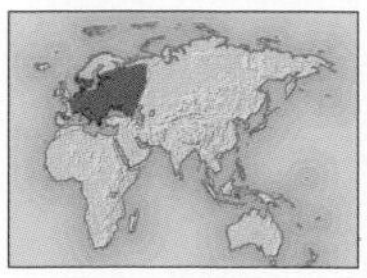

Location W. Europe to W. and C. Asia

Length 3¼–5 in (8.5–13 cm)
Tail 3½–5¼ in (9–13.5 cm)
Weight ⅝–1¾ oz (18–50 g)
Social unit Individual
Status Least concern

The yellow throat of this large, long-tailed mouse contrasts with the brown back and yellowish white underparts. The large, prominent eyes and big ears indicate twilight and nocturnal habits, and the long rear feet allow prodigious jumps. The field mouse climbs trees to 66 ft (20 m), searching for seeds, berries, and small creatures such as caterpillars, spiders, and millipedes. It nests in any suitable hole, among roots, or high in a tree trunk, and aggressively chases away other mice, including similar long-tailed field mice (see right).

Apodemus sylvaticus

Long-tailed field mouse

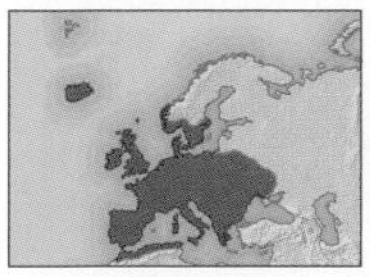

Location Europe, N. Africa

Length 3–4¼ in (8–11 cm)
Tail 2¾–4¼ in (7–11 cm)
Weight 9/16–1 1/16 oz (15–30 g)
Social unit Individual
Status Least concern

Sometimes mistaken for a small field mouse (see left), the long-tailed field mouse may have not only a yellow throat but also an orange-brown chest patch. Its upperparts are gray-brown; the underparts, gray-white. Fast and agile, the long-tailed field mouse consumes many foods, including mushrooms, berries, fruit, worms, and insects. It nests in a burrow or tree-hole, marks its territory with urine, and fights intruders violently.

Micromys minutus

Harvest mouse

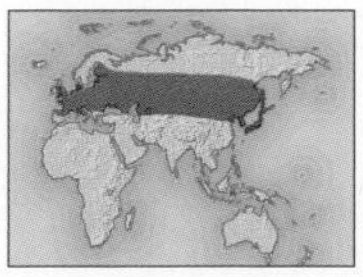

Location W. Europe to E. Asia

Length 2–3 in (5–8 cm)
Tail 1¾–3 in (4.5–7.5 cm)
Weight 3/16–¼ oz (5–7 g)
Social unit Individual
Status Least concern

This tiny mouse is the only Old World rodent with a prehensile tail. It feeds on seeds, including the heads of wheat and other farmed cereals, berries, and small animals such as insects and spiders. In the breeding season, the female gives birth to about 2–6 young, occasionally 12 or more, after a pregnancy lasting 21 days. However, if food becomes scarce, she may eat them—a self-survival strategy that occurs in various rodents.

COLORATION
The harvest mouse has yellowish or reddish brown upperparts and a mainly white underside. The face is rounded but the snout is pointed.

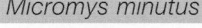

BALL-LIKE NEST

The harvest mouse constructs a spherical nest of finely shredded grass blades and stems, perhaps using an old bird's nest as a base, in a thicket or grassy clump. It is about 3–4¾ in (8–12 cm) across, and is located some 20–51 in (50–130 cm) above ground level. The female's breeding nest is more substantial than the regular nest.

Lemniscomys striatus

Striped grass mouse

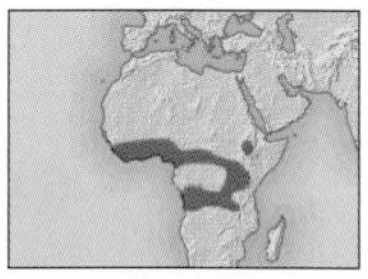

Location W. and E. Africa

Length 4–5½ in (10–14 cm)
Tail 4–6 in (10–15.5 cm)
Weight 11/16–2½ oz (20–70 g)
Social unit Individual
Status Least concern

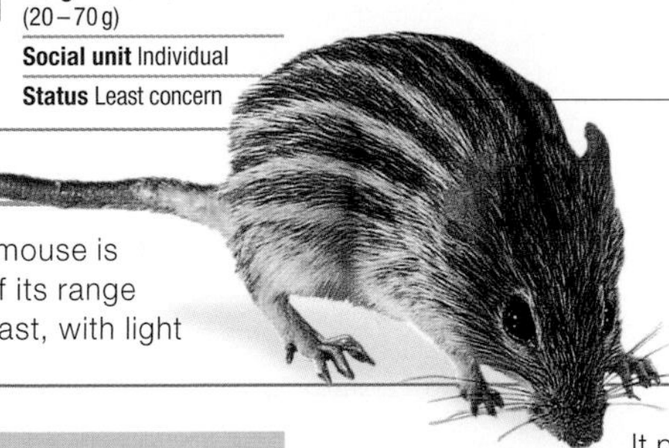

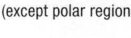

The striped grass mouse is paler in the west of its range compared to the east, with light stripes along the buff or reddish orange back; underparts are brown-tinged white. It lives mainly on the ground with runways leading to feeding areas of grass stems, leaves, farm crops, and the occasional insect. This nervous, jumpy mouse lives alone and may feign death—"play dead"—or shed the skin from its tail when caught by a predator.

Mallomys rothschildi

Smooth-tailed giant rat

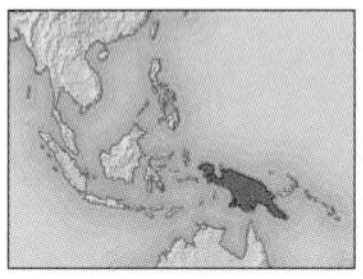

Location New Guinea

Length 13½–15 in (34–38 cm)
Tail 14–16½ in (36–42 cm)
Weight 2–3¼ lb (0.95–1.5 kg)
Social unit Individual
Status Least concern

This rat, also known as Rothschild's woolly rat because of its long, thick fur, has an almost black back, reddish brown upper parts, and pale underside, perhaps with a white band running from the belly up each side. Its scaly tail is brown at the base but turns white about halfway along its length. The smooth-tailed giant rat scampers through the trees in search of shoots, leaves, and other plants, gripping with its sharp-clawed feet. It usually lives in a hollow tree but sometimes inhabits holes in the ground. Females are thought to produce only one offspring per litter.

Rattus rattus

Roof rat

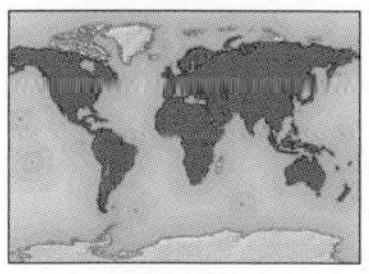

Location Worldwide (except polar regions)

Length 6½–9½ in (16–24 cm)
Tail 7–10 in (18–26 cm)
Weight 5–9 oz (150–250 g)
Social unit Variable
Status Least concern

Also called the ship rat, black rat, or house rat, in early Roman times the roof rat spread around the world from Asia in ships or crates of cargo. It prefers plant matter, such as seeds and fruit, but it can survive on insects, dead animals, feces, and refuse. Roof rats gather in "packs" of 20–60 and may intimidate larger animals such as dogs. The basic coloration is black, but shades of brown also occur, with gray to white on the belly, and whitish or pink feet. The female produces 4–10 young after a gestation period of 20–24 days. This rat can run, climb, and swim very well, and makes a nest of twigs and grass, but often nests in a roof cavity and uses almost any material. The fleas carried by roof rats spread diseases among humans, including bubonic plague, which has killed hundreds of millions of people through the centuries.

MAMMALS

Rattus norvegicus

Brown rat

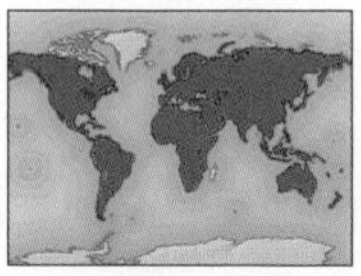

Location Worldwide (except polar regions)

Length 8–11 in (20–28 cm)
Tail 7–9 in (17–23 cm)
Weight 10–21 oz (275–575 g)
Social unit Group
Status Least concern

A hugely varied, opportunistic diet, sharp senses, and great agility have enabled the brown rat, also called the Norway or common rat, to spread all around the world. "Packs" of up to 200, dominated by large males, will attack rabbits, large birds, and even fish. After a gestation of 22–24 days, the female gives birth to 6–9 young in a nest of grass, leaves, paper, rags, or almost any other material. This species is the ancestor of rats bred as pets and for scientific research.

SWIMMING RAT
The brown rat is an exceptional swimmer. It catches small fish and crayfish, and crunches up water snails and aquatic insects.

COLORATION
The brown rat varies from brown to gray-brown or black on its back. It is paler on the underside, and has a long, sparsely haired tail.

Leporillus conditor

Greater stick-nest rat

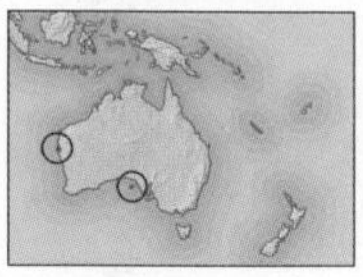

Location W. and S. Australia

Length 7–10 in (17–26 cm)
Tail 5¾–9½ in (14.5–24 cm)
Weight 5½–16 oz (150–450 g)
Social unit Group
Status Vulnerable

Occurring naturally on Franklin Island, W. Australia, and introduced elsewhere, the greater stick-nest rat is almost rabbit sized, with long ears, a rounded nose, and slim, hairy tail. It is gray-brown above and white below, and builds a strong surface nest of sticks and twigs up to 5 ft (1.5 m) high. Large nests may contain many individuals. Active at night, this species feeds on succulent plants and drinks little if any water.

Mus musculus

House mouse

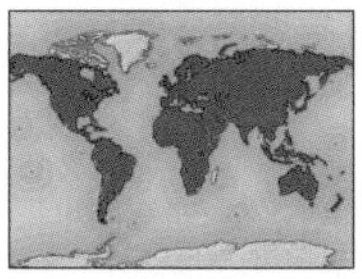

Location Worldwide (except polar regions)

Length 2¾–4¼ in (7–10.5 cm)
Tail 2–4 in (5–10 cm)
Weight ⅜–1 oz (10–35 g)
Social unit Group
Status Least concern

The second most widely distributed mammal, after humans, this mouse survives on a huge range of foods and lives in a family group of dominant male and several females. They communicate by high-pitched squeaks and mark their territory with scent and urine. Maturity is at 8–10 weeks, gestation 18–24 days, and litter size 3–8, with 10 litters in favorable years. The species has been widely bred for pets and scientific research.

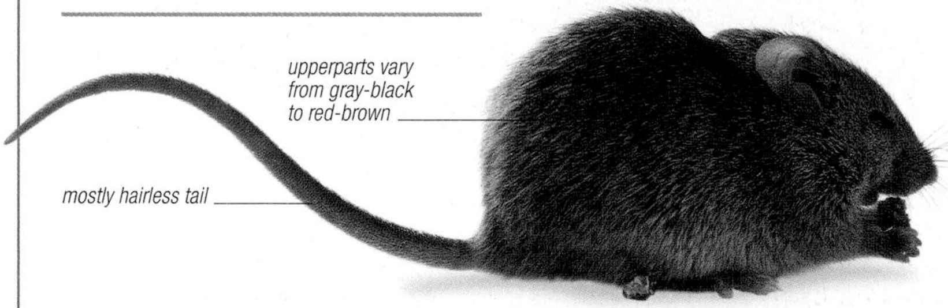

Hydromys chrysogaster

Common water rat

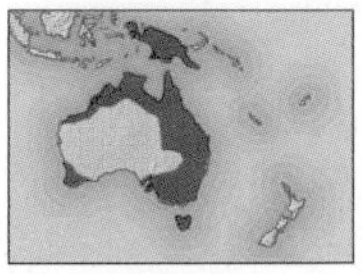

Location New Guinea, Australia (including Tasmania)

Length 11½–15½ in (29–39 cm)
Tail 9–13 in (23–33 cm)
Weight 1½–3 lb (0.65–1.25 kg)
Social unit Individual
Status Least concern

Australia's heaviest native rodent has broad back feet and webbed toes, for swimming. The upperparts vary from brown to gray, the underparts brown to golden-yellow, cream, or even white, and the tail is thick and white-tipped. This water rat is active at dusk and dawn as a powerful predator of shellfish, water snails, fish, frogs, turtles, birds, mice, and even bats.

Notomys alexis

Spinifex hopping mouse

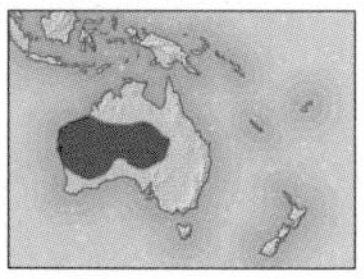

Location W. and C. Australia

Length 3½–7 in (9–17 cm)
Tail 5–9 in (12.5–23 cm)
Weight 11⁄16–1¾ oz (20–50 g)
Social unit Group
Status Least concern

Also called the dargawarra, this large mouse is named after the bushy, spiky desert grass called spinifex, where it often lives. It obtains all its moisture from leaves, seeds, berries, and other plant foods, never needs to drink, and produces some of the most concentrated urine of any rodent. Very sociable, this mouse lives in mixed-sex groups of up to 10, shares nests, and breeds soon after any rainfall.

Acomys minous

Crete spiny mouse

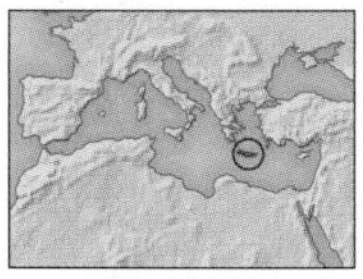

Location Europe (Crete)

Length 3½–4¾ in (9–12 cm)
Tail 3½–4¾ in (9–12 cm)
Weight ⅜–3¼ oz (11–90 g)
Social unit Variable
Status Data deficient

Coarse, stiff hairs on the back and tail give this mouse its common name. Its fur varies from yellow to red, gray, or brown above, and white below. A nocturnal forager, it takes anything edible—mainly grass blades and seeds—and builds only a rudimentary nest. A gregarious species, gestation is 5–6 weeks (long for a mouse). Other females clean and assist the mother at the birth, when the young are already well developed, with open eyes.

MAMMALS

Cavylike rodents

PHYLUM	Chordata
CLASS	Mammalia
ORDER	Rodentia
SUBORDER	Histricomorpha
FAMILIES	18
SPECIES	284

Included in this "umbrella" group are species as diverse as the semiaquatic capybara, the largest living rodent; the New World porcupines, which are arboreal and possess distinctive spines and a prehensile tail; and African mole-rats, which live underground. The defining feature of the suborder is the organization of the jaw muscles (see p.116), and most species are characterized by a relatively large head, a sturdy body, a short tail, and slender legs. Cavylike rodents are found throughout Africa, the Americas, and Asia.

SAFETY IN NUMBERS
By living in a family group, these capybaras greatly increase the chances of spotting a predator before it can attack.

Erithizon dorsata

North American porcupine

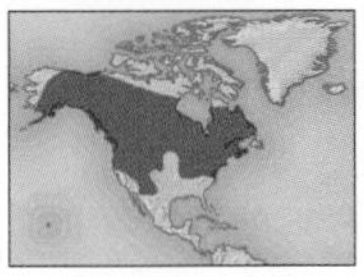

Location Canada, USA

Length 26–32 in (65–80 cm)
Tail 6–12 in (15–30 cm)
Weight 7¾–15 lb (3.5–7 kg)
Social unit Individual
Status Least concern

The large crest of long quills, up to 3 in (8 cm) long on the head, identifies this stocky, short-limbed porcupine. The strong, broad feet have sharp claws and naked soles, to aid grip. The North American porcupine is very vocal, especially during courtship in early winter, when it whines, screeches, grunts, mews, and hoots.

QUILLS AND FUR
The main quills are yellow-white, with black or brown tips. The rest of the body is furred and spined in shades of brown.

CLUMSY CLIMBER

The North American porcupine seems clumsy in trees, yet it climbs extensively for buds, blossom, shoots, leaves, berries, and nuts. It also eats grass and farm crops in summer, and soft bark and conifer needles in winter. Solitary most of the year, it does not hibernate and may share a den in cold spells.

Hystrix africaeaustralis

Cape porcupine

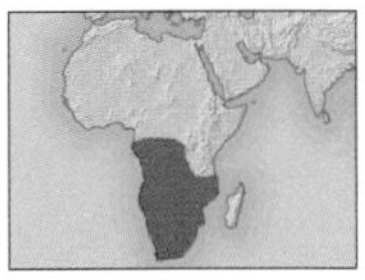

Location C. to southern Africa

Length 25–32 in (63–80 cm)
Tail 4¼–5 in (10.5–13 cm)
Weight 44 lb (20 kg)
Social unit Variable
Status Least concern

These rodents sniff and forage for distances of up to 9 miles (15 km) at night, alone, in pairs, or in small groups, seeking roots, bulbs, berries, and fruit. By day, they rest in caves or rocky crevices. After 6–8 weeks' gestation, the female produces 1–4 young, and the male helps care for them. The Cape (or crested) porcupine communicates by means of quill-rattles, piping squeaks, and grunts.

SPINES AND WHISKERS
The back is covered by banded brown-black and white spines with white tips, intermingled with hairs. The nose has long, stout whiskers.

BARBED DEFENSE

The Cape porcupine cannot shoot out its quills; but, if alarmed, it raises them and charges backward at the enemy. The quills detach easily and their barbed tips work into the aggressor's flesh. These defenses effectively deter attacks by other animals, but these porcupines are killed in large numbers by humans because they damage crops, and for their prized meat.

Coendou prehensilis

Prehensile-tailed porcupine

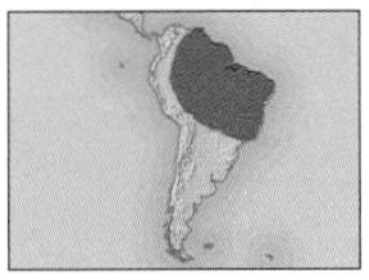

Location N. and E. South America, Trinidad

Length 20½ in (52 cm)
Tail 20½ in (52 cm)
Weight 11 lb (5 kg)
Social unit Individual
Status Least concern

This large, muscular porcupine climbs slowly but surely with curved claws, naked-soled feet, and prehensile tail, which is almost as long as the head and body and lacks fur toward the tip. By day, it sleeps in a hole, in a tree trunk, or in the ground. At dusk, the prehensile-tailed porcupine forages for leaves, bark, fruit, and shoots, and occasionally small animals.

MAMMALS

Cavia aperea

Brazilian guinea pig

Location N.W. to E. South America

Length 8–12 in (20–30 cm)
Tail None
Weight 18–21 oz (500–600 g)
Social unit Group
Status Least concern

A close relative and perhaps ancestor of the domestic (pet) guinea pig (*Cavia porcellus*), this is one of the smallest cavylike rodents. It has a large head, blunt snout, tail-less body, and short legs, with 4-toed front feet and 3-toed rear feet. The long, coarse fur is dark gray-brown to black. It eats leaves, grasses, seeds, flowers, and bark. Guinea pigs live as close neighbors in shrubby grassland, with communal feeding runways but separate nests.

Hydrochoerus hydrochaeris

Capybara

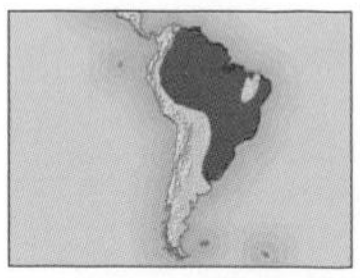

Location N. and E. South America

Length 3½–4¼ ft (1.1–1.3 m)
Tail Vestigial
Weight 77–145 lb (35–66 kg)
Social unit Variable
Status Least concern

The world's heaviest rodent, the capybara has partially webbed toes, and its nostrils, eyes, and ears are set on top of its head, so that it can smell, see, and hear when swimming.It is an excellent swimmer and diver in rivers, lakes, and swamps. It forms various groupings, including male–female pairs, families with young, and larger, mixed herds dominated by one male who mates with all females in his group. They roam a home range, marked with scent, and chase away intruders. After a gestation period of 150 days, 1–8 offspring (usually 5) are born fully furred and can run, swim, and dive within hours of birth. The capybara's numerous predators include humans, who value the meat and hide.

STOUT SWIMMER
The capybara is heavy bodied, with short but sturdy limbs, hooflike claws, and almost no tail.

DAILY ROUTINE

Capybaras rest in the morning, wallow during the midday heat, feed on water plants, buds, and soft tree bark in the evening, rest again around midnight, and resume feeding toward dawn. They move on to find fresh grazing as necessary and may raid farm crops, which makes them pests in some areas.

Dolichotis patagonum

Patagonian mara

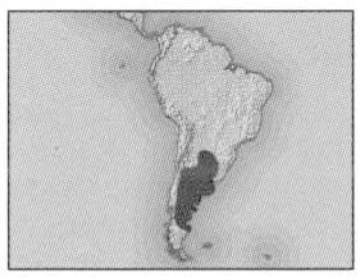

Location S. South America

Length 17–31 in (43–78 cm)
Tail 1 in (2.5 cm)
Weight 4½ lb (2 kg)
Social unit Pair
Status Near threatened

The Patagonian mara or cavy is a large, long-legged rodent that resembles a small deer in looks and behavior. It has a white, collarlike neck patch and

whitish fringe to the short tail. The muzzle is long; the eyes and ears large. The Patagonian mara runs and jumps well, and is mainly a grazer on grass and low shrubs. Male–female pairs stay together for life, digging a large burrow for their 1–3 offspring.

Dasyprocta azarae

Azara's agouti

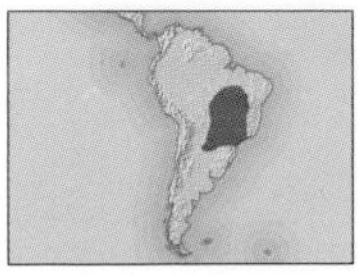

Location E. South America

Length 20 in (50 cm)
Tail 1 in (2.5 cm)
Weight 6½ lb (3 kg)
Social unit Variable
Status Data deficient

This large, social, diurnal species has prominent ears, short legs, and a tiny tail. Its fur is speckled pale to mid-brown, perhaps tinged yellowish

around the underside. The distinctive feet have 5 toes on the front feet, but 3 on the hind feet. Azara's agouti eats a variety of seeds, fruit, and other plant material, and barks when alarmed. People hunt it for meat across much of its range.

Cuniculus paca

Lowland paca

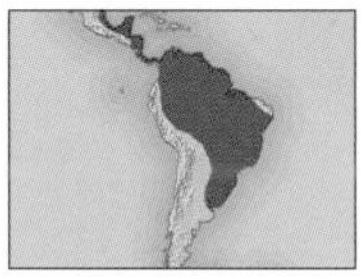

Location S. Mexico to E. South America

Length 23½–32 in (60–80 cm)
Tail ½–1½ in (1.5–3.5 cm)
Weight 13–26 lb (6–12 kg)
Social unit Individual
Status Least concern

Lowland pacas live alone, apart from the mother and single young, which she suckles for 6 weeks. This expert swimmer has a big head, squared-off muzzle, robust body, short, sturdy legs, and tiny tail. The upper body is brown, red, or gray with 4 pale "dotted-line" stripes along each side; the underparts are white or buff. The lowland paca rests in a burrow or hollow tree by day and emerges to eat fruit, leaves, buds, and flowers at night.

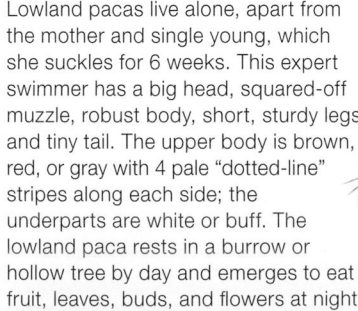

Lagostomus maximus

Argentina plains viscacha

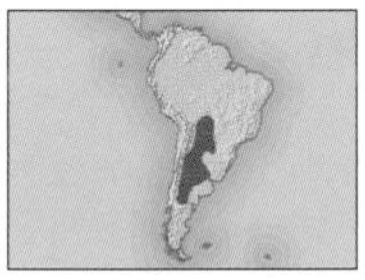

Location C. and S. South America

Length 1¾–2½ in (4.5–6.5 cm)
Tail ½–¾ in (1.5–2 cm)
Weight 15–20 lb (7–9 kg)
Social unit Group
Status Least concern

The largest member of the chinchilla family, the plains viscacha lives in noisy colonies of 20–50. It digs tunnel systems extending more than 980 ft (300 m). At night, it eats grasses, seeds, and roots, and collects sticks, stones, and bones to pile at tunnel entrances. The face is black and white striped, the back is gray-brown fading to white underneath, and the tail is brown-tipped with black. The male is twice as heavy as the female.

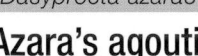

Chinchilla lanigera

Long-tailed chinchilla

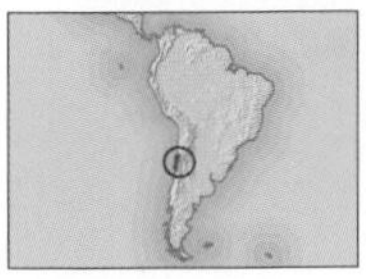

Location S.W. South America

Length 8–9 in (22–23 cm)
Tail 5–6 in (13–15.5 cm)
Weight 14–18 oz (400–500 g)
Social unit Group
Status Critically endangered

This rodent's thick, soft, silky fur, which keeps out the bitter cold of its mountain habitat, has long been prized by humans and, although now protected, has been hunted and farmed in many regions.

Endangered in the wild, the long-tailed chinchilla is common as a pet, with its appealing appearance and usually friendly nature. The fur is silvery gray-blue above and cream or yellowish on the underside, with long gray and black hairs on the tail's upper surface. In the wild, the long-tailed chinchilla forms colonies of 100 or more in rocky areas, sheltering in caves and crevices. It eats most plant foods, especially grass and leaves, sitting up to hold items in the front feet while watching for danger. If threatened, it rears up and spits hard at the aggressor. During the winter breeding season, the female, which is larger than the male, becomes more aggressive toward other females. Up to 4 young, but usually 2–3, are born after a gestation period of 111 days and are suckled for 6–8 weeks.

Octodon degus

Degu

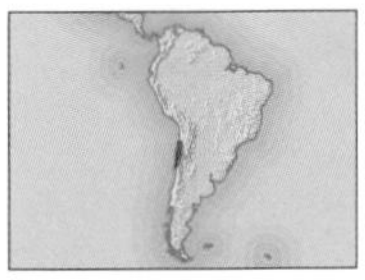

Location S.W. South America

Length 10–12 in (25–31 cm)
Tail 3–5 in (7.5–13 cm)
Weight 6–11 oz (175–300 g)
Social unit Variable
Status Least concern

Resembling a very large mouse, this mountain-dwelling rodent has a stout body. Mainly brown, it has pale underparts, and there may also be yellow furry "lids" above and below each eye, and a yellow neck ring. The long tail, which has a tufted black tip, breaks off easily if grabbed by a predator. The degu lives in colonies and excavates extensive burrow systems. It feeds by day on a varied diet of plant matter and, in the dry season, cattle droppings. It stores excess food in the burrow for winter.

Myocastor coypus

Coypu

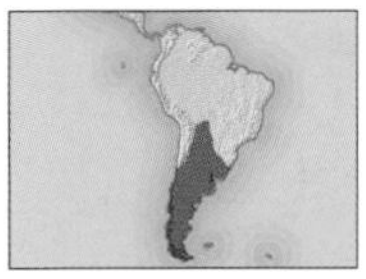

Location S. South America

Length 18½–23 in (47–58 cm)
Tail (13½–16 in 34–41 cm)
Weight 14 lb (6.5 kg)
Social unit Group
Status Least concern

The coypu or nutria was farmed for its dense brown fur, but escapees established colonies in many areas outside its native region. It has a large head, small, high-set eyes and ears, a robust body, arched hindquarters, long, rounded tail, and webbed rear feet for rapid swimming. The coypu eats most vegetation, especially water plants, and lives in bankside tunnels in family groups.

Capromys pilorides

Desmarest's hutia

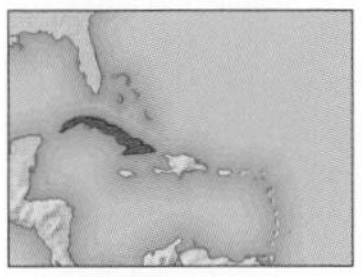

Location Caribbean

Length 22–23½ in (55–60 cm)
Tail 6–10 in (15–26 cm)
Weight 10–15 lb (4.5–7 kg)
Social unit Individual/Pair
Status Least concern

Desmarest's hutia resembles a huge vole with its typically blunt nose, large head, short neck, small ears, stocky body, and short limbs. It has a white nose, red-brown to black upperparts, and gray or yellow-brown underparts.

The strong, tapering, hairy tail and sharp, curved claws are adapted for support and gripping when tree climbing and when foraging for its varied diet of fruit, leaves, soft bark, and, occasionally, lizards. Litter size is 1–4. Hutias live only in the Caribbean, and most species are either severely threatened or already extinct.

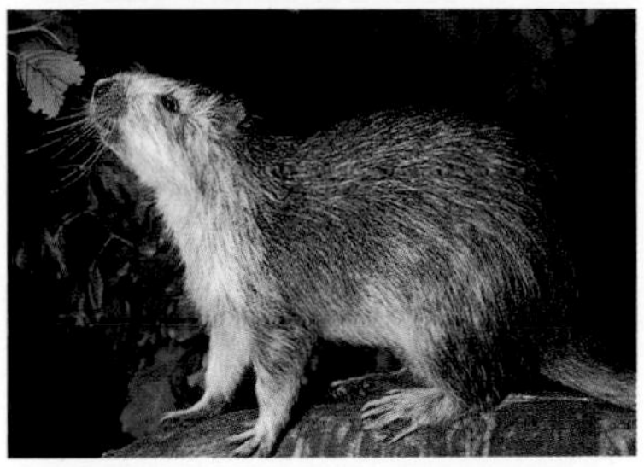

Heterocephalus glaber

Naked mole-rat

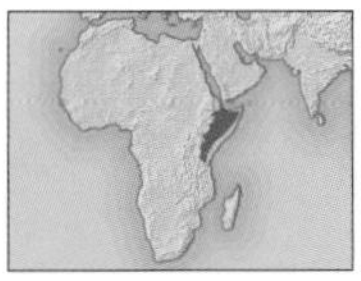

Location E. Africa

Length 3¼–3½ in (8–9 cm)
Tail 1¼–1¾ in (3–4.5 cm)
Weight 1¹⁄₁₆–2⅞ oz (30–80 g)
Social unit Group
Status Least concern

This mole-rat's social system is unique among mammals. Only one dominant female, the "queen," breeds. She may have more than 20 pups per litter and is tended by several nonworkers. The colony's workers form head-to-tail "digging chains" to tunnel and gather food. The naked mole-rat is not truly naked but has pale, sparse hairs over its pinkish gray skin. As in other mole-rats, the massive incisor teeth are for digging and eating, and the eyes and ears are minute. The tail is rounded and the limbs strong, with 5 thick-clawed toes, for digging. Colonies comprise 70–80 individuals that inhabit elaborate tunnel systems. Food-gathering galleries 6–20 in (15–50 cm) deep radiate up to 130 ft (40 m) from the central chamber. New galleries are dug regularly for roots, bulbs, tubers, and other underground plant parts. This rodent surfaces only to travel to another colony.

Cryptomys hottentotus

Hottentot mole-rat

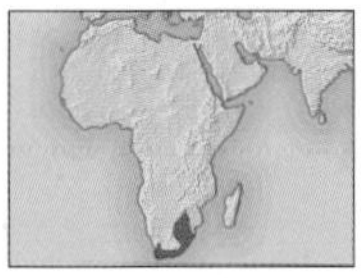

Location Southern Africa

Length 4–7¼ in (10–18.5 cm)
Tail ⅜–1¼ in (1–3 cm)
Weight 3⅝–5½ oz (100–150 g)
Social unit Group
Status Least concern

This mole-rat has a wide head, small eyes and ears, short, sturdy limbs, stocky body, and thick, dense fur. Coloration is pink-brown or gray above, with paler underparts. Extra-large incisor teeth bite through soil as well as bulbs, tubers, roots, and other subterranean plant parts. Colonies of 5–15 individuals dig large, complex tunnel systems and store excess food. Usually only the senior pair breeds.

Bathyergus janetta

Namaqua dune mole-rat

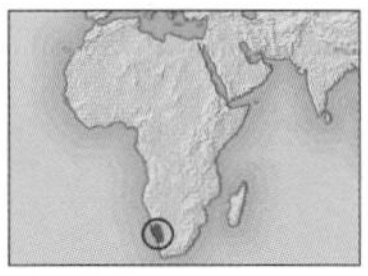

Location S. W. Africa

Length 7–9½ in (17–24 cm)
Tail 1½–2 in (4–5 cm)
Weight Not recorded
Social unit Variable
Status Least concern

This species tunnels in shifting sands, either on the coast or inland. It lives in small colonies in a tunnel system up to 655 ft (200 m) long. The large head has massive protruding incisor teeth, the body is cylindrical, the legs are short but strong, and the tail is tiny. The back is dark brown, the sides gray, and the head and belly dark gray.

MAMMALS

Springhares

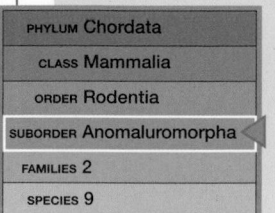

PHYLUM	Chordata
CLASS	Mammalia
ORDER	Rodentia
SUBORDER	Anomaluromorpha
FAMILIES	2
SPECIES	9

This group comprises the springhare and scaly tailed flying squirrels (anomalurids)—2 unusual families of rodents confined to Africa. They have anatomical features that link them with cavylike rodents, but genetic analysis supports their classification in a different suborder. The springhare is a large jumping kangaroo-like rodent of arid habitats, with long hind limbs, wide hooflike rear claws, very long brush-tipped tail, and prominent ears—like those of a hare. Anomalurids have a gliding membrane, although they are not related to flying squirrels of Asia and North America or the colugos. The rough scales on the underside of their tail prevent slippage when they land on tree trunks after gliding.

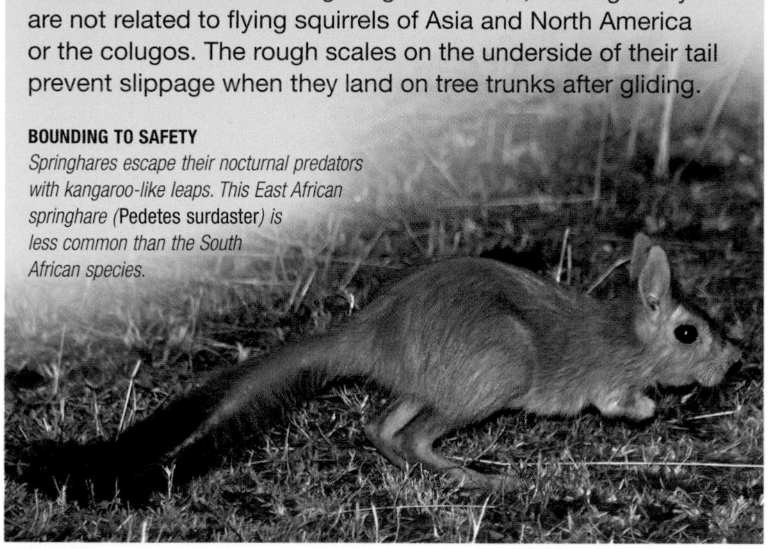

BOUNDING TO SAFETY
*Springhares escape their nocturnal predators with kangaroo-like leaps. This East African springhare (*Pedetes surdaster*) is less common than the South African species.*

Pedetes capensis

South African springhare

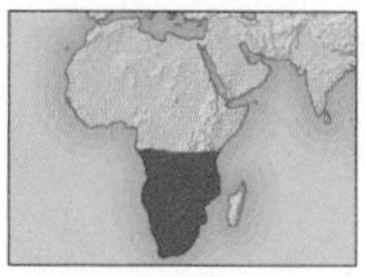

Location C. and E. Africa to southern Africa

Length 10½–16 in (27–40 cm)
Tail 12–18½ in (30–47 cm)
Weight 6½–8¾ lb (3–4 kg)
Social unit Individual
Status Least concern

One of only 2 species of springhares, this species resembles a rabbit kangaroo with a long, bushy tail and large, narrow, upright ears. It makes huge leaps, easily covering 6½ ft (2 m) per bound. The diet consists mainly of seeds, stems, bulbs, and other plant parts, but also includes locusts, beetles, and other invertebrates. When feeding, it tends to bend forward and lope rabbitlike on all fours. The 5 large front claws are specialized for digging. When bounding at full speed, its tail curves up; when sitting up, its tail is used as a support. The springhare sleeps resting on its haunches, head tucked between its hind legs, and the tail wrapped around the whole body. Nocturnal, the South African springhare lives alone or as a male–female pair, and digs several extensive burrows. There appears to be no peak breeding season; the single young is born at any time of year and is suckled in the main breeding burrow for approximately 7 weeks.

Colugos

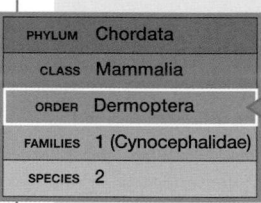

PHYLUM	Chordata
CLASS	Mammalia
ORDER	Dermoptera
FAMILIES	1 (Cynocephalidae)
SPECIES	2

By extending a strong membrane that surrounds their body (the patagium), colugos (also known as flying lemurs) are able to glide more than 330 ft (100 m) between trees, with very little loss in height. The term "flying lemur" is misleading, however, as these mammals are not lemurs and are not capable of true flight. About the size of a domestic cat, they have large eyes, a blunt muzzle, strong claws for climbing, and mottled fur (for camouflage). Colugos, which live in the rain forests of Southeast Asia, strain food (fruit and flowers) through peculiar comb-shaped lower teeth, which are also used to groom the fur.

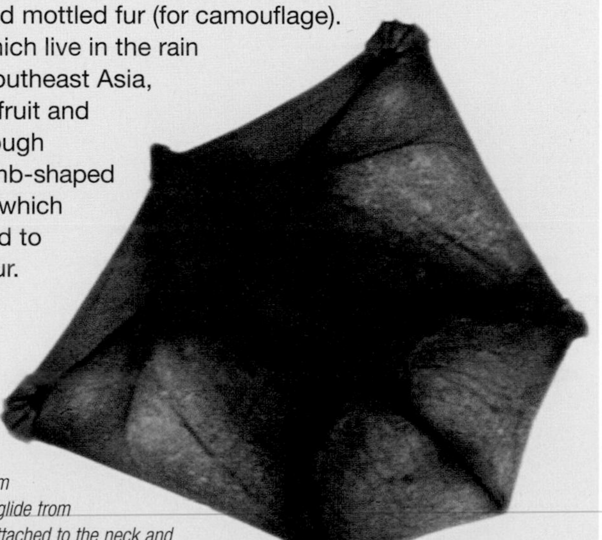

GLIDER
A kite-shaped membrane called the patagium allows colugos to glide from tree to tree. It is attached to the neck and to the tips of the fingers, toes, and tail.

Cynocephalus variegatus

Malayan colugo

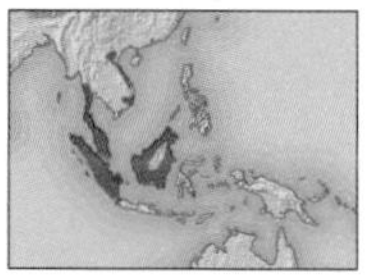

Location S.E. Asia

Length 13–16½ in (33–42 cm)
Tail 7–10½ in (17.5–27 cm)
Weight 2–4½ lb (0.9–2 kg)
Social unit Variable
Status Least concern

Although still common in some areas, numbers of this species have declined rapidly in logged and farmed regions. It has a small head relative to its body with large eyes, little, rounded ears, and a blunt muzzle. The fine, short fur is brownish gray with red or gray on the back, often with lighter flecks to mimic lichen-covered branches, and paler on the underside. The Malayan colugo is active in twilight and at night, eating soft plant parts such as flowers, fruit, buds, and young leaves, and scraping up nectar and sap with its comblike lower incisor teeth. It lives alone or in small, loose groups, inhabiting tree-holes or resting among dense foliage high in the treetops. After a gestation period of about 2 months, the single offspring clings to its mother as she leaps and glides among the trees and is weaned by 6 months.

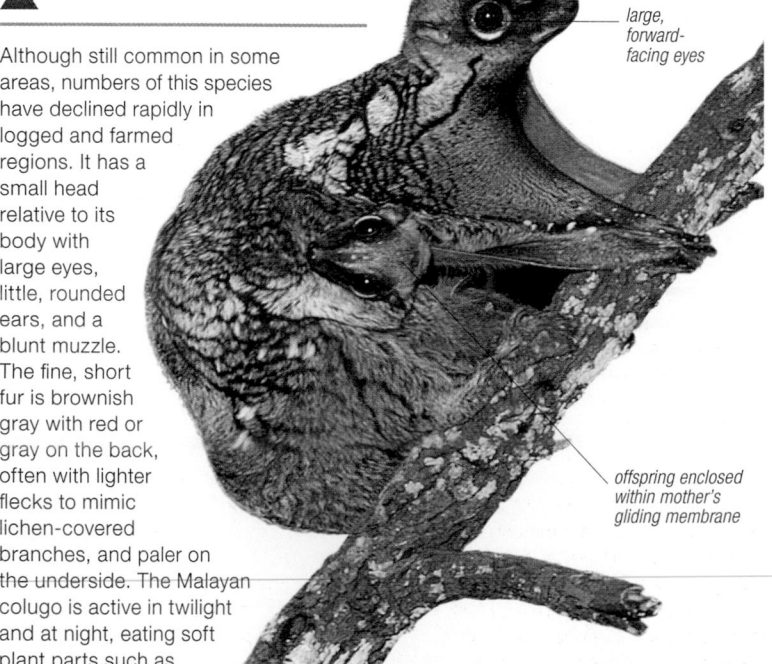

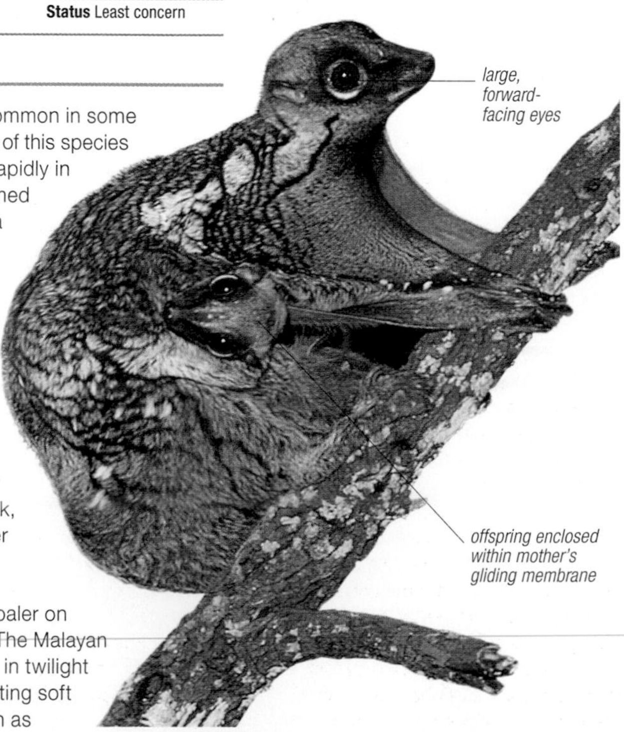

Tree shrews

PHYLUM	Chordata
CLASS	Mammalia
ORDER	Scandentia
FAMILIES	2
SPECIES	20

These small, squirrel-like mammals are not wholly arboreal (they spend much of their time on the ground) and are not true shrews. In fact, they have some features that are associated with primates, such as a large braincase and, in males, testes that descend into a scrotum. Tree shrews lack whiskers and seek prey, such as insects, using well-developed senses of hearing, smell, and vision. Most species have a long, thickly furred tail. Tree shrews inhabit the tropical forests of Southeast Asia.

SEEKING FOOD
Tree shrews are skilful climbers and agile runners. They forage with their hands and pointed snout.

Ptilocercus lowi

Pen-tailed tree shrew

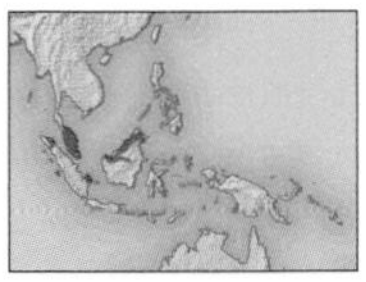

Location S.E. Asia

Length 4–5½ in (10–14 cm)
Tail 5–7½ in (13–19 cm)
Weight ⅞–2⅛ oz (25–60 g)
Social unit Variable
Status Least concern

Named for its mainly naked tail with bushy white hairs at the tip, like a bottle-brush or old-fashioned quill pen, this species is grayish brown above and grayish yellow on the underside. It climbs skilfully with its strong, sharp-clawed limbs, using its tail for balance, and rarely descends to the ground. It makes simple nests in hollow trees or on branches and lives in pairs or small groups. Its varied diet includes worms, insects, mice, small birds, lizards, and fruit.

Dendrogale melanura

Bornean smooth-tailed tree shrew

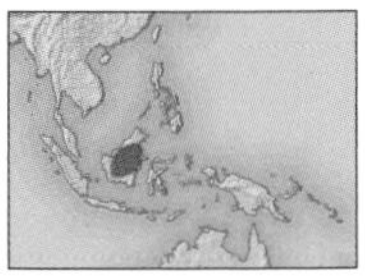

Location S.E. Asia

Length 4–6 in (10–15 cm)
Tail 3½–5½ in (9–14 cm)
Weight 1¼–2⅛ oz (35–60 g)
Social unit Individual
Status Data deficient

This species lives in mountain forests at altitudes of 3,000–5,000 ft (900–1,500 m). Its long limbs and long claws equip it to grasp branches and it spends more time in trees than other tree shrews. The coloration of its upperparts is a mixture of black and buff or cinnamon; the underparts and legs are more red or ocher. It lives alone, feeding by day and sleeping in a leaf-lined tree nest at night. After a gestation period of about 50 days, the female produces 3–4 young.

Anathana ellioti

Indian tree shrew

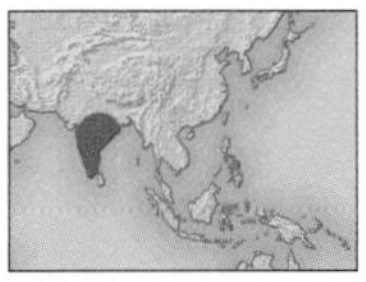

Location S. Asia

Length 6½–8 in (17–20 cm)
Tail 6½–7½ in (16–19 cm)
Weight 5 oz (150 g)
Social unit Individual
Status Least concern

Also known as the Madras tree shrew, this mammal resembles a small, slender gray squirrel. It is speckled with yellow and brown on the upperparts and has a distinctive cream-colored shoulder stripe. The Indian tree shrew has a prominent head with a pointed muzzle, large eyes, and furred ears. It forages actively by day on the ground and among low bushes, as well as in trees, for small edible items, mainly worms and insects but also some fruit. At night, it sleeps in a rock crevice or tree-hole. The species is probably solitary outside the breeding season and does not, as far as is known, defend a specific territory, but information is lacking. The female cares for the young.

Tupaia minor

Lesser tree shrew

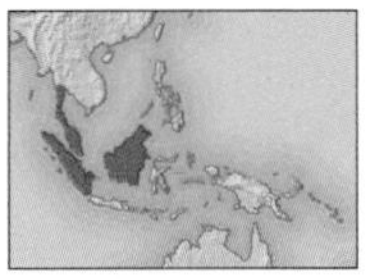

Location S.E. Asia

Length 4¾–5¼ in (11.5–13.5 cm)
Tail 5–6½ in (13–17 cm)
Weight $1\frac{1}{16}$–2½ oz (30–70 g)
Social unit Individual
Status Least concern

This species is more tolerant than most other tree shrews of the loss of natural forest habitat to plantations, parks, and gardens. It is a skilled climber and has a wide diet. It forages by day along branches, in bushes, among fallen logs, and under rocks for small animals, fruit, leaves, seeds, and carrion. Usually 2–3 young are born after a gestation period of 46–50 days. The female leaves the offspring in a leafy nest among the foliage while she feeds, returning sporadically to suckle. Tree shrews fall victim to snakes, mongooses, tree-dwelling cats, and diurnal birds of prey. They often hold food by their front paws while sitting up on their haunches, in the manner of a squirrel, to watch for danger.

GRIPPING FEET

The splayed toes, sharp claws, and pimplelike protrusions on the foot pads of the lesser tree shrew all give an excellent grip on bark and rock. It uses its front legs to pull itself along branches on its belly. This spreads scent produced by abdominal glands. This scent is likely to be recognizable to others of its species, and is probably a means of marking its home range, which it defends against others of the same sex.

SENSES AND BALANCE
The lesser tree shrew's prominent nose, eyes, and ears ensure keen senses; the long tail helps with balance while climbing at speed through the branches.

Primates

PHYLUM	Chordata
CLASS	Mammalia
ORDER	Primates
FAMILIES	15
SPECIES	382

CLASSIFICATION NOTE

Recent research suggests that primates may comprise 2 suborders: Strepsirrhini (lemurs, galagos, lorises, and pottos) and Haplorrhini (tarsiers, apes, and monkeys). Tarsiers have features linking them with both suborders; in this book, they are grouped with Strepsirrhini (traditionally known as the prosimians). To allow adequate coverage, monkeys and apes are presented here as separate groups. Humans do not appear in this book but are classified, with the apes, as hominoids.

Prosimians *see pp.136–9*
Monkeys *see pp.140–9*
Apes *see pp.150–5*

Members of the order Primates—prosimians, monkeys, and apes (but see the Classification note, left)—are a diverse group that form highly complex social units. They are found in South and Central America, Africa and Madagascar, and Southeast and eastern Asia. Primates mostly inhabit tropical rain forests, and their dextrous prehensile (grasping) hands and feet are an adaptation to a largely arboreal lifestyle (some also have a prehensile tail). Traditionally, authorities on primates have recognized about 180–200 species. However, in recent years, there has been a thorough reconsideration of primate species, partly because of the need to identify significant populations for conservation purposes. As a result, many new species have been described since 1990.

Social organization

Only orangutans and a few of the lemurs and galagos lead solitary lives; all other primates exist in social groups. Many species—including most monkeys—live in troops that consist of several females and either one or several adult males. In some species, such as the mandrill, drill, and gelada, huge troops of several hundred individuals periodically split into subgroups of bachelor bands and one male and his harem, both of fixed composition. Chimpanzees and spider monkeys live in large communities of 20–100 individuals that divide into groups of varying composition. Some species, mostly New World monkeys, live in monogamous pairs.

MUTUAL GROOMING
The bonds between members of primate groups are partly maintained by mutual grooming, as shown by these olive baboons. Low-ranking members groom those of higher rank to gain favor and support (in disputes).

Anatomy

Primates form a highly varied group, with members as diverse as mouse lemurs, which may weigh only 1¼ oz (35 g), and gorillas, which can reach over 440 lb (200 kg). Most species have flat nails on the fingers and toes—only a few have claws—and the big toe always has a nail. Most have a tail, except the apes. Apart from some toothed whale species, the higher primates have the largest brain relative to body size of all the mammals, which helps explain their high intelligence. The cerebral hemispheres (which process sensory information and coordinate responses) are highly developed, allowing sight keen enough for accurate tree-to-tree leaps.

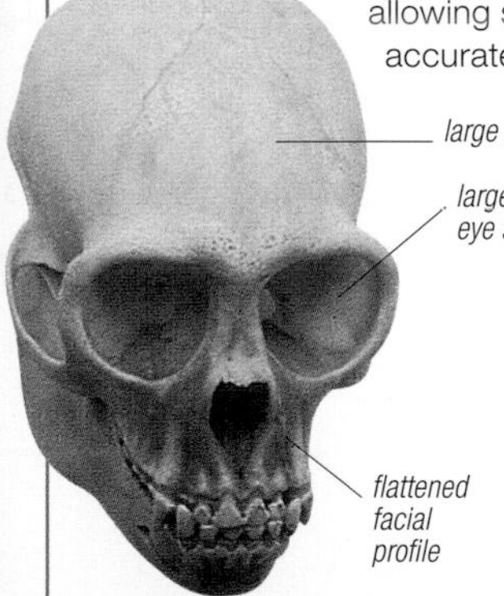

SKULL STRUCTURE
Primates have a large, domed braincase and forward-facing eye sockets. This monkey skull also shows the flattened facial profile that is characteristic of most monkeys and apes.

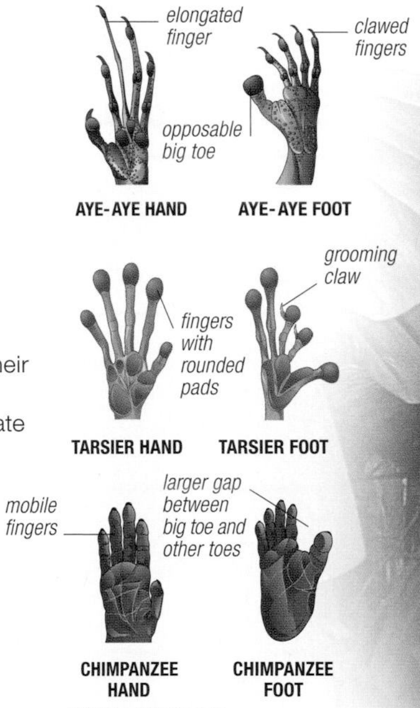

HANDS AND FEET
Hand and foot structure varies according to lifestyle. Claws (aye-aye) and rounded pads (tarsier) improve grip in arboreal species; highly mobile hands and feet (chimpanzee) are vital for species that live on the ground and in trees.

USING A SIMPLE TOOL

SEEKING TERMITES
This common chimpanzee has made a stick tool to obtain food that would otherwise be inaccessible.

DEXTROUS HANDS
The chimpanzee uses its grasping hands to hold and insert the stick into the termite mound.

REMOVING THE TOOL
The termites on guard, provoked by the intrusion, bite the stick with their pincerlike jaws.

SUCCESSFUL ENDEAVOR
As the chimpanzee removes the stick, the termites remain firmly attached.

MAMMALS

Feeding

As a general rule, small primates tend to eat insects, whereas larger species mostly eat leaves and fruit (a large primate cannot sustain itself on insects alone). Small primates have a high metabolic rate and cannot afford the long digestion times needed to process vegetable matter. Some leaf-eating species, such as colobus monkeys and langurs, have a complex stomach containing bacteria to ferment cellulose; other primate species have bacteria in the cecum or in the colon. A few species, including chimpanzees and baboons, hunt vertebrate prey as well as eating vegetable matter. Only tarsiers are entirely carnivorous.

EATING VEGETATION
Of all primates, gorillas probably eat the most plant material. In order to break down cellulose (thereby releasing the cell nutrients), gorillas have large molar teeth and strong jaw muscles for chewing. Their pot belly houses a long digestive tract.

EATING MEAT
Given the opportunity, some primates will hunt (sometimes cooperatively) and kill other animals. This common chimpanzee is eating part of a duiker.

CONSERVATION

Primate numbers are declining severely in the wild due to habitat loss and, more recently, through the illegal hunting of protected species (such as the gorilla) for meat. As a result, many species are now endangered. Primates are also widely used in medical and space research. Although a few species are being reintroduced to the wild from zoos with captive breeding programs, and sanctuaries, such as this chimpanzee orphanage in Zambia (below), provide some refuge, the situation remains bleak.

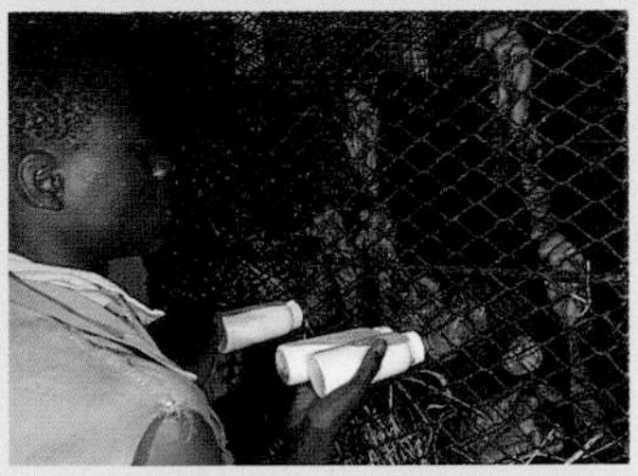

LIFE IN THE TREES
These douc langurs are typical primates in that they are arboreal, have prehensile hands and feet, live in groups, and eat leaves and fruit. Their colorful markings help provide camouflage. Like almost all primates, loss of habitat is a major threat to their survival.

Movement

Most primates spend at least part of their life in trees and have adapted accordingly. To provide a strong hold on branches, the big toe is separated from the other toes in all species except humans, and the thumb is always separated from the fingers, although it is fully opposable (that is, it can turn, face, and touch the other digits in the same hand) only in apes and in some Old World monkeys. The arm and wrist bones are not fused, which increases dexterity. Primates also have "free" limbs—the upper part of each limb is outside the body wall, which allows great freedom of movement. (In other mammals, such as horses, the upper part is inside the body wall—the "armpit" is in fact the elbow joint, which makes movement more restricted.) Some species have a long, prehensile tail, used as a "fifth limb."

SWINGING
Spider monkeys and gibbons, such as this lar gibbon, use their long arms to swing from branch to branch. This is called brachiation.

STANDING
Primates, such as chimpanzees, that are capable of standing and walking on 2 legs (known as bipedalism), tend to have long legs.

CLIMBING
The most common way of moving is on all fours. Quadrupeds, such as this woolly spider monkey, usually have limbs of about the same length.

CLINGING
Vertical clingers and leapers, such as these indri, move with the back held vertically. They have well-developed back limbs for long leaps between trees.

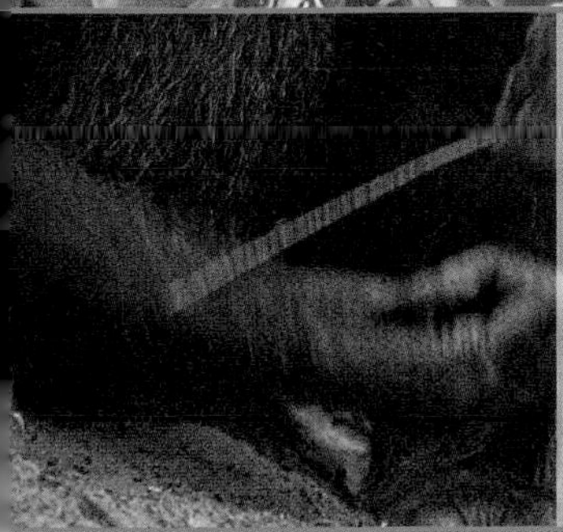

NCING ACT
himpanzee stabilizes the termite-covered y resting it on the opposite forearm.

EATING QUICKLY
Most of the termites, including any on the forearm, are swiftly gobbled before they can escape.

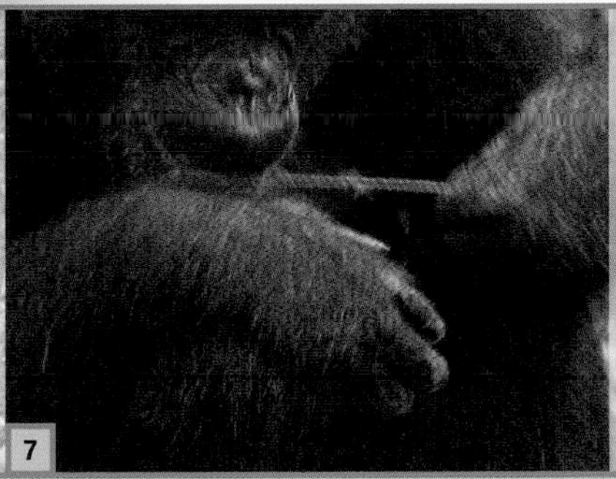

MORE OF THE SAME
Once all the termites have either been eaten or escaped, the chimpanzee will repeat the process.

LEARNED BEHAVIOR
Making and using tools is rare in mammals. Juvenile chimpanzees acquire these skills by copying adults.

Prosimians

PHYLUM	Chordata
CLASS	Mammalia
ORDER	Primates
SUBORDER	Strepsirrhini
FAMILIES	7
SPECIES	90

More primitive than monkeys and apes, prosimians comprise the lemurs of Madagascar, the galagos and pottos of Africa, and the lorises of Asia. Lemurs (which include sifakas, the indri, and the aye-aye) have large ears, an elongated body, long limbs, and most have a long, bushy tail. Lorises, pottos, and galagos are generally smaller than lemurs and tend to have larger eyes. Prosimians mostly inhabit forests and are usually nocturnal (some lemurs are diurnal). Deforestation has endangered many prosimians, especially the lemurs.

Anatomy

Prosimians have a black, doglike snout and a sense of smell that is more highly developed than in other primates. They have large eye sockets and a crystalline layer behind the retina of the eye that reflects light. This increases the amount of light falling on the visual cells and improves night vision. Because most prosimians are arboreal, their hands and feet are adapted for grasping (although they are still less dextrous than monkeys and apes). They have flat nails on all digits except the second toe, which instead has a long claw (the "toilet claw") used for grooming. All species except the aye-aye have a "dental comb"—4 to 6 of the lower front teeth that are pressed together and grow slightly forward. This is used for mutual grooming.

Movement

Most lemurs are quadrupedal and usually run or leap from branch to branch. Weasel lemurs, the avahi, sifakas, the indri, and galagos are vertical clingers and leapers (see below). On the ground, sifakas and indri move on 2 feet, employing sideways hops with their arms in the air for balance. Lorises and pottos usually clamber slowly along branches, clinging tightly at every step (although when startled they can move swiftly).

Communication

Lemurs produce various calls, both to signal alarm (there are often different calls to distinguish aerial and ground predators) and to communicate within or between troops. Pairs of indri occupy treetop territories, and these are marked with loud wailing calls; small sifaka troops define their areas with "shi-fak" calls that sound like hiccups. Male ring-tailed lemurs and bamboo lemurs have a wrist gland with a spur, which they use to mark territory by drawing the wrist sharply across a sapling. This produces a click, creates a scar, and leaves a scent—a gesture that is auditory, visual, and olfactory. Galagos leave scent trails around their territory by placing urine on their feet.

TARSIERS

Not strictly prosimians (see the Classification note on p.134), tarsiers are unusual in that they have many primitive features similar to prosimians, yet they also exhibit characteristics linking them to monkeys and apes, such as a dry, hairy nose. The most striking feature of tarsiers is their enormous eyes: each is slightly heavier than the brain. Other characteristics include a large head and ears, long digits with disklike pads at the tips, very long legs with elongated ankles, and a long tail. The 7 species of tarsiers are found in the forests of Southeast Asia, where they spend much of their time clinging to upright tree stems, scanning the forest floor for prey.

VERTICAL LEAP
When leaping from tree to tree, this Verreaux's sifaka uses its long tail for balance, its muscular legs for propulsion, and its large, grasping hands and feet for a secure landing. These features typify prosimians that keep their bodies vertical when climbing. In midair, a semiupright posture is maintained.

Arctocebus calabarensis

Angwantibo

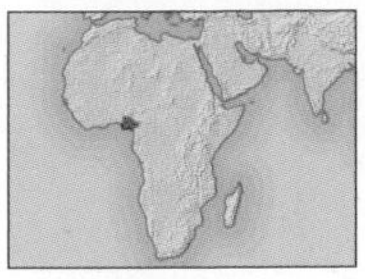

Location W. Africa

Length 9–10 in (22–25 cm)
Tail ⅜ in (1 cm)
Weight 8–17 oz (225–475 g)
Social unit Group
Status Least concern

One of only 2 *Arctocebus* species, this one is orange to yellow on its upperparts and buff beneath. It climbs carefully and deliberately using all 4 equal-length limbs. The second toe is tiny and the first widely separated from the other 3, giving a clamplike grip. Solitary and nocturnal, the angwantibo eats small creatures such as caterpillars, plus a few types of fruit. The other species, the golden angwantibo, lives farther south.

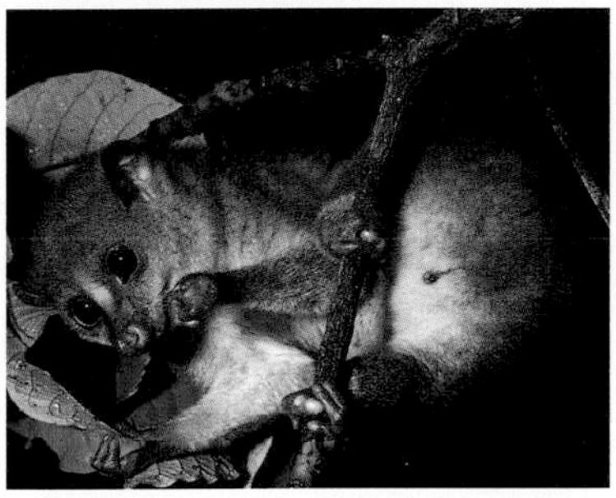

Loris lydekkerianus

Grey slender loris

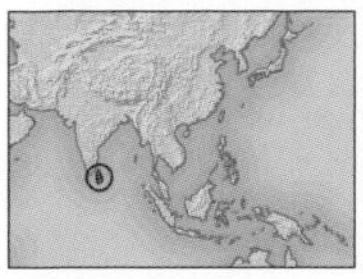

Location S. Asia, Sri Lanka

Length 7–10¼ in (17.5–26 cm)
Tail None
Weight 3–13 oz (85–350 g)
Social unit Individual/Pair
Status Endangered

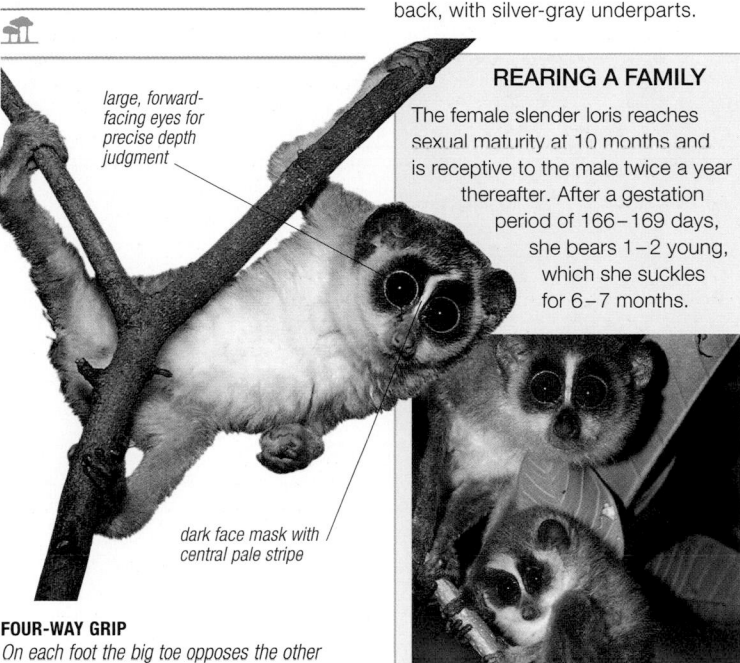

This small, slim primate moves with great deliberation, gripping twigs with all 4 feet—until it suddenly smells, sees, and, with its front legs, snatches a small creature such as an insect or lizard. It also eats soft leaves and buds, fruit, and birds' eggs. By day, the grey slender loris curls up in a tree-hole, dense leaf nest, or a similarly secure place. Its fur color ranges from yellow-gray to dark brown on the back, with silver-gray underparts.

FOUR-WAY GRIP
On each foot the big toe opposes the other 4 toes for a pincerlike grip. The grey slender loris can even sleep holding onto branches.

REARING A FAMILY

The female slender loris reaches sexual maturity at 10 months and is receptive to the male twice a year thereafter. After a gestation period of 166–169 days, she bears 1–2 young, which she suckles for 6–7 months.

Nycticebus coucang

Slow loris

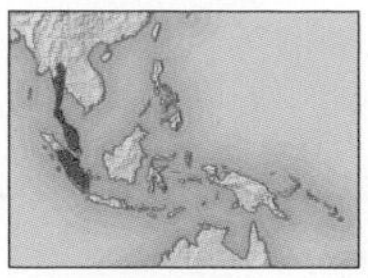

Location S.E. Asia

Length 10¼–15 in (26–38 cm)
Tail ⅜–¾ in (1–2 cm)
Weight 8–23 oz (225–650 g)
Social unit Variable
Status Vulnerable

Aptly named because of its lack of speed, the slow loris has gripping hands like the grey slender loris (see left) and is also a nocturnal tree-dweller. It creeps carefully toward its prey, then lunges with its front limbs. The loris lives alone or in pairs or groups, and the adult male chases other males from its territory, which is marked by urine. The dense, soft fur is brown with a white face and underparts, and dark eye rings and ears.

Perodicticus potto

Potto

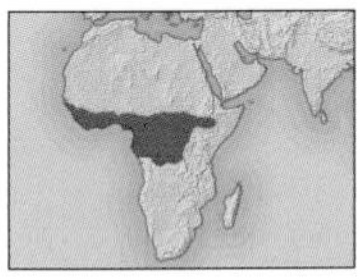

Location W. and C. Africa

Length 12–16 in (30–40 cm)
Tail 1½–6 in (3.5–15 cm)
Weight 1¾–3¼ lb (0.85–1.5 kg)
Social unit Variable
Status Least concern

A careful, nocturnal climber, the potto has very mobile limb joints and reaches out at any angle to bridge gaps between branches. Its fur may be gray, brown, or red; the eyes and ears are small; and the diet consists mainly of fruit, sap, gum, and small animals. A potto can remain immobile for hours to escape attention. If attacked, it will tuck down its head and batter the enemy with the "shield" of horny skin covering spiny bones on the upper back. The gestation period is about 200 days, and the potential life span is over 25 years.

Otolemur crassicaudatus

Brown greater galago

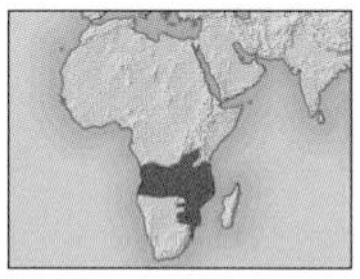

Location C., E., and southern Africa

Length 10–16 in (25–40 cm)
Tail 13½–19½ in (34–49 cm)
Weight 2¼–4½ lb (1–2 kg)
Social unit Variable
Status Least concern

The largest galago, this species locates insects at night using its huge eyes and ears, and snatches them by hand in a split second. It also scrapes gum and sap with its comblike, protruding lower incisor and canine teeth. Deriving its alternative name of bushbaby from its childlike wails, it lives in small family groups of a male–female pair or 2 females, with young. Compared to closely related species, it tends to run on all fours more frequently and to leap in an upright position less often.

Euoticus elegantulus

Western needle-clawed galago

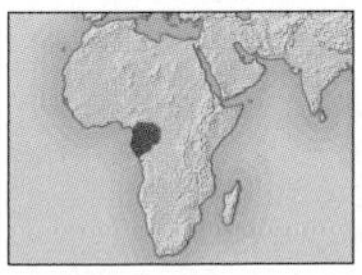

Location W. Africa

Length 4¼–10½ in (10.5–27 cm)
Tail 7½–13½ in (19.5–34 cm)
Weight 10–13 oz (275–350 g)
Social unit Individual/Pair
Status Least concern

Resembling other galagos in form, this species has an orange back, gray underside, pink hands and feet, and a long, gray, white-tipped tail. The oval eyes are rimmed by pale fur, and thin, sharp claws grip well for climbing. Like all galagos and lorises, the second toe on the rear foot has an upward-facing "toilet" claw, for scratching and combing fur. The specially enlarged front teeth scrape wood and bark to obtain gum and sap, which make up three-quarters of the diet, along with a variety of fruit and insects. It may make 1,000 "gum scrapes" per night.

Galago moholi

Moholi bushbaby

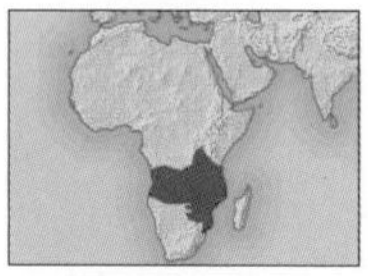

Length 6–6½ in (15–17 cm)
Tail 4¾–10½ in (12–27 cm)
Weight 5–9 oz (150–250 g)
Location E., C., and southern Africa
Social unit Variable
Status Least concern

The moholi bushbaby is also called the lesser bushbaby. It leaps vertically, in enormous jumps of 16 ft (5 m), and its grasping hands and feet are moistened regularly with urine, to maintain grip. It snatches insects from midair by hand or scrapes gum from trees with its comblike lower front teeth. This galago lives in small family groups, which sleep huddled together by day.

Cheirogaleus medius

Fat-tailed dwarf lemur

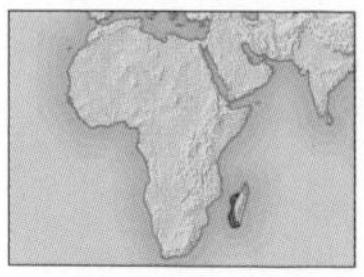

Length 7–10 in (17–26 cm)
Tail 7½–12 in (19–30 cm)
Weight 6 oz (175 g)
Location W. and S. Madagascar
Social unit Individual/Group
Status Least concern

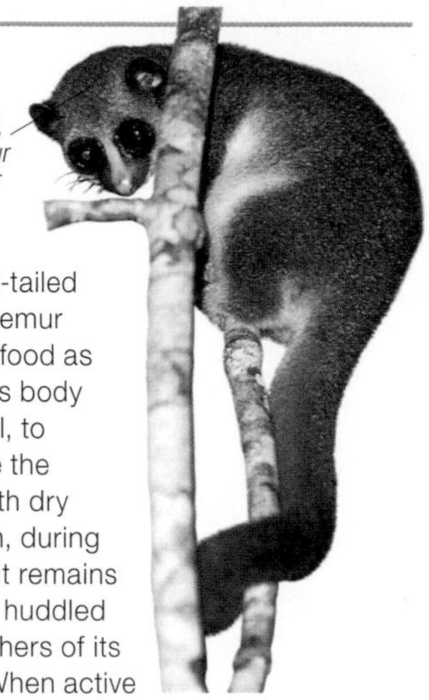

The fat-tailed dwarf lemur stores food as fat in its body and tail, to survive the 8-month dry season, during which it remains torpid, huddled with others of its kind. When active again, it becomes solitary and clambers in trees and bushes at night, seeking flowers, fruit, and insects. It rests by day in a leaf-and-twig nest in a tree-hole or fork.

Varecia variegata

Black and white ruffed lemur

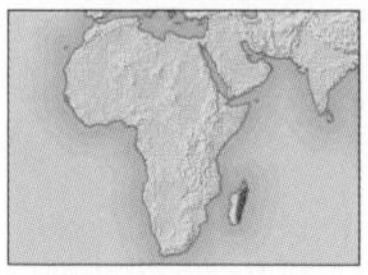

Length 22 in (55 cm)
Tail 3½–4 ft (1.1–1.2 m)
Weight 7¾–10 lb (3.5–4.5 kg)
Location E. Madagascar
Social unit Group
Status Critically endangered

This large species is white or reddish white except for black on the face, shoulders, chest, flanks, feet, and tail.

The black and white ruffed lemur eats a higher proportion of fruit than any other lemur. It builds a leafy nest in a tree-hole or fork for its 2–3 young. The female gives birth after a gestation period of 90–102 days. The offspring remain in the nest for several weeks after which they cling to her. Group size is 2–20, with several dominant females defending the territory.

Eulemur macaco

Black lemur

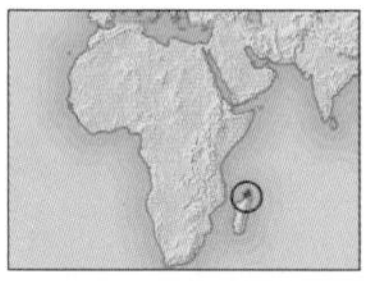

Length 12–18 in (30–45 cm)
Tail 16–23½ in (40–60 cm)
Weight 4½–6½ lb (2–3 kg)
Location N. Madagascar
Social unit Group
Status Vulnerable

A medium-sized species, only the male has the long, soft black fur for which it is named. Females are red, brown, or gray. Unlike most lemurs, the black lemur is also active for part of the night. This behavior may result from hunting and other human disturbance. Led by one female, groups of 5–15 black lemurs forage mainly in trees for fruit, flowers, leaves, and soft bark.

Hapalemur griseus

Grey gentle lemur

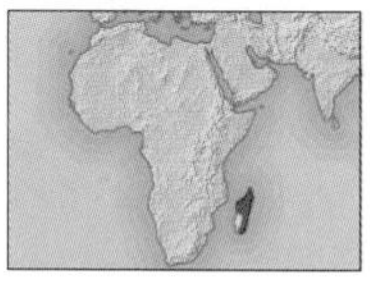

Length 16 in (40 cm)
Tail 16 in (40 cm)
Weight 2¼ lb (1 kg)
Location N. and E. Madagascar
Social unit Group
Status Vulnerable

This species is restricted to humid forests and marshy areas where reeds and the bamboo, on which it feeds, are abundant. It usually lives in groups of 2–5 and is active mainly at dawn and dusk. Females have a single offspring, which is carried underneath the body at first and later on the back. Disturbance of its restricted habitat threatens this specialized lemur.

Lemur catta

Ring-tailed lemur

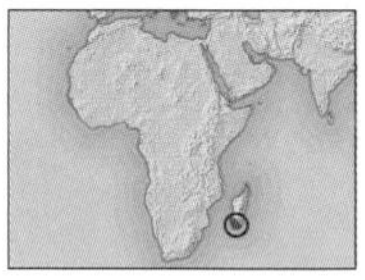

Length 15½–18 in (39–46 cm)
Tail 22–24 in (56–62 cm)
Weight 5½–7¾ lb (2.5–3.5 kg)
Location S. and S.W. Madagascar
Social unit Group
Status Near threatened

This lemur spends more time than its relatives on the ground, even though it is a skilled climber. Very sociable, it forms groups of 5–25, with a core of adult females showing a well-defined hierarchy among themselves and over any males. Young females remain with their mothers and sisters; juvenile males move to other groups. Using their hands, they gather flowers, fruit, leaves, bark, and sap. After 134–138 days' gestation, the female bears one (sometimes 2) offspring. It first clings to the mother's underside and then rides on her back. Like most lemurs, this lemur faces many threats, especially habitat loss.

CATLIKE APPEARANCE
This lemur is catlike both in body proportions and graceful movements. The white face has a dark nose and eye patches; the underparts are whitish gray. The distinctive tail is used for visual and scent signaling.

SUNBATHING

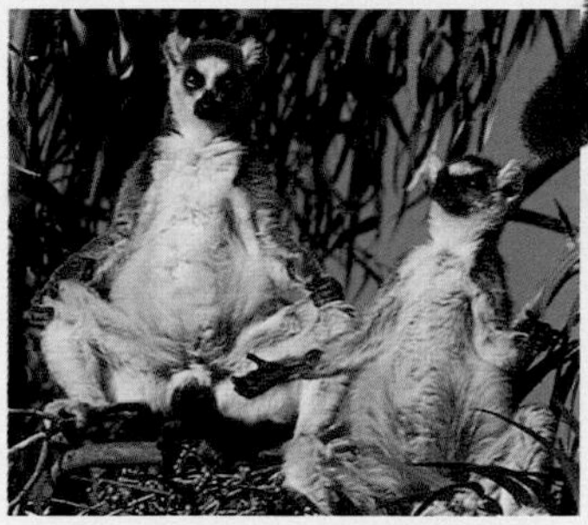

Unlike most other lemurs, the ring-tailed lemur enjoys basking in sunshine, whether on the ground or in trees. It sits upright, with hands on knees in a "sun-worship" posture. Loud alarm calls from others in the group alert it, when necessary, to possible danger—one of the benefits of a highly social lifestyle.

MAMMALS

Microcebus berthae

Madame Berthe's mouse lemur

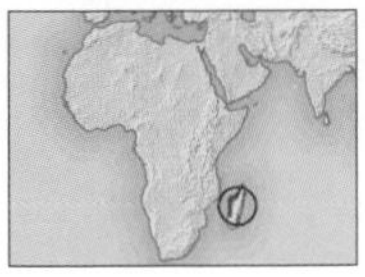

Location Madagascar

Length	3½–4¼ in (9–11 cm)
Tail	4¾–5½ in (12–14 cm)
Weight	1 oz (30 g)
Social unit	Individual
Status	Endangered

The world's smallest primate, this tiny reddish lemur is found only in the Menabe region of western Madagascar.

The species lives in dry deciduous forest, where solitary individuals have bigger home ranges than those of related species in the same area. Males roam much more widely than females and are very promiscuous. Although said to be abundant and adaptable, the total population—perhaps only 8,000 breeding animals—is confined to an area of just 347 square miles (900 square km).

Propithecus verreauxi

Verreaux's sifaka

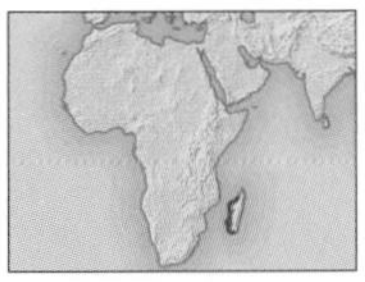

Location W. and S. Madagascar

Length	17–18 in (43–45 cm)
Tail	22–23½ in (56–60 cm)
Weight	6½–11 lb (3–5 kg)
Social unit	Group
Status	Vulnerable

This large, mostly white lemur has brown-black areas on the face, crown, and undersides of the limbs. It uses its powerful legs to move among cactuslike trees with massive, spring-loaded leaps. It eats a wide range of leaves, fruit, flowers, and bark. Living in variable social groupings, it makes its "sifaka" call when two groups dispute territorial boundaries.

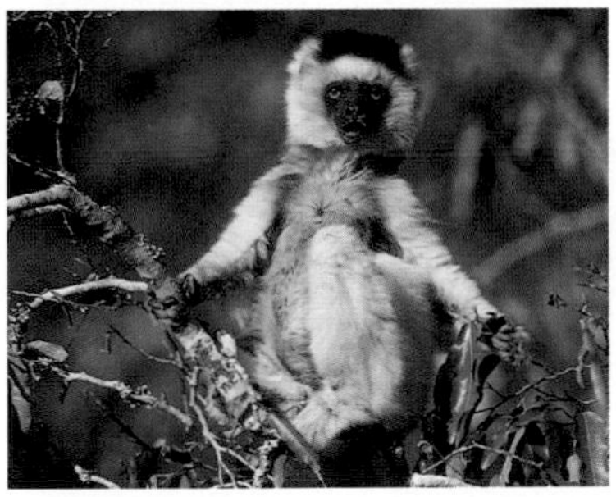

Lepilemur betsileo

Betsileo sportive lemur

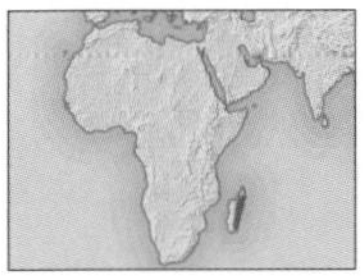

Location E. Madagascar

Length	10 in (26 cm)
Tail	12¾ in (33 cm)
Weight	2¼–2¾ lb (1–1.25 kg)
Social unit	Individual
Status	Data deficient

The nocturnal tree-dwelling sportive lemurs are so-called because of their habit of leaping between vertical tree trunks using powerful hind limbs. On the ground they hop like kangaroos. Named after a Malagasy tribe, the Betsileo sportive lemur is one of more than a dozen species that have been recognized in recent years on the basis of genetic studies. It is grayish brown with a black tail and lives in rainforests of the Fandriana region of eastern Madagascar. Most sportive lemurs eat leaves, supplemented with flowers. They forage alone and defend territories with vocalizations and sometimes physical aggression, but at night small groups may sleep in the same tree hollow.

Indri indri

Indri

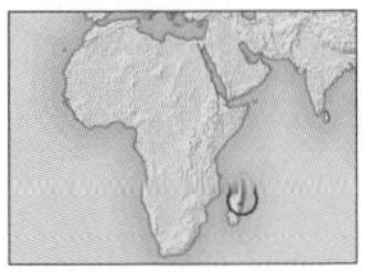

Location E. Madagascar

Length	23½ in (60 cm)
Tail	2 in (5 cm)
Weight	13–15 lb (6–7 kg)
Social unit	Pair
Status	Endangered

Largest of the lemur group, the indri or babakoto ("little father") has very long rear legs, for enormous leaps, but a very short tail. Although diurnal, it is inactive for long periods in the day. The diet is chiefly young leaves, but also fruit, flowers, and seeds. Indris live in pairs with their offspring. The male defends their territory, and the female has first access to food. Indris are mainly black, with white patches.

Daubentonia madagascariensis

Aye-aye

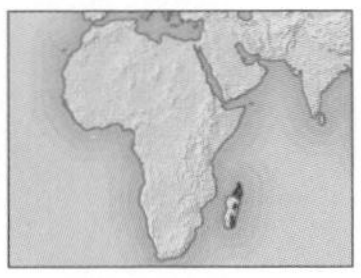

Location N.W. and E. Madagascar

Length	16 in (40 cm)
Tail	16 in (40 cm)
Weight	5½–6½ lb (2.5–3 kg)
Social unit	Variable
Status	Near threatened

The aye-aye has coarse, shaggy black fur with a mantle of white guard hairs. It is specialized as a nighttime primate "woodpecker." It taps trees with its long middle finger, listens intently with its huge ears for wood-boring grubs under the bark, exposes them by gnawing with its rodentlike, ever-growing front teeth, and extracts them with the middle finger. It also eats fruit, including coconut flesh, seeds, and fungi. Aye-ayes share large, stick-made nests but use them in succession, not together. The offspring stays with its mother for 2 years.

CONSERVATION

Believed to be extinct until its rediscovery in 1957, the aye-aye is present in two dozen of Madagascar's national parks and nature reserves, and several small captive breeding colonies have been set up. However, with constant pressure on the island's natural resources, its future is likely to remain entirely conservation-dependent. Here, a one-day-old aye-aye is being weighed.

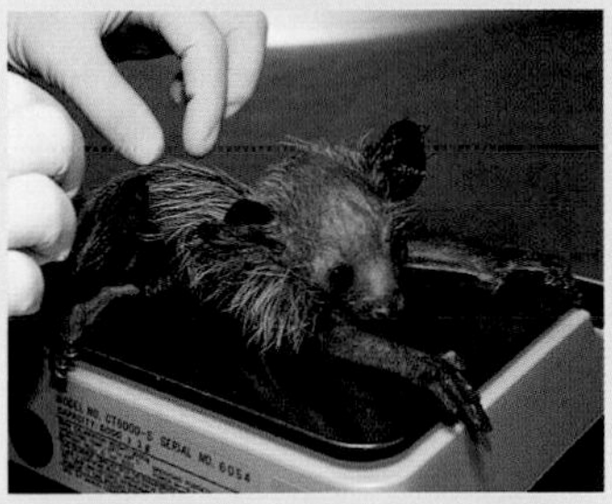

LONG FINGERS
The aye-aye uses its elongated middle finger, which has a double-jointed tip, to extract grubs from under tree bark.

Tarsius bancanus

Western tarsier

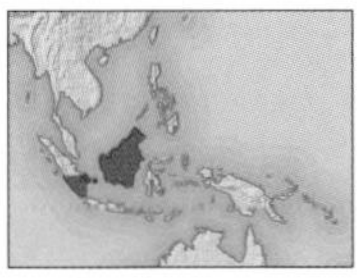

Location S.E. Asia

Length	4¾–6 in (12–15 cm)
Tail	7–9 in (18–23 cm)
Weight	3–4 oz (85–135 g)
Social unit	Individual
Status	Vulnerable

This long-tailed, nocturnal Asian prosimian is small and compact-bodied. Mainly tree-dwelling, the western tarsier has slender fingers and its toes have pads and sharp claws, for gripping branches. It can turn its head to look backward and detect possible predators or prey with its huge eyes and keen ears. Its diet consists mainly of insects. Having spotted its victim, it creeps nearer and then leaps on it and seizes it with its front paws. The female gives birth to one offspring after a gestation period of about 180 days. At first the young is carried by the mother, but it soon learns to cling to her fur.

Monkeys

PHYLUM	Chordata
CLASS	Mammalia
ORDER	Primates
SUBORDER	Haplorrhini (part)
FAMILIES	6
SPECIES	271

This large, diverse group is split into 2 broad, geographically separate subgroups: the Old World monkeys (larger species such as baboons, as well as colobus monkeys and langurs) and the New World monkeys (such as marmosets and spider monkeys), which are distinguished mainly by nose shape. Monkeys are normally found in forests throughout the tropics. Most have short, flat, humanlike faces, although baboons and mandrills have a doglike snout. Many species are endangered by loss of habitat, and the rhesus macaque is one example of a monkey used widely in laboratory research.

Anatomy

Monkeys are characterized by a flattened chest, a hairy nose, a relatively large brain, a deep lower jaw, and sharp canine teeth. Although monkeys are quadrupedal, they are able to sit upright (and will occasionally stand erect), so that the dextrous hands are freed for manipulative tasks (such as picking apart fruit). They have grasping hands and feet, each with 5 digits. Their legs are slightly longer than the arms—much longer in leaping species (such as the red colobus), which also have a long, flexible spine. Monkeys also have a tail that is usually longer than the body, although in some species it is tiny and underdeveloped. A few New World species, such as spider monkeys, have a prehensile tail, sometimes with a bare area at the end with creases and ridges that increase friction for grasping. The tail may be used as a balancing organ and to indicate social gestures.

There are several anatomical differences between Old and New World monkeys. Old World monkeys, which are more closely related to apes than the New World monkeys, have a narrow nasal septum and the nostrils face forward or downward. New World monkeys have a broad nasal septum and nostrils that face sideways. Another major difference is that Old World monkeys have hard sitting pads on the rump, which are absent in New World monkeys.

Social groups

New World monkeys have a great variety of social organizations. Marmosets, for example, usually live in groups consisting of a monogamous pair and subadult offspring that help rear the recent young. Squirrel monkeys, on the other hand, live in very large groups, sometimes over 100, with many females and few males. Spider monkeys live in large communities that split into small groups of varying composition when foraging.

In contrast, Old World monkeys usually exhibit just 2 types of social organization: baboons and macaques live in large, multi-male troops; while the mandrill, the drill, the gelada, guenons, and most langurs live in "harems," with one adult male and several females. Within all monkey social groups, relationships are commonly very close, and grooming is a significant social glue. However, males in some species, such as baboons, will fight furiously with their long, sharp canines for dominance.

Intelligence

Monkeys are intelligent mammals. They are quick to learn, inquisitive, and have an excellent memory. These abilities have helped monkeys succeed in a range of habitats, where they must learn (for example) what they can eat and then remember when and where to find the food again.

AGILE CLIMBER
All monkeys are excellent climbers. This woolly spider monkey has a long prehensile tail, which acts as a "fifth limb" when moving along or resting on branches. The tail has an end section with a naked underside (used to aid grasping) and is strong enough to support both the monkey and its offspring.

Lagothrix cana

Grey woolly monkey

Location C. South America

Length 20–26 in (50–65 cm)
Tail 22–30 in (55–77 cm)
Weight 8¾–22 lb (4–10 kg)
Social unit Group
Status Endangered

Woolly monkeys have thick, soft, close-curled fur, which in this species is gray with black flecks, darker gray on the head, hands, feet, and tail tip. In some individuals, there is a reddish tinge on the underparts. It lives in a mixed troop, which breaks into subgroups to forage primarily for fruit but also for leaves, flowers, sap, seeds, and small creatures. This peaceful, gentle monkey often allows members of another troop into its own troop's territory. After a gestation period of 233 days, the single offspring is born and clings to its mother's underside, moving on to her back at 7 days. It is weaned after 6 months. Males grow larger than females when mature.

CONSERVATION

As well as being threatened by deforestation, adult grey woolly monkeys are frequently hunted, and their young are kept as pets. Conservation depends mainly on hunting restrictions, which in remote parts of the Amazon basin can be difficult to enforce.

MUSCULAR MOVER
Woolly monkeys are stout-bodied and almost pot-bellied, with powerful shoulders, hips, and tail, for hanging from and swinging through trees.

Brachyteles arachnoides

Southern muriqui

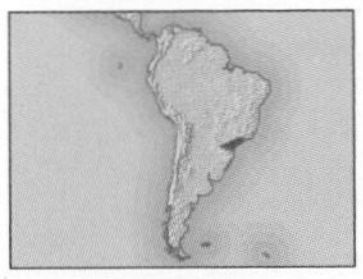

Location C. South America (S.E. Brazil)

Length 22–24 in (55–61 cm)
Tail 26–33 in (67–84 cm)
Weight 21–26 lb (9.5–12 kg)
Social unit Variable
Status Endangered

Also known as the the woolly spider monkey, the southern muriqui has a heavy body, long limbs, and thumbless hands with hooklike fingers. It is the largest New World monkey, and, being a leaf-eater, is endangered due to destruction of Atlantic coastal forests. Just a few hundred are left in 4 or 5 tiny groups on private ranches.

Ateles geoffroyi

Geoffroy's spider monkey

Location S. Mexico, Central America

Length 20–25 in (50–63 cm)
Tail 25–33 in (63–84 cm)
Weight 17–20 lb (7.5–9 kg)
Social unit Group
Status Endangered

Also known as the black-handed spider monkey, this species has black hands, head, and feet, and a cowl-like face surround. Like other spider monkeys, the thumbless hand acts as a simple hook, to swing agilely through trees or to pull fruit-laden branches to the mouth.

Ateles hybridus

Long-haired spider monkey

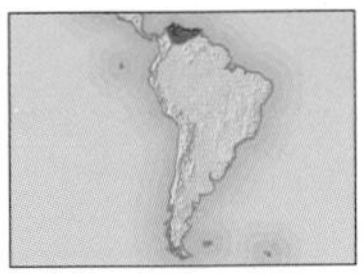

Location N.W. South America

Length 16½–23 in (42–58 cm)
Tail 27–35 in (68–90 cm)
Weight 17–23 lb (7.5–10.5 kg)
Social unit Group
Status Critically endangered

The long-haired spider monkey has a conspicuous triangular white patch on the forehead. It is brown above with pale underparts—giving rise to its other common name of white-bellied spider monkey. Mixed troops of about 20 split into single-sex subgroups of 3–4 to feed on fruit, juicy leaves, and, oddly, soft, decaying wood. They whoop and whinny as they meet up again. Like all spider monkeys, it has long limbs, a slim body, thumbless hands, and a prehensile tail.

Ateles chamek

Black spider monkey

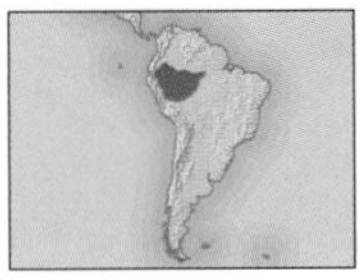

Location W. South America

Length 16–20½ in (40–52 cm)
Tail 32–35 in (80–88 cm)
Weight 21 lb (9.5 kg)
Social unit Group
Status Endangered

Long, black fur and black facial skin identify this otherwise typical spider monkey, which feeds mainly on fruit, berries, flowers, soft leaves, grubs, termites, and honey. They live in large territorial troops, each occupying 370–570 acres (150–230 hectares). The troop splits into variable subgroups to feed, and meets in the evening with greeting calls.

On reaching maturity, females leave to join another troop. After a gestation of 225 days, the single young is born. After 16 weeks, it rides on its mother's back and is weaned by 18 months. Offspring of females high in the dominance hierarchy are more likely to survive to adulthood, when they move to other troops.

Alouatta pigra

Guatemalan black howler

Length 20½–25 in (52–64 cm)
Tail 23–27 in (59–69 cm)
Weight 14–25 lb (6.5–11.5 kg)
Social unit Group
Status Endangered

Location Mexico, Central America

Once considered a subspecies of the mantled howler (*Alouatta palliata*), this monkey is completely black, apart from the male's white scrotal sac. Loud whoops and howls at dawn and dusk proclaim a troop's territory, which may be up to 62 acres (25 hectares). Most troops comprise about 7 members with just one male, who is up to twice as heavy as a female. This monkey eats a plentiful, low-nutrient, leafy diet.

Alouatta seniculus

Venezuelan red howler

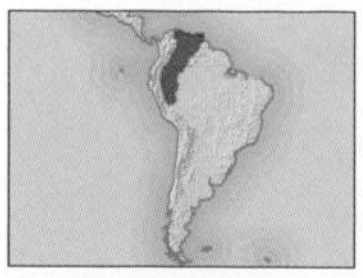

Length 20–25 in (50–63 cm)
Tail 22–27 in (55–68 cm)
Weight 15–20 lb (7–9 kg)
Social unit Group
Status Least concern

Location N.W. South America

Venezuelan red howlers are the largest of the 14 howler monkey species, whose loud howls, whoops, and other calls carry more than 1¼ miles (2 km) through the forest, informing others of a troop's presence or conveying alarm signals. They live mainly in groups of one male and 3–4 females. The male is much heavier than the female. When a new male ousts an existing male from a troop, he may kill the latter's offspring, so that the females become ready to breed with him sooner.

GOLDEN SADDLE
The northern red howler (shown here) has a reddish gold "saddle" on the body, which contrasts with the maroon head, shoulders, and limbs. Other populations are a uniform red color.

TREETOP LIFE

Like all howlers, the Venezuelan red howler is wonderfully adapted to life among trees. Its prehensile tail lacks fur on the underside near the tip, to help it grip and work as a fifth grasping limb. Its climbing ability provides access to a wide variety of leaves and fruit. Howlers need to consume up to 2¼ lb (1 kg) daily of this plentiful, but low-energy, diet. They spend up to three-quarters of the day resting to conserve energy.

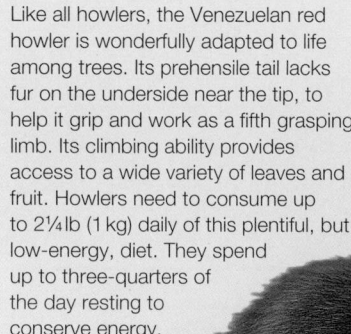

Pithecia monachus

Monk saki

Length 14½–19 in (37–48 cm)
Tail 16–20 in (40–50 cm)
Weight 3¼–6½ lb (1.5–3 kg)
Social unit Group
Status Least concern

Location N. and W. South America

All 9 saki species have broad noses (especially the septum, the central portion between the nostrils), lank fur that falls to the sides from the back and neck, a bushy tail, and extra-long fur around the face, which in the monk, or red-bearded, saki forms a hood or cowl. Compared to many other New World monkeys, this saki is shy and quiet, preferring to stay high in the trees, keep still, and remain unnoticed. It can make a loud alarm call, but under threat its main defense is to bare its teeth. Its diet is fruit and seeds, and it lives in close-knit troops of 4–5, who spend much time grooming each other.

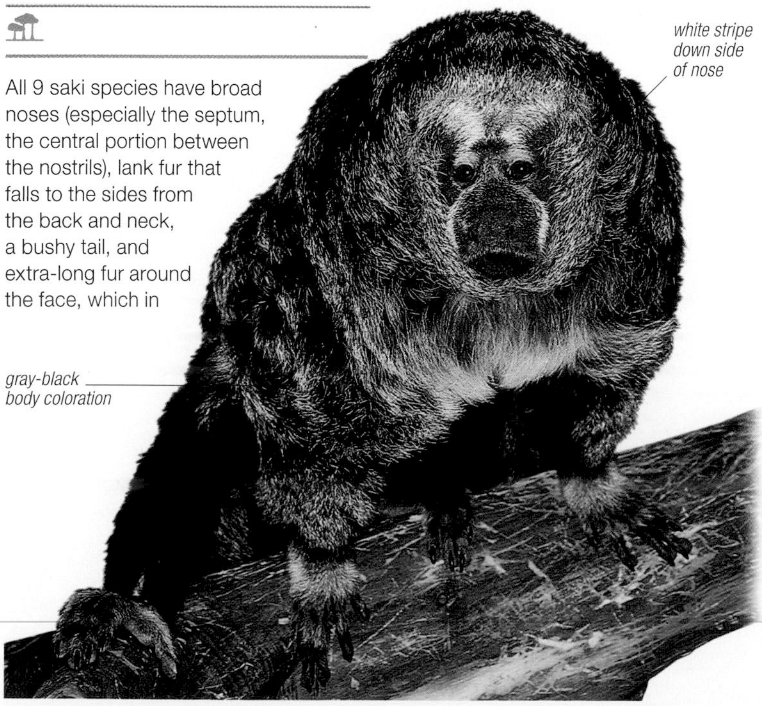

Pithecia pithecia

White-faced saki

Length 13½–14 in (34–35 cm)
Tail 13½–17½ in (34–44 cm)
Weight 4½ lb (2 kg)
Social unit Group/Pair
Status Least concern

Location N. South America

No other New World monkey has such differences between the sexes: the male is black with a white or pale gold face and a black nose, while the female is gray-brown with pale tipped hairs and a dark face. Although it is vegetarian, its teeth have a predatory appearance, with sharp incisors for impaling fruit and long canines to crack seeds and nuts. A typical group is a female, male, and 1–3 young.

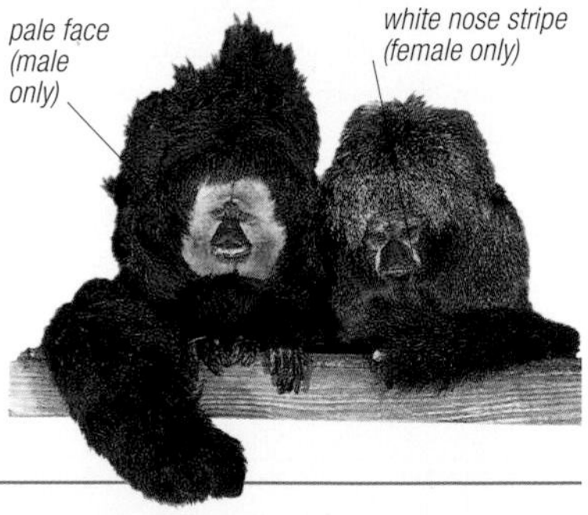

Chiropotes satanas

Black-bearded saki

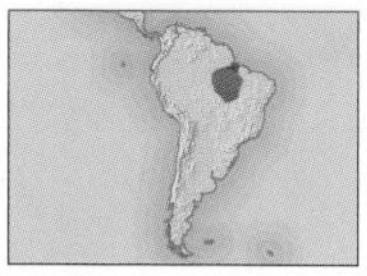

Length 13–18 in (33–46 cm)
Tail 12–18 in (30–46 cm
Weight 4½–8¾ lb (2–4 kg)
Social unit Group
Status Critically endangered

Location N. South America

In this species, the typical saki's long chin fur forms a bushy beard, and the long, dense head fur makes a thick forehead fringe. Both feet and hands grip powerfully and this saki can hang by one limb as it feeds on seeds, hard fruits, and small animals, crushing them with well-developed molar teeth. When excited, it "switches" its tail like a cat and emits a piercing whistle.

Cacajao calvus

Bald uakari

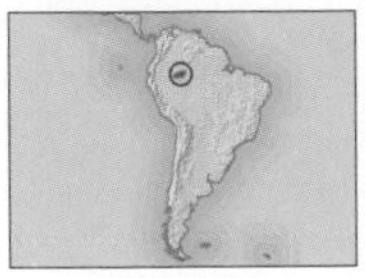

Location N.W. South America

Length 15–22½ in (38–57 cm)
Tail 5½–7½ in (14–18.5 cm)
Weight 6½–7¾ lb (3–3.5 kg)
Social unit Group
Status Vulnerable

Bald uakaris prefer "blackwater," part-flooded forests along small rivers, lakes, and swamps, rather than forests fringing large rivers. They forage by day in trees for seeds, fruit, flowers, and small animals, in a large troop of males and females, usually 10–20 but sometimes up to 100. The troop may mix with similar primates, such as squirrel monkeys, to feed.

COLOR FORMS

Bald uakari subspecies have a variety of fur colors, leading to different names: white uakari (above) in northwestern Brazil; golden uakari on the Brazil–Peru border; red uakari on the Colombia–Brazil border; and pale-backed red uakari further east.

HAIRLESS FACE *The bald uakari has a hairless face and forehead, the skin color varying from pink to deep red.*

Aotus lemurinus

Night monkey

Location Central America to N.W. South America

Length 12–16½ in (30–42 cm)
Tail 11½–17½ in (29–44 cm)
Weight 32–34 oz (900–950 g)
Social unit Pair
Status Vulnerable

Formerly regarded as one species, genetic studies show there may be 11 species of night monkeys, also called douroucoulis, or owl monkeys from their hooting calls in the darkness. They are the only nocturnal monkeys, eating a mixed diet of fruit, leaves, and insects as they clamber cautiously through the branches. They live in male–female pairs and communicate by scents in urine and chest gland secretions. The single young is born after a gestation of 120 days. Weaning takes 8 months and, since juveniles may also stay with parents, close-knit family groups of 4–5 develop.

Callicebus aureipalatii

Golden palace monkey

Location C. South America

Length 31¼ in (80 cm)
Tail 18¾–20½ in (48–52 cm)
Weight 2–2¼ lb (0.9–1 kg)
Social unit Group
Status Least concern

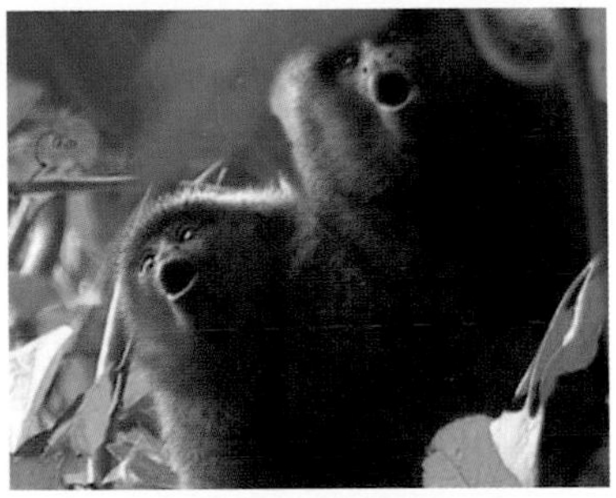

A colorful monkey, with orange-gold coat, white-tipped tail, and reddish hands and feet, this species was named after the organization that maintains the Madidi National Park in Bolivia. Here, the species was discovered in 2004—in the forest of the Andean foothills and floodplain. It is likely that it ranges into southern Peru, too. Groups have a home range of between 42 and 62 acres (17 and 25 hectares) and consume a wide range of rain forest plants. Although much of its population is within a protected area and is not under significant pressure from hunting, the species may be affected by petroleum exploitation, road-building, and hydroelectric projects.

Cebus apella

Brown capuchin

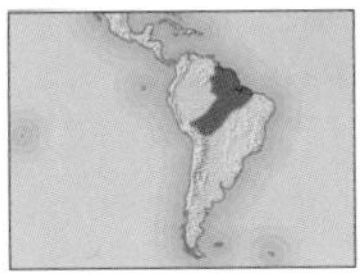

Location N. South America

Length 13–17 in (33–42 cm)
Tail 16–19½ in (41–49 cm)
Weight 6½–10 lb (3–4.5 kg)
Social unit Group
Status Least concern

Capuchins are often regarded as the most adept New World monkeys, and the brown capuchin uses a variety of tools, such as stones to crack hard nuts. It also eats fruit, insects, and some vertebrates, such as frogs, lizards, and even small bats. This species, also known as the tufted capuchin from the furry "horns" above each ear, has the widest range of any New World monkey. Mixed groups of 8–14 are usual as members are not sexually mature until 7 years old—later than most monkeys of similar size.

Callicebus moloch

Red-bellied titi

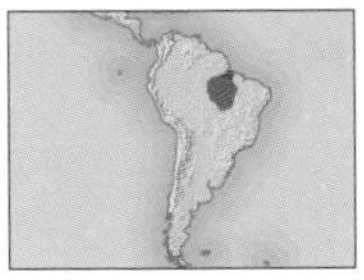

Location N. South America

Length 10½–17 in (27–43 cm)
Tail 14–22 in (35–55 cm)
Weight 1½–2¼ lb (0.7–1 kg)
Social unit Pair
Status Least concern

The 20 or more titi species all have thick, soft fur, chunky bodies, short limbs, and—unlike other New World monkeys of their size—ears almost hidden in fur. They eat mainly fruit, leaves, seeds, and grubs. The red-bellied titi monkey's back is speckled brown, the underparts mainly orange. It relies on its drab coloration and slow movements for camouflage in trees near swamps and pools. Female and male form a close pair-bond and defend a territory of 15–30 acres (6–12 hectares). They intertwine tails and sing a "duet" just after dawn to maintain their family and pair-bonds, and to proclaim their territory.

Cebus olivaceus

Weeper capuchin

Location N.E. South America

Length 14½–18 in (37–46 cm)
Tail 16–22 in (40–55 cm)
Weight 5½–7¾ lb (2.5–3.5 kg)
Social unit Group
Status Least concern

Like the other 12 or so capuchin species, the weeper has a robust build with relatively short limbs, and a prehensile tail. There is no bare skin on the tail's underside, as in many other South American monkeys. The main color is brown, becoming paler on the arms, and even gray or yellow on the face. Seeds, fruit, and small creatures, especially snails and insects, comprise the main diet. The weeper capuchin forms bands of 30 or more, containing mostly females and young, with several males, but only one dominant male breeds. Mothers may look after each other's offspring (known as "allomothering"). Troop members keep in contact with a plaintive, "weeping" cry.

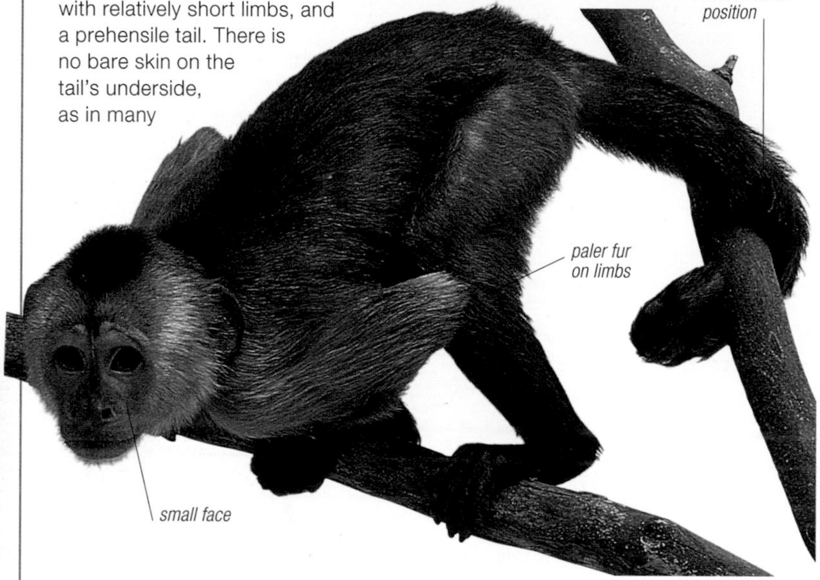

tail habitually held in curled position

paler fur on limbs

small face

Saguinus imperator

Emperor tamarin

Location W. South America

Length 9–10 in (23–26 cm)
Tail 15½–16½ in (39–42 cm)
Weight 16 oz (450 g)
Social unit Group
Status Least concern

Marmosets and tamarins form a distinct group of about 42 American primate species. They are similar to other New World monkeys but differ in body chemistry, have claws rather than nails, and bear 2 rather than 1 offspring. Identified by its flowing white mustache, this species feeds on fruit in the wet season, flower nectar and tree sap in the dry season, and insects, especially crickets, all year round. It often forms a mixed troop with related species such as the saddleback tamarin. Each species responds to the other's alarm calls if a predator is detected nearby.

MUSTACHE
The white, curly-ended mustache of both males and females contrasts with the black face, speckled red- or gray-brown body, and red-orange tail fur.

TAMARIN BABYCARE

Tamarins and marmosets have relatively long gestation periods for such small mammals, the emperor tamarin's being 140–145 days. There are almost always 2 offspring, and the father carries these, as here, except when they are being suckled by their mother.

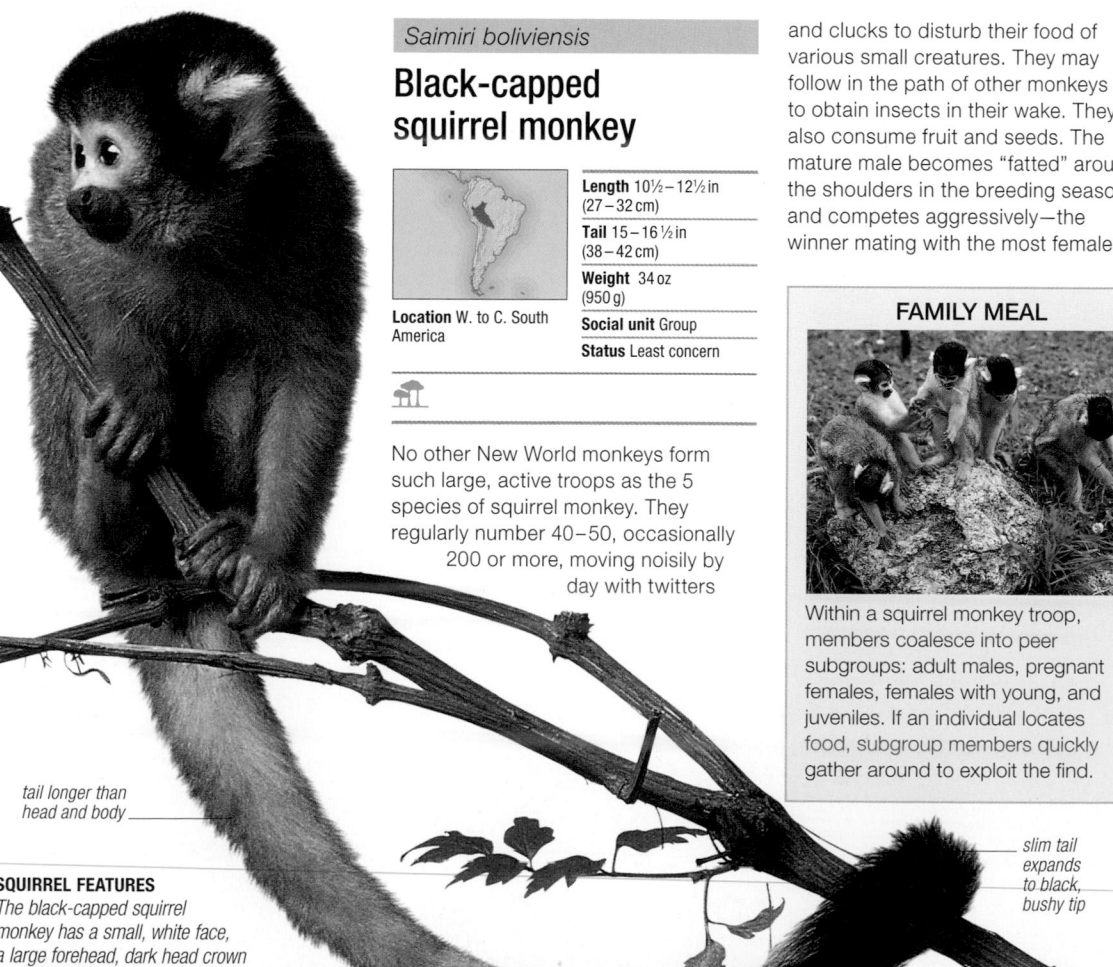

tail longer than head and body

slim tail expands to black, bushy tip

SQUIRREL FEATURES
The black-capped squirrel monkey has a small, white face, a large forehead, dark head crown above, black nose tip and muzzle, and well-furred ears.

Saimiri boliviensis

Black-capped squirrel monkey

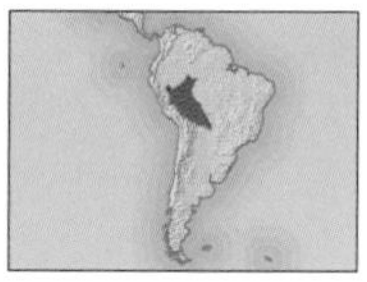

Location W. to C. South America

Length 10½–12½ in (27–32 cm)
Tail 15–16½ in (38–42 cm)
Weight 34 oz (950 g)
Social unit Group
Status Least concern

No other New World monkeys form such large, active troops as the 5 species of squirrel monkey. They regularly number 40–50, occasionally 200 or more, moving noisily by day with twitters and clucks to disturb their food of various small creatures. They may follow in the path of other monkeys to obtain insects in their wake. They also consume fruit and seeds. The mature male becomes "fatted" around the shoulders in the breeding season and competes aggressively—the winner mating with the most females.

FAMILY MEAL

Within a squirrel monkey troop, members coalesce into peer subgroups: adult males, pregnant females, females with young, and juveniles. If an individual locates food, subgroup members quickly gather around to exploit the find.

Callimico goeldii

Goeldi's marmoset

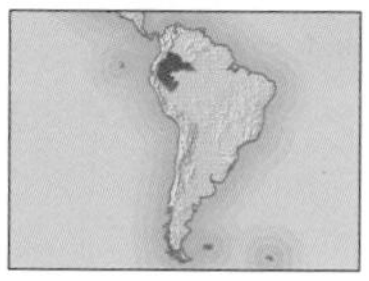

Location N.W. South America

Length 9 in (22–23 cm)
Tail 10–12½ in (26–32 cm)
Weight 21 oz (575 g)
Social unit Group
Status Vulnerable

Goeldi's marmoset is larger than most marmosets and tamarins. Its long fur is black, and it has a "cape" of longer hair on its head and neck. Unlike similar species, it has wisdom teeth. It eats sap and gum (digging its incisor teeth into bark to make these flow), and fruit, insects, and small vertebrates such as lizards. The species forms stable, close-knit groups of up to 10, mainly male–female pairs and their young, and keeps to dense vegetation such as creeper-tangled bamboo.

MAMMALS

Saguinus oedipus

Cotton-top tamarin

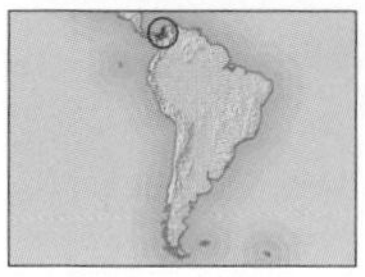

Location N.W. South America

Length 8–10 in (20–25 cm)
Tail 13–16 in (33–40 cm)
Weight 14–16 oz (400–450 g)
Social unit Group/Pair
Status Endangered

The long, white, flowing fur on the head crown distinguishes this tamarin, found in an extremely restricted range in Colombia. Like many marmosets and tamarins, this species has a varied diet and a "helper" system of rearing offspring, in which males and older siblings carry the very young. However, in each troop of 10–12, only one male–female pair breeds; 4 out of 5 births are twins. The cotton-top tamarin has been used in medical research, and currently there are more of these monkeys in captivity than in the wild.

Callithrix argentata

Silvery marmoset

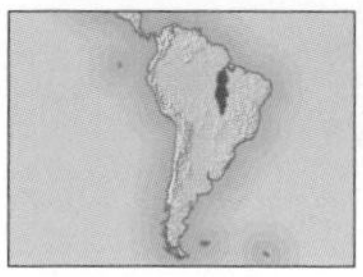

Location C. South America

Length 8–9 in (20–23 cm)
Tail 12–13½ in (30–34 cm)
Weight 12–13 oz (325–350 g)
Social unit Group
Status Least concern

Color is the main distinguishing feature among 23 very similar, closely related marmosets from south of the Amazon. This species has pale silver-gray fur on the back, creamy yellow on the underside, and black on the tail; the face has pink skin and ears. In each small troop of silvery marmosets, only one male and female breed. The rest are siblings, cousins, and other "helpers," who assist in carrying and protecting the infants.

Callithrix pygmaea

Pygmy marmoset

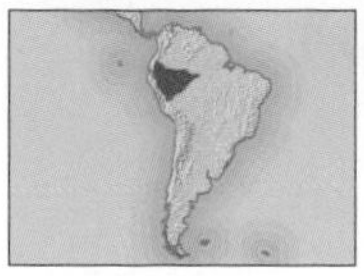

Location W. South America

Length 4¾–6 in (12–15 cm)
Tail 7–9 in (17–23 cm)
Weight 3⅝–4 oz (100–125 g)
Social unit Group/Pair
Status Least concern

The smallest monkey in the world, a curled-up pygmy marmoset fits into a human palm. However, it is long-lived for such a diminutive mammal and may reach 12 years of age. The pygmy marmoset differs from other marmosets in the way it eats gum: it gouges out 10 or more new holes in bark each day, scent-marks them, and returns to these and older holes at intervals to scrape up the sticky, oozing liquid with its long, lower incisor teeth. It also takes flower nectar, fruit, and small creatures such as grubs and spiders. The pygmy marmoset follows the reproductive pattern of other marmosets, with one breeding pair per troop of 5–10. Other troop members, usually older offspring, are childcare "helpers," who carry the 2 newest youngsters after the breeding male has cared for them during the first few weeks.

Callithrix geoffroyi

Geoffroy's marmoset

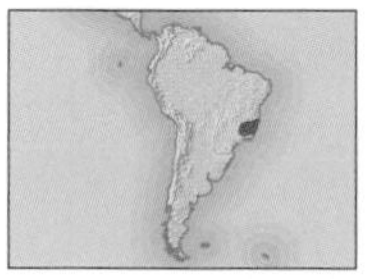

Location E. South America

Length 8 in (20 cm)
Tail 11½ in (29 cm)
Weight 13 oz (350 g)
Social unit Group/Pair
Status Least concern

This marmoset prefers secondary forest, that is, regrown after logging or other disturbance, rather than primary areas. It digs holes in tree bark with its long incisors, marks "ownership" with scent from its perineal gland, and then returns at intervals to scrape up the oozed gum. Like other marmosets, this species also eats fruit and insects. There is only one breeding pair in each troop, but the other members also carry and protect the young.

Leontopithecus rosalia

Golden lion tamarin

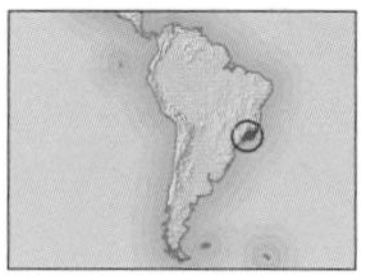

Location E. South America (Rio São João basin)

Length 8–10 in (20–25 cm)
Tail 12½–14½ in (32–37 cm)
Weight 14–29 oz (400–800 g)
Social unit Group/Pair
Status Endangered

This tamarin weighs twice the average for the marmoset and tamarin group. The face is dark gray and the hands, fingers, and claws are long and thin, to probe into bark and crevices for grubs. However some four-fifths of the diet is fruit, supplemented by gum and nectar. It forages by day and sleeps at night in tangled vegetation or, more often, a hole in a tree. There is only one male–female breeding pair in the troop of 4–11. In its "helper" system of rearing young—2 offspring being typical for the group—sexual activity of the helpers is not suppressed, although they do not produce young until they form the dominant partnership.

CONSERVATION

An emblem of conservation, the golden lion tamarin comes from Brazil's Atlantic coast forest—a habitat that has shrunk dramatically. Declared endangered over 30 years ago, its status rose to critically endangered in 1996, when less than 300 wild animals remained. Captive breeding programs have proved very successful. Today, over 1,000 golden lion tamarins live in the wild, many descended from animals raised in captivity. However, with little original habitat left, its future remains precarious.

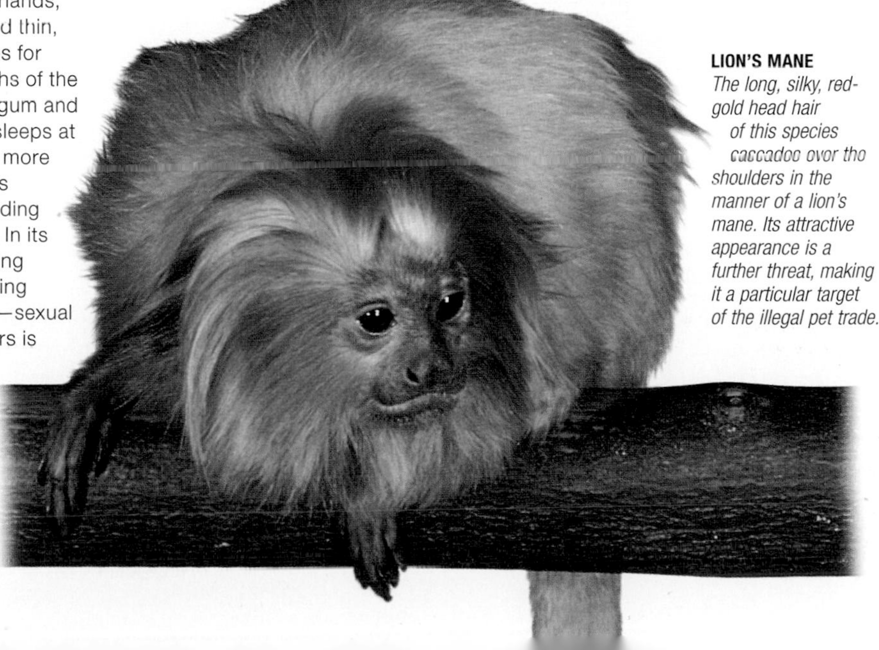

LION'S MANE
The long, silky, red-gold head hair of this species cascades over the shoulders in the manner of a lion's mane. Its attractive appearance is a further threat, making it a particular target of the illegal pet trade.

MAMMALS

Mandrillus sphinx

Mandrill

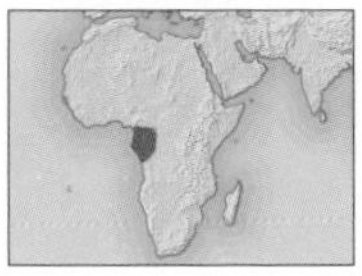

Location W. Central Africa

Length 25–32 in (63–81 cm)
Tail 2¾–3½ in (7–9 cm)
Weight 24–82 lb (11–37 kg)
Social unit Group
Status Vulnerable

A scarlet nose with prominent, bony blue flanges on either side are the unmistakable characteristics of the male mandrill. The female's facial color is much more subdued. She is about a third of the size of the male, which, at 23½ in (60 cm) high to the shoulder, is the largest of all monkeys. Mandrills live in groups in dense African rain forest, spending most of their day on the ground looking for fruit and seeds, as well as eggs and small animals. When night falls, they take to the trees for safety. Troops move over a range of up to 20 square miles (50 square km); they mark their territory with scent and defend it against rivals. Much of what is known about this species comes from studies of captive animals; in the wild, hunted for meat, and with its habitat being destroyed by logging, the mandrill is becoming increasingly scarce.

A WARNING YAWN
When threatened by a predator or approached by rivals, the male will yawn widely to reveal his fearsome teeth, which can be 2½ in (6.5 cm) long.

BOLD ADVERTISEMENT
With his spectacularly colored face—scarlet nose with blue flanges, and yellow beard—the mandrill boldly declares his identity to other animals in the forest. Together with his mauve-blue rump, these colors also announce the male's sex and display his virility to females.

LIFE ON THE GROUND
The mandrill walks and runs on all fours—a form of locomotion made all the more efficient by having fore- and hindlimbs of nearly equal length.

MANDRILL SOCIETY

Mandrills usually live in mixed groups of about 20, which come together to form troops of as many as 250 individuals. There is a strict hierarchy within the group: a dominant male heads each group, mating with fertile females and fathering almost all the infants. Non-breeding males make up the lower ranks.

MALE AND FEMALE
The male is much larger than the female, with a far more colorful face. Some mandrills may live in harems, with one mature male to 20 females.

MOTHER AND CHILD
The mandrill gives birth to one offspring every 18 months or so. At first, the mother will carry the infant on her belly, but as it grows heavier, it will ride through the forest on her back.

FORAGING PARTY
Mandrills often forage in small parties, grunting constantly to stay in touch. When it is time to move on, the dominant male will round up his group with a 2-phase grunt or roar.

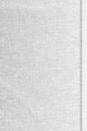

Erythrocebus patas

Patas monkey

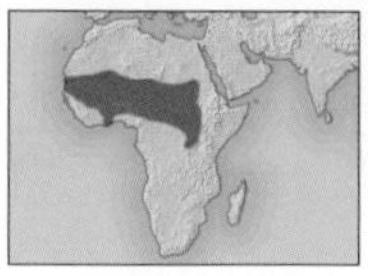

Location W. to E. Africa

Length 23½–35 in (60–88 cm)
Tail 17–28 in (43–72 cm)
Weight 22–29 lb (10–13 kg)
Social unit Group
Status Least concern

This is one of the fastest monkeys when it is running. It has a long, slim body, lengthy limbs, hands, and feet, and short digits. The white moustache and beard contrast with the darker face; the red back contrasts with the white limbs and underside. It lives in troops of up to 10 members, and the single male stays on the periphery, to act as decoy, drawing predators while the females and young hide.

Cercopithecus neglectus

De Brazza's monkey

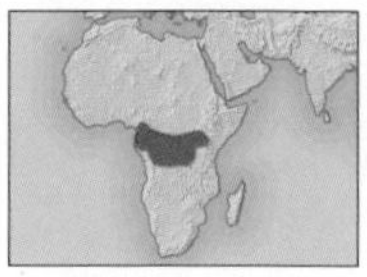

Location C. to E. Africa

Length 20–23 in (50–59 cm)
Tail 23–31 in (59–78 cm)
Weight 15–18 lb (7–8 kg)
Social unit Pair
Status Least concern

One of the most terrestrial of the 20 or so *Cercopithecus* species, De Brazza's monkey is also the only one of the genus that forms male–female pair bonds. Widespread yet inconspicuous over its large range, it marks its territory with saliva and scent, yet avoids rather than challenges any intruders. It has blue-white upper lip and chin fur, and a thin, white thigh stripe. It uses deep, booming calls to communicate, and eats mainly seeds and fruit.

Rungwecebus kipunji

Kipunji

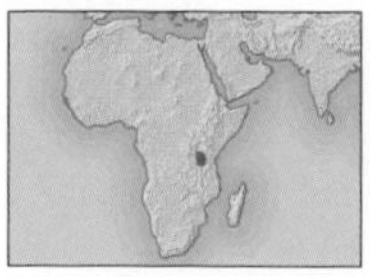

Location E. Africa

Length 33½–35½ in (85–90 cm)
Tail 39½–43½ in (100–110 cm)
Weight 22–35 lb (10–16 kg)
Social unit Group
Status Critically endangered

Discovered in 2003, this omnivorous tree-dwelling primate from the Udzungwa Mountains and Southern Highlands of Tanzania was originally thought to be a mangabey (a type of forest monkey)—but later shown to be more related to ground-dwelling baboons. It may even have originated by crossbreeding between the two groups. It has a barking call, reminiscent of the grunts of baboons. Two isolated populations together inhabit scarcely 38 square miles (100 square km), and the species is threatened by deforestation and hunting.

Papio anubis

Olive baboon

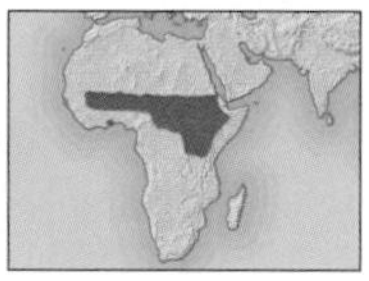

Location W. to E. Africa

Length 23½–34 in (60–86 cm)
Tail 16–23 in (41–58 cm)
Weight 49–82 lb (22–37 kg)
Social unit Group
Status Least concern

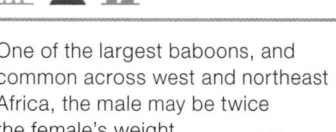

One of the largest baboons, and common across west and northeast Africa, the male may be twice the female's weight. Both sexes have a thick, gray ruff around the cheeks. Average troop size is 20–50, occasionally over 100. They eat fruit, leaves, insects, lizards, and sometimes larger prey such as gazelle fawns.

GROUP ORDER

Young olive baboons are tolerated while in dark "baby fur." As they molt to adult coloration, females take their place at the base of the troop hierarchy. Males are driven off, and must battle their way into another troop.

PRIMATE "DOG" *Typically powerful and doglike, the olive baboon is speckled olive-green, with a black face and rump.*

Macaca munzala

Arunachal macaque

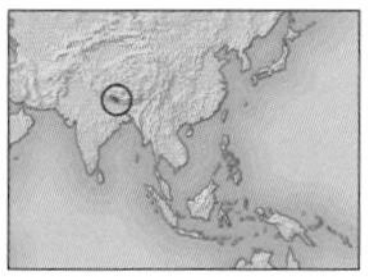

Location S. Asia

Length 23 in (58 cm)
Tail 9 in (23 cm)
Weight 33 lb (15 kg)
Social unit Group
Status Endangered

Restricted to the Indian Arunachal Pradesh state of eastern Himalaya, this macaque was scientifically described in 2005 on the basis of photographs, and differs from related species in having a dark crown patch and pale neck collar. It occurs between 6,650 and 11,483 ft (2,000–3,500 m) in coniferous and broadleaf forests, as well as agricultural land. Like other high altitude monkeys it is robustly built to withstand the cold. Although densely populated in places and relatively tolerant of humans, the Arunachal macaque may be exposed to retaliatory hunting when it raids crops: this could threaten its long-term survival. So far the species is known with certainty from no more than 5 locations within an area of about 1,042 square miles (2,700 square km)—but it may also occur in adjacent areas of Bhutan and Tibet. Field data suggest that there are less than 600 known individuals.

Macaca nigra

Celebes crested macaque

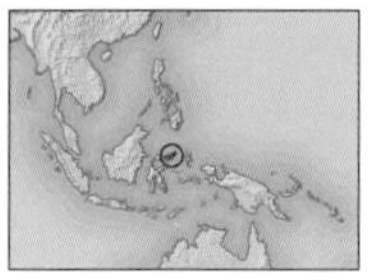

Location S.E. Asia (northern Sulawesi)

Length 20½–22½ in (52–57 cm)
Tail 1¼ in (3 cm)
Weight 22 lb (10 kg)
Social unit Group
Status Critically endangered

The all-black fur, very short tail, and homeland of Celebes (now Sulawesi) give this species the alternative name of Celebes black ape. A crest runs from the forehead back over the crown. Usually flat, it rises when the animal is aroused. This species forms huge, mixed-sex troops of more than 100, yet is generally an inconspicuous, quiet, fruit-eating forest dweller.

midline head crest

Papio papio

Guinea baboon

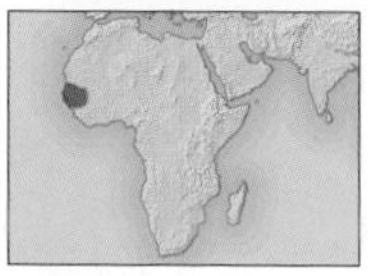

Location W. Africa

Length 27 in (69 cm)
Tail 22 in (56 cm)
Weight 39 lb (17.5 kg)
Social unit Group
Status Near threatened

The male of the smallest baboon species has proportionally the longest mane, reaching almost to his rump. Even for a baboon, the diet is varied: from tough roots to juicy grubs and eggs, and sometimes farm crops. Troops number up to 200, although about 40 is more usual, with several males gathering their own harem of females.

Papio ursinus

Chacma baboon

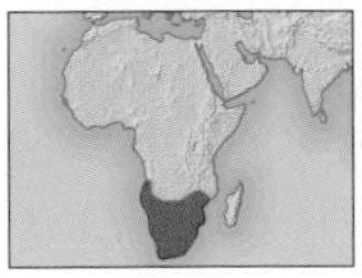

Location Southern Africa

Length 23½–32 in (60–82 cm)
Tail 21–33 in (53–84 cm)
Weight 33–66 lb (15–30 kg)
Social unit Group
Status Least concern

The largest baboon, the chacma has a drooping snout and protruding nostrils. Fur color varies from yellow-gray to black, paler on the muzzle. This intelligent, adaptable primate enjoys very fluid troop composition in some parts of its range. Its wide diet ranges from roots and seeds to insects and young gazelles. In the wild, it is known to use tools such as sticks.

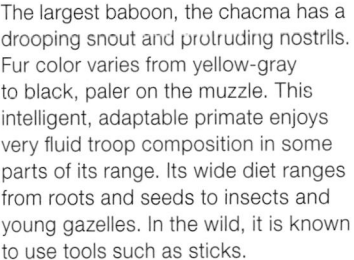

Theropithecus gelada

Gelada

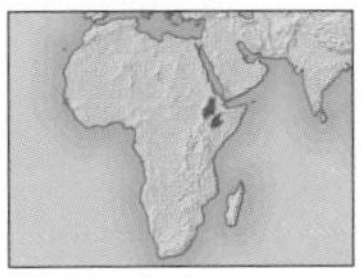

Location E. Africa

Length 28–29 in (70–74 cm)
Tail 18–20 in (46–50 cm)
Weight 42 lb (19 kg)
Social unit Group
Status Least concern

A close cousin of baboons, the pink-chested gelada is restricted to the windy, grassy Ethiopian highlands. Its limited diet of grass blades, stems, and seeds is picked by rapid, dextrous hand movements as it sits and shuffles along. Small groups of females and young led by a male may band into huge but loose troops.

Nasalis larvatus

Proboscis monkey

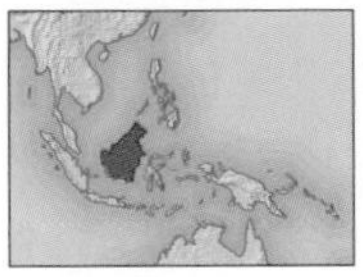

Location S.E. Asia

Length 29–30 in (73–76 cm)
Tail 26–26½ in (66–67 cm)
Weight 46 lb (21 kg)
Social unit Group
Status Endangered

Among the most specialized and distinctive of mammals, the proboscis monkey occupies very restricted habitats near water in lowland rainforest, mangrove swamps, and coasts on the island of Borneo. The average troop is one male with 6–10 females and their young. The male defends his group, honks loudly, bares his teeth, and waves his erect penis at an aggressor.

partly webbed feet

LARGE NOSE
The proboscis monkey's defining feature is its large nose. In mature males, it is long and pendulous, and may play a role in attracting a mate.

CHANGING FACE

At birth, the young have a blue face, dark fur, and a "normal" monkey nose. The coloration changes and the nose grows with age. In females, the nose is much smaller than in the male, but it is still large compared to related species.

Semnopithecus entellus

Northern plains grey langur

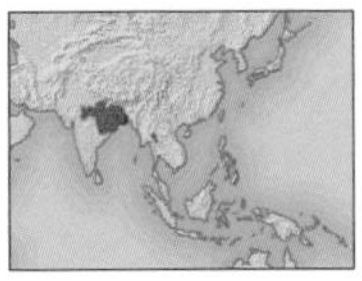

Location India

Length 20–31 in (50–78 cm)
Tail 27–40 in (69–102 cm)
Weight 23–44 lb (10.5–20 kg)
Social unit Group
Status Least concern

This langur is found throughout the Indian subcontinent, except rainforest areas. Coloration varies from dark brown in individuals from the Himalayas to pale fawn in those furthest south. Groups of females and young may be led by one or several males, while other males form bachelor troops.

MIXED DIET

The northern plains grey langur feeds mostly on leaves, fruit, buds, and shoots, which it digests easily in its compartmentalized stomach. Like many wild creatures, these monkeys supplement their diet by licking salt or eating mineral-rich soil. Troops that live around villages also often benefit from leftovers and offerings from local people, who hold the species sacred, identifying it with the Hindu deity Hanuman.

Colobus guereza

Guereza

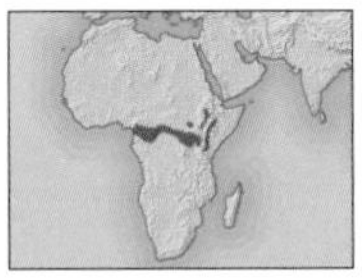

Location C. and E. Africa

Length 20½–22½ in (52–57 cm)
Tail 21–33 in (53–83 cm)
Weight 18–30 lb (8–13.5 kg)
Social unit Group
Status Least concern

Also known as the eastern black and white colobus, this monkey has a black and white face border, "veils" down its flanks and rump, and a bushy tail tip, but is otherwise black. Each male leads a small troop of 4–5 females and young, defending his territory with roars and tremendous leaps. The 3-part stomach houses gut microbes that break down cellulose-based plant food, enabling this species to gain twice as much nutrition than other monkeys from such a leaf-rich diet.

Rhinopithecus strykeri

Myanmar snub-nosed monkey

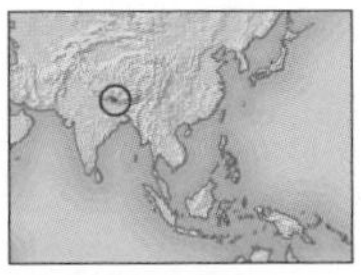

Location S. Asia

Length 21 in (56 cm)
Tail 30½ in (78 cm)
Weight Not known
Social unit Group
Status Critically endangered

This species was described in 2010 from specimens secured from hunters, and lives in the temperate forests of northern Myanmar in the eastern Himalaya. It is black with a white beard and ear tufts, and probably has a total population of only 260 to 330 individuals that are apparently segregated into 3 groups. Five species of snub-nosed monkeys from Asian mountain forests get their name from their stump-noses; reputedly, they avoid getting rain in their nostrils by sitting out rainstorms with their heads facing downward between their knees. Nevertheless, according to local sources, the snub-nosed monkey of rain-soaked Myanmar is often heard sneezing during downpours. Snub-nosed monkeys are entirely vegetarian. In winter, heavy snowfall forces them to descend to lower elevations in search of food—which may bring them closer to villages.

Apes

PHYLUM	Chordata
CLASS	Mammalia
ORDER	Primates
SUBORDER	Haplorrhini (part)
FAMILIES	2
SPECIES	21

Apes are the closest relatives to humans. They are similar to people not only in appearance but also in that they are highly intelligent and form complex social groups. Apes are divided into the lesser apes (the gibbons) and the larger, more humanlike great apes—the orangutan, gorillas, and chimpanzees. They are found in western and central Africa and throughout southern and Southeast Asia, mainly in tropical rain forests. Apes are essentially vegetarian and mostly eat fruit, although some are omnivorous. They are threatened by loss of their forest habitat, hunting and poaching (often for their skins or skulls), and capture for zoos and the pet trade. Chimpanzees were once widely used in medical research.

Anatomy

A shortened spine and a relatively short, broad pelvis lower the center of gravity in apes, thereby facilitating a more upright posture. Apes have a broad chest, with the shoulder blades at the back, which allows an exceptionally wide range of movement in the shoulder joint. A gorilla, for example, can sit on the ground and reach out in any direction to pull in vegetation. Apes also have a flattened face, well-developed jaw, grasping hands and feet, and downward-directed, close-set nostrils. The great apes are very large: the orangutan is the largest arboreal mammal and the gorilla may weigh over 440lb (200kg).

Social organization

The lesser apes form monogamous pairs. They mark their treetop territories with loud, musical songs in which the male and female sing different parts. A maturing young will attempt to establish its own territory by singing alone, until it finds a mate.

The orangutan is the only great ape that is solitary. A mature male controls a large territory with deep, resonant calling. He has access to all the females that enter his domain. The other great apes have well-defined social groups. Gorillas live in troops of 5–10 (occasionally up to 30), consisting of several females, one dominant mature ("silverback") male, and possibly one or 2 other silverbacks (the sons or younger brothers of the dominant male). Chimpanzees live in communities of 40–100. Although there is a dominant male and a social hierarchy, individuals have almost complete freedom to come and go. Foraging occurs in small groups, the composition of which changes daily. Chimpanzees found in West Africa are particularly fond of hunting, and the males cooperate closely to catch several monkeys (for example) at once. Chimpanzees apparently pass on customs and technologies socially, by example, rather than genetically.

Intelligence

Apes are extremely intelligent—even more so than monkeys. They appear to work through problems in the same way that humans do. Chimpanzees, for example, use and sometimes make simple tools, as does at least one population of orangutans in northern Sumatra. The orangutan is one example of an ape that has performed several complex tasks—such as solving puzzles, using sign language, and learning to recognize symbols—in research centers.

MUTUAL GROOMING
Grooming is important in ape societies because it strengthens and maintains bonds between individuals, such as these two chimpanzees. However, it is also a means of gaining favor and is used, for example, by males to ensure support from friends in the event of a challenge to their supremacy.

Symphalangus syndactylus

Siamang

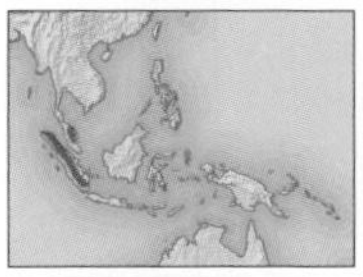

Length 35 in (90 cm)
Tail None
Weight 22–33 lb (10–15 kg)
Social unit Group/Pair
Status Endangered
Location S.E. Asia

The largest gibbon, with a "standing" height of 5 ft (1.5 m), the siamang also has the loudest calls of the group, and the closest-knit families. The female (who is dominant), male, and 1–2 offspring rarely stray more than 100 ft (30 m) from each other, and are usually less than 33 ft (10 m) apart. About three-fifths of their food intake is leaves, and one-third fruits, with a few blossoms and small creatures such as grubs. The family occupies a home range of about 116 acres (47 hectares), but defends only some 60 percent of this as their territory, chiefly using their powerful calls, barks, and screams.

ALL BLACK *The siamang is covered with black, shaggy fur. The male is slightly larger than the female and has a tuft of hair on his genital region that, at a glance, looks like a tail.*

BARKS AND SCREAMS

The siamang's dark gray, elastic throat skin inflates to the size of a grapefruit, to act as a resonator and amplify its amazingly loud calls. The male's screams are thought to discourage other males, while the female's longer, more distinctive series of barklike sounds is associated with territory defense.

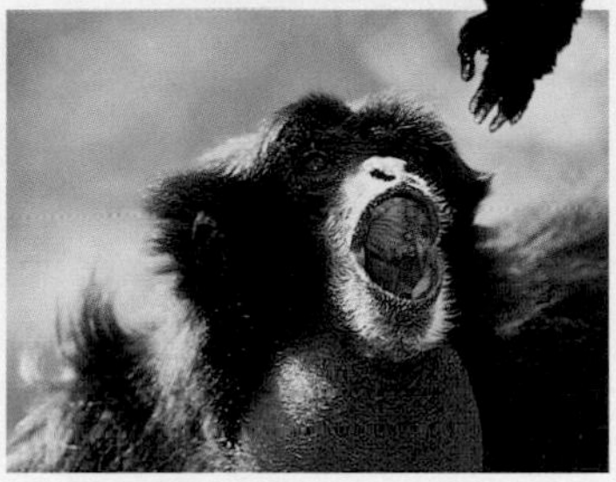

Nomascus concolor

Black-crested gibbon

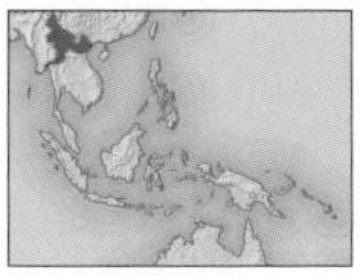

Length 18–25 in (45–64 cm)
Tail None
Weight 10–20 lb (4.5–9 kg)
Social unit Group/Pair
Status Critically endangered
Location S.E. Asia

Unlike most other "flat-headed" gibbons, the black-crested gibbon has long fur on the head crown. The offspring is born with yellow fur, replaced by black fur as it matures; females continue to change to brown or gray. This gibbon eats mainly leaf buds, shoots, and fruit, but rarely animals. It lives in groups, usually a female, a male, and their young.

Nomascus leucogenys

Northern white-cheeked gibbon

Length 18–25 in (45–64 cm)
Tail None
Weight 10–20 lb (4.5–9 kg)
Social unit Group/Pair
Status Critically endangered
Location S.E. Asia

Considered a subspecies of the crested gibbon until 1989, the infants, adult females, and males of this species have similar coloration (see left), and also the "crest" of longer fur on the head. The two species are distinguished mainly by geography: the black-crested gibbon occurs northeast of the Song Ma and Song Bo rivers in Vietnam, while the northern white-cheeked gibbon is found to the southwest. The single young is born after a gestation period of 7–8 months and is dependent on its mother for 18 months.

Hylobates lar

Lar gibbon

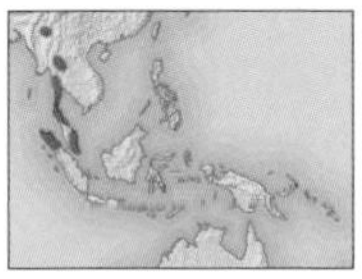

Length 16½–23 in (42–59 cm)
Tail None
Weight 10–17 lb (4.5–7.5 kg)
Social unit Pair
Status Endangered
Location S.E. Asia

The lar gibbon becomes active shortly after dawn, when a female and male "duet" to reinforce their pair-bond. She begins with a series of loud, long hoots that rise to a crescendo; the male responds as these fade, with simpler, more tremulous hoots. Each duet, repeated numerous times, lasts 15–20 seconds according to region. Most of this gibbon's day is spent finding food and eating. Half the diet is fruit, the rest being leaves, insects, and flowers. Some 15 minutes each day is spent in mutual grooming between partners. The lar gibbon rarely moves at night, resting among tree branches or tree forks. Also called the common or white-handed gibbon, the lar gibbon was considered to make lifelong pair-bonds, but recent studies show some serial monogamy with occasional partner changes, and even non-monogamous groupings. Each pair or group defends a territory comprising about three-quarters of its total home range. Gestation is 7–8 months; the single offspring is suckled for 18 months, is adult size by 6 years, and fully mature by 9 years. Lifespan in the wild averages 25–30 years. Deforestation and hunting by humans are the major threats.

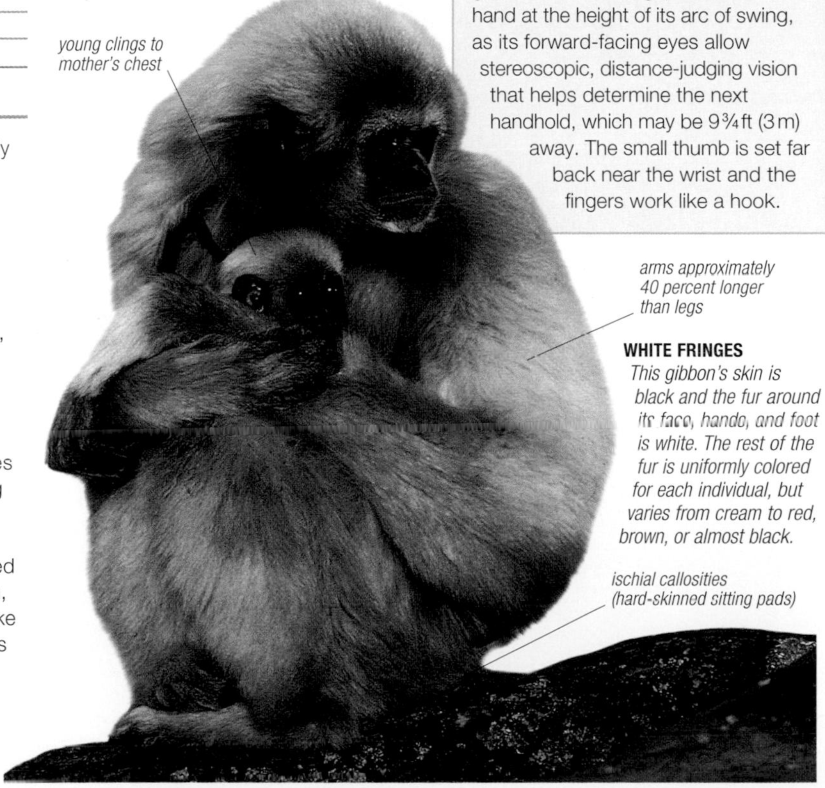

WHITE FRINGES *This gibbon's skin is black and the fur around its face, hands, and foot is white. The rest of the fur is uniformly colored for each individual, but varies from cream to red, brown, or almost black.*

BRACHIATION

The gibbon's arm-hanging, hand-swinging method of movement is called brachiation. It saves energy by maintaining momentum, using the body as a pendulum. The lar gibbon releases its grip with one hand at the height of its arc of swing, as its forward-facing eyes allow stereoscopic, distance-judging vision that helps determine the next handhold, which may be 9¾ ft (3 m) away. The small thumb is set far back near the wrist and the fingers work like a hook.

HANDY FEET

The feet of the lar and other gibbons, like the palms of their hands, have bare, leathery-skinned soles for effective grip. The big toe is able to grasp in opposition to the other toes, so that they can walk upright along branches.

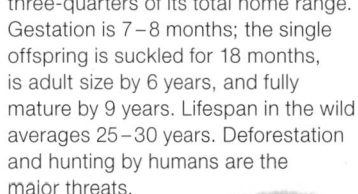

Hylobates moloch

Silvery gibbon

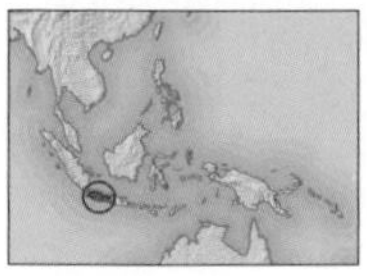

Location S.E. Asia

Length 18–25 in (45–64 cm)
Tail None
Weight 12 lb (5.5 kg)
Social unit Pair
Status Endangered

Pale eyebrows, cheeks, and beard merge with the mainly silvery fur of this gibbon. Its diet is fruit, some leaves, and sometimes nectar and grubs. Like most other gibbons, the family group (female, male, offspring) use calls to defend their territory. Unlike the lar gibbon (p.151), the male and female do not "duet" together.

Pan paniscus

Bonobo

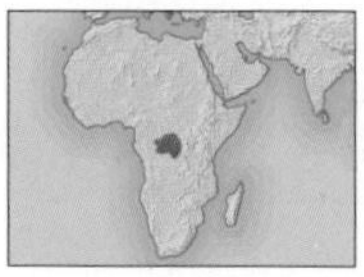

Location C. Africa

Length 28–33 in (70–83 cm)
Tail None
Weight Up to 86 lb (39 kg)
Social unit Group
Status Endangered

The bonobo or pygmy chimpanzee is only slightly smaller than the common chimpanzee (see below), but has a slimmer body and relatively longer and more slender limbs. It feeds chiefly on the ground, mainly on fruit and seeds, but also on leaves, flowers, fungi, eggs, and small animals. This ape can live in groups of up to 80 but is usually found in smaller groups, as it forages and grooms. Sexual relations are common between males, females, and young in various combinations, and may be used to ease social tensions. Females are dominant and leave their family groups when mature; males tend to stay.

FACE AND FUR
The skin of the pygmy chimpanzee is mostly black, even on a juvenile's face. The fur on the crown has a central "parting."

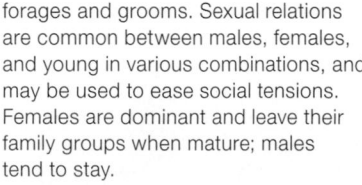

LONG INFANCY

The bonobo is born after 8 months' gestation, and suckled by its mother for up to 3 years. The female continues to protect, groom, and share nests with her offspring for another year or 2. The interval between births is about 5 years.

Pan troglodytes

Common chimpanzee

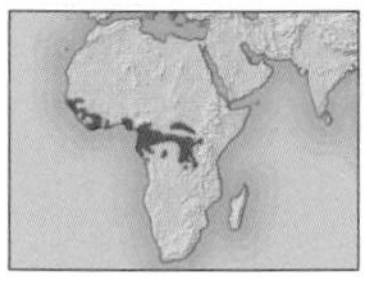

Location W. to C. Africa

Length 25–35 in (63–90 cm)
Tail None
Weight 66–130 lb (30–60 kg)
Social unit Group
Status Endangered

Common chimpanzees live in communities of 15–120. Subgroup composition varies almost hourly for activities such as grooming, feeding, traveling, and defending the territory. This last task is usually carried out by adult male parties, who may attack and kill stray chimps from other communities. Most daylight hours are spent eating—mainly fruit and leaves, but also flowers and seeds. Raiding parties sometimes cooperate to kill and eat animal prey, such as monkeys, birds, and small antelopes. Social bonds may last years, but there are no long-term male–female bonds for reproduction. The single young (rarely twins), born after a gestation period of 8 months, is fed, carried, and groomed by its mother for 3–4 years. It also learns her feeding techniques. Common chimpanzees not only use tools but also make them—for example, stripping side branches from a twig, which it uses to scoop out termites from their nest. The two chimpanzee species are our closest living relatives, and their intelligence, range of emotions, and communication and learning skills have made them valuable to animal trainers, collectors, and researchers. They are also killed for the bushmeat trade.

FACIAL FEATURES
The common chimpanzee uses a wide variety of expressions, clearly visible on the hairless facial area. In particular, it uses its flexible, protrusile lips to make grimacelike "smiles" that actually signify fear.

VOCAL CHIMPANZEES

The common chimpanzee makes more than 30 distinct calls, including the pant-hoot (shown right), which consists of shrieks and roars that can be heard up to 1¼ miles (2 km) away. It is used in many situations and is the most common adult sound. It is thought to identify the caller within its community and solicit information from other members.

NIGHT NESTS

Each adult common chimpanzee builds a new, individual tree nest each night for sleeping. (Rarely, an old nest is renovated or reused.) The common chimpanzee bends over and intertwines many branches to make a firm, leafy platform, usually 10–33 ft (3–10 m) above the forest floor, away from ground-based predators. Infants generally sleep in their mothers' nests until they are about 5–6 years of age.

MAMMALS

Gorilla gorilla

Western gorilla

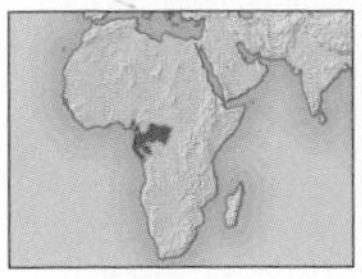

Height 4¼–6¼ ft (1.3–1.9 m)	
Tail None	
Weight 50–430 lb (168–200 kg)	
Social unit Group	
Status Critically endangered	

Location C. Africa

The largest living primates, gorillas are day-active forest dwellers that feed on fruit, leaves, stems, and seeds, as well as a few small creatures such as termites. At night, they bend tree twigs and branches to form a sleeping nest. Living in small, stable groups of 3–20 with strong bonds between the silverback (dominant male) and the females with offspring, the western gorilla uses a wide range of facial expressions, body gestures, and sounds to communicate, including whimpers, grunts, rumbles of contentment, and alarm barks. Its home range of 2,000–4,450 acres (800–1,800 hectares) may overlap with neighboring groups but is not actively defended. The single young is born after a gestation of 8½ months. It clings to its mother's belly for 4–6 months, then rides on her shoulders or back. It first chews vegetation at 4 months and is weaned by 3 years.

SILVERBACK
The adult male gorilla is almost twice the weight of the female, and has a taller bony head crest, longer canine teeth, and a "saddle" of silvery fur.

NERVOUS BEHAVIOR

When nervous, gorillas may "yawn" (above). They usually avoid danger by walking quietly in single file into thick forest—termed "silent flight." In active defense, the silverback barks and stares. If the threat persists, he may charge (see panel, below).

CONSERVATION

The main threats to western gorillas are slash-and-burn clearance of their forest habitat, illegal hunting for the commercial bushmeat trade, and trophy poaching. Almost all gorillas in zoos and parks belong to this species, but reintroductions of captive-bred individuals into the wild are rarely attempted, partly due to the gorilla's complex, close-knit social life. Habitat conservation remains the long-term priority.

Gorilla beringei

Eastern gorilla

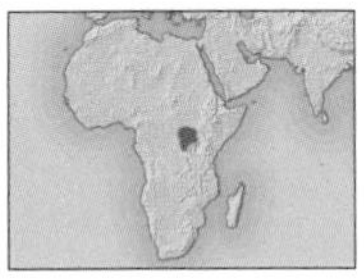

Height 4¼–6¼ ft (1.3–1.9 m)	
Tail None	
Weight 150–460 lb (68–210 kg)	
Social unit Group	
Status Endangered	

Location C. and E. Africa

Previously considered a subspecies of the western gorilla, this species includes eastern lowland gorillas (*Gorilla beringei graueri*), and one or more subspecies of mountain gorillas (such as *Gorilla beringei beringei*). Each group roams a home range of 1,000–2,000 acres (400–800 hectares), which, apart from a central core area, may overlap with neighboring groups. Main foods are leaves, shoots, and stems, especially bamboo; also fruit, roots, soft bark, and fungi. Occasionally, ants are scooped up and swallowed quickly before they bite. At dusk, the group settles to rest—the adult males on the ground, females and young sometimes in nests in trees. Each female shares a nest with her current offspring. The dominant silverback fathers most or all young in the group. He gains the attention of receptive females with mock feeding, hoots, chest-beating, thumping plants, or jump-kicking. A female who has bred, is unlikely to transfer to another male unless her original mate is killed. The offspring remains with its mother until her next birth, after about 4 years. Some eastern gorillas attract tourist income and receive protection, but others are at continued risk from poaching.

GORILLA GROUP
Long-term groups of up to 40 usually comprise one silverback, females, and young. Sometimes brothers, or father and son, stay together, resulting in a multi-male group.

WARM FUR

The mountain subspecies of eastern gorilla has long, shaggy fur to retain body warmth at altitudes of up to 13,200 ft (4,000 m). Only the face, hands, and feet (and the male's chest) lack hair.

DEFENSE DISPLAY

If barking does not deter an attacker (see panel, above), the silverback may begin to hoot. He then stands upright, beats his chest with cupped hands (the bare skin amplifies the sound), and throws vegetation. If this fails, he may charge with a huge roar, and hand-swipe or knock over and bite the invader.

Pongo pygmaeus

Bornean orangutan

Location S.E. Asia (Borneo)

Height 3½–4½ ft (1.1–1.4 m)

Weight 88–175 lb (40–80 kg)

Social unit Individual

Status Endangered

The orangutan is very much a tree-dwelling animal, feeding, sleeping, and breeding in the forest canopy, with only males occasionally coming to ground. It spends most of the day looking for and eating fruit and other food, and at night it builds a sleeping platform by weaving branches together. The female gives birth in a treetop nest, and the tiny infant clings to its mother as she clambers about the canopy. The pair will stay together until the youngster is about 8 years old. Orangutans live in widely scattered communities—probably determined by the availability of food. They are mainly solitary, but may meet up with others at fruit trees, and adolescent females may travel together for 2 or 3 days. All will be aware of neighboring males from their roaring "long-calls." Until fairly recently, orangutans were considered to be a single species—*Pongo pygmaeus*. Genetic research has now led to two distinct species being recognized: the Bornean and the Sumatran (*Pongo abelii*). Loss of habitat presents the main threat to both species as their forest homelands are destroyed by logging and fire. It is estimated that Bornean orangutans now number, 45,000–69,000, while as few as 7,300 Sumatran orangutans survive.

CONSERVATION

Although the orangutan is protected by law, infants are still captured and sold illegally as pets. Projects to rehabilitate rescued orangutans—adults as well as juveniles—have a good success rate, but some animals find it difficult to readjust to life in their natural habitat—and in any case, this is rapidly being destroyed.

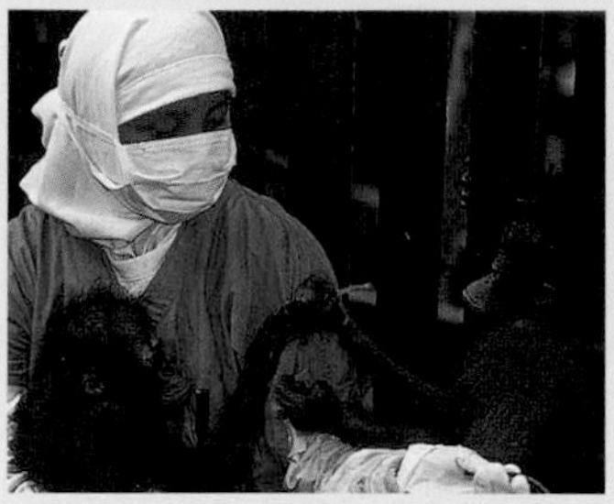

"PERSON OF THE FOREST"
The male orangutan—which means "person of the forest" in Malay—looks strikingly different from the female, with large cheek pads (which grow bigger as the animals age), a long beard and mustache, and a hanging throat pouch. He also has long arm hair, which hangs like a cape when the arms are outstretched.

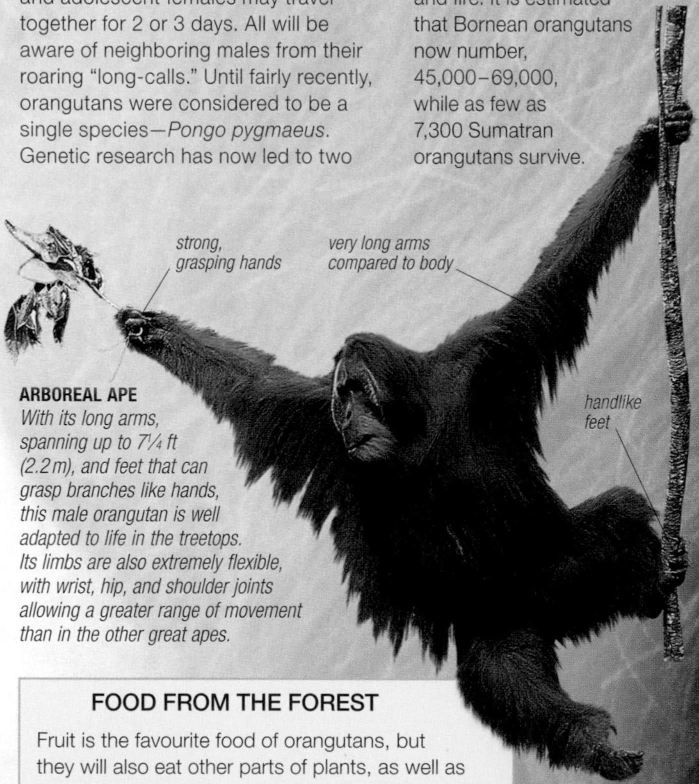

ARBOREAL APE
With its long arms, spanning up to 7¼ ft (2.2 m), and feet that can grasp branches like hands, this male orangutan is well adapted to life in the treetops. Its limbs are also extremely flexible, with wrist, hip, and shoulder joints allowing a greater range of movement than in the other great apes.

FOOD FROM THE FOREST

Fruit is the favourite food of orangutans, but they will also eat other parts of plants, as well as honey, small animals such as lizards, termites and nestling birds, and eggs.

FAMILY GROUP
A mother and her offspring forage together in the forest, plucking fruit and leaves from the trees.

EATING HABITS
Orangutans use their hands and teeth to prepare their food, stripping plants and peeling fruit to expose the succulent flesh.

MAMMALS

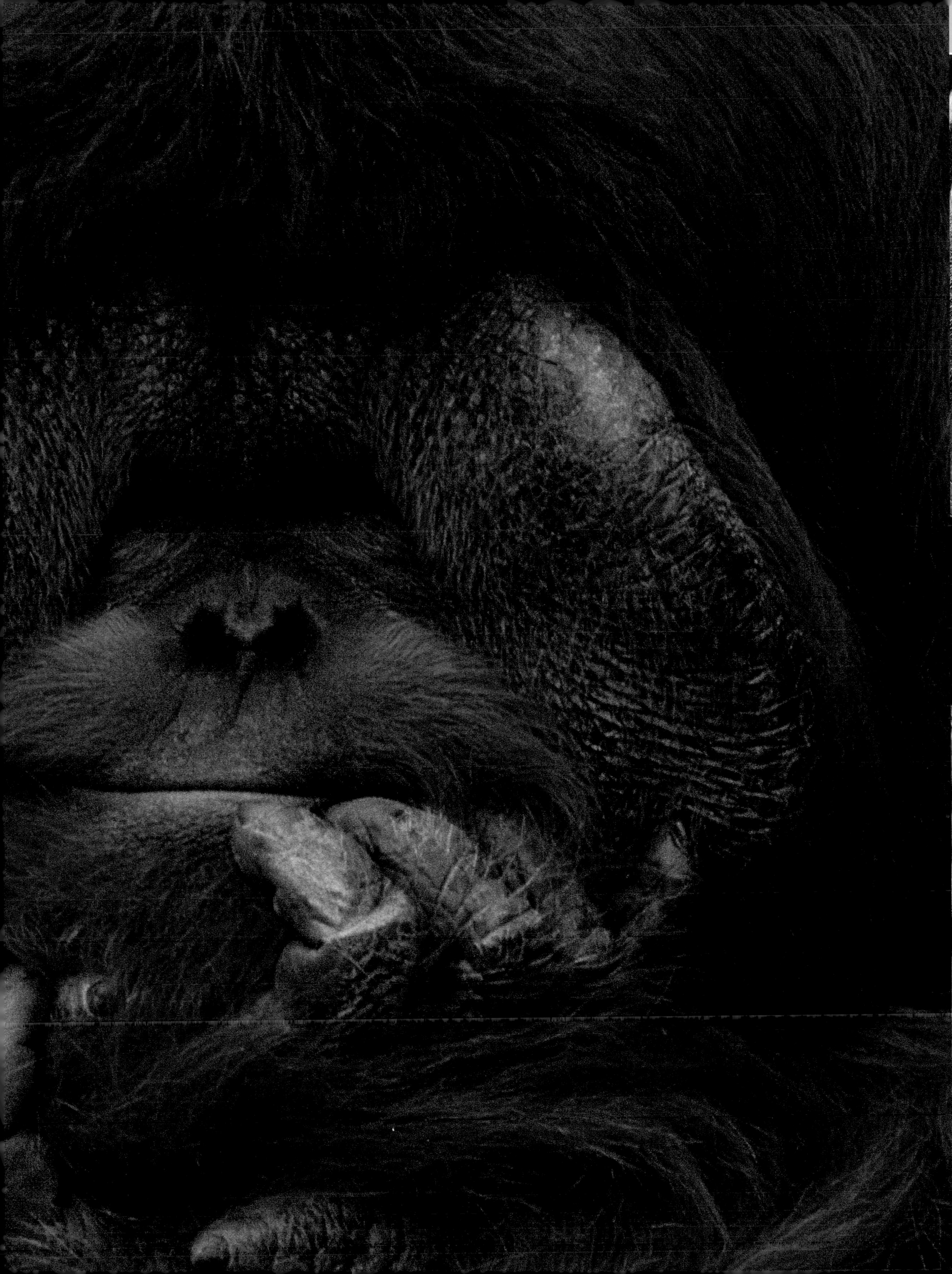

Bats

PHYLUM	Chordata
CLASS	Mammalia
ORDER	Chiroptera
FAMILIES	18
SPECIES	1,117

CLASSIFICATION NOTE

Chiroptera has 2 suborders: Megachiroptera, comprising one family (Pteropodidae), and Microchiroptera (all other bats). Megachiropterans have a foxlike face that lacks features for echolocation (see below).

Bats are the only mammals that possess true, flapping wings and the ability to fly (as opposed to colugos, for example, which glide). Bats' wing membrane (the patagium), an extension of the skin of the back and belly, provides a high degree of maneuverability in flight. Wingspans range from over 5 ft (1.5 m) in the large flying fox to as little as 6 in (15 cm) in the hog-nosed bat. Over half the species echolocate (see below) to capture prey and to navigate at night. Chiroptera is a huge order that comprises nearly a quarter of all mammal species and is exceeded only by rodents in terms of species numbers. Bats are common in tropical and temperate habitats worldwide but are not found in environments that are too cold to support a source of food, such as the polar regions.

Anatomy

Perhaps the most distinctive feature of bats is their wings, which are formed from a double layer of skin stretched between the side of the body and the 4 elongated fingers on each hand. Blood vessels and nerves run between these 2 layers. Extra support is required for the arms to be used as wings, and this is provided by features such as fused vertebrae, flattened ribs, and a strong collarbone. The sternum (breastbone) has a central ridge to which the large muscles used in the downward stroke of the wing are attached. A short, clawed thumb is present in most species at the point where the fingers join, and a cartilaginous spur (the calcar) on the inside of the ankle joint assists in spreading the tail membrane.

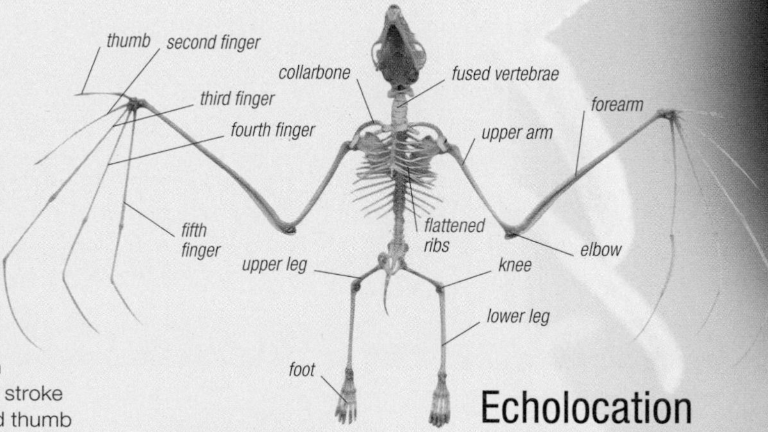

SKELETAL FEATURES
A bat's arms, legs, and greatly elongated fingers provide the framework for the wings. Bats' legs have been rotated 180 degrees so that the knee and the foot bend in the opposite direction to the knee and back foot of other mammals.

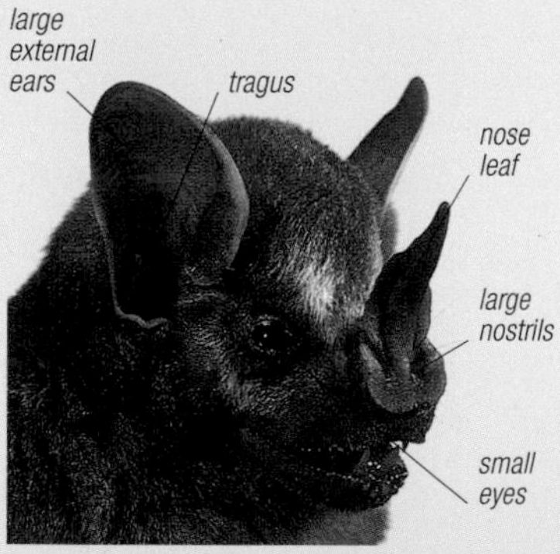

EQUIPPED FOR ECHOLOCATION
Although bats' eyes are well developed, hearing and the sense of smell are more important than sight. Many microchiropterans, such as this neotropical fruit bat, also have a large nose leaf, which assists echolocation. The function of the tragus (a lobe in the front of the ear) is uncertain: it may improve the accuracy of echolocation. Megachiropteran bats generally do not echolocate and have larger eyes (for detecting prey) and smaller ears.

Echolocation

All microchiropterans echolocate. When used in flight, this navigation system makes them formidable hunters. Sounds ("clicks") are produced in the larynx, emitted through the nose or mouth, and directed or focused by the nose leaf (if present). Once the clicks have reflected off an object, the returning echo is picked up via the bat's sensitive ears. The time it takes to receive the echo reveals the size and location of anything in the bat's path.

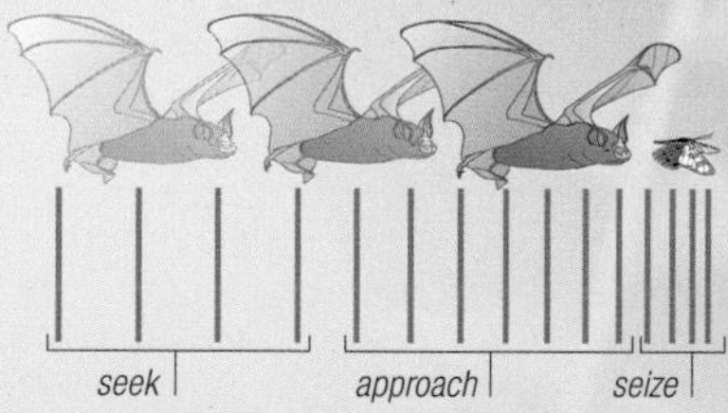

PHASES OF ECHOLOCATION AS A BAT HUNTS AN INSECT

USING SOUND TO FIND PREY
All of the insect-eating bats use echolocation to find airborne prey. When searching for food, such as mosquitoes and moths, the bat emits a series of clicks, represented by red bars on this diagram. As the bat approaches its prey, the time between "clicks" shortens. This helps the bat to pinpoint its target.

CATCHING PREY FROM THE AIR

CLOSING IN ON THE PREY
Using echolocation, this fisherman bat has pinpointed a small surface-swimming fish.

CAPTURE
In one swift movement, the bat catches the fish by raking the water with its long, sharp claws.

HOLDING ON
The bat quickly transfers the fish from its claws to its mouth, so that the fish cannot escape.

FINDING A PLACE TO FEED
In order to eat the fish, the bat must first locate a tree on which to land.

Foraging and diet

Many bats eat insects: some species forage for them among shrubs and trees, while others skim the surface of the forest canopy to catch higher-flying insects. A single bat may eat hundreds of mosquitoes in one night (thereby lessening the incidence of malaria in other animals). Other bats eat fruit, and some use their long tongue to feed on pollen and nectar. Vampire bats use their sharp teeth to make a small incision in the skin of an animal while it sleeps and then drink the blood. Carnivorous bats prey on lizards and frogs; fish-eating bats use the hooked claws on their powerful feet to capture fish (see below).

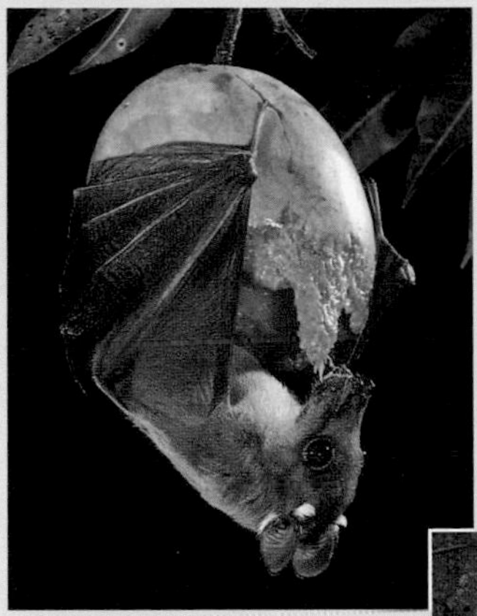

FEEDING ON FRUIT
Most fruit-eating bats use their senses of sight and smell to find food and so lack facial ornaments used in echolocation. This epauletted fruit bat is eating a mango: a group of these bats in a plantation can cause considerable damage. Since fruit bats require a constant supply of ripe fruit, they are found mostly in tropical areas. Fruit bats often feed in groups and fly long distances in search of food.

FEEDING ON BLOOD
Vampire bats are well adapted to feed on blood. They have sharp incisors to cut into flesh, and produce saliva that prevents the blood from clotting. This white-winged vampire bat commonly feeds on the blood of chickens.

CONSERVATION

Despite being rich in terms of species, bats face a huge variety of threats. Globally, one of the most serious is habitat loss, particularly in forested regions. Fruit-eating bats are also persecuted as pests, and are killed by colliding with power lines, while insect-eating bats are threatened by white nose disease—a potentially deadly fungal pathogen that first came to light in North America in 2006. Bats are now routinely fit with microchips and radio collars that allow them to be tracked. This fruit bat (below) will broadcast for several months, showing exactly where it roosts and feeds.

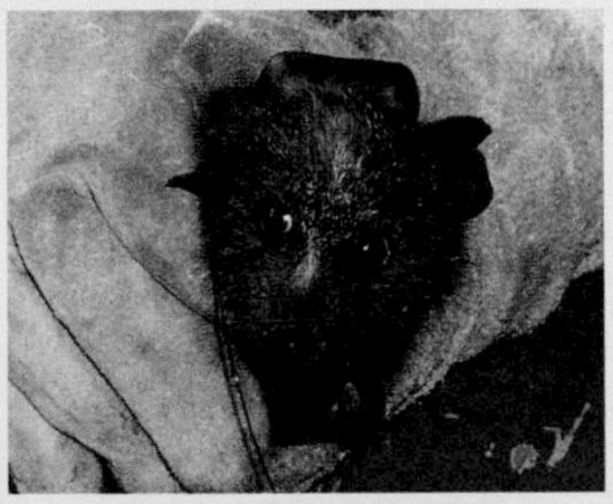

FLYING MAMMAL
Bats, the only mammals that can fly, have structural adaptations that allow them to make up-and-down movements of their wings (in the same way as birds). The open wings of this New World leaf-nosed bat reveal the extent of the wing membranes.

Roosting

Bats often gather in great numbers at a single site, which may be a cave, the roof of an old building, or a hollow tree. All roosting sites must provide a resting place that offers protection from predators, the heat of the sun, the low temperatures of winter (hibernating roosts), and rain. Bamboo bats are small enough to roost in the hollow stems of a plant, while some species of leaf-nosed bats bite into leaf stems so that the leaf droops downward, forming a tent around them. Why bats gather in such large numbers is not fully understood; however, at the end of hibernation, bats living in colonies often weigh more than species that do not.

A PLACE TO REST
Bats, such as these fruit bats, commonly roost in caves during the day, emerging at dusk to feed. Some species use the same roost for many years and gather in groups of many thousands.

...G
...bat attaches itself to a tree (upside down), ...nues to hold the fish firmly in its mouth.

DEVOURING THE PREY
The fish is eaten head first. The bat may use its wings to manipulate the food.

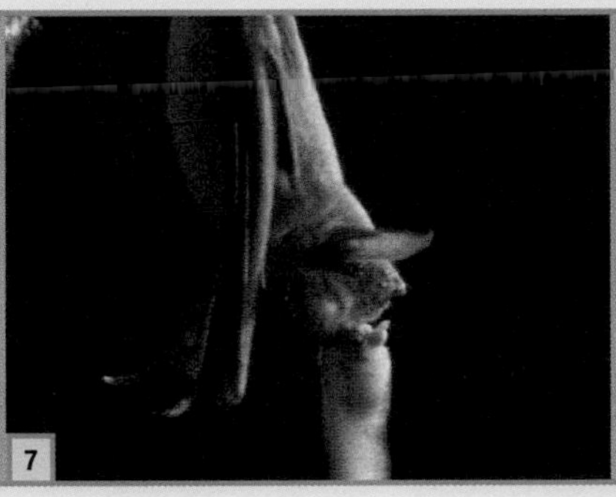

ROOM FOR PLENTY
The bat has highly elastic cheeks, which can be extended during feeding.

ALMOST GONE
With its meal nearly finished, the fisherman bat will soon begin the hunt for more food.

Rousettus aegyptiacus

Egyptian rousette

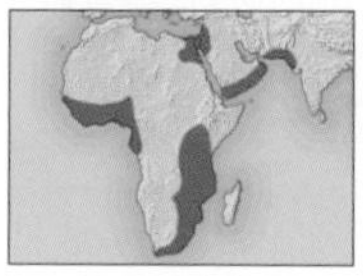

Length 5½–6½ in (14–16 cm)	
Tail ½–¾ in (1.5–2 cm)	
Weight 2⅞–3⅝ oz (80–100 g)	
Social unit Group	
Status Least concern	

Location W. Asia, N. Africa (Egypt), W., E., and southern Africa

Widespread and adaptable, these fruit- and leaf-eating bats are sometimes so common that they reach pest status and damage farm crops. They are also the only fruit bat species to use the high-pitched "clicks" of echolocation. This means they can find their way around and roost in dark caves, rather than sleeping in trees like other fruit bats.

FUR-COVERED FOREARMS
The Egyptian rousette varies from dark brown to slate gray on the back, with lighter, smoky gray underparts. Unusually, its fur extends about halfway along each forearm.

Stylocterium mindorensis

Mindoro stripe-faced fruit bat

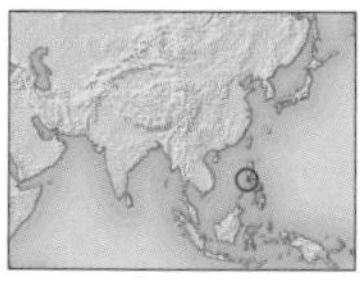

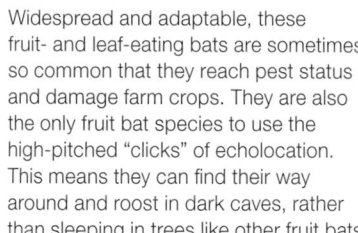

Length 6–7 in (15–18 cm)	
Tail None	
Weight ¼–½ lb (150–210 g)	
Social unit Group	
Status Data deficient	

Location Philippines

With its orange coat and white-striped face, this stunningly colored fruit bat was discovered in 2006 in lowland forest on Philippines' Mindoro Island—following local rumors of its existence. It is found nowhere else though there is speculation that Western Australian aboriginal paintings depict this bat or a closely related species. Currently, the only other member of its genus is known with certainty from Sulawesi. Little is known about Mindoro's stripe-faced fruit bat, and the species is threatened by deforestation and hunting.

Epomops franqueti

Franquet's epauletted bat

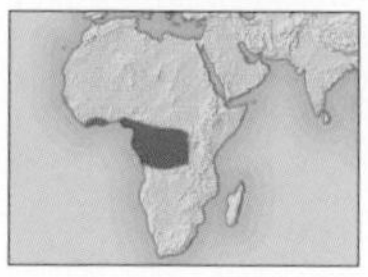

Length 4¼–6 in (11–15 cm)	
Tail None	
Weight 3–3⅝ oz (85–100 g)	
Social unit Group	
Status Least concern	

Location W. and C. Africa

At night, male Franquet's epauletted bats make monotonous high-pitched whistling calls to attract females for mating. The male is slightly heavier than the female and has shoulder patches of long, pale hairs. This species breeds at any time, and twice yearly if guavas, bananas, other fruit, and soft, young leaves are abundant.

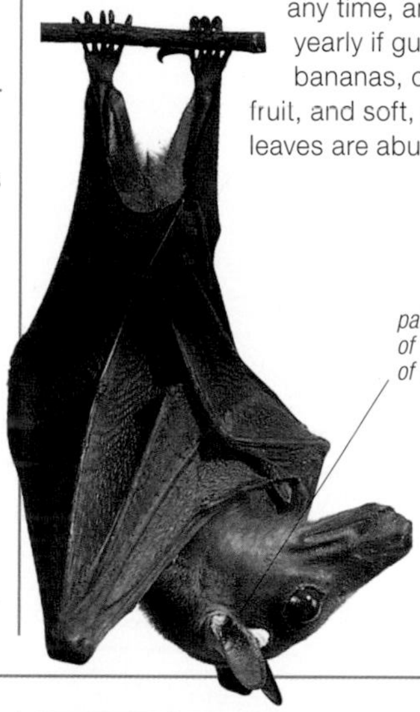

Epomophorus wahlbergi

Wahlberg's epauletted fruit bat

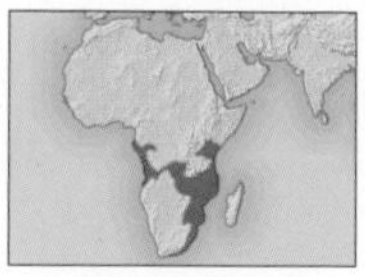

Length 4¾–6¼ in (12–15.5 cm)	
Tail None	
Weight 2⅜–4 oz (65–125 g)	
Social unit Variable	
Status Least concern	

Location E., C., and southern Africa

In addition to white "epaulette" fur patches on the male's shoulders, both sexes have 2 white patches at the base of each ear. These pale tufts may be disruptive camouflage, breaking up the bat's outline when seen from below against dappled leaves. In the breeding season, the male's distinctive call to attract females resembles a squeaky bicycle pump.

Pteropus rodricensis

Rodrigues flying fox

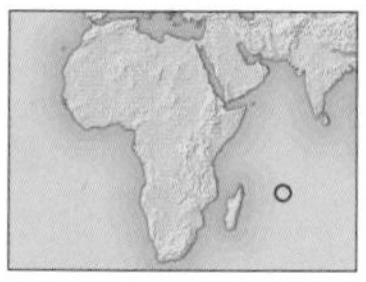

Length 14 in (35 cm)	
Tail None	
Weight 9–10 oz (250–275 g)	
Social unit Group	
Status Critically endangered	

Location Indian Ocean (Rodriguez Island)

Formerly, the daytime roosts or "camps" of this flying fox contained more than 500 individuals. Due to habitat loss, through storm damage and human intervention, and also local hunting for food, the species currently numbers about 4,000 in the wild. Several centers, however, have established successful captive breeding programs. At night, the bats forage in dry woodland for fruit of various trees, such as tamarinds, rose-apples, mangoes, palms, and figs. Like many other fruit bats, they squeeze out the juices and soft pulp, rarely swallowing the harder parts. Observations in captivity show that each dominant male gathers a harem of up to 10 females, with which he roosts and mates. Subordinate and immature males tend to roost in another part of the camp.

Rhynchonycteris naso

Proboscis bat

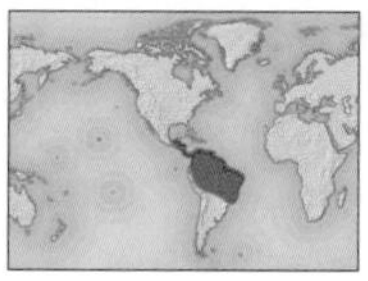

Length 1½–2 in (3.5–5 cm)	
Tail ⅜–½ in (1–1.5 cm)	
Weight ⅛–7/32 oz (3–6 g)	
Social unit Group	
Status Least concern	

Location Mexico to C. South America

Also called the sharp-nosed bat, this species has a long, pointed nose and streamlined appearance. It is a typical small, insectivorous bat, but shows unique roosting behavior as groups of 5–10 (rarely more than 40) rest by day in a line, nose-to-tail on a branch or wooden beam. One adult male may dominate the group and may defend their feeding area—a nearby patch of water where they catch small insects.

Taphozous mauritianus

Mauritian tomb bat

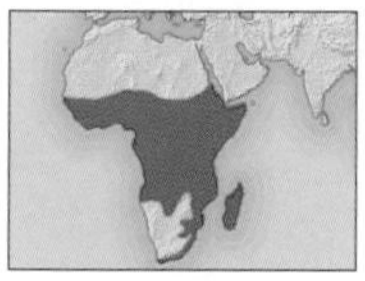

Length 2¾–3½ in (7.5–9.5 cm)	
Tail ¾–1¼ in (2–3 cm)	
Weight 9/16–1 1/16 oz (15–30 g)	
Social unit Group	
Status Least concern	

Location W., C., E., and southern Africa, Madagascar

A member of the sheath-tailed group, the Mauritian tomb bat is known across Africa for its wide range of clicks, squeaks, and other noises just audible to some humans. It also makes ultrasonic sounds for echolocation. This active bat is watchful as it roosts by day in the open on tree trunks and walls, including town buildings. It hunts in clearings, especially over water, for flying insects. The back is grizzled brown-black; the underparts and wings are white.

Macroderma gigas

Ghost bat

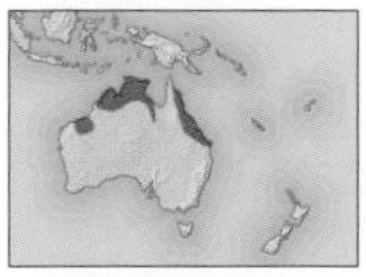

Location W. and N. Australia

Length 4–4¾ in (10–12 cm)
Tail None
Weight 2⅝–5 oz (75–150 g)
Social unit Group
Status Vulnerable

This species is also known as the Australian false vampire bat after the mistaken belief that it feeds on blood. One of the largest microchiropterans, it preys on insects, birds, lizards, and other bats. Its decline may be partly due to the increasing use of its rocky roosting sites for mines and quarries.

Thyroptera tricolor

Spix's disk-winged bat

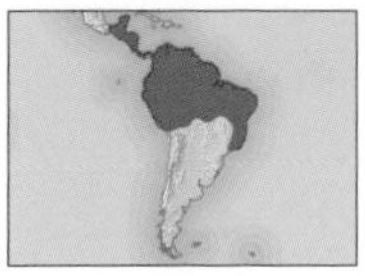

Location Mexico to Central America, South America, Trinidad

Length 1½ in (4 cm)
Tail 1–1½ in (2.5–3.5 cm)
Weight ⅛–3/16 oz (3–5 g)
Social unit Group
Status Least concern

The disk-winged or sucker-footed bat has a rounded, suckerlike structure near each thumb claw in the middle front edge of the wing, and a smaller one on the sole of each foot. Like tiny suction cups, these grip smooth, glossy leaf surfaces so that the bat can shelter within partly furled leaves—roosting head up, unlike nearly all other bats. The grip of a single sucker is sufficient to bear the bat's weight, which averages 0.14 oz (4 g). Alternatively, it rests with others of its kind in small groups of up to 10, among unfurled leaves. Spix's disk-winged bat is the smallest of 4 New World disk-winged species, and feeds on small creatures, many of them nonflying, such as jumping spiders. It is slim and delicate, with a dark or reddish brown back and whitish brown or yellow underparts. Females are slightly larger than the males.

Rhinopoma hardwickei

Lesser mouse-tailed bat

Location W. to S. Asia, N. and E. Africa

Length 2½–2¾ in (5.5–7 cm)
Tail 1¾–3 in (4.5–7.5 cm)
Weight ⅜–9/16 oz (10–15 g)
Social unit Group
Status Least concern

Also called long-tailed bats, the 5 *Rhinopoma* species are the world's only small, insectivorous bats with thin, trailing tails. The tail may be as long as both head and body. This species lives in scrub, semidesert, and tropical forest. When food is plentiful, it may double its body weight, storing fat for several weeks of dry-season inactivity.

Pteronotus davyi

Davy's naked-backed bat

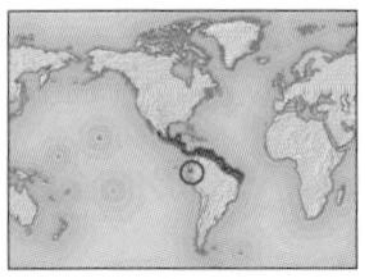

Location Mexico to N. and E. South America

Length 1½–2½ in (4–5.5 cm)
Tail ¾–1 in (2–2.5 cm)
Weight 3/16–⅜ oz (5–10 g)
Social unit Group
Status Least concern

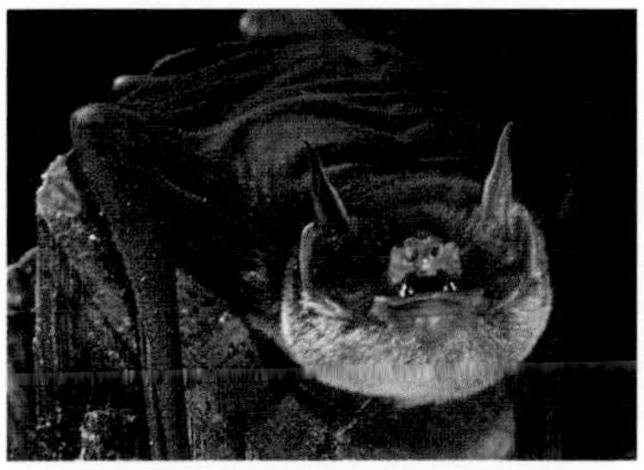

This bat is a common sight at night near towns, as it feeds on flies, moths, and other insects attracted to streetlights. By day, it roosts in large colonies in caves and old mines, often some distance away from its feeding areas. Davy's naked-backed bat has wings that join along the center of the back, obscuring the fur beneath. Bats within this genus are also known as moustached or leaf-lipped bats.

Rhinolophus hipposideros

Lesser horseshoe bat

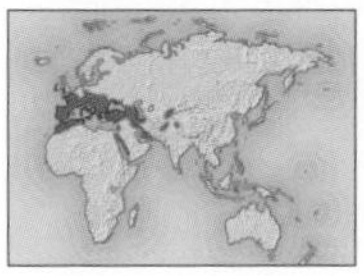

Location Europe, N. Africa to W. Asia

Length 1½ in (4 cm)
Tail ¾–1½ in (2–3.5 cm)
Weight 5/32–⅜ oz (4–10 g)
Social unit Group
Status Least concern

This diminutive bat is widespread in woods and scrub. Despite being classified as least concern, the species is at risk. Its underground, winter hibernation sites, such as deep caverns, have been disturbed, as have summer day roosts in tree holes, caves, chimneys, and mine shafts. Domestic cats take a heavy toll and destruction of woods and hedges has reduced the availability of its prey of small flying insects. The lesser horseshoe is one of the smallest of the 74 species of horseshoe bats.

Hipposideros speoris

Schneider's leaf-nosed bat

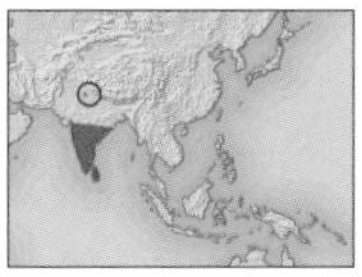

Location S. Asia

Length 1¾–2½ in (4.5–6 cm)
Tail ¾–1¼ in (2–3 cm)
Weight 11/32–7/16 oz (9–12 g)
Social unit Group
Status Least concern

A medium-sized member of this genus of about 69 species, this bat is a typical, small insect-eater. It has a flaplike "leaf" on the upper muzzle around the nostrils, with a U-shaped part below. By day, thousands roost in caves, tunnels, and buildings.

Nycteris grandis

Large slit-faced bat

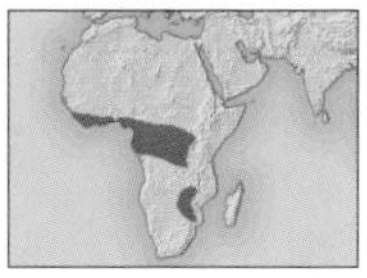

Location W., C., E., and southern Africa

Length 2¾–3¾ in (7–9.5 cm)
Tail 2½–3 in (6.5–7.5 cm)
Weight ⅞–1 7/16 oz (25–40 g)
Social unit Group
Status Least concern

This slit-faced bat has a furrow down the face. This may be partly covered by nose "leaves," so it looks like 2 slits running from nostrils to eyes. A powerful species, it swoops onto other bats, birds, scorpions, sun-spiders, frogs, and even fish near the surface. By day, it roosts in groups of up to 60 in trees, caves, and buildings.

Noctilio leporinus

Greater bulldog bat

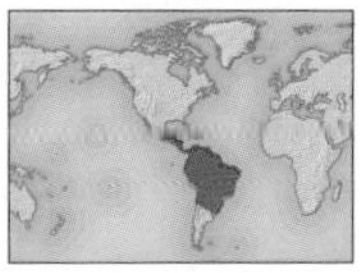

Location Central America, N., E., and C. South America

Length 2½–3 in (6–8 cm)
Tail ½–¾ in (1.5–2 cm)
Weight 9/16–1¼ oz (15–35 g)
Social unit Group
Status Least concern

Also called fisherman bats, the two species of bulldog bats have large nose pads, drooping upper lips, and ridged chins. The greater bulldog bat has velvety fur—orange, brown, or gray—with a distinctive pale stripe along the middle of the back. It roosts by day in hollow trees or caves. At night, it hunts over water or sandy beaches for fish, crabs, and other prey, which it snatches from the ground or water using its large and powerful, sharp-clawed back feet.

Trachops cirrhosus

Fringe-lipped bat

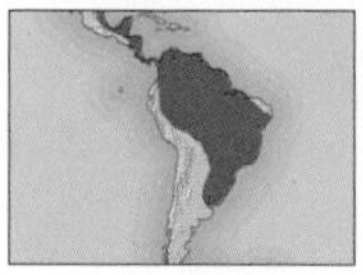

Location Mexico to N. and C. South America

Length $2\frac{1}{2}$–$3\frac{1}{2}$ in (6.5–9 cm)
Tail $\frac{3}{8}$–$\frac{3}{4}$ in (1–2 cm)
Weight $\frac{7}{8}$–$1\frac{1}{4}$ oz (25–35 g)
Social unit Group
Status Least concern

Why this broad-winged, strong-flying bat's lips are studded with papillae (small, wartlike bumps) is not clear. Also known as the frog-eating bat, it hunts along streams, ditches, and similar waterways, killing prey such as insects, frogs, and lizards with its powerful bite. It locates its victims by hearing their sounds, such as the croaking of male frogs, rather than by its own echolocation. In other respects it is a typical bat, roosting in tree holes, hollow logs, and caves.

Anoura geoffroyi

Geoffroy's tailless bat

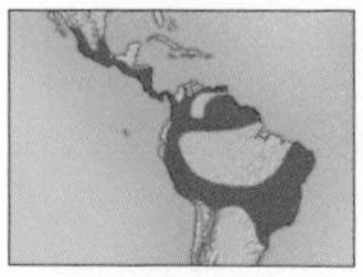

Location Mexico to N. South America

Length $2\frac{1}{2}$–$2\frac{3}{4}$ in (6–7.5 cm)
Tail Up to $\frac{7}{32}$ in (7 mm)
Weight $\frac{7}{16}$–$\frac{5}{8}$ oz (13–18 g)
Social unit Group
Status Least concern

The small, fur-covered tail membrane of this bat gives it the appearance of hairy legs. It also has a small, triangular, upright nose leaf and a long muzzle with protruding lower jaw. These features give it its alternative names of Geoffroy's hairy-legged or long-nosed bat. It hovers in front of night-blooming flowers to sip nectar and gather pollen with its unusual brush-tipped tongue, which is the length of its head. Geoffroy's tailless bat also eats insects, such as beetles or moths. It roosts in caves and tunnels.

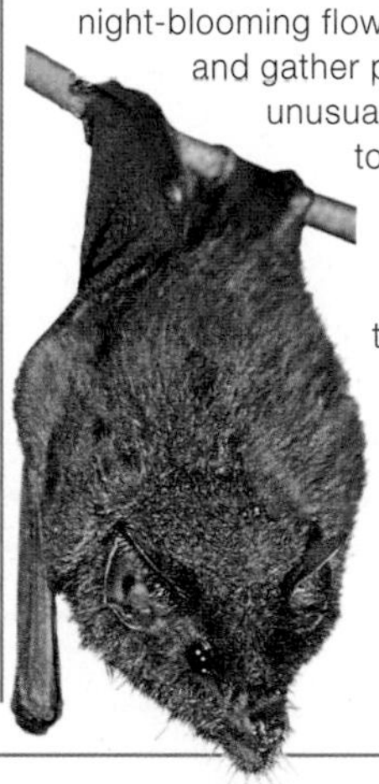

Desmodus rotundus

Common vampire bat

Location Mexico to South America

Length $2\frac{3}{4}$–$3\frac{3}{4}$ in (7–9.5 cm)
Tail $2\frac{1}{2}$ in (6.5 cm)
Weight $\frac{11}{16}$–$1\frac{5}{8}$ oz (19–45 g)
Social unit Group
Status Least concern

The common vampire bat is a strong flier, yet it can also scuttle over the ground with amazing speed and agility, propped up on its forearms and back legs. From dusk, it searches for a warm-blooded victim, such as a bird, tapir, or farm animal—even a seal or a human. The bat lands nearby, crawls closer, bites away any fur or feathers, and laps some 1 fl oz (25 ml) of blood over 30 minutes, its saliva preventing clotting. This bat has a communal roost in a hollow tree, cave, mine, or old building, which it shares with hundreds of others.

RAZOR FANGS

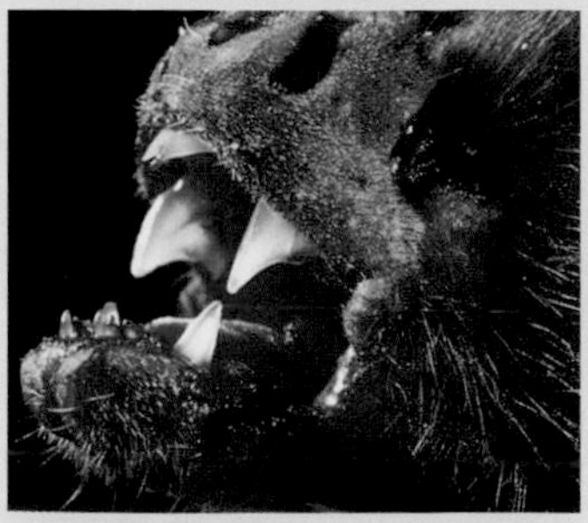

The common vampire bat's thin, pointed, bladelike upper incisors are so sharp that its victim rarely notices as they slice away a piece of flesh about $\frac{3}{16}$ in (5 mm) across.

COLORATION
The fur is dark brownish gray; the underparts are paler, with a buff tinge.

Pipistrellus pipistrellus

Common pipistrelle

Location Europe to N. Africa, W., and C. Asia

Length $1\frac{1}{2}$–2 in (3.5–4.5 cm)
Tail $1\frac{1}{4}$–$1\frac{1}{2}$ in (3–3.5 cm)
Weight $\frac{1}{8}$–$\frac{5}{16}$ oz (3–8 g)
Social unit Group
Status Least concern

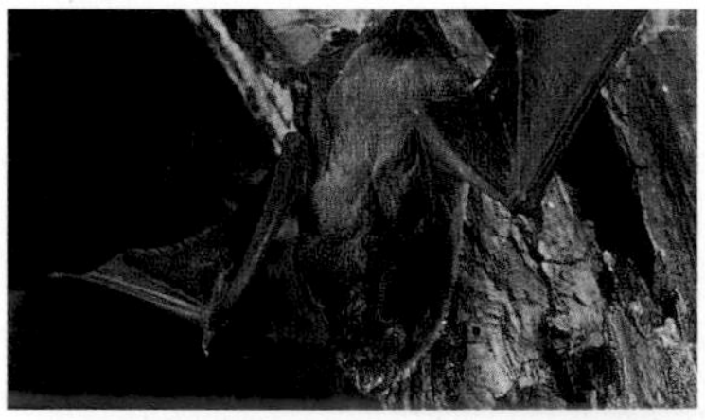

The genus *Pipistrellus* includes almost 62 similar species, of which the common pipistrelle is one of the smallest and most widespread, found in habitats from forest to city parks. It is among the first bats to emerge each evening in pursuit of small flying insects. It roosts by day in crevices, buildings, and bat boxes, and hibernates through winter in similar sheltered places. Nursery colonies may contain up to 1,000 mothers, each with a single young.

Nyctalus noctula

Noctule

Location Europe to W., E., and S. Asia

Length $2\frac{3}{4}$–3 in (7–8 cm)
Tail 2–$2\frac{1}{4}$ in (5–5.5 cm)
Weight $\frac{9}{16}$–$1\frac{3}{4}$ oz (15–50 g)
Social unit Group
Status Least concern

The most widespread of the 8 noctule species, this bat flies high and powerfully before diving steeply to grab flying insects as large as crickets and chafers. By day, it roosts, usually alone, in any available small hollow—for example, in a tree, building, or among rocks. The noctule migrates 1,200 miles (2,000 km) or more between its winter and summer sites. In spring, the female may produce 3 young, in contrast to the single offspring of most small bats.

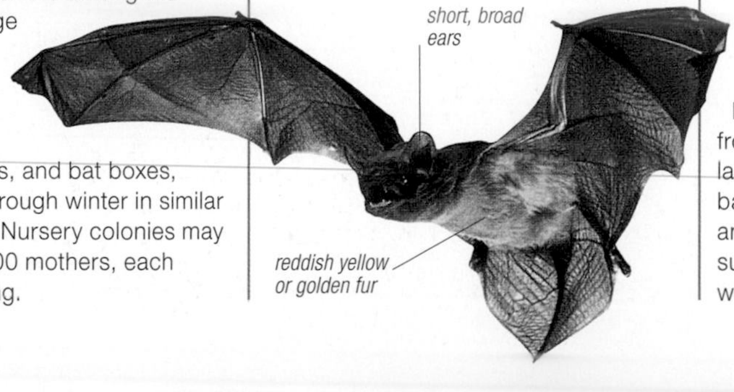

Uroderma bilobatum

Tent-building bat

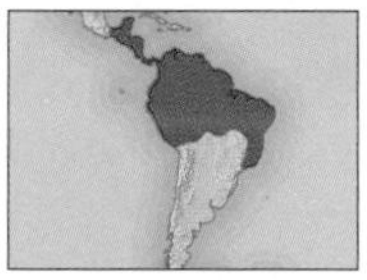

Location Mexico to C. South America, Trinidad

Length $2\frac{1}{4}$–$2\frac{1}{2}$ in (6–6.5 cm)
Tail $1\frac{1}{2}$–$1\frac{3}{4}$ in (4–4.5 cm)
Weight $\frac{7}{16}$–$\frac{11}{16}$ oz (13–20 g)
Social unit Group
Status Least concern

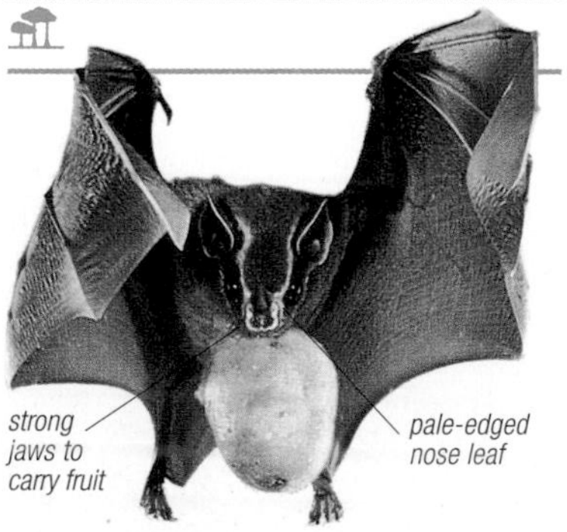

About 15 bat species from the leaf-nosed group roost in "tents"—shelters shaped like umbrellas, cylinders, cones, or flasks, and made by biting leaves such as palm or banana so they droop or fold over. Each tent protects 2–50 or more bats from sunlight, rain, and predators, and lasts up to 3 months. The tent-building bat feeds on a variety of leaves and fruit, chewing them to a pulp and sucking out the juices. It is gray-brown with white stripes on its face and back.

Vampyrum spectrum

False vampire bat

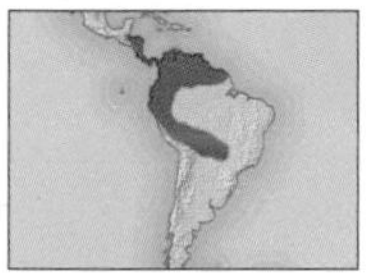

Location Mexico to N. South America, Trinidad

Length $5\frac{1}{4}$–6 in (13.5–15 cm)
Tail None
Weight 6–7 oz (150–200 g)
Social unit Variable
Status Near threatened

Also called Linnaeus' false or spectral vampire, the $3\frac{1}{2}$ ft (1 m) wingspan makes this by far the largest bat in the Americas. As the name suggests, it is not a bloodsucker, but is a powerful predator near the top of the food web and so at particular risk from habitat loss. It hunts other bats, small rodents such as mice and rats, and birds such as wrens, orioles, and parakeets. By day, it roosts in hollow trees in groups of up to 5 individuals.

MAMMALS

Molossus ater

Red mastiff bat

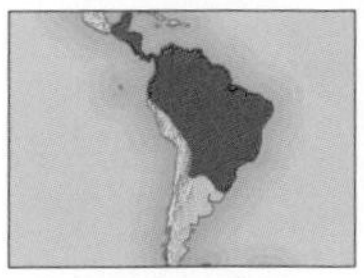

Location Mexico to C. South America, Trinidad

Length 2¾–4 in (7–10 cm)
Tail 1½–2 in (4–5 cm)
Weight 1 1/16–1 7/16 oz (30–40 g)
Social unit Group
Status Least concern

Medium in size, mastiff bats are also called free-tailed velvety bats for their short, soft fur. This species often roosts in buildings and feeds on insects attracted to streetlights. It is more active around dawn and dusk than most bats, and roosts in the middle of the night as well as by day. Huge numbers of insects are stored in its cheek pouches and only chewed and swallowed on return to the roost.

Mops condylurus

Angolan free-tailed bat

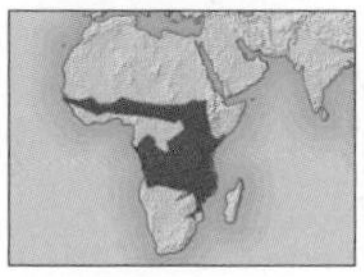

Location W., C., E., and southern Africa

Length 2¾–3 in (7–8.5 cm)
Tail 1½ in (4 cm)
Weight ⅝–1¼ oz (18–35 g)
Social unit Group
Status Least concern

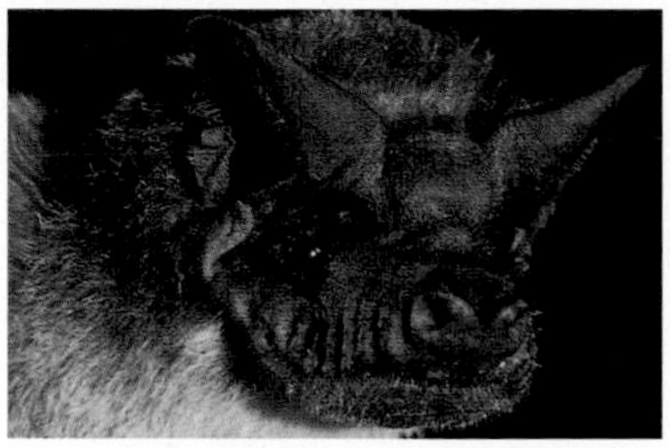

Common and widespread from deserts to rain forests, this bat has a long, mouselike, "free" tail, not enclosed by the tail membranes. It displays the typical bat habit of emerging from its daytime roost in noisy, flapping groups, thereby lessening each individual's risk of being caught by a predator, such as an owl, hawk, or snake. It hunts flying insects, eating them in the air and dropping hard parts such as the legs.

Otonycteris hemprichii

Hemprich's long-eared bat

Location N. Africa to W. Asia

Length 2½–2¾ in (6–7 cm)
Tail 1⅞–2 in (4.7–4.9 cm)
Weight 11/16–1 1/16 oz (20–30 g)
Social unit Group
Status Least concern

Also known as the desert long-eared bat, this is one of the few bat species content in dry, barren habitats. During periods of harsh weather, such as drought, it probably enters a period of hibernation-like inactivity. Its spectacular ears are 1½ in (4 cm) long and often held almost horizontally in flight, such as when swooping onto the ground for insects and spiders. This bat roosts by day in groups of at least 20 in a crevice, cave, or building.

Plecotus auritus

Brown long-eared bat

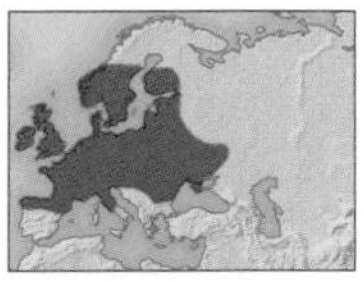

Location Europe, C. Asia

Length 1½–2 in (4–5 cm)
Tail 1½–2 in (4–5 cm)
Weight ¼–½ oz (7–14 g)
Social unit Group
Status Least concern

A compact face and relatively large ears identify the 11 *Plecotus* species of Old World bats. The fur is usually brownish gray, the face darker. This species takes a variety of insects, including moths and beetles, carrying the meal to a perch to eat. Its other habits are typical of small, insect-eating bats, with winter hibernation in caves, mines, and cellars.

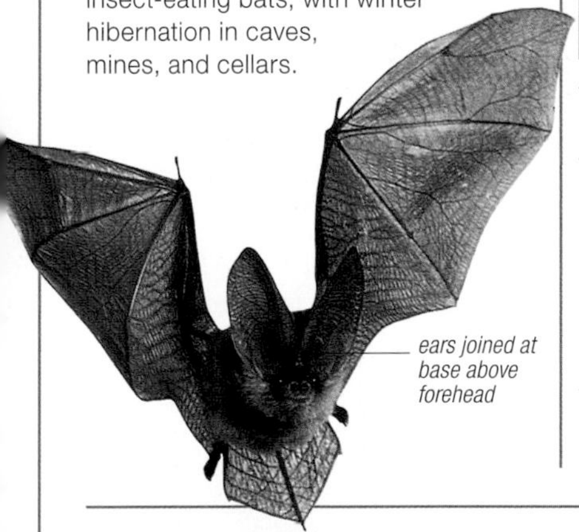

Antrozous pallidus

Pallid bat

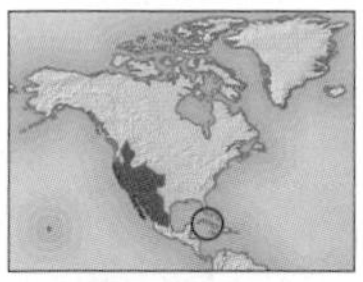

Location W. North America to Mexico, Cuba

Length 2½–3 in (5.5–8 cm)
Tail 1½–2¼ in (3.5–5.5 cm)
Weight ½–1 1/16 oz (14–30 g)
Social unit Group
Status Least concern

This medium-sized, pig-nosed, pale bat tolerates a range of dry habitats, from grassland to scrubby desert and even the intense heat of California's Death Valley. It detects victims by the sounds they make, consuming beetles, crickets, spiders, centipedes, scorpions, lizards, and pocket mice. The pallid bat utters piercing directive cries audible to humans as it "rallies," flying in groups to locate its roost in rocky outcrops, trees, or attics.

Natalus stramineus

Lesser Antilles funnel-eared bat

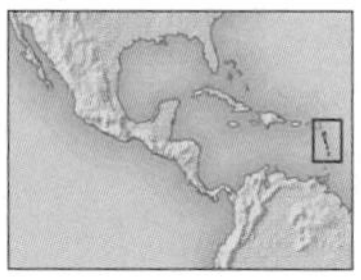

Location Caribbean

Length 1½–1¾ in (4–4.5 cm)
Tail 1¾–2 in (4.7–5.2 cm)
Weight ⅛–3/16 oz (3–5 g)
Social unit Group
Status Least concern

The 5 to 8 species of tropical American funnel-eared bats are tiny and delicate with rounded ears, soft woolly fur, and a tail joined by flight membranes to the legs. The Lesser Antilles funnel-eared bat has a rapid, agile flight, almost like a butterfly. It eats small flying insects and roosts by day in caves.

Myotis daubentonii

Daubenton's bat

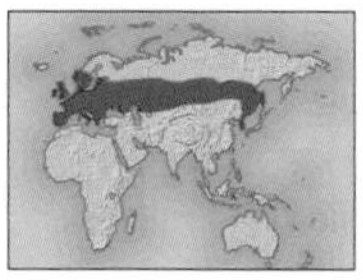

Location Europe to N. and E. Asia

Length 1½–2½ in (4–6 cm)
Tail 1–2 in (2.5–5 cm)
Weight 3/16–9/16 oz (5–15 g)
Social unit Group
Status Least concern

Daubenton's is one of about 102 species in the widespread bat genus *Myotis*—small brown or mouse-eared bats. It flutters 3¼–6½ ft (1–2 m) above water to catch flying insects by mouth or in the pouch of its curled wing or tail membrane—hence its other name of water bat. It also skims the surface and grabs small fish in its large back feet. By day, this bat roosts in trees, buildings, old walls, and bridges. It flies up to 180 miles (300 km) to its winter hibernation site in a cave or mine.

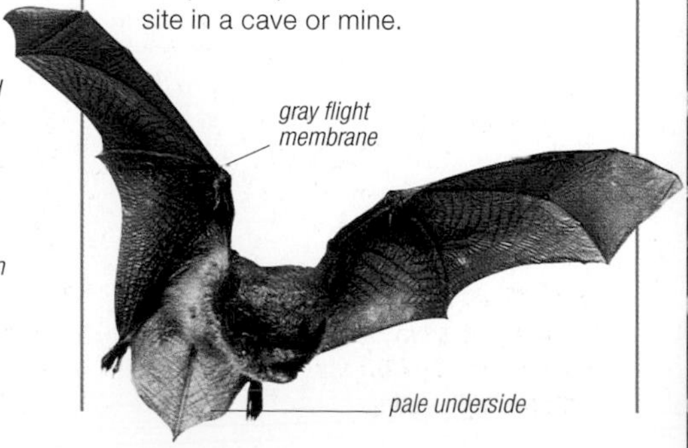

Vespertilio murinus

Parti-coloured bat

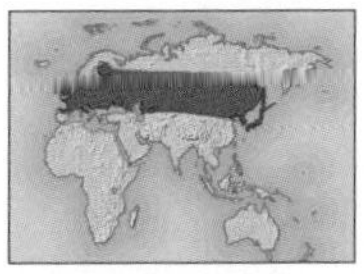

Location Europe to W., C., and E. Asia

Length 2–2½ in (5–6.5 cm)
Tail 1½–2 in (3.5–4.5 cm)
Weight ⅜–⅞ oz (10–25 g)
Social unit Group
Status Least concern

Distinctive coloration of almost black wings and face, pale cream fur below, and brown back hairs tipped with white give the parti-coloured bat its various common names, including frosted bat. It roosts by day in small crevices in cliffs or buildings. In late fall, males fly high near steep rock faces and tall buildings, their courting calls resembling the shrill whine of a high-speed metal grinder.

MAMMALS

Hedgehogs and relatives

PHYLUM	Chordata
CLASS	Mammalia
ORDER	Erinaceomorpha
FAMILIES	1
SPECIES	24

Formerly classified with shrews and moles in the order Insectivora, hedgehogs and moonrats are larger—mostly nocturnal—animals, with proportionately bigger eyes and ears. Their coat has long hair (moonrats), or hairs on the back and sides are modified into protective spines (hedgehogs). When threatened, hedgehogs may roll into a spiny ball, concealing the vulnerable face and underparts. Moonrats are restricted to tropical Southeast Asia, but hedgehogs occur throughout Eurasia and Africa, and some species are adapted to deserts. Both groups are predators of small animals (mostly invertebrates), but hedgehogs take carrion, fruit, roots, and nuts, too.

IMMUNE SYSTEM
Immunity to snake venom allows a hedgehog to take advantage of any snake it comes across as a potential food source.

Erinaceus europaeus

European hedgehog

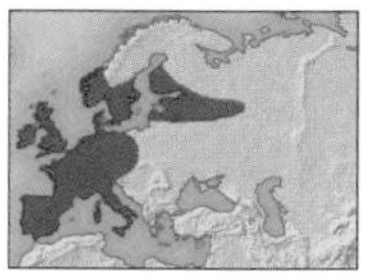

Location Europe

Length 9–10½in (22–27cm)
Tail None
Weight 2–2¼lb (0.9–1kg)
Social unit Individual
Status Least concern

The densely spined European hedgehog roams in urban parks and gardens as well as in hedgerows, fields, and woods at night, nosing—with piglike snuffling (hence "-hog")—for small animals such as worms, insects, and spiders. It also takes birds' eggs and carrion. Its day shelter is a nest of grass and leaves under a bush, log, or outbuilding or in an old burrow. During hibernation, it may wake on mild nights to feed. Mating takes place from May to October and gestation takes 31–35 days. The spines of the 4–5 young appear within hours of birth.

SELF-ANOINTING

The purpose of "self-anointing," when a hedgehog twists around to lick and smear its spines and skin with its own frothy saliva, is not clear. It may be a form of scent marking that helps these normally solitary animals to recognize neighbors on their occasional nightly encounters.

AGILE DEFENSE
The hedgehog runs and climbs with surprising agility. In self-defense it tucks its nonspiny head and legs onto its belly and rolls into a prickly ball.

Echinosorex gymnura

Moonrat

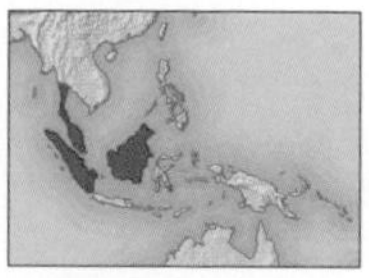

Location S.E. Asia

Length 10–18in (26–46cm)
Tail 6½–12in (16–30cm)
Weight 1–4½lb (0.5–2kg)
Social unit Individual
Status Least concern

Like other moonrats (gymnures), this species makes a territory-marking scent likened to rotting onions. It resembles a combination of hedgehog and small pig with harsh, rough, spiky outer fur, streaked with black and gray-white, and a long, scaly, almost hairless tail. Solitary, the moonrat rests in a burrow or crevice by day; at night it forages for small creatures such as insects, and also swims after fish and other aquatic prey.

Podogymnura truei

Mindanao moonrat

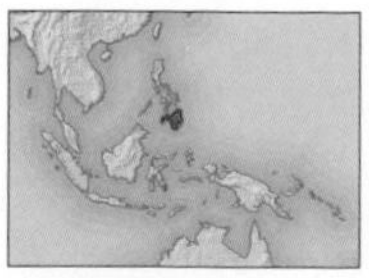

Location S.E. Asia (Mindanao)

Length 5–6in (13–15cm)
Tail 1½–2¾in (4–7cm)
Weight 5–6oz (150–175g)
Social unit Individual
Status Least concern

Inhabiting only one island in the Philippines, this poorly known but locally common species probably forages on the forest floor by day or night, especially around marshes and streams, for varied small animal prey. It lives alone, sheltering in a simple nest of leaves under a rock or log, or in an abandoned burrow. The long, soft fur is mainly gray-red, shading to gray-white on the underside. The distinctively pointed lower snout extends beyond the bottom lip, and the tail is short and coarse furred.

Paraechinus micropus

Indian hedgehog

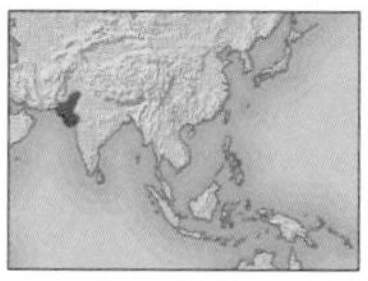

Location S. Asia

Length 5½–9in (14–23cm)
Tail ⅜–1½in (1–4cm)
Weight 11–16oz (300–450g)
Social unit Individual
Status Least concern

Similar to the long-eared hedgehog (see right) in appearance and habits, although slightly spinier, this species has a bare area of skin on the head. It is adapted to desert, dry scrub, and arid grassland, remaining inactive at times of harsh conditions (usually drought). Natural shelters such as rock crevices serve for nests, but the Indian hedgehog may dig a short burrow for this purpose. It hunts at night, seeking out insects, scorpions, and other small creatures. It also eats birds' eggs and scavenges for carrion. This hedgehog may cache (store) food, carrying it back to the nest for later consumption. Coloration is extremely variable with very dark (melanic) and almost white (albino) hues relatively common in the wild, and also brown and yellow banding patterns. Young are born between April and September. There is usually only 1–2 per litter.

Hemiechinus auritus

Long-eared hedgehog

Location W., C., and E. Asia, N. Africa

Length 6–10½in (15–27cm)
Tail ⅜–2in (1–5cm)
Weight 9/16–⅝oz (250–275g)
Social unit Individual
Status Least concern

Similar in appearance to the larger European hedgehog, the long-eared hedgehog has coarse fur on the face, limbs, and belly, but spines elsewhere. Banding on the spines varies from black through brown to yellow and white. It uses natural daytime resting places under rocks and shrubs. It eats various small animals and can become inactive when food or water are scarce.

Shrews and moles

PHYLUM	Chordata
CLASS	Mammalia
ORDER	Soricomorpha
FAMILIES	3
SPECIES	418

Shrews, moles, and solenodons are voracious predators of invertebrates and other small vertebrates. They live frantic lives fueled by a very high metabolic rate; at least a few have toxic saliva for disabling prey. They have small eyes and small (or absent) external ears, often hidden in short, dense fur. Some aquatic forms have webbed or hairy-fringed swimming feet. Moles have broad forefeet and strong claws for burrowing, as well as a long, flexible snout for detecting prey. Shrews are a species-rich family found throughout much of the world except Australasia. Moles are largely confined to the northern hemisphere, and solenodons are exclusively West Indian.

COMPETENT CLIMBER
Shrews have small eyes and well-developed snouts. Despite poor vision, this Eurasian shrew is an adept climber. It also has acute hearing, even though there is no external ear flap.

Solenodon paradoxus

Hispaniolan solenodon

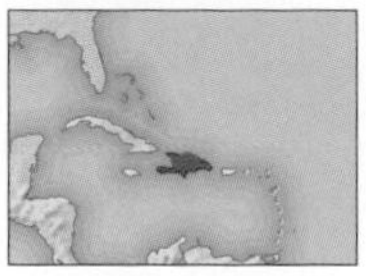

Location Caribbean (Hispaniola)

Length 11–12½ in (28–32 cm)
Tail 6½–10 in (17–26 cm)
Weight 2¼ lb (1 kg)
Social unit Individual
Status Endangered

The 2 species of solenodon—Cuban and Hispaniolan—are large, long-tailed, shrewlike, nocturnal insectivores. Both are under threat. The snout of the Hispaniolan solenodon is long and mobile. Its fur varies from black to red-brown, and its feet, tail, and upper ears are almost hairless. It is fast and agile. It noses and scrabbles on the forest floor with its sharp claws for insects, worms, small lizards, fruit, and other plant matter. Its poisonous bite is used for defense and to stun prey.

Sorex araneus

Eurasian shrew

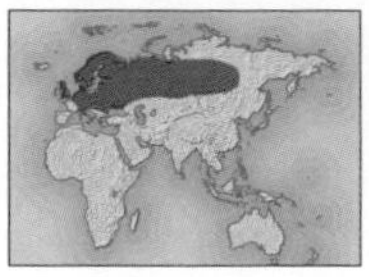

Location Europe to N. Asia

Length 2–3¼ in (5–8 cm)
Tail 1–1¾ in (2.5–4.5 cm)
Weight 3/16–½ oz (5–14 g)
Social unit Individual
Status Least concern

One of the smallest mammals, the Eurasian shrew is adaptable, aggressive, and voracious. It must eat 80–90 percent of its body weight every 24 hours, and it hunts in up to 10 bursts of activity, according to season and conditions. Food includes insects, worms, and carrion. Adults are solitary except for a brief courtship in spring or early fall. After a gestation period of 24–25 days, 6–7 young are born in a special breeding nest made from woven grass and dry leaves. Larger than the usual resting nest, this is similarly sited under a log, root, rock, or in an old burrow. The Eurasian shrew has a pointed, flexible snout and short legs. Its fur is dark brown to black on the back, paler brown on the flanks, and gray-white on the underside. It is territorial, making ultrasonic squeaks, especially when a female gathers her offspring. If cornered, this shrew readily bites.

Megasorex gigas

Giant Mexican shrew

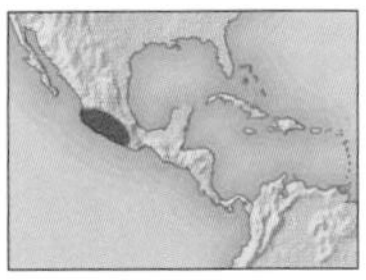

Location S.W. Mexico

Length 3¼–3½ in (8–9 cm)
Tail 1 1/2–2 in (4–5 cm)
Weight 3/8–7/16 oz (10–12 g)
Social unit Individual
Status Least concern

The sole species in the genus *Megasorex,* this large, compact-bodied, but short-tailed shrew prods with its prominent, pointed snout among leaves and loose soil for worms, grubs, millipedes, spiders, and other small prey. The upperparts are dark brown or grayish brown, becoming paler on the underside. This shrew prefers areas of damp soil and moist undergrowth in grassland and forest, ranging from the lowlands to altitudes of 5,600 ft (1,700 m). Only some 20 giant Mexican shrews have been studied and their nesting and breeding habits are not yet known.

Blarina brevicauda

Northern short-tailed shrew

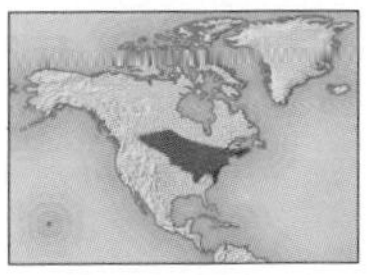

Location S. Canada to N. and E. USA

Length 4¾–5 1/16 in (12–14 cm)
Tail 1¼ in (3 cm)
Weight 11/16 oz (20 g)
Social unit Variable
Status Least concern

Like most shrews, this large, robust species has poor sight but excellent sense of smell, and its poisonous bite (due to toxic saliva) helps disable prey.

It uses mainly scent and touch to hunt its main prey of soil-living animals and—unusually for the shrew group—smaller mammals such as voles and mice, and even some plant matter. This shrew rests and feeds largely underground in runways and old mole or vole tunnels, usually 4–20 in (10–50 cm) deep, and stores items in cold weather. The eyes and ears are tiny, and the snout stouter and less pointed than in other shrews.

Neomys fodiens

Eurasian water shrew

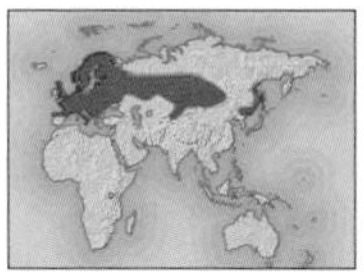

Location Europe to N. Asia

Length 2½–3¾ in (6.5–9.5 cm)
Tail 1¾–3¼ in (4.5–8 cm)
Weight 5/16–7/8 in (8–25 g)
Social unit Individual
Status Least concern

This shrew hunts aquatic insects, small fish, and frogs. It also feeds on land, on worms, beetles, and grubs, and so can survive in damp woods. The small eyes and ears, and the long, pointed snout are typically shrewlike. Solitary but less aggressive than other shrews, the Eurasian water shrew establishes a series of runways and burrows and has a nest of dry grass and old leaves. Here, after 14–21 days' gestation, the female suckles her litter of 4–7 young for approximately 6 weeks.

Suncus etruscus

Etruscan shrew

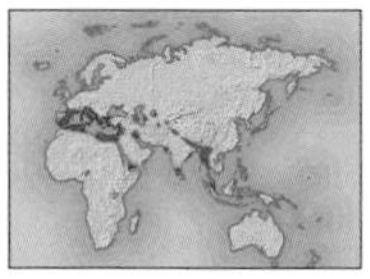

Location S. Europe, S. to S.E. Asia, Sri Lanka, N. to E. Africa, W. Africa

Length 1½–2 in (4–5 cm)
Tail ¾–1¼ in (2–3 cm)
Weight 1/16–1/8 oz (2–3 g)
Social unit Individual
Status Least concern

Also called the pygmy white-toothed shrew or Savi's shrew, this species actively hunts for small prey such as insects, worms, snails, and spiders, then rests for a few hours, through day and night. It nests in a small hole or crevice and is solitary most of the year, forming pairs only in the breeding season. The gestation period is 27–28 days and litter size 2–5, with up to 6 litters per year.

MAMMALS

Suncus megalura

Forest musk shrew

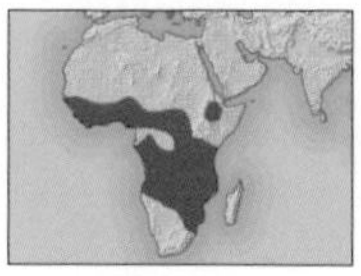

Location W., E., C., and southern Africa

Length 2–2¾ in (5–7 cm)
Tail 3–3½ in (8–9 cm)
Weight 3/16 oz (5 g)
Social unit Individual
Status Least concern

This species is shrewlike in most respects, although it has a relatively long, streamlined body, large and prominent ears, and a tail longer than its head and body. Its soft, velvety fur is brown on the upperparts and almost white on the belly. It hunts in soil, leaf litter, and among branches (using its tail to balance), for invertebrates. Like the other 10 or so *Sylvisorcx* species, it is probably active in bursts through the day and night, and must eat almost its own body weight in food every 24 hours.

Nectogale elegans

Tibetan water shrew

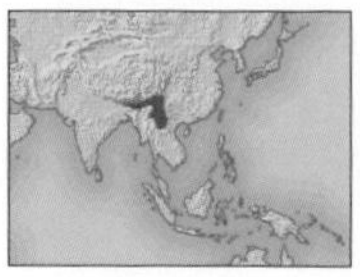

Location S. Asia

Length 3½–5 in (9–13 cm)
Tail 3–4¼ in (8–11 cm)
Weight Not recorded
Social unit Individual
Status Least concern

A small, wary, secretive inhabitant of cold, fast mountain streams in the Himalayas and nearby mountains, this tubby, semiaquatic shrew is slate-gray on the upperparts and silvery white beneath. Its snout is blunt, its eyes and ears tiny, and its black tail is fringed by rows of hairs along each side. It probably eats water insects, fish fry, and other small prey, carrying them to a bank or a midstream rock for consumption. Little is known of its nesting or breeding habits.

Diplomesodon pulchellum

Piebald shrew

Location C. Asia

Length 2–2¾ in (5–7 cm)
Tail ¾–1¼ in (2–3 cm)
Weight ¼–7/16 oz (7–13 g)
Social unit Individual
Status Least concern

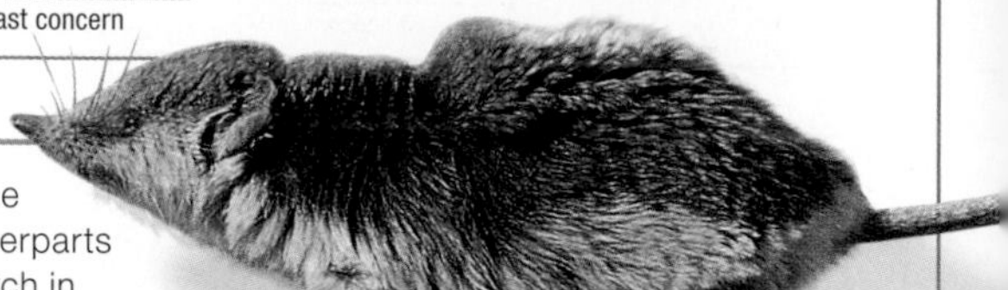

This species derives its name from its coloration: gray upperparts with a distinct oval white patch in the middle of the back, and white underparts, feet, and tail. It has a very pointed snout and long whiskers, even for a shrew. Active at night, when the desert habitat is cooler and prey such as insects and small lizards become more energetic, it hunts mainly on the surface but may also dig in loose sand for grubs and worms. The average litter size is 5, with several litters in a good year.

Scutisorex somereni

Armoured shrew

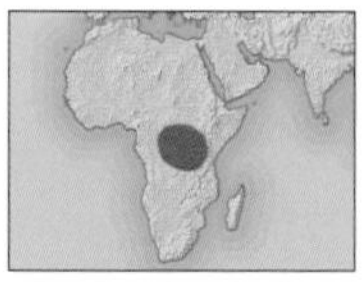

Location C. to E. Africa

Length 4–6 in (10–15 cm)
Tail 2¾–3¾ in (6.5–9.5 cm)
Weight 2½–4 oz (70–125 g)
Social unit Individual
Status Least concern

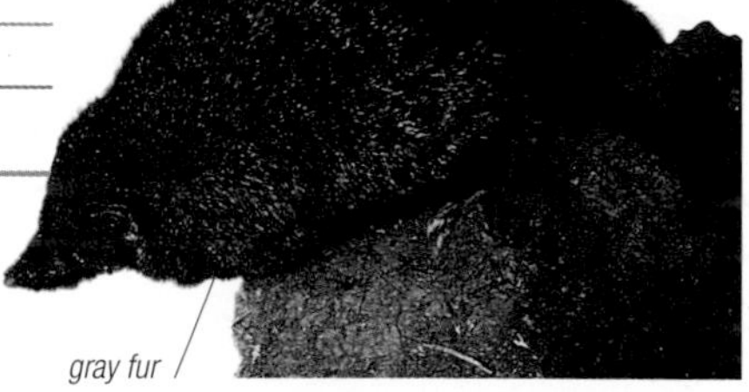

gray fur

This large, woolly coated shrew, also known as the hero shrew, has a distinctively arched and tremendously strong back. This is because its vertebrae have interlocking flanges, or spines, not only along their sides, as in other mammals, but also above and below. The armoured shrew is solitary, and is a skillful climber. Its diet consists of worms, insects, spiders, and carrion.

Crocidura leucodon

Bicoloured white-toothed shrew

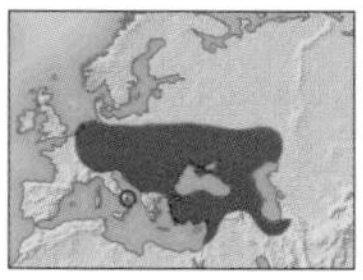

Location Europe to W. Asia

Length 1½–7 in (4–18 cm)
Tail 1½–4¼ in (4–11 cm)
Weight 7/32–7/16 oz (6–13 g)
Social unit Individual
Status Least concern

Sharp demarcation between its gray-white upperparts and whitish yellow underside give this shrew its common name. It also has long, thick whiskers (vibrissae) on its sharp-pointed snout, and a bicolored tail that is less than half the length of the head and body. This adaptable forager in grassland, scrub, wood edge, parks, and gardens eats worms, grubs, and other small invertebrates, hunting mainly by night but also for short periods by daylight. Like similar shrew species, it builds a nest of dry grass in a hole or thick undergrowth. The male produces strong scents from glands on his flanks during the breeding season, generally March to October. Gestation is 31 days, average litter size is 4, and weaning occurs after 26 days.

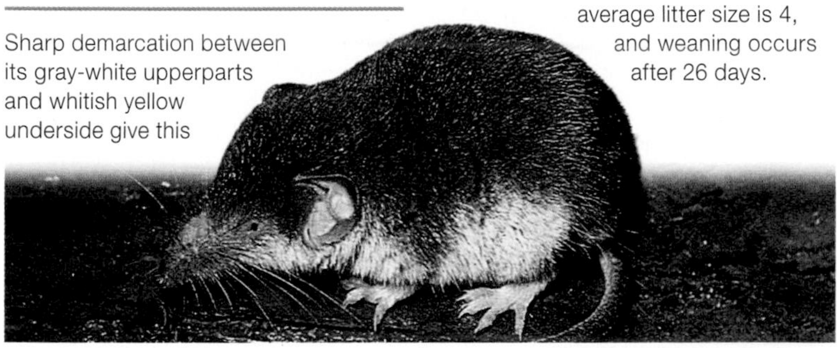

Desmana moschata

Russian desman

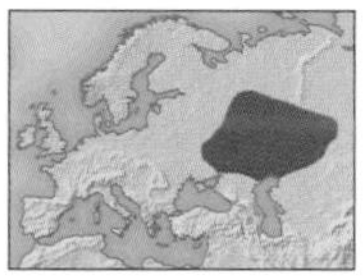

Location E. Europe to C. Asia

Length 7–8½ in (18–21 cm)
Tail 7–8½ in (17–21 cm)
Weight 16 oz (450 g)
Social unit Group
Status Vulnerable

Desmans belong to the mole family, but resemble water shrews. The tail is as long as the head and body, and flattened from side to side for use as both a paddle and a rudder. The rear feet are fully webbed to the toe tips, the front feet partly so. Using its long, sensitive nose, it probes by night for prey in riverbed mud and stones. Unusually for an insectivore, the desman lives in groups and several may share a bank burrow. After 40–50 days' gestation, the 3–5 young are cared for by the female, and weaned by 4 weeks.

CONSERVATION

In addition to being harmed by pollution, the Russian desman faces a new threat: ultrafine plastic nets, which are used for fishing in rivers and lakes. The nets pose a hazard long after they are discarded, by entangling and drowning desmans when they dive. Conservation measures include reintroductions to places where numbers are low, and legal restrictions to limit net use.

OUTER AND INNER COATS
The soft, dense underfur of the Russian desman is covered by long, coarse guard hairs, the coat being rich brown on the head and body, fading to ash-gray on the underside.

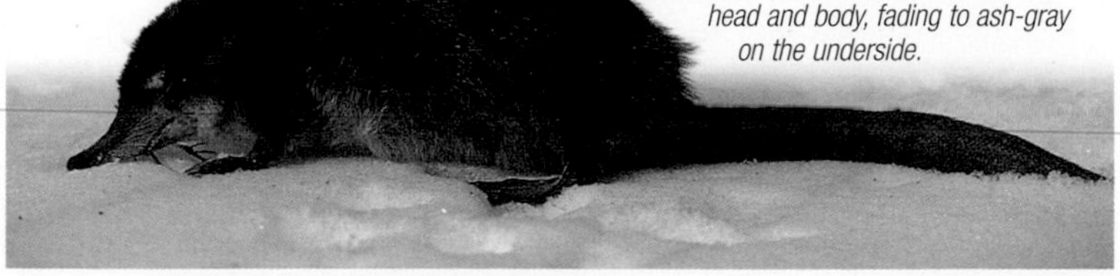

Galemys pyrenaicus

Pyrenean desman

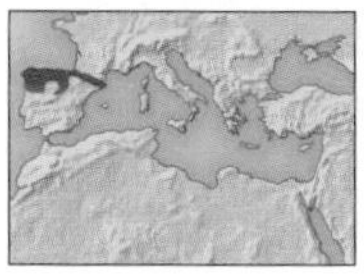

Location S.W. Europe

Length 5 in (12.5 cm)
Tail 5½ in (14 cm)
Weight 1 7/16–1¾ oz (40–50 g)
Social unit Individual/Pair
Status Vulnerable

The Pyrenean desman generally resembles the only other species of desman (see right), although it is smaller, takes lesser prey, such as the aquatic larvae of mayflies and stoneflies, and is more suited to fast-flowing streams. Its long, black snout is almost hairless, its thick fur brown above and silvery below, and its tail slightly flattened from side to side like a rudder. A male and a female may form a loose pair bond, with the male chasing away rivals and the female nesting in a bank burrow, but little more is known of these mammals.

Talpa europaea

European mole

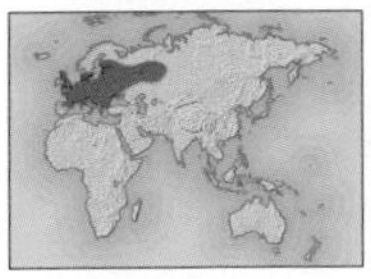

Location Europe to N. Asia

Length 4¼–6½ in (1–16 cm)
Tail ¾ in (2 cm)
Weight 2⅜–5 oz (65–125 g)
Social unit Individual
Status Least concern

Virtually blind, this mole lives mainly underground in tunnels radiating from a central chamber, feeding on worms and other soil animals. When plentiful, the worms are bitten to paralyze them for future use, but, if uneaten, they recover and escape. The female digs a large nest chamber, and after 4 weeks' gestation, gives birth to 3–4 young.

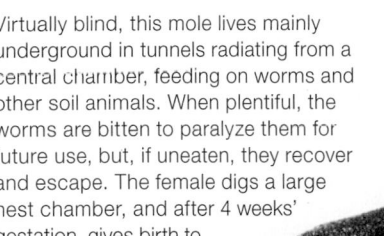

TUNNELING TOOLS

This mole's large front legs have powerful shoulder muscles and broad, outward-facing paws (above) with a strong, spadelike claw on each toe. Anchoring itself with its back feet, it uses its front legs to scoop soil sideways and back, pushing it up as molehills.

REVERSIBLE FUR
Short, dense, black fur that can lie at any angle allows this mole to go forward or backward in its tunnels.

Condylura cristata

Star-nosed mole

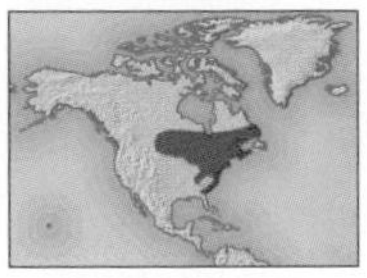

Location E. Canada, N.E. USA

Length 7–7½ in (18–19 cm)
Tail 2½–3 in (6–8 cm)
Weight 1⅝ oz (45 g)
Social unit Variable
Status Least concern

This mole is an expert swimmer. It has a long, sparse-haired, scaly tail, which enlarges in winter with fatty food reserves. In lifestyle and habits it resembles other moles, yet it is less solitary and tolerates meeting others of its kind. Its tunnels are about 1½ in (4 cm) in diameter and 2–24 in (5–60 cm) in depth.

STAR-SHAPED NOSE

An unmistakable snout, with 22 pale, fleshy rays (tentacles) around the nostrils, allows the star-nosed mole to sniff and feel prey in water. It also forages for food among reeds, mosses, and other vegetation. As it hunts, the fleshy rays around its nose wiggle and flex in constant motion.

VARIED DIET
The star-nosed mole eats leeches, snails, small fish, and other aquatic prey, as well as soil animals.

Pangolins

PHYLUM	Chordata
CLASS	Mammalia
ORDER	Pholidota
FAMILIES	1 (Manidae)
SPECIES	8

Similar in shape to armadillos and anteaters, pangolins are covered in overlapping scales, which act as armor and camouflage. Pangolins lack teeth: prey (ants and termites) is collected with the tongue, and powerful muscles in the stomach "chew" the food. Pangolins are found in southern Asia and Africa, in habitats ranging from forest to savanna.

LONG TONGUE
Pangolins use their tongue, which can be extended as far as 10 in (25 cm), to gather ants and termites.

Manis pentadactyla

Chinese pangolin

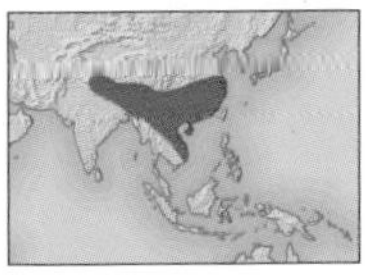

Location E. to S.E. Asia

Length 21½–32 in (54–80 cm)
Tail 10–13½ in (26–34 cm)
Weight 4½–15 lb (2–7 kg)
Social unit Individual
Status Endangered

Bony, pale or yellow-brown scales, up to 2 in (5 cm) across, cover all parts of the Chinese pangolin except for its snout, cheeks, throat, inner limbs, and belly.

When rolled into a ball, no soft areas are exposed. The thin tongue, as long as 16 in (40 cm), scoops up ants and termites. The strongly prehensile tail and long claws make this pangolin surprisingly agile in trees and a powerful burrower.

Manis temminckii

Ground pangolin

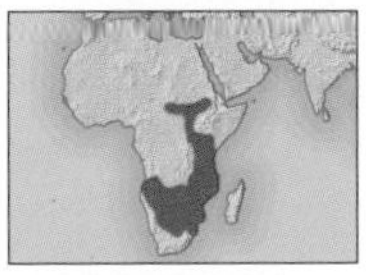

Location E. to southern Africa

Length 20–23½ in (50–60 cm)
Tail 16–20 in (40–50 cm)
Weight 33–40 lb (15–18 kg)
Social unit Individual
Status Least concern

Similar in most respects to the Chinese pangolin (see left), this species rips open termite mounds and ant nests, both in trees and on the ground, with its large claws, and licks up the occupants. This pangolin shows little territorial behavior. The 1–2 young are born after a gestation of about 120 days.

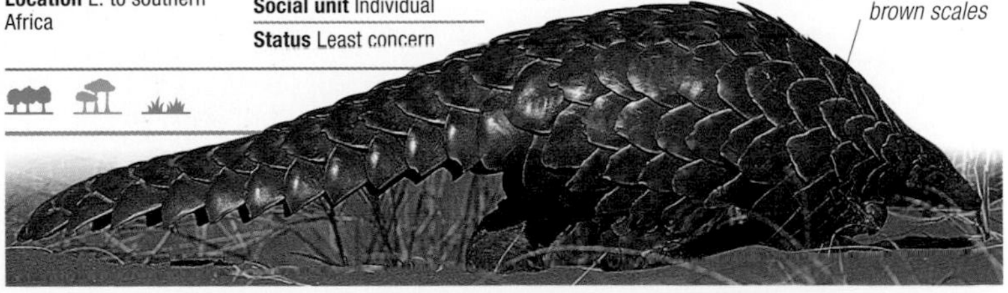

MAMMALS

Carnivores

PHYLUM	Chordata
CLASS	Mammalia
ORDER	Carnivora
FAMILIES	15
SPECIES	285

Although the term carnivore is commonly used to describe an animal that eats meat, it also refers specifically to members of the order Carnivora. While most members of the group eat meat, some have a mixed diet or are entirely herbivorous. Meat-eating carnivores are the dominant predators on land in all habitats: their bodies and lifestyles are highly adapted for hunting. However, there is great variety within the group, which includes species as diverse as the giant panda and the walrus. Uniquely among mammals, carnivores have 4 carnassial teeth. They also have a penis bone (baculum). Indigenous to most parts of the world, carnivores have also been introduced to Australasia.

CLASSIFICATION NOTE

As the evolution of the order Carnivora becomes better known, scientists have subdivided existing families, and included those from the order Pinnipedia (seals and walrus). There are currently 15 families, of which 14 are shown in this book.

Dogs and relatives see pp.168–75
Bears see pp.176–81
Sea lions see pp.182–3
Walrus see p.184
Seals see pp.184–5
Skunks see p.186
Raccoons and relatives see pp.187–8
Red Panda see p.188
Mustelids see pp.189–96
Malagasy carnivores see p.197
Mongooses see p.198
Civets and relatives see p.199
Cats see pp.200–7
Hyenas see pp.208–9

Hunting

Carnivores include some of nature's most skillful and efficient predators. Most use keen senses of sight, hearing, or smell to locate prey, which they catch either by pouncing from a concealed place or by stalking and then running down their quarry in a lengthy chase or swift rush. Many can kill animals larger than themselves. Weasels kill by biting the back of the head and cracking the skull, while cats bite into the neck, damaging the spinal cord, or into the throat, causing suffocation. Dogs shake prey vigorously to dislocate the neck.

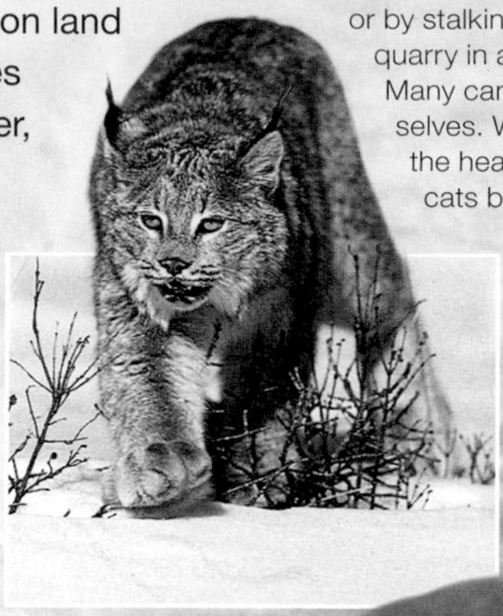

SOLITARY HUNTER
The bobcat, which feeds mainly on small prey such as the snowshoe hare, hunts alone.

HUNTING IN PACKS
Lions generally hunt in groups to capture large animals. The females (males rarely join in) usually stalk to within 98 ft (30 m) and encircle the prey. After a short charge, the animal is brought down with a grab to the flank, then killed by suffocation with a bite to the throat.

Anatomy

Although carnivores vary considerably in size and shape, most share several features that make them well suited to a hunting lifestyle. A typical terrestrial carnivore is a fast and agile runner with sharp teeth and claws, acute hearing and eyesight, and a well-developed sense of smell. Carnassial teeth (see below) are present in predacious living carnivores but are less well developed among omnivorous, herbivorous, and some piscivorous species. Carnivores have either 4 or 5 digits on each limb. Members of the cat family (except cheetahs) have sharp, retractable claws used to rake prey, defend themselves, and climb. Most other carnivores have nonretractable claws, often used for digging.

JAWS AND TEETH
Most carnivores have sharp teeth and powerful jaws for killing and disemboweling prey. The temporalis muscles, which are most effective when the jaws are open, are used to deliver a powerful stab from the sharp canines. The carnassial teeth are sharpened molars in the upper and lower jaws that mesh together perfectly. In combination with the masseter muscles, which can be used when the jaw is almost completely closed, they form a powerful shearing tool for tearing flesh.

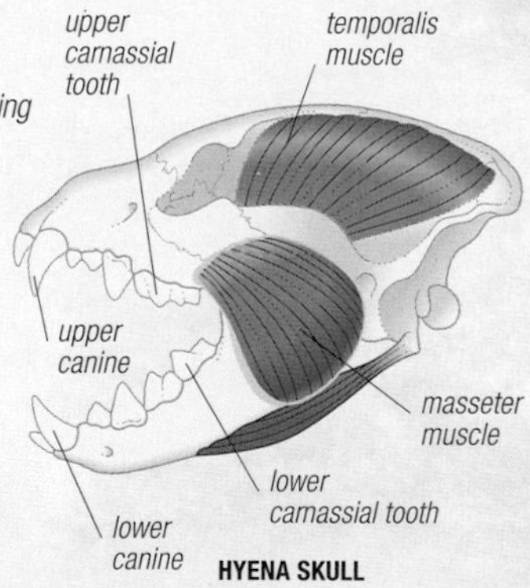

HYENA SKULL

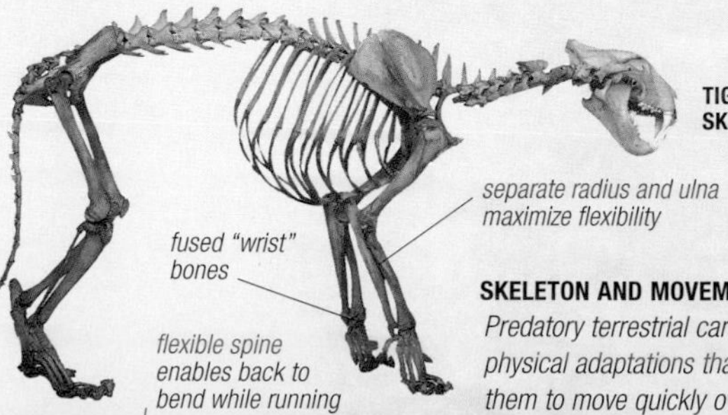

TIGER SKELETON

SIBERIAN TIGER

SKELETON AND MOVEMENT
Predatory terrestrial carnivores have physical adaptations that enable them to move quickly over the ground in pursuit of prey. The spine is generally flexible, the limbs are relatively long, and the collarbone is reduced, maximizing the mobility of the shoulders. To increase the length of their stride and to add to their speed, all carnivores have fused wrist bones, and dogs and cats walk on their toes (rather than the soles of the feet).

Social groups

Although many carnivores live alone or in pairs, others form groups that take different forms, and have complex structures. Lion prides, for example, consist of several related families, although most males leave the pride into which they are born. Lions spend most of their time together, hunt cooperatively, and tend each other's young. In most other groups, individual ties are looser. Red and Arctic foxes live in groups of one adult male and several vixens, but each adult hunts alone in a different part of the group's territory. Elephant seals gather in large numbers only during breeding. The groups are made up of several males and their respective harems of females, which are closely guarded.

PLAY
Young carnivores develop their fighting skills through play. By playing together, these red foxes learn to test another animal's strength without suffering painful consequences.

SHARED PARENTHOOD
These young slender-tailed meerkats are not necessarily the offspring of the adult watching over them. Sharing parental duties is common in many carnivore societies.

MAMMALS

Feeding

Most carnivores live on a diet of freshly killed animal prey, ranging in size from insects, other invertebrates, and small vertebrates to animals as large as buffaloes and reindeer. Carnivores are generally adaptable feeders, seldom restricting themselves to a single food type. However, there are specialists—for example, there are pinniped species that eat only fish. Others, such as bears, badgers, and foxes, eat a mixed diet of meat and plants, while a few, notably the giant panda, are almost entirely herbivorous.

CARRION-FEEDERS
Hyenas feed on live prey and the remains of other animals' kills. Their particularly sharp teeth and strong jaws enable them to break bones and tendons that are too tough for other carnivores.

PLANT EATERS
The diet of the giant panda consists mainly of bamboo shoots and roots, on which it feeds for up to 12 hours a day. Pandas are slow moving compared with other carnivores, and their flat cheek teeth are better suited for grinding than for cutting.

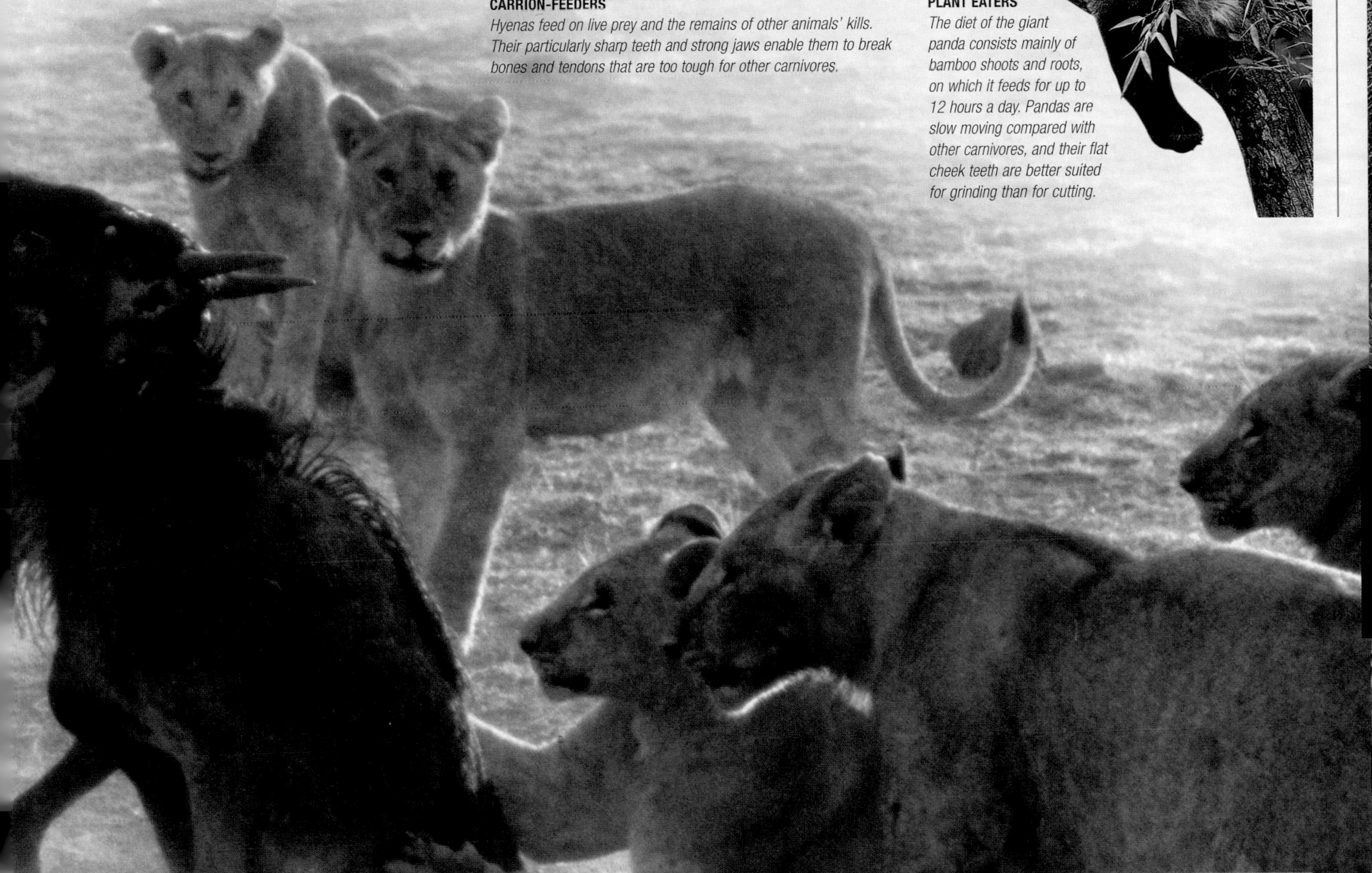

Communication

Carnivores communicate with each other with scent markings, visual signals, and vocalizations. Scent messages, which have the advantage of being persistent, are used to define territory or to find potential sexual partners. They are left by spraying urine or leaving piles of feces, although some animals also rub scent onto objects from glands on their face, between their claws, or at the base of their tail. When animals meet face-to-face, posture, facial expression, and sound are used to pass on a wealth of information, including threats, submissions, advances to partners, and warnings of approaching danger.

GREETING POSTURES
Body language is an important form of communication for African wild dogs, which live in large packs. In this greeting ritual, adult dogs push their muzzles into each other's faces.

MARKING TERRITORY
Bears use trees to leave both scent marks and visual signs. Here, scent is being transferred in saliva and from glands in the bear's feet, while the sharp teeth and claws are being used to rip the bark.

Dogs and relatives

PHYLUM	Chordata
CLASS	Mammalia
ORDER	Carnivora
FAMILY	Canidae
SPECIES	34

Members of the dog family—dogs, wolves, coyotes, jackals, and foxes—are collectively described as canids. They are known for great endurance (rather than sudden bursts of speed) and for opportunistic and adaptable behavior. Dogs are characterized by a slender build, long legs, and a long, bushy tail. Wild canids generally inhabit open grassland habitats the world over, and are absent only from isolated areas such as Madagascar and New Zealand (where the domestic dog is present, however).

Anatomy

Canids have a muscular, deep-chested body covered with a fur coat that is usually uniformly colored or speckled. The lower limbs are developed for strength and stamina: some of the wrist bones are fused, and the front legs cannot be rotated (the bones at the front of the leg are locked). There are 4 digits on the back feet and 5 on the front feet, and each digit has a hard pad. The claws are short, nonretractable, and blunt (other carnivores have sharp claws). Canids also have long jaws, long, fanglike canines (for stabbing prey), and well-developed carnassials (the slashing teeth at the back of the jaws). Canids track their prey by scent, and the long, pointed muzzle houses large olfactory organs. Hearing is also acute, and the ears are large, erect, and usually pointed. Sight is less important, but is still well developed.

Social groups

Smaller species, which usually feed mainly on small rodents and insects, tend to have a flexible social organization but often live either in pairs (for example, jackals) or alone (for example, foxes). However, larger species, such as the wolf and the African wild dog, live in social groups called packs. These packs, which consist of a dominant pair and their offspring, occupy and defend territories, which they mark with urine. The young in a pack are of different ages because older offspring remain in the group for some years, and may help rear new young. Only the dominant pair breed, and the female digs a den in which to give birth. The pack will often perform a bonding ritual, which involves mutual licking, whining, and tail wagging. When hunting, the usual tactic is to track a herd of deer or antelopes (for example) and then cooperatively maneuver so as to separate one animal. This individual is then run down and slashed at by pack members until it falls, exhausted. On returning to the den, the hunters regurgitate meat for the cubs to eat.

Canids and people

Throughout history, canids have proved useful to humankind in many ways. Wild canids are, for example, important controllers of rodent populations, which can spiral quickly if left unchecked. Furthermore, the domestic dog—which descended from the wolf over 10,000 years ago—has always played an important role in a number of human activities. From the tiny chihuahua to the huge St. Bernard (the domestic dog displays more variation between types than any other domestic animal), there are breeds specialized for hunting, herding, guarding, performing, carrying or dragging loads, and companionship.

However, many canid species are considered pests. The wolf, for example, has been hunted and persecuted as a killer of livestock. As a result, this species is now rare throughout its vast range and is extinct in many regions. Other species have fared even worse: the bush dog and the maned wolf, for example, are on the brink of extinction; the red wolf only survived in zoos but has been reintroduced in the wild. On the other hand, the coyote and the red fox—both opportunists—have benefited from the spread of urban environments and are more abundant than ever before.

STRENGTHENING BONDS
Establishing and maintaining bonds between pack members is essential for the survival of dogs that live in social groups. In African wild dogs, bonding behavior, such as licking and whining, frequently occurs before a hunt. It is only by cooperating and hunting as a team that the dogs are able to bring down and kill prey larger than themselves.

MAMMALS

Vulpes vulpes

Red fox

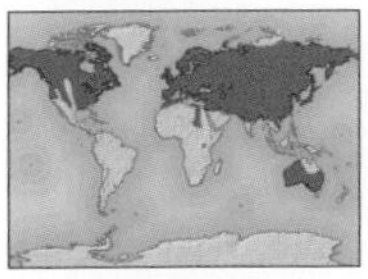

Location Arctic, North America, Europe, Asia, N. Africa, Australia

Length 23–35 in (58–90 cm)
Tail 12½–19½ in (32–49 cm)
Weight 6½–24 lb (3–11 kg)
Social unit Pair
Status Least concern

Active by day and night, the red fox is exceptionally widespread, and adaptable both in habitat, from Arctic tundra to city center, and in diet. Home is an earth (den) in a sheltered place—for example, an enlarged rabbit burrow, a crevice among rocks or roots, or a space under an outbuilding. The basic social unit is a female (vixen) and male (dog), who mark their territory of 0.4–4 square miles (1–10 square km) with urine, droppings, and scent from anal and other glands. Single-male–multifemale groups also occur but only senior females breed. Mating is in late winter or early spring when females make eerie shrieks. Gestation is 49–55 days and litter size up to 12; averages vary from 4–5 in Europe to 6–8 in North America. Both parents and "helper" nonbreeding females care for the cubs and feed them after weaning at 6–12 weeks.

THE UNFUSSY FOX

In grassy or farmed areas, a large part of the red fox's diet comprises lagomorphs, especially rabbits and young hares. The fox stealthily stalks its prey, then makes a dash to catch the victim before it reaches its burrow (rabbit) or accelerates away (hare). The prey is carried by the neck to a secluded spot where the fox can eat at leisure. The red fox also consumes beetles, worms, frogs, birds, eggs, mice, voles, fruit, carrion, and refuse—in fact, almost anything edible.

NOT ALWAYS RED
Coat color varies from grayish or rusty red to almost orange, usually with black on the backs of the ears, sometimes on the lower limbs and feet, and an often black-tinged but pale-tipped tail. All-black and silver-white forms also occur.

bushy tail (brush)

Vulpes cana

Blanford's fox

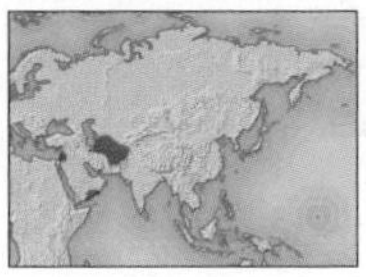

Location W. and S. Asia

Length 16½ in (42 cm)
Tail 12 in (30 cm)
Weight 2–3¼ lb (0.9–1.5 kg)
Social unit Individual
Status Least concern

This small fox has relatively large ears and tail, patchy body coloring in black, gray, and white, white underparts, a dark stripe along the middle of the back, and a stealthy, feline gait. It is a solitary, nocturnal hunter of small creatures, including insects, mainly in barren, rocky hills and grassy uplands. It also consumes appreciable amounts of fruit and is found near orchards and groves. Litter size is 1–3.

Vulpes macrotis

Kit fox

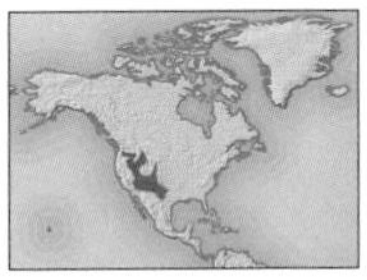

Location W. USA

Length 12½–20½ in (38–52 cm)
Tail 8½–12½ in (22–32 cm)
Weight 3¼–6½ lb (1.5–3 kg)
Social unit Pair
Status Least concern

Similar to the swift fox (see right) in appearance and habits, this species has a more westerly range, but with overlap and perhaps interbreeding in Texas. The kit fox has longer, closer-set ears, a more angular head, and is more heavily built overall. There are 3 color forms: pale gray-brown, dark gray-brown, and intermediate gray. Its habitats vary from grassland to desert, and its diet is omnivorous. Both parents raise the 3–6 young.

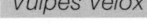

Vulpes velox

Swift fox

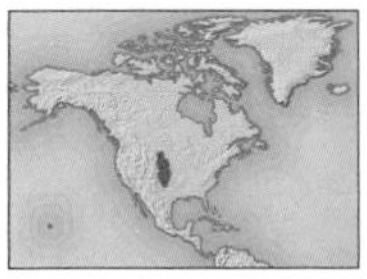

Location C. USA

Length 12½–21 in (38–53 cm)
Tail 7–10 in (18–26 cm)
Weight 3¼–6½ lb (1.5–3 kg)
Social unit Pair
Status Least concern

Recently established as a separate species from the kit fox (see left), the swift fox has a more easterly distribution. Its coloration is similar to that of the kit fox, but it is grayer on the upperparts and buff-orange underneath. It has a bushy, black-tipped tail. Both species dig dens about 3 ft (1 m) deep, with 13 ft (4 m) of tunnels, and mate from December to January—later in northern areas. The gestation period is 50–60 days.

Vulpes zerda

Fennec fox

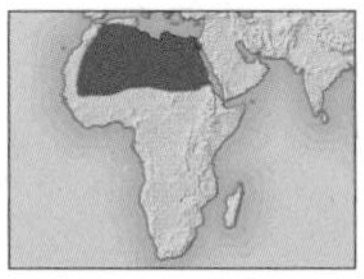

Location N. Africa

Length 9½–16 in (24–41 cm)
Tail 7–12 in (18–31 cm)
Weight 2¼–3¼ lb (1–1.5 g)
Social unit Group
Status Least concern

The smallest fox, the fennec has relatively large ears and a black-tinged tail tip. Its furred soles are adapted for walking on soft, hot sand. Mostly nocturnal, the diet of this fox ranges from fruit and seeds to eggs, termites, and lizards. Unusual among foxes, it associates in groups of up to 10, but relationships are not clear. Each member digs a den several yards into soft earth. Mating occurs in January–February and the 2–5 cubs remain in the den, protected by the female, for 2 months. They are fully mature by 11 months. Hunted for its fur, the fennec fox is also trapped as a pet.

cream to yellowish cream fur

white underparts

Vulpes rueppellii

Rüppell's fox

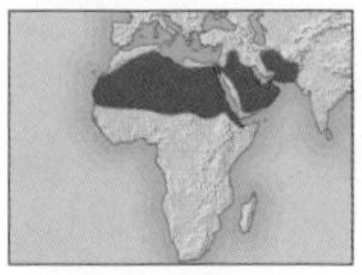

Length 16–20½ in (40–52 cm)	
Tail 10–15½ in (25–39 cm)	
Weight 2¼–6½ lb (1–3.5 kg)	
Social unit Group/Pair	
Status Least concern	

Location N. Africa, W. Asia

white chin, bib, and belly

Rüppell's fox (also called the sand fox) is similar to but slighter in build than the red fox. It has soft, dense, sandy or silver-gray fur to match its arid habitat, black patches on the sides of the muzzle, and a white tail tip. In some regions this species forms monogamous pairs, but in others it gathers in groups of up to 15. It rests by day in a sheltered crevice or burrow, and changes its den every few days. Average litter size is 2–3, born in early spring. It eats a wide variety of foods, from grass to insects, reptiles, and mammals.

Alopex lagopus

Arctic fox

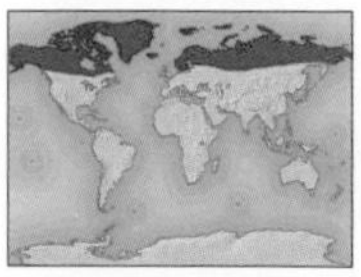

Length 21–22 in (53–55 cm)	
Tail 12 in (30 cm)	
Weight 8¾ lb (4 kg)	
Social unit Group	
Status Least concern	

Location N. Canada, Alaska, Greenland, N. Europe, N. Asia

The Arctic fox has 2 color types, or "phases." Foxes that are "white" phase are almost pure white in winter for camouflage in snow and ice. This phase is associated with the true tundra of open, treeless plains and grassy hillocks. Those of the "blue" phase are more prevalent in mixed coastal and shrubby habitats and are pale gray-brown tinged with blue in winter. The Arctic fox eats a huge variety of foods—mainly lemmings, but also birds, eggs, crabs, fish, insects, seal and whale carcasses, fruit, seeds, and human refuse. Its social group is likewise flexible, with male–female pairs, larger groups of nonbreeders, or a breeding pair plus "helper" females. The den site is extensive, with complex burrow systems for shelter and breeding. Reproduction is closely tied to available food, with more than 15 cubs per litter when lemmings are plentiful, and 6–10 in an average year.

FURRED BUNDLE
The Arctic fox has small ears, a blunt muzzle, and short legs and tail, since these areas lose heat fastest. Every part of its body except its nose is thickly furred.

stout, rounded body under thick fur

SUMMER COATS

The Arctic fox's summer coat is half as thick as its winter one, with less than half of the underfur. In summer, white-phase animals are gray-brown to gray above and gray below; those of the blue phase are browner and darker.

Urocyon cinereoargenteus

Gray fox

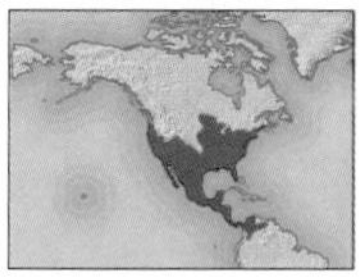

Length 21–32 in (53–81 cm)	
Tail 10½–17½ in (27–44 cm)	
Weight 6½–15 lb (3–7 kg)	
Social unit Pair	
Status Least concern	

Location S. Canada to N. South America

Also called the tree fox, this long-bodied species prefers woodland. It climbs skillfully, leaping up tree trunks and between branches with almost catlike agility. Active at night, it consumes various insects and small mammals, but may rely more on fruit and seeds in certain seasons. The gray fox has a small, dark gray neck mane and central back stripe, and a red tinge to the neck, flanks, and legs, with a buff or white chin and belly. Its den may be in an old burrow or log, but more often in a tree hole up to 30 ft (9 m) above ground, or on a building ledge or in a roof space. Most gray foxes live as breeding pairs.

VULNERABLE CUBS

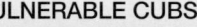

The average litter size for the gray fox is 4 (range 1–10). Each newborn cub is black-furred and, like most foxes at birth, helpless, with eyes closed. Its eyes open at 9–12 days and by 4 weeks it ventures from the den and begins to climb, guarded by a parent. It starts to take solid food 2 weeks later.

GRIZZLED GRAY
The speckled or grizzled coat is due to individual hairs banded in white, gray, and black.

Atelocynus microtis

Small-eared dog

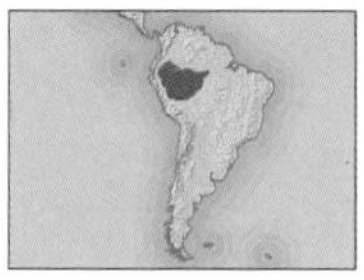

Length 28–39 in (72–100 cm)	
Tail 10–14 in (25–35 cm)	
Weight 20–22 lb (9–10 kg)	
Social unit Individual	
Status Near threatened	

Location N.W. South America

With its rounded, short ears, the small-eared dog resembles a raccoon dog (see p.171) but its fur is much shorter and more velvety, gray to black on the back and varying shades of gray tinged with red-brown on the underside. The black tail is more bushy and foxlike. Also known as the small-eared zorro, this mainly nocturnal and solitary dog is a secretive, little-known inhabitant of tropical forests. It moves with catlike stealth and probably eats mainly small rodents, with some plant matter.

Pseudalopex culpaeus

Culpeo

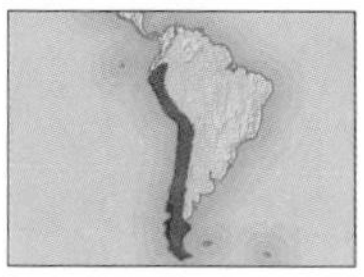

Length 23½–47 in (60–120 cm)	
Tail 12–18 in (30–45 cm)	
Weight 11–30 lb (5–13.5 g)	
Social unit Pair	
Status Least concern	

Location W. South America

A species of open upland and pampas grassland, this large, powerful fox is extensively hunted for its fur and to prevent predation of livestock such as lambs and poultry. Its diet also includes rodents, rabbits, birds and their eggs, and seasonal berries and fruit. Like many foxes, it stores food during times of plenty, burying the excess or wedging it under logs and rocks, for later consumption. The culpeo's coat is grizzled gray on the back and shoulders, more tawny on the head, neck, ears, and legs, with a fluffy, black-tipped tail.

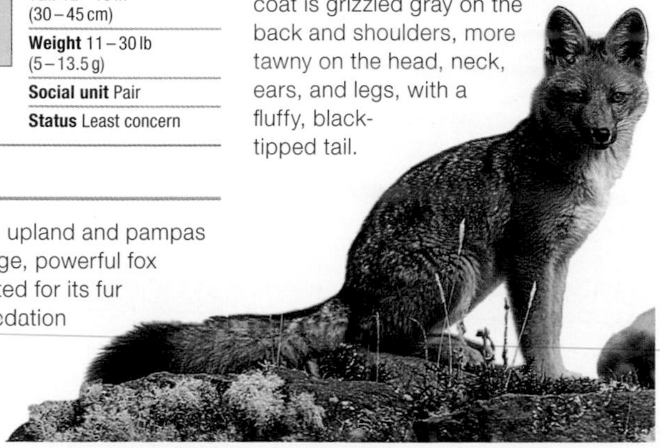

Cerdocyon thous

Crab-eating fox

Location N. and E. South America

Length 25 in (64 cm)
Tail 11½ in (29 cm)
Weight 11–18 lb (5–8 kg)
Social unit Group/Pair
Status Least concern

In addition to crabs—both coastal and freshwater—this medium-sized fox eats much else, including fish, reptiles, birds, mammals, grubs, and fruit. Widespread in many habitats, it shows much variation across its range, although the body is generally gray-brown, with reddish brown face, ears, and front legs, a white underside, and black on the tips of the ears and tail, and the backs of the legs. Active at night, it lives in loose social groups of an adult pair and their offspring.

Nyctereutes procyonoides

Raccoon dog

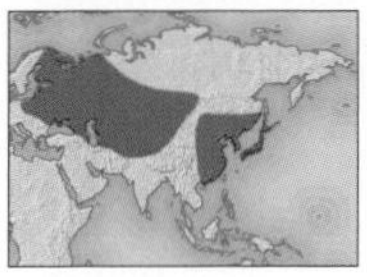

Location Europe, C., N., and E. Asia

Length 20–23½ in (50–60 cm)
Tail 7 in (18 cm)
Weight 17 lb (7.5 kg)
Social unit Group/Pair
Status Least concern

This canid resembles a combination of raccoon and dog, with its black face "mask" and variable black fur on the shoulders and upperside of the tail. It is nocturnal and has a huge dietary range, from fruit to birds, mice, crabs, and fish. It also forages along river banks, lakesides, and the seashore. It lives in pairs or loose family groups; litters average 4–6. The raccoon dog is abundant in Japan, extinct in parts of China, yet spreading rapidly in areas of Europe, where it has been introduced.

READY FOR WINTER

The raccoon dog is an unusual member of the dog family—even youngsters can climb well. It is also unique (for a canid) in that it hibernates in winter. Fall feasting increases body weight by up to 50 percent. A deep sleep follows, usually sheltered in an old fox or badger den.

COLORATION
This raccoonlike dog has long, yellow-tinged, brown-black body fur (especially in winter), black facial patches below the eyes, a white muzzle, short furred legs, and a bushy tail.

Speothos venaticus

Bush dog

Location Central America to N. and C. South America

Length 22½–30 in (57–75 cm)
Tail 5–6 in (12.5–15 cm)
Weight 11–15 lb (5–7 kg)
Social unit Group
Status Near threatened

Long-bodied and short-legged, this day-active predator lives in family-based packs of up to 10. It is a powerful and persistent hunter of ground birds and rodents up to the size of Azara's agouti (see p.130). The pack, however, tackles larger prey, such as rheas and capybaras, swimming efficiently after victims. By night, the group members sleep in dens, in deserted burrows, hollow logs, or under rocks. Average litter size is 4, born after a gestation of 67 days. The male brings food to the suckling female in the den.

Chrysocyon brachyurus

Maned wolf

Location C. and E. South America

Length 4–4¼ ft (1.2–1.3 m)
Tail 11–18 in (28–45 cm)
Weight 44–51 lb (20–23 kg)
Social unit Individual
Status Near threatened

Similar to a red fox (see p.169), but with very long legs, this wolf has long, thick, reddish yellow fur, a black neck crest, central back stripe, and black muzzle. It prefers open, grassy, or low-scrub habitats where it can peer over vegetation for prey and danger. Female and male share a territory, and mate each year, usually in May or June, but otherwise rarely associate. Active at twilight and night, the maned wolf takes a varied diet, including rabbits, birds, and mice, as well as smaller creatures such as grubs and ants, and also appreciable amounts of plant material such as fruit and berries. It is said to kill livestock, especially poultry, and so is hunted as a pest in some areas—yet it is kept as a pet in others. Disease is another major threat. The gestation period is 62–66 days, and the 1–5 pups (average 2) are born in an above-ground den in thick grass or bushes. The mother cares for them alone, suckling them for up to 15 weeks.

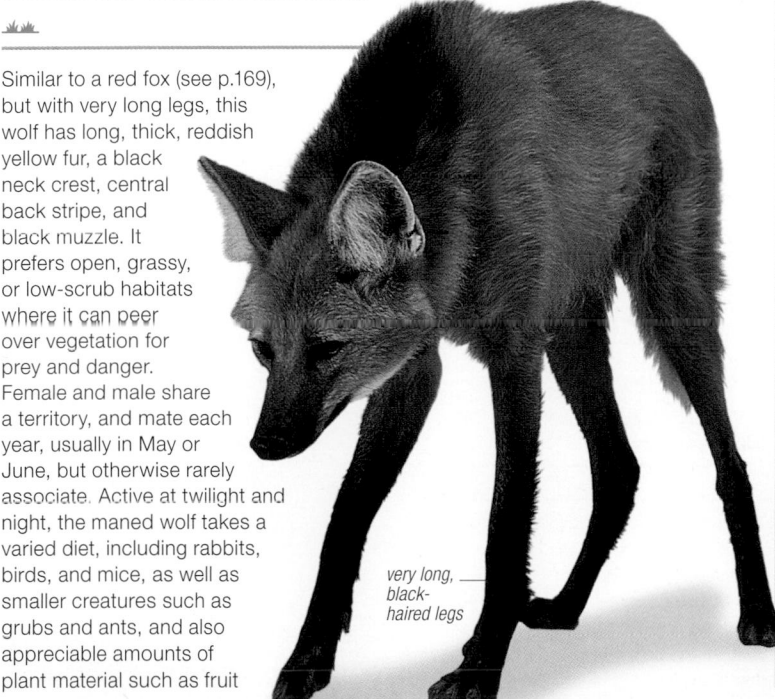

Canis adustus

Side-striped jackal

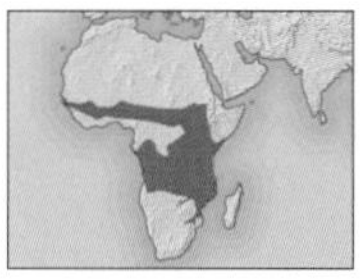

Location W., C., E., and southern Africa

Length 25½–32 in (65–81 cm)
Tail 12–16 in (30–41 cm)
Weight 14–31 lb (6.5–14 kg)
Social unit Pair
Status Least concern

Sometimes sighted foraging at night near city centers, the side-striped jackal is also found in grass, along forest edges, and in mixed farmland. More omnivorous than other jackals, it takes rodents, birds, eggs, lizards, insects and other invertebrates, refuse, carrion, and plant material such as fruit and berries. The basic social group is a female–male pair with their young, which can number up to 6 (average 5 per litter). Offspring are born after 57–70 days' gestation, in a secure den such as an old termite mound or aardvark burrow. Weaned by 10 weeks, they become independent at about 8 months.

Canis aureus

Golden jackal

Location S.E. Europe, N. and E. Africa, W. to S.E. Asia

Length 23½–43 in (60–110 cm)
Tail 8–12 in (20–30 cm)
Weight 15–33 lb (7–15 kg)
Social unit Pair
Status Least concern

These omnivorous, opportunistic jackals usually live as breeding pairs, but in areas with plentiful food, such as refuse dumps near human habitation, they form packs of up to 20. The gestation period is 60–63 days. Average litter size is 5–6 pups (range 1–9), which are cared for in a secure den.

MEALTIME

Golden jackal pups move on from milk to solid food at about 8–10 weeks. At this age they are too young to hunt, so parents, older siblings, and other young adults regurgitate meals for them on return from a successful outing.

SHADES OF GOLD
The coat is mainly pale yellow, gold, or light brown, grayer on the back and gingery on the belly.

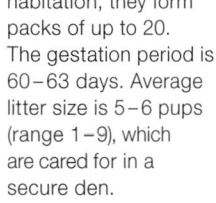

Canis latrans

Coyote

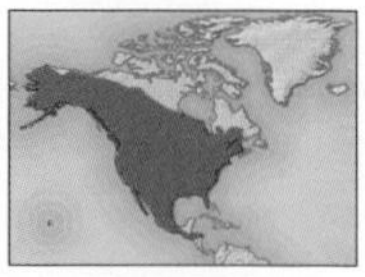

Location North America to N. Central America

Length 27½–38 in (70–97 cm)
Tail 12–15 in (30–38 cm)
Weight 20–35 lb (9–16 kg)
Social unit Variable
Status Least concern

The coyote, like many canids, is highly adaptable in habitat and opportunistic in diet. Once believed to be always solitary, it may also form a breeding pair or, when larger prey is common, gather as a small hunting pack. Food varies from pronghorns, deer, and mountain sheep to fish, carrion, and refuse. The coyote is a rapid sprinter (40 mph/65 kph) and often runs down jackrabbits. Its well-known nocturnal howl usually announces an individual's territory or location to neighbors. Mating occurs from January to March, gestation takes 63 days, and the litter size is 6–18 (average 6). The pups are born in a secure den.

STALK AND POUNCE

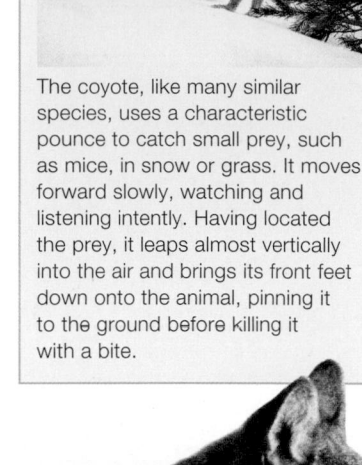

The coyote, like many similar species, uses a characteristic pounce to catch small prey, such as mice, in snow or grass. It moves forward slowly, watching and listening intently. Having located the prey, it leaps almost vertically into the air and brings its front feet down onto the animal, pinning it to the ground before killing it with a bite.

COLORATION
The grizzled buff coat is yellowish on the outer ears, legs, and feet. Underparts are gray or white. The shoulders, back, and tail may be tinged black.

Canis mesomelas

Black-backed jackal

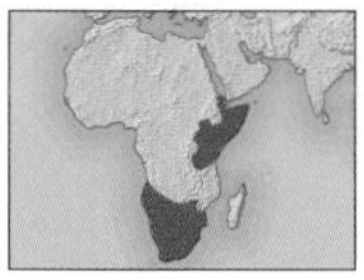

Location E. and southern Africa

Length 18–35 in (45–90 cm)
Tail 10–16 in (26–40 cm)
Weight 13–26 lb (6–12 kg)
Social unit Pair
Status Least concern

This jackal's range extends from city suburbs to the deserts of southern Africa. The main coloration is ginger to red-brown with a distinctive black saddle over its shoulders and back, and a black, bushy tail. Female and male mate for life and hunt together as adaptable omnivores. Their prey includes livestock such as sheep or young cattle. Their breeding habits resemble those of other jackals.

Canis simensis

Ethiopian wolf

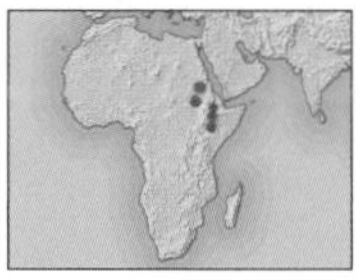

Location E. Africa

Length 3¼ ft (1 m)
Tail 13 in (33 cm)
Weight 33–40 lb (15–18 kg)
Social unit Group
Status Endangered

Formerly known as the Simien jackal, this species' 3 remnant populations (in the Ethiopian highlands) are at risk from habitat loss, competition and diseases from domestic dogs, and overgrazing, which has reduced their prey of hares, rodents, and giant mole rats. Groups of up to 12 wolves congregate noisily at morning, noon, and evening; most hunting is around dawn and dusk. Both parents and young adult "helpers" protect and regurgitate food for the cubs.

Canis rufus

Red wolf

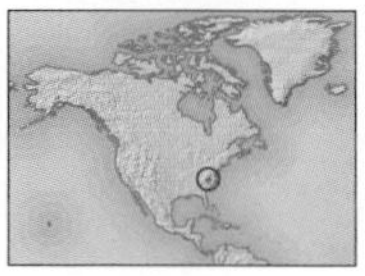

Location Reintroduced to E. USA (North Carolina)

Length 3¼–4 ft (1–1.2 m)
Tail 10–14 in (25–35 cm)
Weight 40–90 lb (18–41 kg)
Social unit Group
Status Critically endangered

By the 1970s, red wolves were believed to be extinct in the wild due chiefly to persecution and interbreeding with coyotes. Reintroduced from 1987 in North Carolina, they have established a population of more than 50. They hunt mammals such as rabbits, coypu, and raccoons, and live in packs with a social organization comparable to that of the gray wolf (see p.174). The coat is tawny-cinnamon mixed with gray and black, and is darkest on the back.

Canis lupus dingo

Dingo

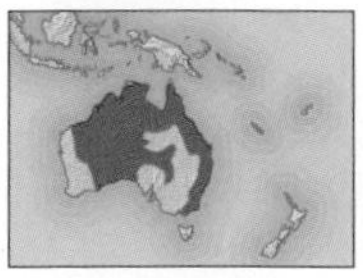

Location Australia

Length 28¼–43 in (72–110 cm)
Tail 8½–14 in (21–36 cm)
Weight 20–47 lb (9–21.5 kg)
Social unit Group
Status Vulnerable

This dog has variously been regarded as a subspecies of the domestic dog; as a subspecies of the domestic dog's ancestor, the gray wolf (see p.174); or as a full and separate species in its own right. It is likely that the dingo descended from the domestic dog within the past 10,000 years, and is now able to survive in the wild in many habitats. Dingoes are found throughout Australia, except for the southwest and southeast, where dingo fences exclude them from livestock; they are classified as pests, both to farm animals and for rabies control. Dingolike dogs also occur wild, semiferal, or semidomesticated, on the mainland and many islands of South and Southeast Asia. Dingoes interbreed readily with domestic dogs, and in parts of Australia one-third of individuals are such hybrids. Prey includes rabbits, rodents, wallabies, small kangaroos, and birds. However, the opportunistic dingo can survive on fruit, plant matter, and carrion. In social behavior and pack system, it resembles the gray wolf.

DINGO OR DOG?
The dingo's coat varies from light sandy to deep red-ginger. Dingo–domestic dog hybrids can look very similar, but may be distinguished by their canine and carnassial tooth shape.

PACK HIERARCHY

Young male dingoes may be solitary and nomadic. Breeding adults usually form settled packs, unless the population is widely spaced, when pairs are likely. About 5 pups (range 1–10) are born after a gestation of 63 days. Senior pack members teach them their place in the hierarchy by nips and other rebuffs.

Otocyon megalotis

Bat-eared fox

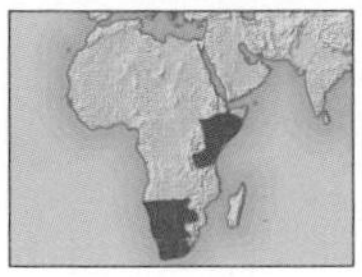

Location E. and southern Africa

Length 18–26 in (46–66 cm)
Tail 9–13½ in (23–34 cm)
Weight 4½–10 lb (2–4.5 kg)
Social unit Variable
Status Least concern

Huge ears and a small face with a pointed muzzle are the bat-eared fox's main external features, but its teeth, too, are very unusual. They are much smaller than those of a typical canid, and with up to 8 extra molars may number 48—more than any other nonmarsupial mammal. Its main diet is insects, especially termites and dung beetles. However, the breeding and social habits of this species are more typically foxlike.

Cuon alpinus

Dhole

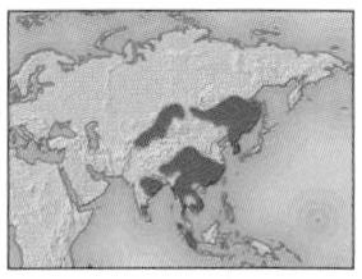

Location S., E., and S.E. Asia

Length 35 in (90 cm)
Tail 16–18 in (40–45 cm)
Weight 33–44 lb (15–20 kg)
Social unit Group
Status Endangered

Also called the Asian red dog, this species is widespread, but with a shrinking overall range and declining numbers. It lives in territorial, day-active packs of 5–12, usually based on an extended family. The fur color is evenly tawny or dark red, with a darker tail and lighter underparts; the legs are relatively short. The main prey is medium-sized hoofed mammals, supplemented by smaller creatures, fruit, and other plant food.

Lycaon pictus

African wild dog

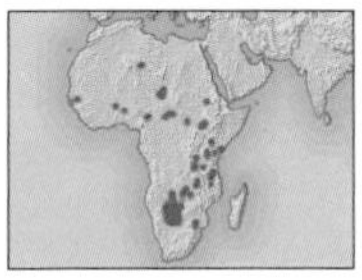

Location Africa

Length 30–43 in (76–110 cm)
Tail 12–16 in (30–40 cm)
Weight 37–79 lb (17–36 kg)
Social unit Group
Status Endangered

Probably the most social canid, the African wild or hunting dog lives in packs of 30 or more adults and young. Only the dominant pair breed, producing a litter of 10–12 (range 2–19) after a gestation of 69–73 days. However, the whole pack cares for and protects the pups, regurgitating food for them until they develop hunting skills by about 12 months. The pack also cooperates to hunt very large prey, such as wildebeest, zebra, and impala. This dog has long legs, and a lean build, with a relatively small head, large ears, and a short, broad muzzle. Unusually for a canid, it has only 4 toes on each foot. Its coat pattern is exceptionally variable, but the muzzle is usually black and the tail tip is white.

CONSERVATION

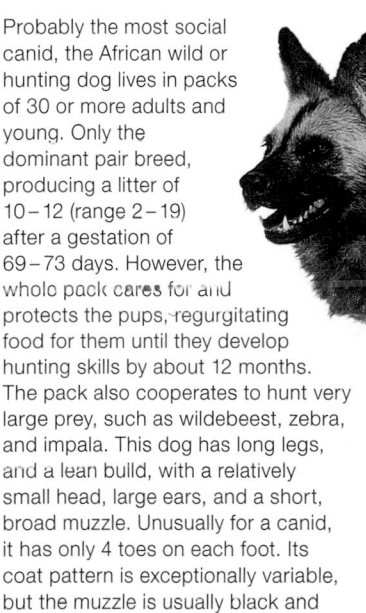

Once widespread across Africa, in many habitats, this wild dog is now reduced to scattered, fragmented populations. It is still persecuted, trapped, shot, and snared, and is also killed accidentally by road vehicles. It suffers both habitat loss and diseases (rabies, distemper) from domestic dogs. Survival depends on active conservation, including tracking pack movements by fitting with radio collars (shown here).

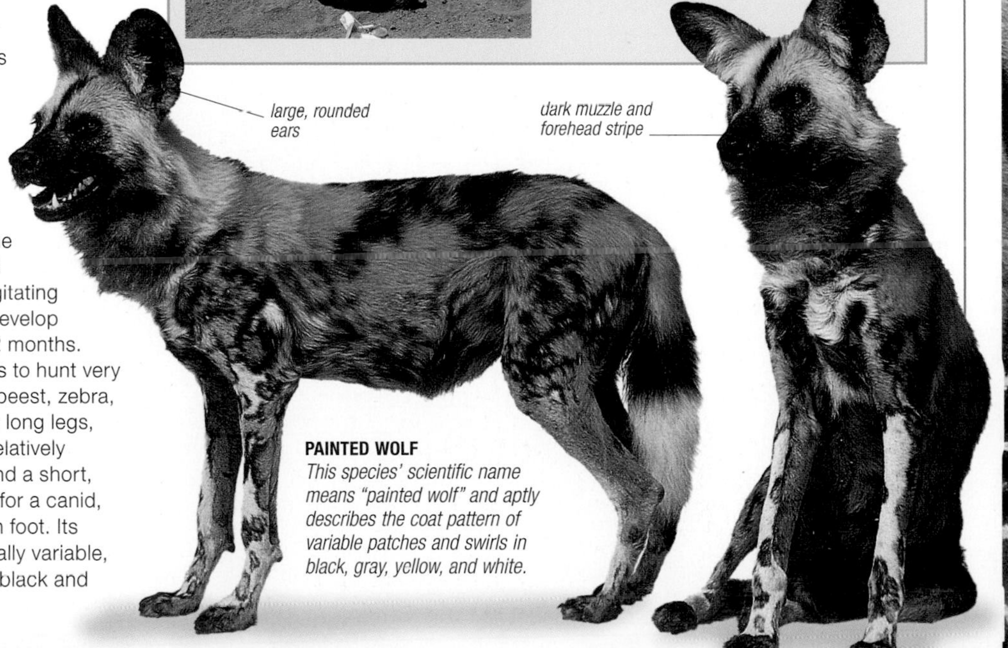

PAINTED WOLF
This species' scientific name means "painted wolf" and aptly describes the coat pattern of variable patches and swirls in black, gray, yellow, and white.

Canis lupus

Gray wolf

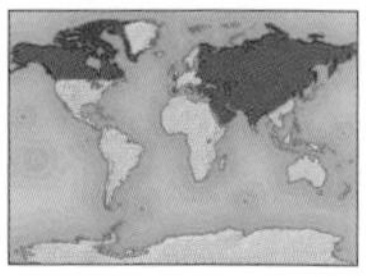

Location North America, Greenland, Europe, Asia

Length 3¼–5 ft (1–1.5 m)
Tail 1–1¾ ft (30–51 cm)
Weight 35–130 lb (16–60 kg)
Social unit Group
Status Least concern

The gray wolf is the largest wild member of the canid family and the ancestor of the domestic dog. Once the world's most widely ranging carnivore, its distribution has since been restricted by widespread human persecution and habitat destruction. An intelligent and social animal, its survival and great success as a predator is dependent on its organization into packs—family groups that commonly consist of 8 to 12 wolves. Packs patrol territories, covering very wide areas, which they maintain by scent markings. The clearly defined hierarchy within a pack centers around a dominant breeding pair that usually mates for life. By hunting in packs the gray wolf is able to take a wide range of prey, including moose and caribou, that may be up to 10 times a wolf's weight.

FEEDING PACK
Having captured their prey, the pack members wait behind the dominant pair for access to the kill.

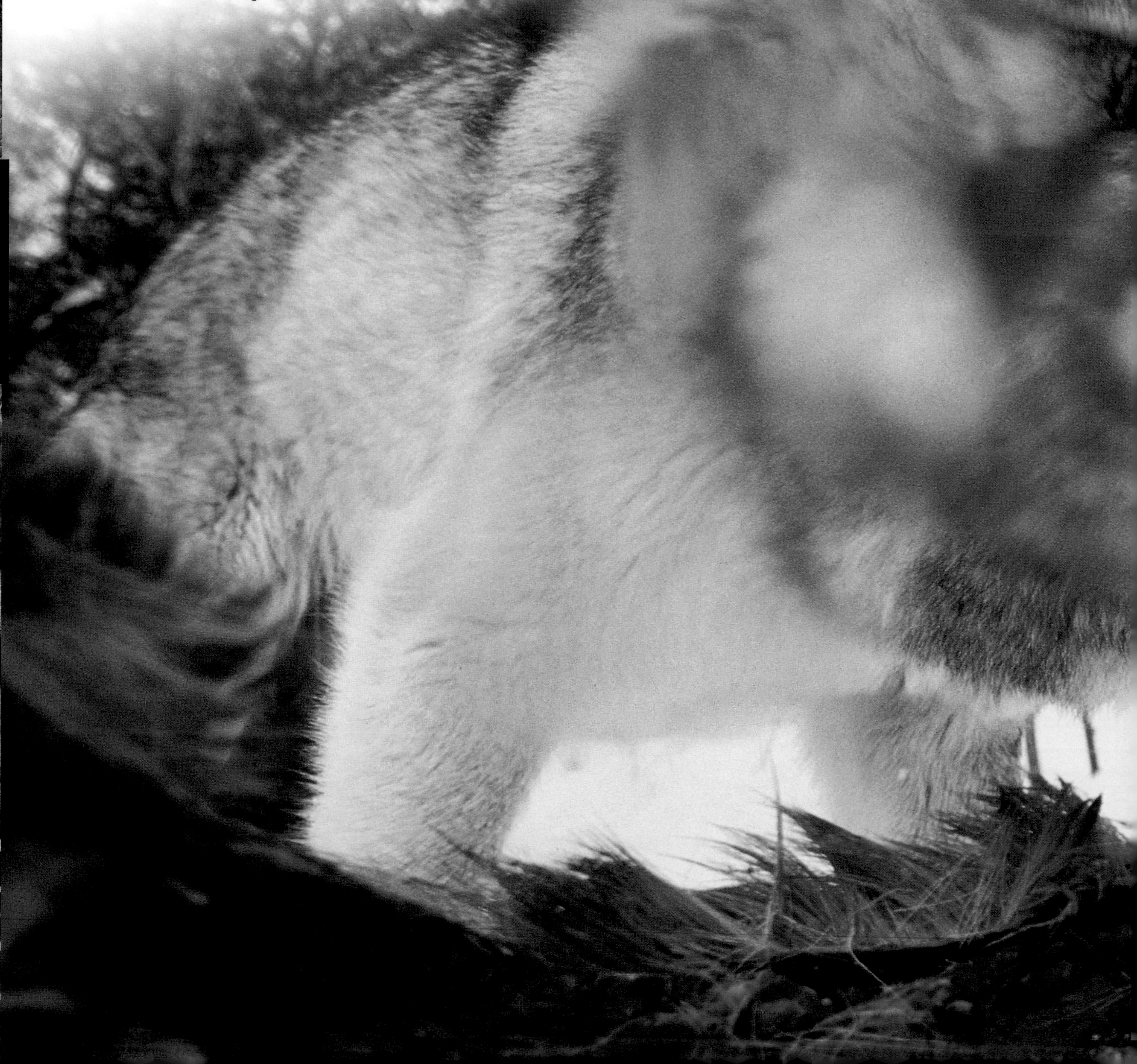

CARE OF THE YOUNG

During the breeding season, which lasts from January to April, the dominant female gives birth to between 4 and 7 pups. After about a month of suckling, pups emerge from the den to receive scraps of food regurgitated by their parents and other pack members. If their food supply has been plentiful, pups will have developed enough to travel with the pack after 3 to 5 months, and by the next breeding season some juveniles will have chosen to leave the pack entirely.

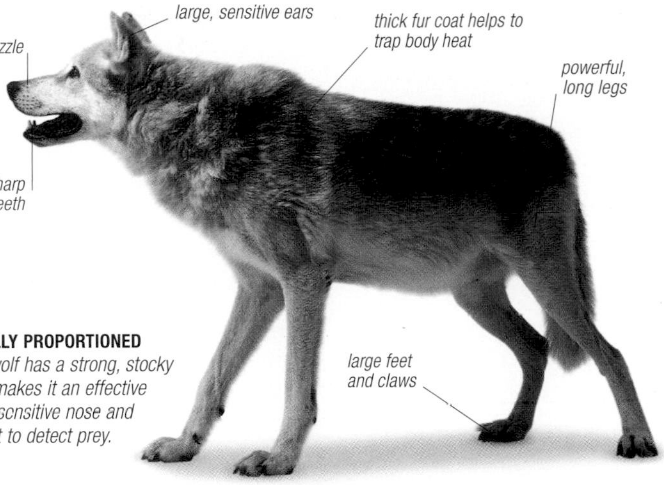

POWERFULLY PROPORTIONED
The gray wolf has a strong, stocky build that makes it an effective hunter. Its sensitive nose and ears help it to detect prey.

PREDATOR AND VICTIM
Although gray wolves can occur close to human settlements, their mythical reputation for ferocity led to their near extermination. Today, most gray wolves live in remote areas, where they hunt herds of large deer or musk oxen.

CALL OF THE WILD
Gray wolves howl to announce their presence and to define and defend their territories. Heard at distances of up to 6 miles (10 km), howling allows rival packs to stay well separated, avoiding confrontation.

Bears

PHYLUM	Chordata
CLASS	Mammalia
ORDER	Carnivora
FAMILY	Ursidae
SPECIES	8

The bear family includes the world's largest terrestrial carnivore, the brown bear, which can stand up to 11 ft (3.5 m) tall. Bears have a heavy build, a large skull, thick legs, and a short tail. They are found throughout Eurasia and North America, and in parts of North Africa and South America, mainly in forests. Unlike most carnivores, bears rely heavily on vegetation as a food source.

Anatomy

Bears are either large or medium-sized, and males are up to 20 percent larger than females. Although the giant panda is one notable exception, most bears have a black, brown, or white coat, and many feature a white or yellow mark on the chest. Despite the fact that they have a keen sense of smell, bears' sight and hearing are less well developed, and this is reflected in their large snout and small eyes and ears.

Most bears have lost the carnassial (shearing) function of the molar teeth. Instead, the molars are flat with rounded cusps, making them effective tools for grinding vegetation. Bears have large, strong paws—a single blow can often kill another animal—and long, nonretractable claws.

Movement

Compared with other carnivores, bears walk slowly and deliberately, with all 5 toes as well as their heels touching the ground (plantigrade gait). They can, however, move quickly if the need arises. When threatened or defending their territory, many bears stand on their back legs to increase their already considerable size. The majority of bears are agile climbers.

Feeding

The diet of most bears consists of a mixture of meat (including insects and fish) and plant material (from roots and shoots to fruit and nuts). Only the polar bear lives exclusively on meat, while the giant panda (which is sometimes classified in a family with the lesser panda and is occasionally grouped with the raccoon family) is almost entirely herbivorous. Because bears depend on plants more than other carnivores, they spend more of their time feeding. Most forage during the day.

Dens and dormancy

Many bears, especially those in cold regions, become dormant in winter. During this time, they retreat to a prepared den and live on reserves of body fat. This state differs from true hibernation (see p.89), which involves a drop in body temperature. Cubs are often born during dormancy. Since they have no fur, the newborn cubs are highly vulnerable and benefit from the snug environment created by their mother's body heat.

CONSERVATION

Of the 8 species of bears, only one—the giant panda—is officially listed as endangered. However, 5 are classified as vulnerable, with numbers that are shrinking, despite efforts to protect them. The reasons for their decline are many. They range from hunting to the melting of the Arctic sea ice, a problem that applies uniquely to the polar bear.

PHYSICAL INTIMIDATION
Bears can be aggressive animals, particularly when competing with each other during the breeding season. When male brown bears come into conflict, they will often try to intimidate one another by making themselves look as large as possible, growling and displaying their teeth. Smaller individuals will usually give way to larger ones, but if a warning is ignored, actual fighting will break out, often leading to serious injury or death.

Ursus maritimus

Polar bear

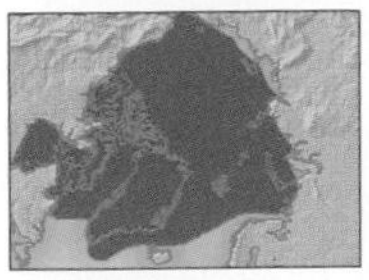

Location Arctic, N. Canada

Length 7–11 ft (2.1–3.4 m)
Tail 3¼–5 in (8–13 cm)
Weight 880–1,500 lb (400–680 kg)
Social unit Individual
Status Vulnerable

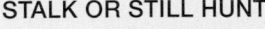

Vying with the brown bear as the largest land-based carnivore, the male polar bear can weigh twice as much as a female. Its favored habitat is a mix of pack ice, shoreline, and open water where seals are found. Some bears move 60 miles (100 km) inland in summer and vary their diet with birds' eggs, lemmings, lichens, mosses, and carrion, such as caribou and musk-oxen. Mating occurs on sea ice in April–May. The pregnant female digs a den in snow or earth and gives birth to 2 cubs (range 1–4) from November to January. The cubs take solid food at 5 months but are not weaned for another 2–3 years.

PAW PADDLES

Polar bears swim readily across open water at up to 6 mph (10 kph). They paddle with the massive forepaws—the rear legs trailing as rudders. The coat's hollow, air-filled guard hairs (see also below) aid buoyancy. When diving, the eyes remain open but the nostrils close as the bear holds its breath for up to 2 minutes, coming up stealthily beneath prey such as seabirds or surface basking seals.

NOT QUITE WHITE
The polar bear's guard (outer) fur is creamy rather than pure white. Hollow and translucent, the guard hairs transmit the sun's heat internally down to their bases, where it is absorbed by the black skin. The dense underfur and thick blubber (fatty layer) under the skin aid insulation.

STALK OR STILL HUNT

The polar bear's chief prey of seals and an occasional walrus are caught by 2 main hunting methods. In the stalk, the bear moves slowly nearer its prey, relying on its camouflaging white coat and "freezing" if the seal looks up. It charges the last 50–100 ft (15–30 m) at up to 34 mph (55 kph). In the still hunt, the bear waits motionless next to a seal's breathing hole, and grabs the prey as it surfaces. The bear bites the seal's head and drags it a short distance for consumption.

Ursus americanus

American black bear

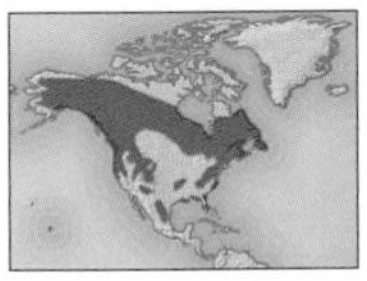

Location North America, Mexico

Length 4¼–6¼ ft (1.3–1.9 cm)
Tail 2¾–6 in (7–15 cm)
Weight 120–660 lb (55–300 kg)
Social unit Individual
Status Least concern

The American black bear is adaptable in habitat, but generally prefers forested country. Its powerful limbs and short claws tear open old logs to search for worms and grubs, and are also excellent for tree climbing, when this bear plucks fruit with its prehensile lips. The ears are larger and more erect than those of the brown bear, and it lacks a prominent shoulder hump. It sleeps in winter, which lasts up to 6 months in the north of its range. This bear may break into outbuildings or vehicles to obtain food left by humans, but it usually flees on confrontation.

BLACK TO BLUE BEARS
In the east, black is the main fur color, but in the west, it may be cinnamon or yellow-brown, and on the Pacific coast, gray-blue.

MAINLY VEGETARIAN

Some 95 percent of this bear's diet is plant-based, including, according to season, roots, buds, shoots, fruit, berries, and nuts, which are often obtained by climbing. It may also become adept at hunting deer fawns and at catching fish.

Ursus thibetanus

Asiatic black bear

Location E., S., and S.E. Asia

Length 4¼–6¼ ft (1.3–1.9 m)
Tail Not recorded
Weight 220–440 lb (100–200 kg)
Social unit Individual
Status Vulnerable

Similar to the American black bear (see left) in appearance and habits, the Asiatic black bear may spend up to half its time in trees. Its main foods include acorns, beech, and other nuts, fruit such as cherries, bamboo shoots and leaves, grasses, herbs, grubs, and insects such as ants. Where its natural forest habitat has been farmed and become fragmented, this bear may raid corn and other crops and, on occasions, has caused human fatalities. Eight months after mating the female gives birth in her winter den to 1–3 (usually 2) cubs. The Asiatic black bear is hunted for its body parts (especially the gall bladder), which are used in Asian cuisine and medicines.

Helarctos malayanus

Sun bear

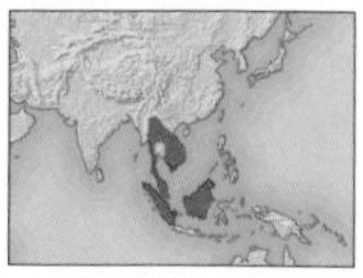

Location S.E. Asia

Length	3½–4½ ft (1.1–1.4 m)
Tail	Not recorded
Weight	110–145 lb (50–65 kg)
Social unit	Individual
Status	Vulnerable

The only truly tropical bear, the sun bear is an elusive, nocturnal, little-known omnivore of hardwood lowland forest. Its sleek, smooth fur varies from black to gray or rusty. It is paler on the muzzle, which is comparatively short. Its stocky, doglike body proportions and small size have led to the local name of dog bear. The sun bear spends much time in trees, even sleeping in a rough nest of bent-over or broken branches. It eats a range of fruit, shoots, eggs, small mammals, grubs, honey (its other name is honey bear), and varied plant food. Habitat loss, as forests are logged and converted to agriculture, is the major threat to the species, and it may raid crops, notably palm-tree plantations for the shoots, leading to persecution by farmers.

white to reddish "sun" chest patch varies from a U-shape to a circle or irregular spot

SMALLEST BEAR
The smallest bear species, the sun bear also has the shortest fur. If seized by a tiger or other predator, the loose skin around its neck allows it to turn and fight.

CLIMBING CLAWS

The sun bear has extremely long, curved claws, an adaptation for tree climbing. It also hugs the trunk with its front limbs and grips with its teeth, to haul itself up. The claws are also used to dig for worms and insects, and to tear up bark or old logs to expose and extract termites and remove honey from wild bees' nests.

LONG TONGUE

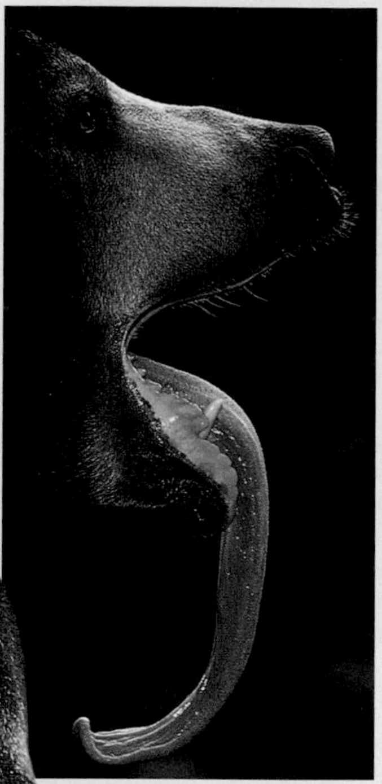

The sun bear's tongue can protrude 10 in (25 cm) to extract grubs, honey, and similar food from holes and crevices. It may also place each front paw alternately in a termite nest; the occupants crawl onto the paw, and the bear then licks them off.

Melursus ursinus

Sloth bear

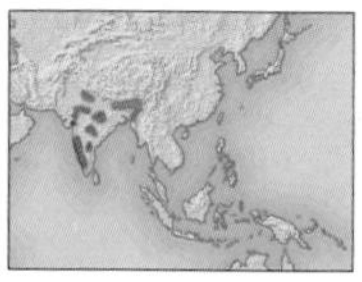

Location S. Asia

Length	4½–6 ft (1.4–1.8 m)
Tail	2¾–4¾ in (7–12 cm)
Weight	120–420 lb (55–190 kg)
Social unit	Variable
Status	Vulnerable

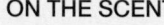

This small to medium-sized bear has a stocky body and short, powerful limbs. It can survive in a variety of habitats, including thorn scrub, grassland, and forest, if its 3 major foods—ants, termites, and fruit—are present (it also eats honey and eggs). Its long, nonretractile foreclaws (see panel above) do not, in this bear, permit effective tree climbing. Like other bears, the adult sloth bear is mainly solitary except during the mating season (June–July). However, brief groupings of 5–7 bears have been observed, even without local concentrations of food to attract them. The female occupies a natural hollow or digs one, and usually has 2 cubs (in November–January). The cubs stay in the den for 2–3 months, ride on the mother's back clinging to her long fur for another 6 months, and become independent after about 2 years. Threats include habitat loss, poaching for body parts (used in traditional medicines), persecution, and cub capture for training and performance.

ON THE SCENT

Bears locate food mainly by smell, sniffing with their long, mobile snout. The sloth bear specializes in ants and termites, tearing open their nests in soil, old logs, or trees with its foreclaws, which are 3 in (8 cm) long. Closing its nostrils and pursing its lips, it sucks the insects through the gap formed by its missing upper incisor teeth. The sucking noises it makes can be heard up to 330 ft (100 m) away.

white chest mark varies from a U-shape to a "Y" or "O"

SHAGGY BEAR
The sloth bear is distinguished by its long, rough fur, especially around the ears, rear neck, and shoulders. Color varies from black to brown or reddish; the muzzle is usually much paler or even white.

STANDING SURVEY

Like other bears, the sloth bear can stand just on its back legs. Once believed to be an aggressive posture, the bear is in fact gaining a clearer view of its surroundings and, more importantly, scenting the air to assess possible food or danger. Injuries to humans, mainly clawings, are usually the result of surprise encounters.

Tremarctos ornatus

Spectacled bear

Location W. South America

Length 5–6½ ft (1.5–2 m)
Tail 2¾–4¾ in (7–12 cm)
Weight 310–390 lb (140–175 kg)
Social unit Individual
Status Vulnerable

South America's only bear, and its largest land mammal after the tapirs, the spectacled or Andean bear is also the most arboreal of the family. Playing a prominent role in the folklore and mythology of many Andean peoples, its multitude of local names include oso achupayero (bromeliad-eating bear), yura mateo (white-fronted bear), yanapuma (black puma), and ucucu (after one of its rare vocalizations). The species once occupied habitats from coastal desert to high-elevation grassland, but human presence increasingly limits it to cloud forests at 3,300–8,900 ft (1,000–2,700 m). Mainly vegetarian, its massive jaw muscles and cheek teeth grind the toughest plants. For example, low-growing plants are simply torn up and chewed, despite cactus spines or leaf barbs on puya bromeliads. A common tree-feeding technique is to edge along a branch, bending other branches to bring fruit within reach. These bears may also create a simple tree platform, 16 ft (5 m) or more across, for feeding and resting. Most matings occur from April to June, but may happen at almost any time so that the birth corresponds with greatest food availability. The pair stay together for 1–2 weeks. The cubs' eyes open at 42 days and they may leave the den, in a hollow among rocks or tree roots, by 3 months. The cubs stay with the mother for 2 years, learning about feeding methods, food types, and threats. As with other bears, the male plays no part in rearing the cubs and, if he encounters them by chance, may attack and kill them.

CUBS IN THEIR DEN
Newborn spectacled bear cubs, like other baby bears, are tiny, each weighing only 12 oz (325 g). Most are born between December and February.

CROPS AND CONFLICT

The diet of the spectacled bear, one of the most herbivorous of bears, includes a huge range of fruit, as well as bromeliads, bulbs of wild orchids and similar flowers, palm shoots and leaf stalks, and, in drier areas, grass stems and cacti. Animal foods include insects, birds, eggs, small mammals, and carrion. Raids on crops, especially corn, and occasional attacks on livestock provoke revenge killings by farmers.

SPECTACLES
The creamy white eye markings vary from complete circles to "eyebrows" above or "teardrops" below, and allow identification of individual bears.

usually black, but occasionally red-brown fur

Ailuropoda melanoleuca

Giant panda

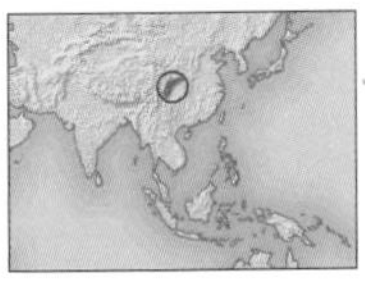

Location E. Asia

Length 5¼–6¾ ft (1.6–1.9 m)
Tail 4–6 in (10–15 cm)
Weight 155–280 lb (70–125 kg)
Social unit Individual
Status Endangered

Instantly recognizable as the worldwide symbol of conservation, the giant panda's own survival is still far from secure. It has a highly restricted diet—99 percent bamboo, using different parts of the 30 or more bamboo species, taking new shoots in spring, leaves in summer, and stems in winter. Carrion, grubs, and eggs are also eaten when available. Normally solitary, the panda feeds mainly at dawn and dusk, and sleeps in a bamboo thicket. It marks its home ranges with anal gland scents, urine, and claw scratches, avoiding confrontation by using overlapping areas at different times. The female, four-fifths the male's size, indicates readiness to mate by moans, bleats, and barks (11 distinct panda calls have been identified). Males gather, chase, and fight each other for the female. One to 2 cubs are born in a den in a hollow tree or rocky cave after 45 days' gestation. Only 6 in (15 cm) long, and 3⅝ oz (100 g) in weight, the newborn cubs are nearly naked.

front limbs more muscled than rear limbs, for climbing

BLACK ON WHITE
The giant panda is white with black ears, oval-shaped eye patches, nose, shoulder "saddle," and limbs. It has erect ears, a broad face, and small eyes.

coarse, oily guard (outer) hairs up to 4 in (10 cm) long

CONSERVATION

More than 150 pandas are kept in captivity worldwide, and about 10 times that number live in the wild. But despite the attention lavished on them, giant pandas are still in decline. One of the reasons for this is the panda's low reproductive rate—only about one in 3 cubs survive their first 6 months. In addition, the wild population lives in isolated fragments. In today's China, demand for land and lumber erode the panda's habitat, contributing to the problems that it has to face.

SIXTH "FINGER"

The giant panda handles bamboo with great dexterity due to an extension of the sesamoid bone in the wrist, which projects as a padlike "false thumb." This can flex and oppose the true thumb (first digit) to grip stems and leaves.

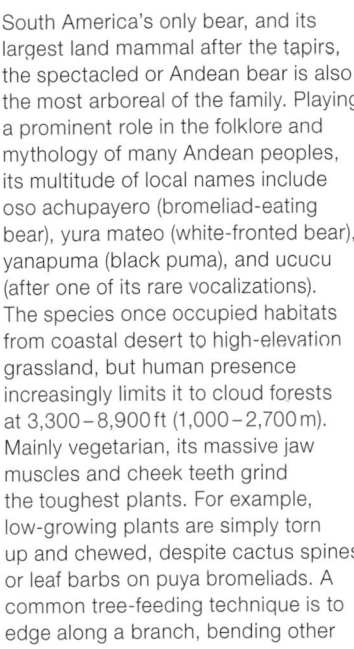

Ursus arctos

Brown bear

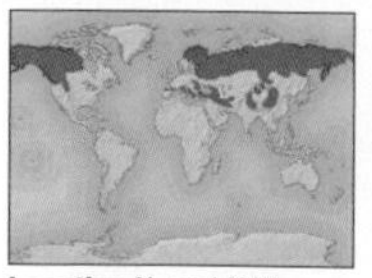

Location N. and N.W. North America, N. Europe, Asia

Length 6½–9¾ ft (2–3 m)
Tail 2–8 in (5–20 cm)
Weight 100–1,000 kg (220–2,200 lb)
Social unit Individual
Status Least concern

The brown bear enjoys the widest distribution of all bear species and varies widely in size depending on its food and habitat. Large areas of open wilderness are important to its survival, which explains why populations can be found in isolated areas such as parts of Alaska and the Yukon, while habitat destruction in the rest of North America and Europe has seen a drastic reduction in its numbers there. The brown bear's distinctive features are its shoulder hump of muscle, and long claws that help it to dig for roots and bulbs. It can stand upright on its hind paws in order to identify a threat or a food source. Although mainly herbivorous, brown bears will readily eat meat when it is available. To avoid winter food shortage, they can den up in dugouts in hillsides, or in brush, for as long as 6 months, during which period the female also gives birth to her cubs. The lifespan of the brown bear is about 25 years in the wild and longer in captivity.

BROWN BEAR SUBSPECIES

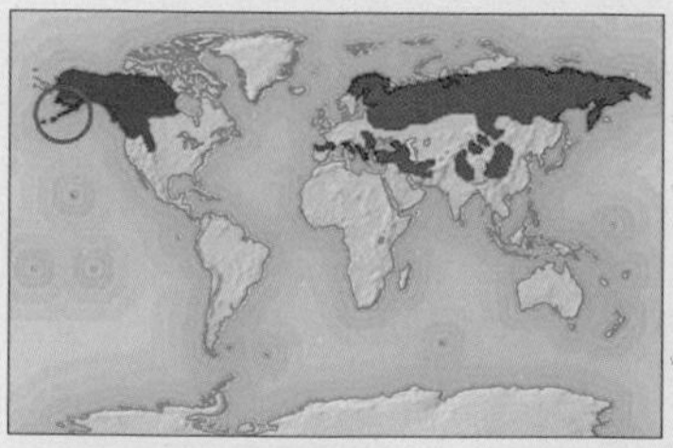

Several subspecies of brown bears are commonly recognized: grizzly bear, Kodiak bear, Alaskan bear, Eurasian brown bear, Syrian bear, Siberian bear, Manchurian bear, and Hokkaido bear. However, their taxonomy is disputed, and rather than being "true" subspecies, they may simply represent size variations due to disparities in food supply.

KODIAK BEAR
Weighing up to 1,720 lb (780 kg), the impressive Kodiak bear, Ursus arctos middendorfi, *is the largest of the subspecies.*

EURASIAN BROWN BEAR
This bear, Ursus arctos arctos, *is the smallest member of the species. Rapid loss of habitat has now restricted it to small pockets of mountain woodland.*

GRIZZLY BEAR
This subspecies, Ursus arctos horribilis, *gets its common name from its "grizzled" coat, the hair being lighter at the tips than at the base.*

POWERFUL BUILD
Brown bears are large, powerfully built animals. Although there is little difference in body length between the sexes, males can be up to twice as heavy as females, which have a smaller, lighter frame. Both sexes feed intensively from spring to fall in order to put on weight in preparation for winter sleep.

SALMON HARVEST

Brown bears are comfortable in water and may wait for hours at waterfalls or in the shallows of a stream ready to dive on their prey. As spawning salmon swim upstream, the brown bear pounces, delivering a crushing bite with its powerful jaws or a stunning blow from one of its large, clawed paws. Salmon is a vital source of protein for coastal populations of brown bear, which are usually the largest of the species; once caught, the fish rarely escape the bear's grasp.

AGGRESSIVE BEHAVIOR
Due to its power, size, and unpredictable behavior, the brown bear has long been considered a threat to humans and livestock. Grizzly bears evolved in open habitat and may act aggressively when defending themselves, since they have little opportunity to find the safety of any cover. A mother with young cubs is a particularly dangerous animal to encounter. Nevertheless, bears will usually avoid coming into contact with humans.

Sea lions, walrus, and seals

PHYLUM	Chordata
CLASS	Mammalia
ORDER	Carnivora
FAMILIES	Otariidae, Odobenidae, Phocidae
SPECIES	35

CLASSIFICATION NOTE

Seals and their allies were traditionally classified in the order Pinnipedia. Today, however, most zoologists believe that these mammals belong to the order Carnivora—the system used here. Current evidence suggests that, among other carnivores, these families are most closely related to bears (Ursidae) and that the sea lions (Otariidae) and the walrus (Odobenidae) are closer to each other than either is to seals (Phocidae).

Although clumsy on land, pinnipeds (seals, sea lions, and the walrus) are supremely agile underwater. They have a streamlined body and powerful flippers and can dive to depths of over 330 ft (100 m). Some species can remain underwater for over an hour. There are 3 families: the Otariidae are the eared seals (sea lions and fur seals), which have small external ears and back flippers that can be rotated forward for movement on land; the walrus (Odobenidae), which has distinctive tusks; and the true seals (Phocidae), which have no external ears. Only eared seals and the walrus can support themselves in a semiupright position on land. Pinnipeds are found worldwide, mostly in temperate and polar seas.

Temperature control

Pinnipeds have several heat-regulating adaptations. In cold water, the blubber insulates the internal organs, and blood flow to the flippers is restricted. In warm conditions, some species wave their flippers to expel excess heat. In addition, true seals and walruses can either contract the blood vessels near the skin's surface (to reduce heat loss in icy water) or they can dilate these vessels to gain heat when basking in the sun. Eared seals, however, will enter the water to avoid overheating.

COLOR CHANGE
The blood vessels in the skin of these walruses are dilated to maximize the amount of heat they can absorb by lying in the sun. As a result, their bodies turn pink.

UNDERWATER ACROBATS
In water, pinnipeds, such as these South American sea lions, are graceful, athletic, and capable of swimming at high speed. While underwater, they can communicate by sounds produced using air retained in the lungs.

Anatomy

Most pinnipeds have a short face, a thick neck, and a torpedo-shaped flexible body. A layer of blubber beneath the skin provides insulation, aids buoyancy, acts as an energy reserve, and protects the organs. All species are covered with hair, except the walrus, which is nearly hairless. Pinnipeds have large eyes for good deep-water vision, excellent hearing, ear passages and nostrils that can be closed underwater, and long whiskers that enhance the sense of touch. Many species display marked sexual dimorphism: elephant seal males weigh 4 times more than females.

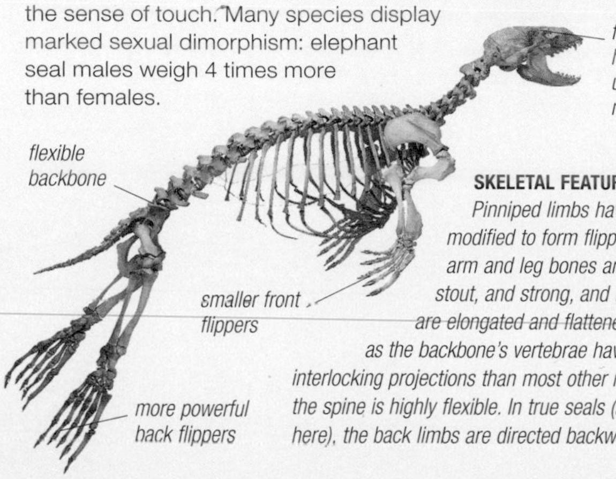

SKELETAL FEATURES
Pinniped limbs have been modified to form flippers: the arm and leg bones are short, stout, and strong, and the digits are elongated and flattened. Also, as the backbone's vertebrae have fewer interlocking projections than most other mammals, the spine is highly flexible. In true seals (shown here), the back limbs are directed backward.

Life cycle

Unlike the other marine mammals (cetaceans and manatees and the dugong), pinnipeds have not abandoned land entirely. In most species, during the annual breeding season, males attempt to set up territories on suitable beaches, fighting savagely for space and excluding weaker males. Females move onto the beaches, sometimes several weeks after the males, and give birth. A few days after a pup is born (usually only one young is produced), the female mates with the male in whose territory she has settled. For the majority of the gestation period, which lasts approximately 8–15 months, pinnipeds are mostly at sea and return to land only when it is time to repeat the breeding process.

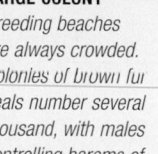

MALE AGGRESSION
In the breeding season, there is fierce competition between males, such as these 2 elephant seals, for mating rights. Only the strongest males are able to secure a breeding territory.

LARGE COLONY
Breeding beaches are always crowded. Colonies of brown fur seals number several thousand, with males controlling harems of 7–9 females.

Callorhinus ursinus

Northern fur seal

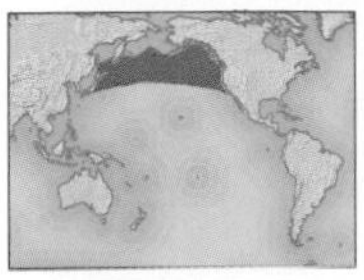

Location North Pacific

Length Up to 7 ft (2.1 m)
Weight 400–600 lb (180–270 kg)
Social unit Variable
Status Vulnerable

The male northern fur seal is brown-gray, while females and juveniles are silver-gray above and red-brown below, with a gray-white chest patch. The front flippers are long and appear "cut off" at the wrist. The diet includes many fish, and also birds such as loons and petrels. Most populations are migratory, with adult males heading south in August. The young stay on land for 4 months and by November follow with their mothers.

Arctocephalus pusillus

Brown fur seal

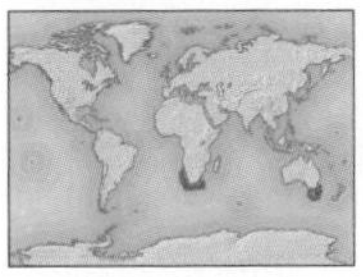

Location Southern Africa, S.E. Australia, Tasmania

Length 6–7½ ft (1.8–2.3 m)
Weight 440–790 lb (200–360 kg)
Social unit Variable
Status Least concern

Brown fur seals from off South Africa tend to be a darker gray-brown than Australian individuals, and they dive twice as deep (to 1,300 ft/400 m). The young are about 28 in (70 cm) long and weigh 13 lb (6 kg) when born, in November–December. They play in "nursery" tidal pools, while the mothers feed at sea for several days at a time.

Zalophus californianus

California sea lion

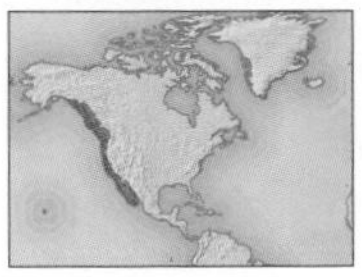

Location W. USA

Length Up to 7¾ ft (2.4 m)
Weight 610–860 lb (275–390 kg)
Social unit Variable
Status Least concern

The "performing seal" of marine parks and aquaria, this species rarely strays more than 10 miles (16 km) out to sea, and often enters harbors and estuaries for food and shelter. The male is dark brown, and females and juveniles are a uniform tan color. Males have a peaklike head crest. The main prey is schooling fish such as herring, squid, and sardines, caught on short (2-minute) dives down to about 245 ft (75 m). In "El Niño years," the California sea lion's diet switches to whiting, salmon, and also birds such as guillemots. During the breeding season (May–July), males fight for small territories on the beaches and rock pools. However, after 2 weeks, they must swim off to feed, and on return, have to battle to regain a territory. The mother cares for her single pup (rarely 2) for 8 days, then enters a cycle of 2–4 days feeding at sea and 1–3 days suckling on land. This usually lasts for 8 months until the next birth.

Phocarctos hookeri

New Zealand sea lion

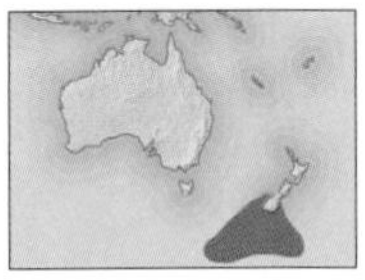

Location Subantarctic islands south of New Zealand

Length 6½–11 ft (2–3.3 m)
Weight 660–990 lb (300–450 kg)
Social unit Variable
Status Vulnerable

Also called Hooker's sea lion, this species is restricted to a few islands south of New Zealand. It forages up to 95 miles (150 km) from land and then retires perhaps ⅔ mile (1 km) inland, to rest among cliffs or trees. Males are dark brown with silver-gray hindquarters and a shoulder mane; females and juveniles are silvery or brown-gray above, yellowish or tan underneath. The diet includes fish, crabs, penguins, and seal pups.

Otaria flavescens

South American sea lion

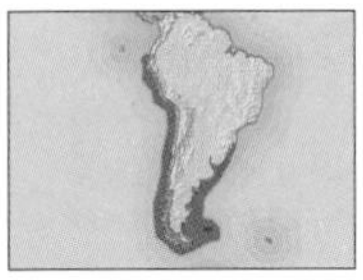

Location W., S., and E. South America, Falkland Islands

Length 7½–9¼ ft (2.3–2.8 m)
Weight 660–770 lb (300–350 kg)
Social unit Variable
Status Least concern

An enormous, heavy head and brown fur that is paler or yellow on the underside identify this powerful sea lion. The male of this nonmigratory species has a copious shoulder and chest mane, and is twice the weight of the female (as in many sea lions). Its breeding (rookery) areas are used year-round for resting. The mothers coax their pups into the water after 1–2 months—a relatively early age for a sea lion.

Eumetopias jubatus

Steller's sea lion

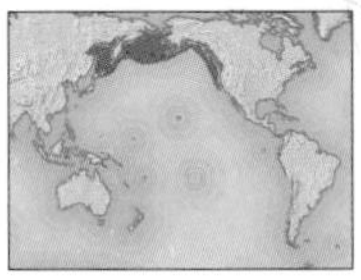

Location North Pacific rim

Length 9¾–11 ft (3–3.3 m)
Weight 1,290–2,500 lb (585–1,120 kg)
Social unit Group
Status Endangered

The largest sea lion, a male Steller's sea lion has a wide muzzle, a huge head, and a thick neck, and may be 3 times the weight of the female. Both are tawny or buff-colored, with black flippers furred only on the upper side. Breeding habits resemble those of other sea lions—colonies numbering over 1,000. Steller's sea lions dive deeply for fish, seals, and otters.

Arctocephalus gazella

Antarctic fur seal

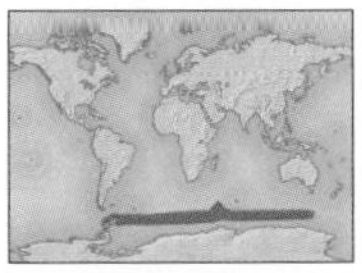

Location Antarctic and subantarctic waters

Length 5¼–6½ ft (1.6–2 m)
Weight 200–460 lb (90–210 kg)
Social unit Variable
Status Least concern

The mane of the male Antarctic fur seal is accentuated by extra muscle and fat, deposited under the skin. He is dark gray-brown, while the female is midgray. Males arrive at their breeding islands in November and compete for territories in which they can mate with about 5 females. Almost exterminated by fur hunting in the 19th century, this species is now recovering despite increased fishing of krill, a major component of its diet.

MAMMALS

Odobenus rosmarus

Walrus

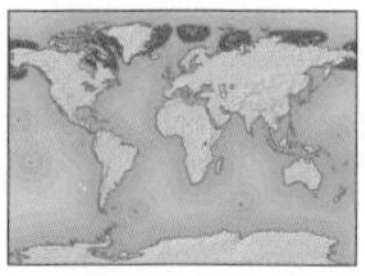

Length 9¾–12 ft (3–3.6 m)
Weight 1.3–2 tons (1.2–2 tonnes)
Social unit Group
Status Data deficient
Location Arctic waters

The male walrus is twice the weight of the female. The walrus's front flippers resemble those of sea lions, and the rear flippers those of seals. Its blunt, thickly whiskered muzzle widens rapidly to the head, neck, and chest, then tapers markedly to the tail, which is embedded in a web of skin. The walrus feeds mainly on seabed dwelling worms, shellfish, sea snails, shrimps, and slow-moving fish. It dives more than 330 ft (100 m) deep, for 25 minutes or more, to find prey using the touch of its whiskers and snout. It then excavates the food with its nose, aided by jets of water squirted from its mouth. Items are eaten mainly by suction with the mouth and tongue, rather than using the teeth. Walruses are social and huddle on land or ice floes in large, mixed groups of hundreds, which split at sea into smaller bands of less than 10; bachelor males tend to form their own gatherings. Courting males make underwater pulses and bellows to attract partners, and mating occurs between January and March. The pup, born after 15 months (which includes 4–5 months' delayed implantation), is up to 4 ft (1.2 m) long and 165 lb (75 kg) in weight. It suckles for 6 months and is weaned over the following 18 months. Mothers are extremely protective of their young, and other females may "adopt" pups that have been orphaned.

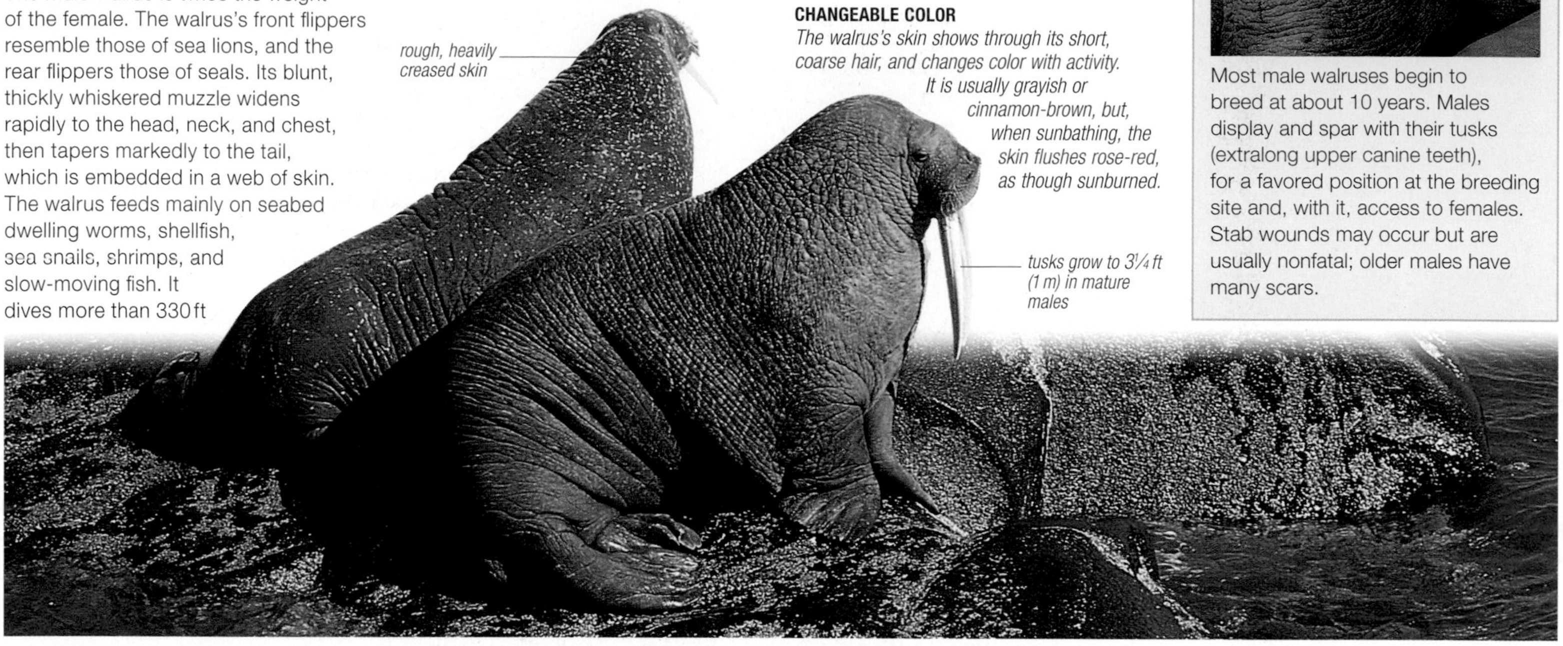

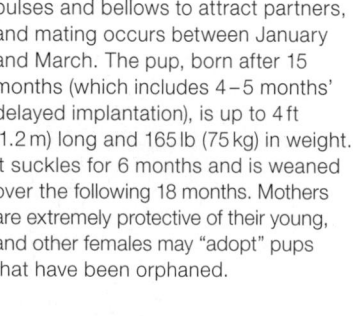

CHANGEABLE COLOR
The walrus's skin shows through its short, coarse hair, and changes color with activity. It is usually grayish or cinnamon-brown, but, when sunbathing, the skin flushes rose-red, as though sunburned.

SPARRING RIVALS

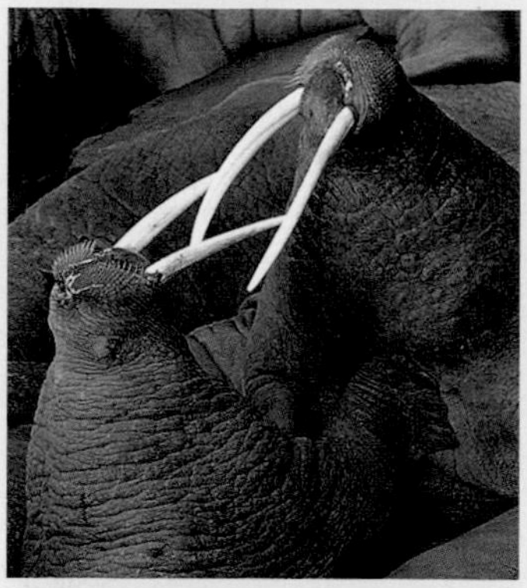

Most male walruses begin to breed at about 10 years. Males display and spar with their tusks (extralong upper canine teeth), for a favored position at the breeding site and, with it, access to females. Stab wounds may occur but are usually nonfatal; older males have many scars.

Monachus monachus

Mediterranean monk seal

Length 7¾–9¼ ft (2.4–2.8 m)
Weight 550–880 lb (250–400 kg)
Social unit Individual/Group
Status Critically endangered
Location Mediterranean, Atlantic (N.W. Africa)

This species has smooth, dark brown fur, paler beneath, and eats mainly fish, such as eels, sardines, and tuna, as well as lobster and octopus. On land, it is less social than most other seals, with the mother–pup pairs or small groups being widely spaced. This extremely rare species is sensitive to human disturbance such as tourism, so hides in sea caves. Cave collapses, pollution, overfishing, and viral infection are also serious threats.

Lobodon carcinophaga

Crabeater seal

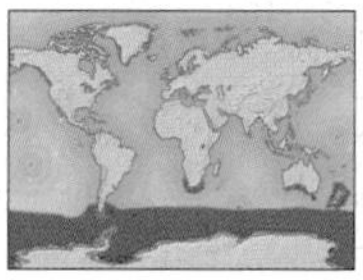

Length 7¼–8½ ft (2.2–2.6 m)
Weight 490 lb (220 kg)
Social unit Variable
Status Least concern
Location Antarctic and subantarctic waters

The long, lithe crabeater seal has oar-shaped, pointed front flippers and a silver-gray to yellow-brown coat with irregular darker spots and rings. One of the most abundant and fastest seals, it swims at 16 mph (25 kph). A typical feeding dive is down to 130 ft (40 m) for 5 minutes. Despite its name, the crabeater seal strains krill using its unusual lobed teeth. Breeding habits are typical for seals, although the male stays with female and pup until weaning takes place at 3 weeks.

Hydrurga leptonyx

Leopard seal

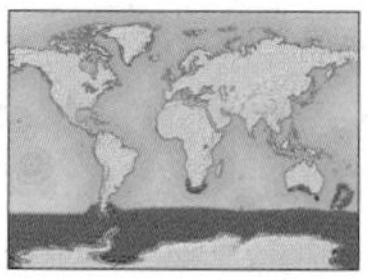

Length 8¼–10 ft (2.5–3.2 m)
Weight 440–1,000 lb (200–455 kg)
Social unit Individual
Status Least concern
Location Antarctic and subantarctic waters

The solitary, sinuous leopard seal is widest at the shoulders and, unusually for a true seal (phocid), swims with its front flippers, which have claws on the fingertips. The head is reptile-like in proportions, with no forehead and a wide, deep lower jaw. The 1-in (2.5-cm) canine teeth are adapted for seizing smaller seals, penguins, and other birds; the diet also includes squid and krill.

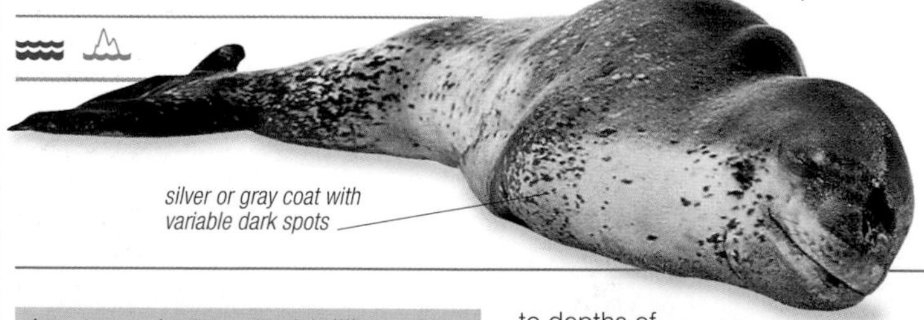

Leptonychotes weddellii

Weddell seal

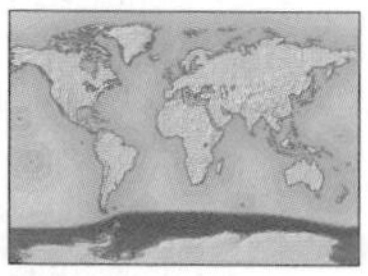

Length 8¼–9½ ft (2.5–2.9 m)
Weight 880–1,320 lb (400–600 kg)
Social unit Variable
Status Least concern
Location Antarctic and subantarctic waters

Bulky, but small-headed and short-flippered, the Weddell seal has a short, blunt muzzle and few, short whiskers. To find fish, squid, and other prey, it dives to depths of 1,600 ft (500 m) for one hour. This seal bites breathing holes in sheet ice with its long, upper incisor teeth.

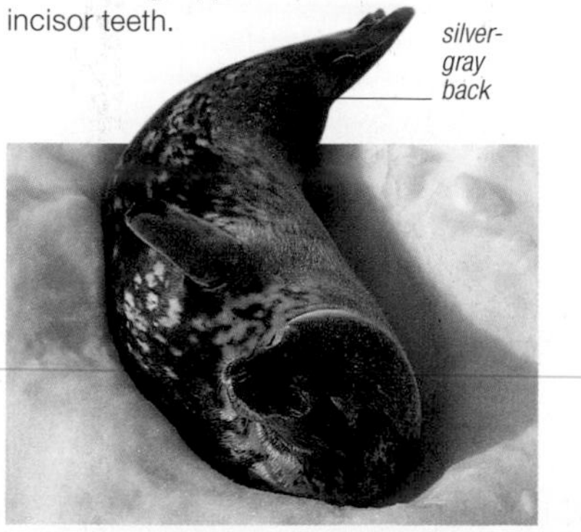

MAMMALS

Ommatophoca rossii

Ross seal

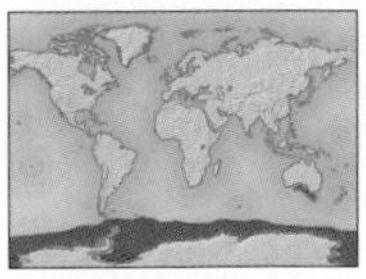

Length 5½–9¾ ft (1.7–3 m)
Weight 290–470 lb (130–215 kg)
Social unit Individual
Status Least concern

Location Antarctic waters

The Ross seal has a distinctively blunt muzzle, a wide head, and long rear flippers. Its fur is the shortest of any seal, and is dark gray to chestnut-brown with a buff underside and broad, dark bands along the body in both adults and pups. Less social than most seals, on ice it lives alone or as a mother–pup pair. The main food is squid, krill, and fish, caught at depths of several hundred yards. In November and December, males battle for territories around breathing holes in the ice used by females.

Mirounga leonina

Southern elephant-seal

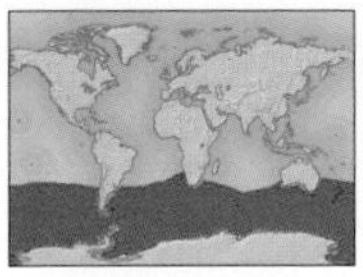

Length 14–20 ft (4.2–6 m)
Weight 2.2–5.5 tons (2.2–5 tonnes)
Social unit Variable
Status Least concern

Location Antarctic and subantarctic waters

The largest pinniped, the male southern elephant-seal is 4–5 times the weight of the female. His huge nose resembles an elephant's trunk, which he inflates when roaring at rivals during the 2-month breeding season. To establish dominance, he also rears up, slaps, and butts. The single pup is born after a gestation of 11 months (including 4 months' delayed implantation) and is suckled for 19–23 days by the ever-present mother, who loses one-third of her body weight during this time. After breeding and molting, these elephant-seals migrate south, feeding on fish and squid and diving to 2,000 ft (600 m) for 20 minutes on average.

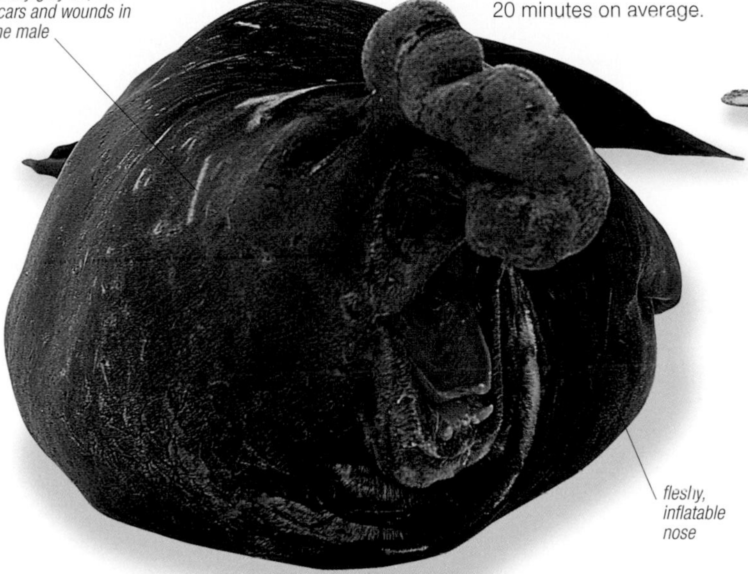

Cystophora cristata

Hooded seal

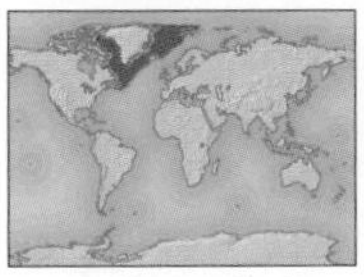

Length 8¼–8¾ ft (2.5–2.7 m)
Weight 660–900 lb (300–410 kg)
Social unit Variable
Status Vulnerable

Location North Atlantic to Arctic Ocean

The hooded seal has a wide, fleshy muzzle that droops over the mouth. The male intimidates his rivals at breeding time by inflating his nasal chamber to form a "hood," which doubles his head size, and also by extruding an internal membrane from his left nostril, which also inflates like a brown-red balloon. Mainly solitary, this seal migrates when not breeding or molting, to follow the pack ice. The pup, born on an ice floe, is up to 3½ ft (1.1 m) long and 66 lb (30 kg) in weight, and is weaned in 4–5 days—the shortest time of any mammal.

Halichoerus grypus

Gray seal

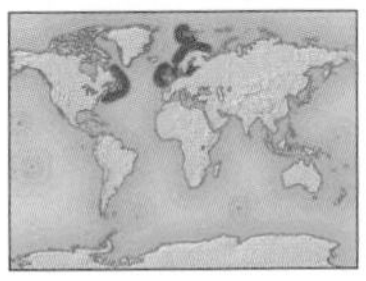

Length 6½–8¼ in (2–2.5 m)
Weight 370–680 lb (170–310 kg)
Social unit Variable
Status Least concern

Location North Atlantic, Baltic Sea

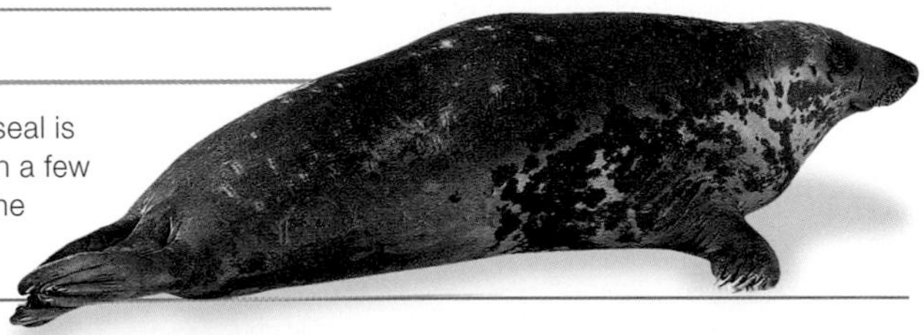

The male gray seal is gray-brown with a few pale patches; the female is paler gray-tan. The face has small eyes, widely separated nostrils, and an angular nose. There are 3 populations: coastal northwest Atlantic, coastal northeast Atlantic, and the Baltic Sea. The first group are 20 percent heavier and breed from December to February; Baltic gray seals breed until April, and those from the northeast, from July to December. The male gray seal does not defend a set territory.

Pagophilus groenlandicus

Harp seal

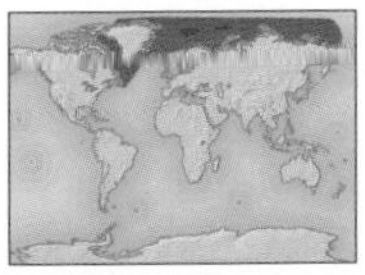

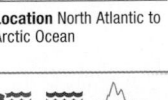

Length 5¼ ft (1.7 m)
Weight 290 lb (130 kg)
Social unit Variable
Status Least concern

Location North Atlantic to Arctic Ocean

The harp seal has close-set eyes in a wide face, black fingertip claws, and silver-white fur with curved dark marks on the back that form a harp shape. It eats cod, capelin, and similar fish, migrating with the edges of pack ice. Social both on ice and in water, harp seals travel in dense, noisy groups. Pups, born February–March on ice, have yellow fur that whitens for 2 weeks before the first molt.

Phoca sibirica

Baikal seal

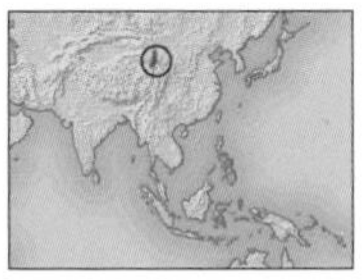

Length 4–4½ ft (1.2–1.4 m)
Weight 175–200 lb (80–90 kg)
Social unit Individual
Status Least concern

Location E. Asia (Lake Baikal)

One of the smaller seals, and the only solely freshwater pinniped species, the Baikal seal resembles its marine cousins in most respects. However, it is mainly solitary, and females tend to mate with the same male over years (serial monogamy). The single pup is born in an ice lair, molts its woolly white coat to the silvery gray adult fur after 6–8 weeks, and may suckle for 10 weeks. At 50–55 years, its lifespan is longer than many other seals.

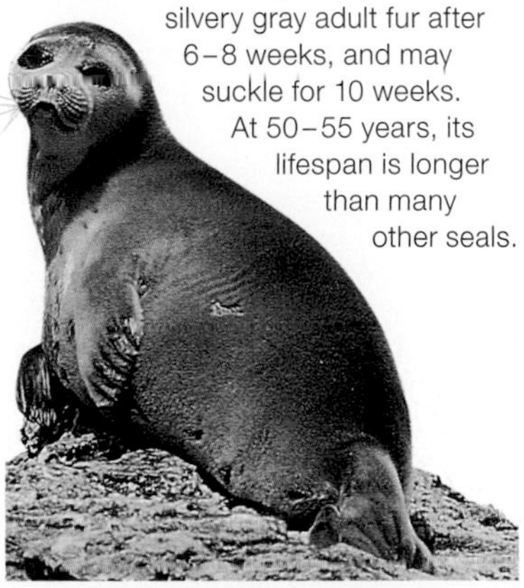

Phoca vitulina

Common seal

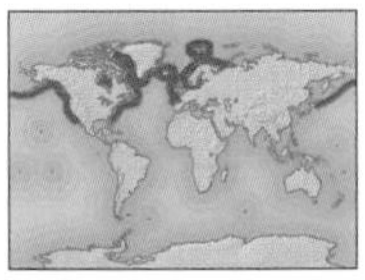

Length 4½–6¼ ft (1.4–1.9 m)
Weight 120–370 lb (55–170 kg)
Social unit Variable
Status Least concern

Location North Atlantic, North Pacific

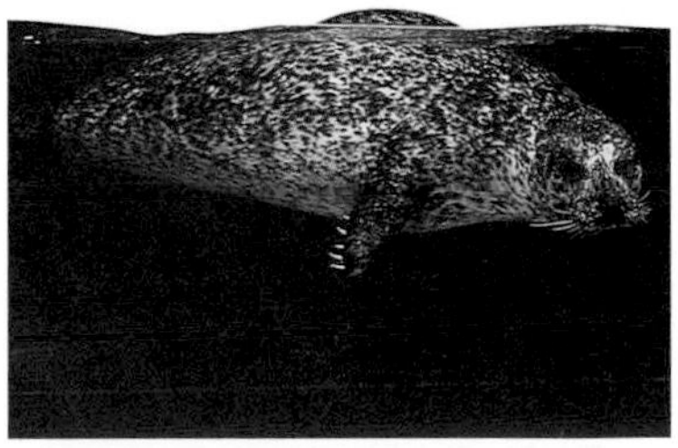

Also known as the harbor seal, this is the most widespread pinniped, with at least 5 subspecies—one of them, the Ungava seal, lives in fresh water in northern Quebec, Canada. Below the large, set-back eyes, the angled, close-set nostrils form a V shape. Color is extremely variable, mainly dark to pale gray-brown with small rings and blotches. A wide-ranging opportunist, the common seal may cause problems in fisheries. Its main prey are herring, sand eels, gobies, hake, and whiting, caught on dives of 3–5 minutes.

Skunks

PHYLUM	Chordata
CLASS	Mammalia
ORDER	Carnivora
FAMILY	Mephitidae
SPECIES	12

Skunks—and the Asian stink badgers— form a group of mammals with an infamous armament of chemical defense. Skunks have not only a more potent repellent than the mustelids, but also the means to fire it at an attacker.

Anatomy

Skunks are ground-dwelling omnivores with a muscular build and strong claws for digging for roots and grubs. They have particularly well-developed anal scent glands armed with muscles, which enable them to squirt foul-smelling sulphurous fluid when threatened or molested. This can repel the largest of predators and a direct facial hit can cause blindness or asphyxiation. Their strikingly black and white pattern serves as a warning of this behavior, and skunks may do "handstands," flashing the tail as a prelude to attack.

Opportunists

Skunks prefer open wooded areas and avoid dense forest. They are mostly active at night or dusk, when they feed on vegetation, fruit, insects, and small vertebrates; certain species may be resistant to snake venom. In some places they raid domestic refuse in the manner of raccoons. They lead solitary lives but some species may gather in communal dens—especially in winter.

TRANSPORTING OFFSPRING
Young striped skunks are carried from place to place in their mother's mouth. This youngster may have wandered off or is simply being moved to another den.

Mydaus marchei

Palawan stink badger

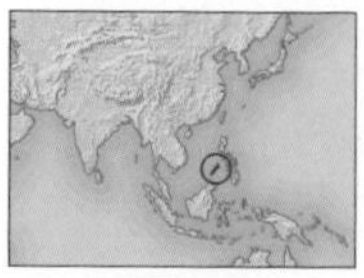

Location Philippines (Palawan and Busuanga islands)

Length 12½–18 in (32–46 cm)
Tail ⅜–1½ in (1–4 cm)
Weight 6½ lb (3 kg)
Social unit Individual
Status Least concern

Slow and ponderous, when attacked the Palawan stink badger accurately squirts an extremely noxious fluid from its anal glands over a distance of 3¼ ft (1 m). It has a short tail, small ears and eyes, a typically badgerlike, stocky body, and a long, flexible, almost hairless snout for rooting out small soil dwellers such as worms, grubs, and slugs. The fur is dark brown with a yellow head cap that tapers to a stripe between the shoulders. The stink badger lives alone in a rocky den or old porcupine burrow.

Conepatus humboldtii

Patagonian hog-nosed skunk

Location S. South America

Length 10–14½ in (25–37 cm)
Tail 12–22½ in (30–57 cm)
Weight 3¼–6½ lb (1.5–3 kg)
Social unit Individual
Status Least concern

Along with the typical small skunk head, stocky body, and fluffy tail, this hog-nosed skunk also has a broad nose pad for rooting up food. Its fur is black or reddish brown with a white stripe along each side; the stripes meet on the head and extend onto the tail. It will feed on virtually anything edible, but mainly eats insects. Like other skunks, it occupies a secure den under a rock, in a burrow, or among bushes, and it can spray enemies with foul-smelling fluid.

Spilogale putorius

Eastern spotted skunk

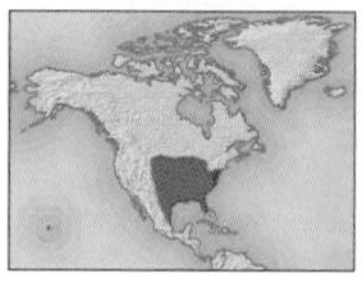

Location E. to C. USA, N.E. Mexico

Length 12–13½ in (30–34 cm)
Tail 6½–8½ in (17–21 cm)
Weight 1–2¼ lb (500–1,000 g)
Social unit Individual
Status Least concern

Heavy-bodied and short-legged, this skunk relies on striking coloration to warn its enemies of the noxious fluid sprayed from its anal glands. The white markings on the body differ in every individual; there is usually, however, a white patch on the forehead and a white tail tip. Food includes a variety of small animals, fruit, and vegetable matter. Generally solitary, up to 8 eastern spotted skunks may share a den in winter.

Mephitis mephitis

Striped skunk

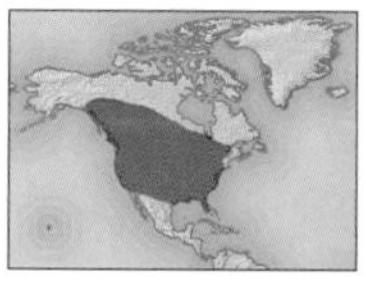

Location C. Canada to N. Mexico

Length 22–30 in (55–75 cm)
Tail 6¾–10 in (17.5–25 m)
Weight 5½–14 lb (2.5–6.5 kg)
Social unit Individual
Status Least concern

As with other skunks, the striped skunk has black-and-white warning coloration. Its diet includes insects, small mammals, birds and their eggs, fish, mollusks, fruit, seeds, and human leftover food. It is solitary but may gather in communal winter dens among rocks, in old burrows, or under outbuildings. The 5–6 young may stay with their mother for more than a year.

NOXIOUS DEFENSE

If threatened, the striped skunk fluffs its fur, arches its back, and lifts its tail. Should the aggressor remain, it stands on its front feet, rear feet in the air, and twists its body. It then ejects a foul-smelling liquid over its head for up to 9¾ ft (3 m), from 2 nozzlelike ducts protruding from its anus.

COLORATION
This skunk is black with a thin, white stripe on the muzzle, and wider, upper-back stripes from head to tail.

MAMMALS

Raccoons and relatives

PHYLUM	Chordata
CLASS	Mammalia
ORDER	Carnivora
FAMILY	Procyonidae
SPECIES	14

CLASSIFICATION NOTE

Some authorities group the lesser panda with the raccoon family, while others place it with the giant panda. However, recent studies suggest that it should be classified in its own family, the Ailuridae and this is how it has been treated in this book.

This family is typified by the raccoons, which are well known for their mischievousness and dexterity. They will boldly approach humans for food and can use their highly mobile hands for opening doors and the like. In addition to raccoons, the family also includes coatis, ringtails, kinkajous, and olingos. Common characteristics include distinctive tail rings and a dark mask pattern on the face. The animals in this family are typically arboreal and have long tails to aid balance. All except the coati are nocturnal. Procyonids are found in the forests of the Americas. The only exception is the lesser panda, which lives in western China and the Himalayas.

Feeding

Most species in this family are omnivorous, and their diet varies with location, season, and the availability of food sources. They will eat fruit, roots, shoots, and nuts, as well as insects and small vertebrates, such as birds, amphibians, and reptiles. Kinkajous mostly eat fruit and are important seed dispersers in their rain-forest habitat. They also have long tongues so they can lap up nectar from flowers. Raccoons use their dextrous hands to reach into streams to feel for crustaceans, fish, and other prey. Their unfussy palate also brings them into urban areas, where they raid garbage cans and fearlessly solicit food from householders. The ringtail and coati are the most carnivorous members of this family and are able to tackle rabbit-sized prey.

Anatomy

Members of the raccoon family are all medium-sized and short-legged, and have a flat-footed (plantigrade), bearlike gait. They commonly have a pointed snout, a relatively long body, a broad face, round or pointed ears, and brown or gray fur. All species have short claws and raccoons feature front paws developed into sensitive, mobile hands. Highly arboreal species, such as ringtails and kinkajous, have the ability to rotate their ankle joints and hang by their feet when feeding or descending tree trunks. Kinkajous have prehensile tails and are therefore able to hang by their tail alone.

OPPORTUNISTIC FEEDING
Raccoons and relatives eat a great variety of food. This common raccoon has used its mobile hands and finely tuned hunting skills to capture a fish.

Bassariscus astutus

Ringtail

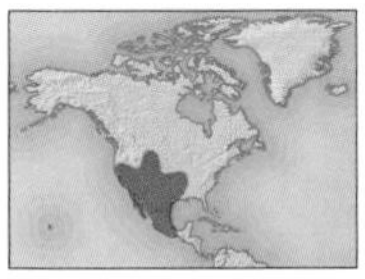

Location C. and W. USA to S. Mexico

Length 12–16½ in (30–42 cm)
Tail 12–17½ in (31–44 cm)
Weight 1¾–3¼ lb (0.8–1.5 kg)
Social unit Individual
Status Least concern

Also called the ringtailed cat or cacomistle, this slender, agile predator has a tail boldly ringed in black and white. The upper body is gray-brown or buff, with black eye rings and white muzzle and "eyebrows." Nocturnal and solitary, this procyonid hunts small birds, mammals, and reptiles, and also forages for grubs, raids birds' nests, and eats fruit and nuts. It marks its territory with droppings and urine and defends it against others of the same sex. Gestation is 51–60 days, with an average litter size of 2–3.

Bassaricyon gabbii

Bushy-tailed olingo

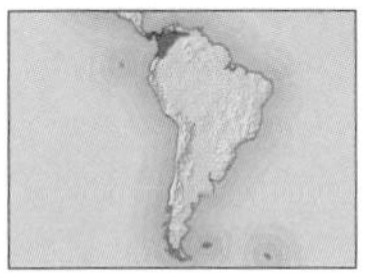

Location Central America to N. South America

Length 14–16½ in (36–42 cm)
Tail 14½–19½ in (37–49 cm)
Weight 2–3¼ lb (0.9–1.5 kg)
Social unit Individual
Status Least concern

This slim, almost catlike procyonid makes its home in most kinds of forest, especially moist, high-altitude cloud forest. It moves with great skill in the branches, rarely coming to the ground, using its strongly grasping hands and feet and its long, fluffy, nonprehensile tail for balance. Night active, the olingo is solitary except for the breeding season when female and male call loudly to each other. Its main diet is fruit, but it also catches grubs and other small creatures.

Nasua nasua

South American coati

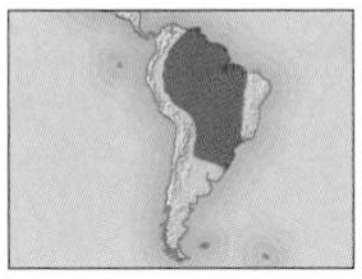

Location South America

Length 16–28 in (41–70 cm)
Tail 12½–28 in (32–70 cm)
Weight 5½–15 lb (2.5–7 kg)
Social unit Individual/Group
Status Least concern

This procyonid's distinctive features include a long, pointed snout. It forages in daytime in noisy groups of 10–20 (rarely over 60), which bustle through vegetation, exploring for anything edible, from fungi and berries to insects and mice, while lookouts around the pack's edge watch for predators. Highly vocal, it uses many different calls, as well as tail movements, to keep in contact. A barking alarm call sends coatis scurrying into dense foliage or the treetops, although they may turn and mob the attacker. The coati also sleeps among the branches. Adult males tend to be solitary and more carnivorous, even cannibalistic on young of their kind. Gestation is 10–11 weeks, and the 2–7 young shelter in a tree nest.

MAMMALS

Procyon lotor

Northern raccoon

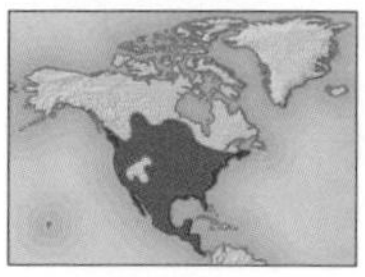

Location S. Canada to Central America

Length 16–25½ in (40–65 cm)
Tail 10–14 in (25–35 cm)
Weight 6½–18 lb (3–8 kg)
Social unit Individual
Status Least concern

Bold and adaptable, with a generalized diet, the raccoon is familiar in many habitats from prairie and woodland to urban sprawl. Active day and night, and normally solitary, raccoons may gather in groups at plentiful food sources such as garbage dumps. Male and female come together briefly for mating with loud chirps and chitters. The female builds a breeding nest in any sheltered site, such as a tree hole, among rocks, or under an outbuilding. Up to 7 offspring (usually 3–4) are born after a gestation of 60–73 days; they venture from the nest after 9 weeks, and are independent by 6 months.

MASKED BANDIT

The "bandit mask" of the Northern raccoon appears to reflect its opportunistic habits. It can climb, dig, and skillfully manipulate doors and latches with its forepaws, and is agile enough to gain access to many livestock enclosures. It may rub dirt off the food before eating it, or rinse it clean if there is water available nearby.

COLORATION
The northern raccoon's long fur varies from pale gray to almost black. The tail has faint dark rings. The ears are short and rounded, and the eyes are small, although the black eye patches make them seem larger.

Procyon cancrivorus

Crab-eating raccoon

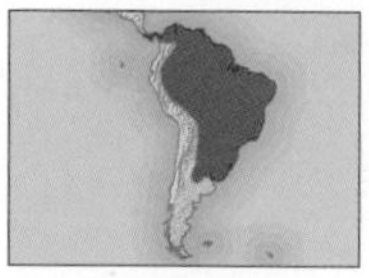

Location Central America to S. South America

Length 18–35 in (45–90 cm)
Tail 8–22 in (20–56 cm)
Weight 4½–26 lb (2–12 kg)
Social unit Individual
Status Least concern

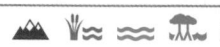

Also called the mapache or osito lavador, this raccoon has short, coarse fur. By night it searches the water's edge in streams, marshes, lakes, and coastlines, feeling with its sensitive, nimble-fingered paws for shellfish, fish, crabs, aquatic insects, worms, and other small prey. The female gives birth to 2–4 (maximum 6) young after an incubation period of 60–73 days, in a den inside a hollow tree, lined with dry leaves and grass. The male takes no part in care of the offspring, which are independent by the age of 8 months.

Potos flavus

Kinkajou

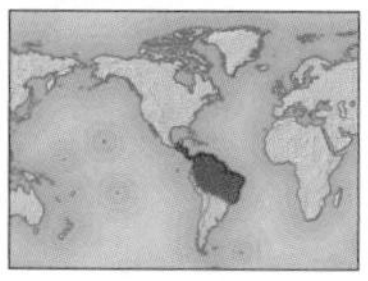

Location S. Mexico to South America

Length 15½–30 in (39–76 cm)
Tail 15½–22½ in (39–57 cm)
Weight 3¼–10 lb (1.5–4.5 kg)
Social unit Individual
Status Least concern

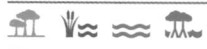

The kinkajou boasts many local names, including mico de noche and martucha. It has a strongly prehensile tail and powerfully grasping feet for great agility in trees. Primarily nocturnal and herbivorous, it also eats grubs, insects, and small vertebrates. The kinkajou makes varied sounds, including squeaks, whistles, grunts, moans, and barks, to proclaim territory, attract a mate, and warn of predators. The single young is born in a tree nest.

Red panda

PHYLUM	Chordata
CLASS	Mammalia
ORDER	Carnivora
FAMILY	Ailuridae
SPECIES	1

The red panda is a cat-sized mammal from the eastern Himalayas with a soft chestnut coat and bushy tail. It climbs well, but also spends time on the ground. On the basis of its anatomy, it has been variously classified with racoons, bears, or mustelids—and it shares the bamboo-eating habit of the giant panda. Recent genetic research suggests that it is not closely related to any of these animals and so is better classified in a family of its own. It lives at cooler, higher elevations than the giant panda and has a more omnivorous diet.

Tree climber

The red panda has partly retractable claws and climbs well. It uses trees not only for feeding but also to escape ground-based predators, and to sunbathe high in the canopy during winter. The female's nest may be in a tree hole, lined with leaves, moss, and other soft plant material, where she rears her 1–5 (usually 2) offspring. Other nesting sites are branch forks, tree roots, and bamboo thickets.

Ailurus fulgens

Red panda

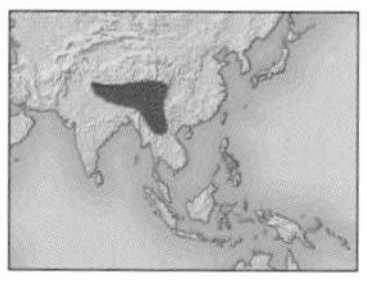

Location S. to S.E. Asia

Length 20–25 in (50–64 cm)
Tail 11–20 in (28–50 cm)
Weight 6½–13 lb (3–6 kg)
Social unit Individual
Status Vulnerable

In addition to bamboo leaves and shoots, the red panda eats other grasses, roots, fruit, and also grubs, small vertebrates such as mice and lizards, and birds' eggs and chicks. It is mainly nocturnal and solitary, but forms pairs during the mating season, and offspring stay with their mother for up to a year. The panda scent marks its territory with droppings, urine, and powerful musklike secretions from the anal glands. It communicates by short whistles and squeaks. Studies in captivity show the gestation period is probably 90 days plus a variable time of delayed implantation. Prime habitats are dense temperate mountain forests, at 6,000–13,200 ft (1,800–4,000 m).

CHESTNUT COLORING
The red or "lesser" panda is red-brown or chestnut with almost white ear rims, cheeks, muzzle, and spots above the eyes. There are also brown, facial "teardrop" stripes.

Mustelids

PHYLUM	Chordata
CLASS	Mammalia
ORDER	Carnivora
FAMILY	Mustelidae
SPECIES	58

Of all the carnivores, the mustelid family is the most diverse and contains the most species. The group includes terrestrial forms (such as ferrets), arboreal species (such as martens), burrowing species (such as badgers), semiaquatic species (such as minks), and fully aquatic species (such as otters). With such a range of lifestyles, the main physical link between species is short legs and an elongated body. Mustelids are found throughout Eurasia, Africa, and the Americas. Although mostly occurring in forest or bush, they have adapted to populate almost every habitat type.

Anatomy

All mustelids have short ears and 5 toes on each foot (most carnivores have only 4 on each back foot); most have a short snout, a long braincase, a long tail, and long, nonretractable, curved claws. Body form tends to be either slender (as in weasels) or heavy and squat (as in badgers). Slimmer forms have a flexible backbone and employ a scampering, bounding gait; stocky forms move with a rolling shuffle. Mustelids have a fur coat that consists of warm underfur and longer, sparser guard hairs. Coat color varies from dark brown or black to spotted or striped. In some species, contrasting patterns—for example, the black-and-white striped heads of badgers—are thought to convey a warning to predators.

Most mustelids have an excellent sense of smell for tracking prey and for communication. All have scent glands in the anus, which generally produce an oily, strong-smelling liquid known as musk. This is secreted into the feces, which are used to mark territory.

Feeding

Reflecting their diverse lifestyles, mustelids have a varied diet. Weasels and stoats, for example, are agile and aggressive and are capable of killing prey larger than themselves, such as rabbits. Some otter species actively hunt fish; others eat mainly shellfish collected by feeling along riverbeds with their sensitive paws. The sea otter cracks open abalone shells by floating on its back on the surface and then hitting the shell against a rock balanced on its chest. Martens, which are arboreal, catch and eat squirrels and birds, while the zorilla, a burrowing species, catches small rodents, lizards, and insects.

Reproduction

Female mustelids do not ovulate automatically. Instead, ovulation is stimulated by copulation, which may last up to 2 hours. This lengthy procedure does expose pairs to predators, but fertilization is almost guaranteed. In many species, the fertilized egg remains dormant and does not implant until conditions are favorable. Therefore, although the gestation period is only 1–2 months, pregnancy may last more than 12 months. Most mustelids are solitary, except during the breeding season.

Fur trade

Mustelid fur is much sought after because it is soft, dense, and water-repellent. Of all the mammal families, the mustelid family has more species with fur valued by humans than any other. Most highly prized is the fur of minks, the sable, and the stoat (when it has its white winter coat). The American mink is farmed for its fur, and in Europe these have escaped in a number of locations and established wild populations, often at the expense of the indigenous European mink. The quest for sea otter fur has brought this species to the brink of extinction, although recently numbers have increased in North America.

AQUATIC HUNTER
The American mink is a voracious hunter, as is typical of mustelids. This species, like all minks and otters, is an excellent swimmer. It can remain submerged for distances of up to 98 ft (30 m) in pursuit of fish, which forms an important part of its diet. Adaptations to a semiaquatic lifestyle include partially webbed feet and a thick, waterproof coat.

Mustela erminea

Stoat

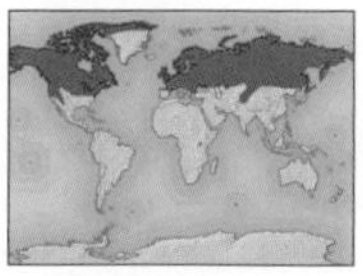

Length 6½–9½ in (17–24 cm)
Tail 3½–4¾ in (9–12 cm)
Weight 2⅛–7 oz (60–200 g)
Social unit Individual
Status Least concern

Location North America, Greenland, Europe to N. and E. Asia

This extremely widespread mustelid adapts to many habitats, and hunts varied prey from mice, voles, and small birds to rats and, often, rabbits. It has the typical mustelid's long, slim, flexible body, pointed muzzle, small eyes and ears, and short legs. In the north of its range, the summer coat of russet to ginger-brown above, demarcated from cream or white below, turns all-white for winter, when the stoat is often called the ermine. However, the tail tip is always black.

Mustela nivalis

Least weasel

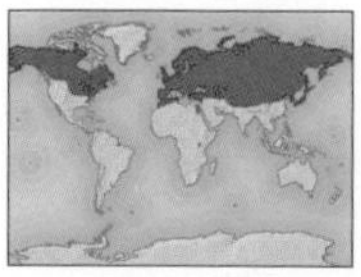

Length 6½–9½ in (16.5–24 cm)
Tail 1¼–3½ in (3–9 cm)
Weight 1¼–9 oz (35–250 g)
Social unit Individual
Status Least concern

Location North America, Europe to N., C., and E. Asia

One of the smallest and most widespread mustelids, this weasel has a small, flattened head hardly wider than the neck, allowing it to enter mouse burrows. It also eats voles, other small rodents, and birds. It is active day and night and must consume one-third of its body weight each day to survive. Overall size varies greatly across the huge range and between sexes, with the male being a quarter as long and up to twice the weight of the female. A small stoat and large least weasel are distinguished by the former's black-tipped tail. Like stoats, least weasels in northern lands turn white in winter, for camouflage in snow. Also like the stoat, this species lives alone, making several nests lined with grass and prey fur or feathers, in a crevice, tree root, or abandoned burrow. After a gestation period of 34–37 days, the female produces a litter of 1–7 (average 5). She cares for them for 9–12 weeks.

russet or chocolate-brown upperparts, legs, and tail

Mustela lutreola

European mink

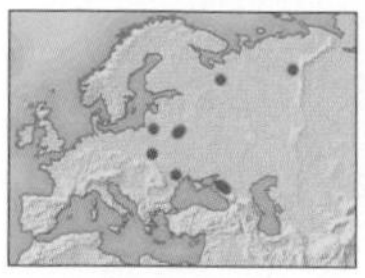

Length 12–16 in (30–40 cm)
Tail 4¾–7½ in (12–19 cm)
Weight 18–29 oz (500–800 g)
Social unit Individual
Status Endangered

Location Europe

The European mink is similar in habits and appearance to the American mink (below), although slightly smaller. It hunts small prey, such as birds, mammals, frogs, fish, and crayfish, on land and in water. The fur is dark brown to almost black with a narrow white edging to the lips. This endangered mink is being bred in captivity for release.

Mustela putorius

European polecat

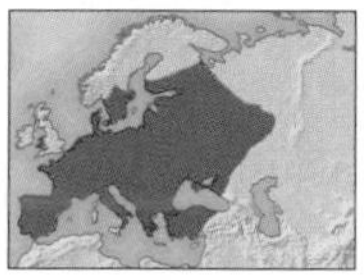

Length 14–20 in (35–51 cm)
Tail 43⁄4–71⁄2 in (12–19 cm)
Weight 1½–3¼ lb (0.7–1.5 kg)
Social unit Individual
Status Least concern

Location Europe

Regarded as the domestic ferret's ancestor, the polecat has long, buff to black hairs with cream or yellow underfur visible between them, and a "mask" across the face. It runs, climbs, and swims well. If threatened, it releases very pungent-smelling anal-gland secretions. As in many similar mustelids, male neighbors defend separate territories, as do females, but male territories usually overlap female ones.

long, sinuous body

Mustela nigripes

Black-footed ferret

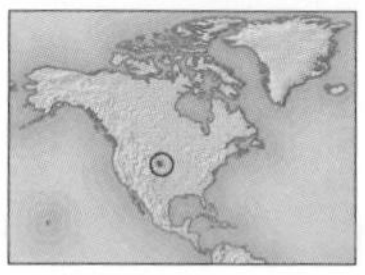

Length 15–16 in (38–41 cm)
Tail 4¼–5 in (11–13 cm)
Weight 32–36 oz (900–1,000 g)
Social unit Individual/Pair
Status Endangered

Location Reintroduced to C. USA

Exceptionally rare, partly due to extermination of its almost exclusive prey of prairie dogs (see p.119), this species was considered extinct in the wild. Captive breeding and release have re-established a few in Wyoming. It chases its prey down burrows and also sets up a nest there, the female giving birth to 3–6 kits.

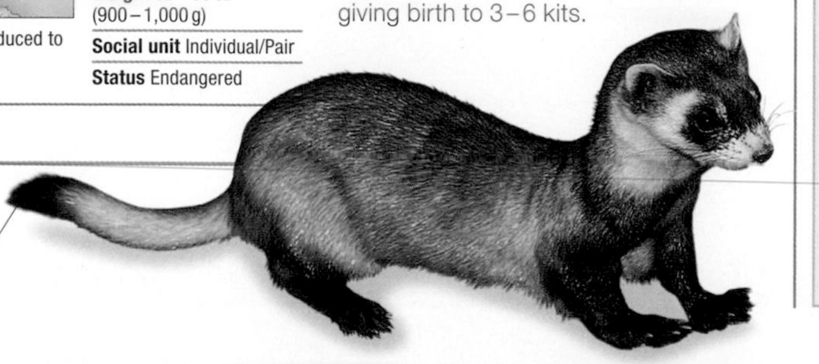

black tail tip

Neovison vison

American mink

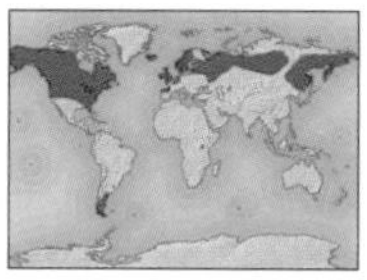

Length 12–21½ in (30–54 cm)
Tail 5½–8½ in (14–21 cm)
Weight 1¼–4½ lb (0.7–2 kg)
Social unit Individual
Status Least concern

Location North America, S. South America, N. and W. Europe, N. and E. Asia

This opportunistic predator uses its partly webbed feet to hunt on land and in water for a variety of small animals, including rats, rabbits, birds, frogs, fish, and crayfish. Since its eyesight is not well adapted for underwater vision, prey is located at the surface and then pursued. In her nesting den among tree roots or rocks, the female suckles 3–6 young for 5–6 weeks. In its natural range, the American mink was once trapped by the thousand; in about 1900 it was taken to South America, Europe, and Russia for fur farming. Escapees established wild populations and are regarded as threats to local wildlife—not only to prey but also to rival predators.

COLOR VARIATION
Most American minks are dark brown to almost black, but approximately 1 in 10 is gray-blue.

MINK TERRITORIES

American mink are territorial, with typical territory sizes of ½–1¾ miles (1–3 km) across for the female and 1¼–3 miles (2–5 km) for the male. Each marks its territory with urine, droppings, and scents from the anal glands. From February to April, males try to mate with females from adjacent territories.

MAMMALS

Martes foina

Beech marten

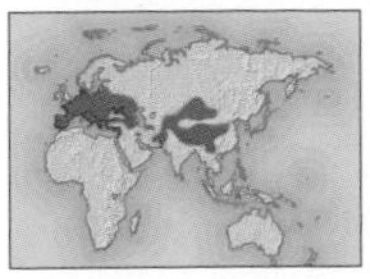

Location Europe, W. and C. Asia

Length 16½–19 in (42–48 cm)
Tail 10 in (26 cm)
Weight 3¼–5½ lb (1.5–2.5 kg)
Social unit Individual
Status Least concern

This marten has adapted to human habitation, and hunts around farms and other buildings for small animals of all kinds. It also scavenges in refuse and eats fruit. It has a relatively short body (for a marten), long legs, and a wide, wedge-shaped head. The fur is brown, with a paler "bow tie" throat patch. Typically solitary, the den is a rocky crevice, tree hole, or old rodent burrow. The territory of up to 200 acres (80 hectares) is marked by droppings. However, near towns, the marten may nest in an outbuilding and be less territorial, foraging with others of its kind. As in many mustelids, there is a period of delayed implantation after mating (230–275 days) followed by 30 days' gestation. Litter size averages 3–4.

Martes martes

European pine marten

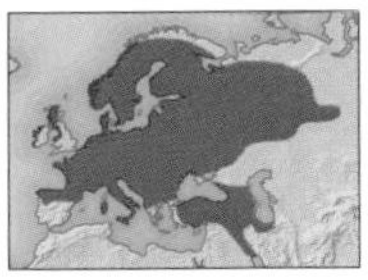

Location Europe to W. and N. Asia

Length 16–22 in (40–55 cm)
Tail 8–11 in (20–28 cm)
Weight 2–4½ lb (0.9–2 kg)
Social unit Individual
Status Least concern

Almost catlike in its movements, this sharp-clawed mustelid climbs well and nests in a tree hole or old squirrel drey (nest). It has a long, slender body and the fur is chestnut to dark brown, with a cream to orange throat "bib." Its bushy tail is used for balance in branches.

Although extremely agile in trees and capable of great leaps, the pine marten takes most prey on the ground, feeding on small rodents, birds, insects, and fruit.

Martes zibellina

Sable

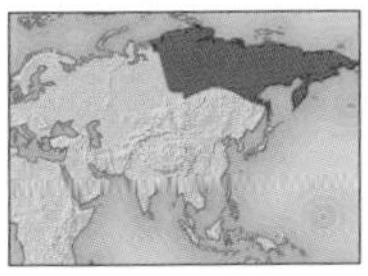

Location N. and E. Asia

Length 12½–18 in (32–46 cm)
Tail 5½–7 in (14–18 cm)
Weight 2–4½ lb (0.9–2 kg)
Social unit Individual
Status Least concern

Few mustelids suffered more hunting for fur than the sable, which is now protected in some regions. It has a brown-black coat, with an indistinct, paler brown throat patch, and the head is wide, with rounded ears.

Compared to other species, its legs are longer, its tail bushier, and its sharp claws partly retractable. Fast and agile on the ground, the sable climbs well but rarely. It has the typical marten diet of small animals and fruit, and takes over an old burrow for its main nest; it also has various temporary dens.

Martes pennanti

Fisher

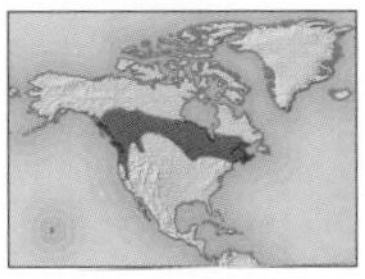

Location Canada to N. USA

Length 18½–30 in (47–75 cm)
Tail 12–16½ in (30–42 cm)
Weight 4½–11 lb (2–5 kg)
Social unit Individual
Status Least concern

Despite its name, the fisher hunts ground prey, from mice to porcupines, as well as scavenging on carcasses. It makes dens in rocks, roots, bushes, and stumps, or high in trees, where it prefers to raise young. Hunted for its long, dense fur, fisher numbers have recovered in some areas but new threats include disturbance by logging.

Martes flavigula

Yellow-throated marten

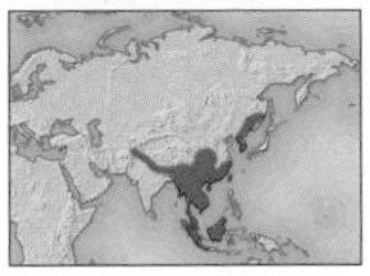

Location E. and S.E. Asia

Length 19–28 in (48–70 cm)
Tail 14–18 in (35–45 cm)
Weight 2¼–11 lb (1–5 kg)
Social unit Individual/Pair
Status Least concern

Similar to the European pine marten (see left) in its climbing agility and long leaps, the yellow-throated marten is larger and has long, dense fur and a bushy tail. Its color varies from dark orange-yellow to brown, with a yellow or white throat patch. It feeds on small rodents, birds, insects, and fruit, on the ground and in trees. Pairs or family groups may catch young deer.

Vormela peregusna

Marbled polecat

Location S.E. Europe to W., C., and E. Asia

Length 13–14 in (33–35 cm)
Tail 4¾–9 in (12–22 cm)
Weight 25 oz (700 g)
Social unit Individual
Status Vulnerable

This polecat is black with variable white or yellow spots and stripes, and the typical black-and-white "face mask." A species of steppes and other dry, open regions, it hunts at twilight and at night for a variety of small animals, especially hamsters. The den is an old, enlarged rodent burrow where the female gives birth to a litter of 4–8 young. When threatened, the marbled polecat arches its head and curls its tail over its body, and may release its pungent anal gland odor.

Poecilogale albinucha

African striped weasel

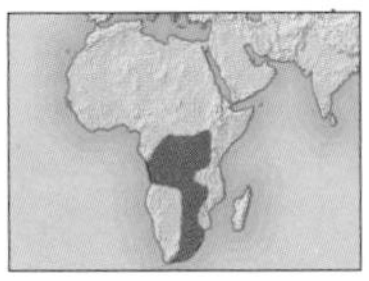

Location C. to southern Africa

Length 10–14 in (25–35 cm)
Tail 6–9 in (15–23 cm)
Weight 8–13 oz (225–350 g)
Social unit Individual
Status Least concern

This exceptionally long, skunklike mustelid is black except for a white patch running from the forehead over the head to the neck, where it splits into 2 white stripes. These divide again into 2, along the back and sides of the body. All 4 stripes unite at the white and bushy tail. The African striped weasel digs well with its long-clawed front feet, and eats almost exclusively mice and similar small rodents, as well as occasional birds and eggs. In defense, it can squirt a pungent spray from its anal glands more than 3¼ ft (1 m).

Galictis vittata

Greater grison

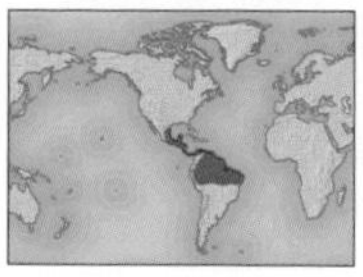

Length 18½–21½ in (47–55 cm)
Tail 5½–8 in (14–20 cm)
Weight 3¼–4½ lb (1.5–2 kg)
Social unit Individual/Pair
Status Least concern

Location S. Mexico, Central and South America

Even for a mustelid, the greater grison is long and sinuous, with a slim, pointed head and flexible neck; the tail, however, is relatively short. Coloration is all gray, other than a white, U-shaped stripe running across the forehead, passing just above each eye and over each ear, tapering toward the shoulder. There is also black below on the muzzle, throat, chest, and front legs. The grison lives alone or in a male–female pair, is an agile runner, swimmer, and climber, and eats various small animals (even worms) and fruit. Its sounds include snorts, growls, screams, and barks.

Ictonyx striatus

Zorilla

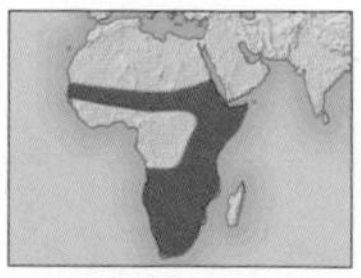

Length 11–15 in (28–38 cm)
Tail 8–12 in (20–30 cm)
Weight 3¼ lb (1.5 kg)
Social unit Individual
Status Least concern

Location W. to E. and southern Africa

Resembling a small skunk (see p.186), the zorilla (striped polecat) is jet black, other than 4 pure white stripes that fan out from the head, along the back and sides, to the tail base. The fluffy tail is mottled white and gray. If threatened, the zorilla hisses, screams, and raises its tail to spray noxious fluid from its anal glands. It digs with its long-clawed, front feet for grubs, mice, and other small creatures.

Gulo gulo

Wolverine

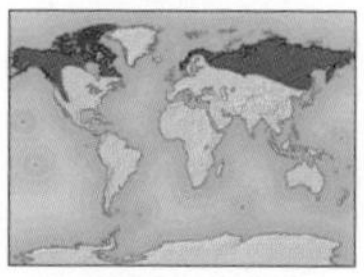

Length 26–41 in (65–105 cm)
Tail 6½–10 in (17–26 cm)
Weight 18–31 lb (8–14 kg)
Social unit Individual
Status Least concern

Location Canada, N.W. USA, N. Europe to N. and E. Asia

The second biggest mustelid after the giant otter, the male wolverine is a third larger than the female. The species is stocky, strong, and bearlike, with extremely powerful jaws that can crunch frozen meat and bone of caribou or similar carcasses in winter; parts of the carcass may be buried for later use. In addition to scavenging, the wolverine runs down victims even on soft snow, with its broad feet and muscular limbs. Also called the glutton, it takes various prey, from deer and hares to mice, as well as birds and eggs, and seasonal fruit. Active all year, it covers up to 31 miles (50 km) daily. The long, dense fur is blackish brown, with a pale brown band along each side, from the shoulder and flank over the rump to the base of the tail. There may be a white chest patch. The wolverine lives in a den among roots or rocks or dug 6½ ft (2 m) into a snowdrift. The average litter is 2–3, and the female suckles her young for 8–10 weeks.

Mellivora capensis

Honey badger

Length 23½–30 in (60–77 cm)
Tail 8–12 in (20–30 cm)
Weight 15–29 lb (7–13 kg)
Social unit Variable
Status Least cocnern

Location W., C., E., and southern Africa, W. and S. Asia

The heavily built honey badger, or ratel, is silver-gray on the upper head, back, and tail, and black or dark brown elsewhere. Its long front claws are well-adapted for digging. The prey of this mustelid includes worms, termites, scorpions, porcupines, and hares. It also cooperates with the Greater honeyguide bird (see p.340), which leads it to bees' nests. The badger opens the nest to provide honey and grubs for both.

Meles meles

Eurasian badger

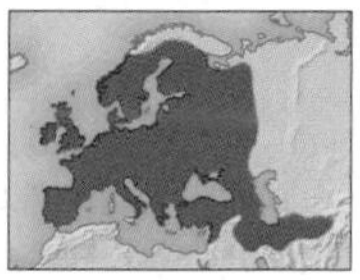

Length 22–35 in (56–90 cm)
Tail 4¾–8 in (12–20 cm)
Weight 22–26 lb (10–12 kg)
Social unit Group
Status Least concern

Location Europe to W. Asia

One of the few group-dwelling mustelids, the Eurasian badger has a small, pointed head and short neck, widening to a powerful body, with short, strong limbs and a small tail. The underparts are black, the main body and tail gray, and the face and neck white, with a black stripe on each side, from the nose over the small eye to the ear. The sight of the Eurasian badger is poor, its hearing better, but it has a keen sense of smell. Nocturnal and omnivorous, this badger varies its diet with season and availability. Earthworms are a staple, supplemented by insects and grubs, as well as frogs, lizards, small mammals, birds and their eggs, carrion, fruit, and other plant matter. Inside the group's sett (see panel, above), the nests are lined with bedding of dry grass, leaves, and moss, which is changed regularly. After up to 10 months of delayed implantation (when the eggs are fertilized but do not immediately implant into the uterine lining), and a gestation period of 7 weeks, the female gives birth to up to 6 cubs. She suckles them for 10 weeks.

COMMUNAL LIVING

Each badger clan averages 6 members: usually a dominant boar (male) with one or more sows (females) and cubs. Their sett—an extensive system of tunnels and chambers—may have 10 or more entrances. Setts are kept clean, and are maintained and enlarged over many generations. Badgers range over a territory of 125–370 acres (50–150 hectares), which they defend against other clans.

VARIABLE STRIPE
The badger's distinctive striped face varies slightly between individuals. It may allow clan members to recognize each other or act as camouflage.

Arctonyx collaris

Hog-badger

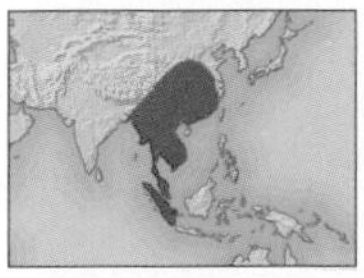

Location S.E. and E. Asia

Length 22–28 in (55–70 cm)
Tail 4¾–6½ in (12–17 cm)
Weight 15–31 lb (7–14 kg)
Social unit Group
Status Near threatened

Named after its looks and its tendency to root and snuffle in the ground like a pig, the hog-badger is an excellent digger, using the very long claws on its front feet. In appearance and habits it resembles the Eurasian badger (see left), being a seasonal omnivore with a very wide range of foods, from fruit and honey to worms and mice. The whitish face has 2 black stripes on either side. The upper body is gray tinged with yellow, and the underparts are black.

Taxidea taxus

American badger

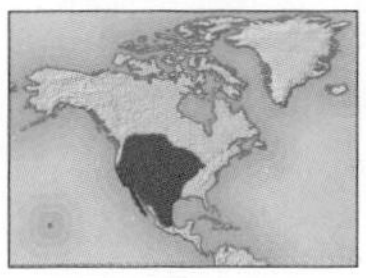

Location S.W. Canada to USA, N. Mexico

Length 16½–28 in (42–72 cm)
Tail 4–6¼ in (10–16 cm)
Weight 8¾–26 lb (4–12 kg)
Social unit Individual
Status Least concern

Smaller but similar in appearance to the Eurasian badger (see left), the American species is, however, solitary. It prefers open country with loose soil, and feeds mainly on small mammals such as prairie dogs and other ground squirrels, which it digs from their burrows. The shaggy fur is grizzled gray on the upperparts, yellow-white below, and black on the limbs.

Melogale personata

Burmese ferret-badger

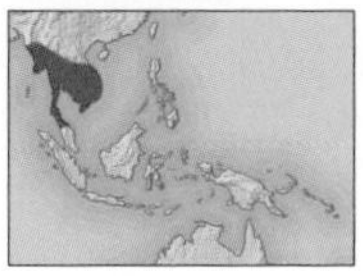

Location S.E. and E. Asia

Length 13–17 in (33–43 cm)
Tail 6–9 in (15–23 cm)
Weight 2¼–6½ lb (1–3 kg)
Social unit Individual
Status Data deficient

Smaller than other badgers, the bushy-tailed Burmese ferret-badger has a long, flexible body resembling a ferret's. It feeds at night on insects, snails, small mammals, birds, and plant matter such as fruit, and sometimes climbs trees. This badger is dark gray or brown, with white or yellow patches on the cheeks and between the eyes. It lives in a burrow and—like most badgers—is fearless if threatened, releasing an offensive anal-gland odor and biting hard.

Melogale moschata

Chinese ferret-badger

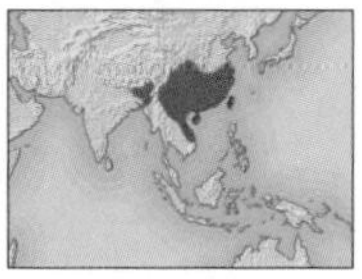

Location E. Asia

Length 13–16¾ in (33–43 cm)
Tail 6–9 in (16–23 cm)
Weight 2¼–6½ lb (1–3 kg)
Social unit Individual
Status Least concern

One of 4 east Asian ferret-badgers that are anatomically intermediate between martens and badgers, this species occurs in woodland and open country centered on southern China. Like other ferret-badgers, its head is boldly patterned with black and white blotches. It is a nocturnal hunter of invertebrates, but also eats carrion, eggs, and fruit, and forages on farmland and close to human habitations. It spends the day in old rodent holes or among rock piles, but does not dig its own burrow. Although it is killed for its pelt and is used in traditional medicine, it is not considered to be globally threatened.

CATCHING PREY
This American badger has caught and killed a ground squirrel, probably by digging it out of its burrow. These badgers do not always eat what they catch and instead store food for later consumption.

MAMMALS

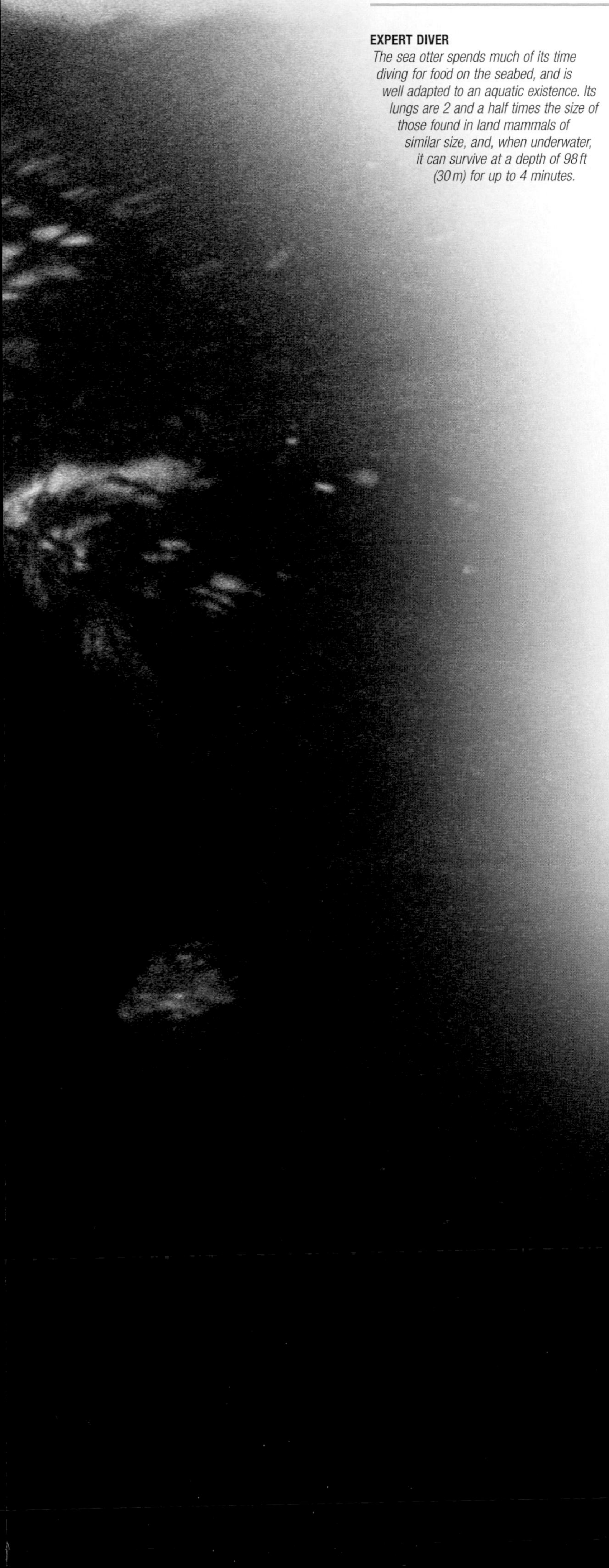

EXPERT DIVER
The sea otter spends much of its time diving for food on the seabed, and is well adapted to an aquatic existence. Its lungs are 2 and a half times the size of those found in land mammals of similar size, and, when underwater, it can survive at a depth of 98 ft (30 m) for up to 4 minutes.

Enhydra lutris

Sea otter

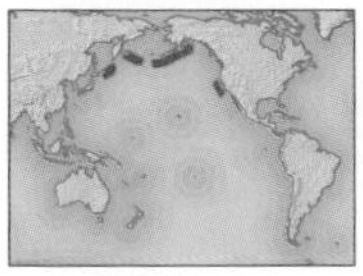

Location North Pacific

Length $1\frac{3}{4}$–$4\frac{1}{4}$ ft (55–130 cm)
Tail 5–13 in (13–33 cm)
Weight 46–62 lb (21–28 kg)
Social unit Group
Status Endangered

The smallest marine mammal, the sea otter lives and feeds in the ocean. It comes ashore only rarely, but is commonly seen close to the coast, particularly near marine kelp forests, floating on its back with its paws out of the water. Superbly adapted to its aquatic way of life, the sea otter has luxuriantly thick fur that keeps it warm in the cold waters it inhabits. It has a strong, flat tail that acts as a rudder, and large, flipperlike hind feet that propel it through the water; the smaller forepaws have retractable claws, like a cat's, which the sea otter uses for holding food and grooming its fur. Excellent eyesight—both underwater and at the surface—a good sense of smell, and sensitive whiskers help it to find food. It forages mainly on the seabed, looking for crabs, clams, sea urchins, and abalone, and has immensely strong teeth for crushing the shells. Before the sea otter sleeps, it may anchor itself by wrapping its body in kelp. Sea otters are social animals, usually found in groups (rafts), with males forming separate rafts from females. In Alaska, hundreds of animals may be found together. Although hunted in the past for its fur—almost to extinction in some areas—the sea otter is now a protected species.

CONSERVATION

Protected from hunting, sea otter numbers have now recovered and stabilized, but only in parts of their original range. Conservation focuses on maintaining healthy kelp forests, so that sea otters have suitable habitats and a reliable source of food. In recent years, sea otters have been successfully reintroduced along the west coast of North America, although the species' range is still much smaller than it was before hunting began.

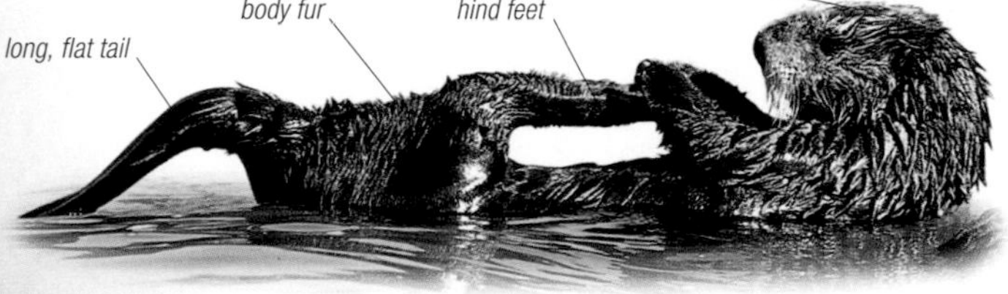

GROOMING TO STAY WARM
The sea otter has the densest fur of all animals—up to 1 million hairs per square inch. Grooming the fur, to keep it clean and waterproof, is essential.

SPECIALIZED BEHAVIOR

The sea otter is a resourceful and adept animal, adapting its behavior in a variety of ways in order to make the most of its environment. Perhaps most remarkably, it has learned to use stones as tools for cracking open shellfish.

TOOL USER
A clam shell can be hard to crack, but the sea otter has devised a reliable method. Lying on its back, it breaks the shell open by hitting it against a rock collected from the seabed.

WATER BABY
The sea otter will sometimes give birth on the rocks, but soon takes her pup into the water. Until the pup learns to groom its fur, the mother keeps it dry by carrying it on her belly, while she floats on her back.

MAMMALS

Lontra canadensis

North American river otter

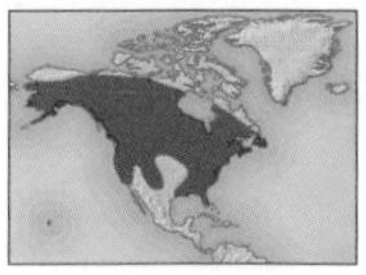

Length 26–43 in (66–110 cm)
Tail 12½–18 in (32–46 cm)
Weight 13–20 lb (6–9 kg)
Social unit Individual
Status Least concern
Location Canada, USA

Probably the most numerous otter, this species is similar to the Eurasian river otter (see below) and is solitary except at mating time, when it makes squeaks, chitters, and whistles. It dwells along riverbanks, lake shores, and coasts, maintaining territories of 3–15 miles (5–25 km). Home is a den in a riverside burrow, under a pile of rocks or a thicket near water, or in a beaver home (lodge).

HUNTER BY DAY OR NIGHT

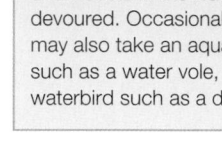

The river otter's main prey is fish, which it catches by day, except in areas disturbed by people, where the otter becomes more nocturnal. Crayfish, frogs, snakes, lizards, and insects in the water are also devoured. Occasionally, this otter may also take an aquatic mammal such as a water vole, or a small waterbird such as a duckling.

SINUOUS AND SILVERY
This long-bodied otter has red- or gray-brown to black velvety fur on the back, paler silvery or gray-brown fur on the underparts, with the cheeks and throat tinged silver or yellow-gray.

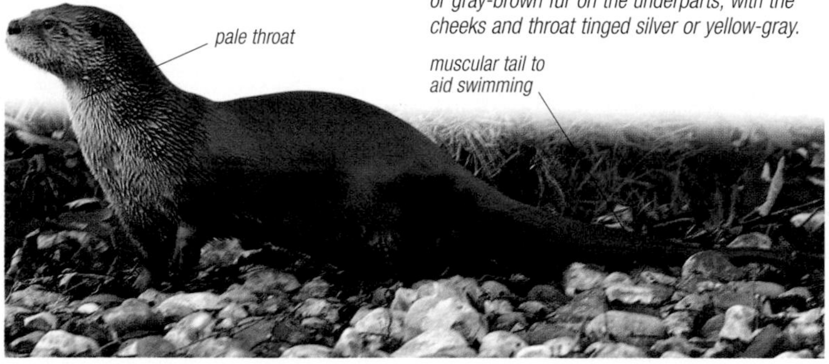

Lutra lutra

Eurasian river otter

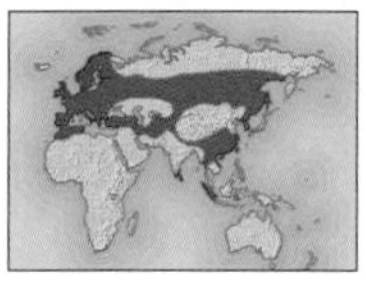

Length 22½–28 in (57–70 cm)
Tail 10–16 in (35–40 cm)
Weight 15–22 lb (7–10 kg)
Social unit Individual
Status Near threatened
Location Europe, Asia

The Eurasian river otter has suffered through hunting for fur, fishery protection, and sports, as well as from water pollution and loss of river habitat caused by bank clearance, irrigation, leisure, and water sports. With its waterproof coat, webbed paws, and stiff whiskers (to feel currents from prey movement), this otter is well adapted to its aquatic habitat. It hunts mainly fish, as well as frogs and other aquatic or amphibious prey. Its coloration is mainly brown, with a paler throat, and the muscular tail is flattened from top to bottom. Inland, the Eurasian river otter hunts chiefly in twilight or darkness, while along coasts, it is more active in daylight. The burrow (holt) is in a bankside territory, 2½–12 miles (4–20 km) long, marked by scent and droppings. The Eurasian river otter is mostly solitary, pairing for 2–3 months in early spring. After 60–70 days' gestation, the 2–3 cubs are suckled for 3 months, and stay with the mother for more than a year.

Aonyx capensis

African clawless otter

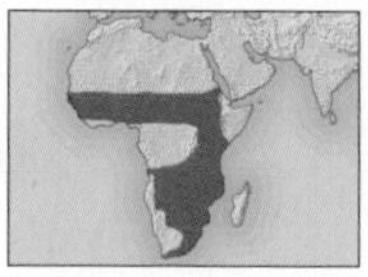

Length 29–37 in (73–95 cm)
Tail 16–26 in (41–67 cm)
Weight 22–35 lb (10–16 kg)
Social unit Pair/Group
Status Least concern
Location W., E., C., and southern Africa

As with other otters, the African clawless otter's long, sinuous body, muscular tail, and short limbs make it well adapted to swim and dive. It catches crabs, frogs, and fish, as well as lobster and octopus along coasts, crushing them with its large teeth. The rear feet are webbed with small claws on toes 3 and 4. The clawless front toes resemble fingers, and are able to feel and hold prey. The clans of paired adults and 2–3 young are exceptionally playful, enjoying mock fighting, mud sliding, and noisy, yapping chases.

Aonyx cinereus

Asian small-clawed otter

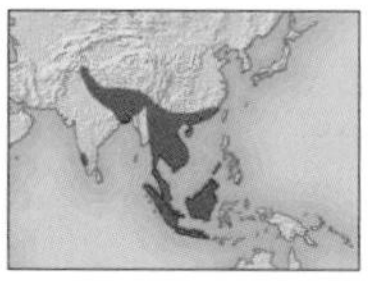

Length 18–24 in (45–61 cm)
Tail 10–14 in (25–35 cm)
Weight 2¼–11 lb (1–5 kg)
Social unit Pair/Group
Status Vulnerable
Location S., E., and S.E. Asia

Smallest of the otters, the Asian small-clawed otter has very short claws that do not extend beyond the fleshy end pads of the partly webbed toes. The cheek teeth are broad, for crushing hard-cased food such as mussels and other shellfish, crabs, and frogs; unusually for otters, fish are relatively unimportant in the diet. The upperparts are brown; the underside is paler, with variable white areas on the lower face, throat, and chest. These otters form loose social groups of about 12, members keeping in touch with noises and scents. Male–female pair-bonds are especially strong. As in many similar species, territories are marked with scent from paired glands at the base of the tail, urine, and droppings (spraints). Litter size varies from 1 to 6, average 2; both parents care for the young.

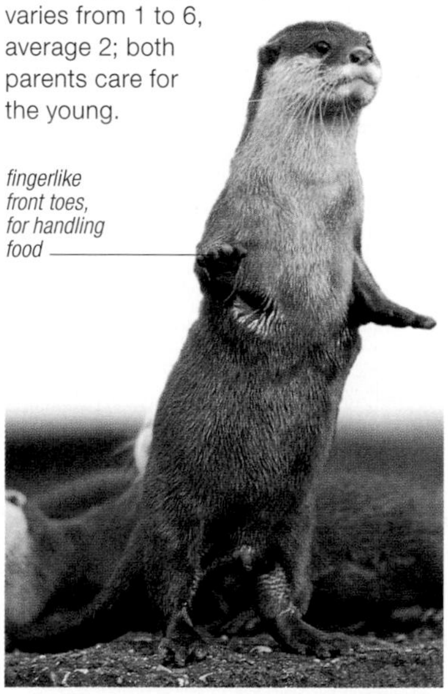

Pteronura brasiliensis

Giant otter

Length 3¼–4½ ft (1–1.4 m)
Tail 18–26 in (45–65 cm)
Weight 49–71 lb (22–32 kg)
Social unit Group
Status Endangered
Location N. and C. South America

The biggest mustelid, this species is similar to a very large river otter. It has short legs, well-webbed toes, and a flattened, wide-based tail, for swimming and diving. Its stout whiskers and sensitive eyes effectively detect prey movements in water. Being so aquatic, it is ungainly on land. The short, dense fur is dark brown, appearing black when wet, with cream spots and patches on the chin, throat, and chest, which may merge into a "bib." The giant otter forms groups of 5–10: usually 2 parents, their offspring, and various younger adults. They live in a communal bank den or burrow and hunt by day for fish, crabs, and other aquatic prey.

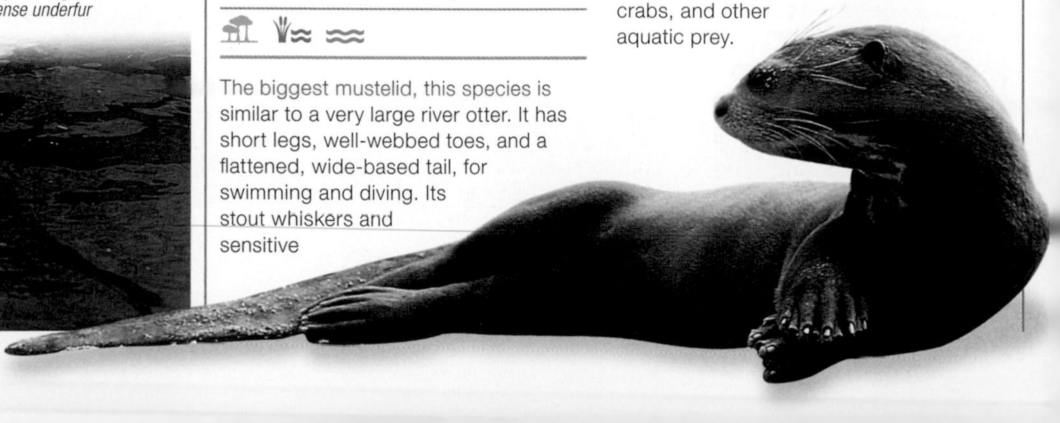

Malagasy carnivores

PHYLUM	Chordata
CLASS	Mammalia
ORDER	Carnivora
FAMILY	Eupleridae
SPECIES	9

This group of carnivores includes civetlike and mongooselike carnivores combined in a single family and entirely restricted to Madagascar. They have had a long isolated evolutionary history and probably originated from African mongooselike ancestors.

CATLIKE APPEARANCE
The catlike fossa is the largest Madagascan carnivore. It hunts mainly lemurs, but it will eat almost any small animal it can catch.

Anatomy

The civetlike species are nocturnal thick-furred carnivores that include the fossa and falanouc. The former is big enough to prey on lemurs in trees and has retractile claws, like those of a cat; the latter is a smaller pointed-headed animal with large, flat feet and bushy tail. Mongooselike species are marked with body stripes, spots, or tail bands, and most have feet with various degrees of webbing.

Predators in isolation

These are the only native mammalian carnivores in Madagascar, together taking a variety of prey appropriate for their body size—ranging from earthworms and insects to mammals and birds. Most are forest dwellers and many—unlike true mongooses—are adept at climbing trees. Malagasy "civets" are nocturnal, but the "mongooses" are mostly active during the day. They sleep in dens consisting of tree hollows or burrows.

Eupleres goudotii

Falanouc

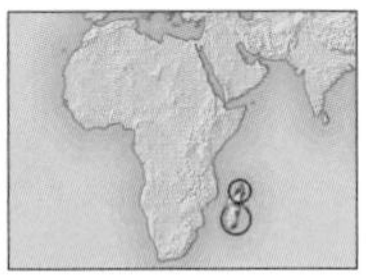

Location E. and N. Madagascar

Length 19–22 in (48–56 cm)
Tail 9–10 in (22–25 cm)
Weight 3¼–10 lb (1.5–4.5 kg)
Social unit Individual
Status Near threatened

Brown on the upperparts, whitish gray on the underside, with a long, slender snout and a short, bushy tail, the falanouc inhabits the Madagascan rain forests and marshes. It uses its long front claws to dig in soil for worms, grubs, insects, slugs, snails, and rodents. The single young, born with eyes open, can follow its mother after only 2 days, and is weaned by 9 weeks. Falanoucs are threatened by habitat loss, humans, dogs, and an introduced competitor, the small Indian civet (*Viverricula indica*).

Fossa fossana

Malagasy fanaloka

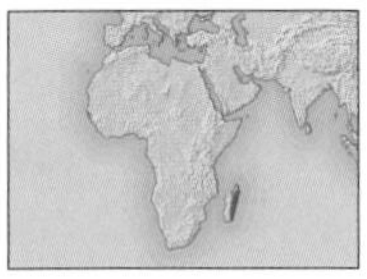

Location Madagascar

Length 16–18 in (40–45 cm)
Tail 8½–18 in (21–30 cm)
Weight 3¼–4½ lb (1.5–2 kg)
Social unit Pairs
Status Near threatened

Found in forests of eastern and southern Madagascar, this spotted carnivore resembles a civet—and used to be classified with them. It lacks the anal scent glands of civets—instead, it probably marks its territory using secretions from its cheek and neck. Pairs of fanalokas defend territories with a range of eerie vocalizations: they spend the day in tree hollows and venture out at night to forage in trees and on the ground for invertebrates, such as worms and crustaceans, as well as frogs, and sometimes fruit.

Salanoia durrelli

Durrell's vontsira

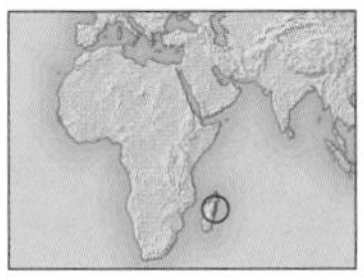

Location Madagascar

Length 12–13 in (31–33 cm)
Tail 7–8½ in (17.5–21 cm)
Weight 1¼–1½ lb (600–675 g)
Social unit Individual
Status Not evaluated

This reddish brown Madagascan mongoose was described in 2010 and named after Gerald Durrell, the conservationist. It is adapted to the marshland area of Lac Alaotra in Madagascar—an important region that harbors many unique species—and has one of the smallest ranges of any carnivore. The species has been observed swimming in Lac Alaotra and it is possible that its broadly padded feet help it move around in the waterlogged habitat: specimens have been trapped on floating mats of vegetation. Here, it is likely that it feeds on crustaceans and mollusks, which it smashes with its strong teeth. It is most closely allied to the insectivorous brown-tailed mongoose of Madagascan dry forests. The long-term survival of Durrell's vontsira is threatened by agricultural encroachment and pollution from fertilizers and pesticides.

Cryptoprocta ferox

Fossa

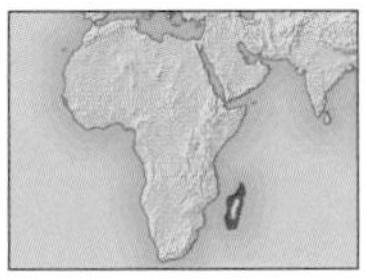

Location Madagascar

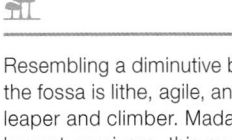

Length 23½–30 in (60–76 cm)
Tail 22–28 in (55–70 cm)
Weight 21–31 lb (9.5–14 kg)
Social unit Individual
Status Vulnerable

Resembling a diminutive brown "big cat," the fossa is lithe, agile, and an excellent leaper and climber. Madagascar's largest carnivore, this muscular, powerful predator hunts by day or night, using the stalk-and-pounce method. It originally specialized in hunting lemurs but now also takes pigs, poultry, and other domesticated animals. A top carnivore, the solitary fossa naturally has a large territory, more than 1½ square miles (4 square km), and thus a low population density. It is threatened by loss of its habitat and persecuted for its attacks on livestock.

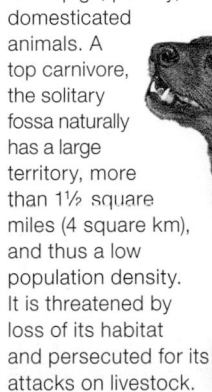

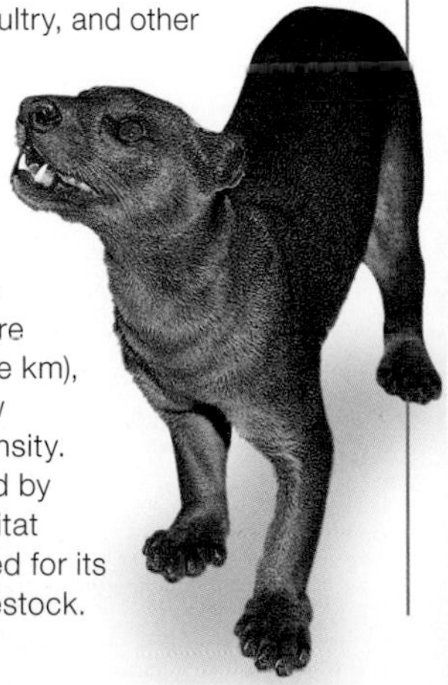

Mongooses

PHYLUM	Chordata
CLASS	Mammalia
ORDER	Carnivora
FAMILY	Herpestidae
SPECIES	33

Native to warm areas of the Old World, mongooses were formerly classified with the civets in the family Viverridae, but are now classified in a separate family—the Herpestidae. Together with the Malagasy carnivores, they are thought to be closely related to hyenas.

LIVING TOGETHER
Mongooses live in complex social groups, called bands, and are generally active during the day.

Anatomy

Most mongooses are smaller than civets; they are shorter tailed and are more uniformly colored. Their anal scent glands are also more complex, with the scent secretion being stored in an external sac. Mongooses have elongated bodies and short legs for movement over the ground.

Snake hunters

These animals live in complex social groups and are generally active during the day. Opportunistic mongooses have a wide ranging diet that includes plant and animal material: they prey on invertebrates, small mammals, frogs, and reptiles—including snakes. It was once thought that mongooses evade envenomation by skillful maneuver when tackling snake prey, but some species have exhibited resistance to snake venom. Their effectiveness at controlling snakes and rodents led to their introduction into places as far afield as Hawaii and New Zealand—where they now pose a threat to native wildlife.

Cynictis penicillata

Yellow mongoose

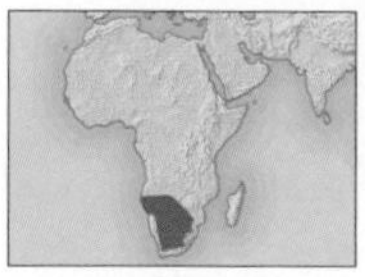

Location Southern Africa

Length 9–13 in (23–33 cm)
Tail 7–10 in (18–25 cm)
Weight 16–29 oz (450–800 g)
Social unit Group
Status Least concern

This mongoose is a yellowish buff color in the south of its range but grayer in the north. The family group (a breeding pair, their offspring, and nonbreeding young adults) occupies and extends a tunnel system taken over from meerkats or ground squirrels. Occasionally, these species all coexist in a large burrow. The main diet is insects, such as termites, ants, beetles, and locusts, as well as birds, eggs, frogs, lizards, and small rodents.

Helogale parvula

Dwarf mongoose

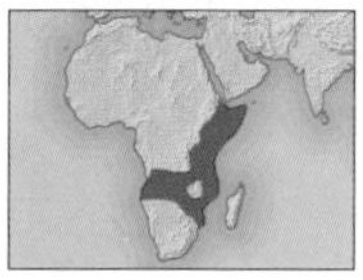

Location E. and southern Africa

Length 7–11 in (18–28 cm)
Tail 5½–7½ in (14–19 cm)
Weight 7–13 oz (200–350 g)
Social unit Group
Status Least concern

The smallest mongoose, this species has thick fur, brown but fine-grizzled in red or black, very small eyes and ears, and long-clawed front feet. It forms packs of 2–20, which "rotate" around the termite mounds of their range. They use the mound as shelter for a few days and feed on insects, lizards, snakes, birds, eggs, and mice. All members of the pack help to care for offspring, which may number up to 6 per female.

Suricata suricatta

Meerkat

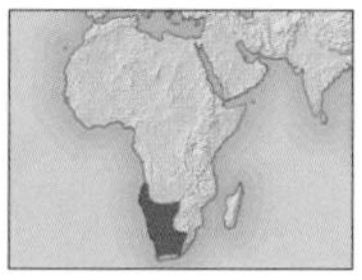

Location Southern Africa

Length 10–14 in (25–35 cm)
Tail 6½–10 in (17–25 cm)
Weight 21–35 oz (600–975 g)
Social unit Group
Status Least concern

EXPERT DIGGERS
The meerkat's long front claws are used to dig its burrow and to find food, mainly insects, spiders, and other small animals, as well as roots and bulbs.

Day-active and social, the meerkat forms colonies of up to 30, which enlarge the former burrow systems of ground squirrels. In early morning, it emerges to sit up and sunbathe, then forages for small prey. Coloration is pale brown on the underside and face, silver-brown on the upper parts, with 8 darker bands on the rear back, dark eye rings, and a dark tip to the slender tail. After a gestation of 11 weeks, the 2–5 young are born in a grass-lined nursery chamber in the burrow.

ON GUARD

While most pack members forage, some act as lookout sentries, especially for hawks and other aerial predators. Sentries stand at vantage points such as on mounds and in bushes, and cheep or cluck warnings. Sharp barks or growls denote more urgent threats and the meerkats dive for cover.

Mungos mungo

Banded mongoose

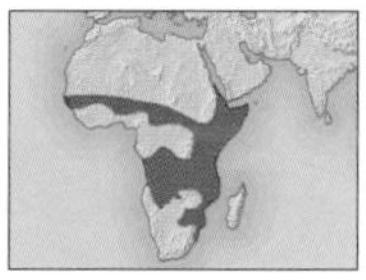

Location Africa

Length 12–18 in (30–45 cm)
Tail 6–12 in (15–30 cm)
Weight 3¼–5½ lb (1.5–2.5 kg)
Social unit Variable
Status Least concern

Common, lively, and opportunistic, this stocky mongoose has distinctive crosswise bands over the rear of the body. The fur is coarse and grizzled, and populations from moist habitats are darker brown than drier-region individuals. Often kept as a pet, the banded mongoose eats varied small items, from termites to birds' eggs. It is often found in packs of 15–20 that include one dominant male.

Civets and relatives

PHYLUM	Chordata
CLASS	Mammalia
ORDER	Carnivora
FAMILY	Viverridae
SPECIES	35

Animals in this family—which includes the civets, genets, and binturong—usually have a slender body and long tail. They are related to cats and hyenas but are more primitive, with a longer snout and additional teeth. Viverrids are distributed throughout Africa, southern Asia, and southeastern Europe, and are found in rain forest, woodland, and savanna. Most are terrestrial, but there are arboreal species (for example, the binturong) and semiaquatic ones (for example, otter genets).

Anatomy

Viverrids typically have a long body and tail, short legs, an elongated neck and head, and a tapered snout. Most civets and genets have spots in longitudinal rows along the body; the binturong has plain black fur. All species have scent glands in the anal region, and in civets these glands produce a substance used to make perfume.

ARBOREAL HUNTERS
Many civets and genets hunt for food in trees as well as on the ground. This large-spotted genet is searching for nesting or roosting birds.

Paradoxurus hermaphroditus

Common palm civet

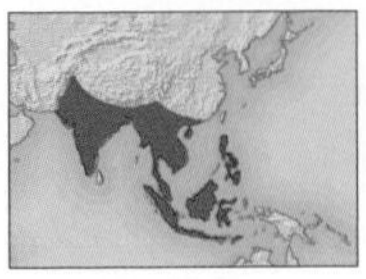

Location S., E., and S.E. Asia

Length 17–28 in (43–71 cm)
Tail 16–26 in (40–66 cm)
Weight 3¼–10 lb (1.5–4.5 kg)
Social unit Individual
Status Least concern

This adaptable, bushy-tailed civet is brownish gray with black stripes on its back, dark flank spots, and a polecat-like "face mask." It stays mainly in trees but may rest by day in a house or on an outbuilding roof. The diet includes much fruit, especially figs, as well as buds, grasses, small animals such as insects and mice, and sometimes poultry. Fermented palm-tree juice is a favorite, giving rise to a local name of "toddy cat."

Viverra tangalunga

Malay civet

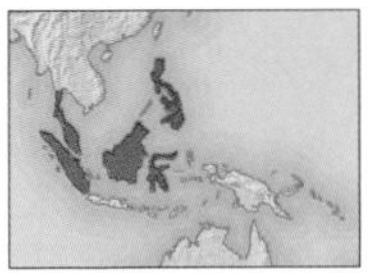

Location S.E. Asia

Length 24–26 in (62–66 cm)
Tail 11–14 in (28–35 cm)
Weight 7¾–10 lb (3.5–4.5 kg)
Social unit Individual
Status Least concern

In addition to the typical civet coat of many dark spots forming lines along the body, this species has a distinctive black-and-white neck collar, white underside, black legs and feet, and about 15 bands along the tail. It climbs only occasionally and feeds mainly on forest-floor creatures including millipedes, giant centipedes, scorpions, and small mammals such as mice. Widespread and common throughout Southeast Asia, the Malay civet is nocturnal and solitary, and lives for up to 11 years.

Genetta genetta

Small-spotted genet

Location W., E., and southern Africa, W. Europe

Length 16–22 in (40–55 cm)
Tail 16–20 in (40–51 cm)
Weight 3¼–5½ lb (1.5–2.5 kg)
Social unit Individual
Status Least concern

This very catlike species, also known as the common genet, has semiretractile claws and is an excellent climber. It takes a variety of small mammals, birds, eggs, grubs, and fruit. In some areas it raids farms for poultry and is considered a pest. The den is a sheltered area under roots or in thick bushes. After the gestation period of 70 days, typically 2–3 young are born with eyes closed.

Prionodon pardicolor

Oriental linsang

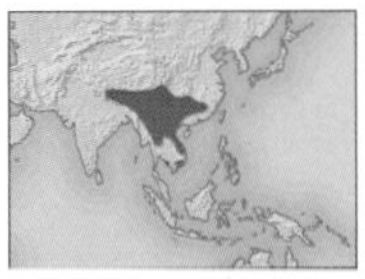

Location S., E., and S.E. Asia

Length 14½–17 in (37–43 cm)
Tail 12–14 in (30–36 cm)
Weight 1¼–3 lb (0.6–1.2 kg)
Social unit Individual
Status Least concern

This small, slender, sinuous linsang seems to "flow" through the branches with amazing grace, using its tail to balance and brake. Both sexes are generally solitary and nocturnal, with large ears and big eyes adapted for night vision. The male is almost twice the size of the female. Its diet consists of small animals, such as frogs, snakes, rats, and mice, as well as carrion. The average litter size is 2–3, with peak births in February and August.

Arctictis binturong

Binturong

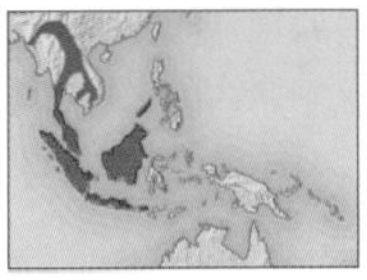

Location S. and S.E. Asia

Length 24–38 in (61–96 cm)
Tail 22–35 in (56–89 cm)
Weight 20–31 lb (9–14 kg)
Social unit Individual
Status Vulnerable

The binturong has shaggy black fur and tufted ears. Its tail is long-haired and prehensile at the tip. It moves slowly and cautiously among branches in search of fruit, shoots, and small animals such as insects, birds, and rodents. By day, it curls up on a secluded branch to rest, but may continue to feed. Individuals mark their territories with scent. The 1–3 young, born after a gestation of 92 days, reach adult size in one year.

Cats

PHYLUM	Chordata
CLASS	Mammalia
ORDER	Carnivora
FAMILY	Felidae
SPECIES	41

With a lithe, muscular body, acute senses, highly evolved teeth and claws, lightning reflexes, and camouflage coloration, cats are model hunters. In fact, cats are the most specialized of the mammalian flesh eaters. They are unusual in that all species appear remarkably similar: the differences between tigers and the domestic cat, for example, are surprisingly small. As a result, classifying smaller cats remains problematic: they can be either grouped together in the genus *Felis* or subdivided into several genera. Cats are found throughout Eurasia, Africa, and the Americas (the domestic cat is found worldwide), from alpine heights to deserts. Many species live in forests. All except the largest cats are expert climbers, and several are excellent swimmers. Most cats are solitary.

Anatomy

Cats have a rounded face and a relatively short muzzle (but a wide gape). The heavy lower jaw helps deliver a powerful bite, and the long canines are used for stabbing and gripping. The carnassials, modified cheek teeth that slice bones and tendons, are highly developed. Cats are covered with soft fur, which is often striped or spotted, and have a tail that is haired, flexible, and usually long. There are 5 digits on the front feet and 4 on the back feet, and each digit has a curved, retractable claw for holding prey. The claws are normally retracted, which helps keep them sharp. However, when required (during climbing, for example), they spring forward via a mechanism similar to a jackknife. The naked pads on the soles of the feet are surrounded by hair, which assists with silent stalking.

Senses

All cats have keen senses. Large, forward-facing eyes enable them to judge distances accurately. The pupils can contract to a slit or to a pinhole (depending on species) in bright light and can dilate widely for excellent night vision. The ears are large, mobile, and funnel-shaped to draw in sounds made by prey, while long, stiff, highly sensitive whiskers aid navigation and night hunting. The sense of smell is also well developed, and in the roof of the mouth is a "smell-taste" organ, called the Jacobson's organ, which detects sexual odors. Secretions from scent glands on the cheeks and forehead, under the tail, and between the claws communicate information such as age and sex.

Hunting techniques

While some cats actively search for prey, others conceal themselves and await passing victims. Many employ a combination of these 2 methods. In either case, the cat's fur usually provides camouflage: a tiger's stripes, for example, blend in with tall grasses, while many forest-living species are spotted, to mimic the effect of sunlight through leaves.

The distance a cat will chase its quarry varies between species. Heavier-built cats, such as tigers, prefer to stalk and pounce; the cheetah uses explosive speed—up to 68 mph (110 kph). Some small cats, such as the serval, hunt in long grass, and use "jack-in-the-box" leaps to surprise and flush out their prey. Cats hunt any animal they can catch and overpower. The big cats specialize in prey larger than themselves, and are capable of dragging a carcass some distance to a safe feeding spot. Smaller cats seek out rodents and birds—some, such as the fishing cat, wade into streams and scoop out fish.

CONSERVATION

The entire cat family is listed by the Convention on International Trade in Endangered Species of Wild Fauna and Flora (CITES), which regulates trade in live animals or their body parts. The main reason for this measure is hunting—either for fur, which has seen a recent resurgence in fashion, or for bones and other remains, which are used in traditional Asian medicine. Habitat destruction has also reduced cat numbers, as many species require large areas to maintain an adequate food supply. In a few species—such as the North American bobcat—the overall population is stable. However, despite intensive action to protect them, many other cats, including the tiger and the lion, are in sharp decline.

TERRITORIAL CONFLICT
All cats are territorial and will fight if their scent marks and vocal warnings are ignored. These male jaguars are threatening one another, and a conflict will ensue if neither of them gives way. The flattened ears show fear, and dilated pupils and bared teeth indicate aggression.

MAMMALS

Prionailurus planiceps

Flat-headed cat

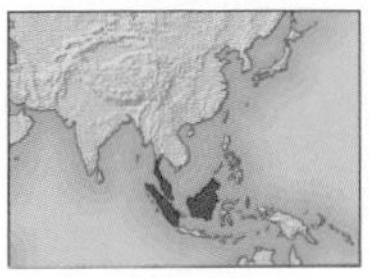

Location S.E. Asia

Length 16–20 in (41–50 cm)
Tail 5–6 in (13–15 cm)
Weight 3¼–4½ lb (1½–2 kg)
Social unit Individual
Status Endangered

Unusually small and low-set ears accentuate the flattened forehead of this semiaquatic fish predator, which also eats shrimps, frogs, rodents, and small birds. Its toes are partly webbed and its claws not fully retractable. The upper premolar teeth are relatively large and sharp to grip slippery food. Slightly smaller than a typical pet cat, the flat-headed cat is usually sighted around rivers, lakes, and swamps, and along irrigation ditches and canals.

Prionailurus viverrinus

Fishing cat

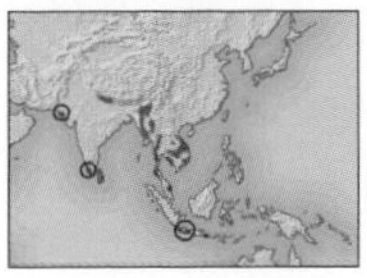

Location S. to S.E. Asia

Length 30–34 in (75–86 cm)
Tail 10–13 in (25–33 cm)
Weight 18–31 lb (8–14 kg)
Social unit Individual
Status Endangered

Olive-gray with black markings and a short tail, the fishing cat is mostly confined to rivers, lakes, marshes, and coastal mangrove swamps. However, its adaptations to water are largely behavioral: its toes are only slightly webbed, and its teeth are not especially suited to grasping slippery prey. Locally common, its dependence on water-edge habitats means its population may suffer as wetlands are affected by drainage, intensive agriculture, human habitation, and pollution.

CIVET-LIKE CAT
This cat's scientific name, viverrinus, *reflects its viverrid- or civet-like proportions: long, stocky body and relatively short legs.*

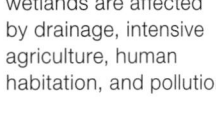

FISH DIET

As its name implies, the fishing cat is a semiaquatic hunter of fish, frogs, snakes, water insects, crabs, crayfish, and shellfish. It scoops prey from the water with its paws, dives in pursuit, sometimes surfacing under a water bird. It also hunts small land mammals such as mice.

Profelis aurata

African golden cat

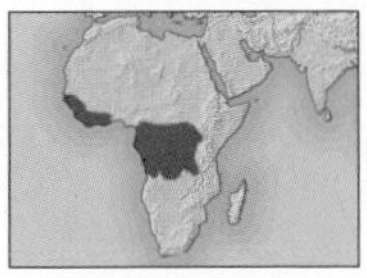

Location W. and C. Africa

Length 24–39 in (61–100 cm)
Tail 6½–18 in (16–46 cm)
Weight 12–35 lb (5.5–16 kg)
Social unit Individual
Status Near threatened

This little-studied, medium-sized cat may vary from gray to red-brown, and it may be faintly spotted or plain. It occurs primarily in tropical rain forest and other forest habitats, especially near rivers. Its prey is mainly rats and other rodents, hyraxes, small forest antelopes, monkeys (perhaps already injured), and similar small mammals. It may also catch birds, mainly on the ground but also in trees.

Lynx pardinus

Iberian lynx

Location S.W. Europe

Length 34–43 in (85–110 cm)
Tail 5 in (13 cm)
Weight 22–29 lb (10–13 kg)
Social unit Individual
Status Critically endangered

Very rare, critically restricted in distribution, and fully protected by law, the Iberian lynx is about half the size of the Eurasian species (see right). Now mainly confined to remote wetlands and uplands, it uses thickets for shelter and open areas that favor its main food of rabbit—with deer fawns, ducks, and other prey in winter.

distinctive spotted coat

Lynx lynx

Eurasian lynx

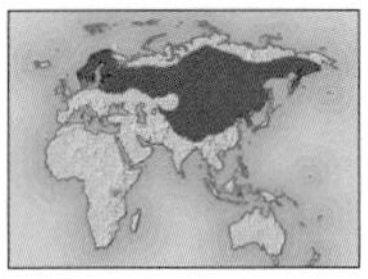

Location N. Europe to E. Asia

Length 2½–4¼ ft (0.8–1.3 m)
Tail 4¼–10 in (11–25 cm)
Weight 18–84 lb (8–38 kg)
Social unit Individual
Status Least concern

Primarily a cat of mixed forest, this lynx has been driven by human presence and persecution to more open woods and rocky mountain slopes. However, it still has one of the widest ranges of all cat species. Its major prey is deer, goats, sheep, and similar hoofed mammals up to 4 times its own size, but, if these are lacking, it hunts hares and pikas.

COAT PATTERNS
The Eurasian lynx has 3 predominant coat patterns: mainly striped, mostly spotted (as here), and plain.

rusty or yellowish gray background color

CONSERVATION

Despite conservation efforts and reintroductions in Western Europe, lynx remain rare because of revenge hunting by livestock farmers, road kills, and mysterious loss of male cubs—possibly caused by a genetic problem.

Lynx rufus

Bobcat

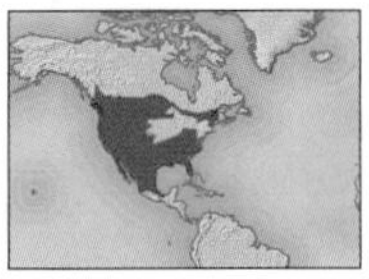

Location S. Canada, USA, Mexico

Length 26–43 in (65–110 cm)
Tail 4¼–7½ in (11–19 cm)
Weight 8¾–34 lb (4–15.5 kg)
Social unit Individual
Status Least concern

Named after its short "bobbed" tail, this medium-sized cat has a ruff-like facial border. Mainly tawny in color, it always has some spots; they may be prominent all over or only on the underside. Ithunts lagomorphs, such as cottontail rabbits in the south of its range and snowshoe hares farther north, but can survive on rodents, deer, and carrion. Its habitat varies from desert to mixed woodland and conifer forest.

variable density of spots on coat

Caracal caracal

Caracal

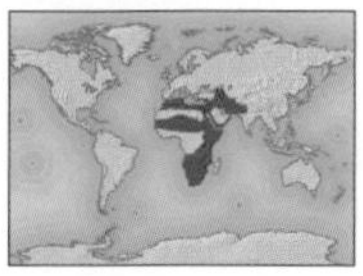

Location Africa, W., C., and S. Asia

Length 23½–36 in (60–91 cm)
Tail 9–12 in (23–31 cm)
Weight 13–42 lb (6–19 kg)
Social unit Individual
Status Least concern

Also called the desert lynx, because of its arid scrubby habitat, the caracal is tawny or reddish, although—as with many cats—occasional black (melanic) individuals occur (see black panther, p.206). It is famed for its ability to spring 10 ft (3 m) vertically and to "bat" flying birds with its paw. Its main foods are rodents, hyraxes, hares, small antelopes, poultry, and other livestock.

Leptailurus serval

Serval

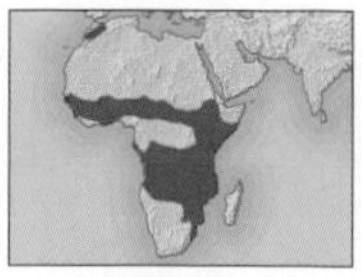

Location Africa

Length 23½–39 in (60–100 cm)
Tail 9½–18 in (24–45 cm)
Weight 20–40 lb (9–18 kg)
Social unit Individual
Status Least concern

Resembling a small cheetah, with its lean body and long limbs, the serval has yellowish fur with dark spots. It prefers to live among reeds and rushes fringing wetlands. Here it hunts rats and similar prey, helping control rodents and thereby aiding farmers (servals rarely attack livestock). The average litter of 2 young is born after 73 days' gestation.

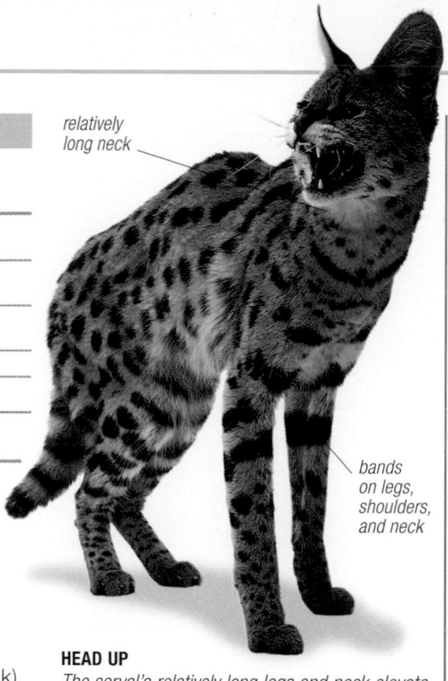

HEAD UP
The serval's relatively long legs and neck elevate its head to 30 in (75 cm) above ground, enabling it to see and hear clearly in long grass.

CAT POUNCE

Having located prey, usually at dusk, and mainly by hearing, the serval excels at the cat pounce. It leaps up to 13 ft (4 m) horizontally and more than 3¼ ft (1 m) high, to strike the victim with its forepaws. This cat eats rats and similar-sized rodents, birds, fish, and large insects such as locusts. Frogs are a favorite with wetland-dwelling servals.

Pardofelis marmorata

Marbled cat

Location S. to S.E. Asia

Length 18–21 in (45–53 cm)
Tail 18½–22 in (47–55 cm)
Weight 4½–11 lb (2–5 kg)
Social unit Individual
Status Vulnerable

A little-known species, found mainly in moist, lowland tropical forest, the marbled cat resembles a small clouded leopard, although they are not related. It is long-tailed, a proficient climber, and probably a nocturnal predator of squirrels, birds, and similar tree dwellers. Data being collected in Thailand should help to shed some light on this species. Gestation is 81 days, litter size 1–4, and sexual maturity is at 21 months.

Felis silvestris

Wild cat

Location Europe, W. and C. Asia, Africa

Length 20–30 in (50–75 cm)
Tail 8½–14 in (21–35 cm)
Weight 6½–18 lb (3–8 kg)
Social unit Individual
Status Least concern

The wild cat resembles a slightly larger, heavier-built, usually longer-furred (especially in winter) version of the domestic tabby cat. The species interbreeds with domestic cats, of which the African subspecies *Felis sylvestris lybica* is presumed to be the ancestor. The wild cat's preferred habitat is mixed broad-leaved woodland, but habitat loss and amenity use of woods has driven it to marginal habitats such as conifer forest, rocky upland, moor, scrub, swamp, and coast. It feeds on rabbits and small rodents, such as rats, mice, voles, and lemmings. This cat climbs well and catches young squirrels or birds in the branches; carrion is also eaten. It mates between January and March and the gestation period is 63–68 days. The female gives birth to an average of 3–4 (range 1–8) cubs in its den in a tree hole, among rocks or tree roots, or in an old rabbit or badger burrow.

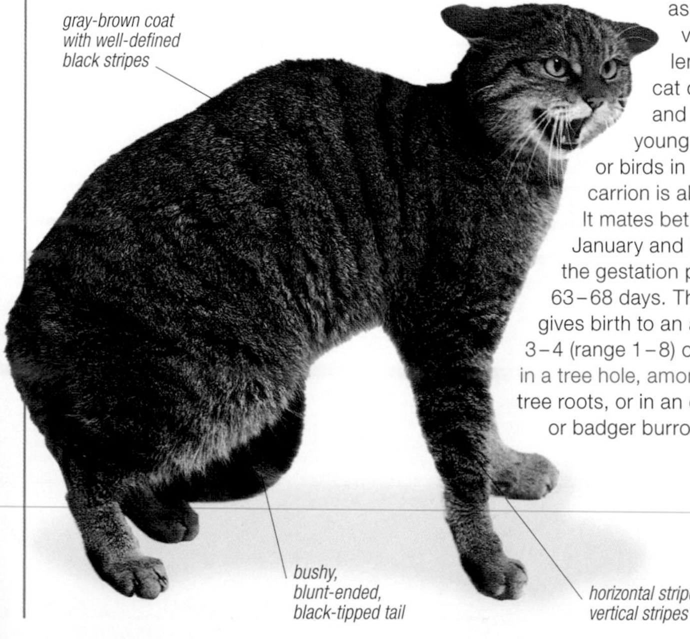

Felis chaus

Jungle cat

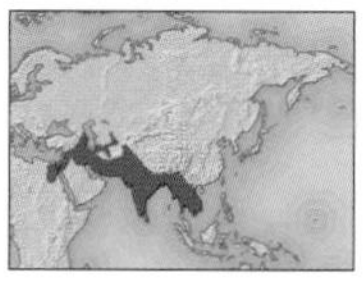

Location W., C., S., and S.E. Asia, N.E. Africa

Length 20–37 in (50–94 cm)
Tail 9–12 in (23–31 cm)
Weight 8¾–35 lb (4–16 kg)
Social unit Individual
Status Least concern

Despite its main common name, this slender, long-legged cat is also known more aptly as the swamp or reed cat. It hunts along marshes, river banks, shores, and also ditches and ponds around human settlements, taking mammals up to the size of coypu, birds (including poultry), reptiles, and, being a strong swimmer, fish and amphibians. Female and male may stay together, and both protect the cubs.

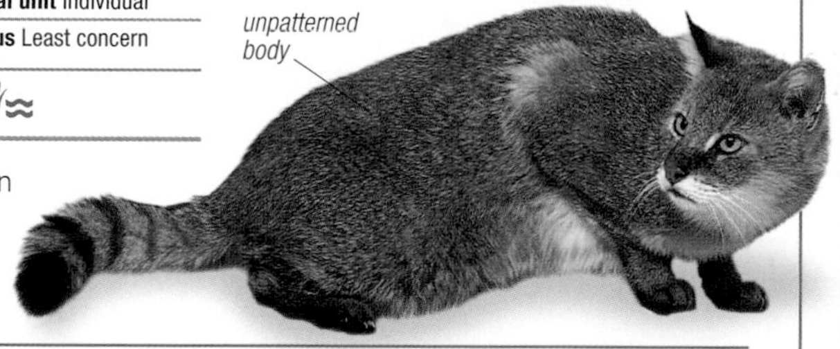

Felis margarita

Sand cat

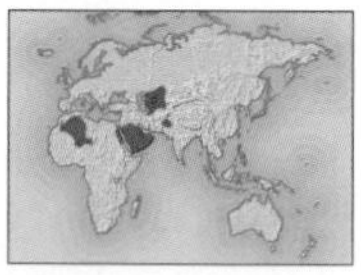

Location N. Africa, W., C., and S.W. Asia

Length 18–22½ in (45–57 cm)
Tail 11–14 in (28–35 cm)
Weight 3¼–7¾ lb (1.5–3.5 kg)
Social unit Individual
Status Near threatened

Surviving on fluids in its food and very little additional water, the blunt-clawed sand cat digs well for its main prey of gerbils and similar rodents, as well as an occasional lizard or snake. It also excavates a den for daytime shelter. The average litter of 3 grows quickly and may be independent in just 6 months.

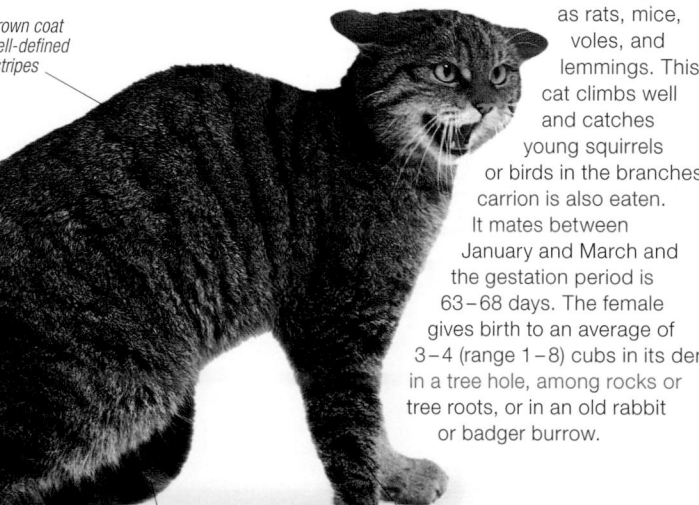

Felis nigripes

Black-footed cat

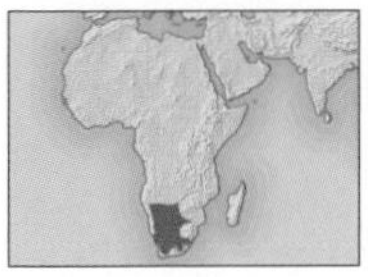

Location Southern Africa

Length 13½–19½ in (34–50 cm)
Tail 6–8 in (15–20 cm)
Weight 3¼–6½ lb (1.5–3 kg)
Social unit Individual
Status Vulnerable

One of the smallest cats, the size of a very small pet cat, this species is pale brown with bold stripes, which thicken on the legs and merge into the all-black undersides of the feet. Prey is likewise small; for example, mice, insects from termites to locusts, spiders, small lizards, and birds. Well adapted to the Karoo, the Kalahari, and other arid regions of southern Africa, it seldom needs to drink water.

Leopardus jacobitus

Andean cat

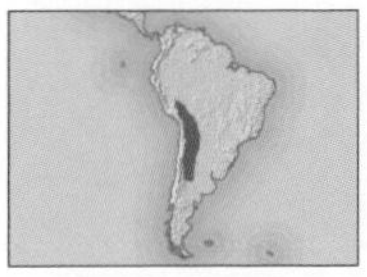

Location W. South America

Length 23–25 in (58–64 cm)
Tail 16–19 in (41–48 cm)
Weight 8¾ lb (4 kg)
Social unit Individual
Status Endangered

Little is known about the habits of the rare Andean, or mountain cat. Small and sturdy, with a long, bushy tail, it has thick, warm, gray-brown fur that is marked with vertical stripes along the upper back, rosette-type spots on the flanks, and bands around the legs and tail. This feline inhabits dry, rocky slopes above the tree line—generally 9,900 ft (3,000 m)—and preys mainly on rodents such as viscachas and formerly chinchillas. Unlike many other cats, this species is not directly at risk from hunting or habitat loss, but hunting threatens some of the prey species on which it depends.

Leopardus pardalis

Ocelot

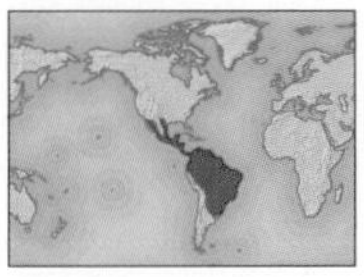

Location S. USA to Central and South America

Length 19½–39 in (50–100 cm)
Tail 12–18 in (30–45 cm)
Weight 25–35 lb (11.5–16 kg)
Social unit Individual
Status Least concern

The chainlike rosettes and spots on this cat's body are highly distinctive. Typically catlike in its nocturnal, solitary, tree-climbing lifestyle, the ocelot is wide ranging and adaptable, living in a variety of habitats from grassland to swamp, as well as most types of forest. It takes a huge variety of prey: chiefly small rodents but also birds, lizards, fish, bats, and larger animals such as monkeys, turtles, young deer, armadillos, and anteaters. The ocelot has a gestation period of 79–85 days and an average litter size of 3 (range 1–2). Females breed from 2 years of age; males from about 2½ years.

CONSERVATION

Ocelots were extensively hunted in the 1960s and 70s, trading at some 200,000 skins per year. The fur trade is now banned by CITES, but ocelot numbers are still on a downward path. As with most cats, one of the main problems the species faces is deforestation, combined with attacks by farmers, and illegal collection for the pet trade.

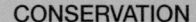

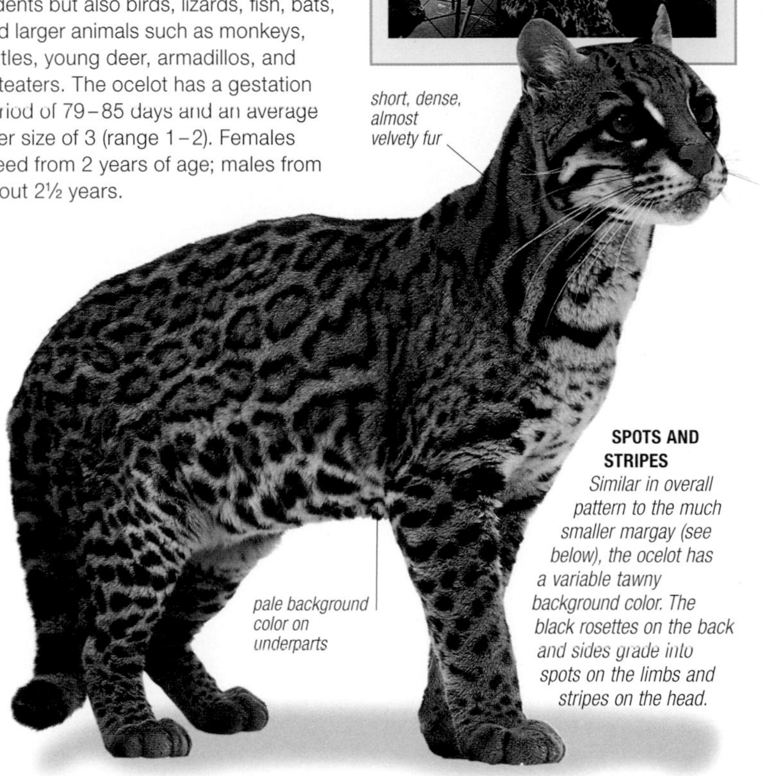

SPOTS AND STRIPES *Similar in overall pattern to the much smaller margay (see below), the ocelot has a variable tawny background color. The black rosettes on the back and sides grade into spots on the limbs and stripes on the head.*

Leopardus guigna

Kodkod

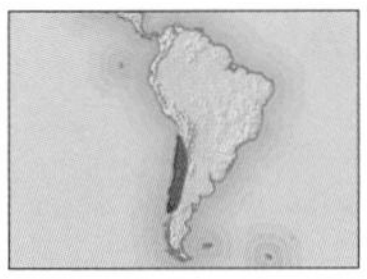

Location W. South America

Length 16½–20 in (42–51 cm)
Tail 7½–10 in (19.5–25 cm)
Weight 4½–5½ lb (2–2.5 kg)
Social unit Individual
Status Vulnerable

The smallest cat in the Americas, the kodkod or güiña closely resembles Geoffroy's cat (see below) but with a thicker tail and smaller head. Its coat is black-spotted gray to ocher, with a ringed tail and a dark throat stripe. Found in moist, cool forests in the Andean foothills of Argentina and Chile, it makes its den in bamboo thickets in the understory. Its main prey are rodents such as mice and rats, and lizards, captured on the ground as well as in trees. It probably hunts by day as well as at night.

Leopardus geoffroyi

Geoffroy's cat

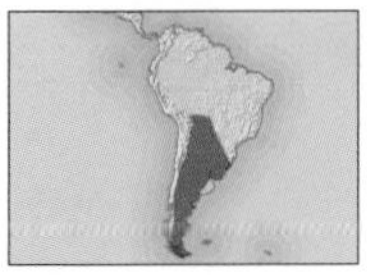

Location C. to S. South America

Length 16½–26 in (42–66 cm)
Tail 9½–14 in (24–36 cm)
Weight 4½–13 lb (2–6 kg)
Social unit Individual
Status Near threatened

Geoffroy's cat, which is sometimes called Geoffroy's ocelot, prefers scrub and shrub to forest and open grassland. It hunts in the branches, on the ground, and in water, for frogs and fish as well as the usual small-cat fare of rodents, lizards, birds, and the (introduced) brown hare. Now protected, this species was hunted for its yellow-brown to silver-gray fur after the trade in ocelot fur (see panel above, right) declined in the 1980s.

Leopardus wiedii

Margay

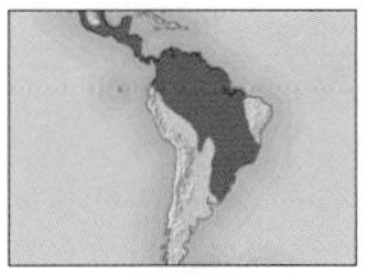

Location S. USA to Central and South America

Length 18–31 in (46–79 cm)
Tail 13–20 in (33–51 cm)
Weight 5½–8¾ lb (2.5–4 kg)
Social unit Individual
Status Near threatened

The margay has exceptional climbing abilities because of its almost "reversible" rear feet, being able to run headfirst down a trunk or hang from a branch by one paw. Its mainly tree-dwelling prey includes rats, mice, squirrels, possums, young sloths, small birds, and invertebrates such as grubs and spiders. It also occasionally eats fruit. The margay is nocturnal, resting in the safety of a tree fork by day. The gestation period is 76–85 days and litter size is one, rarely 2. Following the decline in the availability of ocelot fur (see panel, above), margays became one of the most sought-after small cats for the fur trade. Hunting may still continue illegally in a few areas, but the primary threat to the margay is now forest clearance.

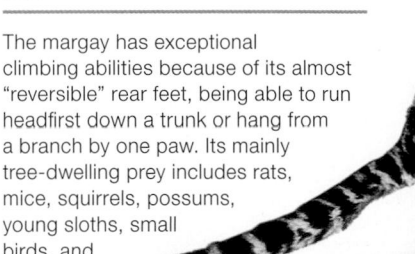

MAMMALS

YOUNG LIFE
A young tiger is dependent on its mother for food for the first year or so of its life. By the time it is 2 years old, it will have enough power, strength, and experience to be able to catch prey for itself. It may start breeding in its fourth or fifth year, and lives on average until the age of 8–10.

TIGER IN ACTION
Tigers in the Ranthambhore reserve in northwest India have been known to charge into lakes after samba deer and become fully submerged with their prey. The tigers in this area have also been reported to have killed and eaten crocodiles.

Panthera tigris

Tiger

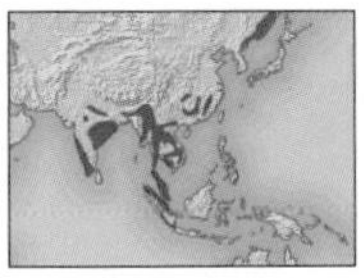

Location S. and E. Asia

Length 4½–9¼ ft (1.4–2.8 m)
Tail 2–3½ ft (60 cm–1.1 m)
Weight 220–660 lb (100–300 kg)
Social unit Individual
Status Endangered

The tiger is the largest member of the cat family, and its orange coat with black stripes and white markings is instantly recognizable. Its size, coat color, and markings vary according to subspecies. Although 6 subspecies are recognized, 3 have become extinct since the 1950s, and the 3 remaining are endangered. The geographical distribution of the tiger once extended as far west as eastern Turkey, but it is now restricted to pockets of southern and eastern Asia. The tiger's habitat varies widely, from the tropical forests of Southeast Asia to the coniferous woodlands of Siberia, but its basic requirements are dense cover, access to water, and sufficient large prey. Hunting mainly by night, it takes mostly deer and wild pigs, and cattle in some regions, but it also eats smaller animals, including monkeys, birds, reptiles, and fish, and readily feeds on carrion. Tigers will also attack young rhinoceroses and elephants. They may eat up to 88 lb (40 kg) of meat at a time and return to a large kill for 3–6 days. Tigers are usually solitary, but are not necessarily antisocial. A male is occasionally seen resting or feeding with a female and cubs, and tigers may also travel in groups.

CONSERVATION

Between 1900 and 2000, tiger numbers fell from an estimated 100,000 to 3,500—approximately the figure today. After decades of poaching, habitat loss, and loss of prey, scattered populations survive only in eastern Russia, China, Sumatra, and in southern Asia from Vietnam to India. Although protected in most areas, tigers are still killed where they threaten local farmers, and a large but clandestine trade operates in their skins and body parts, which are used in Chinese medicine. Programs to save the tiger have met with mixed results, with limited successes set against a trend of overall decline. Radio collars (see below) and antipoaching patrols give these vulnerable predators at least some protection in the wild.

DISTINCTIVE COAT
The Bengal tiger, Panthera tigris tigris, *the most common of the subspecies, exhibits the classic tiger coat: deep orange with white undersides, cheeks, and eye areas, and distinctive black markings. Tiger stripes, which range from brown to jet-black, vary in number, width, and tendency to split. No 2 tigers have the same markings.*

LIVING TIGER SUBSPECIES

The 3 surviving subspecies of tiger differ markedly from one another. In general, those animals from southern areas are smaller and more deeply colored than their larger and paler northern kin. Tigers that live in cold climates also tend to have thicker fur.

SUMATRAN TIGER
The smallest and darkest of the tiger subspecies is the Sumatran tiger, Panthera tigris sumatrae. *Only about 600 are thought to exist today.*

SIBERIAN YOUNGSTERS
The Siberian tiger, Panthera tigris altaica, *is the largest subspecies, and the lightest in color, with the longest coat. Its numbers may be as low as 150–200.*

MAMMALS

Puma yagouaroundi

Jaguarundi

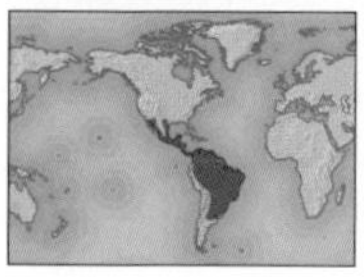

Location S. USA to South America

Length 22–30 in (55–77 cm)
Tail 13–23½ in (33–60 cm)
Weight 10–20 lb (4.5–9 kg)
Social unit Individual
Status Least concern

More mustelid than felid in overall proportions, with a pointed snout, long body, and shortish legs, the jaguarundi has several color forms of unpatterned fur, from black—mainly in forests—to pale gray-brown or red—in dry shrubland. This cat hunts by day, often on the ground, in habitats ranging from semiarid scrub to rain forest and swamp. Its main prey are birds, rodents, rabbits, reptiles, and invertebrates.

Puma concolor

Puma

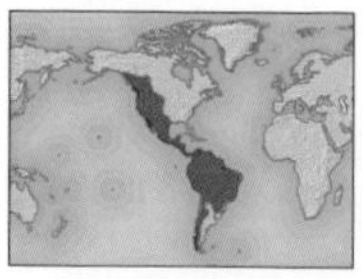

Location W. and S. North America, Central America, South America

Length 3½–6½ ft (1.1–2 m)
Tail 26–31 in (66–78 cm)
Weight 150–230 lb (67–105 kg)
Social unit Individual
Status Least concern

Larger than some "big" cats, the puma—also called panther, cougar, or mountain lion—is probably related more closely to smaller cats. Most of its fur is uniformly buff-colored. It makes many sounds, including an eerily humanlike scream when courting, but it cannot roar. Amazingly adaptable, the puma lives in habitats ranging from tropical rain forest, high mountains, and conifer forest to desert. Small mammals, such as mice, rats, rabbits, and hares, form the staple diet in many areas, as well as occasional sheep, young cattle, moose, and other livestock; the puma rarely scavenges. Births peak from February to September. The 2–3 (range 1–6) spotted cubs are born after an average gestation of 92 days, in a den among rocks or in a thicket. They take solid food from 6–7 weeks.

Neofelis diardi

Bornean clouded leopard

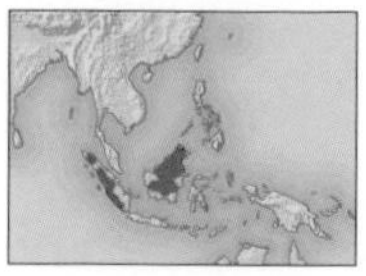

Location S.E. Asia

Length 23¾–41¼ in (61–106 cm)
Tail 21½–35½ in (55–91 cm)
Weight 35¼–50½ lb (16–23 kg)
Social unit Individual
Status Vulnerable

Clouded leopards range throughout Southeast Asia; those from Sumatra and Borneo were recently determined to be a separate species. They have smaller, darker "cloud" markings than those found on mainland Asia. Their numbers are estimated to be larger in Borneo (5,000–11,000) than Sumatra (3,000–7,000). They kill a wide range of prey from monkeys to fish and porcupines, and spend much of their time in trees.

Panthera pardus

Leopard

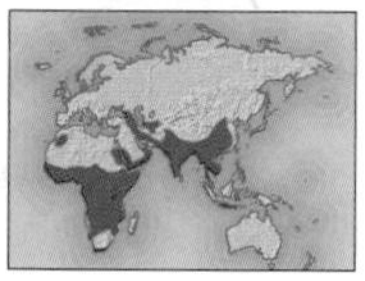

Location W., C., S., E., and S.E. Asia, Africa

Length 3–6¼ ft (0.9–1.9 m)
Tail 24–43 in (60–110 cm)
Weight 82–200 lb (37–90 kg)
Social unit Individual
Status Near threatened

Few other wild cats are as varied in appearance or in prey preference as the leopard, or have a wider geographical range. Its varied diet includes small creatures, such as dung beetles, and large mammals many times its own weight, such as antelopes. A large victim may provide enough food for 2 weeks, although such kills are usually made about every 3 days, twice as often for a female with cubs. The average litter of 2 is born after 90–105 days' gestation, and is cared for by the mother. Weaned by 3 months, they stay with her for a year or more, and siblings may associate for longer. Adaptable to human presence, leopards hunt to within a few miles of big cities, but numbers are falling due to various human activities.

BLACK PANTHER

Like many species of cats and other mammals, leopards may exhibit melanism. As a result of this genetic change (mutation), the skin and fur contain large amounts of the dark pigment, melanin. Most common in moist, dense forests, melanic leopards, known as "black panthers," were once viewed as a separate species. In deserts, leopards are pale yellow; in grass, they are deeper yellow.

CACHING IN TREES

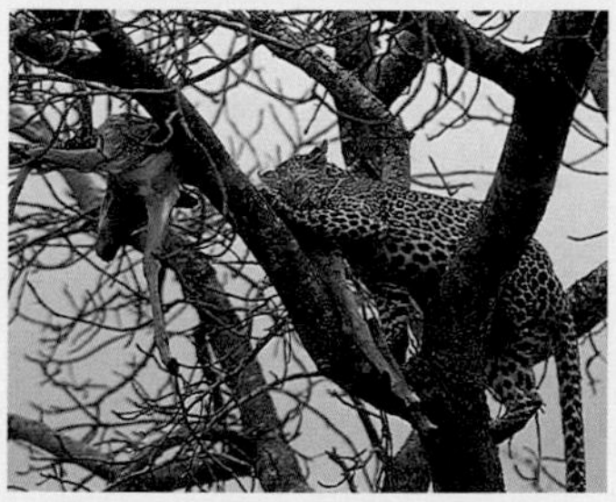

The leopard is an adept climber and uses its immense strength to drag its prey up into trees, for immediate consumption or for caching (hiding for future use). In the branches, it can eat undisturbed, and the meat is safe from scavenging hyenas and jackals.

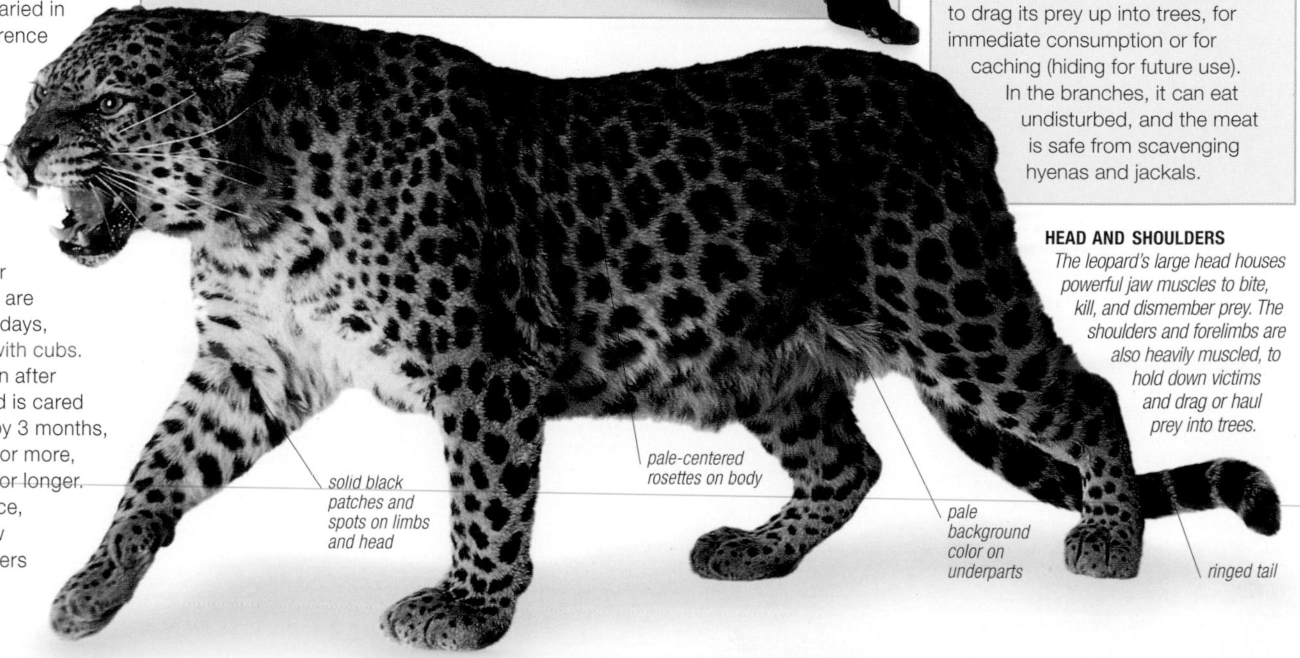

HEAD AND SHOULDERS
The leopard's large head houses powerful jaw muscles to bite, kill, and dismember prey. The shoulders and forelimbs are also heavily muscled, to hold down victims and drag or haul prey into trees.

Uncia uncia

Snow leopard

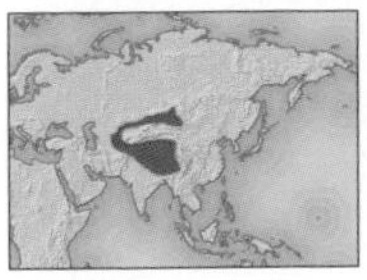

Length 3¼–4¼ft (1–1.3m)
Tail 32–39in (80–100cm)
Weight 55–165lb (25–75kg)
Social unit Individual
Status Endangered

Location C., S., and E. Asia

Resembling the leopard (see p.206) in its wide range of prey, this woolly furred big cat prefers crags and ridges in steppe, rocky shrub, and open conifer forest to altitudes of 16,500ft (5,000m). It can hunt yak or asses, but most prey are smaller—wild sheep, goats, marmots, pikas, hares, and birds. Breeding habits resemble similar-sized felids, although 4–5 cubs may be raised.

Panthera onca

Jaguar

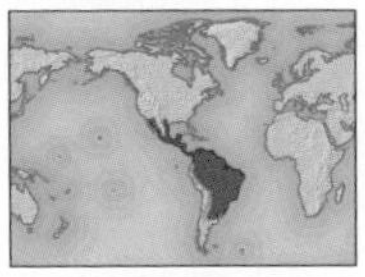

Length 3½–6¼ft (1.1–1.9m)
Tail 18–30in (45–75cm)
Weight 79–350lb (36–160kg)
Social unit Individual
Status Near threatened

Location Central America to N. and C. South America

The New World's only "big" cat, the jaguar resembles the leopard (see p.206) but has rosettes with dark centers, and is more squat and powerful, with a large, broad head and heavily muscled quarters. It prefers a watery environment, such as permanent swamps and seasonally flooded forest, where its main prey are medium-sized mammals such as deer, peccaries, and tapirs. Despite legal protection and reduced hunting for fur, jaguars are increasingly at risk from habitat destruction and their elimination from cattle ranches.

Acinonyx jubatus

Cheetah

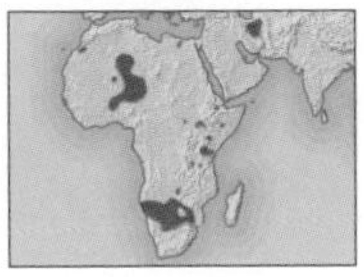

Length 3½–5ft (1.1–1.5m)
Tail 23½–32in (60–80cm)
Weight 46–160lb (21–72kg)
Social unit Individual/Pair
Status Vulnerable

Location Africa, W. Asia

Renowned as the world's fastest land animal, the cheetah can sprint at over 62mph (100kph) for 10–20 seconds, before it begins to overheat. If its prey can stay ahead for longer than this, it invariably escapes. The cheetah eats medium-sized ungulates such as Thomson's gazelle, as well as larger antelopes and smaller animals such as hares. It is more social than any other big cat except lions. Siblings leave their mother at 13–20 months, but may stay together for several more months—indeed, brothers may stay together for years.

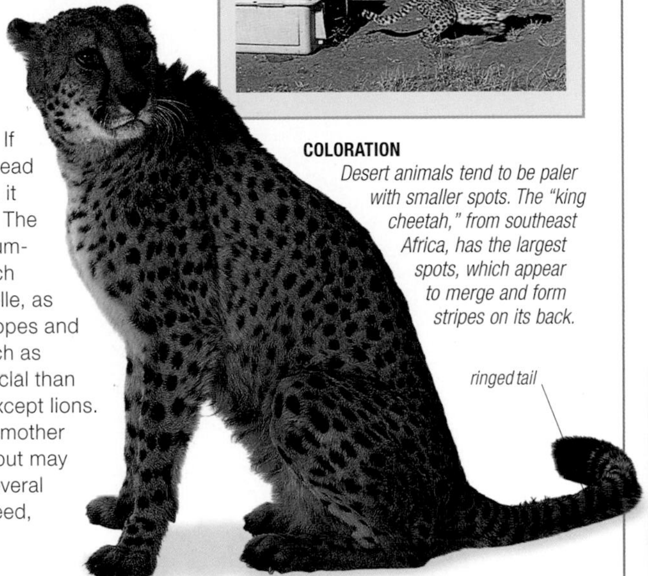

COLORATION
Desert animals tend to be paler with smaller spots. The "king cheetah," from southeast Africa, has the largest spots, which appear to merge and form stripes on its back.

CONSERVATION

In Namibia, in order to monitor the effect of wildlife management and livestock protection measures, some cheetahs are caught and fit with radio collars. Following their release (below), the animal's movements are tracked.

Panthera leo

Lion

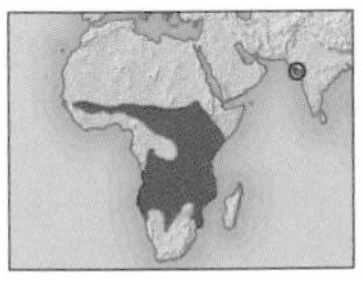

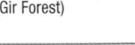

Length 5½–8¼ft (1.7–2.5m)
Tail 3–3½ft (0.9–1.1m)
Weight 280–550lb (125–250kg)
Social unit Group
Status Vulnerable

Location Africa, S. Asia (Gir Forest)

Unique among felids, lions form close-knit, long-term social groups. In females, these are called prides, and average 4–6 related adults and their cubs. Females tend to give birth at the same time and suckle each other's young. Prides occupy home ranges and members cooperate to hunt large prey such as zebra, wildebeest, impala, and buffalo. Individuals also forage alone for small rodents, hares, and reptiles. Adult males live alone, or in coalitions of usually 2–3 unrelated members or 4–5 relatives (originating from the same pride). A coalition defends a large area against other male coalitions, and holds mating rights over prides within it, but this tenure generally lasts only 2–3 years.

MALE AND FEMALE
The lion averages 400lb (180kg) against the lioness's 280lb (125kg). The male's skull is also significantly larger than that of the female.

ROLE PLAYING

Retracted claws and nonexposed teeth show that these 2-year-old lionesses are "play-fighting" to develop skills for the hunt. The tussle helps to determine whether a lioness will be one of those who chase and direct prey—or one who carries out the ambush and kill. Play also helps to establish relative social status within the pride.

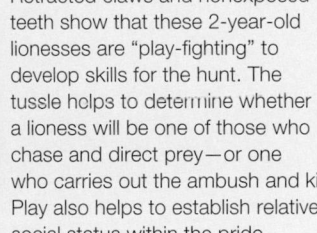

ASIATIC LION

The Asiatic lion (*Panthera leo persica*) survives only in the Gir Forest region of northwest India, with a probable population of about 350. It tends to be smaller than the African lion, with a fold of skin along the central underside of the belly; males have shorter manes. Prides are also smaller, usually 2 related females and young.

Hyenas and aardwolf

PHYLUM	Chordata
CLASS	Mammalia
ORDER	Carnivora
FAMILY	Hyaenidae
SPECIES	4

Although members of this family superficially resemble dogs, they are in fact more closely related to cats and civets and genets. They all feature a distinctive back line that slopes downward from the shoulders to the tail. Hyenas and aardwolves are mainly found in Africa (although the striped hyena extends to southern parts of Asia), in savanna, scrub, and semiarid habitats. They are primarily nocturnal and dig dens that are used to shelter adults and cubs (except in the spotted hyena, where only the cubs seek refuge in dens).

Anatomy

Physical characteristics common to the species in this family include a large head and ears, long front legs and short back legs, a mane on the nape that (except in the spotted hyena) extends down the back, a bushy tail, and short, blunt, nonretractable claws. Hyenas have 4 toes on both the front and back feet; the aardwolf has 5 toes on the front feet and 4 on the back feet. The coat is spotted or striped (the brown hyena has stripes on the limbs only).

Feeding

Hyenas have a broad muzzle and immensely strong jaws (indeed, the jaws are the most powerful of any mammal of comparable size) and teeth for crushing bones. Of the 3 hyena species, the spotted hyena is the most voracious hunter. When hunting cooperatively, spotted hyenas are able to take down large prey, such as zebras. This species is also a highly efficient scavenger—groups of spotted hyenas are capable of driving a lion off its prey. Striped and brown hyenas, on the other hand, scavenge most of the time, although they may also capture small prey. All hyenas are able to digest parts of a kill that other mammals cannot process, such as skin and bone. This means that they occupy a niche not filled by most other mammals. Portions of the meal that they cannot digest, such as ligaments, hair, and hooves, are regurgitated in the form of pellets.

Given these digestive adaptations, it is surprising that one member of this family—the aardwolf—does not eat large prey. Instead, it uses its smaller teeth and sticky tongue to feed on termites. There is little competition for this food source, it requires minimal effort to obtain, and the nutritional value is surprisingly high. An aardwolf may eat as many as 200,000 termites in a single night.

Social groups

While the aardwolf is solitary, and striped and brown hyenas live in pairs or small groups, spotted hyenas live in larger groups called clans. These clans may consist of up to 80 individuals (males and females and their cubs). When cubs are 2–3 months old, they are transferred to a communal den, where all are suckled by any lactating female (brown hyenas employ this system, too). The cubs remain in the den until they are weaned and ready to accompany adults hunting and foraging, which may be at as little as 7 months of age. All hyenas are territorial, and their territories are marked using an anal scent gland (which can be turned inside out). Spotted hyena clans patrol and defend their territory communally.

THE ULTIMATE MEAT EATER
The kill on which this spotted hyena is feeding may have been hunted down by the clan or scavenged from another predator. Because hyenas have powerful, bone-crushing jaws, and eat almost anything, little will be left of this hartebeest carcass when the hyenas leave.

Proteles cristata

Aardwolf

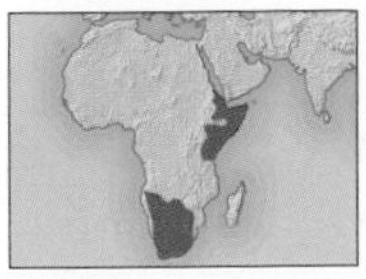

Length 26 in (67 cm)
Tail 9½ in (24 cm)
Weight 20 lb (9 kg)
Location E. and southern Africa
Social unit Individual
Status Least concern

A small relative of the hyena, the aardwolf's specialized diet is termites, particularly surface-foraging nasute (snouted) harvester termites. It also licks up maggots, grubs, and other small, soft-bodied creatures. Its longer front legs and down-sloping body are accentuated by the crestlike back mane, most prominent on the neck and shoulders. This erects under stress so the animal appears larger. The fur is pale buff or yellow-white with 3 vertical stripes on each side, and diagonal stripes across the fore- and hindquarters. The front teeth are hyena-like but the molars are small pegs, the food being ground up by the muscular stomach. The aardwolf is solitary and nocturnal, resting in a burrow by day. It marks its territory with urine, dung, and anal gland secretions. The 2–4 cubs are born after a gestation period of 90 days, emerge from the den at 4 weeks, forage with the mother from 9–11 weeks, and are weaned by 16 weeks.

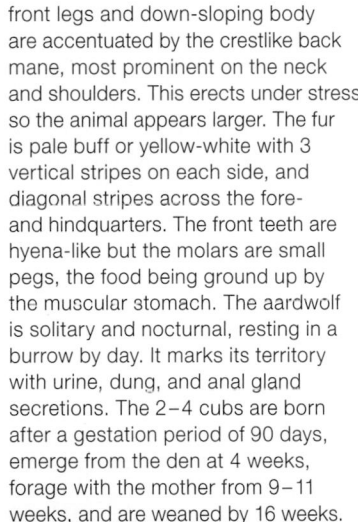

Parahyaena brunnea

Brown hyena

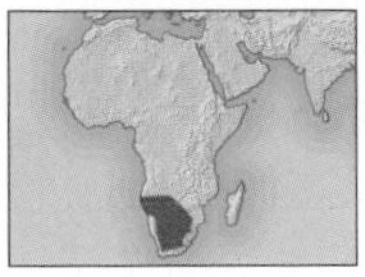

Length 4¼ ft (1.3 m)
Tail 8½ in (21 cm)
Weight 84–105 lb (38–47 kg)
Location Southern Africa
Social unit Variable
Status Near threatened

This species ranges farther into deserts than other hyenas, and can scent carrion from 8½ miles (14 km). It has the typical hyena's powerful jaws and shearing teeth for scavenging on any carcass, including seal pups along the Namib Desert coast. It also catches prey such as springhares. The brown hyena forms loose clans that defend their territory. Its coat is shaggy, dark brown to black, with a pale tawny neck mantle, a gray-patched face, and striped legs.

Hyaena hyaena

Striped hyena

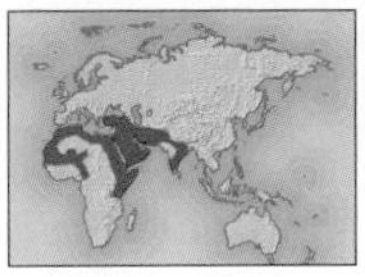

Length 3¼ ft (1.1 m)
Tail 8 in (20 cm)
Weight 77–88 lb (35–40 kg)
Location W., N., and E. Africa, W. to S. Asia
Social unit Individual/Group
Status Near threatened

Preferring savanna and open woodland, this hyena avoids extreme habitats such as deserts. It is gray or pale brown with 5–6 vertical flank bars. The neck mane lessens on the back and merges with the bushy black-and-white tail. Generally solitary, this hyena may form a family group when breeding. It scavenges, hunts prey from insects to hares, and eats fruit and other plant matter.

Crocuta crocuta

Spotted hyena

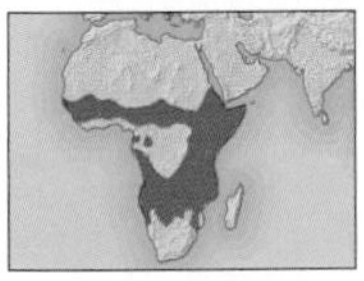

Length 4¼ ft (1.3 m)
Tail 10 in (25 cm)
Weight 135–155 lb (62–70 kg)
Location W. to E. and southern Africa
Social unit Group
Status Least concern

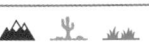

The spotted hyena is the largest hyena species. The female is some 10 percent larger than the male, and her external sexual organs are enlarged so that they are difficult to distinguish from the male's. The social system is female-dominated and based on the clan, which varies from 5 or fewer adults and young in deserts to 50 or more in prey-rich savanna. The clan occupies a communal den, uses communal latrines, and jointly defends its territory of 15–390 square miles (40–1,000 square km), delineated by calls, scent marking, and boundary patrols. The spotted hyena makes many sounds, including whoops, to rally its group or locate cubs, and the famous hyena's "laugh," to signify submission to a senior clan member.

HUNTING HYENA

The spotted hyena is a powerful hunter. Several clan members may form a pack to bring down large prey, such as an adult zebra or wildebeest. When hunting alone, it catches hares, ground birds, and fish in shallows and swamps. It gorges on food, and is able to consume up to one-third of its body weight at one meal.

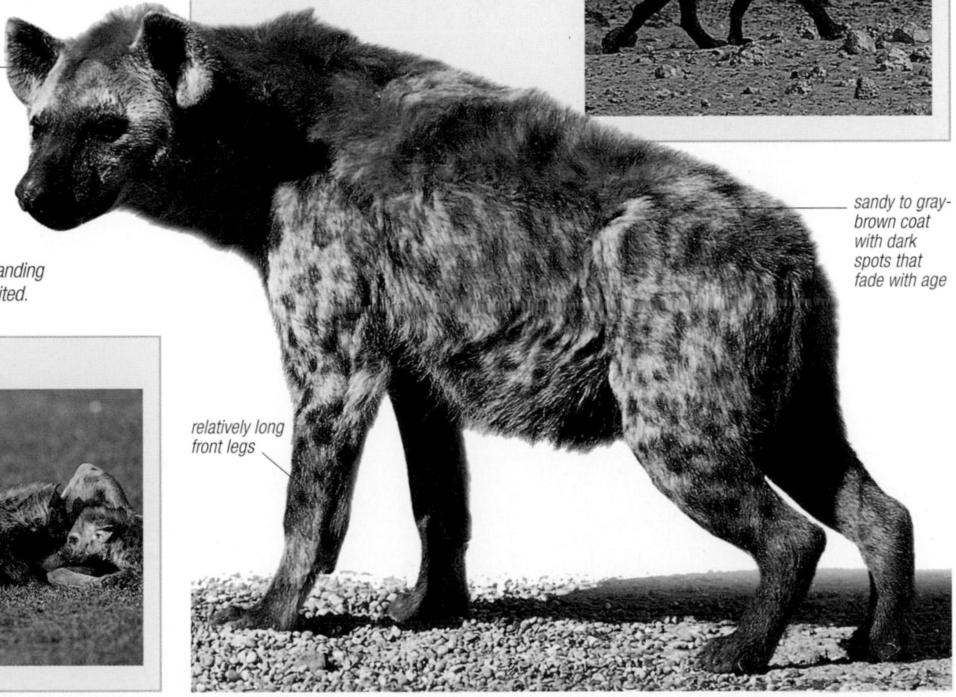

REVERSED MANE
The spotted hyena's neck and back mane is reversed and the erectile hairs slope forward rather than back, standing erect when the hyena is excited.

SIBLING RIVALRY

The spotted hyena mother is solely responsible for cub rearing; the male plays no part. The average litter is 2 (range 1–3), born after 100 days' gestation and weaned at 14–18 months, when nearly full grown. The dominant cub controls access to the mother when suckling and, in times of milk shortage, may kill its sister or brother to improve its own chance of survival.

MAMMALS

Hoofed mammals

PHYLUM	Chordata
CLASS	Mammalia
ORDER	Perissodactyla, Artiodactyla
FAMILIES	13
SPECIES	257

CLASSIFICATION NOTE

Hoofed mammals are classified in 2 orders: Perissodactyla (odd-toed hoofed mammals), and Artiodactyla (even-toed hoofed mammals). Although superficially similar, the 2 orders are not closely related. However, they are often grouped together because they share a range of common traits. There is evidence to suggest that the closest relatives of the artiodactyls are the cetaceans (whales and relatives), and some taxonomists now group them together in the order Cetartiodactyla.

Hoofed mammals are a highly successful group. Their position as the dominant terrestrial herbivores can be attributed largely to their speed and endurance (they are able to outrun most predators), and to the fact that they are well equipped to break down the cellulose in their plant diet. Despite a variety of body forms, most species have a long muzzle, a complex battery of grinding teeth, and a barrel-shaped body. The group consists of odd-toed hoofed mammals (such as tapirs) and even-toed hoofed mammals (such as deer). Wild odd-toed hoofed mammals are found in Africa, Asia, and South and Central America. Wild even-toed hoofed mammals are distributed worldwide (except the West Indies, Australasia, and Antarctica). Hoofed mammals are mostly found in open habitats, such as savanna. Domestic hoofed mammals are found almost anywhere there are humans.

Anatomy

The ability of hoofed mammals to run swiftly for long distances is largely due to the structure of the limbs, which are adapted for simple but powerful forwards and backward movement. Each limb is embedded in the body wall as far down as the elbow or knee joint. Below this joint are the radius and ulna (front limbs) or tibia and fibula (back limbs), then the greatly elongated metapodials (the palm and foot bones in humans). This longer lower limb (and increased movement in the shoulder joint) gives a longer stride length and hence more speed. These animals also have a reduced number of toes, which means fewer muscles and tendons and therefore lower energy demands (which aids endurance). They run on their toes (unguligrade gait), which are encased in hooves.

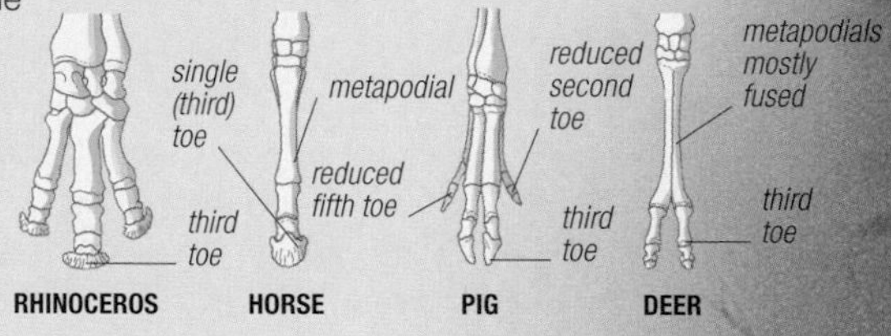

ODD AND EVEN TOES
In odd-toed hoofed mammals (rhinoceros and horse), the leg's weight rests on the central (third) toe. In even-toed hoofed mammals (pig and deer) the weight is borne by the first, third, and fourth toes; the second and fifth toes are greatly reduced (pig) or lost (deer).

REINDEER

RHINOCEROS

HORNS AND ANTLERS
Most hoofed mammals have antlers or horns. Antlers are bony outgrowths of the skull, which in deer are shed each year. Horns are permanent and have a keratinous sheath over a bony core. The horn of the rhinoceros is entirely keratinous.

ESCAPING DANGER
Hoofed mammals must be able to detect danger quickly to survive. To do this they have mobile, tubular ears and acute hearing, an excellent sense of smell, and eyes on the side of the head, giving all around vision. When frightened, they flee at great speed. Antelopes, such as these impala, often make spectacular leaps as they escape.

BIRTH ON THE HOOF

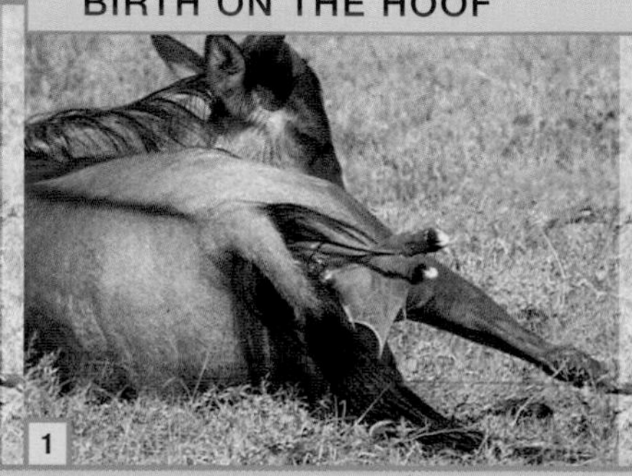

1 **THE BIRTH BEGINS**
During labor, this female wildebeest is vulnerable and restless, alternately lying down and standing up.

2 **THE HEAD APPEARS**
The calf is born head first, but the long front legs are first into the outside world.

3 **A NEW LIFE**
The newborn wildebeest emerges, partially covered by the amniotic membrane.

4 **SYNCHRONIZED CALVING**
Most wildebeests calve in the same 3-week
so that, despite some losses, many calves sur

MAMMALS

Feeding

Almost all hoofed mammals are herbivores. The plants they eat contain indigestible cellulose (the major component of plant cell walls), which is split into digestible carbohydrates via bacterial fermentation. In ruminants (see below), food passes slowly through their system to maximize the nutrition gained. These animals thrive where food is limited but of high quality. In hindgut fermenters (see below), food is not retained in the stomach and passes through the system more quickly. These animals live where food is plentiful but of poor quality. As a result, a larger volume of food must be eaten to obtain enough nutrients.

GRAZING
Grazers feed almost exclusively on grass. To find fresh pasture, some hoofed mammals migrate long distances. The hippopotamus (above) usually feeds at night, using its horny lips to crop the grass.

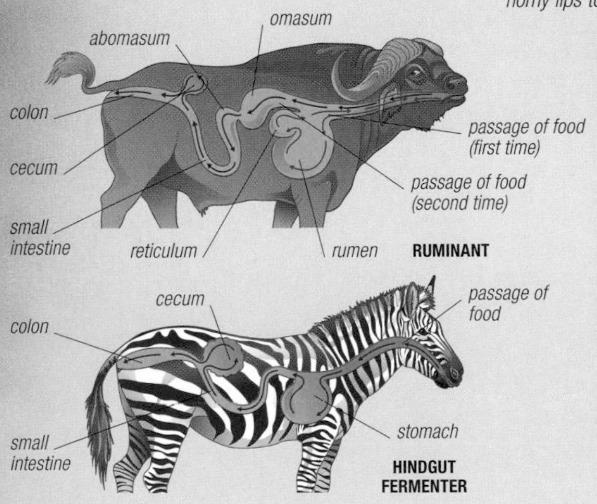

DIGESTIVE SYSTEMS
Ruminants, such as buffaloes, have a complex stomach. In the first chamber, the rumen, bacterial fermentation occurs; food is then regurgitated, re-chewed ("chewing the cud"), and swallowed, this time passing through the digestive system. Hindgut fermenters, such as zebras, have a simple stomach: the fermenting bacteria are in the cecum and at the start of the colon.

BROWSING
Browsers eat almost any plant material. These mountain goats spend much of their time in mountainous areas where grass is scarce—they supplement their diet by eating mosses, lichens, herbs, and woody plants.

MAMMALS

DEFENDING RESOURCES
Hoofed mammals living in arid areas, such as these onagers, often form mixed-sex herds. Males tend to defend resources rather than guard a harem of females.

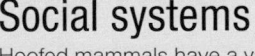

DEFENDING A HAREM
During the breeding season, some male hoofed mammals, such as this red deer, gather a harem, which they defend fiercely. The loud roaring and large antlers of the stag help deter rivals.

Social systems

Hoofed mammals have a variety of social systems, depending on factors such as habitat, body size, whether breeding is seasonal, and whether they migrate. The following examples are typical hoofed mammal social organizations. Tapirs, rhinoceroses, and some forest antelopes are mostly solitary, and a male's territory covers that of several females. Dik-diks live in pairs that occupy small territories. In gazelles, males set up small territories, and female herds wander in and out of these areas. Hartebeests and most zebras live in harems consisting of a male and a number of females. Red deer, on the other hand, form separate-sex herds except during the breeding season (which is known as the "rut"). Males fight with each other, and the most successful will collect a large harem of females.

TIOUS MEAL
other eats the afterbirth. Once this is done, ill lick the calf to stimulate it.

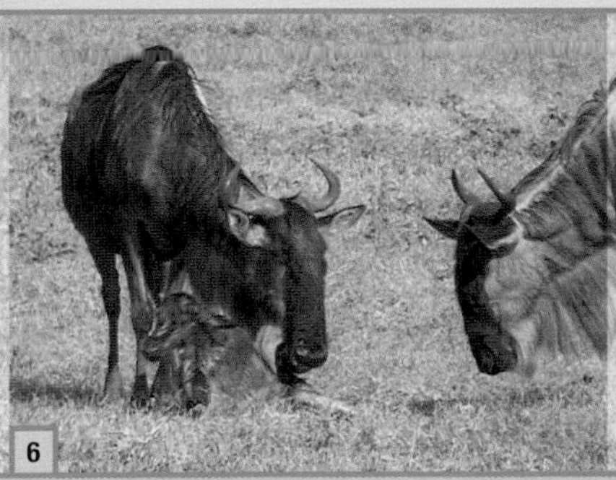

6

PROTECTIVE INSTINCTS
A newborn calf is vulnerable, and the mother is reluctant to let even another wildebeest approach.

7

STANDING UP
The calf attempts to stand. In less than 45 minutes, it will be able to run, reducing the risk of predation.

8

REJOINING THE HERD
Mother and calf quickly rejoin the herd: in a group, the young calf is less conspicuous.

Horses and relatives

PHYLUM	Chordata
CLASS	Mammalia
ORDER	Perissodactyla
FAMILY	Equidae
SPECIES	8

An enduring symbol of grace and freedom, horses and their relatives (equids) are the ultimate odd-toed hoofed mammals—they have just a single toe on each foot. Equids – horses, asses (including the onager and the kiang), and zebras—have a long neck and head, and long, slender legs. They possess great stamina and can run at speed: the swiftest wild equid, the onager, can attain 43 mph (70 kph) for short periods. They are found in grasslands and deserts of Africa and Asia, and various species have been widely introduced across the world.

Anatomy

Equids are characterized by a deep chest, a mane on the neck, a tufted or long-haired tail, a solid hoof on each foot, areas of hard, thickened skin (called chestnuts) on the inside of the front legs above the knee, and mobile lips and nostrils. The eyes, which have oblong pupils, are at the sides of the head for good, all around sight (to help detect predators). Day and night vision is excellent. The ears are long and can twist to locate the sources of sounds, without having to move the body. Hearing is acute.

All species have a heavily haired coat, which is usually uniform in color in horses and asses. Zebras have striking black-and-white stripes; the function of these has had many explanations from social recognition to temperature regulation, or to create a "dazzle" effect to confuse predators.

Feeding

Horses and relatives eat mainly grass (they have a battery of hard-wearing cheek teeth for shearing grass), although they will also feed on desert vegetation and may browse on bark, leaves, buds, and fruit. Unlike cattle (for example), they do not ruminate but instead employ a hindgut fermentation system (see p.211). This allows them to take in large amounts of food, which passes rapidly through the digestive tract. Quality of food is therefore less important than quantity, which means equids can survive in arid habitats. They usually rest during the heat of the day and forage in the morning, evening, and night.

Social groups

Wild horses, plains zebras, and mountain zebras live in groups consisting of mares and their young, led by a "harem" stallion, who protects and herds them. This stallion also defends the group's territory and attempts to prevent other stallions mating with his mares. Young females may remain in the same group as their mothers or they may join a different group; young males leave at maturity and try to collect their own harem. Wild asses and Grevy's zebras, on the other hand, have a different social organization, without long-term associations. Breeding stallions defend large territories—up to 6 square miles (15 square km)—and these are marked by dung piles. The stallion mates with mares that range through his territory.

Equids communicate with each other by whinnying or braying, and these vocalizations vary depending on species. To assess the sexual condition of mares, stallions sniff mares' urine. To analyze the scent in detail, they roll back their upper lip to induce the inhaled air into the Jacobson's organ, a special pouch in the roof of the mouth. This is called the "flehmen" response.

Horses and people

The donkey, a descendant of the African wild ass, was domesticated in the Middle East before 3000 BCE. The domestic horse, derived from the Eurasian wild horse, later replaced the donkey for a number of purposes – including transportation, agriculture, warfare, and recreation—although the donkey remained a popular beast of burden. Domesticated equids continue to serve people.

Most of the wild equids are highly endangered due to habitat loss and hunting: only plains zebra and the kiang are relatively numerous.

FIGHTING FOR DOMINANCE
Conflicts between equids are common throughout the mating season. During this fight, the plains zebra stallions may bite, rear up and strike out with their front feet, and kick with their back feet. The loser is driven off, leaving the victorious stallion with control of the harem, usually between 1 and 6 mares.

Equus asinus africanus

Somali wild ass

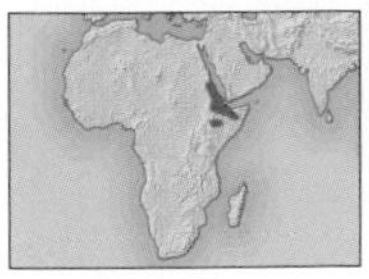

Location E. Africa

- **Length** 6½–7½ ft (2–2.3 m)
- **Tail** 18 in (45 cm)
- **Weight** 440–510 lb (200–230 kg)
- **Social unit** Group
- **Status** Critically endangered

The Somali wild ass lives in rocky deserts where the ground temperature exceeds 122° F (50° C). It eats almost any plant food, from grasses to thorny acacia bushes, and goes without water for several days. The upperparts are buff-gray in summer and iron-gray in winter; the mane is sparse but erect. Females mate only with mature males that hold territories.

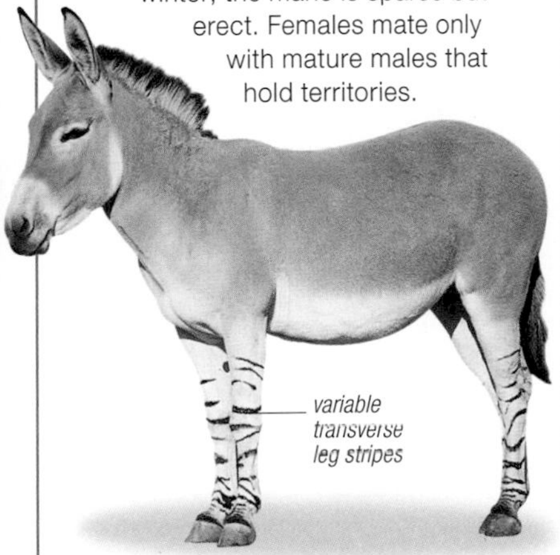

Equus hemionus

Onager

Location W., C., and S. Asia

- **Length** 6½–8¼ ft (2–2.5 m)
- **Tail** 12–19½ in (30–49 cm)
- **Weight** 440–570 lb (200–260 kg)
- **Social unit** Group
- **Status** Endangered

This Asian wild ass eats a variety of vegetation, including grasses and succulent plants. Females and young onagers form loose, wandering herds, while immature males gather in bachelor groups. Solitary mature males kick and bite rivals to occupy the territory they need for breeding. Mainly buff, tawny, or gray in coloration, the onager is white underneath, with a dark mane, back stripe, ear tips, and feathered tail tip.

Equus caballus przewalskii

Przewalski's horse

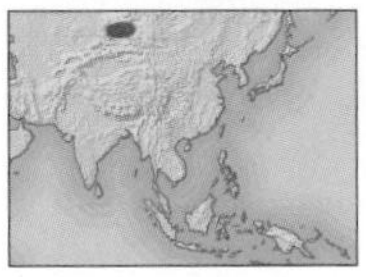

Location Re-introduced to C. Asia

- **Length** 7¼–8½ ft (2.2–2.6 m)
- **Tail** 32–43 in (80–110 cm)
- **Weight** 440–660 lb (200–300 kg)
- **Social unit** Group
- **Status** Critically endangered

This horse now survives in zoos, parks, and field stations, although there have been several attempts to re-introduce it in Mongolia. It lives in cohesive, long-term herds that wander great distances for grass, leaves, and buds. A typical herd is led by a senior mare and has 2–4 other mares, their offspring, and one stallion who stays on the periphery. A single foal is born after a gestation of 333–345 days.

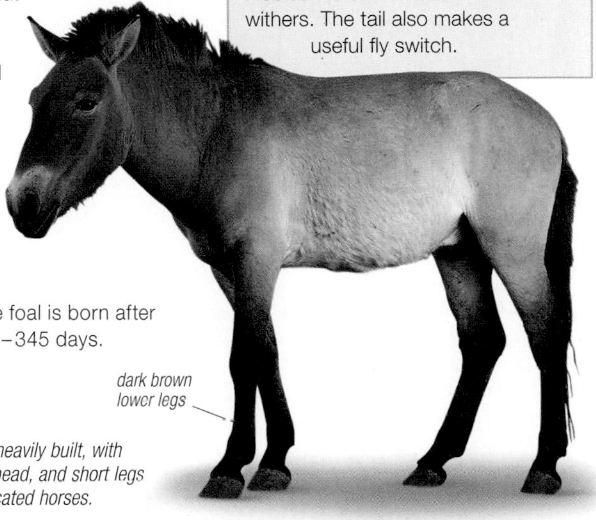

SMALL BUT STOCKY
Przewalski's horse is heavily built, with a thick neck, a large head, and short legs compared to domesticated horses.

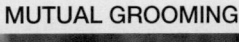

MUTUAL GROOMING

Social grooming is important for this species, as for most wild equids, serving to reinforce herd bonds. Usually 2 animals stand nose to tail so that they can look out for danger in both directions, and each nibbles the other's shoulder and withers. The tail also makes a useful fly switch.

Equus grevyi

Grevy's zebra

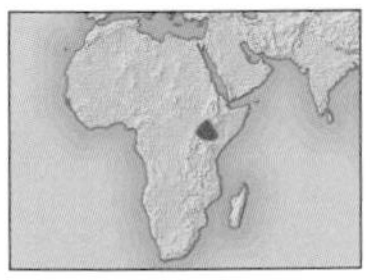

Location E. Africa

- **Length** 8¼–9¾ ft (2.5–3 m)
- **Tail** 15–23½ in (38–60 cm)
- **Weight** 770–990 lb (350–450 kg)
- **Social unit** Group
- **Status** Endangered

The largest zebra and biggest wild equid, Grevy's zebra has dense, narrow stripes that remain distinct all the way down to the hooves; the belly and tail base are white. Males occupy huge territories, up to 4 square miles (10 square km). Females and foals roam freely, perhaps gathering in small, loose herds to graze, but there are no long-term herd bonds.

Equus burchelli

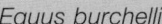

Plains zebra

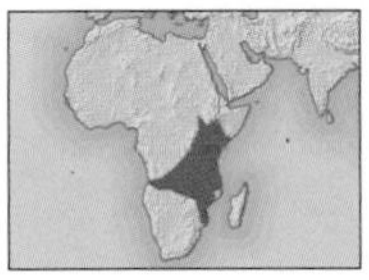

Location E. and southern Africa

- **Length** 7¼–8¼ ft (2.2–2.5 m)
- **Tail** 18½–22 in (47–56 cm)
- **Weight** 390–850 lb (175–385 kg)
- **Social unit** Group
- **Status** Least concern

This successful and widespread zebra has a diet of nine-tenths grass; the remainder is leaves and buds. The main, long-term social unit is a stallion, his harem of one or several mares, and their offspring. Maturing stallions form loose bachelor herds and may challenge for the harem with fierce fights of biting and kicking. The single foal can stand within a few minutes of birth and is grazing after a week.

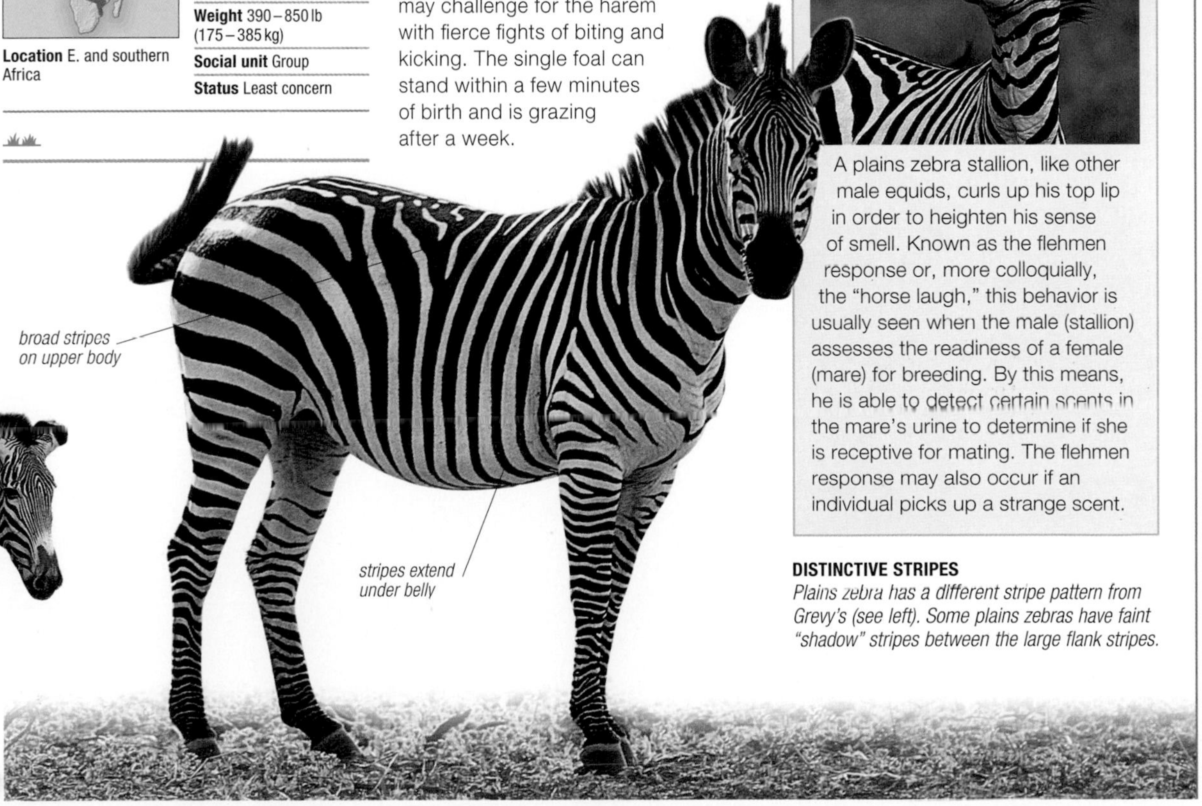

FLEHMEN RESPONSE

A plains zebra stallion, like other male equids, curls up his top lip in order to heighten his sense of smell. Known as the flehmen response or, more colloquially, the "horse laugh," this behavior is usually seen when the male (stallion) assesses the readiness of a female (mare) for breeding. By this means, he is able to detect certain scents in the mare's urine to determine if she is receptive for mating. The flehmen response may also occur if an individual picks up a strange scent.

DISTINCTIVE STRIPES
Plains zebra has a different stripe pattern from Grevy's (see left). Some plains zebras have faint "shadow" stripes between the large flank stripes.

Rhinoceroses

PHYLUM	Chordata
CLASS	Mammalia
ORDER	Perissodactyla
FAMILY	Rhinocerotidae
SPECIES	5

With their huge size, bare, sometimes pleated skin, relatively short limbs, and horned snout, rhinoceroses are almost dinosaur-like in appearance. They live in the savannas of Africa, and in the tropical and subtropical forests and swampy grasslands of Asia. Although rhinoceroses are often considered aggressive—they will charge to scare off an intruder—they are generally timid. All 5 species are endangered, 3 of them critically so. All species require a large daily intake of food (either grass or stems, branches, and leaves) to support their massive body.

Anatomy

Rhinoceroses are large, heavily built animals—the white rhino may weigh up to 2½ tons (2.3 tonnes). The species found in Asia support their bulky frame with thick legs, but African species have surprisingly slim legs and are capable of running at speeds of up to 28 mph (45 kph). Each foot has 3 toes, each with a hoof. The large head features one or 2 horns (depending on species) on the snout. Instead of a bony core (as in the horns of cattle and their relatives, for example), the horns are composed entirely of keratin—a tough protein also found in hair and nails—and the horn perches on a roughened area of the skull (rather than being "rooted" in the skull). Rhinoceroses also have skin up to ¾ in (2 cm) thick, and body hair is usually inconspicuous (although all species have a tail tuft and ear fringes). Asian rhinoceroses have heavily folded skin, giving the appearance of plates of armor.

Of all the senses, smell is the strongest, although mobile, tubular ears provide good hearing. The eyes, however, are small, and rhinoceroses have poor vision.

CONSERVATION

The entire rhino family is listed by CITES, mainly because rhino horn is worth more than its weight in gold. In the Far East, it is powdered and used in medicine, while in Yemen, it is carved to make dagger handles. Both practices are now illegal, although an undercover trade still goes on. However, rhino conservation has had some successes. The Indian rhino population is now climbing steadily, thanks to strict protection, while in Africa, the southern white rhino is also doing well, with about 18,000 animals in reserves and in the wild. The black rhino is also surviving from a catastrophic decline.

Social systems

Although rhinoceroses are mostly solitary, subadults may travel in pairs, female white rhinoceroses sometimes form groups, and Indian rhinoceroses will share a bathing pond without aggression. Adult males of all species tend to be territorial; however, females are not. Stronger male Indian rhinoceroses have a home range that overlaps with the ranges of several females, and is marked by dung piles up to 3¼ ft (1 m) high. Weaker males share the ranges of stronger males but do not attempt to mate. When 2 strong Indian rhinoceros males meet, they may fight using their tusklike lower incisors. Many such conflicts end in the death of one of the combatants. The white rhinoceros, like the Indian species, also employs a "strong" male / "weak" male system. The strong males actively herd females into their territories, and then prevent them from leaving. Black rhinoceroses, however, have less well-defined territories. Little is known about the social behavior of either the Javan or the Sumatran rhinoceros.

HIGH-SPEED CHARGE
Rhinoceroses have a heavy, awkward appearance but are immensely strong and well muscled. If disturbed, an individual such as this black rhinoceros is capable of charging at speeds of 28 mph (45 kph). Even at high speed, it can make rapid changes in direction.

MAMMALS

Dicerorhinus sumatrensis

Sumatran rhinoceros

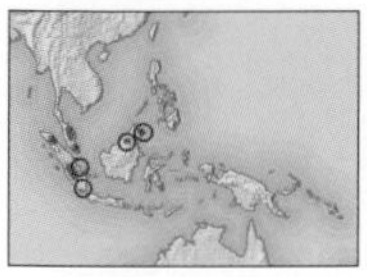

Location S. and S.E. Asia

Length 8¼–10 ft (2.5–3.2 m)	
Tail Not recorded	
Weight Under 1,760 lb (800 kg)	
Social unit Individual	
Status Critically endangered	

Solitary and secretive, this is the smallest and hairiest rhinoceros. It rests in wallows by day and browses at night on twigs, leaves, and fruit, also felling saplings for tender shoots. Its high-altitude habitat, once relatively safe, is now being lost to loggers; horn poachers are also a major threat (the front horn may reach 16 in/40 cm in length). A single calf is born after 7–8 months' gestation and stays with its mother for 18 months, until she next gives birth.

WALLOWING

The Sumatran rhinoceros, like all rhinoceroses, hippopotamuses, and similar sparsely haired mammals, wallows in mud, which dries onto the skin. This is cooling and also protects its delicate surface areas from flies and other biting insects.

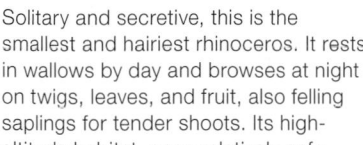

FEW WRINKLES
This small rhinoceros has relatively few skin wrinkles, except around the neck. Sparse hairs cover the skin surface.

Rhinoceros sondaicus

Javan rhinoceros

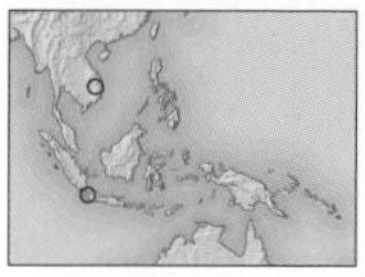

Location S.E. Asia

Length 9¾–11 ft (3–3.5 m)	
Tail Not recorded	
Weight Up to 1½ tons (1.4 tonnes)	
Social unit Individual	
Status Critically endangered	

One of the most rare large mammals, this species is hairless other than on its ears and tail tip. Its thick, gray skin is divided by deep folds to make a "saddle" over the neck with lumps or nodules giving an armor-plated effect. The single horn rarely exceeds 10 in (25 cm) long and is lacking in some females. A solitary, nocturnal browser, it eats a wide range of plants. The Javan—or lesser one-horned—rhinoceros was decimated by lowland forest removal. Two remnant populations survive, taking advantage of coastal mangrove and bamboo marshes. The male is probably territorial, marking his area with dung piles and urine pools; he encounters potential mates at suitable muddy wallows. After 16 months' gestation, one calf is born and stays with its mother for 2 years, possibly longer.

Rhinoceros unicornis

Indian rhinoceros

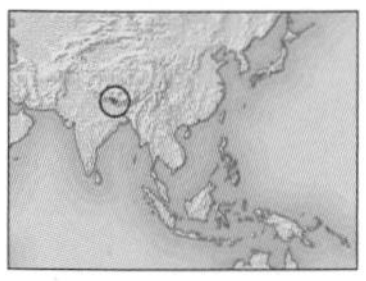

Location S. Asia (Brahmaputra Valley)

Length Up to 12 ft (3.8 m)	
Tail 28–32 ft (70–80 cm)	
Weight Up to 2.4 tons (2.2 tonnes)	
Social unit Individual	
Status Vulnerable	

Like other rhinoceros species, the Indian rhinoceros is generally solitary except for temporary male–female associations when mating, and a mother with her calf. Both male and female have a single horn, up to 24 in (60 cm) long. Males have larger, sharp, tusklike incisors for fighting rivals at breeding time. Otherwise they usually tolerate intruders into their ranges, which vary from ¾–3 square miles (2–8 square km) depending on the quality of habitat. The gestation period is 16 months and the calf may remain with its mother until her next offspring is born, which may be 3 years later. Despite protection projects and an encouraging recovery in numbers, remaining populations of the Indian rhinoceros are scattered and fragmented, and therefore still at some risk. These animals are also still subject to poaching for their horns and other body parts.

PLATES AND RIVETS

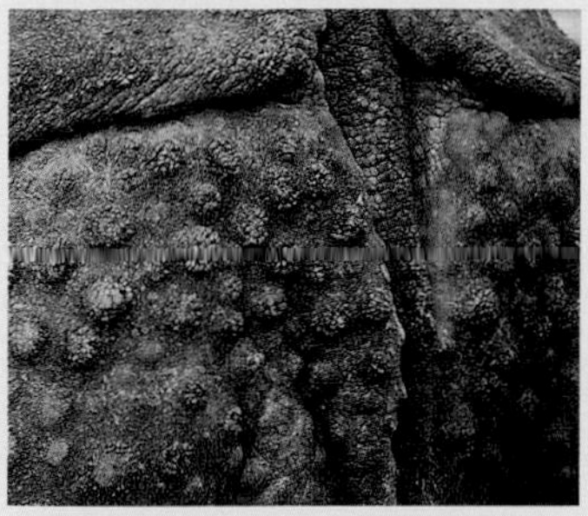

The Indian rhinoceros's skin has deep folds, its armour-plated appearance accentuated by tubercles (lumps), especially on the sides and rear, which resemble rivets. The pink skin within the folds is prone to parasites that are removed by egrets and tick birds, who also act as sentinels.

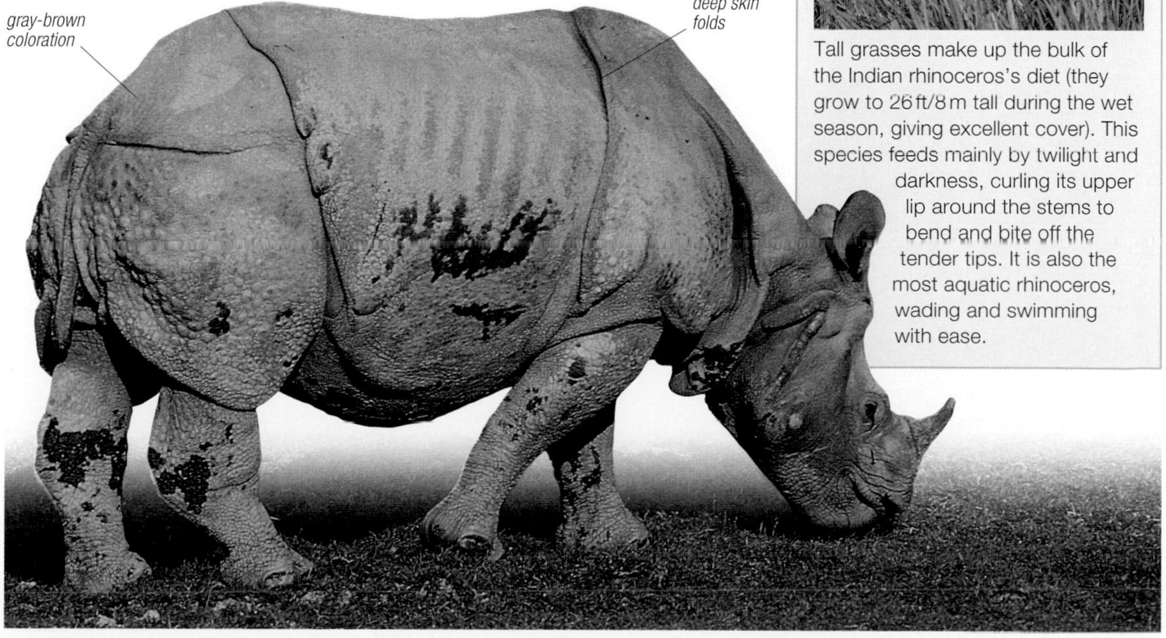

IN THE TALL GRASS

Tall grasses make up the bulk of the Indian rhinoceros's diet (they grow to 26 ft/8 m tall during the wet season, giving excellent cover). This species feeds mainly by twilight and darkness, curling its upper lip around the stems to bend and bite off the tender tips. It is also the most aquatic rhinoceros, wading and swimming with ease.

Diceros bicornis

Black rhinoceros

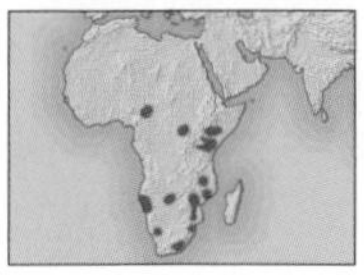

Length 9½–10 ft (2.9–3.1 m)
Tail 23½ in (60 cm)
Weight 1–1¼ tons (0.9–1.3 tonnes)
Social unit Individual
Status Critically endangered

Location W., C., E. and southern Africa

Browsing on a variety of bushes and low trees, the black rhinoceros occupies a variety of habitats but mainly wooded savanna with mosaics of grass and trees. As in other rhinoceros species, its sight is poor but hearing and smell excellent. It feeds mainly by twilight and at night; days are spent dozing in shade or wallowing in mud. The black rhinoceros is solitary, and marks its home range with piles of dung and copious squirts of urine. It may tolerate intruders (of its own species, or human), but this unpredictable rhinoceros may suddenly charge or jab with its horns. Two black rhinoceroses together are usually a mating pair, associating for just a few days, or a mother and calf. The female gives birth after 15 months' gestation. The newborn weighs 88 lb (40 kg), begins to take solid food after a few weeks, and is weaned at around 2 years.

PREHENSILE LIP

The black rhinoceros, also known as the hook-lipped rhinoceros, has a pointed, prehensile upper lip. This curls around new twigs and shoots to draw them into the mouth, where they are bitten off by the molar teeth.

CONSERVATION

Intense demand for rhinoceros horn for traditional Chinese medicines and for dagger handles and similar items in the Middle East led to a massive decline in black rhinoceros numbers from 65,000 in 1970 to just 2,500 in the mid-1990s. Since then, its numbers have increased to over 4,000, with another 600 in captivity. Many of the wild animals are protected from poachers by 24-hour armed guards.

GRAY SKIN, BLACK APPEARANCE
The black rhinoceros's skin is gray with hairs only on the eyelashes, ear tips, and tail end. Its dark appearance is the result of mud dried on the skin.

Ceratotherium simum

White rhinoceros

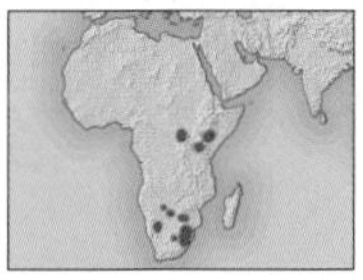

Length 12–13 ft (3.7–4 m)
Tail 28 in (70 cm)
Weight Up to 2.5 tons (2.3 tonnes)
Social unit Group
Status Near threatened

Location E. and southern Africa

The largest and most numerous rhinoceros, the white rhinoceros rivals the hippopotamus as the biggest land animal after the elephant. Males weigh up to half a ton more than females, and have longer horns and a more pronounced nuchal crest. The front horn may reach 4¼ ft (1.3 m) in length, the rear one 16 in (40 cm). This is also the most social rhinoceros, generally placid, with mother–calf pairs staying together for long periods and up to 7 juveniles forming small herds. Mature males, however, tend to be solitary.

An almost exclusive grazer, the wide, straight upper lip—giving the alternative name of square-lipped rhinoceros—and hard lip pads crop grass extremely close. Populations of southern white rhinoceroses (*Ceratotherium simum simum*) are reasonably secure, numbering nearly 18,000, although still conservation-dependent. There may be fewer than 30 of the critically endangered northern white rhinoceroses (*Ceratotherium simum cottoni*).

URINE-MARKING

A male rhinoceros's penis faces the rear, so urine sprays out between the back legs. Male white rhinoceroses mark home ranges in this way. Each needs a territory of about ½ square mile (1 square km) to be selected for breeding.

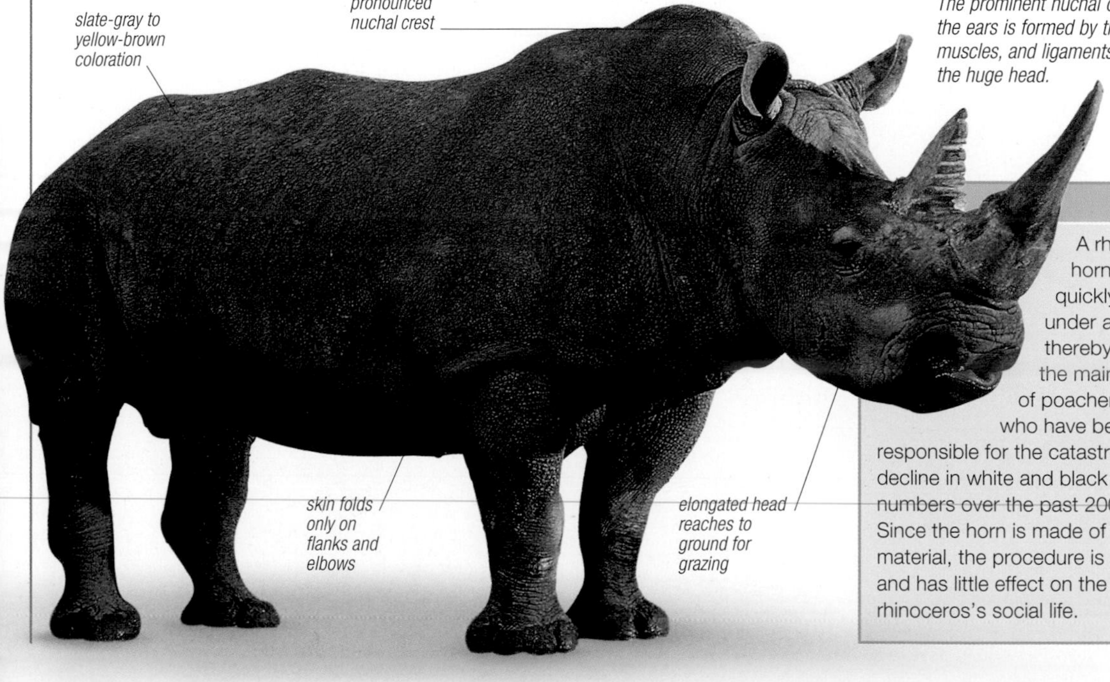

HUMPED SHOULDER
The prominent nuchal crest behind the ears is formed by the bones, muscles, and ligaments to support the huge head.

CONSERVATION

A rhinoceros's horns can be quickly cut off under anesthetic, thereby removing the main target of poachers, who have been largely responsible for the catastrophic decline in white and black rhinoceros numbers over the past 200 years. Since the horn is made of a hairlike material, the procedure is painless, and has little effect on the rhinoceros's social life.

MAMMALS

Tapirs

PHYLUM	Chordata
CLASS	Mammalia
ORDER	Perissodactyla
FAMILY	Tapiridae
SPECIES	4

These mammals can truly be called "living fossils" because as a group they have changed little over the past 35 million years. They are medium-sized animals with a piglike body on relatively high, slender legs, and a short, extensible trunk. Tapirs live in forest areas (never far from water) in Southeast Asia and South and Central America. Much of their time is spent in water with just their trunk exposed (like a snorkel), to escape predators and to keep cool. Malayan, mountain, and Baird's tapirs are all endangered by habitat destruction and hunting.

Anatomy

Tapirs have a streamlined shape, that allows them to move more easily through dense undergrowth. They have a very deep face because their nasal passages are greatly enlarged, with nostrils positioned at the tip of the snout. The sense of smell is acute and is vital in finding food and smelling danger and other tapirs. These animals also have large, erect ears (providing good hearing) and small eyes that are deep in the socket, protected from thorns and sharp branches. Their 3 toes spread out on soft ground, which helps support their weight and prevent sinking. Tapirs have a short, broad tail and hard skin (except on the soles of the feet, which are soft and sensitive). The body hair is usually sparse, except in the mountain tapir, which has a thick coat. Baird's tapir and the South American tapir have a short, bristly mane (which provides protection if a jaguar bites). Most species are entirely brown, gray, or blackish, except for white ear rims in some. The mountain tapir has striking white lips, while South American and Baird's tapirs have light patches on the cheeks, throat, and chin. Only the Malayan tapir has extensive white body markings. Newborn tapirs have white spots and stripes in rows along the flanks and limbs, which provide good camouflage.

JUVENILE COAT
A red-brown coat with horizontal white stripes and spots is typical of young tapirs. The adult coloration begins to appear at about 6 months.

Tapirus terrestris

South American tapir

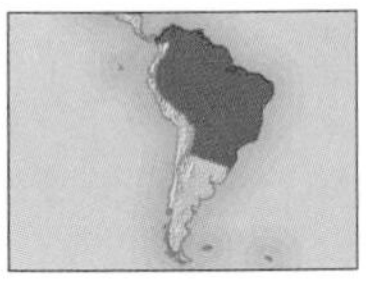

Location N. and C. South America

Length 5½–6½ft (1.7–2m)
Tail 18–39in (46–100cm)
Weight 500–550lb (225–250kg)
Social unit Individual
Status Vulnerable

This bristly coated tapir has white-tipped ears and a short, narrow mane. It favors waterside habitats, swims well, and dives to escape predators such as pumas and jaguars. Browsing selectively by night, it feeds on a wide range of grasses, reeds, fruit, and other vegetation.

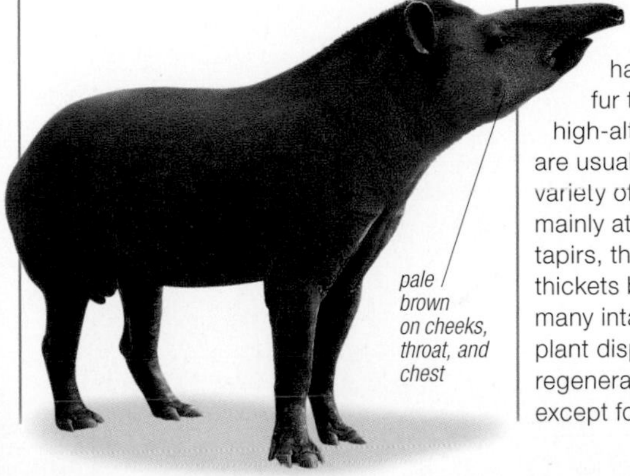

Tapirus pinchaque

Mountain tapir

Location N.W. South America

Length 6ft (1.8m)
Tail 1½ft (50cm)
Weight 330lb (150kg)
Social unit Individual
Status Endangered

The furriest of the 4 tapir species, the mountain tapir has thick, dark brown to black fur that keeps out the cold of its high-altitude habitat. The lips and ears are usually white-fringed. It eats a variety of dwarf trees and shrubs, mainly at dawn and dusk. Like other tapirs, the mountain tapir hides in thickets by day. Its droppings contain many intact seeds, thereby assisting plant dispersal and consequently forest regeneration. They are solitary animals except for mothers with young.

Tapirus bairdii

Baird's tapir

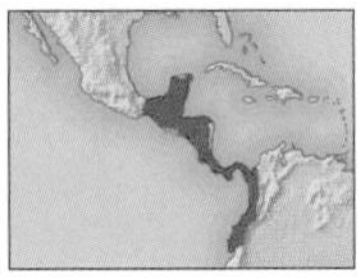

Location S. Mexico to N. South America

Length 6½ft (2m)
Tail 2¾–5in (7–13cm)
Weight 530–880lb (240–400kg)
Social unit Individual
Status Endangered

The largest American tapir, Baird's is dark brown, with pale gray-yellow cheeks and throat, and white-edged ears. It eats a variety of plant parts, from buds and leaves to fallen fruit. The gestation period is about 390–400 days and the female produces one young (very rarely 2), which weighs 11–18lb (5–8kg). Baird's tapir uses shrill whistles to communicate with its young or warn other adults away from its territory.

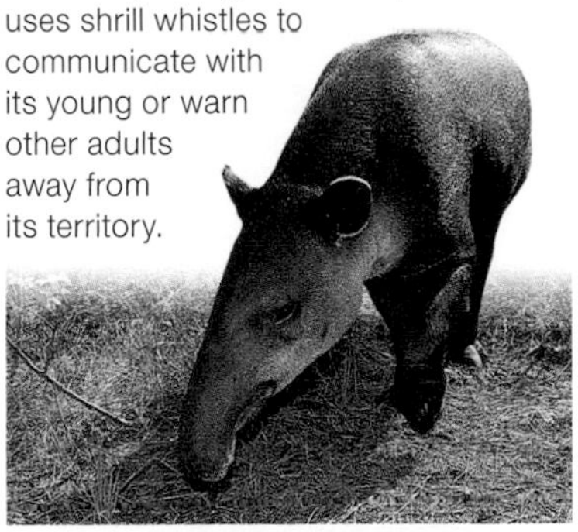

Tapirus indicus

Malayan tapir

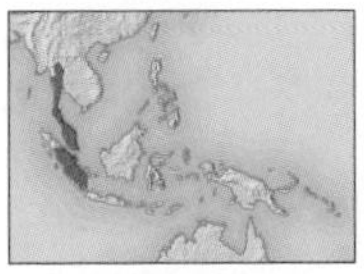

Location S.E. Asia

Length 6–8¼ft (1.8–2.5m)
Tail 2–4in (5–10cm)
Weight 550–1,190lb (250–540kg)
Social unit Individual
Status Endangered

A striking, 2-tone pattern distinguishes the largest and only Old World tapir. It is black with a white "saddle" over the back and rump, which helps break up its body outline in shady forests. The Malayan tapir feeds on soft twigs and young leaves of bushes and saplings, and also on fallen fruit. The male's average range of 5 square miles (13 square km) overlaps the ranges of several females.

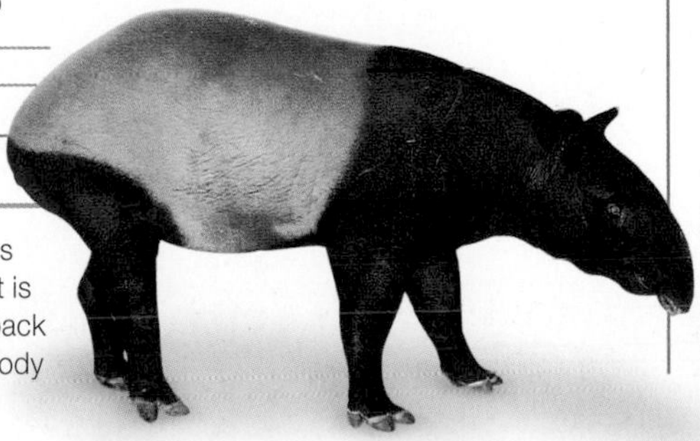

MAMMALS

Pigs

PHYLUM	Chordata
CLASS	Mammalia
ORDER	Artiodactyla
FAMILY	Suidae
SPECIES	19

Despite a reputation for gluttony, wild pigs rarely overeat and are intelligent, adaptable animals. Members of the pig family—which also includes hogs, boars, and the babirusa—are omnivorous (rather than purely herbivorous) and are characterized by a barrel-shaped body, curiously slender legs, a short neck, and a large head. They are found in forests and grasslands throughout Africa and Eurasia and have also been successfully introduced to Australia, New Zealand, and the Americas. The babirusa and the pygmy hog are both endangered because of habitat destruction. Almost all domestic pigs are descended from the wild boar.

Anatomy

One of the most interesting features of a pig's anatomy is its snout, which has a cartilaginous disk at its tip, enclosing the nostrils. The disk is supported by a small bone (the prenasal), not present in other mammals, and is used like a bulldozer when foraging for food. Most species also have upper and lower canines that grow outwards and upward to form tusks. The tips of the lower canines wear against the underside of the upper canines. In the babirusa, however, the male's upper tusks grow up through the skin of the face and then curve backward. Female pigs have smaller canines (female babirusas have no canines).

Pigs also have cloven feet. Two large, flattened hooves bear the animal's weight, but on soft ground the 2 shorter, lateral hooves may touch the ground and help spread the weight. Pigs have thick skin, with hair that is either long and bristly (as in the wild boar) or sparse (as in the babirusa). Most species have a mane down the back of the neck. The tail is thin, twisted, mobile, and usually sparsely tufted.

Fighting

The long tusks of males are used for defense against predators and for fighting other males for social status or mates. Pigs demonstrate 2 distinct fighting styles: lateral and head-to-head. Lateral conflict involves the combatants slashing at each other's shoulders. This fighting style is practiced by pigs, such as the wild boar, that have a long, narrow face, no facial warts, and small tusks. Pigs with a broad head, thick skull, long tusks, and facial warts (for protection against wounds)—such as warthogs and giant forest hogs—tend to fight head-to-head.

Family groups

Pigs live in sounders (families of a sow and her offspring), which communicate by squeals and grunts. Boars join sounders during the mating season. Pigs are the only hoofed mammals to have litters rather than one or 2 young (only the babirusa has twins).

PECCARIES

Although the 3 species of peccary together form a separate family (Tayassuidae) from pigs, the 2 groups are closely related. Peccaries differ from pigs by having short canines that point downward instead of upward, only one lateral hoof (instead of 2) on each back foot, an extremely short tail, a special gland on the upper surface of the rump, and a more complex stomach (3 chambers). Peccaries live in very large, mixed-sex sounders. These units are quite fearless and will attack even big cats. Peccaries are found mostly in forest areas from southwestern USA through Central America to northern Argentina. The Chacoan peccary is heavily hunted and is in danger of extinction.

HEAD-TO-HEAD FIGHTING
When warthogs clash, head-to-head, they are, in fact, only trying to push each other off balance. Lateral fighting is far more damaging.

Hylochoerus meinertzhageni

Giant forest hog

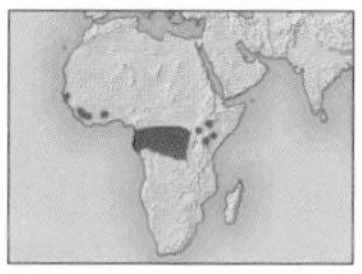

Location W., C., and E. Africa

- **Length** 4¼–7 ft (1.3–2.1 m)
- **Tail** 12–18 in (30–45 cm)
- **Weight** 290–610 lb (130–275 kg)
- **Social unit** Group
- **Status** Least concern

Largest of the pig family, this massive-headed hog has 2 large, wartlike skin growths (excrescences) below and behind each eye, and canines that grow horizontally from the jaw. The straw-colored piglets turn brown then black as they mature, and the long, coarse hair becomes more sparse. Unlike its relatives, this pig does not root; instead, it grazes and low-browses on grasses, sedges, and shrubby vegetation.

Sus salvanius

Pygmy hog

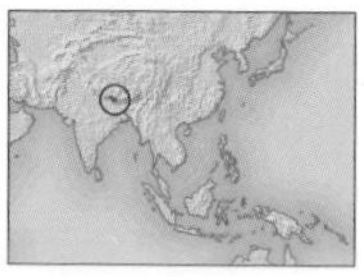

Location S. Asia

- **Length** 20–28 in (50–71 cm)
- **Tail** 1¼ in (3 cm)
- **Weight** 14–21 lb (6.5–9.5 kg)
- **Social unit** Group
- **Status** Critically endangered

This chunky, short-legged pig, the smallest species in the pig family, has a tapering snout and head to push through dense undergrowth. Its overall color is dark brown; the male's upper canines poke slightly out of the sides of the mouth. Both sexes dig large troughs and line them with grassy layers to form nests. Although legally protected, this species is still at risk from poaching and continued removal of its grassy, riverside habitat.

Sus scrofa

Wild boar

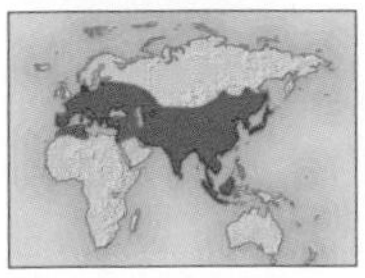

Length 3–6 ft (0.9–1.8 m)
Tail 12 in (30 cm)
Weight Up to 440 lb (200 kg)
Social unit Individual/Group
Status Least concern

Location Europe, Asia, N. Africa

The wild boar, or Eurasian wild pig, is one of the most widely distributed terrestrial mammals, and is also the main ancestor of domestic breeds. It occupies a wide variety of habitats, eats almost any food, runs fast, and swims well. Males live alone except for the mating season, when they join with females and fight rival males for harems. Females are very protective of their young and may band together into groups (sounders) of 20 or more.

FADING STRIPES

Wild boar piglets are typically pale brown with paler stripes along the back and sides of the body. This provides camouflage in their nest of grass, moss, and leaves in a dense thicket. The mother rarely leaves her litter (usually 4–6 piglets), for the first 1–2 weeks. Gradually, she and her young venture from the nest to forage. From 2–6 months, as the piglets become less vulnerable and camouflage is no longer so important, their stripes fade. They become independent at 7 months.

BRISTLY PIG *The wild boar has thick, coarse hair and a narrow mane of longer hair along its spine. Compared with some other wild pigs, it has small eyes and tusks. It has no facial warts.*

Potamochoerus porcus

Bush pig

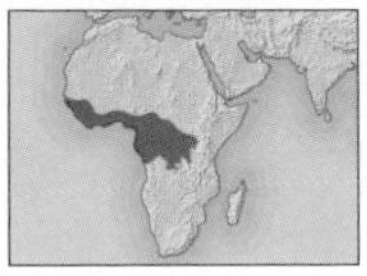

Length 3½–5 ft (1–1.5 m)
Tail 12–17 in (30–43 cm)
Weight 100–290 lb (46–130 kg)
Social unit Group
Status Least concern

Location W. to C. Africa

By far the reddest pig, this species—also known as the red river hog—has long, pointed ears with prominent tufts, a narrow, white stripe along the back, and white facial stripes. It is omnivorous and nocturnal, like other pigs. Highly social, the male stays with his harem of females and offspring, and helps to defend them. Sometimes these family parties of 4–6 congregate into wandering bands of 50 or more.

Babyrousa babyrussa

Babirusa

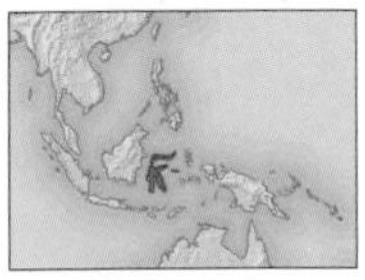

Length 3–3½ ft (0.9–1.1 m)
Tail 10½–12½ in (27–32 cm)
Weight Up to 220 lb (100 kg)
Social unit Individual/Group
Status Vulnerable

Location S.E. Asia (Sulawesi, Togian, and Mangole islands)

The distinctive upper canines of the male babirusa grow through the muzzle and curve back towards the face. Up to 12 in (30 cm) long, they are also loose-socketed and brittle. The almost hairless hide varies in color from brown to gray. Males are primarily solitary, while females and their young travel in groups of about 8. The gestation period of 155–158 days is fairly typical for the pig family; the litter size is only 1–2.

Phacochoerus africanus

Warthog

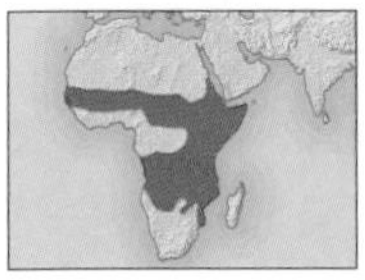

Length 3–5 ft (0.9–1.5 m)
Tail 10–20 in (25–50 cm)
Weight 110–330 lb (50–150 kg)
Social unit Group
Status Least concern

Location Africa (south of the Sahara)

The warthog is a long-legged pig, with a large head. When running, its tail is held straight and upright. Generally active by day, it lives in mixed groups of 4–16 young males or females with young. They shelter and raise their young in grass-lined burrows, dug by themselves or by aardvarks.

GRAZING

The warthog is the only pig adapted for grazing in grassland. Typically, it kneels on its padded "wrists" to nip off the growing tips of grass, using its lips or its incisor teeth. In the dry season, it feeds on underground stems (rhizomes), rooting for them with its toughened snout.

MANED PIG *The warthog's long, dark mane extends from the nape of the neck to the middle of the back, where there is a gap; it then continues on the rump.*

Pecari tajacu

Collared peccary

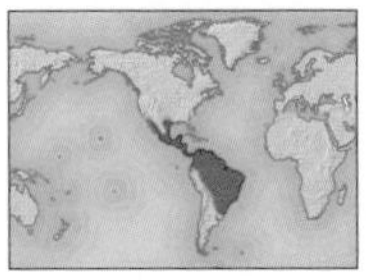

Length 30–39 in (75–100 cm)
Tail ½–2¼ in (1.5–5.5 cm)
Weight 29–66 lb (14–30 kg)
Social unit Group
Status Least concern

Location S.W. USA to S. South America

Also called the javelina, this is the smallest of the 3 peccary species. It is generally dark gray with a whitish, often indistinct, neck collar; the young are reddish with a narrow black stripe along the back. Found in a variety of habitats, the collared peccary eats mainly plant matter, such as berries, shoots, tubers, and bulbs, but also grubs, worms, and small vertebrates, such as snakes and lizards. Peccaries are notably gregarious and form bands of up to 15, of mixed age and sex, which cooperate to repel enemies. Members stand side by side and rub faces in mutual grooming.

Hippopotamuses

PHYLUM	Chordata
CLASS	Mammalia
ORDER	Artiodactyla
FAMILY	Hippopotamidae
SPECIES	2

Members of this family have a semiaquatic lifestyle and are thought to be more closely related to whales than to other even-toed hoofed mammals. They can float and swim, and may remain submerged for over 5 minutes. The hippopotamus lives along rivers and lakes in Africa, while the smaller, less aquatic pygmy hippopotamus is restricted to swampy forest areas in West Africa. The hippopotamus is abundant, but the pygmy hippopotamus is endangered due to habitat destruction and hunting.

Anatomy

Hippopotamuses have a long, heavy body with short, surprisingly insubstantial-looking legs. The enormous head features jaws that allow a huge gape (up to 150 degrees) and carry long, tusklike canine and incisor teeth. The nose is wide and covered with sensitive bristles. The tail is short, tufted, and flattened.

Adaptations for life in the water include webbed toes; eyes, ears, and nostrils located at the top of the head (these are often the only part of the animal protruding above water); and the ability to close the nostrils underwater.

The skin is gray with a pinkish tinge on the underside and around the eyes and skin folds. It is almost hairless in most parts, and extremely thick and fatty. Hippopotamus skin is unusual in that there are no sebaceous glands. Instead, there are mucous glands (which are modified sweat glands) that produce a viscous fluid to keep the skin moist when exposed to air. This fluid, which is pink due to the presence of a red pigment, may also protect against infection and prevent sunburn.

While the hippopotamus weighs about 1⅜ tons (1.4 tonnes), the pygmy hippopotamus, which has relatively longer legs, a much smaller head, darker skin, and eyes at the side of the head, averages only about 550 lb (250 kg).

Feeding

The hippopotamus moves inland at night to graze, generally following trails—marked by dung piles—that lead to its feeding grounds. Each night, an individual will eat about 88 lb (40 kg) of grass. The food is digested in a compartmentalized stomach (the fore-stomach contains bacteria that ferments cellulose). Although this system is slow, the hippopotamus requires less food than animals of a similar size because much of its life is spent supported in water. The pygmy hippopotamus eats roots, grasses, shoots, and fallen fruit, although little is known about its feeding habits.

Social groups

The pygmy hippopotamus is usually found in small groups of up to 3. Female hippopotamuses and their young, however, form groups of usually 10–20 (but sometimes of up to 100) during the day (night-foraging is a solitary affair). They communicate via staccato grunts and deep rumbles that carry some distance in the water. Each group occupies a home range along a section of riverbank or lakeshore, within the territory of a dominant male. This male marks his territory with heaps of dung that he scatters by furiously wagging his tail. Other males enter the territory, but they are tolerated only if they behave submissively and do not attempt to mate. Mating occurs in the water, but the calves are usually born on land.

TERRITORIAL BATTLE
Territorial conflicts between male hippopotamuses are not uncommon, particularly where population densities are high. If, after a period of roaring and ritualized displaying, neither male gives way, a fight will ensue. Using their lower canines as weapons, a battle may last for hours and can result in serious injury.

Hippopotamus amphibius

Hippopotamus

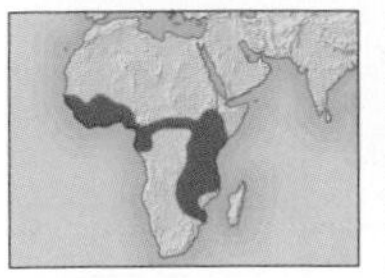

Location Africa

Length 9 ft (2.7 m)
Tail 22 in (56 cm)
Weight 1.5–1.6 tons (1.4–1.5 tonnes)
Social unit Individual
Status Vulnerable

Despite its massive bulk, the hippopotamus swims and walks underwater with grace, and trots with surprising rapidity on land on its short legs. A truly amphibious mammal, its skin has a thin outer layer (epidermis), which dries out easily and is sensitive to the bites of pests such as flies. Despite specialized mucus-producing skin glands, the hide soon cracks unless moistened regularly in water or mud. However, the skin's inner layer (dermis) is up to 1½ in (3.5 cm) thick and formed of a dense mat of fibers that provides great strength. The hippopotamus's main diet is grass, grazed at night, although they have been observed eating small animals or scavenging. The dominant male mates with females in his territory, and the usually single calf is born, generally underwater, after a gestation of 240 days—short for such a large mammal. The mother is fiercely protective, and the calf has few natural predators apart from big cats and hyenas. Hippopotamuses have been known to attack humans, if they feel threatened.

BUOYANCY
The density of the hippopotamus's body is slightly greater than that of water, so it sinks gently and can walk light footed along the bottom. However, if it keeps its lungs well inflated when breathing at the surface, the extra air reduces this density and it can stay afloat with minimal effort.

SENSES UP TOP

The hippopotamus's nostrils, eyes, and ears are all on top of its head, so it can be almost totally submerged yet breathe easily and remain receptive to its surroundings. The nostrils and ears are closed to water entry when diving.

TEMPORARY HERDS

During the dry season, hippopotamuses must wander to find grazing. Instead of each animal returning to its home area by day, some use a nearby pool as a short-term "stop-over" wallow, thereby extending its grazing range. This leads to large gatherings at certain pools, but these lack long-term social or territorial structures.

MOTHER AND CALVES
Calves remain with their mothers after weaning (at 6–8 months) until about 5 years. In this way, family groups develop.

Choeropsis liberiensis

Pygmy hippopotamus

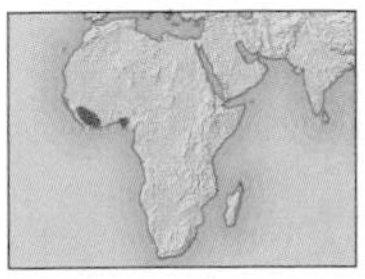

Location W. Africa

Length 4½–5 ft (1.4–1.6 m)
Tail 6 in (15 cm)
Weight 540–610 lb (245–275 kg)
Social unit Individual
Status Endangered

Only one-fifth the weight of its huge cousin (above), the pygmy hippopotamus has a relatively small, less angular head and narrower feet with fewer webbed toes, as adaptations for spending more time on land. It also feeds on a wider variety of plant materials, including shrubs, ferns, and fruit. Pygmy hippopotamuses are usually solitary; although their home ranges overlap, there seems to be little fighting for territory or other form of interaction. They forage at night, following well-worn trails, and spend the day hidden in swamps or sometimes in a riverbank den enlarged from the burrow of some other animal. After a gestation period of 196–201 days, the single calf is born, in water or in the den. The calf risks falling prey to a crocodile or python, but adults have few predators except for leopards and humans. Captive pygmy hippopotamuses have lived longer than many hippopotamuses: 55 compared to about 45 years.

LAND-BASED FORAGER
A pygmy hippopotamus's squat, narrow-fronted form is suited to pushing head down through dense forest vegetation during night-foraging on land.

CONSERVATION

Evidence suggests that the pygmy hippopotamus has always been rare. It survives in dense forests and marshes in Liberia and neighboring West African countries. Despite legal protection, these areas are difficult to patrol and subject to uncontrolled logging and widespread hunting for the bushmeat trade. The flesh is said to resemble pork, yet genetic studies indicate that hippopotamuses are more closely related to whales than to pigs.

Camels and relatives

PHYLUM	Chordata
CLASS	Mammalia
ORDER	Artiodactyla
FAMILY	Camelidae
SPECIES	4

Camelids—camels and their relatives— have long, slender legs and a distinctive gait known as pacing, whereby the front and back legs on the same side move forward together in a rocking motion. Of the Old World camelids—the camels—only one species (the Bactrian camel, from the border area between western China and Mongolia) now survives in the wild. A camel is able to drink up to a quarter of its weight at a time, and can store the water for several days. The New World members of this family, the guanaco and vicuña, are found wild in South America; their domestic descendants, the llama and alpaca, have been bred in the Andes since the time of the Inca civilization. All domesticated camelids are vital to human survival: they provide people with hair, milk, and transportation.

Anatomy

Camelids have a relatively small head, a long, thin neck, and a split upper lip. Camels have either one hump (the dromedary) or 2 humps (the Bactrian camel): these store fat that can be drawn on during lean times. All camelids have a thick coat that provides insulation against daytime heat and warmth during the cooler nights or at altitude. Unlike other hoofed mammals, camelids rest their weight not on their hooves but on the undersides of the 2 digits on each foot, which are cushioned by a fatty pad. This is an adaptation to walking on sandy soil. Camelids are unique among mammals in that they have oval red blood cells, possibly so that these cells can be transported around the body easily, even if the blood is thickened due to dehydration.

Social interaction

In the wild, camelids form groups that consist of one dominant male and a harem of females. "Surplus" males form bachelor bands. While the social systems of the South American camelids have not been studied in great detail, more is known about the way camels interact with each other. Both species go through an elaborate and dramatic ritual when a dominant male is faced with a challenger. First, the harem leader grinds his teeth, rubs a gland on the back of his head against his hump (or front hump in the case of the Bactrian camel), smacks his tail loudly against his rump, and urinates on his back legs, rump, and tail. The 2 then pace side by side, display their tall, humped profiles, and extrude a red bladderlike sac (the dulaa) from the corner of the mouth.

Wild and domestic

The only living Old World wild camelid is the Bactrian camel, although its numbers are now reduced to only 1,000–2,000 individuals. It is taller and slimmer than its domestic counterpart and has more compact, pointed humps. Domestic Bactrian camels are used for transport in cold regions from northern China to Turkey. The dromedary is extinct in the wild, although feral herds now live in central Australia. Domestic dromedaries are found in hot regions in North and northeastern Africa, the Middle East, and northern India through to Kazakhstan. As with other domestic animals, there are different breeds, one of which is kept for its speed and is used in camel racing.

The llama is a domestic animal bred from the wild guanaco and is the traditional pack animal in the Andes. The alpaca is also a domestic species, bred for its fine wool. In the past, the alpaca was thought to be a descendant of the guanaco, but current genetic evidence indicates that it is in fact a descendant of the vicuña, the fine-fleeced, wild camelid of the high-altitude Andes.

DESERT SPECIALISTS
Camels, such as these dromedaries, are well adapted to life in hot climates. Their broad feet provide stability on desert sand, and they have long eyelashes and slitlike, closable nostrils that afford protection during dust storms.

Lama guanicoe

Guanaco

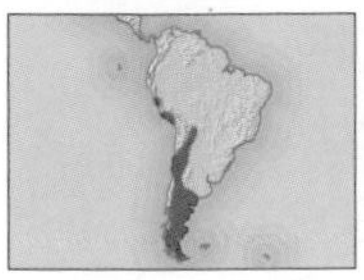

Length 3–7 ft (0.9–2.1 m)
Tail 9½–10½ in (24–27 cm)
Weight 210–290 lb (96–130 kg)
Social unit Group
Status Least concern

Location W. to S. South America

The guanaco prefers cold habitats, mainly grasslands, but also scrub and forest, at heights up to 13,000 ft (4,000 m). It browses and grazes on many grasses and shrubs, as well as lichens and fungi. Typical family groups consist of one male and 4–7 females with young. In the north of its range, the offspring leave their group at about one year of age, compared to closer to 2 years in more southerly populations. Young males form bachelor bands; old males are mostly solitary in their lifestyle.

COLORATION
A typical guanaco is pale to dark brown, with whitish chest, belly, and inner legs, and gray to black head with white-edged eyes, lips, and ears.

LLAMA

Domestic llamas are descended from wild guanacos domesticated 6,000–7,000 years ago. They have been raised by Andean peoples for their wool (fiber), meat, and skins, and have also been used as pack animals.

Vicugna vicugna

Vicuña

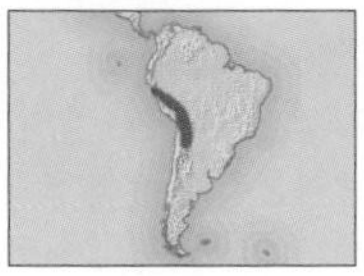

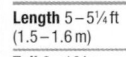

Length 5–5¼ ft (1.5–1.6 m)
Tail 8–10 in (20–25 cm)
Weight 88–120 lb (40–55 kg)
Social unit Group
Status Least concern

Location W. South America

Found in tundralike grasslands at 12,000–16,000 ft (3,600–4,800 m), the vicuña is a strict and selective grazer, grasping perennial grasses with its prehensile, cleft upper lip and snipping off the tips against the tough upper pad. It needs to drink daily. Family groups of one male, 5–10 females, and their young occupy territories, delineated mainly by dung; bachelor males form roving bands.

ALPACA

Alpacas, bred mainly for their thick fiber, were once thought to be domesticated from the guanaco, in the high Andes of central Peru, 6,000–7,000 years ago. New evidence now suggests a vicuña ancestry.

WHITE BIB
The vicuña is pale to dark cinnamon, with a variable whitish chest "bib."

Camelus bactrianus

Bactrian camel

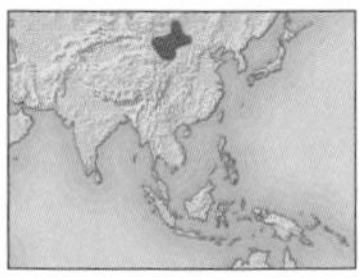

Length 8½–9¾ in (2.5–3 m)
Tail 21 in (53 cm)
Weight 990–1,520 lb (450–690 kg)
Social unit Group
Status Critically endangered

Location E. Asia

Critically endangered in the wild (a domesticated animal is shown), this 2-humped camel can withstand temperatures from –20° F (–29° C) to 100° F (38° C). After a drought, it can drink 24 gallons (110 liters) of water in 10 minutes. It eats grasses, leaves, and shrubs. During the rut, males puff out their cheeks, toss their heads back, and grind their teeth. The winner gains 6–30 females and offspring. One young (rarely twins) is born after 406 days' gestation and is suckled for 1–2 years. Females attain sexual maturity at 3–4 years; males at 5–6 years.

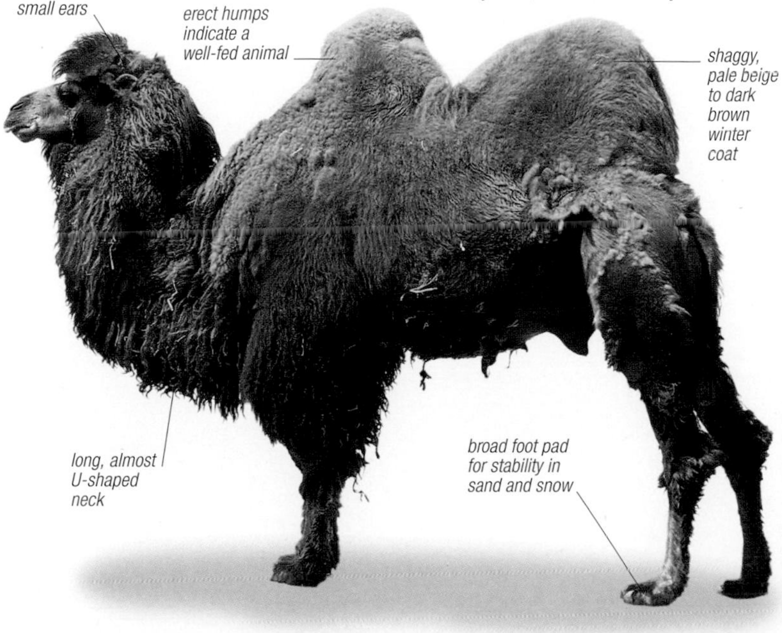

Camelus dromedarius

Dromedary

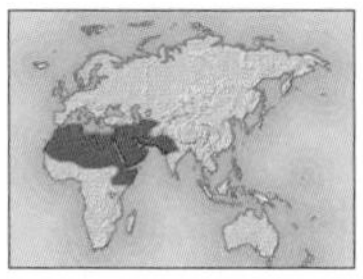

Length 7¼–11 ft (2.2–3.4 m)
Tail 20 in (50 cm)
Weight 990–1,210 lb (450–550 kg)
Social unit Group
Status Not evaluated

Location N. and E. Africa, W. and S. Asia

Extinct in the wild, this one-humped domestic camel shows many adaptations to desert life, losing up to 40 percent of its body weight when food and water are scarce. Allowing its temperature to rise in hot conditions, it reduces sweating to conserve moisture. It feeds on a huge variety of plants, including salty and thorny species, and also scavenges on bones and dried-out carcasses. Dromedaries form small herds of several females and young, and one male, who defends them by spitting, biting, and leaning on opponents. double row of eyelashes and

MAMMALS

Deer

PHYLUM	Chordata
CLASS	Mammalia
ORDER	Artiodactyla
FAMILY	Cervidae
SPECIES	51

Although similar in appearance to antelopes, deer are distinguished by their antlers, which are solid, usually branched, and are shed and regrown each year. Deer are mainly woodland and forest dwellers but can be found in a range of other habitats, from arctic tundra to grassland. They live in northwest Africa, Eurasia, and the Americas. Some species have also been introduced beyond their natural range; for example, to New Zealand and to Britain and mainland Europe.

Anatomy

Most deer have an elongated body, a long neck, large eyes situated at the side of the head, high-set ears, and a small tail. The well-developed third and fourth digits bear the weight, while the second and fifth toes are smaller and usually do not touch the ground. Each year, the coat is molted at least once. In many species, the young have spots on the coat, for camouflage.

The most striking feature of deer, however, is their antlers, which are present only in males (except in reindeer, where both sexes have them). In spring each year, the antlers begin to develop. They grow directly from the skull and are initially covered by finely haired skin (called velvet). As they develop, the velvet dries and is rubbed off so that the antlers are ready to be used in fighting during the breeding season (fall). After breeding (also known as the "rut"), the antlers are shed. It is not certain why antlers are shed and regrown each year, since annual regrowth is metabolically costly—it could be a chance to renew antlers damaged during the rut. Antlers are first grown at 1–2 years of age and are initially simple spikes. As the years progress, the antlers generally become larger and more branched, but regress again in old age. Antler size is an indication of general body condition. Those species that do not have antlers, such as the Chinese water deer, instead possess canine teeth modified into tusks.

Social groups

Social organization depends largely on diet. Smaller species are usually browsers and generally live singly or in small groups. This is because their food occurs in small pockets, which produces competition. Larger species tend to graze more open habitats and therefore compete less for food. These deer often live in herds for protection from predators. Such groups are usually single-sex, except during the rut, when males fight for possession of a harem, using their antlers as weapons and sexual ornaments (to attract mates).

MUSK DEER AND CHEVROTAINS

Musk deer (family Moschidae) and chevrotains (family Tragulidae)—also known as mouse deer—have several features that separate them from other deer (family Cervidae). For example, both have long upper canine teeth instead of antlers. In males, these teeth project below the level of the lower jaw and are used for fighting. The 7 species of musk deer (found throughout Asia) are endangered due to demand for musk, which is secreted from the groin of males during the breeding season and is used as an ingredient in many perfumes. Chevrotains (4 species) live in the rain forests of Africa and Asia.

SURVIVING HARD TIMES
Those deer that inhabit regions where food becomes scarce during winter browse on a variety of plants to survive. These red deer stags will usually feed for longer periods than hinds, possibly because during the fall rut, they must fight for and defend their harem and can therefore spend less time building fat reserves.

Hyemoschus aquaticus

Water chevrotain

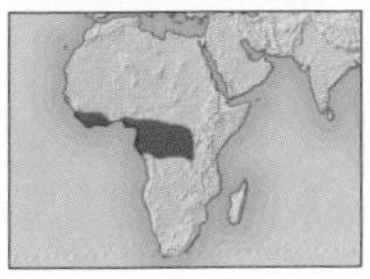

Location W. to C. Africa

Length 28–32 in (70–80 cm)
Tail 4–5½ in (10–14 cm)
Weight 22–26 lb (10–12 kg)
Social unit Individual/Group
Status Least concern

Generally found within 820 ft (250 m) of water, this chevrotain is dark olive-brown, with white markings that include spots on the back, 1–3 stripes along each flank, and chin, throat, and chest bands. It has short legs and ears, a stout body, and swims well, which enables it to escape land-based predators successfully most of the time, but incurs the risk of crocodile attack. Water chevrotains forage for leaves and fallen fruit. Males live singly, while females and their young form small groups.

Muntiacus putaoensis

Leaf deer

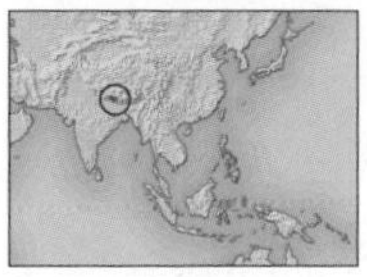

Location S. Asia

Length c. 30 in (80 cm)
Tail c. 3 in (8 cm)
Weight 26–26½ lb (11.8–12.1 kg)
Social unit Individual
Status Data deficient

Initially thought to be the juvenile of another species, specimens of this animal were later shown to belong to an entirely new species of muntjac. The common name comes from local knowledge that the deer is so small it can be wrapped inside a single leaf. It is a primitive species that lives in the dense forests of northern Myanmar and adjacent parts of India. The leaf deer differs from other muntjacs in that both male and female have large canine tusks. The newborns have unspotted coats.

Moschiola meminna

Indian spotted chevrotain

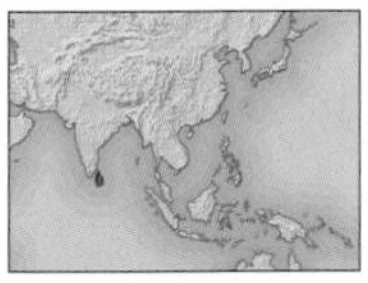

Location Sri Lanka

Length 20–23 in (50–58 cm)
Tail 1¼ in (3 cm)
Weight 6½ lb (3 kg)
Social unit Individual
Status Least concern

Like other chevrotains (mouse deer), this species has 4 fully developed toes on each foot (true deer have 2). It skulks in cover, preferring rocky patches within tropical rain forest. Nocturnal and solitary, it has a spotted back, and striped flanks and throat; however, these pale markings are less distinct than those of the water chevrotain (above) and the main coat color is brown with tiny yellow speckles. Males compete using their sharp, tusklike upper canine teeth. The single young is born after a gestation of 5 months.

Moschus chrysogaster

Alpine musk deer

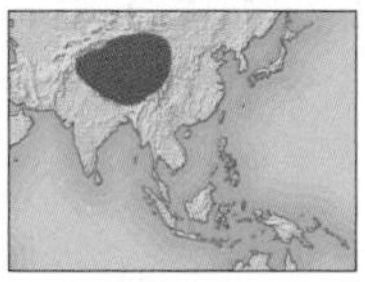

Location S. Asia

Length 28–39 in (70–100 cm)
Tail ¾–2¼ in (2–6 cm)
Weight 15–40 lb (7–18 kg)
Social unit Individual
Status Endangered

An inhabitant of rocky, forested slopes at an altitude of 8,600–11,900 ft (2,600–3,600 m), the musk deer has well-developed side toes to climb rocks and even trees, and to move through soft snow. Its coat is rich, dark brown mottled with gray, paler underneath, with a whitish chin and ear edgings. Its musk secretions are valued in the perfume industry. This has led to hunting and a decline in numbers in the wild.

Dama dama

Fallow deer

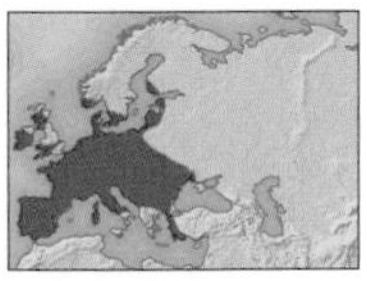

Location Europe

Length 4½–6¼ ft (1.4–1.9 m)
Tail 5½–10 in (14–25 cm)
Weight 77–330 lb (35–150 kg)
Social unit Group
Status Least concern

Fallow deer have long been kept semidomesticated for their beauty and meat, and have been introduced to the Americas, Africa, and Australia. They are active at twilight, consuming many plant foods, from grasses to acorns. Herds may exceed 100. Bucks rut to establish a small patch of land, where they mate.

LYING UP

The fallow deer fawn, like most young deer, "lies up" in thick vegetation or among leaf litter, as the mother feeds. The fawn's instincts are to stay still and silent, camouflaged from predators by its spotted coat. The mother returns at intervals to allow the fawn to suckle, and occasionally leads it to a new lying-up site.

EARTHY HUES *Commonly brown with white spots as here, the fallow deer can also be pale brown, black, or white.*

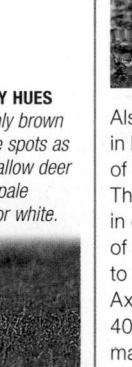

Axis axis

Axis deer

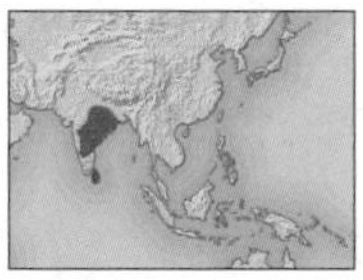

Location S. Asia

Length 3¼–5 ft (1–1.5 m)
Tail 4–10 in (10–25 cm)
Weight 155–175 lb (70–79 kg)
Social unit Group
Status Least concern

Also called the chital, the axis deer lives in large herds of 100 or more, comprised of mixed males, females, and young. They graze in grassland and browse in open woodland, often below troops of langurs (see p.149), who knock fruit to the ground and emit warning calls. Axis deer dash for thick cover at 40 mph (65 kph) when disturbed. The male's antler has one brow tine (prong), and a rear-directed main beam that forks into two points.

Rusa unicolor

Sambar

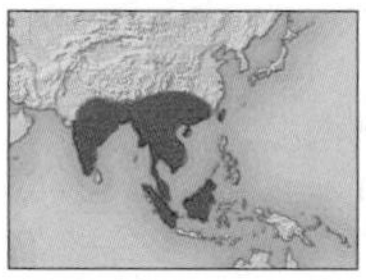

Location S. and S.E. Asia

Length 6½–8¼ ft (2–2.5 m)
Tail 6–8 in (15–20 cm)
Weight 510–770 lb (230–350 kg)
Social unit Individual
Status Vulnerable

The sambar is dark brown except for rusty hues on the chin, inner legs, and tail underside (which has a black tip). The male's 3-point antlers grow to 4 ft (1.2 m) long. Both sexes have a neck mane of thicker fur; this is more prominent in rutting males. Solitary, except for a female with a fawn, and perhaps a yearling, too, these nocturnal deer eat a variety of vegetation.

MAMMALS

Cervus elaphus

Red deer

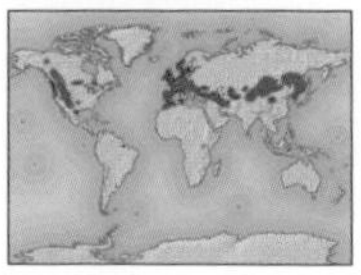

Location Europe to E. Asia, North America

- **Length** 5–6½ ft (1.5–2 m)
- **Tail** 4¾ in (12 cm)
- **Weight** 145–420 lb (65–190 kg)
- **Social unit** Group
- **Status** Least concern

Highly adaptable in habitat and diet, the red deer has been introduced to most continents and is widely farmed for meat, hides, and antler velvet. There is great variation among the 28 or so subspecies, which include the wapitis of China and North America. Females (hinds) form herds led by a dominant hind, with separate male bands except during the fall rut.

MORE BROWN THAN RED
Red deer are red-brown in summer, perhaps with a dark line along the neck and back, and vague flank spots, and then turn dull brown in winter.

RUTTING

Red deer males (stags), like most other male deer, battle during the rutting season. The contest is part display and part physical tussle. Stags roar and bellow, thrash their antlers against bushes and trees, and walk parallel to each other, as they assess whether to fight. If so, they lock antlers, push, twist, and shove. The winner gains a harem.

Cervus nippon

Sika

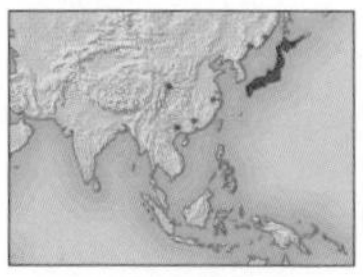

Location E. and S.E. Asia

- **Length** 5–6½ ft (1.5–2 m)
- **Tail** 4¾–8 in (12–20 cm)
- **Weight** 77–120 lb (35–55 kg)
- **Social unit** Group
- **Status** Least concern

The sika has been kept in parks and farmed for centuries, and has been introduced into many regions. Appearance varies among the 14 subspecies, some of which are endangered, but is generally rich red-brown with white spots in summer, and almost black in winter, with perhaps vague spotting on females. The white rump hairs can be flared. The sika eats mainly grasses, including bamboo, twigs, and buds.

Elaphurus davidianus

Père David's deer

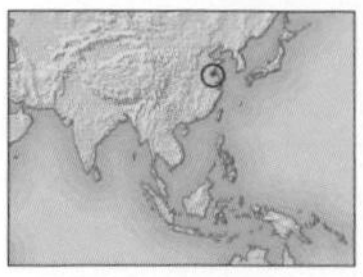

Location E. Asia

- **Length** 7¼ ft (2.2 m)
- **Tail** 26 in (66 cm)
- **Weight** 330–470 lb (150–215 kg)
- **Social unit** Group
- **Status** Extinct in the wild

Distinct from other deer in its body form, this species has a long, horselike face, wide hooves, and a long tail. Also unusual are the male's (stag's) "back-to-front" antlers. The coat is dark gray-fawn in winter, bright red-brown in summer, with a dark central back stripe, and a whorled hair pattern on the rump. Extinct in the wild, the species was saved by captive breeding in England from about 1900. Since the 1980s, it has been reintroduced into the wild in China.

Odocoileus hemionus

Mule deer

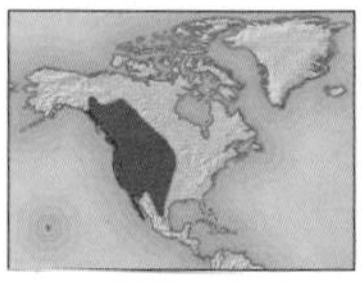

Location W. North America

- **Length** 2¾–7 ft (0.85–2.1 m)
- **Tail** 4–14 in (10–35 cm)
- **Weight** 120–460 lb (55–210 kg)
- **Social unit** Group
- **Status** Least concern

The mule deer is widely distributed in many habitats, and is recorded as eating hundreds of plant species. The main color is gray-brown in winter, and rusty brown in summer. Despite its other name of black-tailed deer, the tail is black on the upper surface only; the rest is white. The face and throat also have variable white areas, with black chin and forehead bands. Rutting is in September–November; 1–2 young are usually born in June.

Odocoileus virginianus

White-tailed deer

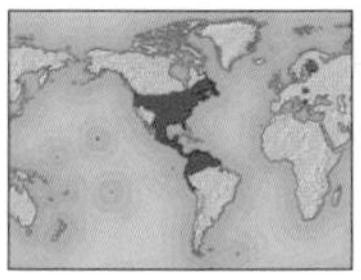

Location S. Canada to N. South America

- **Length** 6–7¾ ft (1.8–2.4 m)
- **Tail** 6–12 in (15–30 cm)
- **Weight** 115–310 lb (52–140 kg)
- **Social unit** Group
- **Status** Least concern

This species is extremely similar in appearance and habits to the mule deer (see left), and in zoos and parks the 2 may interbreed. This rarely happens in the wild, however, even though their ranges overlap. The numerous subspecies become smaller toward the south of the range, shoulder height ranging from 3½ ft (1.1 m) in Canada to 2 ft (60 cm) in Venezuela, where it is known as the venado.

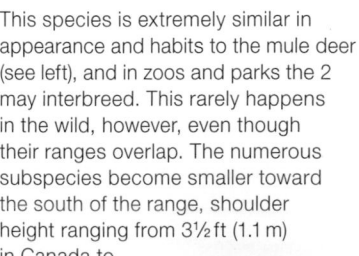

WHITE WARNING
When danger threatens, this deer raises its tail and flashes the bright white underside, as a warning to others in its herd.

YEARLY ANTLER CYCLE

By February, the male white-tailed deer will have cast its antlers. In April–May, they start to grow again, protected by fur-covered skin (velvet), but in September all the velvet will have been rubbed off against trees, leaving the clean bone, before the rut.

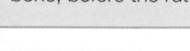

Blastocerus dichotomus

Marsh deer

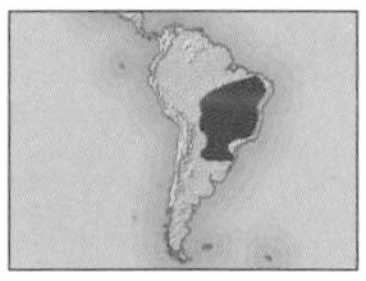

Location C. and E. South America

- **Length** Up to 6½ ft (2 m)
- **Tail** 10 in (25 cm)
- **Weight** 220–310 lb (100–140 kg)
- **Social unit** Individual/Group
- **Status** Vulnerable

Long legs and wide hooves allow the marsh deer to move easily in swamps and floodplains. The largest South American deer, it is reddish brown in summer and darker in winter, with black lower legs, a pale face, and black around the lips and nose. It eats grasses, reeds, water plants, and bushes, and lives alone or in groups of 2–3. It is threatened by habitat loss due to irrigation and conversion to pasture or crops, water pollution, and competition from livestock.

Pudu puda

Southern pudu

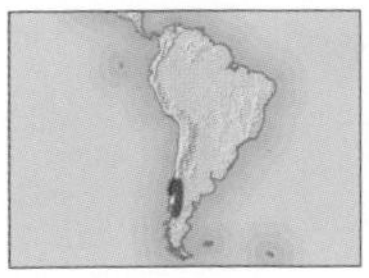

Location S.W. South America

Length 34 in (85 cm)
Tail 2 in (5 cm)
Weight Up to 33 lb (15 kg)
Social unit Individual
Status Vulnerable

The southern pudu is one of 2 small, stocky species of pudu. It is buff to red-brown with rounded ears. The male's antlers are simple spikes, 3¼ in (8 cm) long. Solitary and diurnal, it dwells in moist forests, hiding in understory thickets, where it feeds on bark, buds, fruit, and flowers—but seldom eats grass. It is sexually mature at 6 months.

Rangifer tarandus

Reindeer

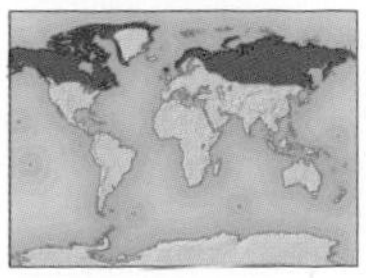

Location N. North America, Greenland, N. Europe to E. Asia

Length 4–7¼ ft (1.2–2.2 m)
Tail 4–10 in (10–25 cm)
Weight 260–660 lb (120–300 kg)
Social unit Group
Status Least concern

Known as caribou in North America, the reindeer has long antlers with a distinctive shovel-like brow tine on one side. It grazes grasses, sedges, and herbs in summer, and mosses, lichens, and fungi in the long winter. One calf is born in May–June after a gestation of 210–240 days.

COLOR VARIATION
American forms have mainly brown coats with darker legs; European and Asian reindeer (shown) are grayer.

MIGRATION

Some reindeer travel 9–40 miles (15–65 km) daily within the same region; others migrate up to 750 miles (1,200 km) twice yearly. In some populations, females and young move to the calving grounds in spring, males following later.

Capreolus capreolus

Western roe deer

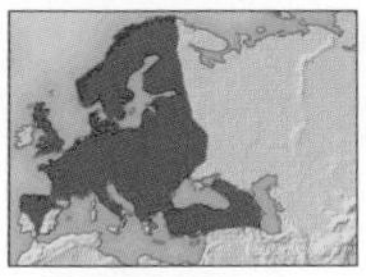

Location Europe, W. Asia

Length 3¼–4¼ ft (1–1.3 m)
Tail 2 in (5 cm)
Weight 44–66 lb (20–30 kg)
Social unit Variable
Status Least concern

This deer has a black muzzle band and variable white chin and throat patches. The white rump patch, which can be fluffed out when the deer is alarmed, is heart-shaped in the female, kidney-shaped in the male (shown). The male has rough-surfaced, 3-point antlers. The sleek, bright red-brown summer coat molts to a longer, denser gray coat in winter.

Alces alces

Moose

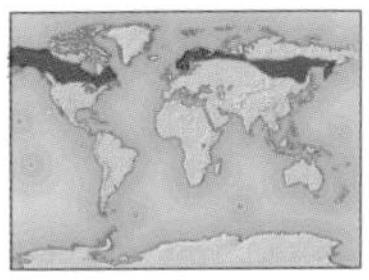

Location Alaska, Canada, N. Europe to N. and E. Asia

Length 8¼–11 ft (2.5–3.5 m)
Tail 4 in (10 cm)
Weight 1,100–1,550 lb (500–700 kg)
Social unit Variable
Status Least concern

The moose is the largest deer, with males up to twice as heavy as females. Found in woods close to swamps, lakes, and other water, it may submerge in summer to reach roots of lilies and other aquatic plants. The winter diet is mainly twigs of trees such as willow and poplar. Moose live alone or in small family groups.

Males rut in September–October. The 1–2 young are born after 242–250 days' gestation, and weaned by 6 months.

COLORATION
The moose is brownish gray in summer, grayer in winter. Its paler-hued, long legs have wide hooves for wading in mud and walking on soft snow.

MOOSE HEAD

The moose's very broad muzzle and flexible lips help to grasp water plants and to strip leaves from twigs. The male's massive antlers may span 6½ ft (2 m); each has up to 20 points, mostly growing from the palm-shaped "beam."

Pronghorn

PHYLUM	Chordata
CLASS	Mammalia
ORDER	Artiodactyla
FAMILY	Antilocapridae
SPECIES	1

Named after the "prong" on its horns, the pronghorn is the only species in its family. The horns are unusual in that they consist of a horny sheath on a bony core (as in antelopes), but are forked and shed yearly (as are deer's antlers).

SWIFT RUNNERS
The pronghorn is one of the fastest mammals—it can reach speeds of over 40 mph (65 kph). In winter, herds of over 1,000 animals may gather.

Antilocapra americana

Pronghorn

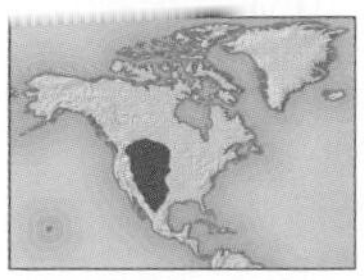

Location W. and C. North America

Length 3¼–5 ft (1–1.5 m)
Tail 3–7 in (7.5–18 cm)
Weight 79–155 lb (36–70 kg)
Social unit Variable
Status Least concern

The pronghorn is red-brown to tan, with white underside, face, rump, and neck bands. The male has a black neck patch, and horns that are longer than the ears, with a forward-facing prong halfway up. The female's horns are shorter than the ears. Herds of over 1,000 form in winter, splitting into smaller groups in summer. Many plants feature in the pronghorn's diet.

MAMMALS

Giraffe and okapi

PHYLUM	Chordata
CLASS	Mammalia
ORDER	Artiodactyla
FAMILY	Giraffidae
SPECIES	2

The giraffe and the okapi are the last surviving representatives of a once highly diverse family. They have long legs, a long, narrow head with small horns, and peculiar, lobed lower canines. The giraffe, with its distinctive long neck, is the tallest living animal—males can reach 18 ft (5.5 m). The 2 species, together known as giraffids, differ in their behavior and ecology because the giraffe lives in woodland savanna (in Africa, south of the Sahara), while the okapi inhabits rain forest (in northeastern Democratic Republic of Congo).

Anatomy

With front legs longer than the back legs, the front parts of the giraffe and the okapi are raised (to facilitate easier browsing). The giraffe has especially long front legs and these are surprisingly sturdy. They are sometimes used in defense: one kick can kill a lion. Both species have a long neck, but what is unexpected about the giraffe's hugely lengthened neck is that it contains only 7 vertebrae, as do almost all other mammals. Each vertebra is greatly elongated, however.

In the giraffe, both sexes have horns, while only the male okapi has horns. The horns are different from those found in other mammals in that they form as cartilage, turn to bone from the tips down, and are covered with skin.

Giraffids have a thick hide to help ward off predators. The okapi has a velvety, dark brown coat with white stripes on the haunches and upper legs (the legs are white below the knees). Giraffes have spotted coats, the pattern of which varies according to location. Both species have a long, copiously tufted tail.

Feeding

Giraffids are browsers with unique 2- or 3-lobed canine teeth, which can be used like a comb to strip leaves from small branches. They also use their thin, mobile lips and long black tongue—which can be extended more than 18 in (45 cm) in giraffes—to gather leaves and shoots. They have a 4-chambered, ruminating stomach (see p.211). Male giraffes tend to be taller than females and are therefore able to feed at higher levels.

Social groups

The giraffe and the okapi have contrasting social systems. Giraffes have home ranges of, on average, approximately 62 square miles (160 square km). When these overlap (which is often), loose associations of up to 25 individuals form. The composition of these herds changes daily. Males are nonterritorial, but a dominance hierarchy is determined in a ritualized fight called "necking"—2 adult males stand side by side, alternately swinging their heads and hitting each other on the neck. Male giraffes have extra bone all over the skull, which provides reinforcement. Only males of high social standing have the right to mate.

Okapis, on the other hand, are mainly solitary animals and are never found in herds. They have much smaller home ranges, and only the dominant males maintain a territory (females move freely from one territory to the next). Male okapis mate with females that wander through their territory.

SWIFT RUNNERS
When danger threatens, a giraffe will take flight, leading others nearby to follow suit. These Angolan giraffes (G. c. angolensis) can reach speeds of over 31 mph (50 kph). Giraffes move directly from a walk to a gallop because their long legs and short body make trotting impossible without tripping.

Okapia johnstoni

Okapi

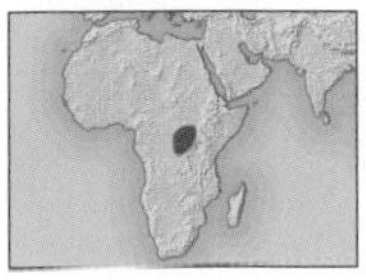

Location C. Africa

Length 6½–7¼ ft (2–2.2 m)
Tail 12–16½ in (30–42 cm)
Weight 440–770 lb (200–350 kg)
Social unit Individual/Pair
Status Near threatened

An elusive browser of thick tropical rain forest, the okapi feeds by day on leaves, soft twigs, shoots, fruit and other plant parts. It relies mainly on hearing in the dense forest and makes a "chuff" sound on meeting another okapi. Rival males "neck fight" like giraffes in the presence of a receptive female, and emit soft moaning sounds during courtship; the female indicates her readiness with similar calls and territorial scent marking. She is slightly taller than the male and 55–110 lb (25–50 kg) heavier. Both sexes look similar: long head and neck, dark muzzle and body (which slopes down from the shoulders), large, rear-set ears, and zebralike stripes on the rump and upper parts of the legs. The female bears a single calf in August–October, after a gestation of 425–491 days. She defends her offspring against predators, but the bond between mother and young is not as strong as in many hoofed mammals.

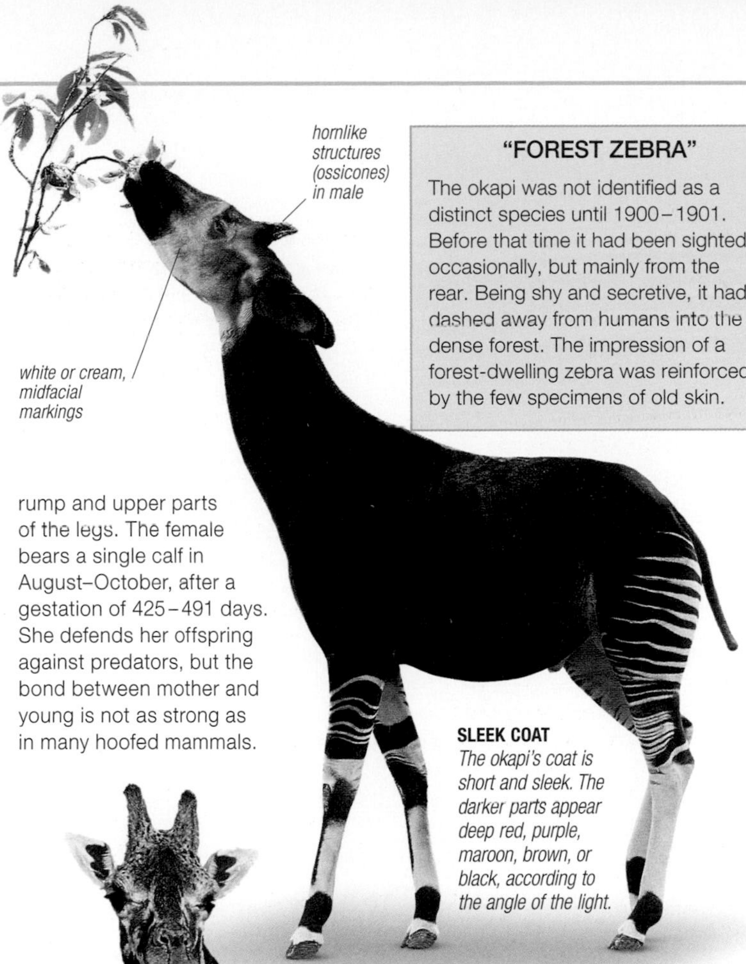

SLEEK COAT
The okapi's coat is short and sleek. The darker parts appear deep red, purple, maroon, brown, or black, according to the angle of the light.

"FOREST ZEBRA"

The okapi was not identified as a distinct species until 1900–1901. Before that time it had been sighted occasionally, but mainly from the rear. Being shy and secretive, it had dashed away from humans into the dense forest. The impression of a forest-dwelling zebra was reinforced by the few specimens of old skin.

LONG TONGUE

The okapi curls its long, black, prehensile tongue around leaves, buds, and small branches, to draw them into the mouth. The tongue is also used for self grooming and, in the female, for cleaning her calf.

Giraffa camelopardalis

Giraffe

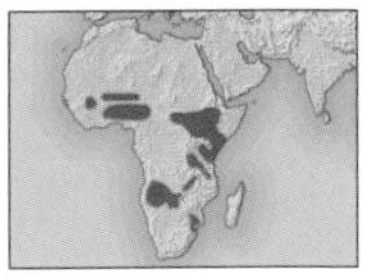

Location Africa

Length 9¾–15 ft (3.8–4.7 m)
Tail 31–39 in (78–100 cm)
Weight 0.66–2 tons (0.6–1.9 tonnes)
Social unit Variable
Status Least concern

A native of dry savannas and open woodland, the giraffe browses higher than any other mammal, mainly for leaves of acacia and wild apricot, but also shoots, fruit, and other vegetation. The combination of greatly elongated tongue, skull, neck, shoulder region (pectoral girdle), and front legs provides the giraffe's great reach when browsing. Usually a small branch is drawn into the mouth with the long, flexible tongue; then the head is pulled away to rake off leaves between the lobe-edged teeth. Among the giraffe's many distinctive features are large eyes and ears; a back that slopes steeply from shoulder to rump; stiltlike legs with large, heavy feet; and a thin tail with a long black tuft for whisking away flies.

The 2–4 specialized horns, called ossicones, are more developed in males than females. Feeding, drinking, and other activities occur in the morning and evening, occupying about 12 hours, with rest (standing up, as in most hoofed mammals) taking place at night, and cud-chewing in the hot midday. Cows mate with local dominant bulls, who have competed with each other by swaying and intercurling necks, and even clashing heads. This activity, "necking," is more a slow-motion ritual than a forceful encounter. It occurs mainly among young bulls, and when a new male arrives in the area. The winner reinforces success by sexually mounting his defeated rival. Cows, after a gestation period of 457 days, give birth to one calf (in rare cases 2), usually in the dry season. The newborn weighs up to 155 lb (70 kg) and stands 6½ ft (2 m) tall. For 10–30 days, the mother keeps it away from the herd; weaning takes place by 13 months. The giraffe's main predators are lions, leopards, and hyenas.

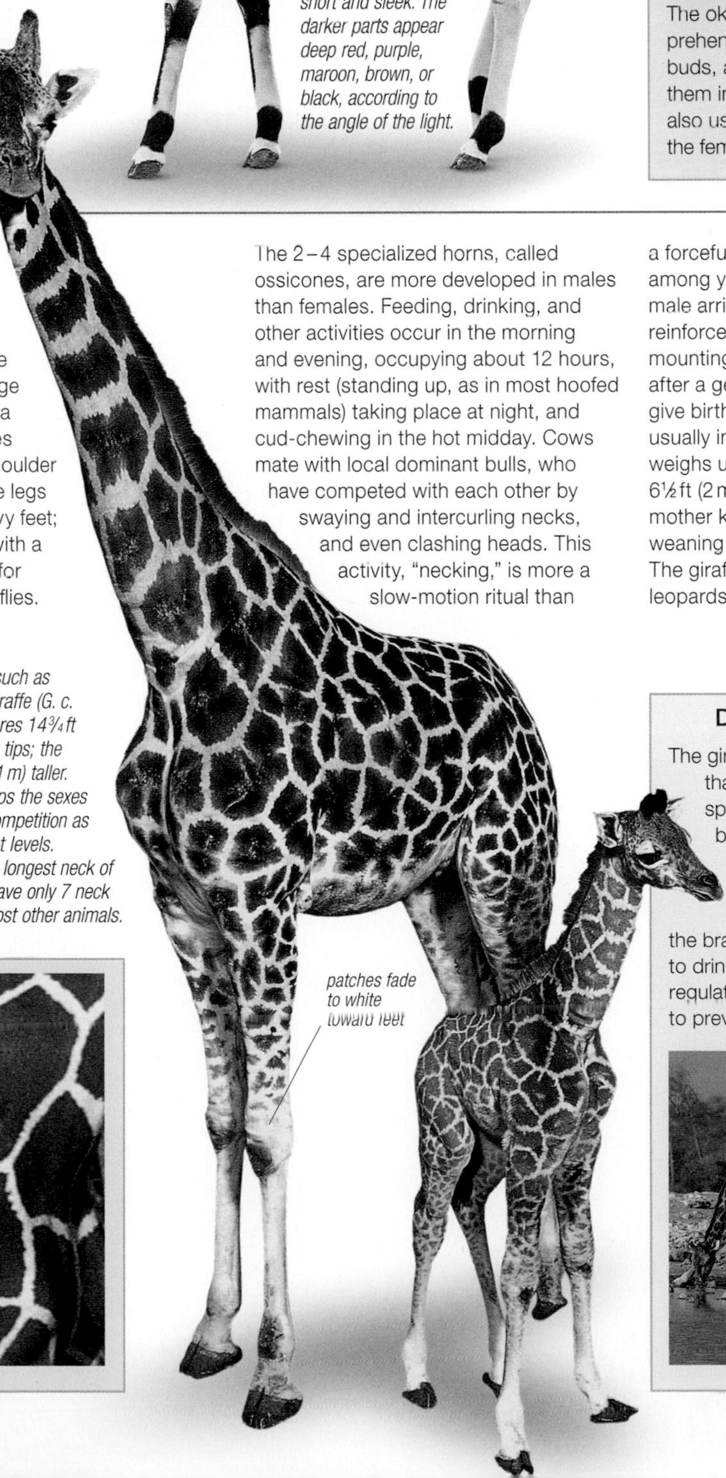

TALLEST ANIMAL
A mature female, such as this Rothschild's giraffe (G. c. rothschildi), measures 14¾ ft (4.5 m) to her horn tips; the male may be 3 ft (1 m) taller. This difference helps the sexes to avoid feeding competition as they utilize different levels. Despite having the longest neck of any animal, they have only 7 neck vertebrae, as in most other animals.

PATCHY DISTRIBUTION

A number of subspecies of giraffe are recognized by skin pattern. The reticulated giraffe (*G. c. reticulata*), pictured here, has very large, sharp-edged, deep chestnut patches separated by fine, white lines. Other subspecies have smaller, irregular, fuzzy-edged patches, which vary from yellow to almost black, with much more white between. Recently, the genetic differences between the subspecies have been thought sufficient to consider them as separate species.

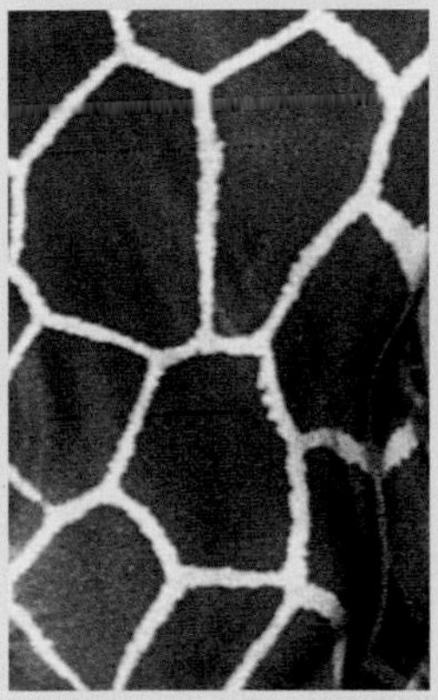

DRINKING PROBLEM

The giraffe's great height means that, to drink water, it must splay its front legs, and even bend at the knees. When upright, its heart has to pump blood upward at enormous pressure to reach the brain, but when the head lowers to drink, a series of one-way valves regulate the blood's force and flow to prevent damage to the brain.

MAMMALS

Cattle and relatives

PHYLUM	Chordata
CLASS	Mammalia
ORDER	Artiodactyla
FAMILY	Bovidae
SPECIES	279

The species that make up this family—the bovids—form a highly diverse group. Members include cattle (wild and domesticated) and their immediate allies (such as bison); sheep and goats and their relatives (such as the chamois); and antelopes (such as the impala), which is a catch-all term for bovids with long, slender limbs. The highest diversity of bovids occurs in Africa, where each species occupies a slightly different niche. They are also found in Eurasia and North America, and a number of species have been introduced to Australasia. Bovids mostly prefer grassland, desert, scrub, and forest habitats.

Anatomy

Although bovids exhibit a wide range of body forms, from sleek, graceful gazelles to the massive, stocky buffalo, all species have unbranched horns consisting of a keratin sheath surrounding a bony core. Unlike the antlers in deer, horns are never shed, and in most species they are present in both sexes. Horns may be straight, curved, or spiraled; keeled, ridged, or smooth; short or long. All have pointed tips.

Bovids have divided ("cloven") hooves: the weight of the animal rests on the 2 central toes of each foot (a pair of small lateral toes is also usually present). The tail varies from small and triangular to long and tufted, and the coat may be smooth and sleek or long and shaggy. Since bovids are often hunted by large predators, they have large, sideways-facing eyes (for good all-around vision); long, mobile ears; and an acute sense of smell. Most species have scent glands located on the face, between the hoofs, and/or in the groin. The glands between the hooves release a scent onto the ground that an isolated animal can follow back to the herd.

Bovids also have a 4-chambered ruminating stomach (see p.211). Food (usually grass or leaves) is pulled in using the dextrous tongue and is then crushed between the lower incisors and a toothless pad in the upper jaw.

Social systems

There is a huge variety of social and breeding systems among bovids. Duikers, for example, are solitary, and dik-diks live in pairs. The impala, however, lives in groups of a male with several females (younger and weaker males form bachelor herds). Male gazelles are territorial: they mate with females that move in groups through their territories. Wild cattle and buffalo, on the other hand, live in less structured groups, although most of the mating is performed by the dominant males.

Bovids and people

Domestic sheep, goats, and cattle are farmed on large and small scales in most countries around the globe and are therefore of huge economic importance to people. Sheep and goats were probably domesticated 8,000–9,000 years ago, in southwest Asia, and their wild ancestors still live in the same region. Cattle, however, were domesticated about 2,000 years after sheep and goats (also in southwest Asia), but the ancestor of most domestic cattle, the aurochs, is now extinct.

Although many wild species are abundant—the wildebeest, for example, numbers in the millions—some, including several species of gazelle, are close to extinction due to hunting.

ESCAPING DANGER
Like most hoofed mammals, the common eland has keen senses, and flees when frightened or chased by predators. Despite their size and massive build, they can gallop at speeds of up to 43 mph (70 kph) and are capable of jumping 5 ft (1.5 m) in the air to clear obstacles in their path.

Tragelaphus spekii

Sitatunga

Location W. and C. Africa

Length 4–5½ ft (1.2–1.7 m)
Tail 8–10 in (20–26 cm)
Weight 110–280 lb (50–125 kg)
Social unit Variable
Status Least concern

Amphibious in habit, the sitatunga occurs in permanent swamps, marshes, and similar watery habitats. Its long, pointed, widely splayed hooves and extremely flexible foot joints are specialized for soft, muddy ground. When in danger from a land-based predator, this antelope retreats to water, and may submerge with only its eyes and nose exposed. At night, males bark warnings or avoidance calls to other males; if they meet, they posture and "horn" the ground. The sitatunga eats many kinds of aquatic and terrestrial plants, including reeds, grasses, and shrubby foliage. While feeding, it may stand in water up to its shoulders. Only the male has ridged, spiraling horns. It also has a grayish tinge to its coat, while the female's is brown to chestnut; both sexes have white around the eyes, and on the cheeks and body. Males are solitary, while females are more social and may live in groups of up to 3. The single calf (rarely twins) is born after 247 days' gestation. There is no specific breeding season.

gray-brown coloration in male

Tragelaphus euryceros

Bongo

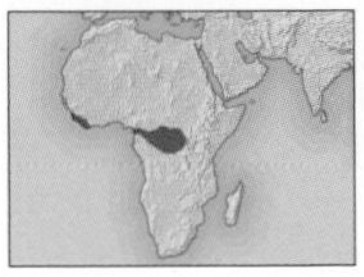

Location W. and C. Africa

Length 5½–8¼ ft (1.7–2.5 m)
Tail 18–26 in (45–65 cm)
Weight 460–890 lb (210–405 kg)
Social unit Individual/Group
Status Near threatened

The largest and most distinctive forest antelope, the bongo has vertical white stripes along the body, and a white chest crescent, cheek spots, nose chevron, and leg bands. The coat is chestnut above, darker below, and it darkens in older males. This selective browser has lyre-shaped horns, longer (up to 37 in/95 cm) in the male, who is solitary. Females form herds of up to 50, perhaps coalescing as their calves associate into nursery bands.

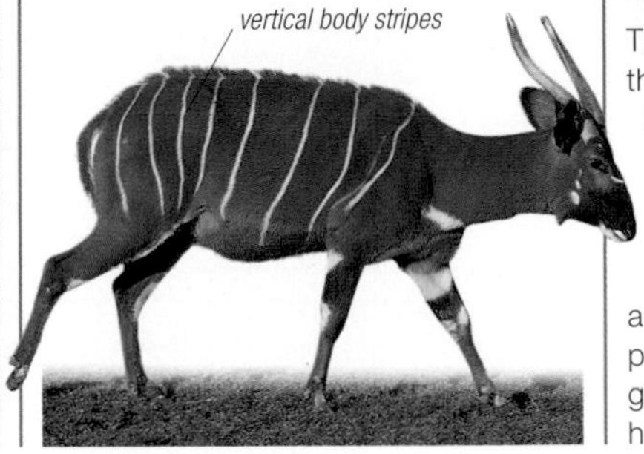

vertical body stripes

Tragelaphus angasii

Nyala

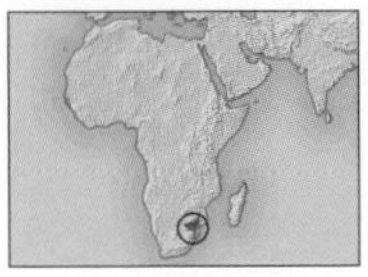

Location Southern Africa

Length 4½–5¼ ft (1.4–1.6 m)
Tail 16–22 in (40–55 cm)
Weight 120–280 lb (55–125 kg)
Social unit Group
Status Least concern

dark, bushy tail

The male nyala is larger and heavier than the female, with a charcoal-gray head and body, indistinct body stripes, tan lower legs, and horns up to 28 in (70 cm) long. Females have no horns and, like juveniles, are red-brown, with a white "V" between the eyes, and vertical white body stripes. Nyalas prefer dense bush near water, and both graze and browse, rearing up to reach higher leaves.

Tragelaphus scriptus

Bushbuck

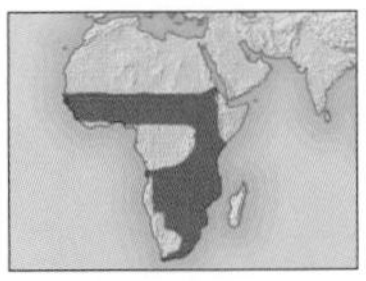

Location W., C., E., and southern Africa

Length 3½–5 ft (1.1–1.5 m)
Tail 8–10½ in (20–27 cm)
Weight 55–175 lb (25–80 kg)
Social unit Individual
Status Least concern

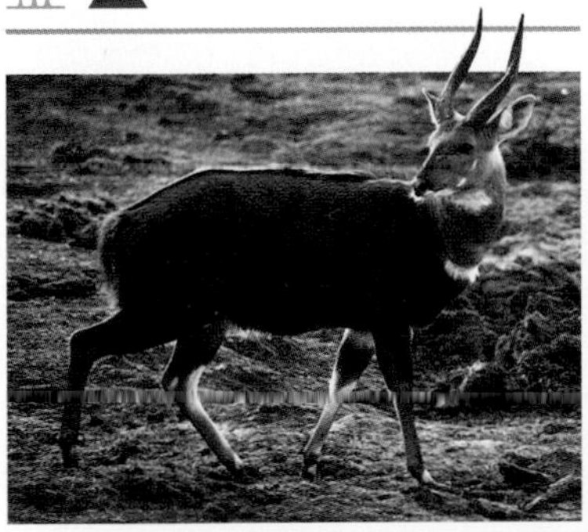

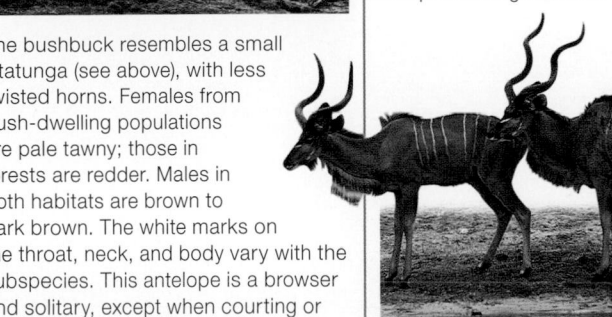

The bushbuck resembles a small sitatunga (see above), with less twisted horns. Females from bush-dwelling populations are pale tawny; those in forests are redder. Males in both habitats are brown to dark brown. The white marks on the throat, neck, and body vary with the subspecies. This antelope is a browser and solitary, except when courting or when a mother is with her young.

Tragelaphus strepsiceros

Greater kudu

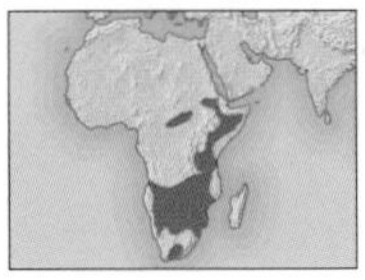

Location E. to southern Africa

Length 6½–8¼ ft (2–2.5 m)
Tail 14½–19 in (37–48 cm)
Weight 260–690 lb (120–315 kg)
Social unit Group
Status Least concern

The male greater kudu is one of the tallest and longest-horned (average 5½ ft/1.7 m) antelopes, and has a long throat fringe. Coloration in both sexes is gray-tinged red or brown, with 6–10 white body stripes, and white nose and cheek marks. This kudu eats leaves, flowers, fruit, herbs, and tubers. Females form groups of up to 6, as do males, except when they compete during the breeding season.

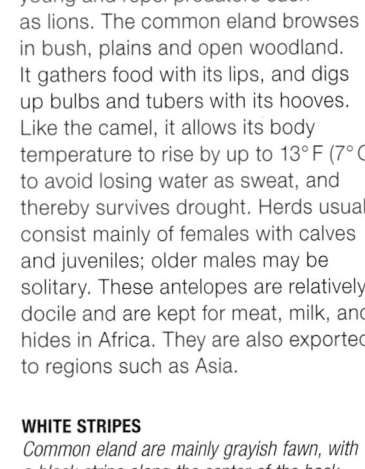

crest along neck and back in male

Taurotragus oryx

Common eland

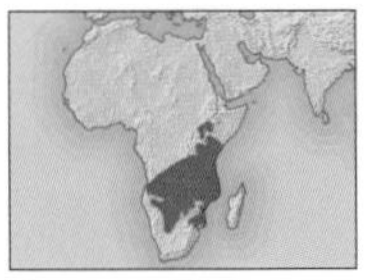

Location C. E, and southern Africa

Length 7–11 ft (2.1–3.5 m)
Tail 23½–35 in (60–90 cm)
Weight 660–2,210 lb (300–1,000 kg)
Social unit Variable
Status Least concern

The two species of eland are the biggest, most cowlike antelopes. Male common elands can weigh up to 2,210 lb (1,000 kg), and have spiraling horns up to 4 ft (1.2 m) long, and a brown-black "topknot" of matted hair on the head. Females weigh up to 1,320 lb (600 kg), with horns half the length of those of the male. Overall body proportions resemble cattle more than antelopes. Also in the manner of cattle, females band together to defend young and repel predators such as lions. The common eland browses in bush, plains and open woodland. It gathers food with its lips, and digs up bulbs and tubers with its hooves. Like the camel, it allows its body temperature to rise by up to 13°F (7°C) to avoid losing water as sweat, and thereby survives drought. Herds usually consist mainly of females with calves and juveniles; older males may be solitary. These antelopes are relatively docile and are kept for meat, milk, and hides in Africa. They are also exported to regions such as Asia.

WHITE STRIPES
Common eland are mainly grayish fawn, with a black stripe along the center of the back, and up to 15 whitish cream vertical body stripes

short mane

shoulder hump

Boselaphus tragocamelus

Nilgai

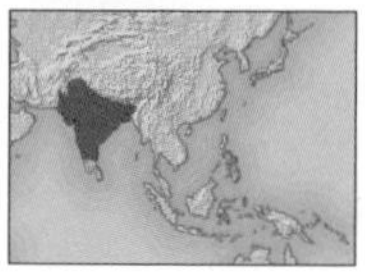

Location S. Asia

Length 6–7 ft (1.8–2.1 m)
Tail 18–21 in (45–53 cm)
Weight Up to 660 lb (300 kg)
Social unit Group
Status Least concern

Also called bluebuck or blue bull, the nilgai is a small-headed bovid with longer front than rear legs, and, in the male, stout, tapering horns 8 in (20 cm) long. The male's coat is gray or bluish gray; the female's is tawny. The nilgai prefers open woods to thick forest, is very wary, with sharp senses, and flees rapidly from predators such as tigers. It grazes and browses on a wide variety of grasses, leaves, and fruit, from early to mid-morning and in the early evening. Males compete for territories—and thus access to groups of 2–10 females—by kneeling in front of each other and lunging with their horns. Breeding occurs during much of the year but most calves, either one or 2, are born in June–October, after a gestation of 243–247 days.

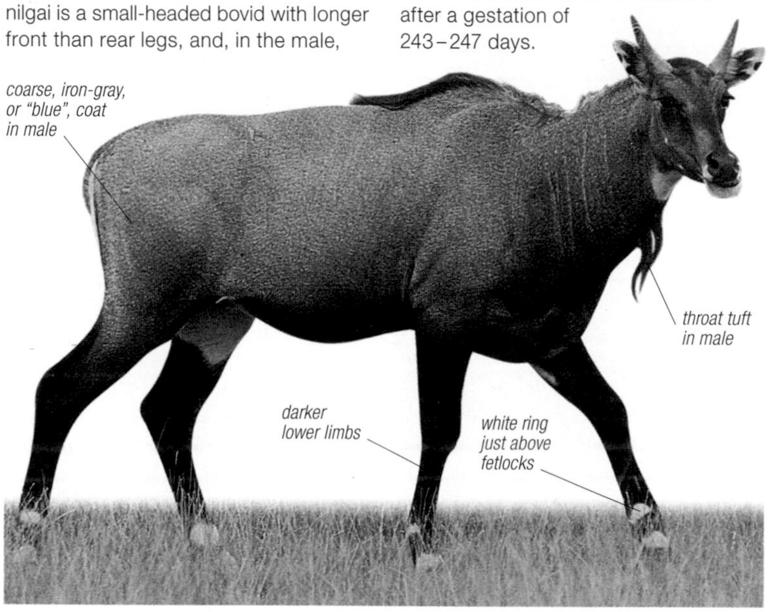

Bubalus depressicornis

Lowland anoa

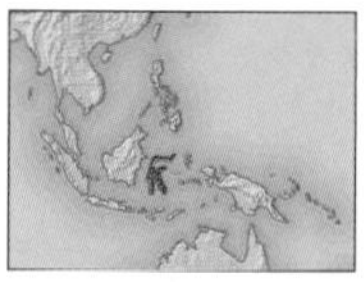

Location Sulawesi

Length 5¼–5½ ft (1.6–1.7 m)
Tail 7–12 in (18–31 cm)
Weight 330–660 lb (150–300 kg)
Social unit Individual
Status Endangered

One of the smallest wild cattle species, this anoa is dark brown to black, with a pale throat bib and facial and leg patches. It has a thick neck, plump body, and short legs, and horns that sweep diagonally backward—adaptations for pushing through dense, swampy forest. Solitary except when breeding, it feeds mainly in the morning on leaves, fruit, ferns, saplings, and twigs. The single calf is born after 9–10 months' gestation.

Bubalus quarlesi

Mountain anoa

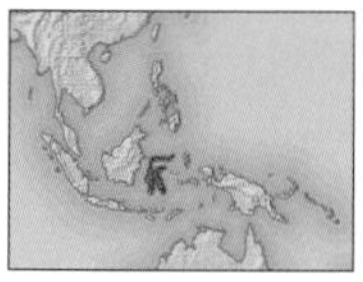

Location Sulawesi and Buton Island

Length 5 ft (1.5 m)
Tail 9½ in (24 cm)
Weight 330–660 lb (150–300 kg)
Social unit Individual/Pair
Status Endangered

Resembling its lowland cousin (see left) in size and overall form, the mountain anoa has a woollier coat, even when fully grown, and fewer white markings, especially on the lower legs and throat. Its horns are smooth surfaced and short, about 6–8 in (15–20 cm) long (those of the lowland anoa being 7–15 in long/18–38 cm). In general, the male is larger and darker than the female, with bigger horns. The mountain anoa is one of the few wild cattle species in Southeast Asia that relies on undisturbed forest habitat, but such terrain is remote and difficult for field studies, and so the detailed habits and population of the species remain unclear. Most sightings are of solitary adults or females with a single young of the year, born after a gestation of 9–10 months. The diet is probably a variety of leaves, and also plentiful mosses, but relatively few grasses.

Tetracerus quadricornus

Chousingha

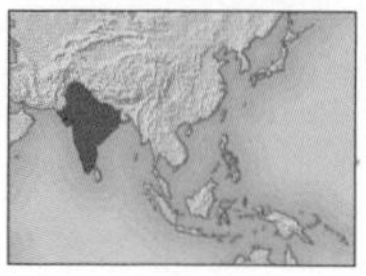

Location S. Asia

Length 32–39 in (80–100 cm)
Tail 4¾ in (12 cm)
Weight 37–46 lb (17–21 kg)
Social unit Individual
Status Vulnerable

Also called the four-horned antelope, the male chousingha has 2 pairs of horns—a feature unique among bovids. The front pair, at only 1¼–1½ in (3–4 cm) long, are half the length of the rear pair. The habits of this shy, fast-moving antelope are little known. It grazes on grasses, sedges, and other plants, usually near water in wooded hills, and communicates by a low whistle for identification or barks for alarm. The brownish coat has a dark stripe on the front of each leg, and a black muzzle and outer ear surfaces.

Bubalus bubalis

Asian water buffalo

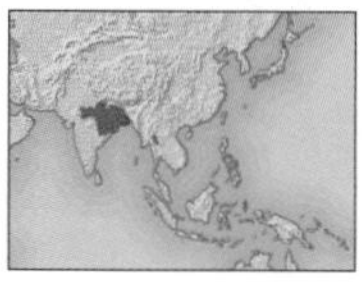

Location S. Asia

Length 7¾–9¾ ft (2.4–3 m)
Tail 23½–39 in (60–100 cm)
Weight Up to 1.3 tons (1.2 tonnes)
Social unit Group
Status Endangered

At more than 1 ton in weight, the Asian water buffalo—or arni as it is also known—is a massive, powerful animal, with the widest horn span of any bovid—more than 6½ ft (2 m). Its large, splayed feet and flexible fetlock joints are suited to the muddy, marshy ground on which this water buffalo grazes in the morning and evening, and occasionally at night, on lush grass and leafy aquatic vegetation. A stable clan of females with young is led by a dominant matriarch (as in elephants), while males form bachelor groups of about 10. Young males spar to assert dominance but avoid serious fighting, and then mix with females at mating time. The Asian water buffalo has been domesticated for thousands of years and spread around the world into various, mostly smaller breeds of less than 1,100 lb (500 kg). The remaining wild populations are scattered and scant, restricted mainly to India, Nepal, and perhaps Thailand.

WALLOWING

During the midday heat, Asian water buffaloes wallow in water or muddy pools, sometimes almost completely submerged, with only their nostrils showing. In addition to keeping them cool, wallowing helps to remove skin parasites, biting flies, and other pests that infest tropical swamps.

BIG-HORNED BOVID
The large horns may curl upward and inward, or point straight out sideways. The face is long and narrow, with small ears. The body is slate black with pale lower legs.

Syncerus caffer

African buffalo

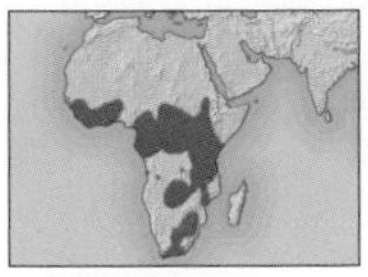

Location W., C., E., and southern Africa

Length 7–11 ft (2.1–3.4 m)
Tail 30–43 in (75–110 cm)
Weight Up to 1,510 lb (685 kg)
Social unit Group
Status Least concern

Africa's only cowlike mammal frequents varied habitats at altitudes of up to 13,200 ft (4,000 m). However, the African buffalo needs a daily drink and is never farther than about 9 miles (15 km) from water. The male may approach twice the weight of the female and has more robust horns that meet in a "boss" on the forehead, a thicker neck, a shoulder hump, and a small hanging fringe of hairs on the throat (dewlap). This buffalo feeds at night and at cool periods of the day, on various grasses, leaves, and other herbage. Males posture for females and dominance, and may fight by pressing or ramming heads. Herd members mutually groom and use mainly sound signals for coordinated actions, such as moving on, flight, or warning. They may also cooperate to mob a predator such as a lion. The single calf (rarely twins) is born after 340 days' gestation, and is fiercely protected by the mother and often by other members of the herd.

HERD BEHAVIOR

African buffaloes are very gregarious and gather at times of plentiful food in herds of 2,000 or more. In the dry season, they split into smaller groups of females and young (including males up to 3 years), or bachelor bands of mature males; older males are more solitary. In any herd, large males dominate smaller ones and also any females.

DARK COLORATION
The African buffalo has a sparse, dark brown coat, big, drooping, hair-fringed ears, a naked muzzle, a long tail, and large feet with rounded hooves.

MIXED BENEFIT

Buffaloes may be infested with lice, ticks, fleas, and similar skin parasites, which birds such as oxpeckers nip off to eat. The bird is fed, and the buffalo cleaned. However, the birds are also known to keep wounds open to feed on the blood, indicating a somewhat parasitic relationship.

Bos javanicus

Banteng

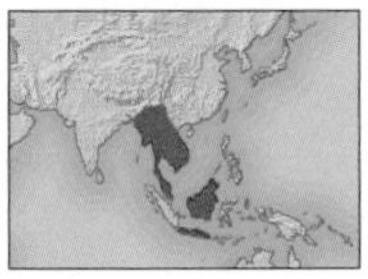

Location S.E. Asia

Length 6–7½ ft (1.8–2.3 m)
Tail 26–28 in (65–70 cm)
Weight 880–1,985 lb (400–900 kg)
Social unit Group
Status Endangered

Ancestor of the domestic banteng and resembling domestic cattle in overall form, the male (bull) banteng is black-brown to dark chestnut; the female (cow) and young are red-brown. All have white undersides, legs, and rump patches. The male's horns are up to 30 in (75 cm) long, angled outward, and then upward; the female's are smaller and crescent-shaped. Banteng live in female–young herds of 2–40 with one adult male, or in bachelor herds. During the monsoon, they move to the hills, returning to the lowlands for the dry season. The wild populations of this bovid are scarce, and their habitat is fast diminishing.

Bos sauveli

Kouprey

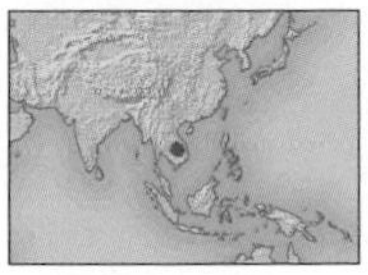

Location S.E. Asia

Length 7–7¼ ft (2.1–2.2 m)
Tail 3¼–3½ ft (1–1.1 m)
Weight 1,544–1,985 lb (700–900 kg)
Social unit Variable
Status Critically endangered

The kouprey (also called Cambodian forest ox) is similar to domestic cattle in general proportions. The male may be black or dark brown with a pendulous dewlap (skin below lower jaw), and L-shaped horns that split at the tips after 3 years' growth. Females may be grayer. Both sexes have off-white legs and pale underparts. They probably form small, loose herds. One of the world's most rare species, the kouprey was identified only in 1937 and is at risk from habitat loss, political instability, poaching, and many other dangers.

Bos grunniens

Yak

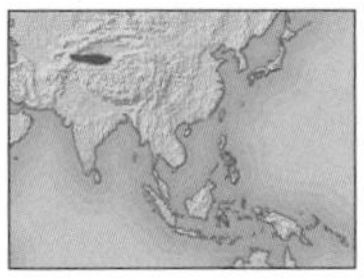

Location C. Asia

Length Up to 10¾ ft (3.3 m)
Tail 23½ in (60 cm)
Weight Up to 1,160 lb (525 kg)
Social unit Individual/Group
Status Vulnerable

Domesticated yak are important to many peoples in South and Southeast Asia for milk, meat, wool, and transportation. The wild yak is larger and quite rare, and is restricted to windy, desolate, bitterly cold steppes at an altitude of up to 19,800 ft (6,000 m), mainly from Kashmir (India) east to Tibet and Qinghai (China). Its dense undercoat of soft, close-matted hair is covered by generally dark brown to black outer hair. The yak grazes grasses, herbs, mosses, and lichens, and crunches ice or snow as a source of water. Females and young form herds, joined by males in the breeding season; otherwise, males form bachelor groups or live alone. A single calf is born every other year, after a gestation period of 258 days.

Bison bison

American bison

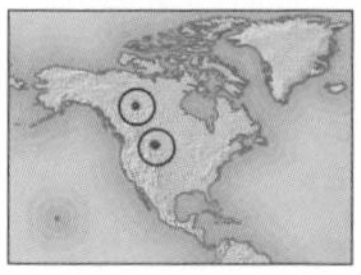

Location W. and N. North America

Length 7–11 ft (2.1–3.5 m)
Tail 1–2 ft (30–60 cm)
Weight 770–2,200 lb (350–1,000 kg)
Population 500,000
Status Near threatened

Also known as the American buffalo, the American bison is massively built, but also deceptively tall, with shoulder heights reaching 6½ ft (2 m). Despite its huge bulk, the bison can run at speeds of up to 37 mph (60 kph). Its sense of hearing and smell are excellent, and essential for detecting danger. Bison spend much of their time grazing and browsing, in loose groups, with rest periods for ruminating. Adult females and their young live in hierarchical herds, which are led by a dominant female. Males usually live apart in bachelor groups and only join females during the mating season. Bison are not territorial, and migration is dictated by season change and adequate food supply. A subspecies, *B. bison athabascae*, is recognized, but the validity of its taxonomy is debated. Over a dozen wild herds now exist, but the species still occupies less than 1 percent of its former range.

MALES BATTLE FOR SUPREMACY

During the mating season, male bison will fight fiercely for the possession of females, usually by head-to-head ramming. Females wanting to mate with the dominant bull gallop about to incite competition between rival males.

CONSERVATION

Despite once numbering around 50 million, the American bison is now virtually extinct in the wild, largely due to widespread commercial hunting since the arrival of European settlers. Although subsequent conservation efforts have led to a significant increase in numbers, most American bison are either captive or have come from captive stock. Yellow Stone National Park (USA) and Wood Buffalo National Park (Canada) are among the few places in which wild herds survive.

WINTER COATS
Although bison are associated with the warm, dry plains of the American West, they are also found in mountainous regions, where they may experience extreme temperatures. In fact, winters offer bison few problems, since their heavy coats and thick manes protect them from the ice and cold.

GRAZING MIGRANT
In the days of the free-roaming herds, bison would make annual migrations of hundreds of miles along traditional routes. However, of a current total population of 500,000 bison, only a very small fraction of these animals roam free in herds, and then only within the 37-mile (60-km) boundaries of the national parks.

MASSIVE FRAME
The bison's massive build is characterized by the towering shoulder hump. The brownish black hair on the neck, head, shoulders, and forelegs is long and shaggy, but the rest of the body is covered in shorter and lighter-colored hair. The large and heavy head sits on a short, thick neck, and features a wide forehead and straggly bearded chin.

Bos gaurus

Gaur

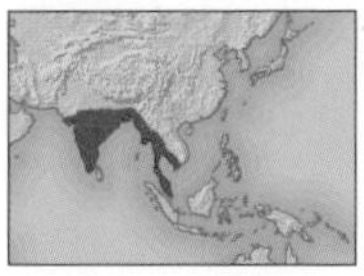

Location S. to S.E. Asia

Length 8¼–11 ft (2.5–3.3 m)
Tail 28–39 in (70–100 cm)
Weight 1,430–2,210 lb (650–1,000 kg)
Social unit Group
Status Vulnerable

Among the biggest cattle, the gaur (seladang or Indian bison) has a huge head and deep body in shades of red, brown, and black, and sturdy, whitish limbs. The S-shaped horns are up to 3½ ft. (1 m) long, and the shoulder hump dips in the lower back before rising to the rump. Rutting males "sing" with bellows that deepen in pitch and carry for 1 mile (1.5 km). The social system and mixed herbage diet are typical of other wild cattle.

Bison bonasus

European bison

Location E. Europe

Length 7–11 ft (2.1–3.4 m)
Tail 12–23½ in (30–60 cm)
Weight 660–2,030 lb (300–920 kg)
Social unit Group
Status Vulnerable

Various lines of evidence, including some genetic studies, suggest that the European bison, or wisent, may be the same species as the American bison (see p.234). Formerly extinct in the wild, the European bison has been park bred and reintroduced into the coniferous Bialoweiza Forest (which straddles the Polish and Belarussian border), where original populations survived until the 1910s. It has a slightly lighter and shorter coat than the American bison, but its habits and herding behavior are similar. The horns, larger in the male, are short and upturned. This bison browses on leaves, twigs, and bark, and grazes on low vegetation. The single calf is born after a gestation period of 260–270 days. It can run within 3 hours and is weaned by one year.

Philatomba walteri

Walter's duiker

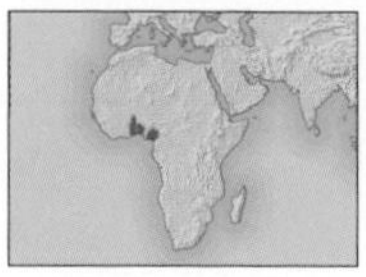

Location W. Africa

Length 30 in (75 cm)
Tail 4½ in (12 cm)
Weight 8¾–13¼ lb (4–6 kg)
Social unit Group
Status Not evaluated

This new species of duiker was described in 2010 on the basis of specimens procured from hunters and markets in the West African countries of Togo, Benin, and Nigeria. It was shown to be genetically and morphologically distinct from other duikers of the region, but so far its habits are virtually unknown. Duikers are nervous, shy antelopes that get their name from their habit of bolting into dense vegetation: duiker means "diving buck." Some—such as Walter's duiker—prefer rain forest to open woodland. Unusually for ungulates, duikers supplement their diet with animal matter—such as insects—and have even been known to kill birds in captivity. The discovery of Walter's duiker highlights the importance of conservation in a part of Africa where exploitation of bushmeat threatens local wildlife. The species is additionally threatened by habitat clearance associated with human population growth.

Sylvicapra grimmia

Common duiker

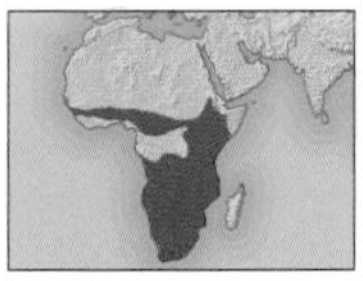

Location W., C., E., and southern Africa

Length 2¼–4 ft (0.7–1.2 m)
Tail 2¾–7½ in (7–19 cm)
Weight 26–55 lb (12–25 kg)
Social unit Individual/Pair
Status Least concern

The common duiker has a tufted forehead, large, pointed ears, and, usually only in males, sharp-pointed horns about 4¼ in (11 cm) long. It is gray to red-yellow haired above, with white underparts, and has a dark nose stripe. Adaptable in habitat, it is a nocturnal browser and also takes small animals and carrion. It lives alone or in pairs, and males defend their territories against rivals.

Kobus ellipsiprymnus

Waterbuck

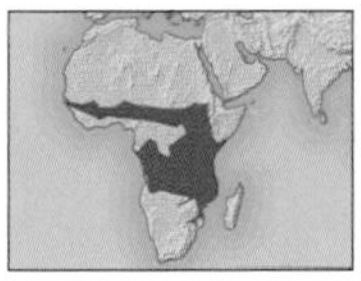

Location W., C., and E. Africa

Length 4¼–7¾ ft (1.3–2.4 m)
Tail 4–18 in (10–45 cm)
Weight 110–660 lb (50–300 kg)
Social unit Individual/Group
Status Least concern

One of the heaviest antelopes, the waterbuck has coarse, long, oily hair that ranges in color from gray to red-brown and darkens with age. There are white markings on the rump, throat, and muzzle, and white "eyebrows," rings above the hooves, and underparts. The horns, normally present only in the male, are up to 3¼ ft (1 m) long. Some 90 percent of the diet is grass, the rest browsed leaves. When threatened, the waterbuck usually dashes to water, where it swims fast or submerges except for the nose. Herds of younger males, usually 2–5 but rarely 50 or more, have a hierarchy based on visual displays, horn length, and frequent fights. Older (6–10 years) breeding males occupy territories. Females are solitary or form loose groups of up to 10.

Kobus leche

Lechwe

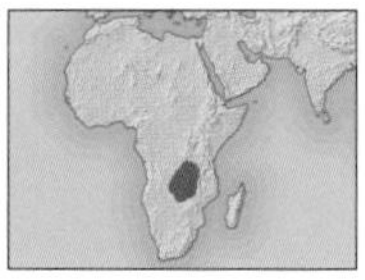

Location C. to southern Africa

Length 4¼–7¾ ft (1.3–2.4 m)
Tail 4–18 in (10–45 cm)
Weight 175–230 lb (79–103 kg)
Social unit Group
Status Least concern

The lechwe, or marsh antelope, eats grasses and aquatic plants exposed by seasonal variations in the weather, responding to water levels across floodplains and swamps. It wades and swims well, forms large herds, and has the lek breeding system (see kob, right). The chestnut to black coat contrasts with the white underparts and black leg stripes. Only the males are horned.

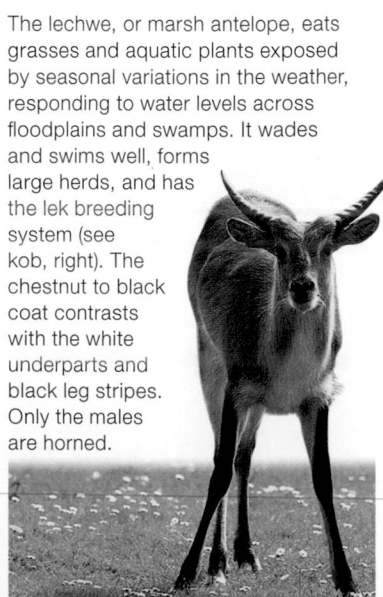

Kobus kob

Kob

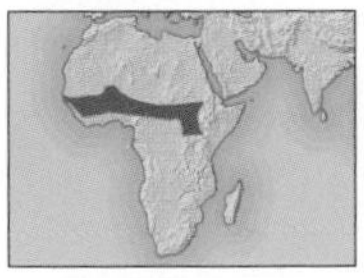

Location W. to E. Africa

- **Length** 4¼–7¾ ft (1.3–2.4 m)
- **Tail** 4–18 in (10–45 cm)
- **Weight** 110–660 lb (50–300 kg)
- **Social unit** Group
- **Status** Least concern

The kob is a graceful but strong grazing antelope, pale cinnamon to brown-black, with white facial and throat markings, and black leg stripes and feet. The male has ringed, lyre-shaped horns. Kobs live at very high densities, and males compete for a patch of ground (known as the lek), perhaps only 50 ft (15 m) across. The winner of this contest gains mating rights over many females.

Kobus vardonii

Puku

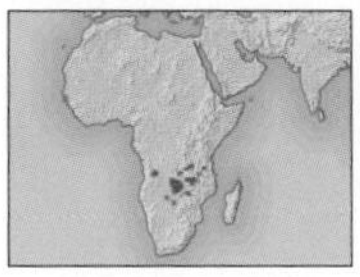

Location W. to E. Africa

- **Length** 4¼–6 ft (1.3–1.8 m)
- **Tail** 7–12 in (18–30 cm)
- **Weight** 145–170 lb (66–77 kg)
- **Social unit** Individual/Group
- **Status** Near threatened

The puku resembles the kob (see left) in its breeding system: it has leks at high population densities; territories at lower ones. The long coat is a uniform golden yellow, and the horns are about 20 in (50 cm) in length. The puku grazes morning and evening and, like other plains antelopes, flees rapidly from danger.

Hippotragus equinus

Roan antelope

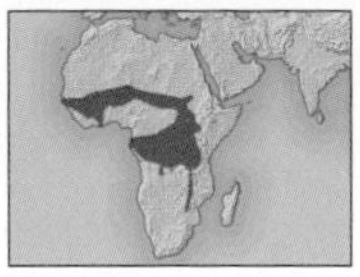

Location W., C., and E. Africa

- **Length** 6¼–8¾ ft (1.9–2.7 m)
- **Tail** 14½–30 in (37–76 cm)
- **Weight** 330–660 lb (150–300 kg)
- **Social unit** Individual/Group
- **Status** Least concern

The roan antelope is red- to brown-coated, with white underneath and black and white facial markings. Both sexes are horned and maned. They survive on poor grass growth and need to drink 2–3 times daily. Herds consist of 12–15 females and young with one dominant male, or younger bachelor males.

Redunca redunca

Bohor reedbuck

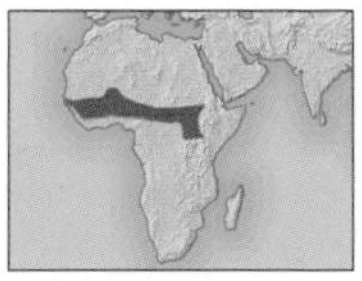

Location W. to E. Africa

- **Length** 3½–5¼ ft (1.1–1.6 m)
- **Tail** 6–18 in (15–45 cm)
- **Weight** 42–210 lb (19–95 kg)
- **Social unit** Variable
- **Status** Least concern

This small, fawn, lightweight savanna antelope has white underparts, throat, and eye-rings. A conspicuous gray patch under the ear marks a scent gland. The bohor reedbuck feeds on grasses and tender reed shoots. Small groups, of a female and young, or horned males, merge into dry-season herds.

Hippotragus niger

Sable antelope

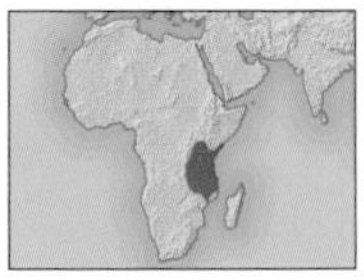

Location E. to S.E. Africa

- **Length** 6¼–8¾ ft (1.9–2.7 m)
- **Tail** 14½–30 in (37–76 cm)
- **Weight** 330–660 lb (150–300 kg)
- **Social unit** Group
- **Status** Least concern

Similar to the roan antelope (see above right) in many respects, the sable antelope gathers in herds of 100 or more during the dry season, when browsing replaces the usual grazing. In the wet season, herds split into bachelor groups of 2–12 males, while dominant males occupy territories and mate with the females there. After an initial concealment, calves form nursery groups, only joining their mothers to suckle.

SABLE CALF
Calf coloration is similar to that of adult females—chestnut or sorrel. It is born after 240–280 days' gestation, and initially is kept away from the herd.

ADULT MALE
Mature males are black but with the same facial pattern as females: white with a central dark blaze and cheek stripes. Both sexes have stout, heavily ringed horns.

Oryx dammah

Scimitar-horned oryx

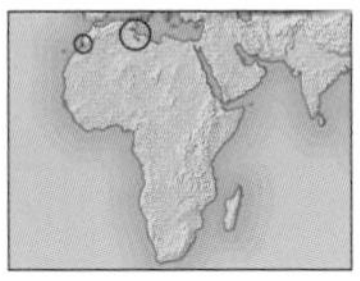

Location N. Africa

- **Length** 5¼–7¾ ft (1.5–2.4 m)
- **Tail** 18–35 in (45–90 cm)
- **Weight** 220–460 lb (100–210 kg)
- **Social unit** Group
- **Status** Extinct in the wild

Specialized for deserts, arid plains, and rocky hills, the scimitar-horned oryx has many physiological adaptations to conserve body water. Its kidneys are very efficient and it sweats only when its body temperature exceeds 116°F (46°C). The large hooves spread the stocky body's weight on soft sand. This oryx feeds on a wide variety of plants in the early morning and evening, and on moonlit nights, resting by day in any available shade. It forms nomadic mixed herds of 20–40, males displaying and tussling to mate with females. After 222–253 days' gestation, the mother leaves the herd to calve, but returns within hours. Young are weaned by 14 weeks and are sexually mature by 2 years. This species was hunted almost to extinction, surviving only on a reserve in north-central Chad. Captive bred animals being held in Tunisia are part of a planned reintroduction program.

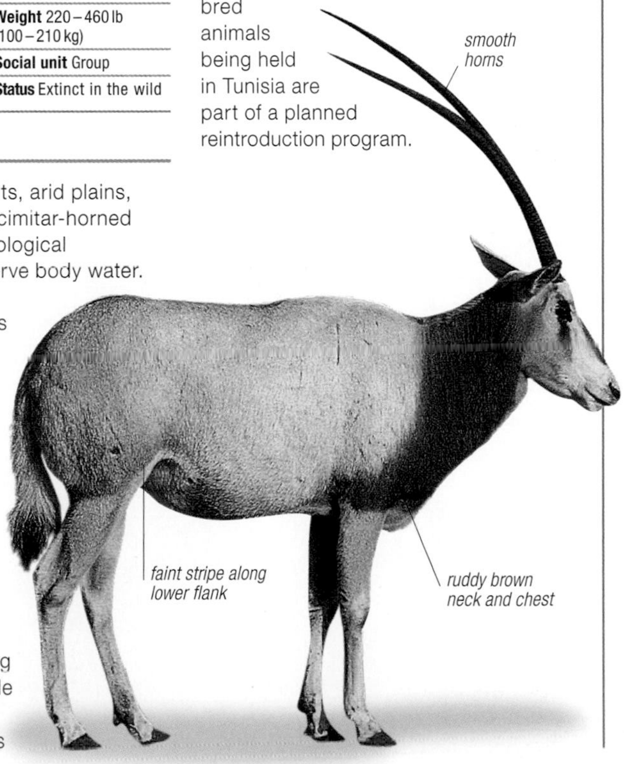

MAMMALS

Oryx gazella

Gemsbok

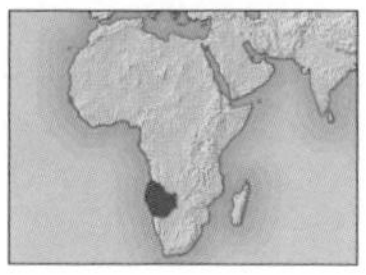

Location S.W. Africa

Length 5¼–7¾ ft (1.6–2.4 m)
Tail 18–35 in (45–90 cm)
Weight 220–460 lb (100–210 kg)
Social unit Group
Status Least concern

The gemsbok, or southern oryx, is a large, distinctively colored antelope of arid grassy scrub and true desert. Its many adaptations for moisture conservation include not panting or sweating until the body temperature exceeds 113° F (45° C), kidneys that make very concentrated urine, and very dry droppings. The main diet is grasses and low shrubs, with wild cucumbers, melons, and similar plants supplying water. Nomadic herds of up to 25 contain females, young, and a few males. Like many desert species, gemsbok breed opportunistically, year-round whenever food is available. The single calf (rarely twins) is born after 260–300 days' gestation, and stays concealed but near the main herd, with occasional visits from the mother to suckle, for up to 6 weeks.

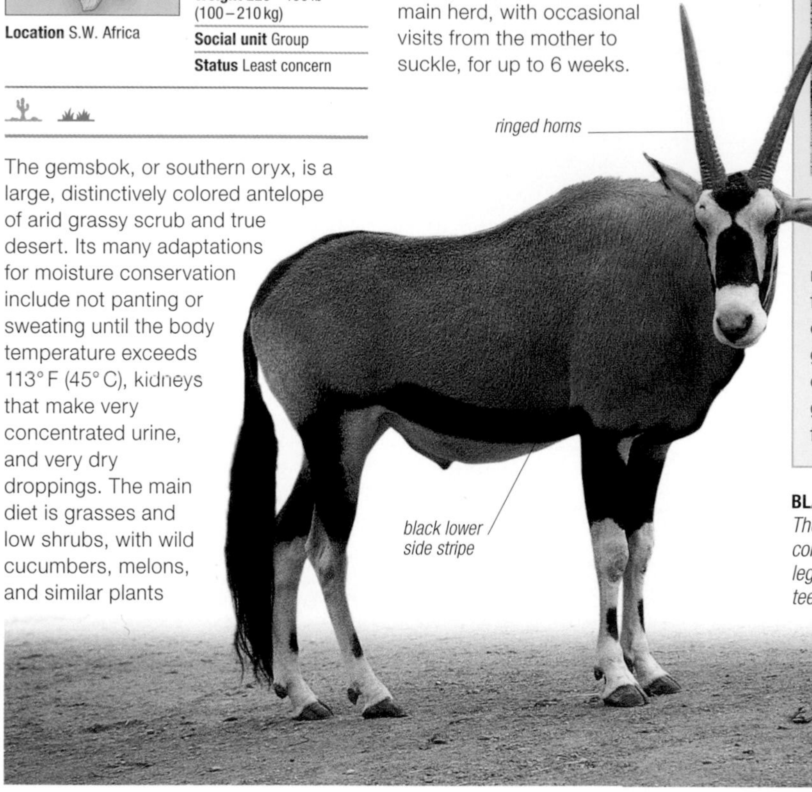

PEACE IN THE SHADE

Gemsbok deal with their hot, dry habitat by grazing at cool times, mainly twilight and at night. They gather in available shade from about 10am to 2–3pm. In most other antelope species, individuals grouped so closely would begin dominance or mating disputes, but gemsbok suspend social interaction for the greater need of survival.

BLACK, WHITE, AND SHADES OF GRAY
The gemsbok has a fawn-tinged gray body and contrasting black and white face, ears, belly, and legs. The broad muzzle and wide row of incisor teeth are adapted to crop coarse grasses.

Addax nasomaculatus

Addax

Location N.W. Africa

Length 5–5½ ft (1.5–1.7 m)
Tail 10–14 in (25–35 cm)
Weight 130–280 lb (60–125 kg)
Social unit Variable
Status Critically endangered

Rare and remote, the addax has similar adaptations to the gemsbok (see left) for desert survival. It wanders in search of almost any vegetation, following rains, and rests in shade at midday. Formerly more numerous, a mixed herd of up to 20 was led by an older male; individuals now live alone or in small bands of 2–4. The coat is gray-brown in winter, sandy to white in summer, with a white facial patch topped by a chestnut forehead tuft, and spiral horns (1½–3 turns).

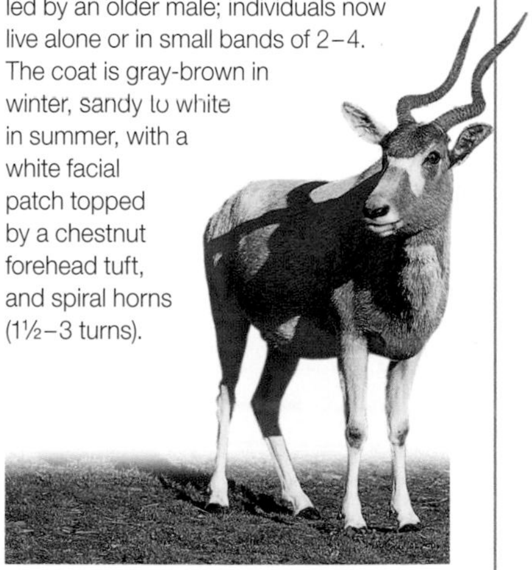

Damaliscus pygargus

Bontebok

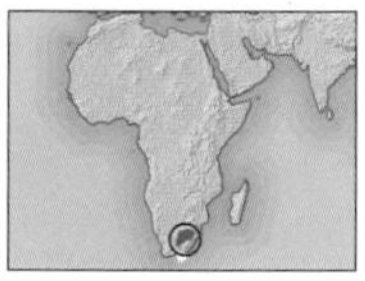

Location Southern Africa

Length 4–7 ft (1.2–2.1 m)
Tail 4–23½ in (10–60 cm)
Weight 150–340 lb (68–155 kg)
Social unit Group
Status Least concern

Also known as the blesbok, the bontebok has a white blaze on its long muzzle, which narrows at the eyes, and continues to the flattened, lyre-shaped horns, which are ringed for most of their 28-in (70-cm) length. Adult males posture and spar with their horns—although rarely fight—to gain a territory. This allows them to dominate a herd of females and young. The male keeps its members together and initiates the herd's travels. Females give birth at traditional calving grounds, after a gestation period of about 8 months; unlike many similar antelopes, they do not isolate or conceal their young. The single newborn can walk within 5 minutes, and soon follows its mother; it is weaned by 6 months. The bontebok grazes mixed grasses and herbs, early and late in the day. It was almost exterminated in the wild by the 1830s, but herds were preserved in parks and reserves where its numbers are now slowly increasing.

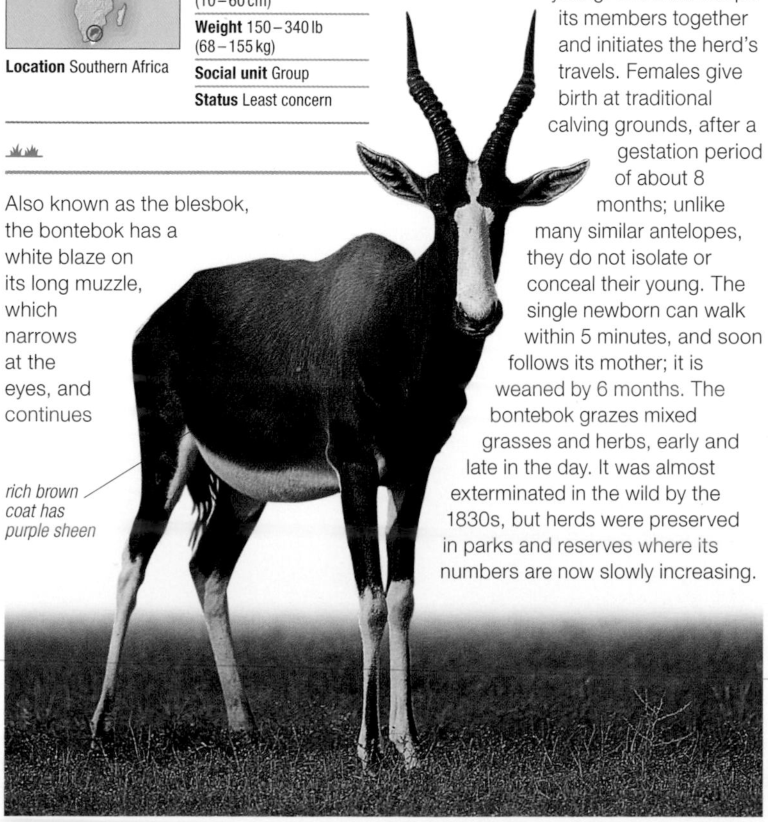

Damaliscus lunatus

Western tsessebe

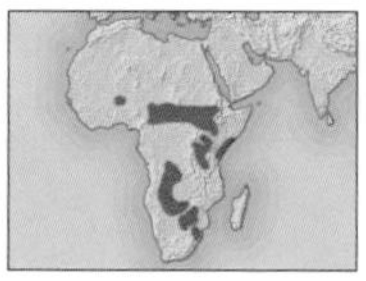

Location W., C., E., and southern Africa

Length 4–7 ft (1.2–2.1m)
Tail 4–23½ in (10–60 cm)
Weight 150–340 lb (68–155 kg)
Social unit Group
Status Least concern

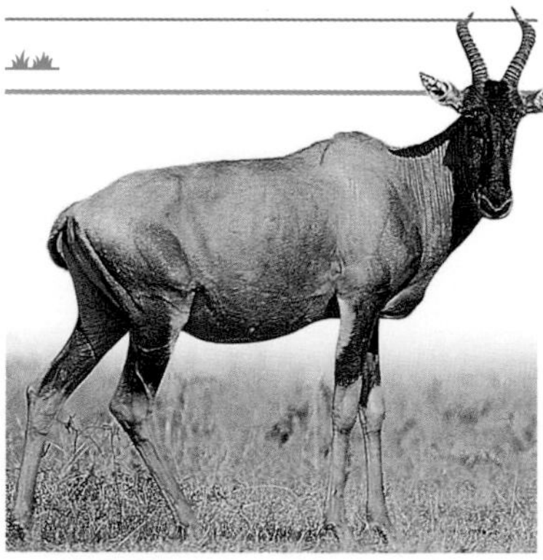

The western tsessebe, or topi, has a long head, a shoulder hump, and a downward-sloping back. Its glossy, red-brown coat is purplish on the upper limbs and darker on the upper muzzle, belly, and lower limbs. It has L-shaped, ringed horns. The western tsessebe lives in seasonally flooded grasslands, and has 2 breeding systems according to conditions: leks (see kob, p.237) in migratory populations; and male territories with harems when resident.

Alcelaphus caama

Red hartebeest

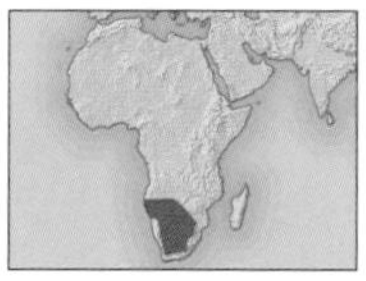

Location Southern Africa

Length 5–8¼ ft (1.5–2.5 m)
Tail 12–28 in (30–70 cm)
Weight 220–500 lb (100–225 kg)
Social unit Group
Status Least concern

Resembling the western tsessebe (see left) in diet and sloping profile, this is the southernmost species of a genus of hartebeest characterized by heavily ringed horns and prominent glands below the eyes. The red hartebeest has a dark tan coat and curving Z-shaped horns. Like other species, there are paler hip patches and black markings on forehead, muzzle, shoulder, and thighs. Historically, hunting constricted its range, but following reintroduction to southern Africa, it is expanding again.

MAMMALS

Connochaetes taurinus

Wildebeest

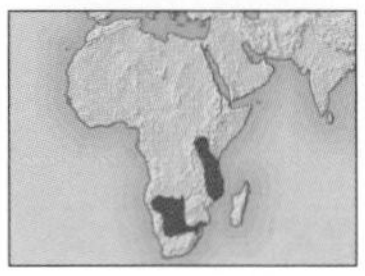

Location E. and southern Africa

Length 5–7¾ ft (1.5–2.4 m)
Tail 14–22 in (35–56 cm)
Weight 260–610 lb (120–275 kg)
Social unit Group
Status Least concern

Also called the brindled gnu or blue wildebeest, this antelope has an unmistakable, large, long-muzzled head, cowlike horns, and high shoulders. The single calf is born after 8–9 months' gestation. It bleats like a lamb, and its fiercely protective mother lows like a domestic cow in reply. Males form bachelor herds at 1–4 years of age, then try to establish solitary territories, with "ge-nuu" calls, ritual posturing, and pushing. Only winning males can mate.

DANGEROUS JOURNEY

Although some wildebeest remain in a home range, most band into large herds and migrate hundreds of miles on an annual journey to find seasonal grazing. The route may be altered by unusual rains stimulating grass elsewhere. At river crossings (above), they are vulnerable to crocodile attack.

LONG MANE
The wildebeest's copious black mane extends from neck to shoulders, and hairs spill over the forehead. The main coat is silver-gray with brown hues fading rearward, and the tail is long and black.

Raphicerus campestris

Steenbok

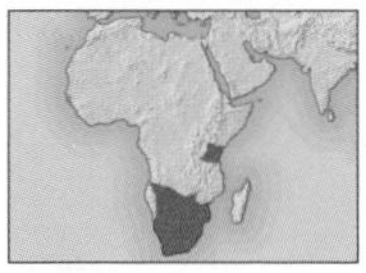

Location E. and southern Africa

Length 24–37 in (61–95 cm)
Tail 1½–3¼ in (4–8 cm)
Weight 15–35 lb (7–16 kg)
Social unit Individual/Pair
Status Least concern

The steenbok, or steinbuck, lives alone or as a pair with largely separate lives, within a territory marked by scents and dung. It both browses and grazes, and digs up roots and tubers with its feet, which lack the 2 lateral toes. Its color is bright rufous-fawn, sometimes tinged silver-gray, pale beneath. There is also a white eye-stripe or ring, and black "finger lines" in the ear, patches on the nose, and between the horns, which are present only in the male.

Oreotragus oreotragus

Klipspringer

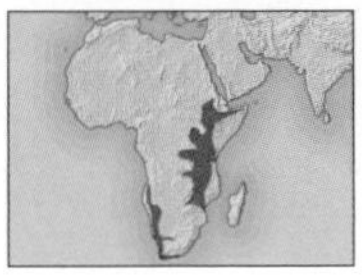

Location E., C., and southern Africa

Length 2½–4 ft (0.8–1.2 m)
Tail 2–5 in (5–13 cm)
Weight 18–40 lb (8–18 kg)
Social unit Pair
Status Least concern

This small, short-muzzled, tiny-hoofed antelope leaps skillfully over the steep, rocky terrain of its native mountains and river gorges. It has a short tail and a dense, glossy olive-yellow coat speckled with yellow and brown, fading to white on the underparts and legs. The male has small, spiky horns. The klipspringer browses on evergreen and other shrubs, and lives in pairs with 1–2 offspring.

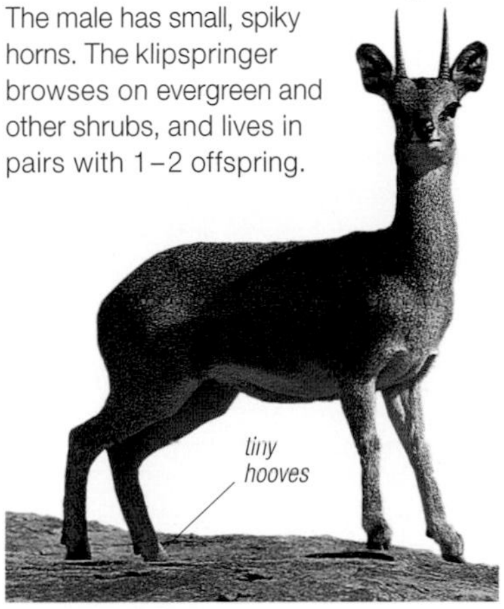

Ourebia ourebi

Oribi

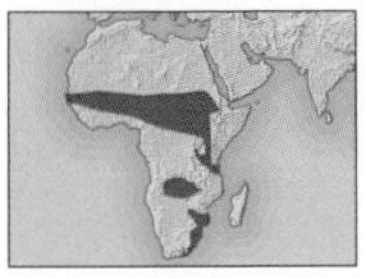

Location W. to E. and southern Africa

Length 3–4½ ft (0.9–1.4 m)
Tail 2¼–6 in (6–15 cm)
Weight 31–46 lb (14–21 kg)
Social unit Variable
Status Least concern

Small, slender, and long-necked, the oribi has fine, silky fur, sandy to rufous above, and white below and on the chin and rump, with longer tufts on the knees. The male has 2 small, spiky, ringed horns. The diet consists of grasses and, in the dry season, bushy leaves. Oribi are socially flexible, with male–female pairs or small herds of 7–8 with 2–3 adult males. The male may help to clean and guard the calf, which is weaned at 2 months.

Madoqua kirkii

Kirk's dik-dik

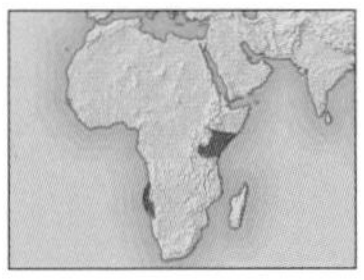

Location E. and S.W. Africa

Length 20½–28 in (52–72 cm)
Tail 14–22 in (35–56 cm)
Weight 6½–15 lb (3–7 kg)
Social unit Pair
Status Least concern

The 4 dik-dik species are named after their alarm call. Kirk's dik-dik has a soft, lank coat, grizzled gray to brown, and more reddish brown on the head, with a forehead crest. The rubbery-bottomed hooves grip effectively on rocks. It eats a wide range of plants and lives as a close-bonded male–female pair and a single offspring, born after 169–174 days' gestation. The newborn remains hidden for 2–3 weeks, and suckles for 3–4 months.

Antilope cervicapra

Blackbuck

Location S. Asia

Length 4 ft (1.2 m)
Tail 7 in (18 cm)
Weight 71–95 lb (32–43 kg)
Social unit Group
Status Near threatened

The blackbuck grazes on grasses, including cereal crops. Females are fawn to yellow, dominant males become black with age, and other males are brown. All have a white underside, rump, muzzle, and eye-ring. The male's horns, up to 27 in (68 cm) long, are ringed at the base and spiral up to 5 turns. When breeding, the male defends his territory and harem.

MAMMALS

Aepyceros melampus

Impala

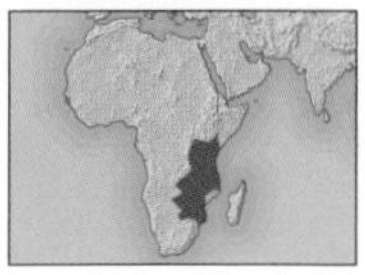

Location E. and southern Africa

Length 3½–5 ft (1.1–1.5 m)
Tail 10–16 in (25–40 cm)
Weight 88–145 lb (40–65 kg)
Social unit Group
Status Least concern

Impala are noisy antelopes. Males (which are horned) make loud, hoarse grunts when rutting; calves bleat; and all emit loud warning snorts as they race off with high leaps, kicking the hind legs out straight and landing on the forelegs. Adaptable grazer-browsers, impala form mixed herds in the dry season. At breeding time, bachelor males compete for territories and females. The coat is reddish fawn with black streaks on the hips and tail.

Litocranius walleri

Gerenuk

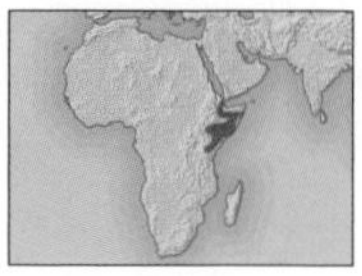

Location E. Africa

Length 4½–5¼ ft (1.4–1.6 m)
Tail 9–14 in (22–35 cm)
Weight 62–115 lb (28–52 kg)
Social unit Individual/Group
Status Near threatened

Also called the giraffe-gazelle, the gerenuk has a very long, slender neck and similar legs. It can curve its spine into an S shape, balancing its weight over its rear legs, in order to stand vertically for long periods. This allows it to browse higher than similar-sized herbivores in open woodland and scattered bush. Seen from the front, its neck, head, and long, wedge-shaped muzzle are extremely narrow, for probing into acacias and other thorny foliage. The gerenuk uses its long, pointed tongue, mobile lips, and sharp-edged incisors to pluck and nip the smallest leaves. It is mainly reddish fawn, with a broad, dark band along the back and upper sides, and white on the underparts, neck, chin, lips, and around the eyes; the tail is black tufted. Only the males have horns, which are 14 in (35 cm) long, relatively thick and curved. Social units are male–female pairs, or small groups of one male and 2–4 females, with offspring. Only territory-holding males breed, from about 3 years old. Younger males form bachelor herds, although one male may attach to the periphery of a female band, and the occasional female may be solitary.

Antidorcas marsupialis

Springbok

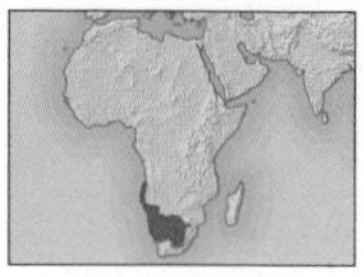

Location Southern Africa

Length 4–4½ ft (1.2–1.4 m)
Tail 6–12 in (15–30 cm)
Weight 66–105 lb (30–48 kg)
Social unit Group
Status Least concern

The springbok is among several bovids that "stott" or "pronk"—leap stiff-legged, high and repeatedly, as if bouncing. This behavior may serve to deter predators. This adaptable herbivore is highly gregarious, but migratory herds once millions-strong now number only 1,500. Breeding habits are as for other gazelle species.

Eudorcas thomsonii

Thomson's gazelle

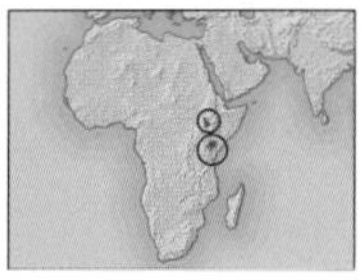

Location E. Africa

Length 3–4 ft (0.9–1.2 m)
Tail 6–8 in (15–20 cm)
Weight 33–66 lb (15–30 kg)
Social unit Group
Status Near threatened

This small gazelle is graceful and speedy. It "stotts" (see springbok, above right) energetically when threatened by, for example, many of the big cats, hyenas, jackals, and similar carnivores, for which it is a staple part of their diet. The most common gazelle in its region, it sometimes forms mixed herds with impala and other gazelle species. It mainly grazes short grasses, yet also browses. Small herds of 10–30 females and young join male bachelor bands, and even lone males, to migrate between grasslands (in the rainy season) and bush (when it is drier). Thomson's gazelles usually produce a single offspring, which is born after 160–180 days' gestation. Although initially mottled darker than the parent, the coat of the young gazelle lightens in 1–2 weeks.

RAPID BREEDING

Thomson's gazelle is one of the few bovids that can breed twice yearly. The first calf is born in January or February, after the rains; the second in July. A newborn is quickly on its feet, but spends the first few weeks lying hidden, until it can keep up with the herd, at 3–4 weeks. It is weaned by 4 months.

DISTINCTIVE MARKINGS
A black flank band separates the sandy fawn back from the white underside. The rufous head has a darker blaze, and white eye-rings that extend along the muzzle, above the black cheek stripes.

Saiga tatarica

Saiga

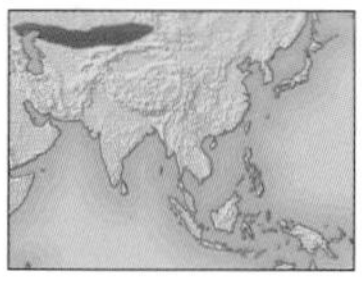

Location C. Asia

Length 3¼–4½ ft (1–1.4 m)
Tail 2¼–4¾ in (6–12 cm)
Weight 57–150 lb (26–69 kg)
Social unit Group
Status Critically endangered

This medium-sized Asian "goat-antelope" has an enlarged nose with down-pointing nostrils, perhaps to control body temperature and/or give a keen sense of smell. The thick, woolly coat is cinnamon-buff above, paler on the underparts, and thickens greatly for winter. The saiga lives on dry steppe and eats varied plants. Smaller breeding groups join to form larger herds for migration. Only males of the species have horns.

MAMMALS

Oreamnos americanus

Mountain goat

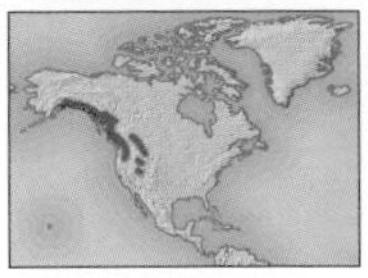

Location W. Canada, N. and W. USA

Length	4–5¼ ft (1.2–1.6 m)
Tail	4–8 in (10–20 cm)
Weight	100–310 lb (46–140 kg)
Social unit	Variable
Status	Least concern

This woolly haired goat survives among ice, snow, rocks, and glaciers. Its large, sharp hooves have hard rims and soft inner pads, to grip slippery surfaces, and its long, yellow-white outer coat and thick, dense underfur conserve body heat. The sharp, back-curved horns are 10 in (25 cm) long, and are slightly thicker in males, which may be 30 percent larger than females. These goats feed on grass, moss, lichens, and twigs. It forms groups of up to 4 in summer, which join to form larger herds in winter.

Rupicapra rupicapra

Alpine chamois

Location S. Europe, W. Asia

Length	3–4¼ ft (0.9–1.3 m)
Tail	1¼–1½ in (3–4 cm)
Weight	53–110 lb (24–50 kg)
Social unit	Individual/Group
Status	Least concern

The Alpine chamois is an agile climber, able to leap 6½ ft (2 m) high, spring 20 ft (6 m) along, and run at 31 mph (50 kph)—its flexible hoof pads giving sure grip on uneven, slippery terrain. Both sexes have slender, black, close-set horns, which are up to 8 in (20 cm) long, and curve back at the tips. It feeds on herbs and flowers in alpine pastures during summer, and moves lower in winter for mosses, lichens, and shoots, as groups disperse.

Capra ibex

Alpine ibex

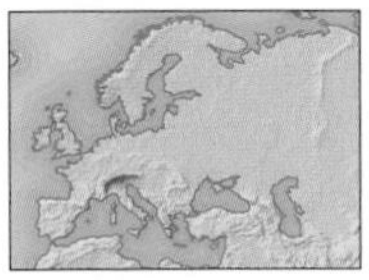

Location S. Europe

Length	4–5½ ft (1.2–1.7 m)
Tail	4–8 in (10–20 cm)
Weight	77–330 lb (35–150 kg)
Social unit	Individual/Group
Status	Least concern

The Alpine ibex dwells at or above the tree line, up to 15,780 ft (4,810 m). The female has a tan coat in summer and the male a rich brown one with yellow-white patches on the back and rump. Both sexes grow a thicker winter coat of more variable color. In spring, they migrate up to alpine pasture, descending in fall to browse buds and shoots. Females and young form stable groups of 10–20; males form single-sex herds.

CLASH OF HORNS

Like many wild (and domestic) goats, male Alpine ibex compete for herd dominance and females by postures, head tosses, and fights. Opponents rear up on their back legs, then lunge forward and clash heads and horns with skull-jarring force.

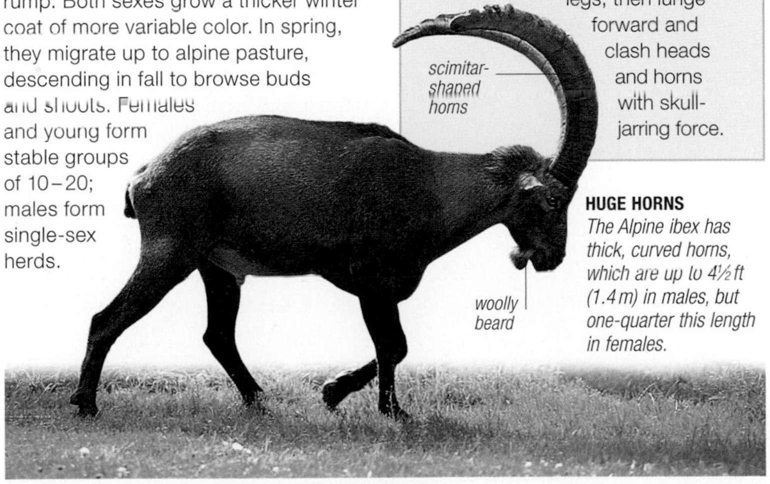

HUGE HORNS
The Alpine ibex has thick, curved horns, which are up to 4½ ft (1.4 m) in males, but one-quarter this length in females.

Hemitragus jemlahicus

Himalayan tahr

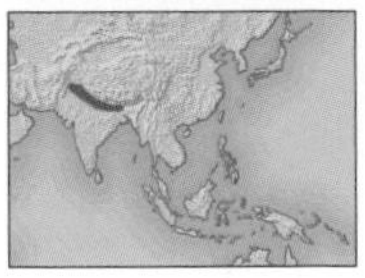

Location S. Asia

Length	3–4½ ft (0.9–1.4 m)
Tail	3½–4¾ in (9–12 cm)
Weight	110–220 lb (50–100 kg)
Social unit	Group
Status	Near threatened

The sure-footed Himalayan tahr has a shaggy, conspicuous mane on its neck and shoulders, extending to its knees; its face and head fur are contrastingly short. The horns, flattened from side to side, reach 16 in (40 cm) in males—twice as long as in females. Like many mountain mammals, the Himalayan tahr migrates high into the mountains in spring, to browse and graze in mixed forests at altitudes of 16,500 ft (5,000 m). It returns to temperate forests as low as 8,250 ft (2,500 m) in fall, when herds of 2–23 females are joined by rutting males, who lock horns and try to topple each other off balance. Young are born the following May or June.

Ovibos moschatus

Muskox

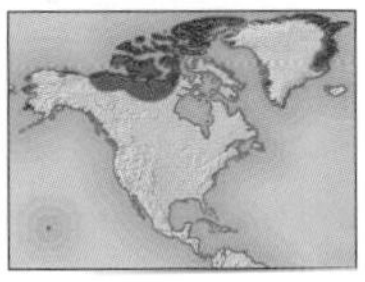

Location N. North America, Greenland

Length	6¼–7½ ft (1.9–2.3 m)
Tail	3½–4 in (9–10 cm)
Weight	440–900 lb (200–410 kg)
Social unit	Group
Status	Least concern

The muskox is named after the strong odor of rutting males, who charge and ram each other for females. Both sexes have broad horns, which nearly meet at a central boss, curving down and then up at the tips. The body is massive; the neck, legs, and tail are short. In summer, the muskox grazes valley sedges and grasses. In winter, it browses on higher ground, where wind keeps the ground clear of snow. Almost exterminated by humans, the muskox has recovered through wildlife management and reintroductions.

DOUBLE COAT
The muskox's outer coat of dark brown guard hairs reaches almost to the ground, to shed rain and snow effectively. The undercoat of fine, soft, pale brown hair provides excellent insulation.

DEFENSIVE CIRCLE

When threatened—for example, from wolves or a polar bear—adult male and female muskoxen gather in a circle, facing outward. The vulnerable youngsters are protected in the middle. Larger herd members may break from the circle to charge or otherwise intimidate the enemy.

MAMMALS

Capra aegagrus

Wild goat

Location W. Asia

Length 4–5¼ ft (1.2–1.6 m)
Tail 6–8 in (15–20 cm)
Weight 55–210 lb (25–95 kg)
Social unit Group
Status Vulnerable

The wild or Bezoar goat, probably the ancestor of domestic goats, grazes and browses in varied habitats, from arid scrub to alpine pasture, at elevations up to 13,800 ft (4,200 m). Females are red-gray to yellow-brown; adult males (a young male is pictured) are bearded and silver-gray with dark markings. Both are horned. Males fight to dominate their bachelor herd and for females.

Capra falconeri

Markhor

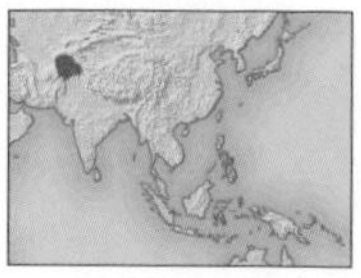

Location C. and S. Asia

Length 4½–6 ft (1.4–1.8 m)
Tail 3¼–5½ in (8–14 cm)
Weight 70–245 lb (32–110 kg)
Social unit Variable
Status Endangered

The markhor occupies various habitats at 2,300–13,200 ft (700–4,000 m), eating tussock grass in summer, and shrubby leaves and twigs on lower slopes in winter, when its short, red-gray coat becomes longer and grayer. The spectacular spiral horns reach 5¼ ft (1.6 m) in the shaggy-throated males, yet only 10 in (25 cm) in the females. The horns are one reason why the markhor is threatened by hunting.

Pseudois nayaur

Greater blue sheep

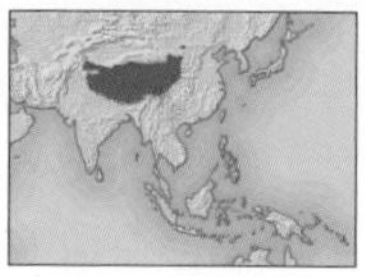

Location S. to E. Asia

Length 4–5½ ft (1.2–1.7 m)
Tail 4–8 in (10–20 cm)
Weight 55–175 lb (25–80 kg)
Social unit Group
Status Least concern

Also called the bharal, the greater blue sheep is camouflaged to survive in rocky, icy alpine zones between tree- and snow-lines. Males are brown-gray, tinged with slate-blue, white below, with a white eyebrow strip, and black flanks and leg stripes. The smooth, 32-in (80-cm) horns splay outward. Females are smaller with shorter horns, and lack most of the black markings. Breeding habits resemble other sheep, with rams competing for harems.

Ammotragus lervia

Barbary sheep

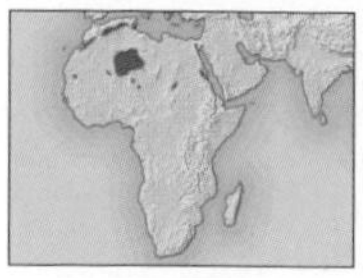

Location N. Africa

Length 4¼–5½ ft (1.3–1.7 m)
Tail 6–10 in (15–25 cm)
Weight 88–320 lb (40–145 kg)
Social unit Individual/Group
Status Vulnerable

The rufous-tawny coat of the Barbary sheep, or aoudad, has a short, upright mane on the neck and shoulders, and a much longer one on the throat, chest, and upper forelegs. Both manes and the crescent-shaped horns, up to 33 in (84 cm) long, are more developed in males. These sheep eat a wide variety of plant matter. Males charge each other, heads lowered, for dominance and access to female groups for breeding.

Ovis ammon

Argali

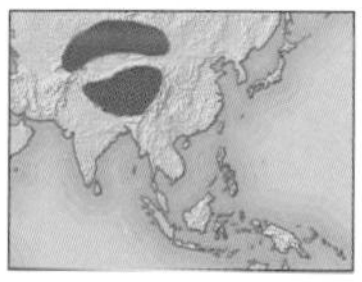

Location C. and S. Asia

Length 4–6 ft (1.2–1.8 m)
Tail 2¾–6 in (7–15 cm)
Weight Up to 440 lb (200 kg)
Social unit Individual/Group
Status Near threatened

Male argali not only charge and clash heads when rutting but also run in parallel and butt the opponent's flank and chest. Also called arkhar, this is the largest wild sheep, with elaborate, ridged horns—up to 5 ft (1.5 m) in males—that corkscrew sideways, twisting 360 degrees or more with age. The coat of this very gregarious bovid is variably pale brown with white legs and rump patch. Like many wild sheep, it is threatened by human hunting, and habitat loss to livestock. Twin young are relatively common.

Ovis gmelini

Anatolian sheep

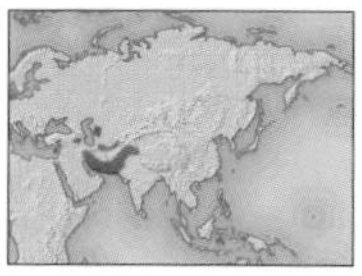

Location W. Asia

Length 3½–4¼ ft (1.1–1.3 m)
Tail 2¾–4¾ in (7–12 cm)
Weight 55–120 lb (25–55 kg)
Social unit Individual/Group
Status Vulnerable

The smallest wild sheep, and probable ancestor of all domestic breeds, the Anatolian sheep frequents uplands and shrubby, grassy plains. The coat is red-brown with a dark central back stripe flanked by a paler "saddle" patch, a short, broad, dark tail, and paler underparts. The curved horns are longer, up to 26 in (65 cm), in males. As in many other wild sheep, females live in small groups with their young, while lone or bachelor-band males compete for dominance and access to females. Success depends on a male's strength in pushing, butting, and ramming, and so most do not begin breeding until 6–7 years old, rutting in late fall.

Ovis canadensis

Bighorn sheep

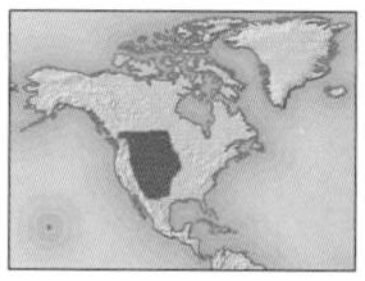

Location S.W. Canada, W. and C. USA, N. Mexico

Length 5–6 ft (1.5–1.8 m)
Tail 2¾–6 in (7–15 cm)
Weight 120–280 lb (55–125 kg)
Social unit Individual/Group
Status Least concern

The bighorn sheep's glossy brown summer coat of brittle guard hairs over crimped gray underfur fades in winter. Before the rut, males display, walk away from each other, turn, advance with a threat jump, and then lunge to head butt with enormous force. This may continue for hours until one gives up. The 1–3 young are born after a gestation period of 150–180 days.

OUT OF REACH

Like many wild sheep, when threatened, bighorn sheep use their gripping hooves and climbing ability to retreat to nearly vertical rocky bluffs and cliffs, where few predators can follow. The young bighorns learn about seasonal pathways and suitable habitats from adults in their group.

BIGGER HORNS
The male's horns curl almost in a circle and may weigh as much as the rest of the skeleton—up to 30 lbs (14 kg). The female's horns are smaller and only slightly curved.

FIGHTING FOR DOMINANCE
During the rutting season, male bighorn sheep fight each other for females with which they will later mate. If neither gives way to the other a fight ensues. The males rear up, as shown here, and then crash their heads together. The weaker individual will eventually be chased away.

Cetaceans

PHYLUM	Chordata
CLASS	Mammalia
ORDER	Cetacea
FAMILIES	11
SPECIES	85

CLASSIFICATION NOTE

This book adopts the traditional method of classifying cetaceans, whereby the order is divided into 2 suborders: baleen whales and toothed whales. It is not certain how they are related to one another, but genetic and morphological evidence suggests the order Cetacea is most closely related to the hippopotamus, which is in the order Artiodactyla.

Baleen whales
see pp.246–9
Toothed whales
see pp.250–261

Cetaceans—whales, dolphins, and porpoises— are perhaps the most specialized of all mammals, with their fish-shaped, hairless body, a flipperlike front limb, and vestigial back limbs (located within the body wall). However, they are true mammals: they breathe air with lungs, and they have mammary glands with which they suckle their young. Cetaceans, which can be divided into baleen whales (such as the humpback whale and the right whale), and toothed whales (such as dolphins and porpoises), are found throughout the world's seas, and some species live in tropical and subtropical rivers. Many species, including the blue whale, have been hunted in such numbers that they are in danger of extinction.

Anatomy

Cetaceans have a hairless, streamlined body to reduce water turbulence. External projections are reduced to the essentials: flippers for steering, a tail with 2 boneless, horizontal flukes (fish have vertical flukes), and usually a dorsal fin for stability. Even the genitals are concealed within folds. Other adaptations to underwater life include a thick layer of blubber (fat and oil) beneath the skin, which conserves body heat, and light, spongy, oil-filled bones. Cetaceans breathe through one (toothed whales) or 2 (baleen whales) blowholes—muscular nostrils usually situated on the top of the head. Toothed whales have a brain that is relatively as large as that of primates, and they are known for their intelligence. Baleen whales have a relatively smaller brain.

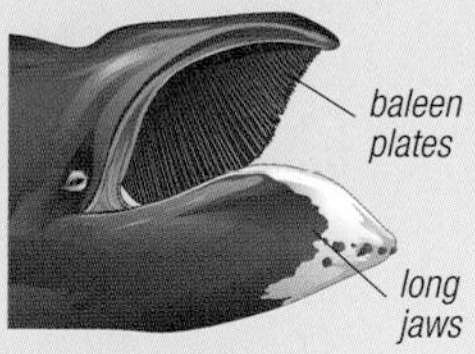

BALEEN WHALE
Instead of teeth, a baleen whale has 130–400 horny (baleen) plates on each side of the upper jaw. The inner edge of each plate has bristles used to sieve food.

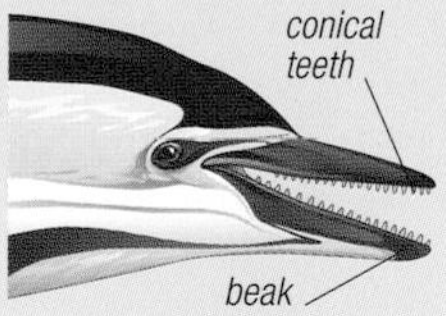

TOOTHED WHALE
The teeth of the toothed whale group are simple, conical, and in most species numerous. The jaws may be extended to form a beak, as seen in dolphins.

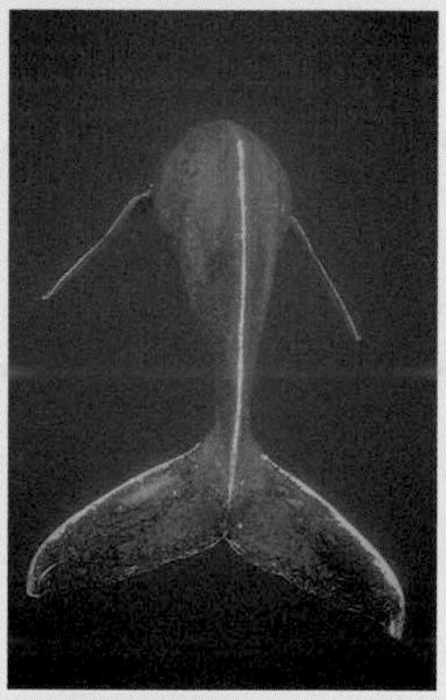

POWERFUL TAIL
In cetaceans, the main propulsive force is the up-and-down movement of the tail. This motion is powerful enough for a whale, such as this humpback, to push two-thirds of its body out of the water when breaching.

Senses

Cetaceans have extraordinarily sensitive hearing. Vision is reasonable—underwater it is excellent up to about 3¼ft (1 m), and in air up to 8¼ft (2.5 m), but color vision is very limited. Some species can focus both eyes ahead, above, or behind them, and some can move their eyes independently. Some freshwater dolphins, however, are nearly or entirely blind. Members of the toothed whale group produce high-frequency clicks for echolocation and can also communicate using a wide range of sounds audible to humans. Other cetaceans employ a variety of vocalizations, but these are not as well studied. Cetaceans have no sense of smell.

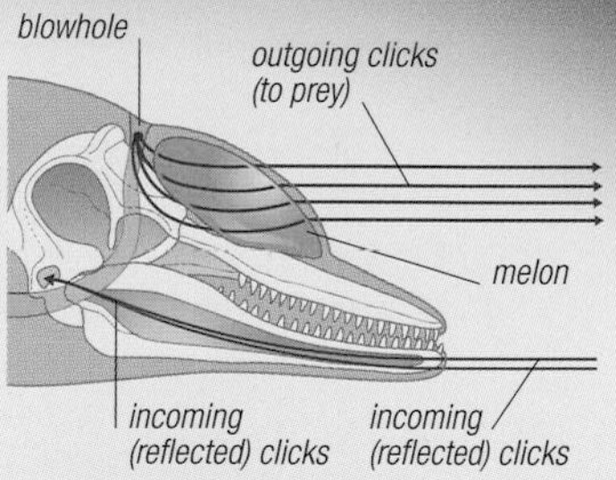

USING ECHOLOCATION
Toothed whales avoid obstacles and catch prey by emitting high-frequency clicks that are reflected off objects in their path. The melon (a fluid-filled swelling) changes shape to focus the clicks. Incoming clicks pass through the jaw.

BREACHING

THE ASCENT
As this southern right whale begins to breach, it surfaces vertically, creating little water turbulence.

THE FLIPPERS APPEAR
The front flippers clear the surface as the whale continues to propel itself upward.

STARTING TO GO DOWN
As the upward surge, which is powered by the tail, is completed, nearly half of the whale is exposed.

ASCENT COMPLETED
As the whale crashes back into the water, it also turns onto its side.

MAMMALS

OCEANIC GIANT
This sperm whale has a typically large, elongated head that is continuous with the long, torpedo-shaped body. Like many cetaceans, sperm whales grow to enormous proportions—in males, up to 59ft (18m) in length and over 55 tons (50 tonnes) in weight.

Reproduction

Some whales, including rorquals (such as the humpback whale), breed during the winter. They migrate from their summer feeding grounds in polar seas to tropical waters (usually around island groups or close inshore), where they give birth and then immediately become pregnant again. In spring, they return to colder waters to feed. Other whales may breed seasonally, but do not migrate to do so. In all cetaceans, copulation is very brief. The male's penis, which is S-shaped, is held inside the body wall. It becomes erect not by filling with blood but by straightening as a result of muscular action. After giving birth, the mother (and in some dolphins, other members of the pod) assists the newborn to the surface to take its first breath.

CALF SUCKLING
Young cetaceans are suckled on milk until they are old enough to eat solid food. The nipples of this Atlantic spotted dolphin are housed in a pouch in the body wall, as is typical of cetaceans.

CONSERVATION

Whales have been hunted by humans throughout history for their meat, bones, and blubber. In the 20th century, with the emergence of intensive hunting and new technology, such as factory ships (above), species numbers declined dramatically. In response, the International Whaling Commission (IWC) banned hunting of certain species in 1966, and these have recovered somewhat. A moratorium on commercial whaling was introduced in 1986. Despite this, several countries continue to whale on scientific grounds, a source of frequent IWC disputes.

Surviving underwater

Even though all cetaceans breathe air, they are able to remain underwater for extended periods, returning to the surface only to exhale ("blow"). In order to remain underwater, the heart rate slows by a half. Also, the water pressure squeezes blood out of the vessels near the skin—the blood can then nourish the vital organs. The water pressure compresses the lungs, which forces air into the trachea and nasal passages where some of the air is absorbed by foamy secretions along the respiratory tract wall. Some toothed whales dive to great depths in search of prey.

SEARCHING THE DEEP
Sperm whales in search of squid dive deeper than other cetaceans. They are known to descend to depths of up to 3,300ft (1,000m) and can remain underwater for 90 minutes.

BLOWING
Cetaceans open their blowholes as they surface and explosively release air and a spray of oil droplets before taking another breath.

EENTERING THE WATER
ıe noise made by a whale striking the surface ay be heard up to 3,300ft (1,000m) away.

GOING UNDER
As the head disappears, only one front flipper and part of the body is visible.

MAKING A SPLASH
The whale's reentry creates a huge splash compared with the initial breach of the water's surface.

THE WHALE DISAPPEARS
With the sequence complete, the whale may breach again. Why whales breach is not clearly understood.

MAMMALS

Baleen whales

PHYLUM	Chordata
CLASS	Mammalia
ORDER	Cetacea
SUBORDER	Mysticeti
FAMILIES	4
SPECIES	14

The most striking feature of baleen whales is their size, which ranges from up to 21 ft (6.5 m) in the pygmy right whale to 110 ft (33 m) in the blue whale. Also characteristic are the baleen plates, which filter prey from the water. The group consists of the gray whale, the rorquals (which include the humpback whale and the blue whale), and right whales. Although commonly found in Antarctic and Arctic regions, they are distributed throughout all the world's oceans, usually in deep water.

Anatomy

All whales in this group have 2 rows of baleen plates (see p.244) that are anchored to either side of the upper jaw. To support these structures, the jaws are elongated. This means that the head, which has 2 blowholes, is large in relation to the body. In right whales, the head comprises up to half the body length, and the jaws are deep (to accommodate long baleen plates). The body is relatively short and stout. Rorquals, on the other hand, have a long, slender body, and shorter baleens.

Feeding

While the gray whale feeds on small crustaceans called amphipods (which are found on the sea bottom), right whales and rorquals eat planktonic crustaceans, which live near the surface. Right whales feed by swimming slowly through dense schools of prey with their mouths open, skimming the plankton out of the water. Rorquals are more active hunters and usually surround their prey. As they feed, the grooves in the throat are relaxed, and the lower jaw becomes a vast sac into which water is gulped. When the water is expelled, the prey remains on the baleen plates. The larger species of rorquals have finer bristles on their baleen plates, which trap crustaceans such as krill and sometimes small fish. Smaller species have coarser bristles to catch larger crustaceans and small fish. The gray whale feeds by stirring up sediments with its relatively short, firm snout, then gulping down its prey, often together with sand, silt, and pebbles.

Migration

Most baleen whale species are migratory. Rorquals, which are known for traveling long distances, feed in the Arctic and Antarctic (where krill is abundant) during the summer. In autumn and winter, they migrate to tropical waters. Here, although they feed little, they give birth and mate again immediately. In spring, they then slowly move back toward higher latitudes, with their suckling calves. Although humpback whales usually migrate along coastlines, other rorquals prefer deeper water. Gray whales migrate farther than any other mammal: approximately 12,450 miles (20,000 km) each year.

Communication

Baleen whales communicate using a variety of sounds, from squeals to rumbles. The most famous whale sounds are the "songs" of male humpbacks. These are produced during winter breeding and consist of a repeated series of high and low notes that are gradually varied as the season progresses. The songs are vital for communication in the vast oceans. Other species employ different sounds; the fin whale, for example, produces a call that is below the range of human hearing and travels huge distances through the ocean.

UNUSUAL FEEDING TECHNIQUE
Baleen whales feed by taking large quantities of small prey. This humpback whale, having circled upward through a school of small fish, is gulping a large mouthful of water and fish. As the mouth closes, the water presses against the rigid upper jaws and flows out over the baleen plates, leaving the fish behind.

Eschrichtius robustus

Gray whale

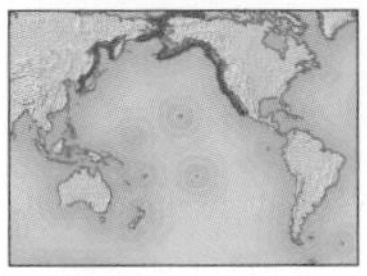

Location North Pacific

Length 43–49 ft (13–15 m)

Weight 15–38½ tons (14–35 tonnes)

Social unit Group

Status Least concern

Gray whales filter-feed like other baleen whales. Unusually, however, they also rely on diving to the shallow sea bed, scooping up huge mouthfuls of mud, and filtering worms, starfish, shrimps, and other small creatures with their short, coarse baleen. This whale's sounds include grunts, wails, moans, and clacking knocks, but the function of these sounds is little understood.

CONSERVATION

Gray whales feed near shore so thay are accessible to watchers, especially in the east Pacific. These whales migrate in groups of up to 10, north to the Arctic for summer feeding, and south to warm-water lagoons to rest and produce calves in winter. They make the longest migrations of any mammal, up to 12,500 miles (20,000 km) yearly. Populations in the east Pacific have risen since legal protection was introduced in 1946. However, west Pacific numbers are still low.

SPY-HOPPING

Many baleen whales, including the gray whale, "spy-hop," becoming vertical in the water with the head well out. They may be detecting other whales, viewing landmarks, or checking water currents for migration. Gray whales swim in a coordinated way, staying in line or arching out of the water together.

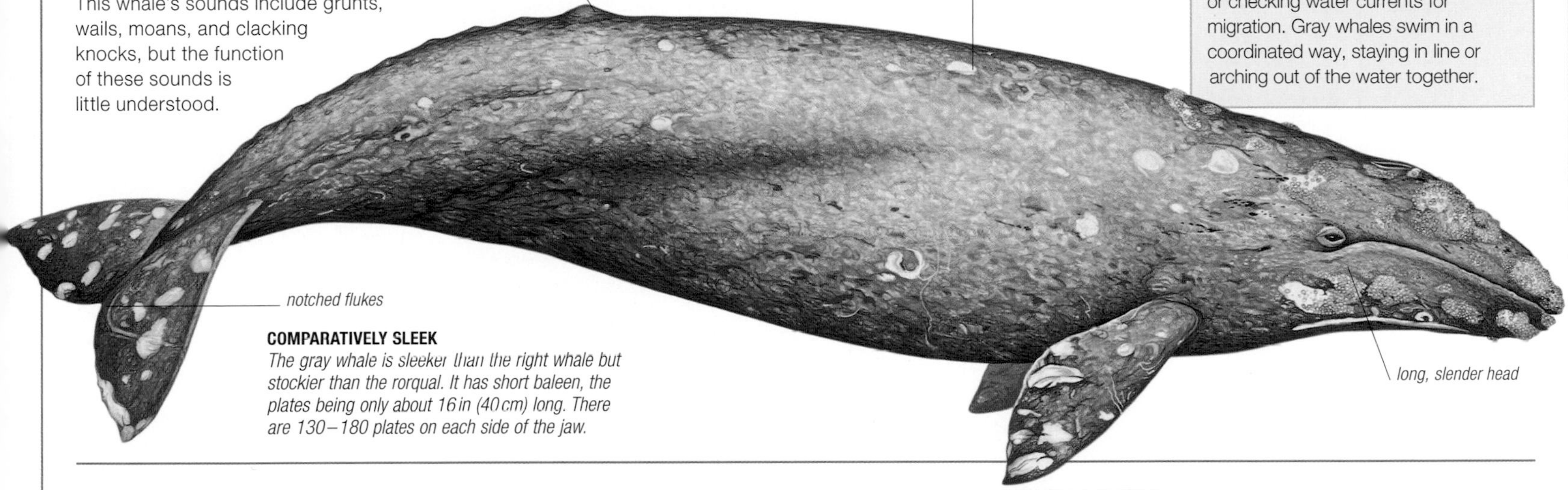

COMPARATIVELY SLEEK
The gray whale is sleeker than the right whale but stockier than the rorqual. It has short baleen, the plates being only about 16 in (40 cm) long. There are 130–180 plates on each side of the jaw.

Eubalaena japonica

North Pacific right whale

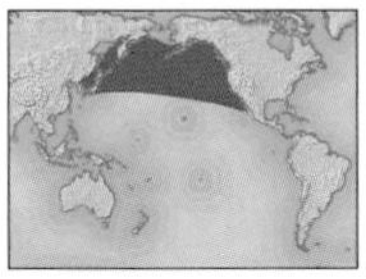

Location Temperate and subpolar waters

Length 43–56 ft (13–17 m)

Weight 44–88 tons (40–80 tonnes)

Social unit Variable

Status Endangered

The most endangered of the large whales, this massive whale is at risk from collisions with ships and from fishing equipment because it swims slowly, dives for only a few minutes, and feeds near the surface. It migrates to the far north or south in summer and returns to warmer, mid-latitude waters in winter, when the female produces one calf, 13–20 ft (4–6 m) long. Twin blowholes give a bushy, V-shaped blow. Sounds include flipper-slaps and breaching.

FILTER FEEDING

The North Pacific right whale feeds alone or in small groups, swimming with mouth open to filter plankton as water rams into its mouth. The narrow baleen plates, mainly blue-black but sometimes white, are up to 9¾ ft (3 m) long. They number 200–270 on each side of the characteristically down-curved upper jaw.

Balaena mysticetus

Bowhead whale

Location Arctic and subarctic waters

Length 46–59 ft (14–18 m)

Weight 55–66 tons (50–60 tonnes)

Social unit Group

Status Least concern

The bowhead whale has a massive head, around one-third of the total weight. The bowhead stays near spreading polar ice all year, enduring weeks of darkness when it may echolate to navigate between and under floes. It surface-skims with its mouth slightly agape. Dives of 5–15 minutes also allow it to feed in midwater or even grub on the bottom. The newborn calf is 13–15 ft (4–4.5 m) long and is suckled for 5–6 months.

SMOOTH SURFACE
The mainly black body is clear of whale lice, barnacles, skin callosities, and similar growths

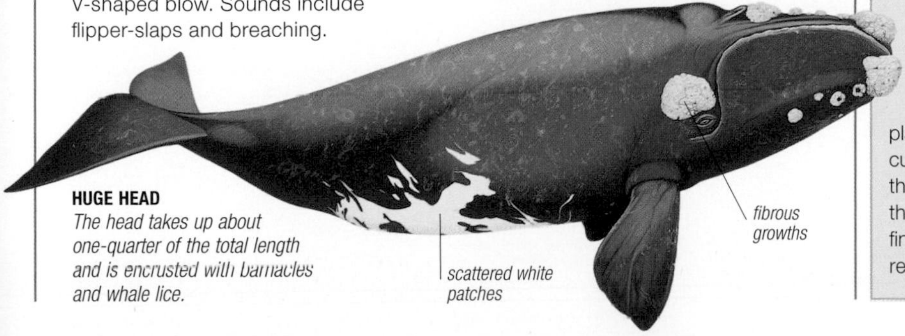

HUGE HEAD
The head takes up about one-quarter of the total length and is encrusted with barnacles and whale lice.

BIGGEST BALEEN

The bowhead has the longest baleen of any whale. The brown- or blue-black baleen plates reach 15 ft (4.6 m) in length, with 240–340 plates on each side of the strongly curved or "bowed" upper jaws (hence the name bowhead). This whale has no throat grooves and no dorsal fin either, giving its body surface a remarkably uncluttered appearance.

Balaenoptera physalus

Fin whale

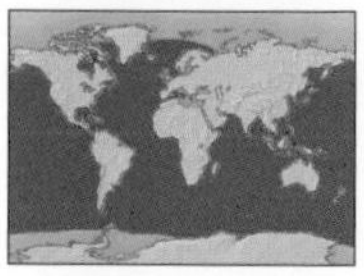

Length 62–72 ft (19–22 m)
Weight 49½–82½ tons (45–75 tonnes)
Social unit Variable
Status Endangered
Location Worldwide (except E. Mediterranean, Baltic, Red Sea, Arabian Gulf)

The fin whale is the second-largest whale, and one of the fastest. In addition to some hums and squeals, it produces an immensely loud, deep moan that can be heard hundreds of miles away. Like other great whales, the fin whale undertakes long migrations from high latitudes in summer, where it feeds on fish and krill, toward tropical regions for winter and breeding. The gestation period is 11 months, the newborn calf 21 ft (6.4 m) long, and weaning takes place at 9–10 months. This means each female has only one offspring every 2 years, so low populations take decades to recover.

FIN, FLIPPERS, AND FLUKES
The fin whale's back, flippers, and flukes are gray. The dorsal fin, set two-thirds of the way along the back, has a concave rear edge.

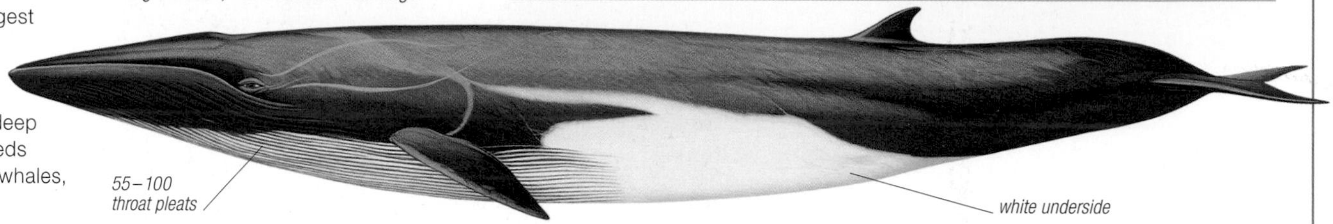

RIGHT-SIDED FEEDING

Fin whales carry out high-speed lunge-feeding, on krill and fish such as capelin or herring. The whales synchronize attacks and take in huge volumes of water, close the mouth, and force out the water to trap fish on the baleen. The whale swims on its right side, which may be why the left part of the mouth is black but the right is white. Such asymmetry in color is very unusual in mammals.

Balaenoptera musculus

Blue whale

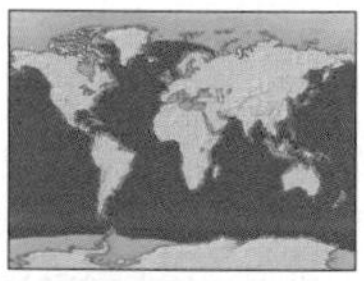

Length 66–98 ft (20–30 m)
Weight 110–176 tons (100–160 tonnes)
Social unit Individual
Status Endangered
Location Worldwide (except Mediterranean, Baltic, Red Sea, Arabian Gulf)

The biggest animal on the planet, the blue whale can consume more than 6 5/8 tons (6 tonnes) of small euphausiid crustaceans, of which its diet almost exclusively consists. It lunges into a school of prey and its throat swells to 4 times the normal width. It closes its mouth, expels the water, and swallows the thousands of food items retained by the baleen. Feeding occurs mainly in summer, in and near rich polar waters. The blue whale is thought to migrate to warmer, lower latitudes for winter, when the females give birth. The calf is 23 ft (7 m) long, 2 3/4 tons (2.5 tonnes) in weight, and is suckled for 6–8 months. Blue whales are usually solitary or in mother–calf pairs, although they may gather as loose groups to feed. They make grunts, hums, and moans, which at volumes greater than 180 decibels are the loudest of any creature sounds.

STREAMLINED GIANT
The blue whale has a slim outline, especially in winter, although it fattens in summer. The tiny dorsal fin is set well to the rear, near the tail. Its coloration is mainly pale blue-gray.

ANATOMY OF A DIVE

When making a deep dive, the whale "headstands," exposing its distinctively wide tail flukes, then descends steeply to a maximum depth of some 655 ft (200 m). The long, narrow flippers play no part in propulsion, the thrust coming from the powerful back muscles that swish the rear body and flukes up and down.

Balaenoptera omurai

Omura's whale

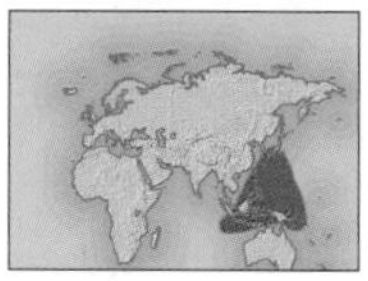

Length 3½–37¾ in (9.6–11.5 m)
Weight 22 tons (20 tonnes)
Social Not known
Status Data deficient
Location Pacific Ocean

Previously considered a pygmy form of Bryde's whale, this new species was recognized by Japanese researchers from studies of material collected by research whaling ships in the Pacific in the 1970s and 1990s, and was subsequently determined to be a more primitive baleen lineage. It is known from only 9 specimens—one a stranding, and others caught in deep waters—between Japan, New Guinea, and Indonesia, but the full extent of its range is not yet well known. An estimate of global population size (1,800 individuals) is likely to be unreliable. Studies on stomach contents of these whales suggest that—like other baleen whales—they feed primarily on krill.

Balaenoptera edeni

Bryde's whale

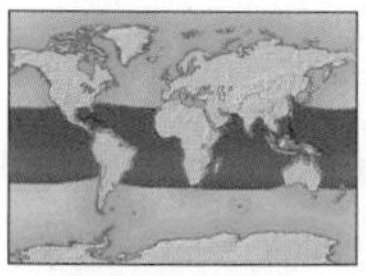

Length 30–39 ft (9–12 m)
Weight 17½–27½ tons (16–25 tonnes)
Social unit Variable
Status Data deficient
Location Tropical and warm temperate waters worldwide

One of 2 similar species with this name (pronounced "Broodah's"), Bryde's whale is mainly coastal in the Eastern Indian and West Pacific oceans. The larger, more offshore Bryde's whale (*Balaenoptera brydei*) is found in the Atlantic and Indian oceans, and parts of the North, East, and South Pacific. Apart from their size difference, both species have blue-gray coloration that is paler on the underside, 40–70 throat grooves, coarse-bristled baleen plates, and a small dorsal fin set two-thirds of the way along the stocky body. These whales dive for up to 20 minutes to feed mainly on schooling fish and krill. They usually live alone or in small, loose herds, but may gather in dozens where prey is plentiful. Both species of Bryde's whale are rapid swimmers, changing speed and direction frequently. They surface steeply to reveal the head first, arch over, and expose the rear of the body but not the tail flukes as they dive. The female reaches sexual maturity at 8–11 years and produces a single calf after a gestation of 12 months.

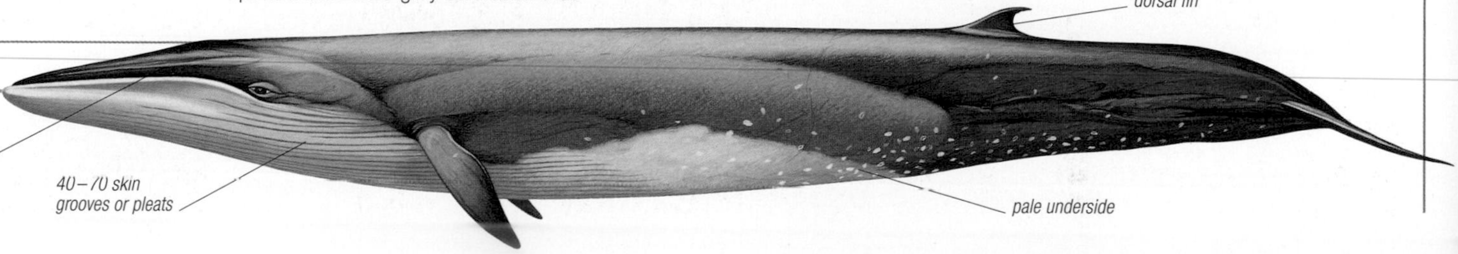

MAMMALS

Balaenoptera borealis

Sei whale

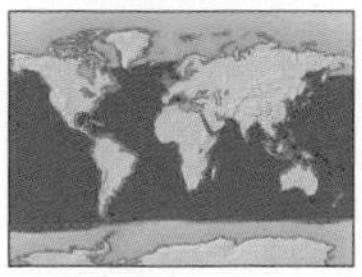

Length 45–52 ft (14–16 m)
Weight 22–27½ tons (20–25 tonnes)
Social unit Group
Status Endangered

Location Worldwide (except Mediterranean, Baltic, Red Sea, Arabian Gulf)

Sei whales frequent more temperate waters (46–77° F/8–25° C) than their more widespread close cousins, the blue and fin whales. The sei whale is slender with a long, slim, pointed head and slightly downcurved jaw line. Its upper surfaces are dark gray, with a relatively sharp transition to the white or pale gray underside. There are 320–340 close-spaced, delicate-fringed baleen plates hanging from each side of the upper jaw. These enable sei whales to take a variety of plankton, from copepods (water-flea-like crustaceans) less than ⅜ in (1 cm) across to schooling fish and squid 12 in (30 cm) long. Sei whales are usually found in schools of 2–5 individuals. They rarely dive deeper than 1,000 ft (300 m), staying underwater for up to 20 minutes. Most births among sei whales are single, although twins occur occasionally.

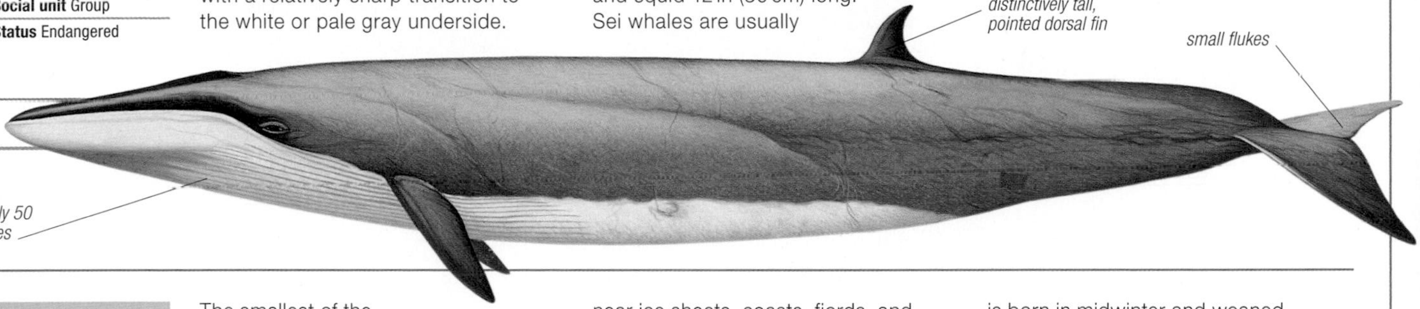

Balaenoptera acutorostrata

Minke whale

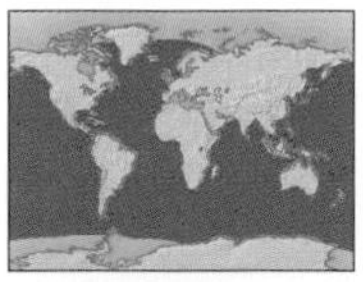

Length 26–33 ft (8–10 m)
Weight 8⅞–14¼ tons (8–13 tonnes)
Social unit Individual
Status Least concern

Location Worldwide (except E. Mediterranean)

The smallest of the rorqual whales, the minke has small baleen plates, 12 in (30 cm) long, with 230–360 in each side of the upper jaw. It has no universal migration pattern and is found in a variety of waters, including open ocean and near ice sheets, coasts, fjords, and estuaries. It feeds alone by gulping in tight-knit swarms of krill or fish. Yet it is not shy, and will approach stationary boats. The 9¾-ft (3-m) calf is born in midwinter and weaned after about 4 months. A smaller minke whale was recently recognized as a separate species—the Antarctic minke (*Balaenoptera bonaerensis*).

SMOKY PATTERNS
The minke has gray, "smoky" patches where the black back meets the white belly. The white band on the flipper may extend to the chest.

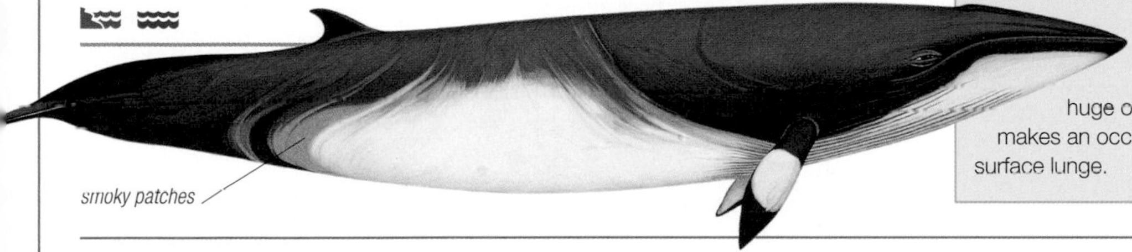

DOLPHINLIKE WHALE

The minke has a dolphinlike shape, with a sharp snout, pointed head, and a ridge from the snout up to the paired blowholes, which are set well back on the forehead. It is a speedy, agile swimmer, able to "surf" huge ocean breakers, and makes an occasional spectacular surface lunge.

Megaptera novaeangliae

Humpback whale

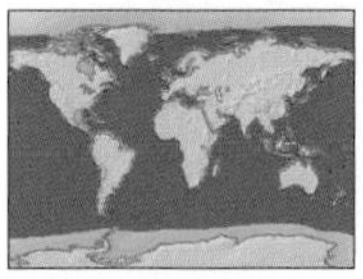

Length 43–45 ft (13–14 m)
Weight 27½–33 tons (25–30 tonnes)
Social unit Variable
Status Least concern

Location Worldwide (except Mediterranean, Baltic, Red Sea, Arabian Gulf)

The humpback whale is dark blue-black above, with paler or white patches below. Color variations, particularly on the underside of the tail, can be used to identify individual humpbacks, as can the knobbly leading edges to the longest flippers of any animal. In spite of its size, it is graceful and athletic, being able to leap out of the water (see below). The humpback whale migrates from cold, food-rich summer waters near the poles to warmer, coastal shallows at lower latitudes for winter calving. Pregnant females spend longest in the feeding areas. It has a wide array of food-gathering methods and small groups cooperate in blowing "curtains" or "cylinders" of underwater bubbles to herd fish. An extremely vocal whale, it also emits numerous sounds, perhaps partly to coordinate its feeding. In winter calving areas, solitary males produce a long, complex series of sounds of amazing variety (see panel below).

BODY AND FINS
The humpback whale has a dumpy body and a fatty pad at the base of the dorsal fin, which is variable in shape, from almost flat to tall and triangular. The span of the tail flukes is about one-third the length of the head and body.

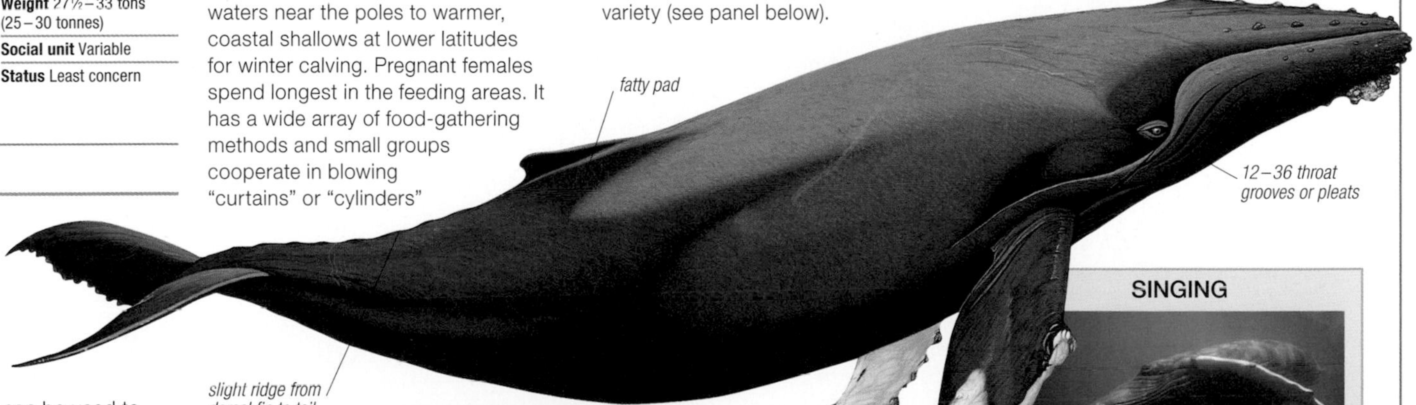

BREACHING

The humpback whale generates sufficient upward force with its tail to lift almost all of its 20-plus tons out of the water. It then twists in the air and falls on its back with a tremendous splash. It is not clear why whales "breach," as such a movement out of the water is called. It may be to create massive sound waves or to ease irritation from skin parasites.

SINGING

The song of the male humpback whale develops from year to year. Lasting up to 30 minutes, it may attract females, warn off males, or be a form of sonar to detect other whales. Above, a male humpback moves into a singing posture, hanging vertically with head downward some 33–130 ft (10–40 m) below the surface.

MAMMALS

Toothed whales

PHYLUM	Chordata
CLASS	Mammalia
ORDER	Cetacea
SUBORDER	Odontoceti
FAMILIES	7
SPECIES	71

A much more diverse group than baleen whales, toothed whales make up almost 90 percent of all cetaceans. The group consists of dolphins (which include the killer whale), river dolphins, porpoises, white whales, sperm whales, and beaked whales. Most are medium-sized—although the sperm whale grows to up to 59 ft (18 m)—and all possess teeth instead of baleen plates. On the forehead, there is a fluid-filled swelling called the melon, in front of which there is usually a beak. Most species are found around the world in deep water and coastal shallows, although a few live in freshwater. Some toothed whales migrate, but only the sperm whale travels long distances.

Anatomy

Toothed whales have simple, conical, pointed teeth that are not divided into incisors, canines, premolars, and molars (as in most other mammals). Each tooth has a single root and is either straight or slightly curved. One set of teeth lasts the whale's lifetime. The number of teeth present varies from more than 40 pairs in each jaw in some dolphins, to a single pair (in the lower jaw) in beaked whales. Unlike baleen whales, toothed whales have only one blowhole, which means that the skull is asymmetrical. The single blowhole usually opens at the top of the head (except in the sperm whale, where the opening is at the front of the skull). Another feature of many toothed whales is the streamlined head and the long, narrow beak seen in most species.

Feeding

While baleen whales trap their prey en masse, toothed whales capture victims individually. The even, conical teeth found in this group are perfect for catching slippery fish, which form the diet of most species. Sperm whales, however, mostly eat squid (but also other prey, such as octopus), and beaked whales have a mixed diet of fish and squid. Killer whales eat other whales (which they attack in packs), fish, and seals, which they sometimes seize from land, using waves to slide onto the beach. They also upset pack ice to knock prey into the water.

Fish-eating species (including killer whales) usually have numerous teeth to hold wriggling prey, while those that eat squid or octopus, for example, have fewer teeth. Sperm whales have functional teeth only in the lower jaw (which is very narrow), and prey is held firmly between these teeth and the roughened palate in the upper jaw.

All toothed whales use echolocation (see p.244) to help find prey (and to avoid objects in their path).

Social behavior

Most toothed whales live in groups called pods or schools, which vary in size from less than 10 to over 1,000 (as seen in some dolphin species). The exact organization of these associations is poorly understood, although it is thought that subgroups form to perform independent tasks, such as feeding. This suggests the presence of complex social structures. Some species, especially killer whales, appear to practice cooperative hunting, whereby prey is "herded" until trapped or surrounded. When swimming in formation, dolphins often leap out of the water, which may simply be playful behavior, or it may function as a form of communication.

STREAMLINED FOR SPEED
Atlantic spotted dolphins are fast, energetic swimmers, and, like several other toothed whale species, they often travel in large pods. The torpedo-shaped body is propelled through the water by powerful thrusts of the tail. The mottled coloration camouflages them from predators and prey in shallow, sunlit waters.

Platanista gangetica

Ganges river dolphin

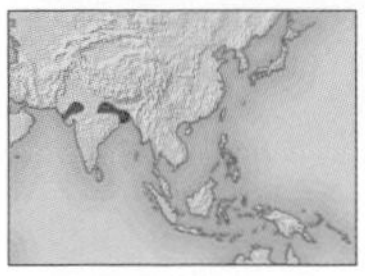

Location S. Asia (Indus and Ganges-Brahmaputra river systems)

Length 7–8¼ ft (2.1–2.5 m)
Weight 185 lb (85 kg)
Social unit Variable
Status Endangered

Also known by the local names susu and bhulan, this exclusively freshwater species has distinctively broad flippers and a long, narrow beak armed with 26–39 pairs of upper, and 26–35 pairs of lower, sharp teeth. The front teeth extend outside the beak's tip, to form a catching cage for fish, shellfish, and similar prey on or near the bottom. The flexible neck allows the head to bend at right angles as the dolphin grubs in the mud or "scans" the area with echolocating sound pulses. The Ganges river dolphin lives in small groups, usually 4–6 but occasionally up to 30. However, its social life and breeding habits are largely unknown. There are 2 subspecies: *Platanista gangetica minor* in the Indus river and its tributaries, and *Platanista gangetica gangetica* in the Ganges-Brahmaputra river system. Both are extremely rare and face numerous threats from humans.

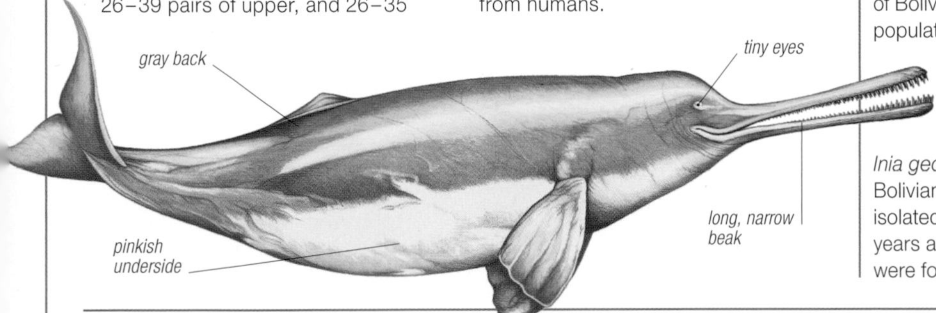

Inia boliviensis

Bolivian river dolphin

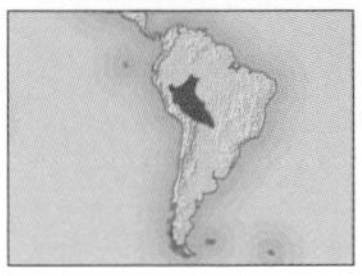

Location C. South America

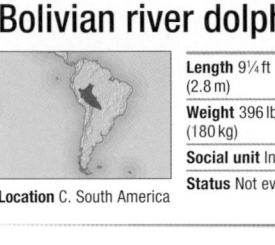

Length 9¼ ft (2.8 m)
Weight 396 lb (180 kg)
Social unit Individual/Pair
Status Not evaluated

Once recognized as a smaller, gray subspecies of the pink Amazon river dolphin, this dolphin is now considered to be a distinct species. It is found in the Beni, Mamoré, and Guaporé rivers of Bolivia—and isolated from other *Inia* populations by 240 miles (400 km) of falls in the Madeira river. Various studies have demonstrated that it is genetically distinct from its more widespread cousin *Inia geoffrensis*. They also indicate that Bolivian river dolphins probably became isolated in the Beni region about 5 million years ago—when Andean mountains were forming and before Madeira's waterfalls arose. Like the Amazon river dolphin, the Bolivian river dolphin preys on bottom-dwelling fishes while swimming, which it snatches using a long beak—helped by a flexible neck. It is more abundant in lagoons and confluences, where prey becomes concentrated by eddies and whirlpools and so is more easily trapped. These dolphins usually travel alone or in pairs. They are less likely to be seen during high water season, perhaps because they then can disperse into flooded forest.

Inia geoffrensis

Amazon river dolphin

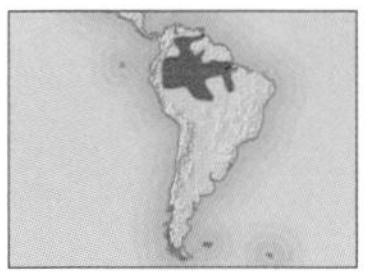

Location South America (Amazon and Orinoco basins)

Length 6½–8½ ft (2–2.6 m)
Weight 220–350 lb (100–160 kg)
Social unit Variable
Status Data deficient

Also called the boto or boutu, this species ranges through the Amazon and Orinoco river systems. It possesses a long, slim beak and flexible neck, for poking in mud for prey during short dives of 1–2 minutes. Being small-eyed and living in murky water, it probably finds its way and food mainly by echolocating sound pulses. The Amazon river dolphin has 25–35 pairs of teeth in both upper and lower jaws. At the front, the teeth are peglike, for seizing prey, but at the rear, they are flatter with peaks or cusps, and thus suited to crushing freshwater crabs, river turtles, and armored catfish. The single calf, 32 in (80 cm) long, is born between May and July.

CAUTIOUS APPROACH

Known for its slow, apparently lethargic lifestyle, the Amazon river dolphin usually lives alone or in twos, occasionally in groups of up to 20. In some areas, it may approach boats or swimmers out of curiosity, but in regions where it is hunted, it has become more cautious.

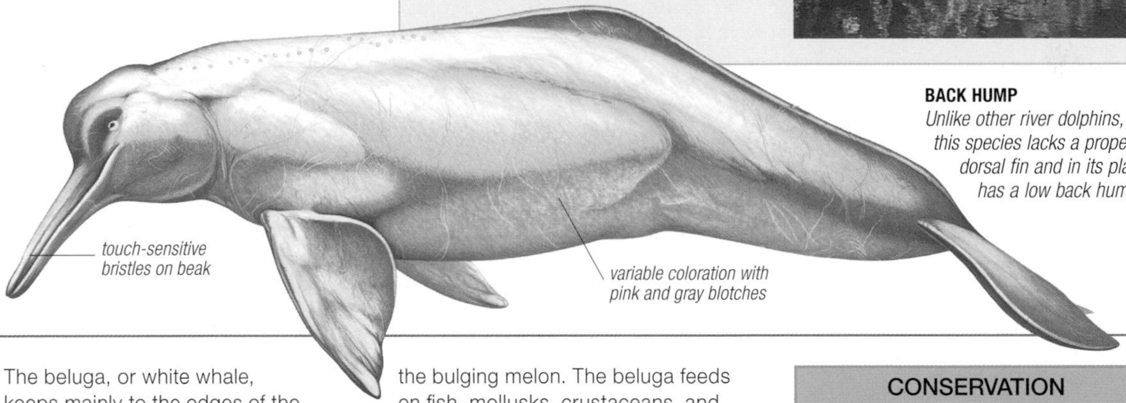

BACK HUMP
Unlike other river dolphins, this species lacks a proper dorsal fin and in its place has a low back hump.

Delphinapterus leucas

Beluga

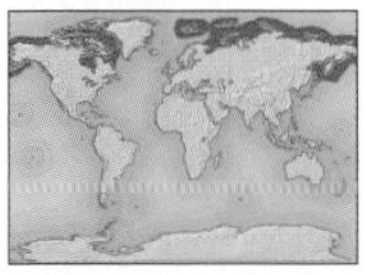

Location Arctic Ocean

Length 13–18 ft (4–5.5 m)
Weight 1⅛–1⅝ tons (1–1.5 tonnes)
Social unit Group
Status Near threatened

The beluga, or white whale, keeps mainly to the edges of the Arctic ice-fields and has been radio-tracked diving to 985 ft (300 m), presumably navigating by echolocation to find prey and breathing holes. This vocal whale emits varied calls, including squeaks, whistles, mews, clicks, and hums. These are audible through the hulls of boats, giving the beluga its nickname of sea canary. The outgoing echolocation sounds are focused by the bulging melon. The beluga feeds on fish, mollusks, crustaceans, and other prey, which it crushes with its 8–11 pairs of teeth in the upper jaw and 8–9 pairs below. The calf is dark gray and 5 ft (1.5 m) long at birth. It is paler by 2 years old, and blue-tinged white by the age of 5.

ALL-WHITE WHALE
The only all-white cetacean, the adult beluga blends with the Arctic ice floes and icebergs. Its skin may be tinged yellow before the summer molt.

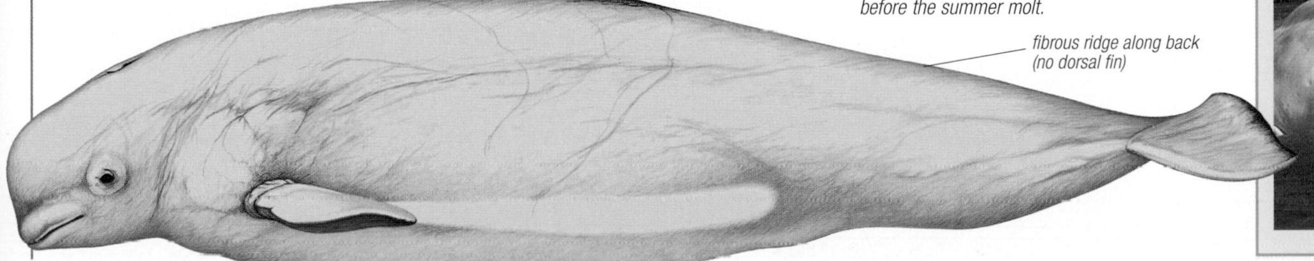

CONSERVATION

As global warming opens up Arctic sea routes, belugas are increasingly threatened by shipping and oil exploration, both of which can cause pollution. Belugas are also targeted by subsistence hunters, but commercial hunting is banned by conservation treaties.

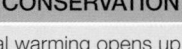

MAMMALS

Monodon monoceros

Narwhal

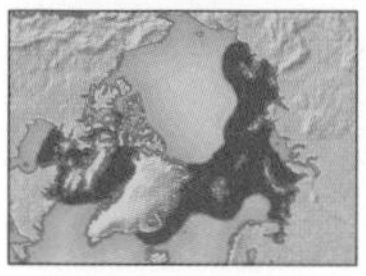

Length 13–15ft (4–4.5m)
Weight ⅞–1¾ tons (0.8–1.6 tonnes)
Social unit Group
Status Near threatened

Location Arctic Ocean

The narwhal has the most northerly range of any mammal, among ice-fields and floes of Arctic waters. It has one ever-growing tooth, the upper-left incisor, which forms a long tusk (see panel, right). Fish, mollusks, crustaceans, and other prey are probably sucked into the mouth by the narwhal's powerful lips and tongue. Like belugas—indeed, often with them—narwhals can sometimes form large schools of thousands, which may be segregated by age and sex. They communicate with a wide variety of sounds, including humlike tones that may be for individual recognition.

COLORATION

The narwhal (female pictured here) is speckled gray and black on a pale background, with more patches on the back merging into larger areas of dark gray.

MALE NARWHAL'S TUSK

The narwhal's tusk grows with age, through the upper lip, spiraling clockwise, to reach 9¾ft (3m) long. It may be a swordlike weapon to "fence" rival males at breeding time. Any use in feeding seems unlikely, since most females lack tusks, but survive well. An occasional male has 2 tusks, or a female has one.

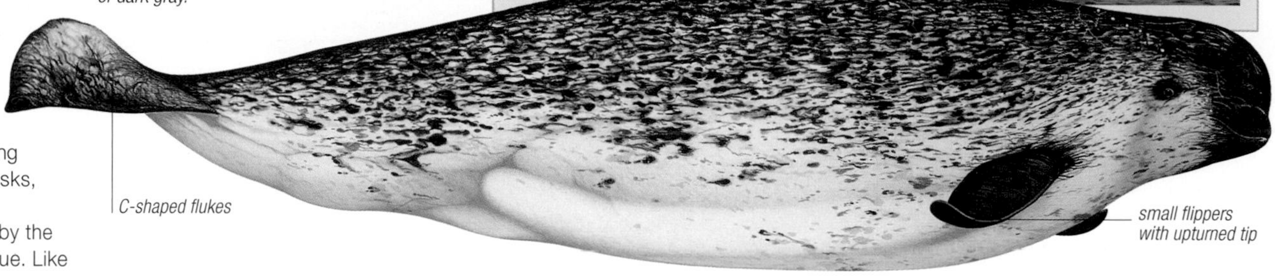

Phocoena phocoena

Harbor porpoise

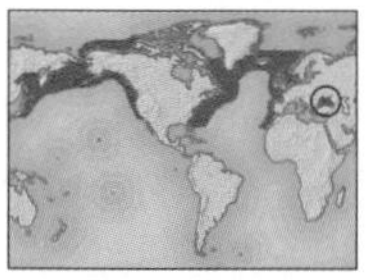

Length ½–6½ft (41.4–2m)
Weight 110–200lb (50–90kg)
Social unit Variable
Status Least concern

Location North Pacific, North Atlantic, Black Sea

The most numerous cetacean over much of its range, the harbor porpoise has nevertheless suffered, like other inshore sea creatures, as a result of human activity. A major danger is being snared in underwater fishing nets when, as an air-breather, it asphyxiates. Apart from humans, its main predators are killer whales, bottlenose dolphins, and large sharks. The harbor porpoise generally forages alone, in waters down to 655ft (200m), for seabed dwellers such as fish and shellfish. It uses very high-pitched echolocation and grips prey with its spade-shaped teeth, numbering 22–28 pairs in the upper jaw and 21–25 in the lower. Groups usually form because prey such as schooling fish are particularly rich in small areas. The single calf is born in early summer and cared for by the mother for up to 12 months.

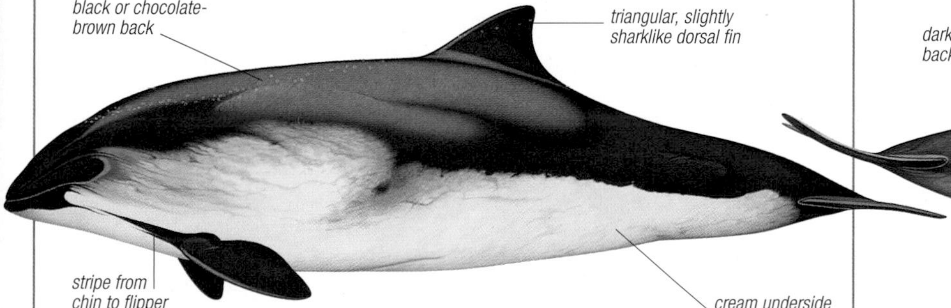

Phocoena sinus

Vaquita

Length 5ft (1.5m)
Weight 105lb (48kg)
Social unit Variable
Status Critically endangered

Location Gulf of California

This small porpoise's very restricted range, in shallow waters less than 130ft (40m) deep at the northern end of the Gulf of California, makes it one of the most vulnerable of all sea mammals. Also called the Gulf of California porpoise, its habits are little known. It is often solitary but may also form small groups of up to 7 individuals. It takes a mixed diet of small fish, squid, and other prey on or near the sea bed, and uses high-frequency clicks for echolocation. The main body color is gray, darker above than below and also around the eyes and mouth. The calf is only 28–32in (70–80cm) long at birth and is suckled by the mother for several months. Such threats as entanglement in fishing nets (leading to over 30 deaths a year), pollution, boat noise, and oil exploration mean that the vaquita's future is bleak.

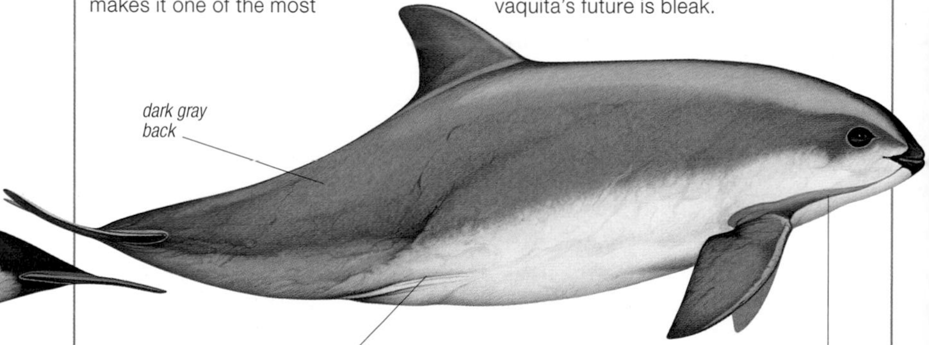

Neophocaena phocaenoides

Finless porpoise

Length 5–6½ft (1.5–2m)
Weight 160lb (72kg)
Social unit Variable
Status Vulnerable

Location Indian Ocean, W. Pacific

Lack of a dorsal fin makes this medium-sized porpoise difficult to identify, or even see, because it surfaces briefly and rolls gently while taking breaths. The vaguely dolphinlike bulging forehead and a slightly beaked snout make it distinctive among the 6 porpoise species. The finless porpoise frequents coastal waters around the Indian and West Pacific oceans, including estuaries and up rivers. Like other porpoises (and unlike dolphins), it leaps from the water only rarely, yet it may "spyhop" in the manner of whales, holding its body vertically, half out of the water, as though standing up to look around. It feeds alone or in small groups of 3–5, occasionally 10 or more. Small fish, mollusks, and crustaceans on or near the seabed are caught using the 13–22 pairs of spade-shaped teeth in both upper and lower jaws. They sometimes follow prey on small-scale, seasonal migrations.

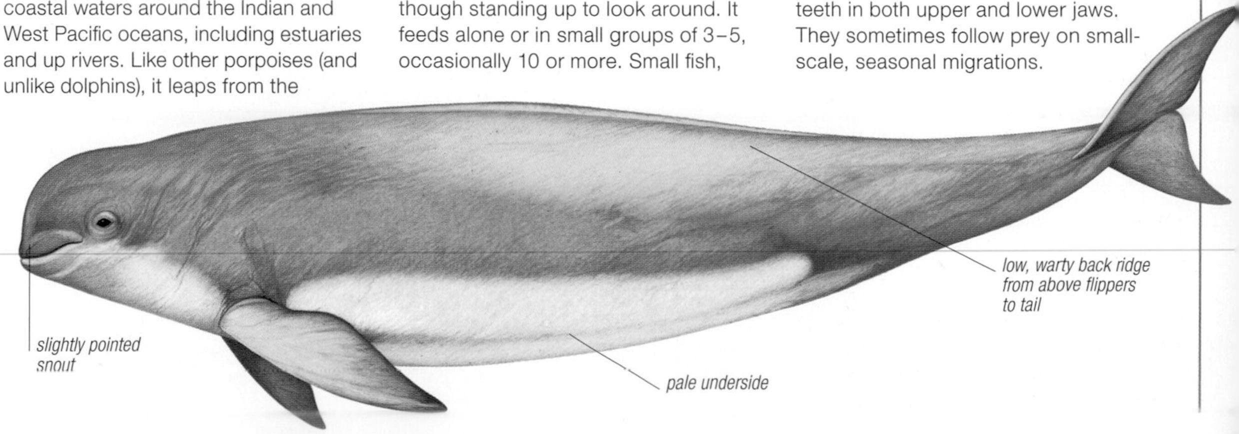

Phocoenoides dalli

Dall's porpoise

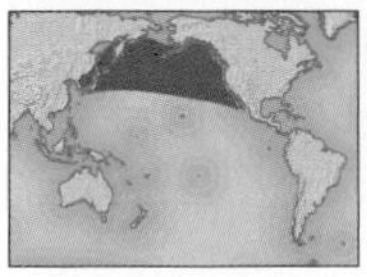

Length 7¼–7¾ ft (2.2–2.4 m)
Weight 380–440 lb (170–200 kg)
Social unit Group
Status Least concern

Location North Pacific

This large, robust-bodied, rapid-swimming porpoise is mainly black apart from a white patch along each flank and perhaps white tips to the flukes and dorsal fin. The head and flippers are small compared to the body, and there are 23–28 pairs of teeth in the upper jaw and 24–28 in (61–71 cm) the lower jaw. Small schools of Dall's porpoises sometimes merge to form vast groups numbering thousands. They make a variety of clicking sounds, and feed on fish and squid—from pilchards at the surface to lanternfish at middle depth. They can swim up to about 34 mph (55 kph).

Sousa teuszii

Atlantic humpback dolphin

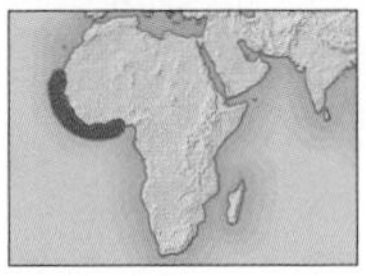

Length 7¾–9¼ ft (2.4–2.8 m)
Weight 620 lb (280 kg)
Social unit Group
Status Vulnerable

Location E. Atlantic

The Atlantic humpback dolphin inhabits shallow coasts, reefs, mangrove swamps, estuaries, and rivers. It swims slowly and feeds mainly on schooling fish. It has 27–38 pairs of stubby teeth in each jaw. This dolphin forms schools of up to 25 and also associates with humans, following shrimp boats for disturbed fish or herding fish toward shore-based nets in return for a share of the catch. It is very similar to *Sousa plumbea*, which frequents the Indian Ocean, and *Sousa chinensis* in the coastal West Pacific. The Atlantic humpback dolphin, however, is slate gray with pale underparts. Those in the Indian Ocean are larger and darker, perhaps with blue-black flecks. Individuals around China are smallest, almost pink with gray flecks around the head and eyes, and the dorsal fin lacks the inward curve on its trailing edge.

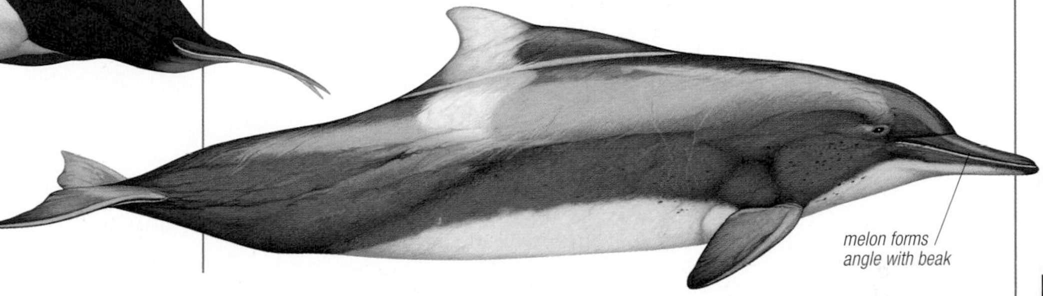

Sotalia fluviatilis

Tucuxi

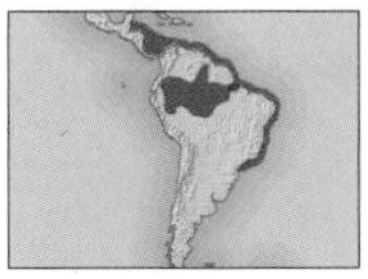

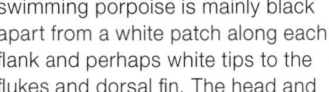

Length 4¼–6½ ft (1.3–2 m)
Weight 77–88 lb (35–40 kg)
Social unit Variable
Status Data deficient

Location Central America, N.E. South America

The tucuxi (pronounced "too-koo-shee") is one of the smallest dolphins, stocky with large flippers, flukes, and dorsal fin. It occurs as a marine form around the coasts and river estuaries of northeast South America, and as a river form in the lower reaches and lakes of the Amazon system. It is a different species to the solely freshwater Amazon river dolphin (see p.251). The tucuxi lives alone or in twos, although larger schools occur, up to 10 in rivers and 30 around the coast. It may leap, somersault, and surf waves—but the reason for this is not clear. Prey up to 14 in (35 cm) long is swallowed whole, including fish such as anchovies and catfish, and squid. In some places, local tradition forbids hunting these dolphins. However, many are accidentally asphyxiated in fishing nets, and some are deliberately killed for meat or fishing bait; the eyes and other parts have been valued as love charms.

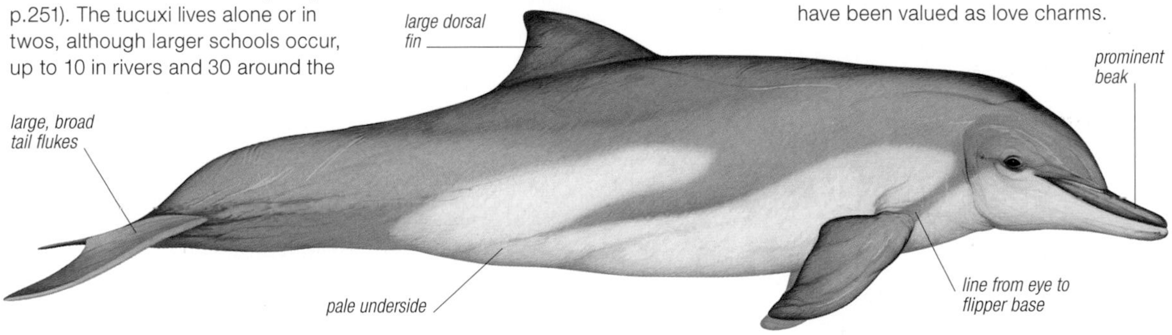

Lagenorhynchus obscurus

Dusky dolphin

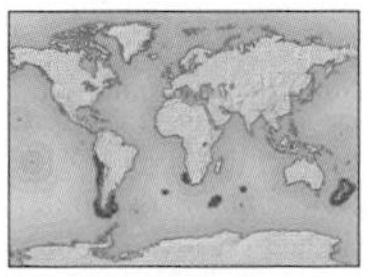

Length 5½–7 ft (1.7–2.1 m)
Weight 155–185 lb (70–85 kg)
Social unit Variable
Status Data deficient

Location S. Atlantic, Indian Ocean, S. Pacific

The dusky dolphin is mainly dark gray to blue-black on its upperside, and pale gray or white on the underside. These areas are separated by a tapering gray stripe from the face along the flank to near the tail base. The head has a smooth profile that widens gradually from beak to blowhole. The species occurs as 3 subspecies based respectively around South America (*Lagenorhynchus obscurus fitzroyi*), southern Africa (*Lagenorhynchus obscurus obscurus*), and New Zealand (unnamed). It prefers waters of 50–64°F (10–18°C) and less than 655 ft (200 m) deep. Off South America, it feeds by day on schooling fish, such as anchovies, and squid; off New Zealand, feeding is mainly by night at middle depths. Schools of dusky dolphins change rapidly in size and composition, varying in number from 2 up to 1,000. Groups often engage in much leaping, chasing, and rubbing.

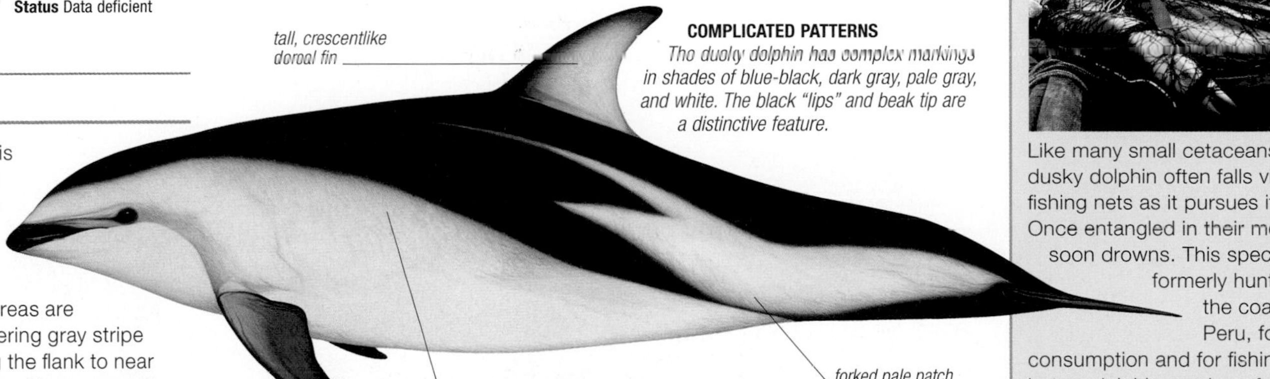

COMPLICATED PATTERNS
The dusky dolphin has complex markings in shades of blue-black, dark gray, pale gray, and white. The black "lips" and beak tip are a distinctive feature.

CONSERVATION

Like many small cetaceans, the dusky dolphin often falls victim to fishing nets as it pursues its prey. Once entangled in their mesh, it soon drowns. This species was formerly hunted off the coast of Peru, for human consumption and for fishing bait, but as dolphin numbers fell, this practice was banned in 1996.

MAMMALS

Lagenorhynchus obliquidens

Pacific white-sided dolphin

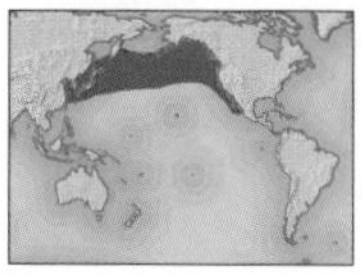

Location North Pacific

Length 7–8¼ ft (2.1–2.5 m)
Weight 65–200 lb (175–90 kg)
Social unit Group
Status Least concern

Distinguished by its tapering head, coloration, and tall dorsal fin, this dolphin often rides ship bow waves in the North Pacific. It has a dark back, a gray-white belly, and a lower flank patch from the beak to level with the dorsal fin. A pale stripe from the tail base extends forward to level with the dorsal fin, then it may fork, arching over the shoulder region to near the eye. This dolphin eats a variety of fish and squid, using 23–36 pairs of small teeth in both the upper and lower jaws. It forms fast-changing schools numbering tens to thousands, and often associates with other dolphins and whales. After a gestation period of 10–12 months, the female gives birth to a single calf, which is 3 ft (90 cm) long. This dolphin makes up a sizable part of the catch in some fishing industries of the northwest Pacific Ocean.

Lagenorhynchus albirostris

White-beaked dolphin

Location North Atlantic

Length 9¼ ft (2.8 m)
Weight 770 lb (350 kg)
Social unit Group
Status Least concern

This large, stocky dolphin is an acrobatic swimmer. Its stubby beak meets the bulging melon of the forehead at a distinct angle. Coloration is mainly dark gray or black with variable gray or white patches extending along the upper and lower flanks; the underside, including the beak, tends to be white. It has 22–27 pairs of robust, cone-shaped teeth in both upper and lower jaws, for feeding on open-water shoaling fish such as herring, although sea bed-dwelling flatfish and squid are also taken. The single calf, 3½–4 ft (110–120 cm) long, is born, like many other dolphins, in summer. White-beaked dolphins congregate in rapidly changing schools numbering from 5 to more than 1,000. They produce machine-gun-like bursts of clicks and a range of squeals to communicate with each other and also to navigate and find prey.

Grampus griseus

Risso's dolphin

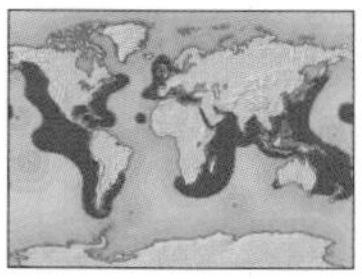

Location Pacific, Atlantic, Mediterranean, Indian Ocean

Length 12½ ft (3.8 m)
Weight 880 lb (400 kg)
Social unit Group
Status Least concern

This distinctive dolphin is readily identified by its large size, blunt and "beakless" head, central crease down the bulging melon, tall dorsal fin, and overall gray, white-scarred coloration. It forms schools of 10–15, but these may aggregate into groups of several hundreds mixed with other cetaceans, especially smaller dolphins and pilot whales. In common with many other dolphins, Risso's dolphin suffers from human activities. It is asphyxiated in fishing nets, accumulates pollutants in the body through the food chain, and swallows pieces of plastic and other refuse.

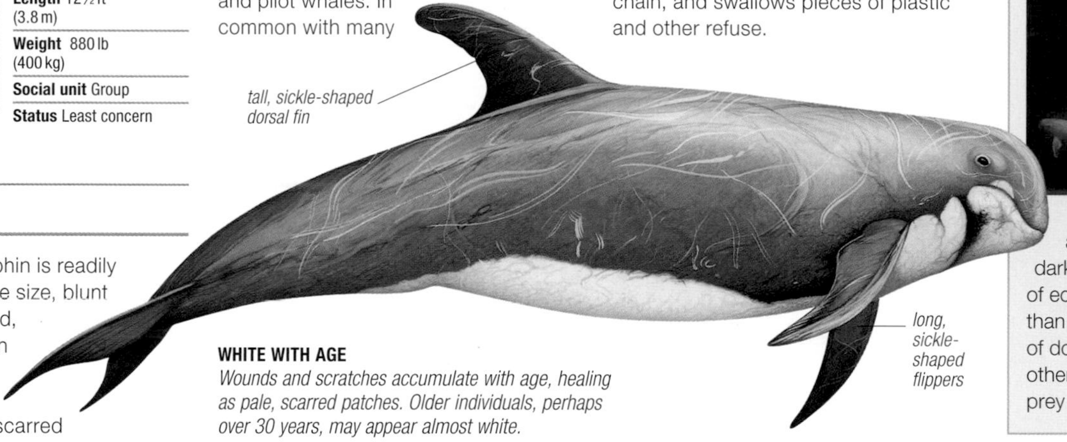

WHITE WITH AGE
Wounds and scratches accumulate with age, healing as pale, scarred patches. Older individuals, perhaps over 30 years, may appear almost white.

COMMUNICATION

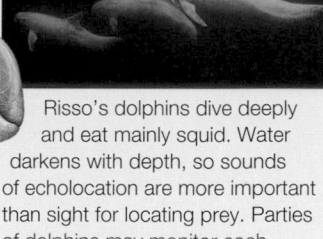

Risso's dolphins dive deeply and eat mainly squid. Water darkens with depth, so sounds of echolocation are more important than sight for locating prey. Parties of dolphins may monitor each other's clicks and echoes, to find prey more efficiently.

Tursiops truncatus

Bottlenose dolphin

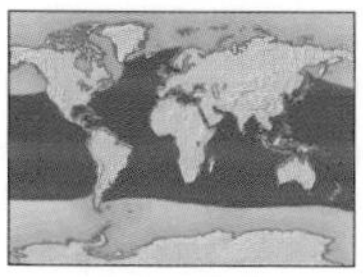

Location Worldwide (except polar regions)

Length 6¼–13 ft (1.9–4 m)
Weight 1,100 lb (500 kg)
Social unit Variable
Status Least concern

This largest of the beaked dolphins is the "performing" species of marine life centers. In fact, there are probably 2 species: *Tursiops truncatus*, which is found worldwide, and the smaller *Tursiops aduncus,* which has a more limited range off the coast of India and in the West Pacific. There may even be more species, considering the wide variation across regions and habitats. Around tropical coasts, the bottlenose dolphin is an average of 6½ ft (2 m) long, and has relatively large flippers, flukes, and dorsal fin. In colder, open oceans they are almost twice as long, with proportionately smaller extremities. Social groupings and feeding methods are similarly varied over its range, as are the leaps, water-slaps, sounds, and calls of the bottlenose dolphin.

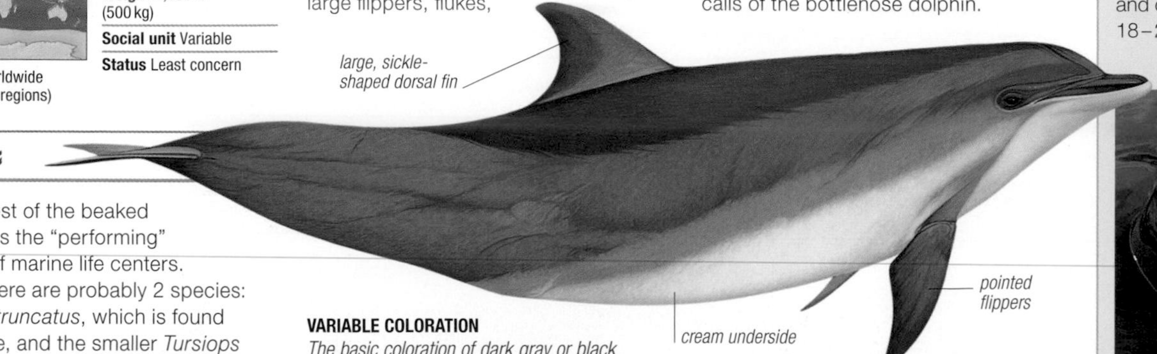

VARIABLE COLORATION
The basic coloration of dark gray or black back, fading to cream underneath, varies in hue and pattern between individuals.

ADAPTABILITY

The bottlenose dolphin, named for its short and robust beak, is an adaptable, successful generalist. It eats many kinds of fish, mollusks, and crustaceans, which it grips with 18–27 pairs of small, conical teeth in both jaws.

MAMMALS

Stenella longirostris

Spinner dolphin

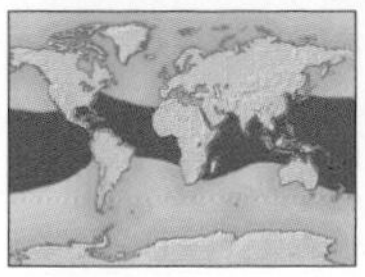

Length 4¼–6½ ft (1.3–2 m)
Weight 99–165 lb (45–75 kg)
Social unit Group
Status Data deficient

Location Tropical waters worldwide

No other dolphin, or possibly cetacean, varies so greatly in body proportions, color, and pattern across its range. There are at least 3 spinner subspecies: one worldwide and 2 in the eastern Pacific, with possibly a fourth, dwarf subspecies in shallow coral reefs in the Gulf of Thailand. There are 45–65 pairs of sharp teeth in each of the upper and lower jaws. Coloration is black or dark gray, fading to a pale or white underside, with black edging to the eyes and lips, although this is very variable. The spinner dolphin dives deep to eat mid-water fish and squid. It forms massive schools of hundreds or thousands, which often associate with other cetaceans, and even with predatory fish such as tuna—although the reason for this is not clear.

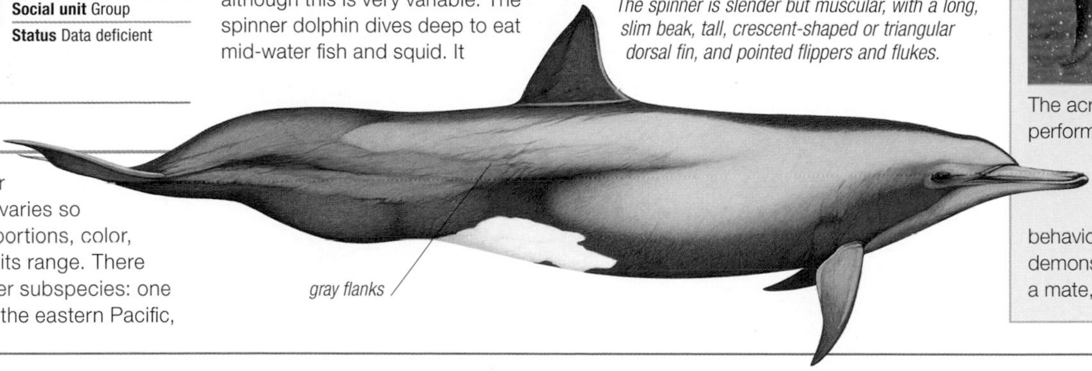

BODY FORM
The spinner is slender but muscular, with a long, slim beak, tall, crescent-shaped or triangular dorsal fin, and pointed flippers and flukes.

SPINNING ACTION

The acrobatic spinner dolphin performs high leaps when it twists in the air or spins on its vertical axis several times. The purpose of this behaviour is unclear. It may be to demonstrate fitness and/or attract a mate, or to dislodge parasites.

Stenella attenuata

Pantropical spotted dolphin

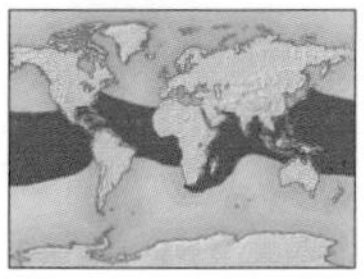

Length 5¼–8½ ft (1.6–2.6 m)
Weight Up to 260 lb (120 kg)
Social unit Group
Status Least concern

Location Tropical and temperate waters worldwide

One of the commonest cetaceans, this dolphin is usually found in waters warmer than 22° C (72° F). It has a slender, streamlined body and slim beak with 40 pairs of teeth in both upper and lower jaws. An elongated, oval, dark gray "cape" extends from the forehead to just behind the dorsal fin. The flanks are lighter gray and the underside is pale. Large schools of thousands are often segregated into mothers with young, older juveniles, and other subgroups. These all associate with other cetaceans, especially spinner dolphins and tuna fish. Pantropical spotted dolphins eat mainly mackerel, flying fish, squid, and other near-surface prey.

Stenella frontalis

Atlantic spotted dolphin

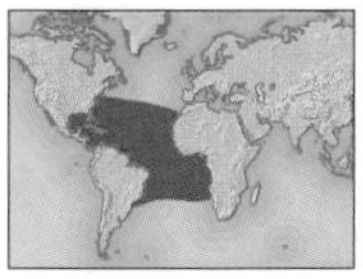

Length 5½–7½ ft (1.7–2.3 m)
Weight 310 lb (140 kg)
Social unit Variable
Status Data deficient

Location Atlantic

The Atlantic spotted dolphin digs with its beak in the sandy sea bed, poking so deep that the whole head is immersed. It also uses various techniques to catch fish, squid, and other prey from surface and mid-waters. It has 32–42 pairs of teeth in the upper jaw and the same in the lower jaw. The Atlantic spotted dolphin forms schools of up to 15 individuals near coasts, but these may gather into larger groups to follow seasonal food. Like its pantropical relative (above), the newborn calf lacks spots. These develop with age, starting on the belly and extending to the sides and back over several years.

STOUTER SPECIES
The Atlantic spotted dolphin is distinguished from its pantropical cousin mainly by its stouter body and beak.

SPOT THE HABITAT

In several spotted dolphin species, including the Atlantic and pantropical, the dark patches and flecks increase in extent with age but may also vary with habitat. These marks are generally less prominent in oceanic groups. In populations that stay near coasts, the spots may be so extensive as to obscure the background color.

Stenella coeruleoalba

Striped dolphin

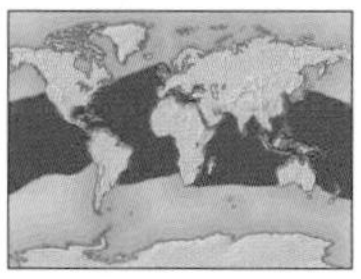

Length 6–8¼ ft (1.8–2.5 m)
Weight 240–360 lb (110–165 kg)
Social unit Group
Status Least concern

Location Tropical and temperate waters worldwide

This species lives in areas of changeable water temperature and may dive as deep as 655 ft (200 m) to seize small fish and squid, using the 40–55 sharp-pointed pairs of teeth present in each jaw. It derives its name from the complex pattern of black and gray stripes along its back and flanks. This active dolphin is capable of a wide variety of acrobatic leaps and spins. Although it is relatively common, its numbers have declined in recent years. In particular, infection by a morbilli virus drastically reduced Mediterranean populations in the early 1990s.

SURFING SCHOOL

Schools of 50–500 striped dolphins race across the ocean and may gather in thousands, leaping high and whistling to keep in contact. They ride pressure waves just in front of migrating great whales, and the man-made equivalent—ship bow waves, especially in Mediterranean and Atlantic waters.

COLORATION
The thin black and wider gray stripes branch or fork and overlay a dark, blue-gray back, blue-gray sides, and pale cream or pink underside. It has a black beak and eye patches.

MAMMALS

Orcinus orca

Killer whale

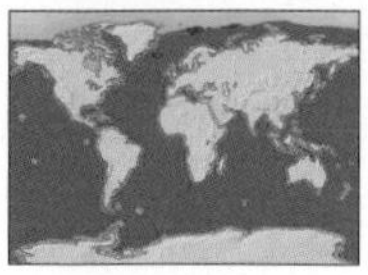

Location Worldwide

Length Up to 30 ft (9 m)
Weight Up to 11 tons (10 tonnes)
Social unit Group
Status Data deficient

Distinctive black-and-white markings make the killer whale—which is also commonly known as the orca—the most easily recognized of the toothed whales and dolphins. It is a highly social whale, living in long-lasting family groups called pods, which consist of adult males and females, and calves of various ages. Pods typically number up to 30 individuals, but groups of as many as 150 whales occur when pods come together to form superpods. Pods are matriarchal and both male and female calves tend to stay with their mother for life. When the young reproduce, their offspring remain to build up multigenerational groupings around the original mother.

Generally, the killer whale's diet is as diverse as its hunting techniques. Fish, ranging from herring to great white sharks, and marine mammals, including whales and seals, are taken, as well as turtles and birds. In some regions, such as the northeast Pacific, however, there appear to be 2 forms of killer whale: transients, who feed on mammals, turtles, and birds; and residents, who eat only fish. Despite its name, the killer whale is approachable and very inquisitive. It has a variety of elaborate surface habits, including spy hopping (rising slowly vertically, until its head is above the water), tail and flipper slapping, and breaching.

PARENTAL CARE
The whole pod may provide parental care, but a newborn calf stays closest to its mother—a bond that will remain strong for the rest of its life.

POD FORMATIONS
Pods may travel in tight formations, with the females and calves at the center, and the males on the fringes, or spread across distances of up to $\frac{2}{3}$ mile (1 km). They communicate using sets of highly distinctive cries and screams, which also act as social signals that reinforce group identity.

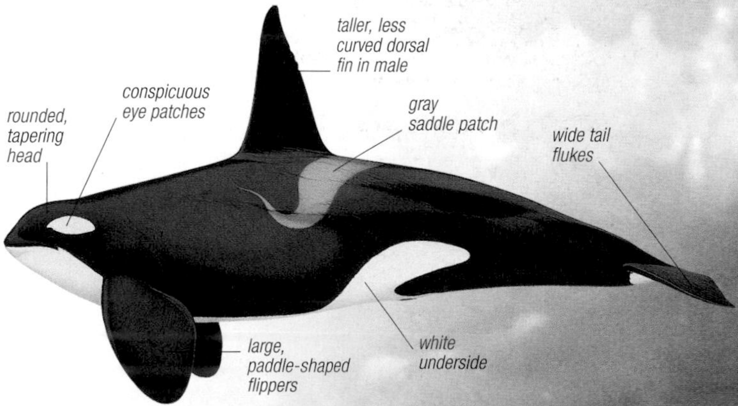

BUILT FOR HUNTING
The killer whale has a powerfully built, stocky body ideally suited to hunting. The broad tail flukes help propel the whale at high speeds, and the dorsal fin—which reaches up to 6 ft (1.8 m) in males—and paddlelike flippers provide stability. Underwater, its markings have the effect of camouflaging the whale from above and below.

INGENIOUS HUNTER
Killer whales are versatile hunters, using a number of different techniques that give them one of the most varied diets in the oceans. They often hunt in pods, chasing down prey or herding fish together before attacking from different angles. Those living off southern South America have perfected the technique of catching sea lions by intentionally beaching themselves to pursue the sea lions in the shallows (shown right), while other methods include tipping over ice floes to unbalance seals and penguins, and breaching next to rocks to wash birds into the sea.

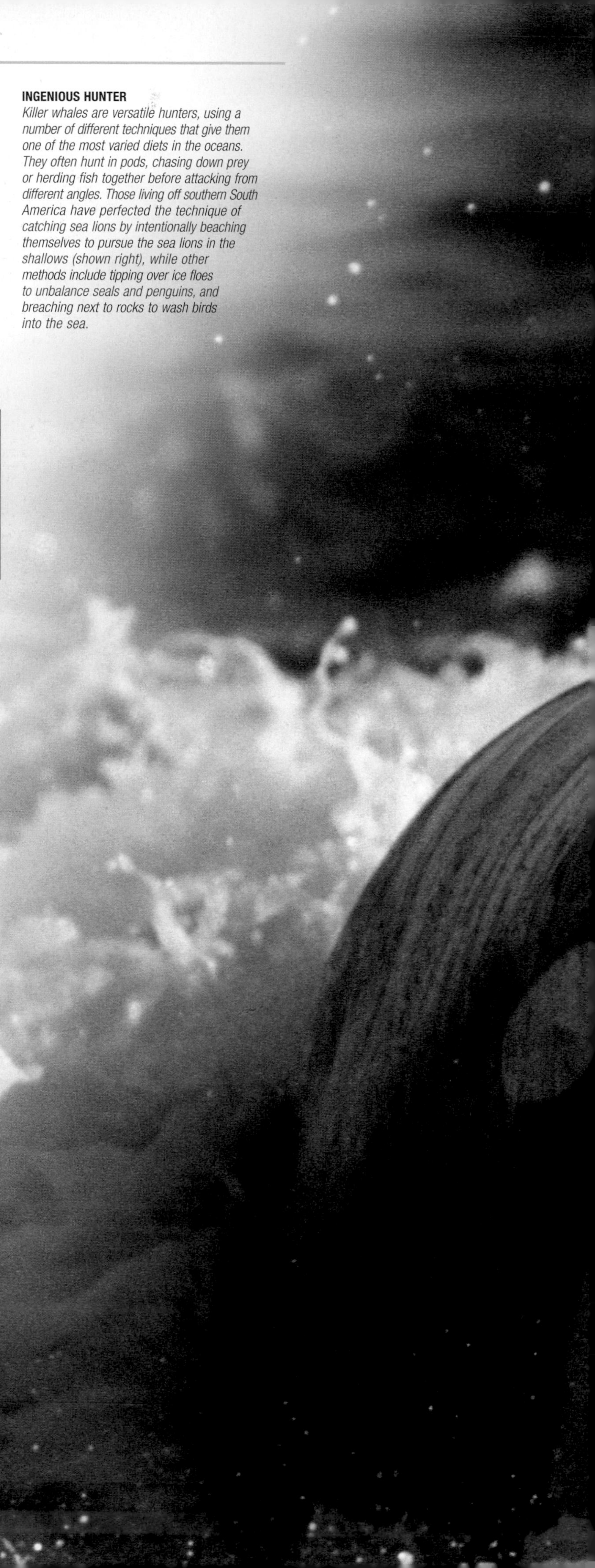

Delphinus delphis

Short-beaked common dolphin

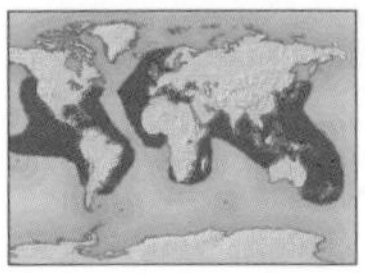

Length 7½–8½ ft (2.3–2.6 m)
Weight 175 lb (80 kg)
Social unit Group
Status Least concern

Location Temperate and tropical waters worldwide

This species has a yellow to buff blaze along the flank, from the face to below the dorsal fin. It tapers to a point, then widens again toward the tail, but as pale gray. There are also dark, narrow stripes from mouth corner to eye and chin to flipper. The short-beaked common dolphin has 40–55 pairs of small, sharp teeth in the upper jaw, and the same in the lower jaw. It lives offshore in the deep ocean; the inshore form is sometimes regarded as a separate species, *Delphinus capensis*. Both species hunt schooling fish and squid to a depth of 985 ft (300 m).

HOURGLASS PATTERN
The flanks have yellow and creamy gray areas that form a distinctive hourglass shape.

SCHOOLING

Short-beaked common dolphins are very social, forming fast-swimming schools of thousands. They leap and tumble, ride waves from ships and great whales, and make many sounds such as clicks, squeaks, and creaks. Their whistles are loud enough to be heard from nearby boats.

Lissodelphis borealis

Northern right-whale dolphin

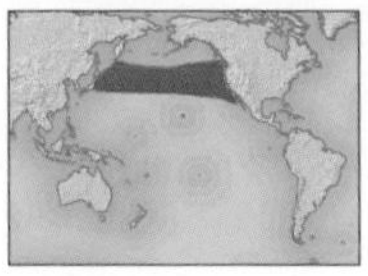

Length Up to 9¾ ft (3 m)
Weight Up to 250 lb (115 kg)
Social unit Group
Status Least concern

Location North Pacific

This sociable dolphin forms schools of 100–200, which merge into gatherings of thousands. It makes varied sounds and associates with other cetaceans, leaping high and riding ship bow waves. Its main foods are midwater fish and squid, down to 655 ft (200 m). A similar species, *Lissodelphis peronii*, is found in southern oceans.

CONSERVATION

Due in part to its schooling habits, the northern right-whale dolphin is at high risk from drift nets set for fish and squid. In the 1980s, over 20,000 died every year. A UN moratorium on drift-netting has since slashed this figure, although some deaths from by-catch still occur.

SLIM AND FAST
The body is slender, with relatively small flippers and tail flukes, indicating that this dolphin is a speedy swimmer. It has no dorsal fin.

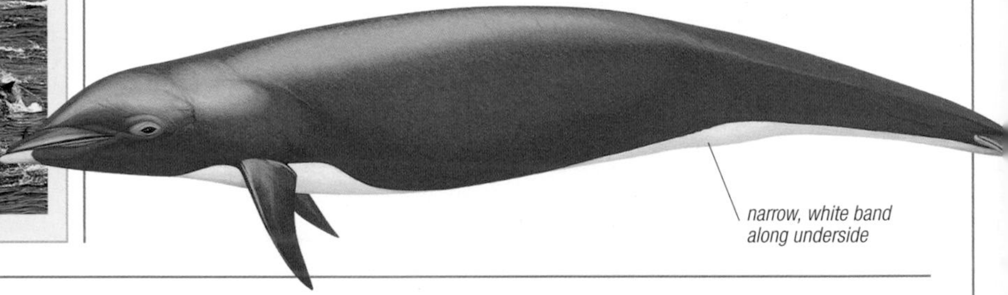

Orcaella brevirostris

Irrawaddy dolphin

Length 7–9¼ ft (2.1–2.8 m)
Weight 200–330 lb (90–150 kg)
Social unit Variable
Status Vulnerable

Location S.E. Asia and N. Australia

This river dolphin's distinctive features include a bulging forehead, no proper beak but ridgelike lips, and "creases" demarcating a neck region between head and body. The head muscles allow a wide range of facial expressions, giving this dolphin an animated appearance to human observers; however, equating expressions to ours is extremely conjectural. Some Irrawaddy dolphins live solely in fresh water, roaming almost 930 miles (1,500 km) inland in the major waterways of the Irrawaddy (Burma) and Mekong (Vietnam). But overall, this is an estuarine and coastal species frequenting muddy, silt-laden river mouths and deltas. It swims slowly in schools of 15 or fewer, feeding on fish, squid, octopus, prawns, and similar prey on or near the seabed, using its 15–20 pairs of teeth in each of the upper and lower jaws. Gestation is estimated at 14 months, and the single calf is about 39 in (100 cm) long and 26 lb (12 kg) at birth. In some regions, these dolphins traditionally work with people to herd fish into nets, receiving reward of food for their cooperation, and even being revered as sacred. However, in other places, Irrawaddy dolphins are viewed as pests at river fisheries or killed for their meat.

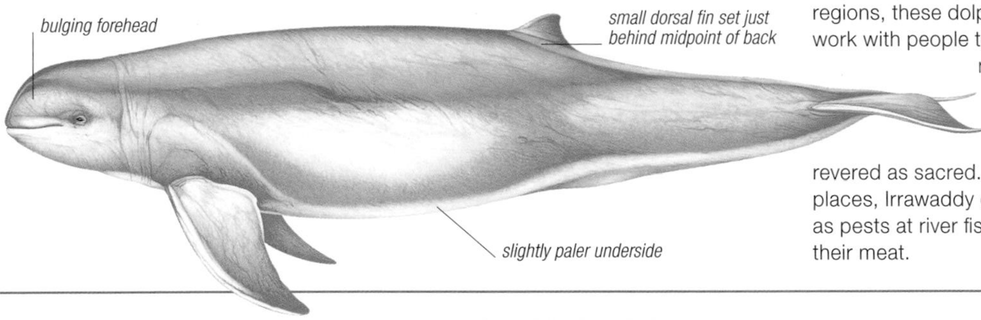

Cephalorhynchus commersonii

Commerson's dolphin

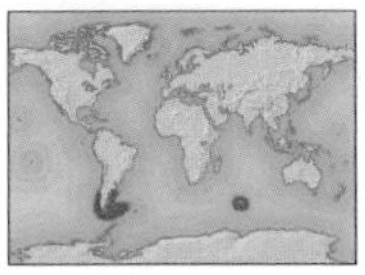

Length 4½–5½ ft (1.4–1.7 m)
Weight Up to 190 lb (86 kg)
Social unit Group
Status Data deficient

Location S. South America, Falkland Islands, Indian Ocean (Kerguelen Islands)

Commerson's dolphin has similar coloration to that of the killer whale. Its forehead slopes smoothly from the snout, merging with its stocky body. The newborn calf, 26–30 in (65–75 cm) long, is gray and becomes two-tone with age. The species forms schools of less than 10 that sometimes expand up to 100. It feeds on seabed dwellers, such as fish, crabs, starfish, and squid. There are 2 populations (probably subspecies), separated by more than 4,970 miles (8,000 km)—those around South America being 10–12 in (25–30 cm) shorter than those in the Indian Ocean.

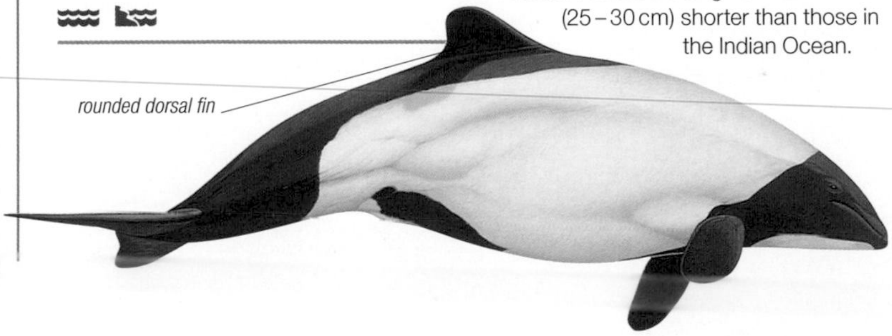

Cephalorhynchus hectori

Hector's dolphin

Length 4–5 ft (1.2–1.5 m)
Weight Up to 125 lb (57 kg)
Social unit Group
Status Endangered

Location New Zealand

One of the smallest dolphins, this species is similar to a porpoise in outline, with a smoothly tapering snout and no distinct beak or melon bulge. It is gray with black flippers, dorsal fin, and tail. The white underside extends a projection up each flank toward the tail. Active and sociable, it forms small schools of up to 5, and spends much time chasing, touching, flipper-slapping, and generally interacting with others. It feeds at various depths, mainly on fish and squid. As an inshore species, Hector's dolphin is at particular risk from entanglement in fishing nets and from pollution.

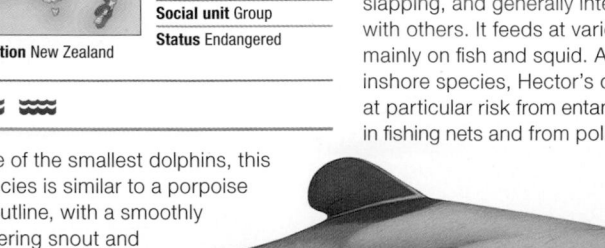

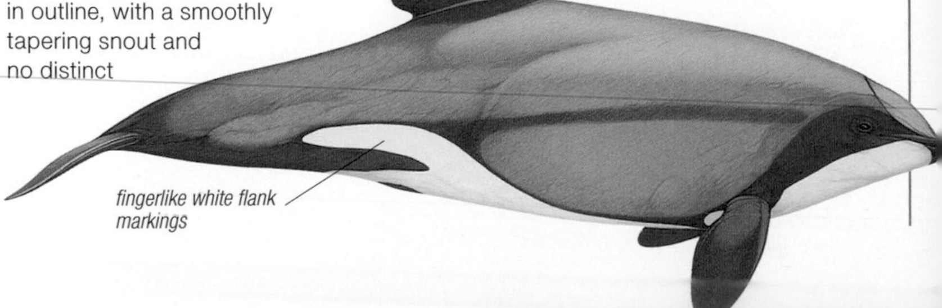

MAMMALS

Pseudorca crassidens

False killer whale

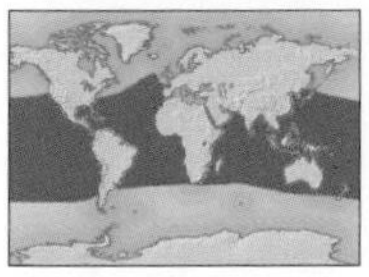

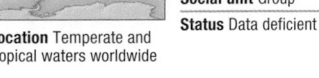

Length 16–20 ft (5–6 m)
Weight 1⅜–1½ tons (1.3–1.4 tonnes)
Social unit Group
Status Data deficient

Location Temperate and tropical waters worldwide

One of the largest dolphins, this rapid swimmer has a long, slim body and a tall, sicklelike dorsal fin. Coloration is uniform black or slate-gray, with a paler underside patch between the flippers, and perhaps pale patches on the sides of the head. It prefers deep oceans but appears occasionally off oceanic islands, in schools of 10–20, rarely up to 300. Equipped with 8–11 pairs of large, conical teeth in both the upper and lower jaws, this formidable hunter pursues large oceanic fish, such as salmon, tuna, and barracuda. It also takes squid and even smaller dolphins.

BULL-NOSED WHALE
The pronounced, rounded melon combined with the absence of a beak gives the false killer whale a bull-nosed profile.

CONSERVATION

The false killer whale makes a wide variety of echolocating and communicating sounds, such as clicks and whistles. It also leaps with agility and skillfully surfs breakers and ship bow waves. However, this confident navigator is often stranded on beaches in vast groups of up to 1,000. Why this happens is not known but, with help to return to the water, some survive.

Globicephala macrorhynchus

Short-finned pilot whale

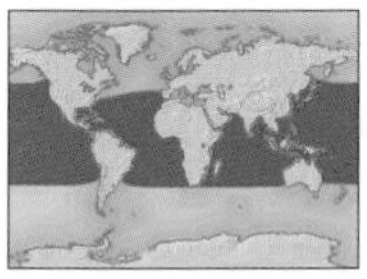

Length 16–23 ft (5–7 m)
Weight 1⅛–2 tons (1–1.8 tonnes)
Social unit Group
Status Data deficient

Location Temperate and tropical waters worldwide

Nocturnally active, the pilot whale feeds mainly on deep-water squid and octopus, diving below 1,600 ft (500 m) for more than 15 minutes. There are 2 very similar species: the short-finned (*Globicephala macrorhynchus*) has smaller flippers (pectoral fins) than the long-finned pilot whale (*Globicephala melas*). In both, there is an anchor-shaped pale patch on the throat and chest, and white streaks behind the dorsal fin and each eye. Females are about half the weight of males, but live 15 years longer, up to 60 years. Along with skin scarring, these features suggest competition between males with battles for females. After a gestation of almost 15 months, the female gives birth to a single calf 4½–6 ft (1.4–1.8 m) long. Pilot whales are still hunted, by driving them into the shallows for slaughter.

BONDING

Pilot whales form schools of tens to hundreds and also associate with other cetaceans such as bottlenose and common dolphins, and minke whales. Within a pilot whale school, adults form long-term bonds and probably recognize each other by individual "signature" whistles. However, many offspring are not closely related genetically to the school's males, suggesting that mating occurs between schools. Females past reproductive age may suckle calves that are not their own.

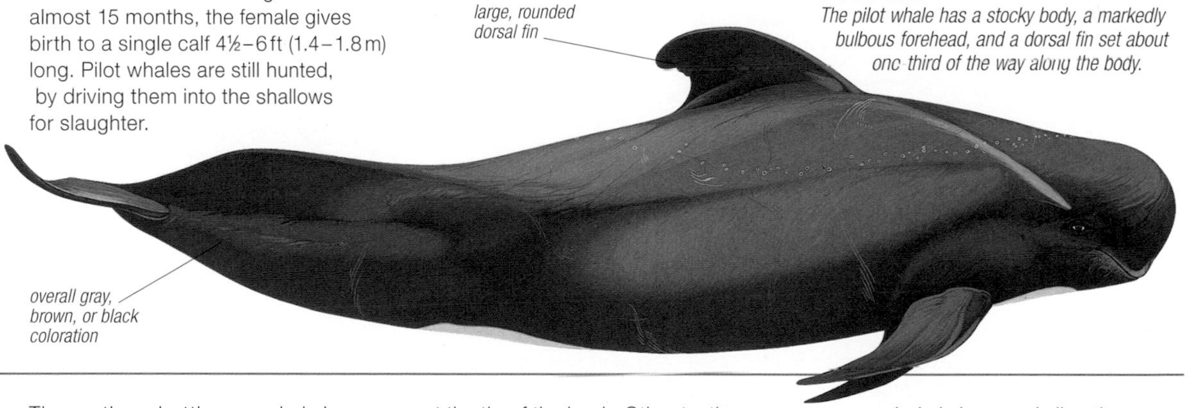

STOCKY BODY
The pilot whale has a stocky body, a markedly bulbous forehead, and a dorsal fin set about one third of the way along the body.

Hyperoodon ampullatus

Northern bottlenose whale

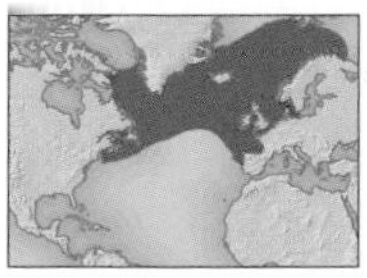

Length 20–33 ft (6–10 m)
Weight Not recorded
Social unit Individual/Group
Status Data deficient

Location North Atlantic

The northern bottlenose whale is one of about 19 species of beaked whales—mostly medium-sized, open-ocean cetaceans that make long, deep dives for squid, starfish, fish, crab, and other food. This species has a bulging forehead and dolphinlike beak. Males, which grow larger than females, have 2 tusklike teeth growing at the tip of the beak. Other teeth are present but the degree of growth above the surface of the gums (eruption) is very variable. Feeding is probably by suction, using the tongue as a piston to draw in water, seabed mud, and prey. The body of the northern bottlenose whale is long and slim, the flippers small, and the erect dorsal fin is set about two-thirds of the way to the tail. Like most other beaked whales, this species congregates in small schools of up to 10, usually containing individuals of the same sex and of similar ages.

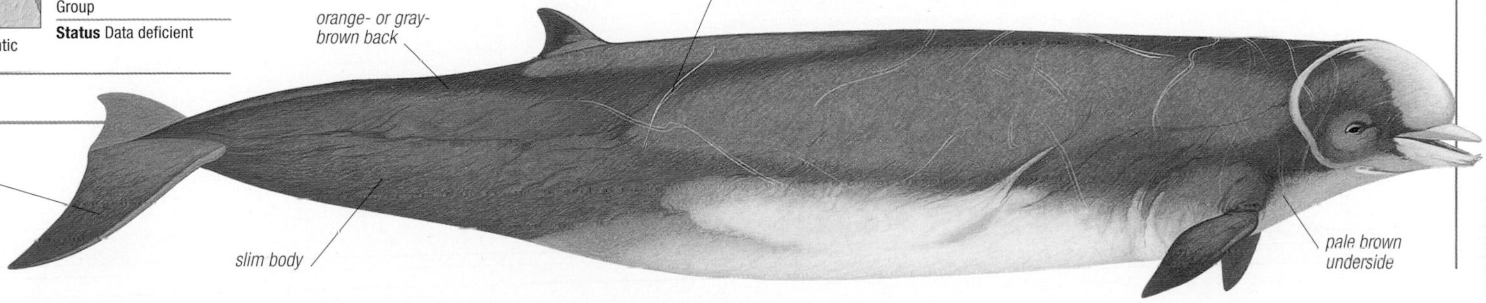

MAMMALS

Ziphius cavirostris

Cuvier's beaked whale

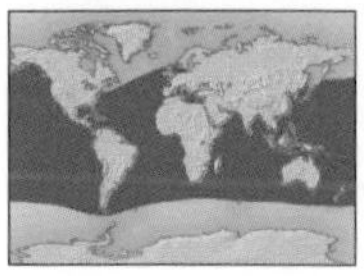

Length 23–25 ft (7–7.5 m)
Weight 3¼–4⅜ tons (3–4 tonnes)
Social unit Variable
Status Least concern

Location Temperate and tropical waters worldwide

This species has the long, slim body and small dorsal fin set well to the rear typical of the beaked whale family. It feeds on deep-living squid and other creatures using suction (see northern bottlenose whale, p.259). The jawline curves up at the tip of the snout and then down. Along with the relatively smooth forehead, this leads to the alternative name of goose-beak whale. The small flippers fit into indentations in the body so that the whale is streamlined for fast swimming, using only its tail flukes, down to great depths. The pale brown to blue-gray body is scarred by parasites and also, in males, by bites from males of the same species. These wounds probably occur during dominance battles at breeding time. Older male Cuvier's beaked whales tend to live alone. Most males have 2 cone-shaped teeth, which project like tusks from the lower jaw. Females and young are toothless. Younger males, females, and offspring, which are nearly 9 ft (2.7 m) long at birth, form schools of rarely more than 10.

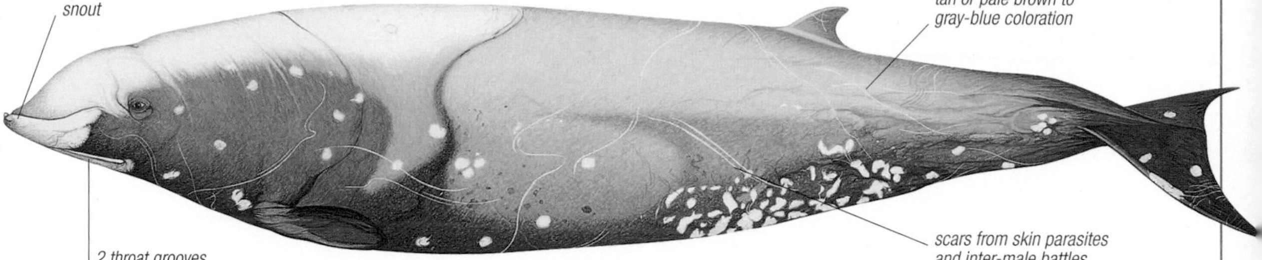

Kogia simus

Dwarf sperm whale

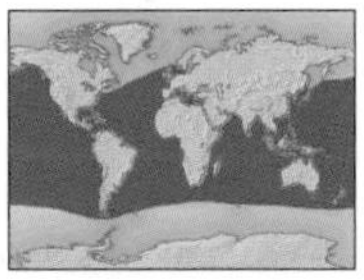

Length 8¾ ft (2.7 m)
Weight 300–600 lb (135–270 kg)
Social unit Variable
Status Data deficient

Location Temperate and tropical waters worldwide

Smallest of the 3 species of sperm whales, this whale dives to nearly 1,000 ft (300 m) for fish, squid, crustaceans, and mollusks. The lower jaw has 7–13 pairs of sharp teeth, and is slung almost sharklike under the large, bulbous head; the upper jaw has only 3 pairs of teeth. Back, fin, flippers, and flukes are blue-gray, shading to cream below. Just behind the mouth and eye is a contrasting, pale crescent, which, in size and position, resembles the gill slit of a fish. A shy creature, the dwarf sperm whale lives alone or in small schools (fewer than 10), and releases a cloud of feces to repel predators. Little is known about its breeding habits: the gestation period appears to vary between 9 and 11 months, and the single calf, about 3 ft (1 m) long, is usually born in autumn. These whales seem prone to group strandings.

Physeter catodon

Sperm whale

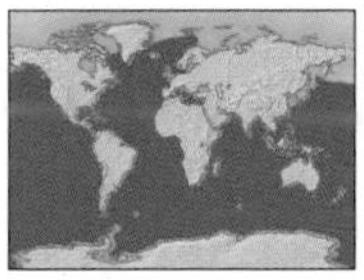

Length 36–65 ft (11–20 m)
Weight 22–63 tons (20–57 tonnes)
Social unit Individual/Group
Status Vulnerable

Location Deep waters worldwide

One of the world's largest carnivores, this massive cetacean makes extremely deep dives for food, mainly squid and octopus, but also fish and, sometimes, giant squid. It can stay submerged for almost 2 hours and has been tracked by sonar to nearly 4,000 ft (1,200 m) deep, with indirect evidence—such as types of bottom-dwelling fish found in its stomach—of dives below 10,170 ft (3,100 m). The ability to make such deep dives may be in part due to its spermaceti organ (see panel, right). Between dives, the whale lies loglike at the surface, a plume of misty air emerging with each exhalation at 45 degrees from the single blowhole. The long, narrow lower jaw is slung under the huge head and bears 50 pairs of conical, round-tipped teeth; there are no visible upper teeth. The main color is dark gray or brown, paler on the underside, with white or cream around the lower jaw. Bull (male) sperm whales are twice the weight of females (cows) and tend to migrate farther north and south, into colder waters, for summer feeding. They form loose, bachelor pods when young but then become more solitary. Females stay nearer the tropics and form mixed groups with their young and juveniles to about 10 years old. In summer or autumn, after a gestation period of 14–15 months, the single calf is born, 13 ft (4 m) long, and suckles for 4 years or more.

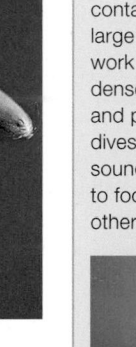

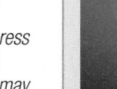

KEEPING IN TOUCH
Sperm whales often swim near, touch, and caress others in their school. They also produce loud, rhythmic clicking and banging sounds, which may aid individual recognition.

SPERMACETI ORGAN

The sperm whale's enormous head contains a spermaceti organ—a large mass of waxy oil. This may work as a buoyancy aid, becoming denser with changing temperature and pressure, to assist very deep dives. It may also function as a sound-lens, like a dolphin's melon, to focus the whale's clicks and other rhythmic noises.

LARGE HEAD
The sperm whale's tall, narrow, boxlike head comprises up to one-third of its total length. The low dorsal fin is set well to the rear.

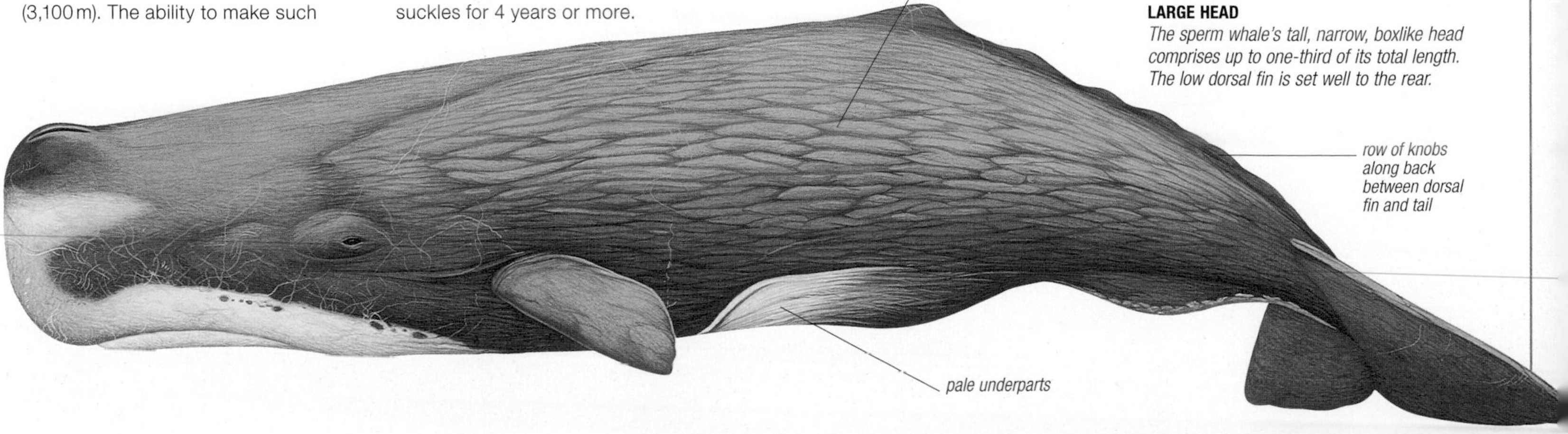

MAMMALS

DIVING DEEP
A sperm whale leaves the sunlit upper levels of the ocean and heads into the dark ocean depths in search of food. Diving to greater depths than any other mammal, it descends at a rate of up to 10 ft (3 m) per second. Away from the light, its eyes are of little use and the whale relies on echolocation.

BIRDS

BIRDS

PHYLUM	Chordata
CLASS	Aves
ORDERS	29
FAMILIES	About 196
SPECIES	About 10,200

Birds are the most accomplished of all flying animals. Their ability to fly has allowed them to spread throughout the world, often to places, such as remote islands and Antarctica, that are beyond the reach of many other animals. Like mammals, birds are endothermic (warm-blooded) vertebrates. However, unlike most mammals, they reproduce by laying eggs. Birds have several adaptations for flight, including wings, feathers, a light but strong skeleton, and a highly efficient respiratory system.

TAKING OFF
The power of flight has enabled birds to exploit aerial niches. In many habitats, birds of prey (such as this tawny eagle), with their powerful wings, acute vision, and sharp bill and talons, are the top predators.

Evolution

Birds evolved from reptilelike ancestors, possibly from tree-dwelling dinosaurs that fed on insects. A lifestyle of arboreal hunting would have promoted the development of such birdlike characteristics as large eyes, grasping feet, and a long snout, later to evolve into a bill. It might also have resulted in the transition from being cold- to warm-blooded—an advantage for animals that feed on insects, which become inactive and slow-moving in cold conditions. Feathers, which were derived from reptilian scales, probably first evolved to provide insulation, although they were no doubt put to the purpose of flight very early on.

One of the earliest-known bird fossils is about 150 million years old, dating from the Jurassic Period (205–142 million years ago). Named *Archaeopteryx lithographica*, this animal was about the size of a crow and showed a combination of reptilian and avian features: it had wings and feathers like a bird but also had a snout, rather than a bill, and the toothed jaws of a reptile. There is some doubt about whether *Archaeopteryx* could fly or just glide, since it lacked the keeled breastbone that provides attachment sides for the muscles needed for powered flight.

During the Cretaceous Period (145–65 million years ago), birds diversified and their anatomy evolved to make possible increasingly efficient flight. It was in this period that the ancestors of living birds appeared. Toward the end of the Cretaceous, a wave of mass extinctions saw the end of the age of the dinosaurs. It is not clear why birds survived—perhaps being warm-blooded helped them withstand a prevailing climatic catastrophe. But having survived, birds as a group flourished into the diversity of forms alive today.

ARCHAEOPTERYX FOSSIL
Archaeopteryx lithographica is thought to represent a link between reptiles and birds: the jaws, snout, and tail (which is supported by vertebrae) are reptilian, but the wings and feathers are birdlike.

Anatomy

A bird has several physical adaptations for flight. Its body is short, strong, and compact, with powerful muscles for moving the wings, and strong legs to launch it into the air and cushion the impact of landing. Its feathers form the flight surface (see panel, below, and p.266); they also provide protection and insulation.

The skeleton of a bird combines remarkable lightness with strength, attributes that are essential for powerful flight. To restrict the bird's

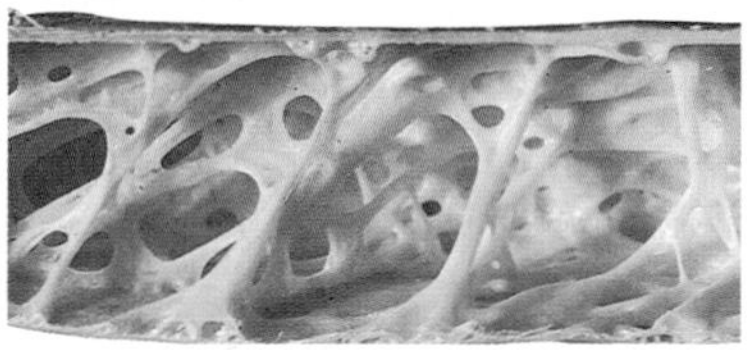

CROSS SECTION OF A BIRD'S BONE
Many bones in a bird are hollow, including the major limb bones, parts of the skull, and pelvis. This reduces the bird's weight, to conserve energy for flight. The bones are strengthened by internal struts.

FEATHERS

Feathers are highly complex structures that are unique to birds. However, they are formed from the same material, keratin, that is found in the hair of mammals and the scales of reptiles. Feathers are subject to considerable wear and tear. To keep them in good condition, birds regularly clean, oil, and reshape their plumage using their bill, an activity known as preening. Other forms of maintenance include scratching, bathing, and sunning. Feathers are shed (molted) and replaced at least once a year. Flying birds have 4 different kinds of feathers, each of which is modified to serve a distinct function: there are 2 types of flight feathers (wing and tail feathers), as well as down feathers and contour feathers.

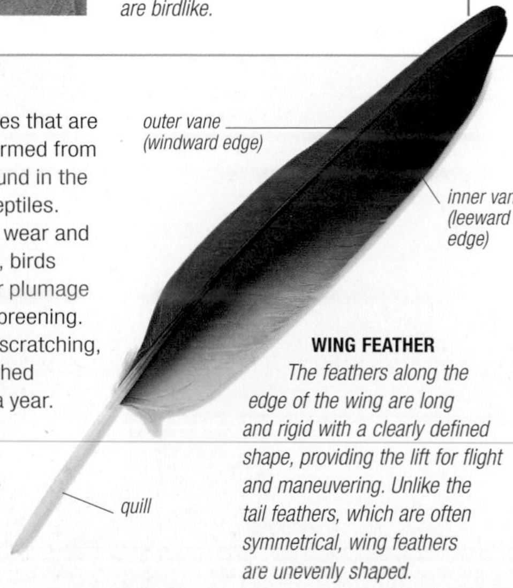

WING FEATHER
The feathers along the edge of the wing are long and rigid with a clearly defined shape, providing the lift for flight and maneuvering. Unlike the tail feathers, which are often symmetrical, wing feathers are unevenly shaped.

FEATHER TYPES
Down feathers form an insulating underlayer, while small contour feathers provide a streamlined covering over the body. The long tail feathers are used for flying and steering.

FEATHER STRUCTURE
The structure of feathers is extremely complex. Most feathers have a central shaft (rachis), from which closely spaced branches (barbs) project outward to form a continuous, flat surface. Minute side branches (barbules) lock the barbs together.

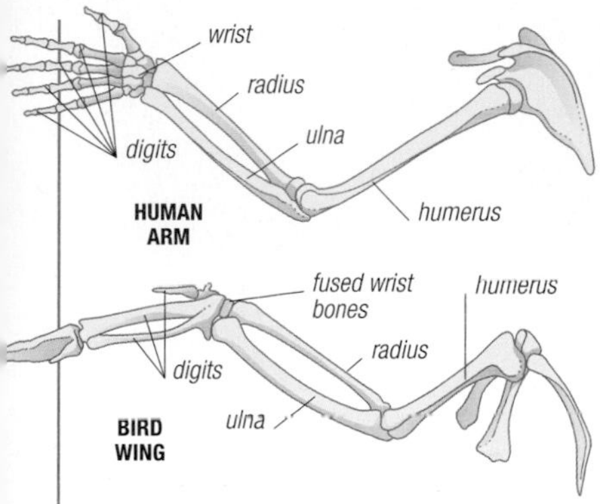

ARM AND WING BONES
In comparison with a human arm, a bird's forelimb (or wing) has undergone a dramatic reduction in the number of bones, especially around the wrist and hand.

weight, a number of bones are greatly reduced in size, and many have fused together, making a rigid frame without the need for large muscles and ligaments to hold the bones together. Most birds' bones are also hollow, lacking marrow. To compensate for their lightness, they are supported by a network of internal struts (trabeculae) at points of stress, giving them great strength. Several bones, including the humerus, contain air sacs, which are connected to the respiratory system.

A bird's wing is the anatomical counterpart of the human arm, although birds have relatively few digits, and some of the "hand" bones are fused together, which contributes to the general rigidity of the structure. Similarly, the "elbow" and "wrist" joints of the wing bones are inflexible in the vertical plane. As a result, when the bird is flying, the wings are held straight out and flapped only at the shoulder, providing rigidity and saving energy by preventing unnecessary movement. The powerful wing muscles are attached to a keel, a large projection that extends at right angles from the breastbone.

Another feature unique to birds is their light, flexible bill (or beak). In some birds, such as parrots, both jaws (or mandibles) move, giving a wider gape. In mammals, only the lower jaw moves. The bill itself has an external covering of keratin, which is light and strong and allows for great variety in form. The bill shape is always adapted to a bird's particular method of feeding, and reflects its diet—for example, finches have a strong, conical bill for cracking seeds, whereas herons have a pointed, daggerlike bill for seizing fish. Birds that feed on insects, such as warblers, have a small, slender bill, while in flesh-eaters, such as hawks, the bill is sharp and hooked for tearing open prey.

Birds have between 2 and 4 toes on their feet, the arrangement and shape varying according to the bird's way of life. Most have 4 toes, of which 3 point forward and one back, a good system for perching (see p.342). Many swimming birds have webs between their toes.

Like humans, birds rely primarily on their sense of sight, followed by hearing. Indeed, many birds have exceptional vision, particularly owls and birds of prey. With their high-speed aerial lifestyle, birds rarely (and in some cases, never) use their sense of smell.

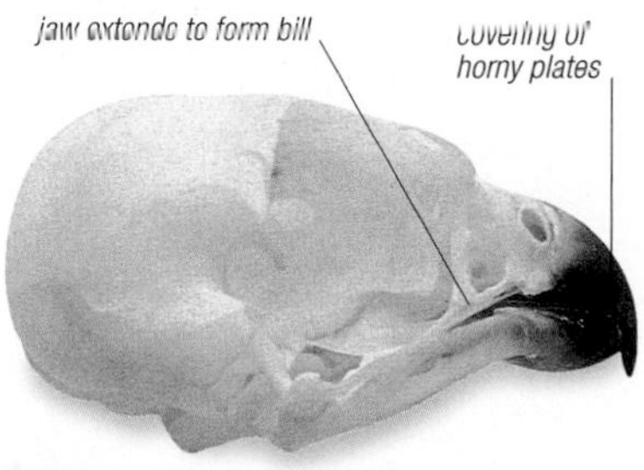

SKULL
The skull of a bird is very light, with many of the bones fused together. The bill consists of an upper and lower mandible, with a highly variable horny sheath on both.

DIGESTIVE SYSTEM

Since they lack teeth, birds are forced to break up particles of food farther down the alimentary canal, in the stomach. The lower part of the stomach (gizzard) grinds down material, often aided by an abrasive paste containing ingested grit or stones, while the upper part (proventriculus) secretes gastric juices. The food is often stored in the esophagus for later assimilation. In some birds, the esophagus expands near its base to make a saclike crop, a further storage vessel. Food can be transferred there very quickly, allowing for large quantities to be ingested in a relatively short time. This is valuable for birds that must risk exposure to predators while feeding.

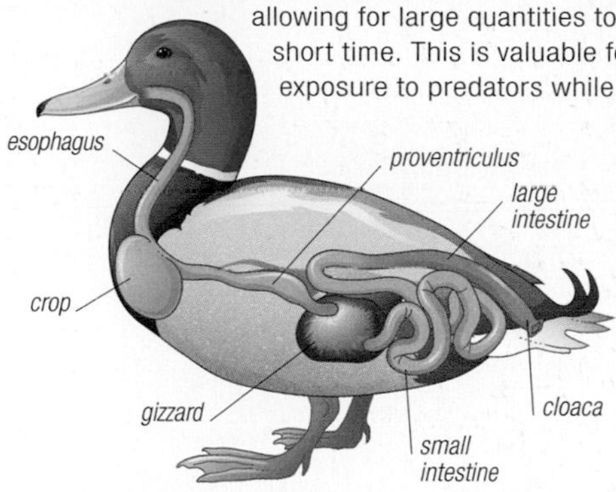

FEEDING AND STORING FOOD
A bird's digestive system is adapted for a dynamic lifestyle. Food is ingested without chewing, and can be stored in the esophagus or the crop. This allows the bird to feed quickly, but to digest in a safer place.

CROP FEEDING
Adult pigeons are unusual in producing a milky secretion from the crop to feed to their young. Here, a turtle dove gives crop milk to a nestling.

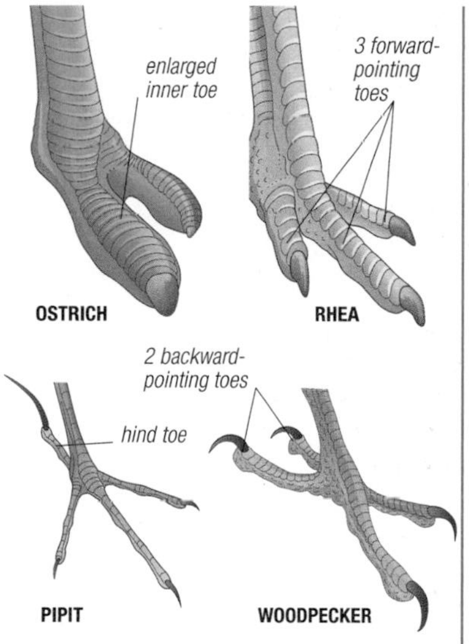

FEET AND TOES
Ostriches and rheas have fewer toes than most birds, enabling them to run quickly. The pipit has the perching foot typical of passerines, with 3 toes pointing forward and 1 back. The arrangement of a woodpecker's toes allows it to climb easily in all directions.

Respiration and circulation

Birds are active animals that have a high metabolic rate. They have an efficient respiratory system that extracts large amounts of oxygen from the air, and a circulation system that can move the oxygen rapidly around the body. Efficient oxygen extraction is also necessary so that birds can remain active at high altitude, where oxygen is less abundant. Although the lungs are small, they are connected to a series of air sacs found throughout the body, which help inflate and deflate the lungs. Instead of air flowing alternately in and out, as it does in mammals, it flows in a single direction. Working in tandem with this efficient oxygen-extraction system, birds have a large heart that pumps at a relatively rapid rate.

Birds need to maintain a constant internal temperature of about 104°F (40°C) to remain active. To achieve this, they control their rate of heat loss (see below). In some species, when there is not enough energy to sustain the metabolism, the internal temperature falls and the bird enters an inactive state called torpor. Some birds that live at high altitude, such as certain hummingbirds, become torpid overnight; others, such as some swifts and nightjars, can stay in this state for days or even weeks.

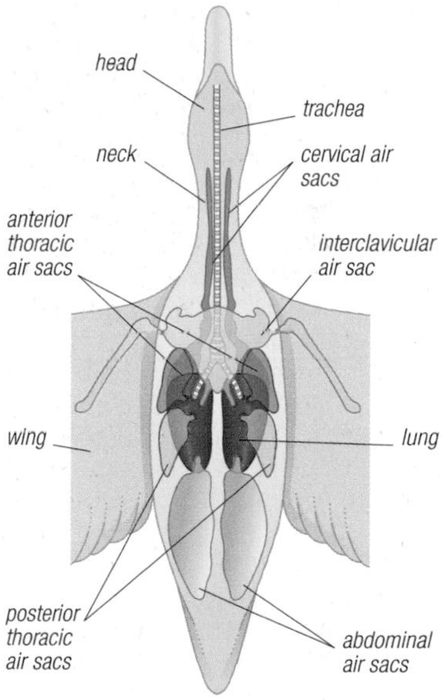

AIR SACS
A bird's lungs are only part of its air-intake system. Connected to the lungs is a system of air sacs, which have thin, nonmuscular walls and perform a similar function to the diaphragm in humans. In all, there are 9 air sacs throughout the body, 8 of which are arranged in pairs.

RESTRICTING HEAT LOSS
To control the rate of heat loss from the body, birds (here, a fieldfare) adjust the volume of insulating air trapped between the skin and the outermost feathers by erecting, or fluffing, the feathers. The greater the volume of air, the less heat is lost.

FAST FLYER
A blue and yellow macaw in flight displays its highly colorful plumage and long, elegant tail. The relatively narrow, tapering wings of macaws enable these birds to fly remarkably quickly, despite their large size.

Flight

Birds are not the only animals capable of powered flight (bats and insects are similarly able), but they include among their number the largest and most powerful of all airborne animals.

A bird's wing achieves lift in the same way as does the wing of an aircraft. As the bird moves forward, air flows more quickly over the upper surface of the wing than the lower one, creating a pressure differential between the surfaces. Since it is the slower-moving air that exerts the greater pressure, the resulting force is upward. The strength of this upward force (called lift) increases with the size of the bird's wing and its forward speed (see panel, below).

There are 2 main modes of flight: flapping and gliding (or soaring). Flapping does not create lift directly; instead, it generates the horizontal motion needed to increase airflow over the surface of the wing and produce lift. This is why most birds have to flap their wings to take off. When gliding, birds hold their wings outstretched and steady to the wind. A gliding bird consumes little energy, but it loses height as its speed, and hence the amount of lift, drop (unless it is riding a current of rising air).

Birds exhibit a great variety of wing shapes, reflecting the way they fly and their general lifestyle. Those that rely on a rapid takeoff and a short burst of speed to escape from predators tend to have broad, rounded wings, since these provide the required acceleration. Species that fly for long periods and need to save energy have long wings. Fast, powerful fliers, such as swifts and falcons, have long, curved wings with pointed tips to reduce drag. Tail shape is also important. Birds that make sudden changes of direction in midair, such as swallows and kites, often have forked tails.

Although most birds can fly, some are flightless—these include the ostrich, rheas, cassowaries, kiwis, the emu, and penguins. It is generally thought that these birds evolved from ancestors that once had the power of flight, since flightless birds have many adaptations for flight, such as hollow bones and wings (although the wings of many flightless birds have since become reduced in size). Flightlessness seems to have come about in one of several ways. For some birds in environments with relatively few predators, flight seems to have become unnecessary—many birds of isolated islands, such as the weka and takahe of New Zealand (which are closely related to cranes), are in this category. Alternatively, flightlessness may have arisen in situations where size and strength on the ground were more important for survival than flight. This is the case for the ostrich, which can run quickly, using its wings for balance; its long legs can also deliver a powerful kick to fend off predators. In penguins, the wings have become similar to flippers and are used by the bird to propel itself through water.

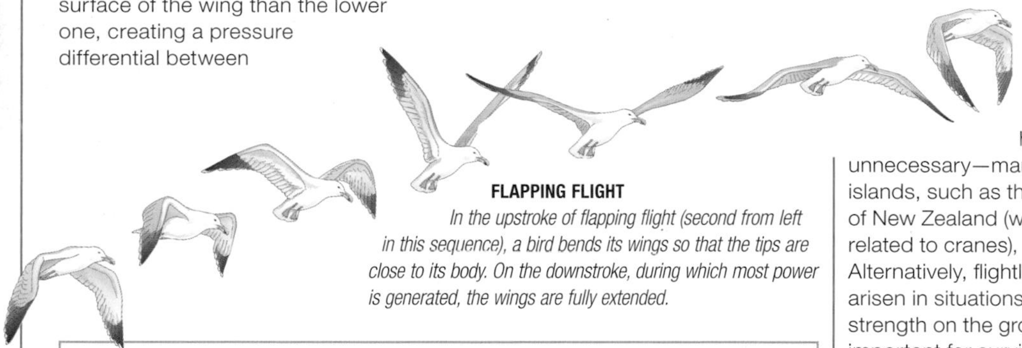

FLAPPING FLIGHT
In the upstroke of flapping flight (second from left in this sequence), a bird bends its wings so that the tips are close to its body. On the downstroke, during which most power is generated, the wings are fully extended.

HOW BIRDS FLY

Birds have a large, keeled breastbone, to which massive flight muscles are attached. When the muscles contract, they bring about the powerful downstroke of the wing, which produces forward propulsion. When the muscles relax, the wing is pulled back up. Feathers (see p.264) also enable birds to fly efficiently. Those on the trailing edge of the wings and on the tail are designed to provide lift and aid maneuverability, while the visible body feathers (contour feathers) streamline the body in flight. In most birds, flying consists of wing-flapping or gliding. Several birds are capable of hovering in midair; hummingbirds can also fly backward.

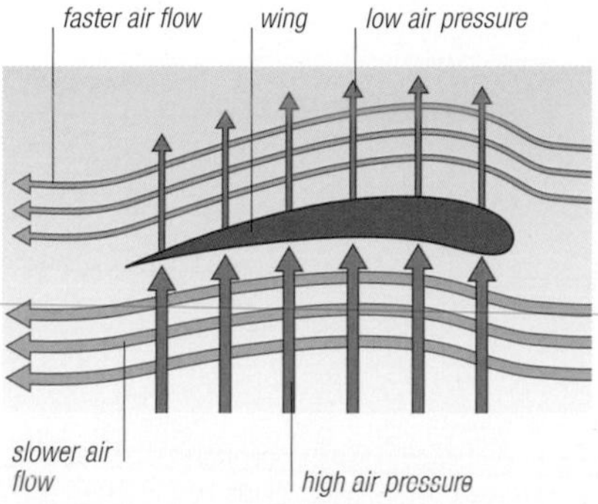

AERODYNAMICS OF THE WING
A bird's wing is curved outward along the upper surface. As a result, air passing over it has to travel a slightly longer distance at a faster pace than the air traveling along the lower surface. The slow-moving air passing under the wing exerts greater pressure, producing an upward force known as lift.

AERIAL MANEUVERING
Frigatebirds are highly distinctive in outline, soaring and gliding above the open sea in search of food. Their narrow, angular wings enable them to fly with great speed and agility, and the characteristic long, forked tail helps them steer.

Reproduction

Most birds are monogamous, breeding in pairs. In general, courtship consists of a male attracting a female with a visual display or by singing. Some birds perform spectacular displays: they may show off physical attributes, evident in the dazzling plumage displays of peacocks and birds of paradise; engage in dancing, as, for example, in cranes (see below); or make aerobatic flights, such as the dramatic swoops and dives of eagles. Many birds also draw attention to themselves using songs. The song, which is unique to the species, serves to repel male rivals and establish a breeding territory; it may also be used to attract a mate.

All birds reproduce by laying eggs, and they go to great lengths to ensure the survival of their clutch. Most do this by placing their egg (or eggs) in a specially built nest, which is either hidden or placed out of reach of predators. The eggs are incubated by one or more adults, almost always including the female. The number of eggs laid in a clutch varies greatly

COURTSHIP DANCE
A male and female red-crowned crane take part in their elaborate dance at the onset of the breeding season. One or both birds bow or bob their heads and leap into the air.

EGGS AND NESTING

Although many birds lay their eggs directly onto the ground or some other surface, most build a structure, known as a nest, in which to put them. Nests provide safety, insulation, and a fixed point for the adults to concentrate on nurturing their eggs and young. An important part of nest-building is choosing a suitable site—one that will provide concealment or inaccessibility from predators. Nests are made from various materials, depending on what is available—usually vegetation, but also animal hair, feathers, or even shed snake skin or human artifacts.

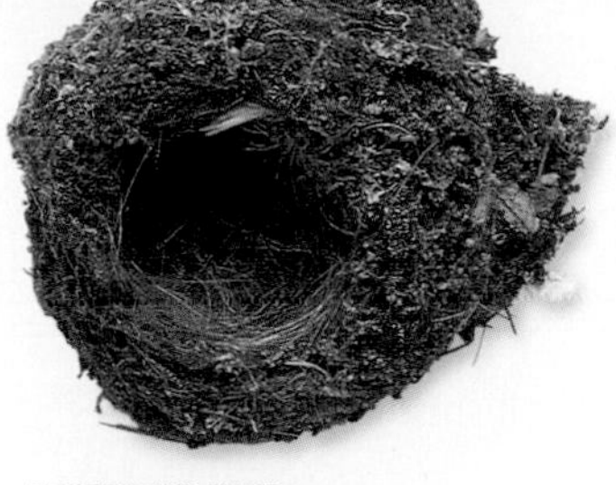

PROTECTIVE NEST
The most common type of nest is an open cup. Most nests are made of more than one material: this chaffinch's nest is bound by spiders' webs and built up with moss, lichen, grass, and feathers.

BLACK-HEADED TROGON
BOUBOU SHRIKE
NATAL CHAT
CETTI'S WARBLER
BLUE SHORTWING

EGG COLOR
Birds' eggs have a wide range of colors. However, all shell colors are the product of just 2 pigments, one derived from hemoglobin and the other from bile. These are added as the egg moves down the female's genital tract.

COMMUNAL FEEDING
Flamingos are among the most sociable of all birds, living in vast groups (sometimes flocks of a million or more) throughout their lives and performing all living functions in company with others. Even their courtship displays are communal, synchronized affairs.

between species, and some birds produce more than a single clutch in a year.

A bird's egg is contained by a light but strong shell that protects the developing embryo and acts as a barrier against bacteria. The shell is made of calcium carbonate, which the female absorbs from her food. Despite appearing to be hard, the shell is porous, allowing for the free exchange of oxygen and carbon dioxide across its surface. Inside, the embryo is nourished by a large reservoir of nutritive material that promotes its growth. Even so, when young birds hatch out, they are often poorly developed. In most species, the hatchlings are blind and without feathers; they also lack the ability to regulate their internal temperature, so they must be brooded. These dependent chicks are referred to as nidicolous or altricial young and are completely reliant on their parents for warmth and food. In contrast, some groups of birds, such as waterfowl and gamebirds, hatch nidifugous or precocial young, which are covered with down and are able to feed themselves within hours of hatching.

BLUE TIT HATCHLINGS

DUCK HATCHLING

THE FIRST DAYS OF LIFE
Most young birds, including passerines (such as the blue tit), depend on adults for food and warmth. Their eggs are laid in well-protected nests. Waders, waterfowl, and gamebirds produce more independent young that can walk and fend for themselves within hours.

Social groups

Birds vary greatly in the way that they relate to their own kind. Some are solitary or form pairs within a defined territory. Many meet up with others for specific activities, such as roosting, feeding, or breeding. Others live in groups throughout their lives.

A social lifestyle carries both advantages and disadvantages. Birds in groups benefit from the collective effort put into searching for food, particularly where supplies are sparsely distributed. Birds in flocks are also less at risk from predators: to be one among many reduces the chances of being singled out and caught, and individuals are more likely to have early warning of a predator's approach. This state of collective vigilance allows birds in a group to spend more time feeding or sleeping than they would if they were alone. In addition, some birds huddle together while roosting to share body heat and reduce heat loss. However, there are also disadvantages in being part of a flock. If food is scarce, competition between group members can cause weaker individuals to become marginalized, unable to feed properly and vulnerable to predators; such birds would be better off on their own. Another disadvantage of social living is the threat of diseases, which spread more rapidly through a group than they do among a population of solitary individuals.

FLOCKING TOGETHER
European starlings feed and roost together, and most also breed in small colonies. Starling roosts are famous for their spectacular aerial maneuvers prior to settling down. Up to 2 million birds may share the same roost site.

Migration

Many birds undertake seasonal migrations. Most migratory species breed in spring and summer in high latitudes, taking advantage of the relatively long days, but move in winter to lower latitudes, where temperatures are higher. Many species show complex patterns of movement, with only certain populations—or in some cases, just females—leaving the breeding areas while others remain behind. Birds that behave in this way are referred to as partial migrants. Not all birds migrate, partly because migration consumes a lot of energy; those that do not migrate at all (including many tropical species) are referred to as sedentary.

The urge to migrate is triggered by a combination of internal physiological cycles (such as hormone levels) and changes in day length. As the time to migrate approaches, birds lay down reserves of fat to sustain them on their journey and show signs of restlessness. Some birds have an impressive ability to navigate, traveling thousands of miles to arrive at a destination with pinpoint accuracy (see panel, below).

MIGRATION ROUTES

Birds have various ways of navigating. They orient themselves mainly using their efficient internal body clock, which enables them to measure changes in day length, as well as the position of the sun and, at night, of the moon and stars. Many species, if not all, can also detect variations in Earth's magnetic field, which they use as a compass. Birds that have already made several migratory journeys recall landmarks and may use clues such as smells or ultrasounds.

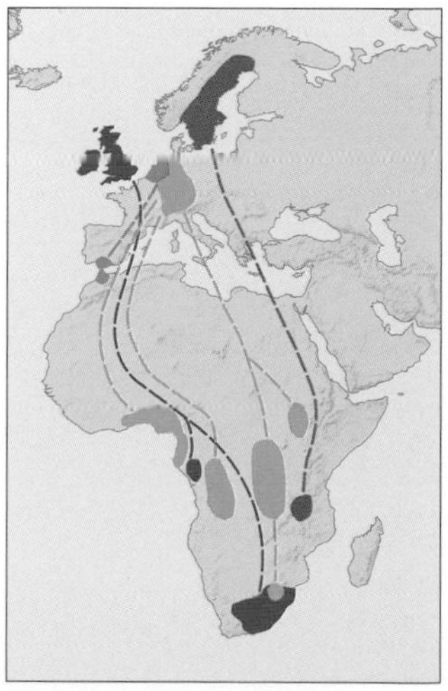

LONG-DISTANCE MIGRANTS
Swallow populations that winter in different parts of Northern Europe follow a range of migration routes. Some travel the relatively short distance to southern Spain or North Africa. Others cover the length of the African continent.

BIRDS

Tinamous

PHYLUM	Chordata
CLASS	Aves
ORDER	Tinamiformes
FAMILIES	1 (Tinamidae)
SPECIES	45

These ground-living birds, which look similar to grouse and partridges, have a plump body and small wings. When running or flying, they tire easily because they have a small heart relative to their size. Tinamous are common in woodland, scrub, and grassland throughout Central and South America.

CAMOUFLAGE
When threatened, this Andean tinamou remains motionless, protected by its excellent camouflage.

Eudromia elegans

Elegant crested tinamou

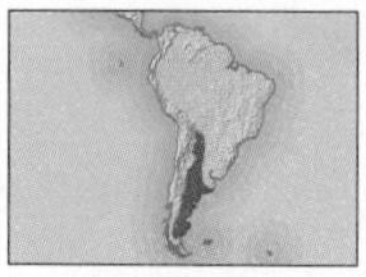

Location S. South America

Height 14½–16 in (37–41 cm)
Weight 14–29 oz (400–800 g)
Plumage Sexes alike
Migration Nonmigrant
Status Least concern

The long, usually forward-curving crest is the distinguishing feature of this tinamou. Light to dark brown, the elegant crested tinamou has wings that are spotted white. One light stripe runs behind the eye, and another runs below it. This relatively shy bird usually moves in small to moderately large groups through woodland, grassland, and brush.

Kiwis

PHYLUM	Chordata
CLASS	Aves
ORDER	Apterygiformes
FAMILIES	1 (Apterygidae)
SPECIES	3

The strange, flightless kiwis all look much the same. They have a stout body covered in soft, hairlike plumage, a slender, curved bill, and no tail. Unlike larger flightless birds, kiwis have 4 toes on each foot. These nocturnal birds are native to New Zealand.

SENSES
Although kiwis have poor eyesight, they have acute senses of hearing and smell and a highly touch-sensitive bill.

Apteryx mantelli

North Island kiwi

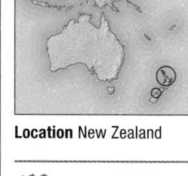

Location New Zealand

Height 20–26 in (50–65 cm)
Weight 3¼–8¾ lb (1.5–4 kg)
Plumage Sexes alike
Migration Nonmigrant
Status Endangered

Once widespread throughout New Zealand's North Island, this brown bird has been severely affected by deforestation and introduced predators such as pigs, dogs, cats, and stoats. When feeding, it walks slowly, tapping the ground with its bill and sniffing. It may push its entire bill—up to 6 in (15 cm) long—into the ground to get at food, which consists of earthworms, cicadas, beetle larvae, centipedes, and fallen fruit. Females lay one to 2 eggs that are very large in proportion to the bird's size.

Cassowaries and emus

PHYLUM	Chordata
CLASS	Aves
ORDER	Casuariiformes
FAMILIES	2
SPECIES	4

These large, flightless birds, from Australia and New Guinea, have a long neck, long legs, and small wings hidden under loose, hairlike plumage. As well as being able to run quickly, they can also swim. Cassowaries and emus have 3 toes on each foot; in cassowaries, the innermost toe bears a sharp claw, up to 4 in (10 cm) long, which can inflict lethal wounds. Cassowaries have a protective casque on their head, and neck wattles that can change color according to the bird's mood.

PATERNAL CARE
The male emu (shown here) does not eat or drink while incubating. After hatching, he cares for the young for up to 8 months.

Casuarius casuarius

Southern cassowary

Location New Guinea, N.E. Australia

Height 4¼–5½ ft (1.3–1.7 m)
Weight 37–155 lb (17–70 kg)
Plumage Sexes alike
Migration Nonmigrant
Status Vulnerable

All cassowaries inhabit dense tropical forest, and as a result are rarely seen. The southern cassowary—the largest of 3 cassowary species—is the only one found in Australia as well as in New Guinea. During the breeding season, males make a low booming sound to attract females, but when not breeding, the adults are solitary. Their diet consists chiefly of fallen fruit.

Dromaius novaehollandiae

Emu

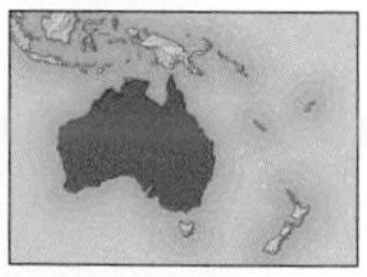

Location Australia

Height 5–6¼ ft (1.5–1.9 m)
Weight 66–130 lb (30–60 kg)
Plumage Sexes alike
Migration Nonmigrant
Status Least concern

Australia's largest native bird, the emu has shaggy, drooping, gray-brown feathers, large legs, but tiny wings. It is highly gregarious, and lives in loose flocks that can contain dozens of birds. It feeds mainly on seeds and berries, and will travel long distances when food is hard to find. Emus are now extinct in Tasmania, but on the Australian mainland, they have benefited from cereal farming, and are now a serious pest in some areas.

Ostrich

PHYLUM	Chordata
CLASS	Aves
ORDER	Struthioniformes
FAMILIES	1 (Struthionidae)
SPECIES	1

The largest of all birds, the ostrich is unmistakable in appearance. It has a long, bare neck and small head, a massive body, and long, muscular legs. The wings are small and covered with loosely packed feathers. The ostrich is unique among birds in having only 2 toes on each foot. Although too heavy to fly, it is capable of running with remarkable speed and stamina—it can travel up to 45 mph (70 kph) for as long as 30 minutes. The single species of ostrich is placed in a separate order from other birds.

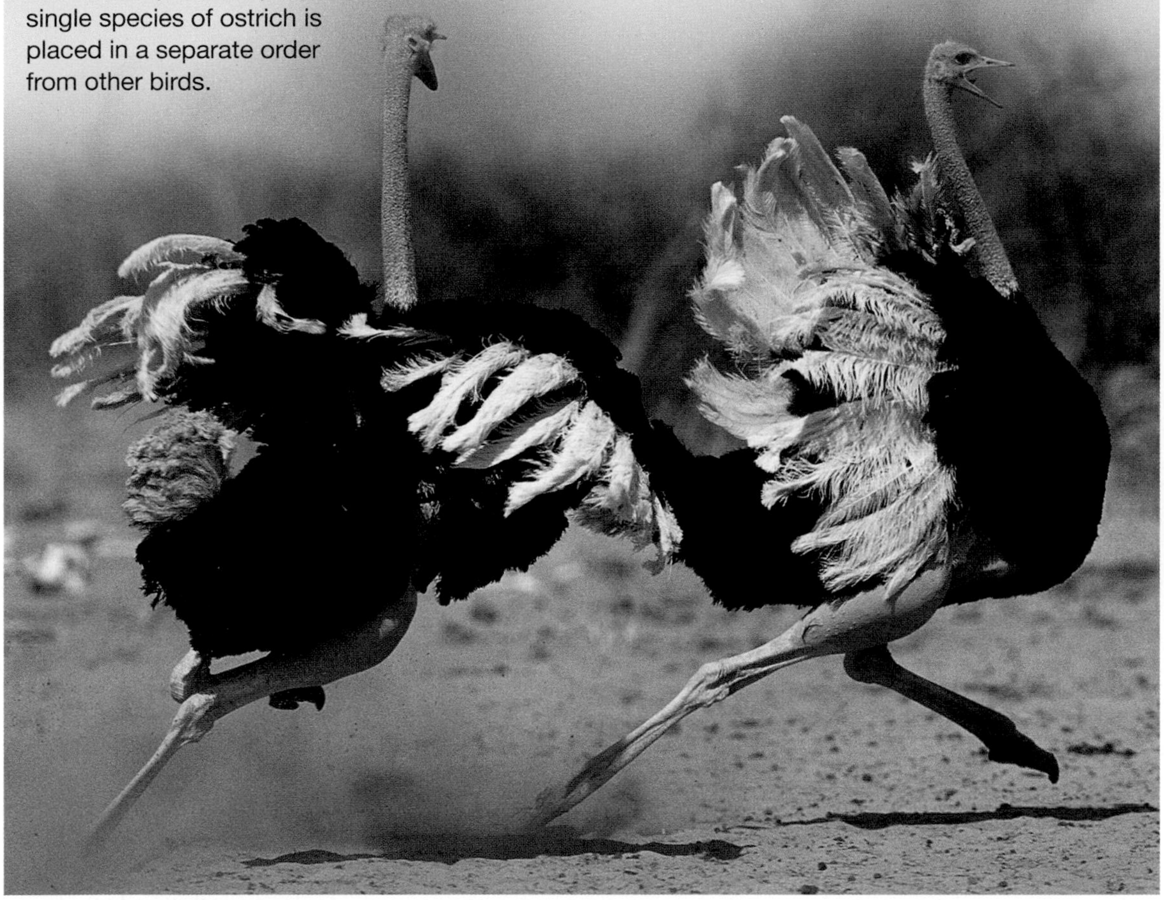

FIGHTING MALES
Male ostriches compete for territory and social status with aggressive displays and occasionally by fighting. The winner of a contest acquires a territory and several females, although only one female, the "major hen," remains during incubation and chick-rearing.

Struthio camelus

Ostrich

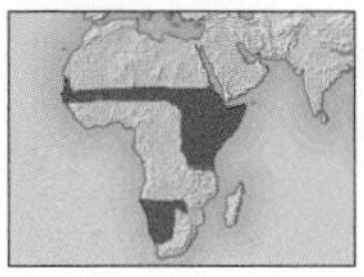

Location W. to E. Africa (south of Sahara), southern Africa

Height 7–9¼ ft (2.1–2.8 m)
Weight 220–350 lb (100–160 kg)
Plumage Sexes differ
Migration Nonmigrant
Status Least concern

Once widespread in Africa and West Asia, ostriches are now restricted largely to eastern and southern Africa, although they are also farmed in other parts of the world. In the wild, they are seminomadic, traveling long distances to find grass and other plant food. They typically form mixed-sex herds, and are rarely found on their own. During the breeding season, males make loud, booming calls and perform elaborate displays. Several females often lay in a single nest, producing a joint clutch of up to 30 eggs. The male takes part in incubation; once the eggs hatch—after about 40 days—he is usually in sole charge of the young.

Rheas

PHYLUM	Chordata
CLASS	Aves
ORDER	Rheiformes
FAMILIES	1 (Rheidae)
SPECIES	2

Although rheas look similar to the ostrich, they are much smaller and have 3 (rather than 2) toes on each foot. The head, neck, and thighs are all covered with feathers, and each wing ends in a claw, which is used to fight predators. When running, rheas use their wings for balance. They occur on upland and lowland plains in most parts of South America.

CARE OF THE YOUNG
Only the male bird incubates the eggs and watches over the young (lesser rheas are shown here). The female goes off in search of other males.

Rhea americana

Greater rhea

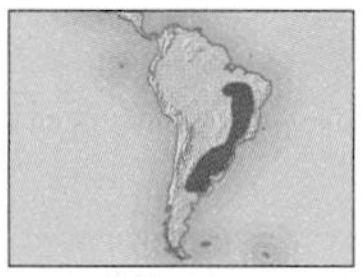

Location E. and S.E. South America

Height 3–5 ft (0.9–1.5 m)
Weight 33–66 lb (15–30 kg)
Plumage Sexes differ
Migration Nonmigrant
Status Near threatened

This long-legged, flightless bird lives in grassland and semiarid scrubland. It is generally gray to brown, and white, the plumage offering effective camouflage against tall grass and scrubby vegetation. The male has a dark collar at the base of its neck during the breeding season. A sociable bird, it lives in groups, although breeding males are solitary for part of the year. In the mating season, males produce a booming call to attract females, then court them with an impressive wing display. The male mates with up to 12 females and then scrapes out a nest, in which the females lay up to 60 eggs between them. After the eggs are laid, the male takes sole charge of incubation and protecting the young chicks. This rhea has suffered through hunting and habitat loss, although it is still locally abundant.

BIRDS

Gamebirds

PHYLUM	Chordata
CLASS	Aves
ORDER	Galliformes
FAMILIES	5
SPECIES	290

This group of mainly ground-dwelling species includes some of the birds most useful to humans. In their domestic forms (including the chicken), they provide an important food source, and many other species (such as pheasants, partridges, and grouse) are hunted for sport or food. Also in this group are the spectacular curassows and the less conspicuous guineafowl and megapodes. Gamebirds are found almost worldwide (even within the Arctic Circle) in a wide range of habitats, including dense forest and high mountains.

DISPLAY
This is the display of the Palawan peacock-pheasant. The "eyes" on the fan are thought to attract the female.

Anatomy

Most gamebirds are plump with a small head and short, rounded wings. Their powerful flight muscles are ideal for rapid escape, but are usually unable to support their heavy body for long distances. The short bill is slightly curved, and the feet are stout and strong for scratching and digging for food. Many species have bare areas of colored skin or long and spectacular tails or crests.

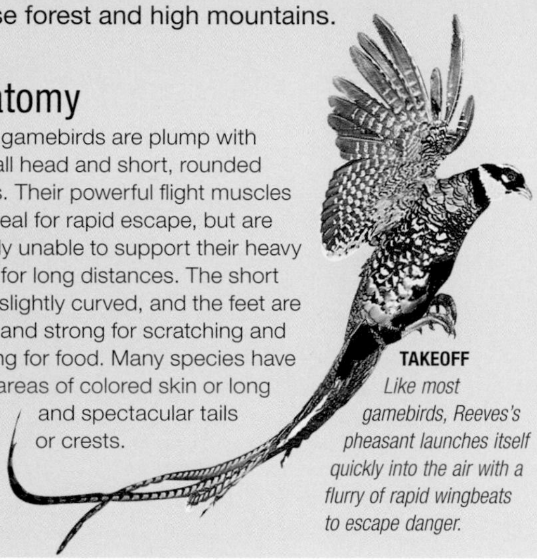

TAKEOFF
Like most gamebirds, Reeves's pheasant launches itself quickly into the air with a flurry of rapid wingbeats to escape danger.

CAMOUFLAGE
Ground-dwellers, like this female black grouse, need cryptic plumage to avoid detection. Another grouse species, the ptarmigan, changes color with the seasons.

Reproduction

Most gamebirds nest in a shallow depression in the ground. Megapodes are unusual in that, instead of incubating their eggs, they store them in mounds or burrows, allowing the sun's heat, microbial activity, or geothermal energy to keep them warm. Young gamebirds fly very soon after hatching—often within a week or, in the case of megapodes, within a few hours.

CLUTCH SIZE
Compared with other birds, many gamebirds lay large clutches. Some produce up to 20 eggs (a pheasant's eggs are shown here).

Leipoa ocellata

Malleefowl

Location W. and S. Australia

Length 24 in (61 cm)
Weight 4½ lb (2 kg)
Plumage Sexes alike
Migration Nonmigrant
Status Vulnerable

The malleefowl and its close relatives—known as megapodes—are the only birds that do not directly incubate their eggs. Instead, eggs are laid in a giant heap of leaves, sticks, and bark, up to 5 ft (1.5 m) high and 15 ft (4.5 m) across, and the heat given off as this decomposes incubates the eggs. During the lengthy incubation period, lasting up to 11 weeks, the parents stay close to the mound and monitor its temperature with their bills. If it gets too hot, they take some of the vegetation away; if too cool, they add more to it. When the young hatch, they dig their way out—fully feathered and able to look after themselves. Male and female malleefowl form pairs that last for years; they live fairly separately much of the time, but come closer together during the summer breeding season. The malleefowl lives on a largely vegetarian diet consisting of fruit, buds, and seeds; however, it will also eat invertebrates such as ants, beetles, spiders, and cockroaches.

Ortalis motmot

Little chachalaca

Location N. South America

Length 15 in (38 cm)
Weight 21 oz (600 g)
Plumage Sexes alike
Migration Nonmigrant
Status Least concern

This species is the smallest member of the Cracidae—a family of gamebirds from the Americas that also includes the curassows. Compared to these, chachalacas are slim-bodied and plain in color, and they do not have crests. There are about 12 species and they are all noisy birds—their name comes from their call. They live mostly in trees, and feed on berries and other fruit.

Crax daubentoni

Yellow-knobbed curassow

Location N. South America

Length 35 in (90 cm)
Weight Not recorded
Plumage Sexes alike
Migration Nonmigrant
Status Near threatened

Like other curassows, this large forest bird feeds mainly on the ground, but it flies up into trees if threatened. Its most striking features are its crest, made of feathers that curl forward, and the fleshy yellow knob at the base of its bill. It eats fruit, leaves, seeds, and small animals. Unusually for gamebirds, curassows nest off the ground, with both sexes helping in the construction. The female lays just 2 eggs—a tiny clutch compared to those of many ground-nesting gamebirds.

Meleagris gallopavo

Wild turkey

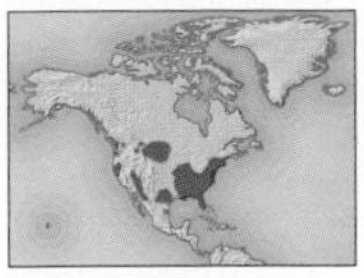

Location North America

Length 4 ft (1.2 m)
Weight 22 lb (10 kg)
Plumage Sexes differ
Migration Nonmigrant
Status Least concern

The wild turkey is a large gamebird with bronze, iridescent plumage and a naked head with conspicuous blue and red fleshy ornaments. The male has a "beard" of hairlike feathers on the upper breast. For much of the year, this turkey is seen in groups of around 20, but in the breeding season, the males set up individual territories. Each male mates with a number of females, courting them with tail fanned, wings spread low, and head held high, giving the characteristic gobbling call. The wild turkey is omnivorous: chicks take up to 4,000 insects per day; adults eat seeds, herbs, roots, buds, and flowers, as well as insects. To defend itself, it pecks with its bill, scratches with its claws, and buffets with its wings.

Callipepla californica

California quail

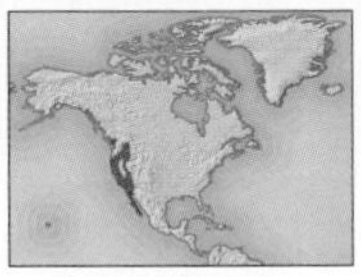

Location S.W. Canada, W. USA, N.W. Mexico

Length 10 in (25 cm)
Weight 6 oz (175 g)
Plumage Sexes differ
Migration Nonmigrant
Status Least concern

A long, black, teardrop-shaped crest, black and white facial feathers, and scaly plumage on the belly are distinctive features of this elegant gamebird. The female is smaller than the male, has a smaller crest, and is generally drabber in appearance. Shy and elusive, the California quail is more often heard than seen. It lives in small flocks of usually 25–30, and eats a variety of seeds and bulbs, as well as leaves and buds.

Tetrao urogallus

Western capercaillie

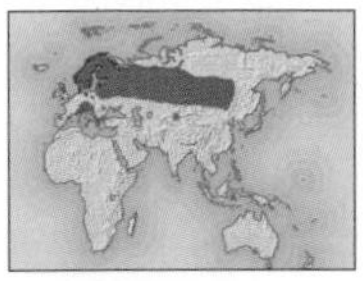

Location N., W., and S. Europe, W. to C. Asia

Length 32–45 in (80–115 cm)
Weight 8¾–10 lb (4–4.5 kg)
Plumage Sexes differ
Migration Nonmigrant
Status Least concern

Males of this large-bodied grouse congregate at display grounds (leks) in their native northern forests, where they strut to display to females; their calls sound like a bottle being uncorked. In summer, it feeds on leaves, buds and berries; in winter, it eats almost exclusively pine needles.

Centrocercus minimus

Gunnison sage-grouse

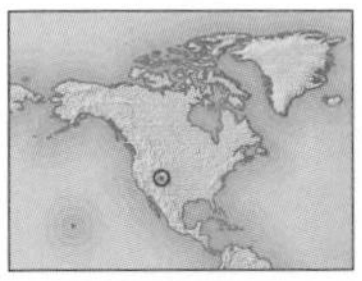

Location W.C. USA

Length 12½–20 in (32–51 cm)
Weight 2–4½ lb (1–2 kg)
Plumage Sexes differ
Migration Nonmigrant
Status Endangered

The smaller of 2 species of North American sage-grouse, this species was named in 2000 and is restricted to southwestern Colorado and southeastern Utah. It is—so far—the only new species of bird to be described from the USA since the 1800s. Male sage-grouse strut in groups with inflated chest sacs and make drumming sounds to attract mates; they favor open country dominated by sagebrush.

Lagopus lagopus

Willow ptarmigan

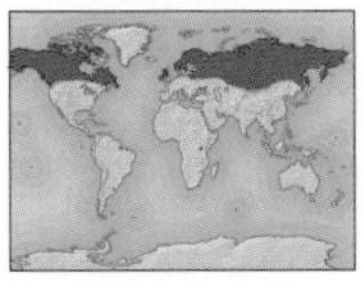

Location N. North America, N. Europe, N. Asia

Length 15 in (38 cm)
Weight 20–25 oz (550–700 g)
Plumage Sexes differ
Migration Nonmigrant
Status Least concern

This exceptionally hardy gamebird—of which there are 20 subspecies—is well adapted to life in the harsh conditions of the northern winter. Like other grouse, its legs and nostrils have feathers for insulation. It tunnels in snow to keep warm, and—except in the British subspecies—the normally reddish brown plumage turns white in winter, giving the bird excellent camouflage.

Tetraogallus caspius

Caspian snowcock

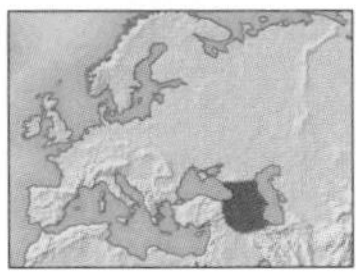

Location S.E. Europe, W. Asia

Length 23½ in (60 cm)
Weight Not recorded
Plumage Sexes differ
Migration Nonmigrant
Status Least concern

Typical of the 5 species of snowcocks, this bird has mottled brown, gray, and white plumage that provides good camouflage against bare rock. It feeds in small flocks, and migrates vertically with the seasons, often descending below the treeline in winter.

Perdix perdix

Gray partridge

Location Europe, W. and C. Asia

Length 12 in (31 cm)
Weight 11–16 oz (300–450 g)
Plumage Sexes differ
Migration Nonmigrant
Status Least concern

The gray partridge is a farmland bird, feeding and nesting among crops as well as in pasture. It has a tawny head and grayish breast, and the male has a conspicuous chestnut horseshoe mark on the abdomen. Coveys of 15–20 birds live together in winter, but late in the season these begin to break up as the males become more aggressive, even fighting with one another. Although changes of mate are common early on, stable pairs soon form, usually of males and females from different coveys.

Coturnix coturnix

Common quail

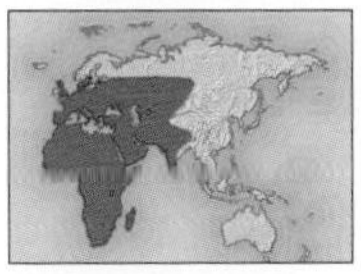

Location Europe, Asia, Africa, Madagascar

Length 7 in (18 cm)
Weight 2½–5 oz (70–150 g)
Plumage Sexes alike
Migration Migrant
Status Least concern

A small and secretive gamebird, this quail is heard much more than it is seen. The first sign of its presence is most often the male's repeated "whit wit-wit" call. During the breeding season, the male and female establish contact by calling, which means they can locate a mate while remaining safely hidden. The common quail roosts on the ground at night in tight groups. Its diet is remarkably varied, and includes seeds, flower buds, leaves, small fruit, and insects and other invertebrates. This is one of the few gamebirds that migrate long distances; those that breed in Europe arrive in spring from Africa.

Tragopan temminckii

Temminck's tragopan

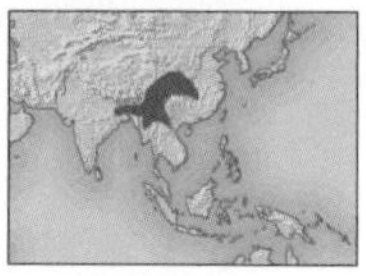

Length 25 in (63 cm)
Weight Not recorded
Plumage Sexes differ
Migration Nonmigrant
Status Least concern
Location S.E. Asia

Male gamebirds often have flamboyant plumage, and this species is no exception. However, with Temminck's tragopan, the male's most conspicuous feature is a blue and red throat wattle, which looks like a multicolored bib. Like the other 4 species of tragopans (all from central and southern Asia), Temminck's is a forest bird, often nesting in bushes or low down in trees, where it makes a simple platform out of sticks. It lives at altitudes of up to 14,850 ft (4,500 m), and feeds mainly on plants—including young shoots and berries—but also eats insects scratched up from the forest floor.

INFLATED DISPLAY

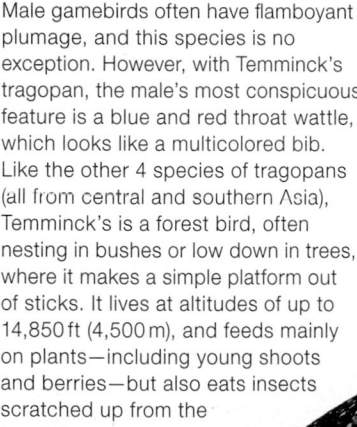

During his courtship display, the male inflates his colorful throat wattle until it covers his breast, and then shakes it to attract the female's attention. If the female is sufficiently impressed, she allows the male to mate.

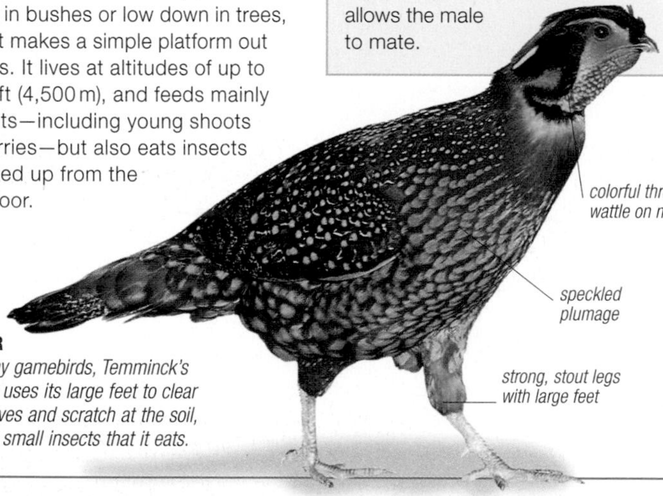

FORAGER
Like many gamebirds, Temminck's tragopan uses its large feet to clear away leaves and scratch at the soil, exposing small insects that it eats.

Pavo cristatus

Indian peafowl

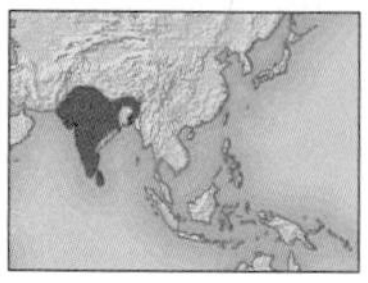

Length 6–7½ ft (1.8–2.3 m)
Weight 8¾–13 lb (4–6 kg)
Plumage Sexes differ
Migration Nonmigrant
Status Least concern
Location S. Asia

The male Indian peafowl, or the peacock, is one of the world's most spectacular gamebirds, with an iridescent blue body and a long train that is spread out like a fan when courting females. The train is not the true tail, but consists of elongated tail coverts, each ending in a colorful "eye." The female—the peahen—is relatively drab, with a shorter train lacking the "eyes." The female chooses a mate on the basis of his appearance, and he usually turns to face her, shaking the train, erect and fanned, to reinforce its effect. He mates with many females, and gives a loud, "kee-ow" call to advertise his presence. The male plays no part in building nests or raising the young. Like almost all gamebirds, the peafowl feeds on the ground, but roosts in tall trees at night, safe from most predators.

Gallus gallus

Red jungle-fowl

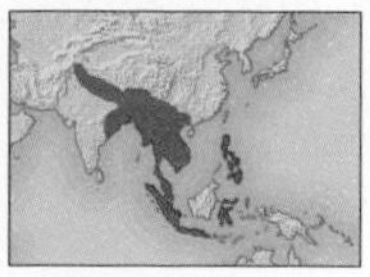

Length 32 in (80 cm)
Weight 1–3¼ lb (0.5–1.5 kg)
Plumage Sexes differ
Migration Nonmigrant
Status Least concern
Location S. and S.E. Asia

This bird is the original ancestor of the domesticated chicken, first raised in captivity at least 5,000 years ago. The male is brightly colored, with fleshy red wattles and comb. The female is smaller and drabber. Hens and chicks use calls to keep in contact and signal danger; the male's "cock-a-doodle-doo" is used to attract females and advertise his presence to rival males.

Argusianus argus

Great argus

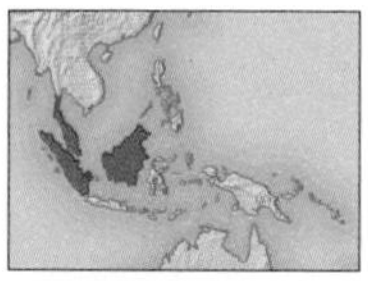

Length 6¼ ft (1.9 m)
Weight Not recorded
Plumage Sexes differ
Migration Nonmigrant
Status Near threatened
Location S.E. Asia

One of the world's largest pheasants, the male of this species has very large secondary flight feathers, decorated with egg-shaped "eyes," and a tail much longer than that of the female. To attract a mate, he calls loudly, raises his tail, and fans out his wings. The females with which he mates raise the young on their own.

Phasianus colchicus

Ring-necked pheasant

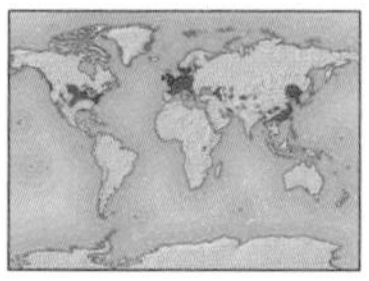

Length 35 in (89 cm)
Weight 1¾–4½ lb (0.75–2 kg)
Plumage Sexes differ
Migration Nonmigrant
Status Least concern
Location North America, Europe, Asia

Male ring-necked pheasants have a dark head with a purple and green gloss and red facial wattles; many also have a white neck ring and maroon breast. Compared to the drab brown female, the male is very colorful and larger, with a long tail. With more than 30 subspecies, this pheasant has been widely introduced, chiefly so that it can be hunted for food.

Numida meleagris

Helmeted guineafowl

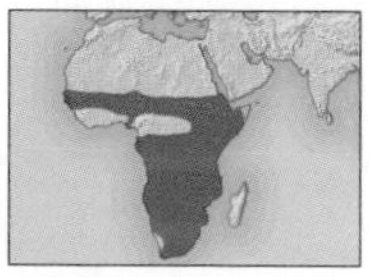

Length 22 in (55 cm)
Weight 2¼–3¼ lb (1–1.5 kg)
Plumage Sexes alike
Migration Nonmigrant
Status Least concern
Location Africa (south of Sahara)

Domesticated long ago as a source of food, the helmeted guineafowl is a flock-forming gamebird whose natural habitat is the open grassland of tropical Africa. Its most conspicuous features are its spotted plumage and bony, hornlike helmet on top of its naked head. Males have a characteristic lateral, hump-backed posture during encounters with other males and while courting. Compared to most gamebirds, it is well able to defend itself and its young—by pecking and scratching, and buffeting with its wings. A very noisy bird, when alarmed it gives a staccato "kek-kek-kekkek-kekkekkek" call.

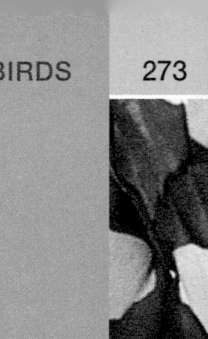

GREATER PRAIRIE CHICKENS
Typical of other members of the grouse family, the males of this North American bird have an elaborate courtship display to impress the females. Males gather on mating grounds called leks—but the tension can lead to aggressive attacks that involve much leaping into the air.

Waterfowl

PHYLUM	Chordata
CLASS	Aves
ORDER	Anseriformes
FAMILIES	3
SPECIES	174

Also known as wildfowl, members of this group include ducks, geese, and swans, as well as a family of 3 South American species known as screamers. Aided by waterproof plumage and webbed feet, they are among the dominant birds of freshwater wetlands. They are also found in estuaries and inshore coastal waters, while a few species are entirely marine. Waterfowl are strong swimmers. They feed mostly from the surface of water, although many ducks dive in search of food, while some species (notably geese, swans, and screamers) graze on land. Waterfowl are powerful fliers, and some species undertake annual migrations of thousands of miles between their wintering and breeding areas.

Reproduction

The reproductive cycle of waterfowl is suited to a life spent on water, which often involves exposure to cold and predators. Unusual among birds, the males have an organ similar to a penis, which is inserted inside the female's cloaca, making it possible for them to mate on the surface of water. Waterfowl usually nest near water, most often among ground vegetation, although some species use holes in trees or rock crevices. The females of some species, such as eiders, pluck their own down to make an insulating cover for their eggs.

YOUNG
Waterfowl hatch in a well-developed state, with their eyes open and a covering of down that dries quickly. Hours after hatching, the chicks (such as these Canada geese) can walk or swim. Leaving the nest, they follow the adult to the relative safety of water, and soon begin to search for their own food.

TAKING OFF
Although some waterfowl can spring almost vertically from water, the heavier species have to resort to a running takeoff. Swans, in particular, need a long runway, gaining speed by pattering with their feet across the surface of the water.

Anatomy

Waterfowl have a plump, buoyant body, with a small head and, usually, a short tail. Most species have a broad, flattened bill (see below) and a long neck for reaching down to feed underwater. Many species are gaudily colored. When molting, most lose all their flight feathers simultaneously. To hide them from predators during the period when they are unable to fly, the males (drakes) acquire a drab coloring, known as an eclipse plumage. Waterfowl are insulated from cold by a layer of down and a layer of fat beneath the skin. Screamers look different from other waterfowl in several respects: they have longer legs, spurred wings, a chickenlike bill, and toes webbed only at the base.

BILL
Like most waterfowl, the common shelduck has a wide, flattened bill. The edges have small ridges, called lamellae, which are used to grip prey or filter edible particles out of the water. The tip of the upper bill may be hardened into a nail for tearing vegetation.

PREENING
Plumage care is important for waterfowl such as this greylag goose, because it is essential that their feathers remain waterproof. A gland on the bird's rump secretes a water-repellent oil. The bird stimulates the gland with its bill and then spreads the oil around its body by rubbing and preening its feathers.

FEET
Waterfowl use their webbed feet to help propel themselves through water. Only the front 3 toes are webbed; the smaller hind toe is raised. This enables waterfowl to walk on land, although they do so with a waddling gait.

Movement

Several adaptations help waterfowl move easily through water. The smooth outline of their body reduces water resistance, while their webbed feet (see left) act as powerful paddles. Thick plumage adds to their buoyancy, although many duck species that dive to find food sleek their plumage to expel air and reduce their buoyancy before submerging. With their huge wings and exceptionally light bones, screamers can fly for hours on end, often soaring. Other species of waterfowl have smaller wings that must be flapped constantly when in flight. Once airborne, however, they are fast and powerful fliers, with some species capable of exceeding 60 mph (100 kph) in level flight.

FLYING FORMATION
Flocks of waterfowl (here, snow geese) fly in a characteristic V-formation. By staying in the slipstream of the leading bird, those further back meet with less turbulence and thereby save energy. The position of hard-working leader is changed regularly.

Anhima cornuta

Horned screamer

Length 33 in (84 cm)	
Weight 4½–6½ lb (2–3 kg)	
Plumage Sexes alike	
Migration Nonmigrant	
Status Least concern	

Location N. South America

Although related to other waterfowl, screamers have many distinctive features, including turkeylike bodies, narrow bills, and long legs ending in partly webbed feet. The largest of the 3 species in the screamer family, the horned screamer is black and white, with a remarkable slender spike, up to 4 in (10 cm) long, that curves upward and forward from its forehead. Like its relatives, it often feeds on land rather than in water. Females lay 4–6 eggs in a nest on the ground, and these take 6 weeks to hatch.

Dendrocygna eytoni

Plumed whistling duck

Length 16–23½ in (40–60 cm)	
Weight 1–3¼ lb (0.5–1.5 kg)	
Plumage Sexes alike	
Migration Partial migrant	
Status Least concern	

Location N. and E. Australia

Like all 8 species of whistling ducks, this species has long legs and large, webbed feet, with a netlike pattern on the ankle and heel, a characteristic that is more typical of a goose than a duck. The plumed whistling duck also has some gooselike behavioral traits: it mostly feeds on land, pulling and clipping grasses. It is believed to remain monogamous for life. The male also cares for the young.

Cygnus olor

Mute swan

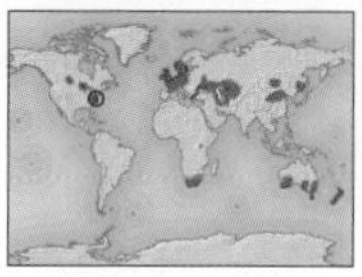

Length 5 ft (1.5 m)	
Weight 26 lb (12 kg)	
Plumage Sexes alike	
Migration Partial migrant	
Status Least concern	

Location North America, Europe, Africa, Asia, Australia

Originally from Europe and Central Asia, this extremely elegant bird has been introduced as an ornamental species in many other parts of the world. When young, the mute swan is grayish brown, but the adult has pure white plumage, an orange-red bill, and black legs and feet. One of the world's heaviest flying birds, it runs or paddles across the water to take off, but once airborne, is a powerful flier, making a distinctive pulsating sound with its wings. Mute swans feed mainly on water, often up-ending to reach plants and small animals in underwater mud. They mate for life, and nest by the water's edge or on small islands, making a mound of vegetation often more than 3¼ ft (1 m) across. Females lay up to 8 eggs in a clutch, and take sole charge of incubation. Once the young have hatched, both parents look after them, until they become independent at about 5 months. Young swans often remain with their parents for longer than this, but they are driven away by the male at the onset of the following breeding season. It takes 3–4 years for the young to become fully mature and able to raise young of their own.

Cygnus atratus

Black swan

Length 3½–4½ ft (1.1–1.4 m)	
Weight 13 lb (6 kg)	
Plumage Sexes alike	
Migration Nomadic	
Status Least concern	

Location Australia (including Tasmania), New Zealand

This is the only swan that is almost entirely black. The innermost flight feathers are curiously twisted, and when threatened, the swan raises these feathers and exposes the white primaries. Exclusively vegetarian, it feeds on aquatic plants; occasionally, it may graze on land. Compared to most other swans, black swans are highly sociable, and also seminomadic. After breeding, they sometimes gather in flocks thousands strong.

LONG-NECKED SWAN *The black swan has the longest neck of all swans, making up much more than half the total length of the bird in flight.*

FAMILY TIES

Like other swans, this species is strictly monogamous. The male and female have a strong bond and both participate in building the nest and caring for the young.

Anseranas semipalmata

Magpie goose

Length 30–35 in (75–90 cm)	
Weight 6½ lb (3 kg)	
Plumage Sexes alike	
Migration Partial migrant	
Status Least concern	

Location S. New Guinea, N. Australia

The only species in its family, this gooselike bird is unique among waterfowl in many ways: its toes are only slightly webbed; the hind toes are unusually long, enabling it to perch easily on small branches; and it is the only waterfowl with legs so long that the ends of the feet extend beyond the tail during flight. It is also the only waterbird to form breeding groups of one male and 2 females. The male is larger, with a more distinctive bony "bump" on his crown.

Anser anser

Greylag goose

Length 30–35 in (75–90 cm)	
Weight 6½–8¾ lb (3–4 kg)	
Plumage Sexes alike	
Migration Migrant	
Status Least concern	

Location Europe (including Iceland), Asia

The greylag goose is the wild ancestor of barnyard geese, originally bred in Central Europe, and the basis for almost all European goose folklore. The name "greylag" derives from its gray coloration and the fact that it undertakes late migrations, or "lags" behind other geese. Males swim in a haughty posture to attract the female, and mating is often preceded by mutual head-dipping.

Branta canadensis

Canada goose

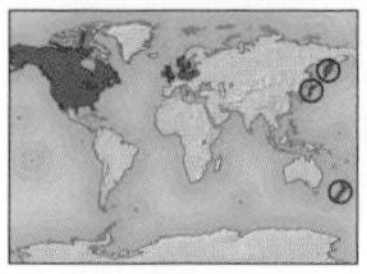

Location North America, N. Europe, N.E. Asia, New Zealand

Length 22–39 in (55–100 cm)
Weight 4½–18 lb (2–8 kg)
Plumage Sexes alike
Migration Migrant
Status Least concern

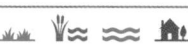

Originally from North America, the highly sociable Canada goose has been introduced widely to Northern Europe and New Zealand. It is becoming increasingly common, a result of its ability to adapt to a wide range of climates, tolerate ecological change, and feed on grass as well as other low vegetation. One of the most variable of all geese in size, subspecies range from the small Arctic forms, up to 4½ lb (2 kg), to the large, most southerly populations, which may occasionally weigh as much as 18 lb (8 kg).

COLORATION
This goose has a black head and neck, white cheeks, and a brown body, with blackish wing-tips and lighter underparts. The bill and feet are black.

GRAZING

Canada geese are diurnal feeders, foraging on dry land as well as on water for plants. Now a common sight in parks outside their native range, flocks leave the water at dawn to feed on grasses, seeds, and grains.

Cereopsis novaehollandiae

Cape Barren goose

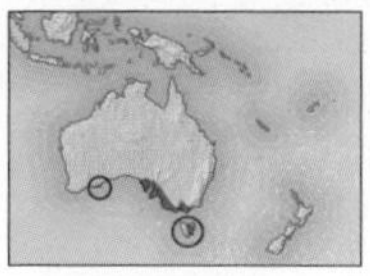

Location S. Australia (including offshore islands and Tasmania)

Length 30–36 in (75–91 cm)
Weight 8¾–13 lb (4–6 kg)
Plumage Sexes alike
Migration Nonmigrant
Status Least concern

With reduced webbing on its toes, this dove-gray goose has adapted for life mainly on land. Its black, stubby bill is mostly hidden by a pale green cere—a fleshy pad at the base of the upper part of the bill. Aggressive and territorial, it performs several threat displays, and may bite or strike with its knobbly wings. Mating occurs on land and is followed by a "triumph" ceremony by pairs after aggressive encounters with rival males.

Alopochen aegyptiaca

Egyptian goose

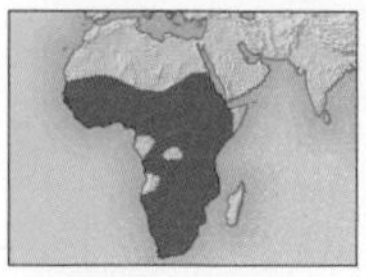

Location Africa (south of Sahara)

Length 25–29 in (63–73 cm)
Weight 5½ lb (2.5 kg)
Plumage Sexes alike
Migration Nonmigrant
Status Least concern

A close relative of the common shelduck (see below, left), this long-legged bird lives in lakes, rivers, and subtropical wetlands, but also spends much of its time on land. Its "bullying" behavior toward smaller species is notable, as is its aggressive behavior within its own species. The female chooses the most aggressive male as a mate, which means there is much fighting during the pairing season. Males have husky calls, while females make loud, nattering sounds.

Tadorna tadorna

Common shelduck

Location Europe, Asia, N. Africa, E. North Atlantic, Mediterranean, W. Pacific

Length 23–26½ in (58–67 cm)
Weight 2¼–3¼ lb (1–1.5 kg)
Plumage Sexes differ
Migration Partial migrant
Status Least concern

An inhabitant chiefly of marine or saline waters, this duck has iridescent, greenish black, white, and chestnut plumage. The male has a knob at the top of its red bill, absent in the female (shown below). Like all shelducks, it is intermediate in behavior between a goose and a typical duck. It forages at the tide-line, usually standing in shallow water, probing for molluscs and other marine invertebrates. Nesting often takes place in holes, such as the abandoned burrows of other animals.

Anas platyrhynchos

Mallard

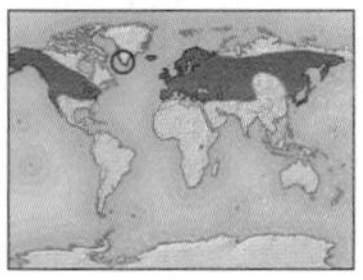

Location North America, S. Greenland, Europe, Asia

Length 20–25½ in (50–65 cm)
Weight 2¼–3¼ lb (1–1.5 kg)
Plumage Sexes differ
Migration Partial migrant
Status Least concern

The mallard has a wide distribution in the Northern Hemisphere, its success being due to its great flexibility—it can adapt to almost all types of aquatic habitats for breeding, even urban environments, and is able to feed in a number of ways (see panel, right). The male has a low-pitched quack and a sharp whistle, while the female is more vocal and has many distinctive quacking calls.

BREEDING MALE
The breeding male has a green head with a white neck ring, a rufous breast, gray flanks, black rump, and a white tail with curled central feathers. The bill is yellow.

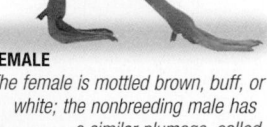

FEMALE
The female is mottled brown, buff, or white; the nonbreeding male has a similar plumage, called eclipse plumage.

UP-ENDING

Mallards often feed by up-ending to reach submerged plants and invertebrates. They also eat by dabbling, grazing or, rarely, diving in shallow water. At certain times of the year, they are a common sight in fields of grain crops.

Anas crecca

Green-winged teal

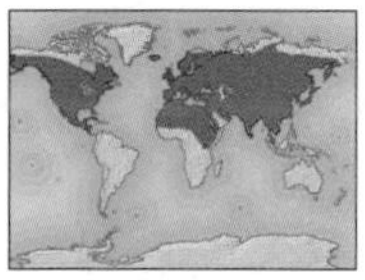

Location North America, Europe (including Iceland), Asia, N. to C. Africa

Length 13½–15 in (34–38 cm)
Weight 13 oz (350 g)
Plumage Sexes differ
Migration Partial migrant
Status Least concern

The smallest dabbling duck of North America and Europe, the green-winged teal breeds in tundra, grassland, and in relatively small bodies of water. Its tiny bill helps it feed on small seeds of aquatic plants; invertebrates also form part of its diet. The male has bright breeding plumage, and utters a cricketlike call that is a double-note whistle, while females produce high-pitched quacks. Both sexes have a green wing patch called "speculum." The North American subspecies is known as the green-winged teal.

Aix galericulata

Mandarin duck

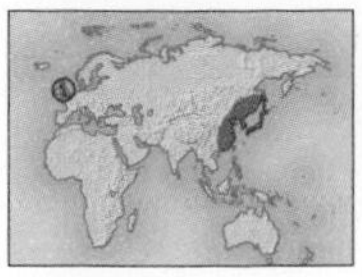

Length 16–19½ in (41–49 cm)
Weight 22 oz (625 g)
Plumage Sexes differ
Migration Partial migrant
Status Least concern

Location N.W. Europe, E. Asia

The male mandarin duck in its breeding plumage is among the most ornate and beautiful of all birds. It has a prominent crest on its head, golden hackles, and a pair of usually bright yellow, sail-shaped feathers on each inner wing. These can be erected above the flanks—their function is purely ornamental. Females and nonbreeding males are mostly olive-brown. This is one of the most arboreal of all ducks, often roosting or perching on branches and nesting in tree cavities; ducklings have to jump from the nest hole to the ground at only one day old. Like other perching ducks, its claws are sharp, to help it cling to branches, and its tail is long and broad, acting as a brake to slow the duck down before it lands in the trees. It also has very large eyes, which help it see at night. The mandarin duck feeds on land, in the trees, and in water, where it dabbles, up-ends, and only rarely dives. Its diet consists of seeds and nuts, as well as some invertebrates, such as insects and land snails. Courtship is highly social, with several males competing for a female's attention. Originally from East Asia, introduced birds have established themselves in Western Europe.

Cairina moschata

Muscovy duck

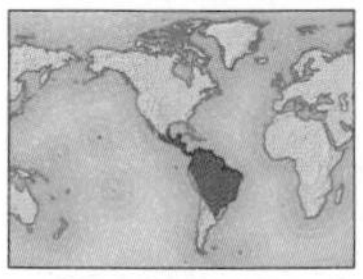

Length 26–33 in (66–84 cm)
Weight 4½–8¾ lb (2–4 kg)
Plumage Sexes alike
Migration Nonmigrant
Status Least concern

Location Central America to C. South America

This large, heavy-bodied duck is mostly blackish as an adult (brownish black when young), and has a long tail, broad wings, and sharply clawed toes—all adaptations for perching in trees. The adult male also has a slight crest on the top of the head, a knob above the nostrils, and bare, warty skin around the eyes. Domesticated birds (below) show many color variants. Unlike most ducks, both sexes are nearly mute.

Polysticta stelleri

Steller's eider

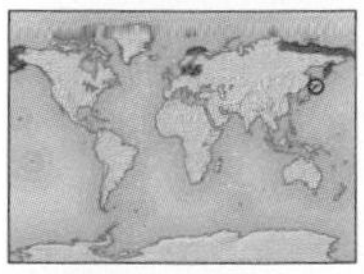

Length 17–18½ in (43–47 cm)
Weight 23–32 oz (650–900 g)
Plumage Sexes differ
Migration Migrant
Status Vulnerable

Location N.W. North America, N. Europe, N.E. Asia

This is the smallest of the 4 eider species, and the only one with an orange-rust breast and underparts (breeding males only). Its gray bill, which has soft, flaplike margins on each side, is probably adapted for scraping invertebrates off rocks, or may serve as a tactile device when foraging in deep, dark waters. It lives entirely in the open sea except during the breeding season, when it may be observed in coastal waters, just before building its nest in low, grassy tundra.

Hymenolaimus malacorhynchos

Blue duck

Length 21 in (53 cm)
Weight 28–32 oz (775–900 g)
Plumage Sexes alike
Migration Nonmigrant
Status Endangered

Location New Zealand

Unique among waterfowl for its slate-blue plumage, this duck is highly adapted for life in cold, clear, rapidly flowing streams, where few other birds can survive. Its main food is the aquatic larvae of insects such as caddisflies, mayflies, stoneflies, and midges, which it obtains by diving. It also feeds by dabbling at the surface, up-ending, and foraging around the riverbed rocks and boulders, where it uses its unique bill to scrape wet algae from the rocks.

Merganetta armata

Torrent duck

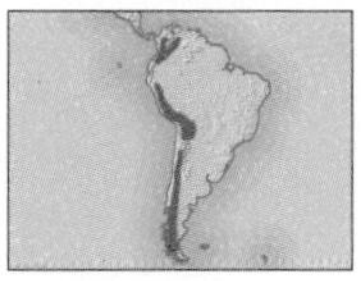

Length 17–18 in (43–46 cm)
Weight 16 oz (450 g)
Plumage Sexes differ
Migration Nonmigrant
Status Least concern

Location W. South America

This highly streamlined, Andean duck never leaves rapid mountain streams, a habitat that few other ducks are able to exploit owing to the swiftness of the currents—the blue duck of New Zealand (see above) is its closest ecological counterpart. It dives and swims upstream, remaining very close to the bottom, probing under rocks and among stones for insect larvae and pupae. Females are rufous, while males are black and white.

Melanitta deglandi

White-winged scoter

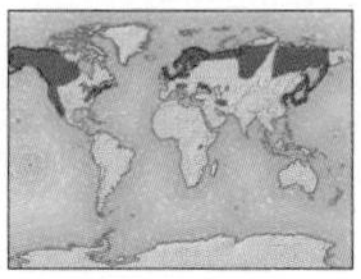

Length 20–23 in (51–58 cm)
Weight 3¼–4½ lb (1.5–2 kg)
Plumage Sexes differ
Migration Migrant
Status Least concern

Location N.W., W., and E. North America, Europe, W., N., and E. Asia

Closely related to the velvet scoter, the white-winged scoter spends much of the year at sea, although it breeds on freshwater inland, and sometimes winters on large lakes. It forages by diving to depths of up to 23 ft (7 m), often remaining underwater for nearly a minute, or longer in deeper waters. It is a stout duck, with a heavy bill adapted for eating mollusks. Its large nostrils may be related to well-developed glands that excrete salts from any seawater that it swallows.

Mergus merganser

Common merganser

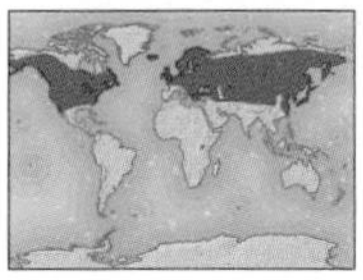

Length 23–26 in (58–66 cm)
Weight 3¼–4½ lb (1.5–2 kg)
Plumage Sexes differ
Migration Partial migrant
Status Least concern

Location North America, Europe (including Iceland), Asia

One of the swiftest of all waterfowl, and the largest of the 7 species of fish-eating mergansers, this duck has a streamlined body that is adapted for rapid underwater swimming and flight—its air speed may approach 60 mph (100 kph). It has a long, narrow, serrated bill to catch and hold slippery fish, which form the main part of its diet. Females are mostly gray above, while males have black backs.

Penguins

PHYLUM	Chordata
CLASS	Aves
ORDER	Sphenisciformes
FAMILIES	1 (Spheniscidae)
SPECIES	17

These distinctive flightless seabirds are adapted for swimming and surviving life in extreme cold. Penguins spend most of their lives in water, propelling themselves with flipperlike wings in pursuit of fish, krill, and squid. They have a thick coat of short, stiff, overlapping feathers that streamlines the body, repels water, and conserves heat. Most species come ashore during the warmer months to breed, usually forming large colonies. On land, they do not perch but instead stand upright and move with a waddling gait. Penguins are confined to the seas of the Southern Hemisphere. Although most common in cold climates, several species are also found where cold currents flow north into tropical regions. Only 2 species, the emperor and Adelie penguins, spend winter in Antarctica.

Swimming

Penguins use 3 different swimming techniques. When idling, they swim slowly at the surface, paddling with their wings, and with their head and tail raised. When hunting, they dive below the surface and effectively fly underwater, flapping their wings to provide power. Most dives last about a minute, but dives of 20 minutes have been recorded. The third form of movement is called porpoising, in which penguins swimming near the surface periodically leap out of the water to breathe.

DIVING
Penguins (here, a king penguin) move much more efficiently in water than they do on land. Some species can swim at speeds of 9 mph (14 kph).

Anatomy

Penguins have a plump body with short legs and webbed feet that are set so far back that they must stand upright on land, balancing on their feet and short, stiff tail. When walking, they put their weight on the soles of their feet, hence their awkward gait. On snow or ice, they may toboggan on their bellies, using their feet and flippers for propulsion. A penguin's body is streamlined in water, and is covered with extremely short feathers that form a sleek, friction-free surface. The wings are specially flattened into flippers (see below). When underwater, they use their feet and tail as a rudder. Penguins have 3 layers that provide waterproofing and insulation: a dense mass of overlapping, oil-tipped feathers; a thick layer of fat under the skin; and, in between, a layer of air that is warmed by the body. The plumage of penguins is black or gray above and white below. Any coloration or ornamentation (such as crests and eye tufts) is confined to the head and neck.

COMMUNICATION IN COLONIES
Apart from the yellow-eyed penguin, all penguins (such as these king penguins) form colonies. When gathered in large numbers, they use calls and visual displays to locate their mates and young.

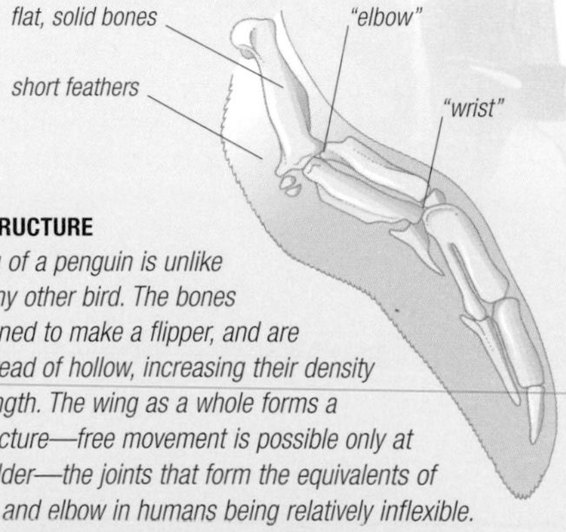

WING STRUCTURE
The wing of a penguin is unlike that of any other bird. The bones are flattened to make a flipper, and are solid instead of hollow, increasing their density and strength. The wing as a whole forms a rigid structure—free movement is possible only at the shoulder—the joints that form the equivalents of the wrist and elbow in humans being relatively inflexible. Well-developed muscles, as large as those of flying birds, help power the flapping of the wings.

Colonies

Most penguins breed in colonies, which can consist of hundreds of thousands of birds. Nests are made of grass, feathers, or pebbles, and the female lays either one or 2 eggs. In some species, once an egg is laid, the female leaves the nest to feed, while the male keeps the egg warm, holding it on top of his feet and beneath folds of skin on his belly. Penguins are still able to walk when the egg is in this position. The first period of incubation can last weeks or even months, during which time the male does not feed but lives on stored fat. When the female returns, the pair take turns to guard the eggs while the off-duty bird feeds at sea.

BURROWING
Not all penguins nest above ground. In open terrain, Magellanic penguins nest in shallow burrows to protect themselves from the elements and from predators, including mammals and other birds.

HUDDLING
In harsh conditions, emperor penguins and chicks huddle together for warmth. The center of a huddle can be about 18°F (10°C) warmer than the edges. Birds take turns to occupy the outermost positions.

Aptenodytes forsteri

Emperor penguin

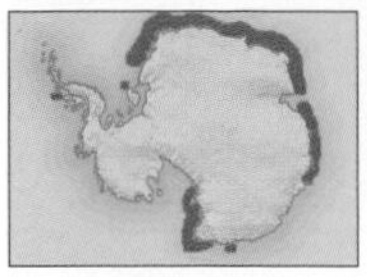

Location Circumpolar around Antarctica

Height 3½ ft (1.1 m)
Weight Up to 82 lb (37 kg)
Plumage Sexes alike
Migration Nonmigrant
Status Least concern

The largest of all the penguins, this bird has highly unusual breeding habits. In winter, the female lays one egg and goes to sea, not returning until the spring. The male carries the egg on his feet and protects it with a "pouch" of feathery skin, huddling with other incubating males to keep warm. He fasts until his mate returns, just as the chick hatches, and then goes to sea, later returning to help rear the chick. This species can dive to depths of 1,750 ft (530 m) for as long as 20 minutes, and travel up to 625 miles (1,000 km) on foraging trips.

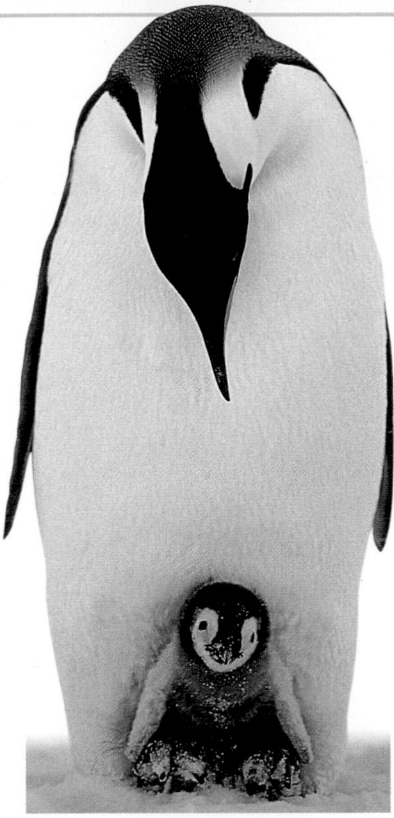

Pygoscelis adeliae

Adelie penguin

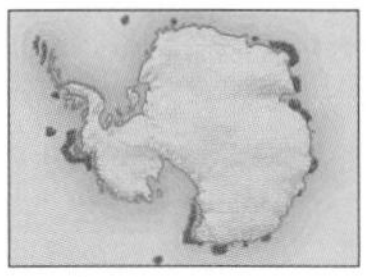

Location Circumpolar around Antarctica

Height 18–24 in (46–61 cm)
Weight 8¾–12 lb (4–5.5 kg)
Plumage Sexes alike
Migration Migrant
Status Least concern

One of the few penguins that nests on mainland Antarctica, along shores that are free of ice in summer, the Adelie penguin is mainly blue-black with pure white underparts and a distinctive white ring around its eye. It breeds in summer in vast colonies of 200,000 or more synchronized pairs, each female laying 2 eggs roughly 2 days apart. Both the male and the female incubate the eggs in shifts. The Adelie penguin is relatively aggressive, and adults are often observed stealing rocks from their neighbors' nests.

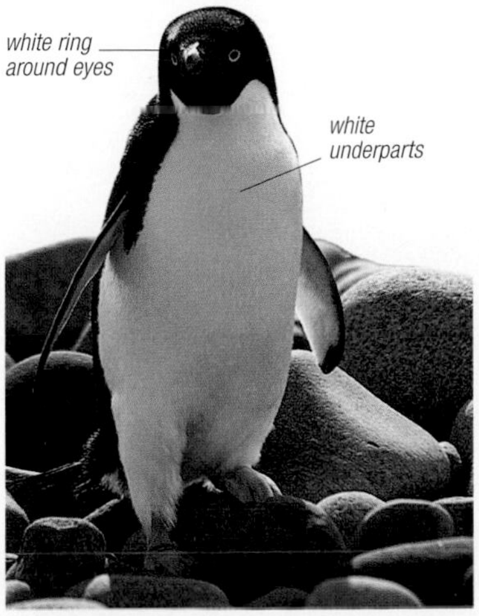

Pygoscelis antarcticus

Chinstrap penguin

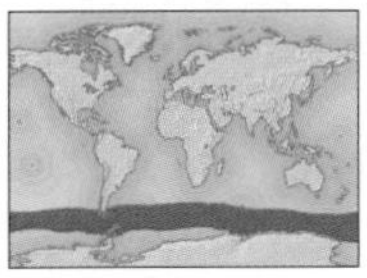

Location Circumpolar around Antarctica

Height 28–30 in (71–76 cm)
Weight 6½–10 lb (3–4.5 kg)
Plumage Sexes alike
Migration Nonmigrant
Status Least concern

Characterized by a thin, black line that runs from ear to ear under the chin, the chinstrap penguin is mostly blue-black with white underparts, cheeks, chin, and throat. It prefers areas of light pack ice, and breeds in high-density, sometimes large, colonies on ice-free areas of coasts. The nests comprise a circular platform of small stones, with a shallow nest cup, and often include bones and feathers. The breeding success of chinstrap penguins is highly variable: it is lower in years when sea ice persists close to colonies, since this restricts access to the sea for foraging adults.

Spheniscus humboldti

Humboldt penguin

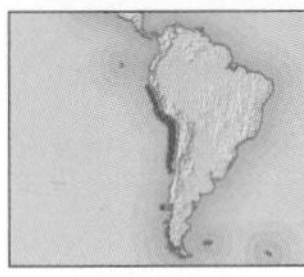

Location W. South America

Height 22–26 in (56–66 cm)
Weight 10–11 lb (4.5–5 kg)
Plumage Sexes alike
Migration Nonmigrant
Status Vulnerable

Small colonies of this penguin occur along the west coast of South America, in the cold but fish-rich Humboldt Current. Coloration is mostly blackish gray with white underparts, but adults have a distinctive black, horseshoe-shaped breast band and a white head stripe. The penguins usually hunt in groups in shallow water, pursuing small, schooling fish such as anchovies and sardines, and they nest in underground burrows, caves, or crevices between boulders. Direct hunting and over-fishing by humans have led to a decline in their population.

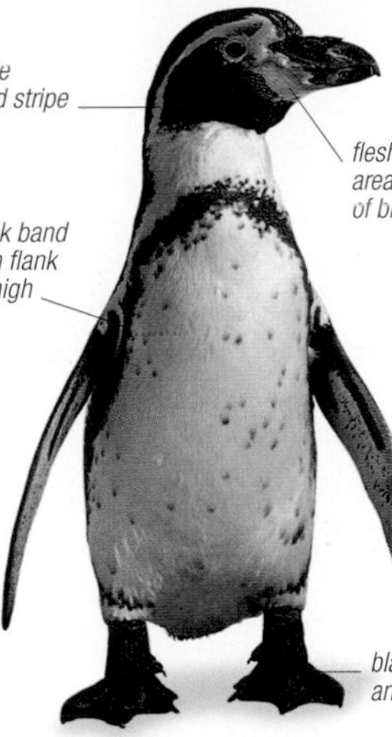

Eudyptula minor

Little penguin

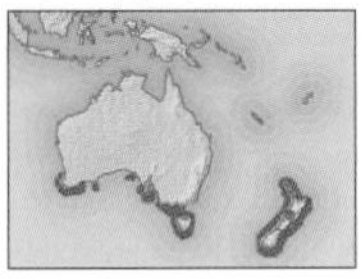

Location S. Australia (including Tasmania), New Zealand

Height 16 in (41 cm)
Weight 2¼ lb (1 kg)
Plumage Sexes alike
Migration Partial migrant
Status Least concern

This mainly blue-gray bird is the smallest species of penguin, and also one of the few that remain fully active after dark. During the day it forages at sea, but after sunset in the breeding season, it returns to land, coming ashore under the cover of darkness. A highly vocal bird, it calls at sea and on land. Little penguins normally nest in underground burrows, but they also breed in caves, crevices, under vegetation among rocks, and sometimes under houses. They lay 2 eggs on a bed of sticks or grass, and the parents take turns at incubation. Chicks are brooded for 7–10 days and guarded for a further 13–20 days. At sea, the little penguin feeds alone or in small groups, catching small fish and swallowing them beneath the surface.

Eudyptes chrysolophus

Macaroni penguin

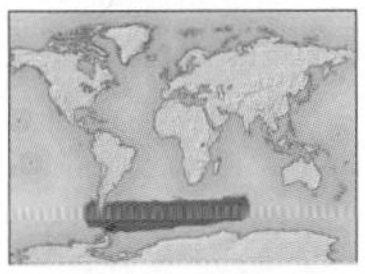

Location S. Chile, South Atlantic, S. Indian Ocean

Height 28 in (71 cm)
Weight 7¾–14 lb (3.5–6.5 kg)
Plumage Sexes alike
Migration Nonmigrant
Status Vulnerable

Like all 6 species of crested penguins, the macaroni penguin has a conspicuous crest of golden plumes on its head. It also has a large, bulbous, orange-brown bill, often ridged in adults. Noisy, aggressive, and demonstrative, it makes raucous braying sounds in its colonies on land and gives short barks at sea. Although 2 eggs are laid, only one chick ever survives to the fledgling stage—usually the chick from the second, larger egg. Both sexes share incubation duties but, unlike most other penguins, the female takes the first shift.

Albatrosses and petrels

PHYLUM	Chordata
CLASS	Aves
ORDER	Procellariiformes
FAMILIES	4
SPECIES	133

Albatrosses and their relatives are oceanic birds that occur throughout the world. They are usually encountered far from land, flying low over waves or dipping into the water to feed on fish, plankton, or various other marine animals. Besides the very large albatrosses, this group includes the smaller fulmars, gadfly petrels, and shearwaters, as well as the tiny storm petrels and diving petrels. All members of the group have tubular nostrils on their upper bill—a unique feature among birds; as a result, they are often known as tubenoses.

Reproduction

All birds in this group breed on land, mostly on inaccessible islands or cliffs. They return each year to established breeding sites, where they form large colonies of up to a million pairs. The female always lays a single egg, often in a burrow dug in soft soil or in rock crevices. After a long incubation, the parents feed the chick on their highly nutritious but foul-smelling stomach oil.

BREEDING COLONY
Like other birds in the group, fulmars often form dense colonies in safe, inaccessible places. Many species visit the colony only after dark, as protection against predators.

SOARING
The long wings of this black-browed albatross are adapted for extended soaring, often for hours on end without a wingbeat.

Anatomy

Albatrosses and their relatives have a short neck, tail, and legs. The front 3 toes are joined by webbing. Most species have very long wings; the wandering albatross, with a wingspan of 11 ft (3.5 m), has the longest wingspan recorded for any bird. Another notable feature of birds in this group is their exceptionally acute sense of smell, which they use to detect food and locate nest sites in the dark. It is thought that each bird may exude its own particular scent.

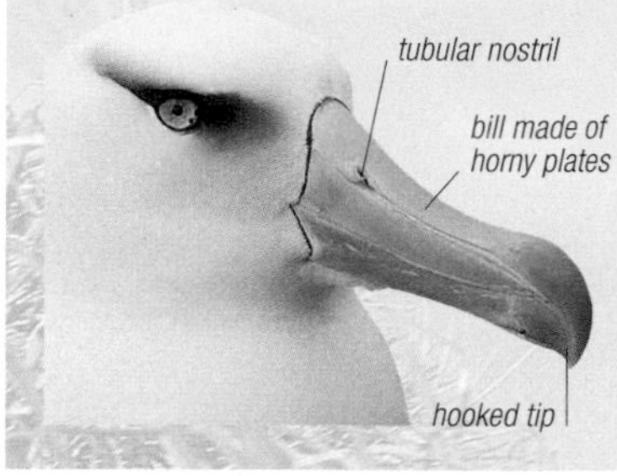

ALBATROSS BILL
Birds in this group have a bill with a hooked tip and sharp blades for dealing with slippery prey. Albatrosses differ from other species in having nostrils positioned on either side of the upper bill rather than being fused into one on the top (see left).

Flight

These ocean-going birds are adapted to continuous flight, often traveling great distances and riding out powerful storms. Their flight method varies according to the bird's size and wingspan. The larger species, including the albatrosses, typically glide rather than fly, making use of the updrafts of wind over waves to give them lift for long periods without expending too much energy (a practice known as dynamic soaring). The tiny, lightweight storm petrels fly with a combination of wingbeats and glides.

FINDING FOOD
Storm petrels fly low over water in search of food. Some species patter on the surface with their feet to alarm their prey or draw it toward the surface.

Diomedea exulans

Wandering albatross

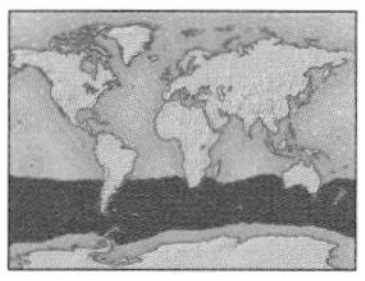

Location Circumpolar around Antarctica

Length 3½ ft (1.1 m)
Weight 18–25 lb (8–11.5 kg)
Plumage Sexes alike
Migration Migrant
Status Vulnerable

This albatross has a wingspan of up to 11 ft (3.5 m), one of the largest of any bird. It feeds mainly at the surface, primarily on squid. It breeds only every other year (at best), since rearing a chick takes a year, and molts in the intervening year. These birds are in decline because they often get caught on baited hooks used in long-line fishing for fish such as tuna.

NESTING

This albatross' nest is a crude but substantial mound of grasses and moss, built on the ground. The parents take turns to incubate the solitary egg. Once it has hatched, they tend their chick for 9 months or more.

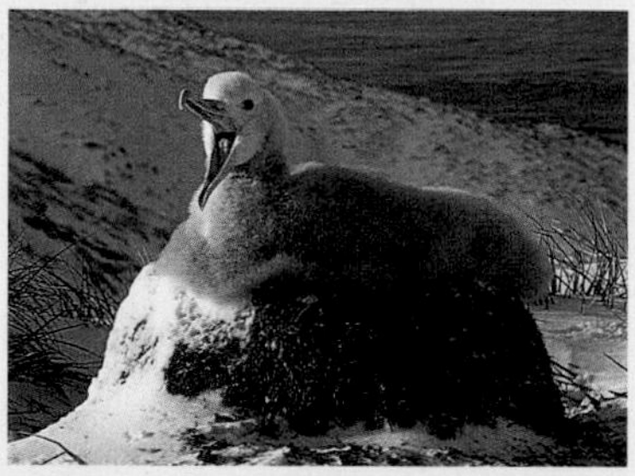

PAIRED FOR LIFE
Albatrosses are monogamous and the male–female bond is very strong. An elaborate courtship display, involving much posturing, head shaking, beak snapping, and ritualized preening is usually initiated by the female.

Thalassarche chlororhynchos

Yellow-nosed albatross

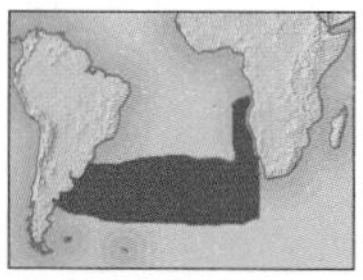

Location South Atlantic

Length 30 in (76 cm)
Weight 5½ lb (2.5 kg)
Plumage Sexes alike
Migration Migrant
Status Endangered

The smallest albatross in the southern ocean, this black and white species takes its name from the yellow ridge and orange tip on its black bill. Squid and fish make up the bulk of its diet, but it also follows ships, feeding on offal thrown overboard. Both parents rear the chick, which is able to fly relatively quickly for an albatross—within about 4 months.

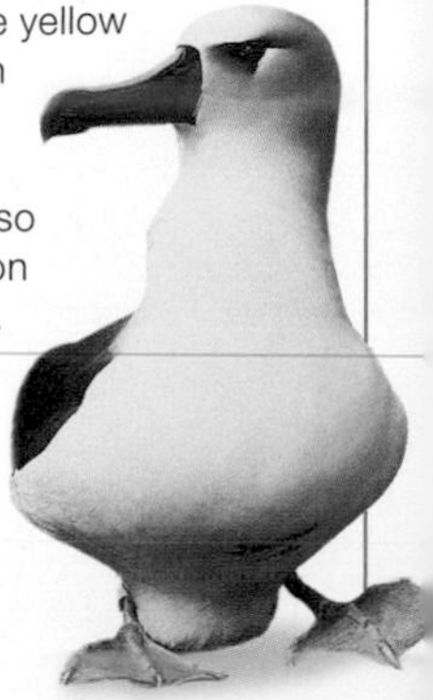

Macronectes giganteus

Southern giant petrel

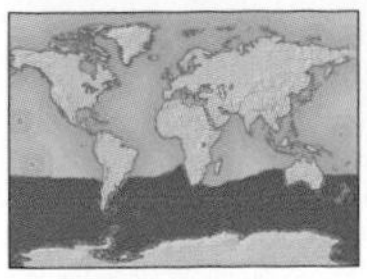

Location Circumpolar around Antarctica

Length 36 in (92 cm)
Weight 11 lb (5 kg)
Plumage Sexes alike
Migration Migrant
Status Least concern

Large and aggressive, with a large bill capable of opening intact carcasses, this bird is one of a few petrels to obtain a significant amount of its food on land. Males in particular prey at seal and penguin colonies and feed on the bodies of whales washed ashore. The adult usually has mottled, grayish brown plumage, but some individuals may be white with black flecks.

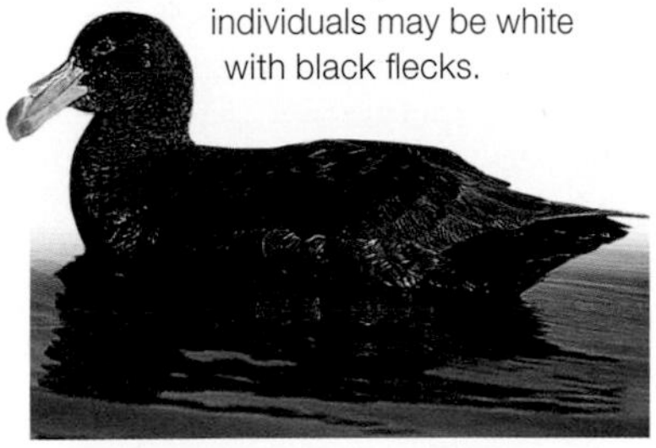

Fulmarus glacialis

Northern fulmar

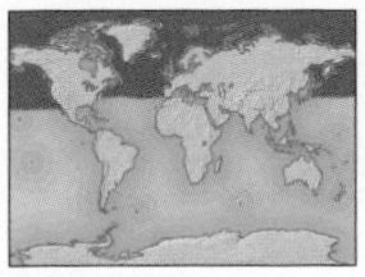

Location Arctic, North Pacific, and North Atlantic oceans

Length 18–20 in (45–50 cm)
Weight 25–30 oz (700–850 g)
Plumage Sexes alike
Migration Migrant
Status Least concern

The fulmar is a common bird of northern waters, flying on characteristic stiff, straight wings. Its numbers have increased dramatically in the past 200 years, particularly in temperate waters of the North Atlantic, possibly due to the increase in offal available from trawlers gutting fish at sea. As well as scavenging from ships, it eats fish, squid, and animal plankton, seizing most of its prey at the surface but sometimes plunging into the water. Most petrels return to breed at the colony where they were born, but the northern fulmar rarely does so. It lays its egg in a barely lined hollow on an earthy or grassy ledge, usually on a cliff, but it will nest on flatter ground where there is no danger from predators. The adults go off in search of food once the chick is about 2 weeks old. If threatened, the youngster defends itself by vigorously spitting an unpleasant smelling oil.

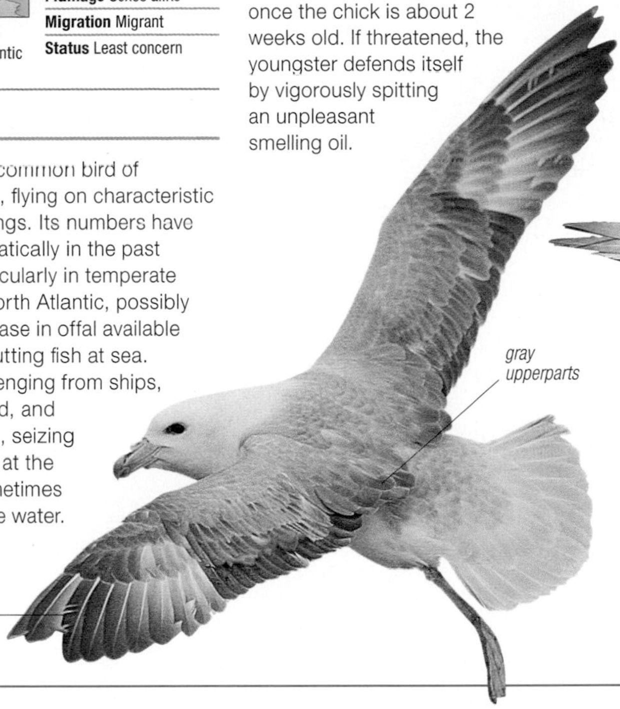

Pagodroma nivea

Snow petrel

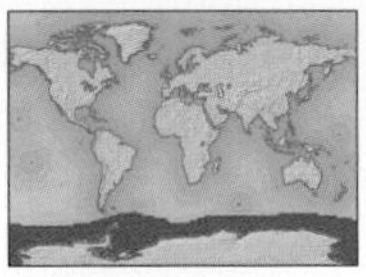

Location Circumpolar around Antarctica

Length 12½ in (32 cm)
Weight 9–16 oz (250–450 g)
Plumage Sexes alike
Migration Migrant
Status Least concern

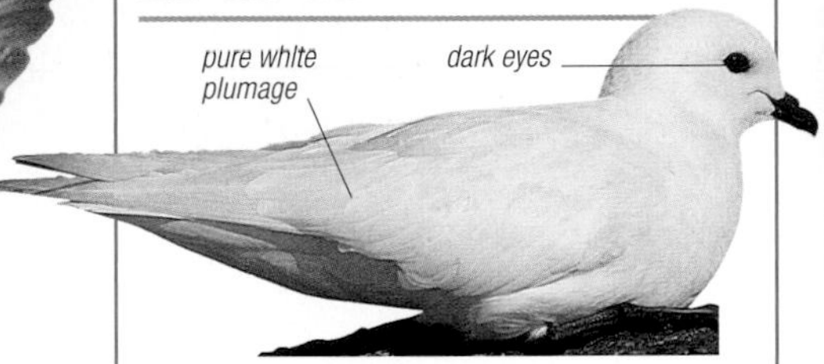

One of the few birds, apart from penguins, that breed on the Antarctic continent, the dovelike snow petrel is rarely seen away from pack ice. It may nest inland, up to 185 miles (300 km) from open water, but breeding is much affected by snowfall, and in some years only one in 5 nesting sites may be occupied. It fiercely defends its nest from other petrels, spitting a foul-smelling oil at them.

Pterodroma cahow

Bermuda petrel

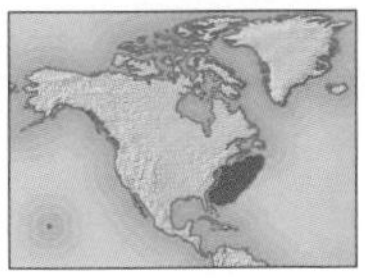

Location Bermuda (ocean range 625 miles/1,000 km)

Length 13 in (33 cm)
Weight 9 oz (250 g)
Plumage Sexes alike
Migration Migrant
Status Endangered

One of the world's rarest seabirds, the Bermuda petrel was brought to the brink of extinction as long ago as the 17th century, as a result of predation by introduced mammals and competition for nesting sites with the white-tailed tropicbird. Conservation measures are helping this black, gray, and white petrel to stage a gradual recovery.

Pachyptila vittata

Broad-billed prion

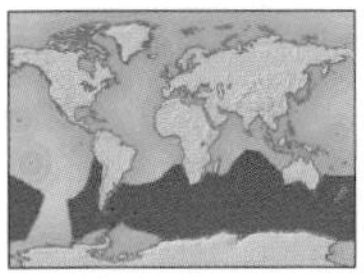

Location South Atlantic, South Pacific, S. Indian Ocean

Length 11 in (28 cm)
Weight 5–8 oz (150–225 g)
Plumage Sexes alike
Migration Migrant
Status Least concern

This medium-gray bird has a darker "M" across both wings, and white underparts. Its bill is broad, with comblike plates used for filtering planktonic prey from the sea. It feeds mainly by "hydroplaning"—with wings outstretched and feet paddling, pushing the sieving bill through the surface water to trap its prey.

Oceanites oceanicus

Wilson's storm petrel

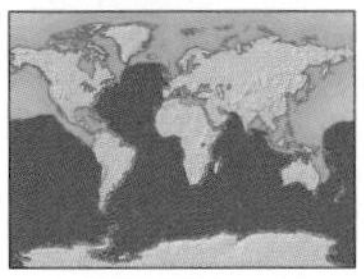

Location Pacific, Atlantic, Indian Ocean

Length 6½ in (17 cm)
Weight 1$\frac{7}{16}$ oz (40 g)
Plumage Sexes alike
Migration Migrant
Status Least concern

A small, soot-black bird with a conspicuous white rump, the storm petrel is often said to be one of the world's most numerous seabirds, with populations running into millions. It breeds around Antarctica, but during the southern winter, it flies north, especially to the northern Indian and Atlantic oceans. It rarely alights on the sea, but catches small fish and crustaceans when pattering on or hovering just above the surface. It can detect prey by smell, and it has been suggested that prey are attracted when the bird stirs its yellow-webbed feet around in the water. When faced with predators, this petrel may squeak and eject stomach oil at them.

Puffinus puffinus

Manx shearwater

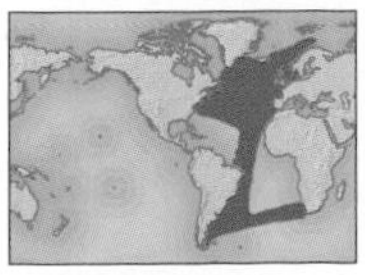

Location Atlantic

Length 12–14 in (31–36 cm)
Weight 13–18 oz (375–500 g)
Plumage Sexes alike
Migration Migrant
Status Least concern

This shearwater breeds in the North Atlantic, forming colonies on offshore islands—mainly Skomer and Skokholm off the western coast of Wales, and Rhum in the Outer Hebrides—filling the air with calls that sound like screams and wails. Although it feeds at sea by day, it is active at the colonies only after dark, to avoid attack by gulls. This bird usually nests in an earth burrow, but it sometimes lays its single egg under rocks. Both parents incubate the egg, working in shifts of 6–7 days, and they feed the chick for up to 70 days, leaving it about a week before it starts to fly. As winter approaches, the shearwater flies south to warmer waters off Brazil.

HOOKED BILL
The Manx shearwater has a relatively slender and hooked bill, with which it seizes fish at the water surface. It may swim a few yards underwater to pursue its prey.

COLORATION

Black above and white below, this bird produces a characteristic black, then white flash as it flies over the sea, looking for small schooling fish such as sprat.

BIRDS

Loons

PHYLUM	Chordata
CLASS	Aves
ORDER	Gaviiformes
FAMILIES	1 (Gaviidae)
SPECIES	5

Also known as divers, loons are superbly adapted for underwater swimming. With their streamlined bodies, legs set far back on the body for efficient propulsion, and strong, webbed feet, they can reach depths of 250 ft (75 m) and stay underwater for several minutes. Dense plumage insulates them in the Arctic and sub-Arctic waters where they live. Despite having small, pointed wings and a pinlike tail, loons fly well. But with feet set so far back, they are almost incapable of walking on land.

PLUMAGE
Loons spend a lot of time taking care of their feathers and can often be seen rolling on their side to reach their belly feathers. A varied breeding plumage (seen here in the Pacific loon) is replaced by simpler, less colorful patterns in winter.

Gavia stellata

Red-throated loon

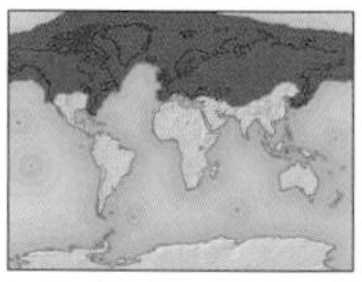

Location North America, Greenland, Europe, Asia

Length 22–28 in (55–70 cm)
Weight 2¼–5½ lb (1–2.5 kg)
Plumage Sexes alike
Migration Migrant
Status Least concern

Distinguished from other loons by the reddish brown throat patch at the base of its thick, long neck, this bird has a plump, oval body that is dark grayish to black with white spots; the underparts are whitish. An inhabitant of coastal bays and inlets, it moves inland to lakes and marshes in northern forests and Arctic tundra during the breeding season, where it emits a loud yodel or wail to attract a mate or establish territory. Courtship behavior involves splashing dives, dipping and shaking of bills, and rushing across the water in pairs. Nests are usually a simple platform of reed, rushes, and grass.

Gavia immer

Common loon

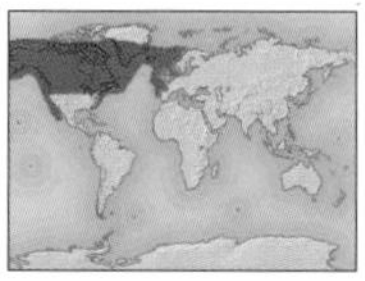

Location North America, Greenland, W. Europe

Length 28–35 in (70–90 cm)
Weight 6½–10 lb (3–4.5 kg)
Plumage Sexes alike
Migration Migrant
Status Least concern

This gray to black and white bird is usually solitary or found in pairs, although large groups of up to 300 may be seen feeding along the coast when not breeding. Like other loons, this species can dive to great depths when hunting for fish and aquatic invertebrates, which it seizes with its bill, sometimes spearing them. It may also dive to avoid predators such as mammals, hawks, and eagles. Both sexes care for the young, and the chicks stay with their parents until fledged, at about 10–11 weeks. Parents sometimes swim with the young on their backs.

Grebes

PHYLUM	Chordata
CLASS	Aves
ORDER	Podicipediformes
FAMILIES	1 (Podicipedidae)
SPECIES	22

Found in sheltered waters throughout the world, grebes are strong swimmers and accomplished divers. They are ideally suited to aquatic life: their small head and thin neck enable them to dive easily when hunting for food, and their feet—which are set well back on the body—have lobed toes with highly flexible joints that give these birds great agility when swimming. Their soft, dense plumage is very water-resistant. Grebes are remarkable for their often elaborate courtship rituals.

"WEED DANCE"
During the "weed dance," a courting pair (here, great crested grebes) dive for vegetation, then rise out of the water to face each other, holding the plants up high.

Tachybaptus ruficollis

Little grebe

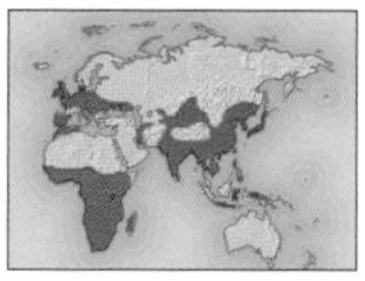

Location Europe, Asia, Africa, Madagascar, New Guinea

Length 10–11½ in (25–29 cm)
Weight 4–8 oz (125–225 g)
Plumage Sexes alike
Migration Partial migrant
Status Least concern

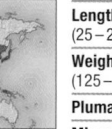

The smallest grebe in Europe and northern Asia, this short-billed bird is brown with a bright chestnut throat and cheeks. It stands and walks on land more easily than other grebes, and is also more likely to fly outside of migration. Both sexes build nests of aquatic vegetation, anchoring them to surface water plants or branches of bushes growing in water.

Podiceps nigricollis

Black-necked grebe

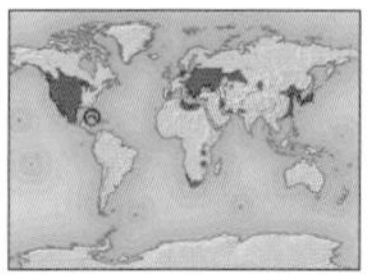

Location North America, Europe, Asia, N. and southern Africa

Length 12–14 in (30–35 cm)
Weight 9–21 oz (250–600 g)
Plumage Sexes alike
Migration Migrant
Status Least concern

This small, dark grebe (non-breeding plumage shown), with a slender, slightly upturned bill, is highly social all year. During the non-breeding season, it prefers saline waters, with thousands gathering on some saltwater lakes. However, it breeds in freshwater lakes and marshes, where it makes floating nests. Adults carry young on their backs for the first few weeks, but little parental care is required after 3 weeks.

Podiceps cristatus

Great crested grebe

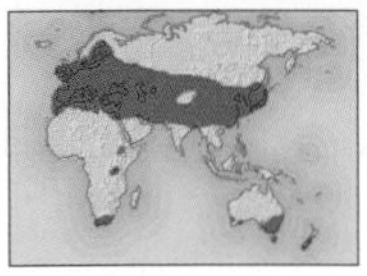

Location Europe, Asia, Africa, Australia, New Zealand

Length 18–20 in (46–51 cm)
Weight 1¼–3¼ lb (0.6–1.5 kg)
Plumage Sexes alike
Migration Partial migrant
Status Least concern

The largest grebe in Europe and northern Africa, the great crested grebe is famous for its remarkably elaborate courtship displays, which involve complex, ritualized postures, dives, and head-shaking; during the displays, its crest is raised and the tippets are flared. It often peers into the water while swimming on the surface to locate fish, then dives up to 100 ft (30 m) deep. Populations of the great crested grebe declined in the mid-19th century, but have recovered with protection and an increase in man-made habitats, such as gravel pits and reservoirs.

PARENTAL CARE

Parents take turns carrying the young on their backs and bringing them food. Each parent develops a preference for particular young, resulting in the adults dividing the brood in order to care exclusively for their favorites.

CREST AND TIPPETS
This elegant bird is distinguished by the ornamental black crest on its crown and its elongated, chestnut and black cheek feathers (tippets).

Podiceps gallardoi

Hooded grebe

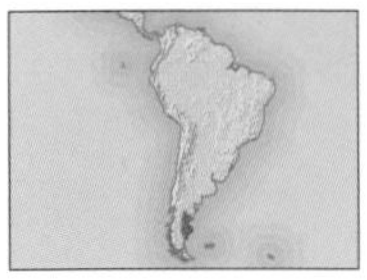

Location S. South America

Length 13½ in (34 cm)
Weight 19 oz (525 g)
Plumage Sexes alike
Migration Nonmigrant
Status Endangered

First described in 1974, this medium-sized, black and white grebe has a cinnamon and black crest and a small, pointed bill. Courtship involves a unique "sky jabbing" display in which males and females come face to face and vigorously move their heads up and down. Populations are small, occurring on isolated lakes in remote areas of Patagonia; the species winters in sheltered bays along the Patagonian coast. It is endangered because of effects of pollution and habitat disturbance during breeding.

Podilymbus podiceps

Pied-billed grebe

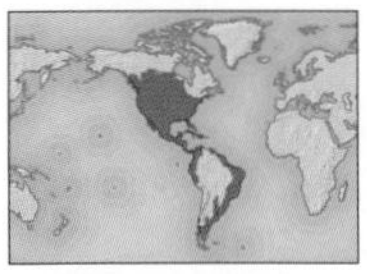

Location North America, Central America, N.W. and S.E. South America

Length 12–15 in (31–38 cm)
Weight 9–21 oz (250–575 g)
Plumage Sexes alike
Migration Partial migrant
Status Least concern

Small and stocky, with a short, arched bill, this brown and white bird is found in freshwater bodies in the breeding season, and winters in areas with open water or in brackish estuaries. Strongly territorial and aggressive, it threatens and chases birds of the same species as well as other waterbirds. Mates may perform a "triumph ceremony" after chasing off an intruder; this involves the pair rising into an upright posture facing each other, and turning back and forth.

Aechmophorus occidentalis

Western grebe

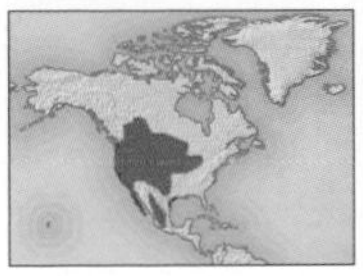

Location C. and S. North America

Length 22–30 in (55–75 cm)
Weight 2¼–4½ lb (1–2 kg)
Plumage Sexes alike
Migration Migrant
Status Least concern

A large grebe with black and white plumage, this bird has a long, slender neck. Mates are in close contact from pairing to taking care of the young. Both sexes build nests on water, anchored to surface plants. Parents take turns carrying young on their backs for the first 2–4 weeks, and feed them until they are 8 weeks old. At the end of the 19th century, tens of thousands of these grebes were hunted for their silky belly feathers, used in coats and hats. Populations have recovered, although they are still in danger from pollution, oil spills, habitat loss, and disturbance by humans.

COURTSHIP RUSH

This species engages in an energetic courtship, following the same rituals as the great crested grebe. The most famous display is the "rush" (shown above), in which birds run across the water side by side. If they continue to be interested in each other, they then perform the "weed dance" (see opposite), in which each bird dives for aquatic vegetation, then holds it high while turning from side to side. In established pairs, the male brings fish to his mate.

PLUMAGE FOR ALL SEASONS
Unlike its smaller North American relatives, the western grebe does not develop any special plumage in the breeding season, and keeps its black crown all year round.

Rollandia micropterum

Short-winged grebe

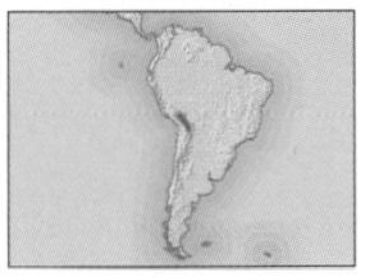

Location W. South America (Lakes Titicaca and Poopo)

Length 15½–18 in (39–45 cm)
Weight 22 oz (625 g)
Plumage Sexes alike
Migration Nonmigrant
Status Endangered

From a distance, this medium-sized South American grebe looks like many of its relatives, with a crested head, brown back and flanks, and white underside. However, its wings are so small that it cannot fly. If threatened, the short-winged grebe patters across the water at speed, flapping its wings rapidly without being able to take off. This unusual bird is found primarily on 2 large lakes of the Central Andes, at an altitude of about 11,900 ft (3,600 m).

Poliocephalus poliocephalus

Hoary-headed grebe

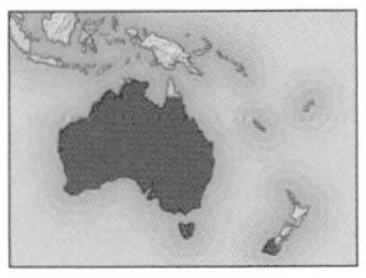

Location Australia (including Tasmania), S. New Zealand

Length 11½–12 in (29–31 cm)
Weight 8–9 oz (225–250 g)
Plumage Sexes alike
Migration Nomadic
Status Least concern

Distinguished by the "brushed back" white, hairlike plumes on its head and upper neck, this bird differs from other grebes in numerous ways. It flies more than other species of grebes, and is probably the least vocal and most social, often found in huge flocks. Courtship is relatively simple compared with that of other grebes, and the nests are typically built in shallower water. The hoary-headed grebe has a dark bill with a pale tip; the males have longer bills.

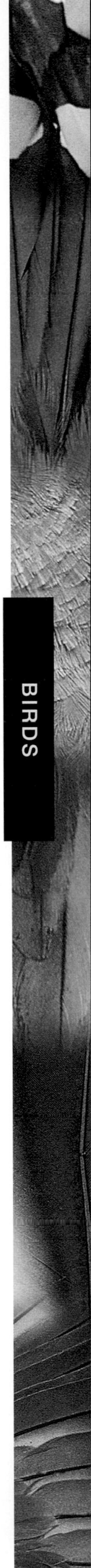

Flamingos

PHYLUM	Chordata
CLASS	Aves
ORDER	Phoenicopteriformes
FAMILIES	1 (Phoenicopteridae)
SPECIES	6

With their extremely long legs and neck, and colorful pink or red plumage, these tall wading birds are striking and instantly recognizable. They are found in the tropics and subtropics, usually in the shallows of salt or brackish water or alkaline lakes, sometimes massed in vast flocks of up to one million birds. Despite their fragile appearance, flamingos occur in many parts of the world where few other animals can survive. They are frequently found in conditions of extreme salinity or alkalinity, and are remarkably tolerant of changes in temperature. Flamingos are specialized feeders, using their unusual downturned bill to filter tiny plants and animals from water.

Colonies

Flamingos are among the most social of birds, even performing their courtship displays in groups. Thousands of individuals open their wings or lift up and turn their heads in one vast, synchronized movement. These group displays seem to bring all the birds of the colony to the same readiness to mate, to ensure rapid, synchronized egg-laying as soon as conditions allow. Breeding colonies often form when the water level of a salt lake drops—the surface is suddenly covered with piles of raised mud, which form the flamingos' nests. The parents care for their young for the first week or two. Once a young flamingo is able to walk and swim, it joins a "creche" (see right).

GROUP DISPLAY
During courtship displays, flamingos may engage in "head-flagging," raising their neck and bill, and turning their head from side to side.

CRECHES
Young flamingos form large groups of up to 300,000 birds (here, lesser and greater flamingos). They are still fed by their parents but are guarded by an unrelated adult.

FLIGHT
Flying flamingos, with their thin neck and legs outstretched, are unmistakable. Large flocks often form long, curving lines. Their long wings and light body enable a relatively easy takeoff.

Anatomy

Flamingos have a slender body set on greatly elongated legs, which are longer in relation to body size than those of any other bird. The limbs are completely bare, allowing the birds to wade deep in highly saline or alkaline water without soiling their plumage. Flamingos have a long, flexible neck that, in the case of the larger species, is also remarkably thin. The head is small, and the bill has a characteristic downward bend. The way in which the upper and lower parts of the bill fit together, combined with the comblike plates on the bill's edges, make it a useful tool for sieving food from water (see right). The flamingo's unique pink or red coloration arises from a dye extracted from food such as algae and shrimps.

STANDING IN WATER
Flamingos (here, a greater flamingo) often stand on one leg for long periods, even when asleep, with the head laid on the body and the other leg tucked under the abdomen. This posture cuts heat loss through the legs and feet.

FEET
Compared with their long legs, the feet of flamingos are relatively small. The front 3 toes are webbed, and the back one is either tiny or missing. The webbing is useful for walking over mud or other soft surfaces.

Feeding

A flamingo usually feeds while wading in shallows, using its feet to stir up mud on the bottom. To feed, it puts its head to the surface (so that its bill is upside down and the tip points backward) and sweeps its head from side to side. By a rapid action of the tongue, it pumps water in and out of the slightly opened bill. Along the inside edge of the bill are rows of plates, known as lamellae, some of which have tiny hairs on them. The flamingo filters out food particles by sieving them past the lamellae and their hairs. The size of particles taken differs between species. Larger flamingos tend to feed on crustaceans, mollusks, and worms, and the smaller species on algae.

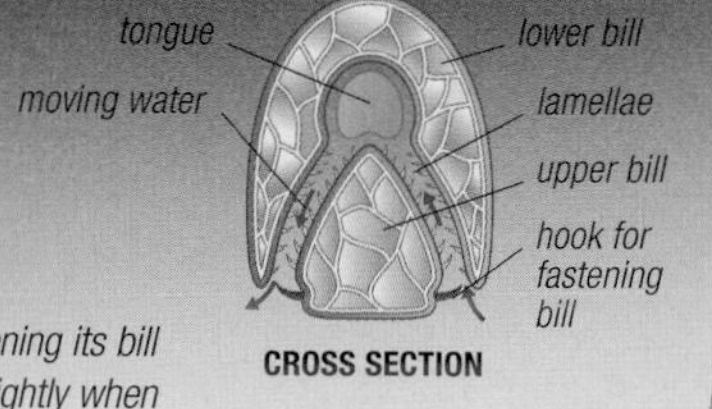

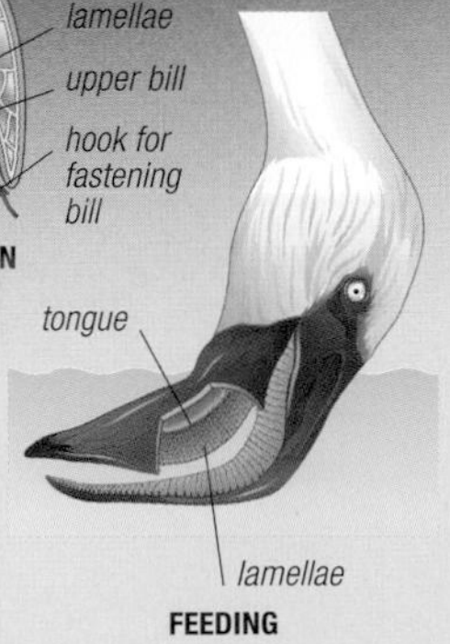

BILL
By opening its bill only slightly when feeding, a flamingo filters out unwanted items. A second filtration happens inside the bill, where rows of tiny plates (lamellae), often with minute hairs attached, act as a sieve to trap the smaller particles that the flamingo needs.

FEEDING ON ALGAE
Lesser flamingos (left) and other smaller species usually feed from the water surface. Larger flamingos immerse their heads completely to feed.

Phoenicopterus roseus

Greater flamingo

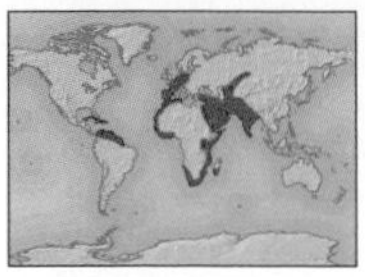

Height 5 ft (1.5 m)
Weight Up to 8¾ lb (4 kg)
Plumage Sexes alike
Migration Partial migrant
Status Least concern

Location Central America, South America, Caribbean, S.W. Europe, Asia, Africa

With its exceptionally long neck and legs, the greater flamingo is the largest member of the flamingo family. It is found in a wide variety of freshwater and saline habitats, particularly salt lakes, estuaries, and lagoons. Greater flamingos outside the tropics often migrate to warmer regions for winter, and in all regions they fly long distances—sometimes up to 300 miles (500 km)—to find food, traveling at night. The greater flamingo's large size enables it to wade out into relatively deep water, whereas other flamingos are restricted to the shallows; it also sometimes swims, up-ending like a duck in order to reach food. This flamingo usually feeds with its head fully immersed, sometimes keeping it underwater for up to 20 seconds. Unlike smaller flamingos, it rarely takes food from the surface, which reduces competition with them. Diet is varied, and includes insects, worms, microscopic algae, and pieces of vegetation. The greater flamingo feeds mainly during the day, even in the hottest conditions. Highly gregarious, it breeds in colonies of up to 200,000 monogamous pairs. Courtship involves complex, synchronized dances—neck stretching, ritualized preening, loud honking—performed by large numbers of males and females. It has a quieter contact call while feeding.

MASS BREEDING

For a social species, the greater flamingo shows a remarkable range of colony size. In some parts of the world, breeding colonies contain huge numbers of birds, but in others—such as southern Europe—they may contain only a few dozen. This variability is one of the reasons why greater flamingos, unlike some of their relatives, are relatively easy to breed in captivity.

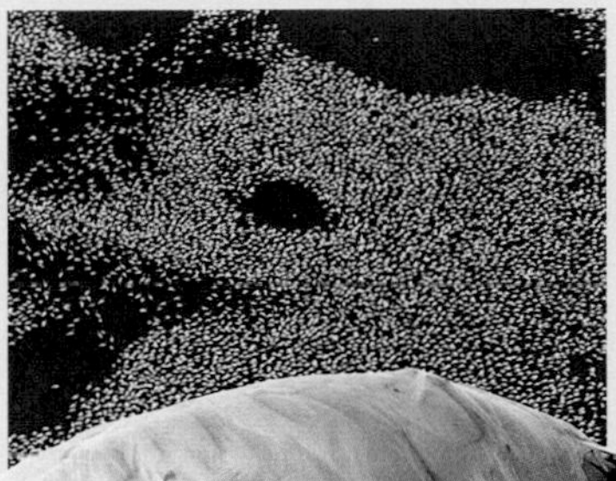

NESTS AND CHICKS

As with all flamingos, nests of greater flamingos are flattened cones of mud, often surrounded by a shallow "moat," and spaced about 5 ft (1.5 m) apart, just beyond pecking distance. Both parents incubate the single egg. They defend the nest when breeding, but are otherwise nonterritorial. Once the chicks can walk, they gather in large creches under the supervision of a small number of adults.

PINK OR PALE
The greater flamingo is usually paler than other species, including the American flamingo, which is similar in habits and in size.

Phoenicoparrus jamesi

Puna flamingo

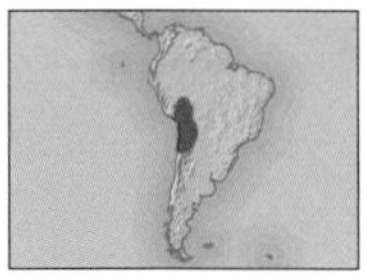

Height 3½ ft (1.1 m)
Weight 4½ lb (2 kg)
Plumage Sexes alike
Migration Nonmigrant
Status Near threatened

Location W. South America

This flamingo derives its common name from its habitat—"puna" is the local word for a high Andean plateau. Its diet consists of diatoms—microscopic algae that abound in salt lakes. It feeds by day in shallow water, walking slowly forward with its bill dipped just under the water surface. Diatoms are filtered out of the water in the bill and swallowed. This flamingo has a variety of calls, both when feeding and on the wing. It is one of 3 South American flamingos, the other 2 being the Chilean flamingo (*Phoenicopterus chilensis*) and the Andean flamingo (*Phoenicoparrus andinus*). All 3 species can be found together in salt lakes, usually at altitudes over 9,900 ft (3,000 m). Most puna flamingos migrate to lower altitudes in winter, but some remain at lakes where there are hot springs. The courtship and breeding habits of this bird are similar to those of the greater flamingo (see above).

CONSERVATION

With its small size and localized distribution, the puna flamingo is vulnerable to anything that affects its reproductive success. In the past, its meat and eggs were widely used for food, and during the first half of the 20th century, it suffered a marked population decline. Today, the major breeding colonies are protected, and its numbers have responded well. In 2005, the total population was estimated to stand at about 1 million birds.

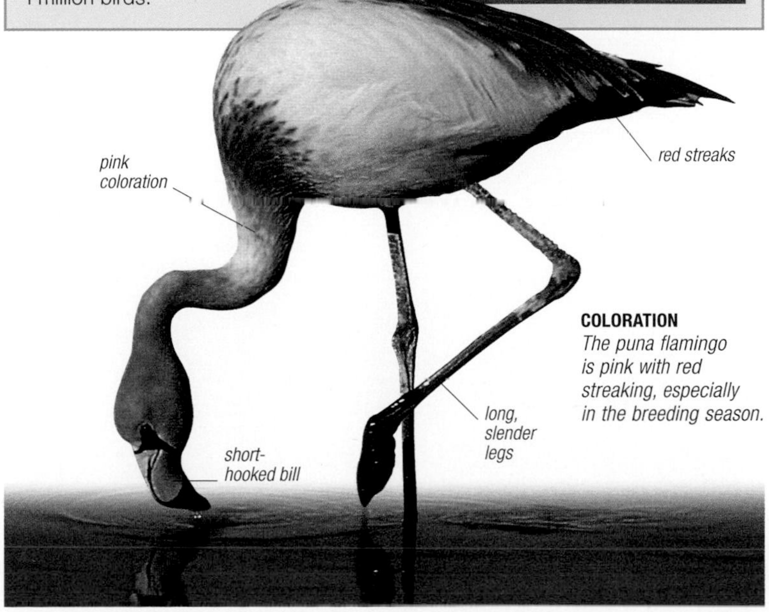

COLORATION
The puna flamingo is pink with red streaking, especially in the breeding season.

Phoeniconaias minor

Lesser flamingo

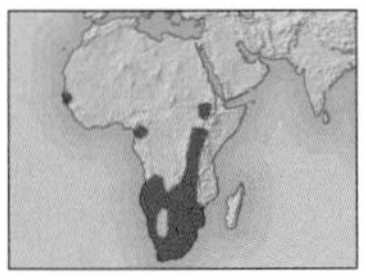

Height Up to 3¼ ft (1 m)
Weight Up to 4½ lb (2 kg)
Plumage Sexes alike
Migration Nomadic
Status Near threatened

Location W., C., and southern Africa

The smallest but most numerous flamingo, the lesser flamingo may be light to dark pink, with a relatively long, dark-colored beak. It is nonmigratory, but will readily move to a new site in search of food. Its diet consists almost entirely of microscopic blue-green algae, abundant in soda lakes. It feeds in shallow water at dusk and after dark, avoiding strong daytime winds. Some colonies are over a million strong—among the largest bird aggregations in the world. Courtship rituals often involve hundreds of birds moving and displaying in synchrony.

Herons and relatives

PHYLUM	Chordata
CLASS	Aves
ORDER	Ciconiiformes
FAMILIES	3
SPECIES	121

This group of wading birds includes herons, egrets, bitterns, storks, ibises, and spoonbills. All have large, plump bodies with long necks and powerful bills. Their long legs help them wade in shallow water, where they feed on fish, amphibians, snails, and crabs. Most feed alone to avoid other birds disturbing their prey, but at night they gather in groups to roost, and many species breed in colonies. These birds are found in freshwater habitats throughout the world, mostly in warm regions not subject to winter freezing.

FEEDING INLAND
The cattle egret finds most of its food away from water. Flocks typically gather around cattle and other large mammals to feed on insects disturbed by their movements.

Anatomy

Herons and their relatives have several adaptations for walking and feeding in shallow water. Long legs allow the bird to keep its plumage dry while it extends its flexible neck to lower its head toward the water. The feet have 4 widely spaced toes, the front 3 joined by webbing that spreads the bird's weight as it walks on mud or marshy vegetation. All birds in this group have broad wings. During flight, herons, bitterns, and egrets retract their neck, whereas most storks and all ibises and spoonbills hold it straight out.

Feeding

Most herons and their relatives use sharp eyesight to find their prey, scanning the water for signs of movement below the surface. Once sighted, a fish or frog will be grabbed (but rarely speared) with a thrust of the bill. Some birds refine the technique—for example, the black heron spreads its wings to cast a shadow on the water and lure fish into the shade. Not all of these birds feed on live prey: the marabou stork is primarily a scavenger that feeds on carcasses. The opportunistic feeding behavior of many species has meant that they are no longer tied to a wetland habitat. Thus, while marabou storks will invade arid habitats—and are even attracted to human refuse tips—in search of carrion, cattle egrets will habitually enter pasture to catch animals disturbed by livestock. As a group, the order is represented on most wetland, coastal, and open terrestrial regions worldwide, with storks generally tolerating drier habitats than herons—and ibises generally favoring warmer latitudes. Those species found in the Northern Hemisphere are migratory, but species in the tropics are sedentary.

Breeding

While many herons are gregarious and may breed in large colonies, bitterns—their close allies—lead more secretive solitary lives. Storks nest alone or in groups depending upon species, whereas ibises and spoonbills are colonial. All members of the order are monogamous and form tight pair bonds that may involve elaborate courtships displays. Their nests range from low-lying platforms among reedbeds to large, untidy stick-nests in trees. Both sexes incubate the eggs and care for their young—which hatch with at least a partial coat of down but are nevertheless helpless and must remain in the nest for a number of weeks.

The hamerkop and shoebill were formerly classified with storks, but are now thought to be allied to pelicans. More recent studies of DNA now suggest that herons, bitterns, ibises, and spoonbills are more closely related to pelicans, too—suggesting that this order should be split into Ciconiiformes (storks) and Ardeiformes (herons, ibises, and relatives).

BILL SHAPES
A long bill is a useful tool for pulling animals from water or mud. Ibises have a long, thin, downcurving bill. Herons have a long, straight bill that tapers to a point. A stork's bill is similar in shape but often curves up or down at the tip, while a spoonbill's flat bill broadens at the end to form a shape similar to a spatula.

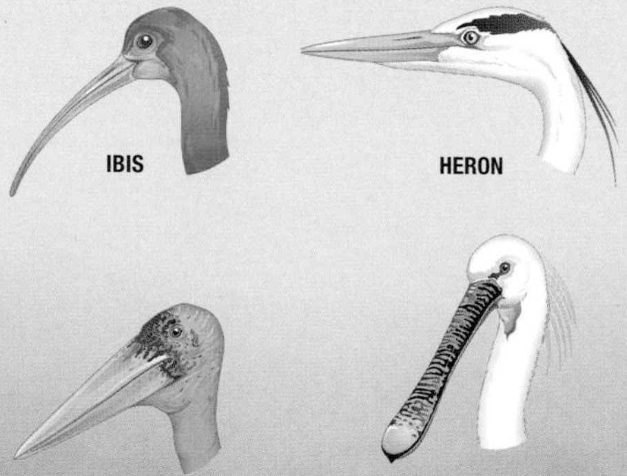

STRIKING AT PREY
When hunting, a heron bends its neck into a characteristic "S" shape. On seeing its prey, the heron thrusts its head forward with lightning speed. Once the prey is caught, it may be subdued by a stab from the bill.

WADING
Herons and their relatives wade slowly and deliberately. Spoonbills (seen here) sweep the water with their bills, using touch-sensitive cells to find small fish and crustaceans.

Ardea cinerea

Grey heron

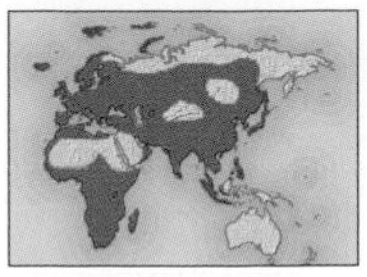

Length 35–39 in (90–98 cm)	
Weight 2¼–4½ lb (1–2 kg)	
Plumage Sexes alike	
Migration Partial migrant	
Status Least concern	

Location Europe, Asia, Africa

The most common and widespread heron in Europe, this bird can be found in all kinds of shallow, freshwater habitats from reedbeds to lakes. It flies with slowly flapping, deeply bowed wings. Grey herons pair for life; the male and female build the nest together, often in a tall tree, but sometimes on the ground. The female usually lays 3–5 eggs. Like other herons, they use ritual courtship and defense displays, such as stretching upward with an arching neck.

APPEARANCE
A large and distinctive bird, the grey heron has a long neck and long legs. Its plumage is gray, white, and black.

FISHING BY SIGHT

The grey heron relies on stealth and rapid reactions to catch its prey. When fishing, it stands close to the water's edge, monitoring the movement of any fish close by. If a fish comes within striking range, the heron tips forward and partly extends its neck, before suddenly stabbing with its bill. It swallows small fish whole, but takes larger ones back to land.

Butorides virescens

Green heron

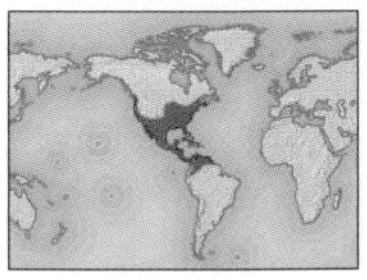

Length 16–19 in (40–48 cm)	
Weight 4–8 oz (125–225 g)	
Plumage Sexes alike	
Migration Partial migrant	
Status Least concern	

Location North, Central, and N. South America

A common bird of American wetlands, the green heron is one of the few birds known to use a tool to acquire food: it drops bait—worms, twigs, or stolen bread—into shallow water in order to lure fish. Tropical and subtropical populations are sedentary, but in temperate North America, these herons migrate—overwintering in warmer coastal regions, especially favoring mangrove swamps.

Nycticorax nycticorax

Black-crowned night heron

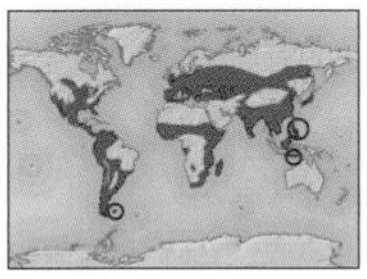

Height 23–25½ in (58–65 cm)	
Weight 18–29 oz (500–800 g)	
Plumage Sexes alike	
Migration Partial migrant	
Status Least concern	

Location North, Central, and South America, Africa, Asia

Unlike most other herons, this small bird feeds at night or at dusk, using its exceptional vision to pinpoint its prey in dim light, and occasionally also responding to sounds. It grasps food with its bill while standing still or walking slowly in the water. It usually feeds singly, but nests and breeds colonially. A good climber, it is often seen clambering over roots or branches close to the water's edge.

Botaurus lentiginosus

American bittern

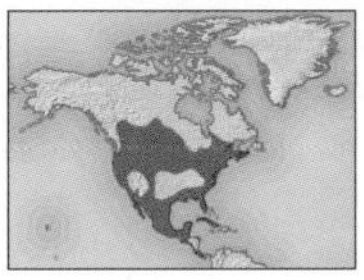

Height 23½–34 in (60–85 cm)	
Weight 18–32 oz (500–900 g)	
Plumage Sexes alike	
Migration Migrant	
Status Least concern	

Location North and Central America, Caribbean

With its streaked and mottled brown plumage, the American bittern is superbly camouflaged for life in dense reedbeds and overgrown marshes. This daytime feeder preys primarily on fish and frogs, stabbing with a lightning-fast lunge of its daggerlike bill. Like other bitterns, when disturbed it assumes a peculiar defensive posture, with its bill upturned and its neck extended, and may sway in the wind with the surrounding grass in order to remain inconspicuous. This bird is widespread across North America, but northern populations—such as in Canada—migrate southward to overwinter in southern USA and the Caribbean. Its nest is a well-concealed platform of vegetation placed in rushes just above the water level or on solid ground.

Ardea alba

Great egret

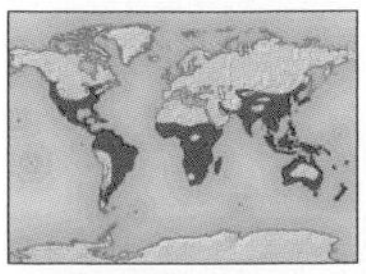

Length 34–39 in (85–100 cm)	
Weight 34–36 oz (950–1,000 g)	
Plumage Sexes alike	
Migration Partial migrant	
Status Least concern	

Location North, Central, and South America, Africa, Asia, Australia

The most widespread of all herons, the great egret is found in all kinds of wetlands from the Americas to Asia and Australia. Its plumage is entirely white, and its legs and feet are black. In the breeding season, long plumes develop on the back, trailing over the relatively short tail. These egrets stalk their prey alone or in loose groups, but within groups individuals defend small territories. Their young in the nest are extremely aggressive, often resulting in the death of weaker chicks.

Mycteria americana

Wood stork

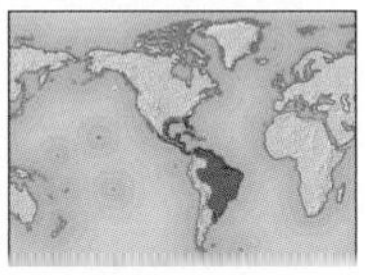

Height 34–43 in (85–110 cm)	
Weight 5½ lb (2.5 kg)	
Plumage Sexes alike	
Migration Partial migrant	
Status Least concern	

Location North, Central, and South America, Caribbean

The wood stork locates food by both sight and touch, enabling it to feed by day or night, and in murky water. The scaly head and neck are bald; feathers would be soiled when dipped in mud. While feeding by touch, the bird walks through the water with its long, thick bill open, moving it from side to side. When it encounters prey, its reactions are extremely quick. White, with black wingtips, it has a dark gray head and neck. The wood stork feeds either alone or in groups, and breeds colonially, building a nest often over water at treetop level, and very occasionally on the ground where there are no ground predators.

Ciconia ciconia

European white stork

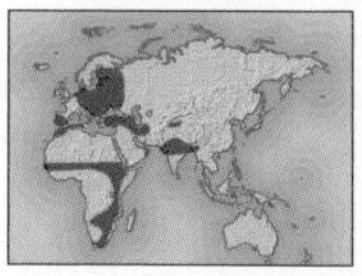

Height 3¼–4¼ft (1–1.3m)
Weight 5½lb (2.5kg)
Plumage Sexes alike
Migration Migrant
Status Least concern

Location Europe, Africa, Asia

In parts of Northern Europe, this elegant, black and white bird is a harbinger of spring, migrating from as far away as southern Africa, where it spends the winter. Today, largely due to habitat change, far fewer storks breed in Northern Europe, but their rooftop nests are still considered good luck. The white stork feeds primarily in shallow water and grassland, but also at the edges of crop fields. It migrates mainly over land, where thermals help its soaring flight.

Leptoptilos crumeniferus

Marabou stork

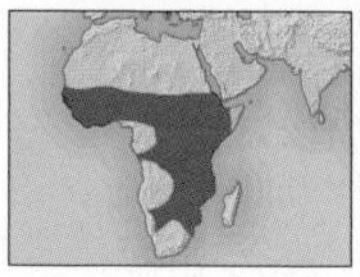

Height 4ft (1.2m)
Weight 11–17lb (5–7.5kg)
Plumage Sexes alike
Migration Nonmigrant
Status Least concern

Location Africa (south of Sahara)

This huge, ungainly looking stork is slaty gray, black, and white, and has a wingspan of nearly 9¾ft (3m), one of the largest of any land bird. Its elegant, soaring flight contrasts markedly with its hunched and unappealing aspect when on the ground. Because of its feeding habits (see panel below), the marabou stork has done very well, and is increasing in numbers throughout its range.

UNGAINLY AIR
Often described as "ugly," this large bird has a nearly featherless head and neck and a massive, wedge-shaped bill. Its pinkish to pale magenta pouch, or wattle, may extend to 14in (35cm) from the base of the bill.

SCAVENGING

Although it is a stork, the marabou often behaves more like a vulture. It soars high up in search of food, and uses its large beak to tear rotting meat from carcasses. Like a true vulture, it has an almost featherless head and neck—an adaptation that helps to prevent it from getting soiled with blood and gore when it inserts its head in carrion.

Threskiornis aethiopicus

Sacred ibis

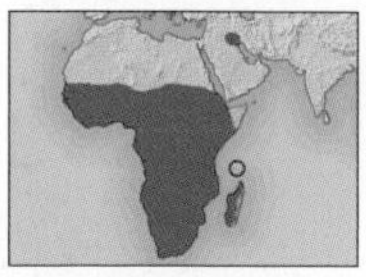

Length 65–89cm (26–35in)
Weight 1.5kg (3¼lb)
Plumage Sexes alike
Migration Migrant
Status Least concern

Location Africa (south of Sahara), Madagascar, Aldabra Island, W. Asia

This medium-sized white ibis, with its strikingly black, featherless head and upper neck, was revered in ancient Egypt. A generalized feeder, it easily cohabits with people and consumes a variety of foods—refuse and offal around human habitations, insects in grassland areas, and aquatic animals in shallow pools.

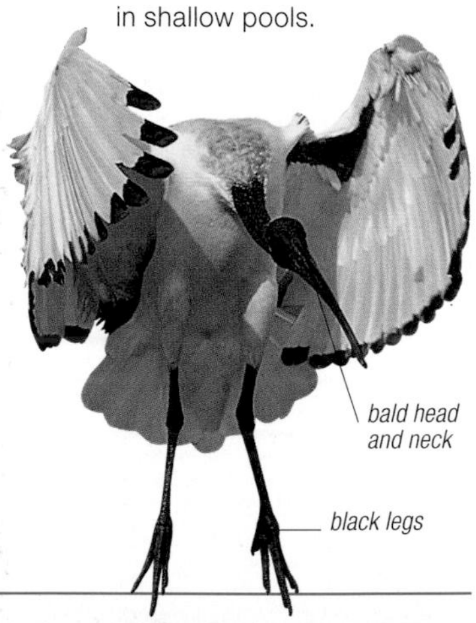

Eudocimus ruber

Scarlet ibis

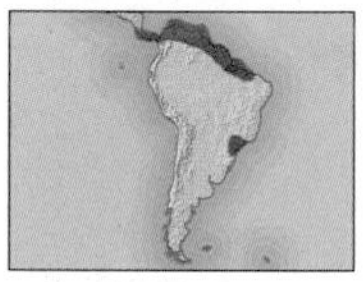

Length 22–27in (56–68cm)
Weight 28–33oz (775–925g)
Plumage Sexes alike
Migration Nonmigrant
Status Least concern

Location S. Central America, N. and E. South America

This vividly colored bird gathers in large flocks—sometimes of tens of thousands of birds—in the breeding season, on the coastal wetlands of northern South America, in swamps, lagoons, and mangroves, as well as on tidal rivers. Birds of a colony pair off for breeding (although they often mate with other partners, too), building their nests in trees close to water. Like other ibises, the scarlet ibis finds its food primarily by touch instead of by sight, probing into soft mud with its long, gently curving bill, usually while walking along. It may also feed like a spoonbill, sweeping its bill from side to side in the water. Crabs, shellfish, and aquatic insects are the major items of prey.

TREETOP ROOST

The scarlet ibis feeds on the ground during daytime, but at dusk, it flies up into waterside trees to roost. This behavior—common among ibises, herons, and their relatives—reduces the danger of being attacked by predators at night.

VIVID COLORATION
With its bright red plumage and black-tipped wings, the scarlet ibis is one of the world's most strikingly colored birds.

Geronticus eremita

Northern bald ibis

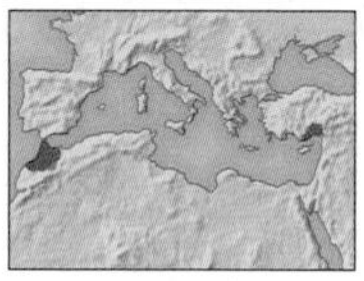

Length 32in (80cm)
Weight Not recorded
Plumage Sexes alike
Migration Migrant
Status Critically endangered

Location N.W. Africa, W. Asia

Once widespread throughout southern Europe, North Africa, and West Asia, this ibis is now extremely rare. It has a bare red head, a straggly black crest on the nape of its neck, and brownish black iridescent plumage. It feeds near pools and dry river beds, using its long, curved bill to probe for insects and other small animals in loose ground. It nests in colonies on steep cliffs, where its young are relatively safe from attack.

Platalea alba

African spoonbill

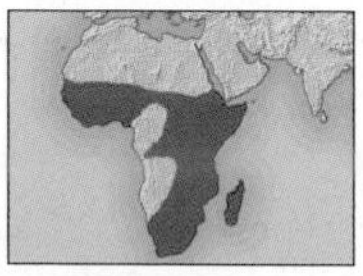

Length 30–35½in (75–90cm)
Weight ½lb (42kg)
Plumage Sexes alike
Migration Partial migrant
Status Least concern

Location Africa (south of Sahara), Madagascar

This graceful, red-legged white bird belongs to a group of 6 species that get their name from the broad, spoon-shaped tips of their bills. Like its relatives, it swings its bill from side to side in the water, creating currents that bring fish within range, which the spoonbill speedily snaps up. It may also feed by night, looking for fish and running after them in a sprint.

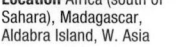

BIRDS

Pelicans and relatives

PHYLUM	Chordata
CLASS	Aves
ORDER	Pelecaniformes
FAMILIES	8
SPECIES	67

This group includes large waterbirds, such as pelicans, cormorants, tropicbirds, frigatebirds, and gannets. Most are strong swimmers, being the only birds with webbing between all 4 toes. Most species have broad wings, and frigatebirds and tropicbirds spend much of their lives on the wing, while gannets, cormorants, and darters are also capable of sustained flight over open sea. All the birds in this group feed on fish. Some species catch their food by making spectacular plunge-dives from a considerable height, and have several adaptations to protect themselves when they hit the water at high speed. Pelicans and their relatives are found in most of the world's seas. Darters, some cormorants and pelicans, the hamerkop, and shoebill all occur on inland water.

Anatomy

A range of physical features helps pelicans and their relatives to catch their prey underwater. The most aquatic members of the group, cormorants and darters, have surface feathers that become soaked easily, reducing buoyancy and making diving easier. The inner feathers remain waterproof and provide insulation. Other members of this group have water-repellent plumage and their bones contain sufficient air space to help them float and fly. Plunge-divers even have extra air sacs beneath the skin to cushion the force of impact as they hit the water.

NOSTRILS
Pelicans and their relatives have small or closed nostrils that prevent water from being forced into their airways when they dive below the surface. Cormorants, boobies, and gannets (shown here), with completely sealed nostrils, breathe through their mouth.

Reproduction

Often choosing remote islands or cliffs that are inaccessible to predatory mammals, pelicans and their relatives nest in colonies. Some pelicans, darters, cormorants, and frigatebirds nest in trees. Males use spectacular displays to attract females—for example, boobies point their head and tail skyward and raise their wings (known as sky-pointing). Nests are often large and usually built by both members of a pair. The young, which are helpless at birth, are fed on regurgitated food. Some young, notably those of frigatebirds, are looked after for many months after fledging.

NESTING
Most of the birds in this group (such as these king cormorants) build substantial nests. Collecting enough material can require many journeys back and forth from the nest site.

Feeding

Pelicans and their relatives use a range of methods for catching fish. Cormorants swim underwater in pursuit of prey. Darters also hunt below the surface but wait in ambush for up to a minute before spearing fish with their bill. Gannets, tropicbirds, boobies, and the brown pelican dive from heights up to 100 ft (30 m), dropping onto fish at speeds up to 60 mph (95 kph). Frigatebirds steal fish from other birds, harrying them in flight until they give up their catch. In contrast, pelicans often hunt cooperatively, forming lines and herding fish into shallows.

SURFACE FISHING
Pelicans have a large throat pouch, which they use to scoop fish from just below the surface. The pouch is also used to collect rainwater for drinking and to dissipate heat in hot weather.

BIRDS

Scopus umbretta

Hamerkop

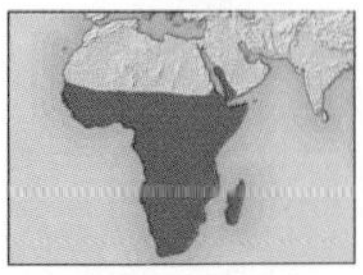

Location Africa (south of Sahara), Madagascar, S.W. Asia

Length 16–22 in (40–56 cm)
Weight 15 oz (425 g)
Plumage Sexes alike
Migration Nonmigrant
Status Least concern

From the German "hammerkopf" meaning hammerhead, this bird's common name is a good description of its unusual shape, with its relatively short bill and heavy crest on the back of its head. Dull brown, it has darker brown primary feathers, and a paler chin and throat. While many waterbirds feed primarily on fish, the hamerkop's main diet consists of amphibians. It feeds in shallow water, using its bill to rake the bottom for frogs and fish; it also flies above groups of tadpoles and snatches them up. The hamerkop builds the largest roofed nest of any bird. Made of twigs, mud, and grass, it is oven-shaped with an entrance tunnel, up to 6½ ft (2 m) across and deep, and sited high up in a tree. While the hamerkop breeds solitarily, nests are often found near each other.

Balaeniceps rex

Shoebill

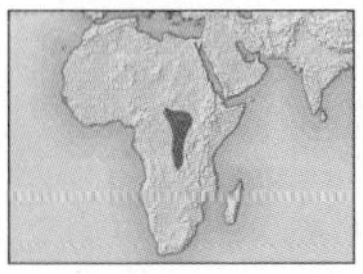

Location Central Africa

Height 3½–4½ ft (1.1–1.4 m)
Weight 10–14 lb (4.5–6.5 kg)
Plumage Sexes alike
Migration Nonmigrant
Status Vulnerable

The shoebill, or whale-headed stork, has a highly distinctive, large, broad, clog-shaped bill. It often hunts for prey in pools that are drying out, feeding mainly on lungfish, and also on frogs and small mammals. The shoebill frequently adopts an unusual method of hunting: it plunges on prey as if falling on it and then cuts up the food with a scissorlike action before swallowing. The shoebill breeds in almost inaccessible papyrus swamps, and usually nests on floating vegetation. In hot weather, it pours beakfuls of water over the eggs to cool them.

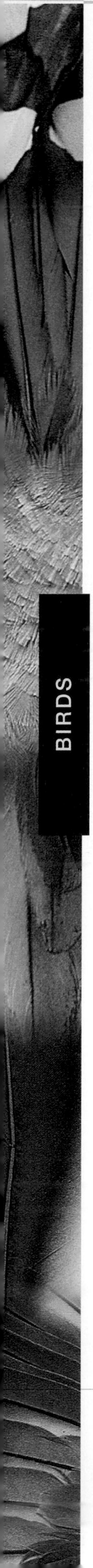

Phalacrocorax auritus

Double-crested cormorant

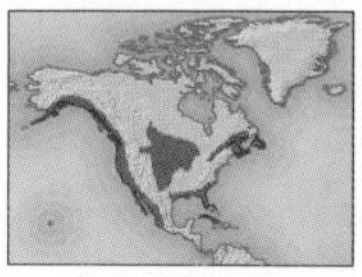

Location North America

Length 30–36 in (76–91 cm)
Weight 3¼–4½ lb (1.5–2 kg)
Plumage Sexes alike
Migration Partial migrant
Status Least concern

This bird is the only cormorant in North America that is widespread inland as well as on the coast. As with other cormorants, it has a streamlined body, a serpentine neck, and large, webbed feet. Unlike other cormorants, for a brief period in spring, it has a double crest on the head. Its increasing population has made it a serious pest of catfish farms on the Mississippi.

Phalacrocorax carbo

Great cormorant

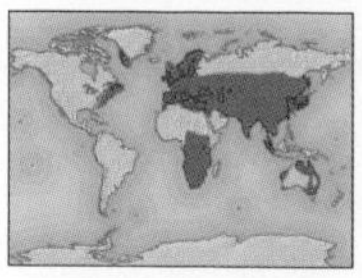

Location E. North America, S. Greenland, Europe, Asia, southern Africa, Australia

Length 32–39 in (80–100 cm)
Weight Up to 7¾ lb (3.5 kg)
Plumage Sexes alike
Migration Partial migrant
Status Least concern

One of the world's most widespread coastal seabirds, the great cormorant is also common inland and can be found on almost any large area of water—fresh, brackish, or saline, natural or artificial—in Europe, Asia, and Africa. It makes its nest in a wide range of habitats, from cliffs and bare rocks to trees and reedbeds. The great cormorant breeds in colonies, in some cases returning to the same location year after year, although pairs of birds usually remain together only for a single season.

LONG AND SLEEK
The great cormorant has a spare, streamlined body, a flexible, serpentine neck, and a stout, hooked bill to help grasp fish.

POWER SWIMMING

Although the great cormorant prefers to hunt in shallow water, it can dive to depths of 100 ft (30 m) or more. When underwater, it swims under its prey with its wings pressed closely to its body. Sturdy legs with large webs make for powerful propulsion to catch fast-moving fish.

Anhinga melanogaster

Darter

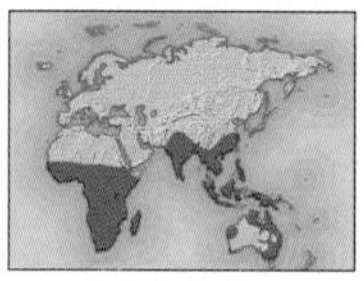

Location Africa (south of Sahara), S. and S.E. Asia, Australia, New Guinea

Length 34–38 in (85–97 cm)
Weight 2¼–4½ lb (1–2 kg)
Plumage Sexes differ
Migration Partial migrant
Status Near threatened

One of 4 similar species, this cormorant-like bird usually swims with only its head and neck above the waterline, the rest of the body being underwater. It has specially adapted vertebrae in its neck that form a Z-shaped kink, which straightens explosively and enables the darter to spear its prey with its pointed bill. It builds a nest of twigs sometimes as high as 16 ft (5 m) above water, where it lays 3 to 6 eggs.

Phalacrocorax atriceps

Imperial shag

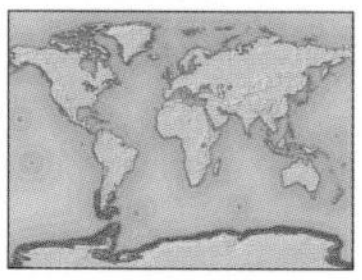

Location Circumpolar around Antarctica, S. South America, Falkland Islands

Length 27–30 in (68–76 cm)
Weight 5½–7¾ lb (2.5–3.5 kg)
Plumage Sexes alike
Migration Partial migrant
Status Least concern

This distinctive, black and white shag, with blue eye-rings, is from the Southern Hemisphere and found mostly on rocky coasts and islands. A highly gregarious bird, it forms dense winter flocks that forage offshore. In summer, it mainly feeds alone. There are several subspecies of the imperial shag, differing slightly in color and markings.

Phalacrocorax harrisi

Flightless cormorant

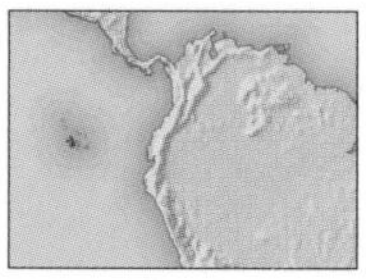

Location Galapagos Islands (Isabela and Fernandina)

Length 35–39 in (89–100 cm)
Weight 5½–8¾ lb (2.5–4 kg)
Plumage Sexes differ
Migration Nonmigrant
Status Endangered

This large cormorant has only the tiniest of wings and long since lost the ability to fly—the absence of terrestrial predators on the Galapagos Islands possibly having made flight unnecessary. It produces very little oil from its preen gland, but the soft, dense body plumage, more like hair than feathers, traps air, which prevents the bird from becoming waterlogged. The flightless cormorant has a long, strong bill that is useful for flushing octopus and fish from the sea bottom. It breeds in small groups. The male rears the young, continuing to feed them for several months, while the female deserts them to find a new mate.

Fregata minor

Great frigatebird

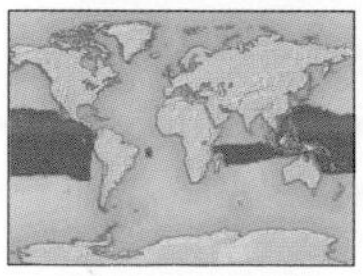

Location Tropical Pacific, Atlantic, and Indian oceans

Length 34–41 in (85–105 cm)
Weight 2¼–3¼ lb (1–1.5 kg)
Plumage Sexes differ
Migration Partial migrant
Status Least concern

With their remarkably light bodies and immense, slender wings, great frigatebirds spend their lives gliding effortlessly over the sea, taking food from the surface or from other birds. Their plumage is not waterproof so, when feeding, they dip only their bill into the water, holding the wings up to avoid getting them wet. This species—one of 5 in the family—has a typically long, hooked black bill and a long, forked tail. The male is black, with a glossy green sheen and long, pale wingbars, and when courting, exhibits a scarlet, balloonlike throat pouch. The female is black and white. The great frigatebird breeds on small, mostly uninhabited islands, forming colonies of up to several thousand pairs.

Phaethon aethereus

Red-billed tropicbird

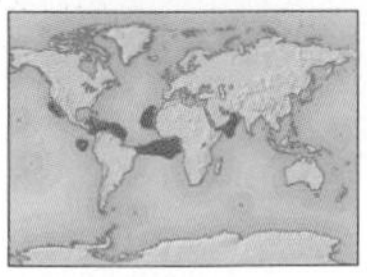

Location E. Pacific, C. Atlantic, and N. Indian oceans

Length 31–32 in (78–81 cm)
Weight 21–29 oz (600–825 g)
Plumage Sexes alike
Migration Migrant
Status Least concern

Although it is a poor swimmer, spending little time on the water, this highly aerial seabird is often found hundreds of miles from land. It feeds by plunge-diving, sometimes from a considerable height, to catch squid and fish, especially flying fish. An elegant bird, with silky white, pink-flushed plumage, it has extremely long tail streamers that it switches from side to side in an elaborate, airborne courtship display. The red-billed tropicbird is one of only 3 species in its family, all restricted to the tropics.

Pelecanus crispus

Dalmatian pelican

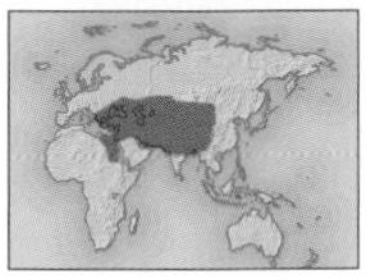

Location S.E. Europe, S. and S.W. Asia, N.E. Africa

Length 5¼–6 ft (1.6–1.8 m)
Weight 22–29 lb (10–13 kg)
Plumage Sexes alike
Migration Partial migrant
Status Vulnerable

This pelican is the largest found in Europe. Despite its size and weight, it is a strong flier, soaring high up during migration on its very large, broad wings. Its plumage is essentially silvery white, with black tips to the wings, and it has an orange-red pouch beneath a pale yellow, orange-tipped bill. The Dalmatian pelican feeds on a wide variety of fish, needing on average 2¼ lb (1 kg) a day. It swims on the surface, up-ending to seize prey. Sometimes a number of birds feed cooperatively, forming a semicircle and driving fish into shallow water, where they can then scoop them up. These pelicans often breed in large colonies, each pair usually rearing one young. Chicks are helpless when first hatched and entirely dependent upon parents; when they are 3 or 4 weeks old, they live in "pods" largely unattended by their parents, which do, however, continue to bring them food. At about 6 weeks, the chicks begin to catch fish, and a month or so later they make their first proper flight.

Pelecanus occidentalis

Brown pelican

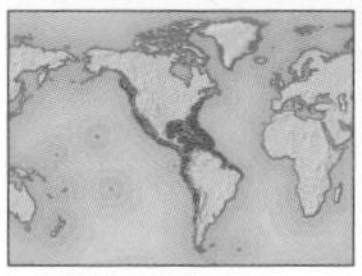

Location North, Central, and South America, Caribbean

Length 3¼–5 ft (1–1.5 m)
Weight Over 7¾ lb (3.5 kg)
Plumage Sexes alike
Migration Partial migrant
Status Least concern

The brown pelican is mainly silver-gray and brown, with a white or white and yellow head and chestnut mane. The greenish skin of the face and throat pouch becomes much more vivid in the breeding season. The only pelican that feeds by plunge-diving instead of by swimming and fishing at the surface, it glides low over the water on outstretched wings, and on sighting a fish, flies up as high as 30 ft (10 m) before folding back its wings and plunging into the sea. It usually nests in trees or shrubs—often in mangroves. However, the Peruvian pelican—a subspecies of the brown pelican—nests on the ground.

RAISING THE YOUNG

Both parents incubate the eggs in shifts of several hours. One or 2 young—very seldom 3—may be raised. Initially, the adults regurgitate food onto the nest floor, but when the chicks are about 10 days old, they take fish directly from the parent's bill.

GROUP LIFE
Highly gregarious, the brown pelican roosts, migrates, and often feeds in groups. It also breeds colonially, on bare, low-lying islands.

Sula nebouxii

Blue-footed booby

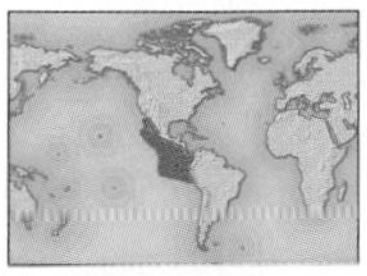

Location W. Mexico to N.W. South America, Galapagos Islands

Length 32–34 in (80–85 cm)
Weight 3¼ lb (1.5 kg)
Plumage Sexes alike
Migration Partial migrant
Status Least concern

This distinctive seabird, with vivid blue legs and feet, is one of 9 species that make up the gannet and booby family. Like its relatives, it has a cigar-shaped body adapted for plunge-diving, and narrow wings that angle back just before it hits the surface. The male is much smaller and lighter than the female and so especially adept at diving into very shallow, inshore water—even in rock pools. Small groups of birds sometimes dive together for food—mainly fish such as flying fish, sardines, anchovy, and Pacific mackerel, or squid. The blue-footed is one of the rarer boobies, with a limited distribution. Even its dispersal is relatively restricted; for example, the Galapagos birds move only to nearby Ecuador.

Sula leucogaster

Brown booby

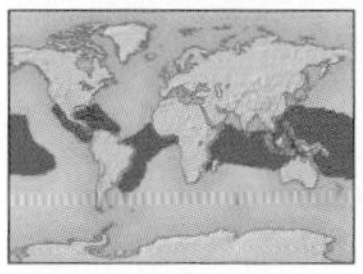

Location Tropical Pacific, Atlantic, and Indian oceans

Length 25–29 in (64–74 cm)
Weight 1½–3¼ lb (0.7–1.5 kg)
Plumage Sexes alike
Migration Partial migrant
Status Least concern

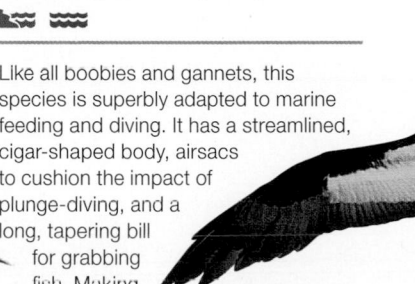

Like all boobies and gannets, this species is superbly adapted to marine feeding and diving. It has a streamlined, cigar-shaped body, airsacs to cushion the impact of plunge-diving, and a long, tapering bill for grabbing fish. Making spectacular dives from 100 ft (30 m) or more, it penetrates the water to great depths, bombarding shoals of squid and fish such as mullet. With its long, sturdy wings, the brown booby is also a graceful flier, and the only member of the family in which the male advertises for a mate in flight.

Morus bassanus

Northern gannet

Location N. Atlantic, Mediterranean

Length 32–35 in (80–90 cm)
Weight 5½–6½ lb (2.5–3 kg)
Plumage Sexes similar
Migration Partial migrant
Status Least concern

The northern gannet is a very streamlined seabird, with a torpedo-shaped body, long, narrow wings, and a daggerlike bill, perfectly adapted to plunge-diving for fish. It nests in densely packed colonies, or "gannetries," on steep cliffs and raised slopes. Highly developed pair-bonding behavior persists throughout the nesting season and, once paired, gannets will remain together for years, returning to the same nest season after season. First breeding occurs between the ages of 3 and 5 years.

NESTING GANNETS
Within their colonies, gannets space their nests at a distance of 2 birds' reach. Gannetries are so densely populated that viewed from afar the high slopes and cliffs can have the appearance of being covered in snow.

One bluish white egg is laid and then incubated by both parents. Young gannets typically leave the nest around 3 months after hatching and migrate without their parents. Juveniles achieve full adult plumage within 5 years.

DAZZLING PLUMAGE
The northern gannet is a very distinctive seabird. Most of its plumage is dazzling white, except for its jet-black wingtips, and the head and back of the neck which have a buff-colored hue, contrasting with its icy blue eyes and bill.

SHOW OF AGGRESSION
The northern gannet can be exceptionally aggressive when defending its nest site, using its powerful bill to stab and grip when fighting.

BEHAVIORAL PATTERNS

Few seabirds are as sophisticated in their behavioral patterns as the northern gannet. Both its plunge-diving technique and pair-bonding behavior are highly dramatic, while the level of parental care it offers is advanced.

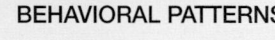

COURTING COUPLE
During spectacular bonding displays, partners fence with their bills, as the female aims to allay the male's aggression.

DRAMATIC PLUNGE-DIVERS
From up to 150 ft (45 m) above the water, the gannet uses its binocular vision to track fish before plunging at speeds of up to 60 mph (100 kph).

FEEDING A NESTLING
The young gannet reaches deep into its parent's throat to feed on regurgitated fish.

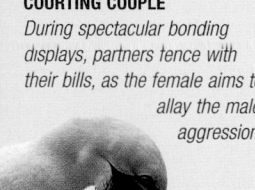

Birds of prey

PHYLUM	Chordata
CLASS	Aves
ORDER	Falconiformes
FAMILIES	3
SPECIES	319

Of all birds, these are the chief carnivores and the most accomplished predators. Although many other birds feed on living animals, birds of prey are set apart by their remarkably acute eyesight, muscular legs, and sharp bill and talons—as well as by their aerial ability and sophisticated hunting techniques. This large group (sometimes also referred to as raptors) includes eagles, hawks, vultures, buteos, buzzards, ospreys, and falcons, among others. Most birds of prey hunt by day, which distinguishes them from owls. As a group, raptors feed on a large assortment of living animals, from worms and snails to fish, reptiles, amphibians, mammals, and other birds. They also eat carrion. Birds of prey are found almost worldwide but are most common in open country in warm parts of the world.

Hunting

Catching live prey takes much practice and often has a high failure rate. Adult birds are usually better hunters than young ones, but even adults miss more often than they strike. Some species vary their technique according to their prey—buzzards, for example, hover over small mammals but search for earthworms on the ground. Others, such as the osprey, are more specialized (see below). Most birds of prey hunt alone, but in a few species (including some eagles) pairs work together, one bird flushing out the prey and the other striking.

FISHING
The osprey feeds almost entirely on fish and has a unique catching technique. It approaches in a low-angled dive and throws its long legs forward to grab a fish with its talons. To help it grip, the osprey may reverse one of its 3 forward-pointing toes.

Anatomy

Birds of prey range in size from falconets, no larger than a sparrow, to condors, which have a wingspan of up to 10 ft (3.2 m). Some are broad-winged and heavy, while others are slight and streamlined. Most species have a large head and short neck, although vultures have a long, bare neck that allows them to reach inside carcasses. One of the most distinctive features of the group is the bill: in almost all species, it is powerful and hooked, with sharp edges for tearing flesh. The exact form of the bill varies, reflecting differences in diet (see below, right). A bird of prey's other main tool is its feet, which are strong and muscular with long, sharp claws known as talons (see below, left). The plumage of most species is in subdued colors (such as brown, gray, black, or blue), often combined with white.

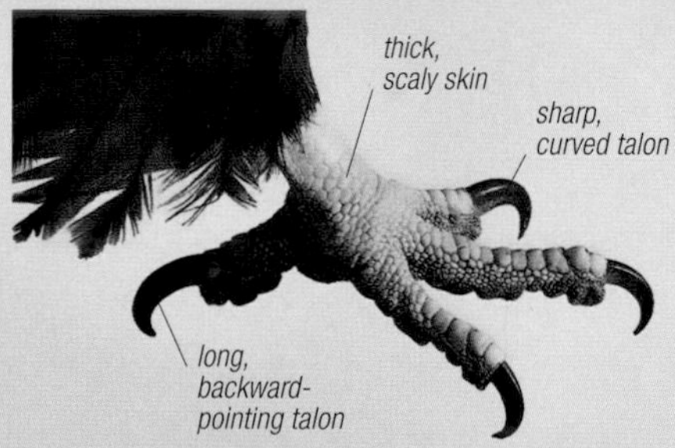

TALONS
Most birds of prey kill with their talons rather than their bill. The talons pierce vital organs and break slender bones. Three of the 4 toes point forward, and one points backward. In many species (including this white-tailed sea eagle), the backward-pointing claw is the longest and sharpest. It gives the foot a powerful pincerlike action for holding and crushing prey that can be difficult for the bird to release.

BALD EAGLE

HAWK

SNAIL KITE

BILLS
The distinct hook and sharp edges of a bald eagle's bill are typical of many birds of prey. Variations on this shape can be seen in other species. Hawks and falcons have a toothlike ridge in the upper bill for breaking the spine of their prey, while the snail kite uses its long, hooked bill to pull snails from their shells.

Senses

Birds of prey hunt mainly by sight, and they are renowned for their acute vision. It is estimated that they can see at least 4 times as much detail as a human, an ability that helps them to pick out prey from a great distance. This arises from having a high concentration of cone cells in the retina, and also from having relatively large eyes. Some species have excellent hearing, notably the harriers, which hunt over thick vegetation. A few vultures also have an exceptional sense of smell, by which they can locate hidden carrion.

EYE PROTECTION
Like many birds of prey, the goshawk has a ridge over each eye, which provides shade and protects the eye from struggling prey. In many species, the eye is also protected by a transparent third eyelid (nictitating membrane).

BIRDS

Feeding

Most birds of prey feed on live animals; the larger the species, the larger the prey that they are able to carry. The bald eagle, for example, is capable of taking a deer fawn if it is not too heavy. Many birds of prey are specialized feeders, including the honey buzzard (which eats wasps and their grubs), and the secretary bird (which eats snakes). As well as live prey, many species also eat carrion from time to time. A few, such as vultures and kites, feed almost entirely on dead animals. One carrion-feeder, the bearded vulture, eats bones, breaking the larger ones by dropping them onto rocks. The palm-nut vulture, named after its favorite food, is unusual in that it feeds mainly on plants.

CARRION-FEEDERS
Up to 6 species of vultures may gather around a carcass at one time, each feeding on different parts of the body. For example, some may feed on soft body parts, while others eat the skin and hide. The lappet-faced vulture (left) is one of the largest species and is often dominant over others.

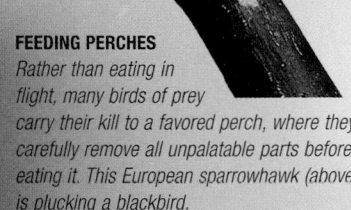

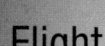

FEEDING PERCHES
Rather than eating in flight, many birds of prey carry their kill to a favored perch, where they carefully remove all unpalatable parts before eating it. This European sparrowhawk (above) is plucking a blackbird.

AERIAL KILLER
One of the most distinctive features of birds of prey is the way they kill with their feet, seizing prey from the ground, water, or air. Steller's sea eagle (seen here) uses its feet to catch Pacific salmon. It also hunts geese, hares, and young seals.

Flight

All birds of prey are able fliers. They use different patterns of flight to suit the way they hunt. Using their large, broad wings to ride thermals or updrafts next to cliffs, vultures and condors can stay aloft for hours, barely flapping their wings and using their high vantage point to search for carrion. In contrast, the slim-bodied falcons attack other birds in midair, maneuvering on long, thin, pointed wings and sometimes diving (or stooping) from a great height (see below). Hawks plan their strikes and then use a burst of speed to surprise or ambush prey. Harriers fly slowly forward over the ground and then drop onto unsuspecting prey. Kestrels and buzzards can hover in one place, watching for movement below and then lunging downward, talons first.

SOARING
Condors and vultures have the largest wings of all birds of prey (the bearded vulture, above, has a wingspan of up to 9¾ ft (3 m)). Both are common in mountainous areas, where rising currents of air help keep them aloft.

STOOPING
Most large falcons live in open country with little cover to conceal them from their prey. To catch birds that could evade them in flapping flight, they often climb to a high altitude and then dive, or stoop, on their target. When diving, the peregrine falcon (right) can reach a speed of 125 mph (200 kph).

CONSERVATION

Many birds of prey are threatened by human activity. Most are listed by CITES, and they include some of the world's most endangered animals, such as the California condor, with a current population of under 400. Even widespread species, such as the peregrine falcon (left), are vulnerable because they are at the top of food chains, and therefore easily harmed by factors affecting their prey. They are also routinely persecuted, because of the perceived threat they pose to livestock.

Cathartes aura

Turkey vulture

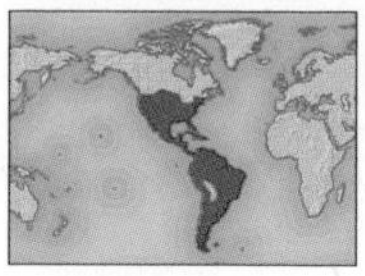

Location S. Canada to S. South America

Length 25–32 in (64–81 cm)
Weight 1¾–4½ lb (0.85–2 kg)
Plumage Sexes alike
Migration Partial migrant
Status Least concern

Often called a "buzzard" in the USA, the turkey vulture is found in an exceptionally wide range of habitats from southern Canada through to Tierra del Fuego. This vulture and its close relatives are the only birds of prey with a good sense of smell, enabling them to locate food even in thick jungle. It soars and glides as soon as the day has warmed up, and often feeds on road kill, possibly explaining its increase in populated areas. Although little is known about its courtship behavior, this bird has been seen performing a ritualized dance on the ground. The female lays 2 eggs and both parents bring food to the young, which are fed by regurgitation from the parents' crops. Once past their first year, the young birds are likely to live for 12–17 years.

Vultur gryphus

Andean condor

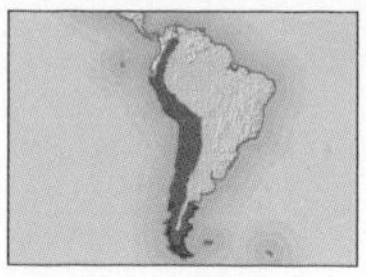

Location W. South America

Length 3¼–4¼ ft (1–1.3 m)
Weight 24–33 lb (11–15 kg)
Plumage Sexes differ
Migration Nonmigrant
Status Near threatened

The Andean condor has the largest wing area of any bird. Because of its immense size, it relies on updrafts from mountains and coastal cliffs to remain airborne, and can travel huge distances with only the occasional flap of its wings. From the ground, its silhouette is highly distinctive, with its large flight feathers spreading out like the fingers of a hand. The males are larger than the females—unusually for birds of prey—and while both sexes have bald heads, the males also have a characteristic white ruff at the base of their necks. Andean condors feed mainly on carrion, and soar at altitudes of up to 18,000 ft (5,500 m) to search for food. Their diet is wide-ranging, and includes not only mountain animals, but also the remains of stranded marine animals, such as seals and whales. In some places, they also feed at seabird colonies, plundering large numbers of eggs. They breed on inland cliffs and reproduce slowly, typically laying a single egg every 2 years.

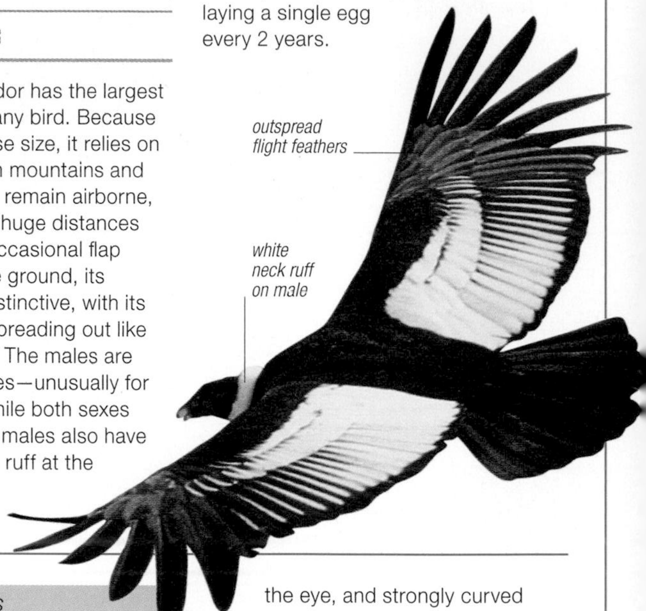

Gymnogyps californianus

California condor

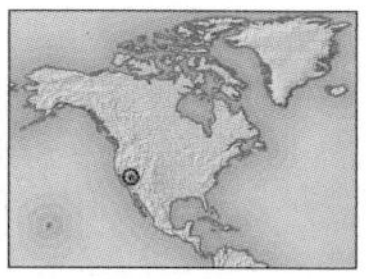

Location W. USA (California, Arizona)

Length 4–4¼ ft (1.2–1.3 m)
Weight 18–31 lb (8–14 kg)
Plumage Sexes alike
Migration Nonmigrant
Status Critically endangered

The California condor is the largest flying bird in North America, and also one of the most endangered. Once found from California to Florida, by 1987 the species no longer existed in the wild, but a captive-breeding program (see panel, below) is slowly helping it recover. California condors are expert at soaring and spend much of their time circling high in the air. At one time, they fed mainly on the remains of bison and pronghorns, but with the arrival of Europeans, this abundant food source rapidly declined. Today, captive-bred condors that have been released feed mainly on dead cattle and deer.

BLACK PLUMAGE
The California condor has mainly black plumage with white tips to the smaller wing feathers and a black neck ruff. The head and the upper part of the neck lack feathers.

CONSERVATION

Through the California Condor Recovery Program, a captive-bred population of condors numbering just 27 in the 1980s has steadily grown to nearly 400 birds today. About half of them are in captivity, and the remainder in the wild. The aim of the program is to re-establish a number of separate breeding colonies, which exist independently of human help, finding their own food.

Pandion haliaetus

Osprey

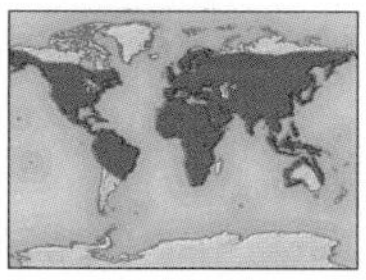

Location Worldwide (except Antarctica)

Length 21½–25 in (54–64 cm)
Weight 3¼–4½ lb (1.5–2 kg)
Plumage Sexes differ
Migration Migrant
Status Least concern

One of the most widespread birds of prey, the osprey is found on every continent except Antarctica. It has white underparts and chocolate-brown upperparts, with a dark stripe through the eye, and strongly curved talons and bill. It plucks live fish from the water during a spectacular, feet-first dive, sometimes totally submerging itself. Its feet have evolved to hold onto slippery prey, with spiny, dry scales on the undersides and a reversible outer toe for carrying fish head-first through the air.

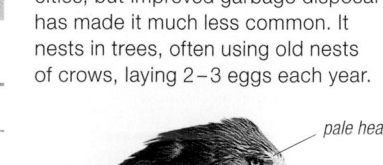

Milvus milvus

Red kite

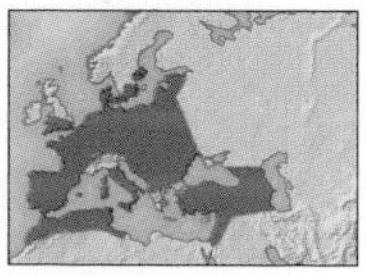

Location Europe, W. Asia, N. Africa, Canary and Cape Verde islands

Length 24–26 in (61–66 cm)
Weight 27–36 lb (750–1,000 g)
Plumage Sexes alike
Migration Partial migrant
Status Near threatened

Kites are agile fliers, typically with long legs and forked tails. The red kite is the largest species, feeding mainly on small mammals, carrion, and young birds. It flies with its wings partly bent, and uses its tail like a rudder, constantly tilting it as it searches for food. Two centuries ago, the red kite was a common scavenger in European towns and cities, but improved garbage disposal has made it much less common. It nests in trees, often using old nests of crows, laying 2–3 eggs each year.

BIRDS

Pernis apivorus

European honey buzzard

Length 20½–23½ in (52–60 cm)
Weight 16–36 oz (450–1,000 g)
Plumage Sexes alike
Migration Migrant
Status Least concern
Location Europe, W. and C. Asia, Africa

Compared to similarly sized birds of prey, the European honey buzzard has small feet, relatively undeveloped talons, and a shallow, curved bill. This reflects its unusual lifestyle as a predator of wasps and other insects. It catches its food in midair, and also follows insects back to their nests, to feed on their developing grubs. Because it depends on insects, it has to migrate southward in winter.

Rostrhamus sociabilis

Snail kite

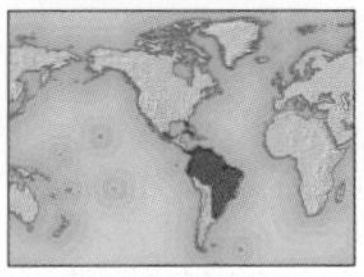

Length 16–18 in (40–45 cm)
Weight 13–14 oz (350–400 g)
Plumage Sexes differ
Migration Partial migrant
Status Least concern
Location S.E. USA (Florida), Cuba, Central America, South America

Formerly called the Everglades kite, this bird is a specialist feeder on aquatic snails. It flies slowly over shallow marshland, snatching up surface-feeding snails. It inserts its fine, heavily curved, hooked bill into the snail and cuts the muscle that attaches it to its shell, allowing it to remove the animal's body. Adult males are dark gray with black primaries; females are brown with streaked buff underparts. Rare in its south Florida range, it can be abundant elsewhere, as on the Argentine pampas.

Haliaeetus vocifer

African fish eagle

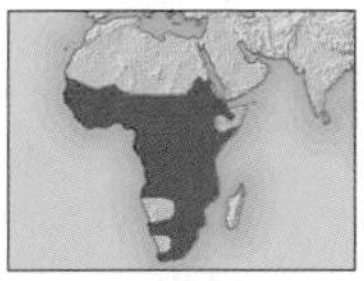

Length 25–29 in (63–73 cm)
Weight 4½–7¾ lb (2–3.5 kg)
Plumage Sexes alike
Migration Nonmigrant
Status Least concern
Location Africa (south of Sahara)

Known as the "Voice of Africa," this eagle throws its head back and gives a loud, yelping cry. With its distinctive white head, it is easy to spot perching on high branches beside rivers. When it locates a fish, it swoops down to catch it with a backward swing of the feet. It will also eat small mammals, birds, and carrion, as well as steal fish from other eagles and fishermen.

Gypaetus barbatus

Bearded vulture

Length 3¼–4 ft (1–1.2 m)
Weight 10–15 lb (4.5–7 kg)
Plumage Sexes alike
Migration Nonmigrant
Status Least concern
Location Europe, Asia, N., E., and southern Africa

The bearded vulture is one of the largest Old World vultures, with huge wings, a conspicuous wedge-shaped tail, and distinctive beardlike feathers at the base of its bill. It eats carrion, but it also specializes at feeding on bones, carrying large ones high into the air and then dropping them onto rocks below. Once the bones have shattered, the bearded vulture lands to feed on the marrow inside.

Neophron percnopterus

Egyptian vulture

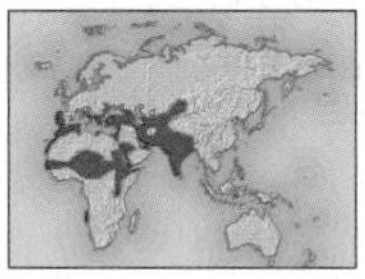

Length 23–28 in (58–70 cm)
Weight 3¼–4½ lb (1.5–2 kg)
Plumage Sexes alike
Migration Partial migrant
Status Endangered
Location Europe, Africa, Asia

The smallest Old World vulture is now endangered because of pollution and decline of large ungulates. Adults have yellow faces and off-white plumage, apart from their black flight feathers; juveniles are speckled brown (shown here). Egyptian vultures scavenge all kinds of food, but are renowned for using stones to break open the eggs of ostriches and other birds. Because of their small size, they have difficulty competing with other vultures at carcasses, and are often the last to feed.

Gyps africanus

African white-backed vulture

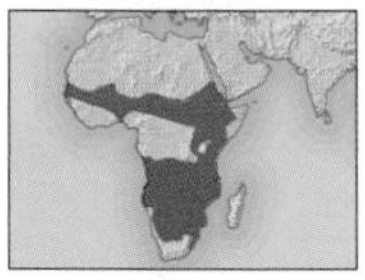

Length 37 in (94 cm)
Weight 8¾–15 lb (4–7 kg)
Plumage Sexes alike
Migration Nonmigrant
Status Near threatened
Location Africa (south of Sahara)

This widespread African scavenger is a griffon vulture—one of a group of 7 species that have long necks that appear to be bald, but that are actually covered in fine down. This absence of large neck feathers allows griffon vultures to reach deep into carcasses without becoming soiled. The African white-backed vulture gets its name from a collar of white feathers at the top of its back, which contrasts with its gray neck. This species is one of Africa's most common vultures, and is often seen in large numbers where food can be found. It makes a variety of hissing and cackling sounds as it jostles for a chance to eat.

FEEDING

White-backed vultures are legendary for their ability to find food. They have a poor sense of smell, but extremely good eyesight, enabling them to spot dead remains from high in the air. Vultures also keep an eye on each other. If one bird sees food and makes a sudden descent, others quickly follow suit.

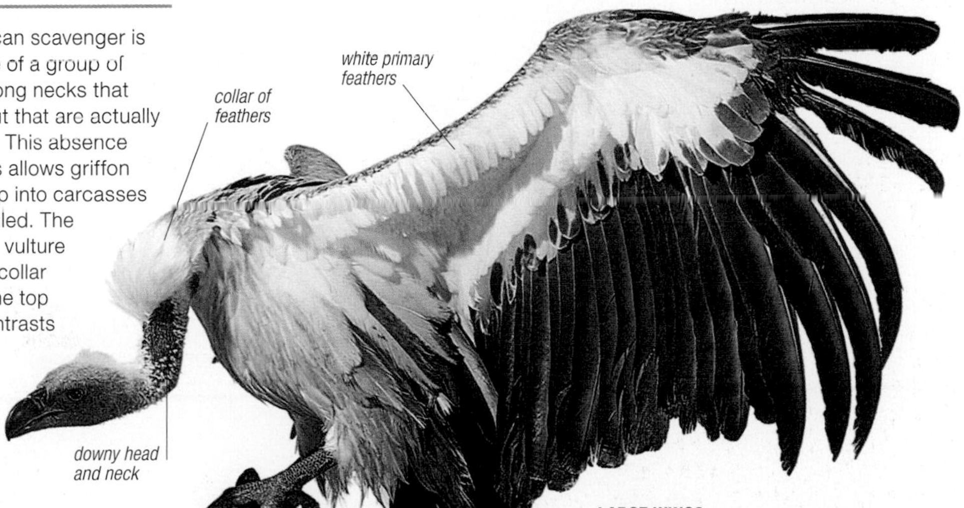

LARGE WINGS
The large, broad wings of the African white-backed vulture enable it to soar and circle on thermals for hours, looking for carrion.

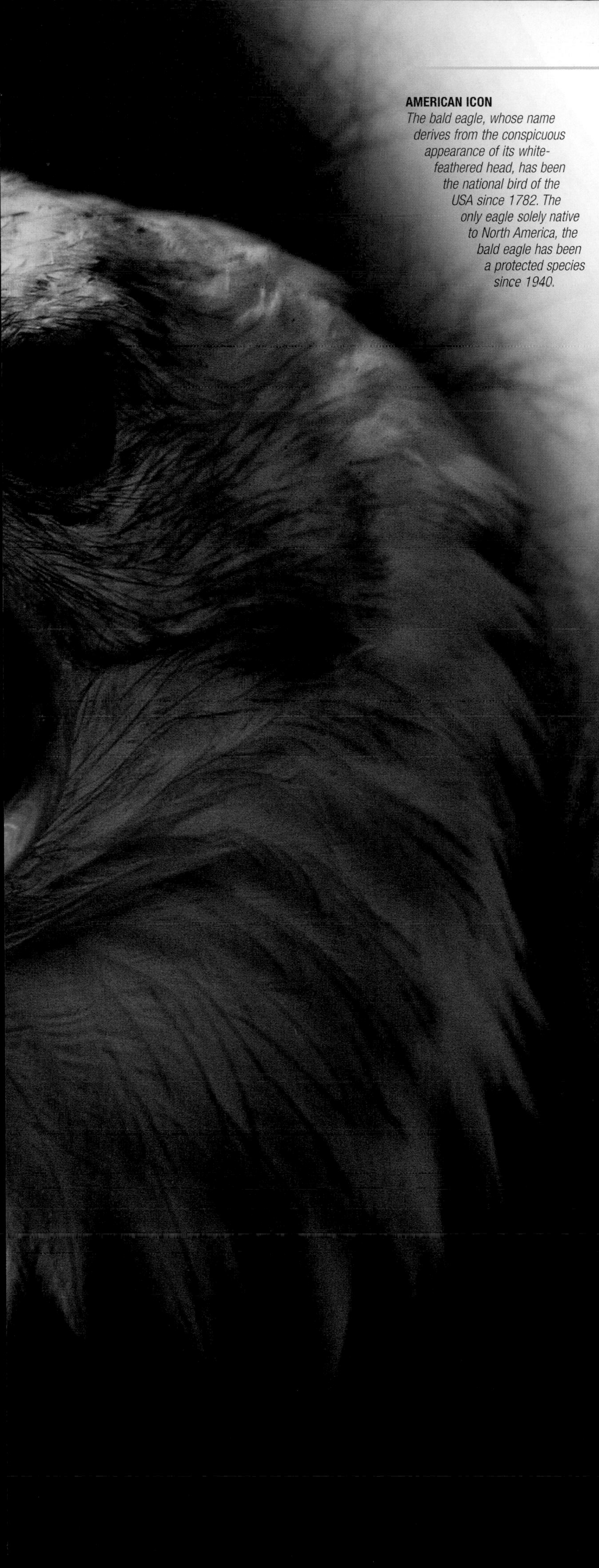

AMERICAN ICON
The bald eagle, whose name derives from the conspicuous appearance of its white-feathered head, has been the national bird of the USA since 1782. The only eagle solely native to North America, the bald eagle has been a protected species since 1940.

Haliaeetus leucocephalus

Bald eagle

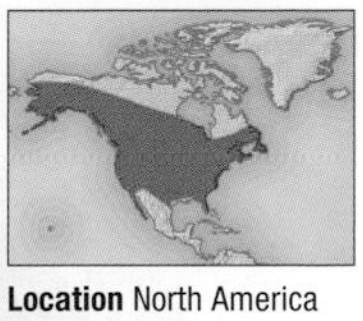

Location North America

Length 28–38 in (71–96 cm)
Weight 6½–14 lb (3–6.5 kg)
Plumage Sexes similar
Migration Partial migrant
Status Least concern

The majestic bald eagle is a large, powerfully built bird, with a wingspan of up to 8¼ ft (2.5 m). Although often found well away from water when wintering, bald eagles are commonly seen close to lakes, rivers, and coastal areas, where they have ample access to fish. Bald eagles pair for life, and while sedentary birds stay together throughout the year, those that migrate separate then come together again at breeding sites. They reinforce their pair bond through often spectacular flight displays involving undulating flight, swooping at one another, and cart-wheeling through the air with clasped feet. Together, they build a very large nest (in a tree or sometimes on the ground), reaching up to 13 ft (4 m) tall. Bald eagles start to breed at about 5 years of age, and usually 2 or 3 eggs are laid. Although the young are cared for by both parents for many weeks, a high percentage do not survive their first year.

DISTINCTIVE FEATURES
The bald eagle is easily spotted by its pure white head and tail, its broad, brown-black wings, and its large, yellow bill. Juveniles take up to 5 years to gain full adult plumage.

FEEDING HABITS

The bald eagle takes food from a range of sources, live and dead, including small birds, carrion (especially during the winter months), and fish. When hunting for fish, it does not usually enter the water like the osprey (see p.296), but instead searches for dead or dying fish or those that live near the surface.

FISHING TECHNIQUE
The bald eagle uses its strong feet equipped with sharp claws to snatch fish from the water's surface.

FOOD FIGHT
As well as piratically stealing food from other predatory birds such as ospreys, bald eagles often fight between themselves over prey items.

WINTER DIET
In winter, bald eagles sometimes collect in large groups close to where salmon come to spawn.

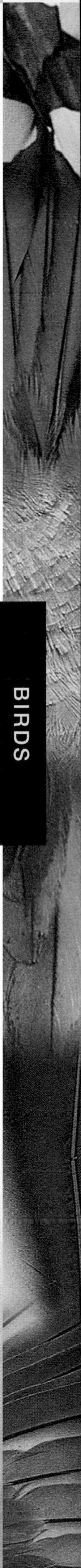

BIRDS

Circaetus pectoralis

Black-breasted snake eagle

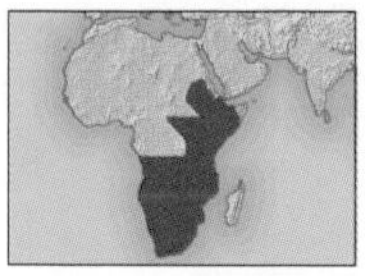

Location E. to southern Africa

Length 26 in (65 cm)
Weight 2¼–5½ lb (1–2.5 kg)
Plumage Sexes alike
Migration Nonmigrant
Status Least concern

A specialized hunter of reptiles, the black-breasted snake eagle has tightly meshed scales on its legs and toes that protect it from biting snakes. It has a large head with large, yellow eyes. The bird has strong legs and small feet, and is bare of feathers. It is often found soaring over open hill slopes as it searches for prey, and will occasionally hover. It feeds primarily on snakes—usually avoiding poisonous ones—but will also take lizards, birds, bats, and even fish. The female lays one egg and incubates it for 48 days. Both parents care for the young; once fledged, the young is dependent for up to 6 months.

Terathopius ecaudatus

Bateleur

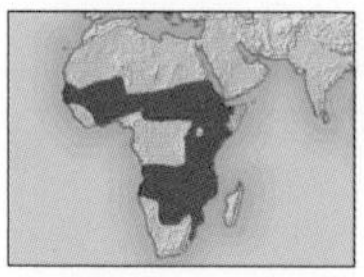

Location Africa (south of Sahara)

Length 23½ in (60 cm)
Weight 4½–6½ lb (2–3 kg)
Plumage Sexes differ
Migration Nonmigrant
Status Near threatened

The bateleur is a colorful eagle with a chestnut-red mantle on its back, a black body and head, and a bare, red or orange face. Females have gray shoulders and black-edged, white secondary flight feathers; males have black secondaries. The long, pointed, almost falconlike wings and the short tail make this bird unmistakable in flight—it tilts from side to side like a tightrope walker trying to balance. The bateleur is mainly a scavenger, but will also take small mammals, birds, reptiles, fish, eggs, and insects. Groups of these eagles may gather at termite mounds when the insects fly from their nests to breed. During courtship, the bird displays a rocking and rolling flight, almost stopping in midair with the wings held open. It builds its large stick nest in an open-branched tree.

Circus cyaneus

Northern harrier

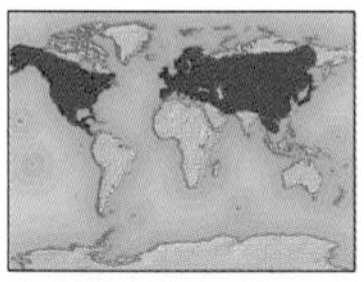

Location North America to N. Central America, Europe, Asia

Length 17–20½ in (43–52 cm)
Weight 13–19 oz (350–525 g)
Plumage Sexes differ
Migration Migrant
Status Least concern

Instead of soaring, the northern harrier flies close to the ground with its wings held in a shallow "V." It has keen eyesight and good hearing, which it uses to locate small animals hidden in vegetation. The female (shown above) is larger than the male, and is brown, with a white rump, whereas the male is gray. Unusually for a bird of prey, the northern harrier nests on the ground, making a nest of sticks and grass.

Accipiter gentilis

Northern goshawk

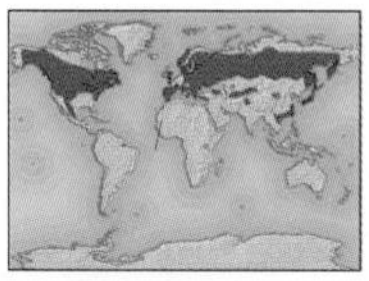

Location Canada to Mexico, Europe, Asia

Length 19–28 in (48–70 cm)
Weight 2¼–3¼ lb (1–1.5 kg)
Plumage Sexes alike
Migration Partial migrant
Status Least concern

A high-speed hunter of woods and forests, the northern goshawk varies considerably across its very wide range. Asian birds are usually pale, while North American ones often have dark heads. Females are always larger than males—sometimes weighing up to half as much again—and young birds are brown, eventually turning gray. Northern goshawks feed on birds up to the size of crows and pigeons, and mammals up to the size of small hares. They often hunt from a perched position at the forest edge, and are shaped for pursuit at close quarters, with short, rounded wings, and a long tail that is used for steering and braking. Apart from these sudden sorties after their prey, they are secretive, and often go unseen. When courting, prospective partners call loudly as they soar into the air, and once their nest is built, they are extremely territorial, driving any intruding birds away. In cold winters, when food is scarce, they are sometimes found well to the south of their normal range.

Buteo buteo

Eurasian buzzard

Location Europe, Asia, N. and E. to southern Africa

Length 20–22½ in (50–57 cm)
Weight 19–36 oz (525–1,000 g)
Plumage Sexes alike
Migration Partial migrant
Status Least concern

This medium-sized raptor, also known as the common buzzard, has large, broad wings and a shortish tail, and is built for soaring on thermals. It is very variable in color. Although large, the Eurasian buzzard catches relatively small prey, such as voles, mice, and insects. It is often seen on the ground, especially in winter, feeding on insects and earthworms. During courtship, the buzzard displays a high, soaring flight with spectacular climbs and stoops, and the male passes nesting material to the female in midair.

Buteo galapagoensis

Galapagos hawk

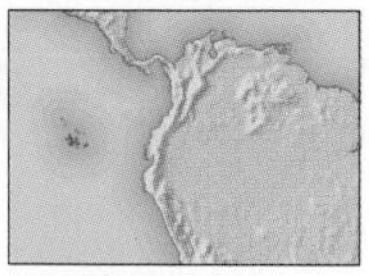

Location Galapagos Islands

Length 22 in (55 cm)
Weight 23–30 oz (650–850 g)
Plumage Sexes alike
Migration Nonmigrant
Status Vulnerable

This is the only diurnal bird of prey on the Galapagos Islands. It is also one of the few bird species that have cooperative breeding habits—females that have extra males helping them are more successful in producing young. The Galapagos hawk is sooty brown all over, with a gray-barred tail. The female is considerably larger than the male. This hawk hunts mainly from the air, gliding in to take prey, but can also hover. It feeds on small mammals, birds, reptiles, and insects. Both parents care for the young—usually one—and will allow it to remain close for up to 4 months before driving it away.

Aquila chrysaetos

Golden eagle

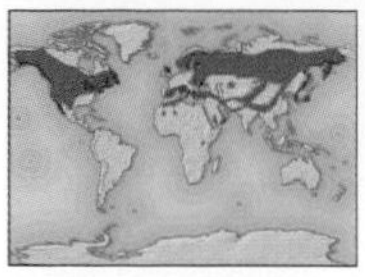

Length 30–35 in (75–90 cm)
Weight 6½–14 lb (3–6.5 kg)
Plumage Sexes alike
Migration Partial migrant
Status Least concern

Location North America, Europe, Asia, N. Africa

With a wingspan of up to 7½ ft (2.3 m), this striking bird is one of the largest land eagles in the Northern Hemisphere. Its plumage is generally dark brown, but it gets its name from the tawny or gold feathers on its nape and crown. The golden eagle is skilled at soaring, but it generally catches its prey by cruising low down across suitable terrain. It breeds on cliff ledges and in tall trees, making platform nests up to 6½ ft (2 m) across. Widely persecuted in the past—through the mistaken belief that it attacks livestock—the golden eagle is now protected in many countries.

FEEDING THE YOUNG

This eagle has a wide-ranging diet, including carrion and living prey, such as rabbits, squirrels, and grouse, and even tortoises in the southern parts of its range. Parents feed their chicks, by tearing food up into strips, for several months after they have fledged. Young birds take 4–5 years to mature, but mortality is high, especially in the first 12 months.

BROAD WINGS
This eagle has broad wings for soaring, and "slotted" flight feathers, spread like the fingers of a hand, for increased lift. It soars at a considerable height, flapping its wings as little as possible, surveying the land below for food.

Pithecophaga jefferyi

Philippine eagle

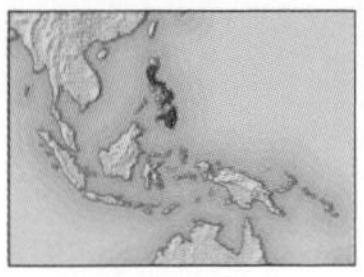

Length 34–39 in (86–100 cm)
Weight 10–18 lb (4.5–8 kg)
Plumage Sexes alike
Migration Nonmigrant
Status Critically endangered

Location Philippines

Threatened by forest clearance and by hunting, this gigantic eagle is one of the world's rarest birds of prey. It feeds on a wide variety of animals, including monkeys, snatching them in a low-level attack. Attempts are being made to breed the eagle in captivity, but with a rapidly shrinking habitat, its future looks uncertain.

Aquila audax

Wedge-tailed eagle

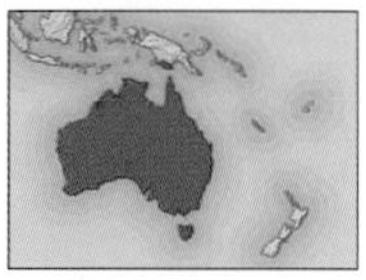

Length 32–39 in (81–100 cm)
Weight 4½–12 lb (2–5.5 kg)
Plumage Sexes alike
Migration Nonmigrant
Status Least concern

Location S. New Guinea, Australia (including Tasmania)

This eagle is Australia's largest bird of prey, with dark brown plumage, and a long, graduated tail that gives it a distinctive silhouette in flight. It feeds on a wide range of animals, including other birds and rabbits, and it also eats carrion, taking the place of vultures in a continent that has no vultures of its own. Wedge-tailed eagles usually nest in trees, lining their nests with leaves.

Harpia harpyja

Harpy eagle

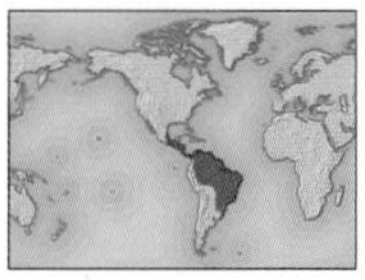

Length 35–39 in (89–100 cm)
Weight 8¾–20 lb (4–9 kg)
Plumage Sexes alike
Migration Nonmigrant
Status Near threatened

Location S. Mexico to C. South America

In habitat and lifestyle, this immense bird closely matches the Philippine eagle (see above), although it lives in a quite different part of the world. Despite its great size, it is unusually agile, which enables it to steer through the treetops to catch its prey. Sloths make up about a third of its food by weight, but it also eats a wide range of other animals, including snakes, lizards, and macaws. Harpy eagles make stick nests high up in old trees, and they often perch on exposed branches, watching the forest canopy below for signs of prey.

POWERFUL PREDATOR
The harpy eagle has gray, black, and white plumage and a long, barred tail. It has a double crest, a strong bill, and very thick, powerful legs and feet.

Polemaetus bellicosus

Martial eagle

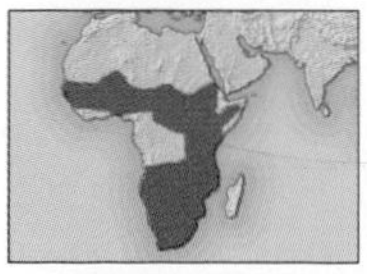

Length 31–34 in (78–86 cm)
Weight 6½–13 lb (3–6 kg)
Plumage Sexes alike
Migration Nonmigrant
Status Near threatened

Location Africa (south of Sahara)

This is the largest of the African eagles, and one of the biggest eagles in the world. It can be found in a wide variety of open habitats, from semidesert and steppes to foothills, and also in moderately forested areas. Its prominent brow gives it a menacing expression. The abdomen is white with gray to black speckling. The martial eagle soars for extended periods in good weather, searching for prey such as gamebirds, hares, hyraxes, small antelopes, monitor lizards, and other medium-sized vertebrates.

CONSERVATION

The population of the harpy eagle is declining as a result of habitat destruction. It needs a very large area to hunt, which makes it vulnerable when a continuous stretch of forest is divided into isolated tracts. Young birds are now being radio-tracked via satellite, to determine their range and the space they need to survive.

BIRDS

Sagittarius serpentarius

Secretary bird

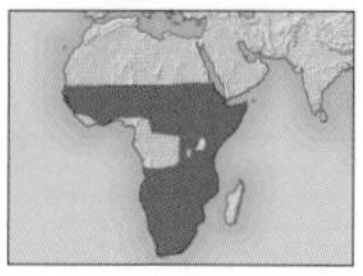

Length 4¼–5 ft (1.3–1.5 m)
Weight 5½–10 lb (2.5–4.5 kg)
Plumage Sexes alike
Migration Nonmigrant
Status Least concern

Location Africa (south of Sahara)

With its storklike legs, long, wedge-shaped tail, and crest of black feathers on the back of its head, the secretary bird is unlike any other raptor. The crest resembles a number of quill pens, as used in the past by secretaries. The flight feathers are black, and the rest of the body plumage is gray. The powerful legs are used for striking prey and running after faster quarry. The secretary bird walks up to 15 miles (24 km) a day through grassland searching for prey. Food includes grasshoppers and other large insects, small mammals, frogs, snakes, lizards, and tortoises. The bird flushes out the prey by stamping on tufts of grass. It then runs after and catches moving items, dealing them repeated blows with its strong feet. The wings act as shields when the bird attacks snakes.

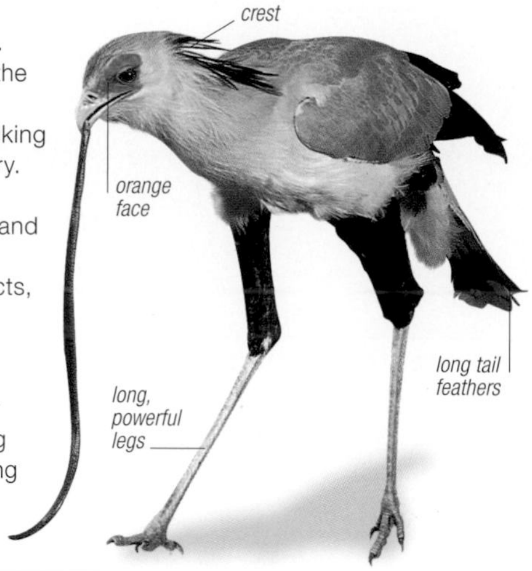

Falco punctatus

Mauritius kestrel

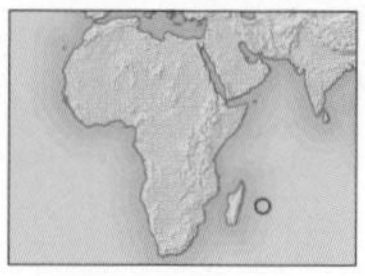

Length 8–10 in (20–26 cm)
Weight 6–8 oz (175–225 g)
Plumage Sexes alike
Migration Nonmigrant
Status Vulnerable

Location Mauritius

This chestnut-brown, cream-fronted bird nearly became extinct because of destruction of its natural habitat of evergreen primary forest, but is now being raised in zoos and encouraged to inhabit secondary forest and scrub. In the 1970s, there were thought to be only 4–8 birds left, but, due to a successful breeding program, there are nearly 1,000.

Caracara plancus

Southern caracara

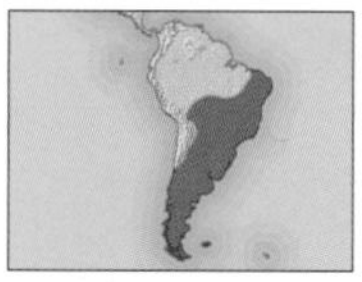

Length 19½–23 in (49–59 cm)
Weight 1¾–3¼ lb (0.85–1.5 kg)
Plumage Sexes alike
Migration Nonmigrant
Status Least concern

Location South America

One of 2 similar American crested caracaras, this striking bird of prey is dark brown with a cream head, breast, and upper back, all finely barred, and a bare, orange face. It spends much of its time walking on the ground. The southern caracara is an opportunistic scavenger, and will often dig for food or chase other birds, including raptors and vultures, to steal it. It eats almost anything, from carrion, eggs, chicks, and frogs to road kill, rotting vegetables, dead and dying fishes, worms, and insects.

Micrastur ruficollis

Barred forest falcon

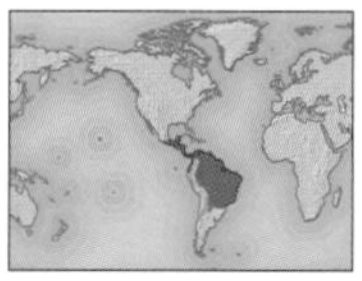

Length 13–15 in (33–38 cm)
Weight 5–8 oz (150–225 g)
Plumage Sexes alike
Migration Nonmigrant
Status Least concern

Location S. Mexico to N. South America

Compared to other falcons, forest falcons have relatively short, rounded wings—a shape that helps them maneuver as they hunt among trees.

This species has long and slender legs, with small feet and sharp talons. It hunts from trees, with short dashes out to seize lizards, and also catches army ants on the ground.

Falco eleonorae

Eleonora's falcon

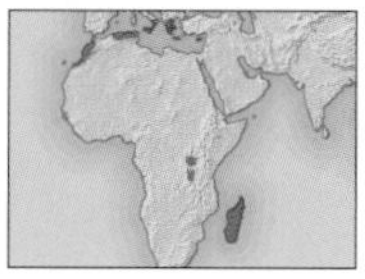

Length 14–16½ in (36–42 cm)
Weight 13–14 oz (350–400 g)
Plumage Sexes alike
Migration Migrant
Status Least concern

Location S. Europe, N. and E. Africa, Madagascar

This fairly small falcon has the typical streamlined falcon shape, but its wings are longer and more swiftlike. It has 2 color forms—normal and melanistic. The normal plumage is dark above and cream with flecks below, while the melanistic form is dark all over. This bird breeds in autumn, usually on cliffs of remote islands, feeding mainly on small birds.

Falco tinnunculus

Common kestrel

Length 12½–15½ in (32–39 cm)
Weight 4–12 oz (125–325 g)
Plumage Sexes differ
Migration Partial migrant
Status Least concern

Location Europe, Asia, Africa

The common kestrel is a small, chestnut-brown falcon with a black-tipped tail as well as black bars and spots. It is one of the few medium-sized birds that can hover for extended periods. This ability allows it to live and hunt in a variety of habitats, especially areas not often frequented by other species, such as the edges of major roads. Its diet consists mainly of small mammals (including voles and mice), insects, and amphibians.

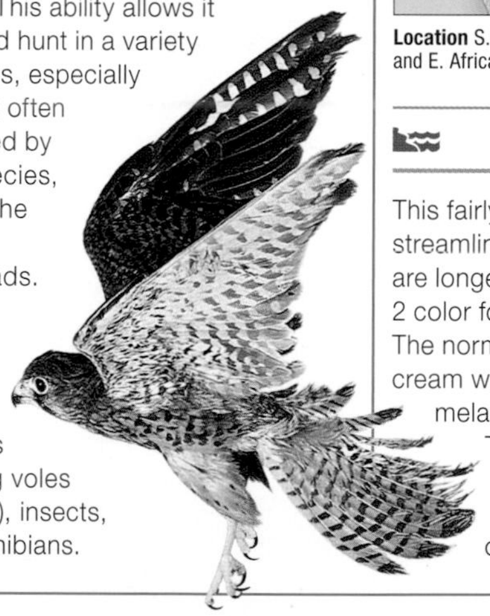

Falco peregrinus

Peregrine falcon

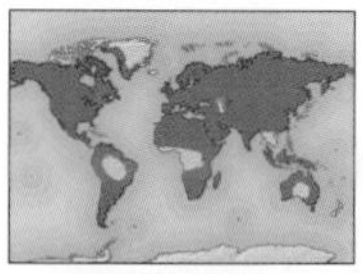

Length 13½–20 in (34–50 cm)
Weight 1¼–3¼ lb (0.55–1.5 kg)
Plumage Sexes alike
Migration Partial migrant
Status Least concern

Location Worldwide (except Antarctica)

The peregrine falcon is one of the world's fastest birds. It also has the widest distribution of any day-flying land bird, being found on every continent, except Antarctica, and on many oceanic islands. Females can be up to 30 percent larger than males. The bird's wings have sharply pointed tips, making them both fast and highly maneuverable; it flies largely by flapping rather than soaring. This falcon usually attacks its prey in a steep, powerful dive, or stoop, during which it may reach a speed of 145 mph (230 kph). Courtship involves aerial displays accompanied by noisy calling. The peregrine falcon has been used in falconry for centuries. In the 1950s and 60s, it was badly affected by DDT pollution, particularly in Europe and the USA, but it is now staging a gradual recovery.

AERIAL PURSUIT

The peregrine falcon may chase its prey—especially doves and pigeons—in order to exhaust them quickly. It usually strikes the prey with its talons, and then follows it to the ground.

PLUMAGE
This falcon has white to cream to rufous underparts, and black, gray, or blue upperparts. The young are brown with buff edging on contour feathers, and vertical stripes on the breast.

Cranes and relatives

PHYLUM	Chordata
CLASS	Aves
ORDER	Gruiformes
FAMILIES	11
SPECIES	228

Long-legged cranes and their relatives are an assortment of birds that look outwardly different from one another but are united by aspects of their internal anatomy—such as the lack of a crop in their digestive system. In addition to cranes themselves, this group includes rails, bustards, trumpeters, and sun-bitterns, among others. By far the largest family is formed by the rails (which include the familiar coots and moorhens). Rails are found worldwide but many of the other families are less widely distributed.

Migration

Many cranes are migratory, traveling vast distances between their breeding and wintering areas. Along the way, large flocks stop at long-established staging posts, where they may stay for a few days before moving on. Cranes breed in wetlands, typically seeking remote places such as Arctic tundra, steppes, highland plateaus, and forested swamps. Some crane species and most other birds in the group do not migrate at all.

NESTING CORNCRAKE
The corncrake is one of the few long-distant migrants among the rails, breeding in Europe in summer and wintering in sub-Saharan Africa.

MOVING NORTH
Some sandhill cranes make long-distance annual migrations between their winter feeding grounds in Texas and their summer breeding areas in Alaska, which are almost 4,000 miles (6,500 km) apart.

Anatomy

Most members of this group have long legs, a slender bill, rounded wings, and modest, cryptic plumage. However, there is otherwise great variation in appearance between species, depending on their habitat and lifestyle. Those that wade in wetlands (cranes), or walk on floating vegetation (the limpkin), have long, slender toes to distribute their weight. The strictly ground-dwelling bustards have shorter toes and strong legs for running over arid land. Aquatic members of the group (coots and finfoots) have lobed feet for swimming. Cranes fly great distances on long, broad wings, while others (mesites, buttonquails, and some rails) have small, rounded wings that are poor or useless for flying.

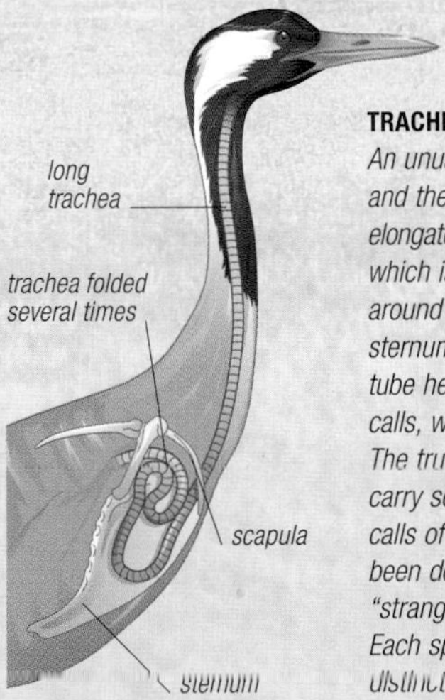

TRACHEA
An unusual feature of cranes and the limpkin is their greatly elongated windpipe, or trachea, which is coiled like a hosepipe around the region of the sternum in the chest. This long tube helps amplify the bird's calls, which are very loud. The trumpeting of cranes can carry several miles, and the calls of the limpkin have been described variously as "strangled" or "shrieking." Each species makes its own distinct, recognizable sounds.

Courtship

Several members of this group have elaborate courtship rituals. Some of the most elegant displays are performed by cranes, which mate for life. Males and females form pairs, and later maintain the bond, with carefully orchestrated dances (see below). Birds that perform similar courtship displays include rails, some bustards (see right), and the sun-bittern, in which the birds hold out their wings, revealing large eye-spots, and fan their tail, often while jumping and running in a circle. In addition to their courtship displays, rails also attract mates by making loud and distinctive calls.

BREEDING PLUMAGE
Breeding male bustards inflate their throat pouch, puff out some of their feathers to make themselves look larger, and call loudly. Many species, such as the great bustard (above), have a ruff around the neck.

CONSERVATION

Cranes and their relatives have a higher proportion of endangered species among their number than any other group of birds, and several have recently become extinct. Threats include habitat destruction, introduced ground predators, and hunting. Among the true cranes, 11 of the 15 species are vulnerable or endangered, including the Siberian crane and whooping crane, with more likely to join them. Protection of cranes is made particularly difficult by their long migrations, requiring the cooperation of many countries.

COURTSHIP DANCE

1 **JUMPING** *Two grey crowned cranes raise their wings and jump into the air.*

2 **RITUAL FLIGHT** *One of the pair flaps its wings and leaps toward the other.*

3 **WING-FLAPPING** *As the leaping bird approaches, the other raises its wings.*

4 **STEPPING BACK** *The leaping bird retreats slightly and begins to bow.*

5 **BOWING** *The 2 birds now lower their wings and bow to one another.*

Mesitornis variegata

White-breasted mesite

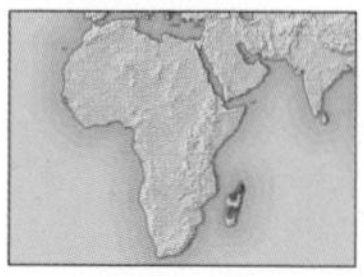

Location Madagascar

Length 12 in (31 cm)
Weight 3⅝ oz (100 g)
Plumage Sexes alike
Migration Nonmigrant
Status Vulnerable

This small forest bird is one of 3 species in the mesite family, all unique to Madagascar, and threatened by the destruction of their natural habitat. Its short, rounded wings and stout legs are typical of a ground-living bird. It feeds mainly on insects and spiders, flicking through fallen leaves or lifting up larger ones in order to find prey. A tangled mass of vegetation piled into a tree or bush, 3¼–9¾ ft (1–3 m) above the ground, serves as a nest.

Turnix varius

Painted buttonquail

Location S.W. and E. Australia, New Caledonia

Length 6½–9 in (17–23 cm)
Weight 2–3⅜ oz (55–95 g)
Plumage Sexes differ
Migration Nonmigrant
Status Least concern

With their compact bodies and camouflage plumage, buttonquails look very much like true quails but they are not closely related. This species is relatively large, reddish in color, and heavily spotted. When foraging for seeds and insects, it scratches with one foot while pivoting on the other, leaving circular marks on the ground.

Anthropoides paradiseus

Blue crane

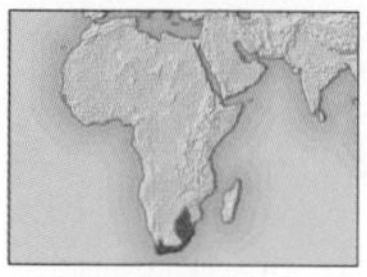

Location S. Africa

Length 39–43 in (100–110 cm)
Weight 9–13 lb (4–6.2 kg)
Plumage Sexes alike
Migration Nonmigrant
Status Vulnerable

This arid country crane has long wing plumes that appear like a tail from a distance. It is restricted to South Africa (where it is the national bird) and Namibia. Although it may gather into large flocks outside the breeding season, overall numbers of this species have sharply declined since the 1970s. This has been caused by factors such as insecticide poisoning on agricultural land and conversion of its grassland habitat into forest; birds are also killed by power-line collisions. Conservation programs are now assisting this species.

Grus japonensis

Red-crowned crane

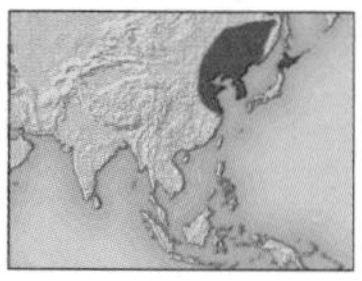

Location E. Asia

Length 5 ft (1.5 m)
Weight 15–26 lb (7–12 kg)
Plumage Sexes alike
Migration Migrant
Status Endangered

With its intricate courtship dances and lifelong partnerships, this elegant bird has long been a symbol of happiness and good luck. The heaviest member of the crane family, it is mainly white, with black flight feathers and a black face and neck. Large flocks congregate on feeding grounds in winter, but in the breeding season, pairs of birds establish territories, defending them vigorously against other cranes.

COURTING DANCES
During courtship, the red-crowned crane performs elaborate dances involving head bobbing, bowing, pirouettes, jumping, and tossing material in the air.

CONSERVATION

The total population of red-crowned cranes stands at about 2,750 birds. About 1,000 of these live in Japan—a spectacular increase from the 1920s, when numbers dropped to just 20. However, the species is still in danger, particularly during its long migration flights.

Balearica regulorum

Grey crowned crane

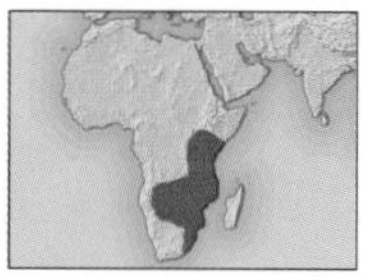

Location E. and southern Africa

Length 3¼–3½ ft (1–1.1 m)
Weight 6½–8¾ lb (3–4 kg)
Plumage Sexes alike
Migration Nonmigrant
Status Vulnerable

African crowned cranes are easily identified by their golden crests, or “crowns,” and by their ability to perch in trees—something that other cranes cannot do. Found in both marshland and cultivated fields, this bird has a relatively short, versatile bill, unlike more aquatic cranes. It stamps its feet while foraging to flush out potential prey, and travels with large mammals to feed on the insects frightened by their movement. Another species, the black crowned crane, occurs in tropical Africa.

Grus grus

Common crane

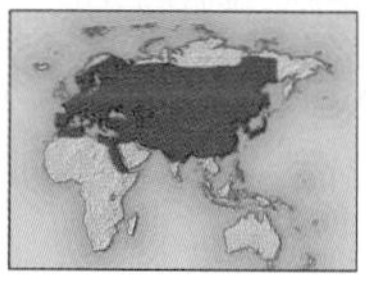

Location Europe, Asia, N. Africa

Length 4 ft (1.2 m)
Weight 11–13 lb (5–6 kg)
Plumage Sexes alike
Migration Migrant
Status Least concern

Like other members of its family, the common crane is a large, elegant bird, with a long bill and long legs. Gray, with a black head and neck, it has a white stripe down the nape, and a red spot on the crown. It gives a loud, trumpeting call, the sound being amplified by an enlarged windpipe that is fused with the breastbone.

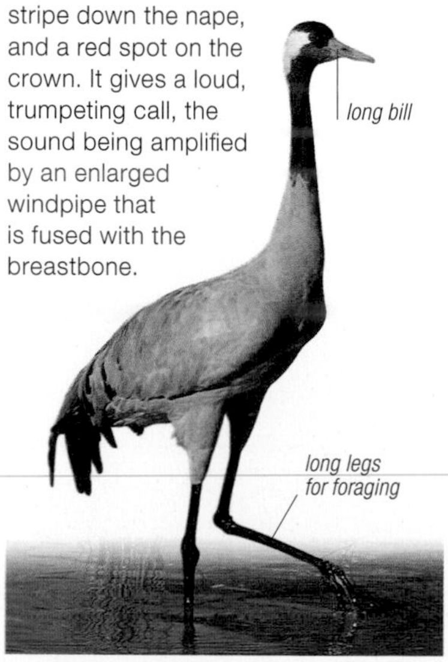

Grus canadensis

Sandhill crane

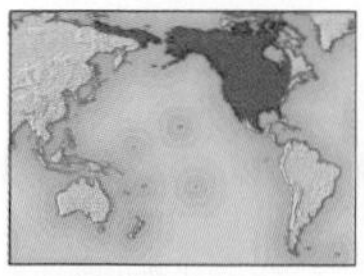

Location North America, N.E. Asia

Length 4 ft (1.2 m)
Weight 8¾–12 lb (4–5.5 kg)
Plumage Sexes alike
Migration Migrant
Status Least concern

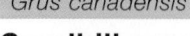

The smaller of the 2 species of cranes native to North America, this is also by far the most common. A gray bird with a red forehead, its relatively short bill enables it to deal with a wide range of foods—plants, insects, and small animals such as mice. Northern populations move south in winter and as many as 40,000 cranes can gather at a single stop-over site, their loud calls carrying across great distances.

Aramus guarauna

Limpkin

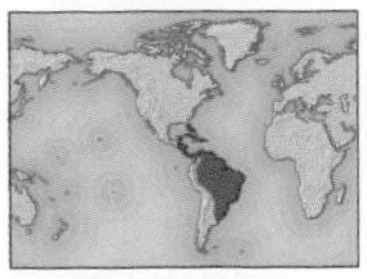

Length 22–28 in (56–71 cm)
Weight 2¼–3¼ lb (1–1.5 kg)
Plumage Sexes alike
Migration Nonmigrant
Status Least concern
Location S.E. USA (Florida), Central and South America

An ibislike bird with a striking brown and white spangled neck, the limpkin has a long, slender bill that is sharpened and slightly twisted at its tip—features that enable it to extract snails from their shells with ease. Breeding pairs often stay together year after year, and the male defends the nest by charging intruders, fighting with its feet, and uttering an unearthly, mournful call.

Crex crex

Corncrake

Length 10½–12 in (27–30 cm)
Weight 4–7 oz (125–200 g)
Plumage Sexes alike
Migration Migrant
Status Least concern
Location Europe, W. to C. Asia, S.E. Africa

Heard far more easily than seen, this cryptically colored bird has a loud and rasping, 2-syllable call, which the male makes both to advertise its territory and to attract females. The call becomes a growling, piglike squeal before mating, or during encounters with other males. The corncrake is declining in many parts of its range as a direct result of changes in farming practices and because the damp pastures it inhabits are increasingly drained and plowed.

Gallinula chloropus

Common moorhen

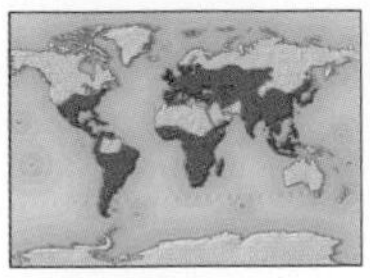

Length 12–15 in (30–38 cm)
Weight 6–18 oz (175–500 g)
Plumage Sexes alike
Migration Partial migrant
Status Least concern
Location Worldwide except Australia

The moorhen is one of the world's most widespread freshwater birds. A member of the rail family, it has distinctive white markings along the flanks and a conspicuous red "shield" and yellow-tipped red bill. Although not as shy as other rails, and often seen moving in the open, if threatened, the common moorhen will flee into dense cover.

Psophia crepitans

Gray-winged trumpeter

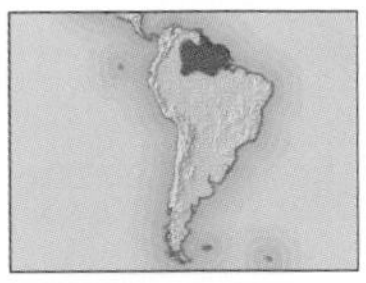

Length 20 in (50 cm)
Weight 2¼–3¼ lb (1–1.5 kg)
Plumage Sexes alike
Migration Nonmigrant
Status Least concern
Location N. South America

This black, humpbacked bird is one of 3 species in the trumpeter family, all found in South America. Small groups of 6–8 live on the forest floor, bathing and roosting together, and foraging for the fallen fruit that are their main food. They use a variety of calls, some loud and trumpeting, to warn of danger, beg for food, mark territory, and threaten intruders. Unusually for birds, several males mate with one female.

Porzana porzana

Spotted crake

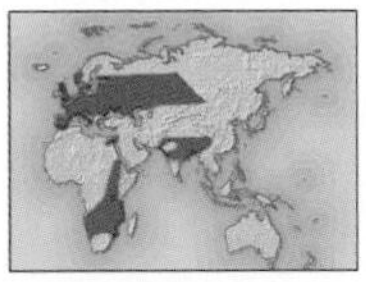

Length 9–9½ in (22–24 cm)
Weight 2–5 oz (55–150 g)
Plumage Sexes alike
Migration Migrant
Status Least concern
Location Europe, W. to C., S., and S.E. Asia, Africa

The spotted crake is a small and secretive wetland bird, with a narrow body that helps it move about among dense reeds and other waterside plants. It walks with a distinctive jerking gait, often bobbing its head and flicking its tail. Most active at dawn and dusk, it feeds mainly on insects, foraging both on land and in water. This extremely wary bird will run for cover or fly off if disturbed. It is generally quiet, but the male calls to attract a female.

Fulica atra

Common coot

Length 14–15½ in (36–39 cm)
Weight 11–43 oz (300–1,200 g)
Plumage Sexes alike
Migration Partial migrant
Status Least concern
Location Europe, Asia, Australia, New Zealand, N. and W. Africa

Compared to some of its relatives, the common coot is an assertive and even aggressive bird. In conflicts with rivals, it displays its white frontal "shield," which contrasts starkly with its heavy black body. If threatened, it will either gather in tight flocks and splash water, or turn on its back and kick out with its feet. It has lobed toes that aid swimming and diving, and it presses air from its plumage before diving to reduce buoyancy.

BIRDS

Porphyrio porphyrio

Purple swamphen

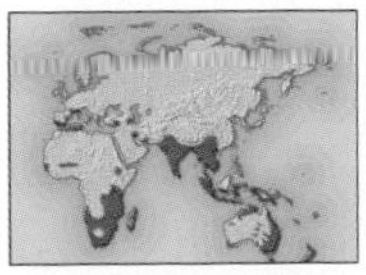

Length 15–20 in (38–50 cm)
Weight 18–46 oz (500–1,300 g)
Plumage Sexes alike
Migration Nonmigrant
Status Least concern
Location S. Europe, W., S., and S.E. Asia, Australia, Africa

One of the biggest members of the rail family—almost as big as a chicken—the purple swamphen is a powerful, robustly built bird with purple and black plumage, a reddish orange bill, and long legs and toes. It feeds on all manner of vegetation, pulling it up with its bill, as well as on aquatic and terrestrial invertebrates. It will also eat the eggs and young of other waterbirds, climbing if necessary to reach the nest. The purple swamphen has a very large range, with several local races differing in plumage and in behavior: some stay in one place, while others move long distances in order to find food. In many areas, males and females pair off for breeding, but in New Zealand, complex breeding groups form, consisting of one or 2 females and 2–7 males. The eggs are laid in one nest and all members of the group help incubate them and rear the young.

Rhynochetos jubatus

Kagu

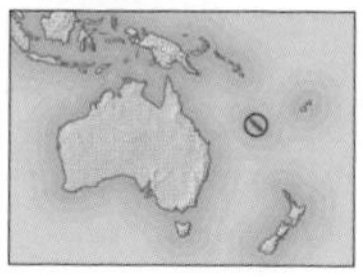

Length 21½ in (55 cm)
Weight 32 oz (900 g)
Plumage Sexes alike
Migration Nonmigrant
Status Endangered

Location New Caledonia

With its stocky body, weak wings, and large, erectile crest, the kagu is a typical island species that has evolved a distinctive appearance and way of life. It lives on the ground, and often stands motionless on one leg, watching and listening for prey. If it strikes and misses, it will use its bill to dig out the prey; unique flaps of skin over the nostrils keep out debris while the kagu forages in soil. When threatened, it raises its shaggy crest and spreads its wings to expose the shieldlike pattern on its flight feathers.

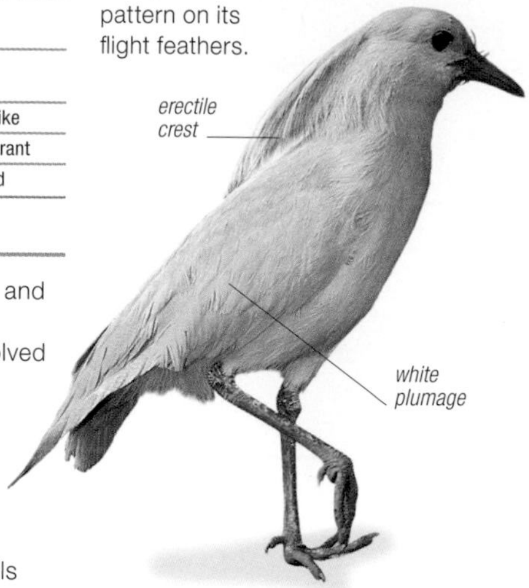

Heliornis fulica

Sungrebe

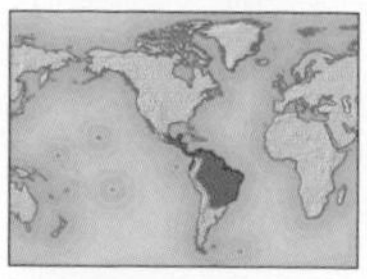

Length 10–13 in (26–33 cm)
Weight 4–5 oz (125–150 g)
Plumage Sexes differ
Migration Nonmigrant
Status Least concern

Location Central and South America

This grebelike bird belongs to the finfoot family, 3 species named for the lobes on their feet that enable them to swim well and move easily on land. The sungrebe is unique in that the male has skin pouches in the "armpit" of each wing in which he carries the young, even in flight. This bird has a slender body and a narrow tail.

Gallirallus calayanensis

Calayan rail

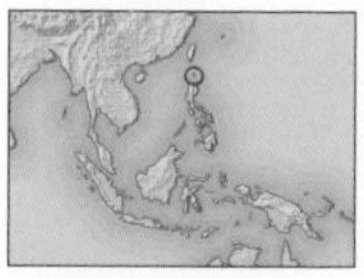

Length c.12 in (30 cm)
Weight 8½ oz (245 g)
Plumage Sexes alike
Migration Nonmigrant
Status Vulnerable

Location Calayan Island, N. Philippines (Calayan Island)

Discovered in 2004—and restricted to Calayan in the far northern Philippines—this is one of several species of rails that has lost the power of flight after evolving on an island. It has uniformly blackish brown plumage, red legs, and a red bill, and has been seen singly or in small groups—roaming on tropical forest floor underlain by coralline limestone.

Eurypyga helias

Sunbittern

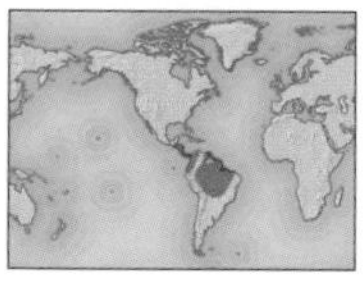

Length 17–19 in (43–48 cm)
Weight 7 oz (200 g)
Plumage Sexes alike
Migration Nonmigrant
Status Least concern

Location Central America, N. South America

An inhabitant of shady rainforest streamsides, the sunbittern is a wary bird that stalks its prey slowly and deliberately, then quickly stabs it with its slender bill. If threatened when in the nest, it moves its neck backward and forward, and hisses like a snake. At other times, it will turn to face its predator, and fan its tail and spread its wings (like the kagu, above) to expose large, eyelike patches. The large "eyes" and the spread of the wings and tail make the sunbittern look very big and imposing. Its mottled, cryptic plumage is soft, enabling it to fly silently. A solitary bird, even paired adults are seldom seen together.

Chlamydotis undulata

Houbara bustard

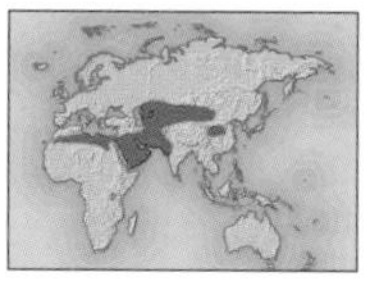

Length 26–30 in (65–75 cm)
Weight 3¼–6½ lb (1.5–3 kg)
Plumage Sexes differ
Migration Nonmigrant
Status Vulnerable

Location Canary Islands, N. Africa, W., C., and E. Asia

This stout, long-legged bird is a typical member of the bustard family—a group of about 22 species that are found only in the Eastern Hemisphere. Like other bustards, it has cryptically colored plumage, and although it can fly well, it spends most of its time on the ground. The houbara bustard searches for food as it walks through its desert homeland, eating seeds and shoots, insects, and small reptiles such as lizards. Of all bustards, it is the one best adapted to the desert, and seldom drinks, obtaining most of the water it needs from food. The male uses a traditional courtship arena for mating, adopting a curious method of attracting the females' attention: it trots blindly about its display ground, with its ruff and crest feathers raised over its head. Although not globally threatened, the houbara bustard is declining fast across its range, and in many parts of West Asia, it is the focus of intensive conservation measures. This species is a highly valued gamebird in the region and is hunted with falcons.

Cariama cristata

Red-legged seriema

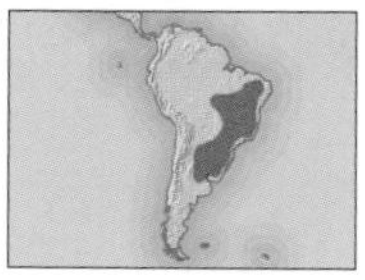

Length 30–35 in (75–90 cm)
Weight 3¼ lb (1.5 kg)
Plumage Sexes alike
Migration Nonmigrant
Status Least concern

Location E. South America

In both shape and behavior, this species resembles the secretary bird (see p.302): both have long legs and crests, and both hunt by striding across the ground. The red-legged seriema often captures large prey in its bill, beating it on the ground to break it into pieces. It usually runs from predators, but may lie down and rely on its cryptic coloring to escape detection. Out of the breeding season it is highly vocal, and its call is one of the most characteristic sounds of South American grasslands. Although generally solitary, this species may also be found in groups at times.

Ardeotis kori

Kori bustard

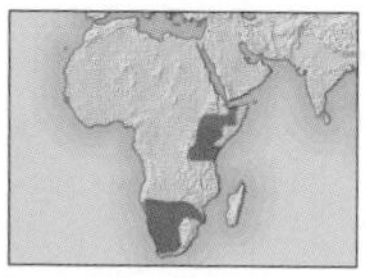

Length 4 ft (1.2 m)
Weight 24–42 lb (11–19 kg)
Plumage Sexes alike
Migration Nonmigrant
Status Least concern

Location E. and southern Africa

Weighing up to 42 lb (19 kg), the kori bustard is one of the world's heaviest flying birds. Like its relatives, however, it lives on the ground and is reluctant to fly unless in serious danger. It often associates with large, herding animals, feeding on insects frightened by the herd as it moves. It also inhabits areas that have recently been burned, eating the new shoots of grass and insects exposed by the lack of vegetation.

DANCING CRANES
Like other members of their family, red-crowned cranes (Grus japonensis) *are built for ostentatious courtship: long legs for dancing, long wing feathers for display, and an elongated coiled windpipe for amplifying their calls.*

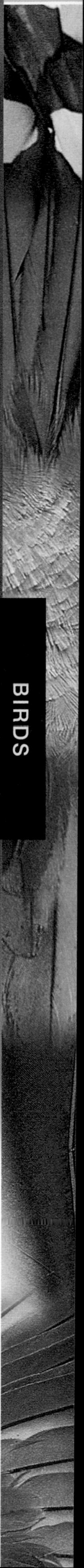

Waders, gulls, and auks

PHYLUM	Chordata
CLASS	Aves
ORDER	Charadriiformes
FAMILIES	19
SPECIES	379

In many parts of the world, these birds are a common sight at sea, along shorelines, and in wetlands. Most are strong fliers that feed on other animals in or near water. Waders or shorebirds (which include sandpipers, plovers, avocets, stilts, snipes, curlews, and jacanas) are long-legged birds that feed by the water's edge. Gulls (which embrace terns, skuas, and skimmers) use their flying skills to catch prey. Auks (including puffins and murres) dive underwater for food. Auks look strikingly similar to penguins but are able to fly and are confined to the Northern Hemisphere. Waders and gulls occur worldwide, but many species are threatened by habitat destruction, oil pollution, or hunting.

Anatomy

Most of the birds in this group have subdued black, white, brown, or gray plumage, but some have colorful bare parts, such as the bill, eyes, legs, and mouth linings. Many go through radical plumage changes, both between seasons and in maturing to adulthood. The 3 groups exhibit major differences, especially in the legs, which are long in the waders but short and web-footed in the rotund, upright auks. Most birds in the group have salt glands above the eyes that enable them to extract the fluids they require from seawater and expel the excess salts through the nostrils.

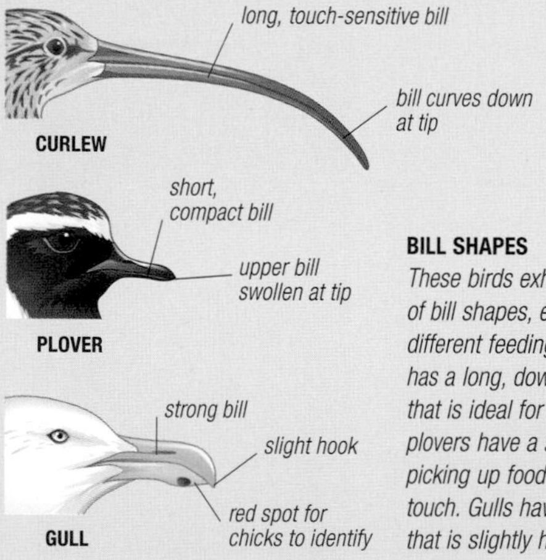

BILL SHAPES
These birds exhibit a wide variety of bill shapes, each adapted for a different feeding method—the curlew has a long, downcurved, touch-sensitive bill that is ideal for probing deep into mud, while plovers have a short, pigeonlike bill used for picking up food that is detected by sight, not touch. Gulls have a robust, multipurpose bill that is slightly hooked for tearing food.

NESTING TERNS
Most terns (including these elegant terns) breed in colonies, seeking an isolated place such as an island or reef for protection. The colonies are usually found on flat, open ground and are often densely populated.

CATCHING FISH

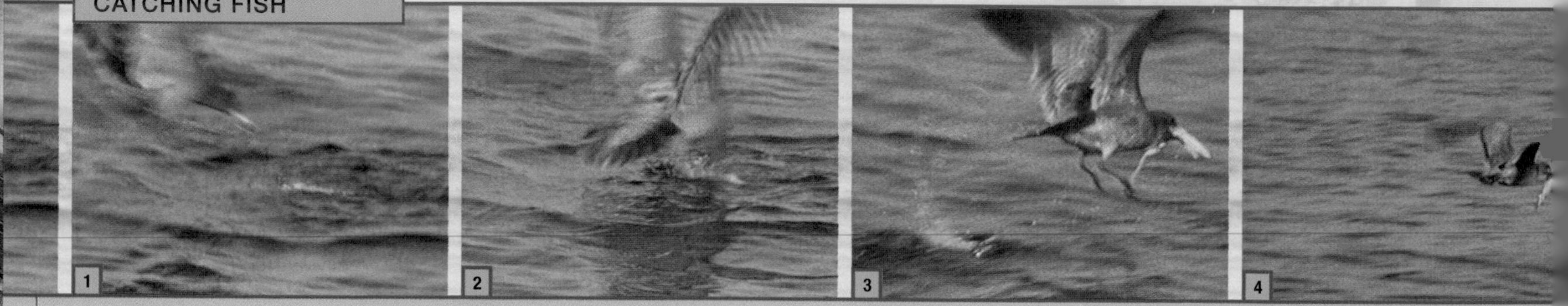

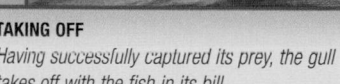

1 FINDING PREY
A western gull detects a fish beneath the surface of the water and swoops down toward it.

2 CAPTURE
With rapid wingbeats, the gull plunges its stout bill into the water and plucks out the fish.

3 TAKING OFF
Having successfully captured its prey, the gull takes off with the fish in its bill.

4 STEADY FLIGHT
The gull flies low over the water, heading toward a suitable feeding site.

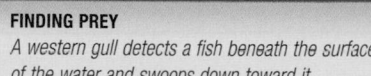

Reproduction

Most birds in this group nest on the ground, laying between one and 6 camouflaged eggs. Other species nest on cliffs, and a few, such as some terns and auks, prefer trees. Many nest colonially, in great numbers (over a million pairs in some auks), and some at extraordinary densities: incubating guillemots virtually rub shoulders with their neighbors. Numerous waders have unusual mating systems, in which either females or males have several mates. In jacanas, painted snipes, and phalaropes, the usual roles of male and female are reversed, the more colorful female initiating display, the male alone incubating the eggs and tending the young.

CLIFFSIDE COLONY
Auks often nest in extremely dense colonies. Murres (left) crowd together on cliff ledges at up to 20 pairs per square yard, their bodies virtually defining the extent of their territory. These birds lay a single egg, each with its own pattern—a way of detecting it among the many similar-looking nest sites.

Flight

The auks, with their short, pointed wings, can fly rapidly in air, but only for limited distances. They use their wings to propel themselves underwater as they swim. The waders have long, pointed wings to help them fly fast and efficiently, but although they are adept at rapid changes of pace and direction, they cannot soar. This skill can be seen in the gulls and their relatives, which spend hours on end on the wing, often traveling extremely long distances over water. Every year, the Arctic tern, which breeds north of the Arctic Circle and winters in Antarctica, travels a distance of about 10,000 miles (16,000 km) each way.

LONG-DISTANCE MIGRANTS
The Arctic tern spends much of its life in flight, feeding on the wing by catching insects in midair and flying low over water to take fishes. It breeds and migrates in large colonies. All terns perform spectacular aerial courtship displays.

CARRYING PREY
The Atlantic puffin has an extremely large, colorful bill. As with all puffins, the upper bill and tongue are ridged with spikes that enable the bird to hold a remarkable number of fishes at a time—up to 62 has been recorded.

Feeding

Apart from seedsnipes, which eat plant matter, all members of this group feed on other animals. The waders probe for invertebrates on beaches, estuarine mudflats, and leaf litter. Gulls are opportunists, picking up fish, eggs, small birds, and mammals where they can, even scavenging at garbage dumps. Gulls and skuas sometimes feed by piracy, intimidating smaller birds into giving up their food. Terns plunge-dive for fish, while skimmers fly low over the water, the lower bill just touching the surface and snapping shut when prey is touched. Auks pursue their prey underwater.

WALKING ON WATER
The jacanas (here, a comb-crested jacana) walk on floating vegetation, their long toes distributing their weight and preventing them from sinking. They eat insects and seeds that they find under the leaves.

FEEDING WADERS
These American avocets are catching their prey by swishing their bill from side to side through water or very soft mud, a technique known as scything. As the bill moves, it comes into contact with edible particles in suspension.

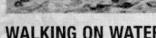

TILE RIVALS
denly, 2 more birds appear, harrassing the by squawking and flapping their wings.

6

CHANGING COURSE
Turning on its side, the gull heads off in a different direction to escape the pursuers.

7

MIDAIR SNATCH
One of the chasing birds has caught up with the gull and is pulling the fish out of its bill.

8

ESCAPE
Having successfully grabbed the fish, the "pirate" gull flees, pursued by the other 2 birds.

BIRDS

Jacana jacana

Wattled jacana

Location S. Central America, South America

Length 6½–10 in (17–25 cm)
Weight 3¼–5¼ oz (90–150 g)
Plumage Sexes alike
Migration Nonmigrant
Status Least concern

The 8 species of jacanas are notable for their long legs and very large feet. They also have a distinctive spur on the leading edge of each wing. The wattled jacana feeds mostly on insects and other aquatic invertebrates, but sometimes eats seeds from rice plants. The female, much larger than the male and weighing about 5¼ oz (150 g), mates with up to 3 males. Each male builds a nest, incubates a clutch of eggs, and rears the young. The female defends the territory from intruders.

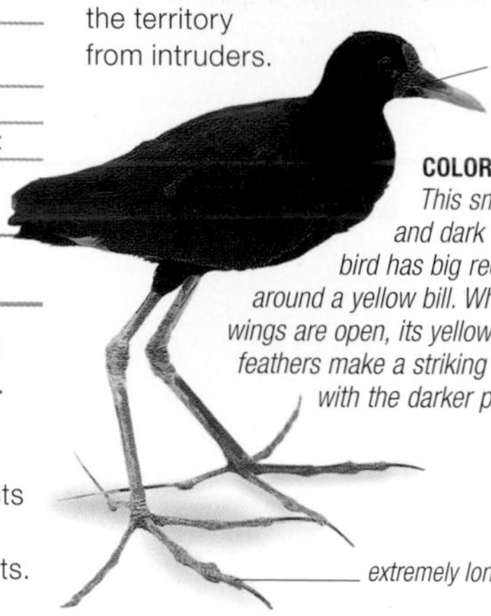

COLORS
This small, black and dark chestnut bird has big red wattles around a yellow bill. When the wings are open, its yellow flight feathers make a striking contrast with the darker plumage.

WALKING ON WATER

As with the other species of jacanas, the wattled jacana has extraordinarily long toes that spread its weight as it walks, enabling it to move easily over the leaves of floating waterplants. Because of this behavior, jacanas are also called lily-trotters.

Dromas ardeola

Crab plover

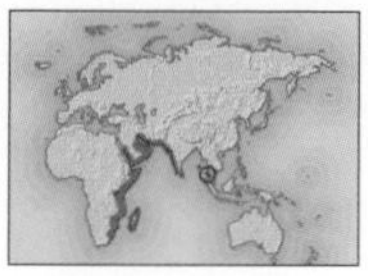

Location E. Africa, S.W., S., and S.E. Asia

Length 15–16 in (38–41 cm)
Weight 8–12 oz (225–325 g)
Plumage Sexes alike
Migration Partial migrant
Status Least concern

This thickset, black and white wader has an exceptionally large, powerful black bill, which it uses to crush the shells of crabs before swallowing them whole. The crab plover is also unusual among waders in that it digs a long burrow, in which it lays a single egg.

Rostratula benghalensis

Greater painted-snipe

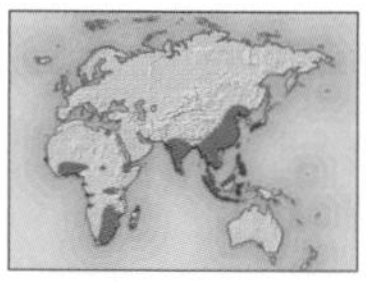

Location Africa, S., E., and S.E. Asia

Length 9–11 in (23–28 cm)
Weight 3¼–7 oz (90–200 g)
Plumage Sexes differ
Migration Nonmigrant
Status Least concern

A medium-sized wader of freshwater marshes, the greater painted-snipe has a distinctive broad, white patch around each eye. It spends much of its time in deep cover, emerging—very

cautiously and mainly at night—to probe the mud for food. The greater painted-snipe has a rather weak, fluttering flight, with legs dangling. The male usually incubates the eggs and rears the young, while the female goes off to find another mate.

Haematopus ostralegus

Eurasian oystercatcher

Location Europe, N.W., N., and E. Africa, S.W., C., E., and S. Asia

Length 16–19 in (40–48 cm)
Weight 14–29 oz (400–800 g)
Plumage Sexes alike
Migration Migrant
Status Least concern

A widespread coastal and freshwater wader, the Eurasian oystercatcher is recognizable by its noisy calls and

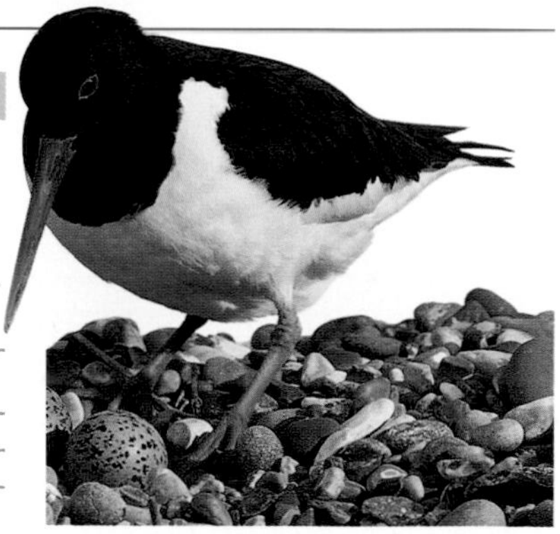

bright orange bill. It feeds mainly on mussels, limpets, and cockles—either by cutting the muscle that holds the 2 halves of the shell together and stabbing the prey inside, or by hammering the shell open on rocks or hard sand.

Ibidorhyncha struthersii

Ibisbill

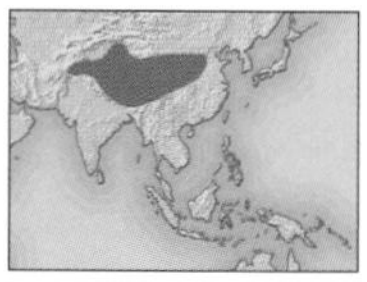

Location C. and E. Asia

Length 15–16 in (38–41 cm)
Weight 10–12 oz (275–325 g)
Plumage Sexes alike
Migration Nonmigrant
Status Least concern

A unique wader with a long, thin, downcurved red bill, the ibisbill is gray-brown above, and has a bluish gray breast, neck, and head, with a black face, and red legs. In spite of its striking coloration, the bird blends remarkably well with stony river banks. It feeds in mountain rivers, raking through the surface and probing under stones to find prey.

Himantopus himantopus

Black-necked stilt

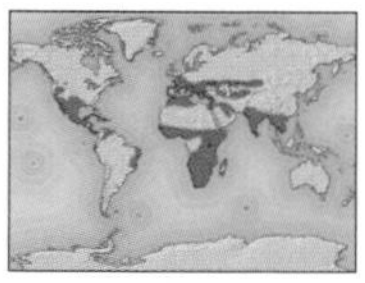

Location Europe, Asia, Africa, North, Central, and South America

Length 14–16 in (35–40 cm)
Weight 5–7 oz (150–200 g)
Plumage Sexes differ
Migration Migrant
Status Least concern

In relation to its body size, this wader has the longest legs of any bird, which enables it to feed in deeper water than most other waders. The black-necked stilt hunts by sight or touch, scything its fine, straight bill through the water to detect prey. This bird also chases insects on and above water, often twisting and leaping to catch them.

Recurvirostra avosetta

Pied avocet

Location Europe, Asia, Africa

Length 16½–18 in (42–45 cm)
Weight 8–14 oz (225–400 g)
Plumage Sexes alike
Migration Migrant
Status Least concern

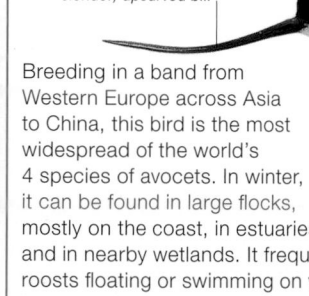

Breeding in a band from Western Europe across Asia to China, this bird is the most widespread of the world's 4 species of avocets. In winter, it can be found in large flocks, mostly on the coast, in estuaries, and in nearby wetlands. It frequently roosts floating or swimming on water, and groups of birds can form large "rafts," looking from a distance rather like gulls. During the breeding season, it tends to move inland, to slightly brackish or saline marshes. The avocet defends its territory very aggressively, calling loudly if it senses danger and chasing off intruders.

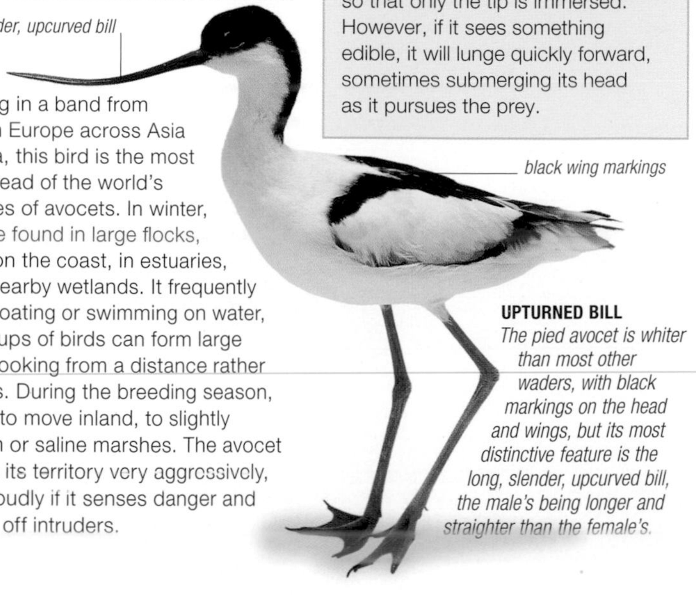

UPTURNED BILL
The pied avocet is whiter than most other waders, with black markings on the head and wings, but its most distinctive feature is the long, slender, upcurved bill, the male's being longer and straighter than the female's.

FEEDING IN WATER

When feeding, the avocet sweeps its upcurved bill from side to side so that only the tip is immersed. However, if it sees something edible, it will lunge quickly forward, sometimes submerging its head as it pursues the prey.

BIRDS

Cursorius rufus

Burchell's courser

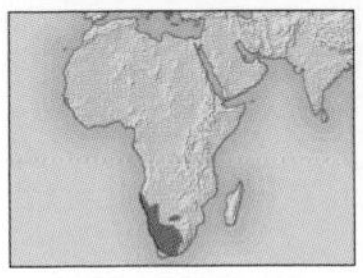

Length 8–9 in (20–23 cm)
Weight 2½–2⅞ oz (70–80 g)
Plumage Sexes alike
Migration Nonmigrant
Status Least concern

Location Southern Africa

Unlike most waders, the 8 species of coursers live in dry habitats, and get their name from their rapid run. Burchell's courser is sandy brown, with a slight rusty tone (as in the juvenile pictured), which provides excellent camouflage in the barren country in which it lives. When threatened, the bird stands still and upright to avoid detection, and then runs away very quickly. If forced to, it will fly far and fast. It feeds in a jerky, stop-start manner, taking insects from the ground.

Vanellus armatus

Blacksmith plover

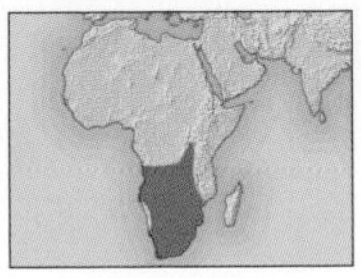

Length 11–12 in (28–31 cm)
Weight 4–8 oz (125–225 g)
Plumage Sexes alike
Migration Nonmigrant
Status Least concern

Location Southern and E. Africa

Widespread in southern and East Africa, this plover is a medium to large wader with very striking black, white, and gray plumage and long, black legs; its eyes are an intense red. Its diet consists of a wide range of invertebrates, frequently taken at night, especially when the moon is bright. Both the male and female are very protective of their young, and will chase or challenge intruders. The blacksmith plover gets its name from its sharp, metallic call.

Pluvialis dominica

American golden plover

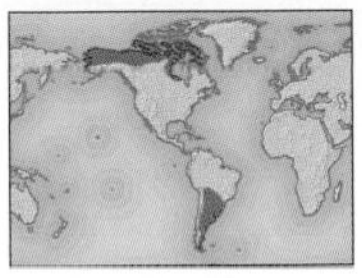

Length 9½–11 in (24–28 cm)
Weight 4–7 oz (125–200 g)
Plumage Sexes differ
Migration Migrant
Status Least concern

Location N. North America, S. South America

This medium-sized plover has a remarkable migration pattern, flying northward over land, but southward largely over the sea. It travels very long distances, wintering on coastal marshes and inland grassland in South America and breeding in drier areas of the Arctic tundra. Although mottled brown in winter, it has beautiful breeding plumage—a gold-spangled black back, and a black face bordered by white above the eyes and on the sides of the neck.

Stiltia isabella

Australian pratincole

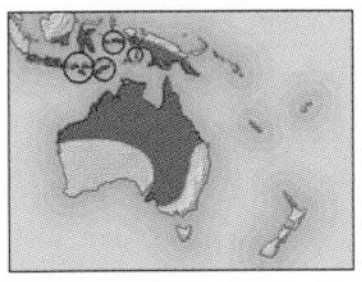

Length 8½–9½ in (21–24 cm)
Weight 2⅛–2½ oz (60–70 g)
Plumage Sexes alike
Migration Partial migrant
Status Least concern

Location S.E. Asia, Australia

This Australian bird is an exception among the 8 species of pratincoles in having long legs and unusually long wings. It is pale cinnamon-brown above and pale tan underneath except for a large, dark chestnut patch on the belly. Its downcurved bill has a red base and black tip. The Australian pratincole runs very fast to catch insects on the ground, but will also feed in the air, sometimes forming aerial flocks that hunt at considerable heights.

Charadrius hiaticula

Common ringed plover

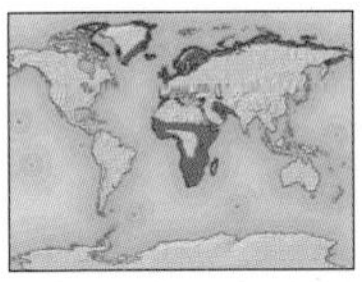

Length 7–8 in (18–20 cm)
Weight 1¾–2½ oz (50–70 g)
Plumage Sexes alike
Migration Migrant
Status Least concern

Location N. North America, Greenland, Europe, Asia, Africa, Madagascar

A short bill, a white neck collar, and black bands on the head and chest are distinctive features of this compact plover. It breeds over a wide range of habitats, but in winter is found mostly on sandy coasts and inland wetlands. It feeds on a variety of invertebrates, foraging for them at night, especially when the moon is full. Both male and female rear the young and will perform a striking "broken wing" display to lure predators from the nest.

Anarhynchus frontalis

Wrybill

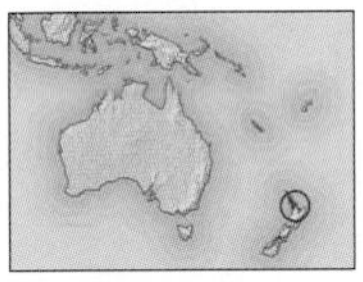

Length 8–8½ in (20–21 cm)
Weight 1⁷⁄₁₆–2½ oz (40–70 g)
Plumage Sexes differ
Migration Migrant
Status Vulnerable

Location New Zealand

The wrybill is the only bird whose bill has a sideways curve. From the side, its bill looks relatively normal—if slightly long for a plover—but when seen from above or head-on, its unusual shape is very evident. The sideways curve is useful when feeding on gravel beaches, allowing the plover to flick stones aside as it hunts for insects and worms. The plumage is soft gray above, and white below, with a narrow, black breastband, more strongly defined in males than in females.

Numenius arquata

Eurasian curlew

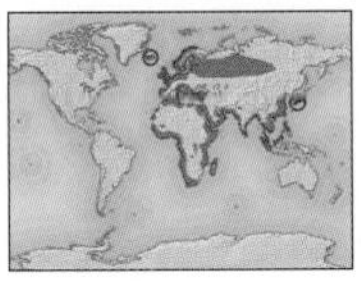

Length 20–23½ in (50–60 cm)
Weight 16–48 oz (450–1,350 g)
Plumage Sexes alike
Migration Migrant
Status Near threatened

Location Europe, Asia, Africa

Famous for its beautiful rising call, the Eurasian curlew is a large, brown-streaked wader with an extremely long, downward-curving bill. Up to 7½ in (19 cm) long, the bill is ideal for extracting animals buried deep in mud or sand. Inland, the bird eats mostly insects and earthworms; on the coast, it feeds on a wide range of worms, shellfish, and crustaceans, especially shrimps and crabs (which the bird usually swallows whole, after removing the legs). It breeds in a variety of habitats, from sand dunes, bogs, and fens to upland heaths and grassland, and becomes strongly territorial at this time. Both male and female defend their territory with a distinctive undulating flight, planing on wings held in a shallow "V" while giving a whistled, bubbling call. Some males will even come to blows with their wings. The Eurasian curlew is one of 8 related species that are found in different parts of the world. They gather in very large flocks for roosting, but feed in smaller groups. Most are strongly migratory—this particular species breeds as far north as the Arctic Circle, but overwinters in sandy and muddy estuaries from Western Europe to East Asia. It is a long-lived bird that can survive for up to 37 years.

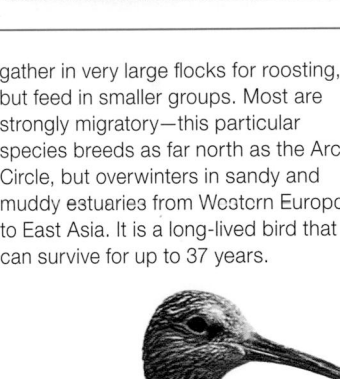

Tringa nebularia

Greenshank

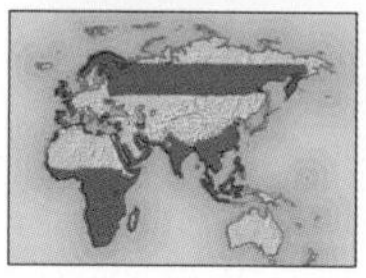

Length 12–14 in (30–35 cm)
Weight 4–11 oz (125–300 g)
Plumage Sexes alike
Migration Migrant
Status Least concern

Location Europe, Asia, Africa (mainly south of Sahara)

Like its 9 other relatives in the genus *Tringa*, this long-billed wader is rarely found far from damp ground. It has dark plumage, which is paler below, and long, gray-green legs. When feeding, the greenshank uses the "walk-and-peck" technique used by other members of its genus. It often runs at its prey or scythes through the water with its bill until it finds food, which includes fish, crustaceans, insects, and invertebrates. Its call, a ringing "tu-tu-tu," is sharper when the bird is disturbed.

Actitis macularius

Spotted sandpiper

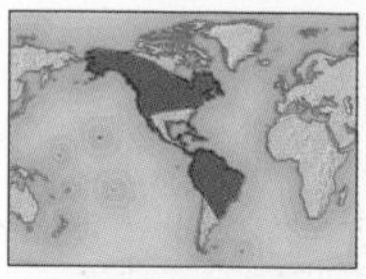

Length 7–8 in (18–20 cm)
Weight 1¼–2½ oz (35–70 g)
Plumage Sexes alike
Migration Migrant
Status Least concern

Location North, Central, and South America

This small sandpiper has a horizontal posture, short legs, and a stout, 2-toned bill. Greenish brown above, it has pale underparts that are boldly spotted with brown in summer; the female has larger, blacker spots. On the ground, it bobs or teeters in its search for food. As in the red phalarope (see right), the female courts the male, with fanned tail and quivering wings. Some mate with up to 4 males in a season, and show little maternal care.

Phalaropus fulicarius

Red phalarope

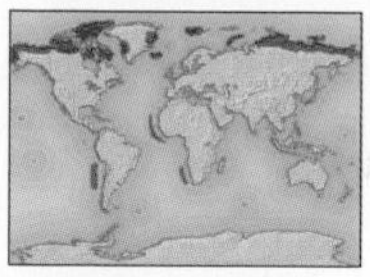

Length 7¾–8½ in (20–22 cm)
Weight 1⁷⁄₁₆–2⅞ oz (40–80 g)
Plumage Sexes differ
Migration Migrant
Status Least concern

Location Circumpolar around Arctic, W. South America, W. Africa

Unlike most waders, which live at or near the water's edge, the red phalarope is an excellent swimmer and spends much of its time either at sea or on muddy pools. It feeds on small animals at the surface and frequently spins around in tight circles to stir up its prey. The red phalarope breeds in marshy ground in the high Arctic tundra, and is remarkable for its breeding behavior. The female takes the lead in courtship, while the male incubates the eggs and rears the young.

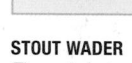

ROLE REVERSAL

The red phalarope's breeding system shows an almost complete reversal of the male and female's normal roles. The female—shown above right—is more assertive than the more drably colored male. She establishes a territory and displays to attract a mate, and he takes care of the young.

STOUT WADER
The red phalarope is a small, rather pot-bellied wader with very short legs and a fairly stout bill. The fleshy lobes between the toes help the bird swim.

Gallinago media

Great snipe

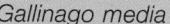

Length 10½–11½ in (27–29 cm)
Weight 5–8 oz (150–225 g)
Plumage Sexes alike
Migration Migrant
Status Near threatened

Location N. Europe, N.W. Asia, Africa

The 16 species of snipes are long-billed and beautifully camouflaged waders that feed mostly under cover of darkness. Like most snipes, the great snipe has a

pointed, 2-toned bill. There is extensive dark barring on the underparts, and the tail corners have more white than most other snipes. In the breeding season, male birds gather at a display ground (lek) and compete with each other, using a variety of sounds. After mating, the female nests alone.

Calidris canutus

Red knot

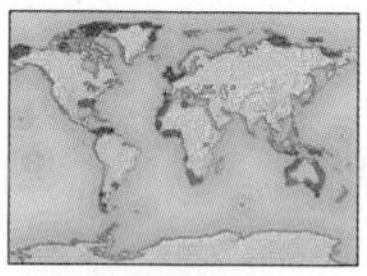

Length 9–10 in (23–25 cm)
Weight 3⅝–8 oz (100–225 g)
Plumage Sexes alike
Migration Migrant
Status Least concern

Location Worldwide, except Antarctica

One of the largest of all sandpipers of the genus *Calidris*, the red knot has gray plumage in winter, but during the breeding season, its face and underparts turn brick-red, while the upperparts are dark and spotted with pale chestnut. Red knots travel up to 7,500 miles (12,000 km) to breed, and form large flocks when feeding and roosting.

Chionis albus

Snowy sheathbill

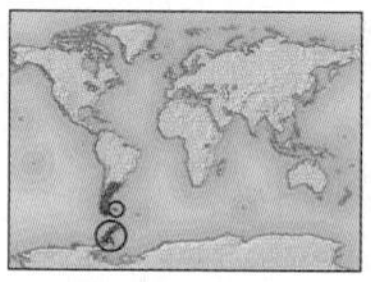

Length 13½–16 in (34–41 cm)
Weight 16–28 oz (450–775 g)
Plumage Sexes alike
Migration Partial migrant
Status Least concern

Location S.E. South America, Falkland Islands, Antarctic Peninsula

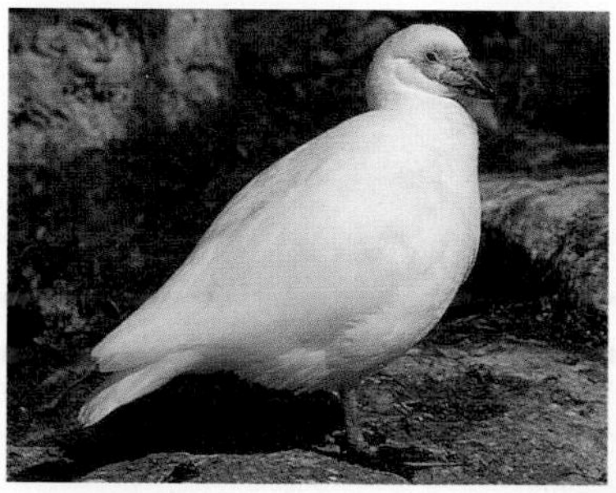

The 2 species of sheathbills constitute the only bird family whose breeding range is entirely confined to the Antarctic and sub-Antarctic region. This species scavenges for food at penguin and seal colonies, feeding on eggs, chicks, and seals' afterbirth, and harassing penguins into regurgitating the food they bring for their young. It makes a cup-shaped nest from grass, bones, pebbles, shells, and feathers.

Stercorarius antarcticus

Brown skua

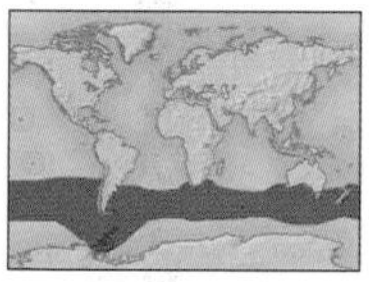

Length 20½–25 in (52–64 cm)
Weight 2¼–4½ lb (1–2 kg)
Plumage Sexes alike
Migration Partial migrant
Status Least concern

Location Circumpolar around Antarctica

Skuas breed in high latitudes, at times close to polar ice. The brown skua is one of 4–6 species in its genus and the most widespread species in the far south. This gull-like scavenger preys on penguins, shearwaters, and other seabirds of the southern oceans, particularly on their eggs and young, and scavenges around fishing boats and ships.

Stercorarius longicaudus

Long-tailed jaeger

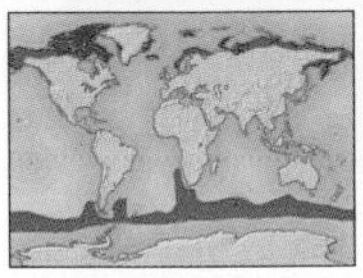

Location Circumpolar around Arctic and Antarctica

Length 19–21 in (48–53 cm)
Weight 8–13 oz (225–350 g)
Plumage Sexes alike
Migration Migrant
Status Least concern

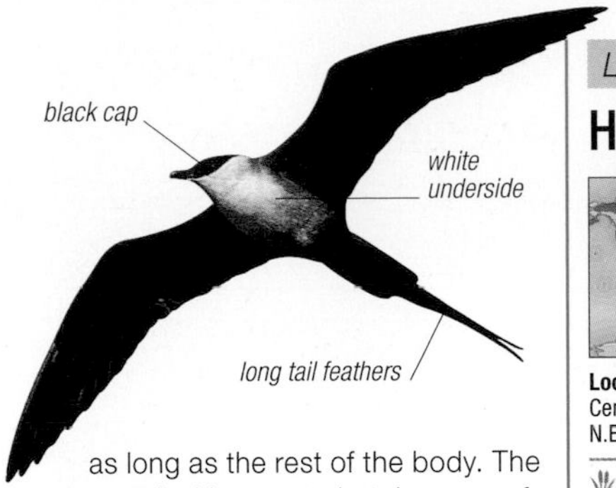

This gray and black jaeger differs from other skuas in having very long central tail feathers, which are clearly visible when it flies. These feathers may be half as long as the rest of the body. The long-tailed jaeger undertakes one of the longest of all bird migrations—from the Arctic to the southern oceans. Lemmings constitute its diet on land, whereas it feeds on fish and robs other birds while at sea.

Pagophila eburnea

Ivory gull

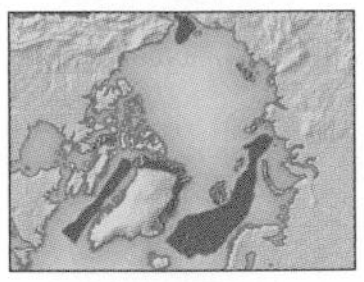

Location Almost circumpolar around Arctic

Length 17½–19 in (44–48 cm)
Weight 19–25 oz (525–700 g)
Plumage Sexes alike
Migration Partial migrant
Status Near threatened

The gull family contains about 50 species, about 35 of which are found in the Northern Hemisphere. Many are familiar birds—both on coasts and inland—but the ivory gull is much less well known. Its entire life is spent in the high Arctic, mostly on the edge of pack ice. This gull feeds chiefly on fish and invertebrates, but also follows polar bears to feed on the scraps they leave behind.

Larus argentatus

Herring gull

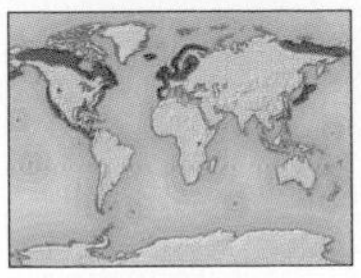

Location North and Central America, Europe, N.E. and E. Asia

Length 22–26 in (55–67 cm)
Weight 1½–3¼ lb (0.73–1.5 kg)
Plumage Sexes alike
Migration Partial migrant
Status Least concern

JUVENILE
Young herring gulls have brown-streaked feathers and take 4 years to acquire adult plumage.

The herring gull is the most abundant coastal bird in much of North America and Europe, and its numbers have grown exponentially in the Northern Hemisphere in the past century. Adaptability is the key to its success, for it will eat almost anything, scavenging from garbage dumps, landfill sites, and sewage outflows. It also steals the eggs and young of other gulls, terns, and birds that nest in open country. Very vocal, especially in breeding colonies, adults tend to return to the same site to breed every year.

GRAY ADULT
One of the larger gulls, this pink-legged bird has a gray back and wings, with black and white tips, a white head and underside, and a yellow bill with a red spot on the lower half.

Larus dominicanus

Kelp gull

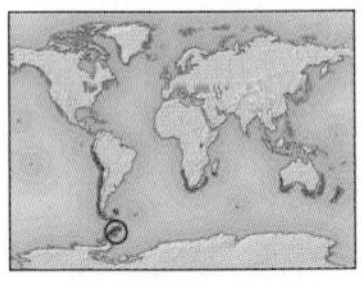

Location South America, Antarctica, southern Africa, S. Australia, New Zealand

Length 21½–26 in (54–65 cm)
Weight 2–3¼ lb (0.9–1.5 kg)
Plumage Sexes alike
Migration Partial migrant
Status Least concern

This large coastal bird is one of the most widespread gulls south of the equator. Its back and wings are black, while the head, tail, and underside are white. Its large, yellow bill has a red spot on the lower half. This opportunistic feeder catches invertebrates stirred up by whales, steals food from terns, and kills birds as large as geese. It also feeds on termite swarms, and scavenges at fish factories and slaughterhouses.

Chlidonias niger

Black tern

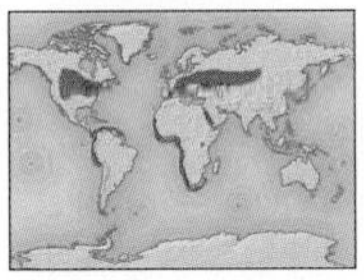

Location North and Central America, N. South America, Europe to C. Asia, Africa

Length 9–11 in (22–28 cm)
Weight 2⅛–2⅝ oz (60–75 g)
Plumage Sexes alike
Migration Migrant
Status Least concern

Terns are graceful birds, with slender bodies and forked tails. Most of them are white with a black cap, but the black tern has dark plumage. Unlike most terns, it does not breed on the coast, but on inland lakes, marshes, and bogs. Here, it feeds mainly on insects plucked from plants, off the water surface, or caught on the wing. Out of the breeding season, when back in coastal waters, small marine fish make up the bulk of its diet.

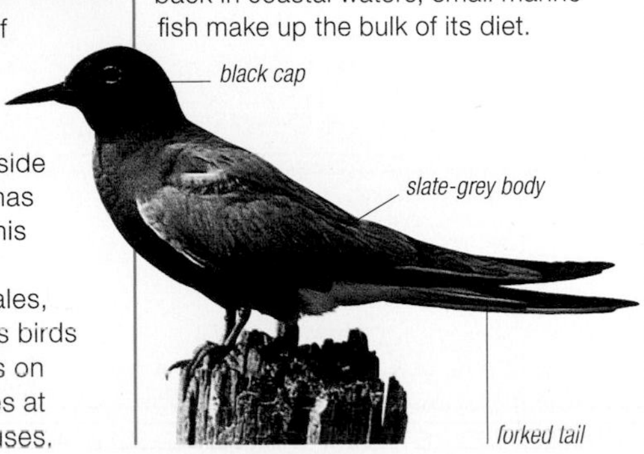

Sterna caspia

Caspian tern

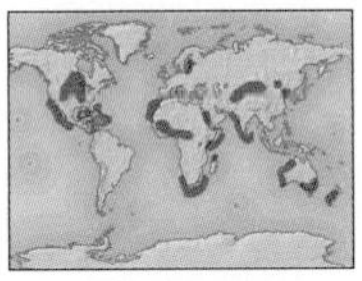

Location North to Central America, Europe, Africa, Asia, Australia, New Zealand

Length 19–22 in (48–56 cm)
Weight 21–28 oz (575–775 g)
Plumage Sexes alike
Migration Partial migrant
Status Least concern

With a wingspan of up to 4½ ft (1.4 m), this is the largest species of tern, approaching the size of larger gulls. It spends winter on lakes, coasts, and estuaries, and in the breeding season is likely to be found in freshwater habitats. It catches fish by plunge-diving, swallowing them head-first while in flight.

Sterna paradisaea

Arctic tern

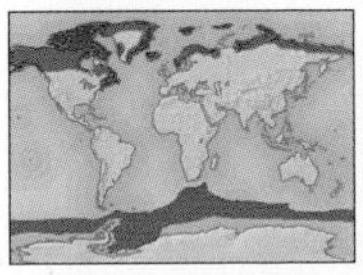

Location Arctic, N. North America, Antarctica

Length 13–14 in (33–35 cm)
Weight 3⅜–4 oz (95–125 g)
Plumage Sexes alike
Migration Migrant
Status Least concern

The Arctic tern makes one of the longest migrations of any bird—a journey of at least 10,000 miles (16,000 km), twice a year. Each autumn, it flies from its nesting place in the north to Antarctica for the southern summer, making the most of the daylight in both hemispheres. It probably spends more time in daylight than any other creature. It feeds mainly on fish, hovering then plunge-diving, or dipping into the water.

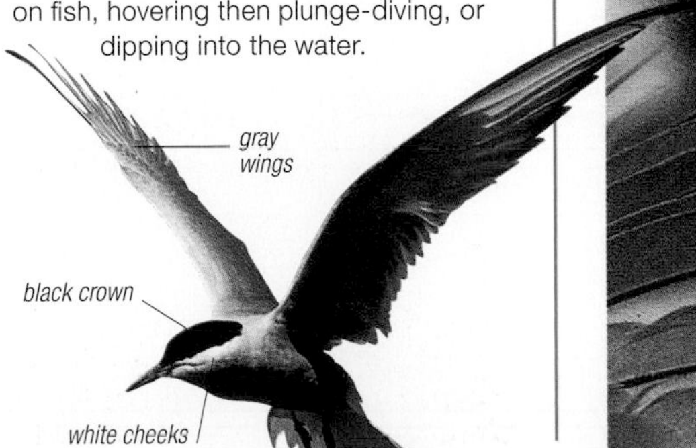

Sterna fuscata

Sooty tern

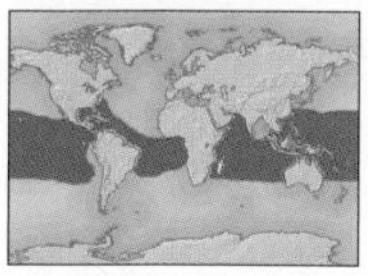

Length 14–18 in (35–45 cm)
Weight 5–9 oz (150–250 g)
Plumage Sexes alike
Migration Partial migrant
Status Least concern

Location Worldwide, in tropical seas

This wide-ranging, tropical tern has white undersides, but is almost completely brownish black above. A bird of the open ocean, it is often seen in large flocks, hundreds of miles from land. Unlike many of its relatives, it does not plunge-dive for its food; instead, it swoops down close to the water, snapping up fish and other small animals near the surface, rarely getting wet. Sooty terns nest in remote islands throughout the tropics, often in enormous, noisy colonies.

Larosterna inca

Inca tern

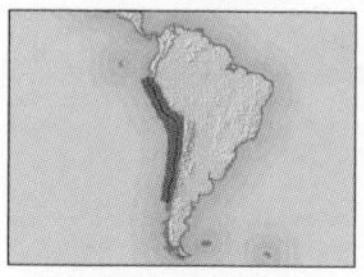

Length 15½–16½ in (39–42 cm)
Weight 6–7 oz (175–200 g)
Plumage Sexes alike
Migration Nonmigrant
Status Near threatened

Location W. South America

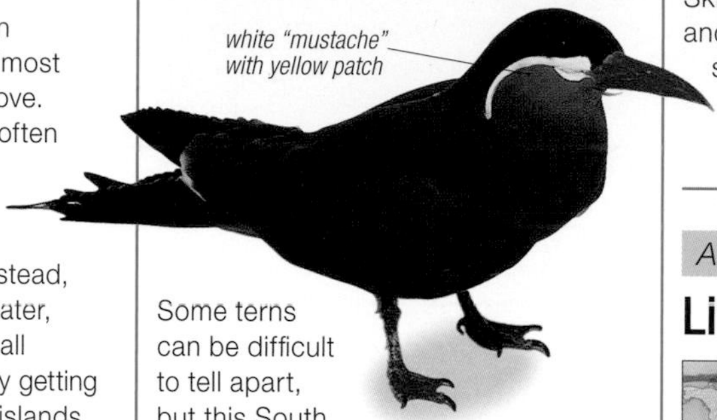

Some terns can be difficult to tell apart, but this South American species is instantly recognizable by its slate-gray color and its remarkable white "mustache" plumes, which trail from its cheeks for 2 in (5 cm) or more. This bird occurs in the Humboldt Current region, feeding on the abundant small anchovies found in the nutrient-rich water. Large numbers of Inca terns congregate over feeding sea lions and humpback whales, diving for fish scraps left by these animals.

Rynchops flavirostris

African skimmer

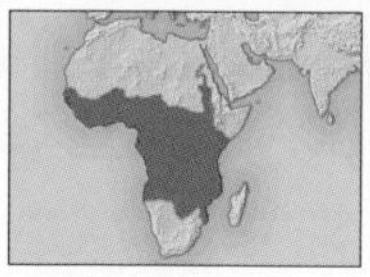

Length 14–16½ in (36–42 cm)
Weight 3⅝–7 oz (100–200 g)
Plumage Sexes alike
Migration Nonmigrant
Status Near threatened

Location Africa

Skimmers are similar to terns in shape and general coloration, but their bill structure is unique, with the lower half much longer than the upper half, and flattened sideways rather like scissor blades. There are 3 species, and they all feed by flying low over water, plowing the surface with the lower "blade" and snapping the bill shut when it touches a fish. The birds feed mostly at dawn and dusk. This species, like the other 2 skimmers, is black with a white underside, and its feet, legs, and bill are bright orange-red.

Alle alle

Little auk

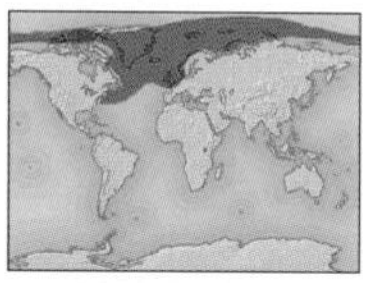

Length 7½–9 in (19–23 cm)
Weight 5–6 oz (150–175 g)
Plumage Sexes alike
Migration Migrant
Status Least concern

Location Circumpolar around Arctic

Also known as the dovekie, the smallest auk in the Atlantic is a black and white bird with a short bill and an upright posture. Like other members of the auk family, it is a pursuit diver, using its stubby wings like flippers to speed after its prey. It feeds mainly on plankton, its bill being specially adapted so that it can catch small prey. The little auk breeds in colonies on steep cliffs along the Arctic coast, but usually winters out at sea.

Aethia pusilla

Least auklet

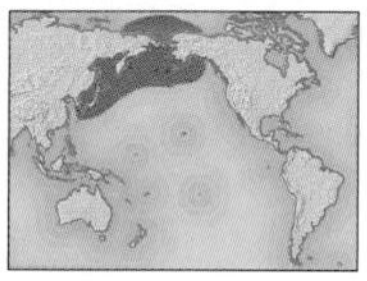

Length 6 in (15 cm)
Weight 3 oz (85 g)
Plumage Sexes alike
Migration Partial migrant
Status Least concern

Location North Pacific, Arctic Ocean

The smallest of all auk species, the least auklet is the same length as a house sparrow, although its body is more heavyset. Breeding colonies may contain as many as one million birds, which set off in insectlike swarms to feed. Plankton is its only food—mainly copepods but also young crustaceans and fish larvae; the bird dives from the surface and swims through the water to find the drifting schools. It lays just one egg a year, in a crack or crevice on bare rock.

Fratercula arctica

Atlantic puffin

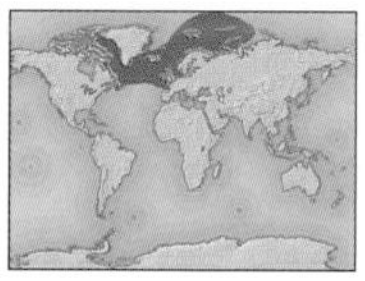

Length 11–12 in (28–30 cm)
Weight 14 oz (400 g)
Plumage Sexes alike
Migration Migrant
Status Least concern

Location North Atlantic, Arctic Ocean

The puffin is the most colorful auk in the North Atlantic, with a multicolored bill and bright orange legs and feet. It sometimes dives to 200 ft (60 m) to find the schooling fish that are its main food. These include sand eels, capelins, herring, and sprats, which are supplemented by animal plankton in winter. When nesting, the puffin digs a burrow 3¼ ft (1 m) or more deep in the ground; the burrow is often lined with feathers and plant matter. Both parents incubate the solitary egg and rear the young, carrying many small fish, packed crosswise into their bills, back to the burrow for it to eat.

PUFFIN SOCIETY

Like most members of the auk family, the Atlantic puffin is a highly social species, nesting in colonies on rocky coasts and offshore islands. On land, puffins stand in groups, but when feeding, they gather in "rafts" out to sea. Feeding areas are usually within 6 miles (10 km) of a colony.

BREEDING COLORS
During the breeding season, the Atlantic puffin's triangular bill is red, yellow, and blue. It fades in late summer when its outer scales are shed.

Uria aalge

Common murre

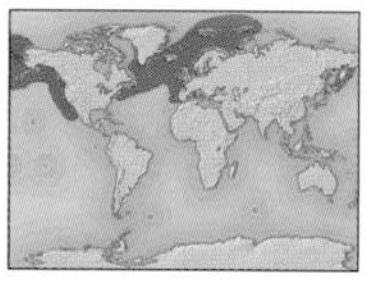

Length 15½–16½ in (39–42 cm)
Weight 30–39 oz (850–1,100 g)
Plumage Sexes alike
Migration Migrant
Status Least concern

Location North Atlantic, North Pacific

A relatively large, blackish brown and white auk, the common murre is gregarious, and often forms dense breeding colonies. Each pair produces a solitary egg, laid directly on the rock. The egg is sharply pointed so that it rolls around in a circle if disturbed, rather than falling off the nesting ledge. The chick leaves the ledge when about 3 weeks old and completes its development at sea.

BIRDS

Pigeons

PHYLUM	Chordata
CLASS	Aves
ORDER	Columbiformes
FAMILIES	2 (Columbidae)
SPECIES	321

The familiar pigeons seen in cities and on arable farms throughout the world are only a small part of a diverse group. Tropical forests are home to an immense variety of species, many of them brilliantly colored, living on the ground or in trees. The larger kinds are generally known as pigeons and the smaller ones as doves. Adult pigeons and doves produce nutritious "crop-milk," which they feed to their young.

ROOSTING
Pigeons and doves often roost communally, either in spaced out groups (like these crested pigeons), or huddled together. When disturbed, several roosting birds will take off with sudden wingbeats, a shock tactic designed to alarm a predator.

Anatomy

Pigeons and doves are plump, full-breasted birds with a small head and bill; the head bobs as the bird walks, to keep it in a constant position relative to the body. These birds are strong fliers, their broad wings driven by powerful breast muscles that enable them to travel long distances at considerable speed. The plumage is thick and soft, although most species have a patch of bare skin around each eye.

IRIDESCENT FEATHERS

BRIGHT COLORATION

PLUMAGE
Many tropical pigeons and doves, such as the magnificent fruit pigeon (left) have bright and varied plumage. Although other species (such as the stock dove, far left) are less strikingly colored, they have small iridescent patches.

Feeding

Pigeons and doves eat mainly plant material. They can be broadly divided into 2 groups: seed-eaters and fruit-eaters. All have a specially adapted gut, with a well-developed crop and a strong, muscular gizzard; the latter is used for grinding food, often aided by ingested grit or stones. Fruit-eaters, taking more easily digestible food, have shorter guts than seed-eaters.

FEEDING YOUNG
These young turtle doves are being fed on "milk" secreted by their mother's crop. Crop-milk is produced by both sexes and is rich in proteins and fats.

BIRDS

Columba livia

Rock pigeon

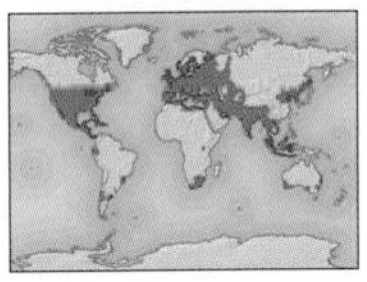

Location North, Central, and South America, Africa, Europe, Asia, Australia

Length 12–13½ in (31–34 cm)
Weight 7–11 oz (200–300 g)
Plumage Sexes alike
Migration Nonmigrant
Status Least concern

Originally from Southern Europe, Asia, and North Africa, the rock pigeon is the wild ancestor of the town pigeon—one of the world's most widespread feral birds. Whereas the wild morph is generally gray, with iridescent highlights on the neck and upper breast, town pigeons are highly variable. Paradoxically, although some feral strains have remarkable homing abilities, wild populations are largely sedentary. The wild rock pigeon nests on sea cliffs and among rocks; for town pigeons, bridges and window ledges provide perfect nest sites.

Columba palumbus

Common wood pigeon

Location Europe, N.W. Africa, W. and C. Asia

Length 16–18 in (41–45 cm)
Weight 10–25 oz (275–700 g)
Plumage Sexes alike
Migration Partial migrant
Status Least concern

The largest pigeon found in Europe, this species has prospered as a result of farming and has become an agricultural pest. It often feeds in flocks on the ground, and is also an agile feeder in trees, clambering to branch tips to feast on fruit and seeds. Sexually active males perform a display flight, accompanied by wing claps, to attract a mate. The female usually produces only a single brood of one or 2 eggs per season. The young are cared for by both parents in the nest.

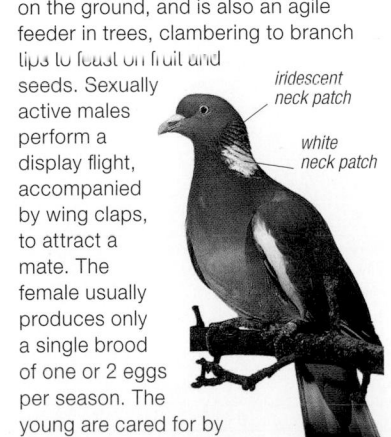

Nesoenus mayeri

Pink pigeon

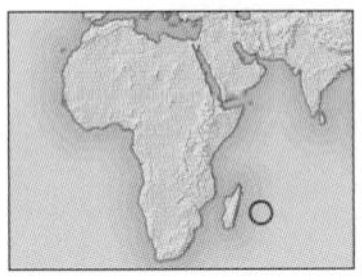

Location Mauritius

Length 12–16¾ in (30–40 cm)
Weight 11–12 oz (300–325 g)
Plumage Sexes alike
Migration Nonmigrant
Status Endangered

One of the world's rarest birds, the pink pigeon is one of many island species that have been harmed by introduced predators and habitat loss. The wild population had diminished to fewer than 20 birds in the 1980s; however, successful captive-breeding and release programs have led to an increase in its numbers. This large pigeon has a soft pink body, a white face and forehead, usually brown wings, and a moderately long, strong bill with a hooked tip.

Streptopelia decaocto

Eurasian collared-dove

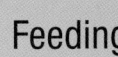

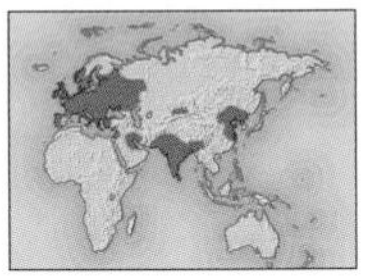

Location Europe, Asia, N.E. Africa

Length 12–13½ in (31–34 cm)
Weight 5–7 oz (150–200 g)
Plumage Sexes alike
Migration Partial migrant
Status Least concern

This slim, pinkish buff dove, with its distinctive black collar, underwent a dramatic and still unexplained expansion during the 20th century, and has become a common sight through much of Europe. It feeds mainly on the seeds and fruit of grasses (including cereals) and herbs, and occasionally on invertebrates and the green parts of plants, typically pecking food from the ground. Highly vocal in the breeding season, its specific name "decaocto" indicates the rhythm of the male's courtship call.

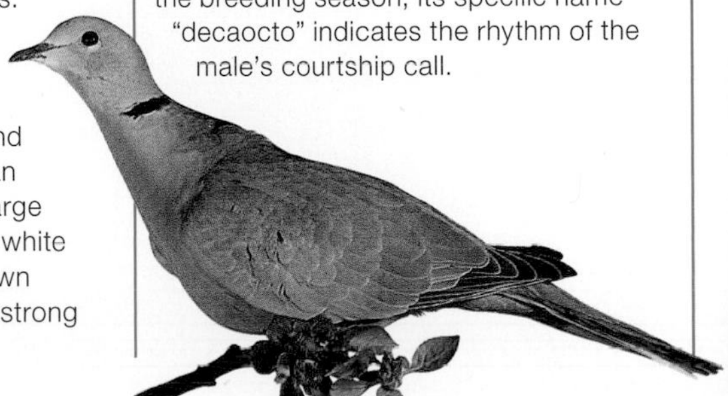

Zenaida macroura

Mourning dove

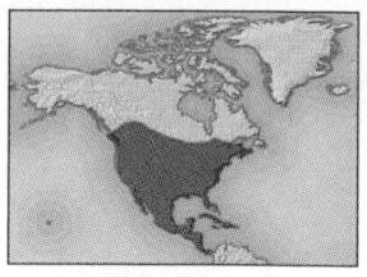

- **Length** 9–13½ in (23–34 cm)
- **Weight** 3⅝–6 oz (100–175 g)
- **Plumage** Sexes alike
- **Migration** Partial migrant
- **Status** Least concern

Location North and Central America, Caribbean

This widespread North American dove gets its name from its mournful, 4-syllable call. Small and slender, it has long, narrow wings and a long, pointed tail. The female is slightly paler than the male. The latter's courtship display consists of gliding, spiraling, and flying above the female, with his wingtips held below the body. A very rapid breeder, this dove may even reproduce in the season of its birth.

Goura scheepmakeri

Southern crowned-pigeon

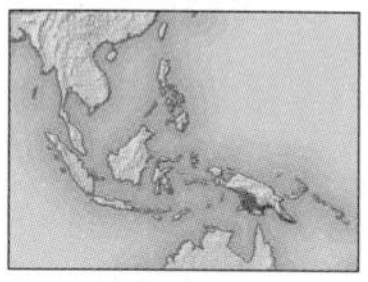

- **Length** 26–29 in (66–74 cm)
- **Weight** 5½ lb (2.5 kg)
- **Plumage** Sexes alike
- **Migration** Nonmigrant
- **Status** Vulnerable

Location S. New Guinea

One of the world's largest pigeons, this species has a distinctive lacy, fan-shaped crest on top of its head. Generally blue-gray, it has a purple-red breast and pale gray wing patches, edged with dark purple. It lives on the ground during the day, feeding on fallen fruit, and roosts in forest trees at night. This species is now vulnerable as a result of loss of its forest habitat and humans hunting it for food and its beautiful plumes. Its rarity has also made it a target for collectors.

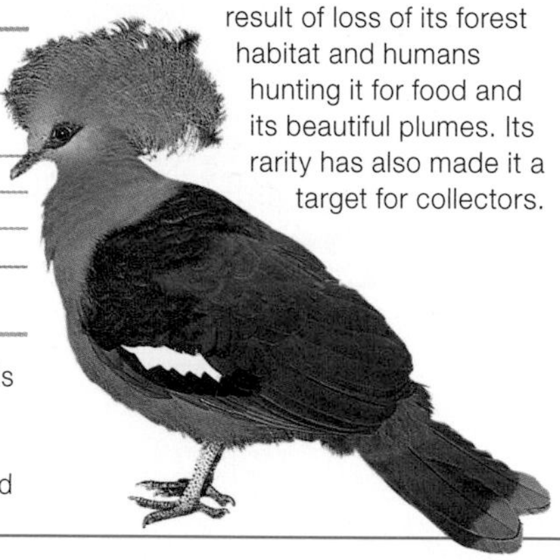

Ptilinopus magnificus

Wompoo fruit dove

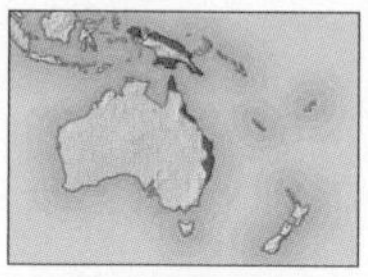

- **Length** 11½–22 in (29–55 cm)
- **Weight** 9–18 oz (250–500 g)
- **Plumage** Sexes alike
- **Migration** Partial migrant
- **Status** Least concern

Location New Guinea, N.E. and E. Australia

Named the wompoo fruit dove after its "wompoo" call, this large, heavy, long-tailed fruit pigeon is threatened by loss of habitat. Despite its richly colored plumage—mostly yellow and green, with a deep purple breast and upper belly, and a gray or greenish gray head and neck—this bird is inconspicuous and surprisingly well camouflaged in the forest canopy in which it lives. It feeds on fruit (usually figs), taken from trees, and very rarely comes to the ground.

Hemiphaga novaeseelandiae

New Zealand pigeon

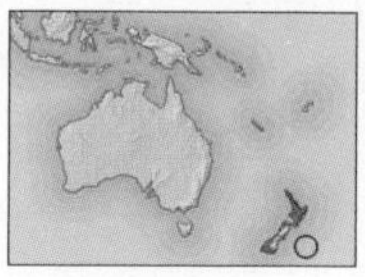

- **Length** 18–20 in (46–50 cm)
- **Weight** 21–29 oz (600–800 g)
- **Plumage** Sexes alike
- **Migration** Nonmigrant
- **Status** Near threatened

Location New Zealand

The largest pigeon in New Zealand, this is the only species that is native to the islands, rather than introduced. It is dark, with iridescent, bronze and green highlights on the upperparts and breast, and a white belly and lower breast. This bird feeds on a wide range of plants and fruit, which it is able to eat whole because of its extendable gape. Several native New Zealand trees are almost totally dependent on this pigeon for their seed dispersal.

Sandgrouse

PHYLUM	Chordata
CLASS	Aves
ORDER	Pteroclidiformes
FAMILIES	1 (Pteroclididae)
SPECIES	16

These intricately patterned birds live in arid areas of Africa and Asia. They spend most of their time on the ground, where they are well camouflaged by their brown or gray, spotted or barred plumage. Sandgrouse resemble both grouse and pigeons. Like grouse, they have a small head, a squat body, and feathers on their legs, yet when they take to the air, they fly strongly on long, pointed wings, with fast, steady wingbeats like those of pigeons. Their thick neck and short legs are also pigeonlike.

RETAINING WATER
Many sandgrouse (including these Burchell's sandgrouse) need to drink once or twice a day. Their belly feathers hold water that is used to supply chicks in the nest, which is normally far from water sources.

Pterocles coronatus

Crowned sandgrouse

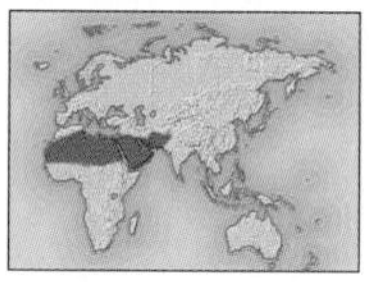

- **Length** 10½–12 in (27–30 cm)
- **Weight** 9–11 oz (250–300 g)
- **Plumage** Sexes differ
- **Migration** Nonmigrant
- **Status** Least concern

Location N. Africa, W. to S. Asia

This well-camouflaged bird lives in some of the most extreme desert areas, helped by its low energy and water requirements, its ability to tolerate air temperatures exceeding 122° F (50° C) for several hours, and its tolerance of water with a high salt content. The male has a sandy-orange crown and a black mask at the base of his bill, while the female is grayer and more barred.

Pterocles namaqua

Namaqua sandgrouse

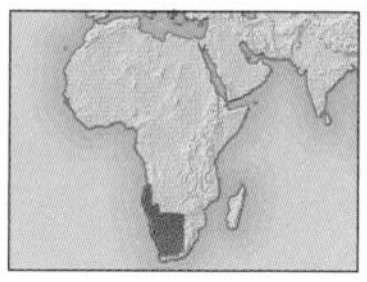

- **Length** 11 in (28 cm)
- **Weight** 6–7 oz (175–200 g)
- **Plumage** Sexes alike
- **Migration** Nonmigrant
- **Status** Least concern

Location Southern Africa

The Namaqua sandgrouse shares many desert-specific adaptations with the crowned sandgrouse (see left): it needs little food or water, and its thick plumage insulates it from high and low temperatures. A pale brown head, white breastband, and mottled brown wings provide excellent camouflage against the stony ground and sand. This bird travels to areas that have received good rainfall in search of seeds, which are its sole source of food.

BIRDS

Parrots

PHYLUM	Chordata
CLASS	Aves
ORDER	Psittaciformes
FAMILIES	1
SPECIES	375

These conspicuous, brightly colored birds have populated most of the world's warmer areas, and are particularly abundant in tropical forests. In addition to true parrots, this group includes the familiar macaws, parakeets, cockatoos, cockatiels, lorikeets, and budgerigars, among others. Noisy and social in the wild, parrots have long been popular as pets, prized for their beauty, intelligence, and impressive learning skills, evident in their remarkable ability to mimic human sounds. Although they may range some distance to forage, very few parrots are truly migratory.

Anatomy

Parrots are easily recognized by their large head, short neck, and strongly hooked bill. They have hard, distinctively glossy plumage, usually predominantly green for camouflage in the forest foliage, with patches of other bright colors. Their feet, consisting of 2 toes pointing forward and 2 backward, are used for climbing trees, and the bill is frequently used as a third limb for climbing or holding. The wings are typically narrow and pointed, enabling parrots to fly with great speed and maneuverability.

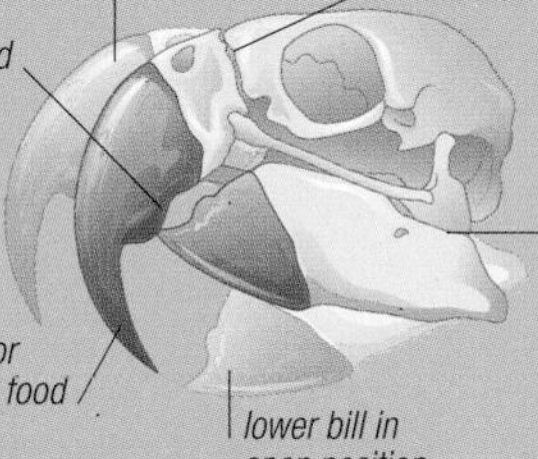

PARROTS' BILL
A parrot's hooked bill is remarkably flexible; both the upper and lower mandibles are hinged against the skull for independent movement.

NECTAR FEEDERS
Lories and lorikeets (here, a rainbow lorikeet) feed on nectar. They have a brushlike tip to their tongue to help them sweep the liquid into their mouth.

Feeding

Virtually all parrots feed exclusively on plant material, including seeds, nuts, fruit, nectar, and flowers. Feeding is often an exercise in coordination between bill and foot, the latter grasping a food item and holding it up for the bill to work over. Although most parrots feed in the trees, many species also readily descend to the ground to forage.

SOCIAL GROUPS
Parrots (here, green-winged macaws and scarlet macaws) often form large, noisy flocks that communicate by making various calls, from harsh squawks to piercing shrieks. Most species form long-term pair-bonds.

CONSERVATION

Nearly a third of all parrot species are under threat in the wild, and a few have actually become extinct. One of the greatest threats is from the caged-bird trade, which accounts for hundreds of thousands of captures and deaths every year. Other threats to their numbers include deforestation.

Trichoglossus haematodus

Rainbow lorikeet

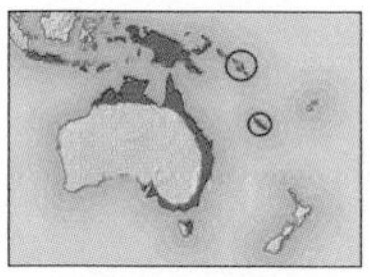

Length 12 in (30 cm)
Weight 5 oz (150 g)
Plumage Sexes alike
Migration Nonmigrant
Status Least concern

Location New Guinea, S.E. Asia, S.W. Pacific, Australia (including Tasmania)

This brightly colored lorikeet has a stocky body and pointed tail. It is highly variable in appearance, with 22 subspecies that differ in size or coloration, or both. Juveniles have dark brown bills, while in adults they are orange or red. Like most other lorikeets and lories, this lorikeet has a brush-tipped tongue adapted for gathering pollen and nectar from flowers. In flight, the rainbow lorikeet screeches continuously while displaying its brightly colored underwings. However, when feeding, its subdued chatter and cryptic coloration make it difficult to detect among foliage. This species causes damage in orchards and vineyards, especially in Australia.

BRIGHTLY COLORED
This startlingly bright bird varies greatly in color. Most subspecies have green upperparts, a purple-blue streaked head, and an orange to red breast with dark scalloping.

FEEDING IN A GROUP

Flocks of rainbow lorikeets congregate to feed in flowering trees and are often found in the company of other nectar- or fruit-eating birds. This bird uses its bill to crush the flesh of fruit to extract the juice and the seeds. The rainbow lorikeet also feeds on insects and their larvae.

Eos reticulata

Blue-streaked lory

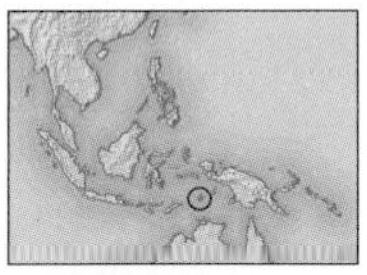

Length 12¼ in (31 cm)
Weight 6 oz (175 g)
Plumage Sexes alike
Migration Nonmigrant
Status Near threatened

Location Indonesia (Tanimbar, Kai, and Damar Islands)

This brightly colored lory is mainly red, with violet-blue streaking on its hind neck and back, a narrow, orange-red bill, and gray legs. A noisy, conspicuous bird, it is most commonly seen in swift overhead flight, emitting shrill call-notes, but also spends much time resting or feeding in trees. Like other lorikeets, it has a brush-tipped tongue to gather pollen and nectar.

BIRDS

Probosciger aterrimus

Palm cockatoo

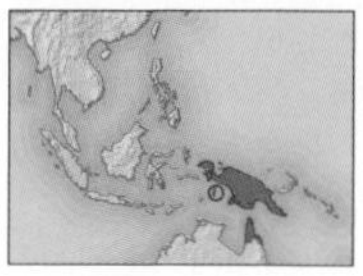

Length 23½ in (60 cm)
Weight 2¼ lb (1 kg)
Plumage Sexes alike
Migration Nonmigrant
Status Least concern

Location New Guinea, N.E. Australia

The largest of about 18 species of cockatoos, the palm cockatoo is also distinguished by its entirely black plumage, huge, hooked bill with which it can crack open palm nuts and other hard-shelled seeds, and a piercing, whistlelike call. When excited or alarmed, the palm cockatoo raises its striking crest of backward-curving, narrow feathers, and its prominent, bare crimson cheek patches deepen in color, causing it to "blush."

Eolophus roseicapillus

Galah

Length 14 in (35 cm)
Weight 12 oz (325 g)
Plumage Sexes alike
Migration Nonmigrant
Status Least concern

Location Australia (including Tasmania)

The most widespread and numerous species of cockatoo, the galah is a familiar sight throughout Australia. Flocks of these noisy and gregarious birds are found in arid scrubland as well as in city parks. Farming has increased this parrot's food supply, leading to a steady rise in population. The galah has a relatively small crest, and its gray back and pink underparts create an alternating effect as it veers through the air.

Nymphicus hollandicus

Cockatiel

Length 12½ in (32 cm)
Weight 3¼ oz (90 g)
Plumage Sexes differ
Migration Nomadic
Status Least concern

Location Australia

The cockatiel is the smallest species of cockatoo and the only one to have a long, pointed tail. The tapering crest is lowered when the bird is resting, or sometimes while feeding. In flight, the cockatiel emits a distinctive warbled chirrup, but is quieter while feeding on the ground or drinking at water holes. This species is very popular as a pet and numerous color variants are found in captive-bred birds.

Nestor notabilis

Kea

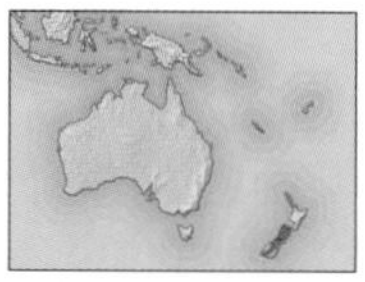

Length 19 in (48 cm)
Weight 29 oz (825 g)
Plumage Sexes alike
Migration Nonmigrant
Status Vulnerable

Location New Zealand (South Island)

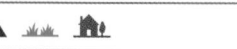

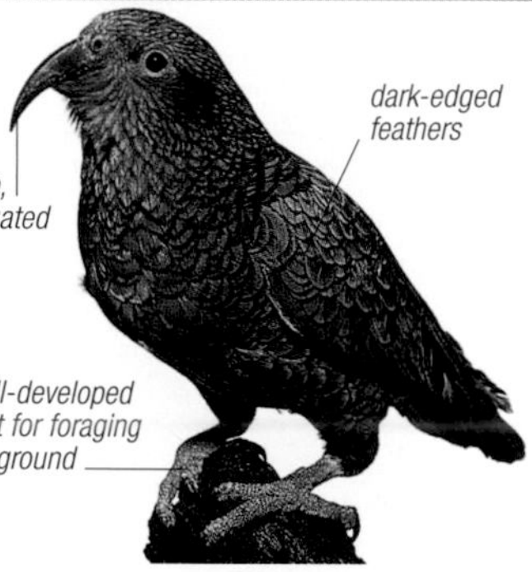

This large, stocky mountain bird is remarkable among parrots for its insatiable curiosity and wide-ranging diet. A natural opportunist, it will examine anything that appears to be edible, and often feeds on carrion, tearing it up with its unusually long bill, which resembles that of a bird of prey. Adults are largely olive-green, with orange underwings.

Eclectus roratus

Eclectus parrot

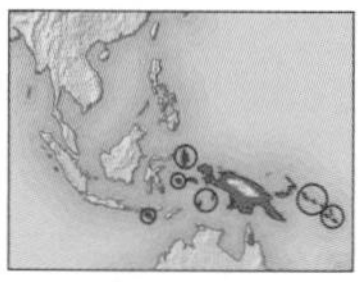

Length 14 in (35 cm)
Weight 18 oz (500 g)
Plumage Sexes differ
Migration Nonmigrant
Status Least concern

Location S.E. Asia, New Guinea, N.E. Australia, S.W. Pacific islands

The male and female eclectus parrots look so different from each other that early naturalists identified them as separate species. Both sexes are large and stocky, with squarish tails and robust bills, but while the male is brilliant green with red flanks and underwings and a yellow bill, the female is red, occasionally with a blue underside and underwings, and has an entirely black bill.

Cacatua galerita

Sulphur-crested cockatoo

Length 20 in (50 cm)
Weight 34 oz (950 g)
Plumage Sexes alike
Migration Nonmigrant
Status Least concern

Location New Guinea, Australia (including Tasmania)

This large, white cockatoo is yellow on the undersides of its wings and tail, and has a yellow crest of narrow, forward-curving feathers. Noisy and active in the morning and late afternoon, it forages in flocks that vary in size from a few dozen birds to several hundred. At night, flocks occupy regularly used roosts, often in trees bordering water courses. This species is very popular as a pet, and both adults and chicks are regularly captured for the live-bird trade, a fact that poses a threat to the survival of some populations.

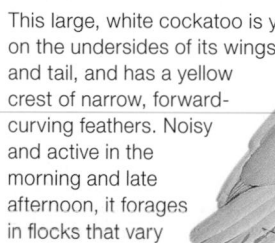

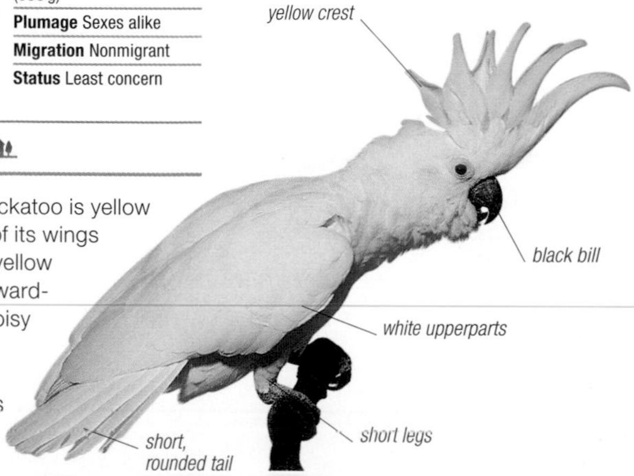

Platycercus elegans

Crimson rosella

Length 14 in (36 cm)
Weight 5 oz (150 g)
Plumage Sexes alike
Migration Nonmigrant
Status Least concern

Location E. and S.E. Australia

This is one of 8 species of rosellas, all of which have similar streamlined bodies and long tails, and are restricted to Australia and its outlying islands.

As its name suggests, the crimson rosella's plumage is red. It has violet-blue cheek patches and outer wing coverts, and a dark blue tail; the back and wings are mottled red and black. The bird tends to become quite tame around farms or in gardens. It has an undulating flight, dropping down toward the ground, gliding upward, and then landing.

BIRDS

Lathamus discolor

Swift parrot

Length 10 in (25 cm)	
Weight 2⅜ oz (65 g)	
Plumage Sexes alike	
Migration Nonmigrant	
Status Endangered	

Location E. and S.E. Australia

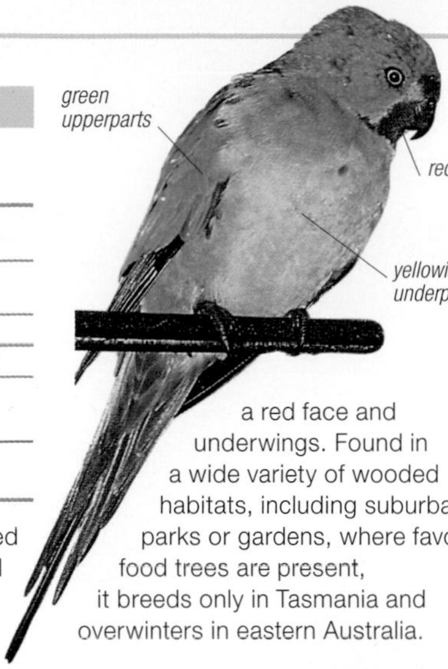

Long, pointed wings and a streamlined shape contribute to making this small parrot a fast flier, hence its name. Its crown is dark blue, contrasting with a red face and underwings. Found in a wide variety of wooded habitats, including suburban parks or gardens, where favored food trees are present, it breeds only in Tasmania and overwinters in eastern Australia.

Melopsittacus undulatus

Budgerigar

Length 7 in (18 cm)	
Weight ⅞ oz (25 g)	
Plumage Sexes alike	
Migration Nomadic	
Status Least concern	

Location Australia

The budgerigar is one of the most numerous of all parrots and is the best known because of its popularity as a pet. Unlike its domesticated counterparts, which may be of different colors, the wild budgerigar is always green, with a yellow face and blue tail. The upperparts of its body are barred black and yellow, providing excellent camouflage while it feeds in vegetation. Often nomadic, this bird is usually seen in large, noisy flocks.

Deroptyus accipitrinus

Red-fan parrot

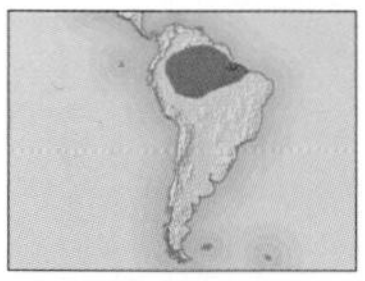

Length 14 in (35 cm)	
Weight 8 oz (225 g)	
Plumage Sexes alike	
Migration Nonmigrant	
Status Least concern	

Location N. South America

This parrot is distinguished by dark red feathers that extend from the nape to the hind neck. When excited or alarmed, it raises its neck feathers to form a spectacular ruff or fan. Its long, rounded tail is also a distinctive characteristic. While perching, the red-fan parrot resembles a bird of prey, with its hawklike eyes. Usually seen in pairs or small groups (rarely of up to 20 birds), this forest-dwelling parrot is inconspicuous but is often heard well before it comes into view.

Psittacus erithacus

Grey parrot

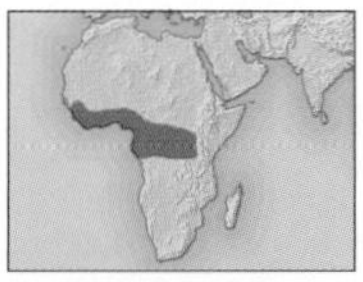

Length 13 in (33 cm)	
Weight 14 oz (400 g)	
Plumage Sexes alike	
Migration Nonmigrant	
Status Near threatened	

Location W. to C. Africa

This stocky, short-tailed parrot is an extremely popular cage bird because of its ability to imitate human speech and perform tricks. Despite being illegally trapped for the pet trade, it is still common in parts of its range. It is found mainly in lowland rain forest, and also in montane rainforest, at forest edges, on plantations, and on farmland or in gardens. It has unusual coloration, its gray plumage contrasting strongly with the bright red or deep maroon of the tail. In flight, members of the flock continually call to each other and also communicate visually by displaying their red tails.

Strigops habroptila

Kakapo

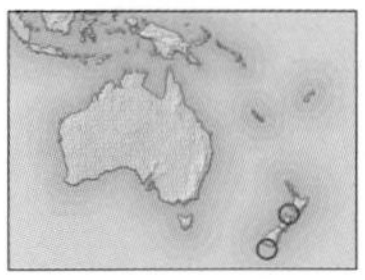

Length 25 in (64 cm)	
Weight 4½ lb (2 kg)	
Plumage Sexes alike	
Migration Nonmigrant	
Status Critically endangered	

Location New Zealand (4 offshore islands)

One of the world's most endangered parrots, the flightless kakapo has highly unusual feeding habits. It is active at dusk and often walks long distances to feeding areas where it chews plants for their juices, leaving balls of fibrous material hanging from the plants. It also digs up or crushes rhizomes with its bill. The males gather at mating grounds (leks), where they dig hollows, and make loud, booming noises to attract females. The adults, chicks, and eggs of the kakapo are highly vulnerable to introduced predators.

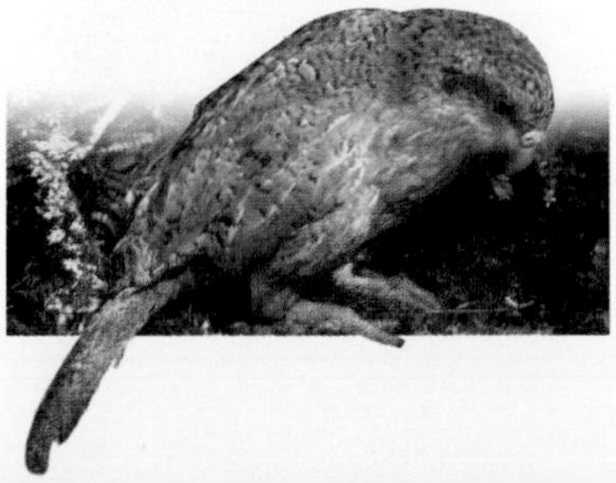

Pionopsitta aurantiocephala

Bald parrot

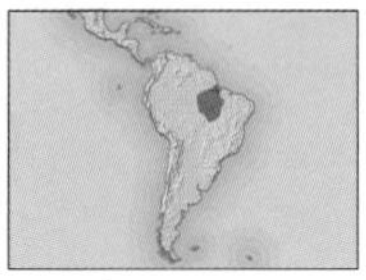

Length 8½–10 in (22–25 cm)	
Weight Not known	
Plumage Sexes alike	
Migration Nonmigrant	
Status Near threatened	

Location C. South America

The bald parrot—so-called because of the bare, bristly head with bright orange skin—occurs in scattered locations in gallery forest of central Amazonian Brazil. When specimens of this remarkable parrot were first collected, they were thought to be juveniles of the allied vulturine parrot *(Pionopsitta vulturina)*, which has a black, bare head and lives in the same region and habitat as the bald parrot. They were later demonstrated to be mature birds of a new species—scientifically described in 2002. It is possible that the bare head of these parrots allows them to feed on sticky fruit without soiling their feathers—like bare-headed vultures are protected from gore while feeding on carcasses. Both bald and vulturine parrots are short-tailed with predominantly green plumage; the bare head is only achieved in adulthood.

Agapornis personatus

Yellow-collared lovebird

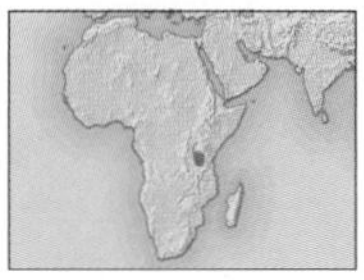

Length 5¾ in (14.5 cm)	
Weight 1¾ oz (50 g)	
Plumage Sexes alike	
Migration Nonmigrant	
Status Least concern	

Location E. Africa (Tanzania)

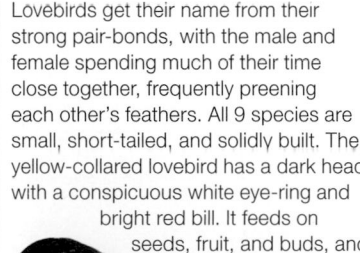

Lovebirds get their name from their strong pair-bonds, with the male and female spending much of their time close together, frequently preening each other's feathers. All 9 species are small, short-tailed, and solidly built. The yellow-collared lovebird has a dark head, with a conspicuous white eye-ring and bright red bill. It feeds on seeds, fruit, and buds, and, unusually for a parrot, builds a nest instead of laying its eggs in an unlined hole.

Amazona aestiva

Blue-fronted parrot

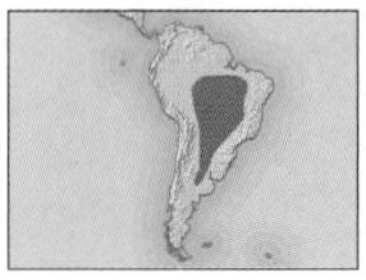

Length 14½ in (37 cm)	
Weight 14 oz (400 g)	
Plumage Sexes alike	
Migration Nonmigrant	
Status Least concern	

Location C. South America

A distinctive blue forehead on a yellow face differentiates this large and stocky parrot from other members of the genus *Amazona*, of which there are about 30 species. Although conspicuous and noisy in flight, the blue-fronted parrot is unobtrusive and quiet while feeding or resting in treetops, where falling pieces of fruit may be the only indication of its presence. Birds gather in flocks at regular nighttime roosts, but mated pairs always stay close together.

Psittacula krameri

Rose-ringed parakeet

Length 16 in (40 cm)
Weight 4 oz (125 g)
Plumage Sexes alike
Migration Nonmigrant
Status Least concern
Location W. to E. Africa, S. and S.E. Asia

This slim-bodied green bird has a rose-pink collar around its hind neck. It has the widest natural distribution of any parrot, stretching from West Africa to Southeast Asia. As a feral bird, it is also found in parts of Europe and North America. The central tail feathers are long and narrow, and the backward-sweeping wings produce a characteristically streamlined flight silhouette.

Myiopsitta monachus

Monk parakeet

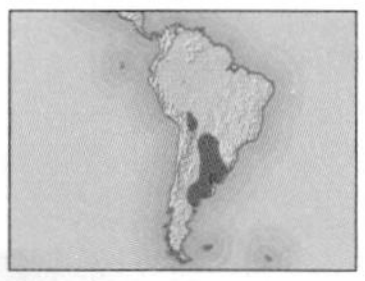

Length 11½ in (29 cm)
Weight 4 oz (125 g)
Plumage Sexes alike
Migration Nonmigrant
Status Least concern
Location C. and S. South America

The nesting habits of this green parakeet are unique among parrots. With several other pairs, it roosts in a communal nest, which forms the center for the birds' daily activities. The colony expands as nests are built alongside or on top of existing ones. This parakeet has a long, green tail, an olive-green abdomen, and a gray-white face and throat. In some parts of South America, it is a serious agricultural pest.

Cyanopsitta spixii

Spix's macaw

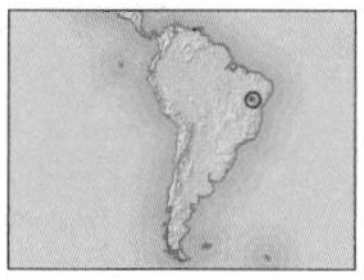

Length 22 in (55 cm)
Weight Not recorded
Plumage Sexes alike
Migration Nonmigrant
Status Critically endangered
Location E. South America

The smallest of all blue macaws, Spix's macaw is a creature of habit, regularly sitting on favored perches atop tall trees and daily following the same flight paths. This parrot feeds on seeds, nuts, and fruit while clambering among branches.

BLUE PARROT
Spix's macaw has a long, pointed tail, and a silvery blue head, with dark gray skin surrounding the eyes. The back and wings are darker blue, and the bill is black.

CONSERVATION

Long one of the world's rarest birds, Spix's macaw vanished from the wild in 2000, when the last free-flying bird disappeared. Two main factors contributed to the species' decline: destruction of river-edge woodland—where it nests in a single species of tree—and the relentless wild-bird trade, which continued until only a handful of birds were left. Fortunately, a population had been built up in captivity, living in seminatural conditions in research stations and zoos. Raised since the 1980s, these now number nearly 100 birds. Captive breeding, together with habitat restoration and controls on illegal collecting, may one day be sufficient to reestablish the species in the wild.

Aratinga jandaya

Jandaya parakeet

Length 12 in (30 cm)
Weight 4 oz (125 g)
Plumage Sexes alike
Migration Nonmigrant
Status Least concern
Location E. South America

One of 30 species of conures, all found in eastern South America, this is a small parrot with a narrow, pointed tail. Its head and neck are yellow, gradually becoming orange-red on the breast and abdomen, and its back and wings are largely green. Usually encountered in noisy flocks of up to 15, this very active bird spends much of its time clambering among branches of trees or shrubs to find fruit. When disturbed, it swiftly takes flight while screeching loudly.

Anodorhynchus hyacinthinus

Hyacinth macaw

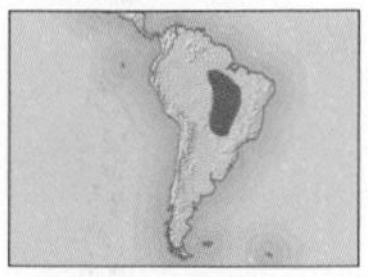

Length 3¼ ft (1 m)
Weight 3¼ lb (1.5 kg)
Plumage Sexes alike
Migration Nonmigrant
Status Endangered
Location C. South America

The largest and probably the most spectacular of all parrots, the hyacinth macaw is distinguished by its rich cobalt-blue plumage, contrasting bright yellow chin patch and ring around the eye, and long, pointed tail. The massive bill is used for crushing hard palm nuts. It feeds mainly in the trees, but comes to the ground to pick up fallen fruit and nuts. Its high value as a cage bird has caused a major decline in its numbers.

Ara chloroptera

Red-and-green macaw

Length 35 in (90 cm)
Weight 2¼ lb (1 kg)
Plumage Sexes alike
Migration Nonmigrant
Status Least concern
Location N. and C. South America

The red-and-green macaw attracts attention by its spectacular, vivid coloration and strident call. A large bird, like some other macaws it has a partly red plumage and a light blue back and rump. The long tail is tipped with blue. Its wings are blue with dark green upper-wing coverts, which give this species its name. The juveniles resemble adults, but have shorter tails. This macaw prefers humid, lowland forest, but in the southern part of its range, where the more common scarlet macaw (*Ara macao*) is absent, it frequents open habitats, including deciduous forest, and is often seen perching on the topmost projecting branches of tall trees. Usually encountered in pairs or small groups, possibly family groups, it sometimes associates with other, more common, macaws, especially at earth banks, where large numbers gather to consume exposed mineral sands. In flight, red-and-green macaws call loudly to each other. This bird feeds on seeds, fruit, and nuts that are often larger or harder than those eaten by other macaws. Adult birds are killed for food or for feathers, while the chicks are taken from the nest for the live-bird trade.

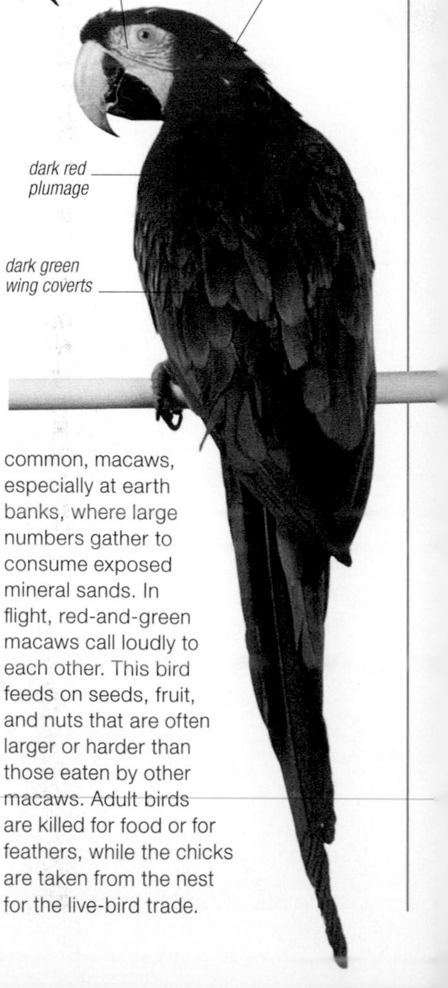

BIRDS

Cuckoos and turacos

PHYLUM	Chordata
CLASS	Aves
ORDER	Cuculiformes
FAMILIES	3
SPECIES	170

Although they have loud voices, cuckoos and turacos are secretive birds. Cuckoos, which occur throughout the world, are generally gray or brown, although some have striking patches or streaks. Turacos, which are exclusive to Africa, are sometimes brilliantly colored, with unique red and green pigments. Both cuckoos and turacos have short wings, a long tail, and 2 pairs of toes on each foot, one pointing forward and the other back, enabling them to climb as well as perch in trees. In addition to cuckoos and turacos, this group includes the hoatzin, placed in a separate family.

BROOD PARASITES
Some cuckoos are brood parasites, laying their eggs in other birds' nests. The eggs often closely resemble those of the host. The cuckoo chick (here, a common cuckoo) may eject the host's eggs or young, and take food brought by its new "parent" (here, a reed warbler).

Clamator glandarius

Great spotted cuckoo

Location S. Europe, W. Asia, Africa

Length 14–15½ in (35–39 cm)
Weight 4 oz (125 g)
Plumage Sexes alike
Migration Migrant
Status Least concern

The largest cuckoo found in Europe, this handsome bird lays its eggs in the nests of crows, magpies, and starlings. Unlike common cuckoos, the young do not evict their nest-mates. However, since they grow faster than the other chicks, they get a larger share of the food brought by the foster parents, with the result that within the first 8 days of their life they reach half their fledgling weight.

Geococcyx californianus

Greater roadrunner

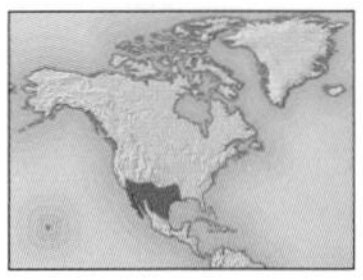

Location S. North America

Length 22 in (56 cm)
Weight 12 oz (325 g)
Plumage Sexes alike
Migration Nonmigrant
Status Least concern

Although it can fly, this long-legged member of the cuckoo family spends most of its time on the ground. Over short distances it can reach speeds of more than 18 mph (30 kph). It walks and runs through the desert, trying to flush out prey—lizards, snakes, birds, and small mammals—and kills whatever it catches with a blow from its strong, pointed bill. This roadrunner also runs to avoid predators, swinging its long tail from side to side like a rudder to make turns while speeding.

Tauraco erythrolophus

Red-crested turaco

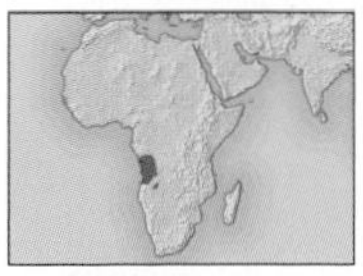

Location S.W. Africa

Length 16–17 in (40–43 cm)
Weight 7–12 oz (200–325 g)
Plumage Sexes alike
Migration Nonmigrant
Status Least concern

Like other turacos, this species derives its bright colors from a copper-based pigment, unique to the turaco family. A tree-dwelling fruit-eater, it spends most of its time hopping from branch to branch, and has a heavy, labored flight. It usually lives in family groups, foraging in the rain forest canopy and vigorously defending both feeding and nesting sites from other turacos as well as other fruit-eating birds.

Cuculus canorus

Common cuckoo

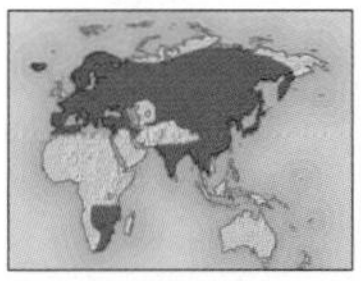

Location Europe, Asia, N.W. and southern Africa

Length 12½–13 in (32–33 cm)
Weight 4 oz (125 g)
Plumage Sexes alike
Migration Migrant
Status Least concern

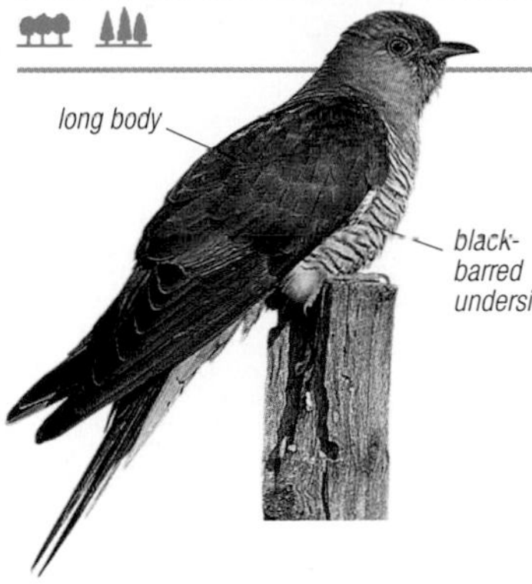

Usually gray with a black-barred, white underside, the common cuckoo breeds mainly in forests throughout Europe and Asia, but travels long distances to spend the winter in Africa and southern Asia. This familiar cuckoo is a classic example of a brood parasite, laying its eggs in the nests of a number of other species (see above).

Crotophaga ani

Smooth-billed ani

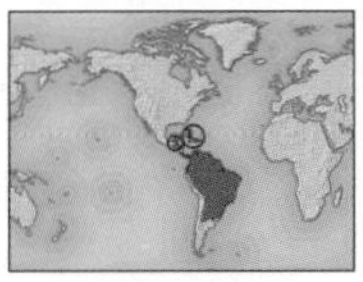

Location S.E. North America, Central America, South America, Caribbean

Length 14 in (35 cm)
Weight 4 oz (125 g)
Plumage Sexes alike
Migration Nonmigrant
Status Least concern

This odd-looking bird has a deep, arched bill and feeds mainly by following cattle, catching the insects they disturb as they move. It lives in groups of up to several pairs, which share the work of incubating the eggs and raising the chicks. These often stay in the nest once they are fledged, helping to raise the next brood.

Opisthocomus hoazin

Hoatzin

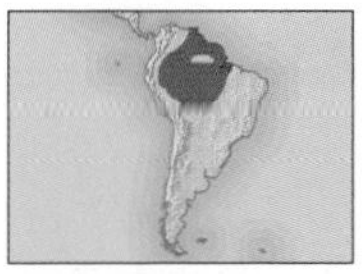

Location N. South America

Length 24–28 in (62–70 cm)
Weight 25–32 oz (700–900 g)
Plumage Sexes alike
Migration Nonmigrant
Status Least concern

This primitive-looking, tree-dwelling bird feeds almost entirely on leaves, a feat few other species can accomplish, because its large stomach enables it to process plant material in a similar way to grazing mammals. The young leave the nest before they can fly and use their tiny wing claws to climb through vegetation; they dive into water if threatened. As many as 8 birds may live together, defending their common territory and helping to rear the young.

Owls

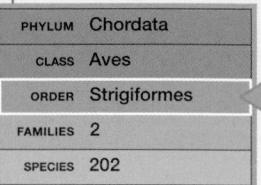

PHYLUM	Chordata
CLASS	Aves
ORDER	Strigiformes
FAMILIES	2
SPECIES	202

Often heard yet seldom seen, owls are hunters that operate mainly at night. They resemble the day-flying predators, hawks and falcons, in having sharp talons and hooked bills for catching and subduing prey, but in addition owls have several adaptations that help them hunt in the dark. Their eyes are very large, to gather all available light, and are forward-facing, to help them judge distance. They have exceptionally acute hearing, and their soft plumage enables them to fly silently. The 2 families of owls—typical owls and barn owls—are found worldwide in most habitats, from dense forest to tundra.

Reproduction

Owls do not build their own nest. Instead, they rely on the previous years' efforts of other birds or simply select a suitable cavity in the ground, a tree, a rocky crevice, or a building. The almost spherical eggs are laid in the nest or straight onto the surface. Most owls lay 2–7 eggs, which usually hatch at 2-day intervals, leading to large discrepancies in the age of the chicks in a brood. If food is scarce, the older chicks take the larger part of the food offered, while the younger chicks may starve.

YOUNG
The youngest owlet in a brood (here, snowy owls) can be 2 weeks younger than the oldest.

NOCTURNAL HUNTER
Under cover of darkness, a little owl closes silently upon its prey. Equipped with sharp talons, acute eyesight, and exceptional hearing, an owl is a formidable nighttime predator.

Anatomy

Owls are highly distinctive, with an upright posture, a large, rounded head, and a short tail. The outer toe is reversible, allowing it to point forward or backward, improving the ability to perch or grab prey. Owls have excellent eyesight, which works as well in daylight as it does at night. They also have exceptional hearing, easily picking up the faint rustle made by a small mammal, even under snow. Some species can hunt in total darkness, their asymmetrical ear openings giving them a three-dimensional perception of sound. All owls have soft, dense plumage, with soft fringes on their flight feathers that muffle the sound of air turbulence.

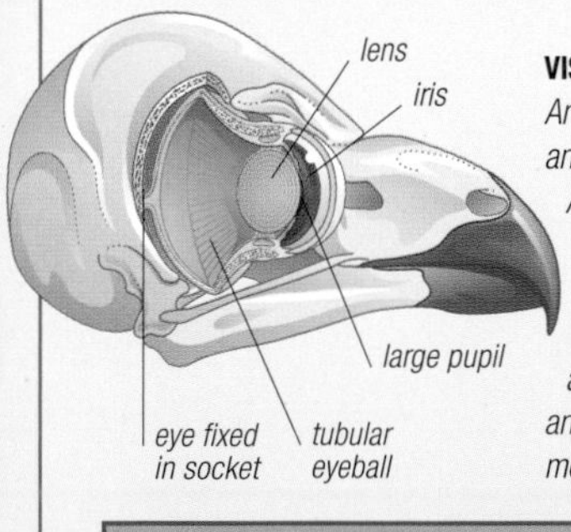

VISION
An owl's eyeballs are tubular and cannot be swiveled. As a result, if the bird needs to look to the side, it must move its whole head. To compensate, an owl can rotate its head and neck about an arc of more than 270 degrees.

Feeding

Owls take a wide variety of living prey, the size of which depends on the owl species. Most feed on insects, birds, or small mammals, and large owls quite commonly catch and eat smaller owls. Owls that live in woods and forests tend to drop from a stationary position onto their prey, but those that occur in open country must hunt on the wing, in slow, low-level, quartering flight.

FISHING
Several species of owls, including Pel's fishing owl (right), feed on fish. They swoop down to pluck fish from the water surface, striking them with their talons and then taking hold with their bristly foot pads.

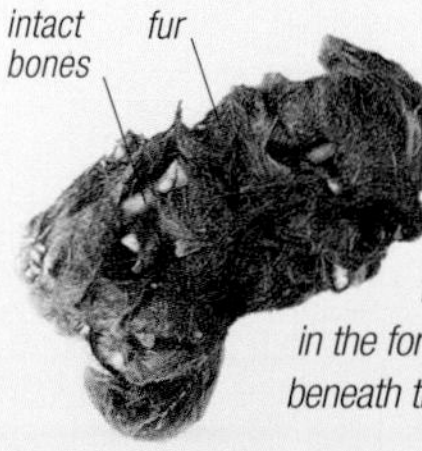

PELLETS
Owls generally swallow their food whole, taking in fur, feathers, bones, and insect chitin. Later, they regurgitate the indigestible parts (see below) in the form of compact pellets, which collect beneath their nest or roosting site.

REGURGITATION

SWALLOWING PREY
A barn owl swallows its prey whole, including much indigestible matter, such as fur and bones.

PREPARING TO CAST
About 6 or 7 hours after eating, the owl is about to regurgitate (or cast) the indigestible parts of its meal.

ROUNDED PELLETS
To ease its passage, the undigested food is rolled into a pellet, with hard parts wrapped in fur and feathers.

CAUGHT IN THE THROAT
Despite being rounded, the pellet causes the owl to strain as it moves up the esophagus.

RELEASING THE PELLET
Finally, a shiny black pellet is cast from the mouth. The owl may cast a second one within 24 hours.

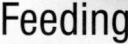

Tyto alba

Barn owl

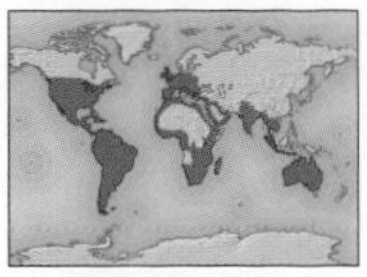

Length 11½–17½ in (29–44 cm)	
Weight 11–23 oz (300–650 g)	
Plumage Sexes alike	
Migration Nonmigrant	
Status Least concern	

Location North, Central, and South America, Europe, Asia, Africa, Australia

ON TARGET
The barn owl is an expert hunter, its excellent vision and hearing enabling it to pinpoint prey in total darkness. It flies low, slowly and silently, then swoops swiftly to the ground. At the last moment, it swings its legs forward, spreading its sharp-clawed toes to grasp and kill its prey.

Found on all continents except Antarctica, the barn owl is the most widespread of all species of owls, and one of the most widely distributed of all land birds. It has a pale, heart-shaped face, long legs covered in white feathers, and a very short tail. The female lays eggs in a hollow tree or an abandoned building. She feeds the young with food brought by the male; she also broods them for up to 3 weeks from hatching, until they have acquired the down they need to keep themselves warm. Changing agricultural practices have reduced the barn owl's food supply; in some areas, this species is now rare.

Otus thilohoffmanni

Serendib scops owl

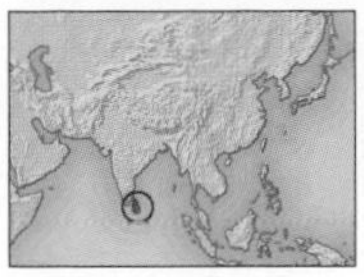

Length 6½ in (17 cm)	
Weight Not known	
Plumage Sexes alike	
Migration Nonmigrant	
Status Endangered	

Location Sri Lanka

First detected in 1995 by its unfamiliar frog-like call in the rainforests of Sri Lanka, this owl was not seen until 2001—when it was photographed and its identity as a new species was confirmed. It is a very small, short-tailed, brownish red owl and has an indistinct facial disk; its eyes are orange in the male, yellow in the female. This species is known only from the lowland wet forests of southwestern Sri Lanka, favoring disturbed areas with tall, dense secondary vegetation and roosting close to the ground. There are probably around 250–1,000 individuals, but the secretive habits of this owl make population estimates difficult. However, this small population is likely to be in decline due to the habitat degradation.

Otus scops

Eurasian scops owl

Length 6½–8 in (16–20 cm)	
Weight 2⅛–4 oz (60–125 g)	
Plumage Sexes alike	
Migration Migrant	
Status Least concern	

Location Europe to C. Asia, Africa

Small and superbly camouflaged, this owl is heard more often than it is seen, its call being a low whistle, repeated every few seconds. The fine, black flecks in its gray or reddish brown plumage make it almost invisible against the bark of a tree. When alarmed, it will stretch itself, and even sway, to imitate a branch and remain hidden. This owl feeds mainly on insects, swooping down on them from a perch, but also eats spiders, earthworms, reptiles, bats, and small birds.

Otus lempiji

Sunda scops owl

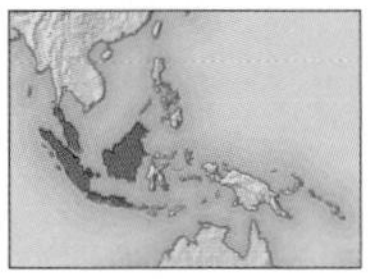

Length 8 in (20 cm)	
Weight 3⅝–4 oz (100–125 g)	
Plumage Sexes alike	
Migration Nonmigrant	
Status Least concern	

Location S.E. Asia

The Sunda scops owl is a small, eared owl, with brown eyes. It occurs in both a brownish gray and a reddish morph. Inhabiting forests, forest edges, plantations, and areas with scattered trees, such as parks and villages, it spends much of the day sitting camouflaged in hiding places, coming out at dusk to hunt mainly for insects. Males and females often call together, giving out a short, mellow hooting call.

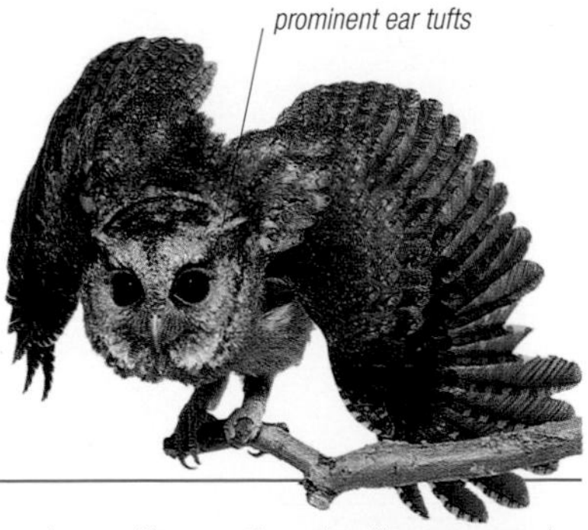

Scotopelia peli

Pel's fishing owl

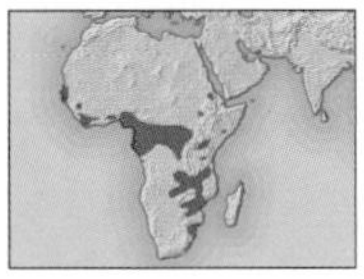

Length 22–25 in (55–63 cm)	
Weight 4½–5½ lb (2–2.5 kg)	
Plumage Sexes alike	
Migration Nonmigrant	
Status Least concern	

Location Africa

This owl is one of 3 African species that specialize in feeding on fish, frogs, and other freshwater animals. Like its relatives, Pel's fishing owl has long, bare legs and curved talons that help it grasp its slippery prey. The plumage of Pel's fishing owl is generally light chestnut, with dark spots and bars, which help camouflage it while it roosts. This large owl lives on the wooded edges of lakes, rivers, swamps, and marshes. It hunts after dark, launching itself from a perch and swooping low over the water. Having seized its prey, it returns to a perch to feed. It makes its nest in a tree hole, usually near water. Females lay one or 2 eggs, but normally only one young owl is reared. Both parents feed and rear the young, which may remain in the nest area for up to 8 months after fledging. A separate group of fishing owls—belonging to the genus *Keputa*—is found in Asia.

Pulsatrix perspicillata

Spectacled owl

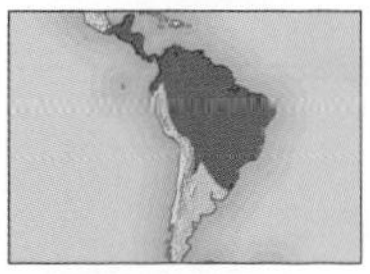

Length 17–20½ in (43–52 cm)	
Weight 21–36 oz (600–1,000 g)	
Plumage Sexes alike	
Migration Nonmigrant	
Status Least concern	

Location S. Mexico to C. South America

This owl, common in the American tropics, gets its name from the ring of white feathers, or "spectacles," around its eyes. Most often found in dense rain forest, it also lives in woodland edges and coffee plantations. Instead of patrolling on the wing like many owls, it hunts from a perch, snatching prey from the ground or foliage. It usually preys on small forest mammals and insects, but also hunts near water, taking crawfish and crabs.

BIRDS

Bubo virginianus

Great horned owl

Location North, Central, and South America

Length 20–23½ in (50–60 cm)

Weight 1½–5½ lb (675–2,500 g)

Plumage Sexes similar

Migration Nonmigrant

Status Least concern

The great horned owl, with its distinctive ear tufts, or "horns," is the largest of the American owls and is found throughout the Americas. It occurs in a broad range of habitats, from forest to desert, and is known to nest at high altitudes, usually in old nests of other large birds, but also in tree cavities and on cliff edges. It is generally a sedentary bird and, especially during the breeding season, very territorial in its nesting and hunting ranges. Usually active between dusk and dawn, its night vision and hearing are extremely acute, making it an effective hunter. Its main prey are small mammals, but insects, reptiles, amphibians, and birds, including other owl species, are also taken. Great horned owls are very vocal, especially during courtship; their loud hoot represents the classic owl call.

POWERFUL BUILD
This very large owl, with its sharp bill, large wings, and powerful talons, is built for hunting. Its piercing yellow eyes and prominent ear tufts add to its intimidating appearance.

SILENT HUNTER
As well as being the largest American owl, the great horned owl is probably also the fiercest. It mainly still-hunts from a series of favored vantage points throughout its territory. It is very swift, and having spotted a target, it glides down, in total silence, to capture its prey in its immensely powerful talons.

PARENTS AND JUVENILES

Great horned owls are very attentive parents, with both males and females tending and feeding their young for at least 6 weeks after fledging. They are also very defensive birds and have been known to drive away intruding humans from their nests.

ON THE NEST
The great horned owl lays between one and 5 eggs, laying a greater number when food is abundant.

JUVENILE OWL
At about 2 months old, the horned owl is almost fully feathered and capable of short flights. At this stage, it can puff itself up and turn its wings forward to look even larger when defending itself.

Nyctea scandiaca

Snowy owl

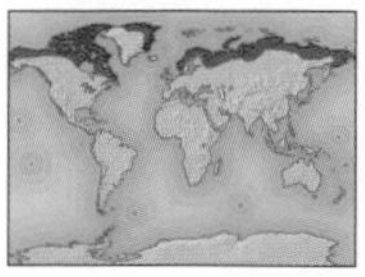

Location Circumpolar around Arctic

Length 22–28 in (55–70 cm)
Weight 2¼–5½ lb (1–2.5 kg)
Plumage Sexes differ
Migration Partial migrant
Status Least concern

With its all-white plumage, the male snowy owl is one of the most distinctive of owls. Long, dense feathers extend right to the toes, and even the bill is largely covered, giving the bird superb insulation against the cold. This owl is most active at dusk and dawn, but becomes diurnal in summer when daylight is continuous. It spends much of the time on the ground or perching on low rocks, and uses its extraordinary eyesight and hearing to locate distant or snow-covered prey, then ambushes it silently. The snowy owl feeds on lemmings, rabbits, hares, and waterfowl. In fact, its breeding cycle is directly related to the abundance of lemmings: the population of the latter tends to rise and fall on a 3- to 4-year cycle, and the snowy owl follows suit.

MOTTLED FEMALE
The female snowy owl is mottled with black—a color scheme that hides her in rocky outcrops when she nests after most of the snow has melted. She is substantially larger than the male.

ON THE GROUND

The snowy owl nests on the tundra, forming only a slight hollow in the ground in which the eggs are laid. The female tends the young, feeding them with food brought by the male.

Strix aluco

Tawny owl

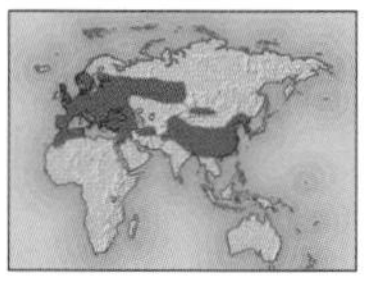

Location Europe, Asia, N.W. Africa

Length 14½–15½ in (37–39 cm)
Weight 6–20 oz (1450–550 g)
Plumage Sexes alike
Migration Nonmigrant
Status Least concern

The tawny owl is found in a broad range of habitats, wherever there is sufficient tree cover to provide a daytime roost. Its usually chestnut-brown plumage is heavily streaked and mottled, providing excellent camouflage among branches and leaves. It has a wide variety of calls, among which the best known is the "twit twoo" call produced during the breeding season. This owl hunts from a perch, and can locate its prey—mainly small mammals, birds, reptiles, and insects—by sound alone.

Glaucidum perlatum

Pearl-spotted owlet

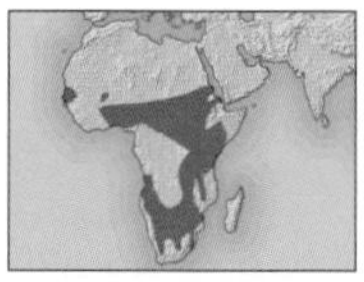

Location Africa (south of Sahara)

Length 6½–8 in (17–20 cm)
Weight 1¾–5 oz (50–150 g)
Plumage Sexes alike
Migration Nonmigrant
Status Least concern

This compact bird is Africa's most diurnal owl, hunting at almost any time of the day or night. Its powerful feet enable it to catch prey even larger than itself. On the back of its head are 2 black patches ringed with white. These "false eyes" serve either to deter predators that might attack from behind, or confuse prey so that they do not know which way to flee.

Ninox novaeseelandiae

Morepork

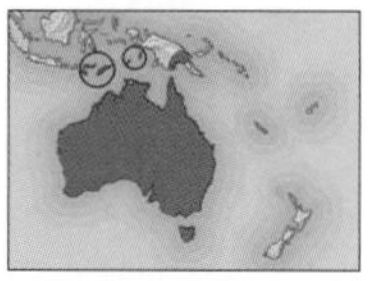

Location Australia (including Tasmania), S. New Guinea, S.E. Asia

Length 2–14 in (130–35 cm)
Weight 5–6 oz (150–175 g)
Plumage Sexes alike
Migration Nonmigrant
Status Least concern

There are several races of this small, stocky owl found in Australia and Asia. One form, found in Queensland, is dark brown, while the form found in Tasmania and New Zealand, and sometimes considered a separate species, is lighter, with contrasting spots. All have a high-pitched, 2-syllable "boobook" call. Wherever it lives, the morepork (also called the boobook owl) roosts in trees during the day, emerging at dusk to feed. It specializes in hunting insects and birds in midair.

Asio flammeus

Short-eared owl

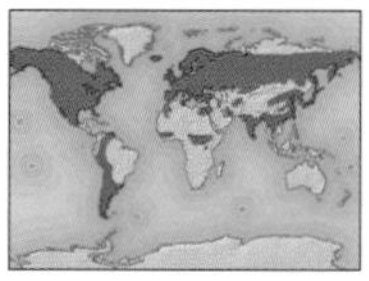

Location North America, W. and S. South America, Europe, Asia, Africa

Length 14½ in (37cm)
Weight 7–18 oz (200–500 g)
Plumage Sexes differ
Migration Partial migrant
Status Least concern

With its long wings and low, flapping flight, this heavily mottled owl can easily be mistaken for a hawk. Usually a solitary bird, it normally roosts on the ground, but in winter, particularly when it snows, it will perch in groups in trees. Groups can also be seen when food is scarce, as large numbers move to richer areas, or when there is an abundance of food in one particular place. This owl finds its prey—mostly small mammals such as mice and voles, and at times birds—mostly by flying, and then pouncing from midair. It lays its eggs in heather, grass, or crops, digging a hollow to hold them safely; unusually for an owl, the female may build a nest from sticks that are lying nearby. It is a fairly quiet owl, with a low, gentle hooting call.

IN BROAD DAYLIGHT

In many parts of its vast range, the short-eared owl—one of the most diurnal of owls—is often seen hunting by day, flying just a few yards above the ground. Its flight is butterflylike, its large wings allowing it to fly slowly without stalling.

ROUND FACED
The short-eared owl has a large, rounded head with a marked facial disk, bright yellow eyes, and 2 short, feathery tufts that look like ears.

Athene cunicularia

Burrowing owl

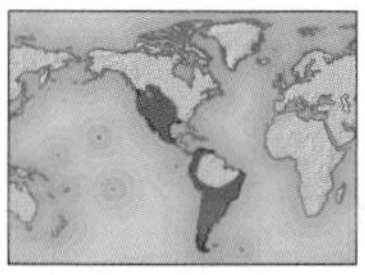

Location North, Central, and South America, Caribbean

Length 7½–10 in (19–25 cm)
Weight 4–9 oz (125–250 g)
Plumage Sexes differ
Migration Partial migrant
Status Least concern

Most owls roost and nest in trees, but the burrowing owl makes its home underground—an adaptation that allows it to live in all kinds of open, treeless areas, from South American pampas to airports and golf courses. When on guard outside the burrow, it adopts a very upright stance and gives a harsh, rattling call, thought to imitate a rattlesnake. It spends much of its time hunting on the ground.

Nightjars and frogmouths

PHYLUM	Chordata
CLASS	Aves
ORDER	Caprimulgiformes
FAMILIES	5
SPECIES	125

These long-winged birds, together with their relatives, the nighthawks, potoos, owlet-nightjars, and oilbirds, are adapted for a life spent mainly in the air; indeed, most cannot walk or hop. They hunt at dusk, dawn, or at night—usually capturing insects on the wing—and roost motionless in trees or on the ground during the day. Many species are known for their loud, distinctive call. Nightjars occur almost throughout the world, in forests and open habitats; their relatives are restricted to the Americas, Asia, or Australasia.

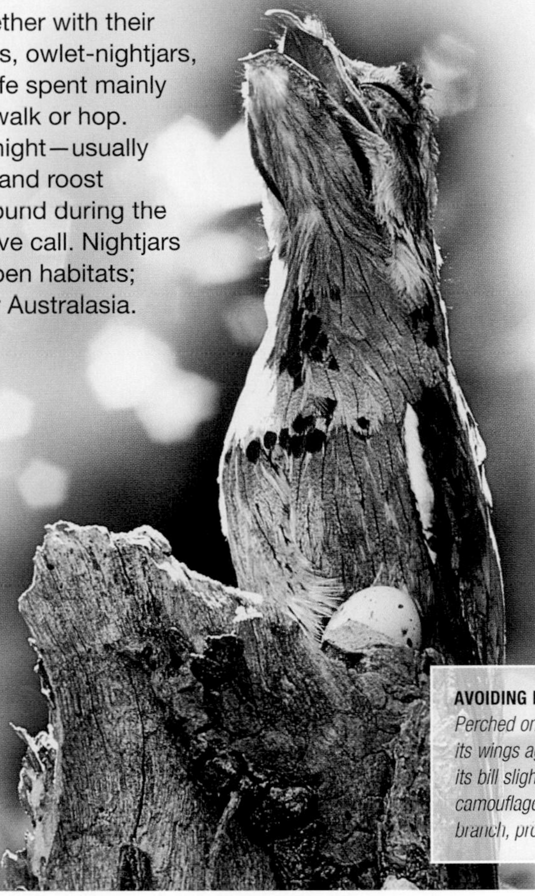

Anatomy

Nightjars and their relatives are round-bodied birds, with a large head and short neck. They have a remarkably large mouth, with an extremely wide gape for trapping insects. Most species have long tails and elongated wings that are ideal for flying swiftly with rapid changes of direction in pursuit of food. Except for owlet-nightjars, these birds have short legs and weak, tiny feet that are unsuitable for walking. Birds in this group usually have brown or gray, cryptically patterned plumage.

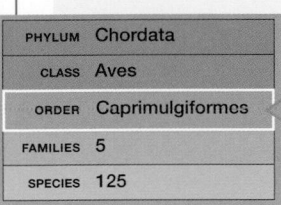

WIDE GAPE
The bill of nightjars and their relatives (here, a red-necked nightjar) is almost as broad as their head and can be opened very wide. In nightjars, this makes an effective trap for flying insects, which are caught in midair, one at a time.

Feeding

Most of the birds in this group feed primarily on insects, which they catch in midair; alternatively, they may swoop down to take them from the ground. Many species fly continually, while others (such as the potoo) make brief but regular sorties from a perch. Frogmouths feed on other birds, mammals, and amphibians, as well as insects. Oilbirds eat only fruit.

BRISTLES
Nightjars (here, the European nightjar) and some of their relatives are equipped with a set of bristles around the edge of their bills. These bristles may be sensitive to touch, and in some species help filter insects into the mouth.

AVOIDING DETECTION
Perched on an upright branch, the common potoo flattens its wings against its body and points its head upward with its bill slightly open. This stance, combined with the camouflage coloring, makes the bird resemble a broken branch, providing protection against predators.

Steatornis caripensis

Oilbird

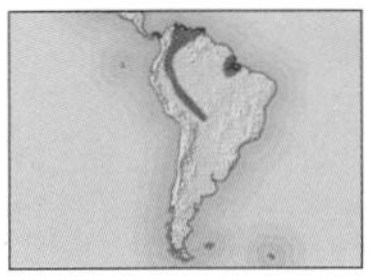

Location N. and C. South America

Length	16–19½ in (40–49 cm)
Weight	13–17 oz (350–475 g)
Plumage	Sexes alike
Migration	Nonmigrant
Status	Least concern

The oilbird is a unique species—the only nocturnal fruit-eating bird in the world. During the day, it remains deep inside caves, sometimes ⅔ mile (1 km) underground. It leaves the cave at night and may travel over 45 miles (75 km) in search of food. Oilbirds use sight and possibly smell to locate the fruit of trees such as laurels and palms, swooping down and plucking them from branches.

ECHOLOCATION

The oilbird nests and roosts in large colonies in caves. It uses echolocation to find its way in the dark; the signals it produces are relatively low-pitched and sound like "clicks" to the human ear. Birds also call to one another to stay in touch, filling the cave with their loud, harsh cries.

LARGE-MOUTHED BIRD
The reddish brown oilbird has white spots on the head, throat, and wings. Its large mouth enables it to carry plenty of food back to the young in the nest—a mound of mud, feces, rotten fruit, and seeds.

Podargus strigoides

Tawny frogmouth

Location S. New Guinea, Australia (including Tasmania)

Length	13½–21 in (34–53 cm)
Weight	6–24 oz (175–675 g)
Plumage	Sexes differ
Migration	Nonmigrant
Status	Least concern

Frogmouths look similar to nightjars, but instead of hunting nocturnal insects on the wing, they pounce on small animals and larger insects on the ground. The tawny frogmouth's large eyes help it to see well in the dark, and it flies through its territory at night, stopping at suitable foraging perches and watching for prey to move nearby. When necessary, it will beat larger animals to death before swallowing them. This species lives in pairs or family groups, and the birds call frequently to maintain contact over the large territory in which they feed and breed. The nest is a simple platform of sticks, placed in the fork of a tree.

DEFENSE POSTURE

If threatened, the tawny frogmouth stands erect, looking like a broken branch on a tree. This posture may sometimes be accompanied by a gaping display in which the bird opens its large, wide mouth.

CAMOUFLAGE
The tawny frogmouth's red- or brown-mottled gray plumage blends in with the color of branches of trees and bushes.

BIRDS

Nyctibius griseus

Common potoo

Length 13–15 in (33–38 cm)
Weight 5–7 oz (150–200 g)
Plumage Sexes alike
Migration Nonmigrant
Status Least concern

Location S. Central America, South America

One of about 5 species of potoos found in Central and South America, the common potoo is a solitary bird that spends the day in trees, perching upright, where it looks like a broken branch. It keeps its conspicuous yellow eyes shut to avoid detection by predators, but creases in the eyelids allow it to continue to watch for danger. This bird comes to life at night, darting from its perch to catch insects in the air. Birds form pairs for breeding, laying a single egg in a knothole or depression in a branch.

Phalaenoptilus nuttallii

Common poorwill

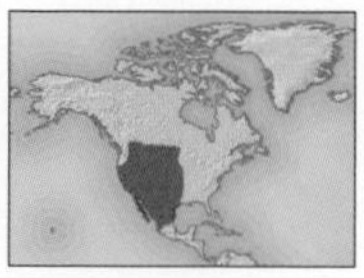

Length 7–8½ in (18–21 cm)
Weight 1¹⁄₁₆–2⅛ oz (30–60 g)
Plumage Sexes differ
Migration Partial migrant
Status Least concern

Location S. Canada, W. and C. USA, N. and C. Mexico

The smallest North American nightjar, the common poorwill gets its name from the male's repetitive 2-note courtship call. It rests on the ground during the day, the intricately patterned plumage providing effective camouflage in its often arid habitat. A night forager, the common poorwill swoops on flying insects from a perch. This bird conserves energy in winter by going into torpor—a hibernationlike state that is rarely found in birds.

Aegotheles cristatus

Australian owlet-nightjar

Length 8½–10 in (21–25 cm)
Weight 1¼–2⅜ oz (35–65 g)
Plumage Sexes differ
Migration Nonmigrant
Status Least concern

Location S. New Guinea, Australia (including Tasmania)

This tree-dwelling bird is a small, nocturnal insect-eater that looks very much like a miniature owl. Compared to other nightjars, it has well-developed feet and legs, and a long, slender tail. There are distinctive dark markings on its face and head. The barred plumage is gray in the male and brown in the female. The Australian owlet-nightjar is an acrobatic flier, with short, round wings permitting it to make brief foraging flights from its perch to capture insects, either on the ground or in the air. It is preyed upon by mammals and monitor lizards.

Caprimulgus europaeus

European nightjar

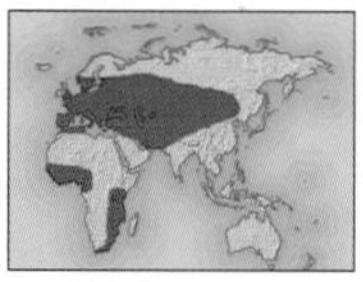

Length 10–11 in (25–28 cm)
Weight 1¾–3⅝ oz (50–100 g)
Plumage Sexes differ
Migration Migrant
Status Least concern

Location Europe, W. to E. Asia, N.W., W., and S.E. Africa

This bird is the only nightjar found in Northern Europe. It takes to the air after dark, catching flying insects on the wing. Males patrol their territories regularly, driving off intruders. During the breeding season, the male has a purring airborne song, which sounds like a piece of machinery. It also claps its wings, to display the white patches on the outer flight feathers. This bird does not build a nest; the eggs are laid directly on the ground, usually close to a tree or shrubs. The female rears the chicks initially, but after about 2 weeks she may leave them with the male and then lay a second clutch.

CAMOUFLAGE

Like other nightjars, this species spends the day resting on the ground or on low branches. Its dull, mottled plumage renders it almost invisible on the ground among vegetation.

BROWN AND MOTTLED
The European nightjar has brown plumage, mottled with black and gray, and long, pointed wings.

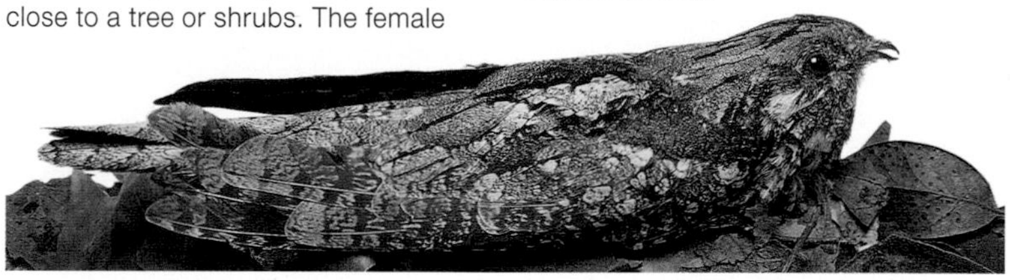

Chordeiles minor

Common nighthawk

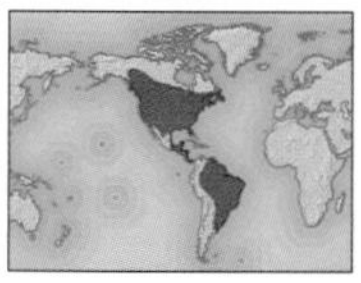

Length 9–10 in (22–25 cm)
Weight 1⅝–3⅝ oz (45–100 g)
Plumage Sexes differ
Migration Migrant
Status Least concern

Location North, Central, and South America

A familiar summer visitor through much of North America, this nighthawk is more active by day than most other species in the nightjar family. It is commonly seen on the wing at dawn and dusk, hawking for insects, and is recognized by its distinctive nasal call. The male has a highly acrobatic aerial display when courting—it swoops down low, making a hollow, booming sound with its wings, and almost colliding with its prospective partner. This bird nests and roosts in open places of all kinds. In some parts of the USA, it has taken to nesting on flat, gravel-covered roofs—a habitat that helps it to avoid some of its predators. The nest is often simply a shallow depression made in the ground, and the female alone cares for the eggs and chicks. When disturbed, the bird often flies up from the nest and settles nearby; occasionally, it will lunge at an intruder. After breeding, large flocks move south for the winter, traveling as far as Argentina.

Macrodipteryx longipennis

Standard-winged nightjar

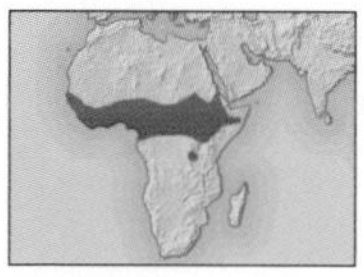

Length 8½–10 in (21–25 cm)
Weight 2⅛–3¼ oz (60–90 g)
Plumage Sexes differ
Migration Nonmigrant
Status Least concern

Location W., C., and E. Africa

Male standard-winged nightjars are instantly recognizable in flight, because their breeding plumage includes 2 elongated wing feathers that can be up to 3 in (78 cm) long. They perform their aerial courtship display at canopy level, often chased by females, but will also select a perch and show off their pennants there. More active during the day than some nightjars, the birds may forage in loose flocks or in pairs, when the female follows the male; they fly very high off the ground, catching insects on the wing.

BIRDS

Hummingbirds and swifts

PHYLUM	Chordata
CLASS	Aves
ORDER	Apodiformes
FAMILIES	3
SPECIES	447

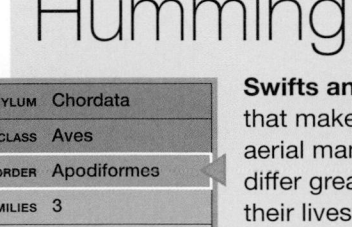

Swifts and hummingbirds share a unique wing structure that makes them acrobatic fliers capable of highly intricate aerial maneuvers. However, their appearance and lifestyles differ greatly. The soberly colored swifts rarely land, spending their lives in midair in search of flying invertebrates; they can sleep, and even mate on the wing. The multicolored hummingbirds hover around flowers to feed, and perch readily. Swifts occur worldwide, but hummingbirds are restricted to the Americas.

FEEDING ON NECTAR
When feeding, a hummingbird (here, a green violetear from Costa Rica) hovers in front of a flower, maneuvering its bill up the tube to draw up the nectar with its long tongue.

Anatomy

Both hummingbirds and swifts have a compact, muscular body, and relatively small feet. Although swifts are dull in color, hummingbirds are remarkable for their dazzling colors and patterns. Hummingbirds have a specialized bill designed for removing nectar from flowers. The bill length and shape are variable, and often match the shape of the flowers at which the birds feed. Swifts have a small bill with a wide gape for trapping tiny insects in flight. The hummingbird family contains many of the world's smallest birds.

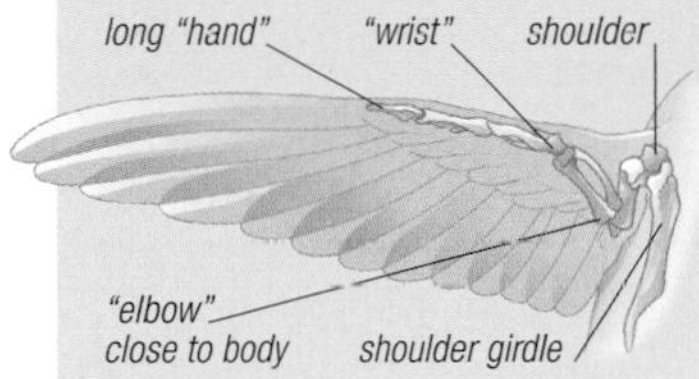

WINGS
In hummingbirds and swifts, the joint between the upper and lower arm, the "elbow" is very close to the body, giving the wings great flexibility and leverage. This feature enables hummingbirds to hover.

Flight

Hummingbirds beat their wings in a figure-eight pattern, which allows them great maneuverability; they are the only birds that can fly backward, and even upside-down. Smaller species may beat their wings 80 times a second. Swifts do not hover, but can vary the speed of their wingbeats to turn sharply.

SWIFT IN FLIGHT
Swifts resemble swallows, but are not closely related to them. They spend more time in the air, and are quicker, more erratic fliers. Swallows perch more readily than swifts.

Eutoxeres aquila

White-tipped sicklebill

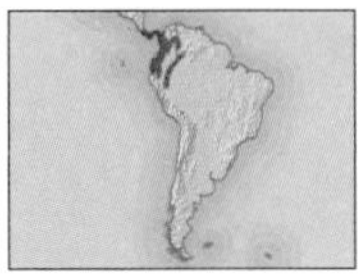

Length 4¾–5½ in (12–14 cm)
Weight ⅜–⁷⁄₁₆ oz (10–13 g)
Plumage Sexes alike
Migration Nonmigrant
Status Least concern

Location S. Central America to N.W. South America

This large hummingbird is green above with streaked underparts, and has white-tipped tail feathers and an unmistakable downcurved bill. The shape of the bill allows it to drink nectar from flowers that have curved throats, especially heliconias. Unlike most hummingbirds, it perches on the flower while feeding from it. It also gleans spiders from their webs and catches insects on the wing. The female lays 2 elliptical eggs in a cup-shaped nest, which is loosely woven from palm fibers and attached by cobwebs to the tip of a hanging leaf.

Eriocnemis isabellae

Gorgeted puffleg

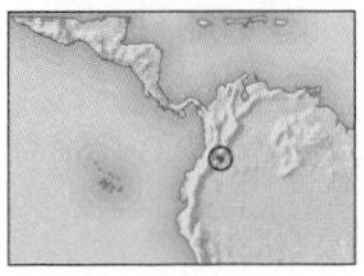

Length 4 in (10 cm)
Weight ¹⁄₁₀–¾ oz (3.9–4.5 g)
Plumage Sexes differ
Migration Nonmigrant
Status Critically endangered

Location N.W. South America (Andes)

Discovered in 2005, this hummingbird is restricted to cold, humid elfin forest (just 19–26 ft/6–8 m high) in a small region—scarcely (9 square miles/10 square km) in area—dominated by rocky outcrops high in the Colombian Andes. It is threatened with extinction by loss of habitat. It has a large, brilliantly colored throat patch with a violet-blue center and iridescent green edges. Known as a gorget, this throat patch along with the white thigh tufts gives rise to its common name.

Ocreatus underwoodii

Booted racket-tail

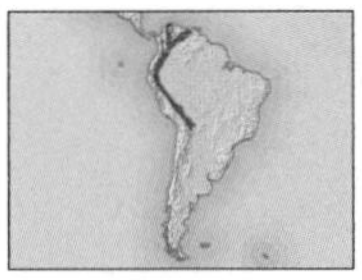

Length 4¼–6 in (11–15 cm)
Weight ⅛ oz (3 g)
Plumage Sexes differ
Migration Nonmigrant
Status Least concern

Location N.W. and W. South America (Andes)

Elongated tail feathers with bare shafts ending in blue-black "racquets" distinguish the male from the female of this tiny species. When courting, the male puts on a spectacular flight display, holding up the fluffy white or brown "puffs" or leg feathers, and flicking his tail feathers up and down to produce a whiplike sound. The male and female fly a regular route when foraging for food, calling to one another to stay in contact. The bird hovers in front of a flower, its rapidly beating wings making a distinct humming noise, and pierces the flower tube with its bill to reach the nectar. Like most hummingbirds, it often becomes torpid at night, its body temperature dropping almost to that of the air in order to save energy. Hummingbirds are so tiny that they would risk starving to death if they maintained their normal body heat throughout the night.

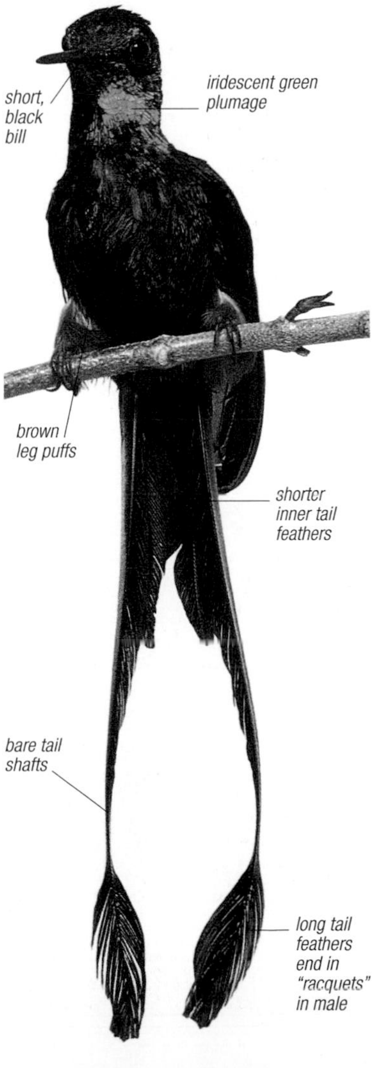

Topaza pella

Crimson topaz

Length 8½–9 in (21–23 cm)	
Weight 3/8–9/16 oz (10–15 g)	
Plumage Sexes differ	
Migration Nonmigrant	
Status Least concern	

Location N. South America

A relatively short, slightly curved bill allows this large hummingbird to feed from a variety of rain forest flowers. Rarely seen on the ground, it lives in the middle and upper levels of the forest, and will sometimes defend a whole treetop when in bloom. The male's plumage is striking—glittering crimson to purple—and it has very long, blackish tail feathers that cross one another toward the tips. The less flamboyant female is generally green in appearance, with a shorter tail.

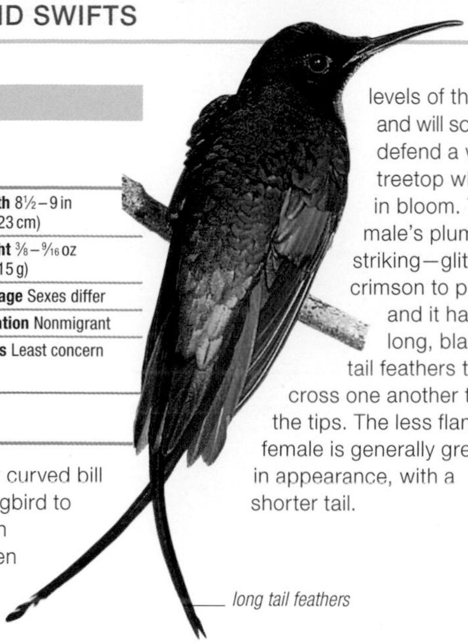

Patagona gigas

Giant hummingbird

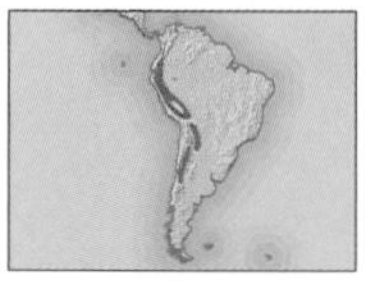

Length 8–8¾ in (20–22 cm)	
Weight 5/8–11/16 oz (18–20 g)	
Plumage Sexes alike	
Migration Nonmigrant	
Status Least concern	

Location W. South America (Andes)

This species is the largest of all hummingbirds—even so, it weighs only 11/16 oz (19 g). Its coloring is rather dull for a hummingbird: brownish overall with pale rump feathers. It often flies more like a swift, sometimes gliding, over dry Andean valleys and arid, steppelike mountain slopes. Commonly found around stands of prickly pears and puyas, it hovers while feeding on the flowers, occasionally clinging to large blossoms.

Ensifera ensifera

Sword-billed hummingbird

Length 6½–9 in (17–23 cm)	
Weight 7/16–9/16 oz (12–15 g)	
Plumage Sexes alike	
Migration Nonmigrant	
Status Least concern	

Location N.W. and W. South America (Andes)

Dark greenish in color, with a deeply forked, blackish tail, this bird has a slender, sword-shaped bill up to 11 cm (4¼ in) long. It is the only bird whose bill is longer than the rest of its body. When resting on a perch it holds the bill almost vertically to reduce the strain on its neck. Apart from flying insects, which it catches by hawking with a wide-open bill, it feeds almost exclusively from flowers with very long tubes, such as brugmansias and daturas. These blossoms usually hang downward and the bird hovers immediately beneath, pushing the bill up into the flower to extract its nectar. This species follows a feeding strategy known as traplining. While some hummingbirds defend clumps of flowers from other birds and even butterflies, trapliners make regular visits to a number of scattered flowers, remembering where each one is and following a set route between them. The intervals between visits mean that the plants have a chance to produce more nectar.

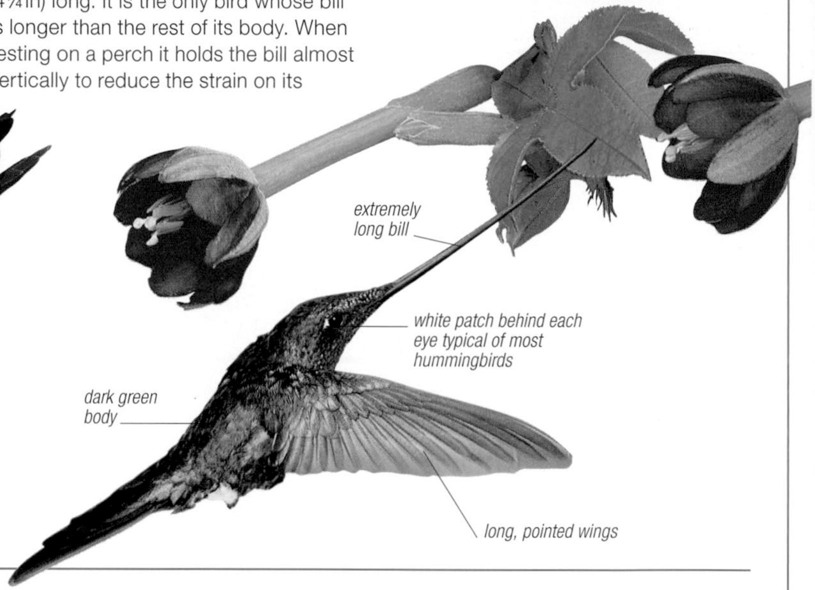

Archilochus colubris

Ruby-throated hummingbird

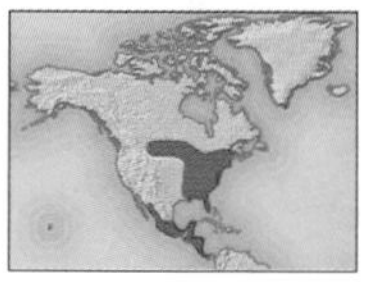

Length 3½ in (9 cm)	
Weight 1/8 oz (3 g)	
Plumage Sexes differ	
Migration Migrant	
Status Least concern	

Location S. Canada, C. and E. USA, Mexico to S. Central America

This is one of the few hummingbirds that migrate as far north as Canada, on an annual journey from Central America that may total over 1,900 miles (3,000 km). For some birds, migration involves a nonstop flight of about 530 miles 530 miles (850 km) across the Gulf of Mexico—a huge distance for such a tiny animal. This hummingbird feeds on nectar, and also uses the holes drilled in tree trunks by sapsucker woodpeckers to feed on tree sap and on the insects this attracts.

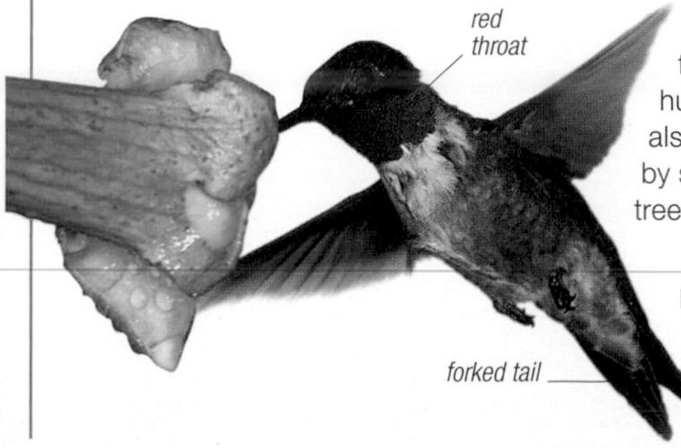

MALE AND FEMALE
A glittering red patch on the throat helps identify this small, iridescent green hummingbird; the female lacks the male's vivid coloring.

FEEDING THE YOUNG

Like other hummingbirds, this bird feeds its young largely on insects, because although nectar is a good source of energy, it does not contain enough protein for growth and development.

Mellisuga helenae

Bee hummingbird

Length 2–2¼ in (5–6 cm)	
Weight 1/16 oz (Up to 2 g)	
Plumage Sexes differ	
Migration Nonmigrant	
Status Near threatened	

Location Cuba, Isle of Youth

The smallest of all birds, the male bee hummingbird weighs less than 1/16 oz (1.6 g). It is green with a grayish white underside, and has an iridescent, fiery red head and bright collar. The female is slightly larger, weighing 1/16 oz (2 g), and has no iridescence on the head or neck. This bird feeds mainly on nectar,hovering with its body held horizontally and pushing its short, straight bill into each flower. The eggs are as little as ¼ in (6 mm) long—smaller than a single pea.

Oreotrochilus estella

Andean hillstar

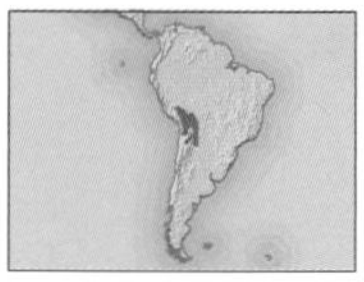

Length 5–6 in (13–15 cm)	
Weight 5/16–11/32 oz (8–9 g)	
Plumage Sexes differ	
Migration Nonmigrant	
Status Least concern	

Location W. South America (Andes)

Common in rocky slopes (with grass and scrub) at altitudes of up to 16,500 ft (5,000 m), this mountain bird does not hover in front of the flowers from which it feeds, but perches on them. Nocturnal torpor is very important for it because nights in the high Andes can be very cold, and the bird would not survive if it tried to maintain its body temperature at daytime level. The male's iridescent emerald-green collar with a black border distinguishes it from the female.

BIRDS

Cypsiurus parvus

African palm swift

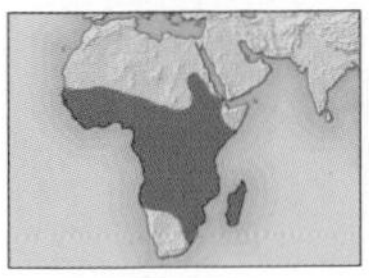

Length 5½–6½ in (14–16 cm)
Weight ⅜–⅝ oz (10–18 g)
Plumage Sexes alike
Migration Partial migrant
Status Least concern

Location Africa, Madagascar

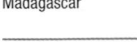

Fan palms are the preferred nesting site of this medium-sized, blackish brown swift, although it will use other palms (as well as artificial sites such as bridges). The nest (seen below) is a pad of feathers and plant fibers, glued with saliva to the vertical side of a drooping palm frond, and the eggs themselves are glued in place with saliva. Nestlings have unusually long claws, helping them cling on as the nest blows about in the wind.

Apus apus

Common swift

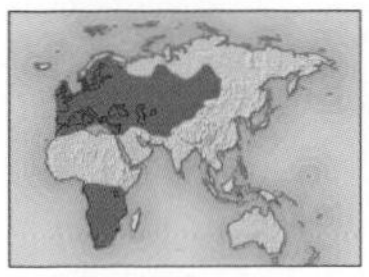

Length 6½–7 in (16–17 cm)
Weight 1¼–1¾ oz (35–50 g)
Plumage Sexes alike
Migration Migrant
Status Least concern

Location N. and southern Africa, Europe, W. to C. Asia

One of the most aerial of birds, the common swift feeds, mates, and even sleeps on the wing. Its nest is a cup-shaped platform formed from plants and feathers that are caught in midair and glued with saliva. It originally nested on rock crevices but now often uses manmade structures. In cold weather, parents often fly hundreds of miles away from the nest. The young survive by becoming torpid; they revive when it warms up, and the parents return with food. After leaving the nest, young swifts remain airborne for up to 3 years, before landing to breed themselves. They spend a relatively short time at their summer breeding grounds.

NARROW WINGS
The common swift has distinctive narrow wings and a forked tail. It has a short bill, small foot, and blackish brown plumage.

ON THE WING

Small, noisy groups of these gregarious birds are often seen feeding and moving over towns. Like other swifts, they are fast and agile, turning in flight by beating each wing at a different rate.

Hemiprocne longipennis

Grey-rumped treeswift

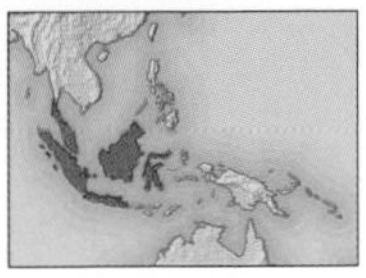

Length 9 in (23 cm)
Weight Not recorded
Plumage Sexes alike
Migration Nonmigrant
Status Least concern

Location S.E. Asia

There are 3 species of treeswifts that belong to a separate family from all other "true" swifts. All of them are found in southern Asia. Like a true swift, the grey-rumped treeswift catches insects on the wing, but it spends less of its time in the air. Between bouts of feeding, it perches on branches—something that true swifts are unable to. This species has long, scythelike wings and a deeply forked tail, but its most distinctive feature is a short, upright crest, attached not to its crown, but to the base of the upper part of its bill. The grey-rumped treeswift makes a minute nest glued to a branch, and lays a single egg. The parents take turns to incubate the egg, covering the nest completely while resting their weight on the branch beneath. Often found in small flocks, nesting pairs of this bird are known to defend their territories aggressively.

Cypseloides niger

Black swift

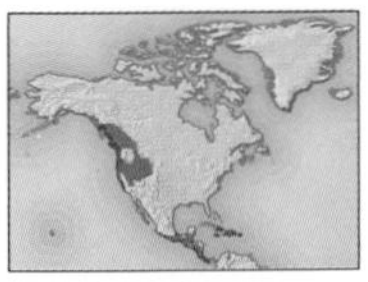

Length 7–8 in (18–20 cm)
Weight 1⅝ oz (45 g)
Plumage Sexes alike
Migration Migrant
Status Least concern

Location W. North America, Central America, Caribbean

A typical, sooty black swift, with long, pointed wings and a slightly forked tail, this fast flier often soars on outstretched wings. Like other swifts, it spends most of its time in the air, and flocks are often seen wheeling through the skies, foraging for insects. Only when breeding does it come down to the ground. Choosing a site next to a waterfall or running water, it builds a half-cup nest formed of living plants—mosses and liverworts—held together with mud.

Collocalia esculenta

Glossy swiftlet

Length Up to 4¾ in (11.5 cm)
Weight Not recorded
Plumage Sexes alike
Migration Partial migrant
Status Least concern

Location C., E., and S.E. Asia, E. Australia

Also known as the white-bellied swiftlet, this highly acrobatic flier is one of a small group of species that nest in caves, often in colonies many thousands strong. It makes its nest entirely from dried saliva, produced by its extra-large salivary glands. Like other swifts, its diet consists of airborne insects, and it is most often seen over forests and gorges, swooping after its prey. In some parts of Southeast Asia, swiftlet nests are harvested, and are used to make birds' nest soup.

Hirundapus caudacutus

White-throated needletail

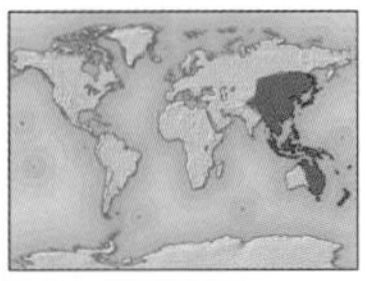

Length 8½ in (Up to 21 cm)
Weight Not recorded
Plumage Sexes alike
Migration Migrant
Status Least concern

Location C., E., and S.E. Asia, E. Australia

Needletails get their name from their tail feathers, which have spiny tips projecting beyond the end of the feathers. There are 4 species, and all of them are extremely fast fliers, reaching speeds of 125 kph (80 mph) during their spectacular courtship displays. The white-throated needletail lives in small flocks or larger groups, and feeds in all kinds of habitats, including farmland and towns. This bird often feeds near the ground, but it also soars high up, catching insects that are sucked high into the sky by columns of rising warm air. The white-throated needletail breeds in Asia and the Himalayas, and builds a shallow, cup-shaped nest in rocky crevices or hollow trees. In the autumn, this bird migrates south across the equator, reaching as far as Tasmania.

Chaetura pelagica

Chimney swift

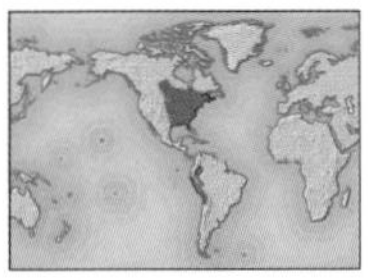

Length 4¾–6 in (12–15 cm)
Weight 11/16–⅞ oz (19–25 g)
Plumage Sexes alike
Migration Migrant
Status Near threatened

Location E. North America, N.W. South America

This small, smoky brown swift is the only one that breeds in eastern North America. It builds a small, half-cup nest of twigs glued together with saliva, mostly in manmade structures such as chimneys and old barns. When not breeding, chimney swifts roost in large flocks; just before and during migration, several thousand birds might roost together. Large chimneys are a favorite roosting site, and birds pour into them at dusk.

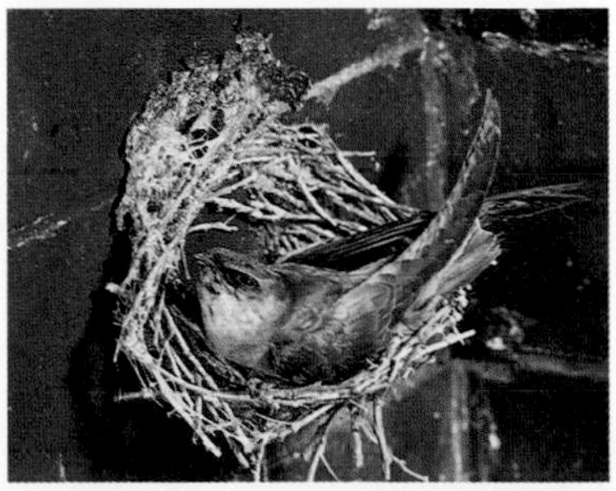

Mousebirds

PHYLUM	Chordata
CLASS	Aves
ORDER	Coliiformes
FAMILIES	1 (Coliidae)
SPECIES	6

Named for their ability to run like rodents, mousebirds scramble about branches in search of buds, leaves, or fruit. Their feet are unique—the 2 outermost toes are reversible, and can point either forward (with the other 2 toes, for hanging), or backward (for grasping). Mousebirds live in flocks and are found only in Africa.

PERCHING
Mousebirds, such as these white-headed mousebirds, have an unusual perching posture. Using their flexible toes, they hold onto branches with their feet held level with their shoulders. They can even perch upside-down.

Colius striatus

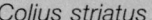

Speckled mousebird

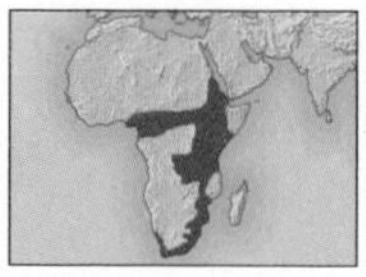

Location C., E., and S. Africa

Length 12–16 in (30–40 cm)
Weight 1⅝–2⅝ oz (45–75 g)
Plumage Sexes alike
Migration Nonmigrant
Status Least concern

This plump bird has a long tail of narrow, stiffened feathers, a stout, softly crested head, and a stubby, downcurved bill. Primarily brown and gray, it has faint barring and speckling on its wings, neck, and throat, and red legs and feet; the bill is dark gray to blackish above and lighter gray below. Like all mousebirds, it often forms groups of 4 to 20—usually consisting of pairs and some immature individuals—that sleep together in a tight cluster hanging from branches. They may preen and even offer food to each other. At times regarded as agricultural pests, mousebirds are often exterminated in farm areas and gardens.

Trogons

PHYLUM	Chordata
CLASS	Aves
ORDER	Trogoniformes
FAMILIES	1 (Trogonidae)
SPECIES	40

These brilliantly colored birds live in tropical forests in the Americas, Southeast Asia, and sub-Saharan Africa. They have short, rounded wings, a long tail, soft, often iridescent plumage, and bright patches of bare skin around the eyes. They grip branches with their small feet, which have 2 toes pointing forward and 2 backward; uniquely, it is the first and second toes that point backward. The short bill has a wide gape for catching invertebrates in flight.

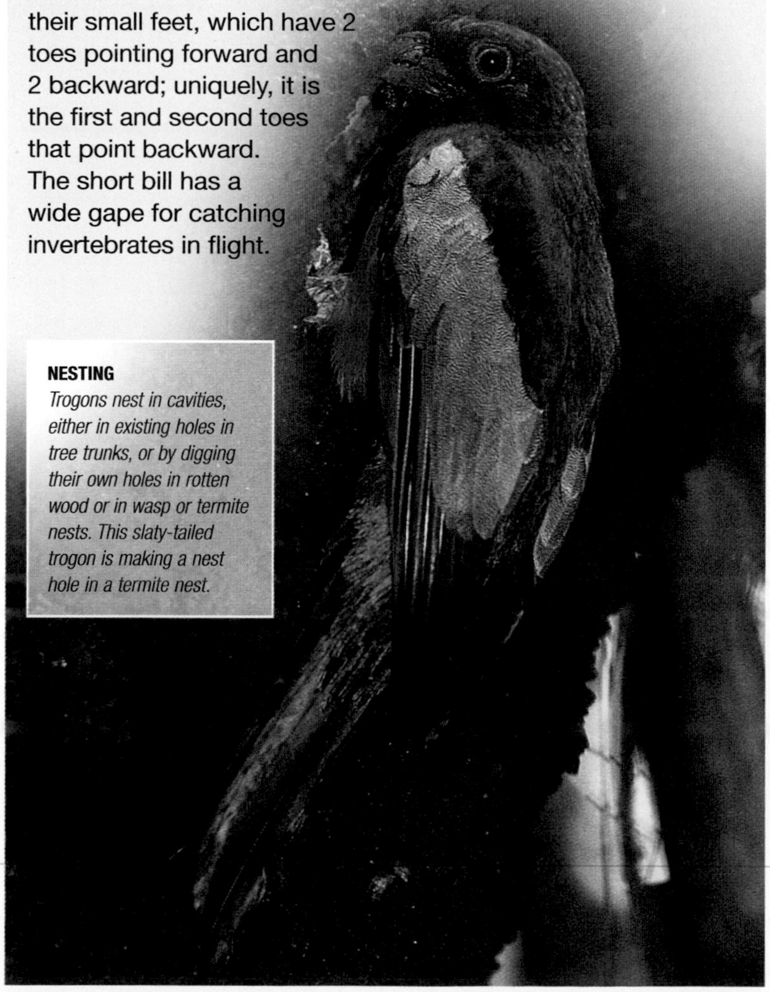

NESTING
Trogons nest in cavities, either in existing holes in tree trunks, or by digging their own holes in rotten wood or in wasp or termite nests. This slaty-tailed trogon is making a nest hole in a termite nest.

Pharomachrus mocinno

Resplendent quetzal

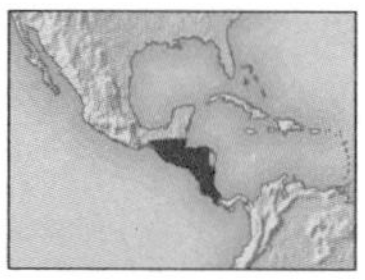

Location Central America

Length 14–16 in (35–40 cm)
Weight 7–8 oz (200–225 g)
Plumage Sexes differ
Migration Nonmigrant
Status Near threatened

The male resplendent quetzal is widely regarded as one of the world's most beautiful birds. It has brilliantly colored plumage—mainly iridescent emerald-green—and a rich crimson-colored breast. Other characteristics, unique to the male, are the exceptionally long tail coverts, which extend beyond the tail to form an elegant train, and the short, bristlelike crest. Both sexes are plump-bodied and have stout heads. This bird's habit of perching motionless for long periods makes it difficult to spot. It flies among trees looking for fruit or insects. Both parents brood the eggs—the tail feathers of the male sticking out of the nest hole as he sits on the eggs—and share in feeding the young.

Trogon violaceus

Violaceous trogon

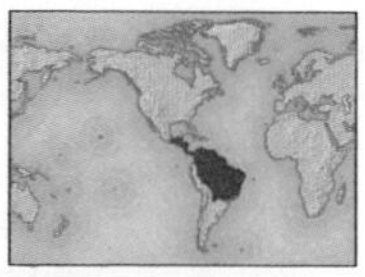

Location S.E. Mexico to C. South America

Length 9–10 in (23–26 cm)
Weight 1⅝–2⅜ oz (45–65 g)
Plumage Sexes differ
Migration Nonmigrant
Status Least concern

Generally solitary, this bird is mainly found in tropical rain forest and woodland (occasionally in dry brushland and farmland), in Central and South America. Males have a black and iridescent violet-blue head, a green back and upper tail, and a yellow-orange breast, while females are primarily gray. The nest consists of an excavated hole, usually in a tree but sometimes built in old termite or wasp nests. This bird feeds on fruit, insects, and other invertebrates.

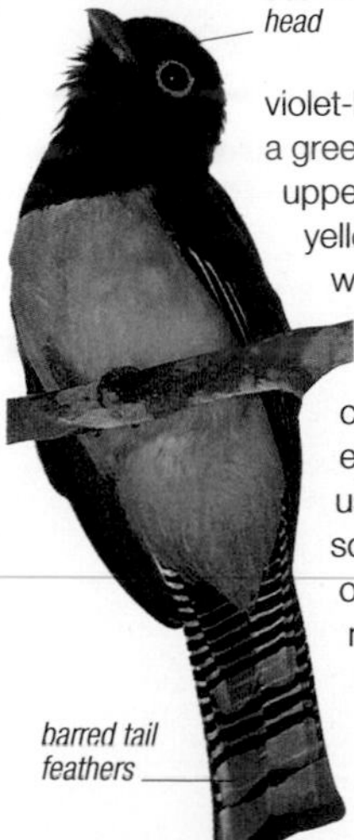

Kingfishers and relatives

PHYLUM	Chordata
CLASS	Aves
ORDER	Coraciiformes
FAMILIES	10
SPECIES	218

Famous for their spectacular dives into water, the kingfishers are arguably the most familiar birds in this group. However, there are 10 very different families in all (including bee-eaters, todies, motmots, rollers, and hornbills), most of which are not waterside birds. They are found worldwide, mainly in woodland, and all nest in holes. These birds range in size from tiny todies, 4 in (10 cm) long, to huge hornbills, up to 5 ft (1.5 m) long.

HUNTING FOR FISH
A kingfisher perches or hovers over water, watching below for signs of movement. When it sees a fish, it plunge-dives head-first and grabs—never spears—its prey.

Anatomy

Most members of this group have a relatively large head and bill, and a compact body. Their legs are frequently short, and their feet tend to be weak. Two of the front toes are usually partially fused together near the base. The wings of most species are broad, but in the elegant and highly aerial bee-eaters, they are relatively long and pointed. Many relatives of the kingfisher, such as the motmots, ground-rollers, bee-eaters, and hornbills, have a long tail, and a large number of species have vividly colored plumage.

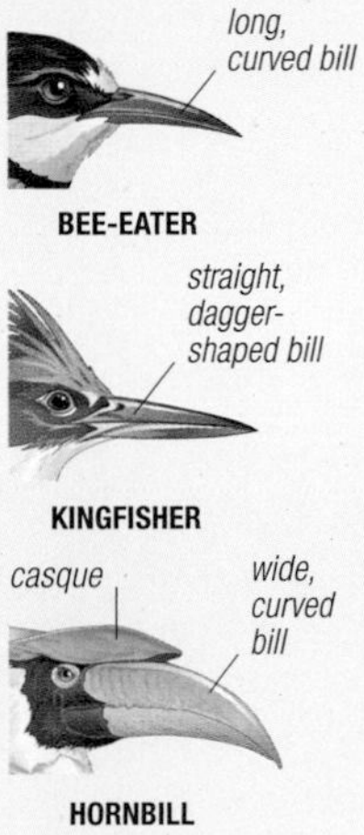

BILL SHAPES
Most members of this group have a strong bill, useful for dealing with animal prey. However, there is great variation in bill shape and size: it is long and curved in bee-eaters and hoopoes; daggerlike in kingfishers and todies; and vastly expanded, often with a horny casque, in hornbills.

CATCHING INSECTS
Rollers, such as this lilac-breasted roller (left), drop from an elevated perch to catch their prey on the ground. A few also catch flying insects in midair.

EATING FRUIT
Like other large hornbills, the great hornbill (right) is mainly a fruit-eater. It uses its long bill to reach fruit on trees and then tosses them back into its gullet.

Feeding

Kingfishers adopt a hunting strategy that is common to many birds in this group. The bird sits still, watching for movement, and then takes off in pursuit—into the water, or down to the ground, or in an aerial chase—before returning to its perch. Since fish are slippery and difficult to control, a kingfisher often stuns its prey by beating it on a hard surface before swallowing it. Most members of this group are meat-eaters, feeding on all kinds of animals, including other birds, fish, mammals, and insects. Hornbills supplement their diet with fruit.

Megaceryle alcyon

Belted kingfisher

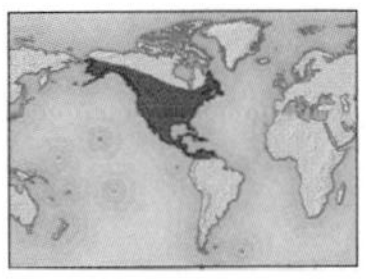

Location North America to N. South America, Caribbean

Length	11½ in (29 cm)
Weight	5 oz (150 g)
Plumage	Sexes differ
Migration	Partial migrant
Status	Least concern

This is one of the few species of kingfishers to be found in North America. It has a short, squarish tail and a craggy, erectile crest. The female belted kingfisher has a prominent reddish brown band across the lower breast. Conspicuous when fishing or sitting on an exposed perch, the belted kingfisher flies off over the water when disturbed, with a loud, rattling call. It feeds predominantly on fish but also eats crustaceans, amphibians, and reptiles. During courtship, these birds circle high overhead and chase each other while uttering shrill cries. The male offers fish to the perching female as part of the courtship ritual. This kingfisher builds its nest in a chamber at the end of a burrow excavated in earth, usually in a river bank devoid of vegetation, but also in man-made earthworks. Birds in the north of the range migrate south when lakes and rivers freeze over.

Ceryle rudis

Pied kingfisher

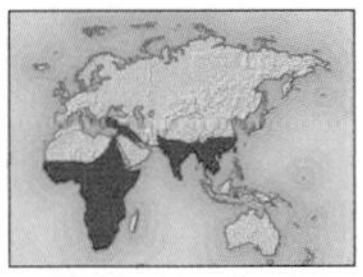

Location Africa, W., S., and S.E. Asia

Length	10 in (25 cm)
Weight	90 g (3¼ oz)
Plumage	Sexes differ
Migration	Nonmigrant
Status	Least concern

This large and boldly marked kingfisher has a wide distribution, stretching from western and southern Africa as far east as China. It has a shaggy crest, and a conspicuous breastband, which is double in the male, and only partly complete in the female. Unlike most other kingfishers, this species is equally at home over fresh and salt water, and sometimes ventures far out over estuaries and shallow coasts. It flies rapidly, but often hovers over the water when locating prey, before diving down to make a catch. The pied kingfisher nests in sandy banks, and often reproduces cooperatively, with up to 4 young nonbreeding adults acting as helpers for a breeding pair. Assertive and vocal, it has a high-pitched call, which it often makes on the wing.

Alcedo atthis

Common kingfisher

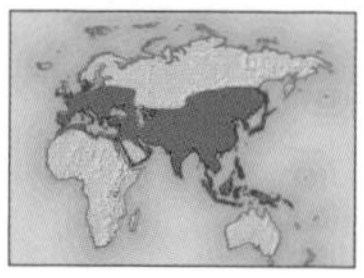

Location Europe, Asia, N. Africa

Length 6½ in (16 cm)
Weight 1¼ oz (35 g)
Plumage Sexes alike
Migration Partial migrant
Status Least concern

The common kingfisher is the only kingfisher encountered in most of Europe, where it is also known as the Eurasian kingfisher. It is a small, swift, and active bird with vivid, distinctive plumage. Size and coloration vary throughout its range, which stretches from Western Europe and northern Africa through to East and Southeast Asia, and 7 subspecies are recognized. In the western sector of its range, it inhabits most aquatic habitats, but prefers lowland, freshwater streams or rivers. In eastern areas, it is more prevalent in coastal or subcoastal habitats, notably estuaries, mangroves, and intertidal pools. At high latitudes, freezing conditions force winter migration, although in warmer areas, birds move locally or are partially resident. Although it supplements its diet with crustaceans, amphibians, and insects, the common kingfisher primarily eats small fish, which it takes by plunge-diving. Each breeding pair occupies a territory of up to ⅔ mile (1 km) along river banks. Pair-bonds are maintained through the breeding season, and both sexes incubate their eggs then care for the chicks for up to 4 weeks, until they are ready to leave the nest.

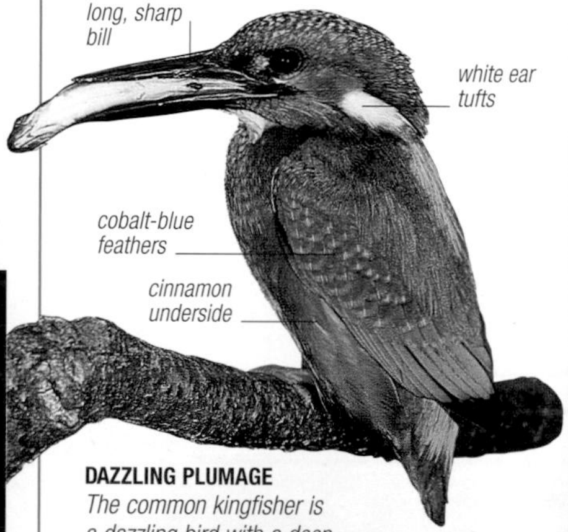

DAZZLING PLUMAGE
The common kingfisher is a dazzling bird with a deep cinnamon underside, greenish blue crown, back, and wings, and dazzling cobalt-blue rump and tail. Its sharp bill is well adapted for striking and grasping fish.

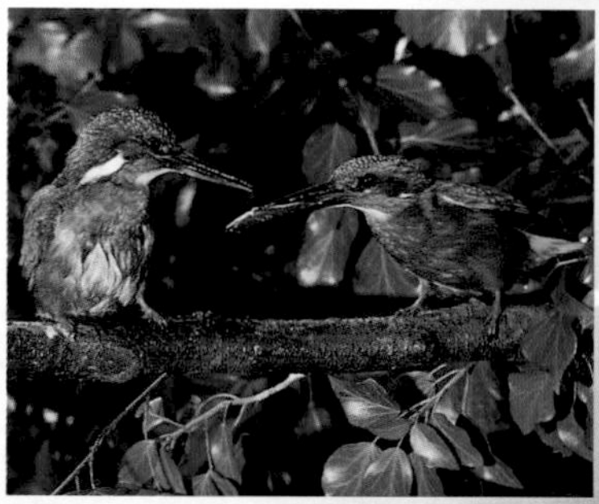

COURTING COUPLE
Courtship behavior is elaborate and includes erratic twisting and turning flights, as well as courtship feeding in which the kingfishers offer each other fish just before copulation.

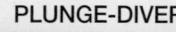

PLUNGE-DIVER

After a steep dive from a favored vantage perch or, less often, from hovering flight, the kingfisher catches its fish usually no deeper than 10 in (25 cm) below the water. Its natural buoyancy and a swift downstroke of its wings send it clear of the water's surface.

LIGHTNING STRIKE
Having captured its prey by a dramatic plunge-dive, the common kingfisher then takes it back to its perch, on which it strikes the fish repeatedly before swallowing it headfirst. Any undigested remains are usually regurgitated as pellets.

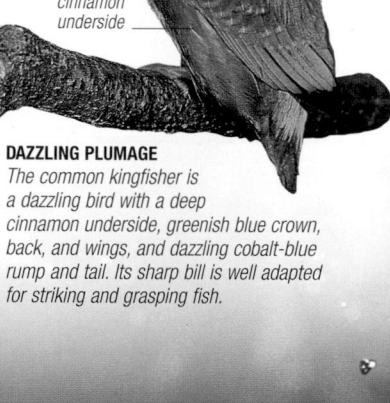

Dacelo novaeguineae

Laughing kookaburra

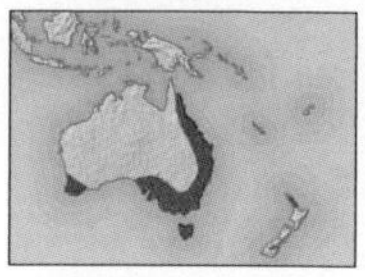

Location E., S.E., and S.W. Australia (including Tasmania), New Zealand

Length 16½ in (42 cm)
Weight 13 oz (350 g)
Plumage Sexes alike
Migration Nonmigrant
Status Least concern

The largest of all kingfishers, the laughing kookaburra is known throughout Australia for its raucous call, which it delivers with partly opened bill pointing skyward and its tail cocked. In urban areas, it becomes quite tame and may even be fed by hand. It is found in strong family groups, which during the day keep within sight or sound of each other, and at night, roost together. This bird has been introduced in southwest Australia, Tasmania, and New Zealand.

EYE-STRIPE
The kookaburra's back and wings are dark brown, while its underparts are a dusky white. Its white head has a distinctive black eye-stripe.

FEEDING

Unlike river kingfishers, this kookaburra captures prey on the ground. It sits on an exposed perch waiting for likely victims—such as insects, snails, frogs, small birds, fish, and reptiles—and then swoops down on them. It crushes small prey in its bill; larger prey, like the snake the bird pictured above has caught, are beaten to death against a branch.

Clytoceyx rex

Shovel-billed kookaburra

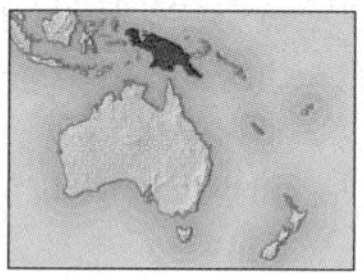

Location New Guinea

Length 13½ in (34 cm)
Weight 11 oz (300 g)
Plumage Sexes differ
Migration Nonmigrant
Status Least concern

A close ally of the better-known Australian kookaburra, the shovel-billed kookaburra is found only in New Guinea, where it is an inhabitant of wet gullies, deep ravines, and heavily shaded, stream-side areas. A large, heavy-bodied kookaburra, it has a dark brown head, mantle, upper back, and wings, while its lower back and rump are a brilliant pale blue. The tail is dull blue in the male and reddish in the female. Juveniles have fine dusky margins on the feathers of their underparts and a collar encircling the hind neck. The peculiarly stubby, broad bill of this bird is adapted to digging for earthworms in the forest floor. The bill is thrust into the soft soil at a slight angle and moved from side to side until it grasps a worm. Little is known of this bird's habits; shy and wary, single birds or pairs are usually encountered only when flushed from the ground to fly up into the forest canopy.

Todus todus

Jamaican tody

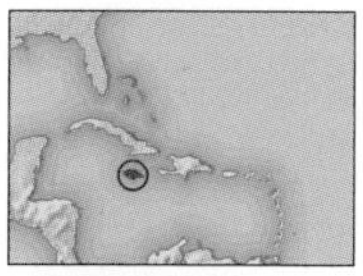

Location Jamaica

Length 4¼ in (11 cm)
Weight 7⁄32 oz (6 g)
Plumage Sexes alike
Migration Nonmigrant
Status Least concern

The Jamaican tody is one of 5 species of small, vivid green, kingfisherlike birds with a large head, a long, flat, bicolored bill, and a prominent crimson bib. It perches on small branches with its bill upturned, looking for insects on the undersides of overhanging leaves and twigs. It usually darts out rapidly to snatch prey, but will sometimes hover. Extremely active from dawn to dusk, it has one of the highest feeding rates known in birds. Its call consists of a loud, nasal "beep."

Momotus momota

Blue-crowned motmot

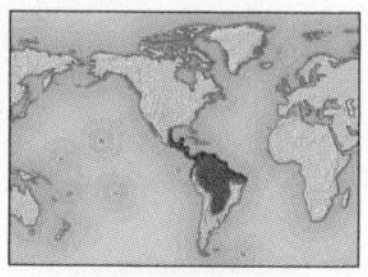

Location Central America to C. South America, Trinidad and Tobago

Length 18½ in (47 cm)
Weight 5 oz (150 g)
Plumage Sexes alike
Migration Nonmigrant
Status Least concern

The blue-crowned motmot is a robust bird with elongated central tail feathers that are bare except for the tips. Its plumage is generally green and it has a black eye-stripe. For most of the day, this quiet, unobtrusive bird sits on a low branch, intermittently swinging its racquet-tipped tail like a pendulum. However, it is very active in the early morning and evening, emitting a far-carrying "hoot-hoot" call. The blue-crowned motmot feeds mainly on insects, either capturing prey on the ground or probing into leaf litter; it also snatches prey from tree trunks.

Merops apiaster

European bee-eater

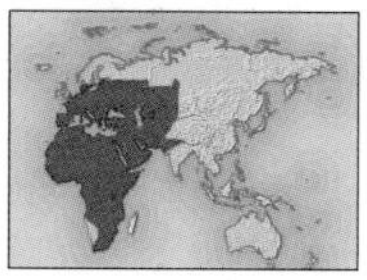

Location Africa, Europe, W., C., and S. Asia

Length 12 in (30 cm)
Weight 2½ oz (70 g)
Plumage Sexes alike
Migration Migrant
Status Least concern

One of the most aerial of all bee-eaters, this medium-sized bird has long wings and a sharply pointed bill. During the day, it perches on telegraph wires, fences, or branches. It preys on stinging insects, which it de-venoms by rubbing the insect's tail-end rapidly against the perch and squeezing it in its bill to expel the venom and sting. It breeds in colonies and nests in burrows.

Merops bullockoides

White-fronted bee-eater

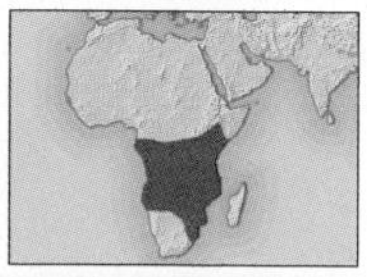

Location Southern Africa

Length 9 in (23 cm)
Weight 1¼ oz (35 g)
Plumage Sexes alike
Migration Partial migrant
Status Least concern

The white-fronted bee-eater lives in a social unit, known as a clan, of up to 16 individuals. These clans associate in nesting colonies of 500 or more birds. Although each clan vigorously defends its own foraging territory, members of different clans greet each other, and regularly visit other burrows at the nesting colony.

BIRDS

Coracias garrulus

European roller

Length 12½ in (32 cm)
Weight 5 oz (150 g)
Plumage Sexes alike
Migration Migrant
Status Near threatened

Location Africa, Europe, W. and C. Asia

The European roller is a heavy-bodied bird with conspicuous, blue and tobacco-colored plumage and black wing-tips. It spends long periods sitting on a prominent perch, such as a bare branch or a power line, looking for prey on the ground. It feeds mainly on insects, but also eats small reptiles, mammals, birds, and fruit. The nest is built in a hollow limb or hole in a tree.

Upupa epops

Hoopoe

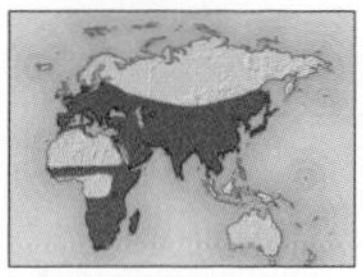

Length 11 in (28 cm)
Weight 2⅝ oz (75 g)
Plumage Sexes alike
Migration Partial migrant
Status Least concern

Location Europe, Asia, Africa, Madagascar

With its bold colors, long, curved bill, and fan-shaped crest, the hoopoe is unmistakable. It spends much of the day on the ground, probing the soil with its bill in search of insects. Its diet also includes earthworms, snails, slugs, and spiders. This bird has an undulating flight with strong wingbeats interspersed by pauses. Hoopoes are notoriously foul-smelling and unhygienic; the fetid state of their nests, arising from the accumulation of excreta, food remains, and the nestlings' body secretions, is thought to deter predators.

Buceros bicornis

Great hornbill

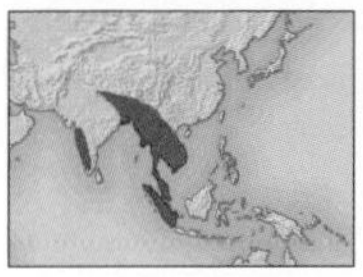

Length 5 ft (1.5 m)
Weight 6½ lb (3 kg)
Plumage Sexes alike
Migration Nonmigrant
Status Near threatened

Location S. and S.E. Asia

The most spectacular of the hornbills, this species is distinguished by its large size, resounding call, and the loud "swooshing" of its wingbeats. It has a massive bill with a prominent casque, and long, rounded wings. The bill, casque, white head, and neck are often stained yellow with oil from the preen gland. Mostly confined to the forest canopy, the hornbill moves easily through it with a series of sideways hops. It follows a regular routine, visiting certain trees at the same time each day. It feeds mainly on fruit, especially figs, but also preys on reptiles, amphibians, small mammals, and birds. Fruit and small prey are grasped with the tip of the bill and moved down the gullet by throwing the head up and back. Larger prey is crushed in the bill and battered against a branch before being swallowed. Like other hornbills, this species builds its nest in the natural hollow of a tree trunk, which it seals with mud, leaving only a narrow opening through which the male passes food to the nesting female. The female breaks out of the nest after almost 3 months, but the entrance is resealed by the chick, which is fed for a further month in the nest.

Phoeniculus purpureus

Green wood hoopoe

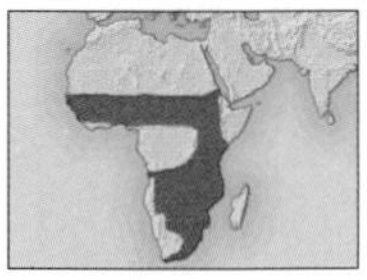

Length 15 in (38 cm)
Weight 2⅝ oz (75 g)
Plumage Sexes alike
Migration Nonmigrant
Status Least concern

Location Africa (south of Sahara)

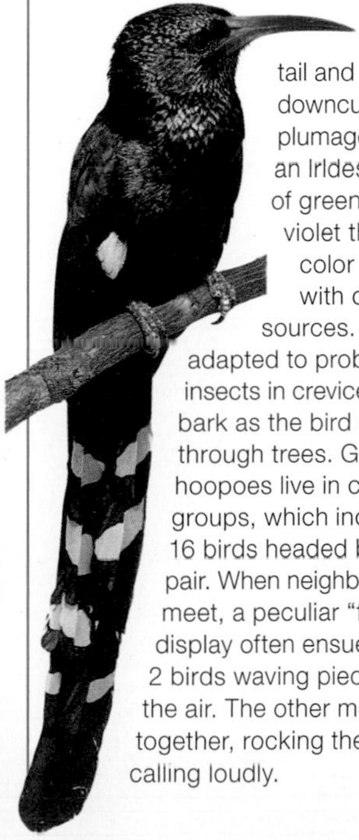

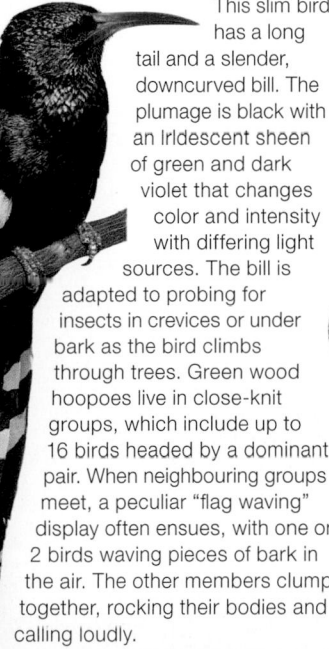

This slim bird has a long tail and a slender, downcurved bill. The plumage is black with an iridescent sheen of green and dark violet that changes color and intensity with differing light sources. The bill is adapted to probing for insects in crevices or under bark as the bird climbs through trees. Green wood hoopoes live in close-knit groups, which include up to 16 birds headed by a dominant pair. When neighbouring groups meet, a peculiar "flag waving" display often ensues, with one or 2 birds waving pieces of bark in the air. The other members clump together, rocking their bodies and calling loudly.

Tockus leucomelas

Southern yellow-billed hornbill

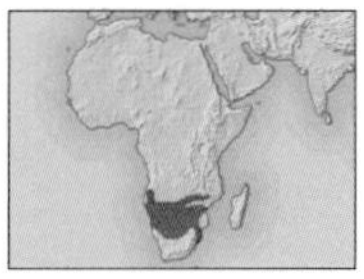

Length 20–23½ in (50–60 cm)
Weight 9 oz (250 g)
Plumage Sexes alike
Migration Nonmigrant
Status Least concern

Location Southern Africa

This small hornbill has a curved, bright yellow bill and black wings with white spots. It feeds on insects and fruit, especially figs. This species sometimes forms cooperative foraging parties with dwarf mongooses, which flush out locusts on which the birds feed. The hornbill, in turn, warns them of approaching danger.

Bucorvus leadbeateri

Southern ground hornbill

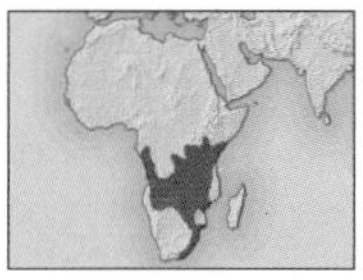

Length 4¼ ft (1.3 m)
Weight Not recorded
Plumage Sexes alike
Migration Not migrant
Status Vulnerable

Location Southern Africa

This turkey-sized bird, and the Abyssinian ground hornbill (*Bucorvus abyssinicus*), are the 2 largest hornbills, and among the few that feed mainly on the ground. Its plumage is almost entirely black, apart from white wing patches, and its most conspicuous feature is a patch of bare, brightly colored skin on its face and throat. The male has a red wattle, whereas the female's is blue. Southern ground hornbills live in groups of up to 8 birds, and feed mainly on small animals. They nest in tree holes, but in each group only the dominant pair breeds—the junior birds help collect food and defend the nest.

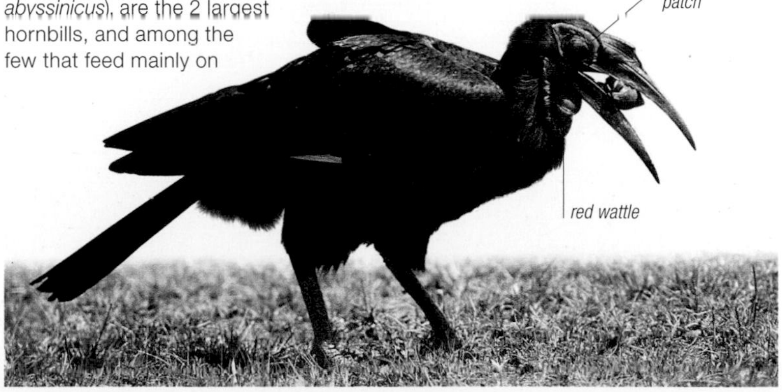

BIRDS

Woodpeckers and toucans

PHYLUM	Chordata
CLASS	Aves
ORDER	Piciformes
FAMILIES	5
SPECIES	411

This group of arboreal birds includes woodpeckers, toucans, barbets, jacamars, honeyguides, and puffbirds. They all have a similar type of foot, with 2 toes pointing forward and 2 backward, which helps them climb easily. All the birds in this group also nest in holes. Woodpeckers and barbets carve out their own nest holes, the woodpeckers using their stout bill as a chisel. These are mainly tropical birds; only woodpeckers are also widespread in temperate areas. Curiously, not a single species occurs in Australia. Toucans and barbets help disperse the seeds of some trees by eating their seeds and then passing them through their body, while woodpeckers limit the spread of some tree diseases by feeding on the insects that carry them.

Reproduction

Although all birds in this group nest in holes, not all use living trees. The smaller woodpeckers prefer to bore into the softer bark of dead and decaying trees. Jacamars, puffbirds, and barbets may dig out burrows, either in the ground or in termite mounds. Unlike the other families, the honeyguides are brood parasites, laying their eggs in the nests of other birds, including woodpeckers. The heels of young toucans are fitted with pads of spikelike projections that protect the ankles from damage against the unlined nest floor.

NESTING
Like most barbets, the black-collared barbet nests in dead trees, sometimes communally. A few species burrow into banks or termite mounds.

Anatomy

The main climbers in this group, woodpeckers and barbets, have strong leg muscles and use their tail for support when holding themselves vertically upright on tree trunks; woodpeckers have specially stiffened tail feathers for this purpose. Chiseling involves delivering powerful blows to the wood surface, so the skull of woodpeckers is unusually thick to absorb the shock; the nostrils are also slitlike to prevent woodchips from entering the airways. Woodpeckers also have a highly distensible, slightly barbed tongue that is coated with a sticky substance for catching insects. Honeyguides have thick skin to protect them from bee stings and specialized bacteria in their gut that enable them to digest beeswax.

WOODPECKER

BARBET

TOUCAN

PUFFBIRD

BILL SHAPE
There is great variation in bill shape in this group. A woodpecker's long, sharp bill is ideal for chiseling out wood, while the stout bill of barbets is used to handle animal prey. The large, serrated bill of toucans is adapted for grasping and tearing fruit. Puffbirds and jacamars have a sharp-tipped bill for catching flies.

FEET
The feet of woodpeckers and all their relatives are described as zygodactylous. Two toes point forward and 2 back, an arrangement that helps them climb and perch on tree trunks.

Feeding

Insects form the greater part of the diet of most birds in this group. Woodpeckers take insects from both the tree surface and deep holes in bark, finding and extracting them with their long tongue. Other species pounce on insects or grab them in midair; jacamars, which take insects in flight, are the only birds to eat the giant morpho butterflies of the American tropics, removing the wings first. Toucans and most barbets feed almost exclusively on fruit. Honeyguides are the only birds that eat beeswax; however, they also feed on insects and fruit.

GUIDING
Honeyguides, such as the lesser honeyguide (right), take their name from their habit of leading large mammals to bees' nests.

EATING FRUIT
Toucans, such as the toco toucan (left), eat mainly fruit. The long bill helps them take fruit from otherwise out-of-reach, slender branches.

NESTING AND FEEDING YOUNG
Woodpeckers use their sturdy bill to bore their own nest holes in tree trunks and defend them vigorously against other hole-nesters. Here, a green woodpecker is feeding its young with ants that it has collected from the ground.

Galbula ruficauda

Rufous-tailed jacamar

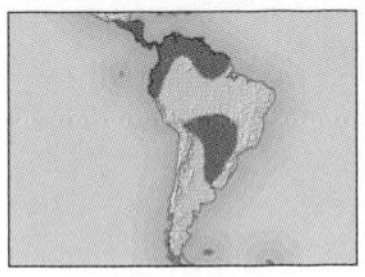

Length 10 in (25 cm)
Weight ⅞ oz (25 g)
Plumage Sexes alike
Migration Nonmigrant
Status Least concern

Location S. Central America, N.W. and C. South America

Found only in Central and South America, jacamars are a family of 17 species of brilliantly colored, iridescent birds that feed on large insects, catching most of their prey in midair. Their diet includes butterflies, bees, winged ants, and dragonflies. The rufous-tailed jacamar has a metallic golden-green head and upper breast, while the lower breast, belly, and tail are rufous. The daggerlike bill is long, thin, and straight. Males have a white chin, the females buff. The chicks hatch with down, unlike most piciform birds. The rufous-tailed jacamar usually forms pairs or family groups, and perches upright on exposed twigs to wait for prey. Both the male and the female incubate the eggs and feed the young. This bird has a sharp call and a distinct loud song: an accelerating "pee-pee-pee" that ends in a trill.

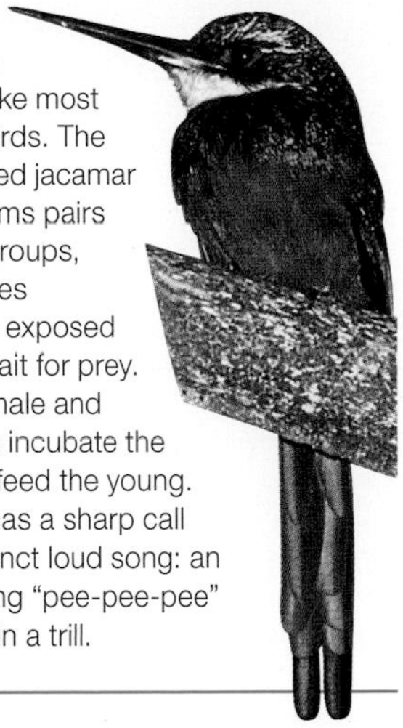

Bucco capensis

Collared puffbird

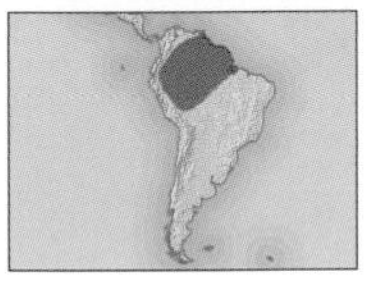

Length 8½ in (21 cm)
Weight 2 oz (55 g)
Plumage Sexes alike
Migration Nonmigrant
Status Least concern

Location N. South America

Puffbirds comprise a family of about 32 species of dull-colored, small to medium-sized, sit-and-wait hunters. Most puffbirds hunt in treetops, and the larger species may catch small reptiles with their strong, hooked bills. The collared puffbird has an orange bill, brown upperparts, and whitish underparts with a large breastband. Its plumage is fluffy, and it has tiny feet. Inhabiting humid, lowland forests, it sits quietly most of the time; only the head moves occasionally as it scans the surroundings for prey. This behavior has earned it names like "sleeper."

Chelidoptera tenebrosa

Swallow-winged puffbird

Length 7 in (18 cm)
Weight 1¼ oz (35 g)
Plumage Sexes alike
Migration Nonmigrant
Status Least concern

Location N. and C. South America

The disproportionately long wings of the swallow-winged puffbird make it look like a bat or a large butterfly in flight. It is an excellent, almost acrobatic, flier and soarer. Black to dark gray, it sits more in the sun, and is easier to spot than other puffbirds. It sits and waits before sallying out to snatch prey on the wing, occasionally hovering in the process.

Lybius dubius

Bearded barbet

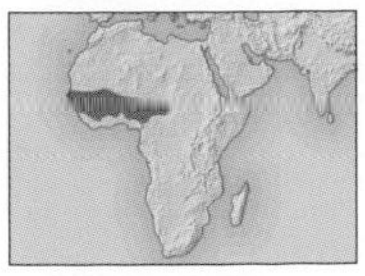

Length 10 in (25 cm)
Weight 2½–4 oz (70–125 g)
Plumage Sexes alike
Migration Nonmigrant
Status Least concern

Location W. Africa

Found in Asia, South America, and Africa, barbets are usually solidly built birds with gaudy plumage. Most have conspicuous bristles at the base of their bill—a feature that gives them their name. The bearded barbet has black upperparts and red underparts, with a black breastband and white flanks. Its bill is large and powerful, and it feeds mainly on fruit, showing a particular liking for wild figs. It also gleans insects from bark.

Megalaima virens

Great barbet

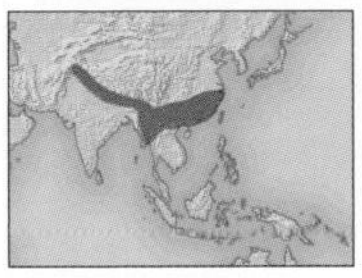

Length 12½ in (32 cm)
Weight 7–11 oz (200–300 g)
Plumage Sexes alike
Migration Nonmigrant
Status Least concern

Location C., S.E., and E. Asia

One of the largest barbets, this clumsy looking bird has a large, pale bill with numerous long bristles. Like other barbets, it often engages in dueting, the male and female singing alternate notes. While calling—a loud, musical sound that is repeated monotonously for several minutes—the bill remains closed and the throat inflates and deflates. Although usually solitary, it may at times be found in feeding flocks in treetops. The bird excavates its nest in a tree hole; both parents feed the young.

Pogoniulus bilineatus

Yellow-rumped tinkerbird

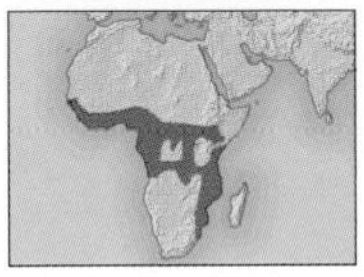

Length 4 in (10 cm)
Weight 7/16 oz (13 g)
Plumage Sexes alike
Migration Nonmigrant
Status Least concern

Location W. to S.E. Africa

A wide-ranging African bird with many regional forms differing in pattern and color, the yellow-rumped tinkerbird has a metallic call consisting of a series of popping sounds. While calling, the bird sits upright, puffing up its throat and making its rump feathers erect. This tiny barbet moves rapidly through canopy, bushes, and vines, picking up fruit and insects. It occasionally hunts insects on the wing. Rather aggressive toward other small barbets, it defends its territory vigorously. Courtship appears to involve flutter flights.

Trachyphonus darnaudii

D'Arnaud's barbet

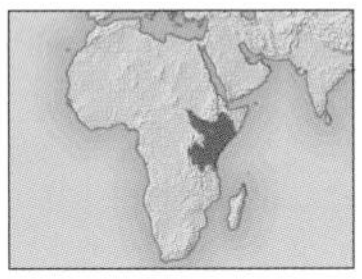

Length 8 in (20 cm)
Weight 1 1/16 oz (30 g)
Plumage Sexes alike
Migration Nonmigrant
Status Least concern

Location E. Africa

Most barbets live in trees, but a small number—particularly in tropical Africa—are equally at home on the ground. D'Arnaud's barbet is one of these species, and is often seen hopping over bare earth or through grass as it searches for food. It may also forage in low bushes and trees for insects, fruit, and seeds. It inhabits flat terrain: grassland with scattered trees, bushland, and open woodland. D'Arnaud's barbet is not shy in the presence of humans, and is best known for the dueting of mating pairs, during which the crown feathers are erect and the tail is cocked every now and then. The bird has various other calls, including alarm calls. While displaying, this yellow-orange, black, and red barbet spreads and swings its tail, bobs its head, and wipes its bill. It is a social bird, living in pairs or small family groups consisting of the breeding pair and subordinate individuals who may assist in raising the young. The mating pair defend their territory aggressively, marking the boundaries vocally by their duets. Both the male and female D'Arnaud's barbet dig the nest vertically into the ground, and line the nest chamber with grass.

BIRDS

Indicator indicator

Greater honeyguide

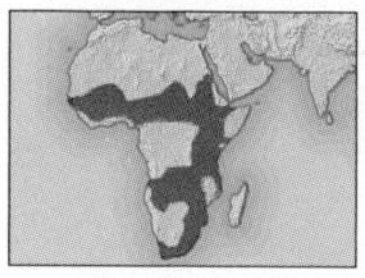

Length 8 in (20 cm)
Weight 1¾ oz (50 g)
Plumage Sexes alike
Migration Nonmigrant
Status Least concern

Location Africa (south of Sahara)

The greater honeyguide is famous for guiding humans and animals—such as honey badgers—to wild bees' nests. Once the nest has been broken open, the honeyguide flies down to it to feed. Apart from beeswax and bees, this bird also feeds on ants, termites, other insect larvae, and even the eggs of other birds. The greater honeyguide is a brood parasite and does not build its own nest. Nestlings have hooked bills with which they kill the nestlings of the host.

Melichneutes robustus

Lyre-tailed honeyguide

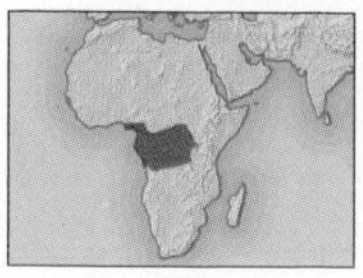

Length 3¾ in (9.5 cm)
Weight 2 oz (55 g)
Plumage Sexes alike
Migration Nonmigrant
Status Least concern

Location C. Africa

The lyre-tailed honeyguide has a lyre-shaped tail and brownish plumage with white markings on its outer feathers. It inhabits lowland and low mountain forests and is also found in coffee plantations. Unlike the greater honeyguide (see left), this species is not known to lead other animals to bees' nests. It is best known for its spectacular flight displays that take place all year round. During these displays, performed by both sexes, the bird flies high above the forest, then descends swiftly in a zigzag or spiral flight. The sound heard during the flights is probably produced by the air passing through the tail feathers. This unusual behavior is associated with feeding sites (bee hives). This honeyguide feeds mainly on beeswax, but also hunts insects in the air. Like the greater honeyguide, it does not build a nest, and probably parasitizes barbets.

Aulacorhynchus prasinus

Emerald toucanet

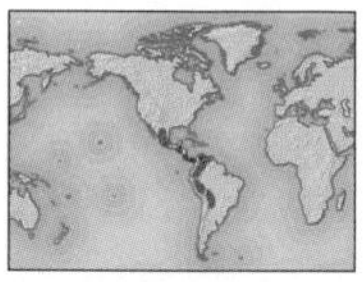

Length 12 in (30 cm)
Weight 5¼ oz (150 g)
Plumage Sexes alike
Migration Nonmigrant
Status Least concern

Location Mexico, S. Central America, W. South America

Toucanets are medium-sized members of the toucan family with relatively modest-sized bills. This species is an upland bird typically found in mountain forests. It lives in pairs and small groups, nesting and roosting in holes. Noisy groups may form in fruiting trees. The call is loud and varied, and includes croaking sounds as well as imitations of many local species of birds. The long, black and yellow bill, white at the base, helps this toucanet reach for fruit and animal prey.

Pteroglossus castanotis

Chestnut-eared aracari

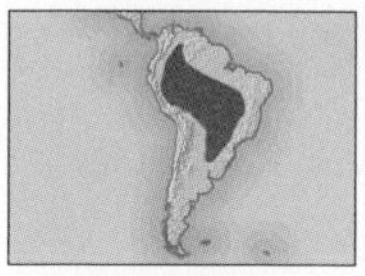

Length 13–16 in (33–40 cm)
Weight 10½ oz (300 g)
Plumage Sexes alike
Migration Nonmigrant
Status Least concern

Location N. to C. South America

Compared to most other toucans, aracaris are smaller and more lightly built, with a slender bill. The long, brightly colored bill of the chestnut-eared aracari has yellow, teeth markings. Living in the forest and forest edge, this bird may be seen restlessly hopping through the tree canopy. It usually occurs in small groups that roost huddled in tree holes and hunt for insects and other small animals together, large flocks sometimes gathering at fruiting trees. Paired birds feed and preen each other. Like other toucans, the chestnut-eared aracari has an undulating flight.

Ramphastos toco

Toco toucan

Length 21–23½ in (53–60 cm)
Weight 20 oz (550 g)
Plumage Sexes alike
Migration Nonmigrant
Status Least concern

Location N.E. to C. South America

The largest of all toucan species, the toco toucan has a spectacular orange bill, up to 7½ in (19 cm) long. The huge bill, white rump, and red undertail coverts are especially conspicuous during the bird's undulating flight, with alternating flaps and glides. The toco toucan is found in woodland, gallery forest, wooded savanna, groves, and even open country. It usually reaches for fruit and prey from a perch, but may descend to the ground for fallen fruit. The large bill helps the bird reach food on the end of thin twigs that cannot support its weight. Once collected, food has to be moved from the tip of the bill into the bird's throat—a task that is achieved with a quick backward toss of the head. Less sociable than other toucans, this bird sometimes participates in periodic invasions of new areas, which occur in certain years. It also migrates locally during some winters. The toco toucan makes its nest in existing holes in dead or living trees and in earth banks; it may also use the nests of terrestrial termites. Its call, a deep, low snore, is often made high up from a branch.

Jynx torquilla

Northern wryneck

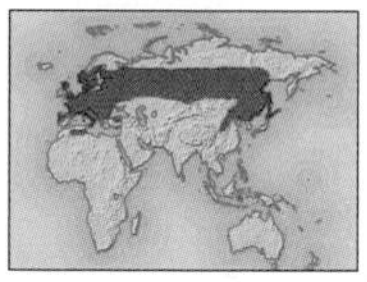

Length 6½ in (16 cm)
Weight ⅜ oz (10 g)
Plumage Sexes alike
Migration Migrant
Status Least concern

Location Europe to Asia, N. Africa

The northern wryneck derives its name from its habit of twisting and writhing its neck, in snakelike fashion, as a form of defensive display. The speckled, brown, gray, buff, and white plumage camouflages it well against bark and the ground. It feeds on the larvae and pupae of ants, opening ant hills with its bill and using its sticky tongue to pick up its food.

Picumnus cirratus

White-barred piculet

Length 4 in (10 cm)
Weight ⅜ oz (10 g)
Plumage Sexes alike
Migration Nonmigrant
Status Least concern

Location N.E. and C. South America

Piculets (about 30 species) are the smallest members of the woodpecker family. Unlike typical woodpeckers, they do not use their tails to brace themselves against trees. The white-barred piculet has a short, soft tail and brownish gray plumage, the male having a distinctive red crown. Found in a wide variety of habitats, including open woodland, savanna, brush, and forest edges, it forages at low heights on thin twigs, vines, and bamboo. This tiny bird feeds on small insects, getting most of its food by vigorous hammering. It nests and roosts in holes in dead branches.

Melanerpes formicivorus

Acorn woodpecker

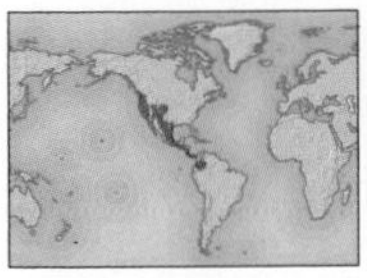

Location W. North America to N. South America

Length 9 in (23 cm)
Weight 2⅜–3¼ oz (65–90 g)
Plumage Sexes alike
Migration Nonmigrant
Status Least concern

Dry acorns may comprise half of the daily food intake of the acorn woodpecker, which has a unique habit of storing acorns (see panel, right). It is also known to feed on other seeds, sap, and fruit. Insects form important food for the nestlings. Both sexes have red crowns, but the forecrown is black in females. The tail and bill are moderately long. Restricted to oak and pine-oak woodland, this sociable woodpecker is often found in a group or breeding unit of 3–12 individuals, who share work on the communal acorn store and on building the nest, and defend the common territory that contains foraging sites, sap trees, the group larder, roosts, and the nest site. The acorn woodpecker is characterized by a complex breeding pattern that may be described as "polygynandrous". Pairs may breed alone, but are typically assisted by males and females that usually belong to previous broods. Helping males participate in reproduction only after their mother has been displaced by another female.

CONTRASTING COLORS
The acorn woodpecker has bold, shiny, black, white, and red plumage that makes it easily recognizable.

STORING ACORNS

The acorn woodpecker drills small, neat holes in tree bark, and jams an acorn firmly into each one. Working together, birds belonging to a single breeding unit can store up to 50,000 acorns in one "granary tree," creating a store of food to last them through the winter.

Sphyrapicus varius

Yellow-bellied sapsucker

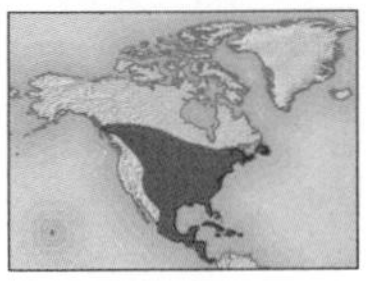

Location North and Central America, Caribbean Islands

Length 8¾ in (22 cm)
Weight 2⅛–2⅞ oz (60–80 g)
Plumage Sexes differ
Migration Migrant
Status Least concern

The yellow-bellied sapsucker drills holes into trees to extract sugary sap. A series of holes are chiseled into the trunks of broadleaved trees such as maple, fruit trees, birch, and poplar. After a short while, the bird returns to feed on the sap oozing out, and on the insects attracted by the sap. This bird has a mewing call-note and drums slowly and irregularly.

Picus viridis

Green woodpecker

Location Europe, W. to C. Asia, N. Africa

Length 12½ in (32 cm)
Weight 7 oz (200 g)
Plumage Sexes differ
Migration Nonmigrant
Status Least concern

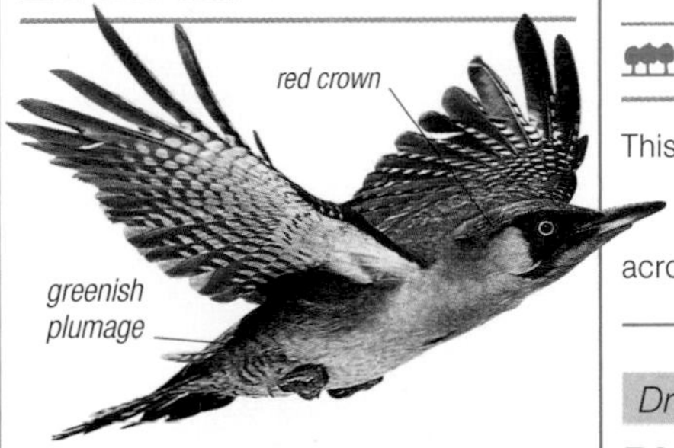

Like the northern flicker (see below), the green woodpecker feeds mainly on the ground, although it uses trees as a refuge, and as places to breed. It has an exceptionally long tongue—even by woodpecker standards—and it uses it to probe into ant nests, extracting them with its sticky tip. The green woodpecker has a loud, laughlike call and steeply undulating flight.

Dendrocopos major

Great spotted woodpecker

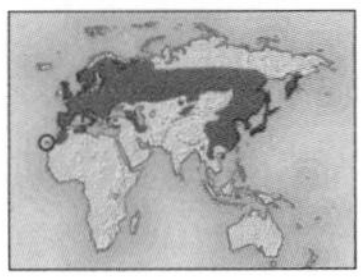

Location Europe, Asia, N. Africa

Length 8–9½ in (20–24 cm)
Weight 2⅜–3⅝ oz (65–100 g)
Plumage Sexes differ
Migration Nonmigrant
Status Least concern

This small, black and white woodpecker has a red lower belly and undertail coverts. The male has a red band across the nape. As is typical in woodpeckers, the strong feet, stiff tail, bill, and tongue of this species are adapted for climbing trees, drilling holes, and probing into cavities. The great spotted woodpecker feeds on fruit, berries, sap from trees, insects and their larvae, and even nestlings of other birds. Early in the breeding season, the bird can be easily detected by the loud drumming sound it makes.

Dryocopus martius

Black woodpecker

Location Europe to Asia

Length 18 in (45 cm)
Weight 13 oz (350 g)
Plumage Sexes differ
Migration Nonmigrant
Status Least concern

This crow-sized bird is Eurasia's largest woodpecker, easily recognized both by its size and by its jet-black plumage and crimson crown. Females have a smaller crown patch than males, placed farther back, but apart from this the 2 sexes are the same. The black woodpecker lives in mature forests and woodland, where it actively chisels out prey that is hidden deep within living and dead wood. It feeds mainly on tunneling insects, such as sawfly grubs and ants, and, when probing for food, it can hack out vertical gashes up to 20 in (50 cm) long. This woodpecker's excavations produce woodchip the size of clothes' pins, which build up in piles on the forest floor. It also pecks at cones, often wedging them in the stumps of fallen trees. This bird lays up to 6 eggs, in holes that it carves out every year. Old holes are very useful to other birds, which take them over when their original owner moves on.

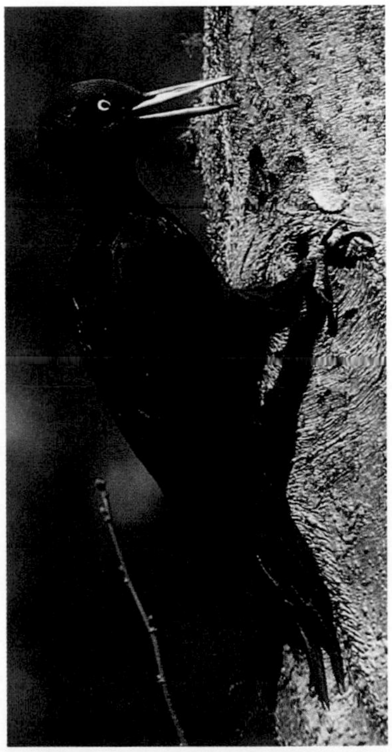

Colaptes auratus

Northern flicker

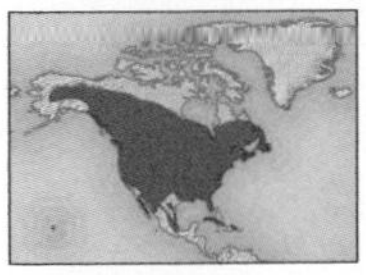

Location North and Central America

Length 12–14 in (30–35 cm)
Weight 3⅝–6 oz (100–175 g)
Plumage Sexes alike
Migration Partial migrant
Status Least concern

The northern flicker has a long, pointed, and slightly curved bill. During confrontations, the bill points forward, and the head and body engage in swinging movements, while the tail and wings are flicked and spread. Often found on the ground, this bird feeds mainly on ants, digging into the earth and pillaging ant nests, with the help of its long tongue. When feeding arboreally, it usually concentrates on the dead parts of a tree. The female, shown here, lacks the black mustachial stripe of the male.

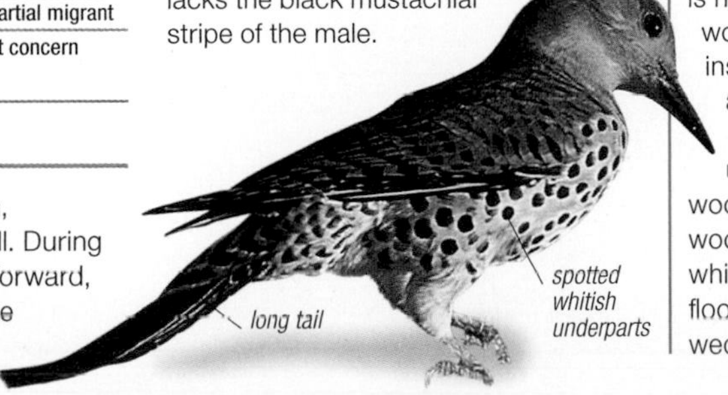

Passerines

PHYLUM	Chordata
CLASS	Aves
ORDER	Passeriformes
FAMILIES	About 96
SPECIES	About 6,000

Most of the world's bird species are passerines. They are sometimes known as perching birds because they have a unique type of foot that enables them to grip even the most slender branches. Passerines are also distinguished by the complex sounds, or songs, made by many species. These are produced using a vocal organ known as the syrinx (which is also found in other birds). Passerines are considered by many to be the most highly evolved of all birds, and many exhibit unusual intelligence. Most live in bushes and trees, but some are adapted to living on the ground; a few (such as swallows) lead an almost entirely aerial existence. Passerines are found in all terrestrial habitats, from arid desert to tropical rain forest. Many species are a familiar sight around buildings and in gardens, and new ones are being discovered all the time.

Perching

Passerines can perch securely on twigs, reeds, and even grass stems. Three of their 4 toes point forward and the other backward. The toes can move independently of each other, and the front toes can oppose the back one, which is especially strong. All 4 toes are level with each other, unlike in many nonpasserines, which have a raised hind toe. When a passerine lands on a perch, its weight causes tendons in the leg to tighten, clamping the toes tightly shut. This grip functions even when the bird is asleep. Despite these advantages, some passerines perch much less often than others. Larks, for example, are essentially terrestrial, and have relatively flat feet to give them balance when running.

PERCHING FEET
Like many passerines, the common waxbill has long, thin toes that are easily wrapped around a small stem. The backward-facing toe is strong and, in contrast to several nonpasserine groups, cannot be reversed. Most passerines have moderately curved, sharp claws that can grip many surfaces.

Anatomy

The passerines' specialized perching foot (see Perching) and well-developed voicebox (or syrinx, see Singing) are among their defining characteristics. These features apart, the members of this group are enormously varied. While a great number of them have subdued coloring, there are also many with spectacularly vivid and bizarre plumage—for example, the remarkable birds of paradise and the multicolored tanagers and finches. Males are often more brightly colored and patterned than females. Most passerines are small birds, but they range from the large, bulky crows and ravens, up to 25 in (65 cm) long, to the short-tailed pygmy tyrant, only 3 in (7 cm) long. Another highly variable feature of passerines is the shape of the bill, which often indicates a specific food preference (see below).

GREENFINCH | WOOD WARBLER | GREAT GREY SHRIKE | SUNBIRD

BILL SHAPE
The shape of a passerine's bill offers important clues to its diet. Species that feed on seeds (such as the greenfinch) often have a short, conical bill, while those that eat invertebrates (such as the wood warbler) tend to have a thin bill. Like many other predatory passerines, the great grey shrike has a hooked and notched bill for subduing relatively large prey. In sunbirds, which feed on nectar, the bill is long and thin, often with a downward curve, to help them reach inside flowers.

Singing

Passerines owe their singing ability to their complex syrinx, which is more highly developed than it is in other birds. Each species has its own distinctive song (or songs), a series of sounds uttered within a defined rhythm and structure. Many passerines have beautiful and complicated songs, among them the larks, wrens, thrushes, nightingales, and lyrebirds. In some species, the chicks learn to sing only by listening to adults of the same species; in others, the songs are innate and do not need to be learned. It is mainly the males that sing, either to claim and keep territory or to attract females.

SINGING MALES
The male nightingale produces some of the most varied of all birdsongs, often delivering them with a characteristic crescendo. It sings both by day and by night, using its daytime songs mainly to mark its territory and deter rivals and its nocturnal songs to attract females.

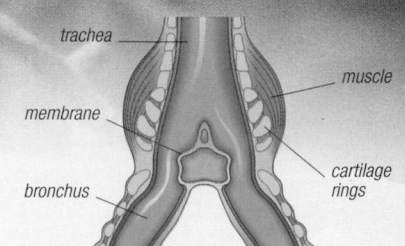

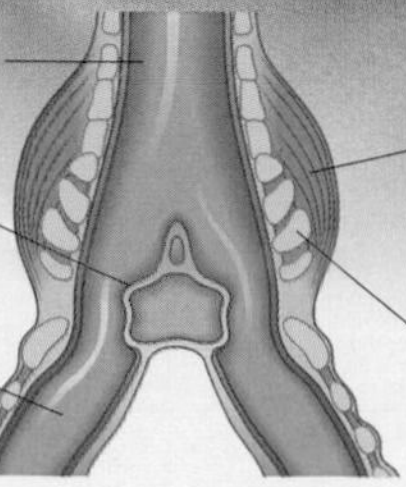

SYRINX
The syrinx is located in the trachea. Membranes in the syrinx vibrate and produce sound as air passes over them. The bird uses muscles, which are attached to rings of cartilage, to modify the sound.

BIRDS

Reproduction

Birds of paradise perform some of the most elaborate courtship displays of all birds, the males attracting females by jumping, hanging upside-down, and making extraordinary noises. However, most passerines have more subtle displays. Many are skilled nest-builders, typically constructing a cup nest from twigs, leaves, and soft material above ground, but they build many more elaborate structures, too. The chicks hatch naked, blind, and helpless, and rely entirely on their parents to feed them until they are ready to leave the nest at 10–15 days old.

ATTRACTING A MATE
Male bowerbirds (such as the spotted bowerbird, left) create elaborate "stages" on which to perform courtship displays. They select a patch of ground, or even build a special structure (the bower), and decorate it with a wide variety of items, such as seeds, grass, and moss, or even human artifacts.

HANGING NEST
Weavers, including the Baya weaver (right), build a highly elaborate domed nest that hangs from a branch, often above water. The nest has an entrance chamber, a nest chamber, and a long tube hanging below. Weavers often nest colonially.

COURTSHIP DISPLAY
Most passerines are monogamous. During the breeding season, a male and female generally form a pair, establish a territory, and build a nest. Among the exceptions to this pattern of behavior are the birds of paradise (such as these Raggiana birds). Males display to females at a breeding arena. After mating, the female builds a nest and cares for her eggs and young alone.

Feeding

Most passerines are small, and therefore require high-energy foods. This means that most species feed either on invertebrates or on seeds, or a combination of the two. Some also take nectar, especially sunbirds. A few rain forest-dwelling families, including birds of paradise, manakins, and cotingas, have an almost exclusive diet of fruit. Among the great variety of passerines are some with more unusual feeding habits. Shrikes are meat-eaters that store their captured prey in a "larder," impaling the bodies of large insects or small vertebrates on thorns or barbed wire before eating. Crossbills feed only on the seeds of coniferous trees, using their crossed mandibles as tweezers to part the scales of a cone in order to remove the seeds.

USING TOOLS
The woodpecker finch of the Galapagos Islands is one of the few animals that uses tools to catch prey. It picks up a small stick or cactus spine in its bill and uses it to pry insects from crevices in the bark of trees.

UNDERWATER FEEDING
Dippers are among the few passerines to venture into water. When feeding, they dive into shallow, fast-flowing streams, propelling themselves downward with their wings, to catch prey as it swims. They also walk along the riverbed, picking up grubs in their bill.

Anting

A few passerines, including starlings and blackbirds in Europe, indulge in an unusual behavior known as anting, in which a bird collects ants in its bill and rubs them over its feathers. Occasionally, the bird will crouch down and let ants run over its plumage. Birds select only ants that exude formic acid, so it is thought that the habit may help cleanse the feathers of parasites. Also unique to passerines is ant following, in which birds follow army-ant columns through tropical forest. The birds do not eat the ants themselves, but catch the small insects fleeing the ants' path.

COLLECTING ANTS
A Eurasian jay lies down among a swarm of ants, allowing the insects to crawl through its plumage. The reason for anting is unknown, but it is thought that acid produced by some ants may act as an insect repellent or insecticide, or as a plumage lubricant.

BIRDS

Order Passeriformes

Family Acanthisittidae

New Zealand wrens

Length 3¼–4 in (8–10 cm)

Species 2

Found only in New Zealand, this family includes the rifleman and rock wren. Both sexes of the rifleman and rock wren incubate eggs and feed the young. One species, the virtually flightless Stephens Island wren, was wiped out—ostensibly by a lighthouse keeper's cat—in the late 1800s; another, the bush wren, was last seen in 1972. Other species are known from subfossils—all were probably weak flyers. Genetic and anatomical evidence suggest that this family represents an ancient lineage of birds that is the sister group to all other passerines.

RIFLEMAN
Abundant and widespread, Acanthisitta chloris *occurs on North and South Islands. It is 3¼ in (8 cm) long; unusually among passerines, the female is larger. She is also more streaked.*

Family Pittidae

Pittas

Length 6–11 in (15–28 cm)

Species 34

Stumpy-looking birds with an acute sense of smell, pittas live on the forest floor in Asia, Australia, and Africa, and feed on insects, spiders, snails, and other invertebrates. They are secretive birds and seldom seen, often blending in with the forest floor despite their highly colorful plumage. The female lays 2–7 eggs and both sexes raise the young.

AFRICAN PITTA
Pitta angolensis*, one of two African species, is a migrant found in thick, evergreen forest. This usually silent bird is 8 in (20 cm) long.*

Family Eurylaimidae

Broadbills

Length 5–11 in (13–28 cm)

Species 15

Broadbills are found across the humid tropics, from West Africa to the Philippines, in habitats including scrub and forest-edge thickets, mangrove swamps, inland forest, and mountain moss-forest. These plump birds have large heads and wide, flattened, hooked bills. Males are usually colored a striking green, red, pink, or blue; females are duller but can be larger than males.

Broadbills feed mostly on fruit and insects; one species eats lizards, and another catches crabs and even fish. Most feed in the midlevel of the forest, filter-sucking insects from leaves and branches. Some broadbills feed alone or in pairs, others gather in foraging flocks of 20–30 when not breeding. These birds build large, pear-shaped nests with a "porched" entrance in the lower half. Woven from rootlets, leaves, and twigs, and often decorated with cobwebs and lichens, these are hung from inaccessible branches. Females lay clutches of 1–8 eggs.

GREEN BROADBILL
The Southeast Asian species Calyptomena viridis *is 8 in (20 cm) long, has long feathers around the bill, nearly concealing it, and 3 black wingbars. Small groups feed in the lower branches on ripening fruit and buds. The green broadbill is also known to feed on insects such as flying termites.*

Family Philepittidae

Asities

Length 4–6 in (10–15 cm)

Species 4

Asities are short, stocky birds found only in dense, evergreen forest mainly in the humid east of Madagascar. Schlegel's asity is also found in the northwest of the island. The velvet asity and Schlegel's asity feed mainly on fruit plucked from bushes, while the 2 sunbird-asities are insect-eaters and also probe flowers, probably for nectar. Only the nest of the velvet asity—a pear-shaped hanging structure with 3 white eggs—has been found. The forest habitat of these birds is fast disappearing with the increase of cultivated land.

COMMON SUNBIRD-ASITY (MALE)
Neodrepanis coruscans *occurs at all levels in the forest, from leaf litter to high canopy. It is 4 in (10 cm) in length and has a downcurved bill.*

Family Dendrocolaptidae

Woodcreepers

Length 5½–14½ in (14–37 cm)

Species 52

Woodcreepers are a family mostly found in South America. They behave in a similar fashion to European and American treecreepers (see p.360). These birds have a long, distinctive tail with stiffened feather shafts that curve inward. Like treecreepers and woodpeckers, they use the tail to support themselves as they climb trees. They also have strong feet and long, sharp claws to help them climb. Generally olive-brown, with rufous wings and tail, many have a sturdy, relatively long, straight or slightly downcurved bill. Some have a thinner bill with a more defined downward curve, and the extreme is reached in the scythebills, which have a long, sickle-shaped bill, up to one-third of their total length. When hunting for insects and invertebrates, woodcreepers work up a tree, either straight up or in a spiral, examining crevices, loose bark, moss, and other plants, flying to the base of the next tree as they finish with one. Larger species also eat lizards and small frogs. Some species follow army ant swarms, feeding on the insects disturbed by the columns of ants as they move. Woodcreepers nest in old woodpecker holes and crevices.

WEDGE-BILLED WOODCREEPER
Found in humid lowland forests, mainly in the Amazon Basin, Glyphorhynchus spirurus *is 5½ in (14 cm) long. Although rarely seen, its sneezing "cheeyf" call often draws attention.*

Family Furnariidae

Ovenbirds

Length 6–10 in (15–25 cm)

Species 246

A large group with many subfamilies, ovenbirds are mostly small brown birds with paler underparts, found in Central and South America. They live in a wide range of habitats, and may be foliage-gleaners, living in the canopy, or leaf-scrapers, which flick leaves in the air and skulk in the densest undergrowth. Many species, such as the true ovenbirds, or horneros, prefer open country, while the cinclodes live along water courses; a few species inhabit marshes. A number of species, such as the Patagonian earthcreeper and brown cachalote, tend to run rather than fly. Ovenbirds are mostly insect-eaters, but some also feed on seeds. Their nesting habits vary widely: while the true ovenbirds build substantial, domed mud-ovens on tree branches, many species nest in natural holes, animal tunnels, or holes they dig themselves. The white-throated cachalote builds a huge stick nest that lasts for many years after the birds have used them. Almost all ovenbirds lay white eggs, usually in clutches of 3–5; some species lay up to 9 eggs. These are mostly incubated over 15–20 days, with the young leaving the nest in 13–18 days. Old ovenbird nests are often used by other birds.

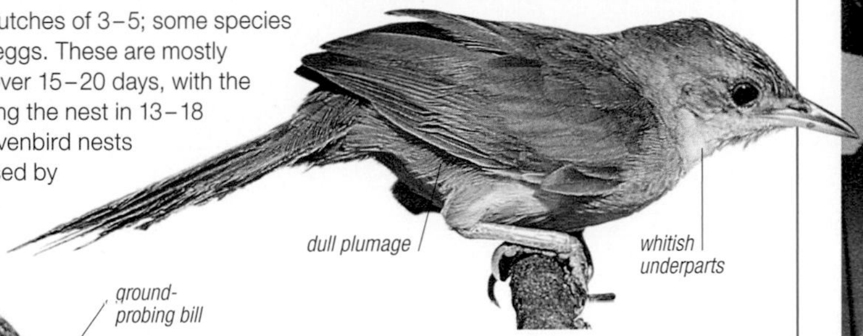

RUFOUS-FRONTED THORNBIRD
Occurring in widely separated areas of semi-arid habitat, Phacellodomus rufifrons *is 6½ in (16 cm) long. It builds a large, hanging stick nest, which has several chambers and is often used by other birds. This South American bird spends much of its time in trees, in pairs or groups.*

RUFOUS HORNERO
Argentina's national bird, Furnarius rufus, *has a misleading common name as it is the least rufous of horneros. Measuring 7–8 in (18–20 cm) in length, it clambers about on the ground or on branches. Its mud-oven nest, which it uses only once, adorns the top of fence posts and telephone poles, lasting for years.*

DARK-BELLIED CINCLODES
Found in southwest South America, Cinclodes patagonicus *is always near fresh water or the sea. It moves along rocky streams and rivers, feeding on small aquatic animals. This bird is 8½ in (21 cm) long, and generally nests in a rock burrow, often in a stream bank. Its call is a sharp "tjit".*

BUFF-FRONTED FOLIAGE-GLEANER
Philydor rufum, *7½ in (19 cm) long, is locally common in humid mountain forests from Costa Rica to Argentina. It searches the canopy for insect prey, hanging upside down athletically.*

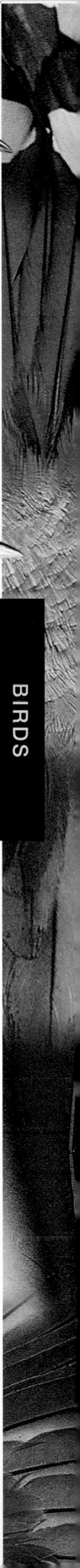

Family Grallariidae

Antpittas

Length 4–9 in (10–22 cm)

Species 51

More often heard than seen, these strong-legged, stubby-tailed birds used to be classified in the same family as antbirds (Thamnophilidae), but they may be more closely related to ovenbirds (see above) and tapaculos (see right). Sexes are alike: both cryptically patterned in browns and grays. They spend more time on the ground than true antbirds, where they move by hopping and feed on invertebrates. The related ant-thrushes (Formicariidae)—which walk, rather than hop—were formerly classified with antbirds, too.

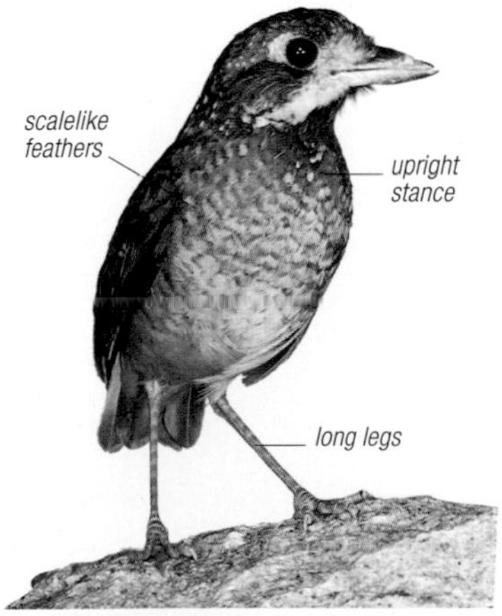

VARIEGATED ANTPITTA
Grallaria varia, *8 in (20 cm) in length, has the long legs and short tail typical of the antpittas. It is found in Guyana, Surinam, French Guiana, and Brazil. This bird flicks leaves aside on the forest floor to probe for earthworms and insects.*

MOUSTACHED ANTPITTA
Many antpittas—such as the Grallaria alleni *from Colombia and Ecuador—are poorly known. It is a skulking bird that lives in dense undergrowth in humid montane forests on the slopes of the Andes.*

JOCOTOCO ANTPITTA
*Discovered in 1997 in southern Ecuador—and still known from only 5 locations—*Grallaria ridgelyi *is a large, boldly patterned antpitta. Like other antpitta species, it has strong legs for hopping about on the ground, where it is known to feed on insects and their larvae, worms, and millipedes. It has only been found in moss-covered montane bamboo forests beside streams, a perpetually wet habitat that is under threat because of deforestation. Its already low population size is undoubtedly in decline because of this.*

Family Rhinocryptidae

Tapaculos

Length 4¼–10 in (11–25 cm)

Species 60

Found principally in southern South America, tapaculos have a distinctive, movable flap of skin covering the nostril. They are long-legged, mainly ground-dwelling birds that feed on invertebrates. Their short, rounded wings and long tails allow them to fly only briefly and weakly. Secretive and rarely seen, tapaculos are identified by their loud calls. They generally build their nest in a hole or crevice at ground level, or in a burrow that they dig themselves, and lay 2–4 white eggs.

CHUCAO TAPACULO
Scelorchilus rubecula *superficially resembles a European robin. It has a striped black and white belly, and is 7½ in (19 cm) long. It is found in south Chilean temperate forest, including bamboo thickets.*

Order Passeriformes *continued*

Family Thamnophilidae

Ant-birds

Length 3¼–14 in (8–36 cm)

Species 220

A high diversity of ant-birds—including ant-shrikes, ant-wrens, and ant-vireos—live in the American tropics, reaching peak diversity in Andean and Amazonian forests. These are largely tree-dwelling birds that get their names from the tendency of certain species to follow swarms of army ants, snatching the small animals flushed out. Many ant-birds have long claws for clinging to vertical saplings as they wait for prey. Few habitually forage on the ground—most snatching insects from foliage. Although otherwise secretive in dense vegetation, ant-birds often betray their presence by their whistling or chattering calls. Most members of the family are sexually dimorphic: males are generally gray, black, or white; females are brown with striking patterns that vary from species to species. A few—such as certain ant-shrikes—have crests. Those that have been studied seem to mate for life and most build a simple cup nest in a tree fork in which they lay 2, rarely, 3 eggs.

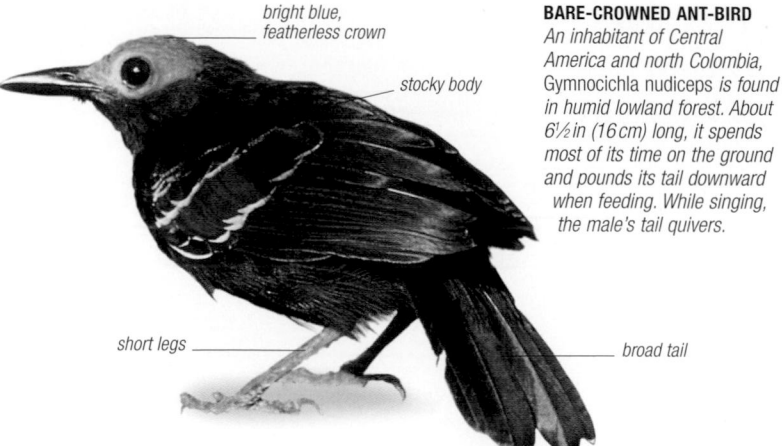

BARE-CROWNED ANT-BIRD
An inhabitant of Central America and north Colombia, Gymnocichla nudiceps *is found in humid lowland forest. About 6½ in (16 cm) long, it spends most of its time on the ground and pounds its tail downward when feeding. While singing, the male's tail quivers.*

Family Tyrannidae

Tyrant flycatchers

Length 2–15 in (5–38 cm)

Species 415

This large family is very diverse and some experts split it into even more species. It contains the world's smallest passerine, the short-tailed pygmy-tyrant, which is just 2 in (5 cm) long with a wingspan of 1½ in (3.5 cm). However, within this large and varied family, body sizes vary widely and include many medium-sized birds. This family is found from the taiga of Alaska, south through North America and the West Indies to the southern tip of South America, and in offshore islands including the Galapagos. However, the vast majority occur in Central and South America, with only 30 species breeding in North America; all the North American species migrate south in the autumn. Tyrant flycatchers have drab plumage overall and the sexes look similar. There are exceptions, however, with some species having colorful crown patches. Most members of the family are insect-eaters, but some catch frogs and lizards, and tropical species regularly eat fruit. Most tyrant flycatchers form monogamous pairs and many have courtship displays that involve wing-stretching or flicking. These birds are not colonial nesters, and their nests vary considerably: many build a simple cup in a tree fork, but some create large, hanging nests suspended from a branch; quite a few nest in cavities. Female tyrant flycatchers in the neotropics lay 2 or 3 eggs, but the clutches can be larger in North America. Females incubate their eggs for 14–20 days and the nestlings fledge in 14–23 days. Usually, both sexes build the nest and feed the young.

EASTERN KINGBIRD
Tyrannus tyrannus, *8 in (20 cm) long, breeds in North America and winters in Central and South America. It eats insects in summer and fruit in winter. Renowned for its vigorous defense of territory, it often attacks other birds, even landing on their backs in flight to inflict blows.*

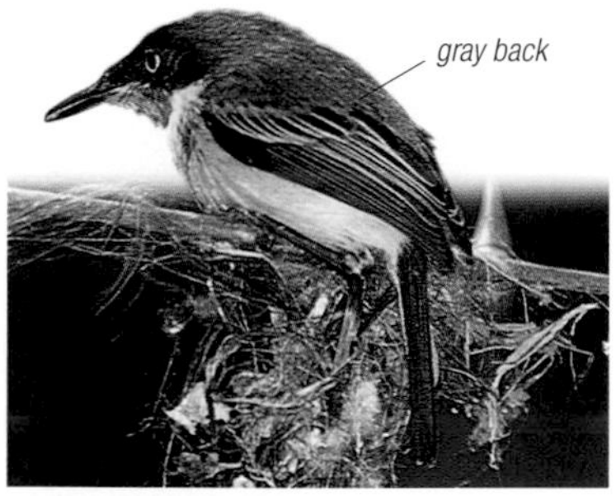

COMMON TODY-FLYCATCHER
The most common and widespread tody-flycatcher, Todirostrum cinereum, *moves quickly among leaves as it catches insects. It can look quite comical as it hops and flutters about, perpetually in motion. It measures 3½ in (9 cm) in length, and occurs in Central America and parts of South America.*

GREAT KISKADEE
Common, widespread, and noisy, Pitangus sulphuratus *has a variety of calls that include "kis-ka-dee." It eats insects and fruit but will also feed on fish and tadpoles, diving for them from a perch. The great kiskadee occurs from Texas, USA, south to Argentina. It is 8½–9½ in (21–24 cm) in length.*

LEAST FLYCATCHER
Empidonax minimus *breeds in eastern Canada and the USA, and winters from Mexico to Panama. This bird, 5–5¾ in (12.5–14.5 cm) long, is the first species of the genus* Empidonax *to migrate north and the last to fly south. It flies out from perches to catch insects and builds a frail-looking cup nest in a tree fork.*

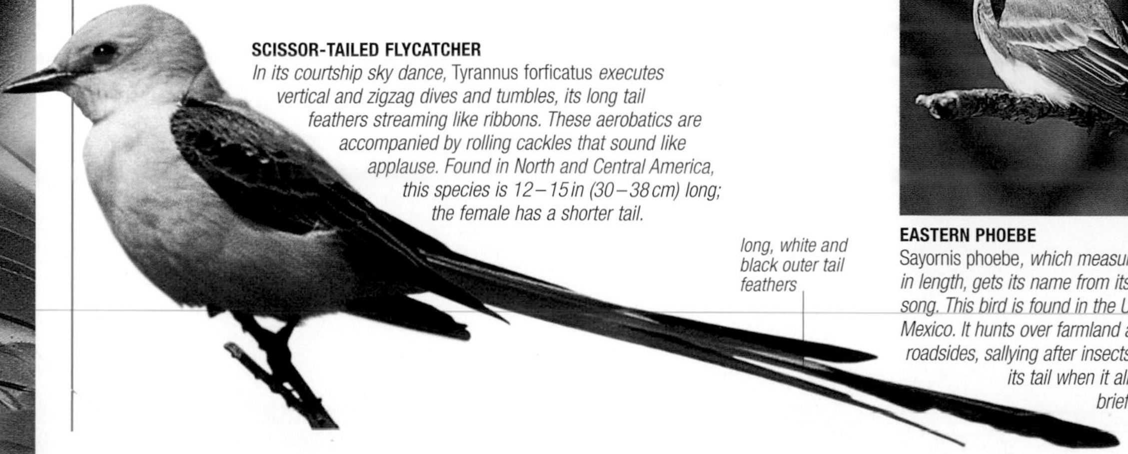

SCISSOR-TAILED FLYCATCHER
In its courtship sky dance, Tyrannus forficatus *executes vertical and zigzag dives and tumbles, its long tail feathers streaming like ribbons. These aerobatics are accompanied by rolling cackles that sound like applause. Found in North and Central America, this species is 12–15 in (30–38 cm) long; the female has a shorter tail.*

EASTERN PHOEBE
Sayornis phoebe, *which measures 7 in (18 cm) in length, gets its name from its hoarse "fee-bee" song. This bird is found in the USA, Canada, and Mexico. It hunts over farmland and wooded roadsides, sallying after insects and waggling its tail when it alights. It may hover briefly over water when chasing flies.*

ROYAL FLYCATCHER
Onychorynchus coronatus *is 6½ in (16 cm) long and has a spectacular, brightly colored, fanlike crest: the male's crest is scarlet with blue tips, while the female's is yellow with blue tips. The crest is normally folded away and sometimes protrudes behind the head, giving the bird a "hammerhead" look. The royal flycatcher is found in Central and South America.*

Family Cotingidae

Cotingas

Length 3½–18 in (9–45 cm)

Species 101

Cotingas vary widely in appearance. They include the 2 bright orange cock-of-the rock species, brilliant blue cotingas, umbrellabirds with their umbrella-like crests, and the bellbirds with their wattles. Most are found in lowland forest, but some occur in mountain forest and shrubbery in the Andes. All cotingas are fruit-eaters, and some also feed on insects. Males of some species display at a lek to attract females. Some species build a tiny platform nest and lay only one egg, which takes 23–28 days to hatch.

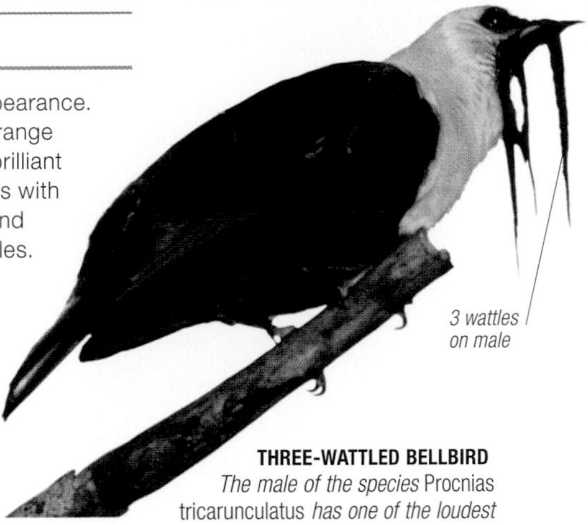

THREE-WATTLED BELLBIRD
The male of the species Procnias tricarunculatus *has one of the loudest bird calls, a deafening "bock" that rings over the forest canopy. This Central American bird is 10–12 in (25–30 cm) long. Its bill opens wide enough to swallow large fruit whole.*

AMAZONIAN UMBRELLABIRD
Found along the Amazon and Orinoco rivers, and in east Andean forest, Cephalopterus ornatus *is 16–19½ in (41–49 cm) long. While courting, males expand their crest and their enlarged wattle as they utter a deep, moaning "boom." Females are smaller.*

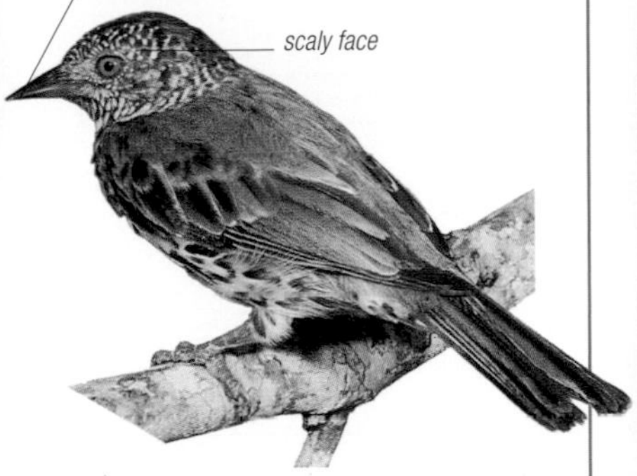

SHARPBILL
The aptly-named Sharpbill (Oxyruncus cristatus)*—ranging from Costa Rica to Paraguay—lacks the bright colors of many birds of this family but can flash a small orange crest. It was previously classified with the tyrant flycatchers, but is genetically closer to certain members of the Cotingidae.*

ANDEAN COCK-OF-THE-ROCK
During courtship, the males of this species, Rupicola peruvianus, *confront each other aggressively at a lek, the display becoming frantic when females appear. This bird is 12 in (30 cm) long, and as its name suggests, is found in the Andes.*

LOVELY COTINGA
Cotinga amabilis, *8 in (20 cm) long, is found in high forest canopies of Central America. The male's outer wing feathers make a tinkling noise during courtship. The female may destroy the nest after her young have flown, to prevent predators from finding the site, which is often reused.*

RUFOUS PIHA
Lipaugus unirufus, *9 in (23 cm) long, inhabits Central America and northwest South America. Males do not display at a lek but may utter loud, whistling calls within earshot of each other.*

WHITE-TIPPED PLANTCUTTER
Phytotoma rutila, *7½ in (19 cm) in length, is found in the dry scrub and acacia groves of Bolivia, Paraguay, Uruguay, and Argentina. This short-crested bird has a mechanical-sounding call.*

Family Pipridae

Manakins

Length 3½–6 in (9–15 cm)

Species 51

Small birds with proportionately big heads, short tails, and broad, slightly hooked bill, manakins are found mainly in the lowlands of Central and South America. Males are generally brightly colored, often black with solid patches of blue, red, yellow, or orange. Females are usually greenish. Males of certain species have modified feathers that can produce a mechanical noise during courtship. The male wire-tailed manakin uses his long, curved, wirelike tail feathers to tickle the throat of a visiting female. Males of some species gather in leks to display to females, either on the ground or on a branch. Successful males mate with several females. The female alone builds the nest—a woven cup in a horizontal tree fork—incubates the eggs, and feeds the young. She generally lays 2 eggs that hatch after 12–15 days. The nestlings feed on insects, but adults feed mainly on fruit, plucking these on the wing. All manakins that have been studied are sedentary and do not migrate.

WHITE-BEARDED MANAKIN
At a lek, the male Manacus manacus, *using a sapling as the display perch, performs stylized jumps and produces noises with his wing feathers. This South American bird is 4 in (10 cm) long.*

BLUE-BACKED MANAKIN
Chiroxiphia pareola, *4¾ in (12 cm) long, is found in South America. Two males often display by jumping over each other on branches; the winner makes short, circular courtship flights.*

Order Passeriformes *continued*

Family Menuridae

Lyrebirds

Length 35–39 in (90–100 cm)

Species 2

Lyrebirds occur in rainforests and scrub in east and southeast Australia. Of the 2 species, Albert's lyrebird is the smaller and has more chestnut plumage, while the superb lyrebird is grayer. Males of both species are known for their extravagant tail, which in the superb lyrebird can be up to 23½ in (60 cm) long, and has 2 distinctive lyre-shaped feathers. Male lyrebirds display from midwinter; while Albert's lyrebird treads down a display platform of vines, the superb lyrebird rakes up numerous low display mounds around his territory. The male's powerful song includes impressive mimicry. While calling, he inverts his tail over his back, shimmering it and turning around slowly.

SUPERB LYREBIRD
Menura novaehollandiae, *34–39 in (86–100 cm) long, is found in southeast Australia and has been introduced in Tasmania. The male of this species does not acquire his full tail until he is 6–8 years old.*

Family Climacteridae

Australasian treecreepers

Length 4–7 in (10–17.5 cm)

Species 7

These birds from Australia and New Guinea are unrelated to true treecreepers of the Northern Hemisphere (Certhiidae) and, unlike them, do not use their tail as support, as they clamber around tree trunks. However, they have evolved to resemble them because of their otherwise similar habits. They have mostly brown plumage, thin, curved bills and strongly clawed feet, and they feed on insects and spiders plucked from tree bark; sometimes they forage on the ground. Their nest—a grassy cup—is usually placed in a tree cavity. Two or 3 eggs are incubated by both sexes.

WHITE-THROATED TREECREEPER
Cormobates leucophaeus, *6½–7 in (16–17.5 cm) long, occurs in eastern and southeastern Australia in a wide range of habitats.*

Family Ptilonorhynchidae

Bowerbirds

Length 8½–15 in (21–38 cm)

Species 20

Male bowerbirds are well known for the elaborate bowers they create to attract females—these are complex structures decorated with brightly colored objects. They also decorate their display grounds (leks) with feathers, flowers, leaves, pebbles, and even clothes pins or colored bits of plastic and paper. They cohabit with one or more females and the latter rear the young. These birds are spread across New Guinea and Australia, where most inhabit damp forest, although some prefer drier areas. Bowerbirds are mainly fruit-eaters, but will also feed on leaves, flowers, seeds, and invertebrates, feeding at all levels from ground to canopy.

SATIN BOWERBIRD
Restricted to northeast and southeast Australia, Ptilonorhynchus violaceus *is 12 in (30 cm) long. The glossy, blue-black male—the most photographed bowerbird—builds an "avenue" bower of sticks, decorating it with bright blue or yellow objects. This bird has a pale yellow bill.*

Family Atrichornithidae

Scrub-birds

Length 6¼–9 in (16–23 cm)

Species 2

The distribution and population of these brown Australian birds have decreased since the mid-1800s. Secretive, they spend much of their time on the ground in thick cover, in wet forest. They feed on insects, lizards, and frogs. Both species of scrub-birds build domed nests.

RUFOUS SCRUB-BIRD
Found in highland rainforest, in a small area of east Australia, Atrichornis rufescens *has a penetrating call but is extremely hard to locate. It is 6½–7 in (16.5–18 cm) long, and is a weak flier that scuttles in leaf litter like a mouse.*

Family Orthonychidae

Logrunners

Length 6½–11 in (17–28 cm)

Species 3

The 3 secretive species of logrunners from forests of New Guinea and eastern Australia are weak-flying birds that live close to the ground. Here, they forage for insects, always preferring to run than fly when threatened. Their tails are spiny due to the stiff bare feather shafts that extend beyond the tips. Logrunners have loud, resonant, voices and one Australian species—the chowchilla—is named after its call. They build large dome nests out of sticks and moss, placed close to the ground. Several other babblerlike birds were formerly classified in this family; Orthonychidae is now thought to be one of several small groups of ancient passerines restricted to Australasia.

AUSTRALIAN LOGRUNNER
Seen in pairs or family parties on the floor of the rainforest in southeastern Australia, Orthonyx temminckii *is 7–8 in (18–20 cm) long.*

Family Pardalotidae

Pardalotes

Length 3½–4½ in (9–12 cm)

Species 4

Pardalotes are small, short-tailed birds found in Australia. Three species are white-spotted to a variable degree (pardalote comes from a Greek word, meaning spotted); the fourth—the striated—has white head and wing stripes. They are insect-eaters and their stumpy bills are adapted for taking sap-sucking scale insects from leaves; like many other birds of the region, they also drink the sweet honeydew exuded by these and related insects—and are often in fierce competition with honeyeaters (see opposite) over sources of this energy-rich food. Pardalotes are typically birds of tall eucalyptus forest, where they feed in the canopy. They descend to earth banks to build their nests in holes, but sometimes nest in tree holes; a clutch consists of 2–5 white eggs. Pardalotes were previously classified with flowerpeckers (p.366), but are now thought to be more closely related to Australasian warblers.

STRIATED PARDALOTE
Endemic to Australia, Pardalotus striatus *is common in eucalyptus forest and other types of woodland. It is about 4 in (10 cm) long. There are at least 5 geographically variable populations, with either black or striped crowns.*

BIRDS

Family Meliphagidae

Honeyeaters

Length 3¾–12½in (9.5–32cm)

Species 183

Found mainly in Papua New Guinea, Australia, Asia, and the Pacific Islands, honeyeaters are mostly nectar- and fruit-eating birds with a characteristic brush tongue adapted to nectar-feeding. This tongue can be extended into nectar about 10 times per second. They are generally dull-colored birds, and the sexes are similar. However, there are exceptions: the regent honeyeater is brighter, with its yellow and black plumage, and the sexes among scarlet honeyeaters differ—males are red and females are brown. Members of this family vary greatly in size. Some resemble warblers or thrushes, others look like hummingbirds, and the larger honeyeaters could almost be mistaken for magpies. Some of these birds do not have feathers on their face and develop wattles. All are found in wooded habitats, with only a few descending to the ground to feed. Outside the breeding season, these birds often migrate in search of flowering trees.

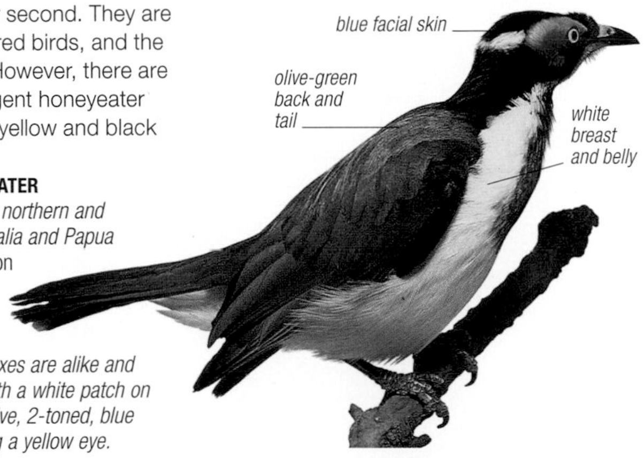

BLUE-FACED HONEYEATER
Distributed across the northern and eastern sides of Australia and Papua New Guinea, Entomyzon cyanotis *is 12in (30cm) long and is found in open forest and farmland. Both sexes are alike and have a black head, with a white patch on the nape, and distinctive, 2-toned, blue facial skin surrounding a yellow eye.*

NOISY MINER
Found only in eastern Australia, including Tasmania, the noisy, gregarious Manorina melanocephala *is locally called the soldierbird. It is 10in (26cm) long and both the sexes look alike. It breeds in colonies from June to December.*

TUI
Endemic to New Zealand, where it is found in native forest and scrub, Prosthemadera novaeseelandiae *has glossy black plumage with a purple-green iridescent color and 2 white throat tufts. An energetic bird with a rich, melodic song, it feeds on nectar, fruit, and insects. The male—12in (30cm) long—is larger than the female.*

LITTLE FRIARBIRD
Philemon citreogularis *is found in Australia, Papua New Guinea, and the Lesser Sundas in Indonesia. At 10½in (27cm) long, it is the smallest of the friarbirds—a group of 16 species within the honeyeater family.*

Family Acanthizidae

Australasian warblers

Length 3½–5in (9–13cm)

Species 66

This family is centered on New Guinea and Australia, but a few species occur westward in the Philippines and Indonesia or eastward on Pacific Islands and New Zealand. Many species have very localized distribution ranges. This fact, however, helps in identification, because similar species can occur in geographically distinct areas. Usually olive, brown, or yellow, the sexes among Australian warblers are similar. Many species are insectivores, while some feed on seeds. They are often seen in pairs, in family parties, and sometimes in mixed-species flocks. Most species do not migrate. The nest made by these birds is a dome of plant fibers, often with a "porch." Females lay 2–4 white, sometimes spotted, eggs. Incubation and fledging each take 15–20 days.

WHITE BROWED SCRUBWREN
A widespread Australian species, Sericornis frontalis *is a dozen or so noisy scrubwrens of dense thickets. The white-browed scrubwren has a rich and varied voice, and can also mimic.*

BROWN THORNBILL
Found in eastern and southeastern Australia and Tasmania, Acanthiza pusilla *is 4in (10cm) long, and has narrow streaks on its gray throat. It forages among the twigs of lower branches and has a variety of calls, among them a scolding note. The nest is built low down in the undergrowth.*

TREEFERN GERYGONE
About 3½in (9cm) in size, Gerygone ruficollis *is found in primary forest in the mountains of New Guinea, at heights of 3,650–10,900ft (1,100–3,300m). Its song is a distinctive and lengthy series of high-pitched whistles, and its nest a pendant, globular structure with a side entrance.*

Family Maluridae

Fairy-wrens

Length 4¾–7½in (12–19cm)

Species 29

Found in Australia, Papua New Guinea, and eastern Indonesia, fairy-wrens are brightly colored birds that hold their tails upright or over their back in the same way as wrens (see p.361). The metallic colors of their plumage include blues, purples, reds, blacks, and whites. Sexes generally differ. Many fairy-wrens are good singers, some being good mimics as well. They hunt for insects and their larvae through the underbrush, and are often found in flocks. Their nest is usually domed, and is made of grass, spiders' webs, and a lining of feathers and plant down. Fairy-wrens lay 2–4 whitish, speckled eggs, which are incubated for 12–15 days; the nestlings fledge in 10–12 days.

SOUTHERN EMU-WREN
Notable for its very long tail, formed of 6 delicate feathers, Stipiturus malachurus *is 6–7½in (15–19cm) long. It occurs along the coastal strip of southern Australia and in Tasmania.*

BLUE WREN
Also known as the superb fairy-wren, Malurus cyaneus *is found in southeast Australia and Tasmania. It is 5–5½in (13–14cm) long, and has a tinny song.*

RED-BACKED WREN
At 4–5in (10–13cm) long, Malurus melanocephalus *is the smallest fairy-wren. Found in northern and eastern Australia, it occupies habitats varying from tall grass to plantations and gardens, and breeds from November to March.*

Order Passeriformes *continued*

Family Callaeatidae

Wattlebirds

Length 10–15 in (25–38 cm)

Species 2

Wattlebirds, confined to New Zealand, are named for the orange or blue wattles at the base of their bill. They are short-winged, weak fliers but are adept at clambering and hopping about branches. The kokako and saddleback are the only living species. Surveys in the 1970s revealed that only tiny colonies of kokako remained in the North Island. Conservationists worked on plans to ensure its survival, which involved preservation of podocarp forests and control over the predation of its eggs and chicks. There are now around 30 separate populations of kokako, but it is still not out of danger. The other extant wattlebird is the saddleback (shown right). The huia, *Heterolocha acutirostris*, last seen in 1907, was unique in that the sexes had different bill shapes: the male had a straight, pointed bill, while the female had a thin, downcurved one. The huia became extinct due to habitat destruction and being hunted for its feathers.

SADDLEBACK
A weak flier, Philesturnus carunculatus*—which is 10 in (25 cm) long—runs and hops through trees, probing the bark for insects. It nests on remote islands and is seldom seen. A small population has recently been introduced to an island in Auckland Harbour where visitor numbers are regulated.*

Family Vangidae

Vanga shrikes

Length 4¾–12 in (12–30 cm)

Species 22

Vanga shrikes are restricted to Madagascar (with the blue vanga also found in the Comoro Islands), but it is possible that other African shrikes belong in this family. All are tree-living birds, some being confined to the evergreen forest in the humid east, while others extend into wooded areas in the western savanna, and some even occur in semidesert scrub. Many of them are boldly patterned in black and white, and have a heavy bill, often hooked at the tip. Although they usually eat insects, some may also feed on treefrogs and small reptiles. They often form loose feeding flocks, sometimes with other species. Some species build a cup of twigs in a tree and lay 3–4 eggs; both parents raise the young.

HOOK-BILLED VANGA
A solitary bird, Vanga curvirostris *is 10–11½ in (25–29 cm) long, and has a heavy, hooked bill. It lives in evergreen forest, brush, plantations, and mangroves across Madagascar.*

Family Malaconotidae

Bush shrikes and relatives

Length 6–12 in (15–30 cm)

Species 58

Confined to Africa, bush shrikes are closely allied to the vanga shrikes of Madagascar, and some—such as the crested helmet-shrikes—may be genetically closer to vangas.

Although many are shy, retiring birds—they are generally more brightly colored than birds of the true shrike family—Laniidae—in which they were previously classified. Some, such as helmet-shrikes, travel in noisy groups, and all have distinctive loud calls that are often delivered from a prominent perch. Tchagras are more somberly colored thicket-dwellers, but have distinctive darker head patterns. Puffbacks are so-called because of the fact they puff out loose feathers on their back during display: displaying males look like fluffy balls. Members of the family occur in forest or dry thorny scrub, where they prey on insects; larger species can also tackle frogs and snakes. They lack the true shrike's habit of impaling victims on thorns and spikes, and tend to be more active in pursuit of prey. Nests are cup-shaped affairs in bushes or trees.

WHITE HELMET-SHRIKE
The most common and widespread of all the species of helmet-shrikes, Prionops plumatus*, which is 6–10 in (15–25 cm) long, is found in bushes in sub-Saharan Africa. It also occurs in areas of cultivation and open woodland. This noisy and gregarious bird gathers in flocks of 5–12, chattering loudly as it forages for food.*

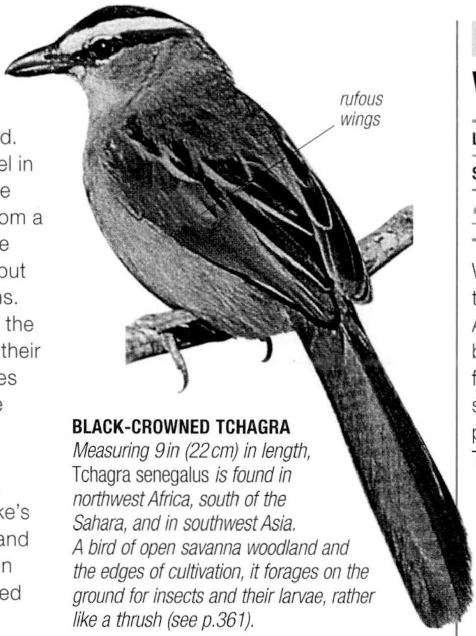

BLACK-CROWNED TCHAGRA
Measuring 9 in (22 cm) in length, Tchagra senegalus *is found in northwest Africa, south of the Sahara, and in southwest Asia. A bird of open savanna woodland and the edges of cultivation, it forages on the ground for insects and their larvae, rather like a thrush (see p.361).*

BOKMAKIERIE
Found in southern Africa, Telephorus zeylonus *occurs in the bush, open savanna, plantations, and gardens. It is 9 in (22 cm) long and, like a thrush, spends much of its time on the ground. This bird performs a beautiful courtship dance in spring. Its name is a transcription of one of its loud, ringing calls.*

Family Artamidae

Woodswallows

Length 4¾–8 in (12–20 cm)

Species 11

Woodswallows occur in parts of Asia, the Pacific Islands, New Guinea, and Australia. Despite the family name, these birds are not closely related to the swallow family, although their flight silhouette is similar to that of martins. Unlike most passerines, they often soar on the wing. They are 4¾–8 in (12–20 cm) long, and have strong bills and short legs and feet, while the wings are quite short and pointed. A characteristic family trait is the birds' habit of huddling together on branches, particularly when roosting. These insectivorous birds catch much of their prey on the wing. They are gregarious and build their grassy, cuplike nests on stumps, in holes or in bushes. All members of the family are colonial nesters. Females lay 2–4 white or cream eggs, with red-brown spots.

BLACK-FACED WOODSWALLOW
Distributed across Australia, New Guinea, the Lesser Sundas, and Timor, Artamus cinereus*—measuring 7½ in (19 cm) in length—is the most common bird in the family. Unlike other woodswallows, it is not very nomadic.*

Family Campephagidae

Cuckoo-shrikes

Length 5½–16 in (14–40 cm)

Species 94

Cuckoo-shrikes have no connection with either cuckoos or shrikes. This is mainly a tropical family consisting of 2 groups of birds ranging from Africa across southern and Southeast Asia to Australia and the islands in the western Pacific; the minivets occur only in Asia. Except for the ground cuckoo-shrike, found on the treeless Australian plains, these are usually tree-living birds. Minivets are colorful birds, the size of wagtails; the male tends to be red and black, whereas the female is yellow or orange and black or gray. The remaining cuckoo-shrikes resemble cuckoos in flight and are often dull gray, ranging from the size of sparrows to that of pigeons. The females of the latter group are often paler, and many species have bristles growing from the base of the bill. The nest is built high in a tree and the female lays 2–5 eggs, usually incubating them alone.

COMMON CICADABIRD
Found in Australia, Papua New Guinea, and eastern Indonesia, Coracina tenuirostris *lives in the canopy of rain forest and other woodland. The female is browner with barred underparts. The name is derived from the male's harsh, descending, cicada-like buzz, which it repeats often. This bird is 9½–10 in (24–26 cm) long.*

YELLOW-THROATED MINIVET
Found from the Himalayas through southern China and Southeast Asia to Borneo, Pericrocotus solaris *lives in open forest. It measures 7½ in (19 cm) in length. The male has a gray throat and an orange belly, but the female is duller with yellow rather than orange underparts.*

Family Cracticidae

Bell-magpies

Length 7–21 in (18–53 cm)

Species 13

Bell-magpies resemble crows with their strong bodies, sturdy bills, and the scales on their legs and feet. However, they are less carnivorous than crows and eat a wide range of food, including fruit. They are found in Asia, Australia, and New Guinea, the most common bird being the Australian magpie. Also included in this family are the butcher birds and currawongs. The lowland and mountain peltops of New Guinea (considered members of the woodswallow family by some) have black plumage with a white face patch and red undertail coverts.

AUSTRALIAN MAGPIE
Found in Australia, New Zealand, and New Guinea, Gymnorhina tibicen *is an opportunist feeder that scavenges on rubbish and also kills smaller birds and rodents. It is 16½ in (42 cm) long and is mostly black-backed, although there are many regional variations.*

Family Platysteiridae

Wattle-eyes and batises

Length 3¼–6½ in (8–16 cm)

Species 38

All of these birds are found in Africa, mostly in forest or forest edges. Wattle-eyes have colorful wattles around their eyes. Batises are small, flycatcher-like birds. Males generally have contrasting black and white plumage—often with a breastband; females are brownish below. They are insectivorous and can catch their prey on the wing. They build cup-shaped nests and lay 2–5 eggs. Various species of wattle-eyes and batises look very similar and are easily confused. However, they can usually be separated on the basis of geographical distribution.

CHINSPOT BATIS
Found in Sudan, Kenya, southwest Africa, and Mozambique, Batis molitor *frequents open woodland, darting out after insects like a flycatcher; it also hovers to search leaves. It is 4 in (10 cm) long, and has a gray crown.*

Family Pachycephalidae

Whistlers

Length 4¾–11 in (12–28 cm)

Species 46

The whistlers make up an Indo-Pacific family of birds, ranging from Indochina to the southwest Pacific. As their name suggests, they are known for their impressively strong calls and songs. They reach their maximum diversity in rain forests on mountains of New Guinea and on tropical islands around the region, where they have evolved into many local species and subspecies—most in the genus *Pachycephalus*. Whistlers are big-headed birds (*Pachycephalus* means "thick-head") and females are usually drabber than males. Plumage varies from plain to colorful plumage: males of many have a blackbreast band so making species identification difficult. They feed on insects, which they glean from foliage—larger species take small vertebrate prey, too. They forage in the mid- to upper levels in forests, and at least one species is a specialist inhabitant of mangrove swamps. Others occur in drier eucalyptus woodland. Information on breeding is limited, but both sexes usually incubate the eggs. Recent evidence suggests that some big-beaked birds from other families—such as shrike-tits (see right) and shrike-thrushes—may be closely related to whistlers and should perhaps be placed in the family Pachycephalidae.

GOLDEN WHISTLER
Found in Indonesia, southern and eastern Australia, Tasmania, and Fiji, Pachycephala pectoralis *shows wide geographical variation: about 73 subspecies have been described. It is 6½–7 in (16–18 cm) long; the gray-brown female lacks the male's black and white and much of the yellow coloring. It may be found in mixed-species flocks.*

Family Falcunculidae

Shrike-tits

Length 5½–7½ in (14–19 cm)

Species 4

Australian shrike-tits have striking plumage: a black and white crested head and yellow underparts. They have large, notched bills, which they use to probe and strip bark in their search for insects and spiders—as well as to "fray" branches and twigs before attaching their nest. Three species are similar in appearance, but widely separated in distribution: eastern birds have gray wings, northern birds have yellowish wings, and the western species has a white belly. All have a variety of whistling and chuckling calls. The remaining member of the family—the wattled ploughbill from New Guinea—is much lesser known. It uses its bill to "plow" bark for insects, but it is debatable whether it is related to the shrike-tits at all.

EASTERN SHRIKE-TIT
Falcunculus frontatus *is found in various parts of Australia, in habitats ranging from rain forests to gardens. It uses its strong bill to tear off loose bark in its search for insects. About 6–7½ in (15–19 cm) long, it has a black crest. The female's throat is olive, the male's is black.*

Order Passeriformes *continued*

Family Corvidae

Crows and relatives

Length 8–26 in (20–66 cm)

Species 129

Habitat All terrestrial

Crows and their relatives are distributed across almost every part of the world, except the Arctic and Antarctic. They are the most highly developed of all birds—intelligent, sociable, and very adaptable, with most of them feeding on a wide range of foods. Many members of this family have mainly black plumage, although some of the jays and magpies are highly colored. Found in open countryside and woodland, in both lowland and upland areas, crows are not migratory birds, although they may move locally between seasons. They usually nest as separate pairs but a notable exception in this regard is the rook, which nests in colonies. A typical crow nest holds 2–8 eggs and is built of twigs, in trees or bushes, or sometimes in holes.

BLACK-BILLED MAGPIE
Similar in appearance to the common magpie, Pica hudsonia *is found in central and western North America. This is a common bird, $18\frac{1}{2}$ in (47 cm) in length, which lives well alongside humans. Gregarious outside of the breeding season, over 100 birds may gather at a winter roost. Its nest is a domed pile of sticks, high up in a tree.*

CLARK'S NUTCRACKER
Found only in the mountains of western North America, Nucifraga columbiana *inhabits juniper and pine woodland. Agile when swooping through canyons, it does not fly long distances and is usually found on treetops. It is 12 in (30 cm) long.*

EURASIAN JAY
The many subspecies of Garrulus glandarius *have slightly different head markings; all are 13 in (33 cm) long. Distributed across Europe and Asia, this bird is a fairly common inhabitant of woodland, but is also found in many other habitats, such as rain forests and taiga. Across much of its range, the Eurasian jay collects and stores acorns as winter food.*

RED-BILLED CHOUGH
Pyrrhocorax pyrrhocorax *is a gregarious bird, particularly outside the breeding season. It forages for insects, especially ants, and also eats berries. During courtship, it performs spectacular aerobatics. The nest is built either on a cliff ledge or in a cave. Measuring 6 in (140 cm) in length, this bird is found in mountain and rocky coastal areas in Europe, North Africa, and South Asia.*

BLUE JAY
Common in woodland and parkland across central and eastern North America, Cyanocitta cristata *is about 12 in (30 cm) long. This bird is found in pairs or small groups, and is always noisy, with a distinctive "peeah peeah" call. It uses mud to build its nest.*

EURASIAN JACKDAW
Widespread across Europe, West Asia, and parts of North Africa, this is a familiar crow that often flies and feeds with rooks. A farmland bird in much of its range, Corvus monedula *also inhabits gorges and sea cliffs, nesting there as well as in buildings and holes in trees. It roosts in flocks of several thousands outside the breeding season. It is about $13\frac{1}{2}$ in (34 cm) long.*

RED-BILLED BLUE MAGPIE
This striking species, 27 in (68 cm) long, has a tail that itself measures $18\frac{1}{2}$ in (47 cm). Found up to a height of 4,900 ft (1,500 m) in the deciduous forests of South and Southeast Asia, Urocissa erythrorhyncha *hunts in flocks at lower levels. Its nest is a rough, flimsy cup in a tree.*

COMMON RAVEN
At 26 in (65 cm), Corvus corax *is the largest crow in the Northern Hemisphere, and is found in North and Central America, Europe, Asia, and North Africa. It lives in open habitats as well as in urban areas in some parts of its range, and up to 20,900 ft (6,350 m) on Mount Everest, often nesting when snow is still on the ground.*

CARRION CROW
Regarded as the archetypal crow by many people in Europe, the all-black Corvus corone *lives in many habitats, including woodland, moorland, farmland, and towns. About 20 in (50 cm) long, it is also found (but with gray in its plumage) through West Asia and in eastern Asia, including Vietnam and Korea. It is a solitary nester, usually making its home in a tree, although in mountainous areas, cliff ledges are often used.*

Family Paradisaeidae

Birds of paradise

Length 4¾–39 in (12–100 cm)

Species 41

Birds of paradise are renowned for the spectacularly beautiful and ornate plumage of the males, which is used in display, although in a few species the sexes are alike. Birds in this family differ in size, and have extremely variable tail lengths. Some species have short, straight bills, whereas others have long, curved ones. Most, however, have rounded wings and strong legs and feet, and use loud calls to attract mates. The majority of birds of paradise are found in New Guinea, mostly in wet montane forests. It is possible that the scarcity of natural mammalian predators on this island enabled these birds to evolve their flamboyant displays; males of some species court females on or close to the ground. The members of this family feed mainly on fruit. Their nests are typically bulky, made of leaves, ferns, and twigs, and are generally placed in tree forks. Their voices vary from soft, drawn-out calls to loud, explosive sounds like a gun being fired.

KING BIRD-OF-PARADISE
In a spectacular courtship display, the male bird expands his feathers and appears almost spherical, waving his green racquets above his head. Cicinnurus regius *is 6½ in (16–17 cm) long, and its 2 thin tail feathers add another 5½ in (14 cm).*

WILSON'S BIRD-OF-PARADISE
Diphyllodes respublica, *6½ in (16 cm) long, is found in Irian Jaya but not Papua New Guinea. Both sexes have areas of bare blue skin on the hind crown, but the male is distinguished by red wings and spiral tail-wires.*

RAGGIANA BIRD-OF-PARADISE
Found only in New Guinea, Paradisaea raggiana, *14 in (35 cm) long, is the best-known bird of paradise. Adult males display communally in traditional lek sites, where up to 20 birds will congregate.*

GREATER BIRD-OF-PARADISE
Found on the Aru Islands and in southern New Guinea, Paradisaea apoda *is 16½–18 in (42–45 cm) long. During courtship, 12–20 males gather in a tree to dance, spreading their wings, while the dominant male stays at the center.*

Family Oriolidae

Old World orioles

Length 8–12 in (20–30 cm)

Species 34

The robust Old World orioles have downcurved bills and long, pointed wings. They inhabit woodland throughout Africa, Asia, New Guinea, and Australia; one species occurs in Eurasia. Most have predominantly yellow plumage, often with black either on the head or the wings; the females are generally dull and more streaked than the males. All of them feed on insects and fruit, and one species, the black-headed oriole, also takes nectar. The distinctive, fluty songs of orioles can be heard over considerable distances. Most of them build a deep cup nest, slung in a horizontal fork on a branch. A number of Old World orioles are migratory.

EASTERN BLACK-HEADED ORIOLE
Found in central, eastern, and southern Africa, Oriolus larvatus *is a noisy and conspicuous species (although it often sits motionless in the canopy), with a loud, fluty call. It is 8¾ in (22 cm) in length, and feeds on seeds, crops, fruit, and caterpillars.*

Family Dicruridae

Drongos

Length 7–28 in (18–72 cm)

Species 26

The drongos, which occur in Africa, Asia, and Australia, have long, forked tails, and in all but 2 cases, are glossy black. Their stout bills are arched and slightly hooked, with a small notch, and the nostrils are often concealed by dense feathers. Drongos are mainly woodland birds, but are also found in open country with scattered trees, up to a height of 10,900 ft (3,300 m). These insectivorous birds also feed on lizards and small birds. They feed like flycatchers—flying out from a branch to snatch prey and returning to devour it—and are well known for chasing birds of prey, which may be several times their own size.

CRESTED DRONGO
Confined to the Comoro Islands and Madagascar, Dicrurus forficatus *is the only drongo in these countries. It is 10 in (25 cm) long, and has a striking crest which sticks up from the base of the bird's bill.*

Family Monarchidae

Monarchs

Length 4¾–12 in (12–30 cm)

Species 95

Found in Africa, Asia, and Australia, these are mostly non-migratory birds of rainforest, open woodland, and shrub savannah. Most monarchs have the broad bill typical of a true flycatcher (Family Muscicapidae) but they are not related to them. They have instead evolved from crow-like ancestors They have evolved this way because of similar foraging habits: catching insects, such as moths, bees, and dragonflies, by making sallies into the air. Other members of the family have chunkier bills for tackling fruit, probing rotten wood—or for exploiting a more opportunistic diet. Their songs are typically weak, but some species have a loud, forceful song. Sexes are alike or strikingly different: males of paradise flycatchers have longer streamerlike tails and in certain species have variable color morphs. Most monarchs build their cuplike nests in tree forks or between twigs, but the magpie-lark is notable for plastering its nest with mud. This behavior led it to be classified with two other Australian mud-nesters (the apostlebird and white-winged chough), before studies of its DNA revealed its true affinities with monarchs. Fantails (Rhipiduridae) were formerly classified with monarchs, before being allocated their own family.

AFRICAN PARADISE FLYCATCHER
Terpsiphone viridis *is 16 in (40 cm) in length, with the male's tail up to 8 in (20 cm) long. The tails of both sexes trail over the side of the small nest cup. This flycatcher is found in forests and riverside woodland, in sub-Saharan Africa.*

MAGPIE-LARK
Common in Australia near surface water, Grallina cyanoleuca *is 11½ in (29 cm) long with a loud, piping, 5-note call. It has adapted to urban spaces and can be seen feeding at roadsides.*

Order Passeriformes *continued*

Family Rhipiduridae

Fantails

Length 5½–7½ in (14–19 cm)

Species 44

Named for their habit of fanning their long tails, these birds belong to the genus *Rhipidura*. They range from the Himalayas in the west to Australasia and the southwest Pacific Islands in the east, and are fiercely territorial. Fantails snatch flying insects with their small, but broad, bills and are very agile when on the wing.

NEW ZEALAND FANTAIL
*Found in New Zealand and satellite islands—such as Chatham Island and Lord Howe Island—*Rhipidura fuliginosa *measures 5½–6½ in (14–17 cm) long. It builds a neat cuplike nest in a tree fork.*

Family Vireonidae

Vireos and relatives

Length 6½–8 in (17–20 cm)

Species 52

The unusual name "vireo" is derived from a Latin word that translates as "to be green." Birds in this family, which includes greenlets, shrike-vireos, and pepper shrikes, weigh from 11/32 oz (9 g) to ⅞ oz (25 g). Their characteristically notched bills are larger than those of the New World warblers (see p.369), the eye-stripes bolder, and the green plumage duller. These birds inhabit broadleaved or mixed forests, and species coexist by foraging in separate areas. Although mainly insectivorous, they also eat fruit, depending on the season. They are migratory birds that breed in the USA and up to northern Canada, but move south in winter to Central or South America. Some vireos are also found in the Caribbean during the breeding season and in winter.

RED-EYED VIREO
About 6 in (15 cm) in length, Vireo olivaceus *lives in deciduous forest in North America, and migrates to South America in winter.*

TAWNY-CROWNED GREENLET
Found in the forests of northern South America, Hylophilus ochraceiceps *is 4½ in (11.5 cm) long and has a rusty-orange crown.*

Family Bombycillidae

Waxwings

Length 6–9 in (15–23 cm)

Species 8

This group consists of 3 waxwings and 5 related species—the 4 silky flycatchers and the hypocolius. The 3 waxwings breed across northern Europe, northern Asia, and much of North America, wintering in the south. They are plump, fawn-brown birds with distinctive crests and silky feathers. Two species have the waxy tips to the secondary wing feathers that give the family its name. Sociable birds, waxwings nest in loose colonies, building their home from twigs and grasses in branches. They feed mainly on berries, flocks turning up even in areas where they are not normally seen and rapidly stripping berry bushes. Like waxwings, the silky flycatchers of southwestern USA and Central America also have silky feathers and prominent crests; however, they catch insects in flight like flycatchers. The hypocolius is a fruit-eating bird with a crest, found only in West Asia.

JAPANESE WAXWING
Bombycilla japonica *breeds in the taiga forest of eastern Siberia and winters farther south in Japan and Korea. Unlike the other 2 waxwings, its secondary feathers lack the waxy drops at their tip. This bird is 6½ in (16 cm) long and is a strong flier.*

Family Laniidae

True shrikes

Length 6–14 in (15–35 cm)

Species 36

These small- to medium-sized birds are the most predatory of all passerines. All have a hooked bill with a toothlike point in the upper part, and strong legs and sharp claws for holding prey. True shrikes feed on insects, but some are known to augment their diet with lizards, small birds, and rodents. Several of them create larders, impaling their prey on thorns or barbed wire, which has given rise to the name "butcher birds" for these shrikes. They are birds of open scrub, and some species are very territorial, even in their winter quarters. They frequently perch on top of a bush, scanning the countryside for prey, and flicking their tail up and down and from side to side. True shrikes are widespread throughout Africa, Europe, and Asia, with 2 species occurring in North America. Bush-shrikes and helmet-shrikes used to be included, but are now classified in a separate family (Malaconotidae). Northern species migrate south in autumn. Shrikes nest in trees or bushes, and the females lay about 2–7 eggs.

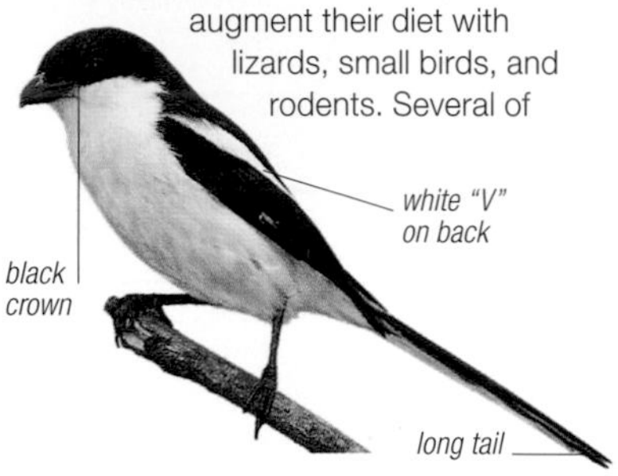

FISCAL SHRIKE
Widespread in Africa, south of the Sahara down to the Cape, in open woodland, parks, and gardens, Lanius collaris *raids other birds' nests and eats the nestlings. About 9 in (22 cm) long, it perches on poles and wires, scanning the ground for insects, mice, and small birds. It will attack birds that enter its territory.*

Family Picathartidae

Rockfowl

Length 15½–20 in (39–50 cm)

Species 2

Rockfowl are unusual in having bare skin on the head. They are found in damp, upland rain forest in West Africa up to 6,950 ft (2,100 m). They nest in a rock cave in their territory, sticking a mud nest to a rock face, at a height of 6½–13 ft (2–4 m). Here, the female lays 2 blotched, brown and gray eggs.

WHITE-NECKED PICATHARTES
A social bird, Picathartes gymnocephalus *is found from Guinea and Sierra Leone eastward to Togo. It is 16 in (40 cm) long, and has a low, croaking call. This bird forages on the ground for insects, frogs, and snails, moving in springy hops.*

Family Dulidae

Palmchat

Length 7 in (18 cm)

Species 1

The single species in this family, *Dulus dominicus*, is found only in Haiti (including the islet of Gonave) and the Dominican Republic. Common and conspicuous in these areas, palmchats are often aggressive and very noisy, chattering loudly in chorus. Like waxwings (see left), with which it is sometimes grouped, the palmchat is a fruit-eater, often seen in flocks. It never comes to the ground, spending all its time in trees, feeding on berries and flowers with its strong, heavy bill. The palmchat differs from waxwings in its nesting habits: it is known to weave a large communal nest around the trunk and lower fronds of palm trees. Woven from twigs and lined with soft bark and grass, this nest sometimes has compartments for 2 to as many as 30 pairs of birds, each pair having a private entrance from the outside. The female lays 2–4 white eggs, which are heavily spotted with gray. Outside the breeding season, palmchats use the communal nest as a nighttime roost.

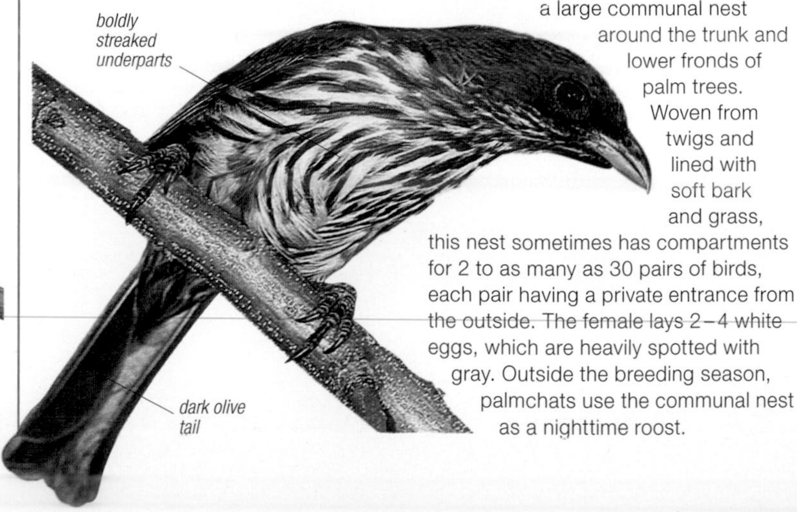

RED-BACKED SHRIKE
Shrikes, such as this female Lanius collurio, *demolish the idea that all passerines are inoffensive little birds. After killing their prey with their stout, hooked bills (including animals that range from insects to small vertebrates, depending on the size of the shrike), shrikes typically impale their victims on thorns. This helps in the dismemberment, but also serves to secure a cache of food for later use.*

Order Passeriformes *continued*

Family Petroicidae

Australasian robins

Length 4¼–7 in (11–18 cm)

Species 46

One group of these birds has red breasts that may have reminded settlers of the European robin, hence their common name. Like the robin, these small birds have an upright stance, and occasionally flick their drooping tail. The 2 sexes are often different, the female being drabber. This tree-living family generally feeds on the ground on insects and their larvae, or sallies forth after flying insects. The nest, a tiny cup of moss bound by spiders' webs, is placed on a branch. The female lays 2–4 pale blue or green eggs with red-brown to violet spots, and is known to incubate them herself for 12–14 days. Many species of Australasian robins are partial migrants.

SCARLET ROBIN
Found in southeast and southwest Australia, and eastern Tasmania, Petroica multicolor *occurs in eucalyptus woods in summer but moves to a more open habitat in autumn. It is 4¾–5½ in (12–14 cm) long, and darts from a stump or low branch to seize prey, such as caterpillars, on the ground.*

CHATHAM ISLAND ROBIN
Petroica traversi *has come back from the brink of extinction on the Chatham Islands, east of New Zealand. Once common on 4 islands, it was nearly wiped out by introduced cats. By 1976, there were only 7 birds left. Their population was boosted by captive-breeding methods.*

Family Paridae

True tits

Length 4¼–9 in (11–22 cm)

Species 58

Many species in this family have adapted to life with humans—they come readily to bird feeders and often use nestboxes for breeding. Found across North and Central America, Europe, Africa, and Asia, these small, acrobatic birds can pluck insects from foliage, or peck at peanuts inside hanging feeders. In autumn and winter, a number of species form mixed-foraging flocks and move along hedgerows, uttering frequent calls, as they search for insects. True tits generally nest in tree holes or nestboxes, lining these with moss before laying a large clutch. The eggs are incubated for 13–14 days and the nestlings fledge in 17–20 days.

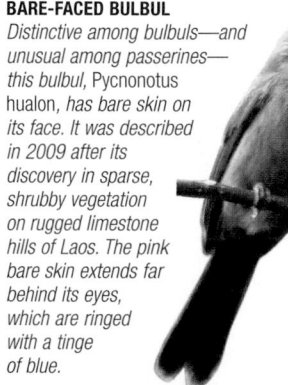

BLACK-CAPPED CHICKADEE
The most widespread North American tit, Parus atricapillus *is one of 6 species named after their "chick-a-dee-dee-dee" call. This bird is 5 in (13 cm) long. In spring, it finds a mate and digs a nest hole in a dead stump, where the female lays 6–8 eggs in a cup of grass, moss, and feathers.*

GREAT TIT
Parus major, *5½ in (14 cm) long, is a dominant tit at feeding stations in Europe. It flocks with other tits in winter, searching for insects, and also eats seeds and fruit. It has a large variety of calls and its song is a sequence of "tee-tee-tu" whistles. The female lays 5–11 white eggs that have reddish spots.*

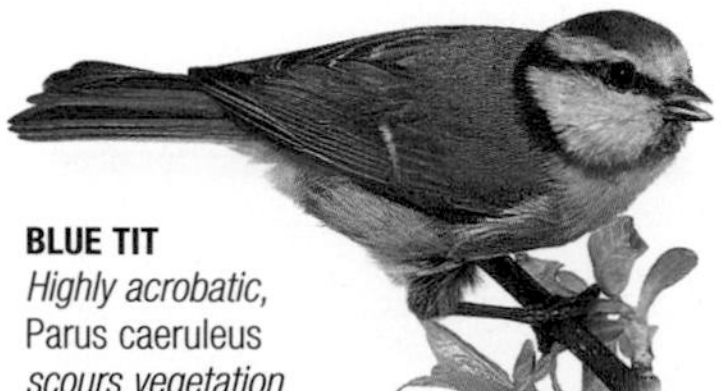

BLUE TIT
Highly acrobatic, Parus caeruleus *scours vegetation for insects and hangs upside-down to feed on seed hoppers. It nests in holes and nestboxes, laying 6–12 reddish-speckled white eggs. It is 4½ in (11.5 cm) long.*

ASHY TIT
Found in southwest Africa, Parus cinerascens *is a bird of dry thorn savanna and Kalahari scrub. It often hangs upside-down when feeding. It is a restless bird, hopping from one branch to the next, and can tear acacia pods in search of insects.*

Family Pycnonotidae

Bulbuls

Length 6–10½ in (15–27 cm)

Species 148

Nearly half the bulbul species occurs in Africa—most making up a group of similar greenish-colored "greenbulls" of the genera *Andropadus* and *Phyllastrephus*—but the majority of the family is found in Asia. These birds primarily inhabit forests, although some have adapted to more open, man-made habitats. The garden species are more gregarious and noisy than forest species. Often located and identified by their calls, few bulbuls have musical songs. Small- to medium-sized birds, many species have dull olive, brown, or gray plumage, sometimes with yellow underparts. Some species have a crest and most have bristles around the base of the bill, which may curve downward toward the tip. Many bulbuls feed on fruit and berries or buds and nectar, some eat only insects, whereas others are omnivorous. The female usually lays 2 eggs per clutch, in a nest that tends to be open in structure to allow the tropical rain to drain through.

BARE-FACED BULBUL
Distinctive among bulbuls—and unusual among passerines—this bulbul, Pycnonotus hualon, *has bare skin on its face. It was described in 2009 after its discovery in sparse, shrubby vegetation on rugged limestone hills of Laos. The pink bare skin extends far behind its eyes, which are ringed with a tinge of blue.*

BLACK-FRONTED BULBUL
Found in Namibia, Botswana, and western South Africa, in forests near rivers and in dry bush near water, Pycnonotus nigricans *is a noisy bird. About 8 in (20 cm) long, it nests from November to March, the female laying 3 pinkish eggs with dark markings.*

RED-WHISKERED BULBUL
The red patch behind the eye in both sexes gives Pycnonotus jocosus *its common name. Its other trademark feature is a black, forward-pointing crest. A common garden bird, about 8 in (20 cm) in length, it occurs from India to Hong Kong.*

ASIAN BLACK BULBUL
Found in South and Southeast Asia, Hypsipetes leucocephalus *is 9 in (23 cm) long. One subspecies from China and northern Vietnam has a white head and throat. This bird feeds on berries and insects in noisy flocks of up to 100. Its call resembles the cry of a kitten.*

YELLOW STREAKED GREENBUL
Found in isolated locations, in mountain rain forests in central and southeast Africa, Phyllastrephus flavostriatus *forages for insects, mainly in trees. Working its way round tree trunks like a woodpecker, it repeatedly flicks one wing at a time, revealing its yellow underwing. This bird is 6–10½ in (15–27 cm) long.*

Family Aegithalidae

Long-tailed tits

Length 4¼–5½ in (11–14 cm)

Species 13

In a separate family from the true tits, this group includes long-tailed tits as well as the North American bush tit. They have short, conical bills and may be more closely related to Old World warblers. They are found in North and Central America, Europe, and Asia. A Himalayan subspecies of the Eurasian long-tailed tit occurs up to 11,200 ft (3,400 m), but descends lower in winter. Groups of long-tailed tits roost together, huddled along a branch. Breeding pairs sometimes have a "helper," a bird that does not mate but helps feed the young. The domed nest is lined with feathers and can take up to 20 days to build. Clutch sizes range from 4 to 12.

LONG-TAILED TIT
About 5½ in (14 cm) long, the tiny Aegithalos caudatus *is found in Europe and Asia, often traveling along hedges in large flocks.*

Family Remizidae

Penduline tits

Length 3¼–5½ in (8–14 cm)

Species 13

These birds, found in North America, Europe, Asia, and Africa, are even more acrobatic than true tits (see opposite), and can climb along the underside of branches. They pick invertebrates off branches and foliage, and also eat seeds and fruit. Small birds with fine, pointed bills, penduline tits forage in groups and nest colonially. Their nest is an unusual pouchlike structure, with a spoutlike entrance tunnel, woven from grass and roots, and hung at the end of a branch. Females incubate 5–10 white eggs for 12 days, and the young fly in 16–18 days.

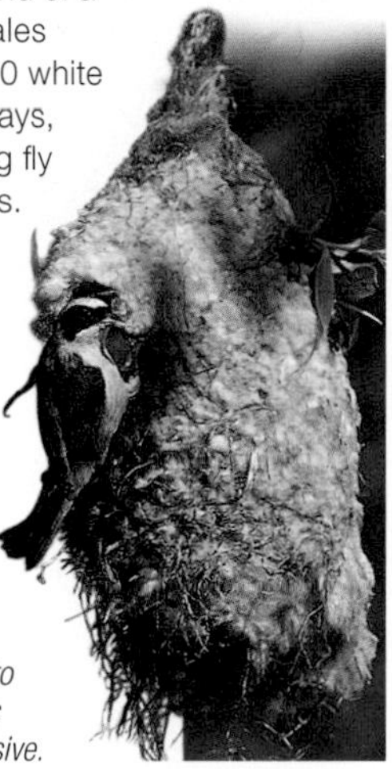

PENDULINE TIT
Remiz pendulinus, *4¼ in (11 cm) long, uses the down from bulrushes that grow in its marshy habitat to build its nest. Its song is unobtrusive.*

Family Cisticolidae

Cisticolas and relatives

Length 4–6 in (10–15 cm)

Species 147

This family of birds is dominated by 3 large genera: the mostly African *Cisticola* and *Apalis*, and the Afro-Asian *Prinia*. These birds lack bright colors, but have distinctive calls and were previously classified with Old World warblers—though they are still considered closely related to them. Cisticolas habitually flick and fan their tails and are found in a variety of open habitats—including grasslands and wetlands. *Apalis* and *Prinia* are longer-tailed, most *Apalis* are found in woodland and *Prinia* in scrub. Their nests are grassy cups, sometimes supported by large leaves. This behavior is taken to a remarkable extent in tailorbirds from Southeast Asia. They are well known for building their nests within a cradle made by sewing the edges of a leaf together with fibers.

TAWNY-FLANKED PRINIA
Found in Asia and southern and eastern Africa, Prinia subflava *has a monotonous "chip, chip" song and a harsh, scolding "sbeee" call. A restless, sprightly bird, it is 4¾ in (12 cm) long, and forages in pairs or small flocks in scrub or vegetation.*

ZITTING CISTICOLA
Found in Europe, Africa, Asia, and Australia, Cisticola juncidis *is only 4 in (10 cm) long, but is disproportionately active and noisy. It has a bouncy, circular song-flight: it beats its wings rapidly to rise, utters a "tzet" call, then drops before rising again. Its fan-shaped tail has black and white markings on the underside.*

Family Sylviidae

Old World warblers

Length 4–8 in (10–20 cm)

Species 305

Old World warblers are found in Europe, Asia, Africa, Australia, and the Pacific Islands—including many restricted in range to oceanic islands. Most have unadorned brown plumage, but have distinctive, sometimes harsh and rasping, calls that are distinctive of the species. Warblers are generally birds of woodland and scrub, and sometimes grassland. The reed warbler group feeds and nests in reedbeds and has colonized islands in the Pacific. One species, *Cettia major*, can be found at 13,200 ft (4,000 m) in bamboo thickets in the Himalayas. Scrubwarblers nest in bushes and woodland throughout Europe. The tiny leaf warblers feed in the tree canopy, picking insects off leaves. Most species are migratory, flying south to warmer climes in autumn after breeding in northern regions. Siberian willow warblers travel 7,500 miles (12,000 km) each way. Old World warblers nest in low cover or on the ground, where females lay 3–7 eggs. Both sexes feed the young.

WILLOW WARBLER
A summer visitor to northern Europe and an inhabitant of Africa during winter, Phylloscopus trochilus *has no particular association with willows. A small warbler, about 4–4¼ in (10–11 cm) long, it can only be separated from the very similar chiffchaff,* Phylloscopus collybita, *by its song: a melancholy verse that ends abruptly, as though cut off. It forages for insects among leaves, often hanging upside-down.*

GRASSHOPPER WARBLER
Increasingly scarce, Locustella naevia *is found in Europe and western Asia and winters in Africa and India. It is 5 in (12.5 cm) long, and favors meadows, ditches, and marshland edges, creeping among weeds and grass, and climbing up stems. Its song is a very long-lasting, insectlike reeling, with only short pauses.*

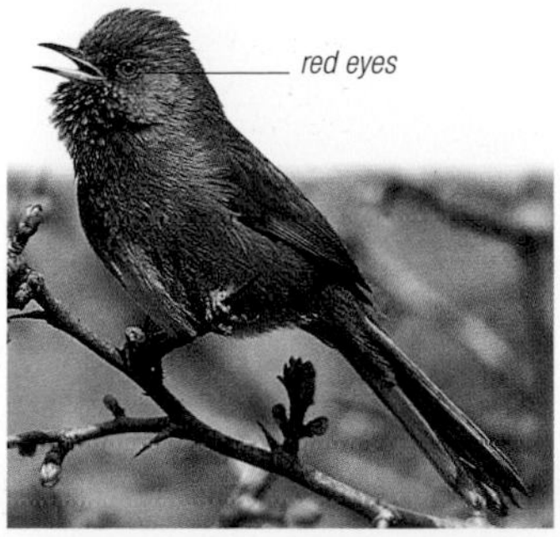

DARTFORD WARBLER
Found in western Europe and northwest Africa, Sylvia undata *is one of the nonmigratory warblers in this region. Winter is a critical time for this species, particularly in the UK, which lies at the northern limit of its range. It is 5 in (13 cm) long, with a long tail, which is often cocked, and short wings.*

CLAMOROUS REEDWARBLER
A marshland warbler found in northeast Africa and Asia, Acrocephalus stentoreus *nests in reeds and other wetland plants, usually over the water. About 7–8 in (18–20 cm) long, it feeds on butterflies, dragonflies, flies, other insects, and small frogs. The female lays 3–6 dark-speckled white eggs. When singing, the male repeats phrases 3 or 4 times.*

BIRDS

Order Passeriformes *continued*

Family Alaudidae

Larks

Length 5–8 in (13–20 cm)

Species 98

Larks are found in virtually all parts of the world—North and South America, Africa, Europe, Asia, and Australia. The skylark, famous for its far-reaching song during flight, is the most well-known member of this family. Larks are ground-dwelling birds, mostly with streaked brown, cryptic plumage that provides good camouflage on the fields and deserts they inhabit; some species, however, have black and white markings. Males and females usually appear similar. Some species, such as the skylark and crested lark, have a crest of feathers that they raise in territorial and courtship displays and when singing. Larks usually have a strong and undulating flight, although they tend to flutter over short distances. The skylark's song-flight is a vertical ascent, high into the sky. Many other species have a climbing or circling song-flight. Larks feed on invertebrates, plant matter, and seeds. Many species flock in autumn and winter and only a few species migrate. The females lay 2–6 eggs, which are generally speckled, in nests made on the ground; the young chicks often leave the nest before they can fly.

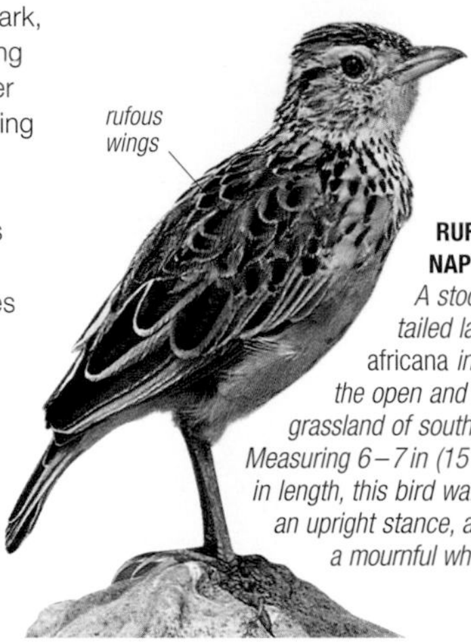

RUFOUS-NAPED LARK
A stocky, short-tailed lark, Mirafra africana *inhabits the open and scrubby grassland of southern Africa. Measuring 6–7 in (15–18 cm) in length, this bird walks with an upright stance, and has a mournful whistle.*

EURASIAN SKYLARK
Alauda arvensis, *7–7½ in (18–19 cm) long, is a streaked, brown lark of farmland and open spaces known for its continuous song when in hovering flight. It is common in North Africa, Europe, and parts of Asia, and has been introduced in Australia and New Zealand. Northern populations migrate south in winter.*

GREATER HOOPOE-LARK
Measuring 7–8 in (18–20 cm) in length, Alaemon alaudipes *inhabits desert and semidesert areas of North Africa (including the Sahara) and West Asia. It often runs away instead of flying. Its striking black and white wing pattern is partly revealed in flight.*

HORNED LARK
Also known as the shore lark, Eremophila alpestris *inhabits tundra and steppe in Europe, and prairie, farmland, and desert in North America. It is also found in North Africa and Asia. About 5½–6½ in (14–17 cm) long, it has little "horns" of feathers.*

Family Hirundinidae

Swallows and martins

Length 4¾–9 in (12–23 cm)

Species 88

Found all around the world, except in the frozen Arctic and Antarctic regions, swallows and martins are often dark blue or green above and paler below. Most have a forked tail, and some have long tail streamers. The bill is short but has a wide gape that is an effective trap for catching insects on the wing. Their dependence on aerial insects makes many species migrants—they breed in temperate areas but fly to the tropics and south temperate areas when the weather gets colder and insect prey disappears. However, many species live in the tropics throughout the year. Swallows and martins have 3 types of nests: natural holes in trees, cliffs, or buildings; tunnels excavated in river banks or sand quarries; and mud cups attached to cliffs or buildings. Some species raise 2 or 3 broods, the clutch size ranging from one to 8 eggs.

CLIFF SWALLOW
Petrochelidon pyrrhonota *breeds from Alaska to Mexico and winters in South America. About 5–6 in (13–15 cm) long, it nests colonially, building gourd-shaped mud nests under the eaves of a building or on a cliff. The female lays 4–6 white eggs.*

PURPLE MARTIN
Progne subis *nests in man-made martin houses and in old woodpecker holes in trees. Colonies can number up to 200 pairs. About 7 in (18 cm) long, it breeds in North America and winters in the Amazon Basin. The male is blue-black all over, but the female is duller above and pale gray below.*

NORTHERN HOUSE MARTIN
A summer visitor to North Africa, Europe, and northern Asia, Delichon urbicum *winters in Africa (south of the Sahara) and Southeast Asia. It builds mud half-cup nests on walls under eaves of buildings, where it nests in groups. This bird is 5 in (12.5 cm) in length.*

PLAIN MARTIN
This martin is resident in Morocco and widespread south of the Sahara in Africa, usually near inland wetlands. Riparia paludicola *is about 5 in (13 cm) long, and most have a white belly but some are entirely brown. It forms very large flocks.*

BARN SWALLOW
A summer visitor to North America, North Africa, Europe, and Asia, Hirundo rustica *winters in Africa south of the Sahara, and in southern Asia, Australia, and South America. It is 7 in (18 cm) long. A lively singer, its nest is a mud half-cup, often on a beam in farm outbuildings. It lays 4–6 red-spotted, white eggs.*

Family Timaliidae

Babblers

Length 4–14 in (10–35 cm)

Species 332

Most babblers are found in Africa, Asia, and Australia—only one species occurs in North America. Most are heavy-bodied birds with a stout bill. They are highly gregarious and maintain contact mainly through quiet calls. Most species are sedentary; only a few migrate. Many babblers are insectivores but some also eat fruit and many are omnivores. The ground feeders rummage in leaf litter, while the more arboreal species find invertebrates on leaves and bark. Nesting habits vary: a number of species build domed nests low down or at ground level; however, song-babblers build an open nest in a bush or tree. Some species even have a cooperative breeding system—birds that are not breeding help the new parents with incubating the eggs and feeding the chicks. Both the incubation and the nestling periods take 13–16 days. Young birds may remain with a group of adults for a year or more.

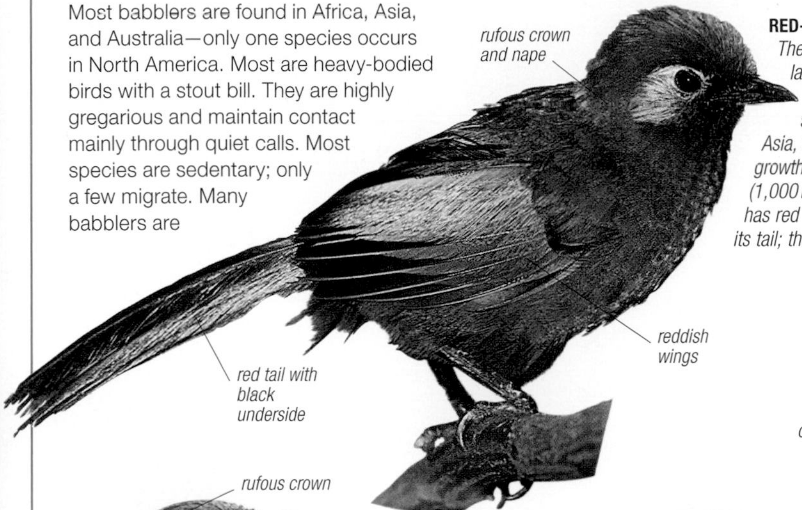

RED-TAILED LAUGHING-THRUSH
There are about 140 species of laughing-thrush, of which Garrulax milnei *is found in south China and Southeast Asia, in scrub, grass, and second-growth forest above 3,300 ft (1,000 m). It is 10 in (25 cm) long and has red wings and a red upper side to its tail; the underside is black.*

COMMON BABBLER
Turdoides caudata *is a cooperative breeder—a small flock of birds helps one pair incubate their eggs and feed the young. This bird, 9 in (23 cm) long, is found from Iraq eastward to Bangladesh, and moves through the undergrowth in forests and gardens foraging for insects, seeds, and berries.*

RED-BILLED LEIOTHRIX
A very popular caged bird, thanks to its melodious, warbling song and attractive plumage, Leiothrix lutea *is found from the Himalayas to south China and Southeast Asia. The female's throat and breast are paler than the male's and she lacks his red coloration in the wings. This bird is 6 in (15 cm) long and forages in bamboo and scrub with other babblers.*

PUFF-THROATED BABBLER
A bird more often heard than seen, the whistled chattering and loud calls of Pellorneum ruficeps *are familiar jungle sounds. This bird is found from India to Southeast Asia and lives in brushwood and bamboo, from sea-level to 4,350 ft (1,300 m). The puff-throated babbler, 6½ in (16 cm) in length, is known to travel in small flocks.*

LARGE SCIMITAR BABBLER
Widespread in evergreen forest and bamboo at heights of up to 7,600 ft (2,300 m), Pomatorhinus hypoleucos *is found on the forest floor, often digging into the soil with its long bill. It is 11 in (28 cm) long, has a high-pitched, mellow hoot, and is found from Pakistan to south China and Southeast Asia.*

NONGGANG BABBLER
*Discovered on limestone highlands straddling the China-Vietnam border, and formally described in 2008, the Nonggang babbler (*Stachyris nonggangensis*), like many of its allies, flies only short distances and prefers to walk—when it searches for invertebrate prey among leaf litter. It forms flocks in winter, but associates in pairs when breeding.*

BEARDED TIT
Found in Europe and Asia, Panurus biarmicus *lacks the unusual bill of its family. A bird of reedbeds, unlike other parrotbills, which prefer thick vegetation, the bearded tit flies through or above the reeds, eating insects, spiders, and reed seeds. It is 6½ in (16.5 cm) long and raises 3 or 4 broods a year; both sexes incubate the eggs and feed the young.*

RUFOUS SIBIA
Also called the black-capped sibia, Heterophasia capistrata *is a Himalayan bird, found from northern Pakistan eastward to southern China. It is 8½ in (21 cm) long and favors deciduous forest, especially oak woodland, but is also seen in coniferous forest. It drinks sweet sap oozing from holes in tree bark, and is also known to feed on insects.*

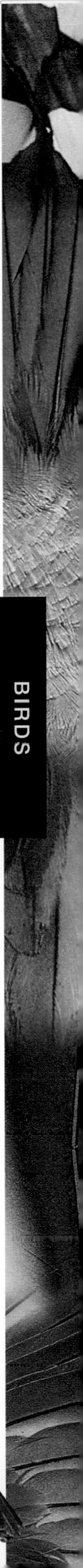

Order Passeriformes *continued*

Family Zosteropidae

White-eyes

Length 4–5½ in (10–14 cm)

Species 117

White-eyes are generally found around forest edges and canopy, from sea level up to 9,900 ft (3,000 m), in Africa and its islands, West Asia, South and Southeast Asia up to Japan, the Indian and Pacific oceans, and Australasia. They have slightly downward-curving, sharply pointed bills, brush-tipped tongues for feeding on nectar, and small, rounded wings. Their most distinctive feature is a white ring around the eye, which varies in size and shape across the group and is the source of their name. Almost the entire family consists of pale green birds that resemble each other. They have a similar "peeuu" call, and the sexes are alike. Highly mobile birds, most white-eyes migrate away from cold weather. However, some nonmigratory populations are found on islands in the Indo-Pacific, where they are vulnerable to freak weather conditions. These birds form distinct subspecies, and they show a lack of genetic variation typical of animals that live in remote places.

JAPANESE WHITE-EYE
Widespread throughout South and Southeast Asia and introduced to Hawaii, Zosterops japonicus*—about 4¼ in (10.5 cm) long—has a yellow throat and chin, and olive-green cap, back, and tail.*

CAPE WHITE-EYE
Common from forests to gardens in Botswana, Lesotho, Mozambique, Namibia, South Africa, and Swaziland, Zosterops pallidus *is about 4¼ in (11 cm) long.*

Family Irenidae

Fairy bluebirds

Length 10–12 in (25–30 cm)

Species 2

These tree-living birds are found in the evergreen or semideciduous forests of Asia: the blue-backed fairy bluebird in wetter parts of India, the Himalayas, southwest China, and Southeast Asia up to Borneo and Palawan in the Philippines; and the black-mantled fairy bluebird in the rest of the Philippines. They are noted for their very long upper and lower tail coverts, extending almost to the tip of the tail. Moving by day through the canopy in search of fruit, fairy bluebirds may also probe tree flowers for nectar. Several individuals may gather in a fig tree in fruit. They have a range of loud calls, including a sharp flight call. The blue-backed fairy bluebird nests in the forks of small trees, where the female builds a platform of twigs and incubates her 2 or 3 eggs. The male helps feed the young.

ASIAN FAIRY-BLUEBIRD
Despite the metallic-blue plumage, the male Irena puella *can be inconspicuous in the canopy—but his repeated, sharp, double whistle often reveals his presence. Females are a duller blue. This bird is 10½ in (27 cm) long, and gathers to feed on fig trees, often with other species.*

Family Reguliidae

Kinglets and goldcrests

Length 3¼–4¼ in (8–11 cm)

Species 6

These small warblerlike birds occur in the Northern Hemisphere, where they are adapted for living and feeding in coniferous forests: they have long claws and grooved feet for grasping needle-leaved twigs and branches as they continually search for minute insect food to fuel their tiny bodies. They get their name from their fiery crown stripe—the feathers of which are erected in display.

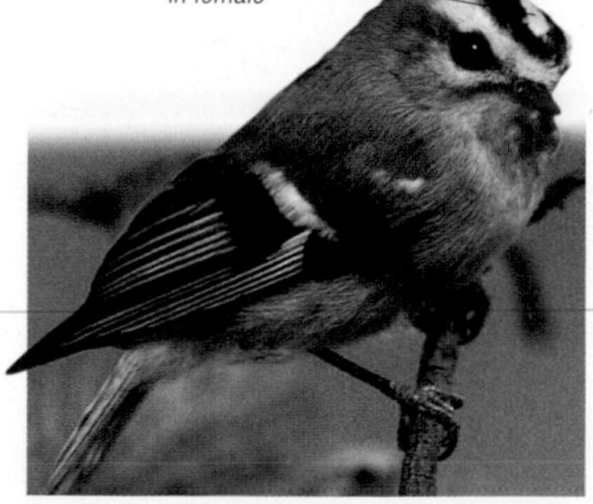

GOLDEN-CROWNED KINGLET
The smallest North American kinglet at 3¼–4¼ in (8–11 cm) long, Regulus satrapa *can withstand the harsh northern winter in the pine woods of New England. It breeds in spruce trees, in a dainty, globelike nest of moss and feathers under a branch. The female lays 8–10 tiny whitish eggs.*

Family Certhiidae

Treecreepers

Length 4–6½ in (10–16.5 cm)

Species 11

Treecreepers are arboreal birds of North and Central America, Europe, Africa, and Asia, with thin, downcurved bills, brown plumage, and pale underparts. They use their stiffened tails for balance as they climb trees in search of insects in the bark, almost always climbing upward unlike nuthatches (see opposite). Their calls and song are high-pitched, and their nest, made of twigs, bark, and moss, is shaped like a loose hammock. They lay 2–9 eggs, which are incubated mostly by the female over 14–15 days. The young, which are fed by both parents, fledge in 14–16 days.

BROWN CREEPER
Once thought to be a subspecies of the treecreeper of Europe, Certhia americana *breeds from Alaska to Newfoundland and south to Nicaragua, wintering in the south of its range. It is 5–5½ in (13–14 cm) long and climbs up trees in a spiral. It flies from the top of a tree to the base of another, searching for insects.*

Family Polioptilidae

Gnatcatchers

Length 4–5 in (10–13 cm)

Species 15

Formerly classified with Old World warblers, but now known to be more closely related to wrens, the small gnatcatchers are confined to the Americas and reach their greatest diversity near the equator. They are active birds with long tails—that are usually cocked and twitching, wren-fashion. However, gnatcatchers are more slender-bodied than wrens and are generally colored gray—not brown—in color. Some have black markings on the head. Most species spend much of their time high in trees, where they forage on insects gleaned from foliage; they often accompany other insect-eaters in mixed flocks in order to drive their prey from cover. Their calls and songs are described as thin, but musical. Gnatcatchers build a delicate nest of petals and plant down, bound with spiders' webs, moss, and lichens, on a horizontal branch. Both sexes incubate the 4–5 eggs.

BLUE-GRAY GNATCATCHER
A gray bird with a long black and white tail, Polioptila caerulea *is 4¼–5 in (11–13 cm) long, and has a thin, nasal call. It is found from southern Canada to Guatemala and Cuba.*

Family Sittidae

Nuthatches

Length 4–8 in (10–20 cm)

Species 29

Birds in this family, found in North and Central America, Europe, North Africa, and Asia, spend their whole life in trees, inching up or down trunks and branches with equal facility. However, the 2 species of rock nuthatches are an exception—they live on rock faces or buildings. Nuthatches do not use their short, square-ended tail for balance, as woodpeckers (see p.338) and treecreepers (see opposite) do. They have long, sturdy toes and claws, and climb by placing one foot higher than the other, balancing on the lower one. They search the bark for insects and invertebrates but also eat seeds and nuts, cracking these with their bill. They nest in holes in trees or rocks, and a few excavate their own holes in rotten wood. Females lay 4–10 red-spotted, white eggs. Many nuthatch species have blue-gray upperparts, although the 3 South Asian species are blue-green. All species, except for some red-breasted nuthatches, are nonmigratory.

EURASIAN NUTHATCH
Bright slate-blue and orange, Sitta europaea, *is 4¼–5 in (11–13 cm) long and has a carrying "peeu, peeu" song. It plasters its nest hole with mud to make the opening smaller. This bird feeds on insects, seeds, and nuts, which it wedges in bark and cracks open with its bill.*

RED-BREASTED NUTHATCH
Sitta canadensis, *4¼–4¾ in (11–12 cm) long, lives in North American coniferous forests as far north as Alaska and Newfoundland. In winter, it feeds on conifer seeds and, when the harvest fails, moves south in large numbers. It smears pine resin around the nest hole, possibly to keep ants out.*

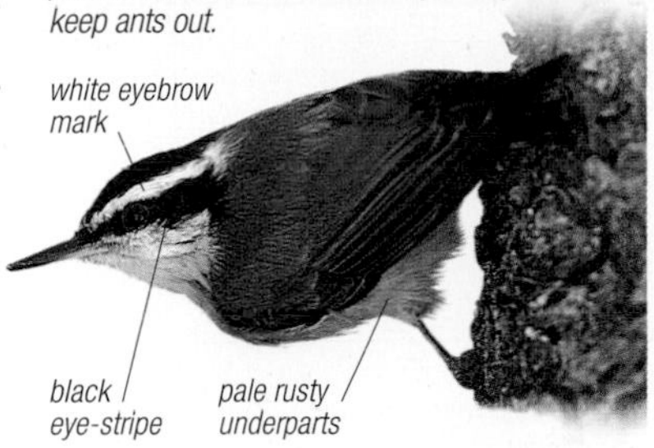

Family Troglodytidae

Wrens

Length 3¼–9 in (8–22 cm)

Species 86

Apart from one species encircling the Northern Hemisphere—the winter wren—this is an exclusively American family. Some are forest birds; others live among cacti in deserts. Wrens are generally brown with darker barred or spotted markings; some species have white patches, too. They spend most of their time in dense undergrowth—often close to the ground, but have loud, harsh calls. Wrens are mostly short-winged, weak fliers; they often have their tail (short in some species, longer in others) cocked over their back. Males build nests in cavities or make domes—depending upon species. Several nests may be constructed, some purposefully used for roosting. The female chooses a nest, which she completes by lining before laying 2–10 eggs. Chicks are either cared for by both parents, or by the female alone; older offspring may act as helpers in raising another brood.

CACTUS WREN
One of the larger wrens, Campylorhynchus brunneicapillus *is 8 in (20 cm) long, and is found in North America from Nevada to Mexico.*

WINTER WREN
Troglodytes troglodytes *is the only wren species that is found outside the Americas. It is 3¼ in (8 cm) long, and is known for its loud, rattling song. In winter, northern birds migrate south.*

Family Turdidae

Thrushes

Length 5–12 in (12.5–30 cm)

Species 185

This family, which includes such familiar birds as the American robin and the Eastern bluebird, is widely distributed in Europe, Africa, Asia, Australia, and North America. In fact, thrushes are native to almost all regions, excluding Antarctica and New Zealand. Many, such as the song thrush, are fine singers, the most notable of them all being the nightingale. Thrushes may form a solid pair-bond, singing to proclaim ownership of territory. Species like the mistle thrush, which nests very early in the year, may be heard singing in the dead of night. Thrushes build cuplike nests of grass and moss, usually in the fork of a bush or tree. Their courtship displays often emphasize their distinctive physical features—for instance, the robin's red breast or the song thrush's speckled breast. During courtship, males of some species, such as the robin, feed the female. However, the female is often solely responsible for nest-building, incubation, and brooding the nestlings. Some species of thrushes, such as the redwing and fieldfare of Scandinavia, are migratory and fly south in winter. Migratory species may form large flocks, but others stick in pairs.

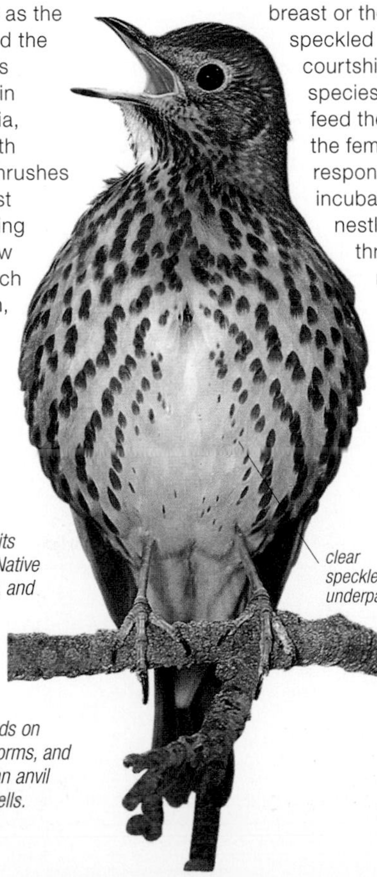

SONG THRUSH
Turdus philomelos *is a fluent singer—often at dusk—and repeats its phrases 3 or 4 times. Native to Europe, North Africa, and northwest Asia, and introduced in Australia and New Zealand, it is a bird of gardens and woodland. About 9 in (23 cm) long, it feeds on berries, insects, and worms, and is noted for its use of an anvil stone to crack snail shells.*

EURASIAN BLACKBIRD
Native to Europe, North Africa, and Asia, and introduced in Australia, New Zealand, and South America, Turdus merula *is 9½–10 in (24–25 cm) long. The male sings melodiously from a prominent perch, particularly in the evenings. Blackbirds are noted for their "chak-chak" calls as they go to roost. They feed on berries, fruit, worms, and insects.*

AMERICAN ROBIN
Turdus migratorius, *called robin in North America, has adapted well to living close to humans. About 10 in (25 cm) long, it feeds on the ground, cocking its head to watch for the movements of worms and other invertebrates. Many populations are migratory, moving north in February; the males leave first. Both sexes build the cup nest and raise 2 or 3 broods.*

EASTERN BLUEBIRD
Found in North America, Sialia sialis *is viewed as the harbinger of spring, arriving in northern areas in late February or March. Males often arrive first, launching into extravagant song-flights from treetops and rising 100 ft (30 m) or more into the air, singing all the way. Nesting in hollow branches or nestboxes, the female lays 3–7 light blue eggs. This bird is 5½–7½ in (14–19 cm) long.*

Order Passeriformes *continued*

Family Sturnidae

Starlings

Length 6½–18 in (16–45 cm)

Species 118

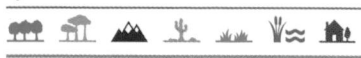

An ability to imitate the sounds they hear is a notable feature of this group of birds, which includes the irrepressible mimics, mynahs. Starlings are found in Africa, Europe, and Asia, but the common starling has been introduced around the world and is now found in 119 nations. These birds have strong legs and bills, and mostly black plumage, often with an iridescent sheen that is particularly noticeable in the African species. Some, such as the long-tailed glossy starling, have long tails, which can be as much as 14 in (35 cm) in length. Some starlings have areas of bare skin on the face, particularly around the eyes, and the wattled starling has loose folds of skin, known as wattles, on its head. Most of these birds are very much at home on the ground, and several species live among human habitation. Most starlings are residents but some are partial migrants, perhaps moving to find food in winter; the group is by and large omnivorous. Starlings lay 2–6 pale blue-green, often brown-spotted, eggs.

HILL MYNAH
Distributed across much of southern and Southeast Asia, Gracula religiosa *is found in moist forest, where its diverse range of piercing calls are a distinctive feature. It is 11–12 in (28–30 cm) long, and is the typical mynah found in pet shops and aviaries.*

SPLENDID GLOSSY STARLING
Distributed across much of Africa, in 28 countries, Lamprotornis splendidus *is a woodland species. Measuring 10½ in (27 cm), this strikingly colored bird is quite shy, and has declined in many places due to large-scale deforestation. Outside of the breeding season, it is known to form flocks.*

EUROPEAN STARLING
Occurring in 119 countries, Sturnus vulgaris *is one of the world's most familiar species, having adapted to living alongside humans. This bird—which is 8½ in (21 cm) long—feeds and roosts in large flocks, often perching on building ledges overnight.*

RED-BILLED OXPECKER
This slender African species, Buphagus erythrorhynchus, *is 8 in (20 cm) long. It feeds on blood that oozes from the wounds of large mammals such as rhinos and buffaloes. It uses its comblike bill to scour the mammal's skin for food, and helps its host by removing ticks and leeches from its body.*

METALLIC STARLING
Australia's only native starling, Aplonis metallica, *arrives annually from New Guinea and nearby islands to breed along the Queensland coast. Hundreds of birds may be found breeding in one tree. This bird is 9 in (22 cm) long, and has glossy green and purple plumage and bright red eyes. It feeds on native and cultivated fruit.*

Family Cinclidae

Dippers

Length 6–8½ in (15–21 cm)

Species 5

These plump, short-tailed birds, found in Europe, Asia, and North and South America, are rather like aquatic versions of wrens (see p.361). Their tail is often cocked in the same way, and they build a similar domed nest of moss, usually lined with grass and leaves, with an entrance hole at the side. The nest is hidden in a rock crevice, among tree roots, under a bridge, or behind a waterfall. The female incubates the 3–6 white eggs for about 16 days, and both parents feed the young. As soon as they leave the nest, the young are able to feed underwater. All 5 species of dippers live near fast-flowing hill or mountain streams, catching mayflies and other insects on the bank and shore, and wading into the water to pick insect larvae, snails, and fish eggs from rocks and pebbles. Their enlarged preen gland—10 times bigger than that of other passerines—gives them extra waterproofing.

WHITE-THROATED DIPPER
One of the few birds able to "walk" underwater, the chubby, pot-bellied Cinclus cinclus *is 7–8½ in (18–21 cm) long and has a rasping "strits" call. Found in Europe, North Africa, and northern Asia, it flies low over the water, its wings making a whirring sound as it flies, and nests in crevices and on ledges under bridges.*

Family Mimidae

Mockingbirds

Length 8–13 in (20–33 cm)

Species 34

Found in North, Central, and South America, the West Indies, and the Galapagos Islands, mockingbirds mostly live on the ground or among brush and thickets. Almost all species are known for their singing and, as the name suggests, some are very good mimics. Many sing at night during the breeding season. Generally long-tailed birds, most mockingbirds are white, gray, or brown; however, 2 species are blue and one is black. They are closely related to thrushes (see p.361), and build similar cup-shaped nests in bushes or trees, where females lay and incubate 2–5 eggs. Mockingbirds are known for their vigorous defence of the nest, and will fearlessly attack intruders.

GALAPAGOS MOCKINGBIRD
Found in the Galapagos Islands, Nesomimus parvulus *often runs instead of flying. It is 10–10½ in (25–27 cm) long, and feeds in competitive groups on eggs, young birds, carrion, and fruit.*

NORTHERN MOCKINGBIRD
Particularly associated with southern USA, Mimus polyglottos *is the state bird of 5 states in the USA. It is 9–11 in (23–28 cm) long, famous for singing on moonlit nights and incorporating other birds' voices in its song.*

Family Rhabdornithidae

Philippine creepers

Length 6 in (15 cm)

Species 3

These insectivorous birds are restricted to the Philippine Islands and are closely related to starlings. The more common stripe-headed creeper is found in the lowlands, while the plain-headed creeper usually lives above 2,950 ft (900 m). With their brown plumage and pale underparts, they resemble treecreepers (see p.360) but have a stouter and straighter bill, and a square-ended tail that is neither stiff nor used for climbing. They forage mainly in the canopy, but will also feed on trunks.

STRIPE-HEADED CREEPER
Rhabdornis mystacalis *feeds on insects picked from twigs and leaves, and on fruit, and probes among flowers, perhaps collecting pollen with its brush-tipped tongue. About 6 in (15 cm) long, it forages in small flocks, sometimes with other species.*

Family Muscicapidae

Old World flycatchers

Length 4–8½ in (10–21 cm)

Species 301

Distributed across Europe, Africa, Asia, and Australia, Old World flycatchers sally out from a perch to catch flying insects in their bill, hence their name. They usually have a broad-based bill, which is surrounded by bristles that help them catch their prey. Species that breed in the northern regions migrate south in autumn when insects begin to get scarce. The plumage of these agile fliers varies: some are brown, others are brightly colored, and the sexes frequently have different plumage. The male is often larger. Some species have a crest, and a few have bright facial wattles. Although some of these birds sing well, most have harsh call notes, and the group as a whole is no match for warblers and thrushes. Most build an open cup in the branches of a tree or bush, where females lay 1–11 mottled eggs.

ASIAN VERDITER FLYCATCHER
Eumyias thalassinus, *measuring 6½ in (17 cm) long, is found from India to south China and south to Borneo. It often perches prominently and delivers a loud, sweet warble. The male is greenish blue, with darker wings and tail; the female is duller and grayer than the male.*

RUFOUS-BELLIED NILTAVA
Niltava sundara *occurs from the Himalayas to southwest China and is resident in Myanmar above 3,300 ft (1,000 m). About 7 in (18 cm) long, it hunts for insects in low scrub and forest undergrowth. The male has a black throat, the female having a white throat patch.*

RUFOUS-GORGETTED FLYCATCHER
Found from Kashmir in north India through the Himalayas to southwest China, and in parts of Southeast Asia, Ficedula strophiata *occurs in forest and forest edges above 3,300 ft (1,000 m). This bird is 5½ in (14 cm) long, the female having a smaller and paler gorget, or throat patch. It has 2 calls: a low croak and a high-pitched "pink."*

NORTHERN WHEATEAR
Oenanthe oenanthe *is the wheatear of Europe, Canada, and Greenland, arriving for the summer from its winter quarters in Africa. Its white rump and black "T" at the tail tip catch the eye in flight. It is 6 in (15 cm) long and eats insects. A bird of open uplands, it nests in crevices, old rabbit burrows, or under rocks; the female lays 5 or 6 pale blue eggs.*

NIGHTINGALE
Luscinia megarhynchos *is renowned for its long and varied song, heard most clearly at night. The song is delivered from within cover; in fact, the nightingale is notoriously difficult to spot. Found in Europe, North Africa, and Asia, it migrates south in autumn. Females lay 4 or 5 olive-brown eggs in a nest that is built in thick cover, near the ground. The nightingale is 6½ in (16.5 cm) long.*

WHITE-STARRED ROBIN
Pogonocichla stellata *has a yellow body and tail, a blue-gray head, and white patches or "stars" over both eyes. It is 6 in (15 cm) long, and reveals a white spot on the throat when it sings its creaking notes. It is found in highland forest in eastern Africa, from the Cape to Malawi.*

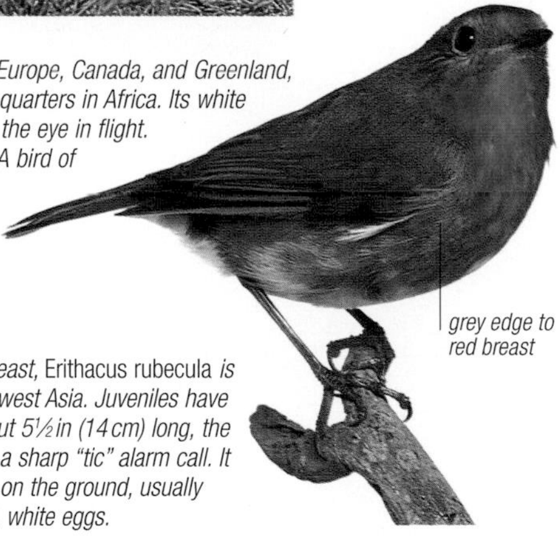

EUROPEAN ROBIN
Perky and bold, with a distinctive red breast, Erithacus rubecula *is found in Europe, North Africa, and northwest Asia. Juveniles have a brown speckled breast and head. About 5½ in (14 cm) long, the European robin has a rippling song and a sharp "tic" alarm call. It feeds on insects and berries, and when on the ground, usually hops. The female lays 4–6 red-spotted, white eggs.*

BLUETHROAT
There are 2 European subspecies of Luscinia svecica*: the Scandinavian form (pictured left) with a red throat patch, breeding in willow and birch forest; and the continental form with a white throat patch, which nests on swampy lakesides and scrubby ditches. There are other subspecies such as* Luscinia svecica magna *in Turkey, which has an all-blue throat. All forms are 5½ in (14 cm) long.*

Family Passeridae

Old World sparrows and relatives

Length 4½–8 in (11–18 cm)

Species 48

Patterned in browns and grays—occasionally with yellow or black patches—sparrows are gregarious seed- and insect-eating birds of open woodland and desert. They often spend much time feeding on the ground and many live alongside human habitations—some species even entering cities. Most lack the weaving skills of the related Ploceidae, but build bulky untidy nests of twigs or grass. Some species are colonial when breeding, others solitary.

HOUSE SPARROW
One of the world's most familiar urban species, Passer domesticus, *is 6 in (15 cm) long and lives easily with humans. It originates from Asia, but is now seen in many countries, although population declines have been noted.*

Family Ploceidae

Weavers and sparrows

Length 5–9 in (13–20 cm)

Species 108

This is a large, mainly African, group of finchlike birds, most of which are known as weavers. They have short, conical bills, short, rounded wings, and plumage that is usually yellow or brown, sometimes both. Some species look confusingly similar. Most weavers build roofed nests, some with long, downward-facing entrance tubes. In some, the males are brightly colored when breeding and have a courtship dance; females may also choose their mate on the quality of their nests.

SOCIABLE WEAVER
Restricted to Namibia, Botswana, and South Africa, Philetairus socius *nests in large colonies of up to 300 birds, the nests adjoining each other in a huge mass on trees or poles. This bird is 5½ in (14 cm) long.*

MALE SPOTTED-BACKED WEAVER
A downward-facing entrance tube built below an intricately woven nest acts as a deterrent to predators, such as larger birds or snakes. The striking black and yellow male of Ploceus cucullatus*—and his fluttering display—entices a female to inspect his handiwork. This African species breeds in noisy colonies.*

Order Passeriformes *continued*

Family Viduidae

Indigo birds and relatives

Length 4¾–13 in (12–33 cm)

Species 20

These African birds are brood parasites of estrildid finches: females lay their eggs in nests of finches and the surrogate parents raise the young. Different species exploit particular species of estrildids—and have evolved a remarkable degree of mimicry to avoid detection. Their eggs are white (like those of estrildids) and their nestlings have the same gape pattern as host nestlings—an arrangement of white spots in the open mouth—to fool the attending parent. These birds occur in open woodland habitat—wherever their host species lives. Males of one group—called whydahs—develop extraordinarily long tail feathers when breeding.

PARADISE WHYDAH
The female of this species is 4¾ in (12 cm), while the male (shown above) is 13 in (33 cm) long—in the breeding season, the magnificent tail of the male adds 8½ in (21 cm) to his length. Vidua paradisaea *does not build nests, but lays its eggs in the nests of the green-winged pytilia. It is a gregarious bird, and common in eastern and central Africa.*

DUSKY INDIGOBIRD
Vidua funerea *mimics the song of the African firefinch and lays its eggs in the firefinch's nest. Found from Kenya to South Africa, and in parts of West Africa, it is 4¾ in (12 cm) long.*

Family Estrildidae

Estrildid finches

Length 3½–5½ in (9–14 cm)

Species 143

Estrildid finches, a subject of much debate with regard to taxonomy, include waxbills, grass finches, parrot finches, and mannikins, and are found in Africa, Asia, open grassland of Australia, and New Guinea. Largely sedentary, these birds move only to find food or water; they feed mainly on grass seeds. Many of these birds pair for life. Most build domed grass nests, while some nest in holes. They lay 4–8 white eggs, which are incubated for 10–21 days. After the breeding season, they generally become sociable, and flock together for safety.

JAVA SPARROW
Although mainly a bird of towns, large flocks of Java sparrows, Lonchura oryzivora, *congregate on agricultural land and can become pests. This species, 5½ in (14 cm) long, is endemic to Indonesia from Java to Sulawesi, where it is now rare. It has also been introduced in Asia, Australia, and Tanzania.*

GOULDIAN FINCH
Restricted to savanna woodland, stony hills, and dense grass in northern Australia, Erythrura gouldiae *is one of the country's most endangered birds, its populations depleted in recent years by infections caused by a mite. It is 5½ in (14 cm) long. Most birds are black-headed, but rarer red- or golden-headed ones also exist.*

ZEBRA FINCH
Taeniopygia guttata *is 4 in (10 cm) long, and has a black, teardroplike line below its eye. Found across mainland Australia, it gathers in flocks near water.*

RED-BILLED QUELEA
Considered by many to be the commonest bird in the world, Quelea quelea, *4¾ in (12 cm) long, is found in 37 African nations. It feeds on rice and corn, and is a crop pest.*

Family Nectariniidae

Sunbirds and spiderhunters

Length 3¼–9 in (8–22 cm)

Species 136

Found throughout Africa, across Asia, and into Australia, almost all male sunbirds have highly iridescent plumage, while females are usually a dull green. After the breeding season, the males molt into a dull, female-type plumage. Sunbirds have long bills for collecting nectar and insects, and some have a long tail that is almost half their body length. The 10 species of spiderhunters, although similar to sunbirds, have much longer bills and stouter bodies, and are found only in Southeast Asia. Sunbirds and spider-hunters resemble hummingbirds (see p.329) and feed in a similar fashion but with slower wingbeats. Their nests, made of fine moss and cobwebs, are oval in shape and are usually suspended from twigs and branches. These aggressive and territorial birds are not migratory, although the pygmy long-tailed sunbird moves annually from Sudan to Congo, and at least 2 other sunbirds move between seasons.

SCARLET-CHESTED SUNBIRD
A dark, sturdy bird with a long, curved bill, Chalcomitra senegalensis *is about 6 in (15 cm) long. It is widespread across open woodland and gardens in West, East, and southern Africa. The male has a brilliant scarlet upper breast.*

PALESTINE SUNBIRD
A scarce bird, Cinnyris osea *(female shown above), is 4–4¼ in (10–11 cm) long. It is found in West Asia and Central Africa.*

VARIABLE SUNBIRD
Widespread across open savanna woodland and gardens of West, East, and southern Africa, Cinnyris venustus *is also known as the yellow-bellied sunbird. It is 4¼ in (11 cm) long. The male is iridescent blue above with a blue-violet upper breast, while the female is olive-brown above with a buff-white breast.*

STREAKED SPIDERHUNTER
Arachnothera magna *is found in dense forest and overgrown clearings from India to China, and south to Malaysia. It is about 6½ in (17 cm) long, and has a long, downward-curving bill.*

COLLARED SUNBIRD
Found in forests and woodland in sub-Saharan Africa, Anthodiaeta collaris *inhabits coastal and riverine woodland. It measures 4 in (10 cm) in length. The male has brighter plumage than the female.*

Order Passeriformes *continued*

Family Dicaeidae

Flowerpeckers

Length 2¾–7½ in (7–19 cm)

Species 48

Flowerpeckers are small, dumpy birds of southern Asia, Australia and New Guinea. They have short tails, long wings and small legs. Males are frequently brightly colored, with bold splashes of red or yellow; females are duller, often with streaked plumage. Most species live in rain forests—but some are frequent visitors to gardens. They feed on insects and fruit, but get their name from their tendency of stabbing into the base of flowers to extract nectar—aided by a forked tongue with tubular tips. Some feed almost exclusively on the fruit of mistletoes. Flowerpeckers may be closely related to sunbirds—and share their habit of building a pouchlike nest suspended between twigs. Females lay 1 to 3 eggs, which they incubate alone or assisted by the male. The insect-eating pardalotes (p.348) were formerly classified with flowerpeckers, but are now thought to be only distantly related.

MISTLETOEBIRD
Dicaeum hirundinaceum *has coevolved with the mistletoe plant. It has a short gut so mistletoe seeds pass through quickly. The sticky seeds then adhere to branches and parasitize new trees.*

ORANGE-BELLIED FLOWERPECKER
Found in Asia, Dicaeum trigonostigma *is 3½ in (9 cm) long and lives in evergreen forests and mangroves. The male has a gray back, head, and upper breast, and an orange-yellow belly and rump, while the female has a gray tinge on its breast and a yellow rump.*

Family Motacillidae

Wagtails and pipits

Length 5½–7½ in (14–19 cm)

Species 66

Wagtails and pipits are ground-dwelling birds, wagtails occurring in Europe, Asia, and Africa, while pipits are found worldwide. They have long legs and long claws, except for the hind claw, which is the shortest in species that perch on trees. Most pipits have streaked brown plumage that helps in camouflage; wagtails have contrastingly colored plumage that is white, black, yellow, or blue-gray. Pipits usually nest on the ground, while most wagtails nest on ledges. Wagtails are generally found near water; some, such as the gray wagtail, are associated with fast-running upland streams, but can also be found near weirs on lowland rivers. Sexes are similar in pipits but there are more marked differences among wagtails. Two to 7 eggs are incubated by both sexes or the female alone.

WHITE WAGTAIL
The subspecies of the European white wagtail that is found in the British Isles, Motacilla alba *is 7 in (18 cm) long, and is common near water and on mown turf. When feeding on the ground, it pauses with tail wagging, then walks rapidly forward, picking up insects in its bill. It nests on buildings and bridges near water, laying 5–6 pale gray eggs.*

RICHARD'S PIPIT
Found in Africa (south of the Sahara), Asia, and Australia, Anthus novaeseelandiae *is also an occasional autumn visitor to Europe. Its song has repeated chirping and trilling phrases and it has a loud flight call that goes "prrreep." This bird nests on the ground, laying 4–6 spotted eggs. It is about 7 in (18 cm) long.*

Family Emberizidae

Buntings, sparrows, and relatives

Length 4–8 in (10–20 cm)

Species 314

This large family is found the world over: buntings in Europe, Africa, and Asia, and New World sparrows and seedeaters in North and South America. They are small birds, with medium-sized legs but large feet that are equipped for scratching the ground to locate food. Their plumage is extremely varied, although no bird in this family has particularly brilliant feathers. The tail is fairly long and sometimes forked, and the wings are long and mostly pointed. Buntings and sparrows have short bills, which are conical and designed for peeling seeds. Most of the species live in open countryside but their habitat preferences vary greatly. They can be found from bleak Tierra del Fuego at the southern tip of South America to the northern tip of Greenland where snow buntings breed, and from the seashore to the high, barren plateaux of the Andes, enduring extremes of climate from very humid to very arid and very hot to very cold. Their nests are cup-shaped and often domed, and are generally built low in a bush or tree or on the ground. The females lay 2–7 eggs, which usually have red, brown, or black markings on a pale background. Most sparrows and buntings are migrants. They can, occasionally, occur well away from their expected wintering grounds, but birders have to be cautious with such records, because this family is particularly popular with aviculturists, and some of the birds observed may have escaped from aviaries.

YELLOWHAMMER
Widespread across Europe (although its numbers are declining in western regions), mainland Asia, and cultivated parts of west Asia, Emberiza citrinella *was also introduced in New Zealand in the 1860s. It measures 6¼ in (16 cm) in length and is common in farmland and open country.*

SONG SPARROW
Melospiza melodia *is found across most of North America, except the tundra in the north and some southern states, and is common to abundant in a wide range of nonforest habitats. It is 6½ in (17 cm) long, has a relatively long, rounded tail, a brown crown with a gray central stripe, and gray cheeks with brown stripes. It also has a grayish breast streaked with black, and a rufous back and tail. This bird breeds from April to July.*

COMMON CACTUS FINCH
Found on most of the Galapagos Islands, Geospiza scandens *is 5½ in (14 cm) long. It specializes in feeding on seeds of* Opuntia *cacti, and has a longer bill and a more pronounced split tongue than other finches, adapted to this purpose.*

SNOW BUNTING
Widespread across the tundra of Europe into northern Asia up to eastern Siberia, and similarly across North America, Plectrophenax nivalis *breeds in the Arctic but migrates to lower latitudes in winter. It breeds from May to August. It is 6¾ in (17 cm) long, and the male, with his white head and wing bar, is brighter than the female.*

DARK-EYED JUNCO
A highly variable species divided into 5 main subspecies, Junco hyemalis *is distributed across North America except the central states. About 6¼ in (16 cm) long, it breeds in forest areas, and winters in many habitats in the USA and Mexico.*

EASTERN TOWHEE
This species, Pipilo erythrophthalmus, *and its western counterpart, the spotted towhee, were originally believed to be one North American species, the rufous-sided towhee. This bird (female shown) is 7½ in (19 cm) long and is found in eastern USA.*

Family Icteridae

New World blackbirds

Length 6–21 in (15–53 cm)

Species 106

New World blackbirds, or troupials, range from the highly colorful orioles, blackbirds, and meadowlarks to the somber-plumaged grackles. It is perhaps surprising that this family also includes the oropendolas, which in many ways appear to be closer to crows (see p.352). All these species are found exclusively in North and South America and the Caribbean. Many live close to human settlements, some benefiting from agriculture and, like the red-winged blackbird, often becoming crop pests. Although highly adaptable, some of the members of this family are extremely rare and local. In some species, males and females differ in size and plumage. Almost all of these birds have conical, unnotched, pointed bills, but in the oropendolas, the top ridge is expanded to form a frontal shield, or casque. This is a noisy family (although not especially musical), and the sounds made by the oropendolas are very impressive. Cowbirds are unusual in that they parasitize the nests of other species to rear their own young. Generally, species in this family are not migratory, although the bobolink undertakes a major migration from North America to Argentina every year.

RED-WINGED BLACKBIRD
Agelaius phoeniceus inhabits marshland in much of North America, and gathers in immense flocks in winter. It is 9 in (22 cm) long. The male, glossy black all over, has bright red shoulder patches that he puffs out when displaying from his song post.

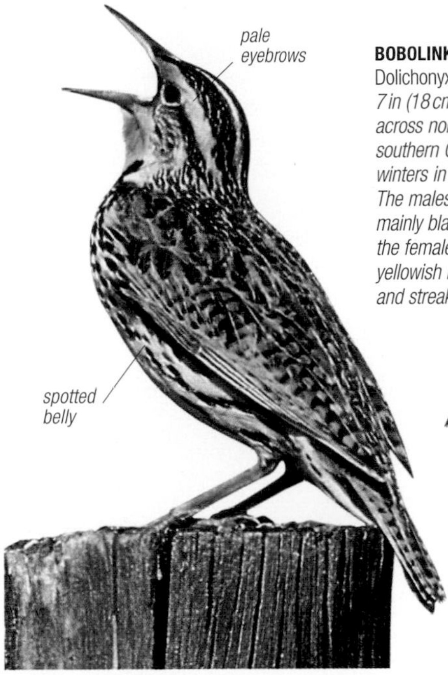

EASTERN MEADOWLARK
A resident of open fields and plains in south and eastern USA, Sturnella magna *is about 9½ in (24 cm) in length and has a distinctive, black V-shaped band on its bright yellow breast. Its back is a mottled brown.*

BOBOLINK
Dolichonyx oryzivorous, *7 in (18 cm) long, is found across northern USA and southern Canada, and winters in Argentina. The males are mainly black, while the females are yellowish brown and streaked.*

CRESTED OROPENDOLA
Psarocolius decumanus *is widely distributed across northern and central South America, and Trinidad and Tobago. About 18½ in (47 cm) long, it is conspicuous for its loud, rasping, gurgling call, ivory bill, blue eyes, and glossy black plumage. This bird nests in colonies.*

COMMON GRACKLE
Found abundantly in farmland, parks, and urban areas of North America, except the western states, Quiscalus quiscula, *12½ in (32 cm) long, is one of the most common birds in its area. The male has a glossy purple head and neck. Both sexes have an unusual twisted tail, which is obvious in flight.*

BULLOCK'S ORIOLE
Icterus bullocki, *8½ in (21 cm) long, inhabits deciduous woodland in southwest Canada, western USA, and northern Mexico, but winters in Central America. In the Great Plains of the USA, it hybridizes with its eastern relative, the Baltimore oriole.*

BROWN-HEADED COWBIRD
Molothrus ater *is about 7½ in (19 cm) long, with the males distinguished by a brown hood. This bird inhabits coastal and southern USA, and is a summer visitor to the central states and Canada.*

Family Prunellidae

Accentors

Length 5½–7 in (14–18 cm)

Species 13

Found in Europe, North Africa, and Asia, accentors are known for their inconspicuous behavior, often shuffling forward or hopping slowly near bushes or rocks or other cover as they scour the ground for food—mainly insects in summer and seeds in winter. They are predominantly gray and brown, sometimes with contrasting orange or black and white markings, and their upperparts are often streaked. The adult males and females and the juveniles all look similar. Accentors have a straight, low flight, usually covering short distances. Some species have a song-flight, while others sing from a perch. In addition to simple pairs, some species of accentors—notably the dunnock—exhibit polygyny (a male with 2 or 3 females) and polyandry (a female with 2 or 3 males). Females lay 3–6 bluish eggs that are incubated for up to 2 weeks. The young fly in another 2 weeks.

DUNNOCK
Native to Europe, Prunella modularis *measures 5½ in (14 cm) in length, and has a whistling call. It is often found under shrubs, hunting for insects. When a dominant male mates with several females there is a subordinate male in attendance; if the subordinate succeeds in mating, he helps feed the chicks.*

Order Passeriformes *continued*

Family Fringillidae

Finches and relatives

Length 4¼–7½in (11–19cm)

Species 186

Found in the Americas, Africa, Europe, and Asia, finches have bills that have evolved for feeding on seeds; some species have particularly strong skulls and large jaw muscles to crack very hard seeds. When feeding, they wedge the seeds in a special groove at the side of their palate, then crush them by raising the lower jaw. The husk is peeled off by the tongue and the kernel swallowed. One group—the Hawaiian honeycreepers—have evolved a diversity of bill shapes for different diets: long and slender for taking nectar, or woodpecker-like for probing bark.

CHAFFINCH
Common in Europe, North Africa, West Asia, and Pakistan, Fringilla coelebs *is found in open country and woodland. It measures 6in (15cm) in length.*

EVENING GROSBEAK
This bulky bird, Hesperiphona vespertina, *is 8in (20cm) long, and inhabits coniferous forest across northern North America, moving southward in winter. It is noisy and gregarious.*

EURASIAN GOLDFINCH
The dainty Carduelis carduelis, *5½in (14cm) long, with its distinctive red face and beige body, is common in the countryside of Europe and Asia.*

RED CROSSBILL
A resident of North America, Europe, and Asia, Loxia curvirostra *is remarkable for its bill. The upper and lower mandibles cross over each other, a feature that allows the bird to extract seeds from ripe pinecones. This bird (female shown) is 6½in (17cm) long.*

BULLFINCH
The attractive bullfinch, Pyrrhula pyrrhula, *6¼in (16cm) long, can quickly strip a pear tree of its buds, making it an unpopular bird with fruit-growers. The male's pink breast combined with its gray back and black and white wings make it easy to identify. This bird is found across Europe, and east into China.*

ISLAND CANARY
The wild ancestor of the common cage bird, Serinus canaria *is 5in (13cm) long and has grayish yellow plumage. It is endemic to the Canary Islands, the Azores, and Madeira, off the west coast of Africa.*

AKIAPOLAAU
Hemignathus munroi *is a critically endangered species restricted to the koa forest above 3,300ft (1,000m) on the slopes of volcanoes of the Big Island of Hawaii. About 5in (12.5cm) long, the male is olive-green above, and yellow below, with a yellow head. The female is smaller and duller.-*

IIWI
A common and noisy species found in native forests above 2,300ft (700m) on Hawaii's Big Island, Kauai, and Maui, Vestiaria coccinea *is rare on other Hawaiian islands. The adult, about 5in (13cm) long, is bright vermilion, with a long, pink, downcurved bill and a yellow ring around the eye.*

Family Thraupidae

Tanagers and relatives

Length 4–11in (10–28cm)

Species 205

Confined to the Western Hemisphere, mostly to the tropics, this family contains some of the most colorful American birds. Relatively few tanagers live in the densest parts of tropical forests; they mostly wander in mixed flocks among the canopy, particularly outside the breeding season. Most birds in this family build open, cup-shaped nests, rarely on the ground. Some feed among low bushes, but few are ground-dwelling birds. They feed on fruit and insects. The euphonias, included within this family, feed on mistletoe berries, while the honeycreepers have bills and tongues designed for sucking nectar from flowers. A few species follow columns of army ants, catching insects and spiders flushed by the ants. Many remain mated throughout the year and therefore hardly sing at all, but a few species, such as the scarlet tanager, do have attractive songs.

SWALLOW TANAGER
A widespread bird in central and northern South America, Tersina viridis *has long, pointed wings and gets its name because it snatches insects in flight—using a short flattened bill that is slightly hooked at the tip. The female (shown here) is green; the male brilliant turquoise blue. Because of its unusual features, it used to be classified in a separate family, outside the tanagers.*

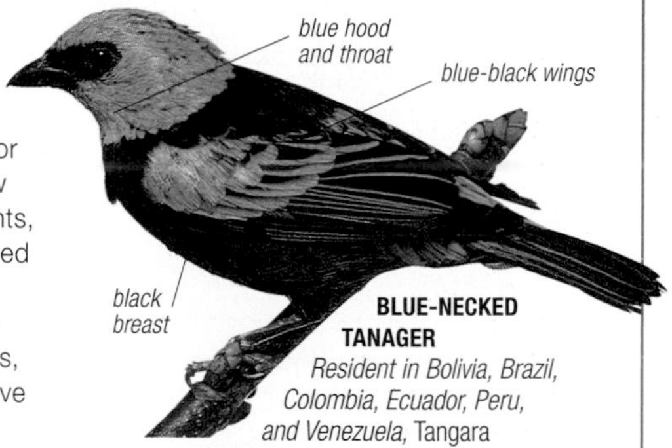

BLUE-NECKED TANAGER
Resident in Bolivia, Brazil, Colombia, Ecuador, Peru, and Venezuela, Tangara cyanicollis *is found between 1,000 and 7,900ft (300–2,400m) in open areas with isolated trees. Both sexes are alike, with a blue hood and throat, and measure 4¾in (12cm) in length. The wings are blue and black, or green and black, depending upon the bird's subspecies.*

MAGPIE TANAGER
Widespread across South America, Cissopis leverianus *inhabits scrub in cloud and rain forests and in cultivated and suburban areas. It is 11in (28cm) long, with a blue-black head, mantle, throat, and breast. Its long, black tail is edged with white, the wings are black and white, while the belly is white.*

SCARLET TANAGER
Found in deciduous forests, Piranga olivacea *breeds from May to July in eastern North America, and winters in South America, from Colombia to Bolivia; it is a rare migrant in the Caribbean islands. This bird measures 6½in (17cm) in length, and the male has a bright scarlet head and body.*

GREEN HONEYCREEPER
Widespread and common in most countries of northern South America, Chlorophanes spiza *is also found in Trinidad and Tobago. About 5½in (14cm) long, it inhabits forest canopies below 4,900ft (1,500m). The male is bright bluish green all over, with a black mask. Both sexes have a yellow bill and red eyes.*

BIRDS

Family Cardinalidae

Saltators, cardinal-grosbeaks, and relatives

Length 5–9 in (12.5–22 cm)

Species 50

Saltators and cardinal-grosbeaks are thick-billed "seed-crunchers" and differ in structure from buntings and sparrows (see p.366), which are "seed-peelers;" the dickcissel is intermediate in this grouping in that it both crunches and peels seeds. These birds are only found in the Americas, particularly within the tropics. Saltators, cardinal-grosbeaks, and New World buntings are included within the group. The saltators are mostly drab-colored birds. Male cardinal-grosbeaks are more brightly colored, as are the New World buntings. It is possible that the *Piranga* tanagers—which share the brilliant red color of cardinals—may belong to this family, too.

DICKCISSEL
Spiza americana *breeds from March to July in central and southern North America and winters in South America. It is 6½ in (16 cm) long, and has a yellow breast, a gray head with a pale eye-stripe, yellowish eyebrow, and grayish back. The male is distinguished by a black bib.*

COMMON CARDINAL
Found along woodland edges and in gardens in southern and eastern North America and into Mexico, Belize, Guatemala, and Bermuda, Cardinalis cardinalis *measures 9 in (22 cm) in length. The male is bright crimson with a black throat and red bill, and the female is buff-olive. Both sexes deliver a loud, liquid, whistling song.*

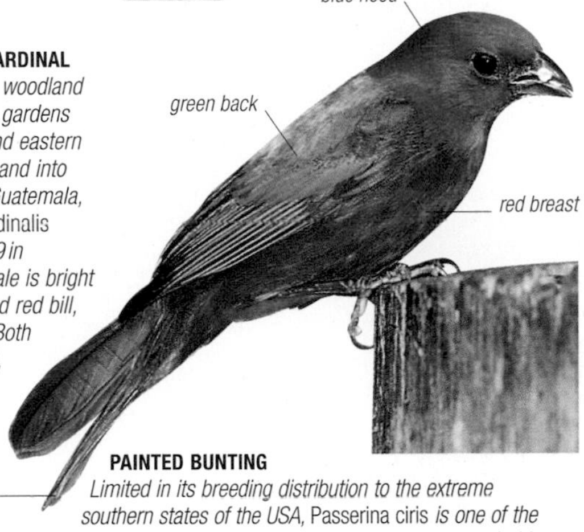

PAINTED BUNTING
Limited in its breeding distribution to the extreme southern states of the USA, Passerina ciris *is one of the most colorful birds in North America. The male has a purple-blue hood and gaudy plumage; the female is bright green above and paler yellow-green below. This bunting is 5½ in (14 cm) long.*

Family Coerebidae

Bananaquit

Length 4¼ in (11 cm)

Species 1

Around 40 subspecies of *Coereba flaveola* have been identified. It is a familiar garden bird and ranges through northern and eastern South America and the Caribbean islands. The bananaquit is common in the northern part of its range, but relatively scarce deeper in the forest. It feeds mainly on nectar but also eats fruit. Both sexes have a black crown and mask, a white eye-stripe, and a vivid yellow breast and belly. It builds a covered domelike nest.

Family Parulidae

New World warblers

Length 4–6½ in (10–16 cm)

Species 122

New World warblers, known in the USA as wood warblers, include many brightly colored birds, with frequent splashes of orange and yellow in their plumage. New World warblers inhabit a wide range of woodland and scrub habitat; many head from North America to Central and South America to spend the winter in tropical latitudes. The pressure of human habitation and agriculture have shrunk these areas, leading to a substantial decline in numbers of breeding birds. Although they are largely insectivorous, these birds also feed on fruit during migration. The songs of different New World warblers are generally well developed but are similar enough to confuse birdwatchers.

HOODED WARBLER
Wilsonia citrina *is an expert fly catcher that inhabits damp forest and wooded swamp, always on, or a few metres above, the ground. It breeds in eastern and central USA, and winters in Central America and the West Indies. This bird is 5½ in (14 cm) long, and both sexes flaunt white tail patches as they move about.*

BLACK AND WHITE WARBLER
Breeding in Canada and eastern USA, Mniotilta varia *winters in the south, as far as northern South America. It forages along branches, probing for insects with its long bill. Its call is a high-pitched whistle. This bird is 4½–5½ in (11.5–14 cm) long.*

KIRTLAND'S WARBLER
Dendroica kirtlandii *breeds only in jack pine woodland in Michigan, USA, and winters in the Bahamas. Habitat management has improved its nesting conditions. It is 6 in (15 cm) long, and both sexes (female pictured) have blue-gray upperparts and white eye-rings.*

COMMON YELLOWTHROAT
Distributed across North America, Geothlypis trichas*—which is 5 in (13 cm) long—winters in southern USA and South America, and prefers low vegetation in fields and swamps. The male has a black face mask bordered with gray, and yellow throat, breast, and undertail coverts. The female is similar but lacks the face mask.*

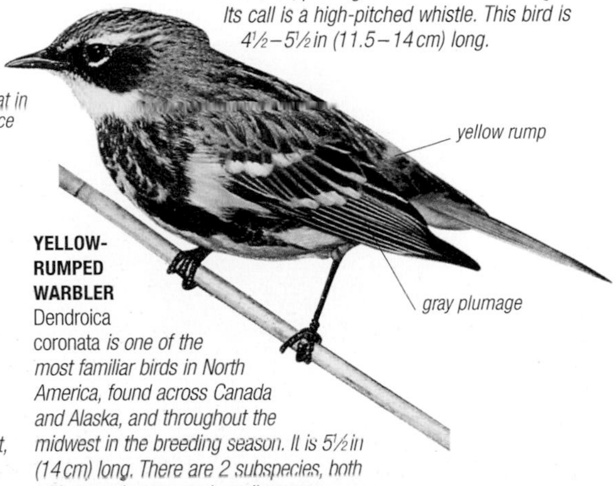

YELLOW-RUMPED WARBLER
Dendroica coronata *is one of the most familiar birds in North America, found across Canada and Alaska, and throughout the midwest in the breeding season. It is 5½ in (14 cm) long. There are 2 subspecies, both with gray plumage and a yellow rump.*

REPTILES

REPTILES

PHYLUM	Chordata
CLASS	Reptilia
ORDERS	4
FAMILIES	64
SPECIES	About 9,400

Reptiles are egg-laying vertebrates that have a tough skin with a covering of scales. These ectothermic (cold-blooded) animals cannot generate internal heat. There are 4 orders of reptiles: snakes, amphisbaenians, and lizards (collectively known as squamates); crocodiles, alligators, and caimans (crocodilians); tortoises and turtles; and tuataras. Most reptiles, including those that live mainly in water, lay eggs on land. The young emerge fully formed without a larval stage.

SCALES
The Gila monster belongs to the largest group of reptiles, the lizards. Unlike amphibians, from which they are descended, all reptiles have scales rather than smooth, moist skin.

Evolution

Reptiles first appeared about 340 million years ago, having evolved from early amphibians. The first reptiles differed from their amphibian ancestors in 2 significant respects: they developed a hard, scaly outer skin that protected them from abrasion and moisture loss; and, more importantly, they evolved a shelled, amniotic egg, in which the embryo developed within a sac of water, protecting it from the environment. Together, these features allowed reptiles to move away from the margins of water bodies, to which the amphibians were restricted, and to colonize land. By the Mesozoic era (230–65 million years ago), reptiles had diversified into an enormous variety of types, and were the dominant land animals. The turtles and tortoises branched off at a relatively early stage, and by 200 million years ago, species similar to those we would recognize today were already present. Some time after this, reptiles began to diversify explosively. Among the orders that appeared at this time were the pterosaurs (flying reptiles) and the dinosaurs, along with 2 orders that survive today, the crocodilians and the tuataras. Later, another evolutionary line led to the appearance of the squamates—the lizards, amphisbaenians, and eventually the snakes. Of the 20 or so orders of reptiles known to have existed in the Mesozoic Era, only 4 survive today.

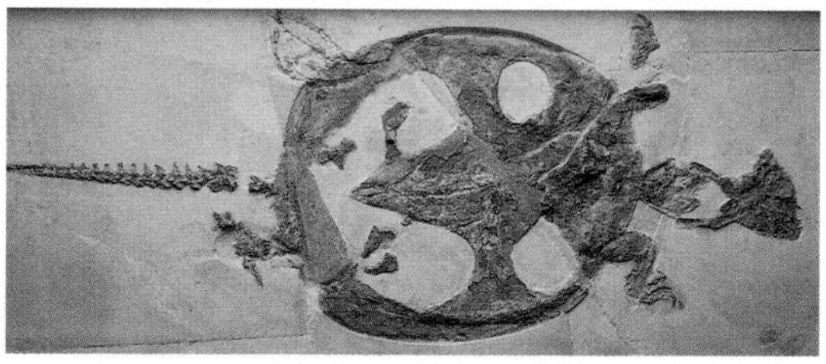

TURTLE FOSSIL
These are the fossilized remains of a turtle, dating back about 200 million years. Turtles and tortoises have changed little since that time, and are the oldest group of living reptiles.

Anatomy

The external anatomy of reptiles shows immense variation, from the long, slender, limbless forms of snakes to the short, stout, shell-covered bodies of turtles. However, all reptiles are characterized by the presence of scales (see panel, below), which form a barrier that protects them from abrasion, attacks from predators and parasites, and dehydration. Scales differ considerably between reptiles, and in some species, scales of various shapes and sizes are found on different areas of the body. Pigment below the scales gives the animal its coloring and markings, which may be for camouflage or for display. In some reptiles, especially male lizards, groups of scales have evolved into dramatic crests, horns, and other features used for display.

LIZARD SKELETON
Lizards are a highly diverse group of animals. Most have a long tail and 4 well-developed legs with long digits, as seen in this skeleton of a monitor lizard. However, some species closely resemble snakes, and do not have any limbs at all.

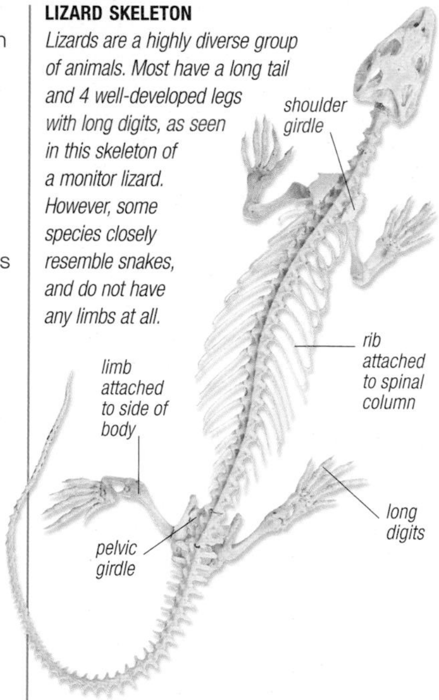

The internal skeleton of a reptile is made up of bony elements and provides a stiff supporting system that is considerably sturdier than that of an amphibian, making reptiles more suited to life on land. Unlike those of mammals and birds, a reptile's limbs support the body from the side, which gives them a sprawling gait when they move. Snakes, along with most amphisbaenians and some lizards, are without functional legs.

Differences in the skull reflect the evolutionary origins of the various reptile groups and are a useful feature for classification. Turtles and tortoises have no openings in their temples, while crocodilians and tuataras have 2 on each side. The squamates also have 2 per side, but they are sometimes joined to form a single, large opening. In many reptiles, bone growth does not

SKIN STRUCTURE

A reptile's skin consists of 2 main layers: the epidermis (outer layer) and the dermis (lower layer). The scales, which are present only on the epidermis, are made of a horny substance, keratin, which is similar in composition to human hair and fingernails. The dermis contains nerves, blood vessels, and cells that support and nourish the epidermis. Unlike a fish's scales, those of a reptile cannot be removed individually. All reptiles replace their scales by shedding their outer skin. This allows room for growth, and also replaces skin that is worn out. Whereas snakes slough their skin in one piece, lizards, crocodilians, turtles and tortoises, and tuataras shed it in chunks or flakes. After a snake has shed its outer skin, it often appears much more brightly colored.

SNAKE SCALES

LIZARD SCALES

CROCODILE SCALES

SCALE TYPES
Reptiles' scales differ greatly in size, shape, and texture. They may be smooth or rough, and may overlap, like roof tiles, butt up against each other, or have layers of stretchy skin in between. In crocodilians, the scales on the back are strengthened by bony plates.

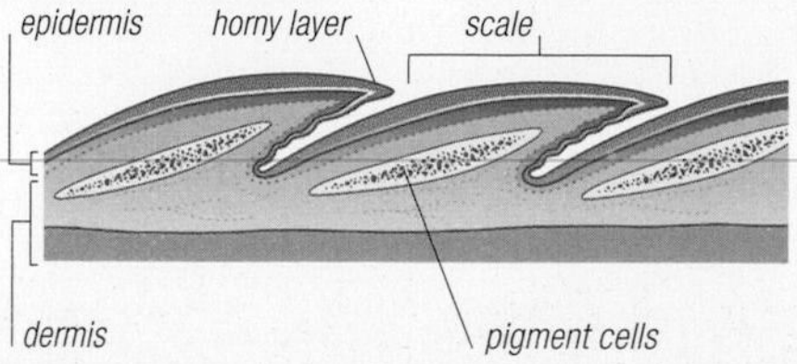

CROSS SECTION OF SNAKE SKIN
The scales of a reptile are made from the thick, horny outer layer of skin. Each scale is joined to its neighbors by flexible, hinged areas, so that the body can move and bend. Pigment cells between the epidermis and the dermis determine the animal's coloration.

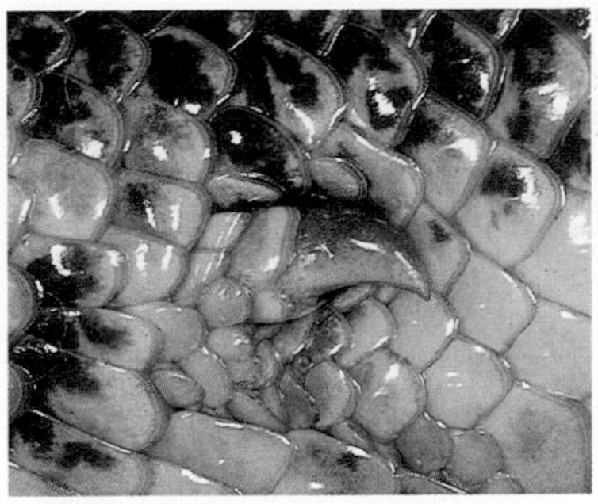

VESTIGIAL LIMBS
In some snake families, such as dwarf boas and some boas and pythons, the vestigial back limbs are still evident, indicating an evolutionary link between snakes and lizards.

end when sexual maturity is reached, which means that some long-lived adults can grow extremely large.

Senses

Reptiles' senses are better developed than those of amphibians, and some have sense organs that are not found anywhere else in the animal kingdom. The eyes are often large and well developed, although many snakes have poor sight, and in some burrowing squamates the eyes are reduced or absent. Turtles and tortoises, crocodilians, and most lizards have mobile eyelids, while in snakes and some lizards they are immovable. Lizards and tuataras have a light-sensitive area on top of their skull, known as the third eye, which is thought to control diurnal and seasonal patterns of activity by measuring day length. Reptiles tend to have poor hearing. Some have no external ear opening or middle ear structure at all, and transmit sounds through the skull bones. Taste is not important to reptiles, but smell is highly developed. Some snakes have heat-detecting organs in their faces, and can detect minute temperature changes to help them to locate prey.

EYE COVERINGS

Crocodilians, turtles, tortoises, and most lizards have 2 movable eyelids (an upper and a lower) as well as a nictitating membrane. This membrane consists of a transparent fold of skin that is drawn over the eye from the side, providing protection while also allowing the animal to see.

In snakes and some lizards, the lower lid, which is transparent, is fused with the upper one, forming a fixed, transparent protective covering over the eye known as the spectacle, or brille.

SHEDDING SPECTACLES
A snake (here, a cross-marked sand snake) sheds the outer layer of its spectacles, or brilles, at the same time as it sloughs the rest of its skin.

ACTIVITY PATTERNS

Maintaining an optimum body temperature is the key to a reptile's survival and lifestyle, as this graph depicting a diurnal lizard's activity patterns shows. At night and in the early morning, the animal shelters from the cold in its burrow. Later, as the temperature rises, it has to bask in the sun in order to obtain the energy required to forage. The lizard must seek shelter around noon to prevent overheating, but re-emerges later, as the air cools, to forage once more.

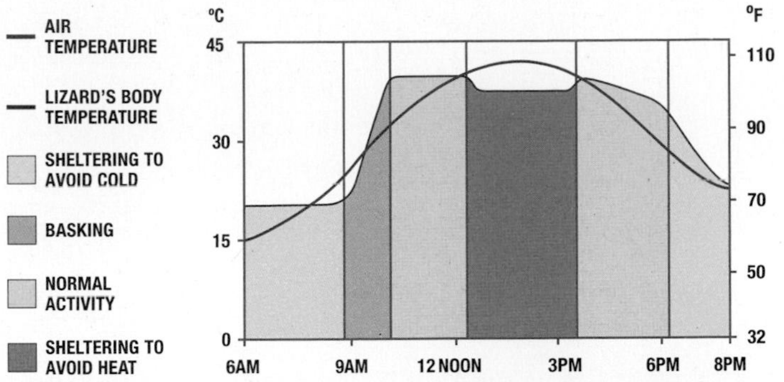

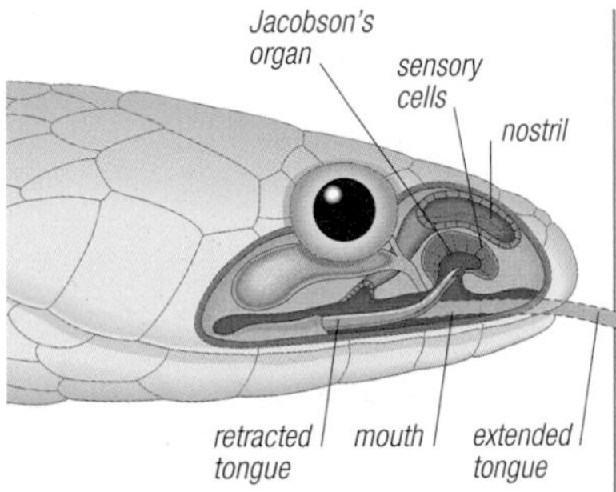

JACOBSON'S ORGAN
Snakes and most lizards extend and flicker their tongue to pick up scent molecules around them. On retracting their tongue, they transfer the molecules to the Jacobson's organ, inside the mouth, where the scents are analysed.

Temperature control

Since reptiles cannot generate heat internally, they depend on external factors to keep their temperature within critical limits. At temperatures below their preferred range (in most species, 86–104° F/30–40° C), they will slow down and may act to raise their temperature, perhaps by basking. This involves flattening or angling their body toward the sun, or pressing their underside against a warm rock.

At very low temperatures, their bodily functions operate at a much reduced rate, although normally this occurs only after the animal has sought shelter, in a burrow or under rocks. In temperate regions, reptiles may shelter for prolonged periods, or hibernate, over winter. Similarly, species from hot, arid climates may shelter for the hottest part of the year, an activity known as estivation. Reptiles that live in tropical regions rarely need to bask.

Reproduction

Little is known about the courtship behavior of reptiles, although chemical communication probably plays a large role. Several reptiles vocalize during the breeding season, and males of many lizards and some other groups indulge in visual displays using bright colors, crests, and flaps of skin (dewlaps) under the chin. These displays serve to establish territories as well as to attract females. In most cases, a female is fertilized by a male, although parthenogenesis, in which a female reproduces without the need for fertilization, occurs in several species of lizards and one species of snake.

Most reptiles lay eggs, usually on land, although a significant number of lizards and snakes give birth to live young. Reptile eggs may have a hard shell, like those of birds, but most have a leathery shell that allows water and oxygen to pass through to the developing embryo. Reptiles hide their eggs in burrows, decaying vegetation, or other similar locations. Incubation periods can last from a few days to several months, with the young of some species overwintering in their nest and emerging nearly one year later. Live-bearing species retain the eggs inside their body and in some cases nourish them through a form of placenta. Hatchling and newborn reptiles are very similar to their parents, although their colors and markings may differ. Parental care is rare except in crocodilians, where it may last for 2 years or more. Some lizards also care for their young after birth.

GIVING BIRTH
The female viviparous, or common, lizard retains her eggs inside her body until they are ready to hatch. The 2–12 young break out of their eggs as soon as they are outside their mother's body and can survive without her care.

EMERGING FROM HIBERNATION
A group of red-sided garter snakes emerge from their communal burrow, where they have been in hibernation since the onset of cold weather in winter.

Tortoises and turtles

PHYLUM	Chordata
CLASS	Reptilia
ORDER	Testudines
FAMILY	13
SPECIES	317

Tortoises and turtles are among the oldest of all living reptiles. They first appeared about 200 million years ago but have evolved little in the intervening time, so that the living species are remarkably similar to those that lived side by side with such animals as dinosaurs. Their most distinctive feature is the hard shell that encloses the soft parts of the body, providing protection and camouflage from predators and the elements. Tortoises and turtles have no teeth and instead use sharp jaws to cut their food. They live on land as well as in freshwater and marine habitats (although all species lay their eggs on land). The terrestrial species are commonly referred to as tortoises, while those that live in fresh water are often called terrapins. The term turtle was traditionally reserved for marine species, but most zoologists now use it to refer to all members of the order. Although they are most common in tropical regions, tortoises and turtles are also found in temperate parts of the world. Some marine species undertake long-distance migrations, either in search of food or to reach their nest sites.

FINDING THE WATER
Upon hatching from their eggs, young sea turtles, such as these green sea turtles, have to find their way to water. It is thought that they either instinctively travel down the slope of the beach or sense small differences in light levels over the land compared with the water (the water being brighter by both day and night). On their way, they must contend with predators such as gulls and crabs.

REPTILES

Anatomy

All tortoises and turtles have a shell, 4 limbs, and a horny, toothless beak in their jaw. The shell consists of upper and lower parts (known as the carapace and plastron, respectively), joined between the front and back legs on each side by a bridge. All parts of the shell have 2 layers: an underlying bony layer and an outer epidermal layer. The outer layer is made of thin, horny plates (scutes), which contain the pigment that gives each species its distinctive coloration. Some species lack scutes and have soft, leathery shells. The shape of the limbs differs between terrestrial and aquatic species: most terrestrial species have short, club-shaped legs, while in aquatic species, they are either webbed or shaped like flippers. Since their ribs are fused to the shell (see below), tortoises and turtles cannot move their ribs to draw air into and out of their lungs. Instead, they use muscles at the tops of their legs to provide the necessary pumping action.

DOMED SHELL

STREAMLINED SHELL

SKELETON
Tortoises and turtles have an unusual skeleton. The ribs and some of the vertebrae are fused to the inner surface of the carapace, and the pelvic and shoulder girdles are in an unusual position inside the rib cage. The skull is heavily built, with no openings behind the eye sockets, as there are in other reptiles. The length of the neck varies greatly between species and determines how the head is withdrawn into the shell (see right).

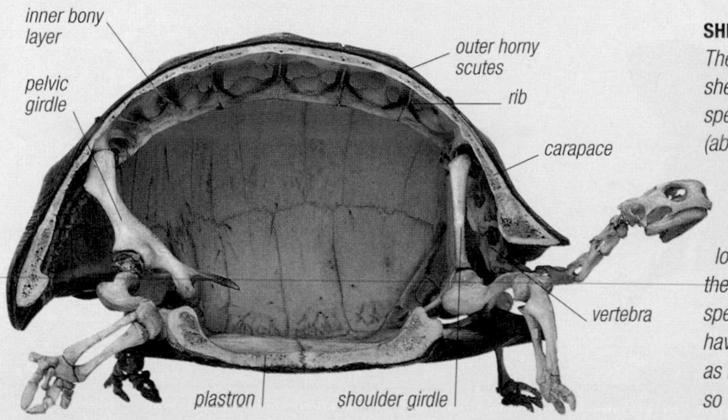

SHELL SHAPES
The shape of a tortoise's or turtle's shell reflects its way of life. Terrestrial species, like the Indian starred tortoise (above top), have a high, domed shell that is difficult for predators to bite or crush. Aquatic species, like the red-eared slider (above), have a low, streamlined carapace that helps them slip easily through water. A few species, such as the pancake tortoise, have a flexible carapace. Others, such as box turtles, have a hinged plastron so that they can close up completely.

SIDE- AND STRAIGHT-NECKED TORTOISES AND TURTLES
Two major subdivisions of tortoises and turtles are recognized, based on the way they withdraw their head into their shell. Side-necked turtles bend their neck sideways, so that the head rests under the edge of the shell; all side-necked turtles are aquatic or semiaquatic, freshwater species. Straight-necked species (which include all terrestrial and some aquatic forms) have a shorter neck, which they bend into a vertical S-shape inside the shell, so that their head appears to go straight back when it is withdrawn.

SIDE-NECKED

STRAIGHT-NECKED

Migration

Some sea turtles migrate long distances from their feeding grounds to the beaches where they lay their eggs. For example, some green sea turtles feed off the coast of Brazil and nest on Ascension Island, in the eastern Atlantic, which involves a return journey of at least 2,800 miles (4,500 km). All the breeding adults arrive at the nesting beaches within a few weeks of each other. Exactly how they navigate is still poorly understood, but they probably use a combination of the earth's magnetic field, the directions of ocean currents, water chemistry, and memory. The greatest distances are traveled by the leatherback turtle, which ranges from tropical seas almost to Arctic waters in pursuit of jellyfish, its preferred food.

SWIMMING
When swimming, sea turtles (such as this hawksbill turtle) use their front limbs for propulsion and their back limbs for steering. The speed at which they swim varies between species, from 1⅘ to about 18 mph (3 to 30 kph).

Reproduction

All tortoises and turtles lay their eggs on land. Tropical species may lay several clutches throughout the year, whereas temperate species lay only once or twice during a breeding season; females can store sperm so that they can continue to lay fertile eggs long after mating. The size of the clutch depends on the size of the species: small turtles lay 1–4 eggs, while the largest species can lay over 100. The green sea turtle is the most prolific species, laying up to 6 or 7 clutches of more than 100 eggs every 2 weeks. Turtles do not care for their young, although at least one species (the Burmese brown tortoise) stays at its nest for a few days after laying to protect the eggs from predators.

EGGS
The eggs of large turtles and tortoises are almost spherical, but those of small species tend to be elongated. The shells may be either hard and brittle or flexible. This leopard tortoise hatchling is breaking out of its shell using an egg tooth that will be shed shortly afterward.

NESTING
Some female tortoises and turtles lay their eggs under rotting vegetation or in the burrows of other animals. However, most species (such as this green sea turtle) use their back limbs to dig a special flask-shaped chamber. Once the eggs have been laid, the sand or soil is carefully replaced and smoothed over to conceal them from predators.

FRESHWATER CARNIVORES
Snapping turtles live in shallow lakes, rivers, and swamps. This common snapping turtle (right) will prey on almost anything that comes within range of its sharp jaws and cavernous mouth, and is small enough to be swallowed.

Feeding

Turtles and tortoises are too slow-moving to pursue active prey, although a few aquatic species—mostly those that live in murky, clouded water—hunt by ambush, remaining motionless in the hope that a fish or crustacean will inadvertently swim past. Tortoises are primarily herbivorous, grazing or browsing on leaves and fruit. Many species also eat animals, such as caterpillars, that appear incidentally in their food, and most will eat carrion given the chance. Freshwater turtles often start life as insectivores, finding enough small aquatic larvae and other small prey to survive but, as they grow, they switch to a diet consisting mainly of aquatic vegetation. Some marine turtles eat only seaweed as adults, while others feed on invertebrates, including jellyfish, sea urchins, and mollusks.

VEGETARIANS
Tortoises such as this Galapagos tortoise (left) graze on grass and other low plants, or browse on bushes and shrubs. They often eat almost continuously when active and are quick to take advantage of more nutritious fallen fruit and even animal carcasses if they are available.

Carettochelys insculpta

Pig-nosed river turtle

Length 28–30 in (70–75 cm)
Breeding Oviparous
Habit Aquatic
Status Vulnerable

Location S. New Guinea, N. Australia

This unusual Australasian turtle lives in fresh water, but has a number of similarities with turtles that live at sea. Its limbs are broad and flipperlike, with relatively few claws, and its gray-green or grayish brown carapace is covered by a layer of soft skin. The shell has a pitted or sculpted surface. Pig-nosed turtles are active foragers, feeding on snails, small fishes, and fruit. The piglike snout gives them their name. They use the snout to breathe while submerged. Females nest in shallow holes on river banks, laying up to 22 thin-shelled eggs.

Chelodina longicollis

Common snake-necked turtle

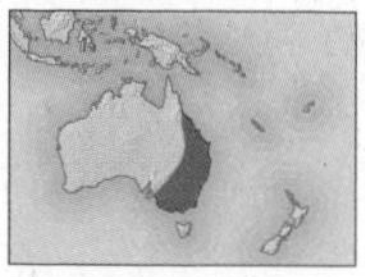

Length 8–10 in (20–25 cm)
Breeding Oviparous
Habit Semiaquatic
Status Not evaluated

Location E. and S. Australia

With its small head and extremely long neck, this turtle is one of Australia's most distinctive freshwater reptiles. Together, the head and neck are often longer than the shell, allowing the turtle to lunge at passing prey, such as fishes, tadpoles, and crustaceans. Its long neck also allows it to "snorkel" while resting on the beds of slow-moving rivers or streams, or in swamps and lagoons. In summer, it may make extensive overland migrations to find water. If disturbed, this turtle can emit a foul-smelling fluid from its musk glands. The female lays 6–24 brittle eggs in a nest dug in grassy or sandy areas at night or after rains.

Chelus fimbriatus

Matamata

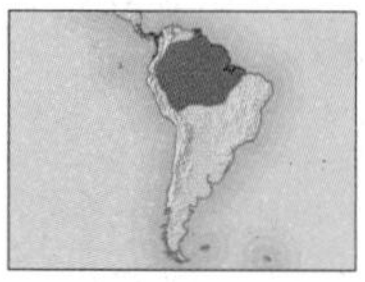

Length 12–18 in (30–45 cm)
Breeding Oviparous
Habit Aquatic
Status Not evaluated

Location N. South America

Most turtles actively seek out their prey, but the knobby shelled matamata lies in wait in shallow, muddy water, using its remarkable camouflage to avoid being seen. Its eyes—set in a triangular head—are small, but it has large external eardrums and sensory skin flaps that help to detect moving prey. It also has a snorkellike nose, which allows it to breathe without rising to the surface. Female matamatas lay up to 28 eggs in a single nesting.

Platemys platycephala

Twist-necked turtle

Length 5½–6¾ in (14–17 cm)
Breeding Oviparous
Habit Semiaquatic
Status Not evaluated

Location N. South America

If threatened, this turtle withdraws into its shell by twisting its head onto its side. Also called the flat-headed turtle, this little-known species has an extremely flat shell that may help it hide under rocks and debris. The carapace is dark or chestnut-brown,or yellow with brown patches, with 2 ridges along its length. A poor swimmer, the turtle remains in shallow pools and puddles, or moves around on the forest floor. It forages for aquatic insects, worms, snails, and tadpoles. Females lay a single egg in a shallow scrape or directly on the surface, and cover it with rotting leaves.

Chelonia mydas

Green sea turtle

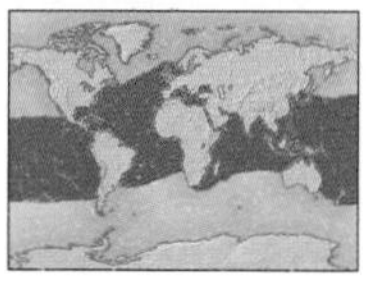

Length 3¼–4 ft (1–1.2 m)
Breeding Oviparous
Habit Aquatic
Status Endangered

Location Tropical, subtropical, and temperate waters worldwide

This graceful, streamlined swimmer is one of the world's most widespread marine turtles found in tropical and subtropical waters across the globe. Despite its name, it is not invariably green, but it always has distinctive, light edging to the scales on its head and limbs, and on scutes around the edge of its shield-shaped carapace. The adults graze on sea grass, mangrove roots, and leaves, but young green sea turtles are more carnivorous, also eating jellyfish, mollusks, and sponges. In order to breed, green sea turtles make their way to isolated, sandy beaches, sometimes traveling across more than 620 miles (1,000 km) of open water. After mating in shallow water, the females crawl ashore after dark, laying 100–150 eggs in deep pits above the high-tide mark. The young hatch in about 6–8 weeks, but before they can reach the water, many are eaten by crabs, gulls, and other seabirds.

LARGE AND STREAMLINED
This large turtle, weighing 145–660 lb (65–300 kg),is a superb swimmer, with a smooth shell and powerful flippers. Adults are greenish brown to black, sometimes streaked with dark brown, red-brown, and yellow.

CONSERVATION

The green sea turtle has been the main quarry of commercial turtle fishing, which has led to its endangered status. With the disturbance of many breeding beaches, suitable nesting sites are also decreasing. Conservation measures for the green sea turtle include legal protection from hunting and egg harvesting, patrols of nesting beaches, and artificial incubation of eggs so that hatchlings can be given a head start, being released into the wild once past their most vulnerable stage.

Eretmochelys imbricata

Hawksbill turtle

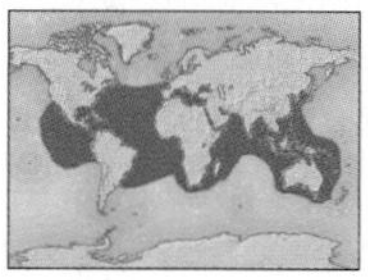

Length 23½–32 in (60–80 cm)
Breeding Oviparous
Habit Aquatic
Status Critically endangered

Location Warmer parts of Atlantic, tropical Indo-Pacific

One of the smallest marine turtles, the hawksbill is easily recognized by its carapace, which has a central keel and serrated edges. Seen closely, its scutes are beautifully marked—the reason why this species has been so widely hunted. Hawksbills are long-lived animals, and are less migratory than other marine turtles. They use their narrow beaks to forage for sponges, mollusks, and other animals on the seabed and among coral reefs.

Caretta caretta

Loggerhead turtle

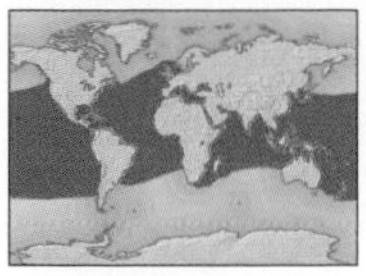

Length 28–39 in (70–100 cm)
Breeding Oviparous
Habit Aquatic
Status Endangered

Location Tropical, subtropical, and temperate waters worldwide

Compared to most other marine turtles, the loggerhead has a large head and powerful jaws, capable of crushing crabs, lobsters, and other hard-bodied prey. In open water, it usually floats near the surface, but it stays near the bottom in estuaries and bays, surfacing only to breathe. Loggerheads breed only every 2 years—or sometimes longer—and lay up to 5 clutches of about 100 eggs.

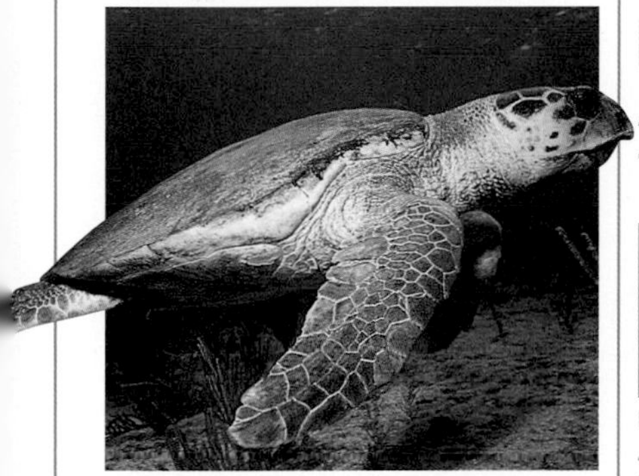

Lepidochelys olivacea

Olive ridley turtle

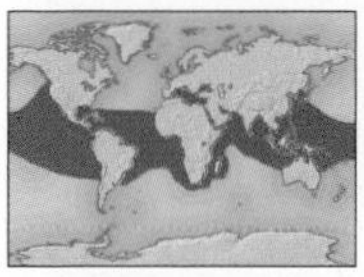

Length 20–30 in (50–75 cm)
Breeding Oviparous
Habit Aquatic
Status Vulnerable

Location Tropical parts of Atlantic, Indian, and Pacific oceans

The smallest sea turtle, the olive ridley turtle has an unmarked, dark or light olive carapace. An active forager, it takes crustaceans, fishes, and squid. It migrates in large numbers to its breeding grounds. Once, huge flotillas of turtles would arrive at sandy beaches to nest; human depredation has put an end to this spectacle.

Chelydra serpentina

Common snapping turtle

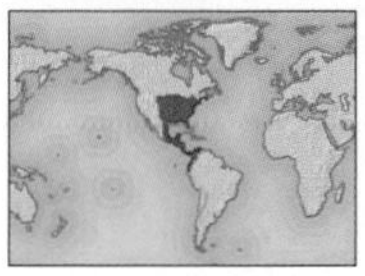

Length 10–18½ in (25–47 cm)
Breeding Oviparous
Habit Aquatic
Status Not evaluated

Location E. and C. North America, Central America, N.W. South America

This large, dark turtle is quick-tempered, biting ferociously when threatened. It has a massive head, a small plastron, and a rough carapace, sometimes covered with algae.

Inhabiting fresh and brackish water, preferably with plenty of plant life, it often lies buried in mud with only the eyes and nostrils exposed. It sits and waits for prey by day but at night is an active forager, lunging with its mouth open to engulf small mammals, birds, fishes, invertebrates, and plants. Mating may occur without preliminaries, but rituals in which the male and female face each other with necks extended have also been seen. Females lay one clutch of 20–30 eggs in a season, in a flask-shaped nest, often in a muskrat lodge; they can retain sperm from one breeding season to the next.

Macrochelys temminckii

Alligator snapping turtle

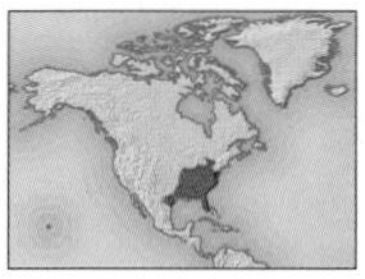

Length 16–32 in (40–80 cm)
Breeding Oviparous
Habit Aquatic
Status Vulnerable

Location S.E. USA

This formidable animal is the world's largest freshwater turtle, with a maximum weight reputed to be over 220 lb (100 kg). Like the matamata (see opposite), it hunts mainly by sitting and waiting, but it is equipped with a lure that entices fishes toward its scissor-sharp jaws. At night, it is an active forager. Males spend their entire lives at the bottom of lakes or rivers, but the females leave the water in spring to lay clutches of 10–50 spherical eggs, buried in mud or sand.

ROUGH AND KEELED
The alligator snapping turtle has a rough shell, with 3 jagged keels, and an extra row of scutes on each side.

FISH LURE

During the day, this turtle lies with its jaws open and lures fishes by wiggling a small, pink, wormlike structure on the floor of its mouth. The hooked upper and lower beaks can deliver a powerful bite.

Emys orbicularis

European pond turtle

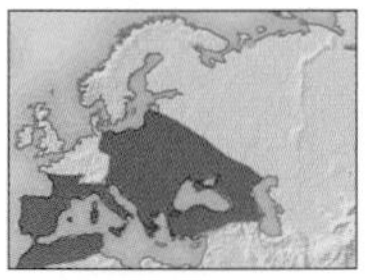

Length 6–8 in (15–20 cm), max 12 in (30 cm)
Breeding Oviparous
Habit Semiaquatic
Status Near threatened

Location N. Africa, Europe, W. Asia

This olive, brown, or black turtle is one of just 2 freshwater species that live in Europe. Its carapace is smooth, and a hinge toward the front of the plastron allows it to be raised when the turtle withdraws its head, although adults cannot close the shell completely. These turtles spend much of their time basking on stones or logs, but they dive at the first sign of danger. They eat frogs, fishes, and other small animals, in water or on land.

Chrysemys picta

Painted turtle

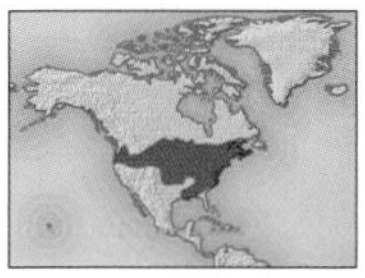

Length 6–10 in (15–25 cm)
Breeding Oviparous
Habit Aquatic
Status Not evaluated

Location North America

This freshwater turtle is one of the most widespread in North America, found in lakes, ponds, and slow-moving streams and rivers from the eastern seaboard to the far Midwest. There are 4 distinct subspecies; all have a flattened, smooth carapace, but some have red margins to their shells, a red stripe on the back, or yellow or red stripes on the neck. This turtle basks for long periods, especially in the morning, several individuals piling up on one another to bask on a favorite log. It is omnivorous and forages actively by day, sleeping on the lake bottom at night. Hunted by birds, fishes, and racoons, it defends itself by hiding its head inside its shell, or by burying itself in mud. The painted turtle hibernates for variable lengths of time, depending on its location. Most females lay 2–20 eggs in 3 clutches, in chambers dug in sandy soil; not all lay eggs every season.

Clemmys guttata

Spotted turtle

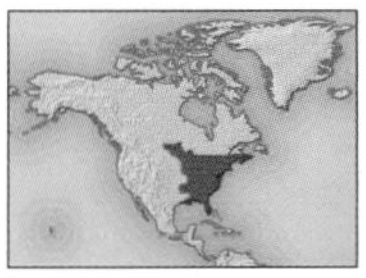

Length 4–5 in (10–12.5 cm)
Breeding Oviparous
Habit Semiaquatic
Status Vulnerable

Location S.E. Canada, E. USA

This small turtle has yellow spots on its black carapace, head, neck, and limbs. An omnivorous active forager, it is itself hunted by birds and small mammals. In summer, the semiaquatic spotted turtle estivates in the muddy beds of lakes and rivers or in muskrat burrows; it may hibernate in similar sites in winter.

REPTILES

Glyptemys insculpta

Wood turtle

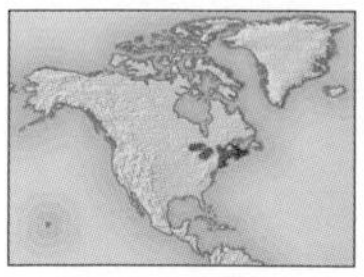

Length 5½–7½ in (14–19 cm), max 9 in (23 cm)
Breeding Oviparous
Habit Semiaquatic
Status Vulnerable

Location S.E. Canada, N.E. USA

Besides being an unusually good climber, the wood turtle is a natural wanderer, and is often found crossing fields or roads, particularly after rain. The scutes on its carapace have concentric ridges, and their somber colors provide it with effective camouflage. Wood turtles are omnivores, eating worms, slugs, insects, and tadpoles, as well as leaves, berries, and other plant food. They mate in water, and the females lay 7–8 eggs a year in flask-shaped nests. At one time, the adult turtles were caught in large numbers for food.

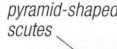

Terrapene carolina

Carolina box turtle

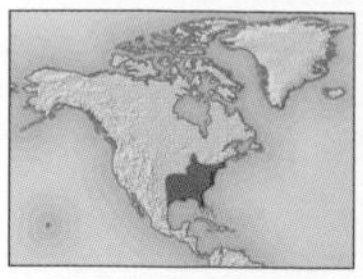

Length 4–8½ in (10–21 cm)
Breeding Oviparous
Habit Semiterrestrial
Status Near threatened

Location E. USA

Known to have lived for as long as 100 years in captivity, in the wild the Carolina box turtle prefers damp habitats, such as meadows, floodplains, or moist, forested areas. Very active in the mornings, particularly after rain, this omnivore forages for slugs, worms, mushrooms, or berries. To escape the heat, it retreats into cover or mud at midday and in midsummer, and survives the cold winter months by hibernating. Females can lay fertile eggs for several years following a single mating. The adults have few natural predators, but the young turtles fall prey to mammals such as raccoons, and birds of prey.

BOXED IN

The front section of the Carolina box turtle's plastron is hinged, so that when the turtle withdraws its head into the shell, the front of the plastron can be raised, effectively shutting the animal into a protective "box."

TAN TURTLE
This individual has a tan shell, but some Carolina box turtles are mainly orange or yellow, and others may be quite dark, with radiating yellow lines.

Trachemys scripta elegans

Red-eared slider

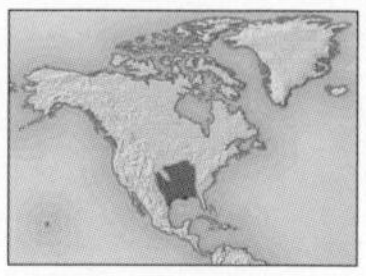

Length 8–12 in (20–30 cm)
Breeding Oviparous
Habit Aquatic
Status Near threatened

Location S. Central USA

Identified by a prominent red stripe behind its eyes, this freshwater turtle is just one subspecies of the many species that live in lakes, rivers, and creeks in North America. Like many of its close relatives, it is fond of basking in sunshine, sometimes forming stacks where basking sites are in short supply, but it is quick to dive into the water if disturbed. As an adult, it lives almost entirely on plant food, but its young eat tadpoles and aquatic invertebrates.

Dermochelys coriacea

Leatherback turtle

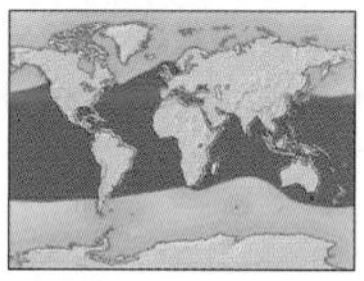

Length 4¼–6 ft (1.3–1.8 m)
Breeding Oviparous
Habit Aquatic
Status Critically endangered

Location Tropical, subtropical, and temperate waters worldwide

Weighing up to 1,770 lb (800 kg), the leatherback is by far the largest of the marine turtles, and one of the animal world's greatest oceanic travelers. Tagged individuals have been known to cross the Atlantic. Leatherbacks also range into cold waters at high latitudes, thanks to a limited ability to generate body heat. Physically, they differ from other marine turtles not only in size, but also in having a narrow, leathery shell, and flippers without claws. Leatherbacks feed primarily on jellyfish, and their throats have backward-pointing spines to prevent their slippery prey escaping. They usually feed near the surface, but they can dive down to 1,300 ft (400 m)—perhaps as much as 3,300 ft (1,000 m)—holding their breath for up to half an hour. Although normally found in the open ocean, leatherbacks gather inshore during the breeding season, when males scramble for females arriving to lay their eggs. As with other marine turtles, the females have a strong homing instinct that guides them to a particular stretch of beach, and they lay eggs on dark, moonless nights to avoid being seen. The hatchlings' shells are covered with small, pearly scales, which soon disappear.

CONSERVATION

Leatherbacks nest on sandy tropical beaches, providing an opportunity to gauge the total population at sea. The results are alarming: in the past 2 decades, leatherback numbers have slumped, with far fewer coming ashore to lay. Measures to protect leatherbacks include a complete ban on their capture; in addition, their eggs are collected and incubated in safe conditions, so that their young can be released into the sea.

EXCURSION ASHORE
Emerging from the sea to lay eggs, a female leatherback laboriously hauls herself up a beach. On average, she may spend less than 2 hours a year on land.

Dermatemys mawii

Central American river turtle

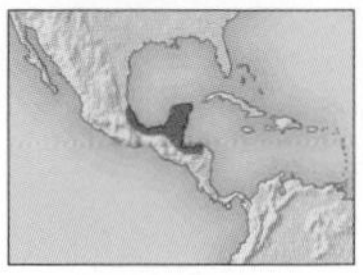

Length 20–26 in (50–65 cm)
Breeding Oviparous
Habit Aquatic
Status Critically endangered

Location S. Mexico to Central America

The webbed feet and streamlined shell of this olive-gray turtle are adaptations for swimming; it is almost helpless on land. Females have olive tops to their heads (yellowish to reddish brown in males), and very short tails. They lay 6–20 eggs in muddy river banks in summer and fall, burying them or covering them with decaying vegetation. While adults are herbivorous, the juveniles also eat mollusks, crustaceans, and, probably, fishes. This turtle is hunted by otters and humans.

Cuora flavomarginata

Yellow-marginated box turtle

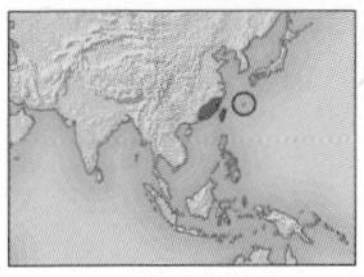

Length 4–4¾ in (10–12 cm)
Breeding Oviparous
Habit Semiaquatic
Status Endangered

Location E. Asia (including Taiwan and Ryukyu Islands)

The yellow-marginated box turtle gets its name from a hinged plastron that allows it to be completely boxed in when withdrawn into its shell, and yellow stripes that run down its back and from each eye to the neck. It spends much of its time in rice paddies, ponds, and streams, but also basks and wanders on land. This turtle feeds on fishes, crustaceans, worms, and fruit. Nesting in soil or sand, females may lay several clutches of one or, occasionally, 2 eggs a year.

Cyclemys dentata

Asian leaf turtle

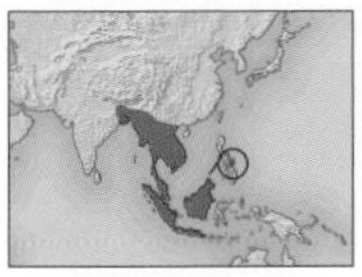

Length 6–9½ in (15–24 cm)
Breeding Oviparous
Habit Semiaquatic
Status Near threatened

Location S.E. Asia

The oval carapace of this turtle has a serrated edge near the tail, and is colored light to dark brown (sometimes mahogany) with faint dark markings. The plastron, which is hinged, is more boldly marked with lines radiating from the center of each scute. As with most turtles, males are smaller than females and have longer, thicker tails. The Asian leaf turtle is an omnivore, feeding on invertebrates and tadpoles, as well as plants. Very active both on land and in water, it prefers shallow streams in either mountains or lowlands. When danger threatens, it may withdraw into its shell or dive to the bottom of the water and hide in the mud. The female digs a chamber in which she lays up to 5 clutches of 2–4 relatively large eggs each year; unusually, in this turtle, the plastron becomes flexible to allow eggs to be laid. Hatchlings are more aquatic than adults and have spines or points—which disappear with age—around the edge of the carapace, perhaps to deter predators.

Kinosternon flavescens

Yellow mud turtle

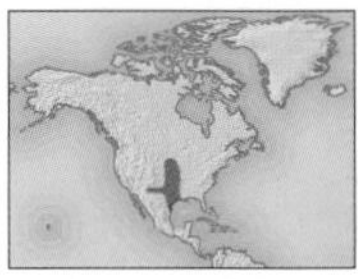

Length 4¾–6½ in (12–16 cm)
Breeding Oviparous
Habit Mainly aquatic
Status Not evaluated

Location S. Central USA, N. Mexico

In addition to withdrawing into its shell for defense, the yellow mud turtle can also release foul-smelling musk from 2 pairs of glands under its carapace, and may occasionally bite. It has especially powerful jaws for crushing its prey, which includes snails, worms, insects, and tadpoles. Primarily diurnal, it spends most of its time in shallow, slow-moving bodies of water in grassland. In midsummer, it shelters from heat in muskrat dens or by burrowing under leaf litter. It hibernates in the cooler parts of its range.

Sternotherus odoratus

Stinkpot

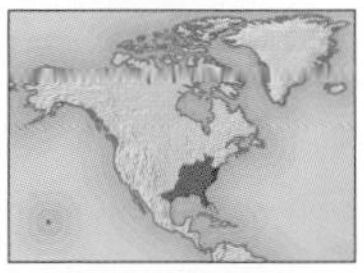

Length 3¼–5 in (8–13 cm)
Breeding Oviparous
Habit Mainly aquatic
Status Not evaluated

Location S. and E. USA

As a form of defense, this turtle expels a foul odor from its musk glands, hence the name "stinkpot." It can also bite viciously. Rarely leaving quiet, shallow, muddy-bottomed waters, it is often covered in algae. It is active by day, but may also hunt at night, and a pair of sensory barbels on the chin may help it to find food—mainly insects, mollusks, plant material, or carrion. It is eaten by bald eagles and red-shouldered hawks, alligators, and fishes such as the largemouth bass. Females nest under tree stumps or in the walls of muskrat lodges, laying 1–5 eggs at a time.

Pelomedusa subrufa

African helmeted turtle

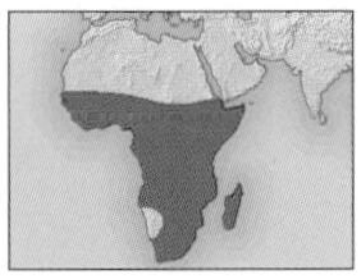

Length 8–12½ in (20–32 cm)
Breeding Oviparous
Habit Semiaquatic
Status Not evaluated

Location Africa (south of Sahara)

The sluggish African helmeted turtle lives in rain pools and watering holes in Africa's open country. During the rainy season, it wanders from pool to pool foraging for frogs, tadpoles, mollusks, invertebrates, and carrion. It may estivate in dry conditions by burying itself in mud. If threatened by crocodiles or other carnivores, it withdraws into its shell, helped by a hinged plastron, and may also discharge a strong-smelling musk and the contents of its cloaca.

Pelodiscus sinensis

Chinese soft-shelled turtle

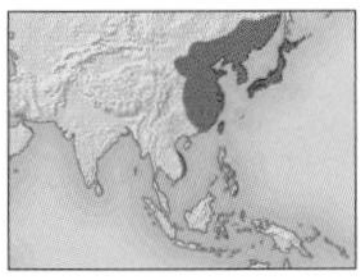

Length 6–12 in (15–30 cm)
Breeding Oviparous
Habit Aquatic
Status Vulnerable

Location E. Asia

With its long snout and tubelike nostrils, the Chinese soft-shelled turtle can "snorkel" in shallow water. It also has webbed feet for swimming. When resting, it lies at the bottom, buried in sand or mud, lifting its head to breathe or snatch at prey. It forages at night, taking crustaceans, mollusks, insects, fishes, and amphibians. If threatened, it cannot withdraw completely into its shell, but it can give a vicious bite.

Chelonoidis elephantopus

Galapagos tortoise

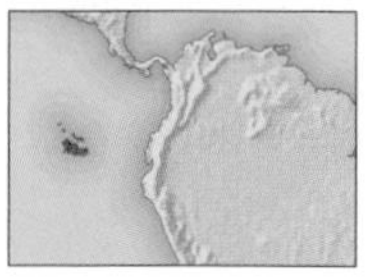

Length Up to 4 ft (1.2 m)
Breeding Oviparous
Habit Terrestrial
Status Vulnerable

Location Galapagos Islands

The Galapagos tortoise, the largest living tortoise, has a huge carapace, massive limbs, and a long neck. There is a great variation in shell shape and overall size depending on which of the Galapagos Islands the tortoise originates. They spend most of their time grazing in small herds and basking in pools or mud wallows. During the breeding season, males become territorial and start looking for mates, while females create nests in chambers in the ground to lay their eggs. Despite some individuals living to well over 100 years of age, the Galapagos tortoise remains vulnerable. The main threat is depredation at a young age by introduced species such as black rats and cats, as well as competition for vegetation from goats and cattle.

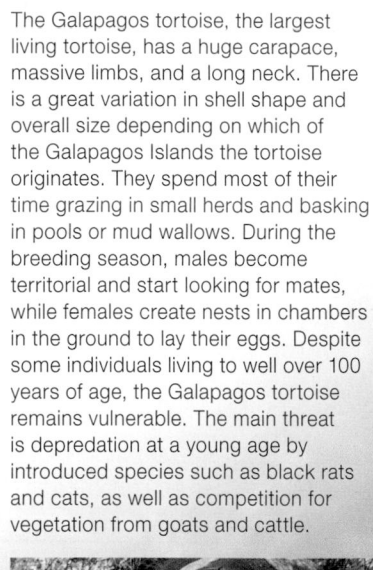

GENERALIZED HERBIVORE
The Galapagos tortoise has strong, toothless jaws well adapted to feeding on whatever type of vegetation it can find, including tough cacti.

SHELL SHAPE
The shell of the Galapagos tortoise varies and is linked to feeding habit. A domed shell is common for subspecies that graze on grasses, whereas a saddle-backed shell is adapted for browsing shrubs.

CONSERVATION

Since 1965, the Charles Darwin Research Station has been running a breeding and repatriation program to boost the dwindling population of Galapagos tortoises. By 2010, the program had released 1,200 tortoises.

UNMISTAKABLE GIANT
The impressive size of the Galapagos tortoise is probably due to its adaptation to living in a difficult environment with an unreliable food supply—the larger the tortoise, the more nourishment it can store.

REPTILES

MATING

The courtship technique of the male Galapagos tortoise is uncompromising. Having located a suitable female, he rams her into submission, nipping her legs to further immobilize her. He then climbs onto her back to mate.

Chelonoidis carbonaria

Red-footed tortoise

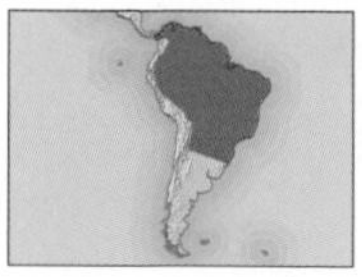

Length 16–20 in (40–50 cm)
Breeding Oviparous
Habit Terrestrial
Status Not evaluated

Location N. to C. South America

The genus *Chelonoidis* includes the world's biggest tortoises, as well as many smaller species, such as this one from South America. Its legs are marked with splashes of red or yellowish orange, and the shell—in adult males—is unusually elongated, with a constriction in the middle. Like most of its relatives, the red-footed tortoise lives mainly on plant food, such as leaves and fallen fruit, but it also scavenges animal remains. Females lay clutches of 2–15 nearly spherical eggs several times a year.

Geochelone elegans

Indian starred tortoise

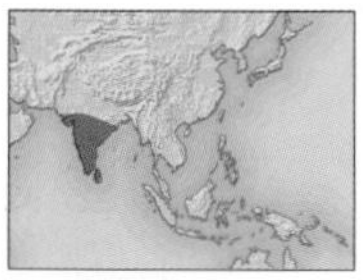

Length Up to 11 in (28 cm)
Breeding Oviparous
Habit Semiaquatic
Status Least concern

Location S. Asia

With its star-shaped markings and knobby carapace, this Asian species is one of the world's most distinctive tortoises. Each of its scutes rises to a rounded point, and the carapace as a whole is strongly domed. The Indian starred tortoise needs plenty of water, and it is most active during the monsoon season. In drier weather, it stirs only in the morning and late afternoon. Each year, the female lays several clutches of up to 10 eggs in a flask-shaped chamber 4–6 in (10–15 cm) deep.

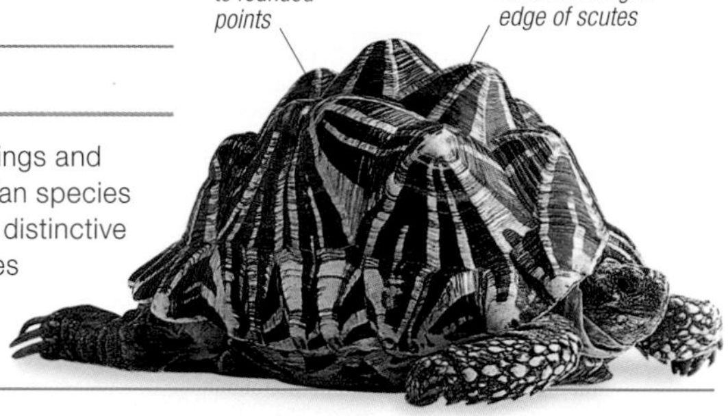

Stigmochelys pardalis

Leopard tortoise

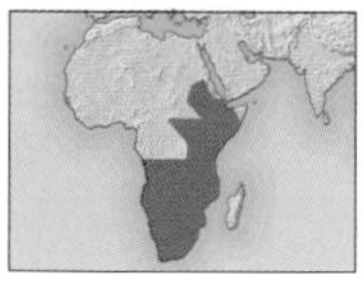

Length 12–28 in (30–70 cm)
Breeding Oviparous
Habit Terrestrial
Status Not evaluated

Location E. to southern Africa

The carapace of the leopard tortoise is yellowish, with scattered dark markings that are bolder, although fewer, in hatchlings. In some individuals, it can grow to 28 in (70 cm) long. This tortoise lives in a variety of habitats, from sandy, coastal scrub to grassland and semidesert, although it is scarce in very dry deserts. It may bury itself to hibernate in cool parts of its range, and to estivate in the hotter parts. The leopard tortoise is herbivorous, grazing on grasses, fallen fruit, fungi, and the leaves of succulents. In the breeding season, the female digs a flask-shaped pit with her hind feet, urinating to soften hard soil, and lays 3–6 clutches of 5–30 eggs at monthly intervals. Hatchlings may have to wait in the nest chamber for several weeks for rain to soften the ground before they can burrow to the surface. The young become prey to monitor lizards, storks, crows, and small mammals.

Gopherus agassizii

Desert tortoise

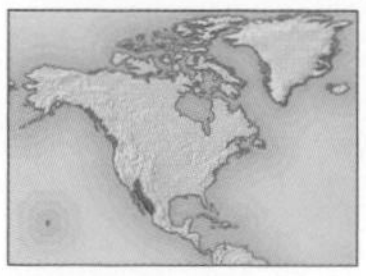

Length 10–14 in (25–36 cm)
Breeding Oviparous
Habit Terrestrial
Status Vulnerable

Location S.W. USA, N.W. Mexico

This tortoise's thick, shovel-shaped forefeet are specially adapted for digging burrows in which to escape the extreme heat of the desert during the day and the cold at night. It shelters individually in burrows in very hot or drought conditions but may gather in large, communal dens dug into gravel banks to escape the cold. It lives for at least 50 years, mainly eating cacti and grasses, but also feeds on insects. Males have longer projections at the front of their plastrons and may attack each other on sight in the breeding season.

Homopus signatus

Speckled padloper

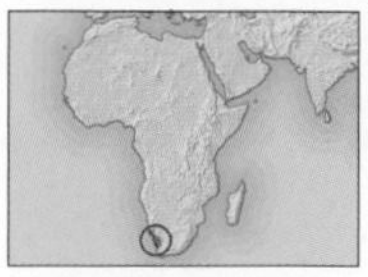

Length 2¼–3¼ in (6–8 cm)
Breeding Oviparous
Habit Terrestrial
Status Near threatened

Location W. South Africa

This African species is the smallest tortoise in the world, with males being particularly tiny. Their small size makes speckled padlopers vulnerable to carnivorous mammals and birds, but it also enables them to squeeze under rocks to escape attack, and to shelter from the sun. The flattened shell is brown, orange-red, or salmon-pink, with fine black markings. Speckled padlopers live in dry habitats, and forage for small, succulent plants.

Kinixys erosa

Serrated hinge-back tortoise

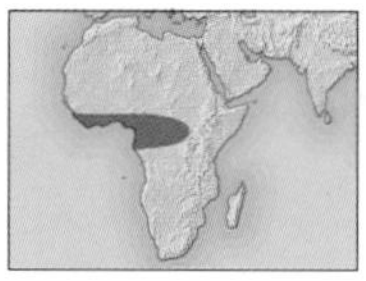

Length 10–12½ in (25–32 cm)
Breeding Oviparous
Habit Semiterrestrial
Status Data deficient

Location W. to C. Africa

A hinge toward the back of the carapace enables this tortoise to shut itself in its shell completely. A fair swimmer, it seeks out marshes and riverbanks in its forest habitat, but when on land, it spends much of the time buried beneath roots and logs. It is an omnivorous species, feeding on plants, fruit, and invertebrates, as well as carrion. In a breeding season, females lay several clutches of up to 4 eggs on the ground, covering them with leaves.

Malacochersus tornieri

Pancake tortoise

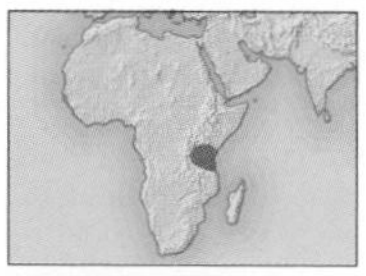

Length 5½–6½ in (14–17 cm)
Breeding Oviparous
Habit Terrestrial
Status Vulnerable

Location E. Africa

The carapace of this tortoise is not only extremely flat but also very flexible, due to openings in the underlying bones. This allows the tortoise to squeeze into narrow crevices in order to escape predatory mammals or birds; it can also wedge itself in, for safety, by digging in its foreclaws and rotating the forelimbs, making extraction difficult. It forages, mainly in the mornings, for grasses, leaves, and fruit, but never moves far from the rocky outcrop in which it spends the night. In summer, this tortoise crawls under a flat stone and estivates to avoid the heat.Females lay several clutches of a single egg throughout the breeding season.

Testudo hermanni

Hermann's tortoise

Length 6–8 in (15–20 cm)
Breeding Oviparous
Habit Terrestrial
Status Near threatened

Location S.E. Europe, Mediterranean islands

Easily confused with the spur-thighed tortoise (see right), this tortoise is the smallest of 3 species that live in southern Europe. It has a domed and slightly lumpy carapace, colored yellow, olive, or brown, with a scattering of irregular dark markings. As with most tortoises, the males are slightly smaller than the females, and have longer tails and concave undersides. Hermann's tortoise is largely a vegetarian, feeding on fruit, flowers, and leaves, although it also eats slugs, snails, and animal remains. It lives in places with dense plant cover, forcing its way through the vegetation on its short but powerful legs. It is not active during the hottest part of the day in summer, and in the extreme south, it becomes dormant. Where winters are cold, it also hibernates for several months each year. As with most tortoises, this species hisses and grunts, particularly during the breeding season. Having mated, females dig flask-shaped nesting chambers, where they lay up to 12 eggs. Like the spur-thighed tortoise, this species was once collected in large numbers for the pet trade, but is now fully protected.

Testudo graeca

Spur-thighed tortoise

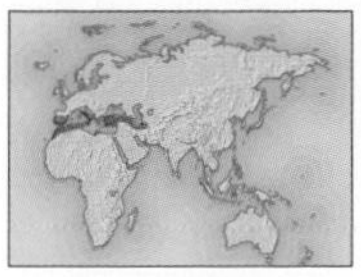

Length 8–10 in (20–25 cm)
Breeding Oviparous
Habit Terrestrial
Status Vulnerable

Location S. Europe, S.W. and W. Asia, N.W. Africa, Mediterranean islands

Conspicuous spurs on the hind legs give this tortoise its name, and help distinguish it from Hermann's tortoise (see left), which overlaps its range. Its shell is similar to the Hermann's but smoother. It lives in grassy places and sand dunes, and feeds on leaves, fruit, and occasionally carrion and mammal dung. The females lay clutches of up to 12 eggs.

Chersina angulata

Angulate tortoise

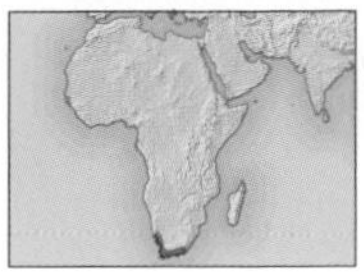

Length 6–8 in (15–20 cm)
Breeding Oviparous
Habit Terrestrial
Status Not evaluated

Location Southern Africa

Also known as the bowsprit tortoise, this species has a high, domed carapace. It eats grasses and succulents—hibernating during the winter—and, in spite of its armor, provides food for monitor lizards, carnivorous mammals, and even birds of prey. Unusually, the males are slightly larger than the females, and fight with rivals at the onset of the breeding season. Females lay one, or rarely, 2 eggs.

Tuataras

PHYLUM	Chordata
CLASS	Reptilia
ORDER	Rhynchocephalia
FAMILIES	1 (Sphenodontidae)
SPECIES	2

Tuataras are the only surviving representatives of a group of reptiles that flourished over 200 million years ago. Found on 2 groups of small islands off the coast of New Zealand, they look similar to lizards. Tuataras live in burrows and are largely nocturnal. Compared to other reptiles, they are remarkably tolerant of cold, remaining active in temperatures as low as 50°F (10°C). They grow slowly, breed infrequently, and live to a great age (possibly over 100 years).

BASKING
Although usually nocturnal, tuataras may occasionally be seen basking on rocks on sunny days. Due to a low metabolic rate, they draw breath infrequently, and when at rest may take only one breath an hour.

Anatomy

Tuataras have a large head, a long tail, and well-developed limbs. They have a different skull structure than lizards (see below), and unlike lizards they have no eardrums, middle ear, or external male sexual organ. A "third eye" is situated on the top of the head, but there is no evidence that this is functional.

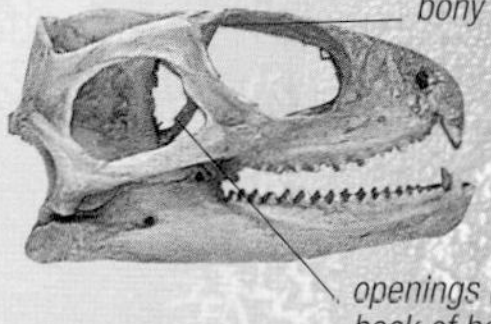

SKULL SHAPE
There are 2 openings towards the back of the tuatara's skull; most lizards have only one such opening. The teeth are not separate structures but serrations along the edges of both the upper and lower jaws.

Feeding

Tuataras are almost entirely insectivorous and feed on the large numbers of scavenging beetles and crickets that are attracted to colonies of seabirds. They also eat other invertebrates, small lizards, and, occasionally, the eggs and chicks of birds whose nest holes they share.

BURROWING
Tuataras either dig their own burrows or inhabit the nest holes excavated by breeding seabirds, such as small petrels. They usually hunt for food just outside the entrance of the burrow.

Sphenodon punctatus

Tuatara

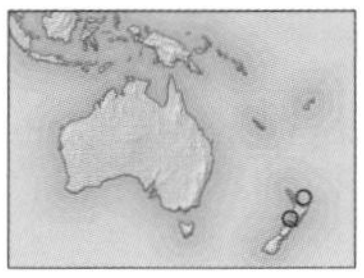

Length 20–23½ in (50–60 cm)
Breeding Oviparous
Habit Burrowing
Status Least concern

Location New Zealand (coastal islands)

With its spiny crest and loose, scaly skin, the tuatara looks deceptively similar to some iguanas (see pp.414–6). It was once widespread in New Zealand, but now survives almost entirely on small, offshore islands, where it is protected from introduced mammals. Adult tuataras feed after dark, with spiders, insects, and worms making up the bulk of their diet. They grow very slowly, and females do not breed until they are at least 20 years old. The eggs spend a year developing inside the female's body, and after being laid, they take at least a year to hatch. Until recently, this species was thought to be the only member of the tuatara family. However, genetic analysis has shown that a group of tuataras on North Brother Island form a distinct species, now known as *Sphenodon guntheri*.

REPTILES

Snakes

PHYLUM	Chordata
CLASS	Reptilia
ORDER	Squamata
SUBORDER	Serpentes
FAMILIES	18
SPECIES	About 3,400

CLASSIFICATION NOTE

Snakes are closely related to lizards and amphisbaenians. There is much debate over the relationships between different types of snakes, and there is no universally accepted system of classification. Most controversy concerns the number of families. While most authorities recognize 18, others consider some of these to be subfamilies and arrive at a smaller number, combining boas and pythons in a single family, for example.

Snakes are formidable and highly evolved predators. Although they have no limbs, no eyelids, and no external ears, these versatile animals move with ease and find their prey using sophisticated senses. All snakes eat other animals, ranging from ants to antelopes, some of which they subdue by constriction or by delivering a venomous bite from specialized fangs. Although snakes cannot chew their food, the bones in their skull are lightly constructed and loosely connected, so that the jaws can be opened wide and the prey swallowed whole. Snakes have established themselves on all the world's major landmasses (except Antarctica) as well as on many oceanic islands. Only about one in 10 are venomous and of these only a small proportion represent a threat to humans.

REPTILES

Anatomy

The shape of a snake's body usually reflects where it lives: climbing species tend to be long and thin; burrowers are often short and stout with short tails and blunt snouts; and sea snakes have flattened, paddle-shaped tails. Unlike other reptiles, snakes have a single row of scales on their underside, the ventral scales, which are usually wide, and smaller scales on the upper surfaces. Some species have large, regular plates on the head, while others have small, fragmented scales. Each eye is covered with a transparent scale, the brille or spectacle, which is replaced when the snake sheds its skin. All scales can be smooth, ridged (keeled), or granular. The internal organs are modified to fit the elongated body. Paired organs are staggered within the body cavity, and only one lung may be functional, with the other reduced in size. Sea snakes have an enlarged lung, part of which forms a buoyancy chamber.

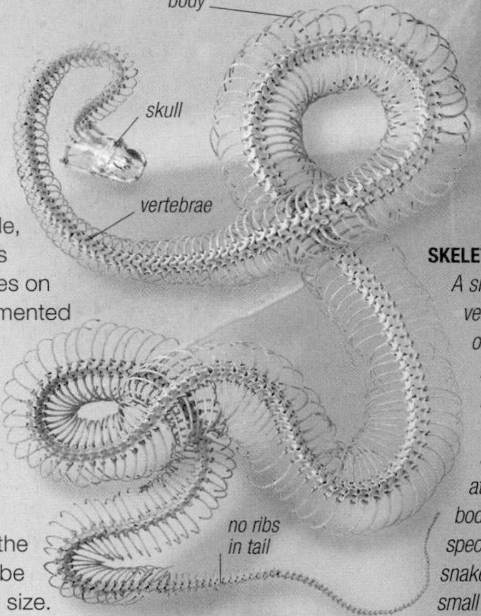

SKELETON
A snake has up to 400 vertebrae that articulate on each other to give a highly flexible skeleton. Each vertebra has 2 winglike processes to keep the spine from twisting. Ribs are attached to the vertebrae in the body but are absent in the tail. Some species have back-limb girdles, and primitive snakes have vestigial back limbs in the form of small spurs, but no snakes have front-limb girdles.

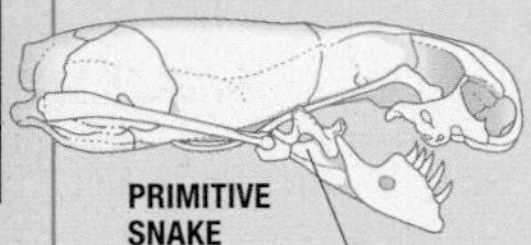

PRIMITIVE SNAKE

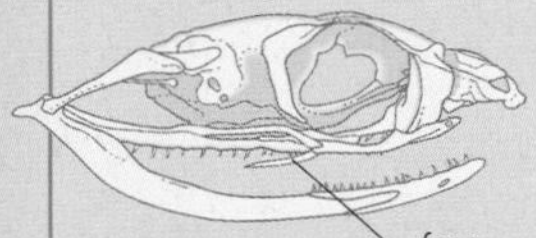

REAR-FANGED SNAKE

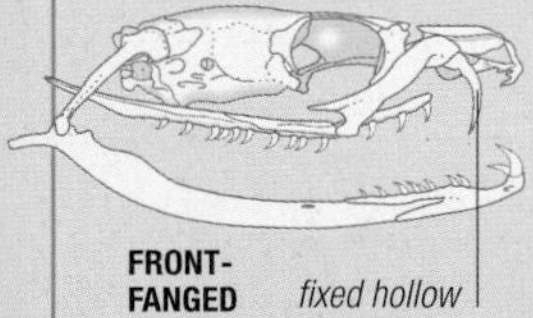

FRONT-FANGED SNAKE

SKULLS AND TEETH
Primitive snakes have a heavy skull with few teeth. Most other snakes have a lighter skull with loosely connected jawbones that can move apart. Teeth are fixed to the upper or lower jaw, or to the roof of the mouth. Venom-injecting fangs may be at the front or rear of the mouth. In some front-fanged snakes, they are hinged to the upper jaw.

SENSES
Snakes have poor eyesight and hearing, and rely instead on other senses. A well-developed sense of smell is supplemented by the Jacobson's organ (see p.373), and active snakes flick their tongue constantly to sample their surroundings. Pit vipers and some boas and pythons can detect small changes in air temperature using organs on their face known as heat pits.

HEAT PITS

FORKED TONGUE

SWALLOWING A LIZARD

1 **STEALTHY APPROACH**
A vine snake slides slowly and carefully along a branch toward an unsuspecting gecko.

2 **THE ATTACK**
In one swift movement, the snake arches over the gecko and, using its rear fangs, injects its venom.

3 **HOLDING ON**
Although the gecko falls from its perch, the snake's tail is wrapped around a branch and it holds on.

4 **SHIFTING GRIP**
Once the gecko is dead, the snake begins to maneuver it into position for swallowing.

Movement

Snakes have evolved several ways of moving around (see right) to compensate for not having legs. The method used involves different uses of their ribs and the muscles attached to them, but also depends on the snake's weight, the speed at which it needs to move, and the type of surface. Most snakes can use different types of locomotion as the need arises. As well as the methods shown here, some vipers use a technique called sidewinding when crossing loose sand (see p.406).

LINEAR PROGRESSION
Heavy snakes and snakes that are moving slowly or stalking prey sometimes progress in a straight line, hooking the trailing edges of the ventral scales over surface irregularities. This occurs in a wavelike sequence, as the snake pulls itself forward.

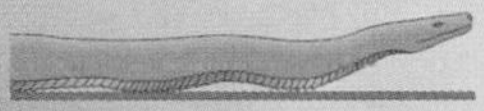

LATERAL UNDULATION
Using each point of its body in turn to push against irregularities in the surface, the snake wiggles from side to side. This method is also used for swimming and sometimes climbing.

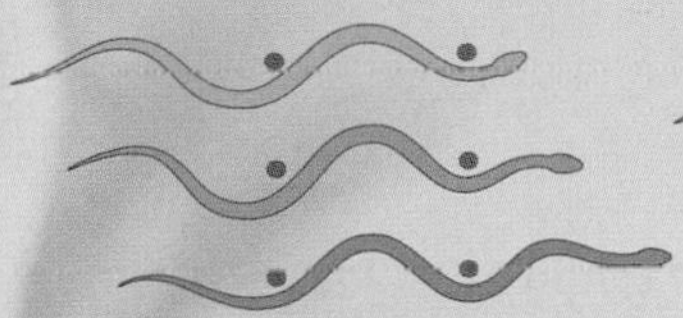

CONCERTINA MOVEMENT
When moving through their tunnels, burrowing snakes expand one end of the body to jam it against the tunnel walls while the other end is thrust forward.

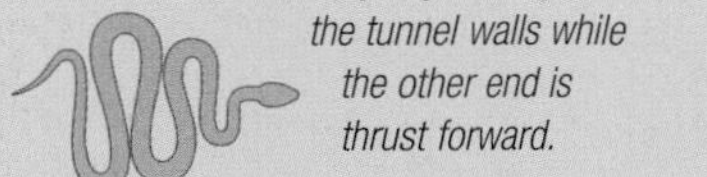

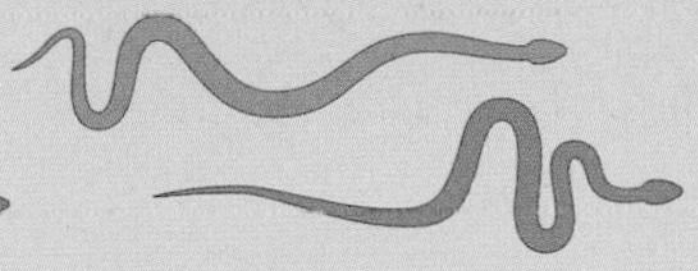

CLIMBING
To enable them to move easily through their habitat, tree snakes have a thinner, more lightweight body than ground-dwelling or burrowing snakes. They have long tails that can be used to grasp branches, and strong vertebrae that help them to cross wide gaps. The pit viper shown here (Schultze's pit viper) is a nocturnal hunter that uses its large eyes and its heat pits to locate prey.

Hunting and feeding

Some snakes are specialized feeders, while others eat a wide range of prey, including other snakes. Small animals or ones that cannot fight back are simply grasped and then swallowed; animals that are larger or better able to defend themselves must be subdued with venom or by constriction before being eaten. A large meal may take a long time to digest, starting in the mouth with powerful salivary juices. Venom itself is a cocktail of modified digestive juices that usually acts quickly on the nervous system or blood tissue of the prey.

SWALLOWING
Flexible jaws and highly elastic skin allow snakes, like this common egg-eater, to eat prey that is larger than their own head. A large meal can take several hours to swallow.

CONSTRICTION
A constrictor kills its prey by suffocation. Each time its victim breathes out, this rock python tightens its coils until breathing stops.

Reproduction

Most snakes lay eggs, but a significant number give birth to live young. Species from temperate regions usually mate in spring, soon after emerging from hibernation, and produce eggs or young in summer. Tropical species may breed in response to rainfall and sometimes have a long breeding season, laying several clutches of eggs each year. Males and females find each other using chemical scent trails, and courtship is rarely elaborate. Snakes show little parental care, although pythons and a few other species coil around their incubating eggs.

EGGS AND LIVE YOUNG
Young snakes, like this monocled cobra, break out of their shell using a sharp, temporary egg tooth. The snake is often coiled tightly inside its shell, and may be up to 7 times longer than the egg.

REPTILES

.OWING HEAD FIRST
ecko is swallowed head first because the follow more easily in this direction.

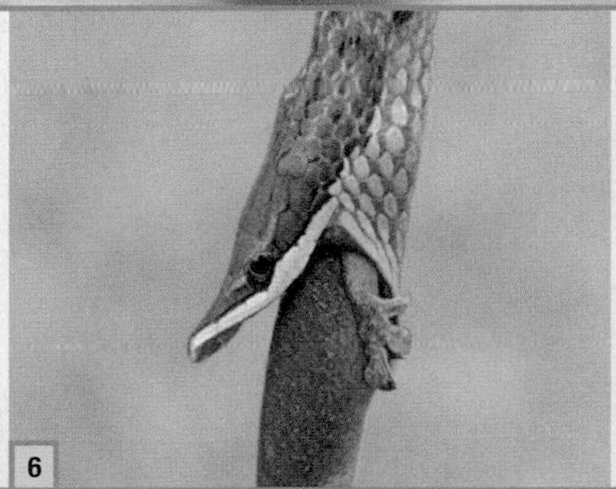

6

EXTENDING THE JAWS
The snake opens its highly flexible jaws and uses its rear fangs to pull the gecko into its mouth.

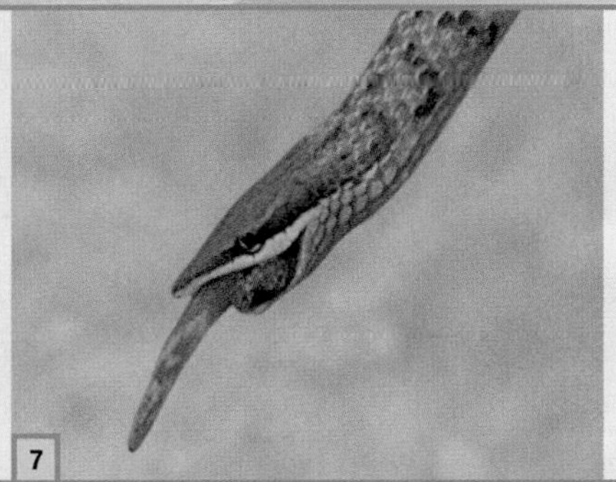

7

MUSCULAR ACTION
When most of the prey is in the throat, wavelike muscle contractions force it into the stomach.

8

DIGESTION
Having completely swallowed its prey, the snake searches for a quiet place to digest its food.

Boas, pythons, and relatives

PHYLUM	Chordata
CLASS	Reptilia
ORDER	Squamata
SUBORDER	Serpentes
SUPERFAMILY	Henophidia
FAMILIES	12
SPECIES	186

Boas and pythons are powerful constricting snakes. They include the world's largest snakes, among them the anaconda (a boa) and the reticulated African and Indian pythons. Boas and pythons do not reproduce in the same way. They also have different distributions: pythons are found in Africa, Asia, and Australasia, while boas occur mainly in the Americas with a smaller number of species in Africa, Asia, and on some Pacific islands. The relatives of boas and pythons, most of which are also constrictors, include the iridescent sunbeam snakes and the aquatic file snakes.

Anatomy

The snakes in this group are often regarded as being relatively primitive. Their skulls are heavier and their jaws more rigid than those of the more advanced snakes (the colubrids, elapids, and vipers). They have also retained several anatomical features from the limbed animals from which they are descended. These include a back-limb (pelvic) girdle and, in most species, the remains of back limbs in the form of small claws or spurs. All species have 2 functioning lungs. Several species of boas and pythons have heat-sensitive pits in the scales bordering their mouths, which they use to locate prey in the dark.

Some of the relatives of the boas and pythons are smaller but otherwise outwardly similar. These include wood snakes and Round Island boas. A few species are brightly colored and others, such as sunbeam snakes, have scales that are highly iridescent. File (or wart) snakes are specialized for an aquatic way of life and are effectively helpless on land. Their scales are granular and rough to the touch, an adaptation for gripping and constricting the fishes on which they feed.

Constriction

Although boas, pythons, and their relatives are not the only snakes that kill by constriction (some colubrids also use this method), most of the constrictors are in this group. When the snake has selected its prey, it throws one or more coils around the animal's body. Each time the victim breathes out, the snake tightens its grip. The victim eventually dies by suffocation, either because it cannot draw breath or because its heart cannot pump blood, rather than by being crushed. Once the prey is dead, the snake loosens its hold and searches for the head. It swallows this end first, gradually releasing the rest of the body from its coils. Constriction is most effective for killing birds and mammals because, being warm-blooded, they have to breathe relatively frequently.

Reproduction

One of the most significant differences between boas and pythons lies in the way they reproduce. Boas (with the possible exception of one species) bear live young—as do wood snakes, pipe snakes, file snakes, and shield-tailed snakes. Pythons, sunbeam snakes, and Round Island boas lay eggs. Pythons are among the small number of snakes that show parental care. The female coils around the clutch of eggs—which can number up to 100 in large species—throughout the 2–3 month period of incubation, to protect them from predators. One or 2 species of python are unique among snakes in that they can regulate the temperature of their eggs by producing metabolic heat.

CONSERVATION

Several species of boas and pythons, especially the largest ones, have been adversely affected by human activities; these large predators require extensive areas of undisturbed land over which to hunt. In addition, many are relatively conspicuous and sluggish and therefore easy targets for hunters, who kill them for their flesh, their skins, or simply out of prejudice. However, some species perform a useful service to humans by controlling the populations of rodents that are attracted to towns and villages.

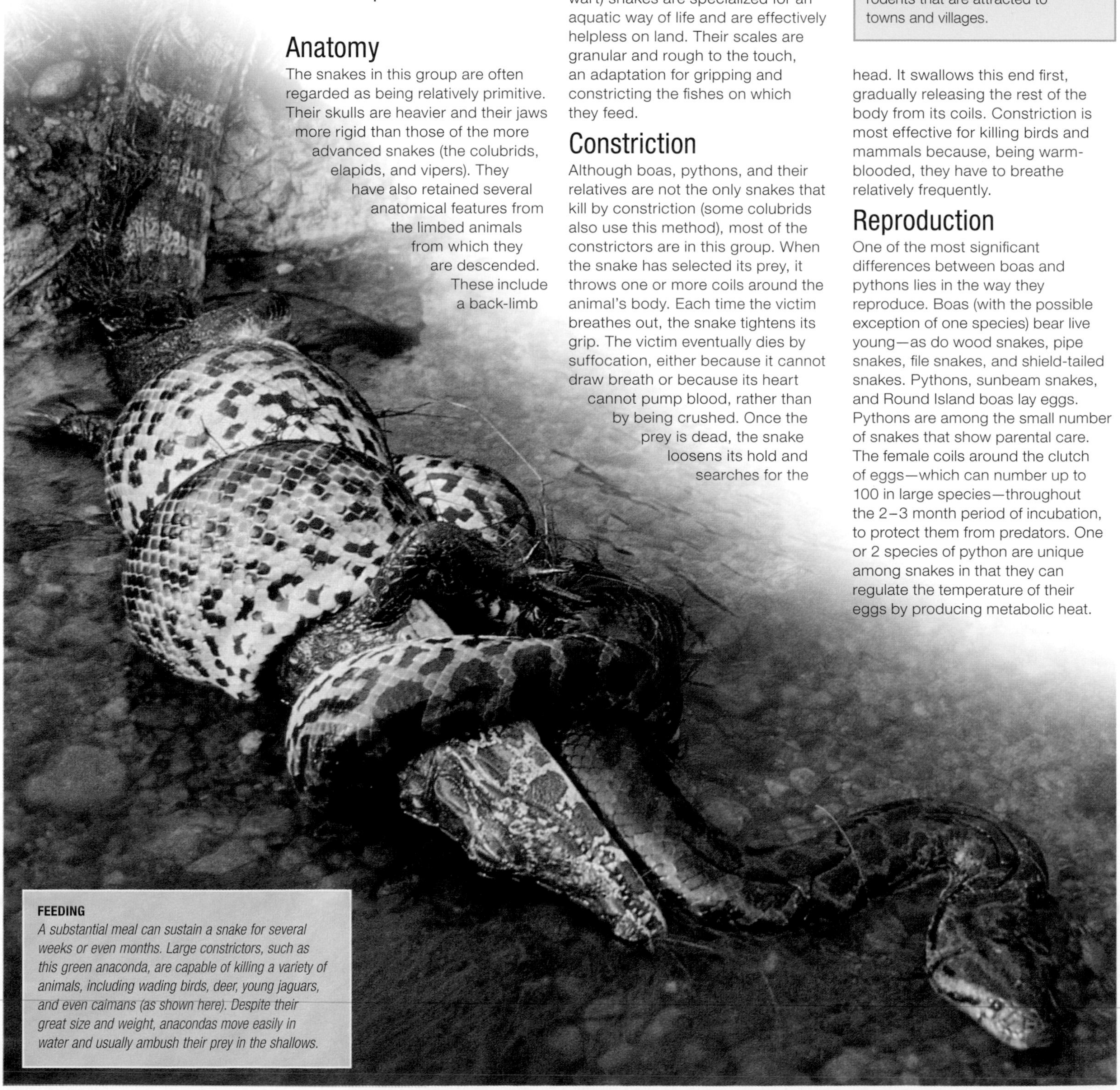

FEEDING
A substantial meal can sustain a snake for several weeks or even months. Large constrictors, such as this green anaconda, are capable of killing a variety of animals, including wading birds, deer, young jaguars, and even caimans (as shown here). Despite their great size and weight, anacondas move easily in water and usually ambush their prey in the shallows.

Anilius scytale

South American pipe snake

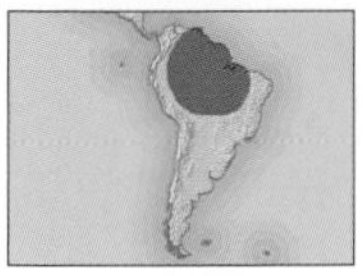

Length 28–35 in (70–90 cm)
Breeding Ovoviviparous
Habit Burrowing
Status Not evaluated

Location N. South America

The sole member of the family Aniliidae, this little-known snake leads a secretive, underground life, its cylindrical body shape adapted to its burrowing habit. The striking red and black coloration

mimics that of venomous coral snakes from the same region, whose defensive behavior the South American pipe snake also imitates. It hunts in burrows, and its diet is thought to consist of small vertebrates, including snakes. It may be preyed upon by other snakes.

Cylindrophis ruffus

Red-tailed pipe snake

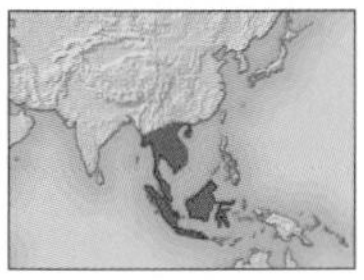

Length 28–39 in (70–100 cm)
Breeding Viviparous
Habit Burrowing
Status Not evaluated

Location S.E. Asia

The underside of this slender snake's tail is red, hence its name. The upper body is black, often with a red mark behind the head. Living mainly underground, this pipe snake occasionally takes to water and may be found in rice paddies and swamps. It actively forages for smaller snakes and eels. Hunted by larger snakes, birds, and mammals, its defense is to raise its tail and expose the red underside, possibly to mimic venomous species such as the krait, or to deflect attack away from its head.

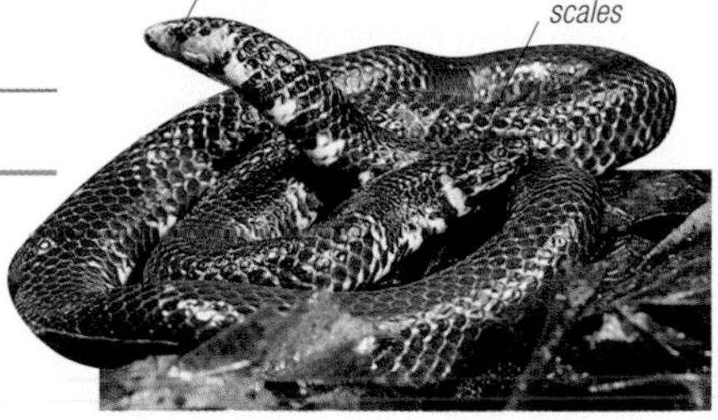

Pseudotyphlops philippinus

Large shield-tailed snake

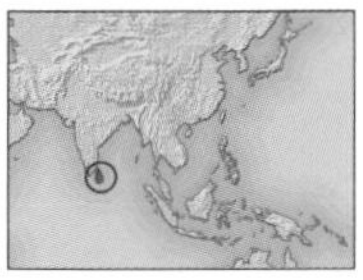

Length 18–20 in (45–50 cm)
Breeding Viviparous
Habit Burrowing
Status Not evaluated

Location Sri Lanka, below 3,300 ft (1,000 m)

This short, stout, mainly brown snake, with yellow edges to its scales, has a tail that ends with a single shield or caudal plate and looks as though it has been chopped off at an oblique angle. The shield, covered with tiny spines that are more concentrated around the edge, is believed to plug the burrow down which the snake moves, keeping it safe from predators. Rarely coming to the surface, the snake has the cylindrical body, pointed head, and smooth scales of a burrowing species, and feeds on the earthworms it finds underground. As with other members of the family Uropeltidae, little is known about this species' social, defensive, or reproductive behavior.

Loxocemus bicolor

Mexican burrowing snake

Length 3¼–4¼ ft (1–1.3 m)
Breeding Oviparous
Habit Burrowing
Status Not evaluated

Location S. Mexico to Central America

This constrictor has the narrow head and cylindrical, muscular body of a burrowing species. Its body is gray, with irregular patches of white scales that develop with age. This snake eats rodents, lizards, and reptile eggs, and is hunted by birds and mammals. Females lay 3–6 large, thick-shelled eggs in chambers dug in soil or leaf litter.

Xenopeltis unicolor

Sunbeam snake

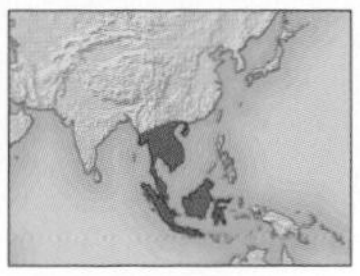

Length 3¼–4¼ ft (1–1.3 m)
Breeding Oviparous
Habit Terrestrial
Status Least concern

Location S.E. Asia

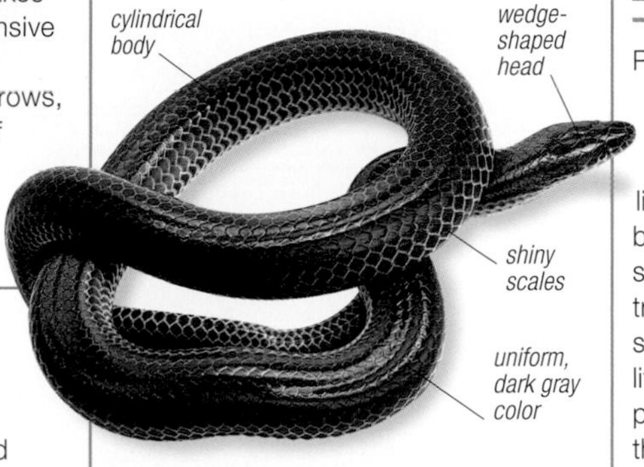

Named after its highly polished, iridescent scales, this snake has the typical body adaptations of a burrowing species. It comes above ground only at night, and eats frogs, lizards, snakes, and small mammals. Females lay 6–9 eggs in undergound nests in soil or leaf litter. The young have a distinct white collar that disappears after 2–3 molts.

Casarea dussumieri

Round Island boa

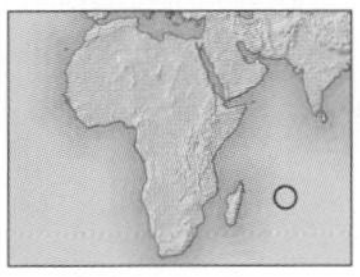

Length 3¼–5 ft (1–1.5 m)
Breeding Oviparous
Habit Terrestrial
Status Endangered

Location Round Island, in Indian Ocean

Poorly known and seldom seen, the Round Island boa is one of the world's most rare snakes, with perhaps only about 1,000 left in the wild. It feeds on lizards, and its narrow head and slender body allow it to squeeze between rocks, so that it can creep up on its prey. Unlike true boas, which bear live young, this species lays eggs, in crevices where leaf litter has accumulated. A breeding program is currently underway to save this snake, but its closest relative, *Bolyeria multicarinata*—also from Round Island—has not been seen since 1975, and is probably extinct.

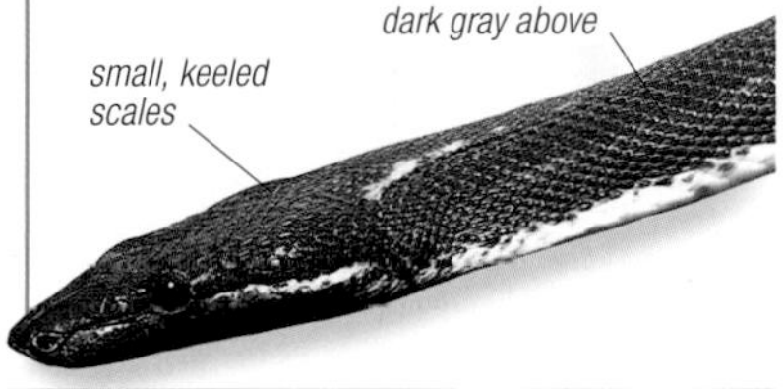

Acrantophis dumerili

Dumeril's boa

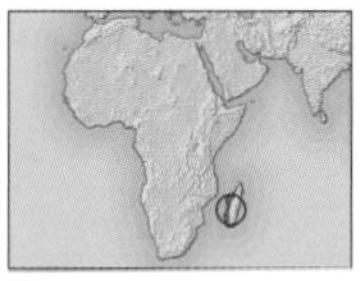

Length 5–6½ ft (1.5–2 m)
Breeding Ovoviviparous
Habit Terrestrial
Status Vulnerable

Location S. and W. Madagascar

Dumeril's boa is intricately patterned in shades of brown, with distinctive, glossy black markings around the mouth. A fairly inactive snake, it prefers to stay hidden in leaf litter on the floor of dry forest, its coloration allowing it to blend in almost imperceptibly. This boa ambushes and constricts its prey, which includes mammals and birds—although it lacks the heat-sensitive facial pits that some boas use to detect warm-blooded prey. The young, typically about 6 or 7, are relatively large at birth.

Calabaria reinhardtii

Calabar ground boa

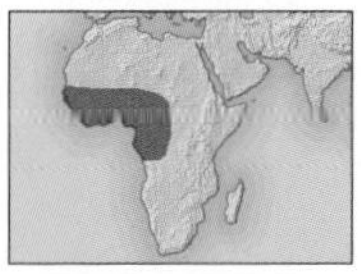

Length 3–3½ ft (0.9–1.1 m)
Breeding Oviparous
Habit Burrowing
Status Not evaluated

Location W. and C. Africa

This boa has a blunt head and a tail that are hard to tell apart. When threatened, it coils into a ball and raises its tail as a "false head" to protect its real head. The scales are black or brown, scattered with red or orange. The Calabar ground boa hunts small mammals in their tunnels and nests, and rarely comes above ground. Formerly classified as a python, it is now considered to be the only egg-laying species of boa.

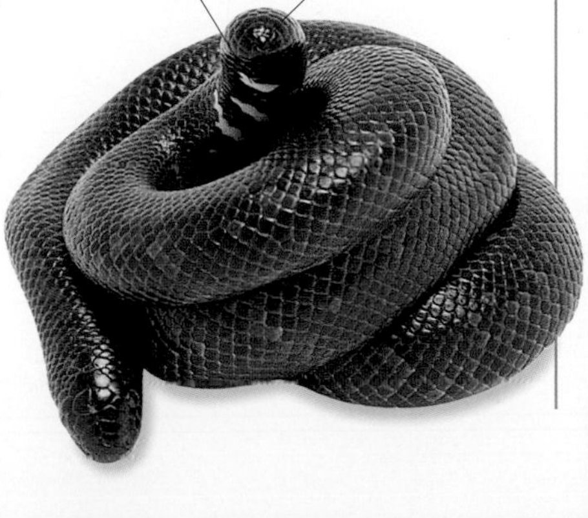

HUNTING BY STEALTH
The common boa will sit and wait, watching its prey from a suitable vantage point. Timing its strike to perfection, it lunges forward and seizes its victim in its jaws before wrapping itself around it. The snake then asphyxiates its prey by tightly squeezing it—a process that can be so quick that small animals may be killed in seconds.

Boa constrictor

Common boa

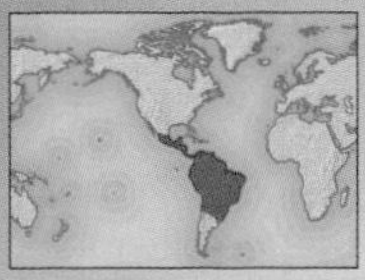

Location Central America, South America, some Caribbean islands

Length 3½–13 ft (1–4 m)
Breeding Ovoviviparous
Habit Terrestrial/Arboreal
Status Not evaluated

A large snake, with a narrow head and a pointed snout, the common boa varies in color (see right) but has dark, saddle markings along the back, sometimes becoming dark red toward the tail. There may be as many as 10 subspecies (see right). The common boa is very adaptable, and uses a huge range of habitats, from tropical forest to dry savanna. It may also be found in urban areas. It climbs well but may equally hunt on the ground, especially in some of the drier habitats in which it occurs (here, it may estivate in order to survive the worst heat of the summer months). It is also an able swimmer and often takes to the water voluntarily. Secretive, and fairly sluggish in behavior, it may be active by day or by night, depending on the climate. It preys on a wide variety of mammals and birds—suffocating them with its muscular coils before swallowing them whole.

COMMON BOA SUBSPECIES

The many subspecies of the common boa vary considerably in size and color. Dwarf island races, such as the Hog Island boa, may be only 3¼ ft (1 m) long, while mainland boas are typically 9¾ ft (3 m)—although in exceptional cases some reach 13 ft (4 m). The color ranges from mainly black (Argentinian boa) or olive-green (Central American boa) to the much lighter, pinkish or silver-gray specimens (Hog Island boa), often with a markedly contrasting tail. All, however, have distinctive saddle markings.

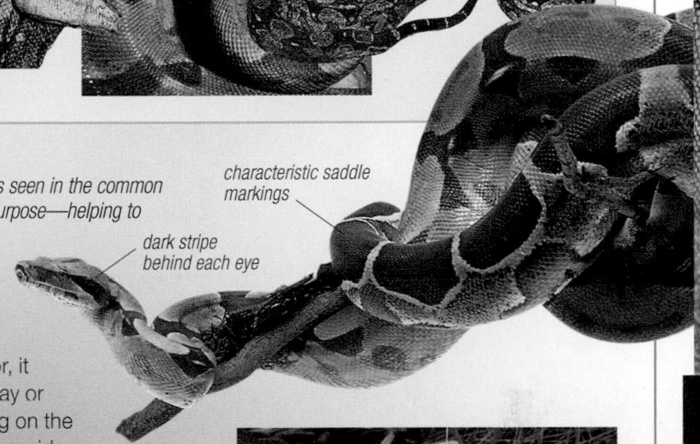

CONCEALING COLORS
The colors and patterns seen in the common boa serve a valuable purpose—helping to break up the snake's outline and enabling it to disappear into its surroundings.

GIVING BIRTH
Depending on the subspecies (here, a red-tail boa), common boas may produce 6–50 young in a litter, each 14–23½ in (35–60 cm) long. The newborn snakes fend for themselves immediately after breaking through the soft membrane that surrounds them at birth.

Candoia carinata

Pacific ground boa

Length 28–39 in (70–100 cm)
Breeding Ovoviviparous
Habit Terrestrial/Arboreal
Status Not evaluated

Location S.E. Asia, New Guinea, Solomon Islands

Also known as the Pacific boa, this variable snake has 2 distinct subspecies. Shown above is subspecies *paulsoni*, which is thick and short-tailed, ranges in color from red or grayish brown to off-white, and is completely terrestrial. Subspecies *carinata* is more slender with a longer tail, is brown or gray with an off-white patch above the cloaca, and is known to climb. Both subspecies feed on lizards, frogs, and small mammals.

Charina bottae

Rubber boa

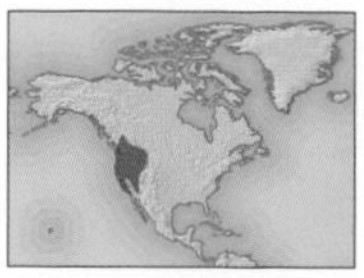

Length 14–32 in (35–80 cm)
Breeding Ovoviviparous
Habit Burrowing
Status Least concern

Location S.W. Canada, W. USA

The common name of this boa derives from its rubbery feel. The head and tail are both blunt, and may be hard to tell apart, as with the Calabar ground boa (see p.387). Its defensive technique is also similar: when a threatening mammal or bird of prey approaches, the snake coils into a ball and raises its tail as a "false head," with which it may pretend to strike in order to deflect attack from its real head. However, the rubber boa is neither venomous nor aggressive. It has the typical body features of a burrowing species, and lives a secretive life below ground or under debris, hunting for birds, reptiles, and small mammals in burrows and tree holes. When winter comes, it hibernates for long periods. While mating, the male stimulates the female with his pelvic spurs and coils his tail around hers. Females give birth to 2–8 young at a time; these measure about 6 in (15 cm) in length, and are pinkish or tan.

Lichanura trivirgata

Rosy boa

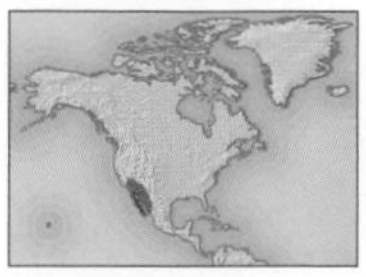

Length 23½–43 in (60–110 cm)
Breeding Ovoviviparous
Habit Burrowing
Status Least concern

Location S.W. USA, N.W. Mexico

This snake usually has a pattern of 3 wide, black, brown, reddish brown, or orange stripes down its cream, buff, or gray body; coloration and size vary between its several subspecies. A burrowing snake, it mostly remains hidden beneath rocks and in crevices, foraging there for birds, reptiles, and small mammals. In the colder parts of its range, it hibernates in winter. While mating, the male stimulates the female with his pelvic spurs as he crawls along her back; 3–5, exceptionally up to 12, young are born.

Corallus caninus

Emerald tree boa

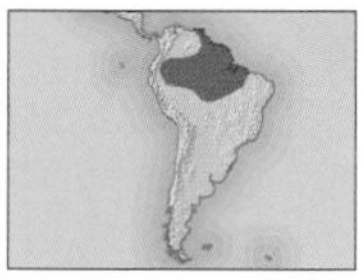

Length 5–6½ ft (1.5–2 m)
Breeding Ovoviviparous
Habit Arboreal
Status Not evaluated

Location N. South America

JUVENILE COLORATION
The emerald tree boa gives birth to 3–15 young each season. They are brick-red, orange, bright red, or yellow, changing to green after about a year.

Like many members of the boa family—particularly the green tree python (see p.392)—this striking South American snake is fully adapted for life off the ground. Its intense green color blends in with the foliage of the rain forests it inhabits, concealing it from predatory birds, while its strong, prehensile tail anchors it firmly to a branch, allowing the boa to lunge out or down to reach a bird or passing mammal. Its eyes have vertical pupils, which help it to sense movement, and it has deep pits in the scales around its mouth, for detecting heat given off by its prey. While mating, the male crawls over the female, and their tails become entwined.

READY TO STRIKE
The emerald boa typically drapes itself around a branch in a series of concentric coils, waiting, with its head hanging down, ready to strike. This boa feeds on small mammals and birds, and has long teeth that help it to grasp its prey firmly.

Corallus hortulanus

Amazon tree boa

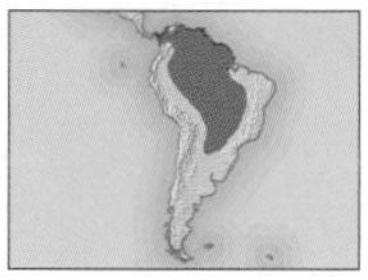

Length 5–6½ ft (1.5–2 m)
Breeding Ovoviviparous
Habit Arboreal
Status Not evaluated

Location N. and C. South America

A flattened body gives this slender, almost entirely arboreal snake rigidity when reaching out across branches; it often hangs down to snatch lizards and birds from lower branches. The Amazon tree boa may come down to the ground at night. Coloration varies from brown or grayish with bars or blotches along the back, to plain orange, yellow, or olive.

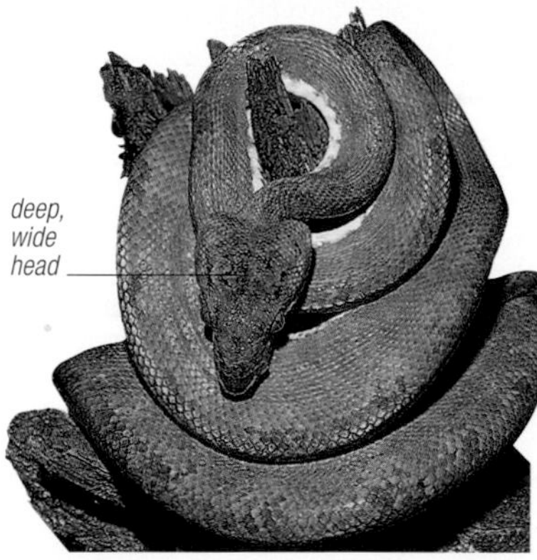

Epicrates cenchria

Rainbow boa

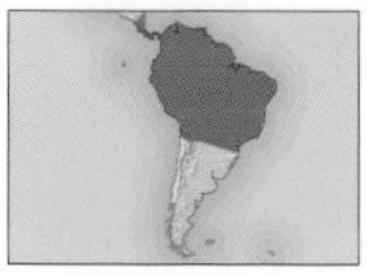

Length 3¼–6½ ft (1–2 m)
Breeding Ovoviviparous
Habit Partly arboreal
Status Not evaluated

Location Central and South America

All subspecies of this boa have smooth, shiny, often highly iridescent scales, with colors varying from dark brown to dark orange. The back has black circles, while the flanks have light-centered spots. This snake rests by day, actively foraging for birds, lizards, and small mammals at night. Only the southern subspecies hibernate. The size and number of young varies greatly between subspecies.

Eunectes murinus

Green anaconda

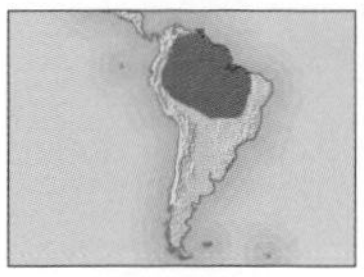

Length 20–33 ft (6–10 m)
Breeding Ovoviviparous
Habit Semiaquatic
Status Not evaluated

Location N. South America, Trinidad

With a maximum weight approaching 550 lb (250 kg), the green anaconda is the world's heaviest snake. It spends most of its life partially submerged in shallow water, and favors areas with thick waterside vegetation, where it can move around unseen. Green anacondas hunt mainly after dark. They lie in wait for mammals such as capybaras and deer, and can even kill fully grown caimans; there are also records of them attacking people, sometimes with fatal results. Males are smaller than females, but have large pelvic spurs, which they use during courtship. Females produce litters of 4–80 young, which take about 6 years to reach maturity. Anacondas may have a lifespan of 25 years or more.

ADAPTED TO WATER

The green anaconda's eyes and nostrils are positioned toward the top of its head, enabling the snake to see and breathe while partially submerged in water.

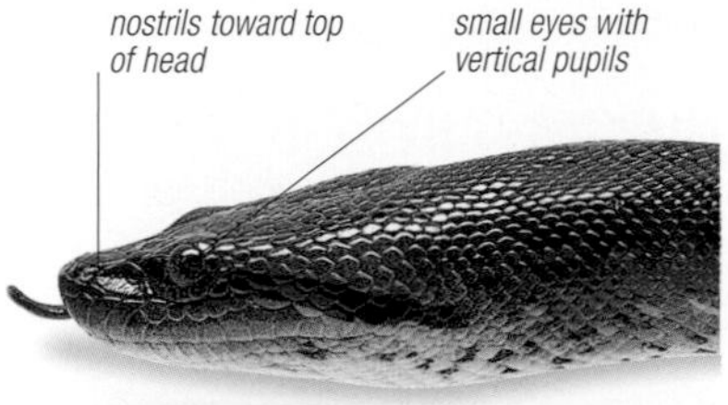

GIANT CONSTRICTOR *The green anaconda's powerful body is strong enough to asphyxiate animals up to the size of a horse. It often ambushes its victims as they arrive to drink.*

Aspidites melanocephalus

Black-headed python

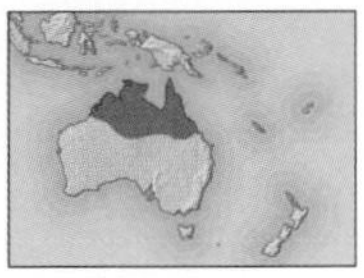

Length 5–8¼ ft (1.5–2.5 m)
Breeding Oviparous
Habit Terrestrial
Status Not evaluated

Location N. Australia

The black-headed python is slender-bodied and has a jet-black head and neck, with irregular brown and cream bars across its body. Not an aggressive snake, it retreats into a burrow or crevice if a predator threatens—usually a mammal or bird of prey. Its diet consists of small mammals, birds, and reptiles, including venomous snakes. Females lay up to 18 eggs at a time, under logs or roots, and in chambers underground. Like most pythons, they coil around the eggs to protect them and regulate their temperature until they hatch.

Antaresia childreni

Children's python

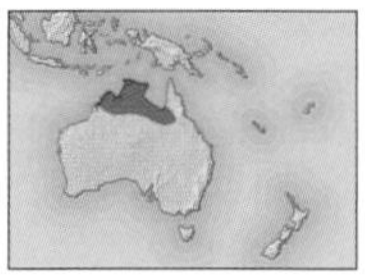

Length 30 in (75 cm), max 39 in (100 cm)
Breeding Oviparous
Habit Terrestrial
Status Not evaluated

Location N. Australia

Named after the Victorian zoologist J.G. Children, this small Australian python spends much of its time hidden away in caves and rock crevices. It specializes in eating lizards, but also ambushes birds and small mammals, including bats. If attacked, it defends itself by striking and biting but, like all pythons, it is nonvenomous. Male Children's pythons have longer and thicker tails than the females. Juveniles have darker markings than adults.

PARENTAL CARE

Seen here coiled protectively around her eggs to regulate their temperature until they hatch, the female Children's python lays a clutch of 8–10 eggs each year, or every other year, in a hollow tree or an underground chamber.

SMALL SNAKE *Very small for a python, this snake is brown or reddish brown, with darker but often faint blotches.*

Morelia spilota

Carpet python

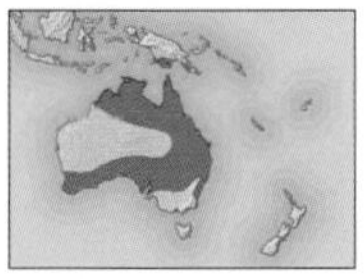

Length 6½ ft (2 m), max 13 ft (4 m)
Breeding Oviparous
Habit Mostly terrestrial
Status Least concern

Location S. New Guinea, Australia

The carpet python is one of Australia's most widespread snakes, found in an unusually wide range of habitats. It is also one of the most variable, with several subspecies, each with a different common name. One feature shared by all subspecies is a bold pattern of irregular markings—which can be brown, gray, reddish brown, or black—laid over variously colored backgrounds. Carpet pythons are active both by night and by day, hunting lizards, birds, and small mammals. Like other pythons, they are nonvenomous, but they can inflict a painful bite. Female carpet pythons lay clutches of about 50 soft-shelled eggs in decaying vegetation or hollow tree trunks, coiling protectively around them until they hatch.

DIAMOND PYTHON

The diamond python, *Morelia spilota spilota*, is the most distinctive subspecies, with an intricate pattern of diamond-shaped blotches from head to tail.

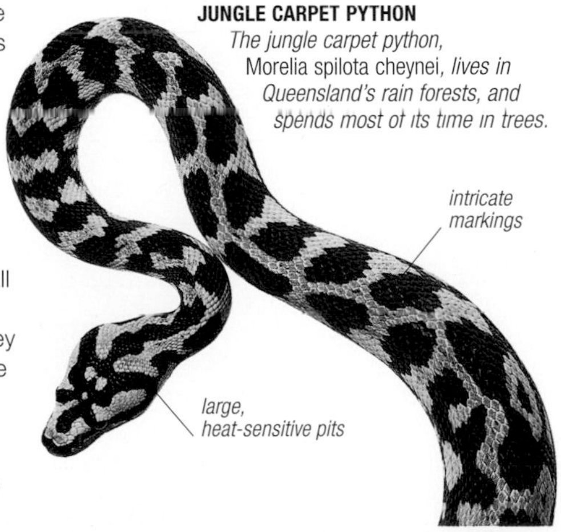

JUNGLE CARPET PYTHON *The jungle carpet python,* Morelia spilota cheynei, *lives in Queensland's rain forests, and spends most of its time in trees.*

Morelia viridis

Green tree python

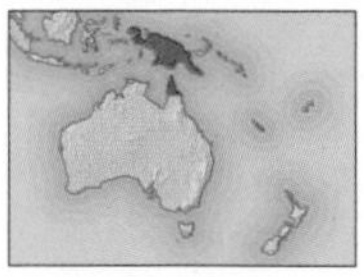

Length 6–7¾ ft (1.8–2.4 m)
Breeding Oviparous
Habit Arboreal
Status Least concern

Location New Guinea and surrounding islands, extreme N. Australia

This bright green python has the same adaptations for an arboreal lifestyle—slender shape, prehensile tail, and green coloration—as the emerald tree boa (see p.390). It also rests and hunts in the same way, coiled around a branch with its head hanging down ready to strike at prey. All the scales around the mouth have prominent heat pits. Females lay 6–30 eggs in a tree hole or among epiphytic plants. The young—which are bright yellow or occasionally red—are hunted by birds of prey.

Python molurus

Burmese python

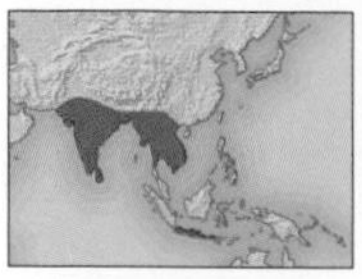

Length 16–23 ft (5–7 m)
Breeding Oviparous
Habit Terrestrial
Status Near threatened

Location S. and S.E. Asia

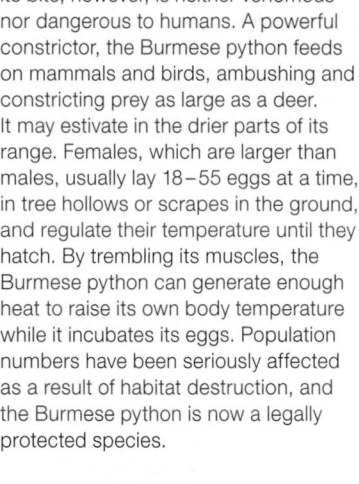

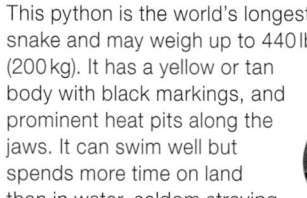

Also known as the Indian python, this very large snake is afforded excellent camouflage by the pattern of interlocking, dark brown blotches on its buff or gray skin. The coloration varies from area to area, but the "arrowhead" marking on the top of the head is constant. As with all pythons, there are heat pits along the jaws. Once adult, this snake has no predators, but it will hiss, strike, or bite if threatened; its bite, however, is neither venomous nor dangerous to humans. A powerful constrictor, the Burmese python feeds on mammals and birds, ambushing and constricting prey as large as a deer. It may estivate in the drier parts of its range. Females, which are larger than males, usually lay 18–55 eggs at a time, in tree hollows or scrapes in the ground, and regulate their temperature until they hatch. By trembling its muscles, the Burmese python can generate enough heat to raise its own body temperature while it incubates its eggs. Population numbers have been seriously affected as a result of habitat destruction, and the Burmese python is now a legally protected species.

Python regius

Royal python

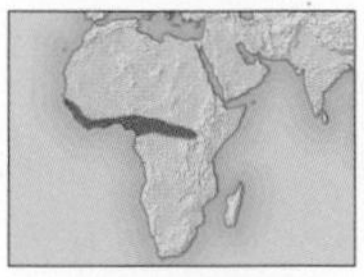

Length 2½–4 ft (0.8–1.2 m)
Breeding Oviparous
Habit Terrestrial
Status Least concern

Location W. to C. Africa

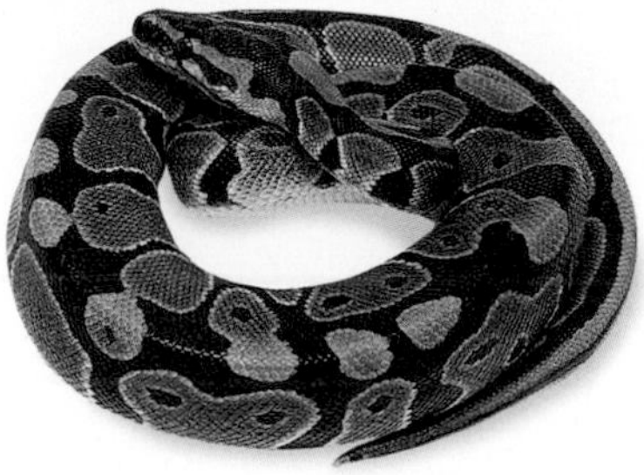

The royal python is short and stocky, with intricate, dark brown markings over a tan or yellowish background, and conspicuous heat pits around its mouth. It defends itself against predators such as mammals and birds by curling up into a ball and hiding its head in the center. In the dry season, this snake may estivate underground. Females lay and incubate a clutch of 3–8 eggs among rocks and in underground chambers.

Python reticulatus

Reticulated python

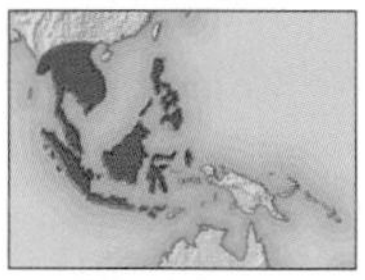

Length 20–33 ft (6–10 m)
Breeding Oviparous
Habit Terrestrial
Status Not evaluated

Location S.E. Asia

This python is the world's longest snake and may weigh up to 440 lb (200 kg). It has a yellow or tan body with black markings, and prominent heat pits along the jaws. It can swim well but spends more time on land than in water, seldom straying far from its den. The reticulated python eats birds, mammals and, very rarely, humans. The female lays and broods 30–50 eggs in hollow trees and underground chambers. This python may live up to 30 years. It is widely hunted for its skin and larger individuals are becoming increasingly rare.

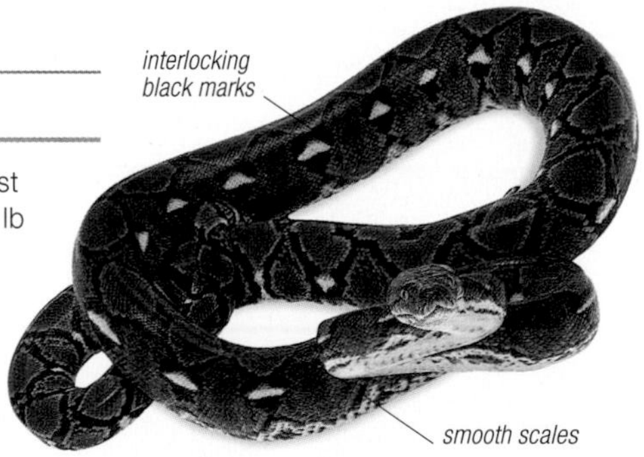

Leiopython hoserae

Hoser's python

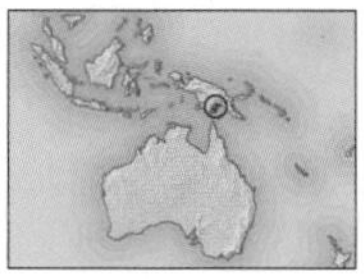

Length 10 ft max (3 m)
Breeding Oviparous
Habit Terrestrial
Status Not evaluated

Location New Guinea

One of several species of "white-lipped" python from New Guinea, Hoser's python has a distinctive pattern of scales on its head, and prefers hotter, more arid habitats than its relatives. White-lipped pythons have an iridescent appearance caused by tiny moisture-repelling ridges on the body scales.

Tropidophis morenoi

Zebra dwarf boa

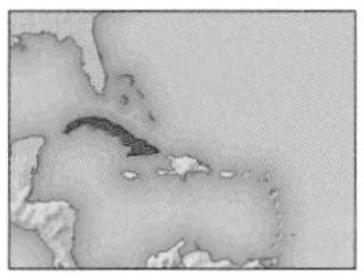

Length 12 in (30 cm)
Breeding Viviparous
Habit Terrestrial
Status Not evaluated

Location Caribbean

So-called because of its banded pattern—formed by fused brown spots on a buff-colored background—this species belongs to a group of tropical American dwarf boas (also called wood snakes), most of which live on the ground. A large number of species have striking color patterns. They have evolved into a great many species within the Caribbean region and reach their highest diversity on the island of Cuba. Other species undoubtedly await discovery. Many—like the Zebra dwarf boa—may be at risk of extinction because of their very restricted island distributions. Dwarf boas stay hidden in low vegetation or even underground during the day and emerge under the cover of night to feed on small vertebrates—including other snakes. They give birth to live young in typical boa fashion.

Tropidophis melanurus

Cuban wood snake

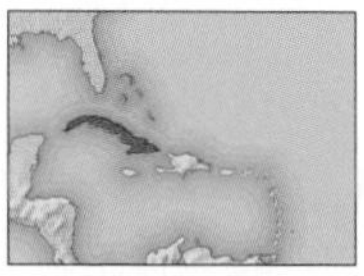

Length 32–39 in (80–100 cm)
Breeding Viviparous
Habit Terrestrial
Status Not evaluated

Location Caribbean

This stocky snake is commonly gray, brown, or buff, with darker markings and a black-tipped tail; it may also be orange, with a yellow-tipped tail. Living mainly on the ground, occasionally climbing into bushes or onto rocks, it defends itself by rolling into a ball or releasing a foul-smelling slime from its cloaca. It feeds on frogs, lizards, smaller snakes, and rodents. Females give birth to around 8 young at a time.

UNUSUAL COLORATION
The lighter orange coloration is much more rare than the darker morph. This snake is the largest in genus Tropidophis.

REPTILES

Colubrids

PHYLUM	Chordata
CLASS	Reptilia
ORDER	Squamata
SUBORDER	Serpentes
SUPERFAMILY	Caenophidia
FAMILY	Colubridae
SPECIES	About 2,100

Sometimes referred to as typical snakes, colubrids form by far the largest family of snakes, accounting for almost two-thirds of all species. They include the familiar garter snakes, grass snakes, and whipsnakes. Colubrids occur in all parts of the world except Antarctica. Measured in both numbers of species and individuals, they are the most numerous snakes on all continents except Australia. Colubrids live in habitats ranging from freshwater lakes, coastal marshes, and estuarine swamps to rain forest and arid desert. Such variation in mode of life and habitat means that they have a wide range of sizes, shapes, and colorations. All the world's rear-fanged venomous snakes are colubrids.

Anatomy

Colubrids have several anatomical features in common. All species are without a functioning left lung and a back-limb (pelvic) girdle. They also lack a small bone in the lower jaw known as the coronoid bone. These 3 characteristics separate colubrids from the more primitive blind snakes, thread snakes, and boas and pythons. Colubrids also differ from these groups in having a more flexible skull, with jaws that can be spread apart to accommodate large items of food. This is possible because the 2 halves of the lower jaw are not connected to one another. With a small number of exceptions, colubrids have large, platelike scales covering their head; this differentiates them from boas, pythons, and most vipers. The arrangement of the head scales is usually consistent within a species and can therefore be a useful feature for identification.

Venom

Some colubrids, notably rat snakes, kill prey by constriction. Although most colubrids lack venom, about a third of all species have a type of venom-producing apparatus known as Duvernoy's gland, which delivers venom to the base of enlarged fangs toward the back of the mouth. Most venomous colubrids have a single pair of fangs, although others have 2 or 3 pairs. The fangs create a wound, into which the venom runs by capillary action. Unlike some front-fanged snakes, colubrids have solid fangs, although the venom may travel along a groove that runs the length of the fang. Most rear-fanged snakes are of little danger to humans because their fangs are too far back to be brought into play unless they chew for a sustained period and because their venom is, in any case, weak. A few species, however, notably the boomslang and twig snakes, are dangerous and have caused human fatalities.

BURROWING ASPS

The burrowing asps are a group of snakes that have been placed in various families, including the Colubridae, sometimes regarded as forming a separate group, the Atractaspidae, which consists of about 62 species. All except one species are found in Africa. These tunneling snakes feed mainly on amphisbaenians. Some have unique fangs that hinge sideways so that the snake can expose them without opening its mouth, injecting its venom with a sideways strike of the head.

Reproduction

Colubrids lay eggs or give birth to live young. In those species that lay eggs, clutch size varies from one to 100 eggs. They are laid in holes or tunnels or under rotting leaves. The young may be similar to the parents but some species have distinctive juvenile coloration and markings.

HUNTING
Compared with many front-fanged snakes, the venomous colubrids have relatively weak venom and inefficient fangs with which to deliver it. Possibly because of this, some colubrids, such as the ratsnake, kill their prey by constriction, causing suffocation. This Everglades ratsnake has caught a barking treefrog. Once the victim is dead, the ratsnake will maneuver its body until the frog's head is in position for it to be swallowed.

Dendrelaphis kopsteini

Kopstein's bronzeback snake

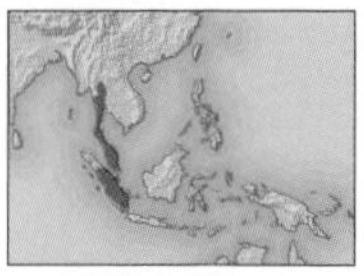

Length 5 ft (1.5 m)
Breeding Oviparous
Habit Terrestrial/Arboreal
Status Not evaluated
Location S.E. Asia

Bronzebacks are arboreal, day-active snakes, named for the bright iridescent sheen of their upper body scales. Thought to be most closely related to the Southeast Asian flying snakes, this species, described in the year 2007, is also known for the bright, brick-red color of its neck, which flares when it expands its neck. It was previously confused with allied species and is known to occur in rain forests throughout Thailand, the Malay peninsula, Singapore, and Sumatra. Like many other bronzebacks, it mainly preys on tree-dwelling lizards, such as geckos, and perhaps frogs.

Boiga dendrophila

Mangrove snake

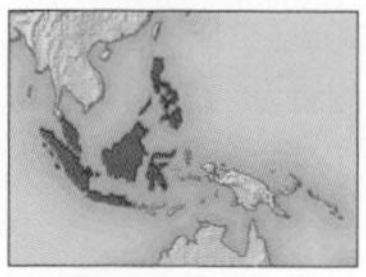

Length 6½–8¼ ft (2–2.5 m)
Breeding Oviparous
Habit Arboreal
Status Not evaluated
Location S.E. Asia

The mangrove snake is mainly glossy black, with vibrant yellow markings on its lips and flanks—these vivid colors may act as a warning to predators. Its body is slightly flattened from side to side, and a distinct ridge runs down the center of its back. If threatened, the mangrove snake draws back its flattened head in readiness to strike, and flares its yellow lip scales. A stealthy nocturnal hunter, this snake preys on lizards, frogs, birds, and small mammals. During the day, it rests in the crook of a tree, or remains coiled among foliage. Females lay 4–15 eggs in leaf litter, rotting stumps, or tree holes.

Bogertophis subocularis

Trans-Pecos ratsnake

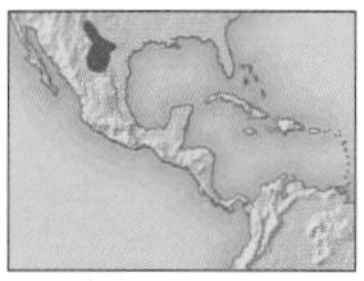

Length 3¼–4¼ ft (1–1.3 m)
Breeding Oviparous
Habit Terrestrial
Status Least concern
Location S. USA, N.E. Mexico

North America is home to a number of ratsnake species, all of them slender-bodied predators that hunt after dark. The Trans-Pecos ratsnake is typical in having variable patterning: one form has large, H-shaped markings, while another—the "blond" form—has paler, circular blotches. Where their distributions overlap, these 2 forms interbreed. Large, prominent eyes indicate that this species is entirely nocturnal. This burrowing snake shelters in cracks and under slabs during the day to escape from excessively high temperatures, and hibernates in winter. When threatened by predatory birds and mammals, it emits a foul-smelling musk and may bite. The female lays 4–8 eggs, which she buries in rotting vegetation, or under rocks.

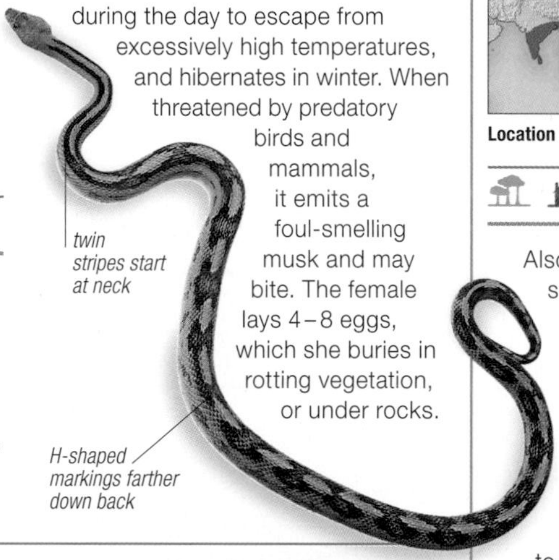

Chrysopelea ornata

Golden flying snake

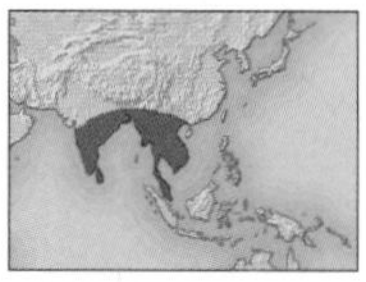

Length 3¼–4¼ ft (1–1.3 m)
Breeding Oviparous
Habit Arboreal
Status Not evaluated
Location S. and S.E. Asia

Also known as the golden tree snake, the golden flying snake can glide among high trees in tropical forests, parks, and gardens, by spreading its ribs to form a concave underside. When threatened—usually by birds of prey or carnivorous mammals—it may launch itself from a high perch in order to escape; if cornered, it will bite. It has a slender body, with green, black-tipped scales, and a narrow head with large eyes. The male has a longer, thicker tail than the female. The golden flying snake frequently rests in trees, among foliage, but always remains alert for possible prey. An active hunter, it grasps and holds its victims—generally small mammals, birds, lizards, and frogs—until the venom takes effect. Although it has a powerful bite, this rear-fanged snake is not particularly harmful to humans. Females lay 6–14 eggs in soil, leaf litter, or rotting wood. The young are 6–8 in (15–20 cm) long when they hatch.

Dasypeltis scabra

Common egg-eating snake

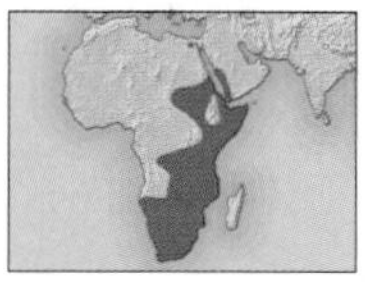

Length 28–39 in (70–100 cm)
Breeding Oviparous
Habit Terrestrial
Status Least concern
Location Africa and W. Asia

Many snakes eat soft-shelled reptile eggs, but this nocturnal African snake specializes in eating the hard-shelled eggs of birds. It feeds heavily during the bird breeding season—swallowing its food whole—and then fasts for the rest of the year. This species and its close relatives are the only snakes that do not have teeth.

A DIET OF EGGS

The very flexible skull, mouth, and throat enable this snake to eat eggs. When an egg reaches the throat, projections of the snake's backbone saw through the shell, which is regurgitated.

MIMICKING THE VIPER

This snake mimics the saw-scaled viper (see p.403) by forming a horseshoe-shaped coil and rubbing the strongly keeled scales on its flanks together, making a rasping sound.

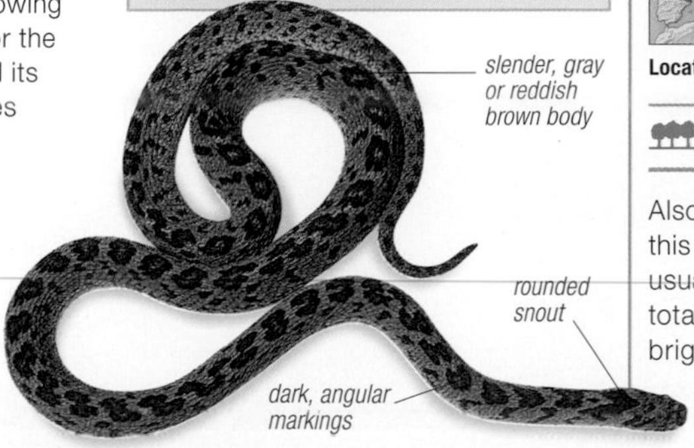

Hierophis viridiflavus

European whipsnake

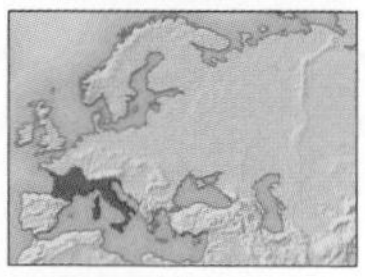

Length 5 ft (1.5 m), rarely 6½ ft (2 m)
Breeding Oviparous
Habit Terrestrial
Status Least concern
Location S. Europe

Also called the western whipsnake, this slender, large-eyed snake is usually greenish yellow, but may be totally black. The young are more brightly colored than the adults.

A fast-moving and agile hunter, it relies largely on sight while hunting lizards, rodents, and smaller snakes. When attacked, it is quick to escape into holes and crevices, and may bite viciously. Several individuals may share a den, resulting in a dense population. Females lay 6–14 eggs in soil, leaf litter, or rotting wood.

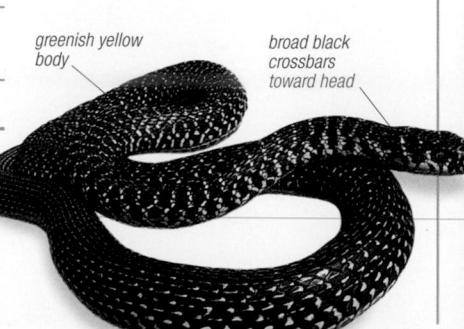

REPTILES

Dipsas indica

Snail-eating snake

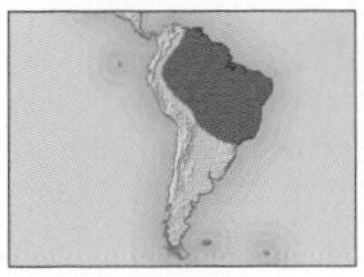

Length 23½–32 in (60–80 cm)
Breeding Oviparous
Habit Arboreal
Status Not evaluated

Location N. South America

A specialized feeder, this snake has jaws that are adapted for extracting snails from their shells—it forces its lower jaw into the shell, hooks its teeth into the flesh, and pulls it out. It hunts at night, resting in tree hollows during the day. The snail-eating snake has a slender body, with a distinctive ridge running along its back, and a rounded head with a blunt snout and large eyes. Males tend to be smaller and have proportionately longer tails.

Dispholidus typus

Boomslang

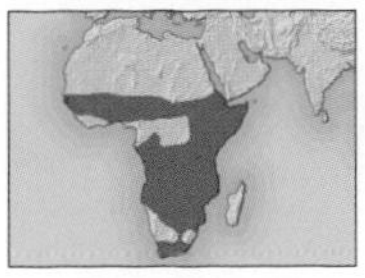

Length 3¼–5½ ft (1–1.7 m)
Breeding Oviparous
Habit Arboreal
Status Not evaluated

Location Africa (south of Sahara)

Deriving its name from the Dutch word for tree ("boom"), this is one of Africa's most venomous arboreal snakes. A highly agile climber, it uses its binocular vision to judge distances as it slips from one branch to another, and when it prepares to strike. Boomslangs feed on lizards, especially chameleons, and also on birds. Their fangs are at the back of their mouths, and they inject their venom by chewing their prey. Adult boomslangs are highly variable in color: they may be green, brown, or almost black, with or without contrasting markings. Males are more colorful than females. When threatened, usually by birds of prey and carnivorous mammals, they inflate their necks to appear larger, and will bite readily. Females lay clutches of up to 14 eggs in tree hollows and dead vegetation.

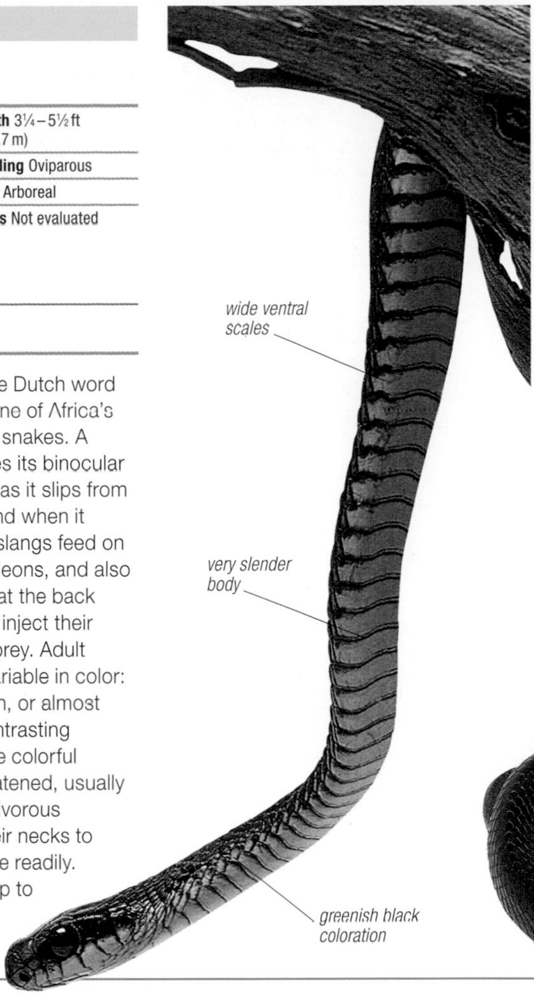

Drymarchon corais

Indigo snake

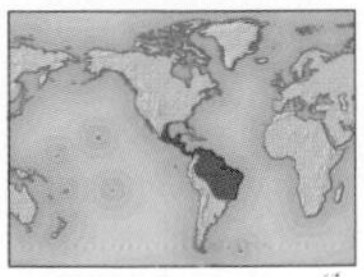

Length 7–9½ ft (2.1–2.9 m)
Breeding Oviparous
Habit Terrestrial
Status Not evaluated

Location S. USA, Central and South America

Also known as the cribo, this highly variable snake is black, brown, tan, or yellowish brown, depending on where it lives (the black Florida form is shown below). It feeds on most animalgroups—fishes, amphibians, lizards, other snakes, birds and their eggs, and small mammals. An opportunistic feeder, it can lunge very quickly at its prey, and sometimes presses it against the side of a burrow or a solid object. Females lay 4–12 eggs in tree stumps, vegetation, or burrows.

Pantherophis guttatus

Corn snake

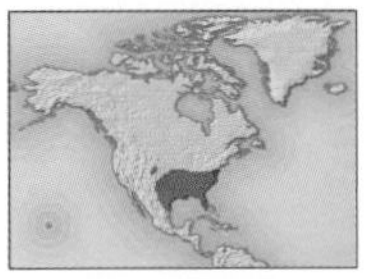

Length 3¼–6 ft (1–1.8 m)
Breeding Oviparous
Habit Terrestrial
Status Least concern

Location C. and S.E. USA

The corn snake is one of the most eye-catching of North America's ratsnakes—a group that includes the gray ratsnake (see right) and the Trans-Pecos ratsnake (see opposite). Like most of its relatives, the corn snake is extremely variable, with 4 different subspecies. The subspecies shown below, *Elaphe guttata guttata*, is the most colorful, and is restricted to southeastern states; the Midwestern subspecies, *Elaphe guttata emoryi*, usually known as Emory's ratsnake, is gray with darker blotches of gray-brown. Corn snakes can be found on the ground, on trees or buildings, and under logs, rocks, and debris. As with other ratsnakes, their flanks meet their undersides at an acute angle, forming ridges that help them to grip bark or walls. Adult corn snakes feed mainly on small rodents, which makes them useful animals on farms. Although they are nocturnal, they are often active during the day at cool times of the year. When threatened, they rapidly vibrate their tails and may rear up, ready to strike. If caught, they excrete a foul-smelling musk. Female corn snakes lay clutches of up to 25 eggs in decaying vegetation.

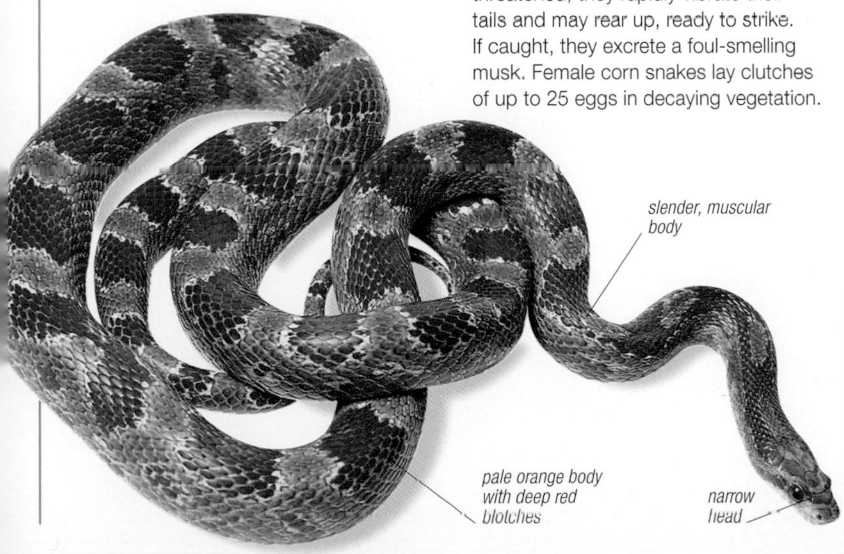

Pantherophis spiloides

Gray ratsnake

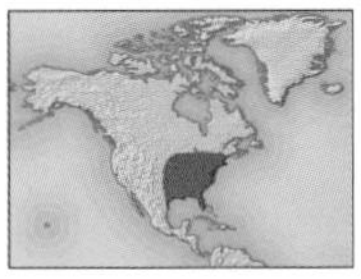

Length 4–6 ft (1.2–1.8 m)
Breeding Oviparous
Habit Terrestrial
Status Least concern

Location S. Canada, C. and E. USA

The gray ratsnake varies greatly in coloration, ranging from bright yellow-orange to much more subdued colors, as in the subspecies *Elaphe obsoleta spiloides*, shown below. Gray ratsnakes favor rocky hillsides with open woodland, often near water, and are excellent swimmers. They feed on rodents and birds, which they kill by constriction. Females lay 5–20 eggs, occasionally up to 40, sometimes in communal sites. A ridge along the flanks helps the snake to wedge itself into the bark when climbing trees.

Zamenis longissimus

Aesculapian snake

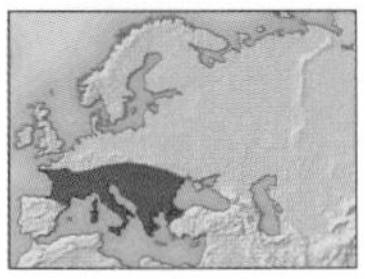

Length 3¼–7¼ ft (1–2.2 m)
Breeding Oviparous
Habit Terrestrial
Status Least concern

Location S. Europe to W. Asia

The name of this snake derives from Aesculapius, the Greek god of medicine. It is thought to be the snake that, coiled around a staff, forms his motif, the caduceus, still used as a symbol of the medical profession today. The Aesculapian snake is olive or brown, occasionally dark gray, with a lighter patch on each side behind the head. The scales are smooth or slightly keeled, giving it a shiny appearance. An agile swimmer and climber, this secretive snake inhabits scrubland, forest edges, and fields, and hibernates during the cooler months.

Zamenis situla

Leopard snake

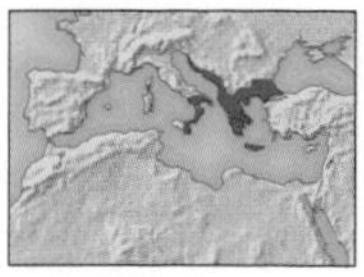

Length 28–39 in (70–100 cm)
Breeding Oviparous
Habit Terrestrial
Status Least concern

Location S. and E. Europe, W. Asia

The leopard snake is cream or gray, with black-edged reddish blotches, or 2 wide, longitudinal reddish stripes, edged with black, along its upper side. It feeds mainly on rodents, which it catches among vegetation or in their burrows, then kills by constriction. If threatened, it may flatten its head and bite. During mating, the male holds onto the female by biting her head or neck and coiling around her. Females lay up to 8 eggs and remain with them for several days.

Erpeton tentaculatum

Tentacled snake

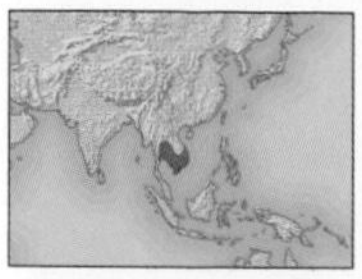

Length 28–39 in (70–100 cm)
Breeding Viviparous
Habit Aquatic
Status Least concern

Location Central S.E. Asia

This unusual snake belongs to a subfamily of colubrids—numbering about 38 species in all—that live in water. It is easily distinguished from its relatives by the unique pair of soft, fleshy tentacles on its snout, which probably have a sensory function. Tentacled snakes are sluggish and nocturnal, and hunt by lying in weed-choked water, waiting for prey to swim by. Females give birth to 5–13 live young underwater.

Gonyosoma oxycephalum

Red-tailed racer

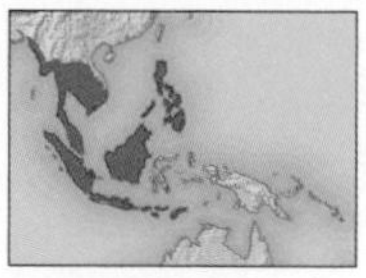

Length 5¼–7¾ ft (1.6–2.4 m)
Breeding Oviparous
Habit Arboreal
Status Not evaluated

Location S.E. Asia

The tail of this long, slender snake may be brown, orange, or gray, but despite its name it is never red. An active hunter, the red-tailed racer forages for birds, bats, and small mammals at night, its bright green color providing excellent camouflage in the trees. When threatened, usually by birds of prey or mammals, it inflates its throat vertically and adopts an S-shaped posture, in readiness to strike. Mating occurs in tree branches.

Heterodon nasicus

Western hognosed snake

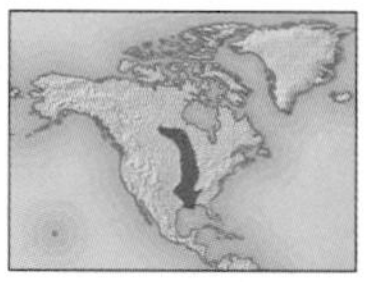

Length 16–32 in (40–80 cm)
Breeding Oviparous
Habit Terrestrial
Status Least concern

Location C. USA to Mexico

With its blotched patterning and thick body, this North American snake looks more like a viper (see p.402) than a member of the colubrid family. Although its venom is toxic, it is not considered dangerous to humans. There are 3 species of hognosed snakes in North America, all of which search through leaf litter for their prey and use their distinctive, upturned snouts for digging and rooting out burrowing toads and other amphibians. If threatened, all hognosed snakes hiss loudly, inflate their bodies, and may make mock strikes; if this fails, they often play dead.

Lampropeltis getula

Common kingsnake

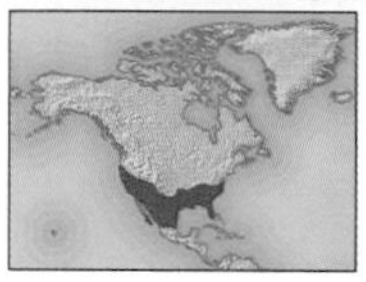

Length 3¼–6½ ft (1–2 m)
Breeding Oviparous
Habit Terrestrial
Status Least concern

Location W. and S. USA, N. Mexico

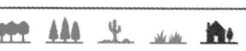

The common kingsnake is a powerful constrictor that actively hunts for prey, frequently entering rodent burrows. It also feeds on birds, lizards, frogs, and snakes, including venomous ones, to whose bite it is immune. Although mainly terrestrial, this snake may climb low vegetation and is also a fine swimmer. If threatened by a predator—mainly birds of prey and mammals—the common kingsnake may bite vigorously and, if grasped, may try to smear its captor with feces. It may be active at night or during the day, depending on the time of year and its distribution, and it hibernates for a longer time in the north of its range than in the south. Mating involves the male crawling along the female's back and biting her neck to hold her steady. The female lays 12 (exceptionally up to 25) eggs in the warmth of decomposing vegetation, rotting wood, or underground chambers. The young measure about 12–14 in (30–35 cm) in length. The common kingsnake can live for over 25 years in captivity.

LAMPROPELTIS GETULA CALIFORNIAE
This Californian subspecies of the common kingsnake may have a dark gray body with cream bands (shown below) or a brown body with a white stripe (shown in panel, right).

A BEWILDERING DIVERSITY

The common kingsnake shows a remarkable variety of colors and patterns. Many herpetologists divide the species into 7 different subspecies, and some believe the total should be as many as 10. To complicate matters further, some subspecies—such as the California kingsnake—have more than one color form.

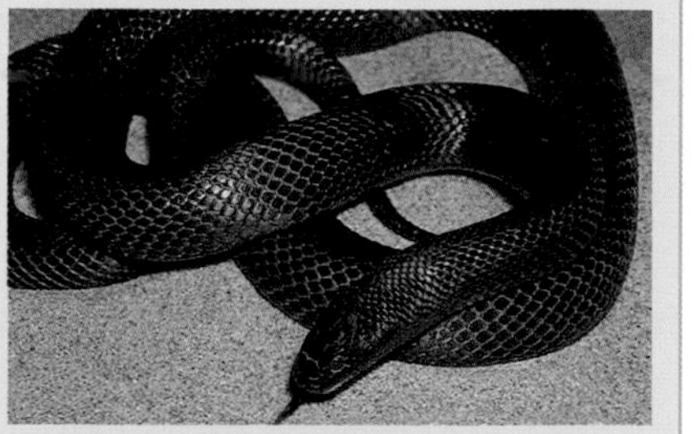

LAMPROPELTIS GETULA NIGRITA
The Mexican black kingsnake is, in its purest form, solid jet black in color, but it may show faint traces of a banded pattern when young. This is one of the most handsome subspecies and a popular pet.

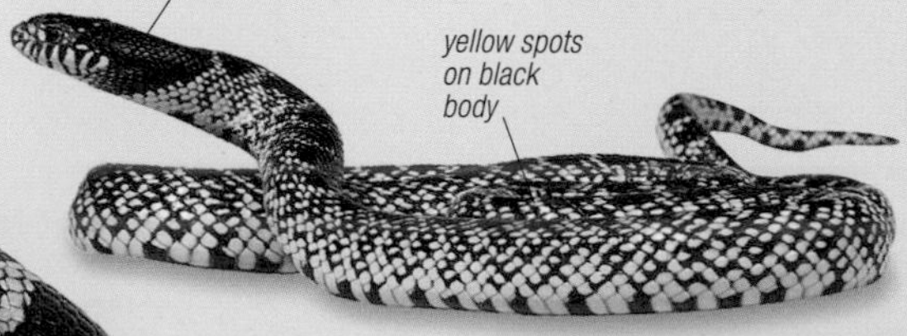

LAMPROPELTIS GETULA SPLENDIDA
This desert snake from Arizona is only active for a short period of the year, when it is neither too cold nor too hot.

LAMPROPELTIS GETULA CALIFORNIAE
The striped form of the Californian kingsnake occurs in some parts of the subspecies' range alongside the banded form (left). Both forms can hatch from the same clutch of eggs.

Lampropeltis triangulum

Milksnake

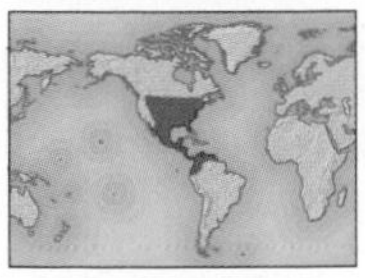

Length 1¼–6½ ft (0.4–2 m)
Breeding Oviparous
Habit Terrestrial
Status Not evaluated

Location North America, Central America, N. South America

The milksnake is one of the world's most widely distributed terrestrial snakes, and also one of the most variable. About 24 subspecies have been identified, with 8 in the United States alone. Most are brightly colored, with red, black, and yellow bands; in some subspecies, the patterning is very similar to that of coral snakes (see p.400)—a feature that probably evolved as a form of defensive mimicry. Milksnakes are secretive, mainly nocturnal, and feed on invertebrates, amphibians, and small rodents, as well as other snakes. Although nonvenomous, they defend themselves by biting, and by discharging a foul-smelling fluid from the cloaca. Females build nests in burrows, under rocks, in tree stumps, or in rotting vegetation, and lay up to 17 eggs.

STARTLING DEFENSE

The milksnake is not venomous. It does, however, mimic the coloration, markings, and even behavior of venomous coral snakes and is easily confused with them in parts of its range. If uncovered from its hiding place, it may thrash around in an attempt to startle its prey with its bright colors.

BODY MARKINGS
The milksnake is usually red, black, and white, the markings arranged as rings around the body or as saddles over the back.

Lamprophis fuliginosus

Brown house snake

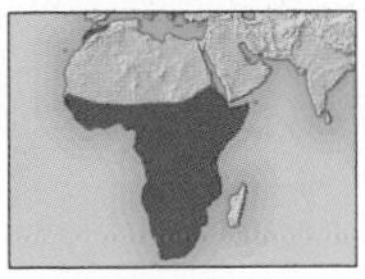

Length 3–5 ft (0.9–1.5 m)
Breeding Oviparous
Habit Terrestrial
Status Not evaluated

Location Africa (mainly south of Sahara)

The brown house snake is a powerful constrictor that actively forages at night for its prey—mainly rodents, but also birds and lizards. It may be tan, brown, orange, or black, with a cream stripe on either side of its head. The male is smaller than the female but has a proportionately longer tail. Females may lay 2 or more clutches of 6–16 eggs per breeding season. This snake may bite if provoked by predators, usually birds of prey or mammals, but it is not venomous.

Langaha madagascariensis

Madagascan leaf-nosed snake

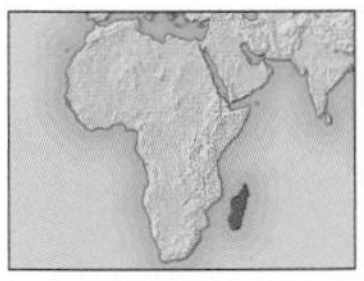

Length 28–35 in (70–90 cm)
Breeding Oviparous
Habit Arboreal
Status Not evaluated

Location Madagascar

This snake is instantly recognizable because of its extraordinary snout. The male has a long, pointed projection on the snout tip; in the female, the appendage is more elaborate and leaf-shaped. Males are light brown with a yellow underside, while females are gray-brown with small, dark markings. Both sexes have a long, slender, vinelike body, which provides excellent camouflage in the vines and branches of their forest habitat. Active during the day and at night, they ambush lizards and frogs.

Natrix natrix

Grass snake

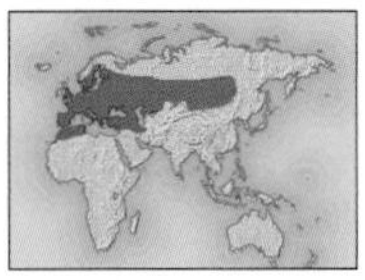

Length 4–6½ ft (1.2–2 m)
Breeding Oviparous
Habit Semiaquatic
Status Least concern

Location Europe to C. Asia, N.W. Africa

The grass snake is a semiaquatic predator, spending much of its time in damp places or in still water. A good swimmer, it can sometimes be spotted rippling its way across the surface of ponds, in search of frogs and fishes. It is one of the most widespread snakes in Europe. Adults are olive-brown, greenish, or gray, usually with a contrasting yellow or white collar just behind the head. Female grass snakes are larger than males, and have longer, thicker tails. After mating, they sometimes lay their eggs at shared nest sites, where the total egg count can reach 200 or more. In places where summers are cool, females often lay their eggs in compost heaps, the warmth of decaying vegetation helping incubation. The grass snake has a lifespan of approximately 15 years in captivity.

PLAYING DEAD

The grass snake ejects a foul-smelling fluid if handled, and sometimes reacts to extreme danger by playing dead. To make the performance as convincing as possible, it turns partly upside down, with its mouth open and tongue exposed.

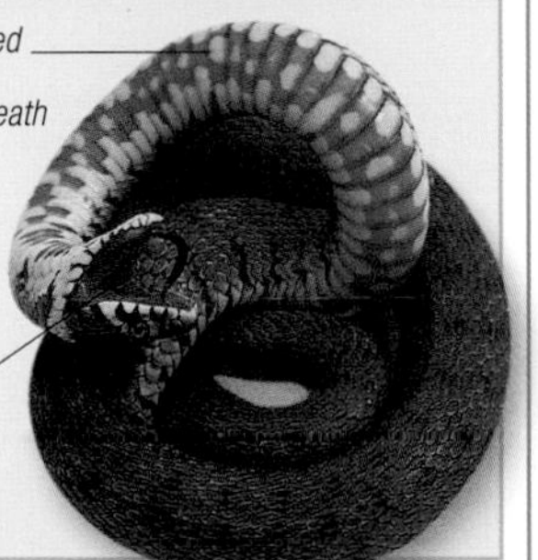

EGG-LAYING
Eggs are laid in decaying vegetation, manure or compost heaps, or other warm locations. Hatchlings measure 5½–8½ in (14–21 cm) long. Males reach maturity in 3 years, females in 4.

Nerodia fasciata

Southern water snake

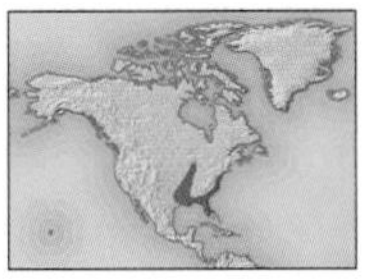

Length 1½–5 ft (0.5–1.5 m)
Breeding Ovoviviparous
Habit Aquatic
Status Least concern

Location C. and S.E. USA

Also known as the banded water snake, this species has a thickset body with keeled scales and highly variable coloration, sometimes with dark bands or blotches. Its eyes are situated toward the top of its head, helping it to see while partially submerged. This snake hunts for frogs and fishes during the day or night, depending on the temperature. If attacked, it may smear its captor with the foul-smelling contents of its anal gland. Many males may successfully mate with the same female. The female gives birth to 2–57 live young.

Spilotes pullatus

Tiger ratsnake

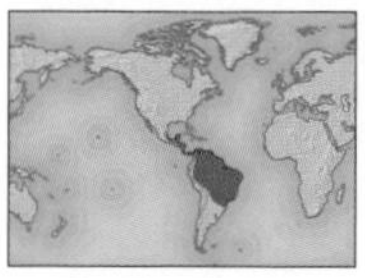

Length 5–6½ft (1.5–2 m)
Breeding Oviparous
Habit Arboreal
Status Not evaluated

Location Central America, N. and C. South America

The tiger ratsnake may be yellow with black bands or spots, black with yellow bars, or black with yellow-centered scales. Its triangular, deep body, which is flattened from side to side, allows the snake to remain rigid while it bridges gaps between branches. Active mainly at night, the tiger ratsnake stalks lizards, frogs, and small mammals, then suddenly strikes out, catching its prey by surprise. When threatened, this snake flattens the front part of its body and forms an S-shaped coil; it may also deliver a strong bite.

Opheodrys aestivus

Rough green snake

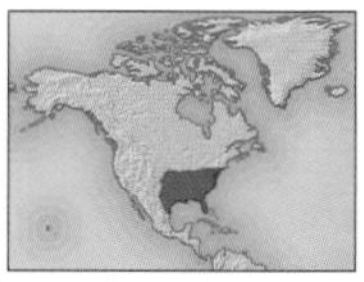

Length 2½–5¼ft (0.8–1.6 m)
Breeding Oviparous
Habit Terrestrial
Status Least concern

Location S.E. USA

Often found among low-growing vegetation, especially near water, this snake is bright green with a white or yellowish green underside. The scales are keeled, hence the rough texture of its skin. Although it typically remains motionless, or moves slowly through vegetation hunting for insects, it is capable of fast movement. The female lays 3–13 elongated eggs.

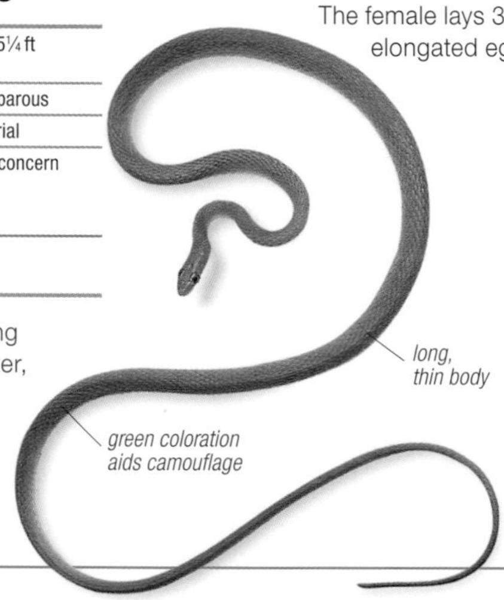

Pituophis melanoleucus

Pine snake

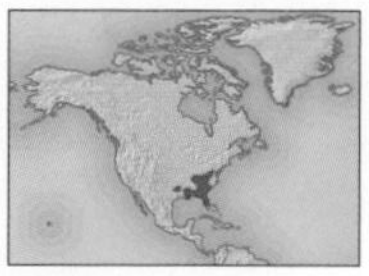

Length 3¼–8¼ft (1–2.5 m)
Breeding Oviparous
Habit Burrowing
Status Least concern

Location S.E. USA

This thickset, powerful constrictor has strongly keeled scales and a slightly pointed snout that is adapted for burrowing. Its coloration is variable, ranging from uniformly black or white to cream or yellowish, with irregular blotches of dark brown or black down the back and flanks. An active forager, it seeks out small mammals, especially burrowing species. The female lays 3–27 eggs in burrows that she often digs herself, using a loop in the front part of her body to remove sand or soil. There are several species of pine snakes, one of which (the Louisiana pine snake) is endangered.

Rhinocheilus lecontei

Long-nosed snake

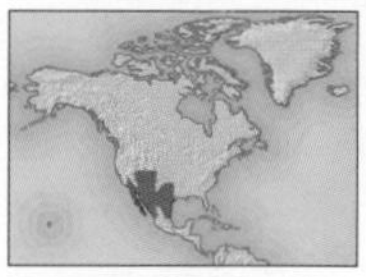

Length 20–39 in (50–100 cm)
Breeding Oviparous
Habit Burrowing
Status Least concern

Location S. USA, N. Mexico

The long-nosed snake lives under rocks and logs, and in rodent holes, and uses its pointed snout for burrowing. It has a markedly underslung jaw—another adaptation that suits it to a semisubterranean way of life. Mainly nocturnal, it feeds on small mammals and birds, as well as lizards and small snakes and their eggs. When threatened, the long-nosed snake hides its head in its coils, vibrates its tail, and discharges a foul-smelling liquid from its anal gland. Females lay one clutch, or sometimes 2, of up to 9 eggs.

Telescopus semiannulatus

African tiger snake

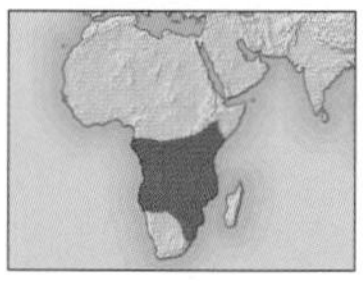

Length 3¼–4 ft (1–1.2 m)
Breeding Oviparous
Habit Terrestrial
Status Not evaluated

Location Southern Africa

A nocturnal hunter, the African tiger snake has large eyes that help it see better in the dark. Its slender, smooth-scaled body is yellowish brown, orange, or pinkish brown, with black blotches that extend across the back as far as the flanks. This slow mover lives mostly on the ground, but also climbs into shrubs, dead trees, and thatched roofs.

Thamnophis proximus

Western ribbon snake

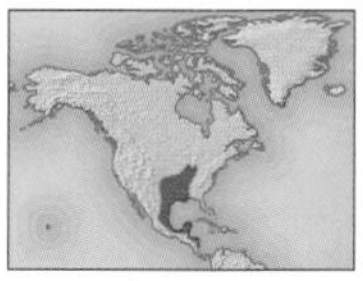

Length 20–39 in (50–100 cm)
Breeding Viviparous
Habit Semiaquatic
Status Not evaluated

Location C. USA to Central America

Like its close relative the common garter snake (see right), this North American snake is fast-moving and agile, as well as being a good swimmer. It feeds mainly on frogs, toads, tadpoles, and small fishes, often pursuing its prey underwater. Western ribbon snakes give birth to up to 24 live young a year, and can reach high densities in suitable habitats.

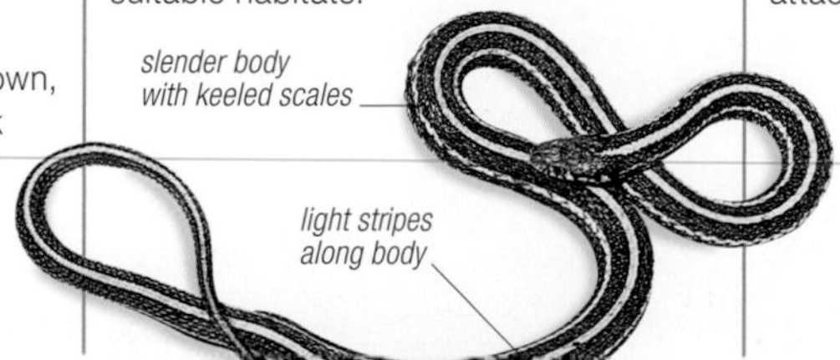

Thamnophis sirtalis

Common garter snake

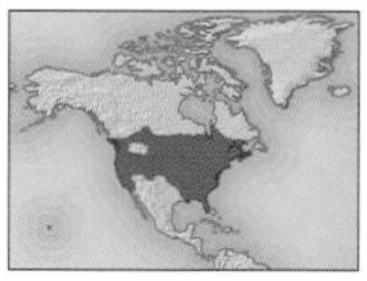

Length 26–51 in (65–130 cm)
Breeding Ovoviviparous
Habit Semiaquatic
Status Least concern

Location North America

The common garter snake invariably lives close to water, but it has the distinction of breeding farther north than any other snake in the Americas. On the fringes of the Arctic, it hibernates en masse, creating an extraordinary spectacle when large numbers of snakes emerge in spring. Common garter snakes feed mainly on earthworms, fishes, and amphibians, and are themselves attacked by mammals and birds of prey.

MATING PATTERNS

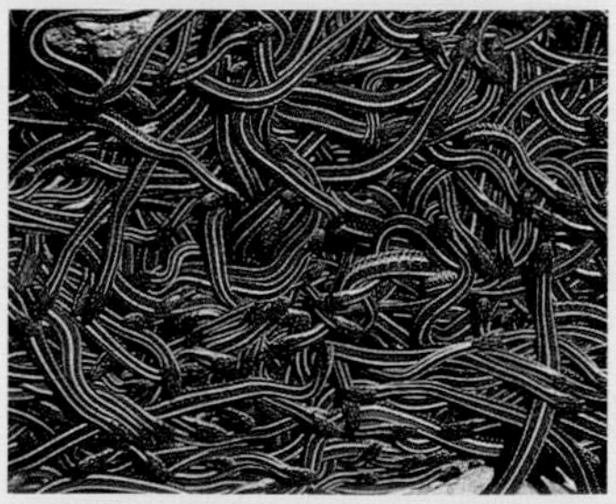

Mating systems vary according to latitude. In the north of the range, males scramble to compete for females, which they often outnumber greatly, as they emerge from hibernation dens. Toward the south, competition is less frenetic.

PHYSICAL APPEARANCE
The 11 subspecies of common garter snakes vary greatly in coloration. Shown here is the vividly striped subspecies T.s. infernalis, *the San Francisco garter snake.*

Elapids

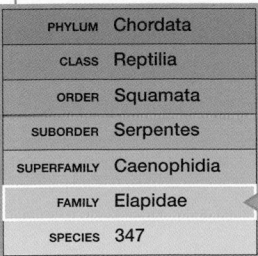

PHYLUM	Chordata
CLASS	Reptilia
ORDER	Squamata
SUBORDER	Serpentes
SUPERFAMILY	Caenophidia
FAMILY	Elapidae
SPECIES	347

Although elapids account for less than one in 10 of the world's snakes, all of them are venomous—many dangerously so. They include cobras, coral snakes, mambas, kraits, and sea snakes, and are found mainly in the tropics and in the Southern Hemisphere. There are ground dwelling and semiburrowing species of elapids, as well as some that live and hunt in trees. Some are fast-moving, diurnal hunters, while others are more secretive, hunting under the cover of darkness. Sea snakes and sea kraits are wholly marine. Elapids are related to colubrids (see pp.393–8) and look superficially similar, but they differ from most of them in having fangs located towards the front of the upper jaw. This allows them to inject venom in a sudden and deadly strike.

Anatomy

Elapids typically have a slender, roughly cylindrical body, with smooth, shiny scales. Sea snakes—sometimes classified in a family of their own—are noticeably different, with a flattened tail for swimming. Most terrestrial elapids have camouflage colors and markings, but coral snakes have bold markings that warn would-be predators to leave them alone. Some cobras intimidate their enemies by raising the front part of their bodies off the ground, and expanding their ribs to form a hood.

Venom

The venom of elapids varies in its potency and the organs and body systems that it affects. Most elapids, including cobras, release venom that acts on the nervous system, paralyzing muscles used in respiration. In some elapids, the venom runs down grooves in the fangs, while in others, the fangs contain an internal venom canal. The elapids include some of the world's most dangerous snakes, such as the taipan. Others, however, are inoffensive and too small to endanger humans.

Reproduction

Most elapids lay eggs but a few are bearers of live young. Sea snakes, for instance, give birth in the water, whereas sea kraits come ashore to lay eggs. A number of Asiatic cobras build nests out of dead leaves and other forest debris. Females guard the eggs fiercely, protecting them from potential predators.

SPITTING COBRA
Some cobras (including the Mozambique spitting cobra shown here), can spray their victims with venom by forcing it out at high speed through apertures in the fangs. Once sprayed into the victim's eyes, it causes great pain and temporary, or sometimes even permanent, blindness.

Acanthophis praelongus

Northern death adder

Length	12–39 in (30–100 cm)
Breeding	Viviparous
Habit	Terrestrial
Status	Not evaluated

Location New Guinea, N. Australia

Despite its name, this snake is not a true adder (see pp.402–3), but an unusually stout-bodied elapid. It is one of Australia's most venomous snakes, hunting by a sit-and wait strategy, and striking at the least provocation. It is normally active at night; during the day, it relies on its camouflage markings to avoid being seen. Adults vary in color from light gray to gray-brown or almost black, with paler rings or bands. Northern death adders feed mainly on small mammals, lizards, and birds, and they use their thin, wormlike tails to lure animals within striking range. They give birth to live young, producing litters of up to 8 each time they breed. A very similar species, the southern death adder (*Acanthophis antarcticus*), is widespread in eastern and southern Australia, and is sometimes found close to urban areas. Like its northern counterpart, it has a bite that is potentially fatal to humans.

ADDER LOOK-ALIKE
A thick body, wedge-shaped head, and narrow neck are all features normally seen in true adders and vipers, rather than in elapids. The northern death adder also has raised "horns" over its eyes.

LIVING LURE

The tip of the northern death adder's tail is very slender, and is considerably paler than the rest of the body. When hunting, the snake lies still, and vibrates the tail tip to lure prey close enough to strike. Rapid-acting venom kills the victim, which the snake can then swallow at leisure.

Austrelaps superbus

Australian copperhead

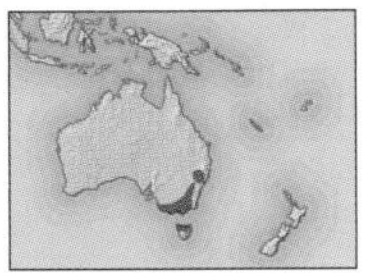

Length	4¼–5½ ft (1.3–1.7 m)
Breeding	Viviparous
Habit	Terrestrial
Status	Not evaluated

Location S.E. Australia, N. Tasmania

This inhabitant of marshes and swamps is active both during the day and at night, and feeds mainly on frogs. It is gray or reddish brown to black above, and cream or yellow on the flanks. Although it hibernates in winter, it can tolerate low temperatures and is active longer than other reptiles.

Dendroaspis angusticeps

East African green mamba

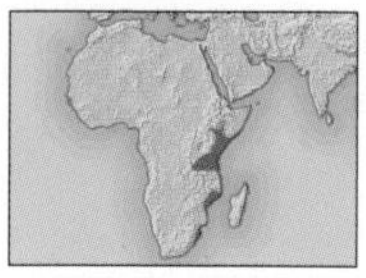

Length 5–8¼ft (1.5–2.5m)
Breeding Oviparous
Habit Arboreal
Status Not evaluated

Location E. and S.E. Africa

The bright green color, slender shape, and long tail of the East African green mamba are all features that are ideally suited to an animal living in trees. Alert and fast moving, this snake chases its prey through the canopy and, if disturbed, it usually moves swiftly to higher branches or passes from the branches of one tree to another; if cornered, it will turn and strike. Although highly venomous, this mamba rarely comes into conflict with humans.

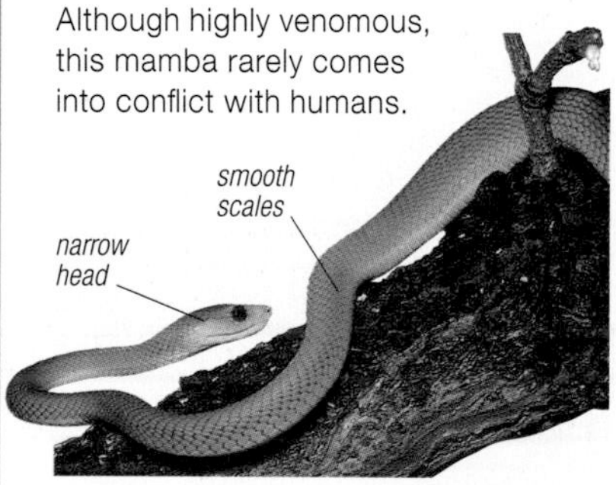

Dendroaspis polylepis

Black mamba

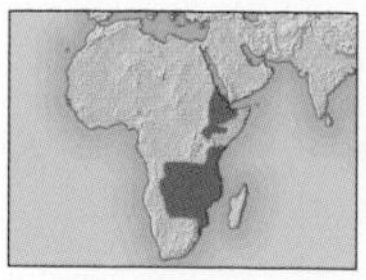

Length 8¼–11ft (2.5–3.5m)
Breeding Oviparous
Habit Terrestrial
Status Least concern

Location E. and southern Africa

In addition to being one of the most poisonous snakes, the black mamba is probably the fastest, making it one of the world's most dangerous predators. In short bursts, it can reach 12½mph (20kph), which is fast enough to overtake someone moving at a brisk run. The black mamba is not actually black, but gray or brown, with a streamlined body covered with large, smooth scales. Highly territorial, it makes its home den in a rock crevice or a hollow tree. It is active during the day, and feeds on birds and small mammals. Females lay 12–17 eggs in chambers underground. The venom of the black mamba is very fast-acting, and bites are often fatal if not treated quickly.

OCCASIONAL CLIMBER
This primarily ground-living snake, mostly found in open woodland, may also occasionally climb. Despite its size, it is highly agile, effortlessly sliding through thorn bushes and trees.

FIGHTING FOR A MATE

During the breeding season, male black mambas fight by raising and intertwining the front half of their bodies, with each snake trying to force its opponent to the ground. However, the fight is ritualized and does not involve any serious harm. Following this test of strength, the winner mates with any females in his territory.

Laticauda colubrina

Yellow-lipped sea krait

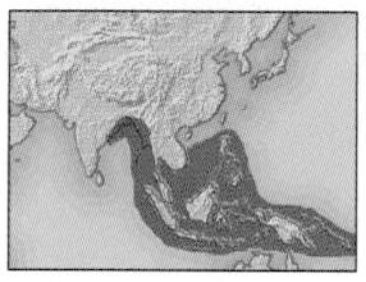

Length 3¼–6½ft (1–2m)
Breeding Oviparous
Habit Aquatic
Status Least concern

Location S. and S.E. Asia

The bluish gray, yellow-lipped sea krait is one of about 40 species of snakes that have flattened, paddle-shaped tails and live and feed at sea. It is found in coastal waters, mangrove swamps, and coral reefs, where it hunts fishes—particularly eels. Like all sea snakes, it is highly poisonous, but poses no threat to humans since it does not bite. Most sea snakes spend their entire lives at sea, and give birth to live young, but the yellow-lipped sea krait and its relatives come ashore to lay their eggs under leaf litter.

Micrurus lemniscatus

South American coral snake

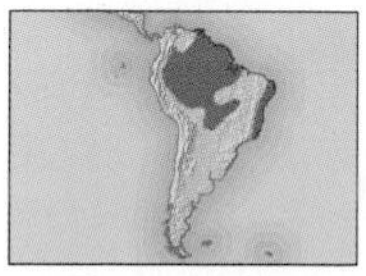

Length 23½–35in (60–90cm)
Breeding Oviparous
Habit Semiburrowing
Status Not evaluated

Location N. and C. South America

This brilliantly colored reptile is one of about 40 species of coral snakes that occur in warm parts of the Americas. Each species has a characteristic pattern of contrasting bands, warning predators that they are poisonous. However, because coral snakes have a semiburrowing lifestyle, generally emerging only at night, their colors are often concealed. Coral snakes feed on small lizards and snakes. They have highly toxic venom, but human fatalities are rare.

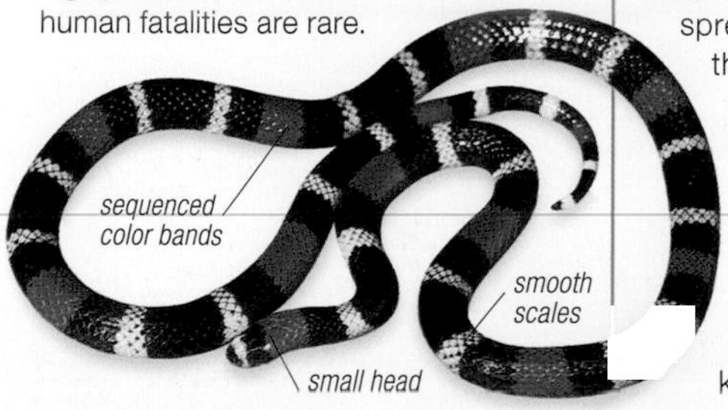

Naja naja

Indian cobra

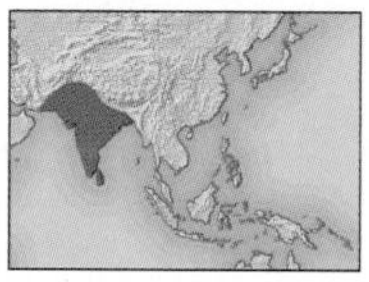

Length 4–5½ft (1.2–1.7m)
Breeding Oviparous
Habit Terrestrial
Status Not evaluated

Location S. Asia

Renowned for its use in snake charming, this animal is one of the most dangerous snakes in India. It is responsible for a relatively small but nonetheless significant proportion of snake-bite mortalities, which total 50,000 per year in India alone. This is partly due to its preference for rice paddies and roadside banks, which are often close to villages and so bring it into contact with humans. Indian cobras are very variable in color, ranging from brown to black, but most individuals have distinctive, pale "spectacle" markings on the back of the hood. As with other cobras, the hood is spread out when the snake feels threatened, but is folded away at other times. Although Indian cobras are sometimes seen basking in the sun, they are most active at night, and are good climbers and swimmers. They feed on small mammals, birds, lizards, and other snakes, killing them within seconds with their potent neurotoxic venom. The principal predators of Indian cobras, other than humans, are carnivorous mammals—notably mongooses—and birds of prey.

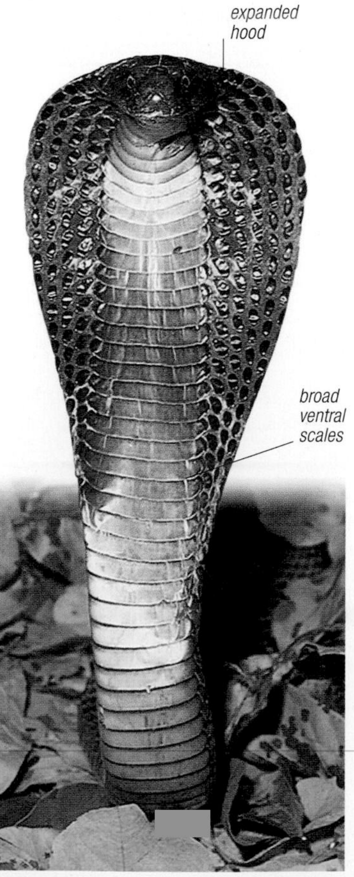

REPTILES

Naja pallida

Red spitting cobra

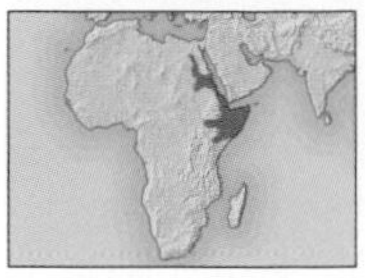

Length 28–47 in (70–120 cm)
Breeding Oviparous
Habit Terrestrial
Status Not evaluated

Location N. and E. Africa

This African snake is one of several cobras that have evolved a unique way of defending themselves. If threatened, they can squirt venom out of small apertures in their fangs, spraying a cloud of droplets 6½ ft (2 m) into the air. Sprayed venom does not kill, but it can cause permanent blindness. The red spitting cobra varies in color from red to gray, and is active at night and early in the morning. Females lay up to 15 eggs in a burrow or in rotting vegetation.

Naja haje

Egyptian cobra

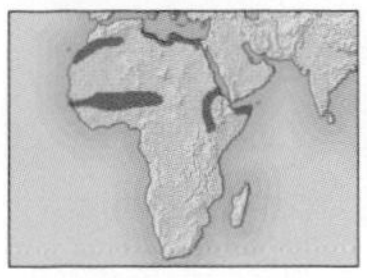

Length 3¼–7¾ ft (1–2.4 m), max 8¼ ft (2.5 m)
Breeding Oviparous
Habit Terrestrial
Status Not evaluated

Location N., W., and E. Africa

Usually brown or grayish, sometimes with black bands, the Egyptian cobra may grow up to 8¼ ft (2.5 m) long. An inhabitant of desert, grassland, and urban areas, it is also found in fields near oases and wadis, and tends to avoid dense forest and the most arid deserts. This cobra is primarily nocturnal, but can sometimes be seen basking in the sun early in the morning. Quick to rear and spread its broad, rounded hood to intimidate an opponent, this highly venomous cobra will hiss and advance on its aggressor if the latter is undeterred, and may eventually strike, delivering a bite that is fatal for humans if an antivenom is not administered promptly. It actively pursues other snakes, small mammals, toads, and birds, including domestic poultry, killing prey with its fast-acting venom; it also feeds on eggs. Individuals are territorial and frequently fight one another. Females lay 8–20 eggs, often in termite mounds, and the eggs hatch after an incubation period of about 60 days.

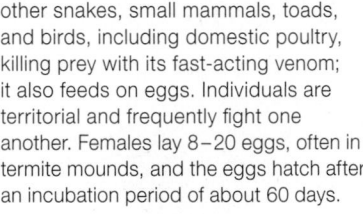

Calliophis bivirgata

Blue coral snake

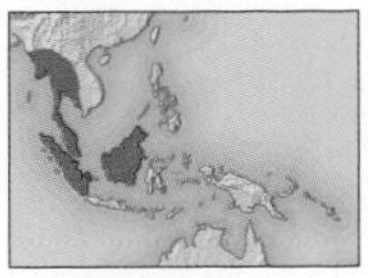

Length 4–4½ ft (1.2–1.4 m)
Breeding Oviparous
Habit Terrestrial
Status Not evaluated

Location S.E. Asia

Bright orange markings on the underside, head, and tail of the blue coral snake act as a warning device to keep potential predators at bay. A highly venomous front-fanged species, its poison glands extend along almost one-third of its entire body length. It is also known as the "100-pace snake," referring to the distance a human is thought capable of traveling before succumbing to the venom. However, it is generally inoffensive, only killing other snakes for food. Only a few cases of human fatalities have ever been recorded.

Notechis scutatus

Australian tiger snake

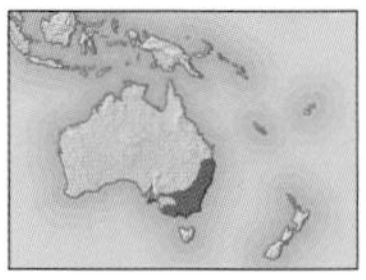

Length 3¼–7 ft (1–2.1 m)
Breeding Viviparous
Habit Terrestrial
Status Least concern

Location S.E. Australia

This extremely venomous species is responsible for many potentially fatal snake bites in Australia. It may be gray, brown, dark brown, or olive-brown, often with narrow, light yellow bands. Although active mainly during the day, the Australian tiger snake may also be active early in the morning and in the evening, and on warm nights; it remains hidden in spells of cold weather. Its prey consists mainly of frogs, which it chases then subdues with its venom. In defense, it flattens its neck and rears its body slightly off the ground. Females give birth to about 30 live young.

Ophiophagus hannah

King cobra

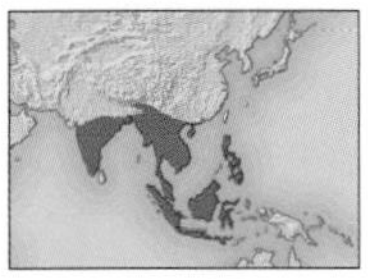

Length 9¾–16 ft (3–5 m)
Breeding Oviparous
Habit Terrestrial
Status Vulnerable

Location S. and S.E. Asia

The king cobra, or hamadryad, is the longest venomous snake and a specialized hunter of other snakes. It sometimes reaches a length of more than 16 ft (5 m), which allows it to overpower and kill other snakes of a considerable size. When threatened, it raises the front third of its body so that it stands 5 ft (1.5 m) tall, erects a narrow hood, and may strike downward, although it rarely attempts to bite. In fact, few cases of king cobra bites on humans have been recorded, because this retiring snake shuns human contact, living mainly in deep forest. Slender and smooth-scaled, the king cobra is a good swimmer and is often found near water. Adults are plain brown, while juveniles are darker and marked with pale chevrons down their backs. Unusually for a snake, this species is at least temporarily monogamous, pairs apparently remaining together during the breeding season. Females lay 21–40 eggs, mostly in piles of dead vegetation, which are guarded by both parents until they hatch. The king cobra may live for more than 20 years in captivity.

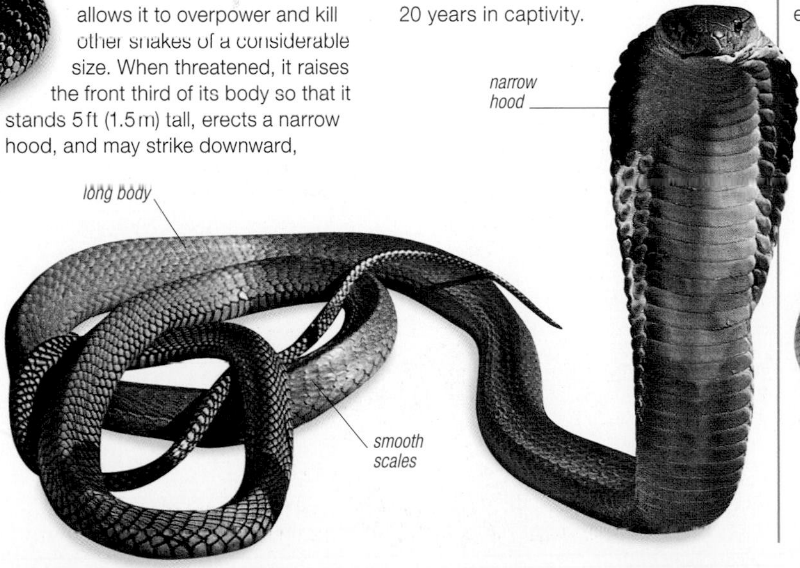

Oxyuranus scutellatus

Coastal taipan

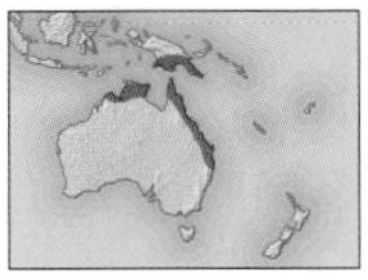

Length 6½–12 ft (2–3.6 m)
Breeding Oviparous
Habit Terrestrial
Status Not evaluated

Location S. New Guinea, N. Australia

Although nondescript in appearance, the plain tan or dark brown coastal taipan is Australia's most venomous snake. It is shy and seldom seen, but when encountered, is capable of striking with deadly speed. It feeds mainly on mammals, but also on birds and lizards, sometimes entering burrows to trap its prey underground. Females lay 3–20 eggs each time they breed. Taipan bites can be fatal to humans, but an effective antivenom means that fatalities are now relatively rare.

Vipers

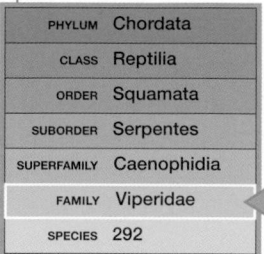

PHYLUM	Chordata
CLASS	Reptilia
ORDER	Squamata
SUBORDER	Serpentes
SUPERFAMILY	Caenophidia
FAMILY	Viperidae
SPECIES	292

Vipers are the most highly evolved of all snakes. They have long, hinged fangs, and some (the pit vipers) also have a pair of heat-sensitive pits between their eyes and nostrils. Vipers are found in a wide range of climates, and compared to other snakes are better equipped to deal with cold conditions. Many live at high elevations or in desert areas with cold winters. Some are even found north of the Arctic Circle. Vipers live on the ground or in trees. Some species use rodent burrows as temporary shelters.

STRIKING POSITION
When preparing to strike, the desert horned viper (shown here) opens its mouth and moves its fangs down from their usual position flat against the roof of the mouth. The jaws are opened wide so that the fangs point forward towards the prey.

Anatomy

Most vipers are short and stocky with wide, triangular heads and rough, keeled scales. The head has long fangs, which are folded against the roof of the mouth when not in use, and large venom glands. Vipers are generally sluggish snakes and rely on camouflage to go undetected: they are often colored to blend in with the surface on which they live and may have intricate geometric markings to break up their outline. Heat detection is more advanced in pit vipers than in any other snakes. Using their heat-sensitive pits, they can not only detect the presence of prey but also gauge its distance and the direction in which it lies. Rattlesnakes, which are a type of pit viper, have a unique warning device in the form of a rattle at the end of the tail.

Venom

Most vipers ambush their prey, some using the brightly colored tip of their tail as a lure. They lunge open-mouthed and stab the victim using their long fangs, penetrating fur and feathers, to reach the vital organs. The venom varies from species to species, but often contains proteins that break down blood cells and cause internal hemorrhaging. The venom being relatively slow acting, vipers deliver large amounts of it, unlike cobras. For the prey, death may not be instantaneous.

Reproduction

Although some vipers lay eggs, most give birth to live young. Females retain the developing eggs in their body and bask during the day to accelerate the embryos' development. Many species breed only every 2, or even 3 years, using the nonbreeding years to feed and recover their body weight.

Atheris mabuensis

Mount Mabu forest viper

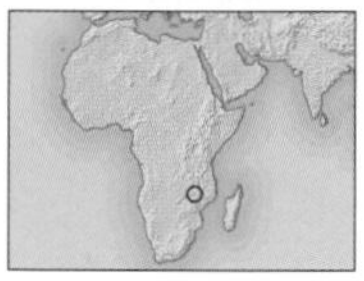

Location E. Africa

Length 15 in (38 cm) max
Breeding Ovoviviparous
Habit Terrestrial
Status Not evaluated

This is the mostly southerly—and smallest—of *Atheris* African forest vipers; it was described as a new species in 2009 and is known only from the midaltitude rain forest of Mount Mabu and Mount Namuli in northern Mozambique. Its highly restricted distribution makes it at risk from habitat destruction. Adults resemble the juveniles of larger allied species and, unlike most members of its genus, are ground dwelling. It may be a specialized hunter—taking small frogs and geckos among leaf litter.

Bitis arietans

Puff adder

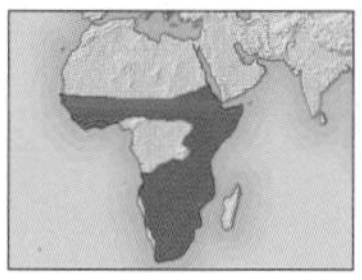

Location Africa (south of Sahara)

Length 3¼ ft (1 m), max 6¼ ft (1.9 m)
Breeding Ovoviviparous
Habit Terrestrial
Status Not evaluated

The puff adder is a thick-bodied snake with a wide, flattened head and a rounded snout. The male is smaller and often more brightly colored than the female. Sluggish and slow moving, larger individuals crawl in a straight line, like a caterpillar. The puff adder hunts in the evening and at night for small mammals and, occasionally, birds and reptiles. It ambushes prey and strikes quickly, injecting large amounts of venom. Males combat and follow the trails of females in the spring. This adder's litter—of 20–40, and in exceptional cases up to 154—is the largest of any snake. The puff adder is highly dangerous to humans, and is responsible for most of the lethal snake bites in Africa.

BODY MARKINGS
The puff adder has a yellowish brown or gray body with white chevrons along its back. Some individuals may also have yellow markings.

DEFENSIVE BEHAVIOR

Highly venomous, the puff adder is ferocious when disturbed, and defends itself by inflating its body and making a long, low hissing sound. If further provoked, it may strike, delivering a fatal bite.

Bitis caudalis

Horned adder

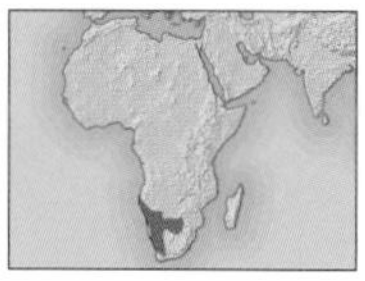

Location Southern Africa

Length 12–20 in (30–50 cm)
Breeding Ovoviviparous
Habit Terrestrial
Status Not evaluated

The horned adder is gray, brown, reddish, or orange, with a single horn over each eye. Its flattened body and rough scales help it shuffle down into loose sand in order to escape the heat and to hide. When crossing loose sand, it may move in a sideways looping motion, known as sidewinding. It hunts mainly in the evening—prey may be attracted to its twitching tail.

REPTILES

Bitis gabonica

Gaboon viper

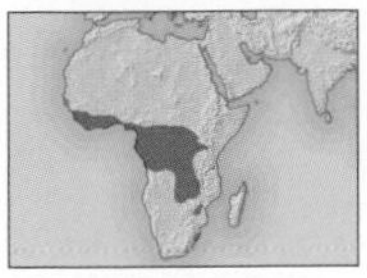

Length 4 ft (1.2 m), max 6½ ft (2 m)
Breeding Ovoviviparous
Habit Terrestrial
Status Not evaluated

Location W. and C. Africa

The immense, thickset body, massive, triangular head with small eyes, and patterning of interlocking, geometric shapes along the back, make the gaboon viper instantly recognizable. When seen away from its natural habitat, it appears to be colorful and conspicuous, with its distinct markings in pale purple, tan, cream, and brown. However, in the wild, its coloration and patterning provide excellent camouflage, making it extremely difficult to detect. A sluggish, usually placid snake, it generally lies motionless on the forest floor, moving only when prey—namely rodents and larger mammals, birds, and frogs—comes within striking range. Large, heavy-bodied individuals normally crawl in a straight line, like a caterpillar. When disturbed, this viper gives a loud, low hiss and may bite, but only as a last resort; the bite is often fatal to humans if not treated at once. Males combat during the breeding season, and females give birth to 16–60 young every 2–3 years.

Echis pyramidum

Northeast African saw-scaled viper

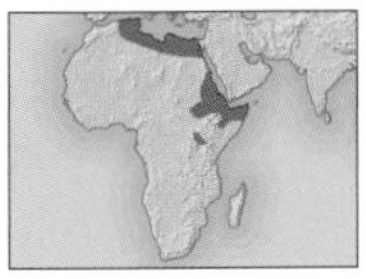

Length 12–23½ in (30–60 cm)
Breeding Oviparous
Habit Terrestrial
Status Least concern

Location N. and N.E. Africa

Despite its small size, this quick-moving snake is highly dangerous, with venom potent enough to kill humans. It belongs to a small genus of vipers (known as saw-scaled or carpet vipers) that have strongly keeled scales that produce a loud rasping sound when they are rubbed together. When threatened, saw-scaled vipers use this sound as a warning signal, in much the same way as rattlesnakes (see p.404) use their tails. Like its relatives, the northeast African saw-scaled viper hunts small mammals, amphibians, reptiles, and invertebrates at night. Females lay clutches of 6–20 eggs.

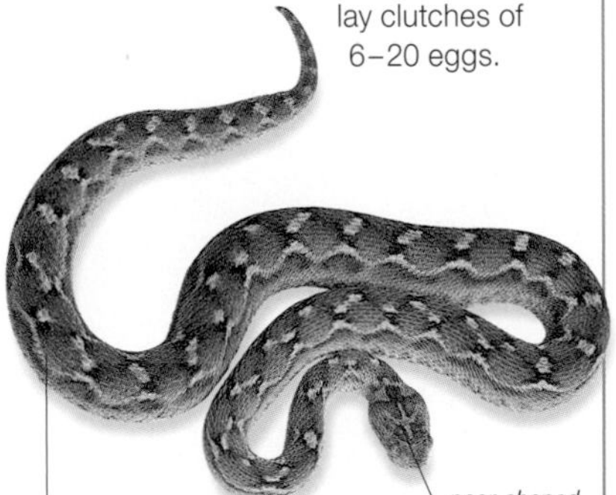

Vipera ammodytes

Nose-horned viper

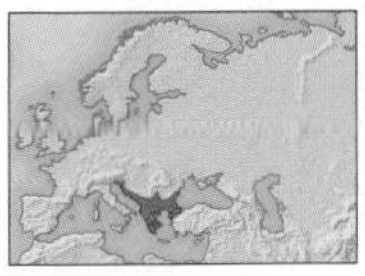

Length 26–35 in (65–90 cm)
Breeding Viviparous
Habit Terrestrial
Status Least concern

Location S.E. Europe, W. Asia

The light brown, reddish brown, or gray nose-horned viper is Europe's most dangerous snake, although its bite is rarely fatal if treated. It has a short, fleshy horn on its snout, the function of which is as yet unknown. It is usually slow-moving, unless provoked, and often coils quietly among rocks on hillsides. Mainly diurnal, it may become nocturnal in midsummer in certain parts of its range.

Cerastes cerastes

Desert horned viper

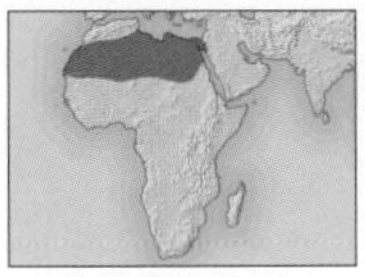

Length 12–23½ in (30–60 cm)
Breeding Oviparous
Habit Terrestrial
Status Not evaluated

Location N. Africa

Also known as the Sahara horned viper, this short, squat snake usually has a thornlike horn over each eye. The strongly keeled scales on its body help the snake shuffle down into the sand, where it remains hidden and protected from the heat. Active at night, it lies in ambush for rodents, lizards, and birds, sometimes partially buried, then launches a rapid strike. When threatened, it forms a coil, rubbing together the scales on its flanks to create a rasping sound.

Daboia russelli

Russell's viper

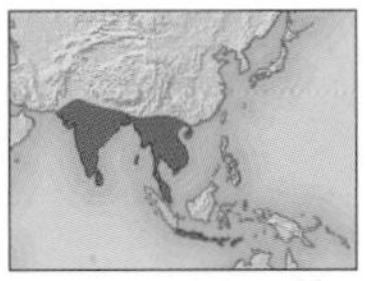

Length 3¼ ft (1 m), max 5 ft (1.5 m)
Breeding Viviparous
Habit Terrestrial
Status Not evaluated

Location S. and S.E. Asia

Although this is a sluggish animal that usually remains coiled and well hidden during the day, Russell's viper is one of the most dangerous snakes to humans in southern Asia. It is distinguished by its light brown body with 3 rows of oval, dark-edged brown blotches down its back. A sit-and-wait predator, it relies on its camouflage to escape detection by unwary prey, then strikes. In defense, it forms a tight coil, hisses, and strikes hard, often lifting its body off the ground in the process.

Vipera berus

Common adder

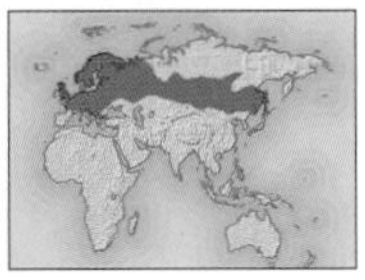

Length 26–35 in (65–90 cm)
Breeding Ovoviviparous
Habit Semiterrestrial
Status Least concern

Location Europe, C. to E. Asia

The common adder—or common viper—is the only venomous snake in northwest Europe, and is the most widely distributed snake in the viper family. It lives in a wide range of habitats, including rocky slopes 9,900 ft (3,000 m) high in the European Alps, and open tundra north of the Arctic Circle. Its color varies from gray, brown, or rust to completely black, but most individuals have a distinctive, zigzag line running down their back. Common adders may occasionally reach as much as 35 in (90 cm) long. They feed primarily on small rodents and lizards, either ambushing or actively pursuing them, but they also spend much of their time basking in the sun. In the north of their range, where winters are colder—for example, in Scandinavia—they hibernate for up to 8 months each year. In spring, the snakes emerge to mate, with males engaging in protracted combat for the chance to pair up with females. Females breed only about once every 2 or 3 years, producing litters of up to 20 young that may have been fathered by several different males. Adders are only mildly venomous. In humans, their bite can cause swelling and pain, but is rarely fatal.

Crotalus atrox

Western diamondback rattlesnake

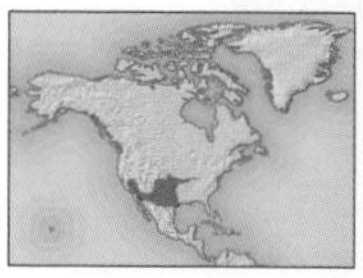

Location S. USA, N. Mexico

Length Up to 6½ft (2 m)

Breeding method Ovoviviparous

Habit Terrestrial

Status Least concern

This large, heavy-bodied snake is North America's most dangerous rattlesnake, with the highest rate of human fatalities. A highly venomous and effective hunter, the western diamondback rattlesnake feeds mainly on small mammals, birds, and lizards, but can be deadly to larger animals or humans, if provoked. The rattle, its most distinctive feature, is a horny section at the end of the tail composed of loosely connected segments, which are added to after each molting. Used to warn and deter intruders, the rattle consists of up to 10 segments in adults. Newborns, measuring about 12 in (30 cm), do not possess a rattle, but instead have a small "button" that develops with molting. However, they are already equipped with needle-sharp fangs and extremely toxic venom. It takes between 3 to 4 years for them to reach sexual maturity.

HEAVY-BODIED SNAKE
This bulky snake varies in color according to its location and habitat, and may be gray, brown, olive-green, or reddish brown. The tail is marked with black and white bands just in front of the rattle.

FATAL STRIKE
The western diamondback rattlesnake usually hunts by ambushing its victim, then striking with its venomous fangs before devouring its stunned prey. It has a high yield of highly poisonous venom containing a mixture of toxins that can kill victims in seconds.

ATTACK AND DEFENSE

The western diamondback rattlesnake is a well-armed and deadly predator that has developed effective methods for hunting prey. However, the solitary snake will often withdraw from confrontation if disturbed, using its rattle to deter intruders.

DEATH RATTLE
The rattlesnake sounds its unmistakable rattle as a means of defense and an ominous warning not to approach.

STALKING PREY
The western diamondback rattlesnake uses the pair of heat-sensitive pits between its eyes and nostrils to detect and locate warm-blooded prey.

EATING ITS PREY
Having stunned a rodent victim with its venomous bite, the rattlesnake waits for it to die before swallowing it whole.

Agkistrodon contortrix

Copperhead

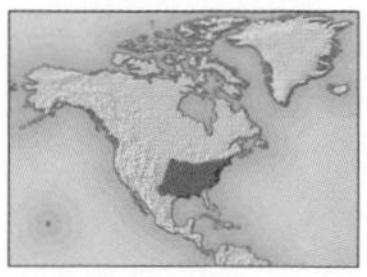

Length 23½–51 in (60–130 cm)
Breeding Viviparous
Habit Terrestrial
Status Least concern

Location C. and S.E. USA

The copperhead is boldly marked with chestnut-brown or orange bands on a gray, tan, or pinkish background. In spring and fall, it basks during the day, but it becomes nocturnal in summer. This viper lies practically hidden among dead leaves or rocks waiting to ambush rodents, frogs, lizards, and invertebrates, using the heat-sensitive pits between its eyes and nostrils to locate prey in total darkness. Its bite is venomous, but rarely serious for humans. The female gives birth to 1–15 young every other year.

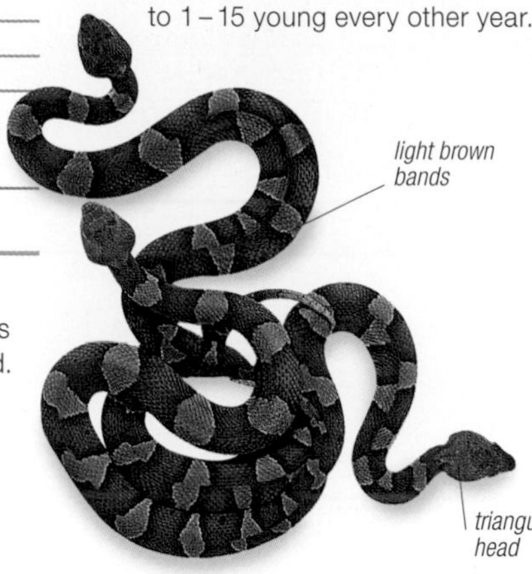

Bothriechis schlegelii

Eyelash pit viper

Length 18–30 in (45–75 cm)
Breeding Viviparous
Habit Arboreal
Status Not evaluated

Location Central America, N. South America

This tropical forest snake is one of a number of pit vipers that live in trees. Like most arboreal snakes, it has a prehensile tail and good binocular vision, but it also has distinctive "eyelashes," formed by raised scales above its eyes. Active at night, it preys on lizards, frogs, and small rodents, and lies in wait near forest flowers to catch hummingbirds as they come to feed. Eyelash pit vipers vary greatly in color, ranging from an intense orange-yellow to grayish green.

Bothrops atrox

Common lancehead

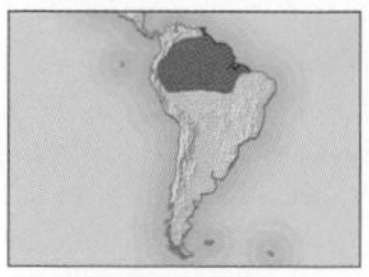

Length 3¼ ft (1 m), max 5 ft (1.5 m)
Breeding Viviparous
Habit Terrestrial
Status Not evaluated

Location N. South America

This is one of the most dangerous snakes in South America because of its rapid strike, potent venom, and its tendency to live near human habitations. Heat-sensitive pits between the eyes and nostrils enable it to detect prey in the darkness. It is also called the fer-de-lance, though some reserve this name for a West Indian species that has a similarly pointed head. Males have longer tails than females, and may fight during the breeding season. Females may give birth to up to 100 young. Juveniles have brightly colored tail tips, which they use to lure prey.

Calloselasma rhodostoma

Malaysian pit viper

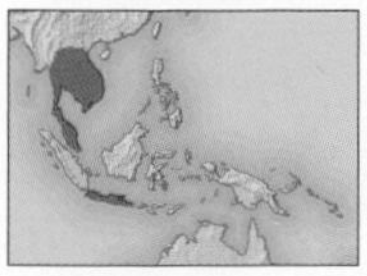

Length 28–39 in (70–100 cm)
Breeding Oviparous
Habit Terrestrial
Status Not evaluated

Location S.E. Asia

This highly venomous snake, usually found at the edges of forests and in clearings, is quick to strike, and is the cause of many snake-bite fatalities in Southeast Asia. It is purplish brown, with light-edged, darker triangles on its flanks, and has a pointed snout. A nocturnal hunter, the Malaysian pit viper uses the heat pits between its eyes and nostrils to help it to locate prey in the dark. Unlike most vipers, this species lays eggs—the female may lay 2 clutches of 13–25 eggs each in a year—and remains coiled around them throughout their incubation.

Crotalus durissus

Tropical rattlesnake

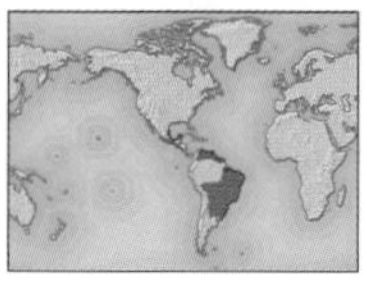

Length 3¼ ft (1 m), max 5 ft (1.5 m)
Breeding Viviparous
Habit Terrestrial
Status Least concern

Location Central America, South America

The bite of the tropical rattlesnake is highly venomous—about 75 percent of untreated bites result in fatalities. There are 14 subspecies recognized, with highly variable coloring, but this is the only rattlesnake over much of its range. Heavily built, it has a broad head and obvious heat-sensitive pits, to detect prey in total darkness. This nocturnal species spends the day coiled, partly hidden by vegetation. The female gives birth to 6–12 young.

Crotalus cerastes

Sidewinder

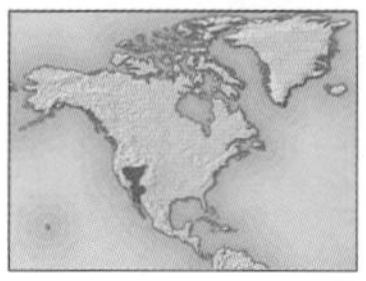

Length 18–32 in (45–80 cm)
Breeding Ovoviviparous
Habit Terrestrial
Status Least concern

Location S.W. USA, N.W. Mexico

The sidewinder is the stoutest of all rattlesnakes and is the only one in which the female is smaller than the male. It has a highly distinctive way of moving—an adaptation to life on windblown desert sand. It feeds mainly on small lizards and rodents, often ambushing its prey from the cover of isolated shrubs. Although sidewinders can be aggressive, by rattlesnake standards their bite is not especially dangerous. Their young are born in late summer, in clutches of 6–12.

BODY COLORATION
The sidewinder varies from cream, gray, tan, or brown to pinkish, depending on the soil on which it lives; markings are poorly defined.

MOVING ON SAND

Sidewinding is a form of movement used by several snake species that live on loose sand. Instead of sliding forward, with its body in contact with the sand, a sidewinding snake throws itself over the surface, following a diagonal path. This kind of movement creates characteristic parallel tracks.

Lachesis muta

Bushmaster

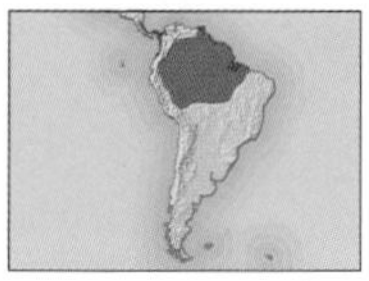

Length 8¼ ft (2.5 m), max 12 ft (3.6 m)
Breeding Oviparous
Habit Terrestrial
Status Not evaluated

Location S. Central America, N. South America

The bushmaster is the longest viper in the world. It is also one of the most venomous, with a 20 percent mortality rate, even if treated; however, it generally avoids contact with humans. A secretive species, it ambushes prey at night—coiled and ready to strike—along a mammal trail. It has a slender, yellow or tan body with black markings. Unusually for a viper, females lay eggs (about 8–15), guarding them until they hatch.

Cryptelytrops albolabris

White-lipped pit viper

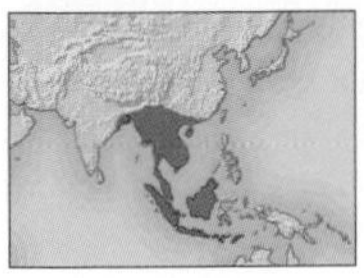

Length 23½–39 in (60–100 cm)
Breeding Viviparous
Habit Arboreal
Status Least concern

Location S.E. Asia

This slender-bodied snake—easily confused with several other green pit vipers from the same part of the world—has a wide, rounded head with prominent heat pits. It lives in trees, hanging head-down at night in a favored hunting place waiting for prey to come within range, then drawing back its head when ready to strike. Its bite is painful, but the venom is mild and rarely fatal to humans. Females give birth to 10–11 young.

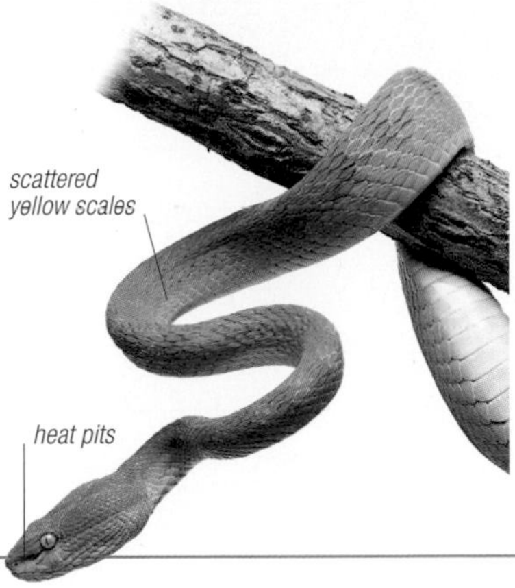

Tropidolaemus wagleri

Wagler's viper

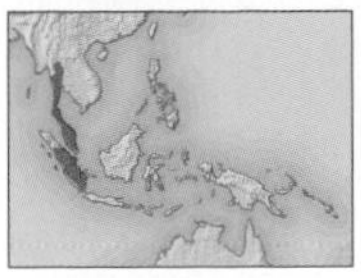

Length 2½–4¼ ft (0.8–1.3 m)
Breeding Viviparous
Habit Arboreal
Status Not evaluated

Location S.E. Asia

Renowned for living in semicaptivity in Malaysia's Penang Snake Temple, this highly variable snake is sluggish during the day, when it rests in trees. At night, it ambushes small mammals, birds, lizards, and frogs by hanging head-down from the branches. It is a thickset viper, with a chunky head, and has various color combinations: usually black, with green or yellow crossbars and scattered yellow spots. Juveniles are green, with yellow and/or red spots arranged into vague crossbands; some individuals retain their juvenile markings throughout their lives. Although its bite is painful, it is seldom fatal to humans.

Blind and thread snakes

PHYLUM	Chordata
CLASS	Reptilia
ORDER	Squamata
SUBORDER	Serpentes
SUPERFAMILY	Scolecophidia
FAMILIES	3
SPECIES	407

These small snakes live underground, emerging above the surface only when driven out by lack of food or by flooding. They have a slender, cylindrical body and are covered with smooth, shiny scales that allow them to move easily through sand and soil. Their skull is more heavily built than in other snakes, and they have jaws that they cannot open widely, with few teeth. Their eyes are covered with scales and are barely functional. Blind and thread snakes are restricted to tropical and subtropical regions. They are often found within ant and termite mounds, and the bulk of their diet is composed of these insects and their larvae.

BURROWER
A thread snake above ground is an uncommon sight. This Peters thread snake from southern Africa has emerged from its burrow in sandy soil.

Rena dulcis

Texas thread snake

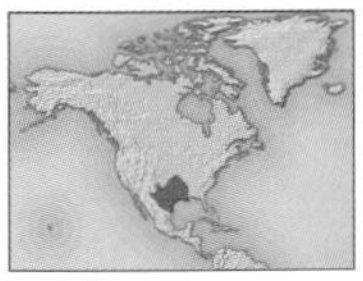

Length 6–10½ in (15–27 cm)
Breeding Oviparous
Habit Burrowing
Status Least concern

Location S. USA, N.E. Mexico

The slender-bodied, pinkish brown Texas thread snake lives almost exclusively below the ground, coming to the surface only at night and after rain. The smooth scales, short tail, and vestigial, scale-covered eyes are all features associated with a burrowing lifestyle. This snake feeds opportunistically on small insects and spiders, and also enters ant and termite nests in search of larvae. When the snake enters a nest, it releases pheromones that imitate that of the insects. The "soldiers," deceived into thinking that the snake is part of the nest, allow it to enter and feed in safety. The chief predators of the Texas thread snake are other snakes and nocturnal birds and mammals. The female lays 2–7 eggs and coils around them until they hatch.

Tetracheilostoma carlae

Barbados thread snake

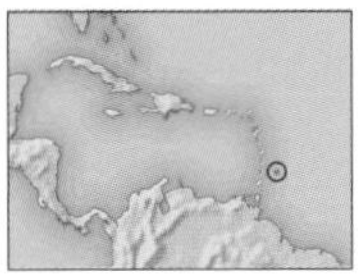

Length Up to 4 in (10 cm)
Breeding Oviparous
Habit Burrowing
Status Not evaluated

Location Caribbean

This tiny snake was described in 2008. It is only found on the upland forest of central Barbados. The larger and recently introduced brahminy thread snake, which breeds more quickly, could be an important competitor and threaten its survival. The Barbados thread snake is the smallest known snake—and has probably evolved to the lowest possible size limit for any snake. It lays a single, highly elongated egg, which hatches into a youngster that is already half the length of an adult snake. A related species found only on St. Lucia island could be the world's second smallest snake.

Ramphotyphlops braminus

Brahminy blind snake

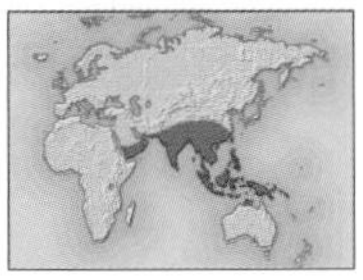

Length 6–7 in (15–18 cm)
Breeding Oviparous
Habit Burrowing
Status Not evaluated

Location W. Asia to N. Australia. Widely introduced elsewhere

This is the only snake known to be parthenogenetic. Males have never been found, and females lay their eggs without having to mate. Each egg is about the size of a peanut. The snake's habit of stowing away in plant pots has resulted in the spread of this species throughout warm parts of the world. Its scale-covered eyes, blunt snout, and smooth scales are all features adapted to burrowing. If in danger, this snake may give off an unpleasant, pungent fluid.

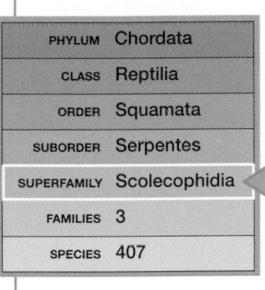

Lizards

PHYLUM	Chordata
CLASS	Reptilia
ORDER	Squamata
SUBORDER	Lacertilia
FAMILIES	24
SPECIES	About 5,500

CLASSIFICATION NOTE

The classification of lizards, like that of snakes, is the subject of much debate, especially the way they are grouped into families. However, their division into the following superfamilies is widely accepted:

Iguanas and relatives *see pp.410–6*
Geckos and snake-lizards *see pp.417–9*
Skinks and relatives *see pp.420–5*
Anguimorph lizards *see pp.426–30*

Lizards are the most successful group of reptiles. They form a large and varied group that is not easy to define concisely: although a typical lizard has a distinct head, 4 well-developed limbs, and a long tail, there are exceptions to all of these generalizations. Some lizards also have an unusual ability to shed their tail when attacked by a predator and then regenerate it. Most species reproduce by laying eggs, although some give birth to live young; a small number actively care for their offspring. Lizards have adapted to survive in habitats all over the world (although they are not found in Antarctica). Most live on the ground, on low-lying rocks, or in trees. There is also a small number of burrowing species and even one that feeds in seawater. A few lizards, notably the geckos, have adapted to live around humans, and in some parts of the world, they are a common sight on the walls and ceilings of buildings.

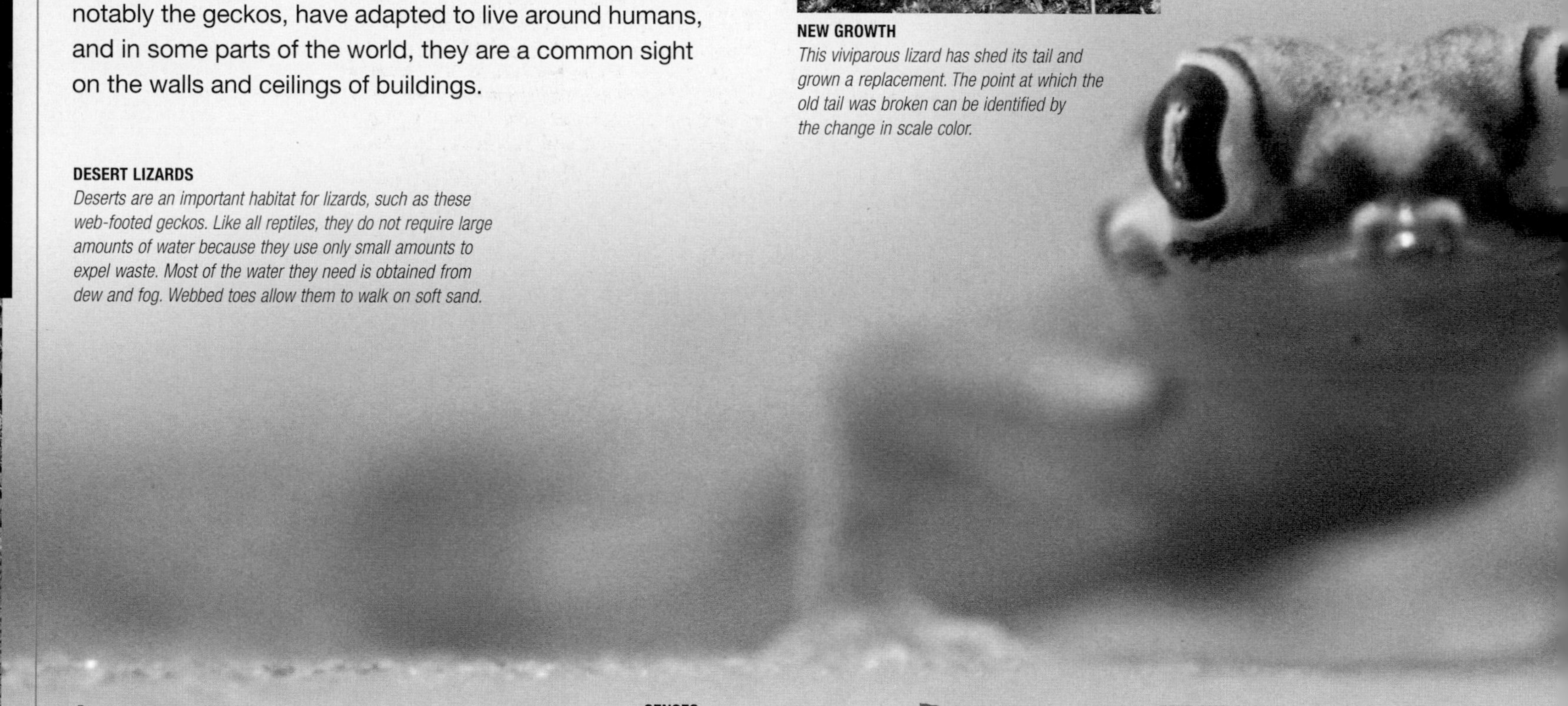

DESERT LIZARDS
Deserts are an important habitat for lizards, such as these web-footed geckos. Like all reptiles, they do not require large amounts of water because they use only small amounts to expel waste. Most of the water they need is obtained from dew and fog. Webbed toes allow them to walk on soft sand.

Regeneration

Some lizards are able to voluntarily break off part of their tail, a phenomenon known as autotomy. This is a defense mechanism that helps the lizard escape when grabbed by the tail. To add to the effectiveness of the strategy, the tail may be brightly colored, often blue, to draw the attacker's attention away from the head. The tail may also continue to move for several minutes after being severed, apparently to distract the predator. The tail breaks at one of several predetermined weak points (see below), and it does not have to be pulled in order to cause a break. Most lizards can regrow a broken tail, but the replacement will not have the same scale arrangement or coloration as the original. A lizard may shed its tail several times during its lifetime.

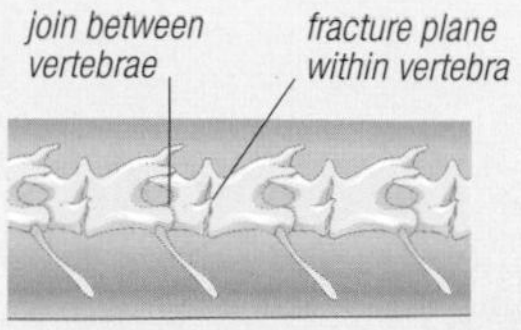

FRACTURE PLANES
A lizard's tail breaks at one of several weak points, called fracture planes, across certain vertebrae. The muscles are arranged in such a way that they, too, will come apart neatly.

NEW GROWTH
This viviparous lizard has shed its tail and grown a replacement. The point at which the old tail was broken can be identified by the change in scale color.

Anatomy

In addition to a distinct head, 4 legs, and a tail, most lizards have external ear openings, and eyelids that can be opened and closed. However, some lizards have short legs or none at all. There are also forms with no eyes or no ears, while others have eyes covered by a transparent scale rather than movable lids. The tongue may be short or long and is usually notched at the tip; forked tongues are found in some species. A lizard's scales may be large, smooth, and overlapping or small, rough, and studlike, in almost any combination. Many have irregular scales or a few large scales scattered among many smaller ones, but none has the single row of wide ventral scales characteristic of snakes (see p.384). Most lizards have sharp teeth along the edges of their jaws.

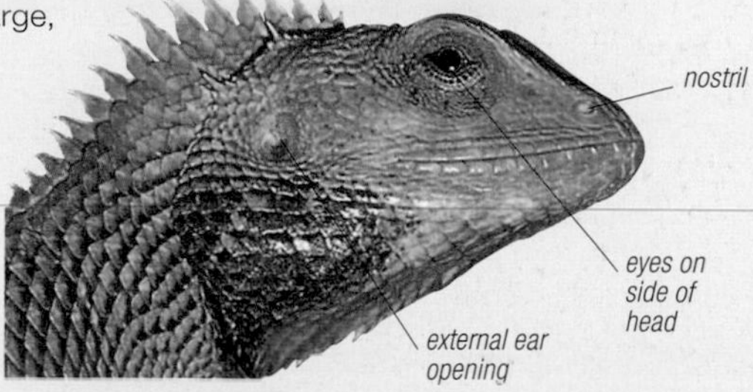

SENSES
A lizard's senses are adapted to its lifestyle. Many species that live underground cannot see, while those living above ground may have good vision, but only a few are able to see colors. Lizards with external eardrums can hear airborne sounds.

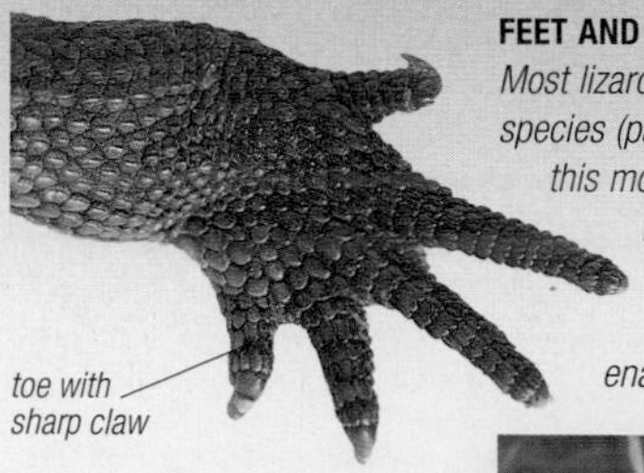

FEET AND CLAWS
Most lizards have 5 digits on each foot. Many species (particularly those that climb, such as this monitor lizard) have sharp claws. Most lizards cannot oppose their digits, although chameleons have their digits fused into 2 opposing groups, enabling them to grip branches.

LEGLESS LIZARDS
Over time, some lizards (such as this Burton's snake-lizard) have lost their limbs. This is an adaptation that helps them to move easily through soil burrows, sand, and leaf litter.

Movement

Most lizards move by walking, climbing, or burrowing. When walking, they usually use all 4 limbs (see right), although some species lift their front legs off the ground while running at high speed; basilisk lizards can even run for short distances over the surface of water in this way, helped by broad feet that spread their weight. Some lizards are agile climbers. To improve their grip, they may have long, sharp claws, while some have a prehensile tail. Among the most impressive climbers are geckos, which can cling to smooth, vertical or even overhanging surfaces using adhesive pads on the tips of their digits (see p.417). Burrowing lizards may use powerful front limbs for digging, although others have reduced limbs or no limbs at all and simply force their way through loose soil and sand with rapid "swimming" movements. A few lizards can glide from high trees (see below).

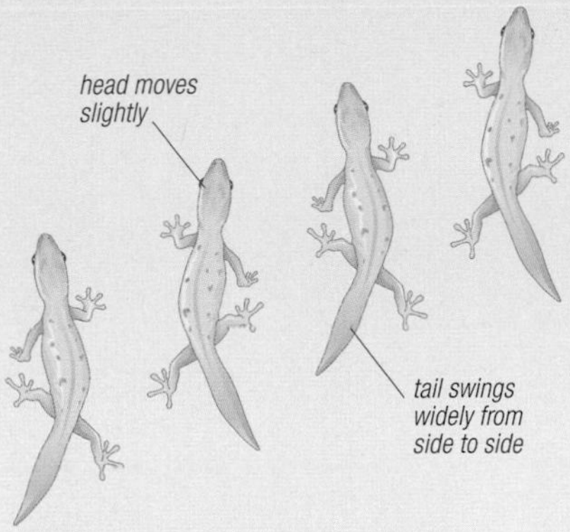

WALKING
A lizard's limbs are held at right angles to the body, rather than supporting it from underneath. This means that as the lizard walks, the weight of the body is thrown from side to side, resulting in a characteristic wiggling motion and causing its tail to swing and provide a counterbalance.

FLYING
The flying lizard (above) has flaps of skin, that act like wings, stretched between elongated ribs. Some other species have similar flaps between their toes. In all cases, the lizards move from tree to tree by gliding rather than powered flight.

CLIMBING TREES
Some large lizards, such as this perentie (right), are surprisingly agile climbers. Perenties usually climb only to escape predators, but other lizards, especially chameleons, spend the greater part of their lives in trees.

CAMOUFLAGE
Chameleons (such as the Cape dwarf chameleon, left) have a remarkable ability to alter the appearance of their skin. Not only does the color change, but markings, such as lines and bars, may appear and disappear.

FIGHTING DRAGONS
The Komodo dragon (right) is the largest of all lizards. When attacking its prey, it usually relies on its jagged teeth and sharp claws. However, when defending itself from another member of its own species, it will often turn its back and thrash its powerful tail.

Defense

Lizards use various active and passive strategies to avoid the many other animals that prey upon them. Other than running away, passive strategies include camouflage and mimicry of inanimate objects, such as sticks and leaves. Some species can alter their coloration, while others make themselves inconspicuous by pressing their bodies close to rocks or tree trunks to avoid casting a shadow. Active strategies include the voluntary loss of the tail (see Regeneration, opposite), and threat displays, which often involve warning coloration (as in blue-tongued skinks) or body enlargement (as in the frilled lizard). Large species, such as monitors, as well as some smaller ones, defend themselves by biting, scratching, or lashing their tail. Two species of lizards, the Gila monster and the Mexican beaded lizard, are venomous.

Iguanas and relatives

PHYLUM	Chordata
CLASS	Reptilia
ORDER	Squamata
SUBORDER	Lacertilia
SUPERFAMILY	Iguania
FAMILIES	3
SPECIES	1,607

Iguanas and their relatives include some of the most colorful and familiar of all lizards. Along with geckos and snake-lizards, they are also the most primitive. They are divided into 3 groups: iguanids, agamids, and chameleons. Many species, notably chameleons, are capable of changing their skin color. Some are highly territorial or have elaborate social rituals. Iguanids and agamids look similar to each other but are found in different parts of the world: iguanids occur throughout the Americas, in Madagascar, and on a few Pacific islands, while agamids are found in Africa, Asia, and Australasia. Chameleons occur in Africa, Madagascar, Asia, and Europe. Iguanas and their relatives occupy various habitats, including desert, grassland, and rain forest. Most species lay eggs but a small number give birth to live young.

Anatomy

Iguanids and agamids look alike and occupy similar ecological niches. The main difference between them is in the teeth: unlike iguanids, the agamids have teeth that are fixed to the jawbone and not replaced if lost. Most species have small scales and a loose fold of skin (or dewlap) under the head. The males are often brightly colored, and many species have ornamentation on their head and a crest down their back. They may also have a brightly colored throat fan, which they use for communication.

Chameleons are adapted to life in the trees. Their narrow body and leaflike profile help make them inconspicuous among foliage, while a prehensile tail and toes that are fused in 2 opposing groups enable them to hold tightly to branches. Chameleons also have a well-known (but often exaggerated) ability to change color (see below). Male chameleons usually have a crest or horns on their head.

Color change

Two types of color change can be seen in the lizards in this group. Chameleons change color by expanding or contracting pigment cells (chromatophores) scattered in their skin. When the cells are expanded, the skin darkens due to the dispersal of melanin pigments in the cells; as the cells are contracted, the skin becomes lighter. A great number of different colors and patterns can be produced in this way. Chameleons change color mainly to control their temperature (making themselves darker when they need to warm up) or to communicate with other members of their species, rather than for camouflage. At night, while they are asleep, chameleons always become paler. Small leaf chameleons and pygmy chameleons, which are brown, have limited ability to change color.

Agamids and iguanids also change color, but this takes place more slowly than in chameleons. Mature males become brighter in the breeding season. Females may change color according to whether or not they are receptive to males: once they have mated, they often develop orange spots (known as gravid coloration) to deter unwanted attention from males.

Communication

In many species of iguanas and their relatives, the males use their bright coloration for display, both to females and to other males. Often occupying a prominent vantage point, they will nod their head vigorously or bob up and down, sometimes exposing brightly colored scales on their throat and chest. Males are highly territorial, and will chase off other males and mate with any receptive female that lives in or strays into their territory.

HUNTING
Chameleons, such as this veiled chameleon, stalk their prey slowly, locating it by sight. Each eye can swivel independently in any direction and is supported on a turret. The muscular tongue is extremely long with a thick, adhesive tip. When prey is within range, the chameleon flicks its tongue out at great speed, catching the prey on the tip and pulling it back into its mouth.

Acanthosaura crucigera

Boulenger's pricklenape lizard

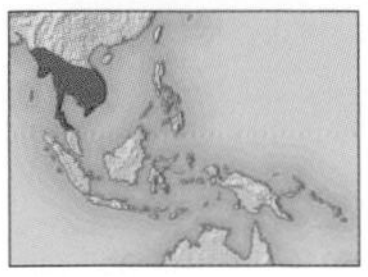

Length 10–12 in (26–30 cm)
Breeding Oviparous
Habit Arboreal
Status Not evaluated

Location S.E. Asia

Relying on its greenish yellow, cryptic coloring, this lizard often escapes detection from predators by remaining motionless, although it may open its mouth wide and bite if cornered. It has long, thin limbs, a long tail, a crest of spines on its back, and larger spines on the nape of the neck and over the eyes. The male is larger, more colorful, and has a more pronounced crest than the female. This lizard lies in wait for insects, then suddenly darts out to catch them. Females lay 10–12 eggs at a time in leaf litter or debris.

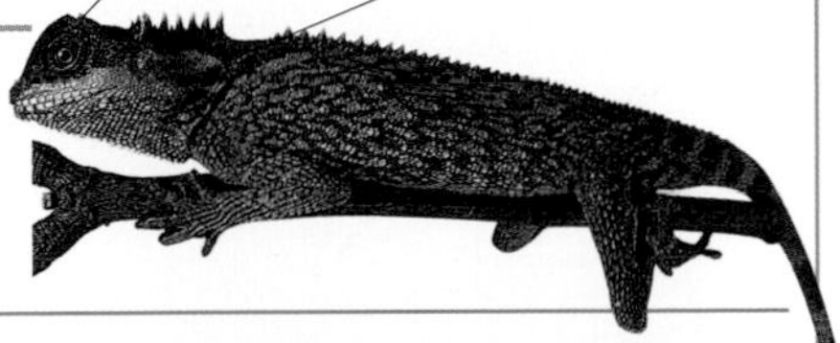

Agama agama

Rainbow lizard

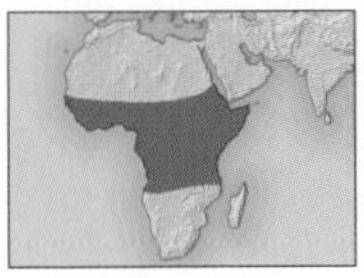

Length 12–16 in (30–40 cm)
Breeding Oviparous
Habit Terrestrial
Status Not evaluated

Location W., C., and E. Africa

Male rainbow lizards are among Africa's most conspicuous reptiles. At night, they are drab and gray, but when they warm up in the sun, they develop a bright orange head, and a blue or turquoise body. Females and juveniles are quite different, with a gray coloration that makes good camouflage. Rainbow lizards live in open habitats, and are often seen around buildings. They feed on insects but become omnivorous when food is short. Males have a head-bobbing courtship display, and females lay 5–7 eggs in holes in the ground.

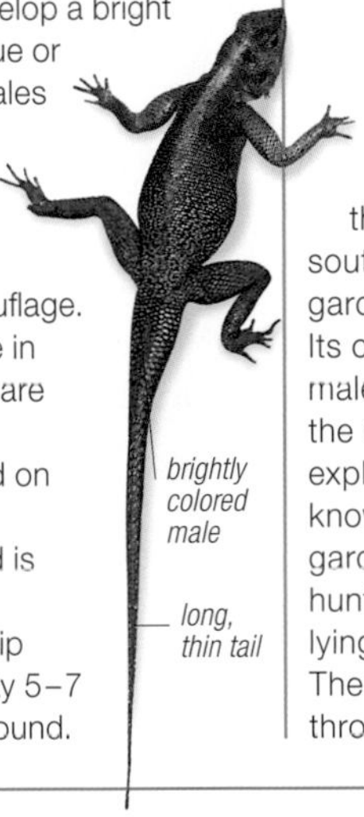

Calotes versicolor

Variable garden lizard

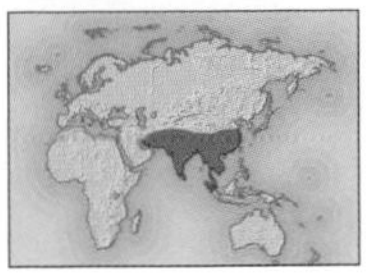

Length 12–14 in (30–35 cm)
Breeding Oviparous
Habit Terrestrial
Status Not evaluated

Location W. Asia to E. and S.E. Asia

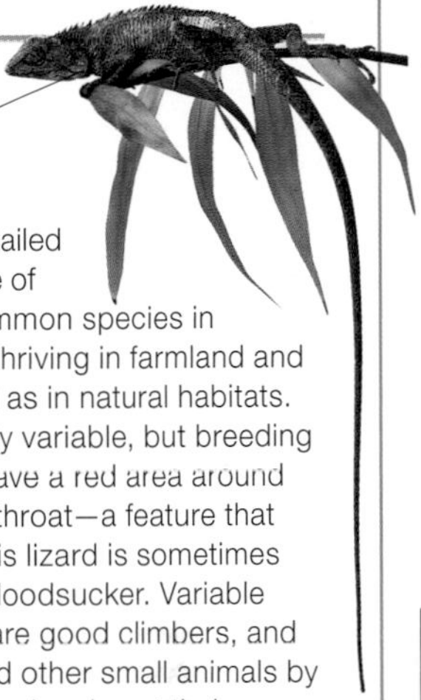

This long- tailed lizard is one of the most common species in southern Asia, thriving in farmland and gardens as well as in natural habitats. Its color is highly variable, but breeding males always have a red area around the mouth and throat—a feature that explains why this lizard is sometimes known as the bloodsucker. Variable garden lizards are good climbers, and hunt insects and other small animals by lying in wait, then lunging at their prey. There are many similar species throughout Asia.

Chlamydosaurus kingii

Frilled lizard

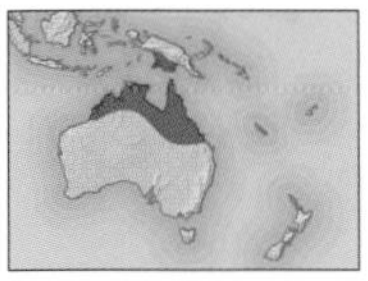

Length 23½–28 in (60–70 cm)
Breeding Oviparous
Habit Terrestrial
Status Least concern

Location S. New Guinea, N. Australia

Australia is home to a wide range of lizards, but this species is the most spectacular. Its name derives from the large, leathery frill around its neck, which is normally folded back over the animal's shoulders. However, if the lizard is threatened, it opens the frill like an umbrella, and at the same time rocks its body and hisses through its gaping mouth. This display is often enough to deter would-be predators, allowing the frilled lizard to make its escape.

CLIMBING TO SAFETY

Frilled lizards forage mainly on the ground. They are good climbers, and after carrying out their defensive display they usually run up a tree trunk, heading for the highest branches.

DEFENSIVE BLUFF
This lizard's frill consists of a flap of skin strengthened by rods of cartilage. It is often brightly colored, contrasting with the animal's gray or brown body.

Draco spilonotus

Flying lizard

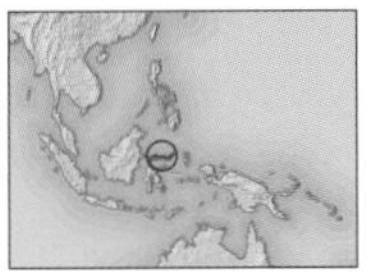

Length 6–8 in (15–20 cm)
Breeding Oviparous
Habit Highly arboreal
Status Not evaluated

Location S.E. Asia

The flying lizard has a pair of "wings," consisting of skin stretched between elongated ribs, which help it glide from one tree trunk to another to avoid predators such as snakes and birds. To take off, the lizard moves its ribs outward and opens the flaps of skin, closing them again on landing. It is generally tan or pale brown, with a throat flap for display that is large and bright yellow in the male, but small and pale yellow in the female. Highly agile, this lizard lives on tree trunks and actively hunts for ants. Females only leave the safety of trees to lay 8–12 eggs, which they bury in sand or soil.

Hypsilurus spinipes

Southern angle-headed lizard

Length 8–13 in (20–33 cm)
Breeding Oviparous
Habit Arboreal
Status Not evaluated

Location Australia (S.E. Queensland and N.E. New South Wales)

This lizard is characterized by a large, angular head, a raised crest (more prominent in males) on the nape and back, and a long tail. It can "freeze" to avoid detection—a useful tactic against predators such as birds, mammals, and monitor lizards—and its exceptionally long toes and claws enable it to climb trees rapidly to escape from danger on the ground. An active forager, the southern angle-headed lizard lunges quickly and suddenly at insects, its sole food source. The female lays eggs in depressions in soil or in leaf litter.

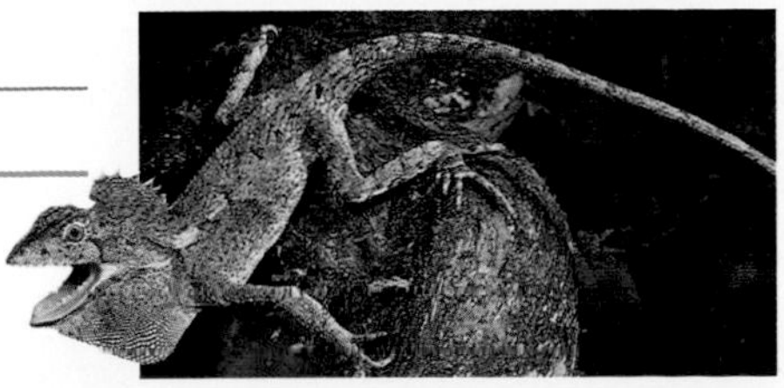

Hydrosaurus pustulatus

Sailfin lizard

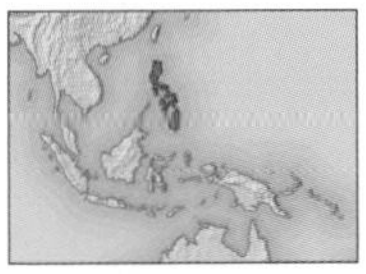

Length 32–39 in (80–100 cm)
Breeding Oviparous
Habit Arboreal
Status Vulnerable

Location Philippines

As an adult, this large, gray or greenish gray lizard has a well-developed crest of toothlike scales from the nape of the neck down the back. The tail also has a crest, up to 3¼ in (8 cm) high, which helps the lizard swim. When in danger from predators, mainly birds of prey, this lizard dashes away at great speed, often diving into water to escape. At other times, it may bask on branches that overhang or are near streams. It forages for leaves and fruit, and feeds on insects opportunistically.

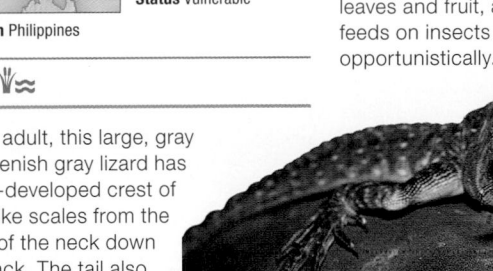

REPTILES

Moloch horridus

Thorny devil

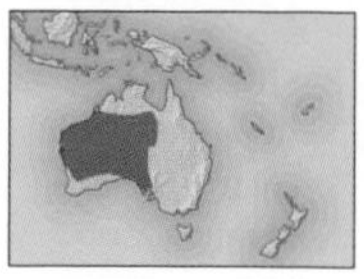

Length 6–7 in (15–18 cm)
Breeding Oviparous
Habit Terrestrial
Status Not evaluated

Location W. to C. Australia

Although it is not often seen—partly because it is well camouflaged—the buff, tan, or gray thorny devil is instantly recognizable by its squat shape, its slow way of moving, rocking backward and forward as it walks, and its armory of sharp spines. These cover the whole of its body, with the largest spines on its head and back, and much smaller ones on its legs and feet. Such protection is vital for the thorny devil because, unlike most lizards, it spends a lot of time foraging in one place on the ground, leaving it vulnerable to attack while feeding. Like the desert horned lizard (see p.416), which shares a similar way of life, the thorny devil feeds during the day, when ants are on the move, and can consume up to 2,500 insects during a single meal, catching them with its projecting tongue. Females lay 3–10 eggs in underground burrows in summer.

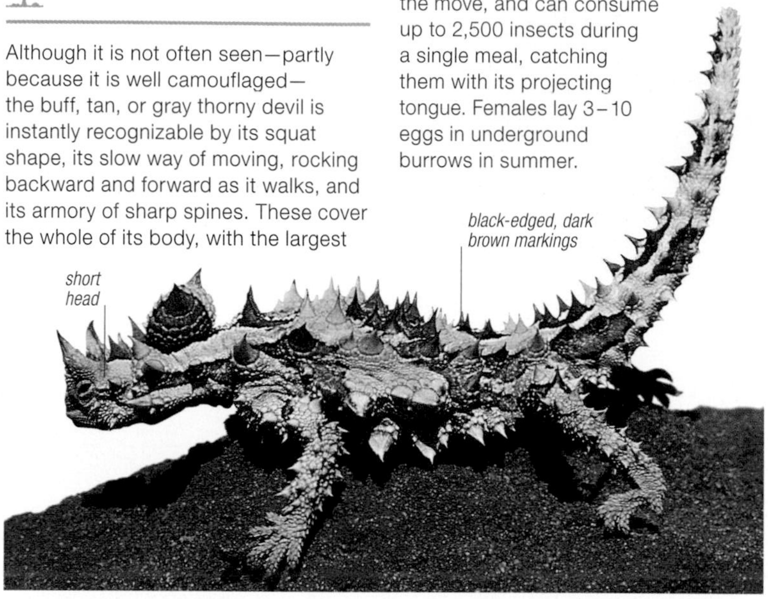

Physignathus cocincinus

Asian water dragon

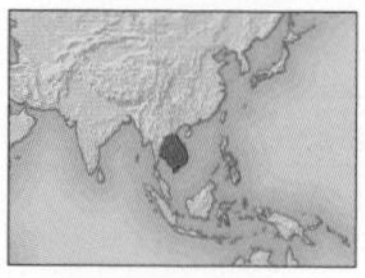

Length 2½–3¼ ft (80–100 cm)
Breeding Oviparous
Habit Terrestrial
Status Not evaluated

Location S.E. Asia (Thailand, Cambodia, and Vietnam)

The Asian water dragon is a large lizard with a crest of prominent, toothlike scales from neck to tail and a raised hump on its neck. It has a green body, but the chin and throat are white, pale yellow, or pinkish. Males become brighter in color during the breeding season. This lizard typically basks near water, into which it may flee for safety, especially when hunted by snakes and birds of prey. It hunts primarily by ambushing its prey, which comprises invertebrates and small vertebrates such as lizards; however, it also forages actively for food, and may occasionally graze upon vegetation. Females lay multiple clutches of 10–15 eggs throughout an extended breeding season, in burrows or shallow scrapes, and cover the eggs with soil and leaves.

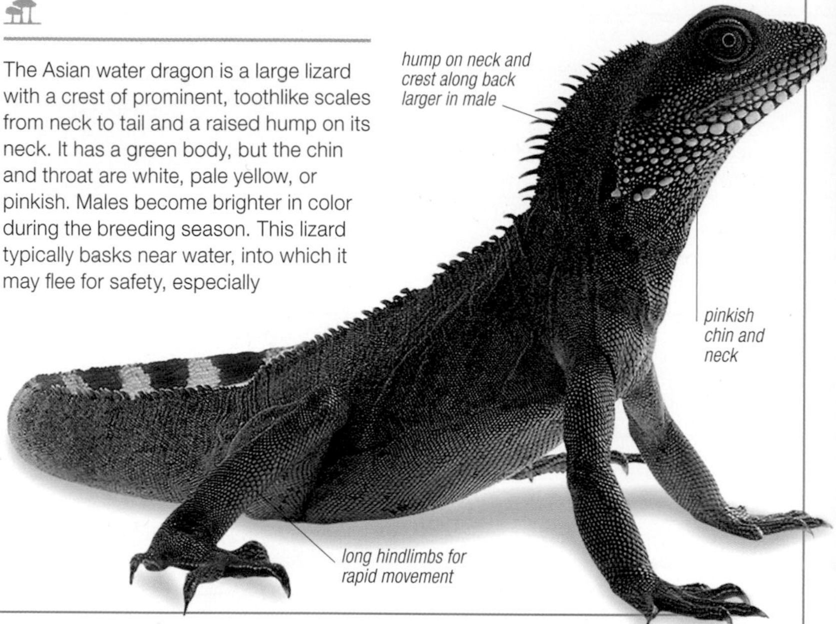

Pogona vitticeps

Central bearded dragon

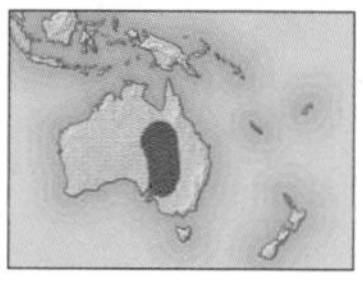

Length 12–18 in (30–45 cm)
Breeding Oviparous
Habit Semiarboreal
Status Not evaluated

Location Australia

This bulky lizard inhabits dry forest, woodland, and scrubland, resting on stumps or raised humps on the ground, and foraging for small insects and vegetation near its regular perch. If confronted by a predator—usually birds of prey, monitor lizards, or snakes—it erects its black throat frill, or "beard," and presents a wide, open mouth. The males, and even the females, are territorial. The female lays several clutches of 15–30, and occasionally more, eggs throughout the breeding season, burying them in sandy soil. The hatchlings' body markings are more distinctive than those of adults, but they do not possess a beard.

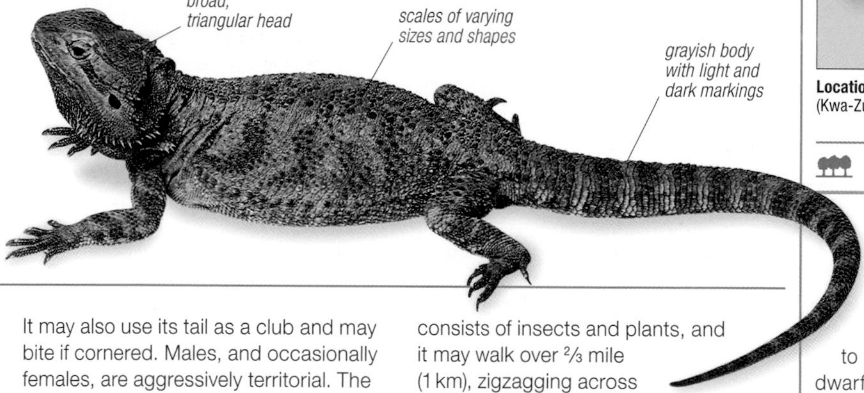

Bradypodion thamnobates

Natal Midlands dwarf chameleon

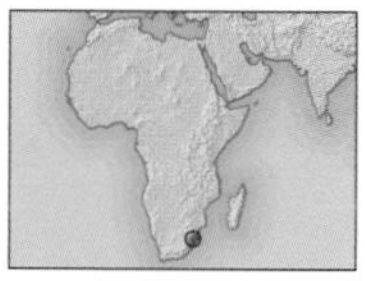

Length 6–7½ in (15–19 cm)
Breeding Viviparous
Habit Arboreal
Status Near threatened

Location Southern Africa (Kwa-Zulu Natal)

Dwarf chameleons are largely restricted to southern Africa, where they live in forest and scrub. They differ from most other chameleons in giving birth to live young. The Natal Midlands dwarf chameleon is a typical species, with a high protuberance, or casque, covering its neck. Its coloration is highly variable, but the long-tailed males, which are often green, are more colorful than the females and juveniles, which are brown or gray. Like other chameleons, it is specially adapted for life in trees, and feeds on insects, which it captures with its long, sticky tongue.

Uromastyx acanthinura

North African mastigura

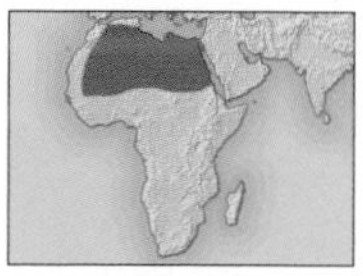

Length 12–16 in (30–40 cm)
Breeding Oviparous
Habit Terrestrial/ Burrowing
Status Not evaluated

Location N. Africa

The North African mastigura has a short, thick, armored tail that it uses to block the entrance to its burrow to deter predatory birds or mammals. Inside the burrow, it inflates its body to wedge itself in. It may also use its tail as a club and may bite if cornered. Males, and occasionally females, are aggressively territorial. The coloration of this lizard varies according to temperature: gray or yellowish brown in cold weather, when it hibernates, changing to orange, red, yellow, or green when active. It emerges only on warm days, basking and orienting its body to the sun, until its body temperature is high enough for foraging. Its diet consists of insects and plants, and it may walk over ⅔ mile (1 km), zigzagging across rocky outcrops in the desert, to find favored plant species. Females lay at least 2 clutches of 20–30 eggs a year, then bury them in side chambers off the main burrow, smoothing over the nest to disguise the site.

Rhampholeon spectrum

Western pygmy chameleon

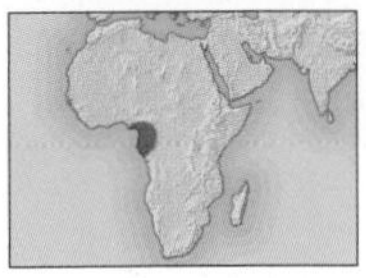

Length 2¾–4 in (7–10 cm)
Breeding Oviparous
Habit Partly arboreal
Status Least concern

Location W. to C. Africa

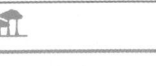

This tiny chameleon—one of the world's smallest—lives in rain forests, where it is camouflaged by its leaflike color and shape. Unlike most chameleons, its ability to change color is limited. It hunts in leaf litter and on low branches, hiding in shrubs at night.

Furcifer pardalis

Panther chameleon

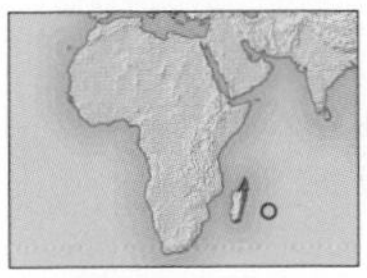

Length 16–20½ in (40–52 cm)
Breeding Oviparous
Habit Arboreal
Status Not evaluated

Location Reunion Island, E. and N. Madagascar

Madagascar and its neighboring islands have over 70 species of chameleons, which is more than in any similarly sized region of the world. The panther chameleon, particularly the adult male, is one of the most brightly colored, with a bewildering array of color schemes, including green, brick-red, turquoise, or any combination of these hues. It can change its color rapidly, but it usually retains its bright lateral stripe. Panther chameleons live in lightly wooded, scrubby habitats, and catch insects and other small animals by shooting out their sticky tongues, which can be longer than the rest of their bodies. They are usually slow-moving, but at the onset of the breeding season, males fight for territories. Each year, females lay several clutches of 12–50 eggs, burying them in moist soil.

CHANGING COLOR

Chameleons are famous for being able to change color for camouflage, but in most species, color is equally important as a means of communication. Male panther chameleons attract females by producing sudden flushes of color, while females change color when pregnant to signal that they are not receptive to mating.

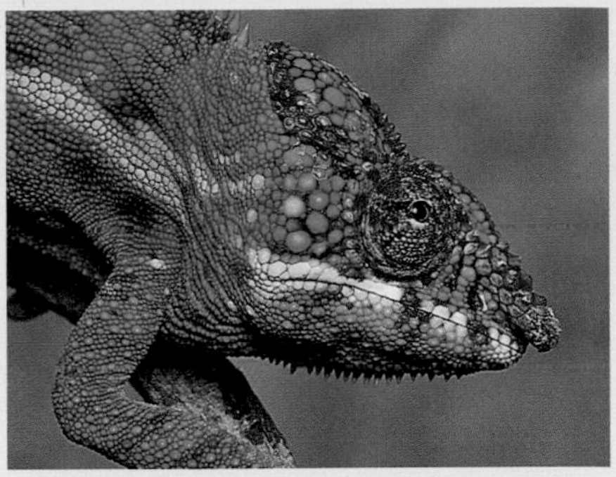

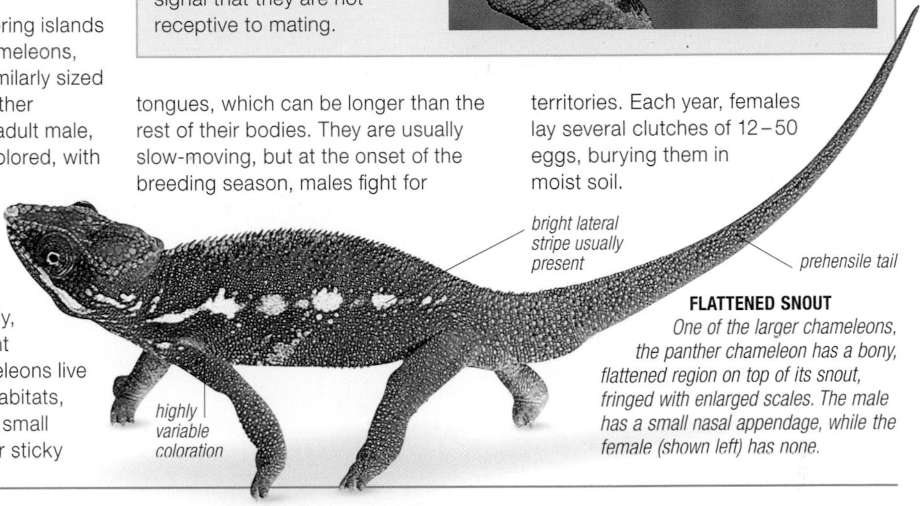

FLATTENED SNOUT
One of the larger chameleons, the panther chameleon has a bony, flattened region on top of its snout, fringed with enlarged scales. The male has a small nasal appendage, while the female (shown left) has none.

Furcifer oustaleti

Oustalet's chameleon

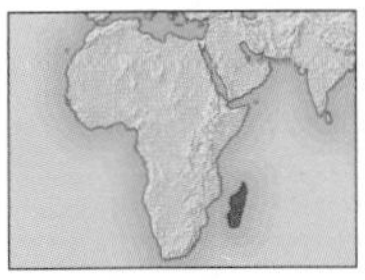

Length 20–27 in (50–68 cm)
Breeding Oviparous
Habit Arboreal
Status Not evaluated

Location Madagascar

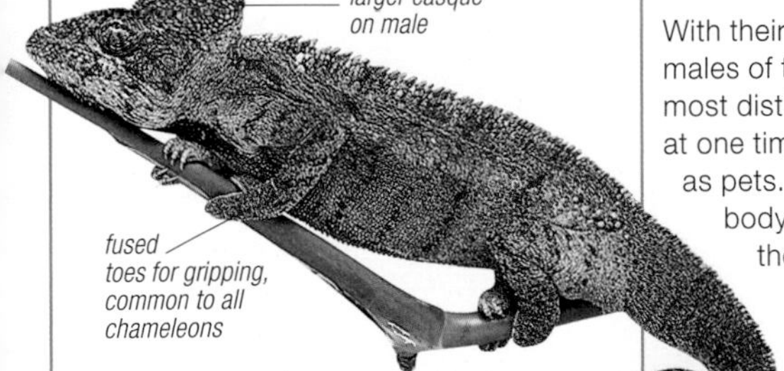

The largest of all the chameleons, this species is rather dull in color—brown with a few yellow, red, or greenish spots—and has no horns. However, it has a large, flattened casque (protuberance) over its head and neck. Unusually for a chameleon, the male is less colorful than the female, but he becomes more flushed with color during courtship. Females lay approximately 50 eggs, in soil or leaf litter. Slow and deliberate in its movements, Oustalet's chameleon carefully stalks prey such as insects and small vertebrates before capturing them with its extendible, sticky tongue. It defends itself against predatory snakes and birds by inflating its body, presenting its side to the enemy, and opening its mouth wide; it may also hiss and bite.

Trioceros jacksonii

Jackson's chameleon

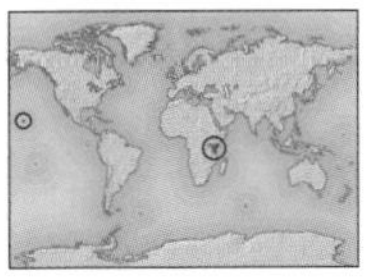

Length 8–12 in (20–30 cm)
Breeding Ovoviviparous
Habit Arboreal
Status Not evaluated

Location E. Africa, introduced in Hawaii

With their 3 forward-pointing horns, males of this species are among the most distinctive of all chameleons, and at one time were often caught and sold as pets. Both sexes have a flattened body and serrated crest along the back, with a bony casque over the neck. Their normal color is green, but specimens that live in East Africa's mountain forests are usually brighter and larger than ones that live lower down. Jackson's chameleon relies primarily on its camouflage to avoid being attacked. However, if cornered, it inflates its body and gives a menacing hiss. Females are ovoviviparous, producing 2 litters of up to 50 young each year. Newborns are light brown, but turn green at about 4 months. In males, horns start to form about 2 months later. In the wild, the lifespan of Jackson's chameleon is probably 2–3 years.

HORNS ON MALE
The male Jackson's chameleon has 3 long, pointed, bony horns on the front of his head, used in fights with other males over territory.

HORNLESS FEMALE

Male and female Jackson's chameleons are easily told apart: the female (shown above) is usually stockier than the male, and the horns are tiny or absent.

Calumma parsonii

Parson's chameleon

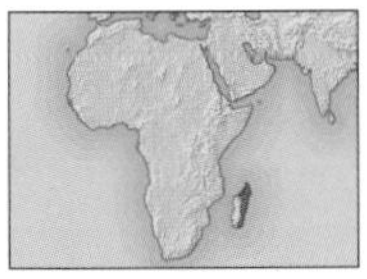

Length 20–23½ in (50–60 cm)
Breeding Oviparous
Habit Arboreal
Status Not evaluated

Location N. and E. Madagascar

This very large, deep-bodied chameleon is usually greenish, but may also be blue. The male is larger and more colorful than the female, with a larger casque (protuberance) on the top of the head and a warty, flattened horn—absent in females—on the snout. Parson's chameleon lives mainly in the trees and is slow-moving, relying on camouflage to escape detection. Defensive behavior includes inflating its body, hissing, or biting, and it may become more brightly colored. Males of this species are highly territorial, and mate with receptive females without preliminaries. The female lays 16–38 eggs, burying them in a hole dug in moist soil.

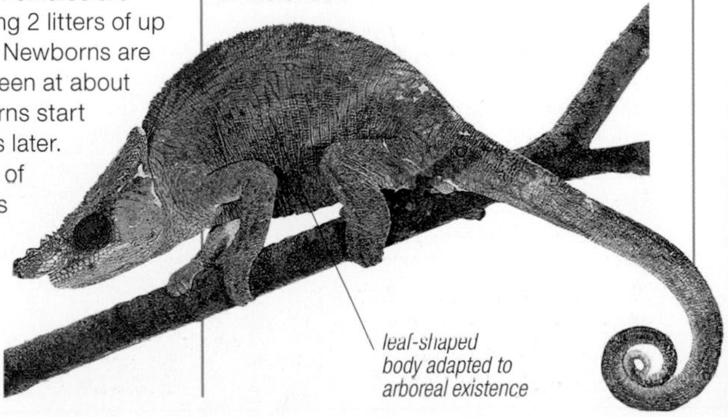

REPTILES

Corytophanes cristatus

Helmeted iguana

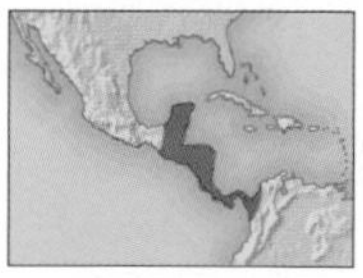

Length 12–16 in (30–40 cm)
Breeding Oviparous
Habit Arboreal
Status Not evaluated

Location S. Mexico to N. South America

The most distinctive feature of the helmeted iguana is the high, narrow casque that extends from the head to the nape of the neck. Other characteristics include long, spindly legs, a narrow body, and a long, thin tail. A crest of small, toothlike scales runs along the back and top of the tail. This iguana relies heavily on camouflage for defense, and runs away when threatened, usually by birds of prey or snakes. It feeds on insects, remaining still until they come within range, then darting out to catch them. Females lay eggs in leaf litter or soil.

KEY FEATURES

The high, narrow casque on the head distinguishes this iguana from all other lizards in tropical America; the precise function of this casque, or "helmet," is unknown. When this iguana is defending its territory, it extends the pouch under its head to ward off other animals.

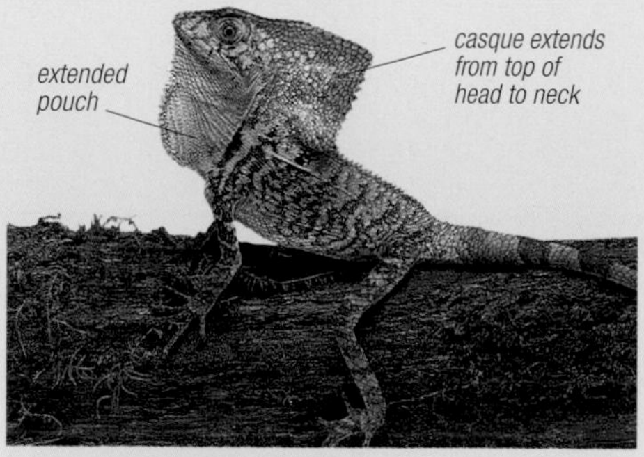

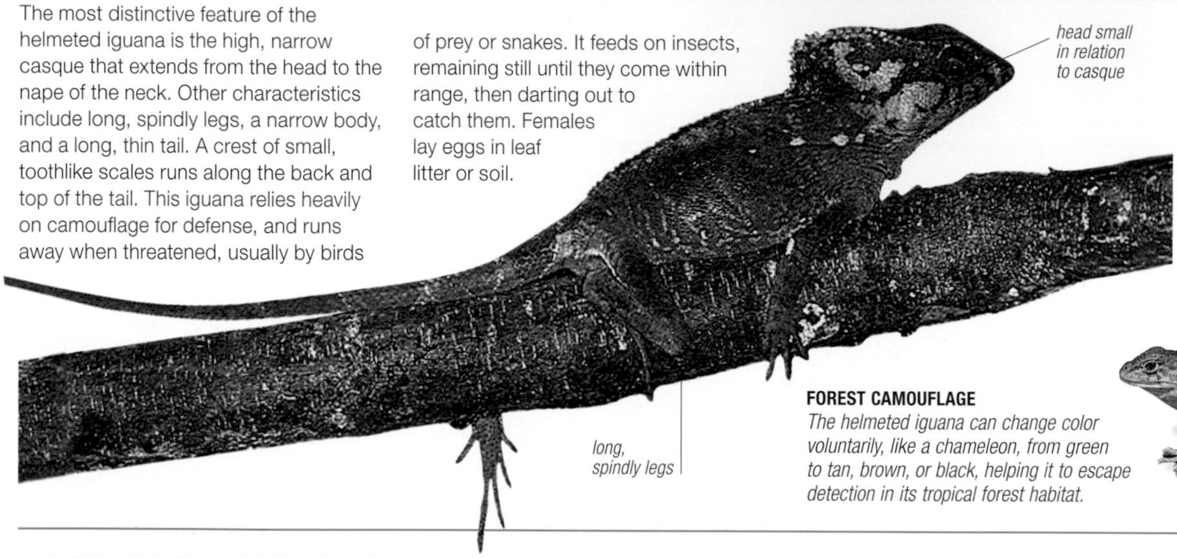

FOREST CAMOUFLAGE
The helmeted iguana can change color voluntarily, like a chameleon, from green to tan, brown, or black, helping it to escape detection in its tropical forest habitat.

Crotaphytus collaris

Collared lizard

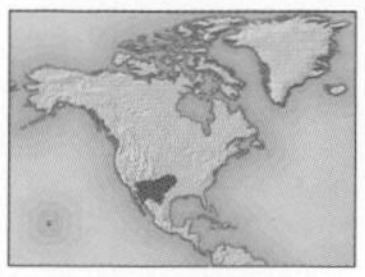

Length 8–14 in (20–35 cm)
Breeding Oviparous
Habit Terrestrial
Status Least concern

Location C. and S. USA, N. Mexico

The brown or greenish brown collared lizard is distinguished by the black and white markings around its neck. It has long hind legs that enable it to flee rapidly from danger, often to retreat to a burrow or beneath a boulder; the front limbs may leave the ground entirely when it runs fast. Active only on hot days, this lizard perches on rocks looking for insects and smaller lizards. Its strong jaws can crush even large prey, and may also be used to bite in defense. Females develop orange patches when carrying their eggs, which they lay in burrows or under rocks.

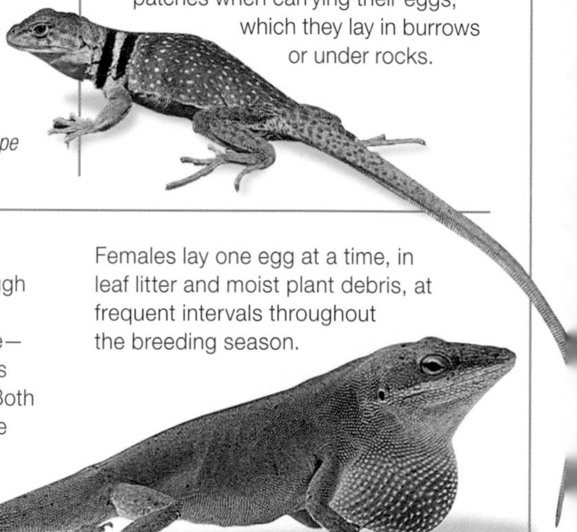

Anolis carolinensis

Green anole

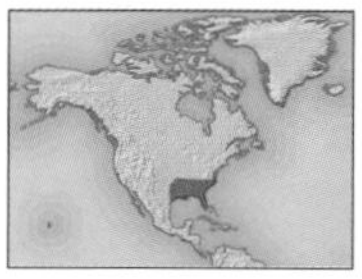

Length 4¾–8 in (12–20 cm)
Breeding Oviparous
Habit Arboreal
Status Least concern

Location S.E. USA

Anole lizards are found only in the Americas. There are over 370 species, and most—including this one—are slender animals with long tails and legs, and distinctive display fans beneath their throats. Like many of its relatives, the green anole is a tree dweller, feeding mainly on insects. It sleeps in bushes at night, but during the day often basks on vertical surfaces such as palm fronds, tree trunks, fence posts, and walls, staying in place with the help of expanded pads on its toes. Although normally bright green in color, it turns brown when it rests in shade—which explains why it is sometimes erroneously called a chameleon. Both sexes possess the pink, extendible throat fan. Males use theirs more often, signaling to rivals and potential mates. Females lay one egg at a time, in leaf litter and moist plant debris, at frequent intervals throughout the breeding season.

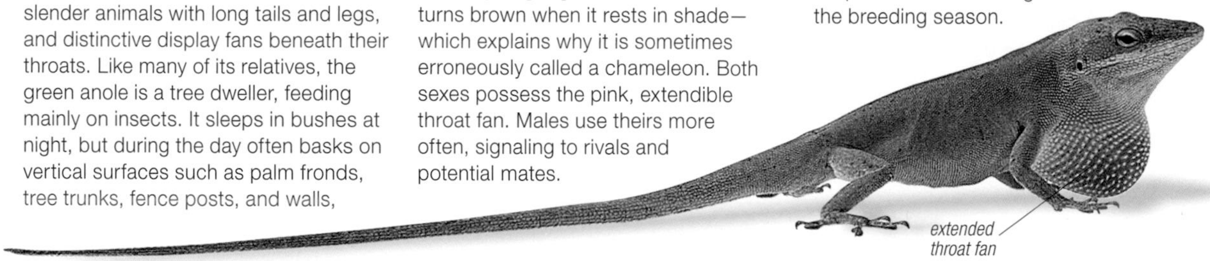

Amblyrhynchus cristatus

Marine iguana

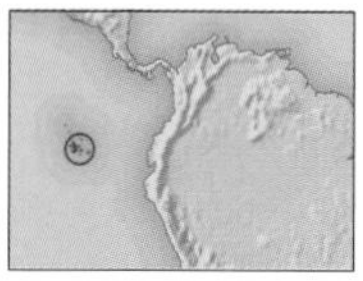

Length 20–39 in (50–100 cm)
Breeding Oviparous
Habit Semiaquatic
Status Vulnerable

Location Galapagos Islands

The marine iguana, with its large, scaly head and back, is the only lizard that forages for food in the sea. It dives for algae in the cold waters off the Galapagos Islands, and its body is specially adapted to cope with both cold and with excess salt. Large adults may dive down to 39 ft (12 m) and can stay underwater for over an hour, but under normal conditions they feed during shallow dives lasting less than 10 minutes. When not feeding, marine iguanas warm up by basking on rocks, and often several thousand can be seen together on the same stretch of shore. During the breeding season, males fight aggressively for the opportunity to mate. Suitable nesting sites are scarce, so thousands of females often nest together, laying 1–6 eggs in sandy burrows. After an incubation period of 2–3 months, the young emerge to feed in the intertidal zone and hide in rock crevices to escape attack from gulls and other seabirds.

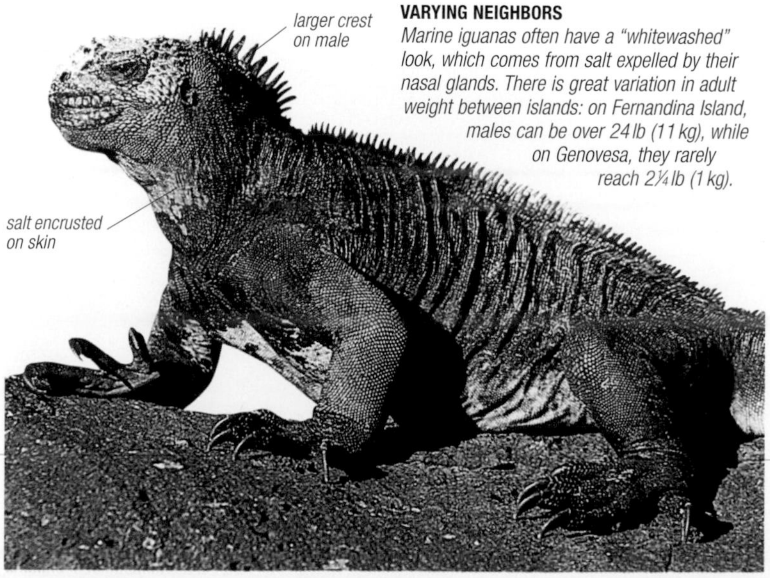

VARYING NEIGHBORS
Marine iguanas often have a "whitewashed" look, which comes from salt expelled by their nasal glands. There is great variation in adult weight between islands: on Fernandina Island, males can be over 24 lb (11 kg), while on Genovesa, they rarely reach 2¼ lb (1 kg).

Conolophus marthae

Galapagos pink land iguana

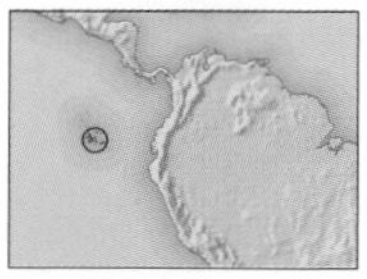

Length 39 in (100 cm)
Breeding Oviparous
Habit Terrestrial
Status Not evaluated

Location Galapagos Islands

Described in 2009, this iguana is found only on Volcan Wolf on Isabela Island and possibly fewer than 100 individuals exist. DNA evidence suggests that it diverged from other Galapagos land iguanas around 5.7 million years ago.

REPTILES

Laemanctus longipes

Casque-headed iguana

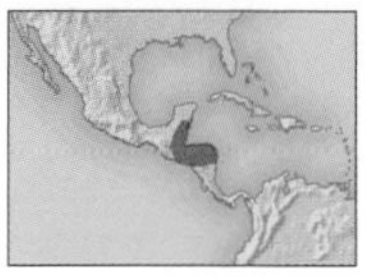

Length 16–28 in (40–70 cm)
Breeding Oviparous
Habit Arboreal
Status Not evaluated

Location S. Mexico to Nicaragua

This extremely slender iguana is green with a yellow face—coloration that provides excellent camouflage among foliage. Its long legs and claws are adapted for climbing, enabling it to escape rapidly up tree trunks when threatened by predatory birds and snakes. It sits and waits for insects to come within range, then darts out to catch them. The female lays 3–5 eggs at a time in shallow scrapes in soil or leaf litter.

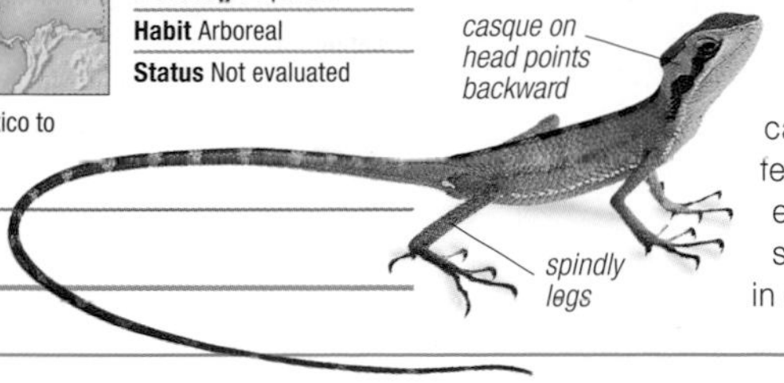

Anolis sagrei

Brown anole

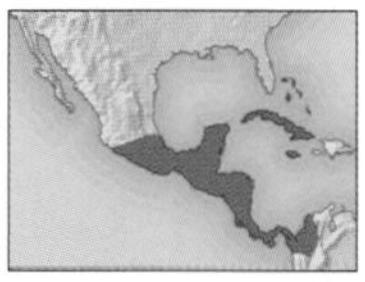

Length 6–8 in (15–20 cm)
Breeding Oviparous
Habit Terrestrial/ Arboreal
Status Not evaluated

Location S. Mexico to Central America, Caribbean

In shape, this lizard (also known as the Cuban anole) closely resembles its relative the green anole (see opposite). However, although it can change body color, its range is limited to brown or gray. Both sexes have an orange throat fan with a white border, which the males use during courtship displays. This anole is common on walls, climbing with the help of adhesive pads on its toes. It forages for flies and other insects, but it is less active in winter, and is hunted by larger lizards, snakes, and birds. Females lay one egg at a time throughout spring and summer. This is a short-lived lizard, with a lifespan of only 2–3 years. However, it is prolific and adaptable, and quick to colonize new areas, often ousting the native anole species.

Basiliscus plumifrons

Green basilisk

Length 23½–30 in (60–75 cm)
Breeding Oviparous
Habit Arboreal
Status Not evaluated

Location Central America

The green basilisk—bright green or bluish green in coloration—rests on branches and in shrubs, waiting for insects and small vertebrates before darting out to catch them. Males are very territorial; a single male may hold territory containing a harem of females with whom he mates. Females lay about 20 eggs in a season. Green basilisks can live up to 10 years in captivity.

WALKING ON WATER

The green basilisk lives in trees bordering ponds, streams, and rivers. When threatened, it drops from its perch and dives into the water to escape. This basilisk has the unique ability to run across the surface of standing water on its hind feet, thus earning itself another common name—the Jesus Christ lizard.

CHARACTERISTIC CRESTS
The green basilisk has 3 crests, located on its head, back, and tail. The back and tail crests aid swimming, helping this lizard escape predators.

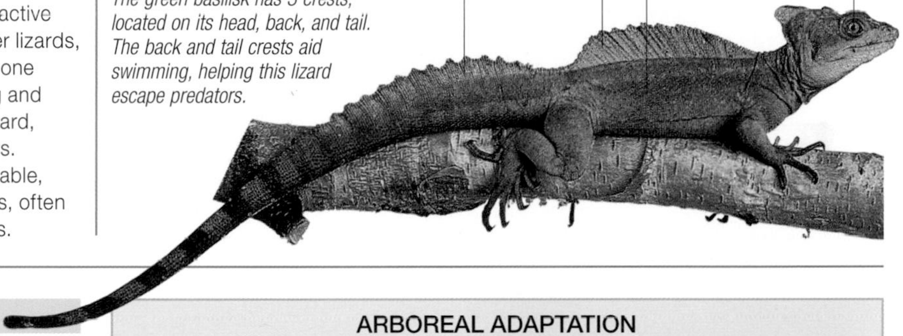

REPTILES

Cyclura cornuta

Rhinoceros iguana

Length 3¼–4 ft (1–1.2 m)
Breeding Oviparous
Habit Terrestrial
Status Vulnerable

Location Caribbean

The massive gray rhinoceros iguana is characterized by several enlarged, raised scales that resemble horns on the snout of the adult. It walks ponderously, with its head held up, foraging for leaves and fruit, but can run very fast when fleeing from danger. When threatened, it may thrash its tail and bite in defense. Males are larger than females, with larger horns and crests, and are highly territorial. Females lay 2–20 eggs in burrows and guard them aggressively. All 8 species of *Cyclura* lizards are endangered due to habitat destruction and predation from introduced mammals such as pigs, dogs, rats, cats, and mongooses.

Iguana iguana

Green iguana

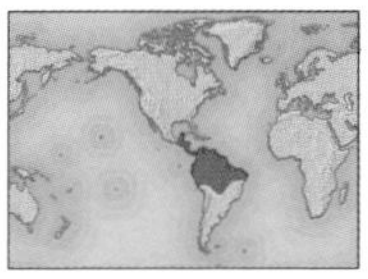

Length 3¼ ft (1 m), max 6½ ft (2 m)
Breeding Oviparous
Habit Arboreal
Status Not evaluated

Location Central America, N. South America

The green iguana is one of the largest lizards in the Americas. Generally grayish or green—although some local populations are orange—it has stout legs, a long tail, and a crest of toothlike scales down its back. Adults also have a fleshy dewlap beneath the throat, which is large in males. Juveniles eat a wider range of food than adults, including insects as well as fruit, flowers, and leaves. Green iguanas defend themselves by lashing out with their tails and claws. Females lay up to 40 eggs in an underground chamber, producing several clutches through an extended breeding season. There is no parental care: newly hatched youngsters have to fend for themselves. Despite being hunted for food and collected for the pet trade, the green iguana is still abundant in parts of its range.

ARBOREAL ADAPTATION

The grayish or green coloration of this large iguana provides excellent camouflage in the tall forest trees in which it lives. It has long legs, and its long claws are specially adapted to aid climbing. Since this lizard usually lives beside rivers, it may drop into the water from overhanging branches to escape from predatory birds or mammals. Shown here are a male (on the left)—easily identified by the large, fleshy dewlap beneath his throat, used for display—and a female green iguana. Adult males are highly territorial.

HEAVY WEIGHT
The average green iguana weighs up to 11 lb (5 kg), but an exceptional specimen may weigh up to 22 lb (10 kg). Despite their predatory appearance, adults are almost entirely herbivorous, often feeding high in trees.

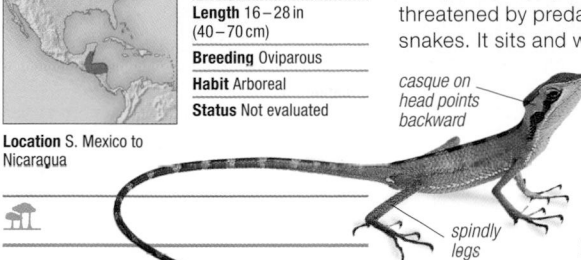

Ctenosaura similis

Black spiny-tailed iguana

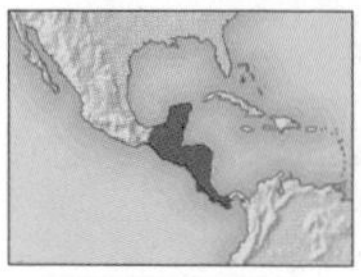

Length 28–39 in (70–100 cm)
Breeding Oviparous
Habit Terrestrial
Status Not evaluated

Location S. Mexico, Central America

The juvenile black spiny-tailed iguana is bright green, lives in trees, and is hunted by birds of prey. The adult, which has few if any predators, is large and gray, with a spikier tail than the juvenile. Adults bask on rocks or stumps, looking out for invertebrates and insects, and also forage for leaves, fruit, carrion, and scraps. Capable of great speed when running away, this iguana may also bite in defense. Females lay 20–30 eggs in burrows or among tree roots.

Sauromalus obesus

Chuckwalla

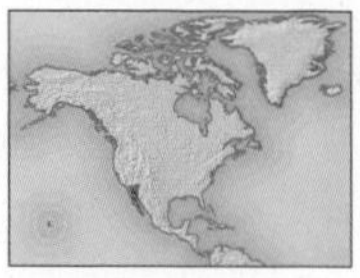

Length 11–16½ in (28–42 cm)
Breeding Oviparous
Habit Terrestrial
Status Not evaluated

Location S.W. USA, N.W. Mexico

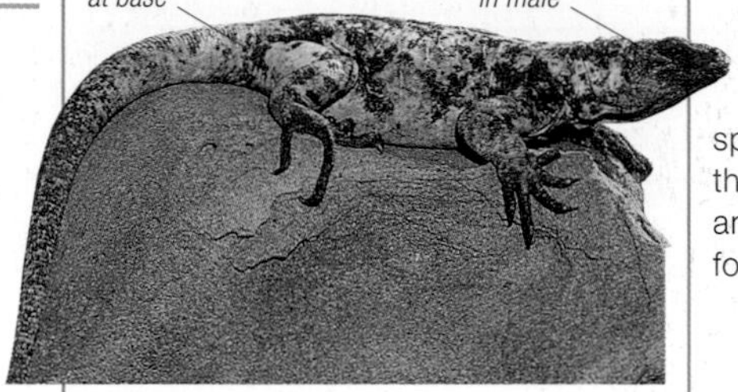

This lizard employs an unusual form of defense: it retreats into rocky crevices and jams itself in by inflating the folds of loose skin on its neck, throat, and flanks, making it hard for predatory birds and mammals to pull it out. It requires very hot conditions before it becomes active, and remains in the crevices in cold weather. In hot weather, it emerges to bask and forage for fruit, leaves, and flowers of succulent plants. Females lay 5–16 eggs, often only in alternate years.

Phrynosoma platyrhinos

Desert horned lizard

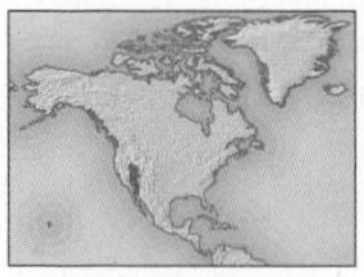

Length 3–5¼ in (7.5–13.5 cm)
Breeding Oviparous
Habit Terrestrial
Status Least concern

Location S.W. USA

The desert horned lizard is one of 16 closely related species that are scattered across dry habitats in North and Central America. Often known as horned toads, because of their spiky scales and round, squat shape, these lizards specialize in feeding on ants, and rely primarily on camouflage for protection. The desert horned lizard is relatively smooth-skinned compared to some of its relatives, but has large "horns" on its head, and smaller spikes on its back and tail. Its coloration—gray, reddish, or tan—helps to conceal it on sandy or gravelly ground, as it moves around slowly searching for food. When it locates a column of ants, it feeds voraciously, picking them up with its extendible tongue. If it is threatened with attack, by birds or carnivorous mammals for example, its first reaction is to "freeze" (it is capable of remaining motionless for long periods). If picked up, it inflates its body with air. Females nest in burrows in sand or loose earth, laying 2 clutches of up to 16 eggs each breeding season. Because these lizards have a round, flat shape, they warm up quickly in the morning sunshine. They hibernate during the winter.

Sceloporus occidentalis

Western fence lizard

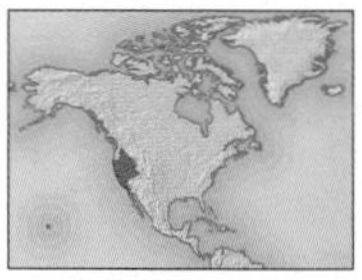

Length 6–9 in (15–23 cm)
Breeding Oviparous
Habit Terrestrial
Status Least concern

Location S.W. USA, N.W. Mexico

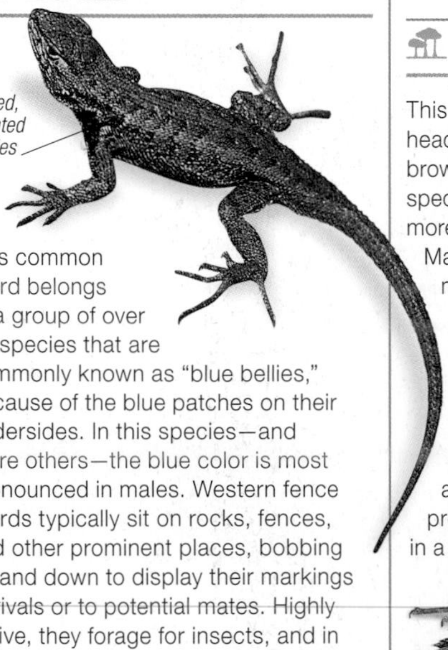

This common lizard belongs to a group of over 60 species that are commonly known as "blue bellies," because of the blue patches on their undersides. In this species—and more others—the blue color is most pronounced in males. Western fence lizards typically sit on rocks, fences, and other prominent places, bobbing up and down to display their markings to rivals or to potential mates. Highly active, they forage for insects, and in between feeding bouts they bask in the sun. Females usually lay a single clutch of 3–14 eggs.

Oplurus cuvieri

Madagascan collared iguana

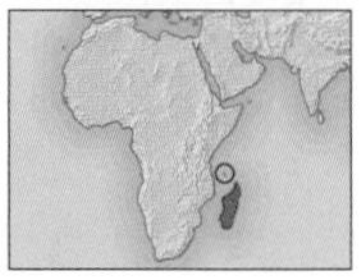

Length 12–14½ in (30–37 cm)
Breeding Oviparous
Habit Arboreal
Status Not evaluated

Location Madagascar, Grand Comoro Island

This is a thickset lizard with a chunky head and spiny tail. Its body is grayish brown, with a distinctive black collar, and speckled brown markings. Males are more intensely colored than females. The Madagascan collared iguana remains motionless for much of the time on a tree trunk, waiting for an insect to come into range before darting out to catch it. When threatened by birds of prey or snakes, it retreats into cracks and crevices in the trunk, using its tail to form a barrier between itself and the predator. Females lay 4–6 eggs in a season.

Liolaemus tenuis

Chilean swift

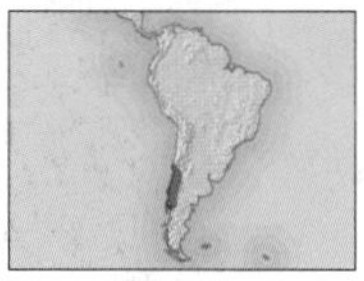

Length 7½–12 in (19–30 cm)
Breeding Oviparous
Habit Mostly arboreal
Status Not evaluated

Location S.W. South America

Also known as the thin lizard, the Chilean swift has a tail that is nearly twice as long as its body. It stalks its insect prey—mainly flies—with the tail held stiffly over its back. The large, green scales on its body overlap like roof tiles. Chilean swifts live in small groups on trees, with one male and several females occupying a territory. Females lay their eggs in a communal site, often under bark; over 400 eggs have been found in a single site. This lizard hibernates during winter.

Tropidurus hispidus

Guianan lava lizard

Length 5–7 in (13–18 cm)
Breeding Oviparous
Habit Terrestrial
Status Not evaluated

Location N. South America

The Guianan lava lizard is dark brown to black above and white below, with a black collar around the neck. Its flat shape allows it to hide in narrow cracks and crevices in rocky outcrops, an effective form of protection against predators such as birds, snakes, and larger lizards. The dark color of this lizard probably helps it to warm up quickly as it basks in the sun, usually on vertical surfaces. It forages actively for insects and spiders. Females lay 3 or more clutches of 4–6 eggs in a year.

REPTILES

Geckos and snake-lizards

PHYLUM	Chordata
CLASS	Reptilia
ORDER	Squamata
SUBORDER	Lacertilia
SUPERFAMILY	Gekkota
FAMILIES	7
SPECIES	1,369

Geckos are small, vocal, usually nocturnal lizards. Some are agile climbers, capable of finding purchase on smooth, vertical, or even overhanging surfaces, including tree trunks, rock faces, and walls and ceilings. There are also many desert species, most of which live in burrows or rock crevices, emerging at night to forage on the ground. Geckos are cosmopolitan lizards, found in most tropical and subtropical countries and on many oceanic islands. This group also includes 2 types of legless lizards that are found only in Australasia. Snake-lizards, which have scaly flaps in place of legs, live on the ground, while blind lizards have eyes that are covered with scales, and live in underground burrows.

Anatomy

Most geckos are small, slender lizards with a relatively large, flat head and large eyes with vertical pupils. Day geckos are often brightly colored, whereas the more typical nocturnal geckos are dull gray or brown to help them blend in with their surroundings. Some also have extraordinary shapes with, for example, leaflike tails to help them with camouflage.

Climbing geckos often have long claws and sometimes flattened pads at the end of each toe. These toe pads, usually absent in ground-dwelling species, are made of thousands of microscopic structures that provide adhesion, allowing them to climb vertical surfaces or even to hang upside down. Some geckos (from the family Gekkonidae) do not have functional eyelids; instead, their eyes are covered with a transparent scale, as in snakes, which they keep clean with their tongue. Others (from the family Eublephaeidae) do have eyelids and can blink—they are mostly ground dwellers from deserts, but some live in caves and one is arboreal. Geckos are unique among lizards in producing an array of sounds that are used to attract mates or to defend territory.

Snake lizards are long and slender, and may have pointed snouts. Like some geckos, they lack functional eyelids, but can be distinguished from snakes by their external ear openings and the several rows of scales along their underside.

Reproduction

Geckos' successful spread around the world may be partly due to their resilient, hard-shelled, and sticky eggs, which can become attached to uprooted trees and drifting debris. Most species lay eggs in pairs but some small geckos lay a single egg at a time. Several females may use a communal egg-laying site, which can contain several dozen incubating eggs as well as the remnants of hatched ones. A few species give birth to pairs of live young. Many demonstrate temperature-dependent sex determination: males are born at higher temperatures, above 88°F (31°C), and females at lower ones.

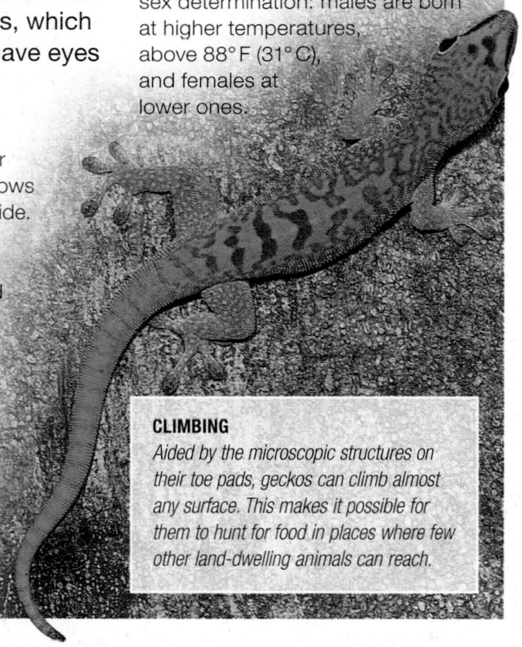

CLIMBING
Aided by the microscopic structures on their toe pads, geckos can climb almost any surface. This makes it possible for them to hunt for food in places where few other land-dwelling animals can reach.

REPTILES

Coleonyx variegatus

Western banded gecko

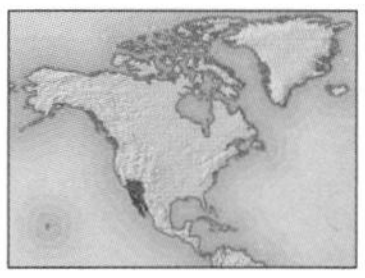

Length 4¾–6 in (12–15 cm)
Breeding Oviparous
Habit Terrestrial
Status Least concern

Location S.W. USA, N.W. Mexico

Unlike typical geckos, the western banded gecko has prominent eyes with movable lids. It appears delicate and translucent, its tiny scales giving a silky texture to its skin. A secretive, nocturnal lizard, it hides during the day in rock crevices, and actively forages at night for small insects and spiders. When captured, it squeaks and may discard its tail. Other defense tactics involve raising its body and curling its tail over its back to look bigger and, possibly, to mimic a scorpion. Females lay 2 soft-shelled eggs at a time.

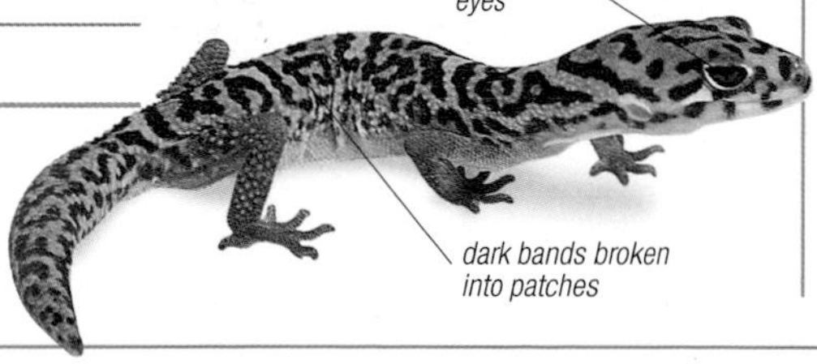

Eublepharis macularius

Common leopard gecko

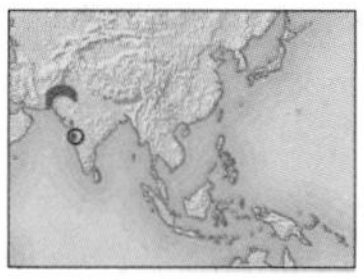

Length 8–10 in (20–25 cm)
Breeding Oviparous
Habit Burrowing
Status Not evaluated

Location S. Asia

The common leopard gecko has a thick tail that is used to store food. In times of plenty, the tail increases in size, but it shrinks during droughts, when the surplus food is metabolized. It may be discarded in defense, but not as readily as in some other lizards. This gecko forages at night. It shelters from extreme heat and cold in burrows or under rocks, hibernates in winter, and may estivate in summer. It has been known to live up to 25 years in captivity.

Hemitheconyx caudicinctus

African fat-tailed gecko

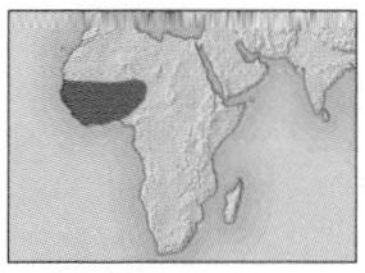

Length 6 in (15 cm), max 10 in (25 cm)
Breeding Oviparous
Habit Terrestrial
Status Not evaluated

Location W. Africa

The tail of the African fat-tailed gecko is used as a food-storage organ; when the lizard is well fed, its tail becomes thicker. It has a tan body with dark brown bands across the back and a broad, dark brown head. Males have a bulge at the base of the tail. A shy animal with a lifespan of about 25 years, it lives in dry savanna, rocky hillsides, or river banks, retreating during the day into burrows and actively foraging for insects at night. It estivates in the dry season. This gecko sometimes bites or discards its tail in defense.

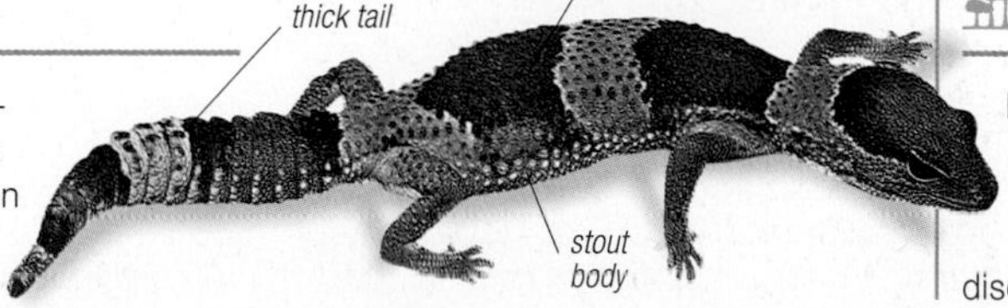

Cyrtodactylus hontreensis

Hon Tre bent-toed gecko

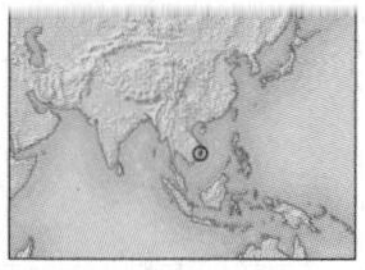

Length 5 in (13 cm)
Breeding Oviparous
Habit Cave-climbing
Status Not evaluated

Location Vietnam

Known only from the cave system of Kien Giang Biosphere Reserve on Vietnam's Hon Tre Island, this distinctively patterned gecko was discovered in 2006. It is one of 5 new gecko species endemic to the island and the second that belongs to the genus *Cyrtodactylus*. However, scientists believe there are many more species of geckos that have yet to be discovered. The Hon Tre bent-toed lizard is pale pink in color, with 5 or so darker, broad, white-edged bands that transverse the length of its slender body from its neck to the base of its tail. It also has 14 irregularly arranged, longitudinal rows of low, smooth tubercles running along the middle region of its body. It belongs to a large genus of "bow-fingered" geckos with delicate birdlike feet armed with claws—enabling it to grip the rocky surfaces of its habitat. Its habits are, so far, little known.

Gekko gecko

Tokay

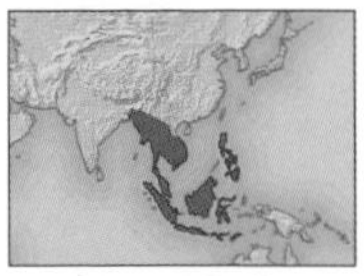

Length 7–14 in (18–36 cm)
Breeding Oviparous
Habit Mostly arboreal
Status Not evaluated

Location S.E. Asia

This gecko is named after the male's loud, explosive call that sounds like "to-kay." It is often found in dwellings, hiding during the day and emerging in the evening to hunt. When threatened by snakes and small, nocturnal mammals, it delivers a hard bite and, if grasped, it discards its tail. The male tokay, which has a larger head than the female, is highly territorial and cannibalistic, eating other tokays as well as insects and other small vertebrates. The male mates with several females. The females then attach a pair of spherical, hard-shelled eggs to a vertical surface, often behind bark, or in a crevice or a cavity in a wall.

TOE PADS

Like most climbing geckos, the tokay has expanded, adhesive toe pads that enable it to gain a firm foothold and move quickly and easily over vertical surfaces such as walls and tree trunks.

NIGHT SIGHT
The tokay is a large, blue or grayish blue nocturnal species with large, yellow eyes that enable it to see well in the dark. The vertical pupils close to a slit with 3 "pinholes" in daylight.

Gonatodes daudini

Union Island gecko

Length 1½–1¾ in (4–4.5 cm)
Breeding Oviparous
Habit Terrestrial
Status Not evaluated

Location Caribbean

Confined to the isolated dry forest of Union Island in the St. Vincent and Grenadines group in the Caribbean, this tiny, red-eyed gecko with enlarged body scales was discovered in rotten wood on Mount Taboi in 2005 at an altitude of 490 ft (150 m). Unlike other tiny members of the genus, this species spends most of its time on the ground. Historically, the species probably had a wider distribution in upland dry forest of the Lesser Antilles—a habitat now cleared through much of this range. Confined to an area of less than 4 square miles (10 square km), the Union Island gecko is now in imminent danger of extinction.

Gonatodes vittatus

Striped day gecko

Length 2¾–3 in (7–7.5 cm)
Breeding Oviparous
Habit Terrestrial
Status Not evaluated

Location N.W. South America

The male striped day gecko is tan, with a black-edged white stripe from the snout to the tail, while the female is gray-brown, with an indistinct white stripe (both are shown below). In both sexes, the eyes lie toward the top of the head. This diurnal gecko lives on the ground among leaf litter, but is also a good climber. It is quick to snap up small insects or spiders. Females often nest together in a communal site, where they lay 2 (rarely single) eggs, each approximately the size of a dried pea.

Hemidactylus frenatus

Common house gecko

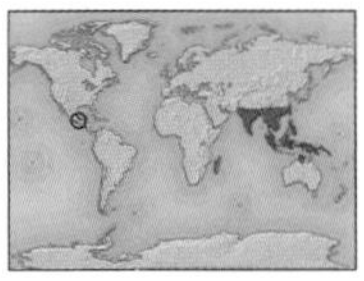

Length 4¾–6 in (12–15 cm)
Breeding Oviparous
Habit Mostly arboreal
Status Least concern

Location Tropical regions worldwide

This widespread, adaptable, gray or brown gecko mostly inhabits buildings, but may also be found on and under rocks and in tree trunks, often in large numbers. It can climb smooth, vertical surfaces and cling upside down to ceilings. Rarely seen during the day, the common house gecko hunts for insects, spiders, or smaller geckos that gather near electric lights after dark; it defends its hunting area. When captured, it squeaks and discards its tail readily to escape.

Pachydactylus rangei

Web-footed gecko

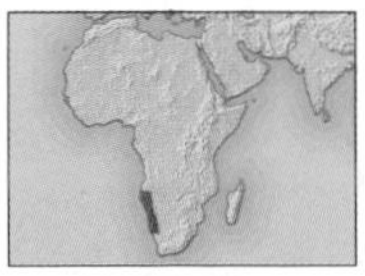

Length 4¾–5½ in (12–14 cm)
Breeding Oviparous
Habit Terrestrial
Status Not evaluated

Location W. southern Africa

This pink gecko is noted for its webbed toes, which enable it to run across desert dunes without sinking into the loose, windblown sand. It spends the day in long tunnels that it digs in the sand, emerging to forage at night. Due to lack of water in the desert, it drinks the water that condenses on its skin on foggy nights. When threatened, it raises itself on stiffened legs to appear larger.

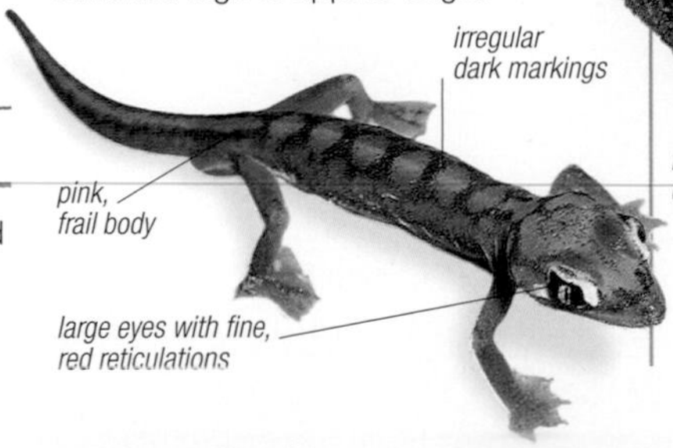

Phelsuma madagascariensis

Madagascar day gecko

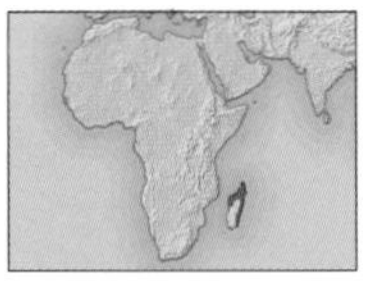

Length 9–12 in (22–30 cm)
Breeding Oviparous
Habit Arboreal
Status Not evaluated

Location N. Madagascar

The largest of the green day geckos, this thickset lizard inhabits tree trunks and walls, perching (often head down) on vertical surfaces with the aid of its toe pads. If it successfully evades attack from birds, it can live 10 years or more. Small groups occupy a single tree, and several females may lay eggs in the same site, under bark or in crevices. Males have a swelling at the base of the tail, and reproductive females have calcium deposits that are visible through the skin of their throat.

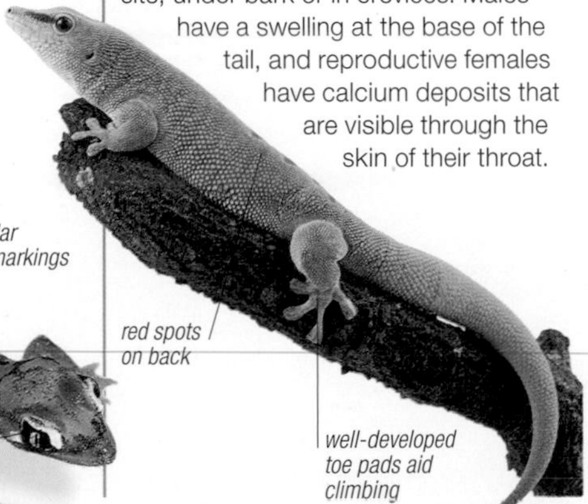

Ptenopus garrulus

Common barking gecko

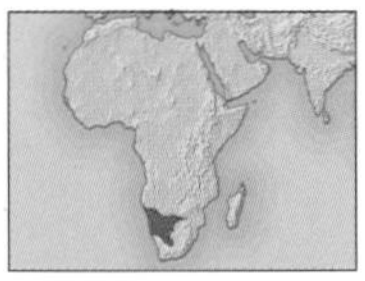

Length 2¼–4 in (6–10 cm)
Breeding Oviparous
Habit Terrestrial
Status Not evaluated

Location Southern Africa

A conspicuous feature of many deserts in southern Africa is the sound of the male common barking gecko calling from the mouth of its burrow after dusk. Both sexes have rounded heads, with a blunt snout, and are yellow or grayish yellow, with darker brown or reddish brown spots; their toes are long and strongly fringed with scales. When in danger, this gecko may "freeze," relying on camouflage to avoid detection by predators. Alternatively, it may retreat into its complex burrow system.

Ptychozoon kuhli

Kuhl's flying gecko

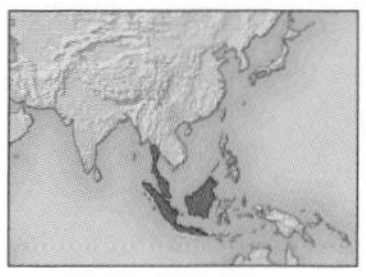

Length 7–8 in (18–20 cm)
Breeding Oviparous
Habit Arboreal
Status Not evaluated

Location S.E. Asia

This tropical forest gecko is one of several lizards that escape danger by jumping from trees and gliding through the air. When not airborne, it often rests head down on tree trunks—a posture that allows rapid takeoff. It is exceptionally well camouflaged—even by gecko standards—and feeds mainly on insects, ambushing or stalking its prey. Females lay 2 eggs each time they breed.

FLYING FEET

Unlike the flying lizard (see p.411), Kuhl's flying gecko glides mainly with its strongly webbed feet. Its fall through the air is also controlled by flaps of skin along its flanks, and its flattened, frilly-edged tail.

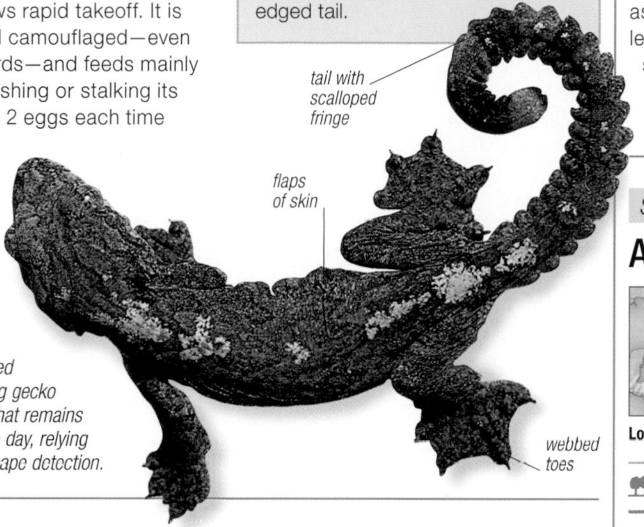

COLORED FOR CAMOUFLAGE
Mainly brown or grayish brown, with indistinct, dark, mottled markings, Kuhl's flying gecko is a nocturnal lizard that remains motionless during the day, relying on camouflage to escape detection.

Rhacodactylus leachianus

New Caledonian gecko

Length 12–14 in (30–35 cm)
Breeding Oviparous
Habit Arboreal
Status Not evaluated

Location New Caledonia

This is an extremely large gecko that can reach as much as 16 in (40 cm) in length, although it is usually smaller. Mottled brown, with a large head and eyes, and a small, thin tail, it lives on tall forest trees, rarely venturing down to the ground. As with most geckos, it eats insects and small vertebrates, but it also feeds on fruit. It uses its powerful jaws to crush food, as well as to deliver a painful bite in defense. New Caledonian geckos appear to live in small family groups that comprise a single male, one or more females, and their young. Females lay eggs in tree holes, behind bark.

Sphaerodactylus elegans

Ashy gecko

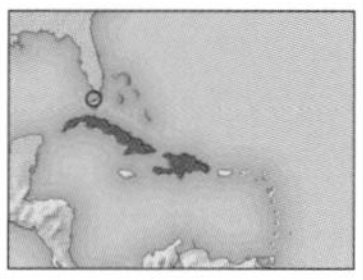

Length 2¾–3 in (7–7.5 cm)
Breeding Oviparous
Habit Terrestrial
Status Not evaluated

Location Caribbean

This tiny gecko is brown, with very fine, speckled white markings and a pointed snout. A lively but short-lived animal, it forages constantly among leaf litter and debris in woods and near houses, catching small insects with a quick dart. Although mainly diurnal, it may also be active at night, often near street lights. Females lay one egg at a time in communal nests—behind bark, among dead leaves, or in crevices.

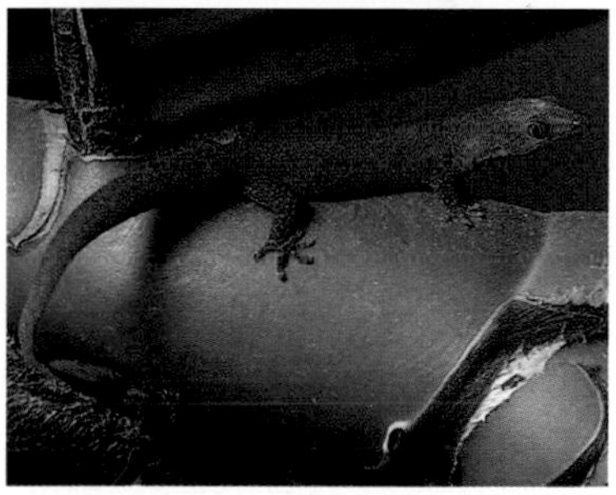

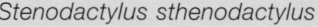

Stenodactylus sthenodactylus

Elegant sand gecko

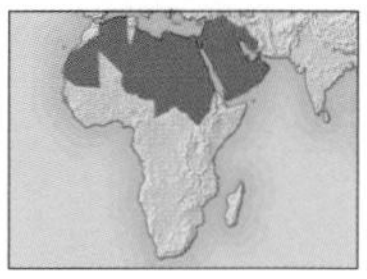

Length 3½–4¼ in (9–10.5 cm)
Breeding Oviparous
Habit Terrestrial
Status Not evaluated

Location N. and E. Africa, W. Asia

This dainty-looking lizard has long limbs, a thin tail, and a slender, beige or sand-colored body, with darker bands and small, pale spots on its back. Its eyes are large, and it has a slightly upturned snout. A skillful digger, the elegant sand gecko lives in holes in the ground or under stones, and buries its eggs in the sand. It becomes active at dusk, when it hunts for termites, ants, small moths, and small beetles and their larvae, walking with a stiff-legged gait to gain a better view of the surroundings. If threatened, usually by birds, this gecko may arch its back to appear larger.

Tarentola mauritanica

Moorish gecko

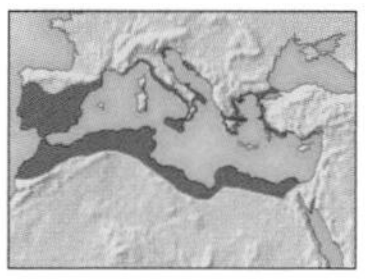

Length 4–6½ in (10–16 cm)
Breeding Oviparous
Habit Terrestrial
Status Least concern

Location S. Europe, W. Asia, N. Africa

The stocky, grayish or brown Moorish gecko often occurs in high densities in rocky scrub, drystone walls, and buildings; it occupies vertical surfaces, and seldom moves far from a crack into which it can retreat. Active mainly at night, in cooler weather it may also be active during the day, and it hibernates in winter. This gecko frequently waits in likely hunting areas, often near lights, and pounces on insects, such as moths, that come near. Females may nest communally, laying their eggs behind flaking rock or plaster, or in crevices.

Saltuarius cornutus

Northern leaf-tailed gecko

Length 6–8½ in (15–21 cm)
Breeding Oviparous
Habit Arboreal
Status Least concern

Location E. Australia

Geckos are generally noted for their camouflage, but this eastern Australian species excels at this form of defense. It has cryptic, gray mottled coloration, a flattened body that is shaped to eliminate shadows, and a remarkably leaflike tail, which can be shed if the animal is attacked (males have a paired bulge at the base of their tail). This gecko is practically invisible when resting on tree trunks and among foliage in its tropical forest habitat, and only gives itself away when it moves. Males are territorial, and females lay 2 eggs at a time in cracks behind bark. Northern leaf-tailed geckos feed mainly at night, actively foraging for insects, which make up most of their diet.

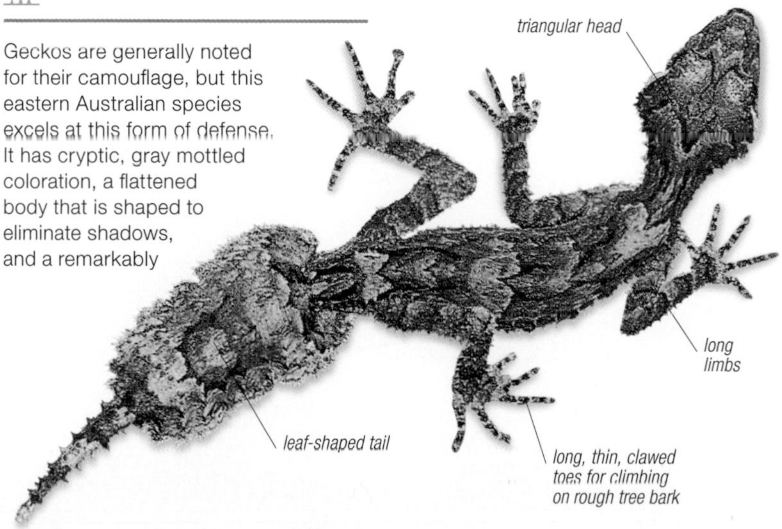

REPTILES

Skinks and relatives

PHYLUM	Chordata
CLASS	Reptilia
ORDER	Squamata
SUBORDER	Lacertilia
SUPERFAMILY	Scincomorpha
FAMILIES	7
SPECIES	2,252

Skinks and their relatives form the largest group of lizards. The skinks are mostly long and slender, and some species are without legs. Some have a method of reproduction that is unique among lizards, the females nourishing their developing young via a placenta. Skinks occur throughout the world but are most numerous in tropical and subtropical areas. Most live on the ground, often in leaf litter, but there are also arboreal and burrowing species. The other families in this group are the girdled and plated lizards, whiptails and tegus, and wall and night lizards. Like the skinks, most of these are ground-dwelling lizards.

Anatomy

Skinks are typically elongated, with a slightly flattened body, a long tail, and a small, wedge-shaped head with small eyes. Burrowing species may have permanently fused, transparent eyelids, which allow them to see while burrowing, and no ear openings.

Some skinks have small legs or no legs at all, an adaptation to burrowing or living in dense vegetation, such as turf. When attacked, skinks readily discard their tail, which is sometimes brightly colored to deflect the attention of predators from their head and body.

Girdled and plated lizards have large, rectangular scales, each of which has a raised keel. This keel may be drawn out into a point in girdled lizards, especially on the scales behind the head and on the tail, which is used to form a barricade when the lizard is pursued into a burrow or crevice, or even as a weapon. When threatened by a predator, the armadillo lizard grasps its tail in its mouth, presenting the attacker with a spiky hoop that it cannot easily attack.

Whiptails and tegus have small scales, a slender body, a pointed snout, and long limbs, although this family also includes some legless forms. Tegus are generally larger than whiptails.

Wall and sand lizards are outwardly similar to whiptails, although they are found in parts of Europe, Asia, and Africa, whereas whiptails are restricted to the Americas. Many members of both families are brightly colored, and they are all prepared to sacrifice their tail if necessary.

Reproduction

Skinks reproduce in various ways. Some species lay eggs, whereas others give birth to live young. In a few species, the developing embryo is nourished by the female via a primitive placenta. In others, the female lays eggs and guards them during incubation by coiling herself around them. She may also move the eggs if danger threatens. When the young hatch, the female licks them clean, and they may remain near her for the first few days.

Some of the relatives of skinks, notably a group of whiptails and another of wall lizards, are parthenogenetic—that is, they can reproduce without males. In these species, each young is an exact replica (clone) of the mother, and is therefore also female, producing eggs of its own as soon as it reaches maturity. Some species are entirely parthenogenetic; in others, there are some populations that include males and reproduce normally.

FEEDING
Most ground-living skinks are secretive animals that spend most of their time under the cover of leaf litter. They generally forage during daylight and take shelter under logs or stones at night. Small skinks feed on invertebrates. The larger species (including the major skink, shown here) also eat plant matter, and some may even eat rodents and birds.

Cordylus cataphractus

Armadillo lizard

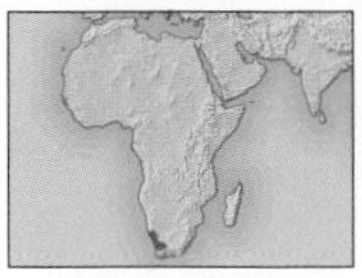

Length 6½–8½ in (16–21 cm)
Breeding Viviparous
Habit Terrestrial
Status Vulnerable

Location Southern Africa

The armadillo lizard is covered with stout, defensive spines on its neck and tail and squarish scales on its back. When threatened, it grasps its tail in its mouth and forms a ring with its body, presenting the spines to deter an approaching predator and making it awkward to attack. Armadillo lizards occur in scrub and rocky outcrops, hiding in large cracks, and hibernating in winter. They live in family groups—the female gives birth to one or 2 large young, and may even feed them—a characteristic that is unusual among lizards, which rarely demonstrate parental care. Insects and spiders make up most of the diet of the armadillo lizard and it, in turn, is hunted by a variety of predators, including birds of prey. It can live for up to 25 years in captivity, slightly more in some exceptional cases.

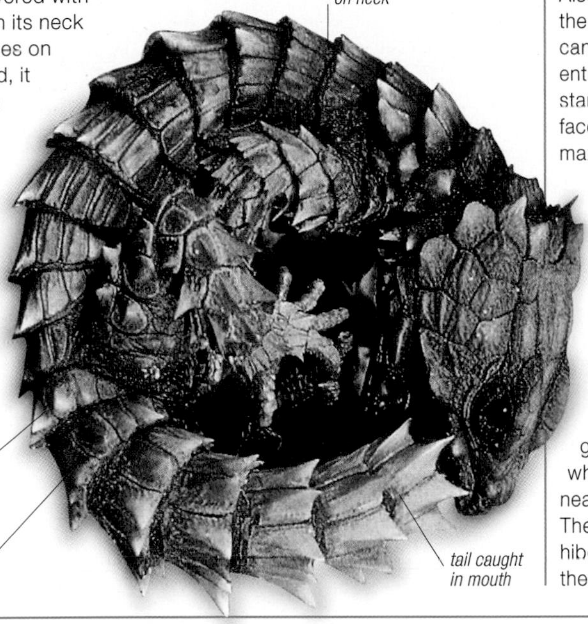

Cordylus giganteus

Sungazer

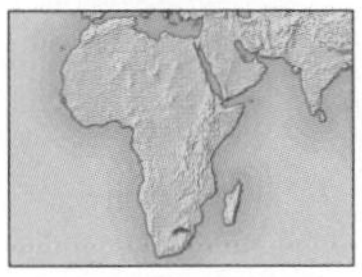

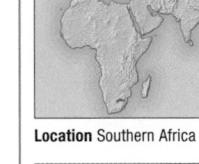

Length 11–15½ in (28–39 cm)
Breeding Viviparous
Habit Terrestrial
Status Vulnerable

Location Southern Africa

Also known as the giant girdled lizard, the yellowish to dark brown sungazer can often be seen basking at the entrance to its burrow, apparently staring at the sun. Burrows usually face north or northwest, providing maximum sunshine throughout the day. This lizard possesses an armored tail, ringed with pointed scales and resembling a pinecone, which it can use to block the entrance to its burrow. If grasped, it wedges the long spines on its head and neck into the burrow roof, making its body difficult to extract. Its diet consists of insects and small vertebrates. Females give birth to one or 2 young, which remain near the adults. The sungazer hibernates during the winter months.

Cordylus rhodesianus

Zimbabwe girdled lizard

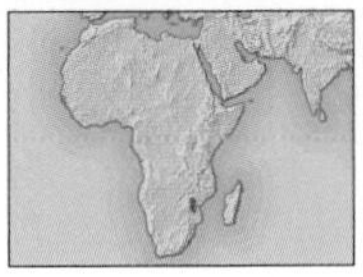

Length 4¾–6¾ in (12–17 cm)
Breeding Viviparous
Habit Terrestrial
Status Not evaluated

Location Southern Africa

This olive-brown lizard is often found in kopjes—the piles of eroded rock that are scattered across southern Africa's grasslands. Its flattened body enables it to hide in narrow cracks or under rocks when threatened, and it uses its tail, which is covered in rings of pointed scales, to block the cracks and deter predators such as snakes, birds, and mammals. The Zimbabwe girdled lizard lives in small family groups—with the young remaining near the parents—and hibernates in winter.

Gerrhosaurus major

Rough-scaled plated lizard

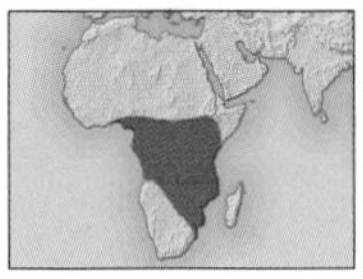

Length 16–19 in (40–48 cm)
Breeding Oviparous
Habit Terrestrial
Status Not evaluated

Location C., E., and southern Africa

The body of the rough-scaled plated lizard is covered with rectangular plates, arranged in rows, that help the reptile wedge itself into cracks in rocky outcrops for protection against birds of prey, snakes, and small mammals. This lizard is light to medium brown, but the male develops a pinkish throat in the breeding season. An active forager, its diet comprises fruit, flowers, invertebrates, and small vertebrates, including smaller lizards. The male is highly territorial, particularly during the breeding season, when he fights with other males. The female lays 2–4 relatively large, oval eggs in rock crevices, in moist soil, and under logs.

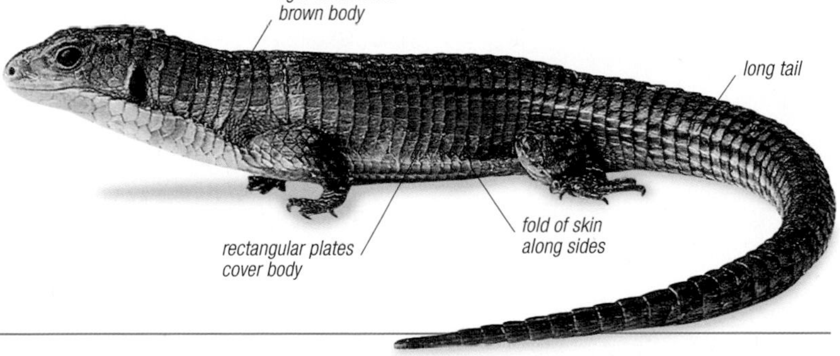

Lacerta agilis

Sand lizard

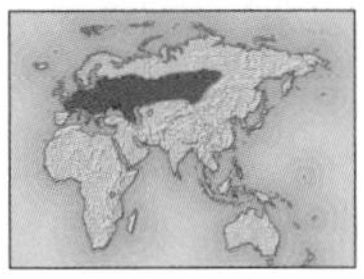

Length 7–9 in (18–22 cm)
Breeding Oviparous
Habit Terrestrial
Status Least concern

Location Europe to C. Asia

This stocky lizard belongs to a group of species that are common in Europe and Asia, sometimes extending north of the Arctic Circle. In Great Britain, it is found in pastures and sand dunes, but farther south, it lives in a much wider range of habitats, including gardens. It has a blunt head and relatively short legs, and its color is very variable, although males are at their brightest during the breeding season. Sand lizards feed on small insects and spiders, and rapidly shed their tails if they are attacked, usually by snakes, birds, or mammals, especially domestic cats. Females lay clutches of 3–14 eggs in the spring and summer, and adults hibernate in winter.

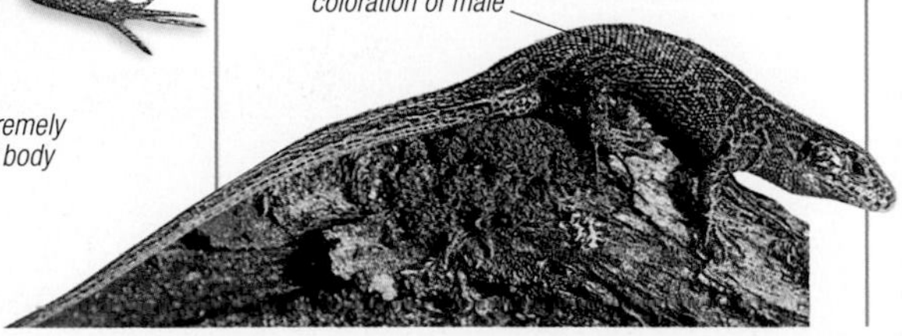

Platysaurus broadleyi

Broadley's flat lizard

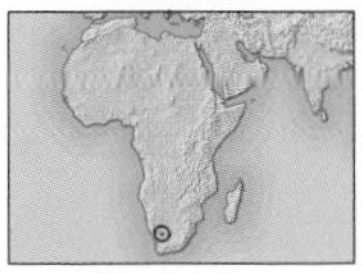

Length 6–8 in (15–20 cm)
Breeding Oviparous
Habit Terrestrial
Status Not evaluated

Location Southern Africa (Augrabies Falls)

The extremely flattened head, body, and tail of this lizard enable it to squeeze into very narrow rock crevices to escape from predatory birds of prey, particularly the rock kestrel. Once wedged in, it is almost impossible to extract. This lizard spends most of the day hunting for insects, especially the small black flies that swarm near waterfalls, darting rapidly to catch the flies when they land, or leaping upward to catch them in flight. It also eats ripe berries. The male has a bluish head, green back, yellow or orange front limbs, and an orange-tan tail. The female and juveniles have dark brown backs with 3 cream stripes and a pale yellow tail. The female lays 2 clutches of 2 eggs each summer, in deep cracks between large boulders and under exfoliating flakes of rock, especially where leaves or debris have accumulated.

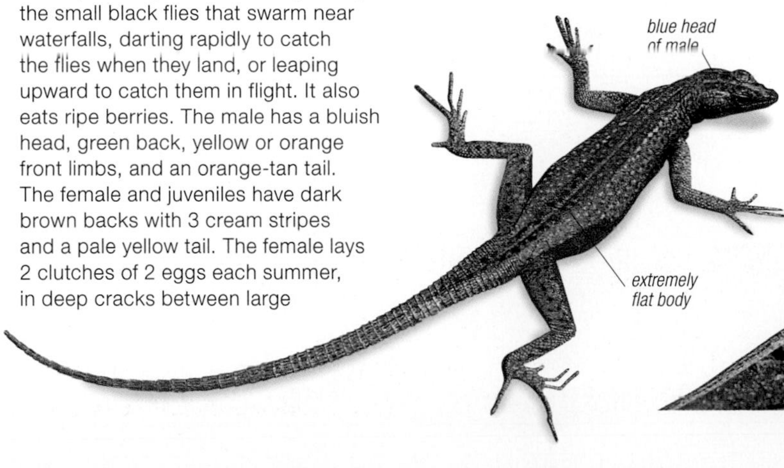

Lacerta schreiberi

Schreiber's green lizard

Length 14 in (36 cm)
Breeding Oviparous
Habit Mostly terrestrial
Status Near threatened

Location S.W. Europe

Schreiber's green lizard is typical of the numerous localized species of "wall lizards" that are found in different parts of Europe. The male is green, with scattered, small black spots, and a blue throat or head (sometimes both) in the breeding season. The female is brown or bright green, with large black markings and a brown head. The young are green with white or yellow, black-edged bars on their flanks. This lizard can be found basking in the open in fields, moorland, or hillsides, often near streams, occupying a position with a good, all-around view. It mostly sits and waits for insects to come within range, but also forages in cracks. When disturbed, it retreats into a burrow or dives into water, hiding under stones on the streambed. If attacked, it can easily discard its tail. This lizard has a lifespan of over 10 years in captivity. The female lays 6–12 eggs at a time.

Zootoca vivipara

Viviparous lizard

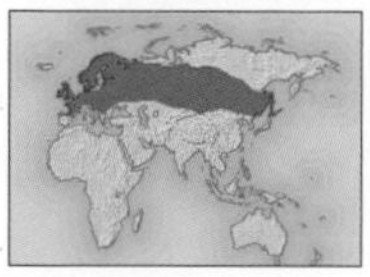

Length 4–4¾ in (10–12 cm)
Breeding Viviparous
Habit Terrestrial
Status Least concern

Location Europe to C. and E. Asia (including Japan)

Also known as the common lizard, this species has one of the largest continuous ranges of any terrestrial reptile, reaching well into the Arctic. It is the only member of its genus that bears live young—an adaptation that helps it survive in regions that are too cold for eggs to develop. In the far north of its range, it can hibernate for up to 8 months at a time and may breed only every second or third summer; however, it makes up for this by having a relatively long lifespan. Viviparous lizards live on the ground, usually in dense vegetation, and feed mainly on insects and spiders. They often bask on a sunny bank, flattening their body and spreading out their legs.

BREAKING OUT

Young viviparous lizards are born in an egg membrane from which they escape almost immediately. They are about 1½ in (4 cm) long and black. In the southern parts of their range—in the Pyrenees—viviparous lizards sometimes lay eggs.

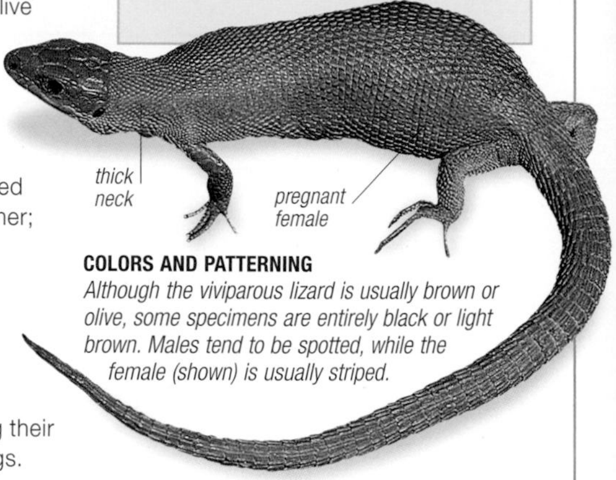

COLORS AND PATTERNING
Although the viviparous lizard is usually brown or olive, some specimens are entirely black or light brown. Males tend to be spotted, while the female (shown) is usually striped.

Timon lepidus

Eyed lizard

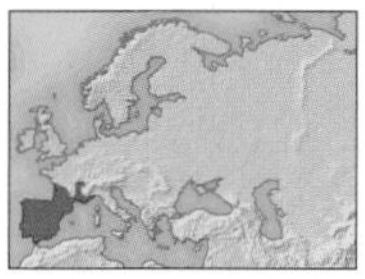

Length 16–32 in (40–80 cm)
Breeding Oviparous
Habit Terrestrial
Status Near threatened

Location S.W. Europe

Also known as the ocellated or jeweled lizard, this thickset green animal, with blue eyespots running along its flanks, is Europe's largest lizard. It inhabits hillsides, fields, and clearings, foraging for large insects, birds' eggs, nestlings, other lizards, and small mammals. Although it hibernates in winter, it may emerge on warm days. If grasped, it bites, discards its tail, and retreats into its burrow. Females lay a single clutch of 6–16 eggs a year.

Podarcis lilfordi

Lilford's lizard

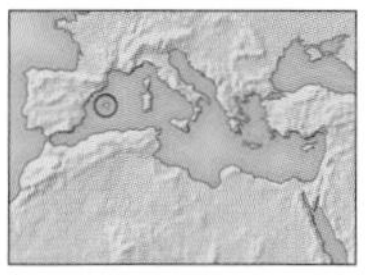

Length 7–9 in (18–22 cm)
Breeding Oviparous
Habit Terrestrial
Status Endangered

Location Balearic Islands

This tough, adaptable lizard can survive in remarkably hostile environments. Although it has few predators (there are no snakes on many of the islands it inhabits), vegetation is sparse and food is scarce, posing the greatest threat to this lizard's survival. It is an active forager, scrambling over rocks and shrubs in a relentless search for food—insects, fishes dropped by seabirds, and food discarded by tourists, as well as flowers and leaves. Some populations are entirely black, while others are brown or green. The female lays about 4 eggs in crevices filled with soil or leaves, or under shrubs.

Podarcis pityusensis

Ibiza wall lizard

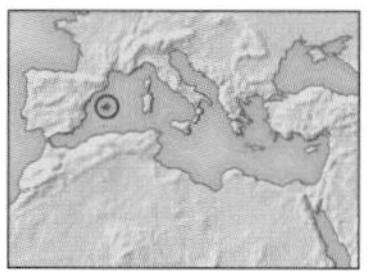

Length 6–8½ in (15–21 cm)
Breeding Oviparous
Habit Terrestrial
Status Near threatened

Location Balearic Islands

With its slender body and long toes, the Ibiza wall lizard is an agile climber. It is most often seen basking on walls, rocky slopes, and fallen trees—sometimes in large numbers—but it invariably scuttles away rapidly if disturbed. Males are generally blue with black markings, while females often have dark stripes, but the colors and patterns of both sexes vary across the species' range. Wall lizards feed on insects, and they lay their eggs in burrows or under stones. The Ibiza wall lizard can live for up to 6 years.

Chalcides ocellatus

Eyed skink

Length 12 in (30 cm)
Breeding Oviparous
Habit Burrowing
Status Not evaluated

Location S. Europe, N. and N.E. Africa, W. Asia

Also known as the ocellated skink, the eyed skink has a stocky, light brown or yellowish brown body with small, white eyespots surrounded by black. Its smooth, glossy scales and small limbs are adaptations to a burrowing lifestyle. Although this lizard is mainly active during the day, it may also be active at dusk in very hot weather. It is often found basking near old ruins or drystone walls near olive groves, never straying far from cover. If threatened by birds, snakes, and cats, it darts into its burrow or into crevices between rocks; if attacked, it may shed its tail. An active forager, the eyed skink searches under stones for small insects and spiders. It lives in loose colonies, and the female lays 3–10 eggs.

Corucia zebrata

Solomon Islands tree skink

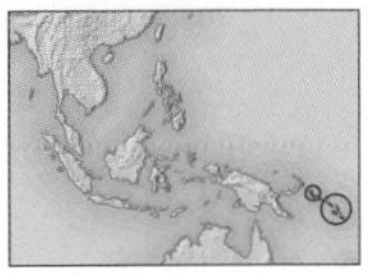

Length Up to 30 in (75 cm)
Breeding Ovoviviparous
Habit Arboreal
Status Not evaluated

Location Solomon Islands

This olive-green or grayish green skink is a highly unusual lizard in many ways. It is very large, with a massive, wedge-shaped head and a long, prehensile tail. Unlike other skinks, the Solomon Islands tree skink is entirely herbivorous, feeding exclusively on leaves and fruit in its forest habitat. Active at night, it shelters in tree holes during the day. If threatened or provoked, it will hiss while raising itself up, and may even bite. This skink is a long-lived species. Although little is known about its reproductive and social behavior, it is likely to be monogamous and appears to live in small family groups or loose colonies. Males and, to some extent, females are highly territorial. A female gives birth to one, occasionally 2, huge young—about one-third of the size of the adult—which stays with its very protective mother for several weeks or even months. The low reproductive potential of this skink makes it vulnerable, although it is not currently a protected species.

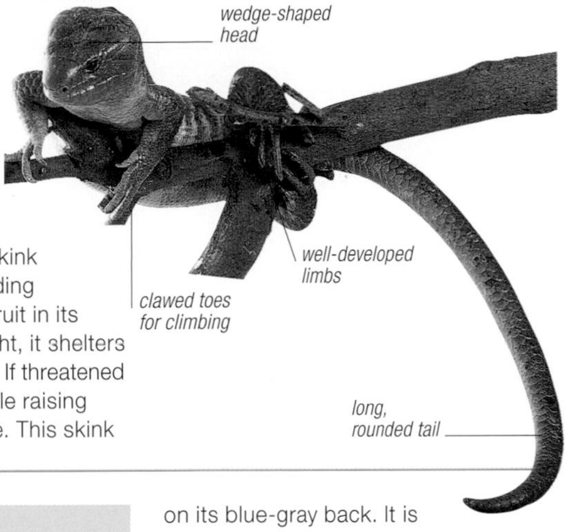

Eumeces schneideri

Berber skink

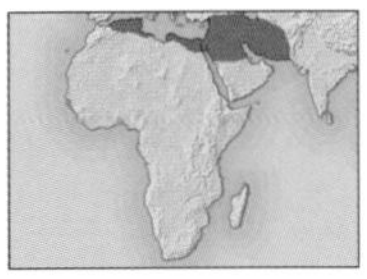

Length 16–18 in (40–45 cm)
Breeding Oviparous
Habit Terrestrial/Burrowing
Status Not evaluated

Location N. Africa to W. Asia

The Berber skink is the largest member of its genus, which is found in North Africa, the Middle East, Asia, and the Americas. Bulky and with a squarish cross section, it has distinctive orange scales on its blue-gray back. It is often found in cultivated land and semidesert, where it digs burrows at the base of shrubs. When attacked, it retreats into its burrow, or may jump into water. Active and quick-moving, the Berber skink forages for insects, spiders, snails, and smaller lizards. It shelters from the sun in the middle of the day and may not emerge on cold winter days. Females lay 3–20 eggs, coiling around them until they hatch.

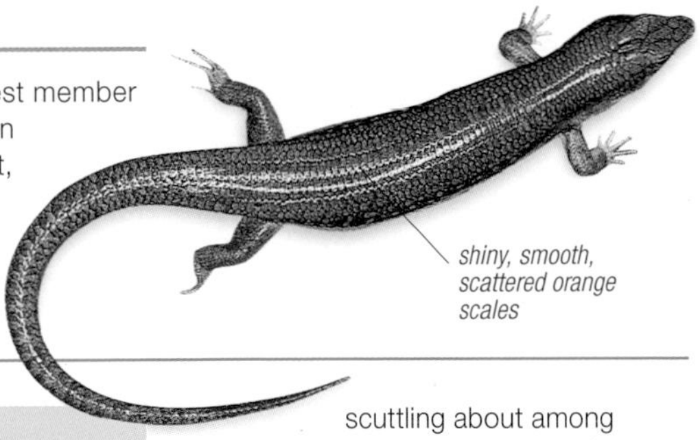

Plestiodon lagunensis

Pink-tailed skink

Length 6½–8 in (16–20 cm)
Breeding Oviparous
Habit Terrestrial
Status Least concern

Location Mexico (Baja California)

Also known as the San Lucas skink, this species is characterized by a long, pink tail that is bright in juveniles but faded in adults. The reason for this coloration is unclear, but it may help to divert predators toward the tail—which can be discarded—instead of toward more vulnerable parts of the body. The pink-tailed skink is a secretive animal, scuttling about among leaf litter or vegetation—where it searches for insects and spiders—or hiding under stones. It is active year-round, although less so during Baja California's brief winters. The female lays 2–6 eggs in tunnels in loose soil. She stays with them until they hatch, and may even move them if they are threatened by unfavorable conditions, such as floods.

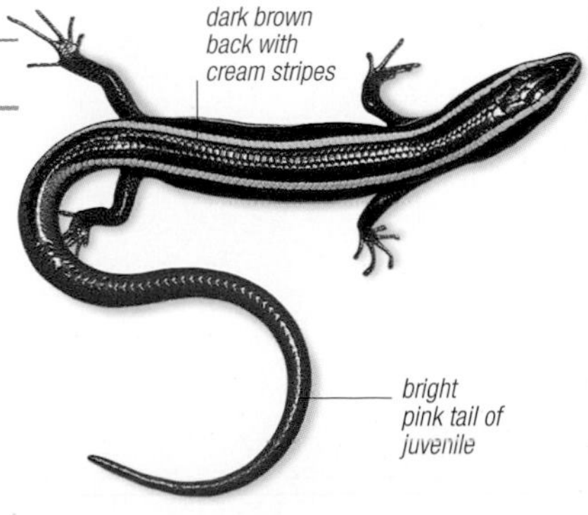

Lamprolepis smaragdina

Emerald tree skink

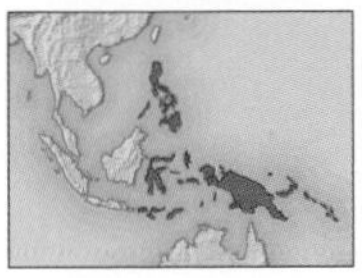

Length 7–10 in (18–25 cm)
Breeding Oviparous
Habit Arboreal
Status Not evaluated

Location S.E. Asia, New Guinea, Pacific islands

The bright green emerald tree skink has a stocky body with a long, tapering tail, a pointed snout, and smooth, shiny scales. It lives on the trunks and among the foliage of large forest trees, and actively forages for insects, which it catches in a rapid, darting movement. If attacked by birds of prey or snakes—its main predators—this lizard will shed its tail in order to escape from its aggressor. Males are territorial and females lay 9–14 eggs.

Bellatorias frerei

Major skink

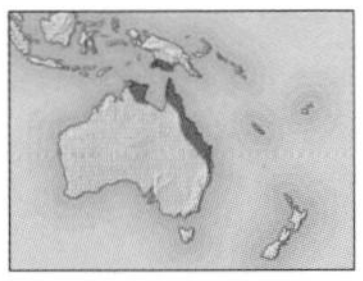

Length 23½–28 in (60–70 cm)
Breeding Viviparous
Habit Terrestrial
Status Not evaluated

Location S. New Guinea, N. and E. Australia

The major skink is one of 20 closely related species that are found in Australia and New Guinea. Like most of its relatives, it has a chunky appearance, with a body that is squarish in cross section, a thick neck, short head, and muscular limbs. Its coloration is dark brown, becoming darker on the flanks, with a white, yellow, or orange underside. The body scales are smooth and glossy, and there are fine lines running down its back. The major skink can be seen basking in the sun or foraging for insects and small vertebrates during the day around the bases of large, buttressed trees, logs, or near rocky outcrops, never too far from cover. Females give birth to up to 6 young, which have scattered white dots on their flanks—these spots disappear as the lizard matures. Within Australia, the major skink is commonly found in forests in northeast New South Wales, islands of the Torres Straits, and Arnhem Land (Northern Territory).

Eutropis longicaudata

Long-tailed skink

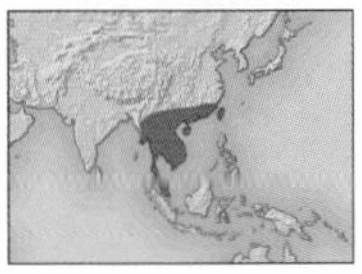

Length 12–14 in (30–35 cm)
Breeding Viviparous
Habit Terrestrial
Status Not evaluated

Location S.E. Asia

The most distinctive feature of this slender skink is its extremely long tail, which can be up to twice the length of the body and head combined. The body scales are shiny and brown, and the flanks are darker brown or black. This lizard lives in clearings and gardens, often near water, and also in coastal scrub. An active forager, it frequently pokes its snout into leaf litter and debris where insects and spiders may be found. Little is known about its reproductive behavior, but most related species give birth to 5–10 young. The genus *Eutropis*, to which it belongs, contains egg-laying as well as live-bearing species.

REPTILES

Trachylepis striata

African striped skink

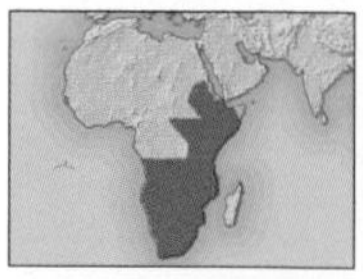

Length 7–10 in (18–25 cm)
Breeding Viviparous
Habit Mostly terrestrial
Status Not evaluated

Location E. and southern Africa

The streamlined body of the African striped skink is highly variable in color, but is often brown, with paler stripes running along the back. It has a pointed head, smooth, shiny scales, and transparent "windows" on its lower eyelids. Males of some populations have an orange head and a yellowish orange throat. This skink lives in a wide variety of habitats, from mangrove swamps to arid areas, and is sometimes found in high densities. Very active during the day, it climbs into shrubs and trees, rocky outcrops, and buildings in search of prey such as small insects and spiders, stopping frequently to bask. When attacked, it can shed its tail. Females give birth to 3–9 young, each about 2¼–3¼ in (6–8 cm) long, which take 15–18 months to mature.

Lepidothryis fernandi

African fire skink

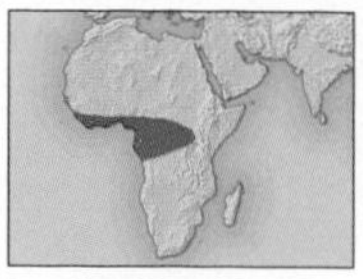

Length 9–14½ in (22–37 cm)
Breeding Oviparous
Habit Terrestrial
Status Not evaluated

Location W. and C. Africa

Also called the fire-sided skink, and formerly known as *Riopa fernandi*, the African fire skink has a short head and stocky body with a brown back, red and black bands on the flanks and face, and blue and black bands on its tail. It lives among leaf litter, and is active at twilight, when it forages for insects and spiders. Like many other lizards, this skink can shed its tail when attacked by a snake or other predator. The female lays up to 8 eggs in leaf litter or rotting wood. Since the African fire skink is rarely seen in the wild, most information comes from observations of captive animals.

Tiliqua rugosa

Stump-tailed skink

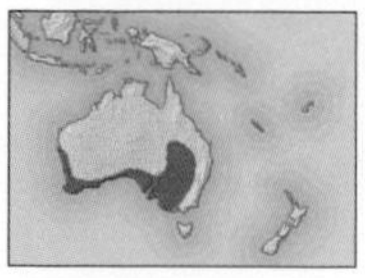

Length 12–14 in (30–35 cm)
Breeding Viviparous
Habit Terrestrial
Status Not evaluated

Location S. and W. Australia

Also known as the shingleback, this skink has a short, stumpy tail that looks like a pinecone, and enlarged, raised dorsal scales. The thick tail acts as a food-storage organ, while the heavily armored body provides protection. Coloration is variable, ranging from dark brown to cream, sometimes with light or dark spots. This slow-moving, omnivorous lizard rests under logs or in leaf litter, and is an active forager. It is temporarily monogamous: pairs remain together for about 8 weeks in the breeding season, then separate, often reuniting the following year. Females give birth to 1–3 relatively large young.

Tiliqua gigas

Giant blue-tongued skink

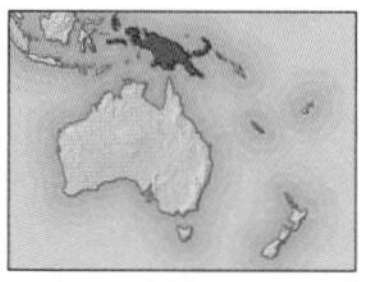

Length 20–24 in (50–62 cm)
Breeding Viviparous
Habit Terrestrial
Status Not evaluated

Location S.E. Asia, New Guinea and satellite islands

A close relative of the eastern blue-tongued skink (see below), this large, New Guinean lizard uses the same defensive technique, displaying its startlingly blue tongue to ward off attack. If this fails, it may bite. This skink constantly pokes its snout about in leaf litter, actively foraging and scavenging for insects, mollusks, fruit, and leaves. It lives in colonies believed to consist of a dominant male and several females and juveniles. Females produce 8–13 young, which often remain close to their mother for the first few days, possibly weeks.

Tiliqua scincoides

Eastern blue-tongued skink

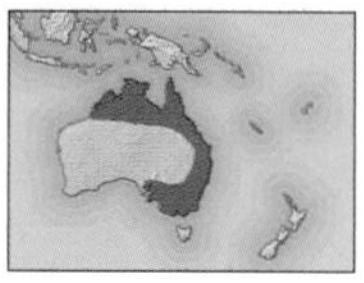

Length 18–20 in (45–50 cm)
Breeding Viviparous
Habit Terrestrial
Status Not evaluated

Location N., E., and S.E. Australia

This large, slow-moving animal is one of Australia's most familiar reptiles. It lives in a wide variety of habitats, and is often seen on roads, where its stubby shape and camouflage coloring make it look like a piece of fallen wood. The eastern blue-tongued skink has a wide-ranging diet, and actively forages for snails, insects, carrion, flowers, fruit, and berries; it will even scavenge for leftovers at picnic sites. During the breeding season, males are territorial; mating takes place after a brief chase, at the end of which the male bites the female on the back of the head. The female gives birth to up to 25 young per litter, following a gestation period of about 150 days. The young take about 3 years to reach maturity. This skink can live for up to 25 years in captivity. Several similar species of skink live in Australia, one of them also occurring in New Guinea and Sumatra.

DEFENSIVE DISPLAY

The eastern blue-tongued skink gets its name from its large, bright blue tongue, which it rolls out of its mouth when threatened. Combined with loud hissing, this visual display is often enough to ward off an attack. This skink is generally harmless, although it has a powerful—if toothless—bite.

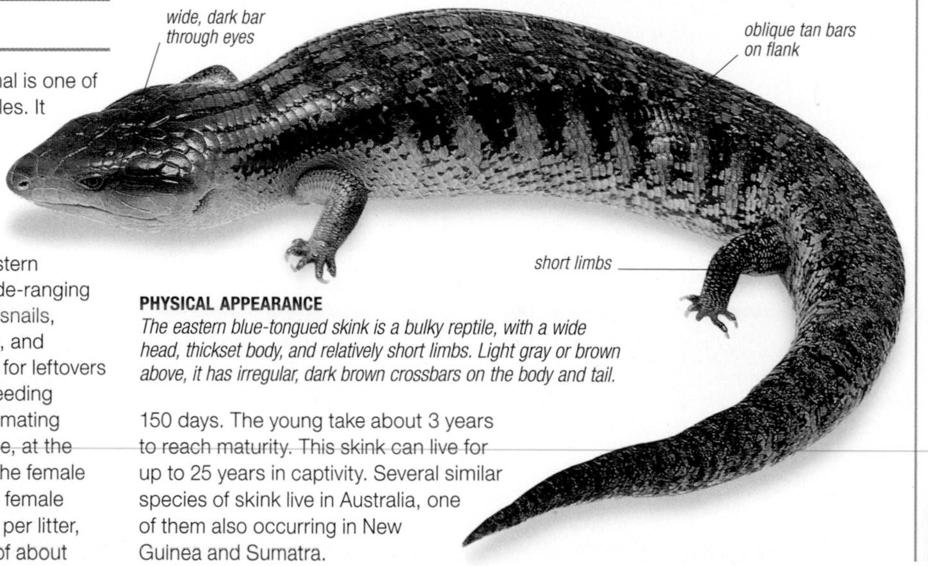

PHYSICAL APPEARANCE
The eastern blue-tongued skink is a bulky reptile, with a wide head, thickset body, and relatively short limbs. Light gray or brown above, it has irregular, dark brown crossbars on the body and tail.

Tribolonotus gracilis

Crocodile skink

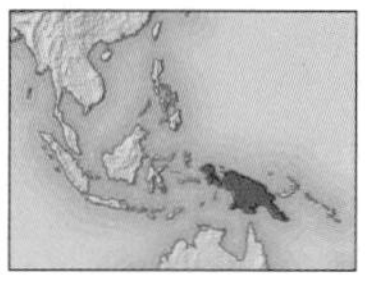

Length 6–8 in (15–20 cm)
Breeding Oviparous
Habit Terrestrial
Status Not evaluated

Location New Guinea

The crocodile skink has a bony, triangular head, 4 rows of spines on the back and tail, and orange, glasseslike rings around its eyes. The glands present on the soles of its feet and on the abdomen are believed to be associated with scent secretion and communication. It is slow-moving and secretive, hiding in leaf litter and debris. If disturbed, it gives out a loud screech. The male usually lives with one or more females, which lay one very large egg in leaf litter or other rotting plant matter.

Tropidophorus grayi

Gray's keeled water skink

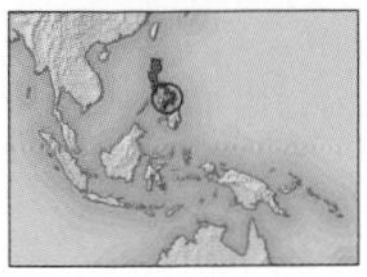

Length 8–10 in (20–25 cm)
Breeding Viviparous
Habit Semiaquatic
Status Least concern

Location Philippines

This slender skink has strongly keeled scales on its back that end in a spine, producing a spiky appearance. Its tail is about the same length as the head and body. It lives in cool conditions, in and around mountain streams, often under rotting logs. If in danger, it may jump into the water and hide under stones on the streambed. Although it is a relatively common skink, little is known about its social and reproductive behavior, except for the fact that the female gives birth to 1–6 young.

Ameiva ameiva

Common ameiva

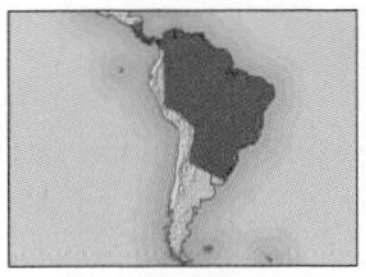

Length 18–20 in (45–50 cm)
Breeding Oviparous
Habit Terrestrial
Status Not evaluated

Location Central America, N. and C. South America

The common ameiva has a streamlined body with a pointed head and a very long tail that may be twice the length of the body. The male, which is more slender and brightly colored than the female, has a bright green area down the center of his back. The female lays several clutches of 2–6 eggs throughout an extended breeding season (March–December). Common ameivas live in loose colonies in forest openings, on the edges of roads and paths, and in cultivated land, usually near human habitation. They bask in patches of sunlight, and shelter under leaf litter or logs on cool days.

Aspidoscelis uniparens

Desert grassland whiptail lizard

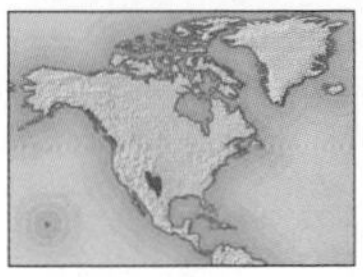

Length 6–9 in (15–23 cm)
Breeding Oviparous
Habit Terrestrial
Status Least concern

Location S. USA, N. Mexico

Found in deserts, grassland, and scrub, this is one of over a dozen very similar lizards that live in North America. It has a long, slender, striped body and whiplike tail, but its most remarkable feature—shared by several other whiptails—is that it is an entirely female species: males are unknown. The desert grassland whiptail lizard is a diurnal hunter that feeds on insects. It also basks in the sunshine for short periods of time throughout the day. Adults bury 1–4 eggs in soil or beneath rocks or logs, and the newly hatched young have bright blue tails.

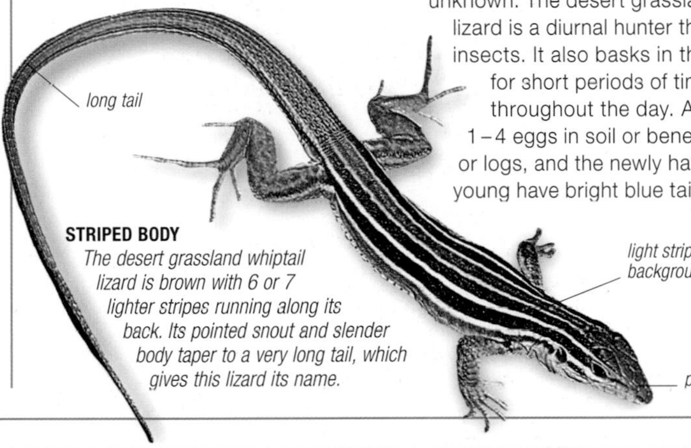

STRIPED BODY
The desert grassland whiptail lizard is brown with 6 or 7 lighter stripes running along its back. Its pointed snout and slender body taper to a very long tail, which gives this lizard its name.

ALL-FEMALE LIZARDS

This lizard reproduces by parthenogenesis, laying fertile eggs without having to mate. Adults and young are always female. Research suggests that this species, and others like it, evolved from single individuals. Each species is a clone—a group of genetically identical animals.

Tupinambis teguixin

Black tegu

Length 32–43 in (80–110 cm)
Breeding Oviparous
Habit Terrestrial
Status Not evaluated

Location N. to C. South America

One of South America's largest terrestrial reptiles, the black tegu shows some striking similarities with monitors (see p.430)—a group of lizards that are found only in Africa, Southeast Asia, and Australasia. It has powerful limbs and long claws for digging, and it uses its long, forked tongue to "taste" the air. Like most monitors, it has a wide-ranging diet, including insects and invertebrates, birds, small mammals, other lizards, and carrion. Black tegus live in forest clearings and on riverbanks, and are good swimmers. When confronted by predators—usually large, carnivorous mammals such as cats, or lizards and snakes when juvenile—they bite and scratch with their claws, and also use their tails as clubs. They walk on their hind legs, and communicate with each other by making loud, snoring noises, which carry a long way in still air. Mating involves the male grasping the female by her neck, and twisting his body under hers. After mating, the female lays 7–12 eggs, sometimes in burrows, but often at the base of a termite mound. The temperature in the mound is kept constant by the termites, which soon repair the mound, sealing in the eggs until they hatch 3 months later. The black tegu is occasionally eaten by humans, and its skin is used in the fashion trade.

Dracaena guianensis

Caiman lizard

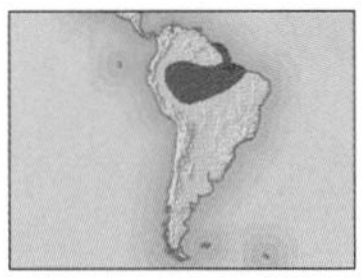

Length 3–3½ ft (0.9–1.1 m)
Breeding Oviparous
Habit Semiaquatic/Arboreal
Status Not evaluated

Location N. South America

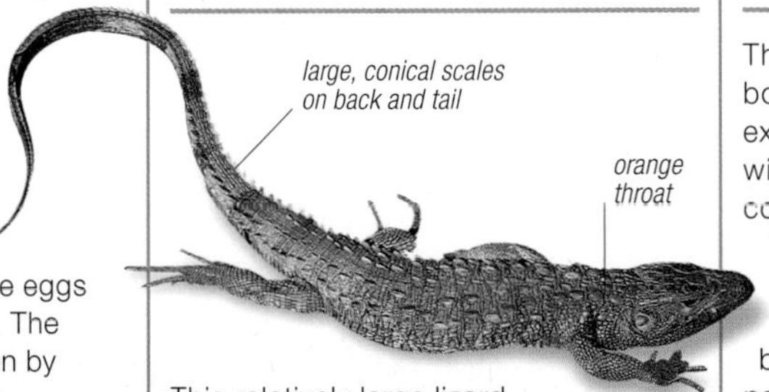

This relatively large lizard resembles a short-headed crocodilian. It is greenish or brown, with yellow or tan blotches on its flanks. The male has an orange and black throat, while the female's is gray. The caiman lizard dives to the bottom of shallow rivers, lakes, and streams to find water snails, which it brings to the surface before crushing the shells with its large, flattened teeth, and swallowing the soft contents. Little is known about this rare reptile, but it is thought to lay eggs.

Xantusia henshawi

Granite night lizard

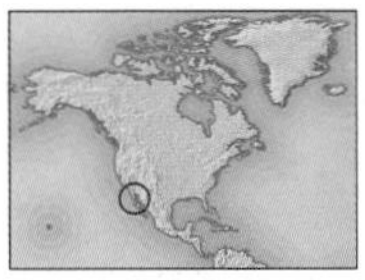

Length 2–2¾ in (5–7 cm)
Breeding Viviparous
Habit Terrestrial
Status Least concern

Location S.W. USA (S. California), Mexico (Baja California)

This nocturnal lizard has a flattened body that enables it to squeeze under exfoliating rock flakes. It is covered with tiny scales that are yellowish in color, with many large, black spots. This lizard actively forages over rock surfaces at night, retreating into narrow crevices when threatened by predators. Although it is not particularly rare, this species is protected because collectors destroy its habitat by removing rock flakes to expose the lizards.

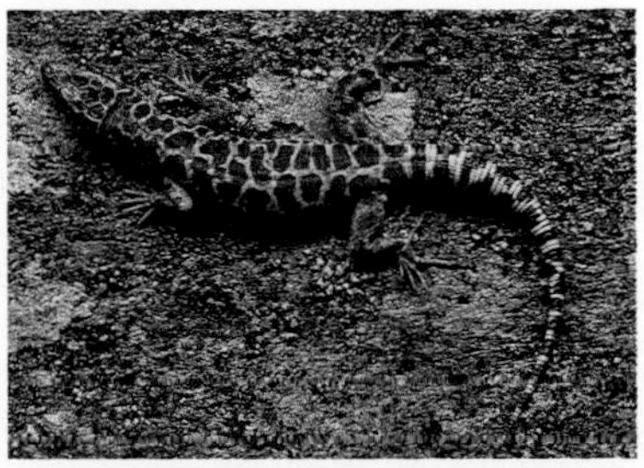

Anguimorph lizards

PHYLUM	Chordata
CLASS	Reptilia
ORDER	Squamata
SUBORDER	Lacertilia
SUPERFAMILY	Anguimorpha
FAMILIES	7
SPECIES	224

The anguimorphs include the largest of all lizards, the monitors, and the venomous lizards, the Mexican beaded lizard and the Gila monster. This diverse group also includes the slow worm and the glass lizards, among others. Anguimorphs are thought to be more evolved than other lizards, and it is likely that they were the ancestors of the snakes—many species have fanglike teeth and a long, forked tongue. Anguimorph lizards are distributed almost worldwide.

Anatomy

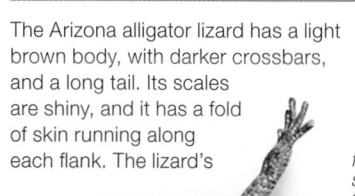

The largest group of anguimorph lizards are the anguids. These slender lizards have smooth scales, and many have small legs or no legs at all (as, for example, in the slow worm).

Monitor lizards are easily recognized by their long neck, narrow head with a pointed snout, powerful limbs, and muscular tail. They have a long, forked tongue, which they use to test their surroundings. This family contains some very large species, including the Komodo dragon.

The venomous lizards, the beaded lizards of North America, are heavily built and have a wide head with a blunt snout, short legs, and a swollen tail. Their venom enters wounds made by sharp fangs in the lower jaw. (Snakes' fangs are in the upper jaw.)

Hunting and feeding

Anguimorph lizards use various methods to find and overpower their prey. Like most lizards, the smaller species feed mainly on insects but some of the larger monitors can subdue large mammals, including pigs and deer. Monitors also feed on carrion and have an acute sense of smell, which they use to locate decaying flesh. The beaded lizards, too, hunt by smell, and will track prey over long distances, using their tongue to follow scent trails left by small mammals and birds. They also feed on the eggs of ground-nesting birds.

Reproduction

Anguimorph lizards may lay eggs or give birth to live young. Several species of monitors incubate their eggs in termite nests, using their strong claws to break into the mounds. When the termites repair the nest, the eggs are entombed in a controlled environment, protected from predators. They often hatch at the start of the rainy season, when the soil softens. Females of some species are thought to return to the mound to help the young break out.

HUNTING
For such large reptiles, monitor lizards can be remarkably agile. The mangrove monitor (shown here) spends a large part of its time in water but also hunts on land. It uses its strong claws to climb trees in search of prey such as this tree snake.

Anguis fragilis

Slow worm

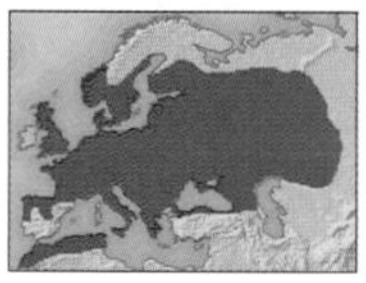

Location Europe to W. Asia, N.W. Africa

Length 12–16 in (30–40 cm), max 20 in (50 cm)
Breeding Ovoviviparous
Habit Terrestrial/Burrowing
Status Not evaluated

With its smooth, scaly body and flickering tongue, this widespread legless lizard looks much more like a snake than a worm. At close quarters, it is easily distinguished from snakes by eyelids that can be closed, and—when threatened—by its ability to shed the end of its tail in order to escape the grip of a predator. Once shed, the tail is very slow to regenerate, leaving many adults with a truncated appearance. Young slow worms are often brightly colored, with a metallic sheen and central stripe. Females tend to keep the stripe as adults, but males are usually a plain coppery brown or gray, although some may have blue spots. The slow worm is a secretive animal, living in habitats that offer plenty of cover, and hiding under logs, flat stones, or piles of trash. It may occasionally bask during the day but is active mainly at dusk, when it emerges to feed on slugs and other invertebrates—a diet that makes it a useful visitor to gardens. Males become fiercely territorial during the breeding season and, after mating, the females give birth to 6–12 live young. Slow worms have a long lifespan, but in the north of their range, up to half of it is spent in hibernation.

Elgaria kingii

Arizona alligator lizard

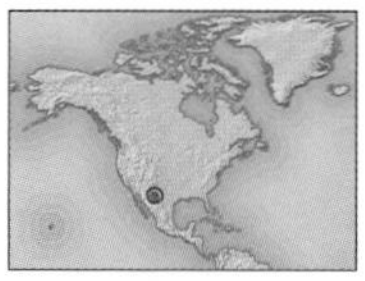

Location S.W. USA, N.W. Mexico

Length 7½–12 in (19–30 cm)
Breeding Oviparous
Habit Terrestrial
Status Least concern

The Arizona alligator lizard has a light brown body, with darker crossbars, and a long tail. Its scales are shiny, and it has a fold of skin running along each flank. The lizard's long, slender body and short limbs are adaptations for moving quickly through grass and other dense vegetation. A secretive species, it is active throughout the day, and sometimes ventures out at dusk. It prefers to remain in moist areas in otherwise dry woodland, and in mountains and upland grassland, often near streams, where it forages among dead leaves and under debris for insects and spiders. It hibernates when food becomes scarce in winter. Preyed upon by snakes, birds, and small mammals, it discards its tail if grasped and may smear its enemy with feces. Alligator lizards live in loose colonies, the males becoming territorial during the breeding season. Females lay 9–12 eggs, buried in moist sand or soil; the hatchlings are boldly banded.

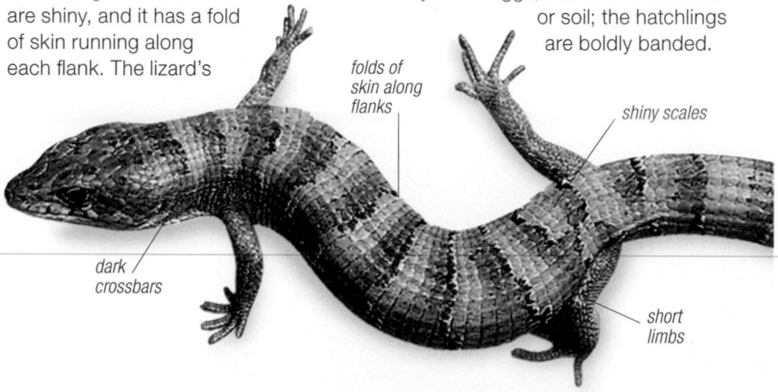

Pseudopus apodus

European glass lizard

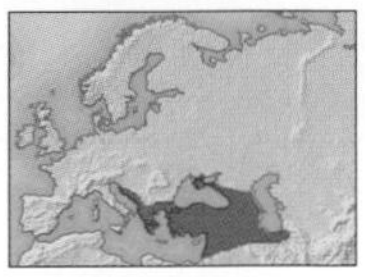

Length 3¼–4 ft (1–1.2 m)
Breeding Oviparous
Habit Terrestrial/Burrowing
Status Not evaluated

Location S.E. Europe, W. Asia

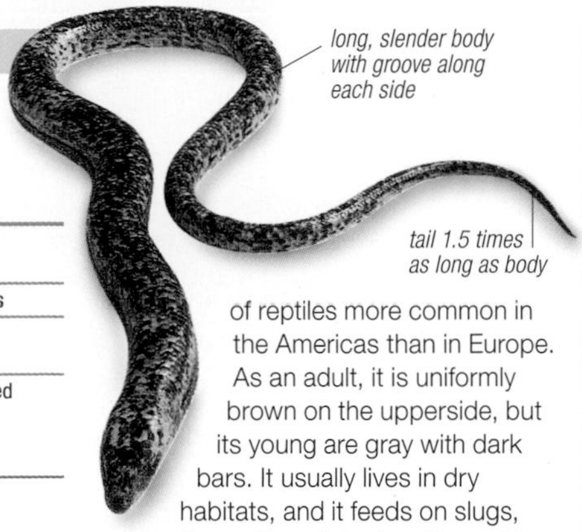

With a length exceeding that of many snakes, this legless lizard is the largest member of the family Anguidae—a group of reptiles more common in the Americas than in Europe. As an adult, it is uniformly brown on the upperside, but its young are gray with dark bars. It usually lives in dry habitats, and it feeds on slugs, snails, and invertebrates. If caught, it may discard its tail or smear its captor with feces. Females lay 8–10 eggs, and bury them in damp sand or soil.

Ophisaurus ventralis

Eastern glass lizard

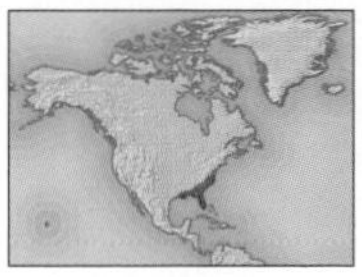

Length 18–43 in (45–108 cm)
Breeding Oviparous
Habit Terrestrial/Burrowing
Status Least concern

Location S.E. USA

This legless lizard is one of 4 closely related species found in North America. Brown with black flanks, it lives mainly in damp meadows—usually on the surface, unlike some of its relatives, which spend most of their time underground. Its tail is over twice as long as the rest of its body, and is easily shed. Active early in the morning and after rain, the eastern glass lizard feeds on slugs, snails, insects, and small vertebrates. The female lays 8–17 eggs in damp soil, and guards them until they hatch.

Anniella geronimensis

Baja California legless lizard

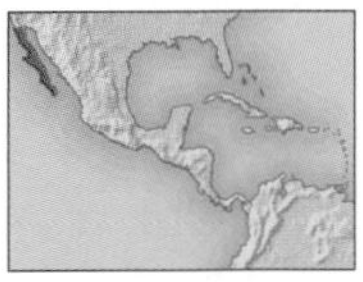

Length 4–6 in (10–15 cm)
Breeding Viviparous
Habit Burrowing
Status Endangered

Location Mexico (Baja California)

The Baja California legless lizard is a small, slender species with a silvery to light brown body. It does not create permanent burrows, but "swims" through loose sand just below the surface, especially around the base of low-growing shrubs. It may come above ground at night or at twilight, leaving behind trails in the sand. This little-studied lizard eats spiders and small insects. If attacked, it can shed its tail in order to escape.

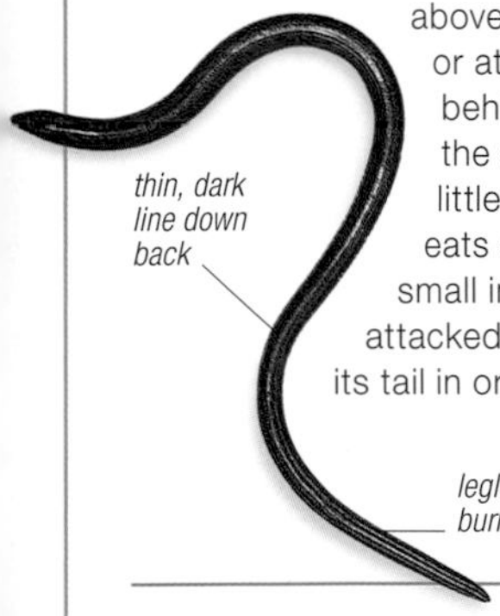

Heloderma suspectum

Gila monster

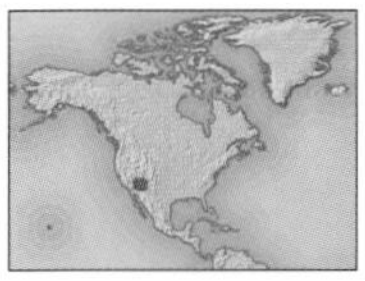

Length 14–20 in (35–50 cm)
Breeding Oviparous
Habit Terrestrial/Burrowing
Status Near threatened

Location S.W. USA, N. Mexico

The Gila monster is one of North America's most distinctive reptiles, and one of only 2 dangerously venomous lizards in the world. Its bulky body is marked with orange, pink, or yellow, contrasting with bands of black warning potential aggressors that it has a toxic bite. Gila monsters live in semidesert or scrub with rocky outcrops, in places where there is ready access to moisture. Compared to most other lizards, they are slow-moving, hunting by day in the spring but at dusk or after dark during the summer heat. The tail stores fat in the same way as a camel's hump; it shrinks when food is hard to find, especially in cold weather when the Gila monster hibernates. Females lay their eggs in late summer, burying clutches of up to 8 in burrows scooped out of damp sand.

FEEDING HABITS

The Gila monster feeds on small mammals and the eggs of quails, doves, and reptiles. It hunts by smell and also "tastes" its surroundings with its tongue. Venom is produced in glands in the lower jaw, not the upper jaw as in snakes. The lizard bites fiercely with its sharp teeth and holds onto its victim while the poison enters the wound. The bite, although painful, is not fatal to adult humans.

POWERFUL BURROWER
This lizard has powerful limbs for burrowing and walks with a slow, lumbering gait. Its coloration may act as camouflage, or as a warning to predators.

Heloderma horridum

Mexican beaded lizard

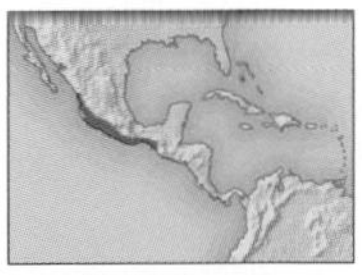

Length 28–39 in (70–100 cm)
Breeding Oviparous
Habit Terrestrial/Burrowing
Status Least concern

Location W. Mexico

A close relative of the Gila monster (see above), the Mexican beaded lizard, too, is venomous, but its colors are more subdued. It is often dark brown with paler blotches, but in the south of its range it is entirely black. This lizard hides in burrows, emerging to forage for birds, especially fledglings, as well as eggs, and small mammals. It bites and injects venom into its prey with the sharp, grooved teeth in its lower jaw. Although painful, its bite is not often fatal to humans. Females lay 4–10 large, elongated eggs in burrows.

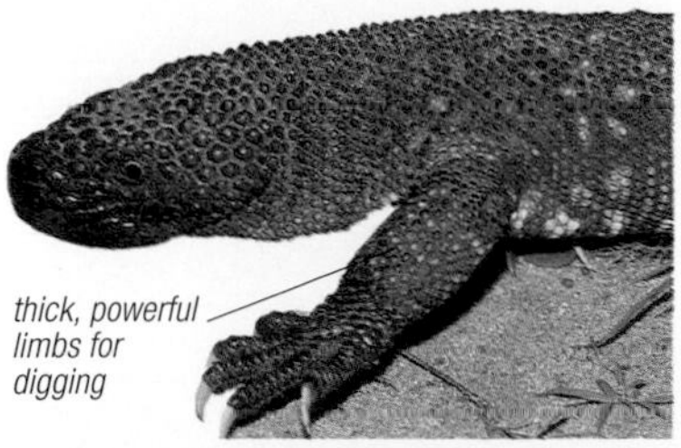

Lanthanotus borneensis

Borneo earless lizard

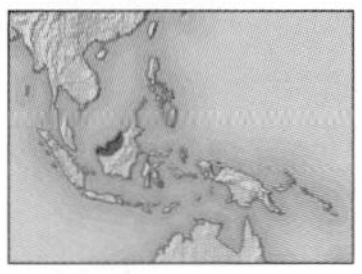

Length 16–18 in (40–45 cm)
Breeding Oviparous
Habit Burrowing/Semiaquatic
Status Not evaluated

Location S.E. Asia

This efficient swimmer inhabits streamside burrows, which may become flooded. Its short limbs and blunt snout are adaptations for burrowing, while the transparent eyelids and nostrils that can be closed by valves are adapted for living in water. Being nocturnal and secretive, it is hardly ever seen. Only about 100 specimens have been collected since it was first described in 1878 and it remains poorly known. Once believed to be related to the heloderms (the Gila monster, above, and the Mexican beaded lizard, left) and, later, the monitors (see p.430), it is now thought not to have any living close relatives.

Varanus komodoensis

Komodo dragon

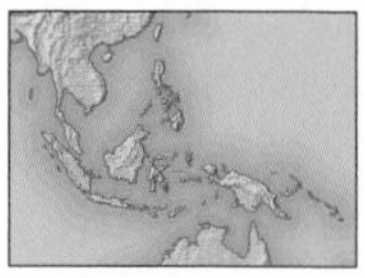

Length 6½–9¾ ft (2–3 m)
Breeding Oviparous
Habit Terrestrial
Status Vulnerable

Location Indonesia (Komodo, Rinca, Padar, W. Flores)

With an average weight of about 155 lb (70 kg), and sometimes double this in captivity, the Komodo dragon is the world's heaviest lizard. It has a long body, well-developed legs, and a deeply forked tongue, which it flicks out as it searches for food. Juveniles are boldly marked, with gray or cream bands, but lose these markings as they mature, becoming uniformly grayish brown, with thickly folded, scaly skin.

Komodo dragons live on scrubby hillsides and in open woodland and dry riverbeds, where they feed entirely on live animals or carrion. They have a keen sense of smell, and can detect the scent of decaying remains from up to 3 miles (5 km). However, they hunt mainly by ambushing prey. Young Komodo dragons attack snakes, lizards, and rodents, but adults target much larger prey, including wild pigs, water buffalo, and deer. They are also cannibalistic—one reason why juveniles spend much of their time in trees.

Adults are mainly solitary, although groups may gather at the site of a kill. During the breeding season, males compete for the chance to mate, wrestling in an upright position with their tails acting as props. After mating, females dig nests in sandy ground, laying clutches of up to 25 eggs. When the eggs hatch—about 9 months later—the young are left to fend for themselves. Komodo dragons take about 5 years to become sexually mature, and in the wild they have a maximum lifespan of about 40 years.

ON THE SCENT OF FOOD
Komodo dragons have good eyesight, but they find most of their food by smell. Like snakes, they "taste" the air with their tongues, which collect scent molecules from the air. They have sharp, serrated teeth, but because they cannot chew, they tear off pieces of food and then throw them backward into their mouths. Komodo dragon saliva is rich in toxic bacteria, and it was thought that when a dragon bites an animal, these bacteria contaminate the wound, thus weakening its prey. In 2009, it was found that Komodo dragons actually produce a toxin, responsible for stunning the prey into shock and preventing its blood from clotting.

CONSERVATION

Komodo dragons are the world's most localized large predators, restricted to a few Indonesian islands. Estimated at 2,500–5,000, the population is a fraction of its size 50 years ago. Its decline is due to hunting, loss of prey, and habitat change. However, Komodo dragons have become increasingly important as a tourist attraction, a practical incentive for their conservation.

A GIANT PREDATOR
Komodo dragons look ungainly, but they are capable of running at up to 11 mph (18 kph) in short bursts. They are also good swimmers. They have a huge capacity for food, eating up to half their own weight in a single meal.

KOMODO DRAGONS AT A KILL

1 FIRST ARRIVAL
Guided by scent, a Komodo dragon homes in on a deer ambushed several hours earlier. The deer escaped, but has since died from the dragon's bite.

2 FEEDING BEGINS
The dragon begins to feed on the deer's body. It has unusually flexible joints between its jaw and skull, enabling it to swallow large chunks of food.

3 QUICK WORK
The dragon feeds rapidly, swallowing flesh, skin, and even bones. Meanwhile, other dragons begin to head for the carcass as its scent drifts downwind.

4 FIGHTING FOR SHARES
As more dragons arrive at the scene, the larger animals try to intimidate the smaller ones into backing away from the kill. If food is scarce, fighting often breaks out.

Varanus albigularis

Savanna monitor

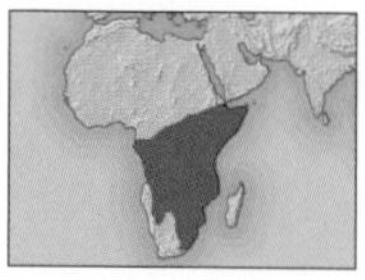

Length 3¼ft (1 m), max 6 ft (1.8 m)
Breeding Oviparous
Habit Terrestrial/Burrowing
Status Not evaluated

Location E., C., and southern Africa

Also known as the white-throated monitor, this lizard has powerful limbs for digging burrows, in which it lives; it also lives in hollow tree trunks. Territorial, with a range of up to 7 square miles (18 square km), it forages for birds, insects, snails, and invertebrates. Its chief predators are martial eagles and ratels, and it is sometimes eaten by humans, despite legal protection. In defense, the savanna monitor puffs up its throat and body, lashes out with its tail, and bites. In a year, a female may lay 2 clutches of as many as 50 eggs.

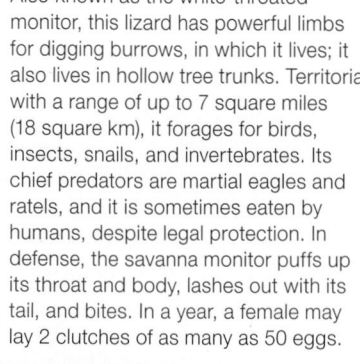

Varanus prasinus

Green tree monitor

Length 30–39 in (75–100 cm)
Breeding Oviparous
Habit Arboreal
Status Not evaluated

Location New Guinea

Well equipped for its arboreal way of life, this species has a light, slender body that enables it to support itself on narrow branches, and a coloration that helps to camouflage it against the trees.

It feeds on insects, small vertebrates, crabs, and birds' eggs. When threatened, it flees through vegetation and may bite if cornered. One of the few monitors that are social by nature, it appears to live in small groups made up of a dominant male, several females, and subordinate males and juveniles. Females lay clutches of up to 6 eggs in a hole in arboreal termite nests or leaf litter.

Varanus varius

Lace monitor

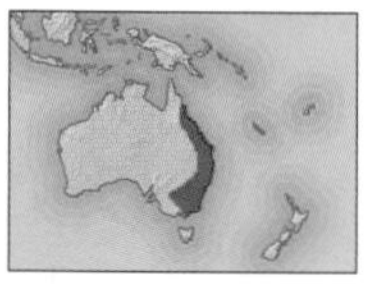

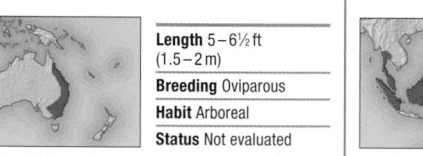

Length 5–6½ ft (1.5–2 m)
Breeding Oviparous
Habit Arboreal
Status Not evaluated

Location E. Australia

Dark gray or bluish, with white or cream markings, the lace monitor has a long neck and tail, and long claws used for scaling trees. It bites and lashes out with its tail if threatened. It forages on the ground and in trees for insects, reptiles, birds and their eggs, and small mammals. Females lay 6–12 eggs in holes in termite mounds, which the termites seal; reportedly, females sometimes return to help the young dig their way out.

Varanus dumerilii

Dumeril's monitor

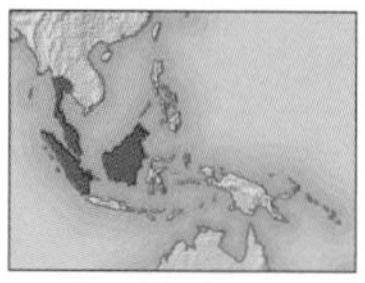

Length 3¼–4¼ ft (1–1.3 m)
Breeding Oviparous
Habit Mostly terrestrial
Status Not evaluated

Location S.E. Asia

Although this slender gray lizard is mostly terrestrial, it is also a good swimmer, and can close its nostrils underwater when in pursuit of crabs to eat. It also forages actively for insects, birds, and bird and turtle eggs. It may occasionally enter the sea to escape from predators, mainly mammals and snakes, which threaten young lizards in particular, or may escape by climbing into trees and shrubs. Males are territorial during the breeding season; after mating, females lay their eggs in soil or leaf litter. This monitor prefers mangrove forests but also inhabits forests away from the coast.

Varanus flavirufus

Sand monitor

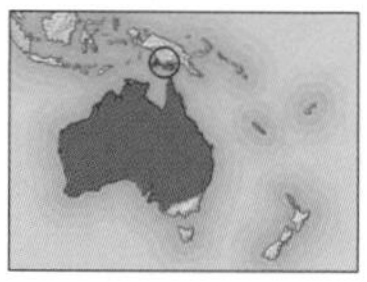

Length 30–39 in (75–100 cm)
Breeding Oviparous
Habit Terrestrial/Burrowing
Status Not evaluated

Location S. New Guinea, Australia

This impressive reptile has a slender, powerful body that varies in color from pale yellow to dark gray, but it almost always has a dark line extending from the back of the eye to the neck. The tail works as a prop when the monitor stands on its back legs, but when threatened, it uses its tail as a whip or club. A voracious feeder, the sand monitor eats mammals, birds, and other reptiles, as well as smaller animals and carrion. It shelters in burrows and in hollow trees, and females often lay their eggs in termite mounds, clawing them open to place the eggs inside.

Varanus bitatawa

Northern Sierra Madre forest monitor

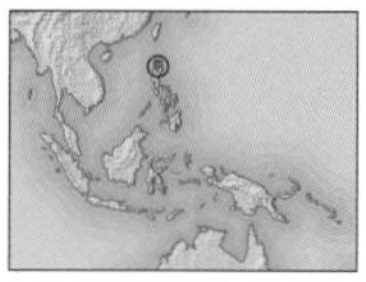

Length 6½ ft (2 m)
Breeding Oviparous
Habit Mostly arboreal
Status Not evaluated

Location Philippines

Discovered in the northern Sierra Madre mountain forests of Luzon, in the Philippines, this spectacular giant monitor was described as a new species in 2010. Longer than a tall man, it has black upper parts blotched with golden spots and a black and green banded tail. The northern Sierra Madre forest monitor spends most of its life high in the trees, some 66 ft (20 m) above the ground, where it feeds almost exclusively on fruit. Its secretive habits and reluctance to cross open spaces probably explain why it eluded detection by scientists for so long. This species of forest monitor is separated from its closest ally in the southern part of the island by a 93-mile (150-km) gap across a low-lying river valley.

Varanus niloticus

Nile monitor

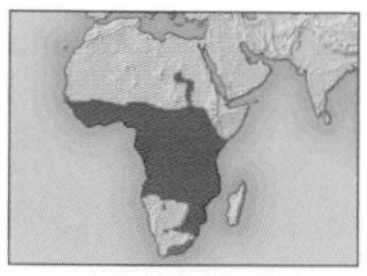

Length 4½–6½ ft (1.4–2 m)
Breeding Oviparous
Habit Semiaquatic
Status Not evaluated

Location Africa (mainly south of Sahara)

The grayish brown Nile monitor is never found far from water. It basks on nearby rocks or tree stumps, and feeds on crabs, mollusks, fishes, and frogs, as well as birds, eggs, and carrion. Hunted by crocodiles and pythons, it uses its tail, claws, and teeth to defend itself. In the cooler part of its range, it hibernates in communal "dens." Females lay 20–60 eggs in termite mounds, and the hatchlings dig their way out after rain has softened the earth.

Shinisaurus crocodilurus

Chinese crocodile lizard

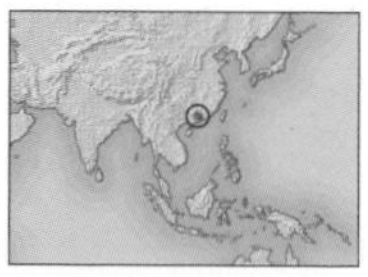

Length 16–18 in (40–46 cm)
Breeding Viviparous
Habit Semiaquatic
Status Not evaluated

Location E. Asia (Kwangsi Province, China)

This rare, little-known lizard is olive in color, with rows of enlarged, bony scales down its back and tail, like those of a crocodile. The male is more colorful than the female but perhaps only in the breeding season. To escape detection, the lizard may "freeze" for several hours at a time, even in midstride—a tactic that, along with its coloration, helps it to merge with its surroundings. It adapts in unusual ways to cool temperatures: it may, for instance, be able to "shut down" its system for hours to conserve energy during very cold nights. The female gives birth to 2–10 young, sometimes over several days.

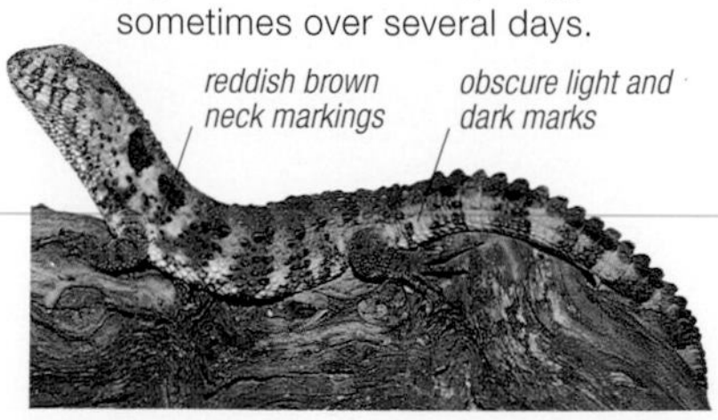

Amphisbaenians

PHYLUM	Chordata
CLASS	Reptilia
ORDER	Squamata
SUBORDER	Amphisbaenia
FAMILIES	6
SPECIES	181

Although they are sometimes called "worm lizards," amphisbaenians are neither worms nor lizards. They are, however, closely related to lizards and belong to the same order as both lizards and snakes. Adapted to an underground life, amphisbaenians are seen above ground only after heavy rain has flooded their tunnels. Their ability to regulate body temperature is limited, which confines them to tropical and subtropical regions.

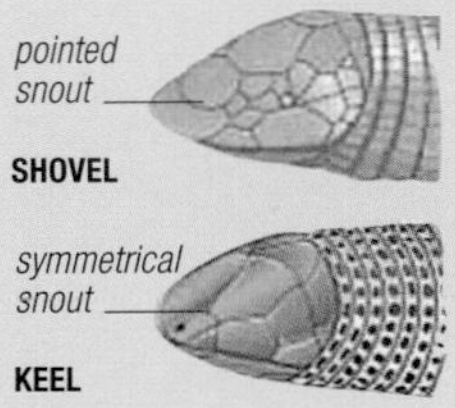

HEAD SHAPE
Amphisbaenians use their thick, heavy skulls to dig tunnels. The shape of the snout varies according to the method of tunneling.

Tunneling

Amphisbaenians construct their own burrows, down which they move in search of worms, insects, and larvae. They make fresh tunnels by using their skulls as rams. Their nostrils point backward, and are thus prevented from being clogged up with soil while burrowing. The lower jaw is recessed, so that, as the head is forced through the soil, the mouth is held tightly closed.

Anatomy

Amphisbaenians are wormlike in appearance with cylindrical bodies and rings of scales that look like body segments. They have rudimentary eyes and smooth heads that look wedge-shaped in side view. They are all without legs, except for 3 species that have front limbs, positioned close to the head and ending in long claws for burrowing. Many amphisbaenians lack pigmentation and are pinkish brown; a few are more colorful.

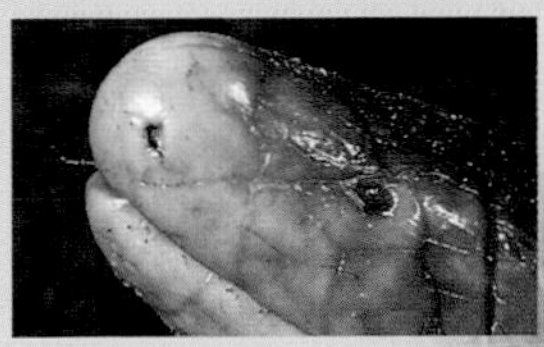

EYES AND EARS
An amphisbaenian's eyes are covered with translucent skin, and like many burrowing animals, they have poor vision. They lack external ear openings.

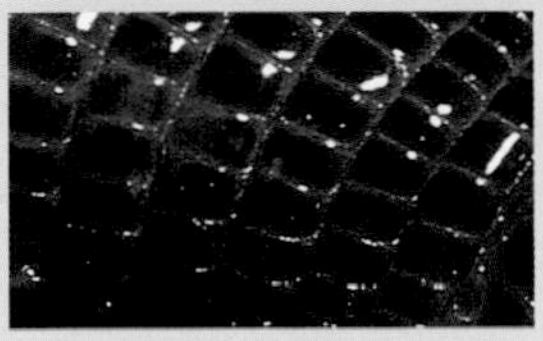

SCALES
Unlike the scales of snakes, which overlap each other, those of amphisbaenians are arranged in concentric rings.

LIMBS
Amphisbaenians spend almost all their lives underground, burrowing in soil, sand, or leaf litter. Most have lost all traces of limbs. However, the members of one family, the Mexican worm lizards (shown here), have retained their forelimbs.

Amphisbaena fuliginosa

Speckled worm lizard

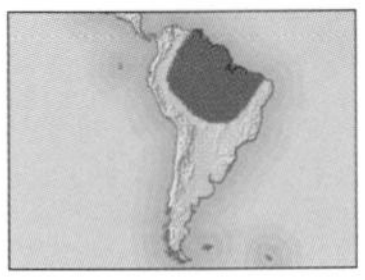

Length 12–18 in (30–45 cm)
Breeding Oviparous
Habit Burrowing
Status Not evaluated

Location N. South America, Trinidad

This South American species is one of the most easily recognized amphisbaenians because of its conspicuous markings. It lives almost exclusively in burrows (although it may come to the surface at night), using its head to ram a tunnel through the ground and its body rings to help it concertina through the tunnel. It feeds on virtually any small vertebrates and insects, attacking them with its strong jaws. If threatened, it may shed part of its tail, which does not regenerate. Like other amphisbaenians, this species is difficult to study since it spends most of its life underground.

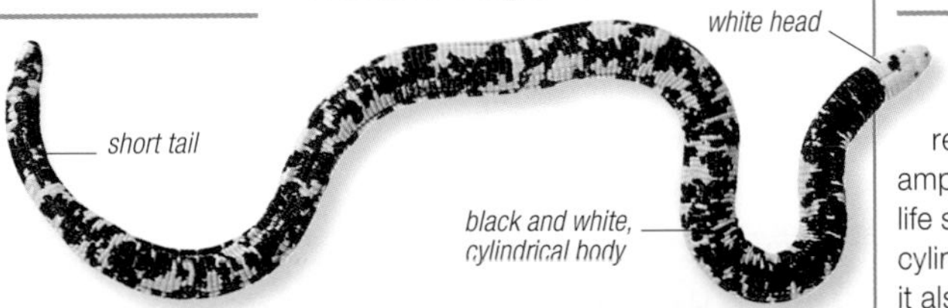

Blanus cinereus

European worm lizard

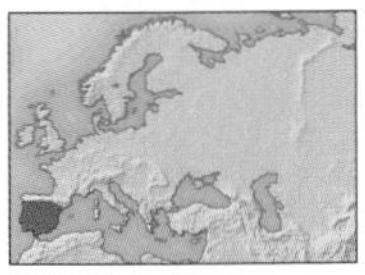

Length 4–8 in (10–20 cm), max 12 in (30 cm)
Breeding Oviparous
Habit Burrowing
Status Least concern

Location S.W. Europe

The European worm lizard has a small, pinkish or purplish brown body, with a short head, pointed snout, rudimentary eyes, and ringlike grooves, all of which help it burrow. Mostly found under rocks and logs, it may appear on the surface at night, especially after heavy rain. It hunts along underground tunnels for earthworms and small insects. *Blanus tingitanus,* found in Morocco and Algeria, used to be considered the same species.

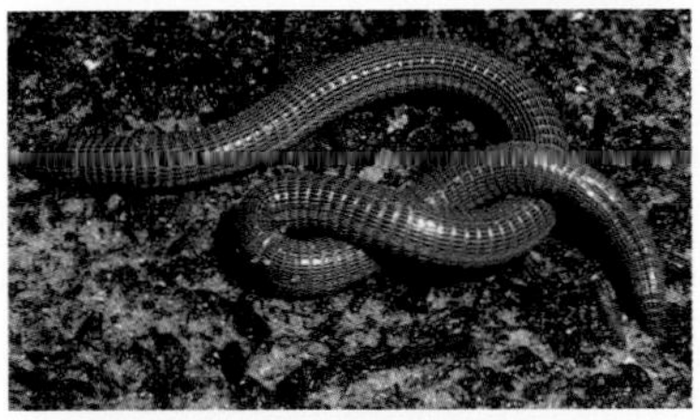

Bipes biporus

Ajolote

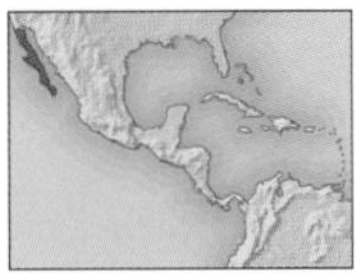

Length 6½–9½ in (17–24 cm)
Breeding Oviparous
Habit Burrowing
Status Least concern

Location Mexico (Baja California)

The ajolote, or mole lizard, is one of the world's most remarkable reptiles. It has the typical amphisbaenian adaptations for a life spent burrowing—a blunt head, cylindrical body, and short tail—but it also has a pair of very short yet powerful front legs. Equipped with clawed feet, they help the ajolote to excavate its way through the ground. The ajolote rarely comes to the surface—and then only after heavy rain—and it feeds by ambushing lizards and other small animals, usually dragging them underground. There are 2 similar species, also found only in Mexico.

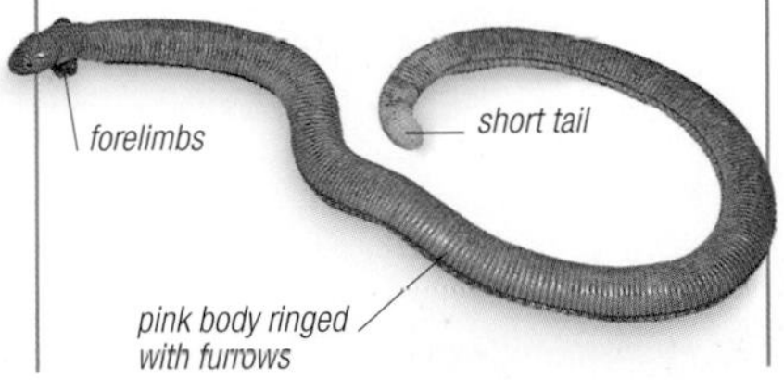

Rhineura floridana

Florida worm lizard

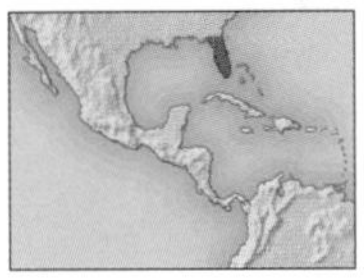

Length 10–14 in (25–35 cm)
Breeding Oviparous
Habit Burrowing
Status Least concern

Location S.E. USA (Florida)

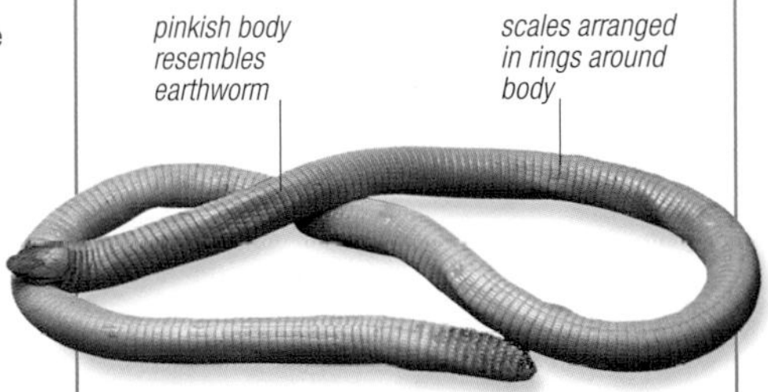

The Florida worm lizard looks like an earthworm, with a pinkish body that has no limbs, no external eyes, and no ear openings. Exceptional specimens can be up to 16 in (40 cm) long. The scales are ill-defined and arranged in rings around the cylindrical body. This worm lizard leads its life almost entirely underground but may be forced to the surface because of heavy rain or ploughing. An opportunistic species, it feeds on invertebrates encountered while burrowing. It is preyed upon by birds—mockingbirds occasionally unearth it when hunting for earthworms.

Crocodiles and alligators

PHYLUM	Chordata
CLASS	Reptilia
ORDER	Crocodylia
FAMILIES	3
SPECIES	24

These large reptiles are among the few survivors from the time of the dinosaurs, having changed little in the last 65 million years. Collectively known as crocodilians, they include crocodiles, alligators, and caimans, and one species of gharial. Crocodilians are formidable, semiaquatic predators. Most of them live in freshwater rivers, lakes, and lagoons, while a few species inhabit tidal reaches and may venture out to sea. Alligators and caimans (with the exception of one species) are found only in North, Central, and South America. Crocodiles occur mainly in Asia, Africa, and Australia, with some species in Central and South America.

Reproduction

Male crocodilians are territorial and mate with several females. Each female lays her eggs near water, either in a mound of vegetation and mud or in an underground chamber. The females guard the eggs and help the hatchlings out of the nest when they hear them calling. Females stay with their young for several months (a year or more in American alligators) to protect them from predators.

PARENTAL CARE
In most species, the female uses her mouth to carry her hatchlings to a shallow pool or backwater.

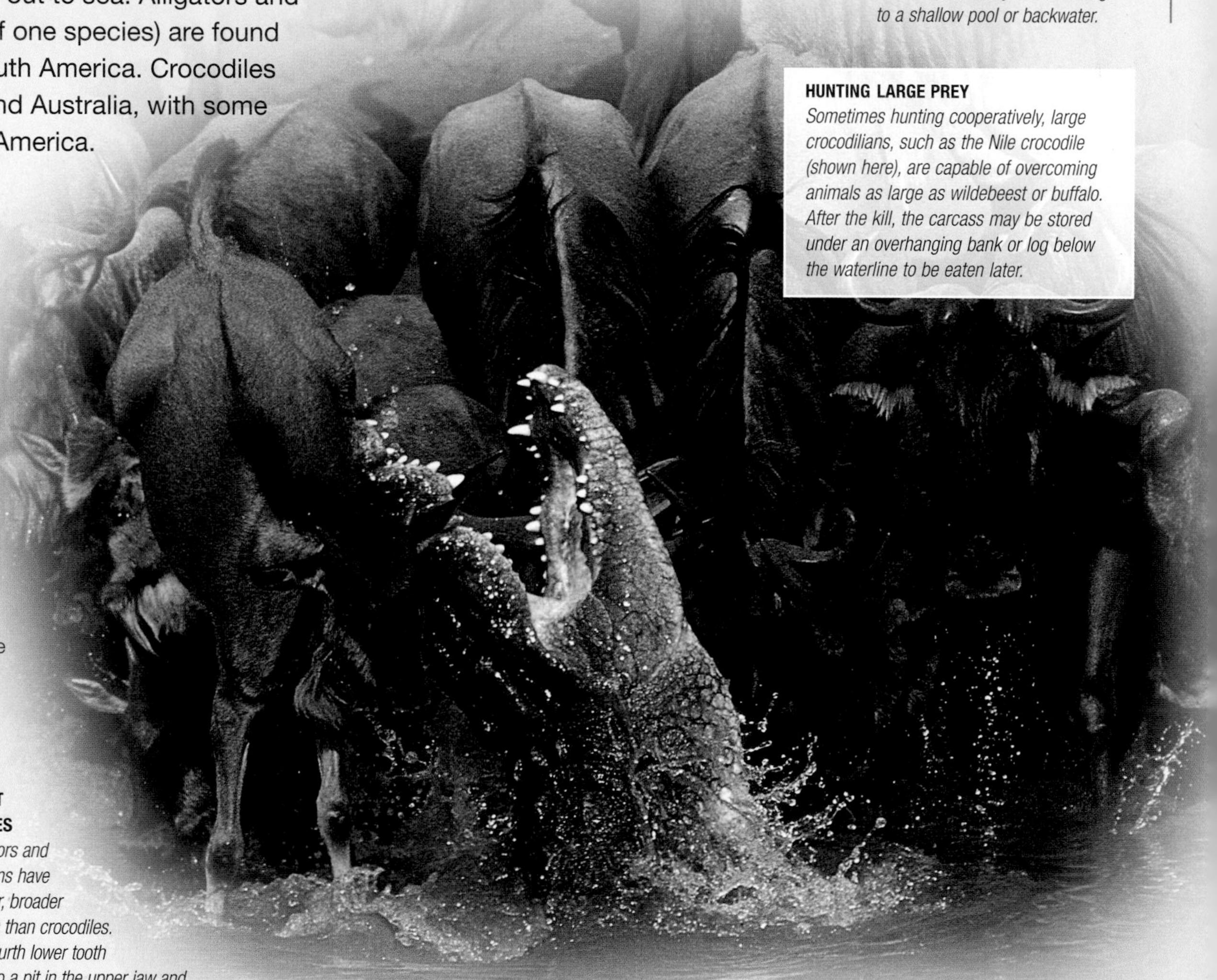

HUNTING LARGE PREY
Sometimes hunting cooperatively, large crocodilians, such as the Nile crocodile (shown here), are capable of overcoming animals as large as wildebeest or buffalo. After the kill, the carcass may be stored under an overhanging bank or log below the waterline to be eaten later.

Anatomy

All species of crocodilians have a wide, slightly flattened body, a long, vertically flattened, muscular tail, powerful jaws, and eyes and nostrils on top of the head so that they can see and breathe while partly submerged. A transparent third eyelid can be closed to protect their eyes when underwater. The large, bony scales have raised ridges, and are shed singly or in pieces, unlike those of other reptiles. Most are dull olive, gray, brown, or black, although juveniles may have lighter markings that disappear with age. The 3 groups of crocodilians differ in the shape of their snout and the arrangement of their teeth (see below).

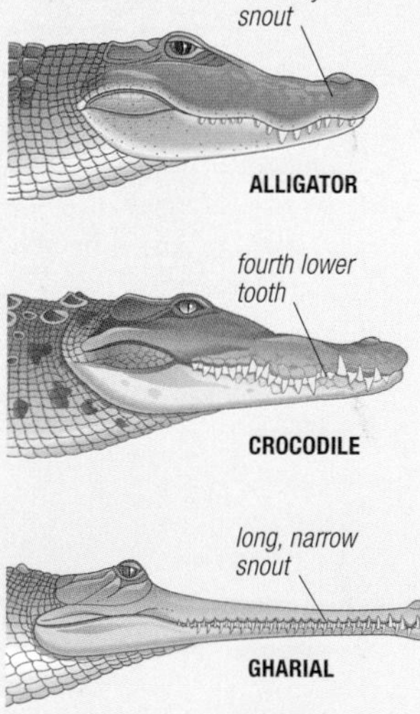

SNOUT SHAPES
Alligators and caimans have shorter, broader snouts than crocodiles. The fourth lower tooth fits into a pit in the upper jaw and is not seen when the mouth is closed. Crocodiles, however, have a notch in their upper jaw, into which the fourth lower tooth fits so that the tooth remains visible when the mouth is shut. The gharial has a long, narrow snout with enlarged front teeth followed by smaller teeth of equal size.

ADAPTATIONS FOR DIVING
Crocodilians have 2 adaptations to stop water entering their lungs when underwater. By closing a flap of skin at the back of the throat, they can open their mouth to catch prey without water passing down the windpipe. Similar flaps can also be closed over the external openings of the nostrils and ears.

Hunting and feeding

Crocodilians are carnivores that eat a mixture of live prey and carrion. They use various techniques when hunting but the most common is to lie in wait at the edges of rivers and lakes for mammals to come within range to drink or try to cross. Another method is to drift toward prey, such as waterfowl, hoping to catch it off guard. Fish-eating species, such as the gharial, take prey with a rapid sideways swipe of their narrow jaws, which offer little resistance to the water. Birds, small mammals, and fishes are swallowed whole, but as crocodilians cannot chew, they must dismember larger prey by spinning violently about their own axis while underwater, holding part of the carcass in their jaws.

HUNTING BY STEALTH
By drifting with only its eyes and nostrils above the water, a crocodilian can move to within striking range without being seen.

STORING FOOD
Having drowned this antelope, the crocodile may eat some of it before securing the remainder in an underwater "larder" to rot, making it easier to dismember.

Alligator mississippiensis

American alligator

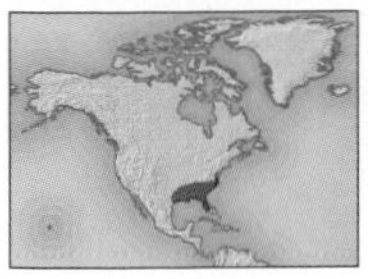

Length 9¼–16 ft (2.8–5 m)
Breeding Oviparous
Habit Aquatic
Status Least concern

Location S.E. USA

Once widely hunted for its skin, the American alligator became seriously endangered in the 1950s, but under legal protection has staged a strong recovery. A large, powerful reptile, it is black with a broad head and rounded snout, and has a heavily plated back. It also has a characteristic large fourth tooth that fits into a socket in the upper jaw. This alligator feeds in or near fresh water, taking animals of all kinds, including birds snatched from low branches. During the breeding season, the males roar to attract females, and mating takes place underwater. Each female lays 25–60 eggs in an enormous nest made of vegetation and mud, digging it open when she hears the calls of the hatching young; she may stay with her brood for as long as 3 years.

ALLIGATOR HOLES

The American alligator prefers to float, partially submerged, in lakes, swamps, and marshes. In midsummer, as the water table falls, it retreats into "alligator holes"—keeping a small body of water open by digging out the sand or mud at the bottom. It feeds on fishes that are trapped in the diminishing pool and animals that visit it.

THE BLACK ALLIGATOR
Both adult and juvenile American alligators are black, but the juveniles have bold, yellow crossbands.

Caiman crocodilus

Spectacled caiman

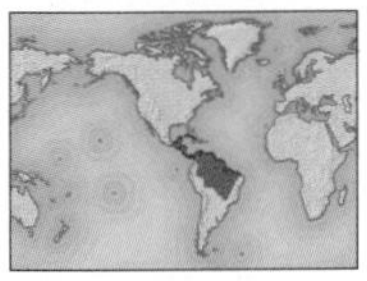

Length 6½–8 ft (2–2.5 m)
Breeding Oviparous
Habit Aquatic
Status Least concern

Location Central America, N. South America

Also called the common caiman, this dull olive crocodile has a bony ridge in front of its eyes. It is found in most freshwater habitats, rarely leaving the water unless driven out by drought, when it will burrow into the mud. It floats on the surface during the day, becoming more active at night. Adults feed on other reptiles, fishes, amphibians, and waterbirds. Males are territorial and establish a dominance hierarchy. Females lay 14–40 eggs in mounds of decaying plants and soil at the edge of water, or on rafts of floating vegetation. Several females may share the same nest and guard it against predators. Although heavily hunted for its skin, this crocodilian has benefited from artificial bodies of water, such as reservoirs, resulting in increases in local populations.

Crocodylus niloticus

Nile crocodile

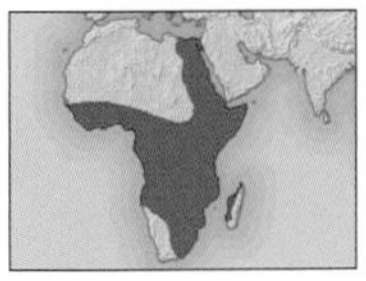

Length 11 ft (3.5 m), max 20 ft (6 m)
Breeding Oviparous
Habit Aquatic
Status Least concern

Location Africa, W. Madagascar

The Nile crocodile has a dark olive to gray body with darker crossbands. It favors larger rivers, lakes, and swamps, and is also found in estuaries and river mouths; it may come ashore to bask on hot days. It feeds on fishes, antelope, zebras, and even buffaloes, leaping up to snatch nesting birds or pull drinking animals into the water. Although solitary by nature, several crocodiles may congregate for feeding, working together to herd fishes into shallow water, where they are easier to catch. Males are territorial and establish a dominance hierarchy. The female lays 16–80 eggs in a hole dug on a bank well above the water level and uses the same nest site for the whole of her life. She guards the eggs throughout the incubation period. The young chirp when they are about to hatch and the female digs them out, gently maneuvering the hatchlings into her mouth, and carrying them into the water in batches. They remain together for 6–8 weeks, then gradually disperse. For the first 4 or 5 years they live in burrows up to 9¾ ft (3 m) long.

JAWS AND TEETH

The Nile crocodile has long jaws, with teeth that are exposed even when its mouth is closed. Like other crocodilians, it can bite but it cannot chew, which presents problems when dealing with large prey such as zebras and buffaloes. Nile crocodiles deal with these by dragging them underwater, and then spinning around to tear off chunks of flesh.

AQUATIC ADAPTATIONS
Like most crocodiles, the Nile crocodile has eyes and nostrils on the top of its head, helping it to see and breathe in water; it also has a flat, powerful tail and webbed hindfeet that help it to swim.

Crocodylus porosus

Saltwater crocodile

Length 16–23 ft (5–7 m)
Breeding Oviparous
Habit Aquatic
Status Least concern

Location S.E. Asia to N. Australia

Also known as the estuarine or Indo-Pacific crocodile, this massive gray, brown, or black animal is the world's largest crocodilian. The biggest specimens on record weighed over 2,200 lb (1 tonne), although decades of hunting mean that such giants are now very rare. Equally at home in freshwater and seawater, these nocturnal reptiles feed on a wide variety of animals, including mammals, birds, and fishes. They present a real threat to humans, and have been responsible for many fatalities. They mate in water, and the females lay about 60 eggs in mound nests on the river bank, high above the watermark. Each female guards her clutch, and stays with her hatchlings for the first few weeks of their life.

WALKING ON LAND
The saltwater crocodile has strong enough legs to raise its body off the ground and so can walk on land.

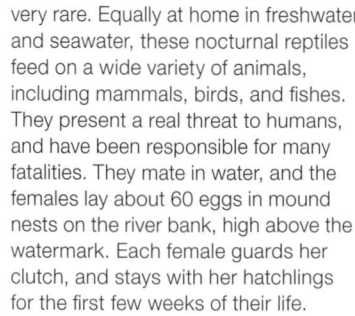

BASKING

The saltwater crocodile may haul itself onto a bank on hot days and bask with its mouth agape. Having its mouth open stops it from overheating, and it allows small birds to peck at parasites and bits of food lodged between its teeth.

Melanosuchus niger

Black caiman

Length 13–20 ft (4–6 m)
Breeding Oviparous
Habit Aquatic
Status Lower risk

Location N. South America

Hunted for its skin, the black caiman has become extinct in many parts of its range, and its numbers are thought to have been reduced by 99 percent in the past 100 years. It has an almost parallel snout tapering sharply at the tip and gray bands on the lower jaw. It floats on the surface by day but may come ashore to feed at night. Fishes, waterbirds, capybara, the occasional domestic animal, and even humans may fall prey to it. The female lays and guards 30–65 eggs, and stays with the young after hatching.

Osteolaemus tetraspis

Dwarf crocodile

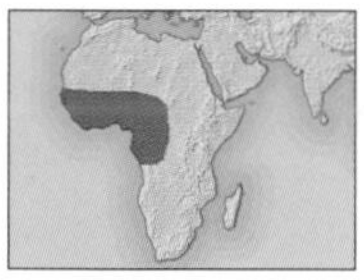

Length 5½ ft (1.7 m), max 6¼ ft (1.9 m)
Breeding Oviparous
Habit Aquatic
Status Vulnerable

Location W. and C. Africa

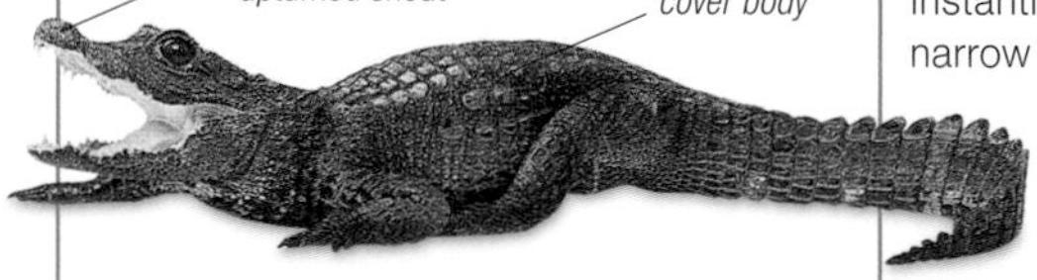

This relatively short, black crocodile lives in permanent pools and swamps in the African rain forest. It hides in burrows and among submerged tree roots during the day, and forages at the water's edge at night, taking its prey with a sideways snap of the jaws. In areas with seasonal flooding, fishes are the main food in the wet season, and crustaceans for the rest of the year. Frogs and toads are also eaten. The female lays usually 10 eggs in a mound of soil and vegetation, guards them, and protects her young in the water. This crocodile is probably common in some parts of its range, but in others, its numbers are depleted because of habitat destruction.

Gavialis gangeticus

Gharial

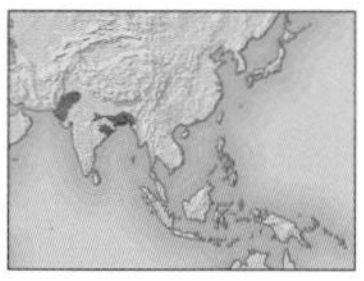

Length 13–23 ft (4–7 m)
Breeding Oviparous
Habit Aquatic
Status Critically endangered

Location N. Indian subcontinent

Instantly recognizable by its unique narrow snout, the gharial—or gavial—is a large but slender crocodilian that spends most of its life in water. Compared to other crocodilians, its legs are relatively weak, and its feet are broadly webbed. Gharials feed primarily on fishes, although they also eat waterbirds. During the breeding season, the male becomes strongly territorial, and assembles a harem of females. The female excavates a nest well away from the water's edge, and lays up to 50 eggs, which at about 5 oz (150 g) each are unusually large. She protects the hatchlings, but does not carry them to water—probably because of the unsuitable shape of her jaws. The gharial almost became extinct in the 1970s as a result of habitat loss, hunting, and fishing; subsequent captive breeding has only had limited success.

A MULTIFUNCTIONAL SNOUT

The gharial's long, slender snout may look fragile, but it is armed with small, extremely sharp teeth. It catches fishes sideways, and then flicks them within its jaws so that it can swallow them head first. The bulbous growth at the end of the nose is present only in males, and is used to produce sounds and bubbles during courtship.

LIFE IN WATER
The highly aquatic gharial inhabits slow-moving backwaters of large rivers, leaving the water only to bask on mud or sandbanks, and to lay eggs. It submerges itself in deep water for safety. Mating, too, takes place in water. The gharial has poorly developed legs and is unable to walk on land, barely managing to push itself over the mud.

REPTILES

EXHALING UNDERWATER
Crocodiles breathe air just as mammals do. This saltwater crocodile will soon have to return to the surface to take in more air.

AMPHIBIANS

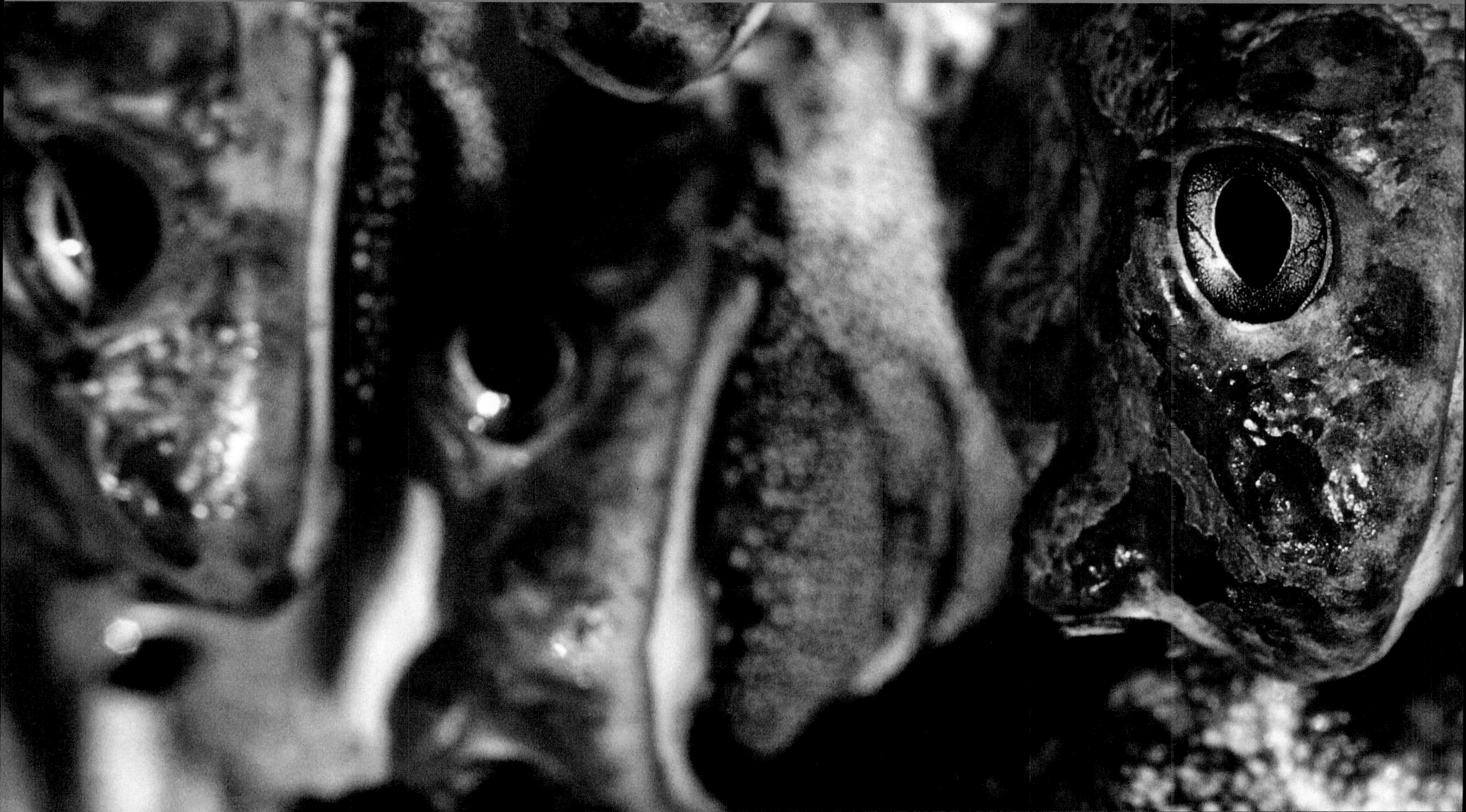

AMPHIBIANS

PHYLUM	Chordata
CLASS	Amphibia
ORDERS	3
FAMILIES	61
SPECIES	6,638

Amphibians are intimately associated with water, most of them spending part of their life in water and part on land. These ectothermic (cold-blooded) vertebrates are divided into 3 groups: newts and salamanders; frogs and toads; and the wormlike caecilians. The life cycle of most amphibians involves a transformation from aquatic larvae (which breathe using gills) to terrestrial adults (which take in oxygen through lungs). Most live close to water in tropical and temperate regions, although some have adapted to survive cold and drought.

TREE FROG
This tree frog, from the rain forests of New Guinea and northern Australia, has moist skin that is well suited to its wet and humid habitat.

Past and present

The first amphibians, which appeared about 370 million years ago during the Devonian Period, were descendants of fleshy-finned fish. They were the first vertebrates to colonize the land, and had scaly skin and a finned tail, but also lungs that could breathe air. They were also the first quadrupeds, evolving 2 pairs of jointed limbs from the fins of their fish ancestors. These enabled them to lift their bodies off the ground, although they were still heavily built animals, with slow and clumsy movements compared to amphibians alive today. By about 280 million years ago, amphibians had become the dominant land animals. However, over the next 70 million years, amphibians were superseded by terrestrial reptiles (which had evolved the ability to lay shell-covered eggs on land) and most species gradually died out.

Much more recently, amphibians have been struck by a far more rapid decline. Many living species are endangered—nearly 500 critically so—and in the last 2 decades, a number have become extinct. The reasons for this are complex, and not fully understood, but they probably involve a range of factors, acting both on a local and global scale. Among them are contamination by waterborne pollutants, infection by chytrid fungi, and climate change. Chytrid disease was first observed in Australia, but it now affects amphibians on several continents, and may eventually spread worldwide.

AMPHIBIAN ANCESTOR
Diplocaulus lived about 280 million years ago and probably resembled a modern newt, but for an unusually shaped head.

Anatomy

Compared to other vertebrates, amphibians have a fairly simple body structure, well adapted for life in water and on land. Most adult amphibians have 4 limbs, with 4 digits on the front limbs and 5 on the back. Newts and salamanders have a relatively small head, a long, slender body, with short limbs of equal length, and a long tail. However, some salamanders have very small limbs with reduced numbers of digits, or lack rear limbs entirely. Most frogs and toads have a distinctive compact body shape, with a large head, short back, small front limbs, large, muscular back limbs, and no tail. Caecilians have a wormlike, limbless body, with a pointed head and a short tail. Other than caecilians, which use smell, most amphibians hunt using sight and have large eyes that help them to locate prey at night. They

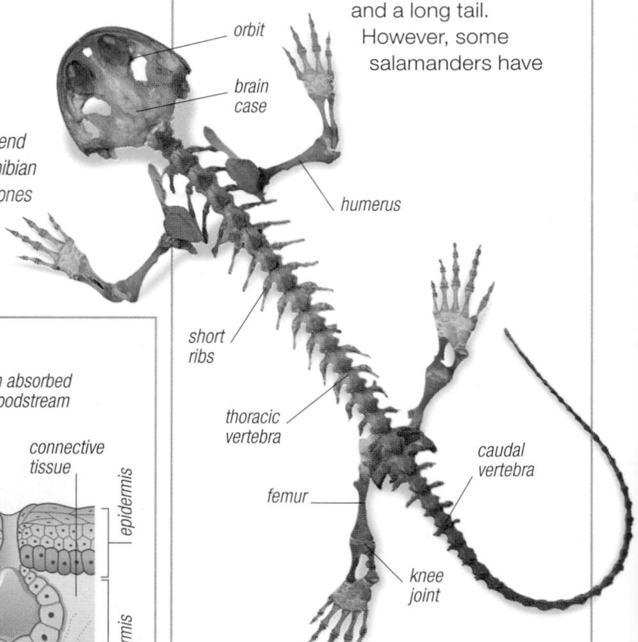

SKELETON
This skeleton of a juvenile Japanese giant salamander highlights an evolutionary trend toward simplifying the amphibian skeleton. It contains fewer bones than other vertebrates, and far fewer than its fish ancestors.

SKIN

The skin of an amphibian is naked and smooth, with no scales or hair to protect it. It is permeable to water and, although numerous mucous glands help to keep the skin moist, most amphibians quickly dry out if they are not in a damp place. All amphibians have poison glands in their skin, and many produce secretions that are distasteful or highly toxic to potential predators. The skin usually contains numerous pigment cells, and many amphibians are brightly colored. This is especially common in poisonous species, in which striking skin patterns serve as a warning. Many amphibians use their skin for breathing, which means that it has to be kept moist at all times. This may also place a limit on how large amphibians can be, since they need to maintain a high ratio of surface area to volume, to absorb an adequate amount of oxygen.

IN SUNLIGHT

IN SHADE

CHANGING SKIN COLOR
The base color of some amphibians' skin changes in response to variations in temperature, light level, or mood. When in sunlight, the skin of the Australian green tree frog (left) is light green but on entering shade gradually turns brown.

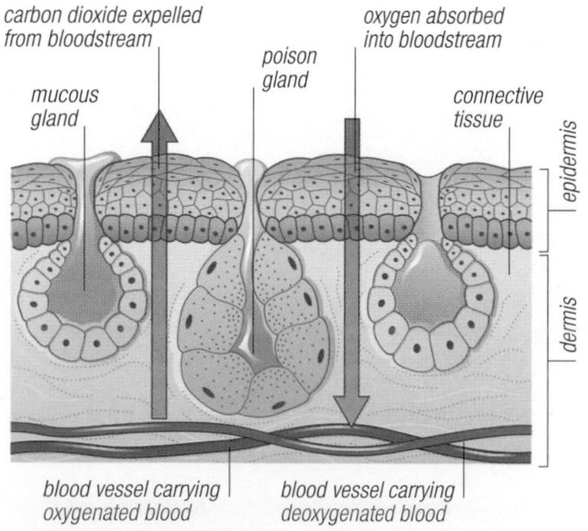

BREATHING THROUGH THE SKIN
Many amphibians can breathe through their skin, absorbing oxygen directly into their bloodstream and expelling carbon dioxide. In addition to being packed with blood vessels—some only a few cells' width from the surface—the skin is very thin, enabling the easy transfer of gases. Mucous glands keep the surface damp, helping the gases to pass through.

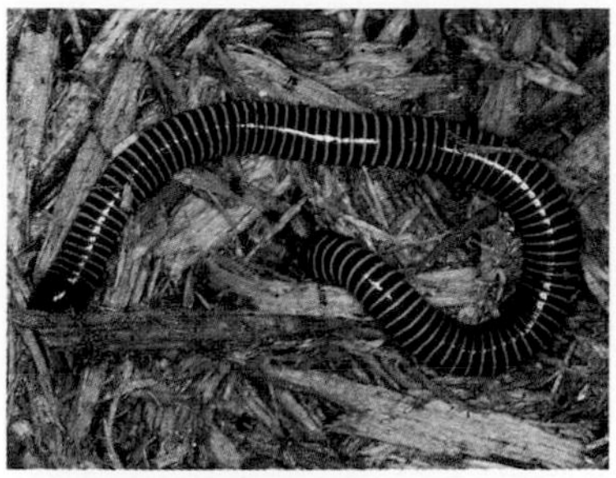

WORMLIKE AMPHIBIANS
With their long bodies, body rings, and lack of limbs, caecilians are often mistaken for large earthworms. Most are found in loose soil or deep leaf litter in tropical forests.

also have an extremely wide mouth, which enables them to consume relatively large prey.

Although most adult amphibians breathe using lungs, they do not have a rib cage and a diaphragm to inflate them. Many amphibians can absorb oxygen into their blood through other organs and tissues, including gills, the skin, and the lining of the mouth. The relative importance of each route varies among different life stages and adult forms. Young amphibian larvae breathe using external gills, as do those salamanders that retain the larval form into the adult stage. Most terrestrial frogs and toads, and some salamanders, have lungs, but also absorb a certain amount of oxygen through the skin and the mouth (see panel, opposite). One group of salamanders, the plethodontids, have no lungs and rely entirely on the skin and the mouth to take in oxygen.

Life cycles

An amphibian's life cycle involves 3 stages—egg, larva, and adult, with the transition from larva to adult involving a radical transformation known as metamorphosis.

Most amphibians deposit their eggs in water. However, some species lay their eggs on land, and others show a variety of ways of retaining them inside their body. In nearly all frogs and toads, fertilization is external, with sperm entering the eggs outside the female's body. During mating, the male clasps the female, usually from above, and sheds sperm into the water at the same moment as the female expels her eggs. In most newts and salamanders, fertilization is internal, the sperm being transferred to the female in small, gelatinous capsules called spermatophores. This requires precise coordination between the movements of male and female, and many newts and salamanders have elaborate courtship behavior to achieve this. In caecilians, the male has a penislike organ, which he uses to insert his sperm directly into the female.

The eggs of amphibians have no shell, but are enclosed in a gelatinous envelope that quickly shrivels up in dry conditions. Eggs may be laid singly, in strings, or in clumps; they may be carefully attached to plants or other solid objects, or just left to drift in the water. Several amphibians show parental care, guarding their eggs against predators and infection from fungal diseases. The number of eggs laid varies enormously: some amphibians lay only one or 2 eggs at a time; others lay up to 50,000. The eggs hatch into minute larvae, which retain sufficient yolk to sustain them for a few days.

The larvae of amphibians (called tadpoles in frogs and toads) are very different in appearance, diet, and lifestyle from adults. Metamorphosis involves a major reconstruction of all parts of the body (see panel, below). The larvae of newts and salamanders are carnivorous, feeding on tiny invertebrates. In contrast, the tadpoles of frogs and toads are typically herbivores, either filter feeding on microscopic plants that live suspended in water, or grazing on algae that cover plants, rocks, and other submerged objects.

The larval stage is an important growth phase for amphibians, but different species vary considerably in the amount of total lifetime growth that is completed in the larval rather than the adult stage. Nearly all amphibians actually become smaller during metamorphosis.

Some amphibians show direct development, in which the larva is not free-swimming but completes its development while still inside the egg and emerges as a tiny version of the adult form. Direct development occurs in a number of species that lay their eggs on land, and also in some that retain their eggs inside the body. Such species, by having no free-swimming larval stage, are not dependent on standing water.

Breeding migrations

Most amphibians live on land as adults but return to water each year to breed. Breeding is usually stimulated by a combination of environmental factors, including temperature, rainfall, and day length. In some species, referred to as explosive breeders, there is a mass migration of adults, all moving toward a breeding pond. Evidence shows that various species locate the same breeding site year after year by using their sense of smell, an ability to navigate by the sun, or by detecting tiny variations in the earth's magnetic field.

TAKING TO LAND
Swarms of newly metamorphosed European common frogs gather by the side of a pond before leaving the water for the first time.

PARENTAL CARE
European newts lay their eggs individually. Wrapping them in leaves may protect them from UV radiation.

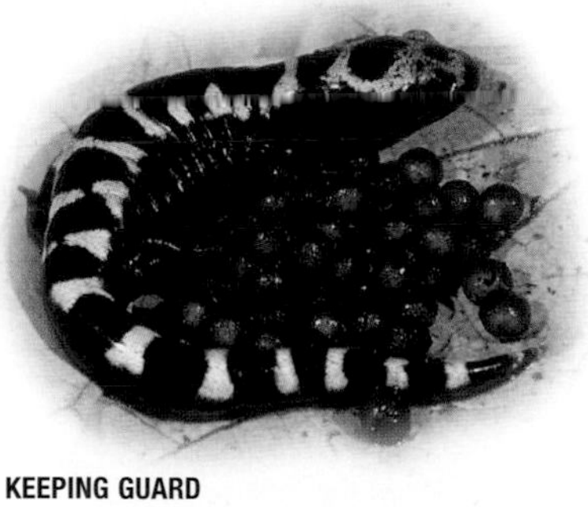

KEEPING GUARD
This marbled salamander guards its clutch of eggs at the edge of a pond. The female curls her body around her eggs to protect them from predators and fungal infection.

GROWTH AND DEVELOPMENT

The 3 stages of amphibian development are clearly illustrated by the European common frog. Embryos start to develop inside the eggs, which are laid in clumps. About 6 days after fertilization, the eggs hatch into tiny tadpoles with a spherical body, well-developed tail, and external gills. Within 4 weeks, these gills have been absorbed and replaced by internal gills. Between 6 and 9 weeks, the back limbs start to develop, the head becomes more distinct, and the body takes on a more streamlined shape. At this stage, the legs are fully functional and the internal gills have been replaced by lungs. By 9 weeks, when the front legs have emerged, the tadpole has the compact body shape of an adult. The tail is then gradually absorbed into the body, until, by about 16 weeks, it has disappeared entirely.

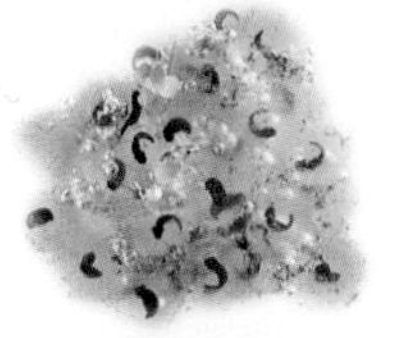

FROG SPAWN

RECENTLY HATCHED

4 WEEKS OLD

6–9 WEEKS OLD

12 WEEKS OLD **9 WEEKS OLD**

EGG TO ADULT
The European common frog's development, from egg to adult, typically takes about 16 weeks, but varies according to temperature levels and food supply.

MIGRATING FROGS
Some species of amphibians, such as these European common frogs, make annual mass migrations to breeding sites, often from great distances and across rough terrain.

Newts and salamanders

PHYLUM	Chordata
CLASS	Amphibia
ORDER	Caudata
FAMILIES	9
SPECIES	597

Newts and salamanders have a slender body with a long tail and, typically, 4 legs of similar size. Of the 3 main groups of living amphibians, they most closely resemble the animals from which all amphibians are descended. They also have some of the most complex and varied life histories. Some salamanders live their entire lives in water, while others are wholly terrestrial. Newts spend most of their lives on land but return to water to breed—other than this, there is no scientific distinction between salamanders and newts. In all but the most primitive species, fertilization is internal (rather than external, as in most frogs and toads). The male does not have a penis but instead passes capsules of sperm to the female during mating. The larvae of newts and salamanders are carnivorous, and have a long, slender body, a deep, finlike tail, and large, feathery external gills. Newts and salamanders are found in damp places and are largely confined to the Northern Hemisphere.

SALAMANDER HUNTING
Salamanders have wider mouths than newts and can catch larger prey like this earthworm. Many have a long tongue that flicks out then opens up to engulf the prey.

Feeding

All newts and salamanders are carnivorous and feed on living prey. They find their prey using a combination of smell and vision; in addition, some aquatic species are very sensitive to water currents made by moving animals. Since they are rarely very active, newts and salamanders do not have a high energy requirement and do not need to feed very often, especially if they have eaten a large amount. When food is abundant, they accumulate fat deposits that enable them to survive periods when it is too dry or too cold for them to feed. The larvae are also carnivorous, eating a wide variety of aquatic invertebrates. In some species, faster-growing larvae become cannibalistic, eating smaller larvae of their own species.

Anatomy

Newts and salamanders differ from frogs and toads in that they retain their tail into adulthood. They also have a relatively small head and small eyes, smell being the most important sense for finding food and social interaction. Most newts and salamanders have 4 legs (with 4 digits on the front limbs and 5 on the back), although some aquatic salamanders have no back legs and only small front legs. Many salamanders have distinct moisture-maintaining "costal" grooves that run around the body. Newts and salamanders vary in the way that they take in oxygen. Some are able to breathe through their smooth, moist skin, while many of those species that spend all their life in water retain the large, feathery external gills characteristic of the larval stage.

LUNGLESS SALAMANDERS
The plethodontids are a group of typically wholly terrestrial salamanders. They have no lungs but instead breathe through their skin and their mouth and throat, which have a rich blood supply.

SIRENS
The sirens, of which there are 4 species, spend their entire lives in water. In addition to having lungs, like some other aquatic salamanders, they take in oxygen through large, feathery external gills on the side of their head. In periods of drought, they can survive for several weeks or months, encased deep inside a cocoon of dried mud. During this time, they feed on their large fat reserves.

COURTSHIP
The courtship of newts and salamanders is varied and often elaborate. Courtship precedes and accompanies the transfer of sperm from male to female in capsules known as spermatophores. The male stimulates the female by secreting compounds known as pheromones. These are usually delivered via the nose but in some species they are passed directly through the skin. European newts—such as these Alpine newts (here the male is on the left)—use visual displays rather than contact to stimulate the female, but in many salamanders the male holds the female in a mating embrace known as amplexus while he stimulates her.

Life cycles

Newts and salamanders have complex life cycles, which typically involve 3 distinct stages: the egg, larva, and adult. The female lays eggs that contain yolk to support the growth of the developing embryo. Some species lay many small eggs, others a few large eggs, each containing a lot of yolk. The eggs hatch into larvae that have feathery external gills and feed on small animals. The larvae grow until they reach the stage at which they undergo metamorphosis into the adult form; this involves the loss of the gills and a switch to breathing air via the lungs and skin. The juvenile stage, which in some is known as the eft stage, is spent on land and may last for several years. In the adult stage, individuals become sexually mature and start to breed; this may involve a brief return to water. There are many variations on this basic pattern, which usually coincide with whether the species is amphibious, terrestrial, or aquatic (see right).

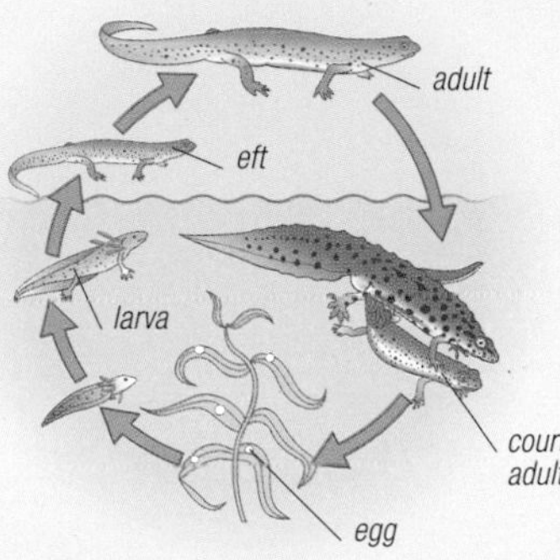

AMPHIBIOUS
This life cycle is typical of American, European, and Asian newts. Adults spend most of their life on land but return to water each spring to breed. This involves a partial reversal of metamorphosis, as individuals develop skin that can absorb oxygen from water and a tail like that of the larva, enabling aquatic newts to swim powerfully.

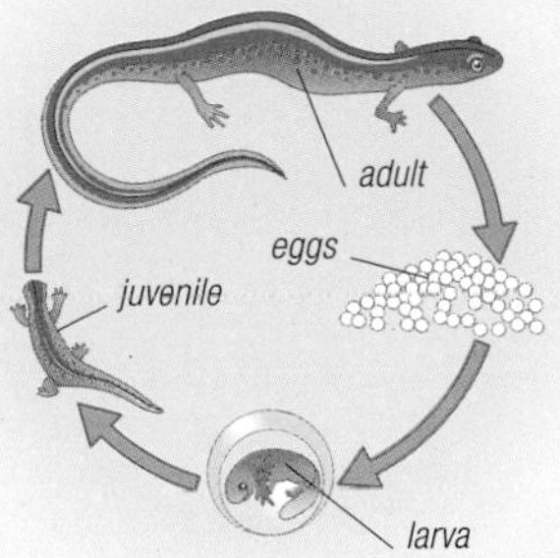

TERRESTRIAL
The eggs of terrestrial salamanders are typically laid on land. There are usually relatively few, large eggs, each containing a lot of yolk, which may be protected by the mother. The larval stage is completed within the egg, which hatches to produce a miniature terrestrial adult. In some species, the eggs may be retained inside the female's body. She gives birth, either to aquatic larvae or to terrestrial juveniles.

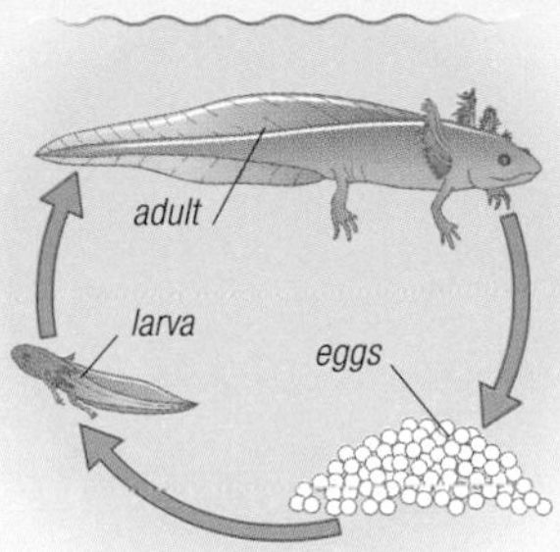

AQUATIC
In the aquatic life cycle, all the life stages are completed in water. The female typically produces a large number of small eggs and does not care for them. The adults become sexually mature while retaining many of the anatomical and physiological features of the larval stage. In many species, there are individual populations that have an aquatic life history, while others have an amphibious one.

NEOTENIC NEWTS AND SALAMANDERS
In neotenic newts and salamanders, individuals become sexually mature while still in the larval stage. They are fully aquatic and have large, feathery external gills. The best known example is the axolotl (shown here), which can be artificially made to metamorphose into a terrestrial adult form by injecting it with the hormone thyroxin.

AMPHIBIANS

Defense

For most newts and salamanders, the risk of being attacked by a predator is greatly reduced by being active only at night. Many species also produce distasteful or toxic secretions from glands in their skin. These glands may be concentrated on the head or the tail. The skin secretions of Californian newts are particularly lethal. Such species are often brightly colored to warn predators, although the effectiveness of this coloration depends on predators learning to associate it with unpleasant consequences. A few species do not produce noxious secretions but mimic the color patterns of those that do. Several species are able to shed their tails when attacked. The detached tail, which twitches after it is dropped, distracts the attention of a predator away from the animal's body.

UNKEN POSITION
Some species of newts and salamanders adopt bizarre defensive positions. This Italian spectacled salamander has taken the contorted "unken" position to alarm predators and expose its bright warning coloration.

COLORATION
Many newts and salamanders have skin patterns and colors that provide camouflage. Others are very brightly colored, usually indicating their capacity to produce noxious or toxic skin secretions. However, in some species (such as this alpine newt), males become more brightly colored to attract females during the breeding season.

Siren lacertina

Greater siren

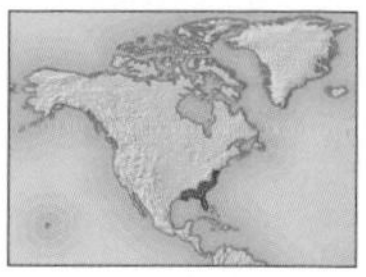

Length 20–35 in (50–90 cm)
Habit Wholly aquatic
Breeding season Early spring
Status Least concern
Location E. and S.E. USA

Sirens are among the largest amphibians in North America. There are 2 species, both with long, eel-like bodies and only one pair of legs, behind their feathery external gills. The greater siren is usually gray or olive-brown, and this color provides it with good camouflage in the muddy water of ponds, lakes, and slow-flowing rivers. This siren spends the day resting on the bottom, but at night forages for food, either crawling over mud or sediment, or swimming with a sinuous movement of its tail. Like most amphibians, greater sirens are almost entirely carnivorous—snails, insect larvae, and small fish making up most of their food. However, during droughts, they encase themselves in mud cocoons, and can survive without food for up to 2 years. Greater sirens lay their eggs in early spring. When their larvae hatch, they are little bigger than most other amphibian tadpoles.

Pseudobranchus striatus

Northern dwarf siren

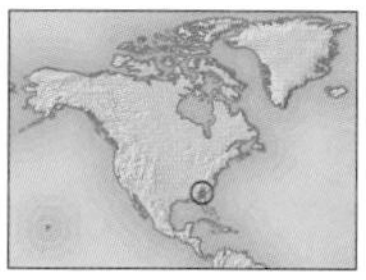

Length 4–9 in (10–22 cm)
Habit Wholly aquatic
Breeding season Not known
Status Least concern
Location S.E. USA

This small, eel-like siren is brown or black above, with one or more yellow or tan stripes along its body. Its forelimbs are small, each with 3 toes, and it has no hind limbs. The feathery external gills help it breathe in water with a low oxygen content. This nocturnal amphibian inhabits swamps, ditches, and marshes, and estivates in mud during the dry season. It feeds on aquatic invertebrates, and is preyed upon by wading birds, water snakes, alligators, and turtles. The female attaches her eggs to water plants.

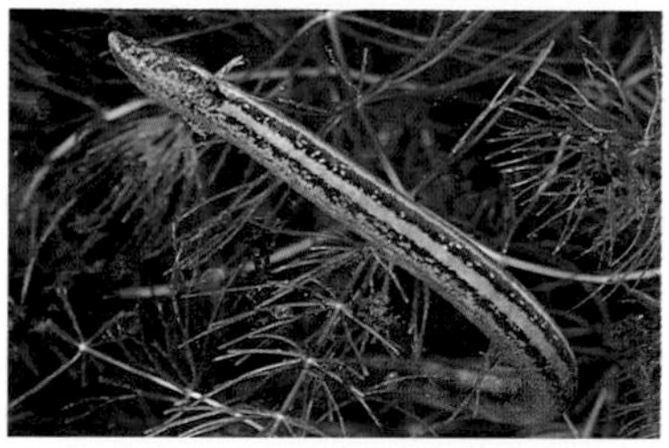

Batrachuperus pinchonii

Western Chinese mountain salamander

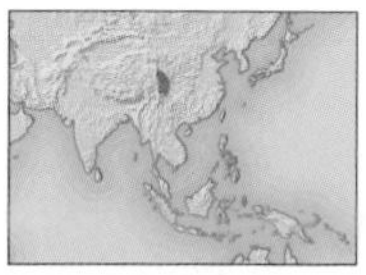

Length 5–6 in (13–15 cm)
Habit Largely aquatic
Breeding season Spring and summer
Status Vulnerable
Location E. Tibet, China (Sichuan Province)

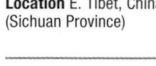

This robust amphibian has a pale brown or olive-green body, usually with darker spots, and about 12 costal grooves along its flanks. It has short legs, with 4 toes on each limb, and a short tail that is flattened from side to side. This nocturnal animal lives in cold, clear, fast-flowing streams at high altitudes. Since there are few fish in these cold water courses, the eggs and larvae of this salamander are relatively safe from predators. The female lays 7–12 eggs in hollows under rocks in the streams.

Cryptobranchus alleganiensis

Hellbender

Length 12–29 in (30–74 cm)
Habit Wholly aquatic
Breeding season Fall
Status Near threatened
Location E. USA

The hellbender is the only North American member of the giant salamander family—a group of 3 species notable for their wrinkled skin and exceptional size. Green, brown, or gray, often with darker spots, it has a head flattened like a spade, small eyes, and a long, paddlelike tail. It is nocturnal, and uses its head to dig out crayfish and other prey. The hellbender has a strong bite, and its skin secretes a noxious slime. Females lay their eggs in fall, and the males guard them until they hatch.

Andrias japonicus

Japanese giant salamander

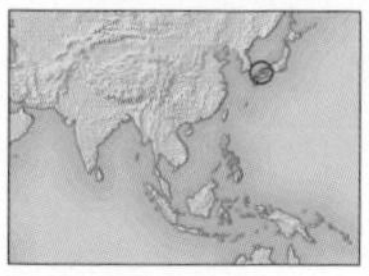

Length Up to 4½ ft (1.4 m)
Habit Wholly aquatic
Breeding season Late summer
Status Near threatened
Location Japan

Along with the Chinese giant salamander (*Andrias davidianus*), the Japanese giant salamander is the largest amphibian in the world. It is gray or brown, with a flattened body, small eyes, and sprawling legs. Its head and throat are warty, and the skin along its sides is deeply folded—a characteristic that helps it to absorb oxygen from the surrounding water. It is active mainly at night, feeding on fish, worms, and crustaceans. During the breeding season, the male digs a spawning pit, in which the female deposits strings of about 500 eggs. The male then guards the eggs, aggressively fending off predators. Japanese giant salamanders can live for over 50 years, but they are a traditional delicacy, and have become increasingly rare.

Necturus maculosus

Mudpuppy

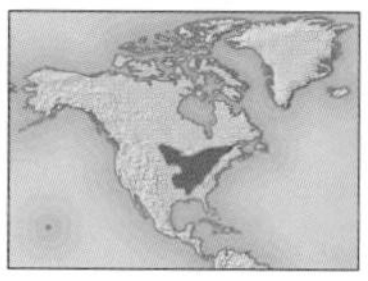

Length 11¾–19½ in (29–49 cm)
Habit Wholly aquatic
Breeding season Fall in north, winter in south
Status Least concern
Location S. Canada, C. and E. USA

The mudpuppy has a long, cylindrical body, a flattened tail, feathery external gills, and 2 pairs of limbs, each with 4-toed feet. It is brown, gray, or black, usually with black spots and blotches. An inhabitant of freshwater habitats, it feeds on small animals, mainly at night. Females lay their eggs in early summer under rocks and logs, then guard them until they hatch, about 2 months later. The mudpuppy has disappeared from many polluted or silted rivers.

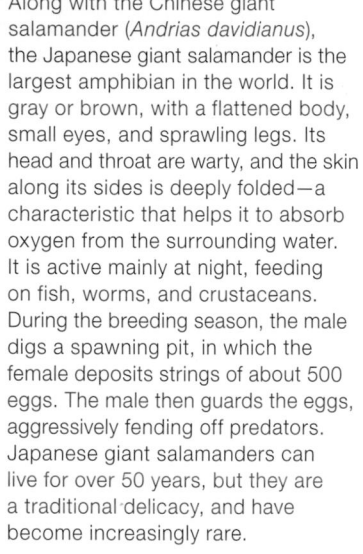

ADAPTABLE GILLS

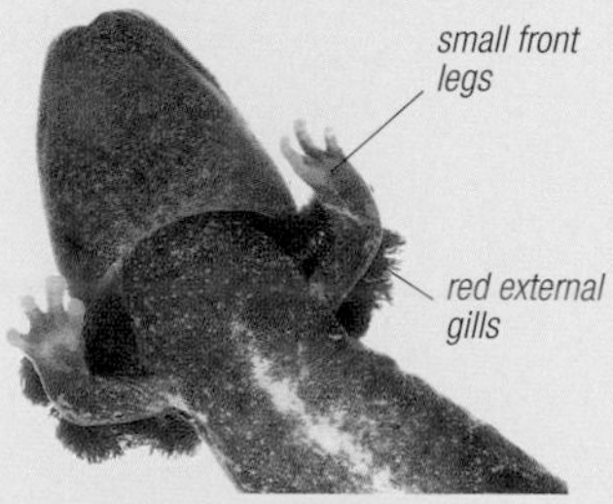

The gills of the mudpuppy adapt to the oxygen content of the animal's habitat. In cool, oxygen-rich water they are usually small, but in stagnant, warm water they tend to be considerably larger.

BODY SHAPE
The flattened, eel-like body shape of the mudpuppy enables it to burrow under rocks and logs.

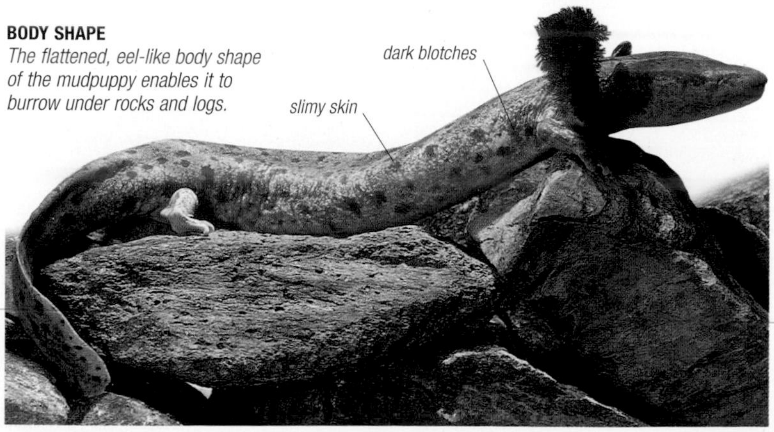

Proteus anguinus

Olm

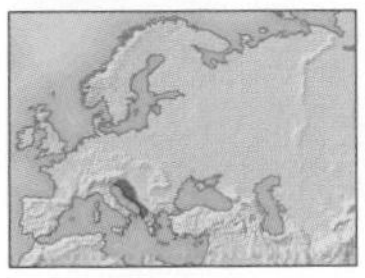

Length 8–12 in (20–30 cm)
Habit Wholly aquatic
Breeding season Not known
Status Vulnerable

Location S. Europe

The olm is one of the few amphibians that has become adapted to life in caves. It has a long and very slender body, a pinkish skin almost devoid of pigment, rudimentary eyes, and 2 pairs of tiny limbs—the front pair with 3 toes, and the rear pair with only 2. It also has red feathery external gills, which it retains throughout its life. Olms inhabit streams in caves, sometimes several miles from a cave opening, and feed on small invertebrates.

Chioglossa lusitanica

Golden-striped salamander

Length 4¾–5½ in (12–14 cm)
Habit Mostly terrestrial
Breeding season Early spring
Status Vulnerable

Location S.W. Europe

The slender body and long tail enable the golden-striped salamander to run very fast, like a lizard. Its coloration is dark brown, with 2 yellow, copper, or golden stripes on its back that merge to form a single stripe on the tail. This salamander has large eyes and a long, sticky tongue for catching insects. If attacked, it may shed its tail. During the breeding season, swellings appear on the upper front legs of males. After mating, females lay their 15–20 eggs under rocks in well-oxygenated but often slightly acidic streams.

Salamandra salamandra

Fire salamander

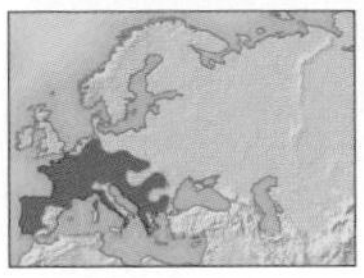

Length 7–11 in (18–28 cm)
Habit Wholly terrestrial
Breeding season Spring, summer, after heavy rain
Status Least concern

Location Europe

Similar in appearance to the tiger salamander (see p.444), this robust amphibian has an eye-catching coloration that warns predators that it is poisonous. It produces a toxic secretion from glands behind its eyes, and this is distasteful to anything attempting to eat it. The fire salamander inhabits forests and woodland on hills and mountains; it spends the winter underground. Breeding involves the male grasping his mate from below and depositing a spermatophore; he then flips to one side so that she falls on it. The eggs develop inside the female until the larval stage, and are then released into ponds or streams. During their development in the oviduct, larvae are occasionally cannibalistic, eating their smaller siblings as eggs or larvae. In some high-altitude populations, the young develop to the adult form inside the mother's body.

FEEDING HABITS

The fire salamander is active mainly at night, especially after rain, when it emerges from beneath logs and stones to feed on worms, slugs, insects, and insect larvae.

BODY COLOR
The body color is variable, depending on geographical location. It can be black and yellow, or yellow with black spots or stripes. In some areas, the yellow may be replaced by orange or red.

Tylototriton verrucosus

Crocodile newt

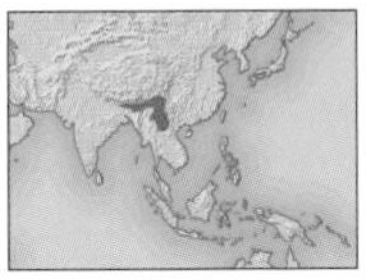

Length 5½–7 in (14–18 cm)
Habit Mostly terrestrial
Breeding season Monsoon
Status Least concern

Location S. and S.E. Asia

Also known as the mandarin salamander, this robust newt has a large head and thickset body, with rough black skin. The prominent orange warts, ridges, and glands indicate that the crocodile newt will exude distasteful skin secretions if attacked. This newt lives most of its life on land—spending winter and dry periods underground—but during the monsoon it migrates to breeding ponds, where it attaches its eggs to water plants. Active at night, the crocodile newt feeds on invertebrates.

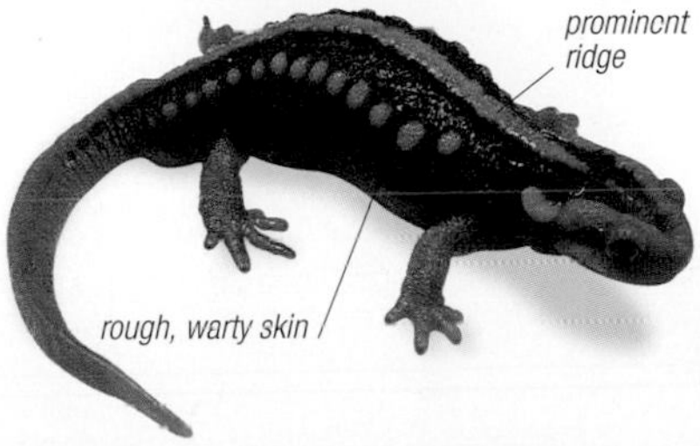

Triturus cristatus

Great crested newt

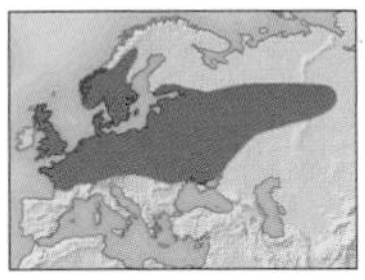

Length 4–5½ in (10–14 cm)
Habit Mostly terrestrial
Breeding season Spring and summer
Status Least concern

Location Europe, C. Asia

Also known as the warty newt, this secretive animal lives mostly on land but spends about 3–5 months of the year in ponds, lakes, and ditches. It is dark brown above, with a characteristically warty skin, and bright orange with black spots below. During the breeding season, the male develops a large, jagged crest on its back, and a white or blue streak along the tail. Like many other newts, the male courts the female with a complex underwater dance, and concludes the display by transferring a spermatophore to the female. The female lays her eggs singly and wraps them in leaves, probably for protection. These newts are widespread throughout Europe, but in many areas their numbers have decreased because their breeding habitat has been drained.

Ichthyosaura alpestris

Alpine newt

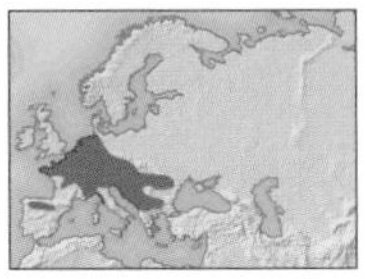

Length 2¼–4¾ in (6–12 cm)
Habit Mostly terrestrial
Breeding season Spring
Status Least concern

Location Europe

The alpine newt has a small body, blue above and bright orange below, and a short tail. In the breeding season, the male develops a low, black and white crest along his back, black and white spots on his sides and tail, and a swollen cloaca. As with most newts, the male attracts the female by fanning his tail and wafting a pheromone toward her snout. In the Balkans, some of these nocturnal animals are permanently aquatic, retaining larval features, such as external gills, into adulthood.

Taricha torosa

California newt

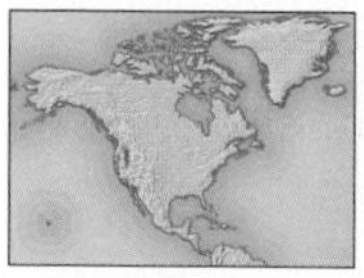

Location W. USA (California)

Length 5–8 in (12.5–20 cm)
Habit Mostly terrestrial
Breeding season Winter and spring
Status Least concern

California is particularly rich in salamanders and newts, and this species is one of over 20 that live in the state. Its rough, warty body is brown or brick-red above, and yellow to orange below. During the breeding season, when these newts carry out mass migrations to ponds and streams, the males develop a smooth, slimy skin, black patches on their feet, and a swollen vent (cloaca). The California newt is protected by a toxic secretion that makes both its skin and flesh extremely poisonous. It is nocturnal and feeds on invertebrates of all kinds.

Pleurodeles waltl

Sharp-ribbed salamander

Location S.W. Europe, N.W. Africa

Length 6–12 in (15–30 cm)
Habit Wholly aquatic
Breeding season Any time, except cold weather
Status Near threatened

This mottled gray-green salamander is one of Europe's largest tailed amphibians. Like most salamanders, it is protected by toxic skin secretions, but it also has sharply pointed ribs that project through its skin when it is held. Together with its chemical defenses, these help deter most predators. Sharp-ribbed salamanders are mainly nocturnal and spend their lives in water, except in times of drought, when they survive by hiding in mud until wetter conditions return. During the breeding season, females lay 200–300 eggs singly or in clumps, fastening them to water plants.

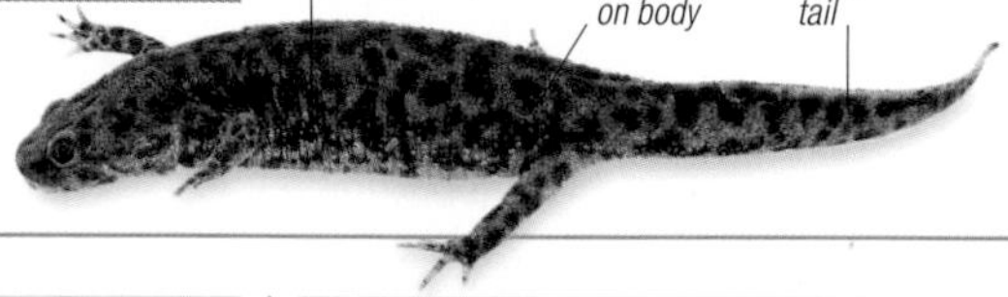

Notophthalmus viridescens

Eastern newt

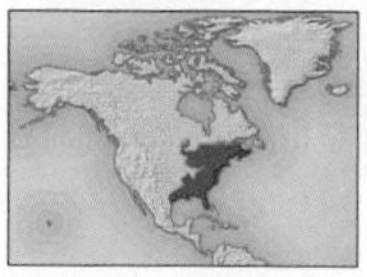

Location E. Canada, E. USA

Length 2½–4½ in (6.5–11.5 cm)
Habit Largely terrestrial
Breeding season Variable
Status Least concern

This common North American newt has a complex life cycle. It breeds in freshwater habitats, but instead of developing directly into adults, its larvae turn into specialized juvenile forms, called red efts. Red efts live on land—usually in grassland, mountain forests, and other damp habitats—and then return to water after 1–4 years, where they reach maturity. Outside the breeding season, adults live on land. During the adult and eft stages, these newts are protected from predators by highly toxic skin secretions.

JUVENILE COLORATION
The young eastern newt, or the red eft, is bright red, orange, or brown all over, with black-edged red spots along its back, and rough skin. This terrestrial juvenile stage lasts for 1–4 years, after which the red eft returns to water to develop into an adult.

ADULT STAGE
This newt completes its development in water. Its upperside turns greenish yellow, with black-edged red spots, and its underside becomes yellow. The breeding male has a very deep tail, a swollen cloaca, and rough pads on the inside of its thighs.

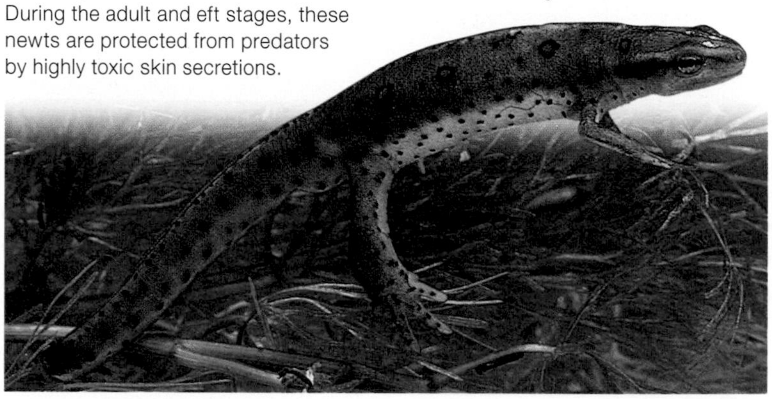

Amphiuma tridactylum

Three-toed amphiuma

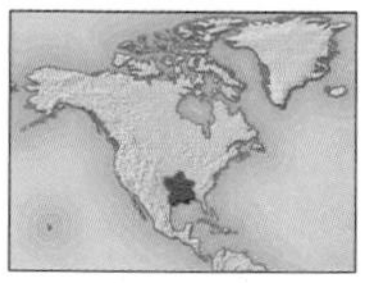

Location C. and S. USA

Length 18–43 in (46–110 cm)
Habit Wholly aquatic
Breeding season Winter and spring
Status Least concern

This large, slimy salamander has a slender, eel-like body that is black, gray, or brown above, and pale gray below. The tail is flattened from side to side, and each of the tiny limbs has 3 toes. It inhabits ditches, swamps, streams, and ponds and, in periods of drought, it can survive buried in mud. Active at night, it feeds on worms and crayfish. This animal can deliver a painful bite.

Dicamptodon tenebrosus

Pacific giant salamander

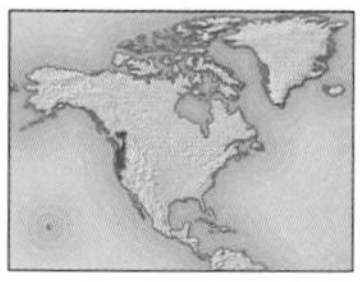

Location S.W. Canada, N.W. USA

Length 6½–13½ in (17–34 cm)
Habit Mostly terrestrial
Breeding season Spring and fall
Status Least concern

Active at night, this dark brown to black salamander, with light brown mottling, lives in wooded habitats in and around streams. It passes through an aquatic larval stage lasting for several years. When individuals reach maturity, they may become terrestrial. However, some never leave the water, becoming aquatic adults. Those that do leave the water are the world's largest terrestrial salamanders. In defense, this animal secretes noxious mucus from its tail. Populations have been adversely affected by the silting of streams.

Ambystoma tigrinum

Tiger salamander

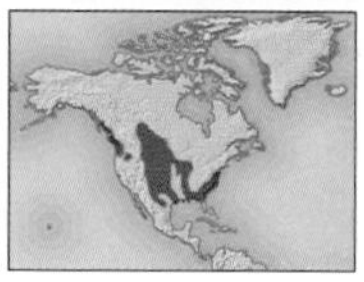

Location North America

Length 7–14 in (18–35 cm)
Habit Mostly terrestrial
Breeding season Variable, after heavy rain
Status Least concern

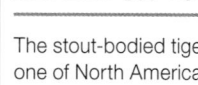

The stout-bodied tiger salamander is one of North America's most colorful amphibians, and is also one of the most locally varied. At least 6 subspecies are known: the eastern subspecies have a dappled pattern of black and yellow markings, but those farther west—including the one shown here—have larger yellow blotches or vertical bars. Adult tiger salamanders generally live on land, and hibernate in burrows dug by other animals. In early spring, once any ice has melted, they migrate to ponds, lakes, pools, and reservoirs to breed, sometimes in large numbers. The females lay up to 7,000 eggs in a single breeding season, and their larvae feed mainly on invertebrates, although some individuals may become cannibalistic, growing larger mouths with extra teeth. In most areas, the young leave the water once they have developed adult bodies, but in some regions—particularly in the west—they may retain their larval form permanently, reproducing without taking up life on land. The chief predators of the tiger salamander are fish, which eat the larvae; however, in recent years, the species has suffered through pollution and habitat loss.

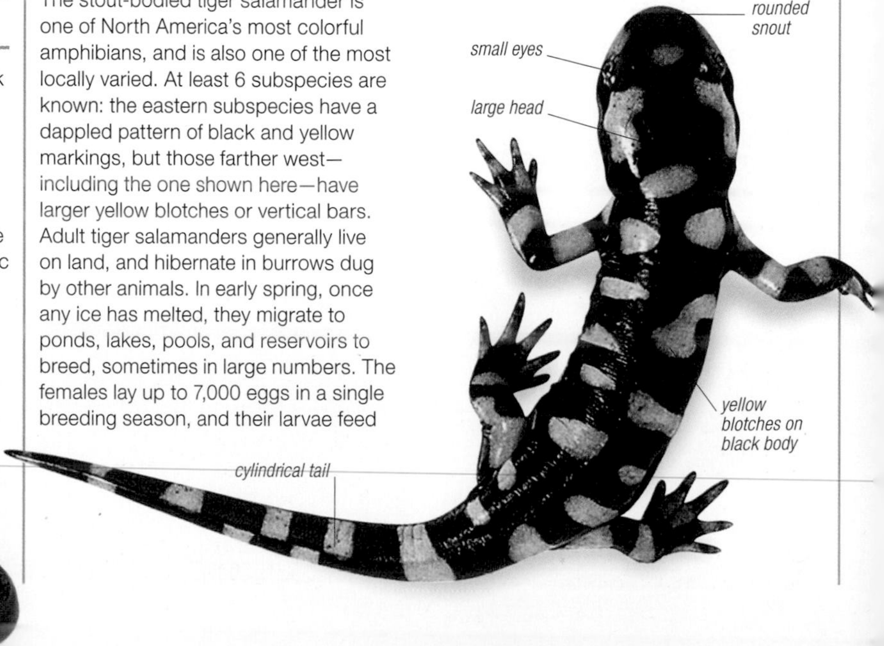

Ambystoma mexicanum

Axolotl

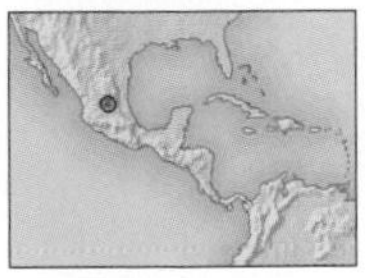

Location Mexico (Lake Xochimilco)

Length 4–8 in (10–20 cm)
Habit Wholly aquatic
Breeding season Summer
Status Critically endangered

This plump amphibian is a classic example of neoteny, or the ability to breed without ever developing a fully adult body. It has a flattened tail and large external gills—features that most salamanders lose when they become mature and take up life on land. Largely nocturnal, the axolotl feeds mainly on invertebrates, and is preyed upon in turn by waterbirds. Axolotls are critically endangered as they live in a single lake, in an area that has seen rapid urbanization in recent years. Once esteemed as a delicacy, they are now legally protected.

CAPTIVE FORMS

Typically black in the wild, many color variants of the axolotl have been bred in captivity. They may be albinos (white with red gills), gray, or mottled black and white. This species is more commonly seen in captivity than in the wild. If the captive animals are injected with a thyroid hormone, they lose their gills and become terrestrial.

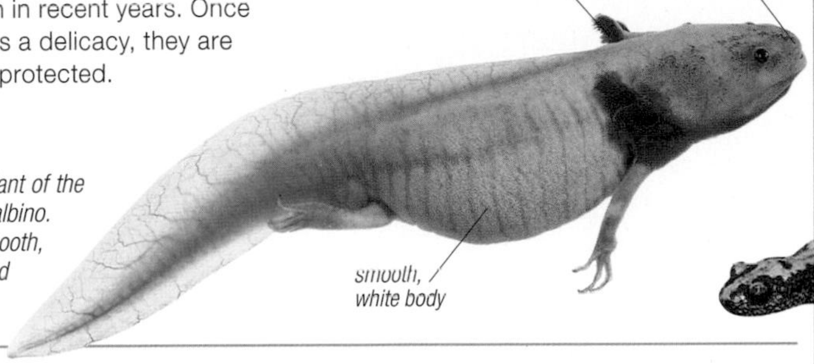

ALBINO
One color variant of the axolotl is the albino. This has a smooth, white body and red gills.

Ambystoma macrodactylum

Long-toed salamander

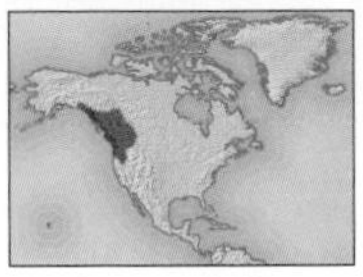

Location S.W. Canada, N.W. USA

Length 4–6½ in (10–17 cm)
Habit Mostly terrestrial
Breeding season Fall and spring
Status Least concern

The long-toed salamander is found in a wide variety of terrestrial habitats, where it spends much of its time underground. However, in spring, large numbers of this nocturnal species migrate to ponds to breed. It has a slender gray or black body, with a yellow, tan, or green dorsal stripe or blotches on the body. When attacked, it coils its body and lifts its tail in the air. Numbers have reduced because of habitat loss and the introduction of fish, that eat the larvae, to its breeding areas.

Plethodon jordani

Jordan's salamander

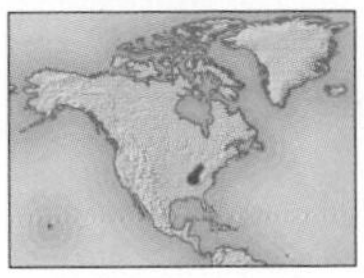

Location E. USA (Southern Appalachian Mountains)

Length 3¼–7½ in (8.5–18.5 cm)
Habit Wholly terrestrial
Breeding season Summer and fall
Status Near threatened

Also known as the red-cheeked salamander, this nocturnal species has a long, slender body and tail, a large head with prominent eyes and, typically, red patches on its legs or cheeks, although this characteristic varies considerably between subspecies. Both males and females are territorial. If threatened, this salamander produces a distasteful slimy secretion from its tail, and may shed its tail if attacked. It mainly feeds on millipedes, beetles, and insect larvae. The male breeds every year, while the female breeds only in alternate years.

Plethodon glutinosus

Slimy salamander

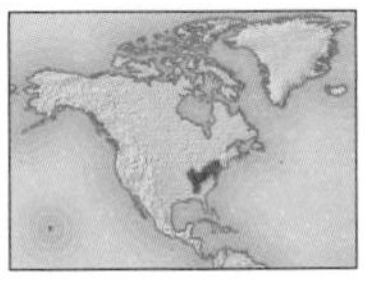

Location E. USA

Length 4½–8 in (11.5–20 cm)
Habit Wholly terrestrial
Breeding season Variable
Status Least concern

This large, slender-bodied salamander gets its name from the gluelike slime that it exudes when handled. The slime is extremely sticky—taking several days to wash off—and it gives the salamander highly effective protection against most predators. Adults are dark blue or black, with many small silver or gold spots. In the breeding season, males develop a swollen vent (cloaca) and a large gland under the chin. The slimy salamander is nocturnal, living in wooded hillsides and ravines.

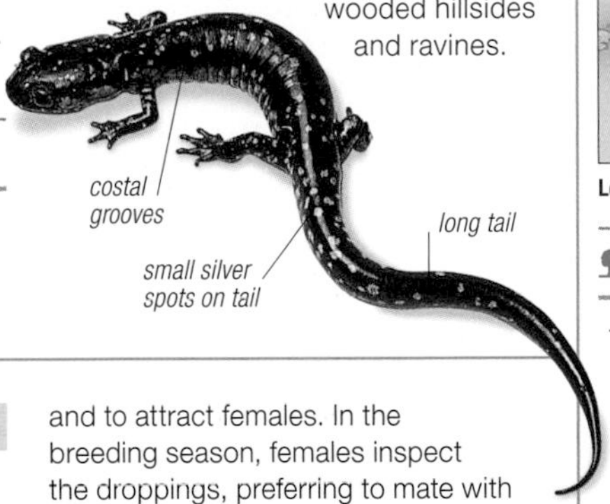

Plethodon cinereus

Red-backed salamander

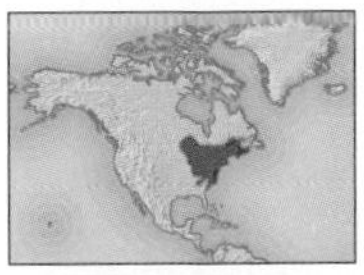

Location S.E. Canada, C. and E. USA

Length 2½–5 in (6.5–12.5 cm)
Habit Wholly terrestrial
Breeding season Fall to early spring
Status Least concern

This small salamander is dark gray or brown, usually with a broad red or red-brown stripe along its body and tail. Males use feces to mark their territory and to attract females. In the breeding season, females inspect the droppings, preferring to mate with males that have eaten a lot of termites. Males can reproduce every year, but females are fertile only in alternate years. The eggs are attached to the roofs of underground chambers.

Desmognathus ochrophaeus

Mountain dusky salamander

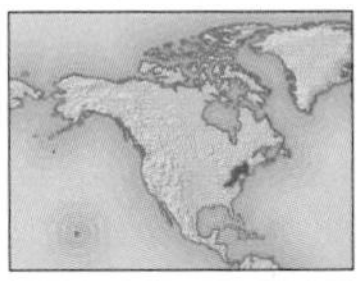

Location E. USA

Length 2¾–4¼ in (7–11 cm)
Habit Wholly terrestrial
Breeding season Spring, summer, and fall
Status Least concern

The mountain dusky salamander belongs to a group of about a dozen North American species that are so similar in their appearance, that it takes expert knowledge to tell them apart. Like most of its relatives, it has a long, slender body, protruding eyes, and a long, cylindrical tail. It is usually mottled dark gray, with legs that vary from brown to red (pictured below is a rare, red-legged variant that may be a mimic of the red-legged Jordan's salamander). The mountain dusky salamander lives on land once it has matured, and spends daylight hours underground. At night, it emerges to feed on insects and other small animals. It is protected by a toxic skin secretion and, if threatened, it sometimes sheds its tail. During the breeding season, adults mate on land. The male stimulates his prospective partner by rubbing a pheromone onto her skin, and abrading the skin's surface with specialized teeth. The female deposits a cluster of 12–20 eggs in a cavity beneath a tree trunk or underground, guarding them until they hatch. The larvae develop in streams. Adults are particularly active after rain, but they spend the winter underground.

Batrachoseps attenuatus

California slender salamander

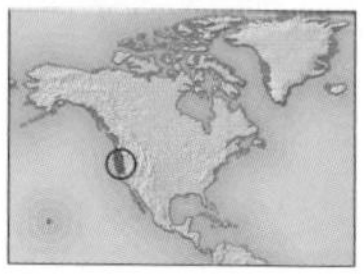

Length 3–5½ in (7.5–14 cm)
Habit Wholly terrestrial
Breeding season Not known
Status Least concern

Location W. USA (California, Oregon)

This salamander lives mainly underground and under logs and rocks, emerging at night after rain. Its extremely long, slender body enables it to enter very small crevices. It is dark brown or black, with a brown, yellow, or red stripe along the back, and tiny, 4-toed limbs. When in danger, the California slender salamander coils and uncoils its tail rapidly, and may shed it if attacked.

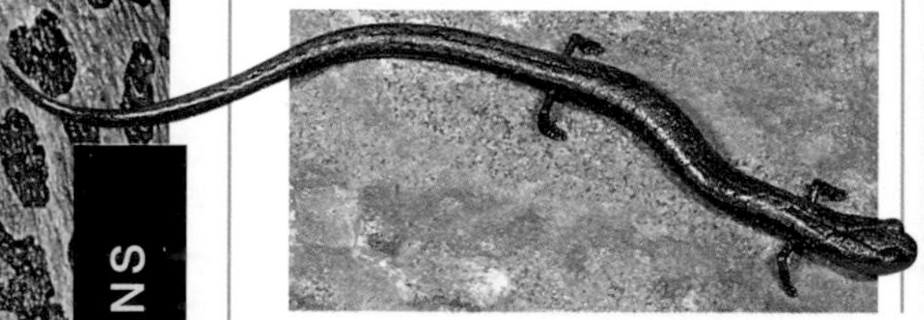

Pseudotriton ruber

Red salamander

Length 4–7 in (10–18 cm)
Habit Mostly terrestrial
Breeding season Any time of year, usually summer
Status Least concern

Location E. and S.E. USA

Brilliantly colored when young, but becoming progressively darker with age, the red salamander is found in damp places, springs, and cool streams in many eastern states of the USA. It has a thick body, a relatively short tail and legs, and a dense scattering of black spots over its red or orange-red skin. Largely nocturnal, it feeds mainly on invertebrates and other salamanders. To ward off predators, the red salamander raises its tail and waves it, but its color may also act as a defense. Although it is harmless, its red markings closely resemble those of the toxic young eastern newts (see p.444)—a similarity that may be a form of protective mimicry. The female lays about 70 eggs, attached to the underside of logs or rocks, and sometimes in water. Adult red salamanders usually spend the winter underground.

Aneides lugubris

Arboreal salamander

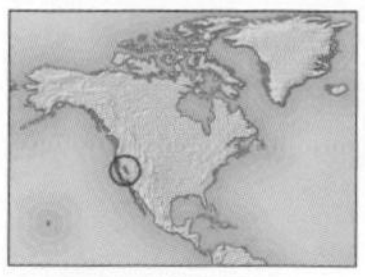

Length 4¼–7 in (11–18 cm)
Habit Terrestrial/Arboreal
Breeding season Not known
Status Least concern

Location W. USA (California)

This long, slender, brown and white salamander has flattened toe tips and a prehensile tail that enable it to climb trees with ease. The relatively large, triangular head contains prominent jaw muscles that are used in fights between individuals, or in defense—for instance, against predatory snakes. It lives below ground in winter and dry weather. The female lays 12–18 eggs in an underground chamber.

AMPHIBIANS

Bolitoglossa mexicana

Mexican mushroom-tongue salamander

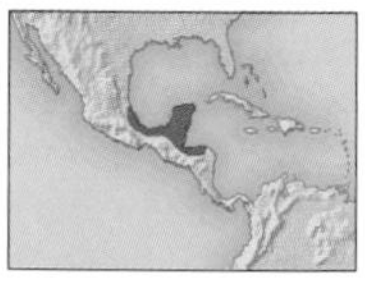

Length 4¾–6 in (12–15 cm)
Habit Arboreal/Terrestrial
Breeding season Any time, except dry/cold periods
Status Least concern

Location S. Mexico to Central America

With its prehensile tail, feet with adhesive pads on the base, and small size, this nocturnal animal is specially adapted for climbing trees and moving around on the smallest of branches. It has a mushroom-shaped tongue, which it flicks out to catch insects. The bronze or pink coloring, with black marbled patterning, acts as a warning that this salamander will emit noxious secretions from its skin if attacked.

Ensatina eschscholtzii

Ensatina salamander

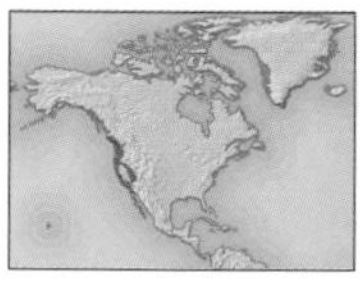

Length 2¼–3¼ in (6–8 cm)
Habit Wholly terrestrial
Breeding season Fall to spring
Status Least concern

Location W. USA

This nocturnal animal lives in damp leaf litter and under logs, and is especially active after rain. It has a short body, long legs, and prominent eyes. Its color depends on the area in which it lives: it may be uniformly brown, or black with white, yellow, orange, or pink blotches. The round tail, which narrows abruptly at the base, is used in defense. The salamander lifts it up and waves it around, emitting a distasteful secretion to deter predators; it may also shed its tail if attacked. During the breeding season, the male has a longer tail and develops a swollen cloaca. The female lays 8–20 eggs in clusters under logs.

Oedipina quadra

Honduran lowland worm salamander

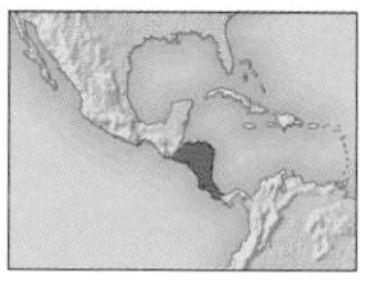

Length 4¾–7½ in (12–19 cm)
Habit Terrestrial
Breeding season Not known
Status Not evaluated

Location Central America

This is one of many new species of rain-forest salamanders discovered in Central America in recent years. It was found while sorting through leaf litter of pristine wet broadleaf forest in Honduras—and gets its name from its elongated body and tiny limbs. It is jet black on upper parts and sides, slightly paler underneath; its tail is nearly rectangular in cross section—hence its scientific name. The species belongs to a family of American lungless salamanders (including Plethodon, Bolitoglossa and allies) that breathe through their moist skin. They have diversified into more than 200 species in the wet forests of Central and northern South America, where they feed on tiny invertebrates, such as springtails. These tropical salamanders do not need free-standing water to breed because they have evolved direct development: eggs laid in the moist habitat hatch into miniature forms of adults, bypassing an aquatic larval stage.

Eurycea guttolineata

Three-lined salamander

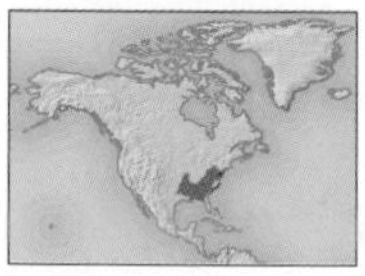

Length 4–7 in (10–18 cm)
Habit Wholly terrestrial
Breeding season Fall and early winter
Status Least concern

Location E. USA

A long and slender body with 13 or 14 costal grooves, prominent eyes, and a very long tail characterize this salamander. It is yellow, orange, or brown, with dark brown or black stripes down the middle of its back and along the sides. It lives under logs and rocks in damp areas near springs and streams; winters and dry periods are spent underground. Spiders, flies, beetles, and ants are the main diet of this predator, which feeds at night, especially after it has rained. During the breeding season, the male develops 2 fleshy projections on the upper lip, 2 protruding teeth, and a swollen cloaca. Eggs are laid in damp cavities.

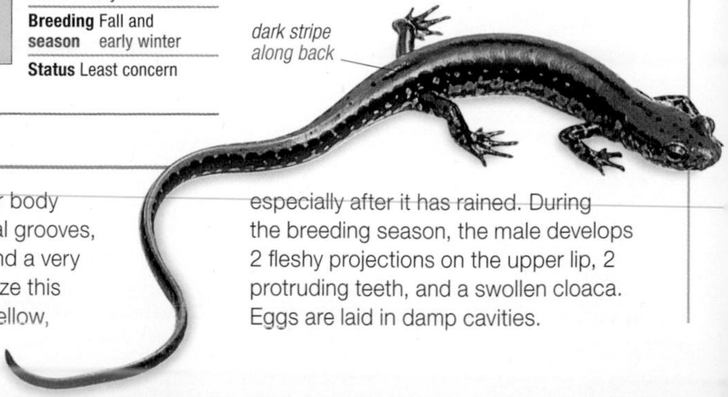

Caecilians

PHYLUM	Chordata
CLASS	Amphibia
ORDER	Gymnophiona
FAMILIES	3
SPECIES	183

These limbless, wormlike animals form the smallest of the 3 major groups of amphibians. They are rarely seen by humans, because they live either in soil burrows or underwater, and are found only in humid, tropical areas. Caecilians have varied life cycles: some species lay eggs, but in others the eggs are retained in the female's body and nourished by secretions produced by the oviduct until the eggs hatch and the young are released by live birth.

Anatomy

Caecilians have long, thin, black or brown to pink bodies and no limbs. They have rudimentary eyes and rely on a good sense of smell to find food and mating partners. Some resemble very large earthworms and burrow into soft soil or mud, using their pointed, bony head as a shovel. All but one species have lungs. The aquatic caecilians resemble eels; they have a fin on the tail that enables them to swim powerfully.

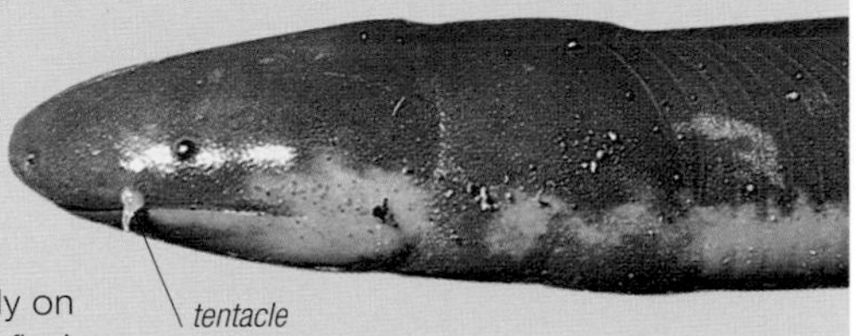

SENSORY TENTACLES
A unique feature of caecilians is the small tentacle located just below each eye. This collects chemical information, which is used to locate prey. Caecilians feed mainly on earthworms, which they catch with their sharp, curved teeth.

Ichthyophis glutinosus

Ceylon caecilian

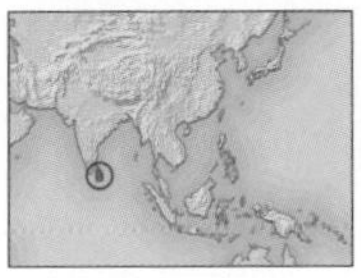

Location Sri Lanka

Length Up to 18 in (45 cm)
Habit Terrestrial/Burrowing
Breeding season Not known
Status Least concern

With its long, thin body covered with well-defined rings, and a short tail, the Ceylon caecilian resembles a large earthworm. It is brown with a bluish sheen, and has a pair of small, retractable tentacles between the nostrils and the eyes that enable it to pick up odors. This caecilian lives underground, mostly in muddy soil and swamps, where it feeds on worms and other invertebrates. The female coils around her eggs to protect them.

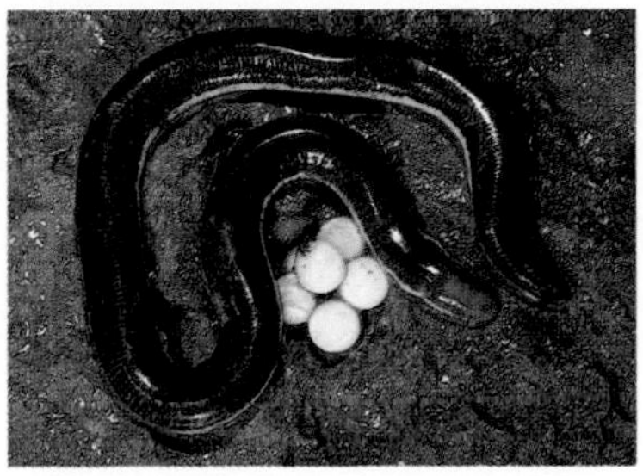

Gegeneophis mhadeiensis

Mhadei caecilian

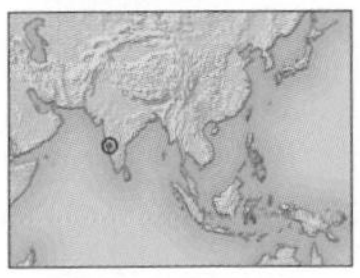

Location India

Length 7–8 in (18–20 cm)
Habit Wholly terrestrial
Breeding season Unknown
Status Data deficient

This caecilian was described in 2008 and is known from only 3 specimens, all collected from 2 adjacent localities in the Western Ghats of India. They were found near human habitations or on agricultural land—a factor that may indicate the species can tolerate human disturbance. It is dark brown with a pinkish head, and with paler rings on its "segments." Its habits are not known, but other members of the genus lay eggs on land and lack a larval stage.

Caecilia tentaculata

Linnaeus' caecilian

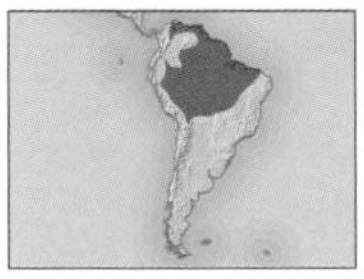

Location N. South America

Length 18–25 in (45–63 cm)
Habit Terrestrial/Burrowing
Breeding season Not known
Status Least concern

This large caecilian has a thick, flattened body, gray or black above and light brown below, and a wedge-shaped snout. It has no tail and, unlike other caecilians, its rear end is covered by a hard shield. Tiny scales are present in the skin in the 100–300 grooves that cover the body, and the tentacles, which are lower than in other caecilians, are not visible from above. Very little is known about its behavior, since it lives buried in loose soil, but it is known to be a solitary animal that lays eggs.

Dermophis mexicanus

Mexican caecilian

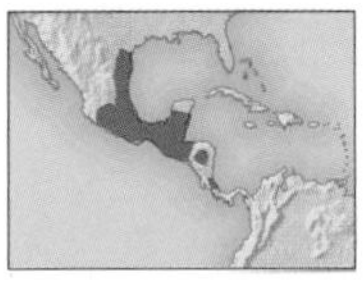

Location Mexico to Central America

Length 4–23½ in (10–60 cm)
Habit Terrestrial/Burrowing
Breeding season Not known
Status Vulnerable

This caecilian lives in the same region of the world as several amphisbaenians (see p.431), and is superficially similar to them, although it lacks legs. Gray, brown, or olive-green above, it has well-defined rings around its body, and a pointed, burrowing snout. It feeds chiefly on invertebrates, but also on lizards, and is completely subterranean, living in a variety of habitats where there is loose soil. The egg and larval stages are completed inside the female's body. After an extended gestation period, the mother gives birth to live young.

Siphonops annulatus

Ringed caecilian

Location N. South America

Length 8–16 in (20–40 cm)
Habit Wholly terrestrial
Breeding season Not known
Status Least concern

Also known as the South American caecilian, this forest-dwelling amphibian lives in soil and resembles a large, thick earthworm. It has a relatively short, stocky, dark blue body, encircled by white stripes and numerous rings. Like other caecilians, it uses its tentacles to pick up the odors of nearby prey—usually worms and other invertebrates. When attacked, it produces a distasteful secretion. The ringed caecilian lays its eggs in soil, where they hatch into miniature adults.

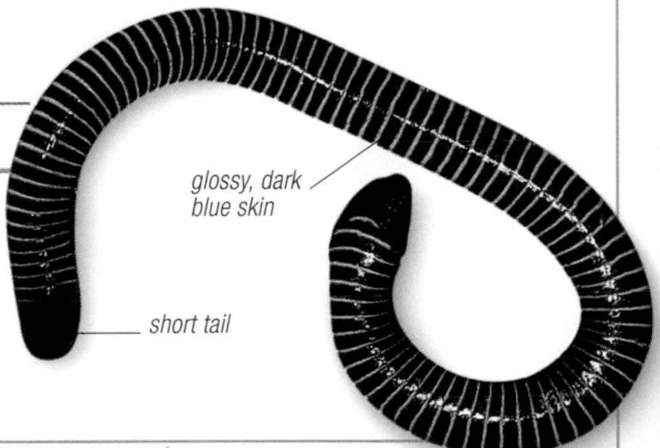

Typhlonectes compressicauda

Cayenne caecilian

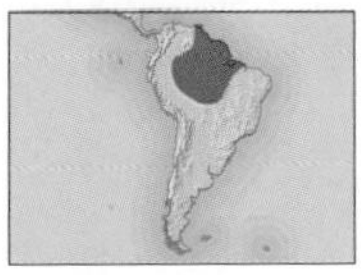

Location N. South America

Length 12–23½ in (30–60 cm)
Habit Wholly aquatic
Breeding season Not known
Status Least concern

An inhabitant of rivers, lakes, and streams, the Cayenne caecilian resembles an eel in shape and the way in which it swims. It has a long, slender, glossy black body, dark gray below, encircled by 80–95 rings. Its tail is flattened from side to side, with a slight fin along the top. The Cayenne caecilian produces toxic secretions when it is attacked, usually by fish. The eggs develop into larvae inside the body of the female, who gives birth to miniature adults.

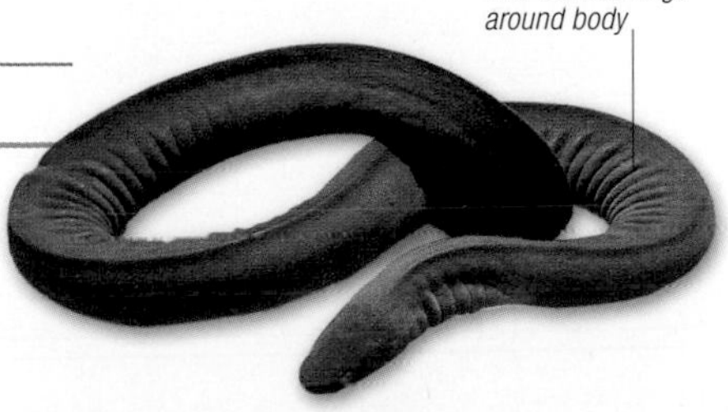

AMPHIBIANS

Frogs and toads

PHYLUM	Chordata
CLASS	Amphibia
ORDER	Anura
FAMILIES	49
SPECIES	5,858

Frogs and toads form much the largest and best known of the 3 groups of amphibians. As adults, they have no tail—unlike other amphibians, their tail is absorbed during metamorphosis from the larval to the adult stage. There is no clear distinction between frogs and toads: the term "toad" is often used to refer only to members of the genus *Bufo*, but it is also used more widely to describe any slow-moving, terrestrial species with a squat body and rough, warty skin. The larvae of frogs and toads, known as tadpoles, mainly feed on algae or plant material and characteristically have a spherical body that contains the long, coiled gut required by such a diet. In many adults, the back limbs are modified for jumping; they are much longer and more muscular than the front limbs. Most frogs and toads live in damp habitats, near the pools and streams in which they breed, but there are several species that live in arid areas. The greatest diversity of species is found in the tropics, particularly in rain forests.

Anatomy

Frogs and toads have a short, rigid body, a large, wide head, and, in most species, front limbs that are much smaller than the back limbs. The back limbs contain an extra section, just above the foot, that is not present in other amphibians. Most species have large, protruding eyes and a conspicuous eardrum (tympanum) on each side of the head, reflecting the importance of the senses of vision and hearing. The mouth is large, and many species have an adhesive tongue that they can flick out at high speed to catch their prey. The female is usually larger than the male, but males of many species have thicker and more muscular front limbs. These enable the male to grip the female securely during mating.

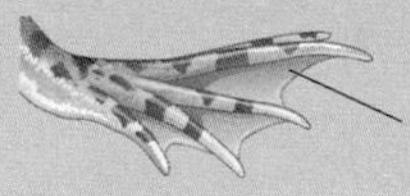

WEBBING

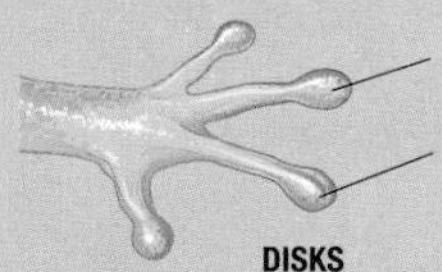

DISKS

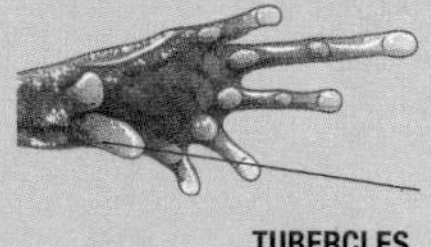

TUBERCLES

FEET
Frogs and toads have 4 digits on the front limbs and 5 on the back. In many species, especially those that spend much of their life in water, the digits on the limbs have webs of skin between them. Many frogs that live in trees have disk-shaped adhesive pads on the ends of their digits, which enable them to gain a secure foothold on smooth, vertical surfaces. Burrowing species have horny protuberances, called tubercles, on their back feet, which they use to excavate soil.

SKIN
The skin of frogs and toads is highly variable. At one extreme, the skin of some toads is so tough that it is used as a substitute for leather; at the other, there are frogs with skin so thin that their internal organs can be seen through it. Toads typically have rough, dry skin that is covered with warts and, in some species, small spines. Frogs generally have smooth, moist skin.

FROG SKIN

TOAD SKIN

Feeding

All frogs and toads are carnivores that feed on live prey rather than carrion. Unable to chew or break up food in their mouth, they have to swallow their prey whole. However, many species can eat large animals (including mice, birds, and snakes), and a single meal can meet their energy needs for a long time so that they do not need to feed often. Few species are active hunters, the most common feeding method being to wait until prey comes within range, either of their long, sticky tongue or of a forward, open-mouthed lunge.

CATCHING

SWALLOWING

SWALLOWING FOOD
Frogs and toads typically shut their eyes when swallowing their food. As their eyelids close, the eyeballs are rolled and pushed downward to increase pressure in the mouth.

SPAWNING SITES
Because bodies of water suitable for breeding are often scarce, many frogs and toads, such as these European common frogs, form large breeding groups, with frogs often converging from a wide area. The collective mass of eggs is also able to retain more heat, enabling the tadpoles to hatch sooner. Ponds and streams that dry up for part of the year are often used, since they do not support other animals that eat eggs and tadpoles.

FINDING A MATE
Many male frogs and toads call to attract females to the spawning site. The calls made by males enable a female to find a mate. More importantly, each species has its own characteristic call, which helps a female to find a mate of the same species in breeding pools used by several species at once. In many species (such as the aquatic frog, right), the effectiveness of the call is enhanced by one or more vocal sacs. The male fills himself with air, which is moved back and forth between the lungs and vocal sacs; sound is produced as it passes over the vocal cords in the larynx.

Reproduction

Once males and females have located one another (see facing page), mating can occur. The male adopts a position known as amplexus, in which he clasps the female from above (see right). Fertilization takes place while the pair are in amplexus, which may last for a period ranging from a few minutes to several days, depending on the species. In all but a small number of species, fertilization is external, the male shedding sperm onto the eggs as they emerge from the female. Each species has a characteristic pattern of egg production and uses a specific type of site for egg deposition. Eggs may be produced singly, in clumps, or in strings. The female usually places them in water, because their development can occur only in moist conditions. Eggs may be dispersed into the water, where they may float or sink, they may be wrapped around vegetation, or they may be glued to plants, logs, or rocks.

AMPLEXUS
Amplexus may take place on land or in water—these tree frogs (right) mate in the branches of trees. Depending on the species, the male may clasp the female just behind her front limbs or just in front of her back limbs. In many species, males outnumber females in breeding groups, and males in amplexus are often attacked by rivals seeking to displace them.

EGGS
The number of eggs produced by a female varies greatly among species, from less than 20 to many thousand. The eggs of the European common toad (left), which are suspended in long strings of jelly, are wrapped around the underwater stems of aquatic plants.

Defense

Frogs and toads typically lack weapons, such as large teeth or claws, that they can use to defend themselves. Most rely on being able to escape by jumping, usually toward open water. Many produce secretions from glands in their skin that make them distasteful or poisonous, sometimes lethally so. Others rely on being well camouflaged. In general, tadpoles lack defenses and are eaten by various predators, notably fish and the larvae of insects, such as dragonflies. Being produced in large numbers means that at least a few survive.

WARNING COLORATION
The poison-dart frog (left) has had to evolve highly poisonous skin secretions because many of its predators (including snakes and spiders) have evolved a resistance to milder toxins. Many species that are distasteful or toxic are vividly colored, which acts as a warning to predators.

CAMOUFLAGE
Most frogs and toads are active only at night and remain entirely motionless by day. Species such as this leaf frog (right) have skin that is colored to match their background and are patterned in such a way that their outline is broken up.

Movement

Frogs and toads move in a wide variety of ways. Most toads are ground dwellers and walk, run, or hop over the ground; few are able to leap large distances. Leaping is rarely used for moving from place to place but is primarily a means of escape from predators. Not all frogs and toads jump: many have back limbs adapted for other types of movement, such as swimming, climbing, and (in a few species) gliding. Burrowing is important for many species, enabling them to hide during the day, or for many weeks or months when it is too cold, hot, or dry for them to be active. They may simply push their way under rocks or logs, or into piles of plant debris, or they may be able to dig deep into the ground. Some desert species can survive being buried for up to 2 or 3 years.

CLIMBING
Some frogs (such as the European treefrog, left) spend their entire life in tree canopies or among other types of vegetation. Their long, slender limbs enable them to jump or straddle gaps between branches, while the adhesive disks on their digits allow them to climb vertical surfaces.

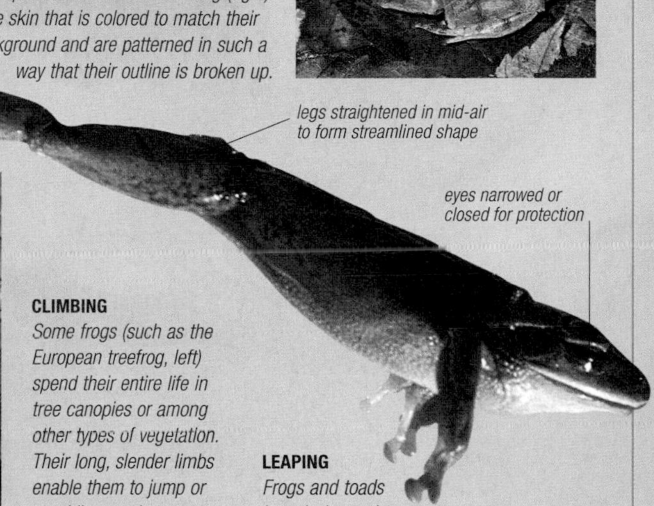

LEAPING
Frogs and toads launch themselves into the air with an explosive burst of energy from their long back legs. On landing, the front legs are pushed forward to cushion the impact.

Ascaphus truei

Tailed frog

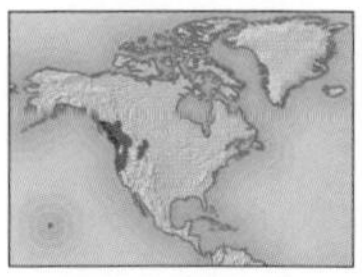

Location S.W. Canada, N.W. USA

Length 1–2 in (2.5–5 cm)
Habit Mostly terrestrial
Breeding season Spring to fall
Status Least concern

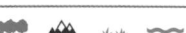

This small green, brown, gray, or reddish amphibian is the only aquatic frog that breeds by internal fertilization. The unique "tail"—found only in males—is an extension of the cloaca, and is used to insert sperm into the female's cloaca. Females lay strings of eggs in cold mountain streams, attaching them to the underside of rocks. The tadpoles take 2 years to mature, and have suckerlike mouths that enable them to cling to rocks in fast-flowing water.

Leiopelma hochstetteri

Hochstetter's frog

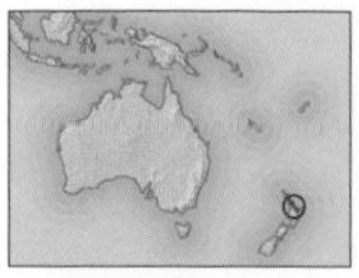

Location New Zealand (North Island)

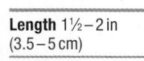

Length 1½–2 in (3.5–5 cm)
Habit Mostly terrestrial
Breeding season Summer
Status Vulnerable

This squat brown frog—an inhabitant of damp ground—is one of just 4 species of frogs that are native to New Zealand. Its eggs, which are enclosed in large, water-filled capsules, develop into nonfeeding tadpoles that stay within the capsule until they mature. When they are fully developed, they emerge as froglets. Hochstetter's frog is still abundant in some localized areas, but 2 other New Zealand species are now very rare, and confined to offshore islands.

Bombina orientalis

Oriental fire-bellied toad

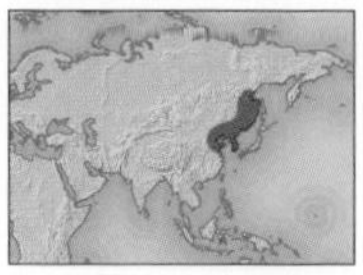

Location E. and S.E. Asia

Length 1¼–2 in (3–5 cm)
Habit Mostly aquatic
Breeding season Spring and summer
Status Least concern

The most distinctive feature of this squat, slightly flattened toad is its bright red underside with black mottling. When threatened, it arches its back, flattens its body, and lifts its limbs over its head to display this vivid coloration. It also produces a distasteful secretion through its skin. The Oriental fire-bellied toad lives in mountain streams in coastal areas, and hides under rocks and logs in winter and during dry spells. The male calls to attract a mate. The female lays small clutches of large eggs under rocks in streams.

Pipa pipa

Surinam toad

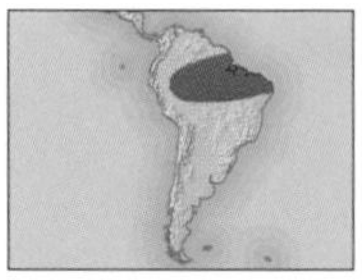

Location N. South America

Length 2–8 in (5–20 cm)
Habit Wholly aquatic
Breeding season Rainy season
Status Least concern

The Surinam toad is gray above and paler below, with a flattened body and flat, triangular head. Like other members of the family Pipidae, it lacks a tongue. An inhabitant of turbid, muddy, and slow-moving water, the Surinam toad is highly adapted to an entirely aquatic existence. Powerful hind limbs aid swimming, sense organs along its sides detect vibrations in the muddy water, tentacle-like projections on its fingers feel for prey, and upward-pointing eyes see above the surface. This toad exhibits unusual mating and breeding patterns: the male clasps the female from above, and the pair turns upside down repeatedly. The eggs are released, fertilized, and trapped in the space between the male's belly and the female's back. They are then absorbed into the skin on the female's back, where they develop in capsules and emerge as miniature frogs.

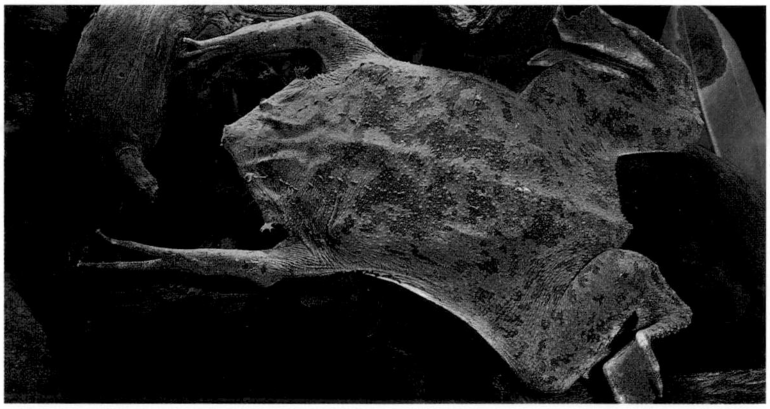

Xenopus laevis

African clawed toad

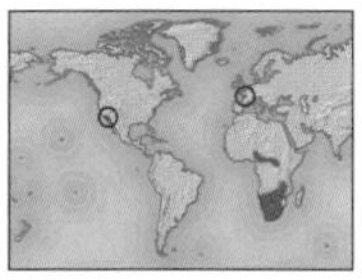

Location C. and Southern Africa. Introduced to USA and Europe

Length 2¼–5 in (6–13 cm)
Habit Mostly aquatic
Breeding season Rainy season
Status Least concern

With its flattened body and large, webbed feet, the African clawed toad (also known as the platanna) is ideally suited to a bottom-dwelling lifestyle in lakes and ponds. Like its close relative the Surinam toad (see right), it has eyes and nostrils positioned on the top of its head, and camouflaged skin to protect it from herons and other predators. During the breeding season, females lay numerous very small eggs, attaching them to various underwater objects. This species is often raised in laboratories for teaching and research.

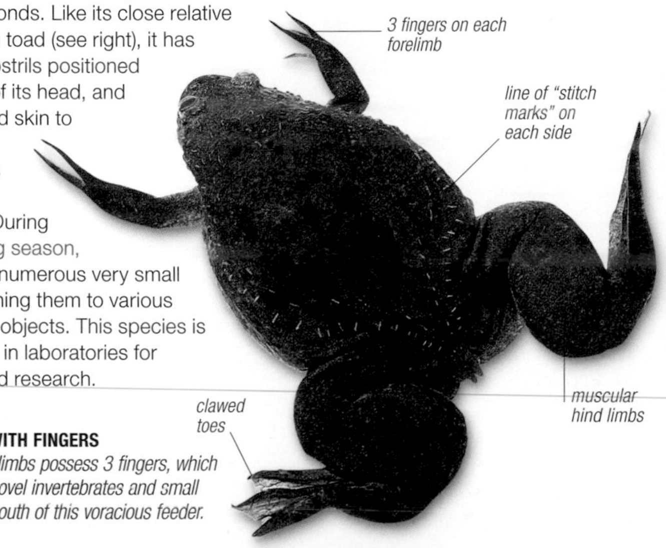

FORELIMBS WITH FINGERS
The small forelimbs possess 3 fingers, which are used to shovel invertebrates and small fish into the mouth of this voracious feeder.

SENSE ORGANS

This toad has a line of white "stitch marks" along its sides. These contain special sense organs that detect vibrations, helping it to locate food and predators in murky water.

Alytes muletensis

Majorcan midwife toad

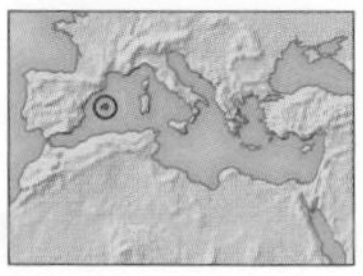

Location Balearic Islands (Majorca)

Length 1½–1¾ in (3.5–4.5 cm)
Habit Mostly terrestrial
Breeding season Spring and summer
Status Vulnerable

One of the world's most rare toads, this tiny species (also known as the ferreret) has been the subject of a very successful conservation program, in which animals bred in captivity are released into suitable habitats on Majorca. It has a yellowish or pale brown body, with dark mottling. Both the males and females call to locate mates. The female transfers a batch of eggs to the male, who carries them until they are ready to hatch. Their tadpoles grow very large, and little growth occurs after they become adults.

Alytes obstetricans

Midwife toad

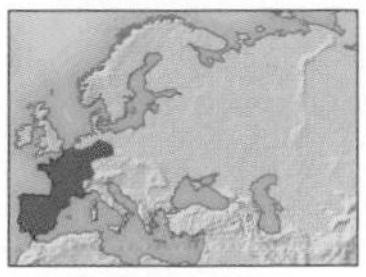

Length 1¼–2 in (3–5 cm)
Habit Mostly terrestrial
Breeding season Spring and summer
Status Least concern
Location W. and C. Europe

The midwife toad is notable for its highly unusual method of breeding. The female lays large, yolk-filled eggs in strings, which are transferred to the male during mating. The male then wraps these strings of eggs around his legs, sometimes carrying 2 strings laid by different females. Once the eggs are ready to hatch, the male deposits them in ponds and pools. This toad is small and plump, with short legs and sandy, pale brown, or gray skin, with darker spots. It inhabits woodland, gardens, drystone walls, sand dunes, and rock slides, hiding in crevices in winter and during dry periods. Active at night, it feeds on invertebrates. Males attract females by calling from a crevice or burrow, making a distinctive, high-pitched "poo-poo" sound.

Pelobates fuscus

Common spadefoot toad

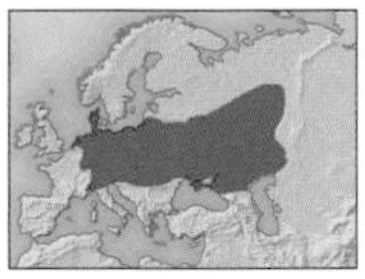

Length 1½–3¼ in (4–8 cm)
Habit Mostly terrestrial
Breeding season Spring
Status Least concern
Location C. and E. Europe, W. Asia

This plump, variable toad has a pale "spade" on each hind foot, enabling it to dig backward into the soil. It lives in sand dunes, and in cultivated areas, preying on insects and other invertebrates. When threatened, it squeals, inflates its body, and raises itself up to appear larger. It emits a garliclike odor, possibly to deter predators.

Megophrys montana

Asian spadefoot

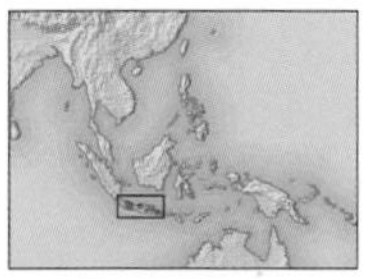

Length 2¾–5½ in (7–14 cm)
Habit Mostly terrestrial
Breeding season Rainy season
Status Least concern
Location S.E. Asia

This tropical toad is one of the world's most effectively camouflaged amphibians. It lives on the forest floor, mimicking dead leaves. Its brown color and its shape are remarkably leaflike, with a sharp snout, pointed "horns" over its eyes, and folds of skin that look like leaf edges. Female Asian spadefoot toads lay their eggs under rocks in streams. After hatching, the tadpoles hang vertically from the water's surface, and feed on very small organisms using their large, umbrella-shaped mouths.

Rhinophrynus dorsalis

Mexican burrowing toad

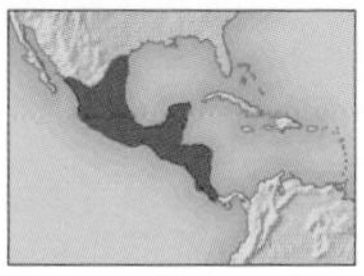

Length 2¼–3¼ in (6–8 cm)
Habit Terrestrial/ Burrowing
Breeding season Rainy season
Status Least concern
Location S. USA to Central America

With its bloated body, pointed snout, and small legs, this unusual-looking toad is adapted to a life spent underground. It lives in low-lying regions where soft soil makes it easy to burrow, emerging only after heavy rain. Adults breed in ponds, and the males have internal vocal sacs that they use to call to females. After the eggs hatch, the tadpoles feed by filtering food from the water, using barbels around their mouths.

Scaphiopus couchii

Couch's spadefoot

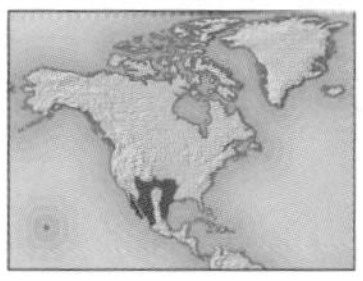

Length 2¼–3½ in (5.5–9 cm)
Habit Mostly terrestrial
Breeding season Spring and summer
Status Least concern
Location S. USA, Mexico

One of 7 similar species, all from North America, this yellow or yellowish green toad, with darker mottling, is a classic example of an amphibian that survives drought by burrowing underground. Like its relatives, it has spadelike black ridges on its hind feet, which it uses to dig its way 3¼ ft (1 m) or more into loose, sandy soil. Once below ground, it encases its body in a watertight cocoon made by shedding several layers of skin, and then remains dormant, waiting for rain. When rain does fall—sometimes many months later—the cocoon breaks down, and this toad makes its way to the surface to breed.

Pelodytes punctatus

Parsley frog

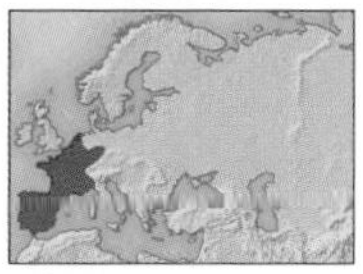

Length 1¼–2 in (3–5 cm)
Habit Mostly terrestrial
Breeding season Spring
Status Least concern
Location W. and S.W. Europe

The small, agile parsley frog lives in damp, vegetated habitats near ponds. It has an unusual method of climbing, which involves clinging onto vertical surfaces such as trees, rocks, and walls, using its belly as a kind of suction pad. Its name derives from the green flecks on its pale skin that resemble chopped parsley. Like most frogs, it is active at night, spends winter and dry spells buried underground, and feeds on insects and other invertebrates. During the breeding season, males call out from underwater to attract females, who call back. Females lay their eggs in broad strips in ponds. The tadpoles sometimes grow to be larger than the adult frogs.

AMPHIBIANS

Notaden bennettii

Holy cross toad

Length 1½–2¾ in (4–7 cm)
Habit Mostly terrestrial
Breeding season After heavy rain
Status Least concern

Location E. Australia

Named after the crosslike pattern of warts on its back, this small, globular toad belongs to a family of over 43 species that are found only in Australia. Like many of its relatives, it spends dry periods underground, and emerges to feed and breed when it rains. This toad feeds on ants and termites, and produces a sticky, defensive secretion when handled. Its tadpoles develop in temporary pools.

Crinia insignifera

Sign-bearing toadlet

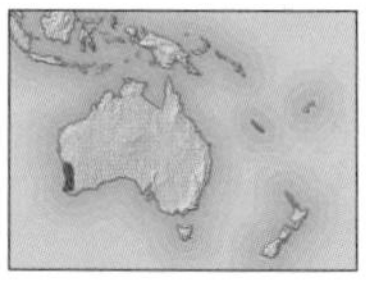

Length ½–1¼ in (1.5–3 cm)
Habit Mostly terrestrial
Breeding season Winter
Status Least concern

Location S.W. Australia

The sign-bearing toadlet is a small Australian frog with a slender body and limbs. It is gray to brown, and often has a dark triangular patch between its eyes; its skin may be smooth or warty. This frog inhabits temporary swamps that form after winter rain, preying on insects and other invertebrates, and spends dry periods underground. During the breeding season, the male calls to attract a mate. Females lay about 60–250 eggs at a time in water, either singly or in clumps; the eggs fall to the bottom of the pond.

Mixophyes carbinensis

Carbine frog

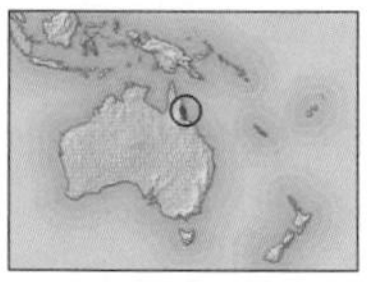

Length 2¼–3 in (6–7.5 cm)
Habit Mostly terrestrial
Breeding season Spring and summer
Status Least concern

Location N.E. Australia

This species was described in 2006 on the basis of specimens collected from Mount Lewis in the Carbine highlands of Queensland—in an area designated as a World Heritage Site. It belongs to a genus of large ground-dwelling frogs restricted to eastern Australia and New Guinea, and lives in montane rain forest near streams and pools. Males call during the breeding season from the leaf-litter floor.

Pseudophryne corroboree

Corroboree frog

Length 1–1¼ in (2.5–3 cm)
Habit Mostly terrestrial
Breeding season Summer
Status Critically endangered

Location Australia (S.E. New South Wales)

This critically endangered animal, also known as the corroboree toadlet, is one of Australia's most colorful amphibians. It lives in moss-covered bogs in the Australian Alps, and lays its eggs in the moss rather than in water. The reason for its vivid colors—yellow or yellow-green, with black stripes—is unclear: unlike most brightly colored amphibians, it does not have toxic skin.

Limnodynastes peronii

Brown-striped frog

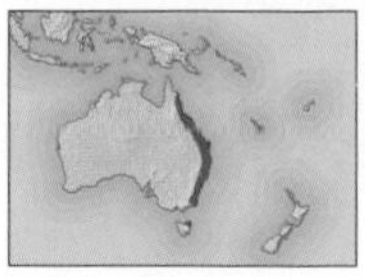

Length 1¼–2½ in (3–6.5 cm)
Habit Mostly terrestrial
Breeding season Spring and summer
Status Least concern

Location E. Australia

Also known as the Australian grass frog, this long-limbed amphibian, with dark longitudinal stripes along its body, lives in and around swamps, feeding at night on insects and other invertebrates. During winter and dry spells, it buries itself underground. The male makes a clicklike call to attract the female. During mating, the pair make a floating foam nest in which 700–1,000 eggs are deposited. The eggs hatch quickly, and the tadpoles develop very rapidly.

Uperoleia lithomoda

Stonemason toadlet

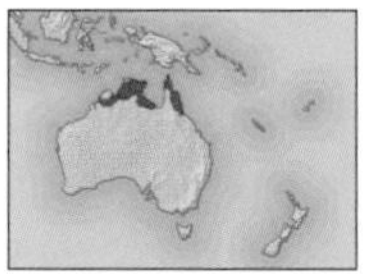

Length ½–1¼ in (1.5–3 cm)
Habit Mostly terrestrial
Breeding season Winter
Status Least concern

Location N. Australia

The clicklike courtship call of this small, squat frog sounds like 2 stones being hit together, hence its name. It is dull brown or gray, with a gold line or row of patches along each side of its body, short limbs, and a horny tubercle on its hind feet that enables it to dig into the soil. It spends dry periods buried underground. The female lays eggs in clumps that fall to the pond floor.

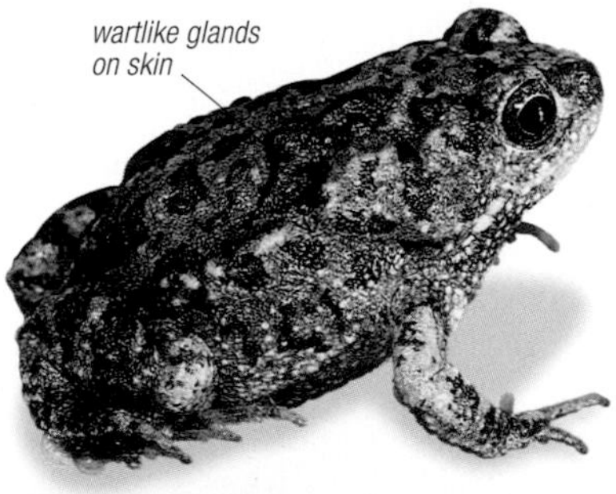

Sechellophryne gardineri

Gardiner's Seychelles frog

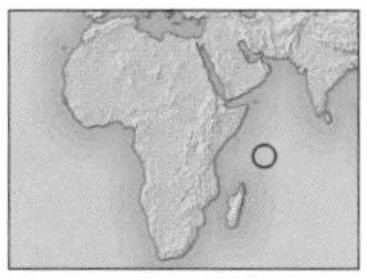

Length ⅜–½ in (1–1.5 cm)
Habit Wholly terrestrial
Breeding season Rainy season
Status Vulnerable

Location Seychelles

This tiny, creamy white to yellowish green frog is notable for its unusual breeding pattern. The eggs, which are laid in clumps on the ground, are guarded by the male and hatch into froglets instead of tadpoles. The froglets crawl onto their father's back and remain there, glued on by mucus, until they have used up their yolk and their legs are fully developed.

Heleophryne purcelli

Cape ghost frog

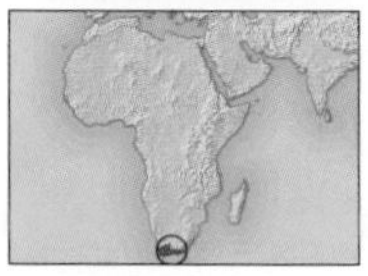

Length 1¼–2¼ in (3–6 cm)
Habit Mostly aquatic
Breeding season Summer
Status Least concern

Location Southern Africa (Cape region)

The Cape ghost frog is ideally suited to its aquatic lifestyle. Its long legs and webbed hind feet enable it to swim in fast-flowing streams, and its flattened toe tips give it a strong grip on slippery rocks and boulders. During the breeding season, males develop tiny spines on their bodies. The tadpoles have large, suckerlike mouths to attach themselves to rocks.

Ceratophrys cornuta

Surinam horned frog

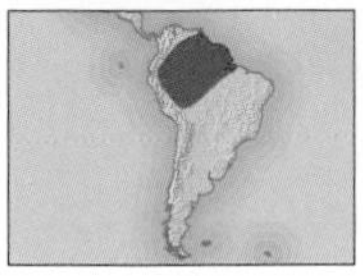

Length 4–8 in (10–20 cm)
Habit Mostly terrestrial
Breeding season Rainy season
Status Least concern
Location N.W. South America

This bulky and intricately patterned frog is a formidable hunter, with an exceptionally large mouth. It ambushes its prey and, if threatened, defends itself ferociously, using its large, sharp teeth. It has a well-camouflaged body, relatively short legs, and a "horn" above each eye. Males are slightly smaller than females and, in the breeding season, they produce a mating call that sounds like cattle lowing. Females lay up to 1,000 eggs, wrapped around aquatic plants. Unlike most tadpoles, Surinam horned frog tadpoles are predatory from the moment they hatch, and attack the tadpoles of other frogs and even each other. They have large, horny beaks and muscular jaws.

HUNTING BY AMBUSH

The Surinam horned frog catches other frogs, lizards, and mice using a sit-and-wait technique. It partly buries itself in soft ground, sometimes with only its eyes projecting above the surface, and then waits for animals to pass by. Like other amphibians, it is triggered into action by movement. Once its prey is within range, it bursts out of hiding, immobilizes its victim with its fangs, and then swallows it whole.

STOCKY BUILD
The Surinam horned frog is a massive, plump-bodied frog, with a huge head and a very wide mouth. It has conspicuous hornlike projections over its eyes.

Telmatobius culeus

Titicaca water frog

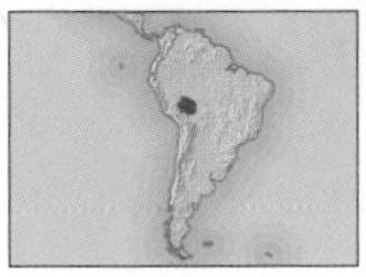

Length 3¼–4¾ in (8–12 cm)
Habit Wholly aquatic
Breeding season Summer
Status Critically endangered
Location South America (Lake Titicaca)

This frog has several adaptations that suit it to a life in the cold, oxygen-impoverished waters of Lake Titicaca. It has long hind legs with fully webbed hind feet for swimming, and very small lungs that enable it to dive. Due to the small size of its lungs, it relies mainly on oxygen taken in through the skin. This uptake is increased by the extensive folds in its skin, which increase the animal's surface area, and the presence of numerous blood vessels and red blood cells.

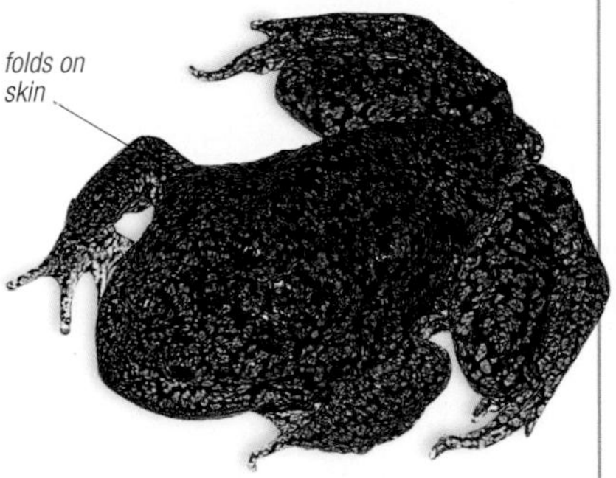

Eupemphix nattereri

Cuyaba dwarf frog

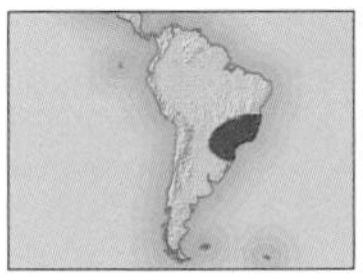

Length 1¼–1½ in (3–4 cm)
Habit Mostly terrestrial
Breeding season After heavy rain
Status Least concern
Location E. South America

The Cuyaba dwarf frog is characterized by the 2 black and white eyespots on its rear. When threatened, it inflates its body and lifts up its rear end, displaying these false eyes to startle the predator. It can also secrete a distasteful substance from glands near its groin to further deter its attacker. This frog has a plump, toadlike body, with short limbs. It is mid-brown, with marbled patterning in darker and lighter shades of brown.

Engystomops pustulosus

Tungara frog

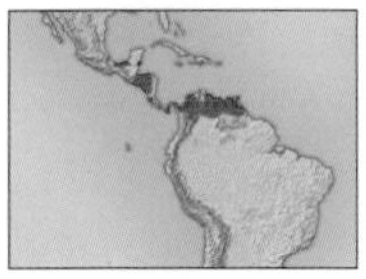

Length 1¼–1½ in (3–4 cm)
Habit Mostly terrestrial
Breeding season Rainy season
Status Least concern
Location Central America, N.W. South America

With its dark gray or brown warty skin, this Central American frog resembles a toad. The male calls loudly to attract females, repeatedly inflating his large vocal sacs to produce a sound that consists of a basic "whine" followed by one or more "chucks." The more chucks he makes, the more likely he is to attract a female; however, his call can also attract predatory bats. The female lays her eggs in floating foam nests in temporary pools.

Leptodactylus pentadactylus

South American bullfrog

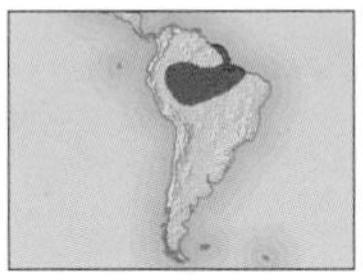

Length 3¼–9 in (8–22 cm)
Habit Mostly terrestrial
Breeding season Rainy season
Status Least concern
Location N. South America

This powerfully built frog is one of the largest members of its genus—a group of 89 species that are mainly confined to the American tropics. It is yellow or pale brown, with dark markings, long legs, large eyes, and conspicuous external eardrums. The male has muscular forearms, with a sharp black spine on his thumbs, which he uses to fight off other males. The female lays eggs near water, depositing them in a foam nest that the male has created by whipping up mucus with his back legs. South American bullfrogs are active at night, feeding on insects and other invertebrates as well as other small animals, and spend dry periods hidden under logs or underground. If picked up, they emit a loud scream to startle the predator into loosening its grip. The hind limbs of this frog are sometimes eaten by humans.

Eleutherodactylus planirostris

Greenhouse frog

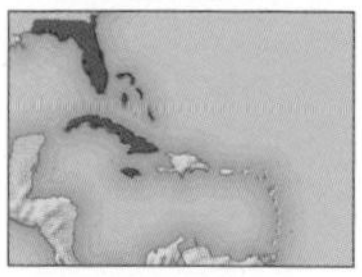

Length 1–1½ in (2.5–4 cm)
Habit Wholly terrestrial
Breeding season Summer
Status Least concern

Location S.E. USA, Caribbean

This tiny frog—one of the world's smallest—belongs to a family of about 201 species that are found throughout warm parts of the Americas. It is brown or tan, with darker mottling or stripes on its back, and well-developed adhesive disks on its fingers and toes. The greenhouse frog usually lives in wooded habitats, but also enters gardens. Like its many relatives, it lays eggs in moist places on the ground, that hatch directly into fully formed frogs.

Bufo bufo

European common toad

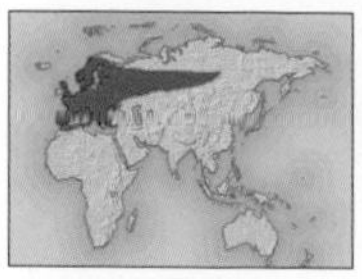

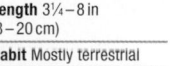

Length 3¼–8 in (8–20 cm)
Habit Mostly terrestrial
Breeding season Spring
Status Least concern

Location N.W. Africa, Europe to C. Asia

One of Europe's most widespread amphibians, the European common toad is a robust animal with a warty or spine-covered skin. Its color ranges from brown or green to brick-red, and females are generally substantially larger than males, particularly in early spring, when they are laden with eggs. At this time of year, males and females congregate in ponds and, after pairing up, the female lays strings of eggs around underwater plants. During the rest of the year, this toad spends much of its time away from water, feeding on insects, slugs, and other small animals. It is mainly nocturnal, hiding by day under logs and in other damp places.

IN SELF-DEFENSE

The European common toad has a large parotoid gland behind each eye, which exudes a distasteful secretion to ward off predators. If threatened, it often swells up by gulping air, and stands on tiptoe to make itself look larger. Despite its menacing appearance, it rarely bites and is harmless to humans.

APPEARANCE
This common toad has large, copper- or gold-colored eyes. During the breeding season, the male develops pads on his front feet for gripping the female.

Anaxyrus americanus

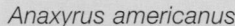

American toad

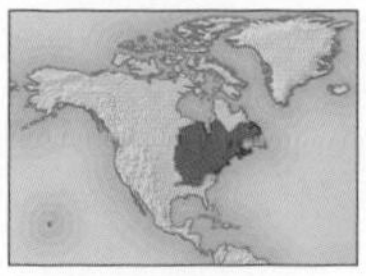

Length 2–3½ in (5–9 cm)
Habit Mostly terrestrial
Breeding season Spring
Status Least concern

Location E. Canada, E. USA

Like the European common toad (see left), the American toad has warty skin, prominent eyes, and a large parotid gland on each side of its head. It shows great variation in color, from brown to brightly colored, and has a spotted chest. An adaptable animal, it is found in all habitats, including gardens and mountains, and is active at night. During courtship, males produce a musical trill lasting 3 to 60 seconds to attract mates. Females lay eggs in long strings in ponds, wrapped around vegetation.

Duttaphrynus kiphirensis

Kiphire toad

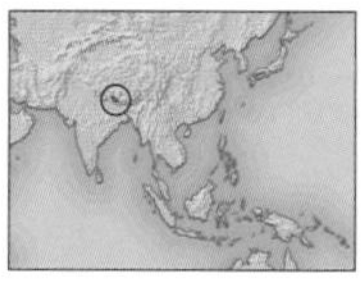

Length 5 in (13 cm)
Habit Terrestrial
Breeding season Probably during the monsoon
Status Not evaluated

Location India

This is one of 30 or so southern Asian *Duttaphrynus* toads previously included in the genus *Bufo* before zoologists split the group to better reflect evolutionary relationships. It was described in 2009 on the basis of specimens collected from a single locality—Kiphire town, in Nagaland, northeastern India, at 3904 ft (1,190 m) above sea level. Its range lies between 2 biodiversity hot spots: the eastern Himalaya and Indo-Burmese lowlands. So far, little is known about its habits—but its closest relatives are mainly nocturnal, feeding on insects and other invertebrates and spending the day hiding under stones and in crevices. These temperate mountain toads are triggered into breeding by the arrival of monsoon rains, when adults lay a double jelly string of eggs in shallow pools along mountain torrents after downpours; during this time, males call with a low "curr, curr" croak. The tadpoles gather in side pools and—in some species—can swarm in considerable numbers.

Amietophrynus gutturalis

Guttural toad

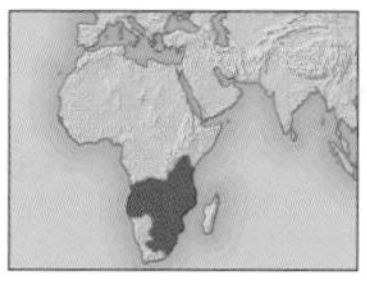

Length 2–4 in (5–10 cm)
Habit Mostly terrestrial
Breeding season Spring
Status Least concern

Location Southern Africa

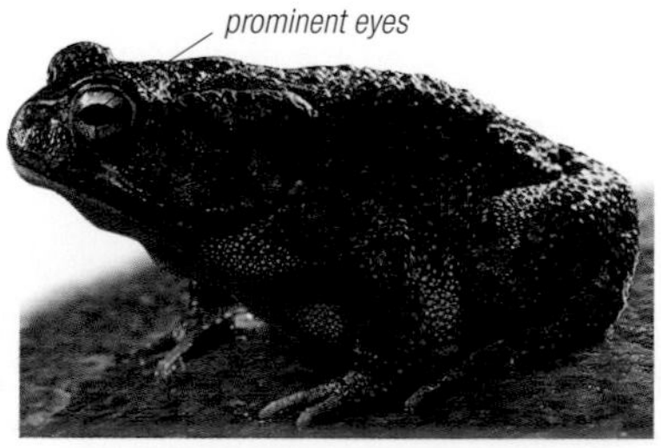

The guttural toad is found in a variety of sites that provide cover, close to ponds, dams, and streams. In urban areas, it can often be seen under streetlights at night, feeding on insects. However, it also preys on other invertebrates and on flying termites as they leave their nest. The female is larger than the male, but the latter has longer forelimbs. Both sexes are pale brown with dark brown patches. The male makes a deep guttural "snore" to attract a mate, and the female lays eggs in water.

Amietophrynus rangeri

Raucous toad

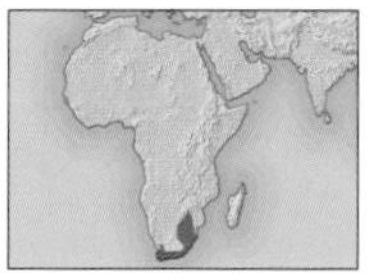

Length 2–4¼ in (5–11 cm)
Habit Mostly terrestrial
Breeding season Spring
Status Least concern

Location Southern Africa

This yellow or pale brown, spotted toad inhabits grassland and woodland close to rivers, streams, and ponds. In cool and dry spells, it hides under logs or rocks, or in rotting vegetation. When cornered by a predator, it produces a sticky, distasteful secretion through its parotid glands. During the breeding season, males let out loud, harsh "quacks" to attract females. Mating activity, such as calling and fighting, is extremely energetic, and males lose weight at this time of year.

Pseudepidalea viridis

European green toad

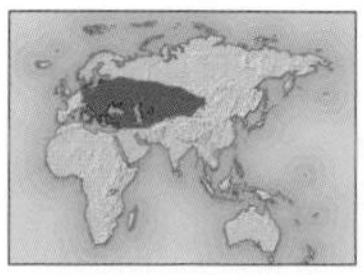

Length 2¼–3½ in (6–9 cm)
Habit Mostly terrestrial
Breeding season Spring and summer
Status Least concern

Location E. Europe, Asia

This robust toad has pale skin with green marbling and red spots. It prefers sandy habitats in lowlands and dry places, but is also found in mountain regions and in urban areas. It spends winter and dry periods under logs, rocks, or buried in rotting vegetation. Active at night, it feeds mainly on insects. During courtship, the male makes a high-pitched call. A female European green toad may lay between 10,000 and 12,000 eggs at a time.

Atelopus zeteki

Golden frog

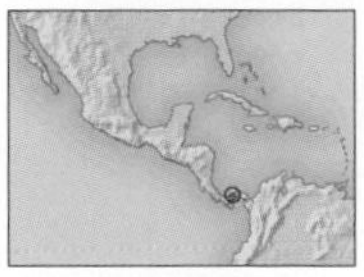

Length 1½–2¼ in (3.5–6 cm)
Habit Mostly terrestrial
Breeding season Rainy season
Status Critically endangered

Location S. Central America (Panama)

This vivid yellow or orange toad, sometimes with black markings, provides a good example of warning coloration. Its skin produces poison when it is attacked and predators learn to avoid it, having experienced the pain once, associating bright color with a noxious stimulus. Like many frogs and toads living in Central America, its numbers have declined drastically in recent years.

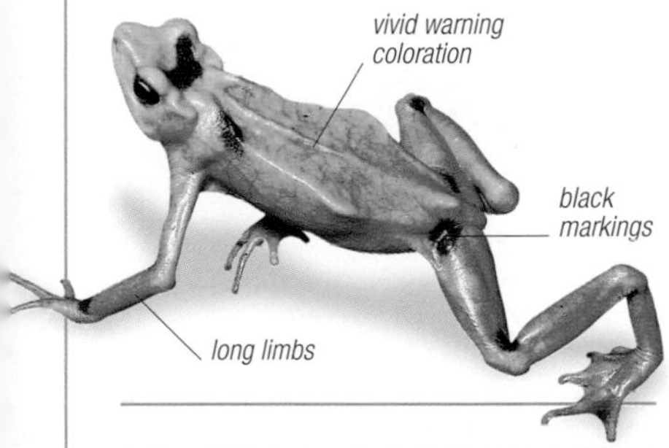

Brachycephalus ephippium

Spix's saddleback toad

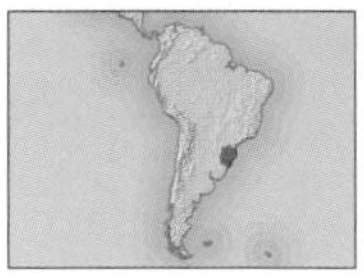

Length ⅜–¾ in (1–2 cm)
Habit Mostly terrestrial
Breeding season Not recorded
Status Least concern

Location E. South America

Also known as the gold frog, this species is an inhabitant of mountain rain forest, where it lives mainly among leaf litter, preying on insects and other small invertebrates. During dry spells, it hides in crevices. The Spix's saddleback toad is a ground dweller—it cannot jump, nor can it climb well, because its stubby toes lack adhesive disks. It has a saddlelike, bony plate embedded in the smooth skin of its back, and is bright orange above, with a yellow underside. The female lays 5 or 6 large eggs on land; there is no tadpole stage, the eggs hatching directly into small frogs.

Nimbaphrynoides occidentalis

Western Nimba toad

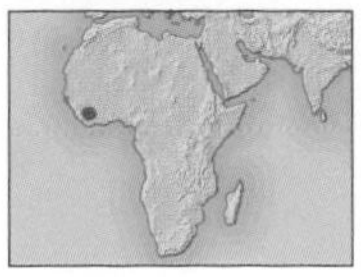

Length ½–1 in (1.5–2.5 cm)
Habit Wholly terrestrial
Breeding season Spring and summer
Status Critically endangered

Location W. Africa (Mount Nimba)

This toad has a highly unusual mode of reproduction: the eggs are fertilized and the young develop inside the mother. After a 9-month gestation period, she gives birth by inflating herself with air and pushing out 2–16 tiny toadlets. The western Nimba toad hides in rocky crevices in the dry season.

Pedostibes hosii

Boulenger's Asian tree toad

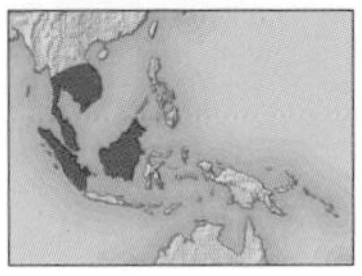

Length 2–4 in (5–10 cm)
Habit Mostly terrestrial
Breeding season Rainy season
Status Least concern

Location S.E. Asia

Unusually for a toad, Boulenger's Asian tree toad is a good climber, with adhesive disks on all its toes. It has warts and small spines on its skin, and varies in color from greenish brown to black, with yellow spots; some females also have a purple tinge. It is usually found along rivers and streams in forest, and is active at night, feeding on ants. The female lays numerous small eggs in strings in flowing water. The tadpoles have suckerlike mouths that enable them to cling to rocks.

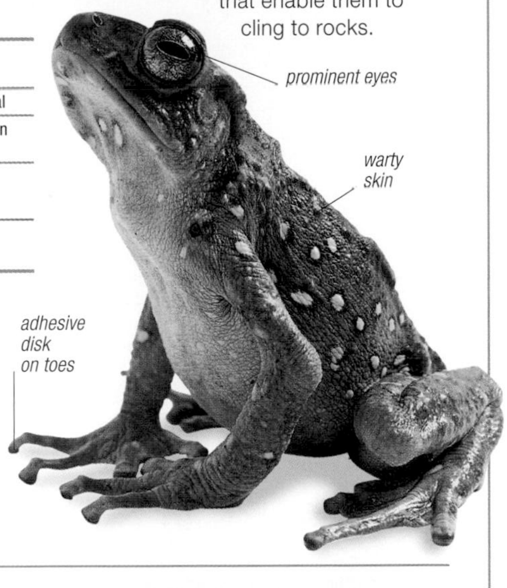

Rhinoderma darwinii

Darwin's frog

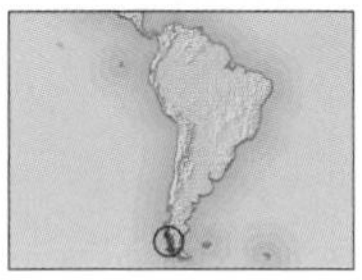

Length 1–1¼ in (2.5–3 cm)
Habit Mostly terrestrial
Breeding season Year round
Status Vulnerable

Location S. South America

An inhabitant of mountain forest in the southern Andes, Darwin's frog is famous for brooding its eggs in its vocal sacs—a form of parental care seen in few other amphibians. Brown or green above with a black underside, it is easily recognized by its small size and sharply pointed snout. Adults live in damp, shady valleys close to flowing water, and are largely nocturnal, feeding on insects and other small animals. During the breeding season, the male attracts the female with a soft, bell-like call. Once the eggs have been laid, the male cares for them until they hatch.

PHYSICAL FEATURES

This small frog has a characteristic fleshy proboscis on the end of its nose. Other features include thin legs, long fingers, webbing on its toes, small eardrums, and eyes with horizontal pupils.

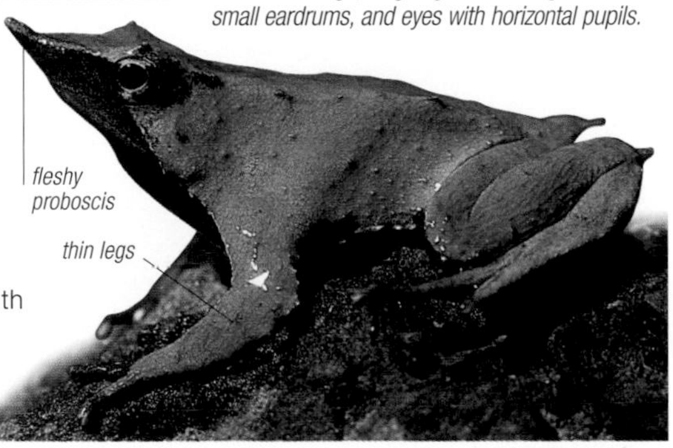

PARENTAL CARE

After a female Darwin's frog has deposited a clump of up to 40 large eggs on the ground, the male guards them for 2–3 weeks. When the embryos begin to move, he swallows up to 15 eggs and retains them in his vocal sacs, where they hatch into tadpoles. The tadpoles require only their own egg yolk in order to develop into froglets. These then emerge from the father's mouth.

Pseudis paradoxa

Paradoxical frog

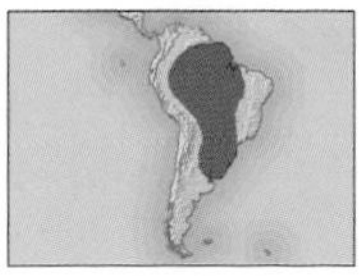

Length 2–2¾ in (5–7 cm)
Habit Wholly aquatic
Breeding season Rainy season
Status Least concern

Location N. and C. South America

In most frogs, adults are considerably larger than mature tadpoles, but in this species the situation is reversed. Paradoxical frog tadpoles have an unusually long life, and can grow up to 10 in (25 cm) long—about 4 times longer than the adult frog. During the final stage of metamorphosis, the animal shrinks, largely by absorbing its very long tail. Most of the features of the adult frog are adaptations for a highly aquatic lifestyle: it has upward-pointing eyes and nostrils that enable it to float just beneath the water's surface, fully webbed hind feet for swimming, and slimy skin.

It lives in ponds, lakes, and swamps, and feeds on aquatic invertebrates, stirring up mud with its feet to disturb prey before catching it. The female lays eggs in a floating foam nest.

AMPHIBIANS

Rhinella marinus

Cane toad

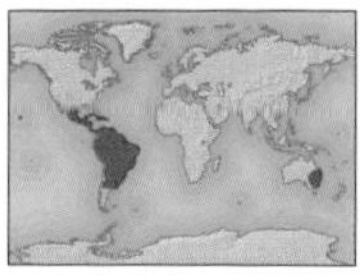

Length 2–9 in (5–23 cm)

Habit Mostly terrestrial

Breeding After rain season

Status Least concern

Location Central America, South America. Introduced to Australia and elsewhere

The cane toad is the world's largest toad. A heavily built creature, it has very tough, warty brown skin that feels leathery to the touch. It has a massive head, with bony ridges above the eyes, and prominent parotid glands that secrete a milky poison whenever the toad is in danger. A hungry feeder, it eats mainly ants, termites, and beetles, but will readily consume a wide variety of other insects and invertebrates as well as other frogs. The cane toad is usually solitary, except in the breeding season. Then males call to females, drawing their attention with a purring trill that sounds like a running motor. The female, which is larger than the male, lays her eggs in long, gelatinous strings, draping them over plants or debris in water. Each clutch may consist of up to 20,000 eggs. In ideal conditions, toadlets will reach adult size within a year of leaving the water. This species has been deliberately introduced into some countries in the hope of controlling insects devastating sugarcane crops—but its predatory nature has devastated native wildlife.

SHORT AND SQUAT
With its heavy, squat body and short limbs, the cane toad is best suited to life on the ground. It usually crawls along or makes short hops.

BIOLOGICAL CONTROL

Introduced into Australia in 1935, as well as into many Pacific islands, the cane toad is the target of eradication programs, rather than of conservation. Most of these programs have met with only limited success, since cane toads are highly adaptable and resilient, with a huge reproductive potential. Biological control may be the best way to limit their spread, but as yet no effective method has been discovered that leaves other amphibians unharmed.

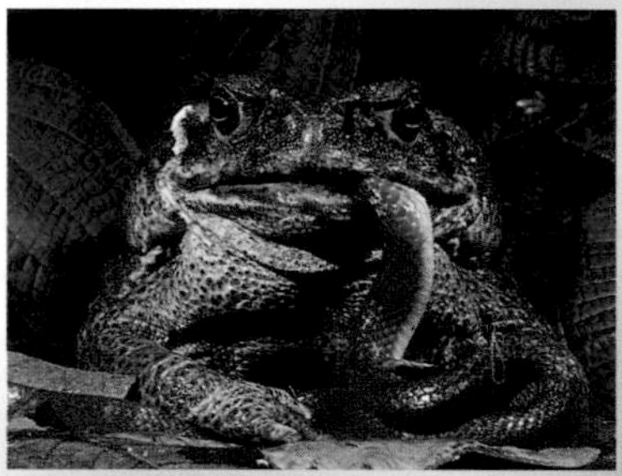

VORACIOUS FEEDER
Emerging at night to feed, the cane toad will eat virtually anything it can fit into its mouth, including snakes. In urban areas, they often gather under streetlights, eating insects attracted by the light.

VENOMOUS DEFENSE

The cane toad is extremely toxic to other animals. If threatened or squeezed, such as when in the mouth of a predator, it oozes venom from the large glands on its shoulders. Smaller glands over the body also produce venom. Anything swallowing this poison may die very quickly—in as little as 15 minutes in some cases.

FORMIDABLE SURVIVOR
The cane toad is highly adaptable, in terms of both survival and reproduction. It is able to thrive in almost any habitat, and can even tolerate brackish water. It feeds on any animal it can catch, and can survive without water for long periods. Breeding, which produces large quantities of eggs, occurs throughout the year.

Litoria caerulea

Australian green treefrog

Location S. New Guinea, N. and E. Australia

Length 2–4 in (5–10 cm)
Habit Terrestrial
Breeding season Spring and summer
Status Least concern

Also known as White's treefrog, this widespread Australasian species is well known for its tame behavior, and its habit of living in or near buildings. Its color varies from green to greenish blue, sometimes with white spots. Like most tree frogs, it has adhesive disks on its toes. In wet conditions, its loose, deeply folded skin allows the frog to take in a large amount of water—an ability that makes it unusually tolerant of drought. The Australian green treefrog is a nocturnal hunter, preying on insects and other invertebrates, and occasionally larger animals, such as mice. Males are smaller than females, and give out a harsh, barklike courtship call. After mating, the female lays 2,000–3,000 eggs in water. These frogs are of considerable interest in medicine: their skin produces several useful antibacterial and antiviral compounds, and a substance that has been used to treat high blood pressure in humans.

Litoria infrafrenata

White-lipped treefrog

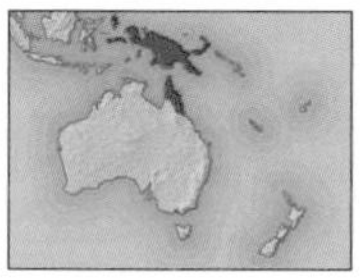

Location New Guinea, N.E. Australia

Length 4–5½ in (10–14 cm)
Habit Mostly terrestrial
Breeding season Spring and summer
Status Least concern

Plain green above, with a distinctive stripe along its lower lip, this tropical species is the largest frog in Australia, and one of the biggest tree frogs in the world. Found in forests and gardens, it feeds after dark on insects and other invertebrates. It has large adhesive disks on its toes, and spends most of its life in trees. During the breeding season, adults descend and females lay up to 400 eggs in water.

Litoria australis

Giant frog

Location N. Australia

Length 2¾–4¼ in (7–10.5 cm)
Habit Terrestrial/Burrowing
Breeding season Rainy season
Status Least concern

This large frog is able to survive underground for very long periods by burrowing into the soil and shedding numerous layers of skin, which create a cocoon around its body. It can retain large amounts of water in its bladder, body cavity, and beneath its skin. A related species, the water-holding frog (*L. platycephala*), is used as a "living well" by aborigines, who dig up the frog and extract drinking water by squeezing it. The giant frog is pale gray, brown, or bright green in color, with a dark brown or black stripe through its eyes. It has a broad head, short, muscular limbs, horny tubercles on its hind feet, and a pair of skin folds on its back. During the breeding season, they are active during the day and can be found beside temporary pools where males call to females from the water's edge. Once they have mated, each female lays up to 7,000 eggs in clumps that sink to the bottom of the pond. The tadpoles are very resilient and can survive water temperatures of up to 109°F (43°C).

Pseudacris crucifer

Spring peeper

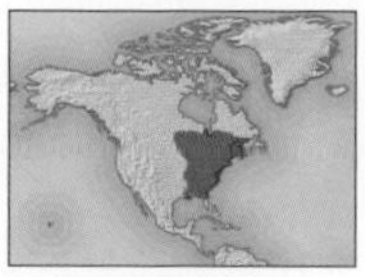

Location S.E. Canada, E. USA

Length ¾–1½ in (2–3.5 cm)
Habit Mostly terrestrial
Breeding season Spring
Status Least concern

Across much of eastern North America, the distinctive call of this tiny but highly agile frog is a sign that spring is under way. It belongs to a group of species known as chorus frogs, which are adept climbers, with reduced webbing, and adhesive disks on their toes. The spring peeper lives in woodland with temporary or permanent ponds, spending winter and dry spells under logs or rotting vegetation.

INFLATED VOCAL SAC

Male spring peepers gather in shrubs and trees next to water in early spring to form choruses, giving out high, piping whistles to attract females. The male's vocal sac amplifies his call as the air is passed back and forth between the sac and lungs over the vocal cords.

BODY COLORATION
The spring peeper is brown, gray, or olive-green, with darker patches that usually form an X-shaped pattern on its back.

Hyla cinerea

Green treefrog

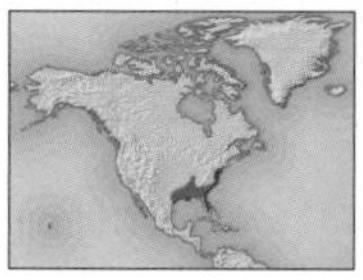

Location S.E. USA

Length 1¼–2¼ in (3–6 cm)
Habit Mostly terrestrial
Breeding season Spring
Status Least concern

This bright green frog can change color quite rapidly, from yellow when calling out for mates, to gray when the weather is cold. In the breeding season, males form choruses around pond edges; some call out to attract mates, while others, called "satellite males," are silent and attempt to intercept females attracted to the calling males. This species is also known as the "rain frog," because it calls just before and during wet weather. It often appears at night at windows, where it feeds on insects attracted to the light.

Hyla arborea

Common treefrog

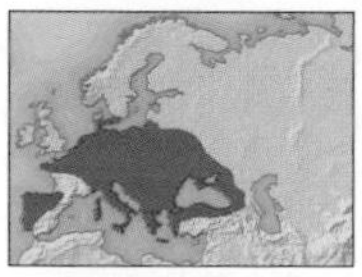

Location Europe, W. Asia

Length 1¼–2 in (3–5 cm)
Habit Terrestrial
Breeding season Spring and summer
Status Least concern

Europe is home to just a few species of tree frogs—this one is the most widespread. Usually bright green, although sometimes yellow or brown, it has a conspicuous dark horizontal stripe on its head and body. It lives in overgrown areas near water, and hibernates in winter. During the breeding season, males make a loud quacking sound, and this can quickly build up into a noisy chorus.

Hyla chrysoscelis

Cope's gray treefrog

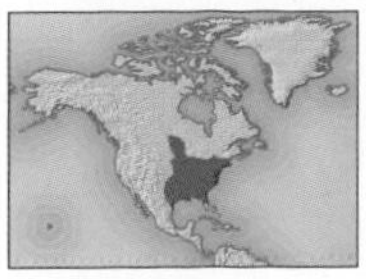

Length $1\frac{1}{4}$–$2\frac{1}{4}$ in (3–6 cm)
Habit Mostly terrestrial
Breeding season Spring
Status Least concern
Location S. Canada, C. and E. USA

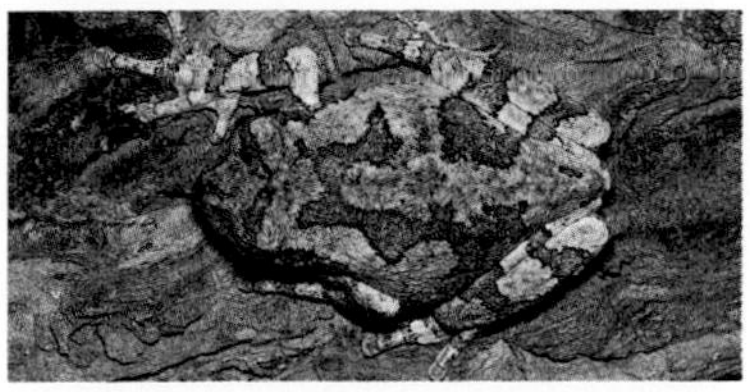

The blotchy gray coloring of this frog enables it to blend perfectly with a background of lichen-covered tree trunks. It can change color in different climatic or lighting conditions: turning darker in cold, dark conditions, and paler in bright light. Its blood contains glycerol, an "antifreeze" that enables it to survive subzero temperatures. Cope's gray treefrog is almost identical to another species, *H. versicolor*, which has a deeper-voiced call and twice the number of chromosomes.

Agalychnis callidryas

Red-eyed treefrog

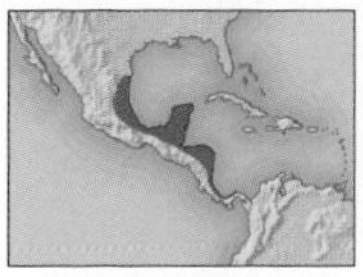

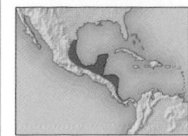

Length $1\frac{1}{2}$–$2\frac{3}{4}$ in (4–7 cm)
Habit Terrestrial
Breeding season Summer
Status Least concern
Location Central America

The jewel-like coloration and red eyes make this one of the most striking frogs. In the breeding season, males congregate on branches overhanging a pond, and call out to females with a series of clicks. The male clambers onto the female's back to mate, and she descends to the pond, carrying the male with her, to take up water, then climbs back up into the tree to lay a batch of about 50 eggs on a leaf over the water. The process is usually repeated several times, the female drawing up water between each new batch of eggs. The eggs hatch after about 5 days, and the tadpoles fall into the pond.

COLORATION

This multihued frog has a green body, with blue and white stripes on its flanks; the insides of its legs are red and yellow. By day, it rests on leaves, with its legs folded along its sides in order to hide all the bright parts of its body, its green back providing camouflage.

GOOD CLIMBER
Like all tree frogs, the red-eyed treefrog has well-developed pads on all its toes to aid climbing.

Phyllomedusa hypochondrialis

Orange-legged leaf frog

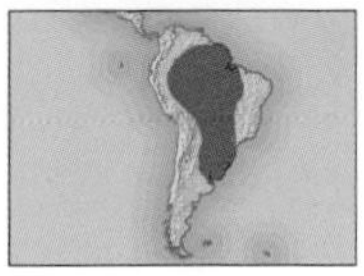

Length 1–$1\frac{3}{4}$ in (2.5–4.5 cm)
Habit Wholly terrestrial
Breeding season Rainy season
Status Least concern
Location N. to C. South America

An inhabitant of arid regions, this long, slender frog secretes a waxy substance from special glands, and rubs it all over its body to reduce water loss. Females surround their fertilized eggs with fluid-filled capsules in order to provide extra water for the developing eggs. The orange-legged leaf frog produces an unpleasant odor to ward off potential predators, and may also feign death.

Scinax ruber

Red-snouted treefrog

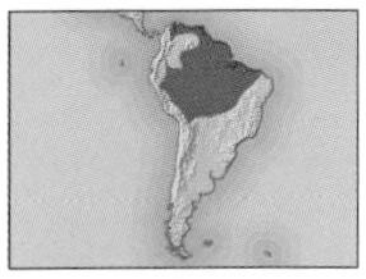

Length 1–$1\frac{1}{2}$ in (2.5–4 cm)
Habit Mostly terrestrial
Breeding season Rainy season
Status Least concern
Location S. Central America to N. South America, Caribbean

The red-snouted treefrog has a slender yellow, silver, or gray body. During the breeding season, males call very loudly to attract mates. The females, which are slightly larger than the males, select their mates on the basis of size, preferring those that are about 20 percent smaller than they are. This ratio allows closest contact with the male cloaca, so that the maximum number of eggs can be fertilized. The female scatters the eggs around the pond to reduce the impact of predation.

Triprion petasatus

Yucatan shovel-headed treefrog

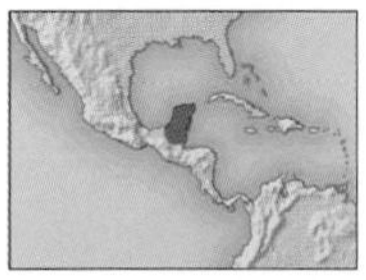

Length $2\frac{1}{4}$–3 in (5.5–7.5 cm)
Habit Mostly terrestrial
Breeding season Rainy season
Status Least concern
Location Mexico (Yucatan region), N. Central America

Characterized by the unique shape of its head, the Yucatan shovel-headed treefrog has a flattened, ducklike snout, and saddlelike flaps behind the eyes, the function of which is unknown. It is gray, brown, or green, with darker spots, and has webbed toes and well-developed adhesive pads on its fingers and toes. This tree frog lives in shrubs and trees that occur in patches in grassland. Active only at night, by day it commonly hides in tree holes, which also provide a refuge during cold and dry weather.

Acris crepitans

Northern cricket frog

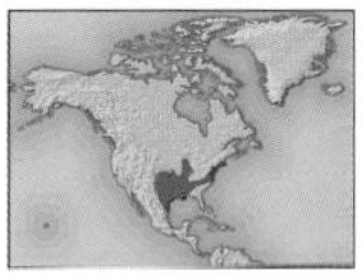

Length $\frac{1}{2}$–$1\frac{1}{2}$ in (1.5–4 cm)
Habit Mostly terrestrial
Breeding season Spring
Status Least concern
Location E. and S. USA

This frog is able to leap large distances due to its long, powerful hind limbs. Its other characteristics include rough skin, and a distinctive, dark, triangular mark between the eyes. It is found among foliage near swamps, ponds, streams, and lakes, and often basks in the sun. In early spring, large numbers converge in ponds, where the males call loudly to attract females, with a series of metallic clicks similar to the sound of crickets. In some areas, tadpoles have black tips on their tails to distract predators.

Gastrotheca monticola

Mountain marsupial frog

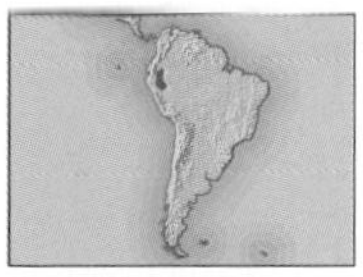

Length $1\frac{1}{2}$–$2\frac{1}{4}$ in (4–6 cm)
Habit Wholly terrestrial
Breeding season Spring and summer
Status Least concern
Location N.W. South America

The female mountain marsupial frog has a brood pouch on her back. During mating, the male places the eggs in the pouch, where they mature. The developing young are joined to the mother's blood system by a link similar to the placenta of mammals. After heavy rain, the mother releases young tadpoles into water (in some closely related species, the young are released as tiny adults). This frog is active at night, and hides under logs during winter and dry periods.

Hyalinobatrachium valerioi

La Palma glass frog

Length ¾–1½ in (2–3.5 cm)
Habit Mostly terrestrial
Breeding season Rainy season
Status Least concern

Location S. Central America to N.W. South America

This small, delicate frog has translucent skin that gives it the appearance of being made from glass. It is an excellent climber, with large adhesive disks on its fingers and toes. Females attach their eggs to foliage overhanging streams. When the pink or red tadpoles hatch, they drop into the water and burrow into the mud or sand on the streambed.

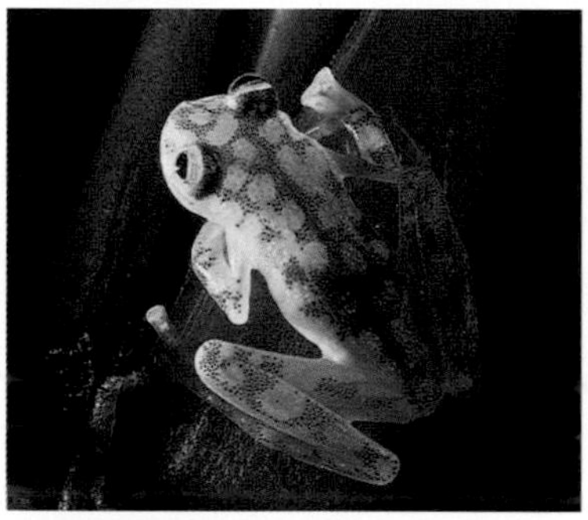

Dendrobates auratus

Green and black poison-dart frog

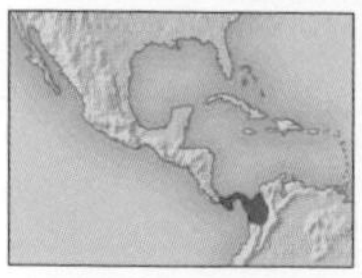

Length 1–2¼ in (2.5–6 cm)
Habit Wholly terrestrial
Breeding season Rainy season
Status Least concern

Location S. Central America to N.W. South America

This brilliantly colored, tropical American frog belongs to a family of about 179 species that includes the most poisonous amphibians in the world. Like its relatives, it lives on the ground, and uses the poison as a defense. Its bright colors show predators that it is toxic, allowing it to hop around unmolested on the forest floor. Courtship among green and black poison-dart frogs is conducted in a highly unusual way; females take the lead, enticing the males to mate by beating their partners on the back with their hind feet.

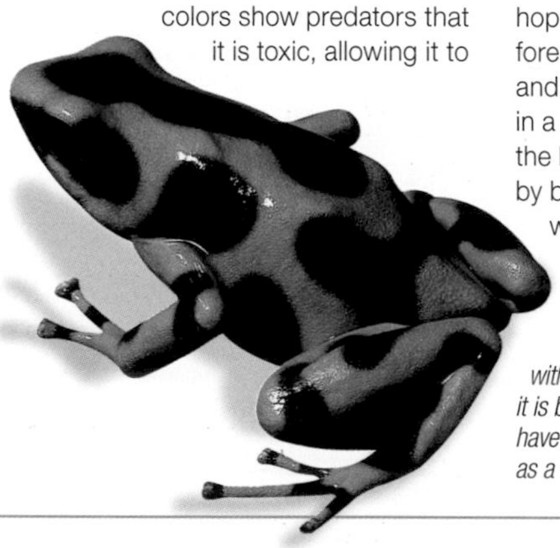

WARNING COLORATION
The green and black poison dart frog is small with a pointed head. A brightly colored species, it is black with green patches, and may sometimes have a golden sheen. This coloring functions as a warning to predators.

PATERNAL CARE

The female green and black poison-dart frog lays clutches of 5–13 eggs in leaf litter, but it is the male that guards the eggs, sometimes caring for more than one clutch at a time. When the eggs hatch, he carries the tadpoles on his back, to small pools in tree holes.

Oophaga pumilio

Strawberry poison-dart frog

Length ¾–1 in (2–2.5 cm)
Habit Wholly terrestrial
Breeding season Rainy season
Status Least concern

Location S. Central America

Although not the most toxic poison dart frog, this Central American species is still highly poisonous. Like other poison dart frogs, it relies on its chemical defense system to keep it out of trouble. Its colors are very variable: in some places it is bright red, but in others it can be brown, blue, or green. Females lay eggs in batches of 4–6, and deposit them singly in water-filled tree holes.

Ranitomeya dorisswansonae

Doris Swanson's poison-dart frog

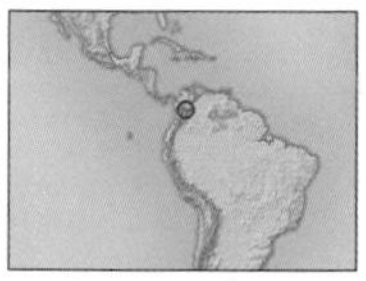

Length ⅔–⁷⁄₁₀ in (1.7–1.8 cm)
Habit Wholly terrestrial
Breeding season Not known
Status Critically endangered

Location N.W. South America

Found in the species-rich central cordillera of Colombia in 2006, this tiny frog is strikingly patterned with red blotches on a black background. It is only known to occur in a forest fragment no bigger than 0.2 square miles (0.5 square km), where it lives on the forest floor and on low-growing bromeliads. It belongs to a genus of small poison-dart frogs recently separated from *Dendrobates* on the basis of genetic evidence. Like all the brightly colored poison dart frogs, the toxins in their skin are derived from insect prey.

Rana temporaria

European common frog

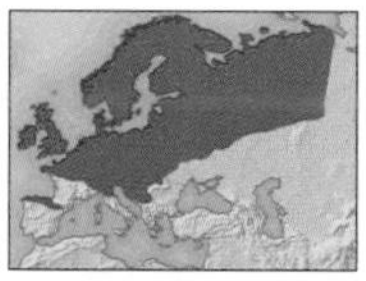

Length 2–4 in (5–10 cm)
Habit Mostly terrestrial
Breeding season Spring
Status Least concern

Location Europe, N.W. Asia

The common frog is one of the most widespread amphibians in Northern Europe, with a range that extends well beyond the Arctic Circle. An inhabitant of damp places, it spends most of its adult life on land, returning to water only to escape attack and to breed. In early spring, large numbers of these frogs make their way to shallow ponds. The males call the females with a deep, growl-like croak, and after mating the females lay floating clumps of several thousand eggs. In warm ponds, the tadpoles reach maturity in 6 weeks, but in cool water they can take 4 months.

Hylarana albolabris

Forest white-lipped frog

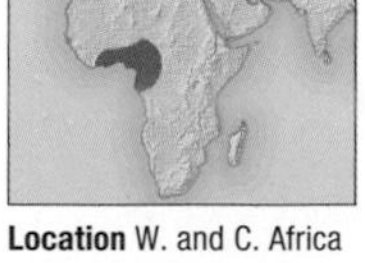

Length 2¼–4 in (6–10 cm)
Habit Mostly terrestrial
Breeding season Summer
Status Least concern

Location W. and C. Africa

Frogs belonging to the family Ranidae, often referred to as the "true" frogs, number more than 340 species, and are found in all the world's major continents. African members are generally poorly known, but those living south of the Sahara belong to the genus *Hylarana*. Characterized by the white stripe on its upper lip, this African species has a pointed snout, prominent folds of skin down each side of its back, and large adhesive disks on its toes. The male has a larger eardrum than the female and, in the breeding season, develops large glands on the upper forelimbs, which enable him to maintain a firm grip on the female while mating. Eggs are laid in still or slow-moving water. This frog lives on low vegetation in secondary rain forest and woodland but is also found in cultivated areas. Its tadpoles have a fringe of skin along their lower lip, the function of which is not known.

Lithobates pipiens

Northern leopard frog

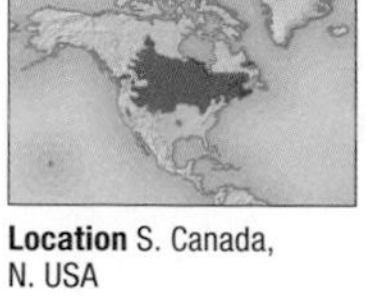

Length 2–3½ in (5–9 cm)
Habit Terrestrial/Aquatic
Breeding season Spring
Status Least concern

Location S. Canada, N. USA

This frog is green or brown with a pointed snout, and 2 or 3 rows of large, irregular, dark spots that have pale borders. Found in meadows near ponds and marshes, it feeds on insects and other invertebrates, and spends winters and dry spells hidden under rocks. The male's call is a loud, deep, rattling "snore," interspersed with grunts. This frog breeds from March to May, depending on the latitude.

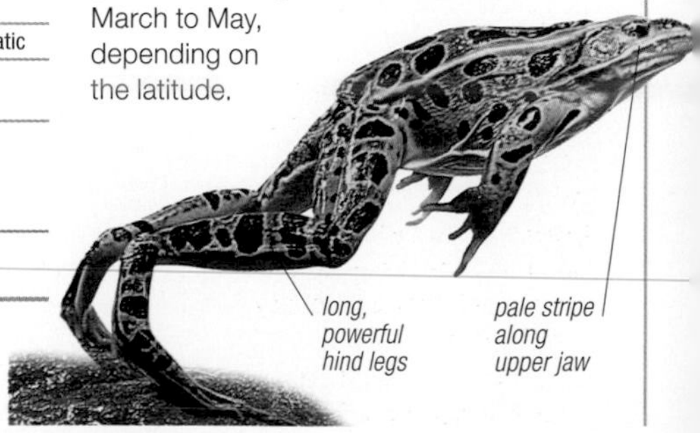

AMPHIBIANS

Lithobates catesbeianus

American bullfrog

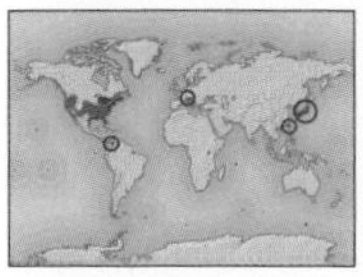

Length 3½–8 in (9–20 cm)
Habit Mostly aquatic
Breeding season Spring and summer
Status Least concern

Location S.E. Canada, W., C., and E. USA. Introduced to Europe, South America, and E. Asia

Renowned for its voracious appetite, the American bullfrog is the largest frog in America. It lives in lakes, ponds, and slow-flowing streams, spending most of its life in water or close to it. In some parts of its range it is plain green above, while in others it is boldly patterned. During the breeding season—in spring and summer—males defend their territories, and produce a deep, sonorous call from a vocal sac under their throats. After mating, females lay several thousand eggs. Their tadpoles can take up to 4 years to become adult.

AMPHIBIAN INVADER

The American bullfrog has a varied diet, including mammals, reptiles, and other frogs. Originally from eastern North America, it has been introduced into several regions farther west, where it has had a harmful effect on local freshwater wildlife.

PHYSICAL APPEARANCE

The most prominent features of this large frog are its long, powerful hind limbs and prominent eardrums. It is green above, with brown markings, and white below.

Lithobates sylvaticus

Wood frog

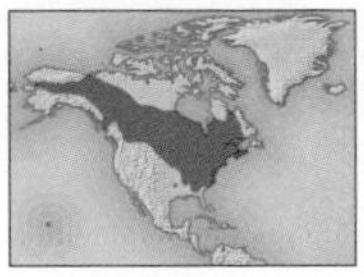

Length 1½–3¼ in (3.5–8 cm)
Habit Mostly terrestrial
Breeding season Spring
Status Least concern

Location Canada, E. USA

Although the wood frog is usually found in temperate woodland in the southern part of its range, it can survive well into the arctic circle, farther north than any other frog in North America. Its blood contains chemicals that act as an "antifreeze," enabling it to live in subzero temperatures. It often breeds while the snow is still on the ground and the breeding ponds are partially frozen. The wood frog varies in color from pink through various shades of brown to black, with darker markings. It has stripes on its legs and a characteristic dark patch, called the "robber's mask," behind its eyes.

Pelophylax ridibundus

Marsh frog

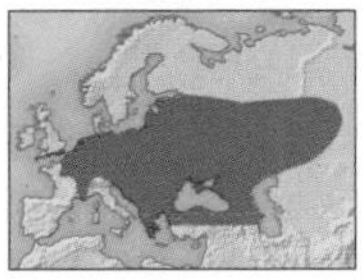

Length 3½–6 in (9–15 cm)
Habit Mostly aquatic
Breeding season Spring
Status Least concern

Location Europe, W. and S.W. Asia

Europe's largest frog, this brown or green species has typically long, powerful hind legs, longitudinal folds of skin on its back, and close-set eyes. An inhabitant of lakes, ponds, ditches, and streams, the marsh frog feeds on fish, lizards, snakes, mice, and other frogs, as well as on insects and other invertebrates. Females lay up to 12,000 eggs in water.

Conraua goliath

Goliath bullfrog

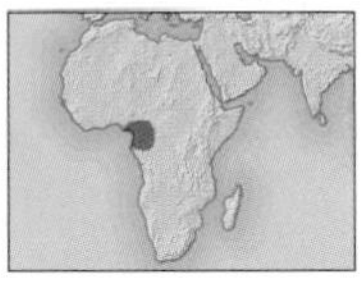

Length 4–16 in (10–40 cm)
Habit Mostly aquatic
Breeding season Rainy season
Status Endangered

Location Cameroon, Equatorial Guinea

Found exclusively in West Africa, this is the world's largest frog. With its powerful hind limbs, long, webbed toes, and smooth, slippery skin, it is well adapted to a life underwater, and is an excellent swimmer and diver. The tips of its toes are dilated, but the function of this feature is not known. The goliath bullfrog lives along streams in jungles, and feeds on other frogs, small reptiles, and mammals. It rarely emerges on land, but when it does it is quick to dive back into water whenever threatened or disturbed. Unusually among frogs, the male of this species is larger than the female and does not call to attract mates. Little is known about the breeding habits of the goliath bullfrog.

Rana dalmatina

Agile frog

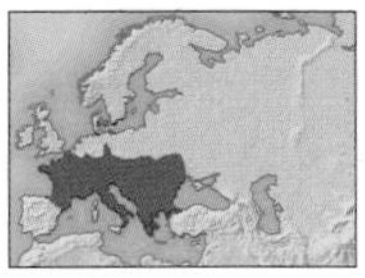

Length 2–3½ in (5–9 cm)
Habit Mostly terrestrial
Breeding season Spring
Status Least concern

Location N., C., and S. Europe

With its exceptionally long hind legs, the agile frog is a remarkable jumper, capable of leaping several yards. It is pale brown, with brown-striped yellow flanks, and has dark brown spots on its back that are usually arranged in an inverted "V" pattern. Males can be distinguished from females by their thicker forelimbs. This frog inhabits open woodland and swampy meadows, spending winter and dry periods under logs and rocks, or buried under vegetation, although males often overwinter under ice in ponds and lakes. They breed as soon as the ice begins to melt in spring, and the eggs are laid in clumps in water.

Ceratobatrachus guentheri

Gunther's triangle frog

Location Solomon Islands

Length 2–3¼ in (5–8 cm)
Habit Mostly terrestrial
Breeding season Rainy season
Status Least concern

This forest-floor animal strongly resembles the Asian spadefoot toad (see p.451), although it is not closely related. It also uses the same kind of camouflage to avoid being seen: its projections, or "horns," help break up the outline of the frog's head, making it difficult to detect against a background of leaves.

Unusually for a frog, the lower jaw of Gunther's triangle frog contains toothlike projections, made of bone, which it uses to hold its prey. These "tusks" are larger in males than in females. The female lays clumps of large eggs, which hatch directly into froglets, in damp soil. The froglets have folds in their skin, which help the absorption of yolk from the eggs.

Platymantis vitiensis

Levuka wrinkled ground frog

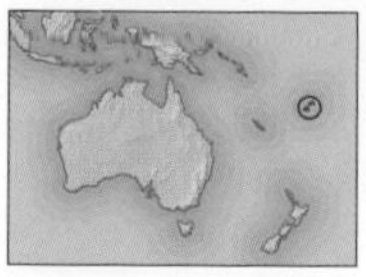

Location Fiji

Length 1½–2 in (3.5–5 cm)
Habit Mostly terrestrial
Breeding season Spring and summer
Status Near threatened

The Levuka wrinkled ground frog is a slender-bodied amphibian that is encountered only in the tropical forests of Fiji. It has a flat head with large eardrums, wrinkled skin, and adhesive disks on its unwebbed fingers and toes. The female lays a small number of large eggs on the ground, which hatch directly into small froglets, the tadpole phase being completed within the egg. Little is known about the behavior and life cycle of this frog.

Ptychadena oxyrhynchus

Sharp-nosed grass frog

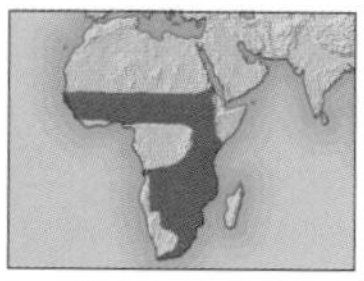

Location W. and C. to southern Africa

Length 1½–2¾ in (4–7 cm)
Habit Mostly terrestrial
Breeding season Spring
Status Least concern

This highly athletic frog can move very fast through grass and in water, its powerful hind legs enabling it to take prodigious leaps. It is yellowish brown above, with brown or black spots and patches, and has a white underside. A pale triangular patch is present on the top of its very pointed snout. This frog inhabits open savanna and thornveld in areas where temporary ponds form after heavy rain, and feeds on insects, especially crickets, and other invertebrates. During dry and cold weather, it remains hidden. The male makes a high-pitched trill to attract the female, and mating takes place on the surface of water after heavy rain. The eggs are fertilized in the air and ejected onto the surface of the water in small batches.

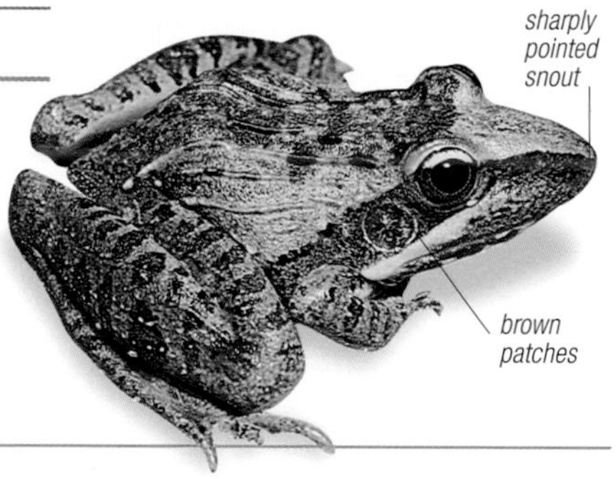

Pyxicephalus adspersus

African bullfrog

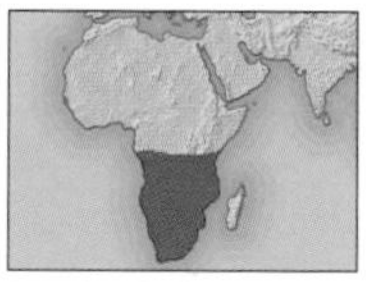

Location C. to southern Africa

Length 3¼–9 in (8–23 cm)
Habit Mostly terrestrial
Breeding season Rainy season
Status Least concern

The African bullfrog inhabits wet and dry savanna, where it may remain buried underground for very long periods, even up to several years if there is no heavy rain. When underground, it encases itself in a cocoon that reduces water loss. It feeds on other frogs as well as insects and other invertebrates. This bullfrog is very large, with a massive head and horny tubercles on its hind feet, for digging. During the breeding season, the male, which is larger than the female, develops a pair of toothlike tusks on the lower jaw, which he uses in fights over territory with other males. He produces a loud, booming call to attract the female. After mating in flooded pools, the male guards both the eggs and the tadpoles. As the tadpoles develop, the male digs channels to enable them to swim away to larger expanses of water.

Mantella viridis

Green mantella

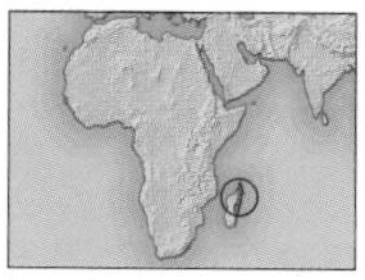

Location N. and E. Madagascar

Length ¾–1¼ in (2–3 cm)
Habit Wholly terrestrial
Breeding season Rainy season
Status Endangered

The mantellas are 16 species of brightly colored frogs, exclusive to Madagascar, which are all threatened with extinction due mainly to the destruction of their forest habitat. Their numbers are further depleted by the worldwide pet trade. The green mantella, which lives near streams in tropical forest, is yellow or pale green on its back and head, black along its sides, and has a white stripe running along its upper lip. Active during the day, it feeds on insects and other invertebrates. The male calls out with a series of clicks to attract the female. After mating, which takes place on land during the rainy season, the female lays eggs close to streams.

Mantella aurantiaca

Madagascan golden mantella

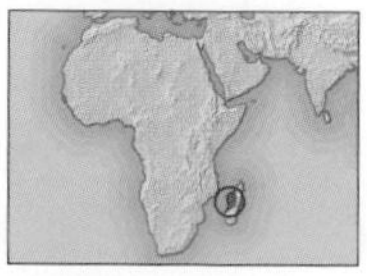

Location W. central Madagascar

Length ¾–1¼ in (2–3 cm)
Habit Wholly terrestrial
Breeding season Rainy season
Status Critically endangered

Strikingly similar to poison dart frogs (see p.460), which live in the American tropics, this Madagascan frog has evolved bright colors to warn predators that it is toxic. Active during the day, it searches for small animals on the forest floor. During the breeding season, adults mate on land, and females lay their eggs in moist leaf litter. Once the tadpoles have hatched, they are swept into pools by rain. Like other mantellas, this species is threatened by deforestation.

Hemisus marmoratus

Mottled shovel-nosed frog

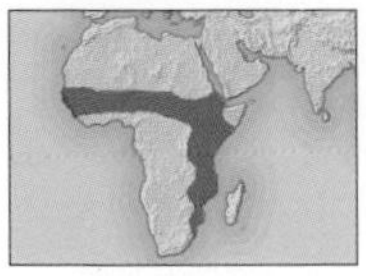

Length 1¼–1½ in (3–4 cm)
Habit Terrestrial
Breeding season Rainy season
Status Least concern

Location W. and C. to S.E. Africa

Unusually for a burrowing frog, this squat species digs into soil head first, using its sharp, tough, spadelike snout, rather than digging backward with its hind feet. It preys on ants and termites. Females lay up to 2,000 eggs in underground chambers close to pools. They guard the eggs until they hatch, then dig a tunnel to release the tadpoles into the water.

Hyperolius marmoratus

Painted reed frog

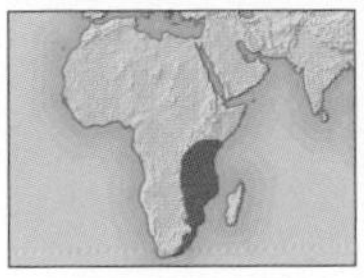

Length 1–1½ in (2.5–3.5 cm)
Habit Terrestrial
Breeding season Spring and summer
Status Least concern

Location E. to S.E. Africa

The painted reed frog is a small, graceful animal with highly variable coloration—it may be striped, spotted, or plain brown. A good climber, its fingers and toes are equipped with adhesive disks as well as webbing. It inhabits areas close to permanent ponds, lakes, and swamps; like most frogs, it spends dry spells hidden under logs and rocks. Active at night, it feeds on insects and other invertebrates. During its long breeding season, which extends through spring and summer, males form choruses close to ponds, and call very loudly to attract mates. Females may come to the chorus more than once in a season, laying a clutch of eggs in small clumps in water at each visit. Males vary in the number of nights they spend in a chorus; those that call on the greatest number of nights get the most matings.

Afrixalus fornasinii

Greater leaf-folding frog

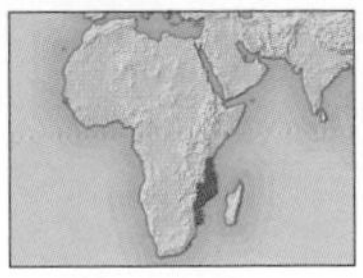

Length 1¼–1½ in (3–4 cm)
Habit Wholly terrestrial
Breeding season Spring and summer
Status Least concern

Location S.E. Africa

Many tree frogs lay their eggs above the ground, but this species has a particularly skillful technique. The female lays her clutch on a leaf overhanging water, and then folds the leaf, fastening it with a glue that she secretes. When the eggs hatch, the tadpoles drop into the water below—a system that gives them a head start in the struggle to avoid being eaten. Greater leaf-folding frogs are brown with dark, longitudinal stripes, and have slender limbs with adhesive disks as well as webbing on their fingers and toes. They live in dense vegetation around pools and swamps, and are active at night. At the onset of the breeding season, large males call from an elevated position in vegetation, making rapid clicks to attract mates. However, smaller males often stay silent, waiting by the callers. When females approach, the noncalling males sometimes manage to intercept them and mate.

Kassina senegalensis

Bubbling kassina

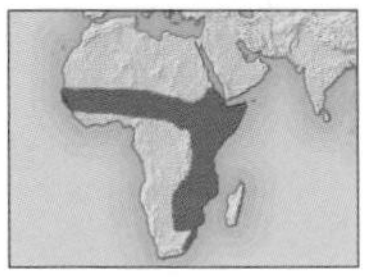

Length 1¼–2 in (3–5 cm)
Habit Mostly terrestrial
Breeding season Rainy season
Status Least concern

Location W. and C. to S.E. Africa

Adapted for walking rather than hopping, the bubbling kassina has slender limbs and slight webbing on the toes of its hind feet. Its plump, elongated body varies in color from beige or yellow to gray, with bold, dark brown or black stripes and blotches. An inhabitant of low-lying savanna and sand dunes, it lives close to ponds and pools. In the breeding season, males begin calling in late afternoon away from water. By nightfall, they have moved closer to call from the pond edge. Mating occurs in water, and the eggs are attached individually to aquatic plants, often in clusters.

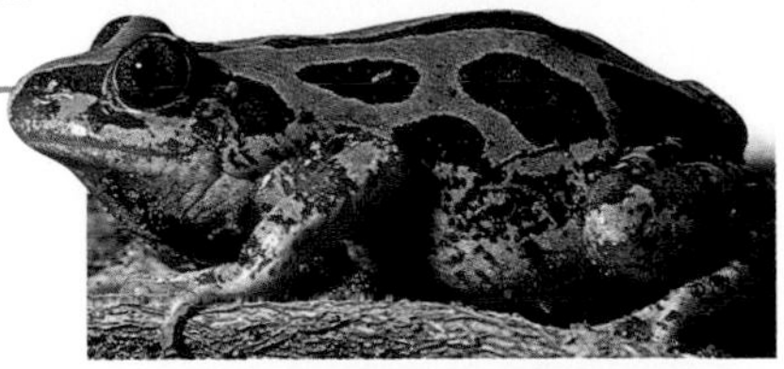

Leptopelis modestus

African treefrog

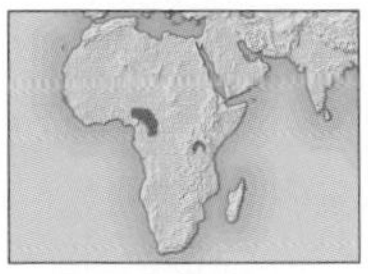

Length 1–1¾ in (2.5–4.5 cm)
Habit Terrestrial
Breeding season Spring and summer
Status Least concern

Location W. Africa, central E. Africa

Also called the modest treefrog, this species has a broad head with a wide mouth, large eyes, and conspicuous eardrums. Gray or pale brown, it has a darker hourglass-shaped patch on its back. Males call to females from trees, producing a deep "clacking" sound while displaying their bright blue or green throat. They will fight to defend their calling positions.

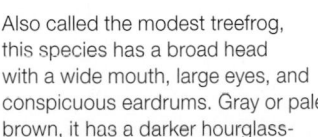

Rhacophorus nigropalmatus

Wallace's flying frog

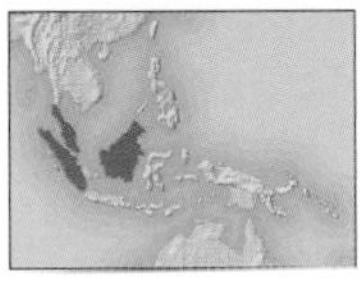

Length 2¾–4 in (7–10 cm)
Habit Terrestrial
Breeding season Rainy season
Status Least concern

Location S.E. Asia

There are over 80 species of *Rhacophorus* frogs, several of which—including Wallace's flying frog—use their feet to glide from tree to tree. The partial webbing between their toes acts as a parachute, enabling them to "fly" across wide gaps to escape predators, steering themselves in midair. Wallace's flying frog is bright green with white spots, and has a flat head with large, protruding eyes and visible eardrums, and large adhesive disks on its fingers and toes for climbing. The male is smaller than the female, and has pads on his thumbs so that he can grasp her during mating. Active at night, Wallace's flying frog is found mostly in brush and forest, and is also frequently found close to human habitation. During the breeding season, the female lays up to 800 eggs in a foam nest attached to foliage overhanging water. When the eggs hatch, the tadpoles emerge and fall into the water below.

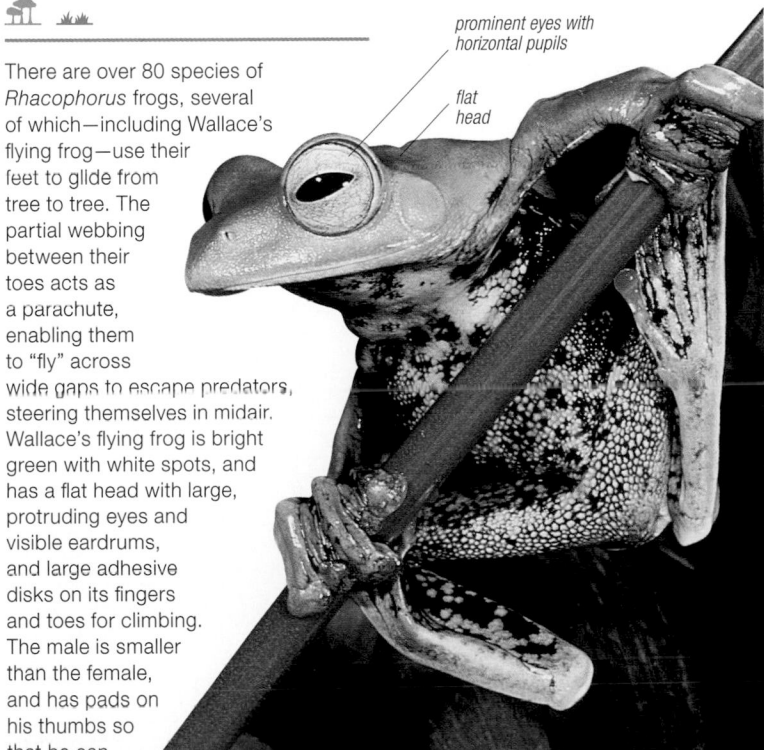

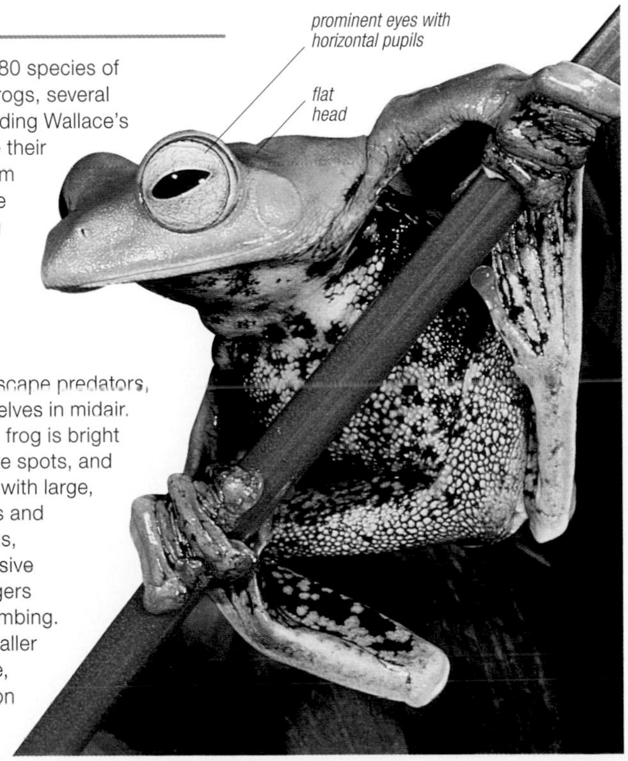

WALLACE'S FLYING FROG
The large webbed feet of Wallace's flying frog act like parachutes as it leaps from tree to tree. This slows its descent so that the frog spends more time in the air and can travel greater distances.

Chiromantis xerampelina

Grey foam-nest frog

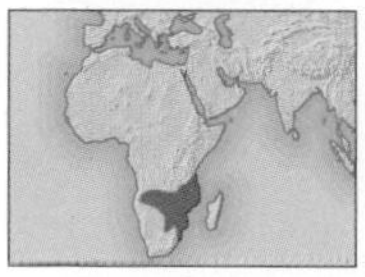

Location Southern Africa

Length 2–3½in (5–9cm)
Habit Mostly arboreal
Breeding season Rainy season
Status Least concern

This tree frog and its close relatives are remarkable for their breeding method and their ability to survive in savanna that has a long dry season. Its skin is adapted to reduce water loss, and it excretes dry droppings rather than watery waste. Usually gray or pale brown, this frog can change color for camouflage—at night it is usually dark-colored, but it can become almost white at midday.

LIVING IN TREES
The grey foam-nest frog has a compact body and prominent eyes with horizontal pupils. It possesses the typical adaptations for an arboreal lifestyle: long, slender limbs, and adhesive disks on all its toes. Its fingers and toes are also webbed.

BUILDING A COMMUNAL FOAM NEST

During the breeding season, the male grey foam-nest frog—which can be distinguished from the female by the pads he develops on the first and second toes of his forelimbs—calls out to females from branches overhanging a pool. He may be joined in this by one or more other males. When a female arrives, one of the males clasps her, they mate, and she produces a secretion that one or several males whip into a foam with their legs. The female then ejects her eggs—up to 1,200 at a time—into the foam, and one or more of the males shed sperm onto them. The exterior of the nest hardens to protect the eggs, but the inside remains moist. After the eggs hatch, the tadpoles burrow out of the nest and fall into the water below.

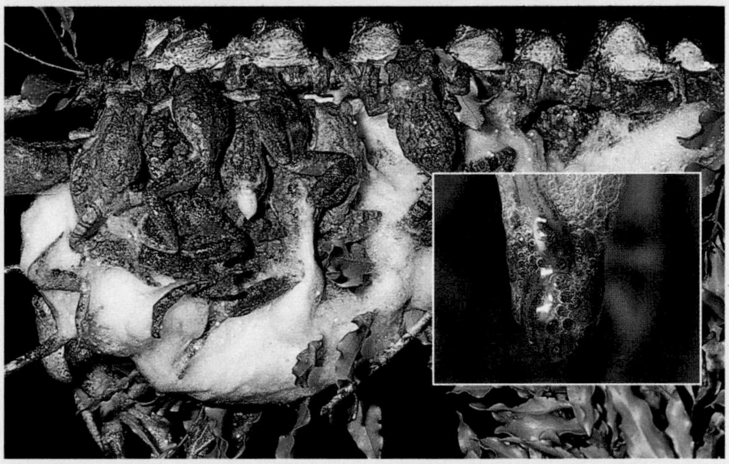

Breviceps adspersus

Bushveld rain frog

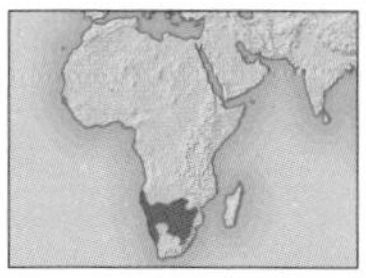

Location Southern Africa

Length 1¼–2¼in (3–6cm)
Habit Terrestrial
Breeding season Rainy season
Status Least concern

The bushveld rain frog lives mostly underground, emerging to feed and mate only after rain. It has a stout body—either light or dark brown, with rows of lighter, yellowish or orange patches with dark borders—and short, stout limbs typical of most burrowing species. When cornered, this frog inflates its body and lodges itself firmly in its burrow. The females are much larger than the males. Since the male is too small to clasp the female during mating, the female emits a secretion from her back that acts as a glue, keeping the mating pair together.

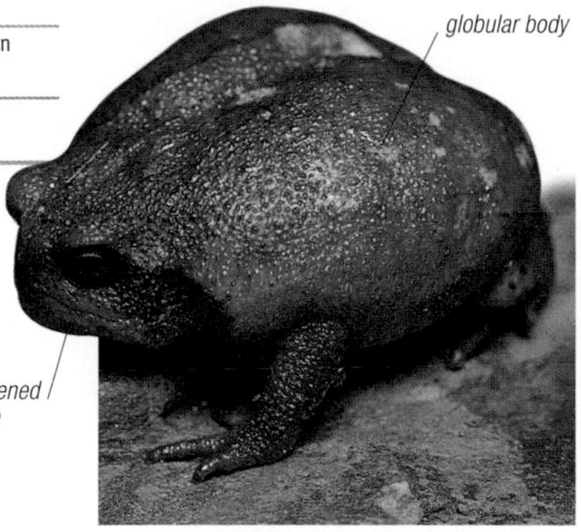

Microhyla ornata

Ornate narrow-mouthed toad

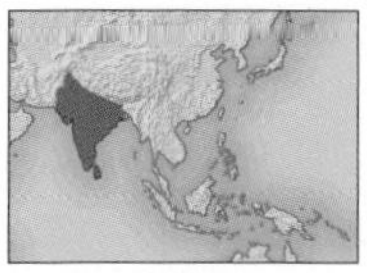

Location S. Asia

Length ¾–1¼in (2–3cm)
Habit Mostly terrestrial
Breeding season Rainy season
Status Least concern

This small, Asian toad has a plump, smooth-skinned body with short forelimbs and long hind limbs, and a small head. It is yellow or ocher, with a darker marbled pattern and dark stripes along its flanks. The ornate narrow-mouthed toad inhabits rain forest and cultivated areas, notably rice paddies, and is also found in grass or leaf litter, preying on insects and other invertebrates. It remains buried underground during dry and cold weather. Breeding takes place in ponds, rice paddies, temporary pools, and slow-moving streams. The male—which is smaller than the female—uses its large vocal sacs to call very loudly from water to encourage females to mate. The female ornate narrow-mouthed toad lays approximately 270–1,200 eggs at a time in water. The eggs, which form a film on the surface, hatch quickly, and the tadpoles develop rapidly, before the breeding pool has time to dry up.

Phrynomantis bifasciatus

Banded rubber frog

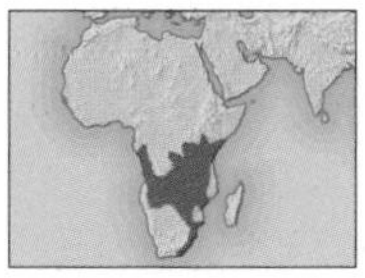

Location Southern Africa

Length 1½–2¼in (4–6cm)
Habit Mostly terrestrial
Breeding season Rainy season
Status Least concern

Unusually for a frog, the short-legged banded rubber frog walks and runs rather than hops or jumps. It is black and glossy, with mainly pink or red spots and stripes down its back; these change color throughout the day, turning paler in bright light. Its head is pointed, and it has small eyes with round pupils. This frog feeds on insects, mainly termites and ants. If threatened, it raises itself on outstretched limbs and inflates its body. It also produces a highly toxic substance from its skin, which may be lethal to humans if ingested. Males produce a trilling courtship call from the pond edge, and females lay clumps of up to 600 eggs, which they attach to aquatic plants.

Gastrophryne olivacea

Great Plains narrow-mouthed toad

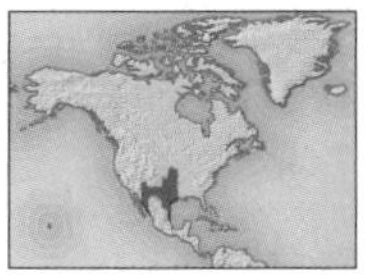

Location C. and S. USA, N. Mexico

Length ¾–1½in (2–4cm)
Habit Mostly terrestrial
Breeding season Spring and summer
Status Least concern

Also known as the Western narrow-mouthed toad, this species has a narrow, pointed head with small eyes, a stout body, and short legs that allow it to burrow and hide in small crevices. Brown or gray in color, it is sometimes distinguished by the dark, leaf-shaped pattern on the surface of its back. In spring, after heavy rain, Great Plains narrow-mouthed toads are known to gather in large numbers around temporary pools. The males call out to the females, making a protracted buzzing sound. After mating, the females lay their eggs in the pool. The tadpoles develop rapidly, before the water dries up.

FISHES

FISHES

PHYLUM	Chordata
CLASSES	Cephalaspidomorphi, Chondrichthyes, Osteichthyes
ORDERS	61
FAMILIES	534
SPECIES	About 32,500

Fishes were the first backboned animals to appear on Earth, and they form the largest group of vertebrates. Unlike the other major divisions of vertebrates, however, fishes are not a natural group; instead, they are an informal collection of 3 classes that are only distantly related to one another. A typical fish breathes using gills, has a body covered with scales, maneuvers using fins, and is ectothermic (cold-blooded). Most species live in either freshwater or the sea, but a few species move between both environments.

UNDERWATER LIFE
Most fishes (such as these butterfly fishes) spend their entire life in water, although a small number of species can survive short periods out of water.

Evolution

Living fishes are usually divided into 3 groups: jawless fishes, cartilaginous fishes, and bony fishes. Each of these groups had a different ancestor and evolved separately from the others.

The first fishes appeared more than 500 million years ago, having probably evolved from soft-bodied, filter-feeding invertebrates. Like living jawless fishes, these fishes had a round, fleshy mouth without jaws, but unlike them, many lacked teeth.

The first fishes (indeed, the first animals) with movable jaws arose around 440 million years ago. Known as acanthodians (or spiny sharks), their jaws evolved from their front gill arches. Linked to the development of jaws was the appearance of teeth. Having jaws and teeth made it possible for acanthodians to feed on a wide range of food. They also had several pairs of spines along the lower sides of their body, from which the paired fins of fishes evolved. Unlike the early jawless fishes, their body was covered with scales.

Fishes with a cartilage skeleton first appeared about 370 million years ago. These were the ancestors of living sharks, skates, and rays.

Some 50 million years before the appearance of cartilaginous fishes, a group of fishes with a bony internal skeleton had appeared. There are 2 main groups of fishes: fleshy-finned fishes, which have fins supported by bony skeletons fringed with rays; and ray-finned fishes, in which the fins are supported only by fin rays. The fleshy-finned fishes almost certainly gave rise to the first 4-limbed land vertebrates. The ray-finned fishes appeared at about the same time but have since become a far larger and more successful group. They gradually developed features that made them more efficient swimmers, including thinner scales, highly flexible fins, and a symmetrical tail. Also important were the evolution of more mobile jaws and gill chambers that could be more easily ventilated.

CEPHALASPIS

STICHOCENTRUS

FISH FOSSILS
Among the first fishes to appear was a form called Cephalaspis *(top). It had a jawless mouth, and its head and gills were protected by a large, bony shield. By contrast, a fossil of the bony fish Stichocentrus (above) reveals jaws, scales, and a much lighter and more flexible internal skeleton.*

Anatomy

Fishes have several physical adaptations for life in water. The body of a fish is covered with smooth scales, and has fins to provide power, steering, and stability. All fishes have gills for extracting oxygen from water, and all have an internal skeleton, although this takes different forms in each of the 3 main groups of fishes. In jawless fishes, the body is supported by a simple rod called the notochord and some rudimentary vertebrae. Sharks, skates, and rays

FIN TYPES
A fish's dorsal, anal, and tail fins occur singly along the midline of the body, while the pectoral and pelvic fins are paired. The common carp (shown here) has only one dorsal fin; however, some other fish have 2 or 3 dorsal fins.

SCALES

Fishes have several types of scales, made of various materials. Cartilaginous fishes have toothlike placoid scales (dermal denticles). Some primitive bony fishes have thick, relatively inflexible scales. They include diamond-shaped ganoid scales, found in gars, and layered cosmoid scales, seen in coelacanths. Like placoid scales, these are made of dentine and a substance similar to enamel. Bony fishes have thin scales made of bone, with one end embedded in the skin and the other exposed: cycloid scales have a smooth exposed surface, while ctenoid scales are rough or spiny.

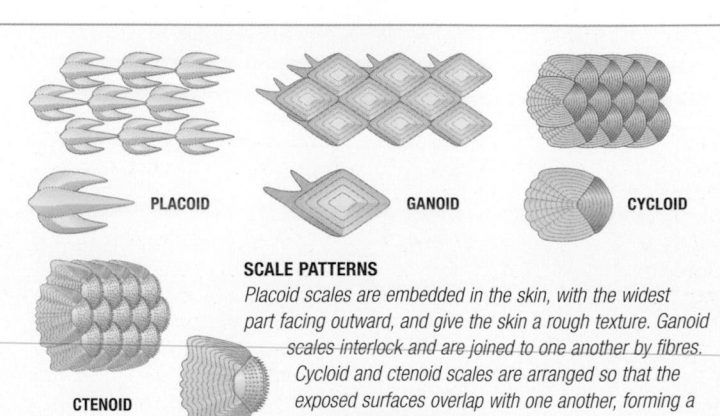

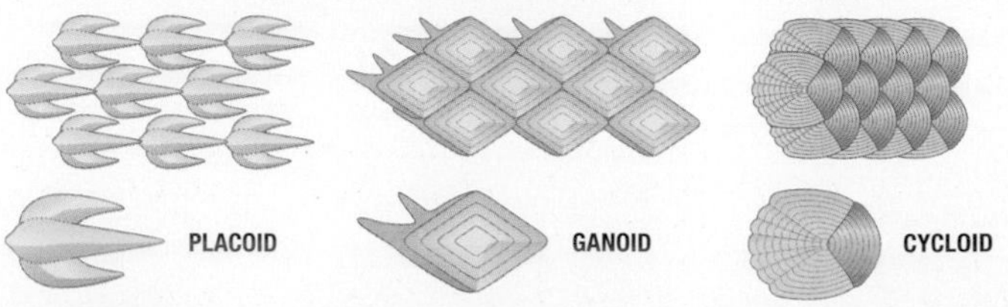

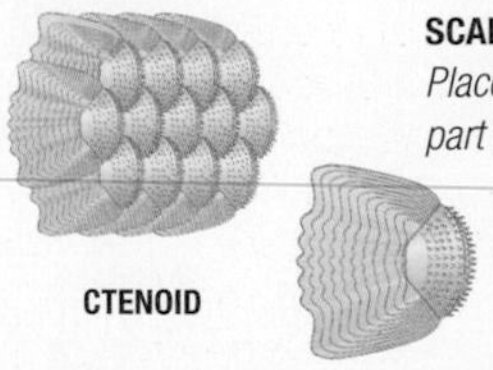

SCALE PATTERNS
Placoid scales are embedded in the skin, with the widest part facing outward, and give the skin a rough texture. Ganoid scales interlock and are joined to one another by fibres. Cycloid and ctenoid scales are arranged so that the exposed surfaces overlap with one another, forming a smooth, flexible covering.

GILLS

A fish's gills are located in internal chambers on either side of the body, just behind the mouth. In most species, each gill consists of a framework of bone or cartilage, supporting tissues that contain densely packed capillaries. As water enters the gill chambers, it is first sieved clean by structures called gill rakers. Behind the rakers are supports called gill arches, which provide attachment points for gill filaments. To increase the efficiency of respiration, the filaments have folds on their surface, known as lamellae. Gases are exchanged across the surface of the lamellae, before being distributed to the rest of the fish's body. In some fishes, the area of the lamellae is 10 times greater than that of the body's outer surface.

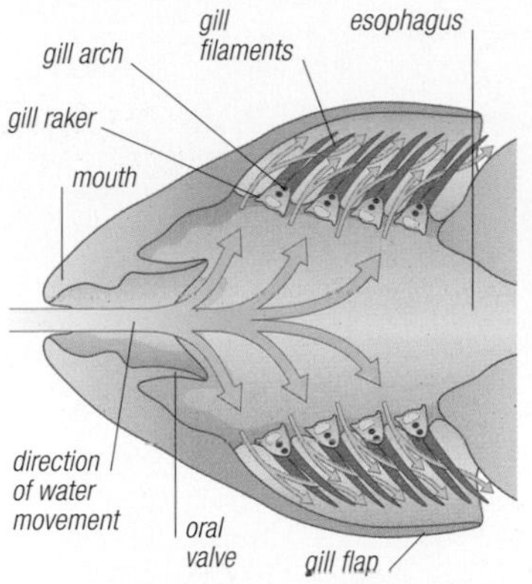

GAS EXCHANGE
Water enters a fish's gills through its mouth and leaves through the gill openings. As it passes through the gills, dissolved oxygen is transferred across the thin outer membrane of the capillaries and enters the bloodstream. At the same time, carbon dioxide is expelled from the capillaries.

BREATHING OUT OF WATER
By keeping their gill chambers moist, mudskippers can survive out of water for long periods. They live in coastal swamps, and at low tide can be found using their well-developed pectoral fins to move around on exposed mudflats.

have a skeleton of cartilage hardened by calcium carbonate, as well as fully developed vertebrae. Most fishes have a skeleton made of bone. In addition to a bony skull and vertebral column, they also have a separate bony fin skeleton.

Although some fishes have smooth skin, most are protected by scales, plates of bone, or spines. Scales, which are the most common covering, provide protection and promote the efficient movement of water over the body, while still allowing the fish to move freely (see panel, opposite). Glands in the skin secrete mucus, which protects the fish from bacteria and, in some species, helps to reduce drag.

Almost all fishes have fins. There are 2 main types: median (or unpaired) and paired fins. Median fins are found singly, on the dorsal or ventral midline of the body. They include the dorsal, anal, and tail (caudal) fins. The paired fins—the pelvic and pectoral fins—are arranged in twos, one on each side of the body. Fins are used mainly for locomotion (see opposite) but they also have other uses: colors and patterns on their surface can be used as signals to warn predators, attract mates, defend territories, or lure prey; the fins of some fishes have spines, which are sometimes poisonous, to protect them from predators.

All fishes have gills, which they use to take oxygen into their bloodstream (see panel, above). Some fishes also have other ways of taking in oxygen. Lungfishes, for example, breathe air using primitive, lunglike organs. Other fishes can absorb oxygen and release carbon dioxide through their skin.

Senses

Fishes collect information about their environment using some sensory systems that are also found in other types of animals—these include vision, hearing, touch, taste, and smell. However, fishes also have some unique sense organs.

Although most fishes have eyes, their ability to perceive light, color, shape, and distance varies greatly between species. This variation can often be related to habitat. Fishes that live in clear water tend to have good vision, while many fishes that live in dark conditions (such as muddy water, caves, or the deep sea) either have poor vision or have lost their eyes altogether. However, some deep-sea fishes have large eyes so that they can collect as much light as possible.

Water is an effective medium for transmitting sound, and most fishes are able to respond to sound waves. Sound vibrations are usually transmitted through the bones and tissues of the head to the inner ear, although in some species they may be amplified by the swim bladder.

In aquatic animals, the distinction between taste and smell is blurred because the same sense organs often respond to chemicals that are dissolved in water and contained in food. Fishes smell using sensory pits (called nares), which are lined with tissues that contain olfactory receptor cells. Most fishes have a good sense of smell, and some can detect chemicals in minute concentration. Taste receptors are usually found in or near the mouth, although some fishes have them on their fins, skin, or whiskerlike barbels, allowing them to taste food on or near the bottom.

Almost all fishes have what is called a lateral-line system, which they use to detect vibrations (including those made by their predators and prey) and changes in water pressure and currents. These changes are detected by organs known as neuromasts, which are usually housed in canals, in the bones of the head and in a canal that runs along the side of the body beneath the scales and is connected to the exterior via pores that run through the scales.

Many fishes are also able to detect electrical waves and impulses. In cartilaginous fishes, signals are received by structures called the ampullae of Lorenzini. These specialized nerve cells, located in small pores at the surface of the skin, contain a conductive gel and can detect the weak electrical currents produced by other animals. Some bony fish can also detect electrical signals. Certain species have organs that produce electrical fields, which they use to detect objects or other fish in low visibility, and to communicate.

TUBULAR EYES

LARGE EYES

EYES
Fishes that live in habitats with little light often have unusual eyes. Spookfishes (top) are deep-water fishes with tubular eyes that point upward. The pinecone soldierfish (above) lives in shallow water but avoids direct sunlight. It feeds at night, using its large eyes to search for food.

TASTE
Catfishes are named after the whiskerlike barbels found in many species. These are covered by taste buds and are used to find food. This is especially useful for species such as the squeaker (shown here), which lives in dark waters at the bottom of deep lakes.

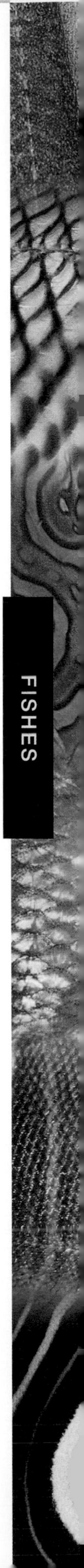

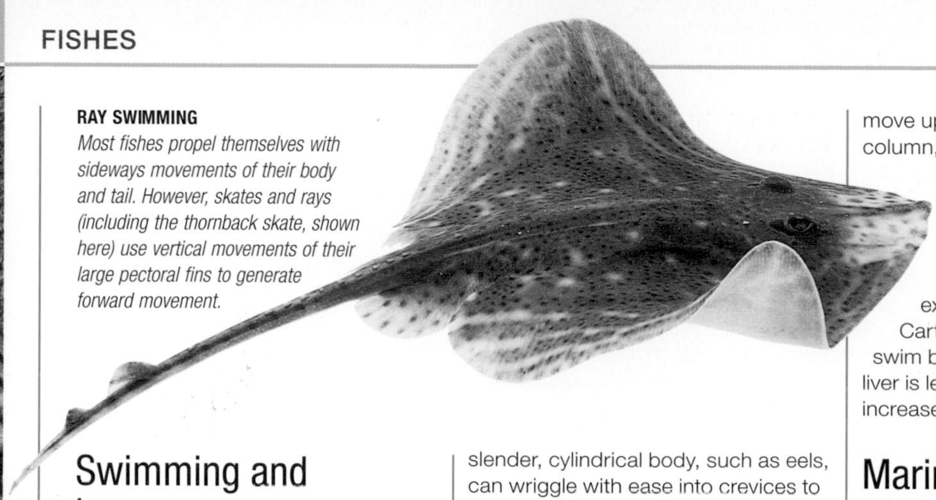

RAY SWIMMING
Most fishes propel themselves with sideways movements of their body and tail. However, skates and rays (including the thornback skate, shown here) use vertical movements of their large pectoral fins to generate forward movement.

Swimming and buoyancy

A fish propels itself through water using the action of its muscles. Along either side of the backbone are bundles of muscle fibers (known as myotomes). By contracting these muscle bundles in sequence, one a fraction of a second after the other, the fish creates a wave motion that travels along its body from front to back and ultimately causes the tail to move from side to side. It is this wave motion that propels the fish through the water. The wave usually begins near the rear third or half of the body so that only the back end moves from side to side. In some fishes (usually fast-moving ones), only the very end of the tail moves. The movement is more pronounced in fishes that have a long, slender body. Some fishes, including eels, can reverse the direction of the wave movement and so swim backward.

Fins play an important part in propulsion and movement. In most fishes, the dorsal and anal fins act in a similar way to the keel of a boat, stabilizing the fish's body. The paired fins have various uses including propulsion. Most fishes use them as control surfaces, adjusting the angle of the fins to move up and down in the water. In sharks and some bony fishes, the paired fins also provide lift as the fish moves forward, while some fishes use their paired fins to "walk" on the bottom and, on rare occasions, on land. The tail fin is used mainly for propulsion but it also has a role in steering.

In general, fishes move relatively slowly, rarely exceeding 3 mph (5 kph). However, some fishes are capable of moving at much greater speed—for example, the wahoo (a close relative of tunas) can reach 50 mph (75 kph) for short periods.

A fish's body shape usually reflects the way it swims. Fishes that need to swim at high speeds in open water for long periods typically have a torpedo-shaped (fusiform) body. A body that is tall and flattened from side to side is less efficient but is common among reef fishes, which need to make sudden turns around dense vegetation or rocky surfaces. Fishes with a slender, cylindrical body, such as eels, can wriggle with ease into crevices to find food or escape from predators. Bottom-dwelling fishes typically have a body that is compressed from top to bottom, helping them remain inconspicuous on the seafloor.

Many fishes are neutrally buoyant. That is, their density is similar to that of the surrounding water, so they can remain suspended at a constant depth in the water column. Most other fishes are slightly negatively buoyant—their fins give them lift as they move, but they will sink if motionless. For bottom-dwelling fishes, such as skates and rays, being negatively buoyant is an advantage. As they move up and down in the water column, most fishes can adjust their buoyancy using a gas-filled organ known as the swim bladder (see p.487). To make themselves more or less buoyant, gas is added to or extracted from this bladder.

Cartilaginous fishes do not have a swim bladder but their large, oil-rich liver is less dense than water and thus increases their buoyancy.

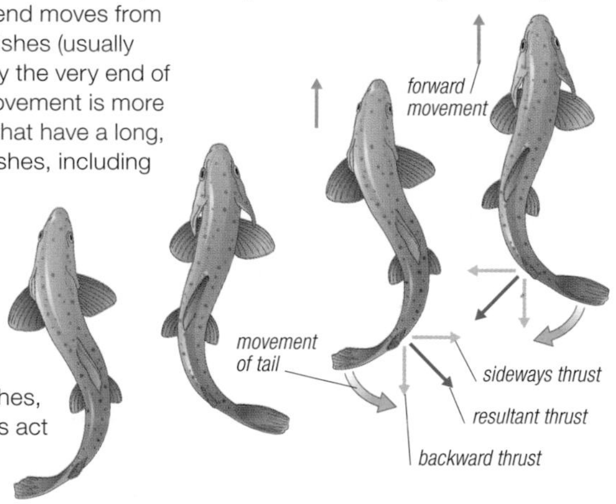

PROPULSION
As a fish's tail moves, it exerts a sideways thrust and a backward thrust on the water. The resultant force acts at an angle halfway between these 2. Movement of the tail from side to side means that the resultant thrusts to left and right produce a net backward thrust. Because every force in nature produces an equal and opposite reaction, this backward thrust propels the fish forward.

Marine and freshwater fishes

Fishes need mechanisms to keep the concentration of water and salts in their body in a stable state. However, this concentration is different from the water in which they live, so they have to counteract osmosis—the tendency of liquids of different concentrations (in their cells and in their environment) to equate by passing through a semipermeable membrane (the cell walls). Marine fishes tend to gain water because the salt concentration in their cells is lower than that of the surrounding saltwater. In contrast, fishes that live in freshwater have a tendency to dehydrate because water always flows from a more concentrated solution to a weaker one to dilute it. Fishes, such as salmon and eels, that move between salt and freshwater are able to cope in both environments. In fishes, water is involuntarily lost and gained through the mouth and gills. To prevent any excesses that could be detrimental to their internal chemistry, fishes "osmoregulate" in different ways.

In marine fishes, the concentration of salts is lower inside their bodies than in the water. Sharks, rays, and coelacanths prevent dehydration by retaining urea, which significantly increases their cell fluid concentration. Other marine fishes counteract water loss by drinking large quantities of seawater and excreting most of the salt. Some have chloride cells in their gills and on the inside the gill cover (operculum) that remove salts from the water that is absorbed and then excrete it. Freshwater fishes control involuntary water intake from the rivers and lakes in which they live by drinking little and producing copious dilute urine. They obtain the salt they need from their food and have cells in their gills that control its loss.

SURVIVING IN FRESH AND SALT WATER
The European eel can survive in both fresh and salt water. It spends most of its adult life in rivers but returns to sea to breed. The opposite is true of many salmon and trout, which move from the sea to fresh water to breed.

Temperature control

Fishes are ectothermic (cold-blooded) animals, which means that their internal temperature is the same as that of the surrounding water. Some species modify their behavior, basking in the sun, seek shade, or rise and fall between warmer and colder depths. Other species regulate their temperature by altering their color, using pigments that change between dark, heat-absorbing colors and light, heat-reflecting shades.

A few fishes—including tuna and the great white shark and its relatives—have some similarities with warm-blooded animals. Some of the heat generated by their large swimming muscles is conserved, so that their internal temperature is higher than that of the water around them. This makes it possible for these fishes to remain active in cold water.

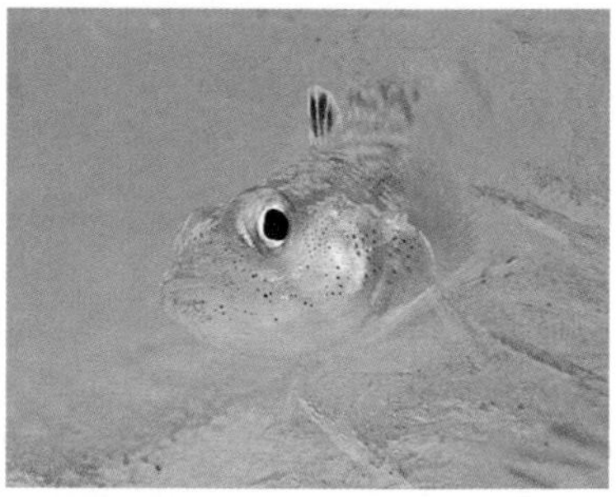

SURVIVING COLD
Seawater freezes at a lower temperature than fish blood. However, the blood of some fishes, such as this icefish, remains liquid even in extremely cold water because it contains proteins that act like antifreeze.

LAYING EGGS
The perch, like many fishes, releases large numbers of eggs to ensure that at least some will eventually produce adult fishes. This is known as broadcast spawning.

Reproduction

The reproductive behavior of fishes is varied. Although in most cases fertilization takes place outside the female's body and the young emerge from the eggs as larvae, there is also a significant number of species in which fertilization is internal and females give birth to live young.

Some fishes breed regularly, often once a year. Others, however, reproduce only once in their lifetime, and may die shortly afterward. The timing of reproduction is controlled by external factors (including changes in temperature, light levels, or day length) or by internal cycles (such as changes in hormone levels).

Some fishes have preferred spawning grounds, laying eggs or releasing their young in places where they are most likely to survive. In order to reach these areas, they may have to undertake migrations of thousands of miles.

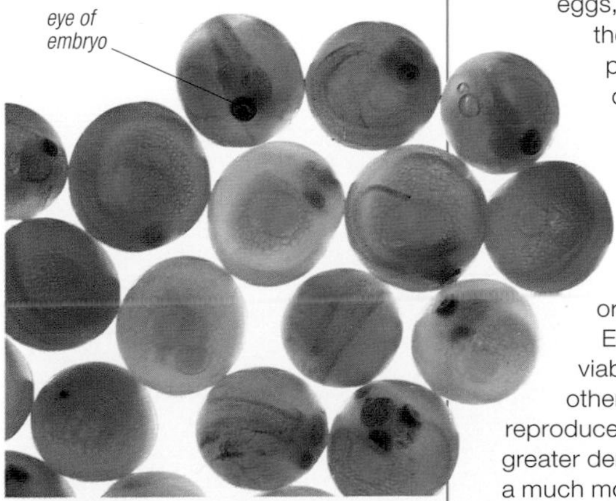

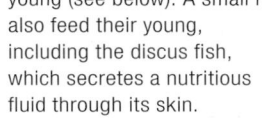

EGGS
Inside its egg, an embryonic fish is nourished by fluids contained in a yolk sac. The egg case is transparent (allowing the eyes of these salmon embryos to be seen from the outside). The time taken for the eggs to hatch often depends on the water temperature.

In some species, large numbers of males and females gather in shoals to breed, without any kind of courtship behavior. However, some fishes conduct complex courtship rituals to increase their chances of attracting a suitable mate. As a prelude to breeding, the male may change color. For example, the male dragonet develops more intense colors, which are displayed to potential mates during elaborate circling movements.

In most fishes (as in many other aquatic animals), fertilization occurs externally. The male releases sperm (milt) over the eggs (roe) as they are released from the female's body. Milt has a thick consistency, which prevents the sperm cells from dispersing too quickly. The eggs of most marine fishes are made buoyant by droplets of oil and float freely as part of the plankton. The eggs of freshwater fishes are typically heavier and have a sticky surface so that they adhere to objects in the water. Most freshwater fishes make a nest for their eggs, and in some species they guard the eggs (see panel, right). In almost all cases, the eggs hatch to release larvae, which are incompletely formed and gradually develop their skeleton, fins, and some of their organ systems.

External fertilization is a viable way for fishes (and other aquatic animals) to reproduce, partly because the greater density of water makes it a much more suitable medium than air for transferring eggs and sperm. Water also provides the developing eggs with nutrients and dissolved oxygen. However, the chances of an embryo surviving to reach adulthood are relatively low, particularly for marine fishes. To compensate for this, females often produce large numbers of eggs (as many as 5 million in some species).

In a small number of species, reproduction begins with internal fertilization. In sharks, for example, the inner edges of the pelvic fins are modified into copulatory organs, called claspers, of which one is inserted into the female's cloaca. Sea water flushes the male's sperm along a groove in this organ and into the female's body.

Where fertilization is internal, the young may be alive when they are released from the female's body. In some species that give birth to live young, the eggs simply hatch inside the mother's body. In others, including some species of sharks and rays, there is a connection between the developing embryo and the female, along which nutrients are passed to the embryo.

Live-bearing females invest a relatively large amount of energy in carrying and nourishing their developing young. However, compared with fishes that reproduce by external fertilization, their young are more developed at birth and have a greater chance of survival. This means that live-bearing fishes can produce and fertilize fewer eggs and yet maintain a stable population.

Compared with other groups of vertebrates, fishes include a relatively large number of hermaphrodites, all of which are bony fishes (see p.487). Some fishes, including certain species of carp and loach, reproduce parthenogenetically, the eggs being fertilized without a male. Other species reproduce by a process called gynogenesis. These include the Amazon molly, which is a hybrid of 2 other species and has a population that consists only of females. Reproduction occurs when a female is fertilized by a male of either of the 2 parent species.

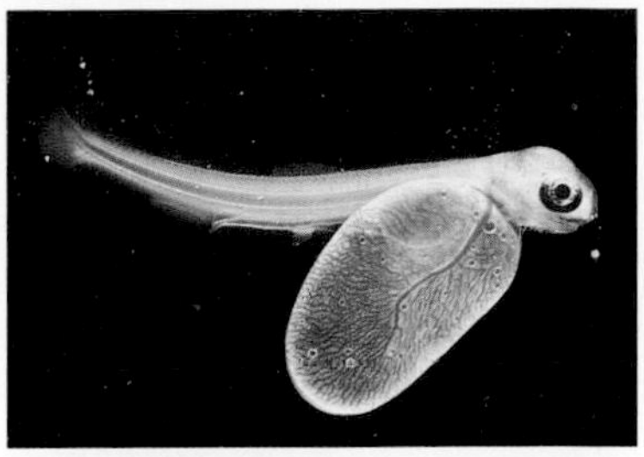

LARVA

ADULT

METAMORPHOSIS
When a fish emerges from its egg, its yolk sac is still attached to its body and provides nourishment during the first weeks of its life. The larva of a brown trout (top) takes 3–4 years to mature to the adult form (above).

PARENTAL CARE

Having released their eggs and sperm into water, many fishes have no further contact with their developing young. However, other species actively care for their eggs or young. Besides protecting their eggs by placing them in a nest, adult fishes may care for them by chasing away predators, cleaning them to prevent infection, or fanning them to provide oxygenated water. When the eggs hatch, the parents may use various ways to protect the vulnerable young (see below). A small number of species also feed their young, including the discus fish, which secretes a nutritious fluid through its skin.

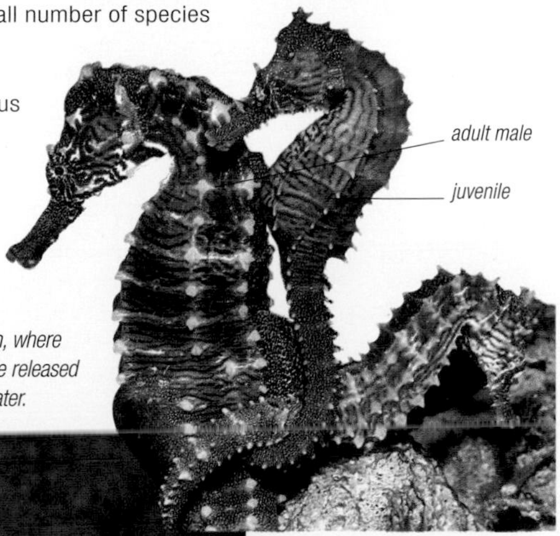

SEAHORSES
Male seahorses (right) play an unusual role in caring for their young. The female places her eggs in a pouch in the front of the male's abdomen, where they are fertilized. The young are released when they hatch, 2–6 weeks later.

MOUTH BROODING
Several hundred species of cichlids brood their eggs in cavities in their mouth and throat. When the eggs hatch, the parents also defend the young from predators by allowing them to shelter inside their mouth.

Jawless fishes

PHYLUM	Chordata
CLASSES	Cephalaspidomorphi
ORDERS	1
FAMILIES	1
SPECIES	About 43

Jawless fishes first appeared more than 500 million years ago, before any other group of living fishes. Although they were once diverse, they are now represented by a relatively small group: lampreys. They have an elongated body, smooth, scaleless skin, and lack jaws. The suckerlike mouth is used to hold and rasp away at their food. Lampreys live in temperate areas, in coastal marine or fresh water as adults and breed in fresh water. Eggs hatch into larvae that remain hidden until they mature into the adult form.

PARASITISM
Adult sea lampreys use their suckerlike mouthparts to attach themselves to other fishes. Once attached, they are difficult to dislodge, and the host often dies as a result of loss of blood or damage to its tissues.

Anatomy

The jawless fishes differ from all other groups of fishes in lacking true, movable jaws; instead, they have an open oral area armed with sharp, toothlike projections that are made of keratin and used in feeding. While they lack a bony internal skeleton, their eel-like bodies are reinforced by a tube of cartilage (notochord) and a few rudimentary vertebrae. They have 7 pairs of porelike gill openings near the head. Like other fishes, their tail ends in a flattened, finlike paddle, and they also have median fins for stabilization, but lack paired fins. The scaleless skin is coated with mucus. Lampreys have moderately good vision and acute senses of smell and taste. They also have lateral line organs that are sensitive to touch.

BODY SECTION
The bodies of lampreys are supported by a notochord that lies below the spinal cord. They also have rudimentary vertebrae made of cartilage.

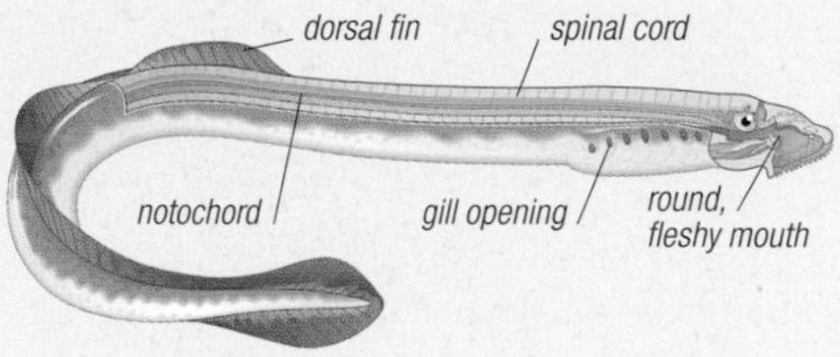

Reproduction

Most lampreys spend their adult lives in marine waters, migrating to fresh waters to breed, but some species spend their entire adult life in fresh water. Spawning pairs excavate shallow nests in gravel beds, lay their eggs, and then die. Young lampreys go through several larval stages before becoming adults and heading out to sea. The larval stages can last up to 6 years, during which time only their heads and mouths are visible protruding from the river bottom. The larvae feed on microscopic organisms and detritus, which they suck in and trap in mucus.

Eudontomyzon graecus

Greek brook lamprey

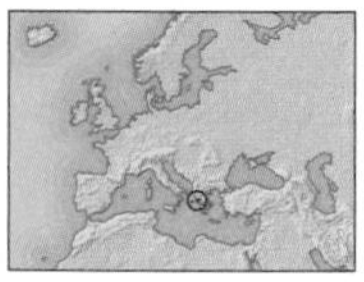

Location E. Mediterranean

Length 7½ in (19 cm) max
Weight c. 1 oz (30 g)
Sex Male/Female
Status Not evaluated

Restricted to the Loúros river drainage of Greece—and described in 2010—this is one of at least 2 nonparasitic species of Greek lamprey. It has a relatively longer tail than its ally, which inhabits an adjacent river system 249 miles (400 km) away. The larvae live in sandy sediments and feed on detritus and small invertebrates. After they metamorphose into the adult form, nonparasitic lampreys do not feed. They breed in highly oxygenated fast-flowing brooks and spring-fed streams with gravelly bottoms—a habitat threatened by pollution and water extraction.

Lampetra fluviatilis

European river lamprey

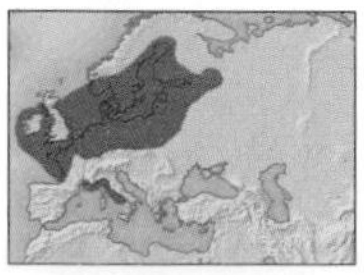

Location North Atlantic, N.W. Mediterranean, Europe

Length 7–19½ in (18–49 cm)
Weight 1–5 oz (30–150 g)
Sex Male/Female
Status Least concern

The European river lamprey, or brook lamprey, spends around a third of its life, between 4 and 7 years, at sea. Unlike other lampreys, which are blood-sucking parasites, adult European river lampreys bite their prey with sharp teeth while holding onto it with their circular sucking mouths. They eat fishes—especially herrings and sprat—and even dead remains, but cease to feed when they become sexually mature and migrate up streams to spawn. Females develop an extra crescent-shaped anal fin at this time. During the larval stage—lasting up to 6 years—this fish is toothless and blind and feeds on algae and particles of organic matter that are filtered from the water that is sucked in through the mouth. Females are longer and heavier than males, but become smaller while spawning. Adults breed just once, in the mid-upper reaches of rivers, and then die.

Petromyzon marinus

Sea lamprey

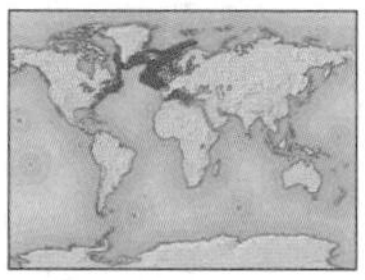

Location North Atlantic, Mediterranean, North America, Europe

Length Up to 4 ft (1.2 m)
Weight Up to 5½ lb (2.5 kg)
Sex Male/Female
Status Least concern

The adult sea lamprey is a parasite that attacks a wide range of salt- and freshwater fishes, including salmon, trout, herring, mackerel, and some sharks. Clinging to prey by its teeth, it uses its rough-surfaced tongue to suck blood. It migrates from the sea to breed in freshwater, but is landlocked in the North American Great Lakes. The opening of the Welland Ship Canal in 1929 allowed lampreys throughout the Great Lakes, and they had a devastating effect on populations of lake trout and other fishes. Since then, the sea lamprey population has been controlled.

SUCKERS

Sea lampreys have oral discs called "suckers" in place of their mouths. The sucker, which is wider than the fishes's body, has a fringed edge and contains many small teeth arranged in concentric rows, with larger teeth surrounding the mouth opening.

EEL-LIKE BODY
Both the adult and the larvae of the sea lamprey have eel-like bodies, which are rounded near the head but flattened at the tail.

FISHES

Cartilaginous fishes

PHYLUM	Chordata
CLASS	Chondrichthyes
ORDERS	12
FAMILIES	51
SPECIES	ABOUT 1,200

Although often described as primitive, cartilaginous fishes include some of the largest and most successful of all marine predators. They are divided into 3 groups: sharks; skates and rays; and a group of deep-sea fishes called chimaeras. All possess a skeleton made of cartilage (rather than bone), specialized teeth that are replaced throughout their lifetime, and skin covered with small, dense, toothlike scales. Most cartilaginous fishes are marine, but some sharks, skates, and rays enter freshwater, and certain tropical species live exclusively in freshwater.

Anatomy

All the fishes in this group have an internal skeleton made of tough, flexible cartilage. This may be strengthened by mineral deposits, and some species have hard, bonelike dorsal spines. However, the skeleton of a cartilaginous fish is much more flexible than that of a bony fish (see p.486), which has a much higher mineral content. On land, a cartilage skeleton would not be rigid enough to support the weight of a large animal, but in water—which has a much higher density than air—cartilage provides an effective skeleton for animals that are up to 33 ft (10 m) long. Most cartilaginous fishes have skin consisting of thousands of interlocking scales, called placoid scales or dermal denticles (see p.468). These have a similar composition to teeth and give the skin a sandpaper-like texture. Some rays have large, thornlike scales but other rays and all chimaeras lack these entirely. Unlike bony fishes, cartilaginous fishes do not have a gas-filled swim bladder.

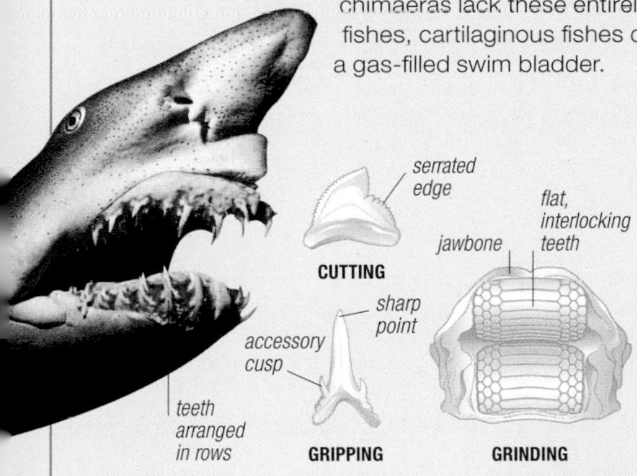

TEETH
Cartilaginous fishes have teeth that are shaped to suit their diet. Serrated teeth are used for cutting, while pointed teeth are used to hold prey. Many rays have flat teeth for grinding food. As teeth are lost, they are replaced by new ones growing behind them.

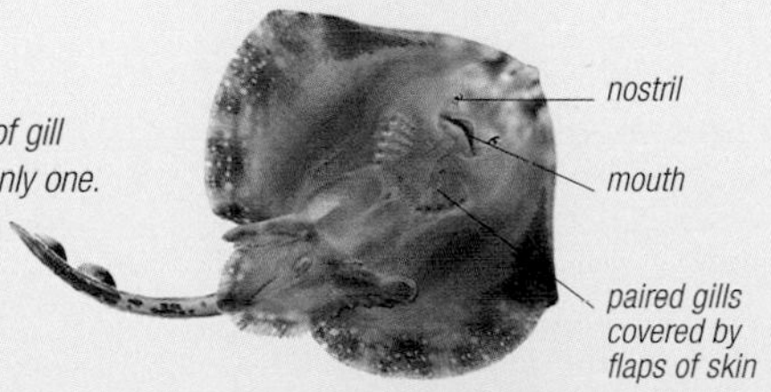

GILL OPENINGS
Sharks and rays have 5–7 pairs of gill openings, while chimaeras have only one. When water enters the mouth, the gill slits are closed. As the water passes out of the open gill slits, the mouth is held closed.

Feeding

All cartilaginous fishes are carnivorous; however, plankton feeders also take in microscopic plants. The greater part of the diet of most species consists of live prey, including other fishes, invertebrates, and occasionally marine mammals. Most species also feed on the remains of dead animals, although only a few species rely solely on carrion. Several of the largest sharks and rays are filter feeders. The teeth of these species have been reduced in size, and stiff projections in the gill arches have developed into sievelike organs for straining small animals out of water.

FILTER FEEDING
Whale sharks drift slowly through water, catching plankton, small fishes, and squid. They live in warm tropical waters, where the food supply is large enough to sustain them.

SWIMMING AND BUOYANCY
Cartilaginous fishes, such as the hammerhead shark, have an oil-rich liver that adds to their buoyancy. However, most are still negatively buoyant and must keep swimming to avoid sinking.

Senses

Cartilaginous fishes have acute senses that can be used to locate prey, even if it is a long way off or buried by sediment. All species have a system of pores, called the ampullae of Lorenzini, which they use to detect the weak electrical signals emitted by other animals. Most also have an effective lateral-line system that responds to very small vibrations, as well as acute vision and a sense of smell that can detect odors in even the weakest solutions.

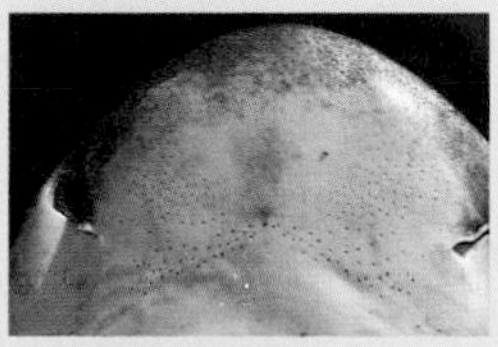

ELECTRICAL SIGNALS
The ampullae of Lorenzini (seen here on the snout of an oceanic whitetip shark) are deep pores connected to electro-receptive nerve endings.

Reproduction

In all cartilaginous fishes, fertilization occurs internally. The male passes sperm into the female's cloaca via a modified pelvic fin. The young are produced by one of 3 processes. The females of some species release leathery egg cases. In other species, the young hatch from the egg inside the female's body and are then released alive. In the remaining species, the young develop inside a placenta-like structure, from which they have a direct nutritional connection with the female. In all cases, there is no larval stage; instead, the young are born as miniature versions of the adults.

LIVE BIRTH
Young lemon sharks hatch inside the female's body before being released tail-first.

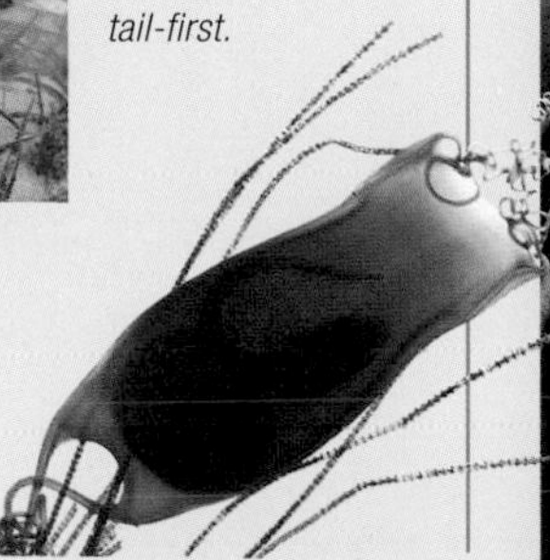

EGG CASE
This egg case contains the embryo of a small-spotted catshark. It is anchored to seaweed by tendrils on the outside.

Sharks

PHYLUM	Chordata
CLASS	Chondrichthyes
SUBCLASS	Elasmobranchi (part)
ORDERS	8
FAMILIES	31
SPECIES	About 500

Sharks are formidable hunters with few natural predators when fully grown. Most have a streamlined body, powerful jaws with several rows of sharp teeth, and acute senses, especially smell. They range in size from the whale shark (the largest of all fishes) to squaloid sharks less than 12 in (30 cm) long. Sharks are most abundant in warm temperate and tropical waters, their habitats ranging from coastal areas to open oceans and deep basins. Some even venture into tropical rivers.

Anatomy

Most sharks have a torpedo-shaped body that allows them to cut easily through water, although several bottom-dwelling species have a flattened body, similar to that of rays. Unlike bony fishes, sharks have rigid fins that cannot be flexed or folded, yet their fins are nonetheless efficient for steering, propeling, and stabilizing the body during swimming. A shark's teeth, which are arranged in rows, are replaced throughout its lifetime. The shape of the teeth varies according to diet (see p.473), although some species have 2 or more types of teeth.

Sharks do not have a swim bladder, but their large, oil-filled liver adds to their buoyancy and helps prevent them from sinking. Many open-ocean sharks need to swim constantly to force water over their gills, but other species can pump water across their gills while motionless. Sharks usually have 5 gill slits on either side of their head; a few species have 6 or 7 pairs.

Food and feeding

Sharks use various tactics to catch prey. Predatory sharks rely on short, high-speed pursuits through open water, or on ambush and surprise. Other species simply search an area for sedentary or slow-moving prey, or excavate burrowing animals. Some sharks are scavengers, using an acute sense of smell to locate carcasses (they can detect a scent in a solution as weak as one part per million). The 3 largest species of sharks are filter feeders, and at least one species, the cookie-cutter shark, is an ectoparasite on other sharks, large fishes, and marine mammals.

Reproduction

Coastal bays, reefs, and atolls are the favored areas for female sharks to lay eggs or give birth (pupping). These areas provide a calm, food-rich environment for young sharks to grow and develop. Baby sharks are born or hatch as miniature adults, even armed with fully functional teeth, and are ready to hunt. Sharks grow slowly to adulthood, often taking a decade or longer to reach maturity. During this time, many young sharks fall prey to larger predators.

CONSERVATION

Over the last few decades, rising human populations and falling stocks of many food fishes have led to an increase in the popularity of sharks as a food source for humans. Compounding this pressure are the growth in the popularity of shark fishing as a sport and a rise in demand for shark fins in Asian cuisine. About 100 million sharks are caught annually, of which 6.2–6.5 million are blue sharks, killed only for their fins. Conservation measures include regulation of catches, protection of threatened species, and protection of pupping grounds.

HUNTING
Sharks of the open oceans (such as the blue shark, shown here) typically have a streamlined body and a powerful tail fin that propels them quickly through the water. The mainstays of their diet are squid and fishes. When hunting, they often circle their prey several times before attacking from below.

Chlamydoselachus anguineus

Frilled shark

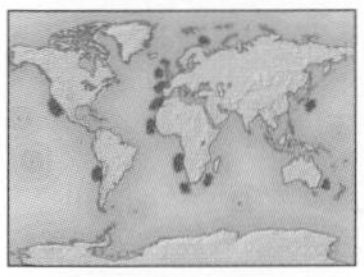

Length 6½ ft (Up to 2 m)
Weight Not recorded
Breeding Ovoviviparous
Status Near threatened

Location E. Atlantic, S.W. Indian Ocean, W. and E. Pacific

This unusual, deep-water species looks more like an eel than a shark, with frilled gill slits, a single small dorsal fin, and a much larger anal fin. It feeds on soft-bodied animals, such as squid, as well as fishes. Similar to some long-extinct species, it has no close living relatives.

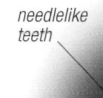

Hexanchus griseus

Bluntnose six-gill shark

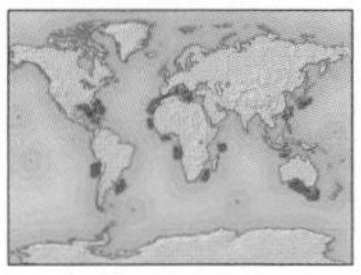

Length 16 ft (4.8 m) or more
Weight 1,320 lb (600 kg) or more
Breeding Viviparous
Status Near threatened

Location Tropical and temperate waters worldwide

This reclusive, primarily deep-water shark has a broad head and cylindrical body. It has 6 gill slits—unlike most other sharks, which have 5—and sharp, comblike teeth. The bluntnose six-gill feeds on rays, squid, bony fishes, and even seals. It also forages opportunistically for slow-moving animals on the seabed by positioning its body, head-down, at an angle of 45–60 degrees to the bottom. With its open mouth directly above the prey, it probably sucks in its food.

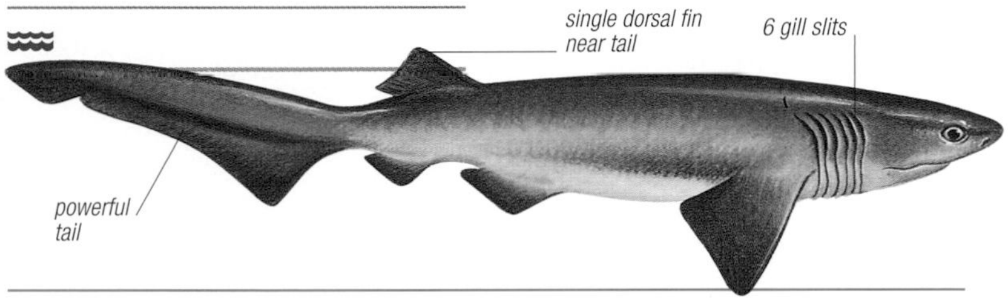

Echinorhinus cookei

Prickly shark

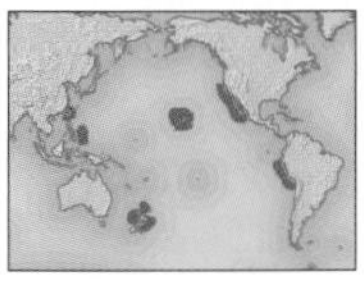

Length Up to 13 ft (4 m)
Weight Over 490 lb (220 kg)
Breeding Ovoviviparous
Status Near threatened

Location W., C., and E. Pacific

The most interesting feature of this shark is the thorny denticles that cover its body. These have star-shaped bases and sharp tips, resembling bramble thorns. Large and slow-moving, the prickly shark spends most of its life foraging on or near the seabed. It feeds on other sharks, ratfishes, bony fishes, cephalopods, and the egg cases of sharks, skates, and rays. It may use its large pharynx and small mouth to create a vacuum effect to suck in prey.

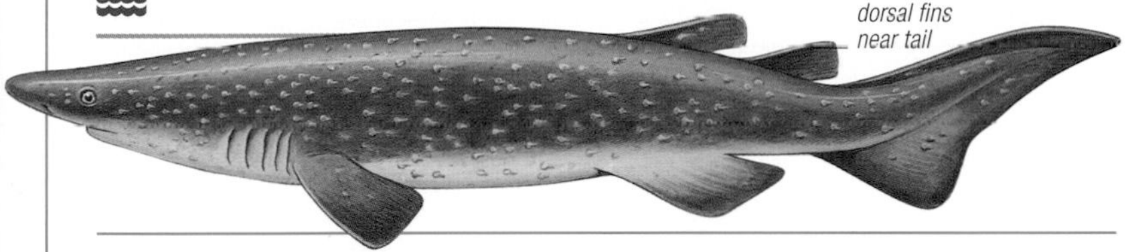

Squalus acanthias

Piked dogfish

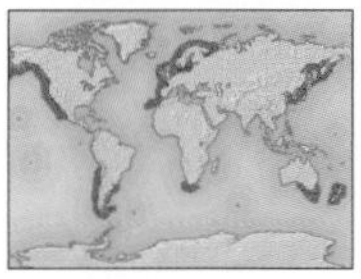

Length 1 m (3¼ ft), max 1.6 m (5¼ ft)
Weight Over 9 kg (20 lb)
Breeding Ovoviviparous
Status Vulnerable

Location Worldwide outside the tropics

Also known as the spurdog, this small, bottom-dwelling shark lives in coastal waters, usually where the temperature is at or below 59°F (15°C). A slow-moving fish, it gets its name from the spines on the front of its dorsal fins, which can inflict painful wounds. It eats crustaceans, sea anemones, and other fishes, and often forms large, single-sex schools. Piked dogfishes give birth after a gestation period of nearly 2 years—a long time for any shark.

Oxynotus centrina

Angular roughshark

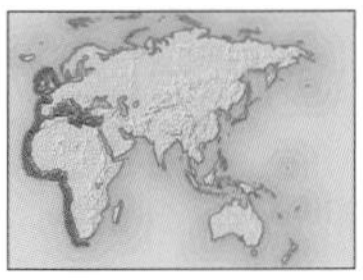

Length Up to 5 ft (1.5 m)
Weight Not recorded
Breeding Ovoviviparous
Status Vulnerable

Location E. Atlantic, Mediterranean

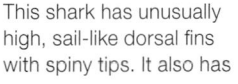

This shark has unusually high, sail-like dorsal fins with spiny tips. It also has a compressed, triangular trunk and furrowed "lips" with lancelike teeth in the upper jaw. There is a large spiracle behind the eye. The expanded body cavity and large, oily liver enable it to maintain buoyancy, allowing it to hover above the ocean floor in search of polychete worms and other invertebrates. It is one of 5 species in the roughshark family.

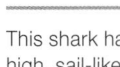

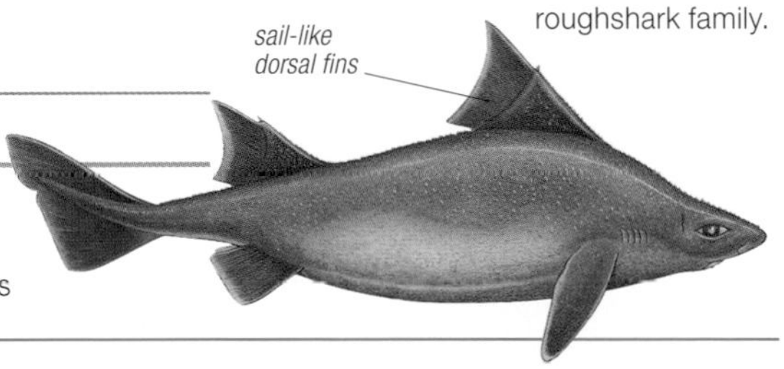

Etmopterus spinax

Velvet belly lantern shark

Length 18 in (45 cm), maximum 23½ in (60 cm)
Weight Not recorded
Breeding Ovoviviparous
Status Least concern

Location E. Atlantic

This small, shoal-forming shark is one of more than 30 similar species. It feeds on invertebrates and small fishes. Tiny luminescent organs on its skin produce light, which help attract potential prey or other velvet belly lantern sharks, frighten predators, and disrupt shadows, making it less conspicuous.

FIRST MEAL

Like many sharks, young lantern sharks are born with an attached yolk sac. This provides them with food during their first few days at sea.

SPINED FINS
The velvet belly lantern shark has a sharp spine at the front of each dorsal fin; its second dorsal fin and fin spine are larger than the first.

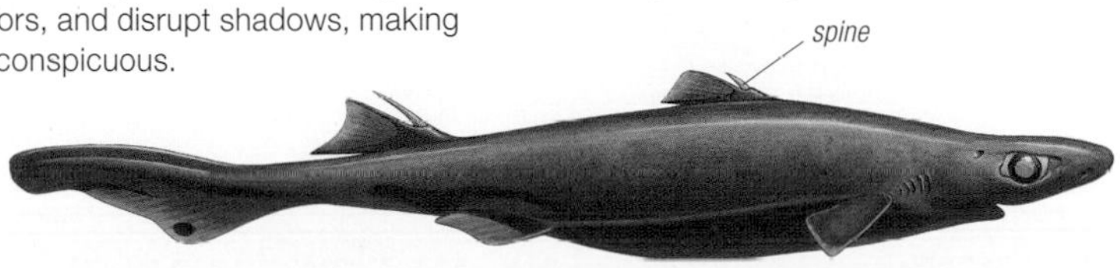

Squaliolus laticaudus

Spined pygmy shark

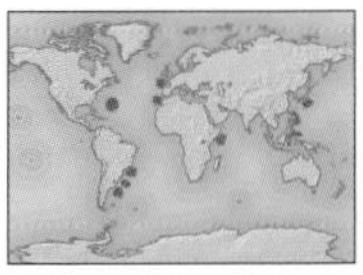

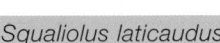

Length Up to 10 in (25 cm)
Weight Not recorded
Breeding Ovoviviparous
Status Least concern

Location North Atlantic, W. South Atlantic, W. Indian Ocean, W. Pacific

One of the world's smallest sharks, this species is unique in having a single dorsal spine. It has well-developed light-producing organs (photophores) on its belly and sides, but hardly any on the back, giving this shark a luminescent underside. The arrangement of these photophores is thought to eliminate shadows, making it harder for predators (such as swordfishes) to see it from below. The first known spined pygmy sharks were caught in 1908 off Japan, but no further specimens were seen until the 1960s.

Pristiophorus japonicus

Japanese sawshark

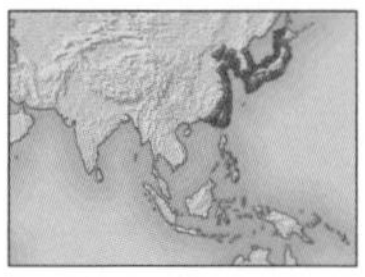

Length Up to 5 ft (1.5 m)
Weight Not recorded
Breeding Ovoviviparous
Status Data deficient

Location W. North Pacific

The Japanese sawshark's body is flattened from top to bottom and has 2 dorsal fins. Unlike rays, it has gills on the side of its head. It has a pair of barbels on a distinctive, saw-shaped snout. Taste sensors on the barbels and snout are used to probe the seabed for small fishes or invertebrates. Its exceedingly sharp teeth are presumably used for destroying prey as well as for defence. Young Japanese sawsharks are born with large teeth, but these remain folded while the young are inside the mother to avoid injuring her. The Japanese sawshark is not to be confused with smalltooth sawfishes (see p.485).

Heterodontus portusjacksoni

Port Jackson shark

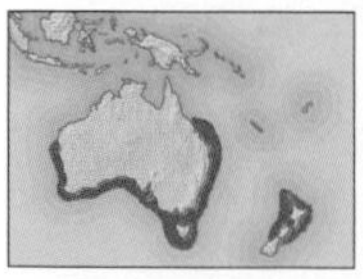

Length Up to 5½ ft (1.7 m)
Weight Not recorded
Breeding Oviparous
Status Least concern

Location Australia (including Tasmania), New Zealand

Like the horn shark (see below), this shark belongs to a family of 8 species that have bulky heads, powerful jaws, and crushing teeth. Its mouth points downward, enabling it to feed on starfish, sea urchins, and other seabed animals. The Port Jackson shark lives in inshore waters, and lays large eggs with spiral cases.

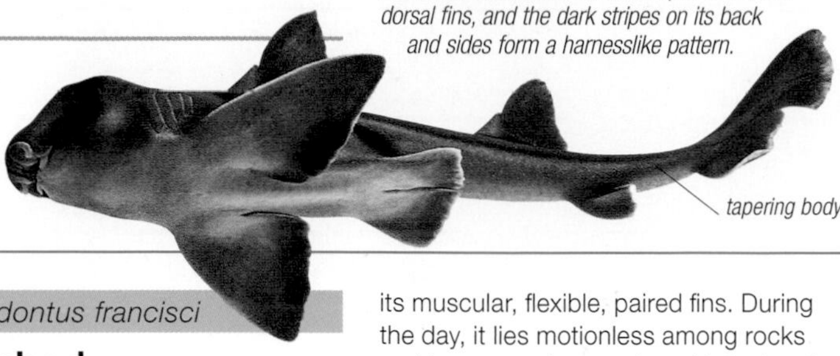

TAPERED BODY
This shark's body becomes progressively narrower from the head to the tail. It has spines on both dorsal fins, and the dark stripes on its back and sides form a harnesslike pattern.

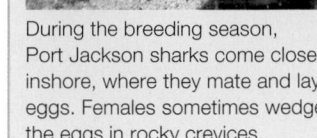

INSHORE MATING

During the breeding season, Port Jackson sharks come close inshore, where they mate and lay eggs. Females sometimes wedge the eggs in rocky crevices.

Heterodontus francisci

Horn shark

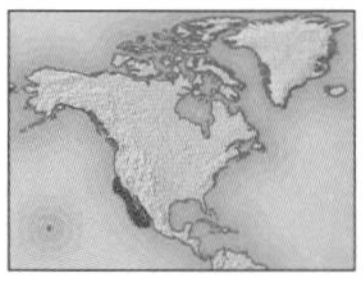

Length 38 in (97 cm), max 120 cm (48 in)
Weight Over 22 lb (10 kg)
Breeding Oviparous
Status Data deficient

Location E. North Pacific

This slow-moving, generally solitary, nocturnal shark is found near rocky or sandy bottoms and kelp beds. It can "walk" across the sea bed using its muscular, flexible, paired fins. During the day, it lies motionless among rocks and in caves, often resting with its head in rock crevices. This shark has a piglike snout and a small mouth. Its enlarged, flattened teeth, located at the back of the mouth, are used to crush prey such as sea urchins or crabs and, sometimes, bony fishes. Horn sharks mate in December or January. The female lays about 30 eggs, which she places under rocks or in crevices. The egg cases have a highly sculptured, spiral shape, which helps to keep them firmly lodged.

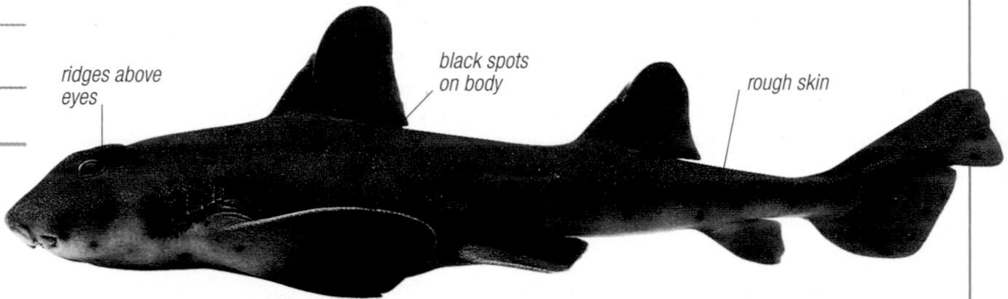

Orectolobus maculatus

Spotted wobbegong

Length 6 ft (1.8 m), max 10 ft (3.2 m)
Weight Not recorded
Breeding Viviparous
Status Near threatened

Location Japan, South China Sea, Australia

A large, slow-moving shark with a flattened body, the spotted wobbegong lives in shallow water close to the shore. Instead of actively hunting for food, it rests on the bottom, and ambushes animals that come within striking range. When the tide falls, it may haul itself from one rock pool to another, partly exposing itself to the air. Although generally unaggressive, spotted wobbegongs can inflict serious injuries if accidentally stepped on.

PERFECT DISGUISE

The spotted wobbegong's mottled coloring and flattened, textured body provide perfect camouflage on the seabed. Lured by the weedlike flaps of skin around its snout, lobsters, crabs, and octopus may swim right up to this shark's waiting mouth.

SPOTTED AND PATTERNED
The spotted wobbegong is elaborately patterned, with light, O-shaped spots and dark saddles on an olive background.

Chiloscyllium plagiosum

Whitespotted bambooshark

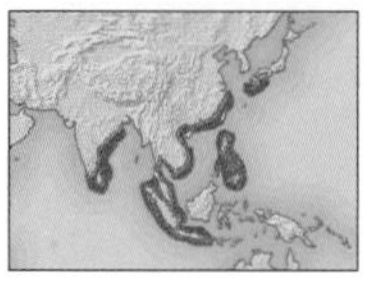

Length Up to 37 in (95 cm)
Weight Not recorded
Breeding Oviparous
Status Near threatened

Location Indo-Pacific

Its small size and attractive appearance have made this shark particularly popular with marine aquarists. It has a slender body, with white spots over a dark background and darker transverse bands. Short barbels on its mouth help it to forage for food; the mouth is located well in front of the eyes. The whitespotted bambooshark uses its thickened pectoral and pelvic fins to clamber onto rocks.

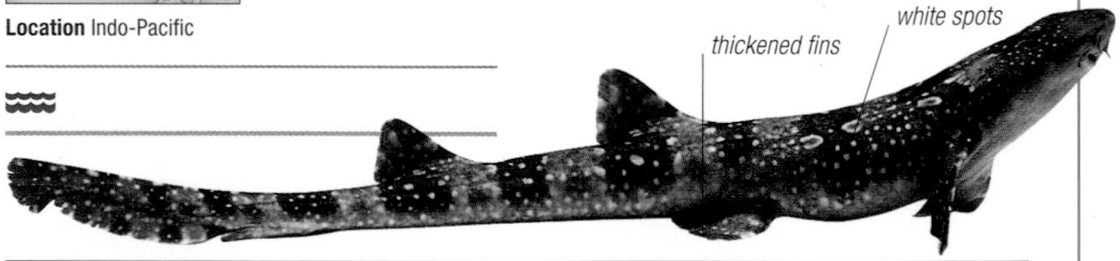

Hemiscyllium galei

Walking shark

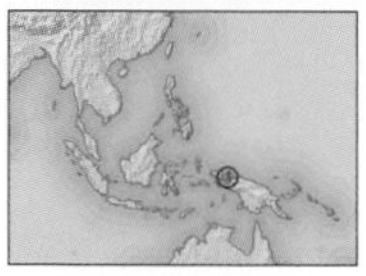

Length 22 in (57 cm)
Weight Not recorded
Breeding Probably oviparous
Status Not evaluated

Location New Guinea

This spectacularly patterned shark belongs to a family of sluggish tropical inshore "bamboo sharks." It was recently discovered on shallow reefs of Geelvink Bay in Indonesian Papua and is marked by dark "saddles" along its back that are dappled with white spots. There are also prominent dark spots running along each side of the body. Little is known about its habits, but it has been observed at night—at depths of 6½–13 ft (2–4 m) – "walking" along the reef bottom using its pectoral and pelvic fins. Like related species, it probably hunts for bottom-living fishes and invertebrates, and spends the day resting under rocks or corals.

FISHES

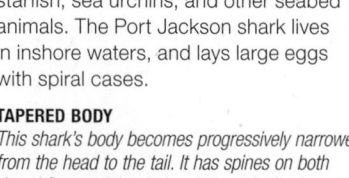

Stegostoma fasciatum

Zebra shark

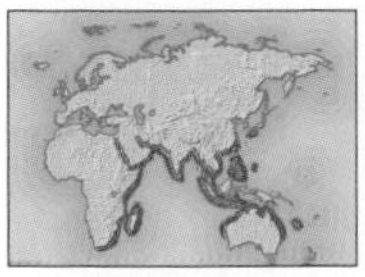

Length 9¼ ft (2.8 m), max 11 ft (3.5 m)
Weight Over 66 lb (30 kg)
Breeding Oviparous
Status Vulnerable

Location Indo-Pacific

Beige with brown spots, the zebra shark has an unusually flexible body that allows it to hunt out shrimps, crabs, and small bony fishes from tight crevices in the coral reefs where it lives. Its mouth, located just behind the snout, points downward, enabling it to feed on mollusks that live on the seabed. During inactive periods, this shark rests on the seabed, propped up on its erect pectoral fins, facing the prevailing current. It is the only species in its family.

Ginglymostoma cirratum

Nurse shark

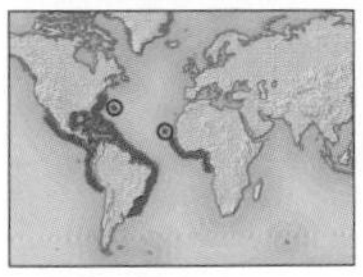

Length 3 m (9¾ ft), max 4.3 m (14 ft)
Weight Over 150 kg (330 lb)
Breeding Viviparous
Status Data deficient

Location E. Pacific, W. and E. Atlantic

The nurse shark has a very tough skin, and a pair of barbels below the mouth for sensing the invertebrates on which it feeds. The small mouth and large pharynx form a powerful suction mechanism, which enables this shark to draw in prey at high speed. During the day, the nurse shark spends its time resting on the seabed, in rocky crevices, or in caves, where it may be found in groups of several dozen individuals, sometimes lying on top of one another. An inhabitant of shallow, inshore areas, it can "walk" along the seabed using its pectoral fins as limbs. The nurse shark is not generally a threat to people but, if provoked, will hold onto its victim with a bulldoglike tenacity.

Rhincodon typus

Whale shark

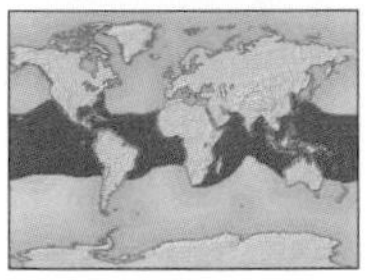

Length 39 ft (12 m), max 46 ft (14 m)
Weight Over 13¼ tons (12 tonnes)
Breeding Viviparous
Status Vulnerable

Location Tropical and temperate waters worldwide

The whale shark is by far the largest fish in the world. However, despite its fearsome appearance, it is quite harmless to humans, and lives almost entirely on plankton, filtering its food with its gills. Its mouth is at the end of its snout—an unusual position for a shark—and although its jaws can be over 3¼ ft (1 m) across, they are armed with the tiniest of teeth. Whale sharks usually feed by cruising slowly near the surface, where they show little alarm at being approached. They have also been reported to feed in a vertical position, using their mouths like giant buckets to trap food. They sometimes gather in large numbers where plankton is abundant, but they otherwise lead solitary lives. Little is known about their breeding behavior, or about their movements across the world's tropical seas.

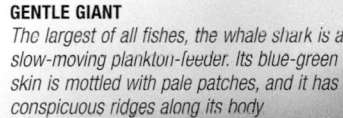

SUCTION FEEDING

In order to feed, the whale shark creates a suction effect by lifting its head above the surface, then sinking quickly to draw water into its open mouth. The water that pours in is filtered for plankton, fishes, and squid by screens on the shark's internal gill slits. This method allows for a large intake of prey.

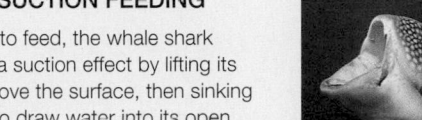

GENTLE GIANT
The largest of all fishes, the whale shark is a slow-moving plankton-feeder. Its blue-green skin is mottled with pale patches, and it has conspicuous ridges along its body

Mitsukurina owstoni

Goblin shark

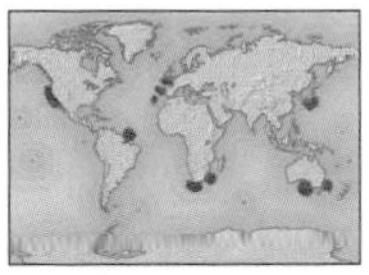

Length 11 ft (3.3 m)
Weight Over 350 lb (160 kg)
Breeding Ovoviviparous
Status Least concern

Location North Atlantic, E. South Atlantic, S. Indian Ocean, W. and E. Pacific

Formally described in 1898, this deep-water shark is white or grayish, with a pointed snout. It has small eyes, and probably detects its prey by sensing electrical fields. It has no close living relatives, and is often described as a living fossil, having changed relatively little for millions of years.

Carcharias taurus

Sand tiger shark

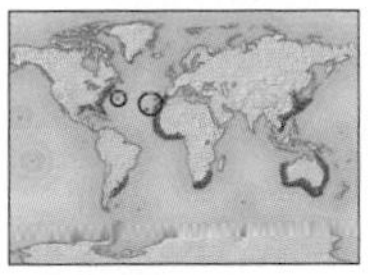

Length Up to 14 ft (4.3 m)
Weight Over 330 lb (150 kg)
Breeding Viviparous
Status Vulnerable

Location W. and E. Atlantic, S. Indian Ocean, W. Pacific

Light brown or beige with darker brown blotches distributed randomly over its body and fins, the sand tiger shark has a distinctive, "snaggle-toothed" appearance. It is a large, cumbersome-looking fish, with a large mouth and dorsal fins situated further back than most sharks. This shark swallows air at the surface and retains it in the stomach to regulate buoyancy. A slow, strong swimmer, it feeds upon a variety of bony fishes, as well as on squid, crabs, and lobsters. Groups of sand tigers have been observed working cooperatively to surround schools of prey. The embryos of the sand tiger shark are known to be cannibalistic. After eating other eggs inside their egg case, surviving embryos hatch, develop teeth, and eat other embryos within the uterus. Only one embryo survives from each of the 2 uteri.

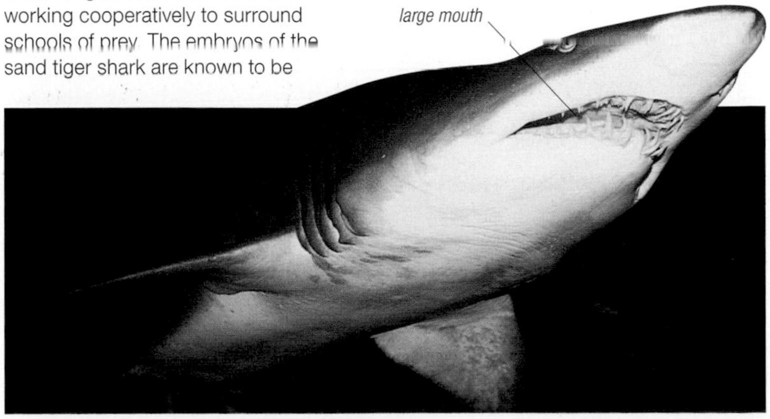

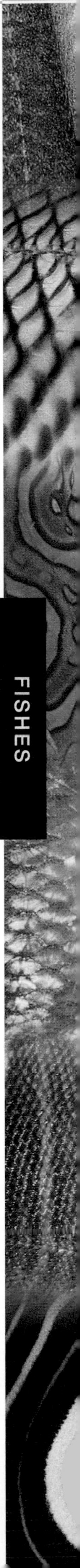

FISHES

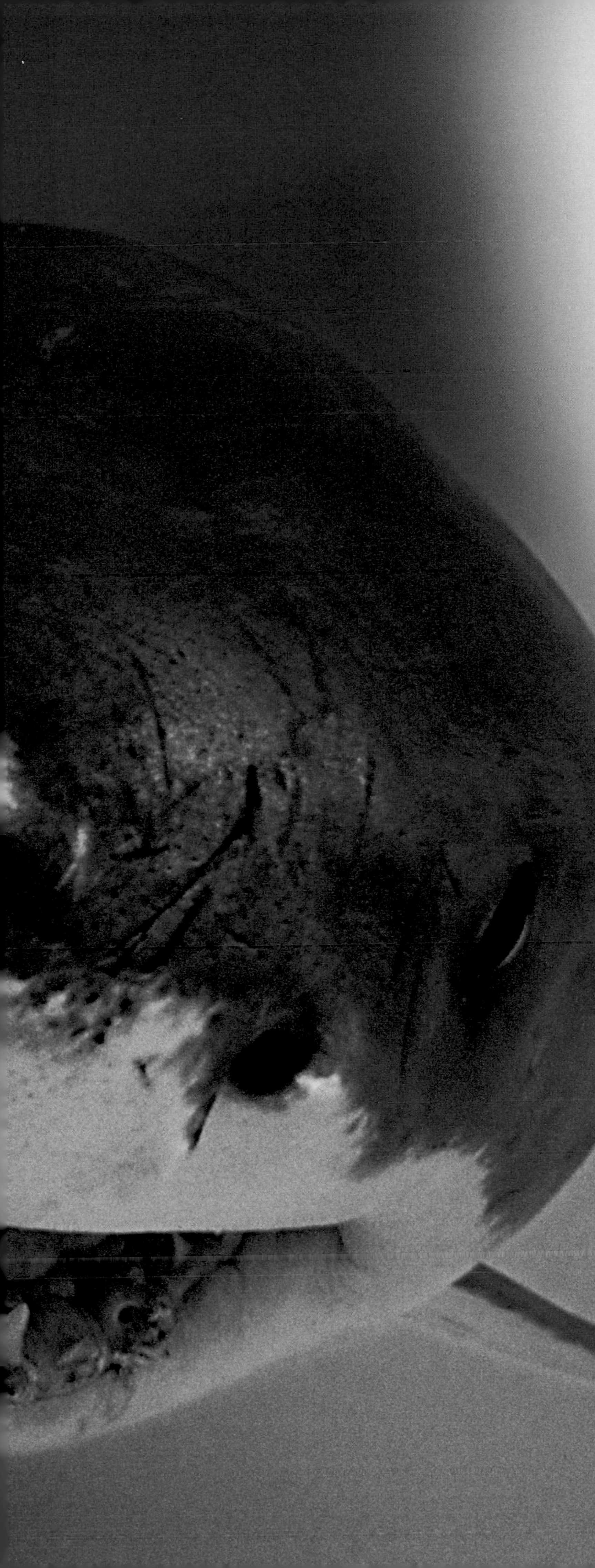

Carcharodon carcharias

Great white shark

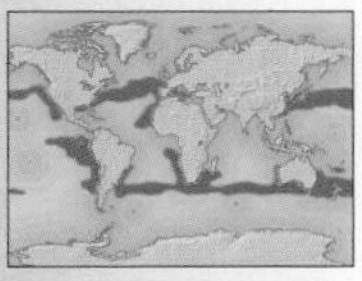

Length 20 ft (6 m), max 26 ft (8 m)

Weight $2^{1}/_{4}$ tons (2 tonnes) or more

Breeding Viviparous

Status Vulnerable

Location Temperate and tropical waters worldwide; occasionally in cold waters

Also known as the white pointer, the great white shark is blamed for more attacks on humans than any other shark—but research shows that most human victims are quickly released. A powerful swimmer, the great white shark cruises through the water, either at the surface or just off the bottom, covering long distances very quickly. It also excels at short, fast chases, and may leap spectacularly out of the water. Like other members of its family, it can maintain its muscles at an unusually high temperature, helping it swim more efficiently. Although usually solitary, it may sometimes be seen in pairs or small groups feeding at a carcass, with larger individuals eating first. The sharks may also swim in a variety of patterns in order to establish their dominance hierarchy. The primary prey is seals, sea lions, dolphins, and large fishes, including other sharks—but this species will feed on any large creature that it can catch. It attacks with great ferocity initially, then retreats while the injured prey weakens, when the shark returns to eat in safety. Females, which are larger than males, give birth to 4–14 live young, 4 ft (1.2 m) or more long. While in the uterus, the young are nourished by eating unfertilized eggs or other embryos. Although this species is protected in some parts of the world, it is a popular sport fish, and its numbers are dropping rapidly.

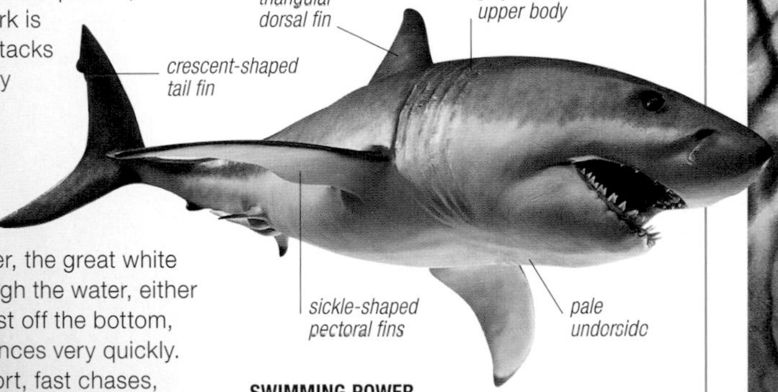

SWIMMING POWER
The great white shark powers through the water with sideways beats of its tail. Its fixed pectoral fins prevent it from nose-diving, and its large dorsal fin aids stability.

AWESOME WEAPONS

Armed with its large, triangular, coarsely serrated teeth, the great white shark is superbly equipped for ripping into the flesh of its unfortunate prey. In some situations, this shark has been known to swim along with its teeth bared, which may serve to warn off competitors for food, or rival sharks intruding on its personal space.

THE GREAT "MAN-EATER"
This very large, formidable-looking shark is greatly feared as a man-eater, partly due to the media fascination that followed the release of the film Jaws. *However, according to the International Shark Attack File, there have been only 321 human deaths attributed to great white shark attacks since records began.*

FISHES

Isurus oxyrinchus

Shortfin mako

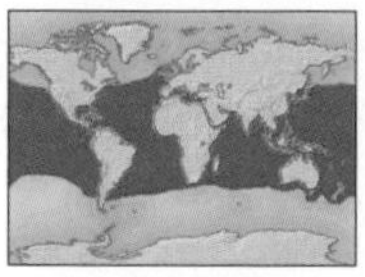

Length Up to 13 ft (4 m)

Weight 1,260 lb (570 kg) or more

Breeding Ovoviviparous

Status Vulnerable

Location Temperate and tropical waters worldwide

Also called the blue pointer, mackerel shark, or snapper shark, this is possibly the world's fastest shark. The tail is adapted for speed, with well-defined keels on either side of the base—a common feature in fast-swimming fishes. This shark has large, unserrated, daggerlike teeth that enable it to stab and grip slippery, fast-moving prey such as mackerel, tuna, bonito, and squids. It can jump several times its length out of the water when pursuing prey, or when hooked.

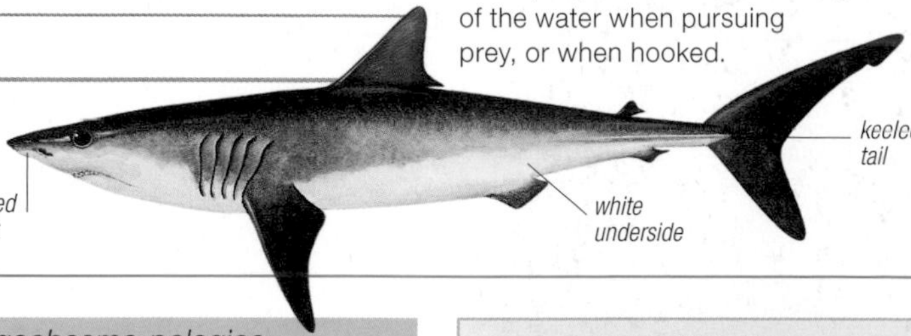

Megachasma pelagios

Megamouth shark

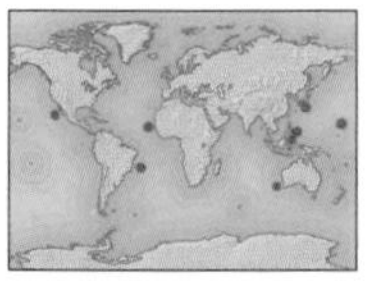

Length Up to 18 ft (5.5 m)

Weight Up to 1,740 lb (790 kg)

Breeding Unconfirmed

Status Data deficient

Location W. and E. Atlantic, S. Indian Ocean, W. and E. Pacific

The megamouth is one of the world's least-known large sharks. It was discovered in 1976 when one specimen became trapped by a ship's underwater equipment off the Hawaiian islands. Around 50 specimens have been seen since. This shark is characterized by an extraordinarily large mouth, which facilitates filter feeding. One of only 3 plankton-eating sharks, the megamouth shark is believed to follow the movement of plankton, rising from deep water to the surface at night and returning to the depths during the day. Its great size means it has few enemies, sperm whales (see p.260) being one of the few predators big enough to tackle this huge fish. However, the megamouth is harmless to humans.

FEEDING AIDS

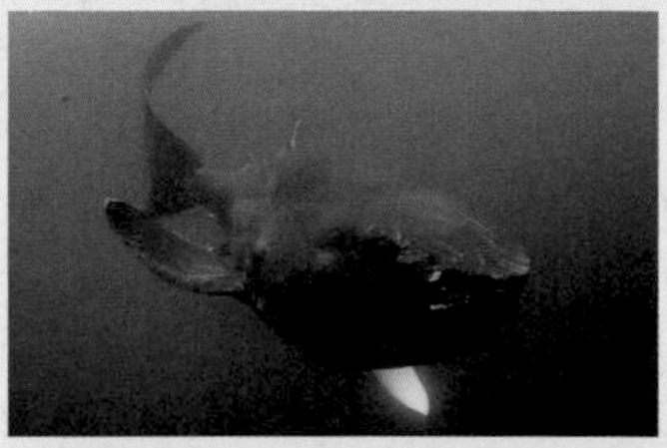

Distinctive coloring in and around this shark's huge mouth may be aids to feeding. A luminescent stripe on the margin of the upper jaw and a silvery roof to the mouth—visible as the shark swims along with jaws open—may serve to attract prey.

COLOR CONTRASTS

The megamouth's back is uniformly gray to blue-black in color, while the underside of the body is whitish. A white stripe runs along the margin of the upper jaw.

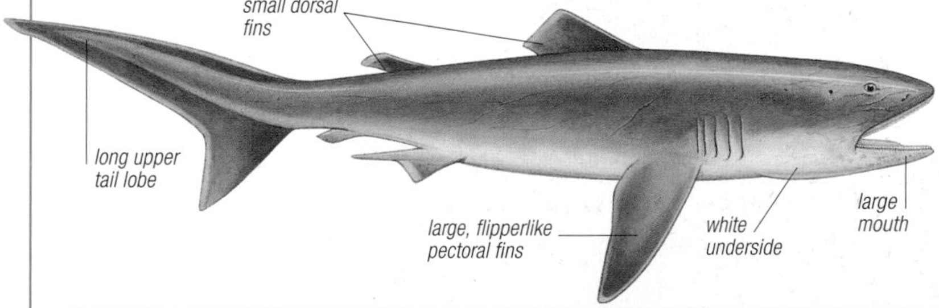

Cetorhinus maximus

Basking shark

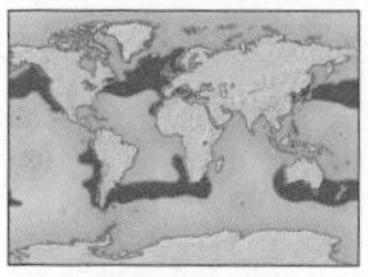

Length 33 ft (10 m), max 49 ft (15 m)

Weight 6⅝ tons (6 tonnes) or more

Breeding Ovoviviparous

Status Vulnerable

Location Cool to warm temperate waters worldwide

The basking shark is the world's second largest fish, after the whale shark (see p.477), and like it, is a harmless, migratory plankton-feeder. It often basks in the sun near the surface of the water, sometimes lazily rolling completely upside down. Although primarily solitary, groups of up to 100 may gather to feed in areas where plankton blooms are dense.

PLANKTON TRAPPER

This filter feeder moves through the ocean with its mouth agape, enabling it to take in large volumes of water and plankton. As the water is expelled through the enlarged gill slits, mucus-laden gill rakers trap the plankton—which the shark quickly swallows.

WATER OUTLETS

This species has very large gill slits that nearly encircle its head. It sheds its gill rakers in winter.

Pseudocarcharias kamoharai

Crocodile shark

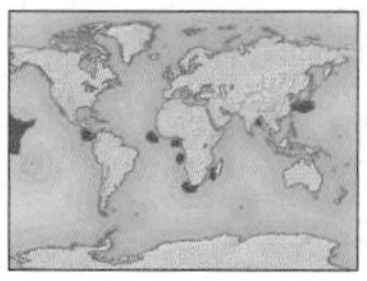

Length 3½ ft (1.1 m) or more

Weight Not recorded

Breeding Ovoviviparous

Status Near threatened

Location Pacific, E. Atlantic, W. and N. Indian Ocean

This small, cylindrical shark is an oceanic species. It has a pronounced snout and large teeth, and is able to extend its jaws unusually far forward to snatch at prey, such as medium-sized bony fishes, squids, or shrimps. The crocodile shark migrates vertically, following its prey toward the surface at night and then sinking again at dawn, its large eyes enabling it to see in dimly lit conditions. Like many sharks, this species is an intrauterine cannibal, with embryos eating each other before the survivors are born. Most sharks that develop in this way produce only one pup per uterus. However, in this shark, cannibalism reduces the number of embryos to 2 per uterus—at this stage, the cannibalism stops, and both embryos normally survive. Pups are about 16 in (40 cm) long at birth and may be born at any time of the year.

Alopias vulpinus

Thresher shark

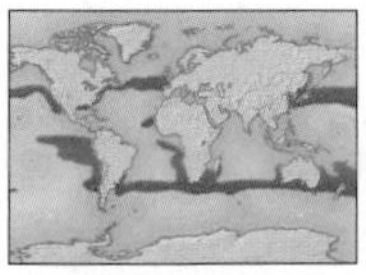

Length Up to 18 ft (5.5 m)

Weight 990 lb (450 kg) or more

Breeding Ovoviviparous

Status Vulnerable

Location Temperate and tropical waters worldwide

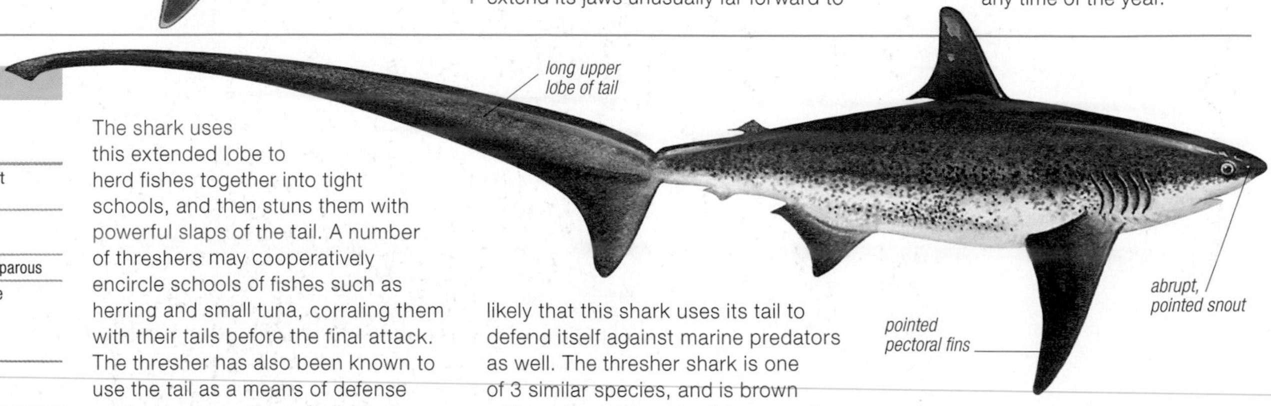

The thresher shark is uniquely shaped, the upper lobe of its tail fin being almost as long as the rest of the body. The shark uses this extended lobe to herd fishes together into tight schools, and then stuns them with powerful slaps of the tail. A number of threshers may cooperatively encircle schools of fishes such as herring and small tuna, corraling them with their tails before the final attack. The thresher has also been known to use the tail as a means of defense against humans. Large individuals caught by fishermen may slap furiously with their tails, one such tail swipe was reported to result in decapitation. It is likely that this shark uses its tail to defend itself against marine predators as well. The thresher shark is one of 3 similar species, and is brown with a white underside. It is generally confined to deeper waters, but the young and some adults may be found in shallower regions around coasts. It may, at times, congregate in schools made up entirely of either males or females.

FISHES

Mustelus mustelus

Smooth-hound

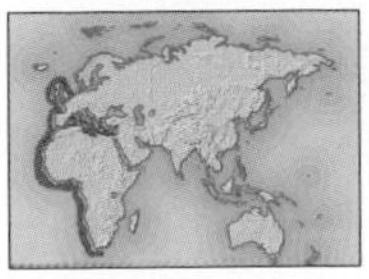

Length Up to 5¼ ft (1.6 m)
Weight 29 lb (13 kg) or more
Breeding Viviparous
Status Vulnerable

Location E. Atlantic, Mediterranean

The relatively smooth skin of the smooth-hound is gray or gray-brown above and lighter below, with few or no markings. This shark's flat, slab-shaped teeth are designed for crushing mollusks, crabs, lobsters, and other invertebrates, as well as bony fishes. It often feeds in schools. Active after dark, the smooth-hound generally swims near the seabed in the shallow, intertidal zone. One of more than 20 similar species, this shark is commercially fished around Europe, the Mediterranean, and West Africa.

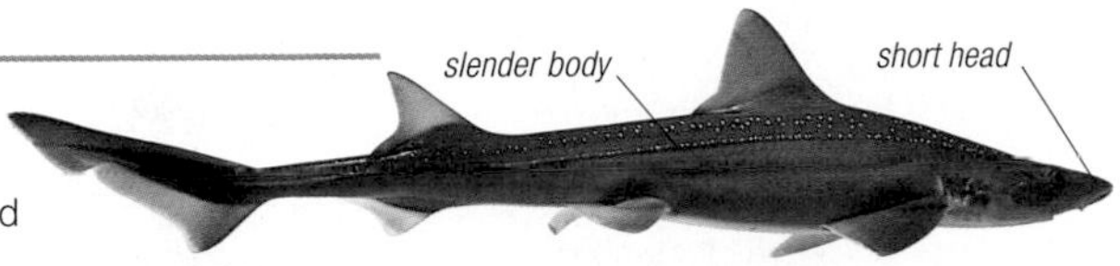

Triakis semifasciata

Leopard shark

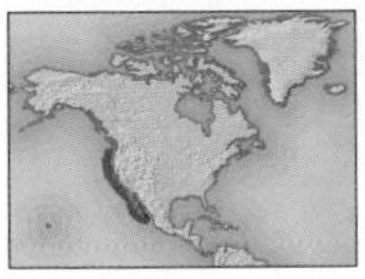

Length Up to 7 ft (2.1 m)
Weight Up to 71 lb (32 kg)
Breeding Viviparous
Status Least concern

Location E. North Pacific

The leopard shark is distinguished by a pattern of conspicuous black markings on its sides and fins, over a light brown background. Its powerful jaws and small, sharp teeth help it to capture a wide variety of prey, including burrowing invertebrates that it pulls out by quickly grasping any exposed parts; the shark is thought to feed both by day and by night. Harmless to humans, this is one of about 6 similar species.

Cephaloscyllium ventriosum

Swellshark

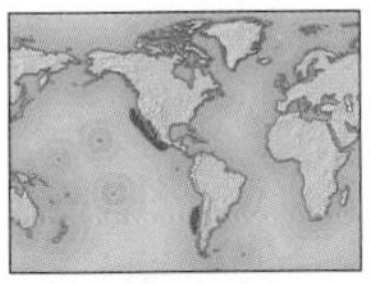

Length Up to 3¼ ft (1 m)
Weight Not recorded
Breeding Oviparous
Status Least concern

Location E. North and South Pacific

When threatened, the swellshark may expand itself by taking in water or air. In addition to making it appear more imposing to potential predators, this defensive technique allows the shark to wedge itself tightly into crevices for safety. This large member of the catshark family has a variable pattern of small spots and dark brown markings. Highly nocturnal, it usually lies motionless in crevices by day, and swims slowly through algae beds or near the seabed at night, presumably preying on diurnal fishes that are less alert at night. The swellshark is one of about 8 similar species.

Scyliorhinus canicula

Small-spotted catshark

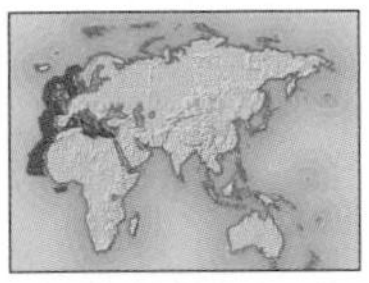

Length Up to 3¼ ft (1 m)
Weight 6½ lb (3 kg) or more
Breeding Oviparous
Status Least concern

Location E. North Atlantic, Mediterranean

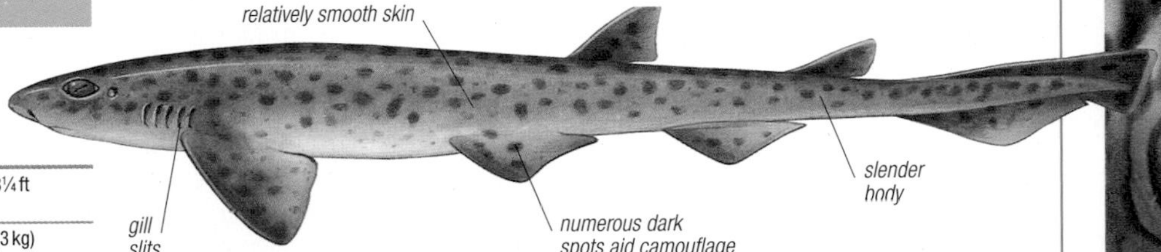

Also called the lesser spotted dogfish, this is the commonest European shark. It is small and slender, with a smooth skin covered with many dark spots and fewer light or white ones. The colors and patterns vary greatly, and help conceal this shark from predatory fishes. Normally solitary, when it does form schools, males and females are rarely found together. This shark is active by both day and night, hunting by scent and electrical sense for worms, mollusks, crustaceans, and small bony fishes, close to the seabed. Females deposit egg capsules on beds of algae; these are often washed up on the shore and are commonly known as mermaid's purses. They are rectangular, with tendrils at each corner that anchor them to the seaweed. In Europe, the small-spotted catshark is commercially fished for its oily liver as well as for food.

Carcharhinus leucas

Bull shark

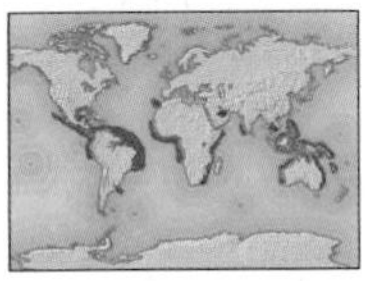

Length Up to 11 ft (3.4 m)
Weight 510 lb (230 kg) or more
Breeding Viviparous
Status Near threatened

Location Tropical and subtropical waters worldwide

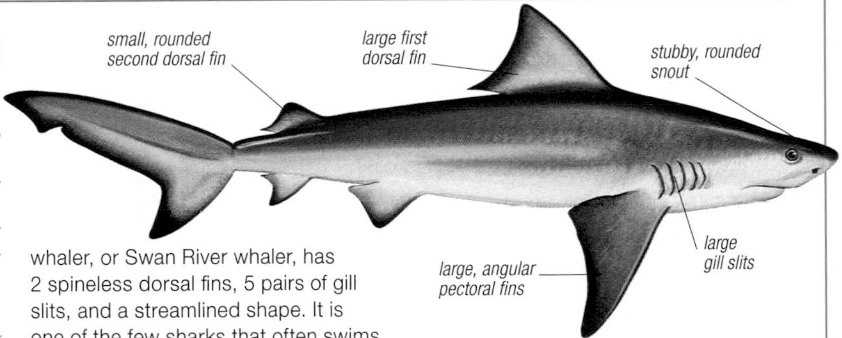

Like all members of the requiem shark family, the bull shark, also known as the river whaler, freshwater whaler, or Swan River whaler, has 2 spineless dorsal fins, 5 pairs of gill slits, and a streamlined shape. It is one of the few sharks that often swims up rivers—it has been found more than 1,870 miles (3,000 km) from the sea in the Amazon. Its diet is unusually wide-ranging and includes bony fishes, invertebrates, mammals, and other sharks, including young bull sharks. One of the world's most dangerous sharks, the bull shark has been involved in numerous provoked and unprovoked attacks on humans.

Galeocerdo cuvier

Tiger shark

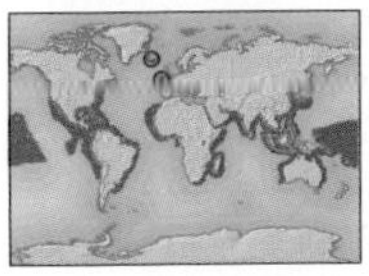

Length 18–25 ft (5.5–7.6 m)
Weight 1,990 lb (900 kg) or more
Breeding Ovoviviparous
Status Near threatened

Location Tropical and warm temperate waters worldwide

A known man-eater and one of the world's most dangerous sharks, the tiger shark has distinctive teeth shaped like a cockerel's comb, and a head that is disproportionately large for its slender, streamlined body. This shark generally prefers coastal waters, often moving inshore to feed at night, and sometimes exploring the very shallow waters of bays and estuaries. Recent studies, however, indicate that it may also cross large expanses of open ocean. It makes forays from the tropics to higher latitudes during the warmer months. The tiger shark is the only ovoviviparous member of the family of requiem sharks; females may secrete uterine "milk" to provide additional nutrition for the embryos. Although not highly valued for its flesh, it is often harvested in commercial shark fisheries.

STRIPED YOUNG
The juveniles and the young adults of this species are conspicuously marked with vertical, tigerlike stripes. These may fade or be absent in adults.

EATING HABITS

A primarily nocturnal hunter, the tiger shark has the most indiscriminate diet among sharks: it employs quick bursts of speed to catch live prey, such as fishes and a variety of marine reptiles, invertebrates, and mammals. It also scavenges on carrion, and may even feed on garbage.

Carcharhinus melanopterus

Blacktip reef shark

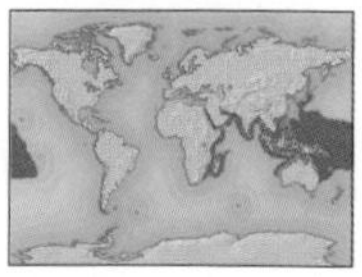

Location Tropical Indo-Pacific

- **Length** Up to 6½ft (2m)
- **Weight** Over 99lb (45kg)
- **Breeding** Viviparous
- **Status** Near threatened

The blacktip reef shark is a familiar species of shark in the tropical Indo-Pacific region. This shark usually occurs in shallow water, and the dorsal and upper tail fin may project above the surface of the water. It is responsible for numerous attacks on humans, both provoked and unprovoked, and is often curious when it spots divers, although it can be driven away. The blacktip reef shark has a streamlined body, which makes it an active and powerful swimmer. As its name suggests, the tips of the fins are black.

Negaprion brevirostris

Lemon shark

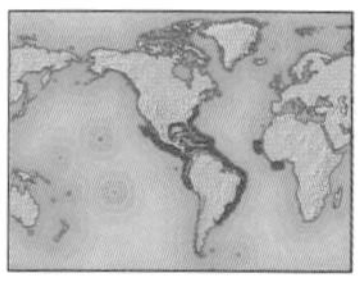

Location E. Pacific, W. and E. Atlantic

- **Length** Up to 11ft (3.4m)
- **Weight** Over 410lb (185kg)
- **Breeding** Viviparous
- **Status** Near threatened

The lemon, or sharptooth, shark adapts well to captivity and hence is one of the most-studied members of the requiem shark family. It is capable of a powerful bite and may attack humans if provoked. Yellowish to light brown in color, this fish has large fins and an abrupt snout. The lemon shark is adapted to survive in brackish waters and areas of low oxygen. It feeds mainly on bony fishes, guitarfishes, and stingrays, but may also eat crustaceans, molluscs, and even seabirds. Adults are most active at night; the juveniles by day.

Triaenodon obesus

Whitetip reef shark

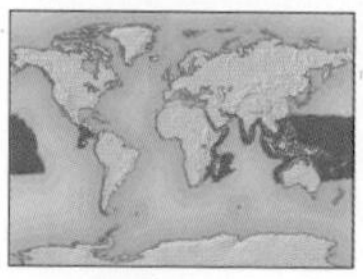

Location Tropical Indo-Pacific

- **Length** 5¼ft (1.6m), max 7ft (2.1m)
- **Weight** Over 40lb (18kg)
- **Breeding** Viviparous
- **Status** Near threatened

While most sharks must swim continuously, mouth agape, to force oxygenated water across their gills, the whitetip reef shark, or blunthead, can pump water across its gills and can therefore rest on the ocean floor. It relies heavily on sound and smell for catching prey—often associating the sound of boat engines with the opportunity to steal fishes, it doggedly pursues speared fishes and may bite fishermen. Nevertheless, it is generally nonaggressive.

CONSERVATION

Despite their reputation for being dangerous, sharks have more to fear from humans than humans do from sharks. Many of the world's inshore species—including the whitetip reef shark—have been heavily overfished in recent years, and are rapidly declining as a result. Sharks are often fished for their fins alone, the rest of the animal being thrown away.

WHITE MARKERS
This slender shark has distinctive white tips on its dorsal fin and upper tail lobe.

Prionace glauca

Blue shark

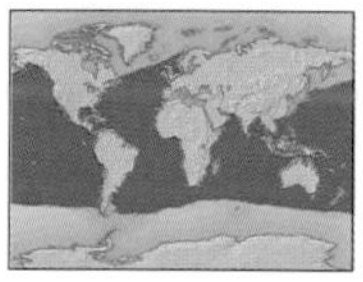

Location Tropical and temperate waters worldwide

- **Length** 12ft (3.8m) or more
- **Weight** Over 440lb (200kg)
- **Breeding** Viviparous
- **Status** Near threatened

Although one of the widest ranging of sharks, this species may be threatened by over-harvesting. It migrates seasonally, moving from cooler to warmer waters. Blue sharks are highly dangerous and are known to attack humans. They sometimes circle potential prey before attacking, and large numbers may gather around a whale or porpoise carcass in a feeding frenzy, biting at chunks of floating debris. They may also be seen following trawlers, and feeding on the trapped fishes.

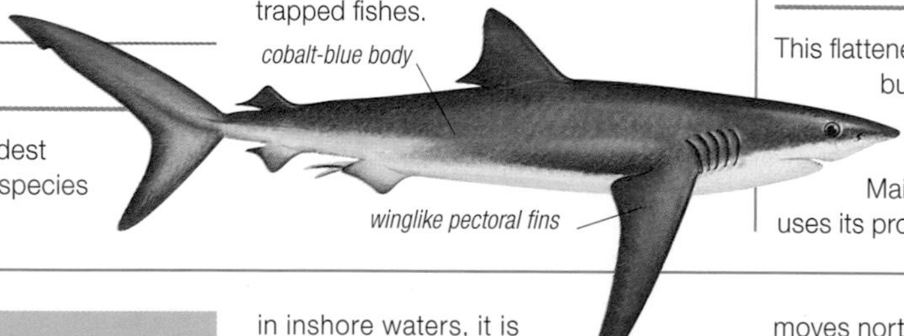

Squatina dumeril

Sand devil

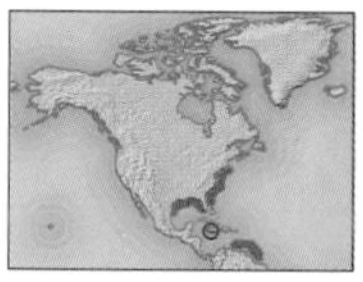

Location W. North Atlantic

- **Length** Up to 5ft (1.5m)
- **Weight** Over 60lb (27kg)
- **Breeding** Ovoviviparous
- **Status** Data deficient

This flattened fish is often found partially buried in sand on the sea floor, its nondescript coloration serving as camouflage. Mainly a sit-and-wait predator, it uses its protrusible mouth to strike with lightning speed at unwary prey, consisting chiefly of crustaceans and other fishes. One of about 14 similar species, the sand devil (or angelshark) is commercially fished for its flesh, skin, and oil.

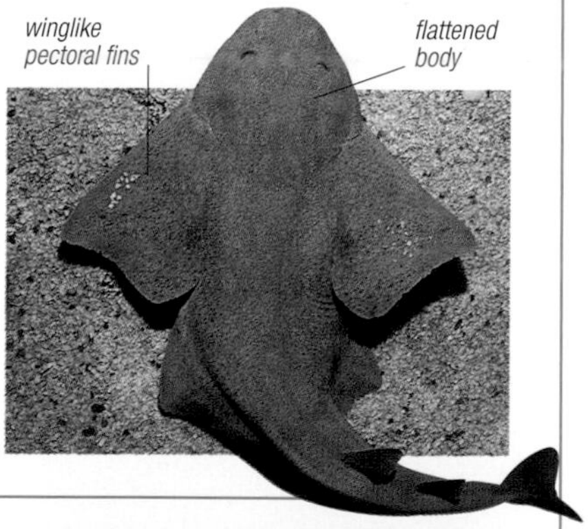

Sphyrna zygaena

Smooth hammerhead

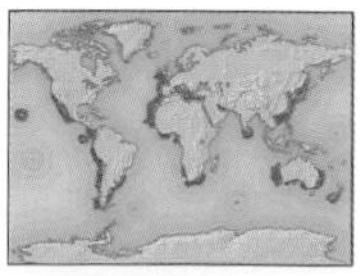

Location Tropical, subtropical, and temperate waters worldwide

- **Length** Up to 13ft (4m)
- **Weight** Up to 880lb (400kg)
- **Breeding** Viviparous
- **Status** Vulnerable

The smooth hammerhead is one of 8 species of hammerhead sharks, instantly identifiable by their bizarre heads, which have eyes at the end of winglike flaps, or "hammers". Found in inshore waters, it is often seen at the surface, with its large dorsal fin slicing through the water. It has an unpredictable temperament, and is known to attack humans. Like its relatives, the smooth hammerhead feeds on a wide variety of fishes, including other sharks. A powerful swimmer, it moves north to cooler waters during the summer. Unlike other sharks, hammerheads often form schools, probably for protection from predators; the young are sometimes observed migrating in huge numbers.

IDENTIFYING FEATURES
Species of hammerhead sharks can be difficult to tell apart. In the smooth hammerhead, the blades of the "hammer" are narrow.

UNIQUE ANATOMY

The reason for the hammerheads' strange anatomy is uncertain. The "hammer" provides these sharks with extra lift as they swim, but it may also help them to track prey, by improving the acuity of their sense of smell, and by giving them a wide visual field.

Skates and rays

PHYLUM	Chordata
CLASS	Chondrichthyes
SUBCLASS	Elasmobranchi (part)
ORDERS	3
FAMILIES	17
SPECIES	About 650

Broad bodies and winglike fins distinguish skates and rays from other cartilaginous fishes. Their flattened shape is generally an adaptation for life on the seabed, although several species, including mantas and eagle rays, swim in open water. Skates and rays are found throughout the world. Rays are more diverse in the tropics, while skates are more widespread in temperate waters. Most skates and rays live along coasts, but some range far into deep-sea basins. Although some rays range between fresh- and saltwater estuaries, there are others that live exclusively in freshwater.

Reproduction

Skates and rays have different modes of reproduction. Most skates breed by releasing large, leathery egg capsules containing one or more developing offspring. Some rays produce eggs that hatch inside the female's body. The young live off the yolk inside the female before being released. In other species, the embryos are further nourished by a fluid produced by the mother and delivered by the membranes of the uterus.

CHIMAERAS

Ratfishes, rabbitfishes, and their relatives—collectively known as chimaeras—make up a small group of cartilaginous fishes that are distinct from sharks, skates, and rays. They form the subclass Holocephali, containing 49 species. Their soft, scaleless bodies are usually long and narrow, and they have a bulky head with large, iridescent eyes. Found in the waters of the Indian, Atlantic, and Pacific, they live at depths up to 26,400 ft (8,000 m).

Anatomy

Skates and rays have flat bodies with large pectoral fins that are used for propulsion. Larger species flap these fins like wings. They have an oil-filled liver like that of sharks, which helps maintain buoyancy, but it is smaller and allows them to remain on the seabed. Most species are colored to match the seabed but they may also partially bury themselves in the sediment. When buried, they breathe through openings called spiracles located behind the eyes. Water enters the spiracles and leaves through the gill openings, bypassing the mouth, which is on the underside. Many skates and rays either lack scales or have large, thornlike scales for protection. Rays have a barbed spine on their tails that is coated with toxic compounds.

Feeding

All species of skates and rays are carnivorous, and most are predators on small, bottom-dwelling fishes and invertebrates, although prey varies considerably between species. These fishes have teeth that are designed to grasp, rasp, or crush food, and interlock in a pattern known as pavement dentition. A few species, such as mantas, are filter feeders that strain small fishes, invertebrates, and plankton with their sievelike gill rakers.

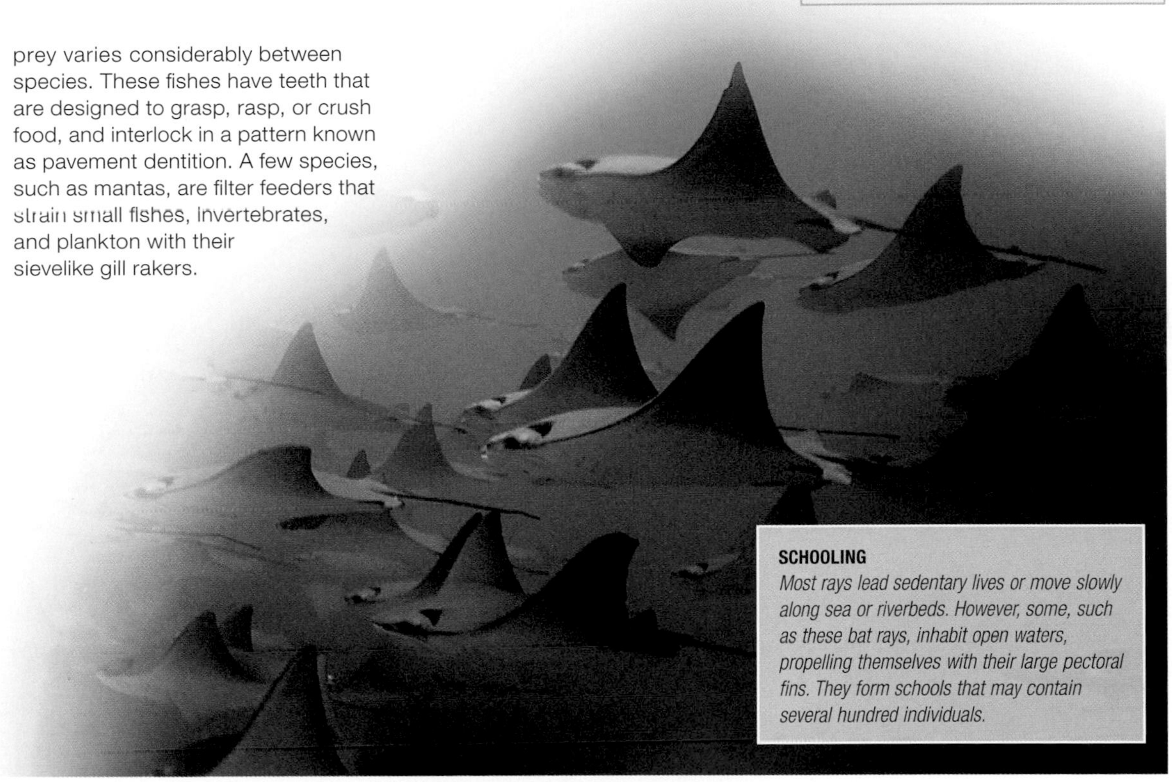

SCHOOLING
Most rays lead sedentary lives or move slowly along sea or riverbeds. However, some, such as these bat rays, inhabit open waters, propelling themselves with their large pectoral fins. They form schools that may contain several hundred individuals.

Raja undulata

Undulate skate

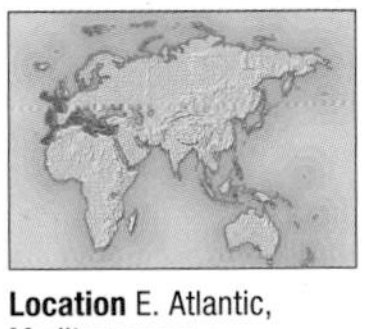

Length Up to 3¼ ft (1 m)
Weight 6½–8¾ lb (3–4 kg)
Breeding Oviparous
Status Endangered

Location E. Atlantic, Mediterranean

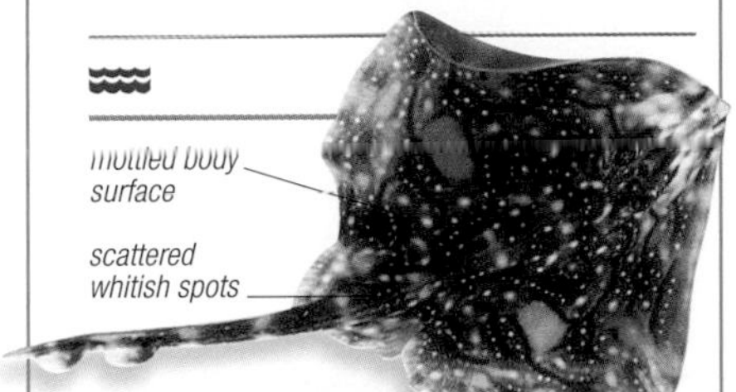

Found on mud or sand at depths of 650 ft (200 m), the undulate skate relies on its intricately patterned, reddish or brown skin to hide from predators. During the breeding season, females lay up to 15 eggs, each in a tough capsule up to 3½ in (9 cm) long.

Raja clavata

Thornback skate

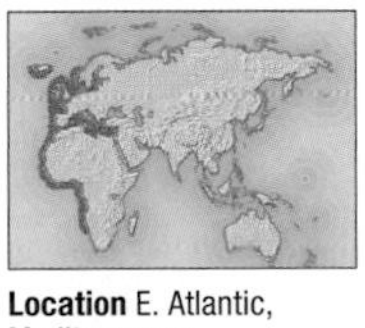

Length Up to 35 in (90 cm)
Weight 4½–8¾ lb (2–4 kg)
Breeding Oviparous
Status Near threatened

Location E. Atlantic, Mediterranean

The thornback skate occurs in varying shades of brownish gray with darker marbling and spotting; its underside is of a lighter color. Both sexes have big spines on their back—hence the common name of this species. As with most rays, it lies camouflaged on the ocean floor.

Dipturus batis

Blue skate

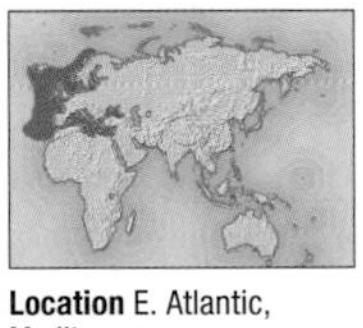

Length Up to 8¼ ft (2.5 m)
Weight 110–220 lb (50–100 kg)
Breeding Oviparous
Status Critically endangered

Location E. Atlantic, Mediterranean

Now severely depleted through over-fishing, this skate is brown and spotted above, and often light gray below. As with the thornback skate (see left), the male tends to be smoother; the female has small prickles on the underside. She lays 10 or more egg cases per year, each of which is about 8 in (20 cm) long, with a dark, thick, flexible shell and a pair of tendril-like thorns on either side. Unlike most other skates, the blue skate is active by day and night. It feeds on seabed animals, especially fishes, worms, and crustaceans.

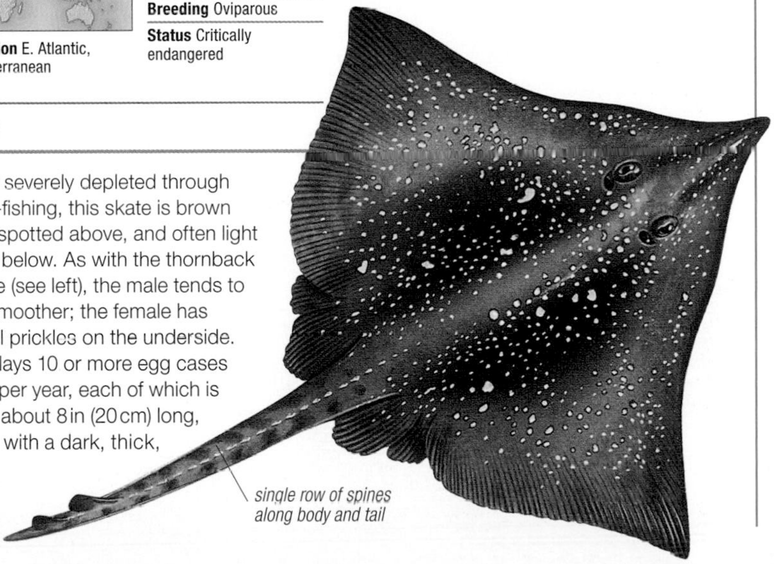

FISHES

Manta birostris

Manta ray

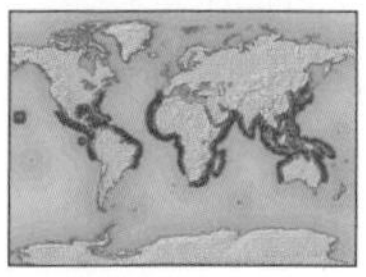

Length 13–23ft (4–7m)
Weight Up to 2 tons (1.8 tonnes)
Breeding Viviparous
Status Near threatened

Location Tropical and sometimes warm temperate waters worldwide

Measuring up to 23ft (7m) across, the manta—also known as the devil ray—is the world's largest ray, swimming through the open ocean like a large bird on slowly flapping wings. However, despite its sinister look, it is a harmless plankton-eater, filtering food with its gills. Its mouth, unlike those of other rays, is positioned in front of its body, allowing it to feed continuously as it moves along. Normally a leisurely swimmer, it can suddenly accelerate if threatened, and can even leap out of the water to avoid large predators, such as sharks and killer whales. Mantas are usually solitary, although they sometimes swim in small, loosely organized schools. They prefer warm waters, and may enter inshore regions in summer. During the breeding season, the male chases the female, swimming beneath her so that their undersides face each other, and inserts his claspers. The female usually gives birth to one or 2 young per year, each about 4ft (1.2m) wide.

FUNNELING FOOD

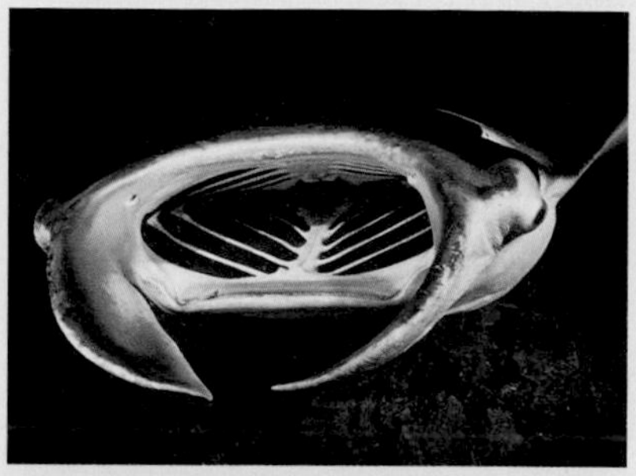

Manta rays have large, flaplike lobes on either side of their heads, and they use these to funnel prey into their mouths. The water passes out through the ray's gills, but the prey is trapped on transverse spongy plates that bridge the gaps between the gill bars. Mantas eat small, schooling fishes, as well as smaller planktonic animals.

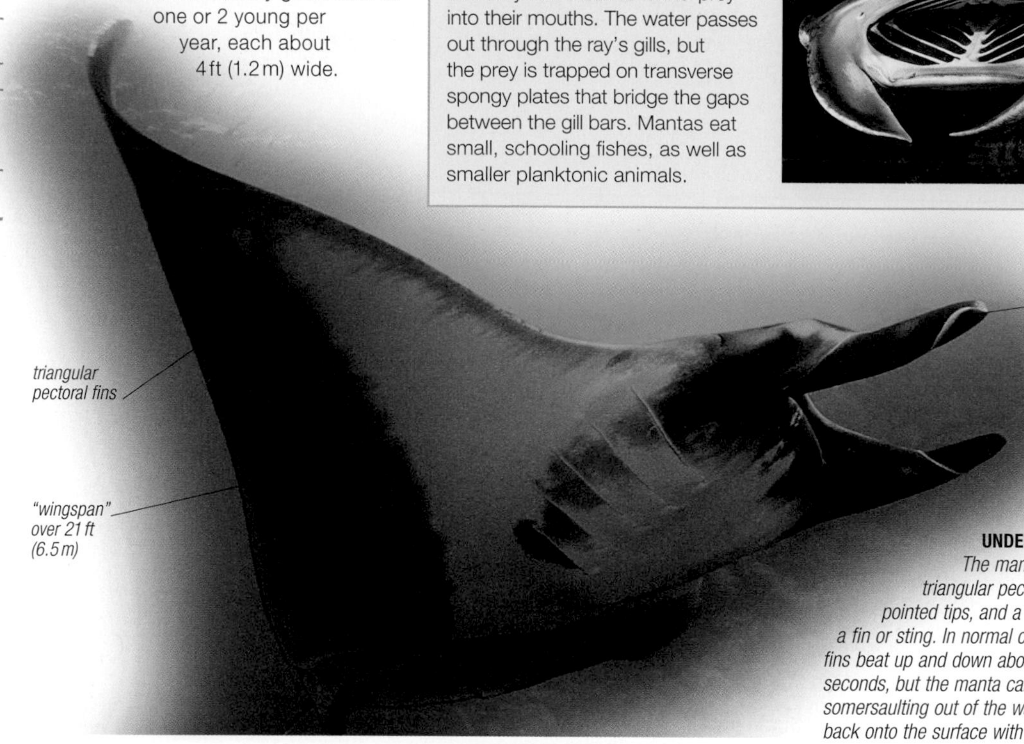

UNDERWATER FLIGHT
The manta has large, triangular pectoral fins with pointed tips, and a short tail without a fin or sting. In normal circumstances, its fins beat up and down about once every 4 or 5 seconds, but the manta can accelerate rapidly, somersaulting out of the water and crashing back onto the surface with a massive impact.

Rhinobatos productus

Shovelnose guitarfish

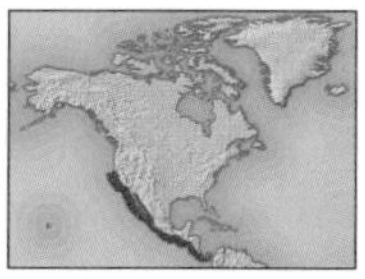

Length Up to 5ft (1.5m)
Weight 33–40lb (15–18kg)
Breeding Viviparous
Status Near threatened

Location E. Pacific

Named for its unusual shape, the shovelnose guitarfish has a broad head with a clear, cartilaginous area on either side of the snout. Its pectoral fins are wide, like those of a typical ray, but the rest of its body is cylindrical, like that of a shark. Found in the warmer areas of the eastern Pacific Ocean, it tends to move inshore in the summer months. The shovelnose guitarfish is generally solitary but sometimes gathers in large numbers, possibly for breeding. It is often found in shallow water around beaches, bays, and estuaries, where it lies partially buried in sand or mud. It feeds on small, bottom-dwelling fishes, and on crustaceans and worms, which it uncovers from the seabed. The female produces 5–25 young, which are about 6in (15cm) long at birth.

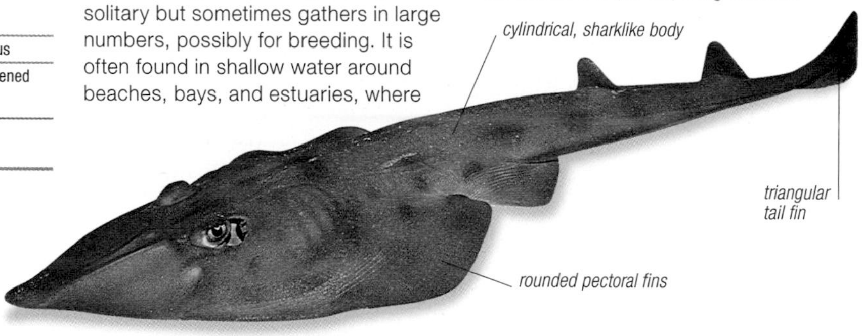

Urolophus halleri

Round ray

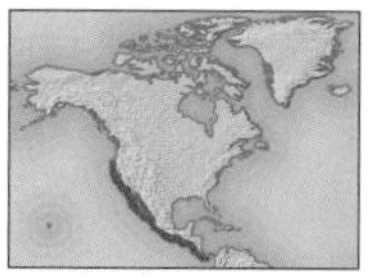

Length 20–23½in (50–60cm)
Weight 4½–8¾lb (2–4kg)
Breeding Viviparous
Status Least concern

Location E. North Pacific

Unlike typical stingrays, the round ray and its relatives, sometimes known as stingarees, have an almost circular body. The body is very flattened, and the tail is relatively short, with a rounded fin at its tip. During warm seasons, large numbers sometimes move into shallow beaches where they may sting bathers who are unfortunate enough to step on them. Although not fatal, the sting of this fish can be intensely painful.

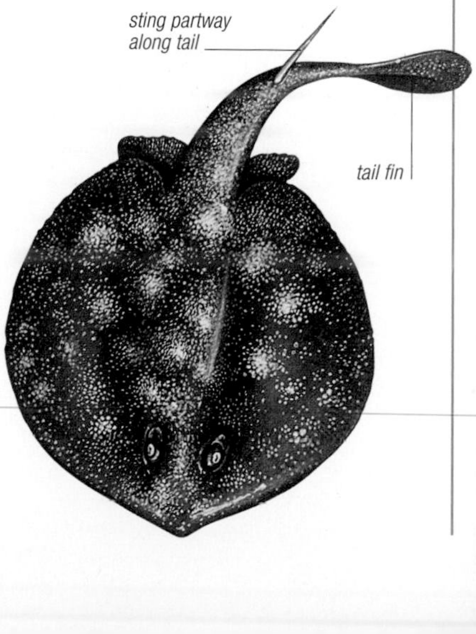

Taeniura lymma

Blue-spotted stingray

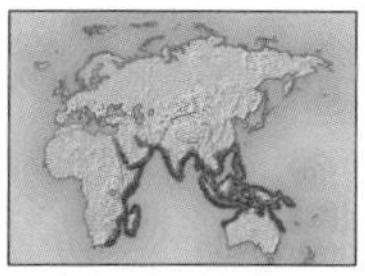

Length Up to 6½ft (2m)
Weight Up to 66lb (30kg)
Breeding Viviparous
Status Near threatened

Location Indo-Pacific

One of the most handsome members of the stingray family, this species has blue spots scattered over its body. In addition, it has blue stripes along the side of its tail and a lighter underside. Like all stingrays, this fish has a toxic spine located at the base of its tail and is capable of inflicting a nasty sting if stepped on or mishandled. Although mainly diurnal, it is sometimes active at night. It moves into shallow areas with the advancing tide to feed on small fishes and invertebrates, especially crustaceans and worms. It is often found in sandy areas next to reefs, and rests in reef caves and crevices when not feeding. The female gives birth to 3–7 young and, like the South American freshwater stingray (see opposite), she nourishes her young by secreting uterine "milk." The colorful appearance of the blue-spotted stingray has made it a popular fish in the aquarium trade and a favorite subject of underwater photographers.

LYING IN WAIT

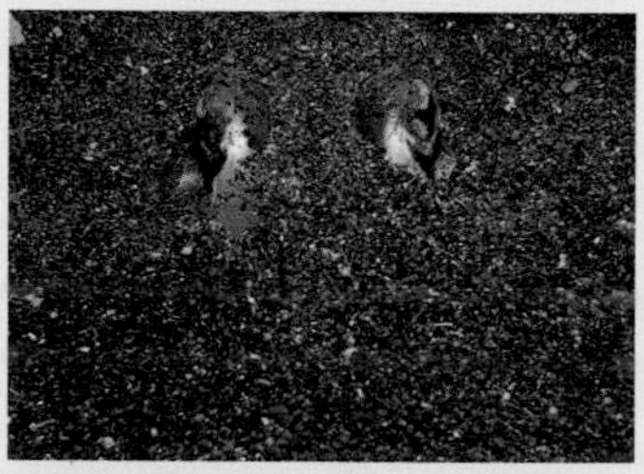

Blue-spotted stingrays normally actively forage for food, but they rest with their bodies largely hidden by mud or sand, with just their eyes projecting above the surface. In shallow water, hidden rays are easily stepped on.

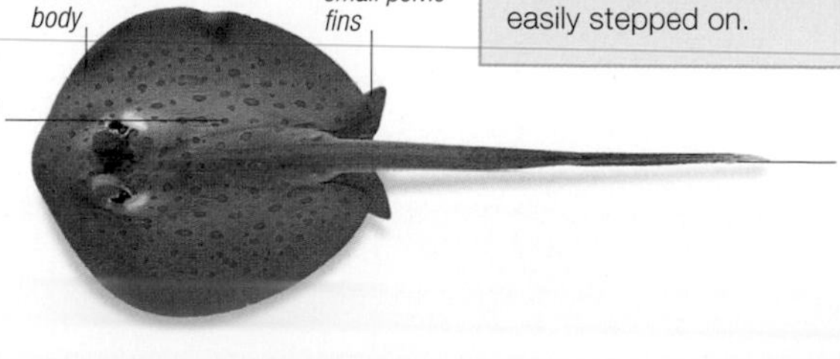

Myliobatis aquila

Common eagle ray

Length 8¼ ft (2.5 m) or more
Weight 44–66 lb (20–30 kg)
Breeding Viviparous
Status Data deficient

Location E. Atlantic, Mediterranean, S.W. Indian Ocean

The common eagle ray has wide but narrow pectoral fins, a blunt snout, and an exceptionally long tail armed with a venomous spine. The body of this fish varies in color from grayish brown to bronze or blackish, and its mouth, positioned on its underside, is equipped with a massive, crushing toothplate. Like its numerous relatives, which are found in oceans all over the world, it feeds on the seabed, but is equally at home in open water. The common eagle ray is believed to prey on a wide variety of small animals, and it excavates these from the sediment by flapping its fins, or by blowing jets of water to clear away mud and sand. A good swimmer, the common eagle ray can leap clear of the water's surface to escape attack from predators. Female common eagle rays give birth to 3–7 live young each year.

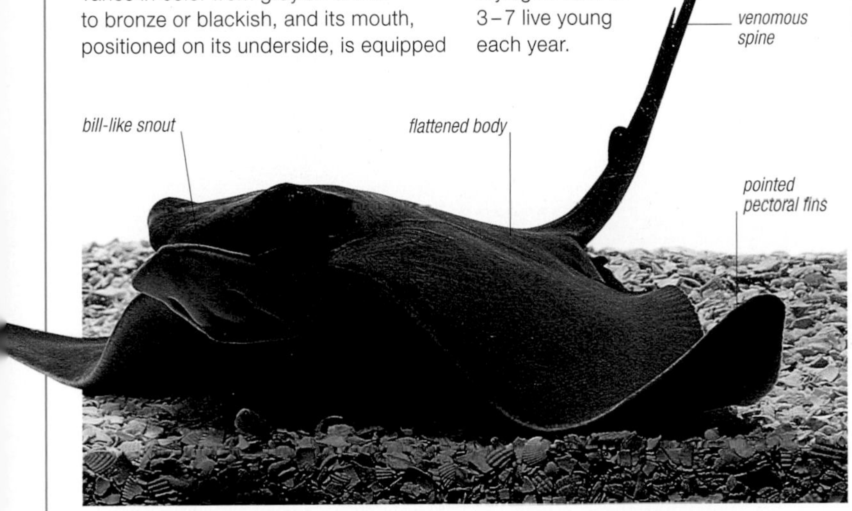

Potamotrygon motoro

South American freshwater stingray

Length Up to 3¼ ft (1 m)
Weight 6½–11 lb (3–5 kg)
Breeding Viviparous
Status Data deficient

Location South America

This round-bodied ray is the most common and widespread freshwater stingray in South America. It is a member of an entire family of stingrays that evolved in freshwater from a marine ancestor. The short tail of the South American freshwater stingray has no fin, and its small pelvic fins are tucked under its pectoral fins. This stingray has few predators, except some larger fishes and caiman, and feeds on small fishes and invertebrates.

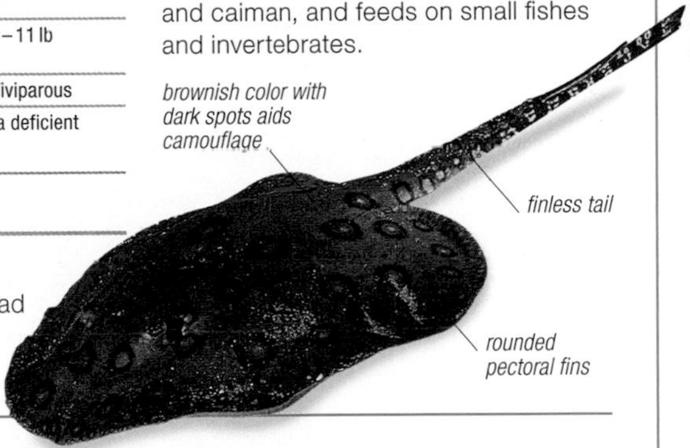

Torpedo marmorata

Marbled electric ray

Length Up to 23½ in (60 cm)
Weight 22–29 lb (10–13 kg)
Breeding Viviparous
Status Data deficient

Location E. Atlantic, Mediterranean

Many cartilaginous fishes use weak electrical signals to detect prey, but this ray can generate shocks, powerful enough to kill other fishes, from specially modified gill muscles at the base of its pectoral fins. Although such shocks are not known to be fatal to humans, contact with this ray can be dangerous. Active during the day and in the evening, the marbled electric ray generally lies on the ocean floor, well camouflaged against the sediment. Females give birth to 5–35 young every year; larger females produce more young.

Pristis pectinata

Smalltooth sawfish

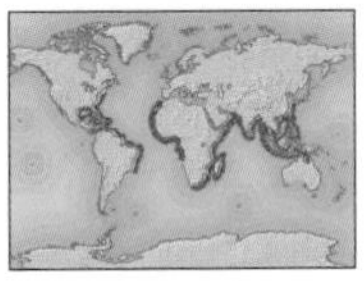

Length 20 ft (6 m) or more
Weight 550–660 lb (250–300 kg)
Breeding Viviparous
Status Critically endangered

Location W. and E. Atlantic, Indo-Pacific

Also known as the wide sawfish, this long-bodied fish has a remarkable snout, which is about one-quarter of its body length and has 24–32 pairs of pointed teeth on either side. Although it is equipped with this sawlike snout, it is not generally aggressive and is dangerous only if mishandled by divers. The smalltooth sawfish has a flattened head with small eyes, which have spiracles situated above and slightly posterior to them. The mouth and gill slits are located on the underside of the head. While typically found in the shallow, sandy, and muddy water of beaches and bays, the smalltooth sawfish may occasionally be found at the mouth of rivers and freshwater streams. It feeds by patrolling the seabed and sucking up small organisms, sometimes using the saw to probe the bottom or excavate buried prey. It also slashes into schools of fishes with its sawlike snout, feeding on dead or maimed fishes that fall to the bottom. The female gives birth to 15–20 young, which are born with a protective membrane over their saws, an adaptation that protects the female while giving birth. At first, the young live in nursery areas, which are usually found in shallow water where there is plenty of vegetation. Loss of these areas due to coastal development has contributed to their decline. One of 6 similar species, the smalltooth sawfish is subject to overfishing in many areas because its flesh is edible and the saw is sold as a souvenir.

Hydrolagus colliei

Spotted ratfish

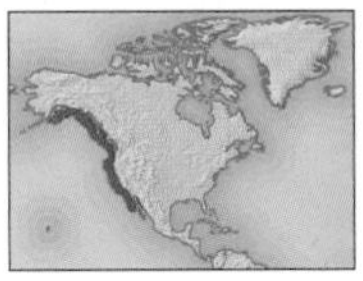

Length Up to 37 in (95 cm)
Weight 4½–13 lb (2–6 kg)
Breeding Oviparous
Status Least concern

Location E. North Pacific

Ratfishes are found in deep, inshore waters in most parts of the world, but the spotted ratfish is one of only 2 species that live on the Pacific coast of North America. Like other ratfishes, it has a disproportionately large head, and a downward-facing mouth that is equipped with large, crushing teeth, and scaleless, slippery skin. Males have a retractable, clublike appendage between their eyes, which is probably used during courtship. Adults are usually dark brown or grayish, with green, yellow, or bluish hues, and they have silvery white spots along their sides. Spotted ratfishes are slow swimmers, and rely on a venomous dorsal spine as their main defense against attack. They patrol the ocean floor in search of prey, feeding on invertebrates and seabed fishes. Females usually lay 2–5 eggs a year; each one is contained in a long, spindle-shaped case. Ratfish liver contains an oil that was once used as a machine and gun lubricant, but is today of little commercial value.

Bony fishes

PHYLUM	Chordata
CLASS	Osteichthyes
ORDERS	48
FAMILIES	482
SPECIES	About 31,000

CLASSIFICATION NOTE

Bony fishes account for more species than any other class of vertebrates. They are split into 2 subclasses: fleshy-finned fishes (Sarcopterygii) have fins that look like flippers, joined to the body by fleshy lobes; ray-finned fishes (Actinopterygii) have fins supported by bony fin rays. Ray-finned fishes are by far the larger group and are divided into 9 superorders. Classification of bony fishes is in a state of constant change as new species are discovered and zoologists learn more about the relationships between them.

Bony fishes form overwhelmingly the largest and most varied group of fishes, accounting for more than 9 out of 10 species. Of the 3 classes of fishes, they evolved most recently and are usually regarded as the most advanced. Although most are small, they vary greatly in size and shape. All bony fishes have a light but strong internal skeleton, made entirely or partly of bone, which supports flexible fins that enable the fish to control its movements with precision. Most also have a gas-filled swim bladder that allows them to adjust their buoyancy within narrow limits. Bony fishes are found in almost all aquatic habitats, including marshes, lakes, rivers, coasts, reefs, and deep oceans. Many have adapted to extreme conditions, and species of bony fishes occur in high-altitude lakes, polar coasts, hot springs, high-salinity ponds, acidic streams, and low-oxygen swamps.

Senses

Most bony fishes have keen senses of vision and hearing. Both of these senses are used in communication and social interaction. They are also thought to contribute to bony fishes' well-developed ability to school. The eyes are generally set on the sides of the head, giving each fish a wide field of view. Rods and cones in the retina give them good color vision, and many species have bright coloration, which helps them attract mates and defend territory. Sound is an effective means of underwater communication. Some species of bony fishes can produce sounds, either using their swim bladder or by rubbing parts of their body together.

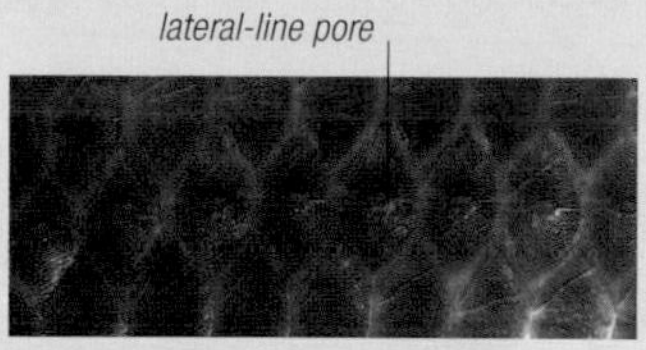

LATERAL LINE
The lateral-line system, which is used to detect movement, is highly developed in bony fishes. This is thought to be another reason why they can move effectively in large schools.

Anatomy

The skeleton of a bony fish consists of 3 main units: the skull, backbone, and fin skeleton (see below). The gills of bony fishes are paired and located behind the head and below the cranium. Unlike in other groups of fishes, the gill openings are covered by a bony flap, called the operculum, and the lower gill chamber contains bony supports, called branchiostegal rays. These 2 structures enable the fish to take a gulp of water into its mouth and then pump it over its gills: the branchiostegal rays help open the mouth and regulate intake, while the operculum acts as a seal over the gills to control the outflow. In this way, a bony fish can respire while stationary. The same 2 structures are also used in seizing and gulping food. Most bony fishes have light, flexible, cycloid or ctenoid scales (see p.468), usually covered by a thin layer of skin that secretes mucus. The mucus repels parasites and disease-causing organisms; in some species, it also prevents moisture loss. Other bony fishes have large, protective scales or no scales at all (although their skin still produces mucus).

SWIMMING
Bony fishes, such as these French grunts, can maneuver precisely. Each fin ray is controlled by a separate set of muscles so that it can move independently of the others. The symmetrical tail typical of bony fishes is an efficient tool for propelling the fish through water.

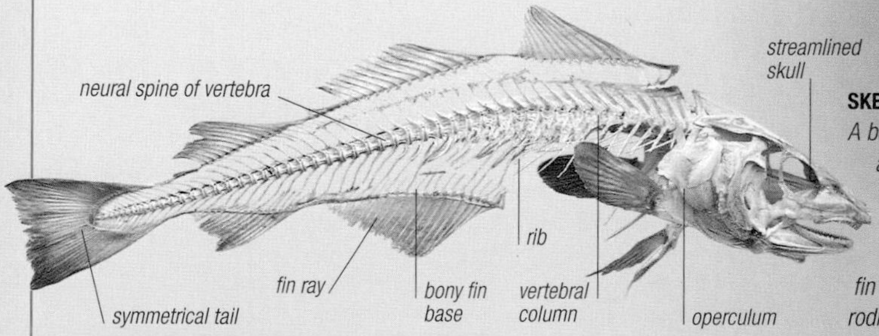

SKELETON
A bony fish's skull encases the brain and supports the jaws and gill arches. Teeth may be found in the jaws, in the throat, on the roof of the mouth, or on the tongue. The backbone, consisting of articulated vertebrae that are linked to the ribs in the abdomen, provides support for the body. Each fin consists of a bony base embedded in the body, from which rodlike structures extend to form the outer fin.

SCHOOLING

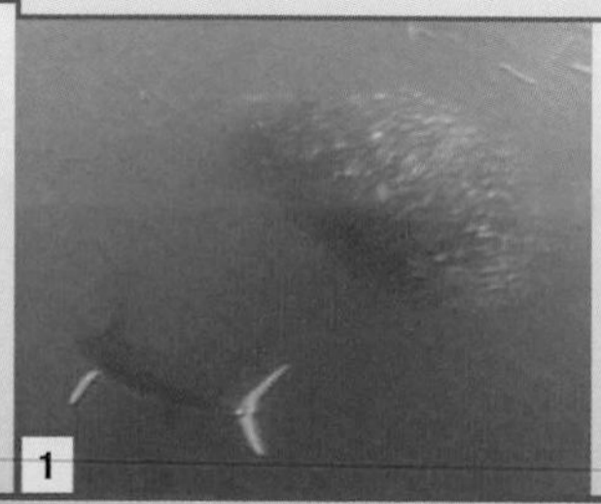

ATTRACTING PREDATORS
A school of fishes is an easy target for predators. These yellowtail horse mackerel have attracted a marlin.

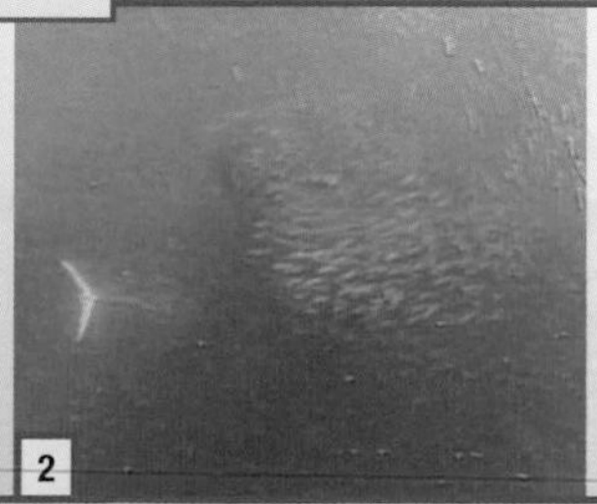

STAYING TOGETHER
As the marlin swims toward the school, the mackerel are able to change direction and stay together.

CLOSING RANKS
While the marlin circles close by, the mackerel pack themselves even more closely together.

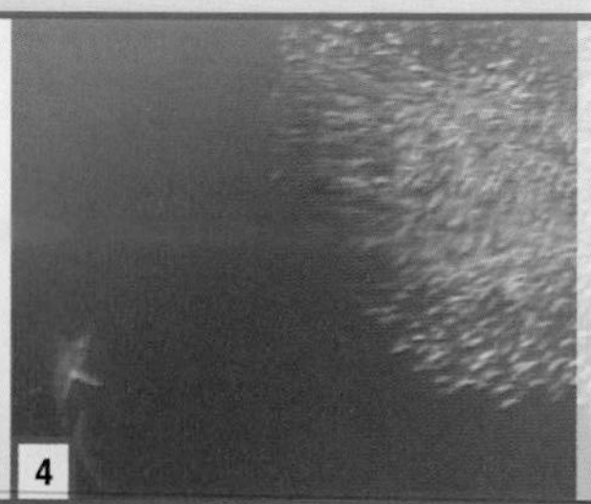

USING THE LIGHT
The marlin stays below the school, where it is easier to pick out the fishes against the light coming from above.

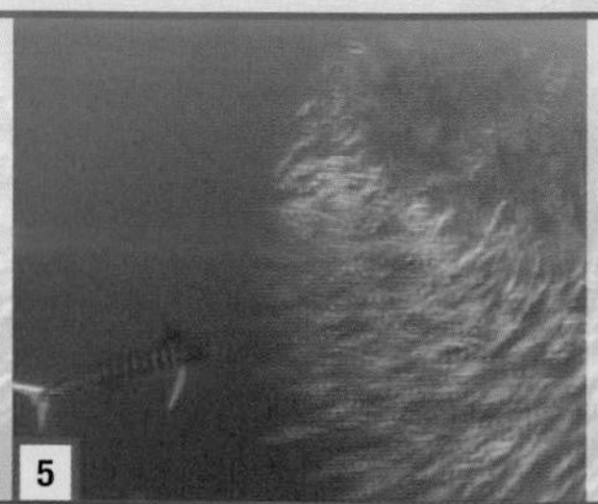

ON THE ATTACK
The marlin attacks again. If the fishes are forced to split into 2 groups, they will try to rejoin each other quickly.

FISHES

Buoyancy

Most bony fishes control their buoyancy using an outgrowth of the intestine known as the swim bladder. This bladder is filled with gas, making it less dense than water. The fish can adjust its buoyancy by regulating the volume and pressure of the gas in the bladder. The swim bladder of a freshwater fish has a greater capacity than that of a marine fish, because freshwater is less dense than sea water and does not give as much support to the fish's body. Having a swim bladder means that bony fishes (unlike cartilaginous fishes) do not need to use their fins for buoyancy control and, instead, can use them entirely for maneuvering.

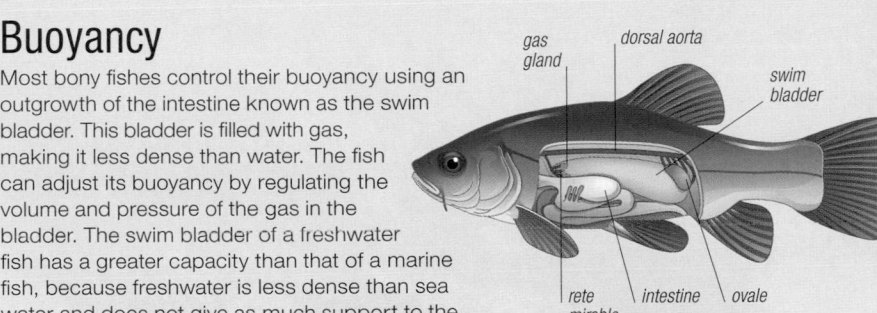

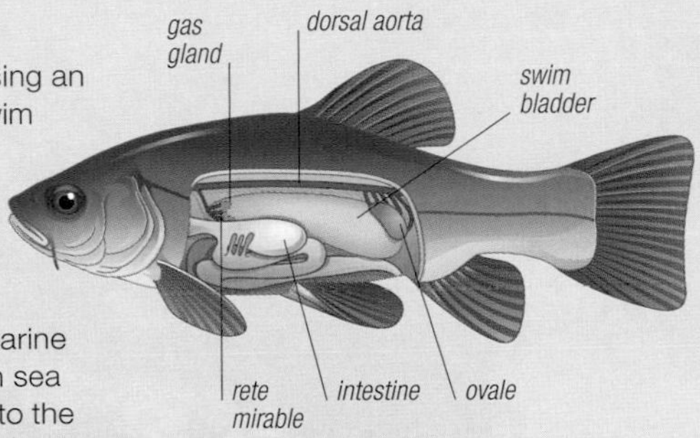

SWIM BLADDER
Apart from some primitive species, most bony fishes adjust their buoyancy by moving gas between the bloodstream and the swim bladder. Gas enters the bladder through the gas gland, which is supplied with blood by a network of capillaries called the rete mirable. Using another structure, known as the ovale, gas is removed from the bladder and reabsorbed into the blood.

Reproduction

In most bony fishes, fertilization occurs outside the body but in a few species it is internal. Marine fishes that live in open water usually produce eggs that float as part of the plankton, often in large numbers because only a small number will actually hatch. Coastal marine and freshwater fishes tend to produce fewer eggs, which are laid in the sediment, in nests, or attached to plants. A few species care for their eggs or even their young. This usually takes the form of protecting them from predators, but some species provide their offspring with food. Bony fishes include some hermaphroditic species, which have both male and female reproductive organs. Others change sex during their adult life: males that become female may become larger and accommodate more eggs (which are larger than sperm); by becoming males, females may assume dominant positions in social groups.

PARENTAL CARE
The male yellowhead jawfish incubates its eggs by storing them inside its mouth. Some other bony fishes (including some cichlids) use their mouth to protect their young from predators.

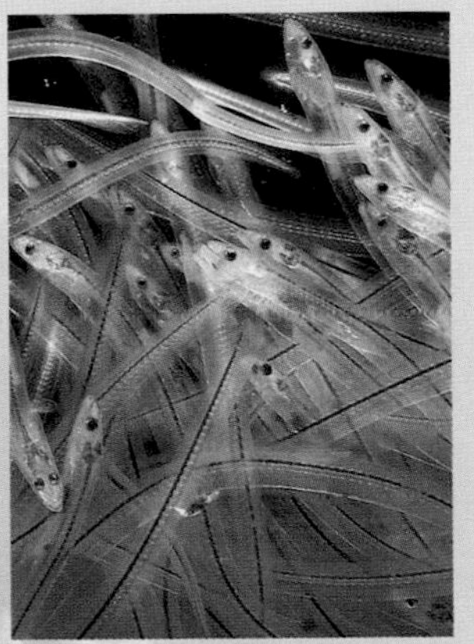

LARVAE
Most bony fishes hatch from eggs and begin life as larvae. These larvae of the European eel are in the process of metamorphosis, during which they are referred to as glass eels.

FISHES

Schooling

Many bony fishes swim in large groups that often contain tens of thousands of individuals. When the fishes are grouped together but behave independently, the group is referred to as an aggregate. In a school, however, the movement of the fishes is much more harmonized, to the extent that the school resembles a single organism. Only bony fishes can swim in this coordinated way. Each fish usually follows a course parallel to its neighbors, maintaining its position using its vision, hearing, and lateral-line system. Fishes in a school are safer than solitary ones, because there are more individuals to look out for danger and it is difficult for a predator to pick out a single fish from a large mass (see left). It is also easier for fishes in a group to find mates and locate food.

STAYING TOGETHER
Schooling fishes pack tightly together, each one typically keeping a distance to its nearest neighbor of no more than one body length. A silvery coloration (seen in these sweepers) helps the fishes see each other's movements because a small change in direction produces a large change in the amount of light reflected.

Fleshy-finned fishes

PHYLUM	Chordata
CLASS	Osteichthyes
SUBCLASS	Sarcopterygii
ORDERS	3
FAMILIES	4
SPECIES	8

Fleshy-finned fishes have an enlarged fin base that consists of muscles connected to an internal skeleton. These fins are used for swimming, or to "walk" on the seabed. In some of the fishes related to the ancestors of this group, the fins developed into limbs for movement on land, and the living fleshy-finned fishes are perhaps the closest relatives of the early terrestrial 4-legged animals. Fleshy-finned fishes are divided into 3 groups. The eel-like lungfish have both gills and one or 2 primitive lungs, and can breathe air directly. They can live in oxygen-poor swamps and even survive out of water. One species, the West African lungfish, can reside in a mud cocoon at the bottom of a dry lake, awaiting the next wet season to replenish the water. The other group, the coelacanths, were thought to be extinct since the time of the dinosaurs until discovered in 1938; they live in deep areas of the Indian Ocean.

AIR-BREATHING FISHES
The adult South American lungfish, which has small gills and 2 lungs, can only breathe air. In the breeding season, the male provides oxygen to its young by releasing it into the water through blood-rich filaments in its threadlike fins.

Latimeria chalumnae

Coelacanth

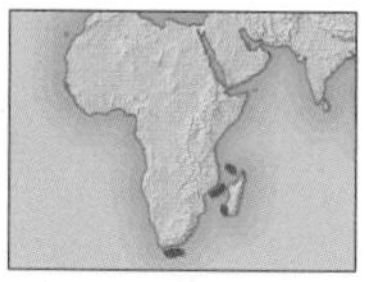

Location W. Indian Ocean

- **Length** 5–6 ft (1.5–1.8 m)
- **Weight** 145–220 lb (65–98 kg)
- **Sex** Male/Female
- **Status** Critically endangered

First seen by scientists in 1938, the coelacanth is a classic example of a "living fossil," because it belongs to a group that was thought to have died out over 65 million years ago. It is found at depths of 500–2,300 ft (150–700 m) along rocky slopes with submarine caverns, swept by strong oceanic currents. It has large, thick, heavy scaling with iridescent white flecks, muscular paired fins, and an unusual tail with an additional central lobe. The pectoral fins of the coelacanth are highly mobile, and it uses these to maneuver into crevices to reach fishes and other prey. Populations are poorly documented and estimates are difficult, but susceptibility to capture by humans and a restricted habitat range have made the coelacanth an endangered species.

Latimeria menadoensis

Indonesian coelacanth

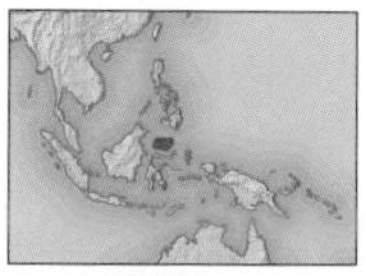

Location Pacific (Celebes Sea)

- **Length** Up to 5¼ ft (1.6 m)
- **Weight** 145–220 lb (65–98 kg)
- **Sex** Male/Female
- **Status** Vulnerable

The Indonesian coelacanth was discovered in the late 1990s, and as yet, relatively little is known about its behavior or ecology. However, since it is physically very similar to the coelacanth of southern Africa (see above), it almost certainly has a comparable way of life. Molecular analysis has shown that the 2 species probably diverged from each other between 4.7 and 6.3 million years ago. Since then, they have been kept apart by the geology of the seabed, and by currents that confine them to their own particular parts of the world. Like its African counterpart, the Indonesian coelacanth appears to be a solitary fish in open water, but it has also been found in groups in caves. So far, population estimates are not available, but it is likely to be endangered. The chief threat to this coelacanth's survival is fishing, which can have a severe effect on a species that is geographically highly isolated.

Neoceratodus forsteri

Australian lungfish

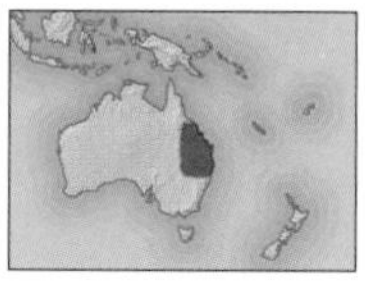

Location E. Australia

- **Length** Up to 6 ft (1.8 m)
- **Weight** Up to 99 lb (45 kg)
- **Sex** Male/Female
- **Status** Not evaluated

Unlike other lungfishes, which often live in pools that sometimes dry up, the Australian lungfish is found in permanent bodies of water associated with dense vegetation, such as large, deep pools, reservoirs, and slow-flowing rivers. A large, freshwater fish with a heavy body, it has a specialized swim bladder that functions as a single lung. Although it normally breathes through its gills, when oxygen levels in the water fall, it gulps air at the surface, breathing through its lung. Unlike African and South American lungfishes, this species has paddle-shaped, paired fins. Its dorsal and anal fins are continuous with its tail fin. The Australian lungfish becomes less active during dry conditions, and can survive for months if kept moist by a covering of damp leaves or mud. It feeds on frogs, crabs, insect larvae, mollusks, and small fishes, crushing its prey between its strong toothplates.

Protopterus annectens

West African lungfish

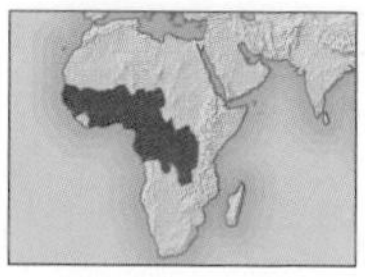

Location W. to C. Africa

- **Length** 6–6½ ft (1.8–2 m)
- **Weight** Up to 37 lb (17 kg)
- **Sex** Male/Female
- **Status** Least concern

The West African lungfish has long, threadlike, paired fins, and is the largest of the 4 lungfishes found in Africa. Like the other 3 species, it breathes through a pair of lungs. At the onset of the dry season, this fish burrows into mud, forming a mucus-filled cocoon. It lives in a variety of freshwater habitats, and is carnivorous, capturing its prey by stealth rather than by rapid pursuit.

FISHES

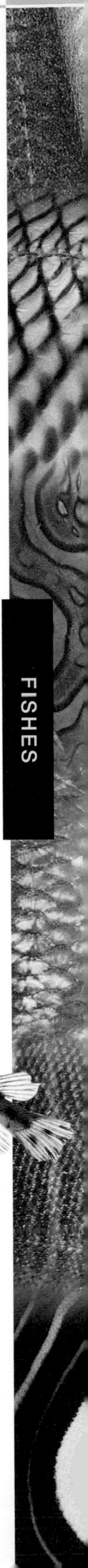

Primitive ray-finned fishes

PHYLUM	Chordata
CLASS	Osteichthyes
SUBCLASS	Actinopterygii
ORDERS	Polypteriformes, Acipenseriformes, Lepisosteiformes, Amiiformes
FAMILIES	5
SPECIES	48

Among the least advanced of all bony fishes, primitive ray-finned fishes have an unusual skeleton composed partly of bone and partly of cartilage. They are divided into 4 groups: sturgeons and paddlefishes, gars, bichirs, and the bowfin (a single species that is placed in a separate order). Primitive ray-finned fishes are among the largest of all freshwater species, with sturgeons reaching lengths of 26 ft (8 m). Unique to the Northern Hemisphere, most species live in freshwater, although some are able to live alternately in fresh and saltwater.

Anatomy

In most primitive ray-finned fishes, the skull and some fin supports are made of bone, while the body and tail are supported by a cartilage notochord with some rudimentary vertebrae. The tail is asymmetrical, with the end of the notochord extending into an upper lobe that is longer than the lower one. Gars, some bichirs, and the bowfin have a swim bladder lined with blood vessels that can be used as a lung when levels of oxygen in the water are low.

Sturgeons are large, bottom-feeding fishes with a flattened snout, a protrusible mouth, and sensitive barbels for locating food. Several rows of large, thickened scales (bony scutes) protect an otherwise scaleless body. Paddlefishes have a long, paddle-shaped rostrum, which may act as a sensory organ or improve the flow of water to the mouth during feeding. These filter-feeding fishes use their gaping mouth and large gill chambers to strain small aquatic organisms from the water.

Gars are cylindrical and highly predatory fishes that catch prey through ambush or bursts of speed. The elongated rostrum and jaws are studded with sharp teeth, resembling the snout of a crocodile. Gars are covered by smooth, thickened ganoid scales (see p.468).

Bichirs have a long, slender body protected by thick scales. Most species (sometimes referred to as the true bichirs) have a row of small finlets on their back and a pectoral fin divided into fleshy lobes.

The bowfin is much smaller than other primitive ray-finned fishes. A predatory fish, it has a blunt head and small teeth.

PADDLEFISHES
There are two species of paddlefishes, both found in large river systems with an ample supply of plankton. The American paddlefish (shown here) lives in the Mississippi river system; it feeds at night and spends the daylight hours resting at the bottom of deep pools. The other species lives in the Yangtze River.

Acipenser sturio

European sturgeon

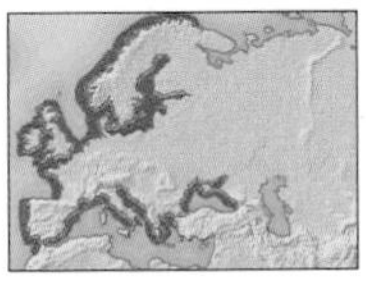

Length Up to 11 ft (3.5 m)
Weight Up to 690 lb (315 kg)
Sex Male/Female
Status Critically endangered

Location North Atlantic, Mediterranean, Europe

Heavily overfished for its flesh and eggs—caviar is made of the unshed eggs of females—the European sturgeon is now extremely rare. One of the largest European fishes that swim up rivers to breed, it is greenish brown and, like other sturgeons, has a downward-facing mouth. It often migrates more than 620 miles (1,000 km) from the sea, laying eggs that stick to the riverbed.

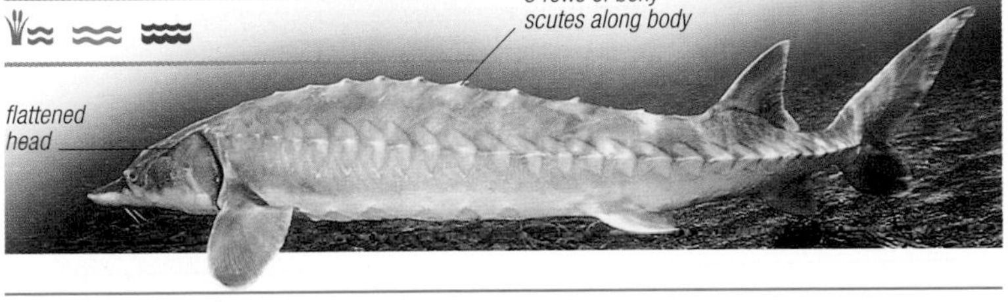

Polyodon spathula

American paddlefish

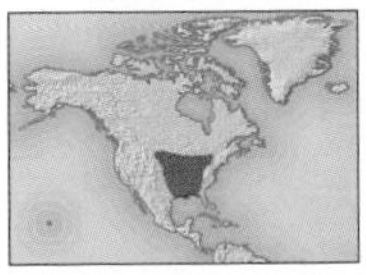

Length 4–6 ft (1.2–1.8 m)
Habit Up to 53 lb (24 kg)
Sex Male/Female
Status Vulnerable

Location C. to S.E. North America

One of the few freshwater fishes that feed by straining plankton from water, the American paddlefish is often described as a living plankton net. It sweeps through the water with its lower jaw dropped and the sides of the head inflated to form a funnel-like opening, filtering large amounts of water and extracting plankton with its gills. The underside of its distinctive paddle-shaped snout is covered with electrosensitive pores. This fish has been heavily exploited for its flesh and eggs.

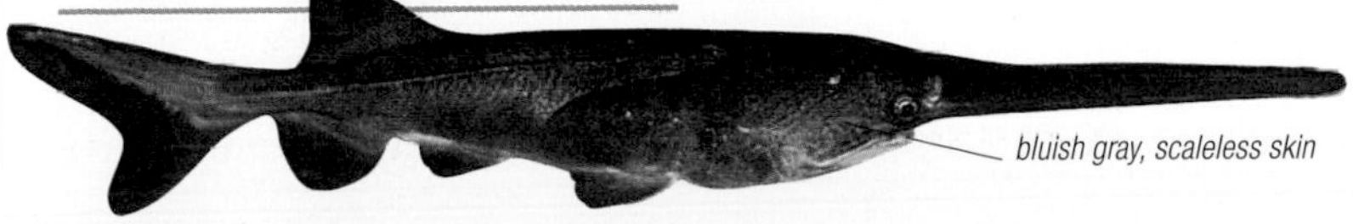

Polypterus ornatipinnus

Ornate bichir

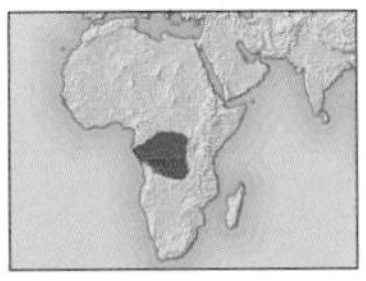

Length 16 in (40 cm) or more
Weight Up to 18 oz (500 g)
Sex Male/Female
Status Not evaluated

Location W. to C. Africa

Bichirs are cigar-shaped fishes with triangular finlets arranged in a row down the rear half of their backs. They have armorlike scales and tubular nostrils, and can breathe atmospheric air since their swim bladder functions much like a primitive lung. The ornate bichir—one of 12 species in the bichir family—is beige, covered by black, netlike markings. It is a slow swimmer that stalks its prey, such as small fishes, amphibians, and crustaceans, quickly sucking them in when in range.

Lepisosteus osseus

Longnose gar

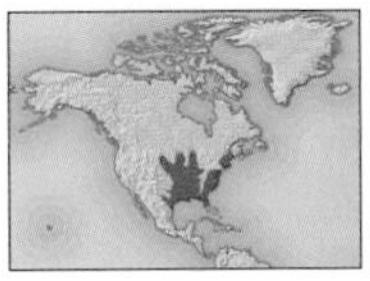

Length 4–6 ft (1.2–1.8 m)
Weight Up to 33 lb (15 kg)
Sex Male/Female
Status Not evaluated

Location C. to E. North America

The longnose gar is one of a group of primitive North American fishes that have long jaws with sharp teeth. Its body is long, and its propulsive fins are set far back—much like those of pikes (see p.502). It is primarily a freshwater fish, but in the southern part of its range, adults are frequently found in brackish water. The longnose gar hangs motionless in the water, hidden by vegetation, and waits for prey to come within striking distance. Then, with a sudden thrust, it takes the prey cross-wise in its mouth, often holding it for several minutes. Its predatory habits make it unpopular among fishermen since it can become entangled in nets and cause a good deal of damage.

Bony-tongued fishes

PHYLUM	Chordata
CLASS	Osteichthyes
SUBCLASS	Actinopterygii
ORDER	Osteoglossiformes
FAMILIES	7
SPECIES	219

The bony-tongued fishes are named after the tooth-studded tongue and palate found in most species. These relatively large fishes—up to 9¾ft (3 m) long—live mainly in freshwater in the tropics. This group contains such varied forms as the large Amazonian arapaima, the peculiar elephantnose fish (which has an extended, downward-curving mouth), the eel-like knifefish, and the delicate and ornate freshwater butterflyfish. Most are carnivores, feeding chiefly on invertebrates, but a few species eat detritus and plant material.

Anatomy

Although some species also have teeth in their jaws, all bony-tongued fishes have many small, sharp teeth on both their tongue and the roof of their mouth, which are used to exert the main biting motion when seizing prey. Most species in this group have a relatively elongated body with prominent eyes and large, hard scales, often with fine ornamentation, although some have reduced scales. The dorsal and anal fins are usually placed far back on the body and are often very long. Some fishes in this group, particularly the elephantnose fish and the knifefish, can produce and detect electrical signals, which they use for navigation and communication. In some species, such as the arapaima and the freshwater butterflyfish, the swim bladder has a lunglike lining so that it can be used as a temporary lung in oxygen-poor water.

ELECTRIC ELEPHANTNOSE
Elephantnose fishes navigate using electrical signals. An electrical field is created around the fish's body by modified muscles, and disturbances in this field are detected by receptor cells.

Chitala chitala

Clown knifefish

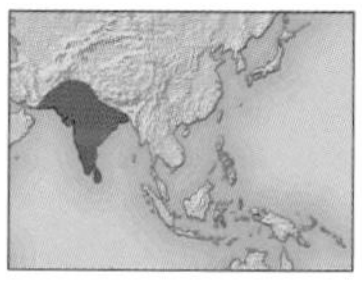

Location S. Asia

Length Up to 34 in (87 cm)
Weight Not recorded
Sex Male/Female
Status Not evaluated

With a body flattened from side to side, adult clown knifefishes are conspicuously hump-backed, whereas the young are relatively slender. The unusually small dorsal fin is featherlike, giving rise to this fish's other common name, the featherback. This fish swims equally well forward or backward, with an undulating motion of its large anal fin—the main propulsive organ. It has a swim bladder modified to form an accessory breathing organ, and frequently rises to the surface to obtain air. It is often seen at the surface, splashing as it rolls over, exposing its silvery flanks.

Pantodon buchholzi

Freshwater butterflyfish

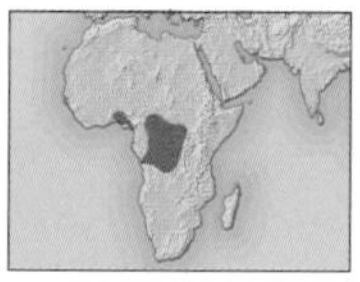

Location C. Africa

Length 4–5 in (10–13 cm)
Weight Up to 18 oz (500 g)
Sex Male/Female
Status Not evaluated

This fish is a surface-feeding insect-eater, able to leap 3¼ft (1 m) or more out of water and glide for short distances using its enlarged, winglike pectoral fins. In addition to hanging motionless at the surface, it is often seen standing for long periods in shallow water, using its pelvic fin rays as stilts. It has a large, lunglike swim bladder, believed to function as an additional air-breathing organ.

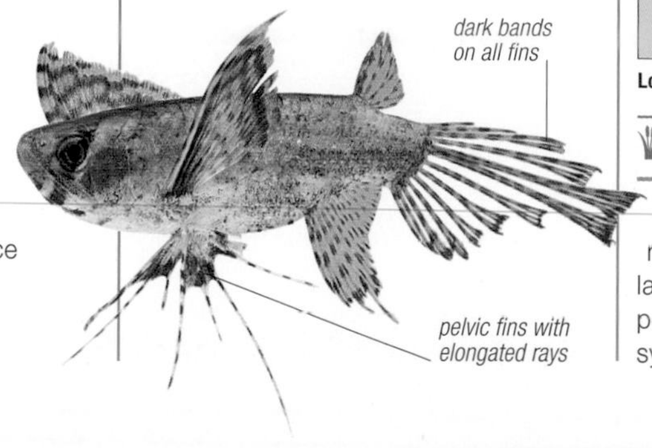

Arapaima gigas

Arapaima

Location N. South America

Length 8¼ft (2.5 m), max 15 ft (4.5 m)
Weight Up to 440 lb (200 kg)
Sex Male/Female
Status Data deficient

One of the largest purely freshwater fishes in the world, the arapaima has a streamlined, gray to dark greenish yellow body with long dorsal and anal fins set back close to its tail, a common feature among fishes that make sudden lunges to catch their prey. Apparently unable to fulfil its respiratory requirements with its gills alone, it takes in air from the surface of the water.

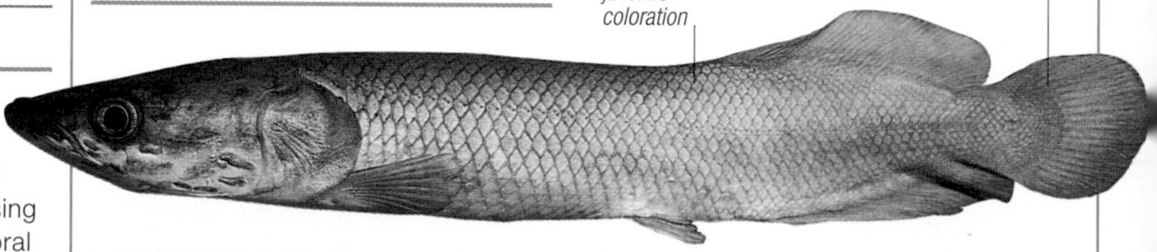

Gnathosomus petersii

Elephantnose fish

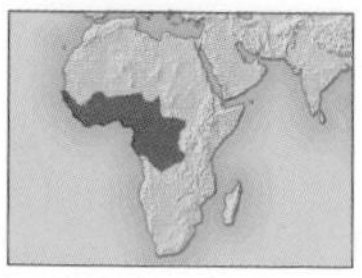

Location W. to C. Africa

Length Up to 9 in (23 cm)
Weight Up to 2¼ lb (1 kg)
Sex Male/Female
Status Not evaluated

Commonly found in still, murky, muddy waters of rivers, lakes, and swamps, this fish possesses a weak electrical system that functions as a kind of radar to detect obstacles, food, and mates. The highly mobile, fingerlike appendage on its chin is used for probing the muddy bottom in search of food. The elephantnose has an unusually large brain that, in relation to body mass, is equalled only by that of humans. Perhaps for this reason, it has a remakable ability to learn, and its playful behavior makes it a popular aquarium fish.

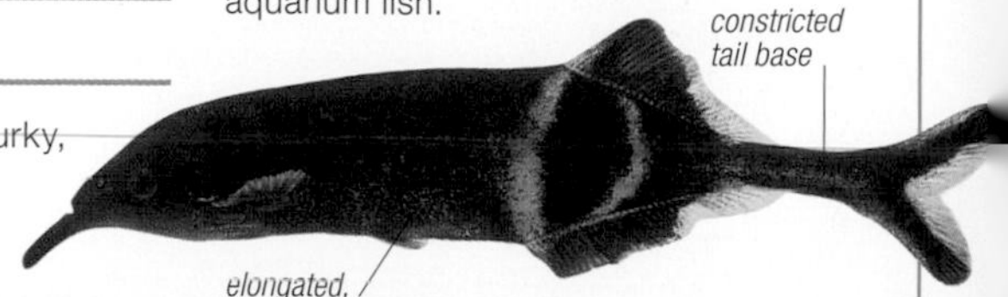

Tarpons and eels

PHYLUM	Chordata
CLASS	Osteichthyes
SUBCLASS	Actinopterygii
SUPERORDER	Elopomorpha
ORDERS	5
FAMILIES	24
SPECIES	About 1,000

This group consists of 3 main lines: eels, tarpons, and halosaurs. Although they look and behave differently as adults, they all start life as a transparent larva (known as a leptocephalus larva). These long, thin larvae may drift in ocean currents for several years before reaching adulthood. Most eels are marine, but a few live mainly in freshwater and return to the sea to breed. Tarpons are largely confined to coastal areas of the tropics, while halosaurs occur in oceans worldwide, down to depths of 6,000ft (1,800m).

Anatomy

Eels are elongated and snakelike. The spinal column consists of more than 100 vertebrae, which makes the body very flexible. Long dorsal and anal fins may extend most of the length of the body, and the pectoral and tail fins are often very small or lost entirely. There are no pelvic fins. Most eels are scaleless, but some species have fine scales embedded in their skin. Tarpons look more like other fishes than like eels. Their bodies are deeper and less elongated, with large, reflective scales, and well-developed and differentiated fins. Halosaurs, however, are more eel-like in shape, with a large head that tapers into a tubular body and elongated tail. They are covered in small scales and have reduced fins, some of which have sharp spines.

Breeding and migration

The breeding cycles of some eels involve remarkable migrations. Most marine eels spawn near where they live, but freshwater eels and some marine species have separate spawning grounds, usually far out to sea. For example, North American and European eels, both of which live in freshwater lakes and rivers, travel over 4,000 miles (6,400km) to different parts of the Sargasso Sea in the western Atlantic. It is thought that during this journey, which takes 4–7 months, they do not eat. Instead, they survive by consuming stored fat and muscle tissue. They may also absorb nutrients through the lining of their mouth. Spawning occurs in deep water, after which the adults die. The leptocephalus larvae are carried passively toward coastal waters by ocean currents. After about 3 years, they develop fins, scales, and pigmentation before beginning their journey upriver to the areas where they grow as adults.

Hunting and feeding

Tarpons, halosaurs, and eels are all predators. The fast-swimming tarpons use speed to catch other fishes. Eels are more likely to ambush their prey, using rocks or reef crevices for cover. Many eels have long, sharp teeth for seizing prey; others have thickened teeth for crushing shells. Some species search the bottom for buried prey or scavenge on dead animals. Deep-sea gulper eels have gigantic mouths and distensible stomachs, and can eat prey much larger than themselves. Halosaurs feed on invertebrates.

HUNTING
Most eels (such as the moray eel, shown here) ambush prey rather than chase it as tarpons do. Moray eels are found in shallow coastal waters of tropical seas. They often hide in crevices, from which they launch surprise attacks when a suitable prey is within reach. The wide mouth is armed with sharp teeth, making it difficult for prey to escape once it has been caught.

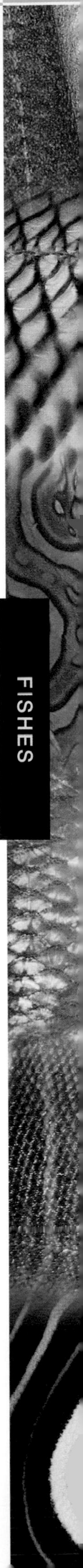

Elops saurus

Ladyfish

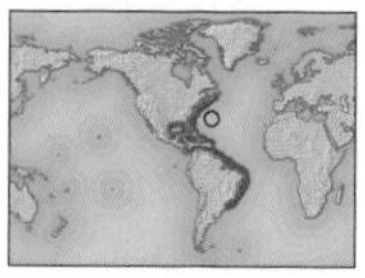

Length Up to 3¼ ft (1 m)
Weight 31 lb (14 kg)
Sex Male/Female
Status Not evaluated

Location W. Atlantic, Gulf of Mexico, Central America, Caribbean

Also known as the ten-pounder, this sleek, blue-gray fish is covered with fine scales. It attacks small fishes in schools, and is well known for its habit of skipping along the surface and for leaping when hooked. Adult ladyfishes move out from the coast into open water to spawn in schools; exact locations are not known, but they are believed to spawn up to 100 miles (160 km) offshore. The transparent, eel-like larvae eventually drift or migrate back to inshore waters.

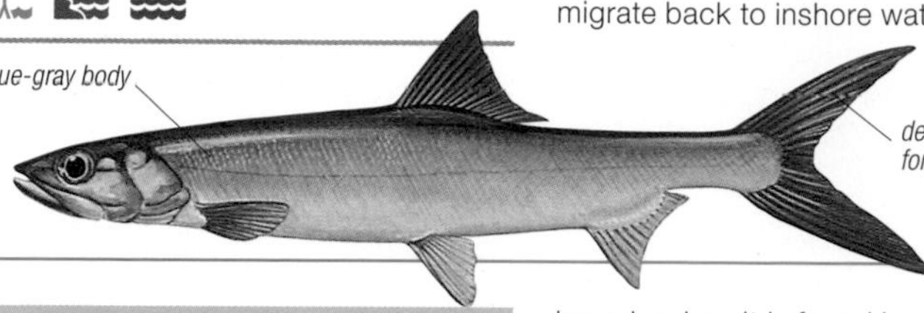

Megalops atlanticus

Tarpon

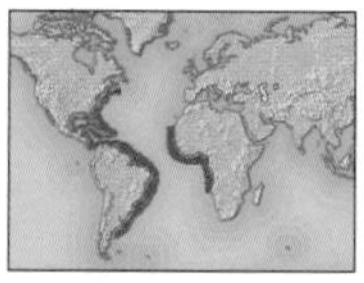

Length ¼–8¼ ft (41.3–2.5 m)
Weight 350 lb (160 kg)
Sex Male/Female
Status Not evaluated

Location W. and E. Atlantic

A large and powerful fish, with a deep body and an upturned mouth that extends far beyond the eyes, the tarpon looks like an unusually large herring. It is found in coastal waters and estuaries, and sometimes also in freshwater, and feeds almost exclusively on schooling fishes such as sardines, anchovies, and mullets. Its lunglike swim bladder enables it to overcome the occasional shortages of oxygen that occur in estuaries and similar kinds of habitat, by breathing air at the surface. It is a popular sport fish, and makes spectacular leaps when hooked. Between late April and August, adult tarpons move into open water to spawn. The transparent, eel-like larvae ultimately end up in estuarine nursery grounds, but also frequently occur in pools or lakes disconnected from the sea.

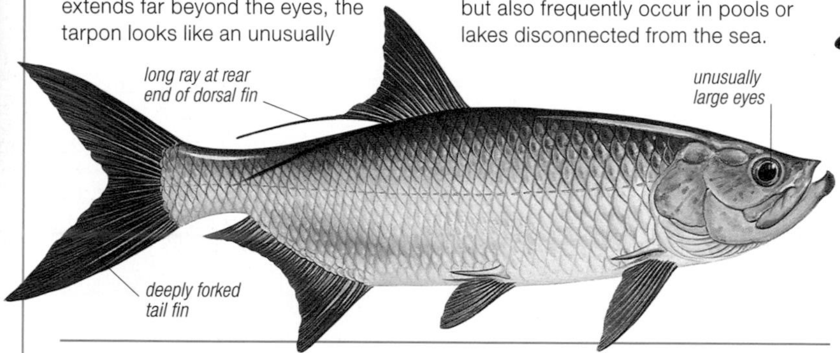

Avocettina infans

Avocet snipe eel

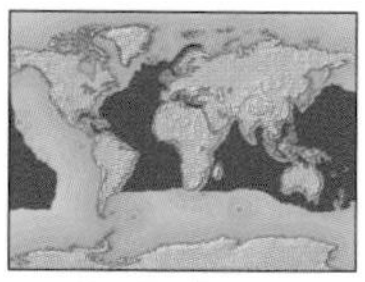

Length Up to 30 in (75 cm)
Weight 1 lb (0.5 kg)
Sex Male/Female
Status Not evaluated

Location Tropical, subtropical, and temperate waters worldwide

A deep-sea eel with a remarkably slender, brown or black body ending in a whiplike tail, the avocet snipe eel belongs to a family of about 9 species. Its long, thin jaws cannot be fully closed and have numerous teeth. This eel probably feeds exclusively on small crustaceans. Immature male and female avocet snipe eels look very similar, but as the males approach sexual maturity, they undergo a marked transformation. Their jaws become drastically shortened, and all their teeth are lost. At this stage of the life cycle, males and females look so different that at one time they were classed in separate families. Very little is known about the avocet snipe eel and its relatives; individuals are usually only captured by specially designed, deep-sea fishing gear during scientific research expeditions.

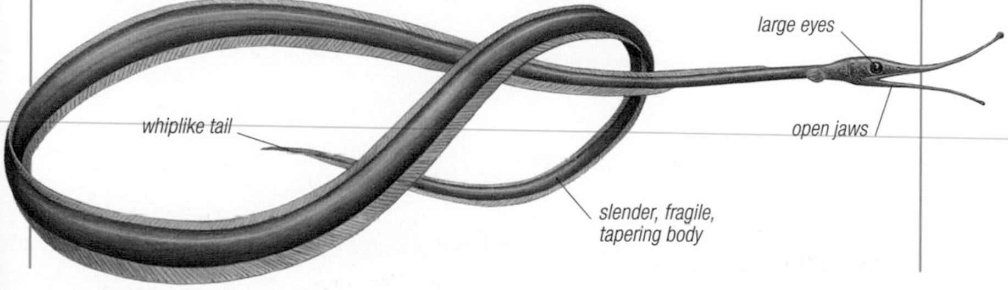

Anguilla anguilla

European eel

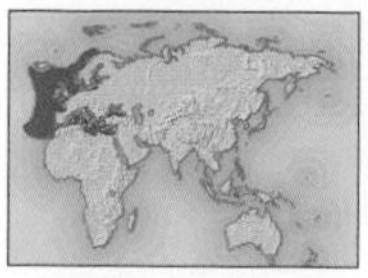

Length Up to 3¼ ft (1 m)
Weight 10 lb (4.5 kg), max 29 lb (13 kg)
Sex Male/Female
Status Critically endangered

Location E. North Atlantic, Mediterranean, Europe

This eel has a remarkable life history. Adults are believed to spawn in the area of the Sargasso Sea, the larvae being transported over a period of 2½ years to European coastal waters by ocean currents. There, they metamorphose into cylindrical, unpigmented "glass eels" before entering rivers in their millions, as silvery, pigmented juveniles called "elvers." The freshwater stage is a feeding and growing phase; as the eel matures sexually, it descends toward the sea. Mature "silver eels" undertake their spawning migration usually from late summer to winter and mainly on dark, moonless, stormy nights. At this stage, the eyes enlarge, the snout becomes more narrow and pointed, and the pectoral fins more lance-shaped—changes that are adaptations to the ocean's depths. Once an extremely common fish, the European eel has declined dramatically in recent decades. Pollution, dams, and overfishing are serious threats, together with climate change.

OUT OF WATER

European eels can survive out of water for several hours in damp, cool conditions. Adults usually travel overland on rainy nights, when the darkness gives them some protection from attack.

ADULT COLORATION
In its freshwater stage, this fish has a yellowish or even golden underside. As it approaches sexual maturity, it becomes silvery below and almost black above (as shown here).

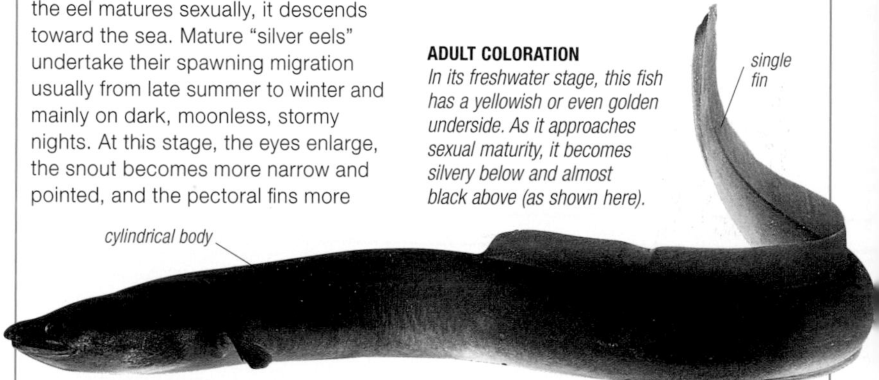

Gymnomuraena zebra

Zebra moray

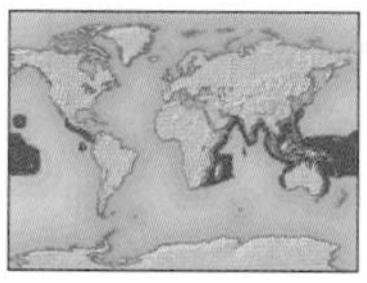

Length Up to 35 in (89 cm)
Weight 22 lb (10 kg)
Sex Male/Female
Status Not evaluated

Location Pacific, Gulf of California, Indian Ocean

Morays are thick-bodied and often powerful eels that typically lurk in clefts in rocks or coral with only their heads showing. These fishes have no pectoral or pelvic fins. There are 200 species, and many, including the zebra moray, are brightly colored. This snakelike fish has a long, cylindrical, highly muscular body, and a large mouth with blunt and pebblelike teeth, adapted for crushing prey. It usually adopts a lie-and-wait feeding technique, and typically lurks in rock or coral crevices, lunging outward to catch its prey. However, it may occasionally leave its hole at night in search of prey. The zebra moray feeds on large and heavily armored prey, particularly crabs and other crustaceans, found around reefs. There is evidence that it can eat larger crabs than other morays of comparable size. An aggressive fish, it is known to defend its territory strongly against intruders. Some morays, although not necessarily this species, are noted for the presence of toxins on their skin.

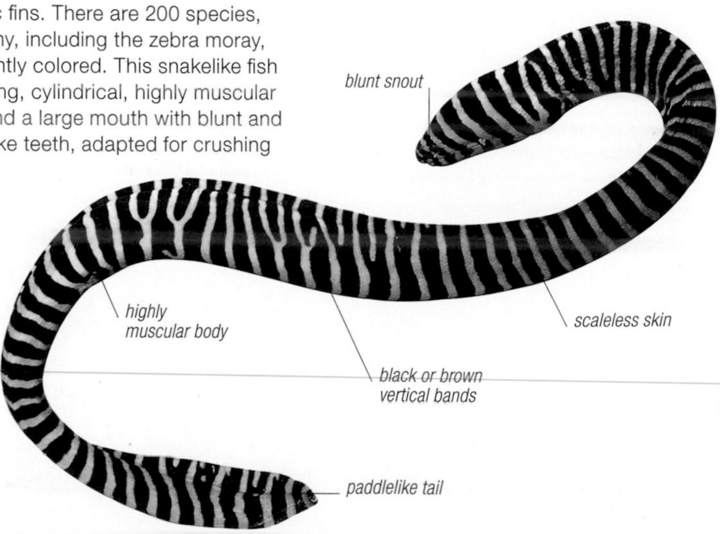

FISHES

Myrichthys colubrinus

Banded snake eel

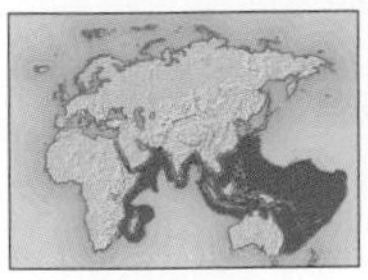

Length Up to 35 in (88 cm)
Weight 6½ lb (3 kg)
Sex Male/Female
Status Not evaluated

Location Indian Ocean, W. Pacific

There are over 300 species of snake eels worldwide and they occur in a wide range of colors and shapes. They are mostly less than 3¼ ft (1 m) long, and burrow in shallow, sandy and muddy bottoms of lagoons, sandy flats, and reefs. This species has a hardened, sharply pointed tail, modified for burrowing tail-first into the sand. Primarily active at night, it feeds on sand-dwelling fishes and crustaceans, and seems to detect its prey by smell.

Heteroconger hassi

Spotted garden eel

Length Up to 14 in (36 cm)
Weight 4½ lb (2 kg)
Sex Male/Female
Status Not evaluated

Location Red Sea, Indian Ocean, W. Pacific

This eel lives in large colonies, with the lower half of its body buried in the sandy seabed. The upper half of its body projects into the water, creating "gardens" that sway gracefully with the current. When a colony is approached by a diver, the eels nearest the diver slowly sink, each into its individual burrow, to hide from the intruder, in perfect gradation from one end of the "garden" to the other. The spotted garden eel feeds by picking up small organisms, such as tiny planktonic invertebrates or larval fishes, one by one, from the water flowing by.

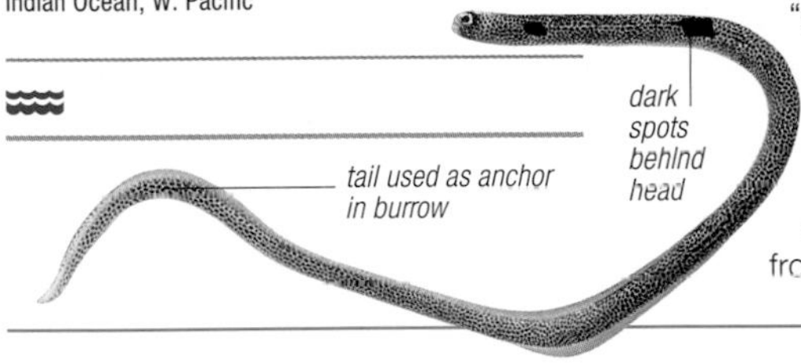

Saccopharynx ampullaceus

Gulper eel

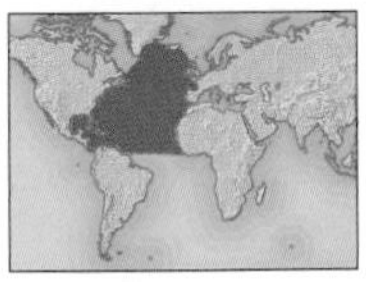

Length Up to 5¼ ft (1.6 m)
Weight 2¼ lb (1 kg)
Sex Male/Female
Status Not evaluated

Location North Atlantic

The gulper eel is a relative of the pelican eel (see right), with a similar small head and large jaws (although smaller than the pelican eel's). It also has a luminescent organ at the tip of its tail which, if dangled in front of its mouth, may be used to entice prey. It is thought that this fish swims slowly, with its enormous jaws swung open so that its prey literally swims into its mouth. A little-studied fish, the gulper eel belongs to a rare family, known from fewer than 100 specimens.

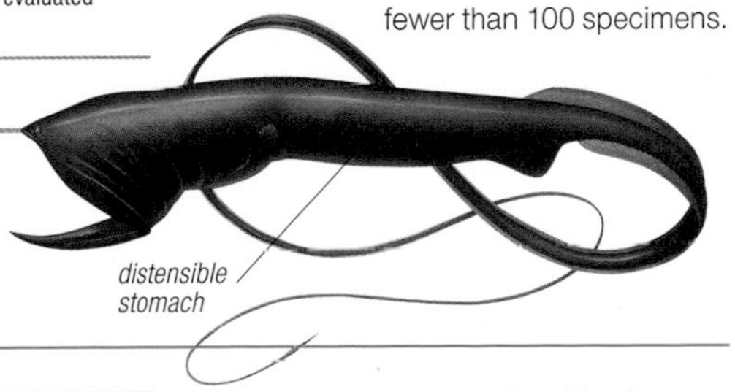

Albula vulpes

Bonefish

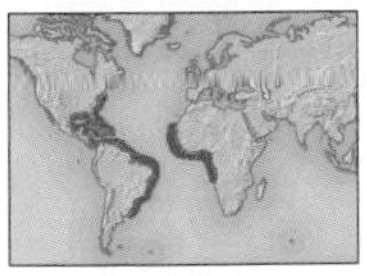

Length Up to 1 m (3¼ ft)
Weight 9 kg (20 lb)
Sex Male/Female
Status Not evaluated

Location Atlantic

With its slender, streamlined body, long head, and high dorsal fin, the bonefish is very much a "standard" fish. Found in shallow, coastal marine habitats, typically over mud flats and mangrove lagoons, it is more important as a sport fish than as food. It feeds on a rising tide, working its way to the muddy bottom with its head down and tail up, blowing jets of water to uncover its prey—usually clams, crabs, and shrimps. Its large size protects it from marine predators.

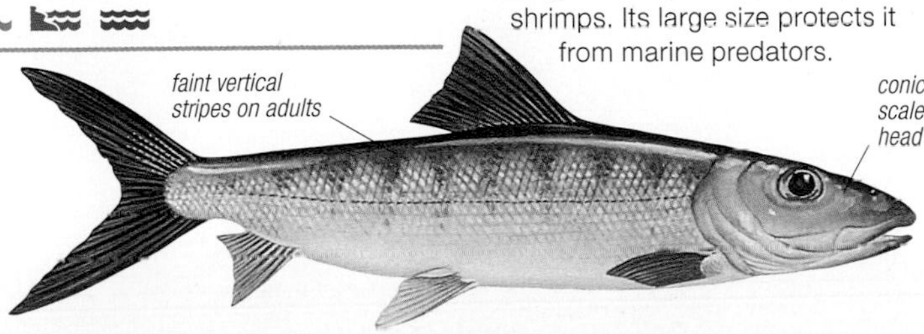

Conger conger

Conger eel

Length 8¾ ft (2.7 m)
Weight 145 lb (65 kg)
Sex Male/Female
Status Not evaluated

Location E. North Atlantic, Mediterranean

This dark gray, thick-bodied fish is one of nearly 200 species in the conger family. As with most eels, it has no scales. With its large size and well-equipped jaws providing protection from most predators, it hunts at night for fishes, crustaceans, and cephalopods. The conger eel is considered an excellent food fish and is caught by anglers in large numbers. It can be found on rocky or sandy bottoms, at depths of up to 330 ft (100 m), but in summer, adult fishes migrate into deeper, offshore waters to spawn. The female conger eel lays between 3 and 8 million eggs, which hatch to produce transparent, eel-like larvae. These drift inshore for 1–2 years before growing into juvenile eels; they reach sexual maturity at 5–15 years.

IN HIDING

Like most of its relatives, the conger eel hides during the day in crevices, emerging only at night to ambush prey. Small conger eels are common in tide pools on rocky shores, as well as in rocky offshore habitats, while large ones often inhabit wrecks of sunken ships.

SNAKELIKE FISH

The conger eel, with its snakelike form, has a long snout, and its gill openings are restricted to a small, crescentlike slit on each side of its body.

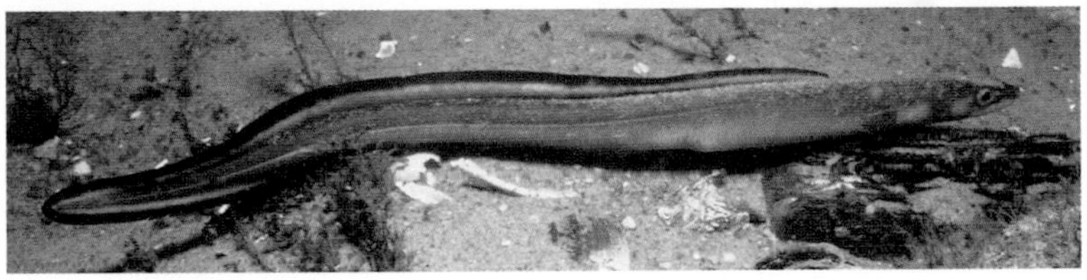

Aldrovandia affinis

Halosaur

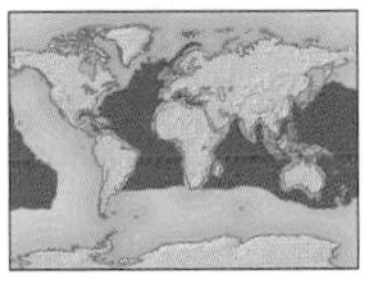

Length Up to 22 in (55 cm)
Weight 13 lb (6 kg)
Sex Male/Female
Status Not evaluated

Location Tropical, subtropical, and temperate waters worldwide

This relatively rare fish is found hovering over the bottom in water between 2,300 and 6,600 ft (700–2,000 m) deep. This species is usually only captured by specially designed, deep-sea fishing gear during scientific research expeditions. Little is known about the halosaur but its well-developed nasal organs, particularly among the males, seems to indicate that sense of smell is a means of communication between the sexes.

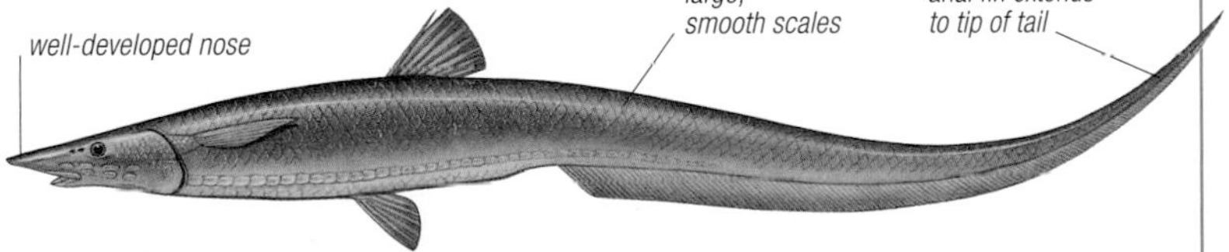

Eurypharynx pelecanoides

Pelican eel

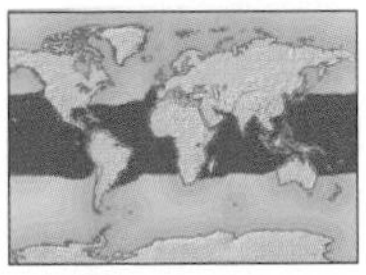

Length 60–100 cm (23½–39 in)
Weight 1 kg (2¼ lb)
Sex Male/Female
Status Not evaluated

Location Tropical and subtropical waters worldwide

This deep-sea fish is one of the most unusual of eels, with a thin, tapering body, but enormous jaws equipped with tiny teeth. Its huge gape and expandable stomach enable it to swallow and digest fishes almost as large as itself. Due to its odd shape and fragile build, the pelican eel is almost certainly a poor swimmer and may not pursue other fishes. Instead, it probably relies heavily on the luminescent organ at the tip of the tail to attract prey.

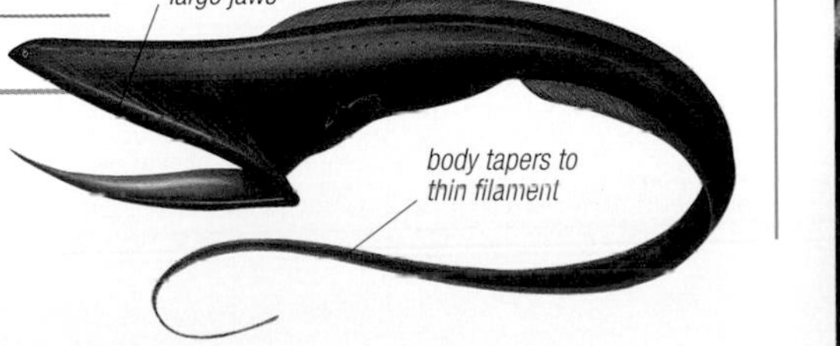

FISHES

Herrings and relatives

PHYLUM	Chordata
CLASS	Osteichthyes
SUBCLASS	Actinopterygii
SUPERORDER	Clupeomorpha
ORDERS	1 (Clupeiformes)
FAMILIES	6
SPECIES	392

These small, streamlined fishes are widespread and abundant. Often forming vast schools, they are an important part of marine food chains and are vital to many coastal economies. The herring family includes sprats, shads, and pilchards; it also includes sardines, although this term is often used to refer to the young or adults of other species. Herrings and anchovies occur from the tropics to temperate zones, in shallow, coastal marine waters and freshwater streams and lakes. Some species live in marine waters but spawn in brackish or freshwaters.

Anatomy

Members of the herring family are similar to one another in overall appearance. They have streamlined bodies that are flattened from side to side, with large scales and well-developed fins without any supporting spines. Most are silver with green or blue hues on the back, becoming progressively lighter on their undersides. Anchovies are generally longer and thinner than herrings, and have a large mouth with a prominent snout. In both anchovies and herrings, the swim bladder is linked to the ear apparatus, which is thought to enhance their hearing. At one time, herrings and their relatives were considered to be related to tarpons and eels because of similarities between their larvae, but now the consensus is that they are sufficiently different to form a group of their own.

CONSERVATION

Herrings have a long history of human exploitation. Easy to catch and to preserve, they are an important source of food and oil. During the days of sail, catches were relatively stable, but the herring harvest soared in the first half of the 20th century, with the introduction of powered fishing vessels and long drift nets. In the 1970s, the Atlantic herring population crashed. It has since recovered as a result of fishing limits and continues to grow.

Feeding

Almost all herrings and their relatives feed on planktonic organisms (mainly crustaceans and their larvae), which they filter out of the water by opening their large mouths to expose their gill rakers. Following the daily vertical movements made by plankton, they usually feed near the surface at night and move to lower depths during the day. Their feeding habits often vary seasonally; herrings and some other species cease to feed during the breeding season. All members of this group feed in dense groups of tens of thousands of fishes, thus attracting predators, including other fishes (such as salmon and tuna), seabirds, and marine mammals. By contributing to the diet of so many other species, herrings and anchovies form a key element in many marine ecosystems.

SCHOOLING
Herring schools are at their largest during the breeding season, which usually falls in the warmer months of the year. Each female may release up to 40,000 eggs, which are attached to the seabed.

Clupea harengus

Atlantic herring

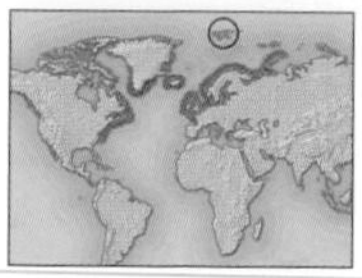

Length Up to 16 in (40 cm)
Weight Up to 25 oz (700 g)
Sex Male/Female
Status Not evaluated

Location North Atlantic, North Sea, Baltic Sea

For centuries, the sleek, silver-scaled herring has been one of the most important commercially fished species in the North Atlantic. Its numbers declined precipitously during the 20th century, due to improved fishing techniques, but are now recovering slowly because of active management. Like most of its relatives, the Atlantic herring is a plankton-feeder, and lives in large, highly mobile schools. It comes to the surface at night, but spends the day in deeper water. Across its wide range, the species is divided into many local races, which differ from each other in habits and in size. Each race uses a number of traditional spawning grounds. Young herrings, which initially resemble tiny eels, hatch on the seabed, but soon swim upward to feed close to the surface. They are eaten in vast numbers by other fishes, and only a tiny proportion of them survive to become adults.

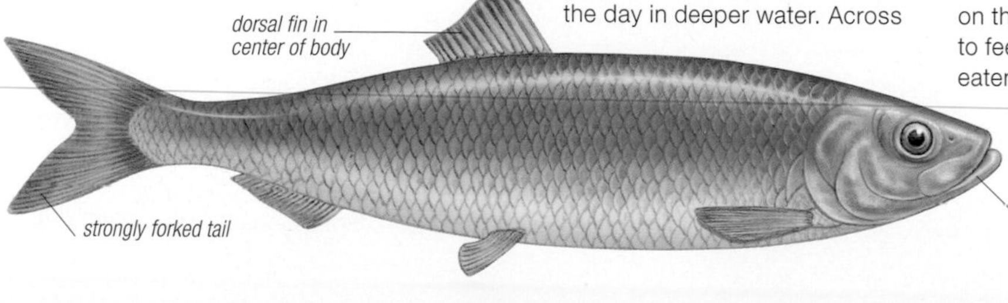

Sardinops sagax caeruleus

California pilchard

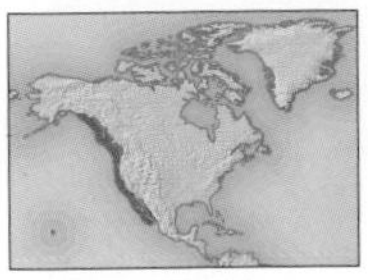

Length 10–14 in (25–36 cm)
Weight 17 oz (475 g)
Sex Male/Female
Status Not evaluated

Location E. North Pacific

This medium-sized silvery fish can be recognized by its distinctive black spots, arranged in rows along its body. In California, it is canned and marketed as the sardine, and is also extensively used to produce oil and fish meal. At one time, schools consisting of an estimated 10 million individuals were found. During the 1930s, over half a million tons a year were caught, but stocks dwindled during the 1950s, and collapsed completely around 1967. Today, the numbers of California pilchard are recovering, but stocks are still far from their original levels.

Alosa sapidissima

American shad

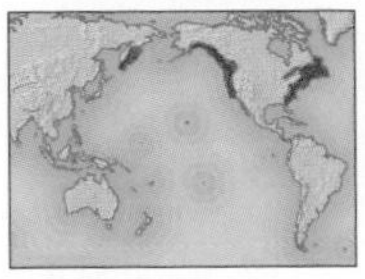

Length 20–23½ in (50–60 cm)
Weight 4½ lb (2 kg)
Sex Male/Female
Status Not evaluated

Location W. North Atlantic, E. North America. Introduced to E. and W. Pacific

With its streamlined silvery body, the American shad looks like many other fishes in the herring order but it is distinctive in that it enters rivers to spawn. Beginning at the age of 4 or 5, American shad migrate to their natal river every year to spawn. Spawning occurs near the shore, after sunset, usually in late summer. Males arrive first, followed by the females, and pairs swim erratically near the surface with their dorsal fins showing. Adults return to the ocean after spawning, and the young shad hatched out in the river migrate to the oceans in autumn. The numbers of American shad have suffered because of dam construction and water pollution. The Pacific populations of this fish have been introduced.

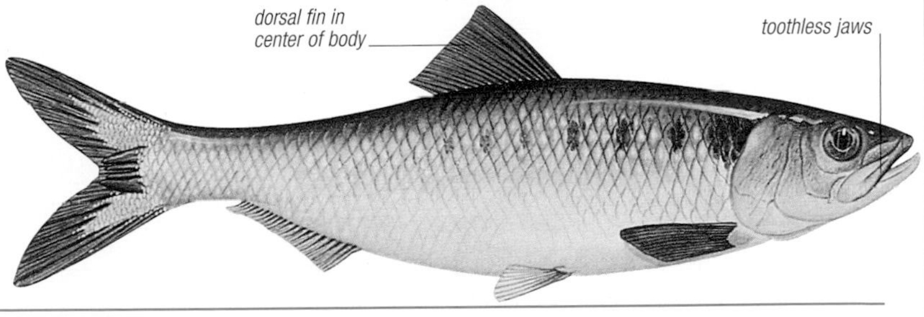

Engraulis ringens

Peruvian anchoveta

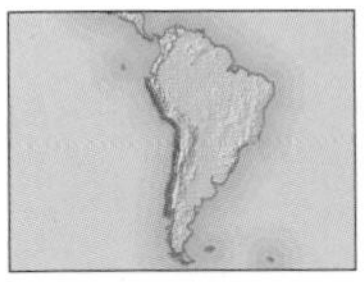

Length Up to 8 in (20 cm)
Weight ⅞ oz (25 g)
Sex Male/Female
Status Not evaluated

Location E. South Pacific

Like its close relative the Californian anchovy (see right), this small, silvery fish feeds on plankton, and forms enormous schools where upwelling currents create a good supply of food. Following the plankton as they migrate, they descend to depths of about 165 ft (50 m) during the day and rise to the surface at night. Peruvian anchovetas also school in huge numbers for safety. This fish spawns largely from winter to early spring.

SLIM FISH
This slender fish, with a pointed snout and deeply forked tail, has a marked resemblance to the Californian anchovy. It swims with an open mouth, filter feeding with the help of its fine gill rakers.

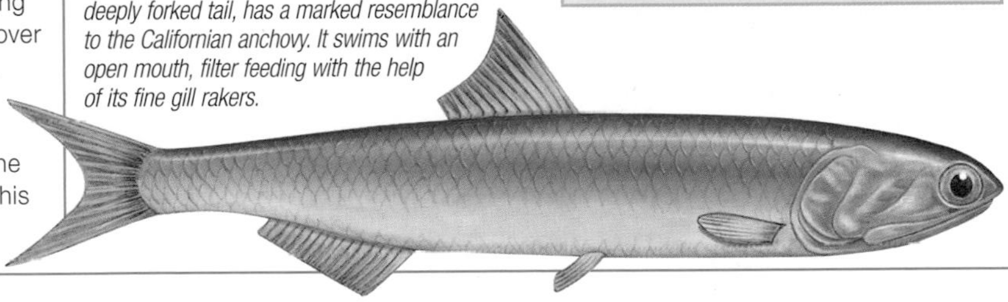

CONSERVATION

Despite its small size, the Peruvian anchoveta is one of the world's most commercially valuable fishes, with up to 11 million tons (10 million tonnes) caught each year. However, its population is prone to sudden crashes, triggered both by El Niño—a periodic temperature rise in the East Pacific—and by overfishing. The last major crash occurred in the early 1980s, with severe repercussions for local people and marine wildlife.

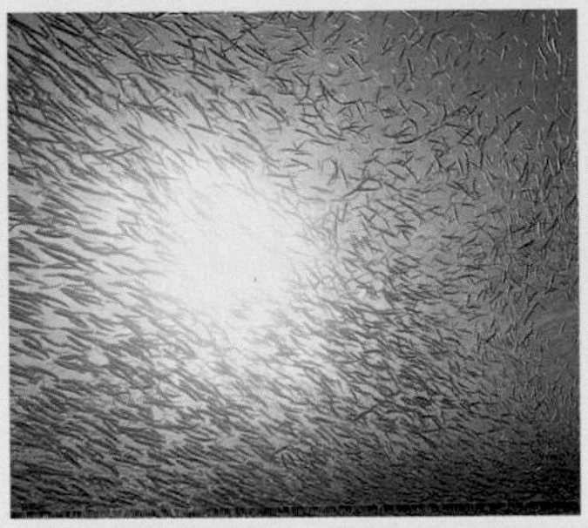

Engraulis mordax

Californian anchovy

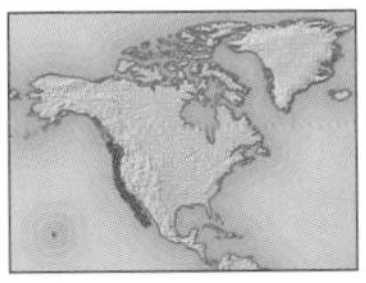

Length Up to 10 in (25 cm)
Weight 1¾–2⅛ oz (50–60 g)
Sex Male/Female
Status Not evaluated

Location E. North Pacific

This small but abundant fish is important commercially and a vital link in the marine chain, providing food for other fishes and fish-eating birds. Slender, with a distinctly pointed snout, it swims with its mouth open, filtering plankton from the water with its gill rakers. The Californian anchovy is one of about 150 species in the anchovy family, many of which are poorly known and difficult to tell apart. Anchovies are found in all the oceans of the world.

Denticeps clupeoides

Denticle herring

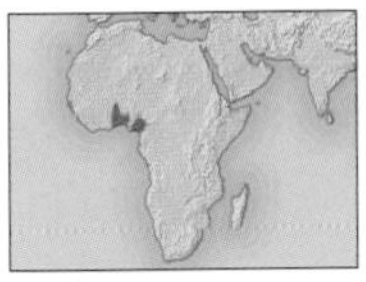

Length 3¼ in (8 cm)
Weight Not recorded
Sex Male/Female
Status Not evaluated

Location W. Africa

This small, silver-colored fish is the most primitive member of the herring group, being virtually indistinguishable from fossil forms. Modified scales along the underside form backward-pointing serrations, and toothlike denticles cover the head. It also has a large anal fin. Unlike most herrings, it lives in fast-flowing, freshwater. It is also the only herring to have a sensory line running along the side of its body. This fish is omnivorous.

Pristigaster cayana

Amazon hatchet herring

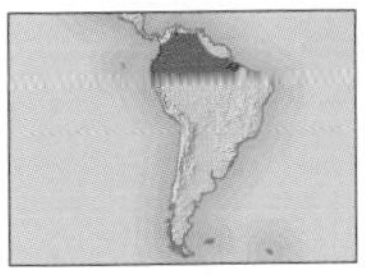

Length 5¾ in (14.5 cm)
Weight Not recorded
Sex Male/Female
Status Not evaluated

Location N. South America

This narrow-bodied, freshwater fish has a distinctive profile, with a deeply curving belly. Like many members of the herring family, its underside has a line of modified scales—or scutes—which have sharp, backward-facing points. Amazon hatchet herrings tend to remain in schools for safety and are omnivorous, filtering plankton from the water through their gill rakers. They inhabit the Amazon river basin as far west as Peru and Colombia. They are also believed to enter the slightly saline waters of the river mouth, but this is not well documented.

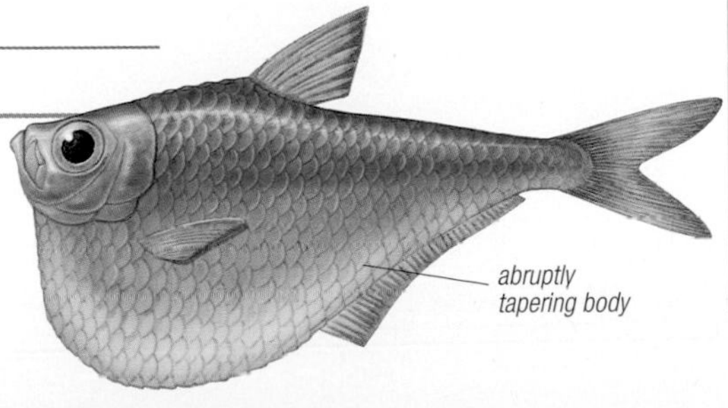

Chirocentrus dorab

Dorab wolf-herring

Length 3¼ ft (1 m)
Weight 14 oz (400 g)
Sex Male/Female
Status Not evaluated

Location Indo-Pacific

This is the largest member of the herring family. It has an elongated, bright blue body and numerous teeth, with 2 fanglike canines protruding outward from its upper jaw. Unlike most herrings, it actively hunts other fishes. Little is known about the behavior of this species, but the fact that it is a large carnivore strongly suggests that it hunts alone or in small groups, rather than in large schools. Along with the closely related whitefin wolf-herring, it is extensively fished in some parts of the world.

FISHES

Catfishes and relatives

PHYLUM	Chordata
CLASS	Osteichthyes
SUBCLASS	Actinopterygii
SUPERORDER	Ostariophysi
ORDERS	5
FAMILIES	73
SPECIES	About 9,600

Catfishes and their relatives account for three-quarters of all species of freshwater fishes. Although they are a varied group, almost all species have a bony internal structure, known as the Weberian apparatus, which is used in sound reception and contributes to their acute hearing. They are divided into 4 groups: milkfishes and relatives, cyprinids (the most widespread of which are carp and minnows), characins (including tetras, tigerfishes, and piranhas), and catfishes themselves. Although mostly confined to freshwater, some species are found in brackish water or the sea.

Anatomy

Catfishes and their relatives vary greatly in size and shape. Most are small fishes, less than 4 in (10 cm) long, although some species reach 16 ft (5 m). The Weberian apparatus that unites most members of this group consists of modified vertebrae that act as a set of levers to transmit sound waves received by the swim bladder to the inner ear. This feature gives catfishes and their relatives an acute sense of hearing, and many species use sound to communicate with each other.

Catfishes typically have a flattened head with a wide mouth that is usually surrounded by whiskerlike sensory barbels; these are used for touch and taste, and they compensate for poor eyesight. Most species have smooth, scaleless skin but in several groups the scales have developed into a dense armor. Nearly all species have sharp or serrated pectoral and dorsal spines, some of which can release venom.

Characins and cyprinids look similar to one another, although most characins are small and often brightly colored, while cyprinids are generally dull-colored and can reach lengths of up to 6½ ft (2 m). Another distinguishing factor is the arrangement of the teeth: characins have teeth in their jaws, often with a replacement row directly behind the functional teeth, whereas cyprinids have toothless jaws but usually have teeth in the pharynx. Characins, unlike cyprinids, also have a small fin (called an adipose fin) between the dorsal fin and tail, which is not supported by rays and stores fat.

Milkfishes and their relatives (which include the beaked salmon) are the only members of this group that do not have a Weberian apparatus. Instead, vibrations are transmitted from the swim bladder to the inner ear by a modified set of ribs.

Feeding

This group includes voracious predators, scavengers, herbivores, and filter feeders. Catfishes are generally predators or scavengers, living close to the bottom in murky water, although a few species are herbivores. Most characins are schooling or free-swimming fishes that eat plants or prey on invertebrates or other fishes. Most species of cyprinids have downward-turned, often suckerlike mouths, with one or more pairs of sensory barbels on either side. They usually feed on detritus or on small invertebrates and plants in the sediment; few are predatory. Milkfishes are filter feeders but some of their relatives feed on small animals.

NIGHT HUNTER
Most catfishes are active at night and feed alone. The barbels around the mouth are covered with taste buds, which help them find food. This long-whiskered (or pimelodid) catfish is found in tropical rivers in South America. The larger members of this group feed on fishes and other animals, including monkeys that fall into the water.

FISHES

Chanos chanos

Milkfish

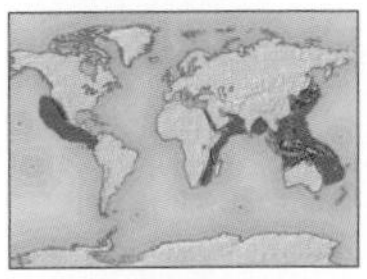

Length Up to 6 ft (1.8 m)
Weight Up to 31 lb (14 kg)
Sex Male/Female
Status Not evaluated

Location E. Pacific, North and South America, Indo-Pacific, Asia, Australia

A large, fast-moving filter feeder, the milkfish has a streamlined silvery body and an unusually large, deeply forked tail. It can tolerate a wide range of salinity but spends most of its life in freshwater, swimming out to sea to spawn. The larvae and juveniles develop in warm, coastal wetlands, in brackish and freshwater environments. Milkfish larvae are collected at sea and cultured in ponds for food in Southeast Asia and the west Pacific; adults are also prized as food fish.

Gonorhynchus gonorhynchus

Beaked sandfish

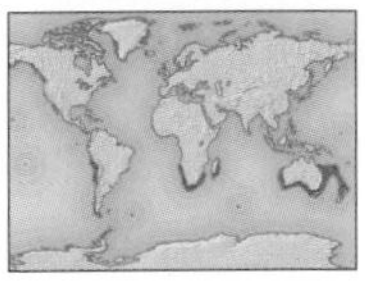

Length Up to 23½ in (60 cm)
Weight Not recorded
Sex Male/Female
Status Not evaluated

Location S.E. Atlantic, Indian Ocean, Pacific Ocean

This fish is adapted for life on the seabed, where it spends most of the daytime hiding in sand or mud, emerging at night to feed. Its upperside is well camouflaged, and its eyes are protected from abrasion by a thin covering of skin. It has silvery flanks, and the tips of its anal and tail fins are tinged with orange or pink. Its dorsal fin is set far back on the body. The beaked sandfish is found on sandy shores across the Southern Hemisphere, at depths of up to 660 ft (200 m). It feeds on invertebrates that live on the seabed, and its small, downward-pointing mouth lacks teeth and is surrounded by small sensory flaps, or papillae, that help it find food. This fish is also called a beaked salmon but this is misleading because it is not a true salmon.

Cyprinus carpio

Common carp

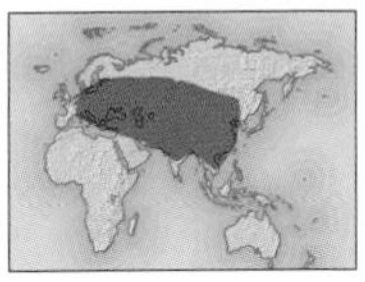

Length Up to 4 ft (1.2 m)
Weight Up to 82 lb (37 kg)
Sex Male/Female
Status Vulnerable

Location W. Europe to S.E. Asia

Originally from Europe and Asia, this deep-bodied, freshwater fish has been raised in semicaptivity for centuries, and introduced to rivers, lakes, and ponds in many parts of the world. It uses its protrusible mouth to grub through the bottom sediment for plant and animal food.

ORNAMENTAL CARP

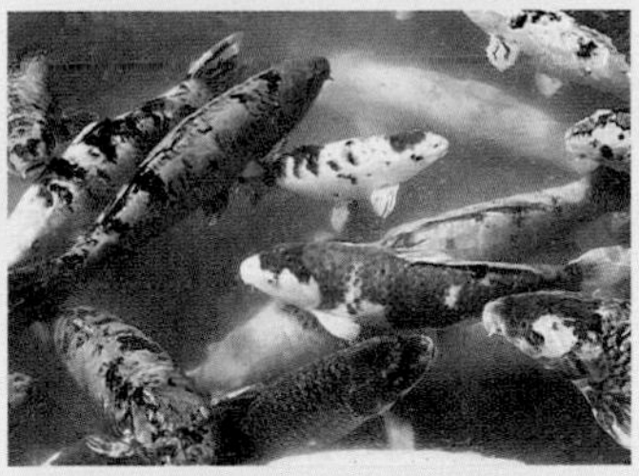

Selective breeding has created different varieties of common carp over the centuries. Mirror carp have large scales, while leather carp have no scales at all. Leather carp are raised for food; the multicolored variety shown here is kept as an ornamental fish.

TOOTHLESS JAWS
The jaws of the common carp have no teeth. It grinds food with the pharyngeal teeth at the back of the throat.

Rhodeus amarus

Bitterling

Length Up to 4¼ in (11 cm)
Weight Not recorded
Sex Male/Female
Status Least concern

Location C. and E. Europe

This small, freshwater fish lives in overgrown waterways, where oxygen levels are often too low for most fishes. The females are silvery, but during the breeding season the males become an iridescent silvery blue. Bitterlings have an unusual way of reproducing: using a long egg-tube, or ovipositor, the female lays her eggs in the mantle cavity of a freshwater mussel, over which the male sheds his sperm. The young grow in this "living nursery," without harming their host.

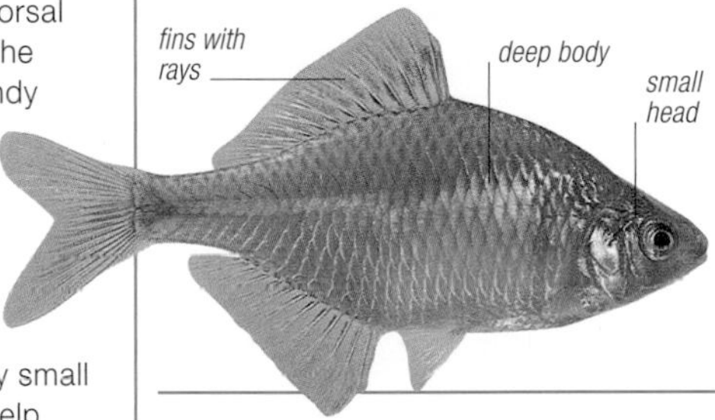

Brachydanio rerio

Zebrafish

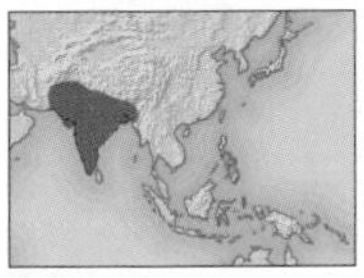

Length Up to 2¼ in (6 cm)
Weight Not recorded
Sex Male/Female
Status Not evaluated

Location S. Asia

Also known as the zebra danio, this highly active, freshwater fish has a slender, silvery yellow body, with a characteristic pattern of 4 horizontal stripes running from its head to the tip of its tail. This tiny fish feeds on freshwater invertebrates, including worms, insect larvae, and crustaceans. The female is slightly larger than the male, and breeds by scattering her eggs in open water, or on the bottom of ponds and pools. Zebrafishes are widely used as a model organism in studies of genetics. Several varieties of this popular aquarium fish have been created by breeders.

Myxocyprinus asiaticus

Chinese sucker

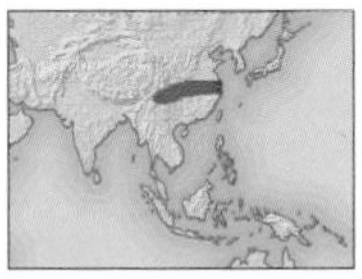

Length Up to 23½ in (60 cm)
Weight Not recorded
Sex Male/Female
Status Not evaluated

Location S.E. Asia

This slow-moving, hump-backed fish is one of the few Asian species in the sucker family. It has protrusible, fleshy lips, which it uses to collect small animals from the riverbed mud. Its body is triangular in cross section, with a pointed back, or hump, and a flat underside. It faces upstream to feed, so that the current pushes it against the riverbed—an adaptation that prevents it from being swept away.

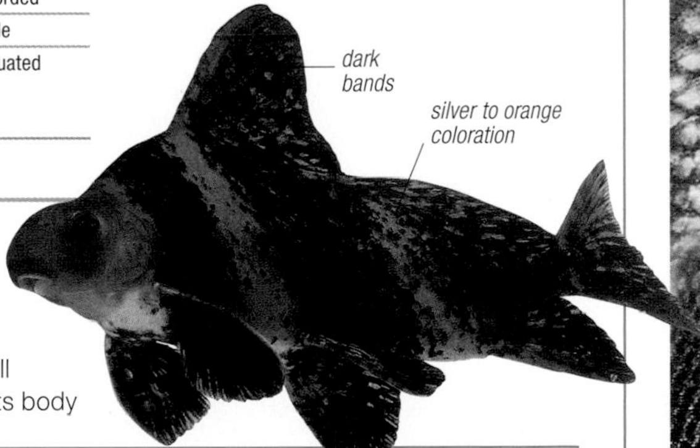

Chromobotia macracanthus

Clown loach

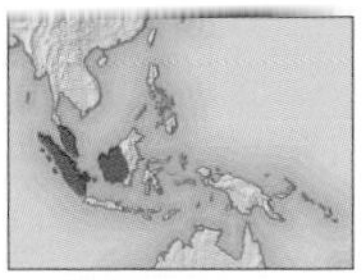

Length Up to 12 in (30 cm)
Weight Not recorded
Sex Male/Female
Status Not evaluated

Location S.E. Asia

A popular aquarium fish, the clown loach is a bottom feeder, feeling for aquatic invertebrates with the 4 pairs of barbels that hang down like a mustache from its mouth. Many loaches—including this one—can defend themselves by flicking out a sharp spine in front of each eye. They use these spines against potential predators, and sometimes against each other. The clown loach spawns in fast-flowing rivers, at the onset of the rainy season.

Gyrinocheilus aymonieri

Siamese algae-eater

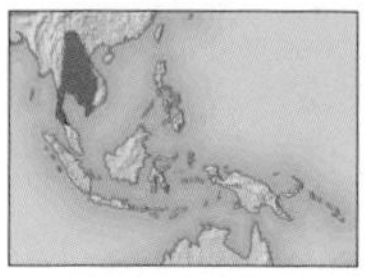

Length 11 in (28 cm)
Weight Not recorded
Sex Male/Female
Status Not evaluated

Location S.E. Asia

Found on solid surfaces at the bottom of flowing water, this slender fish has a modified mouth and lips that it uses to attach itself to rocks and plants, scraping plant material from them. It is golden green above and silvery below, and has a large dorsal fin with 9 branched fin rays. The Siamese algae-eater is commercially exploited for food and the aquarium trade.

Copella arnoldi

Splash tetra

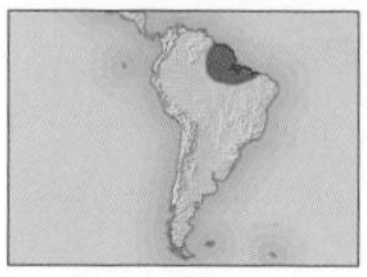

Length 3¼ in (8 cm)
Weight Not recorded
Sex Male/Female
Status Not evaluated

Location N. South America

Also called the jumping characin, this finger-sized fish has a slender, creamy yellow body, with a black stripe on the head and through the eye. It belongs to a group of freshwater species, the lebiasinids, often kept in aquariums. This particular species gets its name from its distinctive reproductive behavior. The female lays her eggs on the underside of overhanging leaves, so that when the water level drops, they are safe from most predators. The male then splashes them with water to keep them alive and the young fishes drop back into the water after they have hatched. Splash tetras feed on worms, insect larvae, and crustaceans.

Hydrocynus vittatus

Tigerfish

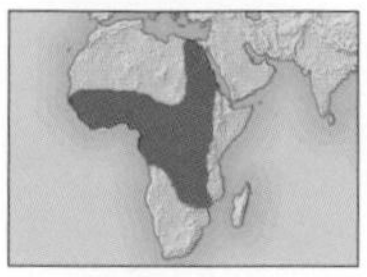

Length 3¼ ft (1 m)
Weight 40 lb (Up to 18 kg)
Sex Male/Female
Status Not evaluated

Location Africa

This African fish is a fierce, freshwater predator, and a well-known gamefish, prized for its habit of fighting back when hooked. It has a sleek silvery body, marked with dark horizontal stripes, a deeply forked tail edged with black, and large, fanglike teeth. Tigerfishes move in schools while feeding, and swallow their prey whole, head first. They can eat fishes up to half their length. Apart from humans, the African fish eagle is their only significant enemy. The giant tigerfish (*Hydrocynus goliath*), a relative of this species and also from Africa, is one of the largest characins, growing up to 6 ft (1.8 m) long.

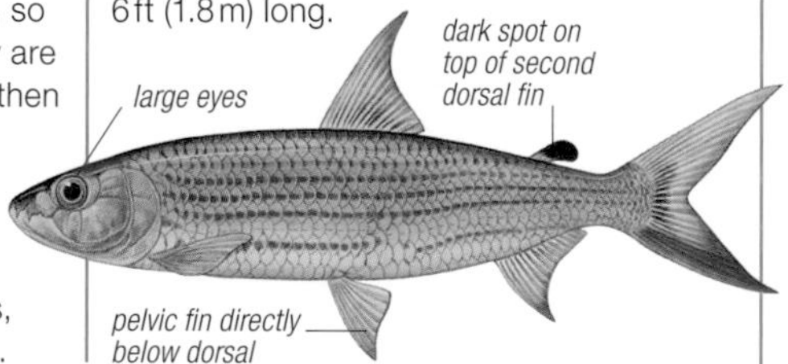

Anostomus anostomus

Striped headstander

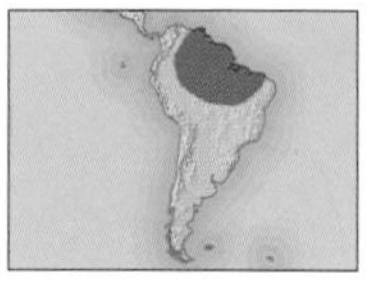

Length 6½ in (16 cm)
Weight Not recorded
Sex Male/Female
Status Not evaluated

Location N. South America

This slender, torpedo-shaped fish is golden yellow, with 3 broad, black horizontal bands running along the body, from the snout to the tail fin. The slender head has an upturned snout, and all fins have reddish bases. This fish gets its name from its habit of swimming head-down, at an oblique angle to the beds of rivers and streams. It swims in small schools (containing up to 40 individuals), often remaining motionless among vegetation. Its upturned mouth enables it to feed from leaves and stems.

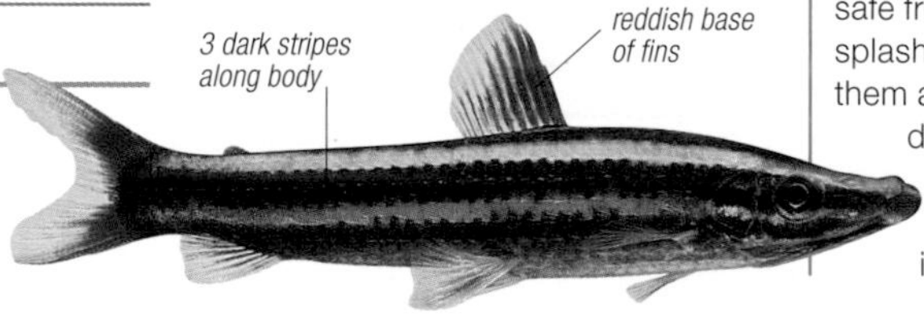

Astyanax mexicanus

Mexican tetra

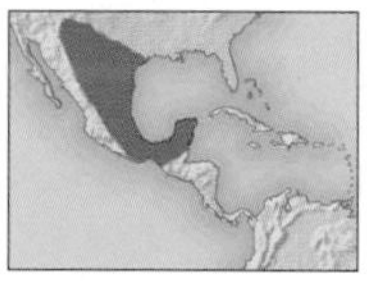

Length 4¾ in (12 cm)
Weight Not recorded
Sex Male/Female
Status Not evaluated

Location S. Central USA to Mexico

Found in springs, creeks, and small rivers, the Mexican tetra typically has a silvery body with red and yellow fins. This fish is of interest to biologists because a variant of the species is found in caves. The subterranean form, called blind cave characin (pictured right), looks very different, with its reduced pigmentation and nonfunctioning eyes covered by skin. Like other cave-dwelling fishes, it finds food by using its pressure-sensing lateral line and its keen sense of smell. Despite the differences in colour and eye development, the 2 forms of Mexican tetra are similar at a genetic level, and can interbreed.

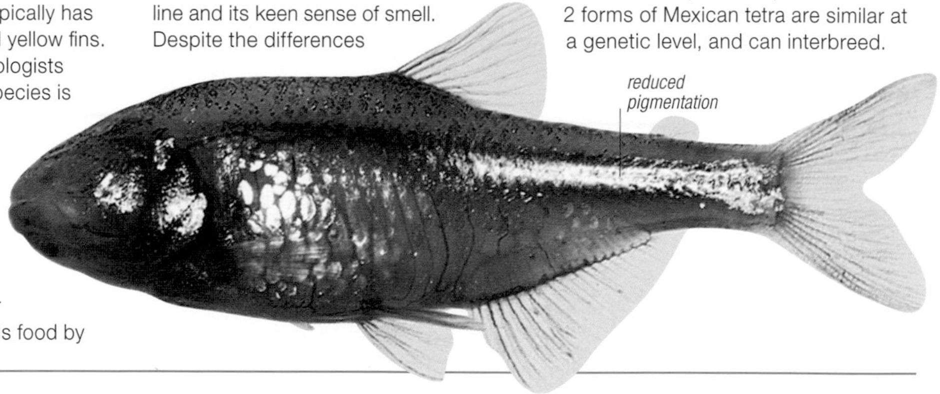

Pygocentrus nattereri

Red piranha

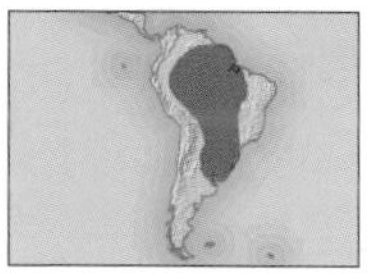

Length 13 in (33 cm)
Weight 2¼ lb (1 kg)
Sex Male/Female
Status Not evaluated

Location N., C., and E. South America

Notorious for its predatory, pack-feeding behavior, the red piranha is extremely variable in appearance, displaying a wide diversity of colour and shape in different parts of its range. Red piranhas typically feed at dawn and dusk, lurking and then dashing at their prey. They generally eat animals smaller than themselves, such as insects, aquatic invertebrates, or other fishes, and will also feed on seeds and fruit. However, given an opportunity, these extremely aggressive, voracious predators can kill much larger prey by hunting as a group, feeding in a frenzied rush reminiscent of the behavior of some sharks. These mass attacks can kill animals as large as capybaras and horses, and have also resulted in some human deaths. During the breeding season, female red piranhas lay up to 1,000 relatively large eggs, and attach them to the roots of trees trailing in the water. Both parents, particularly the male, guard the eggs until they hatch, which takes about 9–10 days.

RED BELLY
The red piranha is usually silvery, with silvery red-flecked eyes, and a head that is dark gray above and orange-red below. In large individuals, the underside is often intensely red—giving this fish its name.

SLICING TEETH

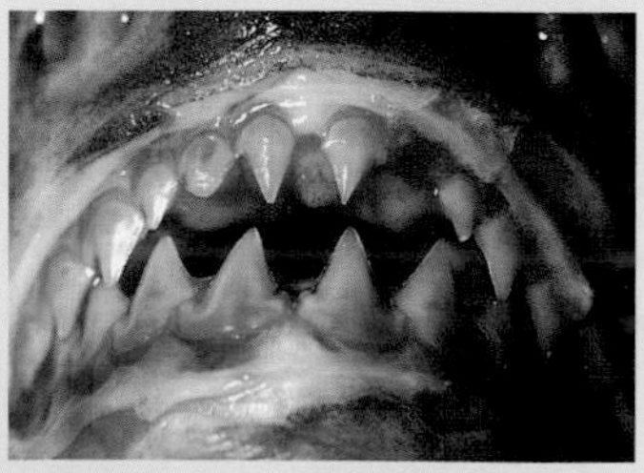

Red piranhas have small but sharp, triangular teeth set in powerful jaws. The teeth interlock when the fish's mouth is closed, enabling it to slice off chunks of flesh. The blunt snout and underslung lower jaw also help it to attack, allowing it to bite with unusual force.

Gasteropelecus sternicla

River hatchetfish

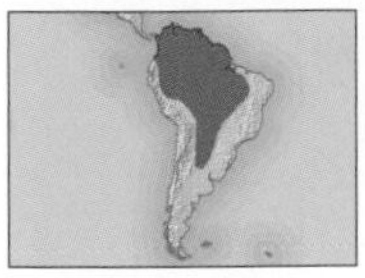

Length 2½ in (6.5 cm)
Weight Not recorded
Sex Male/Female
Status Not evaluated

Location N. to C. South America

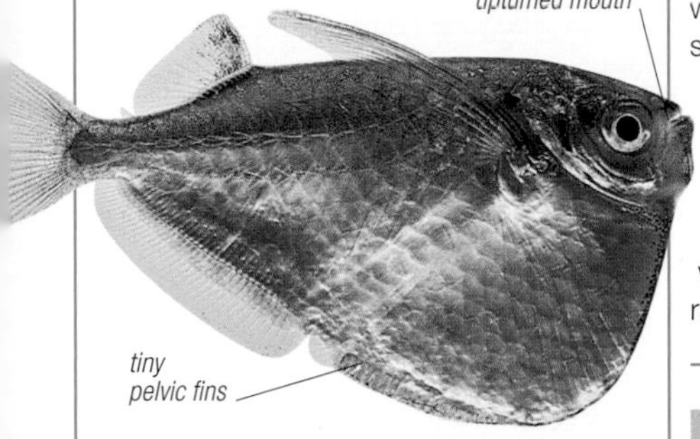

The river hatchetfish gets its name from its narrow shape, and deep, bladelike keel. The most unusual feature of this silver-gray fish is its highly enlarged pectoral fins. When threatened, or when chasing prey, the river hatchetfish uses these fins to gain speed in the water and even to take off into the air. Marine hatchetfishes belong to a different group of fishes, and have a different lifestyle.

Hoplias macrophthalmus

Giant trahira

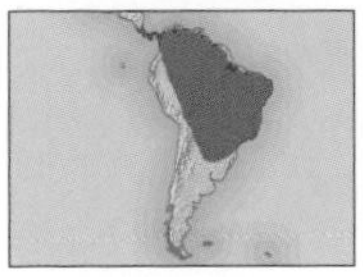

Length Up to 3¼ ft (1 m)
Weight Up to 4½ lb (2 kg)
Sex Male/Female
Status Not evaluated

Location Trinidad, Central America to C. South America

The giant trahira has a thick, stocky body with a blunt snout, very large eyes, and a somewhat rounded stomach. It has 2 or 3 canines in its upper jaw, in addition to a series of conical teeth. This fish can tolerate low oxygen conditions and has been known to move across land at night between bodies of water. An important food fish, it is raised in aquaculture.

Electrophorus electricus

Electric eel

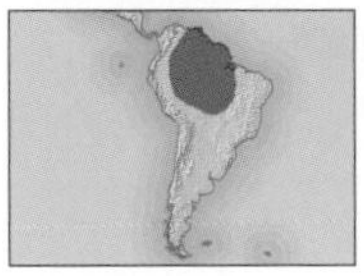

Length Up to 8¼ ft (2.5 m)
Weight Up to 44 lb (20 kg)
Sex Male/Female
Status Least concern

Location N. South America

Despite its name, this fish is not a true eel, but a giant member of the knifefish order with an eel-like body. When fully grown, it can be as thick as a human thigh, and is one of the largest freshwater fishes in South America. It has a continuous fin that runs along its underside, but no fins on its back; its tail tapers to a sharp, finless point. Electric eels have poor eyesight, and find their way by using weak pulses of electricity. However, using modified muscles, or electric organs, that run the entire length of their bodies, they can also produce sudden jolts of up to 600 volts—powerful enough to kill other fishes and potentially lethal to humans.

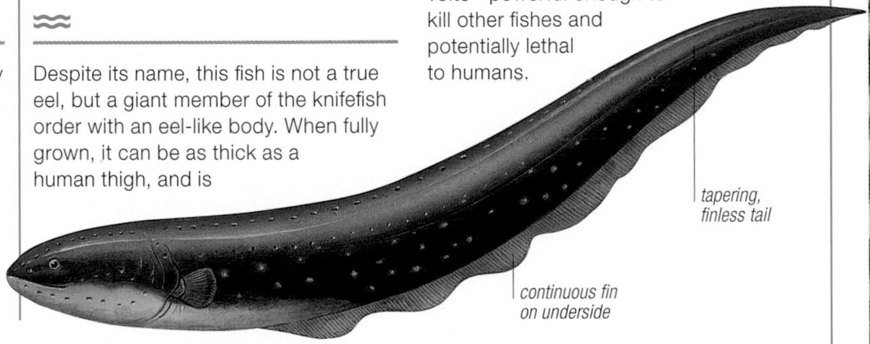

Silurus glanis

Wels catfish

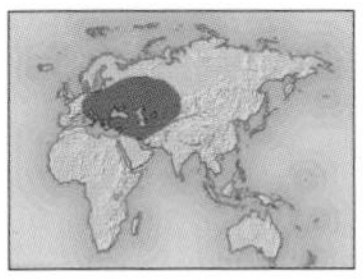

Length Up to 16 ft (5 m)
Weight Up to 660 lb (300 kg)
Sex Male/Female
Status Least concern

Location C. Europe to C. Asia

This huge, bottom-dwelling catfish is one of the largest freshwater fishes in the world. The biggest specimen on record, caught in the 19th century in the Dnieper River in southern Russia, was over 15 ft (4.5 m) long, and weighed over 660 lb (300 kg). However, it is unlikely that any wels catfish of a similar size exist today because they have been heavily fished in most parts of their range. The wels catfish is a solitary fish, lurking at the bottom of rivers and large lakes. It moves around mainly at night, camouflaged by its greenish gray markings that conceal it against riverbed mud. Like most catfishes, it is a voracious predator, eating waterfowl and aquatic mammals, besides crustaceans and smaller fishes. Females lay their eggs when the water temperature rises above 68° F (20° C). The male initially guards both the eggs and the young.

FOOD SENSORS

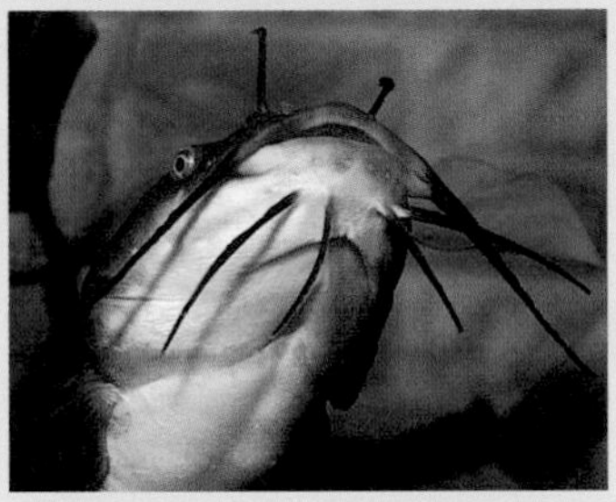

The wels catfish has 2 very long barbels attached to its upper jaw, and 4 smaller ones below its mouth, which help it find food.

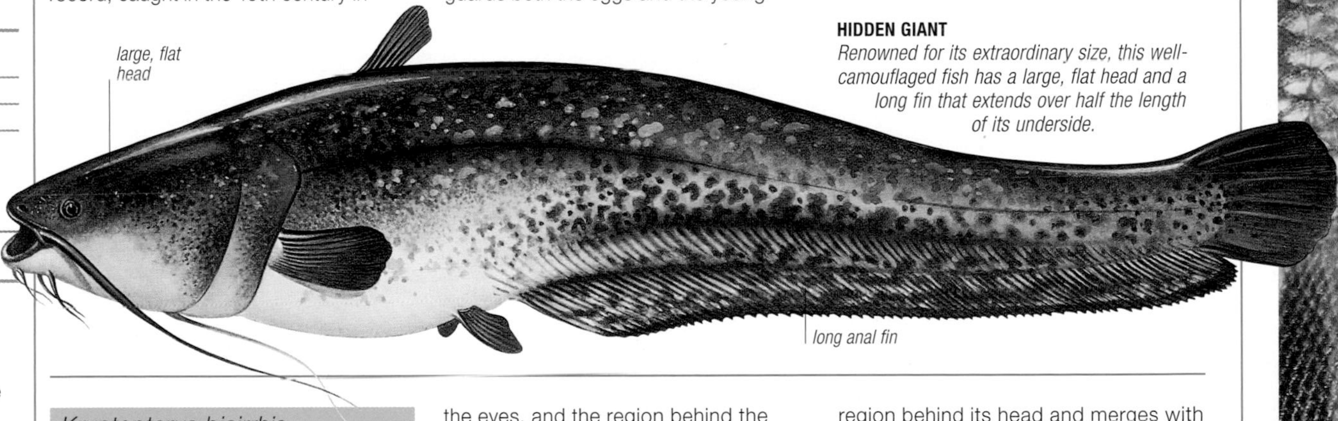

HIDDEN GIANT
Renowned for its extraordinary size, this well-camouflaged fish has a large, flat head and a long fin that extends over half the length of its underside.

Gymnotus carapo

Banded knifefish

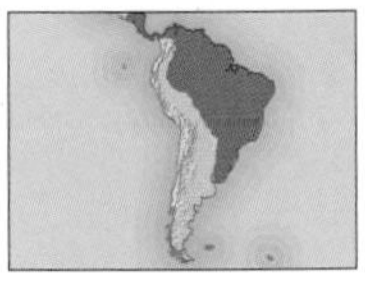

Length 23½ in (60 cm)
Weight Not recorded
Sex Male/Female
Status Not evaluated

Location Central America to C. South America

A close relative of the electric eel (see above right), this freshwater fish has a similar cylindrical body tapering toward its finless, rod-shaped tail. The banded knifefish is a bottom-dwelling species that inhabits murky, still water. It produces weak electric currents with which it senses its environment. Produced by specially adapted muscle tissue close to the tail, the electric currents are also used to communicate with one another and for males to attract mates. It moves by rippling the long fin on its underside, but it is not a strong swimmer.

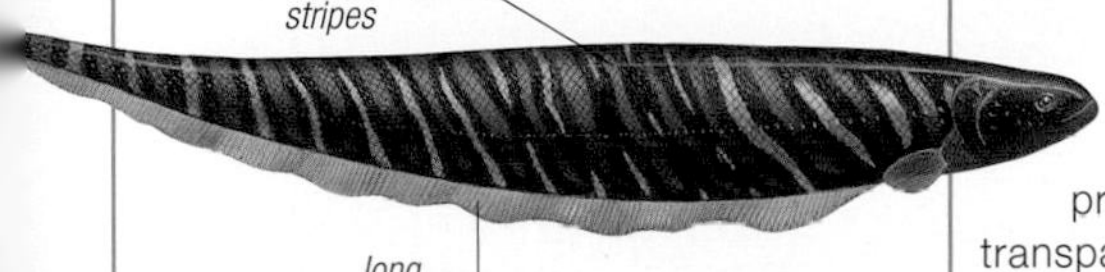

Kryptopterus bicirrhis

Glass catfish

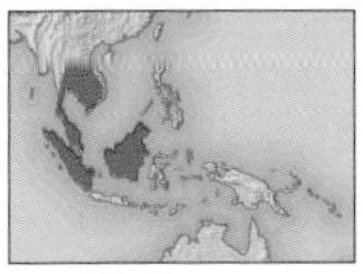

Length 6 in (15 cm)
Weight Not recorded
Sex Male/Female
Status Not evaluated

Location S.E. Asia

Many catfishes are camouflaged by their dark markings, but this slender species is protected by being almost transparent. The only clearly visible parts of its body are the backbone, the eyes, and the region behind the head that contains its digestive and reproductive organs. Its transparency is the result of thin skin, and of body tissues that are suffused by oils, which make its flesh translucent. The glass catfish's body is flattened from side to side, and its dorsal fin is reduced to one or 2 rays, or is missing altogether. As in many other catfishes, it has a very long anal fin, which stretches from the region behind its head and merges with the base of its tail. This fish inhabits lowland flood plains and large rivers with turbid waters. Individuals gather in small groups at an oblique angle to the water surface. Several other groups of fishes contain species that have transparent or nearly transparent bodies. For example, the Indian glassfish, which actually comes from Southeast Asia, is often kept in aquariums.

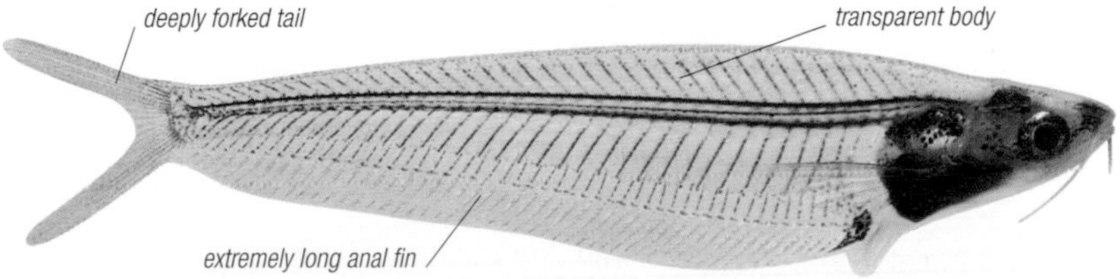

Ancistrus dolichopterus

Bushymouth catfish

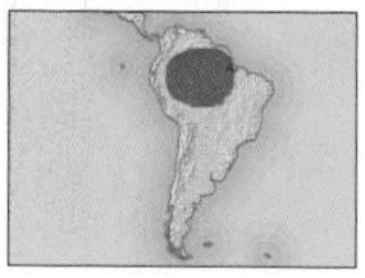

Length Up to 5 in (13 cm)
Weight Not recorded
Sex Male/Female
Status Least concern

Location N. South America

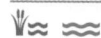

The bushymouth catfish and its close relatives get their name from the distinctive fleshy tentacles on the upperside of their snouts. The males have more tentacles than the females, giving rise to an as yet unproven theory that they are linked to their fitness to breed. When threatened, the bushymouth catfish erects its pectoral and dorsal fin spines, as well as the spines on the sides of its head.

Farlowella acus

Whiptail catfish

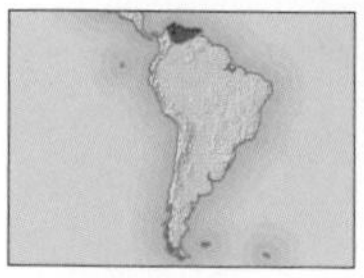

Length Up to 6 in (15 cm)
Weight Not recorded
Sex Male/Female
Status Not evaluated

Location N.W. South America

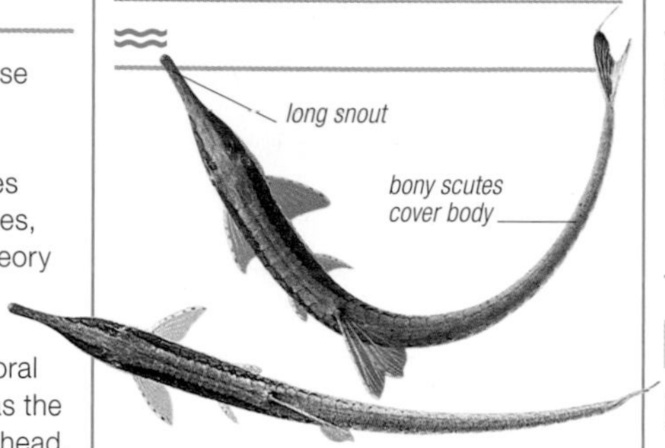

Although they are almost all bottom dwellers, catfishes exhibit a remarkable diversity of body form. The whiptail catfish is unusually long and slender, with a pointed snout, and is camouflaged to resemble a piece of sunken wood. It lives in slow-flowing rivers, and feeds mainly on river-bed algae after dark. During the day, it relies on its cryptic color and shape to protect it from attack. Females lay their eggs on the river bed, and the males guard them until they hatch.

Clarias batrachus

Philippine catfish

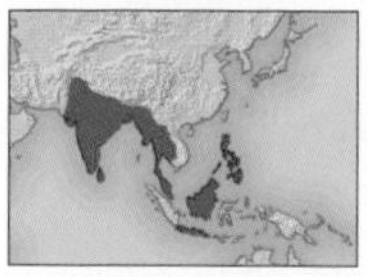

Length Up to 16 in (40 cm)
Weight Not recorded
Sex Male/Female
Status Not evaluated

Location S. and S.E. Asia

This long-bodied catfish has the ability to crawl overland, using its pectoral fins to gain purchase. While on land, it breathes through its specially modified gills, which have strengthened filaments, preventing them from collapsing when exposed to air. The Philippine catfish lives in slow-moving water, walking overland to reach suitable habitats. It is a popular aquarium species; the piebald variety shown here is one of several kinds that have been bred.

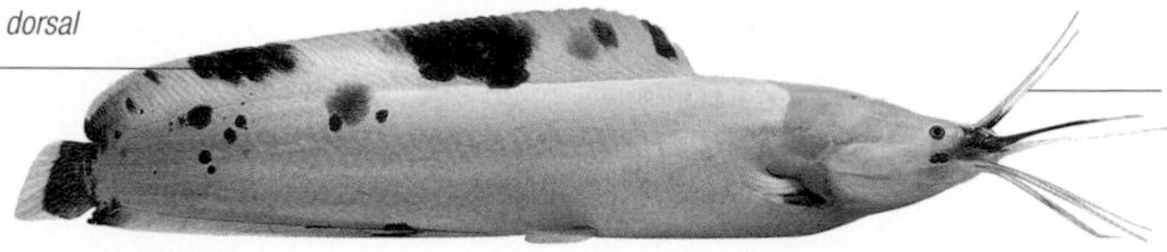

Vandellia cirrhosa

Candiru

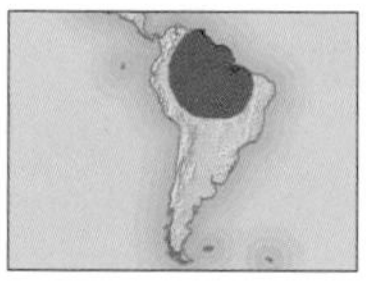

Length Up to 1 in (2.5 cm)
Weight Not recorded
Sex Male/Female
Status Not evaluated

Location N. South America

A tiny catfish with a slender, almost transparent body, the candiru has an unusual way of life. Instead of feeding in open water, it swims into the gill cavities of larger fishes—particularly larger catfishes and characins—using the large, hooked spines of its own gill cover to lodge itself in the gill tissue of its host. It then bites off pieces of gill with its fine, comblike teeth, and feeds on the released blood. Candirus are notorious for swimming up the urethras of mammals, even humans, urinating in rivers. It is thought that, in these instances, the candiru mistakes the flow of urine for the stream of water expelled from the gills of larger fishes.

Synodontis contractus

Upside-down catfish

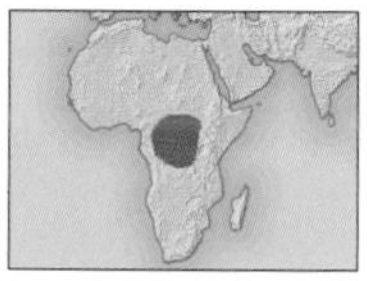

Length 3¾ in (9.5 cm)
Weight Not recorded
Sex Male/Female
Status Not evaluated

Location C. Africa

This African catfish swims upside-down near the surface of rivers and streams, particularly when taking food from the surface or feeding from the undersides of broad leaves and rocks. Its diet includes plant material, aquatic invertebrates, and insect larvae. The upside-down catfish is entirely brown or brown-violet, with a series of small spots scattered across its body. Most fishes are dark above and light beneath—a pattern called countershading that serves as a means of camouflage. In this fish, the arrangement is often reversed, with the underside being darker than the back. The upside-down catfish makes sounds, which are possibly used to communicate with others of its species.

Bagre marinus

Gafftopsail sea catfish

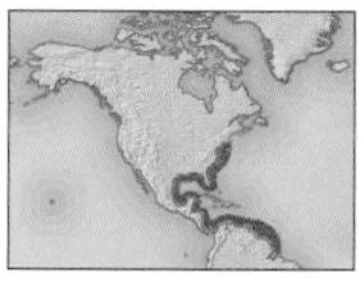

Length Up to 3¼ ft (1 m)
Weight Up to 8¾ lb (4 kg)
Sex Male/Female
Status Not evaluated

Location W. Central Atlantic, Gulf of Mexico, Caribbean

Commercially exploited as a food source, this Atlantic catfish is mainly marine, but sometimes swims into estuaries that have relatively high salinity levels. Its head has a hard, bony shield, but its most conspicuous features are the very long, flattened barbels attached to the sides of its mouth, and the long spines that arch outward from its dorsal and pectoral fins. These serrated spines are also venomous and can inflict painful wounds. The gafftopsail sea catfish erects its dorsal and pectoral spines when threatened. This fish is a bottom-feeder, eating crustaceans and other fishes. During the breeding season (between May and August), the male gafftopsail sea catfish fertilizes the female's eggs and broods them in his mouth until they hatch.

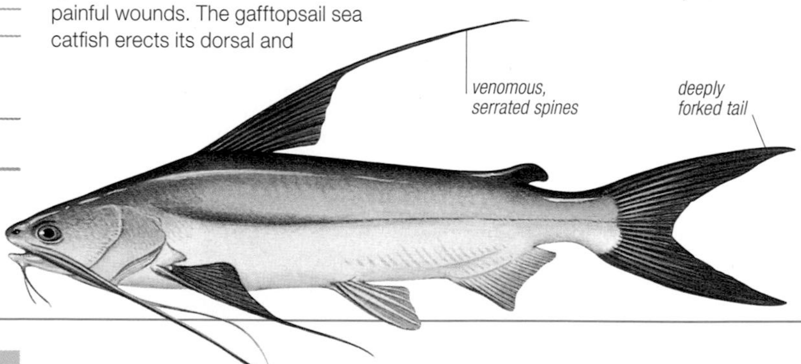

Ameiurus nebulosus

Brown bullhead

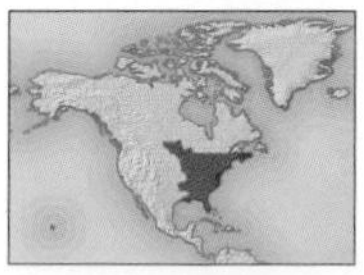

Length Up to 20½ in (52 cm)
Weight 5½–7¾ lb (2.5–3.5 kg)
Sex Male/Female
Status Not evaluated

Location E. North America

This common catfish belongs to a large family of freshwater fishes found in North America. Like other members of the family, it is valued as a food fish. It has a scaleless body, and a large mouth surrounded by 4 pairs of fleshy barbels, which it uses to find food. Its jagged, dagger-like pectoral spines with attached venom glands are a formidable weapon against predators. Both parents guard the eggs in the nest, fanning to increase the air flow around them, and herd the young in tight schools until they reach a length of about 2 in (5 cm).

Salmon and relatives

PHYLUM	Chordata
CLASS	Osteichthyes
SUBCLASS	Actinopterygii
SUPERORDER	Protacanthopterygii
ORDERS	3
FAMILIES	16
SPECIES	538

During their lifetime, many species of salmon and trout make impressive journeys from the oceans (where they live as adults) to freshwater lakes and rivers (where they spawn). Their close relatives include a group of freshwater fishes, the galaxiids. Among their other relatives are the freshwater pikes and a group of marine fishes called argentinoids (which include tubeshoulders and spookfishes, among others). Salmon, trout, and pikes are native to North America, Europe, and Asia, although they have been introduced elsewhere. Galaxiids occur only in the Southern Hemisphere, while argentinoids are found from moderate depths to deep water in all the world's oceans.

Anatomy

Salmon and their relatives have a slim, tapering body, usually with well-developed swimming muscles. The fins are relatively small, except for the tail fin, which is usually large and powerful, enabling the fishes to swim and maneuver quickly. Most species have a small, fleshy fin, known as an adipose fin, near the base of the tail, and pelvic fins that are located far back on the body. Scales in these fishes are either small or entirely absent. As carnivores that feed on a wide variety of prey, most species have a large mouth studded with many sharp teeth. The highly predatory pikes have long teeth for seizing prey and an especially large mouth that allows them to swallow prey almost half their own size. Salmon, trout, and pikes often ambush prey or attack with a burst of speed over a short distance.

Most galaxiids are small fishes, less than 10 in (25 cm) long, with a tubular, scaleless body. They have a square tail and, unlike most other fishes in this group, no adipose fin.

Some of the marine fishes in this group have unusual adaptations for life in deep water. The argentinoids, for example, include some species with large, tubular eyes that point forward or upward (giving good binocular vision). Other argentinoids have light-emitting organs.

Life cycle

Some species in this group complete their entire life in either the sea or in freshwater. However, other species, including many salmon and trout, spend most of their life in the ocean but move to freshwater to breed, a life cycle described as anadromous. Males and females spawn in freshwater rivers. When the young hatch, they grow for a short time in freshwater and then move out to sea. After a period that varies from a few months to several years, the fishes mature into adults and return to freshwater to spawn, usually to the stream where they hatched. To complete the journey, which may be several thousand miles long, they often have to negotiate fast currents or rough terrain, requiring them to leap clear of the water to progress upstream. Breeding adults usually cease feeding before beginning their migration. Having used all their energy on the journey, they usually die soon after spawning. How they find their way to their hatching site is not entirely understood, but the sense of smell and perhaps celestial navigation are thought to be important. Not all salmon and trout follow this type of life cycle: those belonging to landlocked populations spend all their life in freshwater.

Some species of galaxiids are also anadromous. The tiny larvae hatch in freshwater and are swept out to sea. After a few months, they return as juveniles to freshwater, where they mature into adults.

BREEDING MALES
During the migration to their breeding grounds, the males of many species of salmon undergo drastic physical changes. These include the appearance of new colors and markings. Some species also develop a hump on their back and an enlarged lower jaw that bends upward. These pink salmon from the Pacific Ocean are swimming up a stream in Alaska.

Esox lucius

Northern pike

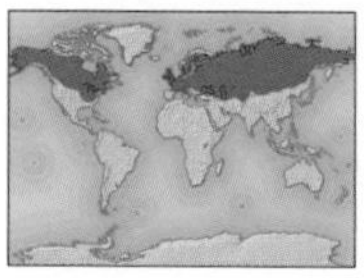

Length Up to 4¼ ft (1.3 m)
Weight Up to 75 lb (34 kg)
Sex Male/Female
Status Least concern

Location North America, Europe, Asia

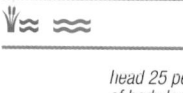

This powerfully built predator is one of the few freshwater fishes found across the whole of the northern hemisphere, except in the Arctic coastal plains. The northern pike is superbly camouflaged with light markings on its body. Like its close relative, the muskellunge (see below), it has a long head with a shovel-like snout, a large, slightly forked tail, and a single dorsal fin. Pike generally spawn during the day, in early spring after the ice has melted. They initially spawn as a pair, but either the male or the female may take on a new partner later. While courting, the male prods the female on her head and pectoral fins with his snout. Female northern pikes grow faster and larger than the males.

STEALTHY HUNTER
The single dorsal fin of the northern pike, set well back near its tail, allows it to approach prey on the surface of lakes without making any warning ripples.

SILENT AMBUSH

The well-camouflaged northern pike hunts by lying in wait among waterweeds, bursting out of cover the instant its prey comes within range. It feeds mainly on other fishes, but also eats insects, frogs, and crayfishes, as well as animals swimming on the surface, such as water voles and young waterfowl.

Esox masquinongy

Muskellunge

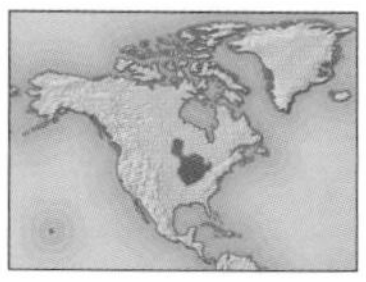

Length Up to 6 ft (1.8 m)
Weight Up to 99 lb (45 kg)
Sex Male/Female
Status Not evaluated

Location E. North America

The muskellunge, or "musky," is the largest member of the pike family, and is notorious among anglers for putting up a ferocious fight when hooked. In common with other pikes, it has a very large mouth armed with sharp teeth, a sloping forehead and beaklike jaws, and a torpedo-shaped body, which can accelerate very rapidly with a sudden flick of the tail. It can be identified by the lack of scales on the lower cheeks and gill covers (opercula). A lie-in-wait predator, it ambushes and feeds on animals as large as muskrats, catching its prey sideways in its mouth, before turning them around and swallowing them head-first. Muskellunges spawn over vegetation in mid-spring, after the ice has melted. A female may pair with more than one male and can lay more than 250,000 eggs in a single season. This fish is found in heavily vegetated lakes and rivers, generally in warmer waters than those inhabited by northern pikes (see above), although the 2 fishes may be found in the same habitat and will occasionally even interbreed. The resulting hybrids are a robust breed with strongly barred markings. The male hybrid is sterile but females are often fertile. The muskellunge is more sensitive to habitat changes than the northern pike, so its numbers decline first should any change occur.

Dallia pectoralis

Alaska blackfish

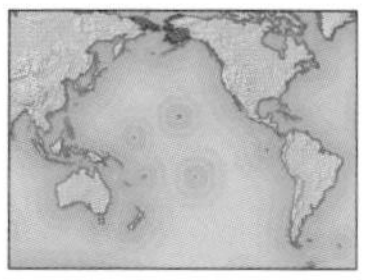

Length Up to 13 in (33 cm)
Weight Up to 13 oz (375 g)
Sex Male/Female
Status Not evaluated

Location N.E. Asia, N.W. North America

The Alaska blackfish belongs to the mudminnow family—a group of fishes that can survive in very cold and stagnant water by gulping mouthfuls of air. They can even survive being partly frozen in the ice for periods of several weeks. This species has a slender body with rounded fins, a blunt head, and a large mouth with a protruding lower jaw. It is dark green or mottled brown, and the males develop a red outline to their fins during the breeding season. This slow-moving fish feeds on insect larvae, snails, and smaller fishes, which it ambushes and hunts with a darting movement.

Stenodus nelma

Sheefish

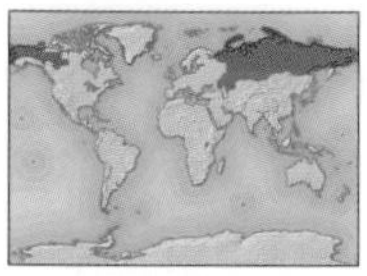

Length At least 5 ft (1.5 m)
Weight 62–88 lb (28–40 kg)
Sex Male/Female
Status Least concern

Location North Pacific, N.W. North America, E. Europe to N.E. Asia

This large member of the salmon family has a long, broad, and shallow head and a large mouth with very fine teeth. As an adult, it feeds mainly on other fishes, but its young eat insect larvae and planktonic animals. Sheefishes live in 2 different ways: some—for example in Canada's Great Slave Lake—are permanent freshwater fishes, which remain in their home range to breed, but most overwinter in brackish water or estuaries, and move upriver in spring. They remain there until autumn, when they spawn. The female is accompanied by one or 2 males during spawning bouts, which take place between dusk and nighttime, and probably occur only every third or fourth year. She lays between 130,000 and 400,000 eggs. Although it is valued as a sport fish, the sheefish is more important in commercial fisheries in Canada, Alaska, and Siberia, where it is commonly found.

Salmo salar

Atlantic salmon

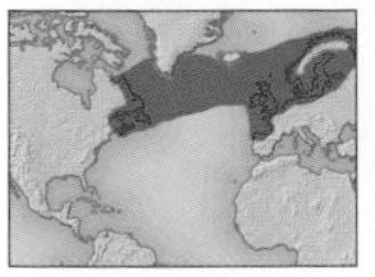

Length Up to 5 ft (1.5 m)
Weight Up to 99 lb (45 kg)
Sex Male/Female
Status Least concern

Location N.E. North America, W. and N. Europe, North Atlantic

The Atlantic salmon is one of the world's most highly prized sport fishes, as well as being widely farmed as a source of food. Wild fishes start their lives as tiny fry—known as alevins—that hatch out among streambed gravel. The fry feed on a diet of insect larvae and other small animals, and after a freshwater life of between 1 and 4 years, they begin their downstream migration to the sea. On entering saltwater, they lose their dark color and develop a silvery sheen, and for up to 4 years they roam widely throughout the North Atlantic, preying on other fishes. At the end of this period, they begin to become sexually mature, and make their way back to the rivers and streams where they originally hatched. Like other salmon, they are powerful swimmers, able to leap over almost all obstacles in their way. They rarely feed on their upstream journey, and by the time they arrive at their spawning grounds, the females are plump with eggs, and the males have developed distinctively hooked jaws. The females are known to use their tails to excavate hollows in gravel, and the males lie alongside while the females lay eggs. Adult Atlantic salmon may die after spawning but, unlike other salmon, a large number of the adults often survive. Emaciated and exhausted, these fishes make their way back to the sea, where they feed and recover. They return to breed again at intervals of one or 2 years.

EARLY DAYS
At the age of a few months, young Atlantic salmon are known as parr. At this stage, they are about 6 in (15 cm) long and have distinctive dark markings, or "fingerprints," along their backs.

READY TO BREED
When they re-enter rivers, Atlantic salmon are silvery, with blue-black backs and black spots along their sides. Their powerful tails enable them to leap up waterfalls and weirs on their way to their breeding grounds.

Oncorhynchus gorbuscha

Pink salmon

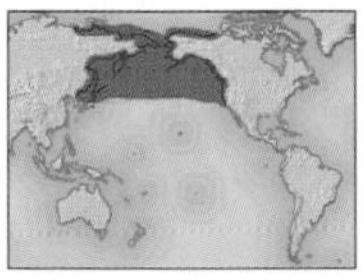

Length Up to 30 in (76 cm)
Weight Up to 14 lb (6.5 kg)
Sex Male/Female
Status Not evaluated

Location N.E. Asia, North Pacific, W. North America

Of all salmon, this species spends the least time in freshwater; it migrates soon after hatching, traveling over 620 miles (1,000 km) at sea. Normally metallic blue, with the approach of the breeding season, it becomes olive-green or yellow, with a distinct red or pink tinge along the sides. The male also develops a high humped back and a hook at the end of the upper jaw. The pink salmon has a lifespan of 2 years, which means that fishes born in even years and those in odd years hardly ever interbreed.

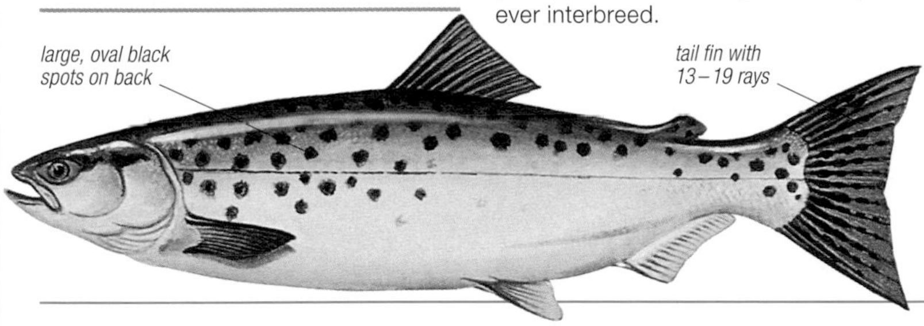

Salvelinus alpinus

Arctic char

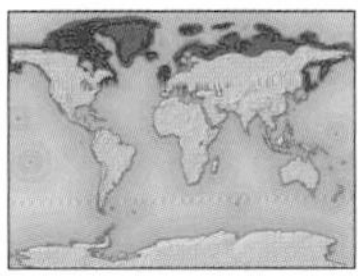

Length Up to 38 in (96 cm)
Weight Up to 26 lb (12 kg)
Sex Male/Female
Status Least concern

Location N. North America, N. Europe, N. Asia, Arctic Ocean

The Arctic char is the most northerly of all freshwater fishes. It is found in lakes in the Arctic tundra, where it spawns during autumn, as well as in mountains farther south, where spawning takes place in winter. Its color varies according to its breeding habits: adults that migrate toward the sea become brightly colored during the breeding season, but landlocked specimens, usually found in lakes, often have bright undersides all year round. The Arctic char is a popular sport fish among anglers in Canada and Scandinavia.

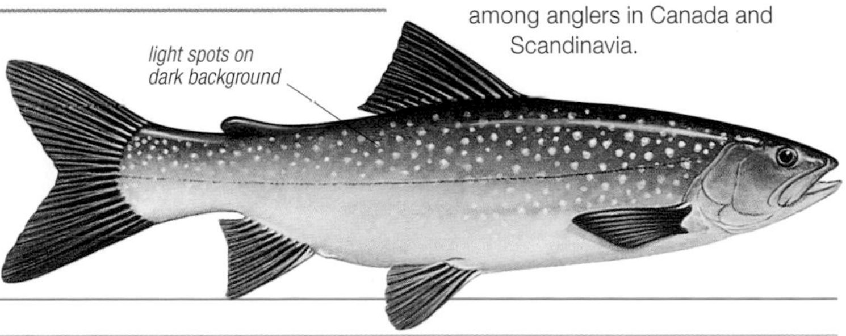

Oncorhynchus mykiss

Rainbow trout

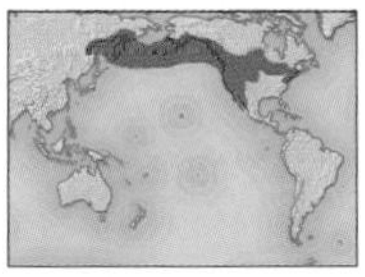

Length Up to 4 ft (1.2 m)
Weight Up to 52 lb (24 kg)
Sex Male/Female
Status Not evaluated

Location N.E. Asia, North Pacific, North America

This black-spotted member of the salmon family is one of the world's most widely introduced freshwater fishes. Originally from rivers and lakes west of the Rocky Mountains, it has been spread eastward in North America, and to countries as far apart as the British Isles and New Zealand, wherever well-oxygenated, cool water can be found. Like all its relatives, the rainbow trout is a predator, feeding on insects, snails, and crustaceans, and—when adult—on other fishes. It takes most of its food from the bottom, but also rises to the surface for flying insects—a habit exploited by anglers using artificial flies. Introduced rainbow trout often spend their entire lives in fresh water, but in their natural habitat, some adults spend part of their time at sea, returning to rivers to spawn. Compared to freshwater rainbow trout, these migratory adults—called "steelheads" in North America—grow faster, live longer, and are more productive.

TROUT FARMING

In many parts of the temperate world, rainbow trout are raised in captivity. Some trout "farms" supply fishes for food but an increasing number raise this trout—a popular sport fish—for angling, using them to stock lakes where they can be caught by line. Farmed trout are fed on protein-rich pelleted food, and grow much more rapidly than they would in the wild.

RAINBOW COLORS
Freshwater forms of rainbow trout range from bluish green to brown on the back and sides, and are white or yellowish on the belly. Spawning males develop a vivid red or pink band on the sides.

Oncorhynchus nerka

Sockeye salmon

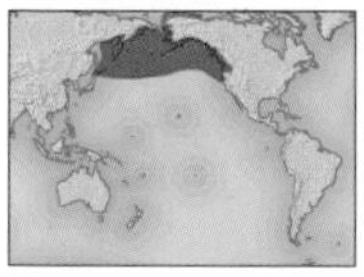

Length Up to 33 in (84 cm)
Weight Up to 15 lb (7 kg)
Sex Male/Female
Status Least concern

Location N.E. Asia, North Pacific, N.W. and W. North America

Also known as the "blueback" and the "red" salmon, the sockeye salmon undergoes a remarkable physical transformation at the time of spawning. Before spawning, sockeye salmon of both sexes have steel-blue heads and backs and silvery sides. However, between June and September, when they leave the ocean and enter their spawning streams, their heads turn bottle-green and their bodies bright red. Of all species of Pacific salmon, the sockeye is unique in requiring a lake to rear in, which makes it dependent on rivers that have a lake in their watershed. The lake-locked form of sockeye salmon—the kokanee—is typically one-sixth the weight of its sea-going kin.

JUMPING THE RAPIDS
The sockeye salmon's determination to reach its native spawning ground is demonstrated by the dramatic way it launches itself over rapids.

SPAWNING

Usually about 4 years after hatching, all adult sockeye salmon eventually return to their native streams to spawn and then die—a final journey that can involve traveling up to 930 miles (1,500 km).

SMALL FRY
When sockeye salmon fry emerge from the gravel, they embark on their first journey to a nearby lake. They will stay there for at least a year before heading out to sea.

NEW LIFE
Having reached its spawning ground, the female sockeye salmon digs a nest in the gravel river bed and lays up to 4,300 bright pink eggs, which are fertilized by the male.

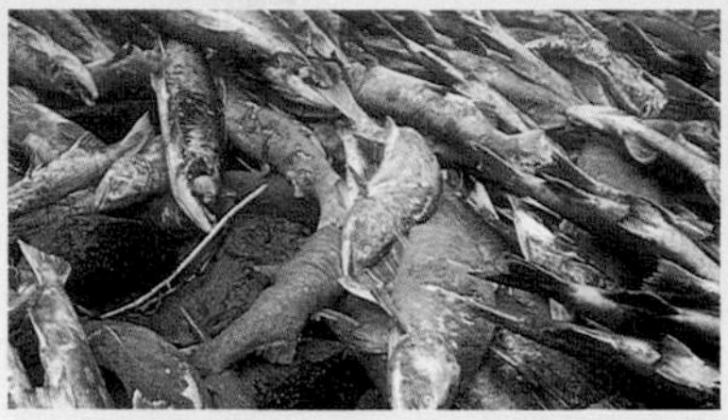

RIVER GRAVEYARD
After spawning, all sockeye salmon die, their bodies littering the same stretch of river in which they themselves were spawned.

DISTINGUISHING FEATURES
The sockeye salmon has a streamlined body and a blunt, conical snout. It is distinguished from other salmon species by its fins, which usually lack definite spots, and its dramatic color change at spawning.

CRIMSON TIDE
The spectacular annual run of the sockeye salmon—sometimes comprising millions of fishes—is a popular attraction, with viewing platforms being built across spawning rivers.

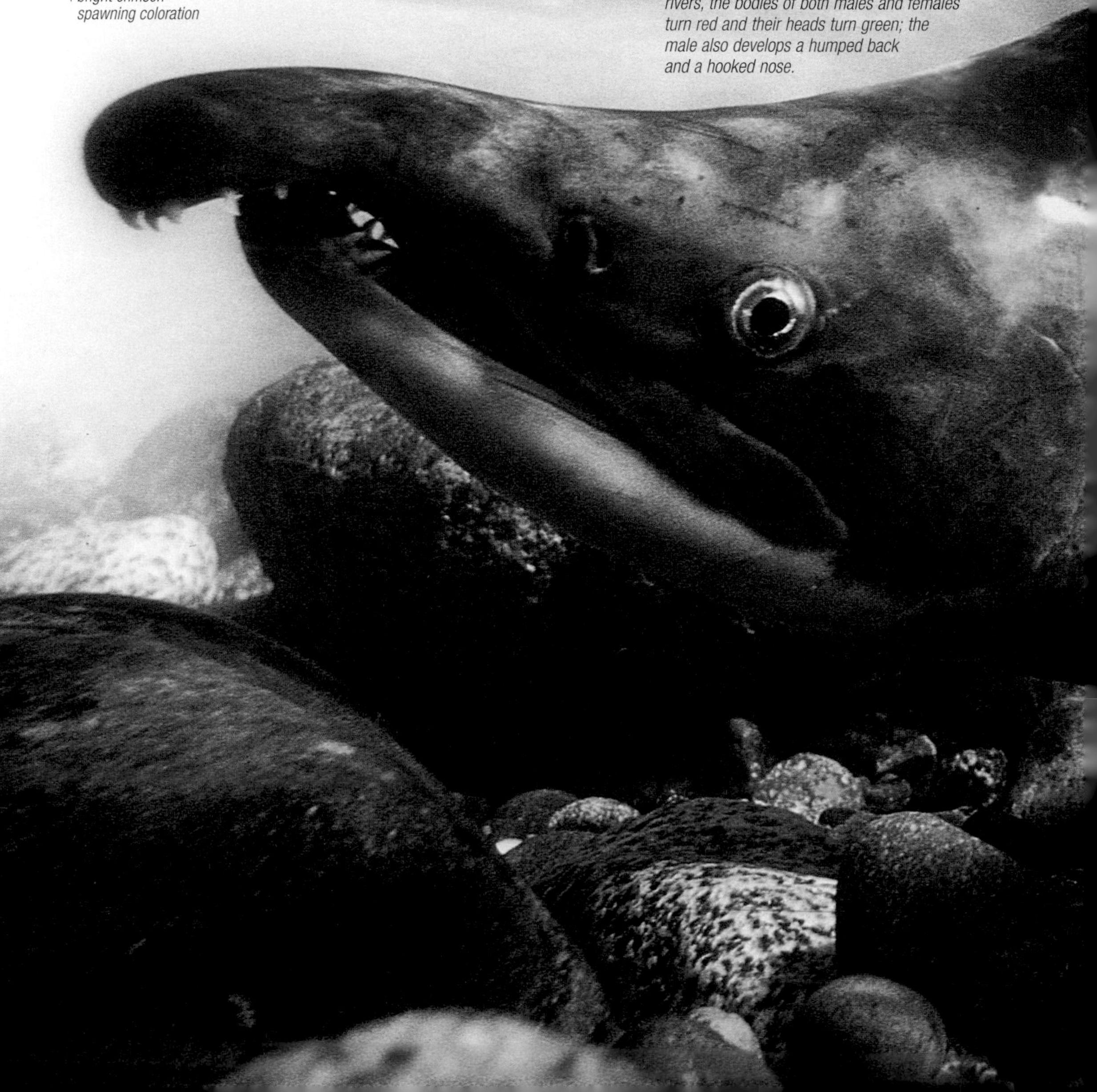

RETURNING SOCKEYE SALMON
The sockeye salmon is anadromous—living in the sea but entering freshwater systems to spawn. As the salmon return to their spawning rivers, the bodies of both males and females turn red and their heads turn green; the male also develops a humped back and a hooked nose.

Coregonus artedi

Cisco

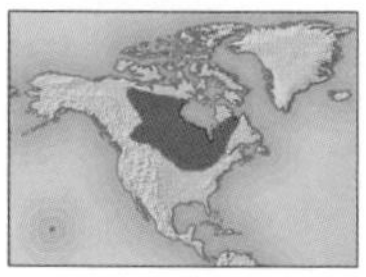

Length Up to $22\frac{1}{2}$ in (57 cm)
Weight Up to $7\frac{3}{4}$ lb (3.5 kg)
Sex Male/Female
Status Least concern

Location N. North America

Essentially a freshwater fish, the cisco, or lake herring, may also be found in brackish water and saltwater. It is usually found in large schools in the middle waters, the depth varying with the seasons. Basically a plankton-feeder, it will also consume a wide variety of other foods. Spawning occurs in autumn and is dependent on the temperature of water. Males are usually smaller than the females and, during the breeding season, the males develop small bumps (pearl organs) that function to stimulate the female to lay her eggs. The young feed on algae and plankton, whereas adults are more predatory and also feed on insects and small crustaceans. The cisco differs from other *Coregonus* species in having upper and lower jaws of the same length and more gill rakers. A major threat to this species is the introduction to lakes of the alewife and lamprey. However, it is also sensitive to pollution and the recent rapid decline of *Diporeia*, a tiny crustacean that is a major food source to lake fishes.

OPEN-WATER FISH
This slender, silvery fish has the characterstic dark upper and light lower body of an open-water schooling fish. It also has the fleshy adipose fin that is typical of the salmon family.

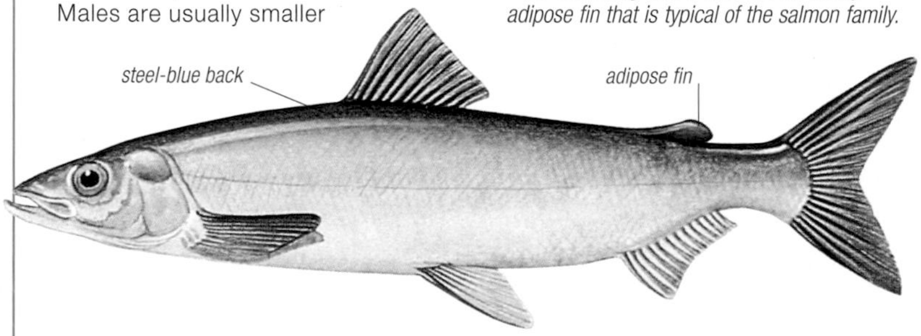

Thymallus thymallus

Grayling

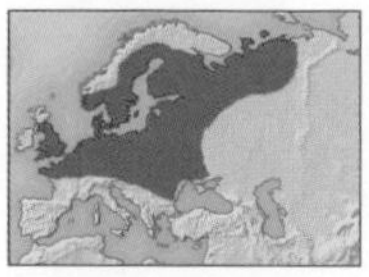

Length Up to $23\frac{1}{2}$ in (60 cm)
Weight Up to 14 lb (6.5 kg)
Sex Male/Female
Status Least concern

Location W., C., and N. Europe

This freshwater member of the salmon family lives in cool lakes and fast-flowing rivers and streams, where it feeds on insects, small worms, and crustaceans. It has an unusually long and tall dorsal fin, which the male curls over the female's back during egg-laying. The grayling is a useful indicator of water quality because it is one of the first fishes to disappear when rivers and streams become polluted.

Thaleichthys pacificus

Eulachon

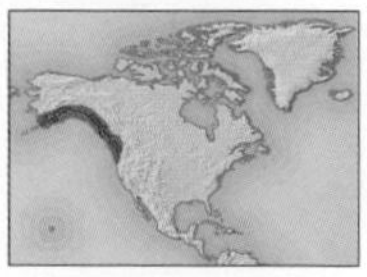

Length Up to 9 in (23 cm)
Weight Up to $2\frac{1}{8}$ oz (60 g)
Sex Male/Female
Status Not evaluated

Location E. North Pacific, W. North America

This is a member of the smelt family—a group of slender, silvery fishes that also includes the capelin (see opposite). Like the capelin, it has a long body and underslung lower jaw, but is blue to blue-brown above. Adult eulachons live at sea, where they feed on plankton, filtering their food from the water. However, they travel upstream in medium- to large-sized rivers to spawn. Few survive spawning, which limits the fish's lifespan to between 2 and 4 years. Eulachons are both a sport and a subsistence resource, and when caught in the mouth of rivers are so oily that they can be used as wicks for candles; hence their common name "candlefish." Their oil used to be traded by native North Americans; the trade routes were often called "grease trails."

Opisthoproctus soleatus

Barrel-eye

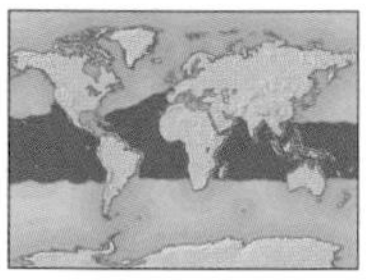

Length Up to $4\frac{1}{4}$ in (10.5 cm)
Weight Not recorded
Sex Male/Female
Status Not evaluated

Location Tropical and subtropical waters worldwide

Barrel-eyes are relatively small fishes named for their tubular, upward-facing eyes, an adaptation that probably helps them hunt other fishes from below. They belong to the same family as the spookfish (see right) and, like them, have most of their fins—apart from the pectoral pair—set close to the tail. This fish is dark above and silvery below, and has a luminescent organ on its rectum that contains light-producing bacteria. The light shines through a lens, and is reflected by the fish's flattened underside, or "sole," which spreads along the length of its body. This diffused light matches the dim light from the surface, making the barrel-eye practically invisible from below. Barrel-eyes are found beneath in tropical waters, down at a depth of about 2,650 ft (800 m).

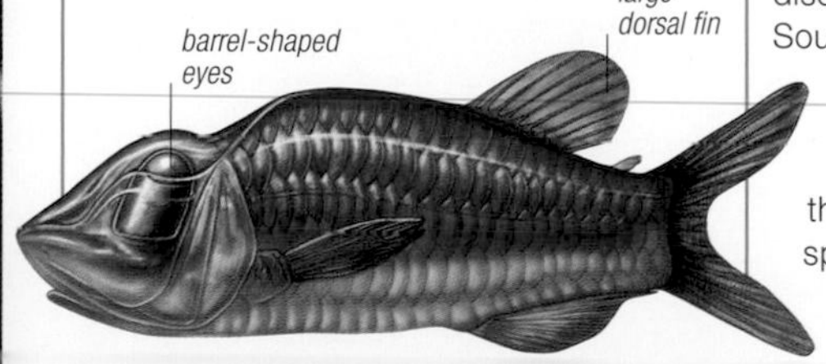

Dolichopteroides binocularis

Spookfish

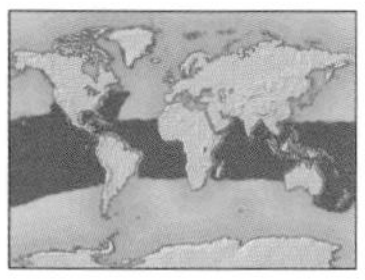

Length At least $3\frac{1}{4}$ in (8.5 cm)
Weight Not recorded
Sex Male/Female
Status Not evaluated

Location Tropical and subtropical waters worldwide, W. North Atlantic

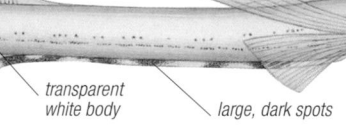

This bizarrely shaped, slow-growing fish is one of about 15 species in the spookfish family. It is distinguished from its relatives by its long, filament-like pectoral fins, which are over half as long as the rest of its body. However, its most striking feature is the highly unusual anatomy of its eyes. Tubular in shape, they point obliquely upward, like a pair of binoculars. This makes light directed at the side of the eyes hard to detect and so, located on the side of each tubular eye is a second eye, complete with retina and lens, making this the only truly 4-eyed fish. Like many deep-sea fishes, the spookfish is very fragile, and prone to damage if recovered from great depths.

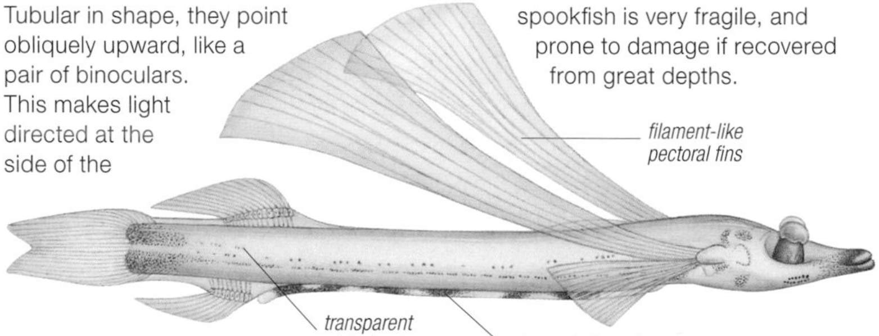

Bathylagus niger

Deep-sea smelt

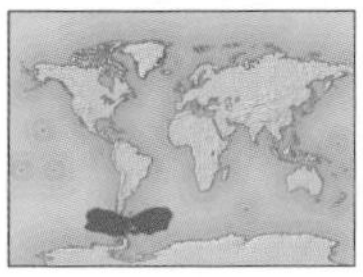

Length $5\frac{1}{2}$ in (14 cm)
Weight Not recorded
Sex Male/Female
Status Not evaluated

Location Southern Ocean

Found between depths of 656–5,118 ft (200–1560 m), this deep-sea smelt was discovered in the Scotia Sea between South America and Antarctica. Its habits are little known but it probably feeds on krill—which makes up nearly 50 percent of the food eaten by related Antarctic species. All *Bathylagus* smelts are small fishes of polar waters.

Dolichopteryx parini

Winged spookfish

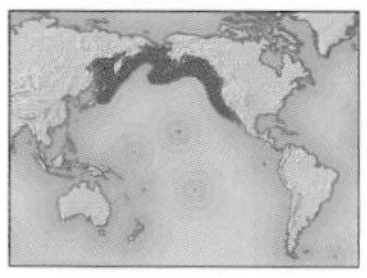

Length $8\frac{1}{2}$ in (22cm)
Weight Not recorded
Sex Male/Female
Status Not evaluated

Location North Pacific

Belonging to a family of remarkable "barrel-eyed" fish, this fish was described in 2001 on the basis of a single specimen from the waters of the North Pacific and is named for its winglike fins. Since then other specimens have been caught, but—like other members of the genus—it remains rare. It feeds on invertebrate planktonic animals, such as free-swimming crustaceans.

Plecoglossus altivelis

Ayu

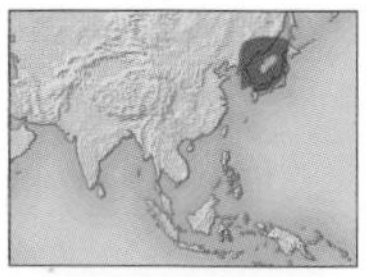

Length Up to 28 in (70 cm)
Weight Not recorded
Sex Male/Female
Status Not evaluated

Location W. North Pacific, E. Asia

The ayu is flattened from side to side, and has a dorsal fin set high on its back and a forked tail. A vegetarian, it feeds on river algae with its modified teeth and digests food with the help of an extra-long gut. The ayu's specialized kidneys enable it to move between sea and freshwater without being affected by the changes in salinity.

FISHES

Platytroctes apus

Legless searsid

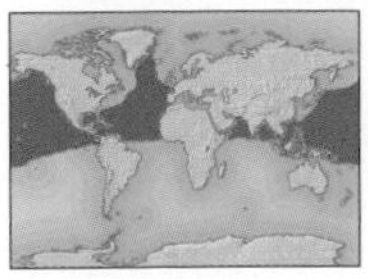

Length Up to 7 in (18 cm)
Weight Not recorded
Sex Male/Female
Status Not evaluated

Location North Pacific, North Atlantic, N. Indian Ocean

The legless searsid belongs to a family of fishes called the tubeshoulders, which get their name from a unique "shoulder" gland, located over the pectoral fin, which emits a luminescent liquid. The liquid forms a glowing greenish cloud in the water, probably helping the fish distract and escape from predators. The legless searsid lacks pelvic fins. It feeds on plankton, possibly using its long gill rakers to sieve them from the surrounding water.

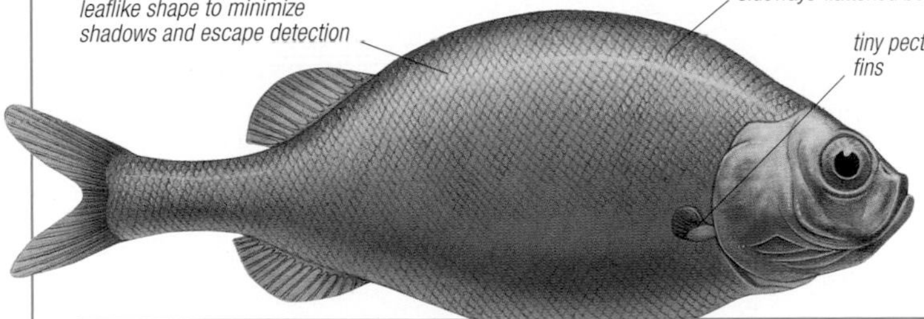

Argentina silus

Greater argentine

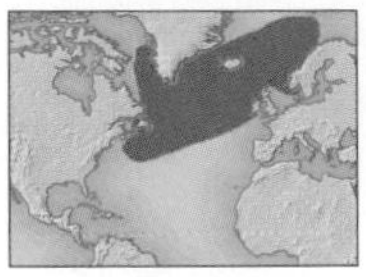

Length Up to 23½ in (60 cm)
Weight Over 16 oz (450 g)
Sex Male/Female
Status Not evaluated

Location North Atlantic, Arctic Ocean

Argentines are open-water or bottom-dwelling fish with slender bodies and large eyes. Their scales are large and have tiny spines. The greater argentine is brown, with an olive-colored back, white belly, and silvery sides that have a gold luster. It feeds on crustaceans, squid, comb jellies, and small fish, using its keen eyesight to spot and ambush prey in the dim conditions that prevail several hundred yards below the ocean surface. The greater argentine moves in schools and can swim away very rapidly to escape bigger fishes that prey on it. The eggs laid by the female float in mid-water currents in the upper regions of the open sea and, after hatching, the young make their way to shallower water, where they are left to survive on their own. The greater argentine is fished for food in Europe and Russia.

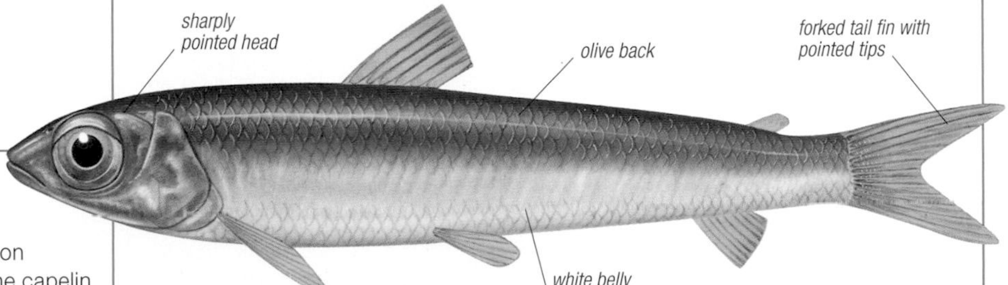

Mallotus villosus

Capelin

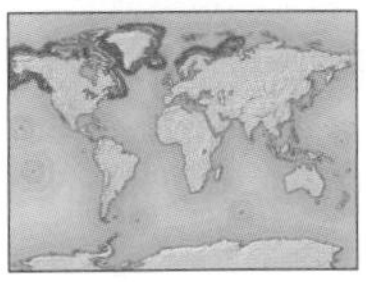

Length Up to 8 in (20 cm)
Weight Up to 1⁷⁄₁₆ oz (40 g)
Sex Male/Female
Status Not evaluated

Location North Pacific, North Atlantic, Arctic Ocean

The capelin lives and breeds at sea, forming large schools that are an important source of food for seabirds, seals, dolphins, and whales; in years when capelin numbers are low, seabird rookeries may entirely fail to reproduce. Slender and with a protruding lower jaw, it is olive-green on the back, merging with silver on its sides and underside, and has a large adipose fin—one of the characteristic features of salmon and their relatives. Although the capelin feeds mainly on zooplankton, using its comblike gill rakers to filter them from the water, it is also known to feed on marine worms and other fishes. During the breeding season, males can be identified by the swollen base of their anal fin and by a velvety band of modified scales (villi) along the sides of their body. When courting, the male stimulates the female by caressing her with his villi, which induces her to spawn. Capelin migrate inshore to spawn on beaches, where the female lays up to 60,000 eggs at high tide. These are buried in the sand and subsequently exposed by the action of the waves. This fish is a major food source for the indigenous populations of Alaska and Canada.

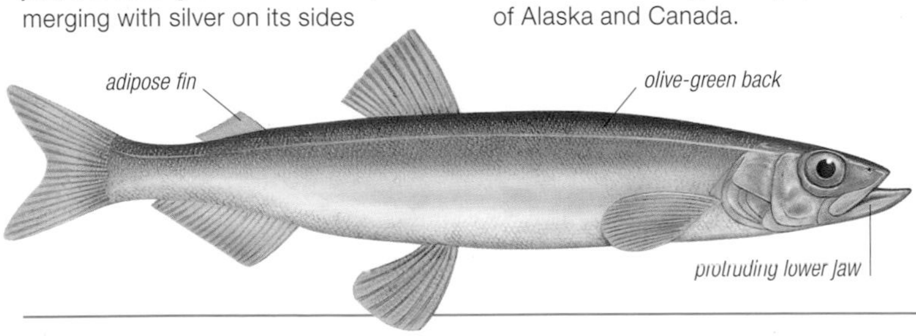

Alepocephalus rostratus

Risso's smooth-head

Length Up to 18½ in (47 cm)
Weight Up to 29 oz (825 g)
Sex Male/Female
Status Not evaluated

Location E. Atlantic, Mediterranean

Smooth-heads—also known as slickheads—are medium-sized, deep-sea fishes. There are nearly 100 species, found in temperate and tropical waters throughout the world. Risso's smooth-head is typical of the family, with its long and compressed body. Its dorsal and anal fins are set well back on the body and accentuate its elongated shape. Dark gray or brown in color, this fish has a slimy and scaleless head, while the rest of the body is covered with smooth scales of moderate size. It is an ambush predator, snapping up its prey, such as invertebrates and crustaceans, from the seabed or from open water.

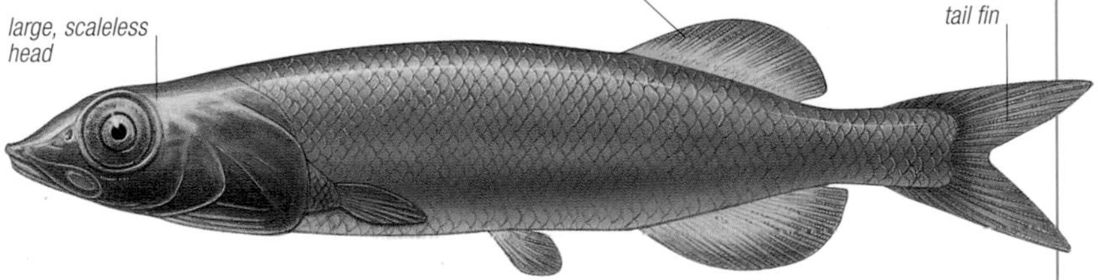

Lepidogalaxias salamandroides

Dwarf pencilfish

Length Up to 2¼ in (6 cm)
Weight Up to ⅞ oz (25 g)
Sex Male/Female
Status Near threatened

Location S.W. Australia

Discovered in 1961, this tiny silver-brown fish is a remarkable example of adaptation to a highly localized habitat. It is found only in the southwest corner of Western Australia, where it lives in small pools of water that lie over peaty sand. Although the water is fresh, it is black and acidic, and dries up entirely in the summer months. Dwarf pencilfishes survive during summer by burrowing up to 23½ in (60 cm) deep. To facilitate this, they have unusually flexible backbones that allow them to wriggle through the damp sand. They remain dormant for up to 5 months, breathing through their skin. Dwarf pencilfishes cannot move their eyes, because they lack external eye muscles, but they can bend their "neck" to look around. At the onset of the breeding season, males mate with females, and the eggs are fertilized inside the female's body. Each female lays up to 100 eggs, and the complete lifespan of the species is often just one year. Dwarf pencilfishes feed mainly on the larvae of aquatic insects.

Galaxias maculatus

Inanga

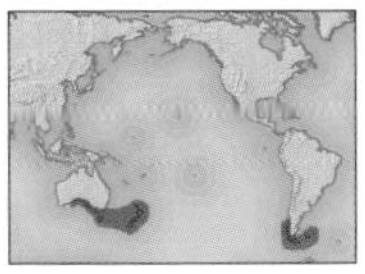

Length Up to 8 in (20 cm)
Weight Not recorded
Sex Male/Female
Status Not evaluated

Location S. Australia, New Zealand, South Pacific, S. Atlantic, S. South America

Also known as whitebait, this widespread fish of the Southern Hemisphere has a small, slender, almost cylindrical body, clear olive-gray to amber above, with a blunt head, and dorsal and anal fins set back close to its tail. It has a high tolerance for sudden salinity changes, linked to its life cycle, which is divided between freshwater and the sea. The eggs are laid on estuarine plants and hatch when the next high tide occurs. The larvae spend up to 7 months at sea, before returning to mature in freshwater.

FISHES

Dragonfishes and relatives

PHYLUM	Chordata
CLASS	Osteichthyes
SUBCLASS	Actinopterygii
SUPERORDER	Stenopterygii
ORDERS	2
FAMILIES	5
SPECIES	About 430

Dragonfishes and their relatives are deep-sea predators with unusual adaptations for catching prey, including large, flexible jaws, sharp teeth, and light-emitting lures. They also have luminous organs on their body known as photophores. This group also includes bristlemouths, viperfishes, slackjaws, and snaggletooths. Although their names reflect their ferocity, these are relatively small fishes whose prey consists of other deep-sea fishes and invertebrates. These fishes are found in all the world's oceans, but their distributions are generally discontinuous.

Anatomy

Most of the fishes in this group have an elongated body and a relatively large head. The head usually contains modified jaws with a very wide gape. For example, slackjaws have no floor to their mouth. Viperfishes have an extra joint in the head that increases their gape; as the jaw opens, organs such as the heart and gills are pushed backward and downward, out of the way of incoming food. The jaws are usually armed with numerous needlelike teeth that are used to seize prey or to prevent food from escaping. Apart from viperfishes, the fishes in this group have a generally scaleless body that is colored black or brown to match the darkened waters. However, some species are silvery or even transparent.

Many species have light-producing organs called photophores. They share this feature with the lanternfishes (see opposite), although in the dragonfishes the photophores are generally larger and more numerous. They are usually located along the sides and underparts of the body or near the eyes. Each is studlike in shape and linked to a nerve. Light is reflected through a silvery backing and focused through a lenslike thickening of the overlying scale. They either shine constantly, or flash on and off. Species that live in deeper water—at depths to about 1,000 ft (300 m)—generally have fewer and weaker photophores, and the light output seems to vary according to the amount of light received from above. Thus, photophores may hide the fishes from predators below them by matching the background illumination from above. Almost all species in this group have a light organ at the end of a chin barbel or a specialized fin ray. This is used as a lure to attract prey within striking distance.

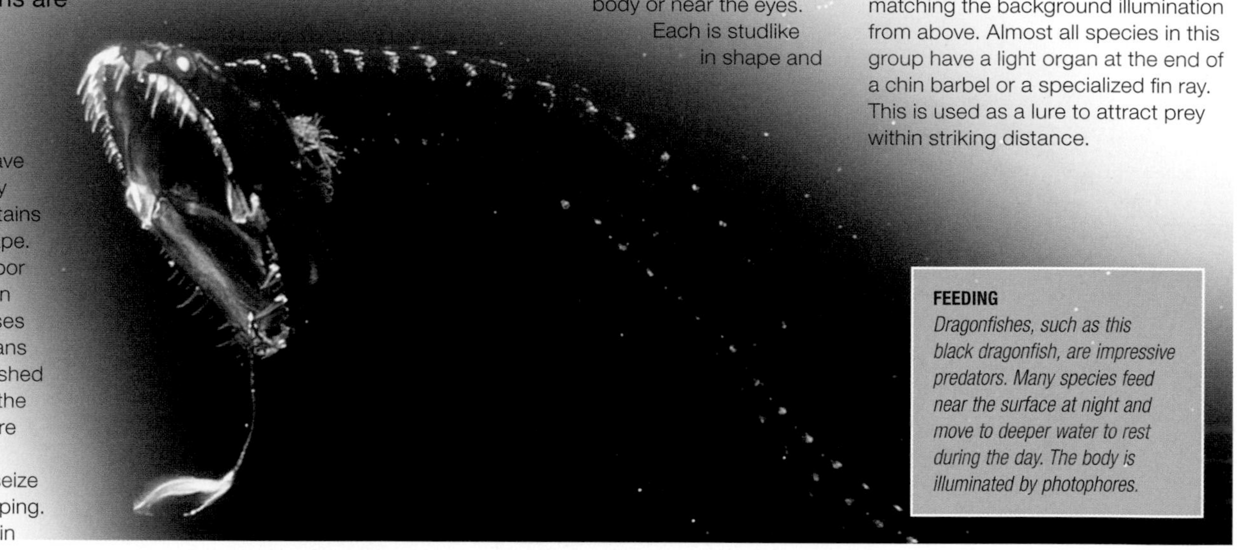

FEEDING
Dragonfishes, such as this black dragonfish, are impressive predators. Many species feed near the surface at night and move to deeper water to rest during the day. The body is illuminated by photophores.

Chauliodus sloani

Sloane's viperfish

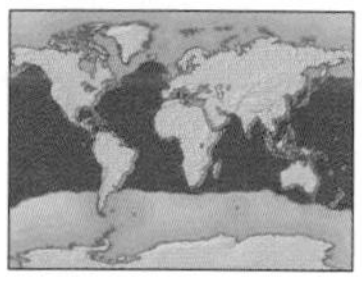

Length Up to 14 in (35 cm)
Weight Up to 1 1/16 oz (30 g)
Sex Male/Female
Status Not evaluated

Location Tropical, subtropical, and temperate waters worldwide

This deep-sea fish has a bluish black, elongated body, a large mouth, and exceptionally long, transparent fangs. The largest fangs are slightly barbed, and are too big to fit into the mouth—instead they protrude outside it when the jaws are closed. The dorsal fin has a long, arching ray that acts as a lure, and the fish also has photophores along its body and around its mouth, which help entice prey toward its jaws. This viperfish spawns all year round. It is one of about 6 species of viperfishes, found at great depths in oceans all over the world.

Idiacanthus antrostomus

Pacific blackdragon

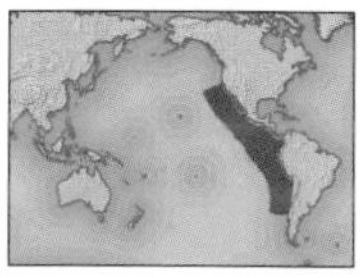

Length Female up to 15 in (38 cm)
Weight Up to 2 oz (55 g)
Sex Male/Female
Status Not evaluated

Location E. Pacific

This bizarre, deep-water marine fish has a black, snake-like body and extremely large teeth that must be rotated to allow the mouth to open and close. Females are about 40 times larger than males. To attract prey, the Pacific blackdragon uses a movable light-producing organ located at the end of a barbel that protrudes from the front of the lower jaw. Other photophores run along the belly. Blackdragons feed by night, moving nearer the surface to prey on small fishes and crustaceans. The larvae are remarkable in having eyes set on long stalks and intestines that extend beyond the tail.

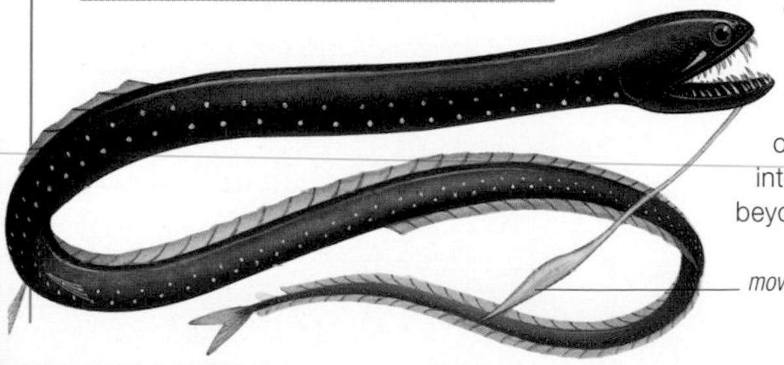

Malacosteus niger

Northern stoplight loosejaw

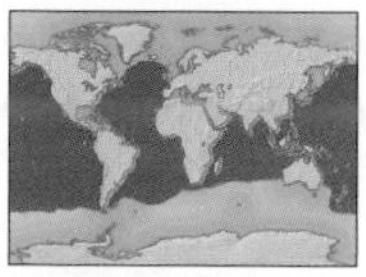

Length Up to 9½ in (24 cm)
Weight Not recorded
Sex Male/Female
Status Not evaluated

Location Tropical, subtropical, and temperate waters worldwide

The Northern stoplight loosejaw is unique in having no membranes joining its jaws and tongue—a feature that may help it make a rapid strike at prey. The head can pivot and the jaws are protractile, which allows it to draw in prey—fishes and crustaceans—much larger than itself. While many species in this family have light-producing organs (photophores) arranged in rows along the body, the Northern stoplight loosejaw has them only around its mouth and is the only fish to produce red bioluminescence.

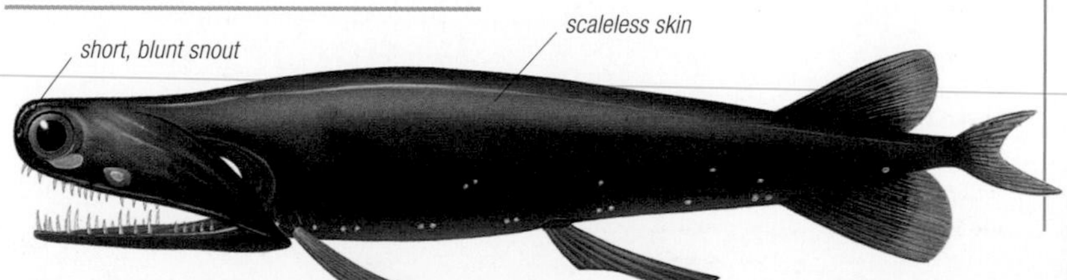

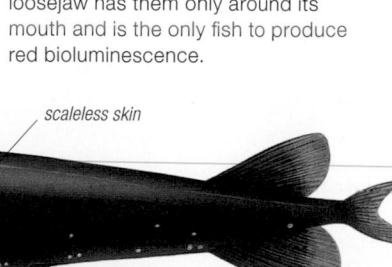

FISHES

Lanternfishes and relatives

PHYLUM	Chordata
CLASS	Osteichthyes
SUBCLASS	Actinopterygii
SUPERORDER	Scolepomorpha, Cyclosquamata
ORDERS	2
FAMILIES	19
SPECIES	About 520

Lanternfishes share with dragonfishes (see opposite) the ability to emit light from organs called photophores on their head and body. They are sometimes regarded as the deep-water equivalent of herrings because they are small, swim in large, dense schools, and form a vital element of deep-water ecosystems by providing food for many other fishes, seabirds, and marine mammals. Lanternfishes are found in great numbers in the middle layers of deep waters off the continental shelves. Their relatives include the slender lancetfish and tripod- and lizardfish, which use their fins to rest on or crawl across the sea floor. These fishes are found in both coastal waters and deep oceans.

Anatomy

All the fishes in this group have thin scales—some silvery and others darkly pigmented—that often fall off when they are caught. Lanternfishes have bioluminescent photophores on their flanks, undersides, and heads, which emit light in shades of green, yellow, or blue. Each species has a unique pattern of photophores, which helps groups maintain cohesion in darkly lit waters. Lanternfishes often make vertical migrations from deep water to surface waters at night to feed on plankton, returning to deep water as the sun rises. They have large eyes and good vision that help them detect changes in light intensity and recognize photophore patterns in dark water. The other fishes in this group are quite different from lanternfishes in anatomy and behavior. Most have no photophores. Lancetfishes are slender deep-water fishes. They have a large mouth containing sharp, pointed teeth and feed on other deep-water organisms. These include lanternfishes, and the lancetfishes make similar daily migrations as they follow their prey between deep water and the surface. Lizardfishes are bottom-living predators that hunt by ambush in tropical and temperate coastal areas. They have many sharp teeth and are camouflaged to match the color and texture of the seafloor. Tripodfishes stand on the tips of 3 elongated fins. They are found on the bottoms of deep ocean basins.

LIGHT PRODUCTION

A lanternfish's photophores resemble small, bright studs (a white-spotted lanternfish is shown here). The males and females of some species have photophores arranged in different patterns, which helps them recognize one another in dark water.

Bathypterois grallator

Tripodfish

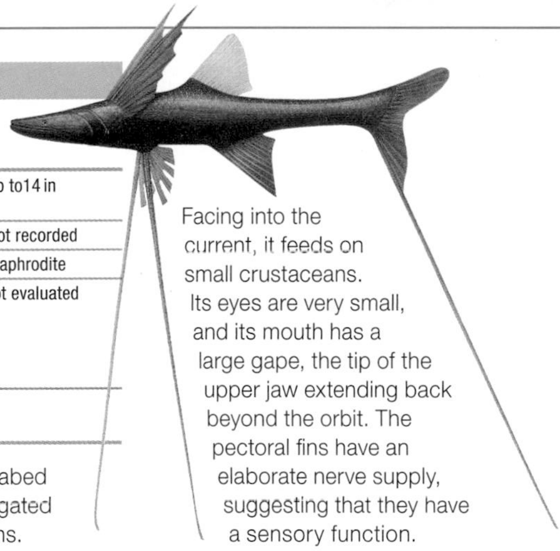

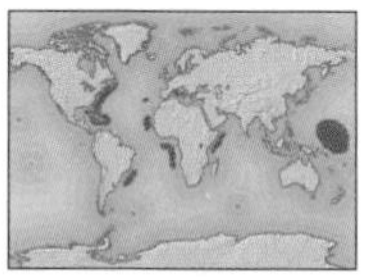

Length Up to14 in (36 cm)
Weight Not recorded
Sex Hermaphrodite
Status Not evaluated

Location Atlantic, Mediterranean, Indian Ocean, W. Pacific

This fish perches over the seabed by forming a tripod with elongated rays from its pelvic and tail fins.

Facing into the current, it feeds on small crustaceans. Its eyes are very small, and its mouth has a large gape, the tip of the upper jaw extending back beyond the orbit. The pectoral fins have an elaborate nerve supply, suggesting that they have a sensory function.

Alepisaurus ferox

Long-snouted lancetfish

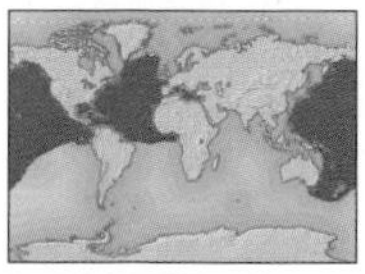

Length Up to 9¼ ft (2.8 m)
Weight 10 lb (4.5 kg)
Sex Hermaphrodite
Status Not evaluated

Location Pacific, North Atlantic, Mediterranean

This large and distinctive deep-sea fish has a spindle-shaped, iridescent body, a deeply forked tail, and a long dorsal fin that stretches for about two-thirds of its length, and is held upright like a sail. The function of this fin is unknown. The large mouth has big, pointed teeth for snagging slippery prey. This fish is a useful source of deep-sea specimens for scientists, because the fish in its stomach are usually intact.

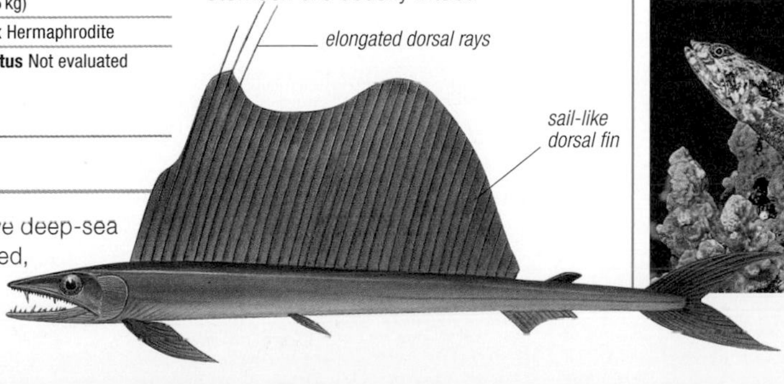

Synodus variegatus

Variegated lizardfish

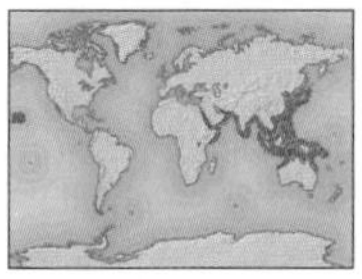

Length 10–13¾ in (25–35 cm)
Weight Not recorded
Sex Male/Female
Status Not evaluated

Location Pacific, Indian Ocean

With its pointed snout and large mouth, this fish's head looks very much like a lizard's, when viewed from the side. It often sits on the seafloor, propped on its pelvic fins, with its head raised in a very lizardlike fashion. It may also bury itself in the sand with only its snout and eyes visible, and lunge at small fishes passing by. The variegated lizardfish has a slender, tubular body and its brown, orange, or reddish coloration effectively camouflages it against the sea floor.

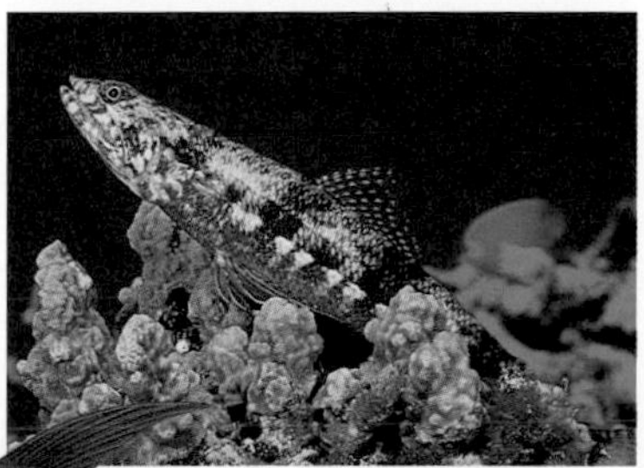

Myctophum asperum

Prickly lanternfish

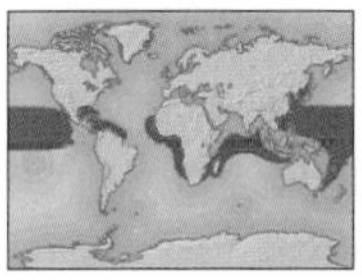

Length Up to 2¾ in (7 cm)
Weight Not recorded
Sex Male/Female
Status Not evaluated

Location North and South Pacific, W. and E. Atlantic, Indian Ocean

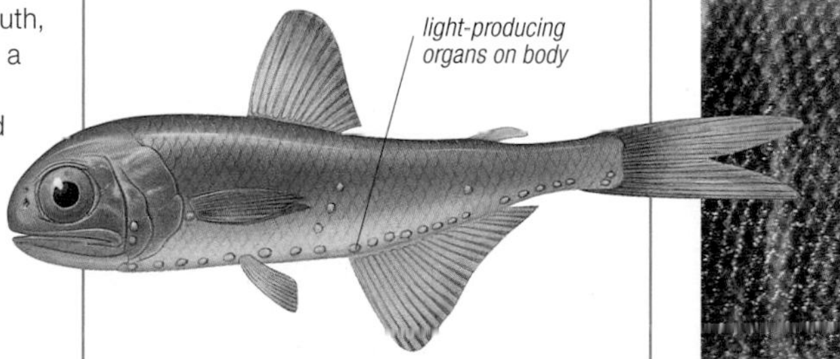

One of the most numerous of the deep-sea fishes and an important link in ocean food chains, lanternfishes eat mainly planktonic crustaceans, and are in turn eaten by seabirds, seals, and larger fishes such as tuna. They derive their name from the light-emitting organs on the head and body. The arrangement of these light organs is an important feature for the identification of the 250 or so species. Differing arrangements of larger light organs near the tail indicate whether the fish is male or female.

FISHES

Cod and anglerfishes

PHYLUM	Chordata
CLASS	Osteichthyes
SUBCLASS	Actinopterygii
SUPERORDER	Paracanthopterygii
ORDERS	5
FAMILIES	37
SPECIES	About 1,600

Cod form a crucial link in marine ecosystems, since they are both predators and prey to many other species. Their close relatives include hake, pollock, and the widespread grenadiers. Anglerfishes are bottom-dwelling or deep-sea predators. They owe their name to a dorsal spine, at the end of which is a lure, which is used to attract prey. This group also contains several other groups, such as cusk-eels (which include pearlfishes and ventfishes) and trout-perches. Apart from the last group, they are mostly marine, with a small number found in fresh or brackish water. Most species live on the seabed but some (such as cod, haddock, and hake) form vast schools.

Anatomy

Cod and their relatives typically have a long, narrow body. The fins, which are supported by relatively weak spiny rays, have a distinctive arrangement: the ventral fins are positioned well forward, and the long dorsal fin is often divided into smaller units. True cod have small, elongated scales, a mouth that opens vertically for feeding on bottom-dwelling animals, and a single sensory barbel on the lower jaw. Hake look similar to cod but have scales on their body and a large mouth with numerous sharp teeth. They are predatory fishes that move in large schools in pursuit of smaller fishes and invertebrates. Grenadiers are covered in rough scales and have a large, bony head that tapers into a thin body and a filament-like tail (giving rise to the alternative name rat-tails). These deep-water fishes scavenge and hunt for small prey on the ocean floor.

The bottom-living anglerfishes are usually colored and patterned to blend into the seabed. They are relatively inactive fishes, with a stout, unstreamlined body and a large head with an enormous mouth. One of the fin rays is modified into a whiplike rod with a lure at the end. The lure often looks like a small marine organism, while in some deep-water species, it has a light-emitting organ at the end.

A typical cusk eel has a small head and a tapered body. The dorsal, anal, and tail fins are often joined to form a single fin. Cusk eels and their relatives have adapted to live in unusual habitats, including caves and areas around volcanic vents on the ocean floor. Some species even live inside the bodies of invertebrates, including molluscs, sea cucumbers, and sea squirts.

True trout-perches are so-called because they have a dorsal fin with spines at the front, like that of perches, and an adipose fin, like that found in trout. Their close relatives include several species that are found only in limestone caves in eastern USA.

Life cycle

In winter and spring each year, cod migrate to established breeding grounds to spawn. They reproduce in great numbers (a single female can produce several million eggs each year), which explains why they have been such an important food for other fishes, marine mammals, seabirds, and humans. In recent years, however, overfishing has greatly decreased the population of many commercial species, and may have even eliminated cod from some of their spawning areas.

Some species of deep-sea angler-fishes overcome the difficulties of finding mates in the darkness of the deep oceans by reproducing in a unique way. Soon after developing from the larval state, the male bites into the body of a much larger female. A permanent attachment is formed, the tissues and circulatory systems of the 2 fishes becoming fused. The male lives on nutrients from the female's circulatory system and releases sperm when she needs to fertilize her eggs. A single female may have one or several males attached to her.

FEEDING
Anglerfishes have a sophisticated feeding method that consumes little energy. The well-camouflaged fish (here, an angler or monkfish) uses its lure to draw prey toward its mouth. Once it is close enough, the anglerfish opens its mouth, creating a strong suction current. The mouth and stomach can both be expanded, so even large prey can be swallowed whole.

Amblyopsis spelaea

Northern cavefish

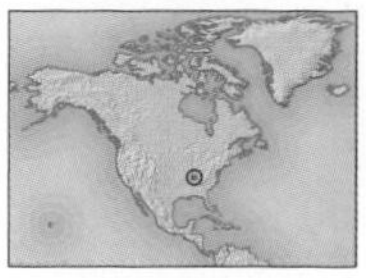

Length Up to 4¼ in (10.5 cm)
Weight Not recorded
Sex Male/Female
Status Vulnerable

Location USA (Kentucky, Indiana)

Like most cave-dwelling fishes, the northern cavefish has pale skin, is completely blind—in this case only vestigial eye tissues remaining under the skin—and has a very localized distribution. It is restricted to the limestone caves straddling the state line between Kentucky and Indiana, an area of less than 40 square miles (100 sq km). It feeds on tiny, cave-dwelling invertebrates, detecting them with the sensory papillae on its head and body.

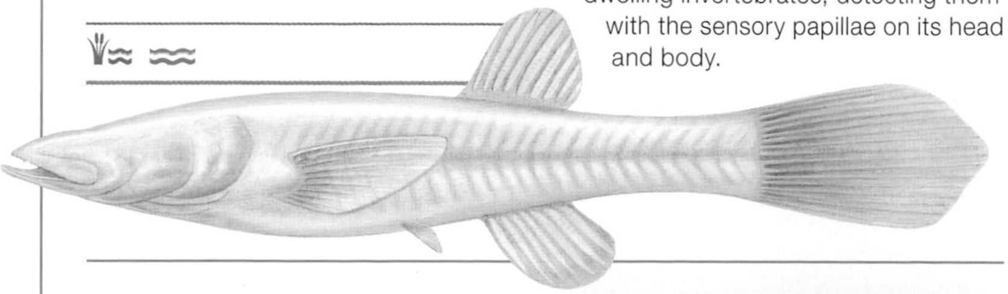

Carapus acus

Pearlfish

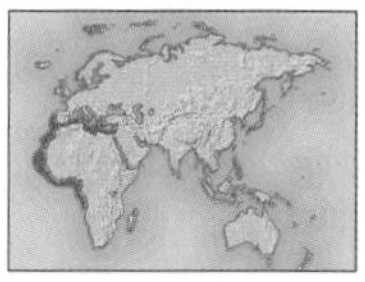

Length Up to 8 in (20 cm)
Weight Not recorded
Sex Male/Female
Status Not evaluated

Location E. Atlantic, Mediterranean

Like most of the other species in its family, this slim, scaleless, silvery white fish spends its adult life inside sea

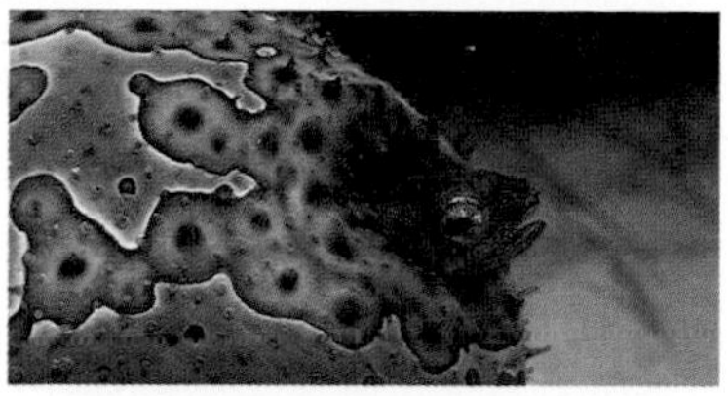

cucumbers. Attracted to them by their shape and the chemicals that they release, the pearlfish enters its host, tail first, through the animal's anus, and takes up residence in the body cavity. After dark, it goes out to feed on other animals, but it may also feed parasitically on its host's internal organs.

Thermichthys hollisi

Ventfish

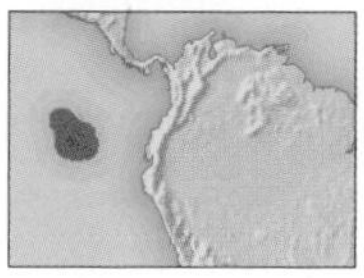

Length Up to 12 in (30 cm)
Weight Not recorded
Sex Male/Female
Status Not evaluated

Location Galapagos Islands

First identified in 1990, this rare fish is found only around hydrothermal vents on the deep-sea bed. It has a short, stout body and a sizable mouth, but small, poorly developed eyes. Like other vent animals, it depends for its survival on the bacteria that thrive in the hot, mineral-laden water that gushes out of vents. It is not known whether it feeds on the bacteria directly or on other animals that use them for food. The ventfish belongs to a family of nearly 200 species, known as live-bearing brotulas, which generally live in shallow water, and sometimes in caves. All give birth to live young.

Ophidion scrippsae

Basketweave cusk-eel

Length Over 11 in (28 cm)
Weight Not recorded
Sex Male/Female
Status Not evaluated

Location E. North Pacific

Cusk-eels are not true eels, but close relatives of pearlfishes (see left) and brotulas (see ventfishes, left). The basketweave cusk-eel normally lives at depths of up to 330 ft (100 m), but some specimens have been found at over 26,400 ft (8,000 m)—the greatest depth for any fish. Like other cusk-eels, this North American species has a blunt head, large mouth, and an elongated body that tapers almost to a point. The scales on its body are arranged in a criss-cross pattern, which give the species its name. Basketweave cusk-eels are active only at night; during the day, they bury themselves in sand with only the tip of their snout showing. At dusk, they partly emerge from the sand, waiting for small fishes or invertebrates to pass by. These fishes may also swim along the ocean floor in search of their food.

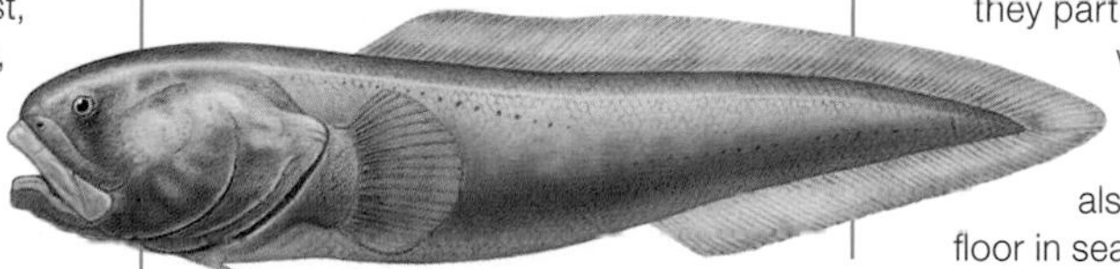

Lota lota

Burbot

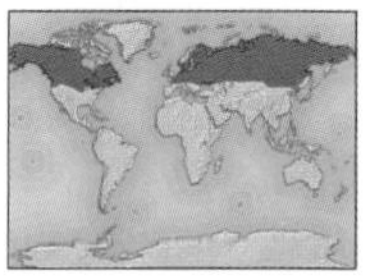

Length Up to 4 ft (1.2 m)
Weight Up to 75 lb (34 kg)
Sex Male/Female
Status Least concern

Location N. North America, Europe, N. Asia

This long, slender fish is the only cod to spend the whole of its life in freshwater. It is usually a dull, dirty green to brown above, with dark mottling, and yellowish to creamy below. The fish hides under tree roots or in crevices by day, and is most active at dawn and dusk, when it hunts for insect larvae, crustaceans, and other fishes. It spawns in winter, and does so—unlike most other fishes—in the middle of the night, and often under the ice.

Merluccius productus

Pacific hake

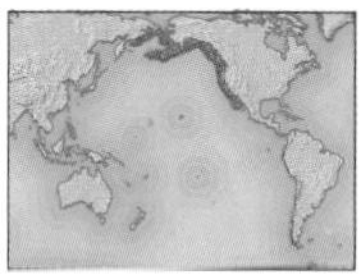

Length Up to 35 in (90 cm)
Weight Up to 11 lb (5 kg)
Sex Male/Female
Status Not evaluated

Location North Pacific

Closely related to the Atlantic cod (see below), the Pacific hake has a similar shape, with an elongated body, large head, and projecting lower jaw. It is bluish gray on its back, and silvery on its sides. Much like the Atlantic cod, it forms large schools and spawns in early spring, when the females lay millions of eggs. These drift on the surface and are widely distributed by ocean currents. Eaten by marine animals such as porpoises, seals, and swordfishes, the Pacific hake has also been heavily harvested by humans in recent years.

Gadus morhua

Atlantic cod

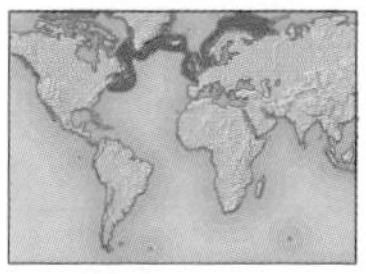

Length Up to 4½ ft (1.4 m)
Weight Up to 55 lb (25 kg)
Sex Male/Female
Status Vulnerable

Location North Atlantic, Arctic Ocean

This large, school-forming fish is one of the world's most commercially important species. It lives in water over the continental shelf, and usually feeds at 100–250 ft (30–80 m) above areas of flat mud or sand. Other fishes, especially herrings, sprats, and capelin, form its main prey. It usually breeds in early spring, with females releasing millions of eggs into the water. Adult cod migrate extensively and keep together, making them relatively easy to catch.

CONSERVATION

The Atlantic cod has been heavily harvested for many years. In the 19th century, specimens of up to 200 lb (90 kg) were sometimes caught, but with today's sophisticated fishing techniques, stocks have shrunk dramatically; fishes weighing over 33 lb (15 kg) are rare.

FULL-BODIED
A long-bodied fish, the Atlantic cod has a bluntly pointed snout, with one barbel hanging from its lower jaw. Coloration ranges from greenish to sandy brown.

Coryphaenoides acrolepis

Pacific grenadier

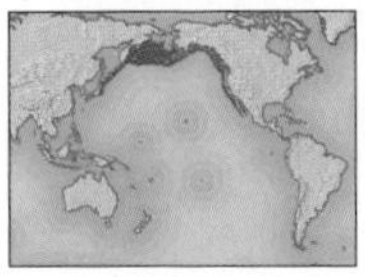

Location North Pacific

Length Up to 34 in (87 cm)
Weight Up to 22 lb (10 kg)
Sex Male/Female
Status Not evaluated

Also known as rat-tails or whip-tails, grenadiers are the dominant fishes on the sloping edges of the world's continental shelves. Like its 400 or so relatives, the Pacific grenadier has a large, bulbous head, a sharp snout, and a body that tapers rapidly into a narrow tail. Its eyes are large, enabling it to see prey on the seabed. The males make surprisingly loud sounds with the muscles in their swim bladder.

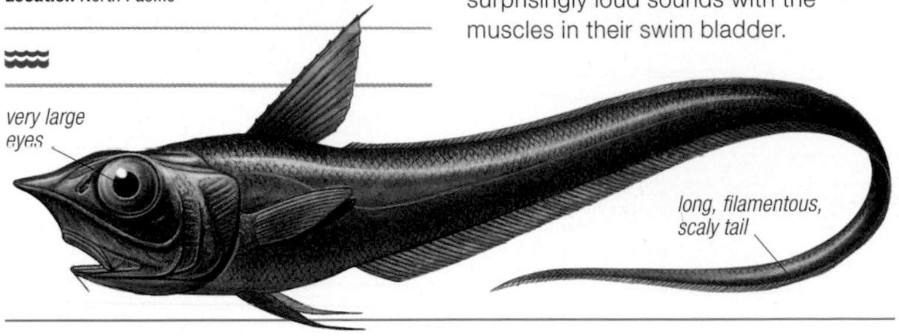

Antennarius maculatus

Warty frogfish

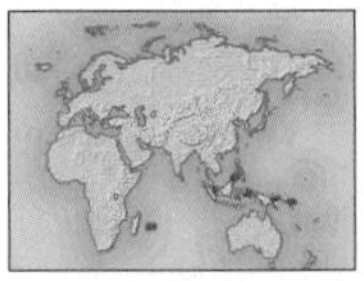

Location W. Pacific, Indian Ocean

Length Up to 4½ in (11.5 cm)
Weight Not recorded
Sex Male/Female
Status Not evaluated

This fish belongs to the anglerfish order—a group noted for their often bizarre appearance. Most anglerfishes live in the deep sea, but frogfishes live in shallow water—at depths of only a few metres—usually in coral reefs. The warty frogfish's dorsal fin has a long spine that ends in a flap of colored skin, which the fish wriggles to mimic a small fish and so entice prey toward its jaws.

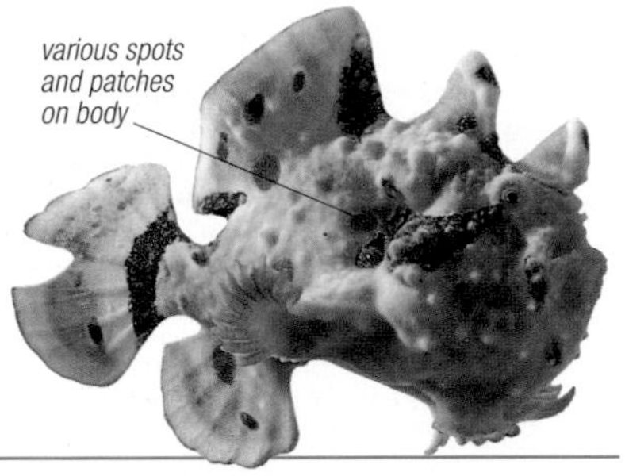

Histrio histrio

Sargassumfish

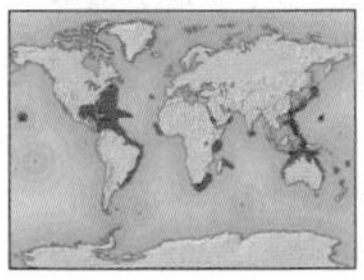

Location Tropical and subtropical seas worldwide

Length Up to 7½ in (18.5 cm)
Weight Up to 14 oz (400 g)
Sex Male/Female
Status Not evaluated

Superbly camouflaged, with highly variable coloring and patterns, this member of the frogfish family lives among drifting weed and is hence very widely distributed. It hunts prey by using its dorsal fin spine as a lure. Sargassumfishes exhibit unusually elaborate courtship behavior, with the males violently nipping and chasing the females. Except in the spawning season, adults are highly cannibalistic.

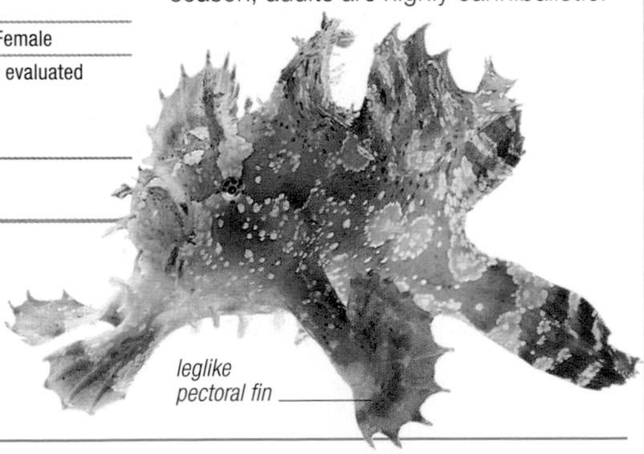

Melanocetus johnsonii

Humpback anglerfish

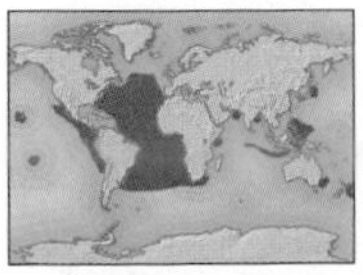

Location Pacific, Atlantic, Indian Ocean

Length Female up to 7 in (18 cm)
Weight Up to 21 oz (600 g)
Sex Male/Female
Status Not evaluated

This deep-sea anglerfish hunts in open water rather than on the seabed. The female is rounded, with an enormous head and jaws, and a lure on her dorsal fin. The male has no lure, and is less than one-fifth as long as his mate. He attaches himself to the female with his jaws but, unlike other deep-sea anglers, swims away after she has spawned.

Lophius piscatorius

Angler

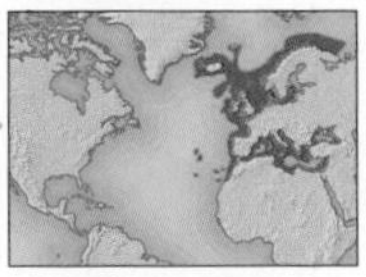

Location E. North Atlantic, Mediterranean, Black Sea

Length Up to 6½ ft (2 m)
Weight Up to 88 lb (40 kg)
Sex Male/Female
Status Not evaluated

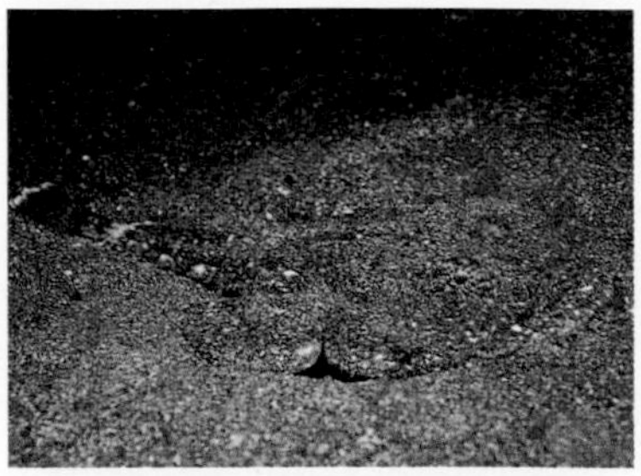

Also known as the monkfish, the angler is dull greenish brown, with a broad and flattened body that tapers rapidly toward the tail. A highly specialized sit-and-wait predator, it is very well camouflaged and has a fleshy lure on its dorsal fin that can be suspended above its mouth. If anything swims close by, the fish opens its jaws and sucks it in. The superbly camouflaged angler is almost invisible against the seabed. Its darkly marbled skin blends into the sediment and the outline of the body is difficult to discern because of the flaps of skin that fringe it. Spawning occurs between May and August depending on the locality. Each female produces up to one million eggs, which are released in a gelatinous sheet that floats at the surface. The bizarre-looking larvae that emerge from the eggs mature at different rates; males after 4 years and females after 6 years.

Histiophryne psychedelica

Psychedelic anglerfish

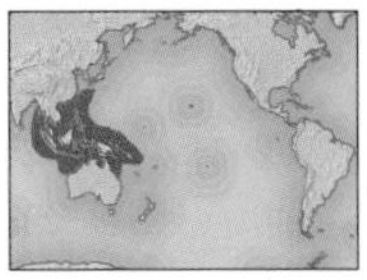

Location S.E. Asia

Length 6 in (15 cm) max
Weight Not recorded
Sex Male/Female
Status Not evaluated

Frogfishes are almost spherical-shaped anglerfishes, characterized by thick folded skin and forward-facing eyes. This distinctive species—pinkish in color, marked with white stripes and with a wide frilly face—was discovered among fishes destined for a public aquarium. In 2008, it was found living among coral rubble in shallows of Indonesian islands. Each fish has a unique pattern of stripes, and the fish "walks" on the seabed using its pectoral fins.

Ceratias holboelli

Kroyer's deep-sea anglerfish

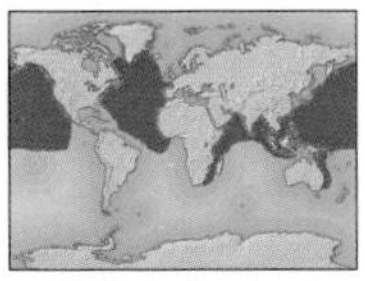

Location Pacific, Atlantic, Indian Ocean

Length Female up to 3¼ ft (1 m)
Weight Up to 110 lb (50 kg)
Sex Male/Female
Status Not evaluated

This fish shows an extraordinary difference between the sexes. The female (pictured) has a large head and mouth, but the male is tiny. When the adult male finds a mate, he bites into her flesh, and then lives parasitically on her. The pair's blood systems merge so that nutrients flow into the male's body; this bizarre lifestyle is found only in deep-sea anglers.

Linophryne arborifera

Bearded angler

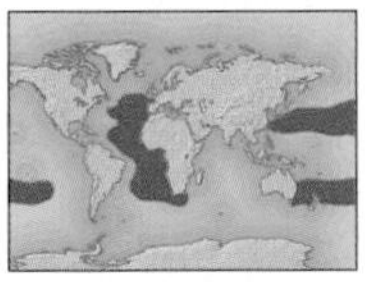

Location Atlantic, W. Pacific

Length Female up to 4 in (10 cm)
Weight Up to 11 oz (300 g)
Sex Male/Female
Status Not evaluated

Female bearded anglers have a luminous lure attached to the tip of their snouts, and also a complex branching barbel, resembling a piece of seaweed, hanging beneath their chins. Like the lure, the barbel gives off light that attracts prey. As in Kroyer's deep-sea anglerfish (see left), the male is much smaller than the female, and lives parasitically on his partner.

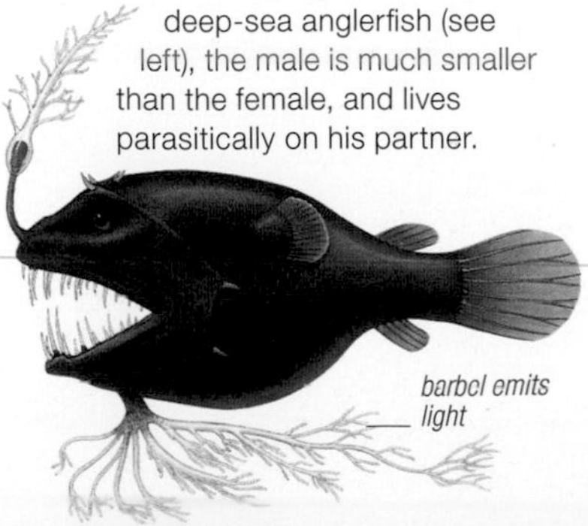

FISHES

Spiny-rayed fishes

PHYLUM	Chordata
CLASS	Osteichthyes
SUBCLASS	Actinopterygii
SUPERORDER	Acanthopterygii
ORDERS	17
FAMILIES	286
SPECIES	About 16,500

Spiny-rayed fishes form the largest and most recently evolved group of fishes, accounting for about half of all species. With such a large number of species they have a remarkable range of shapes, colors, behavior, and specialized adaptations. They vary in size, from tiny gobies ⅜ in (1 cm) long, to huge oarfishes reaching lengths of over 26 ft (8 m). Spiny-rayed fishes are found in almost all aquatic habitats. They dominate inshore marine waters, and are widespread in lakes, ponds, and slow-moving streams, but are also found in the open ocean, including deep waters.

Anatomy

Although spiny-rayed fishes form a large and varied group, they do share many of the same anatomical features. They typically have stiff, bony spines in or near their first dorsal fin. This differentiates them from other ray-finned fishes, which have fins supported by jointed bone segments. However, in some species, the spines are small or missing altogether; in others, such as lionfishes and stonefishes, the spines are armed with venom glands. Most species have ctenoid scales (see p.468), with minute spines on the surface of each scale, giving them a rough texture. In other species, the scales have evolved into bony plates or stout spines, and many have lost their scales completely. Most spiny-rayed fishes have a mobile mouth that can be extended forward, but mouth shape and types of teeth are extremely variable.

Spiny-rayed fishes include a variety of specialized predators, scavengers, and herbivores, which have evolved a remarkable range of body shapes for varying lifestyles. The tuna has a torpedo-shaped body for maximum swimming efficiency. Flatfishes, such as plaice or sole, have an asymmetrical body, with both eyes on one side, and swim with their underside flat against the seabed. Seahorses have a long snout, a head positioned at right angles to the body, and a prehensile tail to grasp marine plants. By far the largest group of spiny-rayed fishes is the perchlike fishes, which form the largest order of vertebrates (with over 9,000 species). As well as perch, this group contains many familiar fishes, such as bass, snappers, mackerel, and butterfly fish.

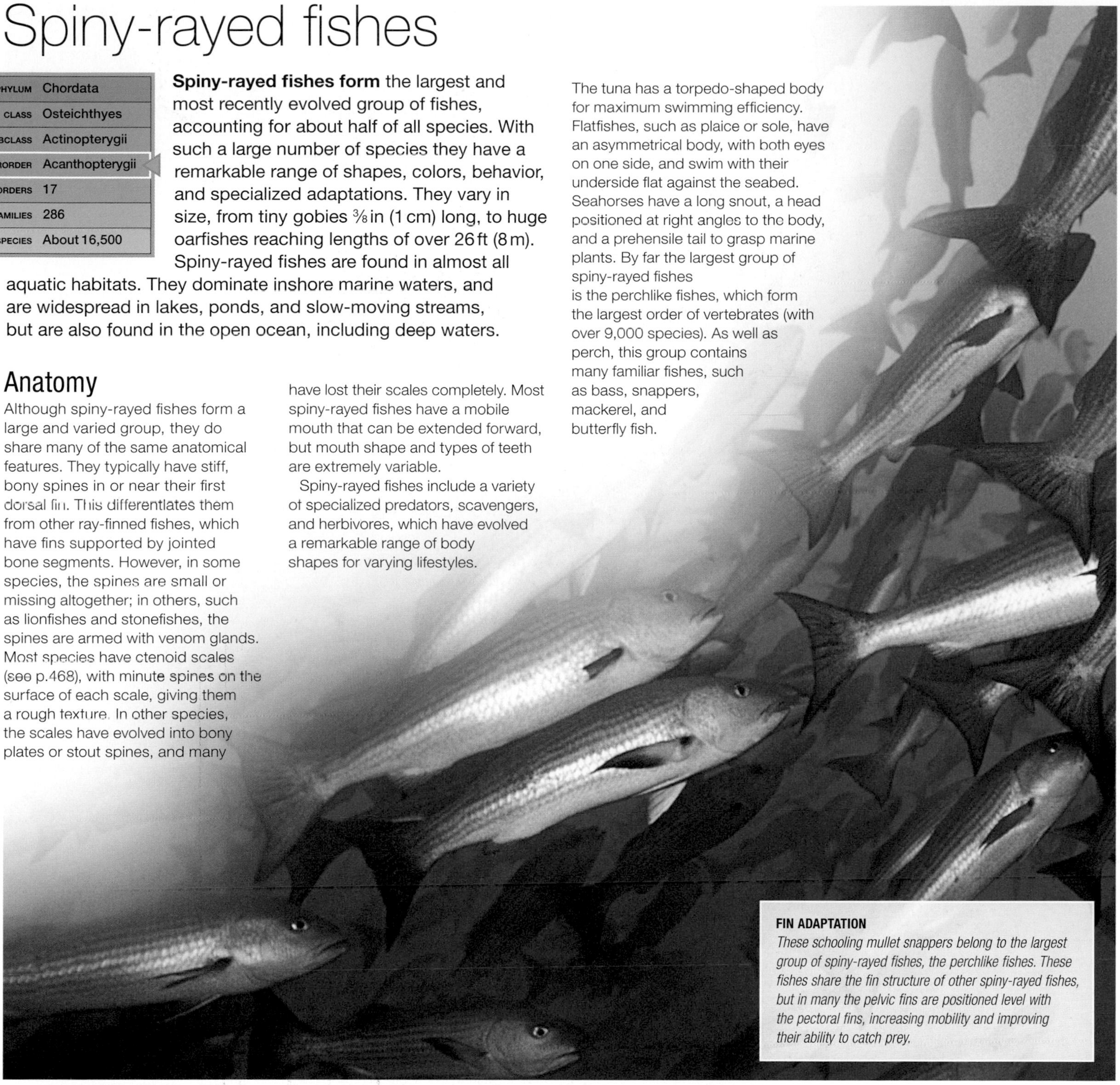

FIN ADAPTATION
These schooling mullet snappers belong to the largest group of spiny-rayed fishes, the perchlike fishes. These fishes share the fin structure of other spiny-rayed fishes, but in many the pelvic fins are positioned level with the pectoral fins, increasing mobility and improving their ability to catch prey.

Leuresthes tenuis

California grunion

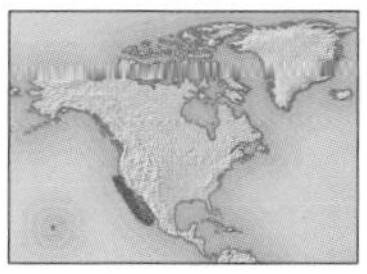

Length Up to 7½ in (19 cm)
Weight Up to 3⅝ oz (100 g)
Sex Male/Female
Status Not evaluated
Location E. North Pacific

The California grunion is one of over 100 species in the American silverside family, and like most of its relatives, lives close to the shore in large schools. It displays remarkable breeding behavior: during spring and early summer, schools of grunion strand themselves en masse on the moist sand of beaches to mate—an extraordinary sight that attracts many spectators. Spawning takes place at night for 2–6 days after a full moon, at high tide (see panel, right). The hunting of spawning grunion is a popular sport in southern California. However, only licensed fishermen are permitted to catch the stranded fishes, using only their hands. The California grunion is thought to feed on plankton and other microorganisms, possibly capturing these tiny organisms by quickly projecting its mouth.

SINGLE STRIPE
A small, slender fish, the California grunion has a silvery blue stripe running the length of its green body, and is silver-white below. It has no teeth, and its mouth is tubelike when projected.

MATING

California grunion beach themselves between waves of surf and mate on the moist sand. The female creates a shallow depression for the eggs, and the male wraps himself around the female, fertilizing the eggs as they are laid.

FISHES

Melanotaenia boesemani

Boeseman's rainbowfish

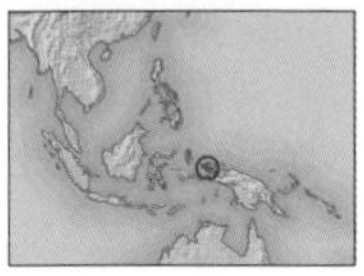

Length Up to 6 in (15 cm)
Weight Not recorded
Sex Male/Female
Status Endangered

Location N.W. New Guinea

Boeseman's rainbowfish is found in the Ajamaru Lakes region in the center of the Vogelkop Peninsula, New Guinea. The color pattern of the males of this species is completely different from most other rainbowfishes. The head and front portion are a brilliant bluish gray, while the fins and the posterior are bright orange-red. There are alternating light and dark vertical bars in between these 2 areas. Females are less brightly colored and have a shorter dorsal fin ray than the males. They lay 100–200 eggs.

Chelon labrosus

Thicklip grey mullet

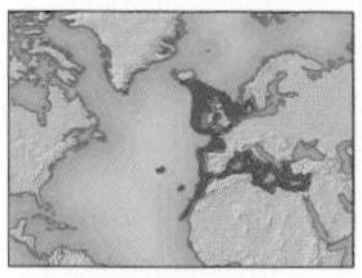

Length Up to 30 in (75 cm)
Weight Not recorded
Sex Male/Female
Status Least concern

Location E. North Atlantic, Mediterranean, W. Black Sea

The common name of this fish is derived from its prominent, thick upper lip. It feeds on organic and algal material found on surface sediments, using its gill rakers to filter the more indigestible material. The muscular stomach and long intestines make digestion more efficient. Found in dense schools in relatively shallow sea water and estuaries, the thicklip grey mullet is popular both as a sport fish and a commercial species. It belongs to a family of about 80 species and was once considered to be a relative of the barracuda because of similarities in the arrangement of the fins.

Cypselurus heterurus

Atlantic flyingfish

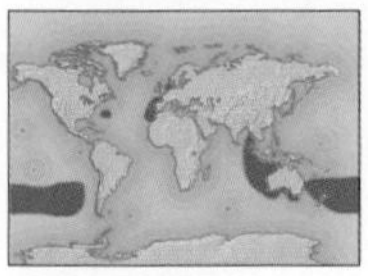

Length Up to 16 in (40 cm)
Weight Not recorded
Sex Male/Female
Status Not evaluated

Location Atlantic, E. Indian Ocean, Pacific

Also known as the four-wing flyingfish, this coastal species lives in the surface waters and feeds on plankton and other small fishes. Unlike other flyingfishes, both the pectoral and pelvic fins are modified for "flight." By opening out these fins, the four-wing flyingfish can glide over long distances—up to 655 ft (200 m)—to escape predators. In order to propel itself clear of the water surface, it may reach speeds of 37 mph (60 kph). The eggs have long, adhesive filaments by which they attach themselves to seaweed to avoid sinking.

Tylosurus acus

Needlefish

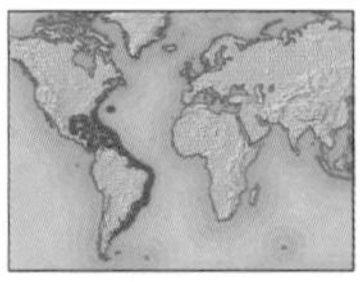

Length 35 in (90 cm), max 5 ft (1.5 m)
Weight Up to 7¾ lb (3.5 kg)
Sex Male/Female
Status Not evaluated

Location W. Atlantic

The largest member of the needlefish family, this species gets its name from its beaklike jaws and slender body. It is dark blue or green above and silvery below, and its jaws contain an impressive array of sharp teeth. To escape predators, this agile fish may jump clear of the water. While leaping, it may inadvertently spear boats or even boaters, causing severe injury due to its size and strength. The young may defend themselves by floating motionlessly on the surface in an attempt to mimic seagrass and escape detection.

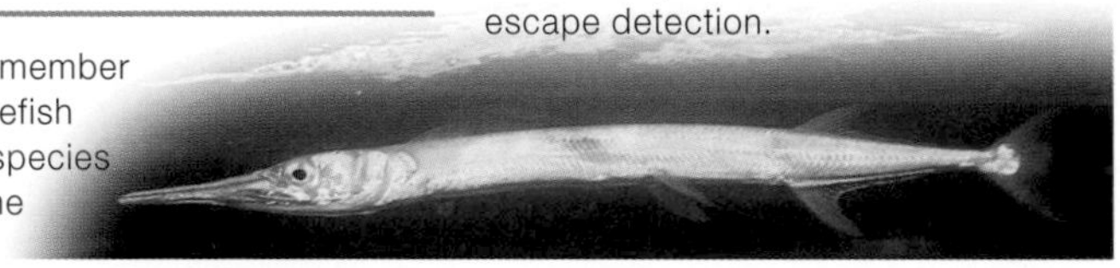

Exocoetus volitans

Tropical two-wing flyingfish

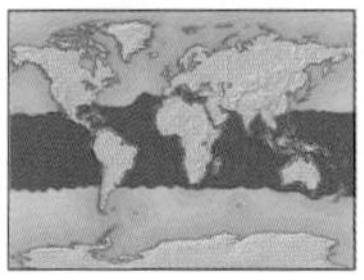

Length Up to 7 in (18 cm)
Weight Not recorded
Sex Male/Female
Status Not evaluated

Location Tropical and subtropical waters worldwide

Despite its name, this fish does not actually fly; instead, it glides over the water on specially enlarged and stiffened fins (see panel, right). Some species of flyingfishes have 2 pairs of "wings," consisting of enlarged pectoral and pelvic fins (see Atlantic flyingfish, above right) but this species glides on its 2 pectoral fins alone. Flyingfishes use this technique to escape predators such as dolphins, billfishes, and tuna, and may achieve speeds of up to 40 mph (65 kph) in the air.

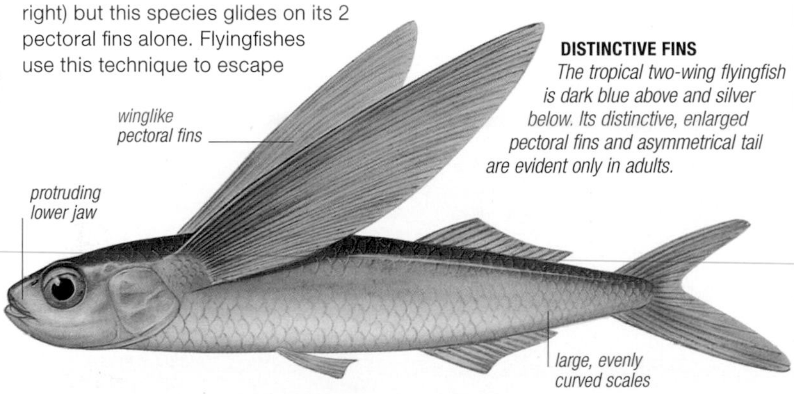

DISTINCTIVE FINS
The tropical two-wing flyingfish is dark blue above and silver below. Its distinctive, enlarged pectoral fins and asymmetrical tail are evident only in adults.

FISH OUT OF WATER

To achieve lift-off, flyingfishes beat their tails rapidly beneath the water for the initial thrust, then extend their pectoral fins to lift them clear of the surface. The lower tail lobe continues to oscillate. At full speed, the fish can glide for up to 12 seconds.

Fundulopanchax amieti

Amiet's lyretail

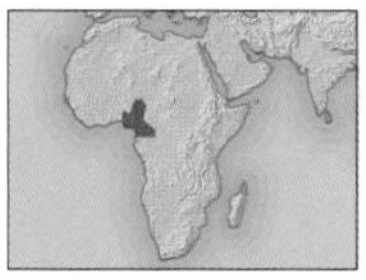

Length Up to 2¾ in (7 cm)
Weight Not recorded
Sex Male/Female
Status Not evaluated

Location W. Cameroon (lower Sanaga system)

Amiet's lyretail, or Amiet's killifish, remained undescribed until 1976, perhaps because of its very limited range. It is found in the stoneless, slow-moving, swampy parts of rivers in the rain forests of Cameroon. The female lays her eggs on the muddy riverbed toward the end of the rainy season. Adult fishes die after only one year, leaving the eggs dormant until the next rainy season. The onset of rains stimulates the eggs to develop and hatch. The fishes mature in only one month and quickly repopulate the river.

Jordanella floridae

American-flag fish

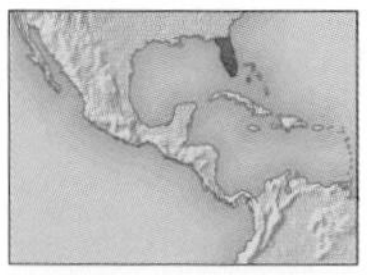

Length Up to 3 in (7.5 cm)
Weight Not recorded
Sex Male/Female
Status Not evaluated

Location S.E. USA (Florida)

This fish belongs to the pupfish family—a group of small freshwater fishes that are particularly common in the Americas. Like most pupfishes, the sexes differ in color, and the male's distinctive pattern and coloration gives the fish its common name. Normally a peaceful eater of algae, the male becomes aggressive during courtship and when guarding eggs. He attracts a female by flashing his red, unpaired fins. After the female has laid her eggs, he drives her away and guards the eggs until they are ready to hatch after about a week.

FISHES

Cyprinodon diabolis

Devils Hole pupfish

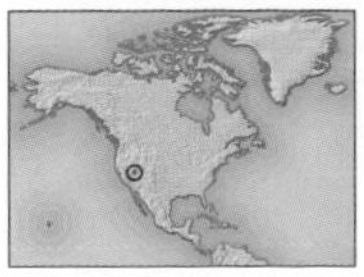

Length Up to 1 in (2.5 cm)
Weight Not recorded
Sex Male/Female
Status Vulnerable

Location S.E. USA (Nevada)

Found only in a single desert spring in Nevada, this tiny fish has one of the most restricted range of any vertebrate. The entire population numbers fewer than 500 individuals and occupies an area of about 215 square ft (20 sq m). Water pollution and fluctuations in the water table threaten its survival. The smallest of all pupfishes, it lacks pelvic fins and the vertical barring seen in most of its relatives. It breeds throughout the year, but rarely lives for more than 12 months.

Anableps anableps

Largescale foureyes

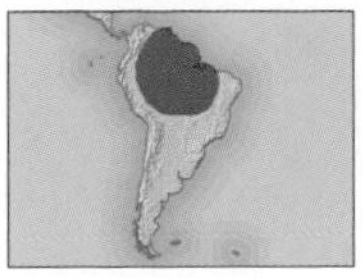

Length Up to 12½ in (32 cm)
Weight Not recorded
Sex Male/Female
Status Not evaluated

Location N. South America

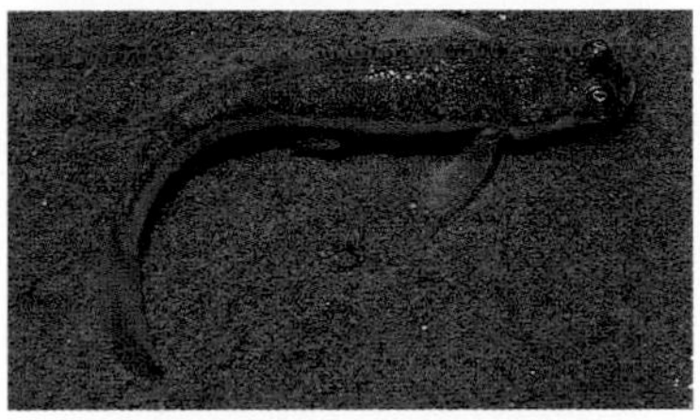

This long-bodied fish has protruding, froglike eyes that are each divided horizontally, allowing it to see above the water and below it at the same time. This enables the fish to hunt equally effectively for aquatic insects underwater or flying insects that have fallen onto the water's surface. It swims rapidly to escape predators like kingfishers and wading birds, and may jump clear of the water to escape predatory fishes. Female largescale foureyes fish can store sperm and so can lay multiple clutches of eggs.

Poecilia reticulata

Guppy

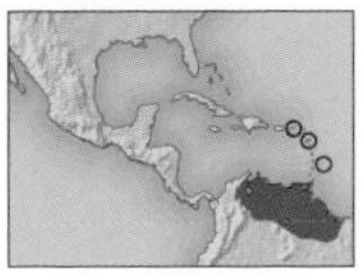

Length 2¼–2¾ in (6–7 cm)
Weight Under 3/16 oz (5 g)
Sex Male/Female
Status Not evaluated

Location Caribbean, N. South America

A popular aquarium fish, the guppy belongs to a family of more than 300 species that are notable for bearing live young. In the wild, the male guppy is very variable in color, often having blotches of blue, red, orange, yellow, and green pigments, as well as occasional spots, bands, or stripes of black. Such variation is greater in the aquarium fish. Very fertile, females may have over 100 young at a time, although the typical brood is 20–40. Adept at feeding on mosquito larvae, the guppy has been widely introduced as an agent of mosquito control.

Poecilia latipinna

Sailfin molly

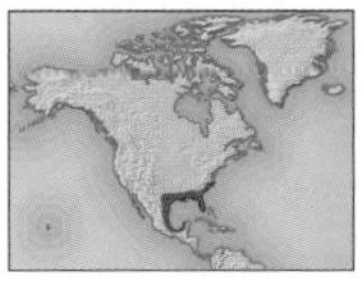

Length ½–2 in (1.5–5 cm)
Weight Not recorded
Sex Male/Female
Status Not evaluated

Location S.E. North America

The sailfin molly derives its name from the large dorsal fin that stands up along almost the entire length of the body. A close relative of the guppy (see above), this fish is also a popular aquarium species. Although it is generally olive-green, many different color forms have been developed. The back, sides, and dorsal fin have rows of spots, which may merge to look like stripes.

Lampris guttatus

Opah

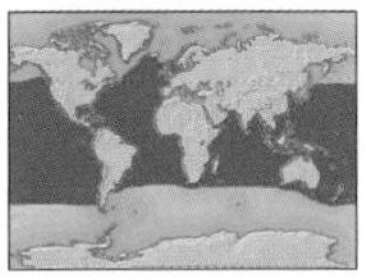

Length Up to 6 ft (1.8 m)
Weight 110 lb (50 kg)
Sex Male/Female
Status Not evaluated

Location Tropical, subtropical, and temperate waters worldwide

Also known as the moonfish, this large fish is an oceanic predator and has large eyes, an oval, silvery blue body—much deeper than it is wide—with strongly contrasting, deep red fins. Unlike most fishes, which swim using their tail, it swims with its pectoral fins, beating them up and down like a pair of wings. The fins are long and stiff, with a narrow profile that minimizes drag as they slice through the water. Despite its toothless mouth and large size—the maximum recorded weight of a specimen is 160 lb (73 kg)—the opah is well known as a very efficient predator, energetically pursuing small fishes and squid. It is found throughout the world but is most common in the North Atlantic and North Pacific oceans. An inhabitant of mid-water levels, the opah is taken on long lines and with gill nets, and is a valuable food fish in Hawaii and on the west coast of the USA.

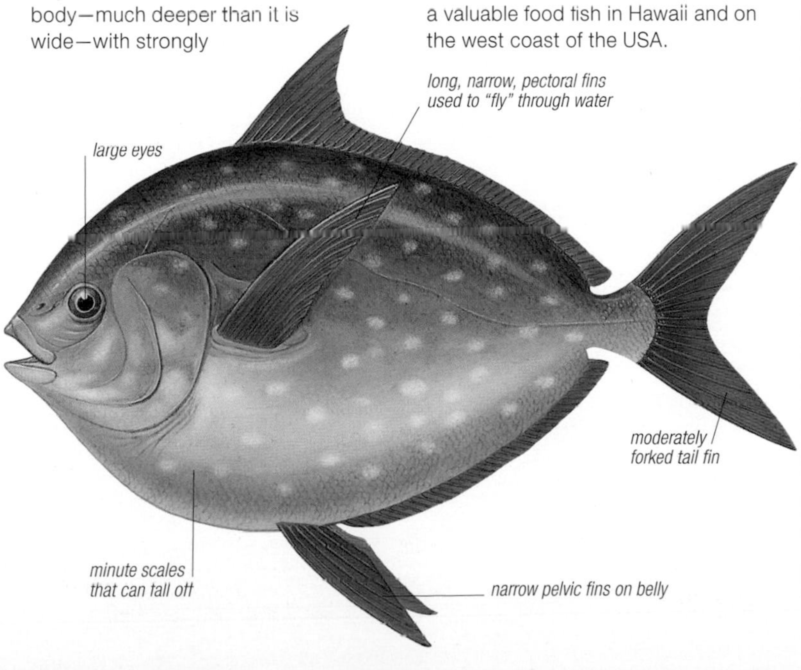

Regalecus glesne

Oarfish

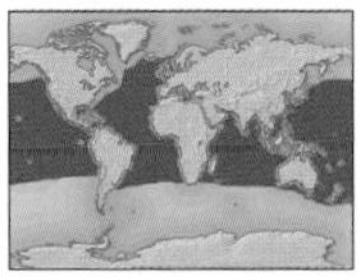

Length Up to 26 ft (8 m)
Weight Not recorded
Sex Male/Female
Status Not evaluated

Location Tropical, subtropical, and temperate waters worldwide

Although it lives worldwide in tropical and temperate seas, this huge, long, ribbonlike deep-sea fish is rarely captured, or even seen alive. Hence little is known about its behavior. Its extraordinary length protects it against most predators and has made it the subject of stories involving sea serpents and marine "monsters." This fish has a silvery body with oblique dusky bars, a short, bluish head, and deep red fins. Its dorsal fin runs along almost the full length of its body, and has long rays forming a crest over the fish's head. The pelvic fins are attached just behind its gill cover, and each consists of a long, single ray, with an expanded tip that looks like the blade of an oar, which is how the oarfish gets its name.

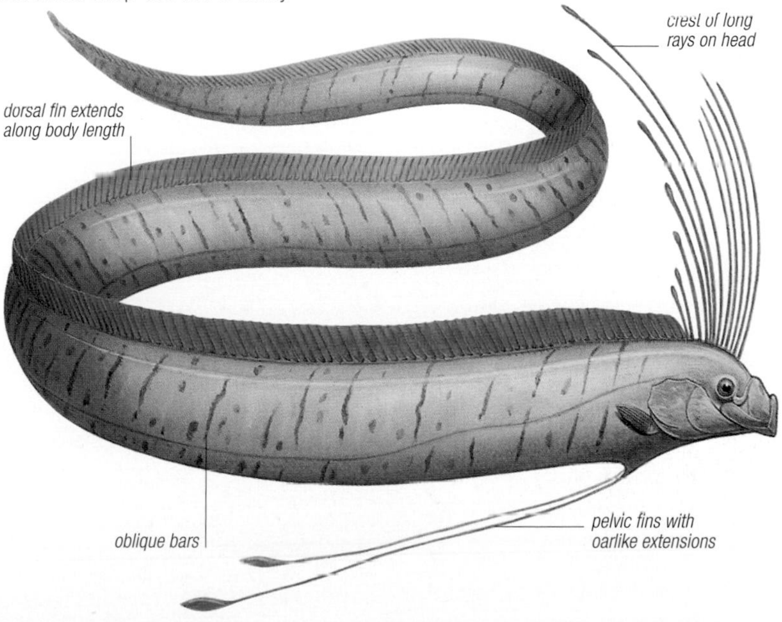

Trachipterus trachypterus

Mediterranean dealfish

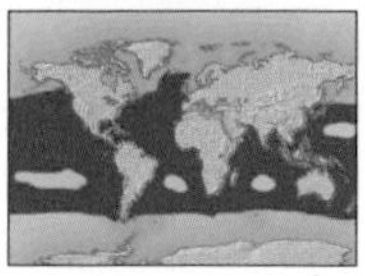

Length Up to 5¼ft (1.6m)
Weight Not recorded
Sex Male/Female
Status Not evaluated

Location Mediterranean, E. Atlantic, Indian Ocean, C. and W. Pacific

A close relative of the oarfish (see p.515), this widespread species differs from it not only in being smaller, but also in having a body that tapers rapidly towards its tail, ending in a tiny, upturned, fanlike fin. It usually has a pattern of 1–4 widely spaced dark blotches on the upper body and one or 2 blotches on the underside; it has no scales. As in the oarfish, its dorsal fin is deep red in color. This carnivore feeds on mid-water fishes and squid, and depends on its large size for protection against predators. Its eggs and larvae are often found in Mediterranean waters but there is little information on its reproductive behavior. The Mediterranean dealfish family contains 11 species; all are rare, and their biology is poorly known.

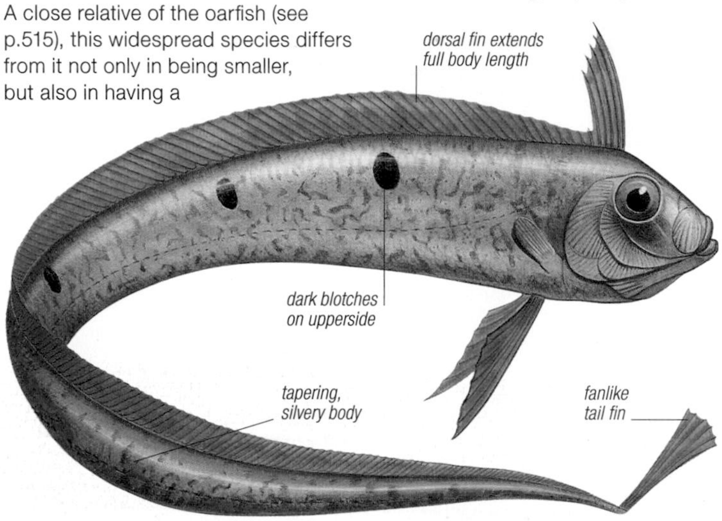

Barbourisia rufa

Velvet whalefish

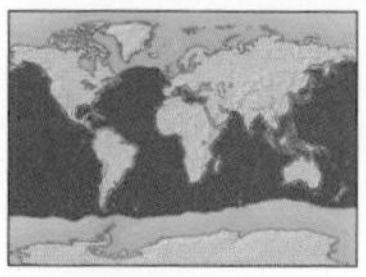

Length Up to 15½in (39cm)
Weight Not recorded
Sex Male/Female
Status Not evaluated

Location Tropical, subtropical, and temperate waters worldwide

This rarely sighted, deep-sea fish has a large mouth, the jaws having many small, depressible teeth. It is reddish orange, appearing black in deep waters, and its elongated body has a flabby feel. The skin of the velvet whalefish is also slightly prickly—a characteristic produced by its scales, each of which has a tiny, central spine. Its fins are placed far back along its body, suggesting that the fish is an ambush predator that attacks with a rapid, headlong rush. The fact that only individual specimens have been collected suggests a solitary lifestyle. This species has not been well studied, as few specimens have been collected, but their size suggests that, like many deep-sea fishes, the velvet whalefish is found at different depths at varying stages of its life. Small young fishes have been found in middle depths, whereas older fishes have been brought up from the seabed, indicating that this is where they spend their adult lives.

Cleidopus gloriamaris

Pineapplefish

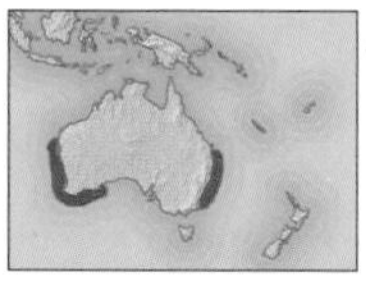

Length 11in (28cm)
Weight 18oz (500g)
Sex Male/Female
Status Not evaluated

Location Indo-Pacific

The pineapplefish gets its name from its shape, and from its large, black-edged scales, which look like the segments of pineapple skin. It belongs to a family of fishes that are protected by bony armor, and by sharp spines on their dorsal and pelvic fins. This particular species has a bioluminescent organ on the lower jaw—orange during the day, glowing blue-green at night—covered by the upper jaw when the mouth is closed. Because this mainly solitary fish ventures out at night from the depths of dark caves, curious small fishes, crustaceans, and other invertebrates are attracted by the glowing organ on its lower jaw, and are caught and eaten. There is only one other species in this family—the pinecone fish—which has its bioluminescent organ on the upper instead of the lower jaw. The bizarre appearance of the pineapplefish makes it extremely popular in the aquarium trade. However, only a skilled aquarist is capable of keeping it alive and well in captivity.

Myripristis murdjan

Pinecone soldierfish

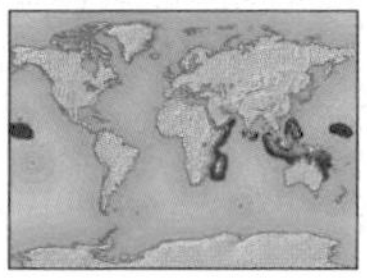

Length Up to 12in (30cm)
Weight Not recorded
Sex Male/Female
Status Not evaluated

Location Pacific, Indian Ocean

The pinecone soldierfish lives in shallow lagoons close to reefs, hiding in crevices during the day and coming out at night to feed on plankton, such as crustacean larvae. It swims with quick, jerky movements, propelled by a deeply forked tail. Like many nocturnal fishes, it has large eyes. The body scales are large, sharp, and very rough, and the spines on the fins and gill covers help deter predators. Its red fins are edged with white. The pinecone soldierfish is capable of producing a chattering noise, but why it does this is not known.

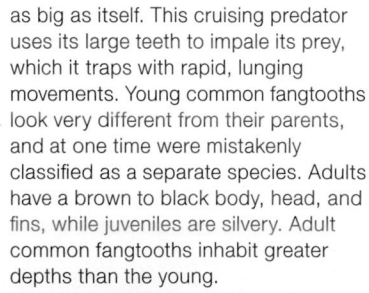

Anoplogaster cornuta

Common fangtooth

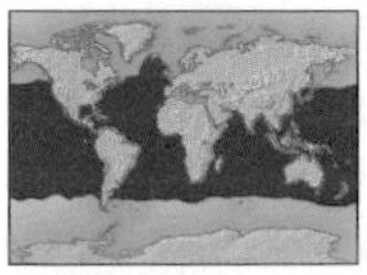

Length 7in (18cm)
Weight Not recorded
Sex Male/Female
Status Not evaluated

Location Tropical, subtropical, and temperate waters worldwide

Like many deep-sea fishes, the common fangtooth has a small body and a disproportionately large head—an adaptation for eating prey that can be almost as big as itself. This cruising predator uses its large teeth to impale its prey, which it traps with rapid, lunging movements. Young common fangtooths look very different from their parents, and at one time were mistakenly classified as a separate species. Adults have a brown to black body, head, and fins, while juveniles are silvery. Adult common fangtooths inhabit greater depths than the young.

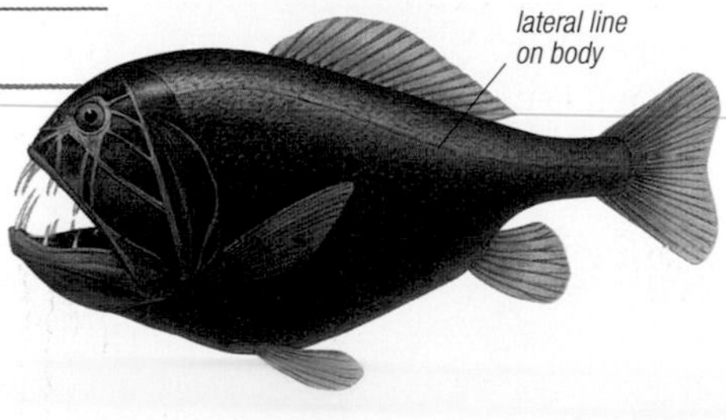

Hoplostethus atlanticus

Orange roughy

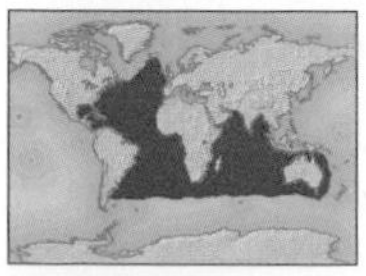

Length 20–23½in (50–60cm)
Weight Not recorded
Sex Male/Female
Status Not evaluated

Location Atlantic, Indian Ocean, W. Pacific

Roughies live mainly at the edges of continental shelves, or on the floor of the open ocean. The orange roughy has a large, blunt head with an upturned mouth, and small teeth arranged in bands. Its body is orange-red, although it appears dark at great depths, which gives it protection from predators. Like its close relative, *Hoplostethus mediterraneus*, the orange roughy is often caught as a food fish, and is very vulnerable to overfishing.

FISHES

Photoblepharon palpebratum

Flashlight fish

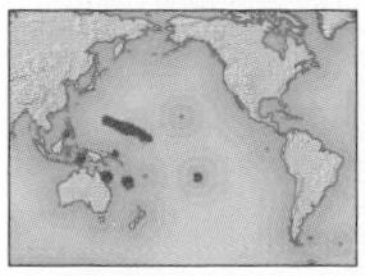

Length Up to 4¾ in (12 cm)
Weight Not recorded
Sex Male/Female
Status Not evaluated

Location W. Pacific

This small, blunt-nosed fish is one of the best examples of bioluminescence—or light production—in the animal world. One of only 8 known species of flashlight fishes, it remains relatively inactive in caves and deeper waters during the day, and moves to shallower areas over reefs at night, where it feeds on small planktonic animals. Individuals defend small territories on the reef. Symbiotic bacteria, contained in an organ under the eye, produce this fish's lime-green light. The light helps it to find prey, startle predators, and communicate with other individuals. Using a black "eyelid" to cover and expose the eye, the fish can synchronize schooling, indicate the presence of another fish, or defend territory.

Gasterosteus aculeatus

Three-spined stickleback

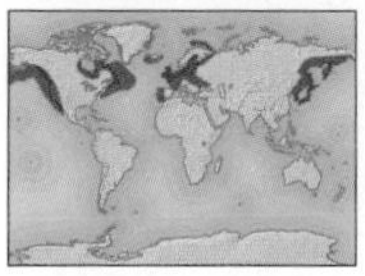

Length 2 in (5 cm), max 4 in (10 cm)
Weight Not recorded
Sex Male/Female
Status Least concern

Location Pacific, Atlantic, Mediterranean, North America, Europe, E. Asia

This widespread, freshwater and coastal fish is easily identified by the 3 movable spines on its back. Its breeding behavior is remarkably complex, with the males building underwater nests for the females to enter. When properly stimulated, the females lay a few eggs in the nests for the males to fertilize, and swim out again. The males aerate the eggs and guard the young.

Hippocampus satomie

Satomi's pygmy seahorse

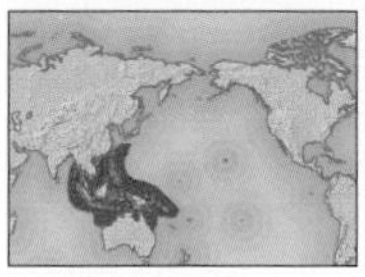

Length ½ in (1.4 cm)
Weight Not recorded
Sex Male/Female
Status Data deficient

Location S.E. Asia

Described in 2008 on the basis of specimens collected in Indonesia, this—the smallest known species of seahorse—is a nocturnal inhabitant of coral reefs. At night, it gathers in small groups on sea fans below reef overhangs at depths of 49–66 ft (15–20 m). Newborn young are black, but become paler and browner as they grow older—providing camouflage in their coral habitat.

Hippocampus guttulatus

Long-snouted seahorse

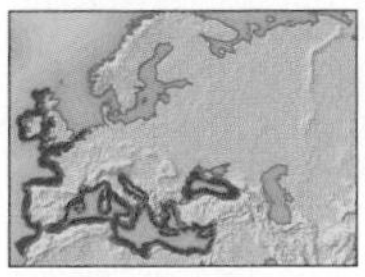

Length Up to 6½ in (16 cm)
Weight Not recorded
Sex Male/Female
Status Data deficient

Location E. North Atlantic, Mediterranean, Black Sea

Long-snouted seahorses are remarkable fish, with pipette-shaped mouths, horselike heads, and bodies covered with bony plates. They feed on planktonic animals, and swim upright, propelled by tiny fins. Most species live among eel grass and seaweed, and anchor themselves with their prehensile tails. Like their close relatives, the pipefish, they have unusual breeding behavior. In the long-snouted seahorse, the female lays the eggs in the male's brood pouch, at the base of his tail. He carries these until the young hatch.

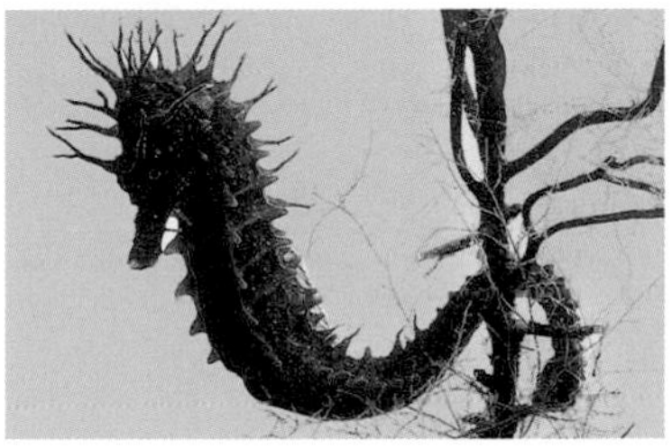

Phyllopteryx taeniolatus

Weedy seadragon

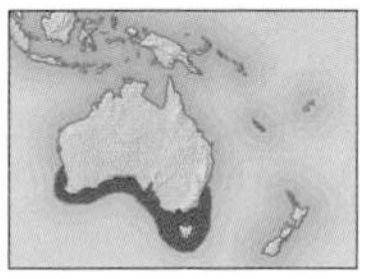

Length Up to 18 in (46 cm)
Weight Not recorded
Sex Male/Female
Status Near threatened

Location S. Australia

The weedy or common seadragon is one of the largest seahorses, and one of the most bizarrely shaped. Like other seahorses, its body is covered with toughened plates, but it also has leaflike flaps, which help provide camouflage. Usually found just above the seabed, the weedy seadragon hides among seaweed and plants growing on rocky reefs. Its bright coloration—brown to orange-red with yellow spotting—does, however, make it relatively easy to spot, and it is occasionally sighted in the open, away from reefs. The leafy seadragon, *Phycodurus eques*, has more elaborate camouflage than this species, and it is less commonly seen. The weedy seadragon searches for small invertebrates, including shrimps, among weeds and sponges near the ocean floor. It also feeds on planktonic animals in mid-water. Numbers of this species are decreasing due to indiscriminate collection.

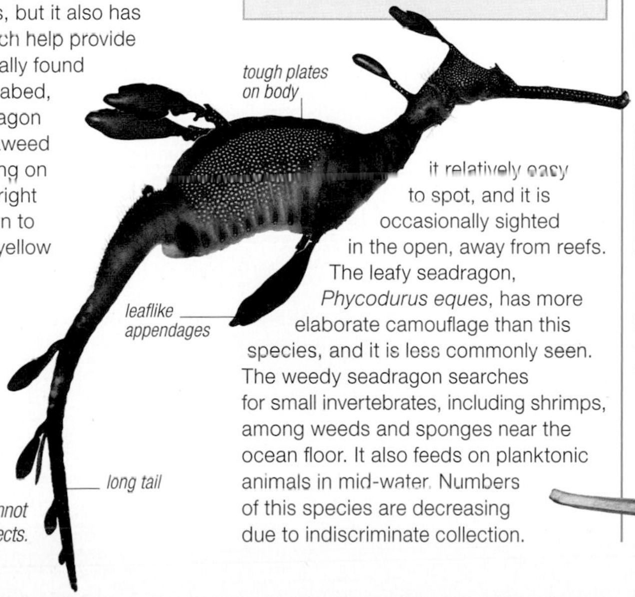

LEAFY PROTRUSIONS
The weedy seadragon is named for the leaflike appendages on its body. Unlike other seahorses, its tail cannot be coiled around objects.

DRAGON'S EGGS

After mating takes place, the fertilized eggs are attached to the wrinkled undersurface of the male's tail and carried around by him. As the eggs hatch, the baby seadragons swim away.

Aulostomus chinensis

Trumpetfish

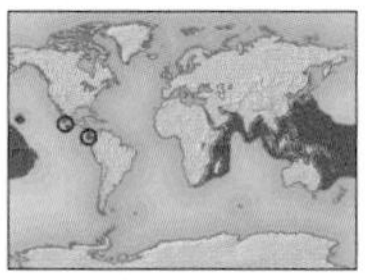

Length Up to 32 in (80 cm)
Weight Not recorded
Sex Male/Female
Status Not evaluated

Location Pacific, Indian Ocean

The trumpetfish is a common reef predator that often conceals itself by floating with its head vertically down among the branches of sea fans and other corals. It has a long, slender, flattened body, and a small mouth at the end of a very long snout. It is generally brown, but some are completely yellow. The trumpetfish hunts by ambush, lying motionless in water like a piece of drifting wood; it is galvanized into sudden action by passing schools of fishes.

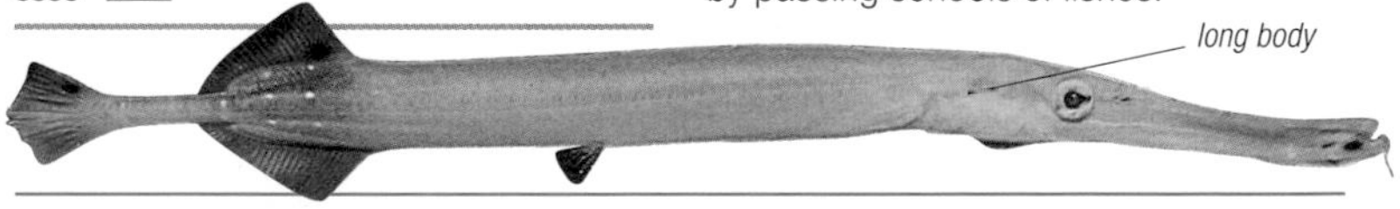

Aeoliscus strigatus

Razorfish

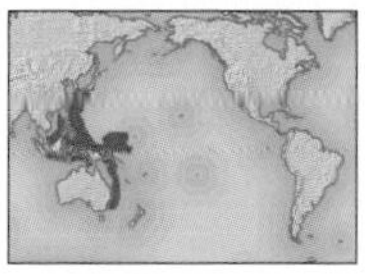

Length Up to 6 in (15 cm)
Weight Not recorded
Sex Male/Female
Status Not evaluated

Location W. Pacific

This fish spends much of its time swimming with its head vertically downwards. Schools of razorfishes move in synchrony while feeding on minute crustaceans that live among the corals and between the spines of sea urchins. The long, narrow body and elongated snout are adaptations to this lifestyle. The belly has a sharp edge, and that gives this fish its common name. A brown stripe runs along its body, which also has scattered spots—the patterning provides effective camouflage in white sandy habitats. The body is almost entirely covered by a transparent armor; only the underside of the tail region is free of this armor to allow for fin movement.

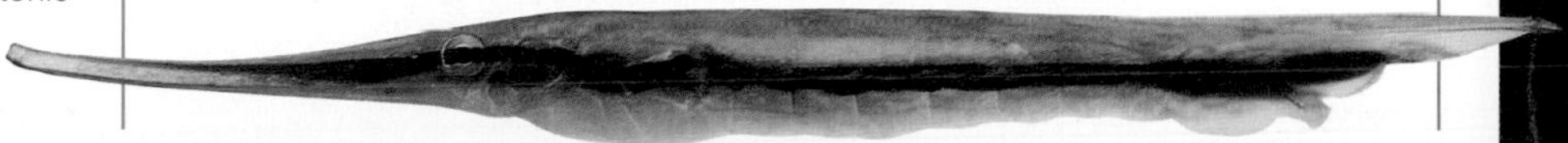

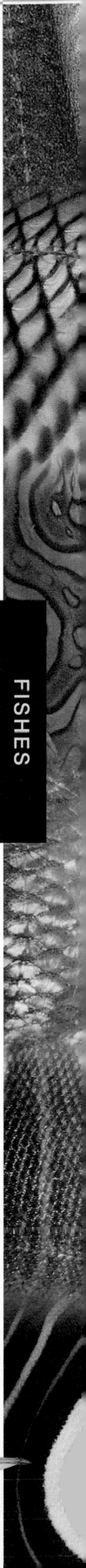

Anoplopoma fimbria

Sablefish

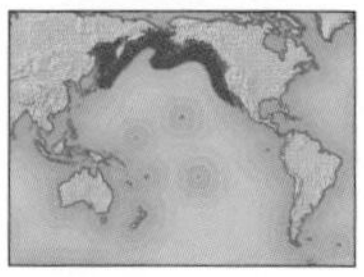

Length Up to 3½ft (1.1 m)
Weight Up to 125 lb (57 kg)
Sex Male/Female
Status Not evaluated

Location North Pacific

Although it is placed in the same order as the scorpionfish, the sablefish belongs to a different family that contains only 2 species. The sablefish is a widespread, North Pacific fish that looks superficially like a cod (see pp.510–1), with its long, streamlined body and widely spaced fins. Adults are blackish to greenish gray above, often with blotches or stripes, and have a white underside; juveniles are blue-black above. Young fishes live near the ocean surface, usually near coasts, where they feed on zooplankton. In contrast, the adults live close to the seabed. The sablefish makes extensive migrations along coasts, moving to deeper water in winter. It is an ambush predator that cruises in groups, chasing prey. It feeds on crustaceans, worms, and small fishes, such as lanternfishes. An important food source, caught in trawls and traps, and on long lines, the sablefish is commercially exploited and widely marketed, mostly in Asia.

Synanceia verrucosa

Stonefish

Length Up to 23½in (60 cm)
Weight Not recorded
Sex Male/Female
Status Not evaluated

Location Indo-Pacific

This fish is a stout, slow-moving predator, with superb camouflage and deadly venom produced by glands at the base of its 13 dorsal spines. The spines are equipped with sharp tips, and these can easily penetrate human feet, sometimes with fatal results. The stonefish hunts by lying in wait for its prey, sometimes half burying itself in the seabed. When partially buried, its upturned head and protruding eyes allow it to locate prey close by.

HIDDEN DANGER

Making the most of its cryptic coloring, the stonefish makes itself even less conspicuous by scooping out a shallow depression on the seabed with its pectoral fins. It then piles up sand or mud around its body, and waits for prey.

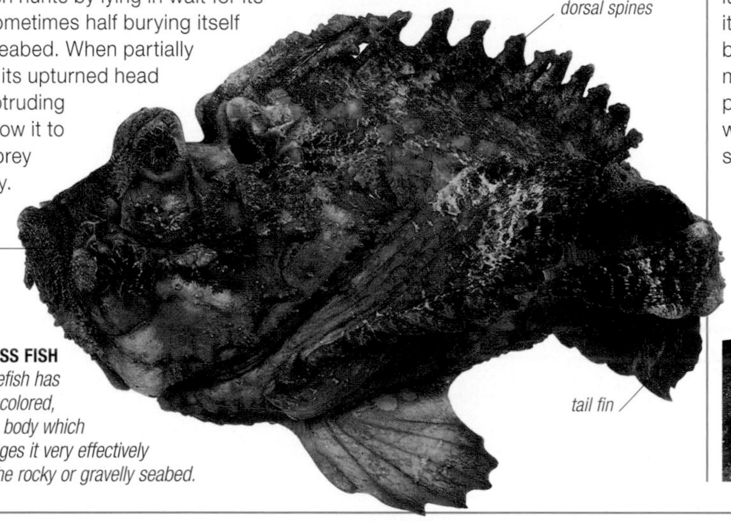

SCALELESS FISH
The stonefish has an oddly colored, scaleless body which camouflages it very effectively against the rocky or gravelly seabed.

Scorpaena porcus

Black scorpionfish

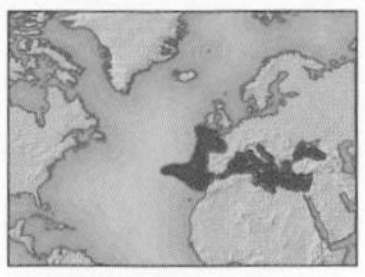

Length Up to 10 in (25 cm)
Weight Not recorded
Sex Male/Female
Status Not evaluated

Location E. North Atlantic, Mediterranean, Black Sea

The black scorpionfish, or rascasse, has a large, bulbous head with large eyes and mouth, a brown body with mottled markings, and several mosslike appendages between its mouth and eyes. It is a sit-and-wait predator that lurks among weed-covered rocks, using its highly effective camouflage to avoid being seen. All of its fins have rounded margins, and the spines of its dorsal, pelvic, and anal fins have poison glands, which, when erect, can inflict a harmful sting on its predators.

Cottus gobio

Bullhead

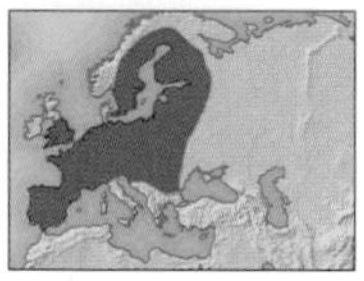

Length Up to 7 in (18 cm)
Weight Not recorded
Sex Male/Female
Status Least concern

Location W. Europe

The bullhead is a solitary freshwater species that lives under stones or in dense aquatic vegetation, where its dark brown, marbled skin provides excellent camouflage. It feeds on the eggs and larvae of fishes and small invertebrates. Spawning occurs between March and May, depending on the locality. The male excavates a small depression under a stone for the female to lay her eggs, and guards the eggs until they hatch.

Chelidonichthys lucerna

Tub gurnard

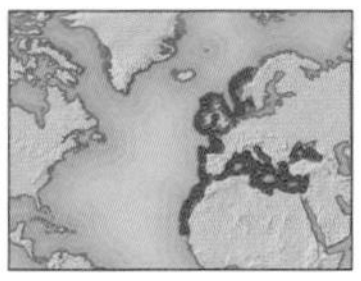

Length Up to 30 in (75 cm)
Weight Not recorded
Sex Male/Female
Status Not evaluated

Location E. North Atlantic, Mediterranean, Black Sea

A reddish colored fish with a bony plated head, the tub gurnard has adapted to life on the seabed.It has 3 isolated, long, and highly mobile rays on each of its pectoral fins that act as legs to support its body on the ocean floor. These fin rays are also highly senstitive and are used to probe the seabed for mollusks and crustaceans, which, along with fishes, form its main diet. The tub gurnard is capable of producing grunting sounds by contracting the muscles associated with its swim bladder. These sounds may help individuals stay together.

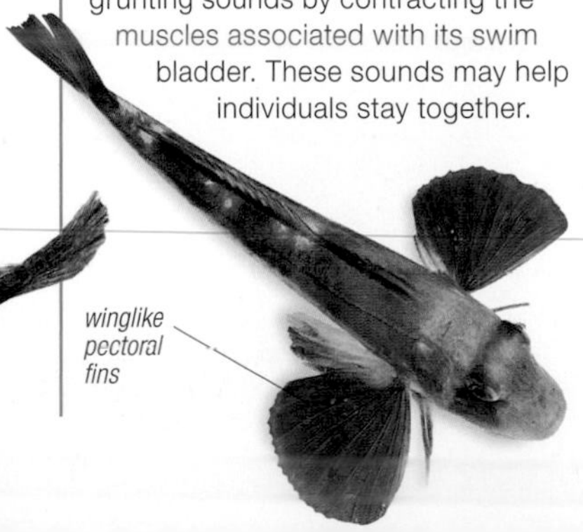

Cyclopterus lumpus

Lumpsucker

Length Up to 23½in (60 cm)
Weight Up to 21 lb (9.5 kg)
Sex Male/Female
Status Not evaluated

Location North Atlantic

Also called the lumpfish, this stout fish has an adhesive disk, formed from modified pelvic fins, on its underside. This acts as a sucker and anchors the slow-moving lumpsucker to rocks and seaweed, allowing it to live in places where the current is strong. Found in water up to 1,000 ft (300 m) deep, it swims into much shallower waters to breed. In the breeding season, the male turns reddish, whereas the female is blue-green. Its eggs are an important source of inexpensive caviar.

PARENTAL CARE

Female lumpsuckers lay large masses of pink eggs, attaching them to rocks in shallow water. The male guards the eggs constantly, and fans water over them to keep them oxygenated. The eggs hatch after 6–8 weeks.

ARMOR PLATING
Lumpsuckers get their characteristic knobbly shape from bony plates scattered over the body. Arranged in rows, these run along the fish's back and sides.

FISHES

Agonus acipenserinus

Sturgeon poacher

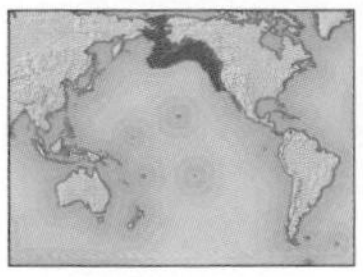

Length Up to 12 in (31 cm)
Weight Not recorded
Sex Male/Female
Status Not evaluated

Location North Pacific

The sturgeon poacher is one of about 50 species of poacher, and derives its name, from its sturgeonlike shape. Like all poachers, it has a large head and elongated snout, and uses its pectoral fins like oars and its tail like a rudder to move along the seabed. All poachers also have a distinctive cluster of finely divided barbels, called cirri, that hang from their downward-pointing mouths. The taste buds on the cirri are used to probe the seabed for crustaceans and small animals. Slow-moving, poachers are easy fishes to catch.

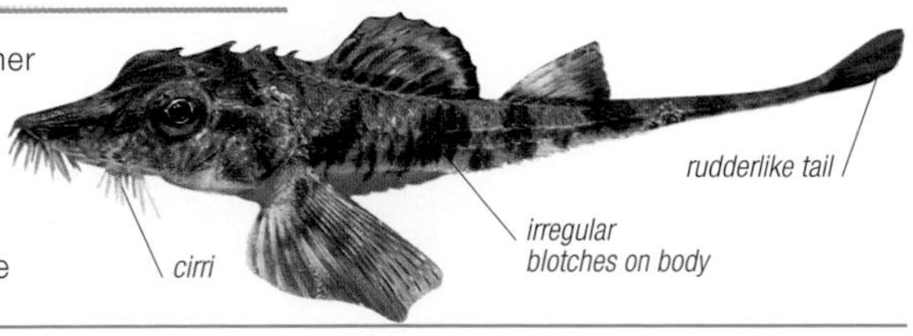

Lopholatilus chameleonticeps

Tilefish

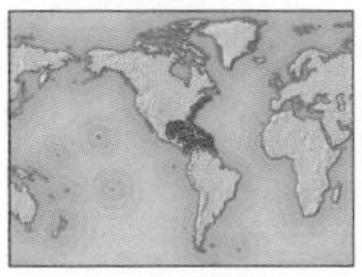

Length Up to 4 ft (1.2 m)
Weight Up to 66 lb (30 kg)
Sex Male/Female
Status Not evaluated

Location W. Central Atlantic

This large, blunt-headed fish belongs to a family of over 40 species, all notable for building mounds or tunneling into the seabed. The tilefish creates funnel-shaped tunnels that can be up to 16 ft (5 m) across and 9¾ ft (3 m) deep. It uses these as a refuge, and also to attract prey, such as crustaceans and small fishes. Each tilefish makes an individual burrow, which it enters head-first, and from which it emerges tail-first.

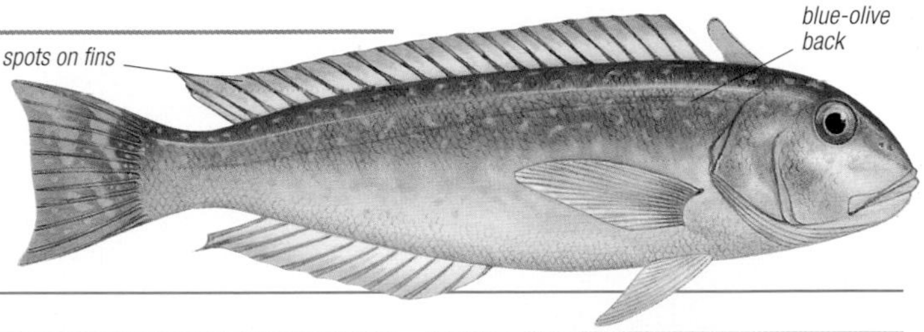

Opisthognathus aurifrons

Yellowhead jawfish

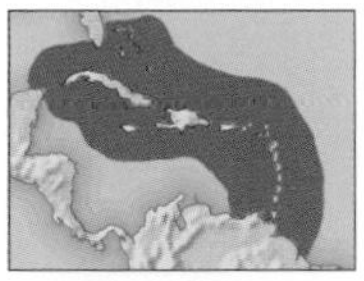

Length 4 in (10 cm)
Weight Not recorded
Sex Not recorded
Status Not evaluated

Location Caribbean to N. South America

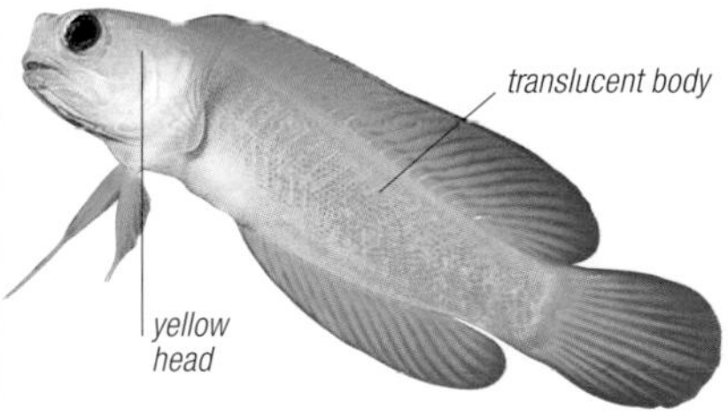

This finger-sized, shallow-water fish lives on the seabed, usually in coral sand and rubble. Found in clear, coastal waters, largely in the tropics, it has a bright yellow head, but its most notable characteristic—shared by other jawfish—is its habit of creating burrows. This fish usually hovers outside the burrow, looking for prey; it mates inside. The male incubates the eggs in his mouth, spitting them out into the burrow before feeding.

Heteropriacanthus cruentatus

Glasseye

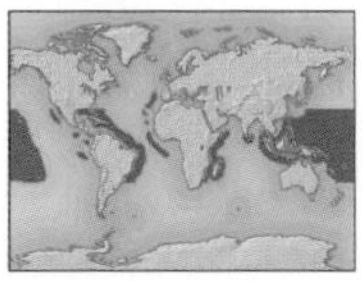

Length Up to 20 in (50 cm)
Weight Up to 5½ lb (2.5 kg)
Sex Male/Female
Status Not evaluated

Location Tropical waters worldwide

This solitary fish gets its name from the brilliant reflective layer—called the tapetum lucidum—in its large, conspicuous eyes. It spends the day under, or near, ridges of rock formations or reefs, emerging only at night to feed on a variety of prey, including small fishes, shrimps, octopus, crabs, and marine worms. Young glasseyes, unlike the adults, spend their time in the open sea.

Anabas testudineus

Climbing perch

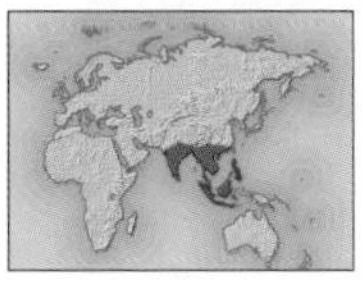

Length Up to 25 cm (10 in)
Weight Up to 150 g (5 oz)
Sex Male/Female
Status Not evaluated

Location S. and S.E. Asia

The climbing perch is a small freshwater fish, adapted to life in stagnant lakes and ponds where the oxygen supply is often poor. It gulps atmospheric oxygen at the surface of the water or climbs out to do so (see panel, right). In some areas, it buries itself in mud in the dry season. The climbing perch is a popular aquarium species, and is also sold as a food fish.

CLIMBING

The climbing perch often uses the stiff spines near the gills and the pectoral fins to climb out and "walk" from one water body to the next on humid nights. It can survive out of water for several days at a time.

SPINY FINS
This small, stocky fish is uniformly gray, olive, or brown. Its fins are short, the dorsal and anal fins having prominent spines.

Caranx hippos

Crevalle jack

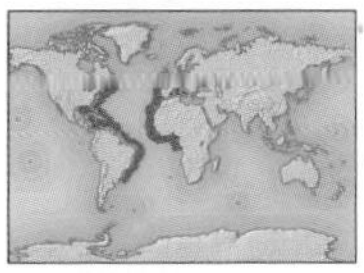

Length Up to 4 ft (1.2 m)
Weight Up to 71 lb (32 kg)
Sex Male/Female
Status Not evaluated

Location E. and W. Atlantic

This fish can be recognized by its large, blunt head, the black spot near the edge of the gill cover, and the narrow-based, deeply forked tail. The largest of all jack fishes, it can tolerate a broad range of salinities. Juveniles, which congregate in large schools, tend to favour brackish water, while older fishes are found in deeper water, often cruising in pairs. It is an active predator of small fishes: when it attacks a school of baitfishes at the surface, the commotion caused can be seen from a distance.

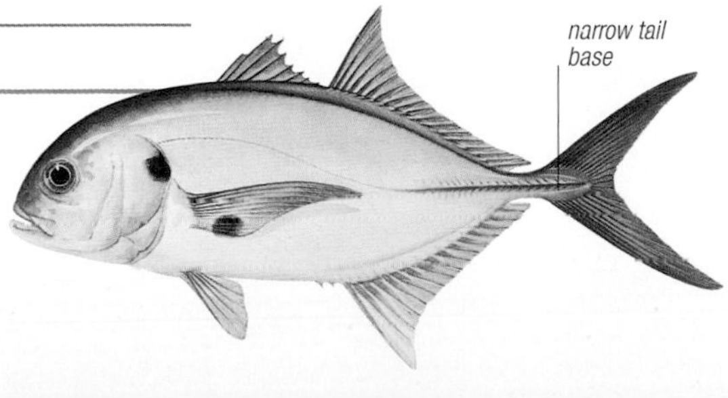

Naucrates ductor

Pilotfish

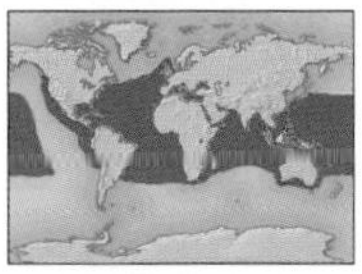

Length Up to 70 cm (28 in)
Weight Not recorded
Sex Male/Female
Status Not evaluated

Location Tropical, subtropical, and temperate waters worldwide

The pilotfish is a small or medium-sized fish with a silvery body marked with 6 or 7 black vertical bands. It derives its name from its habit of swimming just in front of much larger fishes, such as sharks and rays. This arrangement seems to benefit both the pilotfish and its host. The pilotfish feeds off the parasites on sharks and, in turn, benefits from the protection afforded by the larger fish it "pilots"; it also becomes less visible to the smaller fishes on which it preys.

PATIENT HUNTER
The giant grouper usually lurks close to the sea bottom, often lying among coral or rocks, until an unwary crustacean or slow-moving fish swims close by. Capture follows a quick lunge and opening of the cavernous mouth to suck in the prey. Numerous, depressible, needlelike teeth secure the quarry in the mouth.

Epinephelus lanceolatus

Giant grouper

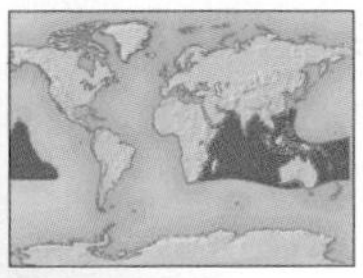

Location Indian Ocean, W. and C. Pacific

Length Up to 8¼ ft (2.5 m)
Weight Up to 880 lb (400 kg)
Sex Sequential hermaphrodite
Status Vulnerable

A dark behemoth of a fish—one of the largest of the groupers—the giant grouper lives in warm, shallow, coastal waters; it may also be seen in estuaries, able to survive in the brackish conditions. Usually solitary, it stays close to coral and rocky reefs, remaining largely in the same area, usually around caves and wrecks. The giant grouper feeds mainly on crustaceans, with a favorite food being the spiny lobster, but it is also known to eat crabs, a variety of fishes such as small sharks and rays, and young sea turtles.

As juveniles, they are preyed upon by other fishes, but as mature adults, they are typically at risk only from humans. Their often massive size and fairly sluggish behavior has made them an easy target of spearfishing. Since their great size requires a large food source and takes many years to develop, their density on reefs is extremely limited. They are easily fished out of entire regions, with little chance of their reappearance unless they are strictly protected. The giant grouper's close relation, the jewfish or Goliath grouper (*Epinephelus itajara*), is now so scarce that, in US waters at least, it is protected, with all harvesting of it prohibited.

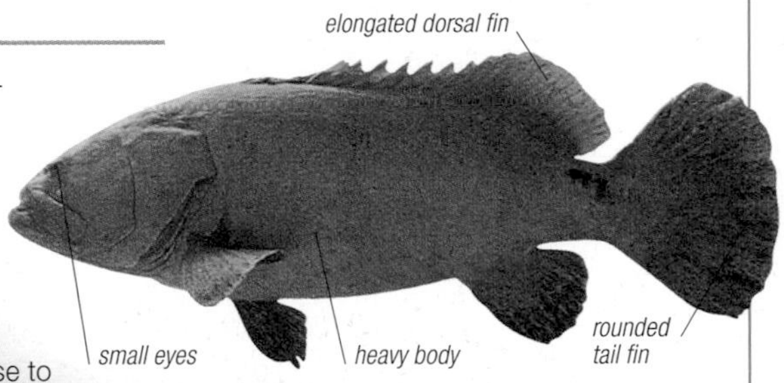

ROBUST BUILD
Also known as the Queensland grouper, this fish has a characteristically heavy body and large mouth. Its typical coloration is brownish, with light mottling.

NOT-SO-GENTLE GIANT
The giant grouper is usually easily approached by divers—but some larger specimens are believed to have attacked and killed humans.

Cromileptes altivelis

Humpback grouper

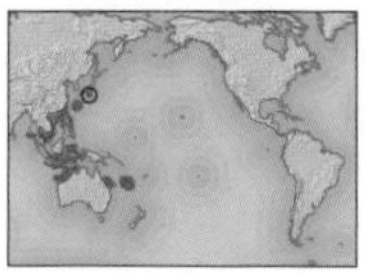

Location Indo-Pacific

Length Up to 28 in (70 cm)
Weight Up to 7¾ lb (3.5 kg)
Sex Sequential hermaphrodite
Status Vulnerable

This small humpback grouper, also known as the Barramundi cod or pantherfish, has very distinctive coloration, with a whitish head, body, and fins all covered with small, widely spaced black spots. Its back rises steeply behind its head in a "hump." Groupers typically change from female to male as they get older. Apart from the fact that it lays eggs, little is known about the breeding biology of this species.

Perca fluviatilis

European perch

Location Europe, Asia

Length Up to 20 in (51 cm)
Weight Up to 10 lb (4.5 kg)
Sex Male/Female
Status Least concern

The European perch is a predominantly freshwater fish that inhabits slow-moving rivers, deep lakes, and ponds. Juveniles feed on small planktonic creatures and invertebrates; adults eat invertebrates and small fishes, such as sticklebacks, minnows, and even other European perch. Smaller fishes tend to group together, while larger individuals are more solitary, preferring to lie close to a large rock or among aquatic vegetation. Important both commercially and as a sport fish, it has been introduced to non-European countries such as Australia. In some places, this has caused significant ecological damage.

EGG STRINGS

Female European perch lay their eggs in lacelike strings, wrapping them around river-bed stones or underwater plants. A single clutch can contain up to 300,000 eggs, connected by their sticky mucus. The eggs hatch in about 3 weeks.

STRIPED
A deep-bodied fish, the European perch has 2 dorsal fins and characteristic dark, vertical stripes, usually 6 in number.

Lepomis cyanellus

Green sunfish

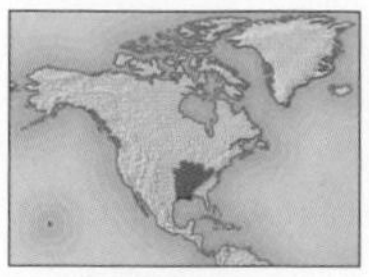

Location E. USA

Length Up to 12 in (31 cm)
Weight 35 oz (975 g)
Sex Male/Female
Status Not evaluated

At home in a wide variety of habitats, from sluggish streams to lakes and ponds, the green sunfish is one of the most common large freshwater fish in North America. It forms relatively large breeding colonies, and is found in groups around aquatic vegetation and shelter. During the breeding season, between late spring and early summer, the male uses his tail to scrape out a shallow nest in which the female lays her eggs. The eggs stick to the gravel bed and the males guard them and the newly hatched young for a few days.

Sphaeramia nematoptera

Pajama cardinalfish

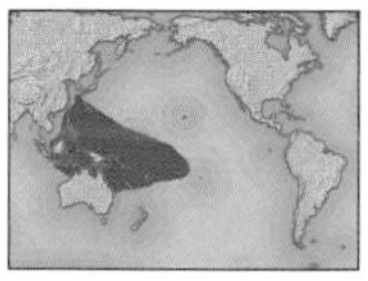

Location Indo-Pacific

Length Up to 3¼ in (8 cm)
Weight Not recorded
Sex Male/Female
Status Not evaluated

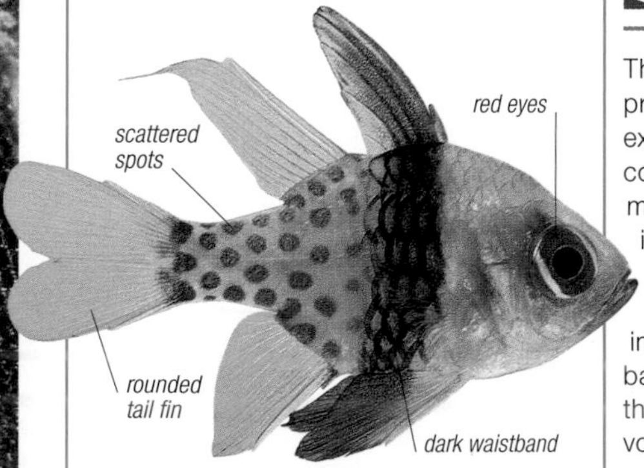

This strikingly marked tropical fish belongs to a family of over 300 species, almost all of which inhabit coral reefs or lagoons. Both sexes have a yellow head, a dark "waistband," and scattered spots toward the tail, but the female is more brightly colored. Males of this species incubate the eggs in their mouths, protecting them until they hatch.

Pomatomus saltatrix

Bluefish

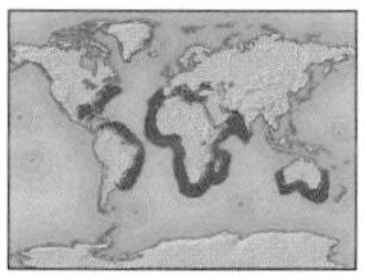

Location Atlantic, Mediterranean, Black Sea, Indo-Pacific

Length Up to 4 ft (1.2 m)
Weight Up to 31 lb (14 kg)
Sex Male/Female
Status Not evaluated

The sole species in its family, this primitive, perchlike fish has an exceptionally wide distribution in coastal and estuarine waters across much of the world. Known as the tailor in Australia, it has strong jaws and rows of razor-sharp teeth, making it a powerful killing machine. Operating in packs, bluefishes attack smaller baitfishes from behind, often herding their prey into tight schools. This voracious feeder may even eat its own young. The bluefish is not territorial, and forms large schools that migrate to warmer waters to mate and spawn. The larvae and young are then carried back to cooler waters by ocean currents.

Remora remora

Shark sucker

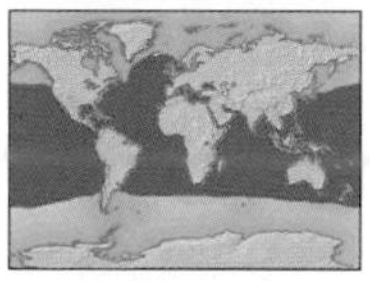

Location Tropical, subtropical, and temperate waters worldwide

Length Up to 34 in (86 cm)
Weight Up to 13 lb (6 kg)
Sex Male/Female
Status Not evaluated

Shark suckers, or remoras, are easily identified by the sucking disk above their head, which they use to fasten themselves onto whales, dolphins, and large fish. This species is most often found attached to sharks. The disk enables it to save energy by using the swimming power of its host for locomotion and gill ventilation. The host protects it from predators, and provides food scraps; in return, the shark sucker gleans parasitic crustaceans from the host's skin and gill cavity.

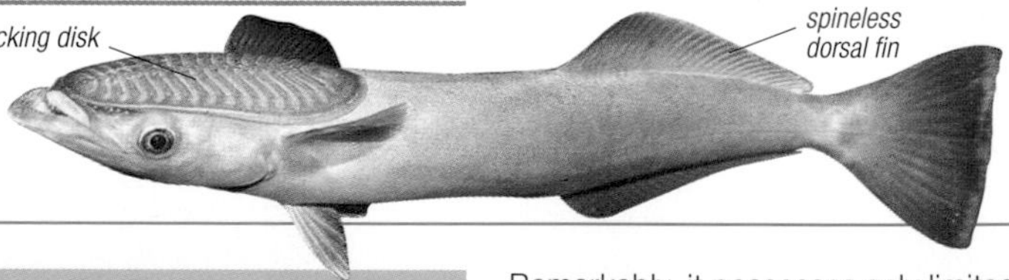

Nomeus gronovii

Man-of-war fish

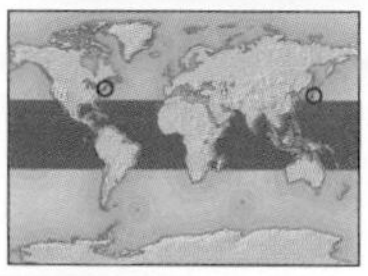

Location Tropical waters worldwide

Length Up to 10 in (25 cm)
Weight Not recorded
Sex Male/Female
Status Not evaluated

During its early life, this blue-spotted fish lives among the stinging tentacles of the Portuguese man-of-war (see p.540)—a habit that keeps it beyond the reach of most predators. Remarkably, it possesses only limited immunity to its host's toxins, and relies on its own swimming agility to avoid being stung. It is known to feed on the tentacles, so it is arguably a parasite of its host. Adults are believed to abandon the relative safety of their hosts for deep waters.

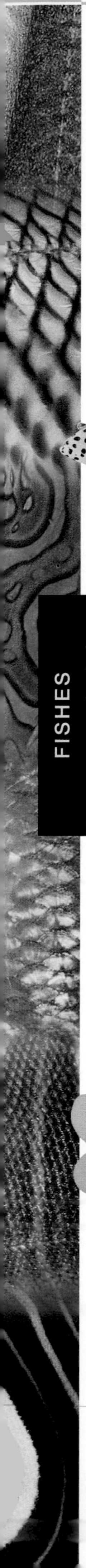
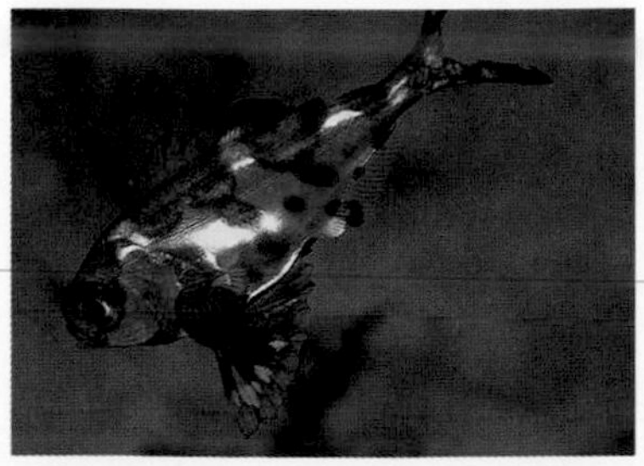

Coryphaena hippurus

Dolphinfish

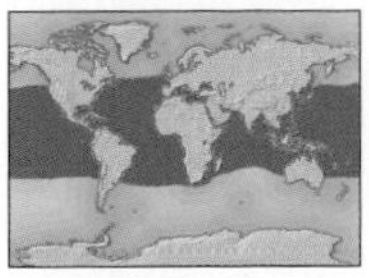

Length Up to 7 ft (2.1 m)
Weight Up to 88 lb (40 kg)
Sex Male/Female
Status Not evaluated

Location Tropical, subtropical, and temperate waters worldwide

Also known as the dorado, this blunt-headed fish has a single dorsal fin stretching along almost the whole length of its elongated body. It also possesses a large, deeply forked tail—a characteristic feature of an oceanic species that relies on speed to catch food and avoid being eaten. It has unusual coloration: a back that is a brilliant metallic blue or green, silvery sides with a golden sheen, and irregular rows of dark or golden spots. An exceptionally rapid swimmer, it can reach speeds of 37 mph (60 kph). This prized food fish is generally restricted to waters warmer than 68°F (20°C) and breeds in the open ocean. Spawning occurs throughout the year, in the evening or at night.

FEEDING

The dolphinfish eats mainly fishes, but also crustaceans and squid associated with floating objects. It follows ships and may form small groups below floating sargassum and other seaweed.

CAMOUFLAGE COLORATION
The countershading on its body helps this fish blend both into the dark depths and the well-lit open sea above, making it invisible to predators approaching from above or below.

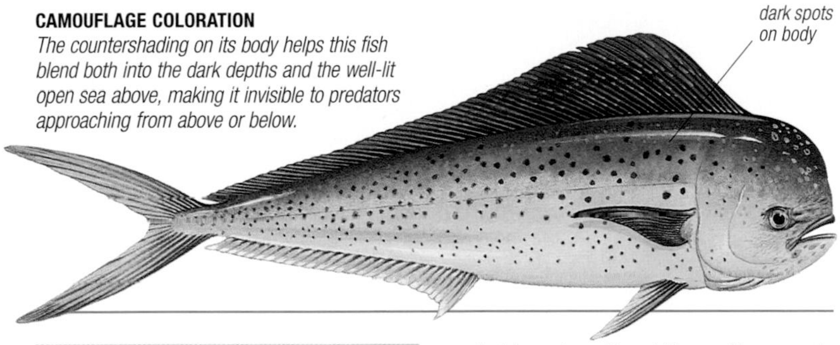

Mullus surmuletus

Surmullet

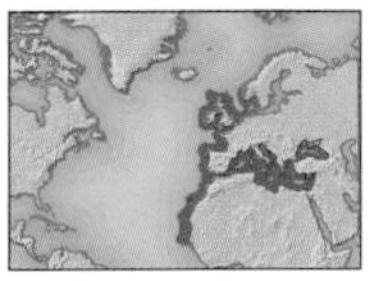

Length Up to 16 in (40 cm)
Weight Up to 2¼ lb (1 kg)
Sex Male/Female
Status Not evaluated

Location E. North Atlantic, Mediterranean, Black Sea

Highly prized as a food fish, the surmullet is reddish in color, and has a darker red stripe along its midline, with several yellowish stripes on its lower flanks. It has a downward-pointing mouth and a pair of barbels under the chin, equipped with sensory organs that help it find food. These barbels, which give the fish its other name—the goatfish—are independently movable. They are thrust into sand and mud to detect small, seabed animals—crustaceans, worms, mollusks, and fishes.

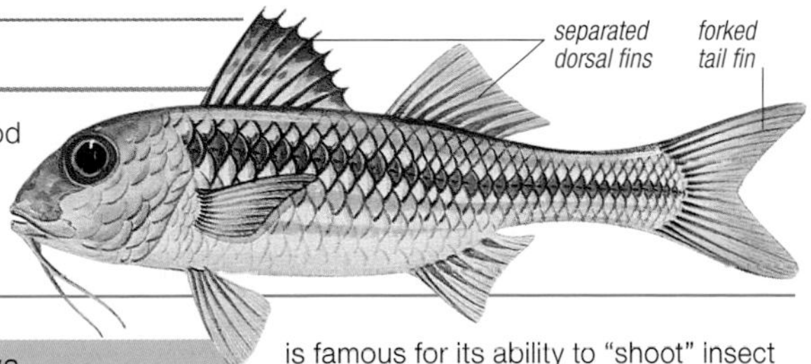

Toxotes chatareus

Spotted archerfish

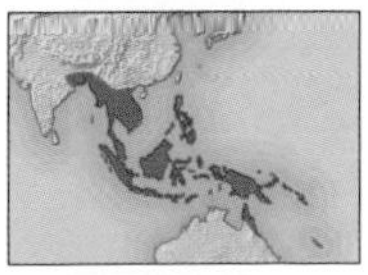

Length Up to 16 in (40 cm)
Weight Up to ¼ lb (21 kg)
Sex Male/Female
Status Not evaluated

Location S.E. Asia, N. Australia

Found in brackish water as well as in freshwater streams and lakes, this deep-bellied, silver and black fish is famous for its ability to "shoot" insect prey off overhanging vegetation, using a precisely aimed jet of water of up to 5 ft (1.5 m). It does this by pressing its tongue against the palate, then rapidly constricting its gill chamber. Little is known about its breeding biology; its eggs float on the surface.

Chiasmodon niger

Black swallower

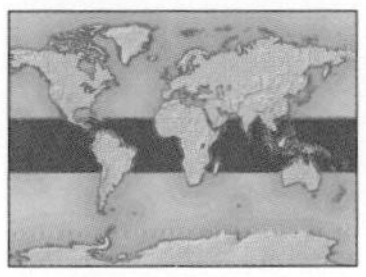

Length Up to 10 in (25 cm)
Weight Not recorded
Sex Male/Female
Status Not evaluated

Location Tropical and subtropical waters worldwide

The black swallower belongs to a small group of marine deep-water fishes that are able to catch and eat fishes larger than themselves. They can do this because their mouths and stomachs can be greatly distended to accommodate prey. Once the contents of the stomach are digested, the fishes regain their normal shape. The black swallower lives in the deep ocean but may move to slightly shallower waters to feed. It has 2 dorsal fins, a long anal fin with a single spine, and a forked tail.

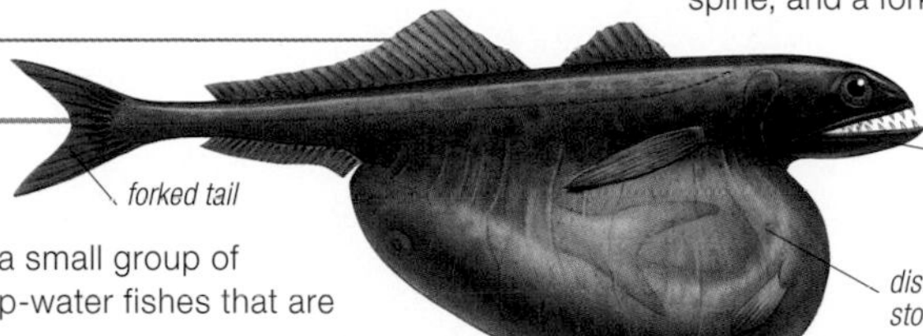

Sciaena umbra

Brown meagre

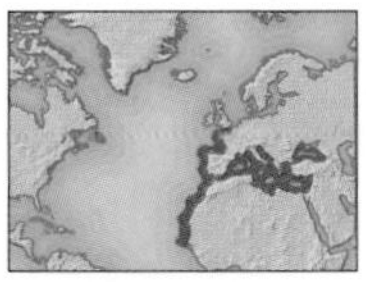

Length 28 in (70 cm)
Weight Not recorded
Sex Male/Female
Status Not evaluated

Location E. North Atlantic, Mediterranean, Black Sea

A deep-bodied fish with a strongly arched back but a nearly straight underside, the brown meagre lives in inshore waters. It belongs to a large family of 290 species. The brown meagre eats small fishes and sea-bed invertebrates, usually after dark. Its most notable feature is its ability to produce loud sounds, used for communication. This fish also has a sharp sense of hearing. The male brown meagre is known to use the muscles on the highly evolved swim bladder to make a drumming sound as a mating cue. This fish seems to have remarkable buoyancy control and appears to swim almost effortlessly.

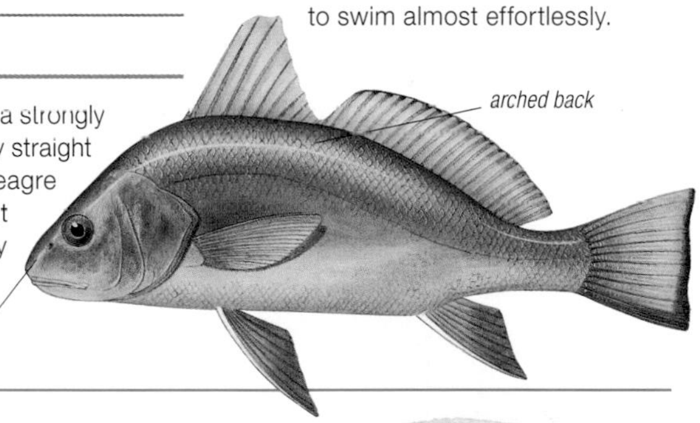

Pomacanthus imperator

Emperor angelfish

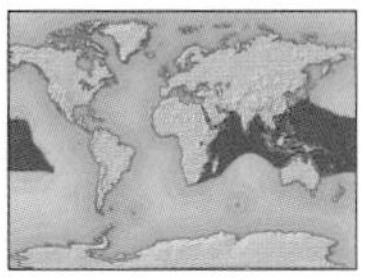

Length Up to 16 in (40 cm)
Weight Not recorded
Sex Sequential hermaphrodite
Status Not evaluated

Location Indian Ocean, Pacific

Angelfishes are among the most flamboyant inhabitants of coral reefs, with intense colors and bold patterns that often differ markedly in juveniles and adults. The emperor angelfish is a typical species, with a deep body flattened from side to side, a small mouth with comblike teeth, and a backward-pointing spine on each cheek. Like other angelfishes, it has the ability, rare among fishes, to digest the tough flesh of sponges, which are plentiful on coral reefs. This fish sometimes produces a loud thumping sound when alarmed. If the male dies, one female of his harem of 2–5 will change sex and assume dominance.

JUVENILES
Young angelfishes have a strikingly different color pattern; they are dark blue to black with many concentric, curved to semicurved white lines on their sides.

STRIPES AND MARKINGS
Adults have alternating, yellow and purple oblique stripes on the body. On the head and around the pectoral fin are light blue, dark blue, and black markings.

Chelmon rostratus

Copperband butterflyfish

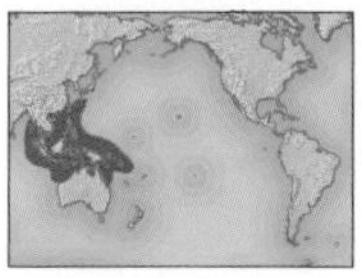

Length Up to 8 in (20 cm)
Weight Not recorded
Sex Male/Female
Status Not evaluated

Location Indo-Pacific

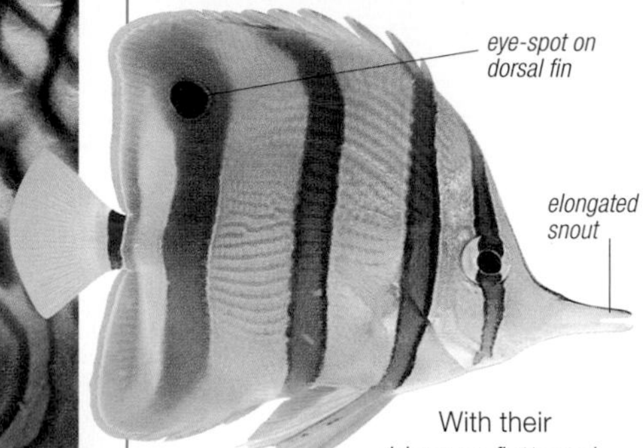

With their sideways-flattened bodies and long, pincerlike snouts, copperband butterflyfishes are ideally shaped for picking small animals from cracks and crevices in coral reefs. This species has 3 orange bands on its sides, and a narrow black band at the base of its tail. These striking markings, and the prominent black eye-spot on the dorsal fin, help confuse predators. At night, the copperband butterflyfish takes on a blotchy coloration to blend in with the background of the reef.

Oxycirrhites typus

Longnose hawkfish

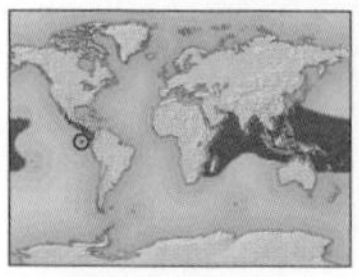

Length Up to 5 in (13 cm)
Weight Not recorded
Sex Sequential hermaphrodite
Status Not evaluated

Location Indian Ocean, Pacific

The 35 species of hawkfishes inhabit tropical waters and are found on coral or rocky reefs. Resting on the seabed as lie-and-wait predators, they ambush unwary invertebrates and small fishes. If threatened, they use their thickened, lower pectoral fin rays to wedge themselves into crevices, making them difficult to extract. The slender longnose hawkfish has whitish and red stripes and spots that form a cross-hatch pattern, and an elongated snout used to pick up prey. This fish is mostly found on its own, but the male is territorial and may maintain a harem within his territory. If a male longnose hawkfish dies, one of the females may change sex and take over as the male of the harem.

Amphiprion frenatus

Tomato clownfish

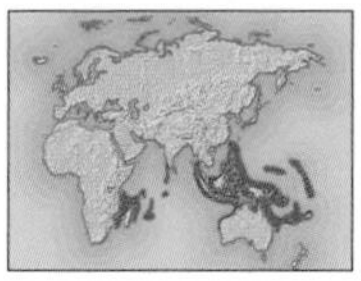

Length 3–5½ in (7.5–14 cm)
Weight Not recorded
Sex Sequential hermaphrodite
Status Not evaluated

Location W. Pacific

This small fish, like other anemonefishes, lives with sea anemones in a symbiotic relationship. It protects itself from the anemone's stings by the mucus covering its body. If the tomato clownfish leaves the anemone, it must, on its return, reestablish its immunity through a series of brief contacts with the anemone, after which it can again immerse itself among the stinging tentacles without harm. It receives shelter from the anemone, and feeds on plankton floating by and algae growing around the anemone. A family unit will usually occupy a single sea anemone. The largest 2 fishes on an anemone are the male and female that do all the spawning; the rest are males, one of which, when the female dies, changes sex and takes her place. Anemonefishes are members of the damselfish family, one of the most colorful groups of fish on coral reefs.

Labroides dimidiatus

Cleaner wrasse

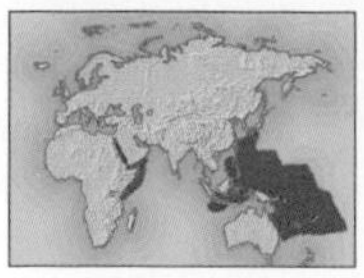

Length Up to 4½ in (11.5 cm)
Weight Not recorded
Sex Sequential hermaphrodite
Status Not evaluated

Location Indo-Pacific

The cleaner wrasse, an important member of coral-reef communities, feeds on parasites infesting other fishes, as well as on their mucus and scales. "Client" fishes queue up at special cleaning stations, which are usually occupied by a pair of wrasses or a group of females with an adult male. Once the dominant male dies, he is replaced not by another male, but by the dominant female, which changes sex.

Maylandia zebra

Zebra mbuna

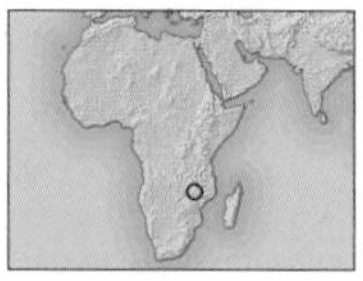

Length Up to 4¾ in (12 cm)
Weight Not recorded
Sex Male/Female
Status Least concern

Location East Africa (Lake Malawi)

This small, metallic blue fish is one of hundreds of cichlid species that have evolved in the relative isolation of Africa's large lakes. It has a small mouth with thickened lips, and scrapes algae from rocks. During breeding, the female picks up her eggs in her mouth, and the male then fertilizes them. After the eggs hatch, the female protects the young for about a week.

Astronotus ocellatus

Oscar

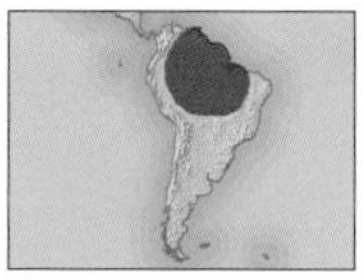

Length Up to 16 in (40 cm)
Weight Up to 3¼ lb (1.5 kg)
Sex Male/Female
Status Not evaluated

Location N. South America

The cichlid family is a huge group—of perhaps more than 1,500 species – of deep-bodied freshwater fish restricted to South America, Africa, and Sri Lanka. The oscar, found in South America, has eye-spots—large, orange-ringed black spots at the tail base—that serve to confuse predators. A single spawning contains up to 2,000 eggs. Like most cichlids, this species tends and guards its young.

Cetoscarus bicolor

Bicolour parrotfishes

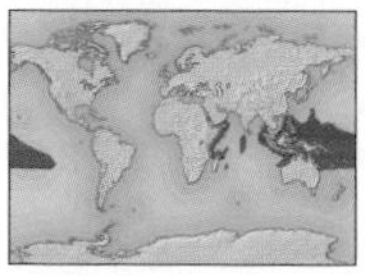

Length Up to 32 in (80 cm)
Weight Not recorded
Sex Sequential hermaphrodite
Status Not evaluated

Location Indian Ocean, Pacific

Parrotfishes get their name from their bright colors and bill-like teeth. As these herbivorous fishes scrape encrusting algae from coral rocks, they also remove and ingest part of the rock. To process this, parrotfishes are also equipped with upper and lower rows of strong, molarlike teeth in the throat. These help in grinding down the rock, transforming it into sand, which is then defecated. Common on coral reefs, parrotfishes are major producers of sand sediments on reefs.

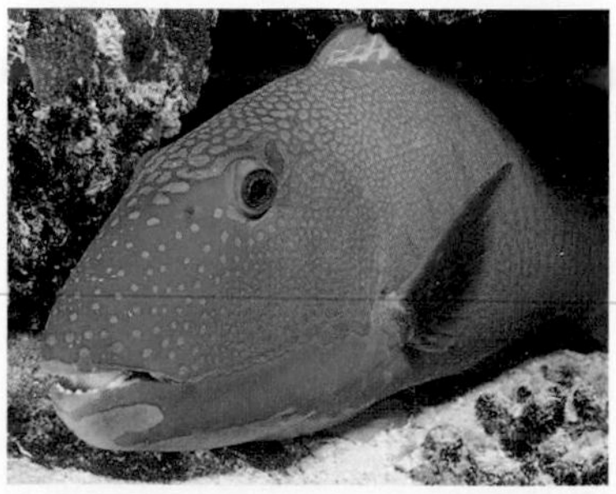

STICKING TOGETHER
Collared butterflyfishes (Chaetodon collare) *live on reef edges in the Indo-Pacific, where they feed on coral polyps. They live in pairs and sometimes gather together in large schools possibly for protection from predators.*

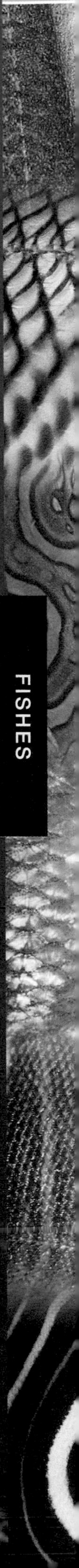

Anarhichas lupus

Wolf-fish

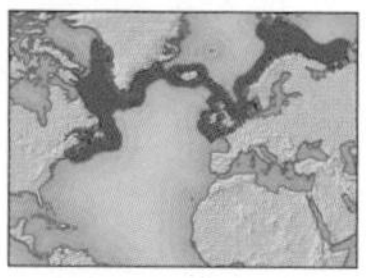

Length Up to 5 ft (1.5 m)
Weight Up to 53 lb (24 kg)
Sex Male/Female
Status Not evaluated

Location North Atlantic, Arctic

The wolf-fish is a solitary, shallow-water predator that has no pelvic fins and only a rudimentary pelvic girdle. In contrast, the teeth are very well developed: they are conical at the front of the jaws, and molarlike at the back. This fish uses its teeth and a bony plate in the roof of the mouth to crush mollusk shells and the carapaces of crabs, lobsters, sea urchins, and other echinoderms on which it feeds. Spawning takes place in deeper water during winter, and the eggs are guarded by the male until they hatch. Although originally caught only as a by-product of trawler fishing, the delicately flavored flesh of the wolf-fish has led to an increase in its commercial importance in recent years.

Chaenocephalus aceratus

Blackfin icefishes

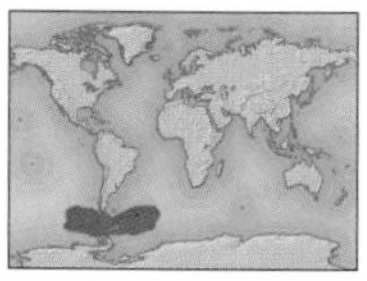

Length Up to 28 in (72 cm)
Weight Up to 7¾ lb (3.5 kg)
Sex Male/Female
Status Not evaluated

Location E. South Pacific, W. South Atlantic

The blackfin icefish produces a natural antifreeze, enabling it to survive in the subzero waters of the Antarctic. It lacks red blood cells and hence looks rather pale, but has excellent blood circulation, and a strong heart that weighs as much as that of a small mammal. Its large, toothy mouth led to it being called the crocodile fish by 19th-century whalers.

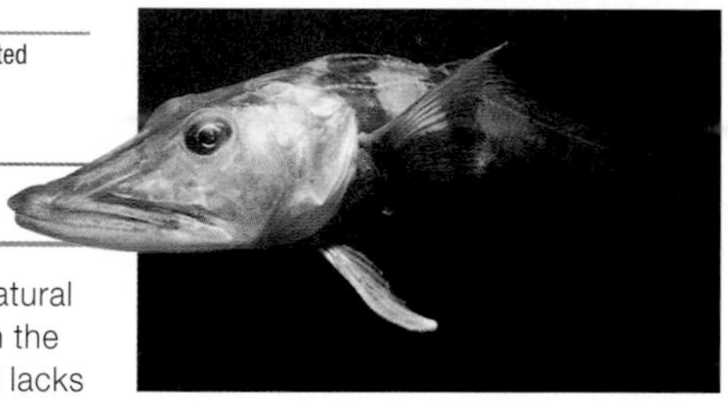

Dissostichus mawsoni

Antarctic toothfish

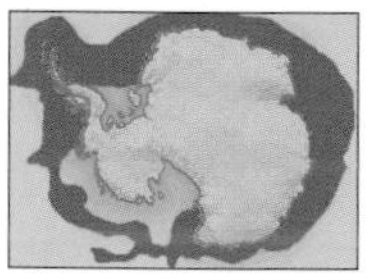

Length Up to 7¼ ft (2.2 m)
Weight Up to 260 lb (120 kg)
Sex Male/Female
Status Not evaluated

Location Antarctic Circle

This bottom-dwelling fish can survive in cold Antarctic waters because it produces special proteins in its tissues and blood that act as an antifreeze. Like many creatures from this region, it grows very slowly, reaching sexual maturity at 8–10 years of age. It may sometimes swim up from the seabed to feed on a range of small fishes, squid, crabs, and prawns. Due to its large size, this fish can be eaten only by a few marine animals, such as sperm whales, killer whales, and elephant seals. The Antarctic toothfish does not have a swim bladder, and achieves neutral buoyancy by having light bones and a high body fat content.

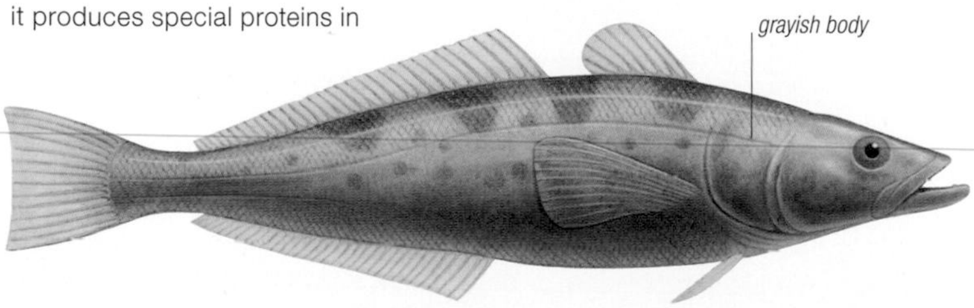

Echiichthys vipera

Lesser weeverfish

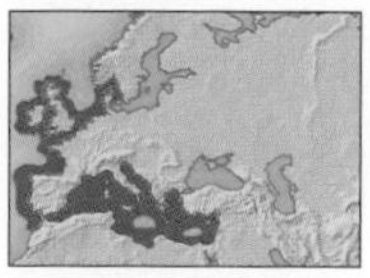

Length Up to 6 in (15 cm)
Weight Not recorded
Sex Male/Female
Status Not evaluated

Location E. North Atlantic, Mediterranean

This venomous fish, frequently found near European beaches, is known to inflict painful wounds if stepped upon. Spines on the first dorsal fin and gill cover have poison glands at their bases; if disturbed, this fish erects these as a warning, or uses them in defense against larger fishes. This species spends much of its time buried in the seabed with only its eyes and the dorsal fin tip exposed.

Nemateleotris magnifica

Firefish

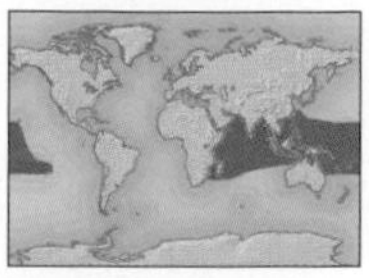

Length Up to 3¼ in (8 cm)
Weight Not recorded
Sex Male/Female
Status Not evaluated

Location Pacific, Indian Ocean

This palm-sized tropical fish is one of about 60 related species that construct burrows in sand or coral rubble on reefs. Its coloration is distinctive, with a pale head and a body that grades smoothly into a deep orange-red near the tail. Adults hover nervously, alone or in pairs, over their burrows, facing the current while feeding on zooplankton. They quickly dart into the burrows when threatened.

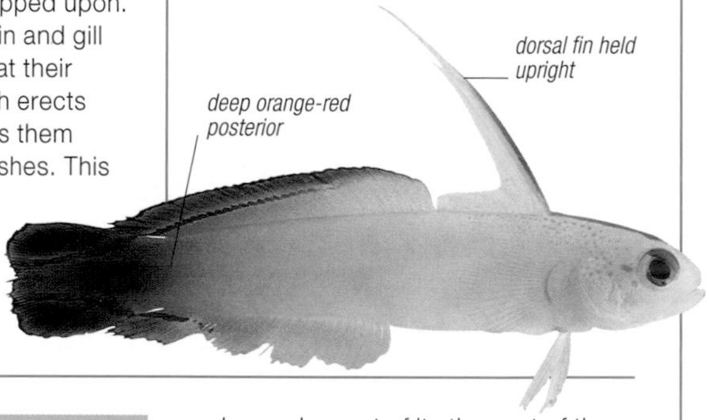

Periophthalmus barbarus

Atlantic mudskipper

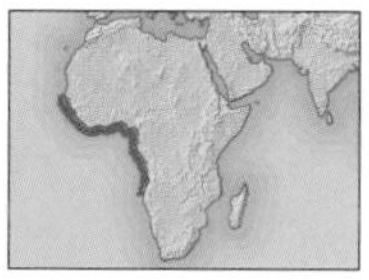

Length Up to 10 in (25 cm)
Weight Not recorded
Sex Male/Female
Status Not evaluated

Location E. Atlantic

Mudskippers have tapering bodies, large heads, and protruding eyes. Like other mudskippers, this species feeds on small animals on the mud surface, and spends most of its time out of the water. It breathes through the skin, which it has to keep moist. It can crawl on its pectoral fins, and also skip and jump across the mud by quick, flexing movements of its body.

Naso lituratus

Orangespine unicornfish

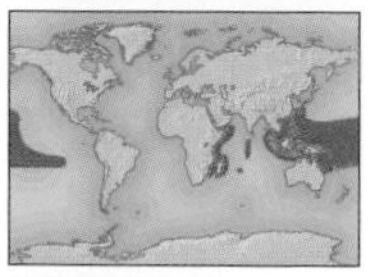

Length 18 in (45 cm)
Weight Not recorded
Sex Male/Female
Status Not evaluated

Location Pacific, Indian Ocean

The orangespine unicornfish belongs to a family of about 82 species known as surgeonfishes because of the knifelike spine or spines on the side of their tail. This species uses its bright orange spines both in defense against predators and as a weapon in fights for dominance, inflicting deep wounds with rapid sideways swipes of the tail. The orangespine unicornfish has a moderately high, compressed body, but no hornlike projection on the forehead, unlike most other unicornfishes. It grazes on leafy algae attached to the seabed, nipping with its incisor teeth. A valued food source, it is caught off coral reefs.

Siganus vulpinus

Foxface

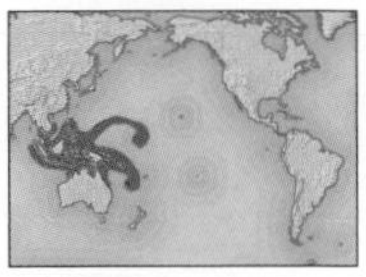

Length Up to 9½in (24cm)
Weight Not recorded
Sex Male/Female
Status Not evaluated

Location Indo-Pacific

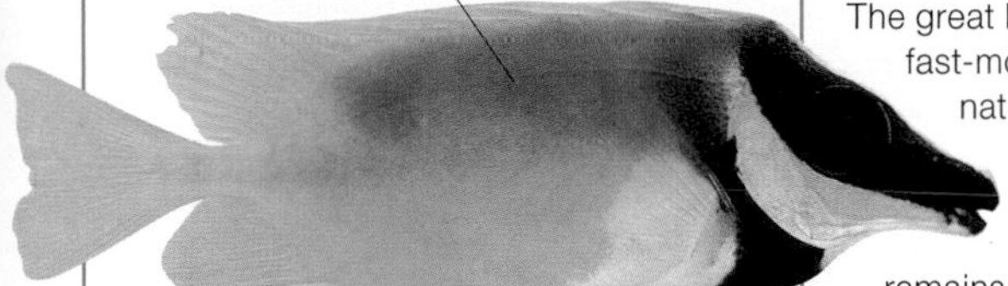

Living close to the bottom in the shallow water above coral reefs, this brightly colored fish feeds primarily on algae that grow at the dead bases of corals. A single row of flat, spadelike teeth enables it to snip off bits of its food. Its single dorsal fin has 13 sharp spines. The conspicuous black, white, and yellow coloration may serve to warn off predators, while venom in its fin spines also forms part of its defensive armor. Males and females apparently mate for life.

Sphyraena barracuda

Great barracuda

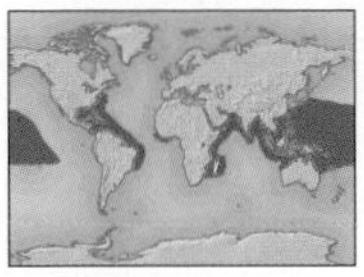

Length Up to 6½ft (2m)
Weight Up to 110lb (50kg)
Sex Male/Female
Status Not evaluated

Location Tropical and subtropical waters worldwide

The great barracuda is a formidable, fast-moving predator, with few natural enemies apart from large sharks. Its large head and powerful jaws conceal knifelike teeth. This fish remains relatively motionless, then lunges forward quickly to grab its prey. It has been known to attack swimmers, often causing serious injury. However, these mishaps have usually occurred in murky waters where a human limb or shiny jewelry can be mistaken for other fishes. Attacks have also been reported in clear waters when the fish has been provoked with a spear. It is curious and will swim up close to divers. The great barracuda is a popular game fish and is sometimes eaten, but large individuals can cause ciguatera poisoning—a potentially fatal condition caused by toxins that barracudas ingest with their food.

SOCIAL JUVENILES

Adult great barracudas are, by and large, solitary cruisers, found typically at or near the surface of open tropical and subtropical seas, in warm and shallow waters. They have been caught occasionally at depths of 330ft (100m). The young, however, sometimes swim in small groups inshore, in shallow, sheltered reefs and mangrove swamps. This allows them to hunt in groups and also provides them with some protection from predators.

LONG BODY
The fish has an elongated body with a long snout. Its body length and associated musculature are specially adapted to produce short, rapid bursts of speed.

Scomber scombrus

Atlantic mackerel

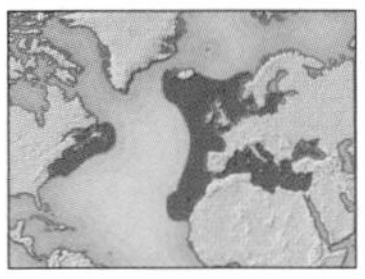

Length Up to 23½in (60cm)
Weight Up to 7¾lb (3.5kg)
Sex Male/Female
Status Not evaluated

Location North Atlantic, Mediterranean, Black Sea

Atlantic mackerel are streamlined predatory fishes that inhabit cold waters and continental shelf areas. In spring, they migrate closer to the shore.

Mackerel group in large schools according to size, feeding on fish larvae and small crustaceans. Spawning occurs from March to June when females lay as many as 450,000 eggs in a season. This commercially important species is extensively fished throughout its range.

Thunnus thynnus

Northern bluefin tuna

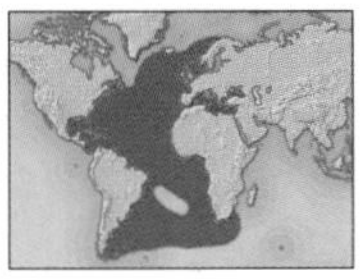

Length Up to 14ft (4.3m)
Weight Up to 2,010lb (910kg)
Sex Male/Female
Status Data deficient

Location Atlantic

The northern bluefin tuna is one of the world's largest and fastest bony fishes, reaching speeds of at least 43mph (70kph). It has a powerful and streamlined body. Its pectoral and pelvic fins and its first dorsal fin can be recessed into grooves to minimize drag. Swimming in schools, the tuna migrates as far as 6,250 miles (10,000km) across ocean basins, feeding on open-water fishes. In recent years, overfishing has lead to a drastic decline in population.

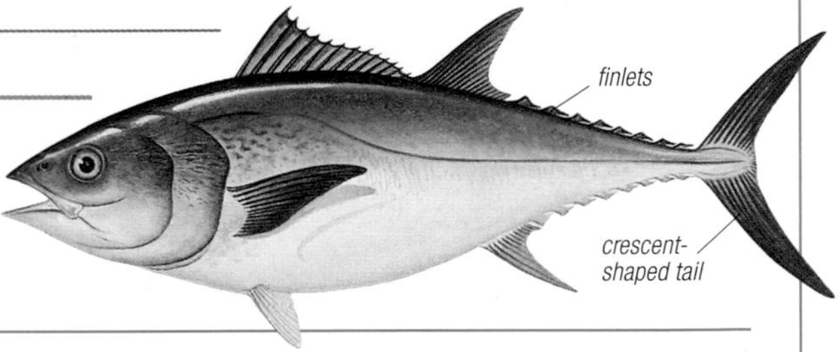

Xiphias gladius

Swordfish

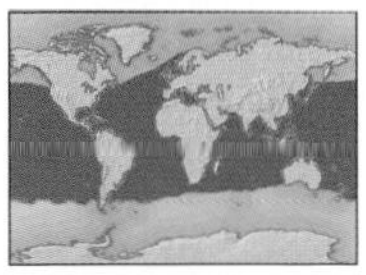

Length Up to 15ft (4.5m)
Weight Up to 1,300lb (590kg)
Sex Male/Female
Status Data deficient

Location Tropical, subtropical, and temperate waters worldwide

A popular food and game fish, this fast-swimming predator gets its name from its long, swordlike snout. Unlike billfishes, such as the blue marlin (see p.528), it certainly uses its snout as a weapon, stunning or impaling smaller fishes and squid. It then slashes the crippled victim into pieces or swallows it whole. The swordfish used to attack small boats occasionally—including those with copper sheathed hulls—with devastating results. Its chief predators are the mako and blue shark, several species of tuna and marlin, and sometimes sea lampreys. Swordfishes spawn in tropical waters throughout the year, peaking between April and September, depending on the area.

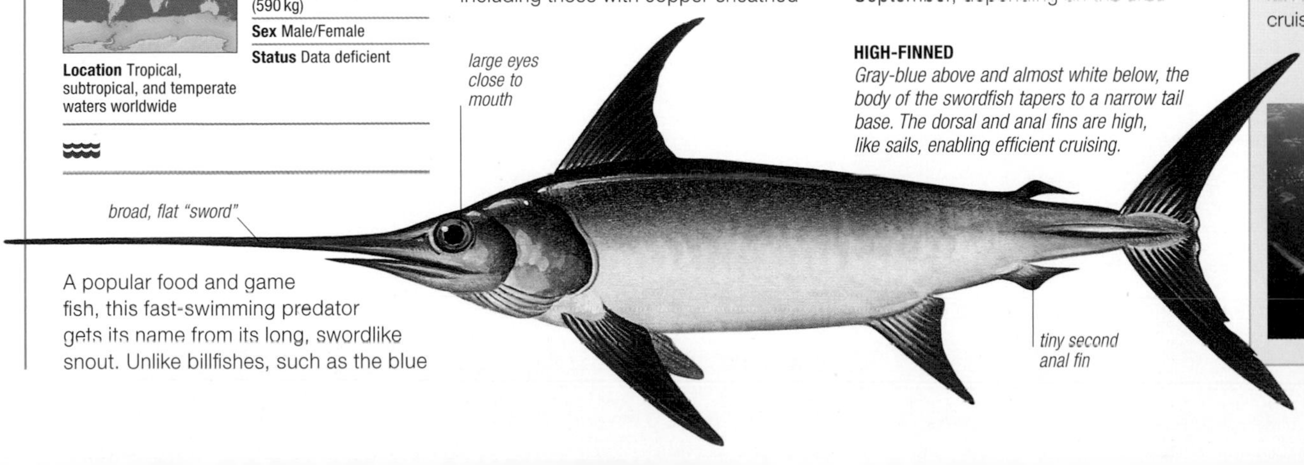

HIGH-FINNED
Gray-blue above and almost white below, the body of the swordfish tapers to a narrow tail base. The dorsal and anal fins are high, like sails, enabling efficient cruising.

SOLITARY HUNTER

Adult swordfishes usually hunt alone. Like tuna and marlin, they have similar proportions of the 2 distinct kinds of swimming muscles: dark red muscles are used for steady cruising, while faster-acting white muscles are used for short bursts of speed.

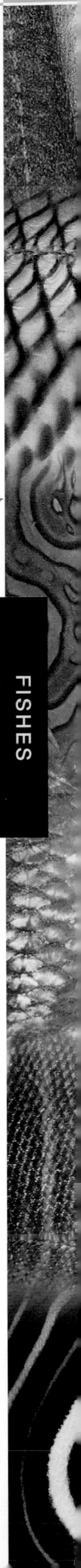

Makaira nigricans

Blue marlin

Location Atlantic

Length Up to 14 ft (4.3 m)
Weight Over 1,985 lb (900 kg)
Sex Male/Female
Status Not evaluated

The blue marlin is renowned among anglers for the way it fights when caught. Hooked blue marlin breach the water surface with extraordinary leaps. The largest member of the billfish family, some individuals can weigh over 1,985 lb (900 kg). The upper jaw forms a large bill, which is apparently used to stun or kill prey. The lateral keels at the base of its tail and the pelvic fins set into grooves make this fish a powerful and fast swimmer, capable of explosive bursts of speed and long-distance travel. Although blue marlin may stray to greater depths, they prefer the warmth of surface waters, where they feed on other fishes such as mackerel, tuna, dolphinfish, and squid.

EGGS AND LARVAE
A single spawning produces millions of eggs, each $\frac{1}{32}$ in (1 mm) in diameter. These possibly hatch within a week. Eggs and larvae are planktonic.

CONSERVATION

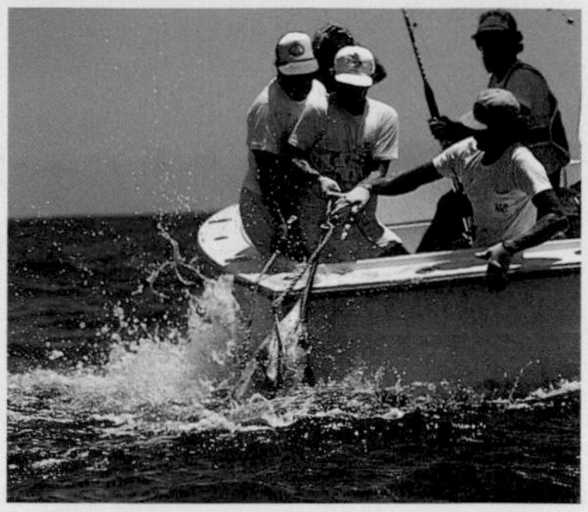

Game fishing can locally affect the numbers of this fish, but there is a much more serious threat in commercial fishing. A number of developing countries, whose offshore waters abound with this species, exploit the blue marlin as a popular source of food. Due to their size, adult marlin are generally safe from predators other than humans, but juveniles fall prey to other sea fishes such as tuna, mackerel, and sharks.

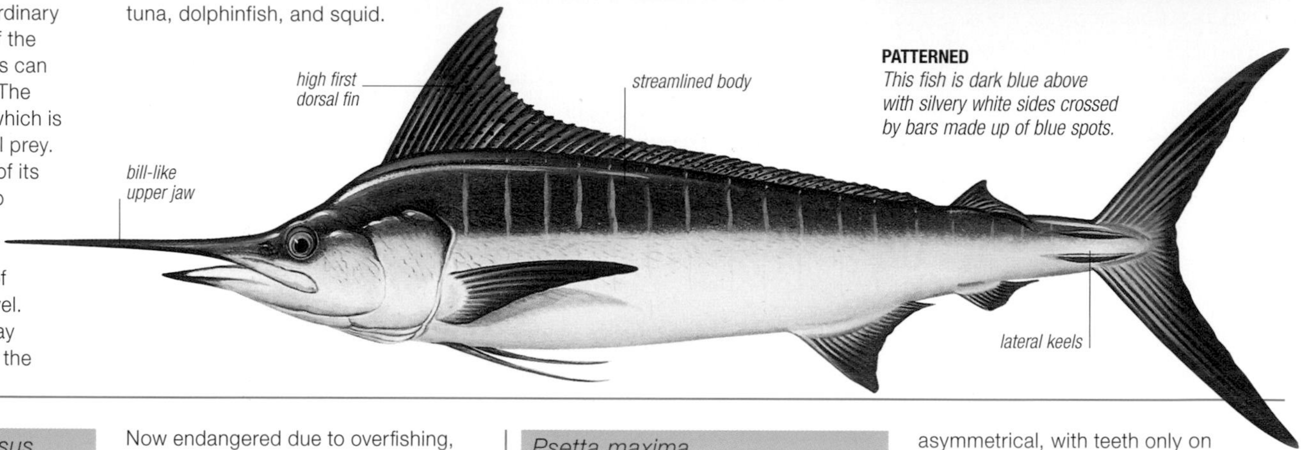

PATTERNED
This fish is dark blue above with silvery white sides crossed by bars made up of blue spots.

Hippoglossus hippoglossus

Atlantic halibut

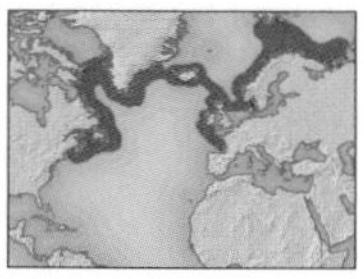

Location North Atlantic, Arctic

Length Up to 8¼ ft (2.5 m)
Weight Up to 690 lb (315 kg)
Sex Male/Female
Status Endangered

Now endangered due to overfishing, the Atlantic halibut is one of the largest flatfishes—a group of over 700 species that have a highly distinctive anatomy, suiting them to life on the seabed. Instead of being flattened from top to bottom, like many seabed fishes, flatfishes are flattened sideways, but habitually lie on one side. The Atlantic halibut lies on its left side, and both its eyes are positioned facing upward, on its right. Compared to most other flatfishes, Atlantic halibuts are unusually active swimmers, and often catch other fishes in midwater.

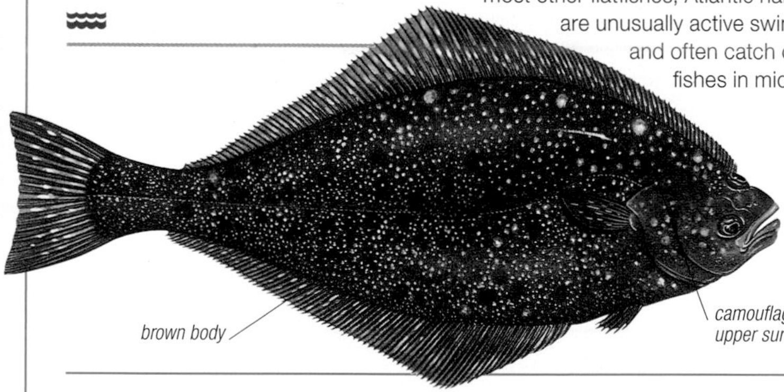

Psetta maxima

Turbot

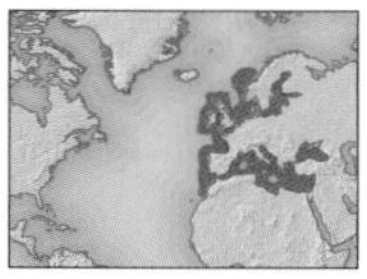

Location E. North Atlantic, Mediterranean, Black Sea

Length Up to 3¼ ft (1 m)
Weight Up to 55 lb (25 kg)
Sex Male/Female
Status Not evaluated

The almost round body of this flatfish has a number of bony tubercles scattered over the surface of the upper side instead of scales. Both the eyes are on the left side, and the mouth is asymmetrical, with teeth only on the blind side. The jaws on the eyed side act like a siphon, drawing in water that is free of sand, thereby making respiration more efficient. Females, which are less numerous than males, produce 10–15 million eggs.

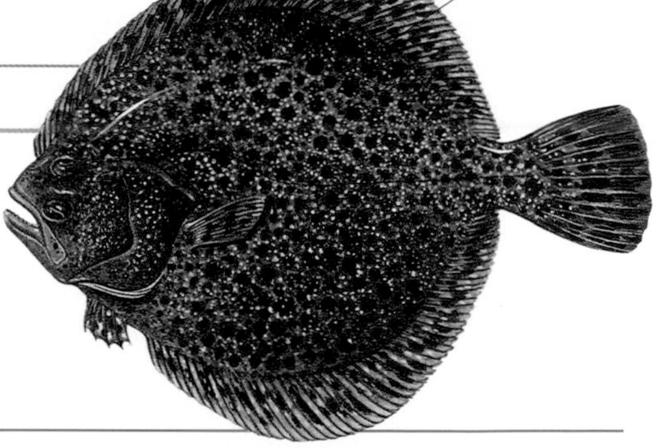

Pleuronectes platessa

European plaice

Location North Atlantic, Arctic, Mediterranean, Black Sea

Length Up to 3¼ ft (1 m)
Weight Up to 15 lb (7 kg)
Sex Male/Female
Status Least concern

As a newly hatched fish larva, the European plaice lives for about 6 weeks near the surface of the sea. As in all flatfishes, a remarkable transformation then occurs that turns it into a bottom dweller. The left eye migrates across the head to the right side, so that the blind left side can lie against the seafloor. There, the European plaice feeds on a diet of thin-shelled mollusks and marine worms. The upper side, capable of rapid and varied color changes, camouflages this fish from potential predators. This is the most important commercial flatfish in Europe.

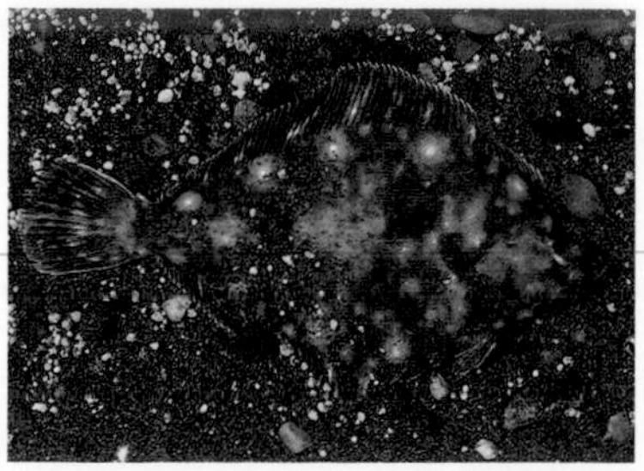

Paralichthys dentatus

Summer flounder

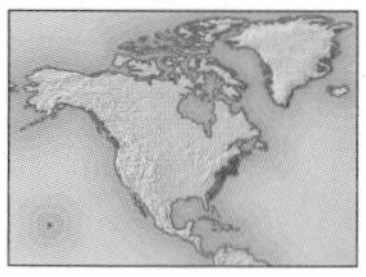

Location W. North Atlantic, E. North America

Length Up to 37 in (94 cm)
Weight Up to 26 lb (12 kg)
Sex Male/Female
Status Not evaluated

Most flatfishes are marine. The summer flounder is unusual in that it is also found in estuaries and in freshwater. It is a bottom-dwelling predator that relies on camouflage, speed, and sharp teeth to catch the small fishes and other sea animals that make up its diet. The eggs of the summer flounder are covered in oil droplets; when the fish spawns, the eggs float up toward the surface, where they hatch into ordinary fish larvae. When a larva reaches about ½ in (1.5 cm) in length, its body becomes flattened and the right eye migrates to the left side. The larva, which lacks a swim bladder, then settles on the bottom.

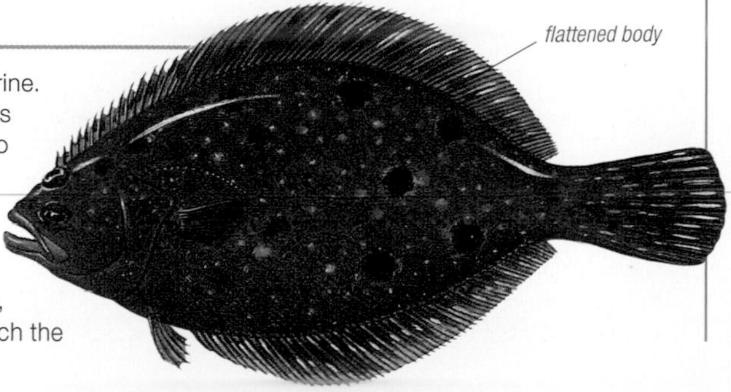

Solea solea

Common sole

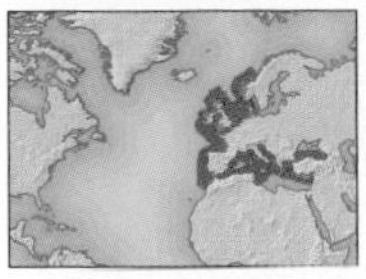

Length Up to 28 in (70 cm)
Weight Up to 66 lb (30 kg)
Sex Male/Female
Status Not evaluated

Location E. North Atlantic, Mediterranean, Black Sea

A commercially important flatfish, the common sole lives in marine and estuarine environments, where it burrows into sandy and muddy floors. Like other flatfishes, it is a highly asymmetrical animal, with both eyes on its camouflaged right-hand side, which always faces upward as it moves over the seabed. The mouth is twisted onto the blind side and has tiny teeth. The common sole is carnivorous, feeding on mollusks and worms, which it locates by smell. The blind side of its head has special filamentous tubercles instead of scales. These are thought to enhance the sole's sense of taste.

Diodon holocanthus

Longspined porcupinefish

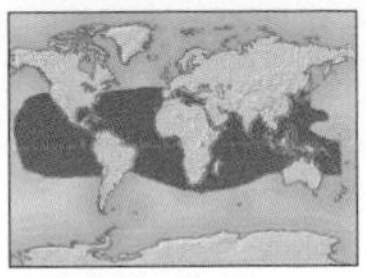

Length Up to 12 in (30 cm)
Weight Not recorded
Sex Male/Female
Status Not evaluated

Location Tropical and subtropical waters worldwide

If threatened, this spine-covered fish swallows large amounts of water, inflating itself until it becomes almost spherical, and making its spines, which are usually folded back, stand on end. But the spiny armor may hinder its ability to swim, and it is often found entangled in nets, much to the annoyance of fishermen. Brown above and yellow below, longspined porcupinefishes are reef dwellers. Their powerful jaws, armed with fused teeth, enable them to crack the shells and skeletons of many marine invertebrates.

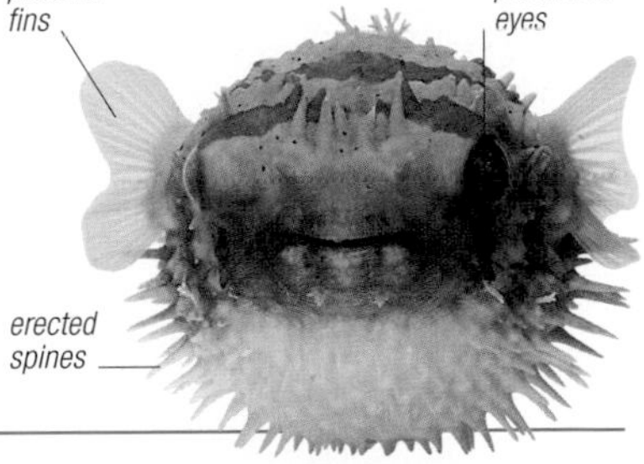

Mola mola

Oceanic sunfish

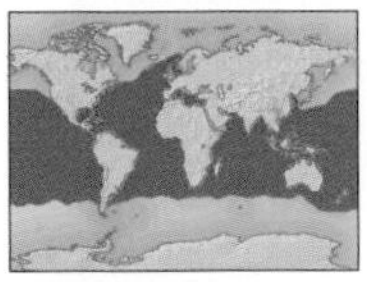

Length Up to 13 ft (4 m)
Weight Up to 2¼ tons (2 tonnes)
Sex Male/Female
Status Not evaluated

Location Tropical, subtropical, and temperate waters worldwide

The world's heaviest bony fish, the oceanic sunfish has a highly distinctive shape. Its scientific name comes from the Latin for millstone, because it looks almost circular when seen from the side. The young are elliptic in shape. The oceanic sunfish feeds primarily on jellyfish, using its teeth to nip pieces off prey. Although generally slow-moving, it may jump clear of the water if alarmed. Sunfishes are occasionally seen basking horizontally at the surface; they probably do this only when they are sick or dying. Hugely prolific, they breed in the open sea, laying up to 300 million eggs, which drift away with the current. Most sightings of sunfishes are in tropical waters, but specimens are occasionally seen at much higher latitudes, where currents keep the temperature relatively warm.

ERECT FINS
The triangular dorsal and anal fins are placed far back, and are held erect like a pair of giant blades. This fish moves by sculling with these fins.

DISK SHAPE

Seen in profile, this fish resembles a huge disk. Its mouth has a padded upper lip, and it has small eyes on either side. The tail is little more than a narrow frill attached directly to the body.

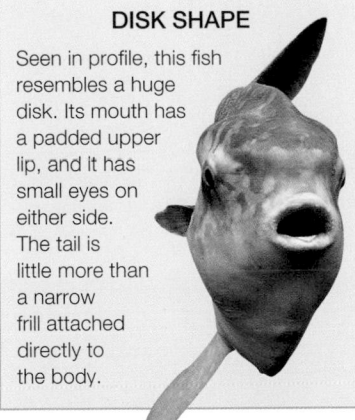

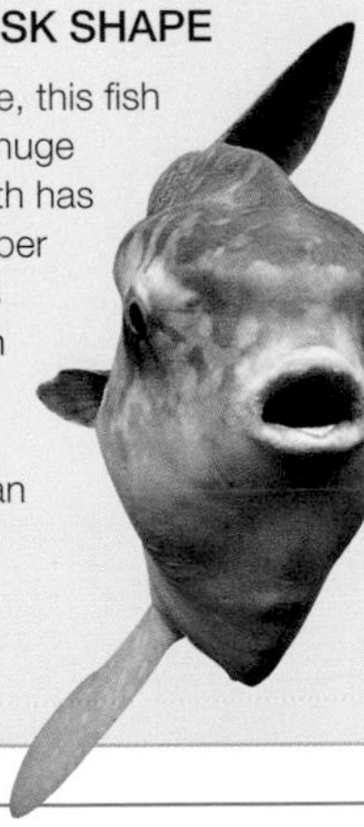

Balistoides conspicillum

Clown triggerfish

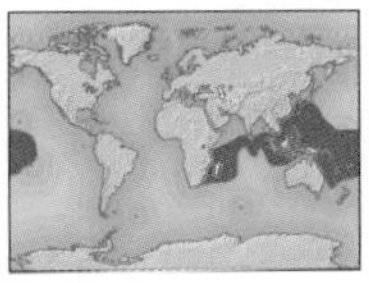

Length Up to 20 in (50 cm)
Weight Up to 4½ lb (2 kg)
Sex Male/Female
Status Not evaluated

Location Pacific, Indian Ocean

This exceptionally colorful, coral-reef species belongs to a family of fishes that can lock their 3 dorsal spines upright, enabling them to wedge themselves into crevices. Its head is covered with tough, protective scales, and its teeth are strong and adapted to crushing the shells of prey such as mollusks and crustaceans. Like other triggerfishes, it can be very territorial, especially during the mating season.

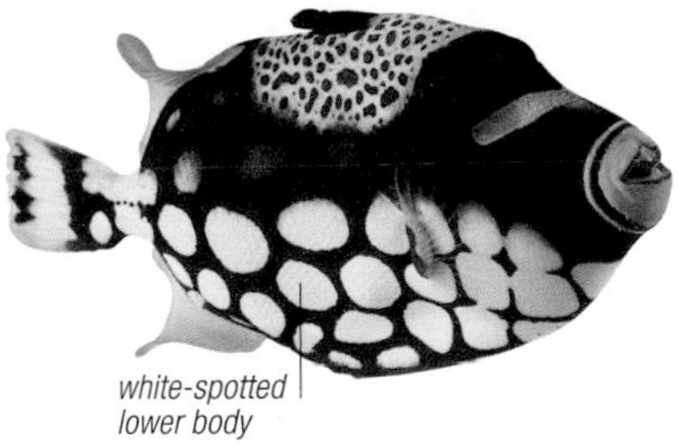

Takifugu niphobles

Fugu

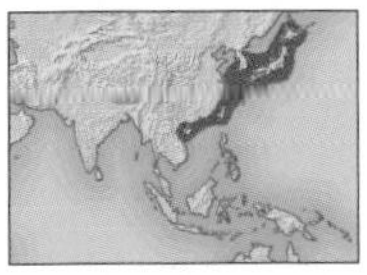

Length Up to 6 in (15 cm)
Weight Not recorded
Sex Male/Female
Status Data deficient

Location W. North Pacific

Despite the fact that the fugu contains a poison more potent than cyanide, it is considered a great delicacy in Japan, where it is prepared with great care before being eaten. To deter predators, this fish may secrete its powerful neurotoxin into the water, or it may inflate its body—by pumping water into an inflatable sac in the stomach—and erect the spines on its sides. The fugu breeds between May and July, 1–5 days after the full and new moons. Thousands of these fishes arrive to spawn on pebbly beaches at high tide.

Lactoria cornuta

Longhorn cowfish

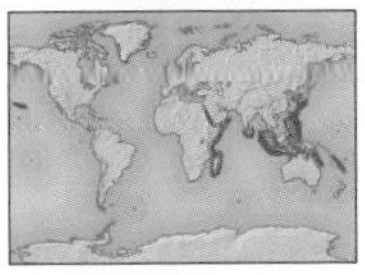

Length Up to 18 in (46 cm)
Weight Not recorded
Sex Male/Female
Status Not evaluated

Location Pacific, Indian Ocean, Red Sea

The longhorn cowfish derives its name from the small, hornlike projections above each eye. Its rigid outer skeleton, composed of fused bony, thick scales, which make it a clumsy swimmer, accounts for its other name, the boxfish. This fish hunts by blowing jets of water into the sand on the ocean floor to expose the bottom-living invertebrates on which it feeds. Longhorn cowfishes are caught and dried to sell as ornaments.

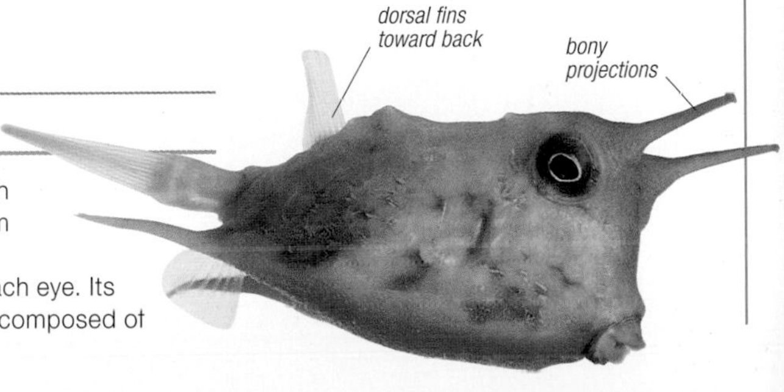

INVERTEBRATES

INVERTEBRATES

Invertebrates are animals that do not have a backbone. Although many of them are small—and therefore easily overlooked—they are immensely varied and widespread, accounting for about 97 percent of all known animal species. While vertebrates form a single phylum, invertebrates are an informal collection of more than 30, and the members of just one phylum—the arthropods—probably outnumber all other animals on earth. Invertebrates are found in every conceivable type of habitat, but they are most plentiful in the oceans, which is where animal life first arose.

LIGHTWEIGHT LIFE
Butterflies are typical invertebrates, with a small body, a high reproductive rate, and a relatively short lifespan. Unlike some vertebrates, invertebrates are ectothermic (cold blooded), which means their activity slows down as temperatures drop.

Evolution

Exactly how multicellular invertebrate animals evolved remains uncertain, but there is little doubt that their ancestors were single-celled organisms similar to living protozoans. One theory, accepted by many biologists, is that loose colonies of these microorganisms formed more permanent partnerships perhaps as hollow balls of cells similar to the living colonial proctotist *Volvox*. Infolding or growth of cells inside the hollow center would produce simple two-layered multi-cellular organisms—the start of invertebrate life.

The earliest signs of animal life are chemical fossils rather than the visible remains of the organisms. Carbon biomarkers specific to sponges are found in rocks dating back about 700,000 million years. The earliest fossils of invertebrates themselves come from late Precambrian times, about 600 million years ago. At the start of the Cambrian Period, about 545 million years ago, there was an acceleration of evolutionary change, known as the Cambrian Explosion. Soft-bodied animals, such as jellyfish, were joined by ones with hard body cases and shells. Hard body parts fossilize well, so from this point onward, the development of invertebrate life is much easier to trace. Although it happened long ago, the Cambrian Explosion brought into being all the invertebrate phyla that exist today. Studies of living species show that these can be divided into 2 groups: the mollusk-annelid-arthropod line and the echinoderm-chordate line. The second of these groups gave rise to animals with backbones.

PAST SUCCESS
Trilobites were among the most successful invertebrates that have ever lived. They existed for about 300 million years, leaving huge numbers of fossils, but died out in a mass extinction about 245 million years ago.

Anatomy

In anatomical terms, invertebrates are united principally by things that they lack: as well as not having a backbone, they do not have a bony skeleton, and they do not have true jaws. As adults, some invertebrates look like plants, and spend their lives in one place; others are instantly recognizable as animals, because they are always on the move.

SYMMETRY
Adult starfish appear to have radial symmetry, which means that equal body parts radiate from a central point. A stag beetle has bilateral symmetry, with a body divided in 2 about a single axis. In bilateral animals, many body parts exist in pairs.

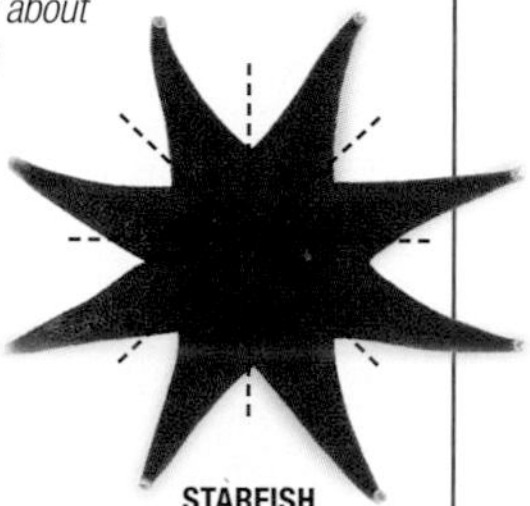

STARFISH

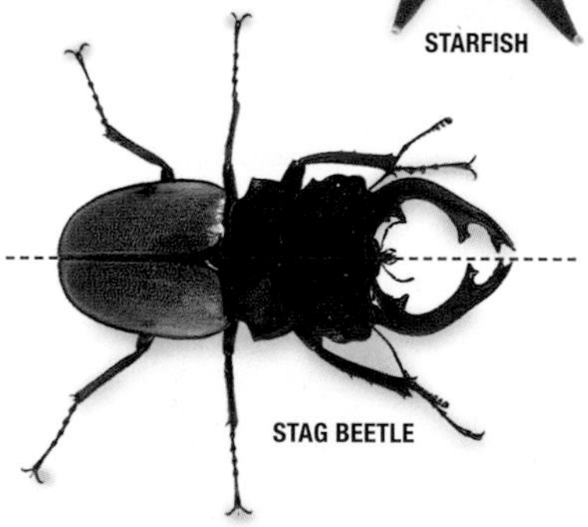

STAG BEETLE

Despite this bewildering variety, invertebrates do follow some underlying patterns, particularly on the cellular level. In the simplest kinds, such as sponges, individual cells are specialized but operate as separate units, rather than in groups. Cnidarians are more complex, with similar cells arranged in tissues. But, in the third and highest grade of organization—seen in most invertebrates—tissues themselves are arranged in organs, and organs into body systems, a pattern that is also followed in the vertebrate world.

This cellular segregation begins as soon as a fertilized egg cell starts to divide. In one species of nematode

SKELETONS

Many invertebrates have a skeleton, which may be internal or external—examples of both types are found in animals that live in water and on land. Some skeletons have evolved chiefly for protection and can withstand powerful impacts, but others are more flexible and exist mainly to keep the body in shape. Skeletons are made of various materials: hard structures often consist of crystalline minerals, while the cuticle of insects is made largely of a plasticlike material called chitin. Once formed, many external skeletons cannot grow and have to be periodically shed and replaced.

SHELLS
A bivalve shell is a 2-part external skeleton. It is relatively heavy but exceptionally hard and grows in step with its owner.

SPICULE SKELETON
Most sponges have an internal skeleton of mineral slivers (spicules), set in a lattice of protein-based fibers. In this glass sponge, they are arranged with almost mathematical precision.

HYDROSTATIC SKELETON
The fluid-filled body cavity of an earthworm is bounded by a flexible wall. The pressurized fluid acts like the air inside a tire.

JOINTED SKELETON
Arthropods, such as this pill millipede, have an external skeleton made of hard plates. Joints between the plates make the skeleton flexible. This kind of skeleton cannot grow.

BACK TO FRONT
Octopuses are bilaterally symmetrical. They have a head with a pair of eyes—one on each side. Although the arms look as if they are radially symmetrical they are not—they are arranged around the mouth. When an octopus swims, it is the rear part of the body that leads.

worm, the adult body consists of exactly 959 cells. Each one is the product of a precise sequence of cell divisions and ends up in a location encoded in the animal's genes.

On a larger scale, one of the fundamental differences between different invertebrates lies in the symmetry of their bodies. Animals in some groups—such as sea anemones—have radial symmetry. Their body parts are arranged in the same way as the spokes or rim of a wheel, usually with the mouth at the center. These animals are often fixed in place, but if they do move, any part of the "wheel" can lead. By contrast, animals with bilateral symmetry usually have a distinct head, which leads the way when they move. These animals have an imaginary central dividing line, which splits them into 2 more or less equal parts.

During the course of evolution, some invertebrates have developed a combination of these body plans. Adult starfish, for example, show almost perfect radial symmetry, except for a small off-center opening (or madreporite) that connects their water vascular system to the outside.

Internally, invertebrates show a wide range of body plans. Sponges—uniquely in the animal kingdom—are full of perforations, or pores. By drawing water through these pores, they filter out tiny particles of food. Some simple invertebrates have a baglike anatomy, with a mouth that leads to a central digestive cavity. This arrangement, seen, for example, in cnidarians and flatworms, has one major drawback: indigestible waste has to be ejected through the mouth. Most invertebrates avoid this problem by having a digestive tract that runs through the body, in effect forming a tube. This plan—which is also found in vertebrates—allows the digestive system to work like a production line, with different parts being specialized for separate tasks.

Although invertebrates do not have bones, many have a skeleton, which protects and supports them. The skeleton may be built of hard parts, but it may also work by fluid pressure (see panel, opposite). Known as a hydrostatic skeleton, this way of supporting a soft body is found only in the invertebrate world.

Senses

For those invertebrates that are permanently fixed in one place, the sensory world is relatively simple. These animals reach out toward anything that gives off the scent of food, and retreat from anything that might be a threat. They are sensitive chiefly to dissolved chemicals, direct contact, or changes in pressure. These animals have specialized nerve cells, or receptors, scattered throughout the body, and a simple nervous system without a brain.

This system works well enough for organisms that rarely move, but mobile invertebrates need more sensory information. They have more elaborate nervous systems, with sense organs to gather information and a brain to process it and trigger the most suitable response. In some of these animals, such as flatworms, the brain is tiny, but in octopuses and other cephalopods, it is highly developed and can be bigger than that of some vertebrates.

Vision is often one of the most important senses for invertebrates that move. Some of them, such as flatworms, have rudimentary eyes without lenses, which simply detect changes in the overall level of light. But many invertebrates have much more complex eyes, equipped with lenses that produce an image. One form, known as a compound eye, is widespread among arthropods, including insects. A compound eye consists of many self-contained units, each with its own receptors and lens system. Each unit responds to part of the field of view, and the total number of units determines how much detail the animal sees. A smaller number of invertebrates—principally cephalopods—have eyes that resemble our own. In these, a single lens focuses light onto a sensitive surface, the retina, which converts the image into signals that travel to the brain.

For invertebrates that live on land, hearing is also a key sense, partly because it is often used for locating a mate. Unlike vertebrates, invertebrate hearing apparatus is not necessarily located on the head. Grasshoppers, for example, have eardrums in their abdomen, while crickets have them on their legs. The same is true of taste organs: flies and butterflies have chemical sensors on their feet, which means they can decide whether something is edible simply by standing on it.

Mobile invertebrates also need to know about their orientation and motion. Most have gravity sensors, that tell them which way up they are, and tension sensors that are triggered when they move body parts such as legs and wings.

TUNING IN
A male mosquito's bushy antennae are used for hearing. This is an example of hearing of a precise kind: the antennae respond to the exact frequency of the female's whining flight. The frequency used varies from one species to another.

MULTIPLE ATTRACTION
Attracted by color and airborne odors, flies congregate on ripe fruit. They taste this sugary food with their mouthparts, and with chemical receptors on their feet.

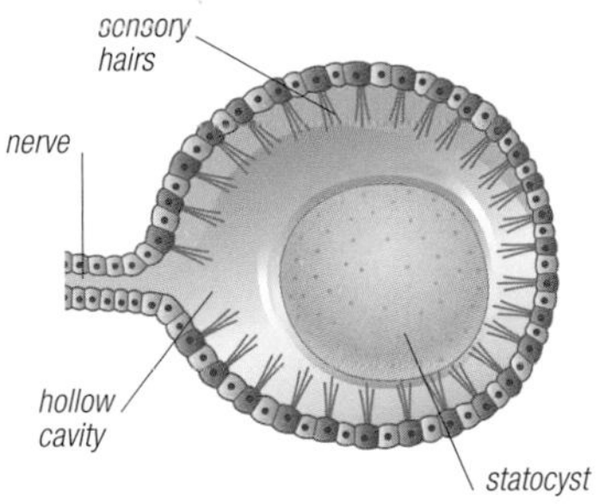

DETECTING GRAVITY
Bivalves have microscopic sensors that respond to the direction of gravitational pull. The sensor consists of a chamber lined with nerve cells that detect the movement of a mineral weight (known as a statocyst).

MIMICRY

Mimicry occurs in the vertebrate world, but invertebrates are the true experts at this form of deception. Invertebrates mimic various objects, from pebbles and twigs to bird droppings, largely to avoid being eaten. They also mimic each other, either as a defense—by resembling species that are dangerous or poisonous—or, more rarely, as a form of camouflage for launching an attack. Mimicry is not only visual: some insects mimic the behavior and even smell of others, so that they can enter their colonies and feed on their young without being attacked.

UNLIKELY MEAL
A postman butterfly caterpillar's first line of defense is its resemblance to bird droppings. If another animal approaches, it waves its menacing-looking "horns," which give off an unpleasant smell.

BUDDING PARENT
An adult hydra can produce young asexually by developing small buds, which grow into new individuals. This parent has grown a single bud, which is now several days old. Eventually, the young animal will break away and become independent.

Reproduction

Invertebrates have a wide range of reproductive techniques, and include animals that reproduce both asexually and sexually. Asexual reproduction usually involves just a single parent. In sexual reproduction, 2 separate individuals are normally needed, although sometimes a single animal can simultaneously take on both male and female roles.

Few invertebrates rely solely on asexual reproduction, because it can lead to problems by creating young that are genetically identical to one another and their parents (see p.30). However, for many species—particularly sap-sucking insects, such as aphids—asexual reproduction is a valuable way of boosting numbers when food is in plentiful supply. Internal parasites, such as flukes, also use asexual reproduction at certain stages of their life cycle. For them, this way of multiplying is particularly important, because their complex life histories mean that each individual offspring has a very small chance of survival. However, with almost all these animals, sexual reproduction intervenes at some point in the life cycle, creating the genetic variations that allow a species to adapt to changing conditions around them.

NEW LIFE
Released by a coral polyp, a parcel of eggs and sperm floats upward. At the sea surface, the parcel will separate, the eggs and sperm will mingle with those of other individuals, and fertilization will occur.

Although asexual reproduction is widespread among invertebrates, most species use sexual reproduction alone. With vertebrates, sexual reproduction almost always involves a male and female animal pairing up, but in the invertebrate world things are often quite different. Animals with both male and female reproductive organs—known as hermaphrodites—are common, and these may either pair up, as in the case of earthworms, or they may fertilize themselves.

Even where individual animals are either male or female, they may reproduce without ever having to meet. This apparently paradoxical situation is common among marine invertebrates, such as sea urchins, and it happens because these animals reproduce by external fertilization. They release their sex cells into the water, and the sperm and egg cells mingle, producing large numbers of fertilized eggs. For this form of reproduction to succeed, the eggs and sperm must be released with precise timing, which is often achieved by coordinating with the changing phases of the moon.

On land, most invertebrates use internal fertilization, with the male and female pairing up, and the male inserting his sperm into the female's body. This can involve complex courtship displays, and sometimes—for example, in spiders—careful behavior on the part of the male, who runs the risk of being eaten by his mate. In the insect world, some male bugs adopt a less ceremonious approach: they use "traumatic insemination," piercing the female's abdomen with an organ that works like a dart. However, internal fertilization does not necessarily involve the direct transfer of sperm. In some scorpions, for example, the male produces a packet of sperm, known as a spermatophore, which it fixes to the ground. The male then presses the female onto the spermatophore, forcing it into her internal genital chamber, where fertilization takes place.

Parental care

Once a female's eggs have been fertilized, most invertebrates leave their young to develop on their own. As a result, their young face an uphill struggle during their first few weeks of life, and a large proportion fail to survive. These losses are offset by the large number of eggs that are produced. A female starfish, for example, can release 2.5 million eggs in 2 hours.

At the other extreme, some invertebrates are attentive parents. Female earwigs, for example, remain with their eggs, periodically cleaning them to ensure that they are not attacked by parasites or by mold. Female octopuses attach their eggs to rocky recesses on the seabed, and remain nearby, squirting jets of water over them to keep them clean and well oxygenated. Invertebrates may also take care of their young after they have hatched. Many arachnids carry their young on their back, while some freshwater crayfish carry theirs on their legs.

METAMORPHOSIS

Most invertebrates change shape as they grow up. This change, known as metamorphosis, allows the young and adults to live in different ways, which can improve their chances of finding food. It also helps species to disperse, because one stage—either the larva (juvenile) or the adult—can have adaptations that enable it to travel. Metamorphosis may be gradual, or rapid and abrupt. In the first type, the young often resemble their parents, but in the second, they look completely different. Metamorphosis is controlled by hormones.

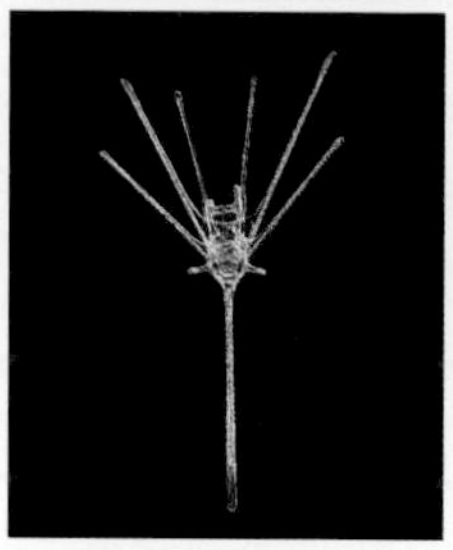

EARLY STAGE

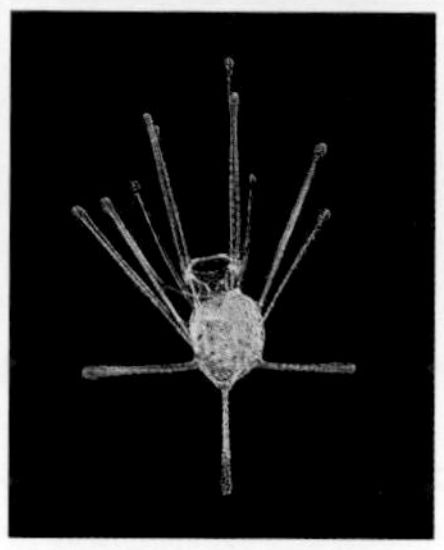

LATE STAGE

DRIFTING YOUNG
Sea-urchin larvae have slender spikes that help them to drift in the plankton. As a larva matures, the spikes are slowly absorbed.

COMPLETE CHANGE
During the pupa stage, a caterpillar's body is broken down and an adult butterfly is formed in its place. The change occurs inside a chrysalis, or pupal case.

SAFE PASSAGE
With her family crowded on her back, a female scorpion guards her brood. At this stage, the young have pale, soft exoskeletons—these darken and harden as they age.

Colonies

Even if they are cared for, most young invertebrates soon disperse to take up life on their own. But with colonial species, animals from the same species remain together, forming permanent groups or colonies. In a colony, the members are usually closely related to one another, and they often divide up the work needed for survival.

Colonial life is a recurring theme in invertebrate evolution, and it can take a variety of different forms. In some colonies—particularly in the sea—the members are physically joined, and may be so highly integrated that they look and behave like a single animal. This kind of organization is seen in

COLONY AFLOAT
A Portuguese man-of-war is an animal colony, formed by a collection of specialized polyps that live and work together.

some hydrozoans, including the Portuguese man-of-war. This animal looks much like a jellyfish but is actually a colony of polyps, one of which forms a baglike float (see opposite). However, floating colonies are comparatively rare: many more invertebrates—including corals, bryozoans, and sea squirts—form fixed colonies, attached to all kinds of objects, from shells to the seabed. Individuals in such colonies are usually connected to one another through joined gut and nerve networks, which allow sensory signals to be passed from one individual to its neighbors.

CENTRE OF ATTENTION
Surrounded by workers, a queen honey bee controls all the activity in her hive. She does this by giving off volatile hormones—called pheromones—that evaporate and spread throughout the nest.

Invertebrate colonies are less common on land, and the members of terrestrial colonies are separate, rather than joined. They reach their greatest development in social insects, which include termites, as well as ants, bees, and wasps. Despite their apparent independence, the members of these colonies rely on each other for survival, just as much as if they were fastened together. In each colony, one individual—the queen—normally lays all the eggs, while the role of the other members is to ensure that the colony is fed, housed, and defended from attack.

Feeding

Collectively, the world's invertebrates eat almost everything that is—or once was—alive. They include herbivores and carnivores, as well as a host of species that feed on dead remains. Their appetites embrace everything that we consume as food, as well as some highly specialized foodstuffs, such as rotting seaweed, feathers, fur, and even animal tears.

In water, a common feeding method among invertebrates involves filtering out edible matter, using body parts that act like a sieve. This way of life supports vast numbers of static invertebrates that live on the sea bed or near the shore. In these animals, evolution has crafted filtering apparatus out of a wide range of body parts: barnacles, for example, use their legs, while bivalves use their gills. Many of these filter feeders actively pump water through their filters, to increase the size of their catch. For these animals, pumping often has the added advantage that it supplies oxygen as well as food. Filter feeding is also used by many planktonic animals, including krill, which in turn provide food for baleen whales—the largest filter feeders of all. Underwater "grazing," in which vegetation is scraped from a solid surface, is also common. It is a technique used by limpets, which have rows of microscopic teeth, and also by sea urchins, which have a set of 5 chalky jaws.

On land, there is no equivalent of filter feeding, although some web-spinning spiders come close to mirroring this way of life, using their webs to trap airborne animals. Instead, most terrestrial invertebrates have to actively forage for food. The smallest herbivores—including many wood-eating species—often burrow through their food, a lifestyle that helps protect them from attack, but most predatory invertebrates move around in the open. In herbivores and carnivores alike, mouthparts are often shaped to deal with particular sources of food. Caterpillars bite off and swallow their food in substantial chunks, but many other insects have a liquid diet. Moths and butterflies, for example, are equipped with a coiled proboscis or "tongue" up to 12 in (30 cm) long, which can be unrolled to drink nectar from flowers or juices from fruit, while horseflies have a pair of bladelike jaws that can slice through skin, allowing them to lap up oozing blood. Despite being carnivorous, spiders are also liquid feeders. They have a tiny mouth, and feed by pouring digestive juices onto their prey or injecting it into its body, and then sucking up the slurry that results.

Given their immense numbers, many invertebrates have an important ecological impact as they feed. Earthworms play a key part in maintaining the fertility of soil, while insects have a major role in pollinating flowers and breaking down dead remains. In many habitats, invertebrates have a much greater effect than larger animals, but unless they are pests, their activity often goes unnoticed by humans.

NATURE'S RECYCLERS
Dozens of dung beetles clamber over a pile of animal droppings. The beetles feed on the dung, and they also roll it away and bury it to provide food for their young.

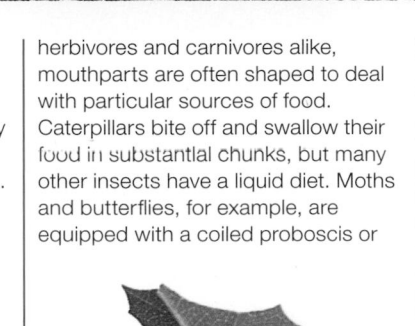

GATHERING FOOD
Dwarfed by its load, a leafcutter ant carries food back to its nest. In some parts of the world, ants eat as much food as all the plant-eating vertebrates in their habitat.

EATING AT LEISURE
Gripping the 2 halves of a mussel's shell, a starfish slowly pries them apart. It can evert its stomach through a paper-thin aperture between the mussel's shells to feed on it.

PARASITES

Invertebrates make up the vast majority of the world's parasitic animals. Some—including ticks and leeches—live on the outside of their hosts. Many others, such as tapeworms and roundworms, live internally, feeding on their host's tissues or on the food that it consumes. External parasites can often survive with a single host, but many internal parasites have complex life cycles, involving several different hosts in sequence. These multiple hosts often include a vertebrate. Most parasites do not kill their hosts, but parasitoids are normally lethal. These animals, which include many insects, lay their eggs in or near the bodies of other animals. When the young hatch, they devour the host from inside.

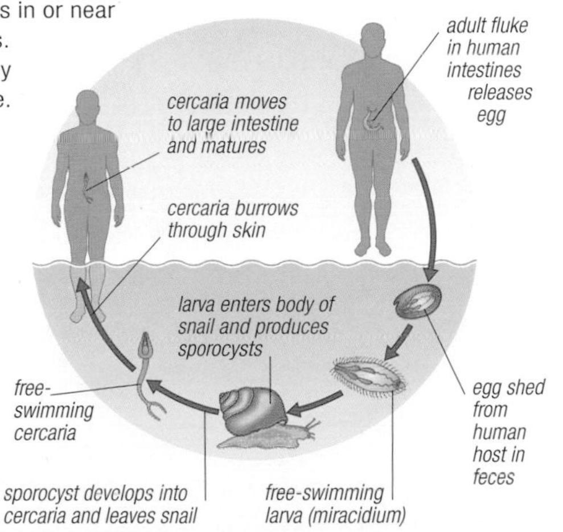

LIFE CYCLE OF A BLOOD FLUKE
The human blood fluke is a parasitic flatworm that uses water snails as intermediate hosts. Its eggs reach water via untreated sewage, and produce free-swimming larvae. These multiply asexually inside snails, producing another swimming stage that attacks humans.

Sponges

PHYLUM	Porifera
CLASSES	3
ORDERS	24
FAMILIES	127
SPECIES	About 10,000

Sponges, which at one time were thought to be plants, are the simplest of all living animals. Unlike most animals, which move through their habitat, sponges are immobile, living attached to solid surfaces and feeding by setting up currents of water from which they filter out particles of food. In other words, their aquatic environment is made to move through their body rather than the other way around. Most sponges are marine animals, although some species have adapted to life in freshwater.

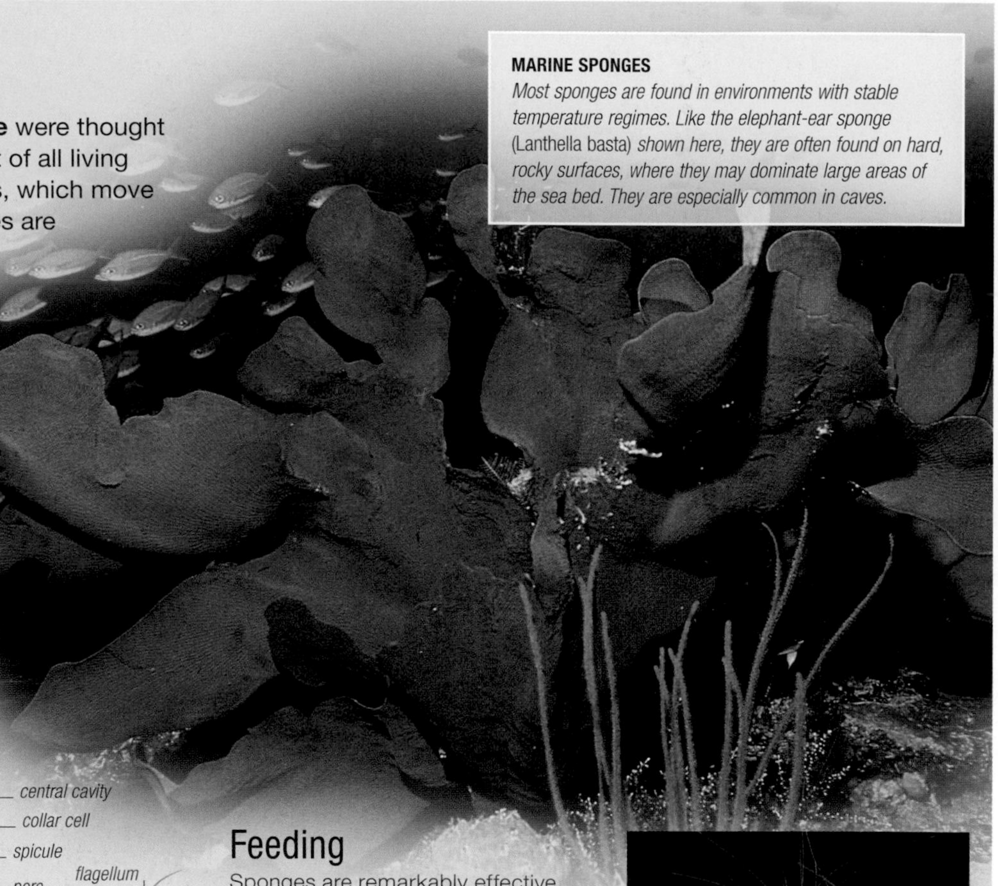

MARINE SPONGES
Most sponges are found in environments with stable temperature regimes. Like the elephant-ear sponge (Lanthella basta) *shown here, they are often found on hard, rocky surfaces, where they may dominate large areas of the sea bed. They are especially common in caves.*

Anatomy

The body of a sponge lacks many features typical of other animals. Although the cells perform specialized functions, they do not form true tissues or organs, and there are no nerve and few muscle cells. However, sponges do have a skeleton—a network of organic collagen fibers and inorganic structures (called spicules) made of silicon dioxide or calcium carbonate. Sponges have no system of symmetry, nor do they have clearly distinct body parts. The simplest sponge is essentially a tube, closed at one end, but complex sponges take on many forms, from crusts, spheres, and barrels to tangled branches.

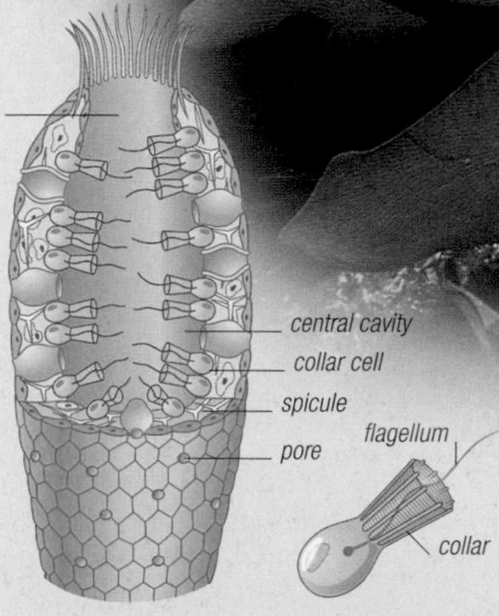

BODY SECTION
Water enters a simple sponge through minute pores in the outer surface. It then passes into a central cavity before being expelled through a larger aperture (the osculum).

Feeding

Sponges are remarkably effective filter feeders. The cells lining the body's inner surface are known as collar cells, each consisting of a ring (or collar) of tentacle-like structures surrounding a whiplike feature called the flagellum. Movement of the flagella creates the current of water that moves through the body. As the water passes through, plankton and particles of organic matter are trapped in the collars.

CARNIVOROUS SPONGE
Found on deep-sea bottoms and in Mediterranean caves, this unusual sponge catches live prey on hooklike spicules at the end of long filaments and digests it externally.

Class Calcarea

Calcareous sponges

Occurrence c.650 spp. worldwide; in shallow water (in crevices, overhangs and caves, on reefs and on seaweed)

These are the only sponges that have skeletal spicules made of calcium carbonate, the mineral that mollusks use to make their shells. Compared to other sponges, they are mostly small—less than 4 in (10 cm) high—lobe- or tube-shaped and drab, but some reef species are brightly colored.

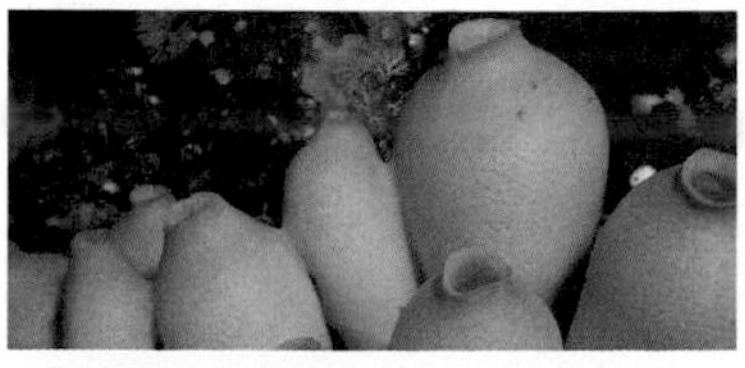

LEUCETTUSA LANCIFER
This flask sponge, less than 2 in (5 cm) high, has the classic compound vase shape of calcareous species. Members of the genus are widespread in shallow waters from the tropics to the subAntarctic.

Class Hexactinellida

Glass sponges

Occurrence c.500 spp. worldwide in deep cold water; mostly at depths of 650–6,000 ft (200–2,000 m) in seas

Most glass sponges are less than 3¼ ft (1 m) high but in some areas many grow together to form "sponge reefs" up to 65 ft (20 m) in height. Their siliceous spicules are usually fused together to form rigid, often beautiful, glasslike masses. These structures are fragile, however, and the skeletons of the species living in deep water tend to collapse when they are dredged up and brought to the surface.

VENUS' FLOWER BASKET
Glass sponges are often objects of remarkable beauty. Species in the genus Euplectella, *such as the one shown here, are found in tropical oceans below a depth of about 500 ft (150 m). They grow up to 12 in (30 cm) tall.*

Class Demospongiae

Demosponges

Occurrence c.7,200 spp. worldwide; in shallow and deep water, on solid surfaces of all kinds

These sponges make up nearly 90 percent of all sponge species, and include the familiar bath sponge, whose populations are now badly affected by overcollection. Demosponges vary in size from less than ⅓ in (1 cm) to well over 3¼ ft (1 m) in height or breadth, and are often brilliantly colored, creating underwater "gardens" that rival corals for their vivid hues. While crusts and mounds are common, these sponges come in a variety of shapes—some even bore into mollusk shells, leaving them riddled with pin-sized holes.

CARIBBEAN PINK VASE SPONGE
Many demosponges form irregular masses, but many others are funnel- and vase-shaped like Niphates digitalis *(above). Up to 12 in (30 cm) tall, it has, like most demosponges, a skeleton of siliceous spicules and collagen fibers.*

STOVE-PIPE SPONGE
Aplysina archeri, *or stove-pipe sponge, can reach a length of over 20 in (50 cm). In sponges with this kind of shape, water is drawn in through pores in the sides of the tube, and expelled through an opening at the top.*

Cnidarians

PHYLUM	Cnidaria
CLASSES	6
ORDERS	24
FAMILIES	300
SPECIES	About 11,000

This group of simple aquatic animals includes sea anemones, corals, jellyfish, and hydroids. The radially symmetrical body is essentially a tube that is closed at one end and open at the other. This tube is either flattened into a bell shape (a medusa) or elongated with the closed end attached to a hard surface (a polyp). In general, medusae swim freely, while polyps live on the seabed. All cnidarians have tentacles around their mouth that contain stinging cells, known as cnidocytes. Most species are marine, although a small number are found in freshwater.

Anatomy

Some cnidarians (such as sea anemones and corals) exist only as polyps. Others (hydroids and some jellyfish) exist as both polyps and medusae at different stages of their life cycle. In both forms, the wall of the body tube consists of just 2 layers of cells: an outer epidermis and an inner gastrodermis, separated by a jellylike matrix (the mesoglea). The central cavity of the tube acts as a gut, and the single opening is used both to take in food and to expel waste. Around the mouth, the body wall is drawn out into one or more whorls of tentacles. These contain the stinging cells, cnidocytes, which are unique to cnidarians. Cnidocytes inject venom into prey and hold it while the tentacles draw it into the mouth.

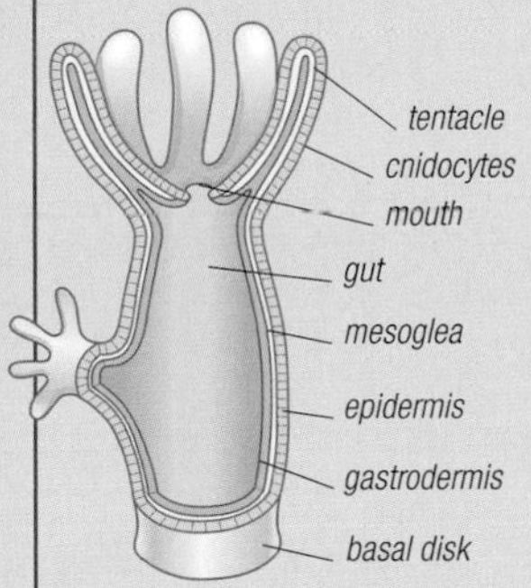

POLYP
The mouth of a polyp faces outward. At the opposite end of the body is a basal disk, which is often attached to a solid surface. Many polyps, such as this hydra, do not have a well-developed mesoglea. Polyps may occur either singly or in colonies.

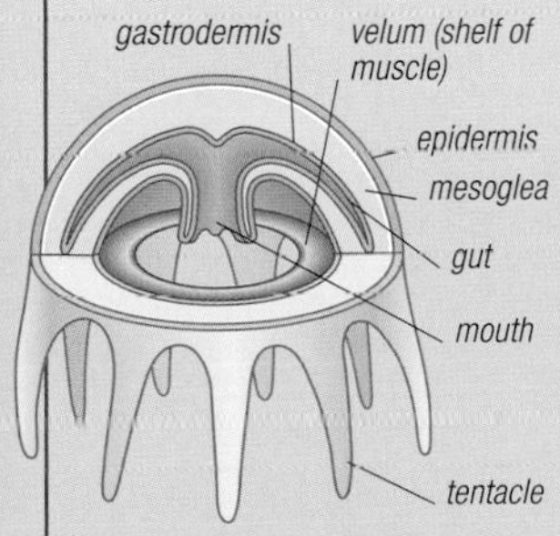

MEDUSA
The body of a medusa looks like a bell, with the mouth at the center of the lower surface and the tentacles at the edges. The mesoglea is thicker than it is in most polyps. Some medusae have a shelf of tissue that is rich in muscle cells and used in locomotion.

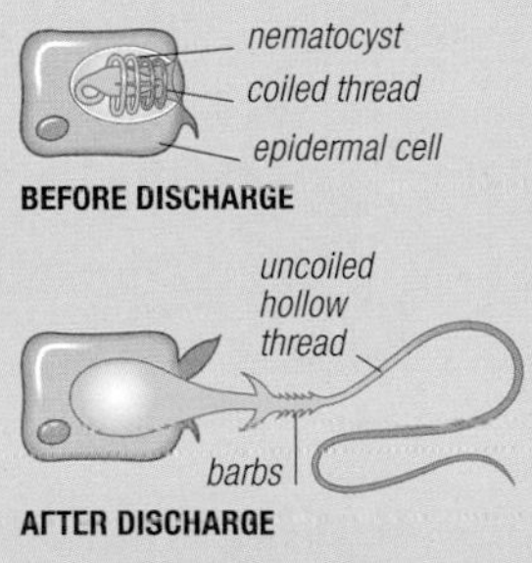

CNIDOCYTES
Each cnidocyte contains a bulblike structure, called a nematocyst, which houses a coiled, barbed thread. When triggered by touch or chemicals, the thread explodes outward and pierces the victim's skin. The tentacles are then used to pull the prey in.

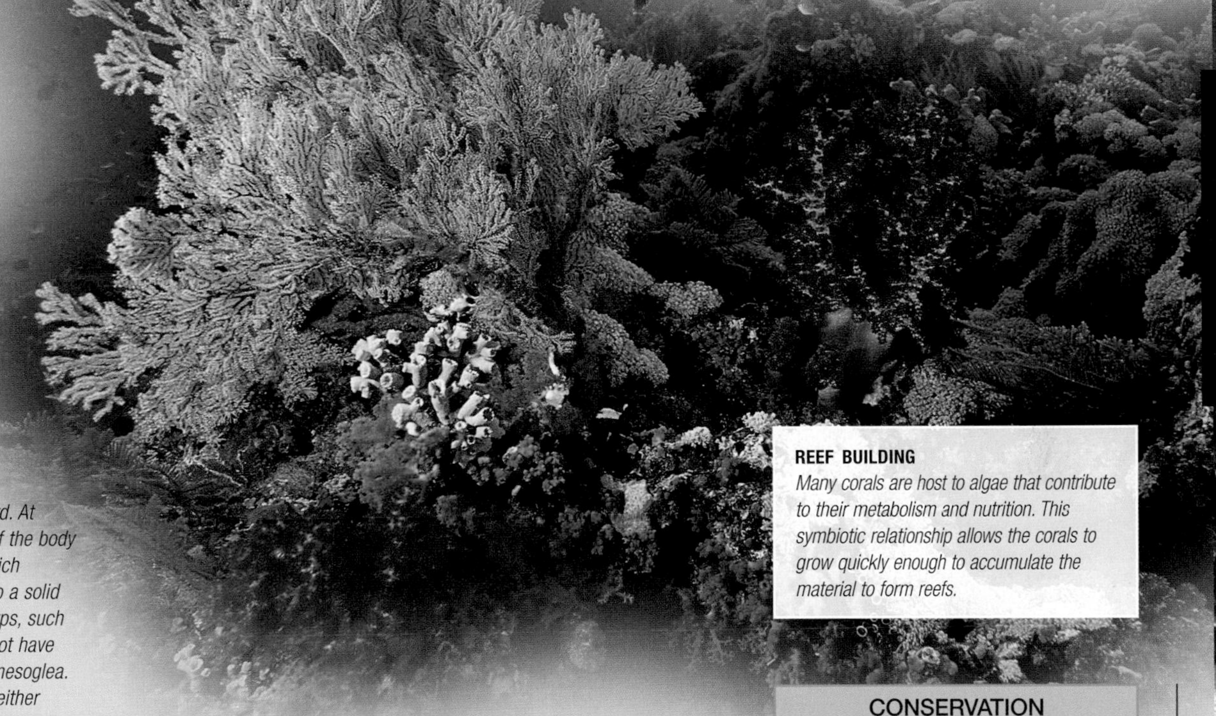

REEF BUILDING
Many corals are host to algae that contribute to their metabolism and nutrition. This symbiotic relationship allows the corals to grow quickly enough to accumulate the material to form reefs.

Locomotion

Cnidarians can change shape by coordinating the contraction of some of their muscle cells. In polyps, the water inside the body cavity can act in a similar way to a skeleton, providing a resistant structure against which the cells are contracted. A polyp can lengthen, shorten, or bend its body—for example, to reach toward potential prey or to retreat from predators. By alternately relaxing and contracting groups of muscle cells, free-living medusae such as the jellyfish (shown right) can move using a weak form of jet propulsion.

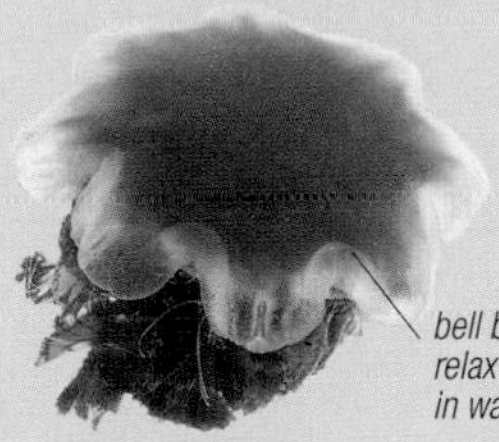

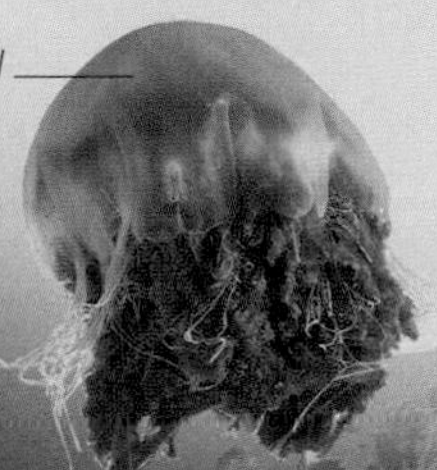

JELLYFISH MOVING
Jellyfish propel themselves through water by contracting their bell, obtaining lift from the water being forced out from under the bell. The bell then relaxes, water re-enters, and the cycle is repeated.

Colonies

Most cnidarians can reproduce asexually (without mating) by producing buds from their own body. In some cases, the new polyp splits completely; in others, the budding is incomplete so that several individuals remain joined together to form a colony. Colonies can reach great size, as, for example, in reef-forming corals. Some planktonic colonies are highly complex. For example, colonies of siphonophores consist of polyps and medusae integrated into what appears to be an individual organism, with specialized polyps forming the equivalent of organ systems.

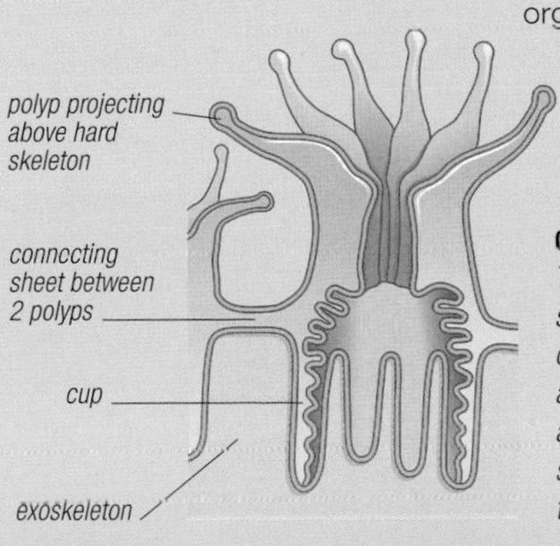

CORAL POLYP
The polyps in a coral colony secrete an exoskeleton of hard calcium carbonate, which may accumulate over time to form a reef. The exoskeleton includes structures called cups, into which the polyps can withdraw.

CONSERVATION

Several dozen cnidarian species (mainly corals) are listed as vulnerable and some that form coral reefs are now endangered. Threats include rising sea temperatures—as a result of global warming—discharge of silt in coastal areas, and collection for souvenirs. If the sea temperature rises above a certain level, the corals' symbiotic algae are expelled, and the corals become bleached and often die.

Class Anthozoa

Anthozoans

Occurrence 7,042 spp. worldwide; on the seabed, in soft sediments

The word "anthozoa" means "flower animal," and with their bright colors, waving tentacles, and fleshy trunks, these animals often look remarkably like plants. However, they are carnivorous—as are all cnidarians—although some species supplement their diet with food produced by microscopic algae harbored in their bodies. Anthozoans vary greatly in size, ranging from minute species less than 1⁄32 in (1 mm) across, to those over 33 ft (10 m) across. When adult, they form polyps—simple, tubular animals with the tube closed at one end and open at the other. The open end is drawn out into one or more whorls of tentacles containing stinging cells, used for feeding and for defense against predators as varied as gastropods, polychaetes, sea spiders, and starfish. Anthozoans multiply asexually and sexually. Many species are solitary, but this group also includes colony-forming hard corals—animals that are responsible for building coral reefs, the biggest structures in the living world. These are formed by the animals' external calcareous skeleton. Other widespread anthozoans include soft-bodied sea anemones, plus sea pens and sea fans that have flexible, horny internal skeletons.

SEA FANS
Growing 9¾ ft (3 m) or more high, sea fans such as Annella mollis *(left) occur widely, but are especially common in tropical seas. They attach themselves to hard surfaces by creeping rootlets, and have a tough but flexible stem, with many branches that bear polyps.*

BUSHLIKE SOFT CORALS
The fleshy masses of most soft corals contain calcareous structures (sclerites) that support them. In the bushlike species (above), the sclerites may protrude when the colony is expanded. A member of the genus Dendronephthya, *this Indo-Pacific coral grows to 3¼ ft (1 m) high. Soft corals like these are either attached to hard surfaces or anchored in soft sediments.*

SEA PENS
With a central axial polyp bearing many lateral polyps, often on regularly arranged side branches, sea pen colonies—found throughout the world—look like old-fashioned feather quills. The lower end of the axial polyp anchors the colony to the seabed. The Pennatula *species on the right grows to 20 in (50 cm) high, but some are double this size. Some of the secondary polyps of sea pens—such as the* Pteroeides *species on the left, which grows to 12 in (30 cm)—pump water through the colony.*

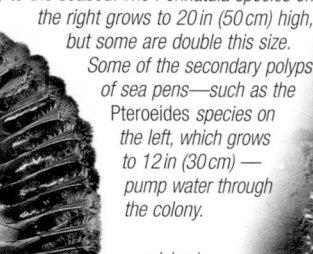

COMMON DEAD MAN'S FINGERS
Soft corals—members of the order Alcyonacea—occur in all oceans. The North Atlantic species Alcyonium digitatum *(above) grows up to 8 in (20 cm) high, and lives in shallow water. Its polyps protrude from a white to pink or orange, fingerlike mass—hence its ghoulish common name. The lobes contract if touched.*

FALSE CORALS
Living in tropical waters worldwide, false corals are intermediate in form between true corals and sea anemones. Their polyps closely resemble those of corals, but lack an external hard framework. Like reef corals, they have symbiotic algae that supply them with nutrients. Amplexidiscus fenestrafer *(above) attains a diameter of at least 12 in (30 cm).*

SEA ANEMONES
Unlike most anthozoans, sea anemones live as solitary polyps; many of these, however, reproduce by dividing and forming clones. The snakelocks anemone, Anemonia viridis *(right), has tentacles up to 4 in (10 cm) long. It needs well-lit conditions because a large part of its food is supplied by symbiotic photosynthetic algae.*

REEF SOFT CORALS
Like dead man's fingers, Sarcophyton *species belong to the order Alcyonacea, and have the same fleshy texture. Members of this genus are most common in warm parts of the Indian and Pacific oceans, and form colonies up to 3¼ ft (1 m) across.*

HORMATHIID SEA ANEMONES
These widespread sea anemones commonly form symbiotic relationships with other marine invertebrates, living on mollusk shells or the carapaces and even claws of crabs. The anemone gains food scraps, while its partner is protected by the anemone's stinging tentacles. The species above, Stylobates aenus, *grows up to 8 in (20 cm) tall.*

ANEMONIA ALICEMARTINAE
Described as a new species in 2001, this bright red anemone is found on the rocky coasts—including tidal pools—of northern Chile. Scientists suggest that such a conspicuous species has not been spotted before because its population has only recently expanded over the last decades.

PORITES SPP.
These massive corals, found in tropical waters, are second only to the staghorns as reef builders. They grow in the form of mounds, sometimes with large, blunt, fingerlike projections. Some Porites *domes on the Great Barrier Reef are almost 33ft (10m) in diameter, and may be 200 years old. The individual polyps, however, are only about 1/8 in (3mm) across.*

MUSHROOM CORALS
While most corals are colonial, mushroom corals consist of a single polyp usually less than 20in (50cm) across. They live in tropical waters and, when young, are attached to the seabed by a stalk. When adult, as in the Fungia *species above, they break away from the stalk and rest on the bottom.*

PAVONA SPP.
This species in the genus Pavona *has a diameter of 10in (25cm) and is a common and conspicuous member of reefs from the Red Sea to the eastern Pacific. It has fine, starlike markings on the surface, and grows in spreading colonies that form columns, plates, or fronds.*

STAGHORN CORALS
The staghorns, belonging to the genus Acropora, *are among the fastest growing corals on reefs. They would quickly swamp most other species if not for the fact that they are fragile and easily damaged in storms. Found in tropical seas, many staghorn colonies resemble antlers or bushes—like the branching species on the left, which may be 6½ft (2m) tall. Some are more spreading and resemble the flat-topped acacia trees on the East African savanna. The platelike growth of the species below—up to 9¾ft (3m) across—intercepts light for its symbiotic algae and shades out potential competitors beneath.*

BRAIN CORALS
The individual polyps of this Lobophyllia *colony are relatively large—1¼in (3cm) in diameter or more. They are arranged in meandering rows, giving the coral a brainlike form. Mature colonies of brain coral may weigh over a ton. Many species inhabit the Pacific Ocean.*

Class Scyphozoa

Jellyfish

Occurrence 234 spp. worldwide; in marine plankton, a few on seaweed or the seabed

With bodies that can be over 6½ft (2m) across and tentacles that may extend many yards, jellyfish are among the largest animals found in plankton. A few species, however, attach themselves to seaweeds, and some spend much of their time on the seabed. A jellyfish's body, or "bell," is shaped like an inverted saucer or bowl, and consists mainly of a thick layer of jellylike mesoglea sandwiched between the outer and inner cell layers (see p.537). The mouth in the center of the underside opens into a simple gut, and stinging tentacles hang down from the bell margin. These simple animals can have complex life cycles. Some (but not all) begin life as small polyps on the seabed. These split into a stack of tiny medusae, or baby jellyfish, which drift away to start life on their own, growing into sexually reproducing adults. Although jellyfish can swim by contracting their muscles, they make little headway against even a gentle current, and are often washed up on beaches after storms. Most are carnivorous, catching prey with their tentacles. This class includes species such as the moon jelly and lion's mane jellyfish.

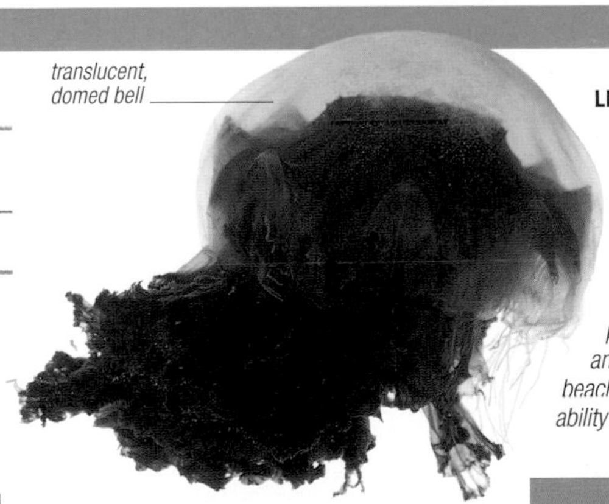

LION'S MANE JELLYFISH
With a bell that can be over 6½ft (2m) across, the lion's mane jellyfish, Cyanea capillata, *which lives in northern waters, is one of the biggest in the world. Its numerous long tentacles pose a danger to swimmers, and even when stranded on a beach, this jellyfish retains its ability to sting for a long time.*

CASSIOPEIA XAMACHANA
A roughly disk-shaped jellyfish, 12in (30cm) across, this species lives in warm, shallow waters, and usually lies quietly pulsating upside down on the seabed. For nutrition, it relies on the symbiotic photosynthetic algae that fill the bladderlike structures around its mouth.

MARIVAGIA STELLATA
Discovered in the Mediterranean in 2010, this species belongs to a group of jellyfish called rhizostomes—characterized by having a complex system of tiny mouths opening along specialized frilly "arms" that hang beneath the bell.

MASTIGIAS PAPUA
These broad-bodied jellyfish, some up to 7in (19cm) across, are most common in the tropics. They are rhizostomes and—like Marivagia stellata *(left)—lack the margin of tentacles found around the bell of many other jellyfish.*

INVERTEBRATES

Class Hydrozoa

Hydrozoans

Occurrence 3,686 spp. worldwide; in plankton, and on sea, river, and lake beds

This group of cnidarians contains a diverse array of animals, with various body shapes and lifestyles, and sizes ranging from as small as 1⁄32 in (1 mm) to as large as 3¼ ft (1 m) across. In the course of their lives, they usually pass through both an attached polyp stage and a free-living medusa stage. Most are colonial, containing tens to thousands of polyps. Some colonies have hard skeletons and resemble corals, while others are often mistaken for jellyfish, drifting through the surface waters. Colonial forms that remain on the seabed mostly have polymorphic polyps, with different polyps being shaped for different tasks. The result is a colony that looks and works like a single animal. Planktonic colonies may also have polymorphic polyps, with some catching prey and others digesting the catch. Mostly carnivorous, hydrozoans use their tentacles to capture organisms ranging from minute planktonic animals to sizable fishes; some are dependent on symbiotic photosynthetic algae. Most animals in this class—which includes the sea fir and by-the-wind-sailor, as well as the freshwater hydras—are harmless to humans, but a few, including the notorious Portuguese man-of-war, have potentially deadly stings. They are preyed on by some mollusks, gastropods, polychaetes, sea spiders, starfish, and fishes. Reproduction occurs both sexually and asexually.

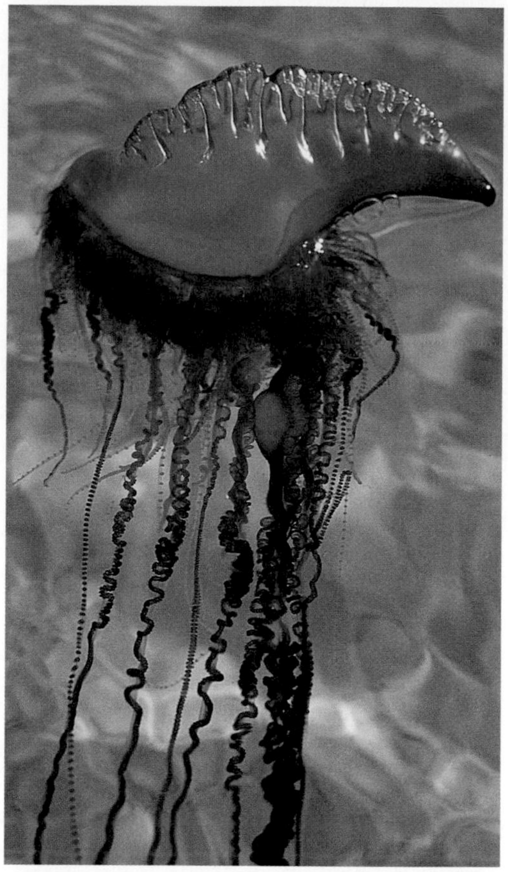

PORTUGUESE MAN-OF-WAR
Found almost worldwide in tropical and temperate seas, the Portuguese man-of-war, Physalia physalis *(right), has a well-deserved reputation for being dangerous. It looks like a single animal, but it is actually a highly integrated colony of polyps, some of which pack a powerful sting. A large, gas-filled float keeps the colony at the surface, and supports trailing tentacles up to 66 ft (20 m) long. These tentacles contain polyps that catch prey, ones that digest food, and others that enable the colony to reproduce.*

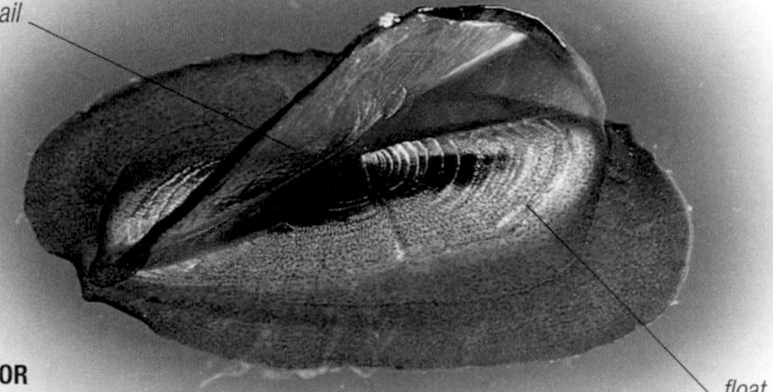

BY-THE-WIND-SAILOR
The coin-shaped Velella velella*—4 in (10 cm) in diameter—is a highly modified hydroid polyp that, like the Portuguese man-of-war, drifts on the ocean surface. It is moved by the wind acting on the small sail attached to the float and, after storms, is often washed up on tropical shores.*

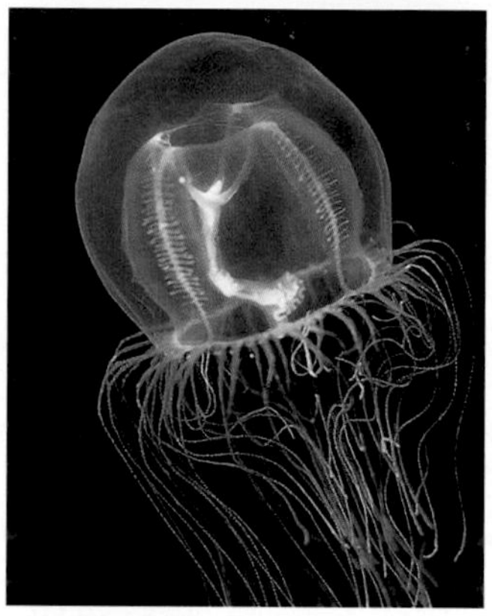

POLYORCHIS SPP.
This delicate hydrozoan medusa has a transparent bell up to 1½ in (4 cm) across, with about 90 tentacles trailing from it. Common on the west coast of North America, it feeds on planktonic animals.

FEATHER HYDROIDS
Species in the widespread genus Aglaophenia *live in shallow water, and their dead remains are often washed up by the tide. Each colony can be up to 23½ in (60 cm) high, and contains numerous feeding and stinging polyps, as well as larger reproductive polyps—the solid yellow structures in the picture (right)—which release floating larvae into the water.*

HYDRACTINIA SPP.
Members of the genus Hydractinia*, distributed worldwide, are small, colonial hydroids, typically ¾ in (2 cm) high, that often form encrustations on animals. The hydroids get improved access to food by being carried around, and in return, protect their host with their stings.* Hydractinia echinata *grows on shells of hermit crabs (see above).*

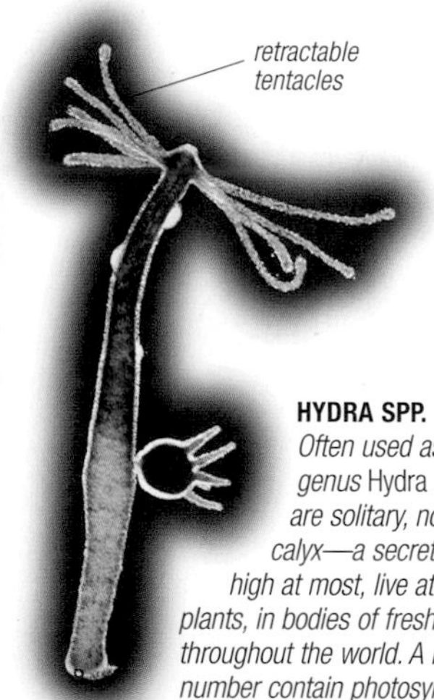

HYDRA SPP.
Often used as a textbook example of a hydroid, the genus Hydra *is, in fact, unusual because its members are solitary, not colonial, and are not surrounded by a calyx—a secreted tube. These tiny animals, 1 in (2.5 cm) high at most, live attached to stones, submerged wood, or plants, in bodies of freshwater throughout the world. A large number contain photosynthetic algae, which give them a green color.*

DISTICHOPORA VIOLACEA
A few hydrozoans resemble corals and have a hard exoskeleton that protects the polyps. Most of these are quite small and brightly colored, such as the Indo-Pacific species Distichophora violacea *(left), which is 2 in (5 cm) high. A few, such as fire "coral"* Millepora*, grow into very large colonies.*

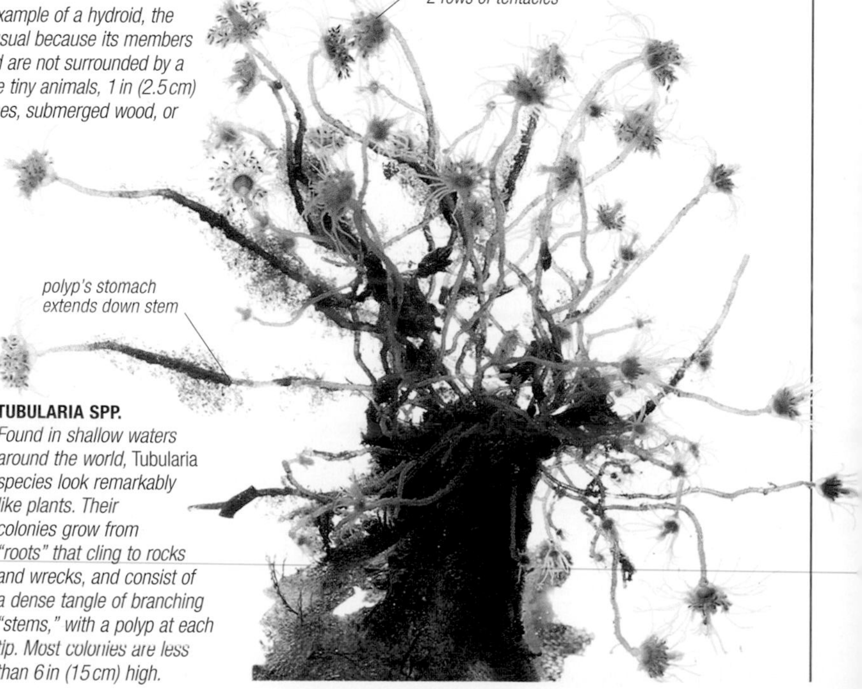

TUBULARIA SPP.
Found in shallow waters around the world, Tubularia *species look remarkably like plants. Their colonies grow from "roots" that cling to rocks and wrecks, and consist of a dense tangle of branching "stems," with a polyp at each tip. Most colonies are less than 6 in (15 cm) high.*

Flatworms

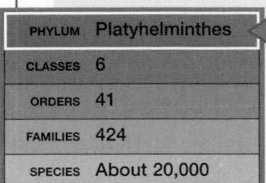

PHYLUM	Platyhelminthes
CLASSES	6
ORDERS	41
FAMILIES	424
SPECIES	About 20,000

Flatworms are the simplest of all the animals that have bilateral symmetry. Their bodies are solid rather than containing an internal cavity, and they have no blood or circulatory systems, and no organs for exchanging gases with the environment. Flatworms include many parasites, such as tapeworms and flukes, as well as a variety of free-living forms that are abundant in freshwater habitats and in the sea, particularly on rocky coasts and reefs.

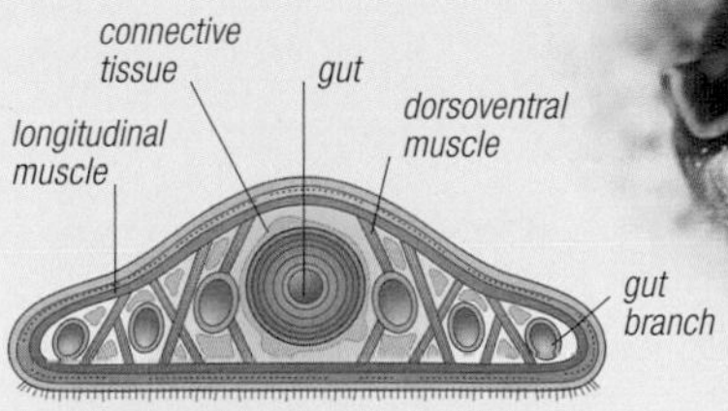

BODY SECTION
Flatworms have a simple body that is not divided into segments. The space between the internal organs is filled with spongy connective tissue.

FREE-LIVING FLATWORMS
Since flatworms rely on diffusion to take in oxygen, their bodies have to be thin. This requirement makes large species—such as this marine turbellarian—much more fragile than other animals of a comparable size.

Class Turbellaria

Turbellarians

Occurrence 5,429 spp. worldwide; on rock and sediment surfaces in water, in damp soil, under logs

Turbellarian flatworms show a variety of different shapes. Although they have a distinct front end, the mouth is positioned some way back on the underside of the body. Most turbellarians are translucent, black, or gray, but some marine species—particularly ones that live in coral reefs—have brightly colored markings. They may be less than 1/32 in (1 mm) to 20 in (50 cm) long. The largest species are paper thin, and this helps them absorb oxygen directly from their surroundings. Unlike other flatworms, which are parasitic, most turbellarians are free living and have simple sense organs to help them negotiate their way around their habitat. Many of them move by creeping, but some of the bigger species swim by rippling their bodies. The majority of turbellarians are predators, feeding on other small invertebrates; some are parasitic, some are commensal, and a few depend on symbiotic photosynthetic algae. Almost all are hermaphrodites.

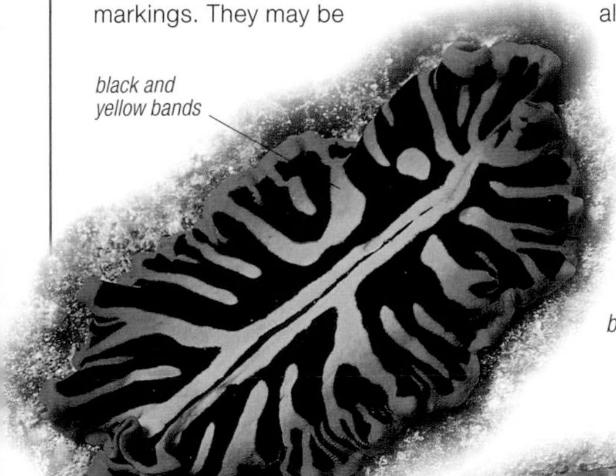

MARINE FLATWORMS
Some marine turbellarians are large and brightly colored. This species— Psuedoceros dimidiatus*—is common on coral reefs in the Indian and Pacific oceans, and grows to 2 in (5 cm) long. It shows a variety of color schemes, including black and yellow traverse bands and a line of black blotches.*

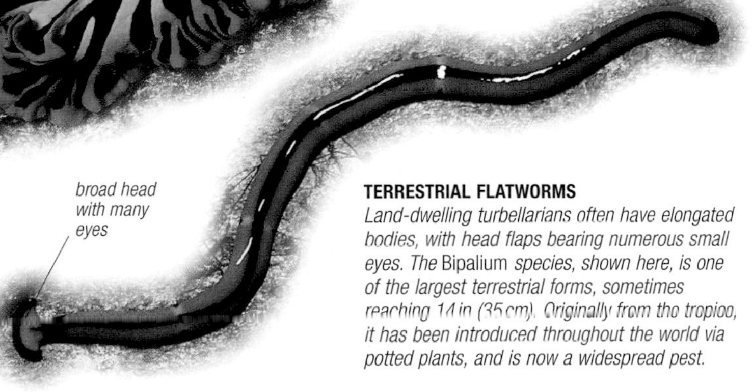

TERRESTRIAL FLATWORMS
Land-dwelling turbellarians often have elongated bodies, with head flaps bearing numerous small eyes. The Bipalium *species, shown here, is one of the largest terrestrial forms, sometimes reaching 14 in (35 cm). Originally from the tropics, it has been introduced throughout the world via potted plants, and is now a widespread pest.*

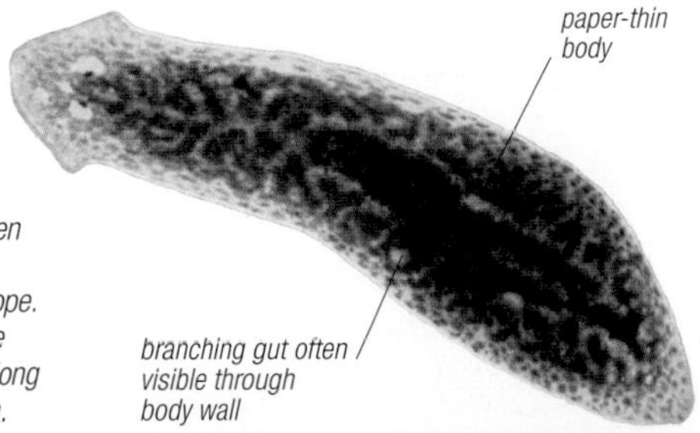

BROWN FRESHWATER FLATWORM
Many turbellarians found in freshwater streams and ponds are drab in color, but their simple eyes—pigmented cups with light-sensitive retina but no lens—are often clearly visible. Dugesia tigrina *is native to North America, but introduced into Europe. Like other turbellarians, they have a single opening to the gut and move by gliding along on a layer of microscopic hairs called cilia.*

Class Cestoda

Tapeworms

Occurrence 5,200 spp. worldwide; within other animals, mostly cartilaginous fishes

Highly specialized for their parasitic way of life, tapeworms are usually less than 1/32 in (1 mm) thick, but can be up to 100 ft (30 m) long. As adults, they live in the intestines of vertebrates—including humans—and they absorb food directly through their body wall. Tapeworms grow from a region just behind the head, forming a long line of segments called proglottids, each of which houses a complete reproductive system. As the proglottids age, they break away from the "tail" and leave the host in the feces, carrying eggs with them. If eaten by an appropriate host animal, the eggs develop—and the cycle begins again.

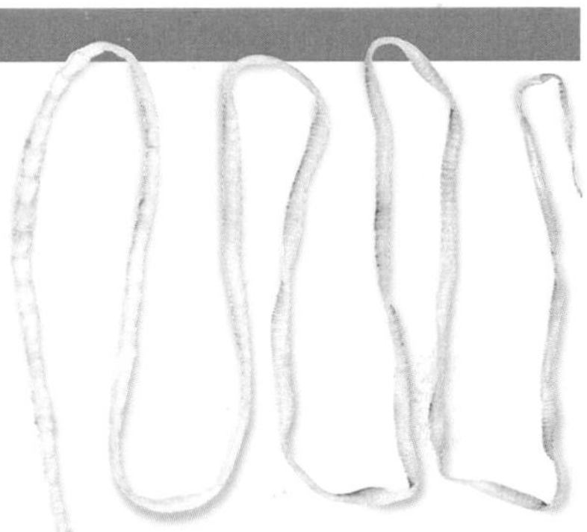

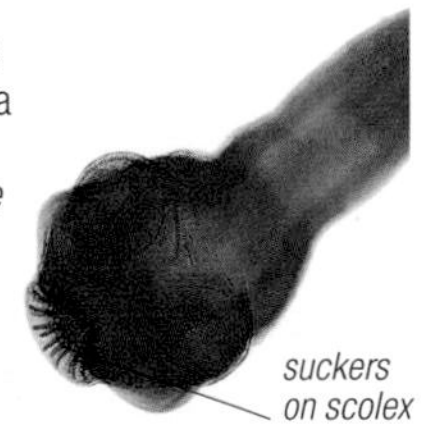

TAENIA TAPEWORMS
The widespread Taenia *species—some over 33 ft (10 m) long—live in humans and other mammals. An adult specimen is shown above, while the close-up on the right shows the head, or scolex. It is equipped with hooks and suckers, with which the worm fastens itself to its host's intestines.*

Class Trematoda

Flukes

Occurrence 1,275 spp. worldwide; within or on other animals

Like tapeworms, flukes are wholly parasitic, but they have a quite different shape. They are typically cylindrical or leaflike, with hooks or suckers at one or both ends, or on the underside. Their length ranges from less than 3/8 in (1 cm) to, in exceptional cases, as much as 20 ft (6 m). Unlike tapeworms, flukes have mouths and digestive systems, and they often burrow through their hosts' tissues. While some attach themselves to internal organs, others are external parasites. Their life cycles are remarkably complex, involving a variety of larval stages, and up to 4 different hosts. The intermediate hosts are often mollusks, but the final host—which harbors the adult stage—is usually a vertebrate. Flukes can cause serious diseases, including bilharzia, which affects millions of people in the tropics.

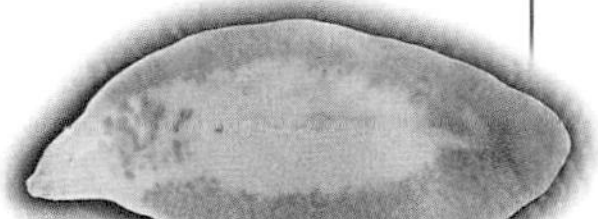

LIVER FLUKES
Several species of flukes live in the livers and bile ducts of mammals, including humans, causing severe damage and even death. The species shown above is the widespread sheep liver fluke, Fasciola hepatica, *which grows to about 1¼ in (3 cm) long. Its intermediate hosts are freshwater snails.*

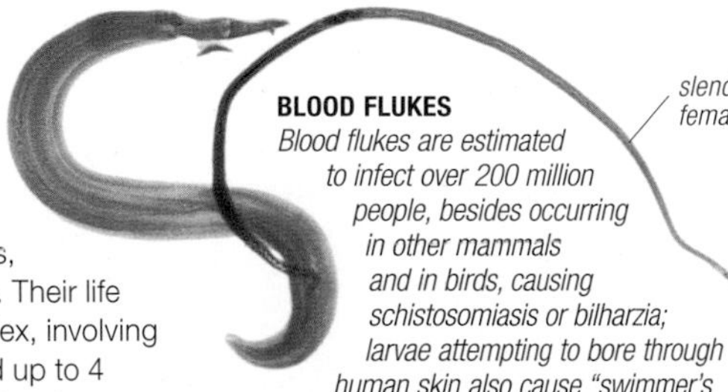

BLOOD FLUKES
Blood flukes are estimated to infect over 200 million people, besides occurring in other mammals and in birds, causing schistosomiasis or bilharzia; larvae attempting to bore through human skin also cause "swimmer's itch." Unusually, the sexes are separate, although the male lives wrapped around the female. This tropical species, Schistosoma mansoni, *is 3/8–1/2 in (1–1.5 cm) long.*

Segmented worms

PHYLUM	Annelida
CLASSES	4
ORDERS	17
FAMILIES	130
SPECIES	About 21,000

The bodies of segmented worms—or annelids—differ from other worms in being divided into a series of linked but partly independent sections, each segment containing the same set of organs. The segments are filled with fluid which, combined with the musculature, gives the body its shape. Segmented worms have various modes of life, many are burrowers, living either in soil or in sediments beneath freshwater or the seawater.

EARTHWORM BURROWS
Feeding mainly on decaying vegetable matter, earthworms are present in soils throughout the world. They are usually found close to the surface but rarely leave their burrows, to avoid being eaten by birds and other animals.

Anatomy

A segmented worm's body consists of a head at one end, a tail at the other, and many trunk segments in between. Each segment is divided from the others by partitions and contains distinctive organs. The mouth is on the first segment, and leads to the gut which passes through all the segments to end in an anus in the tail. On the side of each segment are hooks and bristles (chaetae). Some marine worms have sensory tentacles; others, such as the earthworms, have no obvious projections.

FRESHWATER PARASITES
Leeches are parasitic worms. They use 2 suckers, one at each end of the body, to attach themselves to their hosts.

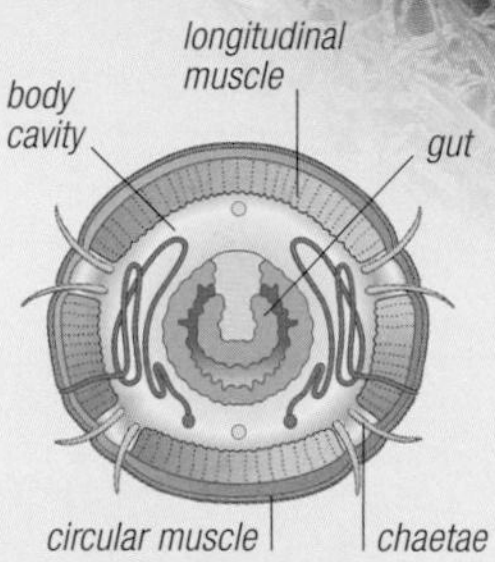

BODY SECTION
The body cavity of an earthworm is filled with fluid and contains a large gut. Short chaetae project through 2 layers of muscle to the body's outer surface.

Movement

An earthworm moves by coordinating the action of muscles in different parts of its body. In each segment, there is one layer of longitudinal muscle and another of circular muscle. When contracted, these muscles make the segment either long and thin or short and fat. As the worm moves through its burrow, short, fat segments anchor part of the body against the wall, while long, thin segments penetrate farther through the soil.

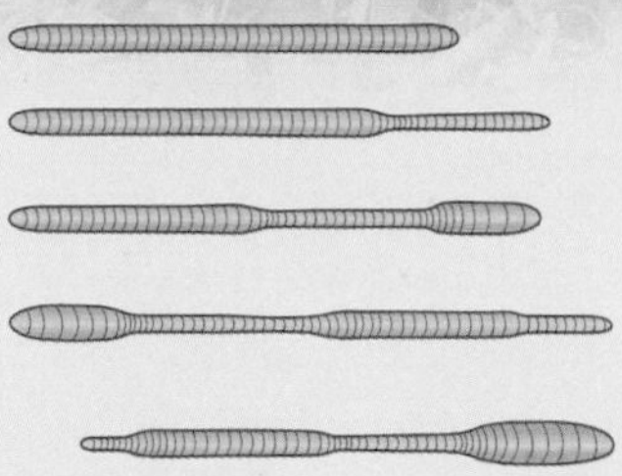

WAVE MOVEMENT
When an earthworm moves, its segments alternately narrow and widen, forming waves of muscular contractions that pass along the body.

Class Polychaeta

Bristleworms

Occurrence 12,220 spp. worldwide; in burrows and tubes, in coral reefs, and amid seaweed and animal colonies on the seabed

Bristleworms, also called polychaetes, form the biggest and most varied class of segmented worms, ranging from less than 1/32 in (1 mm) to 9¾ ft (3 m) long. Lugworms, bloodworms and ragworms, tubeworms, and bamboo worms are all members of this class. Despite being abundant and sometimes brilliantly colored, bristleworms often escape attention because of their burrowing or tube-dwelling lifestyles and their quick reactions when disturbed. The exceptions to this rule are the active hunters, such as ragworms, which can sometimes be spotted in rock pools or under stones. Most bristleworms are long, slender, and worm shaped, but some may be almost circular. They have parapodia—flaps bearing bristles, or chaetae—or at least ridges projecting from the sides of each segment, and heads that bear appendages. Two general types occur: free living and often carnivorous species, which have well-developed parapodia along the body; and sedentary species, which dwell in burrows or tubes, are deposit or suspension feeders, and have very specialized parapodia. A few species are parasitic. Most bristleworms are solitary animals, but some tubeworms may form large colonies or reefs. The sexes are usually separate.

SEA MOUSE
Somewhat inappropriately named after the Greek goddess of love, Aphrodita aculeata *has a feltlike mass of chaetae covering its dorsal surface and scales. This marine scaleworm is up to 8 in (20 cm) long, and is widespread.*

RAGWORMS
Many marine polychaetes, like the ragworm, live in burrows in soft sediment, but can swim well and actively pursue their prey. The predatory lifestyle of the Atlantic species Alitta virens *(right)—which may be up to 20 in (50 cm) long—is reflected in the numerous sensory structures, including eyes and large jaws, at its head end.*

FANWORMS
Some polychaetes live within a tube of their own making, which may be attached to a hard surface or, as in this fanworm, partially buried in soft sediment. Their movement is largely restricted to within the tube, but their feeding tentacles extend out into the water. The widely distributed species shown here— Sabella pavonina*—is 8 in (20 cm) long.*

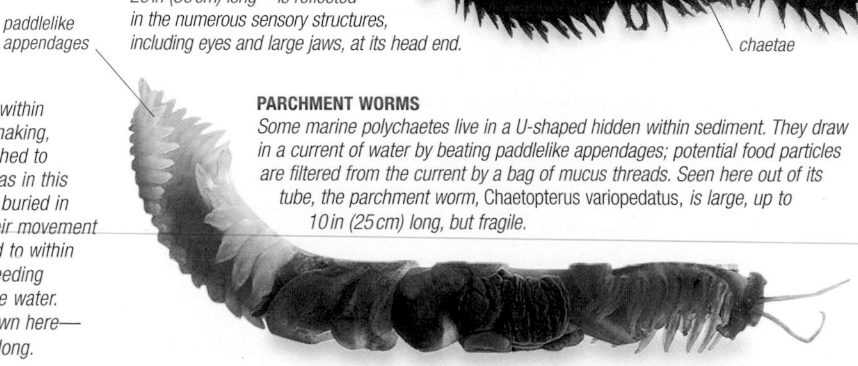

PARCHMENT WORMS
Some marine polychaetes live in a U-shaped hidden within sediment. They draw in a current of water by beating paddlelike appendages; potential food particles are filtered from the current by a bag of mucus threads. Seen here out of its tube, the parchment worm, Chaetopterus variopedatus, *is large, up to 10 in (25 cm) long, but fragile.*

Class Clitellata

Clitellates

Occurrence c.8,000 spp. worldwide; in soil, freshwater or marine sediments, on marine fishes

Habitat All except dry regions

Clitellates are characterized by the presence of the "clitellum," a largely dorsal area of glandular skin that secretes a cocoon where fertilized eggs are placed. The eggs hatch into miniature versions of the adults—an adaptation to life on land where aquatic-style larvae cannot survive. This class contains some of the world's most familiar invertebrates—the common earthworm and its relatives. Earthworms belong to the biggest subgroup, the oligochaetes (comprising about 3,000 species), which live mainly in soil, silt, and freshwater. Leeches form a smaller group (about 500 species), and are either predatory or parasitic. Unlike polychaetes, clitellates lack parapodia and head appendages, and are hermaphrodites. They may be up to 9¾ft (3m) long.

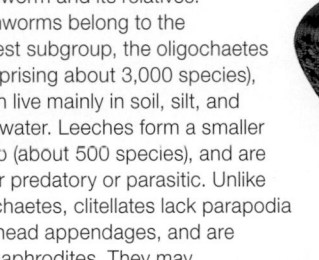

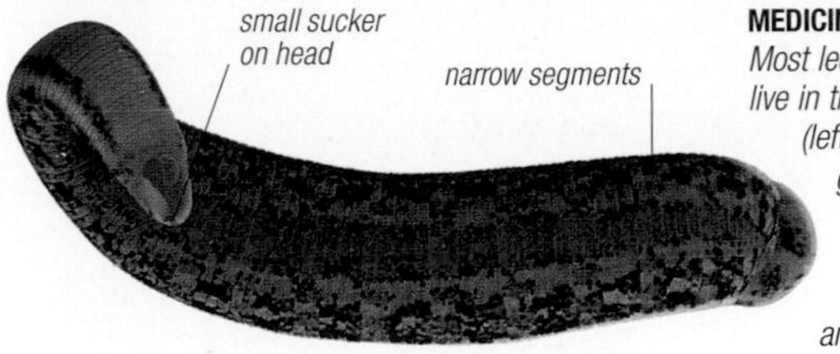

MEDICINAL LEECH
Most leeches capable of penetrating human skin live in tropical rain forests, but Hirudo medicinalis *(left) lives in temperate freshwater, and grows to over 4¾in (12cm) long. It was once commonly used by the medical profession as a bloodletting agent. When it bites, a leech may inject both anticoagulants and anesthetics.*

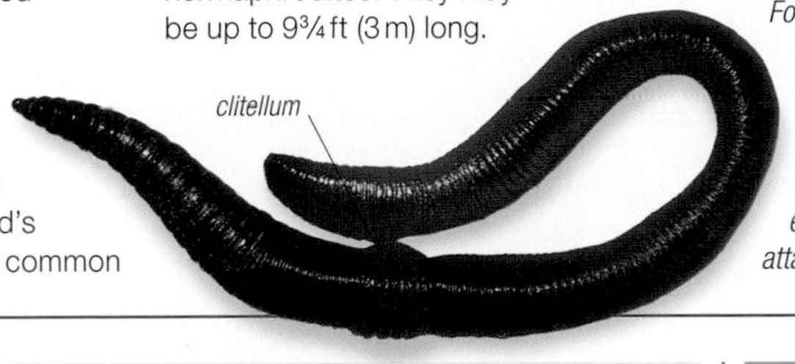

EARTHWORMS
Found almost worldwide, these familiar invertebrates are often seen when they emerge from their burrows in damp conditions, or are pulled from safety by birds. As Charles Darwin pointed out, their importance in aerating and enriching soil is enormous. Lumbricus terrestris *(left) can attain a length of 14in (35cm).*

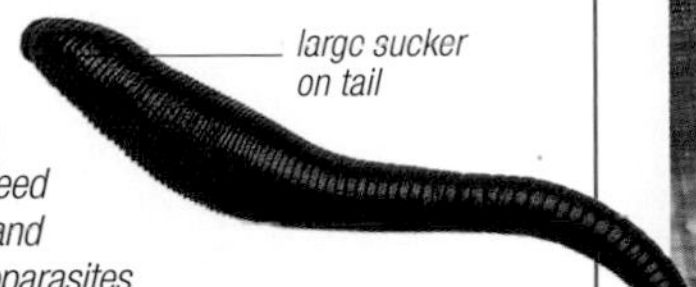

HORSE LEECH
Leeches that feed on mammals and fishes are ectoparasites whose hosts survive the attacks. The horse leech, Haemopsis sanguisuga *(above), found throughout Europe, Asia, and North America, can measure 2¼in (6cm) long when at rest.*

Roundworms

PHYLUM	Nematoda
CLASSES	2
ORDERS	17
FAMILIES	160
SPECIES	About 20,000

Also known as nematodes, these worms are among the most abundant of all animals. Roundworms are either free living or parasitic. The parasitic forms, which include hookworms, pinworms, and threadworms, live inside almost all types of plants and animals, and some are a significant cause of disease. Free-living roundworms are also widespread in all aquatic habitats, including small films of water on land. Their bodies are unsegmented, round in cross section, and tapered toward each end. They have a complex cuticle that is replaced 4 times by molting, between hatching from an egg and reaching sexual maturity.

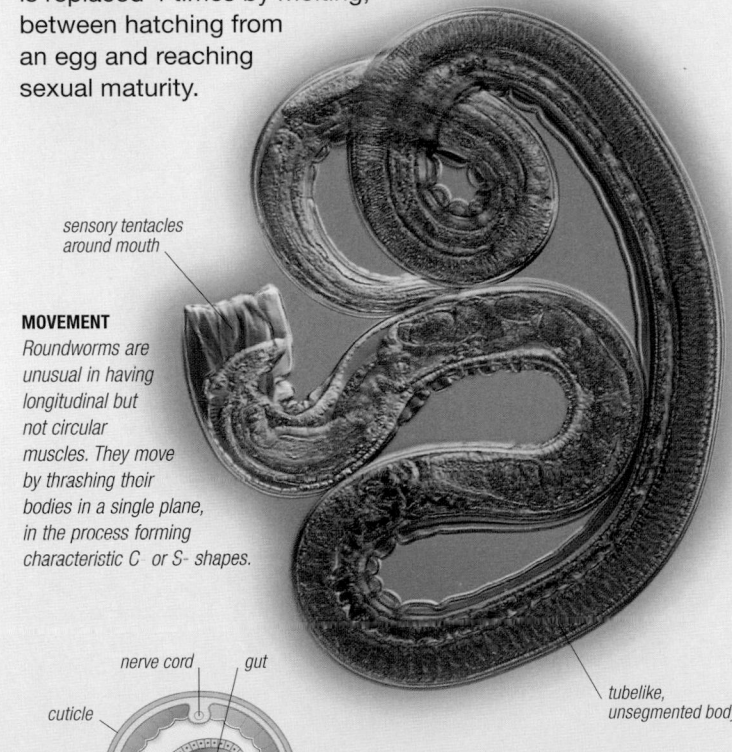

MOVEMENT
Roundworms are unusual in having longitudinal but not circular muscles. They move by thrashing their bodies in a single plane, in the process forming characteristic C- or S- shapes.

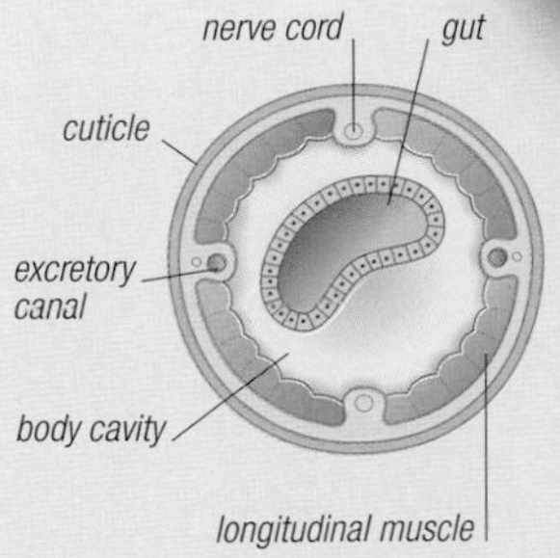

BODY SECTION
The large central cavity in a roundworm's body is occupied by fluid-filled cells. The high internal pressure in this cavity, together with a lattice of fibers in the cuticle, prevents the body from shortening when the muscles contract.

Class Secernentea

Secernenteans

Occurrence c.8,000 spp. worldwide; in water-filled areas around vegetation and in soil, on plants, in animals

The physical differences between secernenteans and other roundworms, such as adenophoreans (see below), are very technical, relating, for example, to types of sense organs. Secernenteans are mostly parasites, some causing serious diseases. They vary greatly in size, from microscopic to several yards long. The largest known secernentean, up to 30ft (9m) in length, lives in the placentas of female sperm whales and is believed to be the largest member of the roundworm phylum. However, the harmful effect of these parasites is not necessarily related to their size: microscopic roundworms, such as pinworms, filarial worms, and hookworms, can multiply rapidly in the host, causing great damage (as in elephantiasis).

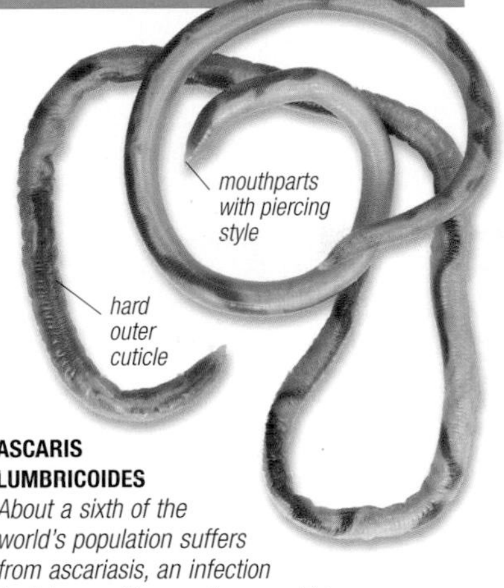

ASCARIS LUMBRICOIDES
About a sixth of the world's population suffers from ascariasis, an infection by this parasitic roundworm, which can grow up to 16in (40cm) long. The infection starts when the eggs are inadvertently taken into the gut with food. The larvae migrate to the lungs, returning to the gut as adults.

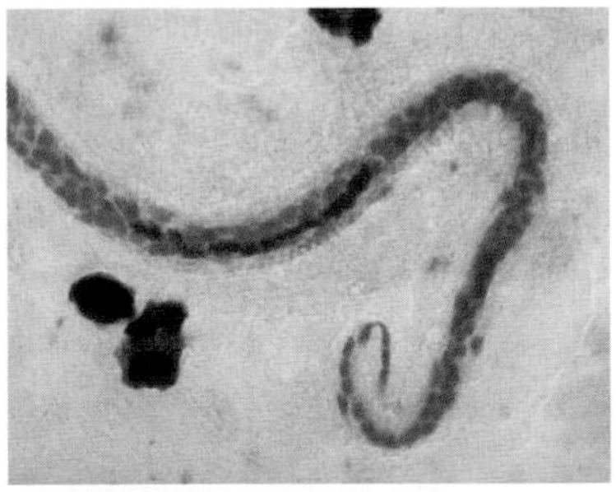

WUCHERERIA BANCROFTI
This microscopic roundworm is the cause of elephantiasis in humans, a tropical disease that produces disabling enlargement of soft tissues, particularly in the legs. The worm lives in the lymphatic system, and is spread by mosquitoes.

Class Adenophorea

Adenophoreans

Occurrence c.12,000 spp. worldwide; in water-filled areas around vegetation and in soils, on plants, animals

Habitat All except desert regions

Adenophorean roundworms are mainly free living rather than parasitic, and are commonly found in most kinds of habitat throughout the world, particularly in marine sediments. They range in size from microscopic to as long as 3¼ft (1m) in exceptional cases. Surveys of seabed mud suggest that vastly more may be awaiting discovery.

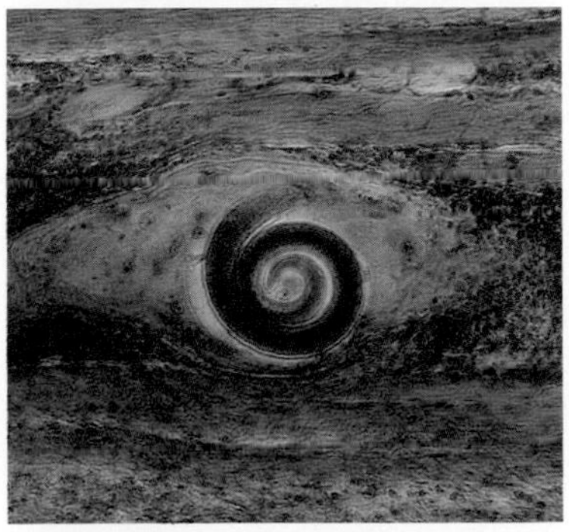

TRICHINA WORM
Trichinella spiralis *is a small roundworm, 1/32in (1mm) long, which lives mainly in rats and other small mammals. Pigs may pick it up while scavenging and, by eating undercooked pork, each year about 40 million humans do the same.*

INVERTEBRATES

Minor phyla

The invertebrate world is classified into about 30 major groups, or phyla. Some contain common and familiar animals but well over half contain organisms that go largely unnoticed—because they are rare, because they are microscopic, or because they live in inaccessible habitats, such as seabed mud. These animals differ from each other as much as invertebrates differ from vertebrates, and they have widely divergent ways of life. Some minor phyla, such as arrow worms, comprise just a handful of species, while a few—such as bryozoans—contain thousands of species, yet both groups are ecologically important. A selection of minor phyla is shown on these 2 pages.

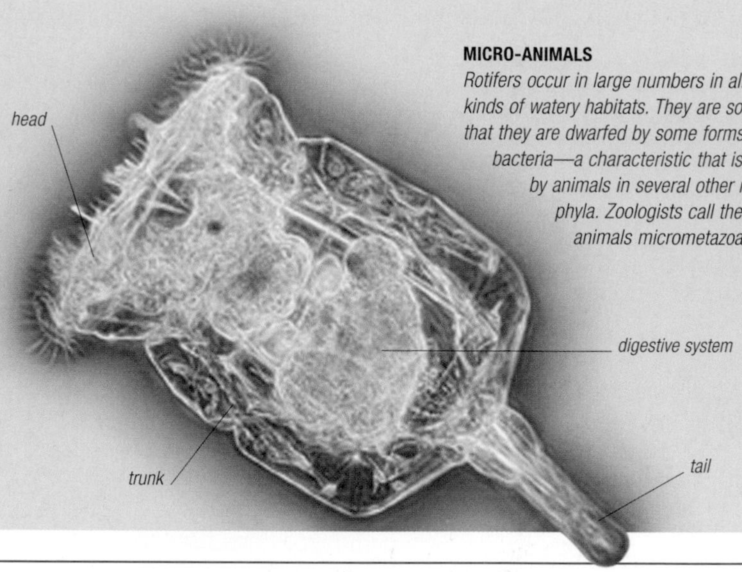

MICRO-ANIMALS
Rotifers occur in large numbers in all kinds of watery habitats. They are so small that they are dwarfed by some forms of bacteria—a characteristic that is shared by animals in several other minor phyla. Zoologists call these tiny animals micrometazoans.

Phylum Ctenophora

Comb jellies

Occurrence c.200 spp. worldwide; planktonic, a few on the seabed

With their tentacles extended, comb jellies, or sea gooseberries, are some of the most attractive and graceful animals in the sea. Although they can swim, by beating their comblike plates of fine cilia (structures similar to hairs), they make little headway against the current, and are often carried along in large swarms by the tide or the wind. As the cilia rows beat, they produce beautiful, iridescent colors. Most comb jellies are small in size, but some can grow to 6½ft (2m) long.

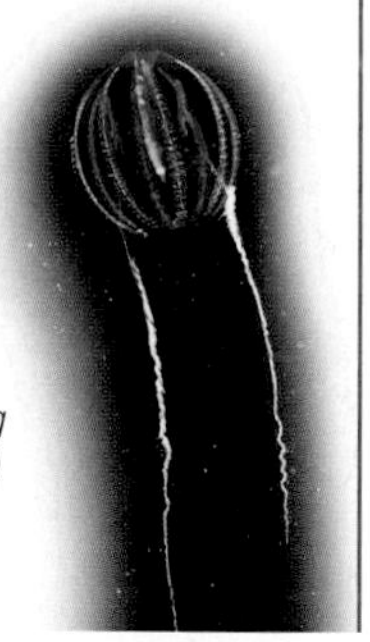

SEA GOOSEBERRY
Comb jellies come in a variety of shapes, from spheres to ribbons and disks. The Atlantic-dwelling Pleurobrachia pileus *(right) is ¾in (2cm) long and has a pair of tentacles 6in (15cm) long.*

Phylum Sipuncula

Peanut worms

Occurrence c.150 spp. worldwide; in soft sediments, mollusk shells, rock crevices, holes in coral or wood

These plump, unsegmented worms, 1/16–28in (0.2–70cm) long, have a peanut- or sausage-shaped body, with walls formed of layers of circular and longitudinal muscles. Their most notable feature is their retractable trunk, known as an introvert, which is extended with the help of hydraulic pressure, and retracted by muscles. Peanut worms are suspension or deposit feeders, and collect organic particles by means of a ring of hydraulic tentacles around the mouth. They appear to be most closely related to mollusks.

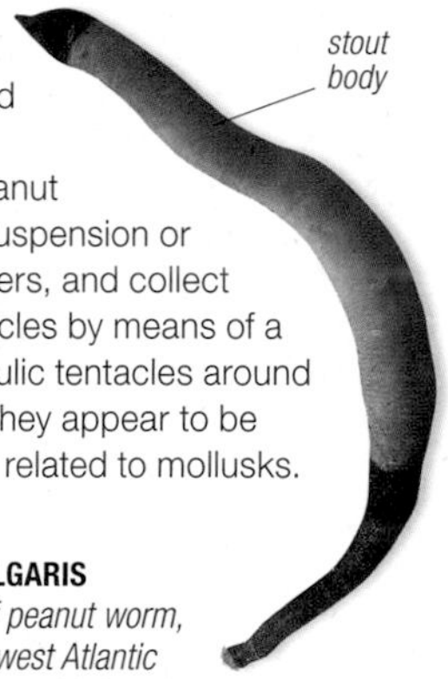

GOLFINGIA VULGARIS
This species of peanut worm, found on northwest Atlantic coasts, lives in mud below the low-tide mark. It is 4in (10cm) long.

Phylum Bryozoa

Bryozoans

Occurrence c.6,000 spp. worldwide; attached to hard surfaces or aquatic environments

Also called sea mats, ectoprocts, or polyzoans, bryozoans are tiny aquatic animals, less than 1/32in (1mm) long, which live inside boxlike cases that are joined together to form colonies. Some bryozoans grow as encrustations on rocks and seaweeds, but others develop a branching, plantlike shape and wave backward and forward in the water. Once dead, these branching colonies are often washed up on the shore, where they may be mistaken for dried-out seaweed. This phylum is ecologically important as a source of food for some marine invertebrates.

ROSS CORAL
Bryozoans bud continuously until thousands of interconnected "zooids" are produced. The North Atlantic Pentapora fascialis *may form colonies up to 6½ft (2m) in diameter.*

Phylum Rotifera

Rotifers

Occurrence c.2,000 spp. worldwide; in vegetation in lakes, rivers, and seas, microfreshwater habitats on land

Rotifers are among the smallest animals: the largest is just 1/8in (3mm) long. They live in water, and are either free-swimming, or attached to solid surfaces. The trumpet-shaped to spherical body usually has a "wheel organ" in front, formed by 2 whorls of hairlike structures (cilia), which is used for feeding and locomotion. Many are transparent, and several are parthenogenetic, males among these being unknown.

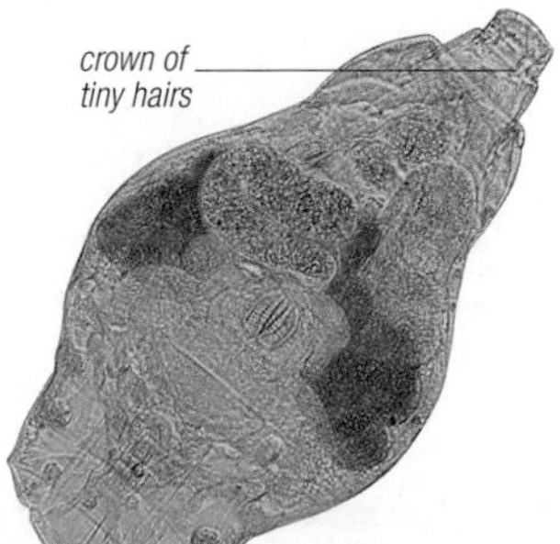

UNIDENTIFIED SP.
Although some rotifers are marine, most live in freshwater habitats, including temporary pools such as those in gutters and water films around leaves and soil particles. Like water bears (see opposite), they are specially adapted to survive drought. They can exist in a dried-out state for many years, coming back to life as soon as they are rehydrated.

Phylum Nemertea

Ribbon worms

Occurrence c.1,400 spp. worldwide; on bed and in surface or middle waters of seas, rivers, and lakes, in forests

A key feature of these mainly marine worms is their unique proboscis. This muscular tube, housed above the gut, is pushed out by hydraulic pressure in order to catch prey; in one large group of species, it bears piercing barbs that allow it to inject toxins. These worms range from less than 1/64in (0.5mm) in length to over 165ft (50m). All have a circulatory system and most have eyes—up to 250in some species. Many are brightly colored. Two groups of these worms are commensal—one within the shells of bivalve mollusks, the other on and within the shell of a variety of crabs.

UNIDENTIFIED SP.
Ribbon worms are noted for their flat and sometimes extraordinarily long bodies. The worm shown here belongs to a group that includes the world's longest animal, which reaches a length of 180ft (55m). This is twice as long as the average adult blue whale.

Phylum Brachiopoda

Brachiopods

Occurrence c.400 spp. worldwide; attached to hard surfaces or buried in soft sediments on the seabed

Encased in bivalved shells, usually with a stalk (pedicle) that may anchor the animal to a hard surface or be used in burrowing, brachiopods look remarkably like bivalve mollusks with stalks attached, but the 2 groups are not related. Also known as lamp shells, brachiopods are especially abundant in colder waters. They are less than 4 in (10 cm) long, and most of the space in the shell is taken up by a loop or spiral of hollow, fringed tentacles (lophophore) surrounding the mouth, with which the animal suspension feeds. Brachiopods are extremely common as fossils: some rocks are made up entirely of brachiopod shells. Once an important group comprising more than 25,000 species, brachiopods appear to have declined as a result of competition with bivalve mollusks.

LIOTHYRELLA UVA
Although brachiopods resemble bivalves, the 2 halves of the shell are dorsal (upper) and ventral (lower), not left and right. The species featured, Liothyrella uva, *is common in Atlantic waters, and has a shell that is up to ¾ in (2 cm) long.*

Phylum Chaetognatha

Arrow worms

Occurrence c.150 spp. worldwide; in plankton, one genus on the seabed

Growing only up to 4¾ in (12 cm) long, arrow worms are relatively small, highly active predators that can eat a third of their own body weight a day. They catch their prey—animal plankton, including larval fishes—with movable spines around their mouths and kill it by injecting tetrodotoxin, a potent poison produced by symbiotic bacteria, and used by very few animals.

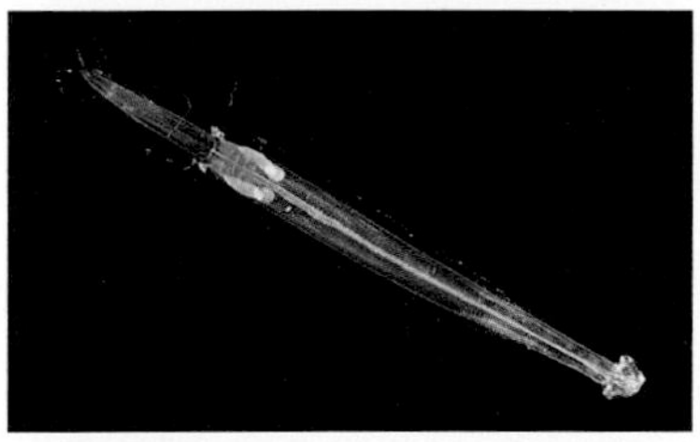

SAGITTID SPP.
The small, torpedo-shaped, transparent arrow worms are carnivores. Like the Sagittid *species above, they live among the plankton in all the oceans—where their large numbers can have great ecological impact; for instance, on fish larvae.*

Phylum Tardigrada

Water bears

Occurrence c.1,000 spp. worldwide; in freshwater, salt water, and damp and wet terrestrial habitats

Habitat All except dry ones

Water bears—or tardigrades—are plump-bodied, microscopic animals with 4 pairs of stubby, clawed legs. Their lumbering movements are remarkably bearlike, but their closest relatives are probably velvet worms. They live in damp habitats such as moss, but can shrivel into a dry husk during drought and survive this way for years. Water bears lay eggs, but some females are parthenogenetic.

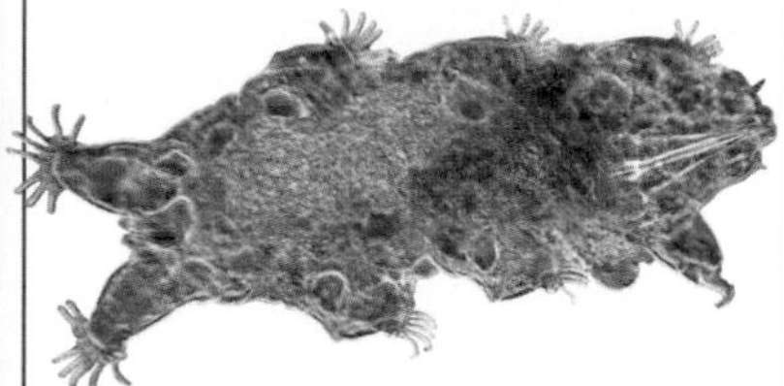

UNIDENTIFIED SP.
Water bears inhabit water films on land and live between sand grains and crevices on the seabed. Many live in clumps of moss on roofs and in gutters. By being dormant for extended periods, they may live over 50 years.

Phylum Echiura

Spoonworms

Occurrence c.200 spp. worldwide; in burrows or other cavities on the seabed

Spoonworms, or echiurans, are sedentary, marine animals with bulbous, sausagelike bodies. Females are up to 20 in (50 cm) long, with an even longer proboscis for sweeping up food. Males are minute and live parasitically inside their partners.

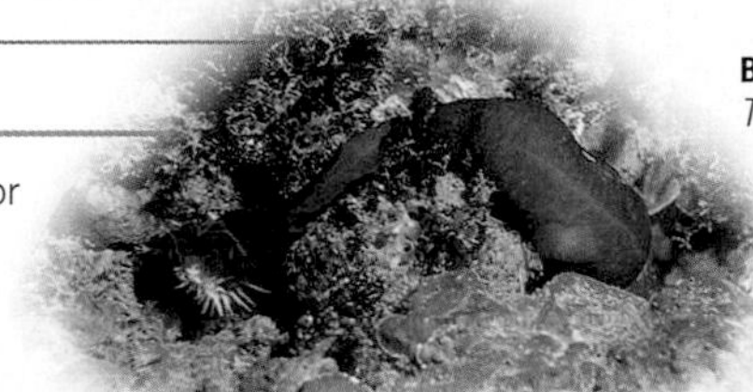

BONELLIA VIRIDIS
This European species has green, toxic skin. Females are up to 6 in (15 cm) long, with a proboscis that can reach out for food over about 3¼ ft (1 m).

Phylum Phoronida

Horseshoe worms

Occurrence c.20 spp. worldwide; in soft sediments or attached to hard surfaces on the seabed

These sedentary marine worms are generally less than 8 in (20 cm) long, and usually remain encased in a secreted chitinous tube, with a horseshoe-shaped crown of as many as 15,000 feeding tentacles around the mouth. The sexes are separate in some species, while others are hermaphrodite. Horseshoe worms are most closely related to brachiopods, or lamp shells (see left).

PHORONIS SPP.
Members of the widespread genus Phoronis *can be 10 in (25 cm) long, but most are much smaller. They have an unsegmented body, usually encased in a chitinous tube buried in sand or mud. The body wall has layers of longitudinal and circular muscles, which enable the worm to move up the tube to feed, and to retreat afterward.*

Phylum Onychophora

Velvet worms

Occurrence c.180 spp. in tropical and southern temperate zones; in moist forests, leaf litter, under stones

These are the only survivors of a group of worms that were abundant in the Cambrian Period. They have cylindrical, wormlike bodies, up to 6 in (15 cm) long, 14–43 pairs of short, unjointed, fleshy legs, each terminating in a pair of claws, and an antennae-bearing head with jaws similar in form to the claws on the legs. These carnivorous animals catch their prey by spraying them with a mucuslike substance from a pair of slime glands opening on either side of the mouth; they use the same technique to defend themselves from their predators.

MACROPERIPATUS TORQUATUS
Velvet worms, such as Macroperipatus torquatus *(below), inhabit warm and humid regions. Their anatomy is basically that of a worm but, like arthropods, they have pairs of legs serially repeated along the body. The legs are fleshy, unjointed, and hydraulically operated.*

Phylum Hemichordata

Hemichordates

Occurrence c.130 spp. worldwide; in marine mud and sand, and other intertidal and subtidal habitats

Hemichordates are unusual among invertebrates in that they have some vertebrate characteristics: a nerve cord that runs along the back, and pharyngeal perforations—structures that have the same anatomical origin as fish gills. The length of the 3-part bodies—comprising a proboscis, a collar, and a long trunk—ranges from less than ½ in (1.2 cm) to 8¼ ft (2.5 m). These animals are found in 2 forms: sedentary, polyplike pterobranchs, whose saclike trunk ends in a stalk that may be prehensile or connect individuals to form colonies; and free living, wormlike forms, in which the trunk is stalkless and bears a terminal anus. The sexes are separate, and reproduction occurs either sexually or asexually, by fragmentation or budding.

ACORN WORMS
Slow-moving and fragile, acorn worms live in U-shaped burrows in intertidal and subtidal habitats in all parts of the world. The large trunk of Balanoglossus australiensis *(above), which may be 3¼ ft (1 m) long, bears numerous gill slits.*

Arthropods

PHYLUM	Arthropoda
CLASSES	19
ORDERS	123
FAMILIES	About 2,300
SPECIES	About 1.2 million

CLASSIFICATION NOTE

Arthropods are divided into 2 subphyla: chelicerates have pincerlike appendages and no antennae, while mandibulates have chewing mouthparts and one or 2 pairs of antennae.

MANDIBULATES
Hexapods *(insects and other 6-legged arthropods) see pp.548–77*
Centipedes and millipedes *see p.578*
Crustaceans *see pp.579–84*

CHELICERATES
Sea-spiders *see p.585*
Horseshoe crabs *see p.585*
Arachnids *see pp.586–93*

Arthropods form the largest phylum of living organisms and account for more than 3 out of 4 known species of animals. They include insects, centipedes and millipedes (myriapods), crustaceans, and arachnids, among others. All arthropods have an exoskeleton that covers a body divided into segments, and all have jointed legs. The first arthropods were marine animals, and many (especially crustaceans) are still found in the sea. However, insects, arachnids, and myriapods have successfully adapted to life on land. Insects, which are by far the most numerous of all arthropods, are the only invertebrates capable of powered flight, having evolved functional wings more than 100 million years before flying reptiles or birds. By providing food for many larger animals, arthropods have become essential to the functioning of most of the world's ecosystems. Krill, copepods, and other crustaceans form the foundation of marine food chains, supporting fish and marine mammals. On land, insects provide food for countless other animal species.

MIGRATION

Arthropods are among the most mobile of invertebrates. Some species make use of this to migrate in response to changes in environmental conditions and food supply. Each year, many spiny lobsters make mass migrations between relatively warm and cool waters. Typically moving in single file, they maintain contact with one another by touch (using their antennae) or sight (picking out the light-colored spots on the abdomen of the individual in front of them). The fall migration is often triggered by violent storms.

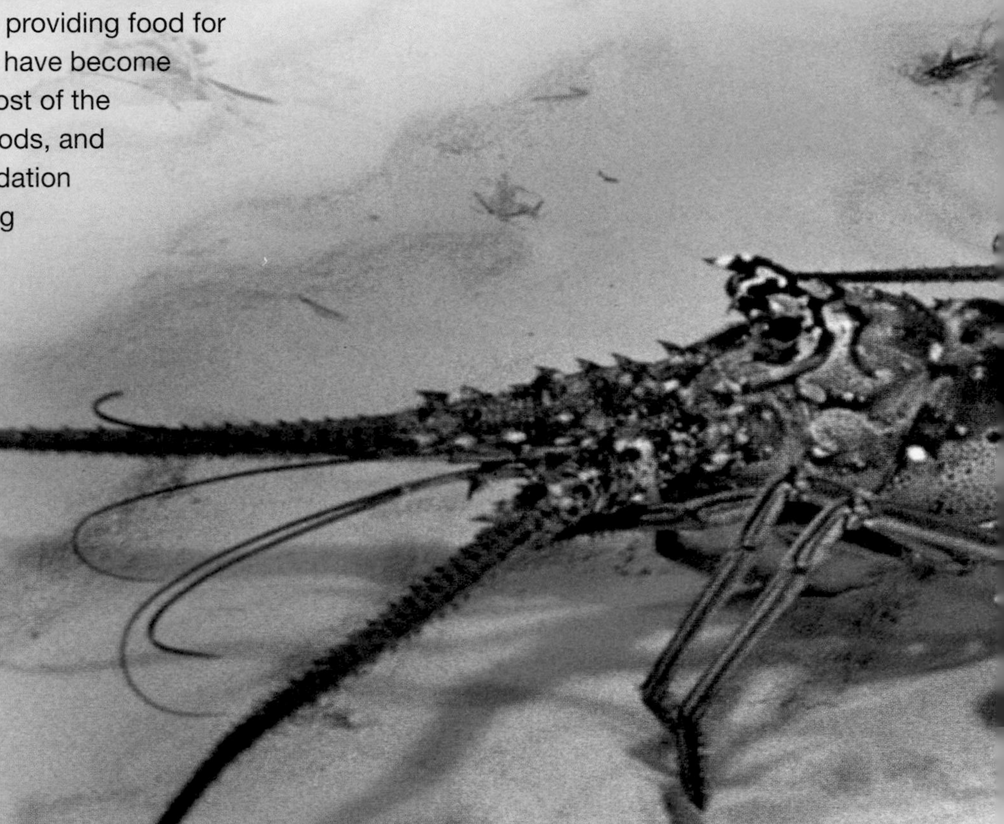

Anatomy

Arthropods share several common features. All species have a bilaterally symmetrical body that is divided into segments (see below). Arising from these segments are several jointed legs, which are arranged in pairs. The body is covered by a tough exoskeleton, which is produced by the epidermis and is composed of protein and a material called chitin. To allow for movement, the exoskeleton articulates at joints and hinges, where the cuticle is soft and flexible. In large marine species, the exoskeleton is strengthened by calcium carbonate, while land species have a thin layer of waterproof wax to stop them from drying out. All arthropods have an open circulatory system: their organs are bathed in a fluid called hemolymph, which is moved around the body by the heart. Gas exchange is carried out by gills, organs known as book lungs, or a system of tracheae. The nervous system consists of a brain that is connected via paired nerve cords to networks of nerve cells in the thorax and abdomen.

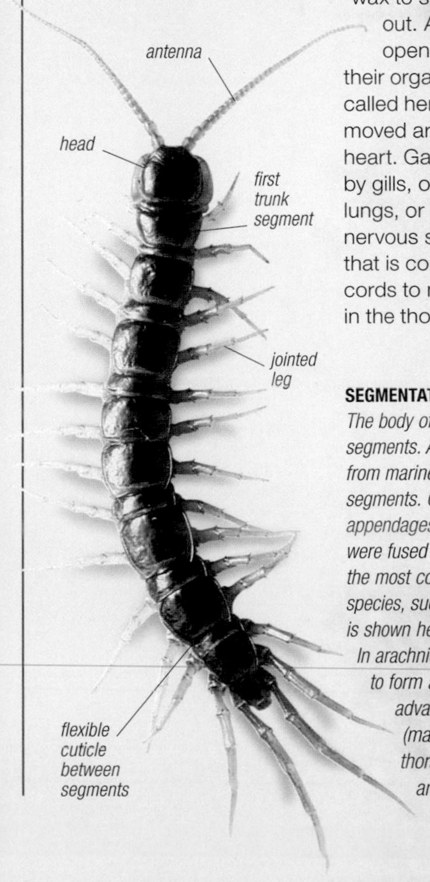

SEGMENTATION

The body of an arthropod is divided into segments. Arthropods probably evolved from marine worms that had unspecialized segments. Over time, eyes, antennae, and appendages appeared, and some segments were fused into functional units, of which the most common is the head. Primitive species, such as myriapods (a centipede is shown here), have a head and trunk. In arachnids, the head and thorax are fused to form a cephalothorax. Insects, the most advanced arthropods, have a head (made of 6 fused segments), a thorax (that has 3 segments), and an abdomen (with 11 segments).

BODY SEGMENTS

Each segment (seen in cross section, below) is essentially a box, with a top (tergum), bottom (sternum), and sides (pleura). These outer surfaces are made from tough, relatively rigid plates (sclerites), which are joined by soft cuticle to allow for movement. Muscles between the tergum and sternum allow the segment to be flattened in shape, while those linking adjacent segments allow the body to flex sideways, curl up, or telescope lengthwise. The nerve cord, digestive system, and heart pass through most segments.

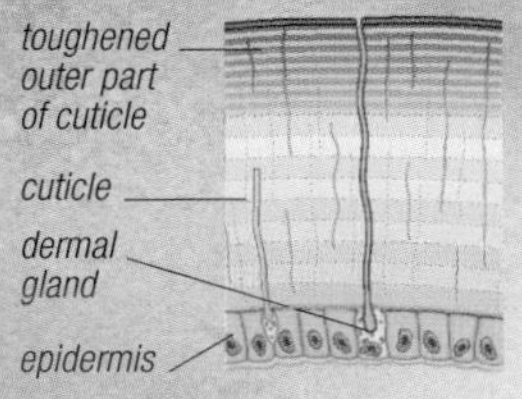

CUTICLE

The cuticle (seen in cross section, left) protects the body of an arthropod from damage and repels pathogens. Although it is a composite of several layers, the cuticle can be repaired: a small wound is quickly sealed; more extensive damage can be repaired by the production of a new cuticle at the next molt. Some species can also regenerate lost limbs.

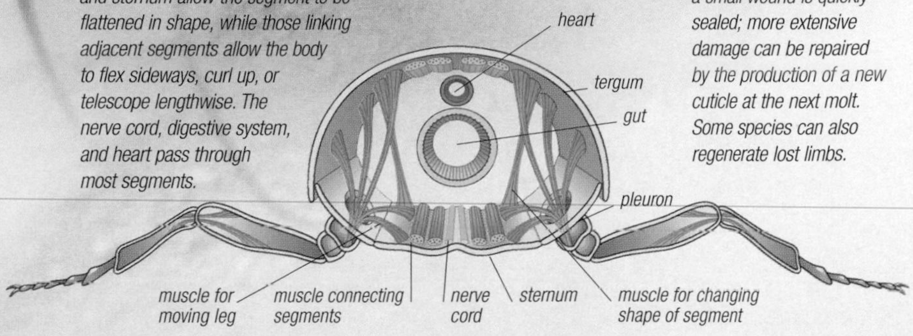

Movement

One of the key features that separates arthropods from most other invertebrates is their jointed limbs. Most arthropods are active animals—moving by walking, swimming, or jumping—which enables them to look for food or mates, colonize new habitats, and escape from predators or adverse local conditions. The number of legs varies from 3 pairs in insects to several hundred pairs in millipedes. Each leg usually ends in a claw of some kind that improves the arthropod's grip on the surface. Many species also have adaptations, such as bristles or adhesive pads, for walking on smooth, vertical surfaces or on water. In some species, the joints between leg segments contain an elastic protein, called resilin. When compressed, this acts as a reserve of energy that can be released when the animal is on the move. In addition to movement, arthropods also use their legs to catch prey, to mate, and to communicate.

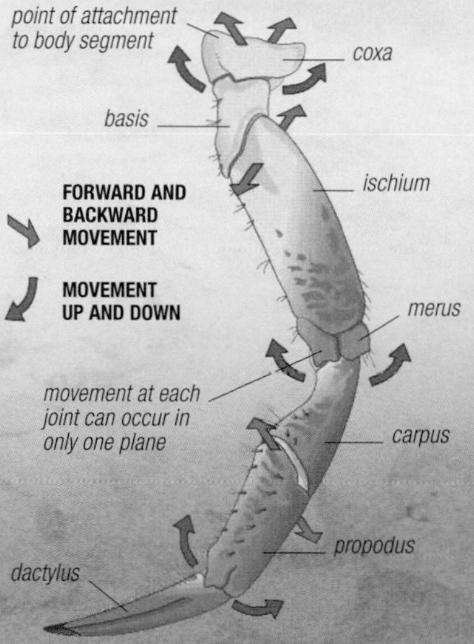

LEG MOVEMENT
An arthropod's limb, such as the first walking leg of a crab (left), consists of a set of hollow, rigid sections. These are joined by areas of flexible cuticle and moved by muscles that connect the segments. The ends of the segments are shaped so that movement at each joint occurs in a different plane to movement at the adjacent joints. The result is a limb that can be moved into almost any position.

WALKING
Most insects, including this beetle, walk with what is known as a tripod gait. As each leg is lifted off the ground, the body is supported by one leg on the same side of the body and 2 legs on the opposite side.

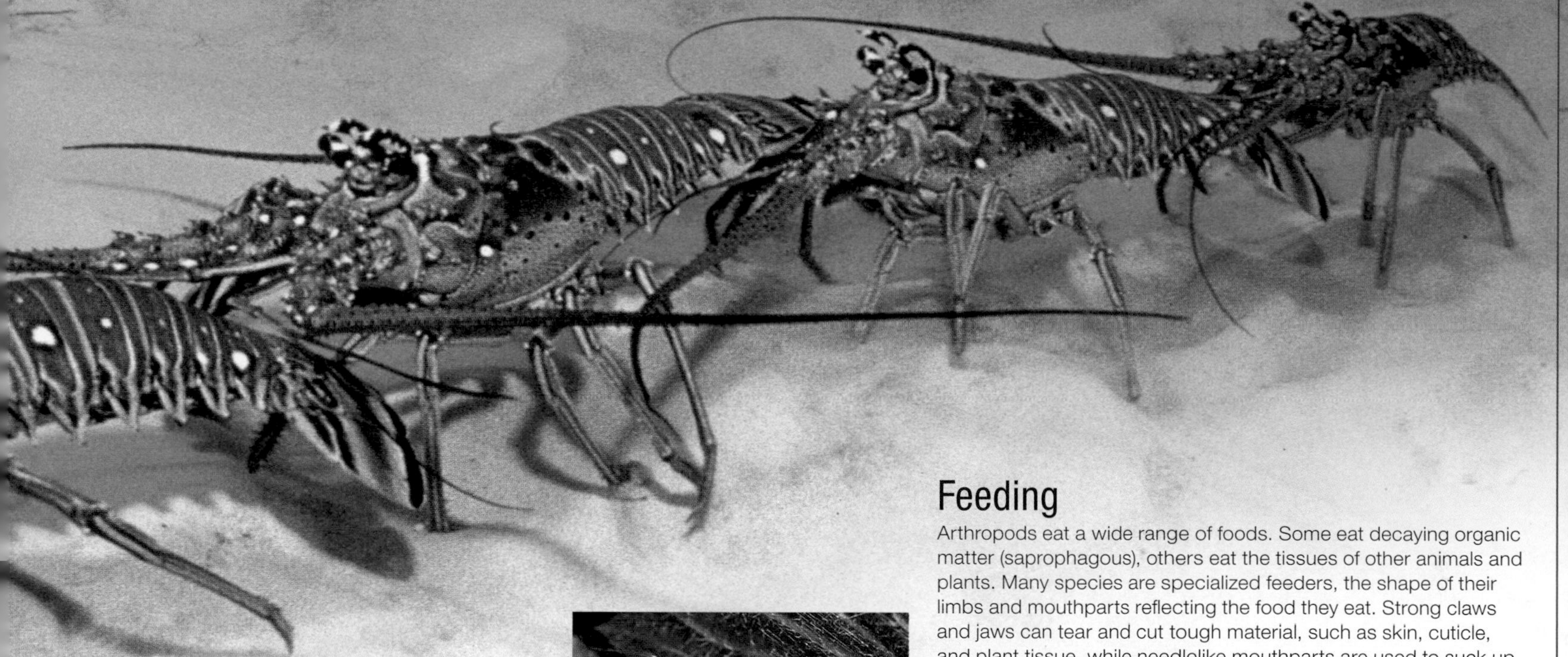

Feeding

Arthropods eat a wide range of foods. Some eat decaying organic matter (saprophagous), others eat the tissues of other animals and plants. Many species are specialized feeders, the shape of their limbs and mouthparts reflecting the food they eat. Strong claws and jaws can tear and cut tough material, such as skin, cuticle, and plant tissue, while needlelike mouthparts are used to suck up liquids, such as sap, blood, and nectar. Many aquatic crustaceans have featherlike structures on their legs or around the mouth that act like strainers to filter minute organisms or organic particles from water. Most arthropods use saliva to lubricate their food as it passes into the digestive system.

PREDATORS
Most arachnids eat other animals. However, they can ingest only liquid food. Having subdued a cockroach, this scorpion will release digestive enzymes over its body and then consume the resulting liquid.

PLANT EATERS
Plant-eating arthropods can eat large quantities of food. A swarm of locusts can consist of up to 50 billion individuals, theoretically capable of consuming 100,000 tons of food in a day.

Growth

One disadvantage of having an exoskeleton is that it must periodically be replaced with a larger one as the animal grows. Arthropods grow and develop in various ways. Crustaceans, myriapods, and arachnids molt throughout their adult life. Larval crustaceans look nothing like the adults. In crabs, for example, the egg hatches into a planktonic zoea larva. This becomes a bottom-dwelling megalopa larva, which in turn develops into the adult. Young myriapods typically look like small adults and become longer by adding trunk segments at each molt, although some centipedes hatch with the adult number of segments. Young arachnids typically look like small versions of their parents. With the exception of bristletails and silverfish, insects stop molting upon reaching sexual maturity.

PARENTAL CARE
It is not unusual for adult insects, centipedes, and arachnids to care for their eggs and young. Female spiders wrap their eggs in silk and carry or guard them until they hatch. Scorpions brood their eggs and carry their young on their back after they hatch.

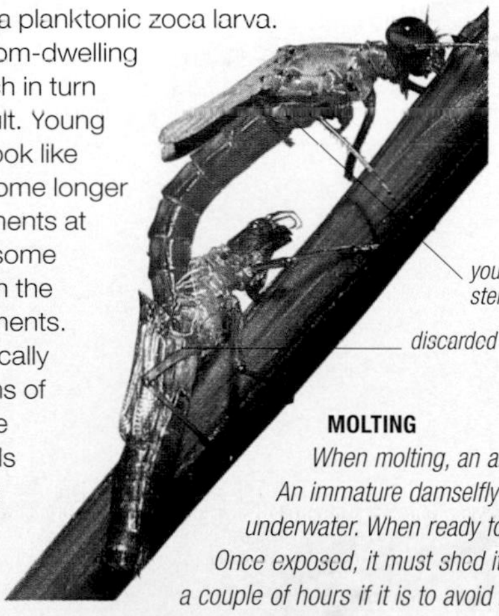

MOLTING
When molting, an arthropod's body is soft and unprotected. An immature damselfly (known as a nymph) molts several times underwater. When ready for the final molt, it crawls out of the water. Once exposed, it must shed its skeleton and take to the air in a couple of hours if it is to avoid being eaten by a predator.

Insects

PHYLUM	Arhropoda
SUBPHYLUM	Mandibulata
SUPERCLASS	Hexapoda
CLASS	Insecta
ORDERS	29
FAMILIES	About 1,000
SPECIES	About 1.1 million

Insects are the most successful animals on earth. They account for more species than any other class: over a million have so far been identified, but it is thought that the true number may be between 5 and 10 million. Insects belong to a group of arthropods called hexapods, all of which have 3 pairs of legs. Many insects also have wings, which makes them the only arthropods capable of powered flight. Combined with their small size and ability to survive in dry environments, flight has enabled insects to colonize a vast range of habitats. Most live on land or in the air, but they are also numerous in freshwater. Without insects, many other forms of life would not exist: the majority of flowering plants are largely dependent on them for pollination, and they form the main component of the diet of many animals. During their lifetime, many insects undergo complete metamorphosis, passing through several physical stages before reaching adulthood.

NON-INSECT HEXAPODS

In addition to insects, hexapods include 3 smaller classes: springtails, proturans, and diplurans. They are collectively known as non-insect hexapods, and a selection are described in this section. All non-insect hexapods are without wings, and many also lack eyes and antennae. Unlike insects, their mouthparts are enclosed in a pouch located on the underside of their head.

SPRINGTAILS
Springtails are small hexapods that occur in great numbers in soil and leaf litter all over the world, including the Arctic and Antarctic.

Anatomy

The body of an insect is divided into 3 segments—the head, thorax, and abdomen—each of which performs different functions. The head supports the mouthparts and much of the sensory apparatus, including the compound eyes and antennae. The thorax, which bears the legs and in many species the wings, is important in locomotion, while the abdomen contains the organs for digestion, excretion, and reproduction. All adult insects breathe air, which enters the body through openings (called spiracles) on the sides of the abdomen and thorax. Immature stages of aquatic species often have gills. The internal organs are bathed in a fluid called hemolymph, which transports nutrients and waste and is pumped around by a tube-shaped heart.

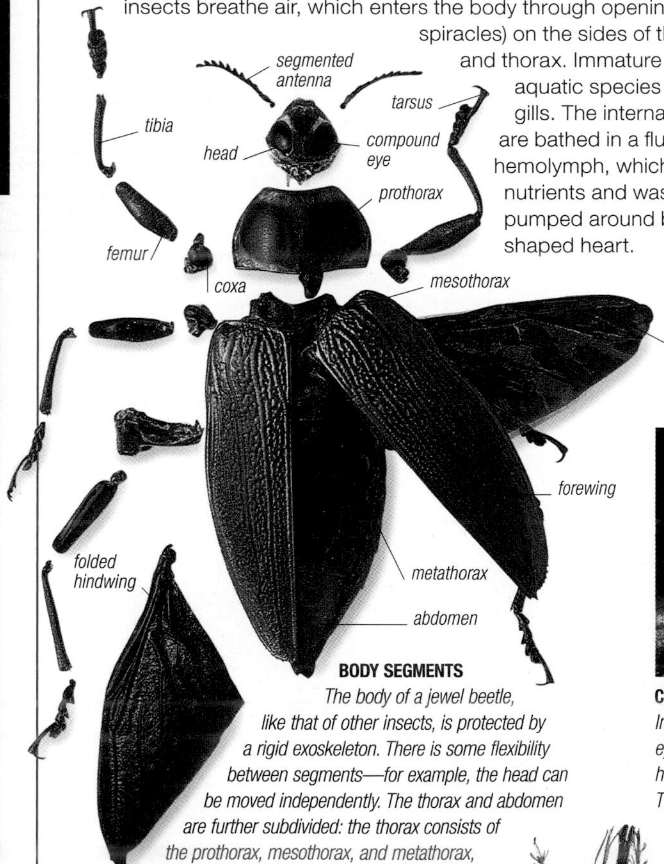

BODY SEGMENTS
The body of a jewel beetle, like that of other insects, is protected by a rigid exoskeleton. There is some flexibility between segments—for example, the head can be moved independently. The thorax and abdomen are further subdivided: the thorax consists of the prothorax, mesothorax, and metathorax, while the abdomen has up to 11 segments.

COMPOUND EYES
Insects have 2 types of eyes: simple eyes (ocelli) and compound eyes. This hornet's ocelli are on top of its head. The forward-facing compound eyes consist of hundreds of light-sensitive units connected by nerves to the brain.

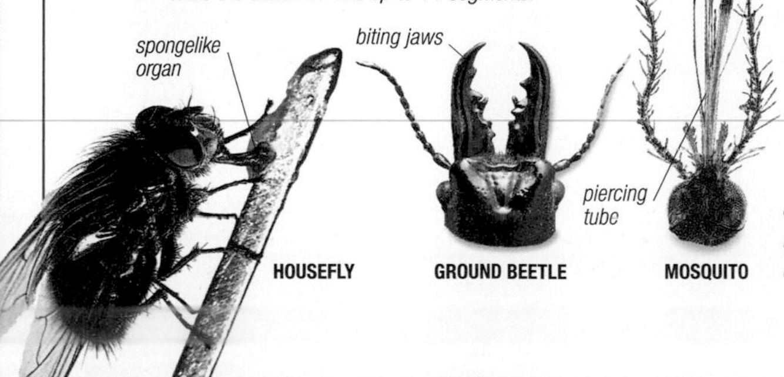

MOUTHPARTS
The mouthparts of insects have evolved into an amazing array of forms. Like most insects, ground beetles have jaws suited for cutting and chewing. Houseflies use a spongelike structure to absorb exposed liquid. Mosquitoes use needlelike mouthparts to pierce the skin of other animals.

Population

There are several reasons why insects have become so successful. Being small, they are able to occupy microhabitats that are inaccessible to other animals. Given the right conditions, they are also able to breed very rapidly, enabling them to respond quickly to increases in the availability of food. For example, a single pair of bruchid beetles have the theoretical potential to produce enough offspring to occupy the entire volume of the earth within 432 days. In practice, this doesn't happen because there are limits to the food supply and because of competition between individuals of the same species and between different species.

SPECIES NUMBERS
High reproduction rates have enabled insects to evolve so rapidly that they make up over half of all animal species alive today. Bugs (such as these firebugs) alone account for 88,000 species. Insects are found on land, in freshwater, and even on the ocean surface.

Life cycles

Insects begin life as an egg. However, once a young insect hatches, it may follow one of several routes to adulthood (see right), varying from a simple series of molts to a complete physical transformation of its body parts. The length of each stage varies greatly between species. For example, a few species of North American cicadas take 17 years to reach adulthood, while some fruit flies may take less than 2 weeks to mature. Once insects reach sexual maturity, they begin to mate. Courtship displays may involve the production of sexual odors, sounds, and even light displays. In almost all insects, fertilization is internal (see pp.30–1). The fertilized eggs are usually laid close to a source of food. However, in parasitic species, they may be laid either on or inside the body of a host animal.

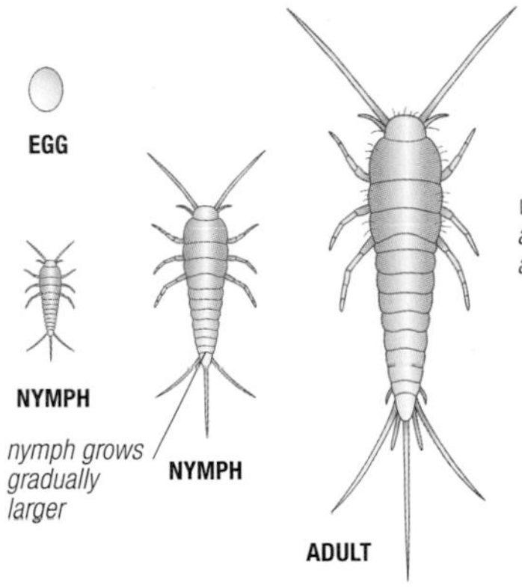

AMETABOLOUS DEVELOPMENT
In some wingless insects, there is little difference in form between young and adults. The nymph develops by shedding its exoskeleton, growing larger after each molt.

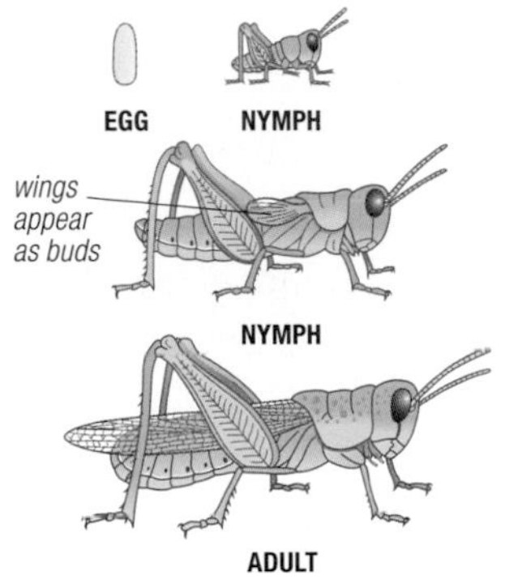

INCOMPLETE METAMORPHOSIS
In some winged insects, such as grasshoppers, the change from the immature stages (nymphs) to the adult is gradual. Nymphs look similar to adults but have no wings or reproductive organs.

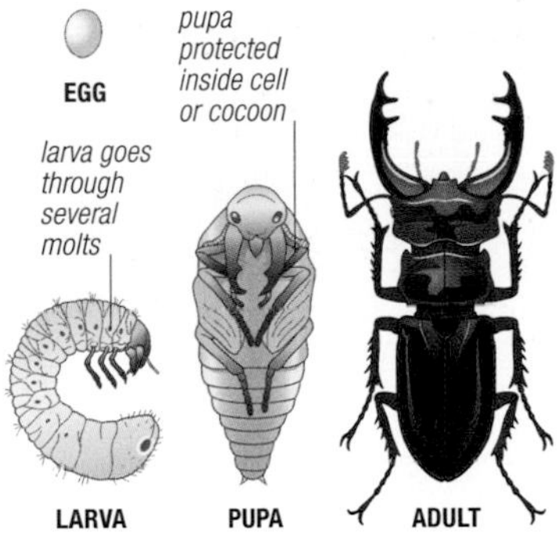

COMPLETE METAMORPHOSIS
In other winged insects, the immature stages (larvae) molt several times. During the last molt, the larvae pupate and the larval tissues are transformed into adult structures.

MIGRATION
Many insects survive low temperatures by becoming dormant. However, some of the larger winged insects are capable of migrating over great distances to avoid winter cold. One of the most celebrated examples is the monarch butterfly (shown here). Each year, tens of millions of monarchs travel up to 2,500 miles (4,000 km) from Canada and the eastern United States to winter in roosting sites in California and Mexico.

TAKING OFF
Beetles protect their delicate hindwings under tough, caselike forewings (elytra). It takes a little time for a beetle to lift its elytra and unfold the hindwings before taking off. If the air is cold, many insects vibrate their flight muscles to warm them up.

Flight and wings

Insects were the first animals to evolve the power of flight, a development that made it easier for them to evade predators and to find food and mates efficiently. An insect's wings, which are made of cuticle, are hinged to the thorax. Most insects have 2 pairs of wings. These are usually hooked together, but in some species, such as dragonflies, they beat alternately, which improves maneuverability. True flies are the only insects that have one pair of wings; they are also the most agile fliers, capable of flying backward, sideways, and even upside down. Almost all flying insects can fold their wings, which means that when they are not flying they can squeeze into small spaces, such as crevices in bark or gaps under stones.

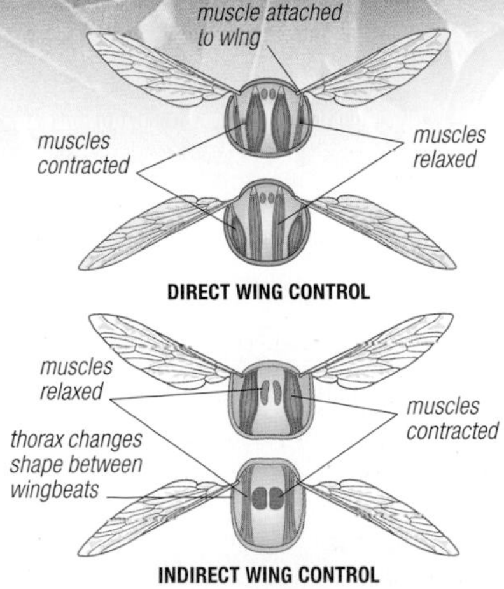

DIRECT AND INDIRECT WING CONTROL
Winged insects use various mechanisms to control their wings. Some control them directly, using muscles attached to the wing bases to produce the upstroke. Other insects move their wings up and down by changing the shape of the thorax. Insects that use this indirect control mechanism usually beat their wings more quickly (100–400 times per second) than those that use direct control (usually about 50 times per second).

NON-INSECT HEXAPODS

Class Collembola

Springtails

Although small and easily overlooked, springtails are present in vast numbers in grassland, leaf litter, and soil. On the underside of the abdomen, they have a ventral tube that allows them to regulate water and grip smooth surfaces, and a fork-shaped jumping organ (furcula), which can be released like a spring to flick them away from predators. There are 32 families (about 8,100 species) of springtails.

Family Poduridae

Water springtail

Occurrence 4 species in the Northern Hemisphere; on the surface of freshwater ditches, ponds, canals, bogs

The most common and widespread species of this family, *Podura aquatica*, is up to $\frac{1}{16}$in (2mm) long. It varies from brown or reddish brown to dark blue or almost black. The furcula is quite flat and long, and reaches the abdominal ventral tube, enabling the springtail to cling to the water's surface and jump effectively. A scavenger, it feeds on small, floating particles of food. Its eggs are laid among vegetation in or around water bodies.

PODURA AQUATICA
This abundant springtail is well adapted to life on water. It may gather in such numbers on freshwater surfaces that the water appears dark.

Family Sminthuridae

Globular springtails

Occurrence 245 spp. worldwide; in leaf litter, fungal fruiting bodies on water surfaces, in caves

Also known as garden springtails, these hexapods are $\frac{1}{32}$–$\frac{1}{8}$in (1–3mm) in length. Ranging from pale to dark brown or green, they have almost spherical bodies. The long, elbowed male antennae in many species are designed to hold the female during mating. Eggs are laid in small batches in soil, and development to sexual maturity may take as little as one month. Several species of these herbivores, such as *Sminthurus viridis*, are pests of crop seedlings.

SMINTHURIDES AQUATICUS
This tiny hexapod—only $\frac{1}{16}$in (2mm) long—is common on the surface of ponds, but does not flock in such large numbers as the water springtail.

Class Protura

Proturans

Like springtails, proturans are sometimes found in leaf litter, but most species live in the soil. These often microscopic animals do not have eyes, tails (cerci), or antennae, and instead they use their front legs as sensory organs. Their mouthparts, which can both pierce and suck, are kept in a pouch and are pushed out during feeding. There are 7 families of proturans, containing about 760 species.

Family Eosentomidae

Eosentomids

Occurrence 340 spp. worldwide; in soil, leaf litter, mosses, humus, decaying wood

These pale and soft-bodied hexapods are $\frac{1}{64}$–$\frac{1}{16}$in (0.5–2mm) long, and have a conical head and an elongated body. They eat dead organic matter and fungi. The eggs, laid in soil or leaf litter, are spherical and have patterns or warts on them. Larvae are identical to adults, but smaller.

EOSENTOMON DELICATUM
This often abundant species is native to Europe, although the genus is found all over the world. It lives in soil, especially chalky soil.

Class Diplura

Diplurans

These blind, elongated, and soft-bodied hexapods are sometimes called two-tailed bristletails due to their 2 terminal cerci. They have a large, distinct head and biting, eversible mouthparts. There are 8 families and about 975 species.

Family Japygidae

Japygids

Occurrence 408 spp. worldwide; in soil

These species are slender-bodied and pale, ranging in length from $\frac{1}{4}$–$1\frac{1}{4}$in (0.6–3cm). The end of the abdomen carries a pair of distinctive, tough, forcepslike cerci which are used to catch small arthropod prey. Eggs are laid in batches in soil.

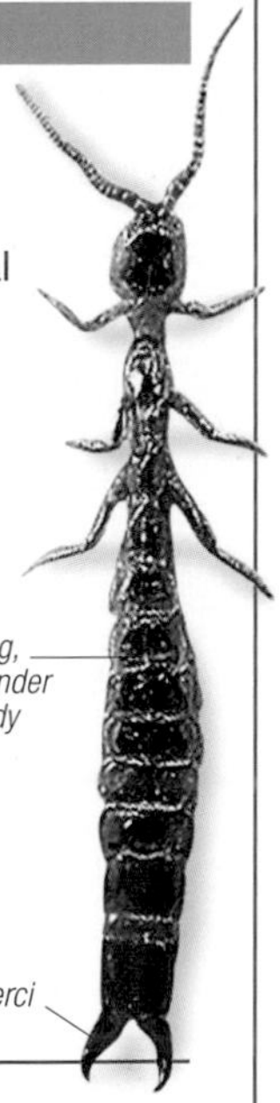

HOLJAPYX DIVERSIUNGUIS
Also called the slender dipluran, this common North American species has stout, terminal cerci. Its abdominal segments are pale around the edges.

INSECTS

Order Archaeognatha

Bristletails

Bristletails are wingless insects that look hump backed in side view. They have simple mouthparts, large eyes that touch each other, and 3 tails (cerci), the middle being the longest. Like silverfish, they do not undergo metamorphosis. There are about 470 species in 2 families.

Family Machilidae

Jumping bristletails

Occurrence 333 spp. worldwide; under stones, in leaf litter, decaying vegetation

These species are elongated and covered with patterns of drab brown or dark gray scales. Up to $\frac{3}{8}$in (1.2cm) long, many can run and jump. They feed on algae, lichen, and plant debris. Eggs are laid in small batches in crevices, and the young mature in 2 years. They may live up to 4 years.

PETROBIUS MARITIMUS
This fast-running insect lives on rocky shorelines in the Northern Hemisphere, often in large numbers.

Order Thysanura

Silverfish

Distinguished from bristletails by their widely separated eyes, and 3 equally sized tails (cerci), silverfish have elongated, flattened bodies, often covered in silvery scales. There are about 570 species in 6 families; some kinds are a familiar sight in houses.

Family Lepismatidae

Lepismatids

Occurrence 295 spp. worldwide, especially in warmer regions; in tree canopies, caves, human dwellings

Lepismatids are brownish, nocturnal insects with compound eyes but no ocelli. They are $\frac{5}{16}$–$\frac{3}{4}$in (0.8–2cm) long. Courtship is simple, and fertilization is indirect. Eggs are laid in small batches in cracks or crevices. Domestic species feed on flour, damp textiles, wallpaper paste, and book bindings. Lepismatids may live up to 4 years.

LEPISMA SACCHARINA
The common silverfish emerges only at night, prefers damp places, and can be found in kitchens, bathrooms, and basements worldwide.

Order Ephemeroptera

Mayflies

The most primitive winged insects, mayflies are unique in undergoing a final molt after their wings have formed. Their underwater nymphs live for 2–3 years, but adults do not feed, and often live for just one day. There are about 3,000 species in 25 families.

Family Siphlonuridae

Primitive minnow mayflies

Occurrence 126 spp. worldwide, but mainly in the Northern Hemisphere; mostly in and near running water

These mayflies have a wingspan of up to 2 in (5 cm). Nymphs are agile swimmers and, when grown, pull themselves out of the water onto a stone or stem.

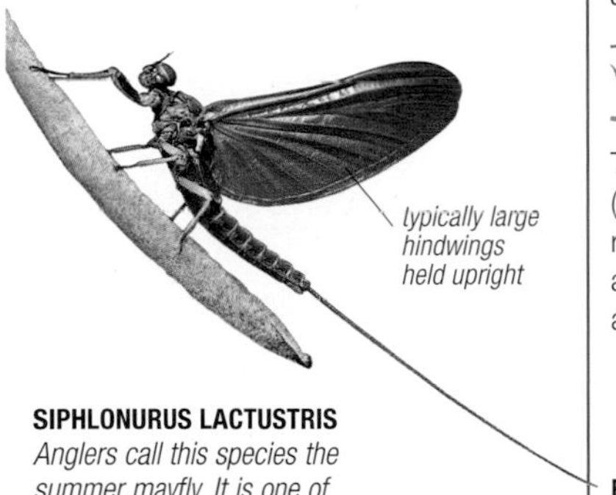

SIPHLONURUS LACTUSTRIS
Anglers call this species the summer mayfly. It is one of the first to colonize a new water course and is tolerant of acidic water.

Family Ephemeridae

Common burrowing mayflies

Occurrence 96 spp. worldwide except Australia; in and on vegetation around freshwater

These are large mayflies, 3/8–1½ in (1–3.4 cm) long, with clear or brownish, rarely dark-spotted, triangular wings and 2 or 3 long tails at the end of the abdomen. Females lay thousands of eggs, which they drop directly into water. The nymphs burrow into the silt at the bottom of the water, using their toothlike mandibles, and eat organic material present in the silt. Adults do not feed. Artificial flies used in trout fishing are modeled on these mayflies.

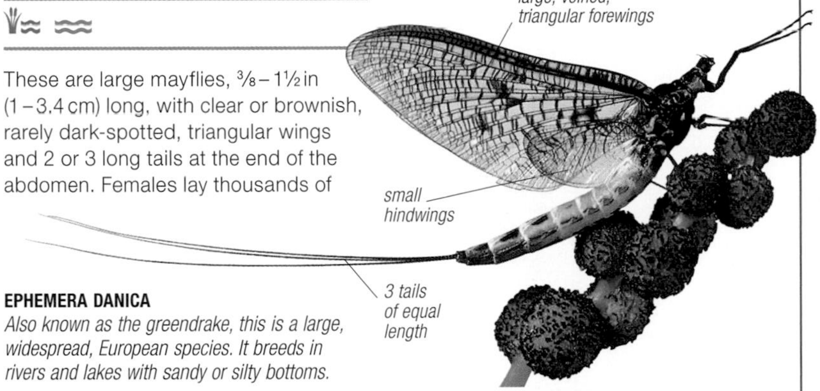

EPHEMERA DANICA
Also known as the greendrake, this is a large, widespread, European species. It breeds in rivers and lakes with sandy or silty bottoms.

Order Odonata

Damselflies and dragonflies

With their large eyes and superb flying ability, these insects are highly effective aerial predators. Damselflies are slender, and tend to rest with wings folded, while the stouter dragonflies rest with wings held out. There are about 5,600 species, in 32 families. All start life as aquatic nymphs.

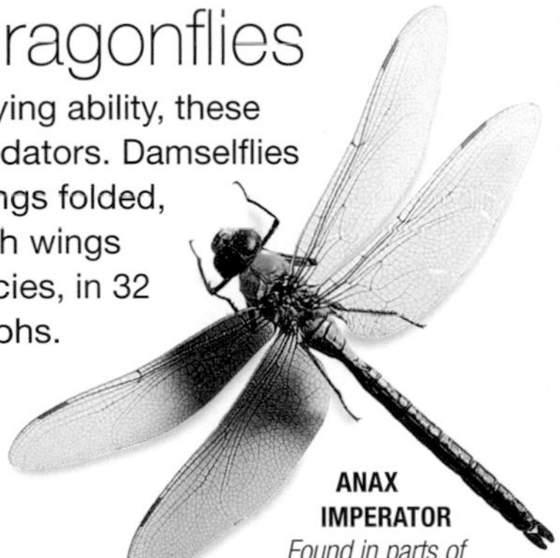

ANAX IMPERATOR
Found in parts of Europe and Asia, the adult emperor dragonfly is a strong flier often seen hunting far from water.

Family Aeshnidae

Darners

Occurrence 444 spp. worldwide; in and near still water, ponds, swamps with vegetation

Also known as hawkers, darners are some of the biggest, most powerful dragonflies, with a wingspan of mostly 2½–3½ in (6.5–9 cm). Their large eyes touch on top of the head. Males hold a territory and patrol it regularly, while females cut slits in plants to lay their eggs. Their nymphs are large and well camouflaged, and hunt animals as large as tadpoles and young fish.

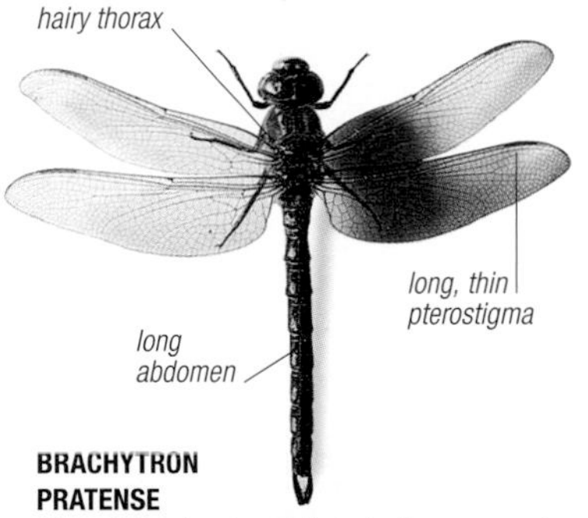

BRACHYTRON PRATENSE
The adult hairy dragonfly typically emerges in May, flying low, with a characteristic zigzag path.

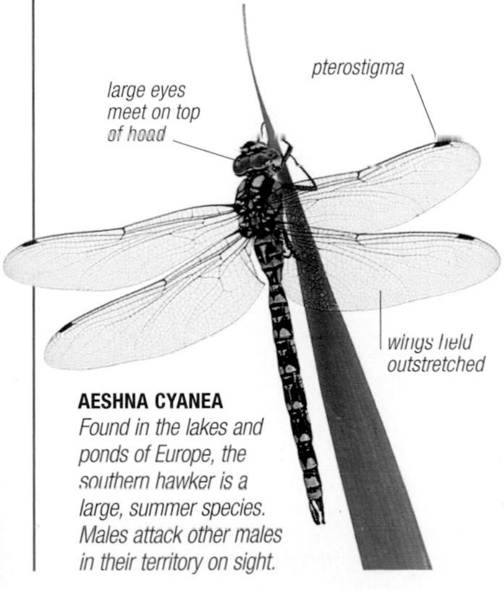

AESHNA CYANEA
Found in the lakes and ponds of Europe, the southern hawker is a large, summer species. Males attack other males in their territory on sight.

AESHNA MIXTA
The migrant hawker, widespread in central and southern Europe, follows along a regular beat to catch prey. It flies in summer and fall, and often migrates across Europe in large numbers.

Family Coenagrionidae

Narrow-winged damselflies

Occurrence 1,142 spp. worldwide, especially in temperate regions; in ponds, bogs, streams, brackish water

These slender-bodied damselflies mostly have a wingspan of ¾–1¾ in (2–4.5 cm). Females make slits in aquatic plants and lay eggs in them. Nymphs hunt on underwater plants.

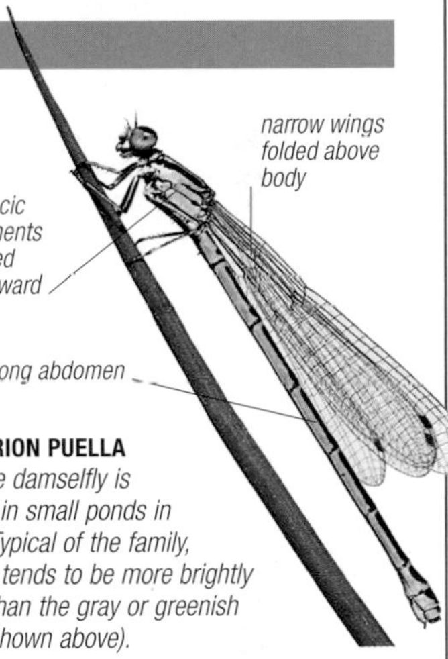

COENAGRION PUELLA
The azure damselfly is common in small ponds in Europe. Typical of the family, the male tends to be more brightly colored than the gray or greenish female (shown above).

Family Lestidae

Stalk-winged damselflies

Occurrence 152 spp. worldwide; in swampy or boggy places, pools, lakes, slow-flowing streams

Also called spread-winged damselflies, these insects have sturdy, bright blue or green bodies, with a metallic sheen. At rest they adopt a vertical pose with the head up and the wings, spanning 1¼–3 in (3–7.5 cm), held out. Males have forceps-shaped claspers at the end of the abdomen. Females lay eggs inside the aerial parts of aquatic plants.

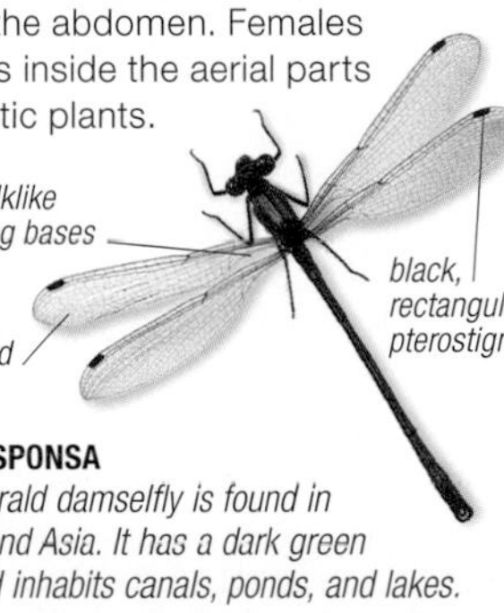

LESTES SPONSA
The emerald damselfly is found in Europe and Asia. It has a dark green body and inhabits canals, ponds, and lakes.

Family Libellulidae

Common skimmers

Occurrence 986 spp. worldwide; in forests near slow-flowing streams, ponds, bogs

Also known as darters due to their unpredictable flight pattern, these stout-bodied, often colorful, insects have a wingspan of mostly 1¾–3 in (4.5–7.5 cm). The broad, flat abdomen makes them highly maneuverable. Females hover over water and dip down to release their eggs. Nymphs hunt for prey in mud and debris.

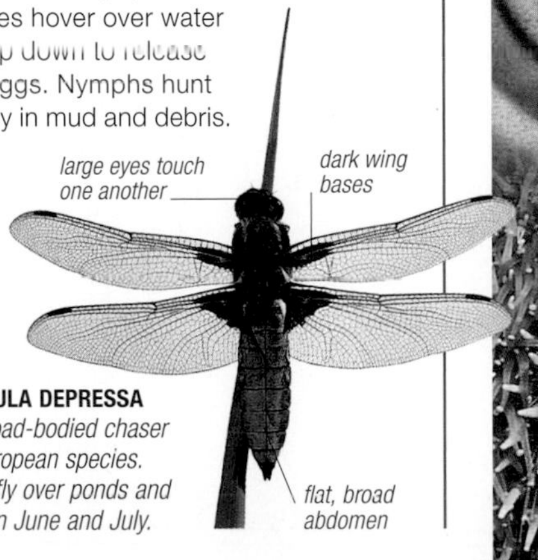

LIBELLULA DEPRESSA
The broad-bodied chaser is a European species. Adults fly over ponds and lakes, in June and July.

INVERTEBRATES

Order Orthoptera

Crickets and grasshoppers

With over 25,000 species in 40 families, crickets and grasshoppers are widespread in all but the coldest parts of the world. Most species have large wings—the front pair, or tegmina, being tough and leathery—but instead of flying away from danger, they often jump away on their powerful hindlegs. They have chewing mouthparts, but their diets vary: crickets, katydids, and their relatives (suborder Ensifera) are predators or omnivores, while grasshoppers and locusts (suborder Caelifera) are entirely vegetarian. All species undergo incomplete metamorphosis, and most adult males stridulate, or "sing," to attract mates.

Family Acrididae

Grasshoppers

Occurrence 10,500 spp. worldwide; on ground, in vegetation

Grasshoppers are ³⁄₈–3¼in (1–8cm) long and have camouflage coloring and patterning. However, some have brightly colored bodies, or wing bands, to warn off predators. Some also produce noxious chemicals. The antennae are always short. Males sing during the day to attract mates. Females are nearly always larger than males and do not have a conspicuous ovipositor. After mating, they lay egg pods containing 15–100 eggs in the ground and protect them by secreting a foamy substance around them. Grasshoppers are agricultural pests; species that periodically form nomadic swarms—known as locusts—destroy crops in some parts of the world.

CHORTHIPPUS BRUNNEUS
The common field grasshopper inhabits grassland from Northern Europe to Spain and Italy. It has a very distinctive song—a series of chirps repeated every 2 seconds or so.

STENOBOTHRUS LINEATUS
The stripe-winged grasshopper lives in dry, sunny grassland and pastures in Europe. It has a high-pitched, alternately loud and soft, song.

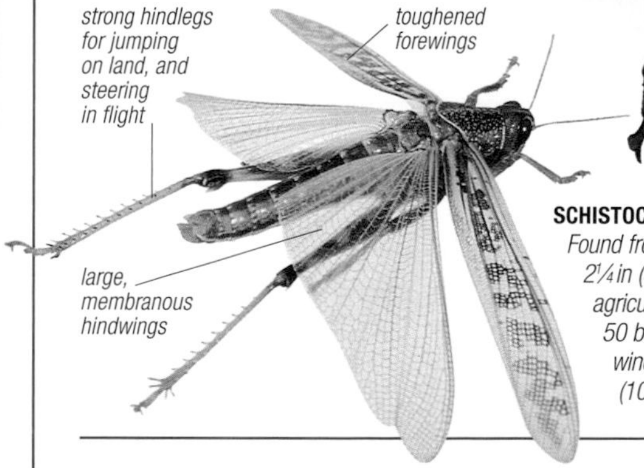

SCHISTOCERCA GREGARIA
Found from Africa to India, the desert locust is 2¼in (6cm) long. It has been a crop pest since agriculture first developed. Swarms of up to 50 billion individuals move with prevailing winds and can devour up to 110,231 tons (100,000 tonnes) of food in a day.

Family Gryllidae

True crickets

Occurrence 4,693 spp. worldwide; on ground, under stones, in leaf litter, on trees

True crickets are ³⁄₁₆–2in (0.5–5cm) long, with slightly flattened bodies. Some species have loud, chirping songs, and are kept as pets. Most true crickets, except the carnivorous tree crickets, are either plant eaters or scavengers.

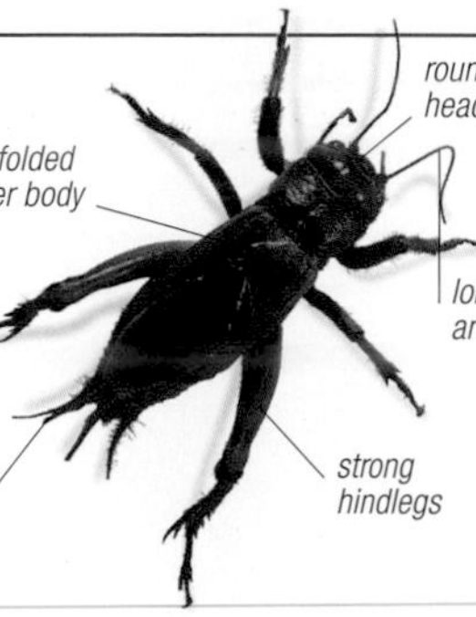

BRACHYTRUPES SPP.
Sometimes called tobacco crickets, these species are distributed across the African savanna. Relatives include pests that attack tea, tobacco, and cotton crops.

Family Gryllotalpidae

Mole crickets

Occurrence 105 spp. nearly worldwide; burrowing in damp sand or soil near streams, ponds, lakes

Like mammalian moles, these crickets are stout and covered with short, velvety hairs. Measuring ¾–1¾in (2–4.5cm) in length, they have short, broad, toothed front legs used for digging. Females dig chambers in the soil to lay their eggs; males make special burrows to amplify their songs, which can be heard up to 1 mile (1.5km) away. Adults and nymphs eat parts of plants as well as small prey, feeding underground during the day and on the surface at night.

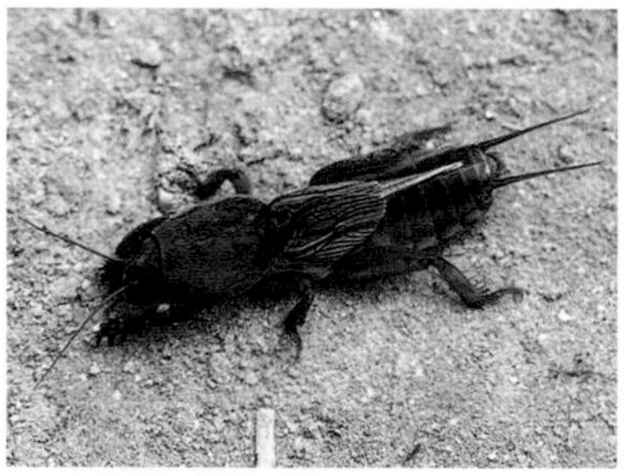

GRYLLOTALPA GRYLLOTALPA
The European mole cricket is attracted to lights after dark. It uses its front legs for digging, and hindlegs for pushing soil along burrows. It can be a crop pest.

Family Rhaphidophoridae

Cave crickets

Occurrence 594 spp. widespread, especially in warmer regions; in caves, humid areas, under logs and stones

Squat, hump backed, and wingless, cave crickets are ½–1½in (1.3–3.8cm) long. They have long hindlegs and even longer antennae that help detect predators. These crickets are drab brown or gray in color. Some cave-adapted species have reduced eyes and soft bodies, and lay their eggs in cave-floor debris. Nymphs start searching for food as soon as they hatch; some scavenge, while others catch living prey.

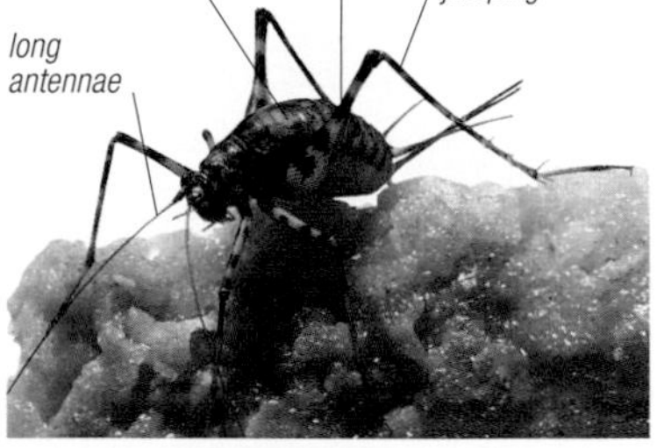

PHAEOPHILACRIS GEERTSI
This African cave cricket is found only in the Democratic Republic of Congo.

Family Tettigoniidae

Katydids

Occurrence 6,728 spp. worldwide, mainly in tropical regions; on vegetation

Also known as bush crickets, these insects get their name from a North American species whose song sounds like "kate-she did." They measure ½–3in (1.5–7.5cm) in length and are brown or green, usually with big wings that slope over their sides. Most species mimic leaves or bark. They may also flash bright, hindwing colors to startle predators. Males sing using a file and scraper system at the base of the forewings. Females lay eggs in plants or soil. Most katydids feed and sing only at night.

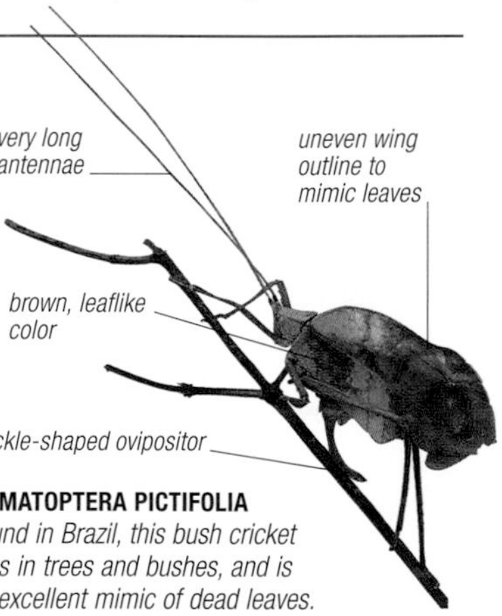

OMMATOPTERA PICTIFOLIA
Found in Brazil, this bush cricket lives in trees and bushes, and is an excellent mimic of dead leaves.

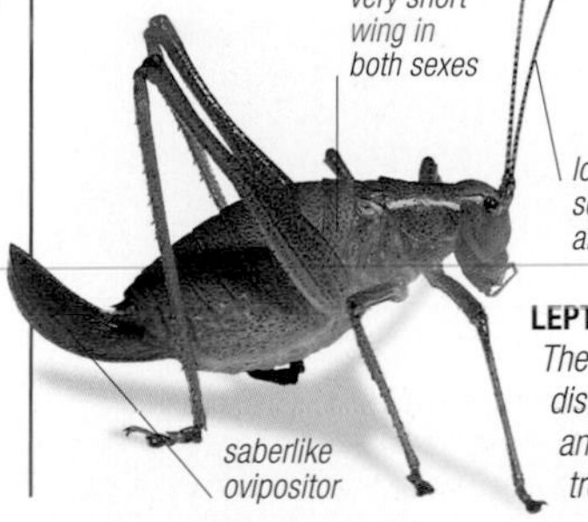

LEPTOPHYES PUNCTATISSIMA
The speckled bush cricket is distributed across Europe, and can be found in bushes, trees, and grassland.

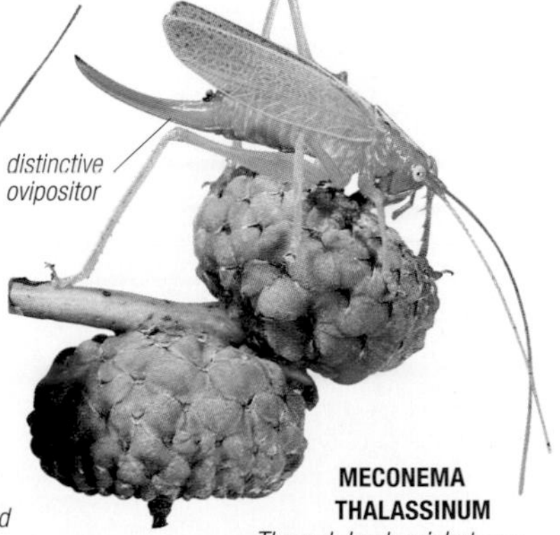

MECONEMA THALASSINUM
The oak bush cricket can be found in Europe, mostly in oak woodland. It is mainly nocturnal and flies readily. Males signal to females by drumming their hindlegs against leaf surfaces.

Order Plecoptera

Stoneflies

Stoneflies are flat, slender-bodied insects that start life in water, developing by incomplete metamorphosis. Adults have 2 pairs of membranous wings, but are reluctant fliers. Most are short-lived, often not feeding at all. There are about 3,000 species in 19 families.

Family Perlidae

Common stoneflies

Occurrence 933 spp. worldwide except Australia; in lakes, rivers, vegetation near running water

As adults, common stoneflies are weak fliers, and often rest on stones close to the water's edge. They are ⅜–2 in (1–4.8 cm) long, and are colored in shades of brown or yellow. The adults do not feed, but the nymphs are often carnivorous.

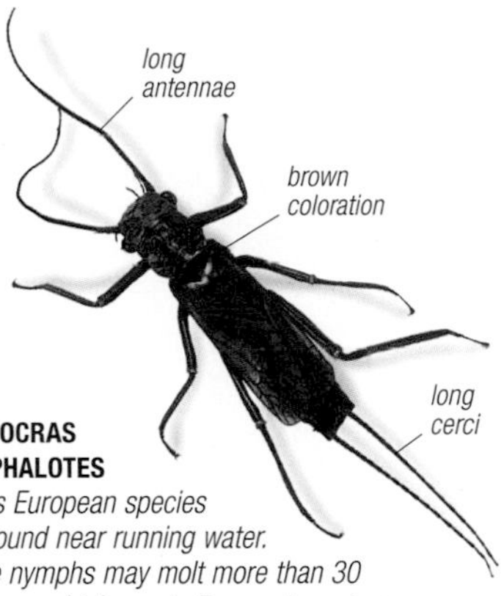

DINOCRAS CEPHALOTES
This European species is found near running water. The nymphs may molt more than 30 times and take up to 5 years to mature.

Order Grylloblattodea

Rock crawlers

Rock crawlers were first discovered in the Canadian Rockies in 1913. Forming one family of 30 species, these small, wingless insects are adapted to cold conditions, and eat dead, windblown, or torpid prey, moss, and other plant matter. Metamorphosis is incomplete.

Family Grylloblattidae

Rock crawlers

Occurrence 30 spp. in cooler regions of the Northern Hemisphere; in limestone caves, decaying wood

Mainly nocturnal, rock crawlers, or ice crawlers, are ⅜–1¼ in (1.2–3 cm) long, have small eyes, and biting mouthparts. Females may lay eggs over 2 months after mating; nymphs take years to develop.

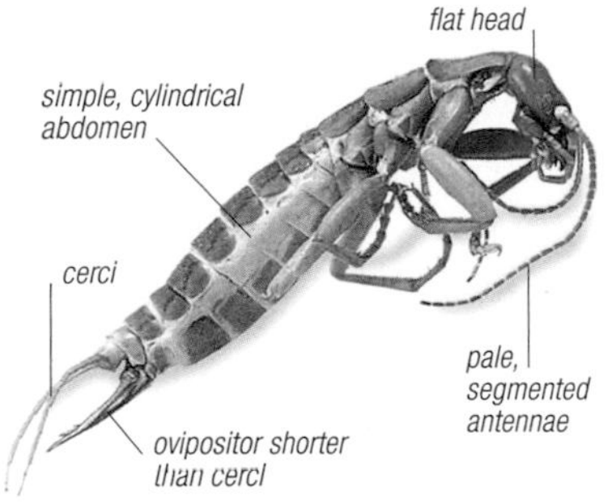

GRYLLOBLATTA CAMPODEIFORMIS
The northern rock crawler is found at high altitudes in the USA and Canada. It lives among rocks near glaciers.

Order Dermaptera

Earwigs

This order contains 11 families (about 1,900 species) of scavenging or plant-eating insects with characteristic abdominal "pincers." Most have short front wings, and fan-shaped hindwings that can be folded. Maternal care is well developed. Metamorphosis is incomplete.

Family Forficulidae

Common earwigs

Occurrence 450 spp. worldwide; in leaf litter and soil, under bark or in crevices

These species are ⅜–1 in (1.2–2.5 cm) long and usually dark brown. Two forcepslike appendages are straighter in females and curved in males. Females guard their eggs and lick them clean.

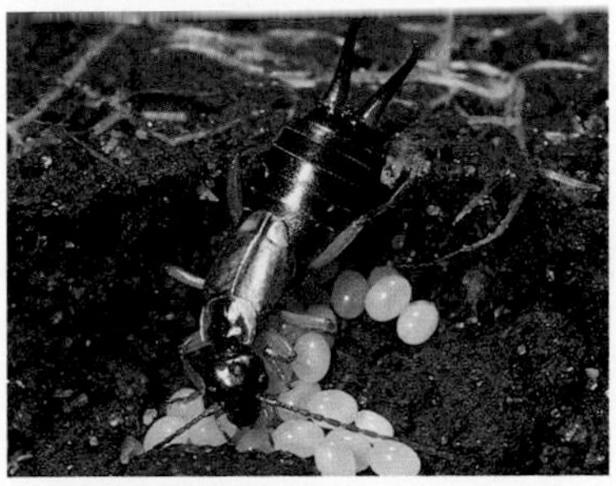

FORFICULA AURICULARIA
The common earwig, now found worldwide, is a pest of vegetables, cereals, fruit, and flowers.

Order Phasmatodea

Stick and leaf insects

These strikingly sticklike or leaflike insects live and feed on plants, and use their remarkably effective camouflage to avoid being eaten. The males—which are rare in some species—have wings, but the females are frequently wingless. They develop by incomplete metamorphosis, and are usually nocturnal. There are about 2,500 species in 3 families.

Family Phasmatidae

Stick insects

Occurrence 2,450 spp. mainly in warm regions; among vegetation or foliage of trees and shrubs

Also called walking sticks, these insects can adopt a twiglike posture that camouflages them superbly; some can even sway in the breeze to blend with the background. Stick insects vary from 1–11½ in (2.5 to 29 cm) in length and have a spiny or warty, brown or green body, with short, tough front wings to protect the large, membranous hindwings. Some species are wingless. When threatened, some stick insects adopt a scorpionlike posture, while others flash their wing colors, or give off noxious chemicals. If attacked, they can shed their legs, which soon grow back. In many species, males occur either rarely or not at all, and the females typically scatter their seedlike eggs on the ground.

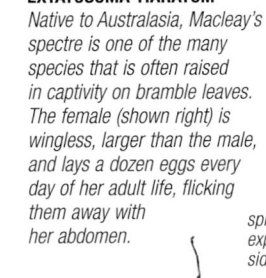

EXTATOSOMA TIARATUM
Native to Australasia, Macleay's spectre is one of the many species that is often raised in captivity on bramble leaves. The female (shown right) is wingless, larger than the male, and lays a dozen eggs every day of her adult life, flicking them away with her abdomen.

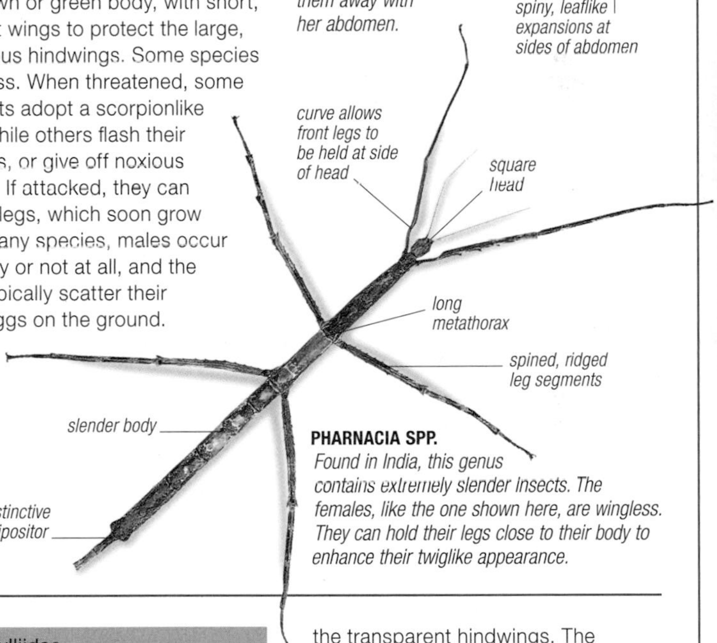

PHARNACIA SPP.
Found in India, this genus contains extremely slender insects. The females, like the one shown here, are wingless. They can hold their legs close to their body to enhance their twiglike appearance.

Family Phylliidae

Leaf insects

Occurrence 30 spp. in Mauritius, Seychelles, Southeast Asia, Australasia; on various plants in well-vegetated areas

Flat, expanded abdomens, extended leg segments, and brown or green coloring all give these insects an uncanny resemblance to leaves. The disguise is helped further by blotches of color and texture, and an ability to sway in the breeze. Some species specialize in looking like dead, wrinkled leaves. Leaf insects range from 1¼–4¼ in (3 to 11 cm) in length. At rest, their veined forewings cover the transparent hindwings. The antennae are short and smooth in females, and long and slightly hairy in males. The eggs are seedlike and are dropped on the ground. Because of their odd appearance, leaf insects are bred as pets all over the world.

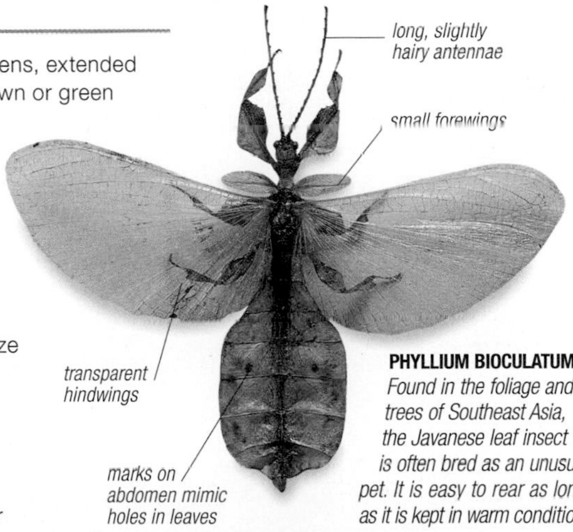

PHYLLIUM BIOCULATUM
Found in the foliage and trees of Southeast Asia, the Javanese leaf insect is often bred as an unusual pet. It is easy to rear as long as it is kept in warm conditions.

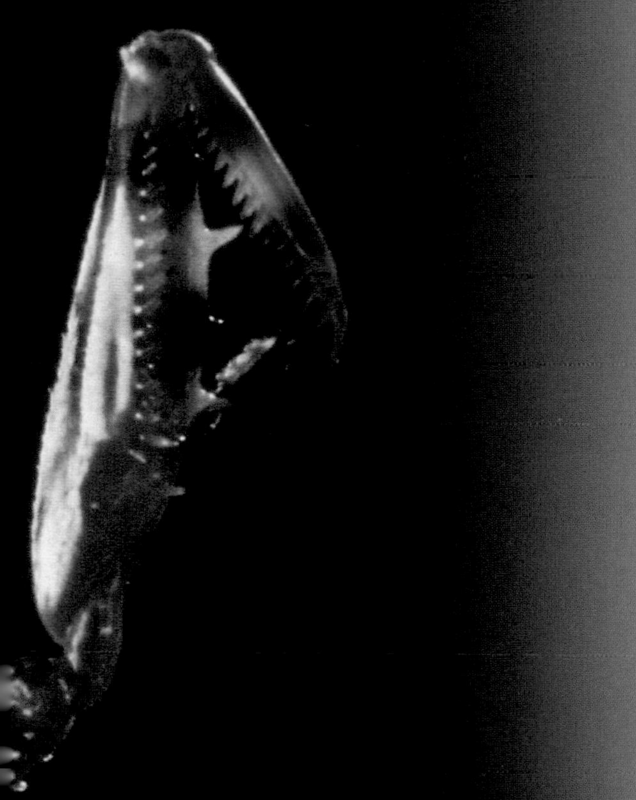

THREAT DISPLAY
When disturbed, mantids like this flower mantid, Creobotra elegans, *often rear up into a defensive pose to display their vivid coloration. In addition to using its excellent eyesight, the mantid avoids attack from bats by using an ultrasonic ear on the thorax between the middle legs.*

Order Mantodea

Mantids

The common name "praying" mantid is derived from the distinctive way in which mantids hold their front legs up and together, as if in prayer. Although variable in shape, mantids are distinguished by their triangular heads and large eyes—they are the only insects able to turn their head around to look behind them—and an elongated prothorax with the front pair of legs distinctively modified for catching live prey. Most mantids are diurnal and eat a large range of arthropods and, occasionally, even vertebrates such as frogs and lizards. Mantids reproduce by laying eggs in a papery or foamlike egg case, which, in some species, is guarded by the female. There are over 2,300 species.

Family Hymenopodidae

Flower mantids

Occurrence 290 spp. in tropical regions, all around the world except Australia; on a wide variety of vegetation

Vivid coloration, including bright reds and greens (see left), allows these species to blend in perfectly with the plants on which they rest, awaiting the arrival of prey. Some parts of the body, such as the legs, often have broad extensions that resemble leaves. The forewings of flower mantids may have colored bands or spirals, or circular marks that look like eyespots. In some species, the colored markings on the front wings can be asymmetrical. The wings of the female flower mantid are sometimes shorter than those of the male. Young nymphs are immediately predatory and are able to tackle prey as soon as their cuticles have hardened. As well as having an ultrasonic ear, used to detect predators, some species have a second ear that is responsive to much lower frequencies, but its function is unknown.

Family Mantidae

Common praying mantids

Occurrence 1,101 spp. in warmer regions, all around the world; on any vegetation where there is ample prey

Most praying mantids belong to this family, and many members vary little in appearance. Most have green or brown coloration and a close resemblance to leaves and twigs. Unlike families such as the flower mantids, the wings of common praying mantids are seldom patterned, although they may have wingspots. Females often have smaller wings than males, and some lack wings entirely. Some species of common praying mantids can grow up to 6 in (15 cm) in length, and the larger species are known to prey on small reptiles, salamanders, and frogs.

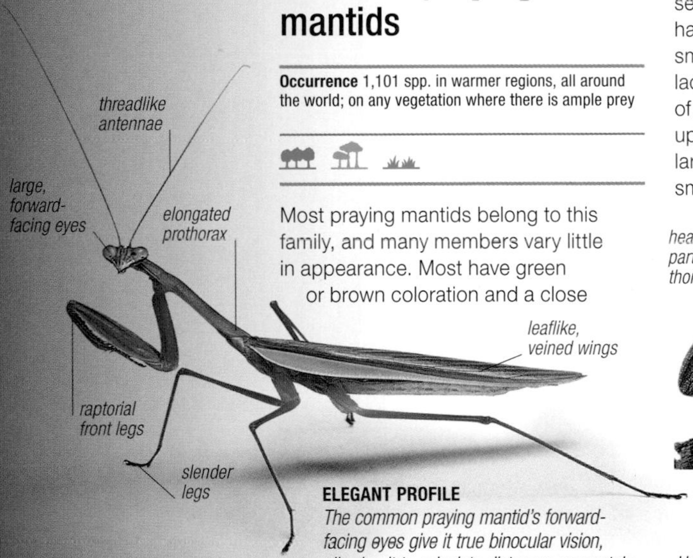

ELEGANT PROFILE
The common praying mantid's forward-facing eyes give it true binocular vision, allowing it to calculate distances accurately. This ability, together with its extendable front legs, makes it an effective hunter.

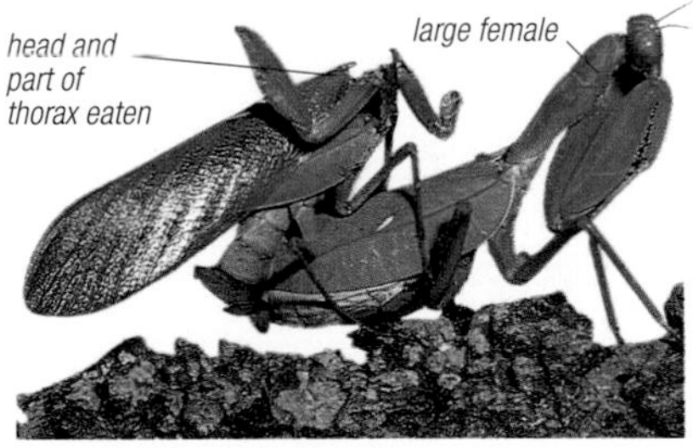

DANGEROUS ENCOUNTERS
Female common praying mantids are famed for eating the male after mating (see above). However, although mating can be dangerous for smaller individuals, males are usually cautious and seldom get eaten by their partner in the wild.

STEALTHY HUNTER

LYING IN WAIT
The mantid lies patiently in wait for a potential meal. Perching on the end of a stick, forelegs together, it focuses on a likely landing site for prey.

THE POUNCE
As an unsuspecting fly comes within range, the mantid acts quickly. It launches itself at its victim, lunging forward and reaching out its long forelegs.

THE CAPTURE
At full stretch, the mantid seizes and impales its victim with its spiky legs before it is even aware of the mantid's presence.

ALL OVER IN SECONDS
Clasped in a strong grip, the helpless fly has no means of escape. Having retreated back to its stick, the mantid begins to eat its prey.

Order Blattodea

Cockroaches

Cockroaches are leathery insects, typically with oval and flattened bodies that enable them to squeeze through tight spaces in search of food or to escape predators. They are sensitive to vibration and can quickly flee from danger. The head is often covered by a shieldlike pronotum and there are generally 2 pairs of wings. The order Blattodea comprises 9 families and about 4,600 species, of which less than 1 percent are pests; the rest are useful scavengers in many habitats. The pest species thrive in warm conditions and where there is poor hygiene and sanitation, and can carry many disease-causing organisms on their bodies.

Adults may produce sexual pheromones to attract mates. The females lay up to 50 eggs in 2 rows surrounded by tough, protective egg cases called oothecae. Metamorphosis is incomplete.

Family Blaberidae

Live-bearing cockroaches

Occurrence 1,189 spp. in tropical regions; in caves, underground

These large insects, usually measuring 1–2¼ in (2.5–6 cm) in length, often have well-developed, pale brown wings with dark markings. In many species, however, the females are wingless and burrow under wood and stones. The pronotum may be very wide. During courtship, some species use sounds to communicate with a mate. All species are ovoviviparous and give birth to live young. The egg sacs are fully extruded from the end of the abdomen, rotated, and then drawn back inside to be brooded within the body of the female. Adults and nymphs are mainly saprophagous or eat dung.

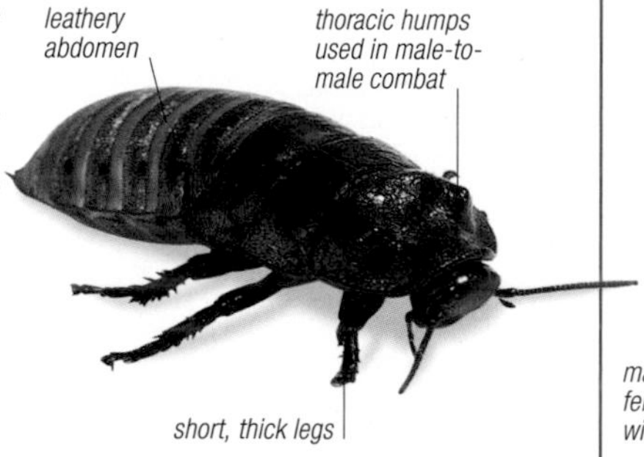

GROMPHADORHINA PORTENTOSA
Better known as the Madagascan hissing cockroach, this insect is 2¼–3¼ in (6–8 cm) long, and startles predators by squeezing air out of its spiracles to produce a loud hiss.

Family Blattellidae

Blattellids

Occurrence 2,240 spp. worldwide in warmer areas; in woodland litter, debris, garbage dumps, buildings

Many of these fully winged species are pale brown but a few are olive-green. Their body length varies from ⁵⁄₁₆ to as long as 4 in (0.8–10 cm). Adults and nymphs are scavengers. Females carry their egg cases, which protrude from their bodies, until the eggs are about to hatch. Thousands of eggs may be produced in a lifetime.

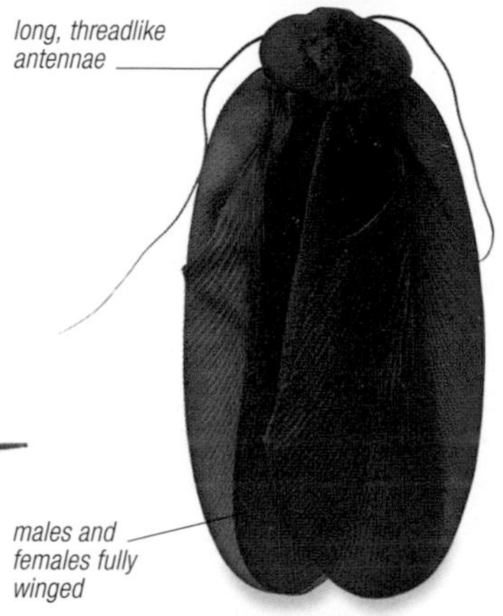

MEGALOBLATTA LONGIPENNIS
The long-winged great cockroach from Peru, Ecuador, and Panama is the largest winged cockroach in the world. It has a wingspan of up to 8 in (20 cm).

Family Blattidae

Common cockroaches

Occurrence 652 spp. mainly in tropical and subtropical regions; common in warehouses, sewers, garbage dumps

Most species of common cockroaches are brown or reddish or blackish brown with markings. They are ¾–1¾ in (2–4.5 cm) long, very active, and can run very fast or fly. Some species emit repellent chemicals that can cause skin rashes. Females can produce up to 50 egg cases, each containing 12–14 eggs, which are stuck to concealed surfaces. Both adults and nymphs are nocturnal and saprophagous.

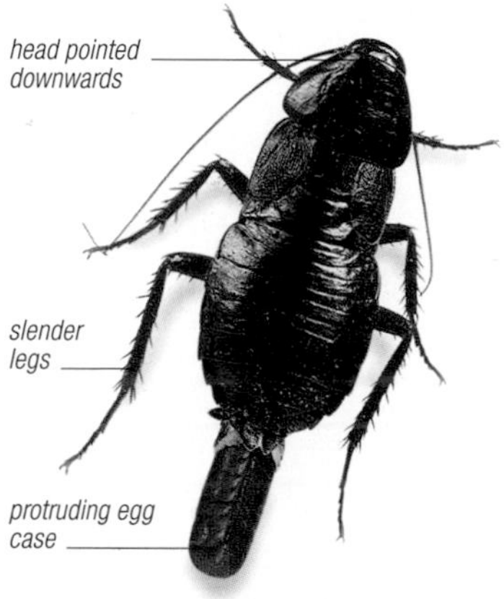

BLATTA ORIENTALIS
The oriental cockroach is a common domestic pest. Like many species, it has spread from its native region by traveling on board ships.

Order Isoptera

Termites

Like ants (see p.576)—with which they are sometimes confused—termites are social insects, living in colonies that can be over a million strong. There are about 2,900 species, in 7 families, and most live in the tropics, where they build impressive nests. They are usually pale and soft, with chewing mouthparts and short antennae, but their body form varies with their role in the colony. Workers are small, while soldiers have large heads and jaws. Queens, which lay all the colony's eggs, can be as big as a human finger. Some termites feed on wood or plants, but others eat fungi that they raise in underground "gardens." Although important in food chains as nutrient recyclers, many are destructive pests. Metamorphosis is incomplete.

Family Rhinotermitidae

Subterranean termites

Occurrence 360 spp. worldwide in warmer regions; in soil or in damp wood touching ground

As adults and nymphs, subterranean termites feed mainly on wood, digesting it with the help of microbes that they harbor in their intestines. They are voracious feeders, and in warm parts of the world are highly damaging pests. Generally cream or light brown, they are typically ¼–⁵⁄₁₆ in (6–8 mm) long.

RETICULITERMES LUCIFUGUS
This termite has distinctive antennae. It makes its nest under the ground, inside damp wood, and is commonly found throughout southern Europe.

Family Termitidae

Higher termites

Occurrence 2,021 spp. worldwide; in trees, soil, underground

Three-quarters of the world's termite species belong to this family. Highly variable, both in appearance and habits, they range in length from ⁵⁄₃₂–½ in (4–14 mm). Queens are much bigger than the workers and soldiers, and may produce thousands of eggs a day. Nests vary from small structures built in trees, to vast earthen towers, with underground "fungus gardens." Many species are crop pests.

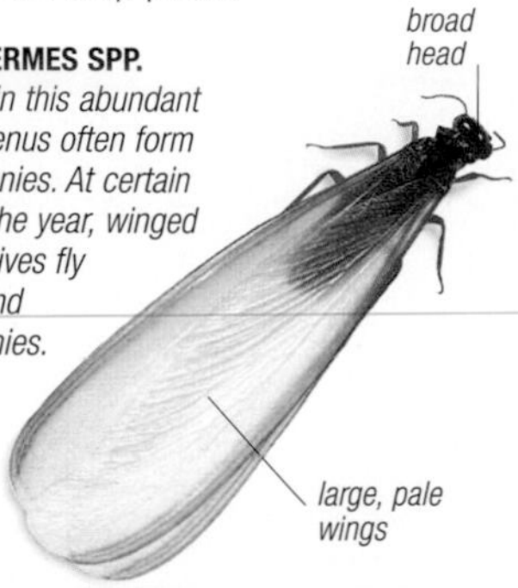

MACROTERMES SPP.
Termites in this abundant African genus often form giant colonies. At certain times of the year, winged reproductives fly off to found new colonies.

GLOBITERMES SULPHUREUS
This Southeast Asian pest makes thick-walled nests partly underground. The nests' soldiers can "explode" to cover intruders in a sticky fluid.

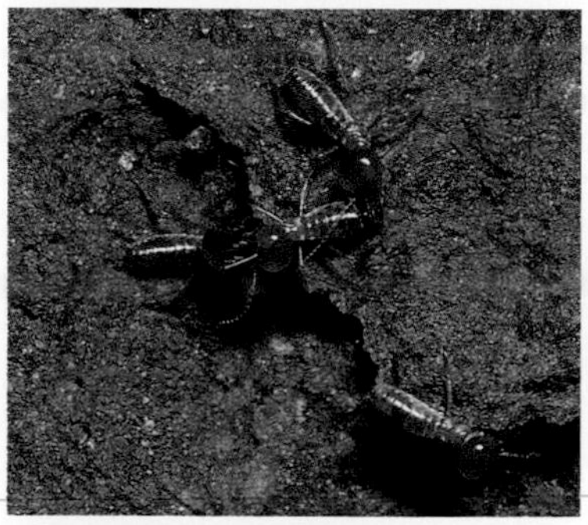

TRINERVITERMES TRINERVOIDES
The snouted harvester termite lives in pastures in central and southern Africa. It is eaten by the aardwolf (see p.209)—an animal that has evolved immunity to its defensive secretions.

Order Embioptera

Web-spinners

This relatively small order consists of 11 families and only about 400 species. Web-spinners are gregarious, and live in soil, leaf litter, and under bark. They make extensive protective tunnels, using the silk made by glands in their front tarsi. Metamorphosis is incomplete.

Family Clothodidae

Clothodids

Occurrence 15 spp. worldwide in tropical and subtropical regions; in soil, leaf litter, under bark and stones

Clothodids are typically elongated insects with short legs. They are 3/16–3/4 in (0.5–2 cm) long, and have small eyes, segmented antennae, and simple, biting mouthparts. As adults, males do not feed but use their mandibles to hold onto the female during mating. Females and nymphs are saprophagous. Females cover their eggs with silk and detritus, and feed the nymphs with pre-chewed food.

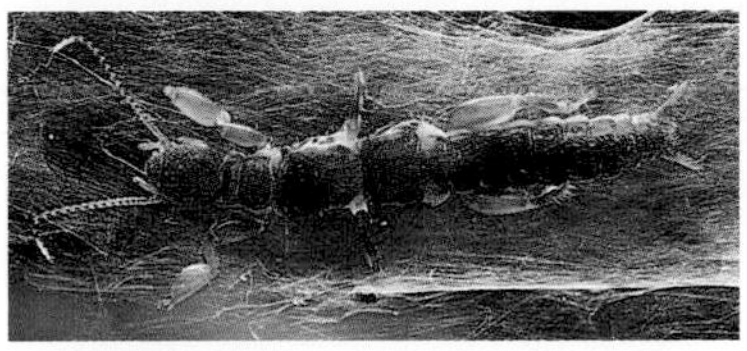

CLOTHODA URICHI
Seen here inside its silken nest, this species is a native of the Caribbean island of Trinidad.

Order Zoraptera

Angel insects

These small, delicate, gregarious, termitelike insects make up a single family of 43 species. They are light straw-yellow to dark brown or blackish in color, with short, abdominal tails (cerci) and unspecialized mouthparts. Metamorphosis is incomplete.

Family Zorotypidae

Angel insects

Occurrence 43 spp. worldwide in tropical and warm temperate regions except Australia; in rotting wood, sawdust

Angel insects are 1/16–1/8 in (2–3 mm) long, with a distinctive triangular head and a pair of antennae. The nymphs and adults eat fungal threads and small arthropods. Sexual behavior may be complex: males of some species give the females mating gifts in the form of secretions from glands in their heads; in other species, the males fight for mates by kicking each other.

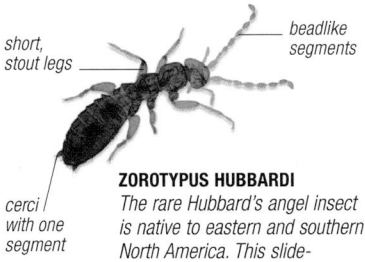

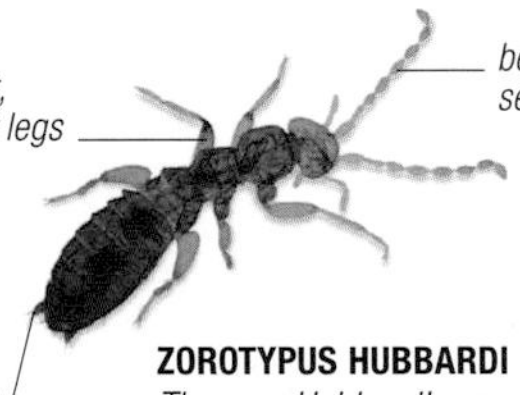

ZOROTYPUS HUBBARDI
The rare Hubbard's angel insect is native to eastern and southern North America. This slide-mounted specimen is stained red.

Order Psocoptera

Barklice and booklice

The order Psocoptera comprises 41 families and 5,600 species of soft-bodied, drably colored insects. The head is large, with bulging eyes, a bulbous forehead, and long, threadlike antennae. Metamorphosis is incomplete, typically with 5 nymphal stages.

Family Liposcelididae

Liposcelid booklice

Occurrence 250 spp. worldwide; in dry leaf litter, under bark, in nests, inside buildings, in food stores

These insects have flattened bodies, with swollen femora on the hindlegs. Most species are wingless. The length of the body ranges from 1/64–3/64 in (0.5–1.5 mm). Some species gnaw on or chew stored flour, cereals, or paper. Eggs are laid in crevices in tree bark, leaf litter, and bird nests. The nymphs look like small adults.

LIPOSCELIS DECOLOR
Found in leaf litter and in a variety of stored, dry produce, this widespread species can become a pest.

Order Phthiraptera

Parasitic lice

Flattened and wingless, the 5,200 species (in 24 families) of parasitic lice are ectoparasites, living permanently on the bodies of birds and mammals without killing them. The mouthparts are used to bite skin, fur, or feathers, or to suck blood. Different lice are linked with specific hosts, and many are restricted to particular areas of the body. Parasitic lice develop by incomplete metamorphosis.

Family Menoponidae

Bird lice

Occurrence 1,153 spp. worldwide; on a variety of bird hosts

Bird lice are 1/32–1/4 in (1–6 mm) long, with a large and roughly triangular head, and biting mandibles. The abdomen is oval, and the legs are short and stout. Each leg has 2 claws. Adults and nymphs of most species feed on feather fragments but some also ingest blood and skin secretions. Eggs are glued in masses to the base of feathers. Some species, such as *Menopon gallinae*, or the shaft louse, can be serious poultry pests.

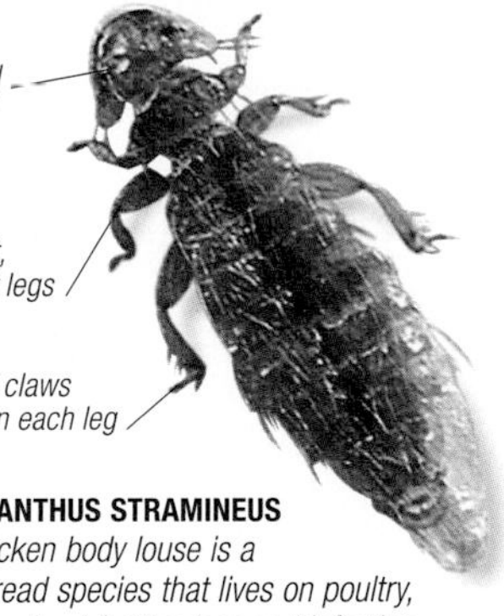

MENACANTHUS STRAMINEUS
The chicken body louse is a widespread species that lives on poultry, bringing about feather loss and infection.

Family Pediculidae

Human lice

Occurrence 7 spp. worldwide; on humans, apes, monkeys

Human lice are small, pale, and elongated, measuring 1/16–1/4 in (2–6 mm) in length, with short, strongly clawed legs for gripping onto their hosts. They have a narrow head and a pear-shaped, flattened body. Adults and nymphs feed on blood. The human louse has 2 subspecies: the body louse, which glues its eggs to clothing, and the head louse, which glues its eggs (nits) to hair. Outbreaks of head lice are common among young schoolchildren. The body louse carries the organism that causes typhus.

PEDICULUS HUMANUS CAPITIS
A female human head louse can lay 9–10 eggs a day, fastening each one separately to a hair. Once cemented in position, the eggs are difficult to dislodge.

Family Pthiridae

Pubic lice

Occurrence 2 spp. worldwide; wherever their hosts (humans and gorillas) live

Also known as crab lice, pubic lice have a squat, flattened body, 3/64–3/32 in (1.5–2.5 mm) long. The middle and hindlegs are especially stout and have very strong, enlarged claws for gripping hair shafts. The nymphs and adults suck the host's blood, leaving bluish marks on the skin. They move very slowly, often hardly at all. A female may lay 30 eggs in her lifetime, and uses a waterproof glue to stick them singly to pubic hair. It takes 4 weeks for the eggs to mature into adults.

PTHIRUS PUBIS
The pubic louse may be found in armpits and beards as well as the groin. Although unpleasant, it is not known to transmit disease.

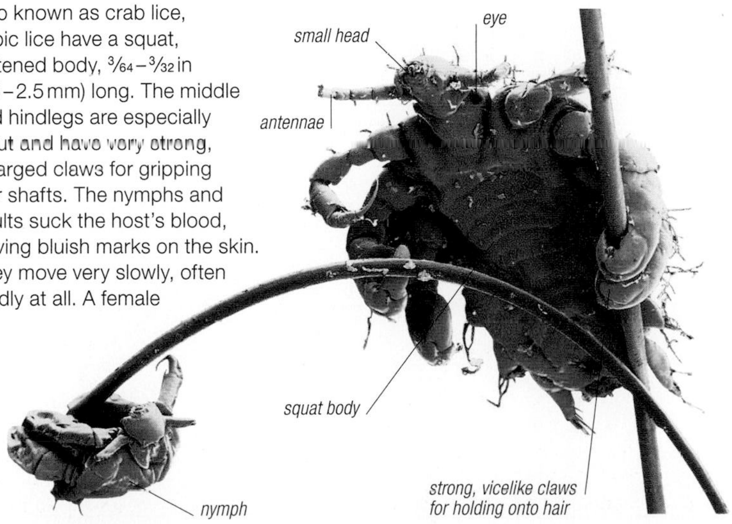

Order Hemiptera

Bugs

Bugs vary in size from species a few millimeters long to giant aquatic predators large enough to catch fishes and frogs. They are found all over the world, in every terrestrial habitat as well as in freshwater. The order comprises 134 families and 88,000 species. There are 4 suborders: Coleorrhyncha, consisting of a single family; Heteroptera or true bugs (including plant bugs); Auchenorrhyncha (including treehoppers, lantern bugs, and cicadas); and Sternorrhyncha (including whiteflies and scale insects). All bugs have 2 pairs of wings; in the Heteroptera, the front ones are typically larger and thickened. The mouthparts, which form a long rostrum, or "beak," pierce and suck, and many sap-sucking species are serious crop pests. Metamorphosis is incomplete.

Family Aleyrodidae

Whiteflies

Occurrence 1,200 spp. worldwide, especially in warmer regions; on a wide range of plants, including crops

These delicate, mothlike bugs are 1/32–1/8 in (1–3 mm) long. They carry a pair of 7-segmented antennae, and 2 pairs of transparent, white, or mottled wings of similar size. Females lay their eggs on tiny stalks on the undersides of leaves. Many of these plant sap-suckers are serious pests.

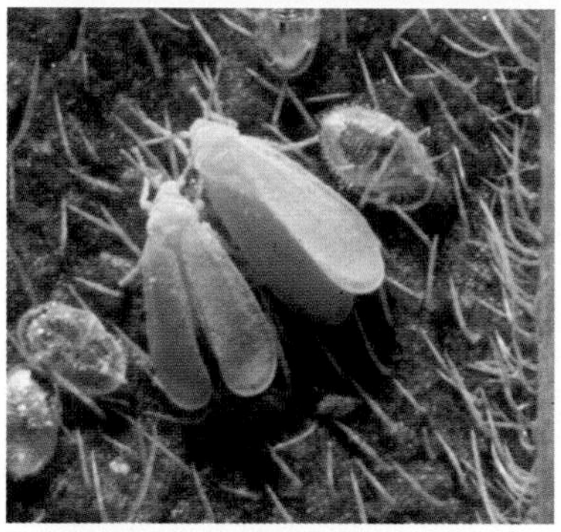

TRIALEURODES VAPORARIORUM
The greenhouse whitefly is a common pest of cucumbers and tomatoes grown under glass. It may also attack field crops in warm conditions.

Family Cercopidae

Froghoppers

Occurrence 2,400 spp. worldwide, especially in warmer regions; on shrubs, trees, herbaceous plants

These squat, round-eyed, plant bugs are very good jumpers. Many species have drab coloration but some are colored in bright red and black or yellow and black patterns. Body length ranges from 3/16–3/4 in (0.5 to 2 cm). Eggs, up to 30 in some species, are laid in soil or in plant tissue. Nymphs feed on plant sap. They surround themselves in foam, which hides them and stops them from drying.

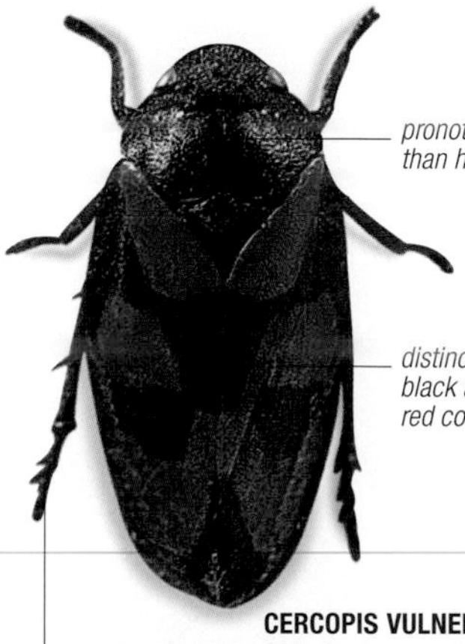

CERCOPIS VULNERATA
This froghopper inhabits grassy places and meadows in Europe and Asia.

Family Cicadellidae

Leafhoppers

Occurrence 16,000 spp. worldwide; almost anywhere with vegetation, abundant in agricultural habitats

Leafhoppers are typically slender-bodied and 1/8–3/4 in (0.3–2 cm) long. Many species are brown or green; some may be brightly striped or spotted. Males attract females with low-amplitude courtship calls, which travel through leaves and stems, instead of being carried through the air. Once a female has mated, she lays hundreds of eggs over a period of 6–8 weeks. Leafhoppers are highly destructive pests, attacking many of the world's staple crops, such as rice, corn, cotton, and sugarcane.

CICADELLA VIRIDIS
Widespread in Europe and Asia, the common green leafhopper can be a pest of fruit, rice, wheat, and sugarcane.

Family Aphididae

Common aphids

Occurrence 2,250 spp. worldwide, especially in northern temperate regions; on most wild and cultivated plants

Their huge reproductive potential makes aphids among the most destructive of all plant-eating insects, affecting almost all crop species by feeding and transmitting diseases. Also called greenflies or blackflies, these are pear-shaped, soft-bodied insects, mostly 3/16 in (5 mm) long, and typically green, pink, black, or brown. The abdomen usually has a pair of cornicles, short tubes that emit a defensive secretion. Reproduction is often parthenogenetic or only seasonally sexual.

MACROSIPHUM ALBIFRONS
The American lupin aphid is a pest of garden flowers, both in North America and Europe. Here, one female (top left) is giving birth.

Family Belostomatidae

Giant water bugs

Occurrence 150 spp. worldwide, especially in tropical and subtropical regions; in standing water, slow-moving streams

Also known as electric light bugs due to their attraction to lights, this family contains the largest bugs, able to eat frogs and fishes. They have a brownish, oval body that is 1/2–4 in (1.5–10 cm) long. The front legs are modified as prey-capturing pincers. Uniquely among insects, females glue their eggs to the male's back, who then broods them.

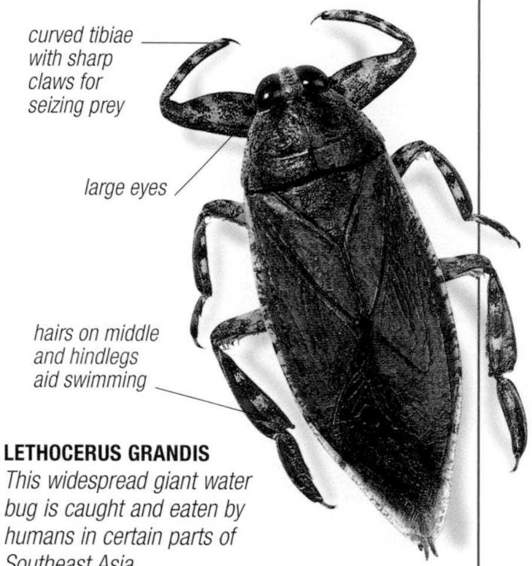

LETHOCERUS GRANDIS
This widespread giant water bug is caught and eaten by humans in certain parts of Southeast Asia.

Family Cicadidae

Cicadas

Occurrence 2,500 spp. worldwide in warmer regions; on suitable host trees or shrubs

Cicadas include some of the world's noisiest insects, producing calls that can be heard up to 1 mile (1.5 km) away. The adults are usually dark brown or green, with squat bodies 3/4–2 1/4 in (2.2–5.5 cm) long. They have short antennae, and prominent eyes set far apart. Only the males sing, using a pair of drumlike tymbal organs located on the sides of the abdomen. Females lay eggs by cutting slits in plants; after hatching, the nymphs feed on roots underground. Some species can live for up to 17 years.

CICADA ORNI
This species, common in southern Europe and the Mediterranean, has a grayish bloom, clear front wings with 11 black spots, and a long rostrum. The adults are eaten by birds such as jays.

PYCNA REPANDA
The Himalayan cicada is commonly found resting and feeding among a variety of tree species, in forests and woodland.

ANGAMIANA AETHEREA
Like all cicadas, the Indian cicada produces songs to signal aggression or to attract mates.

Family Cimcidae

Bed bugs

Occurrence 90 spp. worldwide; on bird and mammal hosts, in nests, caves, crevices in buildings

These flat, oval, reddish brown bugs are $^{5}/_{32}$–$^{3}/_{8}$ in (4–12 mm) long, and have remnant front wings and no hindwings. Their compound eyes have a small number of facets. All bed bugs are blood sucking, temporary ectoparasites that live on birds, bats, and even humans. They live for 2–10 months depending on feeding and temperature. Males practice traumatic insemination, injecting sperm by penetrating the female's body with a sharp sexual appendage. Females may lay more than 150 eggs in their adult life, with 2 or 3 eggs laid per day.

CIMEX LECTULARIUS
The common bed bug feeds at night and senses the presence of a host using temperature, odor, and carbon dioxide concentration as cues.

Superfamily Coccoidea

Scale insects

Occurrence 7,000 spp. worldwide, especially in tropical and subtropical regions; on all parts of many plants

Scale insects are so highly adapted to a sap-sucking life that they hardly resemble insects at all. The females are often wingless and legless, and spend their adult lives permanently attached to plants. Most are under $^{3}/_{8}$ in (1 cm) long, with an oval body covered by a smooth or waxy scale. The males look very different, and often have wings. Many scale insects reproduce asexually, and can be prolific crop pests.

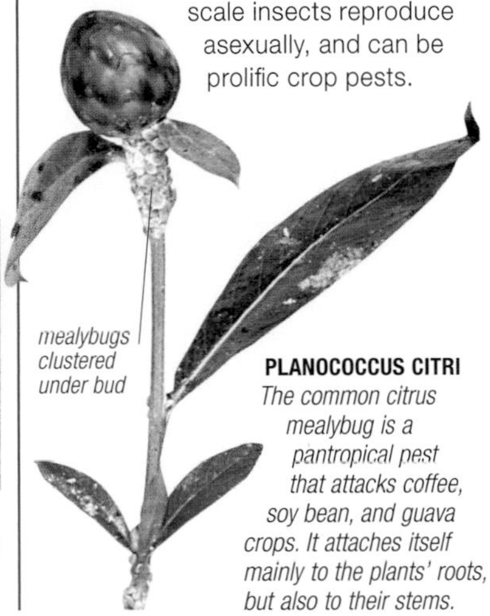

PLANOCOCCUS CITRI
The common citrus mealybug is a pantropical pest that attacks coffee, soy bean, and guava crops. It attaches itself mainly to the plants' roots, but also to their stems.

Family Fulgoridae

Fulgorids

Occurrence 800 spp. in tropical and subtropical regions; mostly on trees and woody shrubs in well-vegetated areas

Many tropical fulgorids are notable for their bizarrely shaped, bulbous heads, and are known as lantern bugs. Their eyes are located at the sides of the head, above the antennae. The thorax is often quite large, as are the wings, which have cross veins. These sap-suckers are $^{5}/_{16}$–4 in (0.8–10 cm) long and have eyespots on the hindwings that can be flashed to deter predators. Adults rest and feed during the day and fly at night. Eggs are laid on host plants and covered by a frothy secretion, which hardens to protect them.

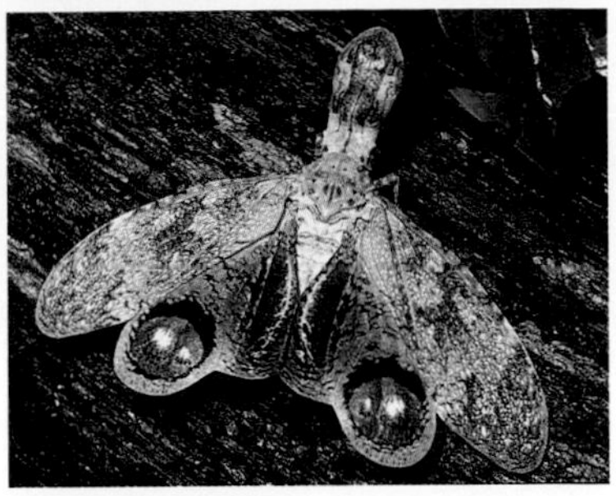

FULGORA SERVILLEI
Found in the forests of South America, this species has a very enlarged head with tooth-like markings. Like most species in the family, it is a sap-sucker.

FULGORA LATERNARIA
Also called the lanternfly or peanut-headed bug, this species has a bulbous head that was once thought to glow. It occurs in South America and the West Indies.

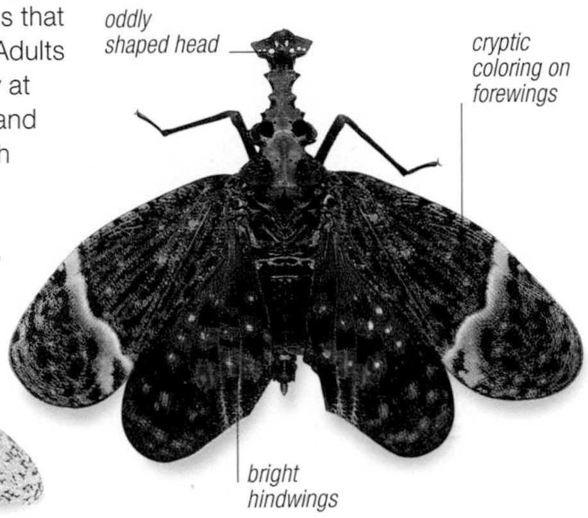

PHRICTUS QUINQUEPARTITUS
The wart-headed bug has colorful wings and an unusual head. It is found in woodland and forests of Panama, Brazil, and Colombia.

Family Gerridae

Pond skaters

Occurrence 500 spp. worldwide; in water bodies ranging from pools, ponds, streams, rivers, and lakes, to warm oceans

Also known as water-striders, these fast-moving bugs are $^{1}/_{16}$–$1^{1}/_{2}$ in (0.2–3.5 cm) long, dark brown or black, and covered with short, velvety hairs. The elongated body has short, stout front legs used for grasping prey, and long middle and hindlegs that help to spread its weight over the water film. These predatory bugs detect prey using ripple-sensitive hairs. Mating follows ripple courtship signals, and eggs are laid on floating objects or embedded in water plants.

GERRIS SPP.
These bugs are widespread and live on water surfaces. Ripples made by prey attract the bugs, which move quickly to attack.

Family Hydrometridae

Water measurers

Occurrence 120 spp. worldwide; in standing water bodies, on floating plants

Like pond skaters, water measurers, also called marsh-treaders, live and feed on the surface film, but they move in a slow and stealthy way, creeping up on their prey. They are $^{1}/_{8}$–$^{3}/_{4}$ in (0.3–2.2 cm) long, with an elongated head and bulging eyes. Most species have either short wings or no wings at all. Water measurers are predators and scavengers in all life stages, using a spearing and sucking technique to feed; they prefer dead or dying prey.

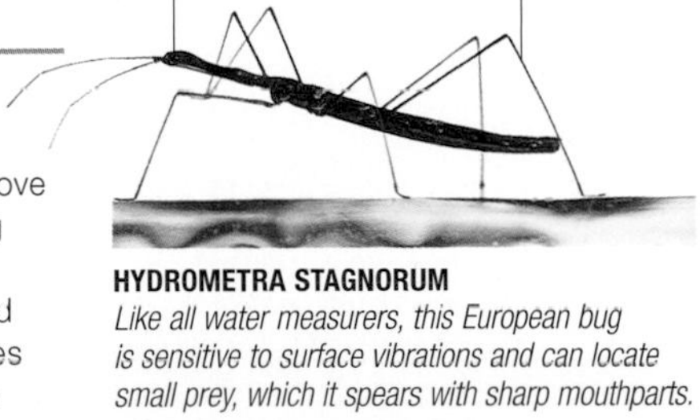

HYDROMETRA STAGNORUM
Like all water measurers, this European bug is sensitive to surface vibrations and can locate small prey, which it spears with sharp mouthparts.

Family Membracidae

Treehoppers

Occurrence 2,500 spp. worldwide, mainly in warmer regions; especially in understory layers in tropical forests

The large, distinctive pronotum, often shaped like a thorn, gives this family the alternative name of thorn bugs. These plant sap-suckers are mostly green, brown, or brightly colored, measuring $^{3}/_{16}$–$^{1}/_{2}$ in (0.5–1.5 cm) in length. Female treehoppers lay egg masses in slits made in the host plant's bark and, depending on the species, a single female may lay up to 300 eggs. The female frequently guards them until they hatch. Ants protect many species in return for honeydew.

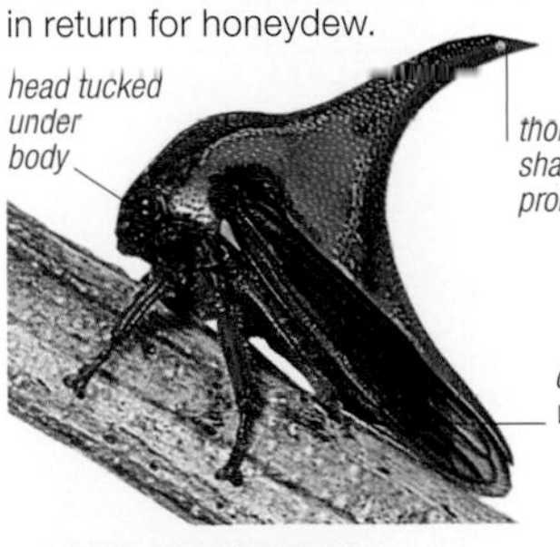

UMBONIA CRASSICORNIS
Members of this genus are found in woodland and forests in North and South America and Southeast Asia. The pronotum is tough and can puncture skin or penetrate shoes.

Family Miridae

Plant bugs

Occurrence 8,000 spp. worldwide; on all aerial parts of a wide range of plants, including crops

Also known as capsid bugs, these fragile, elongated or oval insects form the largest family of true bugs. They show a great variety of colors and markings—some resemble ants—and are generally $^{1}/_{16}$–$^{1}/_{2}$ in (2–16 mm) long. Most are sap-suckers, but some scavenge or hunt live prey. Plant bugs protect themselves by having cryptic coloring, or by giving off a noxious scent when threatened. Eggs are inserted inside plant tissues; a female may lay a total of 30–100 eggs during her adult life. Many species are fruit and crop pests.

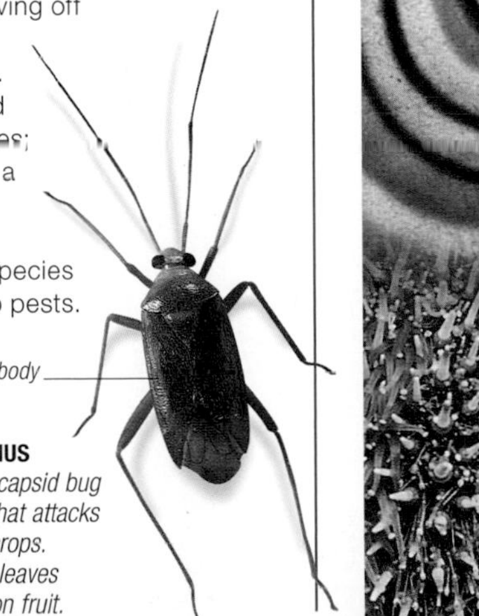

LYGOCORIS PABULINUS
The common green capsid bug is a European pest that attacks fruit and vegetable crops. Feeding by this bug leaves raised, warty spots on fruit.

Order Hemiptera *continued*

Family Notonectidae

Back-swimmers

Occurrence 350 spp. worldwide; in still water, mostly small pools, ponds, edges of lakes

These spindle-shaped bugs, 1/16–1/2 in (2–17 mm) long, are good fliers, but spend much of their lives underwater, swimming upside down. They use their forelegs to catch prey and their hindlegs for swimming.

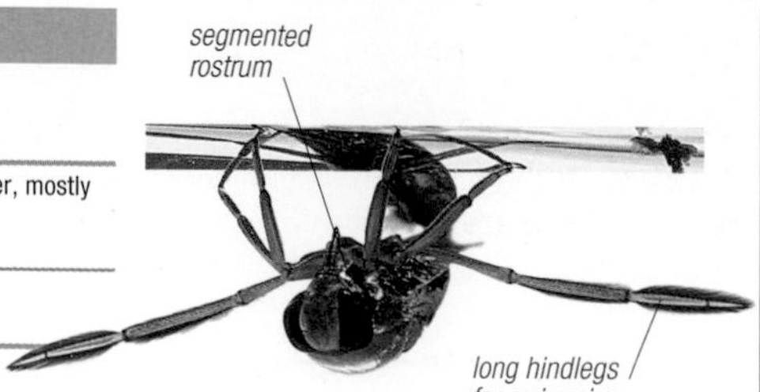

NOTONECTA GLAUCA
Like other back-swimmers, this European species stores air under its wings. It detects its prey by sight, and by sensing surface ripples.

Family Pentatomidae

Stink bugs

Occurrence 5,500 spp. worldwide; on herbaceous plants, shrubs, trees

Also known as shield bugs because of their shieldlike shape, these bugs are 3/16–1 in (0.5–2.5 cm) long. The name stink bugs comes from their ability to give off strong defensive odors from their thoracic glands. Many species are green or brown but some may be brightly colored. The pronotum is broad and may have sharp corners. Females lay clusters of barrel-shaped eggs, up to 400 at a time, on plants. There are 5 nymphal stages. Nymphs are mostly herbivorous to begin with but some turn predatory later, or become mixed feeders. Several species are serious pests of a wide-ranging variety of crops, such as legumes, potato, cotton, and rice.

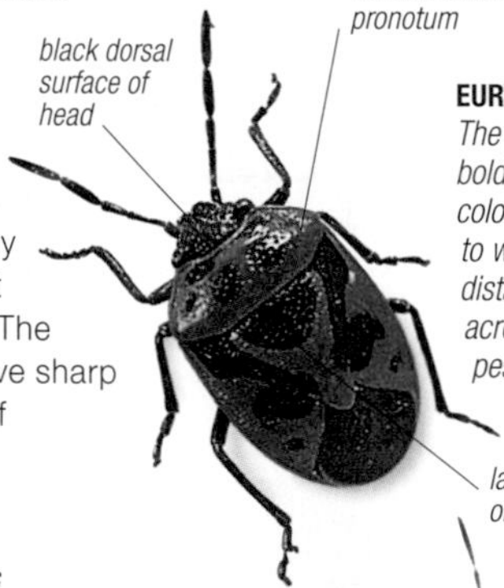

EURYDEMA DOMINULUS
The brassica bug has bold red or orange coloring that serves to warn predators of its distastefulness. Found across Europe, it is a pest of brassica crops.

NEZARA VIRIDULA
The green vegetable bug is abundant in tropical, subtropical, and warm temperate regions. It is a pest of more than 100 plants, mostly vegetables and legumes.

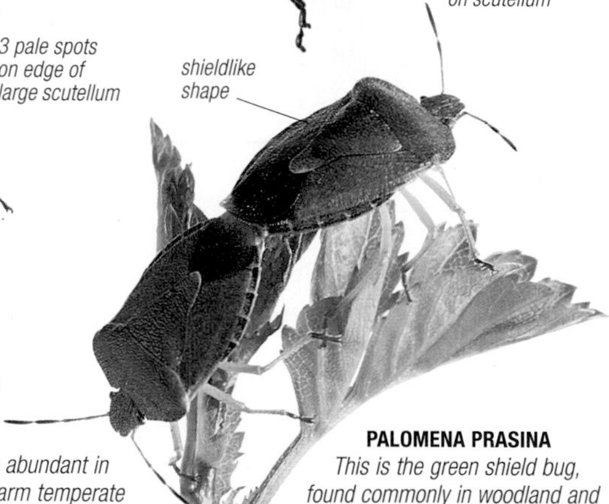

PALOMENA PRASINA
This is the green shield bug, found commonly in woodland and on shrubs across Europe. It is a minor pest of beans and alfalfa crops.

Family Psyllidae

Jumping plant lice

Occurrence 1,500 spp. worldwide; on stems, leaves, and bark of herbaceous and woody shrubs, trees, crops

Looking like small leafhoppers (see p.558) with long antennae, jumping plant lice are 3/64–3/16 in (1.5–5 mm) long, and have 2 pairs of oval wings, held rooflike over the body. They feed on plants, and some species are pests. The females lay stalked eggs on, or in, food plants.

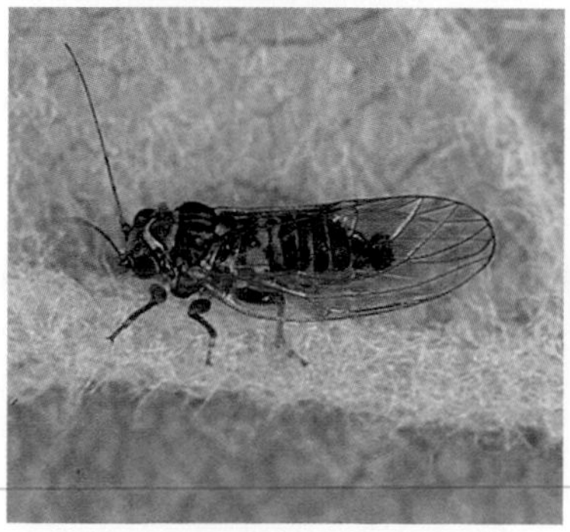

CACOPSYLLA PYRICOLA
Abundant in the Northern Hemisphere, the pear psylla is a major pest of pear trees, from the leaves of which adults and nymphs suck sap.

Family Reduviidae

Assassin bugs

Occurrence 6,000 spp. worldwide, especially tropical and subtropical regions; on ground, in leaf litter, on plants

Assassin bugs get their name from their predatory habits, although some live by sucking blood. Most are 7/32–1 1/2 in (0.7–4 cm) long, with dark, slender bodies. Many attack insect pests; others suck blood of humans—transmitting infection such as Chagas disease.

PLATYMERIS BIGUTTATA
This is a large, highly predatory Afrotropical species, 1 1/4–1 1/2 in (3–3.6 cm) long. Its toxic saliva may cause temporary blindness.

Family Tingidae

Lace bugs

Occurrence 2,000 spp. worldwide; on foliage of herbaceous plants, trees

Lace bugs are small, grayish insects, recognizable by the complex, lacelike patterns on their wings and upper body. They are 1/16–3/16 in (2–5 mm) long. The pronotum can extend, hoodlike, over the head. Females of some species take care of the eggs and nymphs. Many are pests on plants.

TINGIS CARDUI
The spear-thistle lace bug is common in the UK and eats certain types of thistles. A covering of light, powdery wax gives it a pale gray look.

Order Thysanoptera

Thrips

This order consists of 7,400 species within 14 families. Thrips are small and slender, with 2 pairs of narrow, hair-fringed wings. They have large eyes, short antennae, and sucking mouthparts, with one small and one needlelike mandible. A sticky, inflatable structure between the claws aids grip. Metamorphosis is unusual, with 2 nymphal stages and one or more pupalike stages.

Family Thripidae

Common thrips

Occurrence 2,000 spp. worldwide; in leaf litter, on leaves, flowers, and fruit of a wide range of plants

In warm, humid weather, clouds of common thrips, or thunderbugs, form part of the aerial plankton—myriad tiny, winged insects that are blown far and wide by the breeze. Common thrips vary from pale yellow to brown or black, and have very narrow, feathery wings. Typical species are less than 1/16 in (2 mm) long. Some common thrips reproduce asexually, and in most, the female inserts eggs into plant tissues or flowers using a tiny, sawlike ovipositor. The nymphs develop either on plants or in the soil. Adults and nymphs generally live by sucking plant juices, although a few attack other insects. Some species are serious pests of tobacco, cotton, beans, and other crops.

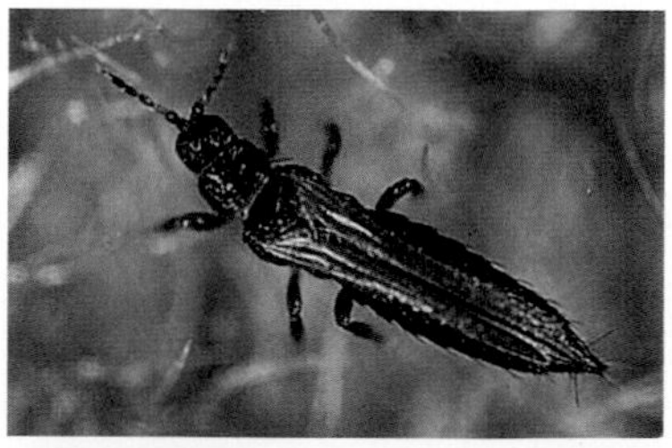

THRIPS FUSCIPENNIS
This dark-colored species, the rose thrips, with distinctive hair on its body, is abundant in the Northern Hemisphere, and is found on a wide range of plant species.

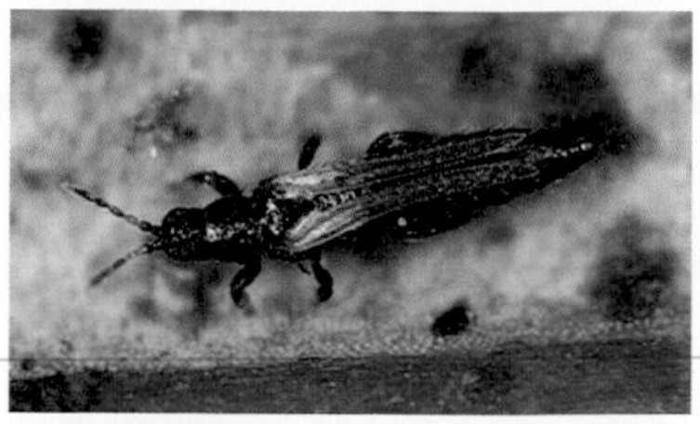

THRIPS SIMPLEX
The gladiolus thrips was originally from South Africa, but is now also widespread in Europe and North America, wherever gladiolus flowers grow.

Order Megaloptera

Dobsonflies and alderflies

These are the most primitive insects to exhibit metamorphosis. Three hundred species in 2 families are weak fliers and never far from water. Larvae are aggressive, aquatic predators. They pupate in soil just above water level. Once metamophosed, the adults never feed.

Family Corydalidae

Dobsonflies

Occurrence 200 spp. worldwide, especially in temperate regions; in running water

Measuring 1–3in (5–7.5cm) in length, with a wingspan of up to 6in (15cm), dobsonflies are large, soft-bodied, gray or brown insects, with 3 ocelli. Males may have large mandibles, used not for feeding but for combat or for holding the female. The latter lays masses of hundreds of thousands of eggs near water.

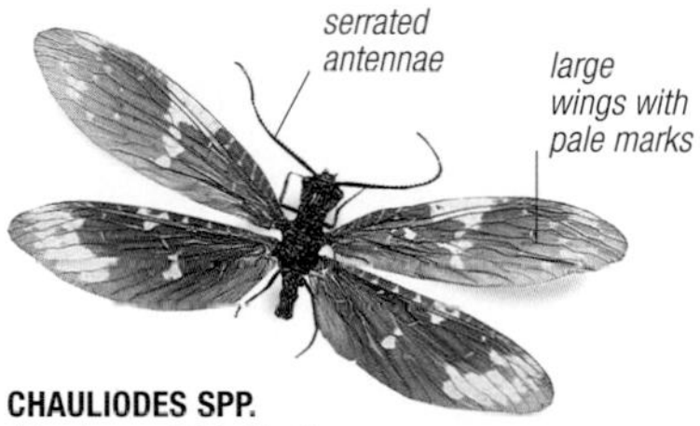

CHAULIODES SPP.
Members of this North American genus called fish flies have distinctly rounded corners on their heads.

Family Sialidae

Alderflies

Occurrence 100 spp. worldwide, especially in temperate regions; in muddy ponds, canals, slow-moving water

Alderflies are dark and smoky-winged; most species are less than ½in (1.5cm) long. They have long, threadlike antennae. Unlike dobsonflies they have no ocelli. Females lay egg masses near water, from which hatched larvae drop. These have fully formed legs for crawling, and often mature in a year.

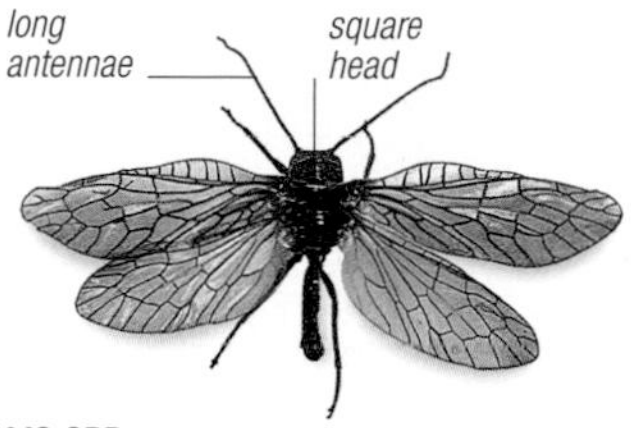

SIALIS SPP.
Common in the Northen Hemisphere, insects of this genus have typically smoky wings with dark veins. Fully grown larvae crawl out of water and pupate in damp soil above the water margin.

Order Raphidioptera

Snakeflies

Snakeflies get their name from the snake-like way in which they catch their prey—raising their head and seizing the prey with a sudden lunge. There are 2 families and about 200 species. Metamorphosis is complete. Larvae hide in bark and litter, and are equally voracious.

Family Raphidiidae

Snakeflies

Occurrence 185 spp. primarily in the Northern Hemisphere; among vegetation

Snakeflies are dark, with a distinct neck formed by a long pronotum. Body length varies between ¼–1¼in (0.6 and 2.8cm). The head is broad in the front and tapered at the rear. Females, slightly larger than males, use their long ovipositor to lay eggs in slits in bark. Larvae are elongate and found under loose bark or leaf litter. Like the adults, they feed on beetle larvae and soft-bodied insects.

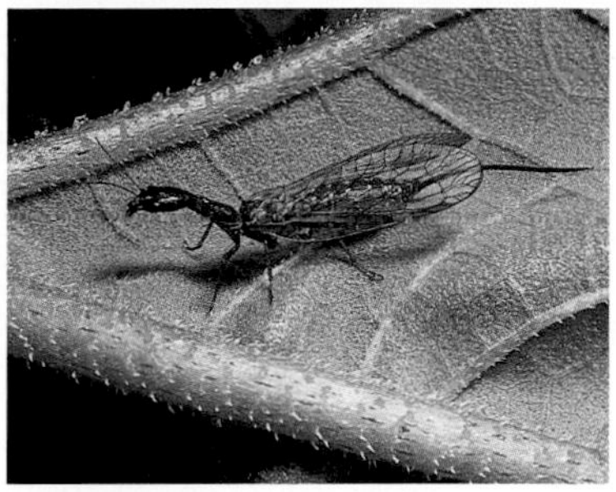

AGULLA SPP.
These snakeflies are found in the woodland of the USA and Canada, from the Rocky Mountains to the Pacific coast.

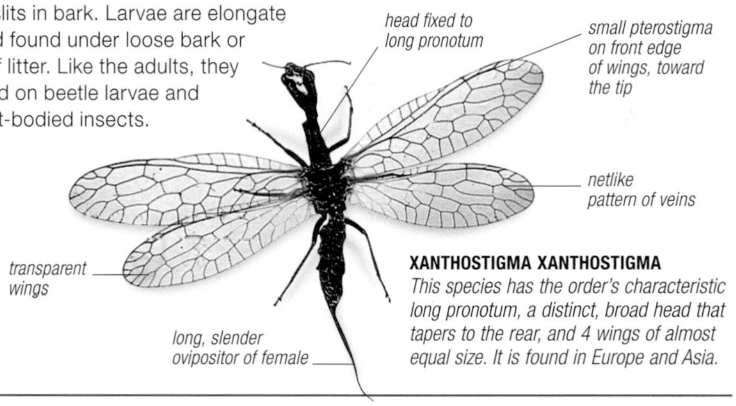

XANTHOSTIGMA XANTHOSTIGMA
This species has the order's characteristic long pronotum, a distinct, broad head that tapers to the rear, and 4 wings of almost equal size. It is found in Europe and Asia.

Order Neuroptera

Antlions, lacewings, and others

This order comprises 14 families and about 11,000 species, characterized by large eyes, chewing mouthparts, long antennae, and 2 pairs of filmy, net-veined wings held over the body when at rest. Metamorphosis is complete. The adults are generally predatory and nocturnal, while the larvae are usually predatory or parasitic.

Family Chrysopidae

Common lacewings

Occurrence 3,059 spp. worldwide; in a range of microhabitats including vegetation, ant nests

Common lacewings are green or brown, with iridescent wings that span ⅜–2in (1–5cm) and carry a complex pattern of veins. The wings also have ultrasound sensors to detect predators such as bats. Adults are nocturnal and prey upon aphids, thrips, and mites.

NOTHOCHRYSA CAPITATA
Like other lacewings, this European species lays eggs on long stalks. The stalks help protect the eggs from attack by ants and other predators.

Family Mantispidae

Mantispids

Occurrence 714 spp. worldwide, mainly in warm temperate to tropical regions; on plants in well-vegetated areas

The front legs of mantispids resemble those of praying mantids (see p.554), and are used in the same way. Their wingspan is ⅜–2¼in (1–5.5cm). Eggs are laid on tree bark, and the larvae of most species feed on spider eggs.

CLIMACIELLA SPP.
Members of this Central and South American genus have bright, wasplike, protective coloration.

Family Myrmeleontidae

Antlions

Occurrence 3,353 spp. worldwide, in semiarid areas in subtropical and tropical regions; in woodland, scrub

Antlions are soft, large, and slender insects that resemble damselflies (see p.551). The long, narrow wings, which span 1½–4¾in (3.5–12cm), may have brown or black patterns and produce a weak, fluttering flight. The head is broader than the pronotum, with large, conspicuous eyes. The club-headed antennae are about as long as the head and thorax together. The abdomen in males has terminal sexual clasping organs that look like earwigs' forceps. Eggs are laid in soil or sand, singly or in small groups. Adults and larvae catch and eat insects and spiders; some species construct conical pits to trap prey. The larvae, sometimes called doodlebugs, live in these pits with only their spiny mandibles showing, and flick sand grains at prey to knock them into the pit. Other larvae live on tree trunks, in soil and debris, or under stones.

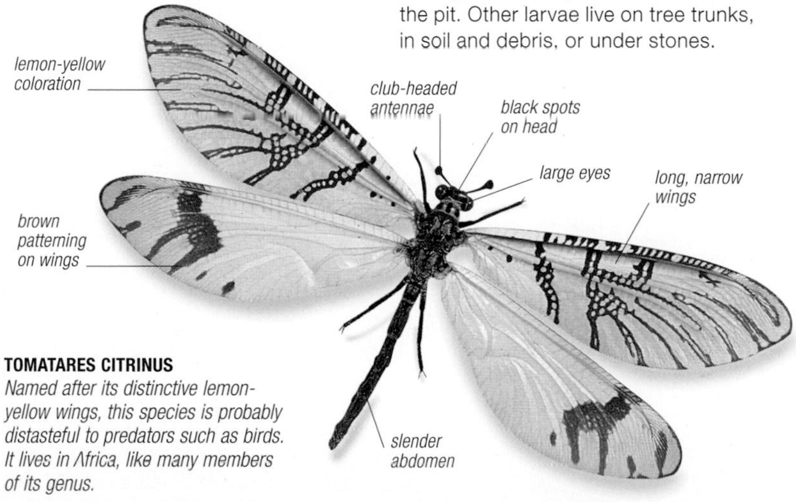

TOMATARES CITRINUS
Named after its distinctive lemon-yellow wings, this species is probably distasteful to predators such as birds. It lives in Africa, like many members of its genus.

Order Coleoptera

Beetles

About one in 3 insects in existence today is a beetle. Collectively they make up an order containing 370,000 known species in 166 families, and have successfully colonized every sort of habitat on land and in fresh water. They range in size from insects that are just visible to the naked eye to tropical giants 7 in (18 cm) long. Beetles are distinguished by their toughened front wings, or elytra, which fold over their membranous hindwings like a case. Because the hindwings are protected, beetles can squeeze into confined spaces. Most species are herbivorous, but the order also includes many predators and scavengers, and some parasites. Beetles develop by complete metamorphosis: there are distinct larva and pupal stages.

Family Anobiidae

Woodworm

Occurrence 1,500 spp. worldwide, especially in temperate regions; in wood, warehouses, stores, houses

These small, elongated to oval beetles are pale or reddish brown to black, and are mostly 1/16–1/4 in (2–6 mm) long. The head is often hooded by the pronotum. Eggs are laid in suitable food, into which the larvae bore tunnels. Several species are pests of cereals, legumes, spices, and tobacco.

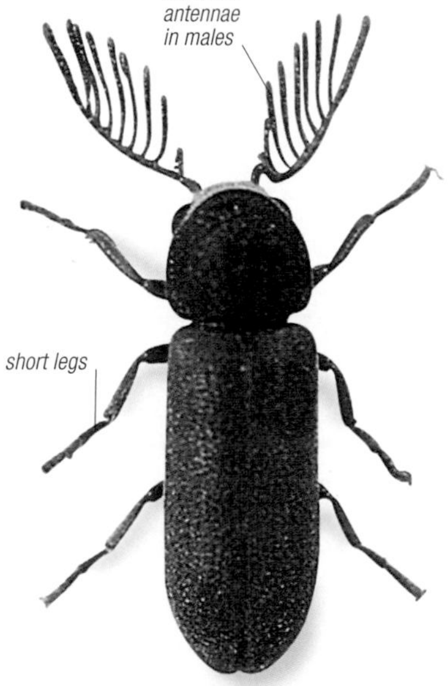

PTILINUS PECTINICORNIS
Found in the UK and central Europe, this species is especially associated with beech trees and may attack wood furniture.

Family Carabidae

Ground beetles

Occurrence 29,000 spp. worldwide; on ground, under stones, logs, in debris, leaf litter, tree foliage

Measuring 1/16–3¼ in (0.2–8 cm) long, these elongated and flattened beetles can be dull or shiny, brown or black, often with a metallic sheen. The head, thorax, and abdomen are clearly differentiated, and the legs are adapted for quick escape from predators. A few species can deter predators using blasts of hot, caustic chemical substances expelled with an audible "pop" from the end of their abdomen. Most species are nocturnal hunters; adults and larvae are mostly predatory. Eggs are laid on the ground, on vegetation, and on decaying wood and fungi.

CICINDELLA CAMPESTRIS
Also known as the green tiger beetle, this fast-running, diurnal species can be found from April to September on sandy soils in Europe.

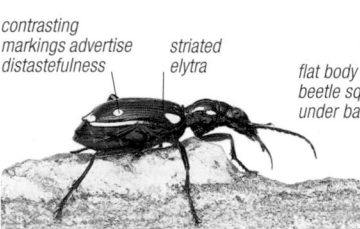

THERMOPHILUM SEXMACULATUM
Commonly called the domino beetle, this Afrotropical species hunts during the day in semiarid areas.

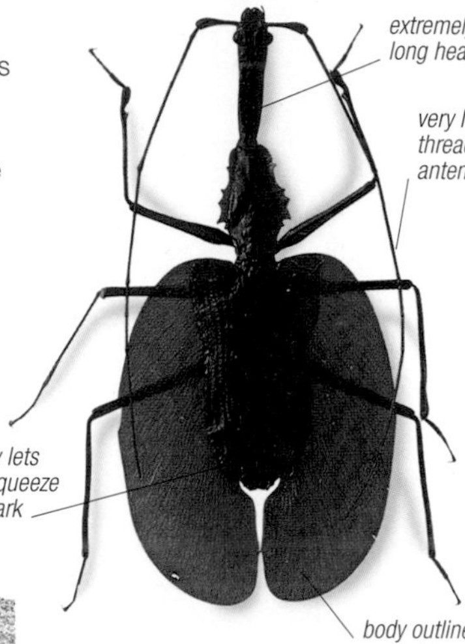

MORMOLYCE PHYLLODES
The violin beetle is found in the tropical forests of Southeast Asia, and feeds on insect larvae and snails.

Family Buprestidae

Jewel beetles

Occurrence 15,000 spp. worldwide, primarily in tropical regions; in sunny glades, on bark or flowers

Most jewel beetles are a brilliant, metallic green, red, or blue, with spots, stripes, and bands. They are 1/16–2½ in (0.2–6.5 cm) in length and taper toward the rear. They lay eggs in wood—some species have heat sensors that can detect freshly burned areas where they may mate and lay eggs—and the larvae are typically wood feeders. Many species are lumber pests. Adult beetles feed on nectar, flowers, and pollen. Their bright wing cases are used in headdresses or as ornaments.

CHRYSOCHROA BUQUETI
This beetle is found in tropical forests of Vietnam and Thailand and is highly prized by collectors. In places, this has led to a decline in local populations.

Family Cerambycidae

Longhorn beetles

Occurrence 30,000 spp. worldwide; on flowers, at sap flows

Also known as timber beetles, these insects are serious pests of wood. Their larvae, responsible for the damage, tunnel through living trees, and also through wood in buildings. Once adult, they feed on sap, pollen, nectar, or leaves. Adult beetles have strong, toothed mandibles, parallel-sided bodies, and are 1/8–6in (0.3–15 cm) long. They are often brightly colored, but their most striking feature is their antennae, which may be up to 4 times as long as the body. Females lay their eggs singly under bark. Like many wood-boring species, the life cycle can take several years to complete.

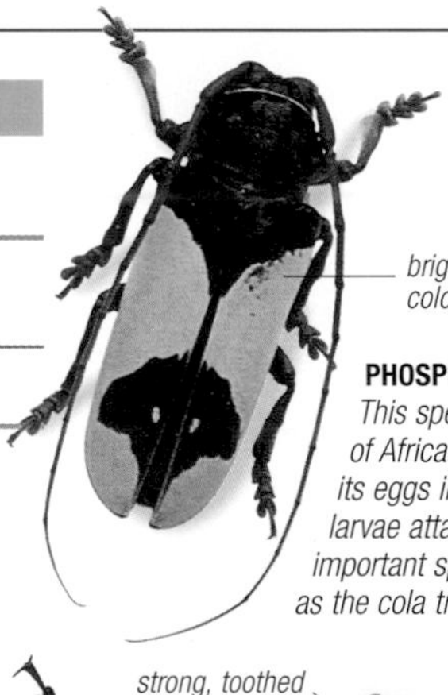

PHOSPHORUS JANSONI
This species is a native of African forests. It lays its eggs in trees, and its larvae attack economically important species such as the cola tree.

STERNOTOMIS BOHEMANNI
This Afrotropical species has the typical longhorn shape. The females use their mandibles to cut through bark, under which they lay eggs.

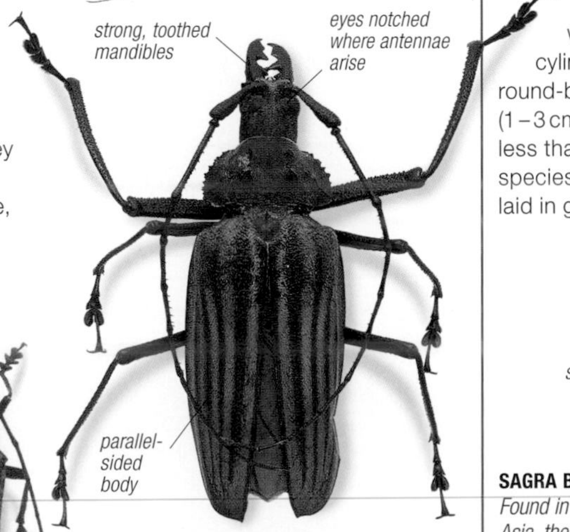

XIXUTHRUS HEROS
This large longhorn beetle is 3¼–4 in (8–10 cm) long and is endemic to Fiji. The males are territorial.

Family Chrysomelidae

Leaf beetles

Occurrence 35,000 spp. worldwide; widespread on most types of plants

Although some eat pollen, adults in this family are mostly leaf eaters—hence the name. This makes many species serious pests, although a few have been used in biological control of weeds. Leaf beetles range from cylindrical and elongated, to round-backed. They measure 3/8–1¼ in (1–3 cm) in length, with the antennae less than half the body length. Many species are brightly colored. Eggs are laid in groups on host plants or in soil.

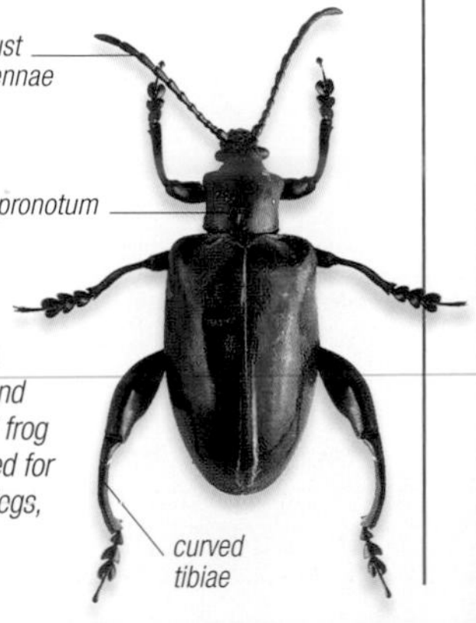

SAGRA BUQUETI
Found in Africa and Asia, the jeweled frog beetles are named for their strong hindlegs, used in male-to-male combat.

Family Coccinellidae

Ladybugs

Occurrence 5,000 spp. worldwide; almost anywhere on foliage where insect prey is present

Also called lady beetles or ladybirds, these short-legged beetles are brightly colored, often with spots or stripes. They are 1/32–1/2 in (1–15 mm) long, rounded and convex, with the pronotum often hiding the head. They are either shiny and smooth, or hairy. Antennae are segmented, with a short, terminal club. Eggs are glued singly or in small groups to plants. Although some species are herbivorous, most adults and larvae are predators of soft-bodied insects. This makes them useful as controlling agents of pests.

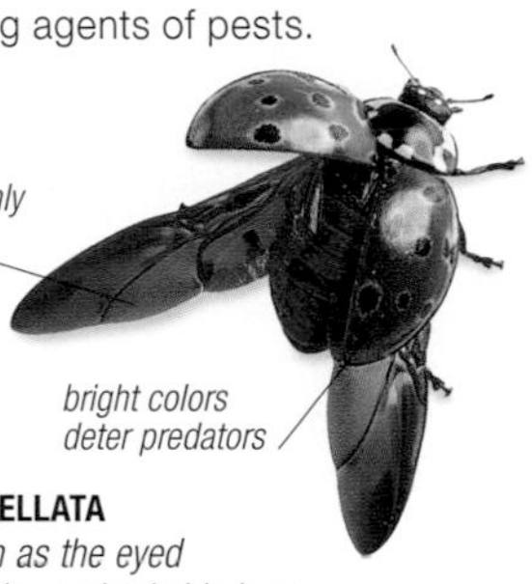

ANATIS OCELLATA
Also known as the eyed ladybird, this unmistakable large European species feeds on jumping plant lice and other small prey.

Family Curculionidae

Weevils

Occurrence 48,000 spp. worldwide; associated with almost every terrestrial and some aquatic plant species

Habitat Terrestrial

Weevils, also known as snout beetles, form one of the largest families in the animal kingdom. Most have a rostrum, which is an extension of the head and carries the mouthparts. The body is 1/32–3½ in (0.1–9 cm) long, and may be cryptically or brightly colored, with patterns and metallic scales. The antennae are usually 11-segmented, and are elbowed or bent. Eggs are laid inside plant tissue or in soil. Mostly diurnal, adults chew and eat soft plants, while larvae mostly feed inside plant tissue. Many weevils are crop pests.

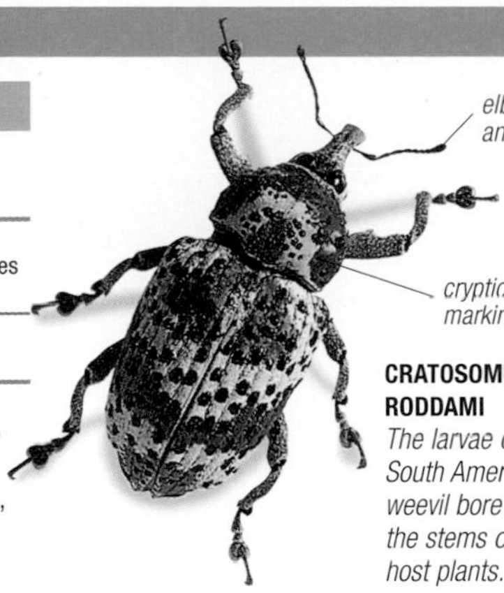

CRATOSOMUS RODDAMI
The larvae of this South American weevil bore into the stems of their host plants.

EUPHOLUS BENNETTI
This brightly patterned insect is found in Southeast Asia and Papua New Guinea.

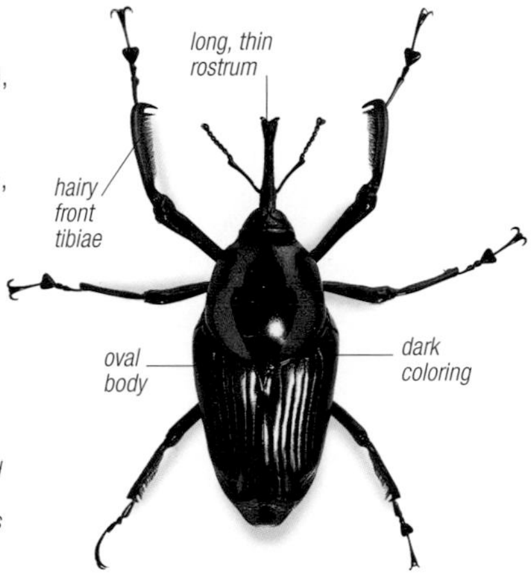

CYRTOTRACHELUS SPP.
The males of this Southeast Asian genus have long, hairy front tibiae, used in courtship displays.

Family Dermestidae

Larder beetles

Occurrence 950 spp. worldwide; mostly associated with dried animal remains

Many species of larder beetles are serious pests of food or animal matter stored in buildings; larvae scavenge on dried meat or fish, and on plant material, woollens, silks, or furs. These beetles are either rounded or slightly elongated, and measure 1/16–3/8 in (2–12 mm) in length. Dull brown or black, they are usually covered with patterns of colored hairs or scales. Females lay up to 150 eggs, in about 2 weeks, on foodstuffs. The hairy larvae feed at a great rate and, in hot countries, mature in a few weeks. The adults feed on pollen and nectar. These beetles are also called skin or museum beetles.

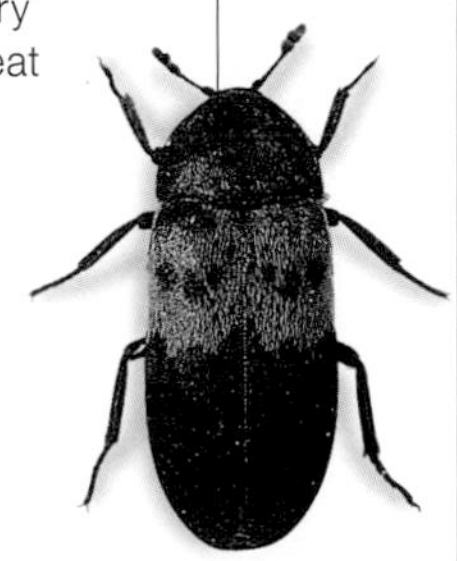

DERMESTES LARDARIUS
Found worldwide, this species inhabits bird nests, warehouses, dry corpses, and animal matter. It feeds on dried meat and cheese in houses.

Family Dytiscidae

Predacious diving beetles

Occurrence 3,500 spp. worldwide; in streams, shallow lakes and ponds, brackish pools, thermal springs

These beetles are 1/16–2 in (0.2–5 cm) long, oval, and mostly black or dark brown. They are fierce predators that attack everything from other insects to mollusks or vertebrates. Larvae are called water tigers. Well adapted to aquatic life, the adults can carry air supplies beneath their wing cases.

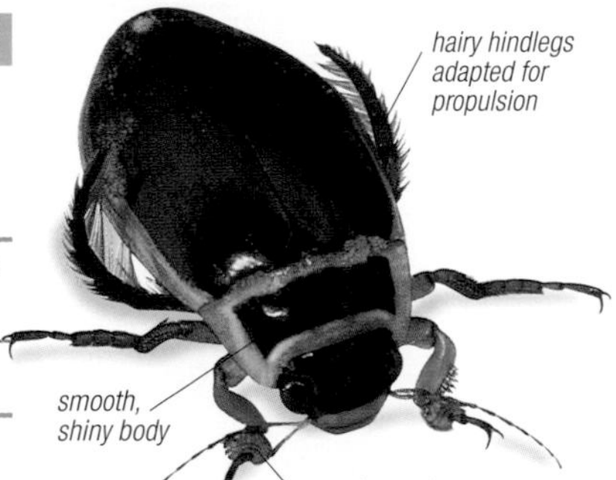

DYTISCUS MARGINALIS
The great diving beetle inhabits ponds and shallow lakes in Europe and Asia. The male has suction pads on its front legs for holding onto the female during mating.

Family Elateridae

Click beetles

Occurrence 9,000 spp. worldwide; around plants, in leaf litter, rotten wood, soil

Click beetles, also called skip jacks, are able to propel themselves into the air, making a loud "click" that frightens predators. Their rapid jump is made possible by powerful thoracic muscles and a unique "peg-and-joint" catch mechanism on the underside of the thorax. They are dull in color, elongated to oval, and parallel-sided, tapering at the rear. Body length ranges from 1/16–2¾ in (0.2 to 7 cm). Eggs are laid in soil and plant matter. The larvae, called wireworms, are generally predatory, although some feed on plant roots and tubers.

CHALCOLEPIDIUS LIMBATUS
Found in South America, this species eats plants as well as insects. Larvae develop in soil, litter, or rotting wood.

Family Histeridae

Hister beetles

Occurrence 3,000 spp. worldwide; in dung, carrion, under bark, in tunnels of wood-boring insects, ant nests

These tough-bodied beetles have a rounded or oval outline; some species may be flattened. Most species are black or metallic, and have striations on the elytra and other body parts. They are mostly less than ½ in (1.5 cm) in length. Eggs are laid in foodstuffs. Adults and larvae are predatory and feed on fly and beetle larvae, and mites.

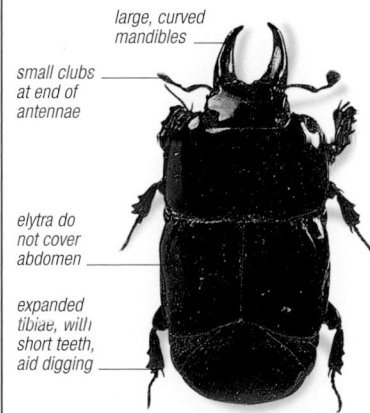

HOLOLEPTA SPP.
Commonly called flat hister beetles, the species in this genus have a very flat and hard body surface. They live under the bark of fallen trees.

Family Lampyridae

Fireflies

Occurrence 2,000 spp. worldwide; on vegetation in woodland and moist grassland, in soil, under stones

Also known as lightning-bugs, many fireflies use species-specific flashes of cold green light made by luminous organs on their abdomen in order to attract mates. The females of some species even imitate the flashing of closely related species to lure males and eat them. Fireflies are 3/16–1¼ in (0.5–3 cm) long. Males are usually fully winged, whereas some females are wingless and may resemble larvae. The larvae eat a small range of snails and other invertebrates. Most adults feed on nectar or dew, although some are predatory.

LAMPROCERA SELAS
This firefly inhabits moist grassland and semiwoodland in tropical South America.

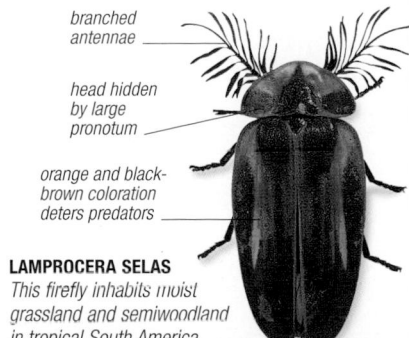

INVERTEBRATES

Order Coleoptera *continued*

Family Lucanidae

Stag beetles

Occurrence 1,300 spp. worldwide, in deciduous woodland and forests; in and on trees

Stag beetles are smooth, black or reddish brown, and mostly ¼–3¼ in (0.6–8.5 cm) long. Males are larger than females and have enlarged mandibles used in male-to-male combat over females. Eggs are laid on decaying tree stumps or roots, and larvae pupate inside a cell of chewed wood fibers. Adults are nocturnal and either do not feed, or take fluids such as nectar.

PHALACROGNATHUS MULLERI
The king stag beetle has a distinctively colored, metallic body. It is found in the rain forests of northern Australia.

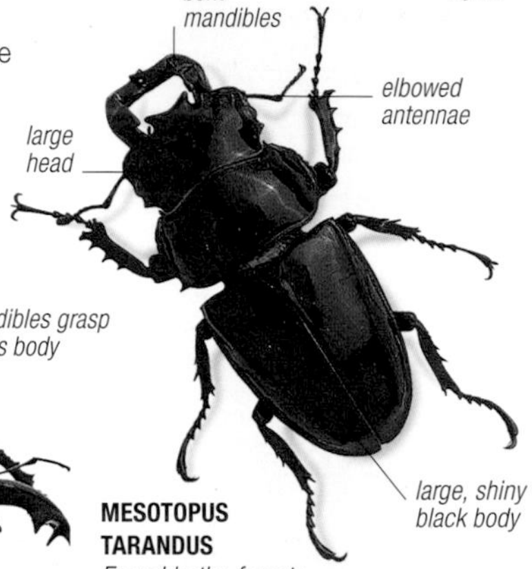

MESOTOPUS TARANDUS
Found in the forests of Africa, this species has large, toothed mandibles, bent almost at right angles.

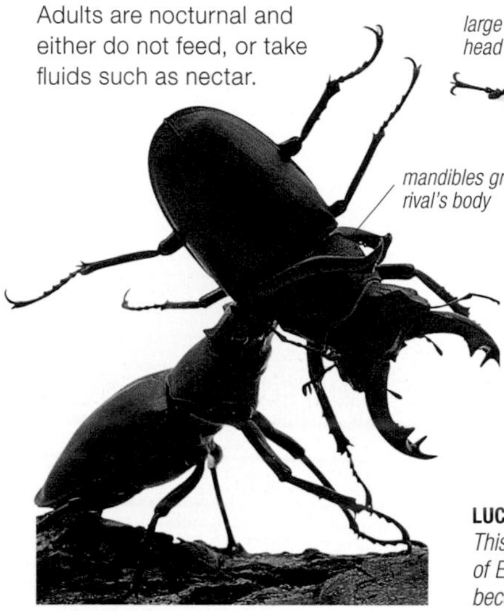

LUCANUS CERVUS
This stag beetle is found in the oak woodland of Europe. It flies after dark. The species is becoming more rare due to loss of habitat.

Family Passalidae

Bess beetles

Occurrence 500 spp. in tropical regions, mainly woodland of South America and Asia; in dead wood

Also called patent-leather or betsy beetles, these insects are shiny black or brown, with a flat body that is ⅜–3¼ in (1–8.5 cm) long. The elytra have noticeable striations, and the head often has a short horn. Adults make high-pitched sounds in aggression or during courtship. Eggs are laid in dead wood.

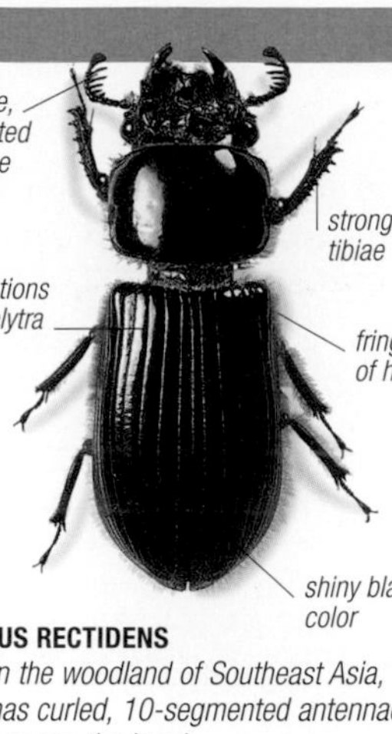

ACERAIUS RECTIDENS
Found in the woodland of Southeast Asia, this beetle has curled, 10-segmented antennae, and a short peg on the head.

Family Scarabaeidae

Scarab beetles

Occurrence 16,500 spp. worldwide; on dung, carrion, fungi, vegetation, under bark, in ant or termite nests

Members of this family are extremely varied in shape, size, and color; subfamilies include dung beetles, Hercules beetles, and leaf chafers. The body length ranges from $\frac{1}{16}$–6½ in (0.2–17 cm). The ends of the antennae have a distinct club made up of 3–7 flat, movable plates. Males of many species have an enlarged head or horns to fight over mates. Eggs are laid—and larvae can be found—in soil, rotten wood, or dung (which dung beetles roll into a ball and bury).

KHEPER AEGYPTIORUM
This is the large, colorful sacred scarab of Africa, probably the first to be revered in Ancient Egypt. It rolls dung into large balls, which it buries and in which it lays eggs.

CHRYSINA RESPLENDENS
Sometimes called precious metal scarabs due to their gold or silver coloring, this rare scarab and its relatives are found in South America.

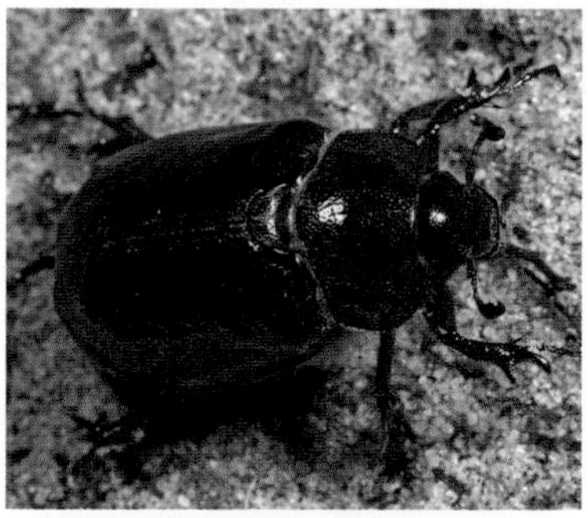

POLYPHYLLA STARKAE
This beetle belongs to a genus of scarab beetles that occur across the Northern Hemisphere. They have their greatest diversity in southern USA. Described in 2009, this species was discovered on the Florida peninsula.

Family Silphidae

Carrion beetles

Occurrence 250 spp. worldwide, mainly in Northern Hemisphere; on ground near carcasses, dung, and fungi

Carrion beetles are usually flat and soft-bodied, $\frac{5}{32}$–1¾ in (0.4–4.5 cm) long. They are black or brown, often with bright yellow or red markings. In carcass-burying species, such as *Nicrophorus* (burying beetles), the elytra are shortened, exposing a few abdominal segments. These beetles bury corpses of small animals and lay eggs on them. Adults often feed larvae on regurgitated carrion.

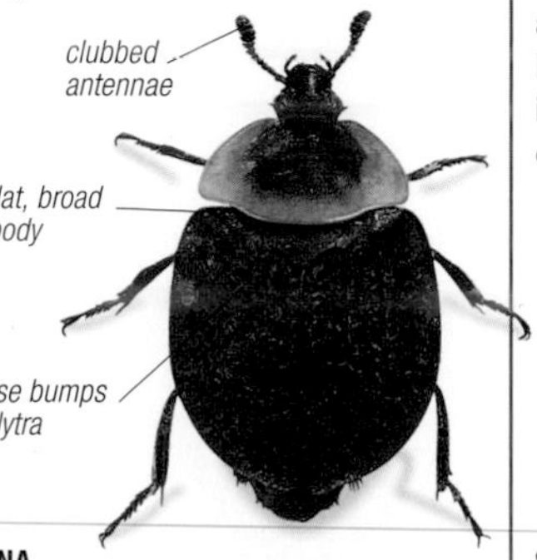

SILPHA AMERICANA
Like all members of its family, this beetle is attracted by the odor of carrion. It is found in North America and has a distinctive broad shape with fairly bright coloration.

Family Staphylinidae

Rove beetles

Occurrence 29,000 spp. worldwide; on ground, in soil, fungi, leaf litter, decaying plants, carrion, ant colonies

Most rove beetles are small—usually under ¾ in (2 cm) in length—smooth, and elongated. They are often brown or black, although some have bright colors and a sculptured body surface. They all have short elytra, with hindwings folded beneath them, and an exposed abdomen. The smaller species tend to be diurnal, while the larger ones are nocturnal. Eggs are laid in soil, fungi, or leaf litter.

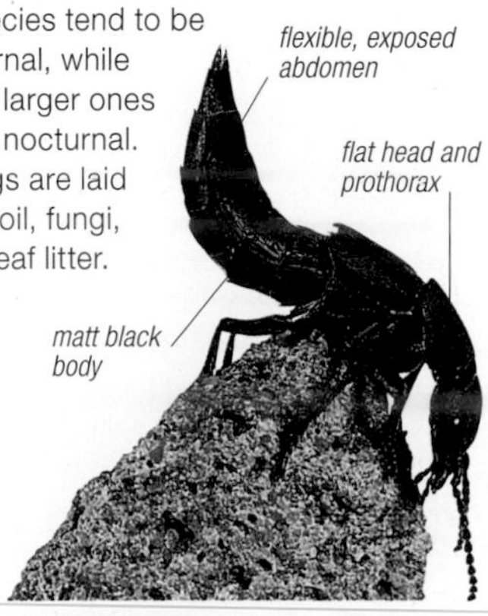

STAPHYLINUS OLENS
Known as the devil's coach horse, this is a large, black European species, ¾–1½ in (2–3.4 cm) long. If threatened, it assumes a scorpionlike posture by curving its abdomen upward.

Family Tenebrionidae

Darkling beetles

Occurrence 17,000 spp. worldwide; on ground in all terrestrial habitats, especially in desert and arid regions

Darkling beetles vary considerably in color (black or brown body, or white elytra), shape (parallel-sided or large and oval), and texture (smooth and shiny or dull and rough). They are $\frac{1}{16}$–2 in (0.2–5 cm) long. Many species have reduced hindwings and do not fly. These beetles eat decaying vegetable or animal matter, and some are pests of stored products. A few secrete a foul chemical repellant.

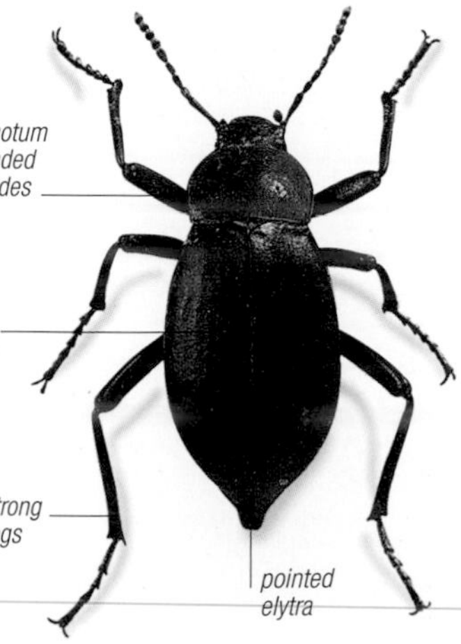

BLAPS MUCRONATA
The cellar, or churchyard, beetle is a nonflying, European species that favors dark places.

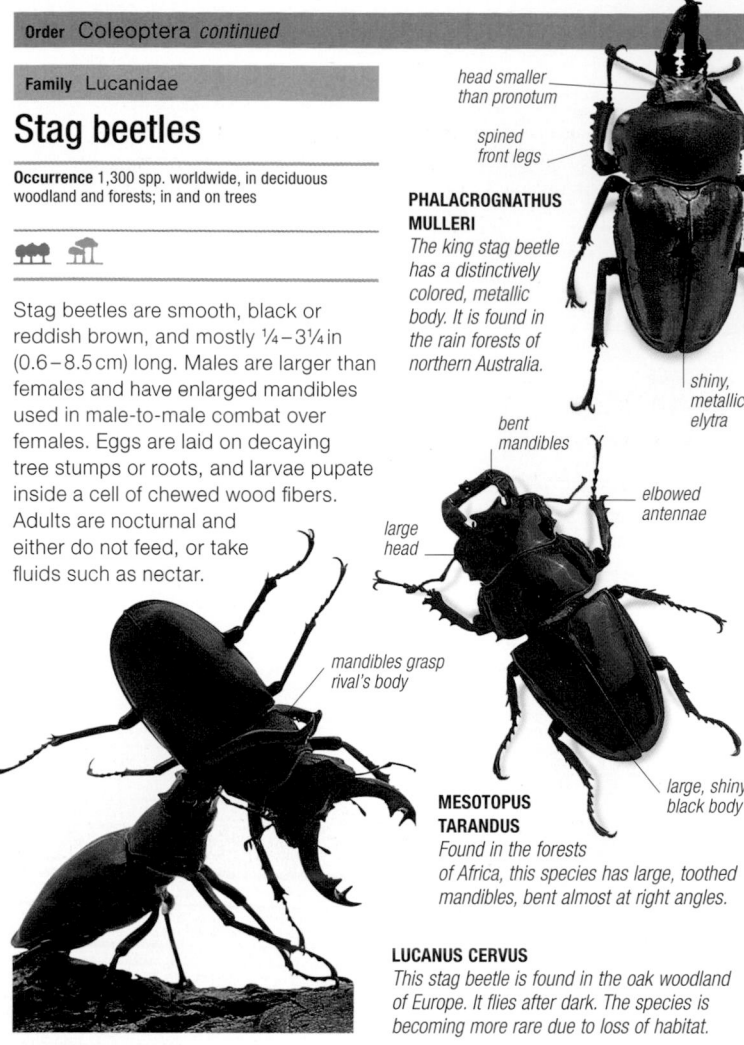

Order Strepsiptera

Strepsipterans

With their small size, parasitic lifestyle, and extreme sex differences, strepsipterans make up one of the most unusual orders of insects. Males usually have wings, the hindwings often being twisted, but females are grublike, and spend their lives lodged in the bodies of larger insects, such as bugs, bees, or wasps. There are about 580 species, in 8 families, and all undergo complete metamorphosis.

Family Stylopidae

Stylopids

Occurrence 268 spp. worldwide; in vegetated habitats where hosts are found, females in bees and wasps

Male stylopids are small and dark, with bulging eyes and bodies that are 1/64–5/32 in (0.5–4 mm) long. Like all strepsipterans, their forewings are tiny and their hindwings relatively large and fanlike. Females have neither legs nor wings, and never leave the body of the host insect. After mating, thousands of eggs hatch inside the female; some species may lay up to 70,000 eggs. The active, tiny, 6-legged larvae (triungulin) leave the female through a special passage and crawl onto flowers to wait for the next host. They may cling to the host's body or be ingested with nectar. In many cases, once inside the host's nest, they leave the body or are regurgitated and proceed to parasitize the eggs or larvae. The sexual organs of the parasitized host also degenerate.

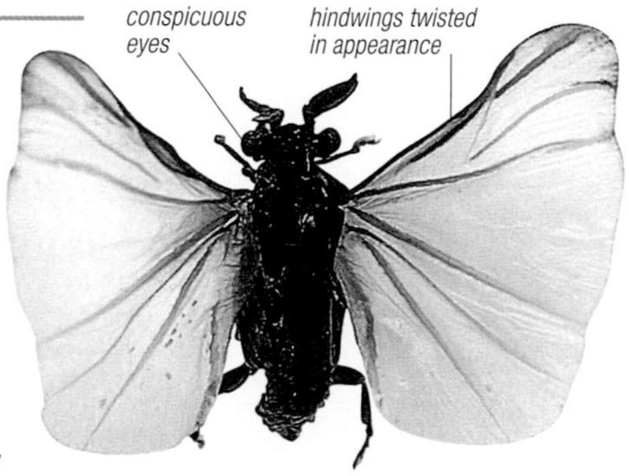

STYLOPS SPP.
Males of these species have a highly distinctive shape, with branched antennae and the order's characteristic berrylike eyes.

Order Mecoptera

Scorpionflies

The males of some species in this order have a scorpionlike, slender abdomen—hence the common name. Metamorphosis is complete. There are 9 families and 550 species of scorpionflies.

Family Bittacidae

Hangingflies

Occurrence 170 spp. in the Southern Hemisphere; in damp, wooded or well-vegetated, shady areas

Hangingflies use their front legs to hang from vegetation and hindlegs to catch prey. The adults are 1/2–1 1/2 in (1.5–4 cm) long. Eggs are laid in soil.

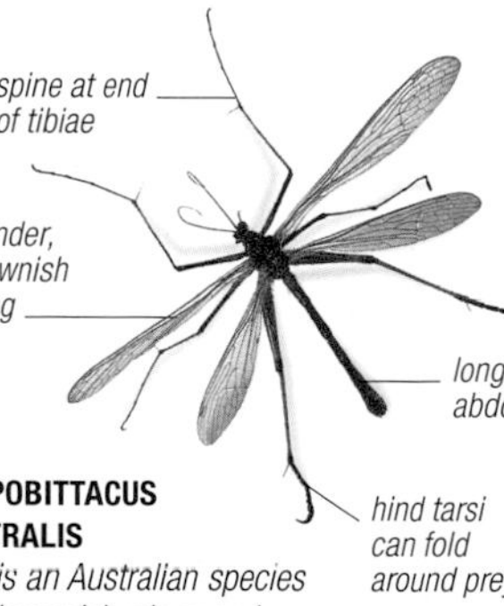

HARPOBITTACUS AUSTRALIS
This is an Australian species with brownish wings and orange-red banded legs.

Family Panorpidae

Common scorpionflies

Occurrence 360 spp. mostly in the Northern Hemisphere; among vegetation in woodland and scrub

These insects are 11/32–1 in (0.9–2.5 cm) long, often with mottled wings. Adults eat nectar, fruit, and insects. They lay eggs in soil, where hatched larvae pupate.

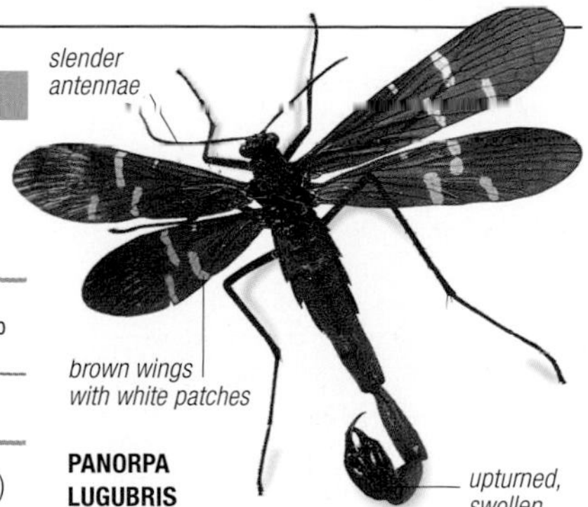

PANORPA LUGUBRIS
This species is common in parts of North America. Males are larger than the females.

Order Siphonaptera

Fleas

Fleas are brown, shiny, wingless insects, superbly adapted for a parasitic lifestyle feeding on mammalian blood. Their bodies are flattened sideways, which helps them to slip through fur, and they are very tough, which makes them hard to kill. They leap onto suitable host animals using their enlarged hindlegs, which store energy in special pads of rubberlike protein. There are about 2,400 species (18 families) of fleas; all undergo complete metamorphosis. Larvae scavenge on debris, dried blood, and adult flea excrement.

Family Pulicidae

Common fleas

Occurrence 200 spp. worldwide; on mammalian hosts, in lairs, burrows

Common fleas parasitize humans and a wide range of other mammals, including dogs, cats, and other carnivores, as well as hedgehogs, rabbits, hares, and rodents. Some of them have a broad host range, occurring on up to 30 different animal species. Common fleas are 1/32–5/16 in (1–8 mm) long and often have combs of backward-pointing bristles on their cheeks—features that are an important aid in species identification. Adult females lay eggs as they feed, dropping them into their hosts' nests, burrows, or bedding, where they hatch to produce pale, wormlike larvae that have tiny, biting jaws. The larvae typically live for 2–3 weeks, before pupating in a minute silk cocoon. Once they have become adult, the fleas remain inside their cocoons for weeks or months until they sense vibrations from a nearby animal. They then emerge from their cocoons and jump aboard their prospective host, attracted by its body heat. However, if they fail to find a host, or become dislodged, they can survive for a long time without food. Many species of common fleas, such as the cat flea, dog flea, and human flea, are of medical importance. Their bite causes severe itching and allergic reactions, as well as spreads diseases and parasitic worms. The bacterium that caused bubonic plague in medieval Europe was carried by various types of rat flea.

SPILOPSYLLUS CUNICULI
Commonly known as the rabbit flea, this species is widespread in the Northern Hemisphere. It carries the rabbit disease myxomatosis.

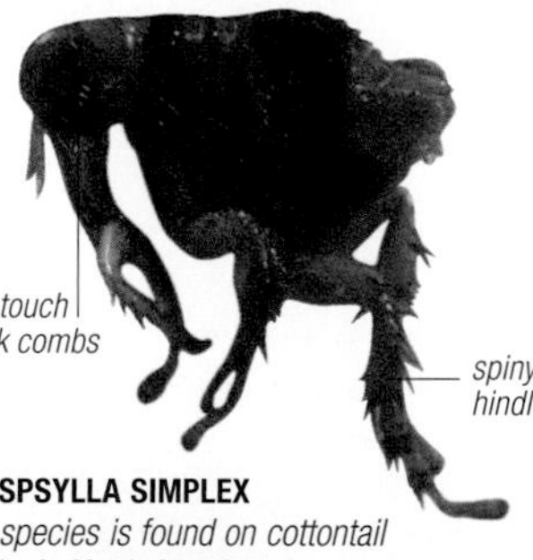

CEDISPSYLLA SIMPLEX
This species is found on cottontail rabbits in North America.

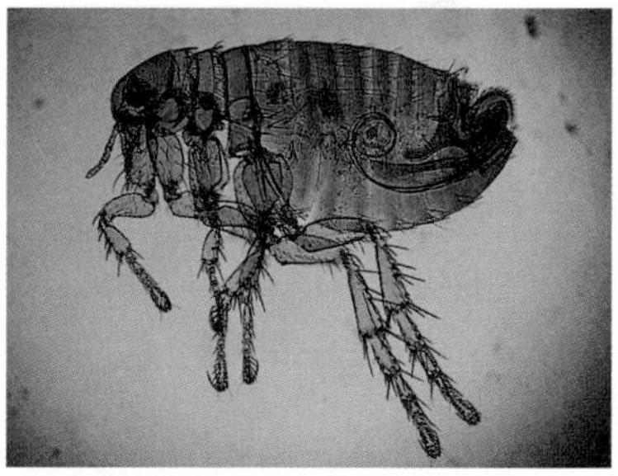

PULEX IRRITANS
This species was once the most common flea attacking humans; today, cat and dog fleas are more dominant. It also attacks pigs, goats, and badgers, and can be a carrier of bubonic plague.

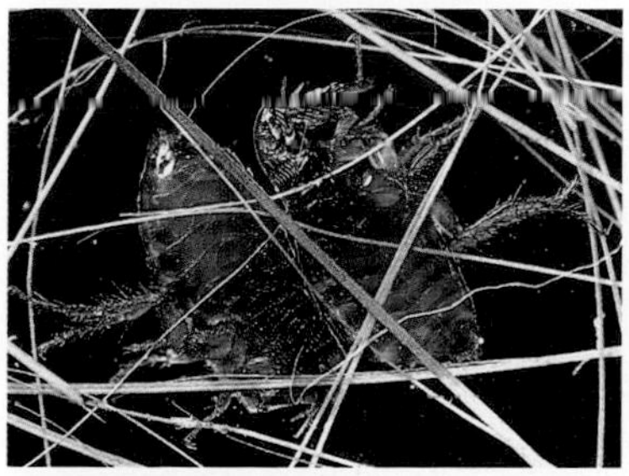

CTENOCEPHALIDES CANIS
Very common in human habitations the world over, the dog flea is reddish brown, with a head that is more domed than that of the cat flea. It is a carrier of the dog tapeworm, which can also affect cats and humans.

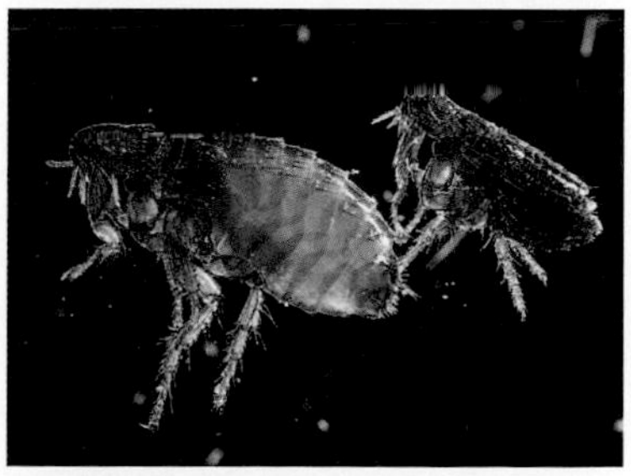

CTENOCEPHALIDES FELIS
The cat flea is a common domestic species found in temperate regions. A cat may have only a few adult fleas feeding on it but its bed supports thousands of larvae. Hungry cat fleas can jump up to 13 1/2 in (34 cm) and readily bite humans.

Order Diptera

Flies

Flies have only a single pair of wings (although some ectoparasitic species are wingless). The hindwings are reduced to small, club-shaped balancing organs called halteres. Flies are ecologically essential as pollinators, parasites, predators, and decomposers, but many damage crops, and the disease-carrying species have a huge impact on wild and domesticated animals, as well as on humans.

There are 130 families and 150,000 species of flies, divided into 2 suborders, Nematocera (26 families) and Brachycera (104 families). The former are delicate, with threadlike antennae and slender bodies, while the latter are more robust, with stout antennae. All have complete metamorphosis: larvae ("maggots") may have a reduced head. Larval feeding habits are among the most diverse in insects.

Family Agromyzidae

Leaf-mining flies

Occurrence 2,500 spp. worldwide; on leaves, stems, seeds, roots of a wide range of plants, including crops

Habitat Terrestrial

This family gets its name from the ability of larvae to chew mines (channels) in leaves. These crop pests also feed inside stems, seeds, or roots. Gray, black, or greenish yellow, they vary from 1/32 to 1/4 in (1–6 mm) in length. Females have a pointed ovipositor, and lay up to 50 eggs a day inside plant tissue. These flies may live for as little as 2 weeks in tropical areas.

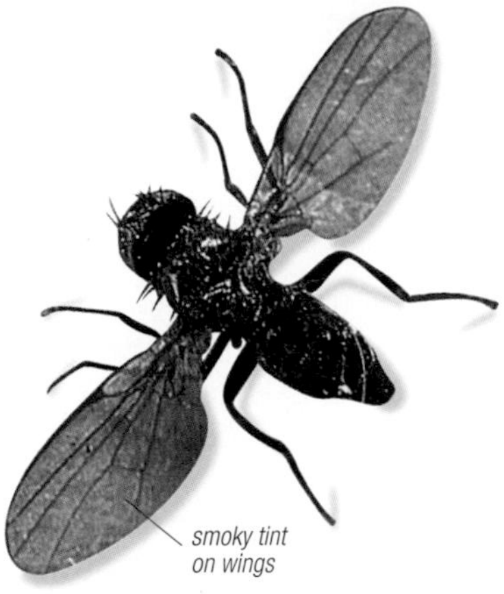

HEXOMYZA SPP.
These species occur throughout the Northern Hemisphere, and in South Africa and Australia. The larvae make galls on the twigs of poplars.

Family Asilidae

Robber flies

Occurrence 5,000 spp. worldwide, especially in tropical and subtropical regions; in a variety of microhabitats

This is a family of predatory flies; both the adults and larvae are predatory, although the latter eat decaying matter as well. Robber flies use their forward-pointed, stout and sharp proboscis to stab prey and inject a paralyzing saliva. They then suck up the body contents of the paralysed insect. They are 1/8–2 in (0.3–5 cm) long, either slender or stout and beelike. Their head is slightly hollow between the eyes, and the face has a tuft of long hairs. Eggs are laid in soil, rotting wood, or inside plants.

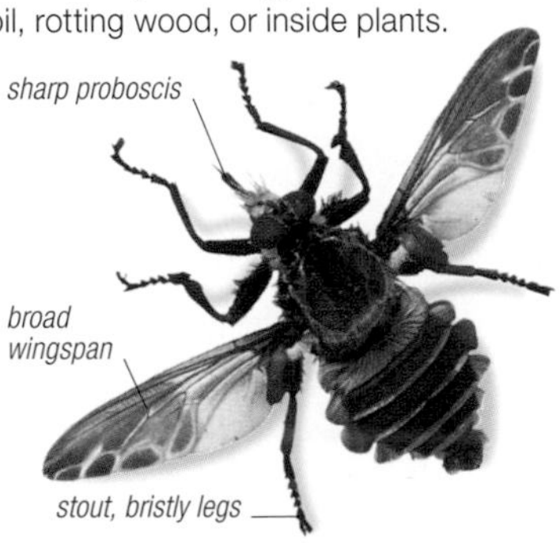

BLEPHAROTES SPLENDIDISSIMUS
This large Australian robber fly has platelike tufts of hair at the sides of its flat abdomen.

Family Bombyliidae

Bee flies

Occurrence 5,000 spp. worldwide, in open, tropical, or subtropical semiarid regions; around flowers, on ground

Bee flies are stout and hairy insects that strongly resemble bumble bees (see p.573). Between 1/16–1 1/4 in (0.2 and 3 cm) long, they are brown, red, or yellow, sometimes with bright markings. They have narrow wings, and when flying produce a high-pitched whine. Like bees, they have a long proboscis that they use to suck nectar from flowers. However, instead of landing on flowers as bees do, bee flies hover in front of them, steadying themselves against the petals with their front legs. Some bee fly larvae eat grasshopper eggs, but most are parasitic and attack beetles, moths, wasps, or other flies.

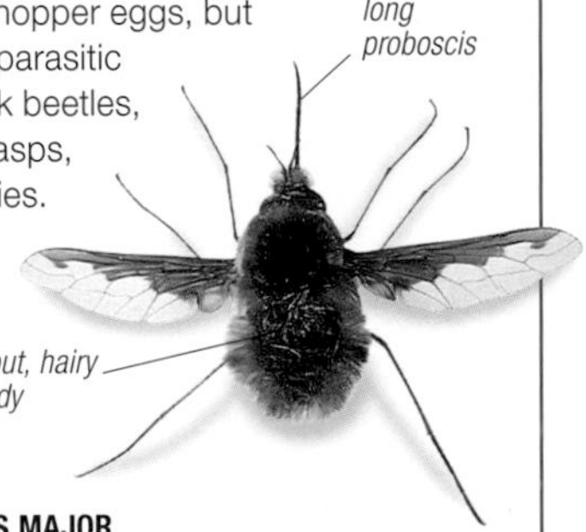

BOMBYLIUS MAJOR
The large bee fly is common and inhabits meadows, hedgerows, and open areas from Europe to China, and North Africa.

Family Calliphoridae

Blow flies

Occurrence 1,200 spp. worldwide; on flowers, vegetation, carcasses, cooked and raw food

Bluebottles and greenbottles are familiar members of this widepread and sometimes troublesome family of flies. Like their many relatives, they have stout, bristle-covered bodies, up to 1/2 in (1.5 cm) long, a metallic sheen, particularly on the abdomen, and a noisy, buzzing flight. Blow flies breed in decaying animal matter, dung, or rotten fishes. Adults feed on liquids from rotting material, nectar, or fruit. The larvae of some species are predators of ants, termites, and the larvae and eggs of other insects. Several species burrow into human flesh and many carry diseases that they can transmit to livestock or humans.

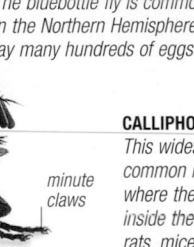

POLLENIA RUDIS
Common in the Northern Hemisphere, cluster flies congregate in large numbers in attics and unheated buildings. In the spring, adult females lay their eggs on earthworms or near fresh worm casts.

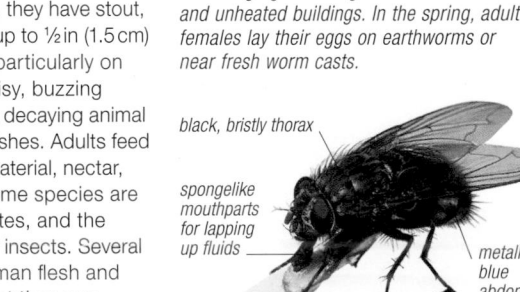

CALLIPHORA VOMITORIA
The bluebottle fly is common in the countryside in the Northern Hemisphere. Females may lay many hundreds of eggs in their lifetime.

CALLIPHORA VICINA
This widespread species is common in towns and cities, where the maggots develop inside the corpses of dead rats, mice, and pigeons.

Family Cecidomyiidae

Gall midges

Occurrence 5,000 spp. worldwide; anywhere near decaying matter, fungi, a wide range of plants

With bodies typically less than 5/32 in (4 mm), gall midges are seldom seen, but their larval homes—plant galls—are much easier to spot. Looking like miniature fruit or malformed buds, the galls are growths triggered by some gall midges, for protection and as a source of food. Like gall wasps (see p.574), gall midges are species-specific, always attacking particular kinds of plants. Adult gall midges have long, slender legs, and wings with a few unbranched veins. A number are serious pests of crop plants.

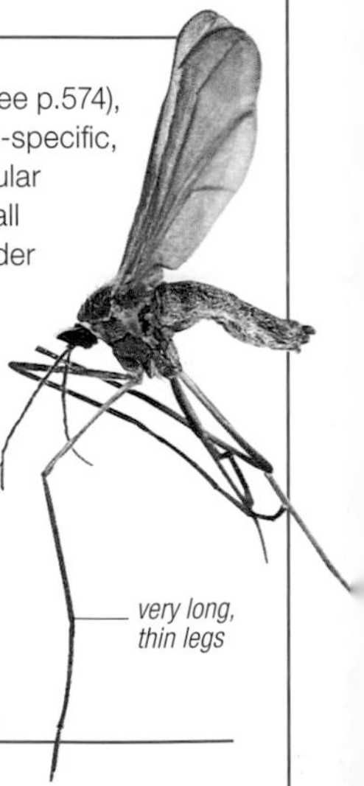

ASPHONDYLIA SPP.
Each species in this genus makes a characteristic gall.

Family Ceratopogonidae

Biting midges

Occurrence 4,000 spp. worldwide, mainly in the Northern Hemisphere; near margins of water bodies

Habitat All

Also called punkies or no-see-ums, these small flies—under 3/16 in (5 mm) long—have a bite that causes severe irritation. They have short, strong legs, dark patterns on their wings, and piercing mouthparts shaped to suck blood and insect body fluids. They mate while flying in a swarm. Eggs are laid in groups of 30–450, and are covered with a protective, jellylike substance.

Some biting midges transmit worm parasites to humans or carry animal diseases. They are pests but are also important pollinators of tropical crops such as cocoa or rubber.

CULICOIDES IMPUNCTATUS
This notorious European species breeds in boggy areas. The bite of the female common midge is extremely itchy and painful.

Family Chironomidae

Non-biting midges

Occurrence 5,000 spp. worldwide; near ponds, lakes, and streams; swarms occur at dusk

These delicate, pale brown or greenish flies resemble mosquitoes, but do not have functional mouthparts. They are 1/32–1/32 in (1–9 mm) long. The males have a slender body and feathery antennae, while the females are stout, with hairy antennae. Most of the 2–3-year life cycle is spent as larvae; adults live no longer than 2 weeks. They mate on the wing in a swarm, and lay their eggs in a mass of sticky jelly on water or plants. The larvae are plentiful in freshwater; up to 100,000 larvae have been recorded in 11 square ft (1 square meter).

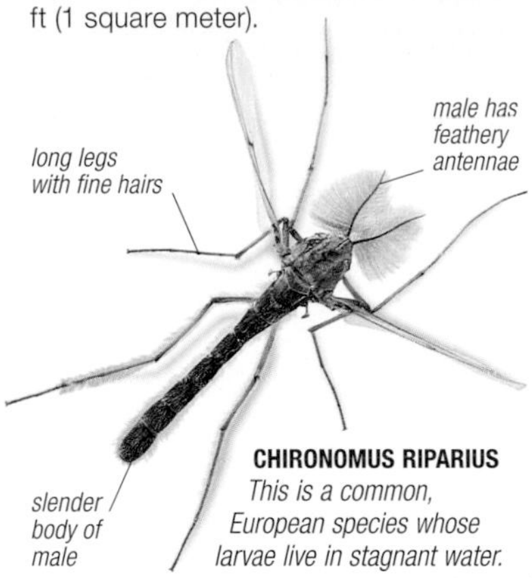

CHIRONOMUS RIPARIUS
This is a common, European species whose larvae live in stagnant water.

Family Culicidae

Mosquitoes

Occurrence 3,100 spp. worldwide, especially in warmer regions; everywhere near water

Habitat All

Mosquitoes are among the world's most dangerous pests, spreading malaria, dengue, and several other potentially fatal diseases. They are 1/8–3/4 in (0.3–2 cm) long, with narrow bodies, a humped thorax, and slender legs. Males often feed on flowers, but females feed on blood—chiefly from mammals and birds—and have syringelike mouthparts for piercing skin. They lay their eggs on stagnant water.

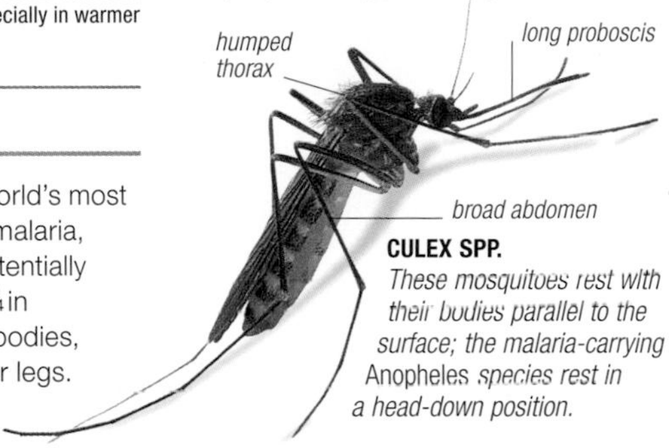

CULEX SPP.
These mosquitoes rest with their bodies parallel to the surface; the malaria-carrying Anopheles *species rest in a head-down position.*

Family Drosophilidae

Pomace flies

Occurrence 2,900 spp. worldwide; near decaying vegetation, fruit, fungi, fermenting liquids

Habitat All

Also known as lesser fruit flies or vinegar flies, these small, yellow, brown to reddish brown, or black species have light or bright red eyes. The thorax and the abdomen may be striped or spotted. Pomace flies are 1/32–1/4 in (1–6 mm) long. Males make a buzzing sound with their wings to attract mates. Eggs are laid near a food source at the rate of 15–25 eggs per day. Larvae feed on bacteria and fungi, while the adults look for fermenting liquids and rotting fruit.

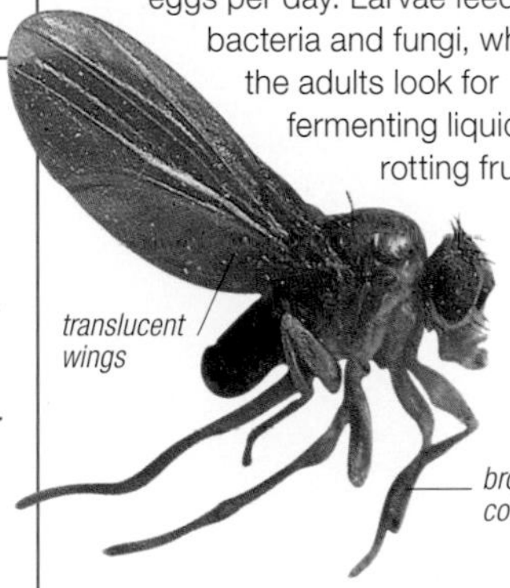

DROSOPHILA MELANOGASTER
The best-known pomace fly, the laboratory fruit fly is often used for genetics studies, as it breeds quickly and has large chromosomes in its salivary glands.

Family Empidae

Dance flies

Occurrence 3,000 spp. worldwide, mainly in the Northern Hemisphere; adults on tree trunks, larvae in rotten wood

The common name of this family comes from the dancing motions of mating swarms. Dance flies are 3/64–1/2 in (1.5–15 mm) long. Most species have a stout thorax, an elongated abdomen, a rounded head with large eyes, and a downward-pointing, sharp proboscis with which they can stab prey. Courtship often involves males offering prey as food to females. Eggs are laid on soil, dung, or plant matter. While adults eat small flies and drink nectar, larvae feed on soft-bodied prey.

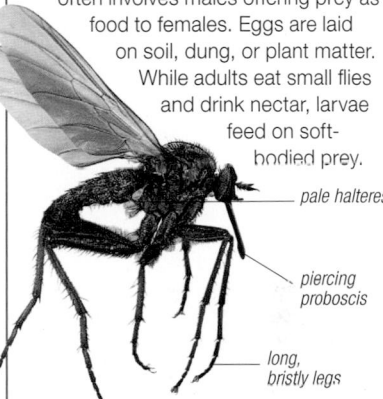

EMPIS SPP.
These flies inhabit damp vegetation. The long proboscis enables the fly to "stab" its prey and extract fluids on which it feeds.

Family Hippoboscidae

Louse flies

Occurrence 200 spp. worldwide, mainly in tropical and subtropical regions; ectoparasites of birds and mammals

These dull brown, stout, flat, and hairy flies, mostly 5/32–9/32 in (4–7 mm) long, have a short proboscis and strong, clawed legs for gripping hairs or feathers. Some have fully formed wings, but in others the wings are vestigial, or absent altogether. Females lay mature larvae (prepupae) one at a time; these develop inside a "uterus," feeding on secretions from special "milk" glands. Adult louse flies are blood-suckers.

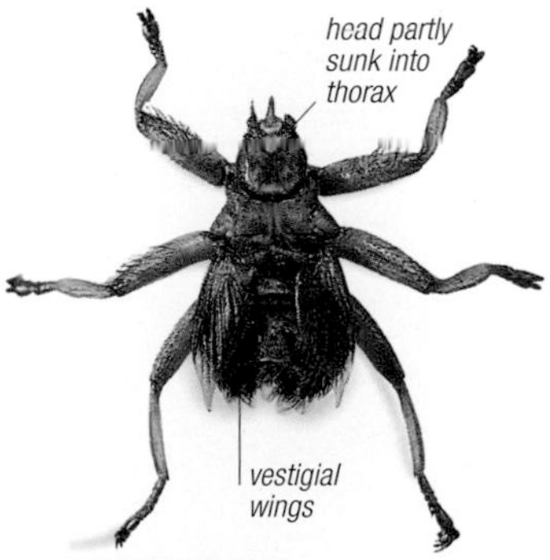

CRATAERINA PALLIDA
Also called the swift louse fly, this European species is parasitic on swifts in their nests. Three-quarters of all louse fly species parasitize birds.

Family Muscidae

Muscids

Occurrence 4,000 spp. worldwide; on flowers, dung, decaying matter, near mammalian hosts

Habitat Terrestrial

The muscid family includes one of the world's most successful insects—the common house fly—together with its many relatives. Muscids are usually drably colored, with bristly bodies up to 3/8 in (1.2 cm) long, and mouthparts that are spongelike for lapping fluids, or piercing for sucking blood. Female houseflies can lay 100–150 eggs per day in rotting matter, dung, fungi, water, or plants. Maggots grow fast and, in ideal conditions, pupate in just over a week. These flies can carry infections such as typhoid and cholera.

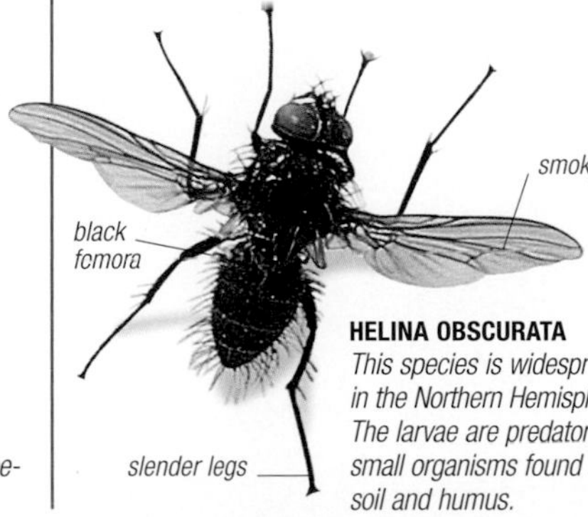

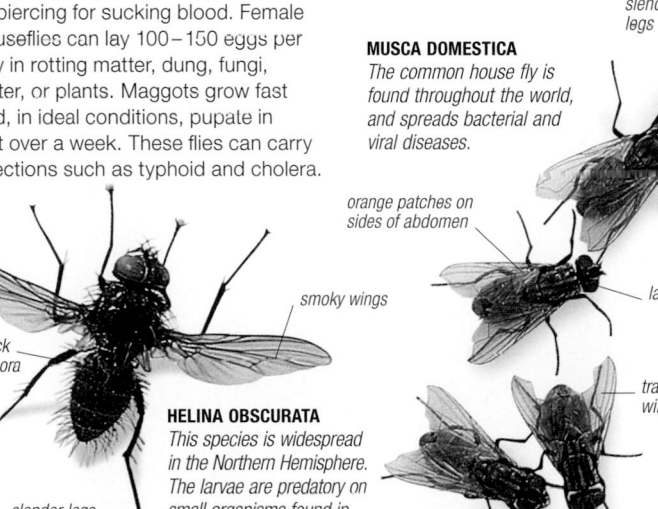

HELINA OBSCURATA
This species is widespread in the Northern Hemisphere. The larvae are predatory on small organisms found in soil and humus.

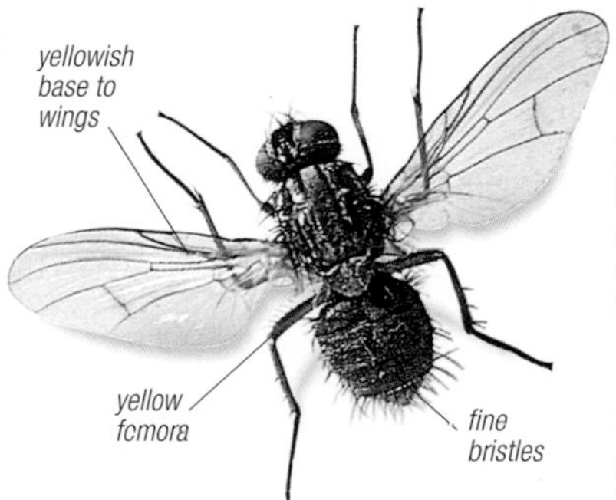

MYDAEA CORNI
This species, found in Europe and Asia, inhabits dung. The larvae are predators on small organisms inside the dung.

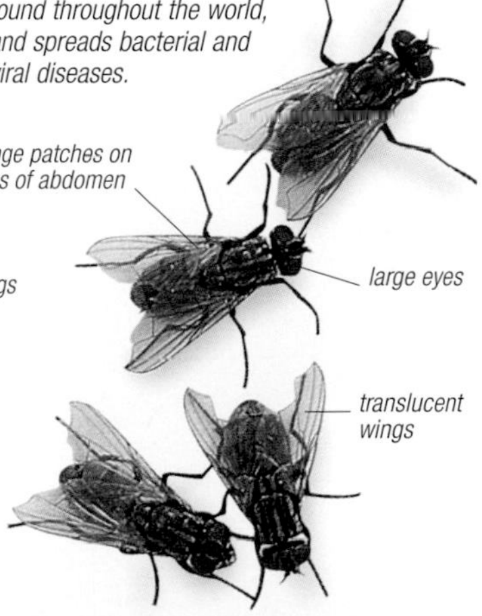

MUSCA DOMESTICA
The common house fly is found throughout the world, and spreads bacterial and viral diseases.

Family Keroplatidae

Fungus gnats

Occurrence 1,000 spp. worldwide; in forests, caves, and other moist places

Usually found in association with fungi in dark, damp forest habitats, keroplatid fungus gnats also occasionally live in caves. Adult keroplatids generally have dark bodies and may have patterned wings; one group mimic wasps. The larvae of some feed on fungi, but others are predators, ensnaring other insects and worms in slimy threads and killing them with oxalic acid produced by mouth glands. Several species, including the New Zealand cave "glowworm" (Arachnocampa luminosa), have luminescent larvae to attract prey.

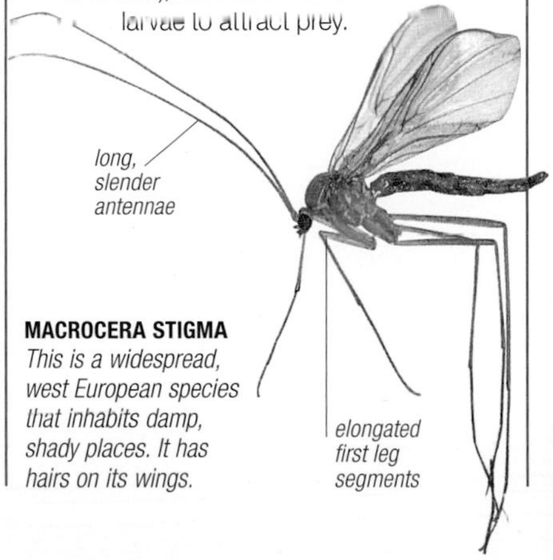

MACROCERA STIGMA
This is a widespread, west European species that inhabits damp, shady places. It has hairs on its wings.

Order Diptera *continued*

Family Oestridae

Bot flies and warble flies

Occurrence 70 spp. worldwide, especially in the Northern Hemisphere and Africa; near mammalian hosts

Many members of this heavy-bodied family look like bees. They are about 5/16–1 in (0.8–2.5 cm) long. Because the adults are short-lived and do not feed, the mouthparts are either tiny or absent. The larvae are all intestinal parasites of mammals, especially horses, goats, and camels. The females of some species lay larvae directly in the nostrils of hosts, where they feed and grow in the sinuses. In warble flies, eggs are laid on the hair of the host, and the larvae burrow under the skin. Fully grown larvae are sneezed out by the host or chew their way out and pupate in the soil.

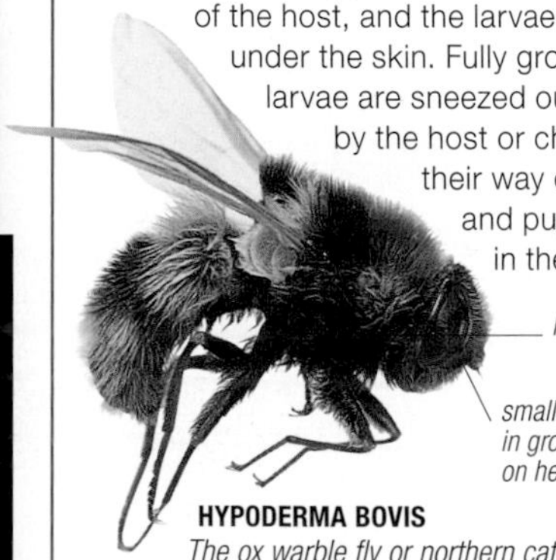

HYPODERMA BOVIS
The ox warble fly or northern cattle grub is common in the Northern Hemisphere, particularly near cattle. It damages hides, and reduces animal vigor and milk production.

Family Phoridae

Scuttle flies

Occurrence 3,000 spp. worldwide; in a wide variety of microhabitats

Habitat All

Also known as hump-backed flies, the brown, black, or yellowish scuttle flies have characteristically fast, jerky movements while running. They are 1/64–1/4 in (0.5–6 mm) long with a strongly down-turned head. The antennae appear to have only one segment and the hind femora are usually very stout. Conspicuous bristles are often present on the body; these look feathery when highly magnified. Adults live from a few days to a week, and may take fluids. Some of the larvae feed on fungi or decaying matter; others feed on insects, snails, and worms.

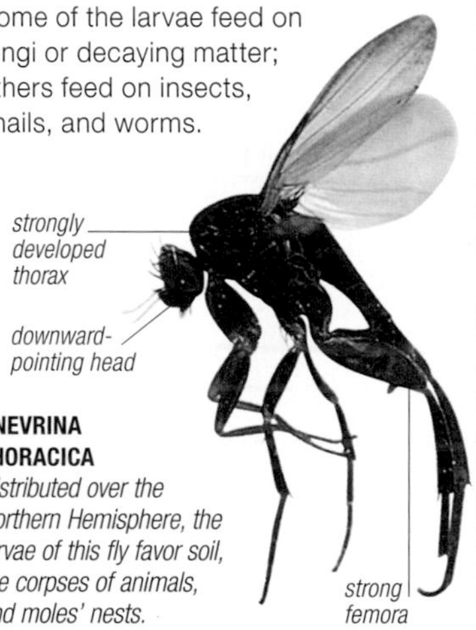

ANEVRINA THORACICA
Distributed over the Northern Hemisphere, the larvae of this fly favor soil, the corpses of animals, and moles' nests.

Family Psychodidae

Sand flies and moth flies

Occurrence 1,000 spp. worldwide, especially in warmer regions; in water, rotting matter, and moist, shady habitats

These flies measure 3/64–3/16 in (1.5–5 mm) in length. Their bodies, wings, and legs are covered with long hairs or scales. The broad wings are held tentlike or partly spread in moth flies, or together above the body in sand flies. Sand and moth flies have a weak, fluttery or hopping flight pattern. Female sand flies are blood-suckers and, particularly in the tropics, several species spread diseases.

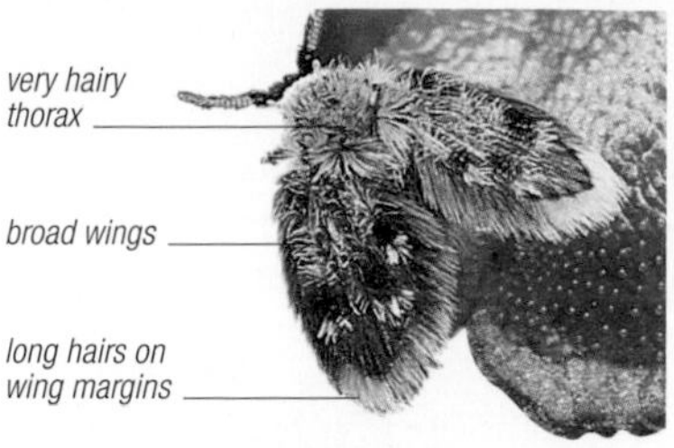

PERICOMA FULIGINOSA
The larvae of this dark, widely distributed moth fly are, like sand flies, saprophagous. They develop in shallow water and in rot holes in trees.

Family Sarcophagidae

Flesh flies

Occurrence 2,300 spp. worldwide, especially in the Northern Hemisphere; foliage, flowers, carrion

Some of these robust-looking flies lay larvae in body cavities or wounds in vertebrates, including humans. They are mostly dull or silvery gray or black, and are 1/16–3/4 in (0.2–2 cm) long. The thorax typically has 3 dark longitudinal stripes on a gray background, and the abdomen appears checkered or spotted. Most females give birth to larvae that are laid or dropped in flight. The larvae of many species feed on carrion, and some are parasitic on other insects, spiders, snails, and worms.

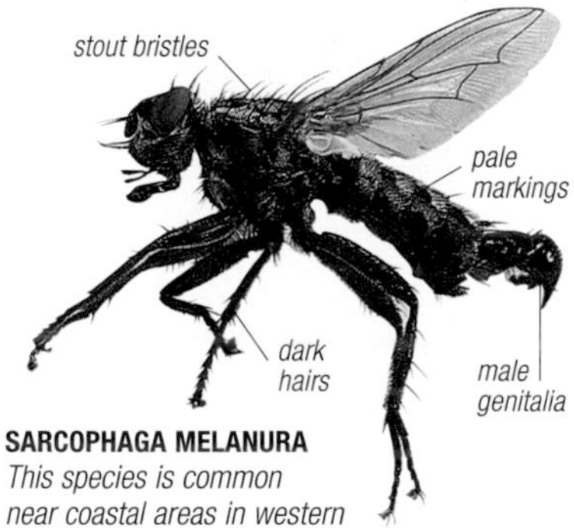

SARCOPHAGA MELANURA
This species is common near coastal areas in western Europe. Its larvae are found in rotting meat and may parasitize snails or insects.

Family Simuliidae

Black flies

Occurrence 1,600 spp. worldwide; larvae in flowing water, adults on plants near water

Black flies, also called buffalo gnats or turkey gnats, are 1/32–3/16 in (1–5 mm) long, and usually black, orange-red, or dark brown. Males drink nectar and water, while females feed on vertebrate blood. They lay their eggs on rocks or vegetation near water. The aquatic larvae, attached to submerged rocks or plants by a hooklike holdfast, filter food from the water with special bristles around the mouth.

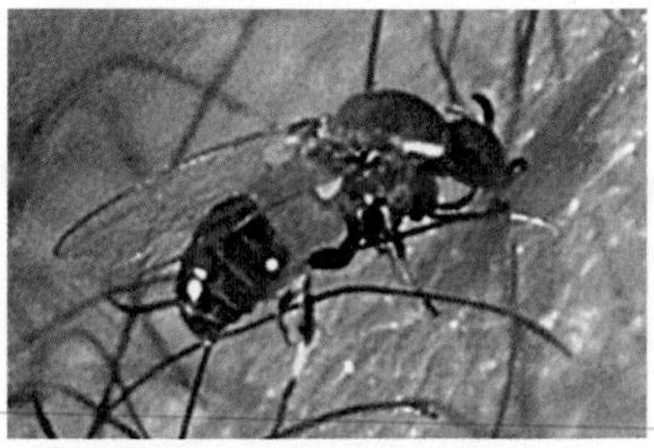

SIMULIUM SPP.
Although known as black flies, some members of this genus are not entirely black. Many species are carriers of the roundworm that causes river blindness in Africa and tropical America.

Family Syrphidae

Hover flies

Occurrence 6,000 spp. worldwide; common on flowers, especially umbelliferous ones

Hover flies include some of the most skilled fliers in the insect world. They feed on pollen and nectar, and can often be seen hovering over flowers, or darting after each other in a high-speed chase. Ranging from 3/8–3/4 in (1–2 cm) in length, they have large eyes and slender wings, and are often colored yellow or orange and black. This coloration—combined with their overall shape—makes them look very much like stinging bees or wasps, and deters many predators from attacking them. Hover fly larvae are very varied, and live on a range of different foods. Some are predators, attacking aphids, sawfly larvae, and other soft-bodied insects, while others feed on rotting vegetation or animal dung. A few live as scavengers in the nests of bees and wasps, where they eat the dead larvae and pupae of their hosts. The larvae of one group, known as drone flies (*Eristalis* spp.), live in stagnant water, and use a telescopic, posterior "snorkel" to breathe.

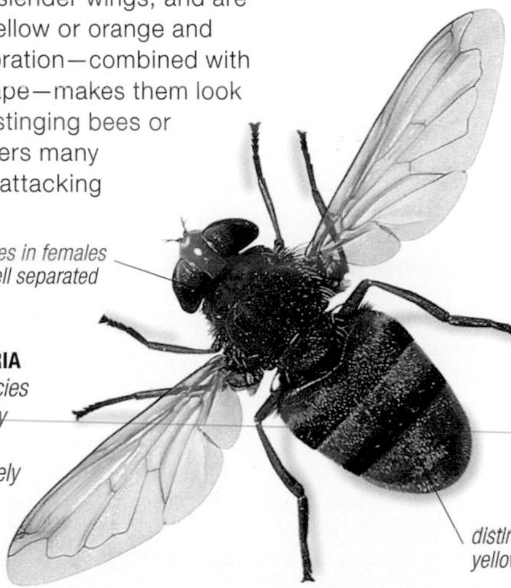

VOLUCELLA ZONARIA
This migratory species is distributed mainly over Europe. It is stout and distinctively banded. Its larvae scavenge inside wasps' nests.

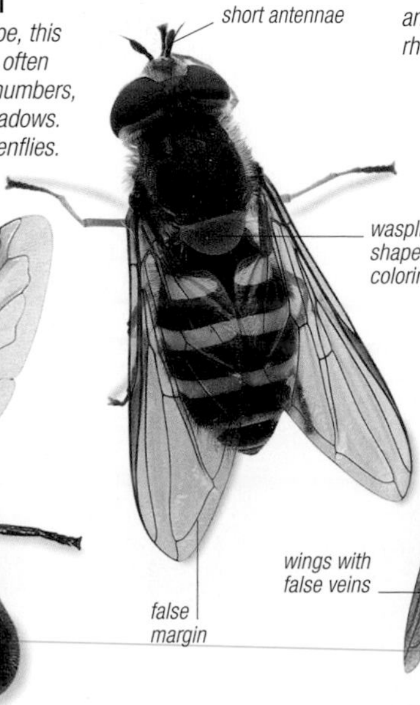

SYRPHUS RIBESII
Common in Europe, this wasplike species often gathers in large numbers, in flower-rich meadows. Its larvae eat greenflies.

SERICOMYIA SILENTIS
The common bog hoverfly is found on acid heathland in Europe. The larvae live in boggy pools, such as those formed after cutting peat, and feed in the decaying rhizomes of reedmace.

INVERTEBRATES

Family Tabanidae

Horse flies

Occurrence 4,000 spp. worldwide, near mammals; many favor damp or wet habitats

Horse flies are stout and hairless with a distinctive, flattened, hemispherical head and colorful, patterned eyes. Most are black, gray, or brown, with brightly marked abdomens, and measure ¼–1¼ in (0.6–2.8 cm) in length. Females have bladelike mouthparts for cutting skin; they feed on the blood of mammals and birds. Eggs are typically laid in a mass of 200–1,000 on debris overhanging water, leaves, or rocks. Larvae are predatory or saprophagous.

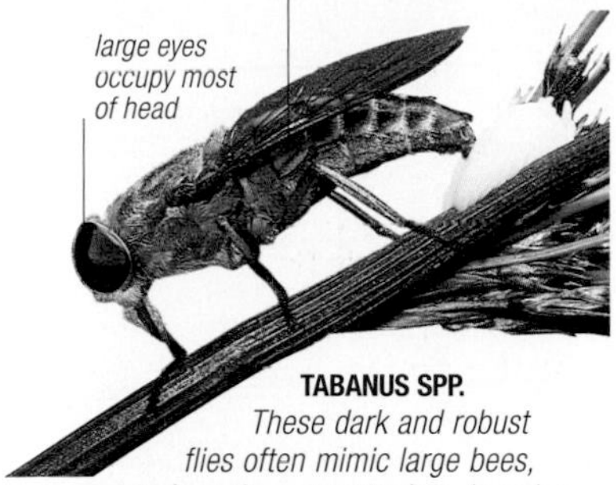

TABANUS SPP.
These dark and robust flies often mimic large bees, such as the carpenter bee, in order to deter predators.

TABANUS SUDETICUS
This giant horse fly has a beelike abdomen with distinctive, pale, triangular marks down the middle. Some species transmit diseases to animals as well as humans.

CHRYSOPS RELICTUS
This twin-lobed deerfly inhabits bogs and ponds in heather moorland in the Northern Hemisphere. Its bite is very painful.

Family Tachinidae

Parasitic flies

Occurrence 8,000 spp. worldwide; where insect hosts are found

Stout and robust, parasitic flies are ³⁄₁₆–¾ in (0.5–2 cm) long, and are very variable in appearance. Most are dark; some are metallic or pale. The abdomen is very bristly, especially toward the rear. Many species look like bristly house flies (see muscids p.567) and some species are almost beelike. Adults feed on sugary fluids, but their larvae are endoparasites of insects. Some species locate crickets to parasitize by listening for their mating songs. Males and females often congregate on hilltops for mating, and the eggs are laid on foliage eaten by hosts, or inside the hosts' bodies. Parasitic flies are mostly diurnal. Many species are used as biological control agents of herbivorous insect pests.

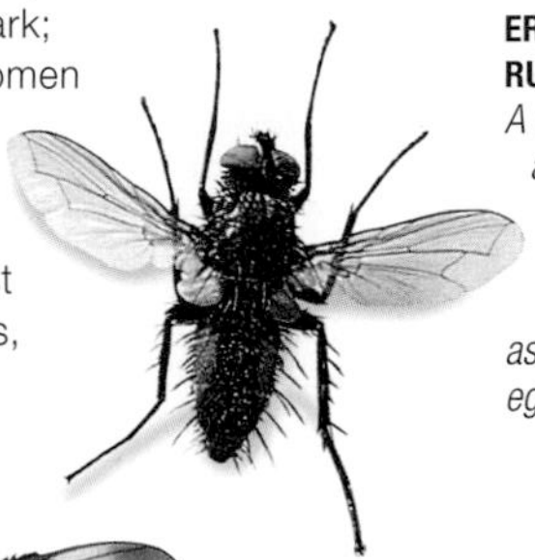

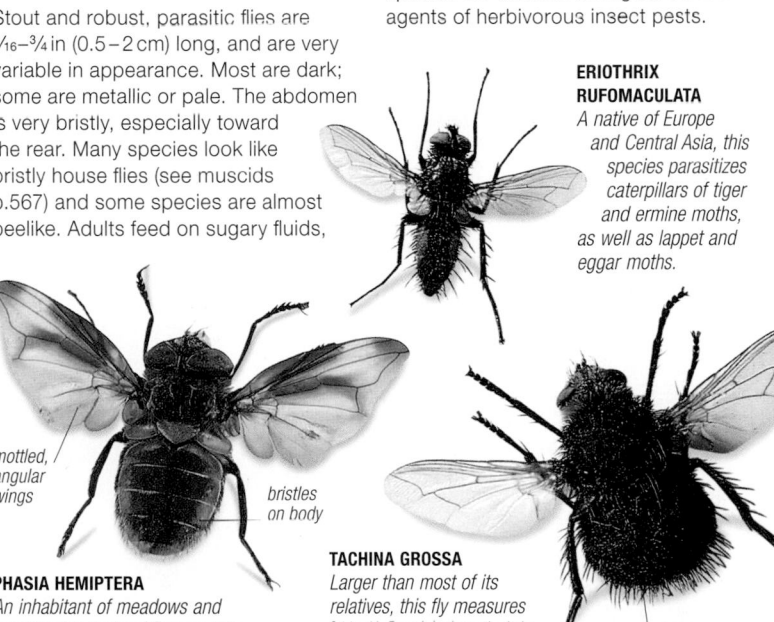

ERIOTHRIX RUFOMACULATA
A native of Europe and Central Asia, this species parasitizes caterpillars of tiger and ermine moths, as well as lappet and eggar moths.

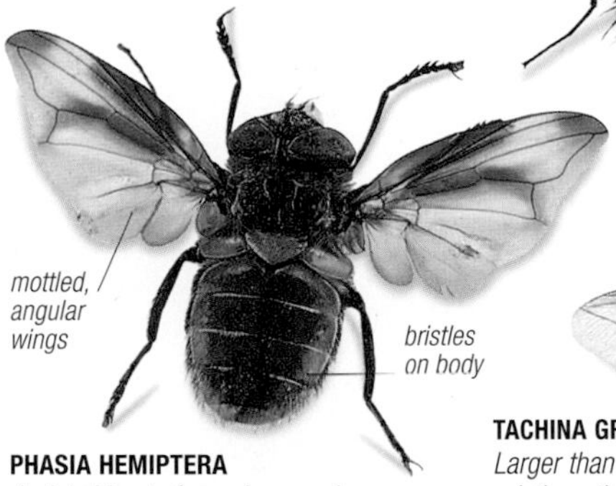

PHASIA HEMIPTERA
An inhabitant of meadows and woodland in parts of Europe, this species parasitizes shield bugs.

TACHINA GROSSA
Larger than most of its relatives, this fly measures ¾ in (1.9 cm) in length. It is found in Europe and Asia.

Family Tephritidae

Fruit flies

Occurrence 4,500 spp. worldwide; on a range of plants, crops, rotting matter

Habitat Terrestrial

Most fruit flies are under ½ in (1.5 cm) long and recognizable by their distinctive wing patterns, which can take the form of bands, patches, and zigzag markings. Females have a pointed ovipositor, which can be longer than the rest of the body. Males display to females by waving their wings. Eggs are laid singly or in groups under the surface of fruit. Adults take fluids, plant sap, nectar, or liquid from rotting matter; the larvae are herbivorous. Many species are crop pests.

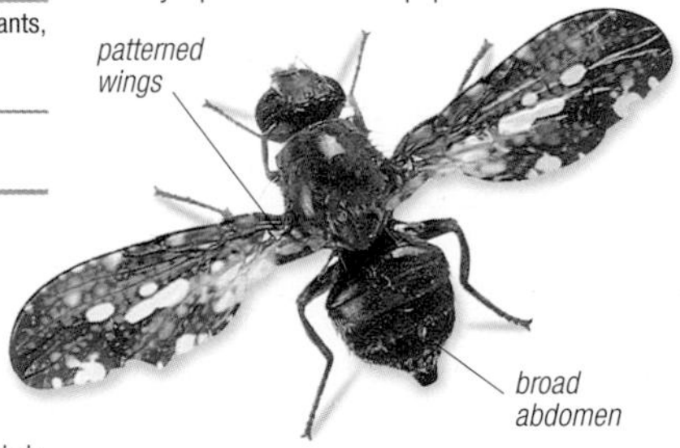

ICTERICA WESTERMANNI
Found in hedgerows, waste ground, and grassy places in Europe, this species of fruit fly feeds on the flowerheads of ragworts.

Family Tipulidae

Crane flies

Occurrence 15,000 spp. worldwide; usually in moist microhabitats

Habitat Terrestrial

Also known as daddy-long-legs, most of these fragile, slender, and elongated flies measure ¼–2¼ in (0.6–6 cm) in length. They shed their legs very easily if caught. The adults of many are short lived and feed on nectar and other fluids. The tough-bodied larvae, called leatherjackets, live in soil, rotting wood, bird nests, and bogs, and are a food source for many animals and birds.

HOLORUSIA SPP.
This genus includes some of the world's largest crane flies. They have a wingspan of 2¼–4 in (6–10 cm).

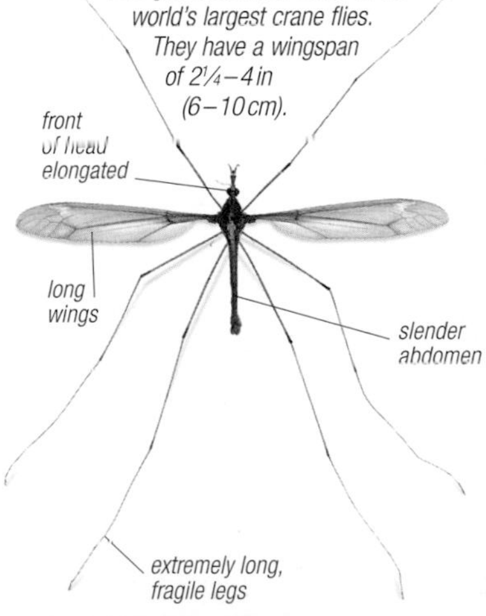

Order Trichoptera

Caddisflies

There are about 10,000 species of caddisflies, in 43 families, and they can be found almost anywhere there is freshwater. The adults are slender, dull-colored insects, with a hairy body and wings, and long antennae. The larvae are soft-bodied, and live in water, often inside a protective, portable case, which each species builds in a characteristic way. They develop by complete metamorphosis.

Family Limnephilidae

Northern caddisflies

Occurrence 1,500 spp. mainly in the Northern Hemisphere; around lakes, temporary pools, ditches

Northern caddisflies are reddish, yellowish, or dark brown, with dark wing markings, and most are under 1 in (2.4 cm) long. Larvae eat organic detritus and algae, and make cases from sand, pebbles, vegetation, or snail shells. Adults may feed on liquids.

LIMNEPHILUS LUNATUS
A widespread species, its larvae are found in a wide variety of freshwater habitats. Coloration ranges between drab shades of black and brown.

Family Phryganeidae

Large caddisflies

Occurrence 450 spp. mainly in the Northern Hemisphere; near ponds, lakes, bogs, slow-moving streams, rivers

Members of this family have light brown or gray markings, and may look mottled. They are ⅜–1 in (1.2–2.6 cm) long and have at least 2 tibial spurs on the forelegs and 4 on the mid- and hindlegs. Larvae make cases out of plant fragments and fibers.

PHRYGANEA GRANDIS
This is the largest caddisfly found in the UK. The male, smaller than the female (shown here), lacks the distinctive dark stripe on the wing.

Order Lepidoptera

Moths and butterflies

With over 165,000 species in 127 families, moths and butterflies are among the most diverse insects in the world. They have minute, overlapping scales on their bodies and wings, and mouthparts in the form of a proboscis. There is no clear-cut difference between butterflies and moths—and no scientific basis for separating them—but as a general rule, butterflies are brightly colored and day flying, with club-ended antennae, while moths are nocturnal and drab. In both, courtship often involves airborne scents and displays. The larvae, called caterpillars, have cylindrical bodies and chewing jaws. They develop by complete metamorphosis, often forming inside a silk cocoon.

Family Arctiidae

Tiger and ermine moths

Occurrence 2,500 spp. worldwide; in well-vegetated areas where host plants occur

Habitat Terrestrial

The hairy and heavy-bodied tiger moths are brightly colored. Ermine moths tend to be pale with black spots or patches and have a wingspan of ¾–2¾ in (2–7 cm). Caterpillars are herbivorous and many are poisonous. Adults take nectar and liquids. Recent evidence groups actiid and noctuid moths together.

ARCTIA CAJA
Also called the garden tiger, this moth is highly unpalatable, and its bright, contrasting coloration advertises this to predators.

Family Bombycidae

Silk moths

Occurrence 100 spp. in tropical areas of Southeast Asia; on mulberry trees, other plants

Long valued as a source of silk, these stout, pale cream, gray, or brown moths have wings that span ¾–2¼ in (2–6 cm). The smooth caterpillars have prolegs on all abdominal segments and usually a short, hornlike structure at the tail. Caterpillars of the commercial silk moth feed on mulberry trees; other species feed on fig and related plants. Pupation takes place inside a dense silk cocoon. Functional mouthparts are absent in adults, which do not feed.

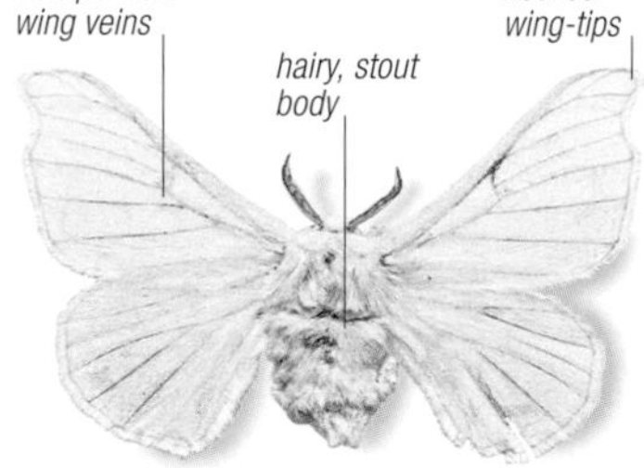

BOMBYX MORI
The silk moth originated in China but is now cultivated in silk farms the world over. It is, however, extinct in the wild.

Family Geometridae

Geometer moths

Occurrence 20,000 spp. worldwide; on deciduous and coniferous trees, woody shrubs, herbaceous plants

The wings of these slender, generally nocturnal moths are rounded, with a span of ½–3 in (1.4–7.4 cm), and complex patterns of fine markings. The females of some species are wingless. These moths are colored brown or green for camouflage; coloration often varies between sexes. Females lay eggs singly or in groups, on bark, twigs, or stems of host plants. The characteristic looping motion of the caterpillars gives them the name "loopers" or "inchworms." Many species are agricultural or forestry pests and can cause severe damage.

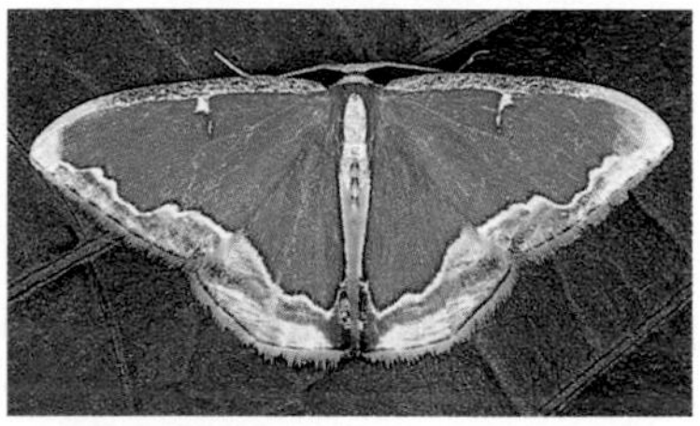

OOSPILA VENEZUELATA
This green species is found mainly in the forests of Venezuela.

Family Hesperiidae

Skippers

Occurrence 3,000 spp. worldwide except New Zealand; in open habitats such as cultivated fields and grassland

Heavy-bodied and mothlike, these butterflies get their name from their rapid, darting flight. Most are drab brown with white or orange markings, but a few can be brightly colored. Their wings span ¾–3¼ in (2–8 cm). The antennae end in a long, curved, and pointed club. Females lay single eggs on host plants. Caterpillars are herbivorous and feed at night.

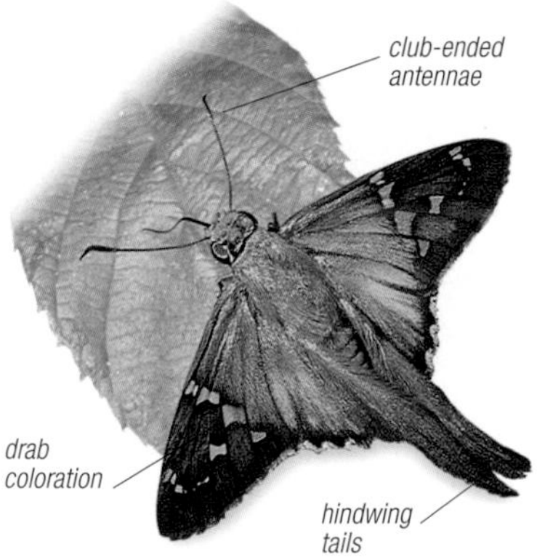

URBANUS PROTEUS
The long-tailed skipper is common in parts of South and North America. It has characteristically long hindwing tails and an erratic flight pattern.

Family Lasiocampidae

Lasiocampids

Occurrence 2,000 spp. worldwide except New Zealand; on foliage of a wide range of deciduous trees

These insects, also known as lappet and eggar moths, are heavy-bodied and very hairy; most are yellowish brown, brown, or gray. The wingspan is mostly under 1½ in (4 cm); females are bigger than males. Eggs are laid in a band of 100–200 around a twig of a host tree. The stout caterpillars may live communally in silk tents or webs that catch the sun's rays. They pupate inside tough, egg-shaped, papery cocoons.

MALACOSOMA AMERICANUM
Also called the eastern tent caterpillar moth, this is an orchard pest. It inhabits forest and woodland areas in the northern USA and southern Canada.

Family Lycaenidae

Blues, coppers, and hairstreaks

Occurrence 6,000 spp. worldwide, particularly in warmer regions; in association with ant nests or host plants

The butterflies in this family are generally small, with a wingspan of ½–2 in (1.5–5 cm), but they are notable for color differences between males and females. Males often have brilliantly colored, iridescent wing surfaces, while females are generally dull. Most species have squat caterpillars, which feed on plants or small insects. Some produce a special fluid that is eaten by ants—in return, the ants guard the caterpillar, and sometimes take it into their nest.

THECLA CORONATA
Also called the Hewitson's blue hairstreak, this species inhabits tropical forests of South America. It is one of the largest and most brilliantly colored members of the family.

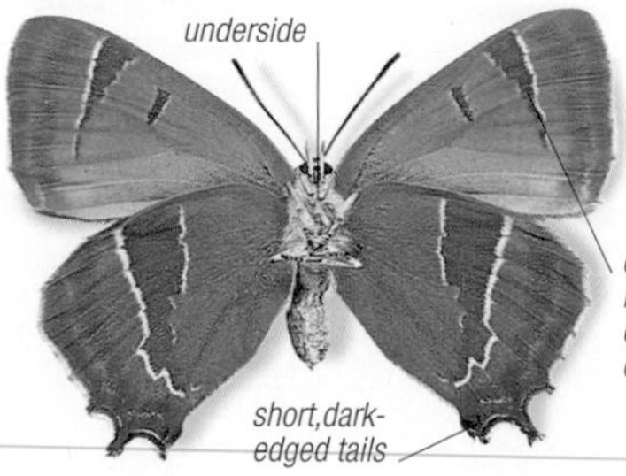

THECLA BETULAE
The brown hairstreak butterfly inhabits woodland in Europe and temperate Asia. The caterpillars eat the leaves of sloe, plum, and birch trees.

LYCAENA PHLAEAS
The small copper butterfly is common in grassy areas up to 6,600 ft (2,000 m) in the Northern Hemisphere. The caterpillars feed on dock leaves.

Family Lymantriidae

Tussock moths

Occurrence 2,600 spp. worldwide; on foliage of deciduous and coniferous trees, shrubs

Tussock moths resemble noctuids (see right) and although most are dull in color, tropical species can be colorful. The wingspan is ¾–2¼ in (2–6 cm); males are slightly smaller than the females, which are sometimes wingless. Females lay eggs in clumps on the bark of host trees and shrubs. Adults lack a proboscis and do not feed. The caterpillars are hairy, often brightly colored, and feed in groups on foliage. They may induce an allergic reaction if touched.

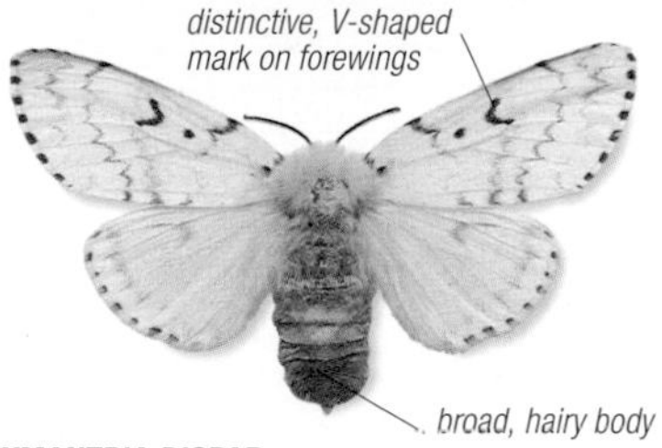

LYMANTRIA DISPAR
Commonly called the gypsy moth, this species is a native of Europe and Asia. Introduced to North America to produce inexpensive silk, it escaped from captivity and has become a pest.

Family Noctuidae

Noctuid moths

Occurrence 22,000 spp. worldwide; on all types of vegetation, including crops

Almost every important crop plant in the world is attacked by one or more noctuid moth species. These moths have narrow forewings and broad hindwings and are mostly dull in color, although the hindwings of some are brightly colored and patterned. The wingspan is usually under 2 in (5 cm). Females lay eggs singly or in groups of 50–300, at the bases of host plants or in the soil. These moths have thoracic hearing organs to detect their main predators, bats.

AGROTIS IPSILON
The dark sword-grass moth inhabits crops and grassy areas across the world. Its caterpillars attack potato, tobacco, lettuce, and cereal crops.

Family Papilionidae

Swallowtails

Occurrence 600 spp. worldwide in warmer regions; in flower-rich, open or shaded areas

This family includes birdwings—the world's largest butterflies, now protected by law. True to their name, many swallowtail species have tails on their hindwings. The wings are typically dark with bands, spots, or patches of white, yellow, orange, red, green, or blue, and wingspan ranges from 1¾–11 in (4.5 to 28 cm). Round eggs are laid singly on host plants and pupation also occurs here. The caterpillars have a forked, fleshy scent gland, which emits an odor that protects them from predators. They are herbivorous and eat a range of foliage. Adult swallowtails feed on nectar and other liquids.

PARNASSIUS APOLLO
Also known as the Apollo, this species is found in mountainous areas of Europe and parts of Central Asia.

PAPILIO GLAUCUS
Commonly called the tiger swallowtail, this North American species inhabits woodland and gardens. Its caterpillars feed among high foliage in many deciduous trees.

ORNITHOPTERA PRIAMUS
Also called the Cairns birdwing, this species inhabits tropical forests in the Moluccas, Papua New Guinea, the Solomon Islands, and northeastern Australia. Females are larger than males, and are drably colored.

Family Nymphalidae

Nymphalid butterflies

Occurrence 5,000 spp. worldwide; in flower-rich meadows, woodland clearings

This family includes some of the world's most eye-catching butterflies, their most characteristic feature being their very short, brushlike front legs, which have a sensory function. These are held off the ground, close to the head. The wingspan is 1¼–6 in (3–15 cm); the upper wing surfaces are brightly colored but the undersides have cryptic coloring to camouflage the butterfly at rest. Males and females may be differently colored, and there may be seasonal variation. Females lay groups of rounded eggs on the foliage of trees, shrubs, and herbaceous plants. The pupae often have warty, conspicuous bumps and hang from the host plant by a group of terminal hooks called cremasters. When young, the caterpillars may feed communally on foliage. Adults feed on nectar and liquids; some are attracted to fruit, dung, carrion, urine, and even gasoline.

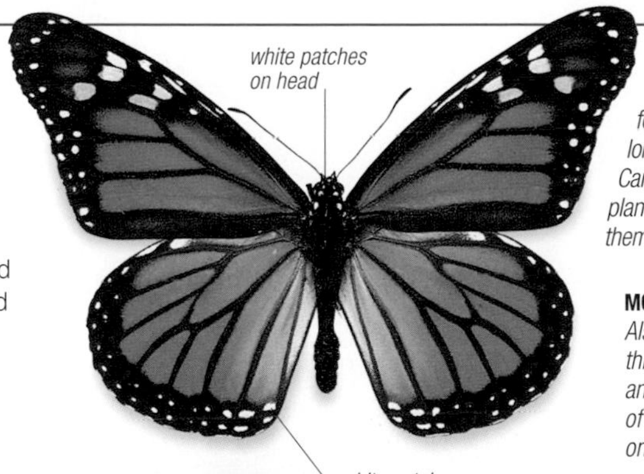

DANAUS PLEXIPPUS
The monarch butterfly is famous for its long-distance migrations, the longest of which extends from Mexico to Canada. Its caterpillars feed on milkweed plants; a substance in these plants makes them unpalatable to predators.

MORPHO PELEIDES
Also called the common blue morpho, this butterfly inhabits forests in South and Central America. With a wingspan of 3¾–4¾ in (9.5–12 cm), it is one of the larger nymphalids. Its caterpillars can produce a noxious defensive odor.

NYMPHALIS ANTIOPA
Called the Camberwell beauty in the UK, and the mourning cloak in North America, this butterfly inhabits grassland in the Northern Hemisphere.

VANESSA ATALANTA
The red admiral is widespread in the Northern Hemisphere. Its caterpillars feed on nettles and hops.

INVERTEBRATES

Order Lepidoptera *continued*

Family Pieridae

Whites and sulphurs

Occurrence 1,200 spp. worldwide; anywhere, often in groups at bird droppings, urine, or puddles in sunshine

The wings of these very common butterflies are usually white, yellow, or orange, with black or gray markings. The wingspan is ¾–2¾ in (2–7 cm) and the pigments in the wing scales are derived from by-products of the food eaten by caterpillars. Eggs are laid singly or in groups of 20–100 on host plants. The pupae, which have a distinctive, spiny projection on the head, are held upright on the host plant by a silk belt.

COLIAS EURYTHEME
Also called the orange sulphur or the alfalfa sulphur, this species inhabits grassy, open areas and cultivated land in North America. Caterpillars feed on clover and can be pests of alfalfa.

ANTHOCHARIS CARDAMINES
The orange tip inhabits damp, grassy meadows and open woodland in Europe and temperate Asia. The undersides of the hindwings are dappled pale green.

PIERIS NAPI
The green-veined white inhabits grassy, open areas and cultivated land in Europe, Asia, and North America.

Family Crambidae

Grass moths

Occurrence 24,000 spp. worldwide, especially in warmer regions; on a range of plants, from aquatic plants to trees

Some species of grass moths have a short "snout" on the front of their head. Most are drably colored. Their wings span ⅜–1¾ in (1–4.6 cm); the forewings may be broad or narrow, but the hindwings are broad and rounded. Eggs are laid on the undersides of leaves.

CHILO PHRAGMITELLA
The Wainscot veneer is found in reed beds, where its caterpillars feed on the stems of reeds belonging to the genus Phragmites.

Family Saturniidae

Saturniid moths

Occurrence 1,200 spp. worldwide, especially in wooded tropical and subtropical regions; on foliage

Variously known as emperor, moon, royal, and atlas moths, these are large and often spectacular insects with broad, often conspicuously marked wings. The wings span 2–12 in (5–30 cm), and the center of each may have an eyespot; some species have hindwing tails. Males have broad, feathery or comblike antennae. Fully grown caterpillars make dense cocoons, attached to the twigs of host plants, and feed on the foliage of a wide range of trees and shrubs. Adults do not feed.

ATTACUS ATLAS
This Southeast Asian moth, commonly called the Atlas moth, has the largest wing area of any insect. It inhabits forest and open, grassy areas. Although easy to raise, it is vulnerable and protected in some countries.

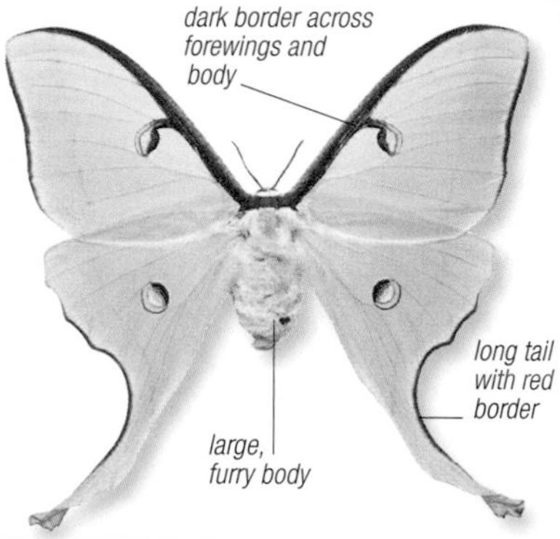

ACTIAS LUNA
The American moon moth is found from Mexico to the Canadian border. The caterpillars feed on the leaves of deciduous trees.

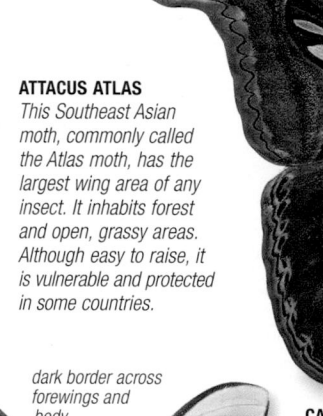

CALLOSAMIA PROMETHEA
The promethea moth inhabits woodland and cultivated areas of southern Canada and the USA. Females are heavy bodied and nocturnal; males are dark, day flying, and resemble Nymphalis antiopa *(see p.571).*

Family Sesiidae

Clear-winged moths

Occurrence 1,000 spp. worldwide, especially in northern temperate regions; around flowers or near host plants

These usually diurnal moths resemble wasps or bees; some even pretend to sting. They are black, bluish, or dark brown, with yellow or orange markings. The wings, large areas of which are transparent, span from ½–1½ in (1.4 to 4 cm) and produce a buzzing sound in flight. Many species are pests of fruit and other trees and shrubs.

SESIA APIFORMIS
With its transparent wings, the hornet moth, from Europe and Asia, convincingly mimics a hornet. Its caterpillars bore inside willows and poplars.

Family Sphingidae

Hawk moths

Occurrence 1,100 spp. worldwide, especially in tropical and subtropical regions; on foliage

With their streamlined bodies and long, narrow wings, hawk moths are among the fastest flying insects, reaching speeds of over 30 mph (50 kph). Their wings measure 1½–6 in (3.5–15 cm) from tip to tip. Adult hawkmoths mostly suck nectar—some hovering like hummingbirds as they do this. They have a long, curled proboscis, up to 10 in (25 cm) long. Although some species are diurnal, most are active after dark. Hawkmoths often migrate long distances to breed. Their caterpillars are known as hornworms, because they have a spinelike process at the end of the abdomen. Some are very serious crop pests.

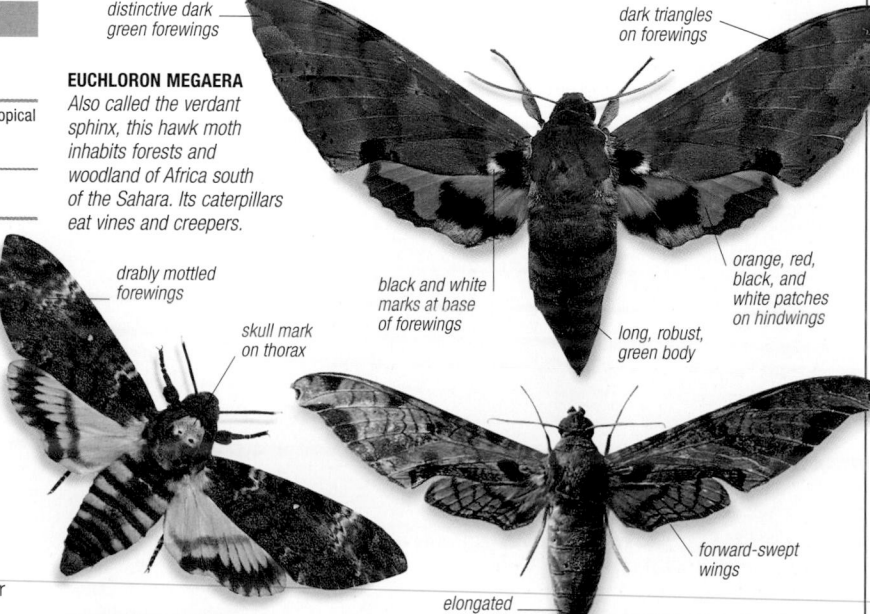

EUCHLORON MEGAERA
Also called the verdant sphinx, this hawk moth inhabits forests and woodland of Africa south of the Sahara. Its caterpillars eat vines and creepers.

ACHERONTIA ATROPOS
The death's head hawk moth is named for the skull-like marking on its thorax. It is found in Africa and Asia, but also migrates to Europe.

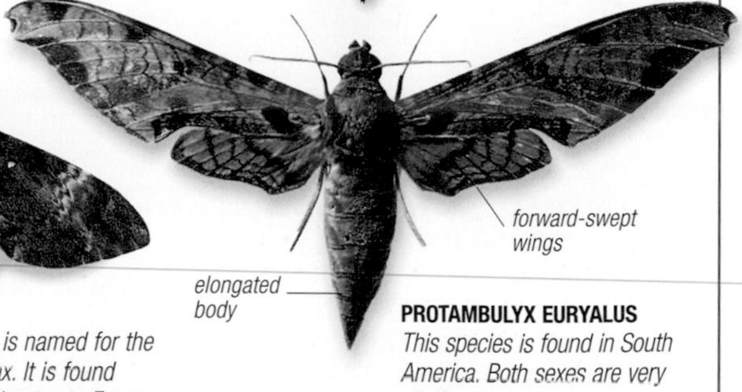

PROTAMBULYX EURYALUS
This species is found in South America. Both sexes are very similar in appearance.

Family Tineidae

Clothes moths and relatives

Occurrence 2,500 spp. worldwide; in rotting wood, fungi, dried organic matter, woollens, fabrics, dry foodstuffs

Many of these small, drab moths are pests, and are variously known as grain moths, case-bearing moths, and tapestry moths. Typically dull brown in color, some have a shiny golden appearance. Their head has a covering of raised, hairlike scales or bristles and the proboscis is short or absent. The wings, $\frac{5}{16}$–$\frac{3}{4}$ in (0.8–2 cm) across, are narrow and are held at a steep angle over the body at rest. Females lay 30–80 eggs over a period of 3 weeks. Caterpillars are saprophagous. Some of them make portable cases of silk and debris, others make a silk web wherever they feed. Adults do not generally feed. Many species will not readily fly, scuttling away from danger instead.

TINEOLA BISSELLIELLA
Also known as the common clothes moth, this widespread species inhabits clothing, carpets, and furs. The female lays eggs in the folds of clothes.

Family Tortricidae

Tortricid moths

Occurrence 5,000 spp. worldwide; on leaves, shoots, buds, fruit

These small moths are also known as bell moths. While most are brown, green, or gray, and patterned to blend in with bark, lichen, and leaves, some can be brightly colored. Their head is covered with rough scales. The wingspan is $\frac{5}{16}$–$1\frac{1}{2}$ in (0.8–3.4 cm) and the forewings, unusually rectangular, are held rooflike over the body at rest. A female can lay up to 400 eggs in a week, singly or in small groups, on the surface of fruit or leaves. The caterpillars feed on leaves, shoots, and buds of plants and often tie or roll leaves with silk. Being fruit- and stem-borers or gall-formers, many species are pests that cause damage to trees and fruit crops.

CLEPSIS RURINANA
Found in woodland and forests of Europe and Asia, the caterpillars of this species roll the leaves of deciduous trees and feed inside.

Family Uraniidae

Uraniid moths

Occurrence 100 spp. in tropical and subtropical regions; on plants, especially euphorbias

This family includes large, long-tailed, colorful, day-flying moths with iridescent wing scales, and dull, nocturnal species without tails. The South American and Malagasy species are so colorful and large that they are often mistaken for butterflies. The wings typically span $2\frac{1}{4}$–4 in (6–10 cm). Adults are often migratory, sometimes forming large swarms, in response to changes in larval food quality. The caterpillars of many species eat plants of the family Euphorbiaceae. They avoid ant predation by dropping from leaves on silk lines.

LYSSA ZAMPA
This large, Southeast Asian moth has a wingspan of 4–$4\frac{1}{4}$ in (10–10.6 cm), a little larger than most species in the family.

ALCIDES METAURUS
This day-flying zodiac moth from the rain forests of North Queensland, Australia, feeds on flowers during the day and flies back to the canopy in the evening.

CHRYSIRIDIA RHIPHEUS
Also known as the Madagascan sunset moth, this is a brightly colored day-flying species that inhabits woodland and forests of Madagascar.

Order Hymenoptera

Bees, wasps, ants, and sawflies

The order Hymenoptera is a vast group of insects that contains about 198,000 species in 91 families. It is divided into 2 suborders: plant-eating sawflies (Symphyta); and wasps, bees, and ants (Apocrita), which—unlike sawflies—have a narrow "waist" and, in females, an ovipositor that may sting. Most species have 2 pairs of membranous wings, joined in flight by tiny hooks. Ecologically, these insects are of tremendous importance in acting as predators, parasites, pollinators, or scavengers. Social Hymenoptera, including ants and bees, are the most advanced insects on earth.

Family Agaonidae

Fig wasps

Occurrence 650 spp. in subtropical and tropical regions; on fig trees

Fig wasps have evolved a symbiotic relationship with fig trees. The female wasps, no more than $\frac{1}{8}$ in (3 mm) long, crawl inside figs, which are lined with tiny flowers. They pollinate some flowers, and lay eggs in others. After developing, the wingless males fertilize the females, which fly away to find new trees.

BLASTOPHAGA PSENES
The common fig wasp is found in Asia, southern Europe, Australia, and California. It pollinates Ficus carica, *the common fig. Females are 20 times more common than males.*

Family Apidae

Honey bees and relatives

Occurrence 1,000 spp. worldwide; in all well-vegetated, flower-rich areas

This family includes one of the world's most useful insects—the honey bee—together with orchid bees and bumble bees. They provide honey and wax, and pollinate most of the world's plants. Typically, they are $\frac{1}{8}$–1 in (0.3–2.7 cm) long. Most females have a venomous sting and a basket to carry pollen on the hindlegs. Honey bees and bumble bees are social insects, living in complex colonies consisting of a queen, males, and sterile female workers. The honey bee nest is a vertical array of wax combs divided into thousands of cells for rearing the young and storing pollen and honey. Queens may lay more than 100 eggs a day. Bumble bee nests are made of grass with wax cells, and the queens raise fewer offspring.

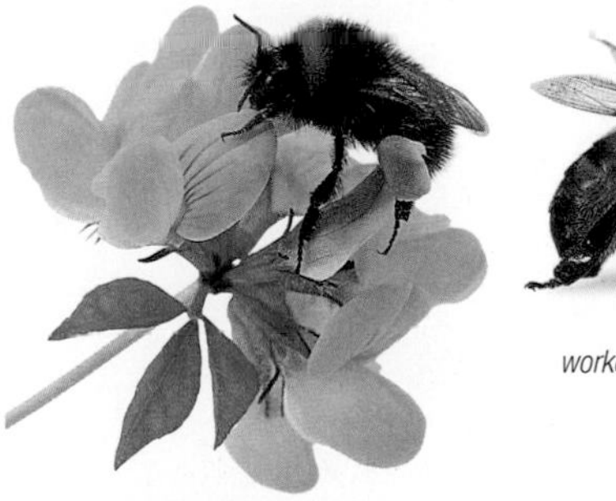

BOMBUS TERRESTRIS
Like all its relatives, the European buff-tailed bumble bee has a body covered by "fur," enabling it to fly in the cool conditions of early spring.

APIS MELLIFERA
Originally from Southeast Asia, the honey bee is now raised all over the world. A honey bee colony has strict division of labor. The queen lays all the colony's eggs, while the workers gather food and tend the larvae. The queen is fertilized by a drone.

Order Hymenoptera *continued*

Family Braconidae

Braconid wasps

Occurrence 25,000 spp. worldwide; on suitable caterpillar hosts

Like ichneumon wasps (see below), braconids are parasitoids—animals that develop inside a living host, eventually killing it. Braconids mainly attack the caterpillars of moths and butterflies, often laying over 100 eggs on or in their victims. The developing larvae feed on the caterpillar's tissues, pupate, and fly away as adults. Adult braconids are typically less than 7/32 in (7 mm) long, and reddish brown or black. Many species are used as biological controls against insect pests.

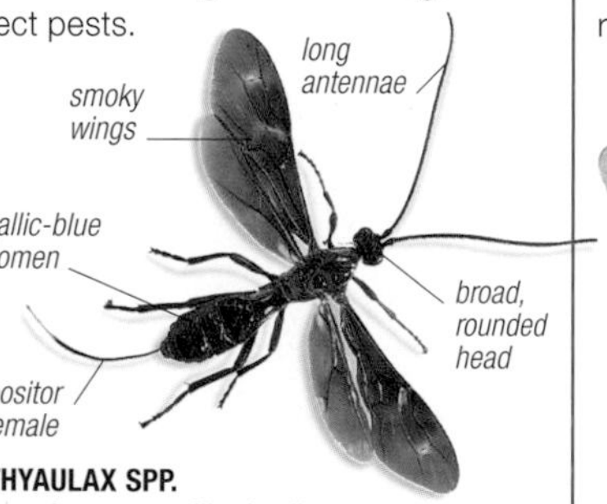

BATHYAULAX SPP.
Larger than most of its family, these braconids measure ½–¾ in (1.5–2 cm), and are found in Africa and Southeast Asia. They are used to control stem-boring caterpillar pests of cereal crops.

Family Chalcididae

Chalcid wasps

Occurrence 1,800 spp. worldwide; on suitable insect hosts

Mostly under 5/16 in (8 mm) long, chalcids make up a small but important family of parasitic insects. Many species lay their eggs in beetle larvae, or in the caterpillars of butterflies and moths, but some are hyperparasites, attacking the eggs or larvae of other parasitic insects. Adult chalcids are typically dark brown, black, red, or yellow, sometimes with a metallic sheen. Their larvae—like those of other wasps—are white and grublike, without legs. Some chalcid species are bred and released to control insect pests.

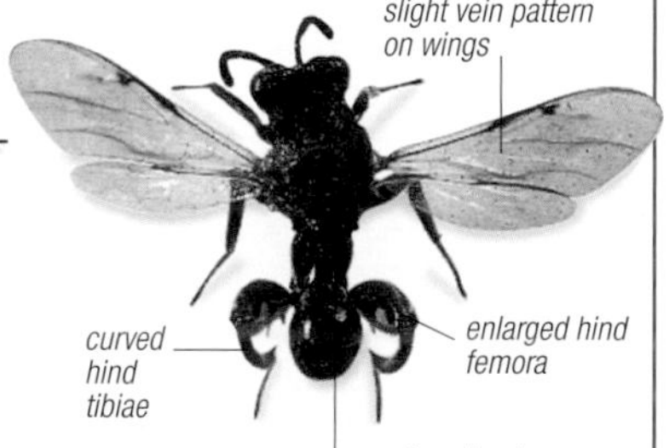

CHALCIS SISPES
This species is native to Europe and Asia. Its larvae parasitize the larvae of soldier flies.

Family Chrysididae

Jewel wasps

Occurrence 3,000 spp. worldwide; on suitable insect hosts

Jewel wasps, also known as cuckoo or ruby-tailed wasps, have very hard and dimpled bodies that protect them from bee and wasp stings. They are 1/16–3/4 in (0.2–2 cm) long. Most species are a very bright metallic-blue, green, or red, or combinations of these colors. The abdomen is hollowed underneath, allowing the wasp to curl up if attacked. Typically, the female lays an egg in the nest of a solitary bee or wasp, and the larva immediately hatches. The jewel wasp larva then feeds on the host larva and also eats the host's provisions. The adults feed on nectar and liquids.

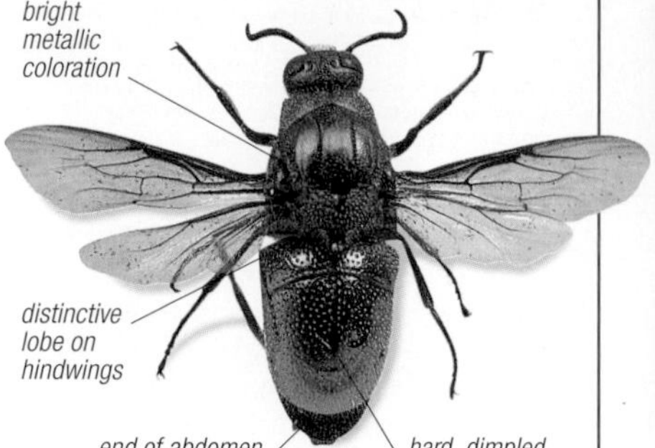

STILBUM SPLENDIDUM
The splendid emerald wasp is a North Australian species that parasitizes solitary mud-nesting wasps.

Family Cynipidae

Gall wasps

Occurrence 1,250 spp. mostly in the Northern Hemisphere; on trees, plants

Gall wasps are shiny, reddish brown or black, and usually fully winged. They measure 1/32–11/32 in (1–9 mm) in length; males are usually smaller than females. The latter have a laterally flattened abdomen and a humped thorax. Many species, by a process not yet fully understood, induce the growth of species-specific galls on oaks and related trees by laying eggs inside the plant tissue. The galls protect and nourish the developing larvae. Life cycles of gall-forming species involve sexual and asexual generations.

ANDRICUS SPP.
These gall wasps are widespread in Europe, and use various oak trees as their hosts.

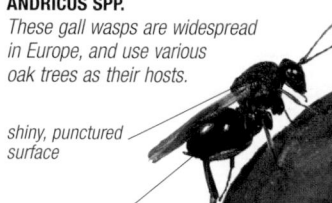

Family Ichneumonidae

Ichneumon wasps

Occurrence 60,000 spp. worldwide, especially temperate regions; wherever suitable insect hosts are found

Ichneumon wasps are of great ecological importance, because they parasitize other insects, including many pests. Using their slender ovipositors, the females lay eggs in the larvae and pupae in a wide range of hosts, sometimes drilling through several inches of wood to reach them. Upon hatching, the ichneumon larvae eat their hosts from inside. Adult ichneumons measure 1/8–1½ in (0.3–4.2 cm) long, and are typically slender, with a narrow waist.

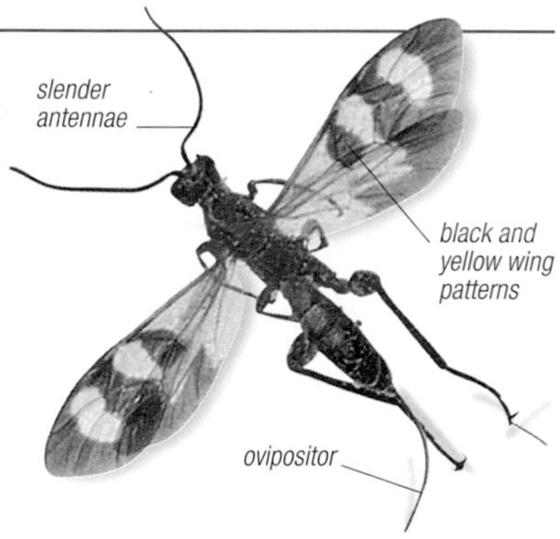

PARACOLLYRIA SPP.
These African wasps have a bright coloring that works as a warning to predators and is mimicked by other insects.

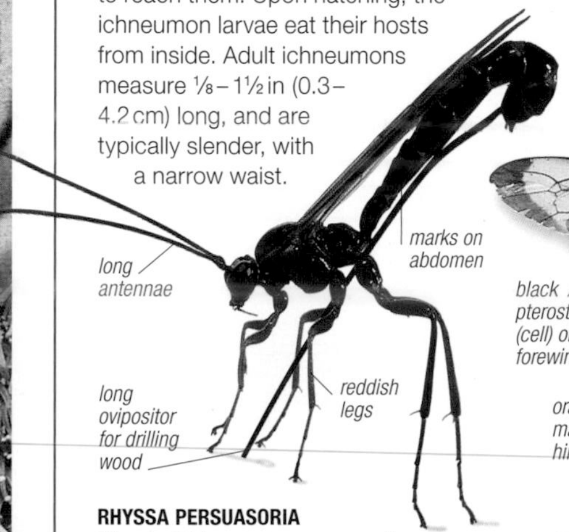

RHYSSA PERSUASORIA
The giant ichneumon wasp uses its long, drill-like ovipositor to reach the larvae of horntail sawflies deep in pine lumber.

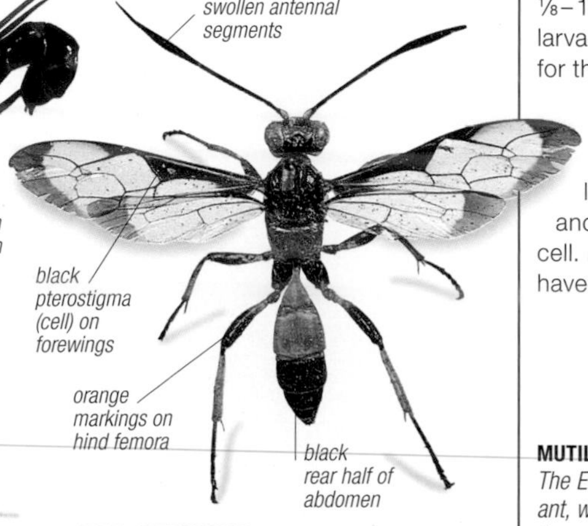

JOPPA ANTENNATOR
This species is characterized by swollen antennal segments and contrasting, wasplike coloration.

Family Mutillidae

Velvet ants

Occurrence 5,000 spp. worldwide, especially subtropical and tropical regions; females found in dry habitats

Despite their name, these are not ants. The name velvet ant comes from the soft, velvety hairs that cover the antlike, wingless females; males have fully developed wings. Velvet ants are 1/8–1 in (0.3–2.5 cm) long. They use the larvae of other wasps or bees as food for their larvae. Females bite open a suitable host cell and, if the larva in it is fully grown, lay their eggs on it and reseal the cell. The larvae then eat the host larva, and pupate inside the cell. Female velvet ants have powerful stings.

MUTILLA EUROPEA
The European velvet ant, widespread in the warmer regions of Europe, parasitizes species of bumble bee.

Family Mymaridae

Fairyflies

Occurrence 1,400 spp. worldwide; in a wide variety of habitats, wherever insect hosts are found

The smallest flying insects on earth belong to this family. Fairyflies are 1/16–3/16 in (0.2–5 mm) long, and are dark brown, black, or yellow. The narrow front wings have a fringe of hairs but no obvious venation. The hindwings appear stalked, and are strap shaped. The females of all species parasitize the eggs of other insects, mostly of plant-hoppers and other bug families. Several fairyfly species are used as biological control agents against insect pests.

ANAGRUS OPTABILIS
This is a specialized parasitoid of the eggs of certain plant-hoppers. Related species have been used to control plant-hoppers that attack rice crops.

Family Pompilidae

Spider-hunting wasps

Occurrence 4,000 spp. worldwide, especially tropical and subtropical regions; wherever spider hosts are found

True to their name, these wasps can paralyze spiders with their strong venom. Females fly or run along the ground in search of spiders, drag the immobilized spider to a nest prepared in mud or in a crevice, and lay a single egg on it before sealing the nest. Spider-hunting wasps have a painful sting. Most species are dark blue or black (although the wings may be dark yellow), and measure 3/16–3¼in (0.5–8cm) long. Males are smaller than the females.

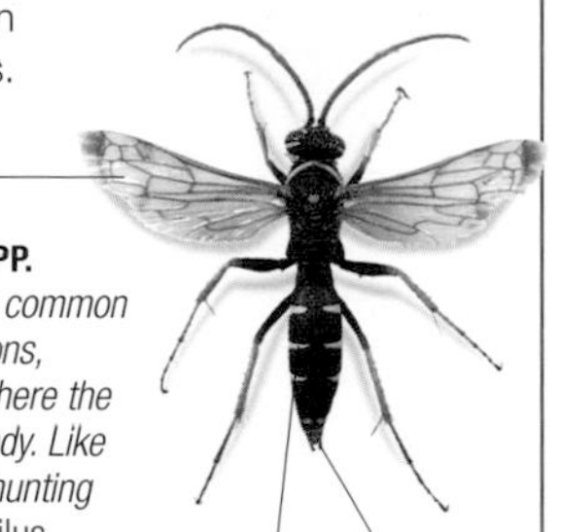

POMPILUS SPP.
This genus is common in warm regions, particularly where the ground is sandy. Like other spider-hunting wasps, Pompilus *wasps are solitary.*

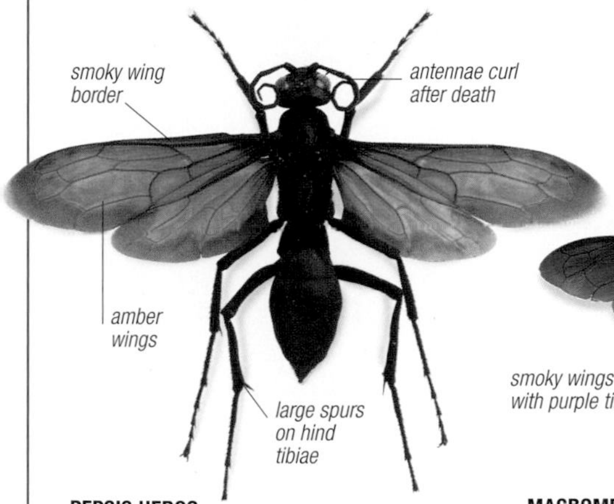

PEPSIS HEROS
The tarantula hawk is the largest spider-hunting wasp, up to 3¼in (8cm) long. As its name suggests, it preys on tarantulas.

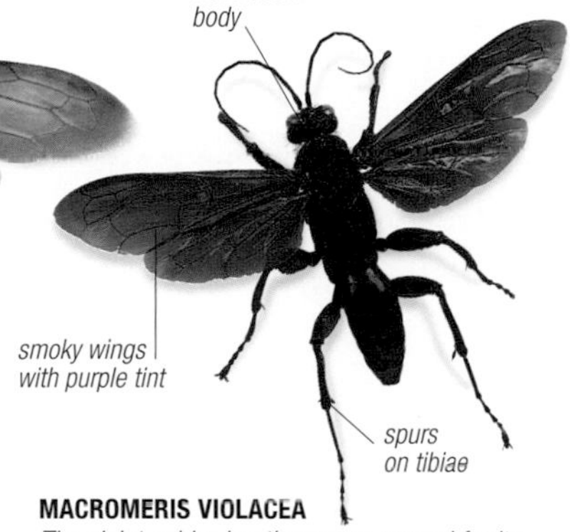

MACROMERIS VIOLACEA
The violet spider-hunting wasp, named for its iridescent purplish wings, is found in Southeast Asia. It is 2–2¼in (5–5.5cm) long.

Family Sphecidae

Digger wasps

Occurrence 8,000 spp. worldwide; in a variety of microhabitats—plant stems, soil, rotten wood

Some species of this family are also known as solitary hunting wasps, sand wasps, or mud dauber wasps. They are 5/32–1¾in (0.4–4.8cm) long, relatively hairless, and often brightly colored. Digger wasps are solitary, and nest in plant stems, soil, or rotten wood. The female catches an insect or a spider, paralyzes it, and buries it in a nest where she lays her egg. In this way, the emerging larva has a ready supply of food.

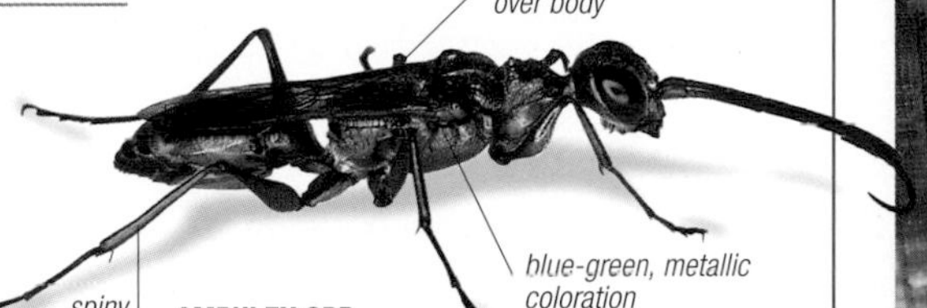

AMPULEX SPP.
These are cockroach-hunters that may enter houses in search of their quarry. This genus is found in tropical parts of the world.

Family Scoliidae

Mammoth wasps

Occurrence 350 spp. worldwide, especially in tropical regions; wherever scarab beetle hosts are found

As their name suggests, these are large wasps, measuring 3/8–2¼in (1–5.6cm) in length, with stout bodies that are densely covered with hair. They are bluish black with reddish brown markings. Males are smaller and slimmer than females, with longer antennae. After mating, females hunt for the larvae of scarab beetles. They then sting, paralyze, and lay a single egg on each larva. The wasp larva pupates in a cocoon that it spins inside the beetle larva's remains.

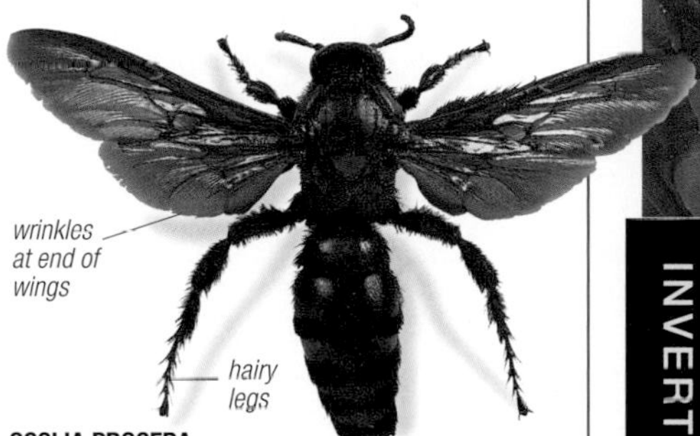

SCOLIA PROCERA
Found in Java, Borneo, and Sumatra, wherever scarab beetle larvae exist; this insect has a painful sting but is not very aggressive.

Family Vespidae

Social wasps

Occurrence 4,000 spp. worldwide; found in a variety of habitats, in nests and foraging widely for prey

The most familiar social wasps are the paper wasps and yellow jackets, which make nests out of chewed wood or other fibers. The queen overwinters, makes a nest in spring, and raises the first brood herself. Workers cooperate with her in caring for subsequent broods. Larvae develop in horizontal "combs," and are fed chewed up insects by sterile female workers. The adult wasps, which measure between 5/32 and 1½in (0.4–3.6cm) in length, roll or fold their wings longitudinally, rather than hold them flat over the body. Nearly all social wasps have warning coloration. They can also have a painful sting.

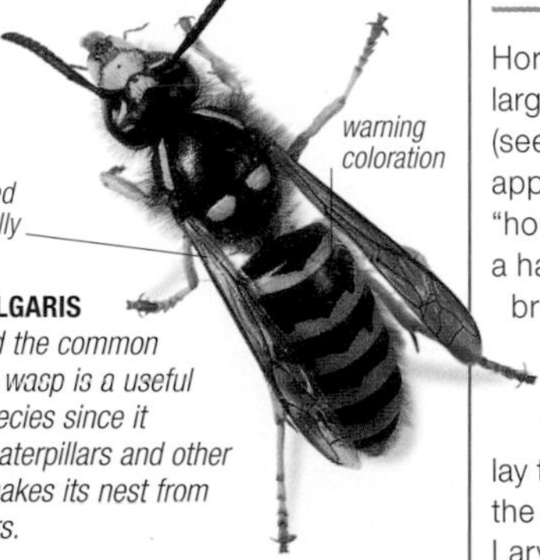

VESPA VULGARIS
Also called the common wasp, this wasp is a useful garden species since it removes caterpillars and other pests. It makes its nest from wood fibers.

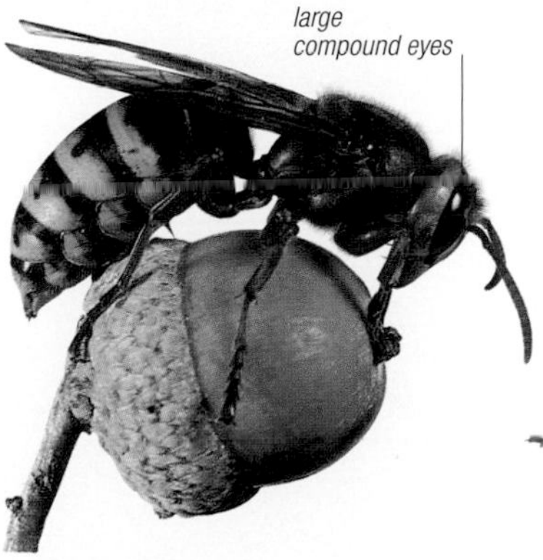

VESPA CRABRO
Commonly known as the European hornet, this wasp makes its nest in hollow trees. Its colonies have only a few hundred workers.

VESPULA GERMANICA
Colonies of the German wasp may be perennial, with more than one queen. This wasp lives mostly in the Northern Hemisphere and was introduced to North America in the late 1800s.

Family Siricidae

Horntails

Occurrence 100 spp. in temperate areas of the Northern Hemisphere; on and near deciduous and coniferous trees

Horntails or woodwasps are the largest members of the sawfly suborder (see right). Despite their menacing appearance, they do not sting—the "horn" at the end of their abdomen is a harmless spine. They are reddish brown, black and yellow, or metallic-purple, and measure ¾–1½in (1.8–4.2cm) in length. Females have a long ovipositor and use it to lay their eggs inside tree trunks, in the process infecting them with fungi. Larvae burrow into heartwood and eat fungus and wood. Pupation occurs in a cocoon of silk and chewed wood.

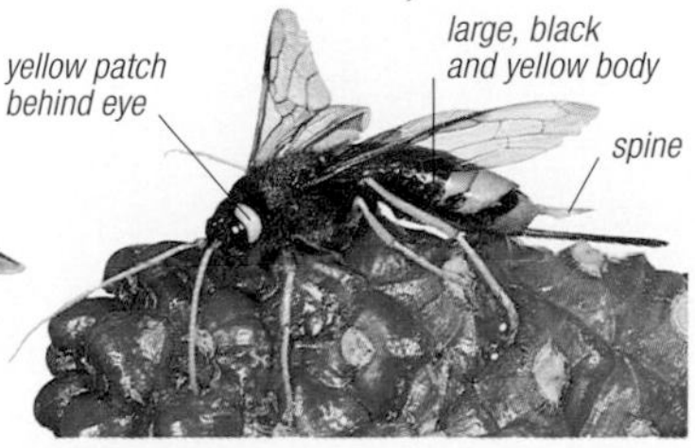

UROCERUS GIGAS
The horntail shown above is a female common horntail or giant woodwasp. The male of this harmless species is elusive.

Family Tenthredinidae

Common sawflies

Occurrence 6,000 spp. worldwide, except New Zealand; on a wide range of plants in gardens, pastures, woodland

Sawflies often look like wasps, but are more primitive insects, classified in a suborder of their own. Unlike wasps, bees, and ants, they do not have a narrow waist, do not sting, and are solitary. Common sawflies are 1/8–¾in (0.3–2.2cm) long. The females use their ovipositor to lay eggs in plants, and the caterpillar-like larvae, which are sometimes brightly colored, usually feed in the open on leaves. Common sawflies often attack fruit, vegetables, and trees, and can be highly destructive pests.

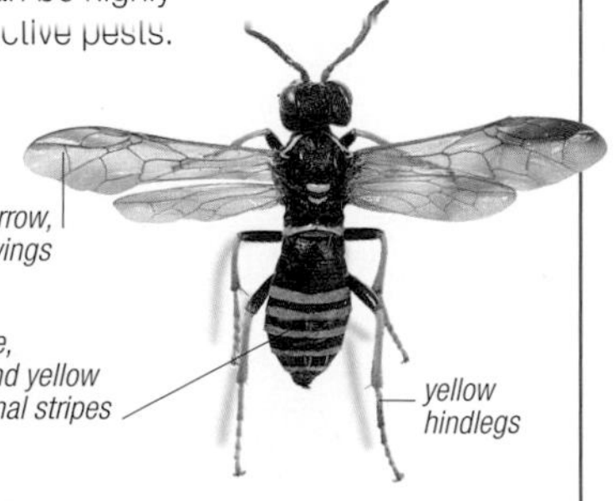

TENTHREDO SCROPHULARIAE
This figwort sawfly is a wasp-mimicking species, found in Europe and Asia. Its larvae feed on mullein and figwort plants.

Order Hymenoptera

Family Formicidae

Ants

Occurrence 9,000 spp. worldwide; in virtually all regions except in Antarctica and on a few oceanic islands

Habitat Most terrestrial habitats

Ants are found almost everywhere on Earth and have an immense impact on terrestrial ecosystems. In many, ants move more earth than earthworms and are vital in nutrient recycling and seed dispersal. Like termites and some bees and wasps, ants are highly social and live in colonies ranging from a handful of individuals sharing a tiny shelter to tens or hundreds of millions inhabiting structures reaching 20 ft (6 m) below ground. Two or more generations of ants overlap at any one time, and adults take care of the young. There are different classes (or castes) of ants within a colony, each with a specific task to perform: some are workers, who also defend the nest; others are dedicated to reproduction. Caste is determined by diet: a rich diet producing reproductives, a poor one producing sterile workers. Mating may take place on the wing or on the ground, after which the males die and the females lose their wings. All ants have glands that secrete pheromones—chemical messages used for communication between colony members as well as for trail making and defense. Numerous species of insects and plants have evolved symbiotic and often highly complex relationships with ants. In some cases, plants provide ants with food or homes; in return, the ants may protect the plants by disturbing potentially damaging herbivores, or stripping leaves from encroaching vines. The ant family is divided into 10 distinct subfamilies, of which the biggest by far are the Myrmicinae and the Formicinae. Some myrmicine ants have stings, while formicines defend themselves by spraying formic acid. The infamous driver and army ants of tropical regions belong to the subfamily Dorylinae. Their colonies move in massive columns, up to several million strong, often raiding termite or ant nests.

EATING HABITS

Some ants are herbivorous seed gatherers or fungus eaters. Other species are carnivorous or omnivorous, while some rely exclusively on the honeydew produced by sap-sucking bugs.

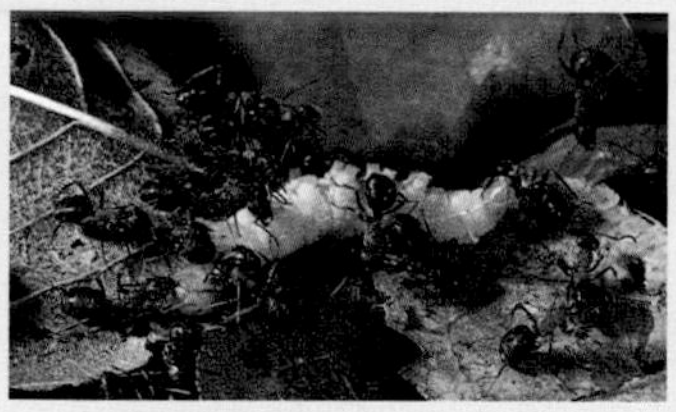

WORKERS WITH PREY
Worker wood ants, Formica rufa, *may cooperate in dragging a large item of prey back to the nest, where it will be cut up.*

LEAFCUTTERS
Leafcutter ants, Atta *spp., chew pieces of leaves into pulp, which they use to fertilize fungus beds—their sole source of food.*

DIFFERENT CASTES

A colony will contain 3 castes of ants: workers, which are always wingless sterile females, queens, and males (which are usually winged and form mating swarms with queens at certain times of the year). The physical differences between the castes of the African driver ant *Dorylus nigricans* (shown here at relative sizes) are typical of most ants, although queens in this species are wingless.

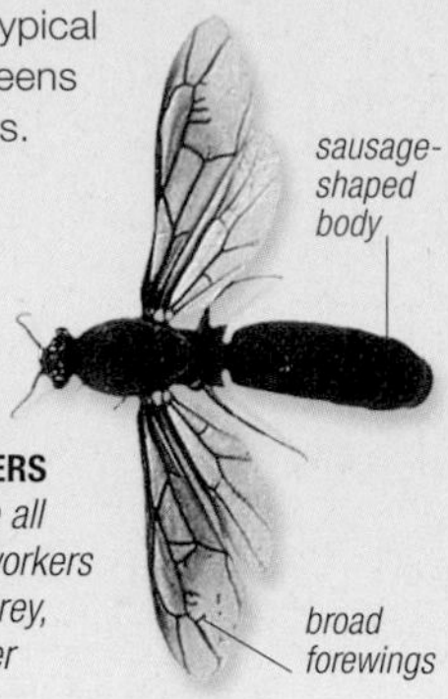

MALE ANT
Because of its body shape, the male is called a sausagefly. Males are often attracted to lights after dark.

FEMALE WORKERS
Worker ants are all female. Large workers carry items of prey, while the smaller workers tend to carry liquid.

QUEEN ANT
The queens of this species are the largest of any ant and can lay 1–2 million eggs every month.

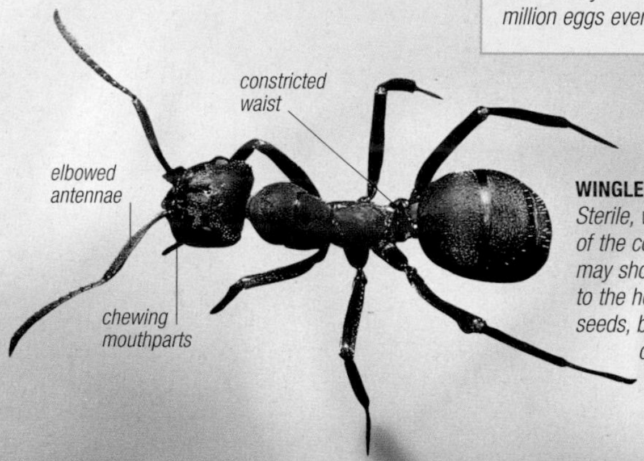

WINGLESS FEMALE
Sterile, wingless females do the work of the colony, and in some species may show amazing adaptations to the head and jaws for crushing seeds, blocking nest entrances, dismembering enemies, and other purposes.

ON THE DEFENSIVE
This worker wood ant, Formica rufa, *has taken a defensive posture with its jaws wide open and abdomen curled under and forward. In this position, it will be able to spray formic acid toward an attacker.*

Centipedes and millipedes

PHYLUM	Arthropoda
SUBPHYLUM	Mandibulata
SUPERCLASS	Myriapoda
CLASSES	Chilopoda Diplopoda
ORDERS	21
FAMILIES	171
SPECIES	About 13,150

Centipedes and millipedes are terrestrial arthropods with many legs, each of their numerous body segments carrying one or 2 pairs. Their heads have biting mandibles and a single pair of antennae. Far less abundant and diverse than insects, they are confined to humid microhabitats because they do not have a waterproof cuticle. However, like insects, they breathe by taking gases directly into the body through openings in the cuticle. Centipedes have a single pair of legs on each trunk segment. They can run fast, although some soil-living species move slower. They are carnivorous and have venomous claws to kill prey. Millipedes are all slow moving, with elongated, cylindrical bodies. Most of their trunk segments are fused together in pairs, called diplosegments, each bearing 2 pairs of legs. They are mostly herbivores or scavengers, using their toothed mandibles to chew rotting organic matter, soil, plants, algae, and moss.

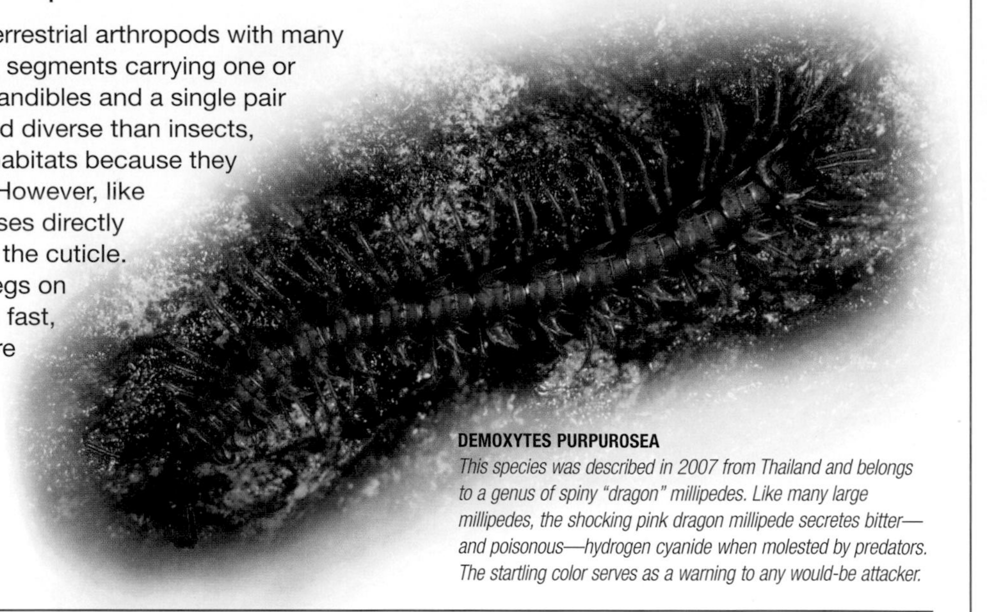

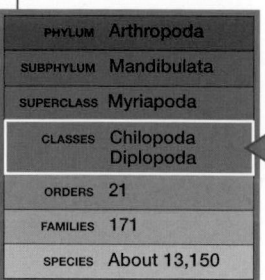

DEMOXYTES PURPUROSEA
This species was described in 2007 from Thailand and belongs to a genus of spiny "dragon" millipedes. Like many large millipedes, the shocking pink dragon millipede secretes bitter—and poisonous—hydrogen cyanide when molested by predators. The startling color serves as a warning to any would-be attacker.

Class Chilopoda

CENTIPEDES

Elongated and flattened, centipedes have a trunk made up of at least 16 segments, most of which carry one pair of legs; the last pair of walking legs is always longer than the rest. Most species are yellowish or brownish, and all have fine, sensory hairs on the body. There are 5 orders, 24 families, and 3,149 species.

Family Lithobiidae

Lithobiids

Occurrence 975 spp. worldwide, especially in temperate regions of the Northern Hemisphere; in cracks, crevices

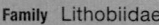

Lithobiids are mostly reddish brown but some can be brightly colored. The tough and flattened body is ¼–1¾ in (0.6–4.5 cm) long. There are 15 pairs of legs, and threadlike antennae with 75–80 segments. The young emerge from the egg with 7 pairs of legs and add a leg-bearing trunk segment at each molt. Lithobiids have a lifespan of 5 to 6 years.

LITHOBIUS VARIEGATUS
The banded stone centipide inhabits leaf litter in Europe's deciduous woodland. It does climb trees in search of food but most often sits and waits on the ground for prey to pass by.

Family Scolopendridae

Scolopendrids

Occurrence 409 spp. worldwide, especially in subtropical and tropical regions; in soil, leaf litter, cracks, crevices

The largest centipede in the world, *Scolopendra gigantea* from South America, belongs to this family. The body length of scolopendrids ranges from 1¼–12 in (3–30 cm). Nocturnal hunters, they are robust enough to catch and kill mice and frogs with their venomous claws, the smaller species being the most deadly. Scolopendrids are typically brightly colored and may be yellow, red, orange, or green, often with dark stripes or bands. They have 21 or 23 pairs of legs, and threadlike antennae with fewer than 35 segments. Usually, there are 4 ocelli, or simple eyes, on either side of the head. Females lay their eggs under soil, rocks, or loose bark.

SCOLOPENDRA HARDWICKEI
Tiger centipedes owe their common name to the markings on their body and their predatory habits. They are found in Southeast Asia, living under stones, rotting wood, and loose bark.

Class Diplopoda

MILLIPEDE

Despite their name, millipedes never have a thousand legs. Some have as many as 750, but even these species are slow runners. To defend themselves, most millipedes rely primarily on their tough exoskeletons, which are reinforced by calcium carbonate. When touched, many also coil up or produce toxic chemicals. This class has 9,973 species, classified in 16 orders and 147 families.

Family Julidae

Cylinder millipedes

Occurrence 903 spp. in the Northern Hemisphere; in soil, leaf litter, caves, under stones and rotting wood

These millipedes have rounded bodies that can range from 5/16–3¼ in (0.8–8 cm) in length. Most are dull-colored but a few have red, cream, or brown spots. The antennae are fairly long and slender. As in all millipedes, the females lay their eggs in nests in soil. There are typically 7 nymphal stages. Cylinder millipedes are common in Europe and Asia.

TACHYPODOIULUS NIGER
This species is commonly known as the black snake millipede. It has a black trunk and white legs, and is up to 2 in (4.8 cm) long. It inhabits woodland in Western Europe and climbs walls, trees, or shrubs to feed after dark.

Family Glomeridae

Pill millipedes

Occurrence 212 spp. in warm and cool temperate regions of the Northern Hemisphere; in soil, caves

Measuring 1/16–¾ in (0.2–2 cm) in length, these broad-bodied millipedes are either small and drably colored or large and brightly marked. The trunk is made up of 13 segments, and the shape of the dorsal plates allows the millipede to roll into a tight ball if threatened. Adults have 15 pairs of legs. Eggs are laid in soil, and when the young hatch, they initially have only 3 pairs of legs. Some members of this family, such as *Glomeris* species, may live up to 7 years.

GLOMERIS MARGINATA
The white-rimmed pill millipede can be found in soil and leaf litter in Europe, and parts of Asia and North Africa. When completely rolled up, it can be confused with the pill woodlouse (see p.584), although it has a shinier body.

Crustaceans

PHYLUM	Arthropoda
SUBPHYLUM	Mandibulata
SUPERCLASS	Crustacea
CLASSES	7
ORDERS	56
FAMILIES	About 1,000
SPECIES	About 70,000

Crustaceans are a very diverse group, ranging from water fleas and copepods—only just visible to the naked eye—to heavy-bodied crabs and lobsters, which include the largest arthropods alive today. Although some crustaceans, such as woodlice, have successfully adapted to life on land, the vast majority of species live in freshwater or in the sea. Aquatic crustaceans include some of the most abundant animals on earth, and they play a key part in many food chains.

Anatomy

Crustaceans differ from insects, centipedes, and millipedes in a number of ways. They have 2 pairs of antennae, compound eyes on stalks, and a cuticle that is often strengthened with calcium carbonate, especially in larger species. The head and thorax are often covered by a shield, or carapace, and the front of this usually extends to form a projection called the rostrum. The thorax has a variable number of segments. The abdomen ends in a tail-like telson. In crabs, the abdomen is short, flat, and tightly curled around to fit under the broad carapace.

Crustacean appendages—what we commonly understand as "legs"—have 2 branches and are specialized for a number of functions such as movement, sensing the environment, respiration, and egg brooding. The first pair of thoracic legs may be enlarged to form chelipeds, with strong claws that are used for defense, handling food, and even sexual signaling. The thoracic appendages, called pereopods, typically have gills. The basal part of some appendages helps in walking, while the abdominal segments usually carry paired swimming appendages called pleopods, or swimmerets.

Feeding

Crustaceans show a wide range of feeding strategies. Most large species capture their prey and kill it by stunning, crushing, or tearing it apart. Many species are filter feeders: they use their thoracic appendages to set up currents in the water and filter small particles of food, which are amassed and pushed into the mouth. Krill feed while swimming and trap particles in hairlike structures called setae on their thoracic legs. Small crustaceans may simply graze on particles of sediment, taking microorganisms from the surface.

Life cycle

Most species of crustaceans have separate sexes. They copulate, lay eggs, and brood the eggs, either within a brood chamber or attached to their legs. Typically, eggs hatch into tiny larvae that float away in the water, feed, and grow, developing trunk segments and legs.

RED CRAB MIGRATIONS
Christmas Island, in the Indian Ocean, has a population of about 100 million red crabs. Each year, triggered by the first rains of the wet season, they migrate from their inland habitats to mate and spawn by the coast. The males, which arrive first, dip themselves in sea water to replenish lost salt and moisture. After mating, females remain by the sea for about 2 weeks, waiting for the eggs to mature before releasing them along the coast.

Class Branchiopoda

Branchiopods

Occurrence c.1,000 spp. worldwide; mainly in freshwater bodies, some in saline or marine conditions

Branchiopods comprise 4 orders: tadpole shrimps (Notostraca), water fleas (Clacodera), clam shrimps (Conchostraca), and brine shrimps (Anostraca). There are 25 families. Although some branchiopods reach a length of 4 in (10 cm), most species are much smaller. The body may be covered by a shieldlike carapace,which, in some species, takes the form of a 2-halved (bivalvate) shell. The first pair of antennae and the second pair of mouthparts (maxillae) are small. The appendages have a flat, leaflike appearance, and carry fringes of fine bristles for filtering particles of food from the water. Branchiopods swim either rhythmically, using their appendages, or jerkily, using the second pair of antennae. Many species possess the respiratory pigment hemoglobin and may appear pink as a result. Females may produce more than one batch of eggs, each consisting of up to several hundred eggs. In some species, these are very drought resistant.

BRINE SHRIMP
Found worldwide, except in the Arctic and Antarctic, brine shrimps (family Artemiidae) live in saline lakes and pools. Their life cycles are rapid, suiting them to habitats that can dry up quickly. Like all brine shrimps, members of the genus Artemia, *such as the one shown here, lay eggs that can be dormant for up to 5 years.*

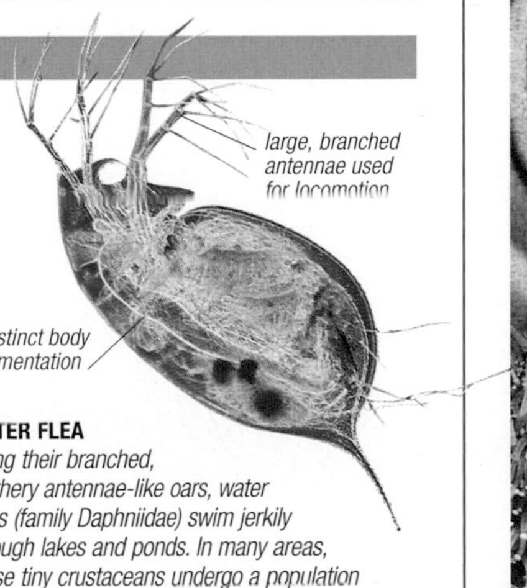

WATER FLEA
Using their branched, feathery antennae-like oars, water fleas (family Daphniidae) swim jerkily through lakes and ponds. In many areas, these tiny crustaceans undergo a population explosion every spring, providing a huge supply of animal food for fishes. The species above belongs to the genus Daphnia.

INVERTEBRATES

Class Ostracoda

Mussel shrimps

Occurrence 5,386 spp. worldwide; in marine, brackish water, freshwater bodies, a few species are terrestrial

Mussel shrimps, also known as seed shrimps, are small, gray, brown, or greenish crustaceans that can measure less than 1⁄32 in (1 mm) to just over 1¼ in (3 cm) in length. They make up 61 families in 6 orders. Their indistinctly segmented body is fully enclosed in a hinged carapace, which can be variously shaped: circular, oval, or rectangular-oval. Its surface may be smooth or sculptured depending upon the species. Mussel shrimps resemble bivalves (see p.595) in having a carapace with 2 halves, strengthened by calcium carbonate, that can be closed by a muscle. The head is the largest part of the body. All species have a single simple eye, the nauplius eye, in the middle of the head; some also have a pair of compound eyes. Most species have separate sexes, which mate in order to reproduce; however, some are parthenogenetic. Some mussel shrimps emit light to attract mates, and the males of some even synchronize their flashing. Eggs are released into water or stuck to plants; they may be very tough and resistant to drying. Mussel shrimps are very important in aquatic food chains and are found worldwide, from shallow to very deep water, as crawlers, burrowers, or free swimmers. They may scavenge, or feed on detritus or particles suspended in the water; some are predatory. There are more than 10,000 ostracod fossil species, and they are important in oil prospecting since different types of fossils indicate different rock strata.

CYPRIDID MUSSEL SHRIMPS
Found in lakes, pools, and swamps, cypridid mussel shrimps (family Cyprididae) include the majority of freshwater species. Some swim, while others crawl on the bottom. They often reproduce parthenogenetically, and in some species males have never been found. Their eggs are resistant to drying.

CYPRIDINID MUSSEL SHRIMPS
The largest mussel shrimps, including the deep-sea genus Gigantocypris, *belong to this widespread, marine group (family Cypridinidae). Cypridinids are good swimmers. They propel themselves with their second antennae, which project from a small notch in the front end of their carapace valves.*

Class Maxillopoda

Maxillopods

Occurrence 17,987 spp. worldwide; in oceans (sediments to deep-sea trenches), freshwater, hot springs, on land

Copepods (Copepoda) and barnacles (Cirripedia) form the most significant subclasses of this large and diverse group of 25 orders and about 310 families. Copepods are abundant everywhere in the ocean—their combined mass exceeds billions of tons, and they are a major source of food for fishes. They are also found in freshwater, from mountain lakes to hot springs. Usually under 3⁄16 in (5 mm) in length, some parasitic species may grow up to 4 in (10 cm). Generally pale, they have no carapace and lack compound eyes. The head is fused with the trunk, and the first pair of thoracic appendages is used for feeding, while the rest help in swimming. Fertilization is external; the male simply attaches a sperm packet to the female's genitals.

Barnacles are entirely marine crustaceans—they may be stalkless, stalked, or parasitic. They have a shell made up of calcareous plates that surround the body, and 6 pairs of thoracic appendages. Most species are hermaphrodite but some are dioecious, with sexes separated in individuals. Barnacles may grow in such numbers on the underside of a ship, that they reduce its speed by about one third.

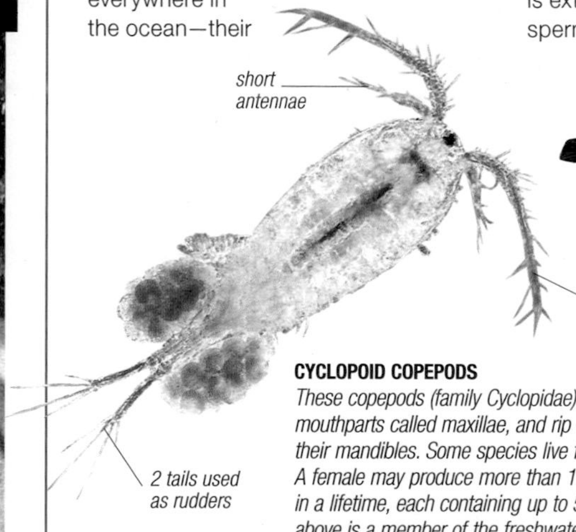

CYCLOPOID COPEPODS
These copepods (family Cyclopidae) grasp prey with mouthparts called maxillae, and rip off small pieces using their mandibles. Some species live for more than 9 months. A female may produce more than 10 pairs of egg sacs in a lifetime, each containing up to 50 eggs. Shown above is a member of the freshwater genus Cyclops.

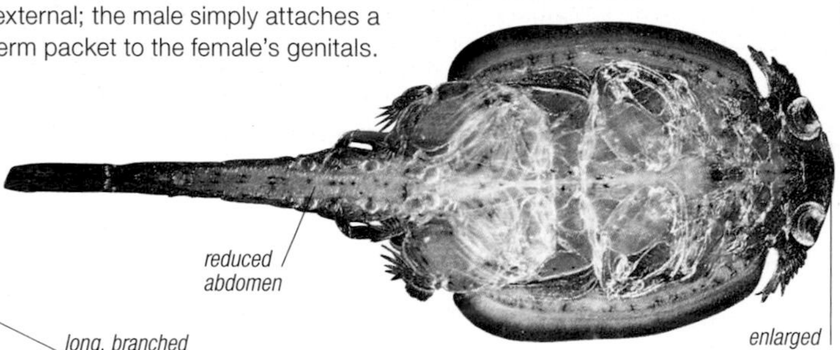

CALIGID COPEPODS
Members of the family Caligidae are ectoparasites on the bodies of marine and freshwater fishes. They have an enlarged head, with modified appendages to grip the host, and a reduced abdominal region. Some are red due to the presence of hemoglobin in their body. Caligus rapax *(shown here) pierces and sucks blood from the skin or gills of the host.*

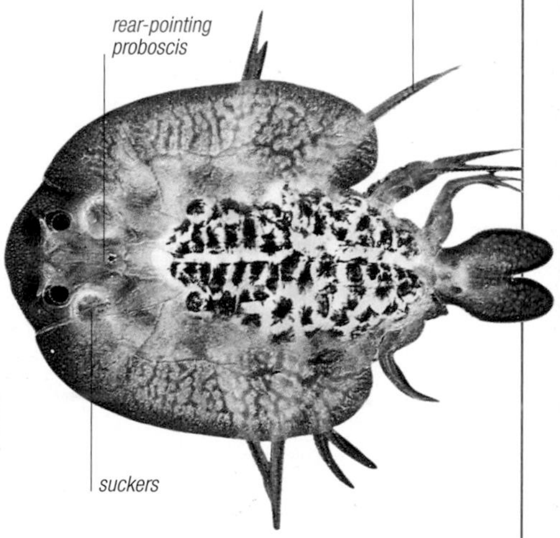

FISH LICE
Belonging to the subclass Branchiura, members of the family Argulidae are commonly called fish lice; their mouthparts are modified for sucking, enabling these parasites to feed on marine or freshwater fishes. A fish infested by them may get fungal infections or even die. Fish lice have a cephalothorax (the head fused with the first thoracic segment), a 3-segmented thorax, and a 2-lobed abdomen. The species shown above belongs to Argulus, *the largest genus in the family.*

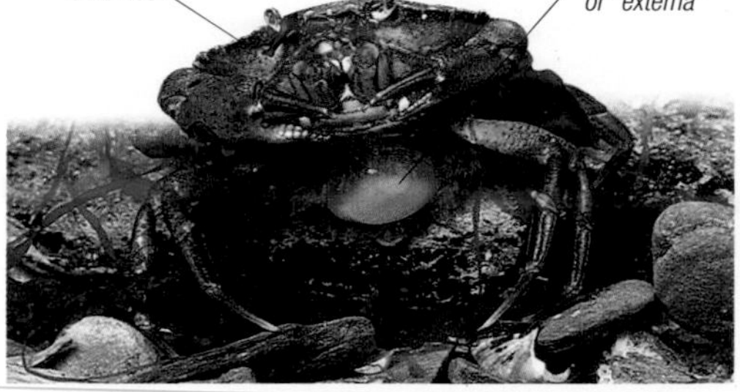

SACCULINIDS
The pink swelling on the underside of this crab is part of a sacculinid (family Sacculinidae)—a parasitic cirripede. Adults have highly branched bodies that spread through the tissues of their hosts, absorbing nutrients. Some, including the species of Sacculina *pictured, can infest half the host population in an area.*

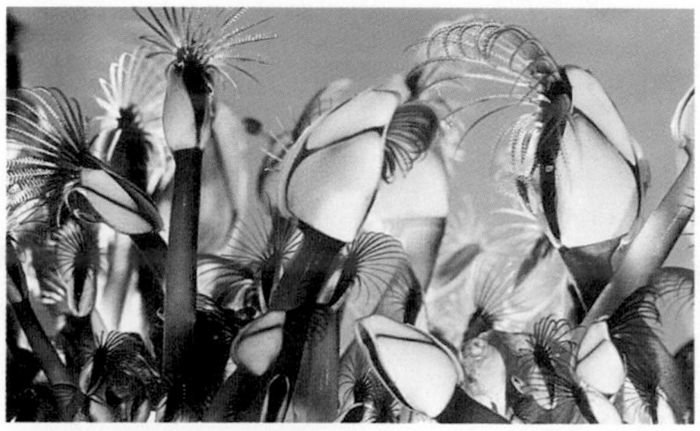

COMMON GOOSE BARNACLES
Goose barnacles (family Lepadidae) inhabit surface waters from polar to tropical oceans. The body has 2 parts: the flexible stalk and the main body. The larvae are free swimming but adults attach themselves to algae, floating wood, turtles, and ships. The largest barnacle, Lepas anatifera, *shown here, is up to 32 in (80 cm) long.*

COMMON ACORN BARNACLES
Members of the family Archaeobalanidae can be up to 4 in (10 cm) in width and height, although most are much smaller. They encrust surfaces in large numbers and filter passing food particles. Semibalanus balanoides *(pictured) withstands freezing in Arctic tidal zones, and can survive out of water for up to 9 hours in summer.*

Class Malacostraca

Malacostracans

Occurrence 38,032 spp. worldwide; in marine intertidal to abyssal zones, freshwater, terrestrial habitats

The world's largest arthropod, the Japanese spider crab (*Macrocheira kaempferi*) belongs to this large and diverse group of crustaceans, which comprises 18 orders and 606 families. It has a legspan of 13 ft (4 m)—in marked contrast to some other malacostracans, which are less than 1/32 in (1 mm) long. Despite this vast difference in size, malacostracans have many features in common. They are often brightly colored, and have tough exoskeletons strengthened with calcium carbonate. Their carapace—if present—acts as a gill chamber, and never covers the abdomen. They often have stalked eyes, and their antennae are prominent. The thorax is made up of 8 segments, while the abdomen has 6 segments and a flattened tail fan (telson) that can be moved rapidly in order to propel the animal away from predators. In many species, the first 1–3 thoracic appendages have become modified into maxillipeds, which are used in feeding; the remaining are used for locomotion. The abdominal appendages help in swimming, burrowing, mating, and egg brooding. Most species belong to the order Decapoda, which contains all the familiar shrimps, crayfish, crabs, and lobsters. They often have an enlarged first pair of appendages equipped with claws. Important in the marine food chain, many of the bigger species are of commercial value.

MANTIS SHRIMPS
Mantis shrimps (family Squillidae) use a unique spearing (sometimes smashing) technique to catch prey. The second pair of thoracic legs is large and specially designed for spearing, and is capable of striking very fast indeed. Mantis shrimps have a flattened, flexible body and stalked, compound eyes. Squilla *species, such as the one pictured right, inhabit tropical and subtropical seas, and are territorial.*

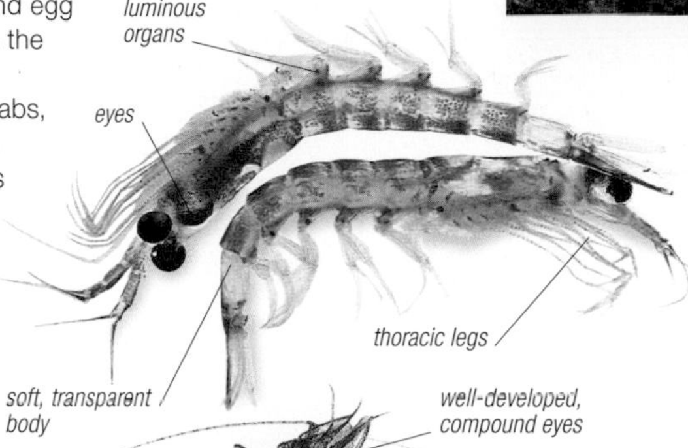

KRILL
Krill (family Euphausiidae) are slender, shrimplike crustaceans that filter food particles from sea water using the long hairs on their thoracic legs. They are highly gregarious, and also luminescent. The species shown left, Antarctic krill (Euphausia superba), *forms very large swarms that sometimes have a collective weight of more than 2 million tons. In the Arctic Ocean, these swarms are an important food for baleen whales.*

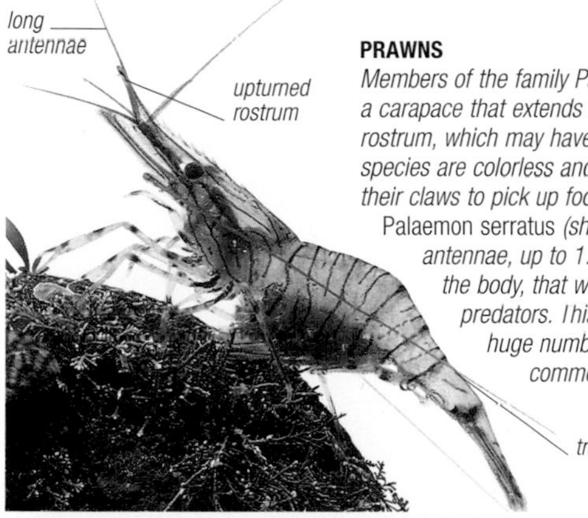

PRAWNS
Members of the family Palaemonidae have a carapace that extends forward to form a rostrum, which may have toothed edges. Most species are colorless and transparent. They use their claws to pick up food. The common prawn, Palaemon serratus *(shown left), has very long antennae, up to 1.5 times longer than the body, that warn it of the presence of predators. This species is caught in huge numbers since it is a common human food.*

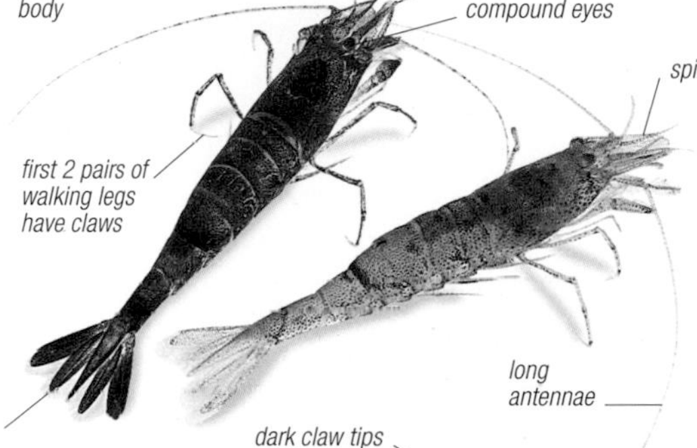

SHRIMPS
Pigment cells (chromatophores) present in their body allow these bottom-dwelling omnivores (family Crangonidae) to change color. The carapace is extended to form a short spinelike, smooth-edged rostrum. The common shrimp, Crangon crangon *(pictured left), is fished commercially along European coastlines.*

BURROWING SHRIMPS
These shrimps (family Callianassidae) make U- or Y-shaped, or branched, networks of burrows in silt and mud in shallow water. They are soft-bodied crustaceans, predatory on small organisms and worms. These shrimps are collected as fish bait in some parts of the world. The picture on the left shows a species from the genus Callianassa. *In this genus, one of the first thoracic legs may be much larger than the other.*

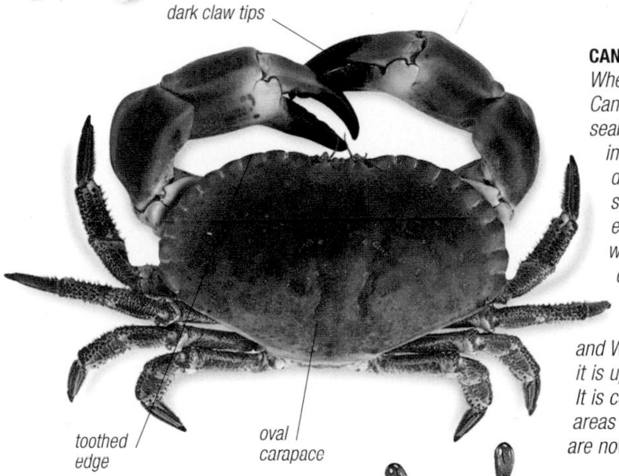

CANCRID CRABS
When adult, cancrid crabs (family Cancridae) typically live on the seabed, where they prey on other invertebrates and scavenge for dead remains. The species shown here is the European edible crab, Cancer pagurus, *which has a distinctive "pie crust" edge to its carapace. Found off the European Atlantic, Mediterranean, and West African coasts, it is up to 12 in (30 cm) across. It is commercially fished in many areas and large specimens are now rare.*

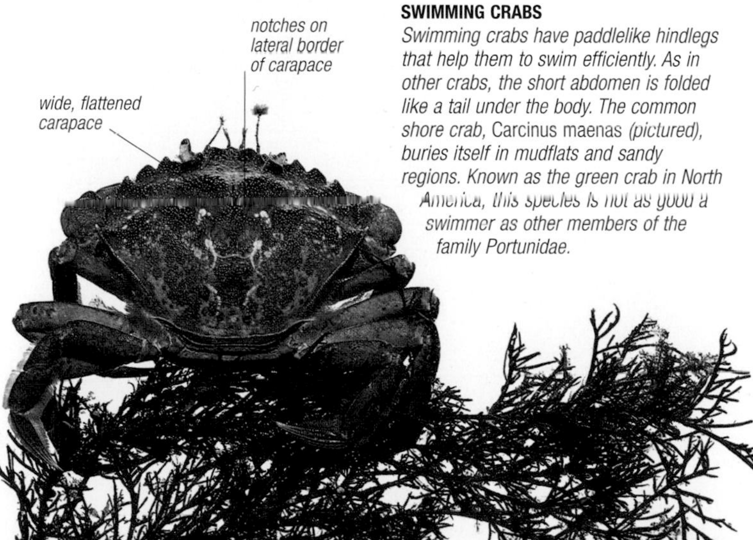

SWIMMING CRABS
Swimming crabs have paddlelike hindlegs that help them to swim efficiently. As in other crabs, the short abdomen is folded like a tail under the body. The common shore crab, Carcinus maenas *(pictured), buries itself in mudflats and sandy regions. Known as the green crab in North America, this species is not as good a swimmer as other members of the family Portunidae.*

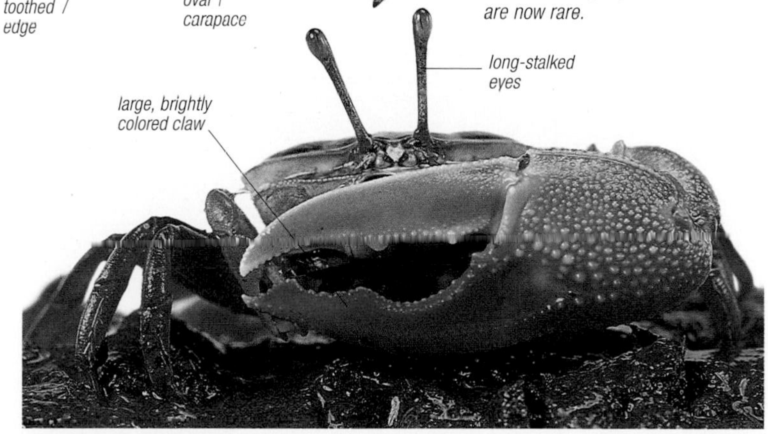

GHOST AND FIDDLER CRABS
In the tropics, ghost and fiddler crabs (family Ocypodidae) abound on low-lying coasts. Ghost crabs live in burrows on sandy beaches, and emerge after dark, feeding on debris washed up by the tide. They have keen eyesight, and are extremely rapid runners, disappearing into their burrows at the first hint of danger. Fiddler crabs, which are smaller, burrow into the mud of mangrove swamps. Males have one large claw, which they wave to attract females and to deter rivals. The claw often makes up half the male's weight. Fiddler crabs feed on organic matter present in silt. There are about 97 species; shown here is the male orange fiddler crab, Uca vocans, *from Africa and Southeast Asia.*

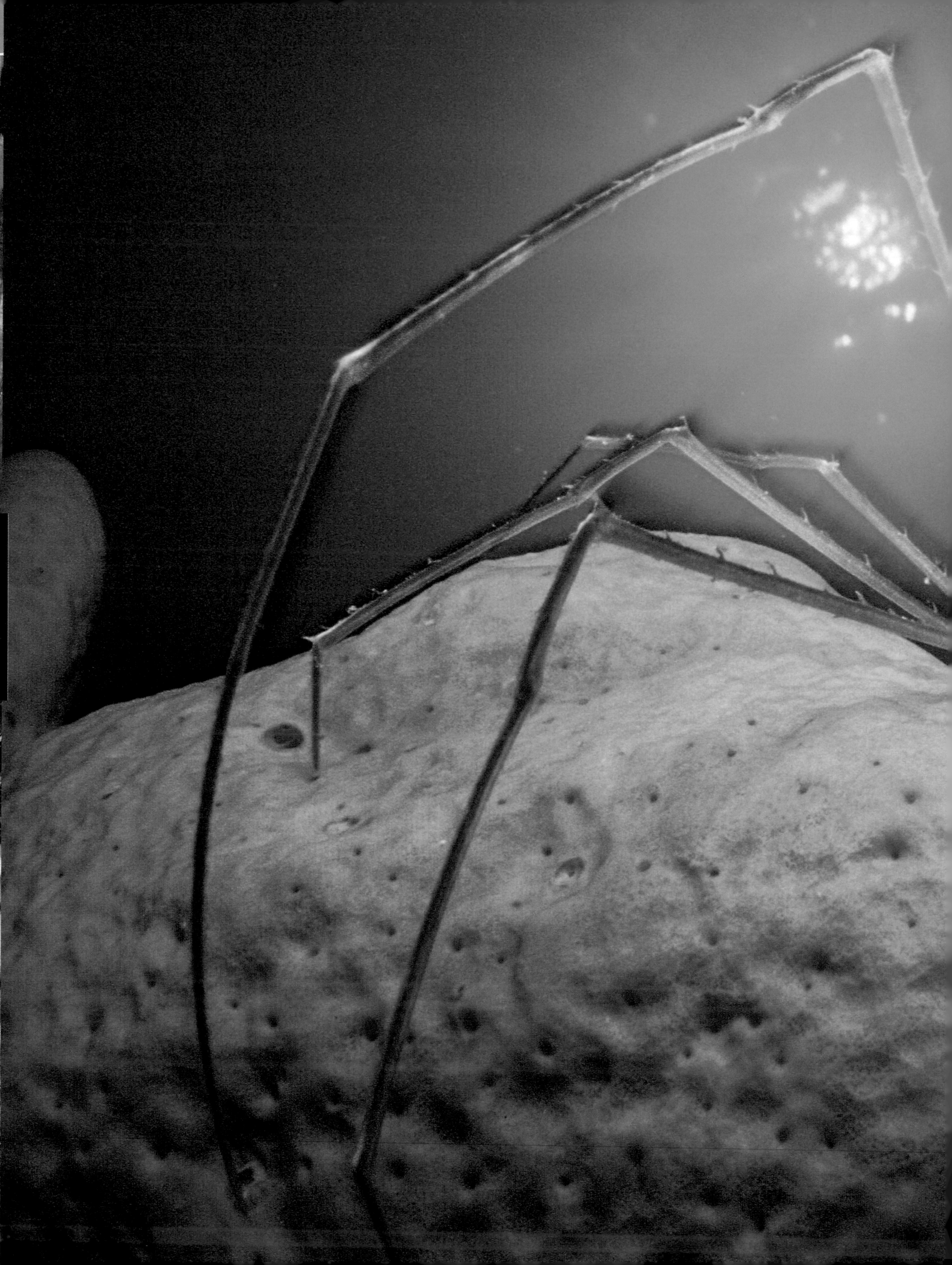

Class Malacostraca

Superfamily Majoidae

Spider crabs and relatives

Occurrence 1,026 spp. in all but polar oceans and seas; from the intertidal zone to waters 6,500 ft (2,000 m) deep

Spider crabs are recognized by their triangular carapace, a shape formed by a narrowing of the front edge, which is often extended as an elongated rostrum. As the common name implies, many species have long, slender legs. In most cases, the pincer-bearing legs (chelipeds) are not much longer than the other legs. Spider crabs use their chelipeds to attach sponges, seaweed, hydroids, and even detritus, such as wood fragments and broken worm tubes, to hooked hairs on their carapace and legs. In some species, the entire body can be covered, making the crab very difficult to see and protecting it from predators. If forced to move to a different location, some will change their camouflage to match their new surroundings. This behavior has given these species the alternative names of decorating or masking crabs. Spider crabs range from small species with a carapace length of around 5/16 in (8 mm) to the massive Japanese spider crab—also known as the giant spider crab—which has a carapace up to 20 in (50 cm) long (see below).

GIANT AMONG ARTHROPODS
The Japanese spider crab, Macrocheira kaempferi, *is the world's biggest living arthropod. Its legs may be up to 5 ft (1.5 m) long.*

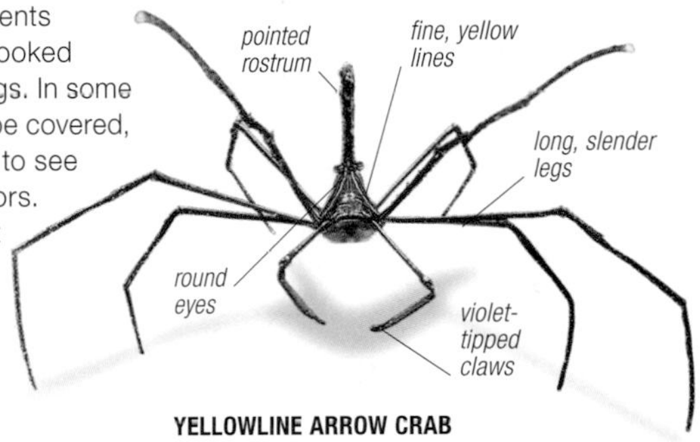

YELLOWLINE ARROW CRAB
Yellow lines on the carapace, running backward from the eyes, and the arrow-shaped rostrum give this species, Stenorhynchus seticornis, *its common name.*

DIMINUTIVE FORM
The dainty yellowline arrow crab—at most, 2½ in (6.5 cm) long—is found in the Atlantic Ocean from the Caribbean to as far south as Brazil. Males can tell how close a female is to laying eggs and, after they have mated, will guard the female from the attentions of other males.

Class Malacostraca *continued*

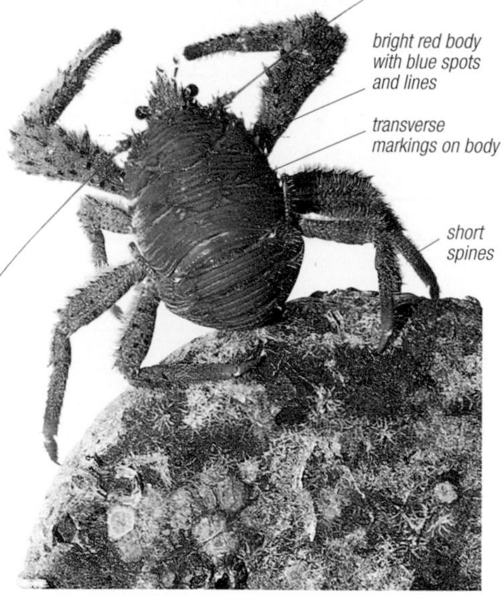

SQUAT LOBSTERS

Squat lobsters (superfamily Galatheoidae) are relatively slow movers but most species use their abdomen to swim rapidly backward to escape predators. Many species are mainly nocturnal and hide under rocks and in cracks during the day. Shown here is the blue-striped squat lobster, Galathea strigosa, *from European shallow seas.*

HERMIT CRABS

The most characteristic feature of hermit crabs (superfamily Paguroidea) is that they carry the shells of gastropods around as homes, changing an old shell for a bigger one as they grow. The white-spotted hermit crab, Dardanus megistos, *shown above, inhabits the lower shores of sandy and rocky, Northern and Western European coasts.*

LAND HERMIT CRABS

Like hermit crabs, land hermit crabs (family Coenobitidae) also carry shells around. An exception is the coconut or robber crab, Birgus latro, *pictured fighting above, which does not have a shell. It is the heaviest terrestrial crustacean, with a weight of up to 8¾ lb (4 kg). It lives in the tropics, and feeds mainly on the fruit of coconut palms. A good climber, its slow speed means it is easy to hunt and is becoming rare in some places.*

CRAYFISH

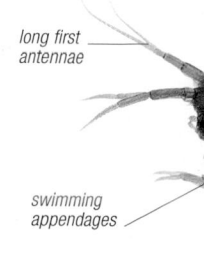

Members of the family Astacidae, freshwater crayfish resemble small lobsters. The first 3 pairs of legs have claws, the front pair being the largest. They make burrows in silt or mud where they hide during the day. The white-clawed crayfish, Austropotamobius pallipes, *shown above, inhabits well-aerated streams in southwest Europe.*

LOBSTERS

The family Nephropidae includes the largest and most commercially important crustaceans, the lobsters. Measuring up to 3¼ ft (1 m) in length, lobsters have a roughly cylindrical carapace with transverse and oblique grooves. The abdomen is elongated with a broad tail fan for rapid backward escape. The common lobster, Homarus gammarus, *shown above, has a blue upper and a yellowish lower surface. Using its large claws, it cracks open mollusks and feeds on them at night.*

PILL WOODLICE

Pill woodlice or pillbugs (family Armadillidiidae) have the ability to roll into a tight ball for protection. They have a rounded back as well as a rounded hind margin. Usually terrestrial, they have 2 pairs of lunglike organs (pseudotracheae) on the abdominal appendages for breathing air. As in many woodlice, the newly hatched young have one pair of legs fewer than they will have as adults. Pictured here is the common pill woodlouse, Armadillidium vulgare.

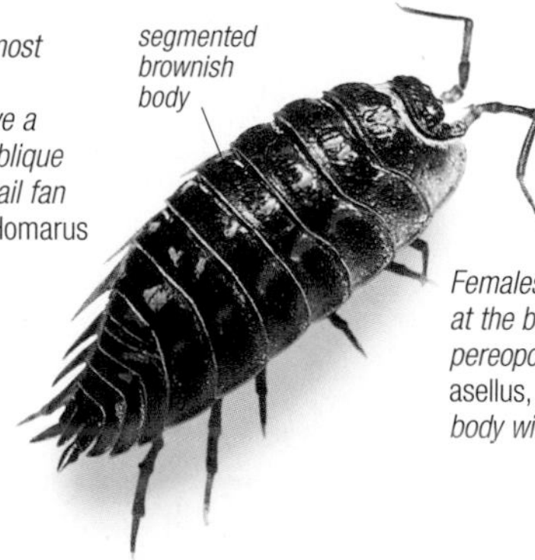

ONISCID WOODLICE

Most oniscid woodlice (family Oniscidae) are terrestrial, living in damp microhabitats, such as rotting wood, under stones, and in caves. They have a flattened body with clear segmentation and no carapace. Females have a brood pouch formed from plates at the bases of the first five pairs of walking legs, or pereopods. The common shiny woodlouse, Oniscus asellus, *pictured left, has a shiny, mottled grayish body with 2 rows of pale patches.*

OPOSSUM SHRIMPS

Shrimps of the family Mysidae, have distinctive movement sensors at the base of the inner pair of flaplike appendages on either side of the tail fan. Most of them live worldwide in estuarine or marine waters (from the coast to the deep sea), and are free swimmers. Elongated and soft bodied, many are pale or translucent but deep-sea species are red. Many form schools and are important food for both fishes and humans.

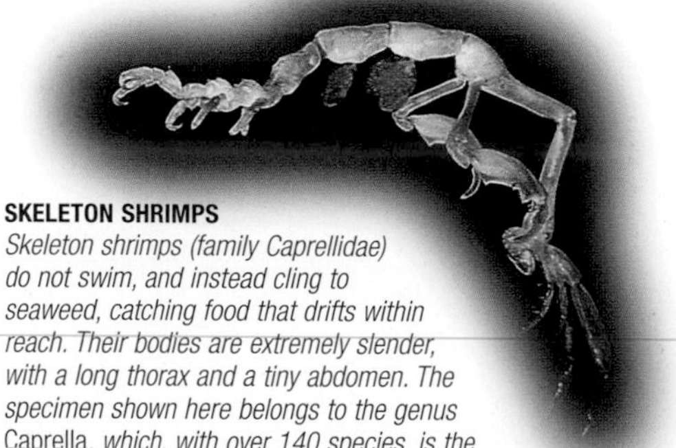

SKELETON SHRIMPS

Skeleton shrimps (family Caprellidae) do not swim, and instead cling to seaweed, catching food that drifts within reach. Their bodies are extremely slender, with a long thorax and a tiny abdomen. The specimen shown here belongs to the genus Caprella, *which, with over 140 species, is the largest in the family. Close relatives of these animals live on whales and dolphins.*

BEACH-HOPPERS

Beach-hoppers and their relatives (family Gammaridae) live on the coast and in freshwater, often among rotting vegetation. Most species—like this one from Europe—have a curved abdomen that can flick the animal away from danger.

Sea-spiders

PHYLUM	Arthropoda
SUBPHYLUM	Chelicerata
CLASS	Pycnogonida
ORDERS	1 (Pantopoda)
FAMILIES	13
SPECIES	About 1,330

Sea-spiders are unusual marine arthropods with a small, cylindrical body and long, slender legs that give them a superficial resemblance to terrestrial spiders. They occur throughout the world's oceans, occupying habitats ranging from shallow, coastal areas to deep ocean waters. Most live on the sea bottom, although some can swim. The greatest numbers of species and the largest individuals—with legspans of up to 30 in (75 cm)—are found in deep water.

Anatomy

Sea-spiders have a small head, or cephalon, 3 trunk segments, and a short abdomen. The head has a raised projection that supports 2 pairs of eyes. It has a proboscis used for feeding, 2 clawed appendages (chelifers) and a second pair known as palps. Most sea spiders have 4 pairs of walking legs arising from the trunk segments but some have 5 or 6 pairs. Most sea-spiders have 4 pairs of walking legs arising from the trunk segments, but some have 5 or 6 pairs. The front pair is used for carrying eggs and grooming. Sea-spiders have no respiratory or digestive organs: gases and dissolved substances are absorbed and released by diffusion.

Feeding

Most sea-spiders are carnivores, feeding on soft-bodied marine animals such as sponges, anemones, corals, hydroids, and lace corals. When feeding, they usually straddle their prey and either suck its tissues up through their proboscis or break off small pieces from its body with their chelicerae and then pass them to their mouth.

STRADDLING PREY
This yellow-kneed sea-spider is feeding on a colony of hydroid polyps. It keeps its legs in one position while moving its body over the polyps.

Family Nymphonidae

Nymphonid sea-spiders

Occurrence 268 spp. in the coastal Atlantic region; in intertidal to shallow water

Slender bodied, and with a stout proboscis, nymphonids are more active and swim around more than other sea-spiders. They measure $\frac{1}{32}$–$\frac{5}{16}$ in (1–8 mm) in length. The chelifers—a pair of appendages at the front—are typically clawed. Both males and females have egg-carrying legs that have 10 segments and may carry masses of up to 1,000 eggs. There may be 4–6 pairs of normal walking legs. These sea-spiders eat soft-bodied invertebrates such as hydroids and bryozoans. Most species inhabit deep water, although some are found in shallow, coastal water.

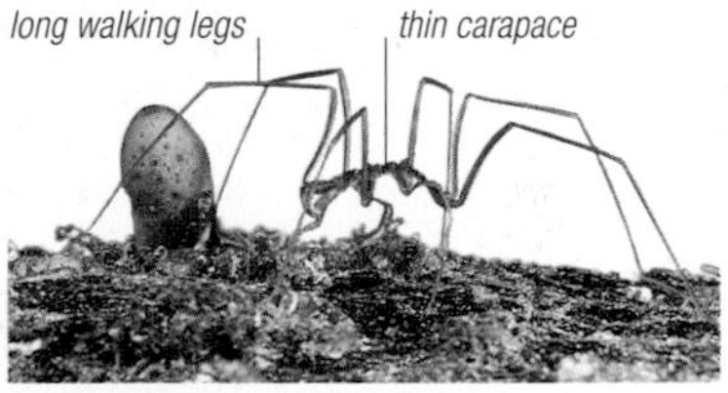

NYMPHON SPP.
These smooth-bodied animals are often overlooked because they are so small. Their legs may be 3 or 4 times longer than their body.

INVERTEBRATES

Horseshoe crabs

PHYLUM	Arthropoda
SUBPHYLUM	Chelicerata
CLASS	Merostomata
ORDERS	1 (Xiphosura)
FAMILIES	1 (Limulidae)
SPECIES	4

Also known as king crabs, these are the only living representatives of a once diverse and successful group of animals that flourished over 300 million years ago. Although they are called crabs, they are not crustaceans, but close relatives of arachnids. They live in shallow waters off the coasts of North America and Southeast Asia, and are most evident in spring, when they emerge from the sea to mate and lay their eggs.

Anatomy

Horseshoe crabs have a fused head and thorax, jointly called the cephalothorax. This is covered by a tough, horseshoe-shaped shell, or carapace, and supports 7 pairs of appendages: one pair of chelicerae, one pair of palps, 4 pairs of walking legs and a chilarium. The abdomen bears a long, spinelike tail and 5 pairs of leaflike gills. Horseshoe crabs have a pair of simple eyes at the front and a pair of compound eyes on either side.

Feeding

Horseshoe crabs are scavengers, burrowing through sediment and mud in search of food such as mollusks, worms, and other marine animals. They catch prey using their chelicerae or their first 4 pairs of walking legs, and then transfer it to the mouth.

REPRODUCTION
Every spring, horseshoe crabs gather along the Atlantic shoreline of the USA to mate. Eggs are deposited in the sand and then fertilized by the males.

Family Limulidae

Horseshoe crabs

Occurrence 4 spp., 1 in North America, 3 in Southeast Asia; on the seabed

The 4 living species of horseshoe crabs all belong to the family Limulidae. They get their name from the horseshoe-shaped shield, or carapace, that covers all of the body except the tail. Colored a drab brown, horseshoe crabs are up to $23\frac{1}{2}$ in (60 cm) long. The abdomen is hinged to the cephalothorax and has leaflike gills on the underside; the tail spine, at the rear, is used to right themselves when overturned. Components of their blue blood have been used in medical research.

LIMULUS POLYPHEMUS
The Atlantic horseshoe crab is not yet an endangered species but is thought to be declining in numbers due to exploitation, habitat loss, and disturbance of egg-laying sites. Its eggs are a major food for many seabirds.

Arachnids

PHYLUM	Arthropoda
SUBPHYLUM	Crustacea
CLASS	Arachnida
ORDERS	12
FAMILIES	661
SPECIES	103,000

This diverse group includes the familiar spiders, scorpions, and harvestmen. However, by far the most common and widespread members of the group are the ticks and mites. All arachnids have 2 main body segments and 4 pairs of legs (unlike insects, which have 3 main segments and 3 pairs of legs). Most are terrestrial animals, although about 10 percent of ticks and mites (and one species of spider) are found in freshwater.

Anatomy

An arachnid's body is divided into 2 sections. The head and thorax are fused to form a cephalothorax, which is joined to the abdomen, in some species by a narrow stalk.

The cephalothorax bears 6 pairs of appendages. The first pair, known as the chelicerae, are pincer- or fang-like and used mainly for feeding. The second pair, called the pedipalps, are leglike or enlarged with claws at the end. The other 4 pairs are walking legs, although in some species, the first pair are long and have a mainly sensory function.

The abdomen may be divided into smaller segments, or have a tail-like extension. It houses the respiratory organs: gases are exchanged via the trachea or organs known as book lungs. In spiders, the abdomen also contains silk glands.

Feeding

Most arachnids are predacious, but a few are scavengers and some mites are parasitic. Large arachnids usually rely on strength to subdue their prey but spiders, scorpions, and pseudoscorpions can also inject venom. Since arachnids have narrow mouths, they cannot ingest large pieces of food. They usually release digestive enzymes over the prey and then suck up the resulting liquid. In scorpions, small pieces of prey are partly digested in a small chamber in front of the mouth.

Some spiders rely on sight and stealth to catch their prey. Others use silk to build traps. For example, some spiders build spirals of sticky silk, called orb webs; those of some tropical species are large and strong enough to entrap small birds. Net-casting spiders make small webs, which they hold in their legs and drop over passing prey.

Reproduction

In most arachnids, the male passes a packet of sperm (or spermatophore) to the female using his pedipalps, chelicerae, or legs. The female releases the eggs, which in some scorpions hatch immediately. In some species, the young develop or hatch inside the female's body. Many arachnids show parental care by guarding their eggs.

HUNTING
Like most arachnids, this wolf spider eats live prey. It has impaled this grasshopper with its chelicerae, which each end in a hinged fang. The adjacent pedipalps are being used to manipulate the prey. Wolf spiders have acute vision and usually hunt at night.

INVERTEBRATES

Order Scorpiones

Scorpions

The most ancient group of all arachnids, scorpions are recognizable by their large, clawlike pedipalps, and their sting-bearing tails. They hunt after dark, often finding prey by touch. Scorpions sometimes use their stings to paralyze prey, but the sting's key function is defense. Females give birth to live young, which they carry on their backs. There are about 1,500 species in 15 families.

Family Buthidae

Buthids

Occurrence 651 spp. worldwide, in warmer regions; in rock cracks, under stones, logs, bark

Despite their relatively small size, 3¼–4¾in (8–12cm) in length, buthids are the most dangerous scorpions. Their potent venom can paralyze the heart and the respiratory system.

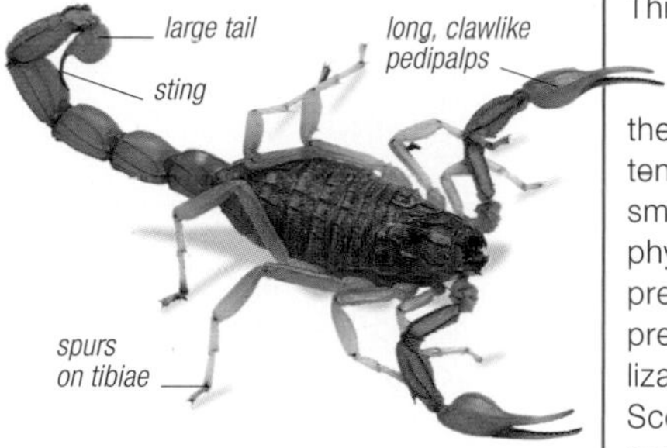

BUTHUS OCCITANUS
An inhabitant of scrubby ground, the common yellow scorpion occurs in parts of North Africa and West Asia. Antidotes to its venom are available.

Family Scorpionidae

Scorpionids

Occurrence 196 spp. in Africa, Asia, and Australia; in cracks, caves, under stones, logs

This family includes the world's largest scorpions, measuring up to 8½in (21cm) long. Although they look menacing, the larger species tend to be less venomous than the smaller ones and rely primarily on physical strength to overpower their prey. Many of these dig to locate prey, which could include spiders, lizards, and even small mammals. Scorpionids sting not only to paralyze their victim, but may also sting mates as part of their complex mating ritual. A litter of 30–35 young stay with the mother for some time after their first molt. Scorpionids have a 5-sided sternum in contrast to the buthids' triangular one.

PANDIUS IMPERATOR
This is the imperial scorpion, some specimens of which can exceed 2⅛oz (60g) in weight. It can be found among leaf litter in the tropical forests of Central and West Africa.

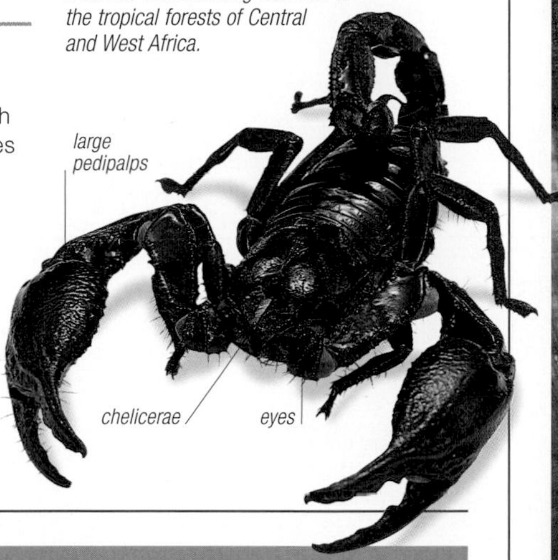

Order Pseudoscorpiones

Pseudoscorpions

Also called false scorpions because they lack a sting and tail, members of this order of 25 families (about 3,300 species) are similar to scorpions in shape. They have venom glands in pincerlike parts in their pedipalps and prepare silk nests in which to molt, brood their young, and hibernate. Eggs are laid in a special brood pouch. Females may cling to other animals to disperse to new habitats.

Family Neobisiidae

Neobisiids

Occurrence 498 spp. worldwide, especially in the Northern Hemisphere; in leaf litter, soil, caves

Habitat Terrestrial

The members of this family are very small, measuring 1⁄32–3⁄16in (1–5mm) in length. They usually have 4 eyes but cave-living species may have fewer eyes or none at all. In color, neobisiids vary from olive- to dark brown, with red, yellow, or cream tinges. The legs are often slightly green, with 2-segmented tarsi. The claw of each pedipalp has one venom gland located in the fixed finger; after paralyzing its prey, the neobisiid shreds it with its large mouthparts.

NEOBISIUM MARITIMUM
Measuring ⅛–9⁄64in (3–3.5mm) long, the maritime pseudoscorpion is found in cracks in rocks and under stones in coastal areas of Ireland, England, and France.

Family Cheliferidae

Cheliferids

Occurrence 292 spp. worldwide, especially in warmer regions; in leaf litter, on tree bark

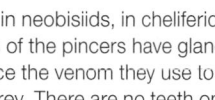

Unlike in neobisiids, in cheliferids both fingers of the pincers have glands that produce the venom they use to paralyze their prey. There are no teeth on the inner surfaces of the pincers. Cheliferids are 3⁄64–3⁄16in (1.5–5mm) long and usually have 2 eyes. Coloration varies from pale to dark brown or black, often tinged with red or olive, with dark markings. As in many arachnids, courtship can be complex; males and females dance holding each other's pedipalps.

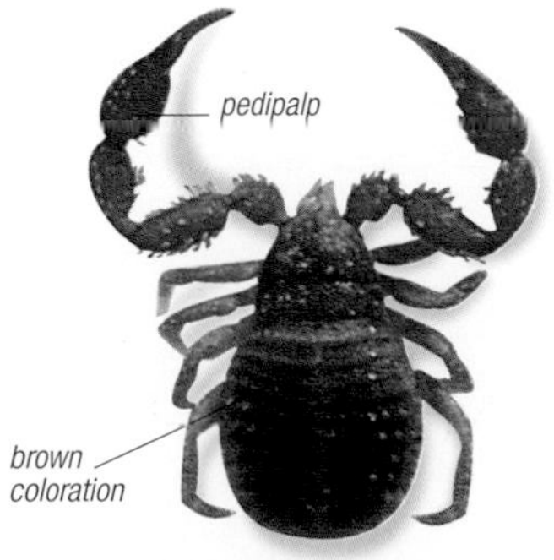

DACTYLOCHELIFER SPP.
While most of these species are distributed across the Northern Hemisphere in a variety of habitats, some are confined to coastal habitats.

Order Solifugae

Sun-spiders or wind-scorpions

Neither true spiders nor scorpions, the animals in this order lack poison glands. Instead, they kill and chew up their prey with their very large, pincerlike chelicerae. Their leglike pedipalps have suction pads that hold down the prey so that it can be eaten. There are about 1,100 species, in 12 families, and most are found either in North America or in the tropics.

Family Solpugidae

Solpugids

Occurrence 190 spp. in Africa, parts of West Asia; on ground, under stones, logs, in burrows in sandy soil

Pale, brown or yellowish, sometimes with bright markings, these sun-spiders are ¼–2¼in (0.6–6cm) in length. Their 4 pairs of legs have a variable number of tarsi and all, except the first pair, have claws with a smooth surface. They have forward facing chelicerae to kill prey; small species and nymphs eat termites. Females lay masses of 20–100 eggs. Some solpugids are active by day; others burrow in sandy soil or hide in cracks, or under stones during the day, and forage in tree canopies at night.

METASOLPULGA PICTA
This species is found in the Namib Desert in Namibia. Like all solpugids, it will bite if handled carelessly, but does not have venom glands.

Family Ammotrechidae

Ammotrechids

Occurrence 83 spp. in the warmer regions of South, Central, and North America; in rotten tree stumps

This family comprises relatively slender species that measure 5⁄32–¾in (0.4–2cm) in length. They are colored in shades of brown. The front margin of the head is rounded, and the tarsi of the first pair of legs do not have claws. Many species are nocturnal and dig into the soil during the day. Some of the smaller ones hide in termite galleries or inside the tunnels of wood-boring insects. These arachnids are predatory on a range of arthropods. Females may lay 2 or more masses of 20–150 eggs in their lifetime.

AMMOTRECHELLA STIMPSONI
This species is found in Florida and other areas in southern North America, where it lives in the bark of rotten, termite-infested trees.

INVERTEBRATES

Order Uropygi

Whip-scorpions

These flat-bodied arachnids have a whiplike tail (telson) at the end of their abdomen and can, in defense, squirt formic and acetic acids from a pair of abdominal glands. Their chelicerae resemble spider fangs. There is 1 family, with 106 species.

Family Thelyphonidae

Vinegaroons

Occurrence 106 spp. in North America and N.E. South America; in leaf litter, under stones, rotten wood

Vinegaroons hunt after dark, hiding in underground burrows during the day. They do not have a sting but can squirt acid over 23½ in (60 cm), and can give a painful pinch with their pedipalps. Excluding the tail, they are up to 3 in (7.5 cm) long.

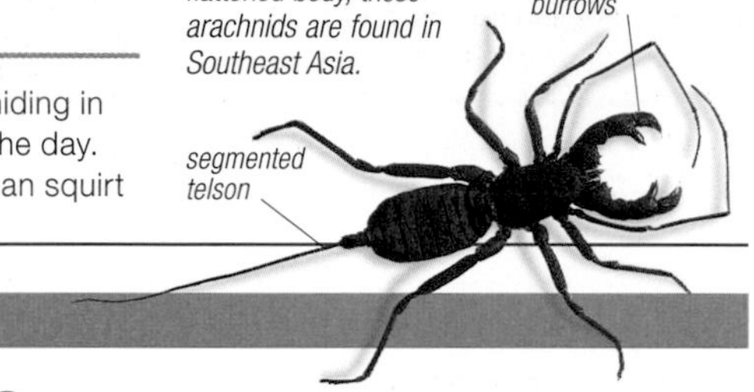

THELYPHONUS SPP.
Dark brown, with a tough, flattened body, these arachnids are found in Southeast Asia.

Order Amblypygi

Whip-spiders

Also known as tail-less whip-scorpions, members of the 5 families (about 160 species) in this order do not sting or bite. They are nocturnal and cave dwelling, with fanglike, segmented chelicerae, large, shiny pedipalps, and an extremely long first pair of legs.

Family Phrynidae

Phrynids

Occurrence 59 spp. in tropical and subtropical regions; under stones, bark, in leaf litter and caves, among rocks

Brown, with dark markings, phrynids have a body 3/16–2¼ in (0.5–6 cm) long. They scuttle rapidly sideways if disturbed. Females carry young for months in a sac under the abdomen.

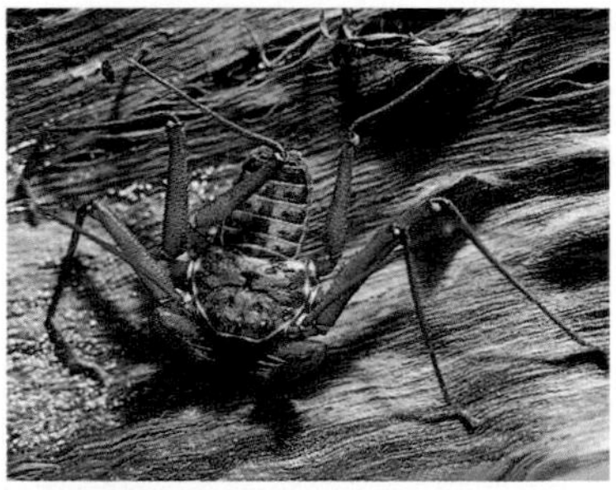

PHRYNUS SPP.
This phrynid, from Central America, has bent its long front legs backward—a typical resting pose.

Order Opiliones

Harvestmen

Harvestmen—or daddy-long-legs—are often mistaken for spiders, but they have rounded bodies, without a slender waist. Their legs are long and slender, and end in microscopic claws. Their mouthparts are small but they produce noxious secretions as a defense. Harvestmen are unusual among arachnids in that fertilization is direct. There are at least 6,125 species, in 48 families.

Family Cosmetidae

Cosmetids

Occurrence 708 spp. mainly in tropical regions of North and South America; under stones, among debris

Like most harvestmen, cosmetids are generally dull in color, although tropical species can be green or yellow, and a few can change color over a matter of weeks to blend with their surroundings. They range from 3/16–3/8 in (5–11 mm) in length. Males tend to be smaller than females but have longer appendages and larger chelicerae. The eyes are located close together on a small "turret" called an ocularium. Females use their ovipositors to lay eggs in cracks and crevices in the soil.

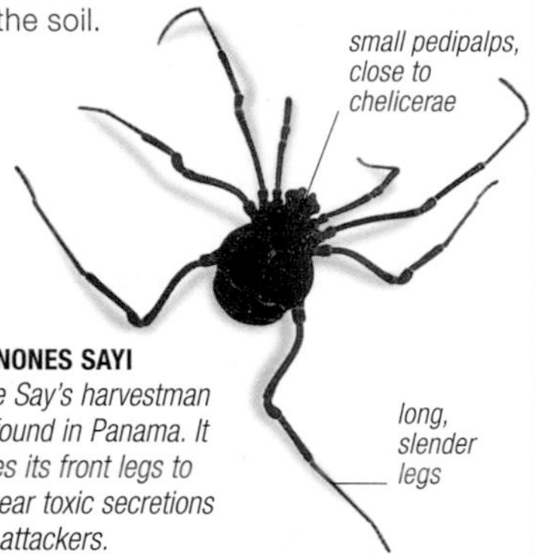

VONONES SAYI
The Say's harvestman is found in Panama. It uses its front legs to smear toxic secretions on attackers.

Family Phalangiidae

Phalangiids

Occurrence 381 spp. worldwide; under stones, among leaf litter in woods, grassy areas

Phalangiids can have a legspan of up to 4 in (10 cm), but their soft bodies are rarely more than 3/8 in (1 cm) long. Like other harvestmen, their eyes are close together, and are on a "turret" above the rest of the body. Males and females differ; the enlarged chelicerae of the males are especially distinct. Females lay 20–100 eggs at a time, burying them in soil or under bark. Most species are nocturnal and prefer humid habitats. Phalangiids are both predatory and saprophagous.

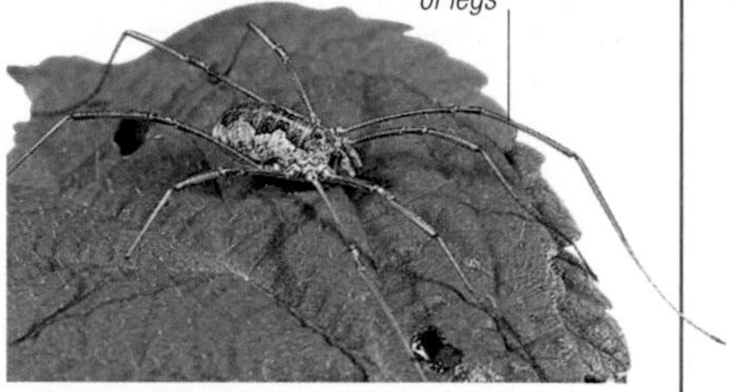

PHALANGIUM OPILIO
Found in the Northern Hemisphere, the diurnal horned harvestman likes sunny spots, and inhabits woods, gardens, and grassland.

Order Acari

Ticks and mites

With over 48,200 species and about 438 families, ticks and mites are present in almost every habitat, particularly on land. They include many pests of crops and stored food, as well as parasites that attack mammals, birds, and reptiles. Most are less than 1/32 in (1 mm) long—although ticks can grow much bigger after feeding on blood—and unlike spiders, their bodies have no distinct divisions.

Family Ixodidae

Hard ticks

Occurrence 686 spp. worldwide; in association with birds, mammals, and some reptiles

Hard ticks have a tough dorsal plate, which fully covers the males, but covers only the front half in females. The abdomen is soft and flexible to allow for large blood meals. The front part of their head projects forward and carries the mouthparts. Hard ticks are 1/16–3/8 in (0.2–1 cm) long but grow larger after feeding. Colors vary from yellow to red or black-brown; some species are highly marked. Some hard ticks carry viral diseases that affect humans, including encephalitis and Lyme disease, as well as diseases affecting cattle and poultry.

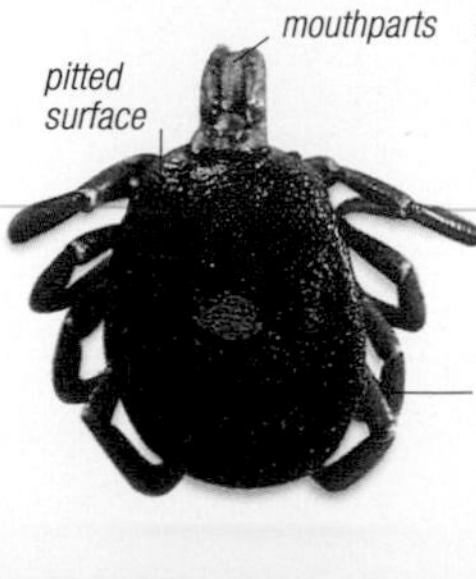

AMBLYOMMA AMERICANUM
The lone star tick is common in the grassland of Central America.

Family Trombidiidae

Velvet mites

Occurrence 279 spp. worldwide, especially in tropical regions; in or on soil, some associated with freshwater

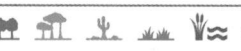

Named after their dense, velvety "fur," these bright red or orange mites eat insect eggs when they are adult, but start life as parasites, attacking insects, spiders, and harvestmen. The adults often emerge from the soil after rain to mate and lay eggs.

TROMBIDIUM SPP.
Trombidium velvet mites are found in many parts of the world. This species, from South Africa, grows up to 3/8 in (1 cm) long.

Family Tetranychidae

Spider mites

Occurrence 689 spp. worldwide; on a range of trees, plants, and shrubs

Habitat All

These soft-bodied mites are orange, red, or yellow, and 1/128–1/32 in (0.2–0.8 mm) long. Many species produce silk. Eggs are laid on leaves, and the young are protected by silk webs. They suck sap and heavy infestations can weaken a plant.

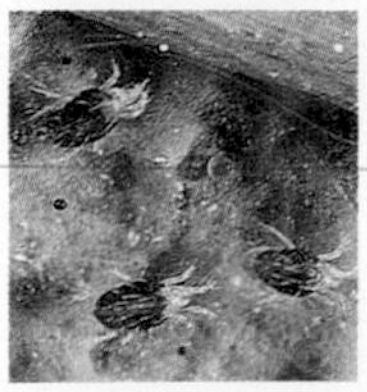

TETRANYCHUS SPP.
Abundant in temperate regions, red spider mites may produce 4–5 generations per year. They are pests of deciduous fruit trees.

Order Araneae

Spiders

Spiders can be distinguished from other arachnids by their distinctive appearance and by their ability to produce silk. A typical spider has 8 eyes, and a body divided into 2 parts—the cephalothorax and the abdomen. The mouth is flanked by a pair of venom-injecting fangs, and a pair of leglike pedipalps. These have a sensory function and, in males, are also used to transfer sperm. All spiders are predatory, injecting venom into their prey. Some use silk to catch prey, but spiders also use silk to protect their eggs, lower themselves through the air, and even to sail on the breeze. There are over 42,000 species, in about 110 families, and they live in every terrestrial habitat, from tropical rain forests to cellars and caves.

Family Araneidae

Orb web spiders

Occurrence 3,006 spp. worldwide; in grassland, meadows, woodland, forests, gardens

Habitat Terrestrial

As the name implies, many of these spiders weave circular webs made up of a central hub with radiating lines and spirals of sticky and nonsticky silk. They typically catch prey in the web, wrap it in silk, cut it out, and take it to a retreat to eat it. Some species spin opaque bands of matted silk in their webs so that birds can see them and do not accidentally destroy the webs by falling into them. However, there are a few species that do not make webs; these prey after dark, using a single thread with a bead of glue at the end to snare moths. The strength of the webs can be such that in Papua New Guinea, the huge webs of the *Nephila* species are used as fishing nets. Orb web spiders often have very large, brightly colored, and patterned abdomens; in tropical species, the abdomen may have unusual, angular shapes. The legs have 3 claws that can be very spiny. Of the 8 eyes, the 4 in the middle often form a square. Orb web spiders are 1/16–1¾in (0.2–4.6cm) long. Males approach females on their webs and pluck threads to attract them. Females lay eggs in silk cocoons that can be camouflaged in webs and litter, or stuck to vegetation or bark.

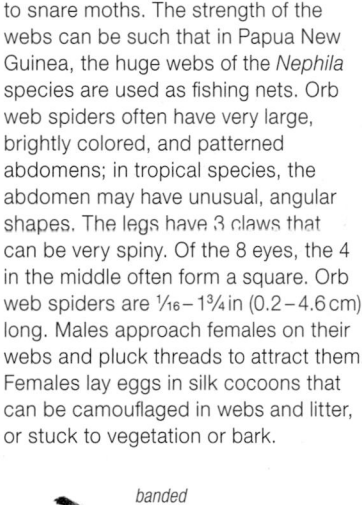

ARANEUS DIADEMATUS
Variably colored (from pale reddish to almost black), the European garden spider, or garden cross spider, inhabits woods and gardens in Europe and North America. Its webs can be nearly 20in (50cm) across.

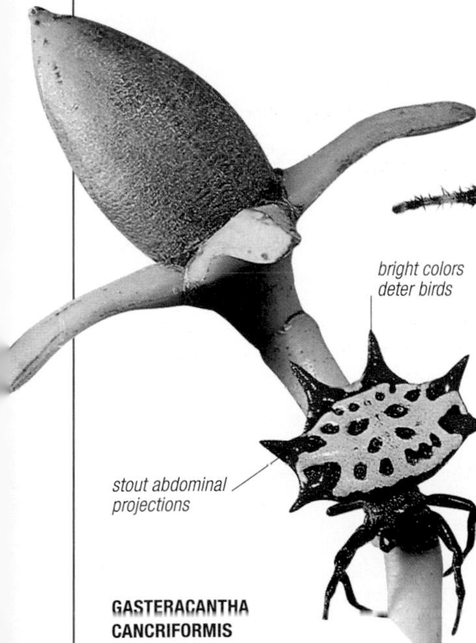

GASTERACANTHA CANCRIFORMIS
These tough-bodied, spiny-bellied orb weavers are found in North and South America. They are often seen hanging in the middle of their webs. They inhabit shrubs and trees, and their webs can also be seen on walls of houses.

MICRATHENA GRACILIS
Called the spined micrathena spider, this species inhabits woodland and forests in North America.

Family Agelenidae

Funnel weavers

Occurrence 1,146 spp. worldwide; in meadows, gardens, bushes, houses, among stones, on walls

Funnel weavers are often long-legged, and have minute, feathery hairs on the body, 8 eyes, and a slender abdomen with dark bars, chevrons, or spots. The body is ¼–¾in (0.6–2cm) long. The funnel weaver's web takes the form of a flat, tangled silk sheet with a tunnel on one side. Females may feed their offspring by regurgitation and, in some species, the spiderlings may eat the mother when she dies.

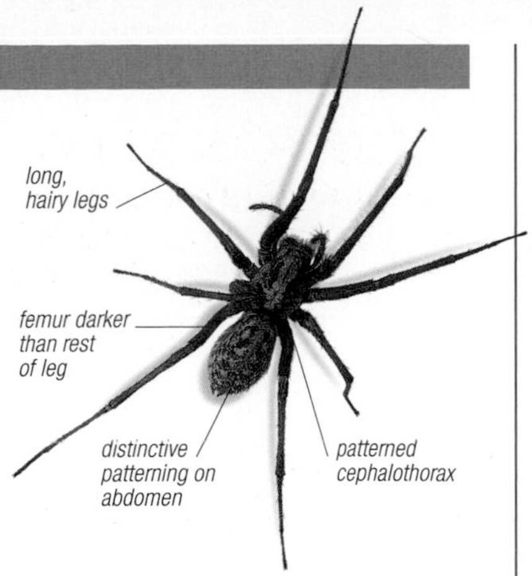

TEGENARIA DUELLICA
This giant house spider can have a legspan of up to 3¼in (8cm), and can run very fast. Members of the genus Tegenaria *are called house spiders. The large spiders that fall into domestic baths usually belong to this genus.*

Family Cybaeidae

Cybaeid spiders

Occurrence 176 species in North America, South America, and Eurasia

Over 100 species of *Cybaeus* make webs in stone crevices or under logs, but this family's best known member is the water spider *(Argyroneta)*: the only spider that lives under water. It ferries air, trapped by its dense abdominal hair, to a silken "diving bell" anchored to pond weed. The bell is used for feeding, courtship, hibernation, and stowing eggs.

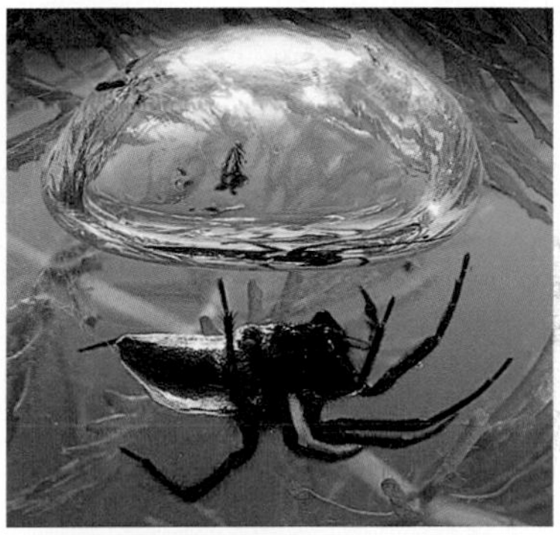

ARGYRONETA AQUATICA
This spider lives permanently in still or slow-moving water, and the air trapped by the body hairs around the abdomen gives it a silvery appearance.

Family Hexathelidae

Funnel-web spiders

Occurrence 178 spp. in tropical regions of the Americas, Africa, Asia, and Australia; in trees, on ground

Funnel-web spiders belong to a group known as mygalomorphs, which also includes tarantulas (see p.593). Unlike other spiders, mygalomorphs are relatively primitive, with fangs that strike downward, instead of closing sideways like a pair of pincers. Funnel-webs are ⅜–1¼in (1–3cm) long and, usually, dark brown. Most make a funnel-shaped retreat that leads into crevices in tree stumps or rocks. Some are highly venomous.

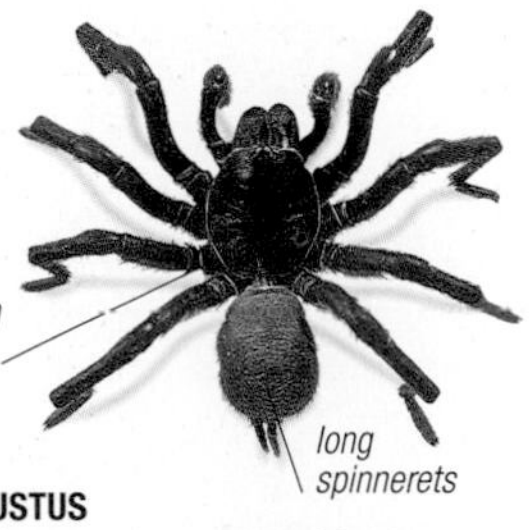

ATRAX ROBUSTUS
The Sydney funnel-web, a native of Southeast Australia, is an aggressive spider that has a painful, venomous bite that may prove fatal.

Family Linyphiidae

Dwarf spiders

Occurrence 4,378 spp. mostly in temperate regions of the north, some in Arctic; on stones, among vegetation

Also known as money spiders, these tiny animals are rarely more than ⅜in (1cm) long. They have large chelicerae with sharp teeth. The legs have strong bristles and the carapace of males may have extensions, which sometimes carry the eyes. Dwarf spiders live in leaf litter or make characteristic, horizontal, nonsticky, sheetlike webs among vegetation. Egg sacs are attached to stones, plants, and other surfaces. These spiders travel over long distances by "ballooning" on long, silk threads.

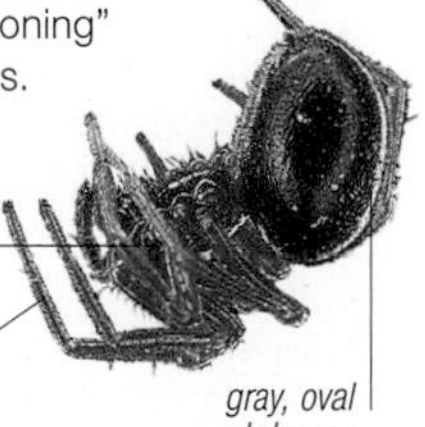

GONATIUM SPP.
These small spiders are widespread in the Northern Hemisphere, inhabiting scrubland and grassland. They prefer low, shaded bushes.

INVERTEBRATES

Order Araneae *continued*

Family Salticidae

Jumping spiders

Occurrence 5,000 spp. worldwide; in grassland, meadows, woodland, forests, gardens

The 5,000 species of jumping spiders (salticids) make up one of the largest of the spider families. Most jumping spiders have stout, hairy bodies, 1/16–1/2 in (2–16 mm) long, and are often drab in appearance, although many tropical species are vividly colored, with elaborate markings. Some species are very antlike, mimicking their prey in both looks and behavior. Although the different species do vary greatly, a distinctive feature common to all is their prominent eyes. At the front of the carapace, which is rectangular, with a rounded rear and square front, sits a row of 4 large eyes. The middle 2 eyes, which are very much larger than the others, resemble old-fashioned car headlights. These eyes enable jumping spiders to judge distance, shape, and movement very accurately, and give them the most keen vision of any spider. The back of the eye capsule can be moved inside the head to keep the image of the prey centered on the retina. The smaller outside eyes and a pair toward the rear of the carapace simply detect movement. Jumping spiders are active during the day, stalking insect prey on the ground, on walls, and in bushes, especially in warm, sunny locations. Their characteristic hunting method of jumping on their prey to seize it gives them their common name. When the weather is bad, they retreat to a small, silken nest they make in a crevice or crack. The spiders use similar shelters over winter and when they molt or lay eggs. The eggs, laid among vegetation, bark, and moss, are wrapped in silk, and the female guards them until they hatch.

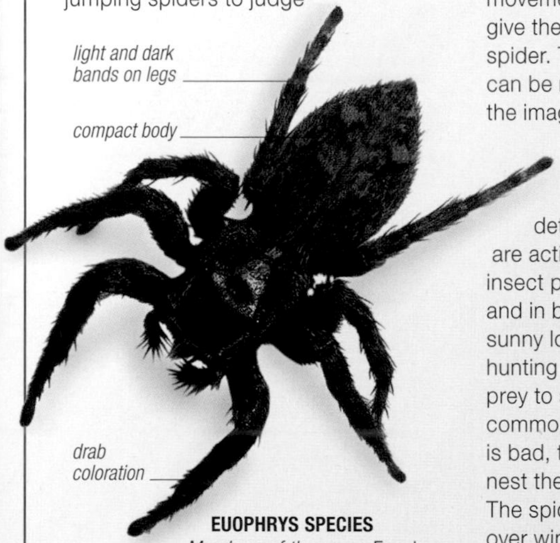

EUOPHRYS SPECIES
Members of the genus Euophrys *are often found on the ground, among grass and low-growing plants, or under stones. As in many species, males fight for access to a female, the larger male usually winning.*

STRIKING APPEARANCE

Some tropical jumping spiders are very brightly colored, with iridescent hairs arranged in patches and spots. In some, the whole of the upper surface of the abdomen may be bright red or orange. Males often have enlarged front legs and ornamented pedipalps, which they use to attract the attention of females.

PHIDIPPUS REGIUS
Males and females of this species, native to S.E. North America, communicate using more than 20 different signals made with their legs and palps. Other spiders in the genus do the same.

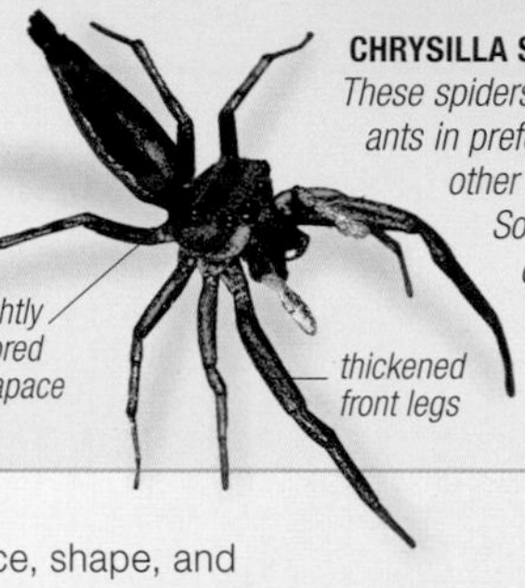

CHRYSILLA SPECIES
These spiders catch and eat ants in preference to any other kind of insect. Some species jump on their prey from behind, while others tackle the ant head first.

JUMPING

Salticids jump not only to catch prey but also to escape from predators. The hind pair or 2 hind pairs of legs are extended rapidly by hydraulic pressure.

ATTACKING PREY
Before it leaps, a jumping spider attaches a safety line of silk in case it gets carried off course. This species is known as the zebra spider on account of its striped markings. The mouthparts (chelicerae) of males are larger than those of the females.

BAGHEERA KIPLINGI
This jumping spider is the only known herbivorous spider. In its native Central American forests, Bagheera kiplingi *lives on acacia trees that produce nutrient-rich nodules to sustain colonies of ants. In turn, the ants clear the acacia of leaf-eating insects. The spider feeds from the nodules, too—but without providing the tree with anything in return.*

ACUTE EYESIGHT
*The 2 outside eyes of this jumping spider (*Salticus *sp.) can detect moving prey up to 10 in (25 cm) away. By turning its body, the spider brings the prey into the field of vision of the 2 middle eyes, which give a large, high-resolution image. The spider then moves toward the prey and jumps at it from a distance of 2–4 in (5–10 cm).*

INVERTEBRATES

Order Araneae *continued*

Family Lycosidae

Wolf spiders

Occurrence 2,374 spp. worldwide, even in Arctic regions; mostly on ground, among leaf litter

Habitat All terrestrial

Wolf spiders are 5/32–1½ in (0.4–4 cm) long and have excellent eyesight, which helps them in their nocturnal hunts. They are pale gray to dark brown, with markings such as bands, stripes, or spots. Females often carry their egg sacs attached to their spinnerets or, in burrowing species, keep the sacs in a silk-lined burrow. They carry spiderlings on their backs. Some wolf spiders are important predators in field ecosystems. This family includes the true tarantula, *Lycosa tarentula*, of southern Europe.

PARDOSA AMENTATA
This European species is variable in appearance—the abdomen, for example, may be brown or gray.

Family Pholcidae

Daddy-long-legs spiders

Occurrence 1,111 spp. worldwide; in caves, leaf litter, dark corners of buildings

Pale legs that are much longer than the body give these spiders a spindly appearance. Colored gray, green, or brown, they have eyes arranged in 2 groups of 3 each, with another pair in between; a few cave-dwelling species, however, are blind. These spiders are ⅛–½ in (3–14 mm) long and have flexible tarsi, with many false joints. Also known as cellar spiders, daddy-long-legs spiders make tangled, irregular webs and quickly wrap prey in silk before biting it. When disturbed, they vibrate their webs rapidly, making themselves appear blurred and difficult to spot. Females carry 15–20 pale, pinkish gray eggs, held with silk, in their chelicerae.

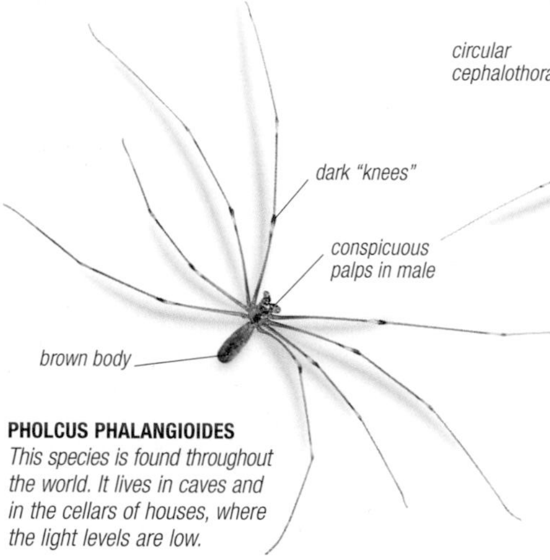

PHOLCUS PHALANGIOIDES
This species is found throughout the world. It lives in caves and in the cellars of houses, where the light levels are low.

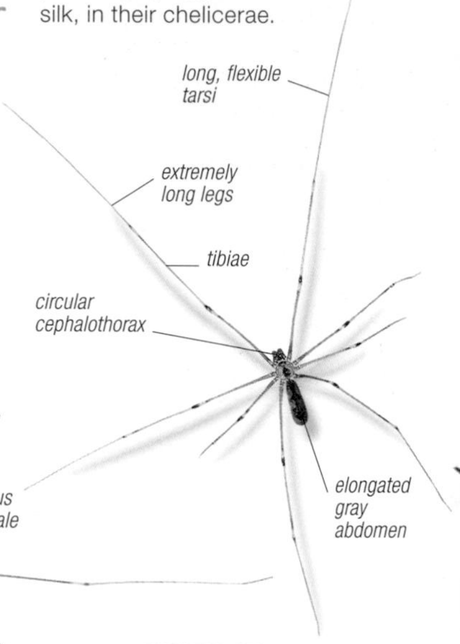

PHOLCUS SPP.
These spiders inhabit caves, rock crevices, and buildings, especially in warmer regions. They all look very similar, and can be separated only by examination of their genitalia.

Family Pisauridae

Nursery-web spiders

Occurrence 333 spp. worldwide; on ground, surface of still water, aquatic plants

Habitat All

These large spiders resemble wolf spiders (see left) in appearance and hunting technique—running and catching prey on the ground rather than using webs. The carapace is oval and has longitudinal markings; the body is ⅜–1 in (1–2.6 cm) long. The females carry the egg sacs in their chelicerae slung below the body, placing them in tentlike nursery webs at hatching time. The sacs and spiderlings are guarded by the females until the second molt, when they leave the web and disperse.

PISAURA MIRABILIS
Found in the pastures and woods of Europe, often among leaf litter, this is a common, ground-living spider.

Family Scytodidae

Spitting spiders

Occurrence 228 spp. worldwide, except Australia and New Zealand; under rocks, in buildings

The common name of these 6-eyed spiders comes from their unique prey-capturing technique. Using a rapid side-to-side movement of the chelicerae, they squirt 2 zigzag streams of sticky, gluelike substance from close range to stick the prey down. Spitting spiders measure 5/32–⅜ in (4–12 mm) long and are cream or yellow-brown with black markings. Females carry the egg sacs under their body until the young emerge.

SCYTODES THORACICA
Often found inside buildings in North America and Europe, this spider can squirt glue over ⅜ in (1 cm). Males are slightly smaller than females.

Family Sicariidae

Six-eyed crab spiders

Occurrence 124 spp. in warm parts of North and South America, Europe, Africa; in shaded areas in rocks and bark

Also known as brown spiders because of their color, these arachnids have 6 eyes, arranged in 3 pairs, and a distinctly hairy body and legs. They are ¼–¾ in (0.6–1.8 cm) long and, in most species, have a violin-shaped mark on the carapace. They make sticky, sheetlike webs, to which some species keep adding as they grow. Females lay 30–300 eggs per sac and keep the sacs at the rear of the web. The bite of these spiders can be very dangerous, causing tissue degeneration.

LOXOSCELES RUFESCENS
Common in Europe and introduced to Australia, the brown recluse spider may bite humans and cause unpleasant lesions that are slow to heal.

Family Sparassidae

Huntsman spiders

Occurrence 1,109 spp. in tropical and subtropical regions; on ground, tree trunks

These drably colored spiders are efficient, nocturnal hunters, able to move sideways with great agility. The larger species, also known as giant crab spiders, can easily tackle lizards. The carapace and abdomen are flattened. Legs may span up to 6 in (15 cm) and the body can be ⅜–2 in (1–5 cm) long. There are 8 eyes of equal size, 4 of which point forward from the front edge of the carapace.

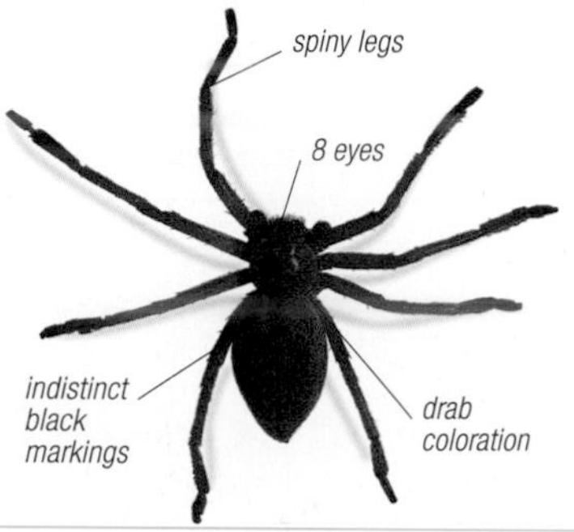

HETEROPODA SPP.
Huntsman spiders are pantropical species. Often found in crates of bananas, they are known to bite warehouse workers but are not dangerous.

Family Theridiidae

Comb-footed spiders

Occurrence 2,3t spp. worldwide; in vegetation, under stones, in leaf litter, cracks, crevices

Also called cobweb spiders, these brown to black arachnids have a comblike row of stout bristles on their hindlegs. The body is 1/16–½ in (0.2–1.5 cm) long and has a very rounded abdomen. Comb-footed spiders construct irregular webs, some of which may have sticky lines extending to the ground. Females are venomous and a few, such as the black widow, are dangerous to humans. They lay up to 1,000 eggs per year, which they often guard in round, silk egg sacs.

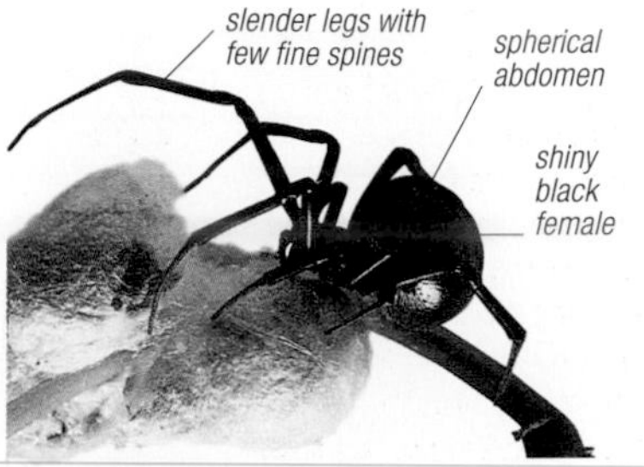

LATRODECTUS MACTANS
The male of this species often dies shortly after mating and may be eaten by the venomous female; hence the common name, black widow.

Family Theraphosidae

Tarantulas

Occurrence 933 spp. worldwide, especially in subtropical and tropical regions; in ground burrows, trees

The name tarantula was originally applied to a wolf spider from southern Europe (see opposite), but nowadays it is far more often used for this family. These are the giants of the spider world, with bodies up to 4¾ in (12 cm) long, and a legspan of up to 11 in (28 cm). Their bodies and legs are covered in bristly hairs, and their coloration varies from pale brown to black, with markings in pink, brown, red, or black. They are mygalomorphs, like funnel-web spiders (see p.589), with fangs that move up and down. When threatened, a tarantula will display its downward-pointing fangs. The large size of tarantulas has led to the belief that their bites are fatal but only some species are venomous; in fact, the venomous species are often relatively small. Tarantulas are nocturnal hunters, relying on their large size to subdue prey, which include small vertebrates such as frogs, lizards, and birds. They crush their prey with their large fangs, pour digestive juices over the body, and then suck up the resulting liquid. Females lay a batch of eggs in a burrow. An egg sac can be the size of a golf ball and contain about 1,000 eggs. Spiderlings stay in the burrow until their first molt, and then disperse to find food and make their own burrows. Several species of this family live for 10–30 years, and some are kept as pets.

BRACHYPELMA VAGANS
The Mexican red-rumped tarantula is a ground-living species that hunts at night. It is found in semidesert and arid regions of Central America.

POECILOTHERIA REGALIS
The tiger or Indian ornamental spider is the world's largest tree-dwelling spider, with a legspan of 6½ in (17 cm). It is found in India and Sri Lanka. It hides in a silk-lined retreat by day and hunts birds, lizards, and arthropods at night.

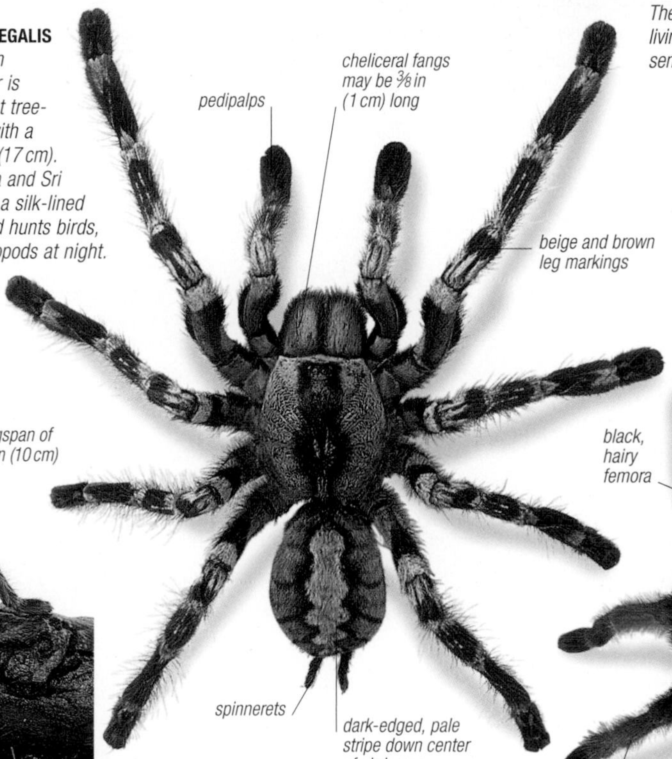

GRAMMOSTOLA ROSEA
Found in South America, especially in Chile, the ground-living Chilean rose is one of the species most commonly kept as a pet because of its docile nature.

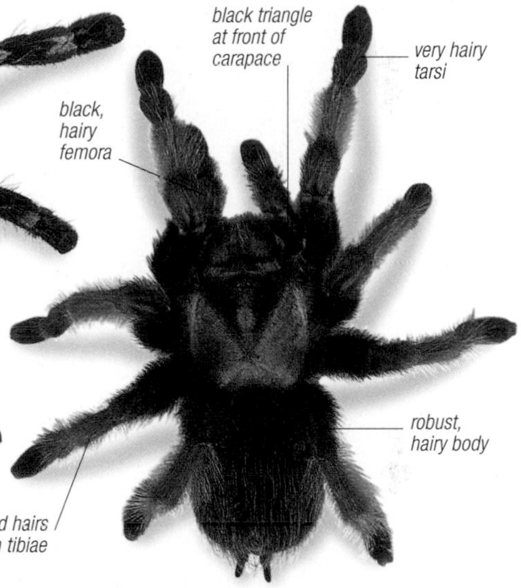

BRACHYPELMA EMILIA
The Mexican red-legged tarantula lives in burrows in desert and semidesert regions of Central America. It kicks up the hairs of its abdomen as a defensive reaction.

Family Thomisidae

Crab spiders

Occurrence 2,146 spp. worldwide; in meadows, gardens, on flowers, other parts of plants, tree bark

The common name derives from these spiders' characteristic crablike sideways movement (although they can also move forward and backward) and usually squat shape; however, some species are elongated. Crab spiders range from 5⁄32 to ½ in (4–14 mm) in length. The carapace is nearly circular, and the abdomen is short, often blunt ended. The first 2 pairs of legs, used to seize prey, are larger and spinier than the other 2 pairs. Crab spiders often ambush insects landing on flowers—species that do so have very powerful venom, which can easily kill a bee much larger than themselves. Many species are colored white, bright pink, or yellow, to match the flowers, and a few can change color. The female keeps her eggs in a flat sac attached to plants, and guards them until she dies.

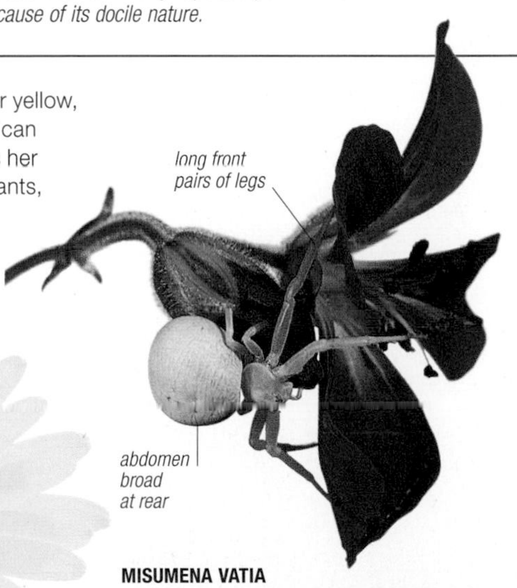

MISUMENA VATIA
This goldenrod crab spider has the ability to change its color in order to merge with the surroundings. Mostly found on white or yellow flowers, it is usually one of these 2 colors (see above and left), but can also be pale green. Common throughout Europe and North America, it can tackle large prey such as butterflies and bees. The female is much larger than the male.

Family Philodromidae

Running crab spiders

Occurrence 600 species worldwide; in meadows, gardens, on flowers, other parts of plants, tree bark

These small to medium spiders are generally colored with browns or grays, and have a body covered in long hairs. They were formerly classified among Thomisidae (left), but are now understood to be more distantly related to them. They are less inclined to ambush their prey—like true crab spiders—and instead catch them while on the run.

INVERTEBRATES

Mollusks

PHYLUM	Molluska
CLASSES	7
ORDERS	53
FAMILIES	609
SPECIES	About 110,000

Of all invertebrates, mollusks have arguably the widest range of body forms. They include gastropods (slugs and snails), bivalves (such as oysters and clams), cephalopods (octopuses, squids, and cuttlefishes, among others), chitons, and several less familiar forms. All mollusks have one or more of the following: a horny, toothed ribbon in the mouth (the radula) that is unique to mollusks; a calcium-carbonate shell or other structure covering the upper surface of the body; and a mantle or mantle cavity, typically with a characteristic type of gill. Mollusks are adapted to life on land and in water; there are also many parasitic species.

Movement

Most mollusks use their broad, flat foot to creep over hard surfaces, often on a thin layer of mucus. Some bivalves are fixed to the surfaces on which they live, and all are relatively inactive, but they often have a well-developed foot, which some species use to dig burrows in sediment. The most mobile and fastest-moving of all mollusks are the cephalopods, especially octopuses and squids. Using undulations of their muscular fins or by flapping them, they can make a slow swimming motion but they are also able to move more rapidly by squirting a jet of water from the mantle cavity.

MUCUS TRAILS
To reduce surface friction, slugs and snails secrete mucus from their foot. Slugs (such as this striped slug) have lost their shells, and generally move more quickly than snails.

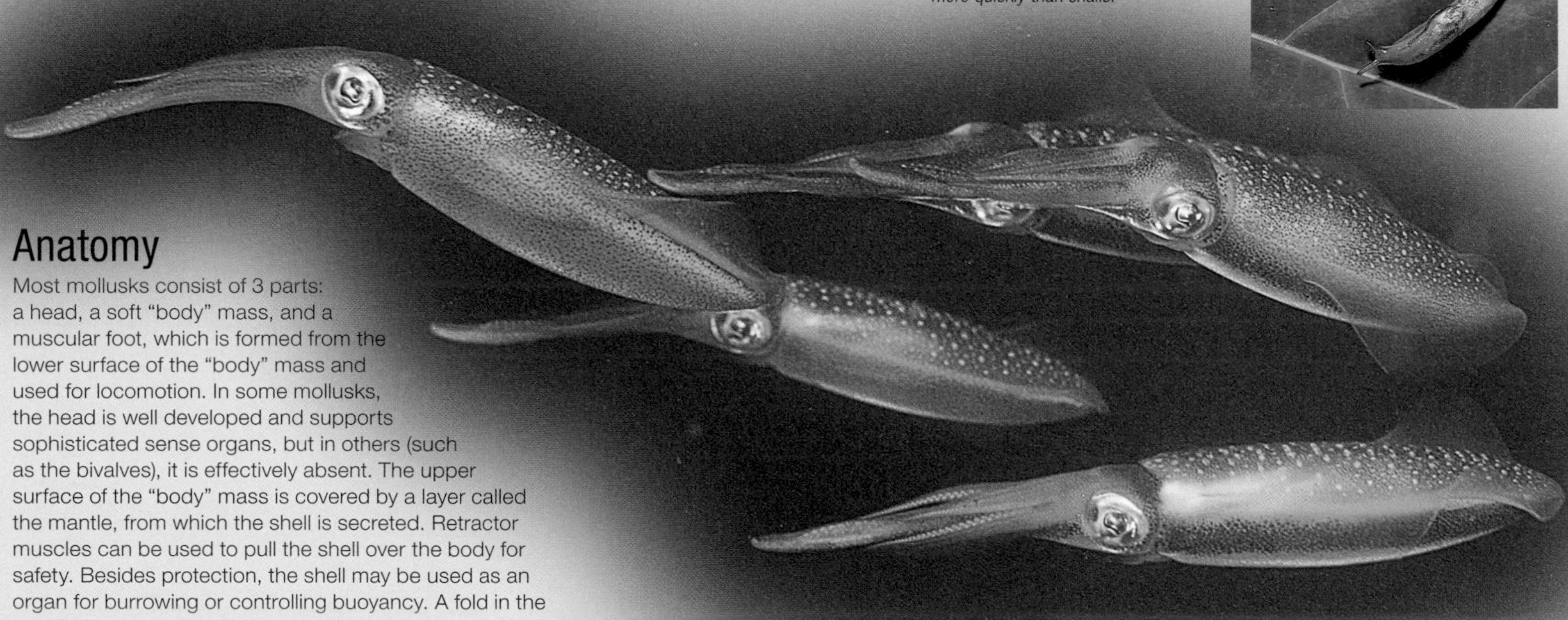

Anatomy

Most mollusks consist of 3 parts: a head, a soft "body" mass, and a muscular foot, which is formed from the lower surface of the "body" mass and used for locomotion. In some mollusks, the head is well developed and supports sophisticated sense organs, but in others (such as the bivalves), it is effectively absent. The upper surface of the "body" mass is covered by a layer called the mantle, from which the shell is secreted. Retractor muscles can be used to pull the shell over the body for safety. Besides protection, the shell may be used as an organ for burrowing or controlling buoyancy. A fold in the mantle forms a cavity that in aquatic mollusks houses the gills.

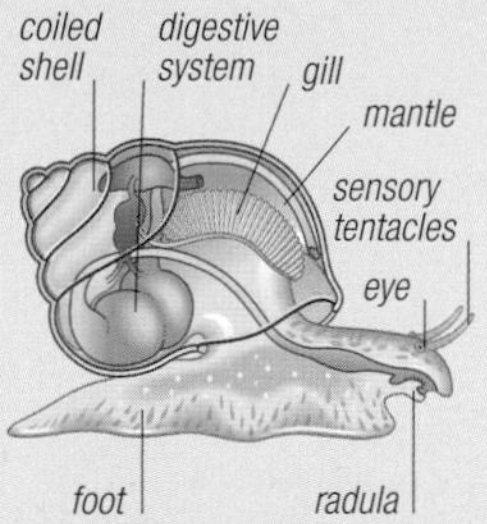

GASTROPOD
Gastropod snails have a coiled shell, a large foot, and a distinct head, with eyes and sensory tentacles. Slugs have no shell.

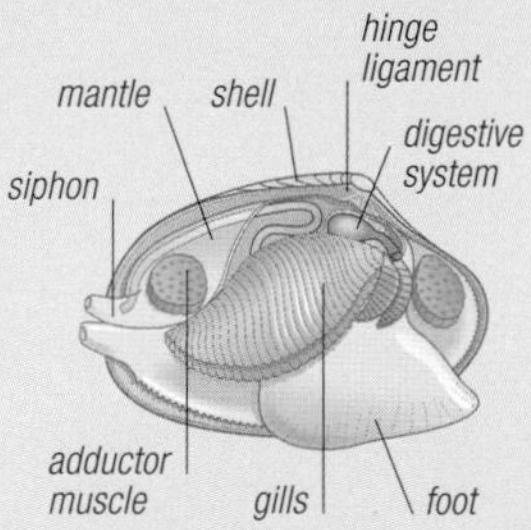

BIVALVE
Bivalves are enclosed within a pair of shells, from which the foot and siphons can be extended. The gills are often large and folded.

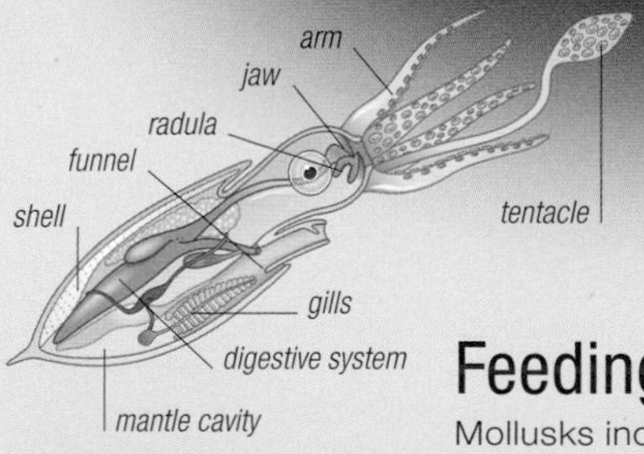

CEPHALOPOD
In cephalopods, part of the head and foot are modified to form arms and a funnel into the gill chamber. The shell is small or absent entirely.

SQUID SCHOOL
All cephalopods are marine. Octopuses tend to be solitary but squids commonly form schools, species of the genus Loligo *being particularly sociable. Cuttlefishes only congregate in the breeding season.*

Feeding

Mollusks include predators, grazers, and browsers, as well as species that feed on deposited or suspended material. Many bivalves and a few gastropods are filter feeders, using their gills to strain particles as small as bacteria from water. Cephalopods are carnivores. They use their long arms, which are often equipped with suckers, to catch fishes, crustaceans, and other prey. These relatively intelligent animals have been known to stalk prey. Besides their radula, cephalopods have a parrotlike beak, which they use to bite their prey. Most other mollusks feed on food ranging from attached algae, plants, and animals to planktonic jellyfish and even slow-moving fishes. Parasitic species obtain nourishment from their host.

HUNTING
Some octopuses are capable of subduing large and formidable prey. This one has overpowered a zebra moray eel.

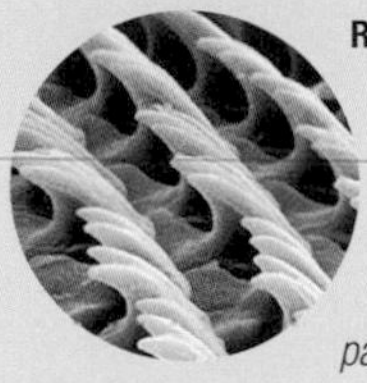

RADULA
The tiny, toothlike denticles on the radula, which may number up to several thousand, are used like a rasp to collect small particles of food.

BIVALVE SHELL
The 2 halves of a bivalve's shell are joined along one edge by elastic material and closed by contracting the adductor muscles.

GASTROPOD SHELL
In most snails, the shell is secreted unevenly from the mantle, which causes it to coil into a clockwise spiral (occasionally counterclockwise).

Class Bivalvia

Bivalves

Occurrence c.15,000 spp. worldwide; in aquatic habitats, mostly on the sea bed, but some in freshwater

Bivalves are the only mollusks that have a shell consisting of 2 parts, called valves, hinged together. In most cases, the valves can be closed tightly to protect all or a large part of the animal's body. Mussels, clams, oysters, and scallops are among the most familiar bivalves, providing a valuable source of food for humans as well as other creatures. Some species, such as the piddock and shipworm, have small, sharp-edged valves, used like drill bits to bore through wood or soft rock. Sizes of bivalve mollusks vary greatly, from less than 1⁄16 in (2 mm) to as much as 3¼ ft (1 m) across. The largest species (giant clam) is about 2 billion times heavier than the smallest. Bivalves are usually static, although some—for example, scallops—can swim. Most are filter feeders, using modified gills to strain food from the water current. The water is drawn in and pumped out through tubes or siphons—especially important for species that live buried in mud or sand. The siphons may be so large as to be permanently outside the shell. Bivalves have tiny heads, and they do not have a radula—the mouthpart that other mollusks use to rasp away at their food. The sexes are separate, although some bivalves may be hermaphrodite.

SCALLOPS
Unusually among bivalve mollusks, scallops can swim using a form of water-jet propulsion, to which they usually resort to escape predatory starfish. This North Atlantic species, Aequipecten opercularis, *may attain a diameter of 3½ in (9 cm).*

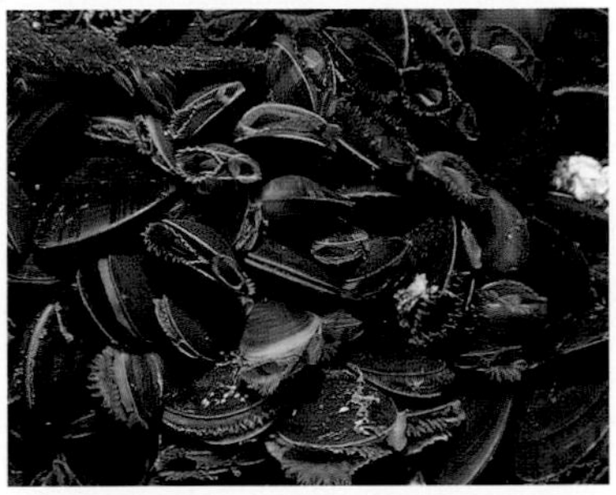

MUSSELS
Mussels attach themselves to hard surfaces by means of small threads, allowing them to hold fast against wave action and water currents. They often live in large clusters, and can cause problems for coastal power stations by blocking their cooling systems. Mussels are often raised in aquaculture systems. Mytilus edulis *(above), a species found in northern waters, may be 6 in (15 cm) long.*

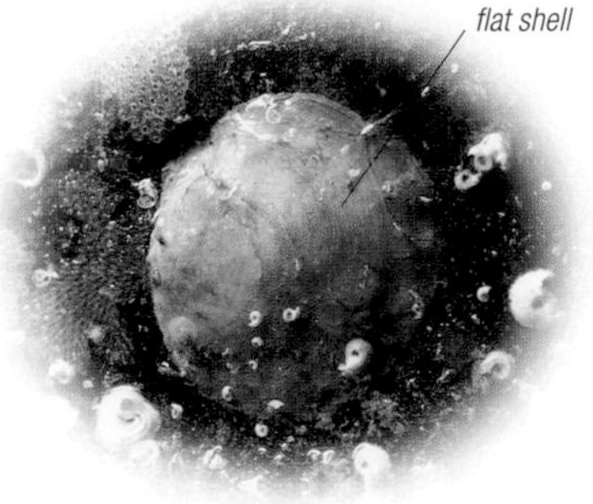

SADDLE OYSTERS
Like true oysters, saddle oysters or jingle shells live with one valve attached to a hard surface. The method of attachment is peculiar: a calcified plug passes from the upper valve through a hole in the lower one onto the surface. The Atlantic Anomia ephippium *(above) is about 2¼ in (6 cm) long.*

THORNY OYSTERS
The tropical thorny oysters, found mostly on rocky outcrops and coral reefs, grow up to 8 in (20 cm) in diameter and have shells covered with elongated spines, as in the Spondylus *species pictured above. Their shells often become covered with algae, which may serve as camouflage.*

EDIBLE OYSTERS
The edible oyster cements its left-hand valve directly onto a surface such as a rock or shell, so that the animal lives lying on its side. A valued food source since prehistoric times, oysters are among the most popular of edible mollusks. This North Atlantic Ostrea *species may grow up to 4 in (10 cm) in diameter.*

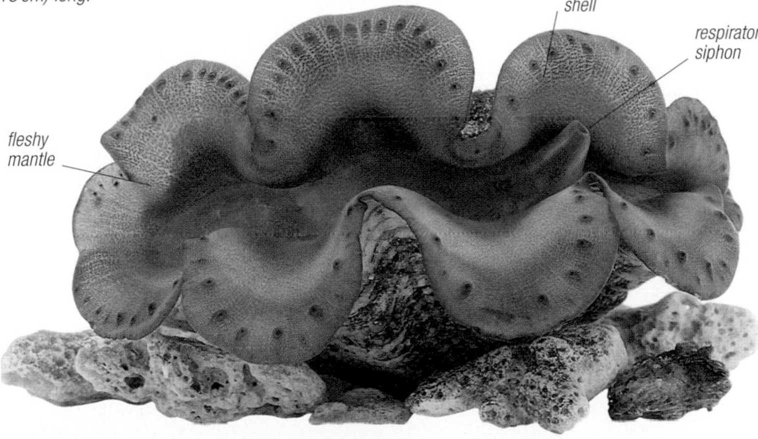

GIANT CLAMS
The tropical, reef-dwelling giant clam, Tridacna sp. *(above), is the largest bivalve; it may measure more than 3¼ ft (1 m) across and weigh over 440 lb (200 kg). Contrary to popular myth, it is harmless to humans. As with all giant clams, most of its dietary needs are met by symbiotic algae that live in its fleshy mantle; using sunlight, the algae produce a supply of energy-rich food. The clam also filters microscopic particles of organic matter from the sea.*

COCKLES
Cockles live in sandy sediments, burrowing just below the surface so the body is half in and half out of the sand. The shell is usually strongly ribbed, stout, and semispherical. Several species, including the Atlantic Cerastoderma edule *(above)—which is 2 in (5 cm) high—are harvested commercially.*

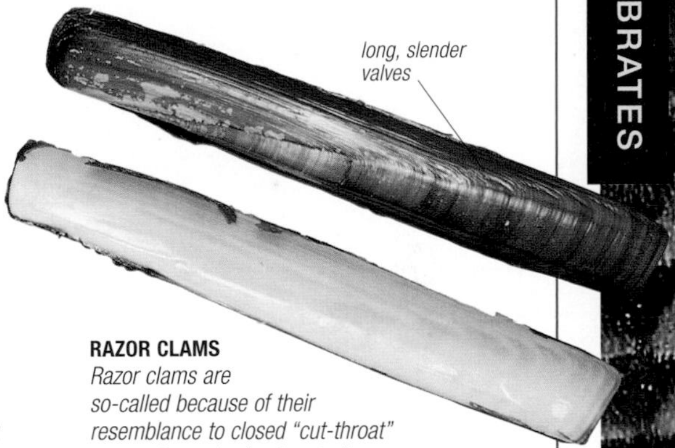

RAZOR CLAMS
Razor clams are so-called because of their resemblance to closed "cut-throat" razors. Occurring in almost all latitudes, they inhabit deep, vertical burrows in sandy sediments, using the shell valves to wedge themselves in. The species whose shell is shown here, Ensis siliqua, *may grow up to 10 in (25 cm) long.*

PIDDOCKS
Using their elliptical shells like drill bits, piddocks bore into materials such as rock, wood, peat, and clay, and live in the excavated burrow. Some piddocks are phosphorescent, glowing with a ghostly blue-green light at night. This North Atlantic species, Pholas dactylus *(left), may be up to 6 in (15 cm) across.*

SWAN MUSSELS
Most bivalves are marine, but some—such as this swan mussel—inhabit freshwater. A species of the genus Anodonta, *it is found in North America and Eurasia, and can grow to 9 in (23 cm) across. Their shells are thin, mainly because calcium is less readily available in fresh-water. In their larval stages, they are parasitic on fishes.*

SHIPWORMS
These wood-boring bivalves occur in all seas, but are particularly common in warm waters. The relatively small shell is used to gouge out a tunnel in which to live, and the shipworm lines it with a chalky tube. Their bodies may extend up to 6½ ft (2 m), although the common shipworm, Teredo navalis *(pictured), rarely exceeds a tenth of this size.*

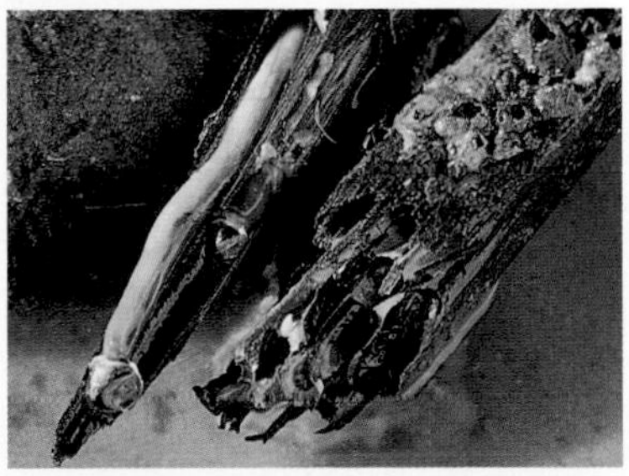

Class Gastropoda

Gastropods

Occurrence c.75,000 spp. worldwide; in aquatic habitats (mostly bottom-dwelling), in wet, damp, and dry regions on land

Habitat All

This vast and variable group of invertebrates includes over four-fifths of the mollusks alive today. The familiar garden snail, slug, winkle, sea hare, sea slug, abalone, pond snail, and conch are all gastropods. Typical species have a spiral shell and move by creeping on a muscular foot, but there are many exceptions to this rule: some,for example, have no shell at all. Other characteristic features include a large head, with eyes and tentacles, and a ribbon-shaped mouthpart called a radula, which is armed with small, often microscopic, teeth and is used to rasp away at food. These mollusks vary widely in size, and can grow up to 3¼ ft (1 m) long. The shape of the shell is highly variable—some form a tall spire, while others flare outward. In some species, an operculum, or "door," shuts the shell opening when the animal withdraws inside. Gastropods are characterized by torsion, a process in which the mantle cavity is rotated counterclockwise up to 180 degrees, until it faces forward and is positioned over the head. This occurs early in larval development and means that the digestive, reproductive, and excretory systems all discharge through the one opening in the shell. Gastropods feed in a wide variety of ways: they may be hunters, browsers, or grazers; some are suspension or deposit feeders; some are parasitic. Although sexes are separate in most gastropods, several groups are hermaphrodites, while in other groups, animals change from one sex to the other during their lifetime.

SEA HARES
Most sea hares have a thin shell (not visible externally), a large, well-developed head with 2 pairs of tentacles, and a large foot. Sea hares inhabit beds of seaweed and sea grasses, where they feed on large, fleshy algae. The North Atlantic species shown above, Aplysia punctata, *may attain a length of more than 12 in (30 cm).*

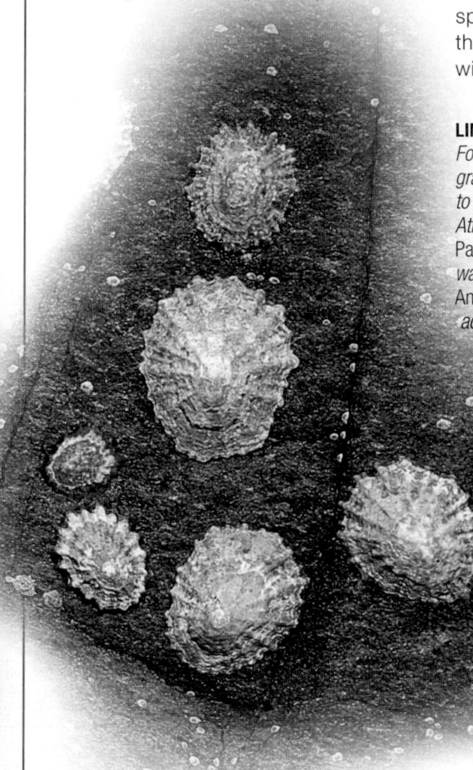

LIMPETS
Found on rocky shores, limpets, like abalones, graze on young algae, using their conical shells to protect themselves from the waves. Of the 2 Atlantic species shown here, the common limpet, Patella vulgata *(left)— 2 in (5 cm) across— wanders over rocks, while the blue-rayed limpet,* Ansates pellucida *(below)—up to ¾ in (2 cm) across—creeps over kelp, feeding on small epiphytic algae and on the kelp itself. Many rock-dwelling limpets are strongly territorial, making circular scars in the rock where they stay locked in position at low tide.*

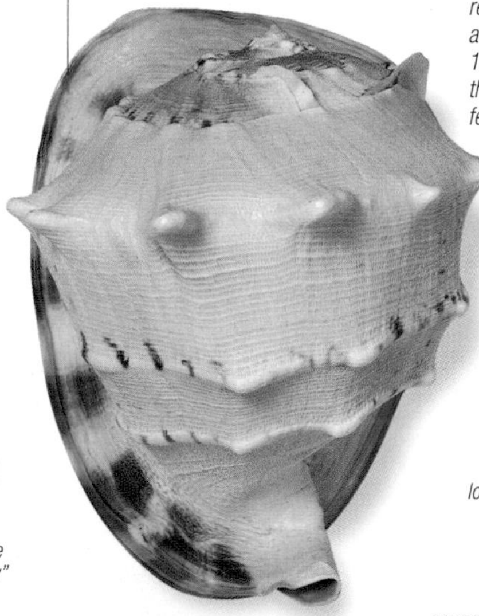

HELMET SHELLS
The stout, colorful, and attractive helmet shells resemble the helmets worn by the gladiators of ancient Rome. The shells may be more than 12 in (30 cm) long. The animals that make them inhabit sandy sediments in the tropics, feeding on other mollusks and sea urchins. The shells of species such as the horned helmet, Cassis cornuta *(left), are much used as ornaments and for carving.*

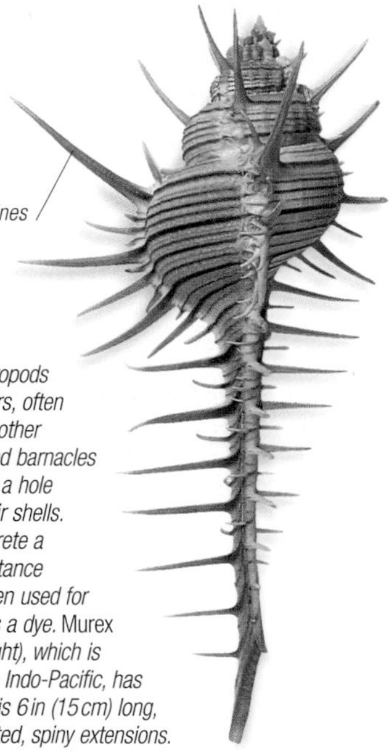

MUREXES
These gastropods are predators, often consuming other mollusks and barnacles after boring a hole through their shells. Several secrete a purple substance that has been used for centuries as a dye. Murex troscheli *(right), which is found in the Indo-Pacific, has a shell that is 6 in (15 cm) long, with elongated, spiny extensions.*

VIOLET SEA SNAIL
The violet sea snail, Janthina janthina, *is one of the few shelled gastropods that inhabit the ocean surface. It does this by secreting a raft of mucus bubbles that keeps it afloat. This warm-water snail is blind and has a paper-thin shell, ¾ in (2 cm) across. It feeds on by-the-wind-sailors (see p.540) and other colonial hydroids that drift through the water.*

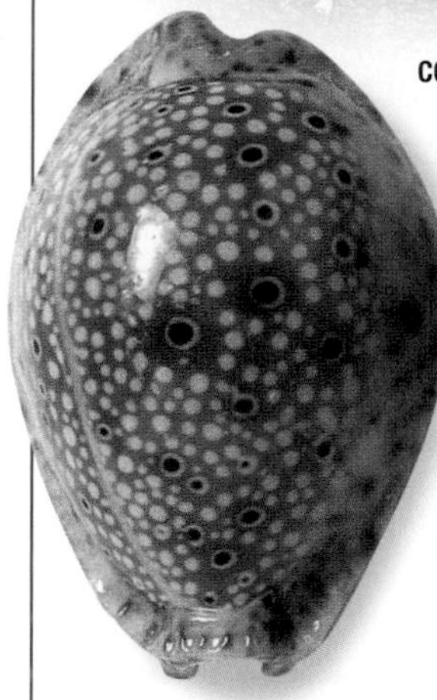

COWRIES
The beauty of their glossy shells ensures that the mostly tropical cowries are particularly sought after by collectors. In life, the shell is covered by a flap of skin, which is a part of its brightly colored mantle. The shell aperture is long and narrow, which helps to protect the cowrie from attacks by predatory crabs. This species, Cypraea ocellata, *can grow up to 2¼ in (6 cm) across.*

HARP SHELLS
The globular harp shells have smooth, regularly spaced ribs, which resemble the strings of a harp. Up to 4 in (10 cm) long, the shells are often brown, red, or pinkish. Species such as Harpa major, *shown here, are mainly nocturnal, and inhabit tropical sandy sediments, into which they burrow. When disturbed, they may discard part of the large foot, in much the same way as a lizard discards its tail.*

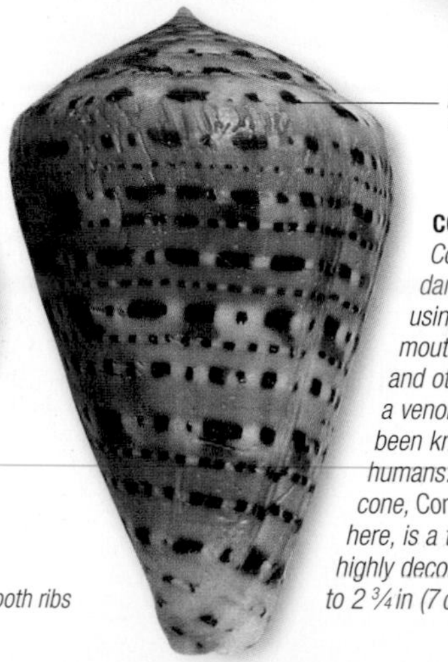

CONE SHELLS
Cone shells are highly dangerous predators: using their harpoonlike mouthparts, they attack fish and other animals, injecting a venom so potent that it has been known to cause death in humans. The West African garter cone, Conus genuanus, *shown here, is a typical example, with a highly decorated, tapering shell up to 2 ¾ in (7 cm) long.*

ABALONES
Also known as ormers, awabis, or ear-shells, abalones are grazing animals that can be over 8 in (20 cm) across. Their shells have a distinctive series of holes, through which the water current drawn into the gill chamber leaves, as do waste products. The iridescent shells of species such as the red abalone, Haliotis rufescens *(above), found off the west coast of North America, are used to make jewelry, and the animals inside are often eaten.*

SACOGLOSSANS
Sacoglossans are sluglike marine gastropods, ⅜ in (1 cm) or so long, that graze on algae, sucking up the contents of their cells. This species, belonging to the widespread genus Elysia, *gets its color from chlorophyll—the bright green pigment that algae use to harness the energy in sunlight.*

SEA SLUGS
True sea slugs lack shells and are often brightly colored with striking patterns. Many species, like Janolus nakaza, *above—which is 3¼ in (8 cm) across and lives on reefs off the southern coast of Africa—have colorful protuberances (papillae) on the back. Most sea slugs live on the seabed, although a few are planktonic.*

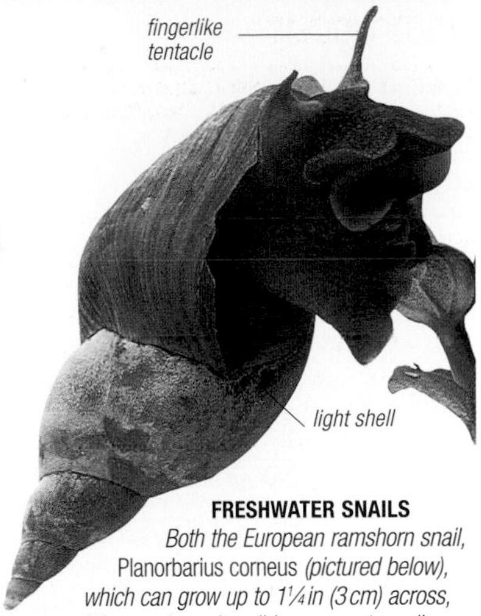

FRESHWATER SNAILS
Both the European ramshorn snail, Planorbarius corneus *(pictured below), which can grow up to 1¼ in (3 cm) across, and the great pond snail,* Lymnaea stagnalis *(above), which is up to 2 in (5 cm) tall, have a single lung instead of gills. Because they breathe air, they can survive in slow-flowing or stagnant water, where oxygen levels are low. The great pond snail and its many relatives are common in freshwater throughout the Northern Hemisphere.*

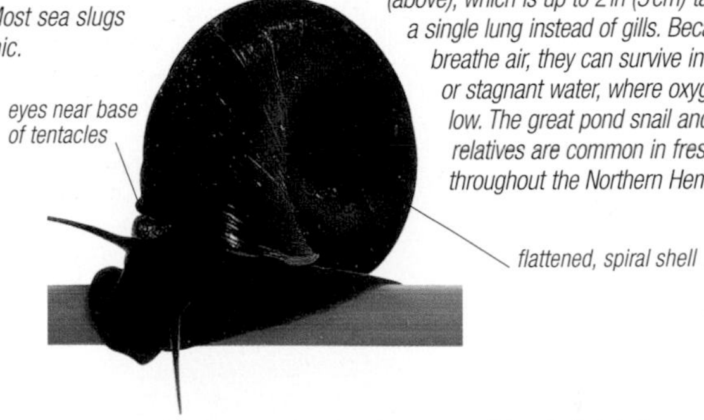

TERRESTRIAL SLUGS
Unlike sea slugs, terrestrial slugs breathe air through their mantle cavity, which functions as a lung. In this European black slug, Arion ater, *usually 6 in (15 cm) long, the opening leading to the cavity can be seen just behind the animal's head.*

GHOST SLUG
This slug avoided detection until 2007—when it was discovered in Wales by a researcher. Selenochlamys ysbryda *prey on earthworms at night:* ysbryd *is Welsh for "ghost." It is unlikely to be native to the UK.*

TERRESTRIAL SNAILS
Land-dwelling snails differ from sea snails in having thin, relatively light shells. Many also lack an operculum—a hard plate that seals the shell's opening. The species pictured right is the giant African snail, Achatina fulicula, *which can reach a length of 12 in (30 cm) with its body extended. Originally from East Africa, it has been introduced to many parts of southern Asia, and in some regions is a serious agricultural pest.*

WINKLES
Winkles are small marine grazers common on rocky shores, in mangroves and salt marshes, and more rarely on mudflats. Some can live at the very top of the shore, where they are wetted only by high spring tides; these species have a gill chamber partly converted into a lung. The European flat periwinkle shown above, Littorina obtusata, *like most species of winkle, is usually less than 1¼ in (3 cm) tall.*

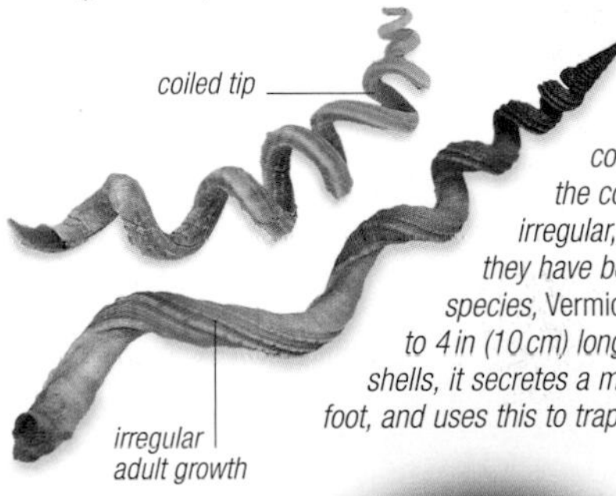

WORM SHELLS
These mainly tropical gastropods start life with tightly coiled shells, but as they grow older, the coils become wider and more irregular, making the shells look as if they have been unwound. This Caribbean species, Vermicularia spirata, *grows up to 4 in (10 cm) long. Like many other worm shells, it secretes a mucus filter from its foot, and uses this to trap small animals.*

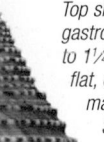

TULIP SHELLS
The spindle-shaped tulip shells or horse conches are predators on other mollusks and worms. Although some live in shallow water and in high latitudes, most prefer warm water and the sublittoral zone (below the low watermark). The body is often red, as in the Fasciolaria *species above, and the shell, up to 10 in (25 cm) long, has a long, tubular projection—the siphonal canal—at the front.*

TOP SHELLS
Top shells are grazing marine gastropods with a conical shell up to 1¼ in (3 cm) high, which has a flat, circular base. They occur on many rocky shores and reefs. Some, like the widely distributed Calliostoma *species shown above, are brightly colored. Males have no penis, and fertilization is external.*

HETEROPODS
The species of the warm-water genus Carinaria *and its relatives are among the few gastropods that have taken up life in the open sea, where they make up part of the plankton. These unusual mollusks, up to 2 in (5 cm) long, have gelatinous bodies and either very thin shells or no shell at all. They appear to live upside down because the shell, if present, is located beneath the body, and the fin—a foot modified for swimming—is uppermost.*

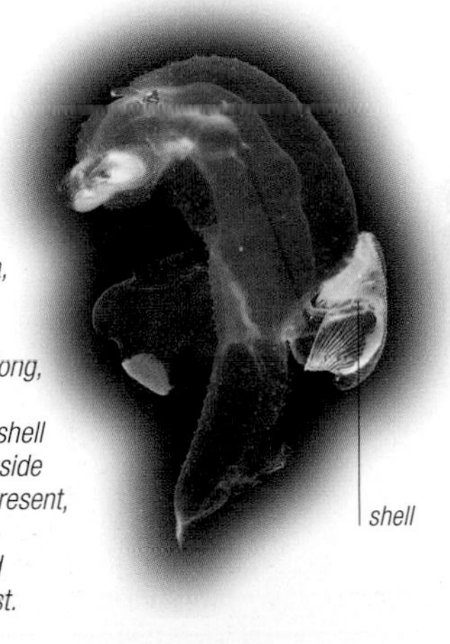

INVERTEBRATES

Class Aplacophora

Aplacophorans

Occurrence 395 spp. worldwide; on the seabed or in association with other animals (especially cnidarians)

With their thin, wormlike bodies, aplacophorans bear very little resemblance to any other mollusks. Typically less than 3/16 in (5 mm) long, they have no trace of a shell, but are covered by a tough cuticle that contains a large number of small, mineral spicules. They have poorly developed heads and no eyes, and move by creeping or by burrowing through sediment. The creeping species feed on cnidarians, while the burrowers feed on tiny animals, or on particles of organic matter. Some species have separate sexes, but most are hermaphrodites. Some keep their eggs until they hatch, but little else is known about their breeding habits.

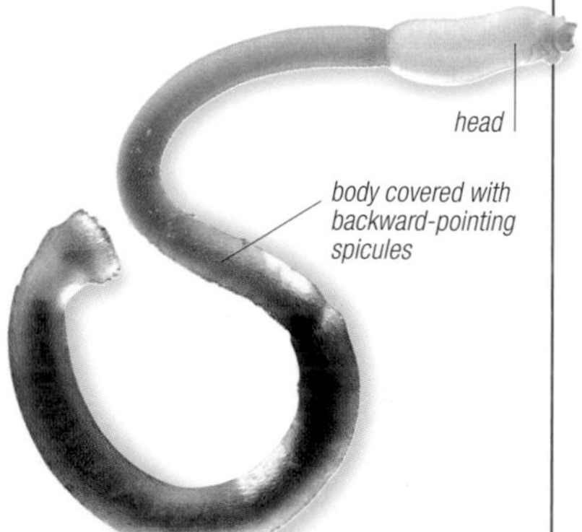

CHAETODERMA SPP.
Members of the genus Chaetoderma, *found in the Northern Hemisphere, are relative giants among aplacophorans, growing up to 3¼ in (8 cm) long. Like other aplacophorans, they lack a foot. They breathe with gills, located in a cavity at the back of the body.*

Class Polyplacophora

Chitons

Occurrence 957 spp. worldwide; on the seabed

Chitons resemble gastropods in having a suckerlike foot, but their shells are quite distinctive, being made of 8 plates. These are arranged in a line, and they arch over the animal's back. These fat, oval to elongated mollusks are covered by a thick "girdle," which often flanks the plates like a lip. It has a rubbery texture, and in many species is covered with bristles or calcareous spicules. Chitons can be up to 16 in (40 cm) long. They have a large, muscular foot, a small head with a well-developed rasping mouthpart (radula), but no eyes or tentacles, and a U-shaped mantle cavity around the foot with 6–88 pairs of gills. In both sexes, there is one pair of gonads (reproductive organs) fused into a single structure. Chitons graze algae from rocks on the seabed. They creep slowly over the surface—often after dark or at high tide—and return to a particular site where they fit snugly against the rock, helping to protect themselves from attack, and from desiccation if exposed to the air.

CHITON STRIATUS
Found on the west coast of South America, the magnificent chiton is large and well arched and surrounded by a broad girdle. It measures 3½ in (9 cm) in length and inhabits intertidal rocks. The central portion of each plate has crowded, straight riblets that are usually worn smooth at the highest point. Head and tail plates have coarse, radiating ribs, and the girdle is covered with polished granules.

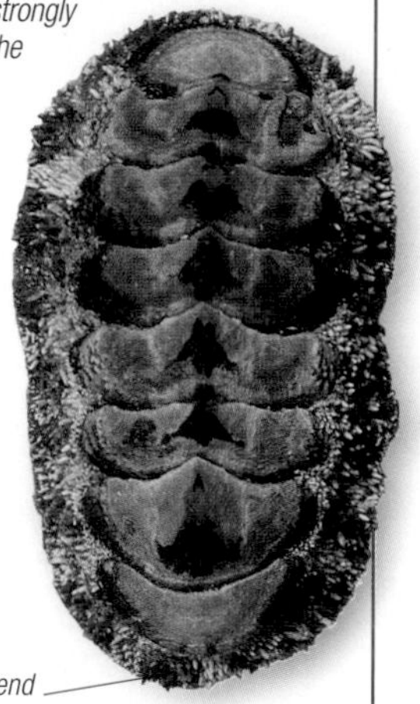

ACANTHOPLEURA GRANULATA
The fuzzy chiton is a strongly humped chiton from the Caribbean. It is surrounded by a thick girdle that is covered with coarse spines. The plates are brown when unworn, but usually appear grayish brown with darker brown crests and sides. Up to 2¼ in (6 cm) in length, it inhabits intertidal rocks.

MARBLED CHITON
Tonicella marmorea, *a North Atlantic chiton up to 1½ in (4 cm) long, has a broad, leathery girdle surrounding the shell plates. It can be found from the intertidal zone to depths of 650 ft (200 m). It occurs on rocks, stones, and pebbles, and in some Scottish sea lochs can attain densities of 50 per square yard.*

Class Monoplacophora

Monoplacophorans

Occurrence 30 spp. in Atlantic, Pacific, and Indian oceans; at depths of 650–23,100 ft+ (200–7,000 m)

Monoplacophoran mollusks were thought to have become extinct more than 300 million years ago, until a handful of specimens were dredged up in the eastern Pacific Ocean in 1952. Since then, they have been found on the deep seabed in widely scattered parts of the world. Monoplacophorans look like limpets (see p.596), but they each have up to 8 pairs of foot retractor muscles, whereas limpets and other gastropods have just one pair. A horseshoe-shaped mantle cavity containing 3–6 pairs of gills surrounds the foot. The shell is conical or cap-shaped, and the head is small, with a large mouthpart (radula). There are no eyes or tentacles. Sexes are separate. These mollusks grow up to 1¼ in (3 cm) long.

Class Scaphopoda

Tusk shells

Occurrence About 576 spp. worldwide; on the seabed, in soft sediments

Widely distributed in all the world's oceans, these burrowing mollusks are easily recognized by the shape of their shells, which are like tapering, hollow tusks, open at both ends, with one opening larger than the other. Tusk shells live head downward in sediment, the head and the foot protruding out of the larger of the shell openings. The small head has numerous fine, ciliated feeding filaments, which contract to transport food particles from the sediment into the mouth. The smaller aperture projects above the sediment surface; respiratory water current is drawn in and let out through this opening. There are no gills; instead, gases are exchanged through the mantle surface.

The sexes are separate, and there is a single reproductive organ. Living tusk shells are rarely seen, but their empty shells, which can be up to 6 in (15 cm) in length, are often washed up on the shore.

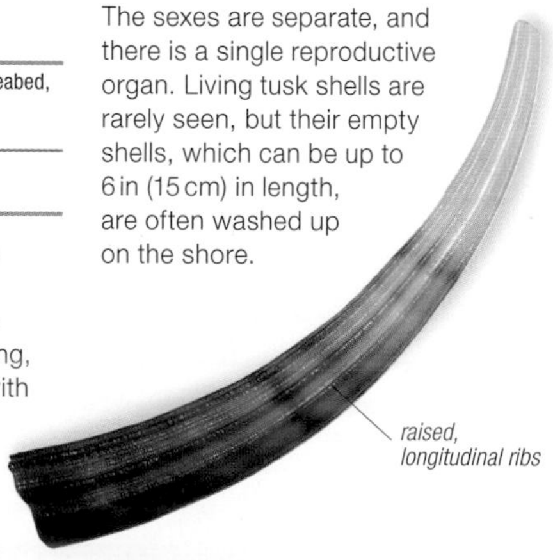

ELEPHANT TUSK
The solid and slightly glossy shell of the species Dentalium elephantinum *has a characteristic notch at its narrower end. This mollusk inhabits offshore sands in the Indo-Pacific, and is 3 in (7.5 cm) long.*

BEAUTIFUL TUSK
This species, Pictodentalium formosum, *found on Indo-Pacific coasts, grows to 3 in (7.5 cm) in length. It differs from other large tusk shells in its head end being only slightly broader than its rear end; it is also straighter than most. Its longitudinal ribs and riblets are crossed by concentric growth lines. A small, notched pipe protrudes from the rear end.*

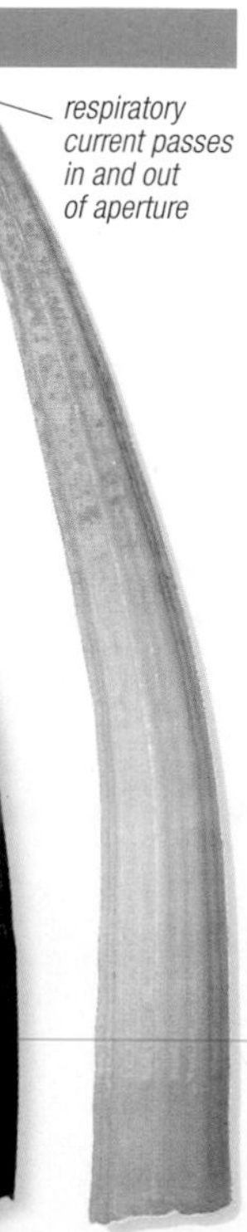

ELONGATED TUSK
Antalis *species are a small group of marine mollusks from the North Atlantic that live in shells 2¼ in (6 cm) long. They feed on microorganisms, using numerous club-shaped tentacles.*

Class Cephalopoda

Cephalopods

Occurrence About 750 spp. worldwide; marine, in open water and on or near seabed

Cephalopods include the largest, fastest, and most intelligent of all invertebrates. Unlike other mollusks, they have arms and tentacles (modified arms), and they can move with speed by a form of jet propulsion, which involves squirting water out of the mantle cavity. The arms and tentacles are used to catch prey. In many species, other modified arms are used by males to transfer spermatophores (sperm packets) to the female. Cephalopods have a rasping mouthpart (radula), a birdlike beak, a large brain, and well-developed eyes. Many early cephalopods had coiled, external shells, but among the existing species, only the nautiluses show this characteristic. In all other cephalopods, such as cuttlefishes, squids, and octopuses, the shell is either internal or absent entirely. Cephalopods vary greatly in size: from tiny *Idiosepius* species, that are just ⅜–¾ in (1–2 cm) long, to giant and colossal squids. Many cephalopods breed only once and then die.

SPIRULAS
These small creatures resembling cuttlefishes are less than 2¾ in (7 cm) long. Their bodies contain a coiled and chambered shell that acts as a float, enabling them to hang head down in the water (see right), at depths of up to 6,600 ft (2,000 m). Using their small arms, they catch planktonic animals that drift within reach. When they die, their shells can float to the surface, and are often carried onto the shore. Shells of the common spirula, Spirula spirula, *shown, are found on tropical beaches.*

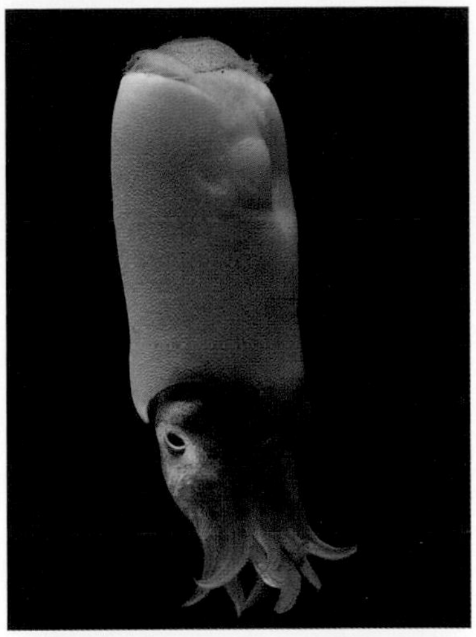

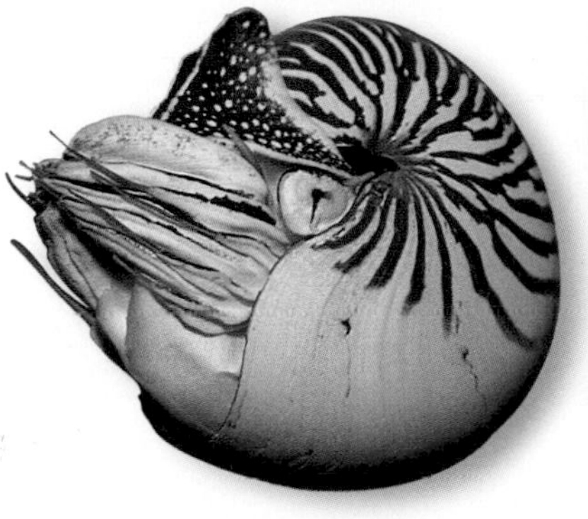

NAUTILUSES
Nautiluses are the only surviving cephalopods with a true external shell, which may be up to 8 in (20 cm) across. It contains a series of gas-filled chambers that act as a buoyancy device, and a large final chamber that houses the animal itself. The pearly nautilus, Nautilus pompilius *(above), lives in the Indo-Pacific.*

GIANT SQUIDS
Members of the genus Architeuthis, *distributed worldwide, giant squids can be over 58 ft (17⅗ m) long. No giant squids has ever been captured alive but they have been photographed. Their bodies are found on beaches, and their undigested beaks in the guts of sperm whales.*

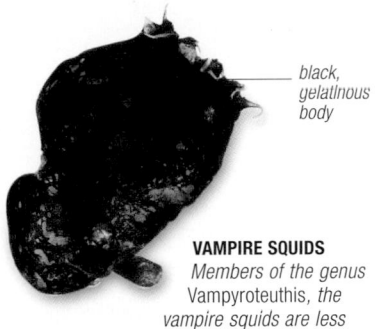

VAMPIRE SQUIDS
Members of the genus Vampyroteuthis, *the vampire squids are less than 2 in (5 cm) long. Like many cephalopods they have 8 arms, plus 2 additional, retractile appendages. They live in tropical waters at depths of up to 9,900 ft (3,000 m).*

PACIFIC SQUIDS
Squids are torpedo-shaped hunters, with internal shells that are reduced to a small, horny "pen." Some hunt alone, but many medium-sized species—such as the Pacific Doryteuthis opalescens, *above, up to 8 in (20 cm) long—feed in large schools. All swim by flapping their fins, or by jet propulsion. Squids have 2 long, prey-catching tentacles that can be retracted, but not completely as in cuttlefishes.*

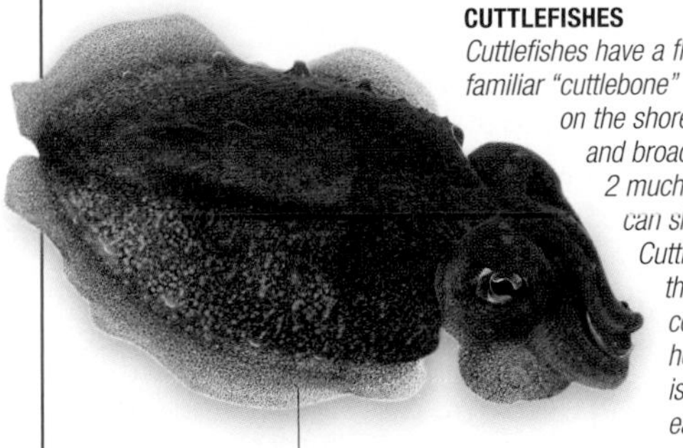

CUTTLEFISHES
Cuttlefishes have a flat, internal shell—the familiar "cuttlebone" that is often washed up on the shore. Their bodies are short and broad, with 8 short arms and 2 much longer tentacles that can shoot out to catch prey. Cuttlefishes are known for their ability to change color. The species shown here, Sepia officinalis, *is common around the eastern North Atlantic and grows up to 12 in (30 cm) long.*

PAPER NAUTILUS
A species in the genus Argonauta, *the pantropical paper nautilus (shown above) is really a type of octopus. It is 8 in (20 cm) long. The breeding female has a superficial, nautilus-like, thin shell—a case secreted by 2 of the arms—in which she broods the fertilized eggs.*

BLUE-RINGED OCTOPUS
The blue-ringed octopus, Hapalochlaena lunulata, *inhabits coral reefs in the Pacific and Indian oceans. It is small—4 in (10 cm) long—and beautiful, but has a venomous bite that can kill humans in 15 minutes.*

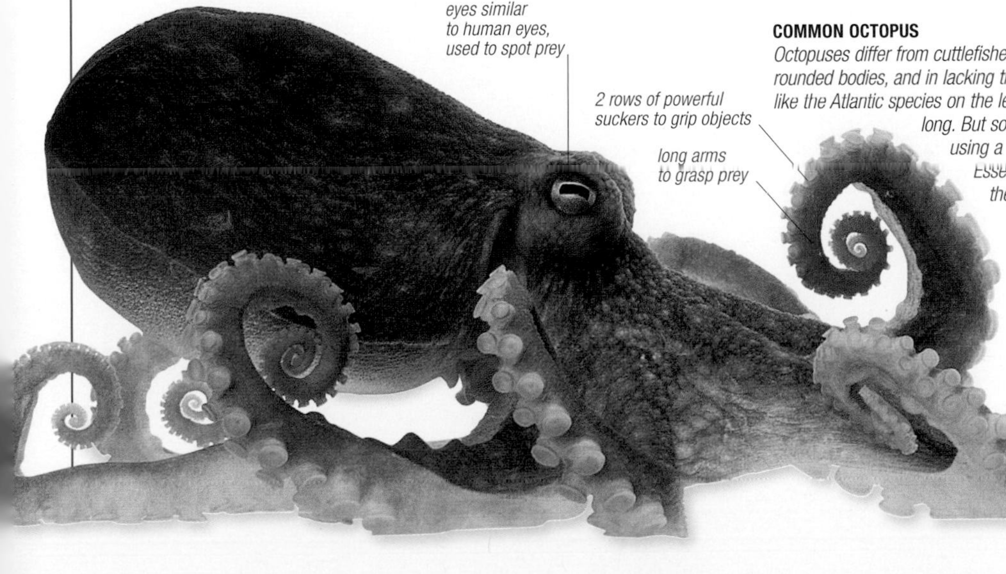

COMMON OCTOPUS
Octopuses differ from cuttlefishes and squids in having more rounded bodies, and in lacking the 2 long, prey-capturing tentacles, like the Atlantic species on the left, Octopus vulgaris, *3¼ ft (1 m) long. But some "sail" bottom currents using a web of skin between the arms. Essentially they have no shell and the body is incredibly flexible.*

WUNDERPUS PHOTOGENICUS
This octopus of Indo-Pacific shallow waters was described in 2006. It has small, stalked eyes and a distinctive white-spotted pattern that becomes more pronounced when the animal is disturbed. It preys on fishes and invertebrates at twilight.

INVERTEBRATES

Enteroctopus dolfeini

Giant Pacific octopus

Occurrence 1 sp. in the north Pacific Ocean, down to depths of 2,500 ft (750 m); mostly on the seabed

The largest species of octopus in the world, the giant Pacific octopus usually grows to about 9¾ ft (3 m) in length and weighs up to 600 lb (272 kg). It lives on the rim of the north Pacific Ocean, from Japan up to the Aleutian Islands and south to California, where it crawls about on the bottom, using its long, sucker-covered arms. It seeks out rocky dens on the seabed, under boulders or in crevices; youngsters will often dig holes under rocks in sand or gravel. Here, the octopus can take refuge from predators—seals, sea otters, sharks, and other large fishes—too big to slip through the den mouth. Foraging mainly at night, this giant octopus looks especially for crabs and lobsters, but also takes shrimps and shellfish, smaller octopuses, and fishes. Often it will return to its den to feed, depositing empty shells and other inedible fragments of prey in piles at the entrance. Like its relatives, this octopus mostly lives alone, except for a brief period (which may occur at any time of the year) when adults come together for mating. The male deposits sperm packets up to 3¼ ft (1 m) long in the female's mantle cavity, by means of a specially modified arm. The female lays her eggs in a den, and will tend them until they hatch, which may take as long as 8 months, depending on the temperature of the water. She will not feed in all this time—and will die soon after her young emerge. Lifespans range up to 4 years.

AGILE GIANT
The giant Pacific octopus is remarkably agile for its size, its suckered arms enabling it to move easily over the ocean floor, and its flexible body squeezing through narrow openings.

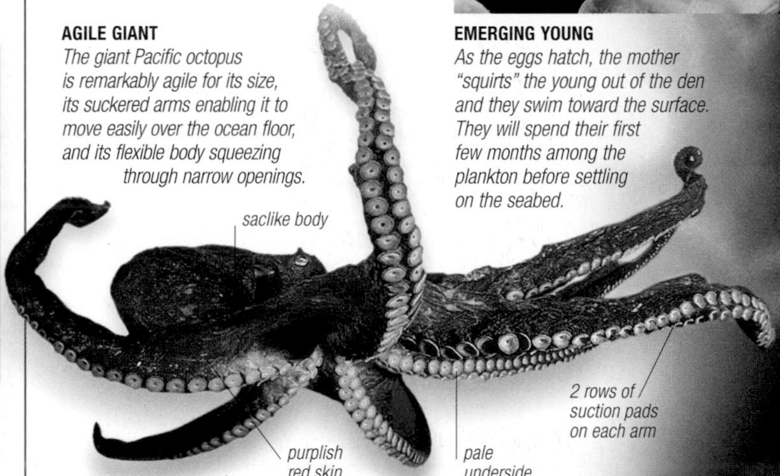

EMERGING YOUNG
As the eggs hatch, the mother "squirts" the young out of the den and they swim toward the surface. They will spend their first few months among the plankton before settling on the seabed.

DEFENSE AND ATTACK

When threatened, the giant Pacific octopus may change color, or flee hidden by an inky cloud. When it catches prey, it grasps it firmly and pulls it apart, bites it open, or breaks through the shell with its beak.

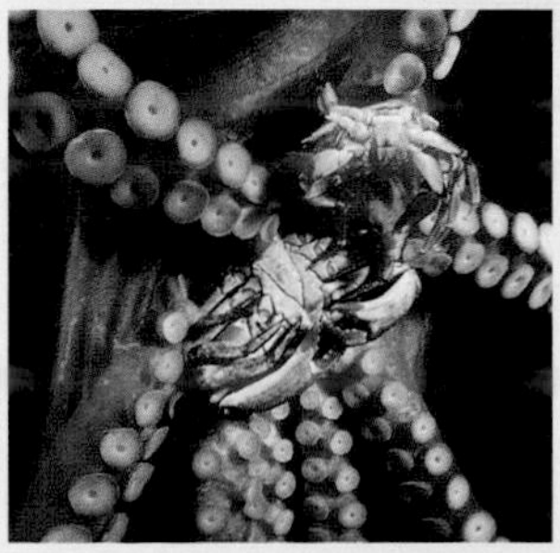

FLEEING FROM DANGER
When a predator gives chase, the octopus ejects an inky substance that forms a dark cloud in the water. Hidden from view, the octopus can swiftly make its escape.

DEALING WITH A CRAB
Holding a crab firmly in its suckers, the octopus makes a hole in the shell with its beak, then injects saliva to liquefy connective tissue so that meat can be extracted.

DEVOTED MOTHER
The female lays many thousands of eggs and stays with them for 5–8 months, guarding them until they hatch. She continuously washes them with a stream of water from her funnel, and grooms them with her arms to keep them free of parasites.

Echinoderms

PHYLUM	Echinodermata
CLASSES	5
ORDERS	38
FAMILIES	173
SPECIES	About 7,000

This group of marine invertebrates includes starfish, brittle stars, sea urchins, and sea cucumbers. Their bodies are typically spiny and usually divided into 5 equal parts arranged symmetrically around a central point. All echinoderms have a skeleton (or test) made of calcium-carbonate plates. They also have a unique internal network, called the water-vascular system, that helps them to move, feed, and take in oxygen. Echinoderms are mostly mobile, bottom-dwelling animals, found on shores, reefs, and the seabed.

Movement

While sea lilies lead stationary lives on the seabed, most other echinoderms are mobile. Starfish and sea urchins move with their tube feet, spines, or the test itself. Sand dollars and sea cucumbers use their feet to burrow into the sediment.

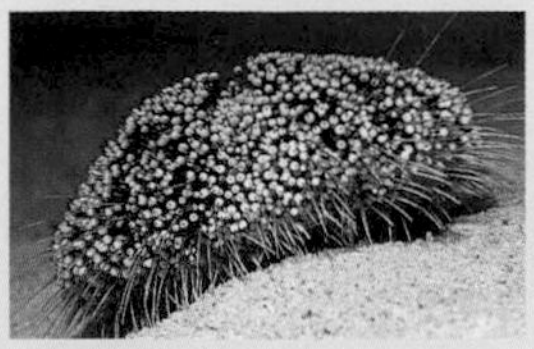

WALKING
Sea urchins walk on the ocean floor by moving water into and out of their tube feet. These extend beyond the spines and grip the surface as the animal pulls itself along.

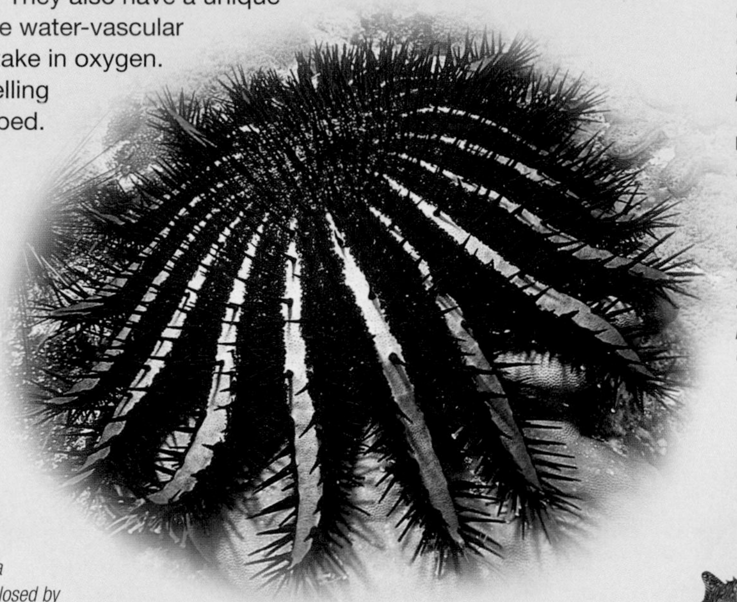

PROTECTIVE SPINES
Many echinoderms are protected from predators by their long, sharp spines. In the crown of thorns starfish shown here, these spines are also venomous. This starfish feeds on coral polyps and periodically destroys large areas of Australia's Great Barrier Reef.

Anatomy

Echinoderms have a wide range of external shapes: their bodies may be drawn out into arms (as in starfish), branched and feathery (as in sea lilies), or spherical to cylindrical (as in sea urchins and sea cucumbers). In most, spiny or knoblike extensions of the skeleton project from the body. The water-vascular system consists of a network of canals and reservoirs, and tentacles that form external appendages called tube feet.

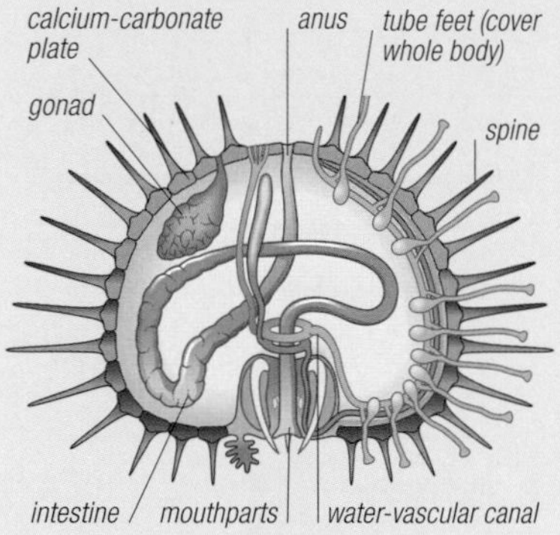

BODY SECTION
An echinoderm's body consists of a central cavity enclosed by calcium-carbonate plates, which either fit together to form a rigid structure (as in sea urchins) or remain separate (as in starfish). Echinoderms have no head and no tail. The mouth is in the center of one surface (the oral end), and the anus, which is sometimes missing, is in the center of the opposite (aboral) surface.

Feeding

Most echinoderms feed on small organisms or particles of organic matter. Some starfish, however, are carnivores, and extend their stomach over the prey, digesting it externally. Most sea urchins are grazers that use elements of their test as a series of teeth.

DEPOSIT FEEDING
Sea cucumbers use small tentacles around their mouths to take in mud and sediment, from which they extract food particles.

Class Crinoidea

Sea lilies and feather stars

Occurrence 426 spp. worldwide; permanently or temporarily attached to the seabed

Sea lilies and feather stars belong to an ancient group of echinoderms, which are abundant as fossils. They have rounded bodies and 5 arms, which often branch repeatedly. All begin life attached to a slender stalk. Sea lilies, which inhabit depths of more than 350ft (100m), remain attached to the stalk throughout their lives; in contrast, feather stars break away from the stalk when young and become free living. Both forms resemble starfish in their basic anatomy, except that their mouths face upward. They feed on plankton, caught by the tube feet on the arms and passed to the mouth. The arms may have a span of up to 28in (70cm), and the stalks can reach 3¼ft (1m) in length. Like some other echinoderms, they are able to regrow lost limbs. The sexes are separate.

arms spread out to collect food

OXYCOMANTHUS BENNETTI
Although stalked sea lilies dominate the fossil record, most surviving forms of crinoids are not permanently attached to the seabed, and some can even swim. Feather stars are filter feeders that occur in all seas, particularly on reefs. This beautiful species has a crown of arms up to 16in (40cm) across.

Class Asteroidea

Starfish

Occurrence 1,853 spp. worldwide; on the seabed

Thanks to their distinctive shape and abundance along coasts, starfish, also called sea stars, are familiar marine invertebrates. Most have a circular body that grades into 5 equal arms, but some have more than 40 arms, and others have arms so short that the animal looks like a cushion. Starfish may swallow small prey whole; some can extend their stomach through the mouth and onto larger animals and digest them externally. When feeding on bivalves, they use powerful tube feet to pull the shell valves apart. Sizes vary from less than ¾in (2cm) to 3¼ft (1m) in diameter. Starfish are famous for their ability to regenerate missing body parts such as lost arms.

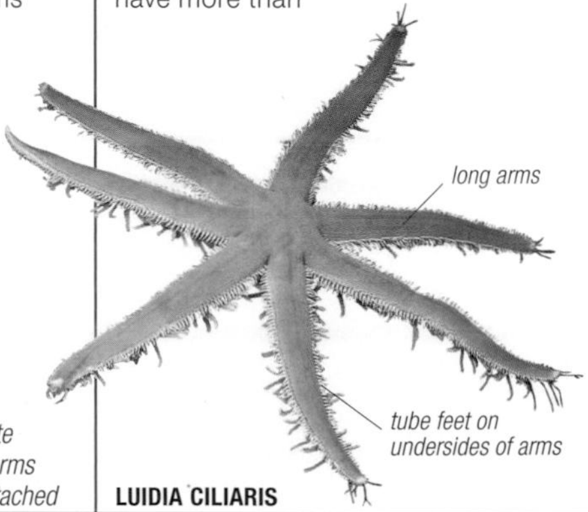

LUIDIA CILIARIS
This seven-arm starfish has 7 long arms instead of the standard 5. Up to 23½in (60cm) across, it lives on coarse sediments in the North Atlantic, where it feeds voraciously on other starfish, sea urchins, and brittle stars.

CULCITA NOVAEGUINEAE
This Southeast Asian species is a typical cushion star, about 10in (25cm) across, with arms that barely project from its body. Cushion stars can be found in a variety of marine habitats, but are particularly common on reefs.

BLUE STARFISH

The blue starfish, Linckia laevigata, *belongs to a group of starfish with elongated snakelike arms. This Indo-Pacific species is not always blue in color; a minority of individuals are purple or orange. It lives just below the intertidal zone and is often found in association with shrimps that feed on detritus and mucus on the body of the starfish.*

Class Ophiuroidea

Brittle stars and basket-stars

Occurrence 2,166 spp. worldwide; on the seabed

Although they are often mistaken for starfish, brittle stars and basket-stars form a separate and distinctive group of echinoderms. They have 5 long, narrow arms which, unlike those of starfish, are sharply demarcated from the small, central disk. The disk measures up to 4 in (10 cm) across, while the entire body can be up to 20 in (50 cm) in diameter. Brittle stars are the most mobile of all echinoderms and, like many of their relatives, can regenerate their arms if these are lost. Unlike starfish, brittle stars are either scavengers, feeding on dead animals, or suspension feeders, collecting small particles of food drifting down to the seabed, helped by a sticky mucus that is strung between the spines on their arms. Brittle stars are extremely abundant on some parts of the ocean floor, forming writhing masses that contain millions of animals. In contrast, basket-stars feed on their own, holding their intricately divided and branched arms at right angles to the current, so that they can catch small animals drifting past.

OPHIOTHRIX SPP.
These brittle stars may occur in dense aggregations on both rocky and sandy areas of the seabed. They have a diameter of up to 8 in (20 cm). Their spine-fringed arms project up into the water and filter feed.

ASTROBOA NUDA
Each arm of a basket-star branches repeatedly to give rise to a mass of armlets up to 30 in (75 cm) in diameter. In the Indo-Pacific species shown above, each armlet ends in tendrils. At night, the basket of arms is held up into the water to catch small swimming animals.

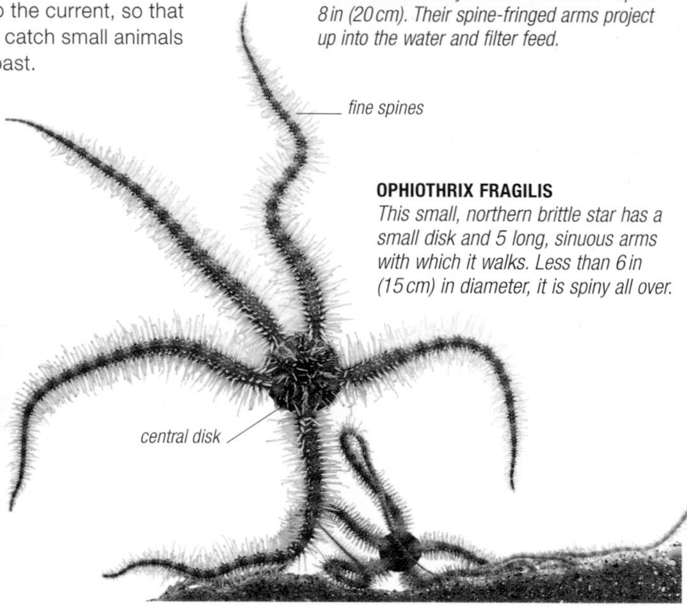

OPHIOTHRIX FRAGILIS
This small, northern brittle star has a small disk and 5 long, sinuous arms with which it walks. Less than 6 in (15 cm) in diameter, it is spiny all over.

Class Holothuroidea

Sea cucumbers

Occurrence 1,800 spp. worldwide; almost all buried in the seabed, or on the seabed surface

With their sausagelike shape and leathery texture, sea cucumbers look very different from other echinoderms, yet their internal anatomy is typical of this group, with organs present in sets of 5. They have 8–30 tube feet around the mouth, arranged in 5 rows; these are modified into finger- or bushlike feeding tentacles. Sea cucumbers vary from less than 1¼ in (3 cm) to 3¼ ft (1 m) in length. Some sea cucumbers discharge sticky, sometimes toxic, threads through their anus; these can entangle a would-be predator.

PSEUDOCOLOCHIRUS VIOLACEUS
This brightly colored, reef-dwelling sea apple cucumber from the Indo-Pacific has a ring of branched tentacles around its mouth. After feeding, each tentacle in turn is inserted into the mouth to be wiped clean. This animal is 4 in (10 cm) long.

Class Echinoidea

Sea urchins

Occurrence 1,090 spp. worldwide; on or buried beneath the seabed

Sea urchins typically have a spherical skeleton (test) of calcareous plates, covered by movable spines. Five rows of tube feet—often hidden by the spines—help these animals to creep over rocks, so that they can graze on algae and small animals. The mouth is on the underside, in the center, and is armed with the "Aristotle's lantern"—a set of 5 large, calcareous plates that work like jaws. After feeding, some sea urchins return to refuges that they have excavated over a period of time in soft rock; in the tropics, pencil urchins use their extra-thick spines to wedge themselves into crevices in coral reefs. The burrowing heart urchins and sand dollars are more disk shaped, and often have very short spines, in some species these look like backswept fur. Burrowing species are deposit feeders, mostly passing food to the mouth using the tube feet and spines. Sea urchins are usually 2¼–4¾ in (6–12 cm) in diameter, but some can grow to 14 in (35 cm).

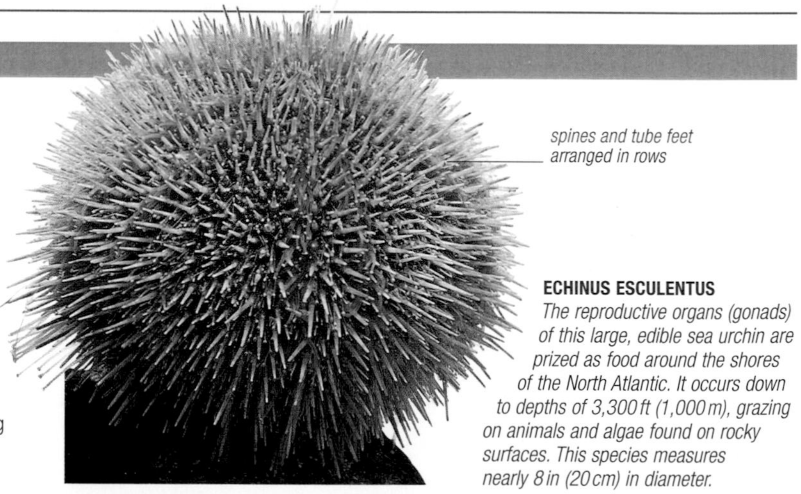

ECHINUS ESCULENTUS
The reproductive organs (gonads) of this large, edible sea urchin are prized as food around the shores of the North Atlantic. It occurs down to depths of 3,300 ft (1,000 m), grazing on animals and algae found on rocky surfaces. This species measures nearly 8 in (20 cm) in diameter.

DIADEMA SPP.
These warm-water sea urchins, 12 in (30 cm) in diameter, have long, thin, brittle spines covered by glandular tissue that secretes a poisonous substance, which can cause severe irritation to human skin. Small fishes and crustaceans often live among the sea urchin's spines, which offer them protection from predators.

SAND DOLLARS
Burrowing sea urchins such as this five-notched sand dollar lack the spherical test and long spines of other species. Like the tropical species pictured, Encope michelini, *they have flattened tests, up to 6 in (15 cm) across, with a slight dome and short spines.*

PENCIL URCHINS
Phyllacanthus imperialis *(right), found in the Indo-Pacific, has only a few blunt but very stout spines, which are often covered by colonies of encrusting organisms. Up to 6 in (15 cm) in diameter, the spherical test is easily visible between the spines.*

Invertebrate chordates

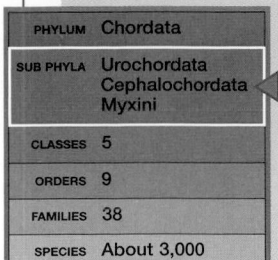

PHYLUM	Chordata
SUB PHYLA	Urochordata Cephalochordata Myxini
CLASSES	5
ORDERS	9
FAMILIES	38
SPECIES	About 3,000

Chordates are animals that have an internal skeletal rod, called a notochord. Most chordates are vertebrates, in which the notochord is largely replaced by interlocking vertebrae. However, there are 3 groups—the tunicates, lancelets, and hagfish—that have no vertebral column and are therefore classified as invertebrates. All are marine. Most tunicates have a bag-shaped body, while lancelets have a long, narrow, rigid body. Hagfish are fishlike and have a cartilaginous skull.

Anatomy

Adult lancelets and lampreys retain their notochord, while in tunicates, it is present only in the larval stage. All species (apart from some tunicates) also have a hollow nerve cord above the notochord, a muscular tail (although in some species this is found only in the larvae), and a perforated throat (or pharynx). In tunicates and lancelets, the pharynx forms an internal filtering mechanism to trap particles of food: water enters through a mouthlike opening (known as the branchial siphon) and leaves via openings that extend through to the body surface.

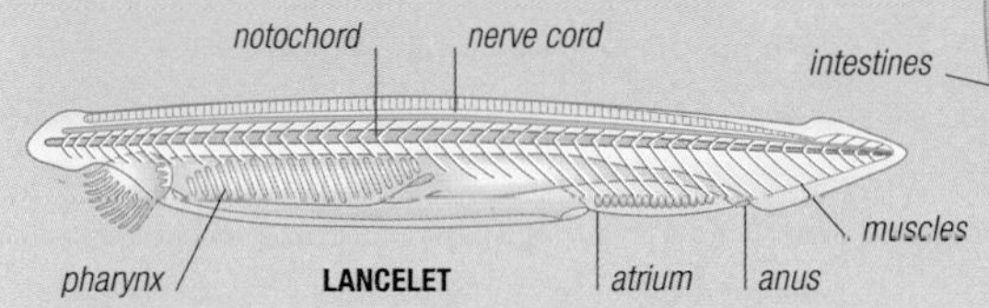

LANCELET

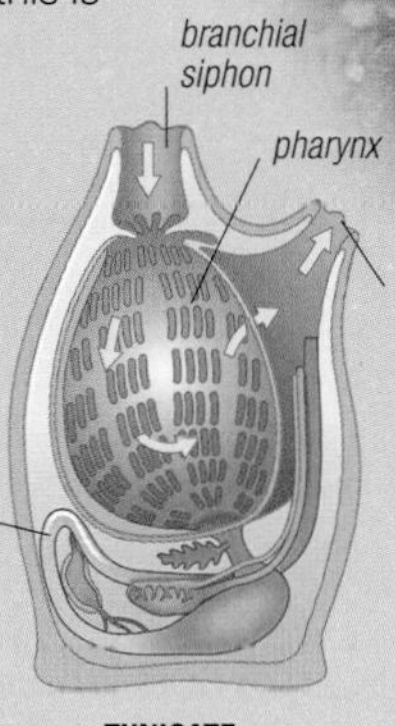

TUNICATE

BODY SECTION
Most tunicates (left) are filter feeders that live attached to the seabed. Water enters through the mouth and leaves through an opening called the atrium. The body of a lancelet (far left) is specialized for swimming: it is long, narrow, flexible, and muscular.

TUNICATE COLONIES
Colonies of tunicates often form a brightly colored mass on the ocean floor. Although the individual members of the colony have separate mouths, they sometimes share an opening for expelling water and waste products.

Class Ascidiacea

Sea squirts

Occurrence 2,813 spp. worldwide; attached to coastal rocks and the seabed

Sea squirts are baglike tunicates fastened to coastal rocks or the seabed. Their spherical or cylindrical tunic surrounds the body. Most filter feed by pumping water through a mesh-lined pharynx, drawing water in through the branchial siphon and out through the atrial siphon. Some live in colonies where the exhalant siphon is shared. Their larvae look like tadpoles and have a tail that contains a notochord—the typical chordate feature.

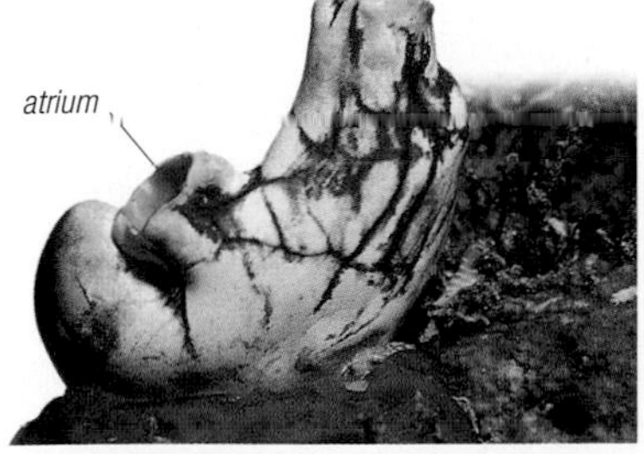

POLYCARPA AURATA
This solitary sea squirt from the Indo-Pacific is covered by a thick, leathery, sometimes brightly colored, protective test or tunic, ¾–6 in (2–15 cm) high. Two siphons project from it, and these are the entry and exit sites of the water current from which the animal filters its food.

Class Thaliacea

Pelagic tunicates

Occurrence 78 spp. worldwide; planktonic in ocean

This class of tunicates contains free-living organisms, drifting among the plankton in the open sea, that have inhalant and exhalant openings at opposite ends of their test. Like sea squirts, pelagic tunicates have a large, perforated pharynx, which occupies most of the body mass. There are 2 basic kinds: pyrosomes, which are colonial, and salps and doliolids, which have alternating phases of solitary and colonial existence. Individuals in the colony (zooids) are up to 10 in (25 cm) long, but some complete colonies may exceed 46 ft (14 m) in length.

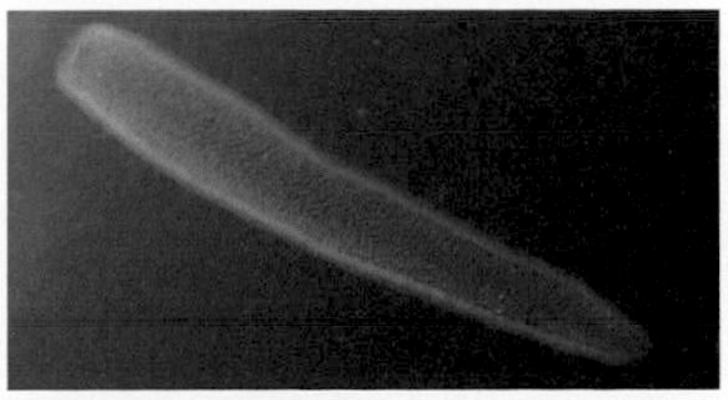

PYROSOMA ATLANTICUM
This colonial, luminescent, warm-water pyrosome, measuring more than 3¼ ft (1 m) in length, resembles a giant planktonic test tube. It filters food from the water, and uses its watery "exhaust" to propel itself slowly through the sea.

Class Leptocardia

Lancelets

Occurrence 30 spp. in tropical and temperate zones; in shallow sand or gravel

Lancelets are small—usually around 2 in (5 cm) long—semi-transparent animals with several vertebratelike features. As well as a notochord, they have gill-like pharyngeal slits, a single dorsal fin, and—like many fish—muscle blocks arranged in V-shaped patterns down the body. Lancelets also have small tentacles around the mouth. They suck in water and food particles by beating microscopic hairs (cilia) that line their pharyngeal slits. The sexes are separate, and they reproduce by external fertilization. In some places, they are caught for food.

BRANCHIOSTOMA LANCEOLATUM
This European species grows up to 2 in (5 cm) long. Like all lancelets, it swims by wiggling its body from side to side. In its normal feeding position, most of its body is buried, with just its head projecting out of the seabed.

Class Myxini

Hagfish

Occurrence 77 spp. in oceans, including deep-sea ocean floors

Despite having a simple cartilaginous skull, hagfish lack an intact vertebral column. They have been classified with the lampreys (see p.472) as "jawless fish," but are now considered not to be closely related. The bodies of hagfish ooze defensive slime, especially when captured. They can knot their body to free the slime and to assist feeding. A knot formed near the tail works its way up to the head, forcing the mouth away with a lump of its preys flesh.

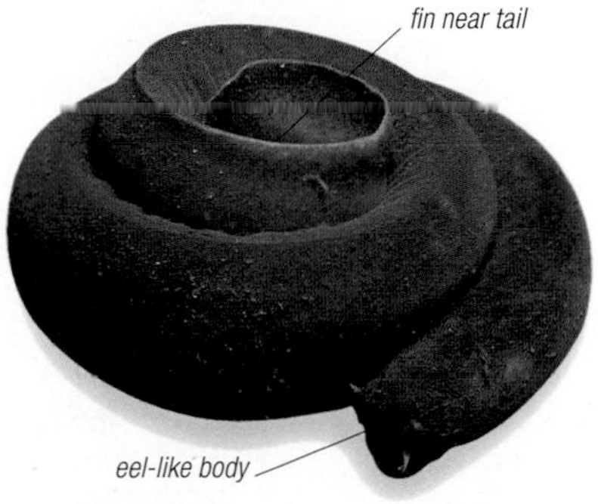

MYXINE GLUTINOSA
This species of hagfish lives in the North Atlantic and Mediterranean where it buries itself in mud and feeds on dead and dying fish.

GLOSSARY

A

ABDOMEN The hind part of the body, lying below the rib cage in mammals, and behind the thorax in arthropods.

ADIPOSE FIN A small fin lying between the dorsal and caudal (tail) fins of some fishes.

AMPLEXUS A breeding position adopted by frogs and toads, with the male using its front legs to hold the female. Fertilization usually occurs outside the female's body.

ANTENNA (pl. ANTENNAE) A sensory feeler on the head of an arthropod. Antennae always occur in pairs, and they can be sensitive to touch, sound, heat, and also taste. Their size and shape varies according to the way in which they are used.

ANTLER A bony growth on the head of deer. Unlike horns, antlers often branch, and in most cases they are grown and shed every year, in a cycle linked to the breeding season.

ARBOREAL Living fully or partly in trees.

ARTICULATION A joint—for example, between adjacent bones.

ASEXUAL REPRODUCTION A form of reproduction that involves just one organism. Asexual reproduction is most common in invertebrates, and is used as a way of boosting numbers quickly in favorable conditions. See also *Sexual reproduction*.

ATRIUM A hollow entrance chamber.

B

BACULUM A bone found in the penis of some mammals.

BALEEN A fibrous substance used by some whales for filtering food from water. Baleen grows in the form of plates with frayed inner edges, which hang from a whale's upper jaw. The baleen plates trap food, which the whale then swallows.

BARBEL A sensory appendage attached to an animal's lips or near its mouth. Many bottom-dwelling fishes have barbels, which they use for finding food hidden on gravel or in mud.

BEAK A set of narrow, protruding jaws, usually without teeth. Beaks have evolved separately in many groups of vertebrates, including tortoises and turtles, and some whales.

BILATERAL SYMMETRY A form of symmetry in which the body consists of 2 equal halves, on either side of a midline. Most animals show this kind of symmetry.

BILL A bird's jaws. A bill is made of bone, with a hornlike outer covering.

BINOCULAR VISION Vision in which the 2 eyes face fully or partly forward, giving overlapping fields of view. This allows an animal to judge depth.

BIOME A characteristic grouping of living things, together with the setting in which they are found.

BIPEDAL Moving on 2 legs.

BLOWHOLE The nostrils of whales and their relatives, positioned on top of the head. Blowholes can be single or paired.

BREACHING Leaping out of the water (usually the sea), and landing back in it with a splash. Breaching is a characteristic form of behavior shown by many large whales.

BREEDING COLONY A large gathering of nesting birds.

BROOD PARASITE An animal—often a bird—that tricks other species into raising its young. In many cases, a young brood parasite kills all its nest-mates, so that it has sole access to all the food that its foster parents provide.

BROWSING Feeding on the leaves of trees and shrubs, rather than on grasses. See also *Grazing*.

C

CALCAREOUS Containing calcium. Calcareous structures—such as shells, exoskeletons, and bones—are formed by many animals, either for support or for protection.

CAMOUFLAGE Colors or patterns that enable an animal to merge with its background. Camouflage is very widespread in the animal kingdom—particularly among invertebrates—and is used both for protection against predators, and for concealment when approaching prey. See also *Mimicry*, *Cryptic coloration*.

CANINE TOOTH In mammals, a tooth with a single sharp point that is shaped for piercing and gripping. Canine teeth are toward the front of the jaws, and are highly developed in carnivores.

CARAPACE A hard shield on the back of an animal's body.

CARNASSIAL TOOTH In mammalian carnivores, a bladelike cheek tooth that has evolved for slicing through flesh.

CARNIVORE Any animal that eats meat. The word carnivore can also be used in a more restricted sense, to mean mammals in the order Carnivora.

CARRION The remains of dead animals.

CARTILAGE A rubbery substance that forms part of vertebrate skeletons. In most vertebrates, cartilage lines the joints, but in cartilaginous fishes—for example, sharks—it forms the whole of the skeleton.

CASQUE A bony growth on the head of an animal.

CAUDAL Relating to an animal's tail.

CECUM A blind-ended part of the digestive tract, which is often used in the digestion of plant food.

CELLULOSE A complex carbohydrate found in plants. Cellulose is used by plants as a building material, and it has a resilient chemical structure that animals find hard to break down. Plant-eating animals, such as ruminants, digest it with the help of microorganisms.

CEPHALOTHORAX In some arthropods, a part of the body that combines the head and the thorax. Animals that have a cephalothorax include crustaceans and arachnids.

CERCUS (pl. CERCI) A paired, hairlike appendage that is attached to the end of the abdomen in some invertebrates, particularly insects.

CHAETA (pl. CHAETAE) A short, stiff hair or bristle, found in many invertebrates. Chaetae sometimes have a sensory function, but in annelid worms, they are more often used in movement.

CHEEK TOOTH see *Molar tooth*, *Premolar tooth*.

CHELICERA (pl. CHELICERAE) In arachnids, each one of the first pair of appendages, at the front of the body. Chelicerae often end in pincers, and in spiders they are able to inject venom. In mites, they are sharply pointed, and are used for piercing food.

CHORDATE An animal belonging to the phylum Chordata, which includes all vertebrates. A key feature of chordates is their notochord, which runs the length of their bodies; it reinforces the body, yet allows it to move by bending.

CHRYSALIS A hard and often shiny case that protects an insect pupa. Chrysalises are often found attached to plants, or buried close to the surface of the soil.

CLASPERS see *Pelvic spurs*.

CLASS A level used in classification. In the sequence of classification levels, a class forms part of a phylum, and is subdivided into one or more orders.

CLOACA An opening toward the rear of the body that is shared by several body systems. In some vertebrates—such as bony fishes and amphibians—the gut, kidneys, and reproductive systems all use this single opening.

CLOVEN-HOOFED Having hooves that look as if they are split in 2. Most cloven-hoofed mammals, such as deer and antelope, actually have 2 hooves, arranged on either side of a line that divides the foot in 2.

COCOON A case made of open, woven silk. Many insects spin a cocoon before they begin pupation, and many spiders spin one to hold their eggs. See also *Pupa*.

COLD-BLOODED see *Ectothermic*.

COLONY A group of animals, belonging to the same species, that spend their lives together, often dividing up the tasks involved in survival. In some colonial species, particularly aquatic invertebrates, the colony members are permanently fastened together. In others, such as ants, bees, and wasps, members forage for food independently, but live in the same nest.

COMMENSALISM A partnership between 2 organisms of different species, in which one benefits, but the other neither gains nor loses.

COMPOUND EYE An eye that is divided into separate compartments, each with its own set of lenses. Compound eyes are a common feature of arthropods. The number of compartments they contain varies from a few dozen to thousands.

CORNICLE In aphids, one of the pair of narrow tubes at the end of the abdomen. The cornicles secrete wax that helps protect aphids from some predators and parasites.

COSTAL Relating to an animal's ribs.

COVERT A feather that covers the base of a bird's flight feather.

CREMASTER A collection of minute hooks on the hind end of a butterfly or moth pupa. The pupa hangs by these hooks from a solid support.

CRYPTIC COLORATION Coloration and markings that make an animal difficult to see against its background.

D

DELAYED IMPLANTATION In mammals, a delay between fertilization of an egg and the subsequent development of an embryo, which allows birth to occur when conditions are favorable for raising young.

DEPOSIT FEEDER An animal that feeds on small specks of organic matter that have drifted down through water, and settled on the bottom.

DETRITIVORE An animal that eats dead plant or animal matter.

DEWLAP A fold of loose skin hanging from an animal's throat.

DIASTEMA A wide gap separating rows of teeth. In plant-eating mammals, the diastema separates the biting teeth, at the front of the jaw, from the chewing teeth at the rear. In many rodents, the cheeks can be folded into the diastema, to shut off the back of the mouth while the animal is gnawing.

DIGIT A finger or toe.

DIGITIGRADE A gait in which only the fingers or toes touch the ground. See also *Plantigrade*, *Unguligrade*.

DIURNAL Active during the day.

DOMESTICATED Relating to an animal that lives fully or partly under human control. Some domesticated animals look identical to their wild counterparts, but many have been bred to produce artificial varieties (breeds), which are not found in nature. See also *Feral*.

DORSAL On or near an animal's back.

DREY The nest of a squirrel, which is built in a tree.

E

ECHOLOCATION A method of sensing nearby objects by using pulses of high-frequency sound. Echoes bounce back from obstacles and other animals, allowing the sender to build up a "picture" of its surroundings. Echolocation is used by several groups of animals, including mammals and a small number of cave-dwelling birds.

ECLIPSE PLUMAGE In some birds, particularly waterfowl, an unobtrusive plumage adopted by the males once the breeding season is over.

ECOSYSTEM A collection of species living in the same habitat, together with their physical surroundings.

ECTOPARASITE An animal that lives parasitically on the surface of another animal's body, often by sucking its blood. Some ectoparasites spend all their lives on their hosts, but many—including fleas and ticks—develop elsewhere, and climb onto the host in order to feed.

ECTOTHERMIC Having a body temperature that is dictated principally by the temperature of the surroundings. Also known as cold-blooded.

ELYTRON (pl. ELYTRA) A hardened forewing in beetles, earwigs, and some bugs. The 2 elytra usually fit together like a case, protecting the more delicate hind wings underneath.

EMBRYO A young animal or plant in a rudimentary stage of development.

ENDOPARASITE An animal that lives parasitically inside another animal's body, either feeding directly on its tissues or stealing some of its food. Endoparasites frequently have complex life cycles, involving more than one host.

ENDOSKELETON An internal skeleton, typically made of bone. Unlike an exoskeleton, this kind of skeleton can grow in step with the rest of the body.

ENDOTHERMIC Able to maintain a constant, warm body temperature, regardless of external conditions. Also known as warm-blooded.

EPIPHYTE A plant that grows off the ground, often in the branches of trees. In the tropics, epiphytes create important microhabitats for many animals, particularly insects and amphibians.

ESTIVATION A period of dormancy in hot or dry weather. Estivation is common in animals that live in the subtropics, where there is a long dry season each year.

EVERSIBLE Capable of being turned inside out.

EXOSKELETON An external skeleton that supports and protects an animal's body. The most complex exoskeletons, formed by arthropods, consist of rigid plates that meet at flexible joints. This kind of skeleton cannot grow, and has to be shed and replaced (ecdysis) at periodic intervals. See also *Endoskeleton*.

EXTERNAL FERTILIZATION In reproduction, a form of fertilization that takes place outside the female's body, usually in water.

F

FAMILY A level used in classification. In the sequence of classification levels, a family forms part of an order, and is subdivided into one or more genera.

FEMUR (pl. FEMORA) The thigh bone in 4-limbed vertebrates. In insects, the femur is the third segment of the leg, immediately above the tibia.

FERAL Relating to an animal that comes from domesticated stock, but that has subsequently taken up life in the wild. Common examples include city pigeons, cats, and horses.

FERTILIZATION The union of an egg cell and sperm, which creates a cell capable of developing into a new animal. In external fertilization, the process occurs outside the body (usually in water), but in internal fertilization, it takes place in the female's reproductive system.

FETUS A developing animal that is partly formed, and approaching the time when it will be born.

FILTER FEEDER An animal that eats by sieving small food items from water. Many invertebrate filter feeders, such as bivalve mollusks and sea squirts, are sessile animals that collect food by pumping water through or across their bodies. Vertebrate filter feeders, such as baleen whales, collect their food by trapping it while they are on the move.

FLAGELLUM (pl. FLAGELLA) A long, hairlike projection from a cell. A flagellum can flick from side to side, moving the cell along. Sperm cells use flagella to swim.

FLEHMEN RESPONSE A characteristic curling of the upper lip shown by many male mammals to enhance a scent by pressing air into the Jacobson's organ.

FLIGHT FEATHERS A bird's wing and tail feathers, used in flight.

FLIPPER In aquatic mammals, a paddle-shaped limb. See also *Fluke*.

FLUKE A rubbery tail flipper in whales and their relatives. Unlike the tail fins of fishes, flukes are horizontal, and they beat up and down instead of side to side.

FOOD CHAIN A food pathway that links 2 or more different species in that each forms food for the next species higher up in the chain. In land-based food chains, the first link is usually a plant. In aquatic food chains, it is usually an alga or other form of single-celled life. See also *Food web*.

FOOD WEB A collection of food chains from a particular habitat. In a food web, each species is often involved in several different food chains.

FORM A depression in the ground used by hares for concealing themselves or their young.

FOSSA (pl. FOSSAE) A pit or depression—for example, in an animal's skull.

FOSSORIAL Burrowing underground.

FRUGIVOROUS Fruit-eating.

FURCULA The forked, springlike organ attached to a springtail's abdomen.

G

GALL A tumorlike growth in a plant that is induced by an animal (typically an insect) or some other organism. By triggering the formation of galls, animals provide themselves with a safe hiding place, and a convenient source of food.

GENUS (pl. GENERA) A level used in classification. In the sequence of classification levels, a genus forms part of a family, and is subdivided into one or more species.

GESTATION PERIOD In mammals, the interval between fertilization and birth, when the developing young is nourished by the mother via the placenta.

GILL An organ used for extracting oxygen from water. Gills are usually positioned on or near the head, or—in aquatic insects—at the end of the abdomen.

GONAD A male or female sex organ.

GRAZING Feeding on grass. See also *Browsing*.

GUARD HAIR A long hair in a mammal's coat, which projects beyond the underfur to protect it and help keep it dry.

H

HALTERES In true flies, the 2 knoblike organs that act like gyroscopes during flight, helping flies maneuver with great precision. Halteres are modified hindwings.

HEAT PITS Heat-sensitive depressions that some snakes have above the front of the upper jaw. These pits detect heat from warm-blooded animals, allowing the snakes to hunt in darkness.

HERBIVORE An animal that feeds on plants or plantlike plankton.

HERMAPHRODITE An animal that has both male and female sex organs, so it can fertilize itself. See also *Sequential hermaphrodite*.

HIBERNATION A period of dormancy in winter. During hibernation, an animal's body processes drop to a low level.

HOME RANGE The area that an animal, or group of animals, uses in its daily activities—for example, when foraging for food. See also *Territory*.

HORN In mammals, a pointed growth on the head. True horns are hollow sheaths of keratin, covering a bony horn core.

HOST An animal on or in which a parasite feeds.

HYPERPARASITE A parasite that attacks another parasite.

I

INCISOR TOOTH In mammals, a tooth at the front of the jaw, used in biting, slicing, or gnawing.

INCUBATION In birds, the period when a parent sits on the eggs and warms them, allowing them to develop. Incubation periods range from under 14 days to several months.

INSECTIVORE An animal that eats insects.

INTERNAL FERTILIZATION In reproduction, a form of fertilization that takes place inside the female's body. Internal fertilization is a characteristic of many land animals, including insects and vertebrates. See also *External fertilization*.

INTRODUCED SPECIES A species that humans have accidentally or deliberately brought into an ecosystem in which it does not naturally occur.

INVERTEBRATE An animal without a backbone.

IRRUPTION A sporadic mass movement of animals outside their normal range. Irruptions are usually short-lived, and occur in response to severe conditions, such as winter cold.

JK

JACOBSON'S ORGAN An organ in the roof of the mouth that is sensitive to airborne scents. Snakes often employ this organ to detect their prey, while some male mammals use it to find females that are ready to mate.

KEEL In birds, an enlargement of the breastbone that anchors the muscles used in flight.

KERATIN A tough structural protein found in hair, claws, and horns.

KINGDOM In classification, one of the 6 fundamental divisions of the natural world.

L

LARVA (pl. LARVAE) An immature but independent animal that looks completely different from an adult. A larva develops the adult shape by metamorphosis; in many insects, the change takes place in a resting stage that is called a pupa.

LATERAL-LINE SYSTEM A sensory system in fishes and some amphibians that detects pressure waves traveling through water, allowing the animal to sense movement. The system consists of sensors arranged in a line along both sides of the body.

LEK A communal display area used by male animals—particularly birds—during courtship. The same location is often revisited for many years.

LOPHOPHORE A crest of short tentacles found in bryozoans and some other marine invertebrates.

M

MANDIBLE The paired jaws of an arthropod, or a bone that makes up all or part of the lower jaw in vertebrates.

MANTLE In mollusks, an outer fold of skin covering the mantle cavity.

MANTLE CAVITY In mollusks, a cavity between the mantle and the rest of the body. The cavity is filled with water in aquatic mollusks, and often contains the gills.

MEDUSA In cnidarians, a bell-shaped body form, which is often free-living. During the complete life cycle, this form may alternate with a form called a polyp, which is often fixed in one place.

MELON A bulbous swelling in the heads of many toothed whales. The melon is filled with a fatty fluid, and is believed to focus the sounds used in echolocation.

METABOLISM The complete array of chemical processes that take place inside an animal's body. Some of these processes release energy by breaking down food, while others use energy by making muscles contract.

METACARPAL In 4-limbed animals, one of a set of bones in the front leg or arm, forming a joint with a digit distally. In most primates, the metacarpals form the palm of the hand.

METAMORPHOSIS A change in body shape shown by many animals—

particularly invertebrates—as they grow from juvenile to adult. In insects, metamorphosis can be complete or incomplete. Complete metamorphosis involves a total reorganization of the organism during a resting stage, called a pupa. Incomplete metamorphosis embraces a series of less drastic changes; these occur each time the young animal molts. See also *Larva*, *Nymph*.

METATARSAL In 4-limbed animals, one of a set of bones in the hindleg, forming a joint with a digit distally. In most primates, the metacarpals form the sole of the foot.

METATHORAX The second segment of an insect's thorax.

MIGRATION A journey to a different region, following a well-defined route. Most animals that migrate do so in step with the seasons, so they can take advantage of good breeding conditions in one place, and good wintering ones in another. See also *Partial migrant*.

MIMICRY A form of camouflage in which an animal resembles another animal or an inanimate object, such as a twig or a leaf. Mimicry is very common in insects, with many harmless species imitating ones that have dangerous bites or stings.

MOLAR TOOTH In mammals, a tooth at the rear of the jaw. Molar teeth may have a flattened or ridged surface for chewing vegetation. The sharper molar teeth of meat-eaters cut through hide and bone.

MOLT Shedding fur, feathers, or skin so that it can be replaced. Mammals and birds molt to keep their fur and feathers in good condition, to adjust their insulation, or so that they can be ready to breed. Arthropods—such as insects—molt in order to grow.

MONOGAMOUS Mating with a single partner, either during the course of one breeding season or throughout life. Monogamous partnerships are common in animals that care for their young.

MORPH A physical variant of a species, sometimes known as a phase. Some species have several clearly defined morphs, which may differ in color or patterning. See also *Polymorphic*.

MUTUALISM Two organisms that interact to the mutual benefit of both.

N

NEUROTOXIN A substance that deactivates nerves or that disrupts the way that they work. Neurotoxins can be lethal in very small amounts. They are found in snake venom and in poisons produced by some other animals, such as frogs, toads, fishes, and insects.

NEW WORLD The Americas.

NICHE An animal's place and role in its habitat. Although 2 species may share the same habitat, they never share the same niche.

NICTITATING MEMBRANE In birds and many reptiles—apart from snakes—a transparent or semiopaque "third eyelid," which moves sideways across the eye. Aquatic birds often use the membrane as an aid to vision when swimming underwater.

NOCTURNAL Active at night.

NOTOCHORD A reinforcing rod that runs the length of the body. The notochord is a distinctive feature of chordates, although in some it is present only in early life. In vertebrates, the notochord becomes incorporated in the backbone during the development of the embryo.

NUCHAL CREST A crest at the nape of the neck.

NYMPH An immature insect that looks similar to its parents but that does not have functioning wings or reproductive organs. A nymph develops the adult shape by metamorphosis, changing slightly each time it molts.

O

OCELLUS (pl. OCELLI) In insects, a small simple eye that detects overall levels of light. Many insects have 3 ocelli, positioned in a cluster on top of the head.

OLD WORLD Europe, Asia, Africa, and Australasia.

OMNIVORE An animal that eats both plant and animal food.

OOTHECA (pl. OOTHECAE) An insect egg case. In some insects—for example, cockroaches—the ootheca has a thin covering, but in others, it is surrounded by a jacket of foam that hardens soon after it has been produced.

OPERCULUM A cover or lid. In some gastropod mollusks, an operculum is used to seal the shell when the animal has withdrawn inside. In bony fishes, an operculum on each side of the body protects the chamber containing the gills.

ORDER A level used in classification. In the sequence of classification levels, an order forms part of a class, and is subdivided into one or more families.

ORGAN A structure in the body, composed of several kinds of tissues, that carries out specific tasks.

OSSICLE A minute bone. The ear ossicles of mammals, which transmit sound from the eardrum to the inner ear, are the smallest bones in the body.

OVIPAROUS Reproducing by laying eggs. See also *Ovoviviparous*, *Viviparous*.

OVIPOSITOR In female insects, an organ at the end of the abdomen, used for laying eggs. The ovipositor often has a tubular or bladelike shape, so that it can insert the eggs in the ground, in plants, or in other animals. In some insects, the ovipositor is modified to form a sting.

OVOVIVIPAROUS Reproducing with eggs that hatch within the female's body so the young one is born alive. The eggs contain all the nutrients that the embryos need, and there is no direct connection between them and the mother. See also *Oviparous*, *Viviparous*.

P

PAPILLA (pl. PAPILLAE) A small, fleshy protruberance on an animal's body. Papillae often have a sensory function—for example, detecting chemicals that help pinpoint food.

PARAPODIUM (pl. PARAPODIA) A leg- or paddlelike flap found in some worms. Parapodia are used for moving, or for pumping water past the body.

PARASITE An animal that lives on or inside another animal, and that feeds either on its host animal or on food that its host has swallowed. The majority of parasites are much smaller than their hosts, and many have complex life cycles involving the production of huge numbers of eggs. Parasites often weaken their hosts, but generally do not kill them. See also *Ectoparasite*, *Endoparasite*, *Parasitoid*.

PARASITOID An animal that is a parasite during its early life but that becomes free-living as an adult. Parasitoids are common among insects. Unlike true parasites, parasitoids usually kill their hosts—typically by eating them.

PARATOID GLAND In amphibians, a gland behind the eyes that secretes poison onto the surface of the skin.

PARTHENOGENESIS A form of reproduction in which an egg cell develops into a young animal without having to be fertilized. Parthenogenesis is common in invertebrates, and it produces offspring that are genetically identical to the parent. In animals that have separate sexes, young produced by parthenogenesis are always female.

PARTIAL MIGRANT A species in which some populations migrate, while others are sedentary. This situation is common in birds with a wide distribution, because the different populations often experience very different climatic conditions. See also *Migration*.

PATAGIUM In bats, the flap of double-sided skin that forms the wing. This term is also used for the parachute-like skin flaps of colugos and other gliding mammals.

PECTORAL FIN One of the 2 paired fins positioned toward the front of a fishes's body, often just behind its head. Pectoral fins are usually highly mobile and are normally used for maneuevring.

PECTORAL GIRDLE In 4-limbed vertebrates, the collection of bones that anchors the front limbs to the backbone. In most mammals, the pectoral girdle consists of 2 clavicles, or collarbones, and 2 scapulae, or shoulder blades.

PEDIPALPS In arachnids, the second pair of appendages, near the front of the body. Depending on the species, they are used for walking, sperm transfer, or attacking prey.

PELVIC FINS The rear paired fins in fishes, which are normally positioned close to the underside, sometimes near the head but more often toward the tail. Pelvic fins are generally used as stabilizers.

PELVIC GIRDLE In 4-limbed vertebrates, the collection of bones that anchors the front limbs to the backbone. The bones of the pelvic girdle are often fused, forming a weight-bearing ring called the pelvis.

PELVIC SPURS Spurs near the rear of the body that some male sharks and rays use during mating. Also known as claspers.

PHASE see *Morph*.

PHEROMONE A chemical produced by one animal that has an effect on other members of its species. Pheromones are often volatile substances that spread through the air, triggering a response from animals some distance away.

PHOTOSYNTHESIS A collection of chemical processes that enable plants to capture the energy in sunlight and convert it into chemical form.

PHYLUM (pl. PHYLA) A level used in classification. In the sequence of classification levels, a phylum forms part of a kingdom, and is subdivided into one or more classes.

PINNA (pl. PINNAE) The external ear flaps found in mammals.

PLACENTA An organ developed by an embryo animal that allows it to absorb nutrients and oxygen from its mother's bloodstream, before it is born. Young that grow in this way are known as placental mammals.

PLACENTAL MAMMAL see *Placenta*.

PLANKTON Floating organisms—many of them microscopic—that drift in open water, particularly near the surface of the sea. Planktonic organisms can often move, but most are too small to make any headway against strong currents. Planktonic animals are collectively known as zooplankton.

PLANTIGRADE A gait in which the sole of the foot is in contact with the ground. See also *Digitigrade*, *Unguligrade*.

PLASTRON The lower part of the shell in tortoises and turtles.

POLYANDROUS Relating to a reproductive system, rare in the animal kingdom, in which females mate with several males during the course of a single breeding season, and the male takes care of the young; seen only in some fishes and some shorebirds.

POLYGAMOUS Relating to a reproductive system in which individuals mate with more than one partner during the course of a single breeding season.

POLYGYNOUS Relating to a reproductive system in which males often mate with several females during the course of a single breeding season.

POLYMORPHIC Relating to a species that occurs in several different forms, known as morphs. The morphs may be geographically separated, so that there is only one in any particular place, or they may occur together, often in set proportions. Morphs persist even though parents from different morphs often interbreed.

POLYP In cnidarians, a body form that has a hollow cylindrical trunk, ending in a central mouth surrounded by a circle of tentacles. Polyps are frequently attached to solid objects by their base. See also *Medusa*.

PREDATOR An animal that catches and kills others, known as its prey. Some predators catch their prey by lying in wait, but most actively pursue and attack other animals.

PREHENSILE Able to curl around objects and grip them.

PREMOLAR TOOTH In mammals, a tooth positioned midway along the jaw, between the canines and the molars.

PREPUPA (pl. PREPUPAE) A resting stage that some insects pass through before forming a pupa.

PREY Any animal that is eaten by a predator.

PROBOSCIS An animal's nose, or a set of

mouthparts with a noselike shape. In insects that feed on fluids, the proboscis is often long and slender, and can usually be stowed away when not in use.

PRONOTUM The upper surface of the front segment in an insect's thorax. In some insects, it is expanded to form a shield.

PROTHORAX The front segment of an insect's thorax.

PTEROSTIGMA A small, colored spot near the leading edge on some insects' wings.

PULMONATE Having lungs. Pulmonate mollusks are air-breathing species that often live on land.

PUPA (pl. PUPAE) In insects, a stage during which the larval body is broken down, and rebuilt as an adult. During the pupal stage, the insect does not feed and usually cannot move, although some pupae may wriggle if they are touched. The pupa is protected by a hard case, which is itself sometimes wrapped in silk. See also *Chrysalis*, *Cocoon*, *Larva*.

QR

QUADRUPEDAL Moving on 4 legs.

RADIAL SYMMETRY A form of symmetry in which the body is shaped like a wheel, often with the mouth at the center.

RADULA A mouthpart that many mollusks use for rasping away at food. The radula is often shaped like a belt, and armed with many microscopic tooth-like denticles.

ROSTRUM In bugs and some other insects, a set of mouthparts that looks like a beak.

RUMINANT A hoofed mammal that has a specialized digestive system, with several stomach chambers. One of these—the rumen—contains huge numbers of microorganisms that help break down the cellulose in plant cell walls. To speed this process, a ruminant usually regurgitates its food and rechews it, a process called "chewing the cud."

RUTTING SEASON In deer, a period during the breeding season when males clash with each other for the opportunity to mate.

S

SAPROPHAGOUS Feeding on dead and decaying matter.

SCAPE The first segment in an insect's antennae, nearest its head.

SCOLEX The head of a tapeworm. The scolex is usually equipped with suckers and/or hooks that enable the worm to fasten itself to its host.

SCUTE A shieldlike plate or scale that forms a bony covering on some animals.

SCUTELLUM A triangular shield behind the pronotum in many insects.

SEBACEOUS GLAND A skin gland in mammals that normally opens near the root of a hair. Sebaceous glands produce substances that keep skin and hair in good condition.

SEDENTARY Having a lifestyle that involves relatively little movement. This term is also used to refer to animals that do not migrate. See also *Migration*.

SEQUENTIAL HERMAPHRODITE An animal with both male and female sex organs that mature at different times.

SESSILE Fastened to a solid surface, instead of being able to move from place to place. Most sessile animals are invertebrates that live in water.

SEXUAL DIMORPHISM Showing physical differences between males and females. In animals that have separate sexes, males and females always differ, but in highly dimorphic species, such as elephant seals, the 2 sexes look very different and are often quite unequal in size.Many other species, particularly among the butterflies, fishes, and birds, show differences in color between the two sexes.

SEXUAL REPRODUCTION A form of reproduction that involves fertilization of a female cell or egg, by a male cell or sperm. This is by far the most common form of reproduction in animals. It usually involves 2 parents—one of either sex—but in some species individual animals are hermaphrodite. See also *Asexual reproduction*.

SILK A protein-based fibrous material produced by spiders and some insects. Silk is liquid when it is squeezed out of a spinneret but turns into elastic fibers when stretched and exposed to air. It has a wide variety of uses. Some animals use silk to protect themselves or their eggs, to catch prey, to glide on air currents, or to lower themselves through the air.

SIMPLE EYE An eye that has a single lens. Simple eyes are found in vertebrates and cephalopod mollusks and, despite their name, are often highly complex. Rudimentary simple eyes are also found in insects. See also *Ocellus*.

SPECIES A group of similar organisms that are capable of interbreeding in the wild, and of producing fertile offspring that resemble themselves. Species are the fundamental units used in biological classification. Some species have distinct populations that vary from each other. Where the differences are significant, and the populations biologically isolated, these forms are classified as separate subspecies.

SPERMATOPHORE A packet of sperm that is transferred either directly from the male to the female, or indirectly—for example, by being left on the ground. Spermatophores are produced by a range of animals, including salamanders (Urodeles) and some insects.

SPICULE In sponges, a needlelike sliver of silica or calcium carbonate that forms part of the internal skeleton. Spicules have a wide variety of shapes.

SPINNERET The nozzlelike organs that spiders use to produce silk.

SPIRACLE In rays and some other fishes, an opening behind the eye that lets water flow into the gills. In insects, an opening on the thorax or abdomen that lets air into the tracheal system.

STEREOSCOPIC VISION see *Binocular vision*.

STERNUM The breastbone in 4-limbed vertebrates, or the thickened underside of a body segment in arthropods. In birds, the sternum has a narrow flap.

STRIDULATE Producing sound by rubbing together 2 parts of the body.

SUBCLASS A level used in classification, between a class and an order.

SUBFAMILY A level used in classification, between a family and a genus.

SUBORDER A level used in classification, between an order and a family.

SUBSPECIES see *Species*.

SUSPENSION FEEDER An animal that feeds on particles of organic matter that are suspended in water. Suspension feeders include brittlestars, some cnidarians, and many annelid worms.

SWIM BLADDER A gas-filled bladder that most bony fishes use to regulate their buoyancy. By adjusting the gas pressure inside the bladder, a fishes can become neutrally buoyant, meaning that it neither rises nor sinks in the water column.

SYMBIOSIS Any partnership between individuals of 2 different species. Animals form symbiotic partnerships with other animals, with plants, and with microorganisms.

T

TARSUS (pl. TARSI) A part of the leg. In insects, the tarsus is the equivalent of the foot, while in vertebrates, it forms the lower part of the leg or the ankle.

TEGMEN (pl. TEGMINA) The leathery forewing of some insects, such as grasshoppers, crickets, and cockroaches.

TERRESTRIAL Living wholly or mainly on the ground.

TERRITORY An area defended by an animal, or group of animals, against other members of the same species. Territories often include useful resources that help the male attract a mate.

TEST In echinoderms, a skeleton made of small calcareous plates.

THORAX The middle region of an arthropod's body. The thorax contains powerful muscles, and bears the legs and wings, if the animal has any. In 4-limbed vertebrates, the thorax is the chest.

TIBIA (pl. TIBIAE) The shin bone in 4-limbed vertebrates. In insects, the tibia forms the part of the leg immediately above the tarsus, or foot.

TISSUE A layer of cells in an animal's body. In a tissue, the cells are of the same type, and carry out the same work. See also *Organ*.

TORPOR A sleeplike state in which the body processes slow to a fraction of their normal rate. Animals usually become torpid to survive difficult conditions, such as cold or lack of food.

TRACHEA (pl. TRACHEAE) A breathing tube, known in vertebrates as the windpipe.

TRACHEAL LUNG In arthropods, a lung that connects to the tracheal system. Tracheal lungs are also found in snakes.

TRACHEAL SYSTEM A system of minute tubes that insects and arachnids use to carry oxygen into their bodies. Air enters the tubes through openings called spiracles, and then flows through the tracheae to reach individual cells. Carbon dioxide flows in the opposite direction.

TUBERCLE A hard swelling on an animal's body.

TUSK In mammals, a modified tooth that often projects outside the mouth. Tusks have a variety of uses, including defense and digging up food. In some species, only the males have them—in which case, their use is often for sexual display.

TYMBAL ORGAN A drumlike organ on a cicada's body that can vibrate to produce a piercing sound. The 2 tymbal organs are on either side of the abdomen.

TYMPANUM The external ear membrane of frogs.

U

UNDERFUR The dense fur that makes up the innermost part of a mammal's coat. Underfur is usually soft, and is a good insulator. See also *Guard hair*.

UNGULIGRADE A gait in which only the hooves touch the ground. See also *Digitigrade*, *Plantigrade*.

UTERUS In female mammals, the part of the body that houses developing young. In placental mammals, the young are connected to the wall of the uterus via a placenta.

V

VENTRAL On or near the underside.

VERTEBRATE An animal with a backbone. Vertebrates include fishes, amphibians, reptiles, birds, and mammals.

VESTIGIAL Relating to an organ that is atrophied or nonfunctional and often reduced in size.

VIVIPAROUS Reproducing by giving birth to live young. See also *Oviparous*, *Ovoviviparous*.

WX

WARM-BLOODED see *Endothermic*.

WARNING COLORATION A combination of contrasting colors that warns that an animal is dangerous. Bands of black and yellow are a typical form of warning coloration, found in stinging insects.

WATER VASCULAR SYSTEM In echinoderms, a system of fluid-filled tubes and chambers that connects with the tube feet, giving them their shape.

WEANING In mammals, the period when the mother gradually ceases to provide milk for her young.

XYLOPHAGOUS Wood-eating.

Z

ZOOID An individual animal in a colony of invertebrates. Zooids are often directly linked to each other, and may function like a single animal.

ZOOPLANKTON see *Plankton*.

INDEX

Page references in italics identify where the main entries for a particular animal appear. Those in bold refer to the main reference to an animal group.

A

C

D

G

H

I

M

N

O

P

Q

R

S

T

ACKNOWLEDGMENTS

DORLING KINDERSLEY would like to thank Daniel Crawford, Emily Newitt, Patricia Woodburn, Richard Gilbert, and Tamara Baillie for administrative help; Steve Knowlden for original design work; Pauline Clarke, Corinne Manches, Janice English, and Emily Luff for additional design assistance; Peter Cross for location research; Clare Double for additional editorial help; Robert Campbell for DTP support; and Andy Samson, Anna Bedewell, Charlotte Oster, Jason McCloud, Martin Copeland, Richard Dabb, and Rita Selvaggio for picture reseach. Thanks also to the Berlin Zoo, Chester Zoo, Dr. David L Harrison, Drusillas Zoo, Exmoor Zoo, Hunstanton Sea Life Centre, IUCN/SSC Canid specialist group, Kate Edmonson at Partridge Films, Marwell Zoological Park, Mike Jordan, Millport Marine Centre, NASA/Finley Holiday Films, the Natural History Museum (London), Paradise Park, Rebecca Tansley at Natural History New Zealand, Robert Oliver, Tropical Marine Centre, Twycross Zoo, Weymouth Sea Life Centre, and World of Birds (South Africa).

For this edition
DORLING KINDERSLEY would like to thank: Ellen Nanney, Smithsonian project coordinator; The IUCN Red List of Threatened Species for the information relating to conservation status of species listed with the book; Laura Palosuo, jacket editor; Caroline Hunt and Dawn Bates for proofreading; Jane Parker for the index; Sam Atkinson, Kathryn Hennessy, Daniel Lord.
DK Delhi would like to thank Jubbi Francis, Ligi John, Monica Saigal, Sakshi Saluja, and Nidhi Sharma for their help on the project.

Picture Credits
The publisher would like to thank the following for their kind permission to reproduce their photographs:

ftl	tl	tc	tr	ftr
fcla	cla	ca	cra	fcra
fcl	cl	c	cr	fcr
fclb	clb	cb	crb	fcrb
fbl	bl	bc	br	fbr

Abbreviations key:
a = above, b = below,
c = center, f = far,
l = left, r = right, t = top

A.N.T. Photo Library: GE Schmida 401frbc; **Prof RJ van Aarde:** 65tc, 312bfr, 454crb; **Kelvin Aitken:** 475cla; **Alamy Images:** Arco Images GmbH 510b, Danita Delimont 243, Images of Africa Photobank 42br, Photoshot Holdings Ltd 435, 464, David Tipling 123t, Ann and Steve Toon 198cr, Whitehead Images 578t; **Gerry Allen:** 476br; **Allofs Photography:** Theo Allofs 214; **David Almquist:** 564crb; **Amgueddfa Cymru – National Museum Wales:** 596cl; **Animals Animals/Earth Scenes:** Anthony Bannister 129crb, Kevin & Suzette Hanle 389tr, Dani-Jeske 178frbc, 233bfr, Johnny Johnson 35tfl, Victoria McCormic 505c, Joe McDonald 27bfr, 32frac, Steven David Miller 32bfr, Charles Palek 325c, Fred Whitehead 583tfr; **APB Photographic:** Alan P Barnes 281ca; **Ardea:** 198crb, 295bfr, 295tc, 409flbc, Alpenmolch 441frbc, Dennis Avon 58cra, 268tl, 327fcfl, Ian Beames 55cl, 288ca, Uno Berggren 120cl, Hans & Judy Beste 98crb, 128bl, 349cfr, 423crb, 430cr, 452cl, 452cb, N. N. Birks 302br, R. M. Bloomfield 338tfr, A E Bomford 471br, J.B. + S. Bottomley 341tfr, John Cancal 82cr, John Cancalosi 325bc, Piers Cavendish 91bl, Graeme Chapman 317crb, John Clegg 14fbl, D Parer & E Parer-Cook 79tr, 434cl, 495crb, Parer & E. Parer-Cook 46bl, John Daniels 372cb, Kevin Deacon 479cfr, Hans D Dossenbach 443bfr, 449c, 449bc, Eric Dragesco 201crb, John S Dunning 339clb, 345tl, 345cr, 347cr, 354cl, 369crb, M D England 237cfl, Jean-Paul Ferrero 34br, 65tl, 93cr, 93crb, 101cra, 125tr, 201bl, 225cl, 259tfr, 275clb, 289tr, 412cl, 420c, 579c, Kenneth W Fink 34c, 40ca, 47bfl, 89flac, 112cr, 112crb, 120tl, 122cl, 152flac, 155, 240bfr, 272tl, 430bfr, Andrea Florence 50bl, 251cfr, Paul Germain 165tr, Bob Gibbons 68tr, M. W. Gillam 47frac, Francois Gohier 39cr, 48tc, 71tl, 107frac, 107tc, 116cfr, 195frbc, 245crb, 248cr, 258cl, 288c, 373bc, 398frbc, Nick Gordon 141cfl, 187br, A. Greensmith 344tc, C. Clem Haagner 166crb, 267tfr, Clem Haagner 333cfl, Chris Harvey 203cl, Masahiro Iijima 204, 205frac, Chris Knights 71flbc, 78cfr, Ferrero Labat 4-5, 27tfr, 43tfr, 433frbc, Keith & Liz Laidler 188tr, 196br, Tom & Pat Leeson 121l, John Mason 89frac, 538bl, E. McNamara 351br, S. Meyers 63tfl, 89frac, Pat Morris 82br, 106cl, 109cra, 123cr, 133cra, 163tr, 164bl, 353cra, 499cfr, 539crb, 583frac, Hayden Oake 90frbc, B. Moose Peterson 45tc, 123bl, 161cl, Moose Peterson 309crb, Moose Peterson / WRP 42frbc, Becca Sounders 476cfl, 476tfr, Peter Steyn 105cl, 327ffrbc, 363fcr, J. Swedberg 299crb, Ron & Valerie Taylor 481br, 484cb, 485br, 523frac, 539bfl, Joanna Van Gruisen 199tr, Adrian Warren 156cra, 400frac, 449tfr, M. Watson 67cfl, 139bl, 175cfr, 192cra, 206frac, 299frbc, Alan Weaving 126clb, 463flac, Wardene Weisser 359cl; **Auscape:** Ferrero-Labat 284frac, Jean-Paul Ferrero 93tfr, 98flbc, 98cra, 173tr, Mike Gillian 96frbc, Neil McDaniel 519cla, D. Parer & E. Parer-Cook 95frbc, T. Shivanandappa 137c, Glen Threlfo 348cl; **Antonio Baeza:** 538br; **Robert E. Barber:** 57c, 183lac, 183bl, 190frbc, 191cra, 314bfl, 332br; **David Barnes:** 183bfl, 184bfr, 185c, 281frac, 290br, 291tcr, 313cl, 536lb, 541frbc, 544bl, 545frac; **Kevin H. Barnes:** 105bl; **Bat Conservation International:** Merlin D Tuttle 156tr, 160bfr, 160crb, 161cra, 161bc; **Fred Bavendam:** 77bfr, 107br, 225frbc, 482cfr, 509bfr, 536bfr, 544cfl; **Julian Bayliss:** 403bl; **Derek A. Belsey:** 40tc, 282br, 283cl, 358tfr, Derek Belsey 277cb, 282cl, 366bfl; **Niall Benvie:** 241cb, 305cra, 323cfr; **William Bernard Photography:** William Bernard 52cb, 111cl, 184ca, 187bl, 226crb; **GK Bhat:** 447tr; **Biosphoto:** 496bc, J.J. Alcalay 101cfr, Pierre Cadiran 164bl, Wigeon Christophe 510bc, Russo Cyril 154bfl, Halleux Dominique 135frbc, Feve Frederic 167frbc, GV-Press / Seitre 162cr, Denis-Huot 41cl, Michel Denis-Huot 17tfr, Vincent Jean 577c, Klein / Hubert 166bfl, Gunther Michael 150, 229tfr, Gasco Nicolas 249cfr, Garguil Philippe 487cfr, Jany Sauvanet 112bl, Rowland Seitre 151fcr, 191crb, 193tr, Montford Thierry 453cla; **Dr Alison Blackwell:** 566frbc; **Dr W.R. Branch:** 452bfr, Dr W. R. Branch 387cfr; **Bruce Coleman Ltd:** Erik Bjurstrom 91bfr, Mark N Boulton 221frac, Bruce Coleman Collection 295frbc, Bruce Coleman Inc 44cl, 70cfr, 205br, Jane Burton 116crb, 162clb, 449cra, 452cr, 454tc, John Cancalosi 153tfr, 211crb, 404cb, Mark Carwardine 260crb, Alain Compost 120b, 133tfr, 165crb, 191frac, 215tl, Gerald S Cubbit 39bfl, 327fc, 383crb, 584tfr, Jack Dermid 436cr, 442br, 445br, Geoff Dore 52cfr, M & P Fogden 31flbc, 50tr, 51tr, 165bl, 343tc, 347tl, 460flac, 465cra, Jeff Foott 54tc, 65c, 186br, 471frbc, 504frac, 504tr, 513frbc, 550ca, 599frac, Christopher Fredriksson 27cl, CB & DW Frith 193tl, Giorgio Guako 409bc, Granville Harris 45cl, Charles and Sandra Hood 375tfr, Johnny Johnson 67tc, 176c, Janos Jurka 227rac, 341bfr, Dr M P Kahl 90br, Dr. M. P. Kahl 451crb, Stephen J Krasemann 274frbc, Felix Labhardt 448c, Gordon Langsbury 67cfr, 312crb, Werner Layer 54bc, 154clb, 180cla, Luiz Claudio Marigo 143tl, 200, 347br, George McCarthy 40cb, Joe McDonald 405c, Bruce Montage 361br, Dr Scott Nielsen 274flbc, 286cfr, 324c, 454frac, Orion Press 266bfr, 304cb, Pacific Stock 15c, 72c, 255tfr, 473tfr, Mary Plage 215cra, Dr. Eckart Pott 79tfl, Allan G. Potts 270ca, Andrew Purcell 569frbc, Hans Reinhard 219tc, 469bfr, 489br, 490crb, John Shaw 459clb, Kim Taylor 28bfr, 91crb, 267flbc, 542flac, 547clb, 548cra, 555crb, 565frbc, 565crb, 569tl, Norman Tomalin 449cr, John Visser 385frac, Uwe Waltz 234cfl, Staffan Widstrans 306bc, Rod Williams 133cr, 137clb, 197br, 202frac, Bill Wood 132br, Kinrad Wothe 138cla, Peter Zabransky 34bfr, Gunter Ziesler 43frac, 111bl, 203bfr, 203cla, 242frbc, 247cl, 278c, 285bc, 317cfl, Christian Zuber 306t; **C. K. Bryan:** 172cr, 232cla; **John Cancalosi:** 95clb, 102cfl, 102cla, 103cra, 141cra, 270bl, 314bfr, 314bl, 368br, 369cl, John Cancalosu 103tfr; **Chris Gomersall Photography:** 361bc, R J Chandler 341tr, 346bl, Brian Chudleigh 311cl, 349frac, 349t, 350tl, 350br, 354t, 356fcl, Bill Coster 347bl, David M Cottridge 289br, 363tr, John Davies 312crb, 316cfr, Paul Doherty 297cfl, Peter Ginn 356bfr, Chris Gomersall 30clb, 281bfr, 300cl, 304flac, 311clb, 312cr, 314cfl, 315clb, 321cla, 326frbc, 352bfr, Michael Gore 33frac, 279bfr, 280bc, 297tr, 339ca, 340cfl, 369fcr, Mark Hamblin 312ca, 321cr, 357crb, J Hollis 326bfr, Barry Hughes 282cfr, Chris Knights 331tc, Mike Lane 310cla, 313tl, 313crb, 351cr, 353br, 362tl, 366cra, Gordon Langsbury 337br, 352flbc, 358bfl, Tim Loseby 357fbr, Mike McKavett 368tr, 369bc, Arthur Morris 354fcl, 358cfr, 366bfr, 369cr, Pete Morris 350tl, 350c, 362br, Carlos Sanchez 304bfl, 312c, 331c, Chris Schenk 64bc, 271cr, Morten Strange 349bl, 351bl, 365fcrb, 366cla, Andy Swash 290cfr, 328tc, 330c, 345cl, 345c, 347crb, 351fcr, 368clb, Roger Tidman 73tc, 277tfr, 277crb, 281frbc, 291tca, 336cfr, 347cl, 358cra, 362cr, David Tipling 271bfr, 275cfl, 276bfr, 277bc, 279tl, 291tbc, 321br, 340c, 340br, 357cr, 358rac, Ray Tipper 316cra, 344tr, Charles Tyler 291br, Cyril Webster 348br, Eric Woods 358cla, Barry E. Wright 268bfr, Steve Young 282frbc; **Prof. Mike Claridge:** 574frbc; **Bruce Coleman Inc:** R. Erite 296frac, Kenneth W Fink 229frac, James Hanken 446flac, Joe McDonald 299bfr, Ivan Polunin 519frac, Norman Owen Tomalin 576bfl; **Wendy Conway:** Wendy Conway 227ca, 283crb, 287br, 311cfr, 367cla; **Corbis:** Yann Arthus-Bertrand 166c, Joe Clark / Tetra Images 20bc, Clive Druett; Papilio 120cra, Natalie Fobes / Science Faction 73br, James Hager 364, Jason Hawkes 81br, Layne Kennedy 234-235, Tannen Maury / epa 32br, Ocean 525, Radius Images 186tl, Visuals Unlimited 14crb, Michele Westmorland / Science Faction 599br; **Cornell University of Ornithology:** J. Surman 29cfl; **Peter Cross:** 57c, 60bc, 102clb, 102crb, 119flac, 129tfr, 130clb, 183cla, 196cfr, 209frbc, 213bc, 216cra, 226tl, 226cla, 237cra, 239cl, 239bl, 240bl, 241cfr, 242frac, 268tr, 275bfl, 276cr, 279cfr, 288bfr; **Dennis Cullinane:** 164cr; **Dr. Melanie Dammhahn:** 139tl; **Manfred Danegger:** 113tfr; **© Tim Davenport/WCS:** 148tr; **Richard Davies:** 556cfr; **David M. Dennis:** 377bfl, 426br, 431frbc, 442cla, 443flbc, 453cr; **Nigel Dennis:** 44tc, 69tc, 106br, 138bc, 139bfr, 139flbc, 139cr, 149flac, 165br, 172bfl, 173frac, 199cra, 209flac, 216tc, 219bc, 233cla, 236flbc, 236bc, 238bfr, 238bl, 239frac, 239cra, 239crb, 240cla, 276frac, 281br, 288flac, 288frac, 297frac, 297cr, 300flac, 300tfr, 305br, 306bfr, 310clb, 311tl, 311cra, 312cfl, 313rbc, 314tfr, 316frbc, 316cl, 323br, 326cl, 336bfr, 337ca, 340cfl, 353clb, 353bl, 354ca, 356tr, 356bc, 360tr, 362fclb, 363br, 365tl, 418frbc, 556frbc, 588br; **R.W. Van Devender:** R. W. Van Devender 442clb, 442bl, 447ca, 450lac; **Dr Frances Dipper:** 540cfr, 540flbc, 594bfl, 594bfr, 604bfl, 604br; **Dorling Kindersley:** NASA / Finley Holiday Films 39br; **James Eaton, Birdtour Asia:** 356cra, 359cr; **Ecoscene:** Sally Morgan 32flbc, Kjell Sandved 27clb; **Edmund D. Brodie, Jr.:** 445bc, Edmund D. Brodie Jr. 441bc; **Elizabeth Whiting & Associates:** 81bl; **Brock Fenton:** 158frac, 158frbc, 159bfl, 159clb, 160cla, 161frac; **David B Fleetham:** 478c; **FLPA:** 118bc, 294cra, 295frbc, Ron Austing 369bl, Fred Bavendam / Minden Pictures 76bl, Rolf Bender 190cfr, B. Borrell 588frbc, Brake / Sunset 111br, Hans Dieter Brandl 63bfl, Robin Chittenden 368ca, M Clark 439bfr, E. Coppolia 283clb, F Di Domerico 184flbc, Yossi Fahhol 333frbc, Tom & Pat Gardner 349fbr, 351tl, Michael Gore 302ca, 328bfr, 353cr, 368bl, D T Gremcock 89frac, Tony Hamblin 59br, David Hosking 145cfr, 503tr, Eric & David Hosking 272cr, 357fcr, Eric Hosking 306frbc, S Jonasson 185frac, 289cl, D. Jones 587flbc, John Karmali 339frac, Gerard Lacz 91br, 135crb, 217bc, Frank W Lane 132tr, 347c, 430rac, Leeson / Sunset 142flac, Donald M. Jones / Minden Pictures 52c (deer), G. Marcoaldi 241bl, K. Maslowski 123br, S & D & K Maslowski 330bfl, S & D Maslowski 170bclb, S Maslowski 369t, Chris Mattison 373tfl, 418clb, 431cl, W. Meinderts 470bfr, W. Meinderts / Foto Natura 158flac, Derek Middleton 68c, 116cra, 443bc, Mikhail 191bl, Hiroya Minakuchi / Minden Pictures 539br, Geoff Moon 319bl, Mark Newman 178cfl, 375br, Philip Perry 216frbc, 238tc, A.A. Riley 29flbc, L. Lee Rue 272bl, Ry Silvestris 142clb, Silvestris 388c, 391cfr, Don Smith 331cfl, Jurgen and Christine Sohns 336cra, Roger Tidman 436tfr, Jan Van Arkel / FN / Minden 545cb, R. Van Nostrand 347tc, John Watkins 321tfr, Weiss & Sunset 303crb, Larry West 461frac, Terry Whittaker 201tr, 201cra, 209cra, 217cra, 221frbc, Roger Wilmshurst 342bl, D. P. Wilson 534cra, W. Wisniewski 31tfl, Martin Withers 332cla, Norbert Wu / Minden Pictures 508t; **fogdenphotos.com :** Michael & Patricia Fogden 229bfr, 329ca, 384crb, 407cr, 408bfr, 409cr, 447clb, 455cfr, 455flac, 465frac, 559tfr; **Foto Natura:** 116bl, Jan van Arkel 576c, Martin Harvey 103clb, 135frac, 135tc, 135crb, 210clb, 222cfr, S. Janssen 175tl, S. Maslowski 122bfr, Wil Meinderts 576flbc, Dietmar Nill 191cl, N. Rosing 122cr; **Fotomedia:** R Dev 236flac, Neeraj Mishra 137frac, E. Hanumantha Rao 137cla, 173bfl, 213cla, 320tl, Thakur Dalip Singh 359tfr, Vivek R. Sinha 215bc, 239bfr, Joanna Van Gruisen 225tl; **Jurgen Freund:** 511ca; **Neil Furey:** 19cb; **Tim Gallagher:** 282tr; **Gabriele Gentile:** 414br; **David George:** 541bl; **Getty Images:** Mark C. Ross / National Geographic 198tl, Kelly Cheng Travel Photography / flickr 69br, Bill Hatcher / National Geographic 112cl, Andrew McConnell 44br, Donald M. Jones / Minden 193b, Birgitte Wilms / Minden 512cr; **Brian Gibbs:** 192flbc; **Michael P. Gillingham:** 114bfl; **Roy Glen:** 349cla; **Francois Gohier:** 246bc, 248tfr; **Jonathan R Green:** 381tr; **Derek Hall:** 34frac, 38ca, 49tl, 53cfl, 53br, 69tl, 278crb; **Howard Hall Productions:** Howard Hall 482cl, 484br, 546c; **Tim Halliday:** 446frac, 446clb; **Derek Harvey:** 145cl; **Martin Harvey Photography:** Martin Harvey 202cra, 207frac, 213tfr, 216tfr, 233frac, 233tfr; **Lawrence Heaney:** 162frac; **Michael Henke:** 517cra; **Daniel Heuclin:** 387c, 401crb, 451bfr; **Dr G. H. Higginbotham:** 363cl; **Jacana Hoa-Qui:** Axel 144crb, Sylvain Cordier 33cfl, Varin Visage 94bfl, Gunter Ziesler 110br, 111tr; **Thomas Holden:** 271bl, 311bfr, 328bl, 341flbc, 346cra, 350bl, 358bc, T. Holden 367crb; **Holt Studios International:** Nigel Cattlin 556clb, 557clb, 560bfr, 560frbc, Peter Wilson 552cra, 569crb; **Hong Chalk Seng:** Chalk-Seng Hong 554c; **Andre van Huizen:** 65ca, 231bfr, 279bfl, 279bc; **Imagestate:** 1, 8-9, 73c, 80c, 324tc, AGE Fotostock 14frbc, 309cfr, Images Colour Library 146c, 180bl, 371c, National Geographic 170bcra, 170btfr, 174b, 182c; **Impact Photos:** Roger Scruton 511frbc; **F. Jack Jackson:** 539tfr, 540bfl, 605fbl; **Jack Jeffrey:** 368fcl; **Mike Johnson:** 514clb, 523tl, 529flbc; **Joint Nature Conservation Committee:** 538frac; **Mike Jordan:** 119bc, 121cfr, 121bl, 122flbc, 123bl, 124bfl, 124bfr, 124ca, 125fbr, 126frbc, 126cr, 128bc, 128br, 162br, 163bc, 242cla, Mike Joran 126br; **Ken Kates:** 100c; **Hiromitsu Katsu:** 114frbc; **Paul Kay:** 539cl, 596flac, 596flac, 596cl; **Kos Picture Source:** Carlo Borlenghi 528tfr; **Mark Kostich Photography:** Mark Kostich 199crb; **Dr C Andrew Henley / Larus:** Dr C Andrew Henley-Larus 94bfr, 95ca, 95cl, 96cra, 99cfr, 99cla, 183bfr; **Vanessa Latford:** 231flbc; **Legend Photography-Andy Belcher:** 520-521, 521; **Ken Lewis:** 283bc, 341bl; **Look:** Konrad Wothe 274cr; **The Mammal Images Library:** P. Myers 94cl; **Andrew Martinez:** 536frbc, 539bl; **Chris Mattison Nature Photographics:** 451clb, 458cr, Chris Mattison 61bc, 271cl, 313cfr, 373frac, 376cfr, 376cr, 377bfr, 378crb, 382frac, 383cfr, 383frac, 391cb, 394crb, 396cfl, 397tr, 401frac, 402flbc, 403frbc, 408c, 408crb, 412bfr, 413tfr, 413cr, 418cfr, 418tr, 419bfl, 419cfr, 421frac, 421br, 422frac, 422cl, 422crb, 423bfr, 423bc, 424bl, 425bfr, 427cfl, 427tfr, 430cl, 431cb, 433tfr, 436cla, 438tc, 445tc, 446ca, 447frbc, 450bfr, 452tr, 453cb, 454tl, 458frbc, 459cfr, 459frac, 459clb, 460frac, 460cla, 463cb, 465cla, 465crb, Geoff Trinder 60ca, 118cr, 225bl, 304br, 305tl, 366cfr, Martin Withers 15bfr, 57cb, 98cla, 102bfl, 103frbc, 114cla, 127cr, 159flac, 190flac, 227bfr, 237cr, 240frac, 240clb, 242br, 275fr, 275crb, 289bl, 296frbc, 300bfr, 301bc, 319tl, 327fbfr, 348cfr, 362crb, 419br; **Dr Luis A. Mazariegos:** 330cr; **McDonald Wildlife Photography:** Joe and Mary Ann McDonald 156c, Joe McDonald 55tr, 110tl, 119tfr, 209bl; **Dr Fridtjof Mehlum:** 314cfr; **Mr W. E. Middleton:** 558cfr; **Minden Pictures:** Fred Bavendam 600bfl, 600cfl, 600ca, Mark Moffet 590c; **Carlos Minguell:** 477frac, 517crb, 519clb, 538tr, 538cra, 605bl; **Natural History Museum:** 538br, 550frac, 557cl; **Natural Science Photos:** C. Blaney 336cl, Cranston 513tc, C. Dani & I. Jeske 44c, Bob Gurr 211frac, David Lawson 411bfr, Jim Merli 52bc, 445bc, W. E. Middleton 336cfl, O. C. Roura 419bc, C & T Stuart 140tr, P. H. & S. L. Ward 439tfr; **Natural Visions:** 26frbc, 469frac, Heather Angel 91cla, 131bl, 372crb, 383clb, 383cb, 469tfr, 503fcrb, 504tc, Tony Martin 251bfr, Paul Ormerod 267cr, Slim Sreedharan 374c, Soames Summerhays 534bfl; **Nature Photographers:** S.C. Bisserot 452cfr, Colin Carver 68c, 461cfr, Don Smith 445cr, Paul Sterry 414cb, 550clb; **Naturefocus, Australian Museum:** C Andrew Henley 97frac; **naturepl.com:** 163tl, 189, 224, Doug Allan 66bfr, 185br, Nigel Bean 82tr, Vanessa Berlowitz 590ca, Peter Blackwell 385cr, Neil Bromhall 460tc, Dan Burton 480cr, 494cb, Jim Clare 54tr, 179cra, Bruce Davidson 153frac, 173crb, Stewart Dawber 337c, G & H Denzau 47flac, Martin Dohrn 83c, Geoff Dore 588cfr, Georgette Douwma 603, Georgette Douwna 74bl, 468tfr, Tim Edwards 319clb, 411cl, Hanne & Jens Eriksen 309cb, 327frac, Geoff Foot 203frac, Jeff Foott 75br, 107frbc, 185bl, 218bl, 234tc, 247tl, 471flbc, 581cfl, Jurgen Freund 107tfr, 375cfr, 540ca, 582c, Nick Garbutt 219crb, Nick Gordon 51bc, David Hall 79bfl, Andrew Harrington 177clb, Tony Heald 88tfr, 269cl, Martha Holmes 247bfr, Charlie Hamilton James 172clb, Paul N. Johnson 225clb, Hans Christopher Kappel 16clb, Kevin J Keatley 192frbc, Simon King 169tc, Brian Lightfoot 293tr, Jeff Lightfoot 504tl, Fabio Liverani 440c, Thomas D Mangelsen 177cla, Nigel Marven 280cfr, Steven David Miller 20tc, 279cr, Steven Miller 430flbc, Naturbiuld 472frbc, Dietmar Nill 121tr, 160cfl, 191cla, 443tc, Ron O'Connor 206frbc, Pete Oxford 35tfr, 106tl, 138frbc, 143cl, 345clb, 430cra, 434cfr, Prema Photos 51cr, 411frbc, Colin Preston 125flac, Rico & Ruiz 443cl, Jeff Rotman 249bl, 473cfl, 475frbc, 483c, 600clb, Phil Savoie 342c, 353ca, 425tfr, Peter Scoones 69bl, 73tl, 92ca, 221tc, Anup Shah 149tfr, 151cbr, 154cfl, 213frbc, 220, 432c, Yuri Shivnev 270frac, John Sparks 16cb, Lynn M Stone 79cfr, 179bfr, David Tipling 15bfl, 278bfr, 293cfr, 307, Tom Vezo 94cr, Gerrit Vyn 271c, 273, Bernard Walton 147rbc, Dave Watts 355, Doug Wechsler 445cfr, 458fr, David Welling 17cb, Doc White 244cl, 248cfr, Mike Wilkes 88bc, 315cfl, 338c, 411tr; **John Nelson:** 157frac; **NHPA / Photoshot:** A.N.T. 17cla, 31bfl, 52ca, 69bc, 95c, 96bfr, 268crb, 344tl, 348fcr, 349fbl, 376cfl, 411bl, 456cla, 462bfr, 488cr, 517cb, 538frac, 168bc, 228bc, 274bl, 294cfl, 295bl, 409frbc, 471tfl, 586bc, B & C Alexander 241bfr, Tom Ang 69crb, J & M Bain 597cfl, Jim Bain 538bfl, Daryl Balfour 286cra, Anthony Bannister 31frac, 74c, 106bl, 192cla, 373flbc, 399c, 400cfr, 465bl, 559flac, A.P. Barnes 30frac, 328frac, G.I. Bernard 545crb, 557cla, 558cla, Joe Blossom 276tl, 284tr, John Buckingham 266tfr, 309frac, N. A. Callow 560frac, G. J. Cambridge 551tl, Laurie Campbell 58cfr, 62bfl, 118tfr, 186bfl, 313tfr, Bill Coster 293crb, 310tr, 585bc, N. R. Coulton 592cl, David Currey 245frac, Stephen Dalton 51bl, 55bc, 128tfr, 159frbc, 266tfl, 329frac, 410c, 415frbc, 415t, 419tl, 542frac, 551clb, Manfred Danegger 58cfr, 113br, 276crb, Nigel Dennis 152cfr, 191bfr, 287tl, 338cb, 358bfr, Dr E. Elkan 565cfr, Robert Erwin 129cr, 442tr, George Gainsburgh 486r, Pavel German 92cla, 92clb, 99cfl, 399bfr, 447frac, 452tl, K Ghani 215frbc, Iain Green 149bfl, Ken Griffiths 336tc, 399cb, E. Hanumantha Rao 343cra, Martin Harvey 71crb, 101tr, 151cr, 285cl, 333cl, 376br, 432bl, Brian Hawkes 326tfr, Daniel Heuclin 47cra, 48frbc, 50cl, 63bc, 96bfl, 131frac, 160tfr, 163br, 296cl, 301cfr, 384c, 387tl, 389frbc, 394bfr, 397flbc, 402tc, 403cr, 406crb, 419frac, 426ca, 427frac, 427bc, 431frac, 443tfr, 447bfl, 456cl, 461br, K. & V. Hurst 70br, T Kitchen & V Hurst 82bl, 174tr, 187t, 188tl, 414tr, 459tr, 461cl, Image Quest 3-D 31frbc, H & V Ingen 547br, Hellio & Van Ingen 282clb, B Jones & M Shimlock 14flbc, 77cb, 537c, 594cfr, 599flac, James Carmichael Jr 291tfr, 590tc, Ralph & Daphne Keller 98bl, 102cfr, 391frbc, 424cb, Rich Kirchner 172tfr, 227bl, Stephen Krassemann 33tfr, 169bl, 297crb, Gerard Lacz 151br, 184tfr, Michael Leach 375frbc, Lutra 471frac, 502tr, Trevor McDonald 76frbc, 77crb, Alberto Nardi 284br, Agence Nature 74tr, 247tfr, 471tfr, 472frac, 522tr, Orion Press 171cra, Haroldo Palo Jnr 94flac, 112bl, 338bl, 340frbc, 344br, Rod Planck 40cr, Dr Ivan Polunin 29cfr, 132bl, Christophe Ratier 284flbc, 286cb, Andy Rouse 171crb, 186frbc, 190br, Jany Sauvanet 129bfr, 142bfr, 161tl, 171cfl, 323cra, 390clb, 447bfr, 453br, 498frbc, 515tl, Kevin Schafer 141cr, 170bfr, 179tfr, 278tfr, 332frbc, 489frac, John Shaw 110tr, 114clb, 115bfl, 210flbc, 328tfr, 338cl, 534frac, Eric Soder 327fcb, R Sorensen & J Olsen 116bfr, Morten Strange 199clb, M. I. Walker 544tr, 579crb, Roy Waller 518frbc, 584bfl, James Warwick 178crb, Dave Watts 97crb, 411clb, Martin Wendler 320bfl, 320bl, 386c, Alan Williams 357bc, Norbert Wu 66flbc, 470cfr, 495bfr, 517cl, 528tl, 539cb, 601, Robert Wu 195frac; **Klaus Nigge:** 294c; **NOAA:** Monterey Bay Aquarium Research Institute 19bl; **Mark Norman:** 599cla; **Gary Nuechterlein:** Gary Nuechierlein 283tfr; **Ocean Eye Films:** Lawson Wood 486c; **OSF:** 208bc, 230, Ida Akachn 589bfr, Doug Allan 252tfr, 526clb, 545tl, Animals-Animals / Richard K. LaVal 111fcr, Kathie Atkinson 92frac, 95cfr, 540flac, 545br, 596cb, 598flac, Eyal Bartov 169frac, G.I. Bernard 115tl, 124crb, 439cl, 580clb, 589frac, Joe Blossom 218bfr, Liz and Tony Bomford 322cfr, Tony Bomford 477fcrb, Clive Bromhall 47c, Mike Brown 316br, Roger Brown 348bl, Robin Bush 315cfl, 450tl, Arthur Butler 309tc, John Chellman / AnimalsAnimals 207frbc, Waina Cheng 555frbc, John Cheverton 30cfl, Martyn Chillmaid 215flbc, Martyn Colbeck 108cl, 152tfr, N.M. Collins 541clb, JAL Cooke 550cla, Daniel J Cox 59bfl, 167tfr, 195br, 304cra, 361fcl, David Curl 315c, Stephen Dalton 284bl, Kenneth Day 293bfr, Richard Day 89frac, 330cb, Mark Deeble & Victoria Stone 196frac, 290tfr, David M Dennis 385frbc, 465bfr, 543frbc, Jack Dermid 440bl, Ajay Desai 149clb, Michael Dick 149cra, Frederik Ehrenstrom 585frac, Paolo Fioratti 336tr, David B Fleetham 77br, 473bl, 475cr, 491c, 524frbc, Michael Fogden 187bc, 305cfl, 331cb, 394clb, 434br, 449frbc, 455flbc, 587cfr, Stephen Foote 545tr, Jeff Foott 111tl, Harry Fox 548clb, 548cb, Frances Furlong 303tfr, C. G. Gardener 541cfr, Max Gibbs 490frac, Nick Gordon 367fcl, Oliver Grunewald 378crb, 378b, Mike & Elvan Habicht 94frac, Howard Hall 73c, 249rbc, 260frbc, 377cfl, 480tfr, David Haring 139frac, John Harris / Survival Anglia 114crb, Terry Heathcote 280tfr, Richard Hermann 73c, 474c, 484cb, 501, 514cfr, 529c, Karen Gowlett-Holmes 75cl, Roger Jackman 455c, Tim Jackson 463bc, Mark Jones 185fl, 380clb, Marx Jones 268cl, Paul Kay 529cla, Isaac Kehimka 119br, Breck P Kent 165cr, Chris Knights 28clb, 283tc, Rudie H Kuiter 489bl, 507br, 522bfr, 522frbc, 538c, 599cfl, Satoshi Kuribayashi 549bfr, 549cfr, 549frac, 549frbc, 590bfl, K & L Laidler 179bc, Lon E. Lauber 67bfr, Zig Leszczynski 377cfr, 439bfl, 450cl, 461tc, 463bfr, 472br, Zig Leszczynski / Animals Animals 597bfr, London Scientific Film 541frac, 541br, Mantis Wildlife Films 452bl, 456bfl, Juan Manuel 327fbl, John McCammon 539bfr, T.S. McCann 31cla, Raymond Mendez 446bfl, Colin Milkins 578frbc, 580tfr, Sean Morris 57cl, 116tr, 336bl, Owen Newman 167bc, 310frbc, G. Synafzsankle Okapia 451bfl, Okapia 232cfr, Stan Osolinski 113tfr, 240crb, 321frbc, Richard Packwood 64cr, John Paling 68br, 70flbc, 605br, Andrew Park 57crb, Peter Parks 534bfr, 545fclb, 580cfl, 580c, 580cr, 580cb, 581tfr, Andrew Plumptre 153bc, Hilary Pooley 210c, Partridge Productions 136cfl, Romford & Borrill / Surval Anglia 164crb, Alan Root 106fcr, 143bl, 343frac, 432br, Norbert Rosing 66ca, 67flac, Tui de Roy 65bc, 301frbc, 380cfl, Kjell B. Sandved 15flbc, 545cfl, DJ Saunders 343bfr, Jany Sauvanet / Okapia 112tr, F. Schneidermeyer 205tc, Krupaker Senani 359cfl, Michael Sewell 35br, 258frac, David Shale 118cl, Chris Sharp 585frbc, Wendy Shattil & Bob Rozinski 39crb, 116c, Alastair Shay 161bl, Jorge Sierra 431cla, Sinclair Stammers 543clb, Marty Stouffer 58tfl, Harold Taylor 61cl, 548bc, TC Nature 353t, 553c, Tony Tilford 359br, Gerald Tompson 343cfr, Robert Tyrrell 330br, Tom Ulrich 96tl, 125cra, 186clb, 331clb, 404flbc, K. G. Vock 548br, P & W Ward 48tr, Babs & Bert Wells 328c, 401flbc, Ian West 422tfr, Kim Westerskov 280cfl, 536bfl, Jack Wilburn 368cr, Eric Wood 293frbc, Conrad Wothe 448bfr, Konrad Wothe 269bl, 277cfr, 312tfr, Norbert Wu 65bl, 72tfr; **Panda Photo:** M. Calandrini 284c; **Otto Pfister:** 225crb, 241cra, 242cr, 305cb, 310bfl, 363crb; **Maslowski Photo:** 287cfl; **Photolibrary:** OSF / Carol Farnet; Foster 590bc; **Photomax:** 488bfr; **Photoshot / Woodfall Wild Images:** Maurizio Biancarelli 29tfl, R. Buchele 60cr, Tom Campbell 90tc, S. Cordier 91fcrb, Mark Hamblin 34flbc, Paul Kay 72c, 493tc, Mike Lane 211tfr, 292c, D. Mason 90bfl, 234flbc, Sean McKenzie 65br, Tom Murphy 32cfl, Tapani Rasanen 300crb, John Robinson 127cfr, Lawson Wood 75tl, Davis Woodfall 61crb; **Mark Picard:** 114bfr, 172bfr, 226bc, Mark Picard 227cfr, 227cb, 271tr, 287cr, 333clb, 367tr, 367crb; **Linda Pitkin / lindapitkin.net:** 77br, 538bfl, 538cfl, 538cla, 538cl, 538bl, 539flac, 539tl, 539tr, 539ca, 539cra, 539cr, 541cfl, 544fcr, 570frac, 604flac; **Planet Earth Pictures:** 274tfr, 274cb, 408cra, 408bl, K & K Ammann 43bfl, Kurt Amsler 518cfr, Ingo Arndt 548c, Pete Atkinson 497cla, Andre Bartschi 142tfr, 317tfr, Gary Bell 177tfr, 476cl, 476bl, 516crb, 585c, Ashley J. Boyd 538frbc, T. Brakefield 205bfr, Jim Brandenburg 115frbc, Robert Canis 592flac, Mary Clay 114tl, 169flbc, Norman Cobley 280cb, Richard Cook 482frac, Richard Coomber 211cb, Peter David 509cr, 523tfr, 597bc, 599tfr, 599cra, Paulo De Oliveira 333frac, 492tfr, 539bl, 545crb, S V Den Nieuwendjik 61c, 153frbc, M & C Denis-Hout 109bfr, Wendy Dennis 316clb, Georgette Douwma 487bfr, 537bfr, 541tfr, 594tfr, 594ca, John Downer 108clb, Geoff du Feu 165tl, Carol Farneti-Foster 93cra, D. Robert Franz 166cra, Robert Franz 226bfl, Nick Garbutt 409cr, Jim Greenfield 512tfr, Roger de la Harpe 337bc, 432tfr, Hans Christian Heap 49br, Robert Hessler 75tr, Paul Hobson 314crb, Steve Hopkin 551frac, 568tfr, Chris Huxley 522flac, Adam Jones 440flbc, Brian Kenney 196tl, 384frbc, 392c, 440tr, 587frbc, Frank Krahmer 47br, John Lee 518tr, Peter Lilja 56cr, Ken Lucas 472cfr, 481crb, 488tfr, 497cl, 517c, 529bl, 588cl, Larry Madin 515frbc, David Maitland 432frbc, Richard Matthews 167tc, 172tl, Neil Mcintyre 113tfr, Pete Oxford 105cr, 270frbc, 391flac, 547frbc, Allen Parker 315frac, Doug Perrine 107bl, 254cfr, 255cfr, 255bl, 477fcla, Carl Roessler 527tfr, G. Van Ryckevorsel 489clb, Peter Scoones 488cl, Angela Scott 108tl, Jonathan Scott 29bfr, 42tr, 43tfl, 62cfr, 109crb, 207cb, 218c, 239tl, John Seagrim 473bfr, Anup Shah 285tfr, Yuri Shibnev 301tl, Marty Snyderman 484frbc, Larry Tacket 144frac, Tom Walker 115bfr, John Waters 130frbc, James D Watt 477fcfr, J. O. Wirminghaus 462cl, Norbert Wu 536tc, 597bfl, Robert Wu 508cr, Andrey Zvoznikov 125br; **Louis W. Porras:** 463cr, Louis W. Parras 427bfr; **Premaphotos Wildlife:** Dr Rod Preston-Mafham 561tfr, 592bfl, Jean Preston-Mafham 570cl, Ken Preston-Mafham 561clb; **Press Association Images:** AP 19br; **John E Randall:** 516tfr, John E Randell 514cr; **Jean Yves Rasplus:** 560clb, 573clb; **Galen B. Rathbun:** 104c, 104fcr, California Academy of Sciences 104bl; **Reinhard - Tierfoto:** Hans Reinhard 280ca, 322c; **Neil Rettig Productions Inc:** Neil Rettig 301bfr; **Rex Features:** Avi Klapper 527bfr; **Dr Gil Rilov:** 539crb; **Mel José Rivera :** 418cra; **Francesco Rovero:** 104bc; **The Roving Tortoise:** Tui De Roy 182crb; **RSPCA Photolibrary:** Adrian Warren 429c; **Save-Bild:** G. Schltz 101frbc; **Kevin Schafer:** Kevin Schafer 129cla, 135frbc, 140rbc, 147tfr, 221tfr, 332clb; **Science Photo Library:** Eye of Science 557frbc, Vaughan Fleming 558tfr, Ted Kinsman 20bl, Dr Kari Lounatmaa 14br (archaea), Will and Deni McIntyre 91tl, Pasieka 14br (bacteria), Dr Morley Read 372clb, Science Pictures 534tfl, Sinclair Stammers 580tr, Andrew Syred 264br, 265bfr, 265cra, M I Walker 541bl, Art Wolfe 141bl; **Sue Scott:** 539flac, 539cr, 540tfr, 580crb, 584crb, 594bc, 596bfr, 598c, 604tr, 604ca; **SeaPics.com:** Daniel W. Gotshall 504cfl, Rudie Kuiter 482bfr, Amos Nachoum 256ca, Doug Perrine 244c, 245cra, 250, 261, 473br, James D. Watt / Seapics.com 487cra, Doc White 194c; **Herb Segars:** 493crb; **shahimages.com :** 6bc; **Wendy Shattil:** 45ca, 206flac, 382cr, Wendy Shattil & Bob Rozinski 130br, 190bl, 289c, 321clb, 346c, 356cl; **C. Andrew Smith:** 114tfr; **Smithsonian Institution:** John E Miller 604clb; **South American Pictures:** Tony Morrison 285cr; **Specialist Stock / Still Pictures:** J.J. Alcalay 64tr, 182tfr, John Cancalosi 404bl, Alain Compost 93bfr, M & C Denis-Huot 134cla, 148frac, Michel Denis-Hout 256cfl, EIA 253frbc, Michael Gunther 152frbc, 216bfr, Yves Lefevre 473cr, Rowland Seitre 97flac, 178tfr, 179frbc, 201, 227tr, 245frbc, 259cfr, Sabine Vielmo 154tc; **Frank Steinmann:** 460clb; **K. Sugawara:** 360tl; **Harold Taylor:** 81tl; **Telegraph Colour Library/Getty Images:** Jan Tove Johansson 28bfl, Paul Losse 26cra; **Jamie Thom:** 2-3; **Claire Thompson:** 206tr; **© Uthai Treesucon / www.arkive.org:** 19crb; **Michael P. Turco:** 195cfr; **USE ngeo:** Michael Nichols 28frac; **Jean Vacelet:** 536cfr; **Eric Vanderduys:** 452clb; **Colin Varndell:** Peter Coe 340cfr, 346crb, Dave Nichols 281flac, Colin Varndell 277bfr, 290bfl; **Vireo:** 352frac, J Dunning 345br, 346t, 347tr, B Schorre 369fcrb, Doug Wechsler 330tl, 340frbc, 354br; **Gernot Vogel:** 394tl; **Prof. Dr. Peter Vogel:** 105cl, 164tr; **Terry Wall:** Terry Wall 272bfr, 290flac; **Bernard Walton:** 147bfr, 147br; **Dave Watts:** 65cb, 92br, 95br, 96frac, 96clb, 97bfr, 98frac, 99bfr, 99cra, 99crb, 102frac, 103crb, 138flbc, 149cla, 184frbc, 241cfr, 242bl, 288cfr, 301lb, 302frac, 348tl, 349br, 356t, 362bl; **WDCS:** Fernando Truijillo 251tr; **Dr Jane Wheeler:** 223tcl, 223tfr; **Wildlife Conservation Society:** Ian Kellett 143bc; **Jack Williams:** 515cl; **Nick Wilton:** 212bc; **Winfried Wisniewski:** 257; **Art Wolfe:** Art Wolfe 90cfr, 90cfr, 90cfr, 130tfr, 134-135c, 153crb, 156frbc, 229bl, 270tfr, 303cr, 308c, 322tfr, 504flbc; **P. A. Woolley and D. Walsh:** P. A. Woolley & D. Walsh 94bl, 99bl; **Norbert Wu:** Brandon Cole 472clb, Bruce Rasner 480cl; **Gunter Ziesler:** 135cra

Additional photography
Max Alexander; Peter Anderson; Irv Beckman; Geoff Brightling; Jane Burton; Peter Chadwick; Andy Crawford; Peter Cross; Geoff Dann; Philip Dowell; Alistair Duncan; Mike Dunning; Andreas von Einsiedel; Ken Findlay; Neil Fletcher; Christopher and Sally Gable; Peter Gardner; Peter Gathercole; Steve Gorton; Frank Greenaway; Peter Hiscock; Colin Keates; Dave King; Cyril Laubscher; Bill Ling; Mike Linley; Kevin Mallet; Maslowski Photo; Chris Mattison; Jane Miller; Tracy Morgan; Gary Ombler; Nick Pope; Rob Reichenfeld; Tim Ridley; Kim Sayer; Tim Shepard; Karl Shone; Harry Taylor; Kim Taylor; M.I. Walker; Mathew Ward; Laura Wickenden; Alan Williams; Jerry Young

Additional illustrations
Martin Camm; Colin Newman